SUBSTITUTED HYDROCARBONS (continued)

Ethers

* $CH_3CH_2OCH_2CH_3$

Diethyl ether

O—CH_2CH_3

Ethyl phenyl ether

Sulfides (thioethers)

* $CH_3CH_2SCH_2CH_3$

Diethyl sulfide

Nitro compounds

$H_3C—N^+(=O)O^-$

Nitromethane

Nitroso compounds

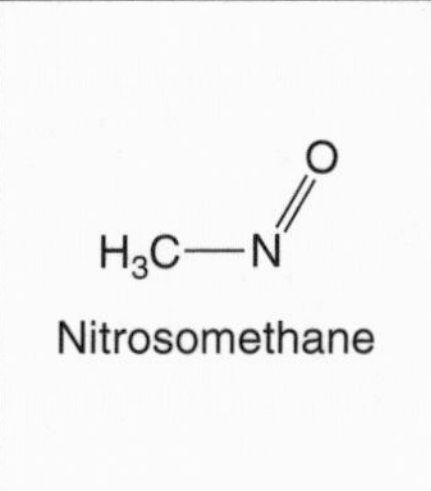

Nitrosomethane

Sulfoxides

$H_3C—S(=O)—CH_3$

Dimethyl sulfoxide

Sulfones

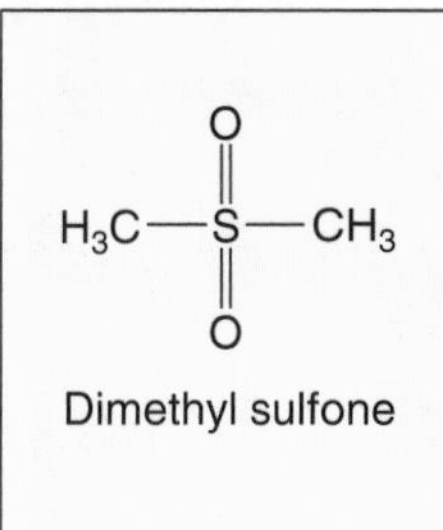

Dimethyl sulfone

Sulfonic acids

$H_3C—S(=O)_2—OH$

Methanesulfonic acid

CARBONYL (C=O) COMPOUNDS

Aldehydes

* $H_3C—C(=O)H$

Acetaldehyde

Benzaldehyde

Ketones

* $H_3C—C(=O)—CH_3$

Acetone (dimethyl ketone)

Carboxylic acids

* $H_3C—C(=O)OH$

Acetic acid

Benzoic acid

DERIVATIVES OF CABOXYLIC ACIDS

Esters

* $CH_3—C(=O)—OCH_2CH_3$

Ethyl acetate

OCH_3

Methyl benzoate

Amides

* $H_3C—C(=O)—NH_2$

Acetamide

NH_2

Benzamide

Acid chlorides

* $H_3C—C(=O)—Cl$

Acetyl chloride

Cl

Benzoyl chloride

Anhydrides

$H_3C—C(=O)—O—C(=O)—CH_3$

Acetic anhydride

Nitriles (cyanides)

* $CH_3CH_2—C≡N$

Propionitrile (Ethyl cyanide)

Organic Chemistry

MAITLAND JONES, Jr.

Princeton University

W · W · NORTON & COMPANY  NEW YORK · LONDON

PRINTED IN THE UNITED STATES OF AMERICA
Composition by TSI Graphics
Manufacturing by R.R. Donnelley & Sons Company
Book design: Jack Meserole
Illustrations: J/B Woolsey Associates
Cover photographs: Nat Clymer
Cover design: Maitland Jones, III, and Susan Hockaday

Library of Congress Cataloging-in-Publication Data
Jones, Maitland, 1937–
Organic chemistry / Maitland Jones, Jr.
p. cm.
Includes index.
ISBN 0-393-97079-5
1. Chemistry, Organic. I. Title.
QD253.J65 1997 96.52944
547—dc21

W.W. Norton & Company, Inc., 500 Fifth Avenue, New York, N.Y. 10110
W.W. Norton & Company, Ltd., 10 Coptic Street, London, WC1A 1PU

1 2 3 4 5 6 7 8 9 0

This book is dedicated to those who oppose tyrannies, large and small, everywhere. For personal reasons I want to single out the intrepid people of Tibet, who continue to resist the cruel occupation of their country, and Wang Dan, Wei Jingsheng, and the others who languish in prison as a result of their support for democracy at Tiananmen in June, 1989.

Contents in Brief

Contents

6

Chapter 6. Rings 191

7

Chapter 7. Substitution and Elimination Reactions: The S_N2, S_N1, E2, and E1 Reactions 233

13

Chapter 13. Conjugation and Aromaticity 567

Chapter 14. Substitution Reactions of Aromatic Compounds 621

Chapter 15. Analytical Chemistry 685

16

Chapter 16. Carbonyl Chemistry 1: Addition Reactions 753

Chapter 20. Derivatives of Carboxylic Acids: Acyl Compounds 1003

Chapter 21. Introduction to the Chemistry of Nitrogen-Containing Compounds: Amines 1075

Chapter 24. Polyfunctional Natural Products: Carbohydrates 1233

Chapter 25. Introduction to the Chemistry of Heterocyclic Molecules 1283

26

Preface

Organic chemistry has long been recognized as an essential ingredient in the education of scientists in a wide range of fields, particularly in the life sciences. Indeed, as more and more disciplines come into contact with organic chemistry, new connections are constantly being forged. By extension, a career in medicine—an applied life science—is also becoming more and more dependent on organic chemistry, and not just because medical schools these days take special notice of the grade in organic! The practice of medicine nowadays increasingly requires a knowledge of the behavior of molecules. The molecular basis of pathogen action and of therapeutic agents are two obvious examples. These are developing areas at the moment, but there can be no doubt that the physician of tomorrow will have to be even more of a scientist—an organic chemist, among other things—than his or her predecessors.

Accordingly, this book is written not just for the chemistry major, but is aimed instead at the person who wants a broad, yet modern introduction to the subject. General principles are stressed, because it is impossible to memorize all the details of this vast subject these days. One can, however, learn to make connections and to apply widely a set of organizing general principles, and that helps to render the prolific detail more manageable.

A Brief Tour of the Book

To understand molecules we must first understand atoms and, especially, the volumes of space occupied by the electrons that orbit the atomic nuclei. Thus, Chapter 1 begins with atoms and electrons. To think sensibly about electrons we need to understand a bit of what quantum mechanics tells us. That does *not* mean that we will all have to become mathematicians, far from it. We only have to understand qualitatively what the mathematicians have to say to us. We also need mathematics because of a problem of scale. We live in a world of certain sizes, and our intuitions are honed by everyday experience. We know how baseballs and basketballs behave. Unfortunately, our experience in the world of such objects poorly equips us to anticipate matters in the large-scale universe—pulsars, quasars, black holes, and other bizarre inhabitants of the macro-universe. What is generally less well appreciated by students is that we operate intuitively just as poorly on the micro scale—the world of atoms and electrons.

Atoms and molecules are really quite strange things; they are not really like the objects that populate our daily experience. For one thing, molecular events take place on a time scale we cannot easily appreciate. Under typical conditions in the gas phase a single molecule undergoes 10^{10}–10^{11} collisions per second. Let's call one of those collisions an ordinary event in a molecule's "life," analogous to an ordinary event in a human life—taking a breath, for example. If you breathe 60 times a minute, you will take over 2.5 billion (2.5×10^9) breaths over an 80-year lifespan. So, in one second, a molecule undergoes more ordinary events than you will take breaths in your entire life. We do not appreciate such differences easily!

The language that makes both the macro scale and the micro scale accessible to us is mathematics. Although we need not do the mathematical operations ourselves, we do need to appreciate some of the things that quantum mechanics has to say to us. Chapter 1 also focuses strongly on Lewis structures—pictorial representations of atoms and ions. Much of this chapter may be review, and if so, this material can be skipped or covered very lightly. However, the ability to write good Lewis structures easily and to determine the locations of charges in molecules with ease is an essential skill. This skill is part of the language of chemistry and will be as important in Chapter 26 as it is in Chapter 1.

Chapter 2 moves on to molecules, and to the regions of space occupied by electrons in molecules: molecular orbitals. Qualitative molecular orbital theory is not too complicated a subject for students, and requires no mathematics. Yet, this simple theory is amazingly powerful in its ability to rationalize and, especially, to predict structure and reactivity. The tutorial in Chapter 2 on qualitative applications of molecular orbital theory is likely to be new to students. This material is very important, as it enables us to emphasize explanations throughout the rest of the book. Not only are traditional subjects such as aromaticity and conjugation (Chapters 12–14) more accessible with the background of Chapter 2, but explanations for the essential, building-block reactions of organic chemistry (Chapters 7 and 9, for example) become possible.

After the two introductory chapters comes a sequence of four chapters on aspects of structure (3–6). Here, some fundamental structural types—some functional groups—are introduced, and stereochemistry is dealt with in depth. This platform allows us to launch a discussion of several archetypal reactions of organic chemistry, and more functional groups, in Chapter 7. Chapter 7 is one of the key chapters, and the reactions of this chapter serve as reference points throughout the book; they are fundamental reactions to which we return over and over again in the later chapters in order to make analogies.

Once we have these basic, reference reactions under control, a general discussion of the role of energetics—kinetics and thermodynamics—becomes appropriate (Chapter 8). Chapters 9 and 10 introduce other building-block reactions, and other functional groups, in the context of a discussion of addition reactions. These provide a foundation for the chemistry of carbonyl compounds, the subject of a series of chapters in the second half of the book (16 and 18–20). Chapter 17 is devoted to a summary of the chemistry of alcohols. As alcohols are exemplars of many of the properties of organic molecules, and are the starting materials or products of many of the reactions of organic chemistry, one may legitimately wonder why we wait so long to bring them up. The answer is that we don't! Chapter 17 appears here as a summary of what has gone before: a collection of all the alcohol chemistry we have discussed in the previous 16 chapters. Alcohol chemistry is scattered throughout the early chapters of the book and is summarized here as a review, and as a prelude to the further chemistry of alcohols that will come in later chapters.

To make an analogy to the study of a language, in the first sequence of reaction chapters (7, 9, and 10) we write sentences constructed from the vocabulary and grammar developed in the early, structural chapters. We will go on in later chapters in the book to more complicated mechanisms and molecules, and to write whole paragraphs and even short essays in organic chemistry. Some of those essays are contained in the "special topics" chapters toward the end of the book (22–26) in which biological chemistry and physical-organic chemistry are further explored with the material of the early chapters as a foundation.

Special Features of This Book

Voice. This book talks directly to the reader, not only about the material at hand, but also about the "how and why" of organic chemistry. It tries to do so in a voice that is personal, not distant. I think it is much easier to enjoy, and learn, organic chemistry if a strong focus on "Where are we and why are we here?" and "What is the best way to do this?" is maintained. Of course, the main arena for such talk must be the lecture, and no book can hope to substitute for the direction and inspiration given by the lecturer. At the same time, it is important for students to hear different voices, to see the path we are all on in different ways. This approach provides at least two views of the material, two different routes to understanding. The sum of these perspectives is likely to be greater than either of them alone.

Flexibility. There is no consensus on the precise order in which to take up many subjects in organic chemistry. This book makes different decisions possible. For example, the spectroscopy chapter is largely free-standing. Traditionally, it comes where it is here, roughly at the midpoint of the book. But cogent arguments can be made that spectroscopy should be introduced earlier, and that is possible, if one is willing to pay the price of delaying the introduction of chemical reactions one more chapter. As in all organic texts, the last few chapters constitute a series of special topics. No one really hopes to finish everything in an organic textbook in one year, and this book provides a number of choices. One might emphasize biological aspects of our science, for example, and Chapters 24–26 provide an opportunity to do this. Alternatively, a more physical approach would see the exciting chemistry of Chapters 22 and 23 as more appropriate.

Extensive Use of Figures and Constructive Use of Color. Organic chemistry is a highly visual subject. Chemists think by constructing mental pictures of molecules, and we talk to each other by drawing pictures. This book favors series of figures over long discussions in the text. The text serves to point out the changes in successive figures. Color is used to highlight change, and track the fates of atoms and groups in reactions. The use of color is judicious, however. There are times when "more is less," and the use of color in organic textbooks can easily fall into this trap. Color is for emphasis. Too much color leads to visual clutter, in which case the emphasis intended by color can be lost.

Cross-Referencing. There is extensive cross-referencing in this book, which is facilitated by the early use of prototypal reaction examples to which analogy is repeatedly made. The reader is consistently reminded that the current material is "not really new—you have seen something like it before, and it's on page xxx." The hope is to supply a unifying treatment in which seemingly different topics are brought together. In turn, this reduces the need to memorize.

Problems. This book incorporates problems in two ways. First, there are many unsolved problems scattered throughout the text, and more, of all degrees of difficulty, are found at the end of each chapter. These range from drill exercises and simple examples designed to emphasize important skills and illustrate techniques, to sophisticated, challenging problems. In the latter cases, we are careful to provide hints and references to material useful for the solution. Answers to all problems of this type can be found in the Study Guide. The Study Guide tries to help students learn how to approach problems, and gives much more than bare-bones answers to most problems. Second, and most critical, is the use of problems as an integral part of the discussion in the text. There are times when a skill must be mastered or a point well understood before further discussion can take place effectively. Rather than simply emphasizing this point in the discussion, a solved, illustrative problem is almost always provided. It is important that these problems be worked. The answer is not there just to be read. These problems are vital to effective progress through the text and must be "read with a pencil."

Molecules in Boxes. Each chapter contains a boxed story of a molecule, which connects the material of the chapter to some aspect of the world around us. Organic chemistry is not a subject confined to the lecture room, or even to the laboratory. It is alive and well all around us. There should be no real surprise about this—after all, we and most of the rest of the biological world are made largely of carbon, and carbon-based creatures and materials affect us greatly. These boxes try to provide some connective tissue between our study of this subject and the rest of the world.

Important Details

Authenticity and Reaction Descriptions. Reactions have been pursued back to the original literature in most cases, and at least to the review literature in all cases. When reactions are discussed, both the general case and at least one specific example are shown. Generality is emphasized, but the particular is not ignored.

Convention Alerts. One factor that can make organic chemistry difficult is that a language must be learned. Organic chemists talk to each other using many different conventions, and at least some of that language must be learned or communication is impossible. In a real sense, this is analogous to learning the vocabulary and grammar of a language. In addition to general treatments of nomenclature at the beginnings of many chapters, we have incorporated numerous "Convention Alerts" in which aspects of the language chemists use are highlighted.

Chapter Summaries. Each chapter ends with a summary of the important concepts, reaction mechanisms, and synthetic procedures discussed therein. These recapitulate and reinforce the material of the chapter. These summaries are further extended with problems and detailed solutions in the Study Guide. In each summary, there is also a section on common errors. What kinds of traps can one easily fall into? How can they be avoided?

Key Terms. Each chapter also contains a glossary of the important terms introduced in the chapter. The Study Guide collects these chapter glossaries into an overall glossary for the book.

"Something More" Sections. Most chapters end with a section in which the material of the chapter is concluded with a special topic. These typically consist of an especially interesting extension or an unusual practical or chemical application of the material in the chapter.

Biological Examples. Throughout the book, reference is made to the connection between organic chemistry and the world of biology. Although many biological reactions are given, these are framed as examples and do not preempt discussions of simple, prototypal reactions. We all have to walk before we can run.

In-Chapter Summary Sections. There are many moments in any course on organic chemistry when it is important to take stock of where we are. What have we done, why did we do it, and where are we going? The introductions to each chapter do some of this, but there are many points in the book where we pause to take stock and summarize. Sometimes these are whole sections, sometimes just a sentence or two. There is even a whole chapter (Chapter 17) that is largely devoted to a summary of the chemistry of alcohols that has permeated the earlier chapters. These sections allow us to focus and keep the direction in which we are going straight in our heads.

Acknowledgments. When I was 13, desperate to spend a summer or two on the tennis circuit, my parents had the wit to see that I was no future Laver or McEnroe, and to insist, over my strenuous objections, that I get a job. That "job" as a volunteer bottle washer and gopher began my lifelong association with William Doering. I went on as an undergraduate to learn organic chemistry from Professor Doering in what was the most exciting (and demanding) course I ever took, and I remained at Yale to do my graduate work with him. I remain his student today, and there is no easy way to acknowledge adequately my debt to him. I can only hope that this book helps others to get off to a similar start, and that they will encounter someone of Doering's intellectual honesty and power along their way.

After serving my sentence as a bottle washer for Doering I began real laboratory work, and was lucky enough to fall under the influence of two masters, Lawrence H. Knox and Gunther Laber. Much of the pleasure of organic chemistry comes from the design and execution of a well-done experiment. From these two patient teachers I learned the great pleasure in carrying out the manipulations of organic chemistry: preparing the exquisite glassware, setting up the apparatus, and finally handling the wonderful liquids and powders that were to be combined to make something new.

I have been lucky enough to have studied with some wonderful teachers both before and after my days as a university student. Stanley Feret and Margaret Frankel taught me in elementary school, and were surely greatly responsible for setting me on my way. Arthur B. Darling, who taught me American History in high school, was particularly inspiring. One could ask for no better example of how to communicate a love for a subject than his. And thirty years ago I spent a postdoctoral year with Jerome Berson, a man who wrote complex papers on carbonium ions that I couldn't yet read sensibly. I learned to read those papers during my year with Jerry, but more important, I discovered an intellectual atmosphere that affected me forever.

Chemists have families, too. Some even have lives. If I have had one during the construction of this book, it is largely the result of the efforts of my family. Adequate thanks to all of them would be far to long to fit here, but their names should be mentioned, and they should know that I understand some of what they put up with. So, thanks to Stephanie and Matt for cheerful, if often crazed chatter by e-mail; to Mait and Perla for delightful dinners in Brooklyn and for introducing me to Tom Waits; and to Hilary for being so splendidly all-Galaxy. My parents really did extract me from the tennis court and get me into a lab for the first time. They had the intelligence to see that a kid who was devoted to picking up frogs might find science a passion. I'm sorry that my father never got to see this book, but he surely had a lot to do with it. I'm glad that my mother has seen it, and I hope she thinks it was worthwhile driving that 13-year-old to and from the lab all those years ago. My wife Susan bore much of the brunt of the writing of this book. Too many absences, too much preoccupation, all coped with. Love and great thanks to her.

Paul Valery said that a poem is never finished, only abandoned. I have found, somewhat to my dismay, that the same is true of textbooks on organic chemistry. Nonetheless, it is surely past time to abandon the writing of this one. But I must point out first that books don't get written by setting an author on his or her way and then waiting for the manuscript to appear. There is a great deal of work to be done; misguided authors need to be straightened out, bribes and threats need to be administered, and commas inserted and deleted. In general, it is an editor's job to make it possible for the author to do the best of which he is capable. Don Fusting and Joe Wisnovsky at W. W. Norton were exemplary in their execution of this role, and my thanks go to them. Jeannette Stiefel, my copy editor, and Rachel Warren at Norton straightened out my prose on many an occasion and were helpful in many other ways. Joel Dubin and the rest of the crew at J/B Woolsey Associates did a great job of producing the art, and coped with great skill and reasonable cheer with my many corrections.

This book also profited vastly from the comments and advice of an army of reviewers, and I am very much in their debt. Their names and affiliations follow: William F. Bailey, University of Connecticut; Ronald J. Baumgarten, University of Illinois at Chicago; John I. Brauman, Stanford University; Donald B. Denney, Rutgers University; John C. Gilbert, University of Texas at Austin; Henry L. Gingrich, Princeton University; Richard K. Hill, University of Georgia; A. William Johnson, University of Massachusetts; Guilford Jones, II, Boston University; Joseph B. Lambert, Northwestern University; Steven V. Ley, Imperial College of Science, Technology and Medicine; Ronald M. Magid, University of Tennessee-Knoxville; Eugene A. Mash, Jr., University of Arizona; Robert J. McMahon, University of Wisconsin-Madison; Andrew F. Montana, California State University-Fullerton; Roger K. Murray, Jr., University of Delaware; Thomas W. Nalli, State University of New York at Purchase; Patrick Perlmutter, Monash University; R. M. Paton, University of Edinburgh; Matthew S. Platz, Ohio State University; Lawrence M. Principe, Johns Hopkins University; Charles B. Rose, University of Nevada-Reno; John F. Sebastian, Miami University; Carl H. Schiesser, Deakin University; Martin A. Schwartz, Florida State University; Jonathan L. Sessler, University of Texas at Austin; Robert S. Sheridan, University of Nevada-Reno; Philip B. Shevlin, Auburn University; Harry H. Wasserman, Yale University; Craig Wilcox, University of Pittsburgh; David R. Williams, Indiana University.

At the risk of slighting some, and delaying unduly the abandonment of this work, I must take a moment to single out two special reviewers, Henry L. Gingrich of Princeton and Ronald M. Magid of the University of Tennessee. These gentlemen read the work line by line, word by word, comma by missing comma. Their comments, pungent at times, but helpful always, were all too accurate in uncovering both the gross errors and lurking oversimplifications in the early versions of this work. The book is much improved by their efforts.

Ron Magid, Peter Gaspar, and Bob Willcott remain steadfast friends from our days together at Yale, many years ago, and are still people with whom I can talk and learn about obscure chemical (and other) subjects. Many other colleagues over the years have also been especially helpful to me; Bob Moss at Rutgers and Matt Platz at Ohio State, as well as Jeff Schwartz and Bob Pascal at Princeton are prime examples.

Despite all the efforts of editors and reviewers, errors will persist. These are my fault only. When you find them, let me know.

Princeton, November, 1996

Organic Chemistry

Introduction

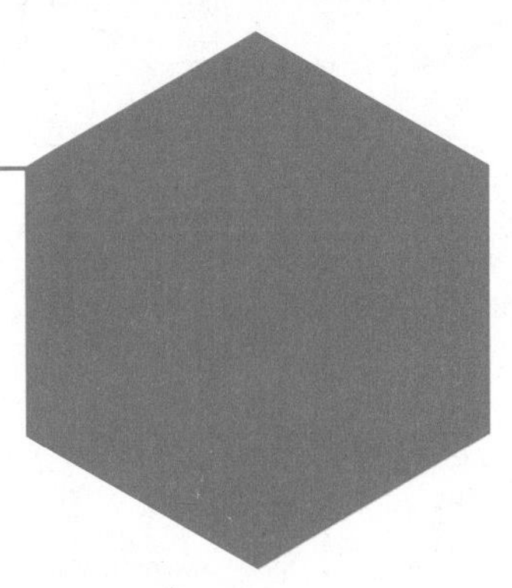

These days, a knowledge of science must be part of the intellectual equipment of any educated person. Of course, that may always have been true, but, I think there can be no arguing that an ability to confront the problems of concern to scientists is especially important today. Our world is increasingly technological, and many of our problems, and the answers to those problems, have a scientific or technological basis. Anyone who hopes to understand the world we live in, to evaluate many of the pressing questions of the present and the future—and vote sensibly on them—must be scientifically literate.

The study of chemistry is an ideal way to acquire at least part of that literacy. Chemistry is a central science in the sense that it bridges such disparate areas as physics and biology, and connects those long-established sciences to the emerging disciplines of molecular biology and materials science. Similarly, as this book shows, organic chemistry sits at the center of chemistry, where it acts as a kind of intellectual glue, providing connections between all areas of chemistry. One does not have to be a chemist, or even a scientist, to profit from the study of organic chemistry.

The power of organic chemistry comes from its ability to give insight into so many parts of our lives. How does penicillin work? Why is Teflon nonstick? Why does drinking a cup of coffee help me stay awake? How do plants defend themselves against herbivores? Why is ethyl alcohol a depressant? All these questions have answers based in organic chemistry. And the future will be filled with more organic chemistry—and more questions. What's a buckyball and how might it be important to my life? How might an organic superconductor be constructed? Why is something called the Michael reaction important in a potential cancer therapy? Read on, because this book will help you to deal with questions such as these, and many more we can't even think up yet.

What Is Organic Chemistry? Organic chemistry is traditionally described as the chemistry of carbon-containing compounds. In the last century, it was thought that organic molecules were related in an immutable way to living things, hence the term "organic." The idea that organic compounds could be made only from molecules derived from living things was widespread, and gave rise to the notion of a vital force being present in carbon-containing molecules. In 1828 Friedrich Wöhler (1800–1882) synthesized urea, a certified organic substance, from heating ammonium cyanate, a compound considered to be inorganic.* Wöhler's experiment really did not speak to the question of vital force, and he knew this. The problem was that at the time there were no sources of ammonium cyanate that did not involve such savory starting materials as horns and blood—surely "vital" materials. The real *coup de gras* for vitalism came some years later when Adolph Wilhelm Hermann Kolbe (1818–1884) synthesized acetic acid from elemental carbon and inorganic materials in 1843–1844 (see structures below).

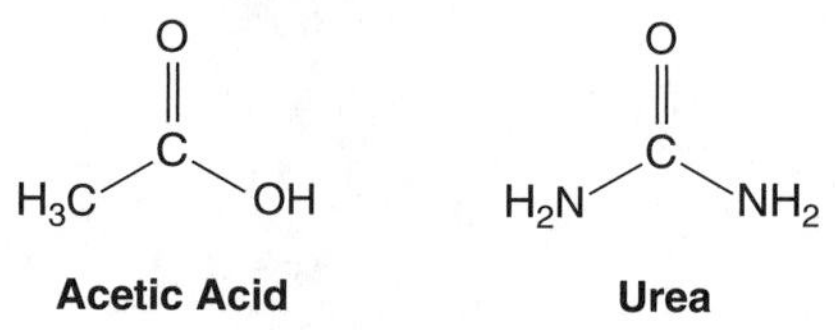

$H_3C-C(=O)-OH$ **Acetic Acid** $H_2N-C(=O)-NH_2$ **Urea**

Despite the demise of the vital-force idea, carbon-containing molecules certainly do have a strong connection to living things, including ourselves. Indeed, carbon provides the backbone for all the molecules that make up the soft tissues of our bodies. Our ability to function as living, sentient creatures depends on the properties of carbon-containing organic molecules, and we are about to embark on a study of their structures and transformations.

Organic chemistry has come far from the days when chemists were simply collectors of observations. In the beginning, chemistry was largely empirical, and the questions raised were, more or less, along the lines of, What's going to happen if I mix this stuff with that stuff?, or, I wonder how many different things I can isolate from the sap of this tree? Later, it became possible to collate knowledge and to begin to rationalize the large numbers of collected observations. Questions now could be expanded to deal with finding similarities in different reactions, and chemists began to have the ability to make predictions. Chemists began the transformation from the hunter–gatherer stage to modern times, in which we routinely seek to use what we know to generate new knowledge.

Many advances have been critical to that transformation; chief among them is our increased analytical ability. Nowadays the structure of a new compound, be it isolated from tree sap or produced in a laboratory, cannot remain a mystery for long. Today, the former work of years can often be accomplished in hours. This expertise has enabled chemists to peer more closely at the "why" questions, to think more deeply about reactivity of molecules. This point is important because the emergence of unifying principles has allowed us to teach organic chemistry in a different way; to teach in a fashion that largely frees students from the necessity to memo-

* Wöhler's urea is an end product of the metabolism of proteins in mammals, and is a major component of human urine. An adult human excretes about 25 g (6–8 level teaspoons) of urea each day. The formation of urea is our way of getting rid of the detritus of protein breakdown through a series of enzymatic reactions. If you are missing one of the enzymes necessary to produce urea, it's very bad news indeed, as coma and rapid death result.

rize organic chemistry. That is what this book tries to do: to teach concepts and tools, not vast compendia of facts. The aim of this book is to provide frameworks for generalizations, and the discussions of topics are all designed with this aim in mind.

We will see organic molecules of all types in this book. Organic compounds range in size from hydrogen (H_2)—a kind of honorary organic molecule even though it doesn't contain carbon—to the enormously complex collections of atoms in biomolecules, which typically contain thousands of atoms and have molecular weights in the hundreds of thousands. Despite this diversity, and the apparent differences between small and big molecules, the study of all molecular properties always begins the same way, with structure. Structure determines reactivity, which provides a vehicle for navigating from the reactions of one kind of molecule to another and back again. So, early on, this book deals extensively with structure.

What Do Organic Chemists Do? Structure determination has traditionally been one of the things that practicing organic chemists do with their lives. Early on, such activity took the form of uncovering the gross connectivity of the atoms in the molecule in question: What was attached to what? Exactly what are those molecules isolated from the Borneo tree or made in a reaction in the lab? Such questions are quickly answered by application of today's powerful spectroscopic techniques, or, in the case of solids, by X-ray diffraction crystallography. And small details of structure lead to enormous differences in properties: codeine, a pain-killing agent in wide current use, and heroin, a powerfully addictive narcotic, differ only by the presence of two acetyl groups (CH_3CO units), a tiny difference in their large and complex structures.

Today, much more subtle questions are being asked about molecular structure. How long can a bond between atoms be stretched before it goes "boing" in its quiet, molecular voice, and the atoms are no longer attached? How much can a bond be squeezed? How much can a bond be twisted? These are structural questions, and reveal much about the properties of atoms and molecules—in other words, about the constituents of us and the world around us.

Many chemists are more concerned with how reactions take place, with the study of "reaction mechanisms." Of course, these people depend on those who study structure; one can hardly think about how reactions occur if one doesn't know the detailed structures—connectivity of atoms, three-dimensional shape—of the molecules involved. In a sense, every chemist must be a structural chemist. The study of reaction mechanisms is an enormously broad subject. It includes people who look at the energy changes involved when two atoms form a molecule, or, conversely, when a molecule is forced to come apart to its constituent atoms, as well as those who study the reactions of the huge biomolecules of our bodies—proteins and polynucleotides. How much energy is required to make a certain reaction happen? Or, how much energy is given off when it happens? You are familiar with both kinds of process. For example, burning is clearly a process in which energy is given off as both heat and light.

Chemists also want to know the details of how molecules come together to make other molecules. Must they approach each other in a certain direction? Are there catalysts—molecules not changed by the reaction—that are necessary? There are many such questions. A full analysis of reaction mechanism requires a knowledge of the structures and energies of all molecules involved in the process, including species called intermediates, molecules of fleeting existence that cannot usually be isolated because they go on quickly to other species. One also must have an idea of the structure and energy of the highest energy point in a reaction, called the transition state. Such species cannot be isolated—they are energy maxima, not energy minima—but they can be studied nonetheless. We will see how.

Still other chemists focus on synthesis. The goal in such work is the construction of a target molecule from smaller, available molecules. In earlier times the reason for such work was sometimes structure determination. One set out to make a molecule one suspected of being the product of some reaction of interest. Now, determination of structure is not usually the goal. And it must be admitted that Nature is still a *much* better synthetic chemist than any human. There is simply no contest; evolution has generated systems exquisitely designed to make breathtakingly complicated molecules with spectacular efficiency. We cannot hope to compete. Why, then, even try? The reason is that there is a cost to the evolutionary development of synthesis, and that is specificity. Nature can make a certain molecule in an extraordinarily competent way, but Nature can't make changes on request. The much less efficient syntheses devised by humans are far more flexible than Nature's, and one reason for the chemist's interest in synthesis is the possibility of generating molecules of Nature in systematically modified forms. We hope to make small changes in the structures and to study the influence on biological properties induced by those changes. In that way it could be possible to find therapeutic agents of greatly increased efficiency, for example, or to stay ahead of microbes that become resistant to certain drugs. Nature can't quickly change the machinery for making an antibiotic molecule to which the microbes have become resistant, but humans can.

What's Happening Now? It's Not All Done. In every age, some people have felt that there is little left to be done. All the really great stuff is behind us, and all we can hope for is to mop up some details; we won't be able to break really new ground. And every age has been dead wrong in this notion. By contrast, the slope of scientific discovery continues to increase. We learn more every year, and not just details. Right now the frontiers of molecular biology—a kind of organic chemistry of giant molecules, I would claim—are the most visibly expanding areas, but there is much more going on.

In structure determination, completely new kinds of molecules are appearing. For example, just a few years ago a new form of carbon, the soccer-ball-shaped C_{60}, was synthesized in bulk by the simple method of vaporizing a carbon rod and collecting the products on a cold surface. Even more recently it has been possible to capture atoms of helium and argon inside the soccer ball. These are the first neutral compounds of helium ever made. Molecules connected as linked chains or as knotted structures are now known. What properties these new kinds of molecules will have no one knows. Some will certainly turn out to be mere curiosities, but others will influence your lives in new and unexpected ways.

The field of organic reaction mechanisms continues to expand as we become better able to look at detail. For example, events on a molecular time scale are becoming visible to us as our spectrometers become able to look at ever smaller time periods. Molecules that exist for what seems a spectacularly short time—microseconds or nanoseconds—are quite long lived if one can examine them on the femtosecond time scale. We are sure to learn much about the details of the early stages of chemical reactions in the next few years. At the moment, we are still defining the coarse picture of chemical reactions. Our resolution is increasing, and we will soon see micro details we cannot even imagine at the moment. It is a very exciting time. What can we do with such knowledge? We can't answer that question yet, but chemists are confident that with more detailed knowledge will come an ability to take finer control of the reactions of molecules. At the other end of the spectrum, we are learning how macromolecules react, how they coil and uncoil, arranging themselves in space so as to bring two reactive molecules to just the proper orientation for reaction. Here we are seeing the bigger picture of how much of Nature's architecture is designed to facilitate positioning and transportation of molecules to reactive positions. We are learning how to co-opt Nature's methods by modifying the molecular machinery so as to bring about new results.

We can't match Nature's ability to be specific and efficient. Over evolutionary time, Nature has just had too long to develop methods of doing exactly the right thing. But we are learning how to make changes in Nature's machinery—biomolecules—that lead to changes in the compounds synthesized. It is likely that we will be able to use Nature's methods, deliberately modified in specific ways, to retain the specificity but change the resulting products. This is one frontier of synthetic chemistry.

The social consequences of this work are surely enormous. We are soon going to be able to tinker in a controlled way with much of Nature's machinery. How does humankind control itself? How does it avoid doing bad things with this power? Those are not easy questions, but there is no hiding from them. We are soon going to be faced with the most difficult social questions of human history, and how we deal with them will determine the quality of the lives of us and our children. That's one big reason that education in science is so important today. It is not that we will need more scientists; rather it is that we must have a scientifically educated population in order to deal sensibly with the knowledge and powers that are to come. So, this book is not specifically aimed at the dedicated chemist-to-be. That person can use this book, but so can anyone who will need to have an appreciation of organic chemistry in his or her future—and that's nearly everyone these days.

HOW TO STUDY ORGANIC CHEMISTRY

Don't Memorize. In the old days, courses in organic chemistry rewarded people who could memorize. Indeed, the notorious dependence of medical school admission committees on the grade in organic chemistry may have stemmed from the need to memorize in medical school. If you could show that you could do it in organic, you could be relied upon to be able to memorize that the shin bone was connected to the foot bone, or whatever. Nowadays, memorization is the road to disaster; there is just too much material. Those who teach this subject have come to see an all too familiar pattern. There is a group of people who do very well early and then crash sometime around the middle of the first semester. These folks didn't suddenly become stupid or lazy; they were relying on memorization and simply ran out of memory. Success these days requires generalization, understanding of principles that unify seemingly disparate reactions or collections of data. Medical schools still regard the grade in organic as important, but it is no longer because they look for people who can memorize. Medicine, too, has outgrown the old days. Now medical schools seek people who have shown that they can understand a complex subject, people who can generalize.

Work with a Pencil. My old professor told me, "Organic chemistry must be read with a pencil." Truer words were never spoken. You can't read this book, or any chemistry book, in the way you can read books in other subjects. You *must* write things as you go along. There is a real connection between the hand and the brain in this business, it seems. When you come to the description of a reaction, especially where the text tells you that it is an important reaction, by all means take the time to draw out the steps yourself. It is not enough to read the text and look at the drawings; it is not sufficient to highlight. Neither of these procedures is reading with a pencil. Highlighting does not reinforce the way working out the steps of the synthesis or chemical reaction at hand does. You might even make a collection of file cards labeled "reaction descriptions" on which you force yourself to write out the steps of the reaction. Another set of file cards should be used to keep track of the various ways you will accumulate to make molecules. At first, these cards will be few in number, and sparsely filled, but as we reach the middle of the course there will be an explosion in the number of synthetic methods available. This subject can sneak up on you and keeping a catalog will help you to stay on top of this part of the subject. We will try to help you to work in this interactive way by interrupting the text with problems, with solutions that follow immediately when we think it is time to stop, take stock, and reinforce a point before going on. These problems are important. You can read right by them of course, or read the answer without stopping to do the problem, but to do so will be to cheat yourself and make it harder to learn the subject. Doing these starred, in-chapter problems is a part of reading with a pencil and should be very helpful in getting the material under control. There is no more important point to be made than this one.

Work in Groups. Many studies have shown that a most effective way to learn is to work in small groups. Form a group of your roommates or friends, and solve problems for each other. Assign each person one or two problems to be solved for the group. Afterward, work through the solution found in the chapter or Study Guide. You will find that the exercise of explaining the problem to others will be enormously useful. You will learn much more from "your" problems than from the problems solved by others.

Work the Problems. As noted above, becoming good at organic chemistry is an interactive process; you can't just read the material and hope to become expert. Expertise in organic chemistry requires experience, a commodity that by definition you are very low on at the start of your study. Doing the problems is vital to gaining the necessary experience. Resist the temptation to look at the answer before you have tried to do the problem. Disaster awaits you if you succumb to this temptation, for you cannot learn effectively that way, and there will be no answers available on the examinations until it is too late. That is not to say that you must be able to solve all the problems straight away. There are problems of all difficulty in each chapter, and some of them are very challenging indeed. Even though the problem is hard or very hard, give it a try. When you are truly stuck, that is the time to take a peek at the Study Guide. There you will find not just a bare bones answer, but, often, advice on how to do the problem as well. Giving hard problems is risky, because there is the potential for discouraging people. Please don't worry if some problems, especially hard ones, do not come easily or do not come at all. Each of us in this business has favorite problems that we still can't solve. Some of these form the basis of our research efforts, and may not yield, even to determined efforts, for years. Part of the pleasure in organic chemistry is working challenging problems, and it would not be fair to deprive you of such fun.

Use All the Resources Available to You. You are not alone. Moreover, everyone will have difficulty at one time or another. The important thing is to get help when you need it. Of course the details will differ at each college or university but there are very likely to be extensive systems set up to help you. Professors have office hours, there are probably teaching assistants with office hours, and there will likely be help, review, or question sessions at various times. Professors are there to help you, and will not be upset if you show enough interest to ask questions about a subject they love. "Dumb questions" do not exist! You are not expected to be an instant genius in this subject, and many students are too shy to ask perfectly reasonable questions.

If you feel uncertain about a concept or problem in the book—or lecture—get help soon! This subject is highly cumulative, and ignored difficulties will come back to haunt you. I know that many teachers tell you that it is impossible to skip material and survive, but this time it is true. What happens in December or April depends on September, and you can't wait and wait, only to "turn it on" at the end of the semester or year. Almost no one can cram organic chemistry. Careful, attentive, daily work is the route to success, and getting help with a difficult concept or a vexing problem is best done immediately.

Atoms, Atomic Orbitals, and Bonding

When it comes to atoms, language can be used only as in poetry. The poet too, is not nearly so concerned with describing facts as with creating images.
—Niels Bohr to Werner Heisenberg*

The picture of atoms that all scientists had in their collective mind's eye at the beginning of this century was little different from that of the ancient Greek philosopher Democritus (~460–370 B.C.), who envisioned small, indivisible particles as the constituents of matter. These particles were called **atoms**, from the Greek word for indivisible. The British chemist John Dalton (1766–1844) had the idea that atoms might have different characteristic weights, but he did not abandon the idea of a solid, uniform atom. That picture did not begin to change dramatically until 1897, when the English physicist J. J. Thomson (1856–1940) discovered the **electron**. Thomson postulated a sponge-like atom with the negatively charged electrons embedded within a positively charged material, rather like raisins in a pudding. Ernest Rutherford's (1871–1937) discovery, in 1909, that an atom was mostly empty space demolished the "pudding" picture and led to his celebrated "planetary" model, in which electrons were seen as orbiting a compact, positively charged nucleus.

It was Niels Bohr who made perhaps the most important modification of the planetary model. He made the brilliant and largely intuitive† suggestion that electrons were required to occupy only certain orbits. As an electron's energy depends on the distance of its orbit from the positively charged nucleus, Bohr's suggestion amounted to saying that electrons can have only certain energies. Thus, the energy states of the electron were (and still are) said to be quantized. Although this notion may seem strange, there are similar phenomena in the everyday world. One cannot create just any tone by blowing across the mouth of a bottle, for example. The tone you hear depends on the size and shape of the particular bottle. If one blows harder, the frequency does not change smoothly; instead a "quantum jump" to an overtone of the original sound occurs.

*Niels Bohr (1885–1962) and Werner Heisenberg (1901–1976) were pioneers in the development of quantum theory, the foundation of our current understanding of chemical bonding.

†Some people's intuitions are better able than others' to cope with the unknown!

In the 1920s and 1930s, a number of mathematical descriptions emerged from the need to understand Bohr's quantum model. It became clear that one must take a probabilistic view of the subatomic world. Werner Heisenberg discovered that it was not possible to determine simultaneously both the position in space and momentum (mass times speed) of an electron.* Thus, one can determine where an electron is at any given time only in terms of probability. One *can* say, for example, that there is a 90% probability of finding the electron in a certain volume of space, but one *cannot* say that at a given instant the electron is at a particular point in space.

The further elaboration of this picture of the atom has given us the conceptual basis for all modern chemistry: the idea of the orbital, the region of space occupied by an electron or a pair of electrons of a certain energy. Both the combining of atoms to form molecules and the diverse chemical reactions these molecules undergo involve, at a fundamental level, the interactions of electrons in orbitals. In atoms these are called **atomic orbitals** and in molecules, they are **molecular orbitals**. Various graphic conventions are used in this book to represent atoms and molecules—letters for atoms, dots for electrons not involved in bonding, and lines for electrons in bonds—but it is important to keep in mind from the outset that the model that most closely approximates our current understanding of reality at the atomic and molecular level is the cloudy, indeterminate—one might even say poetic—image of the orbital.†

1.1 ATOMS AND ATOMIC ORBITALS

In a neutral atom, the **nucleus**, or core of positively charged protons and neutral neutrons, is surrounded by a number of negatively charged electrons equal to the number of protons. If the number of electrons and protons is not equal, the atom or molecule containing the atom must be charged and is called an **ion**. A negatively charged atom or molecule is called an **anion**; a positively charged species is a **cation**. (One of the tests of whether or not you know how to "talk organic chemistry" is the pronunciation of the word cation. It is kat-eye-on, not kay-shun.)

The energy required to remove an electron from an atom to form a cation is called the **ionization potential**. In general, the farther away an electron is from the nucleus, the easier it is to remove, and the lower the ionization potential. Sodium, with a single electron in a relatively high-energy orbital, has an ionization potential of 5.1 electron volts (eV), or 118 kilocalories per mole (kcal/mol). This potential is far lower than that of

*This idea is extraordinarily profound—and troubling. The **Heisenberg uncertainty principle** (which states that the product of the uncertainty in position times the uncertainty in momentum is a constant), seems to limit fundamentally our access to knowledge. For an exquisite exposition of the human consequences of the uncertainty principle see Jacob Bronowski, *The Ascent of Man*, Chapter 11 (Little Brown, New York, 1973).

†In the wonderful quote that opens this chapter, Niels Bohr points out that once we transcend the visible world, all that is possible is modeling or image making. To me, what is even more marvelous about the quote is the simple word "too": It was obvious to Bohr that scientists speak in images, and he was pointing out to Heisenberg that there was another group of people out there in the world who did the same thing—poets.

hydrogen (13.6 eV = 314 kcal/mol),* which has only a single electron in a very low energy orbital. Ionization potential also depends on the effective nuclear charge. For example, the ionization potential of helium is nearly twice that of hydrogen, even though the electrons in both atoms are in the lowest **energy level**, the 1*s* orbital. It is much more difficult to remove one of helium's two electrons than the single electron of hydrogen. Helium's two electrons are attracted more strongly by the doubly positive nucleus than the single electron of hydrogen is by its singly positive nucleus. Figure 1.1 illustrates these points.

		Energy Required Ionization Potential
$_{11}Na$ ⟶	$_{11}Na^+$ + 1 Electron	5.11 eV = 118 kcal
$_1H$ ⟶	$_1H^+$ + 1 Electron	13.6 eV = 314 kcal
$_2He$ ⟶	$_2He^+$ + 1 Electron	24.6 eV = 567 kcal

FIGURE 1.1 The removal of an electron from a neutral atom to form a positive ion requires the application of energy—the ionization potential.

Convention Alert!

Single atoms (in this case, carbon) are often written ^W_ZC, where *W* is the atomic weight and *Z* is the atomic number, the number of protons in the nucleus, which equals the nuclear charge. Here, the superscript is omitted.

The high ionization potential of He has a further explanation: The electrons surrounding the He nucleus completely fill the first energy level. Likewise, the chemical inertness of the other noble gases, which also have high ionization potentials, is due to the stability of their filled outer subshells.

Much of the chemistry of atoms is dominated by gaining or losing electrons in order to achieve the electronic configuration of one of the noble gases (He, Ne, Ar, Kr, Xe, or Ra). The noble gases have especially stable filled shells of electrons: 2 for He, 10 for Ne (2 + 8), 18 for Ar (2 + 8 + 8), and so on. The idea that filling certain shells creates especially stable configurations is known as the **octet rule**. The first shell, called 1*s*, is an exception, as it can hold only two electrons. The second shell can hold eight in two subshells called 2*s* and 2*p*, which can hold two and six electrons, respectively. We will justify these numbers and names shortly, but we need the labels first. Figure 1.2 shows highly schematic pictures of the noble gases He, Ne, and Ar, and uses this terminology.

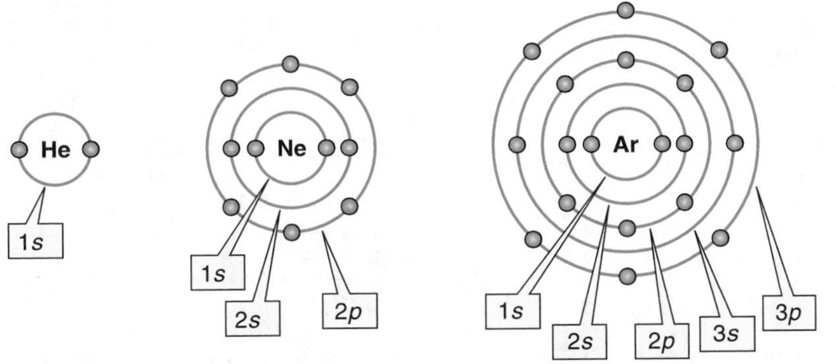

FIGURE 1.2 Highly schematic representations of He, Ne, and Ar. These pictures do not give good three-dimensional representations of these species, but they do show orbital occupancy and electron counting. Better pictures are forthcoming.

*There are several units of energy in use. Organic chemists commonly use kilocalories per mole (kcal/mol); physicists use the electron volt (eV). One electron volt is about 23 kcal. Recently, the International Committee on Weights and Measures has suggested that still another unit, the kilojoule (kJ), be substituted for kilocalorie. So far, organic chemists in some countries, including the United States, seem to have resisted this suggestion (one kilocalorie is equal to 4.187 kJ).

Electrons can be added to atoms as well as removed. The energy that is released by adding an electron to an atom to form a negatively charged anion is called the atom's **electron affinity**, measured in electron volts. The noble gases have very low electron affinities. Conversely, atoms to which adding an electron would complete a noble gas configuration have high electron affinities. The classic example is fluorine: the addition of a single electron yields a fluoride ion, F^-, which has the electronic configuration of Ne. Both F^- and Ne are 10-electron species (Fig. 1.3).

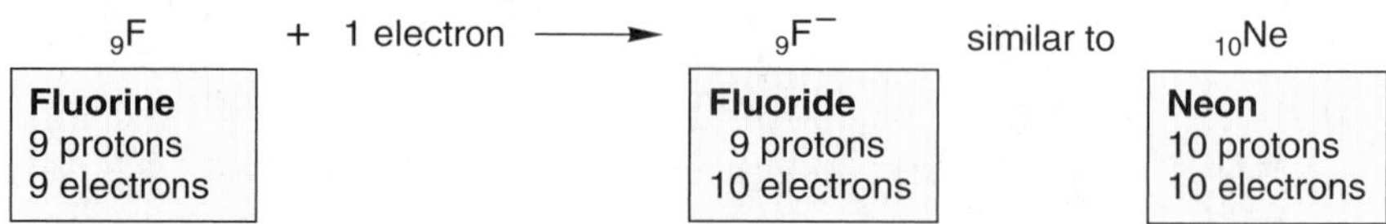

FIGURE **1.3** The ease of adding an electron to an atom to form a negatively charged ion is called the atom's electron affinity. Here addition of an electron to a fluorine atom yields the fluoride ion, which has the same electron configuration as neon.

Table 1.1 shows ionization potentials and electron affinities of some elements arranged as in the periodic table. Notice that atoms with low ionization potentials, whose electrons are easy to remove to give a positive ion, cluster on the left side of the periodic table; atoms with high electron affinities, which accept electrons easily, are on the right side (excluding the noble gases).

TABLE **1.1** Some Ionization Potentials and Electron Affinities

Ionization Potentials and Electron Affinities

Ionization Potentials (eV)

H 13.60							He 24.59
Li 5.39	Be 9.32	B 8.30	C 11.26	N 14.53	O 13.62	F 17.42	Ne 21.56
Na 5.14	Mg 7.65	Al 5.99	Si 8.15	P 10.49	S 10.36	Cl 12.97	Ar 15.76

Electron Affinities (eV)

H 0.75							He ~0
Li 0.62	Be ~0	B 0.24	C 1.27	N ~0	O 1.47	F 3.34	Ne ~0
Na 0.55	Mg ~0	Al 0.46	Si 1.24	P 0.77	S 2.08	Cl 3.61	Ar ~0

Atoms having low ionization potentials often form **ionic bonds** with atoms having high electron affinities by transferring an electron. In sodium fluoride (Na^+F^-), both Na^+ and F^- have filled second shells, and have achieved the stable electronic configuration of the noble gas Ne. In potassium chloride (K^+Cl^-), both ions have the electronic configuration of Ar, another noble gas (Fig. 1.4).

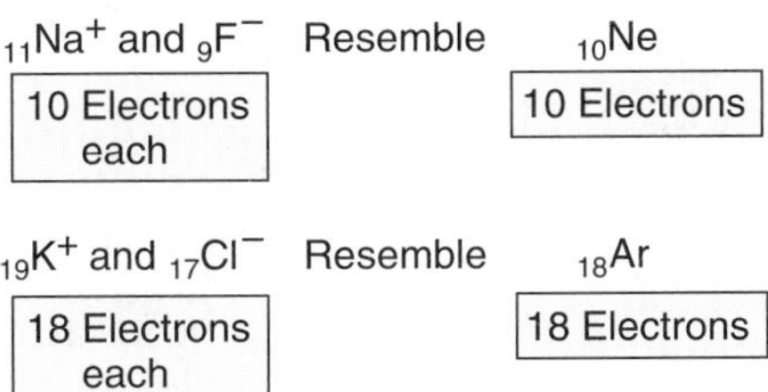

FIGURE 1.4 In the ionic compounds NaF and KCl, each atom can achieve a noble gas electronic configuration with a filled octet of electrons.

Compounds formed by ionic bonding are held together by the electrostatic attraction of opposite charges and are traditionally the province of inorganic chemistry. Nearly all of the compounds of organic chemistry are held together not by the electrostatic attractions present in ionic compounds but by the sharing of electrons in covalent bonds.

1.2 COVALENT BONDS AND LEWIS STRUCTURES

The formation of molecules through covalent bonding will be the subject of much of the remainder of this chapter as well as Chapters 2 and 3. The bond formed by sharing a pair of electrons is called a **covalent bond**. A covalent bond is shown as a pair of dots between the two atoms or as a line joining the two. These two-electron bonds will be prominent in almost all of the molecules shown in this book. For many simple molecules, even the line between the atoms is left out, and the compound is written simply as A_2 or AB. You are left to supply mentally the missing two electrons (Fig. 1.5).

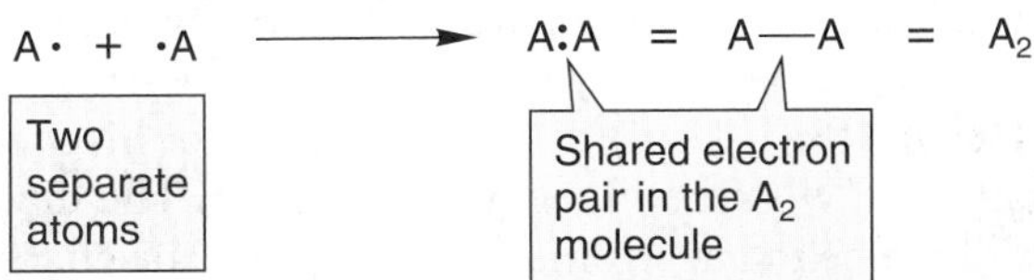

FIGURE 1.5 Bonding through the sharing of electrons is not ionic, but covalent. Covalent bonding can also result in stable electronic configurations.

The idea of covalent bonding was largely the creation of the American chemist Gilbert Newton Lewis (1875–1946) and the molecules formed by covalent bonds are usually written as what are called **Lewis structures**. In almost all covalently bonded molecules, every atom can achieve a noble gas electronic configuration by sharing electrons. In the H_2 molecule, each hydrogen atom shares two electrons and thus resembles helium. In F_2, each fluorine has a pair of 1*s* electrons in the lowest energy shell and is surrounded by eight other electrons, six unshared (**nonbonding** or **lone-pair** electrons) and two shared in the fluorine–fluorine bond. Thus, each fluorine in F_2 has a filled octet and, by virtue of sharing electrons, resembles neon. With the exception of H, electrons in 1*s* orbitals are not involved in bonding, and are not shown in Lewis structures. This omission can make the construction and interpretation of Lewis structures complicated. Always remember to account for the unshown 1*s* electrons. In HF, the hydrogen is helium-like and the fluorine is neon-like (Fig. 1.6).

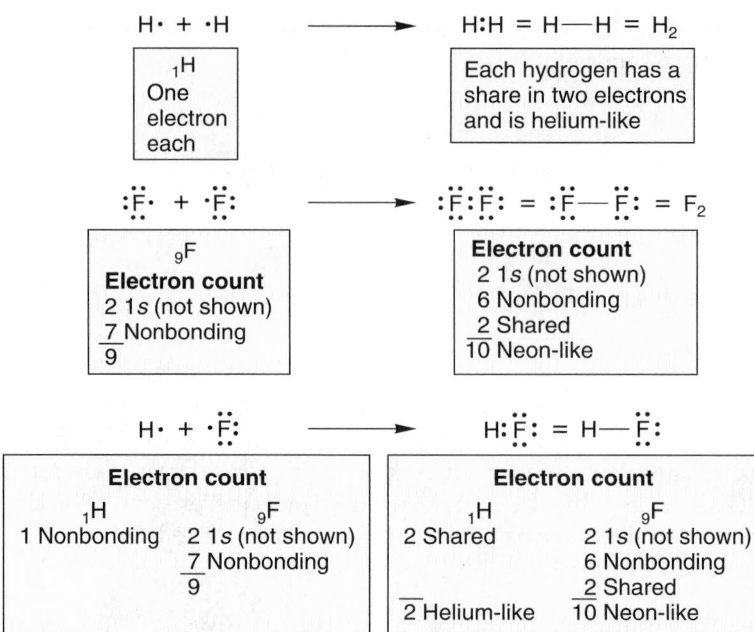

FIGURE 1.6 Lewis dot structures for H_2, F_2, and HF. Every atom has a noble gas electronic configuration.

Hydrogen fluoride is different from any of the molecules mentioned so far in that the two electrons in the bond are shared unequally between the two atoms. In any diatomic molecule made from two different atoms there cannot be equal sharing of the electrons, because one atom must attract the electrons more strongly than the other. The extreme case for this unequal electron sharing is the ionic bonding seen at the end of Section 1.1 in Na^+F^- or K^+Cl^-. Molecules in which electrons are unequally shared contain **polar covalent bonds**. Often a small arrow pointing from positive to negative is used to indicate the direction of the dipole, or δ^+ (partial positive charge) and δ^- (partial negative charge) signs are placed on the appropriate atoms. We speak of such molecules as having **dipole moments**. The dipole moment is measured in debye units after the Dutch physicist and chemist Pieter Debye (1884–1966), who spent much of his working life at Cornell University and won the Nobel prize in 1936. The dipole moment is equal to r, the distance in angstroms (Å) separating the charges, times q, the charge in electrostatic units (esu) (Fig. 1.7).

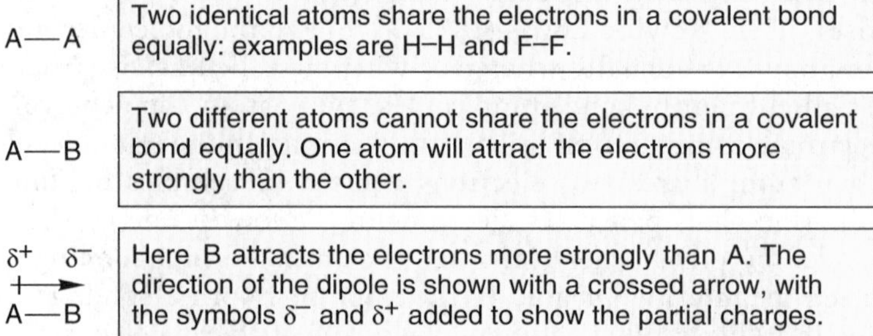

FIGURE 1.7 A polar covalent bond between atoms A and B. There is a dipole moment in this molecule.

The tendency of an atom to attract electrons is called **electronegativity**. Atoms with high electron affinities are the most electronegative and are found at the right of the periodic table. Atoms with low ionization potentials are called electropositive and lie at the left of the table. Table 1.2 gives some electronegativity values.

TABLE 1.2 Some Electronegativities

Electronegativities

1 (IA)	2 (IIA)	3 (IIIA)	4 (IVA)	5 (VA)	6 (VIA)	7 (VIIA)
H 2.3						
Li 0.9	Be 1.6	B 2.1	C 2.5	N 3.1	O 3.6	F 4.2
Na 0.9	Mg 1.3	AL 1.6	Sc 1.9	P 2.3	S 2.6	Cl 2.9
K 0.7						Br 2.7
Rb 0.7						I 2.4

PROBLEM **1.1**

Indicate the direction of the dipole in the indicated bonds of the following molecules:

H—Cl, H—F, Li—F, H_3C—Cl, HO—Br, Br—Cl

Not all molecules containing polar bonds have dipole moments. Many do, including all those shown in Problem 1.1, but consider carbon dioxide (CO_2), a linear molecule of the structure O=C=O (the double lines represent double bonds in which the carbon atom shares two pairs of electrons with each oxygen atom). Oxygen is more electronegative than carbon (see Table 1.2), so there will be two polar bonds as shown in Figure 1.8. The dipole moment of a molecule is the vector sum of all dipoles within the molecule. In O=C=O the two dipoles are equal in magnitude and pointed in exactly opposite directions; they cancel. There are two polar carbon–oxygen bonds in carbon dioxide, but no dipole moment.

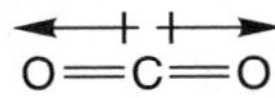

FIGURE **1.8** Even though carbon dioxide contains two polar bonds, there is no net dipole moment in the molecule. The bond dipoles cancel.

In representing three-dimensional structures on a two-dimensional surface such as a blackboard or book page, it is necessary to devise some scheme for showing bonds directed away from the surface, either toward the front (up) or toward the rear (down). In the following figures, the solid wedges (▬) are coming out toward you and the dashed wedges (┅) are retreating into the page. If visualizing these molecules is hard for you at first, by all means make models.

PROBLEM **1.2**

Which of the two molecules in Figure 1.9 has a dipole moment and which does not? Look carefully at the shapes of the two tetrahedral molecules.

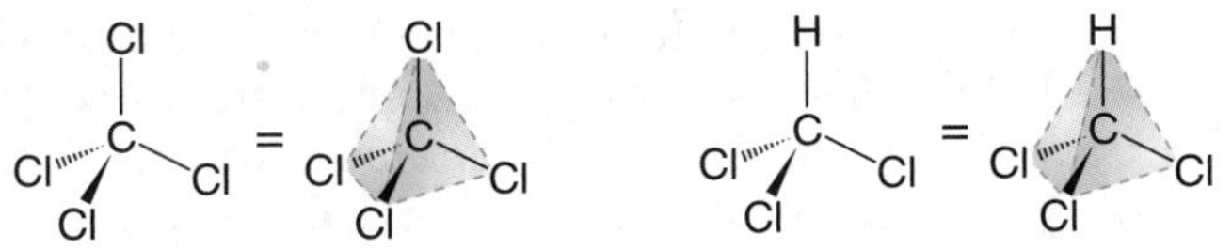

Carbon tetrachloride (CCl_4)

Chloroform ($CHCl_3$)

FIGURE **1.9** The dashed lines show the outlines of the tetrahedra.

Carbon Dioxide

$\ddot{O}=C=\ddot{O}$

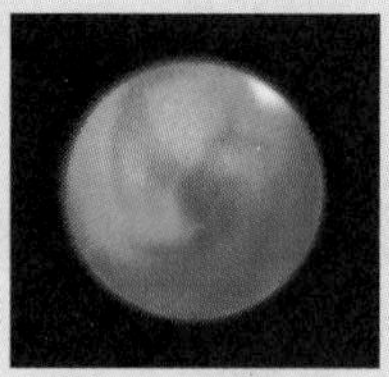

Mars

Carbon dioxide (CO_2) is a colorless, odorless gas making up only 0.03–0.04% of our atmosphere. Carbon dioxide is far more prevalent on Mars and Venus, but apparently was scrubbed from Earth's atmosphere when the oceans formed. Oceans never formed on Venus, as it's far too hot there, and apparently, there was not enough water on Mars to do the job. Carbon dioxide doesn't have a liquid form but sublimes directly from the solid, known as dry ice, to the gas.

Carbon dioxide is a major player in the carbon cycle on earth. The fate of much organic carbon is to be transformed into CO_2 either through burning, as with oil and gas, or through metabolism of dead organic matter. Green plants reverse the process through photosynthesis, in which they use the energy in light to combine CO_2 with H_2O to make sugars. We and other creatures ingest the sugars, and eventually return the carbon to the atmosphere as CO_2. It now seems certain that the amount of CO_2 in the atmosphere is increasing, largely as the result of human activity. The amount has increased by about 15% in the last 100 years, and there has been a special burst in the last decade or so. This increase may have important consequences. As CO_2 builds up in the atmosphere it acts as a "greenhouse gas" (methane, CH_4, is another), trapping infrared (IR) radiation (heat) from the surface of the earth and contributing to global warming. The issue is surely not totally understood. Still, there is a clear correlation between the amount of CO_2 in the atmosphere and temperature. What is not certain is the nature of this correlation. For example, it seems that the rise in CO_2 sometimes follows, rather than anticipates, temperature change. It is an enormously complicated issue and much more research in chemistry (and geology, oceanography, and atmospheric sciences in general) is needed before we will understand the relationship. Still, if the surface temperature increases only a few degrees, there will be substantial melting of the polar ice caps, and it's going to get very wet indeed in a lot of currently coastal areas. It is estimated, for example, that the New Jersey coast will be quite near Princeton if a substantial portion of the ice cap melts.

In order to write Lewis dot structures for molecules, we need to know the number of electrons in an atom. For atoms in the second row of the periodic table, one can tell immediately how many electrons are available by knowing the atomic number of the atom. The number of available electrons is equal to the total number of electrons in the atom, less the two $1s$ electrons. The electrons in the outermost subshell, the least tightly held electrons, are called the **valence electrons**. Electronic configurations are shown in shorthand by writing the name of the orbital and appending the number of electrons as a superscript. For example, the electronic configuration of helium ($_2$He) would be written $1s^2$ and that of beryllium ($_4$Be) as $1s^22s^2$.

For atoms inthe third row, neither the $1s$ pair nor the eight electrons in the second shell ($2s^22p^6$) are shown in the Lewis structure. Instead, only the valence electrons in the third shell are indicated. In complete Lewis structures, every valence electron is written as a dot. In a slightly more abstract form of a Lewis structure, pairs of electrons in bonds are shown as lines and only the nonbonding valence electrons remaining on atoms appear as dots. Some examples of Lewis structures are shown in Figure 1.10.

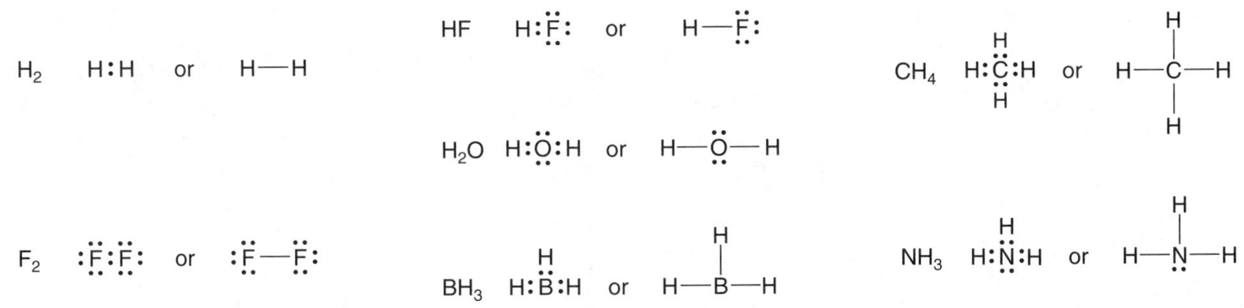

H_2 H:H or H—H

HF H:F: or H—F:

H_2O H:O:H or H—O—H

F_2 :F:F: or :F—F:

BH_3 H:B:H or H—B—H

CH_4 H:C:H or H—C—H

NH_3 H:N:H or H—N—H

FIGURE 1.10 A few Lewis dot structures.

Let's draw Lewis structures for some of these more complicated molecules. To construct the Lewis dot structure of methane (CH_4), first determine that carbon ($_6C$) can contribute four electrons to bonding (six less the two inner-shell 1*s* electrons) and that each hydrogen contributes an electron. Thus four two-electron covalent bonds are formed, with no electrons left over. For ammonia ($:NH_3$), nitrogen ($_7N$) contributes five electrons (seven minus the 1*s* pair), and the three hydrogens contribute one electron each. Therefore three nitrogen–hydrogen bonds can be formed with two electrons, with a nonbonding pair left over (Fig. 1.11).

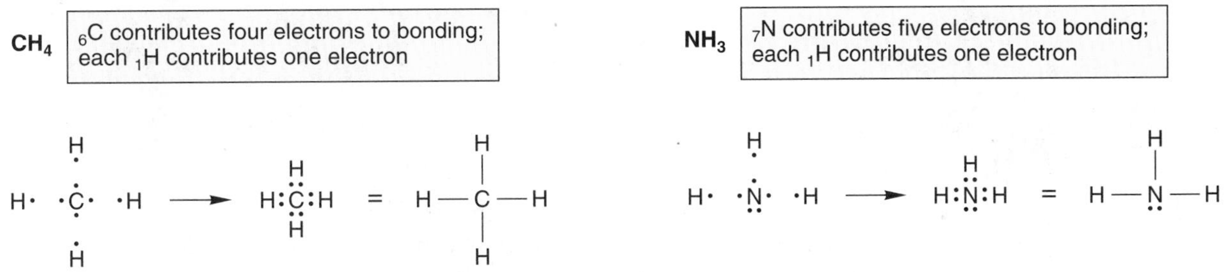

CH_4 $_6C$ contributes four electrons to bonding; each $_1H$ contributes one electron

H· ·C· ·H ⟶ H:C:H = H—C—H

NH_3 $_7N$ contributes five electrons to bonding; each $_1H$ contributes one electron

H· ·N· ·H ⟶ H:N:H = H—N—H

FIGURE 1.11 Construction of a Lewis structure for methane, CH_4, and ammonia, $:NH_3$.

PROBLEM 1.3

Construct Lewis structures for the following neutral molecules:

*(a) BF_3 (b) H_2Be (c) BH_3 *(d) $ClCH_3$ (e) $HOCH_3$ (f) $H_2N–NH_2$

ANSWER

(a) BF_3 The first task is to determine the gross structure of BF_3. Is it B–F–F–F or some other structure with F–F bonds? Start by working out the number of electrons available for bonding in F and B. Fluorine ($_9F$) has seven electrons available for bonding (9 – 2 1*s*) and thus has a single "odd" electron. Boron ($_5B$) has three electrons (5 – 2 1*s*).

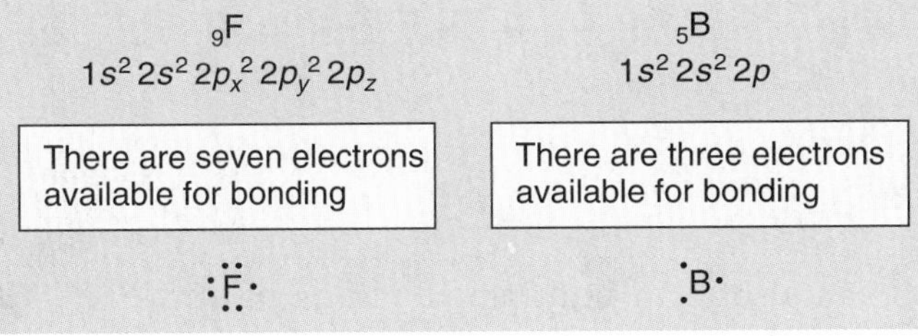

$_9F$

$1s^2 2s^2 2p_x^2 2p_y^2 2p_z$

There are seven electrons available for bonding

:F·

$_5B$

$1s^2 2s^2 2p$

There are three electrons available for bonding

·B·

*Problems marked with an asterisk are answered in whole or part in the text. Detailed answers to all the other in-chapter and additional end-of-chapter problems, many with comments, appear in the *Study Guide*.

ANSWER (CONTINUED)

There is no easy way to form a molecule with both boron–fluorine and fluorine–fluorine bonds. Moreover, B has three electrons available for bonding, and a structure with three boron–fluorine bonds can be nicely accommodated through covalent bonds in which the "odd" electron on each fluorine is shared with one of boron's available three electrons:

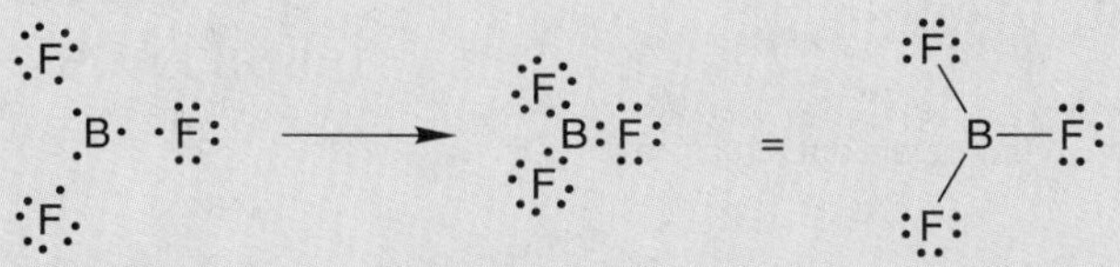

(d) $ClCH_3$ Chlorine ($_{17}Cl$) is in the third row of the periodic table, so we ignore the 10 1*s*, 2*s*, and 2*p* electrons, which leaves 7 electrons for bonding. Like fluorine, chlorine has a single odd electron that can form a covalent bond with one of the four bonding electrons of carbon. Carbon forms three other covalent bonds through sharing electrons with the three hydrogen atoms. This leads to the following Lewis structure:

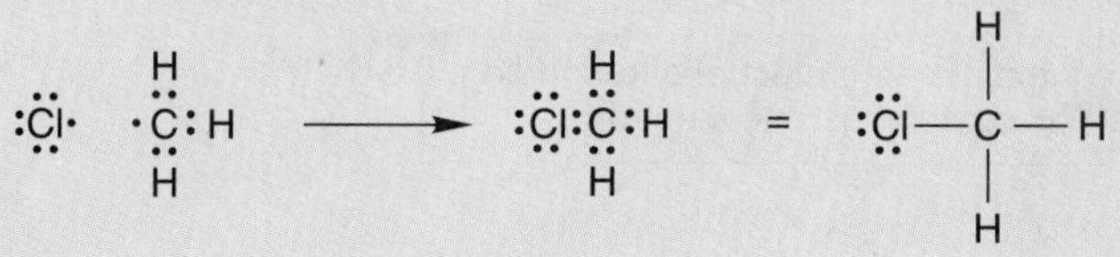

Atoms of elements below the first row of the periodic table can participate in multiple bonding—there can be more than one bond to most other atoms. Usually, the structural formula will give you a clue when this is necessary. Ethane ($H_3C{-}CH_3$) requires no multiple bonds; indeed it permits none, as can be seen from construction of the Lewis structure. Each carbon is attached to four other atoms—all four of the electrons available for bonding to carbon are shared in single bonds to the three hydrogen atoms and the lone other carbon. Therefore, all electrons can be accounted for in single bonds between the atoms (Fig. 1.12).

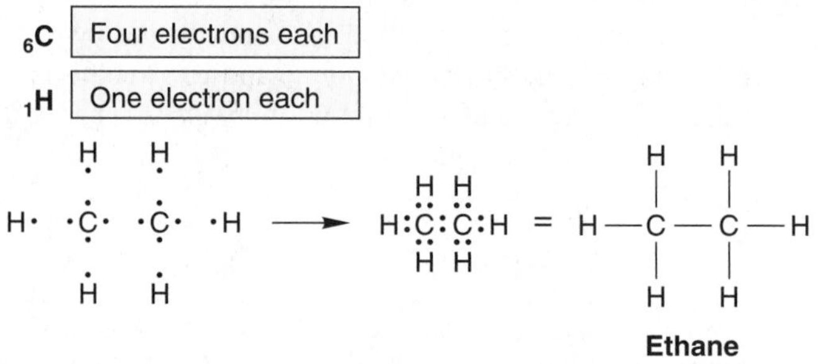

$_6$**C** Four electrons each

$_1$**H** One electron each

Ethane

FIGURE **1.12** Construction of a Lewis structure for ethane, $H_3C{-}CH_3$.

Carbon can also form multiple bonds, as in ethylene (H_2CCH_2) and acetylene (HCCH). In H_2CCH_2, only three of carbon's four bonding electrons are used up in forming single covalent bonds to one pair of hydrogens and the other carbon. There is an electron remaining on each carbon, and these two electrons are shared in a second carbon–carbon bond. Thus, ethylene contains a carbon–carbon double bond (Fig. 1.13).

$_6$C Four electrons each
$_1$H One electron each
A double bond
Ethylene

FIGURE **1.13** Construction of a Lewis structure for ethylene, $H_2C{=}CH_2$.

In acetylene (HCCH), for each carbon, two of the four bonding electrons are used in forming single bonds to one hydrogen and the other carbon. Two electrons are left over on each carbon, and these are shared to create a triple bond between the carbons (Fig. 1.14).

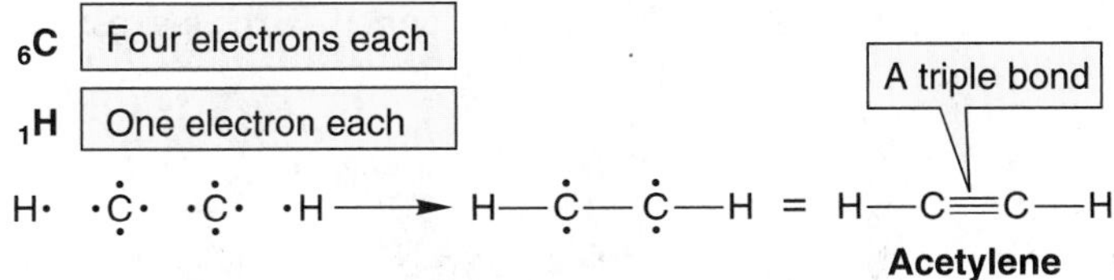

FIGURE **1.14** Construction of a Lewis structure for acetylene, HC≡CH.

PROBLEM **1.4**

Draw Lewis structures for the following neutral compounds. Use lines to indicate electrons in bonds and dots to indicate nonbonding electrons.

*(a) CH_3	(c) Br	(e) NH_2
*(b) CH_2	(d) HO	(f) $H_3C{-}N$

ANSWER

In this problem, you must again determine the number of bonding electrons on each atom. Then make the possible bonds and see how many electrons are left over. As each compound is neutral, the number of bonding electrons will be equal to the atomic number less the 1*s* electrons. For hydrogen, there will always be only the single 1*s* electron.

(a) CH_3 Carbon has four bonding electrons and forms normal covalent bonds with three hydrogens, each of which contributes a single electron. There is an odd electron left over on carbon, which is shown as the usual Lewis dot.

(b) CH_2 Now carbon can form only two covalent bonds because only two hydrogen atoms are available. Two nonbonding electrons are left over.

PROBLEM **1.5** In each of the following compounds, there is at least one multiple bond. Draw a Lewis structure for each molecule. Use lines to indicate electrons in bonds and dots to indicate nonbonding electrons.

(a) F_2CCF_2	(d) H_2CCO	(g) H_3COCOH (with O above the second C)
*(b) H_3CCN	(e) $H_2CCHCHCH_2$	
(c) H_2CO	*(f) H_3CNO	

ANSWER (b) H_3CCN The methyl (CH_3) group should be familiar by now. Its three covalent bonds are constructed by sharing three of carbon's four available electrons with the single electrons of the three hydrogens.

```
  H            H
H:C·    =      |
  H         H—C·
               |
               H
```

The other atoms, the carbon and the nitrogen, bring four and five electrons, respectively ($_6$C, 6 − 2 1*s* electrons = 4; $_7$N, 7 − 2 1*s* electrons = 5). Carbon–carbon and carbon–nitrogen covalent bonds can be formed, which leaves two unshared electrons on carbon and four on nitrogen.

```
   H                      H
   |                      |
H—C· ·Ċ· ·N̈:     =     H—C—Ċ—N̈:
   |   ·   ·              |  ·  ·
   H                      H
```

Formation of a triple bond between carbon and nitrogen completes the picture, and nitrogen is left with a nonbonding pair of electrons.

```
   H                      H
   |                      |
H—C—Ċ—N̈:       =     H—C—C≡N:
   |  ·  ·                |
   H                      H
```

(f) H_3CNO Once again there is a methyl group. Nitrogen has five available electrons, and oxygen six ($_8$O, 8 − 2 1*s* electrons = 6). A carbon–nitrogen single bond and a nitrogen–oxygen double bond can be formed. Nitrogen is left with one pair of nonbonding electrons, and oxygen with two pairs.

```
   H                    H                   H
   |                    |                   |
H—C· ·N̈· ·Ö:    =    H—C—N̈—Ö:     =     H—C—N̈=Ö:
   |     ·  ·           |     ·            |
   H                    H                   H
```

As we examine the nature of the bonding involved in the various structural types of organic chemistry, it is vital to be able to draw Lewis structures quickly and easily. It is also extremely important to be able to determine the charge on atoms in different structures. For single atoms, this is easy as long as you remember to count the two 1*s* electrons, which are never shown except for H and He. The total number of electrons is compared with the nuclear charge, the atomic number. Figure 1.15 gives a few examples.

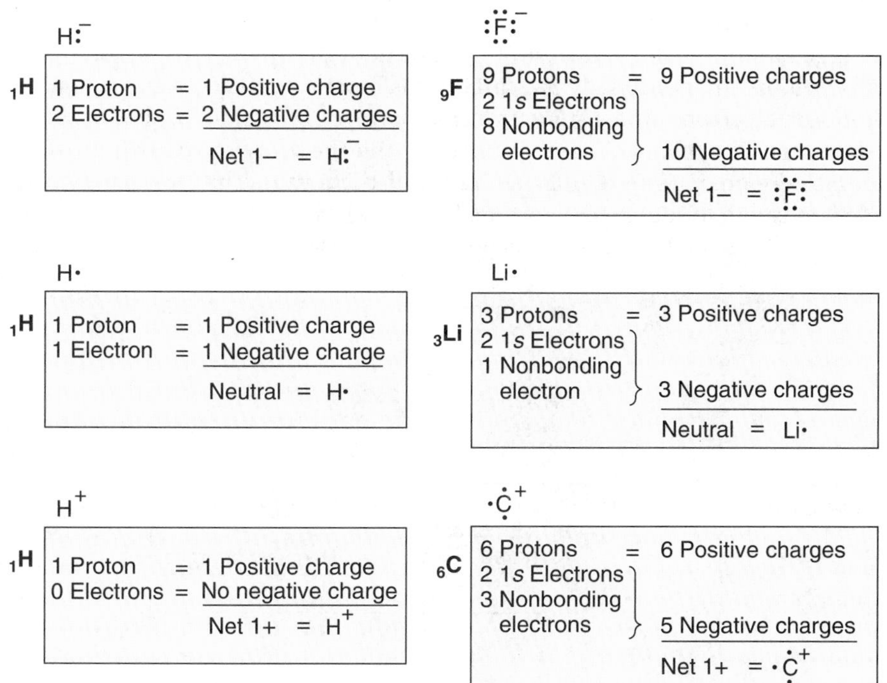

FIGURE 1.15 Lewis structures and charge determinations for some neutral atoms and charged species (ions).

In molecules, the determination of charge is a bit harder, because you must take care to account for shared electrons in the right way. As usual, remember that every atom except hydrogen and helium has a pair of unshown nonvalence 1*s* electrons. Count all nonbonding valence electrons as "owned" by an atom, as well as one electron from each two-electron bond to it. In H_2, for example, each hydrogen has only a share in the pair of electrons binding the two nuclei, and therefore each hydrogen is neutral. In F_2 each fluorine has two 1*s* electrons, which are not shown, six nonbonding (lone pair) electrons, and a share in the single covalent bond between the two fluorine atoms. This count gives a total of nine electrons, exactly balancing the nine positive nuclear charges. In $H_2C{=}CH_2$, each carbon has a pair of 1*s* electrons and a share in four covalent bonds for a total of six. As each carbon has a nuclear charge of +6, both are neutral. Figure 1.16 shows these examples.

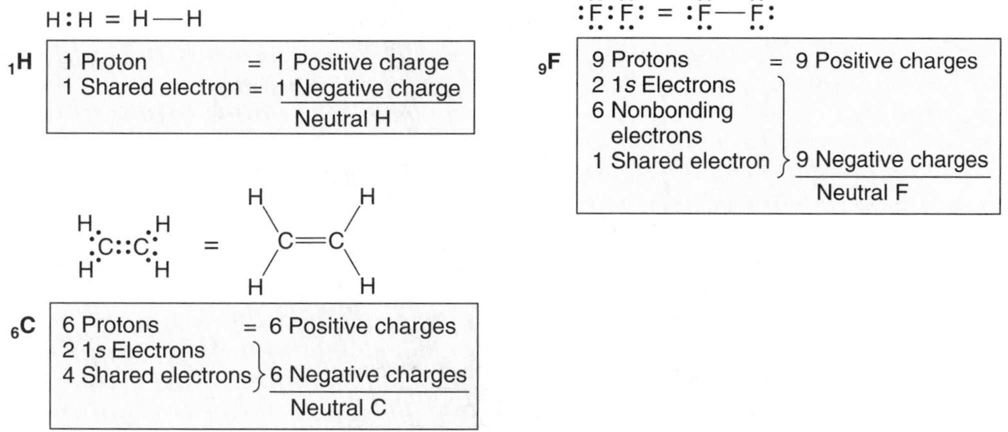

FIGURE 1.16 Electron counting in some simple molecules.

Let's verify that the carbon of the methyl anion ($^{-}$:CH_3) is negatively charged and that the nitrogen of the ammonium ion ($^{+}NH_4$) is positively charged. In the methyl anion, carbon has a pair of 1*s* electrons, two nonbonding electrons, and a share in three covalent bonds, for a total of seven electrons. The nuclear charge is only +6, and therefore the carbon must be negatively charged. The nitrogen atom of the ammonium ion has two 1*s* electrons and four shared electrons for a total of six. The nuclear charge is +7, and therefore the nitrogen is positively charged (Fig. 1.17).

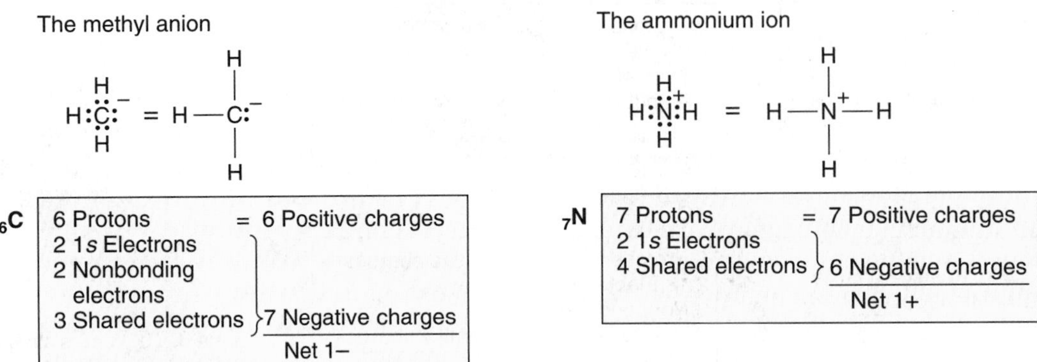

FIGURE 1.17 Examples of the calculation of the charge on an atom in a molecule.

Figure 1.18 shows some more ions and works through the calculations of charge.

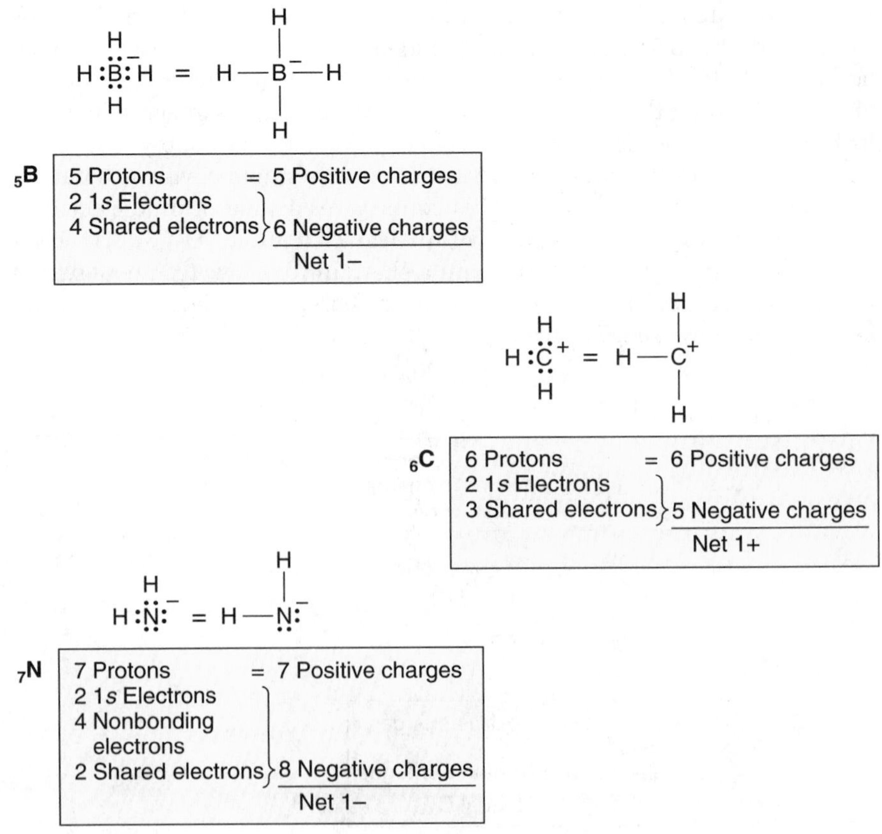

FIGURE 1.18 More calculations of charge in molecules.

PROBLEM 1.6

Draw Lewis structures for the following charged species. The charge is shown closest to the charged atom. For example, in $^{+}NO_2$, it is the nitrogen that is positive, not oxygen.

(a) ^{-}OH
(b) $^{-}BH_4$
(c) $^{+}NH_4$
(d) ^{-}Cl
(e) $^{+}CH_3$
*(f) $^{+}OH_3$
*(g) $^{+}NO_2$ ($ON^{+}O$)

ANSWER

(f) $^{+}OH_3$ Neutral oxygen has six electrons available for bonding ($_8O$, 8 – 2 1*s* electrons), and so positively charged oxygen must have five electrons.

Three single bonds can be made with the three hydrogen atoms.

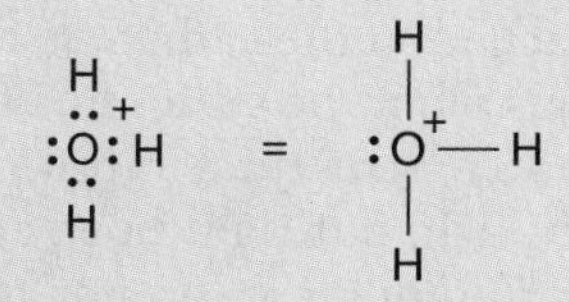

(g) $^{+}NO_2$ Nitrogen has five electrons available for bonding, so ^{+}N must have four. Each oxygen has six. Two nitrogen–oxygen double bonds can be formed, leaving each oxygen with two pairs of nonbonding electrons.

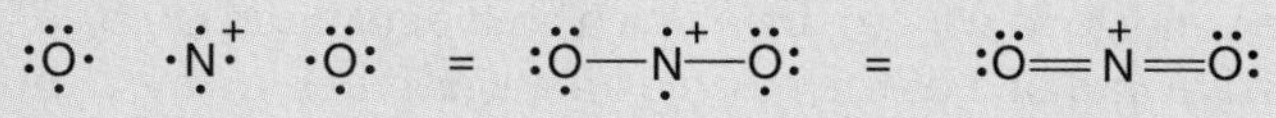

PROBLEM 1.7

Add charges to the following compounds wherever necessary.

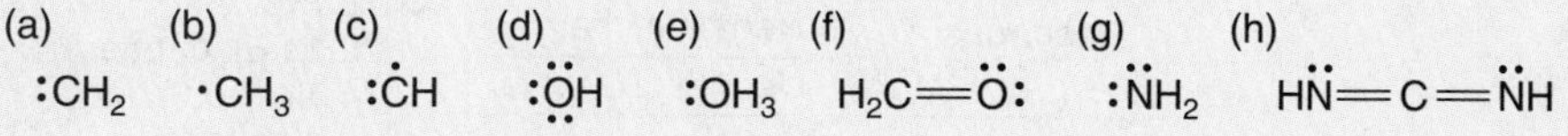

(a) $:CH_2$ (b) $\cdot CH_3$ (c) $:\dot{C}H$ (d) $:\ddot{O}H$ (e) $:OH_3$ (f) $H_2C{=}\ddot{O}:$ (g) $:\ddot{N}H_2$ (h) $H\ddot{N}{=}C{=}\ddot{N}H$

PROBLEM 1.8

Add electrons to complete the following Lewis structures. The charge is placed as close as possible to the charged atom.

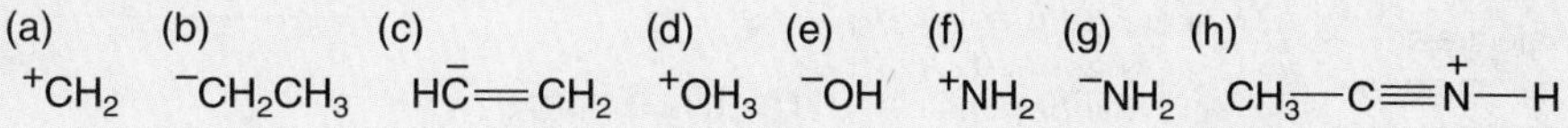

(a) $^{+}CH_2$ (b) $^{-}CH_2CH_3$ (c) $H\bar{C}{=}CH_2$ (d) $^{+}OH_3$ (e) ^{-}OH (f) $^{+}NH_2$ (g) $^{-}NH_2$ (h) $CH_3{-}C{\equiv}\overset{+}{N}{-}H$

1.3 INTRODUCTION TO RESONANCE FORMS

Often, there will be more than one possible Lewis structure for a molecule. This is especially true for charged molecules, but applies to many neutral species as well. (You may already have encountered this phenomenon in comparing your answers to Problems 1.6 and 1.7 with mine.) How do we decide which Lewis structure is the correct one? The answer almost always is that neither Lewis structure is complete all by itself, and that the molecule is best described as a combination of all reasonable Lewis structures, that is, a combination of **resonance forms**. A good example is nitromethane, H_3CNO_2. First of all, draw a good Lewis structure for nitromethane. Unfortunately, the formula H_3CNO_2 is ambiguous. How is the NO_2 part, the nitro group, to be drawn? Is it N–O–O or are both oxygens attached to nitrogen? There really is no way to tell from the formula. This problem exists even for the H_3C part. Why not imagine the structure as H–H–H–C? But carbon has four electrons available for bonding and is tetravalent. Hydrogen has only one electron and is monovalent. So, this kind of linear structure seems impossible. Accordingly, the alternative structure, in which each hydrogen is attached to the carbon, is fairly easy to find. However, it is by no means obvious what the structure of the nitro group is. As you work through the various structural types of organic chemistry, this kind of problem will diminish, but right now, at the very beginning, it can be quite daunting. For the moment, do not hesitate to refer to the end paper at the front of this book, where a compendium of common "functional group" structures is reproduced. In Chapter 9, we will treat resonance forms much more extensively.

Figure 1.19 shows that, even though nitromethane is a neutral molecule, there is no good way to draw it without separated charges. The nitrogen is positive and one of the oxygens is negative. But this is not the only Lewis form possible! We can draw the molecule so the other oxygen bears the negative charge. In the real nitromethane, the two oxygens share the negative charge equally (Fig. 1.20).

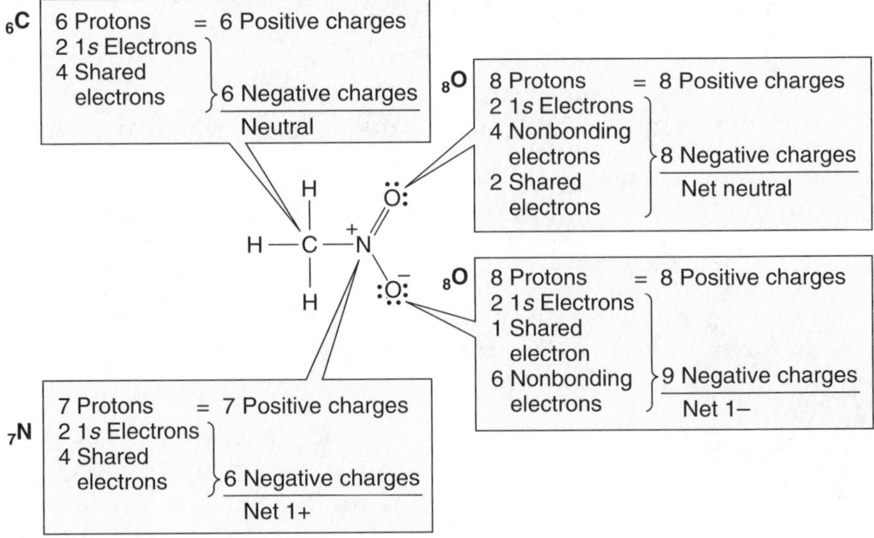

FIGURE 1.19 One electronic representation of nitromethane, $H_3C{-}NO_2$.

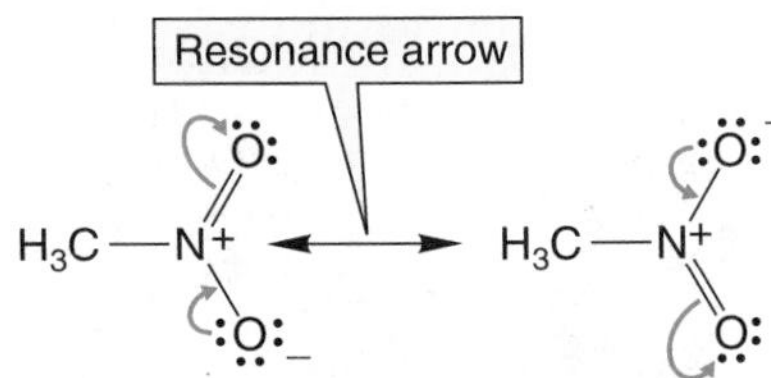

FIGURE 1.20 Two equivalent electronic representations for nitromethane. These are called resonance forms. Notice the special, double-headed arrow between the two forms and the curved arrows used to convert one resonance form into the other.

PROBLEM 1.9

Draw a Lewis structure for nitric acid ($HO{-}NO_2$), and verify that the nitrogen is positive and one of the oxygens is negative (see Fig. 1.19).

Convention Alert!

In Figure 1.20 curved arrows are used to interconvert the two Lewis structures, or resonance forms. Two things must be stressed here. First, although the curved arrows in Figure 1.20 have no physical significance, they do constitute an extraordinarily important bookkeeping device. We will use curved arrows throughout this book, and all organic chemists use them to keep track of electron movement. Being able to draw resonance forms quickly and accurately is an essential skill for any organic chemist. Second, notice that the bookkeeping is accomplished by moving or "pushing" *pairs* of electrons. Be careful when doing this sort of thing not to violate the rules of valence, not to make more bonds than are possible, for this is an easy mistake to make. These curved arrows are two-electron arrows. Figure 1.21 gives some examples of this kind of electron-pair pushing. This device, the arrow formalism, will be discussed extensively in Chapter 7.

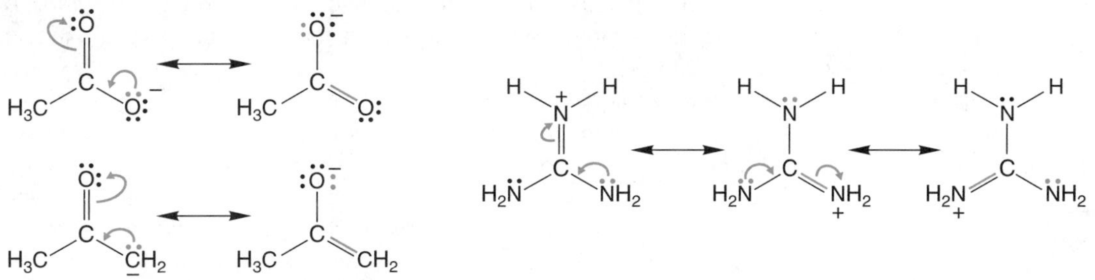

FIGURE 1.21 More molecules best represented as combinations of resonance forms.

PROBLEM 1.10

Use the arrows to convert your Lewis structure for nitric acid ($HO{-}NO_2$, Problem 1.9) into a resonance form.

It is also important to be clear on another point. Nitromethane does not spend half its time as one resonance form and half as the other. *There is no equilibration between the resonance forms.* Resonance involves only the movement of electrons, never atoms. Nitromethane is best described as the combination of the two forms representing different electronic structures.

Convention Alert!

The special double-headed arrow (⟷) used in Figures 1.20 and 1.21 is reserved for resonance phenomena and is never used for anything but resonance. A pair of arrows (⇆) indicates equilibrium, the interconversion of two chemically distinct species, and is never used for resonance (Fig. 1.22). This point is most important in learning the language of organic chemistry. It is difficult because it is arbitrary. There is no way to reason out the use of the different kinds of arrows; they simply must be learned.

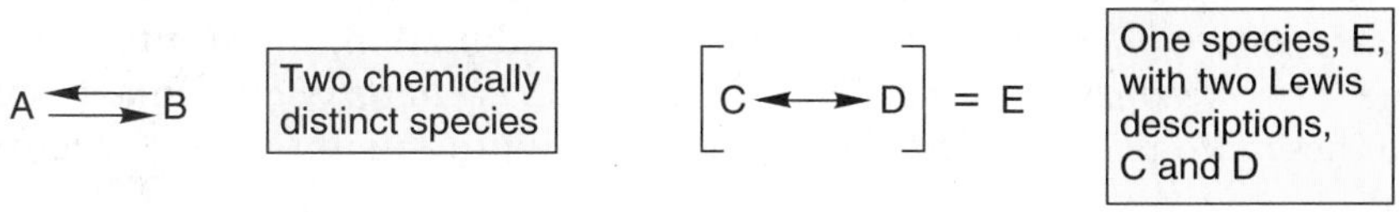

A ⇄ B Two chemically distinct species

[C ⟷ D] = E One species, E, with two Lewis descriptions, C and D

FIGURE 1.22 The difference between equilibrium (two different species, A and B; two arrows) and resonance (different electronic representations, C and D, for the same molecule, E; double-headed arrow).

*PROBLEM 1.11 Draw another structure for nitromethane in which every atom is neutral. *Hint*: There are only single bonds in this structure.

ANSWER You can arrive at the answer by using arrows to push electron pairs in the following way:

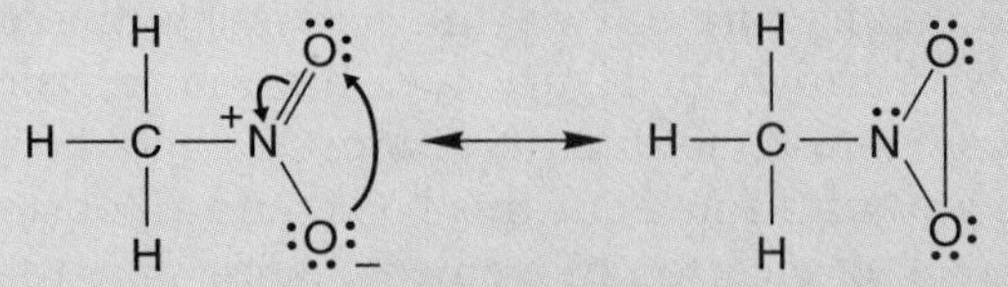

Note that, as a new bond is formed between the two oxygen atoms, the two charges originally on oxygen and nitrogen are canceled. Notice also the long oxygen–oxygen bond. *Remember*: In drawing resonance forms, you move only electrons, not atoms. The oxygen–oxygen distance must be the same in each resonance form.

Is this cyclic form an important resonance form? Is it a good representation of the molecule? The problem is the long bond between the oxygens. This bond must be weak just because it is so long. In the language of organic chemistry, we would say that the cyclic resonance form contributes little to the structure of nitromethane.

Is this cyclic structure a resonance form? That is a tricky question, and the answer is "yes and no." Yes, it is a resonance form if, as above, we have been careful not to move any atoms in the cyclic structure. Resonance involves different electronic structures that may not differ in the positions of atoms. If atoms have been moved, as in the figure below in which there is a normal oxygen–oxygen bond, the two structures are in equilibrium and are not resonance forms.

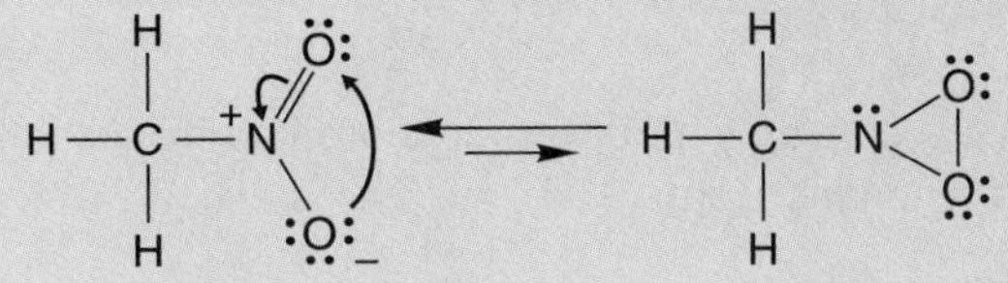

PROBLEM 1.12 Verify that the C, N, and O atoms of your new structure are neutral.

Formaldehyde

FIGURE 1.23 Resonance forms for the molecule formaldehyde, $H_2C{=}O$.

The carbon–oxygen double bond gives us another opportunity to write resonance forms. As an example take the simple compound, formaldehyde ($H_2C{=}O$, Fig. 1.23).

Carbon ($_6C$) has a pair of 1*s* electrons and shares in four covalent bonds; therefore it is neutral. Oxygen ($_8O$) has a pair of 1*s* electrons, four nonbonding electrons, and a share in two covalent bonds, for a total of eight electrons. Oxygen is also neutral. However, we can push electrons to generate a second Lewis structure in which the carbon is positive and the oxygen negative. The real formaldehyde is a combination, *not a mixture,* of these two resonance forms. Charge separation is energetically unfavorable so these two resonance forms will not contribute equally to the structure of the formaldehyde molecule. Still, neither by itself is a perfect representation of the molecule, and in order to represent formaldehyde well, both electronic descriptions must be considered.

PROBLEM 1.13

Verify that the carbon of the polar resonance form of formaldehyde in Figure 1.23 is positive and the oxygen is negative.

*PROBLEM 1.14

There is a third resonance form for formaldehyde, but it contributes very little to the structure and is usually ignored. Can you find it and explain why it is relatively unimportant?

ANSWER

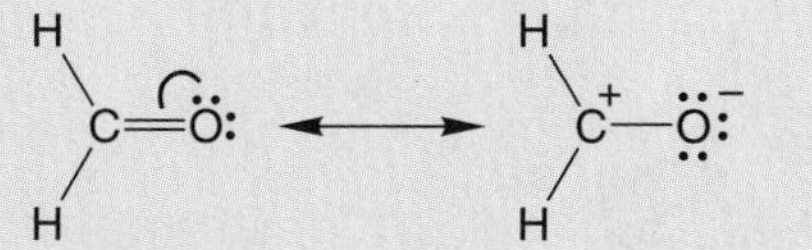

The electrons can be pushed (arrows) in the other direction so as to make the carbon negative and the oxygen positive.

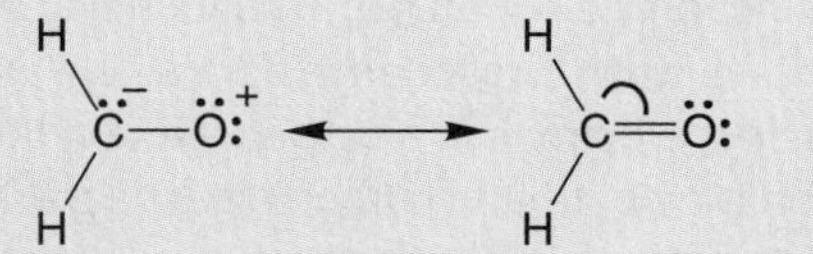

This new resonance form is another valid electronic description of the carbonyl compound, and so must be considered. However, oxygen is more electronegative than carbon, and therefore better able to accommodate a negative charge. In the new resonance form, the more electronegative oxygen is positive and the more electropositive carbon is negative. This description will not contribute very much to the real structure of formaldehyde.

PROBLEM 1.15

Use the arrow formalism to convert each of the following Lewis structures into another resonance form. Notice that the last of these asks you to do something new—to move electrons one at a time in writing Lewis forms.

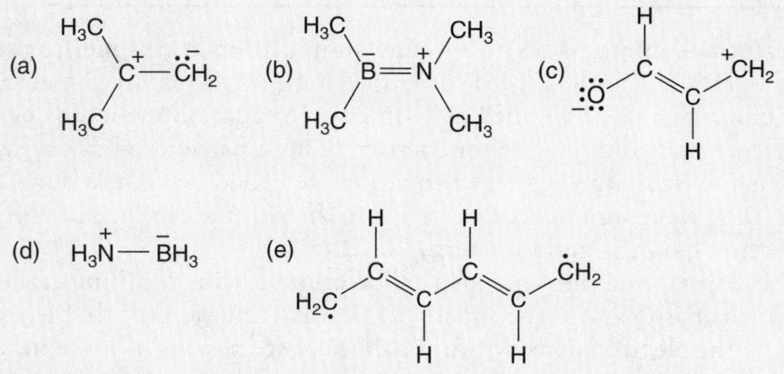

(a) (b) (c)

(d) $H_3\overset{+}{N}—\overset{-}{B}H_3$ (e)

PROBLEM **1.16** Use the arrow formalism to write resonance forms that contribute to the structures of the following molecules:

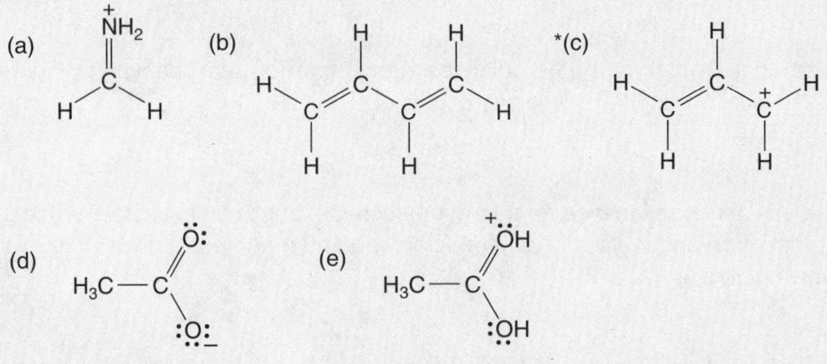

(a) (b) *(c)

(d) (e)

ANSWER (c)

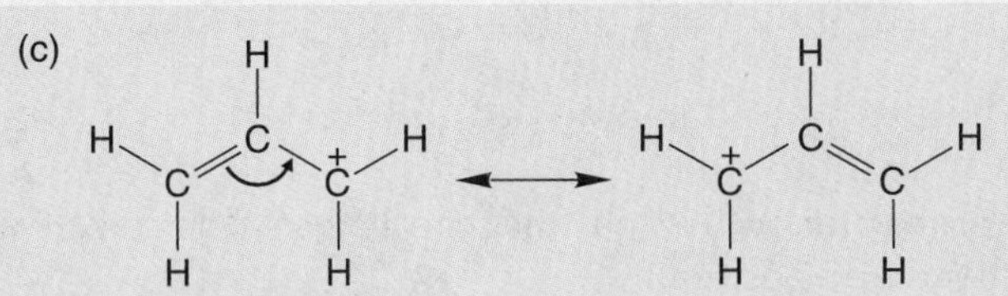

As we have seen, it is not only electron pairs (nonbonding electrons) that can be redistributed (pushed) in writing resonance forms, but bonding electrons as well. In this case, a pair of electrons in a carbon–carbon double bond is moved. The positive charge on the right-hand carbon disappears, but reappears on the left-hand carbon. In the real structure, the two carbons share the positive charge equally, each bearing one-half the charge. The structure is sometimes written with dashed bonds to show this sharing:

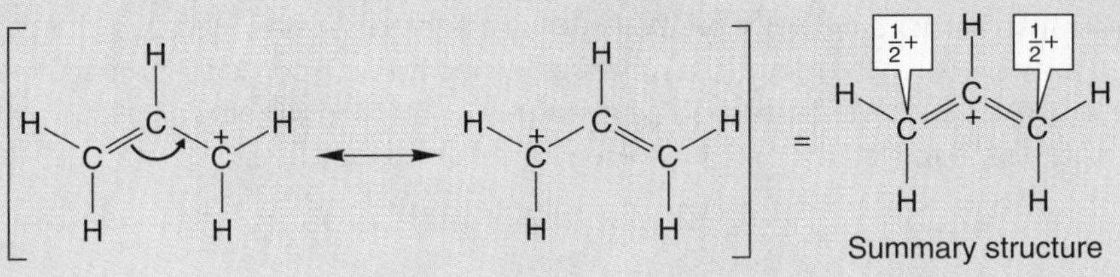

Summary structure

1.4 MORE ON ATOMIC ORBITALS

The electrons in atoms do not occupy simple circular or spherical orbits. Erwin Schrödinger (1887–1961) described a negatively charged electron in the vicinity of a positive nucleus with a wave equation, which recognized that electrons could have the properties of both particles and waves. The solutions to Schrödinger's wave equation are called **wave functions**, written ψ. Each wave function corresponds to the volume of space occupied by an electron having a certain energy.

These wave functions are also called **orbitals**, thus recalling earlier descriptions of atoms. It is the square of the wave function, ψ^2, that is proportional to the electron density. Although there are regions of space in which ψ and $\psi^2 = 0$ (zero electron density = zero probability of finding an electron), ψ does not vanish at a large distance from the nucleus, but maintains a finite value, even if very small.

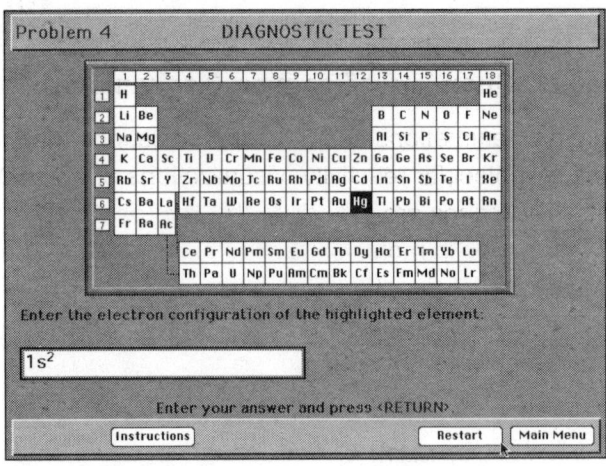

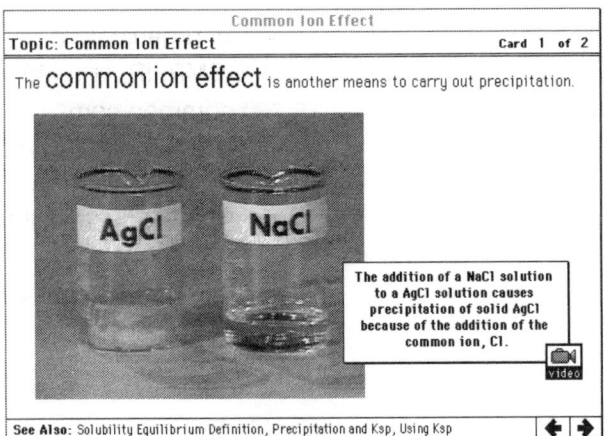

Each wave function is determined by a set of **quantum numbers**. The first quantum number is n, the principal quantum number, which may have the integral values $n = 1, 2, 3, 4, \ldots$. It is generally related to the distance of the electron from the nucleus and hence to the energy of the electron. The higher n, the greater the average distance of the electron from the nucleus, and the greater its energy. The principal quantum number of the highest energy electron of an atom also determines the row occupied by the atom in the periodic table. Hydrogen and helium are in the first row, Li, Be, B, C, N, O, F, and Ne in the second, Na, Mg, Al, Si, P, S, Cl, and Ar in the third, and so on (Table 1.3).

TABLE 1.3 Principal Quantum Number (n) of the Highest Energy Electron

Atom	n
H, He	1
Li, Be, B, C, N, O, F, Ne	2
Na, Mg, Al, Si, P, S, Cl, Ar	3

The second quantum number, l, is generally related to the shape of the orbital and depends on the value of n. It may have only the integral values 0, 1, 2, 3, ... $(n - 1)$ [0 up to $(n - 1)$]. So, for an orbital for which $n = 1$, l must be zero; for $n = 2$, l can be 0 or 1; and for $n = 3$, the three possible values of l are 0, 1, and 2. The quantum number l is always referred to by a letter, s for $l = 0$, p for $l = 1$, d for $l = 2$, and f for $l = 3$, which leads to the common designations of orbitals shown in Table 1.4.

TABLE 1.4 Relationship between n and l

n	l	Orbital Designation
1	0	1*s*
2	0	2*s*
2	1	2*p*
3	0	3*s*
3	1	3*p*
3	2	3*d*

The third quantum number m_l, depends on l as did l on n. It may have the integral values $0, \pm1, \pm2, \pm3, \ldots, \pm l$, and is generally related to the orientation of the orbital in space. Table 1.5 presents the possible values of n, l, and m_l.

TABLE 1.5 Relationship between n, l, and m_l

n	l	m_l	Orbital Designation
1	0	0	1*s*
2	0	0	2*s*
2	1	−1	2*p*
2	1	0	2*p*
2	1	+1	2*p*
3	0	0	3*s*
3	1	−1	3*p*
3	1	0	3*p*
3	1	+1	3*p*
3	2	−2	3*d*
3	2	−1	3*d*
3	2	0	3*d*
3	2	+1	3*d*
3	2	+2	3*d*

Finally, there is s, the spin quantum number, which may have only the two values $\pm½$. All this may look complicated but it's really quite simple. Table 1.6 gives a complete listing of the possible combinations of quantum numbers through $n = 3$.

As shown in Tables 1.4–1.6, orbitals are designated with a number and a letter in the following way:

$$1s\ (n = 1, l = 0) \quad 2s\ (n = 2, l = 0) \quad 2p\ (n = 2, l = 1)$$

For $n = 2$, $l = 1$ (a 2p orbital), m_l may take the values −1, 0, +1; thus, there are *three* 2p orbitals, one for each value of m_l. These equi-energetic orbitals are differentiated by arbitrarily designating them as $2p_x$, $2p_y$, or $2p_z$. A little later we will see that the "x, y, z" notation indicates the relative orientation of the 2p orbitals in space.

TABLE 1.6 Possible Combinations of Quantum Numbers for $n = 1, 2$, and 3

n	l	m_l	s	Orbital Designation
1	0	0	±½	1*s*
2	0	0	±½	2*s*
2	1	−1	±½	2*p*
2	1	0	±½	2*p*
2	1	+1	±½	2*p*
3	0	0	±½	3*s*
3	1	−1	±½	3*p*
3	1	0	±½	3*p*
3	1	1	±½	3*p*
3	2	−2	±½	3*d*
3	2	−1	±½	3*d*
3	2	0	±½	3*d*
3	2	1	±½	3*d*
3	2	2	±½	3*d*

For $n = 3$, we have the following possibilities:

$$3s\ (n = 3, l = 0) \quad 3p\ (n = 3, l = 1)$$

Just as for $n = 2$, m_l may now take the values −1, 0, +1. The three 3*p* orbitals are designated as $3p_x$, $3p_y$, and $3p_z$.

$$3d\ (n = 3, l = 2)$$

As $l = 2$, m_l may now take the values: −2, −1, 0, +1, and +2; thus there are five 3*d* orbitals. These turn out to have the complicated designations $3d_{x^2-y^2}$, $3d_{z^2}$, $3d_{xy}$, $3d_{yz}$, and $3d_{xz}$. Mercifully, in organic chemistry we only very rarely have to deal with 3*d* orbitals and need not consider the *f* orbitals, for which $n = 4$ and which are even more complicated.

For each of these orbitals, the spin quantum number *s* may be either +½ or −½. We may now designate an electron occupying the lowest energy orbital, 1*s*, as either 1*s*, with $s = +½$, or 1*s*, with $s = −½$, *but nothing else.* Similarly, there are only two possibilities for electrons in the $2p_x$ orbital: $2p_x$, with $s = +½$, or $2p_x$, with $s = −½$. That is why it is impossible for more than two electrons to occupy any orbital! When we write $1s^2$ we mean that the 1*s* orbital is occupied by two electrons, which have opposite (paired) spin quantum numbers ($s = ±½$). The designation $1s^3$ is meaningless, as there is no way to put a *different* third electron in any orbital. This rule is called the **Pauli principle**, after Wolfgang Pauli (1900–1958), who first articulated it in 1925. These ideas are summarized in Figure 1.24.

$1s^2$ means the 1*s* orbital contains two electrons:

Electron No.1

$n = 1, l = 0, m_l = 0, s = +\frac{1}{2}$

Electron No. 2

$n = 1, l = 0, m_l = 0, s = -\frac{1}{2}$

FIGURE 1.24

We can now write electronic descriptions for atoms, following what is known as the **aufbau principle** (aufbau is German for building up or construction), which simply makes the reasonable assumption that we should fill the available orbitals in order of their energies, starting with the lowest energy orbital. To form these descriptions of neutral atoms, we add electrons until they are equal to the number of positive charges (thus the number of protons) in the nucleus. Table 1.7 gives the electronic descriptions of H, He, Li, Be, and B.

TABLE 1.7 Electronic Descriptions of Some Atoms

Atom	Electronic Configuration
$_1$H	$1s$
$_2$He	$1s^2$
$_3$Li	$1s^22s$
$_4$Be	$1s^22s^2$
$_5$B	$1s^22s^22p_x$

For carbon, the next atom after boron, there is a choice to be made in adding the last electron. For carbon, the first five electrons are placed as in boron, but where does the sixth and last electron go? One possibility would be to put it in the same orbital as the fifth electron to produce $_6C = 1s^22s^22p_x^2$. In such an atom the spins of the two electrons in the $2p_x$ orbital must be paired (opposite spins).

*PROBLEM **1.17**

Explain why the two electrons in the $2p_x$ orbital of carbon ($_6C = 1s^22s^22p_x^2$) must have opposite (paired) spins.

ANSWER

Because they are in the same orbital ($2p_x$), both electrons have the same values for the three quantum numbers n, l, and m_l ($n = 2$, $l = 1$, m_l = either +1, 0, or −1). If the spin quantum numbers were not opposite, one +½, the other −½, the two electrons would not be different! The Pauli principle, (No two electrons may have the same values of the four quantum numbers), ensures that two electrons in the same orbital must have different spin quantum numbers.

Alternatively, the sixth electron could be placed in another $2p$ orbital to produce $_6C = 1s^22s^22p_x2p_y$. The only difference is the presence of two electrons in a *single* $2p$ orbital ($_6C = 1s^22s^22p_x^2$) versus one electron in each of two *different* $2p$ orbitals ($_6C = 1s^22s^22p_x2p_y$). The three $2p$ orbitals are of equal energy, so how is one to make this choice? **Hund's rule** (Friedrich Hund, b. 1896) states that for orbitals of equal energy (such as the three $2p$ orbitals), the electronic configuration with the greatest number of parallel spins (same spin quantum number) results in the lowest energy overall. Electrons are negatively charged and, as like charges repel each other, placing electrons in different orbitals minimizes destabilization owing to that repulsion. Hund's rule tells us that the second configuration, $_6C = 1s^22s^22p_x2p_y$, is lower in energy than $_6C = 1s^22s^22p_x^2$, and that the spins of the two electrons in the $2p$ orbitals must be the same (parallel). Electron–electron repulsion is minimized if the spins of these two electrons are the same because two electrons with the same spin may not occupy the same region of space (orbital). Figure 1.25 shows these two possible electronic configurations for carbon.

The convention used to designate spin shows the electrons as little arrows pointing either up ↑ or down ↓. Two arrows in opposite directions (↑↓) show paired spins and two arrows in the same direction (↑↑) show parallel, or unpaired, spins.

Convention Alert!

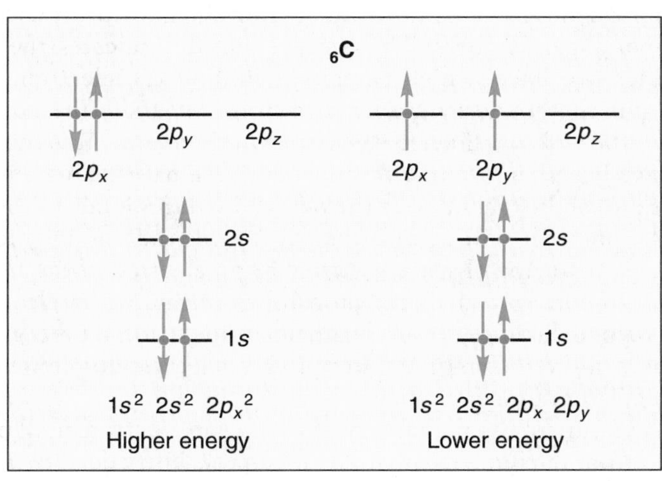

FIGURE **1.25** An application of Hund's rule to the carbon atom. The electronic configuration with the larger number of parallel spins is lower in energy. Note the use of the arrow convention to show electron spin.

PROBLEM 1.18 Explain why the fifth and sixth electrons in a carbon atom may not occupy the same orbital as long as they have the same (parallel) spins.

TABLE 1.8 Electronic Descriptions of the Remaining Atoms in the Second Row

Atom	Electronic Configuration
$_6C$	$1s^22s^22p_x2p_y$
$_7N$	$1s^22s^22p_x2p_y2p_z$
$_8O$	$1s^22s^22p_x^22p_y2p_z$
$_9F$	$1s^22s^22p_x^22p_y^22p_z$
$_{10}Ne$	$1s^22s^22p_x^22p_y^22p_z^2$

Now we can write electronic descriptions for the rest of the atoms in the second row of the periodic table (Table 1.8). Notice that for nitrogen we face the same kind of choice just described for carbon. Does the seventh electron go into an already occupied $2p$ orbital, or into the remaining empty $2p_z$ orbital? The configuration in which the $2p_x$, $2p_y$, and $2p_z$ orbitals are all singly occupied by electrons with the same spin is lower in energy than any configuration in which a $2p$ orbital is doubly occupied. Again, Hund rules!

PROBLEM 1.19 Write the electronic configurations for the next row in the periodic table, $_{11}Na$ through $_{18}Ar$.

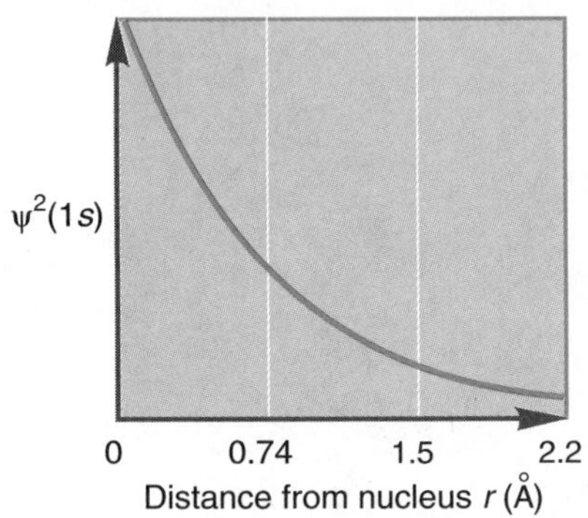

FIGURE 1.26 A plot of $\psi^2(1s)$ versus distance r for hydrogen (1*s*). Note that the value of ψ^2 does not go to zero, even at very large *r*.

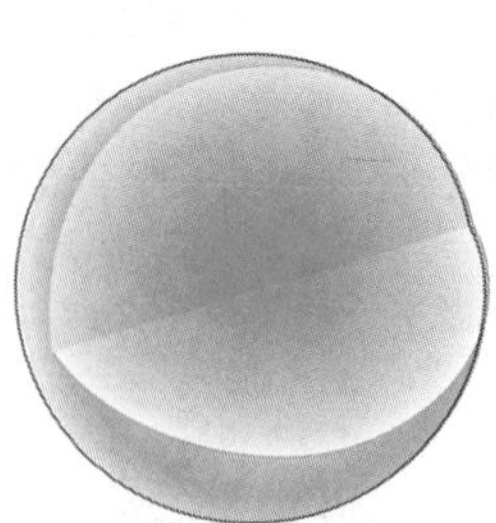

FIGURE 1.27 A three-dimensional picture of the 1*s* orbital. The surface of the sphere denotes an arbitrary cutoff. The wave function continues in all directions, although its value becomes very small indeed far from the nucleus.

Wave functions, or orbitals (solutions to Schrödinger's equation), can be graphed to show the value of ψ as a function of *r*, the distance from the nucleus. As it is ψ^2 that is related to electron density, it is usually this parameter that is plotted against *r*. Figure 1.26 shows this plot for the lowest energy solution to the Schrödinger equation, the 1*s* orbital.

Figure 1.26 shows that the probability of finding an electron falls off rather sharply in all directions as we move out from the nucleus. There is no directionality—the 1*s* orbital is symmetrical in all directions. It is the quantum number *l* that is associated with orbital shape, and all *s* orbitals, for which *l* always equals zero, will be spherically symmetric. Note also that the electron density never goes to zero—there is a finite, though very, very small probability of finding an electron at a distance of several angstroms (or feet, or miles, or light years) from the nucleus. Figure 1.27 translates this two-dimensional graph into a three-dimensional picture of the 1*s* orbital, a spherical cloud of electron density, at its maximum near the nucleus. Although this cloud of electron density is arbitrarily bound by the spherical surface at some distance from the nucleus, it does not really terminate sharply there. This surface simply indicates the volume within which we have a high chance of finding the electron. We can choose to put the boundary at any percentage we like—the 95% confidence level, the 99% confidence level, or any other value. It is this picture that we have been approaching throughout these pages. This representation of the probability function is the one you will remember best (probably) and the one that will be most useful in our study of chemical reactions. This is the region of space occupied by an electron in the 1*s* orbital.

Figure 1.26 correctly gives the probability of finding an electron at a particular point a distance *r* from the nucleus, but it doesn't recognize that as *r* increases, a given change (Δr) in *r* produces a greater volume of space and thus a greater number of points.

To get a picture of the probability of finding an electron at all points at a distance *r* from the nucleus, it is the radial probability density (Fig. 1.28)

that is the most intuitively useful view. This picture takes account of the increasing volume of spherical shells Δr thick as the distance from the nucleus r increases.

To summarize: First, we can do no better than to say that within some degree of certainty the electron is within a certain volume of space. Second, the cutoff is arbitrary—the electron density does not become zero at long distance from the nucleus. Third, an orbital can contain only two electrons, which will have what are called "paired" or "opposite" spin quantum numbers, $s = +\frac{1}{2}$ and $-\frac{1}{2}$. Finally, it is the quantum number l that is related to the shape of the orbital, and all s orbitals have spherical symmetry.

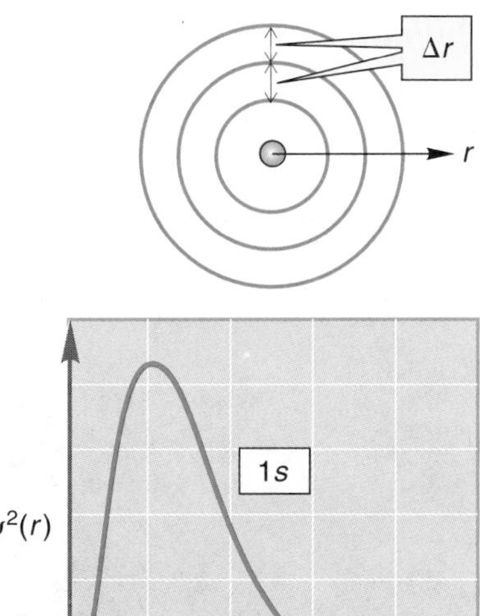

FIGURE **1.28** A slice through the three-dimensional 1*s* orbital. At relatively low values of *r* (short distance from the nucleus) Δr will contain a smaller volume than at higher values of *r* (larger distance from the nucleus). The outer spherical shell contains greater volume than the middle shell even though they are both Δr wide. This plot of ψ^2 (*r*) versus *r* weighted for the increasing influence of Δr as *r* gets large is a radial probability density plot.

The wave function ψ has many of the properties of other kinds of waves. There are, for example, the peaks and valleys evident in waves in liquids and in a vibrating string. Peaks correspond to regions for which the mathematical sign of ψ is positive and valleys correspond to negative regions. In a vibrating string, when we pass from a length of string in which the amplitude is positive to one where it is negative, we find a stationary point called a **node**.* Nodes also appear in wave functions at the points at which they change sign; where $\psi = 0$. Of course, if $\psi = 0$, then $\psi^2 = 0$, and the probability of finding an electron at a node is also zero. Thus, a node is a region of space in which the electron density is zero. The number of nodes in an orbital is always one less than n, the principal quantum number. Thus, the $n = 1$ orbital (1*s*) has no nodes, all $n = 2$ orbitals have a single node, $n = 3$ orbitals have two nodes, and so on. As we turn our attention from the 1*s* orbital to higher energy orbitals, we will have to pay attention to the presence of nodes.

The next higher energy orbital is the 2*s* ($n = 2$, $l = 0$). Because $n = 2$, there must be a single node in this orbital, and we expect spherical symmetry, as with all *s* orbitals. Figure 1.29 shows a plot of ψ^2 versus r, the radial probability density, and a three-dimensional representation of the 2*s* orbital. This figure shows a spherical node, the region at which the sign of ψ and ψ^2 is zero.

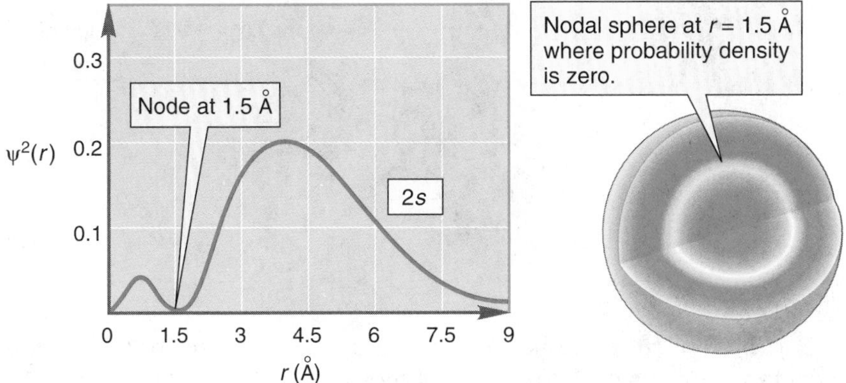

FIGURE **1.29** A plot of $\psi^2(r)$ versus *r* for a 2*s* orbital and a cross section through the orbital.

When $n = 2$, l can equal 1 as well as zero. This combination of quantum numbers gives the $2p_x$, $2p_y$, and $2p_z$ orbitals, each of which must possess a single node. Because l is not zero, the 2*p* orbitals are not spherically sym-

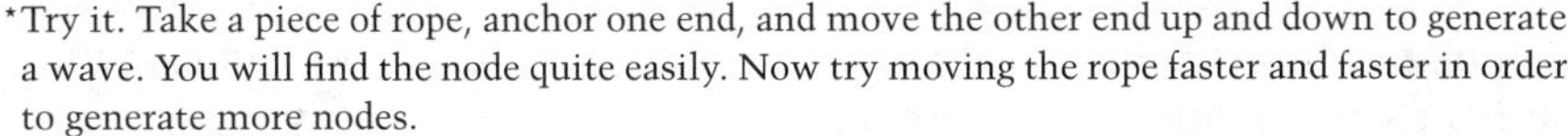

*Try it. Take a piece of rope, anchor one end, and move the other end up and down to generate a wave. You will find the node quite easily. Now try moving the rope faster and faster in order to generate more nodes.

metrical. As you may have guessed from the *x, y, z* designations, the 2*p* orbitals are directed along the *x, y,* and *z* axes, which intersect at 90° angles. The lobes of 2*p* orbitals are slightly flattened spheres, and the orbital as a whole is shaped roughly like a dumbbell (Fig. 1.30). The node is the plane separating the two halves of the dumbbell. In one lobe of the orbital the sign of the wave function is positive; in the other it is negative. These signs are usually indicated by a change of color rather than (+) and (–) to avoid confusion with electrical charge. The three 2*p* orbitals together are shown in Figure 1.31. As you can see, the stylized dumbbells used to represent 2*p* orbitals are easier to draw than the more accurate picture shown in Figure 1.30.

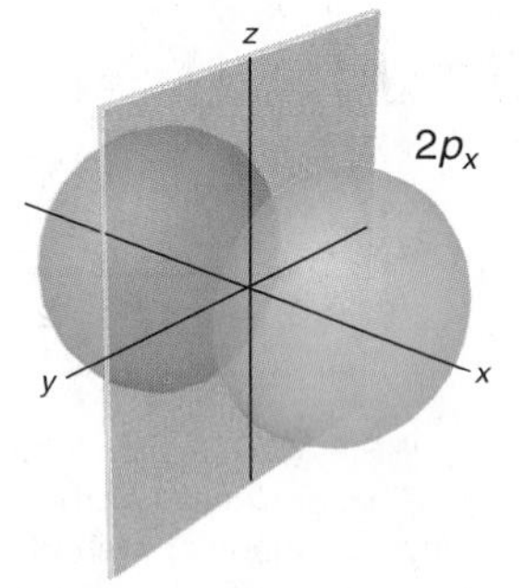

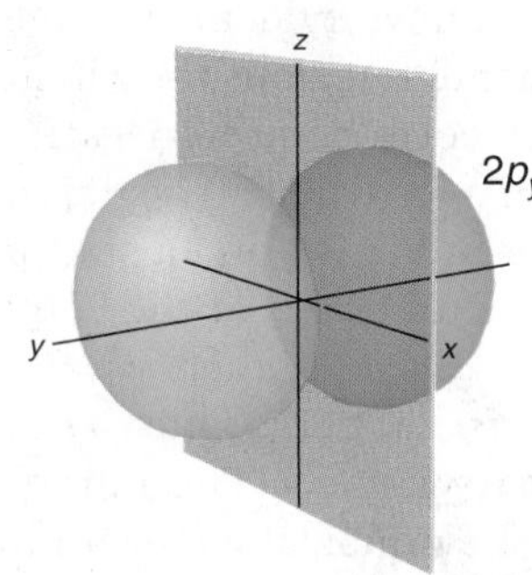

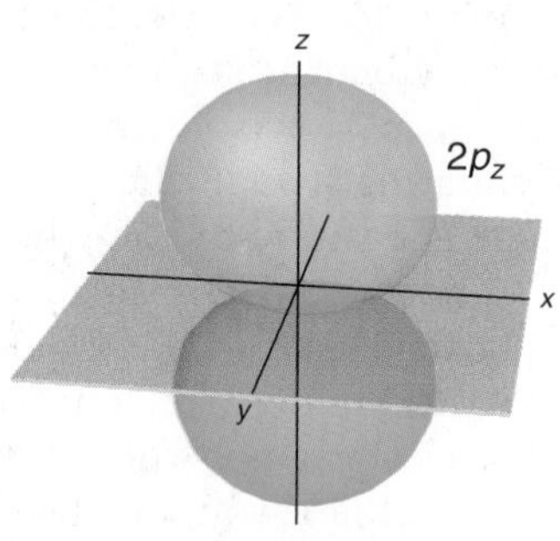

FIGURE **1.30** An accurate three-dimensional representation of three 2*p* orbitals. Note the nodal planes separating the lobes where the sign of ψ differs.

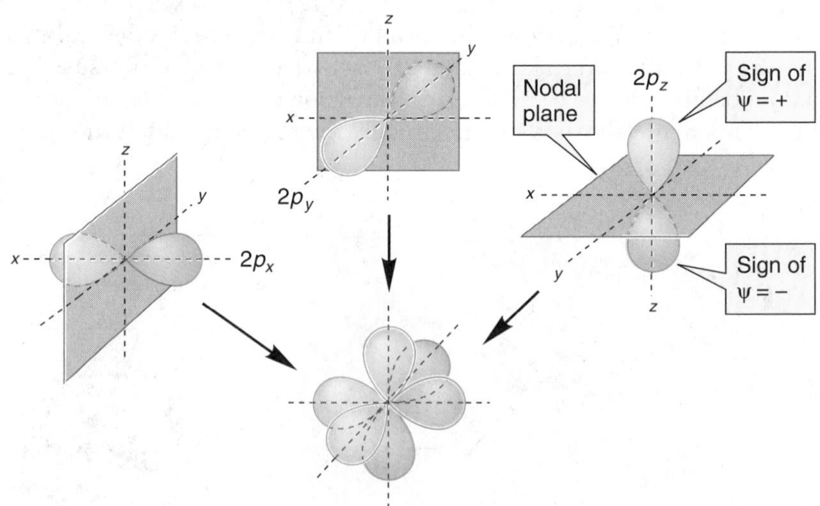

FIGURE **1.31** A schematic three-dimensional representation of three 2*p* orbitals, shown separately and in combination.

For the first time, we begin to get hints of the causes of the complicated three-dimensional structures of molecules: Electrons are the "glue" that holds the atoms of molecules together, and electrons are confined to regions of space that are by no means always spherically symmetrical.

We can now make some quite detailed pictures of atoms using the shapes described here. In the drawings in Figure 1.32, the 1*s* electrons have been omitted in the atoms past He. Structures for these atoms, in which each electron is shown as a dot, also ignore the 1*s* electrons, which are not strongly involved in bonding for atoms other than hydrogen.

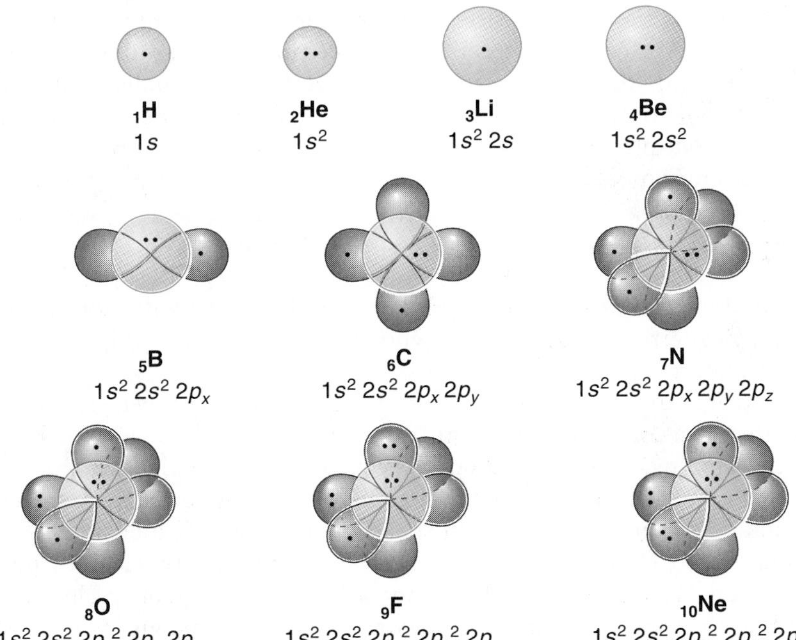

FIGURE 1.32 Some schematic pictures of atoms. The 1*s* electrons are not shown as dots except for H and He.

1.5 SUMMARY

NEW CONCEPTS

Orbitals (wave functions) are defined by four quantum numbers, n, l, m_l, and s. Electrons may have only certain energies determined by these quantum numbers, and therefore may occupy only certain regions of space, called orbitals. Orbitals have different, well-defined shapes: *s* orbitals are spherically symmetric, *p* orbitals are roughly dumbbell shaped, and *d* and *f* orbitals are more complicated. Orbitals may contain a maximum of two electrons.

Some molecules cannot be well described by a single Lewis structure, but are better represented by two or more different electronic descriptions. This phenomenon is called "resonance." Be sure you are clear on the difference between an equilibrium between two different molecules, and the description of a single molecule using a number of different resonance forms.

REACTIONS, MECHANISMS, AND TOOLS

There are no reactions to speak of yet, but we have developed a number of tools in this first chapter.

Lewis structures are drawn by representing each electron by a dot. These dots are transformed into arrows if it is necessary to show electron spin. Exceptions are the 1*s* electrons, which are held too tightly to be importantly involved in bonding except in hydrogen. These electrons are not shown in Lewis structures except in H and He. Somewhat more abstract representations of molecules are made by showing electron pairs in bonding orbitals as lines—the familiar bonds between atoms.

The double-headed resonance arrow is introduced to cope with those molecules that cannot be adequately represented by a single Lewis structure. Different *electronic* representations for the molecule are written. The molecule is better represented by a combination of these so-called resonance forms than by any one form. Be very careful to distinguish the resonance phenomenon from chemical equilibrium. Resonance forms give multiple descriptions of a single species. Equilibrium describes two (or more) different molecules.

COMMON ERRORS

Here, and in similar sections throughout the book, we will take stock of some typical errors made by those who attempt to come to grips with organic chemistry.

Electrons are not baseballs. Nothing is harder for most students to grasp than the consequences of this observation. Electrons behave in ways that the moving objects in our ordinary lives do not. No one who has ever kicked a soccer ball or caught a fly ball can doubt that on a practical level it is possible to determine both the position and speed of such an object at the same time. However, Heisenberg demonstrated that this is *not* true for an electron. We cannot know both the position and speed of an electron at the same time. Baseballs move at a variety of speeds (energies) and a baseball's energy depends only on how hard we throw it. Electrons are restricted to certain energies (orbitals) determined by the values of quantum numbers. Electrons behave in other strange and counterintuitive ways. For example, we have seen that a node is a region of space of zero electron density. Yet, an electron occupies an entire 2*p* orbital in which the two halves are separated by a nodal plane. A favorite question, How does the electron move from one lobe of the orbital to the other? simply has no meaning. The electron is *not* restricted to one lobe or the other, but occupies the *whole orbital* (Fig. 1.33). Mathematics makes these properties seem inevitable; intuition, derived from our experience in the macroscopic world, makes them very strange.

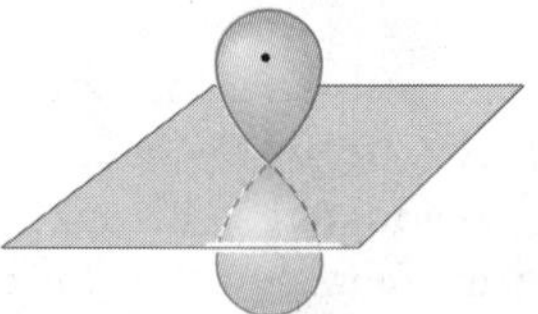

FIGURE **1.33** An electron occupies an entire 2*p* (or other) orbital—it is not restricted to only one lobe.

It is easy to confuse resonance with equilibrium. On a mundane, but nonetheless important level, this confusion appears as a misuse of the arrow convention. Two arrows separate two entirely different molecules (A and B), each of which might be described by several resonance forms. The amount of A and B present at equilibrium depends on the equilibrium constant. The double-headed resonance arrow separates two different electronic descriptions (C and D) of the same species, E (Fig. 1.22).

1.6 KEY TERMS

Anion A negatively charged atom or molecule.

Atom A neutral atom consists of a nucleus, or core of protons and neutrons, orbited by a number of electrons equal to the number of protons.

Atomic orbital One of the energy levels allowed for an electron in an atom. These orbitals result from the solution of Schrödinger's equation describing the motion of an electron in the vicinity of a nucleus. Atomic orbitals have different shapes, which are determined by quantum numbers. The *s* orbitals are spherically symmetric, *p* orbitals roughly dumbbell shaped, and the *d* and *f* orbitals are even more complicated.

Aufbau principle When adding electrons to a system of orbitals, first fill the lowest energy orbital available before filling any higher energy orbitals. Electron-electron repulsion is minimized by filling systems of equi-energetic orbitals by singly occupying all orbitals with electrons of the same spin before doubly occupying any of them. See **Hund's rule**.

Cation A positively charged atom or molecule.

Covalent bond A bond formed by the sharing of electrons through the overlap of atomic orbitals (or molecular orbitals: see Chapter 2).

Dipole moment A dipole moment in a molecule results when two opposite charges are separated.

Electron A particle of tiny mass (1/1845 of a proton) and a single negative charge.

Electron affinity A measure of the tendency for an atom or molecule to accept an electron.

Electronegativity The tendency for an atom to attract electrons.

Energy level See **atomic orbital** and **wave function**.

Heisenberg uncertainty principle For an electron, the uncertainly in position times the uncertainty in momentum (or speed) is a constant. We cannot know the exact position and momentum (speed) of an electron at the same time.

Hund's rule For a set of equi-energetic orbitals, the electronic configuration with the maximum number of parallel spins is the lowest in energy. That is,

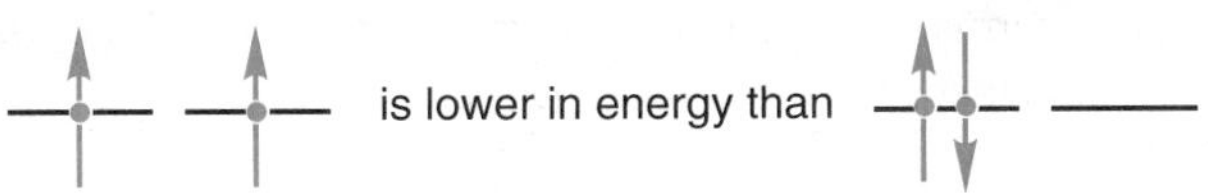

Ion A charged atom or molecule.

Ionic bond The electrostatic attraction between a positively charged atom or group of atoms and a negatively charged atom or group of atoms.

Ionization potential A measure of the tendency of an atom or molecule to lose an electron.

Lewis structure In a Lewis structure, every electron (except the 1*s* electrons for atoms other than hydrogen or helium) is shown as a dot. In slightly more abstract Lewis structures, electrons in bonds are shown as lines connecting atoms.

Lone-pair electrons Electrons in an orbital that is not involved in binding atoms. See **nonbonding electrons**.

Molecular orbital An orbital not restricted to the region of space surrounding an atom, but extending over several atoms in a molecule. Molecular orbitals are formed through the overlap of atomic or molecular orbitals. Molecular orbitals can be bonding, nonbonding, or antibonding. See Chapter 2.

Node The region of zero electron density separating regions of opposite sign in an orbital. At a node the sign of the wave function is zero.

Nonbonding electrons Electrons in an orbital that is not involved in binding atoms. See **lone pair**.

Nucleus The positively charged core of an atom containing the protons and neutrons.

Octet rule The notion that special stability attends the filling of the 2*s* and 2*p* atomic orbitals to achieve the electronic configuration of neon, a noble gas.

Orbital The volume of space in which an electron is likely to be found. Another word for "wave function." These are three-dimensional representations of the solutions to the Schrödinger equation.

Pauli principle No two electrons in an atom or molecule may have the same values of the four quantum numbers.

Polar covalent bond Any shared electron bond between two different atoms must be polar. Two nonidentical atoms must have different electronegativities and will attract the shared electrons to different extents, creating a dipole.

Resonance forms Many molecules cannot be represented adequately by a single Lewis form. Instead, two or more different electronic representations must often be combined to give a good description of the molecule. These different representations are called resonance forms.

Quantum numbers These numbers evolve from the Schrödinger equation and characterize the various solutions to it. They may have only certain values, and these values determine the distance of an electron from the nucleus (n), the shape (l), and orientation (m_l), of the orbitals, and the electron spin (s).

Valence electrons The outermost, or most loosely held, electrons.

Wave function (ψ) A solution to the Schrödinger equation, another word for "orbital."

1.7 ADDITIONAL PROBLEMS

PROBLEM 1.20 Use the arrow formalism to write structures for the resonance forms contributing to the structures of the following ions:

(a) Carbonate ion

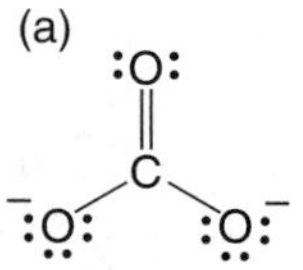

(b) Sulfate ion

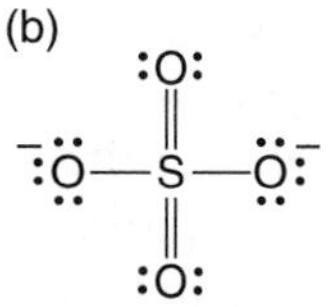

(c) Nitrate ion

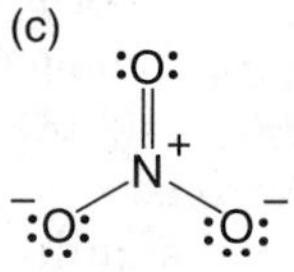

(d) Guanidinium ion

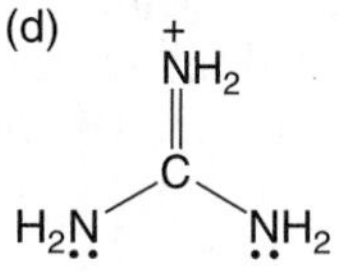

(e) A vinyl ammonium ion

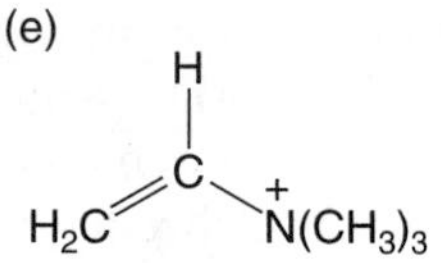

PROBLEM 1.21 Use the arrow formalism to write resonance forms contributing to the structures for the following molecules:

(a) $H_2\bar{C}—N{=}\overset{+}{N}$:

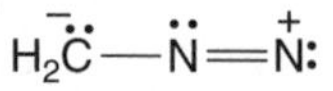

(b) $H_3C—\bar{N}—N{=}\overset{+}{N}$:

(c) $H_3C—\overset{+}{C}{=}N—\bar{N}—CH_3$

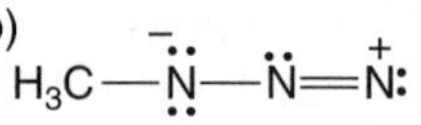

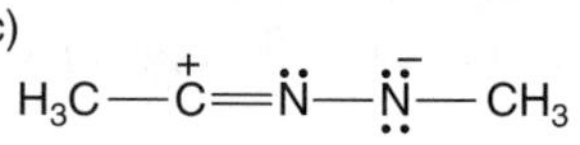

(d) $(H_3C)_2\bar{C}—N{=}\overset{+}{C}—CH_3$

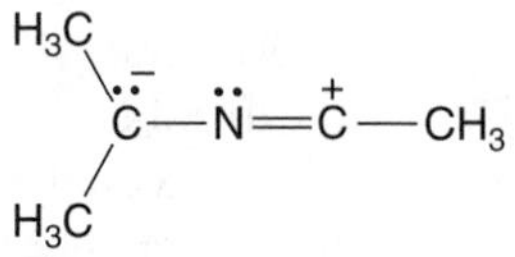

(e) $(H_3C)_2\overset{+}{C}—N(CH_3)—\bar{N}—CH_3$

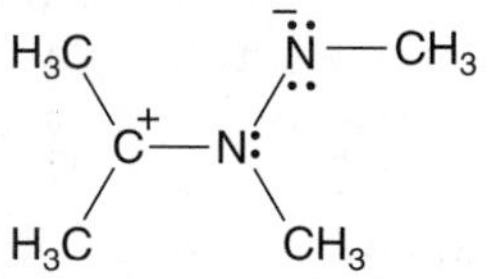

(f) $^{-}$:O—N=$\overset{+}{C}$—CH_3

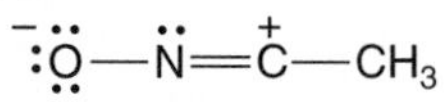

PROBLEM 1.22 Draw resonance forms for the following cyclic molecules:

(a)

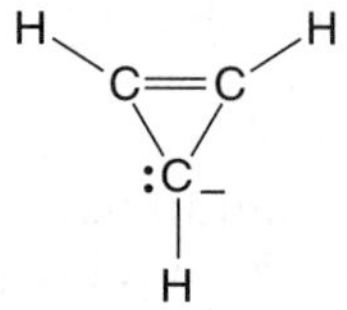

(b)

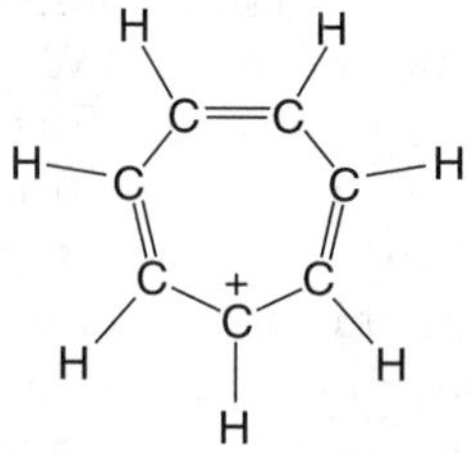

(c)

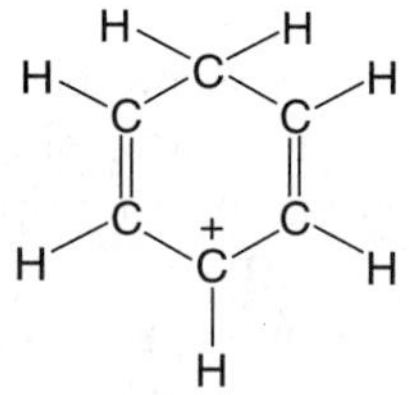

PROBLEM 1.23 Draw resonance forms for the following acyclic molecules:

(a)

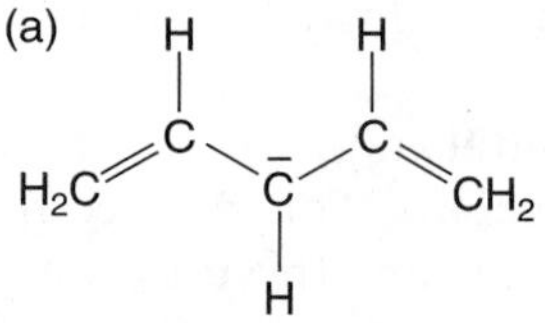

(b)

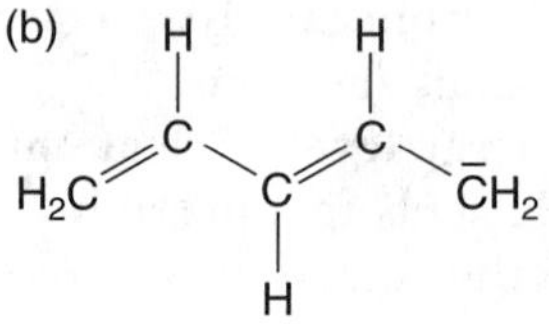

PROBLEM 1.24 Ozone (O_3) is another molecule resembling those in Problem 1.21. These days it has a rather bad press, as it is present in too small an amount in the stratosphere and too great an amount in cities. Write a Lewis "dot" structure for ozone and sketch out contributing resonance forms. Write one neutral resonance form. Be careful with this last part; the answer is tricky.

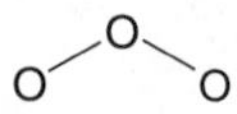

PROBLEM 1.25 Add charges to the following molecules where necessary:

(a) (b) 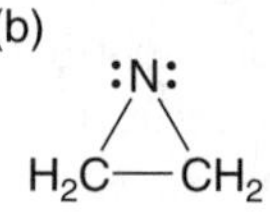 (c) 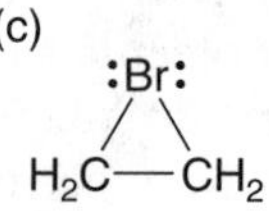 (d)

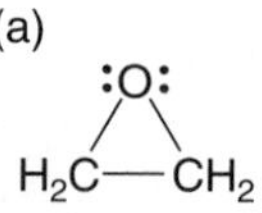

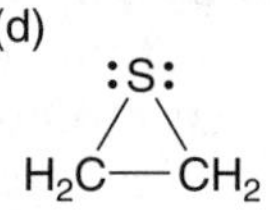

PROBLEM 1.26 Add charges to the following molecules where necessary:

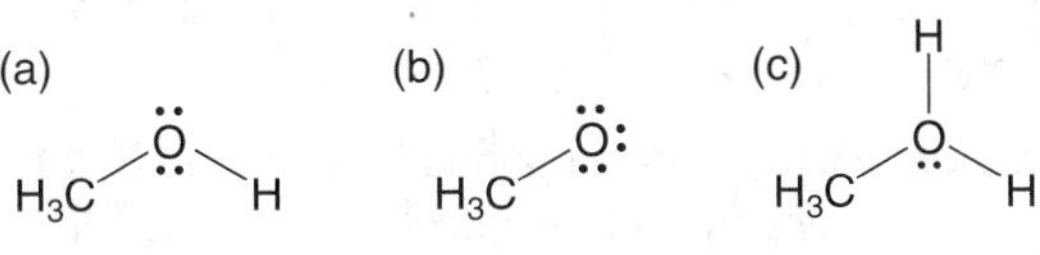

(d) H_3C–S–H (e) H_3C–S (f) H_3C–S(H)–H

(g) H_3C–N–H (h) H_3C–N(H)(H)–H (i) H_3C–N(H)–H

(j) H_3C–P–H (k) H_3C–P(H)(H)–H (l) H_3C–P(H)–H

PROBLEM 1.27 Write Lewis dot structures for the following neutral diatomic molecules. In O_2 and F_2, there is a single bond between the two atoms, but in N_2 there is a triple bond between the two atoms.

O_2 N_2 F_2

PROBLEM 1.28 Atomic carbon can exist in several electronic states, one of which is (of course) lowest in energy and is called the "ground state." Write the electronic description for the ground state and at least two higher energy, "excited states."

PROBLEM 1.29 Write the electronic configurations for the following ions:

(a) Na^+ (b) F^- (c) Ca^{2+}

PROBLEM 1.30 Write the electronic configurations for the atoms in the fourth row of the periodic table, $_{19}K$ through $_{36}Kr$. (*Hint*: The energy of the 3*d* orbitals falls between that of the 4*s* and 4*p* orbitals. Don't worry about the m_l designations of the 3*d* orbitals).

PROBLEM 1.31 Write electronic configurations for $_{14}Si$, $_{15}P$, and $_{16}S$. Indicate the spins of the electrons in the 3*p* orbitals with a small up or down arrow.

PROBLEM 1.32 There is an instrument, called an electron spin resonance (ESR) spectrometer, that can detect "unpaired spin." In which of the following atoms and molecules would the ESR machine find unpaired spin? Explain.

O O^+ O^{2-} Ne^+ F^-

PROBLEM 1.33 For the Lewis structure of carbon monoxide shown below, first verify that both the carbon and the oxygen atoms are neutral.

:C=Ö:

Second, indicate the direction of the dipole moment in this Lewis structure.

As you have just shown, based on this Lewis structure, carbon monoxide should have a substantial dipole moment. In fact, the experimentally determined dipole moment is very small, 0.11 D. Draw a second resonance structure for carbon monoxide, verify the presence of any charges, and indicate the direction of any dipole in this second resonance form. Finally, rationalize the observation of only a very small dipole moment in carbon monoxide.

PROBLEM 1.34 Would you expect formaldehyde, shown below, to have a greater dipole moment than carbon monoxide? Why or why not?

H_2C=Ö:

PROBLEM 1.35 Consider three possible structures for methylene fluoride (CH_2F_2), one tetrahedral (structure **A**), the others flat (structures **B** and **C**). Does the observation of a dipole moment in CH_2F_2 allow you to decide between structures **A** and **B**? What about structures **A** and **C**?

A B C

PROBLEM 1.36 While wandering in an alternative universe you find yourself in a chemistry class and, quite naturally, you glance at the periodic table on the wall. It looks (in part) like this:

$_1H$	$_2He$						
$_3Li$	$_4Be$						
$_5B$	$_6C$	$_7N$	$_8O$	$_9F$	$_{10}Ne$	$_{11}Na$	$_{12}Mg$
$_{13}Al$	$_{14}Si$	$_{15}P$	$_{16}S$	$_{17}Cl$	$_{18}Ar$	$_{19}K$	$_{20}Ca$

Deduce the allowed values for the four quantum numbers in the alternative universe.

PROBLEM **1.37** In a different alternative universe, the following restrictions on quantum numbers apply:

$n = 1, 2, 3, \ldots$
$l = n - 1, n - 2, n - 3, \ldots, 0$
$m_l = l + 1, l, \ldots, 0, \ldots, -l - 1$
$s = \pm \frac{1}{2}$

Call the elements in this universe **1**, **2**, **3**, ... and assume that Hund's rule and the Pauli principle still apply.

(a) For $n = 1, 2,$ and 3 show what orbital subshells are available and label them as *s, p, d,* and so on.

(b) How many electrons can be accommodated in the first three shells ($n = 1, 2,$ and 3) in this universe?

(c) Provide electronic descriptions for elements **1–14** in this universe. (In *our* universe Li would be written $1s^2 2s$.

(d) Construct a periodic table for elements **1–23** in this universe.

Molecules, Molecular Orbitals, and Bonding

Organic chemistry just now is enough to drive one mad. It gives one the impression of a primeval, tropical forest full of the most remarkable things, a monstrous and boundless thicket, with no way of escape, into which one may well dread to enter.

—Friedrich Wöhler*

In chapter 1, we examined atomic structure and began a study of the collections of atoms called molecules. This chapter will enlarge the discussion of atomic orbitals to include *molecular orbitals,* the regions of space occupied by electrons in molecules. We know that electrons on atoms are confined to certain volumes of space and that these spaces are not all alike in shape. A dumbbell-shaped $2p$ orbital is quite different from a spherical $2s$ orbital, for example. Covalent bonding between atoms involves the sharing of electrons. This sharing takes place through overlap of atomic orbitals with other atomic or molecular orbitals, the regions of space containing electrons.

Chemistry is largely the study of the structures and reactivities of molecules. A century ago, when chemistry was a young science, there seemed to be little time to worry too much about the "how and why" of the science; there was too much discovering going on, too much information to be collected. Nowadays, no longer is some vast jungle being explored by brute force; chemists are now aiming more and more for new discoveries. For example, the pharmaceutical industry is moving away from the random testing of every new compound discovered to the application of synthetic, mechanistic, and theoretical chemistry to the design of new drugs. These efforts are in their early stages, but there is every reason to believe that humankind will succeed in this endeavor. The more we understand, and the better our models of nature are, the more likely it is that the current transformation from trial-and-error methods to intellectually driven efforts will be successful.

Further, directed progress in chemistry depends on understanding how molecules react and the application of that understanding in creative ways. These days, the idea that both the structures of molecules and the

*Wöhler (1800–1882) obtained an M.D. degree, but had the good sense to take up chemistry instead of practicing medicine. One of his main contributions to both organic and inorganic chemistry was the notion that certain groups of atoms would behave in more or less the same manner regardless of the details of their molecular environment. This idea began to bring a semblance of order to organic chemistry.

reactions they undergo can be understood by examining the shapes and interactions of atomic and molecular orbitals has gained widespread acceptance in the chemical community, and an important branch of theoretical chemistry embraces these ideas. Although molecular orbital theory can be approached in a highly mathematical way, we will not follow that path. One of the great benefits of this approach is that it can be appreciated quite readily by nonmathematicians, and even very "low level," highly qualitative molecular orbital theory can provide striking insights into structure and reactivity.

In Chapter 1, we constructed Lewis dot structures for molecules. Our first task in this chapter is to elaborate on this theme to produce better pictures of the bonds that hold atoms together in molecules. We'll need to consider how electrons act to bind nuclei together, and we'll take as our initial example hydrogen (H_2), the second simplest molecule.

PROBLEM 2.1 What is the simplest molecule? If H_2 is the second simplest molecule, the answer to this question must be "H_2 minus something." What might the "something" be? The answer will appear further along in the text, so think about this question for a while now.

2.1 THE COVALENT BOND: HYDROGEN (H_2)

We start with two hydrogen atoms, each consisting of a single proton, the nucleus, surrounded by an electron in the spherically symmetrical 1*s* orbital. In principle, the situation is this simple only as long as the two hydrogen atoms are infinitely far apart. As soon as they come closer than this infinite distance they begin to "feel" each other. *Remember*: wave functions do not vanish as the distance from the nucleus increases. In practice, we can ignore the influence of one hydrogen atom on the other until they come quite close together, but as soon as they do, the energy of the system changes greatly. As one hydrogen atom approaches the other, the energy decreases until the two hydrogens are 0.74 angstrom (Å, where 1 Å = 10^{-8}cm) apart, from which point the energy of the system rises sharply, asymptotically approaching infinity as the distance between the two hydrogens declines toward zero (Fig. 2.1).

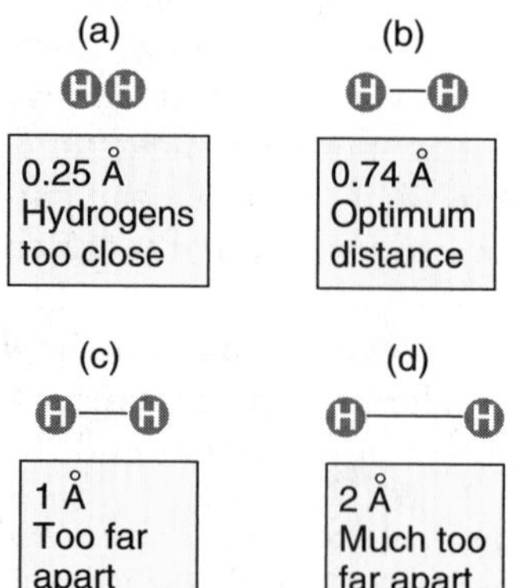

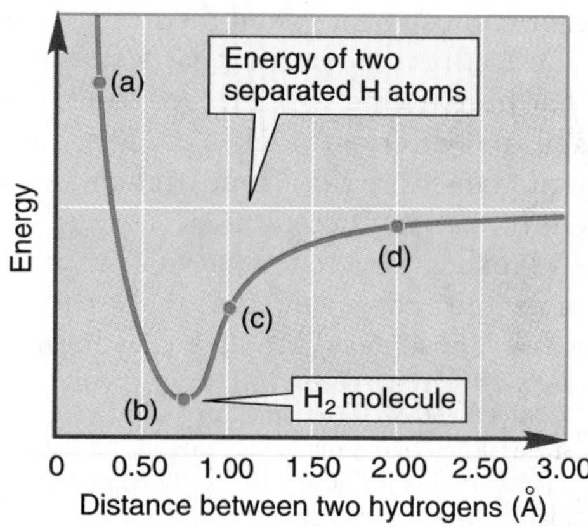

FIGURE 2.1 A plot of the energy for the hydrogen molecule (H_2) as a function of the distance between the two hydrogen nuclei. The point of minimum energy corresponds to the equilibrium internuclear separation, the bond distance.

In the hydrogen molecule, two 1*s* electrons serve to bind the nuclei together—the system H–H is *more stable than two separated hydrogen atoms.* The two negatively charged electrons attract both nuclei and hold the molecule together. Finally, when the atoms approach too closely, the positively charged nuclei begin to repel each other and the energy goes up sharply.

We begin our description of bonding by combining the two 1*s* atomic orbitals (ψ's) of the two hydrogen atoms to produce two new orbitals called molecular orbitals. The first molecular orbital (Φ) results from a simple addition of the constituent atomic orbitals, and it can be written as $\psi(H_a, 1s) + \psi(H_b, 1s) = \Phi_{bonding} = \Phi_B$. This **bonding molecular orbital** is drawn in Figure 2.2, which shows that Φ_B looks much like what one would expect from a simple addition of two spherical 1*s* orbitals.

H_a + H_b ⟶ H_a H_b

$\psi(H_a) = \psi(H_a, 1s)$ | $\psi(H_b) = \psi(H_b, 1s)$ | $\Phi_B(H_2) = \psi(H_a) + \psi(H_b)$

FIGURE **2.2** The bonding molecular orbital Φ_B, formed from the combination of two hydrogen 1*s* atomic orbitals: 1*s* + 1*s*.

Convention Alert!

Orbitals are customarily designated by Greek letters. We have already used ψ for atomic orbitals. Many people continue to use ψ for molecular orbitals as well. We will switch and use Φ. Be careful in your other reading, as you will encounter situations in which ψ is used for all orbitals and even places in which our use of Φ and ψ is reversed.

As the bonding orbital Φ_B is concentrated between the two nuclei, two electrons in it can interact strongly with both nuclei. This attraction explains why Φ_B, in which the electron density is highest between the nuclei, is strongly bonding.

Quantum mechanics tells us another important thing. *The number of wave functions (orbitals) resulting from the combining process must equal the number of wave functions (orbitals) going into the calculation.* There must be another molecular orbital that results from our combination of the *two* hydrogen atomic 1*s* orbitals. It corresponds to $\psi(H_a, 1s) - \psi(H_b, 1s)$ and is called the **antibonding molecular orbital**, Φ_A. It is shown in Figure 2.3.

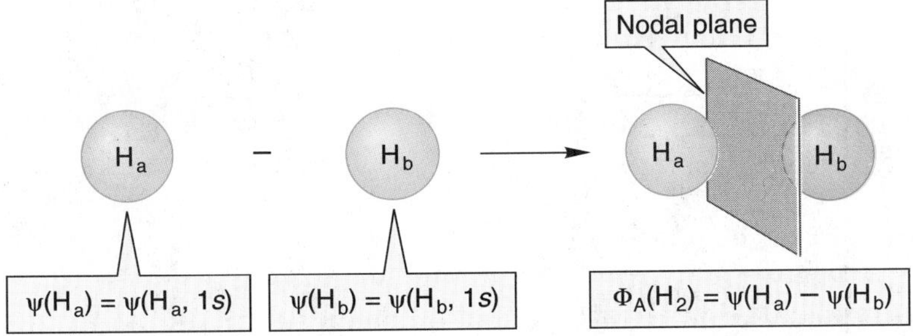

FIGURE **2.3** The higher energy, antibonding molecular orbital Φ_A, formed from the combination of two hydrogen 1*s* atomic orbitals: 1*s* − 1*s*. Notice the node—a plane separating the two lobes of opposite sign.

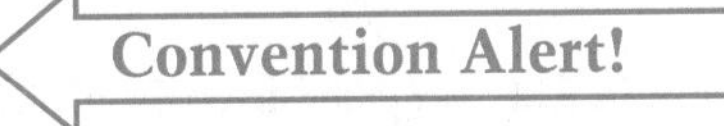

It is easy to confuse the plus and minus signs meant to designate the sign of the wave function with electrical charges or other symbols in the figures. Accordingly, it has become traditional to use color to indicate regions of different sign in orbitals. There are even traditional colors, blue and green.

In Φ_A, unlike Φ_B, there is a nodal plane between the two hydrogen nuclei. Recall that a *node* is a region in which the sign of the wave function is zero. As we pass from a region where the sign of the wave function is positive to a region in which it is negative, the sign of Φ must go through zero. As noted in Chapter 1, the square of the wave function corresponds to the electron density, and if Φ is equal to zero, then Φ^2 is also zero. Therefore the probability of finding an electron at a node must be zero as well.

In contrast to Φ_B, in the antibonding molecular orbital (Φ_A), the electron density is zero between the nuclei, because that is where there is a node. Accordingly, if a pair of electrons were in this orbital, the nuclei would be poorly shielded from each other, and electrostatic repulsion would force the two positively charged nuclei apart.

Thus, the combination of two hydrogen 1*s* atomic orbitals yields two new, molecular orbitals, Φ_B and Φ_A. This seems intuitively reasonable: There are, after all, only two ways in which two 1*s* atomic orbitals can be combined. The signs of the wave functions can be either the same, as in the bonding orbital Φ_B, or opposite, as in the antibonding orbital Φ_A. Figure 2.4 shows a very simple, and most useful graphic device for summing up these relations. The atomic orbitals (ψ's) going into the calculation are shown at the left and right sides of the figure. For molecular hydrogen, these orbitals are the two equivalent hydrogen 1*s* atomic orbitals, which must be of the same energy. They combine in a constructive, bonding way (1*s* + 1*s*) to give the lower energy Φ_B (bottom), and in a destructive, antibonding way (1*s* – 1*s*) to give the higher energy Φ_A (top).

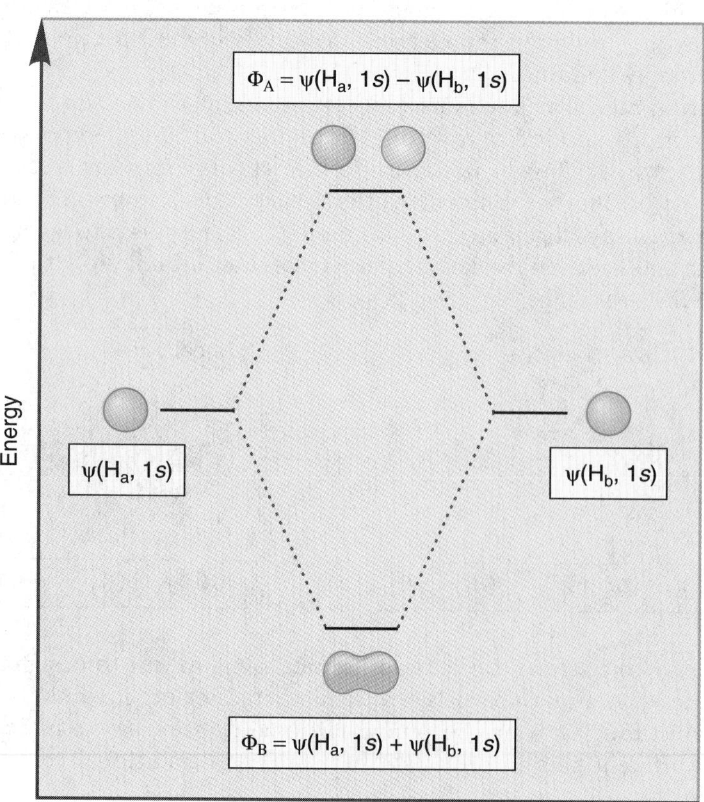

FIGURE 2.4 A graphical representation of the combination of two atomic 1*s* orbitals to form a new bonding and antibonding pair of molecular orbitals, Φ_B and Φ_A.

PROBLEM 2.2

Sketch the orbitals produced through the interaction of two carbon 2*s* atomic orbitals.

*PROBLEM 2.3

Sketch the orbitals produced through the interaction of a carbon 2*s* atomic orbital overlapping end-on with a carbon 2*p* atomic orbital.

ANSWER

The two orbitals can interact in a bonding way (2*s* + 2*p*) or an antibonding way (2*s* – 2*p*):

2*s* – 2*p*

Antibonding orbital — note the new node (bar)

2*s* + 2*p*

Bonding orbital

The analogy to more familiar, real world wave forms works here, too. Waves can reinforce or interfere with each other. If we imagine adding two waves (here representing 1*s* wave functions) as shown in Figure 2.5, we can easily see the result as building up electron density between the two atoms, exactly where it serves to hold the two nuclei together. On the other hand, the combination of waves of opposite amplitude generates a node between the two nuclei.

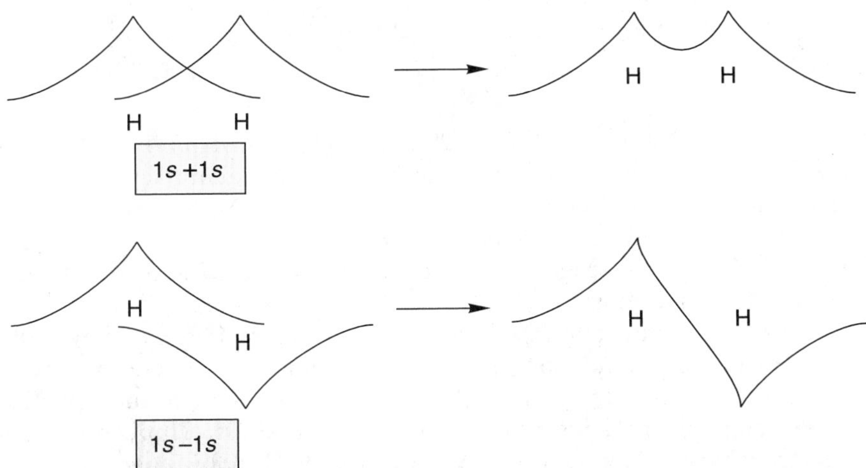

FIGURE 2.5 Two waves interacting in a reinforcing way (analogous to 1*s* + 1*s*) and a destructive way (analogous to 1*s* – 1*s*).

Note that the graphical construction of molecular orbitals takes place without reference to electrons (Fig. 2.4). Only after the diagram has been constructed do we have to worry about electrons. But now let's count them up and put them in. In the construction of H_2, each hydrogen atom brings one electron. In Figure 2.6, these are placed in the appropriate 1*s* orbitals ψ (H_a) and ψ (H_b). In the new molecule, H_2, the two electrons will go into the lower energy Φ_B. Because these two electrons are in the same orbital their spin quantum numbers must be different. The Pauli principle made this point earlier (Chapter 1, p. 30). The antibonding molecular orbital (Φ_A) is empty.

Convention Alert!

We represent electron spin with two small arrows as in Figure 2.6. The arrows point in opposite directions to indicate that the spin quantum numbers are opposite ($s = \pm 1/2$), or in the same direction to show two electrons with the same spin.

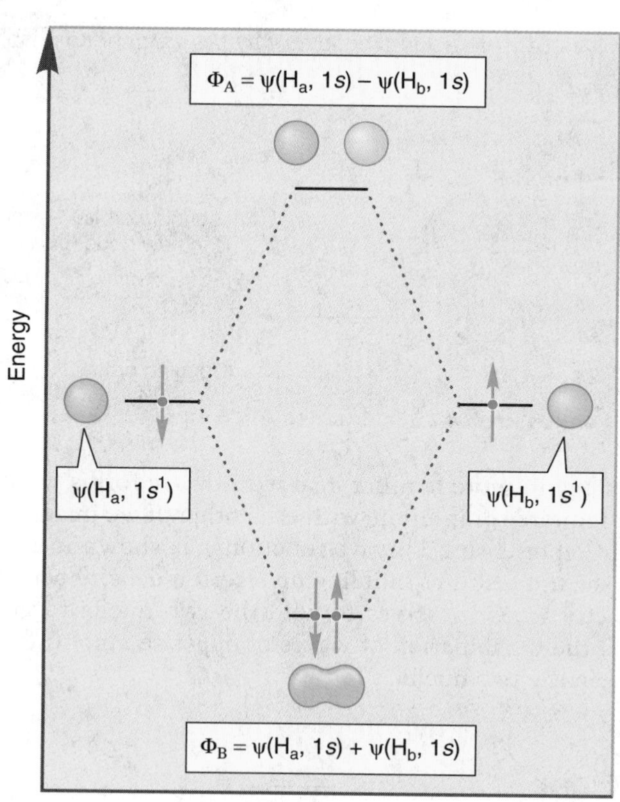

FIGURE 2.6 The electronic occupancy for the molecule H_2. The two electrons can be accommodated in the lower energy bonding molecular orbital (Φ_B) as long as the spins (shown as arrows) are opposite (Pauli principle).

Not all combinations of orbitals are productive. If two orbitals approach each other in such a way that the new bonding interactions are exactly balanced by antibonds, there is no net interaction between the two. Such orbitals are called **orthogonal** orbitals. The way in which orbitals approach each other in space is very important. Figure 2.7 shows two such "net-zero" interactions.

What is the connection between stability and energy? The lower the energy of an orbital, the more stable an electron in it is. A consequence of this stability is that the strongest bonds in molecules are formed by electrons occupying the lowest energy molecular orbitals. Throughout this chapter we will be taking note of what factors lead to low-energy orbitals and, thus, to strong bonding between atoms.

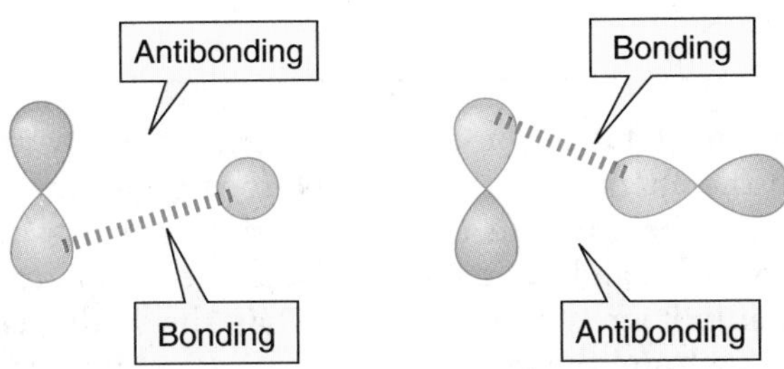

FIGURE 2.7 Two examples of noninteracting (orthogonal) orbitals. In each case, the bonding interactions are exactly canceled by antibonding interactions.

*PROBLEM 2.4

Contrast the interactions between two 2*p* orbitals approaching in the two different ways shown in Figure 2.8.

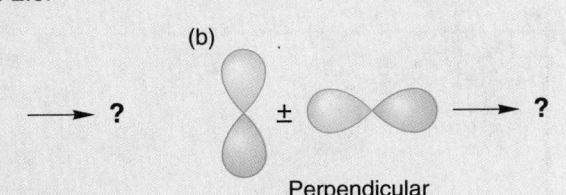

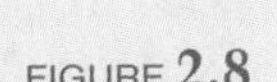

FIGURE 2.8

ANSWER

A pair of 2*p* orbitals interacting side by side produces two new orbitals (a), one bonding (2*p* + 2*p*), the other antibonding (2*p* – 2*p*). On the other hand, interaction end-on involves no net bonding or antibonding. The two exactly cancel, producing no net interaction (b). The two orbitals are orthogonal.

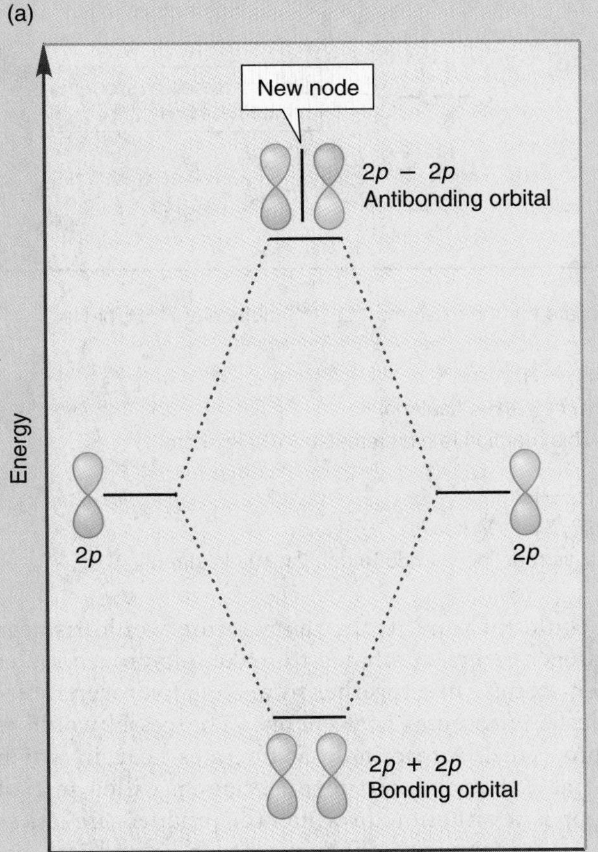

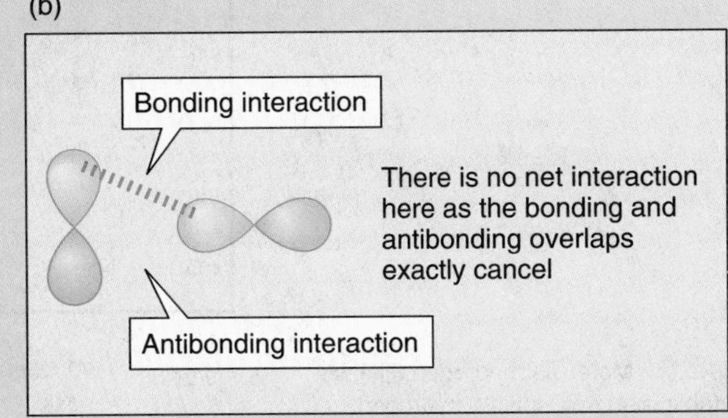

2.2 BOND STRENGTH

How much is the stabilization apparent in Figure 2.6 worth? The difference between the energies of the two electrons in Φ_B and the energies of the two separate electrons in atomic 1*s* orbitals is 104 kcal/mol. This number is very large, and the hydrogen molecule is held together (bound) by this substantial amount of energy. That the hydrogen molecule is bound by 104 kcal/mol means that this amount of energy would be liberated by the creation of a molecule of H–H from separated hydrogen atoms. This same amount of energy must be applied to break the bond and regenerate two hydrogen atoms (Fig. 2.9).

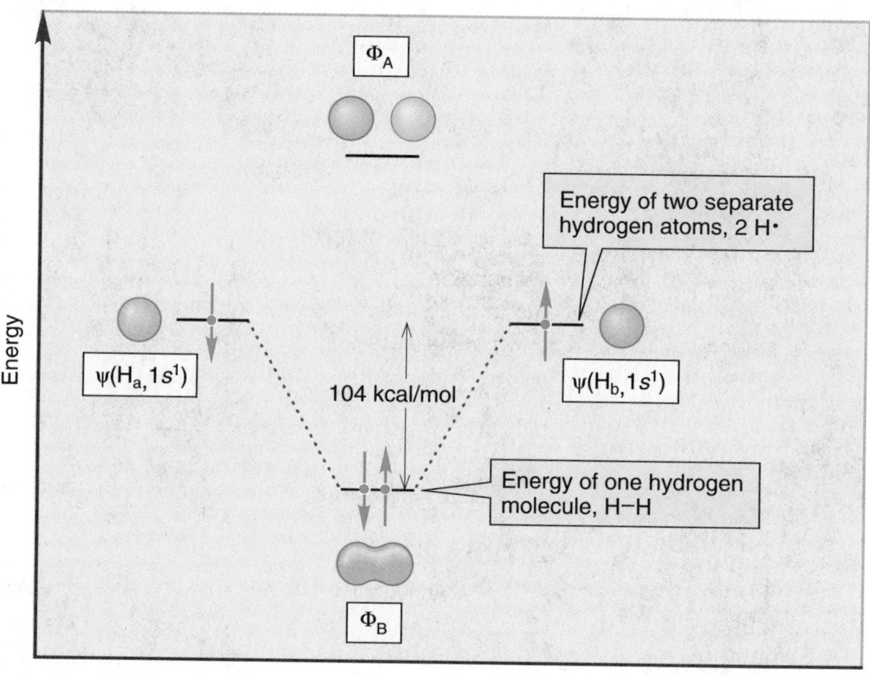

FIGURE 2.9 Molecular hydrogen (H_2) is more stable than two isolated hydrogen atoms by 104 kcal/mol. The combination of two hydrogen atoms to make a hydrogen molecule releases 104 kcal/mol and is called an exothermic reaction. The reverse reaction, the formation of two hydrogen atoms from molecular hydrogen, is called endothermic and requires the application of 104 kcal/mol of energy.

Two separate hydrogen atoms ⇌ molecular H–H, or H_2

H• + H• ⟶ H—H
$\Delta H° = -104$ kcal/mol
This reaction is exothermic by 104 kcal/mol

H• + H• ⟵ H—H
$\Delta H° = +104$ kcal/mol
This reaction is endothermic by 104 kcal/mol

Actually, if we could measure it, the temperature would rise ever so slightly as two hydrogen atoms combined to make a hydrogen molecule. As the two hydrogen atoms came together to make a hydrogen molecule, 104 kcal/mol would be released as heat energy. The vessel would warm up as the process proceeded. A reaction that liberates heat, in which the products are more stable than the starting materials, is called an **exothermic reaction**. The opposite situation, in which the products are less stable than starting material, requires the application of heat and is called an **endothermic reaction**.

The sometimes troublesome sign convention shows $\Delta H°$, the enthalpy (bond energy) change for an exothermic reaction, as a negative quantity: for A + B → C, $\Delta H° = -x$ kcal/mol. For an endothermic reaction, $\Delta H°$ is positive: for A + B → C, $\Delta H° = +x$ kcal/mol. In Figure 2.9, for example, H· + H· → H–H, $\Delta H° = -104$ kcal/mol, is an exothermic reaction, and H–H → H· + H·, $\Delta H° = +104$ kcal/mol, is an endothermic reaction.

Convention Alert!

In graphs such as Figure 2.10, energy is plotted against a quality known as reaction progress or reaction coordinate. Reaction coordinate simply means "progress of the reaction," and monitors some geometrical change taking place as the reaction proceeds. In this reaction, the formation of molecular H_2 from two hydrogen atoms, a suitable reaction coordinate might be the distance between the two hydrogen atoms. It is a fair approximation at this point to take "reaction coordinate" as indicative of the progress of the overall reaction. We will use the term "reaction progress" from now on.

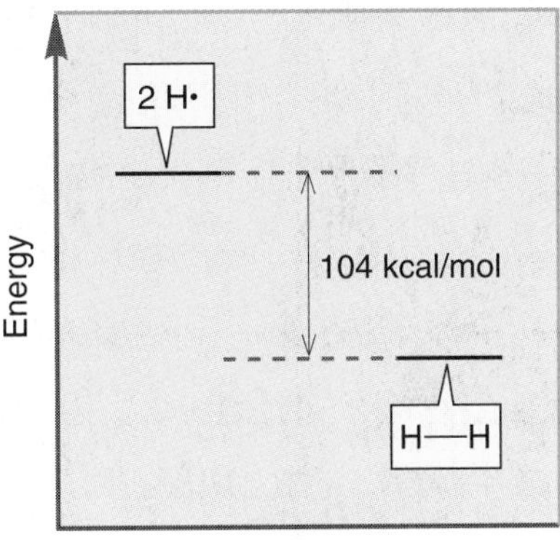

FIGURE **2.10** Another schematic picture of the formation of molecular hydrogen from two hydrogen atoms. This picture clearly shows the release of energy, in this case, 104 kcal/mol, as H–H is formed. If we read the figure from right to left, we also see the endothermic formation of two hydrogen atoms from the hydrogen molecule.

This amount of energy, 104 kcal/mol for hydrogen, is called the **bond dissociation energy** (BDE), and is the amount of energy that must be applied for **homolytic bond cleavage** (i.e., the energy needed to break the bond into a pair of neutral species). We could also imagine breaking a typical bond in another way, **heterolytic bond cleavage**, in which an anion–cation pair would be formed. Normally, this is not the path followed, as it costs even more energy to develop separated charges than to make neutral species.

Remember the arrow convention that keeps track of these two different kinds of bond breaking. Two-electron arrows have standard, double-barbed arrows, whereas single electron (homolytic) bond breakings are shown using pairs of single-barbed arrows (Fig. 2.11).

Convention Alert!

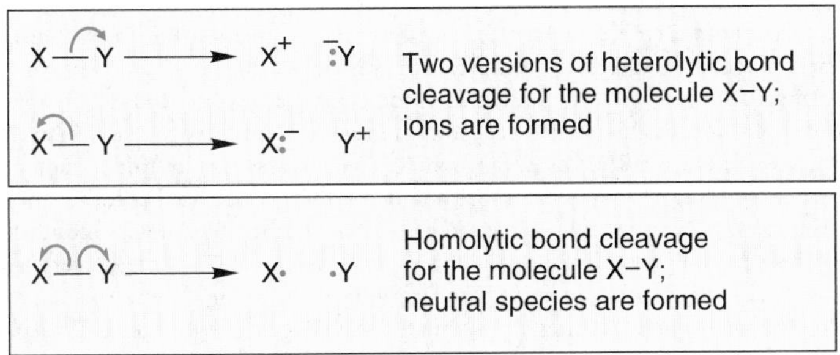

FIGURE **2.11** Three ways to break a bond. In two, an electron pair becomes no longer shared, but associated entirely with one of the atoms. This heterolytic cleavage produces a pair of ions. In the third kind of bond cleavage, the pair of binding electrons is split evenly between the atoms, giving a pair of neutral species. This process is called homolytic cleavage.

*PROBLEM 2.5 Sketch the profile of an endothermic reaction. See Figure 2.10 for the sketch of an exothermic reaction.

ANSWER Aha! A trick question. If you have written an exothermic reaction, you have already written an endothermic reaction as well. You need only read your answer backward. We are psychological prisoners of our tendency to read from left to right. Nature has no such hangups! The formation of molecular hydrogen from two hydrogen atoms (Fig. 2.10, left to right) is exothermic (104 kcal/mol of energy is given off as heat), and the formation of two hydrogen atoms from a single hydrogen molecule is endothermic by the same amount (104 kcal/mol of heat energy must be applied).

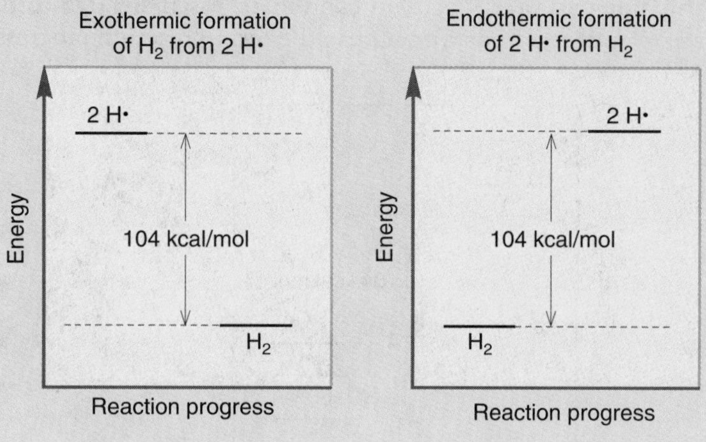

By way of calibration, a 25 °C (77 °F) room provides about 15–20 kcal/mol of thermal energy to the molecules in it. It is important to begin to develop a feeling for which bonds are strong and which are weak, to start to build up a knowledge of approximate bond strengths.* The bond in the hydrogen molecule is a strong one. As we continue our discussion of structure we'll make a point of noting bond energies as we go along. Unfortunately, there is no way to acquire this knowledge except by learning some bond strengths. Fortunately, there are not too many numbers to remember. Table 2.1 gives a few important bond dissociation energies (BDEs). These numbers do not have to be known precisely, but it is important to have a rough idea of the bond strengths of common covalent bonds.

*To describe a bond as strong or weak is an arbitrary function of human experience. A 100-kcal/mol bond is "strong" in a world where room temperature supplies much less thermal energy. It would not be strong on the sunlit side of Mercury (where it's *hot*: the average temperature is about 377 °C, or 710 °F). Similarly, if you were a life form that evolved on Pluto (where it's *cold*: the average temperature is about –220 °C, or –361 °F), you would regard as strong all sorts of "weak" (on Earth) interactions, and your study of chemistry would be very different indeed. One area of research in organic chemistry focuses on extremely unstable molecules held together by weak bonds. The idea is that by understanding extreme forms of weak bonding we can learn more about the forces that hold together more conventional molecules. Chemists who work in this area deliberately devise conditions under which species that are normally most unstable can be isolated. They create very low temperature "worlds" in which other reactive (predatory) molecules are absent. In such a world, exotic species may be stable, as they are insulated from both the ravages of heat (thermodynamic stability) and the predations of other molecules (kinetic stability).

TABLE 2.1 Some Average Bond Dissociation Energies.

Bond	BDE (kcal/mol)	Bond	BDE (kcal/mol)
C–H	96–99	C–I	52
N–H	93	C=C	146–151
O–H	110–111	C=N	143
S–H	82	C=O	173–181
C–C	83–90	C≡C	199–200
C–O	85–91	C≡N	204
C–N	69–75	H–I	71
C–F	116	H–Br	88
C–Cl	79	H–Cl	103
C–Br	66	H–F	136

This table gives averages over a range of different compounds. A few compounds may lie outside these values. In later chapters more precise specific values will appear.

Even the simplest molecule, H_2^+, is bound quite strongly.* The small amount of information already in hand—a molecular orbital picture of H_2 and the bond strength of the H–H bond (104 kcal/mol)—enables us to construct a picture of this exotic molecule and to estimate its bond strength. We can imagine making H_2^+ by allowing a hydrogen atom (H·) to combine with H^+, a bare proton. All the work necessary to create a molecular orbital diagram has already been done in the construction of Figure 2.4. We are still looking at the combination of a pair of hydrogen 1*s* atomic orbitals, so the building of a diagram for H_2^+ begins exactly as did the construction of the diagrams for H_2. Indeed, we come to precisely the same molecular orbital diagram, reproduced in Figure 2.12. The only difference

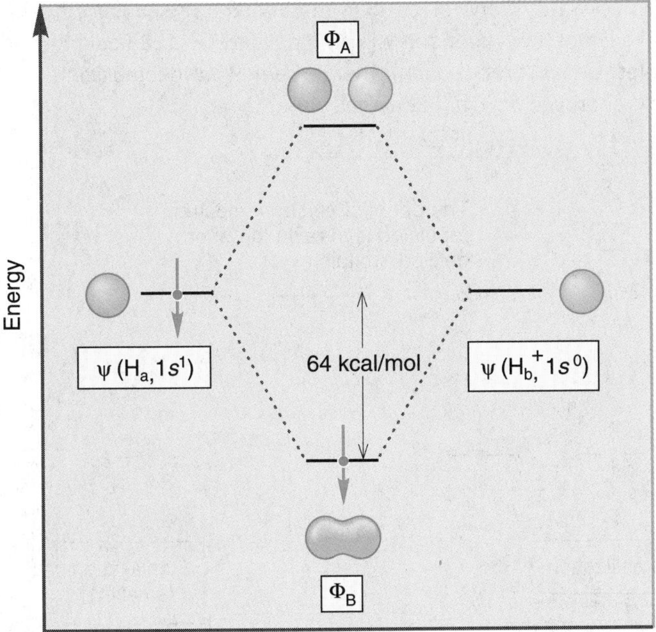

FIGURE 2.12 An orbital interaction diagram for H_2^+. Notice that the diagram is the same as the one for H_2—only the number of electrons is different.

*Here is the answer to Problem 2.1. The simplest molecule is hydrogen (H_2) minus one electron. Another electron cannot be removed to give something even simpler because H_2^{2+} is not a molecule—there are no electrons to bind the two nuclei.

comes when we put in the electrons. Instead of having two electrons, as does H_2, with one coming from each hydrogen atom, H_2^+ has only one. Naturally, it goes into the lower energy, bonding molecular orbital, Φ_B. What would we guess about the bond energy of this molecule? If two electrons in this orbital result in a bond energy of 104 kcal/mol, it seems reasonable to guess first that stabilization of a single electron would be worth about 52 kcal/mol. And we are amazingly close to being correct! The H_2^+ molecule is bound by 64 kcal/mol.

It may seem somewhat counterintuitive that a strong (low energy) bond is associated with a large number. The lower the energy, the higher the number representing bond strength! The H_2 molecule has a bond dissociation energy of 104 kcal/mol, for example, whereas the more weakly bound H_2^+ is held together by only 64 kcal/mol. A strong bond means a low-energy species. The diagram in Figure 2.10 should help keep this point straight. A strong bond does not mean high energy; it means low energy. Remember also the sign convention: 2 H· → H–H, $\Delta H° = -104$ kcal/mol.

*PROBLEM 2.6

Although our guess is quite close to reality, it is a little bit low. The H_2^+ molecule is more stable (lower in energy) than we thought. Why is our estimate of bond strength low? Why might the stabilization of two electrons in an orbital be less than twice the stabilization of one electron?

ANSWER

In making this estimate, we divided the known bond strength of H_2 by 2. The supposition was that if two electrons in the bonding orbital are stabilized by 104 kcal/mol, one electron should be stabilized by 52 kcal/mol. However, we neglected to worry about the repulsive forces between two electrons in the same orbital. The energy of H_2 is *raised* by the repulsive forces between two negatively charged electrons occupying the same orbital. When an electron is removed to form H_2^+, these repulsive forces disappear!

Look at it backward. If one electron in the bonding molecular orbital is stabilized by 64 kcal/mol, two electrons will be stabilized by 128 kcal/mol. However, these two electrons will repel each other, a somewhat destabilizing factor, and the real bond energy is only 104 kcal/mol.

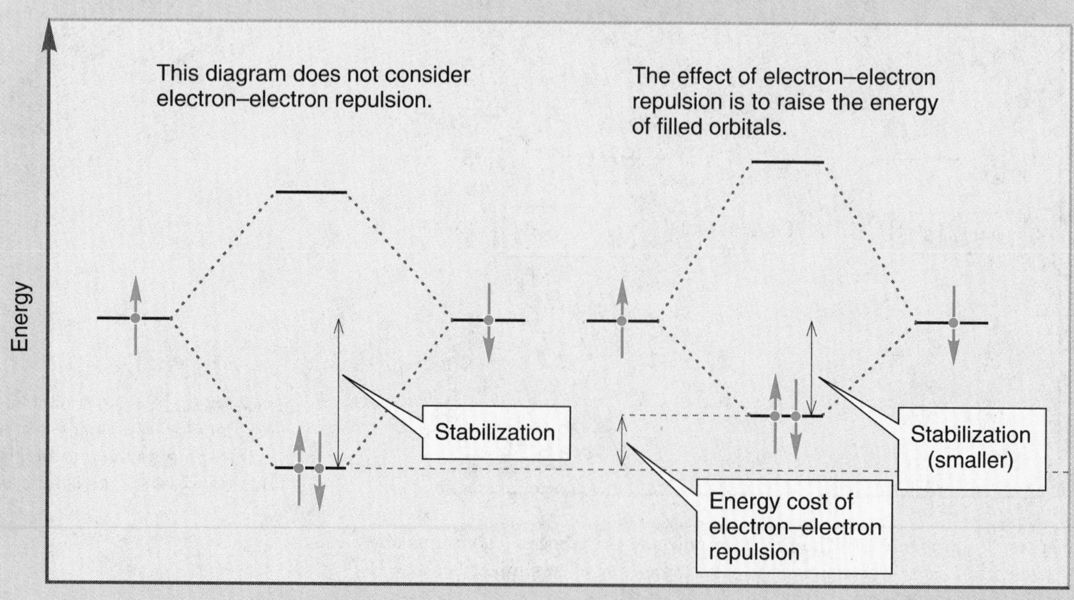
This diagram does not consider electron–electron repulsion.
The effect of electron–electron repulsion is to raise the energy of filled orbitals.
Energy
Stabilization
Stabilization (smaller)
Energy cost of electron–electron repulsion

As in H_2, the antibonding molecular orbital (Φ_A) is empty in H_2^+. Why the emphasis on Φ_A? What is an empty orbital, anyway? The easiest, nonmathematical way to think of Φ_A in H_2 is as the place the next electron would go. As an example, let's construct a molecule of helium, He_2. Each He, like each H in H_2, brings only a 1*s* orbital into the calculation. Just as in the molecules H_2 and H_2^+, the molecular orbital system is created by the combination of two 1*s* atomic orbitals. We can reuse our orbital interaction diagrams of Figures 2.4 and 2.12 to start to draw Figure 2.13.

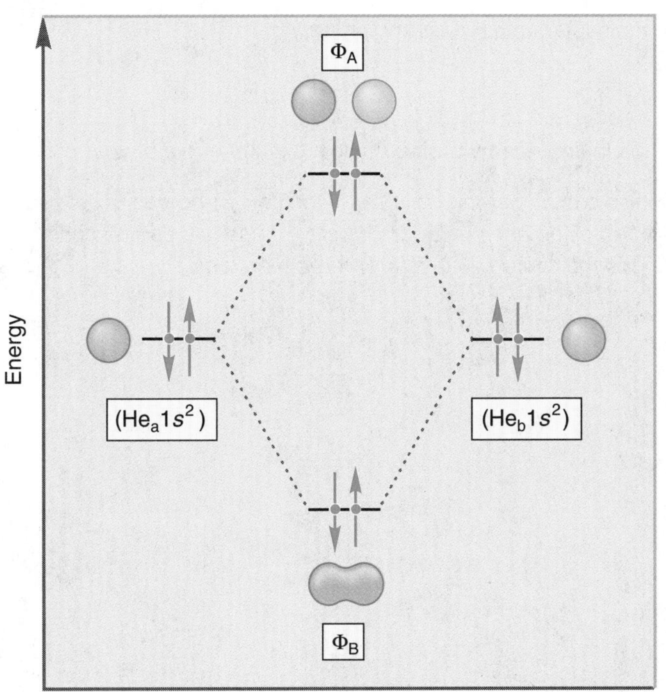

FIGURE 2.13 An orbital interaction diagram for He_2. There is no net bonding because the stabilization owing to the pair of electrons in the bonding molecular orbital is offset by the destabilization from the two electrons in the antibonding orbital.

Helium

Helium (He) is the only substance that remains liquid under its own pressure at the lowest temperature recorded. There are only about five parts per million of helium in the atmosphere, but it reaches substantially higher concentrations in natural gas, from which it is obtained. Helium comes from the radioactive decay of heavy elements. For example, a kilogram of uranium gives 865 L of helium after complete decay. There's not much helium on Earth, but there's a *lot* in the universe! About 23% by weight of the universe is helium, mostly produced by thermonuclear fusion reactions between hydrogen nuclei in stars. So, an outside observer of our universe (whatever that means) would probably conclude that helium was some of the most important stuff around.

Each helium atom brings two electrons, though, so the electronic occupancy of the orbitals will be different from what it is for H_2 or H_2^+. The bonding orbital (Φ_B) can only hold two electrons (Pauli principle, Chapter 1, p. 30) so the next two must occupy Φ_A. An electron in this antibonding orbital is destabilizing to the molecule just as an electron in Φ_B is stabilizing, so there is no net binding in the hypothetical molecule He_2. Molecular helium (He_2) is, in fact, unknown.

PROBLEM 2.7 Derive the molecular orbital diagram for He_2^+.

*PROBLEM 2.8 Estimate the binding energy for He_2^+. How did you arrive at your estimate?

ANSWER The molecular orbital diagram for this molecule can be easily derived from Figure 2.13 by removing one electron. The He_2^+ molecule can be constructed from He + He^+. This molecule will have only three electrons.

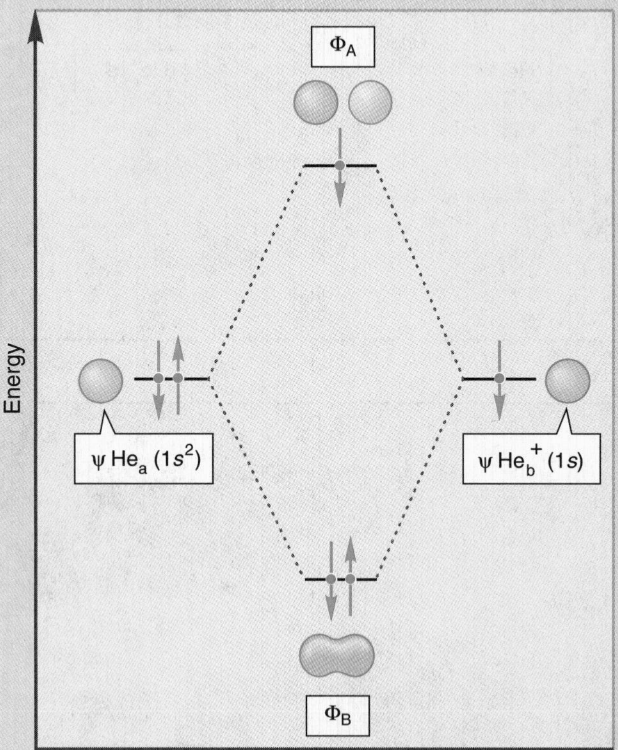

Assuming that an electron in the antibonding orbital Φ_A is destabilized about as much as one in the bonding orbital Φ_B is stabilized, there is net one bonding electron in this molecule. The He_2^+ molecule should be stabilized by about the same amount as H_2^+, another molecule with a single electron in the bonding molecular orbital. This estimate turns out to be correct, and He_2^+ is bound by a little more than 60 kcal/mol.

PROBLEM 2.9

We can use the answer to Problem 2.6 to work out the answer to a more subtle question. In He_2, both the bonding and antibonding orbitals are filled with two electrons. Consider electron–electron repulsion to explain why the stabilization of the two electrons in Φ_B is less than the destabilization of the electrons in Φ_A.

2.3 LINEAR AND SQUARE H_4

Let's use the techniques of this chapter to work out the molecular orbital description of the hypothetical, but nonetheless interesting and important molecule H_4. You may legitimately ask, Why are you inflicting this crazy molecule on me? The answer is that patterns repeat in nature, and H_4 provides the simplest model for the many four-orbital systems encountered later. For example, we will have to deal extensively with combinations of four 2*p* orbitals in considering reactions of doubly bonded carbon–carbon systems, and we need to build up the background here. The molecule H_4 is a useful model for what comes later, the somewhat more complex chemistry of larger molecules.

Up to this point we have made only diatomic molecules and geometry has not been important. For this molecule, questions of geometry certainly do arise. Several H_4 molecules are possible. We will first construct linear HHHH, then square H_4.

The choice of where to start is crucial. We know the molecular orbitals of H_2 and it is surely tempting to try to combine two of these half-made H_4 molecules into the real* thing.

So let's just take the molecular orbitals for H_2 and combine them into the molecular orbitals for linear H_4. There are two easy ways to do this, and each will work. In Figure 2.14, the two H_2 molecules are placed end to end. It would work just as well to place one hydrogen molecule inside the other (see Problem 2.10).

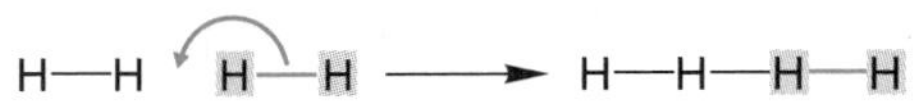

H—H H—H ⟶ H—H—H—H

FIGURE 2.14 The strategy for construction of H_4. The two H_2 molecules will be placed end to end.

When we consider molecules that are larger and harder to manipulate, we must apply simplifying assumptions to keep the business of orbital construction manageable. In the orbital-building process we will look only at the combinations of the two bonding molecular orbitals of H_2 (bonding with bonding) and the two antibonding molecular orbitals (antibonding with antibonding). Why don't we need to look at the interaction of bonding with antibonding molecular orbitals? The strength, and thus the importance, of an interaction between two orbitals depends strongly on two things. One is the degree of overlap—two orbitals at great distance from each other do not overlap significantly. This point is important—bond strength depends on orbital overlap; the more, the better. The strength of a bond between two atoms is very small if they are not quite close together. The second factor is the relative energy of the two orbitals

*Actually, linear H_4 is unstable relative to two molecules of hydrogen (H_2) and is quite unreal. One of the wonderful things about theory is that it can be utterly unhampered by adherence to reality. We can calculate what H_4 would be like if it existed, and this information will be useful later in making analogies to molecules that really do exist. Many marvelous things have been discovered by calculations of molecules that *could not* (and sometimes did not) exist.

(A and B). The strongest interaction is between orbitals of equal energy, and the interaction becomes weaker as the orbitals grow farther apart in energy. This notion is shown graphically in Figure 2.15.

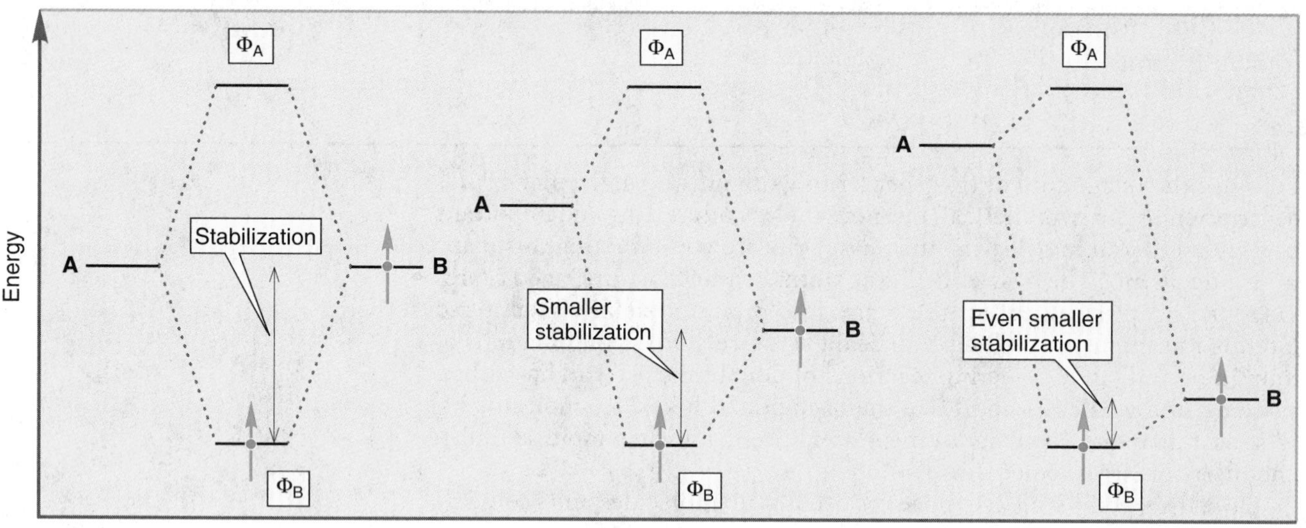

FIGURE 2.15 The interaction of two orbitals **A** and **B** as the energy difference between them increases. The closer the two orbitals are in energy, the greater the stabilization (shown as an arrow) of the orbital created by their bonding interaction, and the greater the destabilization of their antibonding combination.

In a problem such as this one for H_4, we need to consider only the strongest interactions between orbitals. In this case, that means that we must look only at the combinations of two equi-energetic bonding H_2 orbitals and of the two equi-energetic antibonding orbitals. Figure 2.16 shows the results of these combinations.

A set of four new molecular orbitals emerges: $\Phi_B + \Phi_B = \mathbf{A}$, $\Phi_B - \Phi_B = \mathbf{B}$, $\Phi_A + \Phi_A = \mathbf{C}$, and $\Phi_A - \Phi_A = \mathbf{D}$. How do we know how to order these in energy? Simple, just count the nodes; the new molecular orbitals increase in energy as the number of nodes (antibonding interactions) increases. So the ordering is as shown in Figure 2.16.

How easy this is! A very simple procedure has yielded a useful picture of the molecular orbitals of a quite exotic species, the molecule HHHH. In summary: We built the new molecular orbitals from known starting materials, in this case, the molecular orbitals of H_2. First, we allowed two bonding molecular orbitals of H_2 to interact to form the new molecular orbitals we called **A** and **B**. Next, we did the same thing with a pair of antibonding molecular orbitals to generate **C** and **D**. This level of approximation requires only that we examine the strongest orbital interactions, so it was not necessary to look at the combinations of bonding H_2 orbitals with antibonding H_2 orbitals. Finally, we counted nodes in the new molecular orbitals **A**, **B**, **C**, and **D** to order them in energy correctly. Although HHHH may seem exotic, it is an important model for the nearly identical constructions we will make for combinations of 2*p* orbitals. The next step, transforming these molecular orbitals into those for square H_4, is even easier than what we have done already.

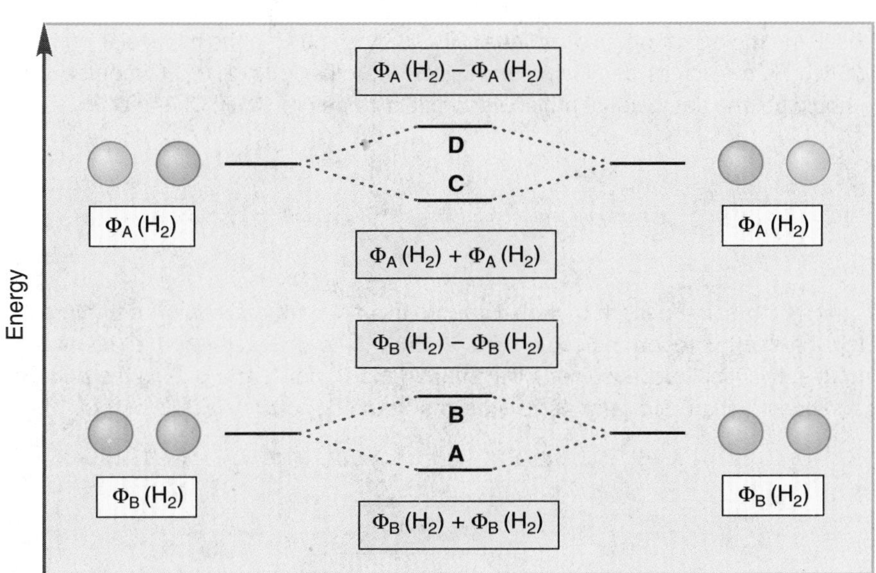

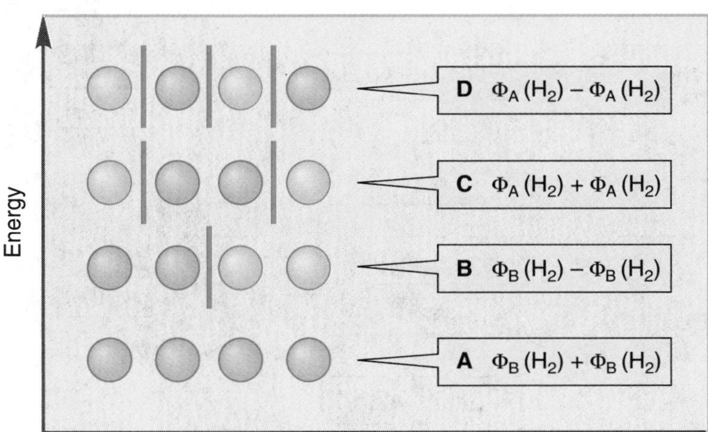

FIGURE 2.16 The set of molecular orbitals for linear H_4 (**A**, **B**, **C**, and **D**) formed from combinations of the bonding and antibonding molecular orbitals of H_2. The molecular orbitals for HHHH are ordered in energy by counting nodes (shown as bars).

Now let's make the molecular orbitals for square H_4 out of the molecular orbitals for linear H_4. To do this we need only transform the molecular orbitals we have for the linear version of H_4 (**A**, **B**, **C**, and **D**) by the appropriate bending motion into the square arrangement. What happens to the shapes and energies of the molecular orbitals as they are changed from linear to square (Fig. 2.17)?

HHHH —bend→ HH
HH

FIGURE 2.17 The schematic formation of the molecular orbitals for square H_4 from the molecular orbitals for linear H_4.

*PROBLEM **2.10** Build the molecular orbitals for linear HHHH by combining the molecular orbitals of two H_2 molecules by placing one inside the other (Fig. 2.18). Of course you should get the same set of molecular orbitals as before, **A**, **B**, **C**, and **D**.

FIGURE **2.18**

H—H H—H ⟶ H—H—H—H

ANSWER Exactly the same pattern of molecular orbitals appears as we get from interacting the two hydrogen molecules end to end. Here the ordering (just count the nodes) is critical, because $\Phi_B \pm \Phi_B$ gives **A** and **C**, not **A** and **B** as in the end-to-end construction. Similarly, $\Phi_A \pm \Phi_A$ gives **B** and **D**, not **C** and **D** as before.

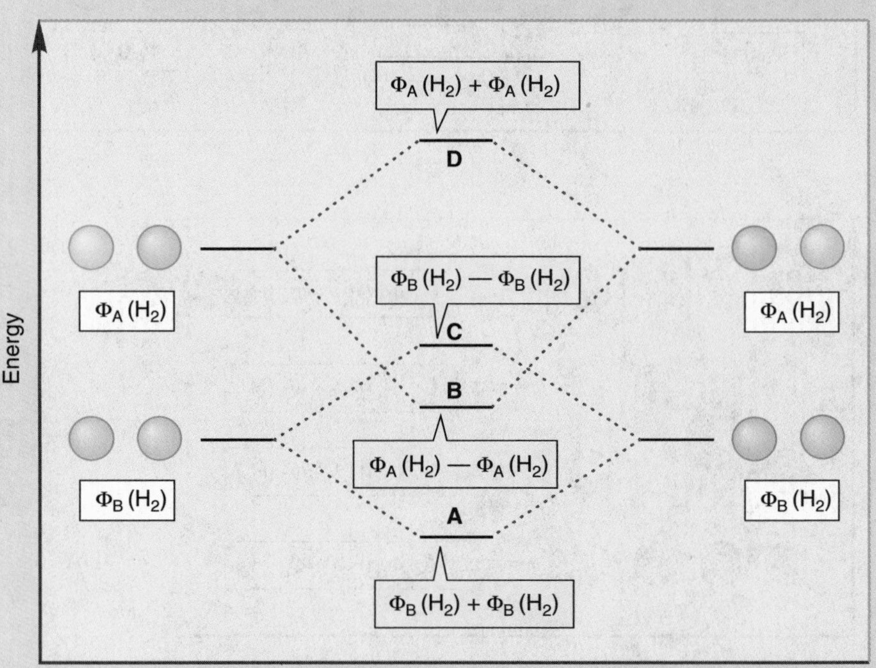

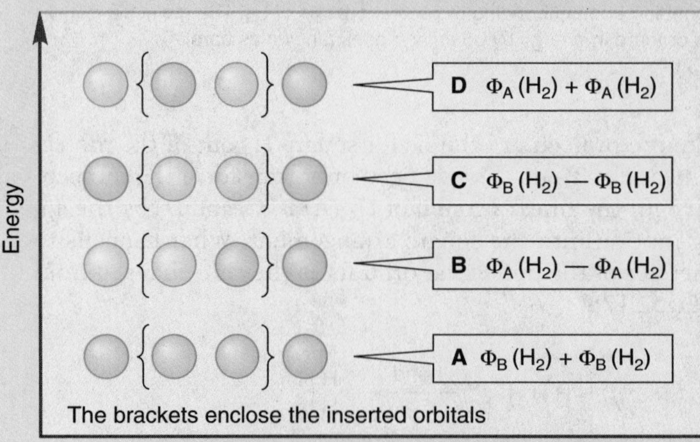

The result is a new set of four molecular orbitals, **E**, **F**, **G**, and **H**. Orbitals **A** and **C** become **E** and **F**; they are lowered in energy through the creation of a new bonding interaction between the ends of the molecule. Orbitals **B** and **D** go up in energy as a new antibonding interaction appears between the ends in **G** and **H** (Fig. 2.19).

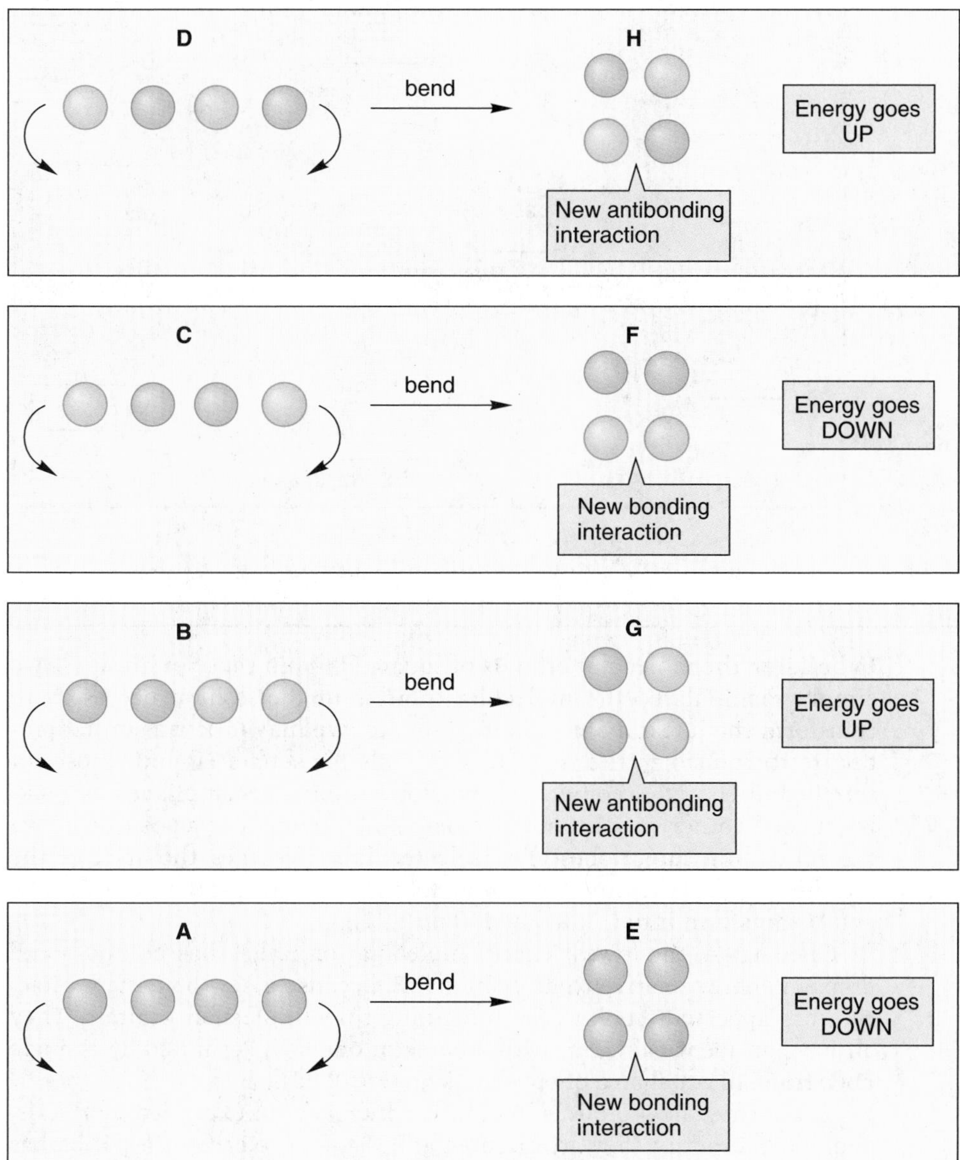

FIGURE 2.19 The result of bending the four molecular orbitals for linear HHHH (**A**, **B**, **C**, and **D**), is the creation of the four molecular orbitals for square H_4 (**E**, **F**, **G**, and **H**). The new interaction between the old terminal hydrogens of HHHH determines whether the energy of the new molecular orbital is higher or lower than that of its ancestor.

Figure 2.20 shows the results of these transformations. The new set of cyclic orbitals, **E**, **F**, **G**, and **H** can be ordered in energy by counting the nodes. Orbital **E** has none, **F** and **G** have one each and so are equal in energy, and **H** has two and is the highest in energy (Fig. 2.20).

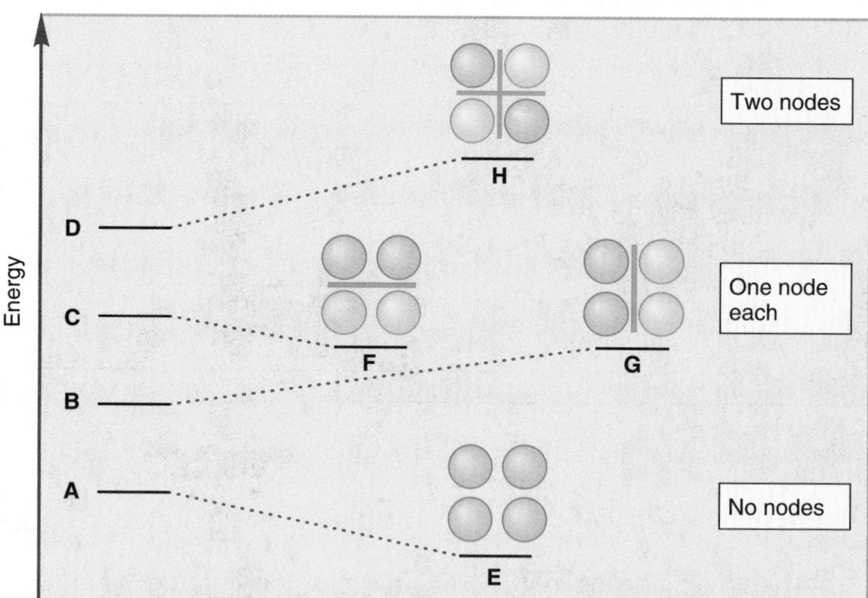

FIGURE 2.20 The energy ordering of the molecular orbitals for square H_4. The **E** orbital has no nodes, **F** and **G** have one node each, and **H** has two.

2.4 SOMETHING MORE: THE MOLECULAR ORBITALS OF TETRAHEDRAL H_4

In building the molecular orbitals of square H_4 from those of linear H_4 we simply manipulated the molecular orbitals in the fashion necessary to transform the linear molecule into a square. We may follow a similar procedure to transform the molecular orbitals for square H_4 into those for tetrahedral H_4. The geometrical transformation is accomplished by picking up one corner, or lobe, of the square and moving it to a position above the remaining three lobes. The selected lobe becomes the apex of the tetrahedron, and the three remaining lobes build the base. Orbitals **E**, **F**, **G**, and **H** transform into **I**, **J**, **K**, and **L** (Fig. 2.21).

Once again, the lowest energy molecular orbital, **I**, has no nodes and drops in energy relative to its "parent," **E**, because a new bonding interaction has appeared. Look at the remaining three molecular orbitals. They all have an identical form, with an area of one sign separated by a single node from an equal area of opposite sign (Fig. 2.22).

These three new orbitals look rather like a set of expanded *p* orbitals, and, like *p* orbitals, they are oriented at 90° angles to each other. Each has a single node, so they all have the same energy, as shown in Figure 2.22.

PROBLEM 2.11 Point out the new bonding and antibonding interactions in orbitals **I**, **J**, **K**, and **L**.

PROBLEM 2.12 Place the appropriate number of electrons in the diagrams for neutral H_4 and H_4^+ using the diagrams in Figures 2.20 and 2.22.

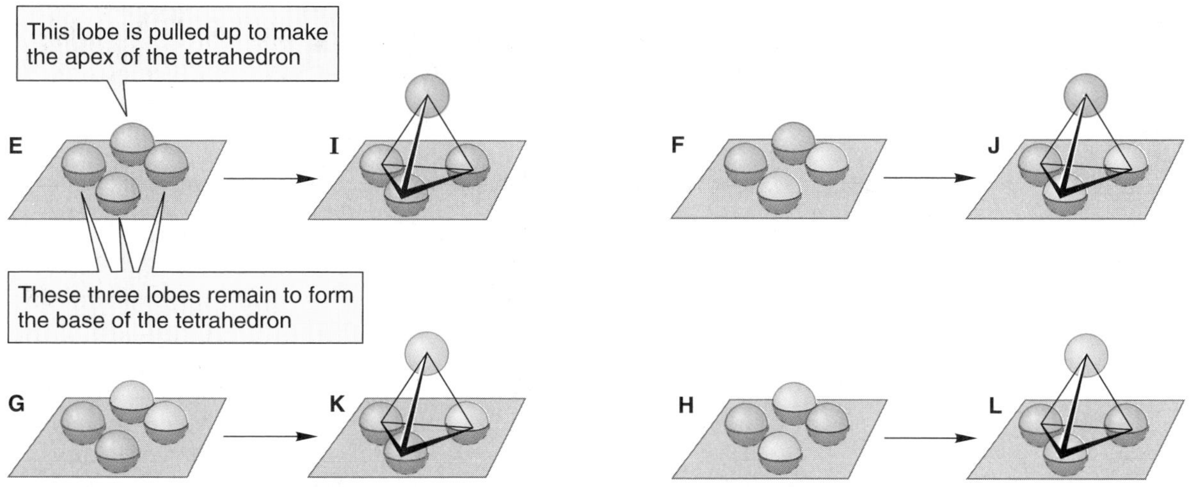

FIGURE **2.21** The method for transforming the molecular orbitals for square H_4 into those for tetrahedral H_4 is to move one lobe to a point in space above the plane of the other three lobes. The molecular orbitals for square H_4 (**E**, **F**, **G**, and **H**) transform into those for the tetrahedral species (**I**, **J**, **K**, and **L**).

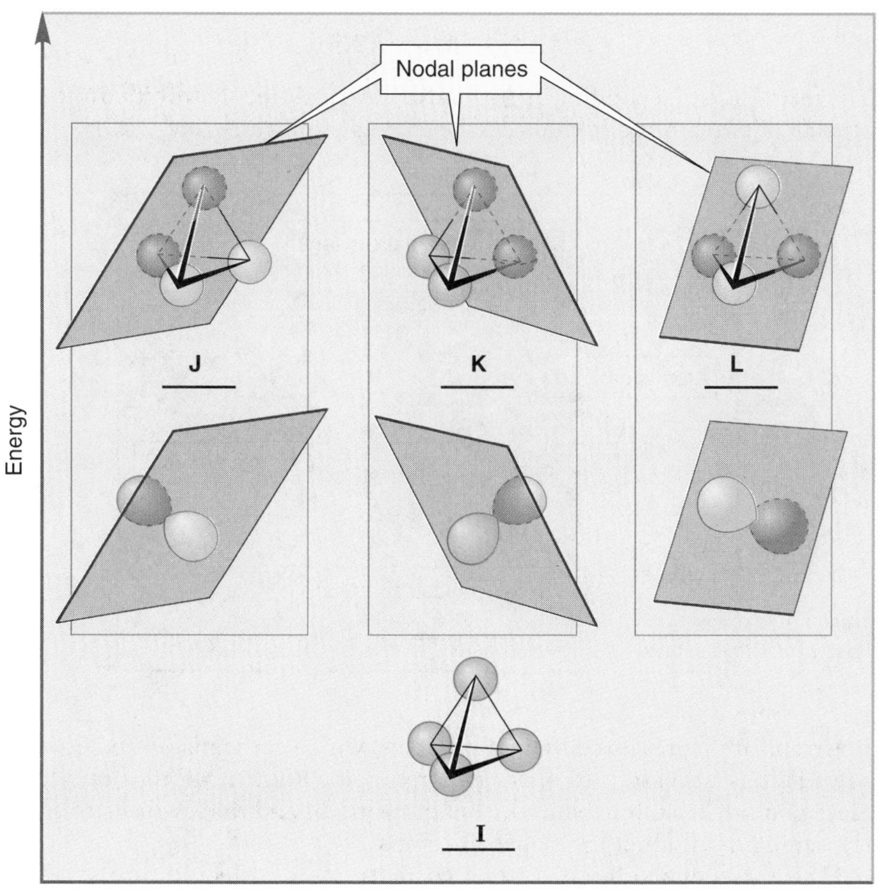

FIGURE **2.22** Each of the three molecular orbitals **J**, **K**, and **L** has a single nodal plane, and therefore the same energy. Notice the similarity to a set of atomic 2*p* orbitals.

PROBLEM 2.13 Use a diagram to show how **F** and **G** move up and **H** moves down in energy as the tetrahedron is created from the square.

PROBLEM 2.14 Which would be most stable: linear, square, or tetrahedral H_4^{2+}?

2.5 SUMMARY

NEW CONCEPTS

When two orbitals overlap, two new orbitals are formed: one lower in energy than the starting orbitals, the other higher. Two electrons can be stabilized in the new, lower energy orbital. The amount of stabilization depends on the amount of overlap of the two orbitals and on the relative energy of the overlapping orbitals. The closer the two are in energy, the greater the stabilization. The overlap of orbitals and the attendant stabilization of electrons compared with those in separated atoms are the bases of covalent bonding.

This orbital-forming process can be generalized: *n* orbitals into the calculation requires *n* new orbitals (solutions) out.

REACTIONS, MECHANISMS, AND TOOLS

The formation of a bonding and an antibonding molecular orbital from the overlap of two atomic orbitals can be shown graphically (Fig. 2.23):

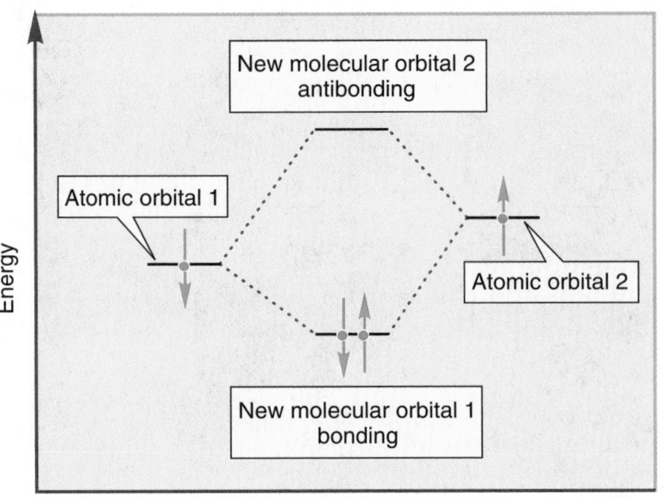

FIGURE 2.23 The overlap of two atomic orbitals produces two new molecular orbitals. One is lower in energy and bonding, the other higher in energy and antibonding.

Electrons are represented by vertical lines, which are transformed into arrows if it is necessary to show electron spin. Remember that only two electrons can be stabilized in the bonding orbital and that two electrons in the same orbital must have opposite spins.

The molecular orbitals for one geometry of a molecule can be transformed qualitatively into the molecular orbitals for a different geometry of the same molecule by applying the appropriate motion. For example, the molecular orbitals for square H_4 can be made from the molecular orbitals for the linear species H—H—H—H by moving the ends of linear H—H—H—H closer together. The resulting change in energy can be gauged qualitatively by noticing whether new bonding or antibonding interactions appear.

COMMON ERRORS

Constructing molecular orbitals through combinations of atomic orbitals (or other molecular orbitals) can be daunting, at least in the beginning. Remember these hints.

1. The number of orbitals produced at the end must equal the number at the beginning. If you start with *n* orbitals, you must produce *n* new orbitals. Here is a way to check your work as it proceeds.
2. Keep the process as simple as you can. Use what you know already, and combine orbitals in as symmetrical a fashion as you can.
3. Pay attention to questions of geometry. The question, Make the orbitals of H_4, really is meaningless. The proper reply is, What H_4 (linear, square, or tetrahedral) is to be made?
4. The closer in energy two orbitals are, the more strongly they interact. At this primitive (but useful) level of theory, you need only interact the pairs of orbitals closest in energy to each other.
5. When orbitals interact in a bonding way (same sign of the wave function), the energy of the resulting orbital is lowered; when they interact in an antibonding way (different signs of the wave function), the energy of the resulting orbital is raised.
6. To order the product orbitals in energy, count the nodes; the more nodes in an orbital, the higher it is in energy.

These methods will become much easier than they seem at first, and they will provide a remarkable number of insights into structure and reactivity not easily available in other ways.

The sign convention for exothermic and endothermic reactions seems to be a perennial problem. In an exothermic reaction, the products are more stable than the starting materials. Heat is given off in the reaction and $\Delta H°$ is given a minus sign. Thus, for the reaction $A + B \rightarrow C$, $\Delta H° = -x$ kcal/mol. An endothermic reaction is just the opposite. Energy must be applied to form the less stable products from the more stable starting material. The sign convention gives $\Delta H°$ a plus sign. Thus, $A + B \rightarrow C$, $\Delta H° = +x$ kcal/mol.

2.6 KEY TERMS

Antibonding molecular orbital Occupation of an antibonding molecular orbital by an electron destabilizes a molecule.

Bond dissociation energy (BDE) The amount of energy that must be applied to break a bond into two neutral species. See **homolytic bond cleavage**.

Bonding molecular orbital Occupation of a bonding orbital by an electron stabilizes a molecule.

Endothermic reaction In an endothermic reaction, the products are less stable than the starting materials. Energy, usually in the form of heat, must be supplied.

Exothermic reaction In an exothermic reaction the products are more stable than the starting materials. Energy, usually in the form of heat, is evolved in such a reaction.

Heterolytic bond cleavage The breaking of a bond to produce a pair of oppositely charged ions.

$$X{-}Y \longrightarrow X^+ + :\bar{Y}$$

Homolytic bond cleavage The breaking of a bond to form two neutral species.

$$X{-}Y \longrightarrow X\cdot + \cdot Y$$

Orthogonal orbitals Two noninteracting orbitals. The bonding interactions are exactly balanced by antibonding interactions.

2.7 ADDITIONAL PROBLEMS

The following problems deal mostly with the construction of molecular orbitals from atomic and/or other molecular orbitals. We have largely exhausted the possibilities for two orbitals interacting, but the first two problems give a bit more practice. We then move on to bigger systems: H_3, H_4, CH_2, NH_3, H_5, and planar methane (CH_4). These problems are demonstrations of what this kind of orbital manipulation can accomplish. The molecules H_3 and H_4 are especially important. Although they may seem to be exotic species (and are!), they serve as reference points for molecules we will deal with extensively in the future.

PROBLEM 2.15 In Problem 2.4 you looked at two possible interactions of a pair of 2*p* orbitals, side by side and perpendicular, or end-on. Now let a pair of 2*p* orbitals interact end to end, and draw the two new molecular orbitals produced.

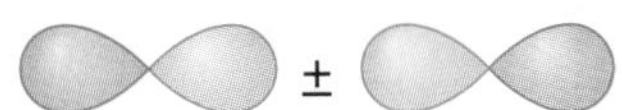

PROBLEM 2.16 In Problems 2.4 and 2.15, you looked at three possible interactions of a pair of 2*p* orbitals, side by side, end-on, and end to end. Two interactions produced a pair of new molecular orbitals, but one, the end-on interaction, was what we have called a net-zero interaction, in which the bonding overlaps were exactly balanced by the antibonding overlaps. The orbitals are orthogonal. There is a fourth possibility for two 2*p* orbitals, shown below. Is this a net-zero interaction?

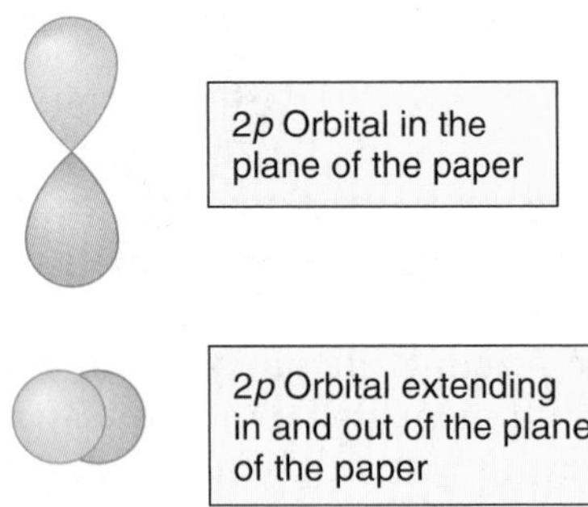

PROBLEM 2.17 Indicate whether the following reactions are exothermic or endothermic. Estimate by how much. Use Table 2.1.

(a) $H_3C\cdot + \cdot\ddot{Br}: \rightleftharpoons H_3C{-}\ddot{Br}:$

(b) $H_3C{-}\ddot{Cl}: \rightleftharpoons H_3C\cdot + \cdot\ddot{Cl}:$

(c) $H_2C{=}CH_2 + H{-}H \rightleftharpoons H_2C(H){-}CH_2(H)$

(d) $H_2C{=}CH_2 + H{-}\ddot{Cl}: \rightleftharpoons H_2C(H){-}CH_2(:\ddot{Cl}:)$

PROBLEM 2.18 Let's extend our discussion of H–H a little bit to make the molecule linear HHH. Use the molecular orbitals for H_2 and the 1*s* atomic orbital of H. Place the new H in between the two hydrogen atoms of H–H. Watch out for net-zero interactions!

H–H ± H ⟶ H–H–H

H_2, Φ_B, Φ_A H, 1*s*

H_2 Molecular orbitals ± H Atomic orbital ⟶ Molecular orbitals for HHH

(a) How many orbitals will there be in linear HHH?
(b) Use the molecular orbitals of H_2 and the 1*s* orbital of hydrogen to produce the molecular orbitals of linear HHH. Sketch the new orbitals.
(c) Order the new orbitals in energy (count the nodes).

PROBLEM 2.19 (a) Use the molecular orbitals for linear HHH to produce the molecular orbitals for triangular H_3. Recall the transformation of linear HHHH into square H_4 you did in the chapter (p. 57).

(b) Place the molecular orbitals of linear and triangular H_3 on the same energy scale. At this level of theory (low, but very useful), orbitals with the same number of nodes can be placed at the same energy level for a given molecule. This does not mean that orbitals of *two different molecules* with the same number of nodes are at the same energy.

PROBLEM 2.20 Which will be lower in energy, linear H_3^+ or triangular H_3^+? What a sophisticated question! Yet, given the answer to Problem 2.19, it is very easy.

PROBLEM 2.21 Now make a set of molecular orbitals very much like the ones you made for linear HHH in Problem 2.18, but this time use $2p$ orbitals, not $1s$ orbitals. Place one $2p$ orbital in between the other two. Use the molecular orbitals $2p + 2p$ and $2p - 2p$ that you constructed in Problem 2.4.

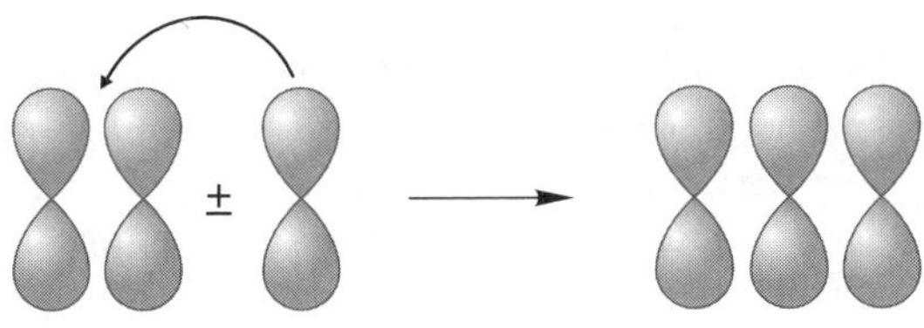

Order the new orbitals in energy.

PROBLEM 2.22 Both in the chapter and in Problem 2.10 you made the orbitals for linear HHHH. Make a related set of molecular orbitals, but use $2p$ orbitals rather than $1s$ orbitals. Again, use the molecular orbitals $2p + 2p$, and $2p - 2p$ that you constructed in Problem 2.4 as your building blocks.

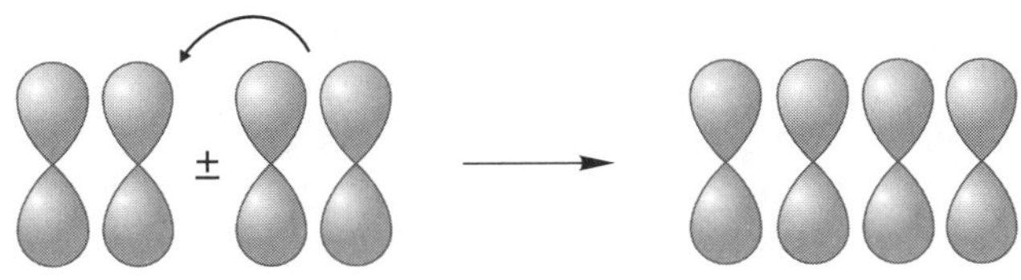

Order the new orbitals in energy.

PROBLEM 2.23 You have already seen how the molecular orbitals of square H_4 can be formed by bending the molecular orbitals of linear H_4 (see Fig. 2.19). Now construct the molecular orbitals for square H_4 by combining two sets of H_2 molecular orbitals (Φ_B, Φ_A) side by side. *Remember*: You need only consider interactions of Φ_B with Φ_B and Φ_A with Φ_A. Of course you should get the same set of molecular orbitals as before, **E**, **F**, **G**, and **H**.

Now comes a set of more challenging problems. It is probably not necessary that you be able to do problems of this level, but they are well within your range now, and they demonstrate the power of these simple orbital manipulations. Besides, this kind of thing is fun. The questions themselves will lead you through the answers.

PROBLEM 2.24 Generate the molecular orbitals for linear methylene, H—C—H, by combining the atomic orbitals of carbon with the molecular orbitals of hydrogen, H_2.

(a) Show clear pictures of the molecular and atomic orbitals you are using.
(b) How many molecular orbitals will linear methylene have?
(c) Draw pictures of the molecular orbitals for linear methylene.
(d) Order these molecular orbitals in terms of energy. Place them on a scale relative to the energy of a lone, nonbonding carbon $2p$ orbital.
(e) Place the appropriate number of electrons in the orbitals, being careful to indicate the spin quantum number for each electron (use a little ↑ or ↓ arrow to show spin).

PROBLEM 2.25 Generate the molecular orbitals for planar ammonia, NH_3. Do this by taking combinations of the molecular orbitals for triangular H_3 (see Problem 2.19 for these orbitals) and the atomic orbitals of nitrogen.

(a) Show clear pictures of the molecular and atomic orbitals you are using to make the molecular orbitals of planar ammonia.
(b) How many molecular orbitals will ammonia have?
(c) Draw pictures of the molecular orbitals for planar ammonia. Show clearly how these are generated .
(d) Order the bonding and nonbonding molecular orbitals in terms of energy. Place them on a scale relative to the energy of a lone, nonbonding $2p$ orbital. You do not have to order the antibonding orbitals.
(e) Place the appropriate number of electrons in the orbitals. Be careful to indicate the spin quantum number for each electron (use a little ↑ or ↓ arrow to show spin).

PROBLEM 2.26 The compound H_5 could have the trigonal bipyramidal structure **A**. Regardless of whether or not it does have this structure (or even if H_5 exists), you can calculate the molecular orbitals for such a species.

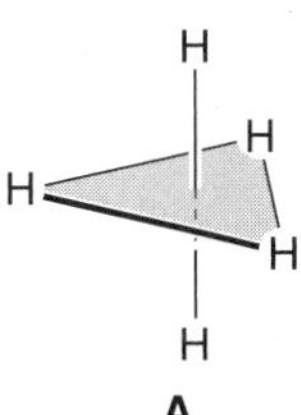

A

(a) Show the molecular orbitals for H_2 and triangular H_3 (Problem 2.19).
(b) Determine how many molecular orbitals **A** will have.

(c) Combine the molecular orbitals for H_2 and cyclic H_3 in an appropriate way to make **A**. Be very careful to show the orbitals you are using and show how you are combining them to make the orbitals of **A**. *Remember*: You need only interact the orbitals closest in energy.

(d) Order the molecular orbitals of **A** in energy.

(e) Place the appropriate number of electrons in the orbitals. Don't forget to take account of electron spin.

PROBLEM 2.27 Methane (CH_4) is tetrahedral. Nevertheless, try to imagine planar methane and construct its molecular orbitals.

Allow the molecular orbitals of square H_4 to interact with the atomic orbitals of carbon to produce the molecular orbitals of planar methane. You may start with the molecular orbitals of square H_4 (p. 59, Fig. 2.19).

(a) Show clearly the molecular and atomic orbitals you are using.

(b) How many molecular orbitals will planar methane have?

(c) Draw the molecular orbitals for planar methane. Show how these are produced by clearly indicating which molecular orbitals are interacting with which atomic orbitals.

(d) Give the energy ordering for all bonding and nonbonding orbitals of planar methane. You need not worry about antibonding orbitals.

(e) Place the appropriate number of electrons in the orbitals.

3

Alkanes

God never saw an orbital.
—Walter Kauzmann*
February, 1964

The chemical reactions of a given compound are a function of its detailed structure. No compound reacts exactly like another, no matter how similar their structures. Yet there are gross generalizations that can be made, and some order can be carved out of the chaos that would result if compounds of similar structure did not at least react in somewhat the same fashion. To a first approximation, the reactivity of a molecule depends on the presence of specific **functional groups**, commonly encountered groups of atoms that generally react in the same way in all molecules.

The endpaper at the beginning of this book presents an overview of some of the functional groups that are important in organic chemistry. This endpaper is not to be memorized, and it is certainly not complete. You will want to work through it, however, being certain you can write good Lewis structures for all the different kinds of molecules, and begin to become familiar with the variety of functional groups important in organic chemistry. These different kinds of molecules will all be discussed in detail in later chapters. We start in this chapter with **alkanes**, the simplest members of the family of **hydrocarbons**—compounds composed entirely of carbon and hydrogen.

Alkanes all have the molecular formula C_nH_{2n+2}. The simplest of the alkanes is **methane**, CH_4. Vast numbers of related molecules can be constructed from methane by replacing one or more of its hydrogens with other carbons and their attendant hydrogens. Linear arrays can be made, as well as branched structures and even rings (ring compounds, called cycloalkanes, have a slightly different formula: C_nH_{2n}). Figure 3.1 shows a few schematic representations. Before we can go farther we need to make sense of these schematics, and as usual, we must start with structure. The first task is to describe the shape of methane. Once we have determined this, we can go on to investigate other members of the alkane family.

*Walter Kauzmann (b. 1916) was chairman of the Princeton University chemistry department in 1964 when I interviewed for a job. He made this comment in response to my typically convoluted molecular orbital answer to some simple question.

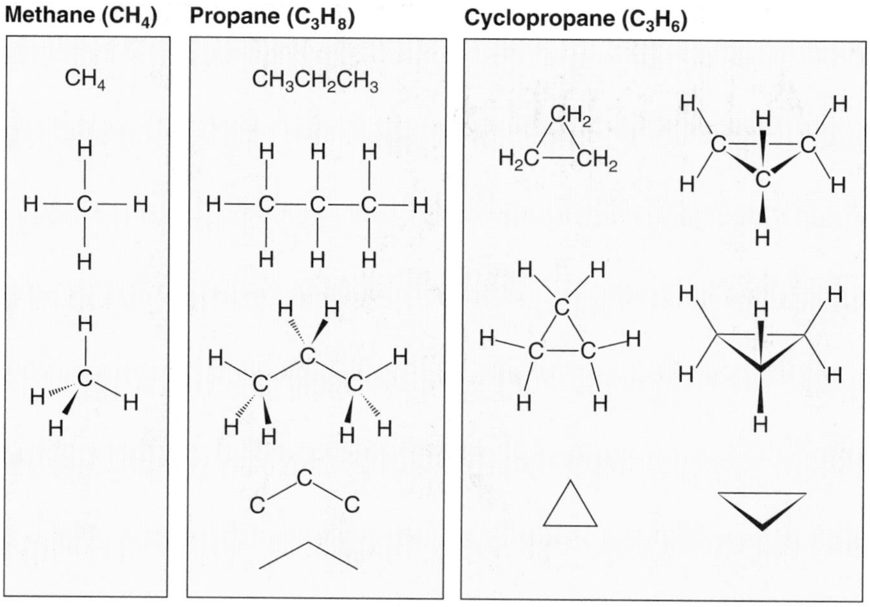

FIGURE 3.1 Several different ways of drawing alkane structures. Notice, but don't worry about, the extremely schematic representations for these molecules. Note the different conventions used in these representations.

PROBLEM 3.1 Draw Lewis structures for the following two linear alkanes (no branching—all carbons in a row): butane (C_4H_{10}) and pentane (C_5H_{12}).

3.1 THE STRUCTURE OF METHANE (CH_4)

As always, we start with what we already know, in this case the structure of a carbon atom. The molecular formula of methane tells us that we need to make four bonds between hydrogen and carbon. This kind of molecular construction was already done, in Chapter 2, when we made H_2 out of two separated hydrogen atoms (p. 42). We will follow a similar process to generate our "first-guess" structure for methane. This first guess will turn out to be (badly) wrong. One may legitimately wonder why one would go through such an exercise. Why not go straight away to the Truth? First, there isn't any Truth—what is here is bound to be flawed. Second, we learn from mistakes. We'll often find out a great deal from our errors, and this is just such an example. Much can be learned from the ways in which our first guess must be adjusted in order to describe Nature better. This process of successive approximation is very common in science. Our development of the structure of methane will serve as a prototypal discussion. We go slowly and in great detail here because we will use the concepts of the next few pages repeatedly as we work through the various structural types of organic chemistry.

A carbon atom has the electronic configuration $1s^22s^22p_x2p_y$. The electrons in the $1s$ orbital are too low in energy to participate strongly in bonding, and the $2p_z$ orbital is empty. As our first-guess model we can imagine forming bonds between each of the four hydrogen $1s$ orbitals and the orbitals on carbon that contain electrons. In doing this, we mimic the construction of the hydrogen–hydrogen bond in H_2 from the overlap of a pair of singly occupied hydrogen $1s$ orbitals. Figure 3.2 shows the overlap of the available carbon $2s$, $2p_x$, and $2p_y$ orbitals with the four hydrogen $1s$ orbitals. Two predictions are clear: first, we should have two carbon–hydrogen bonds formed from the overlap of the hydrogen

1*s*–carbon $2p_x$ and hydrogen 1*s*–carbon $2p_y$ orbitals; second, these bonds must be at 90° to each other, because of the 90° angle between the carbon 2*p* orbitals. The third and fourth bonds are difficult to locate exactly, as in this first model they are formed from the overlap of a carbon 2*s* orbital with two hydrogen 1*s* orbitals. Both the 2*s* and 1*s* orbitals are spherically symmetric, and thus no directionality can be induced by the shape of the orbitals. We might guess that these two new carbon–hydrogen bonds would be aimed so as to keep the electrons in the bonds as far from each other as possible, thus minimizing repulsions.

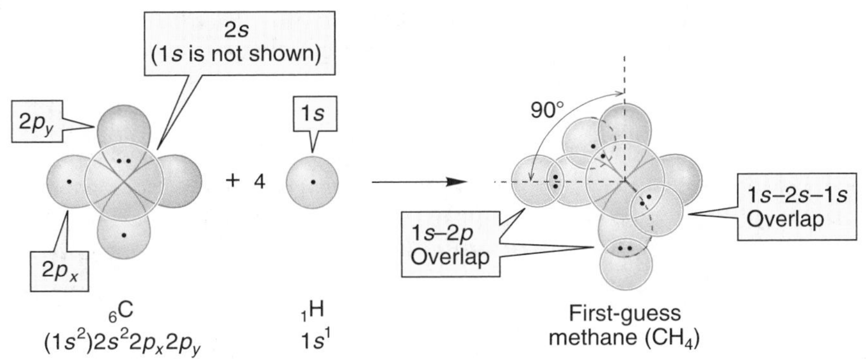

FIGURE 3.2 A first-guess structure for methane, CH_4. Bonds are formed by the overlap of hydrogen 1*s* with carbon 2*p* and 2*s* orbitals. This requires some 90° bond angles in this model. Carbon's 1*s* electrons are shown in parentheses because they are not used in bonding.

*PROBLEM 3.2

Sometimes people see a contradiction in this first-guess model. There appear to be too many electrons in the carbon 2*s*–hydrogen 1*s* system. We have spent some effort in Chapter 1 to explain the Pauli principle (p. 30), and to show why only two electrons may occupy any atomic or molecular orbital. You are used to thinking of two-electron bonds, and our new system is clearly more complicated than that. But there is no violation of the Pauli principle in the model above. Explain.
Hint: Remember that overlap of *n* orbitals produces *n* new orbitals. See Problem 2.18.

ANSWER

This problem asks you to consider the molecular orbital system made up of overlapping 1*s*, 2*s*, and 1*s* orbitals containing four electrons. Where are these four electrons? *Three* overlapping atomic orbitals will produce *three* molecular orbitals. That is all we really need to notice in this problem. There is plenty of room for four electrons. One of the three new orbitals is bonding, one nonbonding, and one antibonding. The four electrons will occupy the lowest two molecular orbitals. There is no violation of the Pauli principle here. See also the discussion in Problem 2.18 (p. 64) of the linear HHH molecule.

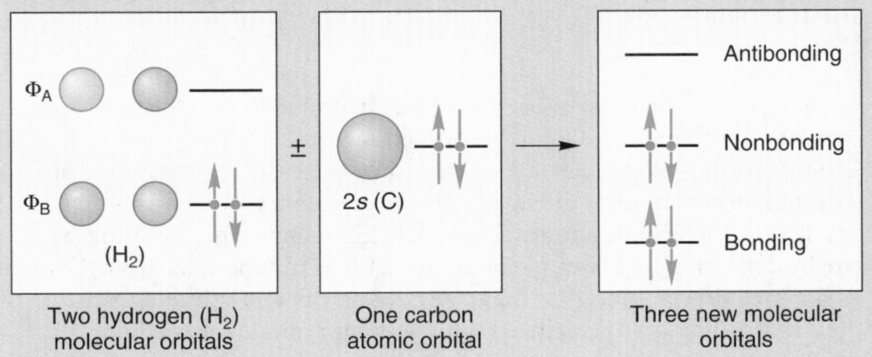

Our first-guess model has a fatal flaw. Modern spectroscopic methods show that there is no 90° H—C—H angle in methane and that all carbon–hydrogen bonds are of equal length! The four carbon–hydrogen bonds are identical. So we're wrong, after all, even though the Pauli principle is not violated. See Problem 3.2. Knowing that we are wrong, we can begin to rationalize why, and try to construct a model that does a better job of describing Nature. Don't be offended by this—science works this way all the time. It's very rare that we'll be smart enough to get it right the first time. The goal is to use what we learn from our wrong guesses to come closer next time. We can already anticipate one problem in our first model—the electrons in some of the bonds are only 90° apart. They could be farther apart, and thus electron–electron repulsion could be reduced.

PROBLEM 3.3 What about this repulsion? To be sure, the negatively charged electrons and positively charged nuclei will repel other like charges. But there is more to it than that. Consider the interaction through space of two bonds, each containing two electrons in a filled orbital, and show that the farther apart they are, the lower in energy the system must be.

There are other, more subtle problems. If you worked out Problem 3.2 you know that the H—C—H, 1*s*–2*s*–1*s* bonding system consists of three orbitals and four electrons. A system of three molecular orbitals emerges: one bonding, one nonbonding, and one antibonding. The two lowest energy molecular orbitals are occupied by the four available electrons (Fig. 3.3).

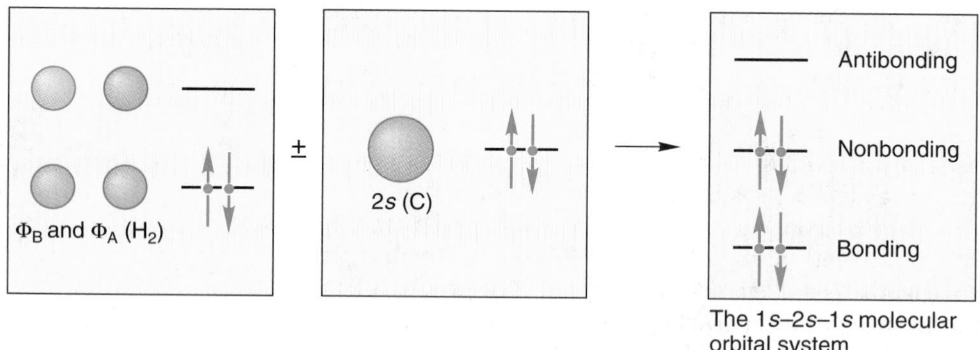

FIGURE 3.3 The orbital system made from the overlap of a carbon 2*s* atomic orbital with two hydrogen 1*s* orbitals consists of three molecular orbitals: one bonding, one nonbonding, one antibonding.

Although the lowest energy orbital is bonding, the middle orbital is essentially nonbonding. This is not an efficient use of electrons in holding atoms together. Bonded systems are more stable than their separated constituent atoms because of the stabilization originally shown in Figure 2.9 (p. 48), and repeated here in Figure 3.4. Electrons in nonbonding orbitals are by definition no lower in energy than those in separated atoms. If there is a better (lower energy) arrangement of atoms and orbitals, Nature will find it. It is our job to construct a model that reproduces Nature.

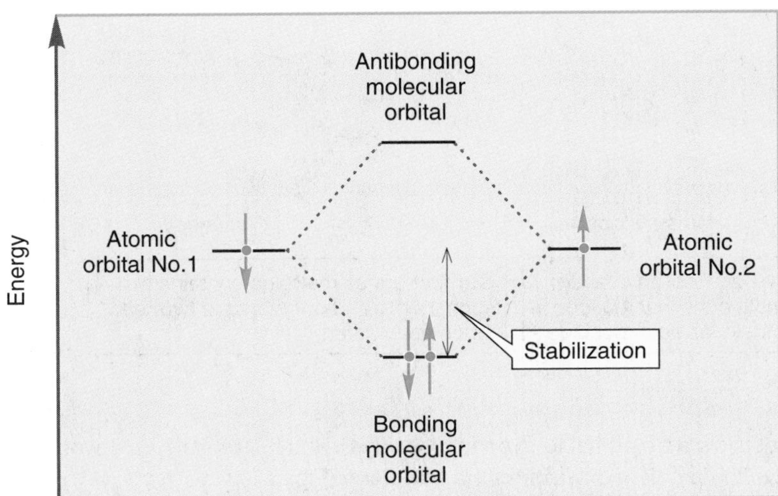

FIGURE 3.4 The interaction of two atomic orbitals produces a bonding and antibonding pair of molecular orbitals. An electron in the bonding molecular orbital will be stabilized (lower energy) relative to an electron in one of the separated atomic orbitals.

A third problem with our first-guess model might be labeled "inefficient overlap," or "wasted orbitals." Recall from Chapter 2 (p. 55) that the strength of a bond (the amount of stabilization) is directly related to the overlap of the constituent orbitals. Efficient overlap of atomic orbitals leads to highly stabilized, low-energy molecular orbitals. Poorer overlap leads to reduced stabilization. One way to reduce overlap (and thus, stability) is to increase the distance between the orbitals. The H_2 system is less stable at a bond distance of 1 Å than at its optimum distance of 0.74 Å (Chapter 2, p. 42; Fig. 3.5).

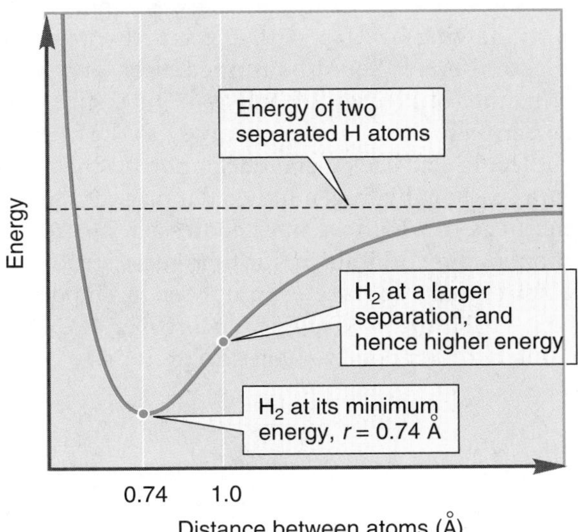

FIGURE 3.5 The formation of hydrogen (H_2) from two separated hydrogen atoms.

Another way to use orbitals inefficiently is to point them away from each other, or otherwise not to take full advantage of potential orbital overlap. Notice that our 2*p* orbitals are quite poorly designed for end-on overlap. Overlap can take place only with one lobe. In our current model for methane, the back lobes of the 2*p* orbitals are unused and are thus wasted (Fig. 3.6).

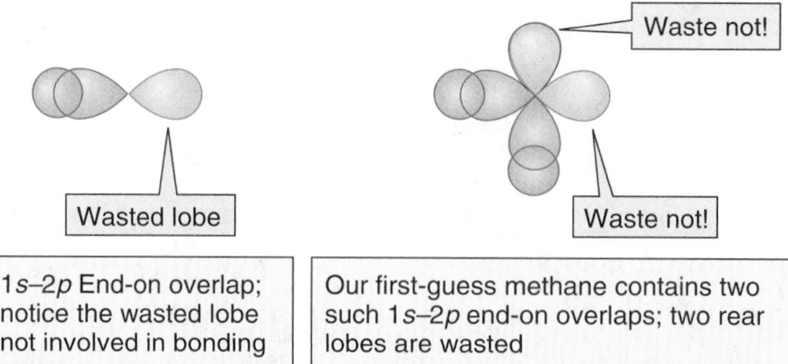
Waste not!
Wasted lobe
Waste not!
1s–2p End-on overlap; notice the wasted lobe not involved in bonding
Our first-guess methane contains two such 1s–2p end-on overlaps; two rear lobes are wasted

FIGURE 3.6 End-on overlap of a $2p$ atomic orbital with a $1s$ orbital. Note the wasted lobe of the $2p$ orbital. Our first-guess model for methane has two such overlapping pairs of orbitals.

PROBLEM 3.4 Why not allow a $2p$ orbital to overlap sideways with an s orbital? Why would this not solve the problem of waste of the back lobe (Fig. 3.7)?

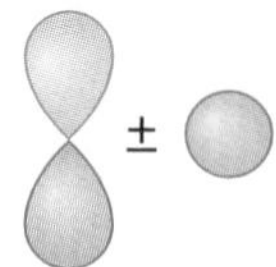
±

FIGURE 3.7 The sideways overlap of an s and a $2p$ orbital. Why is there no interaction between the two?

To summarize: The first model for methane has three problems. First, the bonds, and the electrons in them, are too close together, which leads to electron–electron repulsion and destabilization. Second, there are two electrons in a nonbonding orbital that are not contributing to holding the molecule together. Finally, the model uses $2p$–$1s$ overlap and thus the rear lobes of the $2p$ orbitals are wasted.

One problem may be eliminated by using those electrons in the carbon $2s$ orbital more efficiently. Recall that the empty $2p_z$ orbital was uninvolved in our first formulation of methane. Now imagine that one electron is promoted from the lower energy carbon $2s$ orbital to the higher energy $2p_z$ orbital. There would then be four singly occupied carbon orbitals ($2s$, $2p_x$, $2p_y$, $2p_z$), allowing overlap with the four hydrogen $1s$ orbitals. Then four two-electron bonds could be formed, each of which would be stabilized in the manner of Figures 2.9 and 3.4. To be sure, it "costs" energy to move an electron from the lower energy $2s$ to the higher energy $2p_z$ orbital, but the molecule gets back a great deal more in the stabilization of electrons in bonding carbon–hydrogen molecular orbitals. So now we have a second-guess methane in which we have addressed the problem of wasting electrons in nonbonding orbitals in our first-guess structure. We know that we still are not correct, because even our second model predicts the unobserved 90° H—C—H angles (Fig. 3.8). Moreover, we still have not dealt with the problem of our poorly shaped $2p$ orbitals.

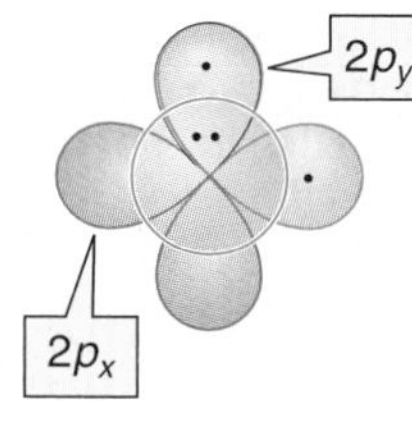
$2p_y$
$2p_x$

Carbon atom
$(1s^2)2s^22p_x2p_y$

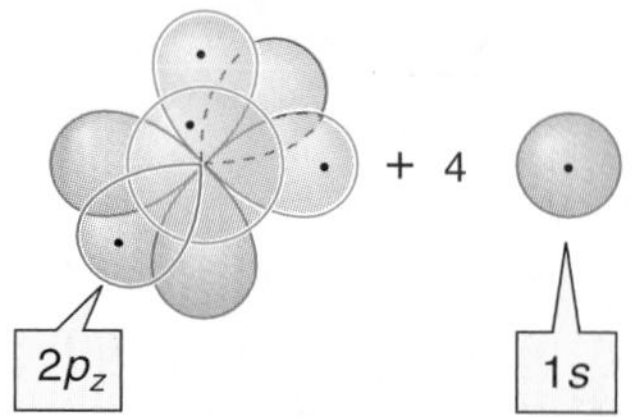
+ 4
$2p_z$
1s

Carbon atom with an electron promoted from the $2s$ orbital to the $2p_z$ orbital
$(1s^2)2s, 2p_x, 2p_y, 2p_z$

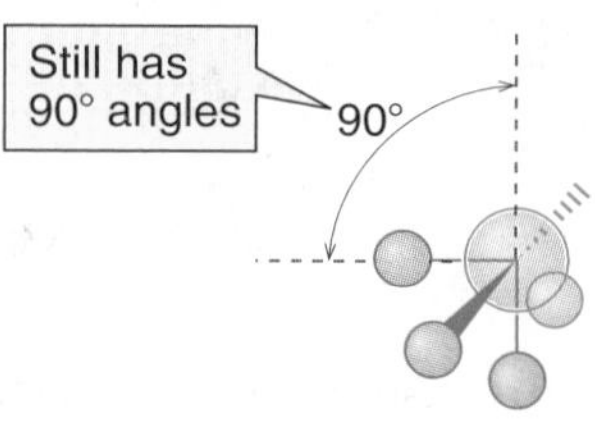
Still has 90° angles
90°

Second-guess methane

FIGURE 3.8 Our second-guess model for methane in which an electron has been promoted from the lower energy $2s$ orbital to a higher energy $2p$ orbital. Now there are four two-electron carbon–hydrogen bonds, but we still predict the wrong (90°) bond angles.

3.2 HYBRID ORBITALS: MAKING A BETTER MODEL FOR METHANE

3.2a *sp* Hybridization

Nature doesn't "worry" about such things; Nature merely *is* the energy minimum situation. We make models to describe Nature, and one of them goes by the name of **hybridization**.

The raw material from which we can make the four carbon–hydrogen bonds of methane consists of the four hydrogen 1*s* atomic orbitals and the 2*s*, $2p_x$, $2p_y$, and $2p_z$ atomic orbitals of carbon. As we've seen, there are difficulties in using these orbitals directly. There's not much that can be done about the hydrogen 1*s* orbitals, but mathematically, at least, there is a way to change the carbon orbitals. If we combine atomic wave functions we get composite (hybrid) orbitals, made up of fractions of the available atomic orbitals. This situation is easiest to see if we begin by combining one 2*s* with one 2*p* orbital, as shown in Figure 3.9.

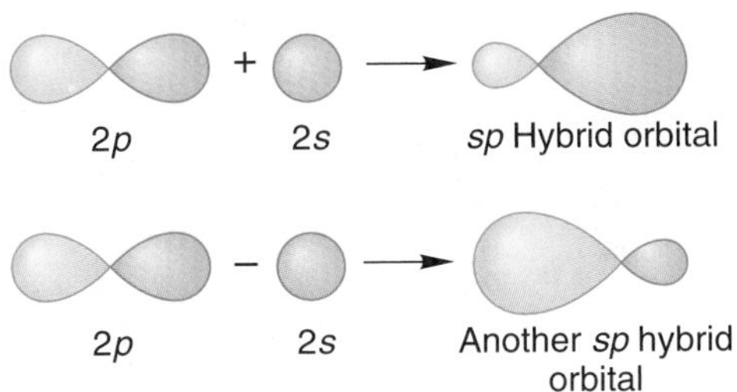

FIGURE 3.9 A pair of *sp* hybrid orbitals formed from the combination of a 2*s* and a 2*p* orbital. Each of the *sp* hybrids has 50% *s* character and 50% *p* character.

The two can be combined in a constructive way (2*p* + 2*s*) or a destructive way (2*p* – 2*s*). The result is a pair of distorted, hybrid orbitals. No longer are there equal-sized lobes, as in a 2*p* orbital. In the first case (2*p* + 2*s*), we get an expansion of one of the original lobes (the one with the same sign as the 1*s* orbital) and a shrinking of the other (the one with the opposite sign) to produce a new orbital that looks very much like a lopsided 2*p* orbital. The second case (2*p* – 2*s*) is also lopsided, but in the opposite sense. These two new orbitals are called ***sp* hybrid** orbitals. The designation *sp* means that the orbitals are composed of 50% *s* orbital (50% *s* character) and 50% *p* orbital (50% *p* character).

Figure 3.10 shows the real shape of one of these orbitals. It is traditional to use the schematic form in drawings, rather than attempt to reproduce the detailed structure of the orbital.

Note that the *sp* hybrids solve most of the problems we saw previously. They are unsymmetrical, as they have a fat and a thin side. The problem of waste is much reduced and these hybrids can form stronger bonds than can pure 2*p* orbitals. They are also aimed 180° from each other. The electrons in the bonds are as remote from each other as possible, and the repulsion problem is minimized. In this model, no longer are electrons wasted in nonbonding orbitals. All electrons are in bonding molecular orbitals (*sp* + 1*s*).

A number of compounds can be found in which the bonding can be nicely described by using *sp* hybridization. The classic example is H–Be–H, a linear hydride of beryllium. Beryllium ($_4$Be) has the electronic configuration $1s^2$, $2s^2$ and thus brings only two electrons to the bonding scheme (*Remember*: The 1*s* electrons are not used). We construct our picture of beryllium hydride (BeH_2) by making two *sp* hybrids from the Be 2*s* atomic orbital and one of beryllium's three equivalent 2*p* orbitals, arbitrar-

$2s \pm 2p_x \longrightarrow$ Two *sp* hybrid orbitals

One *sp* orbital in schematic form ...

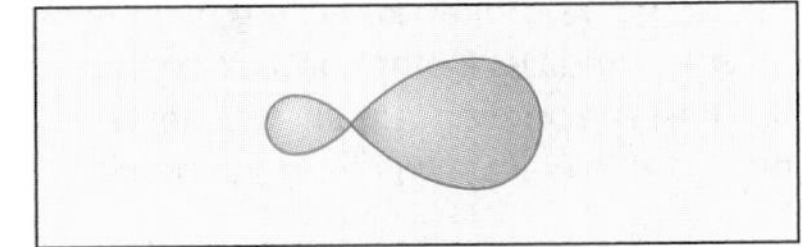

... and more correctly:

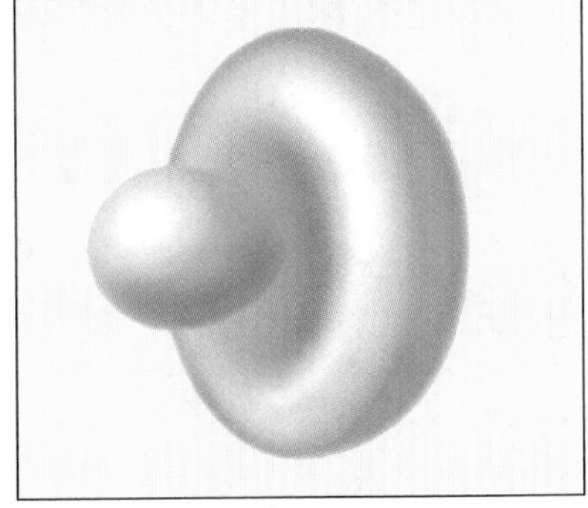

FIGURE 3.10 Schematic and detailed pictures of an *sp* hybrid orbital.

ily taken as the $2p_x$ orbital. The overlap of each of these hybrids with a singly occupied hydrogen 1*s* orbital produces two Be—H bonds (*sp* + 1*s*), directed at angles of 180° with respect to each other. All bonds of this type, with cylindrical symmetry, are called **sigma bonds** (σ bonds). The combination of Be (*sp*) and H (1*s*) orbitals must also yield an antibonding orbital (*sp* – 1*s*), which in this case is empty, as shown in Figure 3.11. This antibonding orbital is also cylindrically symmetrical, and is therefore properly called a σ bond.

Convention Alert!

The antibonding orbital is designated σ*, where the asterisk (*) means "antibonding."

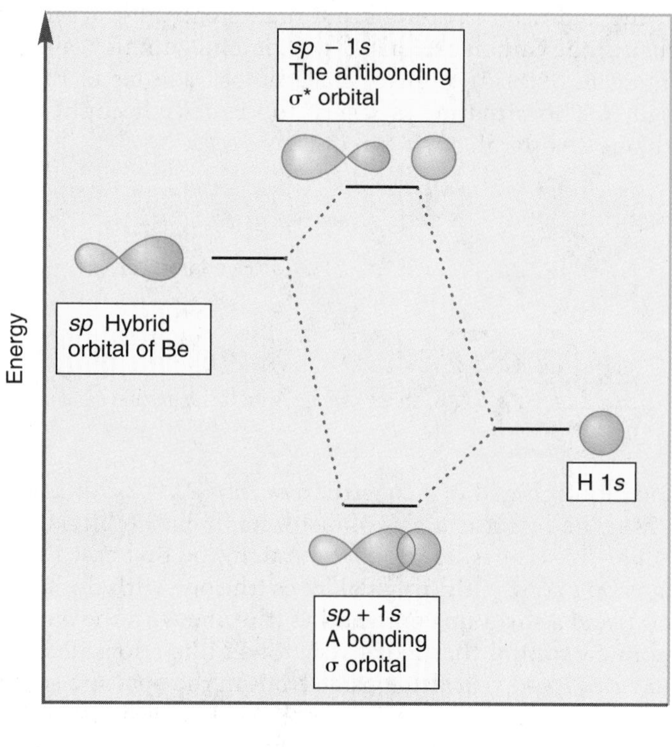

FIGURE 3.11 The formation of bonding (σ) and antibonding (σ*) orbitals from the overlap of an *sp* hybrid orbital with a hydrogen 1*s* orbital. Beryllium hydride is an *sp* hybridized compound. Note that the interaction diagram shows the 1*s* orbital at lower energy than the hybrid *sp* orbital.

It is no accident that we use two atomic orbitals in our description of the bonding in BeH_2. *Two* hybrids are needed for bonding to *two* hydrogens, which means we must combine *two* unhybridized atomic orbitals (2*s* and $2p_x$) to get them.

PROBLEM 3.5 What happens to the $2p_y$ and $2p_z$ orbitals? Elaborate the picture in Figure 3.11 to show the unused, empty, 2*p* orbitals on Be.†

†This kind of thing generally drives physical chemists crazy. As an orbital is nothing more than the electron probability region, how can there be an "empty orbital?" The problem is at least partially avoided if the empty orbital is thought of as "the place the next electron would go."

*PROBLEM 3.6

First, use the hybridization model to build a picture of the linear molecule, H–C–H. Recall that a molecular orbital picture of this linear molecule was constructed in Chapter 2 (Problem 2.24, p. 65). If you didn't do that problem, you might try it now and contrast your "molecular orbital" answer with your "hybridization" answer.

ANSWER

The hybridization part of this problem should be relatively easy. The carbon atom is attached to two other atoms, the two hydrogens. As for BeH_2 (p. 73), the carbon is hybridized *sp*. The 2*s* and $2p_x$ orbitals are combined to form two hybrid *sp* orbitals. Bonds to hydrogen are made through overlap of *sp* and 1*s* orbitals. The $2p_y$ and $2p_z$ orbitals remain unchanged. Carbon uses two of its four bonding electrons in the carbon–hydrogen bonds, leaving two for the remaining 2*p* orbitals. Each contains a single electron.

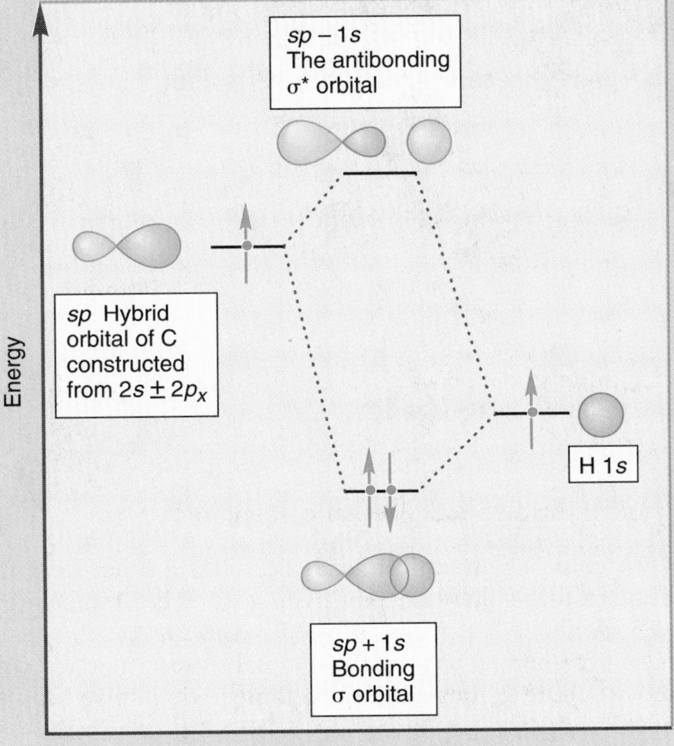

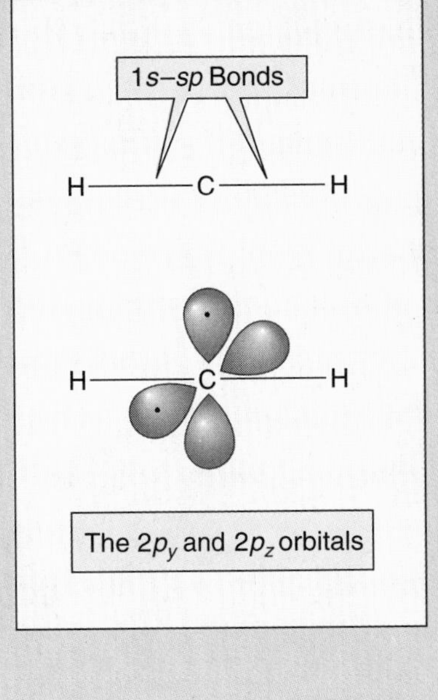

3.2b sp^2 Hybridization

Let's extend this way of thinking to analyze borane (BH_3), a slightly more complicated molecule, before we come to our next attempt at a realistic picture of methane. In BH_3, we must make three bonds from boron to hydrogen. Boron's electronic configuration is $_5B = 1s^2 2s^2 2p_x$, and thus there are three electrons available to form bonds with the three singly occupied hydrogen 1*s* orbitals. This time we combine the boron 2*s*, $2p_x$, and $2p_y$ orbitals to produce the three hybrids we need in our model. These are called, naturally, **sp^2 hybrids** (constructed from one *s* orbital and two *p* orbitals) and have 33% *s* character and 67% *p* character. They resemble the *sp* hybrids, shown in Figure 3.10, which gives both schematic and more accurate representations of an *sp* orbital. As with *sp* hybrids, the lopsided distribution allows for efficient overlap with a hydrogen 1*s*, or any other singly occupied orbital.

$$2s + 2p_x + 2p_y \longrightarrow \text{Three } sp^2 \text{ hybrid orbitals}$$

PROBLEM 3.7 Boron also forms a compound, BF_3. Create a bonding scheme and draw a Lewis structure for this molecule.

How are these new sp^2 hybrids arranged in space? Your intuition may well give you the answer. They are directed so that electrons in them will be as far as possible from each other. The angle between them is 120° and they are aimed toward the corners of a triangle (Fig. 3.12).

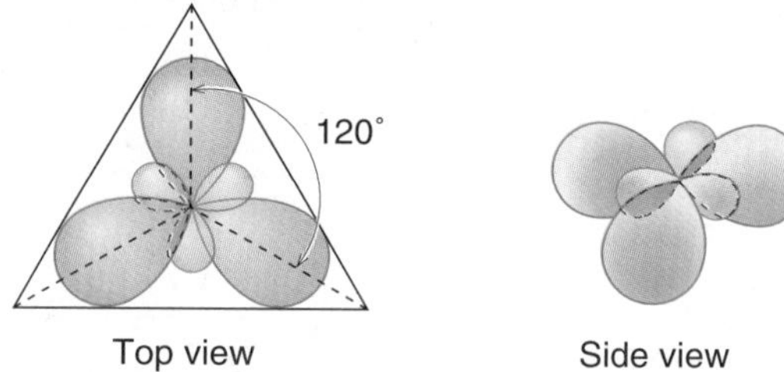

FIGURE 3.12 Top and side views of the arrangement in space of the three sp^2 hybrid orbitals.

PROBLEM 3.8 The $2p_z$ orbital on boron was not used in our construction of BH_3. Sketch it in, using the drawing of Figure 3.12.

3.2c sp^3 Hybridization: Methane, Revisited

Now we come again to methane. For CH_4 we need not two or three hybrids, but four—there are four hydrogens to be attached to carbon. Fortunately, we have just what we need in the 2*s*, $2p_x$, $2p_y$, and $2p_z$ atomic orbitals of carbon. In our mathematical description we combine these orbitals to produce four new, **sp^3 hybrids**, which also look similar to *sp* hybrids (Fig. 3.10).

$$2s + 2p_x + 2p_y + 2p_z \longrightarrow \text{Four } sp^3 \text{ hybrid orbitals}$$

The familiar lopsided shape appears again, and thus we can expect strong bonding to the singly occupied hydrogen 1*s* orbital. By now you will surely anticipate that these orbitals will be aimed in space so as to keep the orbitals, and any electrons in them, as far apart as possible. The previous two examples, BeH_2 and BH_3, were relatively easy, as linear arrangements and triangles are familiar to us all. Now we move into three dimensions. How are four objects arranged in space so as to maximize the distance between them? The answer is, at the corners of a tetrahedron (Fig. 3.13).

The carbon in methane is hybridized sp^3 (25% *s* character; 75% *p* character) and the orbitals are directed toward the corners of a tetrahedron. The four hydrogen atoms, each with an electron in a 1*s* orbital, overlap with the hybrids to form bonding and antibonding pairs (Fig. 3.14).

Methane, a tetrahedral molecule

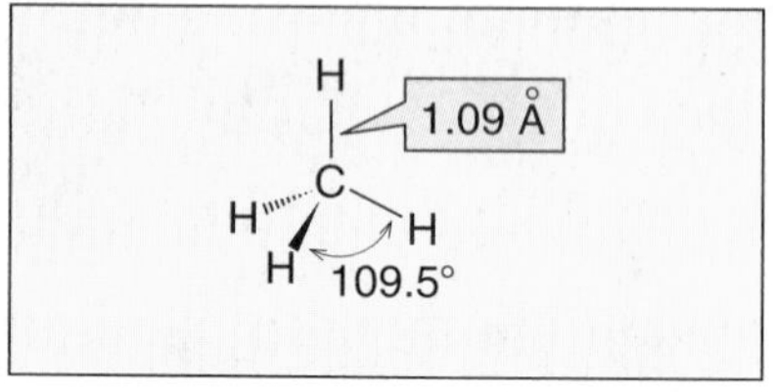

A tetrahedron

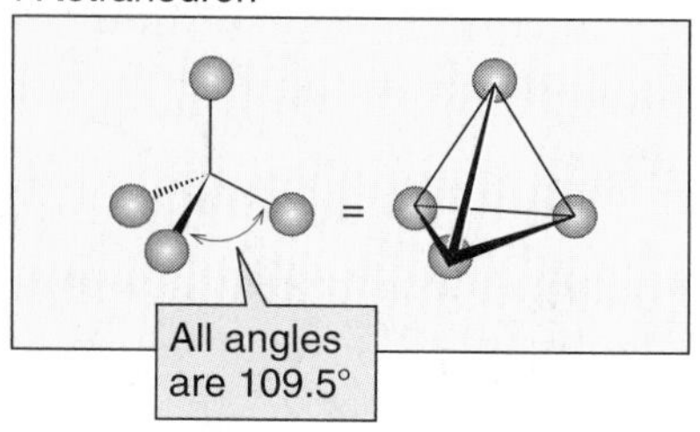

FIGURE 3.13 A schematic view of the tetrahedral arrangement of the four hydrogens surrounding the central carbon atom of CH_4.

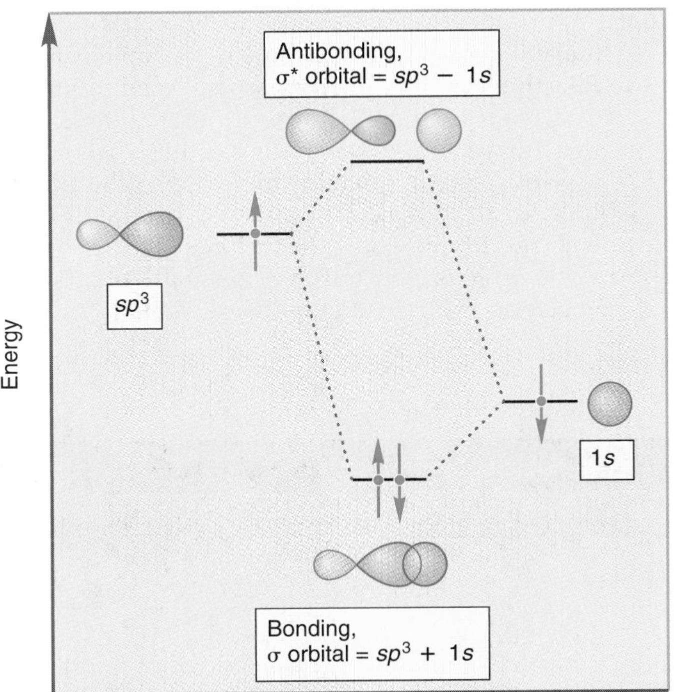

FIGURE **3.14** The formation of a carbon–hydrogen bond through the overlap of an sp^3 hybrid orbital with a hydrogen 1*s* orbital. Note the formation of both the bonding (σ) orbital and the antibonding (σ*) orbital.

The resulting H–C–H angles are 109.5°, and the length of a carbon–hydrogen bond in methane is 1.09 Å (Fig. 3.13).

The carbon–hydrogen bond in methane is very strong; its bond dissociation energy is 104 kcal/mol. This means that 104 kcal/mol must be applied in order to break one of the carbon–hydrogen bonds in methane homolytically (Fig. 3.15).

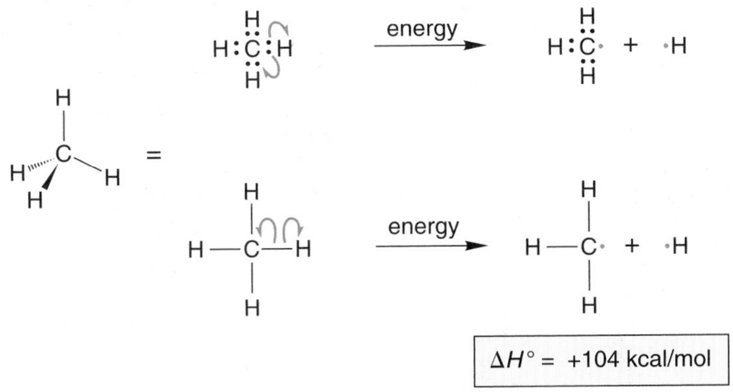

CH_4 (H:C:H with H above and below) —energy→ $\cdot CH_3$ + ·H

H–CH₃ —energy→ H–C(H)(H)· + ·H

$\Delta H° = +104$ kcal/mol

FIGURE **3.15** Homolytic cleavage of one carbon–hydrogen bond in methane. This reaction is endothermic by 104 kcal/mol.

To summarize: The hybridization model solves the three problems listed below, which were identified earlier when the first-guess model for methane was introduced (p. 72).

1. In first-guess methane, the electrons in the bonds were far from being as remote from each other as possible. The hybridization model yields four hybrid orbitals directed toward the corners of a tetrahedron, the best possible arrangement.

2. In first-guess methane, two electrons occupied a nonbonding orbital and, therefore, did not contribute to holding the molecule together. In the hybridization model, all electrons are in bonding orbitals (Fig. 3.14).
3. In first-guess methane, bonds were formed through overlap between carbon $2p$ and hydrogen $1s$ orbitals, thus wasting the rear lobes of the $2p$ orbitals. In the hybridization model, overlap is much improved in the bonds formed using the fat lobes of hybrid sp^3 orbitals and hydrogen $1s$ orbitals (contrast Figs. 3.6 and 3.14). Table 3.1 reviews the properties of sp, sp^2, and sp^3 hybrid orbitals.

TABLE 3.1 Properties of Hybrid Orbitals

Hybridization	Constructed from	Angle between Bonds (°)	Examples
sp	50% s, 50% p	180	BeH_2, linear HCH
sp^2	33.3% s, 66.7% p	120	BH_3, BF_3
sp^3	25% s, 75% p	109.5	CH_4, $^+NH_4$

Methane

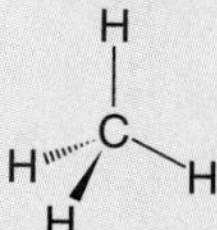

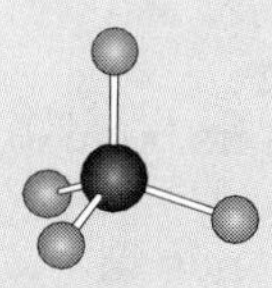

Methane (CH_4), the simplest of all hydrocarbons, is nonetheless important and biologically active. Methane constitutes about 85% of natural gas, and leaks into the atmosphere from wellheads and other pipeline junction points. It is also produced by plant decay and by bacteria in the stomachs and intestines of pigs and cattle. In this case, it enters the atmosphere through belching and flatulence. (In the movie *Mad Max Beyond Thunderdome*, the trading village Bartertown derived its energy from the methane produced by herds of pigs kept in the depths of Undertown.)

The amount of methane in the atmosphere has increased dramatically during industrial times, roughly doubling over the last 100 years. Methane is, alas, an important greenhouse gas, and its accumulation is cause for concern. In the atmosphere, it slowly oxidizes to carbon dioxide and enters the cycle again through photosynthesis.

Methane may have been essential to the formation of life. The early atmosphere on earth was relatively rich in methane and ammonia (as are the current atmospheres of the outer planets). Given an energy source such as ultraviolet (UV) radiation or lightning, methane and ammonia react to form hydrogen cyanide (HCN), a molecule that polymerizes to give adenine, a common component of ribonucleic acid (RNA) and other molecules of biological importance. In the presence of water and an energy source, methane and ammonia react to give amino acids, the building blocks of proteins.

3.3 DERIVATIVES OF METHANE: METHYL (CH_3) AND METHYL COMPOUNDS (CH_3X)

We can now imagine moving to a less symmetrical situation. As soon as one of the four hydrogens surrounding the central carbon in methane is replaced with another atom, pure tetrahedral symmetry is lost. We might well anticipate that one of our newly constructed sp^3 hybrids could overlap in a stabilizing way with any atom X offering a singly occupied orbital. For maximum stabilization we want to fill the new, bonding orbital created from the overlap of an X orbital with one of the hybrids, which requires two and only two electrons (Fig. 3.16). All manner of derivatives of

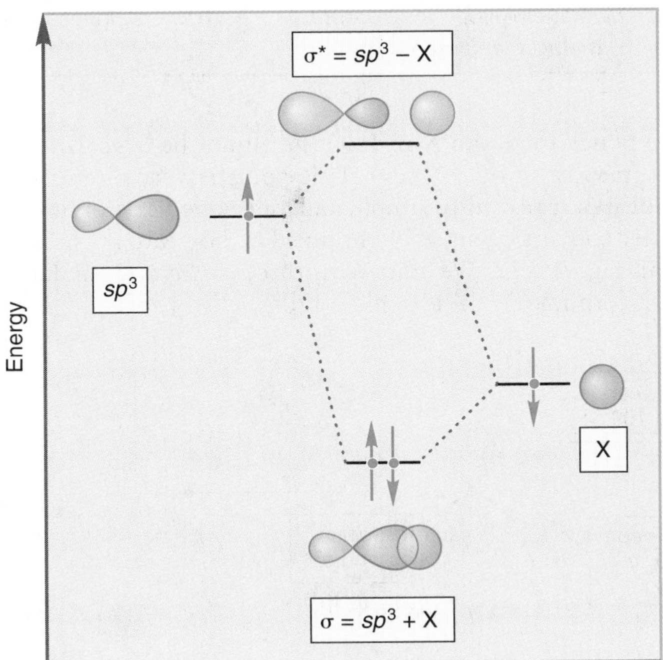

FIGURE 3.16 Formation of a carbon–X bond from overlap of an sp^3 hybrid with an *s* orbital of X. Notice that this overlap offers stabilization to only two electrons.

methane, called **methyl compounds** (CH_3–X) are possible. Their names are formed by dropping the "ane" from the name of the parent, and appending "yl." The generic term for derivatives of alkanes is **alkyl**. Such compounds are often abbreviated R–X, where R stands for a general alkyl group.* Table 3.2 shows a number of methyl derivatives with their melting and boiling points.

The bonding in methyl compounds closely resembles that in methane, and it is conventional to speak of the carbon atom in H_3C–X as being sp^3 hybridized, just as it is in the more symmetrical CH_4. Strictly speaking, this is wrong, although sp^3 is a good approximation. The four bonds to carbon in H_3C–X are not equivalent, as an X is not H. The bond from C to X is not the same length as that from C to H, and the H–C–H bond angle cannot be exactly the same as the X–C–H angle. As sp^3 hybridization yields four exactly equivalent bonds directed toward the corners of a tetra-

*Why R? In the old days R stood for *der Rest* (German for "residue" or "remains") or, in English, radical, a general name for groups that were commonly found in organic chemistry. These days "radical" is better reserved for the neutral species in which a single electron occupies a nonbonding orbital, R• (see Section 3.4 for a more detailed description), but the general use of R to mean "any alkyl group" remains.

TABLE 3.2 Some Simple Derivatives of Methane, Otherwise Known as Methyl Compounds

$H_3C{-}X$	Common Name	mp (°C)	bp (°C)	Physical Properties
$H_3C{-}H$	Methane	−182.5	−164	Colorless gas
$H_3C{-}OH$	Methyl alcohol or methanol	− 93.9	65.0	Colorless liquid
$H_3C{-}NH_2$	Methylamine	− 93.5	− 6.3	Colorless gas
$H_3C{-}Br$	Methyl bromide	− 93.6	3.6	Colorless gas/liquid
$H_3C{-}Cl$	Methyl chloride	− 97.7	− 24.2	Colorless gas
$H_3C{-}CN$	Methyl cyanide or acetonitrile	− 45.7	81.6	Colorless liquid
$H_3C{-}F$	Methyl fluoride	−141.8	− 78.4	Colorless gas
$H_3C{-}I$	Methyl iodide	− 66.5	42.4	Colorless liquid
$H_3C{-}SH$	Methanethiol or methyl mercaptan	−123	6.2	Colorless gas/liquid

hedron, the bonds to H and X in $H_3C{-}X$ cannot be *exactly* sp^3 hybrids, only approximately sp^3 ($sp^{2.8}$, say). This point is very often troubling to students, but it is really quite simple and, once one has seen it, even obvious. Consider converting an sp^2 hybridized carbon into an sp^3 hybridized carbon, as in Figure 3.17. We start with three sp^2 hybrids and a pure, unhybridized $2p$ orbital (H—C—H angle = 120°).

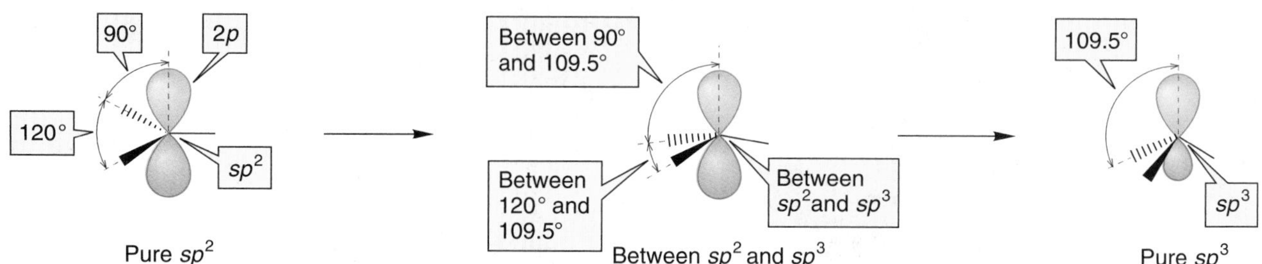

FIGURE 3.17 A thought experiment. The conversion of an sp^2 hybridized carbon into an sp^3 hybridized carbon. What is the hybridization at a point intermediate between the starting point (sp^2) and end point (sp^3)?

As we bend the hybrid orbitals, we will come to a point at which the 120° H—C—H angle has contracted to 109.5°, or sp^3 hybridization. Our orbitals, originally ⅓ *s* and ⅔ *p* (sp^2: 33.3% *s*, 66.7% *p* character) have become ¼ *s* and ¾ *p* (sp^3: 25% *s*, 75% *p* character). *They have gained p character in the transformation from* sp^2 *to* sp^3. Now look at the $2p$ orbital. As we bend, it goes from pure *p* (100% *p* character) to ¼ *s*, ¾ *p* character (sp^3: 25% *s*, 75% *p* character). *It has gained s character in the transformation.* There is a smooth transformation as we pyramidalize the molecules from sp^2 to sp^3, and in between these two extremes, the hybridization of carbon is intermediate between sp^2 and sp^3, something like $sp^{2.8}$. The hybridizations sp^2 and sp^3 are *limiting* cases, applicable only in very symmetrical situations. The hybridization of a carbon (or other) atom is intimately related to the bond angles! If you know one, you know the other.

In general, methyl compounds ($H_3C{-}X$) will closely resemble methane in their approximately tetrahedral geometry. They cannot be perfect tetrahedra, however.

3.4 THE METHYL CATION ($^+CH_3$), ANION ($^-:CH_3$), AND RADICAL ($\cdot CH_3$)

Table 3.2 could be augmented to include $^+CH_3$, $^-:CH_3$, and $\cdot CH_3$, three special cases of H_3C-X. The molecules in Table 3.2 are all stable species, and they can be purchased from any chemical supply company. The three new molecules are members of a class of compounds called **reactive intermediates**. This simply means that these molecules are too unstable to be isolable under normal conditions, and must usually be studied by indirect means, often by looking at what they did during their brief lifetimes, or sometimes by isolating them at very low temperature. These three species are especially important because they are prototypes—the simplest examples of whole classes of molecules to be encountered later when the study of chemical reactions takes over our attention.

In the cation $^+CH_3$, X is "nothing." To make this molecule, one could imagine removing **hydride** ($^-:H$) from methane to leave the positively charged **methyl cation** ($^+CH_3$), the simplest example of a **carbocation**, a molecule containing a positively charged carbon* [Fig. 3.18(a)].

In the **methyl anion** ($^-:CH_3$), X is a pair of electrons. Removal of a proton (H^+) from methane would leave behind a pair of nonbonding or lone-pair electrons in the negatively charged methyl anion, a simple example of a carbon-based anion, or **carbanion**. Both of these ways of producing $^+CH_3$ and $^-:CH_3$ involve the concept of breaking a carbon–hydrogen bond in unsymmetrical fashion, a process known as heterolytic bond cleavage, (Chapter 2, p. 49; Fig. 3.18). Remember the arrow formalism—the arrows of Fig. 3.18 move the *pair* of electrons in the carbon–hydrogen bond to the hydrogen (a) or to the carbon (b).

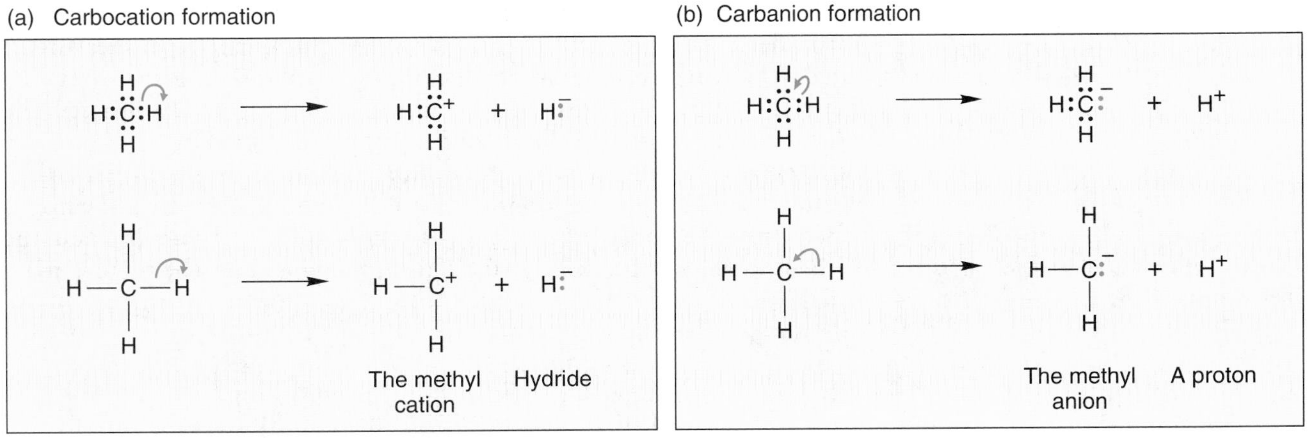

(a) Carbocation formation

The methyl cation | Hydride

(b) Carbanion formation

The methyl anion | A proton

FIGURE 3.18 The formation of (a) the methyl cation ($^+CH_3$) and (b) the methyl anion ($^-:CH_3$) by two different heterolytic cleavages of a carbon–hydrogen bond in CH_4.

Recall that there is another way to think of breaking a two-electron bond, and that is to allow one electron to go with each atom at an end of the breaking bond. This homolytic bond cleavage (Chapter 2, p. 49) in

*A topic of a long and too intense argument in the chemical community, a carbocation is sometimes called a **carbonium ion** (by traditionalists) or a **carbenium ion** by others. The compromise "carbocation" is both aptly descriptive and avoids the emotional reactions of the staunch defenders of the other terms.

methane would give a hydrogen atom (H•) and leave behind the neutral **methyl radical** (•CH_3). Here a single electron plays the part of X in CH_3–X, and the single-barbed "fishhook" arrow convention is used (Fig. 3.19).

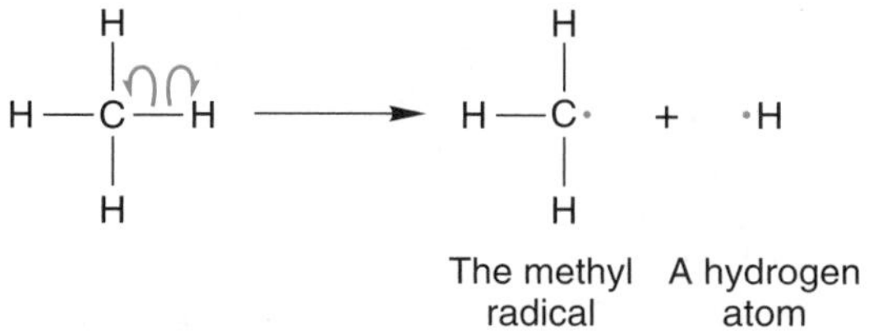

$$H_3C\text{—}H \longrightarrow H_3C\cdot \;+\; \cdot H$$

The methyl radical   A hydrogen atom

FIGURE 3.19 The homolytic bond cleavage of a carbon–hydrogen bond in methane to give a hydrogen atom and the methyl radical.

All three of these species have been observed, although each is extremely reactive. They have structures, though, and we can make some predictions for at least two of them. In the methyl cation ($^{+}CH_3$), carbon is attached to three other groups, suggesting the need for three hybrid orbitals (recall BH_3, p. 75) and that, in turn, suggests sp^2 hybridization (Fig. 3.20).

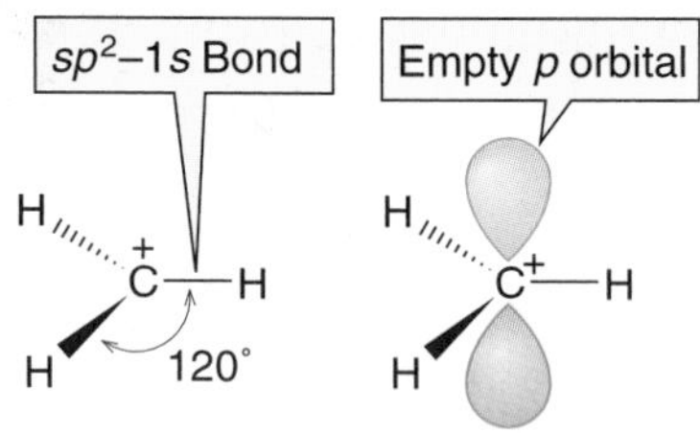

FIGURE 3.20 The sp^2 hybridized methyl cation, $^{+}CH_3$.

On the other hand, the carbon in the methyl anion is attached not only to three hydrogens but also to the pair of nonbonding electrons. In the methyl cation, with no lone-pair electrons, we have a pure *p* orbital (zero *s* character) and the species is as flat as a pancake (C–C–C angle = 120°). The methyl anion has two more electrons than the cation and we will have to consider them in arriving at a prediction of structure. Recall that *s* orbitals have density at the positively charged nucleus. Therefore, it would seem reasonable that a negatively charged electron would be more stable (lower in energy) in an orbital with a lot of *s* character. A pyramidal structure seems appropriate for the methyl anion, although of course the molecule cannot be a perfect tetrahedron. We can't predict exactly how pyramidal the species will be, and it's a difficult quantity to measure in any case, but recent calculations predict the structure in Figure 3.21. Remember that hybridization and bond angles are intimately related. The more *s* character in the orbital containing the nonbonding electrons, the more pyramidal the species will be. In a limiting example in which the electrons are in a pure 2*s* orbital (0% *p* character), the carbon–hydrogen bonds must be made from pure *p* orbitals (0% *s* character), and the C–C–C angle will be exactly 90°. This anion would be a very steep pyramid (Fig. 3.22).

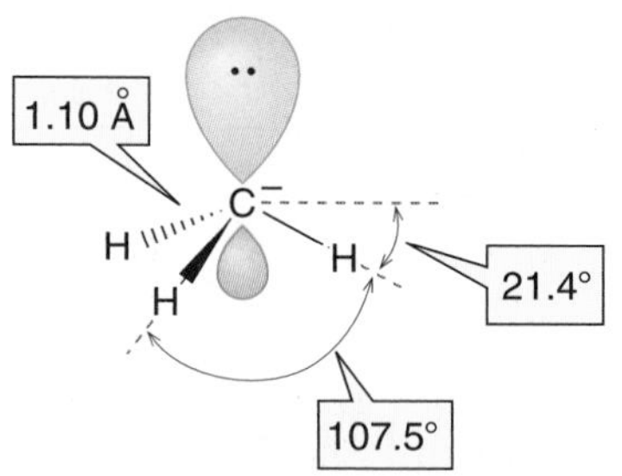

FIGURE 3.21 The structure of the methyl anion, $^{-}$:CH_3.

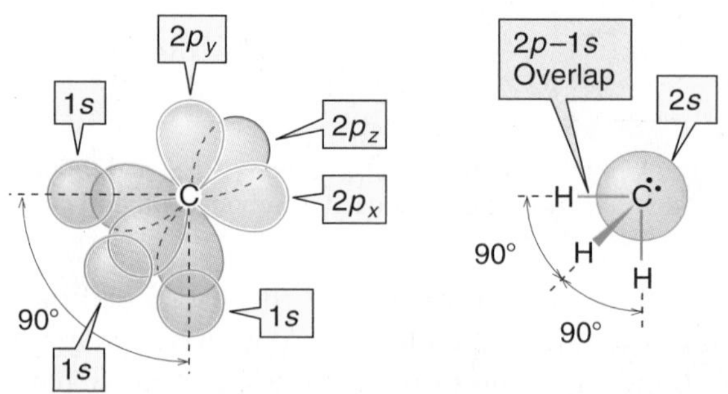

FIGURE 3.22 An unhybridized methyl carbanion. If the "lone-pair" electrons are in a 2*s* orbital, the carbon–hydrogen bonds must be made from overlap of pure 2*p* orbitals with hydrogen 1*s* orbitals. This way of making carbon–hydrogen bonds is very inefficient. Do you remember why?

PROBLEM 3.9 As we saw in Figure 3.21, the methyl anion does not have the limiting, 90° structure of Figure 3.22. Why not?

It is harder to deal with the neutral methyl radical ($\cdot CH_3$) in which there is only a single nonbonding electron. At present it is not possible to choose between a planar species and a rapidly inverting and very shallow pyramid, although it is clear that the methyl radical is close to planar (Fig. 3.23). Do not be disturbed by this! Our understanding of organic chemistry is still incomplete and many simple things are beyond us. There is still lots to do!

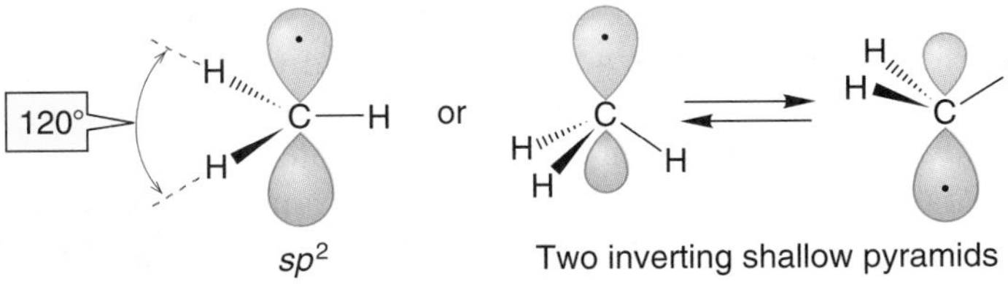

120°

sp[2]

or

Two inverting shallow pyramids

FIGURE 3.23 The methyl radical ($\cdot CH_3$) is either flat or a rapidly inverting shallow pyramid.

PROBLEM 3.10

Draw a structure for the molecule **A** at the half-way point for the inversion shown at the right of Figure 3.23. What is the hybridization of the carbon atom in **A**?

3.5 ETHANE (C_2H_6), ETHYL COMPOUNDS (C_2H_5X), AND NEWMAN PROJECTIONS

There is a special substituent X that produces a most interesting $H_3C{-}X$. In this case, we let X = another methyl group. Now $H_3C{-}X$ becomes $H_3C{-}CH_3$, methylmethane in the methyl-X terminology, but known always as ethane, C_2H_6. This compound is the second member of the class of alkanes (Fig. 3.24). You might try to anticipate the following discussion by building a model of ethane and examining its structure now.

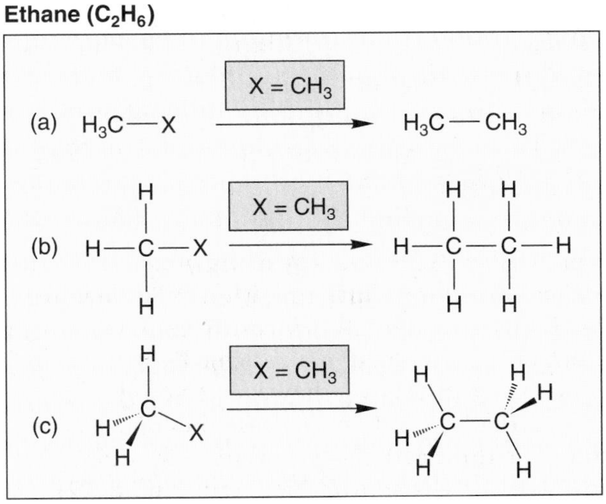

Ethane (C_2H_6)

(a) $H_3C{-}X$ —[X = CH_3]→ $H_3C{-}CH_3$

(b) —[X = CH_3]→

(c) —[X = CH_3]→

FIGURE 3.24 The construction of ethane ($H_3C{-}CH_3$) by the replacement of X in $H_3C{-}X$ with another methyl group. Remember the drawing convention. Solid wedges are read as coming out at you, whereas dashed wedges are seen as retreating into the paper (Problem 1.2; Fig. 1.9, p. 15).

We will soon come to a section (Section 3.6) devoted to the various drawing conventions (codes) but we must take a moment here to repeat one important drawing convention. The solid wedges of Fig. 3.24 are taken as coming forward, out of the paper. The dashed wedges are read as retreating from you into the paper. It takes only a little practice before drawings such as the ones in Figure 3.24 come to life, and appear clearly three dimensional.

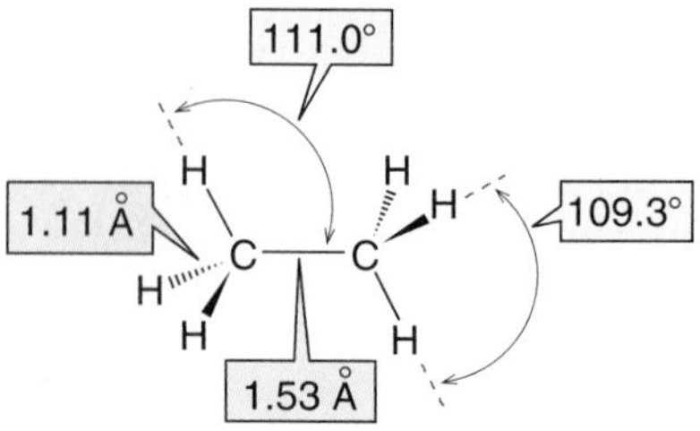

FIGURE 3.25 The detailed structure of ethane.

Even in a molecule as simple as ethane there are a number of interesting structural questions. From the point of view of one carbon, the attached methyl group takes up more room than the much smaller hydrogens. So we would expect the $H_3C{-}C{-}H$ angle to be slightly larger than the H—C—H angle. And so it is $H{-}C{-}CH_3 = 111.0°$ and $H{-}C{-}H = 109.3°$ (Fig. 3.25).

There are more complicated structural questions. How are the hydrogens of one end of the molecule arranged with respect to the hydrogens at the other end? There are two extreme, or limiting structures, called **conformations**, for ethane. Figure 3.24 shows ethane in only one conformation. Conformations are three-dimensional molecular structures interrelated by rotations about bonds. In this case, they are **eclipsed ethane** and **staggered ethane**, each shown in a side view in Figure 3.26. In the eclipsed conformation each of the three hydrogens on one methyl stands directly in front of a hydrogen on the other methyl. There is a **dihedral angle** of 0° between the hydrogens. The dihedral angle is the torsional, or twist, angle between two carbon–hydrogen bonds on adjacent carbons; the angle between the H—C—C and C—C—H planes, and is shown in Figure 3.26 as ϑ. In the staggered conformation, the dihedral angle ϑ is 60° and the hydrogens are maximally far apart. Be sure you see how the two forms are interconverted by rotation about the carbon–carbon bond. Use a model!

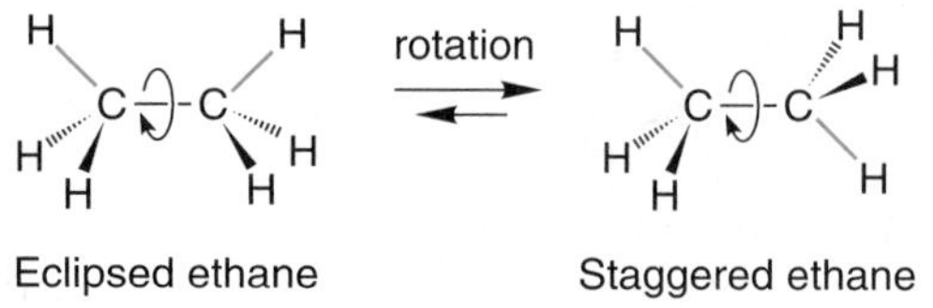

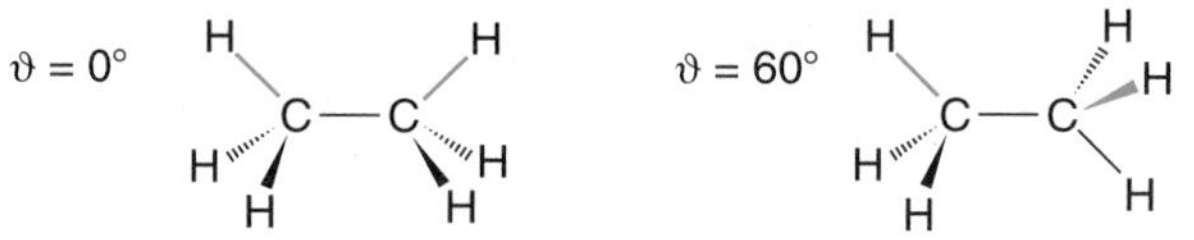

FIGURE 3.26 Two conformations of ethane: the eclipsed and staggered forms.

Convention Alert!

There is a particularly effective way of viewing these structures called a **Newman projection**, after its inventor, Melvin S. Newman (1908–1993) of Ohio State University. Like all devices, it contains arbitrary conventions, which can only be learned, not reasoned out. Its utility will repay you for the effort many times over, however, so do it! One first imagines looking down a particular bond, in this case the lone carbon–carbon bond in the molecule. The three carbon–hydrogen bonds attached to the "front" carbon are now drawn in as shown. Notice in Figure 3.27 that from this end-on view the angle between the carbon–hydrogen bonds on the front carbon is 120°. The rear carbon is next represented as a circle—a kind of exploded view—and the H atoms attached to it are put in. It is easy to construct the representation for the staggered form (Fig. 3.27), but not so easy for the eclipsed molecule. If we were strictly accurate in the drawing, we wouldn't be able to see the hydrogens attached to the rear carbon, because each is directly behind an H on the front carbon (the dihedral angle is 0°). So we cheat a little and offset these eclipsed hydrogens just enough so we can see them (Fig. 3.27). Newman projections are extraordinarily useful in making three-dimensional structures clear on a two-dimensional surface. Problem 3.11 offers a chance to practice visualizing and creating Newman projections. This problem is important.

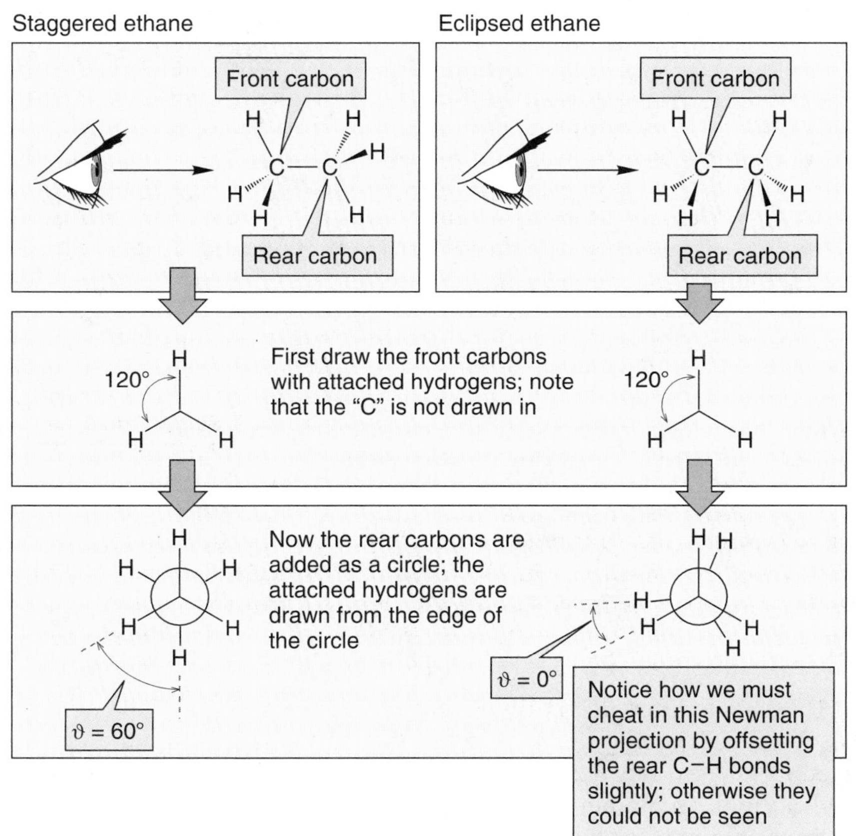

FIGURE 3.27 A Newman projection for staggered ethane. The dihedral angle (ϑ) between two carbon–hydrogen bonds is 60°. A Newman projection for eclipsed ethane. The dihedral angle (ϑ) between two carbon–hydrogen bonds is 0°.

PROBLEM 3.11

Write Newman projections for the staggered conformations found by looking down the carbon–carbon bonds of ethyl chloride (CH_3-CH_2-Cl) and 1,2-dichloroethane ($Cl-CH_2-CH_2-Cl$). In the second case, there are two staggered forms of different energy. Can you estimate which will be more stable?

Which of these forms of ethane represents the more stable molecule? One's first guess would surely be the staggered form, and that would be right. The electrons in the carbon–hydrogen bonds repel each other, and it is energetically best to have them as far from each other as possible, as in the staggered form. This repulsion can be effectively displayed in a plot of dihedral angle as a function of energy, as in Figure 3.28. Indeed, calculations show that the eclipsed form *is not even an energy minimum.* It is an energy maximum, the top of the barrier separating two staggered forms. Such a point is called a **transition state (TS)**.

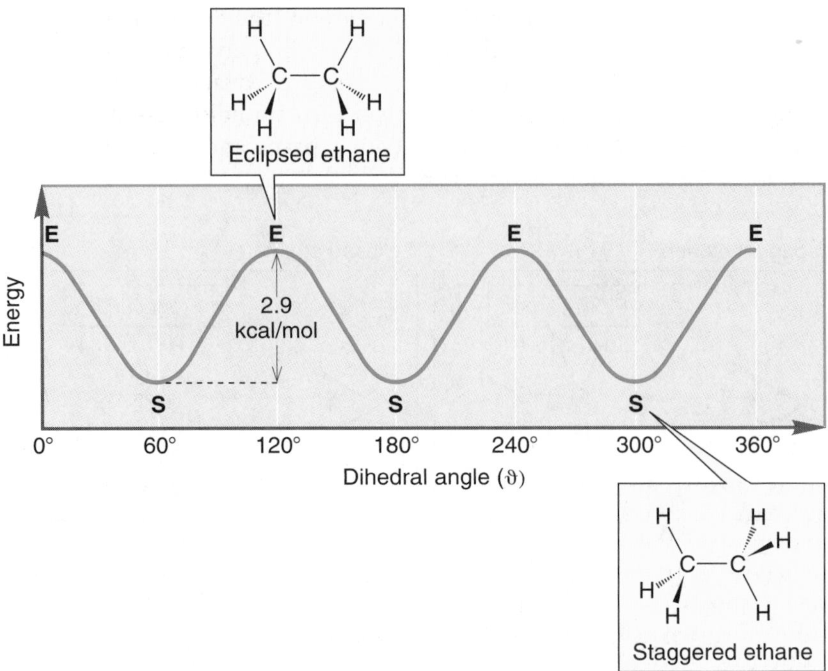

FIGURE 3.28 A plot of the energy change in ethane as rotation around the carbon–carbon bond occurs.

PROBLEM 3.12 You have recently seen a transition state, although the term wasn't used. Where? *Hint*: Where have we described two forms of a molecule that interconvert by passing through a species **A**? This **A** turns out to be a transition state.

How big a difference is there between the two forms? This question asks how high the barrier is between two of the minimum energy staggered forms (see Fig. 3.28). The barrier turns out to be a small number, 2.9 kcal/mol, an amount of energy easily available at room temperature. As there are three eclipsed interactions in eclipsed ethane, we can see that each pair of eclipsed carbon–hydrogen bonds "costs" about 1 kcal/mol **torsional strain**. Under normal conditions, ethane is said to be freely rotating, as there is ample thermal energy to traverse the barrier separating any two staggered forms. In colder places, where there was much less thermal energy available, alien organic chemists would have to worry more about this rotational barrier. Thus, their lives would be even more complicated than ours.

*PROBLEM 3.13 Imagine 1,2-dideuterioethane ($DCH_2{-}CH_2D$), where D = deuterium, on some cold planet such as Pluto. If there is not enough energy available to overcome the barrier to rotation, how many 1,2-dideuterioethanes would there be?

ANSWER This problem can be reduced to, How many isomers of 1,2-dideuterioethane are possible? This problem is actually much tougher than it looks. Part of it is easy, but there is a hidden, and quite subtle, difficulty. By far the easiest way to attack this problem is to use Newman projections. Sight down the carbon–carbon bond and examine all possible staggered conformations.

ANSWER (CONTINUED)

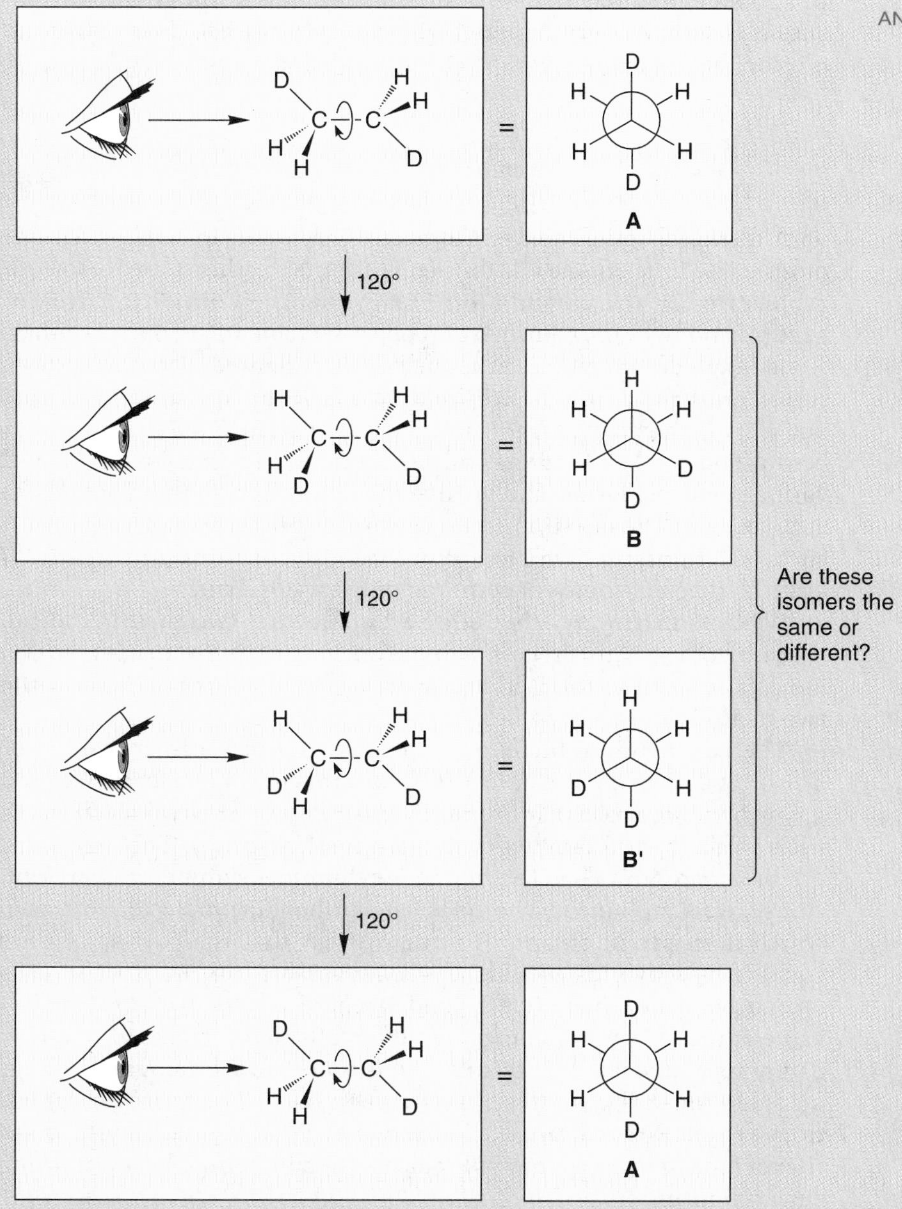

The first 120° rotation takes us from **A** to **B**, and the two are clearly different. A second 120° rotation generates **B′**, and a third 120° rotation takes us back to **A** as a 360° rotation is completed. The hard part of this problem is ferreting out the relationship between **B** and **B′**. Are they the same or not? At this point, this problem is very tough to do without models. Your models will show you that **B** and **B′** are not the same. There are three different possible staggered forms of 1,2-dideuterioethane. Just how **B** and **B′** differ is an interesting question and we will deal with it extensively in Chapter 5.

The chemical and physical properties of ethane closely resemble those of methane. Both are low-boiling gases and both are quite unreactive under most chemical conditions. But not all! One need only light a match in a room containing ethane or methane and air to find that out. This reaction is actually very interesting. If we examine the debris after the explosion, we find two new molecules, water and carbon dioxide (Fig. 3.29).

FIGURE 3.29 The combustion of ethane to give carbon dioxide, water, and energy.

$$2\ H_3C{-}CH_3 \xrightarrow[\text{match}]{7\ O_2} 4\ CO_2 + 6\ H_2O + \Delta \text{ and } h\nu$$

Ethane — Heat — Light

A thermochemical analysis indicates that water and carbon dioxide are more stable than ethane and oxygen. Presumably, this is the reason ethane explodes when the match is lit. Energy is all too obviously given off as heat (Δ) and light ($h\nu$). But why do objects, stable in air, such as ethane or wood, explode or burn continuously when ignited? How are they protected until the match is lit? We'll approach this question in Chapter 8, but it's worth some thought now in anticipation. This matter is serious because our molecules are also less thermodynamically stable than their various oxidized forms, and we live in an atmosphere that contains about 20% oxygen. The question could be turned into, Why can humans live in such an atmosphere without spontaneously bursting into flame? This issue has been of some concern, not only for novelists.*

We can imagine making ethane by allowing two methyl radicals to come together. This process is by no means entirely imaginary; methyl radicals do react to form ethane. Figure 3.30 looks at the combination of two methyl radicals in some detail.

The two hybrid orbitals overlap to form a new carbon–carbon bond. The orbital diagram in Figure 3.30 shows this in graphic form. We produce a new bonding molecular orbital that is filled by the two electrons originally in the hybrid orbitals of the two methyl groups. This new bonding orbital is a σ orbital, as the required cylindrical symmetry is present. Of course, we simultaneously produce an antibonding molecular orbital (σ*), which is empty in ethane. We might guess that the overlap of the two equal-energy hybrids would be very favorable, and it is. The carbon–carbon bond in ethane has a bond strength of about 90 kcal/mol. This value is the amount of energy by which ethane is stabilized relative to a pair of separated methyl radicals, and is therefore the amount of energy required to break this strong carbon–carbon bond. The formation of ethane from two methyl radicals is exothermic by 90 kcal/mol, and the dissociation of ethane into two methyl radicals is endothermic by 90 kcal/mol. In equation form,

$$H_3C\cdot + \cdot CH_3 \rightarrow H_3C{-}CH_3 \qquad \Delta H^\circ = -90\ \text{kcal/mol}$$

or, equivalently

$$H_3C{-}CH_3 \rightarrow H_3C\cdot + \cdot CH_3 \qquad \Delta H^\circ = +90\ \text{kcal/mol}$$

Note once again the sign convention for ΔH°. In the exothermic reaction the product $H_3C{-}CH_3$ is lower in energy than the starting material

*Eliot ... claimed to be deeply touched by the idea of an inhabited planet with an atmosphere that was eager to combine violently with almost everything the inhabitants held dear ... "When you think of it, boys," he said brokenly, "that's what holds us together more than anything else, except maybe gravity. We few, we happy few, we band of brothers—joined in the serious business of keeping our food, shelter, clothing and loved ones from combining with oxygen."

—Kurt Vonnegut
God Bless You, Mr. Rosewater

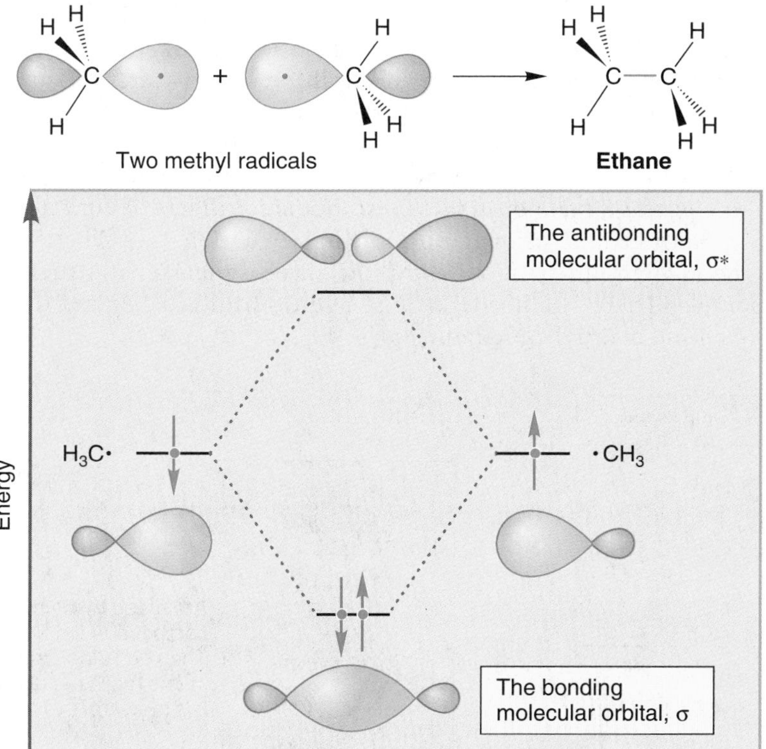

FIGURE 3.30 The formation of ethane through the combination of a pair of methyl radicals. Notice again the creation of both bonding (σ) and antibonding (σ*) orbitals.

(two methyl radicals) by 90 kcal/mol; hence the minus sign. By contrast, in the endothermic reaction, 90 kcal/mol must be supplied to ethane to produce the higher energy pair of methyl radicals, and the sign becomes plus.

Just as we imagined replacing one of the hydrogens in methane with an X group to produce a series of methyl compounds, so we can form **ethyl compounds** (CH_3CH_2–X) from ethane. Table 3.3 shows some common ethyl compounds and a few of their physical properties.

TABLE 3.3 Some Simple Derivatives of Ethane: Ethyl Compounds

CH_3CH_2–X	Common Name	mp (°C)	bp (°C)	Physical Properties
CH_3CH_2–H	Ethane	–183.3	–88.6	Colorless gas
CH_3CH_2–OH	Ethyl alcohol or ethanol	–117.3	78.5	Colorless liquid
CH_3CH_2–NH_2	Ethylamine	– 81	16.6	Colorless liquid
CH_3CH_2–Br	Ethyl bromide	–118.6	38.4	Colorless liquid
CH_3CH_2–Cl	Ethyl chloride	–136.4	12.3	Colorless liquid
CH_3CH_2–CN	Ethyl cyanide or propionitrile	– 92.9	97.4	Colorless liquid
CH_3CH_2–F	Ethyl fluoride	–143.2	–37.7	Colorless gas
CH_3CH_2–I	Ethyl iodide	–108	72.3	Colorless liquid
CH_3CH_2–SH	Ethanethiol or ethyl mercaptan	–144	35	Colorless liquid
$^+CH_2CH_3$	Ethyl cation			Reactive intermediate
$^-{:}CH_2CH_3$	Ethyl anion			Reactive intermediate
$\cdot CH_2CH_3$	Ethyl radical			Reactive intermediate

How many different $CH_3CH_2{-}X$ compounds are there? From a two-dimensional, schematic picture of ethane there appear to be two ethyl compounds with the same formula but different structures. Compounds of this type are generally called **isomers** (from the Greek words meaning "the same parts"). If the two ethyl compounds of Figure 3.31 actually existed, they would be isomers. However, the corners apparent in the structures in Figure 3.31 do not in fact exist, but are artifacts of our attempt to represent a three-dimensional molecule in two dimensions. These corners show how important it is to know the three-dimensional structures of compounds. There is but one kind of hydrogen in ethane, and therefore only one kind of ethyl compound.

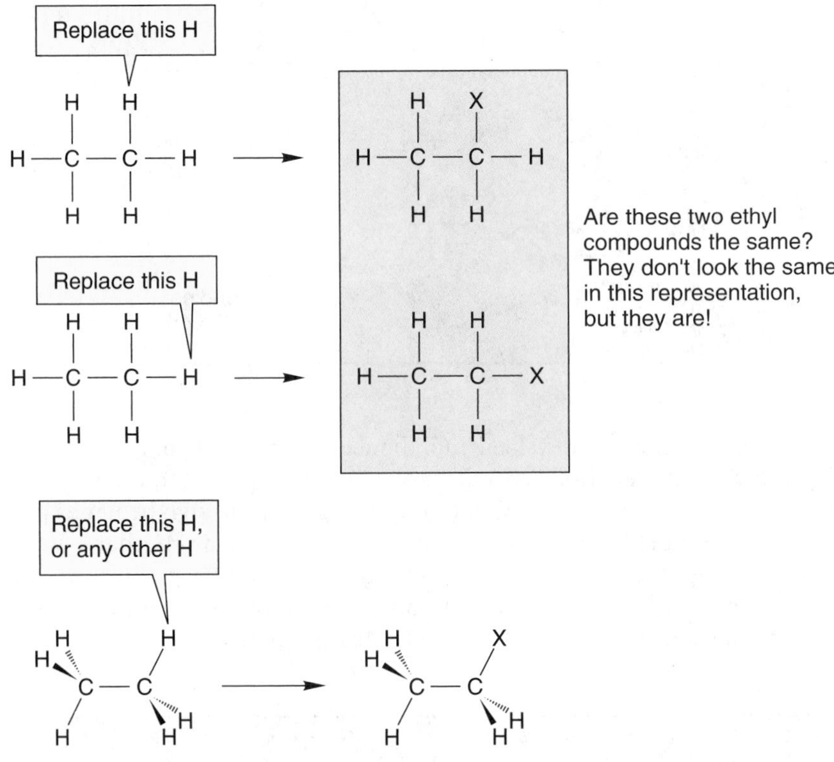
Replace this H
Replace this H
Are these two ethyl compounds the same? They don't look the same in this representation, but they are!
Replace this H, or any other H

FIGURE 3.31 The two-dimensional representation of ethane suggests that there are two different kinds of ethyl compound, $CH_3CH_2{-}X$. We must look at ethane in three dimensions to see why there is only one $CH_3CH_2{-}X$.

3.6 CODED STRUCTURE DRAWINGS

In the preceding sections, there have been many different representations of the very simple molecule, ethane. Figure 3.32 recapitulates them and adds some new ones.

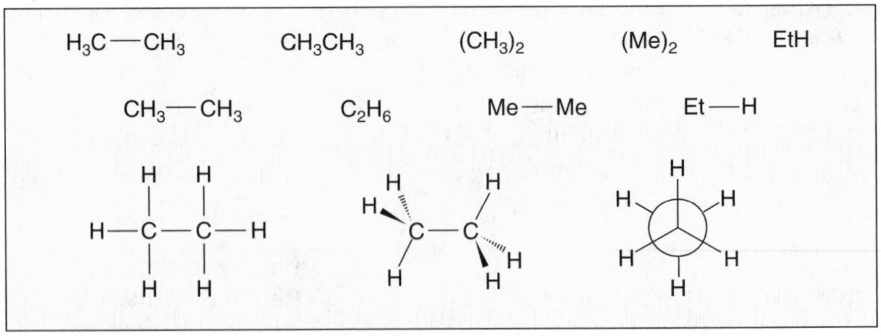

$H_3C{-}CH_3$ CH_3CH_3 $(CH_3)_2$ $(Me)_2$ EtH

$CH_3{-}CH_3$ C_2H_6 Me—Me Et—H

FIGURE 3.32 Several different representations of ethane. Methyl groups are sometimes written as Me and ethyl groups as Et, especially in colloquial use.

The difficulty of representing the real, dynamic, three-dimensional structure of ethane is more apparent now. In the real world, there is often not the time to draw out a good representation of even as simple a molecule as ethane carefully. The solid and dashed wedges of Figure 3.32 are the traditional attempts at adding a three-dimensional feel to the two-dimensional drawing. More complicated molecules can raise horrendous problems. Adequate codes are needed and you must learn to see past the coded structures to the real molecules both easily and quickly. It's worth considering here, at this very early point, some of the pitfalls of the various codes.

Convention Alert!

First of all, it is not simple to represent even ethane in a linear fashion. Clearly, neither C_2H_6 nor Et–H is as descriptive as H_3C-CH_3, as they don't even show the connectivity of the atoms in a general way. The representation H_3C-CH_3 is better, but even this picture lacks three dimensionality. We used this representation in this chapter, but you will see the variant CH_3-CH_3 in other places and, probably, in this book as well. The two formulations are equivalent, even though the second structural formula, CH_3-CH_3, seems to indicate that the bond to the right-hand carbon comes from one of the hydrogens on the left-hand carbon. Of course, it does not; it is the two carbons that are bonded, as H_3C-CH_3 shows. Yet it is easier to write CH_3-CH_3 and you will see it often.

The form $(CH_3)_2$ can also represent ethane, and this requires that you fight through a deeper level of abstraction and know that the two methyl groups are attached through their two carbon atoms. One often sees groups encased in parentheses in shorthand formulas. The groups, here methyl groups, are treated as units. You must learn to expand these formulas and, once again, to see what atoms are attached to each other. It is easier than it looks at first, but there can be no denying that it takes some practice. Now let's see what happens when we let X = CH_3 and create "ethylmethane," best known as propane.

PROBLEM 3.14 Draw three-dimensional structures for the following compounds: $(Me)_2CH_2$, $(CH_3)_4C$, $(CH_3)_3CH$, $EtCH_3$, $(Et)_2$, $CH_3CH_2CH_3$, EtMe, MeEt.

PROBLEM 3.15 Make three-dimensional drawings of the "two" propanes created by replacing the two "different" hydrogens in a flat structure of ethane. Convince yourself that there is but one propane. By all means use your models.

3.7 PROPANE (C_3H_8) AND PROPYL COMPOUNDS (C_3H_7X)

Propane, the third member of the alkane series, has many of the chemical and physical properties of ethane and methane. Figure 3.33 shows several representations for propane, including two new ones.

In one of these representations [Fig. 3.33(d)], only the carbon frame is retained, and all eight hydrogen atoms are implied. Not even the carbons are shown in Figure 3.33(e), and the reader is left to fill in mentally all the atoms. This most abstract representation is the scheme of choice for organic chemists and we will gradually slip into this way of drawing as we move along through the various structural types. Notice that many, but not all of the schematic representations for propane show that it is not a strictly linear species. All carbons are hybridized approximately sp^3, and the C–C–C angle will be close to 109° (in fact, it is 112°).

Convention Alert!

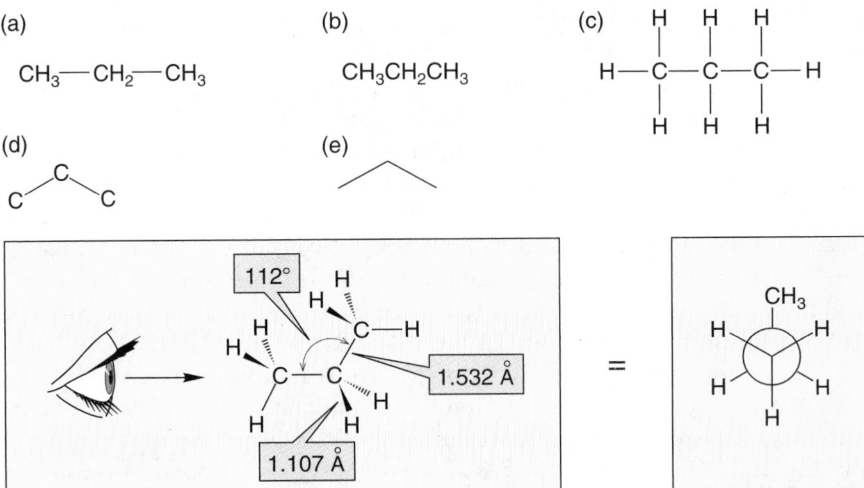

FIGURE 3.33 Several different representations of propane.

The Newman projection (p. 84) of Figure 3.34, constructed by looking down one of the two equivalent carbon–carbon bonds, shows the structure of propane particularly well. A plot of energy versus dihedral angle is similar to that for ethane, except that the rotational barrier is slightly higher, 3.4 kcal/mol (Fig. 3.34). In propane, it is a large methyl group that is eclipsed with a hydrogen at the top of the barrier (the transition state), whereas in ethane it is two small hydrogens that must pass each other.

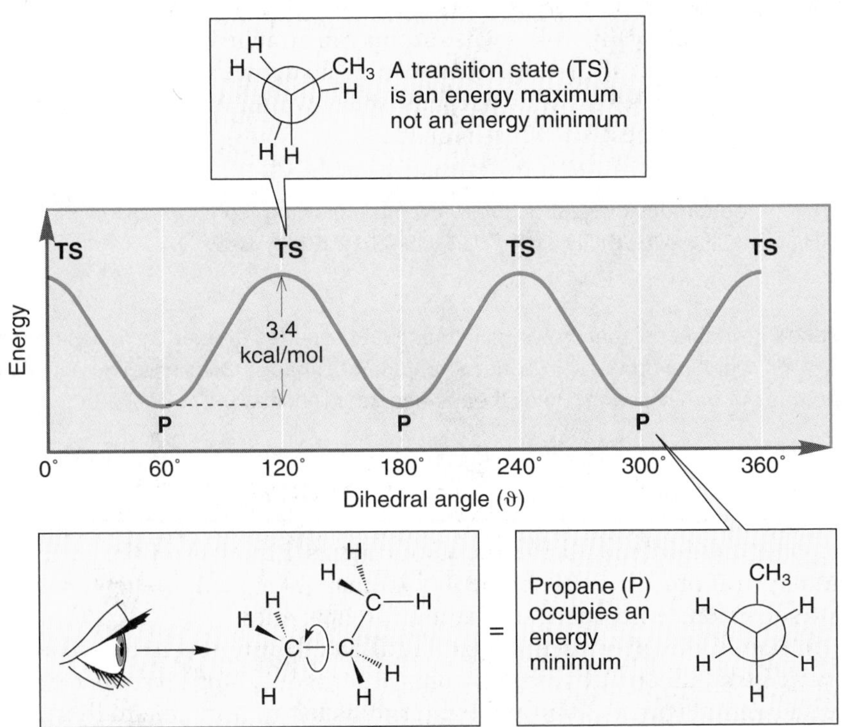

FIGURE 3.34 An energy versus dihedral angle (ϑ) plot for propane.

Replacement of a hydrogen in propane with an X group yields propyl compounds, but the situation is more complicated than it was in methane or ethane. In both methane and ethane all hydrogens were equivalent, and there could be only one methyl or one ethyl derivative. In propane, there are two different kinds of hydrogen. There is a central CH_2 group, called a

methylene group, as well as two equivalent methyl groups at the two ends of the molecule (Fig. 3.35).

Either of these two kinds of hydrogen can be replaced by an X to give a propyl compound (Fig. 3.36). The linear compound $CH_3{-}CH_2{-}CH_2{-}X$, in which X replaces an end methyl hydrogen, was once named normal propyl X (abbreviated ***n*-propyl** X) and you may well still see this nomenclature. Recently, the International Union of Pure and Applied Chemistry (IUPAC) dropped the "*n*" in favor of a simpler name. The branched compound $CH_3{-}CHX{-}CH_3$, in which X replaces a hydrogen on the central methylene carbon, is called **isopropyl** X (and sometimes, *i*-propyl X).

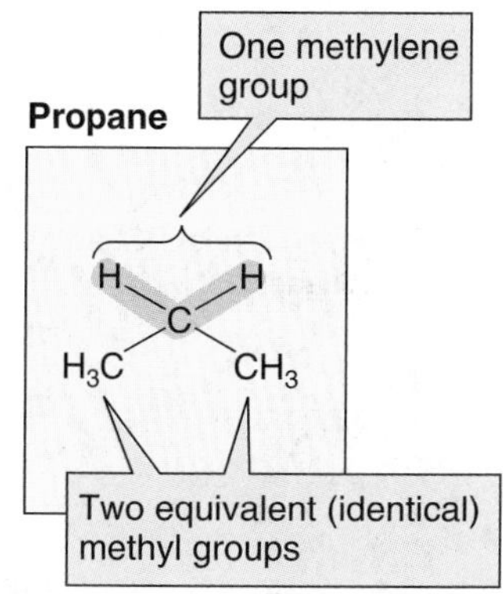

FIGURE 3.35 Propane contains one methylene (CH_2) group and two equivalent methyl groups (CH_3).

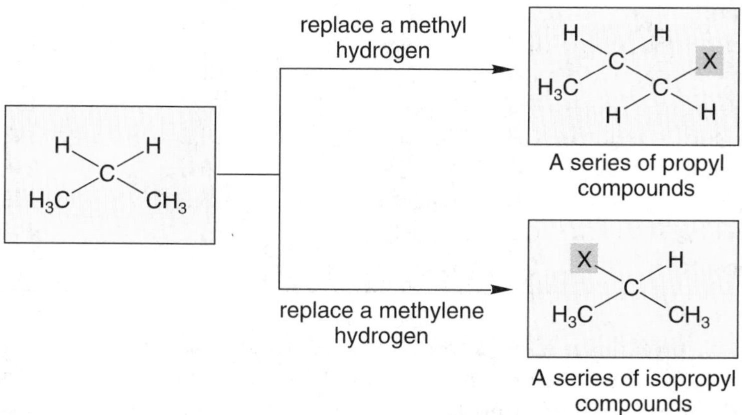

FIGURE 3.36 Propyl and isopropyl compounds.

Make careful three-dimensional drawings of propyl alcohol ($CH_3{-}CH_2{-}CH_2{-}OH$) and the related isopropyl alcohol ($CH_3{-}CHOH{-}CH_3$).

PROBLEM 3.16

3.8 BUTANES (C_4H_{10}), BUTYL COMPOUNDS (C_4H_9X), AND CONFORMATIONAL ANALYSIS

If we once again let X be methyl, we generate butane and isobutane from our two different propyl compounds (Fig. 3.37).

X = CH_3

Butane

X = CH_3

Isobutane

FIGURE 3.37 If X = CH_3, we produce butane and isobutane.

Butane is the next member of a series of straight-chain alkanes. Once we get beyond four carbons, a systematic naming protocol, the IUPAC system, is used. The straight-chain five-carbon alkane is pentane, the six-carbon is version hexane, seven carbon atoms is heptane, and so on. Some straight-chain alkanes are collected in Table 3.4.

TABLE 3.4 Some Straight-Chain Alkanes

Name	Formula	mp (°C)	bp (°C)
Methane	CH_4	−182.5	−164
Ethane	CH_3CH_3	−183.3	− 88.6
Propane	$CH_3CH_2CH_3$	−189.7	− 43.1
Butane	$CH_3(CH_2)_2CH_3$	−138.4	− 0.5
Pentane	$CH_3(CH_2)_3CH_3$	−129.7	36.1
Hexane	$CH_3(CH_2)_4CH_3$	− 95	69
Heptane	$CH_3(CH_2)_5CH_3$	− 90.6	98.4
Octane	$CH_3(CH_2)_6CH_3$	− 56.8	125.7
Nonane	$CH_3(CH_2)_7CH_3$	− 51	150.8
Decane	$CH_3(CH_2)_8CH_3$	− 29.7	174.1
Undecane	$CH_3(CH_2)_9CH_3$	− 25.6	195.9
Dodecane	$CH_3(CH_2)_{10}CH_3$	− 9.6	216.3
Eicosane	$CH_3(CH_2)_{18}CH_3$	36.8	343.0
Triacontane	$CH_3(CH_2)_{28}CH_3$	66	449.7
Pentacontane	$CH_3(CH_2)_{48}CH_3$	92	

Butane provides a nice example of **conformational analysis**, the study of the relative energies of conformational isomers. Moreover, this material anticipates the very important sections in Chapter 6 on the conformational analysis of six-membered rings of carbon (cyclohexanes). Let's start by constructing a Newman projection by looking down the C(2)–C(3) bond of butane (Fig. 3.38).* If we make the reasonable assumption that the

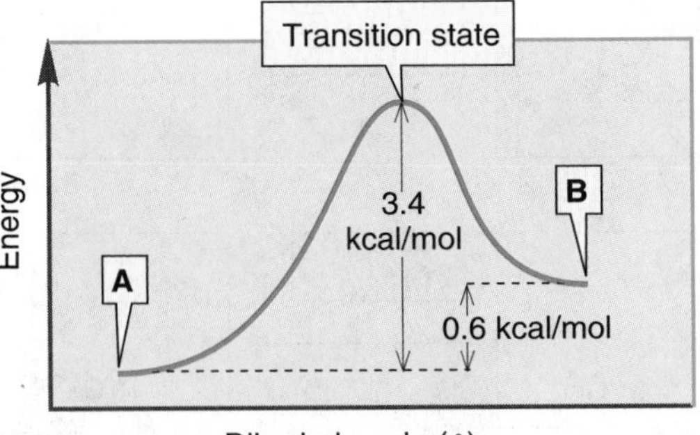

FIGURE 3.38 Different representations of butane. The most stable conformation for butane is the *anti* form **A**. A clockwise rotation of 120° leads to the less stable *gauche* form **B**.

*The notation C(2), C(3), and so on will be used to refer to the carbon atom number.

relatively large methyl groups will best be kept as far from each other as possible, we come to conformation **A**, called the *anti* form. As in ethane (p. 83), a 120° clockwise rotation passes over a transition state (energy maximum) to a new minimum, **B**. Conformation **B** will be less stable than **A** because in **B** the two methyl groups are closer than they are in **A**. Conformation **B** is called *gauche* butane and the methyl–methyl interaction is a *gauche* interaction. *gauche* Butane (**B**) is about 0.6 kcal/mol higher in energy than the *anti* form (**A**). The eclipsed transition state for the interconversion of **A** and **B** lies about 3.4 kcal/mol higher than **A**.

Another 120° clockwise rotation brings us to a second *gauche* butane, **B'**. The transition state for the conversion of **B** to **B'** involves an eclipsing of two $C-CH_3$ bonds and lies 3.8 kcal/mol above **B** and **B'**, a total of about 4.4 kcal/mol above **A**. A final 120° clockwise rotation returns us to **A** (Fig. 3.39).

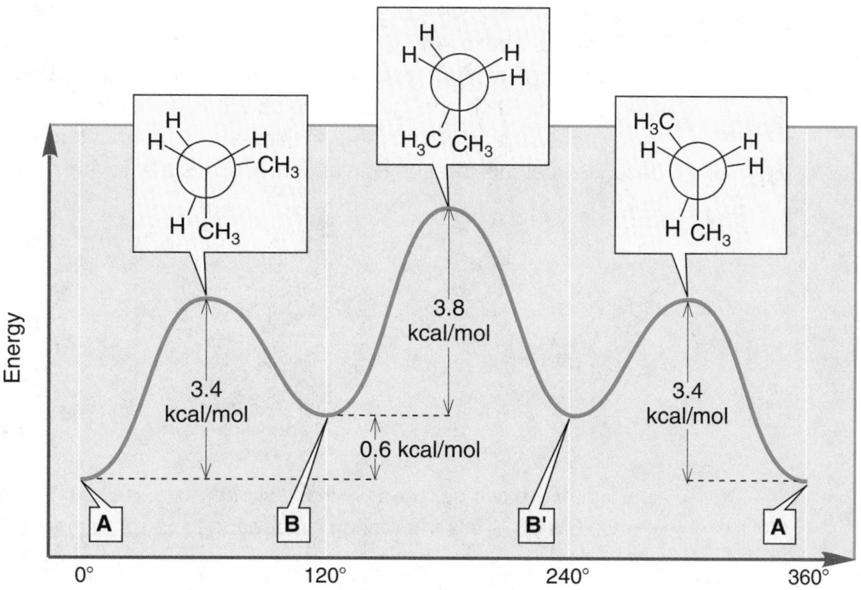

FIGURE 3.39 A dihedral angle (ϑ) versus energy plot for butane.

The branched compounds such as isobutane are more difficult to name systematically than their straight-chain isomers. The first members of this series have common names, but as the number of carbons increases, and complexity sets in, the IUPAC system takes over. We start our analysis of branched compounds with the butyl compounds, C_4H_9-X. Butane contains two kinds of hydrogen atoms in its pairs of equivalent methylene (CH_2) and methyl (CH_3) groups, which should produce two butyl

*PROBLEM **3.17**

Recall how much energy an eclipsed C—H/C—H interaction costs in ethane (p. 86). Given that value, and the information that the transition state for the interconversion of **A** and **B** is 3.4 kcal/mol higher in energy than **A**, calculate how much each eclipsed C—CH_3/C—H interaction in the transition state for the interconversion of **A** and **B** is worth. How much is the eclipsed C—CH_3/C—CH_3 interaction in the transition state for the conversion of **B** to **B′** worth (see Fig. 3.39)?

ANSWER

In this problem you need to calculate the amount of destabilization caused by C—CH_3/C—H and C—CH_3/C—CH_3 interactions. Each C—H/C—H eclipsed interaction costs about 1.0 kcal/mol (p. 86). The transition state for the interconversion of **A** and **B** is an eclipsed form. In the figure below, it is reached through a 60° clockwise rotation of the rear carbon. A second 60° clockwise rotation takes us to **B**.

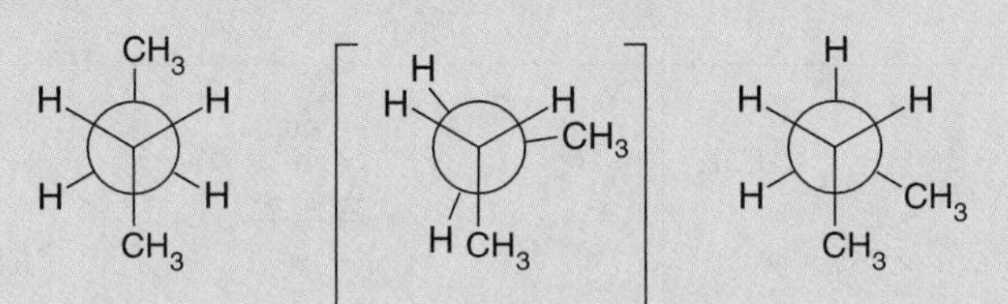

A | Transition state | **B**

In the transition state, there is one eclipsed C—H/C—H interaction and two eclipsed C—H/C—CH_3 interactions. We know that the transition state lies 3.4 kcal/mol above **A** (Fig. 3.39). The C—H/C—H interaction costs 1.0 kcal/mol, and so the two C—H/C—CH_3 interactions must produce the remaining 2.4 kcal/mol. Each one must cause about 1.2 kcal/mol destabilization.

Figure 3.39 shows that the transition state for the interconversion of **B** and **B′** lies 3.8 kcal/mol above **B**. The transition state for this conversion is shown below.

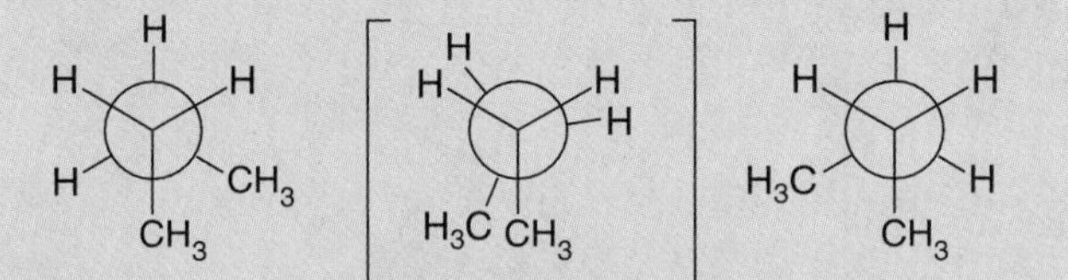

B | Transition state | **B'**

Each of the two eclipsed C—H/C—H interactions costs 1.0 kcal/mol, so the single C—CH_3/C—CH_3 interaction must cause the remaining 1.8 kcal/mol destabilization.

PROBLEM **3.18**

Draw Newman projections constructed by looking down the indicated carbon–carbon bond in the molecules in Figure 3.40. For the second molecule make an energy versus dihedral angle graph.

FIGURE 3.40

compounds, as shown in Figure 3.41. Similarly, isobutane contains a single **methine group** (CH) and three equivalent methyl groups and should also yield two butyl compounds.

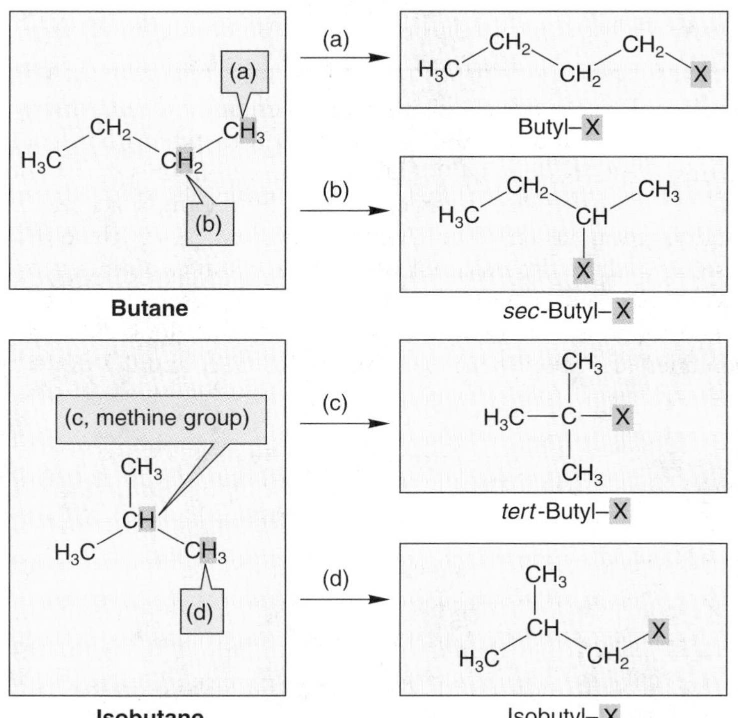

FIGURE 3.41 Replacement of one of the hydrogens of butane with X yields two kinds of butyl derivatives, butyl–X and *sec*-butyl–X. Replacement of hydrogen with X in isobutane yields two more butyl compounds, isobutyl–X and *tert*-butyl–X.

It appears that there should be four derivatives of C_4H_9–X. Indeed there are, and Figure 3.41 shows these butyl–X compounds. The four different types of butyl compound are called **butyl**, **isobutyl**, ***sec*-butyl** (short for *secondary*-butyl), and ***tert*-butyl** (short for *tertiary*-butyl). A new and important terminology has been introduced. A **primary carbon** is attached to only one other carbon atom. Ethane contains two primary carbon atoms. **Secondary carbons** are attached to two carbons and **tertiary** carbons to three. A **quaternary** (*not* quarternary) **carbon** atom is attached to four other carbons. Thus, the names for the butyl groups tell you something about the structure. These commonly encountered groups and names must be learned (Fig. 3.42).

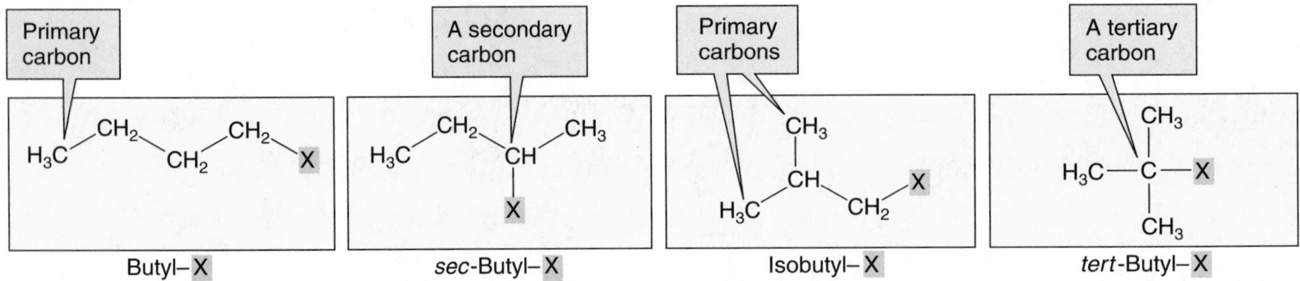

FIGURE 3.42 The four different kinds of butyl group.

PROBLEM 3.19 Draw two compounds containing quaternary carbons.

3.9 PENTANES (C_5H_{12}) AND PENTYL COMPOUNDS ($C_5H_{11}X$)

One would now anticipate that four pentanes would emerge when we transform our four butyl-X compounds into five-carbon compounds by letting X = CH_3 (Fig. 3.43). Yet, only three pentanes exist! Somehow the technique of generating larger alkanes by letting a general X become a methyl group has failed us. The problem lies again in the difficulty of representing three-dimensional structures on a two-dimensional surface. Two of the four structures in Figure 3.43 represent the same compound, even though in this two-dimensional code they look different. The three pentanes are called pentane, 2-methylbutane, and 2,2-dimethylpropane.

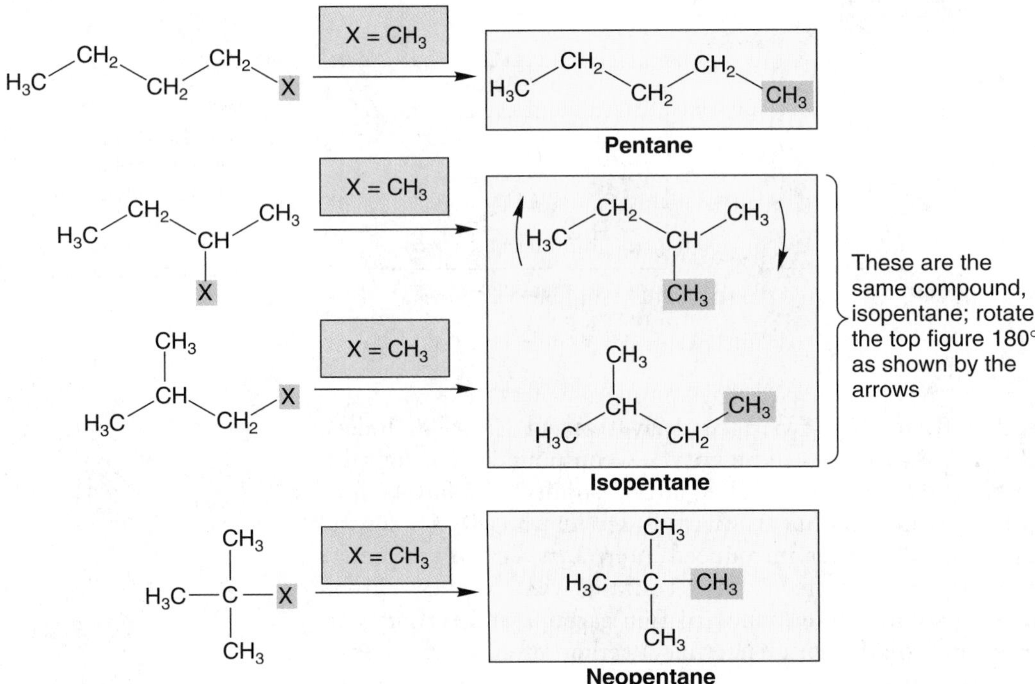

FIGURE 3.43 If we let X = CH_3, it appears as though four pentanes should exist. Yet there are only three pentanes: pentane, isopentane, and neopentane. Two of the pentanes we created are the same. Make models to be sure you understand this point.

These last two compounds are traditionally called "isopentane" and "neopentane." Section 3.10 will describe the naming system for alkanes.

One important lesson to be learned from Figure 3.43 is that the two-dimensional page will lie to you more often than not if you let it. As we have already seen, there is a tension between our need to talk quickly and efficiently to each other and the accurate representation of these complicated organic structures. This tension is at least partly resolved by using codes, but the price of this use is that you *always* must be able to see past the code to the real, three-dimensional structure if you expect to be able to think effectively about organic chemistry. These codes become quite abstract (see p. 90), and it is worth going through the process of developing the abstraction one more time. Consider the representations for isopentane in Figure 3.44. The first level of abstraction is to suggest only vaguely the real molecule, as in Figure 3.44(a). One then goes to the more schematic structure of Figure 3.44(b), and then to a structure in which the hydrogens are left out Figure 3.44(c). Finally, one even removes the carbons, as in Figure 3.44(d), to give the ultimate abstraction for the structure of isopentane. Here nothing survives from the original picture except a representation of the backbone. You are left to supply the carbons, which must be put in at every vertex and terminus, as well as the attendant hydrogens. Most important, you must be able to translate these highly schematic structures into three dimensions, and to see the way these skeletal pictures transform into molecules.

(a) H—C—C—C—C—H (with H atoms on each carbon; CH_3 branch: H—C—H, H)

(b) H_3C—CH_2—CH(CH_3)—CH_3

(c) C—C—C(C)—C

(d)

FIGURE 3.44 Four increasingly abstract representations for isopentane.

Draw schematic representations for the pentanes and hexanes. PROBLEM 3.20

Actually, there is even more to be done. These line drawings can be approximated with ball-and-stick models, in which balls represent atoms and sticks represent the bonds between atoms. Such representations do a relatively poor job of showing the **steric** (spatial) requirements of the real molecules. Chemists have therefore evolved space-filling models, which attempt to show the volumes of space carved out by the atoms. Figure 3.45 shows some representations of isopentane.

$(H_3C)_2CH—CH_2—CH_3$

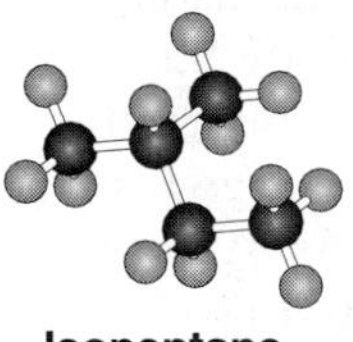

Isopentane

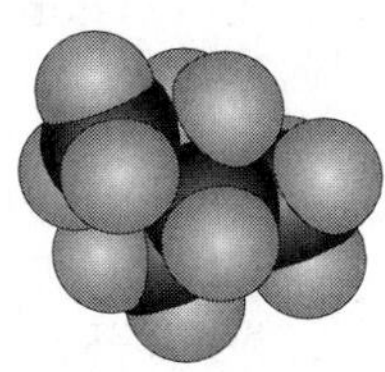

FIGURE 3.45 Views of isopentane.

So there are only three pentanes, but no less than eight pentyl derivatives (Fig. 3.46).

Pentane

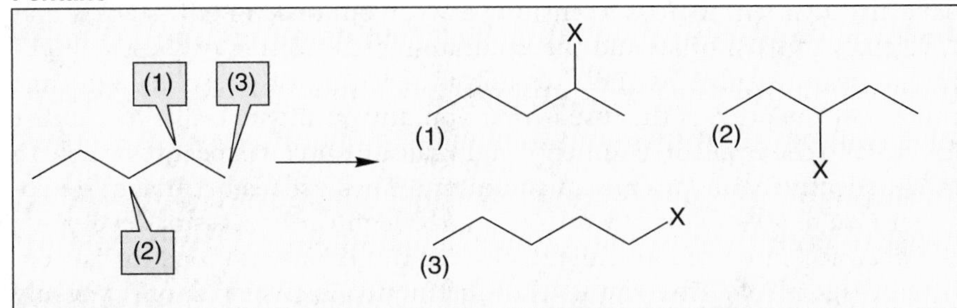

Isopentane

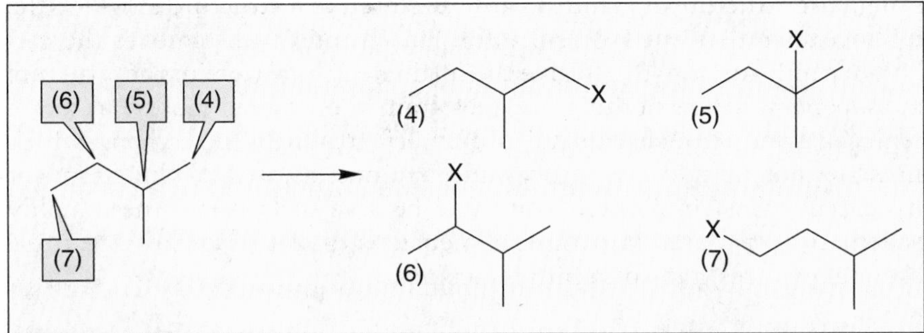

Neopentane

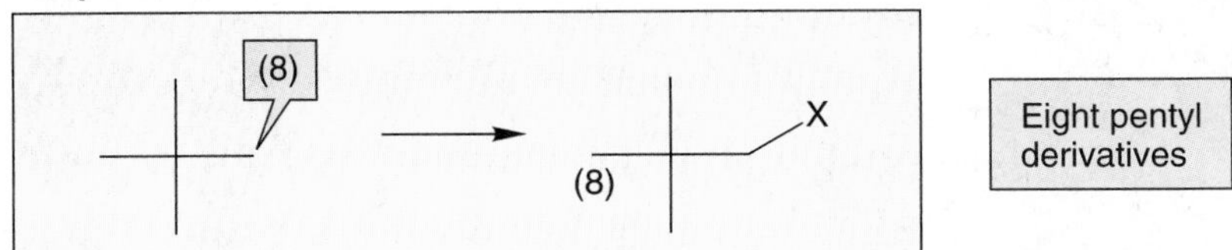

FIGURE 3.46 The eight pentyl derivatives.

3.10 THE NAMING CONVENTIONS FOR ALKANES

Clearly, things are rapidly getting out of hand. Although it is not too difficult to remember the four butyl groups, the eight pentyls constitute a tougher task—and the problem rapidly gets worse as the number of carbons increases. A system is absolutely necessary. In practice, the old, nonsystematic names are retained through the butanes, but once we reach five carbons the IUPAC system largely takes over, although a few old favorite names are still occasionally seen.

Here are the most important rules of the naming convention:

1. Search for the longest straight chain of carbons. The structure is named as a derivative of this parent hydrocarbon. This rule can be a bit tricky at first. *Remember*: The corners apparent in many two-dimensional representations are not present in the real structures. In the examples in Figure 3.47 it is perfectly reasonable to count around the corner in order to find the longest chain. If this is not clear to you, by all means go back to Problem 3.15. Do not be reluctant to use your models. An atom or group attached to the longest straight chain is given a position number. The chain is numbered

from the end that produces the lowest position number for the substituent.

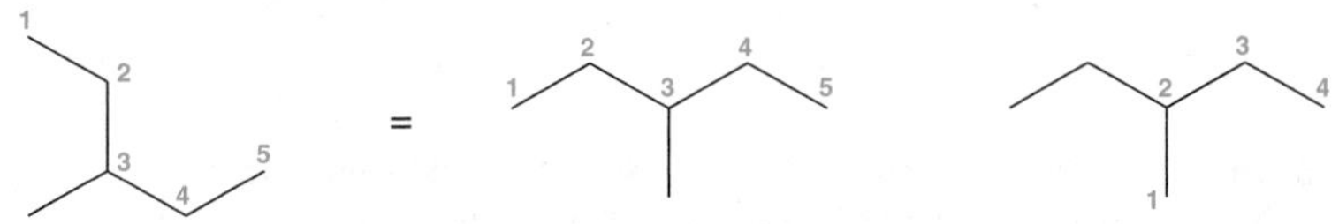

3-Methylpentane *not* **2-Ethylbutane**

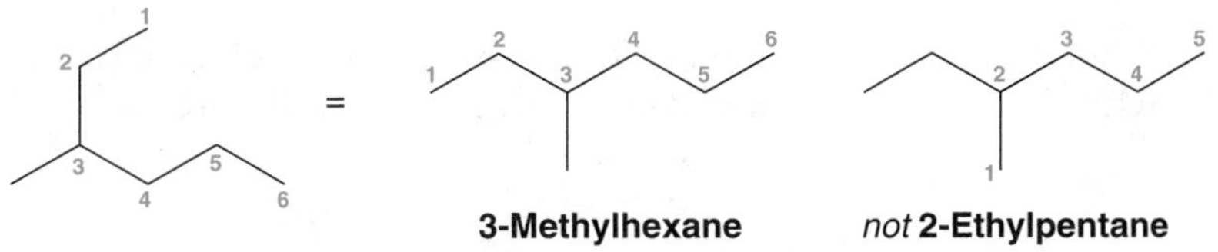

3-Methylhexane *not* **2-Ethylpentane**

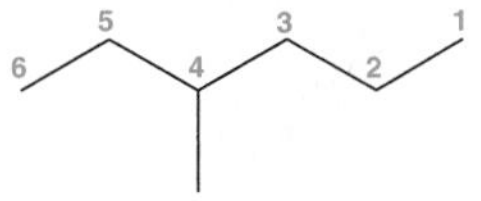

not **4-Methylhexane**

FIGURE 3.47 In finding the longest straight chain of carbons it is permissible, indeed often necessary, to count around corners. Substituents are numbered and the numbering is done from the end that yields the lowest possible number for the substituent.

2. In a substituted alkane, the substituent is given a number based on its position on the parent hydrocarbon. The longest chain is numbered so as to make the number of the substituent position as low as possible. Some examples are shown in Figure 3.48.

2-Chlorobutane *not* **3-Chlorobutane**

3-Fluorohexane *not* **4-Fluorohexane**

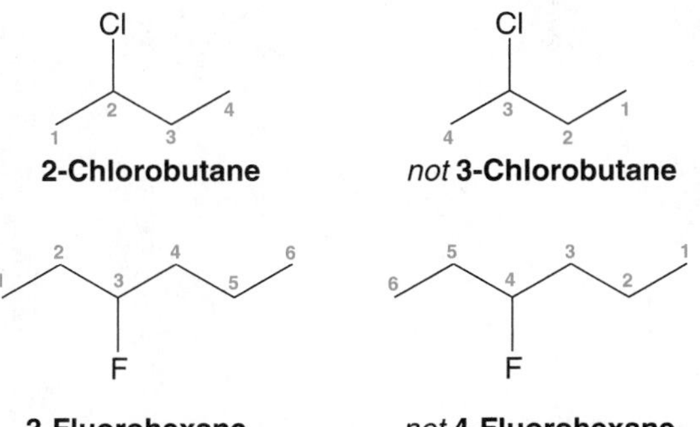

FIGURE 3.48 The chain is numbered so as to produce the smallest number for a substituent.

3. In a multiply substituted alkane, all the substituents receive numbers. If the substituents are identical, the prefixes di-, tri-, tetra-, and so on are used. Figure 3.49 gives some examples.

2,3-Dimethylbutane **2,3,4-Trichloropentane** **1,1,1,5,5,5-Hexafluoropentane**

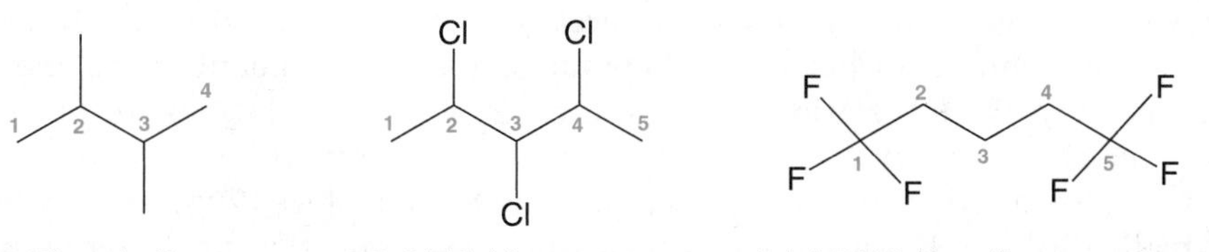

FIGURE 3.49 Examples of polysubstituted alkanes in which the substituents are identical.

If the substituents are not identical, they are ordered alphabetically, as shown in Figure 3.50. Prefixes such as *sec*- and *tert*- are ignored in determining alphabetical order. Thus "*tert*-butyl" would appear in the name earlier than "chloro."

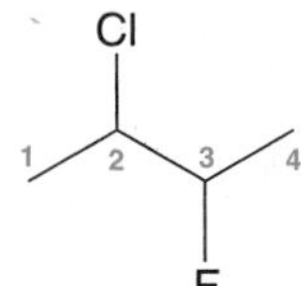

2-Chloro-3-fluorobutane
not 3-chloro-2-fluorobutane
not 2-fluoro-3-chlorobutane

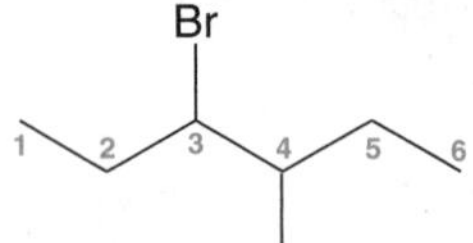

3-Bromo-4-methylhexane
not 4-bromo-3-methylhexane
not 3-methyl-4-bromohexane

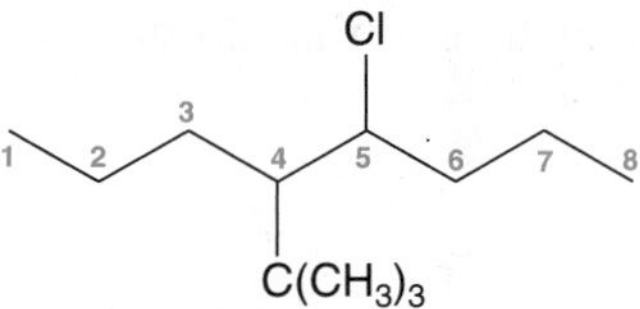

4-*tert*-Butyl-5-chlorooctane
not 4-chloro-5-*tert*-butyloctane

FIGURE **3.50** Nonidentical substituents are incorporated into the name alphabetically.

4. When the rules above do not resolve the issue, the name that starts with the lower number is used, as demonstrated in Figure 3.51.

2,5,5-Trimethylheptane
not 3,3,6-trimethylheptane

FIGURE **3.51** In unresolvable cases, start with the lowest possible number.

PROBLEM **3.21** Write IUPAC names for the following compounds:

PROBLEM **3.22** Write structures for the following compounds: (a) 2-bromobutane, (b) decane, (c) *tert*-butyl chloride, (d) 1,4-difluoropentane, (e) 2,4,4-trimethylheptane.

3.11 WRITING ISOMERS

A favorite question in courses on organic chemistry is, Write all the isomers of pentane (or hexane, or heptane, etc.). Such problems force one to cope with the translation of two-dimensional representations into three-dimensional reality. Figuring out the pentanes is easy; there are only 3 isomers. Even writing the hexanes (5 isomers) or heptanes (9 isomers) is not really difficult. But getting all the octanes (18 isomers) is a tougher proposition, and getting all the nonanes (35 isomers) with no repeats is a real challenge. Success depends on finding a systematic way to write isomers. Thrashing around writing structures without a system is doomed to failure. You will not get all the isomers and there is a high probability of generating repeat structures. Any system will work as long as it is truly systematic. One possibility is the following, shown generating the 9 heptanes in Figure 3.52:

Heptane

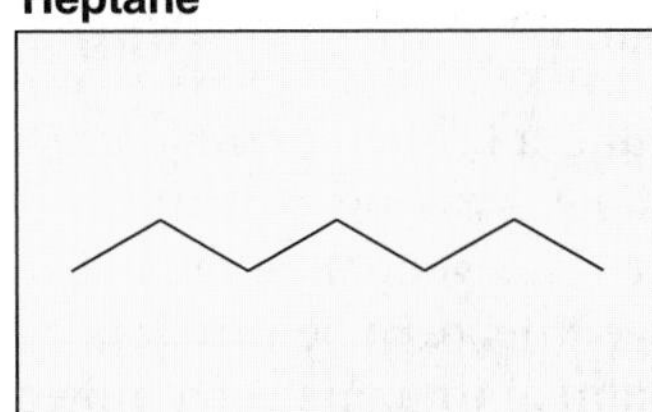

Common repeats

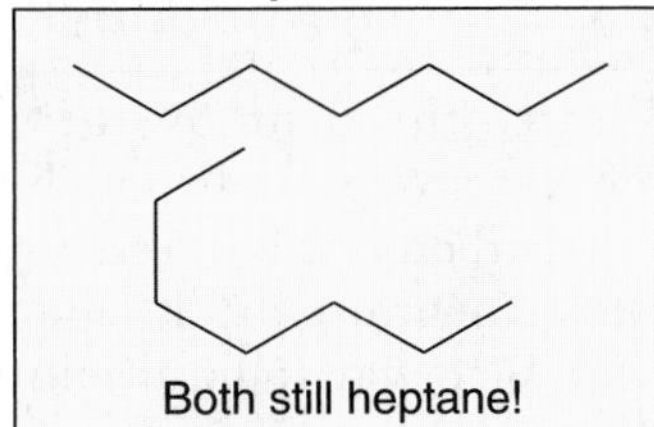

Hexanes —
one C atom must be added

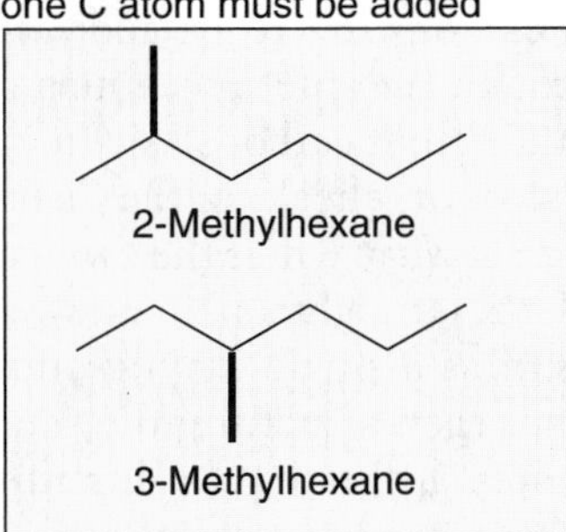

Still 3-methylhexane

Still 3-methylhexane

Pentanes—
two C atoms must be added

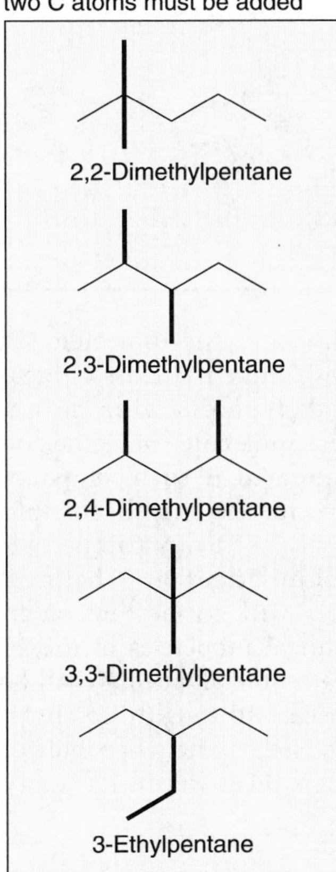

Still 2,3-dimethylpentane

Still 2,3-dimethylpentane

Butane—
three C atoms must be added

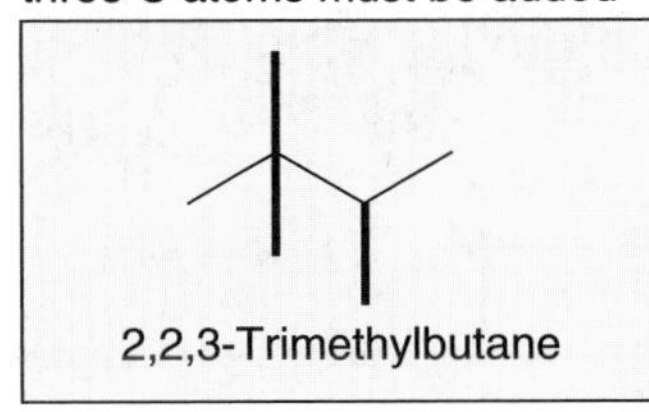

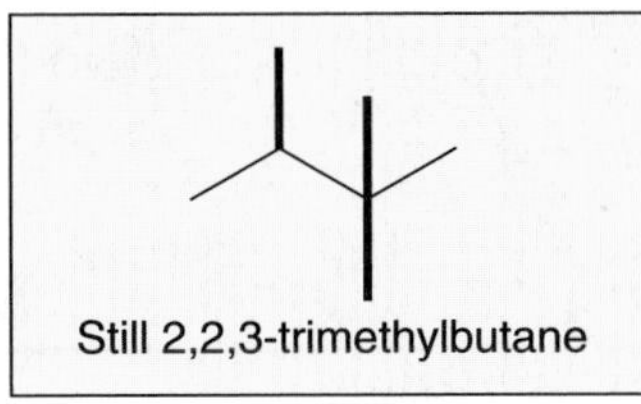

FIGURE 3.52 A method for finding all the isomers of a hydrocarbon. The nine heptanes are used as examples.

1. Start with the longest straight chain possible. This system generates heptane itself.
2. Shorten the chain by one carbon, and add the "extra" carbon as a methyl group at all possible positions, starting with the left-hand side carbon and moving to the right. This generates the methylhexanes. You must check each isomer you make to be sure it is not a repeat. One way to be absolutely certain is to name each molecule as you generate it. If you repeat a name, you have repeated an isomer.
3. Shorten the chain by one more carbon. The longest straight chain is now five carbons (pentanes). Two carbons are to be added as two methyl groups or a single ethyl group. Two methyl groups can be placed either on the same carbon or different carbons. Again, start at the leftmost carbon and move to the right, checking each isomer to be sure it is not a repeat. Be sure to see that when the two carbons are placed on the *same* carbon of the straight chain they can be added either as two methyl groups or as a single ethyl group. Take care to look carefully as you generate each structure to be sure you have not repeated an isomer. This step generates the dimethylpentanes and 3-ethylpentane.
4. Shorten the chain again, and try to fit the three "extra" carbons on a butane chain. There is only one way to do this.

PROBLEM 3.23 Write and name all the isomers of octane, C_8H_{18}.

3.12 RINGS

All the hydrocarbons we have met so far have the molecular formula C_nH_{2n+2}, and are called **saturated alkanes**. There is a class of closely related molecules that shares most chemical properties with the saturated alkanes, but not the general formula. These molecules have the composition C_nH_{2n} and are sometimes called **unsaturated**. In such compounds, we could find a way in principle (and often in practice as well) to attach two more hydrogen atoms to bring us up to C_nH_{2n+2}. Every pair of hydrogens that is "missing" constitutes one degree of unsaturation. What might the bonding in such molecules be? It probably will not deviate much from what we already know, because the chemical properties of these molecules resemble very closely those of the saturated alkanes. Recall how we first thought of making ethane. We allowed two roughly sp^3 hybridized carbons to approach and the overlap of the pair of singly occupied orbitals generated a bonding, filled σ carbon–carbon bond as well as its empty, antibonding counterpart, σ* (Fig. 3.53).

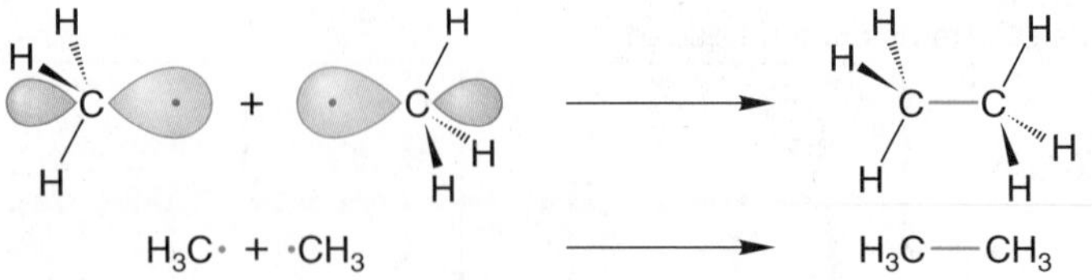

$$H_3C\cdot + \cdot CH_3 \longrightarrow H_3C{-}CH_3$$

FIGURE 3.53 The formation of ethane through the overlap of two singly occupied hybrid orbitals.

Suppose we replace two of the hydrogens of ethane with carbons, here shown as methyl groups. This substitution corresponds to the construction of butane from two ethyl radicals. There is little difference between this process and the dimerization of two methyl groups. Now make one more small change. Instead of using methyl groups to replace the pair of hydrogens in Figure 3.53, use a chain of connected methylene groups to form a ring (Fig. 3.54).

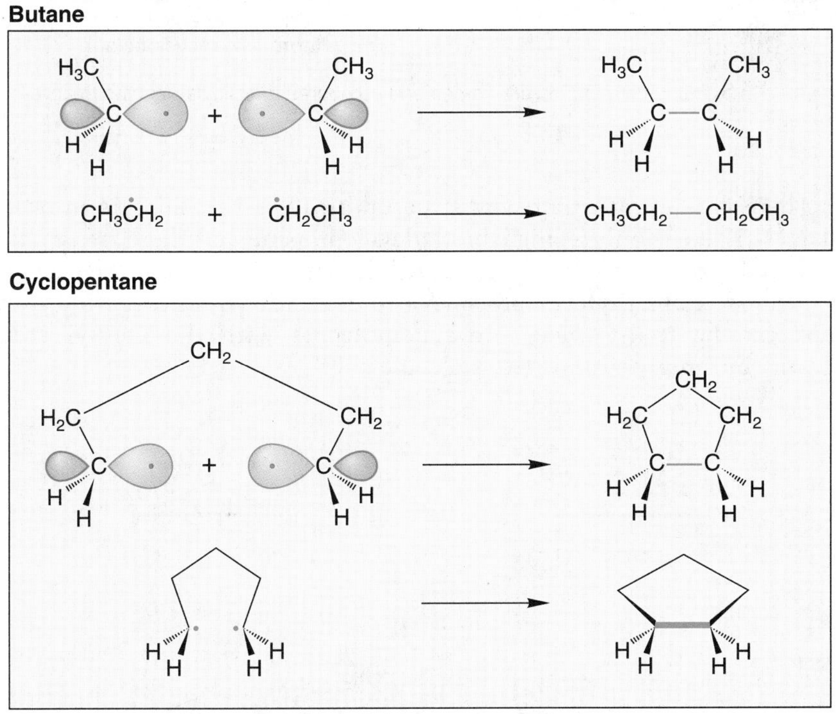

Butane

$CH_3\dot{C}H_2 + \dot{C}H_2CH_3 \longrightarrow CH_3CH_2—CH_2CH_3$

Cyclopentane

FIGURE 3.54 The formation of butane and cyclopentane through the overlap of two singly occupied hybrid orbitals.

There is no essential difference between this construction of a ring compound and the process used to make ethane and butane. Nor would we expect to see big differences in chemical properties. The bonding in pentane is not significantly different from that in cyclopentane (Fig. 3.54). These ring compounds are named by attaching the prefix "cyclo" to the name of the parent hydrocarbon (Table 3.5).

TABLE 3.5 Some Cyclic Alkanes (Cycloalkanes)

Name	Formula	mp (°C)	bp (°C, 760 mmHg)
Cyclopropane	$(CH_2)_3$	−127.6	− 32.7
Cyclobutane	$(CH_2)_4$	− 50	12
Cyclopentane	$(CH_2)_5$	− 93.9	49.2
Cyclohexane	$(CH_2)_6$	6.5	80.7
Cycloheptane	$(CH_2)_7$	− 12	118.5
Cyclooctane	$(CH_2)_8$	14.3	148–149 (749 mmHg)

Monosubstituted cycloalkanes require no numbers. In multiply substituted compounds substituents are assigned numbered positions, and the name is constructed so as to minimize the numbers. Multiple substituents are named alphabetically (Fig. 3.55).

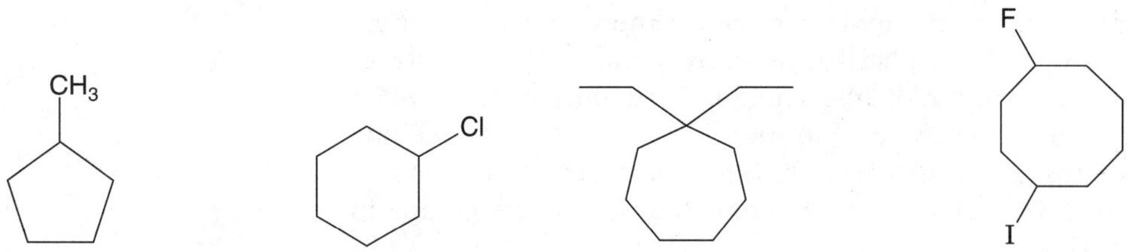

Methylcyclopentane **Chlorocyclohexane** **1,1-Diethylcycloheptane** **1-Fluoro-4-iodocyclooctane**
not 4-Iodo-1-fluorocyclooctane
or 4-fluoro-1-iodocyclooctane

FIGURE 3.55 Ring compounds are named and numbered in a similar fashion to acyclic alkanes.

There is one important new structural feature that appears in cycloalkanes. There is only a single methylcyclopropane, but there are *two* isomers of 1,2-dimethylcyclopropane. As shown in Figure 3.56, there is no "sidedness" to methylcyclopropane. The molecule with the methyl group "up" can be transformed into the molecule with the methyl group "down" by simply turning the molecule over.

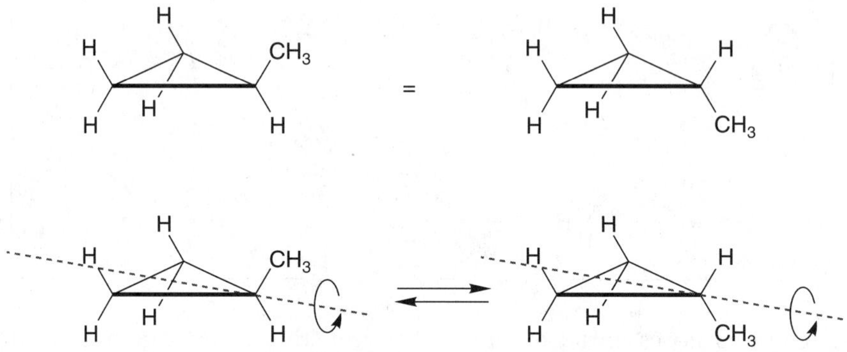

Rotate 180°

FIGURE 3.56 There is only one isomer of methylcyclopropane. In the figure a thicker line is used to show the side of the cyclopropane nearest to you.

However, no number of translational or rotational operations will suffice to change the 1,2-dimethylcyclopropane with both methyl groups on the same side of the ring, called **cis**, into the one with the methyl groups on opposite sides, called **trans**. The two are certainly different. The use of models is absolutely mandatory at this point. Make models of the two compounds and convince yourself that nothing short of breaking carbon–carbon bonds will allow you to turn *cis*-1,2-dimethylcyclopropane into the isomeric *trans*-1,2-dimethylcyclopropane (Fig. 3.57).

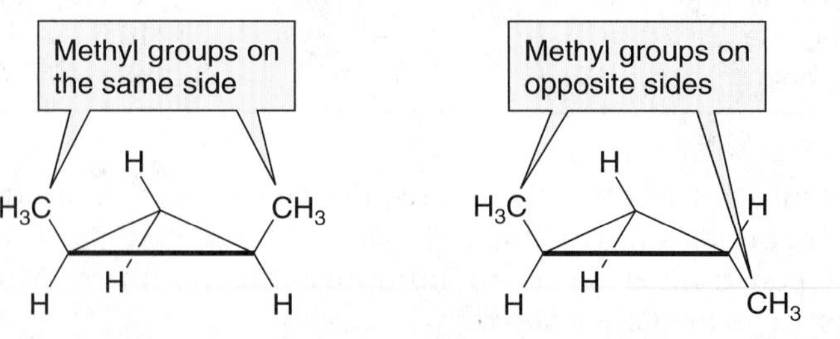

***cis*-1,2-Dimethylcyclopropane** ***trans*-1,2-Dimethylcyclopropane**

FIGURE 3.57 There are two isomers of 1,2-dimethylcyclopropane, called cis and trans (but see Problem 3.24).

PROBLEM 3.24

The situation is even more complicated, as you will see in Chapter 5. There are really two isomers of *trans*-1,2-dimethylcyclopropane! Use your models to construct the mirror image of the *trans*-1,2-dimethylcyclopropane you made and see if it is identical to your first molecule.

3.13 PHYSICAL PROPERTIES OF ALKANES AND CYCLOALKANES

Simple saturated hydrocarbons and cycloalkanes are colorless gases, clear liquids, or white solids, depending on their molecular weight. To many people they smell bad, although I have long thought that these molecules have been the victims of a bad press, and I don't think they smell bad at all. The odor of cooking gas, mostly saturated hydrocarbons, comes from an additive, a mercaptan (R–SH) (Remember: R stands for a general alkyl group, p. 79), put in specifically so that escaping gas can be detected by smell. Tables 3.4 and 3.5 have collected some physical properties of straight-chain and cyclic alkanes. Why do the boiling points increase as the number of carbons in the molecule increases? The boiling point is a measure of the ease of breaking up intermolecular attractive forces. There is a factor that stabilizes the liquid phases of hydrocarbons called **van der Waals forces**. When two clouds of electrons approach each other dipoles are induced as the clouds polarize in such a fashion as to stabilize each other by opposing opposite charges (Fig. 3.58).

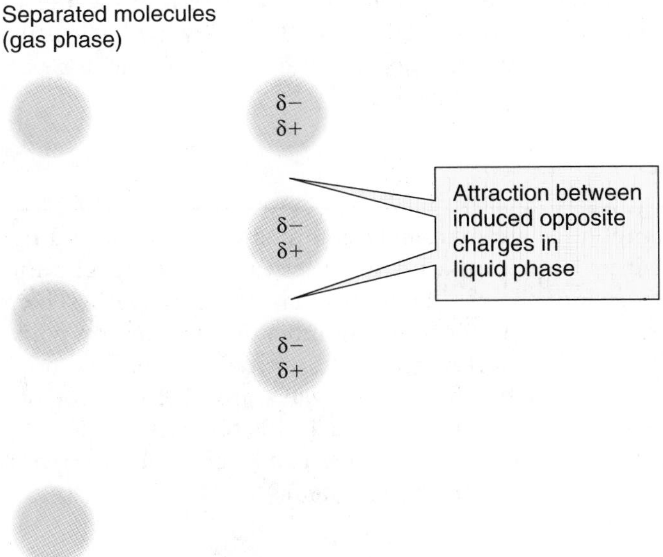

FIGURE 3.58 The stabilization of molecules through van der Waals forces.

Of course, many alkanes have small dipoles to begin with, but they are very small and do not serve to hold the molecules together strongly. Other molecules are much more polar and this polarity makes a big difference in boiling point. Polar molecules can associate quite strongly by aligning opposite charges. This association has the effect of increasing the boiling point.

The more extended a molecule is, the more potent the induced dipole can be. More compact, more symmetric molecules have smaller induced dipoles, and therefore lower boiling points. A classic example is the difference between pentane and neopentane (Fig. 3.59). The more spherical neopentane boils about 25 °C lower than the straight-chain isomer. Isopentane is less extended than pentane but more extended than neopentane, and its boiling point is right between the two.

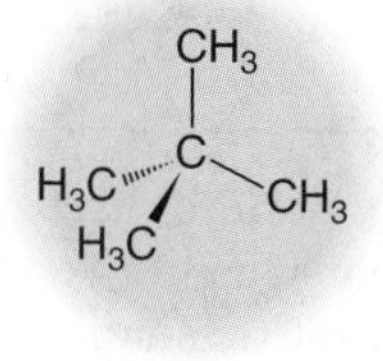

Highly symmetrical neopentane has a nearly spherical cloud of electrons (bp 9.5 °C; mp −16.5 °C)

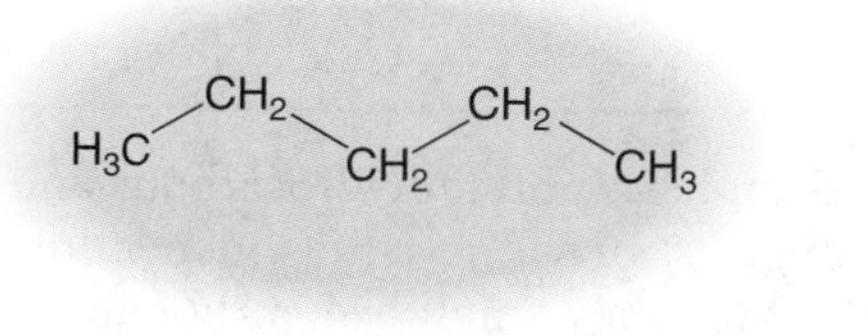

The more extended molecule pentane has a much greater surface and has greater intermolecular interactions (bp 36.1 °C; mp −130 °C)

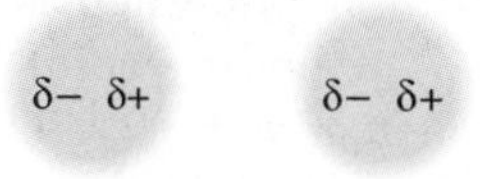

Minimal interaction between two spheres allows for relatively weak van der Waals forces

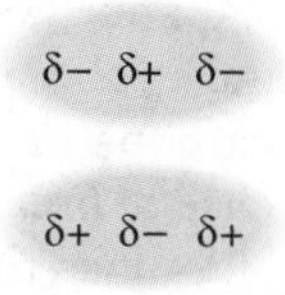

More extensive contact possible in the extended molecule allows more powerful van der Waals interactions

FIGURE 3.59 The more extended pentane boils at a higher temperature than the more compact neopentane does.

Symmetry is especially important in determining melting point, as highly symmetric molecules pack well into crystal lattices. It takes more energy to break up a well-packed lattice. Neopentane, for example, melts 113 °C higher than pentane does.

3.14 SOMETHING MORE: POLYCYCLIC COMPOUNDS

Ring compounds are extremely common. Both small and large varieties are found in Nature, and many varieties of exotic cyclic molecules not found so far outside the laboratory have been made by chemists. Moreover, rings can be combined in a number of ways to form polycyclic molecules. Here is an opportunity for you to think ahead. How might two rings be attached to each other? Some ways are obvious, but others require some thought. We will work through this topic as a series of problems. A number of different structural types can be created from two rings. These problems try to lead you through them.

*PROBLEM 3.25 Find a compound of the formula $C_{10}H_{18}$ composed of two five-membered rings.

ANSWER Two five-membered rings can be joined in a very simple way to make bicyclopentyl. There is no real difference between this process and the formation of ethane from two methyl radicals (p. 104).

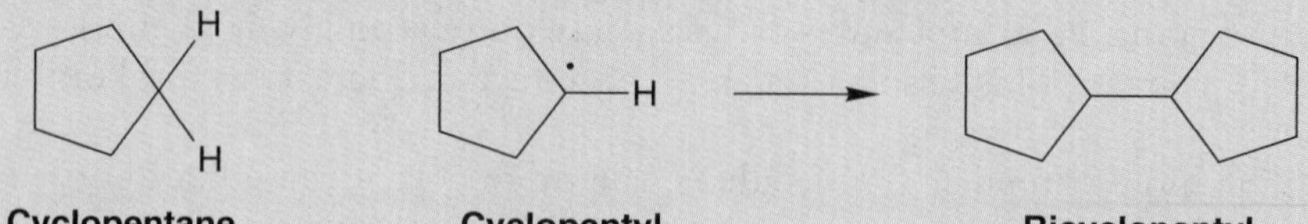

Cyclopentane (C_5H_{10}) **Cyclopentyl** (C_5H_9) → **Bicyclopentyl** ($C_{10}H_{18}$)

Another structural type still containing two five-membered rings has the formula C_9H_{16}. Clearly, we are not dealing with a simple dimer here as we are short one carbon. The two rings must share one carbon somehow. Draw this compound.

*PROBLEM 3.26

ANSWER

In this case, the two rings share a single carbon. The way to do it is to let one carbon be part of both five-membered rings.

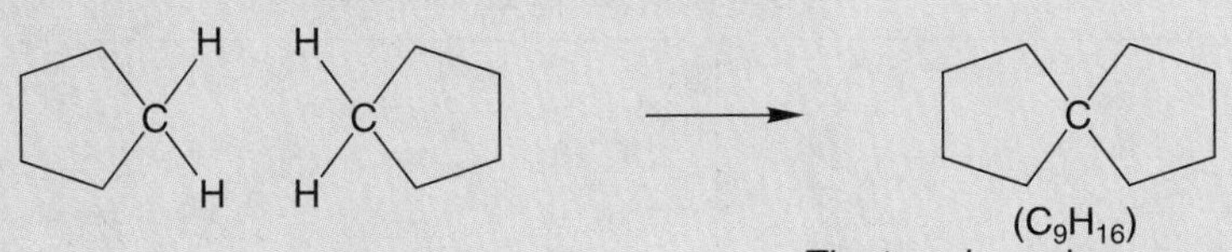

(C_9H_{16})
The two rings share one carbon

Two five-membered rings can share more than one carbon. Find a molecule of the formula C_8H_{14}. There are two similar examples of this kind of "fused" molecule. Try to find both. *Hint*: Focus on the two shared carbons and the attached hydrogens. (Make a model.)

*PROBLEM 3.27

ANSWER

The two rings share two carbons in the following way:

(C_8H_{14})
The two rings share two carbons

But the problem points out that there are two of these compounds. Remember that rings have sides. There are cis and trans forms for this molecule (p. 106). The two rings can be fused together in a cis or trans fashion:

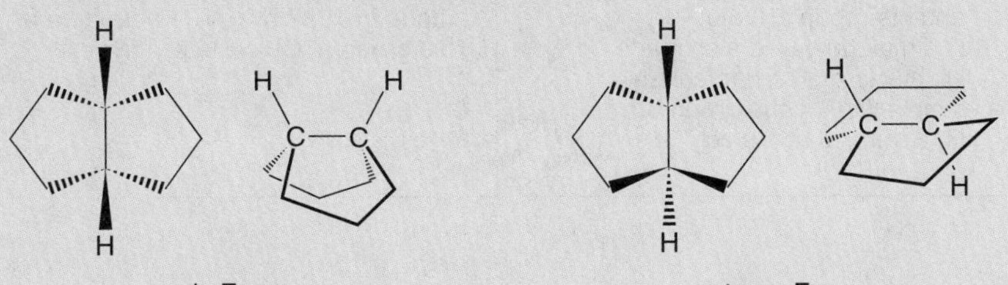

cis Form

trans Form

Finally, and most difficult, there is a molecule, still constructed from five-membered rings, that has the formula C_7H_{12}. In this structural type, the two five-membered rings must share three carbons. Find this molecule.

*PROBLEM 3.28

ANSWER

Finally, we come to C_7H_{12}, in which the two rings share three carbons. A "cage" structure results.

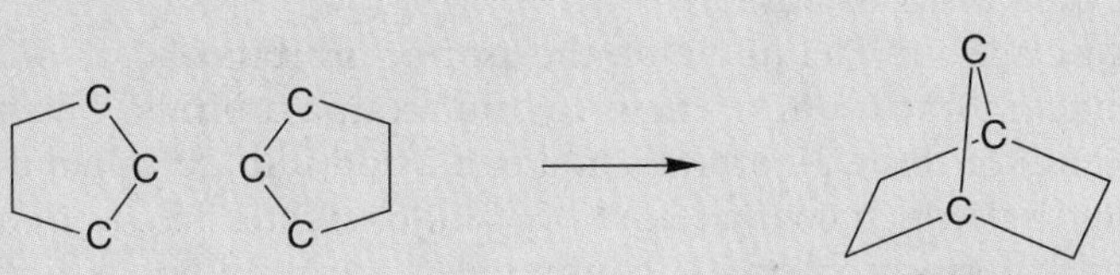

(C_7H_{12})
Now three carbons are shared

Polycyclic compounds can be exceedingly complex. Indeed, much of the fascination that organic chemistry holds for some people is captured nicely by the beautifully architectural structures of these compounds. Figure 3.60 shows four molecules. Two molecules are "natural products" and two are not (yet) found outside the laboratory: [1.1.1]propellane and dodecahedrane.

Aflatoxin B_1
made in 1966 by G. Büchi and his research group at MIT

Progesterone
made in 1967 by G. Stork and his group at Columbia University

[1.1.1]Propellane
made in 1982 by K. B. Wiberg and his group at Yale; this molecule was also made slightly later in a particularly simple way by the group of G. Szeimies at Munich

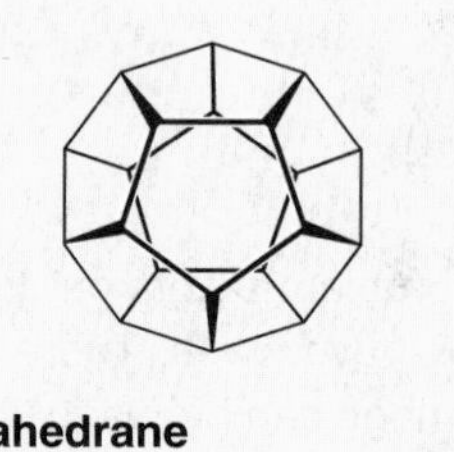

Dodecahedrane
made in 1982 by L. A. Paquette and his group at Ohio State University

FIGURE 3.60 Some polycyclic molecules.

3.15 SUMMARY

NEW CONCEPTS

In this chapter, a bonding scheme for the alkanes is developed. We continue, as in Chapter 2, to form bonds through the overlap of atomic orbitals. Some atomic orbitals, 2*p* orbitals in particular, are not well suited to forming strong bonds. For example, overlap with one lobe of a 2*p* orbital wastes the potential overlap with the other lobe. Moreover, 2*p* orbitals are directed at 90° angles, and this does not keep the electrons in the bonds as far from each other as possible.

We describe a model in which the atomic orbitals of carbon are mathematically combined to form new, hybrid atomic orbitals. The results of a combination of three 2*p* orbitals and one 2*s* orbital are called sp^3 hybrids, reflecting the 75% *p* character and 25% *s* character in the hybrid. These hybrids solve the problems encountered in forming bonds between pure atomic orbitals. The sp^3 hybrids are asymmetric, and have a fat and a thin lobe. Overlap between a hydrogen 1*s* orbital and the fat lobe provides a stronger bond than that between 1*s* and carbon 2*p* orbitals. In addition, these hybrids are directed toward the corners of a tetrahedron, which keeps the electrons in the bonds as far apart as possible, thus minimizing destabilizing interactions. Other hybridization schemes are sp^2, in which the central atom is bonded to three other atoms, and *sp*, in which a central atom is bonded to two other atoms. Molecules hybridized sp^2 are planar, with the three attached groups at the corners of a triangle, and *sp* hybridized molecules are linear.

Even simple molecules have complicated structures. Methane is a simple tetrahedron, but one need only substitute a methyl group for one hydrogen of methane for complexity to arise. In ethane, for example, we must consider the consequences of rotation about the carbon–carbon bond. The minimum energy conformation for ethane is the arrangement in which all carbon–hydrogen bonds are staggered. About three kilocalories per mole above this energy minimum form is the eclipsed form, the transition state, the high-energy point (not an energy minimum, but an energy maximum) separating one staggered form of ethane from another. The three kilocalories per mole constitutes the rotational barrier between the two staggered forms. This barrier is small compared to the available thermal energy at room temperature, and rotation in ethane is fast.

REACTIONS, MECHANISMS, AND TOOLS

In this chapter, new molecules are built up by first constructing a generic substituted molecule such as methyl–X (CH_3–X). The new hydrocarbon is then generated by letting X = CH_3. In principle, each different carbon–hydrogen bond in a molecule could yield a new hydrocarbon when X = CH_3. In practice, this is not such a simple procedure. The problem lies in seeing which hydrogens are really different. The two-dimensional drawings are deceptive. One really must see the molecule in three dimensions before the different carbon–hydrogen bonds can be identified with certainty.

The coding or drawing procedures can get quite abstract. It is vital to be able to keep in mind the real, three-dimensional structures of molecules even as you write them in two-dimensional code. The Newman projection, an enormously useful device for representing molecules three dimensionally, is described in this chapter.

The naming convention for alkanes is introduced. There are several trivial names that are commonly used and which, therefore, must be learned.

COMMON ERRORS

It is easy to become confused by the abstractions used to represent molecules on paper or blackboards. Do not trust the flat surface! It is easy to be fooled by a "corner" that does not really exist, or to be bamboozled by the complexity introduced by ring structures. One good way to minimize such problems (none of us ever really becomes free of the necessity to think about structures presented on the flat surface) is to work lots of isomer problems, and play with models!

3.16 KEY TERMS

Alkanes The series of saturated hydrocarbons of the general formula C_nH_{2n+2}.

Alkyl compounds Substituted alkanes. One or more hydrogens is replaced by another atom or group of atoms.

Butyl group The group $CH_3CH_2CH_2CH_2$.

***sec*-Butyl group** The group $CH_3CH_2CH(CH_3)$.

***tert*-Butyl group** The group $(CH_3)_3C$.

Carbanion A compound containing a negatively charged carbon atom. A carbon-based anion.

Carbenium ion A name suggested for a molecule containing a trivalent, positively charged carbon. It is not widely used. See **carbocation**.

Carbocation The compromise and currently widely used name for a molecule containing a trivalent, positively charged carbon atom.

Carbonium ion The traditional name for a molecule containing a trivalent, positively charged carbon. It has fallen into disuse.

cis "On the same side." Applied to specify stereochemical (spatial) relationships in ring compounds (and for double bonds, as we will see in Chapter 4).

Conformation The three-dimensional structure of a molecule. Conformational isomers are interconverted by rotations about bonds.

Conformational analysis The study of the relative energies of conformational isomers.

Dihedral angle The torsional, or twisting angle between two bonds. In an X–C–C–X system, the dihedral angle is the angle between the X–C–C and C–C–X planes.

Eclipsed ethane The conformation of ethane in which all carbon–hydrogen bonds are as close as possible. This conformation is not an energy minimum, but the top of the barrier separating two molecules of the stable, staggered conformation of ethane.

Ethyl compounds Substituted ethanes; $CH_3CH_2{-}X$ compounds.

Functional group An atom or group of atoms that generally reacts the same way no matter what molecule it is in.

Hybridization A mathematical model in which atomic orbital wave functions are combined to produce new, combination, or hybrid orbitals. The new orbitals are made up of fractions of the pure atomic orbital wave functions. Thus, an sp^3 hybrid is made of three parts p wave function and one part s wave function.

Hydride Hydrogen with a pair of electrons in its 1s orbital, $H{:}^-$.

Hydrocarbons Molecules containing only carbon and hydrogen.

Isobutyl group The group $(CH_3)_2CHCH_2$.

Isomers Molecules of the same formula but different structures.

Isopropyl group The group $(CH_3)_2CH$.

Methane The simplest stable hydrocarbon, CH_4.

Methyl anion $^-{:}CH_3$.

Methyl cation $^+CH_3$.

Methyl compounds Substituted methanes; $CH_3{-}X$ compounds.

Methylene group The group CH_2.

Methyl radical $\cdot CH_3$.

Methine group The group CH.

Newman projection A convention used to draw what one would see if one could look down a bond. The groups attached to the front atom are drawn in as three lines. The back atom is represented as a circle to which its attached bonds are drawn. Newman projections are enormously useful in seeing spatial (stereochemical) relationships in molecules.

Primary carbon A carbon atom attached to one other carbon.

Propyl group The group $CH_3CH_2CH_2$.

Quaternary carbon A carbon attached to four other carbons.

Reactive intermediates Molecules of great instability, and hence fleeting existence under normal conditions. Most carbon-centered anions, cations, and radicals are examples.

Saturated alkanes Alkanes of the molecular formula C_nH_{2n+2}.

Secondary carbon A carbon attached to two other carbons.

Sigma bond (σ) Any bond with cylindrical symmetry.

***sp* Hybrid** A hybrid orbital made by the combination of one s and one p atomic orbital.

sp^2 Hybrid A hybrid orbital made by the combination of one s orbital and two p orbitals.

sp^3 Hybrid A hybrid orbital made by the combination of one s orbital and three p orbitals.

Staggered ethane The energy minimum conformation of ethane in which the carbon–hydrogen bonds (and the electrons in them) are as far apart as possible.

Steric A generic word referring to the arrangement in space of atoms and groups of atoms.

Tertiary carbon A carbon attached to three other carbons.

Torsional strain Destabilization caused by the proximity of bonds (usually eclipsing) and the electrons in them.

trans "On opposite sides." Used to specify stereochemical (spatial) relationships in ring compounds (and on double bonds as we will see in Chapter 4).

Transition state The high-energy point separating two energy mimima.

Unsaturated compound A compound not having the maximum number of hydrogens. The only examples we have seen so far are the ring compounds, which have the molecular formula C_nH_{2n}.

van der Waals forces Intermolecular forces in molecules caused by induced dipole–induced dipole interactions.

3.17 ADDITIONAL PROBLEMS

PROBLEM **3.29** Write Lewis structures for all the starred structural types in the list of functional groups on the endpaper at the beginning of this book. Show nonbonding electrons as dots and electrons in bonds as lines.

PROBLEM **3.30** Draw all the isomers of "bromopentane," $C_5H_{11}Br$. Give proper IUPAC names to all of them.

PROBLEM **3.31** Draw all the isomers of "chlorohexane," $C_6H_{13}Cl$. Give proper IUPAC names to all of them. *Hint*: There are 17 isomers.

PROBLEM **3.32** What is the approximate hybridization of carbon in the following compounds?

(a) H₂C=O (b) benzene (c) H—C≡C—H (d) CCl₄

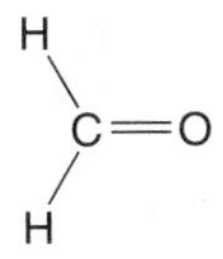

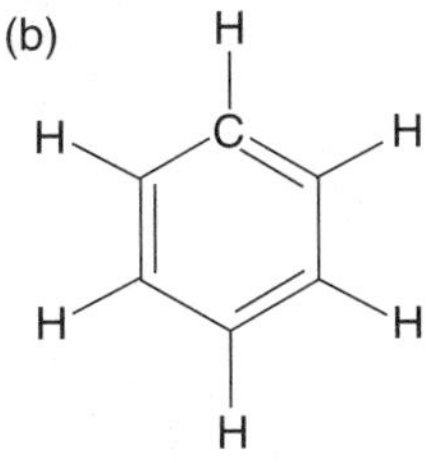

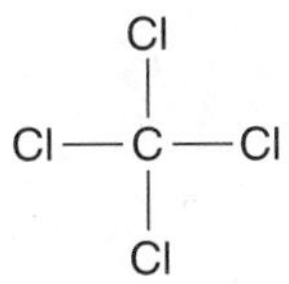

PROBLEM **3.33** Use the boiling point data in Table 3.4 to estimate the boiling point of pentadecane, $C_{15}H_{32}$.

PROBLEM **3.34** Rationalize the differences in the melting points for the isomeric pentanes shown below.

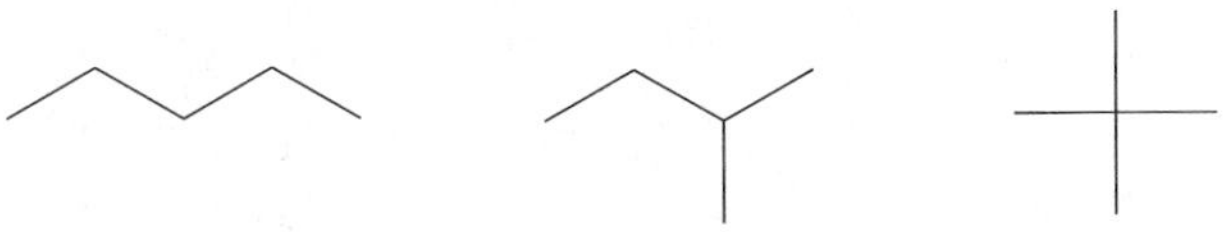

Pentane mp −129.7 °C

Isopentane mp −159.9 °C

Neopentane mp −16.5 °C

PROBLEM **3.35** Imagine replacing two hydrogens of a molecule with some group X. For example, methane (CH_4) would become CH_2X_2. (a) What compounds would be produced by replacing two hydrogens of ethane? (b) What compounds would be produced by replacing two hydrogens of propane? (c) What compounds would be produced by replacing two hydrogens of butane?

PROBLEM **3.36** Draw Newman projections of the eclipsed and staggered conformations of 1,2-dichloropropane by looking down the C(1)—C(2) bond.

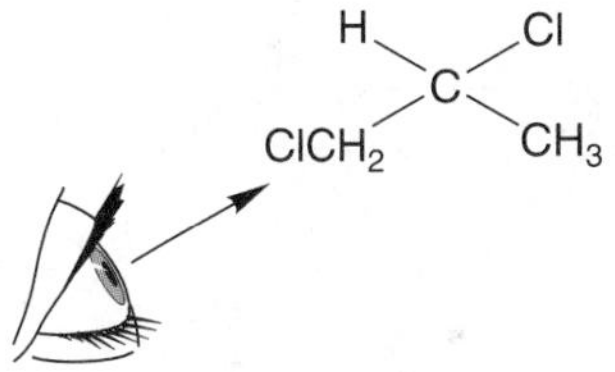

PROBLEM **3.37** Draw Newman projections constructed by looking down the C(1)—C(2) bond of 2-methylpentane. Repeat this process looking down the C(2)—C(3) bond. In each case, indicate which conformations will be the most stable.

PROBLEM **3.38** Although a chlorine is about the same size as a methyl group, the conformation of 1,2-dichloroethane in which the two chlorines are eclipsed is higher in energy than the conformation of butane in which the two methyl groups are eclipsed. Explain. *Hint*: Size isn't everything!

PROBLEM **3.39** Write IUPAC names for all the pentanes and hexanes you drew in your answer to Problem 3.20.

PROBLEM **3.40** Draw and write IUPAC names for all the isomers of nonane (C_9H_{20}). (*Hint*: There is a rumor that there are 35 isomers!) This problem is hard to get completely right.

You may want to continue isomer drawing and naming on your own. But be careful, this is no trivial task. For example, there are 75 structural isomers of decane and over 6500 for tetradecane.

PROBLEM **3.41** Write IUPAC names for the following compounds:

(a) (b)

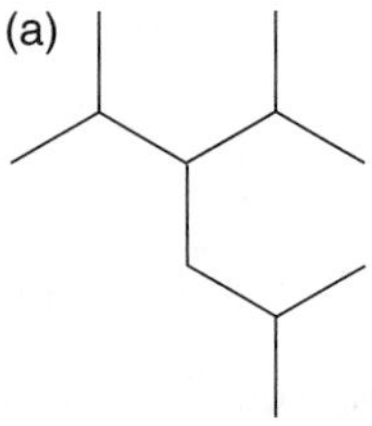

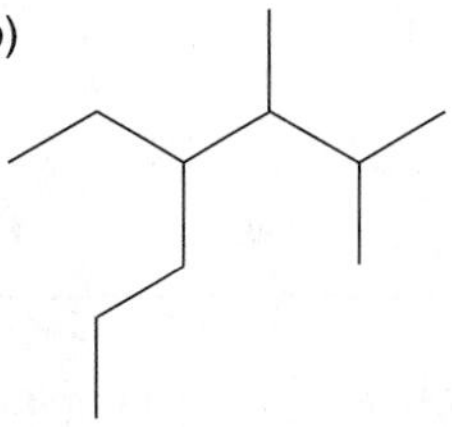

(c)

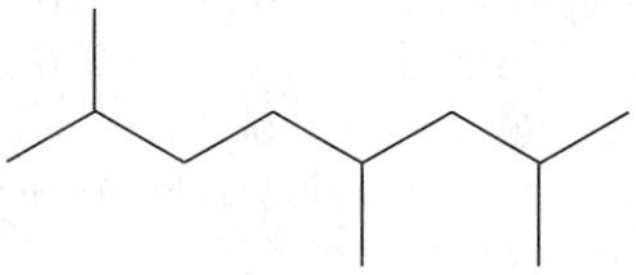

(d)

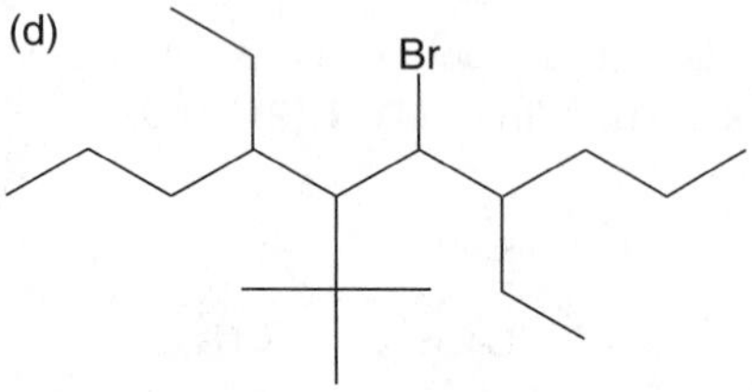

PROBLEM **3.42** Draw structures for the following compounds:

(a) 4-Ethyl-4-fluoro-2-methylheptane
(b) 2,7-Dibromo-4-isopropyloctane
(c) 3,7-Diethyl-2,2-dimethyl-4-propylnonane
(d) 2-Chloro-7-iodo-5-isobutyldecane

PROBLEM **3.43** Although you should be able to draw structures for the following compounds, each is named incorrectly below. Give the correct IUPAC name for each of the following compounds:

(a) 2-Methyl-2-ethylpropane
(b) 2,6-Diethylheptane
(c) 1-Bromo-3-propylpentane
(d) 5-Fluoro-8-methyl-3-*tert*-butylnonane

PROBLEM **3.44** One of the octane isomers of Problem 3.23, 3-ethyl-2-methylpentane, could have been reasonably named otherwise. Draw this compound and find the other reasonable name. The answer will tell you why the name given is preferred.

PROBLEM **3.45** Name the following compound:

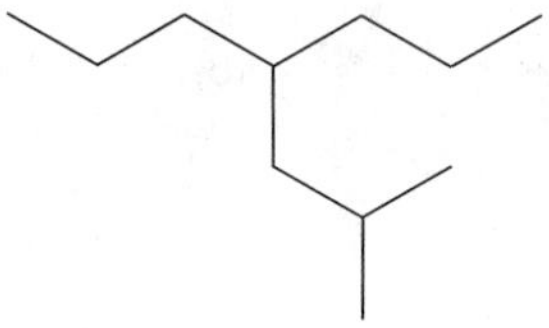

Alkenes and Alkynes

On the atomic and subatomic levels, weird electrical forces are crackling and flaring, and amorphous particles ... are spinning simultaneously forward, backward, sideways, and forever at speeds so uncalculable that expressions such as "arrival," "departure," "duration," and "have a nice day" become meaningless. It is on such levels that magic occurs.

—Tom Robbins*
Skinny Legs and All

Not all hydrocarbons have the formula C_nH_{2n+2}. Indeed, we have already seen some that do not: the ring compounds, C_nH_{2n} (Fig. 4.1).We noted in Chapter 3 (p. 104) that the chemical properties of these ring compounds closely resemble those of the acyclic saturated hydrocarbons. There is another family of hydrocarbons that also has the formula C_nH_{2n}, many of whose chemical properties are sharply different from those of the ring compounds and saturated chains we have seen so far. These compounds are different from other hydrocarbons in that they contain carbon–carbon double bonds. They are called **alkenes** to distinguish them from the saturated alkanes.

There are also hydrocarbons of the formula C_nH_{2n-2}, whose chemical properties resemble those of the alkenes but not those of the alkanes or cycloalkanes. These compounds are the **acetylenes**, or **alkynes**.

The structures and some of the properties of the families of **unsaturated hydrocarbons**, called alkenes and alkynes, are discussed in this chapter.

$CH_3CH_2CH_2CH_3$

Butane
(C_4H_{10})
a saturated alkane

H_2C—CH_2
|　　|
H_2C—CH_2

Cyclobutane
(C_4H_8)
a cycloalkane

FIGURE 4.1 The chemical properties of saturated alkanes, C_nH_{2n+2}, are very similar to those of the cycloalkanes, C_nH_{2n}.

4.1 ALKENES: STRUCTURE AND BONDING

The simplest alkene is properly called **ethene** but is almost universally known by its common, or "trivial" name, **ethylene**.†

Several spectroscopic measurements and chemical reactions show that ethylene has the formula C_2H_4 and is a symmetrical compound composed of a pair of methylene groups, H_2CCH_2. Earlier, when we encountered methane (CH_4), we constructed a bonding scheme designed to reproduce the way four hydrogens were attached to the central carbon. Here we have

* Thomas Eugene Robbins is an American author, born in 1936 in Blowing Rock, NC.

†I like the old names and will use many of them. The newer, systematic IUPAC names are fine when complexity develops—we obviously could not have a trivial name for every compound. Yet some of the flavor of organic chemistry is lost when the system is applied too universally. The trivial names connect to history, to the quite correct image of the bearded geezer slaving over the boiling retort, and I like that.

FIGURE 4.2 Replacement of one hydrogen in methyl (CH_3), with a methylene (CH_2) group leads to the framework of ethylene (H_2CCH_2). Note that each carbon so far has only three bonds. The full bonding scheme for ethylene is not yet in place.

a slightly different problem. Each carbon atom in ethylene is not attached to four other atoms as in methane, but to three. It seems we will need a different bonding rationale with which to describe Nature now. Our strategy will be to develop a bonding scheme for the simplest trivalent compound of carbon, methyl (CH_3), and then to extend it to ethylene, in which each carbon is attached not to three hydrogens as in methyl, but to two hydrogens and the other methylene (CH_2) group (Fig. 4.2).

Three bonds are needed to make attachments to the three hydrogens in CH_3. Using the atomic orbitals of carbon to overlap with the three hydrogen 1*s* orbitals leads to problems, just as it did in our earlier construction of methane (see Section 3.2).

*PROBLEM 4.1 Produce a model for neutral methyl ($\cdot CH_3$) using the unhybridized atomic orbitals of carbon and the hydrogen 1*s* orbitals. Critically discuss the shortcomings of this model. What's wrong with it?

ANSWER There is more than one way to construct such a model. Start by determining how many electrons are available. Carbon ($_6C$) supplies four (six less the two low-energy 1*s* electrons), and each hydrogen has one, for a total of seven. We might form three carbon–hydrogen bonds by overlap of the carbon $2p_x$, $2p_y$, and $2p_z$ orbitals with three hydrogen 1*s* orbitals, for example. This process uses six electrons (two each in the carbon–hydrogen bonds), and would leave the carbon 2*s* orbital to hold the remaining electron.

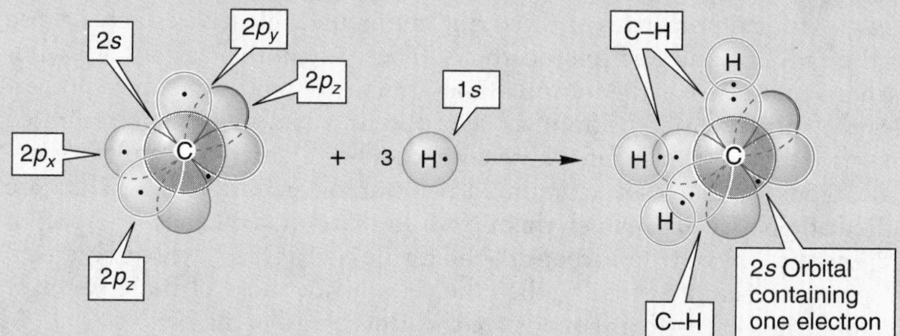

However, this model gives a "methyl" with carbon–hydrogen bonds 90° apart. It is possible for a central carbon to be surrounded by three hydrogens 120° apart, and so we might suspect that this unhybridized model with its 90° angles will be destabilized by electron–electron repulsion. Moreover, overlap of a 1*s* orbital with a 2*p* orbital is inefficient. The rear lobes of the 2*p* orbitals are wasted. Stronger bonds can be constructed from directed, hybrid orbitals.

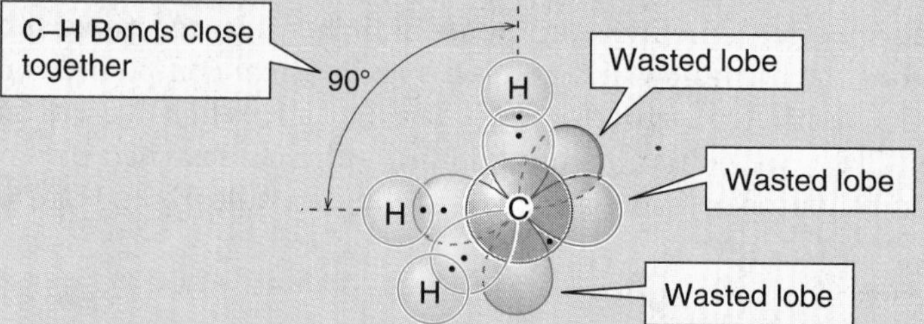

Other possible answers include making two carbon–hydrogen bonds from 2*p*–1*s* overlap and the third from 2*s*–1*s* overlap. This leaves the last electron in a 2*p* orbital. Consequently, this model has the same problems as noted previously.

As in our earlier discussion of sp^2 hybridization (Chapter 3, p. 75), we will create three new hybrid orbitals, which will do a better job of attaching the carbon atom to the three hydrogens than do the pure atomic orbitals. Three hybrids are needed, and our mathematical operations will therefore involve combining three wave functions (atomic orbitals) to produce the three new hybrid orbitals. Recall that in such quantum mechanical calculations the number of new orbitals always equals the number of orbitals originally combined in the calculation, here three. So, let's combine the carbon $2s$, $2p_x$, and $2p_y$ orbitals to produce three new, **sp^2 hybrids**. The $2p_z$ orbital unused in our calculation remains, unhybridized and waiting to be incorporated in our picture (Fig. 4.3). Remember that the choice of $2p$ orbitals to be combined is arbitrary. Any pair of p orbitals can be used, leaving the third left over.

C $(1s^2)\ 2s^2\ 2p_x 2p_y 2p_z$

Combine $2s$, $2p_x$, $2p_y$ ⟶ three sp^2 hybrid orbitals; $2p_z$ is not used

FIGURE 4.3 The combination of three atomic orbitals of carbon ($2s$, $2p_x$, and $2p_y$) leads to three new orbitals called sp^2 hybrids. The $2p_z$ orbital is not involved in the calculation.

The new sp^2 hybrids are generally shaped like the sp^3 hybrids constructed earlier. They are directed orbitals—there is a fat lobe and a thin lobe so they lead to quite efficient overlap with the hydrogen $1s$ orbitals (Fig. 4.4).

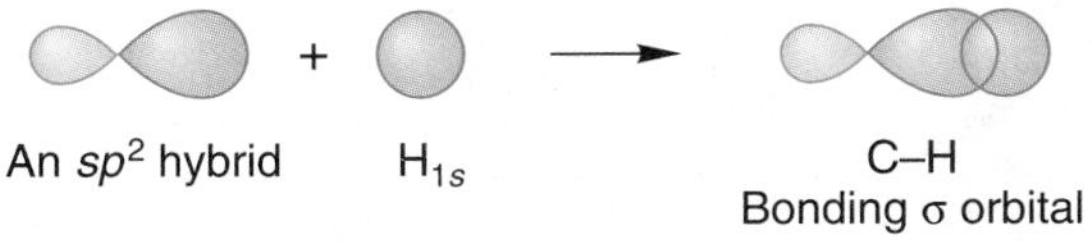

An sp^2 hybrid + H_{1s} ⟶ C–H Bonding σ orbital

FIGURE 4.4 The directed sp^2 hybrid orbital overlaps with the spherical $1s$ orbital on hydrogen to produce a carbon–hydrogen σ bond. The antibonding orbital (σ*) is not shown.

What do you guess is the angle between the hybrid orbitals? Remember that one of the problems with using unhybridized carbon atomic orbitals is that this process produces bonds that are too close together. Repulsions between filled orbitals are not minimized in the unhybridized structure. The hybridization scheme produces strong bonds (good overlap) that are directed so as to minimize repulsions. As Figure 4.5 shows, the best way to arrange three things in space (the hydrogens in this case) surrounding a central object (here, the carbon) so they are as far from each other as possible, is to put them at the corners of an equilateral triangle. And indeed, the angles between the three sp^2 hybrids are exactly 120°. Recall the earlier discussion of BH_3 in Section 3.2b.

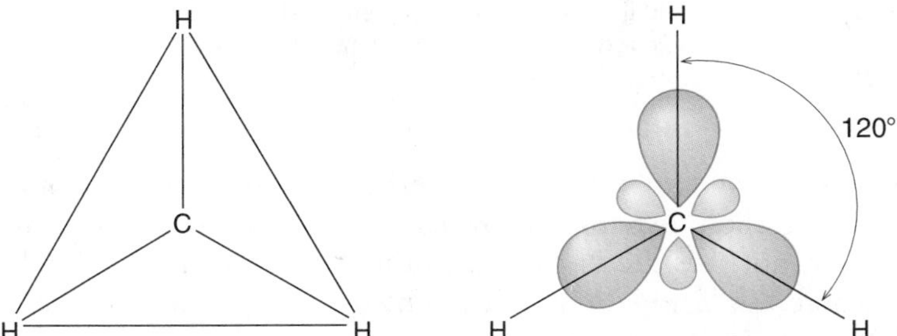

FIGURE 4.5 In planar CH_3 the angle between the carbon–hydrogen bonds is 120°. The hydrogens lie at the corners of an equilateral triangle. Repulsive overlap between the filled carbon–hydrogen bonds is minimized in this arrangement.

Each of the three sp^2–$1s$ overlaps produces a bonding ($\sigma = sp^2 + 1s$) and an antibonding ($\sigma^* = sp^2 - 1s$) molecular orbital. Stabilization will be maximized if we form two-electron bonds, as the stabilized bonding molecular orbital will be filled and the antibonding, high-energy molecular orbital left empty (Fig. 4.6).

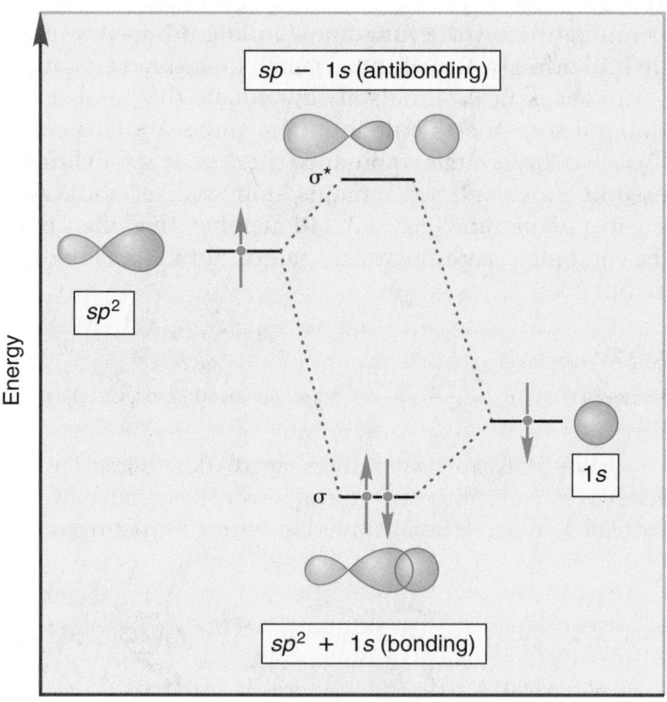

FIGURE 4.6 The overlap of sp^2 and 1*s* orbitals can occur in a bonding, stabilizing way (sp^2 + 1*s*), or an antibonding, destabilizing way (sp^2 – 1*s*).

Each hydrogen supplies a single electron. Therefore, in order to form three, two-electron bonds we need a single electron from carbon for each of the three sp^2 hybrids (Fig. 4.7).

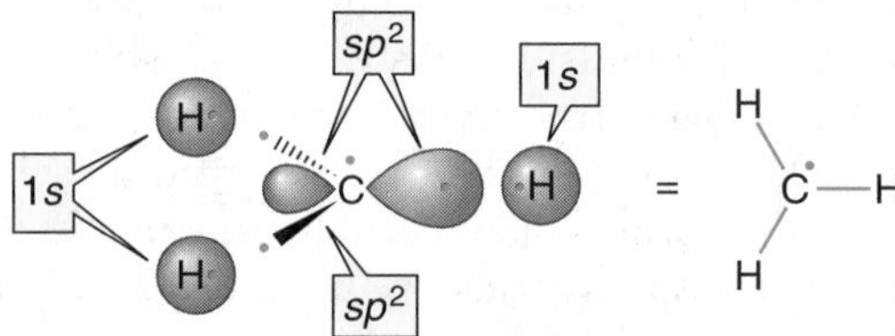

FIGURE 4.7 Three two-electron carbon–hydrogen bonds can be formed using one electron from each hydrogen and three electrons from carbon (shown in red). This leaves one electron on carbon (shown in red) left over. In this figure, two of the sp^2 orbitals are shown schematically as wedges.

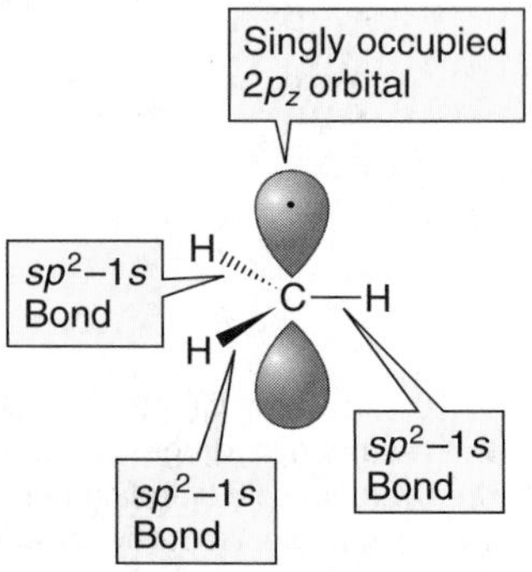

Singly occupied $2p_z$ orbital

sp^2–1*s* Bond

H

C—H

H

sp^2–1*s* Bond

sp^2–1*s* Bond

FIGURE 4.8 A view of sp^2 hybridized CH_3. Don't be confused by the electron in one lobe of the 2*p* orbital. The two lobes are not separate, and the electron occupies the whole orbital, not just one lobe.

Bonding to two hydrogens and the other carbons uses three of the four electrons of carbon, leaving one over. Where is this electron? Remember that we did not use the carbon $2p_z$ orbital in creating the sp^2 hybrids. That's where the last electron of carbon goes. So our sp^2 scheme leads to the picture in Figure 4.8. There is a carbon atom, surrounded by a planar array of three hydrogen atoms, and the unhybridized $2p_z$ orbital extends above and below the plane of the four atoms. The carbon–hydrogen bonds are familiar two-electron bonds (don't forget the empty antibonding orbitals, though) and there is a single electron in the $2p_z$ orbital.

*PROBLEM 4.2

Construct the bonding molecular orbitals for planar methyl (CH_3). Use the molecular orbitals for cyclic H_3 (**A**, **B**, and **C**) given in the problem. Allow them to interact with the appropriate atomic orbitals of a carbon atom placed at the center of the triangle of hydrogens. Next, order the bonding molecular orbitals in energy (Fig. 4.9).

Molecular orbitals for H H H

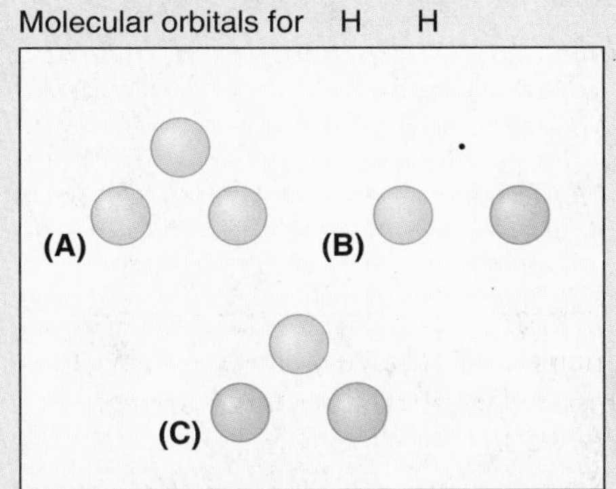

±

Atomic orbitals for carbon

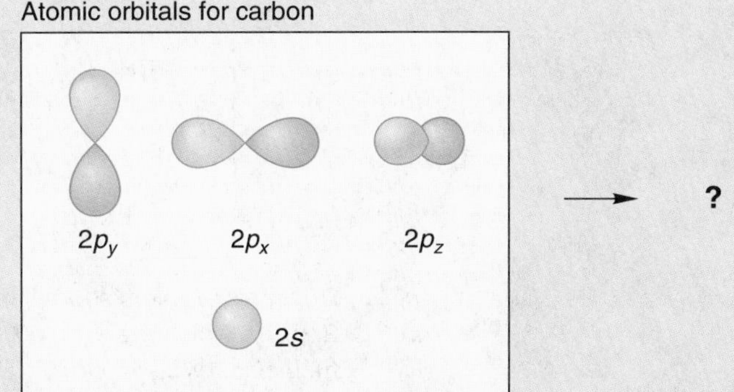

FIGURE 4.9

ANSWER

Geometry is important. In planar methyl, the carbon is symmetrically surrounded by three hydrogens, and we must be careful to overlap the orbitals in this way. The key to this problem is to use the appropriate H_3 molecular orbital to overlap with the proper atomic orbital of carbon. In this case, the lowest energy molecular orbital of H_3 (**A**) is perfectly set up to interact with the lowest energy atomic orbital of carbon, 2*s*. This produces two new molecular orbitals, CH_3 1 and CH_3 2.

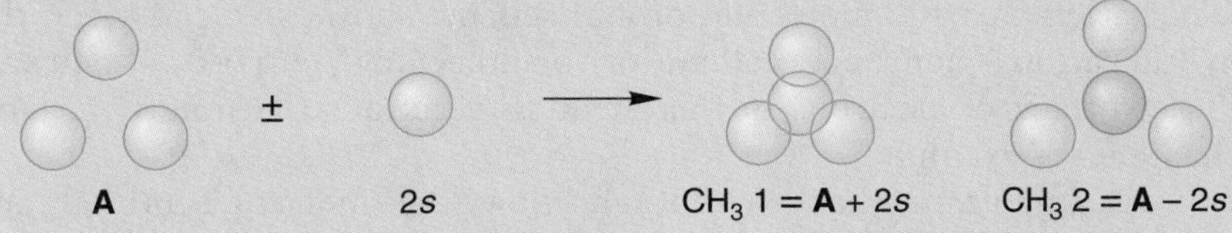

The other two higher energy molecular orbital's of H_3 (**B** and **C**) are aligned so as to overlap well with the $2p_x$ and $2p_y$ orbitals of carbon, respectively. This overlap produces the bonding molecular orbitals CH_3 3 and CH_3 5. The $2p_z$ orbital is unused in this scheme.

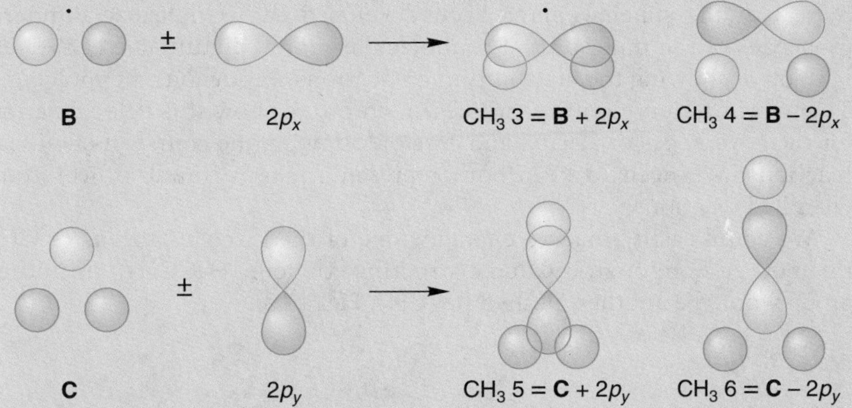

In ordering the new bonding orbitals in energy, be sure to remember to check your answer by counting nodes.

ANSWER (CONTINUED)

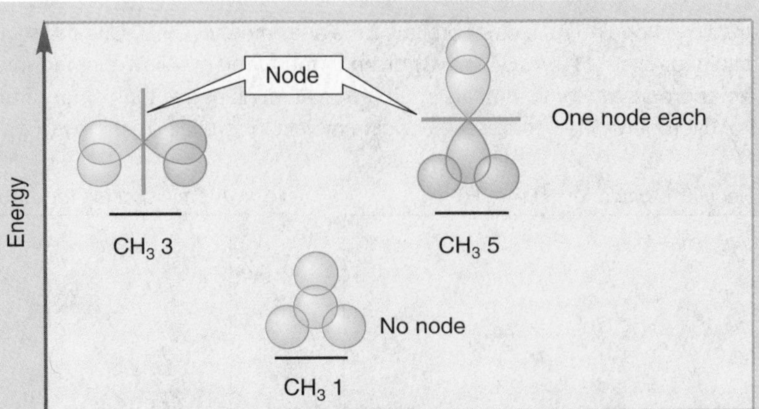

Does this problem seem familiar? It should. We did a very similar one in Problem 2.25, in which we worked out the molecular orbitals of planar ammonia. The molecular orbitals of any XH_3 molecule will look qualitatively the same.

We have seen this molecule, CH_3, before! It is nothing more than a methyl group. In Problem 4.2, we developed a molecular orbital picture of methyl in which the delocalization of electrons throughout the molecule was emphasized. In the text, we have built up a picture that emphasizes localized bonds. There are advantages to both schemes. In one sense the molecular orbital picture is probably more "real," as electrons are not as limited to the regions of space that the hybridization scheme suggests. However, we do not make horrible energetic mistakes if we ignore the delocalization that the molecular orbital scheme shows so clearly, and the hybridization scheme is excellent for "bookkeeping" purposes. It helps us to keep track of electrons in the chemical reactions that follow in later chapters, for example.

It is important that we do not view the molecular orbital and hybridization schemes as being in conflict or as giving substantially different pictures of the molecule methyl, CH_3. Note, for example, that the geometry is exactly the same in the two schemes. We are humans, stuck with our inability to apprehend the properties of electrons easily, and needing approximations in order to represent Nature. Different approximate bonding schemes have been developed that emphasize different properties of the molecules. The molecular orbital picture does an excellent job of showing the distribution of electrons throughout the molecule. The hybridization picture sacrifices an ability to show this delocalization for the advantages of clarity, and ease of following the course of chemical reactions. We need to keep both representations in mind as we study chemical reactions.

We could easily imagine changing one of the three hydrogens of CH_3 for another atom or group. In constructing ethylene (H_2CCH_2), this other group would be another methylene (CH_2, Fig. 4.10).

FIGURE **4.10** We can imagine replacing one H in CH_3 with a CH_2 (methylene) group. This transformation gives us our first orbital picture of ethylene ($H_2C{=}CH_2$), although the carbon–carbon double bond is not yet completely drawn.

So our first picture of ethylene is derived from the joining of a pair of methylene (CH_2) groups (Fig. 4.11).

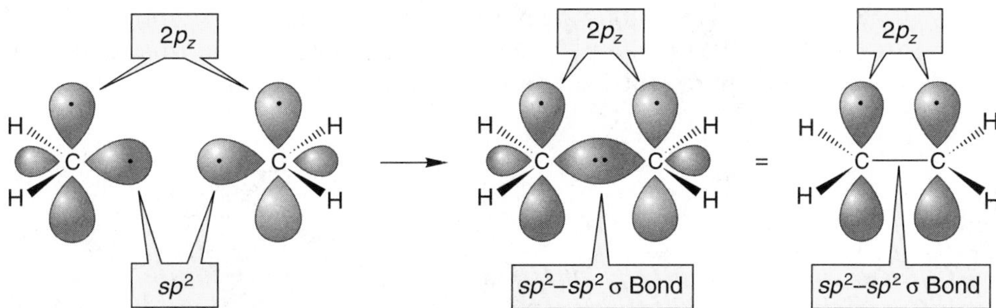

FIGURE 4.11 The two sp^2 hybrids overlap to make an sp^2–sp^2 σ bond joining the two carbons of ethylene. The two $2p_z$ orbitals remain, and extend above and below the plane defined by the six atoms.

This structure is quite analogous to that produced in the construction of ethane from a pair of sp^3 hybridized carbons (Section 3.5). In ethane, two sp^3 hybrids overlapped to form a bonding σ orbital and an antibonding σ* orbital (Fig. 3.30, p. 89). Construction of the carbon–carbon bond in ethylene begins with the similar overlap of a pair of sp^2 hybrids. Once more, a bonding σ orbital and antibonding σ* orbital are produced, this time by the constructive and destructive overlap of a pair of sp^2 hybrids (Fig. 4.12).

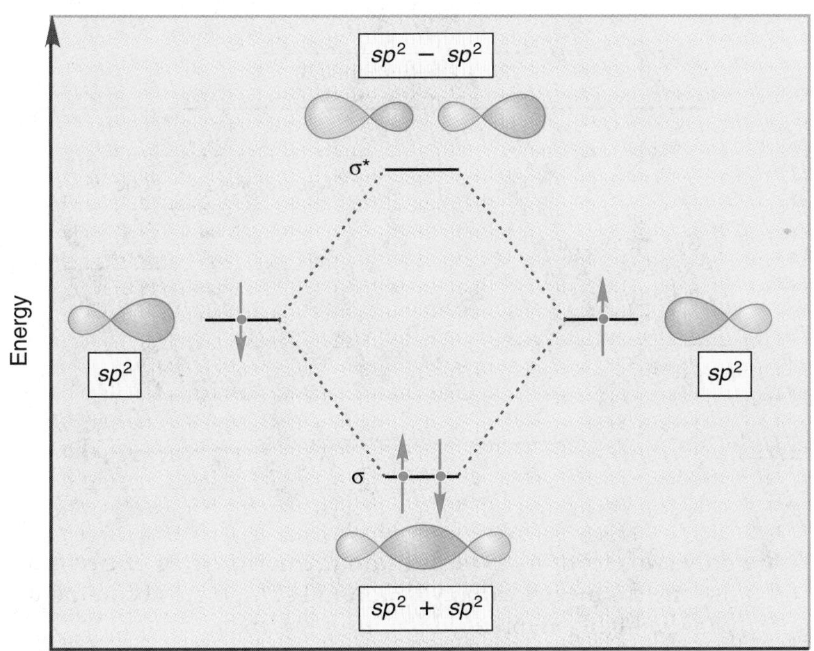

FIGURE 4.12 Overlap of the two sp^2 hybrids produces a filled bonding molecular orbital, $\sigma = sp^2 + sp^2$, and an empty, antibonding molecular orbital, $\sigma^* = sp^2 - sp^2$. As there are only two electrons, only the low energy, bonding molecular orbital will be filled.

So far, we have ignored the $2p_z$ orbital on each carbon. It is time to take them into account. The two carbon $2p_z$ orbitals in question are only a bond apart and will interact quite strongly. Moreover, they are of equal energy and that too will contribute to a strong interaction. (*Remember*: Orbitals of equal energy interact most strongly, Chapter 2, p. 55,56.) We know what happens when atomic orbitals overlap—bonding and antibonding molecular orbitals are created. The new orbitals formed from 2*p*–2*p* overlap are shown schematically in Figure 4.13. These new molecular orbitals are not σ orbitals as they do not have cylindrical symmetry. There is a plane of symmetry instead, and this makes them π **orbitals**, here called π (bonding) and π* (antibonding).

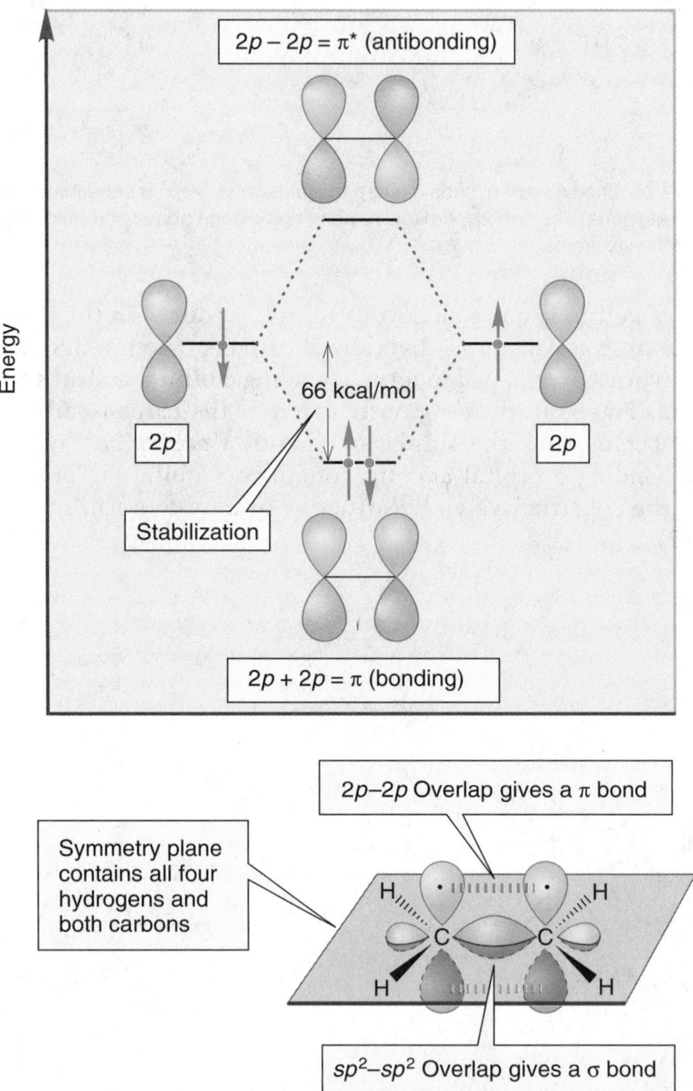

FIGURE 4.13 Overlap of a pair of 2*p* orbitals results in a lower energy, bonding π orbital and a higher energy, antibonding π* orbital. The stabilization amounts to about 66 kcal/mol when the π orbital is filled.

There are only two electrons to be put into the new system of two molecular orbitals, one from each carbon $2p_z$ orbital. These are accommodated very nicely in the highly stabilized bonding π molecular orbital.

So the two carbons of ethylene are held together by two bonds. One is made up of sp^2–sp^2 σ overlap and the other of 2*p*–2*p* π overlap. There is a **double bond** between the two carbons. The convention is to simply draw two lines between the atoms, making no distinction between the two bonds, although, as you have just seen, this approximation is far from justified (Fig. 4.14).

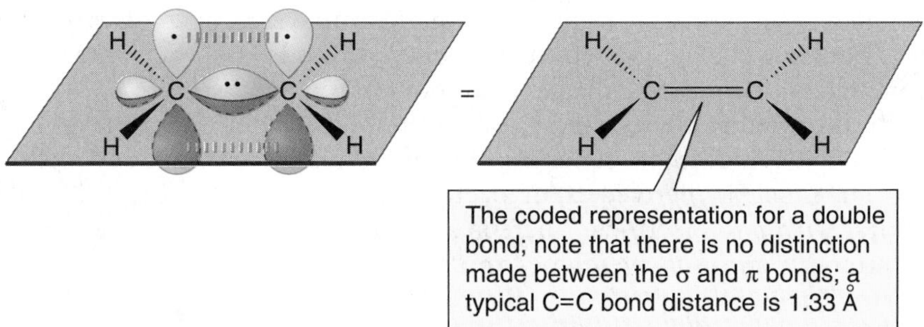

FIGURE 4.14 The π and σ bonds of a carbon–carbon double bond are not differentiated in the coded representation of an alkene.

The overall bonding scheme for a carbon–carbon double bond includes both σ and π bonds (and their empty antibonding counterparts) and is also shown in Figure 4.14. Carbon–carbon double bonds are quite short, with a typical bond distance being 1.33 Å.

In ethane, there was nearly free rotation about the newly formed carbon–carbon bond. Will the same be true for ethylene? Actually, the argument is a bit more subtle. There isn't really free rotation about the carbon–carbon bond in ethane. There is a 3-kcal/mol barrier, produced by the need to pass through a structure in which the carbon–hydrogen bonds are eclipsed (recall Fig. 3.28, p. 86). Our real question should be, How high is the barrier to rotation about the carbon–carbon double bond in ethylene? To begin the answer we construct Newman projections by sighting along the carbon–carbon bond (Fig. 4.15).

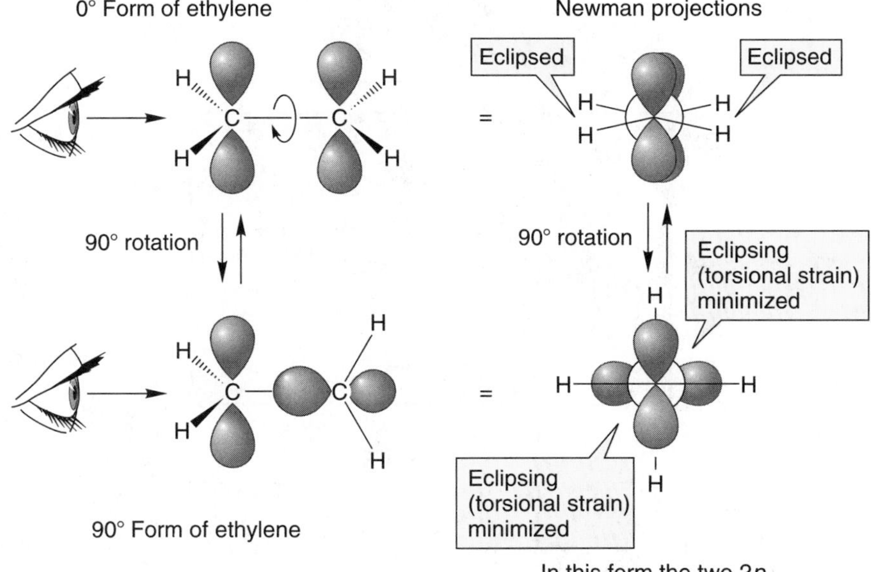

FIGURE 4.15 Although rotation about the carbon–carbon bond relieves torsional strain, it also destroys the overlap between the two $2p_z$ orbitals on the adjacent carbons. This decoupling leads to a high barrier to rotation. When one $2p$ orbital is rotated 90°, orbital overlap is lost.

Note that in the 0° conformation there will be torsional strain induced by the eclipsing of the carbon–hydrogen bonds. This strain is removed in the 90° rotated arrangement. In our discussion of ethane, we found that each pair of eclipsed hydrogens induces approximately 1-kcal/mol destabilization. Therefore our initial guess might be that the 0° form would be about 2 kcal/mol higher in energy than the 90° form. That would be approximately right, *if torsional strain were the whole story.* But it is not, as we have ignored the pair of $2p_z$ orbitals, and their interaction completely overwhelms the small amount of torsional strain.

In the 0° form the two *p* orbitals overlap; in the 90° form they do not (Fig. 4.15). When rotation occurs, the 2*p*–2*p* overlap declines and with it, the stabilization derived from occupying the new bonding molecular orbital with two electrons. There is an enormous stabilizing effect in the 0° arrangement that overcomes the relatively minor effects of torsional strain (Fig. 4.15).

PROBLEM **4.3** Show that in the 90° rotated form there is zero overlap between the pair of carbon $2p_z$ orbitals.

Figure 4.16 plots the energy change as rotation occurs. The barrier to rotation here is no paltry 3 kcal/mol as it is in ethane, but 65.9 kcal/mol, an amount far too high to be available under normal conditions. Alkenes are "locked" in a planar conformation by this amount of energy. We will soon see some consequences of this locking.

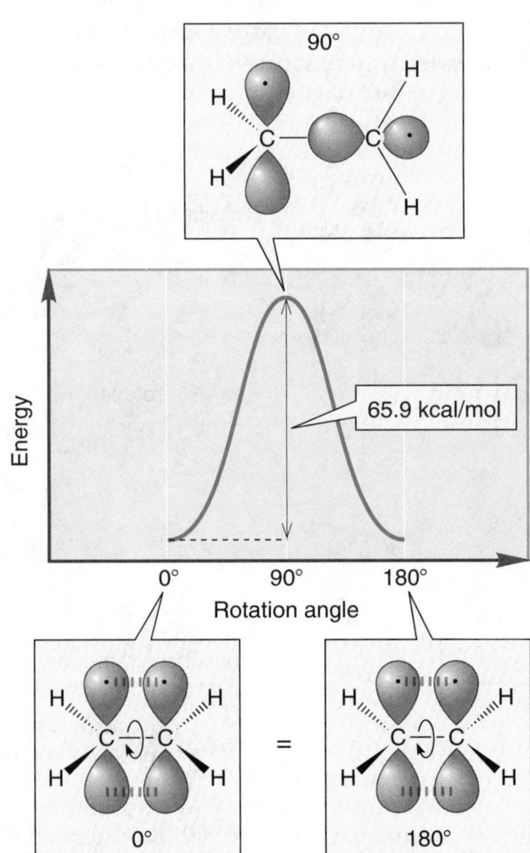

FIGURE **4.16** A plot of potential energy versus rotation angle in an alkene.

To summarize: Two double-bonded carbons are held together by a σ bond made up of overlapping sp^2 orbitals and a π bond made from 2*p* orbitals overlapping side to side. There is no free rotation about a carbon–carbon double bond, as such rotation diminishes overlap between the

$2p$ orbitals making up the π bond and costs energy. A full 90° rotation requires about 66 kcal/mol of energy.

Ethylene

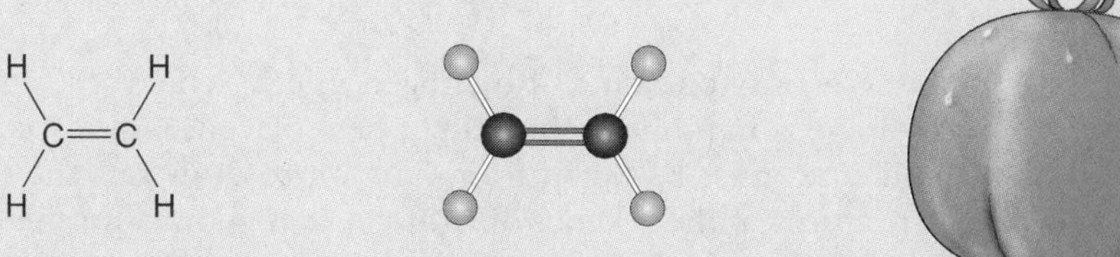

It's amazing, but the simplest of all alkenes, ethylene, is an important plant hormone. Among other functions, ethylene acts to promote ripening of fruit. Moreover, production of ethylene is autocatalytic; that is, a little ethylene induces the formation of more from the amino acid methionine, and its effects are magnified. Tomatoes are now typically shipped green in well-ventilated containers so they will arrive unspoiled. Ripening can then be started by exposure to ethylene.

Ethylene also induces abscission; the falling of leaves or flowers, by promoting the formation of the enzyme cellulase that weakens cell walls by destroying the cellulose from which they are made. A weakened abscission layer forms at the base of the leaf or flower, and wind or rain then breaks the stem. Indeed, it was this phenomenon that indirectly led to the discovery of ethylene as a plant hormone. Trees surrounding gas lamps were often defoliated when gas leaks occurred. A principal component of the illuminating gas was ethylene, and a little investigation showed that this chemical was the cause of the defoliation.

4.2 DERIVATIVES, ISOMERS, AND NAMES OF ALKENES

There is but one ethylene (or ethene, if you insist). We can make derivatives of this molecule by mentally replacing a hydrogen with some "X" group. Figure 4.17 shows a few such compounds. These are not called "ethenyl" or even "ethylenyl" compounds, but instead bear the delightfully trivial name, so far untampered with, of "**vinyl**."

Ethylene ($H_2C{=}CH_2$) ⟶ Vinyl–X ($H_2C{=}CHX$)

X =

Br	Vinyl bromide	OH	Vinyl alcohol
Cl	Vinyl chloride	CN	Vinyl cyanide
F	Vinyl fluoride	SiH_3	Vinyl silane
I	Vinyl iodide	NH_2	Vinylamine

FIGURE **4.17** Some substituted alkenes—vinyl compounds.

If we now let X = CH_3 (methyl), we get a compound, C_3H_6 (propene) or, commonly, propylene (Fig. 4.18). Notice how the root prefix *prop* has been retained for this three-carbon species, and only the vowel changed to note the difference between saturated prop<u>a</u>ne and unsaturated prop<u>e</u>ne.

We can now imagine replacing a hydrogen in propene with an "X" group, but the situation is not as simple as it was with ethylene. Although

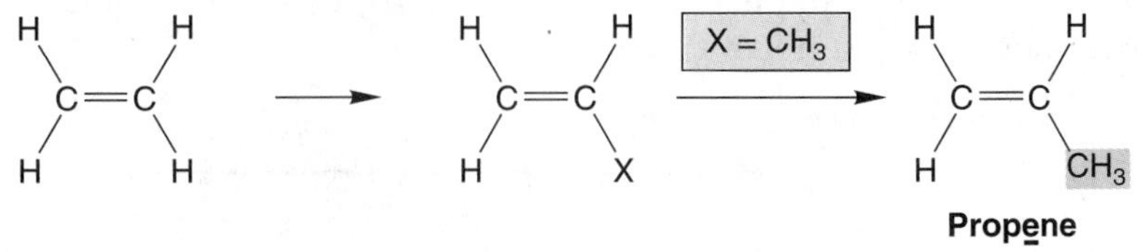

$X = CH_3$

Propene

FIGURE **4.18** When $X = CH_3$ (methyl), we get propene (C_3H_6). Note the change of vowel from prop<u>a</u>ne to prop<u>e</u>ne.

there is only one kind of hydrogen in ethylene, there are several different hydrogens in propene that might be replaced. A quick count is misleading and care must be taken from now on to draw out the three-dimensional structures of the alkenes on which we are doing our mental replacements. In practice, this is quite simple, because alkenes are planar, and most structures are easily drawn. If we just write a shorthand structure, we predict three new kinds of compounds when a hydrogen is replaced with an "X" group. As shown in Figure 4.19, these compounds are all known.

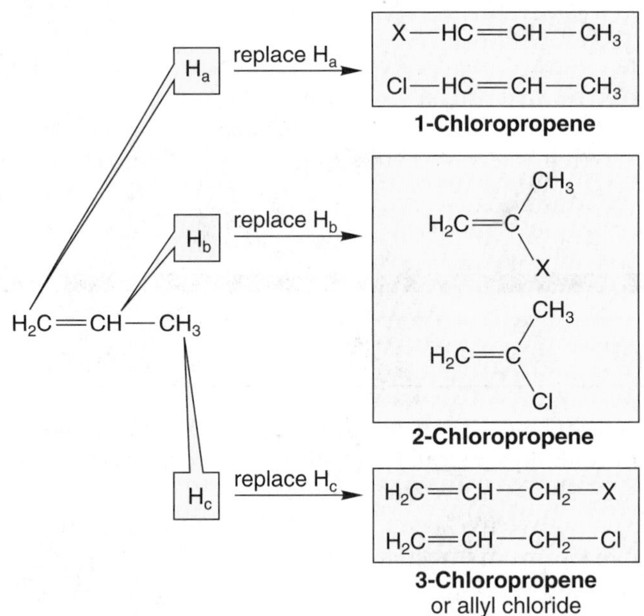

FIGURE **4.19** Substitution of three different hydrogens in propene appears to give three different propenyl–X compounds. In fact, we have missed something, because there are four such compounds, not three.

$H_2C{=}CH{-}CH_2$
The allyl group

$H_2C{=}CH{-}CH_2{-}Cl$
Allyl chloride

$H_2C{=}CH{-}CH_2{-}Br$
Allyl bromide

$H_2C{=}CH{-}CH_2{-}NH_2$
Allylamine

$H_2C{=}CH{-}CH_2{-}OH$
Allyl alcohol

$H_2C{=}CH{-}CH_2{-}CN$
Allyl cyanide

FIGURE **4.20** Some allyl compounds.

Note especially the use of the common name **allyl** for the $H_2C{=}CH{-}CH_2$ group, as this terminology is commonly encountered. Figure 4.20 shows some allyl compounds.

However, there is one severe problem with this analysis. In reality, there are *not* three kinds of substituted propenes, but four. The more detailed drawing in Figure 4.21 shows why. The coded structure of Figure 4.19 is inadequate to show that the two hydrogens on the end (called terminal) methylene group are *not* the same. One, H_a, is on the same side of the double bond as the methyl group, whereas the other, $H_{a'}$, is on the opposite side from the methyl. Replacement of hydrogens H_a and $H_{a'}$ with X gives two different compounds.

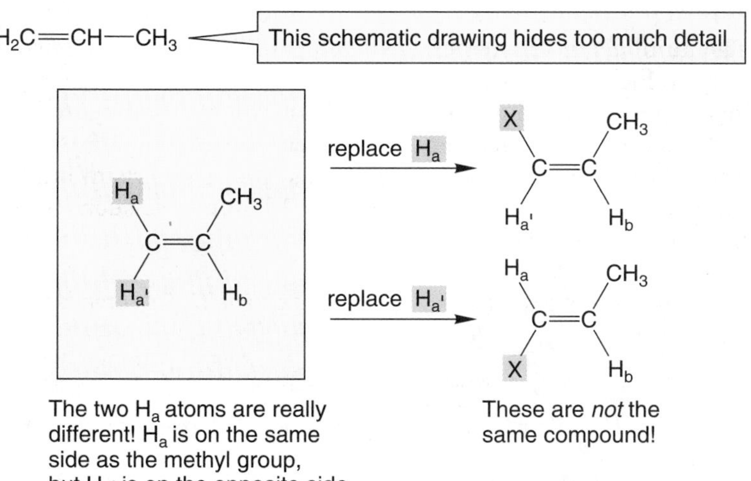

$H_2C{=}CH{-}CH_3$
This schematic drawing hides too much detail
replace H_a
replace $H_{a'}$
The two H_a atoms are really different! H_a is on the same side as the methyl group, but $H_{a'}$ is on the opposite side
These are *not* the same compound!

FIGURE 4.21 A more detailed drawing shows the problem. In propene, the hydrogen cis to the methyl is not the same as the hydrogen cis to the other hydrogen. Replacement of H_a with X gives a compound with the two remaining hydrogens on the same side of the double bond. Replacement of $H_{a'}$ gives a compound with the two remaining hydrogens on opposite sides of the double bond.

These two molecules are interconvertible only by rotation about the carbon–carbon double bond (Fig. 4.22).

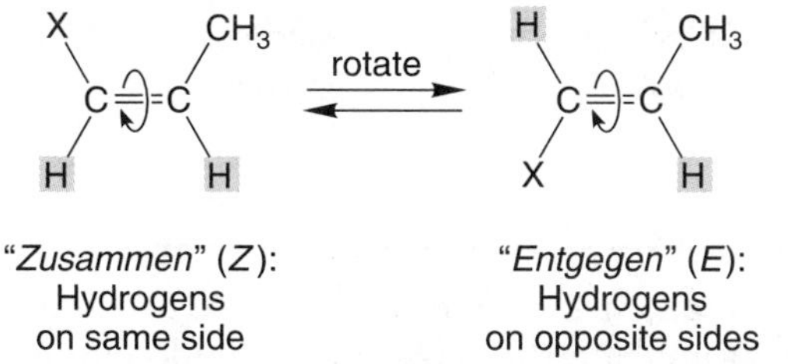

rotate

"*Zusammen*" (*Z*): Hydrogens on same side

"*Entgegen*" (*E*): Hydrogens on opposite sides

FIGURE 4.22 The (hypothetical) interconversion of two isomers of $CH_3CH{=}CH{-}X$ by rotation about the carbon–carbon double bond. This rotation requires about 66 kcal/mol and cannot occur under normal conditions.

We know this rotation would require about 66 kcal/mol, too high an amount of energy for interconversion to be common. The molecule with the two remaining hydrogens on the same side of the double bond is designated as cis or (*Z*) (for *zusammen,* German for "together"). The compound with the hydrogens on the opposite side is called trans or (*E*) (for *entgegen,* German for "opposite"). Recall the use of cis and trans in our discussion of ring compounds (Fig. 3.57, p. 106).

If we let X = methyl we get the butenes (C_4H_8). The four butenes are 1-butene, *cis*-2-butene, *trans*-2-butene, and 2-methylpropene (or isobutene, but also sometimes called isobutylene) and are shown in Figure 4.23.

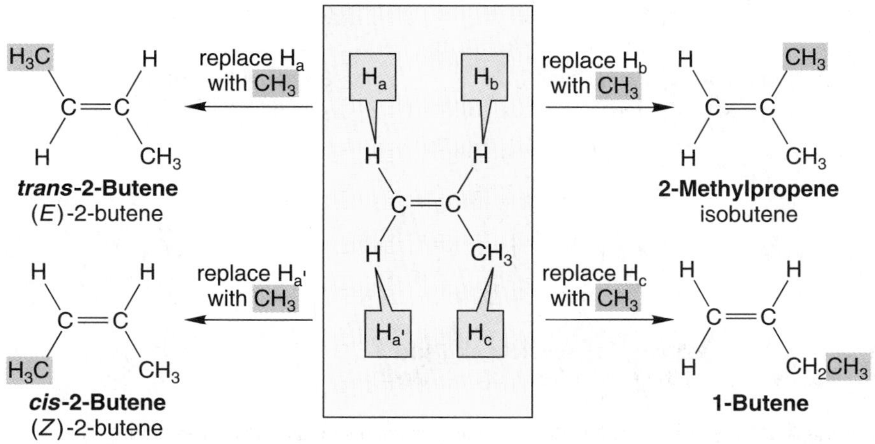

replace H_a with CH_3 → ***trans*-2-Butene** (*E*)-2-butene

replace $H_{a'}$ with CH_3 → ***cis*-2-Butene** (*Z*)-2-butene

replace H_b with CH_3 → **2-Methylpropene** isobutene

replace H_c with CH_3 → **1-Butene**

FIGURE 4.23 Replacement of the four different hydrogens in propene by a methyl group leads to the four butenes.

Convention Alert!

The numbers in the names 1-butene, *cis*-2-butene, and *trans*-2-butene are used to locate the position of the double bond. We will shortly discuss the naming convention for alkenes, but you might try to figure it out here from the names and structures of the butenes.

Not every alkene is capable of cis/trans (*Z*/*E*) isomerism! Ethylene and propene are not, and of the butenes only 2-butene has such isomers. Try Problem 4.4. It is simple, but worthwhile.

PROBLEM 4.4 Which of the following alkenes is capable of cis/trans (*Z*/*E*) isomerism?

At this point, the systematic naming protocol takes over. Five-carbon compounds (C_5) are called pentenes, six-carbon (C_6) compounds are called hexenes, and so on. Be alert for isomerism of the cis/trans (*Z*/*E*) kind. It takes some practice to find it.

PROBLEM 4.5 Write all the isomers of the pentenes (C_5H_{10}) and hexenes (C_6H_{12}). Be alert for isomerism of the cis/trans (*Z*/*E*) kind in these molecules.

4.3 NOMENCLATURE

Most of the rules for alkenes are similar to those used for alkanes, with a number attached to indicate the position of the double bond. Chains are numbered so as to give the double bond the smallest possible number. One important rule that is different in the alkenes is that the name is based on the longest chain containing a double bond, whether or not it is also the longest chain in the molecule. When molecules contain more than one double bond they are named as "dienes," "trienes," and so on. In these cases, the longest straight chain is defined as the one with the greatest number of double bonds. Cyclic molecules are named as "cycloalkenes." These rules are illustrated with the molecules in Figure 4.24.

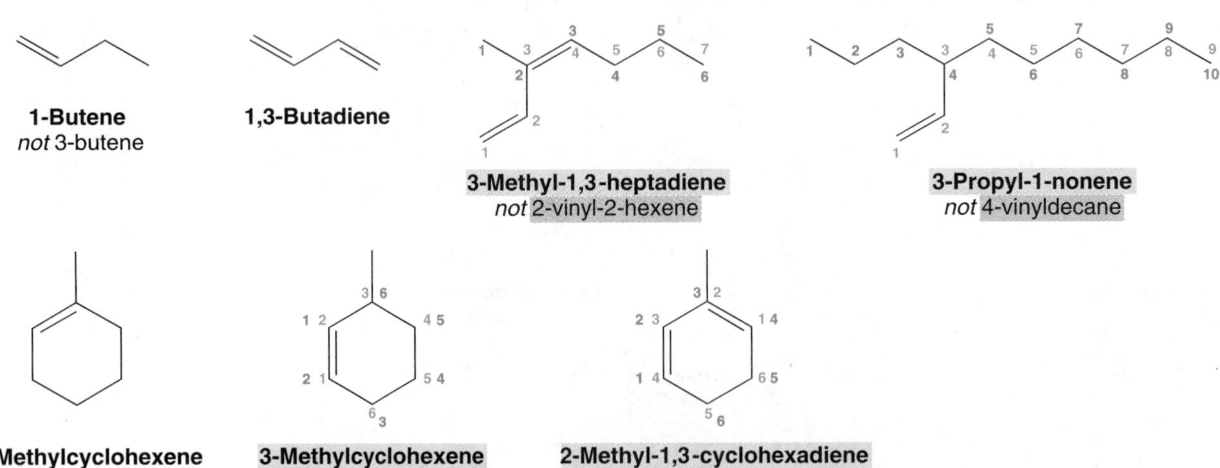

FIGURE 4.24 Some examples of the naming protocols for alkene nomenclature. The convention keeps the number denoting the position of the double bond as low as possible.

Name the pentenes and hexenes of Problem 4.5.

PROBLEM **4.6**

Finally, there are examples in which the numbering of a substituent becomes important. Usually, it is the double bond numbering that must be considered first. The double bond is usually given the lowest possible number [Fig. 4.25 (a)]. An important exception is the hydroxyl (OH) group, which has priority over the double bond, and is given the lower number [Fig. 4.25 (b)]. Only if there are two possible names in which the double bond has the same number do you have to consider the position of the substituent [Fig. 4.25 (c)].

(a)

trans-4-Chloro-2-pentene
not trans -2-chloro-3-pentene

(b)

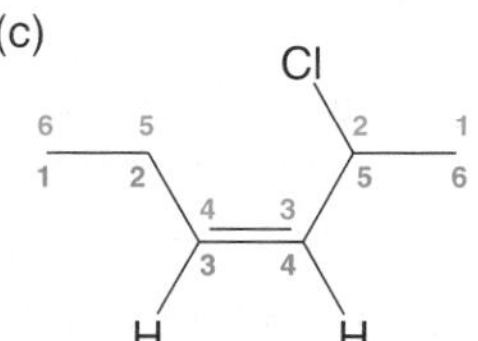

***trans*-2-Butene-1-ol**
not trans-2-butene-4-ol

(c)

***cis*-2-Chloro-3-hexene**
not cis-5-chloro-3-hexene

FIGURE **4.25** The double bond takes precedence over most substituents, but when two names put the double bond at the same numbered position, we use the name in which the substituent has the lower number.

Draw the following molecules: (*Z*)-2-pentene, 2-chloro-1-pentene, (*E*)-3-penten-2-ol, 4-bromocyclohexene, 1,3,6-cyclooctatriene.

PROBLEM **4.7**

In Figure 4.24, the structure of 3-methyl-1,3-heptadiene is not completely specified by the name given. Even if the cis/trans nomenclature is applied, problems remain. What's the difficulty? Can you devise a system for resolving the ambiguity?

*PROBLEM **4.8**

In describing fully the structure of 3-methyl-1,3-heptadiene we need to specify stereochemistry in order to distinguish the two possible forms. In one of the structures, two saturated alkyl groups (methyl and propyl) are cis while in the other they are trans.

ANSWER

trans CH_3 and $CH_2CH_2CH_3$

cis CH_3 and $CH_2CH_2CH_3$

Often, cis/trans will do the job, but here it fails. There is but one H, so you cannot find two hydrogens “on the same side” (cis), and two hydrogens “on opposite sides” (trans). That is the problem; it is up to you to devise a protocol for resolving it. For more about the device used by organic chemists, the (*Z*/*E*) system, read on in the text.

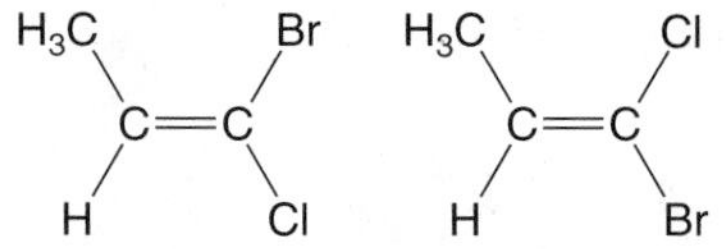

FIGURE 4.26 Sometimes the cis/trans convention is inadequate to distinguish two isomers. Which of these two isomers would you call cis and which trans?

Usually the cis/trans naming system is adequate to distinguish these pairs of isomers, but not always. Figure 4.26 gives an example in which cis/trans is inadequate. In the molecule 1-bromo-1-chloropropene there are clearly two isomers, but it is not obvious which is cis and which is trans. It was to solve such problems that the old cis/trans system was elaborated into the (*Z*/*E*) nomenclature by three European chemists, R. S. Cahn (1899–1981), C. K. Ingold (1893–1970), and V. Prelog (b. 1906). A priority system is set up, which ranks the groups attached to the double bonds. The compound with the higher priority groups on the same side of the double bond is (*Z*), and the other is (*E*). This priority system has other, very important uses to be encountered in Chapter 5, so we will spend some time here elaborating on it, even though we will see it again in another context.

4.4. THE CAHN–INGOLD–PRELOG PRIORITY SYSTEM

Convention Alert!

1. The first choice in setting priorities is made on the basis of atomic number. The atom of higher atomic number has the higher priority (Fig. 4.27). Thus, a methyl group, attached to the double bond through a carbon atom (atomic number = 6), has a higher priority than a hydrogen (atomic number = 1). Similarly, in the second compound, oxygen (atomic number = 8) has a higher priority than boron (atomic number = 5).

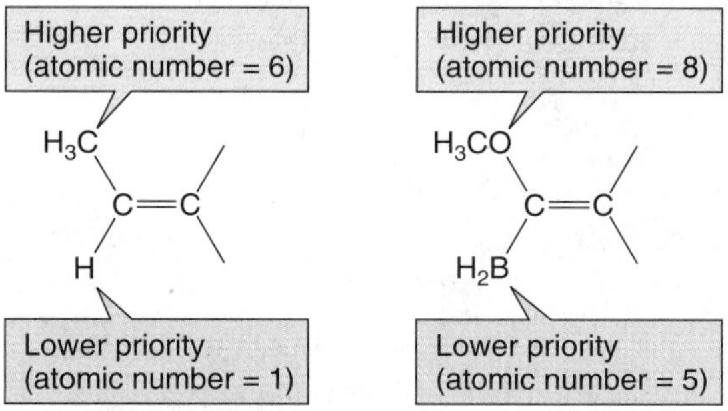

Higher priority (atomic number = 6)
Higher priority (atomic number = 8)
Lower priority (atomic number = 1)
Lower priority (atomic number = 5)

FIGURE 4.27 The substituent with the higher atomic number gets the higher priority.

How is this priority system used in practice? Consider the two isomers first encountered in Figure 4.26. Bromine has a higher atomic number than chlorine, so the isomer in which the higher priority methyl group and the bromine are on the same side is (*Z*), and the isomer in which the lower priority hydrogen is on the same side as bromine is (*E*) (Fig. 4.28).

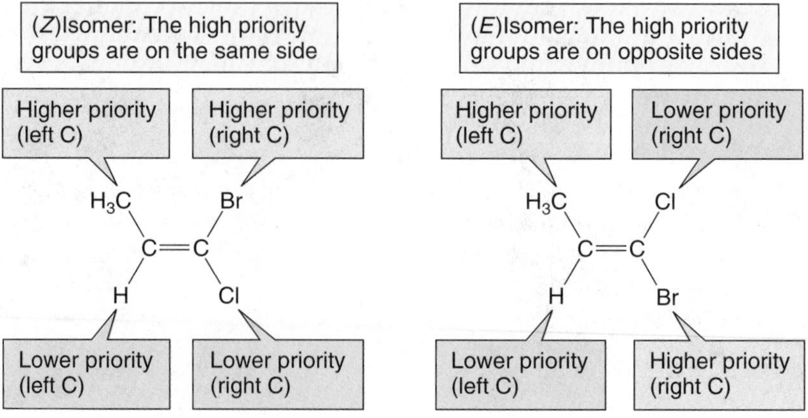

FIGURE 4.28 Use of the Cahn–Ingold–Prelog priority system.

2. For isotopes, atomic weight is used to break the tie in atomic number. Thus deuterium (atomic number = 1, atomic weight = 2) has a higher priority than hydrogen (atomic number = 1, atomic weight = 1) (Fig. 4.29).
3. Nonisotopic ties are broken by looking at the groups attached to the tied atoms. For example, 1-chloro-2-methyl-1-butene has a double bond to which a methyl and an ethyl group are attached. Both these groups are connected to the double bond through carbons, so Rule 1 does not differentiate them (Fig. 4.30).

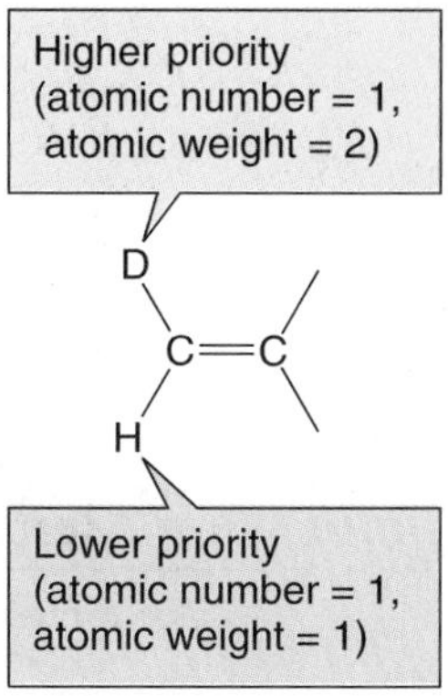

Higher priority (atomic number = 1, atomic weight = 2)

Lower priority (atomic number = 1, atomic weight = 1)

FIGURE **4.29** When the atomic numbers are the same, the heavier isotope gets the higher priority.

1-Chloro-2-methyl-1-butene

Same priority (atomic number = 6)

Same priority (atomic number = 6)

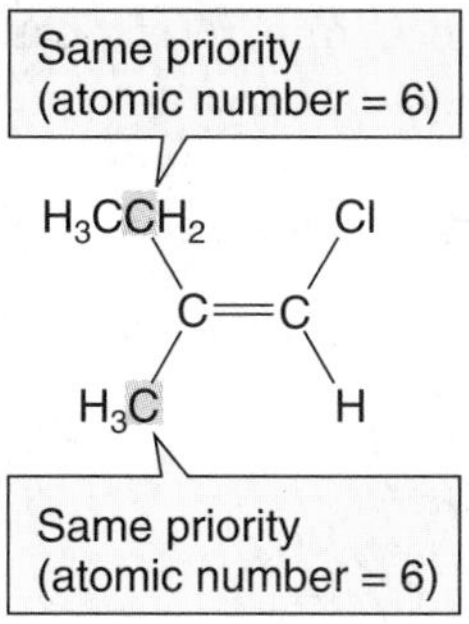

FIGURE **4.30** In this molecule, the two **C** atoms attached to the double bond of course have the same atomic number and, therefore, the same priority according to the Cahn–Ingold–Prelog Rule 1. How is the tie broken?

To break the tie, we look at the groups attached to these carbons. The carbon of the methyl group is attached to three hydrogens, whereas the "first" carbon of the ethyl group is attached to two hydrogens and a carbon. The ethyl group gets the higher priority (Fig. 4.31). If the groups are still tied after this procedure, one simply looks further out along the chain to break the tie, as in Problem 4.9 (Fig. 4.31).

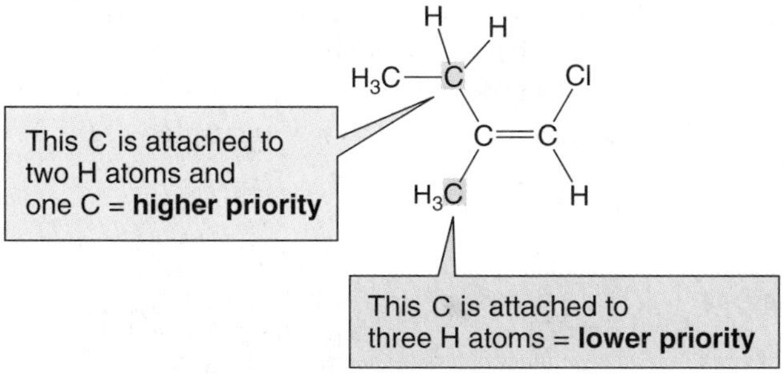

This C is attached to two H atoms and one C = **higher priority**

This C is attached to three H atoms = **lower priority**

FIGURE **4.31** The indicated methyl C is attached to three H atoms. The indicated ethyl C is attached to C, H, and H. The indicated ethyl C gets the higher priority.

PROBLEM **4.9**

Determine which of the molecules in Figure 4.32 is (*Z*) and which is (*E*).

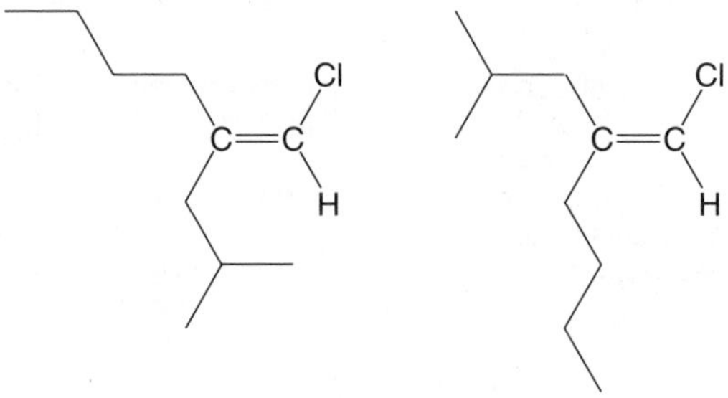

FIGURE **4.32**

PROBLEM **4.10** Draw pictures of the following molecules:

(a) (*E*)-3-Fluoro-3-hexene
(b) (*E*)-4-Ethyl-3-heptene
(c) (*Z*)-1-Bromo-2-chloro-2-fluoro-1-iodoethylene

4. Multiple bonds are treated as multiplied single bonds. A double bond to carbon is considered to be two single bonds to the carbons in the double bond, as shown in Figure 4.33. As we will shortly see, it is also possible to connect two carbon atoms through a triple bond. The priority system treats triple bonds in a similar fashion (Fig. 4.33), which results in an isopropenyl group being of higher priority than a *tert*-butyl group, for example (Fig. 4.34).

FIGURE **4.33** The priority of doubly bonded carbons is determined by adding two single carbon bonds as shown. A triple bond is treated similarly.

tert-Butyl group

Isopropenyl group

Lower priority

Higher priority

FIGURE **4.34** An isopropenyl group has a higher priority than a *tert*-butyl group.

To summarize: The technique of assigning (*Z*) and (*E*) is to determine the priority (1 = high, 2 = low) at each carbon of the double bond. The isomer with the two high-priority groups on the same side of the double bond is (*Z*). The isomer with the two high-priority groups on opposite sides is (*E*). Two examples are given in Figure 4.35.

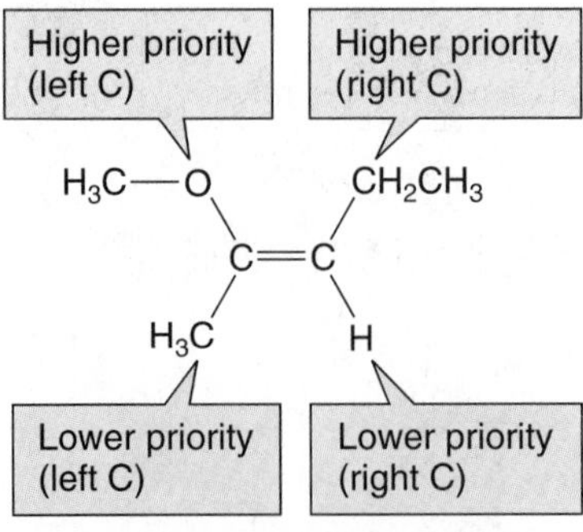

The two high priority groups are on the same side — this molecule is (*Z*)-2-methoxy-2-pentene

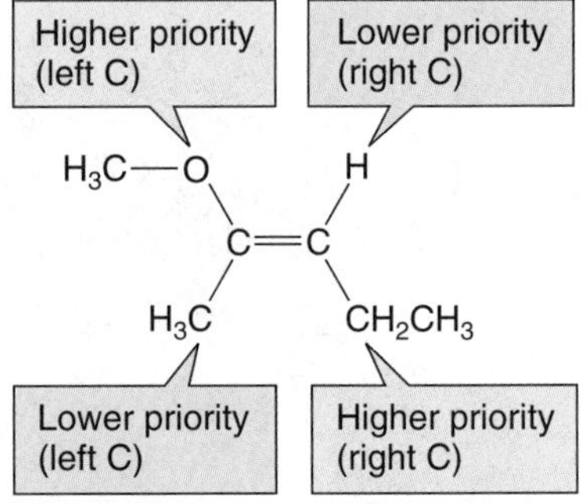

Here the situation is different — the two high priority groups are on opposite sides; this molecule is (*E*)-2-methoxy-2-pentene

FIGURE **4.35** Two examples of the Cahn–Ingold–Prelog priority system at work.

Which of the isomers in Figure 4.36 is (*Z*), and which is (*E*)? PROBLEM **4.11**

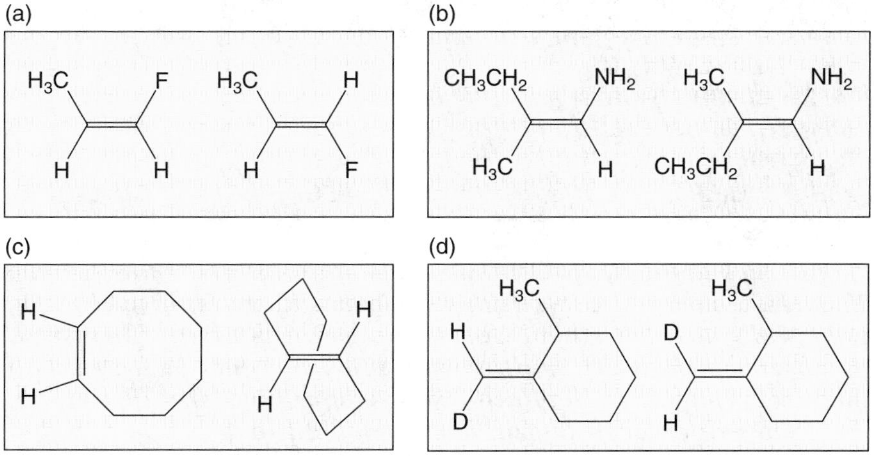

(a) (b) (c) (d)

FIGURE **4.36**

4.5 RELATIVE STABILITY OF ALKENES: HEATS OF FORMATION

The **heat of formation** (ΔH_f°) of a compound is the enthalpy of formation through the reaction of its constituent elements in their standard states. The standard state of an element is the most stable form of the element at 25 °C and 1-atm pressure. For an element in its standard state, ΔH_f° is taken as zero. For carbon, the standard state is graphite. For graphite, as well as simple gases (e.g., H_2, O_2, and N_2) the ΔH_f° is 0 kcal/mol.

The more negative—or less positive—a compound's ΔH_f° is, the more stable it is. A negative ΔH_f° for a compound means that its formation from its constituent elements would be *exothermic*—heat would be liberated. By contrast, a positive ΔH_f° means that the constituent elements are more stable than the compound and its formation would be *endothermic*—energy would have to be applied.

Remember: Bonding is an energy-releasing process. We would expect to find that the formation of a molecule from its constituent atoms is an exothermic reaction. For example, consider the simple formation of methane from carbon and hydrogen. Carbon in its standard state (graphite) and gaseous H_2 have ΔH_f° = 0 kcal/mol, whereas methane's $\Delta H_f^\circ = -17.8$ kcal/mol. The formation of methane from graphite and hydrogen releases 17.8 kcal/mol (is exothermic by 17.8 kcal/mol) (Fig. 4.37).

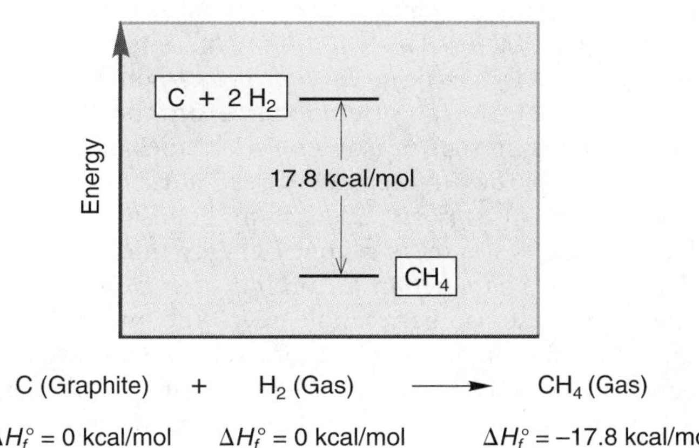

FIGURE **4.37** The formation of methane from graphite and gaseous hydrogen is exothermic by 17.8 kcal/mol; ΔH_f° of methane = – 17.8 kcal/mol.

Table 4.1 gives the heats of formation of a number of hexenes.

TABLE **4.1** Heats of Formation for Some Hexenes

Isomer	ΔH_f° (kcal/mol)	
$CH_2{=}CHCH_2CH_2CH_2CH_3$	−10.0	Least stable
cis $CH_3CH_2CH{=}CHCH_2CH_3$	−11.2	↑
trans $CH_3CH_2CH{=}CHCH_2CH_3$	−12.1	
$(CH_3)_2C{=}CHCH_2CH_3$	−16.0	↓
$(CH_3)_2C{=}C(CH_3)_2$	−16.6	Most stable

First look at the pair of disubstituted isomers, *cis*- and *trans*-3-hexene. The trans isomer is the more stable compound (more negative ΔH_f°). Figure 4.38 shows the reason; in the cis isomer there is a cis ethyl–ethyl torsional interaction, which is absent in the trans compound. In the trans isomer, both alkyls are eclipsed by hydrogen, and the destabilizing ethyl–ethyl repulsion is missing.

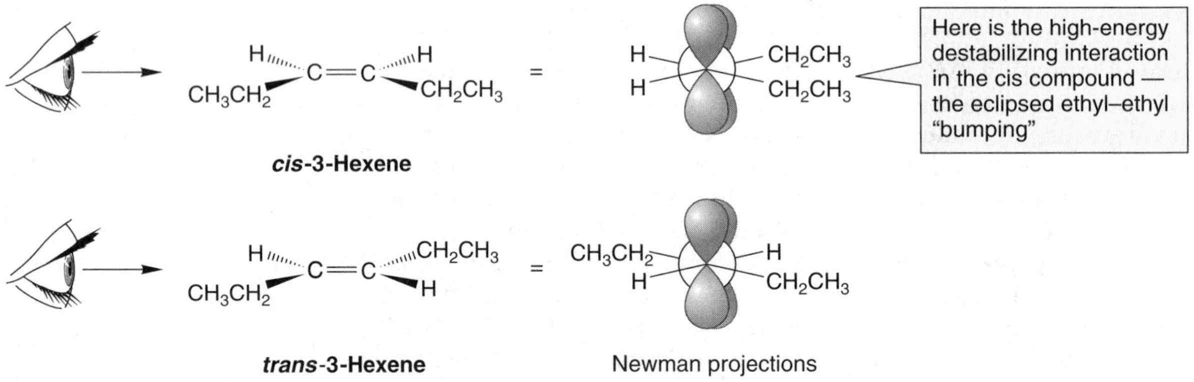

FIGURE **4.38** Newman projections show the unfavorable ethyl–ethyl steric interaction in *cis*-3-hexene.

But both disubstituted isomers are more stable than the monosubstituted 1-hexene, and the trisubstituted isomer is more stable than either disubstituted molecule. Tetrasubstituted 2,3-dimethyl-2-butene has the lowest energy of all (most stable). Evidently, the degree of substitution (the number of alkyl groups attached to the double bond) is important in stabilizing the molecule. In general, the more substituted the double bond the more stable it is (Table 4.1).

The reason for this has been the subject of some controversy. One good way to look at the question focuses on the different kinds of carbon–carbon bonds present in isomeric molecules of different substitution patterns. In all of these molecules, three kinds of carbon–carbon linkages are present, sp^2–sp^2, sp^2–sp^3, and sp^3–sp^3 bonds (Fig. 4.39).

An electron in a 2*s* orbital is at lower energy than an electron in a 2*p* orbital. The more *s* character in an orbital, the more an electron in it is stabilized. In these isomeric molecules, the more substituted the carbon–carbon double bond is, the more relatively low-energy (strong) sp^2–sp^3 bonds are present. By contrast, less substituted alkenes have more relatively high energy (weaker) sp^3–sp^3 bonds. The more strong bonds pre-

Monosubstituted

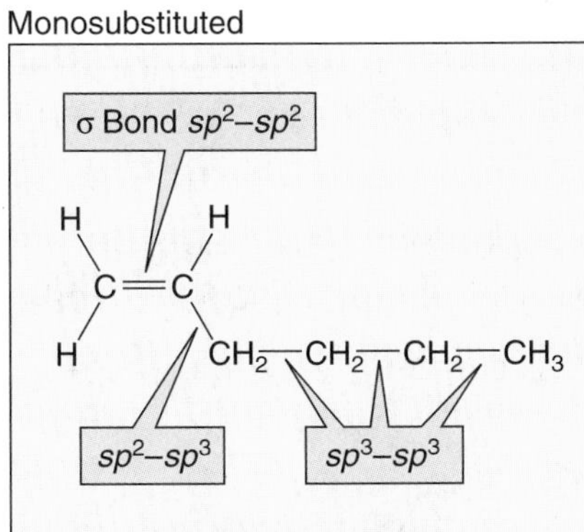

Disubstituted

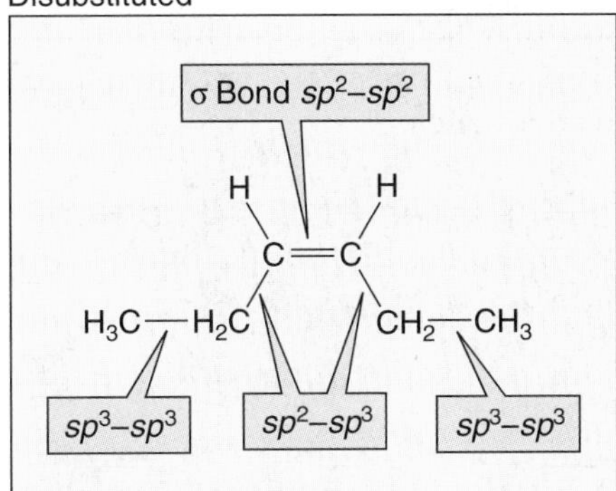

Trisubstituted

Tetrasubstituted

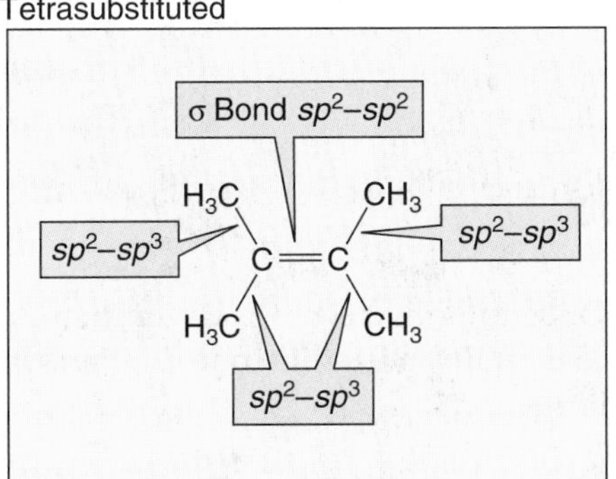

FIGURE 4.39 There are three kinds of carbon–carbon bonds in the hexenes: sp^2–sp^2, sp^2–sp^3, and sp^3–sp^3.

sent, the more stable is the molecule. Figure 4.40 shows the number of sp^3–sp^3, sp^2–sp^3, and sp^2–sp^2 bonds in the isomeric hexenes of Table 4.1. 2,3-Dimethyl-2-butene, the molecule with the strongest bonds, is the most stable isomer of the set, and 1-hexene, with the weakest set of bonds, is the least stable.

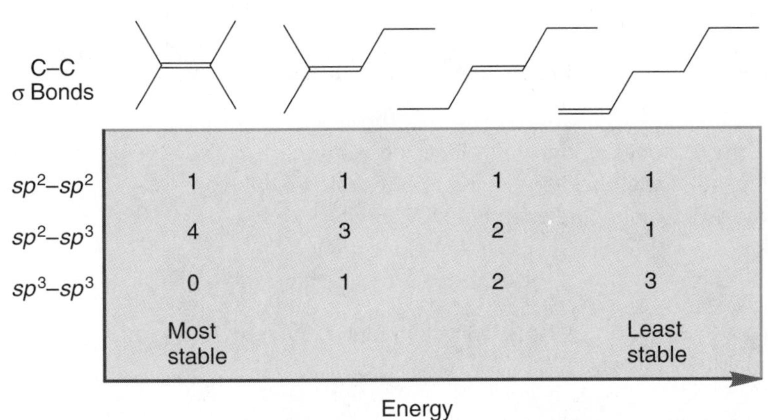

C–C σ Bonds
sp^2–sp^2 1 1 1 1
sp^2–sp^3 4 3 2 1
sp^3–sp^3 0 1 2 3
Most stable
Least stable
Energy

FIGURE 4.40 The isomeric hexenes have different numbers of these three types of carbon–carbon bonds. The more of the stronger sp^2–sp^3 bonds, and the fewer of the relatively weak sp^3–sp^3 bonds, the more stable the isomer.

PROBLEM 4.12

Carry out a similar analysis for the series of molecules: 1-pentene, 2-methyl-2-butene, (*Z*)-2-pentene.

4.6 DOUBLE BONDS IN RINGS

Even small rings can contain double bonds. Figure 4.41 shows a number of **cycloalkenes**.

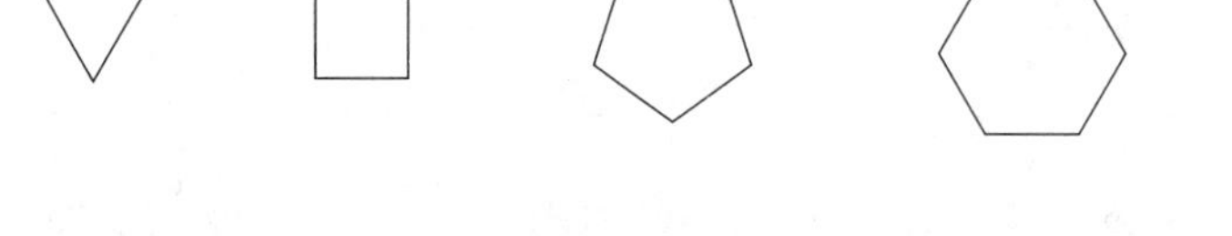

FIGURE **4.41** Some cyclic alkenes.

PROBLEM **4.13** Name all the compounds in Figure 4.41.

Once the ring becomes larger than cyclopropene, it becomes possible for it to contain more than one double bond. Several such molecules are shown in Figure 4.42.

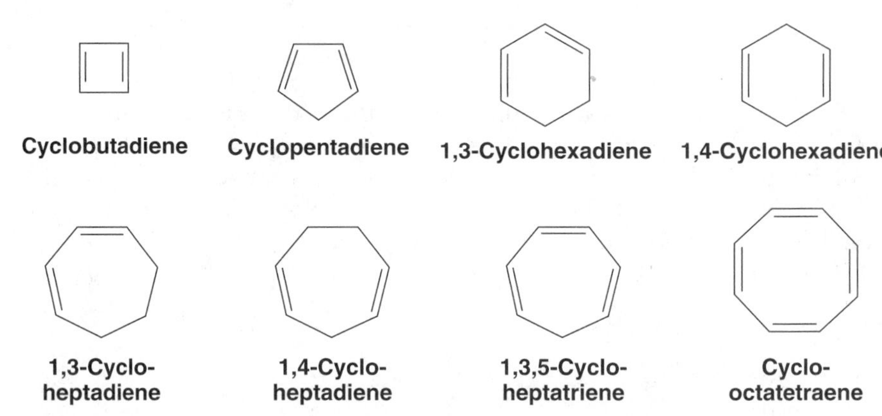

FIGURE **4.42** Some cyclic polyenes.

PROBLEM **4.14** Write the structures for 1,3,5-cyclohexatriene, 1,3,5,7-cyclooctatetraene, 3-methyl-1,4-cyclohexadiene, 2-fluoro-1,3-cyclohexadiene, and 2-bromo-1,4-cycloheptadiene.

*PROBLEM **4.15** Draw both (*E*)- and (*Z*)-1-methylcycloheptene. Which would you expect to be more stable? Explain.

ANSWER In (*Z*)-1-methylcycloheptene, the two higher priority groups will be on the same side of the double bond, whereas in (*E*)-1-methylcycloheptene they will be on opposite sides. On one side, hydrogen is lower priority than CH_2, and on the other side, CH_3 is lower priority than $CH_2{-}C$. So, the cis compound, with the higher priority groups on the same side, is (*Z*), and the trans compound, with the higher priority groups on opposite sides, is (*E*).

ANSWER (CONTINUED)

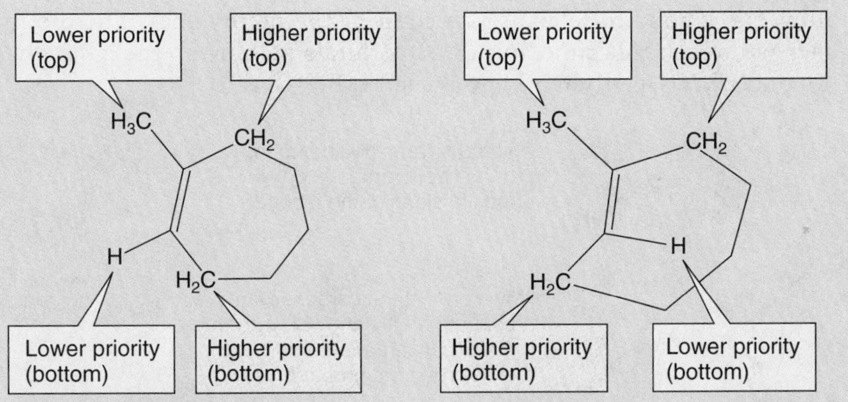

(*Z*)-1-Methylcycloheptene **(*E*)-1-Methylcycloheptene**

Even the drawing gives a clue to the relative stabilities. It is trivial to draw the (*Z*) form, but the (*E*) form requires you to stretch bonds in order to make the necessary connections. If you made a model, you saw that the (*E*) isomer contains a badly twisted double bond, with poor overlap between the 2*p* orbitals making up the π bond. There are no such problems in the (*Z*) isomer, which is much more stable.

All the double bonds in Figures 4.41 and 4.42 are cis. For some reason this is often a sticky point for students. Make sure you understand why all the double bonds shown in the figures are appropriately called cis. To make the point absolutely clear we will use cyclopentene as an example. Note that the two hydrogens on the double bond are on the same side. The two alkyl groups (part of the ring) are on the other (Fig. 4.43). The molecule is properly called cis, or (*Z*).

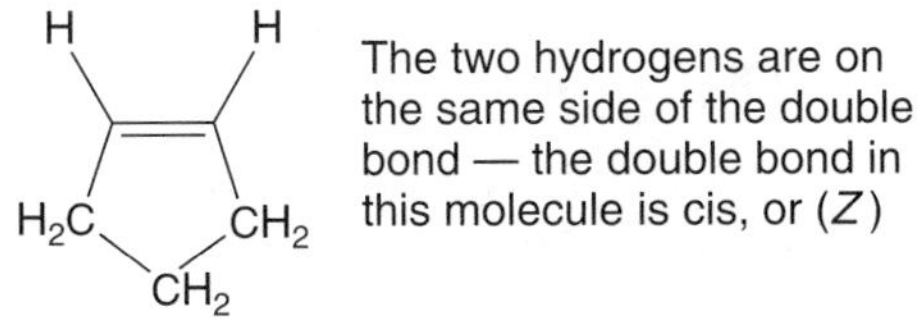

***cis*-Cyclopentene**

FIGURE **4.43** The double bond in cyclopentene is cis.

It is extremely difficult to put a trans double bond in a small ring, and it's worth our time to see why. Double bonds depend on overlap of *p* orbitals for their stability. There is no great problem in a ring containing a cis double bond (Fig. 4.43), but if the cycloalkene is trans, the double bond becomes severely twisted, and highly destabilized. Remember that π bonds derive their stability from overlap of 2*p* orbitals, and that a fully twisted (90°) π bond is about 66 kcal/mol higher in energy than the 0° form (p. 124). For a small, or medium-sized ring, it is not possible to bridge opposite sides of the double bond without very severe strain. Let's use cyclopentene as an example again. A single CH_2 group cannot span the two carbons attached to opposite sides of a trans double bond (Fig. 4.44). This difficulty is easy to see in models. Use your models to try to make *trans*-cyclopentene. Hold the double bond planar, and then try to introduce the bridging methylene groups. They won't reach. Next, allow the π bond to relax as you connect the methylenes. The methylenes will reach this time, but the $2p$ orbitals making up the π bond will no longer overlap efficiently. Make sure you see this.

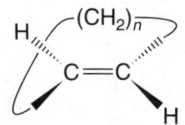

To start a trans double bond, put the two hydrogens on opposite sides of the π bond

Now connect the carbons with a chain of methylene $(CH_2)_n$ groups; if the ring is large enough ($n \geq 6$), there is no great problem

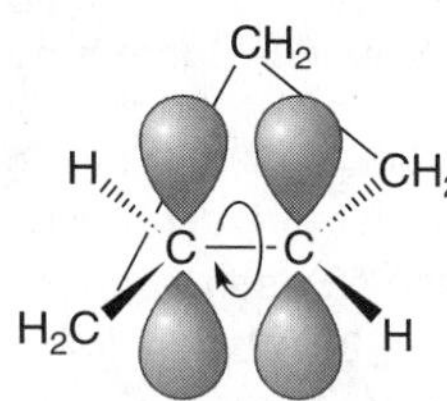

To make *trans*-cyclopentene, the double bond carbons must be bridged by only three methylenes; this cannot be done without twisting the $2p$ orbitals out of overlap; this twisting is very costly in energy terms

FIGURE **4.44** It is difficult to incorporate a trans double bond in a small ring. Either the bridge of methylene groups is too small to span the trans positions or the $2p$ orbitals making up the π bond must be twisted out of overlap.

Of course the larger the ring, the less problem there is because more atoms are available to span the trans positions of the double bond. In practice, the smallest trans cycloalkene stable at room temperature is *trans*-cyclooctene. Even here, the trans isomer is 11.4 kcal/mol less stable than its cis counterpart (Fig. 4.45).

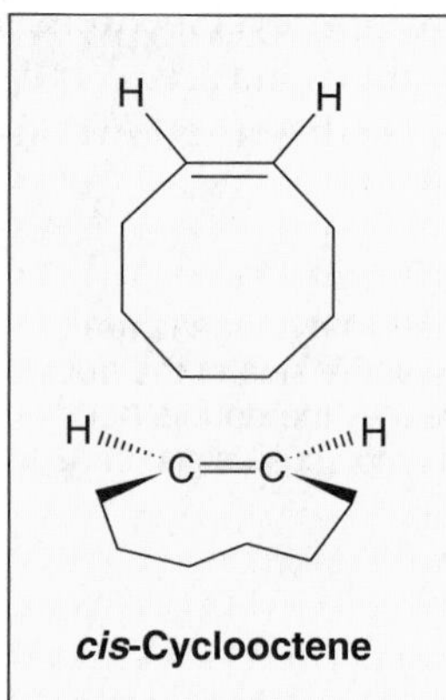

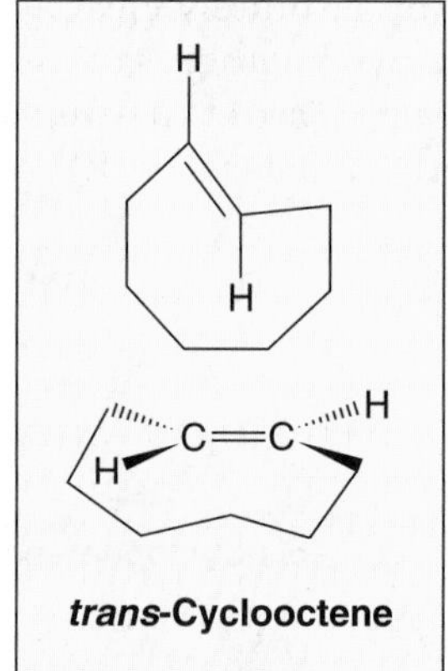

FIGURE **4.45** *trans*-Cyclooctene is the smallest ring in which a trans double bond is stable at room temperature. The cis isomer is more stable by 11.4 kcal/mol.

PROBLEM **4.16** Calculate the equilibrium distribution of *cis*- and *trans*-cyclooctene at 25 °C given an 11.4-kcal/mol energy difference between the two isomers (of course, this question assumes that equilibrium can be reached—something not generally true for alkenes). The relationship between the energy difference (ΔG) and the equilibrium constant (K) is $\Delta G = -RT \ln K$, or $\Delta G = -2.3RT \log K$, where R is the gas constant (1.986 cal/deg·mol) and T is the absolute temperature.

Many polycyclic compounds containing double bonds are known, including a great many natural products. Figure 4.46 shows some common structures.

Limonene

Vitamin A

CH_2OH

Cholesterol

HO H H H

Morphine

HO HO O H N CH_3

FIGURE **4.46** Some natural products incorporating cycloalkenes.

It was noticed in 1924 by the German chemist, Julius Bredt (1855–1937), that there was one structural type that was conspicuously absent among the myriad compounds isolated from sources in Nature. There seemed to be no double bonds attached to the **bridgehead position** in what are called "bridged bicyclic molecules" (Fig. 4.47). You have met such structures before in Chapter 3 (p. 109), and they will reappear in some detail in Chapter 6, but you might make a model now of a simple bicyclic structure. Figure 4.47 shows two kinds of bicyclic molecules, "bridged" and "fused," and briefly introduces some terminology.

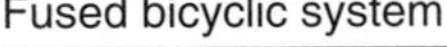

Bridged bicyclic system

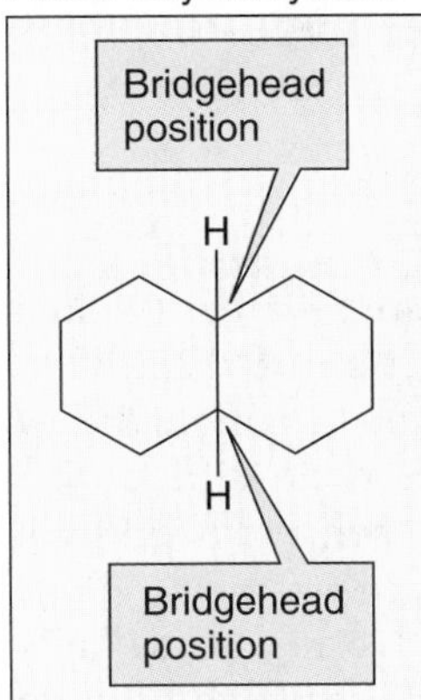

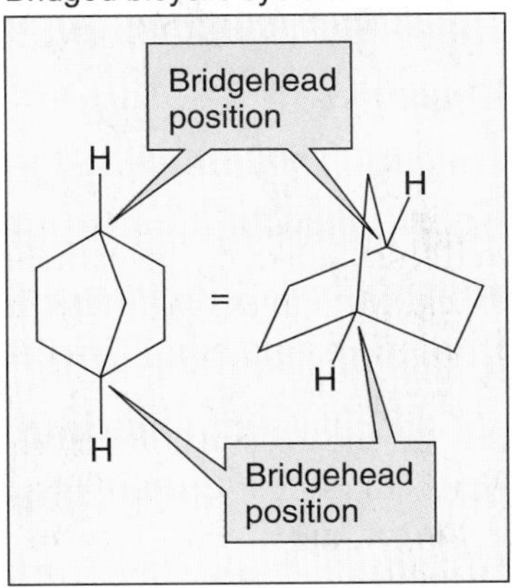

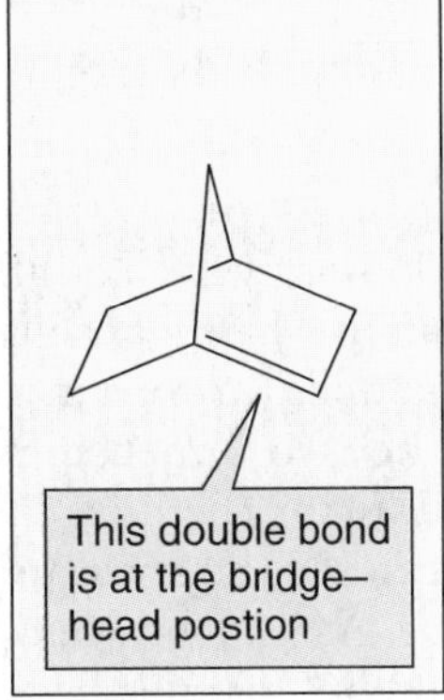

FIGURE **4.47** Fused and bridged bicyclic molecules. Molecules containing double bonds at the bridgehead positions of bicyclic systems are conspicuous by their absence in Nature. Apparently, great instability is caused by the incorporation of a double bond at the bridgehead position. Why are these compounds so unstable?

Bredt was unable to explain this absence of double bonds at the bridgehead position, but was alert enough to note the phenomenon, which has become deservedly known as **Bredt's rule**. What is the reason behind this rule? First of all, the rigid cage structure with its pyramidal bridgehead carbons requires that the π bond, the "double" part of the alkene, be formed not from $2p$–$2p$ overlap, but by $2p$–hybrid orbital overlap. Overlap is not as good as in a normal alkene π system (Fig. 4.48).

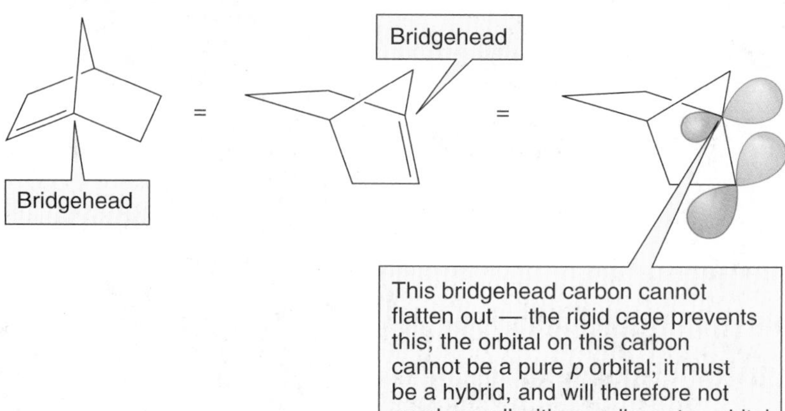

FIGURE 4.48 The orbital at the bridgehead position cannot be a pure $2p$ orbital because of the rigidity of the bicyclic system. Three views of a bicyclic molecule containing a double bond at the bridgehead.

But there is even more to this question, and a careful three-dimensional drawing of the molecule best reveals the story. As is so often the case, the Newman projection is the most informative way to look at the molecule (Fig. 4.49). Note that the orbitals making up the (hypothetical) double bond

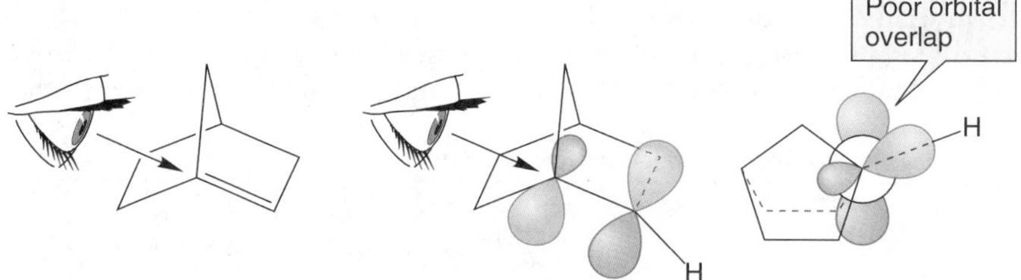

FIGURE 4.49 A Newman projection shows that in these bridgehead alkenes the orbitals making up the "π bond" cannot overlap well.

do not overlap at all well. Just as trans cycloalkenes are severely twisted, so are bridgehead alkenes. Moreover, the rigid structure of these compounds (make a model!) allows for no relief. This is a good example of how the two dimensions of the paper can fool you. There is no difficulty in drawing the lines making up the double bond on the paper, and unless you can see the structure, you will almost certainly be fooled.

As the bridges get longer, flexibility returns and bridgehead "anti-Bredt" alkenes become stable (Fig. 4.50). Clever syntheses have been devised so that even quite unstable compounds can be made and studied. In simple bridgehead alkenes, the limits of room temperature stability are reached with bicyclo[3.3.1]non-1-ene (do not worry yet about the naming system; we will deal with it in Chapter 6), which contains a trans double bond in an eight-membered ring. In the simple trans cycloalkenes, it is also the eight-carbon compound that is the first molecule stable at room temperature. It doesn't seem unreasonable that there should be a rough correspondence between the trans cycloalkenes and the bridgehead alkenes, which also contain a trans double bond in a ring.

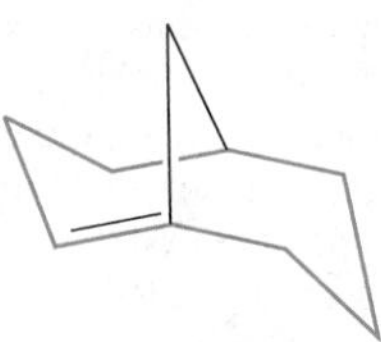

FIGURE 4.50 The smallest bridgehead alkene stable under normal conditions is bicyclo[3.3.1]non-1-ene. This molecule contains a *trans*-cyclooctene, shown in color.

PROBLEM 4.17

Is the bicyclo[3.3.1]non-1-ene shown in Figure 4.50 (*Z*) or (*E*)?

***PROBLEM 4.18**

Draw the other form (*Z* or *E*). Why is it less stable than the one in Figure 4.50? *Hint*: Focus on the rings in which the double bonds are contained.

ANSWER

The form in Figure 4.50 is (*Z*), as the higher and lower priority groups are on the same side of the double bond. On the left-hand carbon of the double bond, the higher priority group is CH_2 and the lower priority group is H. On the right-hand carbon of the double bond the higher priority group is $CH_2–CHC_2$ and the lower priority group $CH_2–CH_2C$. In this form, the two higher priority groups are on the same side of the double bond. Thus the correct label is (*Z*). The other form is the (*E*) structure, with the two higher priority groups on opposite sides.

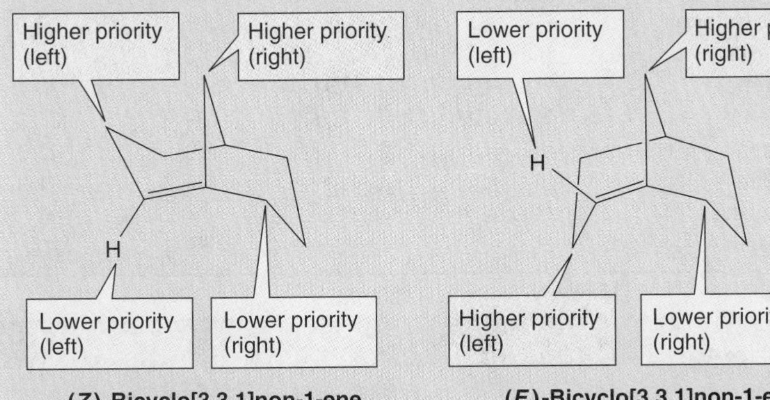

(*Z*)-Bicyclo[3.3.1]non-1-ene **(*E*)-Bicyclo[3.3.1]non-1-ene**

Note that this molecule contains a double bond in two different-sized rings. The double bond of the (*Z*) form is cis in the six-membered ring and trans in the eight-membered ring. In the (*E*) form, the double bond is cis in the eight-membered ring and trans in the six-membered ring. The smaller the ring, the less stable is a trans double bond (why?), so the (*E*) form is less stable than the (*Z*) form.

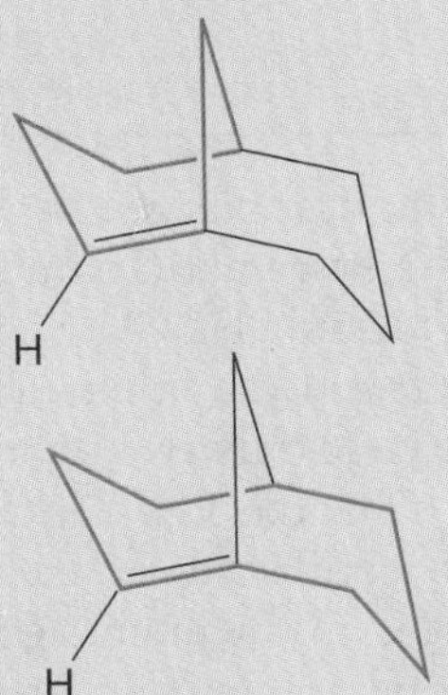

(*Z*) Form: double bond cis in six-membered ring (red, top), trans in eight-membered ring (red, bottom)

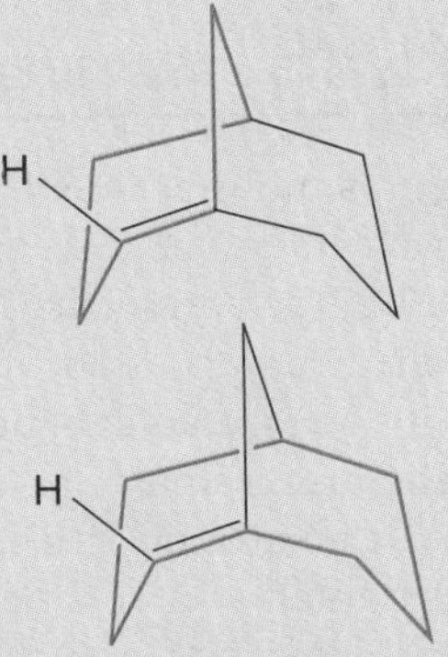

(*E*) Form: double bond cis in eight-membered ring (red, bottom), trans in six-membered ring (red, top)

4.7 PHYSICAL PROPERTIES OF ALKENES

There is little difference in physical properties between the alkenes and their saturated relatives, the alkanes. Their odors are a bit more pungent and perhaps justify being called "evil-smelling." In fact, the old trivial name for alkenes, **olefins**, readily evokes the sense of smell. Tables 4.2 and 4.3 collect some data for alkenes and cycloalkenes.

TABLE **4.2** Some Simple Alkenes

Name	Formula	mp (°C)	bp (°C)
Ethene (ethylene)	$H_2C{=}CH_2$	−169	−103.7
Propene (propylene)	$H_2C{=}CH{-}CH_3$	−185.2	− 47.4
1-Butene	$H_2C{=}CH{-}CH_2{-}CH_3$	−185	− 6.3
cis-2-Butene	$CH_3{-}CH{=}CH{-}CH_3$	−138.9	3.7
trans-2-Butene	$CH_3{-}CH{=}CH{-}CH_3$	−105.5	0.9
Isobutene	$(CH_3)_2CH{=}CH_2$	− 40.7	− 6.6
1-Pentene	$H_2C{=}CH{-}(CH_2)_2{-}CH_3$	−138	30.1
1-Hexene	$H_2C{=}CH{-}(CH_2)_3{-}CH_3$	−139.8	63.3
1-Heptene	$H_2C{=}CH{-}(CH_2)_4{-}CH_3$	−119	93.6
1-Octene	$H_2C{=}CH{-}(CH_2)_5{-}CH_3$	−101.7	121.3
1-Nonene	$H_2C{=}CH{-}(CH_2)_6{-}CH_3$		146
1-Decene	$H_2C{=}CH{-}(CH_2)_7{-}CH_3$	− 66.3	170.5

TABLE **4.3** Some Simple Cycloalkenes

Name	mp (°C)	bp (°C,)
Cyclobutene		2
Cyclopentene	− 135	44.2
Cyclohexene	− 103.5	83
Cycloheptene	− 56	115
cis-Cyclooctene	− 12	138
trans-Cyclooctene	− 59	143
cis-Cyclononene		167–169
trans-Cyclononene		94–96 (at 30 mmHg)

4.8 ALKYNES: STRUCTURE AND BONDING

Like the simplest alkene, ethylene, the smallest alkyne is generally known by its trivial name, **acetylene**, not the systematic name, ethyne. Acetylene is a symmetrical compound of the formula HCCH. Each carbon is attached to two other atoms: one hydrogen and the other carbon. Accordingly, a reasonable bonding scheme must yield only two hybrid bonds. As in the discussion of alkanes and alkenes (Chapter 3, p. 73; Chapter 4, p. 115), the atomic orbitals of carbon are combined to yield hybrid orbitals, which will do a better job of bonding than the unchanged atomic orbitals. Two bonds are needed, one for the bond to hydrogen, and another for the bond to carbon. Therefore, we need to combine only two of carbon's atomic orbitals to give us our hybrids. A combination of the $2s$ and $2p_x$ orbitals will yield a pair of ***sp* hybrids**. We can anticipate that these new *sp* hybrids will be directed so as to keep the bonds, and the electrons in them, as far apart as possible, which produces 180° angles. (Recall our discussion of BeH_2 in Section 3.2, p. 73; Fig. 4.51.)

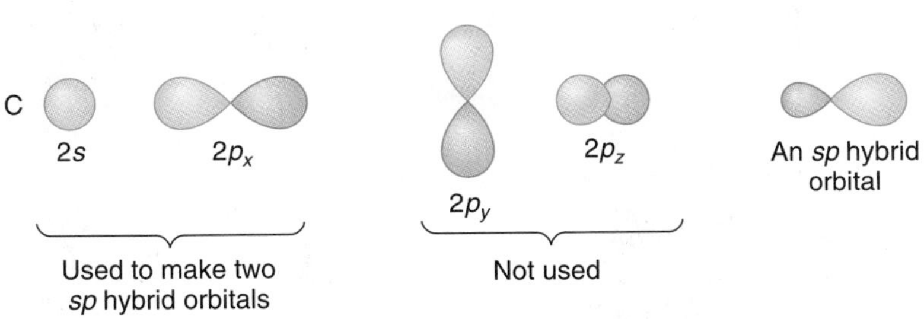

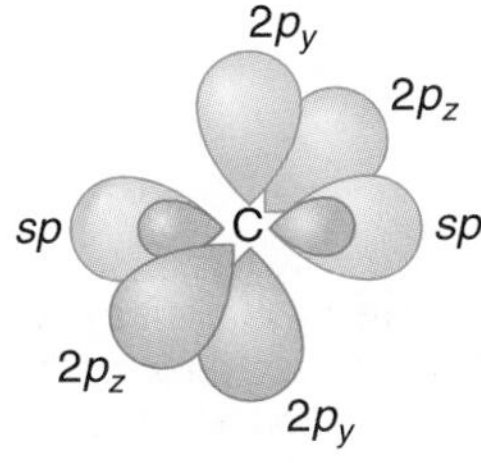

FIGURE 4.51 To make an *sp* hybrid we combine the wave functions for the carbon 2*s* and $2p_x$ orbitals. The $2p_y$ and $2p_z$ orbitals are not used in the hybridization scheme.

These new *sp* hybrid orbitals generally resemble the sp^3 and sp^2 hybrids we made before. As we used only two of carbon's four available atomic orbitals, the $2p_y$ and $2p_z$ orbitals are left over. Figure 4.52 shows an *sp*-hybridized carbon atom. (From now on, we will not show the small lobes).

The bond to hydrogen is formed by overlap of one *sp* hybrid with the hydrogen 1*s* orbital, and the carbon–carbon bond is formed by the overlap of two *sp* hybrids. This forms the σ system of the molecule (Fig. 4.53).

FIGURE 4.52 An *sp*-hybridized carbon atom. Note the left over $2p_y$ and $2p_z$ orbitals.

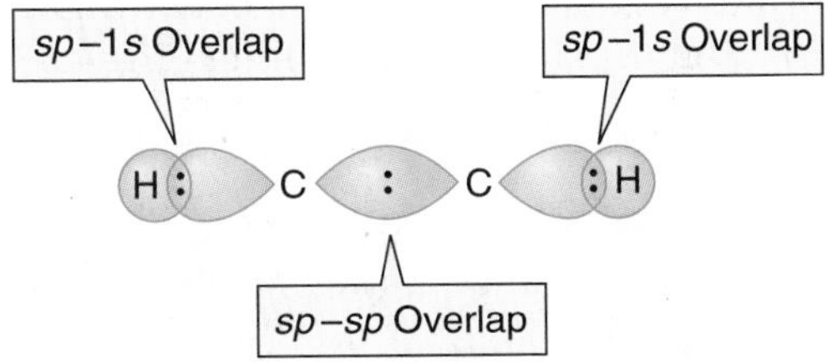

FIGURE 4.53 The σ bonding system of acetylene.

The figure shows the bonds formed by *sp–sp* and *sp*–1*s* overlap, but don't forget that these overlapping hybrid and atomic orbitals create both bonding and antibonding molecular orbitals. The empty antibonding orbitals are not in Figure 4.53, but are shown in the schematic construction of Figure 4.54.

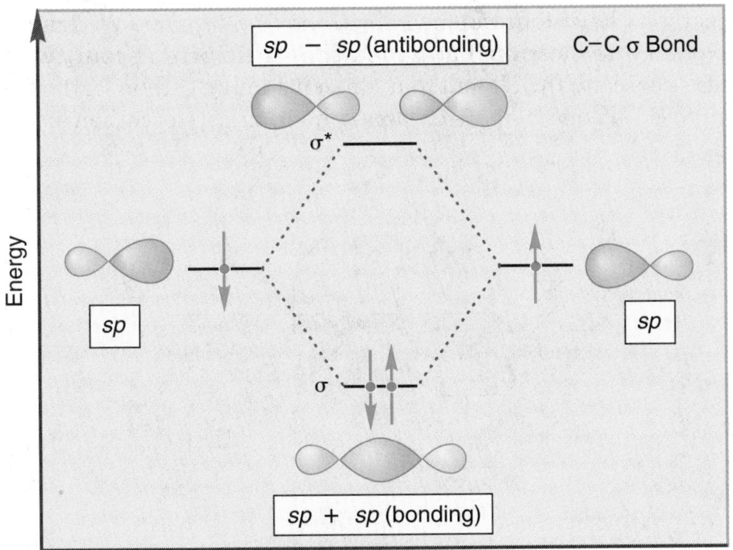

FIGURE 4.54 An orbital interaction diagram showing the bonding and antibonding orbitals formed by *sp–sp* overlap.

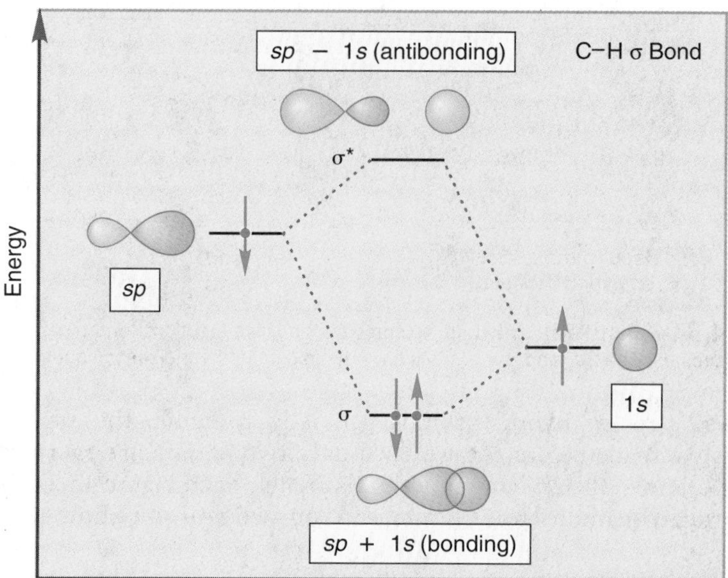

FIGURE **4.54** (CONTINUED) An orbital interaction diagram showing the bonding and antibonding orbitals formed by *sp*–1*s* overlap.

As in the alkenes, the remaining unhybridized 2*p* orbitals also overlap to form π bonds. This time there are two *p* orbitals remaining on each carbon of acetylene, and we can form a pair of π bonds that are directed, as are the constituent 2*p* orbitals, at 90° to each other (Fig. 4.55).

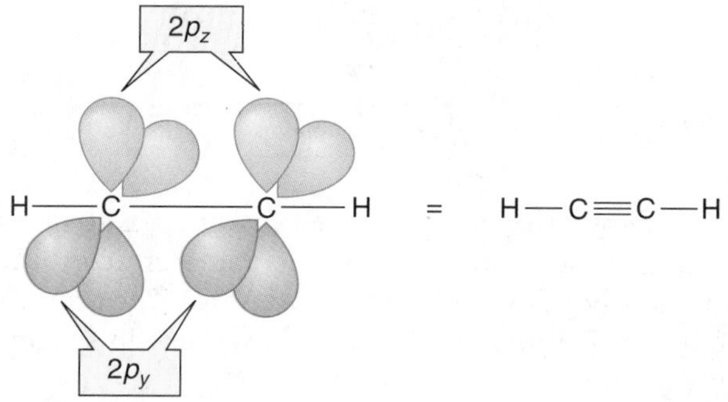

$2p_z$

H—C———C—H = H—C≡C—H

$2p_y$

FIGURE **4.55** Overlap of two $2p_y$ and two $2p_z$ orbitals forms a pair of π bonds in alkynes.

Both carbons have four valence electrons (*Remember*: We ignore the very low energy 1*s* electrons) and each can participate in four two-electron bonds. These are the σ bond to hydrogen and the carbon–carbon **triple bond** composed of one carbon–carbon σ bond and two carbon–carbon π bonds (Fig. 4.56).

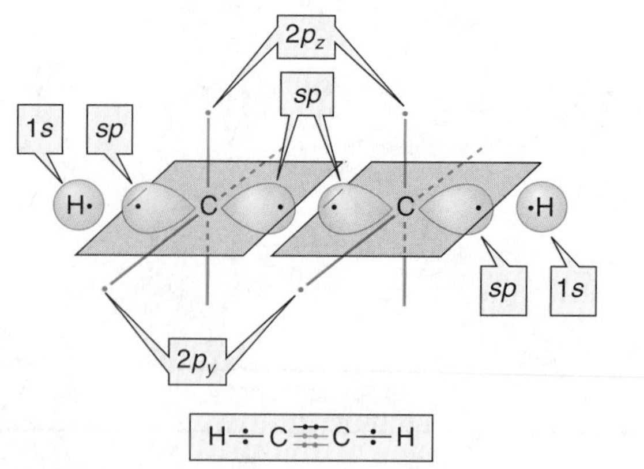

FIGURE **4.56** A highly schematic construction of the π and σ bonds in acetylene.

PROBLEM **4.19**

Draw the orbital interaction diagrams for construction of the π bonds of acetylene from the 2*p* orbitals.

One of the structural consequences of triple bonding is an especially short carbon–carbon bond distance (~1.2 Å), considerably shorter than either carbon–carbon single or double bonds (Fig. 4.57).

$H_3C—CH_3$ 1.54 Å $H_2C{=}CH_2$ 1.33 Å $HC{\equiv}CH$ 1.20 Å

FIGURE **4.57** Bond lengths in angstroms (Å) of simple two-carbon hydrocarbons.

4.9 RELATIVE STABILITY OF ALKYNES: HEATS OF FORMATION

Heats of formation show that alkynes are very much less stable than their constituent elements. Table 4.4 summarizes heats of formation for some alkynes, alkenes, and alkanes. The virtues of using hybrid orbitals for σ bonding rather than *p* orbitals for π bonding are apparent. The more π bonds in a molecule, the more positive the heat of formation, and the more endothermic the formation of the compound from its constituent elements.

TABLE **4.4** Heats of Formation for Some Small Hydrocarbons

Hydrocarbon	ΔH_f° (kcal/mol)	Hydrocarbon	ΔH_f° (kcal/mol)	Hydrocarbon	ΔH_f° (kcal/mol)
Ethane	−20.1	Ethylene	12.5	Acetylene	54.5
Propane	−25.0	Propylene	4.8	Propyne	44.6
Butane	−30.2	1–Butene	−0.1	1–Butyne	39.5
		2–Butene	−1.9(cis)	2–Butyne	34.7
			−2.9(trans)		

The very positive heats of formation for alkynes have their practical consequences. When a very high heat is desired, as for many welding applications, acetylene is the fuel of choice.

Disubstituted alkynes are more stable than their monosubstituted isomers. For example, we can see from Table 4.4 that the heat of formation of 1-butyne is about 5 kcal/mol more positive than that for 2-butyne. This phenomenon echoes the increasing stability of alkenes with increasing substitution (Section 4.5).

4.10 DERIVATIVES AND ISOMERS OF ALKYNES

Of course, we can imagine replacing one of the hydrogens of acetylene with another group X to give substituted alkynes. Figure 4.58 gives an example of an "ethynyl" compound. In practice, two naming systems are used here, and the figure shows them. Ethynyl chloride can also be called chloroacetylene, and both naming systems are commonly used.

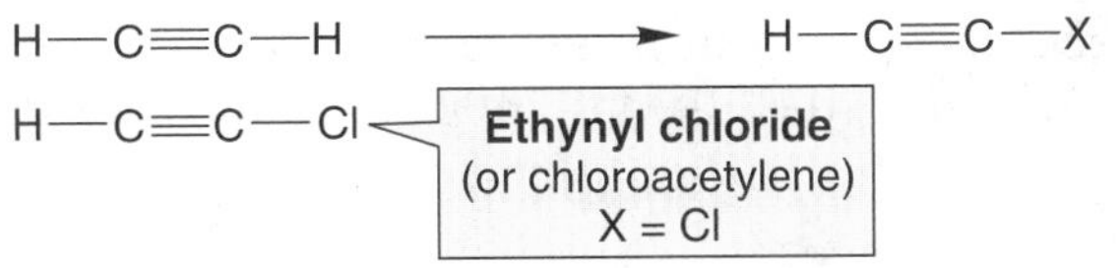

$H—C{\equiv}C—H \longrightarrow H—C{\equiv}C—X$

$H—C{\equiv}C—Cl$ — **Ethynyl chloride** (or chloroacetylene) X = Cl

FIGURE **4.58** An ethynyl compound, chloroacetylene.

We can start a construction of the family of alkynes by replacing X in Figure 4.58 with a methyl group, CH_3. This produces propyne (or methylacetylene), the lone three-carbon alkyne (Fig. 4.59).

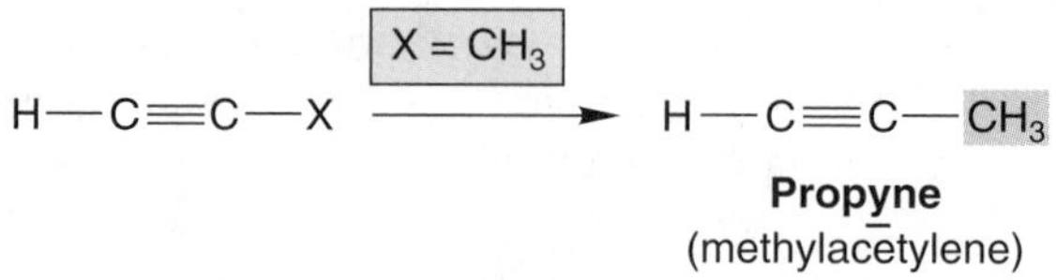

$$H{-}C{\equiv}C{-}X \xrightarrow{X = CH_3} H{-}C{\equiv}C{-}CH_3$$

Propyne
(methylacetylene)

FIGURE 4.59 Replacement of X with a methyl (CH_3) group, gives propyne.

There are two different hydrogens in propyne, and therefore two different ways to produce a substituted compound (Fig. 4.60).

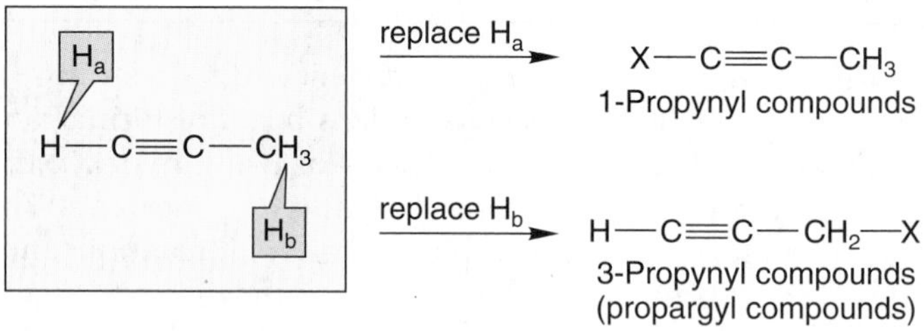

H_a, H_b: $H{-}C{\equiv}C{-}CH_3$

replace H_a → $X{-}C{\equiv}C{-}CH_3$
1-Propynyl compounds

replace H_b → $H{-}C{\equiv}C{-}CH_2{-}X$
3-Propynyl compounds
(propargyl compounds)

FIGURE 4.60 There are two kinds of hydrogen in propyne, and each can be replaced with X to give derivatives.

When the acetylenic hydrogen is replaced, the compounds can be named either as acetylenes or as "1-propynyl" compounds. 3-Propynyl compounds result from replacement of a methyl hydrogen with X. The common name **propargyl** is reserved for the $HC{\equiv}CH{-}CH_2$ group and is often seen. Note the relationship between the allyl (Fig. 4.20, p. 126) and propargyl groups.

When $X = CH_3$ we produce the two four-carbon acetylenes, which must be butynes (Fig. 4.61).

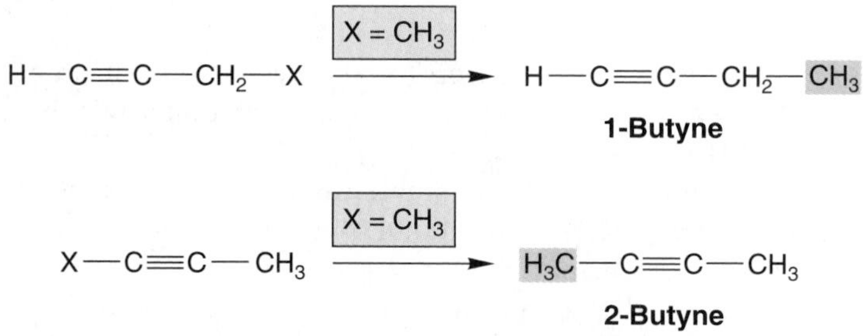

$$H{-}C{\equiv}C{-}CH_2{-}X \xrightarrow{X = CH_3} H{-}C{\equiv}C{-}CH_2{-}CH_3$$

1-Butyne

$$X{-}C{\equiv}C{-}CH_3 \xrightarrow{X = CH_3} H_3C{-}C{\equiv}C{-}CH_3$$

2-Butyne

FIGURE 4.61 If $X = CH_3$, we get the two butynes, 1-butyne and 2-butyne.

The naming protocol is similar to that used for the alkenes. As with the alkenes, the triple bond is designated a position by assigning it a number, which is kept as low as possible. When both double and triple bonds are present, the compounds are named as "enynes," and the numbers designating the positions of the multiple bonds kept as low as possible. If the numbering scheme would produce two names in which the lower number could go to either the "-ene" or the "-yne", the "-ene" gets it (Fig. 4.62).

It was at this point in the alkenes that we encountered cis/trans isomerism, and it's appropriate to search for it here as well. It takes only a short look to see that there can be no such isomerism in a linear compound (Fig. 4.63), and so the alkynes are simpler to analyze structurally than the alkenes.

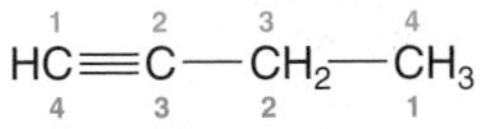

$HC{\equiv}C{-}CH_2{-}CH_3$

1-Butyne
not 3-butyne

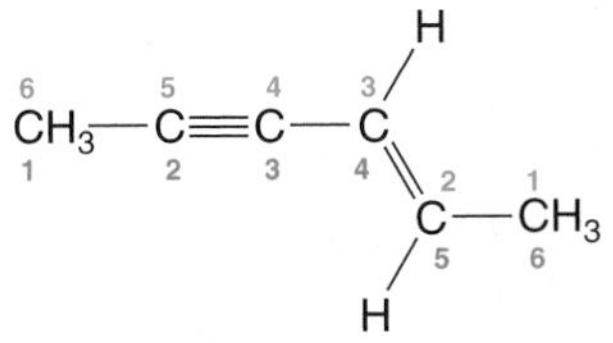

$CH_3{-}C{\equiv}C{-}CH{=}CH{-}CH_3$

***trans*-2-Hexen-4-yne**
not trans-4-hexen-2-yne

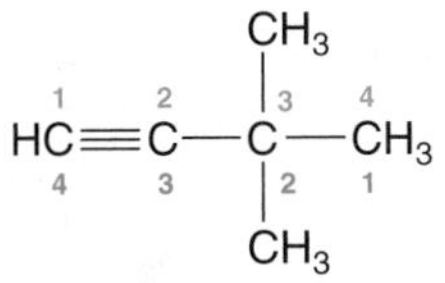

$HC{\equiv}C{-}C(CH_3)_2{-}CH_3$

3,3-Dimethyl-1-butyne
not 2,2-dimethyl-3-butyne

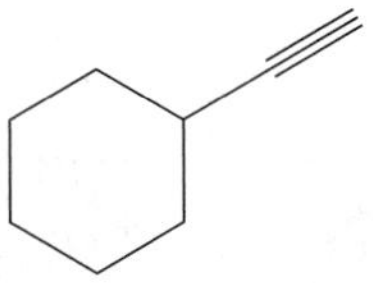

Ethynylcyclohexane
or cyclohexylacetylene

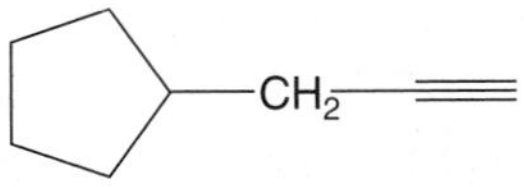

3-Propynylcyclopentane
or propargylcyclopentane

FIGURE 4.62 Some examples of the naming convention for alkynes.

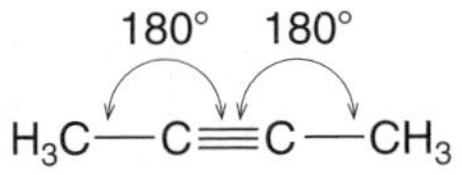

180° 180°

$H_3C{-}C{\equiv}C{-}CH_3$

FIGURE 4.63 The *sp* hybridization requires 180° angles, and therefore there can be no cis/trans isomerism in alkynes.

Draw all the pentynes, hexynes, and heptynes.

PROBLEM 4.20

Name all the isomers in Problem 4.20.

PROBLEM 4.21

4.11 TRIPLE BONDS IN RINGS

Like double bonds, triple bonds can occur in rings, although very small ring alkynes are not known. The difficulty comes from the angle strain induced by the preference for linear, 180° bond angles in an acetylene. When a triple bond is incorporated in a ring, it becomes difficult to accommodate these 180° angles. Deviation from 180° reduces the overlap between the *p* orbitals making up one of the π bonds of the acetylene and that raises the energy of the compound. The same problem exists for the alkenes, but it is less severe (Fig. 4.64).

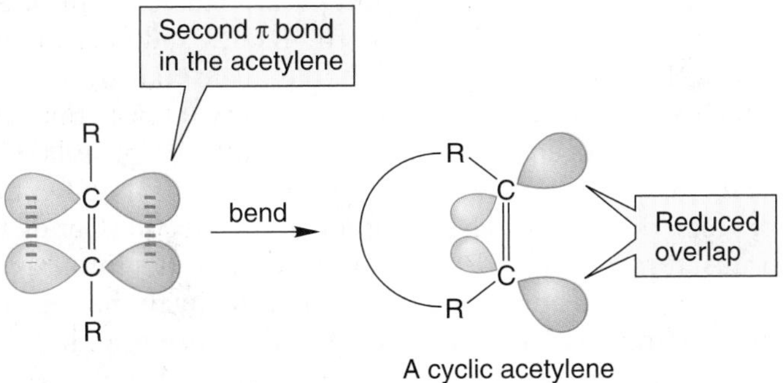

FIGURE 4.64 Incorporation of a triple bond in a ring is difficult because orbital overlap between the 2*p* orbitals of one of the π bonds is reduced by the bending required by the ring.

In practice, the smallest ring in which an alkyne is stable under normal conditions is cyclooctyne. Thus, it and larger cycloalkynes (cyclononyne, etc.) are known, but the smaller cycloalkynes are either unknown or have only been observed as fleeting intermediates (Fig. 4.65).

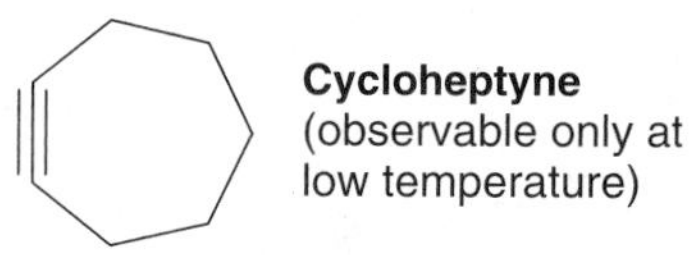

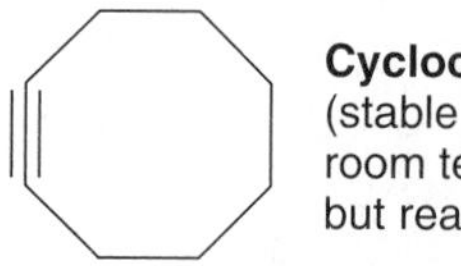

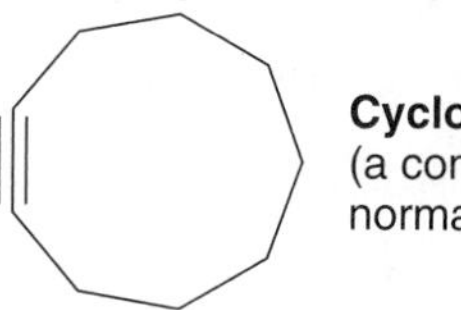

FIGURE 4.65 Some cycloalkynes. Cyclooctyne is the smallest unsubstituted cycloalkyne that is stable at room temperature.

4.12 PHYSICAL PROPERTIES OF ALKYNES

The physical properties of alkynes resemble those of alkenes and alkanes. Table 4.5 collects some data.

TABLE 4.5 Some Simple Alkynes

Name	Formula	mp (°C)	bp (°C)
Acetylene (ethyne)	$HC{\equiv}CH$	− 80.8	−84
Propyne	$HC{\equiv}C{-}CH_3$	−101.5	−23.2
1-Butyne	$HC{\equiv}C{-}CH_2CH_3$	−125.7	8.1
2-Butyne	$CH_3{-}C{\equiv}C{-}CH_3$	− 32.2	27
1-Pentyne	$HC{\equiv}C{-}(CH_2)_2CH_3$	− 90	40.2
1-Hexyne	$HC{\equiv}C{-}(CH_2)_3CH_3$	−131.9	71.3
1-Heptyne	$HC{\equiv}C{-}(CH_2)_4CH_3$	− 81	99.7
1-Octyne	$HC{\equiv}C{-}(CH_2)_5CH_3$	− 79.3	125.2

4.13 MOLECULAR FORMULAS AND DEGREES OF UNSATURATION

How is one to determine the possible structures for a molecule of a given molecular formula? For example, what are the possibilities for C_6H_{10}? Cyclohexene, 1-hexyne, and 1,3-hexadiene are only a few of them. Before we begin to write down the possibilities, we need to know what are the general possible structural types. In this case, cycloalkenes, acyclic alkynes, and acyclic dienes are among the structures whose formula is C_6H_{10}. Given a formula for an unknown hydrocarbon, we are faced with the problem of determining the number of π bonds and rings. Saturated alkanes have the formula C_nH_{2n+2}. Both alkenes and cycloalkanes have the formula C_nH_n. Such molecules are said to have "one degree of unsaturation," which simply means that they contain two fewer than the maximum number of hydrogens, as defined by the alkane formula, C_nH_{2n+2}. Alkynes and cycloalkenes have the formula C_nH_{2n-2}, and contain four fewer hydrogens than the C_nH_{2n+2} number. These compounds have two degrees of unsaturation. The **degree of unsaturation** (Ω) for a hydrocarbon is the total number of π bonds and rings, and can always be determined by calculating the number of hydrogens for the corresponding saturated alkane (H_{2n+2}), and then subtracting the number of hydrogens actually present and dividing by two. This procedure also works if hydrogens are replaced with other monovalent atoms such as the halogens. Such atoms are treated as if they were hydrogens. Later, when we see more complicated molecules containing nitrogen and oxygen, we will have to update this subject. Figure 4.66 works out some examples.

C_6H_{12} (C_nH_{2n})

For C_6H_{2n+2}, there would be 14 hydrogens.

Calculation:
14 – 12 = 2/2 = 1 degree of unsaturation,
which means that there must be one π bond or one ring.

C_9H_{16} (C_nH_{2n-2})

For C_9H_{2n+2}, there would be 20 hydrogens.

Calculation:
20 – 16 = 4/2 = 2 degrees of unsaturation,
which means that there must be a total of two π bonds and rings.

$C_{18}H_{30}$ (C_nH_{2n-6})

For $C_{18}H_{2n+2}$, there would be 38 hydrogens.

Calculation:
38 – 30 = 8/2 = 4 degrees of unsaturation,
which means that there must be a total of four π bonds and rings.

FIGURE 4.66 Some calculations of degrees of unsaturation.

PROBLEM 4.22

Calculate the degrees of unsaturation (Ω) for the following formulas. Give two possible structures for each one. (a) C_5H_6 (b) C_7H_8 (c) $C_{10}H_{10}$ (d) $C_5H_8Br_2$.

4.14 SOMETHING MORE: ACIDITY OF ALKYNES

It is generally most difficult for a base (B:⁻) to remove a proton from a hydrocarbon to give an anion. In other words, hydrocarbons are generally very weak acids. We have already had a brief discussion of the methyl anion in Chapter 3 (p. 81), but methane, the parent of this anion, is an extraordinarily weak acid (Fig. 4.67)

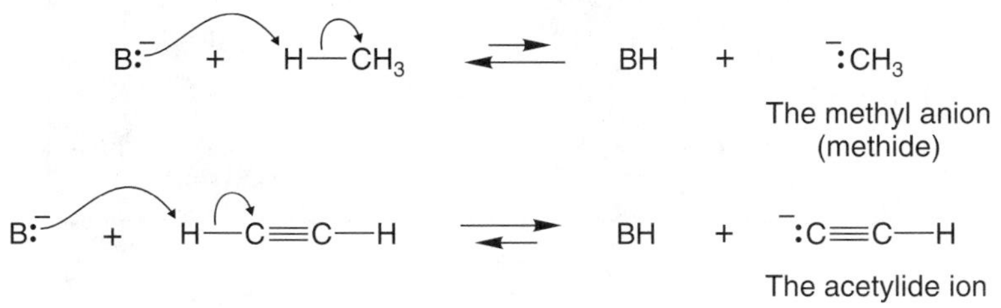

$$B:^- + H—CH_3 \rightleftharpoons BH + {}^-:CH_3$$

The methyl anion (methide)

$$B:^- + H—C{\equiv}C—H \rightleftharpoons BH + {}^-:C{\equiv}C—H$$

The acetylide ion

FIGURE 4.67 Strong bases (B:⁻) can remove the terminal hydrogen of an acetylene. Terminal acetylenes are reasonably strong acids for hydrocarbons.

A remarkable property of terminal acetylenes is that they are reasonably strong Brønsted acids (proton donors). This means that compared to most hydrocarbons it is easy to remove the terminal acetylenic hydrogen as a proton, leaving behind an anion, called an **acetylide**. Acetylenes are stronger acids than alkanes by roughly a factor of 10^{30}! Be sure you are clear on this point! Acetylenes are not strong acids in the sense that hydrochloric acid (HCl) and sulfuric acid (H_2SO_4) are, but they certainly are in comparison to most other hydrocarbons.

Why should acetylenes be relatively strong hydrocarbon acids? Consider the orbital containing the residual pair of electrons in the two cases of Figure 4.67. For the methyl anion, this orbital is hybridized approximately sp^3, with roughly 25% *s* character. In the acetylide, the corresponding orbital is *sp* hybridized, and therefore has 50% *s* character. An electron in an *s* orbital is substantially lower in energy than one in a *p* orbital, and so the more *s* character the orbital has, the lower in energy an electron in it is. The nonbonding electrons in the acetylide ion ($HC{\equiv}C{:}^-$) are in a much lower energy orbital than those in the nonbonding orbital of the methyl anion ($H_3C{:}^-$), also called methide (Fig. 4.68).

Methane
base, B^-
Methide
+ BH
C is hybridized approximately sp^3, therefore the electrons are in an orbital of about 25% *s* character
H—C≡C—H
Acetylene
base, B^-
Acetylide
+ BH
Now, C is hybridized approximately *sp*, and the electrons are in an orbital of approximately 50% *s* character

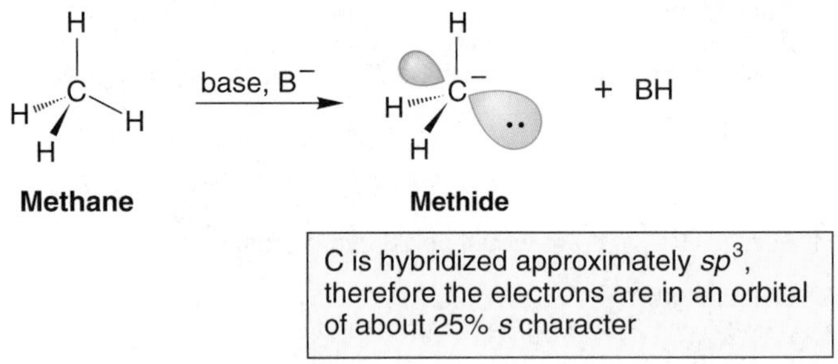

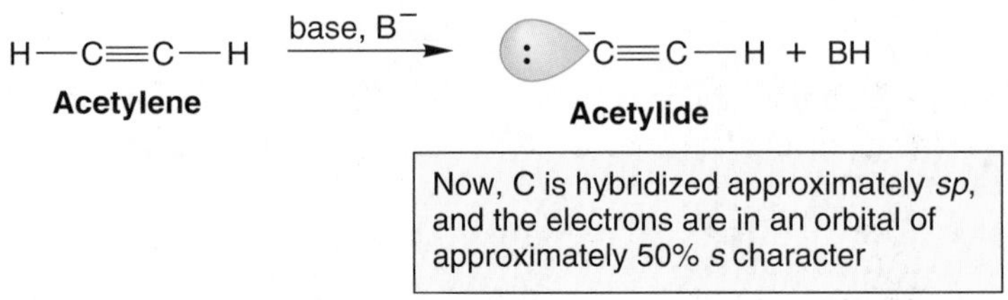

FIGURE 4.68 In an acetylide ion, the two nonbonding electrons are accommodated in an *sp* orbital. An *s* orbital is lower in energy than a *p* orbital, and so the more *s* character in a hybrid orbital the more stable an electron in it is.

If the hypothesis that the hybridization of the anion is critical is correct, then alkenes should be intermediate in acidity between alkanes and alkynes. This hypothesis is exactly right: alkenes are about 10^{10}–10^{12} *more* acidic than alkanes and 10^{18-20} *less* acidic than alkynes (Fig. 4.69).

Very weak acid
Intermediate
H—C≡C—H
Relatively strong acid (for a hydrocarbon)
Acidity

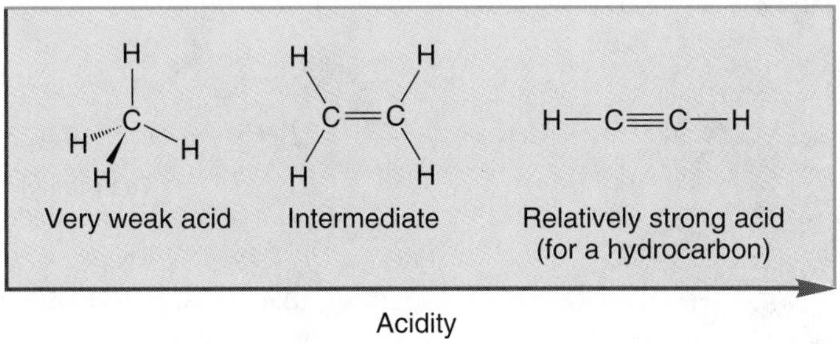

FIGURE 4.69 Alkenes are intermediate in acidity between alkanes and alkynes.

4.15 SUMMARY

NEW CONCEPTS

This chapter deals mainly with the structural consequences of sp^2 and *sp* hybridizations. The doubly bonded carbons in alkenes are hybridized sp^2 and the triply bound carbons in alkynes are hybridized *sp*. In these molecules, atoms are bound not only by the σ bonds we saw in Chapter 3 but by π bonds as well.

These π bonds, composed of overlapping 2*p* orbitals, have substantial impact upon the stereochemistry of the molecule. Alkenes can exist in cis (*Z*) or trans (*E*) forms—they have sides. Unlike the alkanes, which contain only σ bonds with very low barriers to rotation, there is a substantial barrier (~66 kcal/mol) to rotation about a π bond. The linear alkynes do not have cis/trans isomers and the question of rotation does not arise.

Double and triple bonds can be contained in rings, if the ring size is large enough to accommodate the strain incurred by the required bond angles. The smallest trans cycloalkene stable at room temperature is *trans*–cyclooctene. The smallest stable cycloalkyne is cyclooctyne.

More substituted alkenes and alkynes are more stable than their less substituted relatives.

REACTIONS, MECHANISMS, AND TOOLS

The Cahn–Ingold–Prelog priority system is introduced. It is used in the (*Z*/*E*) naming system. We will encounter it again very soon in Chapter 5.

In this chapter, we continue the idea of constructing larger hydrocarbons out of smaller ones by replacing the different available hydrogens with an X group. When X is methyl (CH_3), a larger hydrocarbon is produced from a smaller one.

The removal by a base of a hydrogen from the terminal position of acetylenes to give acetylides is mentioned. Terminal alkynes are moderately acidic molecules.

SYNTHESES

Acetylides can be formed from acetylenes.

$$CH_3{-}C{\equiv}C{-}H \ + \ \underset{\text{Base}}{B{:}^-} \longrightarrow CH_3{-}C{\equiv}C{:}^- \ + \ B{-}H$$

COMMON ERRORS

The concept of cis/trans (*Z*/*E*) isomerism is a continuing problem for students. Errors of omission (the failure to find a possible isomer, for example) and of commission (failure to recognize that a certain double bond, so easily drawn on paper, really cannot exist) are all too common. The idea begins with the notion that, unlike σ bonds, π bonds have substantial barriers to rotation (~66 kcal/mol). The barrier arises because of the shape of the *p* orbitals making up the π bond. Rotation decreases overlap and raises energy. Accordingly, double bonds have sides, which are by no means easily interchanged. Substituents cannot switch sides, and are locked in position by the high barrier to rotation. It is vital to see why there are two isomers of 2-butene, *cis*- and *trans*-2-butene, but only one isomer of 2-methyl-2-butene (Fig. 4.70).

This relatively easy idea has more complicated implications. For example, the need to maintain planarity (*p*–*p* overlap) in a π bond leads to the difficulty of accommodating a trans double bond in a small ring, or at the bridgehead position of bridged bicyclic molecules. In each case, the geometry of the molecule results in a twisted (poorly overlapping) pair of *p* orbitals.

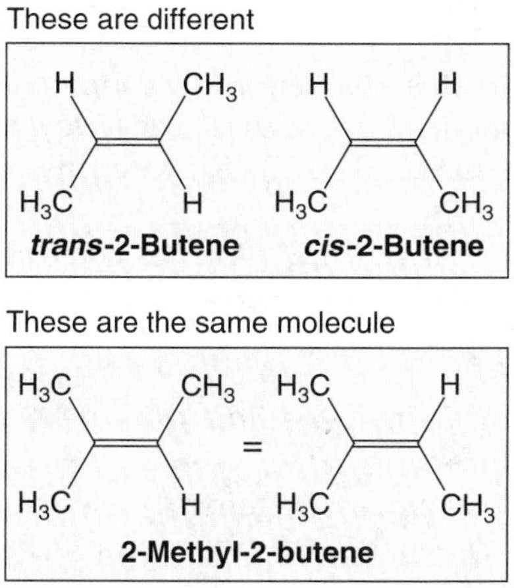

These are different

***trans*-2-Butene** ***cis*-2-Butene**

These are the same molecule

2-Methyl-2-butene

FIGURE 4.70 Both *cis*- and *trans*-2-butene exist, but there is only one isomer of 2-methyl-2-butene.

4.16 KEY TERMS

Acetylenes Hydrocarbons of the general formula C_nH_{2n-2}. These molecules, also called alkynes, contain carbon–carbon triple bonds. The parent compound, HC≡CH, is called acetylene, or ethyne.

Acetylide The anion formed by removal by base of a terminal hydrogen from an acetylene.

Alkenes Hydrocarbons of the general formula, C_nH_{2n}. These molecules, also called "olefins," contain carbon–carbon double bonds.

Alkynes Hydrocarbons of the general formula C_nH_{2n-2}. These molecules, also called acetylenes, contain carbon–carbon triple bonds.

Allyl The group $H_2C{=}CH{-}CH_2$.

Bredt's rule Bredt noticed that there were no examples of bicyclic molecules with double bonds at the bridgehead position.

Bridgehead position The atoms in a bicyclic system from which the bridges emanate.

Cahn–Ingold–Prelog priority system An arbitrary system for naming certain isomers of alkenes. It determines a priority system for ordering attached groups.

Cycloalkenes Ring compounds containing a double bond within the ring.

Degree of unsaturation (Ω) In a hydrocarbon, this is the total number of π bonds and rings.

Double bond Two atoms can be attached by a double bond composed of one σ bond and one π bond.

Ethene The simplest alkene is $H_2C{=}CH_2$. It is usually known as ethylene.

Ethylene The simplest alkene is $H_2C{=}CH_2$. It is more properly known as ethene, a name that is rarely used.

Heat of formation (ΔH_f°) The heat evolved or required for the formation of a molecule from its constituent elements in their standard states. The more negative the heat of formation, the more stable the molecule.

Olefins Hydrocarbons of the general formula, C_nH_{2n}. These molecules, also called alkenes, contain carbon–carbon double bonds.

π Orbitals Molecular orbitals made from the overlap of *p* orbitals. Such orbitals have a plane of symmetry.

Propargyl group The $H{-}C{\equiv}C{-}CH_2$ group.

***sp* Hybridization** The combination of the 2*s* and one 2*p* atomic orbital into hybrid orbitals.

***sp* Hybrids** Orbitals formed through the combination of the 2*s* and a single 2*p* orbital. The two hybrid orbitals are directed 180° from each other.

***sp*² Hybridization** The combination of the 2*s* and two 2*p* atomic orbitals into hybrid orbitals.

***sp*² Hybrids** Orbitals formed through the combination of the 2*s* and a pair of 2*p* atomic orbitals. They are directed toward the corners of an equilateral triangle.

Triple bond Two atoms can be attached by a triple bond composed of one σ bond and two π bonds.

Unsaturated Containing less than the maximum number of hydrogens. Alkenes and alkynes (as well as ring compounds) are unsaturated molecules.

Vinyl group The $H_2C{=}CH$ group.

4.17 ADDITIONAL PROBLEMS

In this chapter, we encounter the determination of molecular formulas. The following three review problems (4.23–4.25) deal with this subject.

PROBLEM **4.23** Alkanes combine with oxygen to produce carbon dioxide and water according to the following scheme:

$$C_nH_{2n+2} + (3n+1)/2\ O_2 \rightarrow (n+1)\ H_2O + n\ CO_2$$

This process is generally referred to as combustion. An important use of this reaction is the quantitative determination of elemental composition (elemental analysis). Typically, a small sample of the compound is completely burned and the water and carbon dioxide produced are collected and weighed. From the weight of water the amount of hydrogen in the compound can be determined. Similarly, the amount of carbon dioxide formed allows us to determine the amount of carbon in the original compound. Oxygen, if present, is usually determined by difference. The determination of the relative molar proportions of carbon and hydrogen in a compound is the first step in deriving its molecular formula.

If combustion of 5.00 mg of a hydrocarbon gives 16.90 mg of carbon dioxide and 3.46 mg of water, what are the weight percents of carbon and hydrogen in the sample?

PROBLEM **4.24** Calculate the weight percents for each element in the following compounds: (a) C_5H_{10} and (b) C_9H_6ClNO.

PROBLEM **4.25** A compound containing only carbon, hydrogen, and oxygen was found to contain 70.58% carbon and 5.92% hydrogen by weight. Calculate the empirical formula for this compound. If the compound has a molecular weight of approximately 135 g/mol, what is the molecular formula?

PROBLEM **4.26** How many degrees of unsaturation are there in compounds of the formula C_5H_8? Write at least eight isomers including ones with no π bonds and ones with no rings.

PROBLEM **4.27** Draw and write the IUPAC names for all of the isomers of 1-heptene (C_7H_{14}). There are 36 isomers, including 1-heptene.

PROBLEM **4.28** Draw and write the IUPAC names for all of the isomers of 1-octyne (C_8H_{14}). There are 32 isomers, including 1-octyne.

PROBLEM **4.29** Draw and write the IUPAC names for all of the isomers of dichlorobutene ($C_4H_6Cl_2$). There are 27 isomers, according to usually reliable sources.

PROBLEM **4.30** Write IUPAC names for the following compounds:

(a)

I

(b)

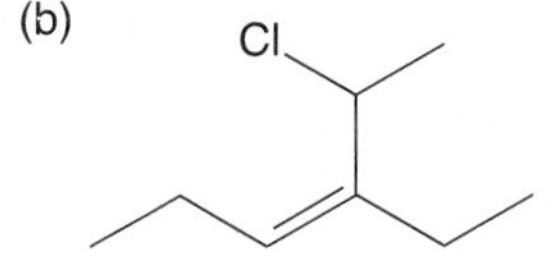

(c)

Br

(d)

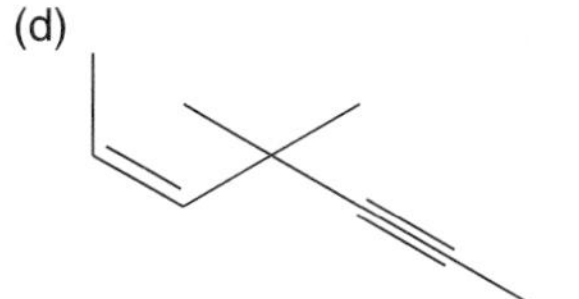

PROBLEM **4.31** Draw structures for the following compounds:

(a) *(Z)*-3-Fluoro-2-methyl-3-hexene
(b) 2,6-Diethyl-1,7-octadien-4-yne
(c) *trans*-3-Bromo-7-isopropyl-5-decene
(d) 5-Chloro-4-iodo-6-methyl-1-heptyne

PROBLEM **4.32** Although you should be able to draw structures for the following compounds, they are named incorrectly. Give the correct IUPAC name for each compound:

(a) 5-Chloro-2-methyl-6-vinyldecane
(b) 3-Isopropyl-1-pentyne
(c) 4-Methyl-6-hepten-1-yne
(d) (*E*)-3-Propyl-2,5-hexadiene

PROBLEM **4.33** Find all the compounds of the formula $C_4H_6Br_2$ in which the dipole moment is zero. It is easiest to divide this long problem into sections. First, work out the number of degrees of unsaturation so that you can see what kinds of molecules to look for. Then find the acyclic molecules. Next work out the compounds containing a ring.

PROBLEM **4.34** Find all the compounds of the formula C_4H_6 in which there are four different carbon atoms. There are only nine compounds possible, but two of them are quite strange at this point. *Hint*: One of these molecules contains a carbon that is part of two double bonds, and the other is a compound containing only three-membered rings.

PROBLEM **4.35** Find all the isomers of the formula C_4H_5Cl. As in Problem 4.33, it is easiest to divide this problem into sections. First, find the degrees of unsaturation so that you know what types of molecule to examine. There are 23 isomers to find. *Hint*: See the hint for the previous problem.

PROBLEM **4.36** Find the compound of the formula $C_{10}H_{16}$, composed of two six-membered rings, that has only three different kinds of carbon atom.

PROBLEM **4.37** Develop a bonding scheme for the compound $H_2C{=}O$ in which both carbon and oxygen are hybridized sp^2.

PROBLEM **4.38** (a) Develop a bonding scheme for a ring made up of six CH groups, $(CH)_6$. Each carbon is hybridized sp^2. (b) This part is harder. The molecule you built in (a) exists, but all the carbon–carbon bond lengths are equal. Is this consistent with the structure you wrote in (a), which almost certainly requires two different carbon–carbon bond lengths? How would you resolve the problem?

PROBLEM 4.39 In this chapter, we developed a picture of the methyl cation ($^+CH_3$), a planar species in which carbon is hybridized sp^2. In Chapter 3, we briefly saw compounds formed from several rings, bicyclic compounds. Explain why it is possible to form carbocation (a) in the bicyclic compound shown below, but very difficult to form carbocation (b). It may be helpful to use models to see the geometries of these molecules.

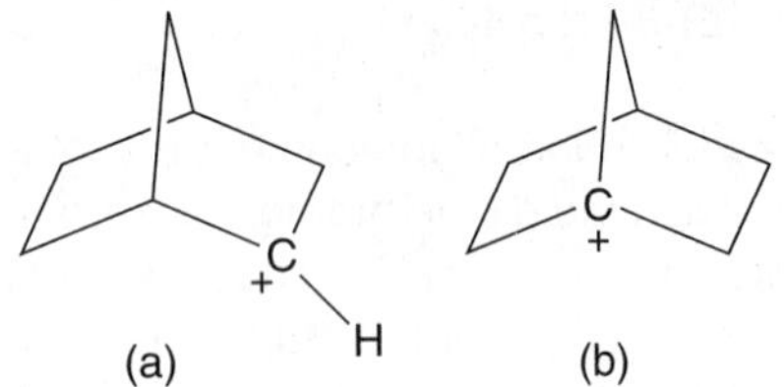

Stereochemistry

I held them in every light. I turned them in every attitude. I surveyed their characteristics. I dwelt upon their peculiarities. I pondered upon their conformation.

—Edgar Allan Poe*
Berenice

In Chapter 3 (p. 107), we found a strange phenomenon. There seemed to be an extra isomer of *trans*-1,2-dimethylcyclopropane. Now we will discover why two *trans*-1,2-dimethylcyclopropanes exist, which leads into a detailed discussion of **stereochemistry**, the structural and chemical consequences of the arrangement of atoms in space. We have already seen several examples of stereoisomeric molecules. Substituents on both ring compounds and alkenes can be attached in cis [or (Z) = same side] and trans [or (E) = opposite side] arrangements (Fig. 5.1).

***cis*-1,2-Dimethylcyclopropane**

Two representations of *trans*-1,2-dimethylcyclopropane

***cis*-2-Butene [or (*Z*)-2-butene]**

***trans*-2-Butene [or (*E*)-2-butene]**

FIGURE 5.1 *cis*- and *trans*-1,2-Dimethylcyclopropane and *cis*- and *trans*-2-butene are pairs of stereoisomeric molecules.

In this chapter, we will examine more subtle questions of stereoisomerism. Most molecules of Nature are **chiral**. The word "chiral" is derived from the Greek and means "handed." Chiral molecules are related to their mirror images in the same way that your left hand is related to your

*Edgar Allan Poe (1809–1849), was an American poet and writer probably most famous for his macabre stories and poems. As the quotation shows, he was also an early devotee of stereochemistry.

right. They are not identical! Indeed many molecules of Nature (for example, the amino acids) are handed in the same way: One might say that they are all left handed. The source of the ubiquity of left handedness is one of the great questions left in chemistry. Was it chance? Did an accidental left-handed start determine all that followed? When we encounter extraterrestrial civilizations will their amino acids (assuming they have any) also be left handed or will we find right-handed amino acids? As it now appears we will find no cohabitants within our own solar system, and as physical contact with our galactic neighbors is unlikely, you might consider how you would convey to a resident of the planet Altair 4 our concepts of left and right.

This topic is a complicated but important one that is absolutely essential to organic and biological chemistry. When we come to an examination of reaction mechanisms—of the way molecules come together to make other compounds—we will often use chiral molecules to provide the level of detail necessary to work out how reactions occur. Moreover, the molecules of Nature are largely chiral, and this first look at chirality provides a base for virtually all of biochemistry and molecular biology.

You are very strongly urged to work through this chapter with models. It is a rare person who can see easily in three dimensions without a great deal of practice.

5.1 CHIRALITY

The phenomenon of handedness, or **chirality**, actually surfaces long before a molecule as complicated as *trans*-1,2-dimethylcyclopropane. We went right past it when we wrote out the heptanes in Figure 3.52 (Chapter 3, p. 103). One of these heptanes, 3-methylhexane, exists in two forms. To see this, draw out the molecule in tetrahedral form using C(3) as the center of the tetrahedron. Next, draw the mirror image of this molecule. Contrast these structures with 3-methylpentane treated the same way, with C(3) still the center of the tetrahedron (Fig. 5.2).

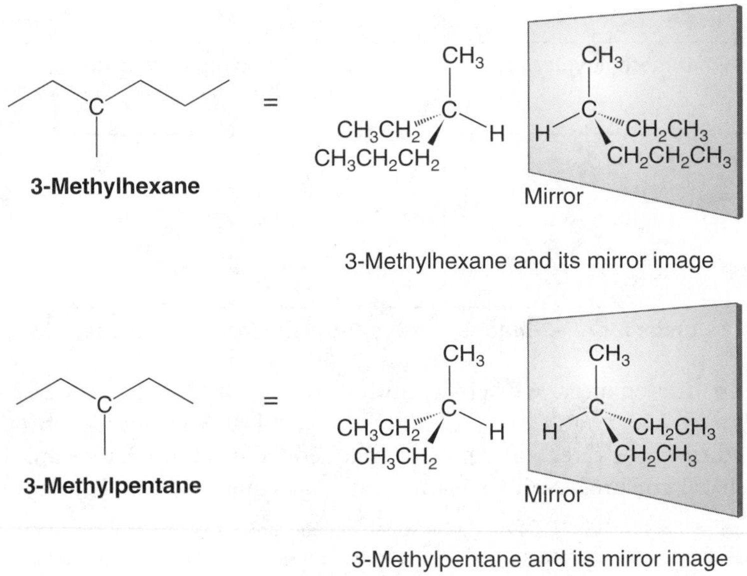

3-Methylhexane

3-Methylhexane and its mirror image

3-Methylpentane

3-Methylpentane and its mirror image

FIGURE 5.2 3-Methylhexane and 3-methylpentane reflected to show their mirror images.

Now see if the reflections are the same as the originals. For 3-methylpentane, one need only rotate the molecule 180° and slide it over to superimpose the two mirror images (Fig. 5.3). The molecule and its mirror image are clearly identical, and hence, not chiral. However, Nature seems to lay a trap for us as there are many motions of the mirror image that will *not* directly generate a molecule superimposable on the original. In such cases, we may be led to think we have uncovered chirality. For example, try rotating the molecule as shown at the bottom of Figure 5.3. The original and the newly rotated molecule are not simply superimposable. But chiral? No, for as long as there is one motion (Fig. 5.3, top) that does generate the mirror image, the molecule is **achiral** (not chiral).*

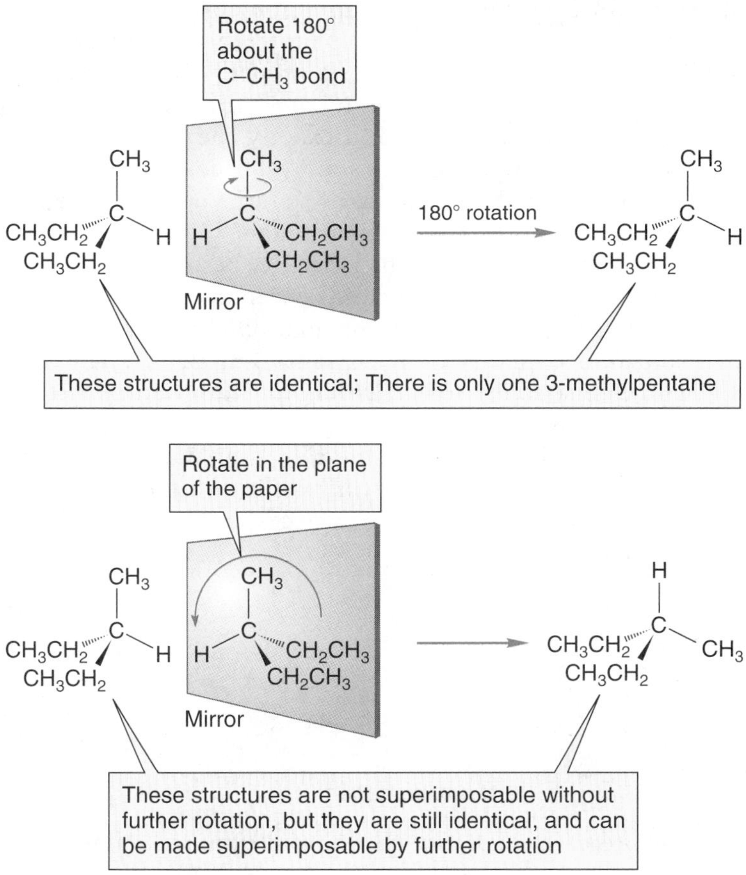

FIGURE **5.3** 3-Methylpentane and its mirror image are identical. Rotate the mirror image at the top of the page 180° about the carbon–methyl bond to see this. Although an unproductive motion tells us nothing about the chirality of the molecule, a motion that produces a superimposable mirror image tells us that the molecule is not chiral (achiral).

Now consider Figure 5.4. Here we apply to 3-methylhexane the same rotation that we applied before to 3-methylpentane. In this case, we cannot superimpose the original onto the newly rotated molecule no matter how many ways we try. The propyl group in one form winds up where the ethyl group is in the other. Indeed, no amount of twisting and turning can do the job: The two are irrevocably different.

*There is a language problem introduced by the choice of Greek for the word (chiral) to describe the phenomenon of handedness of molecules. In Greek, the word for "not chiral" is "achiral." This poses no problem in Greek, but, at least in spoken English it does. It is often difficult to distinguish between "Compound X is a chiral molecule" and "Compound X is achiral."

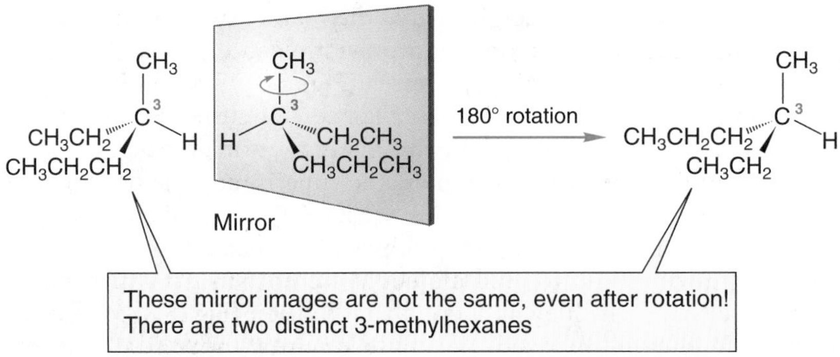

FIGURE 5.4 3-Methylhexane and its mirror image are not identical.

Models are essential now, as you must be absolutely convinced of this point, and certain of the difference between 3-methylpentane and 3-methylhexane. The more symmetric 3-methylpentane is achiral, whereas 3-methylhexane and your hand are chiral. The two stereoisomers of 3-methylhexane are related in exactly the same way your two hands are. A chiral compound can simply be defined as a molecule whose mirror image is not superimposable on the original. The two versions of 3-methylhexane are called **enantiomers**.

Figure 5.5 shows schematically the difference between 3-methylhexane and 3-methylpentane. The difference in symmetry is obvious in these color-coded pictures. In 3-methylhexane, four different groups surround a central carbon atom, which is not the case for 3-methylpentane. Our next job is to see how these stereoisomers differ physically and chemically, and to explore the circumstances under which chirality will appear. What structural features will suffice to ensure chirality? Will, for example, the phenomenon of four different groups surrounding a carbon be a sufficient condition to ensure chirality? Will it be a necessary condition? This chapter discusses such questions.

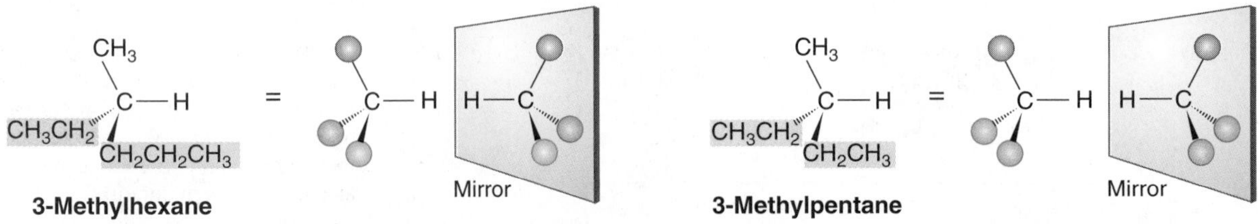

FIGURE 5.5 Color-coded schematic drawings of 3-methylhexane and 3-methylpentane and their mirror images.

PROBLEM 5.1 Are the following molecules chiral? (a) Bromochlorofluoroiodomethane, (b) dibromochlorofluoromethane, (c) 3-methylheptane, (d) 4-methylheptane.

5.2 PROPERTIES OF ENANTIOMERS: PHYSICAL DIFFERENCES

Almost all physical properties of enantiomers are identical. Their melting points, boiling points, densities, and many other physical properties do not serve to distinguish the two. But they do differ physically in one somewhat obscure way: They rotate the plane of plane-polarized light to an equal degree but in *opposite* directions. Rotation of the plane of polarization is called **optical activity** and molecules causing such a rotation are optically active. The enantiomer that rotates the plane in a clockwise direction (as one faces a beam of light passing through the sample) is called **dextrorotatory**. If the rotation is counterclockwise, it is **levorotatory**. The direction of rotation is indicated by placing a (+) for dextrorotatory, or (–) for levorotatory, before the name (Fig. 5.6).

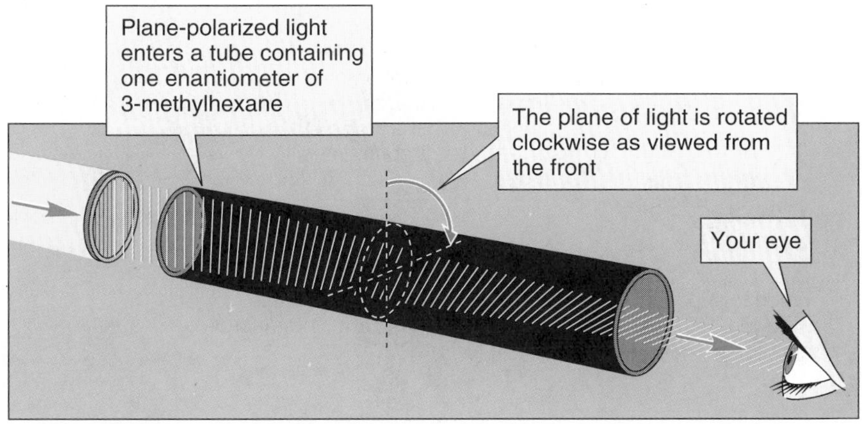
Plane-polarized light enters a tube containing one enantiometer of 3-methylhexane
The plane of light is rotated clockwise as viewed from the front
Your eye

FIGURE 5.6 The (+)-enantiomer rotates the plane of plane-polarized light in a clockwise direction. The (–)-enantiomer—not shown here—will rotate the light an equal amount in the counterclockwise direction.

Very often, equal amounts of the dextrorotatory and levorotatory stereoisomers are present. In such a case, no rotation can be observed as there are equal numbers of molecules rotating the plane in clockwise and counterclockwise directions, and the effects cancel. Such a mixture is called a **racemate**, or **racemic mixture**. The presence of a racemic mixture of enantiomers is often indicated by placing a (±) before the name as in Figure 5.7, which brings up a technical problem.

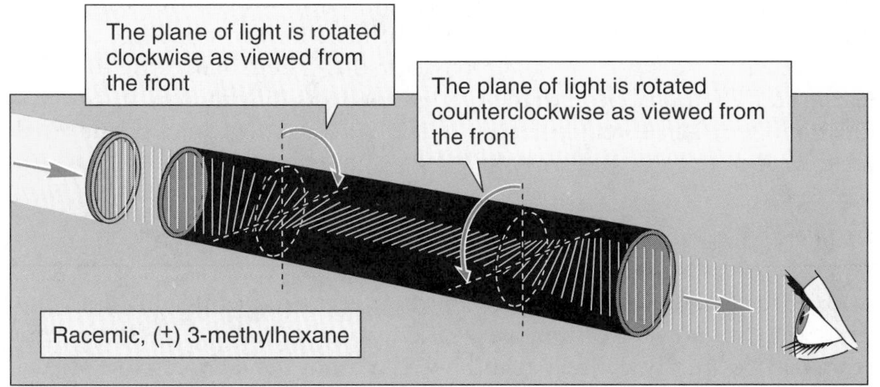
The plane of light is rotated clockwise as viewed from the front
The plane of light is rotated counterclockwise as viewed from the front
Racemic, (±) 3-methylhexane

FIGURE 5.7 There is no net rotation by a racemic mixture of enantiomers. One isomer will rotate the plane of plane-polarized light to the right and the other an equal amount to the left.

When we are talking about a chiral molecule, how do we know whether the subject of our colloquy is a single enantiomer or the racemic mixture? Often the text will tell you by indicating (+), (–), or (±). If no indication is made, you are almost certainly talking about the racemic mixture. Drawings present other problems. If the subject is a single enantiomer, the drawing will certainly point this out. Sometimes an asterisk (*) is used to indicate optical activity.

Convention Alert!

There are times, however, when a three-dimensional representation is important, but the subject is a 50:50 racemic mixture of the two enantiomers rather than a single enantiomer. Unless both enantiomers are drawn, the figure appears to focus only on a single enantiomer. Our molecule, 3-methylhexane, illustrates this in Figure 5.8. If we want to show the three-dimensional structure of this molecule we must draw one of the enantiomers. To be strictly correct, both enantiomers should be drawn in any discussion of this molecule unless it really is a single enantiomer that is under discussion. In practice this is rarely done, and you must be alert for the problem. Unless optical activity is specifically indicated, it is usually the racemate that is meant.

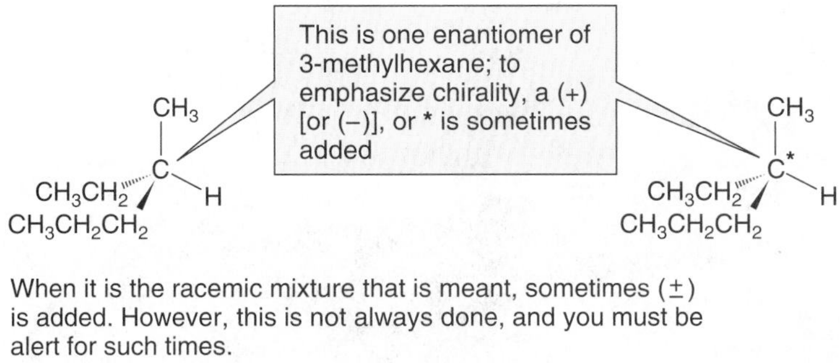

This is one enantiomer of 3-methylhexane; to emphasize chirality, a (+) [or (−)], or * is sometimes added

CH_3 C H CH_3CH_2 $CH_3CH_2CH_2$

CH_3 C* H CH_3CH_2 $CH_3CH_2CH_2$

When it is the racemic mixture that is meant, sometimes (±) is added. However, this is not always done, and you must be alert for such times.

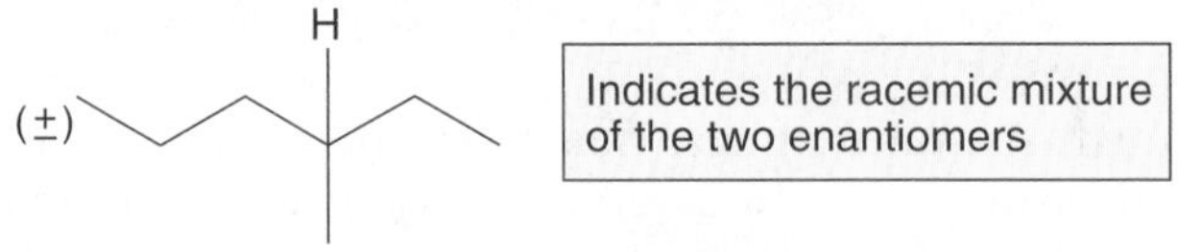

(±) H

Indicates the racemic mixture of the two enantiomers

FIGURE 5.8 A variety of conventions is used to indicate the presence of a single enantiomer.

Clearly, we desperately need a way to identify which enantiomer we mean when we talk to each other. In practice, we cannot take the time to describe in detail the orientation in space of the groups surrounding each atom in a molecule. We need a convention in order to specify the **absolute configuration**, and naturally one has been devised. Conventions are sometimes necessary but are almost always difficult to learn, because by their very nature there is no way to reason them out. There is nothing to do but learn them. You will be rewarded for your efforts by a facility in communication that allows you to specify a detailed structure quickly and (with luck) easily.

5.3 THE (*R*/*S*) CONVENTION

Convention Alert!

In order to specify the absolute configuration of a molecule, one first identifies the **stereogenic center**, very often a carbon atom.† A stereogenic atom can be simply defined as follows: An atom (usually carbon) of such nature and bearing groups of such nature that it can have two nonequivalent configurations.

In our example of 3-methylhexane, the stereogenic carbon, C(3), is the one upon which we have based the tetrahedral structures of Figures 5.2–5.5. Next, the four groups (atoms or groups of atoms) attached to the stereogenic carbon are identified and given priorities according to the Cahn–Ingold–Prelog scheme we first met in Chapter 4 (p. 130) when we discussed cis/trans or (*Z*/*E*) isomerism. The system is applied here so as to

†The word "stereogenic," though officially suggested in 1953, came into general use only very recently when it was urged by Princeton professor Kurt M. Mislow (b. 1923) as a replacement for a plethora of unsatisfactory terms such as "chiral carbon," "asymmetric carbon (or atom)," and many others. These old terms are slowly disappearing, but will persist in the literature (and textbooks) for many more years.

be able to identify any stereogenic atom as either (*R*) or (*S*) from a consideration of the priorities of the attached groups (**1** = high, **4** = low). The application of the Cahn–Ingold–Prelog system is a bit more complicated than that for determining (*Z*) or (*E*) for alkenes, so the system will be summarized again here.

The atom of lowest atomic number is given the lowest priority number, **4**. In 3-methylhexane, this atom is hydrogen (Fig. 5.9). In some molecules, one can give priority numbers to all the atoms by simply ordering them by atomic number. Thus, bromochlorofluoromethane is easy: H is **4**, F is **3**, Cl is **2**, and Br is **1**. The priorities are assigned in order of increasing atomic number so that the highest priority, **1**, is given to the atom of highest atomic number, in this case bromine (Fig. 5.9).

A subrule is illustrated by another easy example, 1-deuterioethyl alcohol (Fig. 5.10). We break the tie between the isotopes H and D by assigning the lower priority number to the atom of lower mass, H. So H is **4**; D is **3**; C is **2**; and O, with the highest atomic number, is priority **1**.

There can be more difficult ties to break, however. The molecule *sec*-butyl alcohol (2-hydroxybutane) illustrates this. The lowest priority, **4**, is H and the highest priority, **1**, is O. But how do we choose between the two carbons? The tie is broken by working outward from the tied atoms until a difference is found. In this example, it's easy. The methyl carbon is attached to three hydrogens and the methylene carbon to two hydrogens and a carbon. The methyl carbon gets the lower priority number **3** by virtue of the lower atomic numbers of the atoms attached to it (H,H,H = 1,1,1) and the ethyl carbon (attached to C,H,H = 6,1,1) gets the higher priority number, **2** (Fig. 5.11).

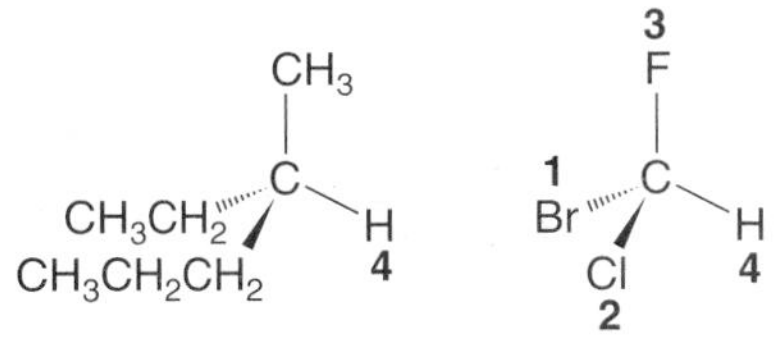

FIGURE 5.9 The atom of lowest atomic number, often hydrogen, is given the lowest priority, **4**. The remaining priorities are assigned in order of increasing atomic number.

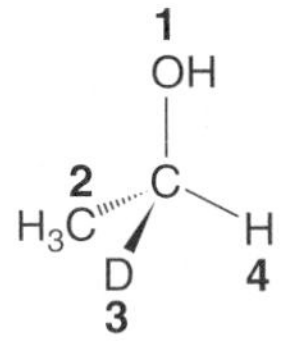

FIGURE 5.10 When the atomic numbers are equal, the atomic weights may be used to break the tie.

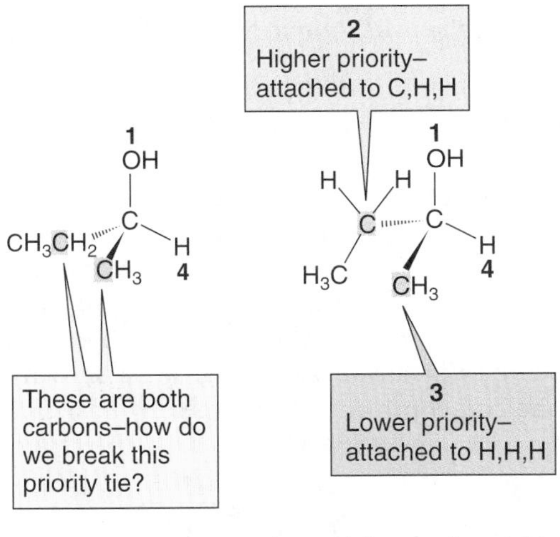

FIGURE 5.11 When both the atomic number and the atomic weights are equal, one looks at the atoms to which the tied atoms are attached and determines their atomic numbers.

Sometimes there will be even more difficult ties to break, and another look at 3-methylhexane illustrates this well. It is easy to identify H as the atom with the lowest atomic number, but the other three atoms are all the same; they are carbons. How do we break this tie? We work outward from the identical atoms until a difference appears. In 3-methylhexane, the methyl group is attached to three hydrogens, but each methylene group is attached to two hydrogens and one carbon. As H is of lower atomic number than C, the methyl group gets priority number **3** (Fig. 5.12).

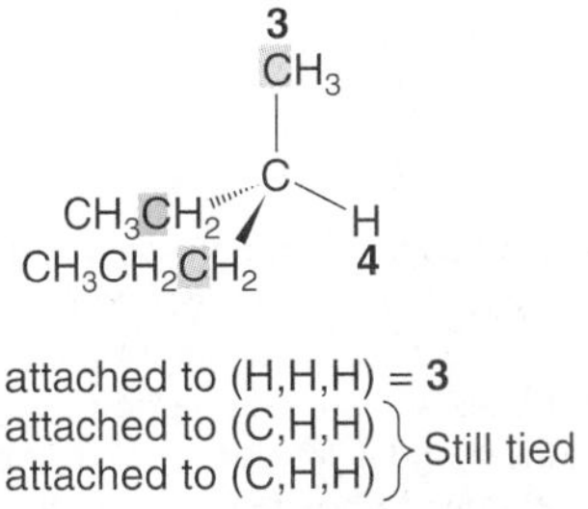

FIGURE 5.12 An example of a complicated tie-breaking procedure. In this case, we start by looking at the atoms attached to the three tied carbons. This procedure does not completely resolve the tie.

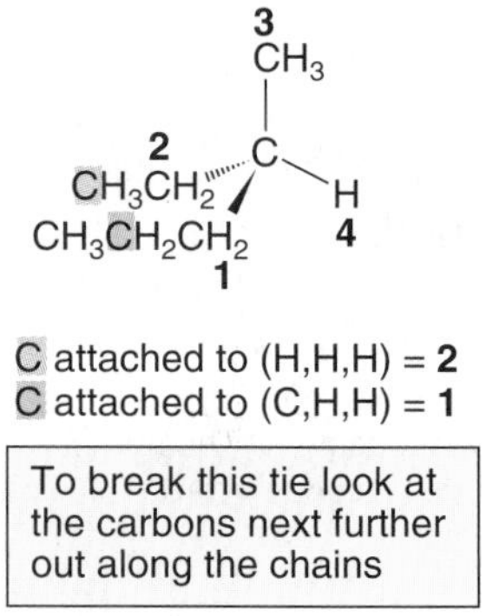

FIGURE 5.13 The final assignment of priority numbers in 3-methylhexane.

Now we are faced with yet another tie, and must find a way to distinguish the pair of methylene groups. Again we work outward. Look at the carbons attached to the two tied methylenes. One is a methyl carbon and is attached to three hydrogens. The other is a methylene and is attached to two hydrogens and a carbon. The methylene attached to the methyl gets priority **2** and the other methylene is left with the highest priority, **1** (Fig. 5.13). In practice, this detailed bookkeeping is often unnecessary.

In this example, it is really quite easy to see that the smaller ethyl group is of lower priority than the larger propyl group, and thus its carbon will get the lower priority number. More difficult examples can be devised, however!

PROBLEM 5.2 Identify the priorities of the compounds in Figure 5.14.

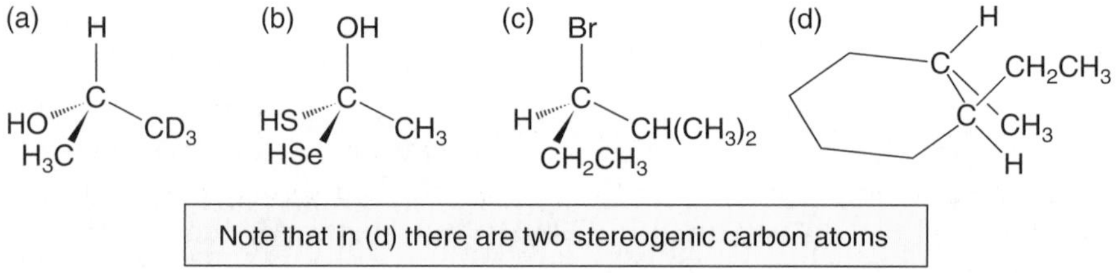

FIGURE 5.14

PROBLEM 5.3 Draw the mirror images of the chiral compounds (a–d) in Figure 5.14.

How do we deal with multiple bonds? Carbon is often doubly or triply bonded to another atom. Carbon–carbon double bonds, carbon–carbon triple bonds, and carbon–oxygen double bonds are treated as shown in Figure 5.15 (recall Fig. 4.33, p. 132). In the example in Figure 5.15, hydrogen will obviously get the lowest priority, **4**, and the bromine the highest priority, **1**. But both the other substituents on the stereogenic carbon are carbons. The tie is resolved by treating the carbon atom in the carbon–oxygen double bond as being attached to two oxygens as shown in Figure 5.15. Therefore it gets the higher priority, **2**, and the carbon attached to only one oxygen gets the remaining priority, **3**.

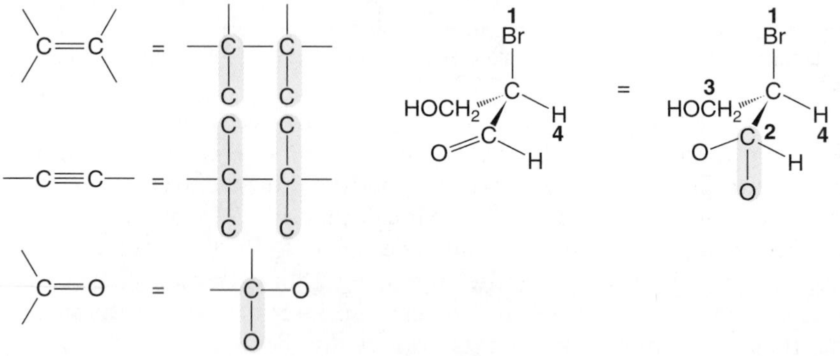

FIGURE 5.15 Carbon–carbon double bonds and carbon–carbon triple bonds are elaborated by adding new carbon bonds as shown. The carbon–oxygen double bond is treated by adding a new bond from carbon to oxygen. Thus, the carbon is treated as being attached to one hydrogen and two oxygens and gets a higher priority number than the carbon that is attached to two hydrogens and one oxygen.

Once priority numbers have been established, we imagine looking-down the bond from the stereogenic atom (usually carbon) toward the atom of lowest priority (**4**, often H). The other three substituents (**1**, **2**, **3**) will be facing you. Connect these three with an arrow running from highest to lowest priority number (**1**→**2**→**3**). If this arrow runs clockwise, the enantiomer is called (*R*) (Latin: *rectus*, "right"); if it runs counterclockwise it is called (*S*) (Latin: *sinister*, "left") (Fig. 5.16). The label (*R*) or (*S*) is added to the name as an italic prefix in parentheses. Figure 5.17 shows the (*R*) and (*S*) forms of some examples from the previous paragraphs.

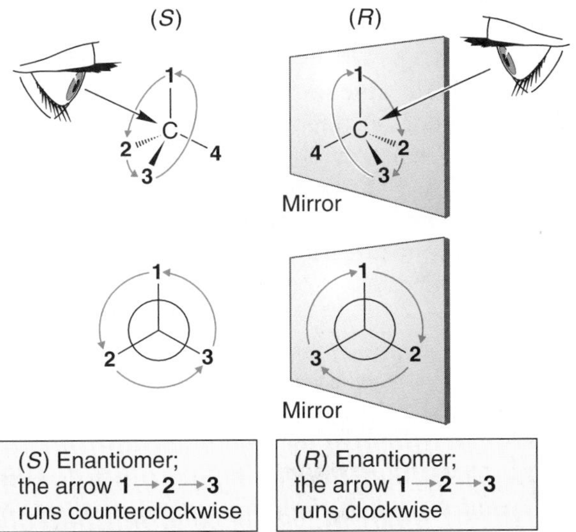

FIGURE 5.16 The convention for naming enantiomers as (*S*) or (*R*).

Note that your eye must look from the stereogenic carbon toward priority **4**; if you look in the other direction, or draw the arrow **3**→**2**→**1**, you will get the convention backward!

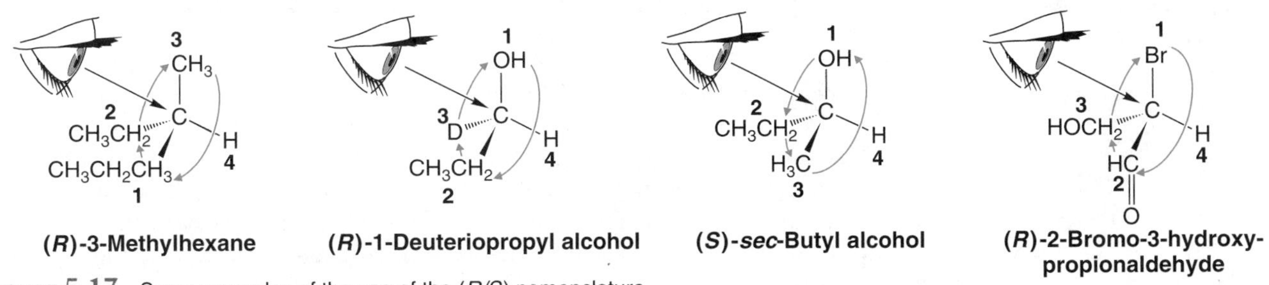

FIGURE 5.17 Some examples of the use of the (*R/S*) nomenclature.

PROBLEM 5.4 Designate the stereogenic carbons of the compounds in Problem 5.2 as either (*R*) or (*S*).

PROBLEM 5.5 Draw out the enantiomers of the molecules in Figure 5.17.

This convention looks a bit complicated. It is easier than it appears right now, but it just must be learned, and cannot be reasoned out. It's quite arbitrary at a number of points. For the price of learning this convention we get the return of being able to specify great detail most economically. Time and again we will use the subtle differences between enantiomers to test ideas about reaction mechanisms—of how chemical reactions occur. The study of the effects of stereochemistry on reaction mechanisms has been, and remains, one of the most powerful tools used by chemists to unravel how Nature works. Practice using the (*R*/*S*) convention and it will become easy. It's well worth it—indeed, it's essential.

5.4. THE PHYSICAL BASIS OF OPTICAL ACTIVITY

We have covered the way in which enantiomers differ physically. Enantiomers rotate the plane of plane-polarized light in opposite directions. Before we go on to talk about how enantiomers differ in their chemical reactivity, let's take a little time to say something about the phenomenon of optical activity. How does a collection of atoms (a molecule) interact with a wave phenomenon (light), and how does the property of chirality (handedness) induce the observed rotation?

Light consists of electromagnetic waves vibrating in all possible directions perpendicular to the direction of propagation (Fig. 5.18). Passing ordinary light through a Nicol prism (or a piece of Polaroid film) eliminates all

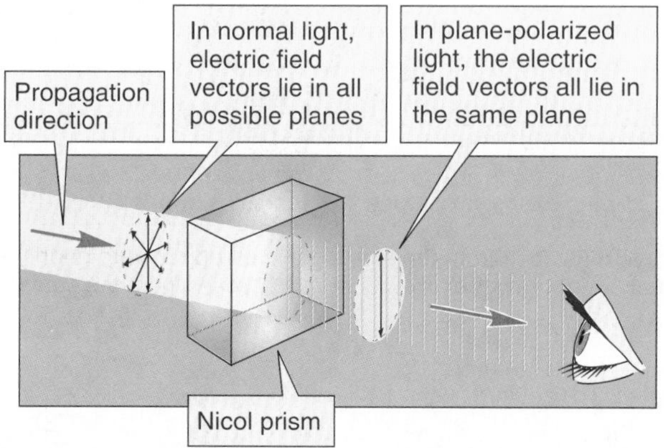

FIGURE 5.18 A Nicol prism transmits only plane-polarized light.

waves but those whose electric vectors are vibrating in a single plane. The classic experiment to demonstrate this uses two pairs of Polaroid sunglasses. If we observe light through one pair, all is still visible, although the light has been plane polarized. The second pair of glasses only transmits light when oriented in the same way as the first pair. If it is turned 90° to the first pair none of the plane-polarized light can pass, and darkness results (Fig. 5.19).

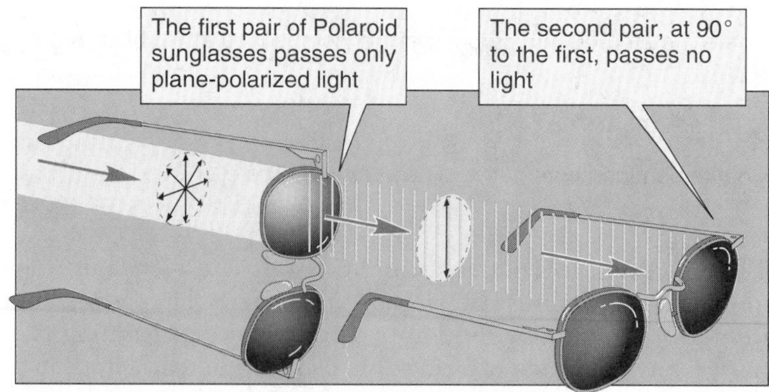

FIGURE 5.19 A classic experiment using Polaroid sunglasses: (a) The first pair of sunglasses acts as a Nicol prism and transmits only plane-polarized light. (b) If the second pair is held in the same orientation, light will still be transmitted. If it is held at 90° to the first pair, no light can be transmitted.

The electrons in a molecule, moving charged particles, generate their own electric fields, which interact with the field of the light. As we have already seen, many molecules are not highly symmetric, and the interactions of their unsymmetric electric fields with that of the plane-polarized light act to change the electric and magnetic fields. Thus, the plane of polarization is rotated slightly. For achiral molecules this rotation is canceled by an equal and opposite rotation induced by a molecule in the mirror-image arrangement (Fig. 5.20).

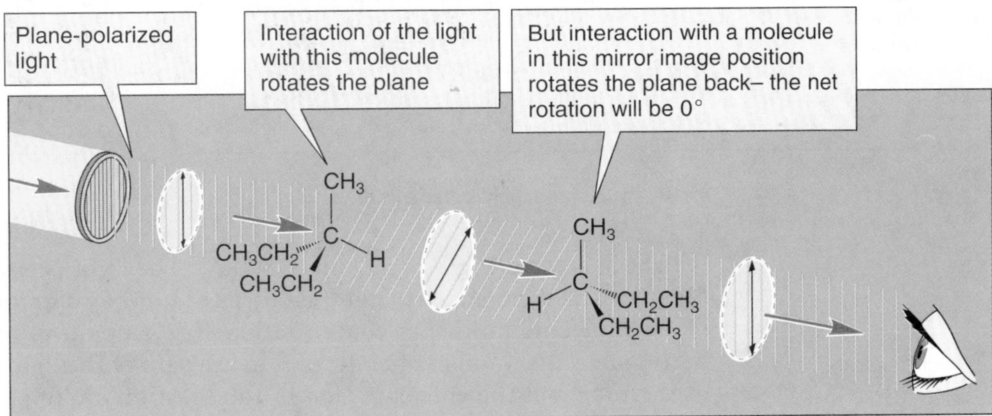

FIGURE 5.20 Achiral molecules such as 3-methylpentane do not give a net rotation of plane-polarized light. Equivalent molecules exist in mirror-image positions, and thus the net rotation must be 0°.

The result is no net rotation of the plane. However, when the interaction is with a chiral molecule, *there are no compensating mirror-image molecules* (Fig. 5.21). There must be a net change in the plane of polarization, and this is generally large enough to be observed by a device called a

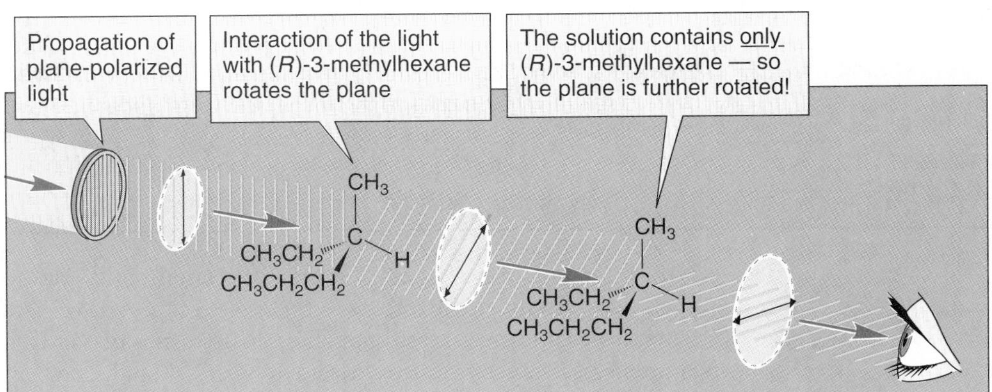

FIGURE 5.21 Optically active molecules such as (*R*)-3-methylhexane will rotate the plane of plane-polarized light. There are no compensating mirror-image molecules for this kind of molecule.

polarimeter (Fig. 5.22). The amount of rotation varies with the wavelength of light, so monochromatic light must be used to produce results easily reproduced in other laboratories. Generally, the sodium D line is used, which

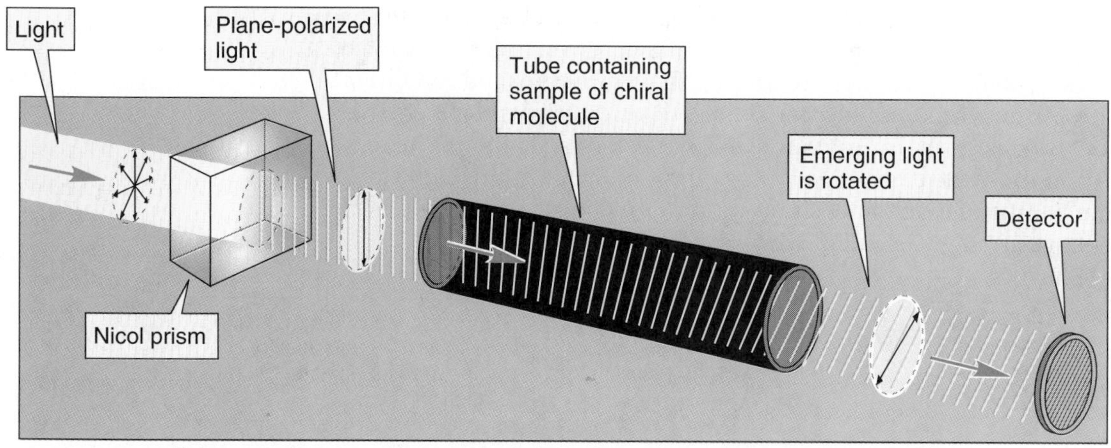

FIGURE 5.22 A highly schematic drawing of a polarimeter.

is light of a wavelength of 589 nm (5890 Å). The light is passed through a Nicol prism to polarize it, and then through a tube containing a solution of the molecule in known concentration. The amount of rotation is proportional to the number of molecules in the path of the light, so the length of the tube and the concentration of the solution are important. A term called the specific rotation, [α], which is the rotation induced by 1-g/mL solution 10 cm long, is reported. The specific rotation [α] is related to the observed rotation α in the following way:

$$[\alpha] = \alpha/cl, \text{ where } c = \text{concentration in grams per milliliter (g/mL), and } l = \text{length of the tube in decimeters (dm)}$$

Sub- and superscripts are used to denote the wavelength and temperature of the measurement. Thus, $[\alpha]_D^{20}$ means the sodium D line was used and the measurement was made at 20 °C.

PROBLEM 5.6 Create a scheme like those in Figures 5.20 and 5.21 to show what happens when plane-polarized light is passed through a solution of racemic 3-methylhexane.

5.5 PROPERTIES OF ENANTIOMERS: CHEMICAL DIFFERENCES

Now let's look at how enantiomers differ chemically. We still can't discuss specific chemical reactions, but we can make some important general distinctions. Note that there is no difference in how the two enantiomers of 3-methylhexane interact with a "Spaldeen," an old-fashioned smooth red rubber ball (Fig. 5.23). Your left and right hands interact in

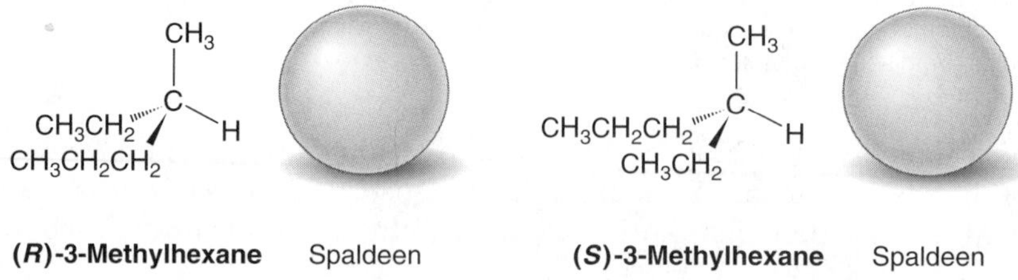

CH_3, CH_3CH_2, $CH_3CH_2CH_2$, C, H — **(R)-3-Methylhexane** Spaldeen

CH_3, $CH_3CH_2CH_2$, CH_3CH_2, C, H — **(S)-3-Methylhexane** Spaldeen

FIGURE 5.23 There is no difference in how (R)- and (S)-3-methylhexane interacts with the featureless smooth rubber ball, once called a "Spaldeen."

*PROBLEM 5.7

Actually, this is only true if the name has been worn off by too much stoop-ball playing. Show that the two enantiomers interact differently with a Spaldeen with the name "Spalding" still visible.

ANSWER

In the approach shown, (*R*)-3-methylhexane interacts with the new Spaldeen to place the ethyl group near the "S" and the propyl group near the "G." These interactions are opposite in the (*S*) enantiomer; the propyl group approaches the "S" and the ethyl group approaches the "G."

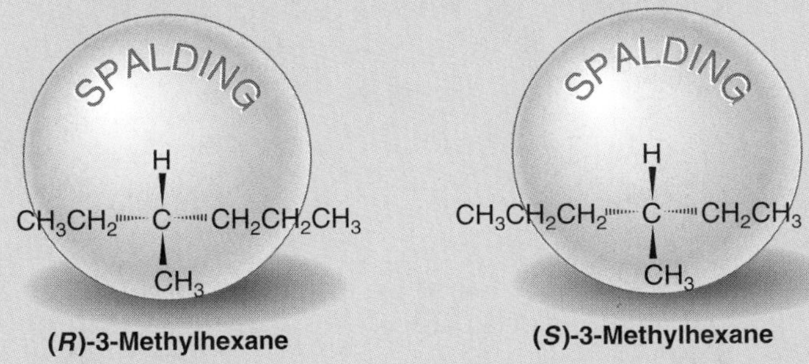

You should be certain that this result is not caused by the particular approach chosen for the two objects. Try interacting 3-methylhexane by bringing it toward the Spaldeen "H and methyl first" or simply turn it over (H down, methyl up) in the figure above.

exactly the same way with the featureless Spaldeen (Fig. 5.24). Chemically, the Spaldeen represents an achiral molecule.

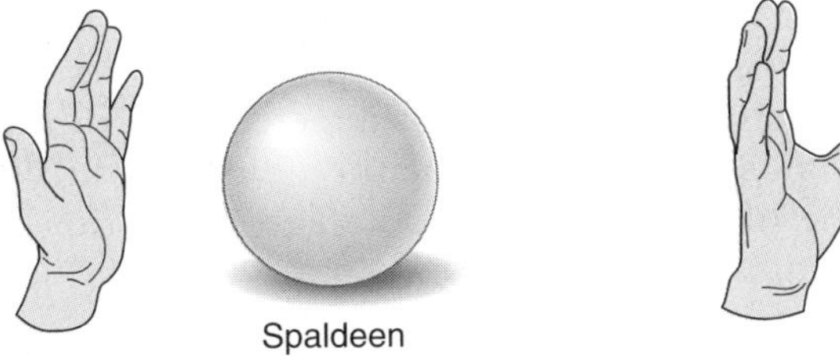

FIGURE 5.24 Right and left hands also interact identically with a Spaldeen. In both the right and left hand, the thumb and little finger are opposed by the same featureless red rubber ball.

Figure 5.25 elaborates this picture by showing the two enantiomers of 3-methylhexane approaching an achiral molecule, propane.

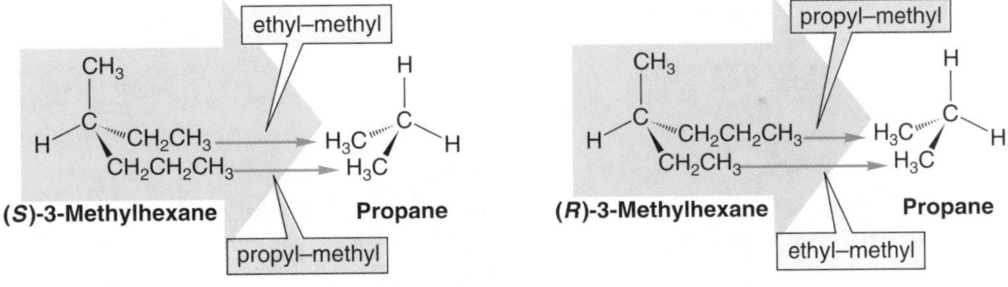

FIGURE 5.25 The two enantiomers of 3-methylhexane interact in exactly the same way with the achiral molecule, propane.

In each instance, the 3-methyl group of 3-methylhexane approaches the hydrogen of propane and both the ethyl and propyl groups are opposite methyl groups of propane. One can *always* find exactly equivalent interactions for the two enantiomers and propane, no matter what approach you pick. Try it with models.

Now imagine your hands, or the two enantiomers of 3-methylhexane, interacting with a chiral object. The Spaldeen with the label will do, but let's use a more common object, such as the shell in Figure 5.26.

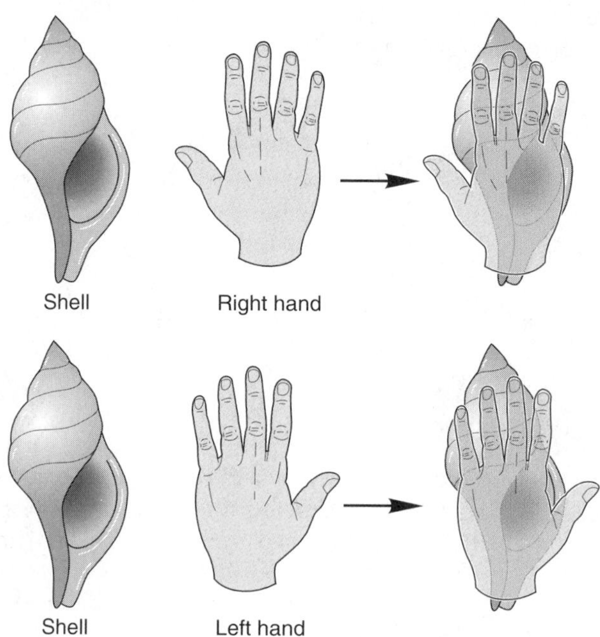

FIGURE 5.26 All sorts of common objects are chiral. A shell is one example. The interactions between a chiral object and two enantiomers—here a shell and two hands—are different. The enantiomeric right and left hand interact differently with the chiral shell.

PROBLEM 5.8 Convince yourself that the shell in Figure 5.26 is chiral.

As your right hand approaches the shell (Fig. 5.26), it is the little finger that opposes the opening, and the thumb that is directed away from it. For the left hand the situation is reversed: It is the thumb that is directed toward the opening and the little finger that is not. The interactions of the two hands with the shell are distinctly different.

PROBLEM 5.9 Convince yourself that turning the shell upside down doesn't change the situation.

For the molecular counterpart of this, look at the approach of the two enantiomers of 3-methylhexane to another chiral molecule, (*R*)-*sec*-butyl alcohol (Fig. 5.27). As the (*S*) enantiomer approaches, propyl is opposite

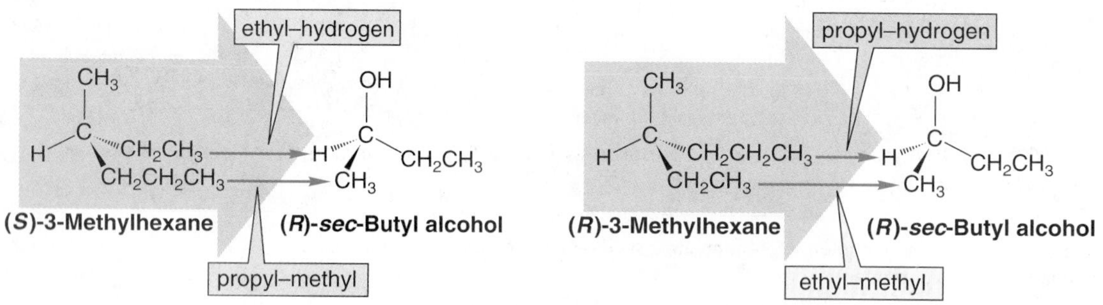

ethyl–hydrogen
CH_3 H C CH_2CH_3 $CH_2CH_2CH_3$ → OH H C CH_2CH_3 CH_3
(*S*)-3-Methylhexane **(*R*)-*sec*-Butyl alcohol**
propyl–methyl

propyl–hydrogen
CH_3 H C $CH_2CH_2CH_3$ CH_2CH_3 → OH H C CH_2CH_3 CH_3
(*R*)-3-Methylhexane **(*R*)-*sec*-Butyl alcohol**
ethyl–methyl

FIGURE 5.27 The approach of (*S*)- and (*R*)-3-methylhexane to (*R*)-*sec*-butyl alcohol. The two approaches are very different and in sharp contrast to the interaction shown in Figure 5.25 with the achiral molecule propane.

*PROBLEM **5.10**

A reader of an early version of this chapter suggested that students would ask, quite correctly, Why can't I turn the picture of (*R*)-3-methylhexane upside down, and then the ethyl–hydrogen and propyl–methyl interactions with (*R*)-*sec*-butyl alcohol would be just like the interactions in the top half? Explain carefully why turning the molecule upside down doesn't produce superimposable mirror images.

ANSWER

It is true that turning the molecule upside down will make two of the interactions the same. Propyl will be opposite CH_3 and C_2H_5 will be opposite H. However, *everything else is wrong!* No longer are CH_3 and OH being opposed, for example (red bonds). Simply turning (*R*)-3-methylhexane over does not produce the same interactions—or even similar interactions to those between (*S*)-3-methylhexane and (*R*)-*sec*-butyl alcohol.

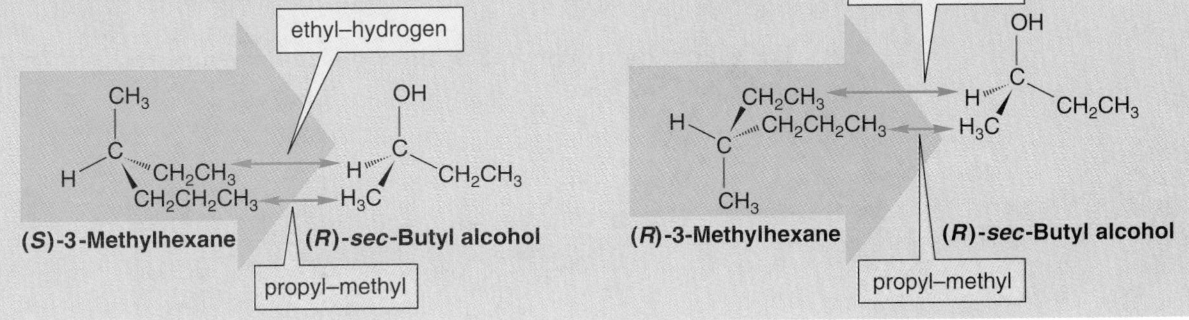

(*S*)-3-Methylhexane **(*R*)-*sec*-Butyl alcohol** **(*R*)-3-Methylhexane** **(*R*)-*sec*-Butyl alcohol**

methyl and ethyl is opposite H. For the other (*R*) enantiomer, the interactions are propyl with H and ethyl with methyl (see Problem 5.10). The two approaches are very different. Indeed, there is no approach to the chiral alcohol that can give identical interactions for the two enantiomeric 3-methylhexanes. Try some. Contrast this situation with the approaches to the achiral propane shown before in Figure 5.25. The two stereoisomeric enantiomers interact *identically* with achiral propane but *differently* with the chiral molecule (*R*)-*sec*-butyl alcohol. This situation is general: Enantiomers have identical chemistries with achiral reagents but different chemistries with chiral ones.

A practical example of the consequences of chiral interactions is the different smells of enantiomeric compounds. One smells a molecule by binding it in a decidedly chiral receptor. There are many examples in which one enantiomer binds differently from the other or not at all. For example, (*R*)-carvone smells like spearmint and (*S*)-carvone like caraway (Fig. 5.28).

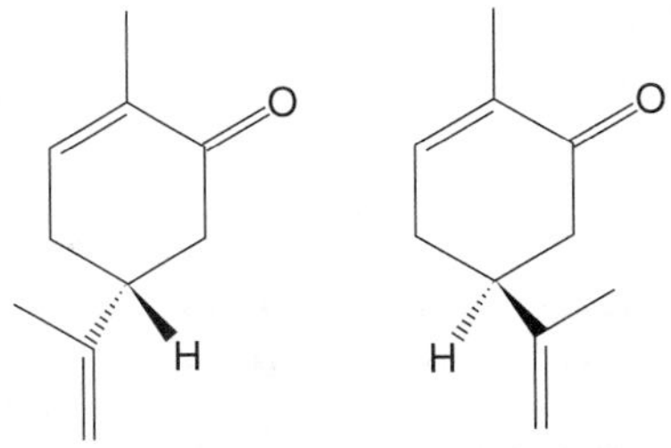

(*R*)-(–)-Carvone (spearmint) **(*S*)-(+)-Carvone** (caraway)

FIGURE **5.28** The interaction of chiral molecules with chiral receptors in your nasal passages can be very different. For example, (*S*)-(+)-carvone smells like caraway, whereas the (*R*)-(–)-carvone enantiomer smells like spearmint.

PROBLEM **5.11**

Convince yourself that the assignment of (*R*) and (*S*) in Figure 5.28 is correct.

5.6 INTERCONVERSION OF ENANTIOMERS BY MOBILE EQUILIBRIA: *GAUCHE* BUTANE

In Chapter 3 (Figs. 3.40 and 3.41, p. 97), you analyzed the conformations of butane, finding an *anti* form lower in energy than a *gauche* form (Fig. 5.29).

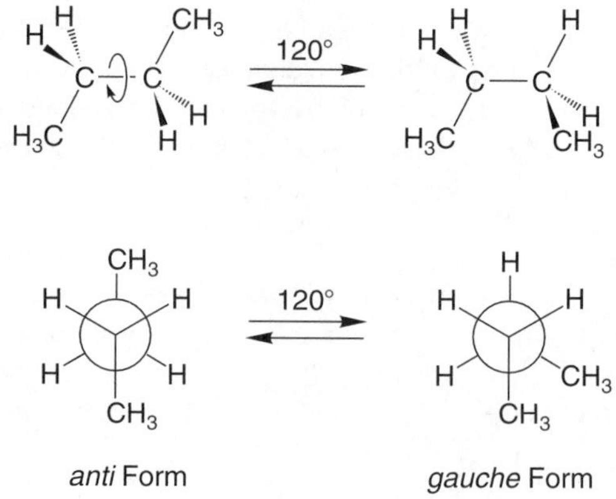

FIGURE 5.29 The *anti* and *gauche* forms of butane shown both in framework and Newman projections. The Newman projections clearly show the *gauche* methyl–methyl interaction.

The *gauche* form of butane is chiral. It cannot be superimposed on its mirror image (Fig. 5.30).

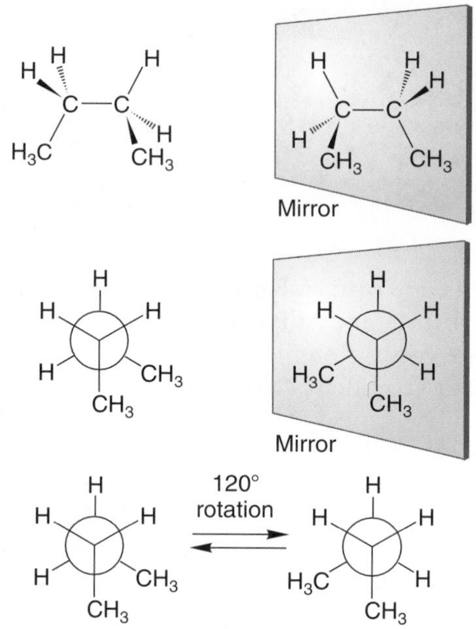

FIGURE 5.30 The *gauche* form of butane is chiral! The mirror images are *not* superimposable. However, rapid rotation around carbon–carbon single bonds interconverts these "conformational enantiomers."

Why then is butane itself achiral? No matter how hard we try, no optical rotation can be observed for butane. The nearly free rotation about the carbon–carbon bond in butane serves to interconvert the two enantiomers of the *gauche* form. If we were able to work at the extraordinarily low temperatures required to "freeze out" the equilibrium between the two *gauche* forms we could separate butane into enantiomers (Fig. 5.30). The two *gauche* forms are **conformational enantiomers**. We will encounter this phenomenon again when we study the stereochemistry of cyclic compounds.

5.7 DIASTEREOMERS: MOLECULES CONTAINING MORE THAN ONE STEREOGENIC ATOM

We have just looked at the different interactions between a single enantiomer of a chiral molecule and an enantiomeric pair of molecules (Fig.

5.27). Let's take this interaction business to its limit and see what happens if we actually form bonds to produce new molecules. We can't worry yet about real chemical reactions—we are still in the tool-building stages—but that doesn't matter. We can imagine some process that will make an attachment, replacing the two X substituents with a chemical bond, shown in all succeeding figures with a blue screen (Fig. 5.31).*

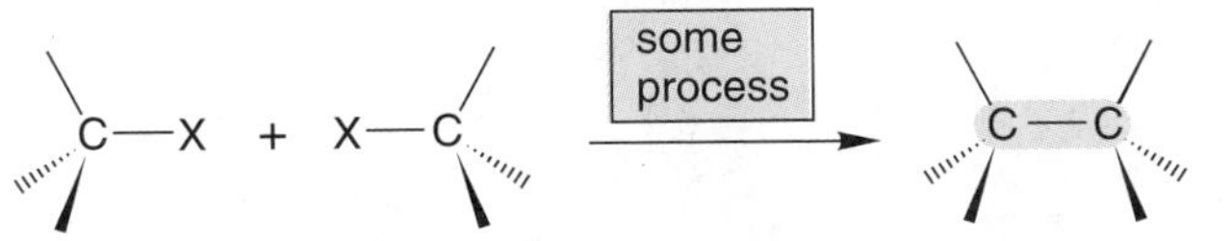

FIGURE **5.31** A hypothetical creation that removes the two X substituents and joins the carbons. The new connection is shown in blue.

If we follow this hypothetical process (which mimics some very real chemical reactions), we generate a new molecule containing two stereogenic carbons. First, let's see what happens when a single chiral molecule reacts with each isomer of a pair of enantiomers. It's worth taking a moment to see if we can figure out what's going to happen before writing out real molecules. Let's assume we are using the (***R***) form of our single enantiomer. It will react with the (*R*) form of the enantiomeric pair to generate a molecule we might call (***R***–*R*), and with the (*S*) form to generate (***R***–*S*) (Fig. 5.32).

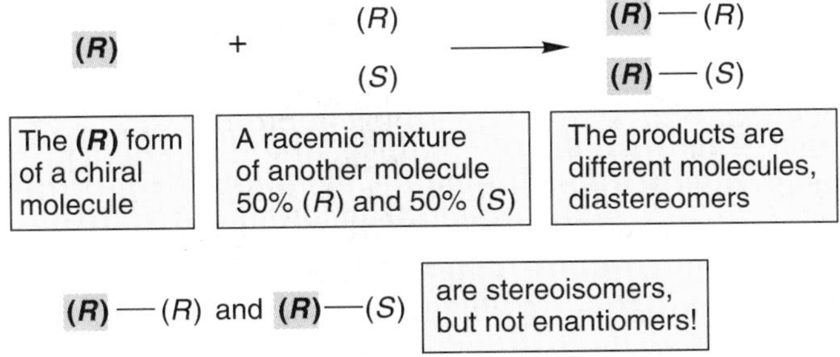
(R) + (R) (S) → (R)—(R) (R)—(S)
The (R) form of a chiral molecule
A racemic mixture of another molecule 50% (R) and 50% (S)
The products are different molecules, diastereomers
(R)—(R) and (R)—(S)
are stereoisomers, but not enantiomers!

FIGURE **5.32** A schematic analysis of the reaction of one enantiomer with a racemic mixture of another molecule. The products of this reaction are stereoisomers but not enantiomers. They are called diastereomers.

We can tell a lot even from the crudely schematic coded picture of Figure 5.32. First of all, (***R***–*R*) and (***R***–*S*) are clearly neither identical nor enantiomers (nonsuperimposable mirror images). For example, the mirror image of (***R***–*R*) is (*S*–***S***), not (***R***–*S*).

PROBLEM **5.12** What is the mirror image of (*R*–*S*)? This question sounds insultingly simple, but most people get it wrong!

Yet these molecules are stereoisomers—they differ only in the arrangement of their constituent parts in space. There is a new word for such molecules—they are **diastereomers**, "stereoisomers that are not mirror images."

*In this and many succeeding figures, the molecules are drawn in eclipsed forms for clarity. As you know, these are not stable arrangements and will be converted into staggered forms by rotation about the central carbon–carbon bond (Chapter 3, p. 83). None of the arguments presented is changed by this rotation, and in the staggered arrangements it is more difficult to visualize the relationships involved.

*PROBLEM 5.13 If the two *gauche* forms shown in Figures 5.29 and 5.30 are enantiomers, what is the relationship between one *gauche* form and the *anti* form?

ANSWER The *gauche* forms are mirror images and, therefore, enantiomers. Each *gauche* form is a stereoisomer of the *anti* form, but not its mirror image. Stereoisomers that are not mirror images are diastereomers.

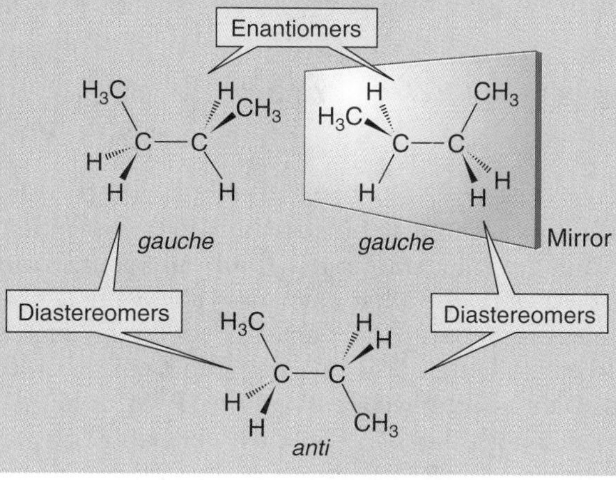

PROBLEM 5.14 We have seen diastereomers twice before. Show two different kinds of diastereomeric molecules. *Hint*: Start with a good definition of diastereomer.

Now let's make all this theory a little more real, and look at some molecules. Imagine a hypothetical reaction (X = Cl) in which the (*S*) enantiomer of 2-chlorobutane (*sec*-butyl chloride) reacts with the two enantiomers of 1,1,1-trifluoro-2-chloropropane (Fig. 5.33).

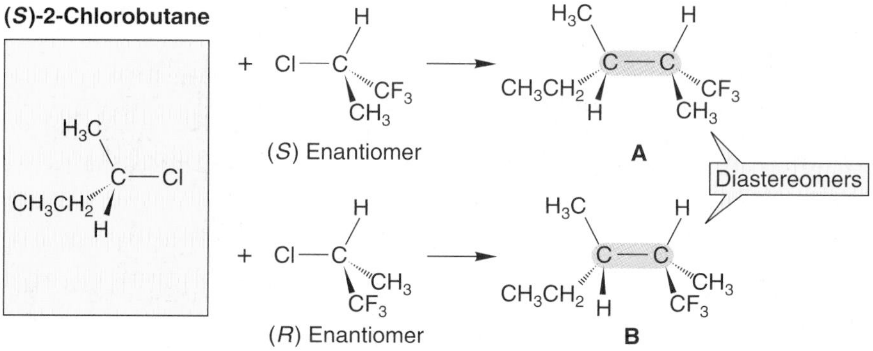

FIGURE 5.33 The hypothetical reaction of (*S*)-2-chlorobutane with the two enantiomers of 1,1,1-trifluoro-2-chloropropane. Two stereoisomeric products, **A** and **B**, are formed. The products **A** and **B** are not mirror images (enantiomers) so they must be diastereomers.

Two new products, **A** and **B**, are produced from our hypothetical reaction. Look carefully at them. Are they identical? Certainly not. Are they enantiomers? No, again. They are stereoisomers, though, so they must be diastereomers.

PROBLEM 5.15 Verify the claims of the preceding paragraph. Are **A** and **B** diastereomers?

The (*R*) form of 2-chlorobutane would also produce two new products, **C** and **D** (Fig. 5.34), which are neither identical nor enantiomeric, and thus must also be diastereomers.

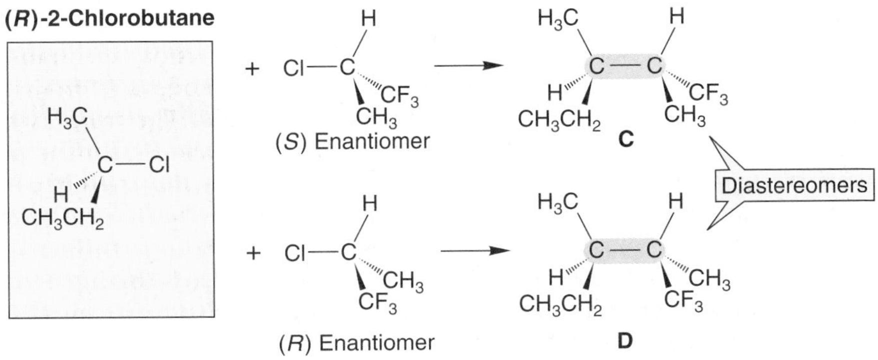

FIGURE 5.34 The hypothetical reaction of (*R*)-2-chlorobutane with the same mixture of enantiomeric 1,1,1-trifluoro-2-chloropropanes to give another pair of diastereomers, **C** and **D**.

Now the question is, What are the relationships among **A**, **B**, **C**, and **D**? As Figure 5.35 shows, **A** and **D** are enantiomers and **B** and **C** are enantiomers. So all possible combinations of the two molecules have produced two pairs of enantiomers, or four stereoisomers in all. The (*R*) and (*S*) relationships are also shown in Figure 5.35. As a simple theoretical analysis would predict (Fig. 5.36), we have formed all possible isomers in this combination [(*S–R*), **A**, (*S–S*), **B**; (*R–S*), **D**; and (*R–R*), **C**].

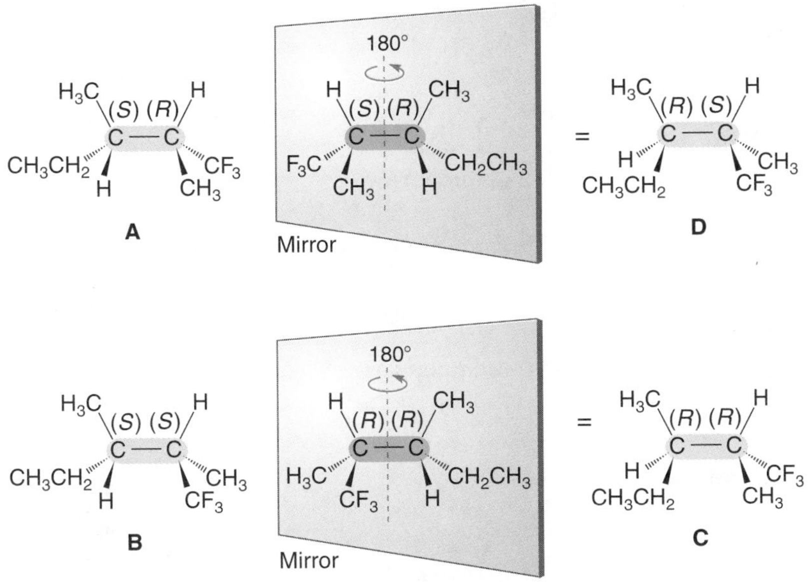

FIGURE 5.35 Two pairs of enantiomers, **A** and **D** and **B** and **C**, are formed from this hypothetical reaction. Combinations **A** and **B**, **A** and **C**, **B** and **D**, and **C** and **D** are diastereomeric pairs. Each of these pairs is composed of two stereoisomers that are not mirror images.

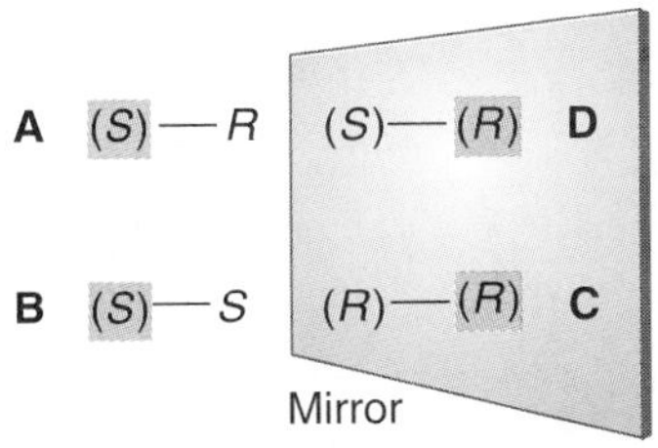

FIGURE 5.36 (*R*)- and (*S*)-2-chlorobutane reacting with (*R*)- and (*S*)-1,1,1-trifluoro-2-chloropropane should give four isomers: two pairs of enantiomers.

PROBLEM 5.16

(*S*)-2-Chlorobutane and (*S*)-2-chloro-1,1,1-trifluoropropane (Fig. 5.33) are used to produce isomer **A**, which is properly designated as the (*S*,*R*) isomer (Fig. 5.35). Check that there is no mistake in this analysis, and then explain how the two (*S*) compounds can produce the (*S*,*R*) isomer, **A**.

We might be able to generalize even at this early stage. A molecule with one stereogenic center has two possible stereoisomers, a single pair of enantiomers, the (*R*) and (*S*) forms. A molecule containing two stereo-

(*RRR*)	All (*R*)
(*RRS*) (*RSR*) (*SRR*)	Two (*R*), one (*S*)
(*SSR*) (*SRS*) (*RSS*)	One (*R*), two (*S*)
(*SSS*)	All (*S*)

FIGURE 5.37 The eight possible stereoisomers for a molecule containing three stereogenic atoms.

genic centers gives two pairs of enantiomers, or four stereoisomers in all. It seems that for n stereogenic carbons, we get 2^n isomers. We can immediately predict that, for $n = 3$, stereogenic atoms will produce 2^3 or 8 isomers. This is easy to check—what isomers are possible? The easiest way to tell is not to examine some complicated molecule and thus incur the task of drawing a large number of stereoisomers, but to analyze the problem in a simpler fashion. A stereogenic carbon can only be (*R*) or (*S*). Figure 5.37 shows the possibilities for three such atoms.

The eight possible isomers appear easily in this coded form. Note that there is nothing in this analysis that requires the stereogenic atoms to be adjacent—they need only be in the same molecule.

This procedure reliably gives you the *maximum* number of possible stereoisomers. There are some interesting (and important) special cases, however, and this very general method won't spot them. Let's imagine using our hypothetical chemical coupling reaction again, but this time combining two molecules of *sec*-butylchloride. This procedure is exactly the same as the one we used before to generate the four stereoisomers of Figures 5.33 and 5.34, **A**, **B**, **C**, and **D**. We allow our (*R*) and (*S*) starting materials to combine in all possible ways (Fig. 5.38).

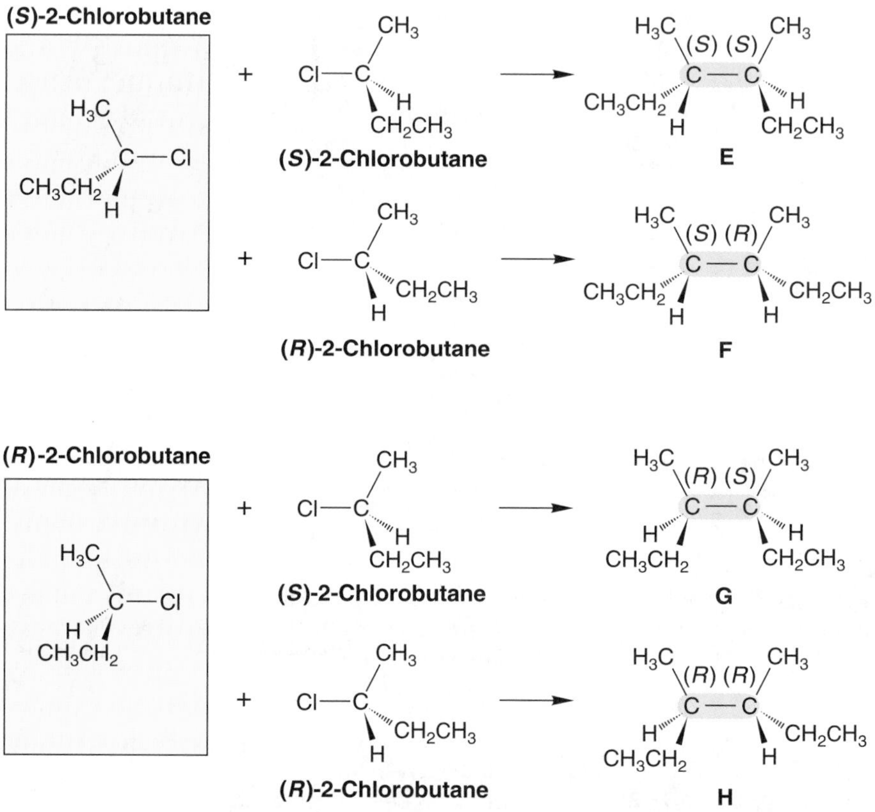

(*S*)-2-Chlorobutane

+ **(*S*)-2-Chlorobutane** ⟶ **E** (*S*) (*S*)

+ **(*R*)-2-Chlorobutane** ⟶ **F** (*S*) (*R*)

(*R*)-2-Chlorobutane

+ **(*S*)-2-Chlorobutane** ⟶ **G** (*R*) (*S*)

+ **(*R*)-2-Chlorobutane** ⟶ **H** (*R*) (*R*)

FIGURE 5.38 All possible combinations of (*R*)- and (*S*)-2-chlorobutane (*sec*-butyl chloride) produce four isomers, **E**, **F**, **G**, and **H**.

In the previous example, we found that we had made two pairs of enantiomers, (*S*–*R*) and (*R*–*S*), **A** and **D**, and (*S*–*S*) and (*R*–*R*), **B** and **C** (Figs. 5.35 and 5.36). This time one pair of enantiomers, **E** and **H**, does appear (Fig. 5.39).

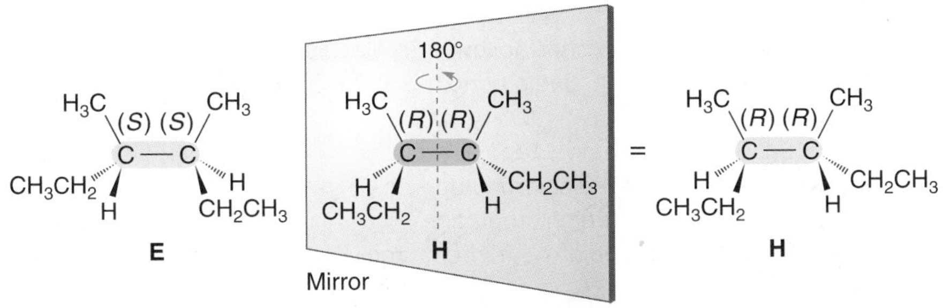

FIGURE 5.39 Here **E** and **H** are enantiomers.

However, the second potentially enantiomeric pair of molecules, **F** and **G**, are in fact identical species (Fig. 5.40)!

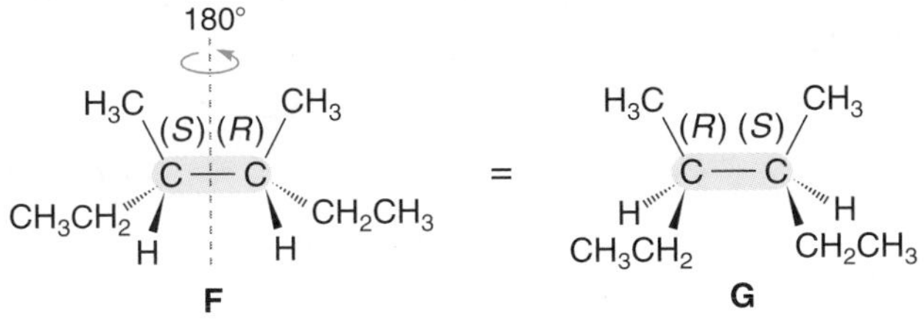

FIGURE 5.40 The **F** and **G** molecules are identical. This compound is meso.

So in this example we get only three stereoisomers, not the maximum number of $2^2 = 4$. Clearly, we are one stereoisomer short because of the identity of **F** and **G**. This molecule, **F** = **G**, contains stereogenic carbons but is not chiral. Such a molecule is called a **meso compound**.

Twice before we have seen molecules that for different reasons give no observable rotation of plane-polarized light. Achiral molecules contain no stereogenic atoms and cannot induce optical rotation. A racemic mixture of chiral molecules contains equal numbers of right- and left-handed compounds, which rotate the light equally in clockwise and counterclockwise directions, respectively, for a net zero rotation. Although a meso compound does contain stereogenic atoms, it is achiral and induces no optical rotation.

We just answered the second part of our earlier (p. 158) question. No, "four different groups attached to a stereogenic carbon" is not a sufficient condition for optical activity. Our meso compound **F** = **G** certainly has carbons to which four different groups are bonded, but it is equally clearly not chiral. Even though searching for carbons attached to four different groups may be a good way to begin a hunt for chirality, one must be careful. Finding such a carbon does not guarantee we have found a chiral molecule, unless we are certain that the molecule contains one and only one such carbon. Alas, it is also the case that not finding such a carbon does not guarantee that there will be no chirality (for examples, see p. 185). The ultimate test remains the superimposability of mirror images. If the mirror images are superimposable, the molecule is not chiral; if the mirror images are different, nonsuperimposable, the molecule is chiral.

When will meso compounds appear? Meso compounds occur when there is a plane or point of symmetry in the molecule. In effect, this divides the molecule into halves, which contribute equally and oppositely to the rotation of plane-polarized light. Compound **F** = **G** provides an example of a molecule with a plane of symmetry in the eclipsed forms we have been using, and a point of symmetry if we draw the molecule in its energy minimum, staggered arrangement (Fig. 5.41).

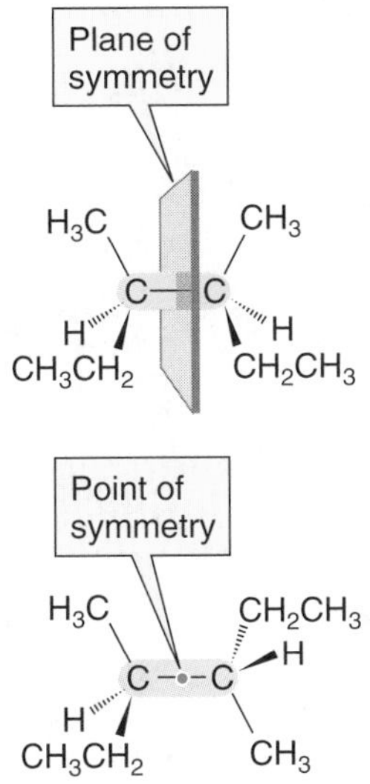

FIGURE 5.41 Two views of the meso compound **F** = **G**. A plane of symmetry exists in the eclipsed form and a point of symmetry in the stable, staggered arrangement.

PROBLEM 5.17

Draw all the stereoisomers of the following molecules in which there is free rotation about carbon–carbon single bonds. Some may be achiral:

(a) 2,2,3,3-Tetrabromobutane
(b) 2,2-Dibromo-3,3-dichlorobutane
(c) 2,3-Dibromo-2,3-dichlorobutane
*(d) 2,3-Dibromo-2-chloro-3-fluorobutane

Hint: It is easier to analyze these molecules in their eclipsed forms.

ANSWER

(d) This molecule contains stereogenic carbons and yields the full complement of $2^2 = 4$ stereoisomers. The molecules are shown in eclipsed forms for clarity and ease of analysis, but these are not the minimum energy arrangements, which will have staggered bonds.

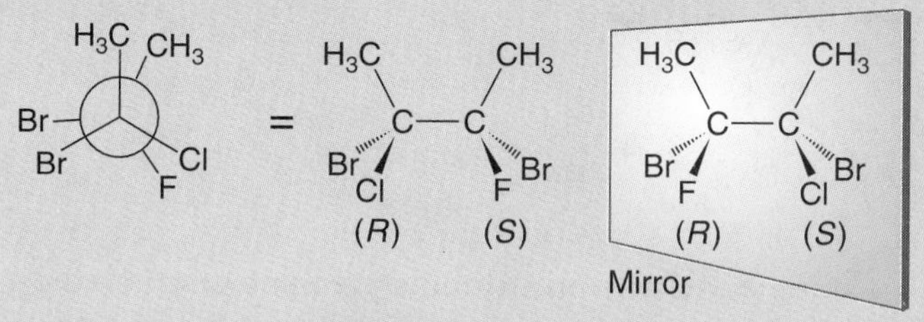

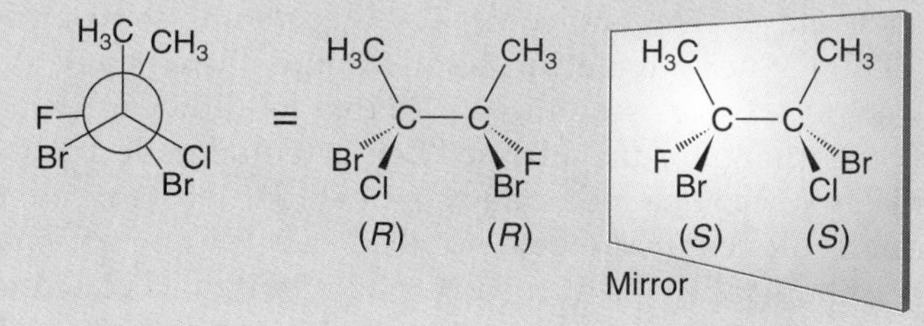

(*R*) (*S*) (*R*) (*S*)

Mirror

(*R*) (*R*) (*S*) (*S*)

Mirror

*PROBLEM 5.18

How does the situation change if there is *no* free rotation about carbon–carbon single bonds for 2,3-dibromo-2-chloro-3-fluorobutane (see Problem 5.17d)?

ANSWER

If rotational barriers are high, then each staggered form would be isolable. The number of isomers increases dramatically, as each isolable staggered form will have an enantiomer. Now there are 12 total isomers, in six pairs of enantiomers. The answer is given in Newman projections.

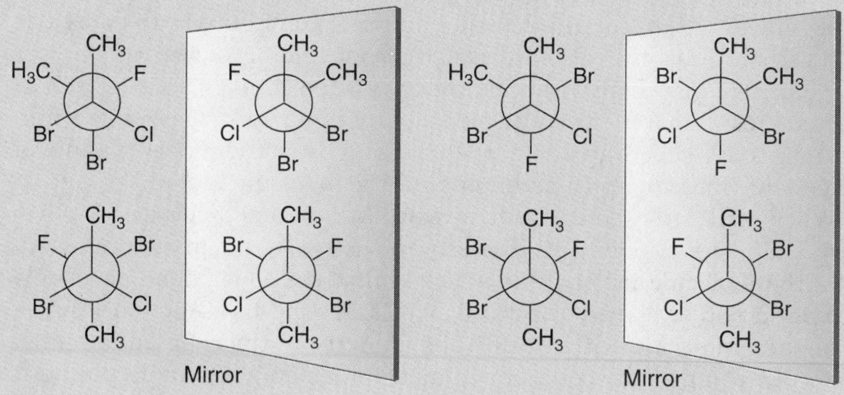

Mirror

Mirror

ANSWER (CONTINUED)

Mirror

Mirror

PROBLEM 5.19

Identify each stereogenic carbon in the compound of Problem 5.17c as either (*R*) or (*S*).

5.8 PHYSICAL PROPERTIES OF DIASTEREOMERS: OPTICAL RESOLUTION

The formation of diastereomers allows the separation of enantiomers. Separation of enantiomers, called **resolution**, is a serious experimental difficulty. So far we have ignored it. Enantiomers have identical physical properties (except for the ability to rotate the plane of plane-polarized light), and one might legitimately wonder how in the world we are ever going to get them apart. At several points we used a single enantiomer without giving any hint of how a pair of enantiomers might be separated. The key to this puzzle is that diastereomers, unlike enantiomers, have different physical properties—melting point, boiling point, and so on.

One general procedure for separating enantiomers is to allow them to react with a naturally occurring chiral molecule to form a pair of diastereomers.* These can then be separated by taking advantage of one of their different physical properties. One typically can separate such a pair by crystallization, because the members of the diastereomeric pair will have different solubilities. Then, if the original chemical reaction can be reversed, we have the pair of enantiomers separated. Figure 5.42 outlines the general scheme and begins with a schematic recapitulation of Figure 5.33, which first described the reaction of a single enantiomer with a racemic mixture to give a pair of diastereomers. Be sure to compare the two figures.

FIGURE 5.42 Resolution is a general method for separating the constituent enantiomers of a racemic mixture. A single enantiomer of a chiral molecule is used to form a pair of diastereomers, which can be separated physically. If the original chemical reaction can be reversed, the enantiomers can be isolated.

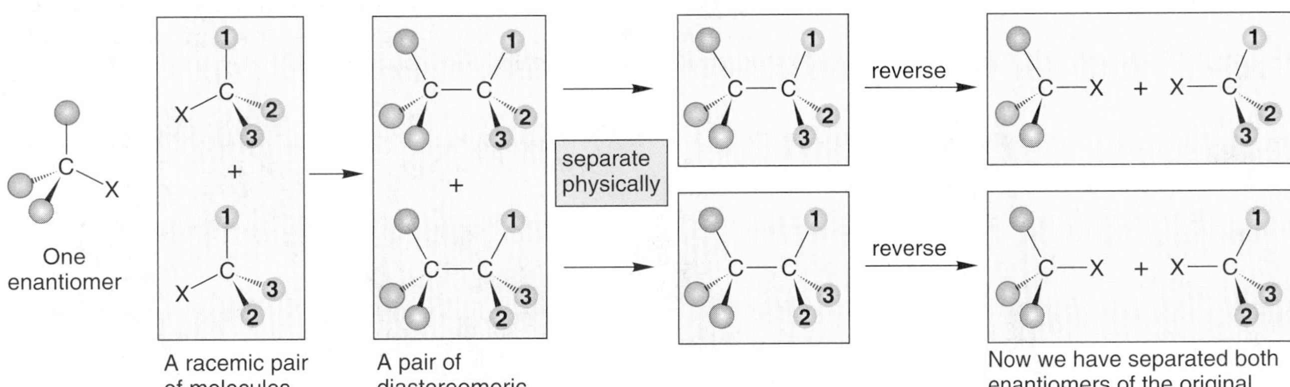

*Of course, there is no magic in using a molecule derived from natural sources. One made in the laboratory will do as well. However, many molecules found in Nature are easily isolable as pure enantiomers, and it is often convenient to use them.

It is not even necessary to form covalent bonds. For example, in the traditional method for separating enantiomers of organic acids, optically active nitrogen-containing molecules, called alkaloids, are used to form a pair of diastereomeric salts, which can then be separated by crystallization. These alkaloids have wondrously complex structures (for more on these fascinating molecules, see Chapter 21, p. 000). Two examples, brucine and the notorious strychnine, are shown in Figure 5.43 along with the general procedure for this kind of resolution.

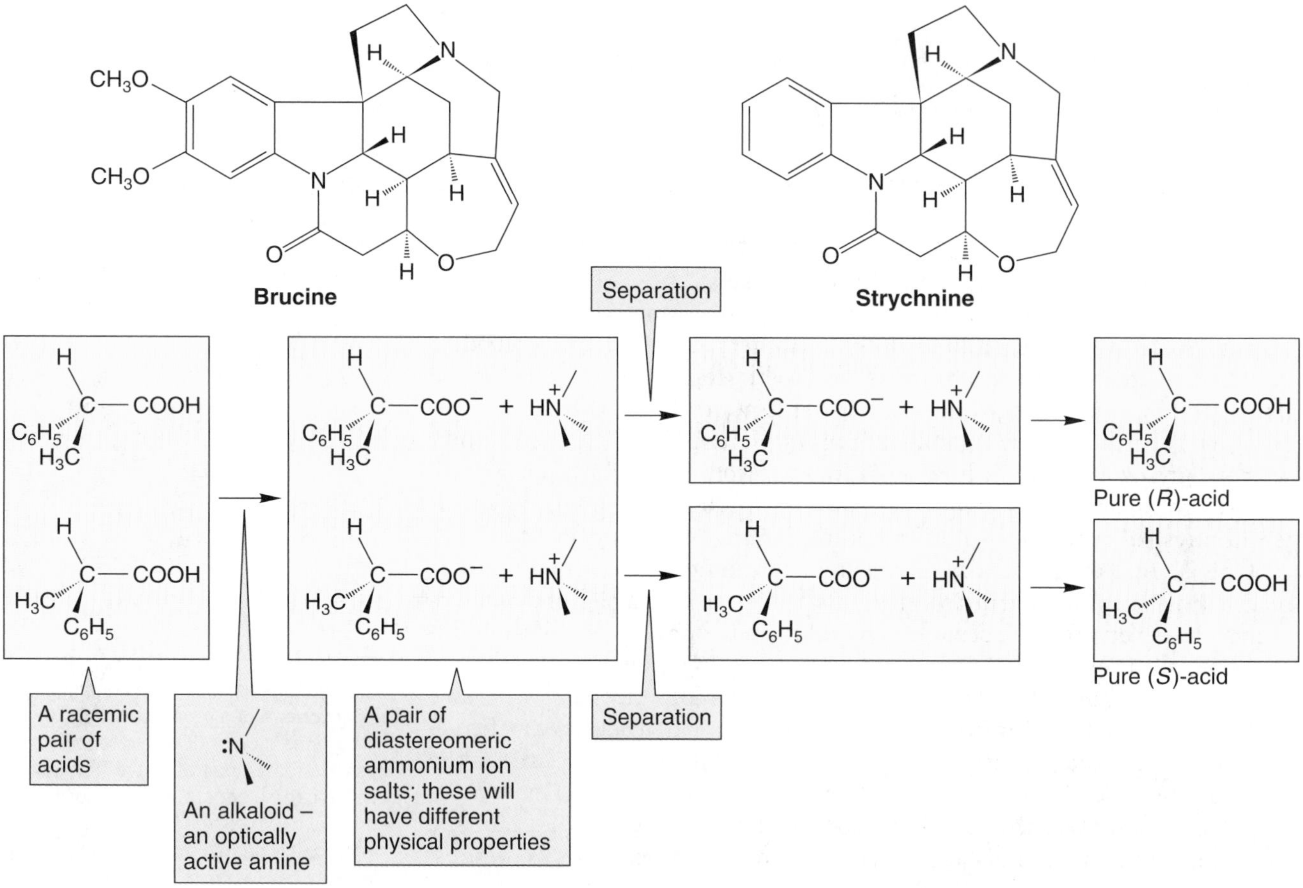

Brucine

Strychnine

A racemic pair of acids

An alkaloid – an optically active amine

A pair of diastereomeric ammonium ion salts; these will have different physical properties

Separation

Separation

Pure (*R*)-acid

Pure (*S*)-acid

FIGURE 5.43 Two alkaloids, brucine and strychnine, are commonly used to separate the enantiomers of chiral organic acids. Diastereomeric salts are first formed, then separated by crystallization, and the individual enantiomeric acids are regenerated.

*PROBLEM 5.20 Identify with an asterisk (*) all the stereogenic carbons in brucine.

ANSWER

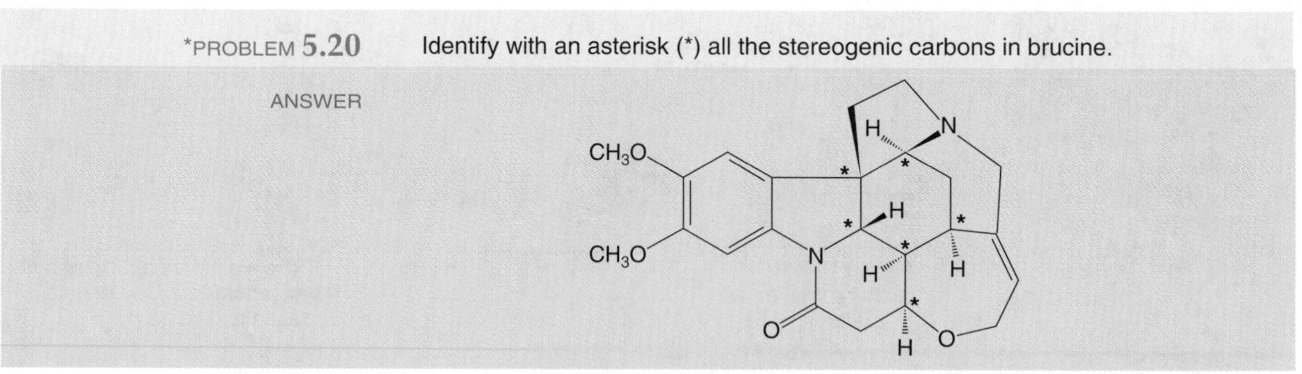

PROBLEM 5.21 Identify each stereogenic carbon in brucine as (*R*) or (*S*).

These days, this general procedure has been extended so that all manner of enantiomeric pairs can be separated by chromatography. In such a technique, covalent chemical bonds are not formed, as they are not in the salt formation shown in Figure 5.43. Rather, advantage is taken of the formation of partial bonds—complexes—as the pair of enantiomers passes over an optically active substrate. The complexes are diastereomeric and thus one will be more stable (contain a stronger partial bond) than the other. One enantiomer will be held more tightly than the other and will pass through the chromatography apparatus more slowly (Fig. 5.44).

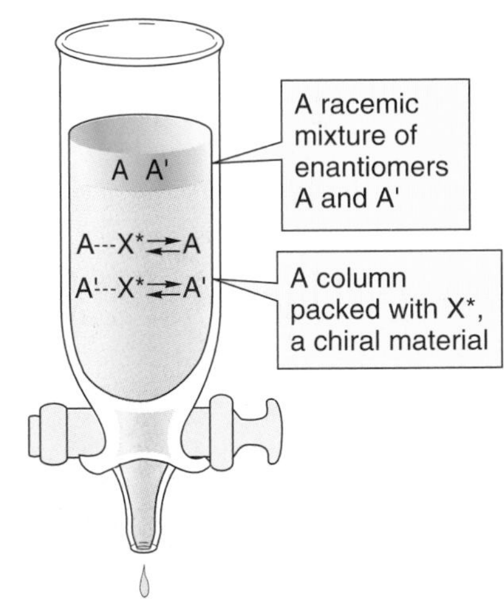

FIGURE **5.44** A chromatography column for separating enantiomers. The column is packed with an optically active substrate (X*) that forms a complex with the enantiomers A and A'. These complexes are diastereomers and have different physical properties, including bond strengths of A ··· X and A' ··· X. Both AX and A'X are in equilibrium with the free enantiomers, and these equilibria will be different for the two diastereomeric complexes. Therefore, A and A' will move through the column at different rates and emerge at different times.

Strychnine

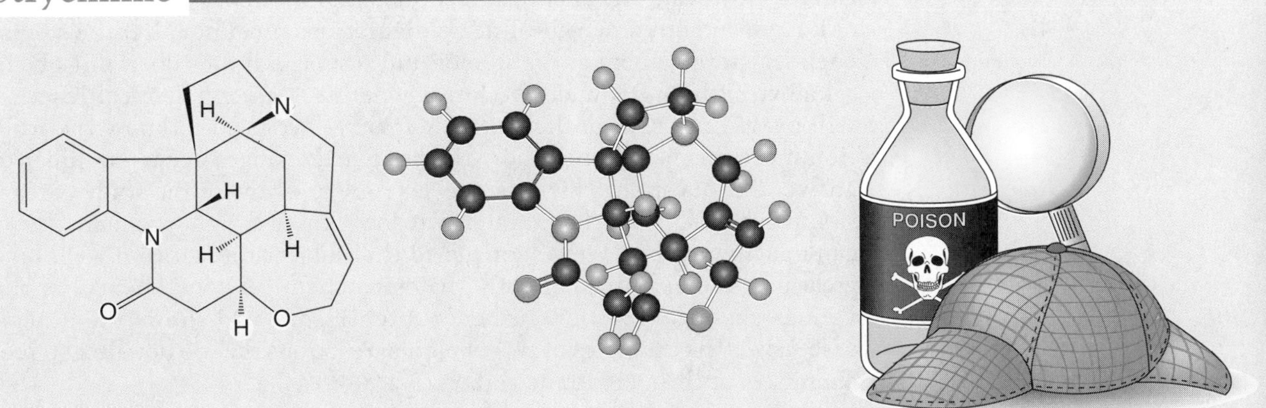

The notorious poison strychnine was first isolated from the beans of *strychnos ignati Berg* by Pelletier and Caventou in 1818. It constitutes about one-half of the alkaloids present in the beans and makes up 5–6% of their weight. Its structure is obviously complicated and was only determined correctly by Sir Robert Robinson in 1946. It was a mere two years more before a physical determination of the structure by X-ray diffraction was reported by Bijvoet, confirming Robinson's deductions, and presaging the demise of chemical, as opposed to physical, structure determination as a viable enterprise. Robert B. Woodward provided the first synthesis of strychnine in 1954 in a landmark paper that begins with the exclamation "Strychnine!" (hardly the usual dispassionate scientific writing). The introduction to Woodward's paper makes good reading, and provides an interesting defense of the art of synthesis in the face of critics who thought that the profession would surely be rendered obsolete by the increasingly powerful physical methods. The ensuing years have shown Woodward's defense to be correct. (See *Tetrahedron,* **1963**, *19,* 247; your chemistry library probably has it.)

Of course, much interest in strychnine centers on its pharmacological properties. It is a powerful convulsant, lethal to an adult human in a dose as small as 30 mg. Death comes from central respiratory failure and is preceded by violent convulsions. Strychnine is the lethal agent in many a murder story, real and imagined. One example is Sir Arthur Conan Doyle's Sherlock Holmes mystery, *"The Sign of the Four,"* in which Dr. Watson suggests the lethal agent to be a "powerful vegetable alkaloid ... some strychnine-like substance."

5.9 DETERMINATION OF ABSOLUTE CONFIGURATION (*R* OR *S*)

Now that we have achieved the separation of our racemic mixture of enantiomers into a pair of optically active stereoisomers, we face the difficult task of finding out which enantiomer is (*R*) and which is (*S*). This problem is not trivial! Indeed, in Chapter 24, when we deal with sugars, we'll find that until rather recently, there was *no* way to be certain, and one just had to guess (correctly in the case of the sugars, it turns out). One would like to peer directly at the structures, of course, and under some circumstances this is possible.

X-ray crystallography can determine the relative positions of atoms in a crystal, and a special kind of X-ray diffraction called "anomalous dispersion" can tell the absolute configuration of the molecule. But this is not a generally applicable technique—one needs a crystalline compound, for example. It does serve to give us some benchmarks, though. If we know the absolute configuration (*R* or *S*) of some compounds, we may be able to determine the absolute configurations of other molecules by relating them to the few compounds of known absolute configuration. We must be very careful, however. The chemical reactions that interconvert the molecules of known and unknown absolute configurations must not alter the stereochemical arrangement at the stereogenic atoms, or if they do, it must be in a known fashion. How do we know whether a given chemical reaction will or will not change the stereochemistry? We need to know the reaction mechanism—to know how the chemical changes occur—in order to answer this question. This reason is just one of many for the study of reaction mechanisms. At the moment we don't know any mechanisms so we can't go any further. We'll soon get to this subject though, and we'll take great pains to examine carefully the stereochemical consequences of the various chemical reactions we encounter. Figure 5.45 shows the general case, as well as examples of two chemical reactions that do not alter stereochemistry at the stereogenic carbon (Fig. 5.45).

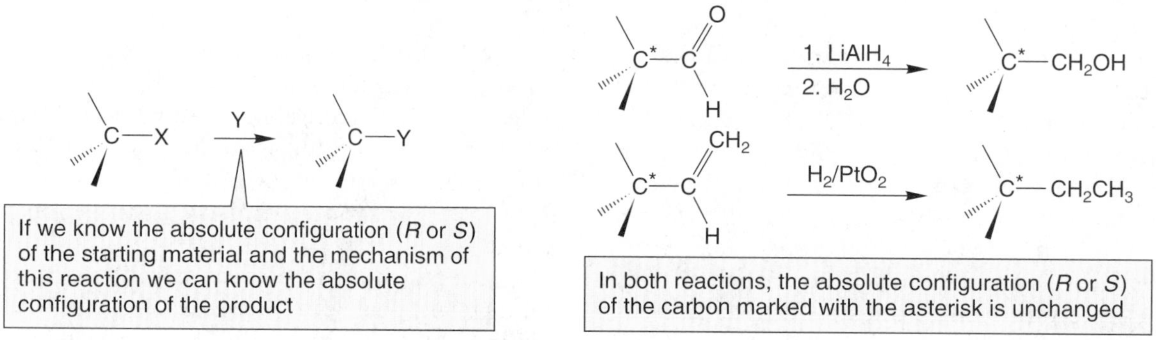

C—X →(Y) C—Y

If we know the absolute configuration (*R* or *S*) of the starting material and the mechanism of this reaction we can know the absolute configuration of the product

C*—CHO →(1. $LiAlH_4$; 2. H_2O) C*—CH_2OH

C*—CH=CH_2 →(H_2/PtO_2) C*—CH_2CH_3

In both reactions, the absolute configuration (*R* or *S*) of the carbon marked with the asterisk is unchanged

FIGURE 5.45 To know the absolute configuration of the product in this reaction you must know two things: the absolute configuration of the starting material *and* the mechanism of the reaction with Y. Only if these two things are known will the absolute configuration of the product be known with certainty.

5.10 STEREOCHEMICAL ANALYSIS OF RING COMPOUNDS (A BEGINNING)

Most of the stereochemical principles in the previous sections apply quite directly to ring compounds. We'll start with a simple question. How many isomers of chlorocyclopropane exist? That looks like an easy question, and it is. There is only one (Fig. 5.46).

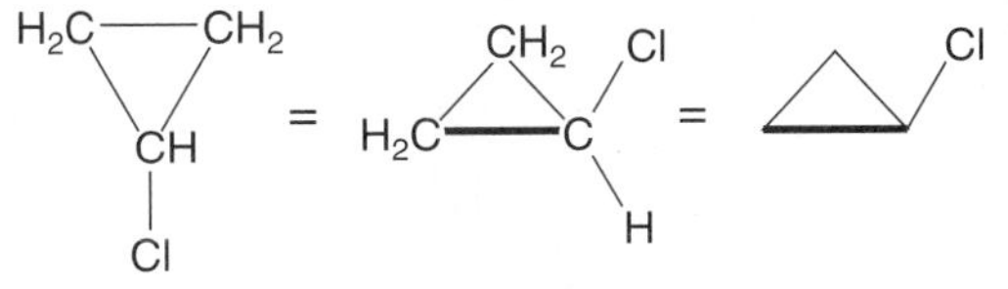

FIGURE 5.46 Different representations of the single isomer of chlorocyclopropane.

To be certain, however, we should examine this compound for chirality. Again this process is simple. As Figure 5.47 shows, the mirror image is easily superimposed on the original, once again through a 180° rotation. Chlorocyclopropane is an achiral molecule.

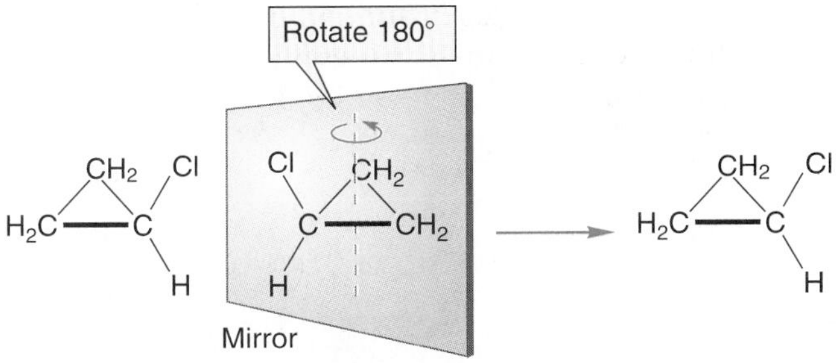

FIGURE 5.47 Chlorocyclopropane is superimposable on its mirror image and is, therefore, an achiral molecule.

How many isomers of dichlorocyclopropane exist? This question is much tougher than the last one. There are four isomers of dichlorocyclopropane. Most people don't get them all at this point, so it's worth going carefully through an analysis. Where can the second chlorine be placed in the single monochlorocyclopropane? As Figure 5.48 shows, there are only two places, the same carbon as the first chlorine, or one of the two equivalent adjacent carbons. This analysis yields two compounds called **structural isomers**.

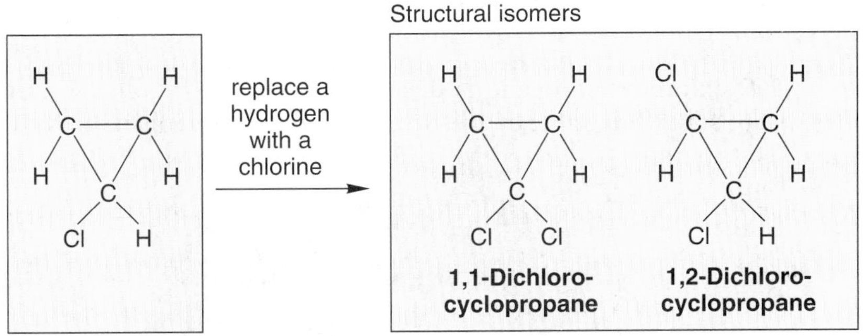

FIGURE 5.48 Replacement of one hydrogen of chlorocyclopropane with another chlorine can occur in only two places, the position to which the first chlorine is attached or one of the other two equivalent carbons. This procedure generates the structural isomers 1,1-dichlorocyclopropane and 1,2-dichlorocyclopropane.

But that only brings us to two isomers; we're still two short of reality. As first mentioned in Chapter 3 (p. 106), rings have sides. We can find one more isomer if we notice that the second chlorine can be on the same side (the cis isomer) or the opposite side (the trans isomer) of the ring (Fig. 5.49). These compounds are stereoisomers, more precisely, diastereomers (stereoisomers that are not mirror images). Now we are just one isomer short.

FIGURE 5.49 1,2-Dichlorocyclopropane can exist in cis and trans forms. In the figure, all the ring hydrogens have been drawn in to help you see the difference between these two molecules. Be absolutely certain you see why these two molecules cannot be interconverted. Use models.

Diastereomers

1,2-Dichlorocyclopropane = ***cis*-1,2-Dichlorocyclopropane** **trans-1,2-Dichlorocyclopropane**

Once we are sure we have found all the different positions on which to put the two chlorines it's time to look for chirality. Figure 5.50 shows that neither 1,1-dichlorocyclopropane nor *cis*-1,2-dichlorocyclopropane is chiral. The mirror images are easily superimposable on the originals. Note that *cis*-1,2-dichlorocyclopropane is a meso compound. It contains stereogenic carbons but is superimposable on its mirror image, and therefore is achiral.

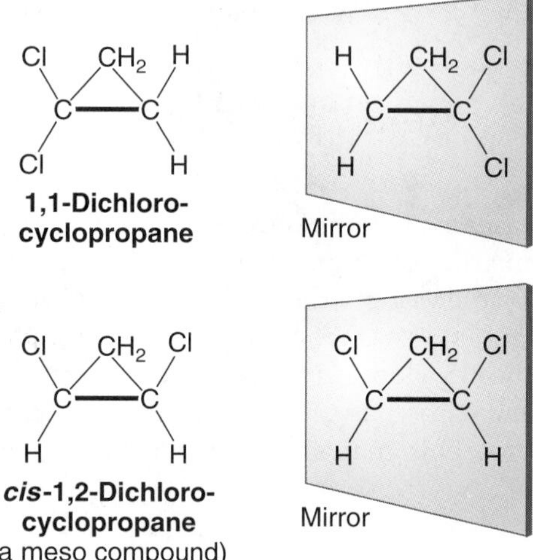

FIGURE 5.50 Both 1,1-dichlorocyclopropane and *cis*-1,2-dichlorocyclopropane are superimposable on their mirror images and, therefore, are achiral.

trans-1,2-Dichlorocyclopropane is different. In this molecule the mirror image is not superimposable on the original and this stereoisomer is chiral (Fig. 5.51).

FIGURE 5.51 *trans*-1,2-Dichlorocyclopropane is *not* superimposable on its mirror image. It is a chiral molecule and, therefore, two enantiomers exist. This stereochemical relationship is probably the most difficult you've encountered so far, so be sure you can see why no manipulation of one enantiomer will make it superimposable on the other.

***trans*-1,2-Dichlorocyclopropane** Mirror

So now we have the four isomers: 1,1-dichlorocyclopropane, *cis*-1,2-dichlorocyclopropane, *trans*-(1*S*,2*S*)-dichlorocyclopropane, and *trans*-(1*R*,2*R*)-dichlorocyclopropane (Fig. 5.52).

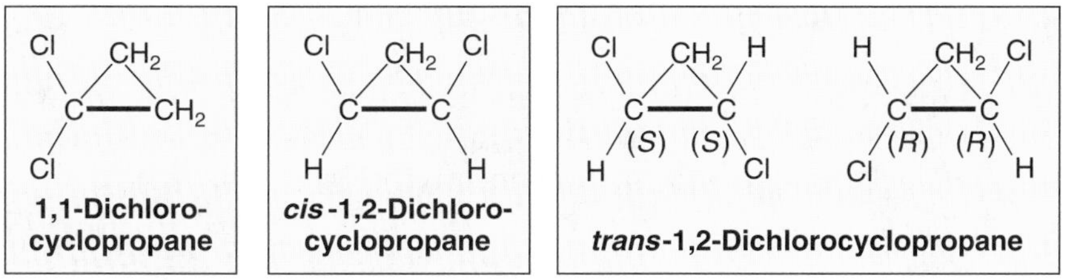

Cl–C(Cl)–CH$_2$–CH$_2$ ring	Cl, H / C–C / Cl, H (CH$_2$)	(S)(S) and (R)(R)
1,1-Dichlorocyclopropane	*cis*-1,2-Dichlorocyclopropane	*trans*-1,2-Dichlorocyclopropane

FIGURE 5.52 The four stereoisomers of dichlorocyclopropane.

PROBLEM 5.22 Verify the assignment of absolute configuration in the two enantiomers of *trans*-1,2-dichlorocyclopropane in Figure 5.52.

PROBLEM 5.23 Identify the stereochemical relationships between the four isomers of Figure 5.52. Find all the pairs of enantiomers and diastereomers.

This section illustrates one procedure for working out an isomer problem, but any rational technique will work. The important thing is to devise some systematic way of searching. I happen to like the one above in which one first finds all the structural isomers (in this case 1,1- and 1,2-dichlorocyclopropane), then searches for stereoisomerism of the cis/trans type, and finally examines each isomer for chirality. But it doesn't matter what procedure you use as long as you are systematic. What will not work is a nonrational scheme. So don't start a "find the isomers" problem by just writing out compounds without thinking first. No one can do it that way.

PROBLEM 5.24 There are six isomers of dichlorocyclobutane. Find them all and name them carefully.

We started with small rings for a reason—the inflexibility of small rings makes them flat, or very nearly so. Three points determine a plane, so cyclopropane must be planar. Cyclobutane need not be absolutely flat, and we can see that it is not if we take the time to make a model. At the same time, it cannot be far from planar. Larger rings are more complicated though, and we'll defer an examination of their stereochemistry until Chapter 6, which looks at a number of structural questions about rings.

5.11 REVIEW OF ISOMERISM

The words isomer and stereoisomer are often tossed around quite loosely, and it's worth a bit of review. Structural isomers (often called *constitutional isomers*) are molecules of the same formula, but differing connectivity. Their constituent parts may well be different (but need not be). Butane (two methyl groups and two methylene groups) and isobutane (three methyl groups and one methine group) are typical examples of structural isomers (Fig. 5.53).

Stereoisomers have the same connectivity, but differ in the arrangement of their parts in space. Two kinds of stereoisomers exist: enantiomers and diastereomers. Enantiomeric molecules are nonsuperimposable mirror images of each other. Simple examples are (*R*)- and

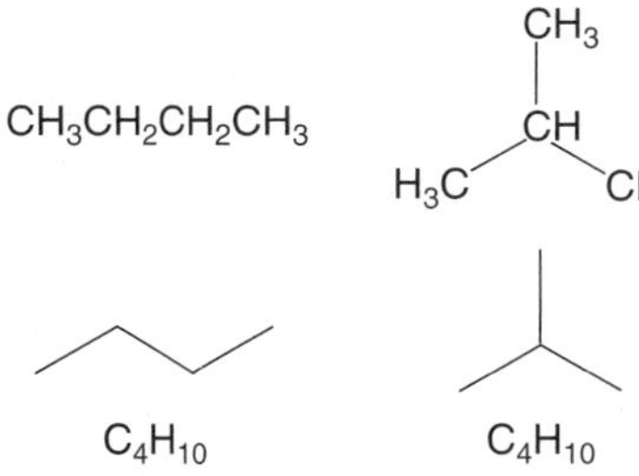

$CH_3CH_2CH_2CH_3$	CH_3–CH(H_3C)–CH_3
C_4H_{10}	C_4H_{10}

FIGURE 5.53 Butane and isobutane are structural isomers.

PROBLEM 5.25 Find a pair of structural isomers whose constituent parts are not different.

(*S*)-3-methylhexane. Slightly more complicated are (2*S*,3*S*)-dichlorobutane and its enantiomer, (2*R*,3*R*)-dichlorobutane. These pairs are not structural isomers, since they have the same connectivity (Fig. 5.54).

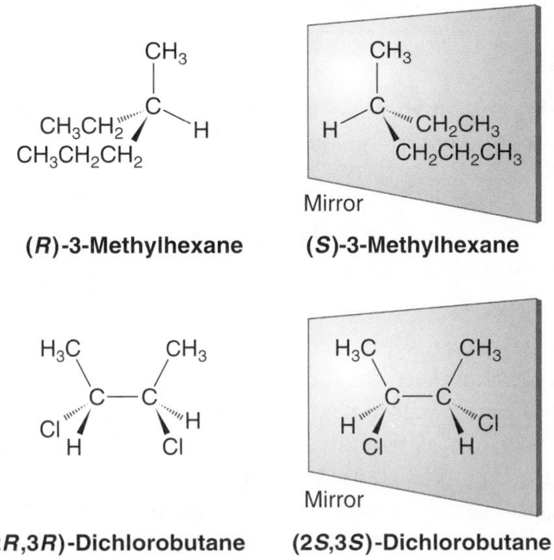

(*R*)-3-Methylhexane **(*S*)-3-Methylhexane**

(2*R*,3*R*)-Dichlorobutane **(2*S*,3*S*)-Dichlorobutane**

FIGURE 5.54 Some examples of stereoisomers.

Diastereomers include all stereoisomers that are not mirror images; *cis*- and *trans*-2-butene are examples, as are *meso*-2,3-dichlorobutane and either the (*R*, *R*) or (*S*, *S*) isomer of 2,3-dichlorobutane in Figure 5.54 (Fig. 5.55).

Stereoisomers also include conformational isomers, in which different isomers are generated through rotations about bonds. Eclipsed and staggered ethane are typical examples. Note that a conformational isomer need not be an energy minimum—the eclipsed conformation of ethane is an energy maximum, for example. Conformational isomers can be either enantiomeric or diastereomeric. The two *gauche* forms of butane are conformational enantiomers, but the *gauche* and the *anti* forms of butane are conformational diastereomers (Fig. 5.56).

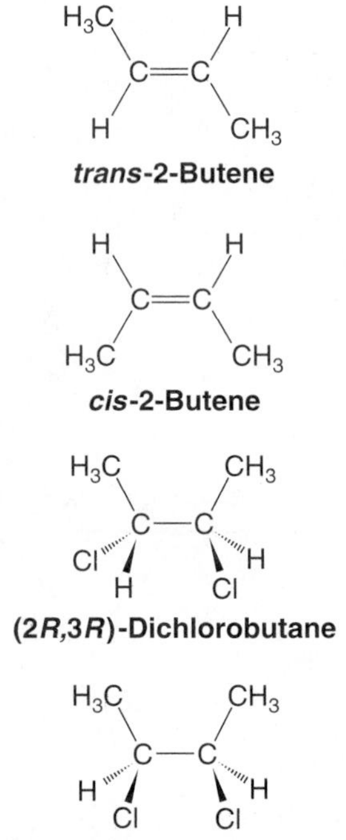

***trans*-2-Butene**

***cis*-2-Butene**

(2*R*,3*R*)-Dichlorobutane

***meso*-2,3-Dichlorobutane**
(2*S*,3*R*)-dichlorobutane

FIGURE 5.55 Some examples of diastereomers.

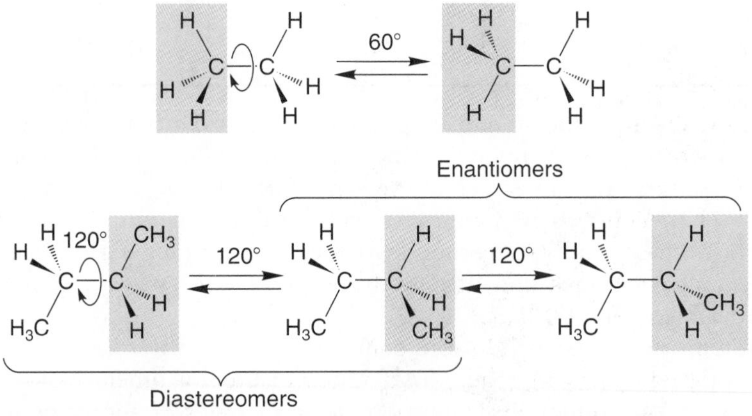

Enantiomers

Diastereomers

FIGURE 5.56 Some examples of conformational isomers.

Show that the two conformers labeled "enantiomers" in Figure 5.56 really are mirror images.

*PROBLEM **5.26**

ANSWER

The trick is to find a way to make an easy comparison. In this case, it takes only a simple 180° rotation to put the molecules into the proper orientation. Now it is easy to see that the two molecules from Figure 5.56 [labeled (a) and (b)] are indeed mirror images.

(a) 180° Mirror (b)

5.12 SOMETHING MORE: CHIRALITY WITHOUT "FOUR DIFFERENT GROUPS ATTACHED TO ONE CARBON"

Is the presence of a carbon atom attached to four different groups a necessary condition for chirality? No. We asked this question first on page 158. This section answers it by giving a number of examples of chiral molecules without this kind of atom.

trans-Cyclooctene is chiral, yet it contains no carbon atom attached to four other groups. Figure 5.57 gives one representation of the nonsuperimposable mirror images of this molecule, but you may need to use models to convince yourself that two enantiomers really do exist. Be careful! This molecule is racemized by rotation around the carbon–carbon double bond, and your models are likely to be eager to do exactly that.

Two double bonds can be connected directly to each other. We will see these **allenes** again in detail in Chapter 12, but you are well equipped after this chapter to understand the bonding and geometry of an allene (Fig. 5.58).

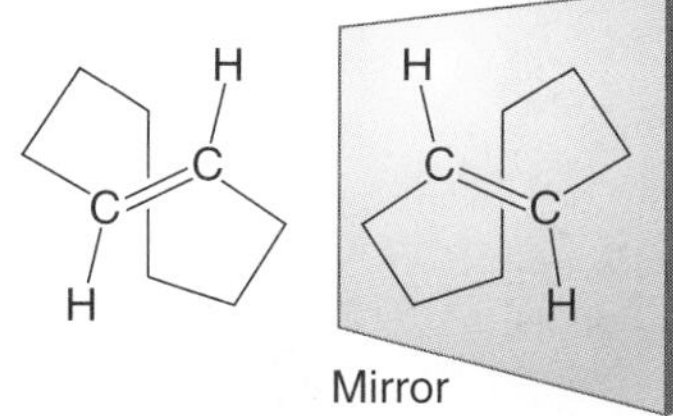

FIGURE **5.57** *trans*-Cyclooctene is a chiral molecule.

$H_2C{=}C{=}CH_2$

FIGURE **5.58** Allene is a molecule containing two double bonds that share one carbon.

What is the hybridization of each carbon of allene?

*PROBLEM **5.27**

ANSWER

$H_2C{=}C{=}CH_2$
Allene

The end carbons are attached to three groups (two hydrogens and the other carbon), and are therefore hybridized sp^2. The central carbon is attached to two other groups (the two methylene groups), and therefore is hybridized *sp*.

Make a careful, three-dimensional drawing of 1,3-dimethylallene. Pay special attention to the geometry at the ends of the molecule.

*PROBLEM **5.28**

ANSWER

The key to this problem is to notice that overlap of the 2*p* orbitals on the end sp^2 carbons with the 2*p* orbitals on the central, *sp* hybridized carbon requires that the

ANSWER (CONTINUED) groups on the end carbons be in perpendicular planes.

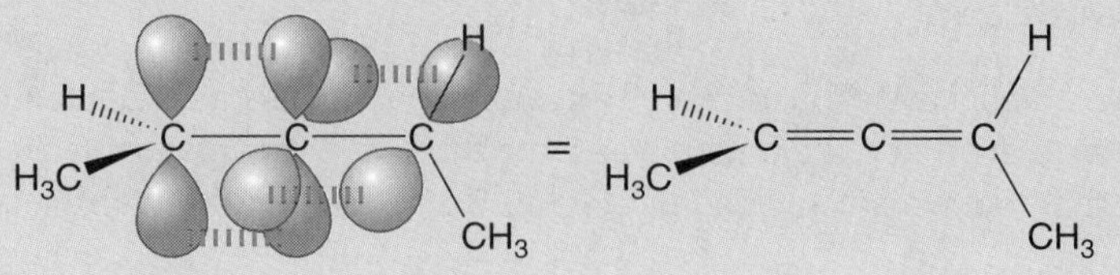

*PROBLEM 5.29 Draw the mirror image of 1,3-dimethylallene and verify that the two enantiomers are not superimposable.

ANSWER The mirror images of 1,3-dimethylallene are not superimposable, although it may take some work with models before you are certain!

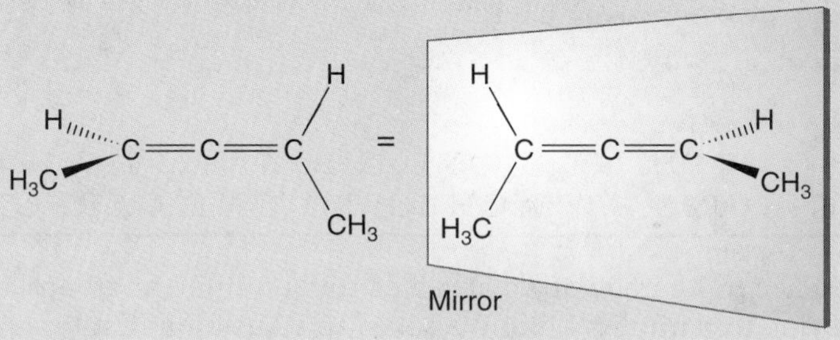

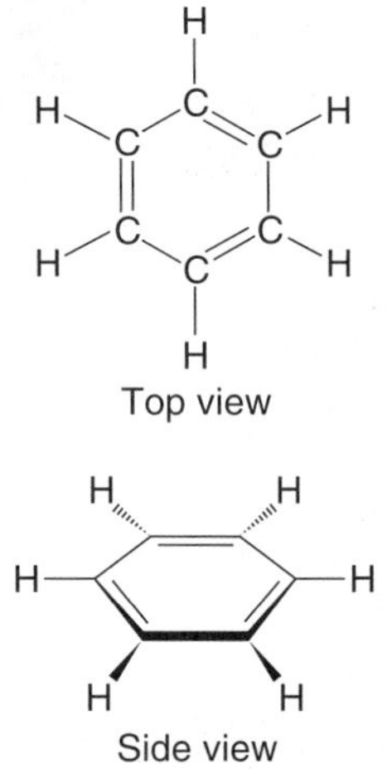

Mirror

In short, substituted allenes can be chiral, although these molecules have no "carbons attached to four different groups."

There exists a flat molecule, C_6H_6, called benzene. Benzene consists of a six-membered ring of carbons to which hydrogens are attached, one at each vertex (Fig. 5.59). The key point is that benzene is flat as a pancake (for much more on benzenes see Chapters 13 and 14). Two benzenes can be attached to each other to make a molecule called "biphenyl" (Fig. 5.60). If we now substitute the four carbons directly adjacent to the junction between the two rings in the correct way, a chiral molecule can be achieved! It is necessary that the substituents be large enough so that rotation about the junction between the two rings is restricted (Fig. 5.60).

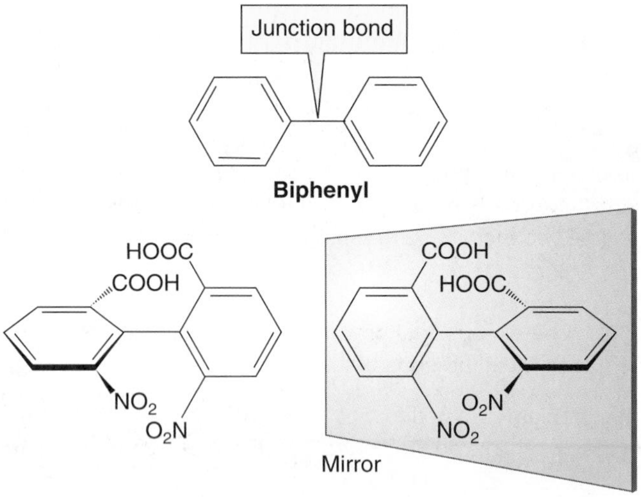

Top view

Side view

FIGURE 5.59 Two views of benzene.

Junction bond

Biphenyl

Mirror

FIGURE 5.60 Biphenyl and a substituted (chiral) biphenyl.

Show that rotation about the junction bond will racemize the two enantiomers of Figure 5.60.

PROBLEM 5.30

Finally, we can use benzene in another way to construct an odd chiral molecule. Two benzenes can be attached in what is called a "fused" fashion (Fig. 5.61).

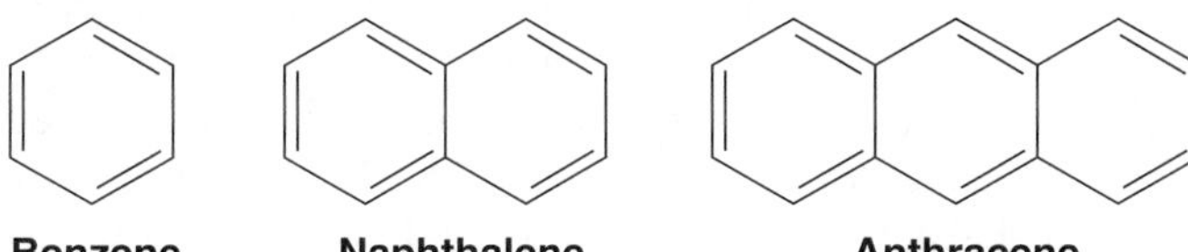

FIGURE 5.61 Some fused benzenes.

We can imagine continuing this process in several ways. We might simply continue to add benzene rings in a linear fashion, as in Figure 5.61. Or, we might add the restriction that new rings be added so as to produce a curve (Fig. 5.62).

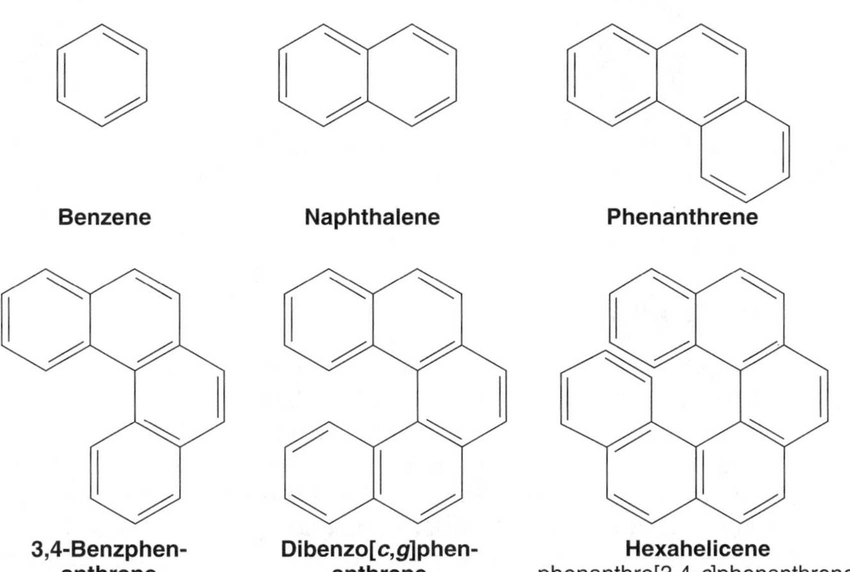

FIGURE 5.62 Hexahelicene is chiral. Why?

If we follow the latter course, the addition of the sixth ring produces a chiral molecule!

Explain why hexihelicane is chiral. You may need models to do this.

PROBLEM 5.31

5.13 SUMMARY

NEW CONCEPTS

This chapter deals almost exclusively with the concept and consequences of the "handedness" of molecules—chirality. Some molecules are related to each other as are your left and right hands. These are nonsuperimposable mirror images, or enantiomers.

An absolutely safe method of finding chiral molecules is to examine the mirror image to see if it is superimposable on the original. In particular, finding a carbon attached to four different groups is neither a necessary nor sufficient condition for chirality. A meso compound is an example of an achiral molecule containing carbons attached to four different groups.

Enantiomers have identical physical properties except that one stereoisomer rotates the plane of plane-polarized light by some amount to the right, whereas the other rotates the plane by the same amount but in the opposite direction. Enantiomers have identical chemistries with achiral molecules, but interact differently with other chiral molecules.

An examination of the consequences of the presence of more than one stereogenic carbon in a single molecule leads to the discovery of diastereomers; stereoisomers that are not mirror images.

REACTIONS, MECHANISMS, AND TOOLS

The arbitrary, but important (*R/S*) convention is based upon the Cahn–Ingold–Prelog priority system and allows us to specify absolute configuration of molecules.

Resolution, or separation of enantiomers, is generally accomplished by allowing the racemic mixture of enantiomers to react with a single enantiomer of a chiral agent to form a pair of diastereomers. Diastereomers, unlike enantiomers, have different physical properties and can be separated by crystallization or other techniques. If the original chemical reaction that formed the diastereomers can be reversed, the pure enantiomers can be regenerated.

COMMON ERRORS

This chapter continues the journey into three dimensions begun in Chapter 4. Most commonly encountered problems have to do with learning to see molecules in three dimensions and in particular, with the difficulty of translating from the two-dimensional page into the three-dimensional world. There can be no hiding the difficulty of this endeavor for many people, but it will yield to careful work and practice.

There are a few small errors to be careful to avoid. There are many arbitrary parts to the (*R/S*) convention, and this ensures that there is no way to figure them out. The (*R/S*) convention is something that must be memorized.

When a single enantiomer is drawn, be sure to ask yourself whether this is done deliberately, in order to specify *that particular enantiomer*, or simply as a matter of convenience to avoid the work of drawing both enantiomers of the racemic pair. Usually some indication will be made if it is indeed a single enantiomer that is meant.

Remember: Finding a carbon with four different groups attached is a fine way to *start* looking for chirality, but it is neither a sufficient nor necessary condition. Problem makers love to find exotic molecules that prove this point!

5.14 KEY TERMS

Achiral Not chiral.

Absolute configuration The arrangement in space of the atoms in an enantiomer.

Allene A 1,2-diene. A compound containing a carbon atom that is part of two double bonds.

Chiral A chiral molecule is not superimposable on

its mirror image.

Chirality The ability of a molecule to exist in two nonsuperimposable mirror-image forms; "handedness."

Conformational enantiomers Enantiomers interconvertable through (generally easy) rotations around bonds within the molecule.

Dextrorotatory The rotation of the plane of plane-polarized light in the clockwise direction.

Diastereomers Stereoisomers that are not mirror images.

Enantiomers Nonsuperimposable mirror images.

Levorotatory The rotation of the plane of plane-polarized light in the counterclockwise direction.

Meso compound An achiral compound containing stereogenic atoms.

Optical activity The rotation by a molecule of the plane of plane-polarized light.

Polarimeter A device for measuring the amount of rotation of the plane of plane-polarized light.

Racemic mixture (racemate) A mixture containing equal amounts of two enantiomeric forms of a chiral molecule.

Resolution The separation of a racemic mixture into its constituent enantiomeric molecules.

Stereochemistry The physical and chemical consequences of the arrangement in space of the atoms in molecules.

Stereogenic center An atom, usually carbon, of such nature and bearing groups of such nature that it can have two nonequivalent configurations.

Structural isomers Molecules of the same formula but with different connectivities among the constituent atoms.

5.15 ADDITIONAL PROBLEMS

PROBLEM **5.32** Find all the chiral isomers of the heptanes (Fig. 3.52, p. 103). Designate the stereogenic atoms with an asterisk (*).

PROBLEM **5.33** Draw the (*S*) enantiomers of these compounds.

PROBLEM **5.34** Find all the chiral isomers of the octanes (Problem 3.23, p. 104).

PROBLEM **5.35** Draw the (*R*) enantiomers of all of these compounds whose name must end in "pentane."

PROBLEM **5.36** Now find all the chiral isomers of the nonanes (Problem 3.40, p. 97). Indicate those with more than one stereogenic carbon.

PROBLEM **5.37** Draw three-dimensional pictures of all the steroisomers of 3,4-dimethylheptane and 3,5-dimethylheptane. It is probably easiest to draw them in the eclipsed arrangement, even though this is not a low-energy conformation. Determine the absolute configuration (*R* or *S*) of each stereogenic carbon. Designate the stereogenic carbons with an asterisk.

PROBLEM **5.38** Find all the chiral isomers of the hexenes (Problem 4.5, p.147) and heptenes (Problem 4.27, p. 153).

PROBLEM **5.39** Find all the chiral isomers of the hexynes (Problem 4.20, p. 147), heptynes (Problem 4.20, p. 147), and octynes (Problem 4.28, p. 153).

PROBLEM **5.40** In Problem 3.31, you drew all the isomers of "chlorohexane" $C_6H_{13}Cl$. Which of these isomers contains stereogenic carbons? Be sure to be alert for molecules containing more than one stereogenic carbon. Designate the stereogenic carbons with an asterisk.

PROBLEM **5.41** Draw three-dimensional pictures of the stereoisomers of the molecules in your answer to Problem 5.40 that contain more than one stereogenic carbon.

PROBLEM **5.42** Which of the following molecules can exist as cis/trans (*Z/E*) isomers? Draw the (*E*) and (*Z*) forms. Which molecules possess stereogenic carbons? Designate the stereogenic carbons with an asterisk.

(a) $(CH_3)_2C{=}CHCH_2CH_2CH(CH_3)CH_2CH_2OH$

(b) $(CH_3)_2C{=}CHCH_2CH_2C(CH_3){=}CHCH_2OH$

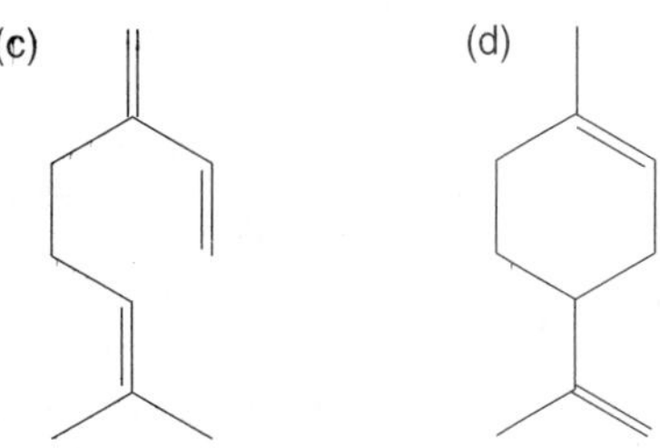

PROBLEM **5.43** Draw the (*R*) enantiomer for each molecule containing a stereogenic carbon in Problem 5.42.

PROBLEM **5.44** Draw all the stereoisomers of the following molecule. Identify all pairs of enantiomers and diastereomers.

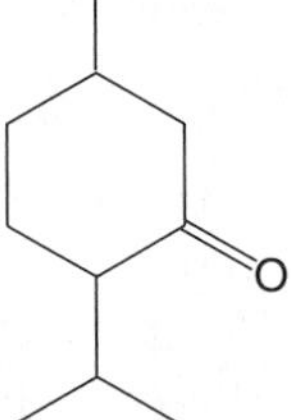

PROBLEM **5.45** In Problem 4.35 (p. 153), you worked out all the isomers of the formula C_4H_5Cl. Find all the chiral isomers of the ring compounds of this formula. Indicate the stereogenic carbons with an asterisk.

PROBLEM **5.46** Draw the four chiral isomers found in Problem 5.45, show the mirror images, and indicate the absolute configuration (*R* or *S*) of the stereogenic carbons.

PROBLEM **5.47** In Problem 4.33 (p. 153), you found nine cyclopropanes of the formula $C_4H_6Br_2$. Find the chiral ones. Indicate the stereogenic carbons with an asterisk.

PROBLEM **5.48** One of the isomers in the answer to Problem 5.47 has a pair of bromines on one carbon. Draw the stereoisomers of this molecule, and use the (*R/S*) convention to show the absolute configuration of the stereogenic carbons.

Rings

Clearly the ring had an unwholesome power...
—J. R. R. Tolkien*
Lord of the Rings

In these early chapters, we have used our powers of imagination to picture the three dimensionality of organic molecules. But nowhere is thinking in three dimensions more important than in depicting the myriad structures formed from chains of atoms tied into rings. It's not always easy to see these structures clearly, so do not be reluctant to work with models, especially at the beginning. All organic chemists use them. Polycyclic compounds are especially complicated. No one can see all the subtlety of these compounds without models. Much of chemistry involves interactions of groups in proximity, and the two-dimensional page is often quite ineffective in showing which atoms are close to others. As you go on in this subject, your ability to use the two-dimensional surface to depict the three-dimensional molecular world will increase, but none of us will ever outgrow the need for models. Rings were first encountered in Section 3.12, and Figure 6.1 recalls some simple and complex ring compounds.

Two schematic views of cyclobutane

$H_2C—CH_2$
$H_2C—CH_2$

A complicated, polycyclic compound, bicyclo[2.2.1]heptane

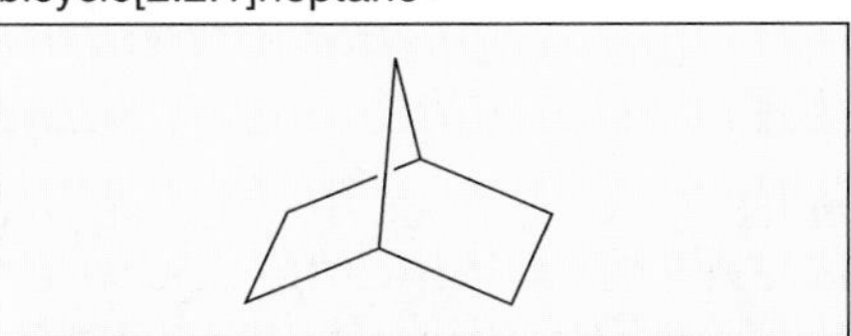

Two schematic views of cyclopentane

H_2C, CH_2, CH_2, H_2C, CH_2

A really complicated, polycyclic compound that nobody tries to name systematically — it's called "dodecahedrane"

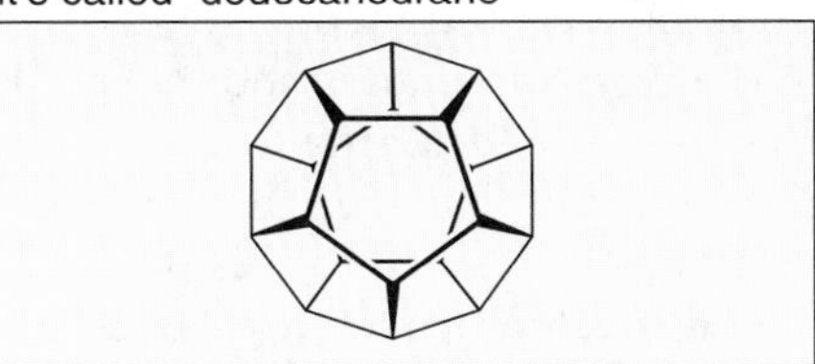

FIGURE 6.1 Some simple and complex ring structures.

*John Ronald Reuel Tolkein (1892–1973) was Merton Professor of English language at Oxford University.

Spectacular new molecules containing rings are always appearing. In 1984, Professor Kenneth B. Wiberg (b. 1927) of Yale University reported the construction of the polycyclic molecule, tricyclo[1.1.1.$0^{1,3}$]pentane (also known as [1.1.1]propellane, Fig. 6.2), a marvelously exciting molecule. You are now only six chapters into your study of organic chemistry, yet it is possible for you to appreciate why the chemical world was knocked out by this compound. So that becomes Problem 6.1. (How many introductions contain problems? Sadism abounds, but at least this molecule probably won't appear on an exam.)

Tricyclo[1.1.1.$0^{1,3}$]pentane ([1.1.1]propellane) **Bicyclo[1.1.1]pentane**

FIGURE 6.2 The strange ring compound, tricyclo[1.1.1.$0^{1,3}$]pentane ([1.1.1]propellane), and its more prosaic cousin, bicyclo[1.1.1]pentane.

*PROBLEM 6.1 Why is tricyclo[1.1.1.$0^{1,3}$]pentane unusual? Would you expect it to be especially stable or unstable with respect to its cousin bicyclo[1.1.1]pentane? Why?

ANSWER This molecule has all sorts of problems. Of course, it contains no less than three cyclopropane rings, and that alone will introduce severe strain. However, the factor that differentiates this molecule from the "merely" severely strained is that all four valences of the "bridgehead" carbons are aimed in the same direction! What kind of hybridization can be involved? That's no easy question, and the chemical world has spent quite some effort at puzzling out just why this molecule is as stable as it is. Although that's not a reasonable subject for an introductory book, two things about this problem are important. First, we can see why this molecule is so remarkable. Second, the isolation of this molecule shows how imperfect our understanding of chemistry still is! However, one must admit that it is a pleasure to see challenging curiosities continue to appear. There is much to learn and many molecular marvels still hiding from us.

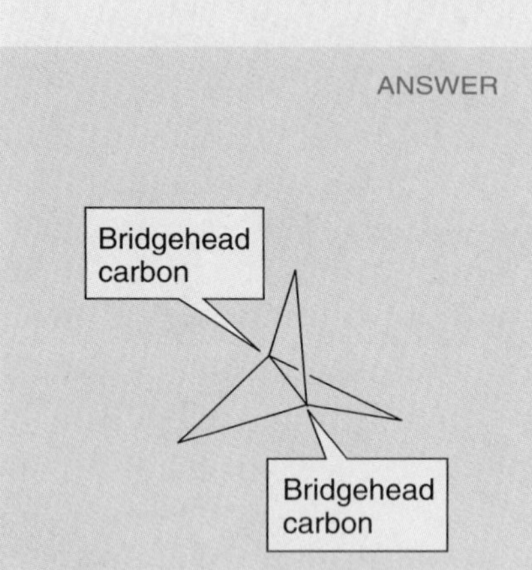

6.1 RINGS AND STRAIN

Much can be learned from the simplest of all rings, cyclopropane, $(CH_2)_3$. Indeed, even at this early stage we can make some preliminary judgments about this compound. In doing so, we would be recapitulating the thoughts of Adolph von Baeyer (1835–1917) who in 1885 appreciated the importance of the deviation of the internal angles in most cycloalkanes from the ideal tetrahedral angle, 109.5°. He suggested that the instabilities of the ring compounds should parallel the deviations of their internal angles from the ideal; the further the angle from tetrahedral, the less stable the molecule. The idea that **angle strain** is an important factor in ring stability has continued to this day (Fig. 6.3). In fact, this smallest ring compound, cyclopropane, represents an extreme example of the effects of angle strain. The strain induced by the reduction of the ideal tetrahedral angle of 109.5° to 60° is so great as to make other bonding arrangements

109.5° <109.5° >109.5°

A normal tetrahedron with a bond angle of 109.5°

A small ring will constrict this angle to less than 109.5°

A large ring will expand this angle to more than 109.5°

FIGURE 6.3 Ring compounds will have angle strain if angles are forced to be significantly smaller or larger than the ideal 109.5°.

possible. One way of looking at cyclopropane* views the bonds between the carbons as "bent." The *internuclear* angle in an equilateral triangle is, of course, 60° and cannot be otherwise. However, the *interorbital* angle is apparently quite a bit larger (Fig. 6.4) and is estimated at 104°.

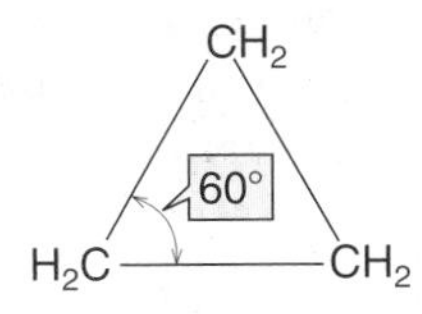

Cyclopropane
(internuclear angle is 60°)

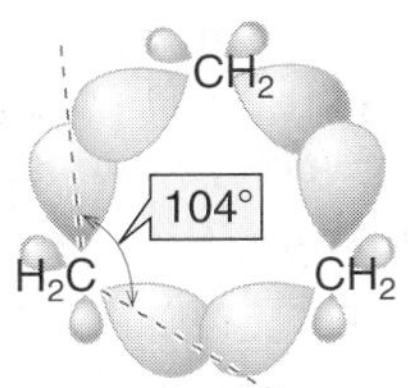

Cyclopropane
(interorbital angle is 104°)

FIGURE 6.4 The internuclear angle (60°) and the interorbital angle (~104°) are different for cyclopropane.

Of course, the carbon skeleton of cyclopropane is flat—three points determine a plane—and planarity causes still other problems. Figure 6.5 shows a Newman projection looking down one of the three equivalent carbon–carbon bonds of cyclopropane.

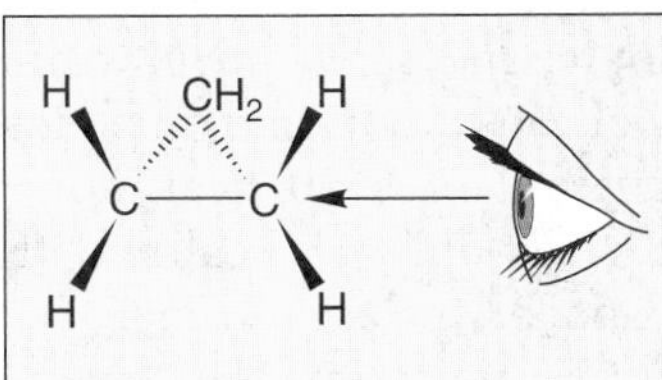

This Newman projection attempts to show the eclipsed C—H bonds; in cyclopropane all the C—H bonds are eclipsed

FIGURE 6.5 In cyclopropane, all carbon–hydrogen bonds are eclipsed. *Remember*: The solid wedges are coming toward you and the dashed wedges are retreating from you.

Note that the four carbon–hydrogen bonds shown in Figure 6.5 are eclipsed. We have seen this before in the eclipsed form of ethane (and other alkanes). Ethane has a simple way to avoid this destabilizing opposing of atoms and electrons; it adopts the staggered conformation and minimizes the problem. There is no way for cyclopropane to do this. Its hydrogens are stuck in the eclipsed arrangement and there is no possible release. We can even make an estimate of how much damage this **torsional strain** (eclipsing strain) is causing. The rotational barrier in ethane is the result of three pairs of eclipsed hydrogens (Chapter 3, p. 86), and amounts to 3 kcal/mol (Fig. 6.6).

*Whenever you see words such as "One way of looking ..." you should know that someone is simplifying, sometimes to excess. At best the author is uncertain and at worst he or she doesn't believe what follows.

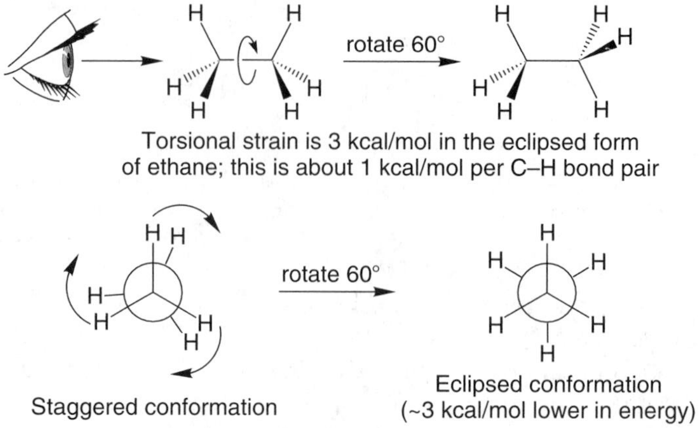

rotate 60°

Torsional strain is 3 kcal/mol in the eclipsed form of ethane; this is about 1 kcal/mol per C–H bond pair

rotate 60°

Staggered conformation

Eclipsed conformation (~3 kcal/mol lower in energy)

FIGURE 6.6 The destabilization introduced by a pair of eclipsed carbon–hydrogen bonds is about 1 kcal/mol.

As mentioned in Chapter 3 (p. 86), a reasonable guess would put the energy cost of each pair of eclipsed hydrogens at about 1 kcal/mol. The total strain of cyclopropane is composed of torsional strain (~6 × 1 = 6 kcal/mol) plus angle strain, which is much more difficult to estimate without the help of calculations or experiments (Fig. 6.7).

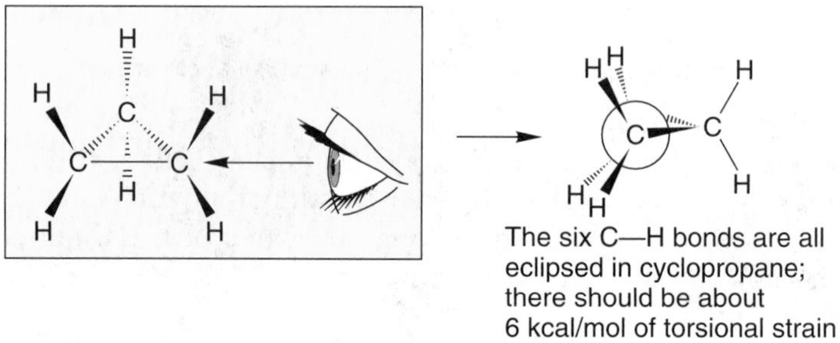

The six C—H bonds are all eclipsed in cyclopropane; there should be about 6 kcal/mol of torsional strain

FIGURE 6.7 The six pairs of eclipsed carbon–hydrogen bonds should produce about 6 kcal/mol of torsional strain in cyclopropane.

The effects of high strain in cyclopropane show up in a number of ways, including the unusual reactivity of the carbon–carbon bonds. Breaking the carbon–carbon bond in ethane requires 90 kcal/mol, but the carbon–carbon bond in cyclopropane needs only 65 kcal/mol (Fig. 6.8). This difference allows a first estimation of the strain energy of cyclopropane (see p. 200 for two more quantitative estimates). Strain destabilizes cyclopropane by about 25 (90 – 65) kcal/mol.

Ethane
Bond dissociation energy about 90 kcal/mol

Cyclopropane
Bond dissociation energy about 65 kcal/mol

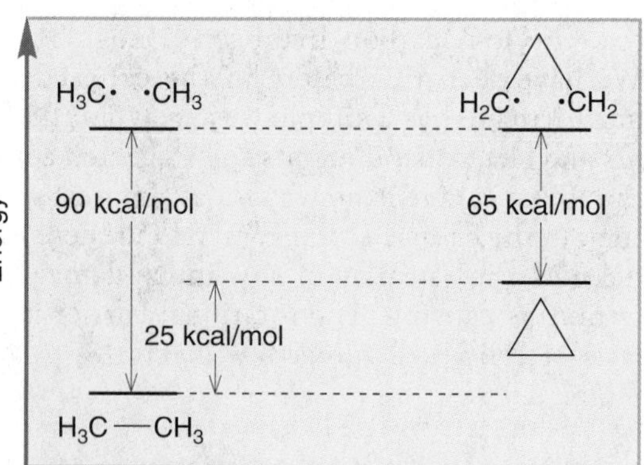

FIGURE 6.8 The bond dissociation energy of a carbon–carbon bond in cyclopropane is only 65 kcal/mol. Cyclopropane is strained by about 90 – 65 = 25 kcal/mol.

At first glance, the other cycloalkanes would seem to have similar difficulties. In cyclobutane, for example, angle strain is less of a problem as cyclobutane's 90° is a smaller deviation from the ideal 109.5° than is cyclopropane's 60°. Along with this relief, however, comes increased torsional strain, as there are four pairs of hydrogens eclipsed as opposed to cyclopropane's three pairs (Fig. 6.9). One might estimate the torsional strain in planar cyclobutane at 8 kcal/mol (1-kcal/mol strain for each pair of eclipsed carbon–hydrogen bonds).

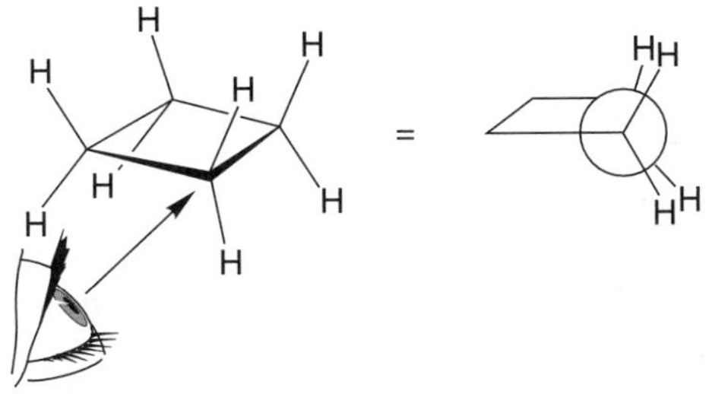

FIGURE 6.9 There are eight pairs of eclipsed carbon–hydrogen bonds in cyclobutane, and therefore about 8 kcal/mol of torsional strain.

Unlike cyclopropane, cyclobutane has a way to balance torsional and angle strain. The four-membered ring need not be planar, but can distort or "pucker" somewhat if this will result in an energy lowering. Let's look at the consequences of puckering the ring—of moving one methylene group out of the plane of the other three (Fig. 6.10).

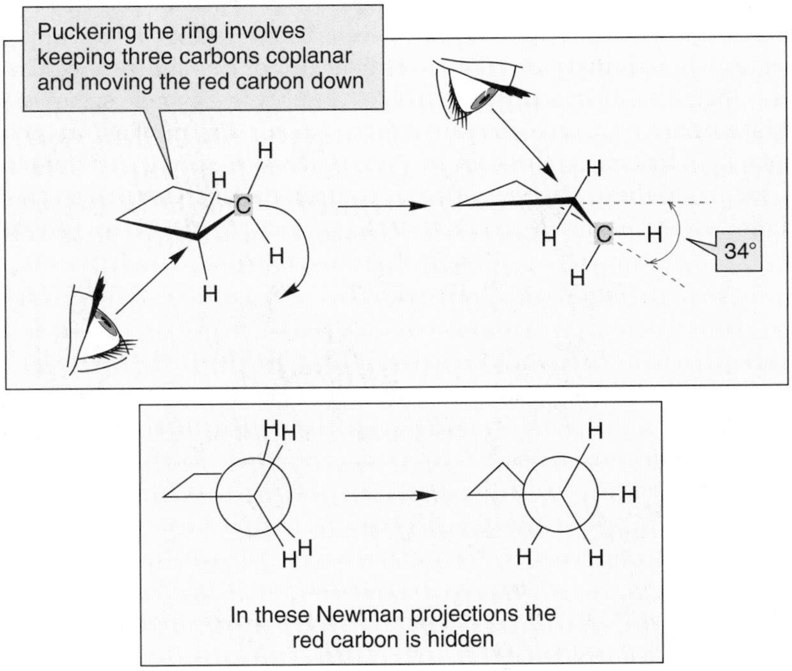

Puckering the ring involves keeping three carbons coplanar and moving the red carbon down

34°

In these Newman projections the red carbon is hidden

FIGURE 6.10 Puckering relieves torsional strain in cyclobutane.

As the ring puckers, torsional strain is reduced, but angle strain is increased. A balance between the two effects is struck in which the ring puckers about 34° and the C–C–C angle closes to about 88°. This form is not static, however; the ring is in motion, much as are the acyclic alkanes. The planar form of cyclobutane is like the eclipsed form of ethane and lies not in an energy well, but at the top of an energy barrier separating a pair of puckered cyclobutanes. This barrier is very small, about 1.4 kcal/mol (Fig. 6.11).

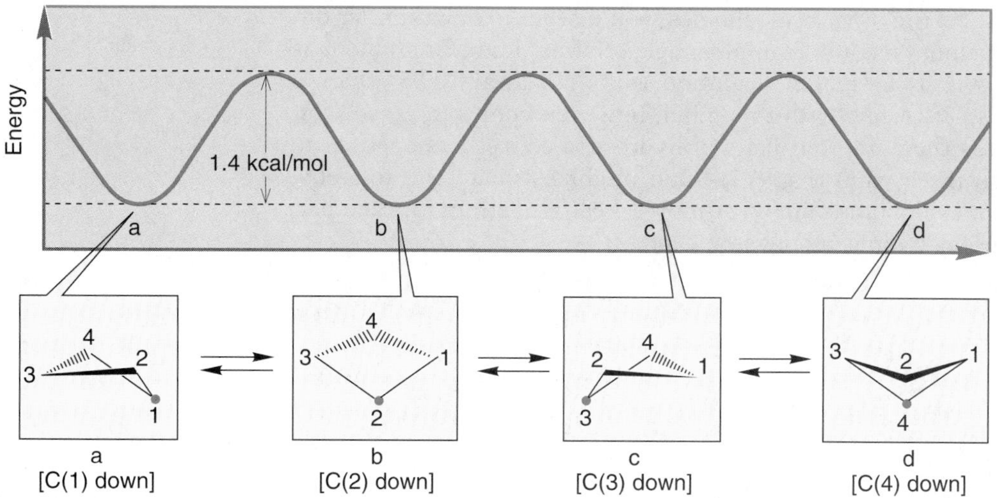

FIGURE 6.11 Cyclobutane is a mobile, not static, molecule in which different nonplanar forms rapidly interconvert. The planar form is the high-energy point (transition state) separating the energy minima, a, b, c, and d, which are equivalent forms of nonplanar cyclobutane. In this figure, the bending is exaggerated—the flap is really only 34° out of plane.

Let's now look at cyclopentane, $(CH_2)_5$, the next larger cycloalkane, and see how it distorts in a similar way. Were cyclopentane planar, it would suffer the torsional strain induced by five pairs of eclipsed hydrogens. As each pair of eclipsed hydrogens contributes about 1 kcal/mol, this amounts to the rather large figure of 10-kcal/mol destabilization. The internal angle in a planar pentagon is 108°, however, so angle strain would be quite small. As in cyclobutane, the ring distorts from planarity, relieving some eclipsing at the cost of increased angle strain. For nonplanar cyclopentane, there are two forms of comparable energy, the "envelope" and the "twist" form (Fig. 6.12).

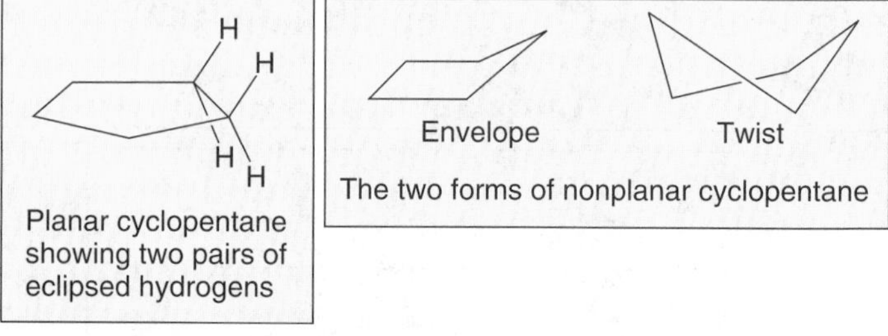

FIGURE 6.12 Cyclopentane is also nonplanar. In this case there are two forms that achieve nearly the same energy.

PROBLEM 6.2 Draw a Newman projection looking down one of the carbon–carbon bonds in planar and envelope cyclopentane. Use models!

Neither the puckered envelope form nor the twist form of cyclopentane is static. In the envelope form the "flap" moves around the ring, generating the five possible puckered isomers. This motion requires only a series of rotations around carbon–carbon bonds and is summarized in Figure 6.13. A model will help you to visualize this motion. Hold two adjacent carbons, sight down the bond that connects them, and convert one form of the envelope into another.

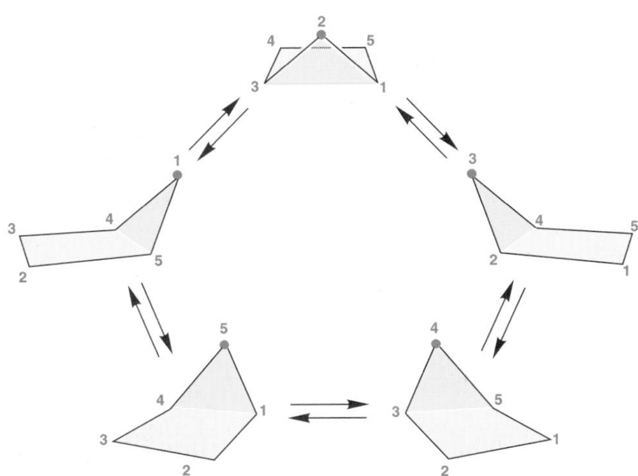

FIGURE **6.13** The five equilibrating nonplanar envelope cyclopentanes. The out-of-plane carbon in each form is indicated with a red dot.

By now you must be expecting that a similar distortion from planarity will be present in cyclohexane, $(CH_2)_6$, but you are probably prepared neither for the magnitude of the effect nor its consequences for the structure and stability of cyclohexanes. In contrast to the small rings, distortion from planarity in cyclohexane relieves both the angle and torsional strain of the planar structure. The internal angle in a planar hexagon is 120°, larger, not smaller, than the ideal sp^3 angle of 109.5°. Deviation from planarity will *decrease* this angle and thus *decrease* angle strain. Torsional strain from the six pairs of eclipsed hydrogens in planar cyclohexane can be expected to contribute about 12 kcal/mol of strain, which is also decreased in a nonplanar form. So in cyclohexane, *both* angle and torsional strain will be relieved by relaxing from a planar structure. Remarkably, this relaxation produces a molecule in which essentially all of the torsional and angle strain is gone. The formation of the energy minimum cyclohexane, called the "chair" form, is shown in Figure 6.14. The internal angle in cyclohexane (111.5°) is close to the ideal C–C–C angle in a simple straight-chain alkane (112° in propane) and the carbon–hydrogen bonds are nicely staggered. Newman projections looking down the carbon–carbon bonds show this, as will (even better) a look at a model.

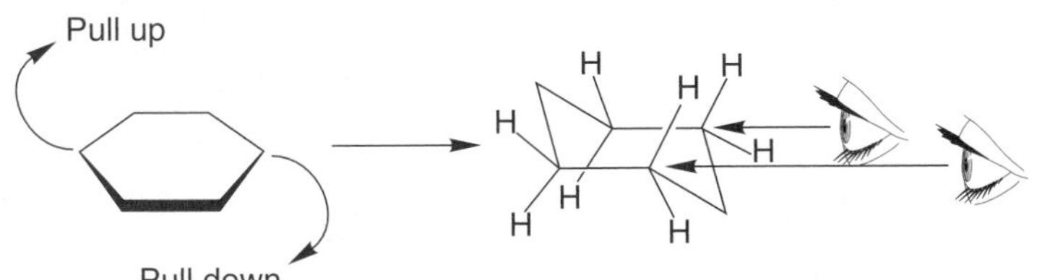

Planar cyclohexane: 120° C–C–C bond angles and about 12-kcal/mol torsional strain from C–H bond eclipsing; there will also be angle strain from the too wide 120° C–C–C internal angles

Chair cyclohexane 111.5° C–C–C bond angles; about 0 kcal/mol torsional strain

A double Newman projection of the chair form of cyclohexane; note the perfect staggering of the C–H bonds

FIGURE **6.14** Planar and chair cyclohexane.

The details of cyclohexane stereochemistry are important enough to warrant a lengthy discussion (Section 6.3, p. 204), but we can do a few things here in preparation. First of all, it is necessary to learn to draw a decent cyclohexane. No person can truly be described as educated unless he or she can do this, and anyone can, regardless of artistic ability. So learn how to do it and the next time your roommate mentions some obscure European writer, impressing you with his or her erudition and calling into question your sophistication, confound your tormenter with a perfectly drawn cyclohexane!

There are a few tricks that make the construction of a perfect drawing easy. First of all, within the ring, opposite bonds are parallel (Fig. 6.15). With a little practice, keeping the proper bonds parallel should let you easily draw the carbon framework of cyclohexane.

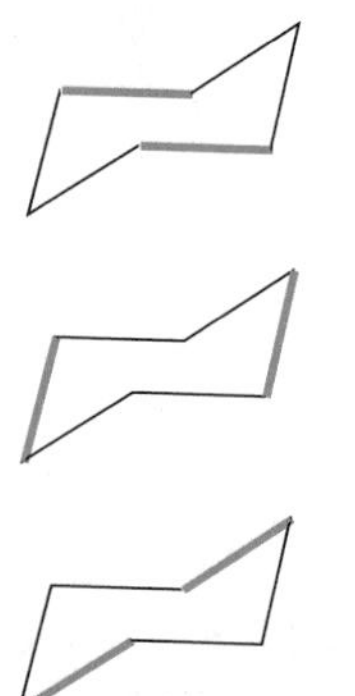

FIGURE 6.15 In chair cyclohexane, there are three pairs of parallel carbon–carbon bonds.

It is a bit more difficult to get the hydrogens right. It helps to tip the ring a little and not to draw it with the "middle" bonds flat (Fig. 6.16). Now we see the molecule as it would rest on a flat surface. Six of the carbon–hydrogen bonds are easy to draw; three of them point straight up and three straight down. These are called the **axial hydrogens**.

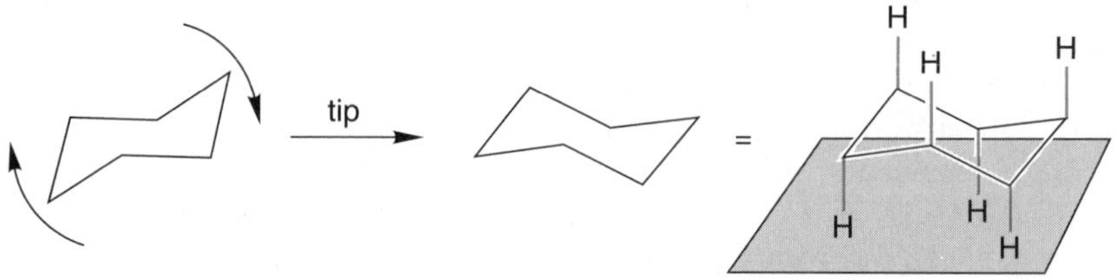

FIGURE 6.16 The set of six "straight up and down" or "axial" carbon–hydrogen bonds of chair cyclohexane are shown in red. All axial carbon–hydrogen bonds are parallel.

So far, so good, but it is the positioning of the last six hydrogens that gives people the most trouble. To get them right, take advantage of the parallel carbon–carbon bonds encountered before in drawing the cyclohexane ring. Each member of this second set of six **equatorial** carbon–hydrogen bonds is parallel to the two carbon–carbon bonds one bond away. Like the six axial carbon–hydrogen bonds, the equatorial bonds also alternate up and down but this set points only slightly up or slightly down (Fig. 6.17).

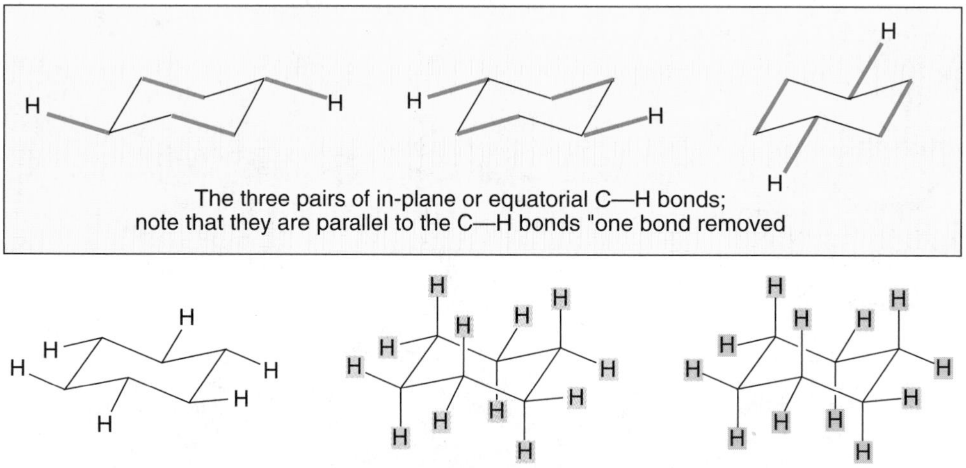

The three pairs of in-plane or equatorial C—H bonds; note that they are parallel to the C—H bonds "one bond removed

The six equatorial hydrogens of chair cyclohexane

All 12 C—H bonds of chair cyclohexane; axial hydrogens are shown in red, equatorial hydrogens in blue

Up hydrogens in green, down hydrogens in gray

FIGURE 6.17 The equatorial carbon–hydrogen bonds are parallel to the ring carbon–carbon bonds one bond away. The drawing of a perfect chair cyclohexane is now complete.

Now you have a perfect cyclohexane ring in its energy-minimum chair form. This exercise in drawing lets us see one important thing immediately. Note again that there are two different kinds of hydrogen. There is the "straight up and down" set of six, which are called the axial hydrogens, and the other set of six, which lie roughly along the equator of the molecule and are called the equatorial hydrogens (Fig. 6.18).

The six axial hydrogens of chair cyclohexane

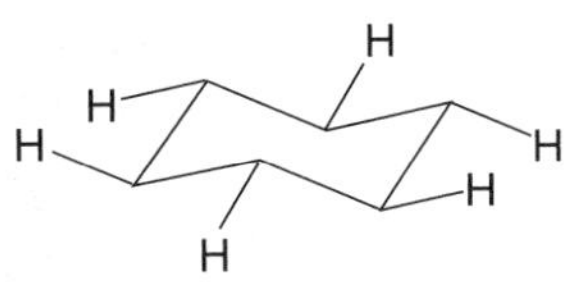

The six equatorial hydrogens of chair cyclohexane

FIGURE 6.18 The two sets of hydrogens in chair cyclohexane: axial and equatorial.

*PROBLEM 6.3

What happens to a set of axial hydrogens when you convert one chair into another? Try it with models. This problem is important and simple.

ANSWER

The set of axial hydrogens becomes the set of equatorial hydrogens! An *up* hydrogen always remains *up*, however. Similarly, a *down* hydrogen is always *down*, whether axial or equatorial.

Up axial becomes up equatorial

ring flip

Down axial becomes down equatorial

Larger rings will also be nonplanar, but no more spectacular surprises such as cyclohexane's energy minimum form remain. Planar heptagons and octagons have internal angles of 129° and 135°, respectively, and thus have substantial angle strain. Both seven- and eight-membered rings would have severe torsional strain were they to remain planar. There is relaxation to nonplanar forms but some amount of strain always remains in these medium-sized rings. It is a complicated business to analyze medium-sized rings completely, and we will not embark upon it. Rings with minimal angle strain will have severe interactions between the carbon–hydrogen bonds, and thus substantial torsional strain. Relieving those eclipsings usually induces increased angle strain. As in cyclopentane, there is a compromise to be made between minimizing angle and torsional strain, and Nature does the best it can, finding the lowest energy structural compromise. Unlike cyclohexane, however, in the medium-sized rings there is no obvious low-energy solution. Often there are several somewhat different minima, close to each other in energy, and separated from each other by quite low barriers.

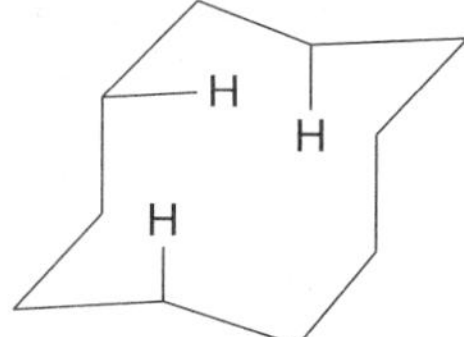

FIGURE 6.19 The structure of cyclodecane shows partial cyclohexane rings. A number of destabilizing interactions between carbon–hydrogen bonds are apparent.

Fragments of chair cyclohexanes are sometimes seen in the structures of medium-sized ring compounds. Cyclodecane is an example, as Figure 6.19 shows. One can also see some of the destabilizing interactions between carbon–hydrogen bonds in this structure.

As rings get even bigger, the strain energy decreases. The limit is an infinitely large ring, which resembles an endless chain of methylene groups. In such a species, it is possible to stagger all carbon–hydrogen bonds, and, of course, in an infinite ring there is no angle strain. So after cyclohexane, the next strain-free species is reached when the ring size has grown large enough to approximate an infinitely large ring.

6.2 QUANTITATIVE EVALUATION OF STRAIN ENERGY

Strain is an important factor in chemical reactivity. The more strained a compound, the higher its energy. The higher in energy a compound, the more likely it is to react. We have just seen how a combination of torsional and angle strains can act to change the energy of a ring compound, cyclopropane. Now we move on to some quantitative measures of the energies of ring compounds. We introduce some general techniques and look ahead to our detailed consideration of energy in Chapter 8.

6.2a Heats of Formation

To recapitulate material from Chapter 4 (p. 133), the heat of formation (ΔH_f°) of a compound is the enthalpy of its formation by the reaction of its constituent elements in their standard states. The standard state of an element is the most stable state at 25 °C and 1-atm pressure. For an element in its standard state, ΔH_f° is taken as zero. The more negative (or less positive) a compound's ΔH_f°, the more stable it is. A negative ΔH_f° for a compound means that its formation from its constituent elements would be *exothermic*—heat would be liberated. By contrast, a positive ΔH_f° means that the constituent elements are more stable than the compound and its formation would be *endothermic*—energy would have to be applied (Fig. 6.20).

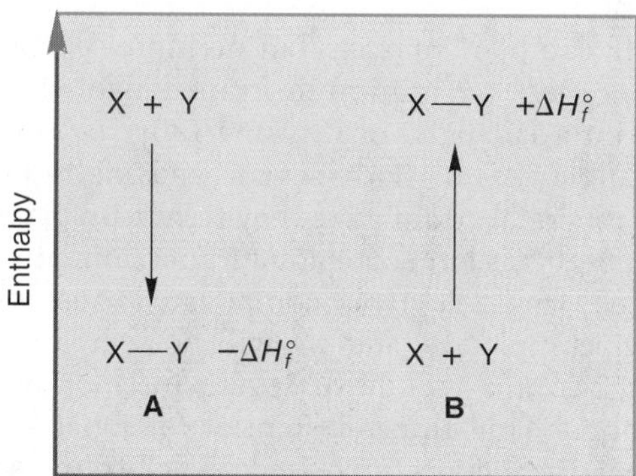

FIGURE 6.20 For a compound more stable than its constituent elements, ΔH_f° is negative. In **A** the compound X–Y is more stable than its constituent elements X and Y; therefore ΔH_f° of X–Y is negative. For a compound less stable than its constituent elements, ΔH_f° is positive. In **B** the constituent elements X and Y are more stable than the compound X–Y, therefore ΔH_f° of X–Y is positive. The more negative (less positive) the heat of formation, the more stable the compound.

Table 6.1 collects the heats of formation for some cycloalkanes. It also calculates the ΔH_f° *per methylene group* for these hydrocarbons.

TABLE **6.1** Heats of Formation for Some Cycloalkanes

Molecule	ΔH_f° (kcal/mol)	ΔH_f° per CH_2 (kcal/mol)
Cyclopropane	+12.7	+4.2
Cyclobutane	+ 6.8	+1.7
Cyclopentane	−18.7	−3.7
Cyclohexane	−29.5	−4.9
Cycloheptane	−28.3	−4.0
Cyclooctane	−29.7	−3.7
Cyclodecane	−36.9	−3.7
Cyclododecane	−55.0	−4.6
$(CH_2)_\infty$		−4.9

PROBLEM **6.4**

How would one get the value for a CH_2 group in an infinite chain, $(CH_2)_\infty$? Use the following data to estimate the value of ΔH_f° of an individual methylene group in kilocalories per mole: ΔH_f° (hexane) = −39.9, (heptane) = −44.8, (octane) = −49.8, (nonane) = −54.5, (decane) = −59.6 kcal/mol.

Note that the ΔH_f° for a methylene (CH_2) group in a strain-free straight-chain alkane, $(CH_2)_\infty$, is −4.9 kcal/mol. This value is exactly the ΔH_f° we calculate for a methylene group in cyclohexane. Cyclohexane really is strain-free. We can now use these data to calculate some strain energies for the cycloalkanes. First, calculate (Table 6.2) the ΔH_f° for the ring constructed from strain-free methylene groups. The difference between the calculated, *strain-free* ΔH_f° and the real, *measured* ΔH_f° is the strain energy (strain E in the table). For example, cyclopropane has a measured ΔH_f° of +12.7 kcal/mol. A strain-free cyclopropane would have a ΔH_f° of 3×-4.9 kcal/mol = −14.7 kcal/mol. The difference between these two values (27.4 kcal/mol) is the strain energy. The strain energy per CH_2 is (27.4)/3 = 9.1 kcal/mol.

TABLE **6.2** Strain Energies for Some Cycloalkanes

Molecule	ΔH_f° (kcal/mol)	ΔH_f° per CH_2 (kcal/mol)	Calcd ΔH_f° (kcal/mol)	Strain E (kcal/mol)	Strain E per CH_2 (kcal/mol)
Cyclopropane	+12.7	+4.2	−14.7	27.4	9.1
Cyclobutane	+ 6.8	+1.7	−19.6	26.4	6.6
Cyclopentane	−18.7	−3.7	−24.5	5.8	1.2
Cyclohexane	−29.5	−4.9	−29.4	0.1	0
Cycloheptane	−28.3	−4.0	−34.3	6.0	0.9
Cyclooctane	−29.7	−3.7	−39.2	9.5	1.2
Cyclodecane	−36.9	−3.7	−49.0	12.1	1.2
Cyclododecane	−55.0	−4.6	−58.8	3.8	0.3

Now look at the other rings. Cyclopropane and cyclobutane are about equally strained, although the strain per methylene is significantly higher for cyclopropane. Cyclopentane and cycloheptane are slightly strained, and the strain gets worse in the medium-sized rings until we reach a 12-membered ring. Strain declines in the larger rings until (with some aberrations) we reach the hypothetical strain-free infinite ring.

PROBLEM 6.5 Use models to look for the sources of strain in medium-sized rings.

6.2b Strain Analyzed by Heats of Combustion

Strain energies can also be determined from an analysis of heats of combustion (ΔH_c°). This method is attractive for analyzing strain because the measurements can be made and compared directly. Combustion of a hydrocarbon is an exothermic reaction. The products, H_2O and CO_2, are more stable than the starting hydrocarbon and oxygen. This difference, the exothermicity of the reaction, appears as heat and light (Fig. 6.21), which is easily apprehended by looking at the light and feeling the heat evolved by a propane stove.

$$C_3H_8 + 5\ O_2 \rightarrow 3\ CO_2 + 4\ H_2O + \text{Energy}$$

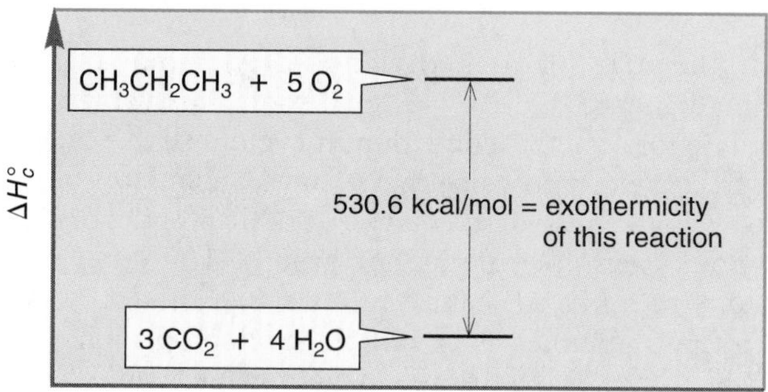

FIGURE 6.21 The combustion of propane to give CO_2, H_2O, and energy.

Convention Alert!

The conventions are confusing here. When writing energy as the product of a reaction it is conventional to show it as "+ energy." The enthalpy of this exothermic reaction, ΔH, is negative, however! So,

$$C_6H_{12} + 9\ O_2 \rightarrow 6\ CO_2 + 6\ H_2O \qquad + 937\ \text{kcal/mol}$$

but,

$$C_6H_{12} + 9\ O_2 \rightarrow 6\ CO_2 + 6\ H_2O \qquad \Delta H = -937\ \text{kcal/mol}$$

This method can be used to establish energy differences between molecules. Compare, for example, the heats of combustion for octane and an isomer, 2,2,3,3-tetramethylbutane (Fig. 6.22).*

*A reviewer of this book, RMM, who sat with me in chemistry classes at Yale, many, many years ago, reminds me that we were once told that this molecule, the only isomer of octane with a melting point above room temperature (mp 104 °C), was known as "solid octane." We were then told that this was likely to be the *only* fact we retained from our organic chemistry course! Now you are stuck with this knowledge.

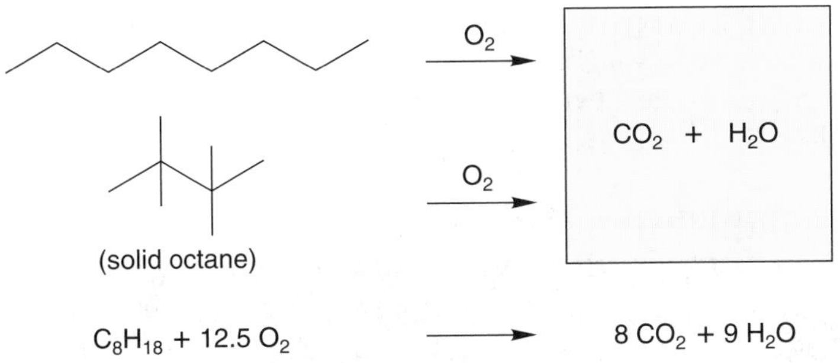

O_2

CO_2 + H_2O

O_2

(solid octane)

C_8H_{18} + 12.5 O_2 ⟶ 8 CO_2 + 9 H_2O

FIGURE 6.22 Combustion of two isomeric octanes.

The experimentally measured heat of combustion of octane is 1307.60 kcal/mol, and that of 2,2,3,3-tetramethylbutane, 1303.04 kcal/mol. Figure 6.23 graphically shows the relationship between these values and the energies of the molecules. As these isomeric molecules produce exactly the same products on burning, the difference in their heats of combustion is the difference in energy between them. Of course, this analysis also tells you which isomer is more stable, and by how much.

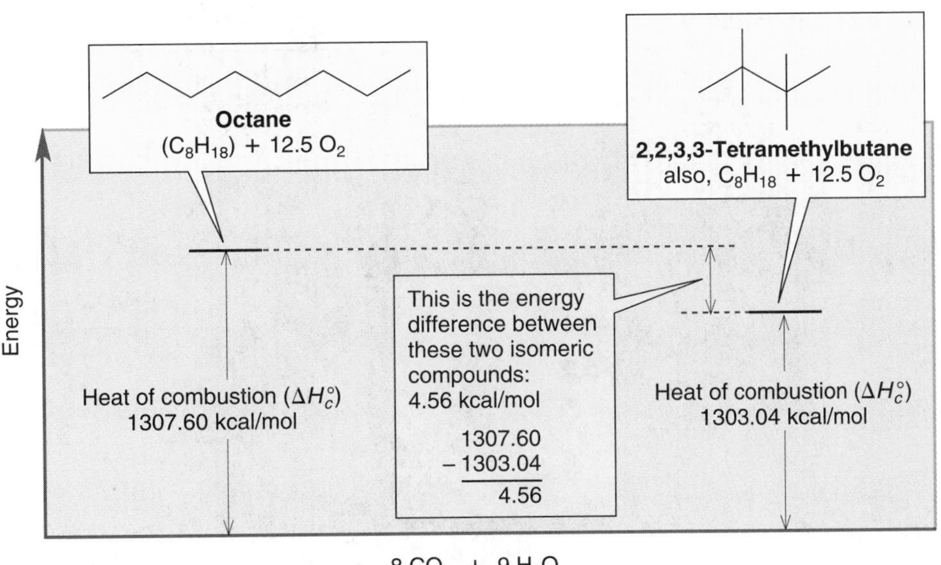

FIGURE 6.23 A quantitative picture of the combustion of two octanes.

PROBLEM 6.6

The heats of combustion of heptane, 3-methylhexane, and 3,3-dimethylpentane, respectively, are 1149.9, 1148.9, and 1147.9 kcal/mol. Carefully draw a diagram showing the relative stabilities of these molecules.

We can also use heats of combustion to measure the relative strain energies of the cycloalkanes. The higher the ΔH_c°, the less stable the compound (Fig. 6.24). Once more we need a baseline, and we use the ΔH_c° of a strain-free methylene group in an infinite chain of methylenes, 157.5 kcal/mol. This value is determined in much the same way as was used to get the ΔH_f° for a strain-free methylene (p. 201). Then we measure the ΔH_c° values for the series of cycloalkanes, finding the ΔH_c per methylene for each ring. Subtraction of 157.5 kcal/mol, the ΔH_c° of a strain-free methylene group, gives the strain energy per CH_2 and, therefore, the strain energy of the ring compound. As Table 6.3 shows, there is a good correspondence between the strain energies measured from ΔH_f° and ΔH_c°.

TABLE 6.3 Strain Energies for Some Cycloalkanes

Molecule	Experimental ΔH_c° per CH_2 (kcal/mol)	Strain E per CH_2 (ΔH_c°–157.5) (kcal/mol)	Strain E from ΔH_c° (kcal/mol)	Strain E from ΔH_f° (kcal/mol)
Cyclopropane	166.3	8.8	26.4	27.4
Cyclobutane	163.9	6.4	25.6	26.4
Cyclopentane	158.7	1.2	6.0	5.8
Cyclohexane	157.4	–0.1	–0.6	0.1
Cycloheptane	158.3	0.8	5.6	6.0
Cyclooctane	158.6	1.1	8.8	9.5
Cyclodecane	158.6	1.1	11.0	12.1
Cyclododecane	157.8	0.3	3.6	3.8
$(CH_2)_\infty$	157.5			

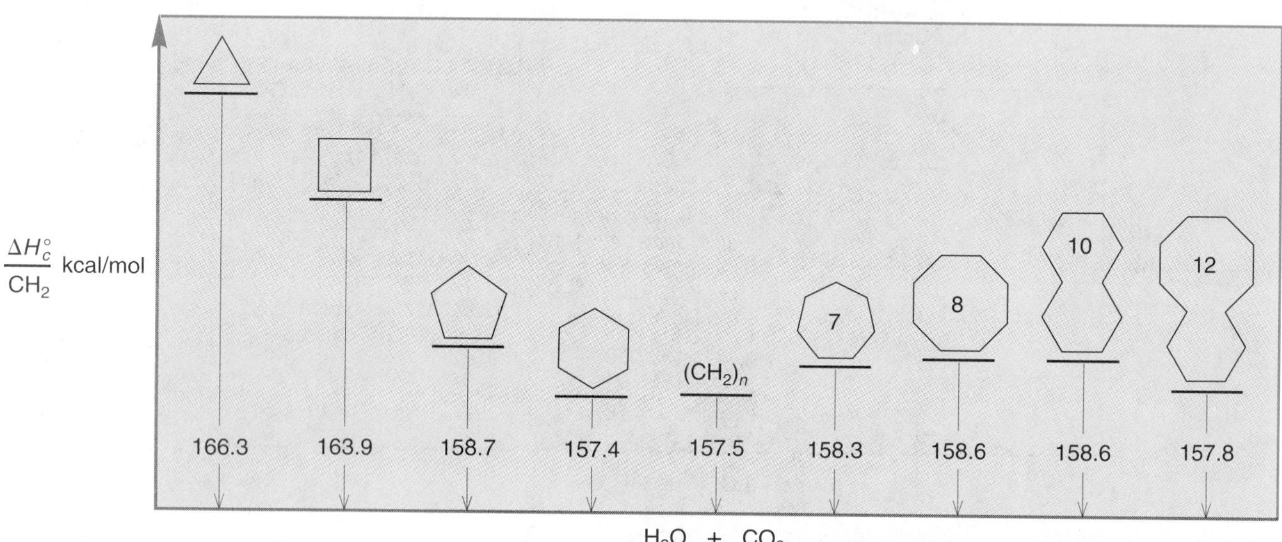

FIGURE 6.24 The heats of combustion per CH_2 for a series of cycloalkanes and a strain-free sequence of methylene (CH_2) groups. The higher the heat of combustion, the less stable the compound.

6.3 STEREOCHEMISTRY OF CYCLOHEXANE: CONFORMATIONAL ANALYSIS

Earlier, when we were considering the ways in which ring compounds distort so as to minimize strain, we saw that cyclohexane adopted a strain-free chair conformation. Now it is time to look at cyclohexane in detail. Six-membered rings are most common in organic chemistry, and a great deal of effort has been made over the years at understanding their structure and reactivity. One chair form of cyclohexane can easily be converted into another. We are now going to look in detail at this transformation and estimate the energies of the various intermediates and transition states along the path from one chair to the other. This process is an example of **conformational analysis**.

A little manipulation of a model will show the overall conversion, but it's not easy to come to the actual pathway for the process. If we move carbons 1, 2, 3, and 4 into one plane, with carbon 6 above the plane and carbon 5 below the plane, we come to a "half-chair" structure (Fig. 6.25).

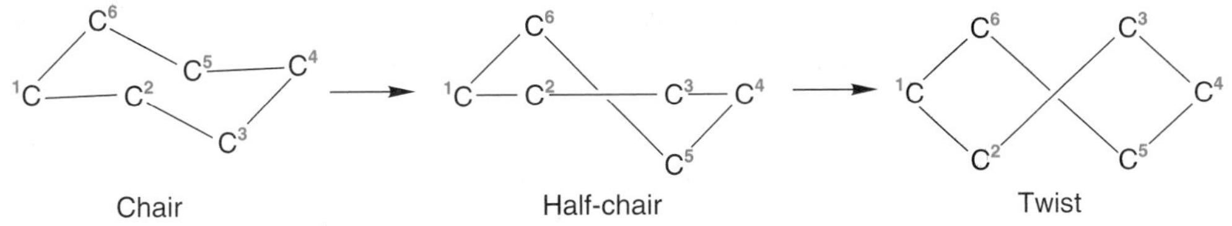

Chair → Half-chair → Twist

FIGURE 6.25 Conversion of the energy minimum chair cyclohexane into the half-chair and then the twist form.

The half-chair conformation does not represent a stable molecule, but is instead a picture of the top of the energy mountain pass (a transition state) leading to a lower energy molecular valley called the twist conformation or, sometimes, twist-boat (Fig. 6.25). The half-chair contains many eclipsed carbon–hydrogen bonds, which become staggered somewhat in the twist conformation, and angle strain is partially relieved as well. The twist arrangement can pass through a second half-chair to give the other chair cyclohexane. Kinetic measurements allow an evaluation of the energies involved (Fig. 6.26). The half-chair and the twist lie 10.8 and 5.5 kcal/mol above the chair, respectively.

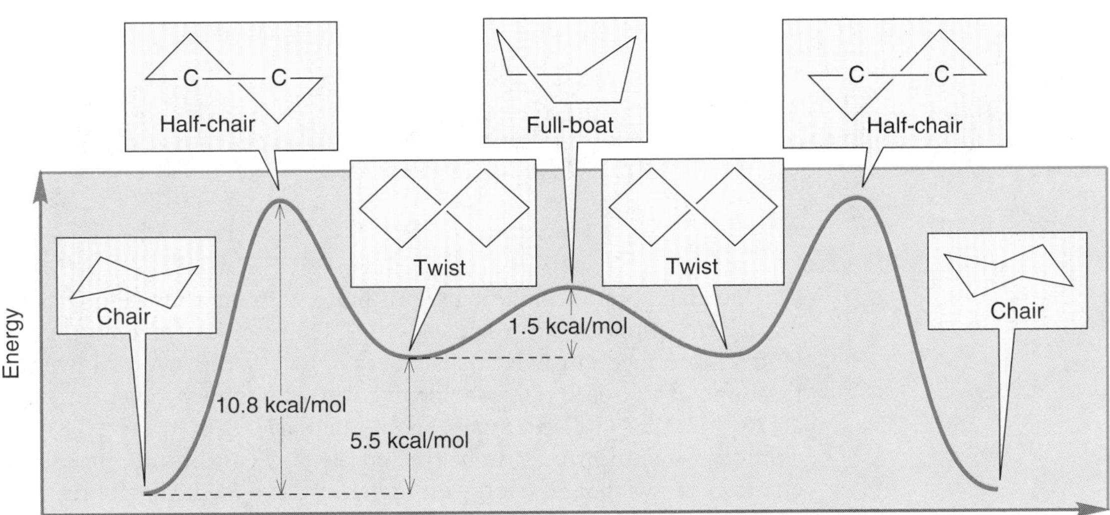

FIGURE 6.26 The interconversion of two chair cyclohexanes. The two chairs and the two twist forms are intermediates (energy minima), and the two half-chairs and the full-boat are transition states (energy maxima).

Do not confuse the twist with the full-boat (Fig. 6.26). The full-boat is not an energy minimum but is, like the half-chair, an energy maximum. It lies at the top of the barrier (like the half-chair, it, too, is a *transition state*) separating two twist forms. The boat is only 1.5 kcal/mol higher than the twist (7.0 kcal/mol higher than the chair), and can never be isolated because it is not an energy minimum.

Both the half-chair and the full-boat suffer from the kinds of strain we have seen before. In the half-chair, much of the ring is planar and there will be both angle and torsional strain. The full-boat also has torsional strain, but there is a further hydrogen–hydrogen interaction that is destabilizing. This interaction is between the "prow" and "stern" carbons and between the two "inside" hydrogens at the prow and stern of the boat. This new kind of strain is induced when two atoms come too close to each other and is called **van der Waals strain** (Johannes Diderik van der Waals, 1837–1923) (Fig. 6.27).

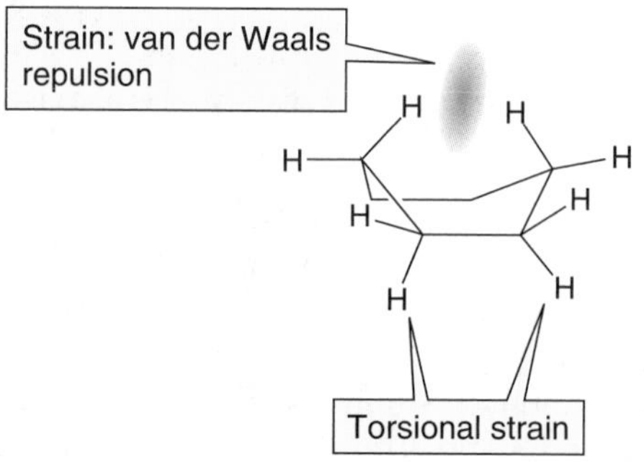

FIGURE 6.27 The full-boat form.

*PROBLEM 6.7 Draw a Newman projection looking down the "side" $H_2C{-}CH_2$ bond of the full-boat cyclohexane.

ANSWER This task is relatively easy. Once again, set your eye slightly to one side so as to see the carbon–hydrogen bonds on the rear carbon.

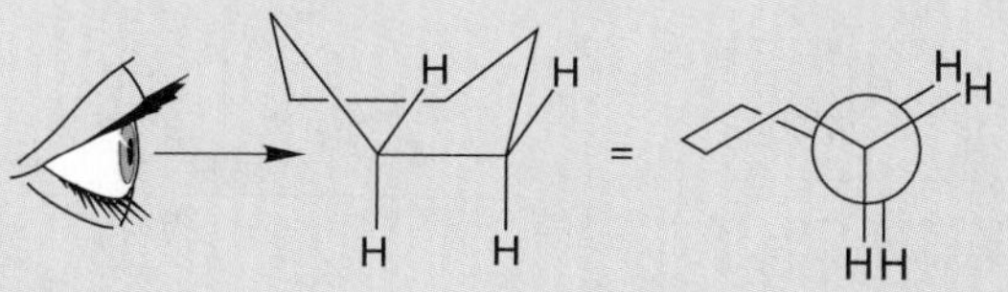

One chair cyclohexane will equilibrate with another as long as the environment supplies the requisite 10.8 kcal/mol to traverse the barrier. Under normal conditions this is an easy process, but at very low temperature, one can "freeze out" one chair form. How much twist form is there at 25 °C? We will see many such calculations in Chapter 8, but the chair–twist equilibrium can be treated like any equilibrium process. A calculation shows that an energy difference of $\Delta G = 5.5$ kcal/mol results in an enormous preference for the more stable isomer of the pair, about 9999:1. The "take home lesson" here is that small energy differences between two molecules in equilibrium result in a very large excess of the more stable isomer.

6.4 MONOSUBSTITUTED CYCLOHEXANES

Recall (Problem 6.3, p. 199) that "flipping" from one chair to the other interconverts the set of six axial hydrogens with the set of six equatorial hydrogens (Fig. 6.28). The danger of relying too strongly on coded two-

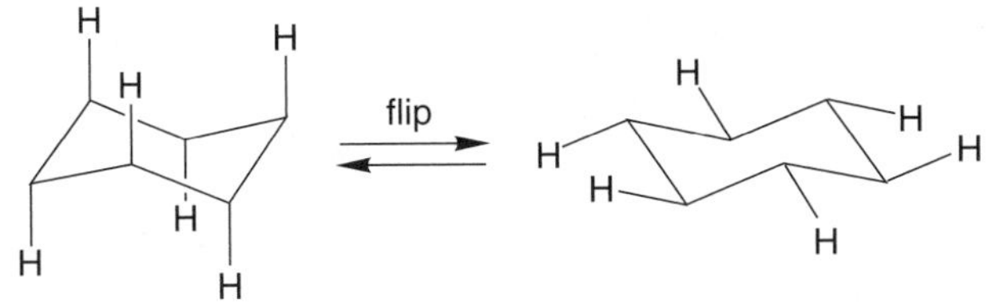

FIGURE 6.28 In cyclohexane, a ring flip interconverts axial and equatorial hydrogens.

dimensional representations of molecules is shown nicely by methylcyclohexane. The coded structures give no hint of the richness of the complicated, three-dimensional structure of the cyclohexane; they even hide the presence of two conformational isomers of methylcyclohexane (Fig. 6.29). The methyl group, or any substituent, can be either in an axial or

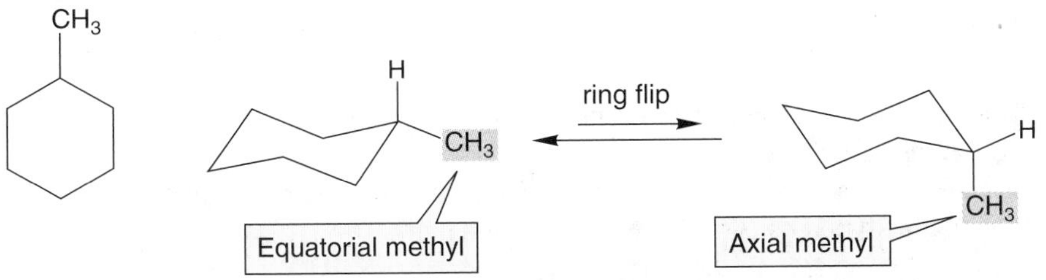

FIGURE 6.29 The uninformative skeletal structure hides the existence of two conformational isomers of methylcyclohexane. In one isomer the methyl group is equatorial, and in the other it is axial.

equatorial position, and these two molecules are diastereomers. (*Remember*: Diastereomers are stereoisomers that are not mirror images—see Chapter 5, p. 170).

Given that two isomers of methylcyclohexane exist, our next job is to determine which is more stable. As it turns out, we can even make a reasonable guess at the magnitude of the difference in energy between the two diastereomers. First of all, notice that axial and equatorial methylcyclohexane are interconverted by a chair flip (Fig. 6.29).

What turns out to be the crucial factor in creating the energy difference between the two forms is shown by the Newman projections made by sighting down the bond attaching a ring methylene group to the carbon bearing the methyl group. In the axial form, there are two *gauche* interactions between the methyl group and the nearby axial carbon–hydrogen bonds (Fig. 6.30). Note the destabilizing "1,3-diaxial interaction" between the axial hydrogen and the axial methyl group.

Although it is easy to see one of the *gauche* interactions, the perspective of the drawing hides the other unless you are careful. No amount of

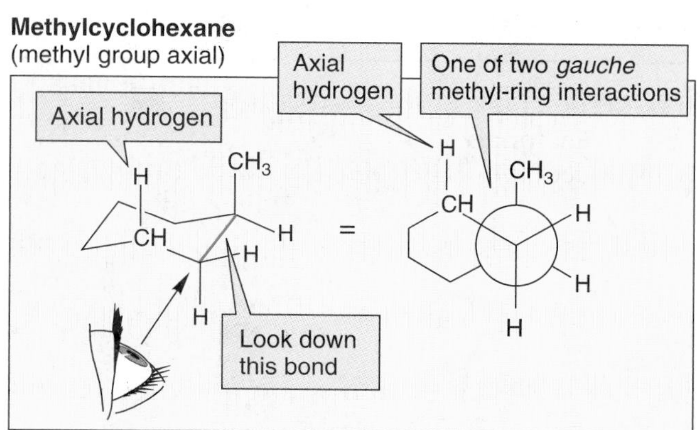

***gauche*-Butane**

A *gauche* methyl–methyl interaction

FIGURE 6.30 Methylcyclohexane with the methyl group axial. The Newman projection shows the *gauche* interaction with one of the two equivalent ring methylene groups. Each of the two *gauche* interactions in this compound resembles a *gauche* interaction in butane.

words can help you here as much as working out this stereochemical problem yourself. It is very important to try Problem 6.8. The answer follows immediately.

Recall our conformational analysis of butane in Chapter 3 (p. 93). The *gauche* form of butane was about 0.6 kcal/mol less stable than the *anti* form, in which the terminal methyl groups of butane were as far apart as possible.

*PROBLEM 6.8 Draw the Newman projection of methylcyclohexane looking from the other adjacent methylene group toward the methyl-bearing carbon. Be sure that you see the second *gauche* methyl-ring interaction. Next, compare this *gauche* interaction to that in butane. Are the two exactly the same?

ANSWER It is difficult to place your "eye" correctly for this view, but if you look along the indicated carbon–carbon bond, the appropriate Newman projection can be constructed.

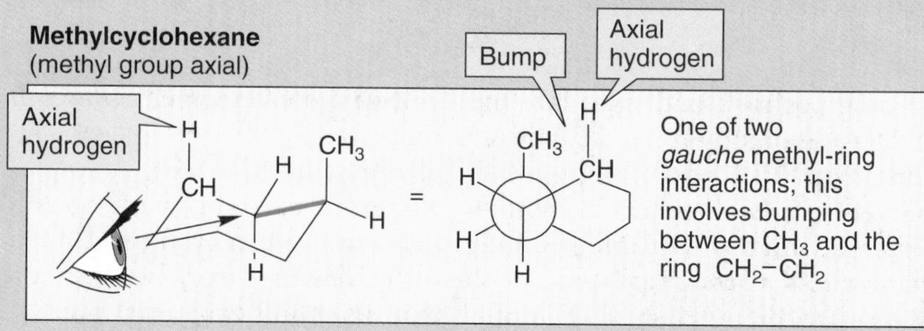

***gauche*-Butane**

Bump

Here, the *gauche* interaction is a methyl–methyl bumping

So, the two interactions are very similar, but not exactly the same. In butane two methyl groups bump, whereas in axial methylcyclohexane a methyl group bumps with a CH_2-C.

A Newman projection of the isomer with the equatorial methyl group made from the same perspective reveals none of these *gauche* interactions—this arrangement resembles the *anti* form of butane (Fig. 6.31).

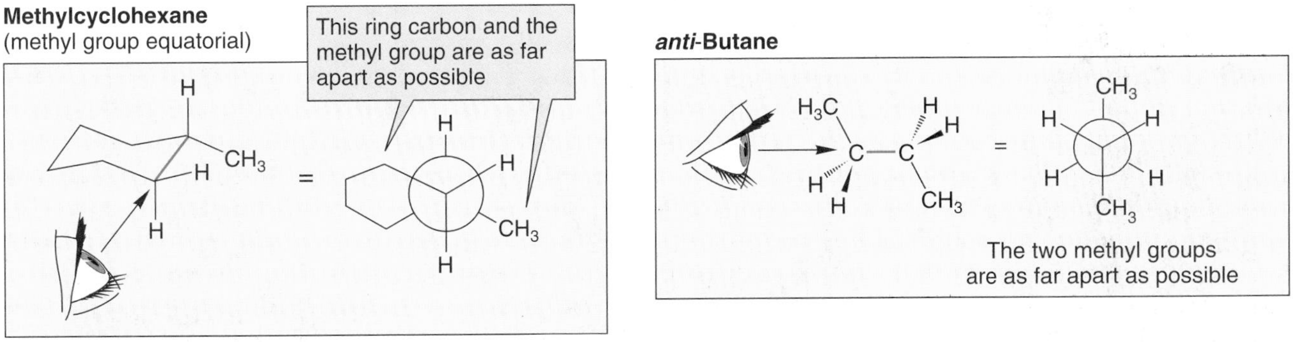

Methylcyclohexane
(methyl group equatorial)

This ring carbon and the methyl group are as far apart as possible

***anti*-Butane**

The two methyl groups are as far apart as possible

FIGURE 6.31 Methylcyclohexane with an equatorial methyl group. There are no *gauche* methyl-ring interactions. This isomer is 1.74 kcal/mol more stable than the molecule with the methyl group axial. Note the resemblance to the *anti* form of butane.

Accordingly, we might make a guess that the isomer with the methyl group equatorial would be more stable than the axial isomer, and that the energy difference between the two would be about twice the difference between *gauche*- and *anti*-butane, $2 \times 0.6 = 1.2$ kcal/mol. We'd be close. The real difference is 1.74 kcal/mol.

PROBLEM 6.9

Why isn't the difference exactly 1.2 kcal/mol? Point out some of the differences between a *gauche* interaction in butane and that in axial methylcyclohexane.

Calculations of the percentage of isomers in the equatorial and axial forms of methylcyclohexane will become clearer in Chapter 8. For now it's fair to say that about 95% of the molecules will have the methyl group equatorial at any given instant. Once more, a small energy difference (1.74 kcal/mol) has resulted in a large excess of the more stable isomer. Don't make the mistake of thinking that methylcyclohexane consists of a mixture of 95% molecules locked forever in the conformation with the methyl group equatorial, and 5% of similarly frozen molecules with their methyl groups axial. These molecules are in rapid equilibrium. At any given moment 95% will have their methyl groups equatorial, but all molecules are equilibrating between equatorial and axial methyl isomers.

PROBLEM 6.10

The formula for doing these calculations is $\Delta G = -RT \ln K$, where R is about 2 cal/deg mol, and T is the absolute temperature (see Chapter 8 for more details). Calculate the amounts at equilibrium at 25 °C of two compounds differing in energy by $\Delta G = 2.8$ kcal/mol.

A group larger than methyl would result in an even greater dominance of the equatorial form over the less stable axial arrangement. The 1,3-diaxial "bumpings" responsible for the destabilizing *gauche* interactions will be magnified, and reflected in the equilibrium constant (Fig. 6.32; Table 6.4).

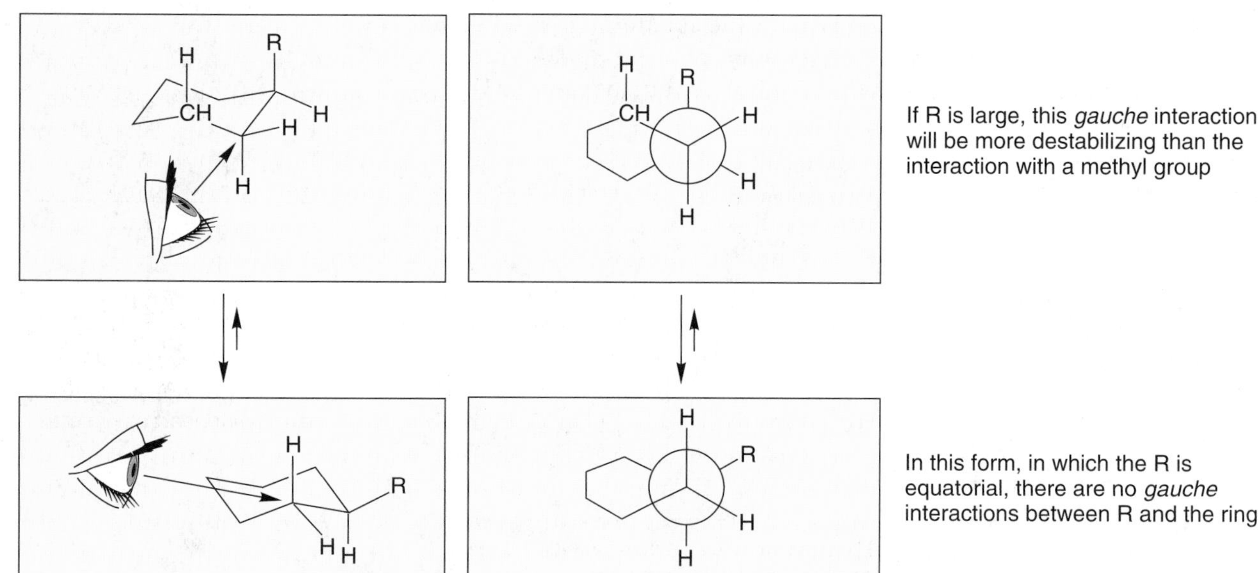

FIGURE 6.32 The larger R, the more destabilizing will be the *gauche* interactions with the ring methylene groups, and the more favored will be the equatorial isomer.

TABLE 6.4 Axial–Equatorial Energy Differences for Some Alkylcyclohexanes at 25 °C

Compound	ΔG (ax/eq) (kcal/mol)	*K*
Methylcyclohexane	1.74	19.5
Ethylcyclohexane	1.79	21.2
Propylcyclohexane	2.21	43.4
Isopropylcyclohexane	2.61	86.0
tert-Butylcyclohexane	5.5	11,916

Note the large axial–equatorial (ax/eq) energy difference for the *tert*-butyl group. The value is approximately 5–5.5 kcal/mol, which translates into an equilibrium constant of almost 12,000 at 25 °C. The situation is complicated for the *tert*-butyl group, because the conformation with an axial *tert*-butyl group is so greatly destabilized that a twist conformation, with an energy about 5 kcal/mol higher than the conformation with an equatorial *tert*-butyl group, may be lower in energy than the pure chair with an axial *tert*-butyl group (Fig. 6.33). Regardless, the conformation of *tert*-butylcyclohexane with the *tert*-butyl group equatorial is favored enormously.

FIGURE 6.33 The large *tert*-butyl group distorts the equilibrium far toward the much more stable equatorial form. The geometrical relationships can be estimated with confidence.

*PROBLEM 6.11

It is often said that the *tert*-butyl group locks the molecule into the form with the *tert*-butyl group equatorial. Is this a good way to put it? Why not?

ANSWER

No! The molecule is not locked at all. The forms with the large *tert*-butyl group axial and equatorial are in rapid equilibrium, with the equatorial form greatly predominating. Very few molecules are in the axial form at any one time, but the equilibrium is mobile. This molecule is not locked into one form and immobilized forever.

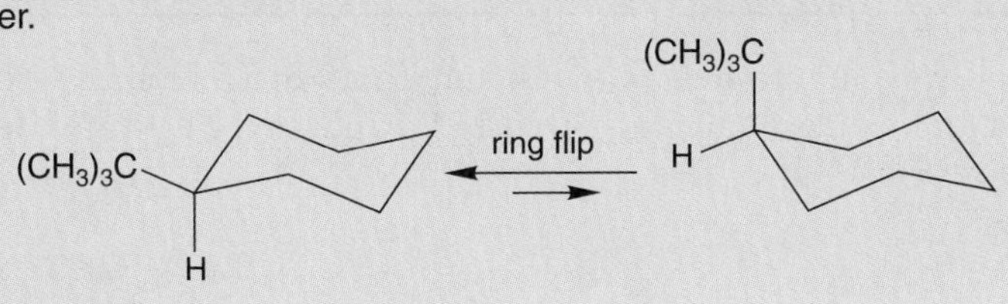

$(CH_3)_3C$ — H — ring flip — $(CH_3)_3C$ — H

As we will soon see, in studies of reaction mechanisms it is often very helpful to know with some precision the spatial relationships between groups. Ring compounds are quite helpful in this regard, as, for example, the rigid frame of cyclopropane allows for quite accurate estimations of angles and distances to be made. But cyclopropanes are so strained as to be sometimes too reactive for mechanistic work. In fact, the ring is often opened in chemical reactions. For example, one wouldn't want one's rigid framework disappearing as the reaction occurred, but this is all too likely for cyclopropanes. Other rings are flexible and don't allow us the firm predictions of angles and distances we want. However, a large group such as *tert*-butyl enables us to prejudice the mobile equilibrium between cyclohexane rings so strongly in favor of the form with the *tert*-butyl group equatorial that we can determine with confidence the positions of other atoms in the molecule (Fig. 6.33). We will see later that this is a frequently used technique.

Wonders still abound! In 1990, an Israeli research group reported that the all-trans isomer of 1,2,3,4,5,6-hexaisopropylcyclohexane was more stable in the conformation in which all six isopropyl groups were axial, not equatorial (Fig. 6.34). Surely, if one isopropyl group prefers the equatorial position, six isopropyl groups would do so as well. Not so, because for a group the size of isopropyl, adjacent trans groups bump badly when they are both equatorial. These 1,2-dialkyl interactions overwhelm the usual 1,3-diaxial interactions, and the all-axial form is preferred. It is not hard to see this effect once it is pointed out, but it must be admitted that the Israelis' observation is deliciously counterintuitive.

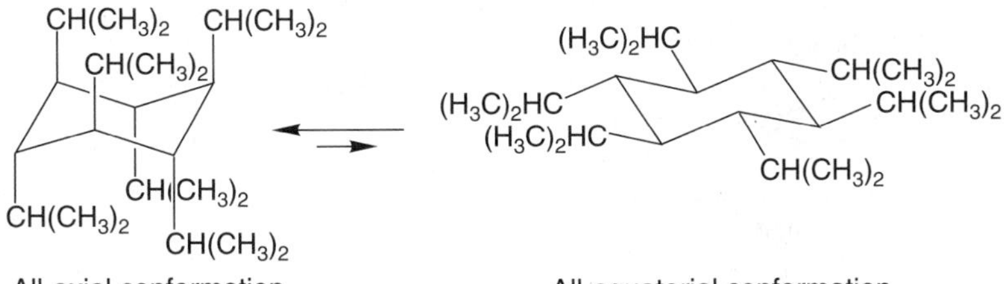

$CH(CH_3)_2$ $CH(CH_3)_2$ $CH(CH_3)_2$ $CH(CH_3)_2$ $CH(CH_3)_2$ $CH(CH_3)_2$

All-axial conformation

$(H_3C)_2HC$ $(H_3C)_2HC$ $(H_3C)_2HC$ $CH(CH_3)_2$ $CH(CH_3)_2$ $CH(CH_3)_2$

All-equatorial conformation

FIGURE 6.34 *All-trans*-1,2,3,4,5,6-hexaisopropylcyclohexane prefers the conformation in which all isopropyl groups are axial.

To summarize: All cycloalkanes except cyclopropane distort from planarity so as to minimize strain. The amount of strain can be measured in a variety of ways including measurements of heats of combustion or heats of formation. Cyclohexane adopts an energy-minimum chair form in which there are six hydrogens in the axial position and six in the equatorial position. Cyclohexane is a mobile system, as two possible chair forms interconvert. This interconversion exchanges substituents in the axial and equatorial positions.

6.5 DISUBSTITUTED RING COMPOUNDS

We have already mentioned that cis and trans diastereomers exist for dichlorocyclopropanes (Chapter 5, p. 182), and that the trans form is chiral (Fig. 6.35).

Mirror

***cis*-1,2-Dichlorocyclopropane**
(an achiral molecule, a meso compound)

***trans*-1,2-Dichlorocyclopropane**
(there is a pair of enantiomers)

FIGURE 6.35 *cis*-1,2-Dimethylcyclopropane is achiral, but there is a pair of enantiomers for *trans*-1,2-dimethylcyclopropane, a chiral molecule.

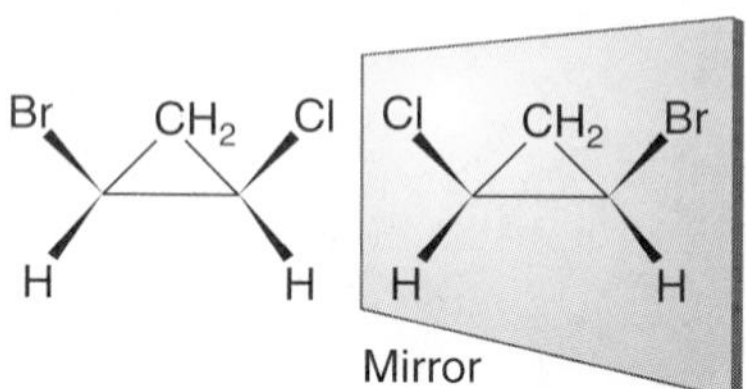

cis-1-Bromo-2-chlorocyclopropane is a chiral molecule—a pair of enantiomers exists

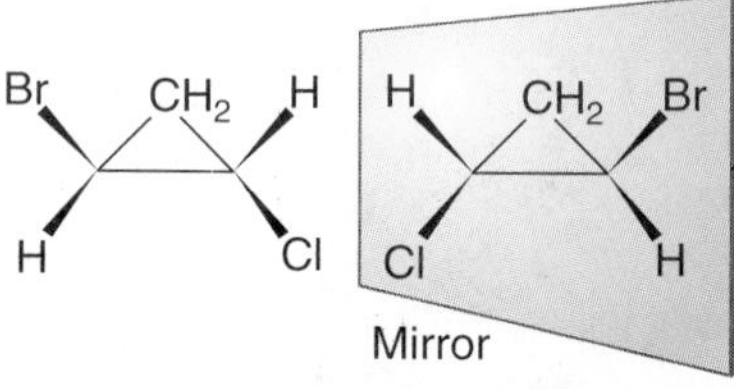

trans-1-Bromo-2-chlorocyclopropane is also chiral, and, of course, there is a pair of enantiomers

FIGURE 6.36 Both *cis*- and *trans*-1-bromo-2-chlorocyclopropane are chiral.

When the two substituents on the ring are different, as in 1-bromo-2-chlorocyclopropane, two pairs of enantiomers are possible, and both the cis and trans forms of this molecule are chiral (Fig. 6.36).

This relatively simple situation becomes harder to see in cyclohexanes where the molecules are not planar, but it doesn't really change. We will first look at 1,1-dialkylcyclohexanes and then at the somewhat more complicated 1,2-dialkylcyclohexanes.

6.5a 1,1-Disubstituted Cyclohexanes

1,1-Dimethylcyclohexane is a simple, achiral molecule in which the cyclohexane axial–equatorial equilibration induced by flipping one chair to the other interconverts equivalent molecules, each of which has one methyl group axial and one equatorial. As for cyclohexane itself, the equilibrium constant for the equilibration of the two equivalent forms of the 1,1-dimethyl isomer must be 1 (Fig. 6.37).

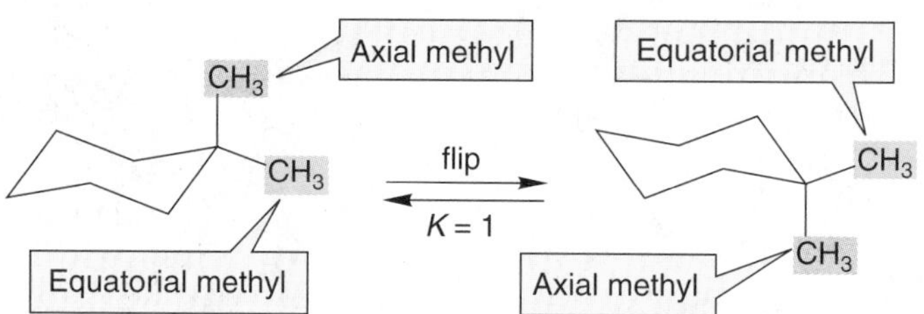

FIGURE 6.37 1,1-Dimethylcyclohexane is not chiral. The ring flip converts the axial methyl group into an equatorial methyl group and vice versa.

1-Isopropyl-1-methylcyclohexane is only slightly more complicated. There are two isomers, nearly equal in energy, neither of which is chiral (Fig. 6.38).

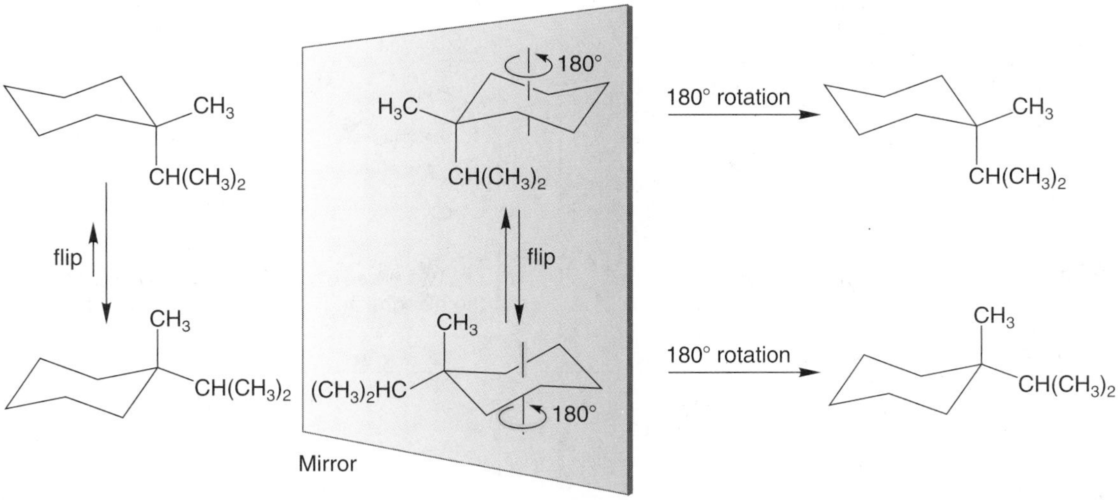

FIGURE 6.38 Both isomers of 1-isopropyl-1-methylcyclopropane are achiral—the mirror images are superimposable.

Use the data in Table 6.4 to calculate the energy difference between the possible isomers of 1-isopropyl-1-methylcyclohexane.

PROBLEM 6.12

6.5b 1,2-Disubstituted Cyclohexanes

Just like 1,2-dimethylcyclopropane, 1,2-dimethylcyclohexane can exist in cis and trans forms (as can any 1,2-disubstituted ring compound), but the chair form of cyclohexane makes the compounds look somewhat different from the more familiar rigid isomers of cis and trans disubstituted cyclopropanes (Fig. 6.39).

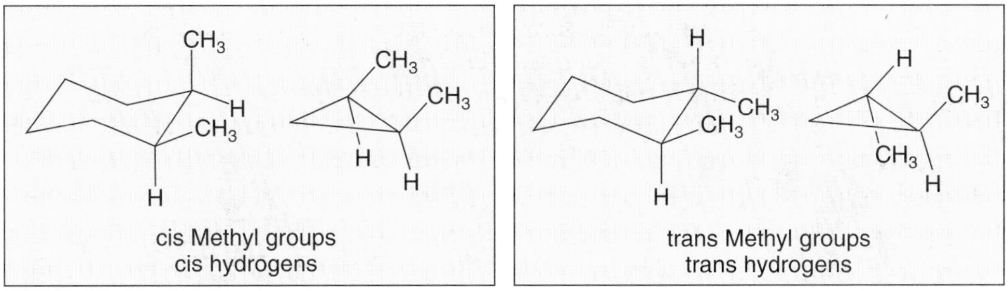

cis Methyl groups
cis hydrogens

trans Methyl groups
trans hydrogens

FIGURE 6.39 Three-dimensional representations of *cis*- and *trans*-1,2-dimethylcyclohexane contrasted with *cis*- and *trans*-1,2-dimethylcyclopropane.

The words cis and trans refer to the "sidedness" of a molecule and do not depend strictly on the angles between the groups. In a cis disubstituted cyclopropane, the dihedral angle between the cis groups is 0°, whereas it is 60° in the disubstituted cyclohexane. The methyl groups are refered to as cis in either case. In *trans*-1,2-dimethylcyclopropane, the dihedral angle between the methyl groups is 120°, whereas in the six-membered ring it is 60° (Fig. 6.40).

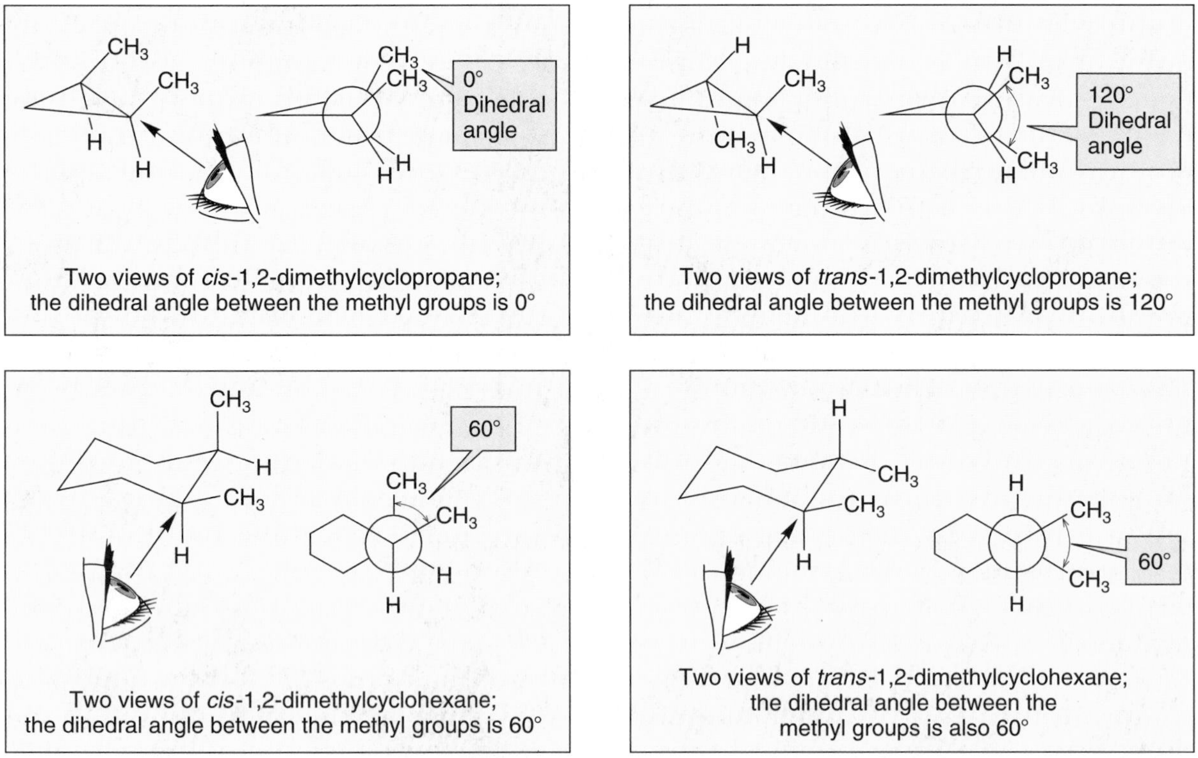

Two views of *cis*-1,2-dimethylcyclopropane; the dihedral angle between the methyl groups is 0°

Two views of *trans*-1,2-dimethylcyclopropane; the dihedral angle between the methyl groups is 120°

Two views of *cis*-1,2-dimethylcyclohexane; the dihedral angle between the methyl groups is 60°

Two views of *trans*-1,2-dimethylcyclohexane; the dihedral angle between the methyl groups is also 60°

FIGURE **6.40** cis and trans Isomerism in the planar, inflexible cyclopropanes and the non-planar, flexible cyclohexanes.

PROBLEM **6.13** Figure 6.40 only shows the equatorial, equatorial form of *trans*-1,2-dimethylcyclohexane. What is the dihedral angle between the methyl groups in the much less stable axial, axial dimethyl isomer?

In *cis*-1,2-dimethylcyclohexane, one methyl group is axial and the other is equatorial (Fig. 6.41). The conformational ring flip does not alter the structure—the axial methyl becomes equatorial, and the equatorial methyl becomes axial. The equilibrium constant between these two equivalent compounds must be 1.

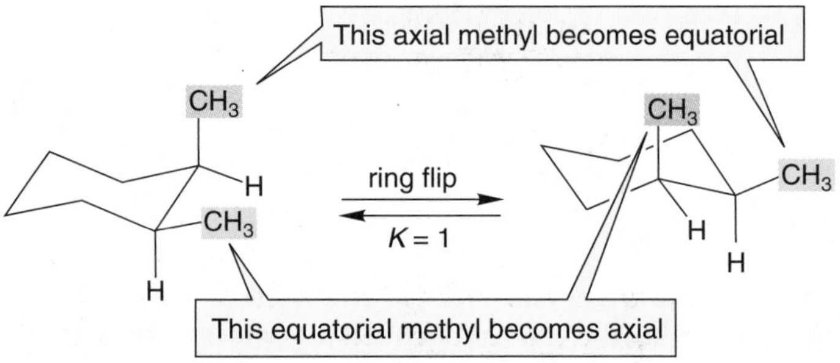

This axial methyl becomes equatorial

ring flip

$K = 1$

This equatorial methyl becomes axial

FIGURE **6.41** When the ring flip occurs in *cis*-1,2-dimethylcyclohexane, the axial methyl group becomes equatorial and vice versa.

Note that the two conformational isomers of *cis*-1,2-dimethylcyclohexane are enantiomers. The equilibration between the two conformations produces a racemic mixture of the two enantiomers. Have we seen a similar situation before? Indeed we have; just recall our discussion of the *gauche* forms of butane (Chapter 3, p. 93). The equilibration of the two *cis*-1,2-dimethylcyclohexanes is very similar to the equilibration of two *gauche* forms of butane (Fig. 6.42).

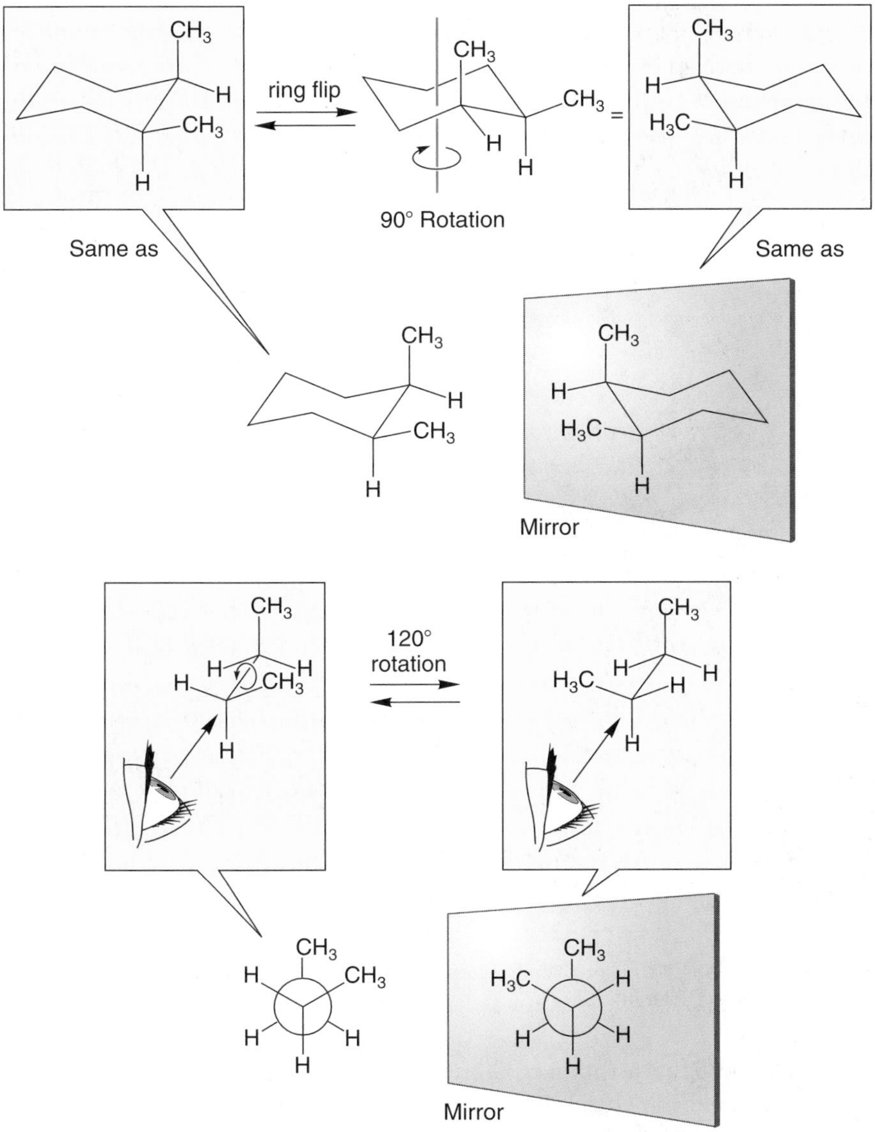

FIGURE 6.42 The ring flip in *cis*-1,2-dimethylcyclohexane converts one enantiomer into the other. A very similar process converts one enantiomer of *gauche* butane into the other.

The trans form of this molecule presents a different picture. *trans*-1,2-Dimethylcyclohexane can either have both methyl groups axial or both equatorial (Fig. 6.43). The ring flip converts the diaxial form into the

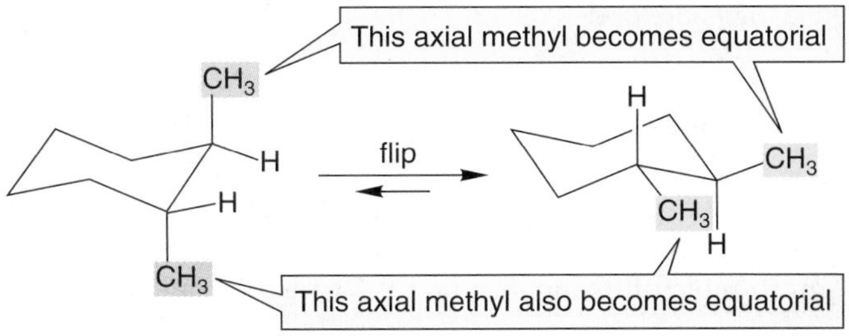

FIGURE 6.43 The ring flip in *trans*-1,2-dimethylcyclohexane interconverts a molecule with two axial methyl groups and one with two equatorial methyl groups.

diequatorial form. Let's apply conformational analysis to predict which of these two diastereomers will be more stable. As usual, Newman projections are vital, and in Figure 6.44 we sight along the carbon–carbon bond attaching the two methyl-substituted carbons. In the diequatorial isomer, there is one *gauche* methyl–methyl interaction costing 0.6 kcal/mol. In the diaxial isomer, the two methyl groups occupy *anti* positions and their interaction with one another is not destabilizing. However, for each axial methyl group we still have the two *gauche* interactions with the ring methylene groups (see p. 209), and this will cost the molecule 2×1.74 kcal/mol, or 3.48 kcal/mol. We can predict that the diequatorial isomer should be $3.48 - 0.6 = 2.88$ kcal/mol more stable than the diaxial isomer. In turn, this predicts that at 25 °C there will be more than 99% of the diequatorial form present.

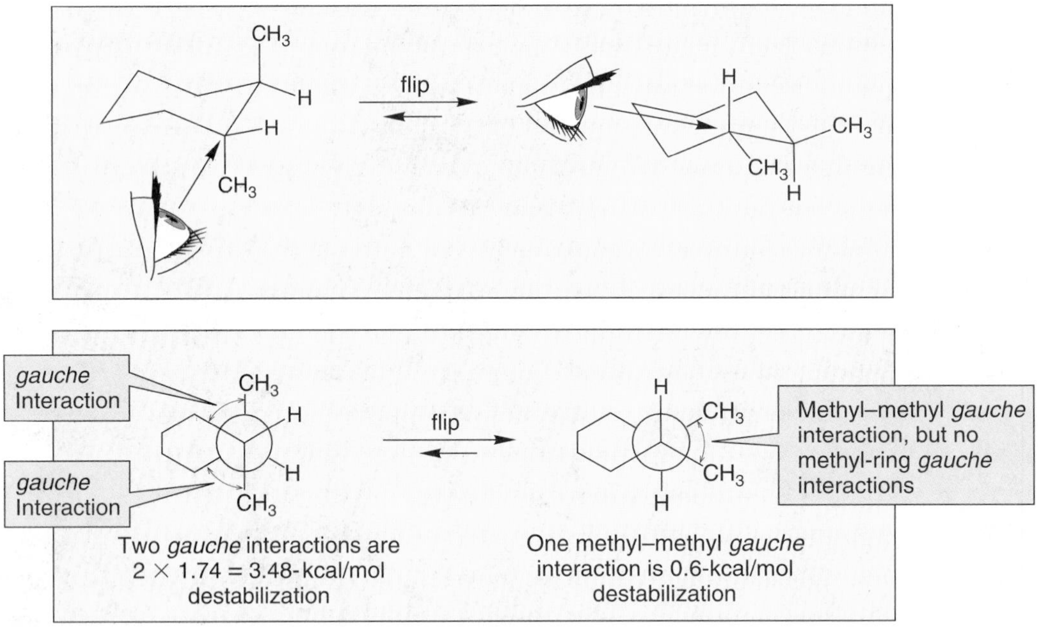

FIGURE **6.44** The diequatorial form of *trans*-1,2-dimethylcyclohexane should be more stable than the diaxial form.

In principle, both the diaxial and diequatorial forms should be resolvable—they are both chiral molecules. In practice, we cannot isolate *trans*-diaxial-1,2-dimethylcyclohexane; it simply flips to the much more stable diequatorial form. The diequatorial form can be isolated and resolved.

PROBLEM 6.14 Show that the ring flip of the diequatorial form to the diaxial form will not racemize optically active *trans*-1,2-dimethylcyclohexane. What is the relation between the equatorial–equatorial and axial–axial forms of *trans*-1,2-dimethylcyclohexane?

Let's stop for a moment and summarize: Neither *cis*-1,2-dimethylcyclopropane nor *cis*-1,2-dimethylcyclohexane can be resolved, as the cyclopropane is a meso compound (Chapter 5, p. 175), and the cyclohexane flips into its mirror image. However, both *trans*-1,2-dimethylcyclopropane and *trans*-1,2-dimethylcyclohexane *can* be resolved; they are not superimposable on their mirror images. The point is that in a practical sense the planar cyclopropanes and nonplanar cyclohexanes behave in the same way. This finding has important consequences. In deciding questions of stereochemistry, we can treat the decidedly nonplanar 1,2-dimethylcyclohexanes *as if* they were planar. Indeed, all cyclohexanes can be treated as planar for the purposes of stereochemical analysis, because the planar forms represent the average positions of ring atoms in the rapid chair–chair interconversions.

Now let's look at some slightly more complicated molecules in which the two substituents on the ring are different. In 1-isopropyl-2-methylcyclohexanes, cis and trans isomers exist, and once more we will have to worry about the effects of flipping from one chair form to the other. In the cis compound, there are two different conformational isomers (Fig. 6.45). In one, the methyl is axial and the isopropyl group is equatorial. When the ring flips, the axial methyl group becomes equatorial and the equatorial isopropyl group becomes axial.

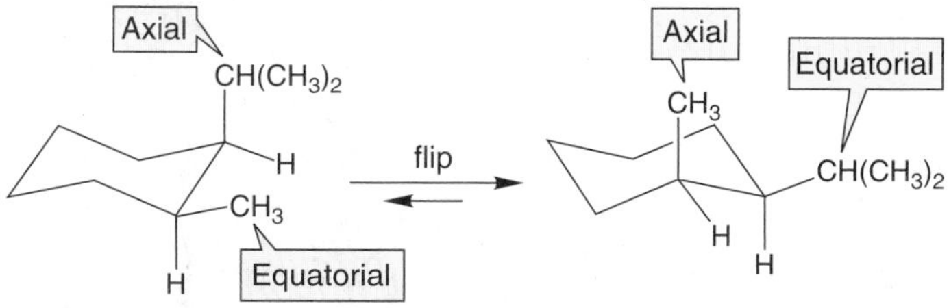

FIGURE 6.45 *cis*-1-Isopropyl(ax)-2-methyl(eq)cyclohexane flips to *cis*-1-isopropyl(eq)-2-methyl(ax)cyclohexane.

Unlike the dimethyl case, these two compounds are not enantiomeric. They are conformational diastereomers—different conformational isomers with different physical and chemical properties. In principle, each could be isolated and resolved. So there is a total of four possible cis isomers, the two shown in Figure 6.45 and their mirror images (Fig. 6.46).

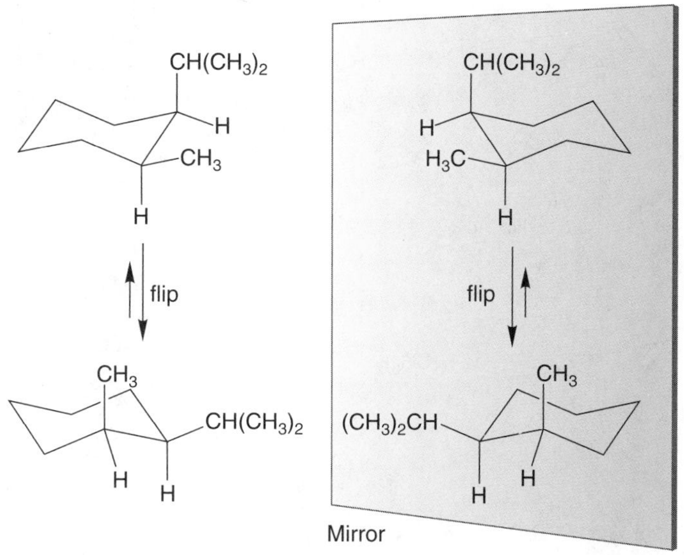

FIGURE 6.46 The four stereoisomers of *cis*-1-isopropyl-2-methylcyclohexane. Two pairs of enantiomers exist.

The rigid molecule *cis*-1-isopropyl-2-methylcyclopropane cannot undergo any ring flip and there is only a single pair of enantiomers (Fig. 6.47).

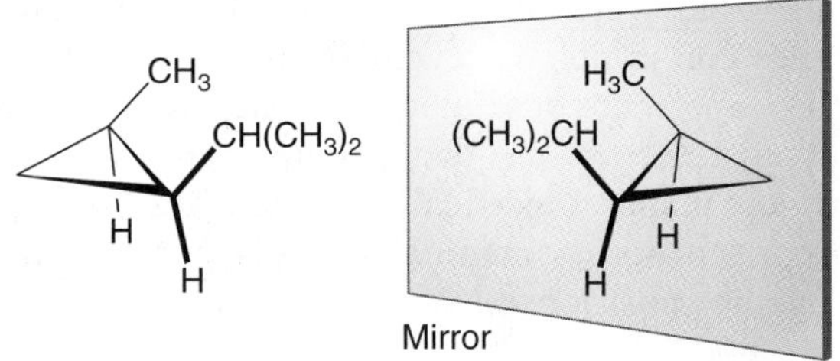

FIGURE 6.47 The two enantiomers of *cis*-1-isopropyl-2-methylcyclopropane. The lack of a possible ring flip reduces the total number of possible stereoisomers.

PROBLEM 6.15 Label the stereogenic carbons of Figures 6.46 and 6.47 as (*R*) or (*S*).

The situation is similar in the case of the trans form. Although the inflexible cyclopropane ring has but a single pair of enantiomers (Fig. 6.48), there are two pairs of enantiomers in the flexible cyclohexane (Fig. 6.49).

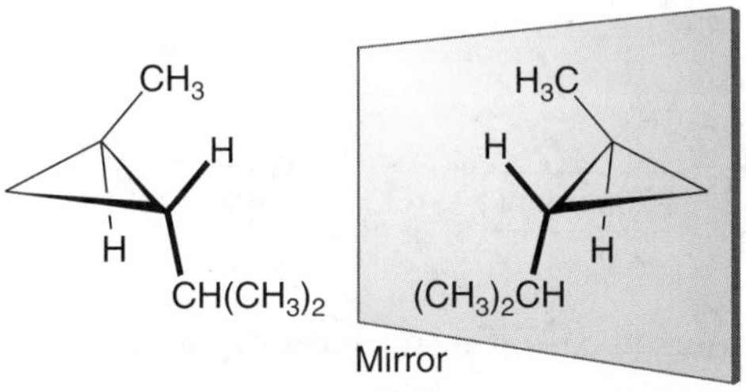

FIGURE 6.48 The single pair of enantiomers of *trans*-1-isopropyl-2-methylcyclopropane. The rigid cyclopropane ring prevents any ring flip.

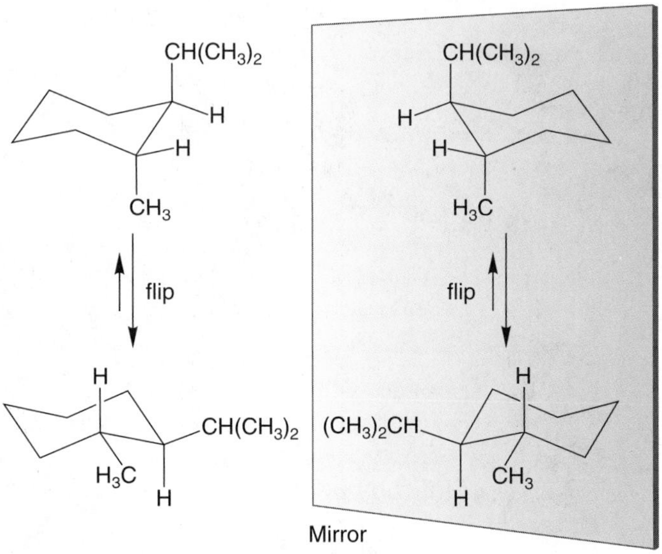

FIGURE 6.49 The four stereoisomers of *trans*-1-isopropyl-2-methylcyclohexane. Two pairs of enantiomers exist.

PROBLEM 6.16 Point out the pairs of enantiomers and diastereomers in Figure 6.49. Assign all the stereogenic carbons as (*R*) or (*S*).

*PROBLEM 6.17

Use the data of Table 6.4 to estimate the energy difference between *cis-* and *trans*-1-isopropyl-2-methylcyclohexanes.

ANSWER

For the cis compound, ring flipping converts **A** with an axial isopropyl group and an equatorial methyl group into **B** in which the isopropyl group is equatorial and the methyl is axial. Table 6.4 tells you that an isopropyl group is more stable in the equatorial position by 2.61 kcal/mol. A methyl group is more stable in the equatorial position by only 1.74 kcal/mol. The more stable conformation will be **B** by 2.61 − 1.74 = 0.87 kcal/mol.

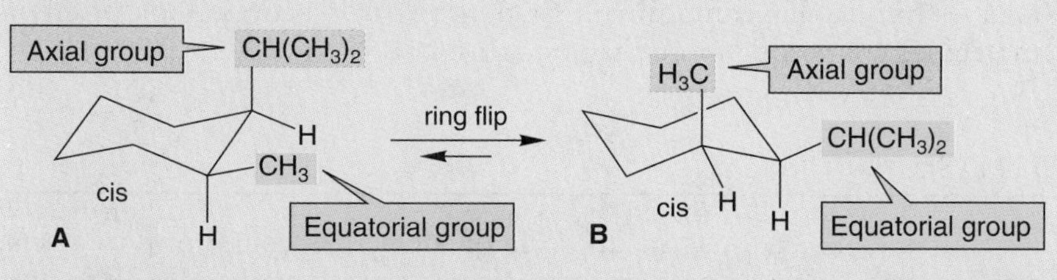

Axial group $CH(CH_3)_2$ H CH_3 cis **A** H Equatorial group ring flip H_3C Axial group $CH(CH_3)_2$ cis H H **B** Equatorial group

The trans compound has both groups axial (**C**) or both groups equatorial (**D**). Conformation **D** will be preferred by 2.61 + 1.74 = 4.35 kcal/mol. However, **D** suffers a methyl–isopropyl *gauche* interaction that will be destabilizing by somewhat more than 0.6 kcal/mol. So, our final answer would be approximately 4.35 − 0.6 = 3.75 kcal/mol.

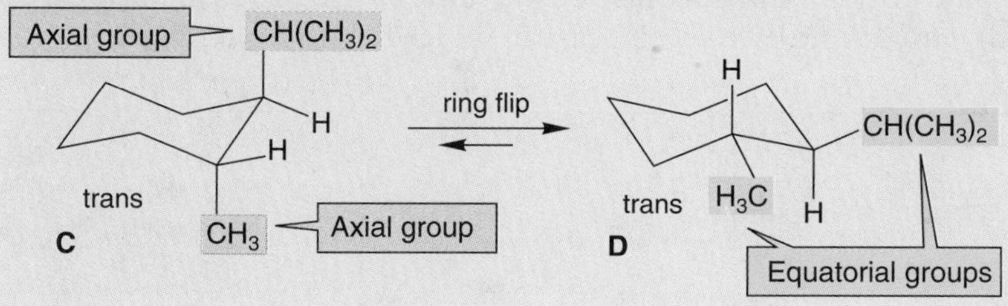

Axial group $CH(CH_3)_2$ H H trans **C** CH_3 Axial group ring flip H $CH(CH_3)_2$ trans H_3C H **D** Equatorial groups

PROBLEM 6.18

Estimate the energy difference between the two conformational isomers of *cis*-1,3-dimethylcyclohexane.

*PROBLEM 6.19

Does ring flip racemize *trans*-1,3-dimethylcyclohexane, as it does for *cis*-1,2-dimethylcyclohexane (Fig. 6.42, p. 215)?

ANSWER

No. Figure 6.42 shows that the ring flip of *cis*-1,2-dimethylcyclohexane converts one enantiomer into the other. Ring flip in *trans*-1,3-dimethylcyclohexane, which, like *cis*-1,2-dimethylcyclohexane, has one axial methyl group and one equatorial methyl group, *regenerates the same enantiomer.* Ring flip cannot racemize this compound.

***trans*–1,3-Dimethyl–cyclohexane**

CH_3 H H CH_3 ring flip CH_3 H H 180° CH_3 turn over CH_3 H H CH_3

These are the same thing!

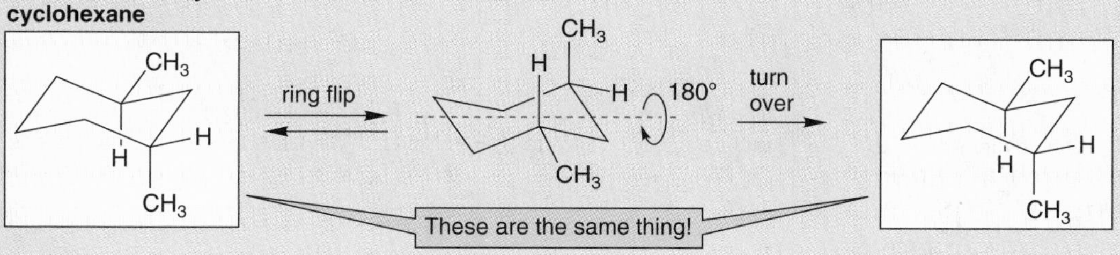

The "take home lesson" here is that it takes a careful analysis of the stereoisomers formed by ring flipping of substituted cyclohexanes to see all the possibilities. First, be sure you have made a *good* drawing (follow the procedure on p. 198). Next, determine whether or not the isomer is chiral by drawing the mirror image and seeing if it is superimposable on the original. Now do a ring flip of the original isomer. Is the ring-flipped isomer the same as the original? Is it the mirror image? Is it completely different? Differently substituted cyclohexanes give all three possibilities.

Now that we have dealt with both simple and complicated cycloalkanes (molecules containing a single ring), it is time to look briefly at molecules containing more than one ring.

6.6 BICYCLIC COMPOUNDS

Of course, one can imagine all sorts of molecules containing two separated rings. There is nothing special about such molecules. However, there are other molecules in which two rings share a carbon or carbons, and these are more interesting. There are three general ways in which two rings can be connected by sharing carbons: They can share either a single carbon or two or more carbons. These modes of attachment are called **spiro** (one carbon shared), **fused**, and **bridged** (more than one carbon shared). Fused is really a special case of bridged, with $n = 0$ (Fig. 6.50).

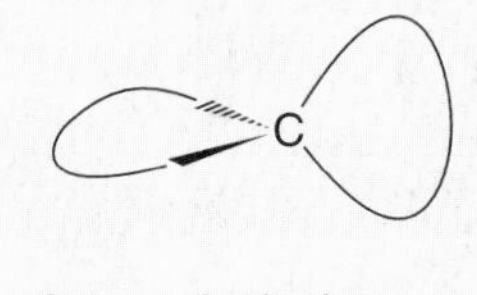

Spiro substitution (one carbon is shared)

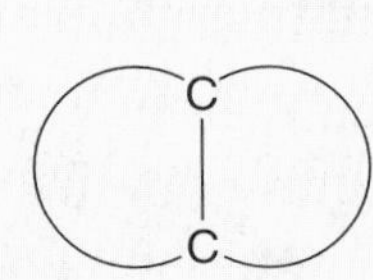

Fused substitution (two carbons are shared)

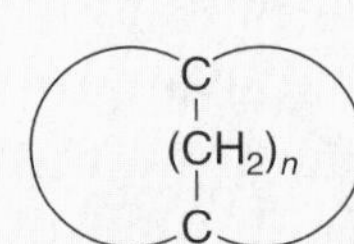

Bridged substitution (more than two carbons are shared; if $n = 0$, the molecule is fused)

FIGURE **6.50** Rings can share a single carbon (spiro), two carbons (fused), or more than two carbons (bridged).

Spiro substitution is common in both "natural products" and in molecules so far encountered only in the laboratory. Note in Figure 6.51 how poorly a two-dimensional picture represents the real shape of such molecules. Even the simplest spirocyclic compound, spiropentane, is badly served by the flat page. The central carbon is approximately tetrahedral, and we really must use wedges to give a three-dimensional feel to this kind of molecule.

The two-dimensional picture of spiropentane is highly misleading ...

... as the central carbon must look like this ... nearly tetrahedral

... this means that the two rings must lie in different, perpendicular planes

FIGURE **6.51** A two-dimensional picture of spiropentane is misleading. The two rings lie in perpendicular planes.

Figure 6.52 shows some "natural" and "unnatural" spiro compounds.

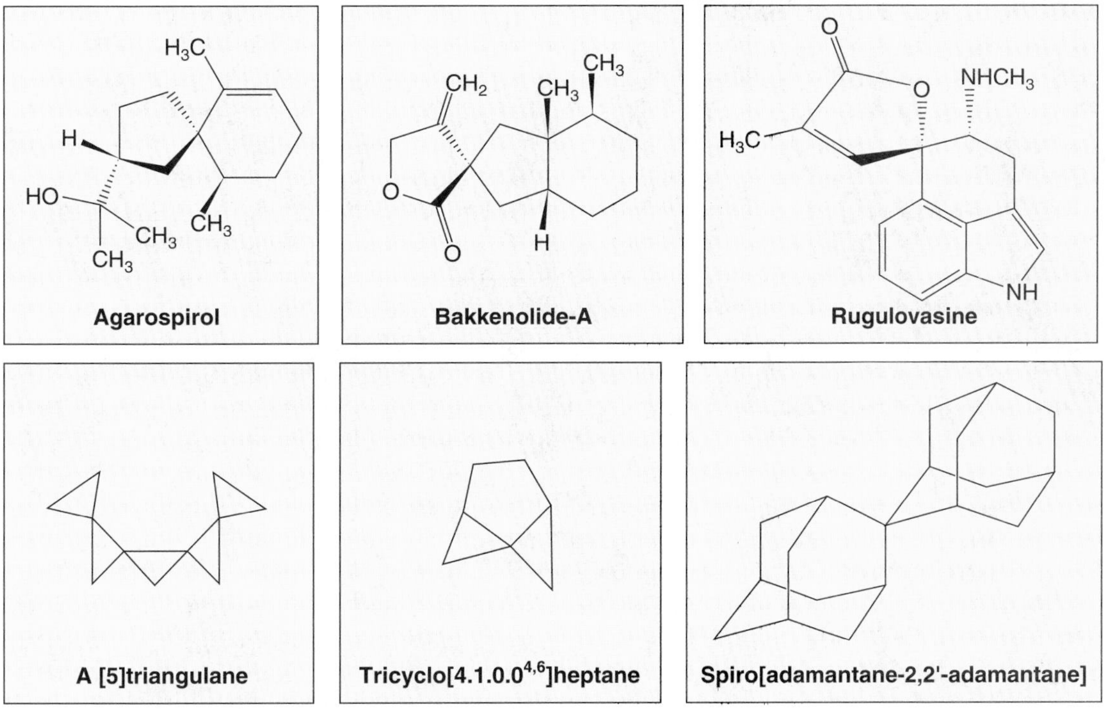

Agarospirol

Bakkenolide-A

Rugulovasine

A [5]triangulane

Tricyclo[4.1.0.0^{4,6}]heptane

Spiro[adamantane-2,2'-adamantane]

FIGURE 6.52 Some spiro compounds. Don't worry about the exotic names.

Bicyclic molecules in which two rings share two or more carbons are even more important than the spirocyclic compounds. The simplest way in which two rings can share more than one carbon is for two adjacent carbons to be shared in a fused structure (Fig. 6.53).

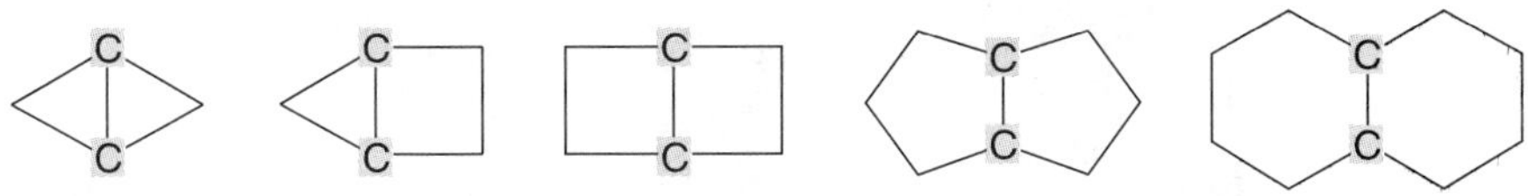

FIGURE 6.53 Some schematic, two-dimensional structures for fused bicyclic compounds in which two rings share a pair of adjacent carbon atoms. The "fusion" positions, or "bridgeheads," are shown in red.

The hydrogens attached at the ring junction positions can be either on the same side (cis) or on opposite sides (trans). In practice, both stereochemistries are possible for larger rings, but for the small rings only the cis ring junction is stable (see Problem 6.20). Figure 6.54 shows a number of compounds that contain this kind of ring fusion.

Aflatoxin B_1

cis Ring fusion

Cortisone

trans Ring fusion

trans Ring fusion

6-Demethyl-6-deoxytetracycline

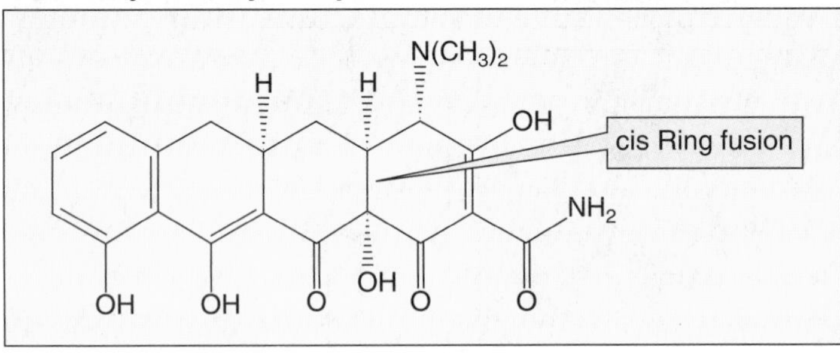

FIGURE 6.54 Some fused polycyclic molecules. Note the cis and trans ring fusions.

PROBLEM 6.20 Use your models to convince yourself that the trans stereochemistry is not possible for the molecules at the left of Figure 6.53.

Particularly important are *cis*- and *trans*-decalin, the compounds formed by the fusion of two cyclohexanes. Although Figure 6.55 shows these compounds in schematic form, by now we can certainly guess that the real shape will be much more intricate.

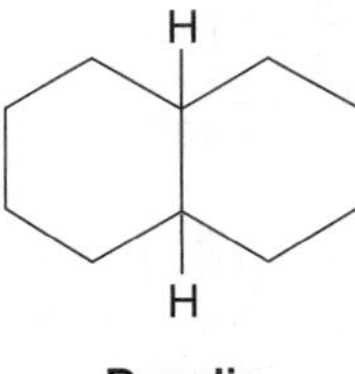

Decalin

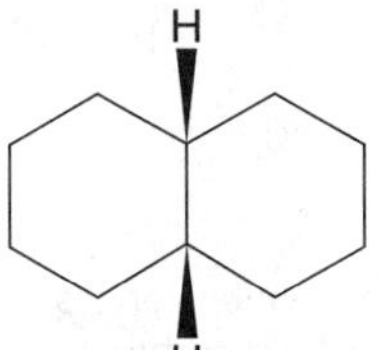

Schematic picture of *cis*-decalin

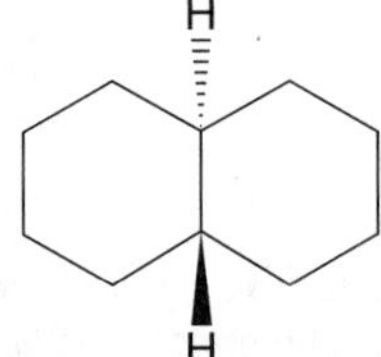

Schematic picture of *trans*-decalin

FIGURE 6.55 In decalin, the compound formed by two fused cyclohexanes, the two hydrogens at the ring fusion can be either on the same side (cis) or on opposite sides (trans) of the rings.

It's easy enough to draw them out though, if one goes about it carefully. First of all, draw a single, perfect chair cyclohexane, and put in the axial and equatorial bonds at two adjacent carbons. As for any trans disubstituted cyclohexane (p. 215), the two rings of *trans*-decalin will be attached either through two equatorial or two axial bonds. As shown in Figure 6.56, attachment through two axial bonds is impossible—the distance to be bridged is too great (try it with models).

So we are left with only the two equatorial bonds for trans attachment of a second ring, and this time the new chair fits in easily (Fig. 6.57).

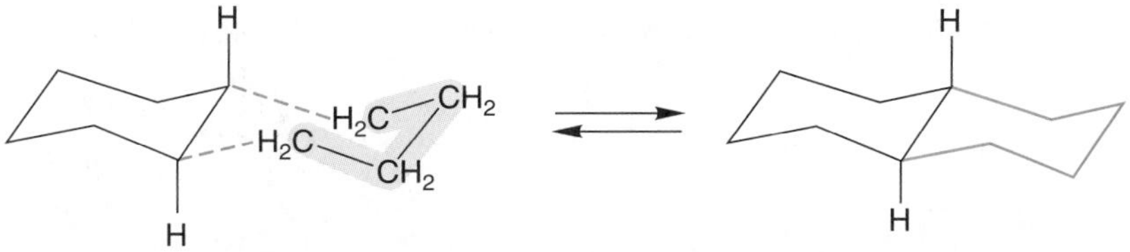

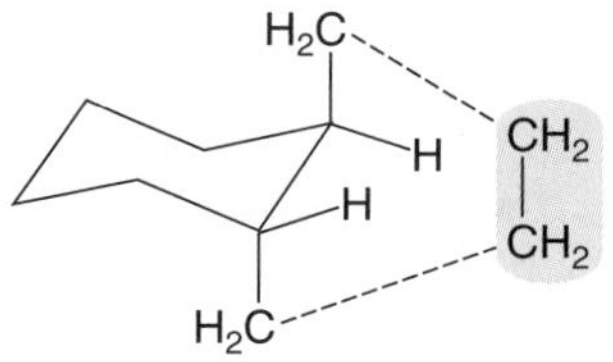

FIGURE 6.56 A futile attempt to construct *trans*-decalin by connecting two axial carbons with a pair of methylene groups. It is not possible to span the two axial positions with only four methylene groups; two methylene groups occupy axial positions leaving only two others (blue) to complete the ring.

It is easy to connect two equatorial positions with a chain of four methylenes; the hydrogens at the fusion positions (bridgeheads) must be axial

trans-Decalin: two fused chair cyclohexanes

FIGURE 6.57 *trans*-Decalin can be easily made by connecting two adjacent equatorial positions through a chain of four methylene groups.

Cis substitution must involve one axial and one equatorial bond in the attachments to the second ring (Fig. 6.58). If we pay attention to the rules for drawing perfect chairs, it's easy to produce a good drawing for this molecule. Remember that every ring bond in a cyclohexane is parallel to the bond directly across the ring (p. 198).

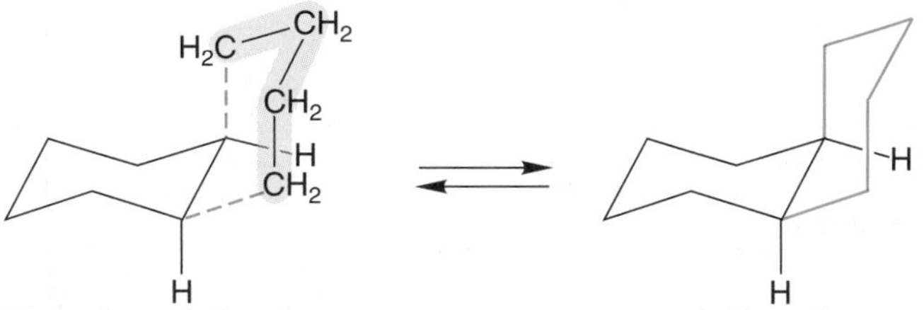

To make a cis junction we must connect one axial and one equatorial position with a four-carbon chain

***cis*-Decalin**
The two bridgehead hydrogens occupy one axial and one equatorial position

FIGURE 6.58 *cis*-Decalin can be constructed by connecting one axial position with one equatorial position.

PROBLEM 6.21

Use models to convince yourself that *cis*-decalin is a mobile molecule, undergoing easy double chair–double chair interconversions, whereas *trans*-decalin is rigidly locked.

Now imagine two rings sharing three carbons. This sharing produces a different kind of structure, called a bridged bicyclic molecule (Chapter 3, p. 109). Two cyclopentanes, for example, can share two or three carbons. We have seen the first case before (Fig. 6.53) and the second is drawn out in Figure 6.59. In each case, the carbons at the fusion points are called the **bridgehead positions**.

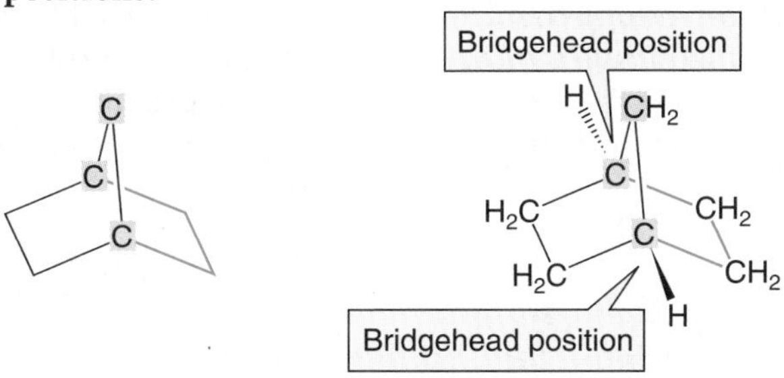

Bridgehead position

Bridgehead position

FIGURE 6.59 Two five-membered rings sharing three carbons. The shared carbons are shown in red, the carbon–carbon bonds in blue complete one cyclopentane. A more complete structure for this bicyclic hydrocarbon is also shown.

*PROBLEM 6.22 Design a molecule in which two rings share four carbons.

ANSWER One sure-fire way to do a problem such as this is to start by drawing the carbons to be shared (four, in this example). Then add the remainders of the other rings.

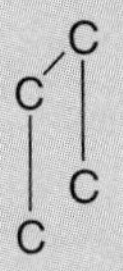

The four carbons to be shared

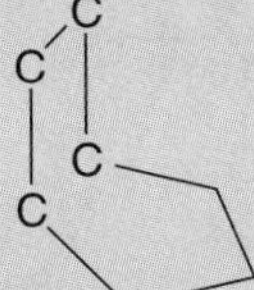

Here are those four carbons incorporated into one ring

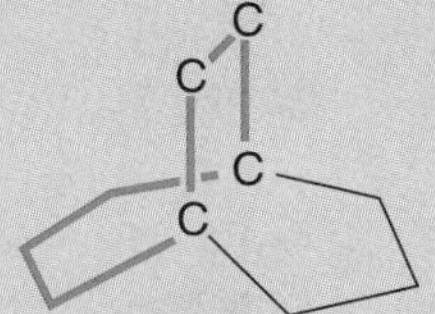

Here they are incorporated into another ring

Such molecules are called "bridged" and are extremely common. Three compounds found in Nature are shown in Figure 6.60.

β-Pinene

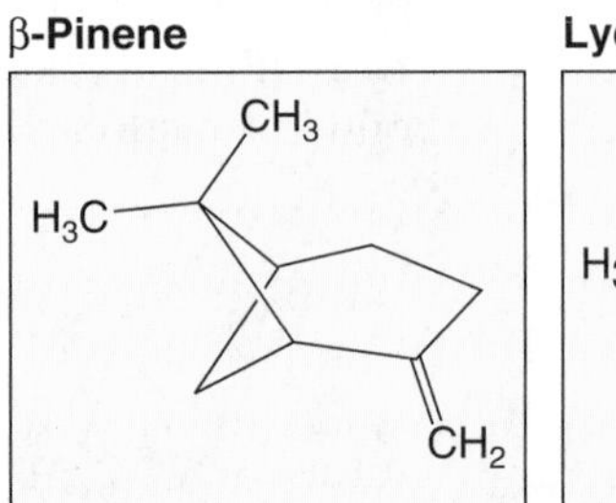

Lycopodine

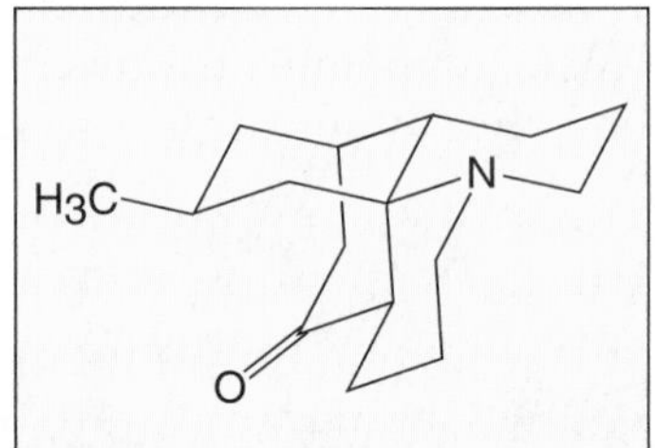

Calicheamicin

FIGURE 6.60 Three naturally occurring bicyclic molecules.

Convention Alert!

Bicyclic compounds are named in the following way: One first counts the number of carbons in the ring system. The molecule in Figure 6.61 has eight ring carbons. Thus the base name is "octane." The molecule is numbered by counting from the bridgehead carbon [carbon number 1, C(1)] around the largest bridge first, proceeding to the other bridgehead. One then continues counting around the second longest bridge and finally numbers the shortest bridge. Any substituents can now be assigned a number. So the compound in Figure 6.61 is a 3,3-dimethylbicyclooctane.

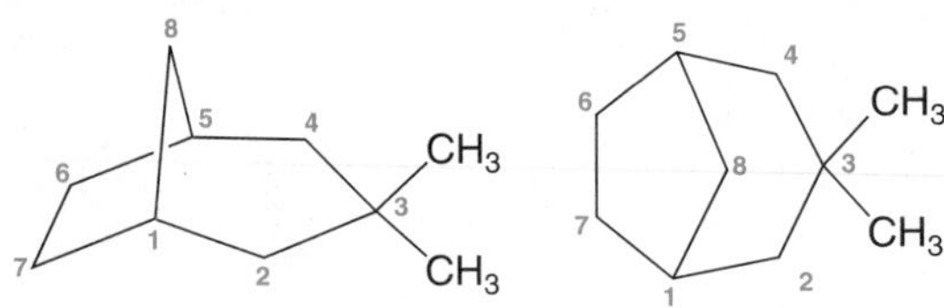

FIGURE 6.61 Part of the naming protocol for a typical bicyclic compound.
1. Count from one bridgehead around the longest bridge to the other bridgehead (1, 2, 3, 4, 5).
2. Next, count around the second longest bridge (6,7).
3. Last, number the shortest bridge (8). This compound is a 3,3-dimethylbicyclooctane.

The bridges are counted from the bridgeheads and assigned numbers equal to the number of atoms in the bridges, *not counting the bridgehead atoms themselves.* These numbers are enclosed in brackets, largest first, in between the designation "bicyclo" and the rest of the name (Fig. 6.62). So this compound is called 3,3-dimethylbicyclo[3.2.1]octane.

1 Atom bridge

3 Atom bridge

2 Atom bridge

CH_3

CH_3

3,3-Dimethylbicyclo[3.2.1]octane

FIGURE 6.62 More of the naming protocol for bicyclic compounds. This molecule is 3,3-dimethylbicyclo[3.2.1]octane. The red dots show the bridgehead atoms. These are not counted in sizing the bridges, but are counted in numbering the compound.

PROBLEM 6.23 Make drawings of the following three compounds: (a) *cis*-bicyclo[3.3.0]octane, (b) 1-fluorobicyclo[2.2.2]octane, (c) *trans*-9,9-dimethylbicyclo[6.1.0]nonane.

PROBLEM 6.24 Name the compounds in Figure 6.63.

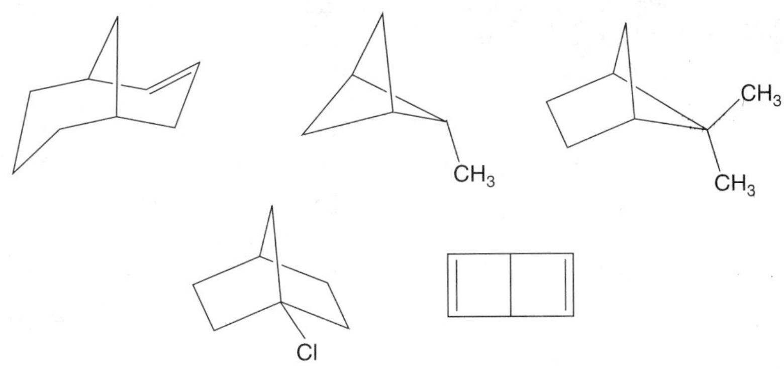

FIGURE 6.63

As with the 1,2-fused systems we must worry about the stereochemistry at the bridgehead positions. At first, this may seem trivial—where could these hydrogens be but where they are shown in the figures? The answer is, "inside the cage!" In the compounds we have already examined, it is not easy for hydrogens to occupy the inside position. As shown in Figure 6.64, all four valences (bonds) of the bridgehead carbons of bicyclo-[3.2.1]octane would be pointing in the same direction if one or two bridge-

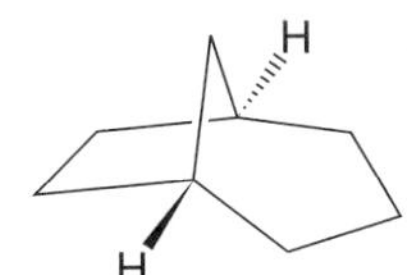

"Out, out" bicyclo[3.2.1]octane

"In, out" bicyclo[3.2.1]octane

"In, in" bicyclo[3.2.1]octane

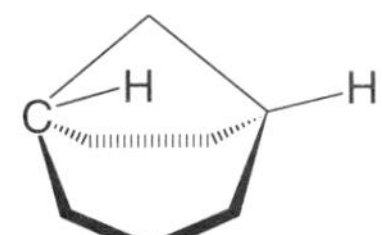

Side view of "in, out" bicyclo[3.2.1]octane as seen by the eye

FIGURE 6.64 Three stereoisomers of bicyclo[3.2.1]octane; "out, out," "in, out," and "in, in." Only the first isomer is known. This side view of "in, out" bicyclo[3.2.1]octane shows the problem—the four valences at one bridgehead carbon are pointed in the same direction, This position is hopelessly strained. The "in, in" isomer is even worse. Neither of these molecules has been made. The "out, out" molecule is quite normal.

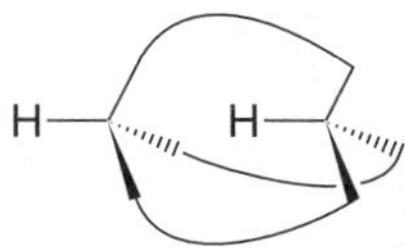

FIGURE 6.65 If the bridges are long enough, there is no angle strain in an "in, out" isomer.

head hydrogens were inside the cage. But now imagine increasing the size of the bridges. In principle, it should be possible to make the bridge chains long enough so that a normal, nearly tetrahedral, arrangement can be achieved with one or both bridgehead hydrogens inside the cage (Fig. 6.65). And so it is. The molecules shown in Figure 6.66 are both known, as are several other examples.

FIGURE 6.66 Two known "in" or "in, out" compounds.

6.7 POLYCYCLIC SYSTEMS

We needn't stop here. More rings can be fused or attached in fused or bridging fashion to produce wondrously complex structures. The naming protocols are complicated, if ultimately logical, and won't be covered here. Many of the compounds have common names that are meant to be evocative of their shapes. Prismane and basketane are examples. The molecular versions of the Platonic solids, tetrahedrane, cubane, and dodecahedrane are all known, although the parent tetrahedrane still evades synthesis (Fig. 6.67).*

Cholesterol
a polycyclic compound
a steroid

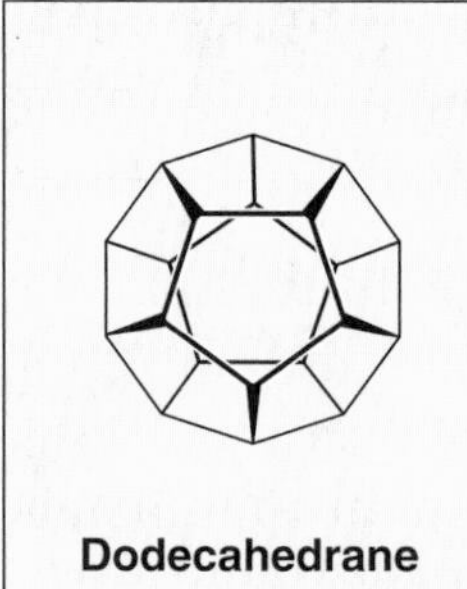

Dodecahedrane

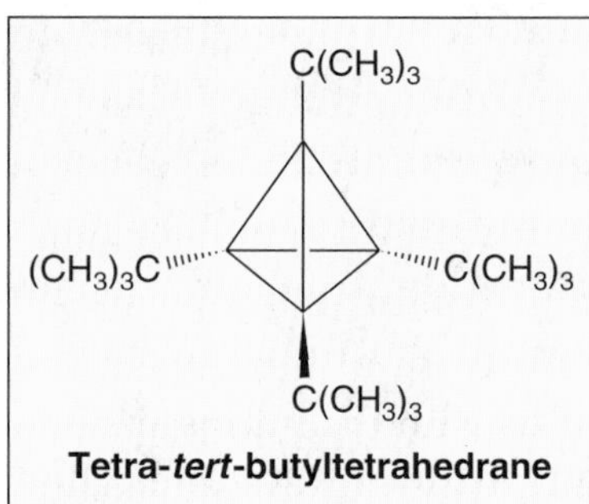

Tetra-*tert*-butyltetrahedrane

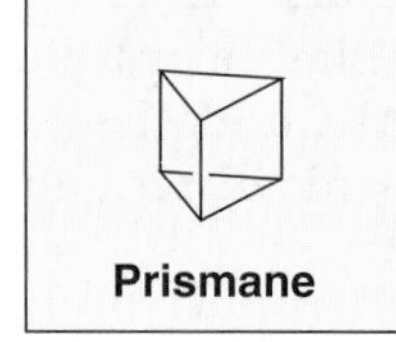

Prismane

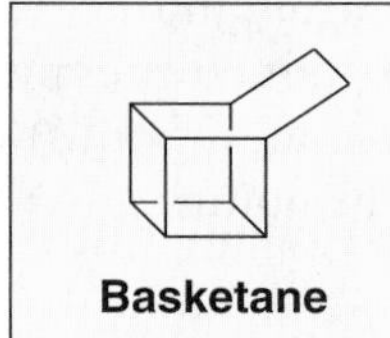

Basketane

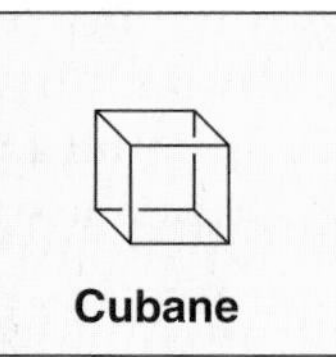

Cubane

FIGURE 6.67 Some polycyclic molecules.

*The last Platonic solid, the icosahedron, has no appropriate carbon version, as the atoms at the vertices have six, not four, neighbors. In a classical "icosahedrane," carbon would have to be hexavalent! There are icosahedral molecules known, such as the extraordinarily stable carboranes ($C_2B_{10}H_{12}$) shown in Figure 6.68, but the cage bonding in such compounds is not made up of normal two-center, two-electron bonds. In the carboranes, neither carbon nor boron is hexavalent, but rather "hexacoordinate." The lines are not conventional two-electron bonds but mere devices to map the connectivity in the molecule. We will see more examples of this kind of bonding in Chapter 23.

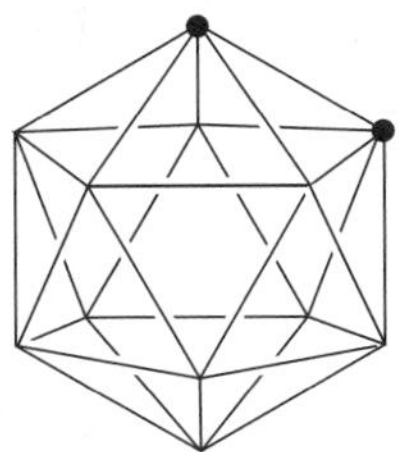
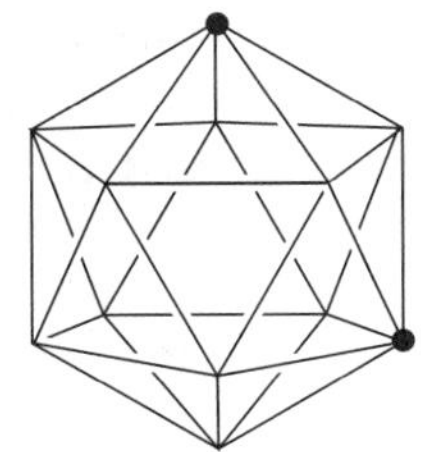
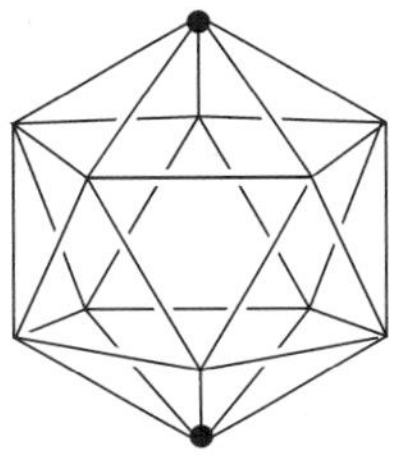

FIGURE **6.68** Three icosahedral molecules. The dots represent CH, and the other vertices are BH. The formula of each of these molecules is $C_2B_{10}H_{12}$. The lines between the atoms cannot represent normal two-electron bonds. (Why not?) These molecules are amazingly stable.

6.8 SOMETHING MORE: ADAMANTANE AND DIAMOND

Consider constructing a polycyclic molecule by expanding a chair cyclohexane. First, connect three of the axial bonds to a cap consisting of three methylene (CH_2) groups all connected to a single methine (CH) group. This process produces the molecule, tricyclo[3.3.1.1^{3,7}]decane, better known as "adamantane," $C_{10}H_{16}$ (Fig. 6.69).

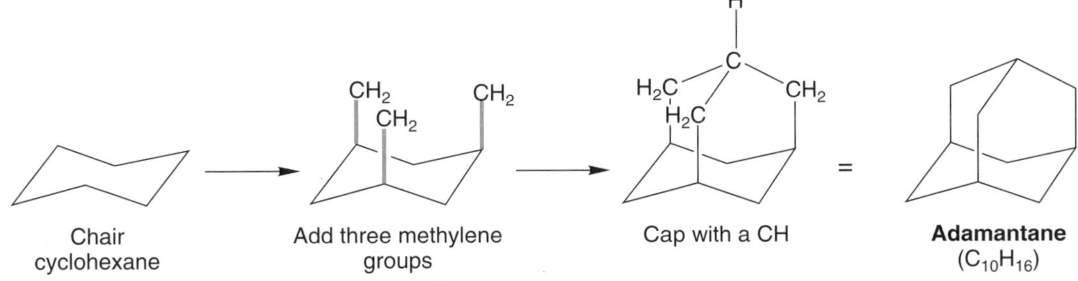

FIGURE **6.69** A schematic construction of the polycyclic molecule adamantane. If we cap a chair cyclohexane with three methylene (CH_2) groups all attached to a single methine (CH), we produce adamantane.

1-Aminoadamantane

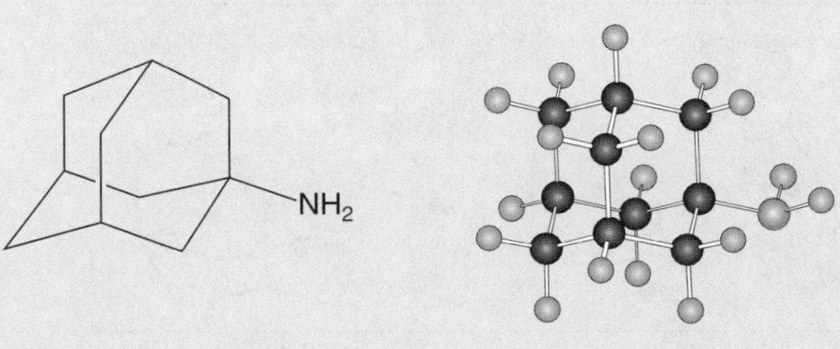

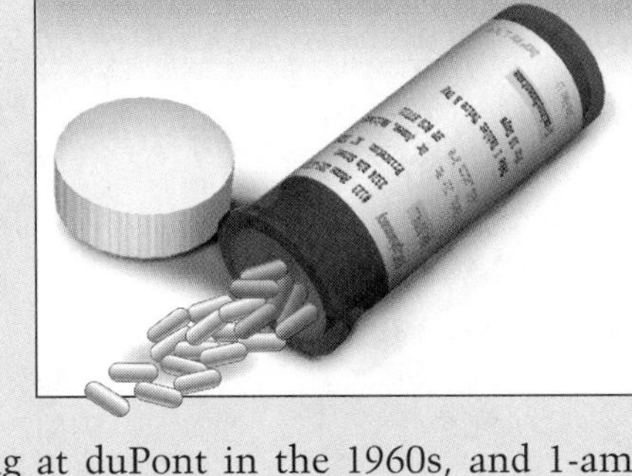

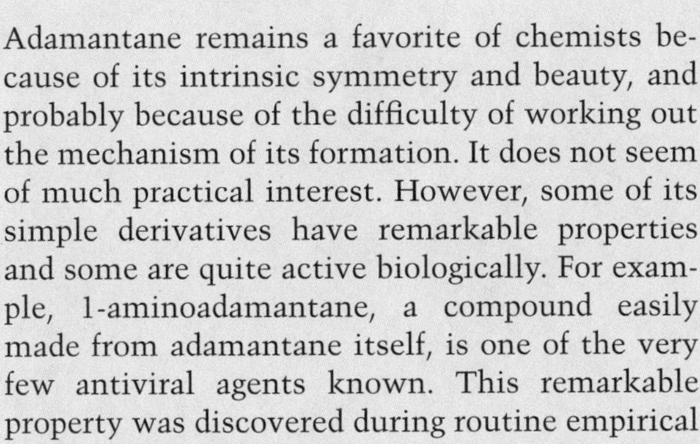

Adamantane remains a favorite of chemists because of its intrinsic symmetry and beauty, and probably because of the difficulty of working out the mechanism of its formation. It does not seem of much practical interest. However, some of its simple derivatives have remarkable properties and some are quite active biologically. For example, 1-aminoadamantane, a compound easily made from adamantane itself, is one of the very few antiviral agents known. This remarkable property was discovered during routine empirical screening at duPont in the 1960s, and 1-aminoadamantane has since been marketed, mostly as an agent against influenza A and C. It apparently works by migrating through the cell membrane to attack the virus within. It is the adamantane cage, acting as a molecular ball of grease, that helps in the membrane penetration. It has been speculated that other, large, symmetrical hydrocarbon "blobs" might act in the same way. Unfortunately, they are not so easy to make.

As Figure 6.70 shows, adamantane is composed entirely of perfect chair six-membered rings. As you would expect, this molecule is nearly strain-free, and constitutes the thermodynamic minimum for all the $C_{10}H_{16}$ isomers.

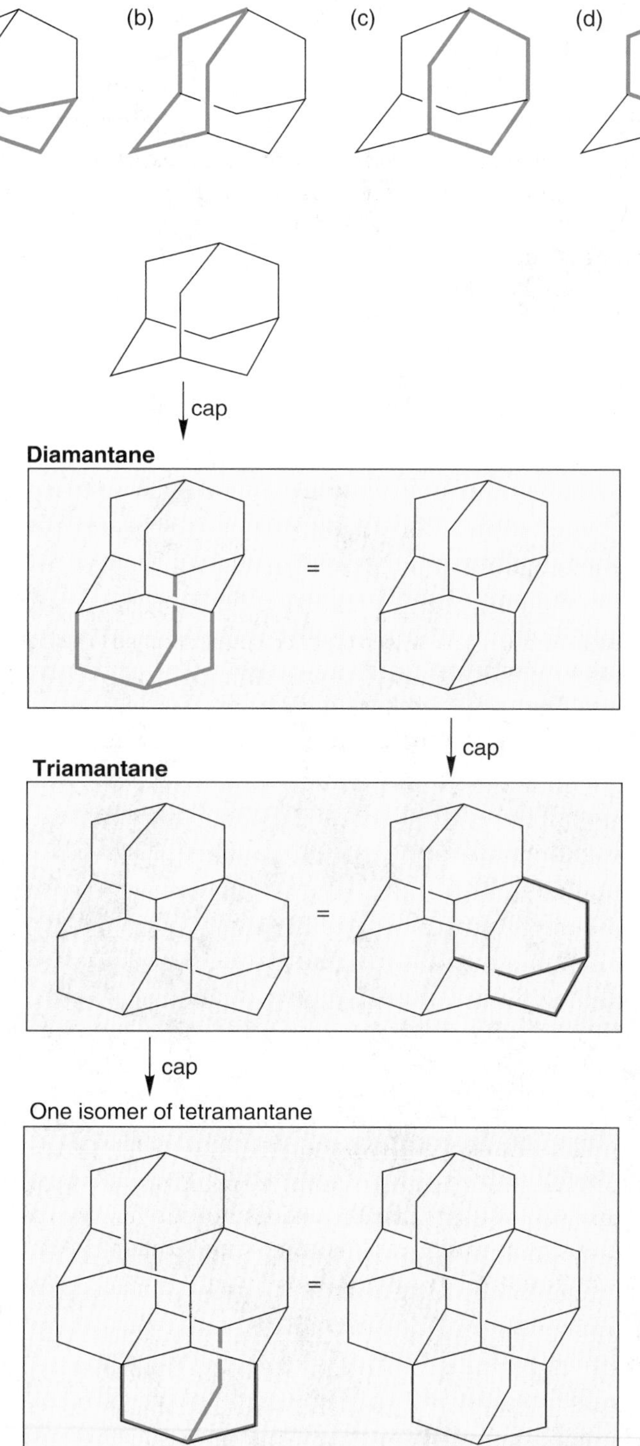

FIGURE 6.70 Adamantane is composed entirely of chair cyclohexanes. The first two (a and b) are easy to see as chairs, but (c) and (d) may require the use of models.

FIGURE 6.71 The continuation of this capping process leads to polyadamantanes and, ultimately, to diamond.

Adamantane was first found in the 1930s in trace amounts in petroleum residues by the Czech chemist Stanislav Landa (1898–1981), but can now be made easily in quantity by a simple process discovered by the American chemist Paul von R. Schleyer (b. 1930).

Consider what happens when we continue the capping process begun in our transformation of chair cyclohexane into adamantane in Figure 6.69. Adamantane itself is composed of only chair cyclohexanes, so we have a number of possible places to start the process (Fig. 6.71). Addition of one more four-carbon cap produces "diamantane," two caps, "triamantane," four, "tetramantane," and so on.

The ultimate result of the capping process is a molecule composed of a network of adamantanes, and this is the structure of diamond.

PROBLEM 6.25 What is the empirical formula of diamond? For a large polymeric molecule such as diamond, the edges of the molecule are insignificant in figuring out the formula.

PROBLEM 6.26 Can you draw other isomers of tetramantane?

Adamantane chemistry remains full of astonishing surprises. For years it has been known that synthetic diamonds could be made from another polymeric form of carbon, graphite, by treatment at high temperature (~2300 K) and very high pressure (7×10^4 kg/cm^2). However, in the 1980s in the Soviet Union and Japan, it was discovered that diamond films could be grown at low temperature and pressure by passing a stream of hydrogen containing a few percent methane through an electric discharge (Fig. 6.72). The mechanism of this reaction remains obscure. Any ideas?

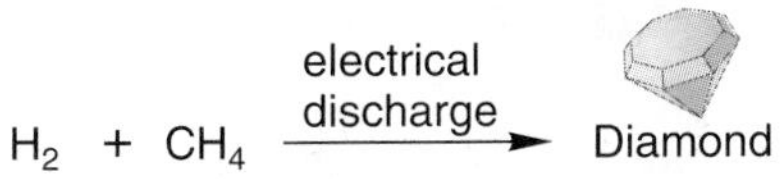

$$H_2 + CH_4 \xrightarrow{\text{electrical discharge}} \text{Diamond}$$

FIGURE 6.72 A simple (and astonishing) synthesis of diamond.

6.9 SUMMARY

NEW CONCEPTS

This chapter deals exclusively with the structural properties of ring compounds. Two kinds of destabilizing effects on rings are discussed. Torsional strain, the destabilizing effect of eclipsed carbon–hydrogen bonds, has been mentioned before when the acyclic hydrocarbons were discussed. The planar forms of ring compounds are particularly subject to torsional strain. In addition, rings contain varying amounts of angle strain, depending on the ring size. Any deviation from the ideal tetrahedral angle of 109.5° will introduce angle strain. Rings adopt nonplanar forms in order to minimize the combination of torsional and angle strain.

One nonplanar ring deserves special mention. Cyclohexane avoids torsional and angle strain by adopting a chair conformation in which a nearly ideal tetrahedral angle is achieved and all carbon–hydrogen bonds are perfectly staggered. In cyclohexane, there is a set of six axial hydrogens that is converted into the set of six equatorial hydrogens through rotations about carbon–carbon bonds called a ring flip.

There are many possible quantitative evaluations of strain. This chapter uses both heats of formation (ΔH_f°) and heats of combustion (ΔH_c°) to arrive at values for the strain energies of various rings. As rings increase in size, strain initially decreases, reaching a minimum at the strain-free cyclohexane. Strain then increases until large ring sizes are reached.

By combining rings in spiro, fused, and bridged fashion, very complex polycyclic molecules can be constructed.

REACTIONS, MECHANISMS, AND TOOLS

The few reactions encountered here are not really new. There is, however, a somewhat more detailed treatment of the formation of carbon dioxide and water from hydrocarbons and oxygen (combustion), and a discussion of the effects of strain on the bond energy of the carbon–carbon bond.

COMMON ERRORS

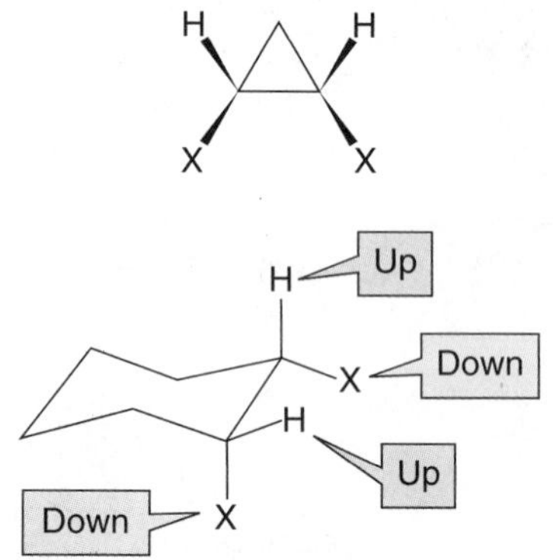

FIGURE 6.73 In both these compounds the two hydrogens shown are cis, as are the two X groups. This relationship is easy to see for the flat cyclopropane, but harder in the nonplanar cyclohexane.

You are presented in this chapter with the most difficult challenge so far in visualizing molecules. Moreover, the difficulty in translating from two dimensions to three is compounded by the mobile equilibria present in many ring compounds. In analyzing complicated phenomena, such as the equilibrations of disubstituted cyclohexanes, be sure to use models, at least at first.

Two small points seem to give students trouble. First, one hears constantly, I'm no artist, I can't draw a good chair cyclohexane! Nonsense. It does not take an artist, merely an artisan. Go slowly, do not scribble, and follow the simple procedure outlined in this chapter (p. 198). Drawing a perfect chair cyclohexane is one thing that everyone can do.

Determining which groups are cis and trans in ring compounds causes problems. *Remember*: cis means "on the same side," and trans, "on opposite sides." Bonds to cis groups do not have to be parallel, merely both up or both down. Figure 6.73 gives two common examples, one straightforward, the other not so obvious.

6.10 KEY TERMS

Angle strain The increase in energy caused by the deviation of an angle from the ideal demanded by the hybridization.

Axial hydrogens The set of six, straight up and down hydrogens in chair cyclohexane. Ring flip interconverts these hydrogens with the set of equatorial hydrogens.

Bridged rings In a bridged bicyclic molecule, two rings share more than two atoms.

Bridgehead position The bridgehead positions are shared by the rings in a bicyclic molecule. In a bicyclic molecule, the three bridges emanate from the bridgehead positions.

Conformational analysis The determination of the minimum energy form of a molecule through an analysis of steric interactions.

Equatorial hydrogens The set of six hydrogens, also "up and down," but more or less in the plane of the ring in chair cyclohexane. These hydrogens are interconverted with the set of axial hydrogens through ring "flipping" of the chair.

Fused rings Two rings sharing only two carbons.

Spiro rings Two rings that share a single carbon.

Torsional strain The destabilization induced in a molecule through eclipsed bonds.

van der Waals strain When two atoms are too close together, the energy of the system increases. This increase is called van der Waals strain.

6.11 ADDITIONAL PROBLEMS

PROBLEM **6.27** By now you have encountered most structural types and have some experience with rings. Write and name all the isomers of the formula C_5H_8. Stay alert for isomerism of the cis/trans (*Z/E*) and (*R/S*) types. Indicate chiral molecules with an asterisk. This problem is very hard at this point.* There are all sorts of odd structural types among these isomers. It should not be hard to get most of them, but getting them all is really tough. Think in three dimensions and *organize*. *Hint*: The total number of isomers, not including enantiomers, is 28.

PROBLEM **6.28** Write out the pairs of enantiomers for the chiral isomers in Problem 6.27.

PROBLEM **6.29** Assign (*R*) and (*S*) configurations for the versions of the chiral cyclic molecules shown on the left in the answers to Problem 6.28; given in the Study Guide.

PROBLEM **6.30** Draw the two possible chair forms of *cis*- and *trans*-1,4-dimethylcyclohexane. Are the two forms identical, enantiomeric, or diastereomeric? In each case, indicate which chair form will be more stable and explain why. Is either of these molecules chiral?

PROBLEM **6.31** Draw the two possible chair forms of *cis*- and *trans*-1-isopropyl-4-methylcyclohexane. Are the two forms identical, enantiomeric, or diastereomeric? In each case, indicate which chair form will be more stable and explain why. Is either of these molecules chiral?

PROBLEM **6.32** Draw the two possible chair forms of *cis*- and *trans*-1,3-dimethylcyclohexane. Are the two forms identical, enantiomeric, or diastereomeric? In each case, indicate which chair form will be more stable and explain why. Is either of these molecules chiral?

PROBLEM **6.33** Draw the possible chair forms of *cis*- and *trans*-1-isopropyl-3-methylcyclohexane. Are the two forms identical, enantiomeric, or diastereomeric? In each case, indicate which chair form will be more stable and explain why. Is either of these molecules chiral?

PROBLEM **6.34** Use the data in Table 6.4 (p. 210) to calculate the energy difference between the possible isomers of *cis*- and *trans*-1,4-dimethylcyclohexane.

PROBLEM **6.35** Use the data in Table 6.4 (p. 210) to calculate the energy difference between the possible isomers of *cis*- and *trans*-1-isopropyl-4-methylcyclohexane.

PROBLEM **6.36** In Section 6.5b (p. 213), you saw that for purposes of stereochemical analysis you could treat the decidedly nonplanar *cis*- and *trans*-1,2-dimethylcyclohexanes as if they were planar. The planar forms represent the average positions of ring atoms in the rapid chair–chair interconversions. Use planar representations of the following cyclohexanes to determine which molecules are chiral.

(a) *cis*-1,3-Dimethylcyclohexane
(b) *trans*-1,3-Dimethylcyclohexane
(c) *cis*-1-Isopropyl-3-methylcyclohexane
(d) *trans*-1-Isopropyl-3-methylcyclohexane
(e) *cis*-1,4-Dimethylcyclohexane
(f) *trans*-1,4-Dimethylcyclohexane
(g) *cis*-1-Isopropyl-4-methylcyclohexane
(h) *trans*-1-Isopropyl-4-methylcyclohexane

PROBLEM **6.37** Draw both chair forms for the following cyclohexanes. Indicate the more stable chair form in each case. Not all examples will be obvious; some choices may be too close to call.

(a) CH_3, $CH(CH_3)_2$, Cl

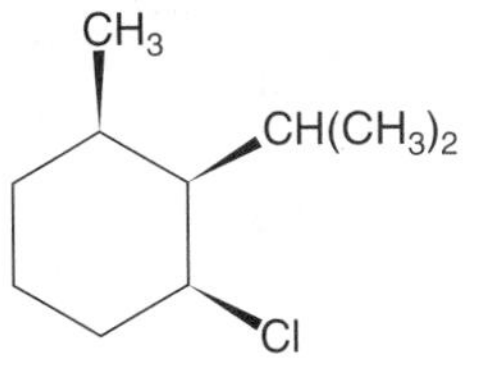

(b) CH_3, $CH(CH_3)_2$, Cl

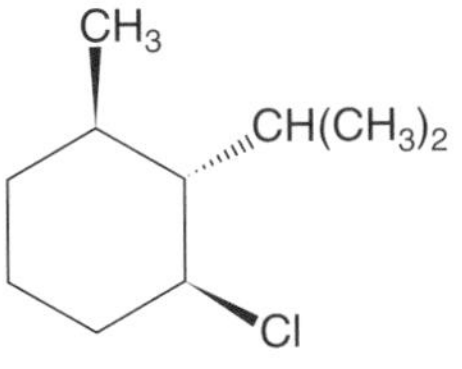

(c) CH_3, $CH(CH_3)_2$, Cl

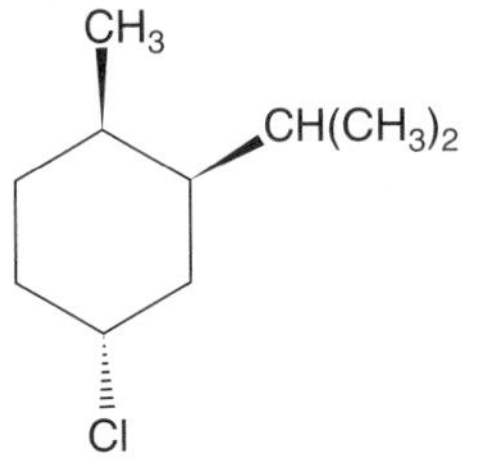

(d) CH_3, $CH(CH_3)_2$, Cl

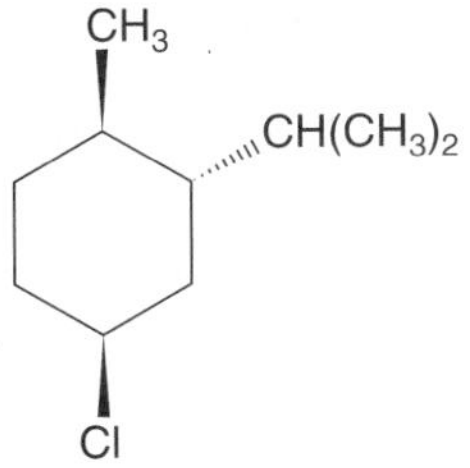

*Indeed, this problem constituted the *entire* first hour examination in my first course in organic chemistry.

PROBLEM **6.37** (CONTINUED)

(e) (f)

CH_3 CH_3

Cl $CH(CH_3)_2$ Cl $CH(CH_3)_2$

PROBLEM **6.38** Name the following compounds:

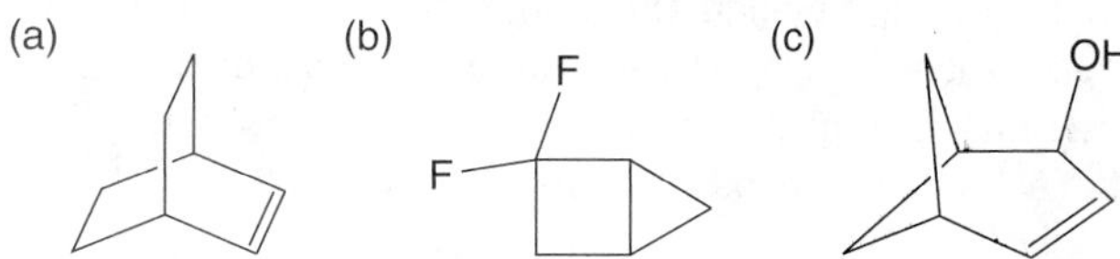

PROBLEM **6.39** Write structures for the following compounds:

(a) 2,2-Dibromo-3,3-dichloro-5,5-difluoro-6,6-diiodobicyclo[2.2.1]heptane
(b) Bicyclo[1.1.1]pentan-1,3-diol
(c) Hexamethylbicyclo[2.2.0]hexa-2,5-diene

PROBLEM **6.40** The sugar glucose is a six-membered ring containing an oxygen atom. There are groups attached at each of the five ring carbons as shown. The red "squiggly" bond merely means that the hydroxyl group at the 1-position may be either up or down. That is, there are two structures for glucose. Draw these two molecules in three dimensions.

HOH_2C 6 O 1 OH 5 2 HO 4 3 OH OH

PROBLEM **6.41** One of the isomers of methylbicyclo[2.2.1]heptane can exist in two diastereomeric forms (we are not counting mirror images, which increases the total number of stereoisomers to four). Write the possible isomers of methylbicyclo[2.2.1]heptane and explain how one of them can exist in two forms.

PROBLEM **6.42** (a) Can one of the isomers of methylbicyclo[2.2.2]octane also exist in two diastereomeric forms? That is, does it behave just like the methylbicyclo-[2.2.1]heptane mentioned in Problem 6.41? (b) Can one of the isomers of methylbicyclo[2.2.2]octane exist in two enantiomeric forms?

Substitution and Elimination Reactions: The S_N2, S_N1, E2, and E1 Reactions

Don't look back, something might be gaining on you.
—Leroy Robert "Satchel" Paige*

We are about to embark on a study of chemical change. The last six chapters covered various aspects of structure, and still more analysis of structure will appear as we go along. Without this structural underpinning no serious exploration of chemical reactions is possible. The key to all studies of reactivity—reaction mechanisms—is an initial, careful analysis of the structures of the molecules involved. We have been learning parts of the grammar of organic chemistry. Now we are about to use that grammar to write some sentences and paragraphs. Let's begin with a reaction central to most of organic and biological chemistry, the replacement of one group by another (Fig. 7.1).

The general case

$$R{-}L + N \rightleftharpoons R{-}N + L$$

R =

H_3C Methyl group	$(CH_3)_2CH$ Isopropyl group
CH_3CH_2 Ethyl group	$(CH_3)_3C$ *tert*-Butyl group

FIGURE 7.1 This generic reaction presents an overall picture of the substitution process. A leaving group (L) is replaced by a new group, N, to give R–N. The box shows a standard set of alkyl groups that will be used throughout this chapter.

This conversion of one alkyl halide into another is an example of the **substitution reaction** (Fig. 7.2). This reaction looks simple, and in some ways it is. The only change is the replacement of one halogen by another.

A specific example

$$\underbrace{C_6H_5C(=O)OCH_2CH_2}_{R}{-}\underbrace{Cl}_{L} + \underbrace{Na^+\ I^-}_{N} \xrightarrow[24\ h]{100\ °C} \underbrace{C_6H_5C(=O)OCH_2CH_2}_{R}{-}\underbrace{I}_{N} + \underbrace{Na^+\ Cl^-}_{L}$$

(80%)

FIGURE 7.2 The replacement of one halogen by another is an example of the substitution reaction.

Yet, as we examine the details of this process we find ourselves looking deeply into chemical reactivity. This reaction is an excellent prototype

*Satchel Paige (1906–1982) was a pitcher for the Cleveland Indians and later the St. Louis Browns. He was brought to the major leagues by Bill Veeck after a long and spectacular career in the Negro Leagues. Some measure of his skill can be gained from realizing that Paige pitched in the major leagues after his 50th birthday.

that nicely illustrates general techniques used to determine reaction mechanism.

One reason that we spend so much time on the substitution reaction is that it *is* so general; what we learn about it and from it is widely applicable. Not only will we apply what we learn here to many other chemical reactions, but these processes have vast practical consequences as well. Substitution reactions, commonly, (if sometimes inaccurately) called displacement reactions, are at the heart of many industrial processes and are central to the mechanisms of many biological reactions including some forms of carcinogenesis, the cancer-causing activity of many molecules. The molecules of which we are composed are bristling with substituting agents. It seems clear that when some groups on our DNA (deoxyribonucleic acid, see Chapter 26) are alkylated in a substitution reaction, tumor production is initiated or facilitated. Beware of alkylating agents.

This chapter is long, important, and possibly challenging. It represents a first introduction to reactivity, the central area of organic chemistry. In it, we cover four of five building block reactions, two of them substitution reactions (S_N1 and S_N2) and two of them elimination reactions (E1 and E2). The other fundamental reaction, additions, will appear shortly, in Chapter 9. From now on, you really cannot memorize your way through the material; you must try to generalize. Understanding concepts and applying them in new situations is the route to success. This chapter contains lots of problems, and it is important to try to work through them as you go along. I know this is repetition, and I don't want to insult your intelligence, but it would be even worse not to warn you of the change in the course that takes place now, or not to suggest ways of getting through this new material successfully. You cannot simply read this material and hope to get it all. You must work with the text and the various people in your course, the Professor and the teaching assistants, in an interactive way. The problems try to help you do this. When you hit a snag, or can't do a problem, find someone who can and get the answer. The answer is more than a series of equations or structures and arrows; it also involves finding the right approach to the problem. There is much technique to problem solving in organic chemistry; it can be learned, and it gets much easier with practice. *Remember*: Organic chemistry must be read with a pencil in your hand.

7.1 ALKYL HALIDES: NOMENCLATURE AND STRUCTURE

Alkyl iodides, bromides, chlorides, and, occasionally, fluorides are common substrates for the substitution reaction and are collected under the name of **alkyl halides**. Alkyl halides have the formula $C_nH_{2n+1}X$, where X = F, Cl, Br, or I. Quite appropriately, alkyl halides are named as fluorides, chlorides, bromides, and iodides. Both common and systematic names are used, although as the complexity of the molecule increases, the system naturally takes over. The important rules to remember for naming saturated halides are (1) minimize the number given to the substituent, and (2) name multiple substituents in alphabetical order. If a carbon–carbon double bond is present, the double bond takes precedence over the halogen when positional numbers are assigned (see the last example in Table 7.1). Table 7.1 gives some examples.

TABLE 7.1 Some Alkyl Halides, Names, and Known Properties

Compound	Names	bp (°C)	mp (°C)
CH_3—F	Methyl fluoride	–78.4	–141.8
CH_3CH_2—Br	Ethyl bromide	38.4	–118.6
$(CH_3)_2CH$—I	Isopropyl iodide, or 2-iodopropane	89.4	– 90.1
$(CH_3)_3C$—Cl	*tert*-Butyl chloride, or 2-chloro-2-methylpropane	51.0	– 25.4
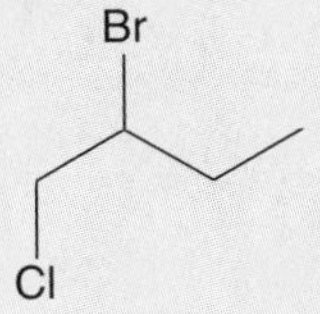 Br, Cl	2-Bromo-1-chlorobutane	147	
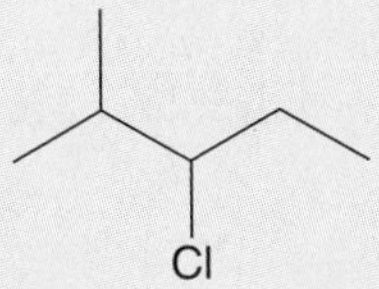 Cl	3-Chloro-2-methylpentane	117	
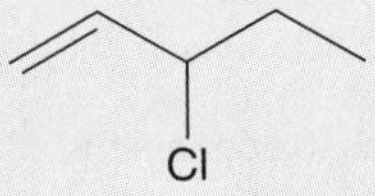 Cl	3-Chloro-1-pentene	94	
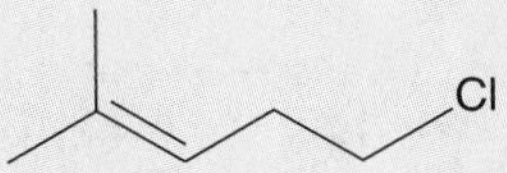 Cl	5-Chloro-2-methyl-2-pentene *not* 1-chloro-4-methyl-3-pentene		
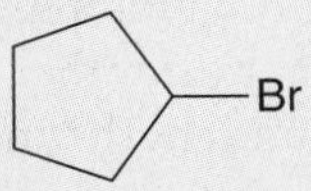 Br	Bromocyclopentane	138	
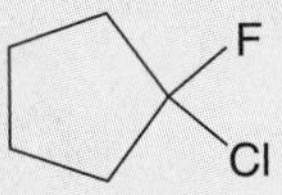 F, Cl	1-Chloro-1-fluorocyclopentane		
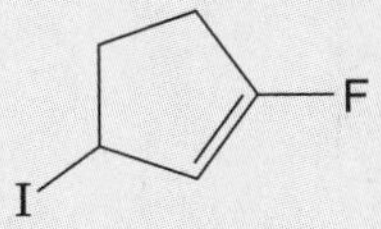 F, I	1-Fluoro-3-iodocyclopentene *not* 3-fluoro-1-iodocyclopent-2-ene		

Halides are classified as methyl, primary, secondary, tertiary, and vinyl (Fig. 7.3). Chapter 3 (p. 97) introduced this nomenclature. *Remember*: A primary carbon is attached to only one other carbon, a secondary carbon to two others, and a tertiary carbon to three others. In vinyl halides, there is always a double bond to which the halide is directly attached.

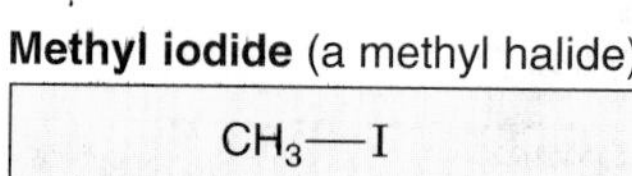

Ethyl iodide (a primary halide)

CH_3CH_2—I

Isopropyl iodide (a secondary halide)

$(CH_3)_2CHI$

tert-Butyl iodide (a tertiary halide)

$(CH_3)_3C$—I

Vinyl iodide

H_2C=CH

FIGURE 7.3 Some alkyl and vinyl halides at various levels of abstraction.

TABLE 7.2 Dipole Moments of Some Simple Alkyl Halides

Alkyl Halide	Dipole Moment (D)
Methyl fluoride	1.85
Ethyl fluoride	1.94
Methyl chloride	1.87
Ethyl chloride	2.05
Methyl bromide	1.81
Ethyl bromide	2.03
Methyl iodide	1.62
Ethyl iodide	1.91

As pointed out in Chapter 3 (p. 107), substituted alkanes are polar molecules. The halogens are electronegative atoms and strongly attract the electrons in the carbon–halogen bond. Alkyl halides have dipole moments of approximately 2 D (Fig. 7.4; Table 7.2).

means $\overset{+}{\delta}$—$\overset{-}{\delta}$

FIGURE 7.4 A substantial dipole moment exists in the polar carbon–halogen bond, X=F, Cl, Br, and I.

Simple alkyl halides are nearly tetrahedral, and therefore the carbon to which the halogen is attached is approximately sp^3 hybridized. The bond to halogen involves an approximately sp^3 carbon orbital overlapping with an orbital on the halogen for which the principal quantum number varies from $n = 2$ for fluorine to $n = 5$ for iodine. The C–X bond lengths increase and the bond strengths decrease as we read down the periodic table. Some typical values for methyl halides are shown in Figure 7.5.

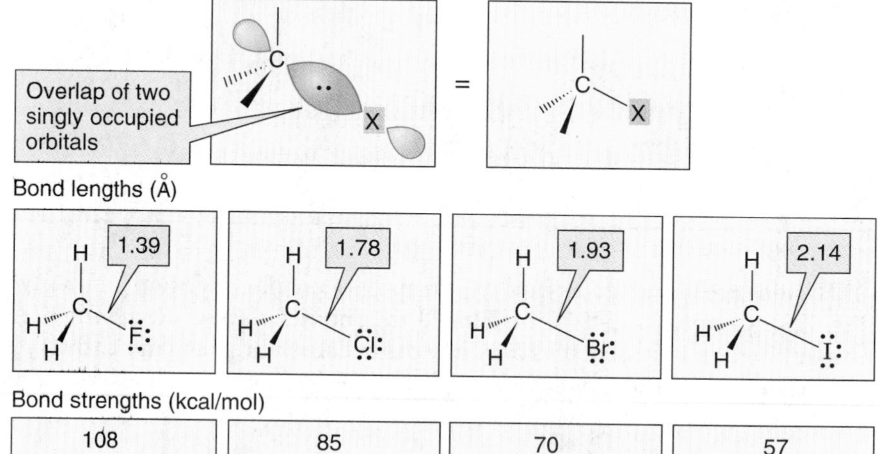

FIGURE 7.5 The C–X bond of an alkyl halide is formed by the overlap of a singly occupied carbon sp^3 orbital with a singly occupied halogen orbital. The structures of alkyl halides at the bottom show both C–X bond lengths (in angstom units) and C–X bond strengths (in kilocalories per mole). *Remember*: These values refer to homolytic bond breaking (formation of two neutral radicals, Chapter 2, p. 49; Chapter 3, p. 77).

Now it is time to see how these species take part in the substitution reaction. Our discussion of reaction mechanism starts with acids and bases.

7.2 REACTION MECHANISM: BRØNSTED* ACIDS AND BASES

A **Brønsted acid** is any compound that can donate a proton. A **Brønsted base** is any compound that can accept a proton. Let's look first at a very simple process, the reaction of solid potassium hydroxide (KOH) with gaseous hydrochloric acid (HCl). This reaction is nothing but a competition of two Brønsted bases (Cl^- and HO^-) for a proton, H^+. The stronger Brønsted base (HO^-) wins the competition (Fig. 7.6).

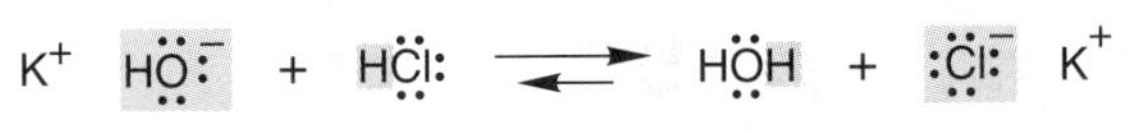

$$K^+ \; HO^- + HCl \rightleftharpoons HOH + Cl^- \; K^+$$

FIGURE 7.6 The two Brønsted bases, HO^- and Cl^-, compete for the proton, H^+.

Hydrochloric acid (HCl) is called the **conjugate acid** of Cl^-, and Cl^- is the **conjugate base** of hydrochloric acid. Conjugate bases and acids are related to each other through the gain and loss of a proton (Fig. 7.7).

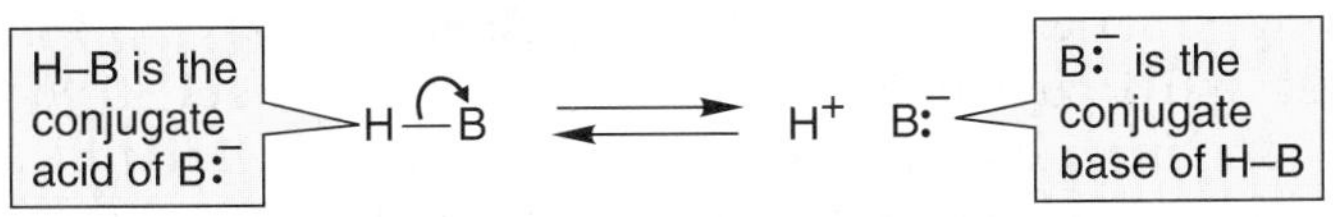

H–B is the conjugate acid of B:⁻ → H—B ⇌ H^+ B:⁻ ← B:⁻ is the conjugate base of H–B

FIGURE 7.7 Conjugate acids and bases are related by the gain and loss of a proton. **Caution!** This is a formalized picture and does not imply the reaction of bases with a bare proton, H^+. Outside of the gas phase, bare protons do not exist.

PROBLEM 7.1

What are the conjugate acids of the following molecules?
(a) H_2O (b) ^-OH (c) NH_3 (d) CH_3OH (e) $H_2C{=}O$ (f) $^-CH_3$

PROBLEM 7.2

What are the conjugate bases of the following molecules?
(a) H_2O (b) ^-OH (c) NH_3 (d) CH_3OH (e) CH_4 (f) $HOSO_2OH$ (g) $^-OSO_2OH$

A molecule may be *both* a Brønsted acid and a Brønsted base! Consider water, for example. Water can both donate a proton (act as a Brønsted acid) and accept a proton (act as a Brønsted base). Figure 7.8 shows two reactions, the protonation of water by hydrogen chloride, in which water acts as a base, and the protonation of hydroxide ion by water, in which water is acting as the acid.

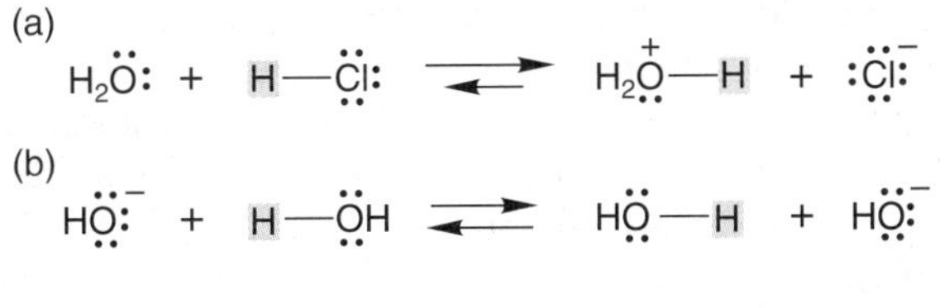

(a)

$$H_2O: + H—Cl: \rightleftharpoons H_2\overset{+}{O}—H + :Cl:^-$$

(b)

$$HO:^- + H—OH \rightleftharpoons HO—H + HO:^-$$

FIGURE 7.8 (a) Here, water acts as a Brønsted base, accepting a proton. The bases water (H_2O) and chloride (Cl^-) compete for the proton. (b) Water can also act as a Brønsted acid, donating a proton. Two hydroxide ions (HO^-) compete for the proton.

Many substitution reactions begin with a protonation step. As strong acids are better able to donate a proton than weaker acids, it will be important for us to know which molecules are strong acids and which are

*These acids and bases were named after Johannes Nicolaus Brønsted (1879–1947).

TABLE 7.3 Some pK_a Values for Assorted Molecules[a]

Compound	pK_a
HI	–10
H_2SO_4	–9
HBr	–9
HCl	–7
ROH_2^+	–2
H_3O^+	–1.7
HNO_3	–1.4
HF	3.2
RCOO<u>H</u>	4–5
H_2S	7.0
H_4N^+	9.2
$CH_3O\underline{H}$	15.2
H_2O	15.7
RO<u>H</u>	16–18
HC≡CH	25
NH_3	38
$H_2C{=}CH_2$	45
$(CH_2)_3$ (cyclopropane)	46
CH_4	59
$(CH_3)_3C\underline{H}$	>70

[a]When there is a choice, it is the underlined H that is lost.

weak. The dissociation of an acid in water can be described by the equation

$$HA + H_2O \rightleftarrows H_3O^+ + A^-$$

This leads to an expression for the equilibrium constant, K.

$$K = [H_3O^+]\,[A^-]/[HA]\,[H_2O]$$

Because it is present as solvent, in vast excess, the concentration of water remains constant in the ionization reaction. This equation is usually rewritten to give the acidity constant, K_a,

$$K_a = [H_3O^+]\,[A^-]/[HA]$$

where the square brackets indicate concentration.

By analogy with pH, we can define a quantity **pK_a**.

$$pK_a = -\log K_a$$

The stronger the acid, the lower the pK_a. Table 7.3 gives pK_a values for some representative compounds. This short list spans about 80 powers of 10 (pK_a is a log function). In practice, any compound with a pK_a lower than about +5 is regarded as a reasonably strong acid; those with pK_a values below 0 are very strong. As we discuss new kinds of molecules more pK_a values will appear. Should you memorize this list? I think not, but you will need to have a reasonable idea of the approximate acidity of different kinds of molecules.

7.3 THE ARROW FORMALISM

There is a most important bookkeeping technique called the **arrow formalism**. It's arbitrary and, at best, only sketches the broad outlines of what happens during a chemical reaction. In particular, it does not constitute a **reaction mechanism**, a description of how the reaction occurs. If a reaction mechanism is not an arrow formalism, what is it? A short answer would be a description, in terms of structure and energy, of the intermediates and transition states separating starting material and product. Implied in this description is a picture of how the participants in the reaction must approach each other.

Only rarely will the curved arrow formalism be able to do more than offer clues to an understanding of a reaction. Despite these limitations, it is a powerful device of enormous utility in helping to keep track of bond makings and breakings. In this formalism, chemical bonds are shown as being formed from pairs of electrons flowing from a donor toward an acceptor, often, but not always, displacing other electron pairs. Figure 7.9 shows two examples. In the first, hydroxide ion, acting as a Brønsted base, displaces the pair of electrons in the hydrogen–chlorine bond. The reaction of ammonia (:NH_3) with hydrogen chloride is another example in which the electrons on neutral nitrogen are shown displacing the electrons in the hydrogen–chlorine bond.

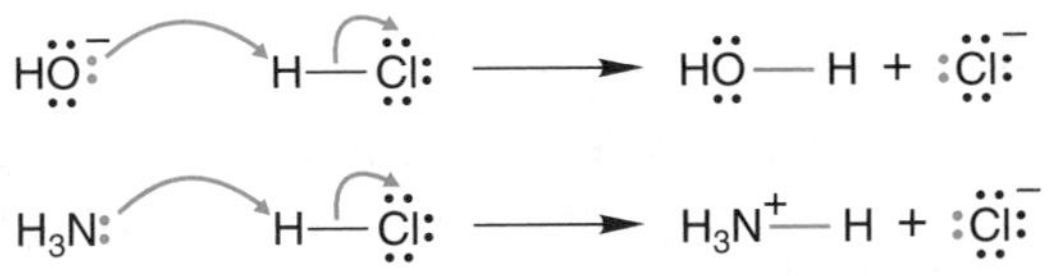

$HO^{-} + H{-}Cl \longrightarrow HO{-}H + Cl^{-}$

$H_3N: + H{-}Cl \longrightarrow H_3\overset{+}{N}{-}H + Cl^{-}$

FIGURE 7.9 Two demonstrations of the application of the curved arrow formalism. We will see many other examples.

*PROBLEM 7.3

Draw an arrow formalism for the reversals of the two reactions in Figure 7.9.

ANSWER

This problem may seem to be mindless, but it really is not. The key point here is to begin to see all reactions as proceeding in two directions. Nature doesn't worry about "forward" or "backward." Thermodynamics dictates the direction a reaction will run in a practical sense.

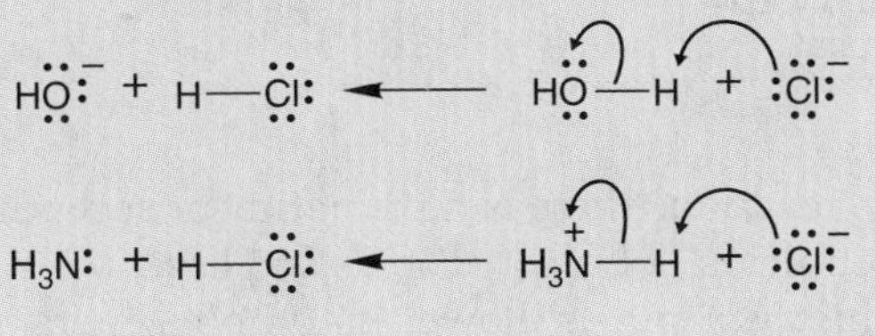

$HO^{-} + H{-}Cl \longleftarrow HO{-}H + Cl^{-}$

$H_3N: + H{-}Cl \longleftarrow H_3\overset{+}{N}{-}H + Cl^{-}$

Let's consider the mechanistic deficiencies of this venerable device, the curved arrow formalism. To me, the arrows evoke the coming together of pairs of electrons: the interaction of filled orbitals. We know this cannot be a good mechanistic representation of a successful reaction because the interaction of two filled orbitals is not stabilizing (Fig. 7.10)! We will need better orbital descriptions of mechanisms than the ones the arrow formalism implies. Yet the arrow formalism remains a powerful mapping technique, and we will use it repeatedly to show which bonds are made and which are broken in a reaction. All organic chemists use it; it is an important part of the language, but we must be clear as to what it is (an essential mapping device), and what it is not (a reaction mechanism). We will use it extensively, beginning with the description of the reactions of acids and bases.

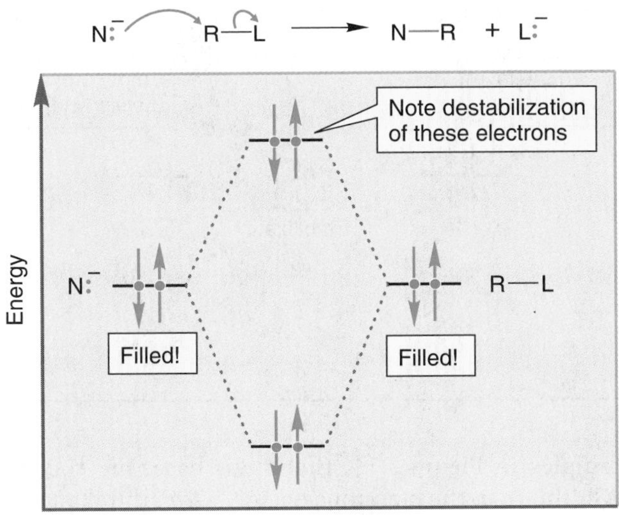

FIGURE 7.10 Although extraordinarily useful, the curved arrow formalism cannot be applied without critical thought. Here, the arrow formalism evokes the coming together of two pairs of electrons—one attached to N and the other in the R–L bond. But this combination yields the orbital diagram at the bottom, which shows a destabilizing interaction.

7.4 LEWIS ACIDS AND BASES

Now consider two conceptually related processes: the reaction of boron trifluoride with fluoride ion, F^-, and the reaction of the methyl cation with hydride, $H:^-$ (Fig. 7.11).

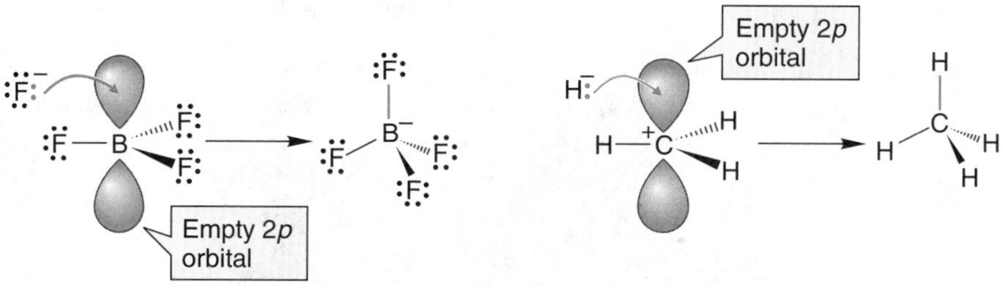

FIGURE 7.11 The fluoride and hydride ion are Lewis bases reacting with the Lewis acids BF_3 and $^+CH_3$.

In these two reactions there is no donating or accepting of a proton, and so by a strict definition of Brønsted acids and bases we are not dealing with acid–base chemistry. There is, however, a more general definition of acids and bases that deals with this situation. A **Lewis base** is defined as any species with a reactive pair of electrons. This definition may seem very general and indeed it is. It gathers under the "base" category all manner of odd compounds that don't really look like bases. The key word that confers all this generality is "reactive." Reactive under what circumstances? Almost everything is reactive under some set of conditions. And that's true; compounds that are extraordinarily weak Brønsted bases can still behave as Lewis bases.

If the definition of a Lewis base seems general and perhaps vague, the definition of a **Lewis acid** is even worse. A Lewis acid is anything that will react with a Lewis base. Therein lies the power of these definitions. They allow us to generalize—to see as similar reactions that on the surface appear very different. To describe Lewis acid–Lewis base reactions in orbital terms, we need only point out that a stabilizing interaction of orbitals involves the overlap of a filled orbital with an empty orbital (Fig. 7.12).

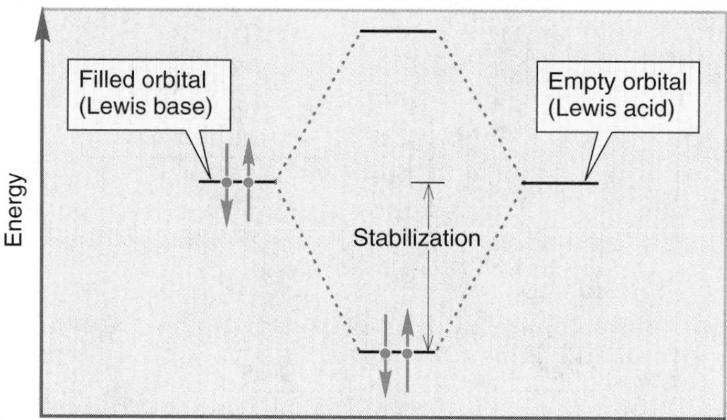

FIGURE 7.12 The interaction between a filled orbital (Lewis base) and an empty orbital (Lewis acid) is always stabilizing. The only question is by how much.

In the examples in Figure 7.11, the Lewis bases are the fluoride and hydride ions. In fluoride, the electrons occupy a 2*p* orbital and in hydride a

1*s* orbital. The Lewis acids are boron trifluoride and the methyl cation, each of which bears an empty 2*p* orbital. A schematic picture of these interactions is shown in Figure 7.13.

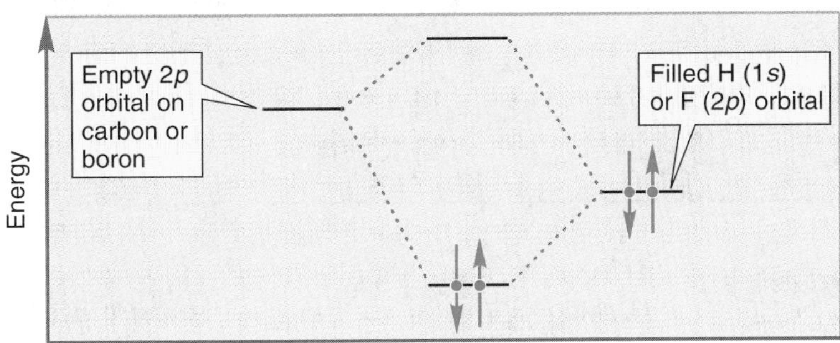

FIGURE 7.13 A schematic interaction diagram for the reactions of Figure 7.11.

7.5 HOMO–LUMO INTERACTIONS

It is possible to elaborate a bit on the generalization that "Lewis acids react with Lewis bases" by remembering that the strongest interactions (most stabilizing) between orbitals are between the orbitals closest in energy (Chapter 2, p. 55, 56). For stabilizing filled-empty overlaps, this will almost always mean that we must look at the interaction of the **H**ighest **O**ccupied **M**olecular **O**rbital (**HOMO**) of one molecule (**A**) with the **L**owest **U**noccupied **M**olecular **O**rbital (**LUMO**) of the other (**B**) (Fig. 7.14).

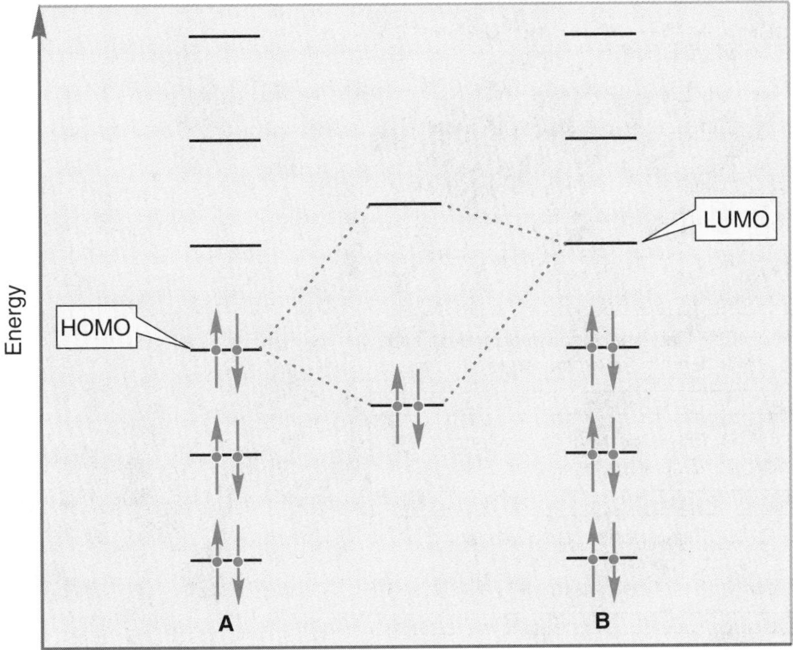

FIGURE 7.14 In two typical molecules, the strongest stabilizing interaction is likely to be between the HOMO of one molecule (**A**) and the LUMO of the other molecule (**B**), because the strongest interactions are between the orbitals closest in energy.

In the reaction of BF_3 with fluoride, it is the filled fluorine 2*p* orbital that is the HOMO and the empty 2*p* orbital on boron that is the LUMO (Fig. 7.15). *All* Lewis acid–Lewis base reactions can be described in similar terms.

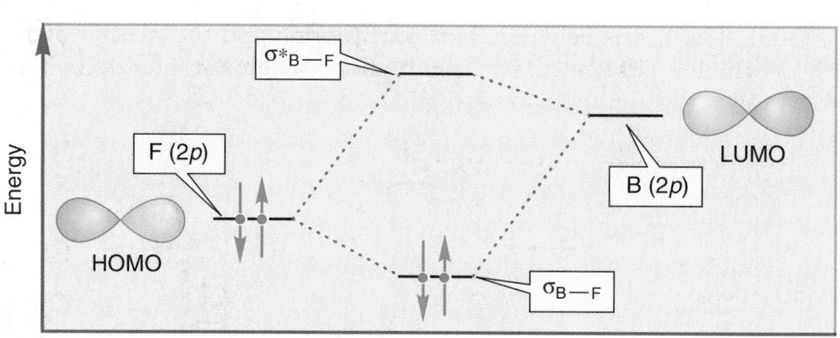

FIGURE 7.15 The interaction of the LUMO for BF_3 (an empty 2*p* orbital on boron) and the HOMO of F^-, a filled 2*p* orbital on fluorine, produces two new molecular orbitals: an occupied bonding σ orbital and an unoccupied antibonding σ* orbital.

PROBLEM 7.4 Identify the HOMO and LUMO in the reaction of the methyl cation with hydride in Figure 7.11.

*PROBLEM 7.5 Point out the HOMO and LUMO in the following reactions.

(a) $H\ddot{O}:^- + Li^+ \longrightarrow H\ddot{O}Li$

(b) $H_2C{=}CH_2 + BH_3 \longrightarrow {}^+CH_2{-}CH_2{-}\overset{-}{B}H_3$

(c) $\bar{:}CH_3 + H{-}\ddot{Cl}: \longrightarrow CH_4 + :\ddot{Cl}:^-$

(d) $H_3N: + H{-}\ddot{Cl}: \longrightarrow {}^+NH_4 + :\ddot{Cl}:^-$

(e) $H\ddot{O}:^- + H_3\overset{+}{O}: \longrightarrow 2\,H_2\ddot{O}:$

ANSWER

(a) HOMO: Filled nonbonding orbital on HO^-
LUMO: Empty 2*s* orbital on Li

(b) HOMO: Filled π orbital of ethylene
LUMO: Empty $2p_z$ orbital on BH_3

(c) HOMO: Filled nonbonding orbital of the methyl anion
LUMO: Empty σ* orbital of H–Cl

(d) HOMO: Filled nonbonding orbital of ammonia, $:NH_3$
LUMO: Empty σ* orbital of H–Cl

(e) HOMO: Filled nonbonding orbital of hydroxide
LUMO: Empty σ* orbital of an O–H bond of H_3O^+

7.6 REACTIONS OF ALKYL HALIDES: THE SUBSTITUTION REACTION

A great many varieties of substitution reactions are known, but all involve a competition between a pair of Lewis bases for a Lewis acid. Figure 7.16 shows five examples with the displacing group, the **nucleophile*** (from the Greek, "nucleus loving") and displaced group, the **leaving group**, identified. *Both* the nucleophile (usually Nu) and leaving group (usually L) are Lewis bases.

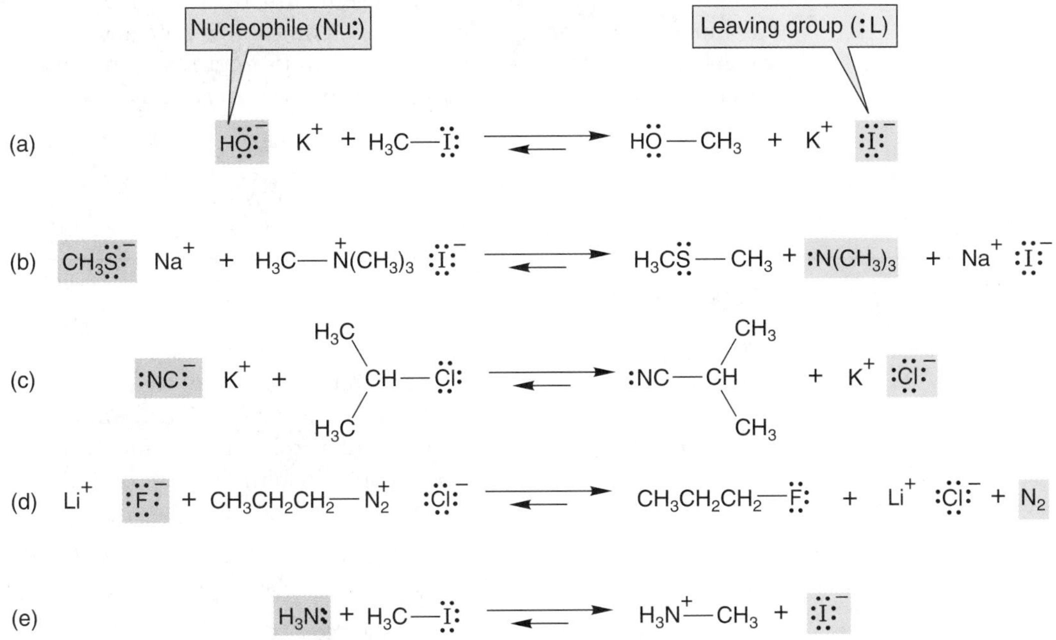

Nucleophile (Nu:) Leaving group (:L)

(a) $HO^- \; K^+ + H_3C{-}I \rightleftharpoons HO{-}CH_3 + K^+ \; I^-$

(b) $CH_3S^- \; Na^+ + H_3C{-}\overset{+}{N}(CH_3)_3 \; I^- \rightleftharpoons H_3CS{-}CH_3 + :N(CH_3)_3 + Na^+ \; I^-$

(c) $:NC^- \; K^+ + (H_3C)_2CH{-}Cl \rightleftharpoons :NC{-}CH(CH_3)_2 + K^+ \; Cl^-$

(d) $Li^+ \; F^- + CH_3CH_2CH_2{-}\overset{+}{N_2} \; Cl^- \rightleftharpoons CH_3CH_2CH_2{-}F + Li^+ \; Cl^- + N_2$

(e) $H_3N: + H_3C{-}I \rightleftharpoons H_3\overset{+}{N}{-}CH_3 + I^-$

FIGURE 7.16 Five examples of the substitution reaction. The nucleophiles for the reactions reading left to right are (a) hydroxide (HO^-), (b) methyl mercaptide (CH_3S^-), (c) cyanide (NC^-), (d) fluoride (F^-), and (e) ammonia ($:NH_3$). The leaving groups are iodide (I^-), trimethylamine [$N(CH_3)_3$], chloride (Cl^-), nitrogen (N_2), and iodide (I^-) again.

Identify the HOMOs and LUMOs in the reactions of Figure 7.16. PROBLEM 7.6

Notice that these are all equilibrium reactions. We will consider equilibrium in detail in Chapter 8, but we need to see here that these competition reactions are all, in principle at least, reversible. A given reaction may not be reversible in a practical sense, but that simply means that one of the competitors has been overwhelmingly successful. There are many possible reasons why this might be so. One of the equilibrating molecules might be much more stable than the other. If one of the nucleophiles were much more powerful than the other, the reaction would be extremely exothermic *in one direction or the other.* Reactions (a–c) of Figure 7.16 are examples of this. We are used to thinking of reactions running from left to right, but this prejudice is arbitrary and even misleading. Reactions run *both ways.* There are other reasons why one of these reactions might be irreversible. Perhaps the equilibrium is disturbed by one of the products being a solid that crystallizes out of the solution, or a gas that bubbles away. If so, the reaction can be driven all the way to one side, even against an unfavorable equilibrium constant. Under such circumstances, an endothermic reaction can go to completion. In reaction (d) of Figure 7.16 both

*Nucleophiles are always Lewis bases.

propyl fluoride (bp 2.5 °C) and nitrogen are gases, and their removal from the solution will drive the equilibrium to the right.

The second point to notice is that there is no obligatory charge type in these reactions. The displacing nucleophile is not always negatively charged [reaction (e)], and the group displaced does not always appear as an anion [reactions (b) and (d)].

If we examine a large number of substitution reactions, two catagories emerge. The boundaries between them are not absolutely fixed, as we will see, but there are two fairly clear-cut limiting classes of reaction. One is a kinetically second-order process, common for primary and secondary substrates. The other is a kinetically first-order reaction, common for tertiary substrates. We will take up these two reactions in turn, examining each in detail.

7.7 SUBSTITUTION, NUCLEOPHILIC, BIMOLECULAR: THE S_N2 REACTION

7.7a Rate Law

An analysis of many substitution reactions shows that the rate is proportional to the concentrations of both the substrate, R–L, and the displacing agent, called the nucleophile, Nu:⁻ (Fig. 7.17).* This bimolecular reaction is second order overall, and first order in both R–L and Nu:⁻, which leads to the name **S**ubstitution, **N**ucleophilic, **bi**molecular, or **S_N2** reaction.

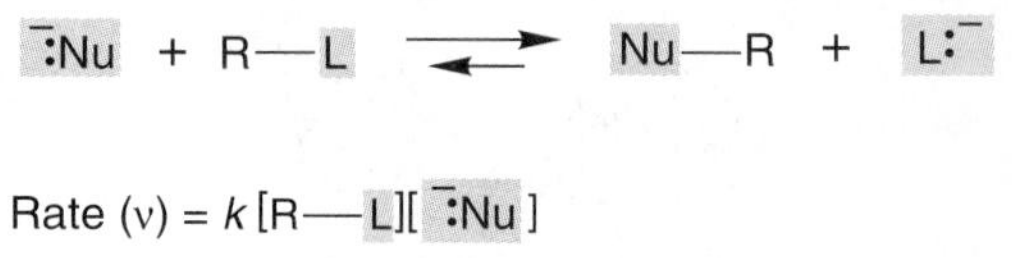

$$\overline{:}\text{Nu} + \text{R—L} \rightleftharpoons \text{Nu—R} + \text{L}\overline{:}$$

$$\text{Rate } (v) = k[\text{R—L}][\overline{:}\text{Nu}]$$

FIGURE 7.17 For many substitution reactions, the rate v is proportional to the concentrations of both the nucleophile, [Nu:⁻], and the substrate, [R–L]. This bimolecular reaction is first order in Nu:⁻ and first order in R–L. The rate constant, *k*, is a fundamental constant of the reaction.

7.7b Stereochemistry of the S_N2 Reaction

One of the most powerful of all tools for investigating the mechanism of a reaction is a stereochemical analysis. In this case, we need first to see what's possible, and then to design an experiment that will enable us to determine the stereochemical results. If we start with a single enantiomer (Chapter 5, p. 160) (*R*) or (*S*), there are several possible outcomes of a substitution reaction. We could get **retention of stereochemistry**, in which the stereochemistry of the starting molecule is preserved. The entering nucleophile occupies the same stereochemical position as did the departing leaving group (Fig. 7.18).

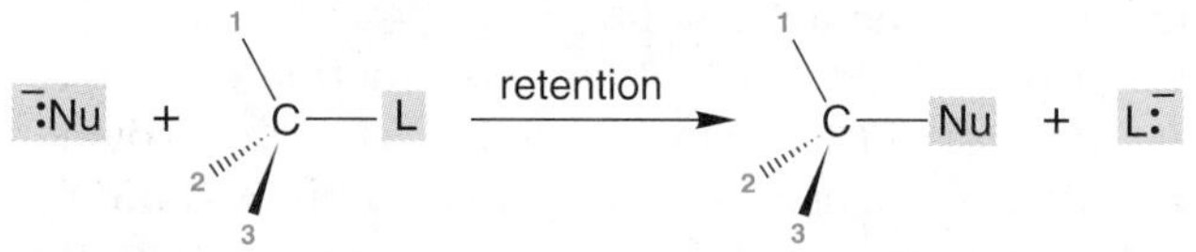

FIGURE 7.18 One possible stereochemical outcome for the S_N2 reaction. In this scenario, the product is formed with *retention of configuration*—the product has the same handedness as the starting material.

A second possibility is **inversion of stereochemistry**, in which the configuration of the starting material is reversed; (*R*) starting material going to (*S*) product and (*S*) starting material going to (*R*) product. In this kind of

*Here, R stands for a generic alkyl group. Don't confuse this R with the (*R*) used to specify one enantiomer of a chiral molecule.

reaction, the pyramid of the starting material becomes inverted in the reaction like an umbrella in the wind (Fig. 7.19).

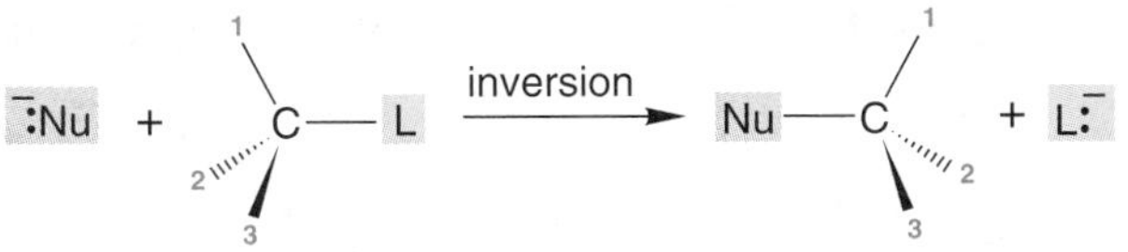

FIGURE 7.19 Another possible stereochemical result for the S_N2 reaction. The handedness of the starting material could be reversed in the product, in what is called *inversion of configuration.* In this and Figure 7.18, notice how the numbers enable us to keep track of inversions and retentions.

It is also possible that optical activity could be lost. In such a case, (*R*) or (*S*) starting material goes to a racemic mixture—a 50:50 mixture of (*R*) and (*S*) product (Fig. 7.20).

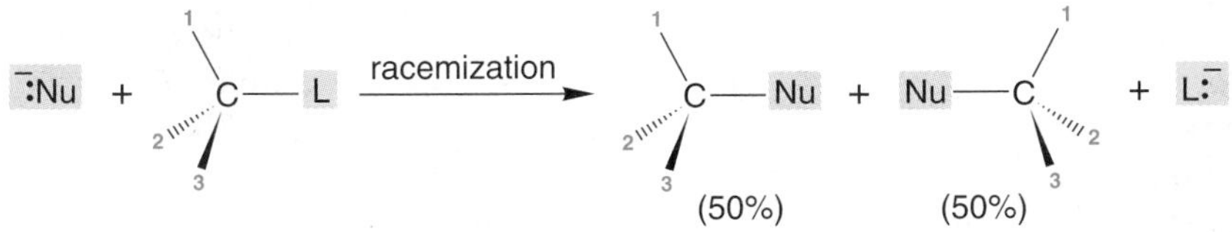

FIGURE 7.20 A third possible stereochemical result for the S_N2 reaction. In this case, optically active starting material [either (*R*) or (*S*)] is converted into a 50:50 racemic mixture of enantiomeric products.

Finally, it is possible that some dreadful mixture of the (*R*) and (*S*) forms is produced. In such a case, it is very difficult to analyze the situation, as several reaction mechanisms may be operating simultaneously.

*PROBLEM 7.7

Sometimes matters are even more complicated. It is possible, for example, that an (*R*) starting material could produce an (*R*) product by a process involving inversion. Here is a conceivable example. Explain.

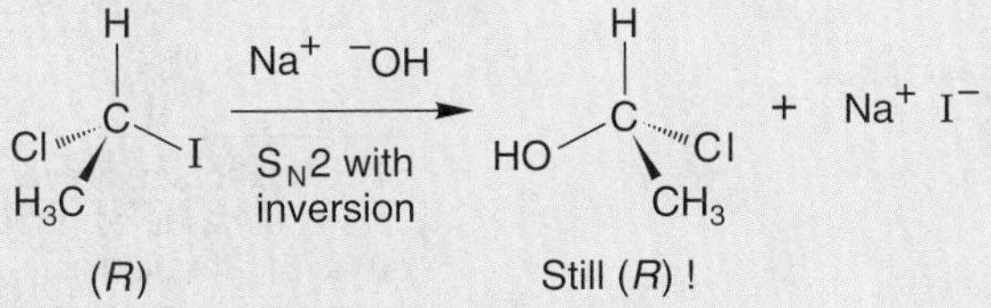

ANSWER

The designations (*R*) and (*S*) are determined for each molecule through the Cahn–Ingold–Prelog priority system (Chapter 4, p. 130; Chapter 5, p. 160). *If the groups attached to carbon change,* as in the example shown, there is no guarantee that inversion will change an (*R*) carbon to an (*S*) carbon. In the figure, the "umbrella" is inverted, but at the same time the substituents attached to the stereogenic carbon have also changed from (C, Cl, H, I) to (C, Cl, H, OH). Application of the priority system shows that both molecules are (*R*) despite the inversion.

For most secondary and primary substrates, substitution occurs with complete inversion of configuration. We now are faced with three questions: (1) How do we know the reactions go with inversion? (2) What does this tell us about the reaction mechanism? (3) Why is inversion preferred to retention or racemization?

The answer to the first question is easy; we measure it. Suppose we start with the optically active iodide (R–I) in Figure 7.21 and monitor the reaction with radioactive iodide ion ($^-$I*).† We measure two things: the in-

†In the following discussion we have indicated radioactive iodine with an asterisk in both the text and the figures.

corporation of radioactive iodide and the optical activity of R–I*. If radioactive iodide reacts with R–I with retention of configuration there will be no loss of rotation at all (I* and I are not significantly different in their contributions to the rotation of plane-polarized light) (Fig. 7.21).

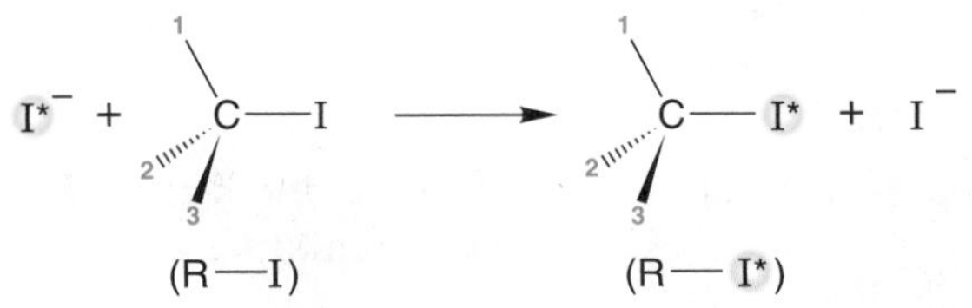

I^{*-} + (R—I) ⟶ (R—I*) + I^-

FIGURE 7.21 A mechanism involving retention predicts that incorporation of radioactive labeled iodide will induce no change in optical rotation.

What happens if the reaction goes with inversion? The observed optical rotation of R–I is certainly going to decrease as, for example, (*R*)-iodide is converted into (*S*)-iodide. But it is usually not obvious how the rate of loss of optical activity will compare with the rate of incorporation of I*. If every replacement occurs with inversion, the rate of loss of optical activity must be twice the rate of incorporation of I*. Every conversion of an (*R*)-iodide (R–I) into an (*S*) iodide (I*–R) generates a molecule, I*–R, whose rotation exactly cancels that of a molecule of starting R–I. In effect, the rotation of two molecules goes to zero for every inversion (Fig. 7.22), and this is exactly what is observed experimentally. Every incorporation of I* is accompanied by the inversion of one (*R*)-iodide to an (*S*)-iodide.

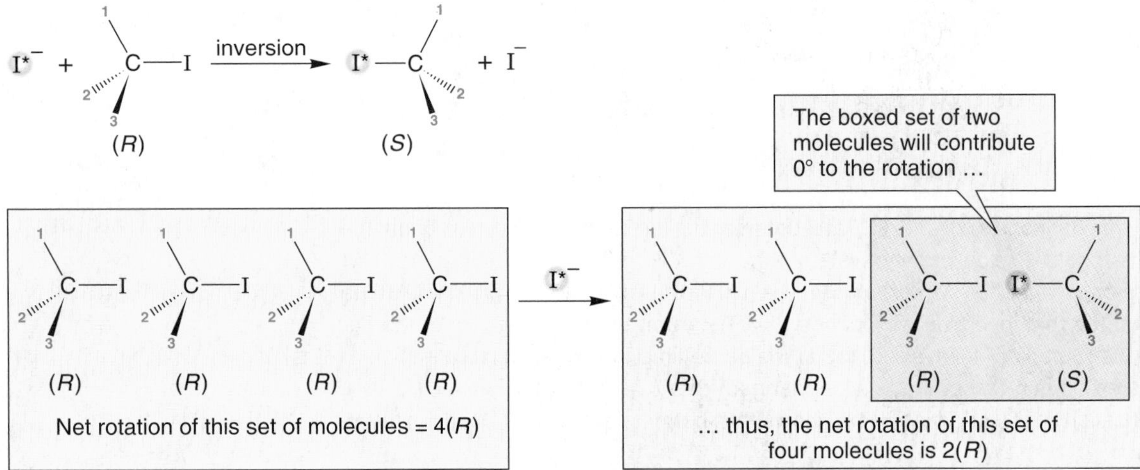

FIGURE 7.22 A mechanism involving inversion. For the incorporation of one atom of radioactive iodide, the rotation of the set of molecules shown will be cut by a factor of 2.

Now, what does this result tell us about the mechanism? The observation of inversion tells us that the nucleophile, the incoming atom or molecule, must be approaching the substrate from the rear of the departing atom. Only this path can lead to the observed inversion. Figure 7.23 shows the reaction, along with an arrow formalism description of the bonds that are forming and breaking.

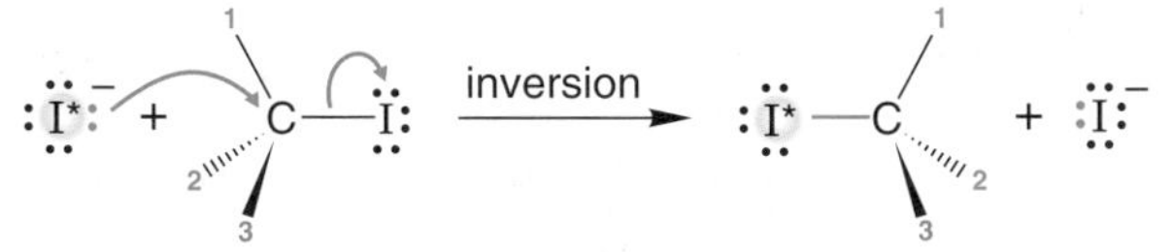

FIGURE 7.23 The observation of inversion in the S_N2 reaction means that the entering nucleophile, radioactive (or labeled) iodide, must enter the molecule from the side opposite the departing leaving group, here unlabeled iodide. The arrows show the path of approach of the entering nucleophile and the departing leaving group. The tetrahedral "umbrella" is inverted during the reaction.

Had the reaction occurred with retention, a frontside approach would have been demanded (Fig. 7.24).

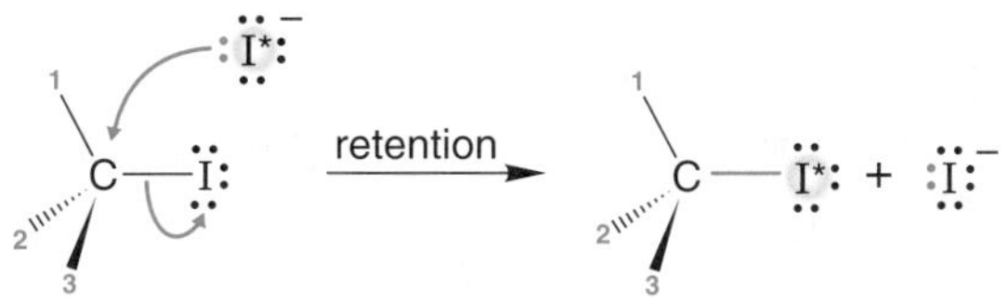

FIGURE 7.24 If retention had been the experimentally observed result, the incoming nucleophile would have had to enter the molecule from the same side as the departing leaving group. The tetrahedral "umbrella" would have retained its configuration. This result is *not* observed.

All this is useful; we now know the path of approach for the nucleophile in the S_N2 reaction, and so we know more about the mechanism of this process. But our new-found knowledge only serves to generate the inevitable next question. *Why* is inversion preferred? Why is frontside displacement—retention—never observed? Here comes part of the payoff for all the work we did in Chapters 1 and 2. One of the key lessons of those early chapters was that interactions of *filled* and *empty* orbitals will be stabilizing, as the two electrons originally in the filled orbital can be accommodated in the new bonding orbital, and no electrons need occupy the antibonding molecular orbital. Now let's look at this reaction in orbital terms. A filled nonbonding orbital "*n*" on the nucleophile interacts with an orbital on the substrate, R–L, involved in the bond from R to the leaving group, L. As the interaction of two filled orbitals is not stabilizing, the filled orbital *n* must overlap with the empty, antibonding σ^* orbital of R–L. This overlapping is nothing more than the interaction of the nucleophile's HOMO, *n*, with the sigma bond's LUMO, σ^* as is shown schematically in Figure 7.25.

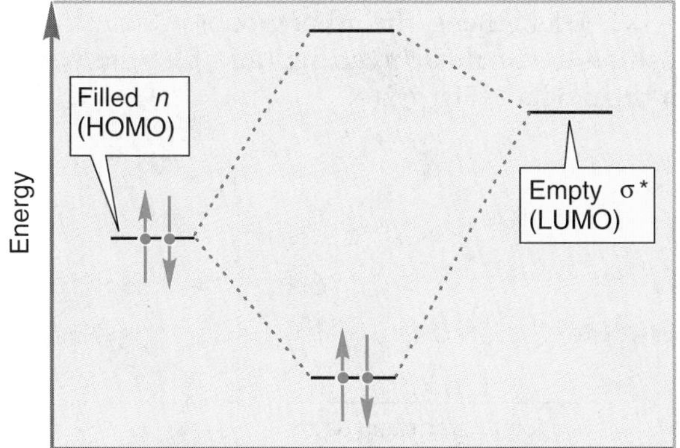

FIGURE 7.25 An orbital interaction scheme for the stabilizing interaction of a filled nonbonding orbital (*n*) on the nucleophile overlapping with the empty, antibonding (σ^*) orbital on the substrate.

What does σ* look like? Figure 7.26 shows us. Note that σ* has one of its large lobes pointing to the *rear* of R—L. That is where overlap with the filled *n* orbital will be most efficient.

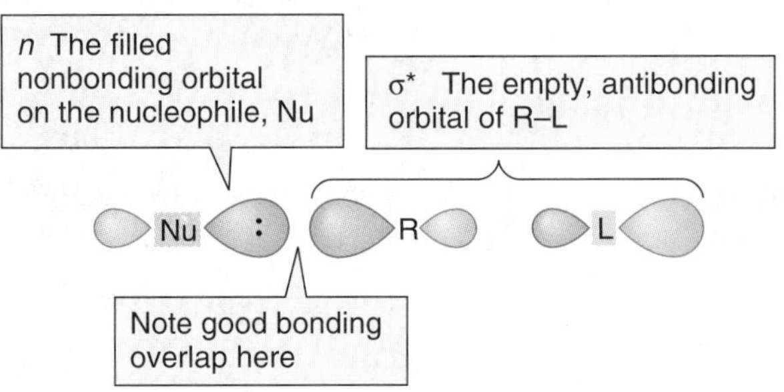

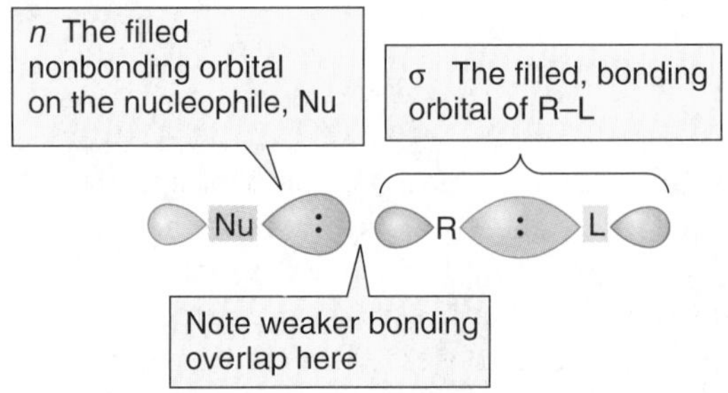

FIGURE 7.26 (a) The overlap of the filled nonbonding orbital (*n*) with the "outside" lobe of the empty σ* orbital of R—L takes advantage of good overlap with the large lobe of σ*. (b) Contrast (a) with the destabilizing interaction of *n* and σ, the two filled orbitals. The interaction of two filled orbitals is destabilizing, but the overlap of the lobes involved, one fat, the other thin, is poor. This interaction is not important.

σ* The empty antibonding orbital of R–L
R
L
Antibonding
Bonding
Nu
n The filled nonbonding orbital of the nucleophile, Nu

FIGURE 7.27 Frontside substitution (retention) requires overlap with the inside part of σ*. There is poor overlap with the small lobes of σ* as well as offsetting bonding and antibonding interactions.

Now look what happens if the orbitals overlap in a way that would lead to retention through a frontside substitution. Not only is the magnitude of interaction reduced by the relatively poor overlap with the small lobe of σ*, but there is both a bonding and antibonding interaction, and these will tend to cancel each other (Fig. 7.27). Frontside displacement is a very unlikely process indeed.

We can now map out the progress of the bond making and breaking in the S_N2 reaction. The reaction starts as the nucleophile Nu:⁻ approaches the rear of the substrate. As the HOMO, the nonbonding orbital *n* on Nu:⁻, overlaps with the LUMO, the antibonding σ* orbital of C—L, the bond from C to L is weakened as the antibonding orbital begins to be occupied. The bond length of C—L increases as the new bond between Nu and C begins to form. As C—L lengthens, the other groups attached to C bend back, in umbrella fashion, ultimately inverting (just like the wind-blown umbrella) to form the product (Fig. 7.28).

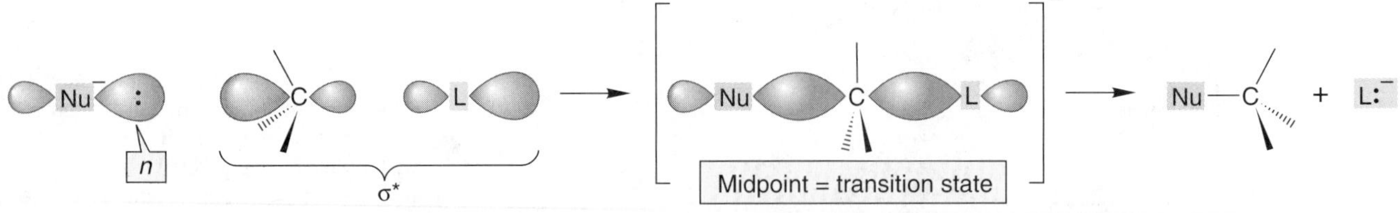

FIGURE 7.28 As the S_N2 reaction proceeds, the bond from Nu to C forms, the bond from C to L breaks, and the umbrella inverts.

Look at the midpoint of the reaction. For a symmetrical displacement such as the reaction of radioactive $^-$I* with R—I discussed before, this will be the point at which the groups attached to C are exactly half-way inverted (Fig. 7.29).

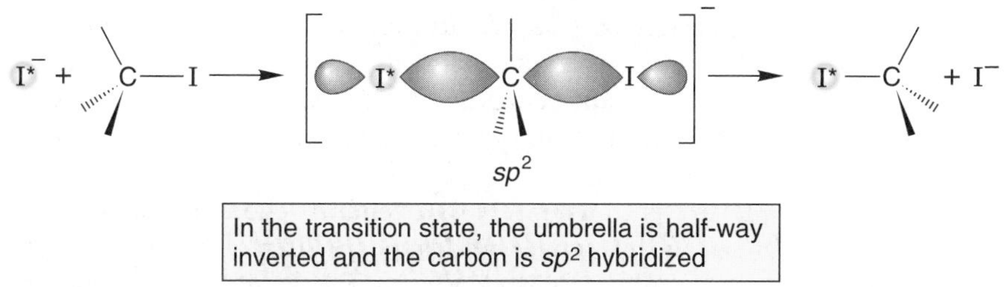

sp^2

In the transition state, the umbrella is half-way inverted and the carbon is sp^2 hybridized

FIGURE 7.29 In a symmetrical inversion process, the midpoint, called the transition state, is the point at which the umbrella is exactly half-inverted.

What is the hybridization of the central carbon at this midpoint? It must be sp^2, as this carbon is surrounded by three coplanar groups. There are partial bonds to the incoming nucleophile, $^-$:I* and the departing leaving group, $^-$:I. No picture shows the competition between the two nucleophiles ($^-$:I* and $^-$:I) as clearly as this one. What Lewis acid are they competing for? A carbon $2p$ orbital, as the figure shows. What is this structure? At least in the symmetrical displacement of $^-$:I by $^-$:I*, it is the half-way point between the equi-energetic starting material and product. This structure is the **transition state** for this reaction. Is this transition state something one could potentially catch and isolate? No! It occupies not an energy minimum, as stable compounds do, but it sits on an energy maximum along the path from starting material to product. We have seen such structures before: Recall, for example, the eclipsed form of ethane or the transition state for the interconversion of two pyramidal methyl radicals (Chapter 3, p. 83; Fig. 3.23; Problem 3.12; Fig. 7.30).

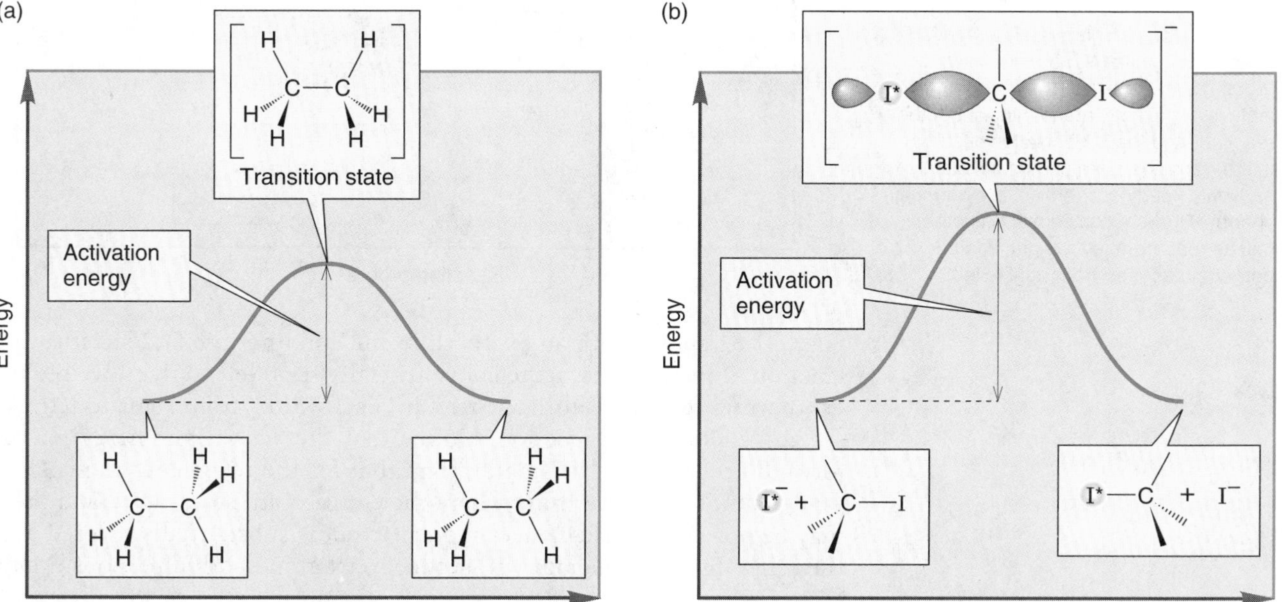

FIGURE 7.30 Two analogous diagrams. (a) Shows the rotation about the carbon–carbon bond in ethane; (b) shows the displacement of $^-$I by $^-$I*. In each case, two equi-energetic molecules are separated by a single energy barrier, the top of which is called the *transition state,* shown in brackets.

Now, let's consider the change in energy as the S_N2 reaction proceeds [Fig. 7.30 (b)]. The starting material and products are separated by the high-energy point of the reaction, the transition state. The energy difference between starting material and transition state is related to the rate of the reaction. The larger this difference, the higher the barrier, the slower the reaction. So, a quantity of vital importance to any study of a chemical reaction is the energy difference between starting material and the transition state, called the **activation energy**. This diagram is symmetrical only if the substitution reaction is symmetrical; if starting material and product are of equal energy.†

This specific picture is transferable to a general second-order substitution reaction, but some differences will appear. First of all, usually the starting material and product will not be at the same energy, and the transition state cannot be *exactly* half-way in between starting material and product. It will lie closer to one than the other (Fig. 7.31).

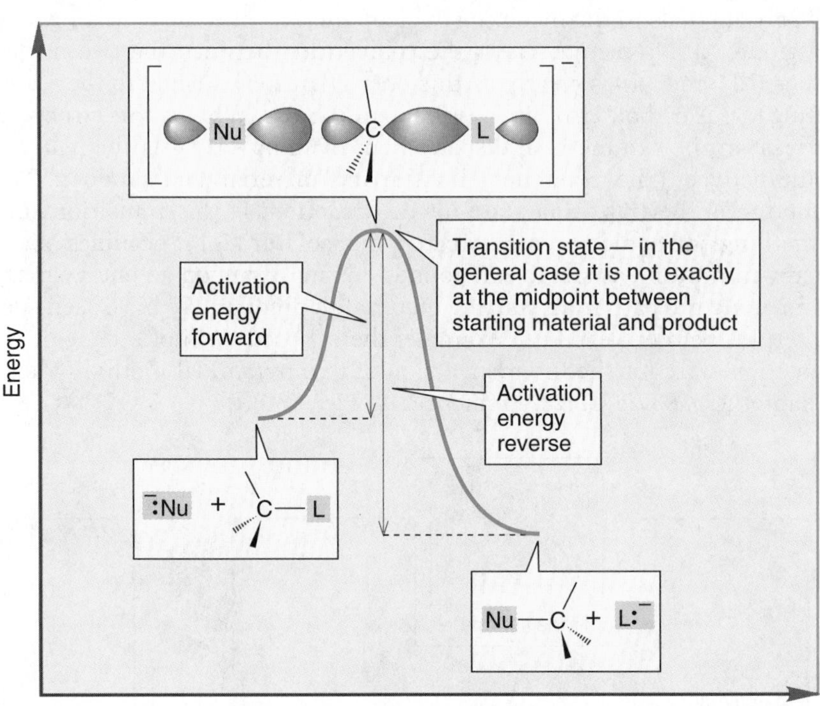

FIGURE 7.31 A general S_N2 reaction. If the nucleophile (Nu) and the leaving group (L) are different, the transition state is not exactly at the midpoint of the reaction, and the hybridization of carbon in the transition state cannot be exactly sp^2, only nearly sp^2 (notice its pyramidal shape). The reaction shown is exothermic as read from left to right, and endothermic as read from right to left.

Figure 7.31 shows both an exothermic and endothermic S_N2 substitution reaction. If we read the figure conventionally from left to right, we see the exothermic reaction, but if we read it "backward," from right to left, we see the endothermic reaction. Notice that the activation energy for the forward process must be smaller than that for the reverse reaction. Thermodynamics will determine where the equilibrium settles out, and there may be practical consequences if one partner is substantially favored over the other, but one diagram will suffice to examine both directions. Many

†Actually, even our picture of the displacement of $^-$:I by $^-$:I* is slightly wrong. Normal iodine (I) is not *exactly* the same as radioactive iodine (I*), and even this reaction is not truly symmetrical. We can ignore the tiny differences in isotopes, however, as long as we keep the principle straight. If you want to be more correct, insert the words "except for a tiny isotope effect" wherever appropriate.

of us are prisoners of our language, which is read from left to right, and we tend to do the same for these energy diagrams. Nature is more versatile, or perhaps less prejudiced than we are.

(*S*)-Malic acid

CH_2COOH

H— —OH

COOH

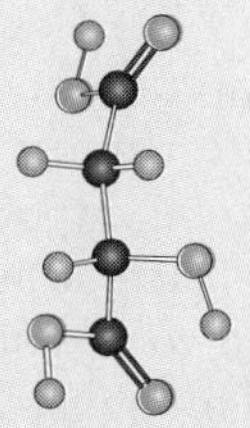

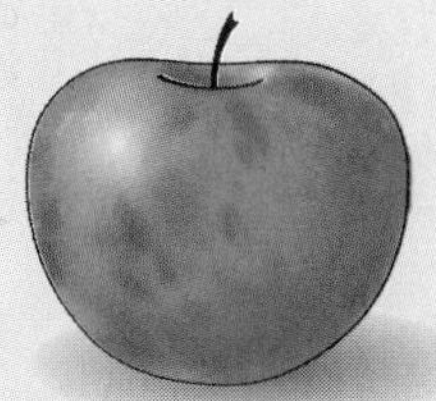

This chiral acid figured strongly in the original discovery of inversion of configuration in the S_N2 reaction. Malic acid is sometimes called "apple acid" because of its high concentration in apples and other fruits. It is a molecular carrier of the carbon dioxide absorbed by plants. The CO_2 appears in the CH_2COOH group of malic acid. The principal actor in the early mechanistic work was Paul Walden, a chemist born in 1863 in what is now Latvia and who died in Tübingen, Germany in 1957. Indeed, the inversion now known to occur universally in the S_N2 reaction is sometimes called "Walden inversion."

All this lets me tell a wonderful story about Paul Walden told to me by my professor, William Doering. Professor Doering was visiting Tübingen in 1956 to talk about his work and overheard a discussion by a now-famous German chemist who was lamenting the fact that Tübingen had not been bombed during the war. It seems that most of his colleagues in other "more fortunate" cities had newly built laboratories courtesy of Allied bombing. There was an old gentleman sitting in the corner who had been wheeled into the department from his nursing home to hear Doering's seminar. Paul Walden—for indeed it was he—was 92 at the time. He overheard the remark and replied, scathingly, "Es ist nicht der Käfig, sondern der kleine Vogel darin!" (It is not the cage that matters, but the little bird inside.)

Now let's examine the effects of structural change in the various participants in this reaction, the R group, the nucleophile, the leaving group, and the solvent.

7.7c Effects of Substrate Structure: The R Group

The structure of the R group makes a huge difference in the rate of the S_N2 reaction. We anticipated this result when we mentioned earlier that the practically useful S_N2 reaction was restricted to methyl, primary, and secondary substrates. By implication, the rate of the S_N2 reaction with tertiary substrates is zero, or at least negligibly small. Table 7.4, which gives average relative rates of S_N2 reactions for a variety of R groups, shows that the preceding statement is correct. Why should this be? The simple answer seems to be that in a tertiary substrate the rear of the C—L bond is guarded by three alkyl groups, and the entering nucleophile can find no unhindered path along which to approach the fat lobe of σ^* (Fig. 7.32). So the S_N2 reaction is disfavored for tertiary substrates in which steric hindrance to the approaching nucleophile is especially severe. Another reaction, favorable for tertiary substrates, becomes possible. It is called the S_N1 reaction, and we will deal with its mechanism in Section 7.8.

TABLE 7.4 Average Rates of S_N2 Substitution Reactions for Different Groups

R	Average Rate
$CH_2{=}CHCH_2$	1.3
CH_3	1
CH_3CH_2	0.033
$CH_3CH_2CH_2$	0.013
$(CH_3)_2CH$	8.3×10^{-4}
$(CH_3)_3CCH_2$	2×10^{-7}
$(CH_3)_3C$	~0

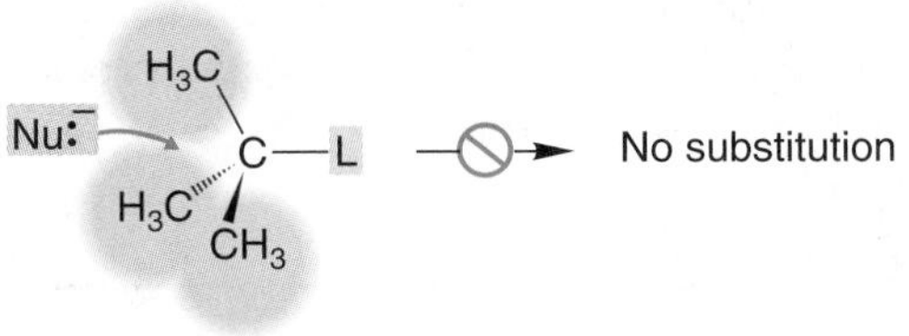

Nu:⁻ → H_3C, H_3C, CH_3 C—L —⊘→ No substitution

FIGURE 7.32 For tertiary substrates, approach from the rear is hindered by the alkyl groups, here all shown as methyls. This steric effect makes the S_N2 reaction impossible.

If this steric argument is correct, secondary substrates should react more slowly than primary substrates, and primary substrates should be slower than methyl compounds. In general, this is the case (Table 7.4). In practice, the S_N2 reaction is usually useful as long as there is at least one hydrogen attached to the same carbon as the leaving group (methyl, primary, and secondary substrates). The small size of hydrogen opens a path at the rear for the entering nucleophile.

PROBLEM 7.8 Explain the following change in rate for the S_N2 reaction:

$$R{-}O^- + CH_3CH_2{-}I \longrightarrow R{-}O{-}CH_2CH_3 + I^-$$

Rate for R = CH_3 is much faster than for R = $(CH_3)_3C$

This picture, which emphasizes steric effects, allows us to make a prediction. In principle, there must be some primary group so gigantic that the S_N2 reaction would be unsuccessful (Fig. 7.33).

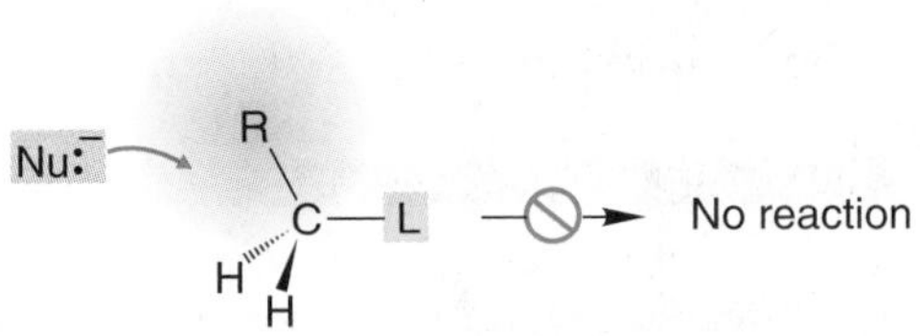

Nu:⁻ → R, H, H C—L —⊘→ No reaction

FIGURE 7.33 Even for primary substrates, usually very reactive in the S_N2 reaction, an R group so large as to prevent access to the rear of the C—L bond can be imagined.

In practice, it is rather easy to find such groups. Even the neopentyl group, $(CH_3)_3CCH_2$, is large enough to slow the bimolecular displacement reaction severely, because the *tert*-butyl group blocks the best pathway for rearside displacement of the leaving group (Fig. 7.34; Table 7.4).

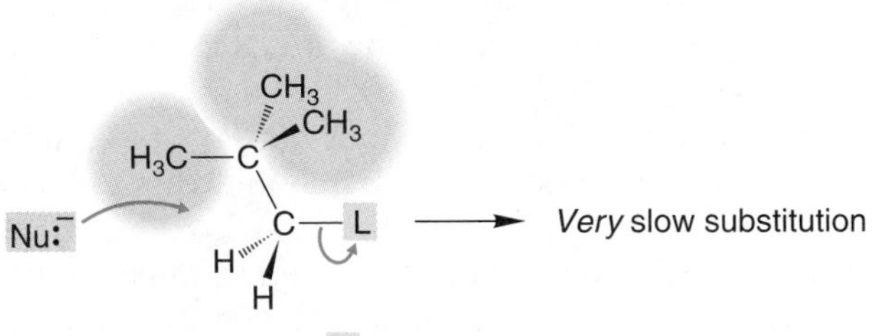

Nu:⁻ → H_3C—C(CH_3)(CH_3)—C(H)(H)—L ⟶ *Very* slow substitution

Neopentyl – L

FIGURE 7.34 Even neopentyl compounds, in which a *tert*-butyl group shields the rear of the C—L bond, are hindered enough so that the rate of the S_N2 reaction is extremely slow (see Table 7.4).

The S_N2 reaction also takes place when ring compounds are used as the substrates. Although the reaction proceeds normally with ring compounds, there are some interesting effects of ring size on the rate of the reaction (Table 7.5; Fig. 7.35).

Compound	Relative Rate
Cyclopropyl bromide	$< 10^{-4}$
Cyclobutyl bromide	8×10^{-3}
Cyclopentyl bromide	1.6
Cyclohexyl bromide	1×10^{-2}
Isopropyl bromide	1.0

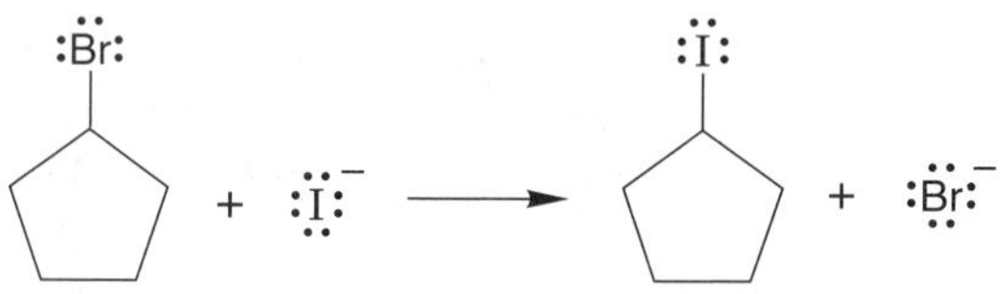

TABLE 7.5 Relative Reactivities of Cycloalkyl Bromides in the S_N2 Reaction

:Br: + :I:⁻ → :I: + :Br:⁻

FIGURE 7.35 The S_N2 reaction takes place normally with cyclic substrates.

*PROBLEM 7.9

Use a ring compound to design a test of the stereochemistry of the S_N2 reaction.

ANSWER

This problem asks something new of you; here you are required to design an experiment. Keep in mind what you are trying to do. You want to probe the stereochemical requirements of the S_N2 reaction, and you must use a ring compound. There are many possibilities. Basically, the ring provides a rigid framework that allows determination of the stereochemical results of the S_N2 reaction. Here is one possibility that uses a cyclopentane ring:

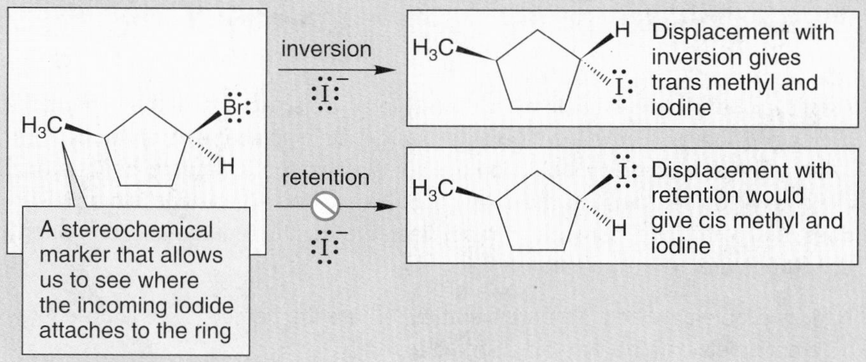

H_3C Br: H

A stereochemical marker that allows us to see where the incoming iodide attaches to the ring

inversion :I:⁻

H_3C H :I: Displacement with inversion gives trans methyl and iodine

retention :I:⁻

H_3C :I: H Displacement with retention would give cis methyl and iodine

Why should small rings be so slow? As with any question involving rates, we need to look at the structures of the transition states for the reaction to find an answer. In the transition state, the substrate is hybridized approximately sp^2, and that requires bond angles close to 120°. The smaller the ring, the smaller the C—C—C angles. For a three-membered ring, the optimal 120° must be squeezed to 60°, and for a cyclobutane, to 90°. This contraction introduces severe angle strain in the transition state, raises its energy, and slows the rate of reaction (Fig. 7.36).

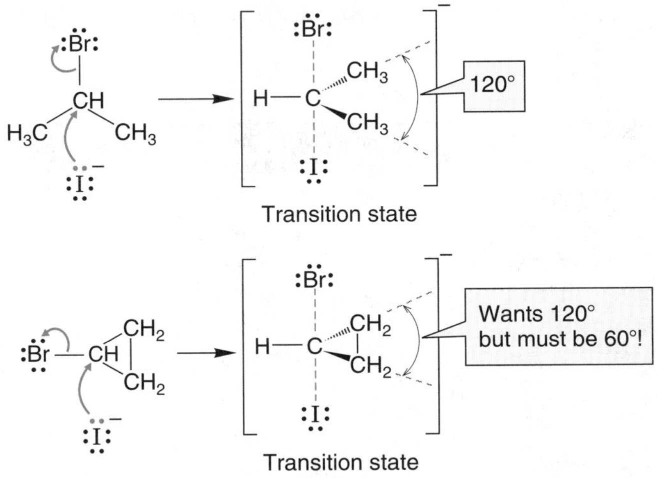

FIGURE 7.36 The transition states for S_N2 reactions of isopropyl bromide and cyclopropyl bromide.

PROBLEM 7.10 Wait! The argument presented above ignores angle strain in the starting material. Cyclopropane itself is strained. Won't that strain raise the energy of the starting material and offset the energy raising of the transition state? Comment. *Hint*: Consider angle strain in both starting material and transition state—in which will it be more important?

In cyclohexane, the rate is apparently slowed by steric interactions with the axial hydrogens. There can be no great angle strain problem here (Fig. 7.37).

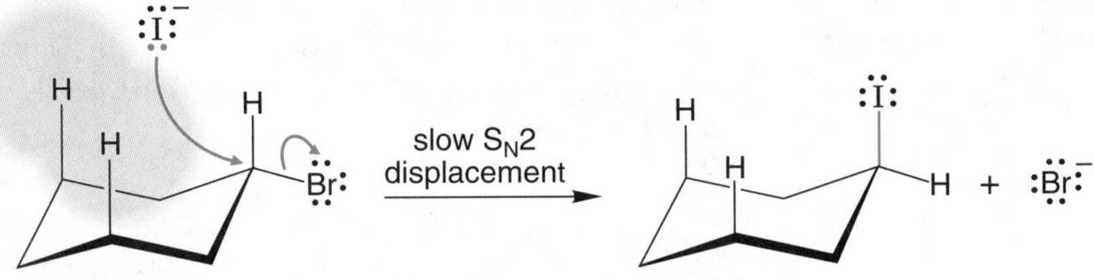

FIGURE 7.37 The slow S_N2 displacement of bromide by iodide in cyclohexyl bromide. Incoming iodide is apparently blocked somewhat by the axial carbon–hydrogen bonds.

*PROBLEM 7.11 Axial cyclohexyl iodide reacts more quickly in S_N2 displacements than equatorial cyclohexyl iodide. Draw the transition states for displacement of iodide from axial cyclohexyl iodide by iodide ion and for the analogous reaction of equatorial cyclohexyl iodide. What is the relation between these transition states? Why does the axial iodo compound react faster than the equatorial compound? **Caution!** Hard, tricky, question.

ANSWER A tricky problem indeed. The transition states are the same! There is no energy difference between the two transition states.

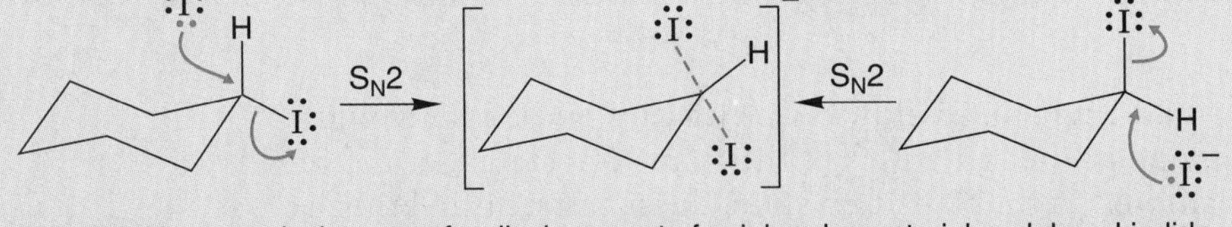

The transition state is the same for displacement of axial and equatorial cyclohexyl iodide.

Therefore, the reason that axial cyclohexyl iodide reacts more quickly than equatorial cyclohexyl iodide cannot lie in the transition state energies. However, the axial iodo group raises the energy of the *starting material*, thus lowering the activation energy.

ANSWER (CONTINUED)

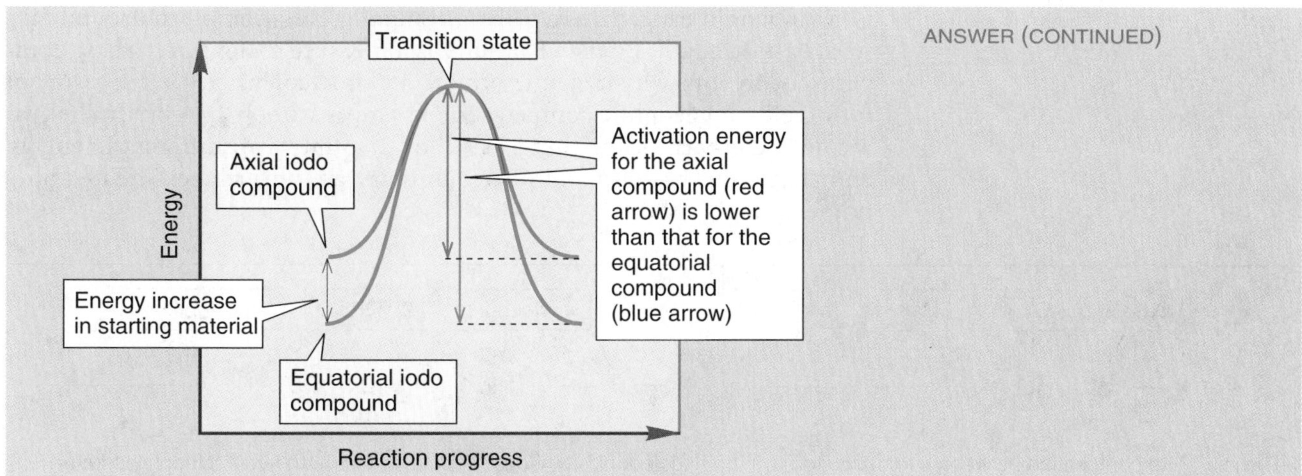

7.7d Effect of the Nucleophile

Some displacing agents are more effective than others. They are better players in the competition for the carbon 2*p* orbital, the Lewis acid. In this section, we examine what makes a good nucleophile; that is, what makes a powerful displacing agent.

Remember: Size can be important. We have already seen how the S_N2 reaction can be slowed, or even stopped altogether, by large R groups (Figs. 7.33 and 7.34). Presumably, large nucleophiles will also have a difficult time in getting close enough to the substrate to overlap effectively with σ^* (Problem 7.8, p. 252). Figure 7.38 shows two nucleophiles, carefully chosen to minimize all differences except size.

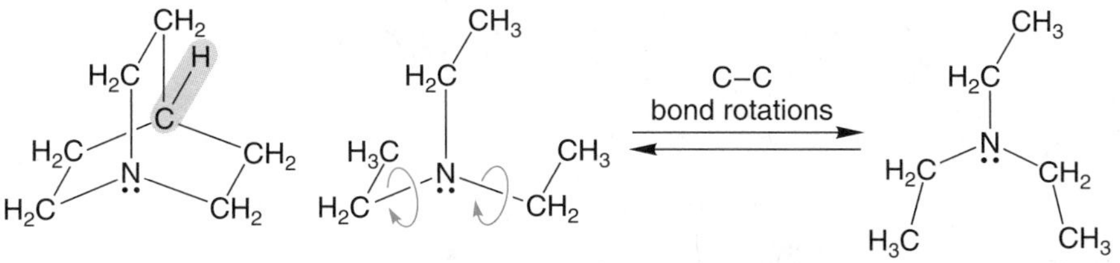

C–C bond rotations

FIGURE 7.38 In both these molecules, the nucleophilic nitrogen atom is flanked by three two-carbon chains. In the cage compound, they are tied back by the CH group shown in red, and are not free to rotate. In triethylamine, they are free to rotate and effectively increase the bulk of the nucleophile.

The cage molecule, in which the three ethyl-like groups are tied back out of the way, is a more effective displacing agent than is triethylamine in which the three ethyl groups are freely rotating. Triethylamine is effectively larger than the cage compound in which the alkyl groups are tied back (Fig. 7.39).

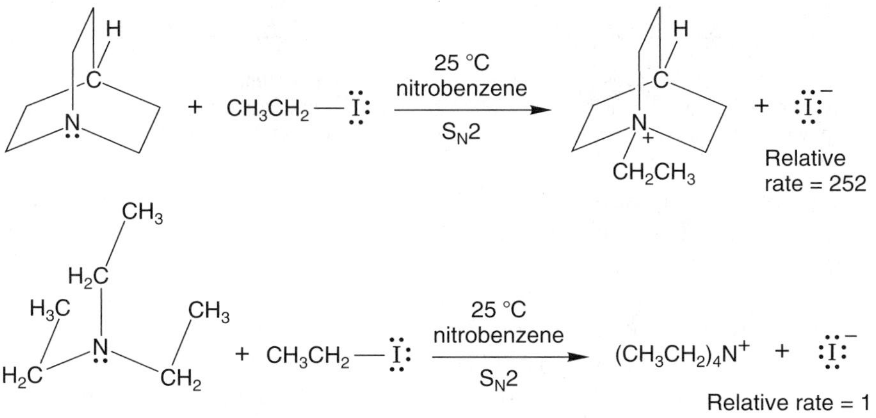

+ CH_3CH_2—I: $\xrightarrow[S_N2]{25\ °C,\ \text{nitrobenzene}}$ + :I:$^-$

Relative rate = 252

+ CH_3CH_2—I: $\xrightarrow[S_N2]{25\ °C,\ \text{nitrobenzene}}$ $(CH_3CH_2)_4N^+$ + :I:$^-$

Relative rate = 1

FIGURE 7.39 The tied-back cage compound reacts faster with ethyl iodide than does triethylamine, which is effectively the larger compound. The larger nucleophile has a more difficult time in attacking the rear of the carbon–iodine bond in ethyl iodide.

One would expect there to be a general correlation of nucleophilicity with base strength. Brønsted basicity is a measure of how well a base competes for an empty hydrogen 1*s* orbital, and nucleophilicity is a measure of how well a nucleophile competes for an empty carbon 2*p* orbital. The two are not exactly the same thing, as 1*s* and 2*p* orbitals are different in energy and shape, but basicity and nucleophilicity are nevertheless related phenomena (Fig. 7.40).

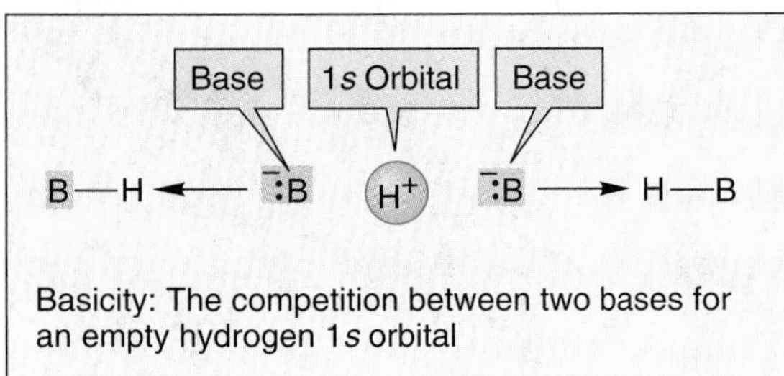

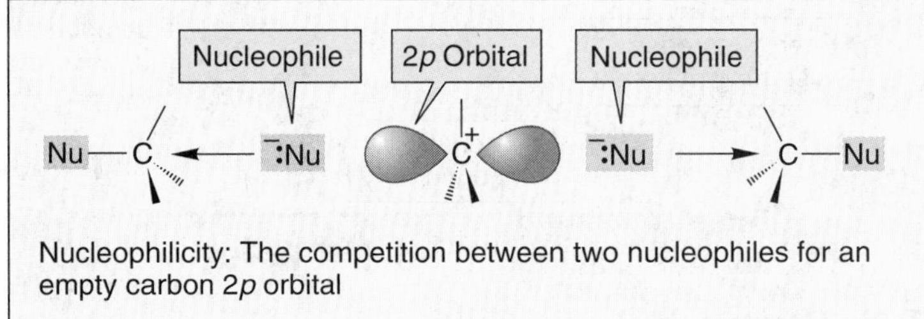

FIGURE 7.40 Basicity and nucleophilicity are not the same, but they are related phenomena.

Actually, it is simple to refine our discussion of good nucleophiles. What do we mean by "a good competitor for a carbon 2*p* orbital"? We are talking about the overlap of a filled and empty orbital, and the resulting stabilization is a measure of how strong the interaction is. *Remember*: The strongest orbital interactions, and hence the greatest stabilizations, come from the overlap of orbitals close in energy (Fig. 7.41).

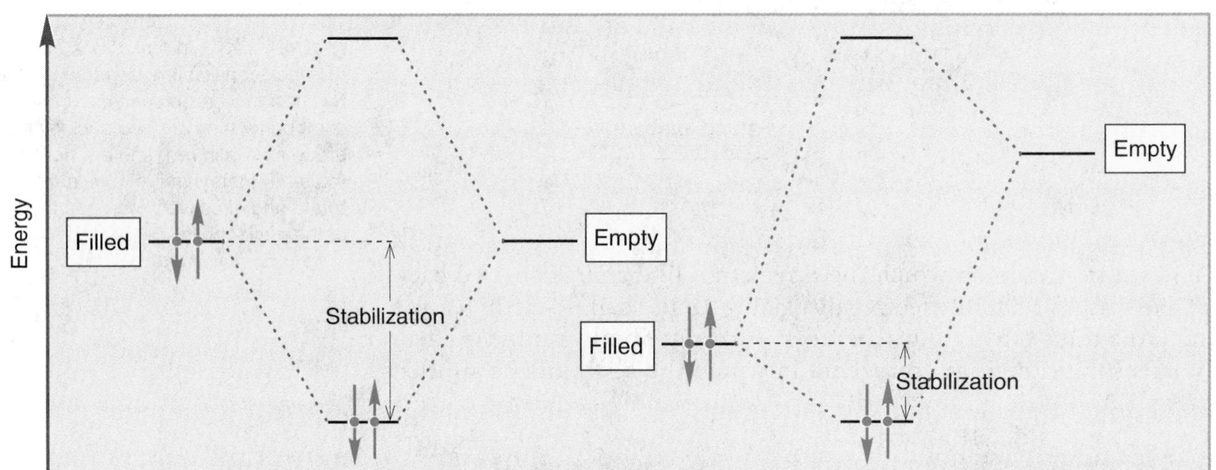

FIGURE 7.41 Stabilization is greater when filled and empty orbitals of equal or nearly equal energy interact than it is when orbitals that differ greatly in energy interact. The further apart two orbitals are in energy, the smaller the stabilization resulting from their overlap.

So we can anticipate that energy matching between the orbital on the nucleophile and an empty carbon orbital will be important in determining nucleophilicity.

Table 7.6 and Figure 7.42 show some nucleophiles segregated into general catagories. It is not possible to do very much better than this, as nucleophilicity is a property that depends on the reaction partner. A good nucleophile with respect to carbon may, or may not, be a good nucleophile with respect to displacement on some other atom. Energy matching is important in the reaction. If we change the reaction partner, and thus the energy of the orbital involved, we will change the stabilization involved as well (Fig. 7.41).

TABLE 7.6 Relative Nucleophilicities of Some Common Species

Species	Name	Relative Nucleophilicity
	Excellent Nucleophiles	
NC^-	Cyanide	126,000
HS^-	Thiolate	126,000
I^-	Iodide	80,000
	Good Nucleophiles	
HO^-	Hydroxide	16,000
Br–	Bromide	10,000
N_3^-	Azide	8,000
NH_3	Ammonia	8,000
NO_2^-	Nitrite	5,000
	Fair Nucleophiles	
Cl^-	Chloride	1,000
CH_3COO^-	Acetate	630
F^-	Fluoride	80
CH_3OH	Methyl alcohol	1
H_2O	Water	1

$^-NH_2 > NH_3$ $\quad$ $^-NH_2 > {}^-OR > {}^-OH$ $\quad$ $I^- > Br^- > Cl^- > F^-$

$^-SH > SH_2$ $\quad$ $H_2Se > H_2S > H_2O$

$^-OH > OH_2$ $\quad$ $R_3P > R_3N$

FIGURE 7.42 Some relative nucleophilicities. Be careful! Nucleophilicity (Lewis basicity) is a hard-to-categorize quantity. Relative nucleophilicity depends, for example, on the identity of the reaction partner, the Lewis acid, as well as the nature of the solvent.

Table 7.6 and Figure 7.42 reveal some spectacular exceptions to the "a good Brønsted base is a good nucleophile" rule. Look at the halide ions, for example. As we would expect from the pK_a values of the conjugate acids, HX (where X is a halide), the basicity order is $F^- > Cl^- > Br^- > I^-$. The nucleophilicity order is exactly opposite the basicity order (Fig. 7.43). Iodide is the *weakest* base but the *strongest* nucleophile. Fluoride is the *strongest* base but the *weakest* nucleophile. Why?

To answer this question, we need more data. It turns out that the effectiveness of these (and other) nucleophiles depends on the so-often neglected solvent. It is easy to forget that most chemical reactions are run in "oceans" of solvent, and it is perhaps not too surprising that these oceans of other molecules are not always without substantial effects on the reaction! Figure 7.44 examines rates for typical displacement reactions by halides in solvents of differing polarity and in the total absence of solvent (the gas phase). Note that iodide is a particularly effective nucleophile, relative to other halides, in water, a **protic solvent**. Protic solvents are polar solvents containing protons (for example, water or alcohols). In polar solvents that do not contain hydroxylic protons, such as acetone, $(CH_3)_2C{=}O$, the reactivity order is reversed. In the absence of solvent, in the gas phase, the order is also reversed.

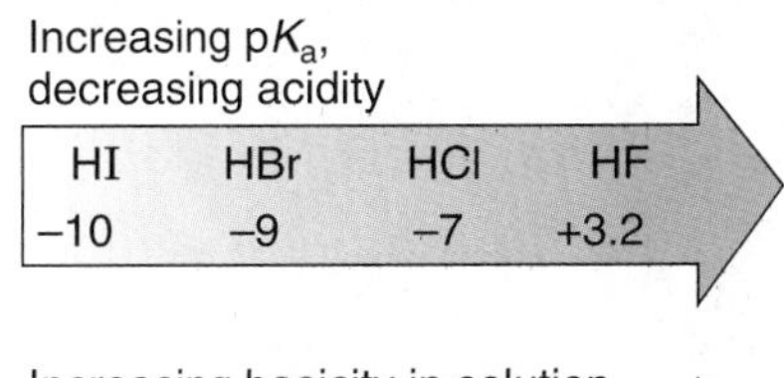

Increasing basicity in solution

I^- Br^- Cl^- F^-

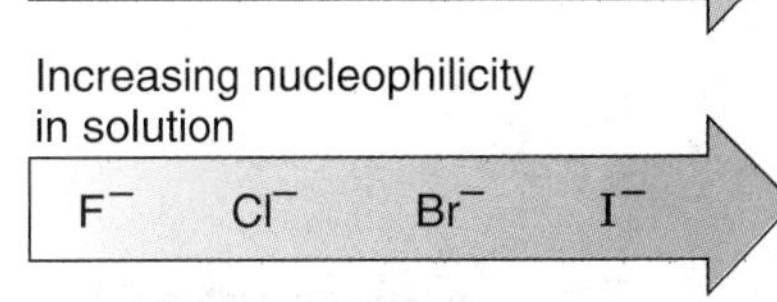

FIGURE 7.43 For the halide ions in solution, the orders of basicity and nucleophilicity are opposite.

Reaction	I	Br	Cl	F
	Relative rates			
$X^- + CH_3—I \xrightarrow[H_2O]{S_N2} X—CH_3 + I^-$	160	14	1	---
$X^- + CH_3—Br \xrightarrow[\text{acetone solvent}]{S_N2} X—CH_3 + Br^-$	1	5	11	---
$X^- + CH_3—Br \xrightarrow[\text{gas phase, } no \text{ solvent}]{S_N2} X—CH_3 + Br^-$	---	<0.015	0.02	1

X = Halogens

FIGURE 7.44 In the highly polar and protic solvent water, iodide is the best nucleophile (fastest reacting). In the polar, but aprotic solvent $(CH_3)_2C{=}O$ (acetone), the order of nucleophilicity is reversed. In the gas phase, where there is no solvent, fluoride emerges as the best nucleophile! Read the table across not down. The dashed line means "not measured."

In a protic solvent, a halide can act as a *nucleophile* and react with the substrate through interaction with σ* of the C–X bond. Alternatively, a halide can behave as a *base* by interacting with one of the protons of the solvent molecule. Chloride and fluoride are by far the better bases and thus hydrogen bond to protonic solvents much more strongly than iodide.

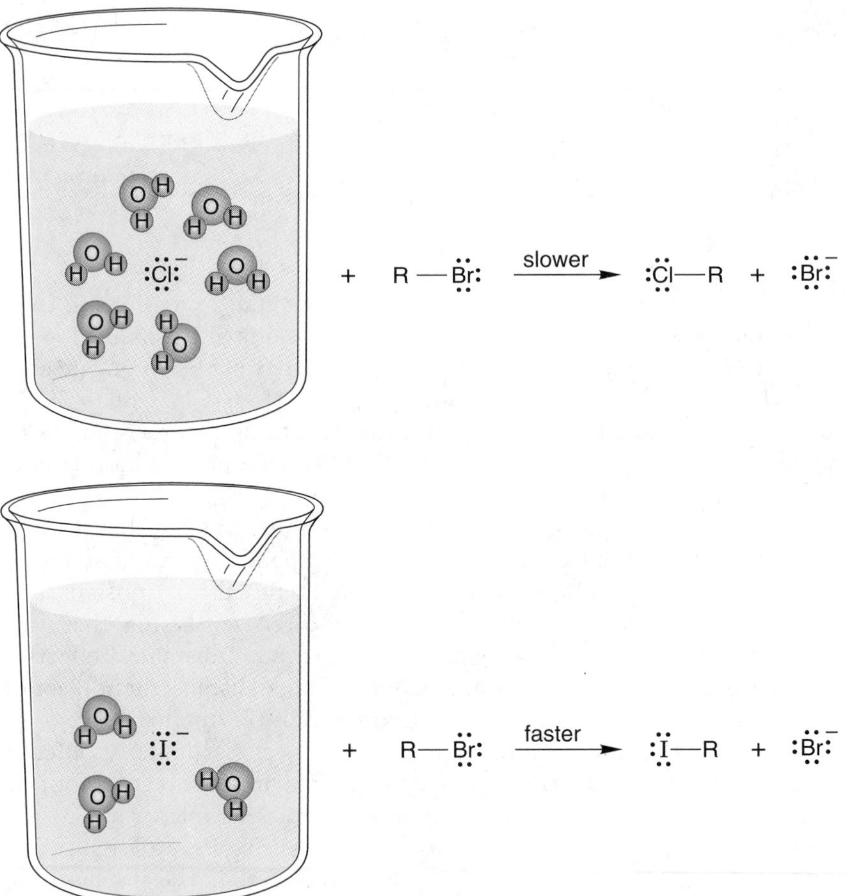

FIGURE 7.45 In a hydroxylated solvent, the highly solvated chloride ion reacts slower with R–Br than the less encumbered iodide ion.

The halides become highly solvated in protonic solvents and their relative sizes increase dramatically. They become more encumbered, and therefore less effective nucleophiles. Iodide, the poorer base, is less hydrogen bonded to protonic solvents and thus more free to do the S_N2 reaction (Fig. 7.45).

As the polarity of the solvent decreases, fluoride becomes more competitive with iodide. What's the limit of this effect? A reaction with *no* solvent, of course. Pure nucleophilicity, with no competing effects of solvent, can only be examined in the absence of other bulk molecules. We can do that by running the reaction in the gas phase. It is no trivial matter to examine reactions of ions in the gas phase, as the stability of ions depends heavily on solvation. Yet, in the last two decades such experiments have become possible. In the gas phase, fluoride, the better base, is also the better nucleophile (Fig. 7.44).

7.7e Effect of the Leaving Group

Most leaving groups depart as anions.

$$\text{Nu:}^- + \text{R–L} \rightarrow \text{Nu–R} + \text{L:}^-$$

It seems obvious that the more stable $L:^-$ is, the easier it will be to displace it. The stability of the anion $L:^-$ is related to the ease of dissociation of its conjugate acid, H–L, and thus to the pK_a of H–L. Acids (H–L) with low pK_a values (strong acids) are related to good leaving groups, $L:^-$ (weak bases). Acids with high pK_a values (weak acids) are related to poor leaving groups (strong bases). Figure 7.46 shows some good and bad leaving groups and the pK_a values of their conjugate acids, H–L.

$$\text{H–L} + H_2O \rightleftarrows H_3O^+ + \text{L:}^-$$

Acid	pK_a	Leaving Group	Name
		Good Leaving Groups	
HI	−10	$^-$I	Iodide
HBr	− 9	$^-$Br	Bromide
HCl	− 7	$^-$Cl	Chloride
$HOSO_2R$	− 6.5	$^-OSO_2R$	Sulfonate
H_3O^+	− 1.7	OH_2	Water
		Bad Leaving Groups	
HF	+ 3.2	$^-$F	Fluoride
H_2S	+ 7.0	$^-$SH	Thiolate
HCN	+ 9.2	$^-$CN	Cyanide
H_2O	+15.7	$^-$OH	Hydroxide
HOR	+16–18	$^-$OR	Alkoxide

FIGURE 7.46 Some good and bad leaving groups and their conjugate acids.

Many times it would be convenient to convert an alcohol (R–OH), which is generally cheap and available (in the chemical industry, those two terms are usually synonymous) into another compound, R–Nu. One naturally considers an S_N2 substitution reaction of R–OH (Fig. 7.47) with some nucleophile Nu:⁻.

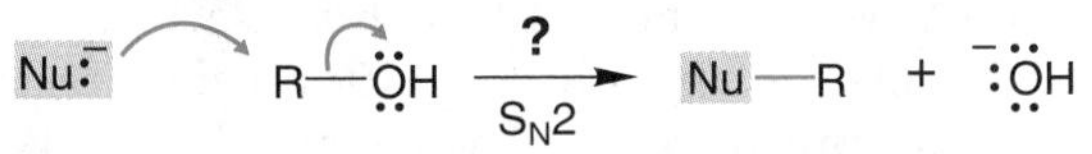

$$Nu{:}^- + R{-}\ddot{O}H \xrightarrow[S_N2]{?} Nu{-}R + {:}\ddot{O}H^-$$

FIGURE 7.47 A hypothetical S_N2 reaction converting an alcohol, R–OH, into a new compound, Nu–R.

The problem is that ^-OH is an extraordinarily poor leaving group (Fig. 7.46). The pK_a of water is 15.7. Water is a rather weak acid, and its conjugate base, hydroxide ion, is a strong base. There are several ways to circumvent the problem, the simplest of which is to transfer a proton to the alcohol (protonate the alcohol) with a strong acid to give the conjugate acid of the alcohol, $R{-}\overset{+}{O}H_2$ (Fig. 7.48).

$$R{-}\ddot{O}H + H{-}B \longrightarrow R{-}\overset{+}{\ddot{O}}H_2 + B{:}^-$$

FIGURE 7.48 The first step in the reaction of an alcohol with the acid H–B is protonation of the Lewis base oxygen. This converts the leaving group from the very poor leaving group, hydroxide, into the excellent leaving group, water.

PROBLEM 7.12 What *will* happen when an alcohol is treated with a strong base? Think "simple."

This reaction converts the leaving group from hydroxide into water. Water is a weak base; its conjugate acid is H_3O^+, a strong acid with a pK_a of −1.7. Therefore, water is easily displaced by a good nucleophile. In the reaction of an alcohol with HBr, the strongest available nucleophile is Br^-, the conjugate base of the HBr used to protonate the alcohol. So, if we want to convert an alcohol into a bromide we need only treat it with HBr. The reaction and its mechanism are shown for the conversion of ethyl alcohol into ethyl bromide in Figure 7.49.

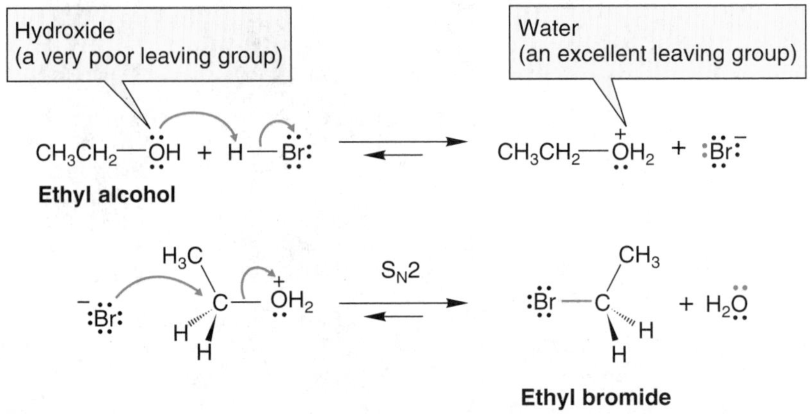

Hydroxide (a very poor leaving group)

Water (an excellent leaving group)

$$CH_3CH_2{-}\ddot{O}H + H{-}\ddot{B}r{:} \rightleftharpoons CH_3CH_2{-}\overset{+}{\ddot{O}}H_2 + {:}\ddot{B}r{:}^-$$

Ethyl alcohol

$${:}\ddot{B}r{:}^- + H_3C(H)(H)C{-}\overset{+}{\ddot{O}}H_2 \overset{S_N2}{\rightleftharpoons} {:}\ddot{B}r{-}C(CH_3)(H)(H) + H_2\ddot{O}$$

Ethyl bromide

FIGURE 7.49 The mechanism for the conversion of ethyl alcohol into ethyl bromide by treatment with HBr. Note especially the technique of changing the leaving group from the very poor hydroxide to the excellent water.

This procedure is a general one—the conversion of a poor leaving group into a good one—and we will see it often. Another procedure uses a two-step process in which the alcohol is first converted into a sulfonate. Sulfonates are excellent leaving groups (Fig. 7.46) and can be displaced by all manner of nucleophiles to give desirable products (Fig. 7.50).

The general case

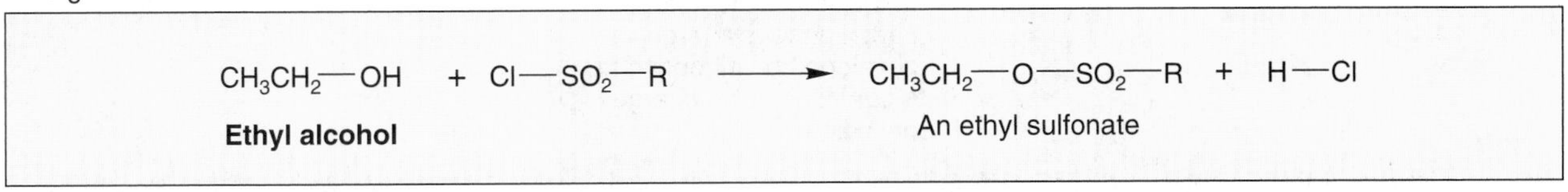

$$CH_3CH_2{-}OH + Cl{-}SO_2{-}R \longrightarrow CH_3CH_2{-}O{-}SO_2{-}R + H{-}Cl$$

Ethyl alcohol — An ethyl sulfonate

A specific example

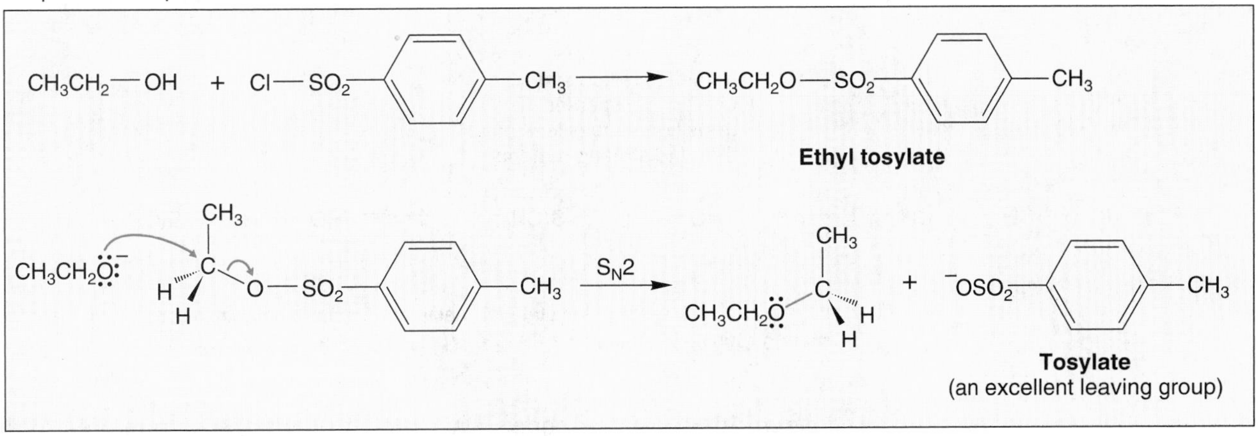

$$CH_3CH_2{-}OH + Cl{-}SO_2{-}C_6H_4{-}CH_3 \longrightarrow CH_3CH_2O{-}SO_2{-}C_6H_4{-}CH_3$$

Ethyl tosylate

$$CH_3CH_2\ddot{O}{:}^- + CH_3CH_2{-}O{-}SO_2{-}C_6H_4{-}CH_3 \xrightarrow{S_N2} CH_3CH_2\ddot{O}{-}CH_2CH_3 + {}^-OSO_2{-}C_6H_4{-}CH_3$$

Tosylate
(an excellent leaving group)

FIGURE 7.50 The conversion of ethyl alcohol into another ethyl derivative. In this sequence of reactions the poor leaving group, OH, is first converted into the good leaving group, sulfonate. The leaving group is then displaced through an S_N2 reaction with a nucleophile, Nu:$^-$.

PROBLEM 7.13

Explain why sulfonates should give rather stable anions (weak nucleophiles) and are therefore good leaving groups.

7.7f Effect of Solvent

At first, the behavior of the S_N2 reaction as the polarity of solvent is changed is confusing. As the solvent polarity is increased, some S_N2 reactions go faster, some slower. Two examples are shown in Figure 7.51. In the first reaction (a) a change to a more polar solvent results in a faster reaction, but in the second reaction (b), the rate decreases when the solvent is made more polar.

There is a way to attack almost every problem that poses a question having to do with rates. The rate of a reaction is determined by the energy of the transition state—the high point in energy as the reaction proceeds

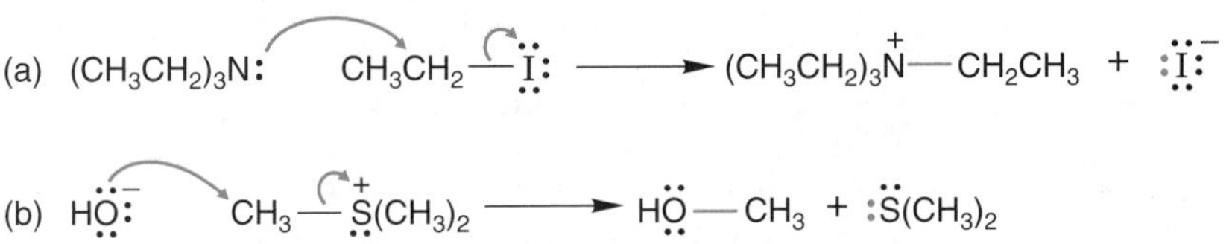

(a) $(CH_3CH_2)_3N{:} \quad CH_3CH_2{-}\ddot{I}{:} \longrightarrow (CH_3CH_2)_3\overset{+}{N}{-}CH_2CH_3 + {:}\ddot{I}{:}^-$

(b) $H\ddot{O}{:}^- \quad CH_3{-}\overset{+}{S}(CH_3)_2 \longrightarrow H\ddot{O}{-}CH_3 + {:}S(CH_3)_2$

FIGURE 7.51 Opposite effects of a change in solvent polarity on the rates of two S_N2 reactions. Reaction (a) goes faster when the solvent is made more polar, whereas Reaction (b) goes more slowly.

from starting material to products. In order to solve a question involving rates, we must know the structure of the transition state of the reaction. Figure 7.52 shows the structures of the transition states for the two S_N2 reactions of Figure 7.51.

In the first example (a), a more polar solvent will act to stabilize the charged products and the polar, charge-separated transition state. There will be very little effect on the uncharged starting materials. The energy

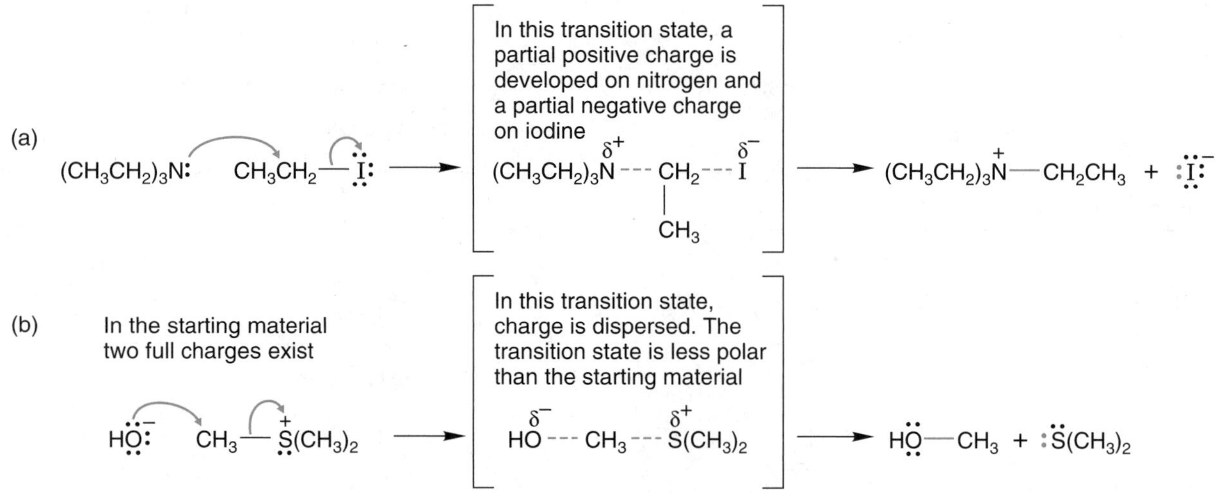

(a) $(CH_3CH_2)_3N:$ $CH_3CH_2—I:$ ⟶ [In this transition state, a partial positive charge is developed on nitrogen and a partial negative charge on iodine; $(CH_3CH_2)_3\overset{\delta^+}{N}---CH_2(CH_3)---\overset{\delta^-}{I}$] ⟶ $(CH_3CH_2)_3\overset{+}{N}—CH_2CH_3$ + $:I:^-$

(b) In the starting material two full charges exist

$HO:^-$ $CH_3—\overset{+}{S}(CH_3)_2$ ⟶ [In this transition state, charge is dispersed. The transition state is less polar than the starting material; $\overset{\delta^-}{HO}---CH_3---\overset{\delta^+}{S}(CH_3)_2$] ⟶ $HO—CH_3$ + $:S(CH_3)_2$

FIGURE 7.52 The transition state for reaction (a) is more polar than the starting material. The transition state for reaction (b) is less polar than the starting material.

diagram in Figure 7.53 shows what this differential solvation does to the rate of this reaction. The height of the transition state is decreased relative to the starting material and the reaction will go faster. It takes less energy for the molecules to traverse the barrier when the polarity of the solvent increases.

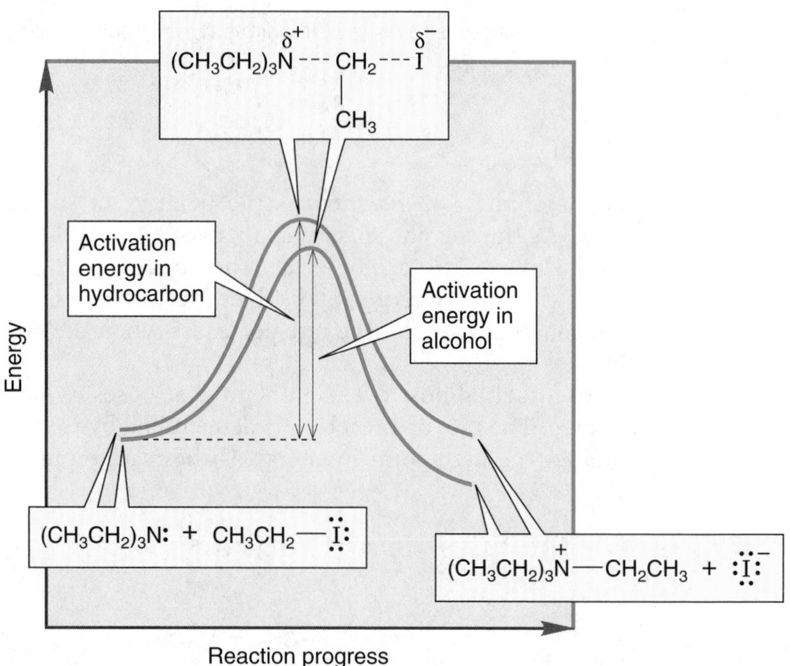

FIGURE 7.53 In this reaction, the polar transition state will be stabilized by a change to a more polar solvent more than will the neutral starting materials. The result is a decreased activation energy for the reaction and a faster rate. Hydrocarbon solvent (nonpolar, blue line), alcohol solvent (much more polar, red line).

The second example (b) is different. Here the starting materials bear full charges. This charge is dispersed as the reaction goes on. A change to a more polar solvent will stabilize the fully charged starting materials more than the transition state or products. The activation energy increases and the reaction slows (Fig. 7.54).

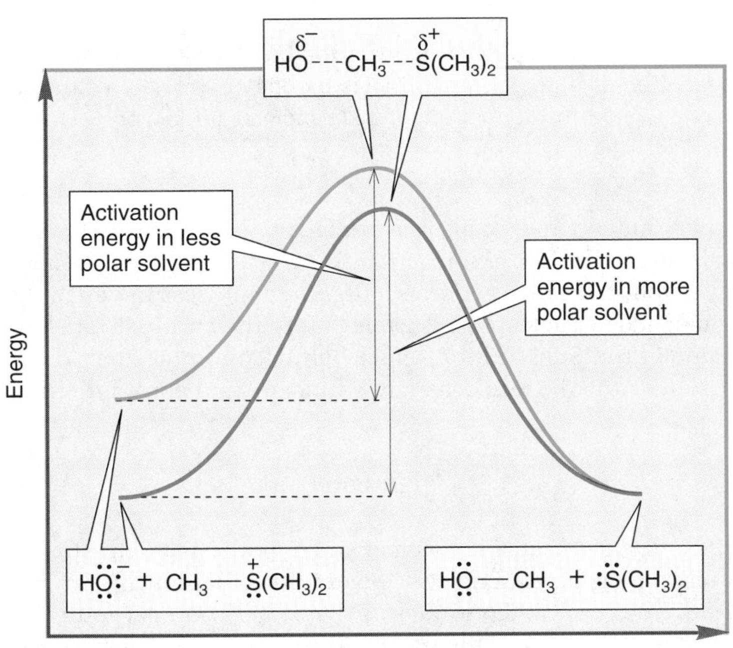

FIGURE 7.54 In this reaction, the transition state will be stabilized by a change to a more polar solvent less than will the charged starting materials. The result is an increased activation energy for the reaction and a slower rate. Alcohol solvent (blue line), alcohol and water solvent (more polar, red line).

To summarize: One limiting case of substitution is the S_N2 reaction, which is bimolecular and common for primary and secondary substrates. The other limiting example is the S_N1 reaction, which is a unimolecular process that is common for tertiary substrates.

7.8 SUBSTITUTION, NUCLEOPHILIC, UNIMOLECULAR: THE S_N1 REACTION

7.8a Rate Law

tert-Butyl bromide does not undergo a substitution reaction when treated with the good nucleophiles listed in Table 7.6. The rate of the second order S_N2 displacement on *tert*-butyl bromide by hydroxide ion is effectively zero. Yet, *tert*-butyl bromide does react with the much less powerfully nucleophilic molecule water to form the product of substitution, *tert*-butyl alcohol (Fig. 7.55).

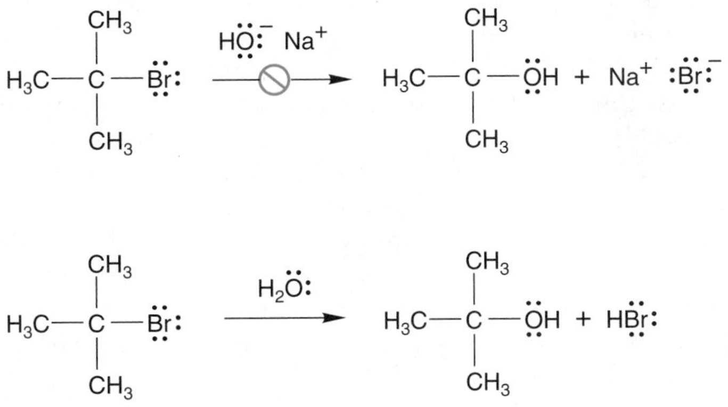

$$(CH_3)_3C{-}Br \xrightarrow[\text{(no reaction)}]{HO^-\ Na^+} (CH_3)_3C{-}OH + Na^+\ Br^-$$

$$(CH_3)_3C{-}Br \xrightarrow{H_2O} (CH_3)_3C{-}OH + HBr$$

FIGURE 7.55 Despite the far greater nucleophilicity of hydroxide ion, the reaction of *tert*-butyl bromide with water is successful in producing *tert*-butyl alcohol, whereas reaction with hydroxide ion is not.

PROBLEM 7.14 Although *tert*-butyl bromide does not give *tert*-butyl alcohol on reaction with hydroxide ion, an alkene product (C_4H_8) is produced. Provide a structure for the product and an arrow formalism for this bimolecular reaction.

The first clue to what's happening in Figure 7.55 comes from an examination of the rate law for the reaction. It turns out that the formation of *tert*-butyl alcohol in this reaction is a first-order process in which the concentration of added nucleophile is irrelevant to the rate. The rate law leads to the name of the Substitution, Nucleophilic, unimolecular, or the **S_N1 reaction**, where k is the rate constant, a fundamental constant of the reaction.

$$\text{Rate} = v = k[\textit{tert}\text{-butyl bromide}]$$

At this point, a reasonable mechanistic hypothesis might simply view the S_N1 reaction as beginning with the unimolecular ionization shown in Figure 7.56. The reaction continues as the carbocation is captured by the solvent water, in what is called a **solvolysis reaction**. Solvolysis simply means that the solvent, here water, plays the role of nucleophile in the reaction. In the first step of the S_N1 reaction, *tert*-butyl bromide ionizes to a tertiary carbocation, which then is captured in a second step by the nucleophilic solvent water to give the protonated alcohol. Any base present, here likely to be bromide or water, can remove a proton (deprotonate the protonated alcohol) to give the alcohol itself (Fig. 7.56).

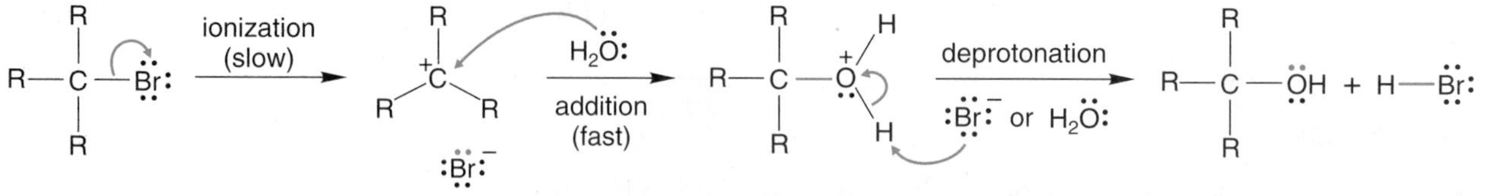

FIGURE 7.56 A schematic mechanism for the S_N1 reaction. A slow ionization is followed by a relatively fast capture of the carbocation by a nucleophile.

We can describe the reaction with an energy diagram as we did for the S_N2 reaction (Figs. 7.30 and 7.31, p. 249, 250). This diagram is more complicated, because of the formation of the intermediate carbocation. The presence of the intermediate means that there will be two transition states separating it from the starting material and the product. The transition state for the slow step, the initial endothermic ionization, is higher than the transition state for the second, faster step, the capture of the ion by the nucleophile (Nu:$^-$) (Fig. 7.57).

This second step determines the structure of the product *but not the rate of the reaction.* If there are several nucleophiles present they will compete with each other in the capture of the carbocation to produce different products, but the rate of the reaction will not be affected (Fig. 7.58). Note that the reverse of the ionization, recapture of the carbocation by bromide, is nothing more than another example of a reaction of this intermediate with a nucleophile.

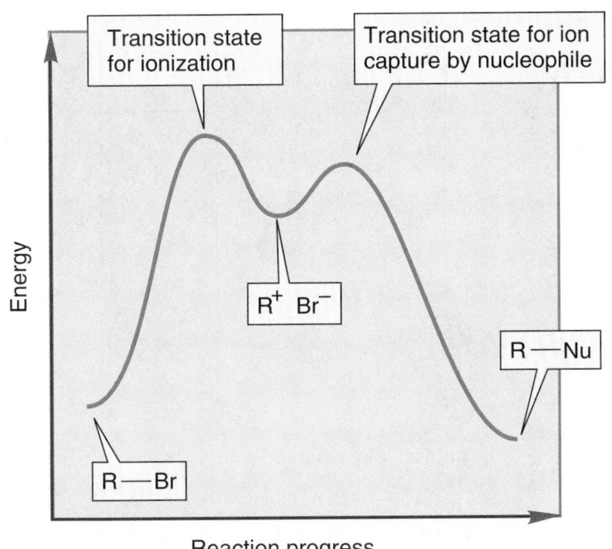

FIGURE 7.57 An energy diagram for the S_N1 reaction. Notice that the transition state for the slow step in the reaction—the ionization—is higher in energy than the transition state for the capture of the ion by a nucleophile.

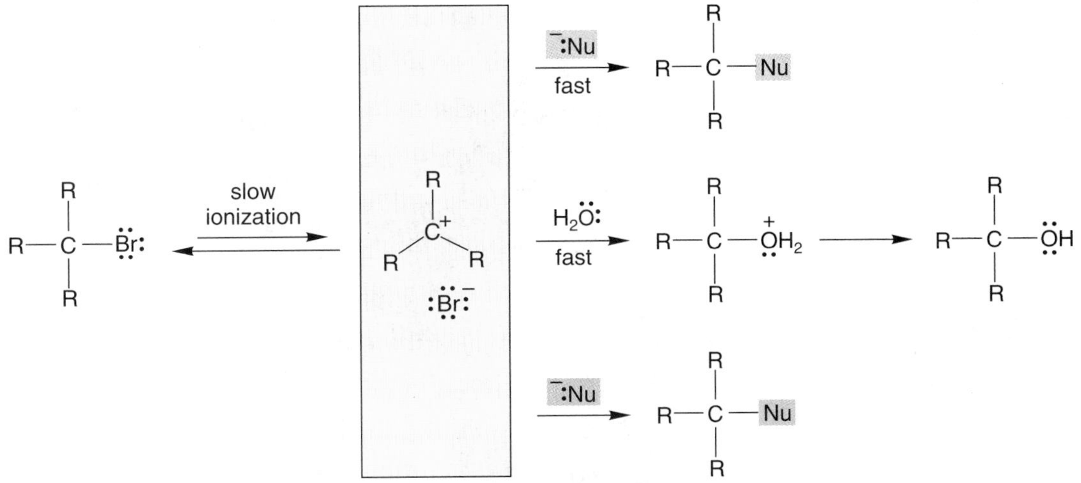

R_3C—Br: ⇌ (slow ionization) R_3C^+ + :Br:⁻

⁻:Nu (fast) → R_3C—Nu

$H_2\ddot{O}$: (fast) → R_3C—$\overset{+}{O}H_2$ → R_3C—ÖH

⁻:Nu → R_3C—Nu

FIGURE 7.58 The nucleophile determines the structure of the product as it captures the intermediate carbocation. This figure shows the capture of the carbocation by four nucleophiles (⁻:Nu, ⁻:Nu′, ⁻Br, and H_2O). However, the nucleophiles present have no effect on the rate of formation of the carbocation, the slow step in the reaction.

We will now see if this simple hypothesis allows a reasonable prediction of the stereochemical course of the reaction.

7.8b Stereochemistry of the S_N1 Reaction

If the S_N1 reaction were as straightforward as suggested by Figures 7.56–7.58, what would be expected if an optically active starting material were used? The crucial intermediate in the reaction is the planar carbocation, an achiral molecule! Once the free, planar carbocation is formed, optical activity is inevitably lost. Addition of the nucleophile must take place at both faces of the cation forming equal amounts of (*R*) and (*S*) product. This simple mechanism demands racemization, as Figure 7.59 shows.

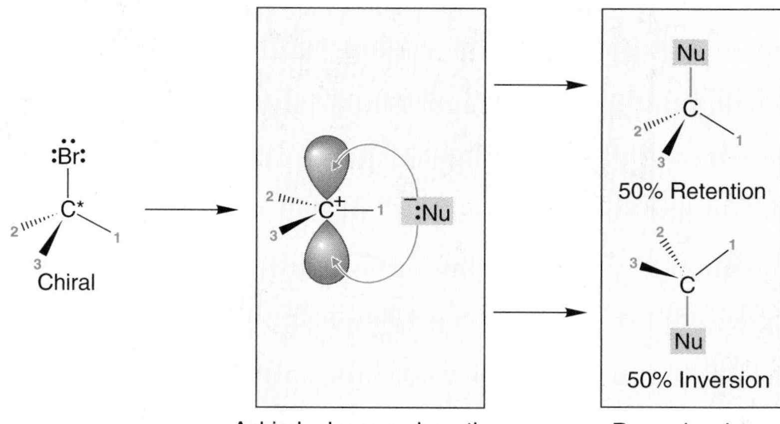

FIGURE 7.59 Loss of optical activity occurs as soon as the free, planar carbocation is formed. Addition of any nucleophile will occur equally at the two available lobes of the 2*p* orbital to give a racemic mixture of products.

Sometimes racemization *is* the actual result. However, usually the stereochemical results are not so clean, and an excess of inversion is found. For example, the two optically active molecules of Figure 7.60 give 21 and 18% excess inversion in the S_N1 reaction with water.

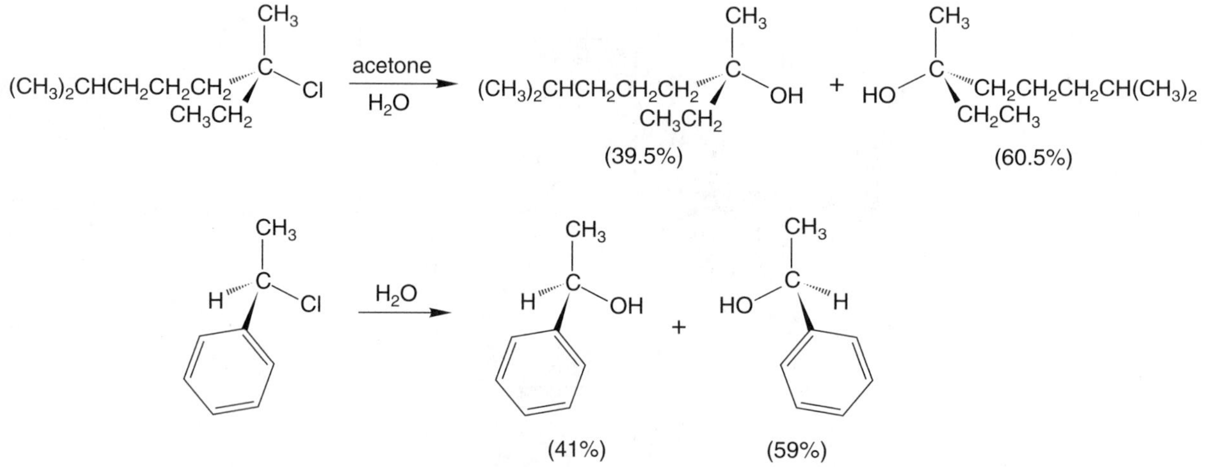

$(CH_3)_2CHCH_2CH_2CH_2$, CH_3CH_2, CH_3–C–Cl $\xrightarrow[H_2O]{acetone}$ $(CH_3)_2CHCH_2CH_2CH_2$, CH_3CH_2, CH_3–C–OH (39.5%) + HO–C(CH_3)(CH_2CH_3)$CH_2CH_2CH_2CH(CH_3)_2$ (60.5%)

H, CH_3, C_6H_5–C–Cl $\xrightarrow{H_2O}$ H, CH_3, C_6H_5–C–OH (41%) + HO–C(CH_3)(H)C_6H_5 (59%)

FIGURE 7.60 Usually, complete racemization is not the result in S_N1 substitution. In most reactions, there is an excess of the product resulting from inversion of configuration.

What is the source of residual optical activity, and why is excess inversion (as opposed to retention) observed? These two questions are vitally important, as satisfactory answers must be found if we are not to be forced to abandon the proposed mechanism for the S_N1 reaction. The answers turn out to be simple. In the mechanism outlined in Figures 7.56–7.58, we assumed that the leaving group had diffused away from the cation, leaving behind a *symmetrical* ion. What if reaction with solvent occurs before the leaving group diffuses completely away? What if reaction occurs with the *pair* of ions shown in Figure 7.61? The leaving group guards the side of the cation from which it departed, and makes addition by solvent water more difficult at this "retention" side. The result is usually an excess of inversion.

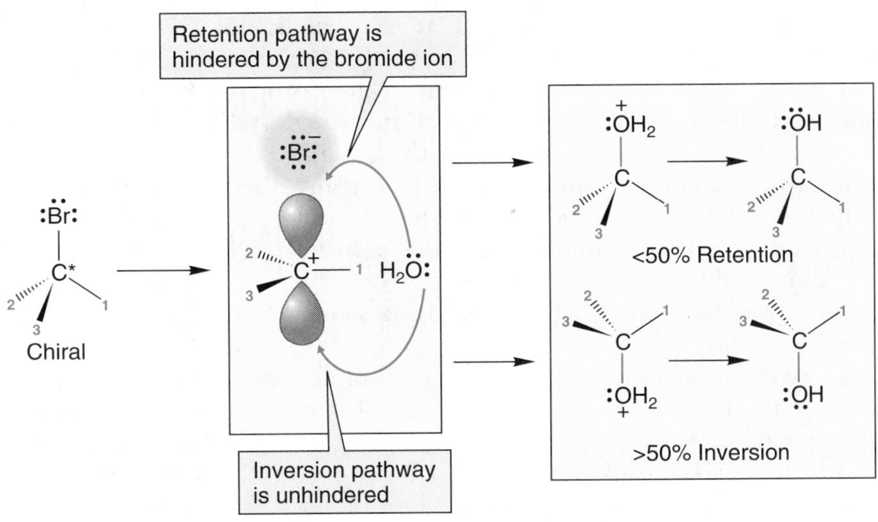

FIGURE 7.61 Unless the leaving group, here bromide ion, completely diffuses away, it shields the carbocation along the retention pathway and this results in an excess of inversion in most S_N1 reactions.

PROBLEM 7.15

The second reaction of Figure 7.60 shows an S_N1 reaction of a secondary halide. Can you see why the S_N1 reaction should be especially favorable for this secondary halide?

***PROBLEM 7.16**

Consider the role of solvent on the stereochemistry of the S_N1 reaction shown in Figure 7.61. A more polar solvent favors racemization. Why?

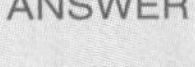

ANSWER

Polar solvents stabilize ions. The more stable an ion is, the less reactive it is, and the longer it lives. Long life allows for movement of the two ions away from their positions in the original ion pair. So, the more polar the solvent, the more likely it is that the symmetrical ion of Figure 7.59 will be formed from the unsymmetrical ion pair of Figure 7.61. A polar solvent favors racemization, the inevitable result of formation of the symmetrical carbocation of Figure 7.59.

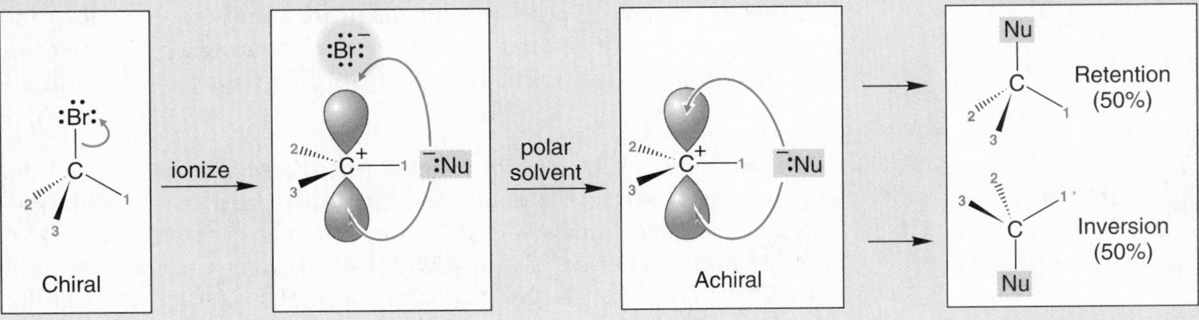

7.8c Effects of Substrate: The R Group

The rate of the S_N1 reaction is much greater for tertiary than for secondary substrates, and simple primary alkyl and methyl halides do not react at all by the S_N1 process. Our mechanistic hypothesis shows us why. If carbocation formation is the key step, only those compounds that can form relatively stable ions can react in S_N1 fashion.

Modern measurements show that the stability of carbocations increases with substitution (we will see why in Chapter 9). Table 7.7 shows the heats of formation of carbocations in the gas phase. *Remember*: The more positive a species' heat of formation the less stable it is (Chapter 4, p. 133). We can't make too much of the *quantitative* data, as these gas-phase measurements cannot take into account the effects of solvent, which are very large indeed. Nonetheless, they do give a good qualitative picture of carbocation stability: tertiary carbocations (three alkyl groups) are more stable than secondary ones (two alkyl groups), and secondary carbocations are more stable than primary ones (one alkyl group). The methyl cation ($^+CH_3$), with no alkyl groups, is least stable of all. Do not confuse these relative stabilities with absolute stabilities. Carbocations are almost all very unstable species. A small atom such as carbon does not bear a positive charge with any grace, and a carbocation will react with available Lewis bases (for example, halide ion) very quickly. Do not say, "The *tert*-butyl cation is very stable," but rather, "*For a carbocation*, the *tert*-butyl cation is quite stable."

TABLE 7.7 Heats of Formation for Some Carbocations

Cation	Substitution	ΔH_f° (kcal/mol)
$^+CH_3$	Methyl	261.3
$^+CH_2CH_3$	Primary	215.6
$^+CH_2CH_2CH_3$	Primary	211
$^+CH_2CH_2CH_2CH_3$	Primary	203
$H_3C\overset{+}{C}HCH_2CH_3$	Secondary	183
$(CH_3)_2\overset{+}{C}H$	Secondary	190.9
$(CH_3)_3C^+$	Tertiary	165.8

There are many reactions in which tertiary and secondary carbocations are intermediates, but no simple reactions for which a primary or methyl cation is required. These intermediates are simply too unstable. In practice, this means that when you find yourself writing a mechanism using a primary carbocation, think again, it's probably wrong!

*PROBLEM 7.17 We have avoided for the moment a detailed discussion of *why* more substituted carbocations are more stable than less substituted ones.* Nevertheless you can see the outlines of why this is so. Draw out a three-dimensional picture of the ethyl cation. (*Remember*: The positively charged carbon and the three surrounding atoms are coplanar—this part of the ethyl cation is flat.) Now consider the overlap of one of the carbon–hydrogen bonds of the methyl group with the empty $2p_z$ orbital. Will this be stabilizing? Why?

ANSWER Interactions between filled and empty orbitals are stabilizing. This fundamental tenet of chemistry can be applied in this case. If a filled carbon–hydrogen bond overlaps with the empty 2*p* orbital of the carbocation, the result is a stabilizing interaction.

*If avoiding the question drives you crazy, and I think it should at least make you a little uncomfortable, be sure to do this problem—the answer falls right out. See also Chapter 9.

ANSWER (CONTINUED)

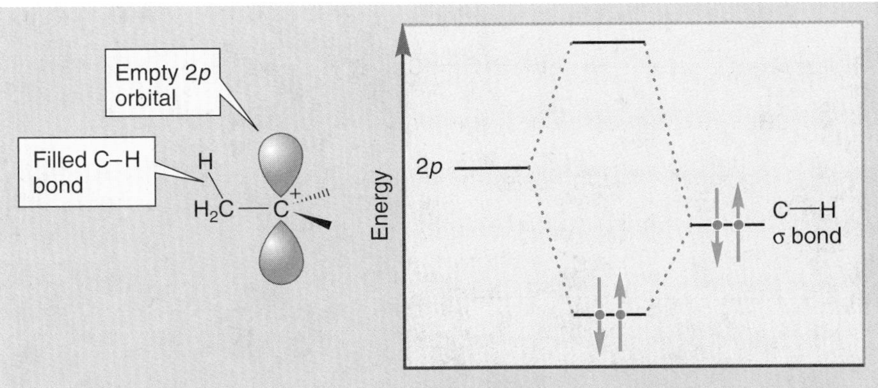

This interaction can only occur if the appropriate carbon–hydrogen bonds exist. The more alkyl groups, the more stabilization there is. For example, in

H—C^+(H)(H)

no such stabilization is possible without the carbon–hydrogen bonds of an alkyl group

So, tertiary compounds, which form the most stable, *tertiary* carbocations, commonly react through the S_N1 mechanism, and methyl and primary compounds, which would have to form extremely unstable carbocations, do not. Naturally enough, secondary compounds react faster than primary ones in the S_N1 reaction, but more slowly than tertiary compounds (Fig. 7.62).

R_3C—L	>	R_2CH—L	>>>	RCH_2—L	>	H_3C—L
Tertiary		Secondary		Primary		Methyl

FIGURE 7.62 Relative reactivities in the S_N1 reaction.

7.8d Effects of the Nucleophile

We have already seen that the nucleophile does not affect the rate of the S_N1 reaction—the reaction is first order in substrate. Yet, the nucleophile does determine the product structure. Note the distinction between the **rate-determining step** and the **product-determining step**. For example, reaction of *tert*-butyl bromide in water containing a small amount of an added nucleophilic ion such as cyanide leads to substantial amounts of *tert*-butyl cyanide. Even though water is present as solvent, and thus has an immense advantage over cyanide ion because of concentration, the more powerfully nucleophilic cyanide still competes. Cyanide is many times more reactive toward the carbocation than the solvent water (Table 7.6, p. 257). In this reaction, the rate-determining step (the slow ionization) and the product-determining step (the faster capture of the intermediate ion by nucleophiles) are sharply separated (Fig. 7.63).

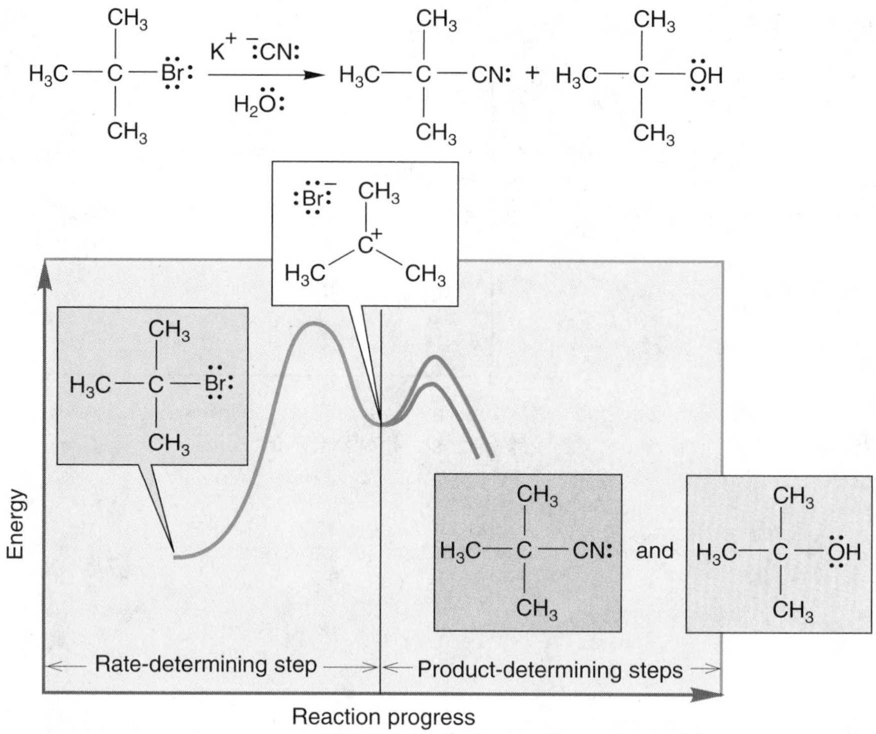

FIGURE 7.63 The rate-determining and product-determining steps are sharply distinguished in the S_N1 reaction.

7.8e Effect of the Leaving Group

The identity of the leaving group has a large influence on the rate of the S_N1 reaction because the structure of the departing group affects the ease of ionization. We have already discussed how the structure—and hence stability—of the carbocation affects the rate of ionization, but obviously the stability of the negatively charged counterion will be important as well. The more stable the anion, the more easily it will be lost (Fig. 7.64). Some good leaving groups were shown in Figure 7.46 (p. 259).

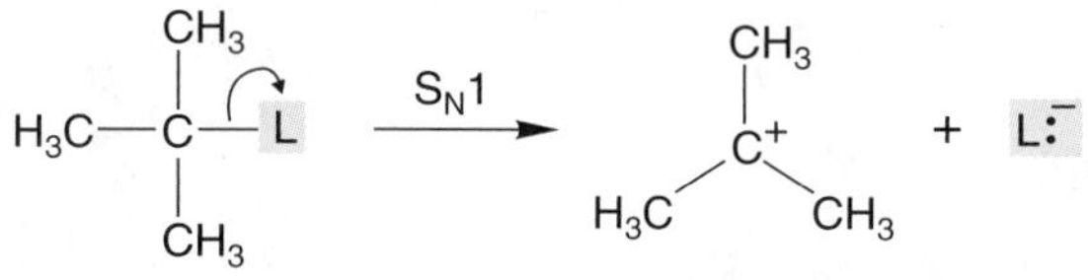

FIGURE 7.64 The rate of ionization is affected by the stability of the leaving group, $L:^-$. The more stable the leaving group, the faster the ionization.

7.8f Effect of Solvent

For the S_N2 reaction the effect of solvent was rather complicated and depended on the charge type of the reaction. Some S_N2 reactions were accelerated by a change to a more polar solvent; others were slowed down (p. 261). For most S_N1 reactions the situation is simpler. Ions are usually formed in the slow step of the reaction. This process will be favored by a polar solvent that will stabilize the polar species generated by the ionization (Fig. 7.65).

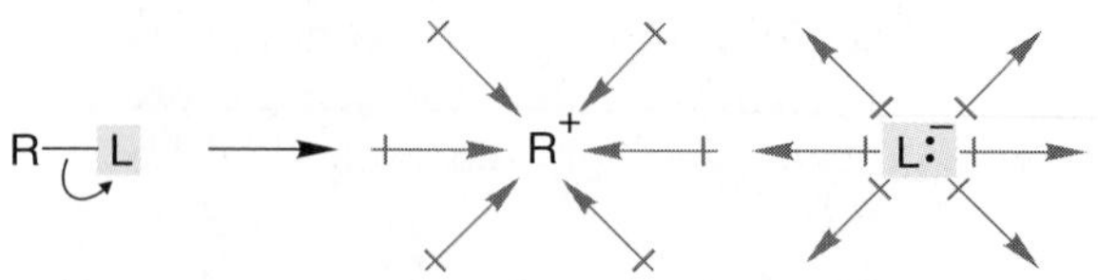

FIGURE 7.65 In a polar solvent, represented by the dipole arrows in the figure, the charged species formed in the S_N1 reaction are stabilized. A nonpolar solvent cannot do this as well. One can expect that the S_N1 reaction will usually be favored by a polar solvent.

7.9 OVERVIEW OF THE S_N2 AND S_N1 REACTIONS

These two substitution reactions are interconnected by more than the ultimate formation of a substitution product. They complement each other—when one is ineffective the other is likely to work well. For example, the S_N2 reaction works best for methyl and primary substrates, which is exactly the point at which the S_N1 reaction is least effective. When the S_N2 reaction fails, as with tertiary substrates, the S_N1 reaction is most effective (Fig. 7.66).

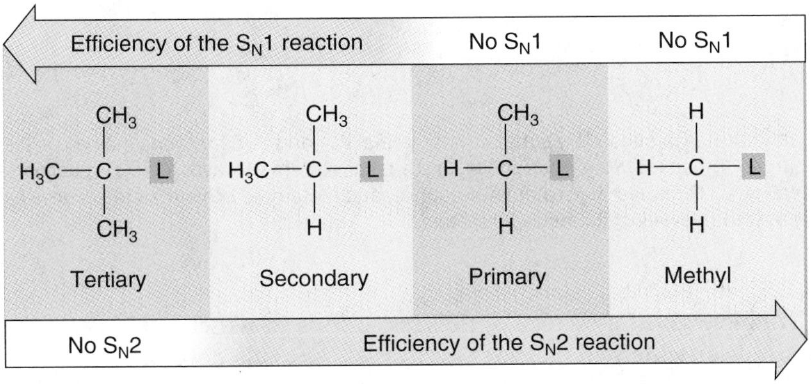

Efficiency of the S_N1 reaction — No S_N1 — No S_N1

Tertiary: $(CH_3)_3C-L$ | Secondary: $(CH_3)_2CH-L$ | Primary: CH_3CH_2-L | Methyl: H_3C-L

No S_N2 — Efficiency of the S_N2 reaction

FIGURE 7.66 The S_N2 and S_N1 reactions are complementary. When one is ineffective, the other works well.

So, at the extremes the situation is clear. The S_N2 dominates for methyl and primary substrates because there is little steric hindrance to the displacement of the leaving group and because the competing S_N1 reaction is disfavored by the necessity of forming the highly unstable methyl or primary carbocations (Fig. 7.67).

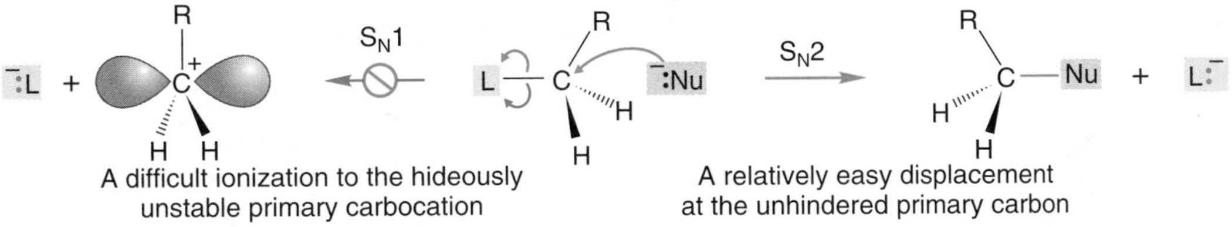

S_N1 (blocked) — S_N2

A difficult ionization to the hideously unstable primary carbocation

A relatively easy displacement at the unhindered primary carbon

FIGURE 7.67 A comparison of the S_N1 and S_N2 reactions for primary substrates. The S_N2 reaction is greatly favored.

The S_N1 dominates for tertiary substrates because ionization can produce the relatively stable tertiary carbocation and because the competing S_N2 reaction is strongly disfavored by the steric hindrance to the nucleophile approaching from the rear (Fig. 7.68).

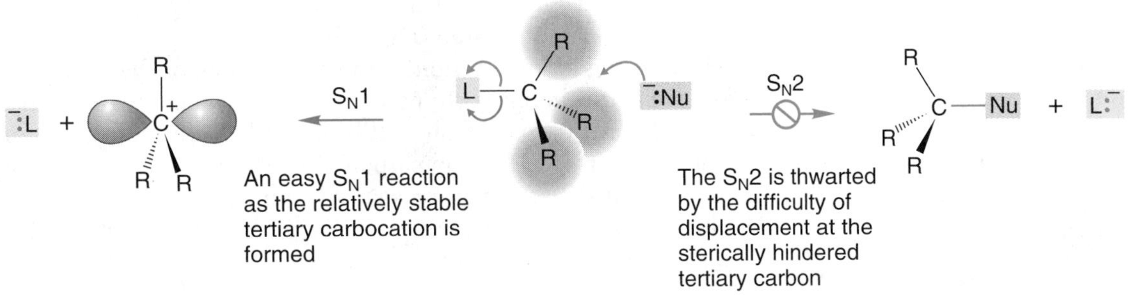

S_N1 — S_N2 (blocked)

An easy S_N1 reaction as the relatively stable tertiary carbocation is formed

The S_N2 is thwarted by the difficulty of displacement at the sterically hindered tertiary carbon

FIGURE 7.68 For tertiary systems, the S_N1 reaction dominates; the ionization produces the relatively stable tertiary carbocation and the S_N2 reaction is hindered by the three R groups that thwart approach of the nucleophile from the rear.

In the middle ground—secondary substrates—we might expect to be able to find both reactions. Small R groups, use of a powerful nucleophile, and a solvent of proper polarity for the charge type of the reaction can favor the S_N2 reaction. Large R groups, a weak nucleophile, and a polar solvent will favor the S_N1 reaction (Fig. 7.69).

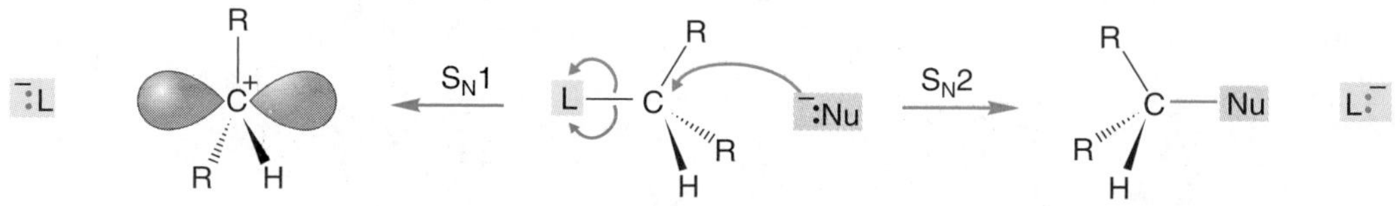

FIGURE 7.69 For secondary substrates, both the S_N1 and S_N2 reactions may be possible. Large R groups, a weak nucleophile, and a polar solvent will favor the S_N1 reaction, whereas small R groups, a powerful nucleophile, and the proper solvent for the charge type of the reaction will select for the S_N2 reaction.

We have given a picture of these reactions in which extremes are emphasized; in which an "either S_N2 or S_N1" view is developed. The situation is messier than this simple explanation would suggest, especially for secondary compounds. Imagine beginning to ionize a bond to a leaving group. As the bond stretches, positive charge begins to develop on carbon (δ^+) and rehybridization toward sp^2 begins as the groups attached to the carbon flatten out (Fig. 7.70).

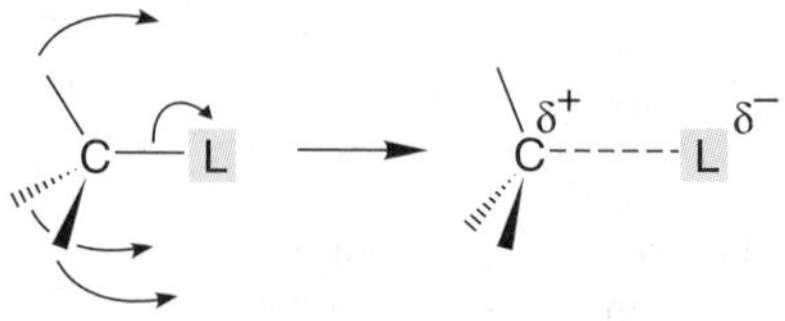

FIGURE 7.70 The beginnings of an S_N1 reaction. The bond to the leaving group $:L^-$ has lengthened and the tetrahedron has begun to flatten out. The hybridization of the central carbon is changing from nearly sp^3 to the sp^2 of the planar carbocation.

If we simply continue this ionization process we have the pure S_N1 reaction. If we let a nucleophilic solvent molecule interact with the rear of the departing bond and ultimately form a bond to carbon, we have the S_N2 reaction with inversion (Fig. 7.71).

How strong an interaction with the rear lobe of the orbital is needed before we call it bonding? In polar solvents, ions are highly solvated and in an S_N1 reaction solvation will begin to be important as soon as a δ^+ charge begins to build up on carbon. Solvation will be especially easy at the rear of the departing group as the leaving group shields the frontside. We have already seen that the leaving group can hinder reaction at one side of the developing cation and produce some net inversion in the reaction (Fig. 7.61).

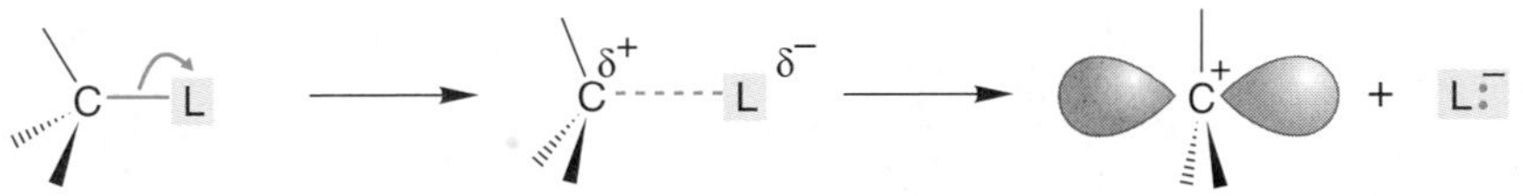

In the classic S_N1 reaction, this ionization continues to produce the planar carbocation. Notice, however, that the carbocation when first formed is not symmetrical. The leaving group lurks on one side, partially shielding the cation from solvent and other nucleophiles.

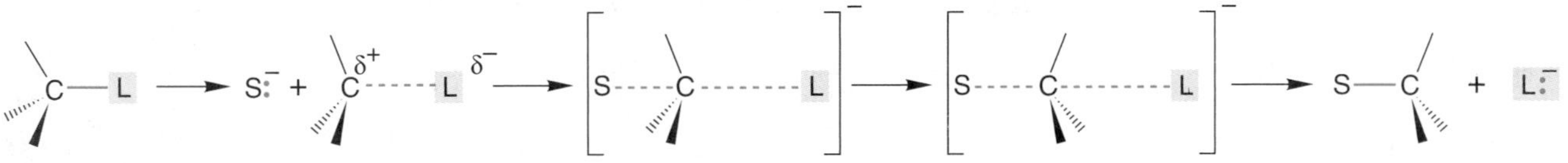

However, solvation at the side away from the old leaving group can be converted into nucleophilic attack (S_N2) to give inverted product.

FIGURE 7.71 Especially for secondary substrates, the difference between S_N2, displacement by solvent, and preferential solvation at the rear of the departing bond is a subtle one.

Now we begin to see the difficulty of telling the two mechanisms apart in some situations. When does "preferential solvation from the rear" become "displacement from the rear"? It becomes very hard to sort out in some cases. Organic chemistry is frustratingly, if delightfully, messy.

This messiness raises its head in other ways. Almost all substitution reactions are accompanied by some formation of alkene. In organic chemistry few reactions proceed in 100% yield, quantitatively producing one molecule from another. A large part of the business of organic synthesis is the search for specificity. Synthetic chemists seek reagents and reaction conditions that greatly favor one reaction pathway over all others. In particular, attempts to synthesize molecules through substitution reactions must take account of competitive reactions called eliminations. The elements of H–L are lost (eliminated) from the substrate to give an alkene (Fig. 7.72). Sections 7.10 and 7.11 will examine two general mechanisms for these elimination reactions.

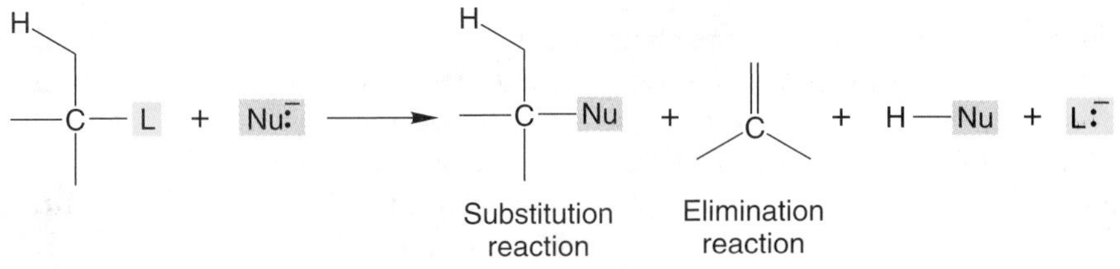

Substitution reaction

Elimination reaction

FIGURE 7.72 In almost all S_N1 and S_N2 reactions, substitution is accompanied by some elimination of H–L to give an alkene.

7.10 THE UNIMOLECULAR ELIMINATION REACTION: E1

7.10a Rate Law

If we measure the rate law for the formation of the alkene accompanying an S_N1 reaction, we find that this elimination reaction is also a first-order process. As in the S_N1 reaction, the slow step is the ionization of the substrate to give a carbocation. The intermediate carbocation can either add a nucleophile to finish off the S_N1 reaction, or lose a proton to produce an alkene in the **E1 reaction**. The carbocation is an intermediate in *both* processes (Fig. 7.73).

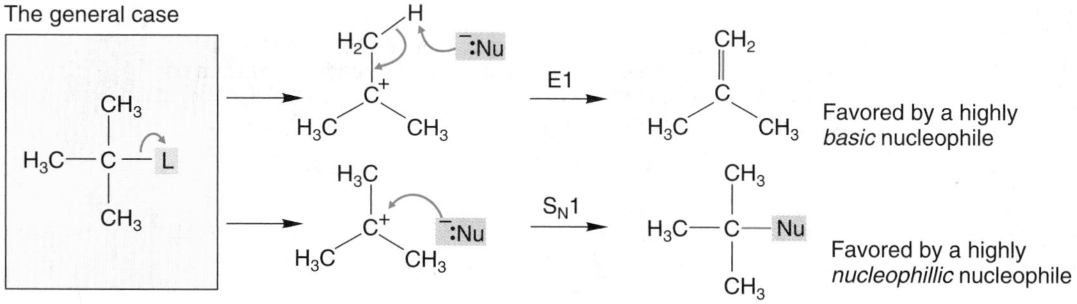

The general case

Favored by a highly *basic* nucleophile

Favored by a highly *nucleophillic* nucleophile

FIGURE 7.73 The carbocation intermediate in the S_N1 reaction can do two things: it can continue the S_N1 reaction by adding a nucleophile ($Nu:^-$) or, if a proton is available, an alkene can be formed in what is called the E1 reaction. Note that in the elimination phase the nucleophile is acting as a Lewis base.

The mixture of E1 and S_N1 products depends primarily on the nucleophile's basicity. Strong Brønsted bases will be more effective at removing the proton than weak Brønsted bases. That's exactly what we mean by a strong Brønsted base—something that has a high affinity for a proton. Good nucleophiles will be less effective in the E1 reaction, but more effective in the S_N1 reaction (Fig. 7.74).

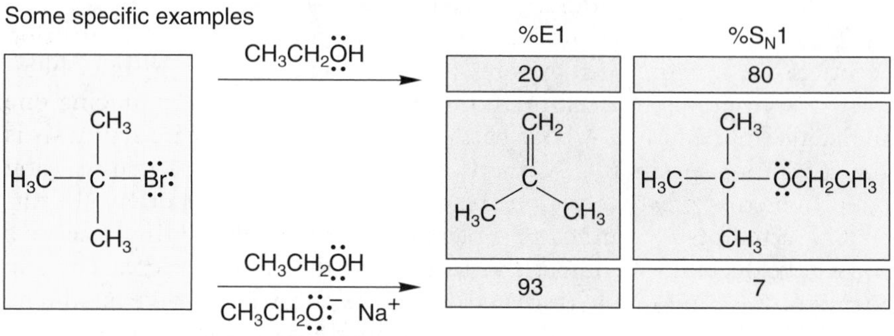

Some specific examples

Conditions	%E1	%S_N1
CH_3CH_2OH	20	80
CH_3CH_2OH, $CH_3CH_2O^-\ Na^+$	93	7

FIGURE 7.74 The E1 and S_N1 reactions compete. A strong base (ethoxide) favors the E1 reaction. For pure ethyl alcohol, the ratio of E1 product (isobutene) to S_N1 product (*tert*-butyl ethyl ether) is 0.25. In the presence of the much stronger base sodium ethoxide, this ratio increases to about 13.

7.10b Orientation of the E1 Reaction

In many E1 reactions, more than one hydrogen can be removed from the carbocationic intermediate to give an alkene. When there is a choice, the major product is the more substituted, more stable alkene (Chapter 4, p. 133). Of course, statistical effects will also influence the product mix, but the overwhelming factor is alkene stability. The more substituted alkene is favored strongly in the E1 reaction. For example, 2-bromo-2-methylbutane reacts in a solution of ethyl alcohol and water to give a mixture of 2-methyl-2-butene and 2-methyl-1-butene in which the more substitued alkene predominates by a factor of 4 (Fig. 7.75).

To see why the more substituted alkene is favored, look at the transition state for proton removal. In the transition state the carbon–hydrogen bond is lengthened, and the double bond of the alkene will be partially formed (Fig. 7.76).

PROBLEM 7.18 Draw careful arrow formalisms for the reactions of Figure 7.75.

PROBLEM 7.19 The reaction in Figure 7.75 also produces two products of the S_N1 reaction: one $C_5H_{12}O$ and the other $C_7H_{16}O$. What are they and how are they formed?

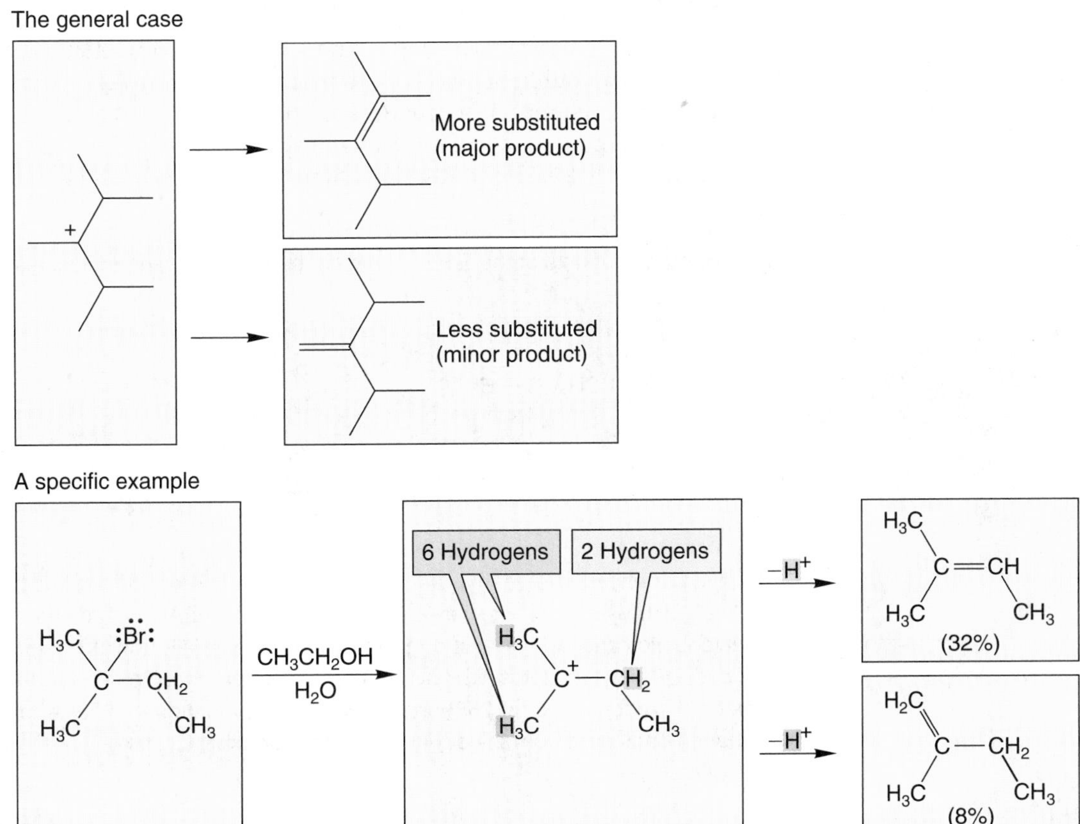

FIGURE 7.75 The more substituted (more stable) alkene is favored in the E1 reaction, even when statistical factors favor the less substituted isomer. This reaction is known as Saytzeff elimination.

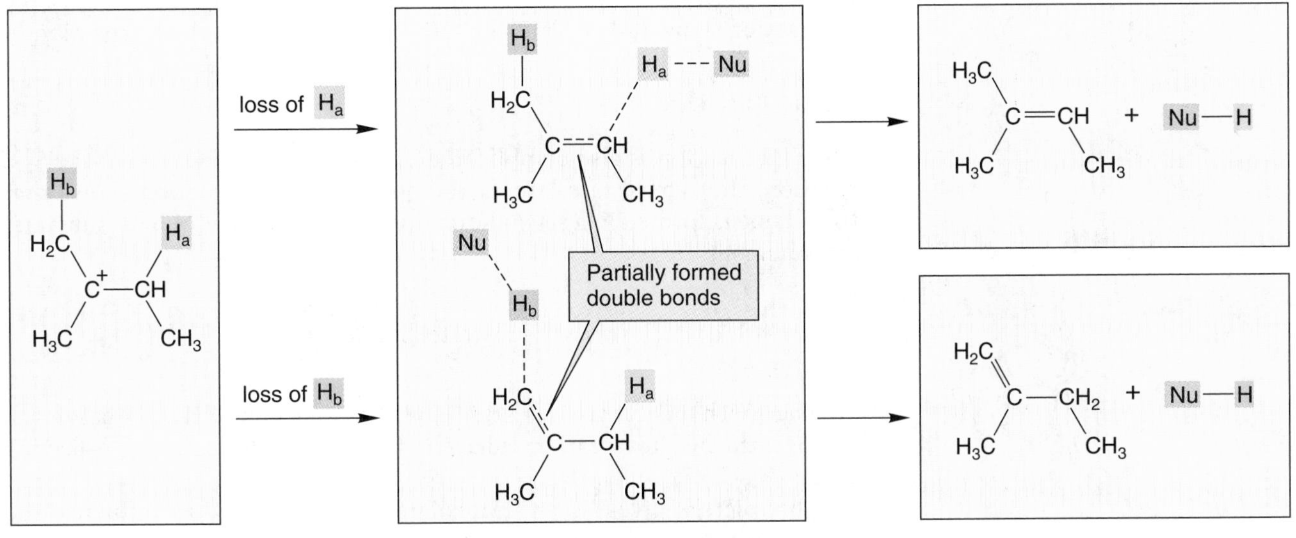

FIGURE 7.76 In the transition state, a double bond is partially formed. The increased stability of more substituted double bonds will be felt, and the transition state for formation of the more substituted alkene (loss of H_a) will be lower in energy than the transition state for formation of the less substituted alkene (loss of H_b).

The factors that operate to stabilize alkenes with fully formed double bonds will also be operative in the transition states in which there are partially formed double bonds. So, the more alkyl groups the better, and the E1 reaction preferentially produces the most substituted alkene possible. This phenomenon is called **Saytzeff elimination**, after Alexander M. Saytzeff (1841–1910), the Russian chemist who discovered it.*

PROBLEM **7.20** Predict the products of the E1/S_N1 reactions of the following molecules in water:

(a) $(CH_3CH_2)_3C—Br$

(b) [structure with Br]

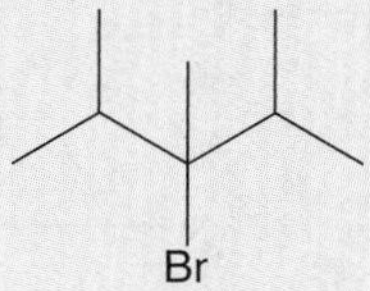

(c) [bicyclic structure with Br]

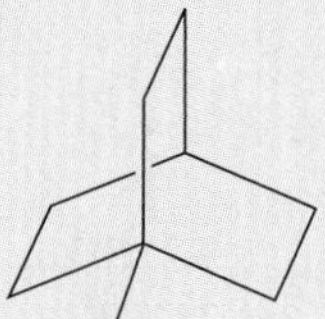

PROBLEM **7.21** It was noticed as early as 1862 that treatment of isobutyl halides, $(CH_3)_2CHCH_2–X$, with silver acetate, Ag^{+} ^{-}OAc (^{-}OAc is "acetate," a weak nucleophile, and Ag^+ is a strong Lewis acid, especially toward chloride), produced *tert*-butyl acetates, $(CH_3)_3C–OAc$, and isobutene. Remember the stability order of carbocations (p. 268) and provide an explanation for this behavior. *Hint*: Reason backward from the structure of the product.

7.11 THE BIMOLECULAR ELIMINATION REACTION: E2

As with the substitution reaction, there are two common mechanisms for the elimination reaction. We have just explored the first-order process, the E1 reaction. Now let's look at the bimolecular reaction, the E2 reaction.

7.11a Rate Law

The S_N2 reaction is also complicated by alkene formation. As for the S_N2 reaction, the typical rate law for the bimolecular **E2 reaction** shows that the alkene is formed in a second-order process, first order in both substrate and nucleophile.

$$v = k[\text{substrate}]\,[\text{nucleophile}]$$

The mechanism involves attack by the nucleophile (Lewis base) at a proton attached to a carbon adjacent to the carbon bearing the leaving group (Fig. 7.77).

This picture gives us an immediate clue as to how to influence the product mixture from competing E2 and S_N2 reactions. As in the E1–S_N1 competition (p. 273), strong Brønsted bases (high affinity for hydrogen 1*s* orbital) should favor the E2 and strong nucleophiles (high affinity for carbon 2*p* orbital) the S_N2 (Fig. 7.78).

*The name Saytzeff can be transliterated as Zaitzev, Saytzev, or Saytseff.

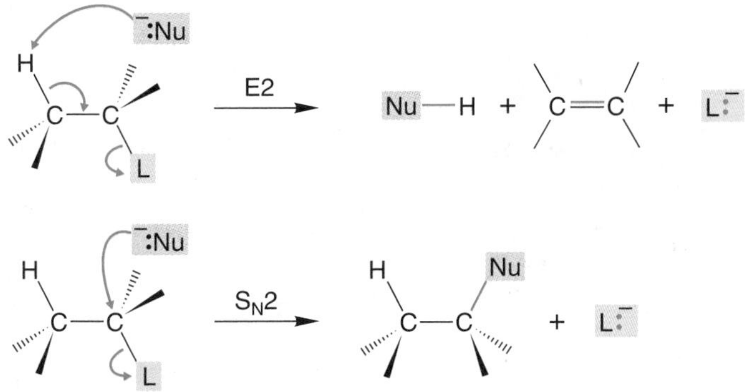

FIGURE 7.77 Arrow formalism pictures of the competing E2 and S_N2 reactions.

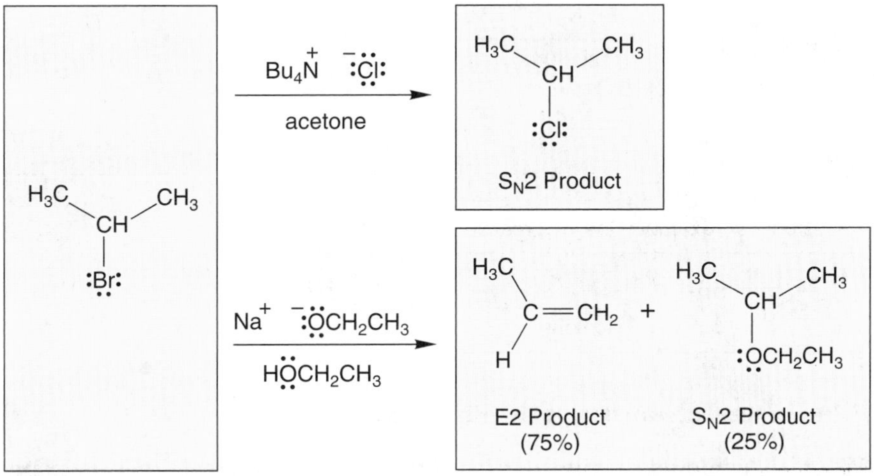

FIGURE 7.78 The E2 and S_N2 reactions compete. In the specific examples shown, the relatively good nucleophile, chloride, gives exclusively the S_N2 reaction, whereas the much more basic ethoxide ion gives mostly the product of the E2 reaction.

7.11b Effect of Substrate Branching on the E2–S_N2 Mix

The rate of the S_N2 reaction depends strongly on the structure of the substrate. The more highly branched the substrate, the more hindered is the obligatory approach from the rear of the leaving group and the slower the S_N2 reaction. This effect reaches a limit with tertiary substrates in which the S_N2 rate is effectively zero. The slower the S_N2 reaction, the more important is the competing E2. With a typical tertiary substrate such as *tert*-butyl bromide, the E2 reaction is the only effective reaction under E2–S_N2 conditions. The steric requirements for deprotonation are far less severe than for attack at the rear of the carbon–bromine bond of *tert*-butyl bromide (Fig. 7.79).

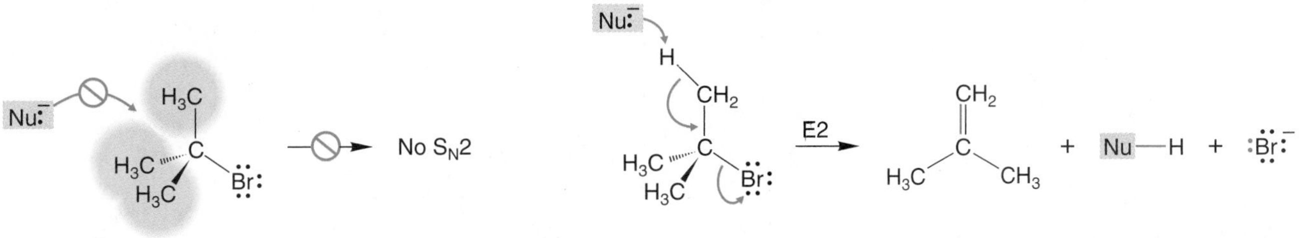

FIGURE 7.79 For tertiary substrates, the S_N2 reaction does not proceed, but the E2 reaction is relatively unhindered. Attack from the rear in S_N2 fashion is blocked by the three methyl groups. The E2 reaction is relatively easy.

With less hindered substrates the S_N2 reaction can compete much more effectively with the E2, and with primary substrates the S_N2 is dominant (Fig. 7.80).

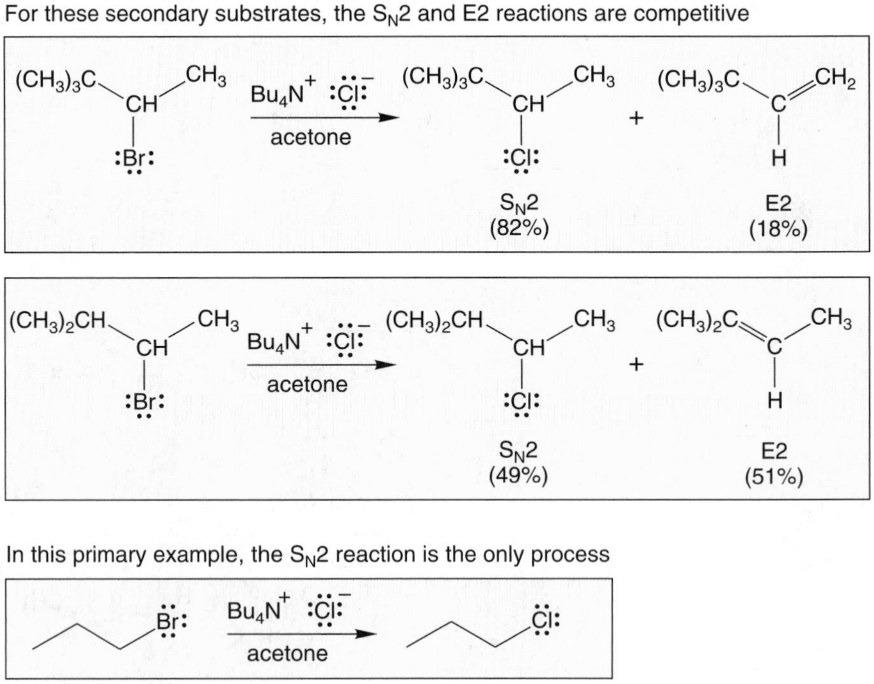

For these secondary substrates, the S_N2 and E2 reactions are competitive

$(CH_3)_3C{-}CH(Br){-}CH_3 \xrightarrow[\text{acetone}]{Bu_4N^+ \ :\ddot{Cl}:^-} (CH_3)_3C{-}CH(Cl){-}CH_3$ (S_N2, 82%) + $(CH_3)_3C{-}C(H){=}CH_2$ (E2, 18%)

$(CH_3)_2CH{-}CH(Br){-}CH_3 \xrightarrow[\text{acetone}]{Bu_4N^+ \ :\ddot{Cl}:^-} (CH_3)_2CH{-}CH(Cl){-}CH_3$ (S_N2, 49%) + $(CH_3)_2C{=}C(H){-}CH_3$ (E2, 51%)

In this primary example, the S_N2 reaction is the only process

1-bromobutane $\xrightarrow[\text{acetone}]{Bu_4N^+ \ :\ddot{Cl}:^-}$ 1-chlorobutane

FIGURE 7.80 For secondary substrates, the E2 and S_N2 reactions are competitive, and for primary substrates the S_N2 reaction dominates.

7.11c Stereochemistry of the E2 Reaction

When we explored the mechanism of the S_N1 and S_N2 reactions, an examination of the stereochemical outcome of the reactions was crucial to our development of their mechanisms. Stereochemical experiments are common in organic chemistry, and are one of the reasons we spent so much time and effort earlier on stereochemical analysis. In the E2 reaction, there are four possible extreme orientations. They can be classified by noting the dihedral angle between the two groups that are leaving, the hydrogen and the leaving group L (Fig. 7.81).

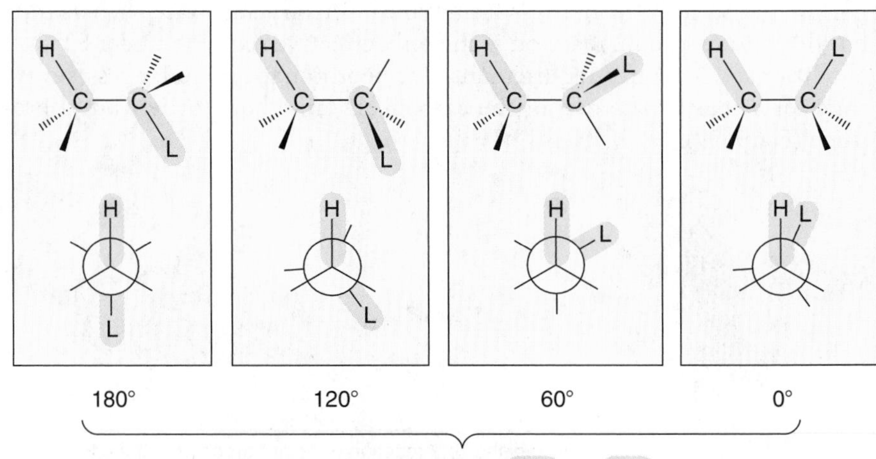

FIGURE 7.81 Four possible arrangements for the E2 reaction. The Newman projections emphasize the dihedral angle between the C–H and C–L bonds involved in the reaction.

As the C–H and C–L bonds lengthen in this reaction, the molecule flattens out as the $2p$ orbitals that will constitute the π system of the alkene are formed. *Remember* (Chapter 4, p. 123): In the product alkene the $2p$ orbitals are perfectly lined up, eclipsed, with a 0° dihedral angle (Fig. 7.82).

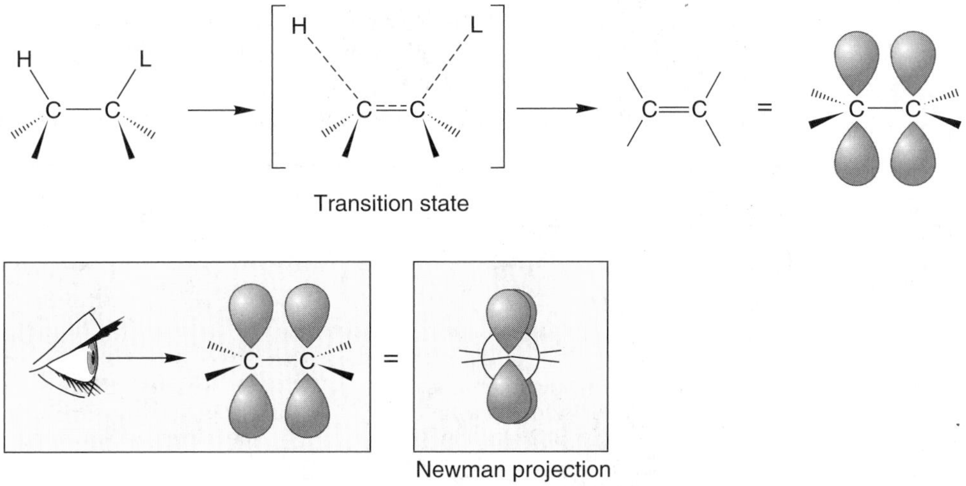

Transition state

Newman projection

FIGURE 7.82 The creation of a π bond as the E2 reaction occurs. In the E2 reaction, the C–H and C–L bonds lengthen as the π bond between the carbons is created. Note especially the requirement for orbital line-up as the π bond is generated. The dihedral angle between the two p orbitals in the product alkene is 0°.

Note that only the syn (0°) and anti (180°) arrangements of Figure 7.81 can lead smoothly to an alkene with perfectly overlapping $2p$ orbitals. The 120° and 60° forms cannot produce alkene without an additional rotation to line up the orbitals (Fig. 7.83).

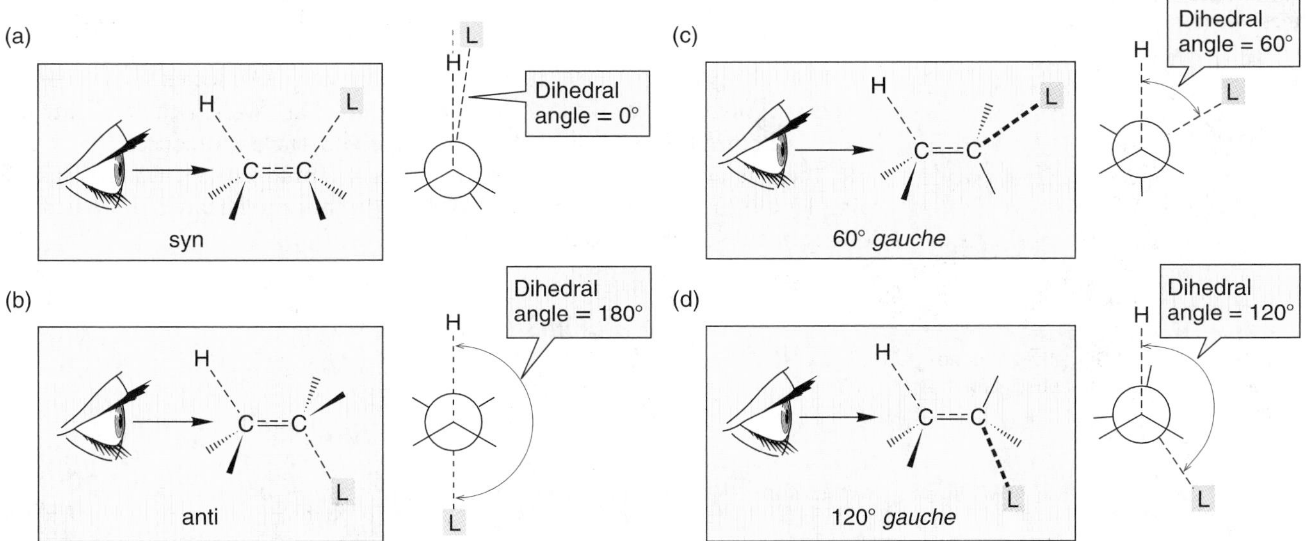

FIGURE 7.83 Four views of the transition state for the E2 reaction. In examples (a) and (b), where the dihedral angle is 0° or 180°, the π bond can be generated without an extra rotation. In examples (c) and (d), where the dihedral angle is 60° or 120°, an extra rotation must take place for the π bond to be made.

Thus we can anticipate that the 60° and 120° *gauche* arrangements for elimination should be disfavored. It is much more difficult to decide which of the remaining two possibilities, syn (0°) or anti (180°) E2 will be preferred in the E2 reaction.

The E2 is a kind of remote S_N2 reaction. The Lewis base (Nu:$^-$) attacks the hydrogen and the pair of electrons in the carbon–hydrogen bond does the displacing of the leaving group. If we keep this analogy in mind we can see that the anti (180°) elimination involves displacement from the rear, as in the S_N2 reaction, but syn (0°) elimination requires frontside displacement. We might therefore expect the anti E2 to be favored (Fig. 7.84).

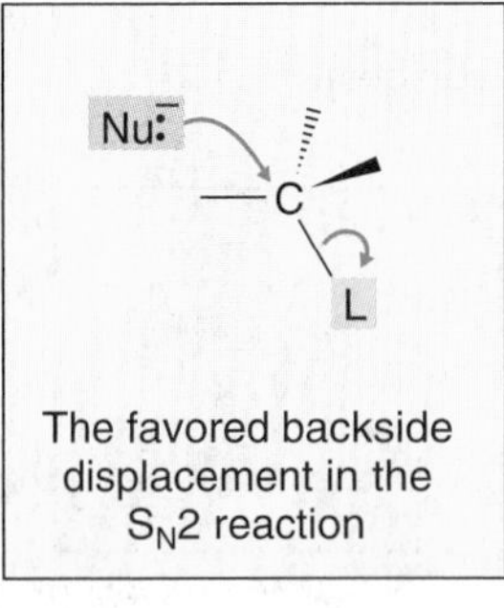

The favored backside displacement in the S_N2 reaction

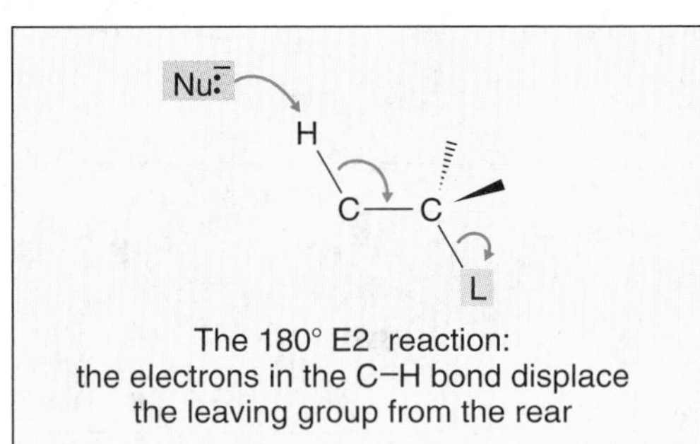

The 180° E2 reaction: the electrons in the C–H bond displace the leaving group from the rear

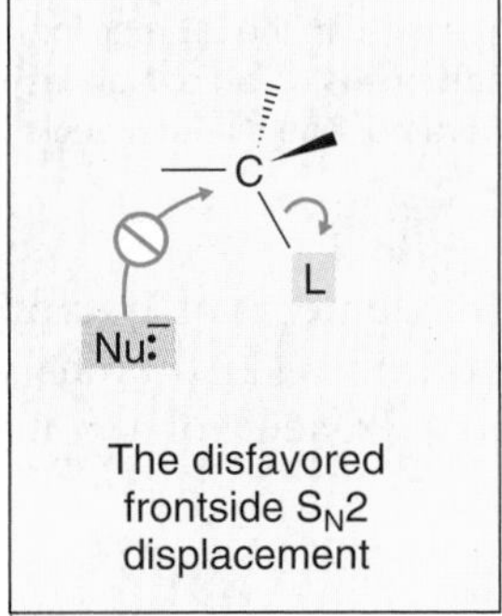

The disfavored frontside S_N2 displacement

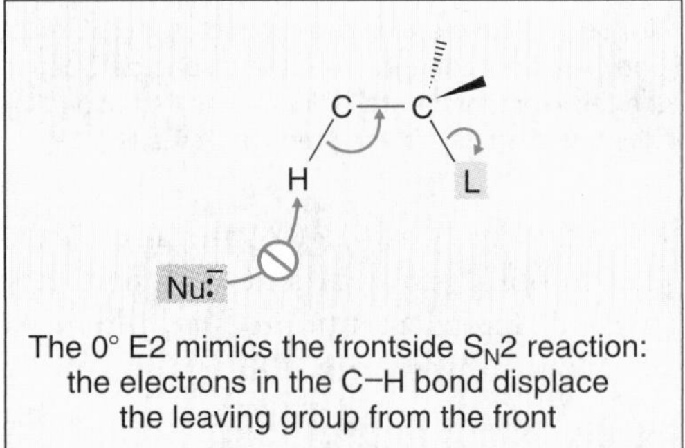

The 0° E2 mimics the frontside S_N2 reaction: the electrons in the C–H bond displace the leaving group from the front

FIGURE 7.84 The 0°, syn E2 reaction resembles a frontside S_N2 reaction, whereas the 180°, anti E2 reaction looks like the favored, backside S_N2 displacement reaction.

Now the problem is to find an experiment that will let us see this level of detail. As usual, we will use a stereochemical analysis, and this requires a certain complexity of substitution in the substrate molecule. For example, there is no use in examining the stereochemistry of the E2 reaction of a molecule as simple as 2-chloro-2,3-dimethylbutane because there is only one tetrasubstituted alkene possible (Fig. 7.85).

$$H\text{—}C(CH_3)_2\text{—}C(CH_3)_2\text{—}Cl \xrightarrow{Nu:^-} (H_3C)_2C\text{=}C(CH_3)_2 + Nu\text{—}H + :Cl:^-$$

FIGURE 7.85 2,3-Dimethyl-2-butene is the only possible tetrasubstituted alkene in this reaction.

PROBLEM 7.22 What other alkene could be formed in the reaction of Figure 7.85?

More highly substituted molecules can reveal the stereochemical preference, however. For example, consider the molecules of Figure 7.86. In diastereomer **A**, syn (0°) elimination will lead to a product in which the methyl groups are cis [this is, however, the (*E*) alkene], whereas anti, (180°) elimination will give the compound with trans methyls [the (*Z*) alkene]. In **B**, these preferences are reversed: syn elimination produces the

(*Z*) alkene, and anti elimination gives the (*E*) alkene. Many tests of this kind have been carried out, and it is certain that in acyclic molecules anti (180°) elimination is favored over the syn process (0°) by very large amounts, typically greater than 10^4 (Fig. 7.86).

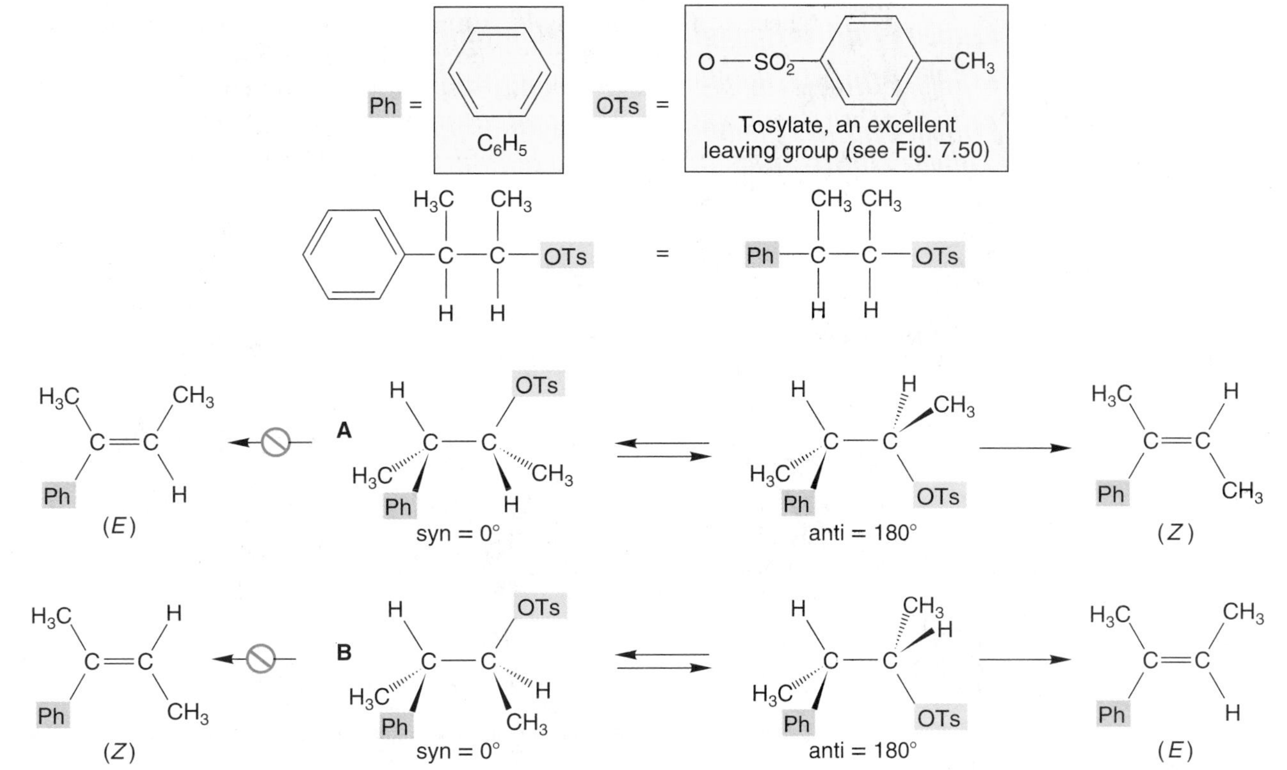

FIGURE 7.86 The possibilities for 0° and 180° E2 reactions for the two diastereomers shown. In each case, 180°, anti elimination wins out over 0° syn elimination by a large factor.

*PROBLEM 7.23

In chemistry, it is customary to report one's research results by submitting papers to the chemical literature. As a kind of quality control, these papers are sent to experts in the field for comment and review before publication. Imagine that you are such a referee looking over the experiment described in Figure 7.86. Do the data absolutely require the conclusion that anti (180°) eliminations are preferred to syn (0°)? Obviously, I think not. What do you think?

ANSWER

There is a problem. The analysis of Figure 7.86 shows a 180° elimination proceeding from a staggered, energy minimum form. By contrast, the 0°, syn elimination of Figure 7.86 proceeds from a fully eclipsed arrangement, which is a transition state, not an energy minimum (Chapter 3, p. 86; Fig. 7.30, p. 249). This comparison is scarcely fair! In order to make a real comparison of the syn and anti E2 processes, we will have to be more clever.

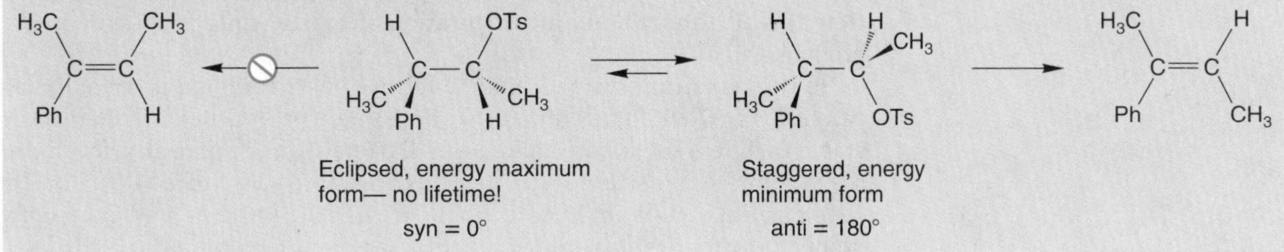

It is important to remember that orbital overlap is vital in this reaction, and a syn elimination in which the orbitals overlap well may compete effectively with an anti elimination in which the perfect 180° dihedral angle cannot be attained. For example, some polycyclic molecules give syn elimination. In the example shown in Figure 7.87, the 2-bicyclo-[2.2.1]heptyl or "2-norbornyl" system, the *ideal* trans (180°) elimination (loss of HOTs) is impossible, as only a 120° arrangement of the departing groups can be achieved. By contrast, the **syn elimination** (loss of DOTs) has a perfect 0° dihedral angle between the two departing groups. The result is that elimination is very slow and that the perfect syn elimination is preferred to the 120° **anti elimination** (Fig. 7.87).

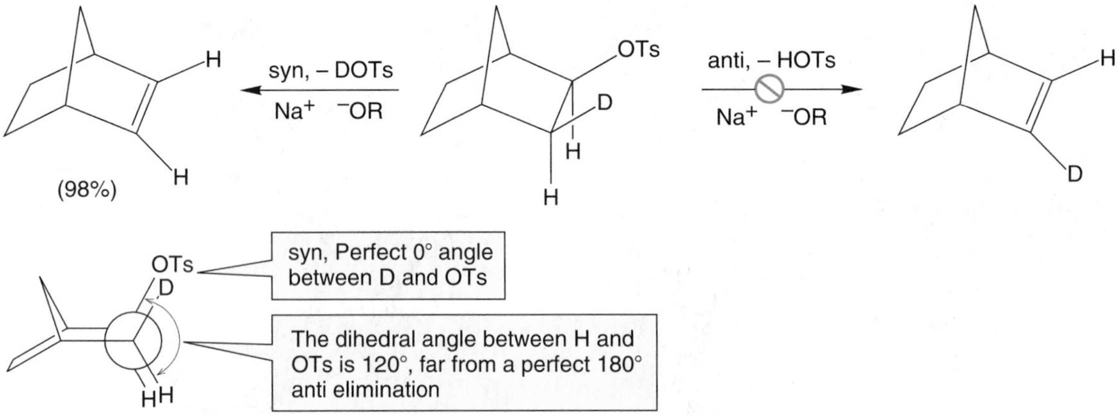

FIGURE 7.87 In some cases, syn elimination can be favored over an imperfect anti elimination. In this example, syn elimination can take place through an ideal, 0° arrangement, but the anti elimination must involve an unfavorable dihedral angle of 120°.

PROBLEM 7.24 Draw all the possible stereoisomers of 1,2,3,4,5,6-hexachlorocyclohexane. First, draw the molecules flat using some schematic device to indicate the stereochemistry of the chlorine atoms. Next, make three-dimensional drawings of these molecules.

PROBLEM 7.25 One of these isomers reacts much more slowly in the E2 reaction than all the others. Which one is it and why is it so slow to undergo an E2 reaction? *Hint*: Remember the steric requirements for the E2 reaction.

7.11d Orientation of the E2 Reaction

Generally, Saytzeff elimination is the rule. The most substituted, and therefore most stable alkene is the major product in this regioselective reaction (Fig. 7.88). **Regiochemistry** simply refers to the outcome of a reaction in which more than one orientation of substituents is possible in the product.

In the transition state for alkene formation, the π bond is partially developed (Fig. 7.89). The factors that make more substituted alkenes more stable than less substituted alkenes will operate on the partially developed alkenes in the transition states as well. The more substituted the partially developed alkene is, the more stable it will be, and this will inevitably produce the more substituted alkene as the major product (Fig. 7.90).

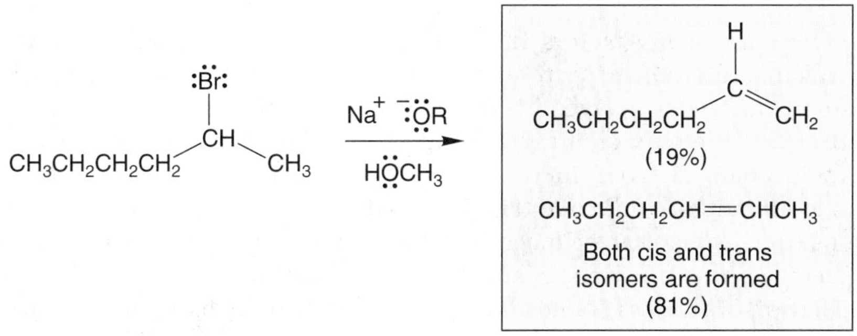

$CH_3CH_2CH_2CH_2CH(Br)CH_3 \xrightarrow[HOCH_3]{Na^+\ ^-OR}$ $CH_3CH_2CH_2CH_2C(H)=CH_2$ (19%) + $CH_3CH_2CH_2CH=CHCH_3$ Both cis and trans isomers are formed (81%)

FIGURE 7.88 In this E2 reaction, it is the more substituted alkene that is usually formed preferentially.

Nu:⁻
H
C—C
:Br:
Nu
H
C==C
:Br:
Transition state for the E2 reaction
Nu—H
C=C
:Br:⁻

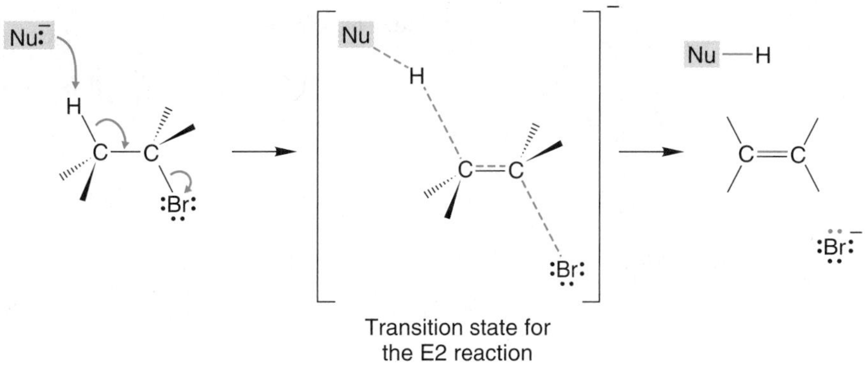

FIGURE 7.89 In the transition state for the E2 reaction, there is a partially formed double bond. More substituted partial double bonds will be more stable than less substituted partial double bonds, just as is the case in fully formed double bonds.

:Br:
Na⁺ ⁻:OCH₃
HOCH₃
:Br:
H
CH₃O:
In this transition state for the formation of the major product, there is a partially formed disubstituted double bond
Major product
Na⁺ ⁻:OCH₃
HOCH₃
:Br:
H
CH₃O:
In this transition state for the formation of the minor product, there is a partially formed monosubstituted double bond
Minor product

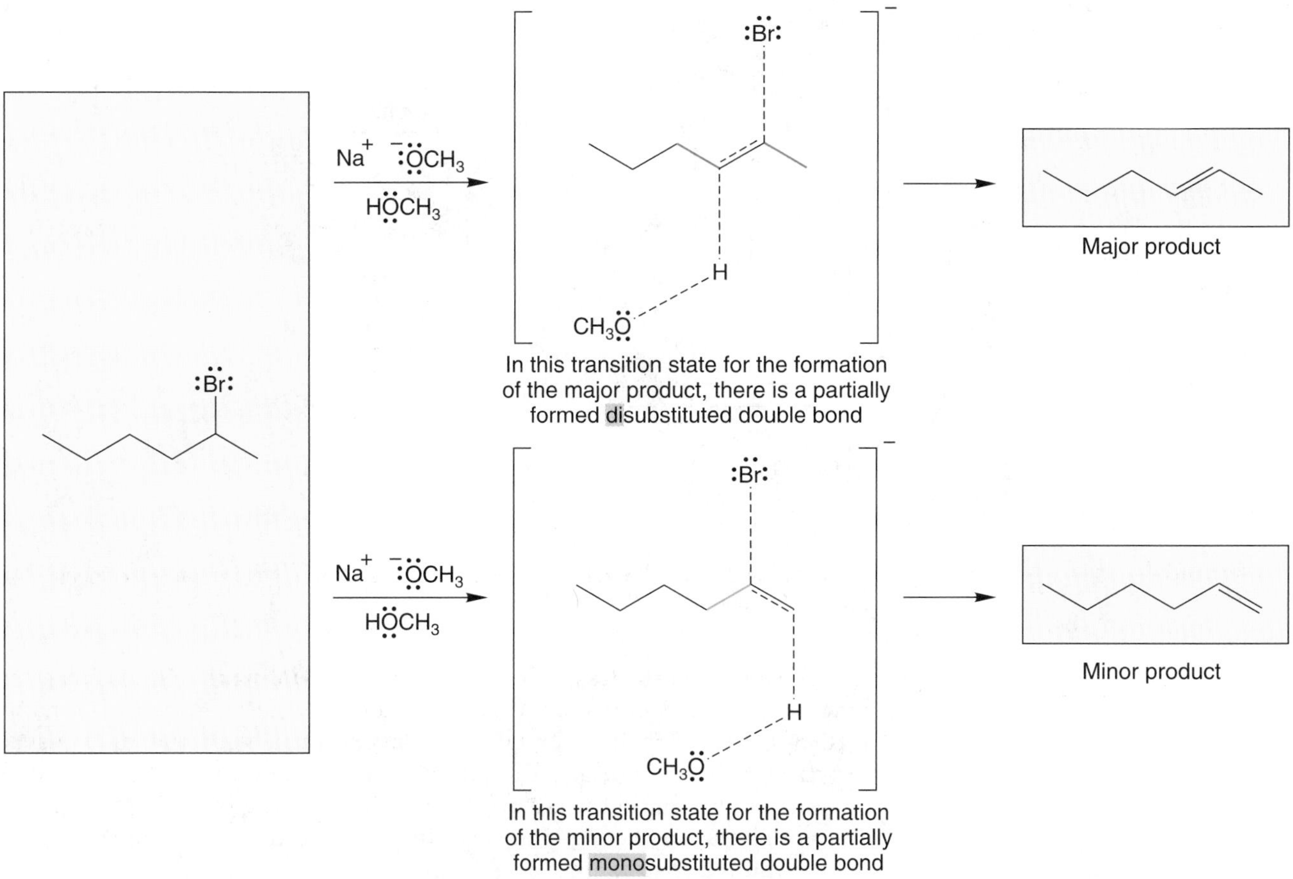

FIGURE 7.90 In the E2 reaction, the more stable transition state leads to the major product, the more substituted alkene.

There are many factors that influence the ratio of alkene products, however, and some are quite surprising at first. The size of the base makes a small difference. Larger bases favor the less stable alkene. Although there is only a very small change in the product distribution when the base is changed from methoxide to *tert*-butoxide in the reaction of 2-iodobutane with alkoxide (Fig. 7.91), when the base is made truly enormous these effects can be magnified. In order to form the more substituted alkene the base must penetrate further into the middle of the substrate. Formation of the less stable alkene requires attack by base further toward the periphery of the molecule and thus has lower steric requirements.

The general case

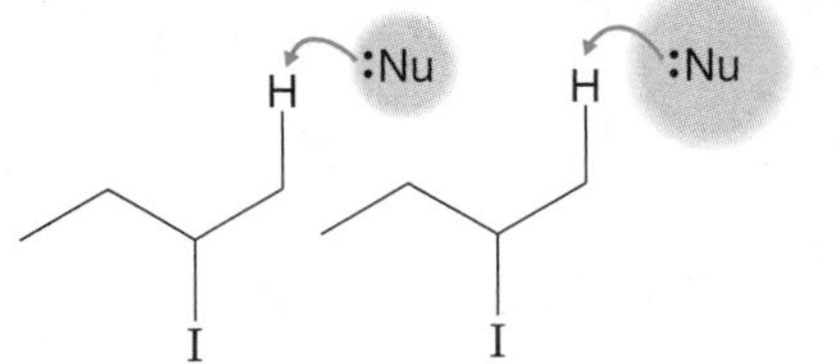

Size makes little difference in the approach to one of the hydrogens whose removal leads to the less substituted alkene

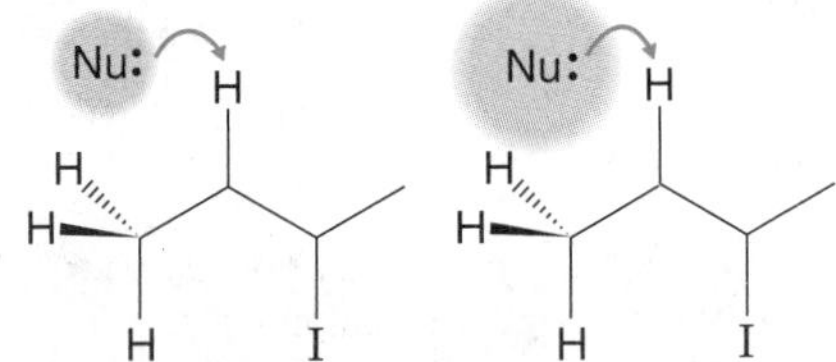

There are greater steric problems in approaching one of the hydrogens in the interior of the molecule; thus size of the base operates to favor the less substituted alkene (Hofmann elimination)

A specific example

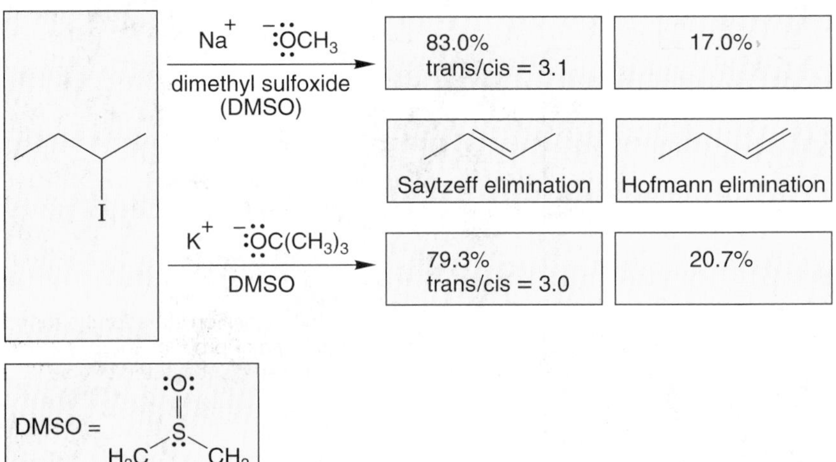

FIGURE 7.91 The effect of increasing base size is to favor the less stable "Hofmann" product somewhat.

Formation of the less stable alkene is called **Hofmann elimination** after the man who discovered this preference, August W. von Hofmann (1818–1892). This kind of elimination contrasts with the more usual Saytzeff process in which the more stable alkene is the major product. Hofmann elimination is favored by certain leaving groups. Typical examples of Hofmann leaving groups are fluoride and the 'onium ions, such as ammonium and sulfonium (Fig. 7.92).

L

Na^+ $^-$:ÖR

HÖCH$_3$

(cis and trans isomers)

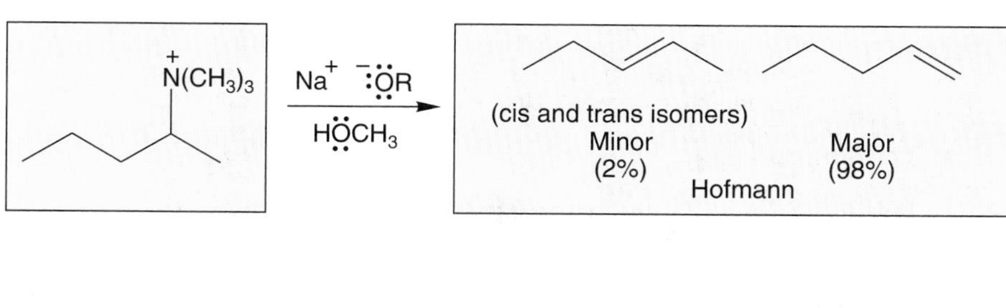

:Br:

Na^+ $^-$:ÖR

HÖCH$_3$

(cis and trans isomers)
Major
(69%)
Saytzeff
Minor
(31%)

L	2-Alkene (%)	1-Alkene (%)
I	81	19 Saytzeff preferred
Br	72	28 Saytzeff preferred
Cl	67	33 Saytzeff preferred
F	30	70 Hofmann preferred

FIGURE 7.92 The regioselectivity of the E2 reaction depends on the identity of the leaving group. Leaving groups such as fluoride, ammonium (R_3N^+), and sulfonium (R_2S^+) lead to predominant Hofmann elimination (the less stable isomer is the major product).

*PROBLEM 7.26

How can the nature of the leaving group determine the structure of the alkene? Consider the possible products of an E2 reaction from the compound in Figure 7.93. First, show what the possible products are and identify the different hydrogens that must be lost to produce them. Then, draw Newman projections for the arrangements leading to the two products. Remember the preference for 180° anti elimination in the E2 process. Finally, use the Newman projections to explain why a large leaving group might favor Hofmann elimination.

$\overset{+}{N}(CH_3)_3$

FIGURE 7.93

ANSWER

The possible products are

$^+N(CH_3)_3$

H_a H_b

loss of H_a → +

loss of H_b →

ANSWER (CONTINUED) Here are the Newman projections for loss of H_a and H_b.

In the 180° arrangement for an E2 elimination there is a bad steric interaction between ethyl and the large trialkylammonium ion

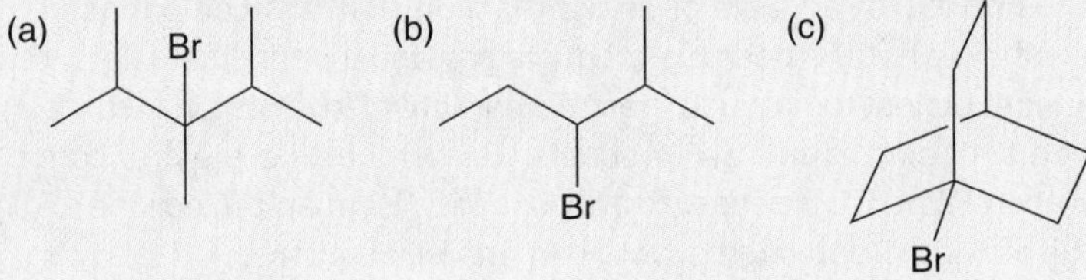

Here, the large group is flanked by two small hydrogens — a more stable arrangement

The larger the leaving group, the worse will be the steric interaction with the ethyl group in the 180° arrangement giving the more stable, more substituted product. By contrast, the arrangement leading to the less stable, less substituted product is relatively unhindered.

PROBLEM 7.27 Predict the products of reaction of the following compounds with hydroxide ion. Be careful with the molecule in (c). The answer here is tricky.

7.12 WHAT CAN WE DO WITH THESE REACTIONS? SYNTHETIC UTILITY

Your ability to make molecules has just increased enormously. Starting from rather easily accessible compounds, the alkyl halides, and inorganic materials, myriad other structural types are suddenly available. Figure 7.94 summarizes the situation in a general way. Not all reactions will work for all R–L molecules and elimination reactions will surely also occur.

The figure adds several reactions to the ones we have discussed specifically. For example, it sneaks in a brand new synthesis of substituted acetylenes by using the acetylide ion as a nucleophile (usually only effective with primary R–L compounds). We mentioned the acidity of acetylenes at the very end of Chapter 4 (p. 149) and even warned that more was coming about this reaction. Monosubstituted (terminal) acetylenes are decent acids (pK_a ~ 25), which means that treatment of a terminal acetylene with a strong base can result in removal of a proton and formation of the acetylide. A favorite base for doing this is the amide ion, $^{-}NH_2$ (Fig. 7.95). There are some other new reactions in Figure 7.94. Go over each of them carefully and be sure you see how each works.

Keeping track of synthetic methods can be difficult, especially when learning from a book mostly devoted to mechanism. You are right at the beginning of synthetic chemistry now, and here is a hint on how to keep track of it. Make a 4 × 6-in. file card for every structural type and then keep a short list of the available synthetic reactions on the card. To help in ordering things, and to keep as far as possible from the dreaded mindless memorization, you might make a mechanistic notation for each reaction as well. The reaction of acetylides with alkyl halides to make new acetylenes might get an "S_N2 with primary substrates" beside it, for example. You can start your catalog with the reactions in Figure 7.94. Be sure to keep up to date; synthetic ability is the kind of thing that sneaks up on you, like whatever it was that worried Mr. Paige.

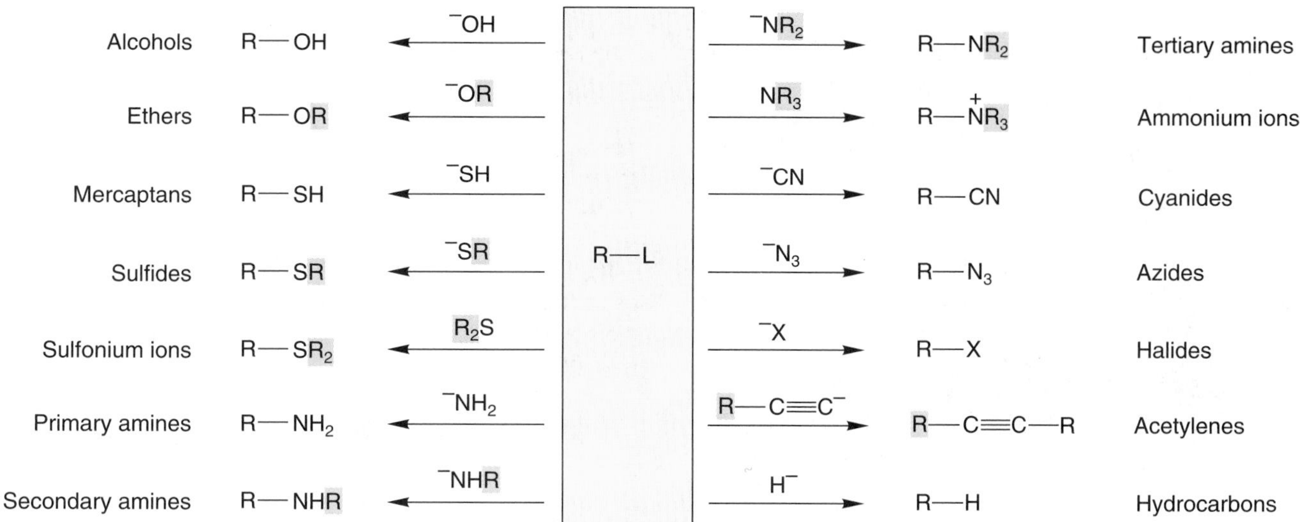

FIGURE 7.94 The synthetic potential of the substitution reaction.

Figure 7.94 also fails to show the potential complications we have talked about. *Remember*: There is always a competition between the substitution reactions, S_N1 and S_N2, and the E1 and E2 reactions. Under some circumstances the major products of the reactions summarized in Figure 7.94 will be alkenes. Your file card catalog should include the synthesis of alkenes through the E1 and E2 reactions. The identity of the leaving group, so conveniently avoided by the L in the figure, makes a difference, and the structure of R, also deliberately vague in the figure, is vital. In

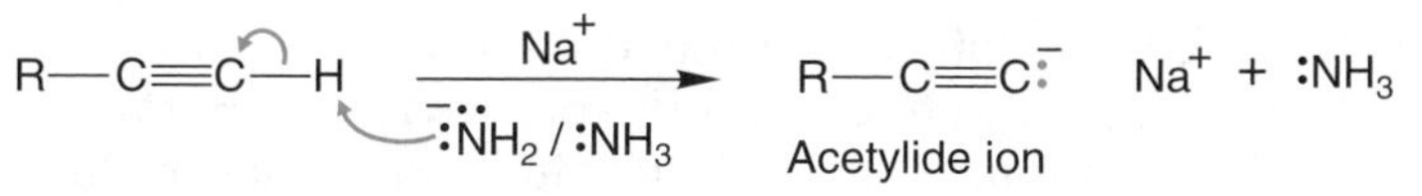

$$R{-}C{\equiv}C{-}H \xrightarrow[\,:\ddot{N}H_2^- \,/\, :NH_3]{Na^+} R{-}C{\equiv}C{:}^- \quad Na^+ + :NH_3$$

Acetylide ion

FIGURE 7.95 Formation of an acetylide from a terminal alkyne.

PROBLEM 7.28 Use the pK_a data in Table 7.3 (p. 238) to decide if sodium amide would be effective at producing vinyl anions.

order to design a successful synthetic transformation, you have to think about all the components of the reaction. A sample file card is shown in Figure 7.96.

Alkyl Bromides		
Reaction	Mechanism	Notes
R—L + Br^-	S_N2	Needs good leaving group
		Primary or secondary R
		E2 competes
R—L + Br^-	S_N1	*tert* R group, polar solvent
		E1, E2 compete

FIGURE 7.96 A sample file card for alkyl bromides.

PROBLEM 7.29 Predict the products of the following reactions:

$HO^- + CH_3Br \rightarrow$
$HO^- + (CH_3)_3CBr \rightarrow$
$H_2O + (CH_3)_3CBr \rightarrow$

PROBLEM 7.30 Devise syntheses for the following molecules. You may use any inorganic reagent, organic iodides containing up to four carbons, and the special reagents potassium cyanide and propyne.

(a) N_3

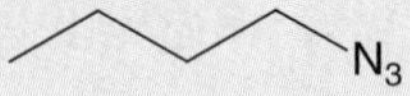

(b)

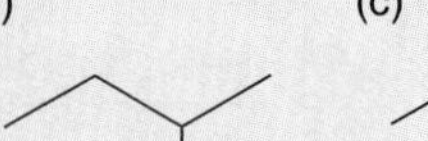

(c)

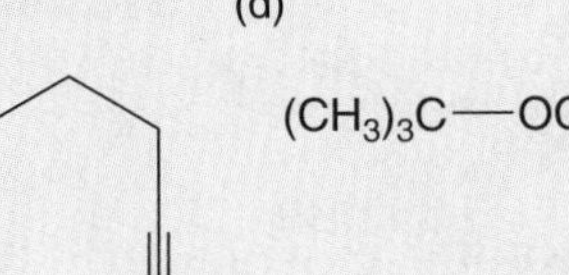

(d) $(CH_3)_3C—OCH_3$

(e) $(CH_3)_3C—N_3$

PROBLEM 7.31 Indicate which member of the following pairs of compounds will react faster in (1) the S_N1 reaction and (2) the S_N2 reaction. Explain your reasoning.

(a) 1-Bromobutane and 2-bromobutane
(b) 1-Chloropentane and cyclopentyl chloride
(c) 1-Chloropropane and 1-iodopropane
(d) *tert*-Butyl iodide and isopropyl iodide
(e) *tert*-Butyl iodide and *tert*-butyl chloride

7.13 SOMETHING MORE: THE E1cB REACTION

When we discussed the Hofmann elimination reaction, a steric effect was used (Problem 7.26, p. 285) to explain the preference for Hofmann elimination of alkyl ammonium and sulfonium ions. But what of fluoride? A fluorine is just about the same size as hydrogen, so we can't make much of a case for steric factors here. Before we can talk sensibly about fluorine we have to mention another kind of loss of H—L, the **E1cB reaction** (elimination, unimolecular, conjugate base). This reaction is in essence the opposite of the E1 reaction. It is a reaction in which the hydrogen is lost first to generate an anion. Subsequent ejection of the leaving group as an anion generates the alkene (Fig. 7.97).

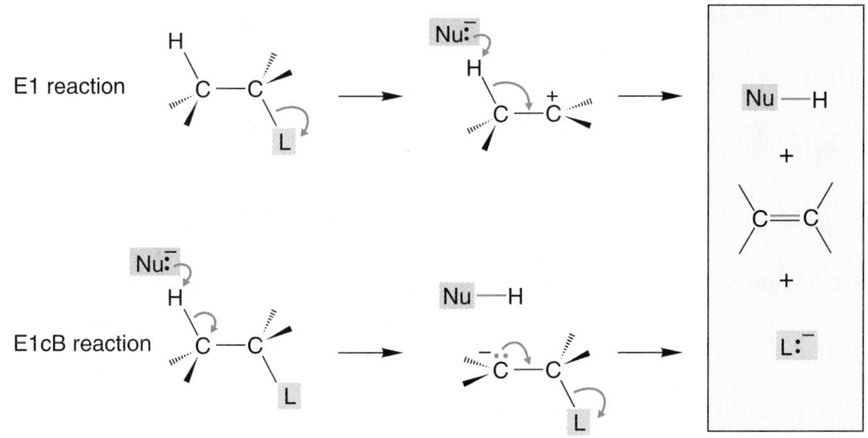

E1 reaction

E1cB reaction

FIGURE 7.97 The E1 and E1cB reactions contrasted. In the E1 reaction, the slow step is ionization to give the carbocation. In the E1cB reaction, a proton is first removed to give an anion that subsequently eliminates the leaving group.

This reaction is relatively uncommon , but some examples are known. The requirements for the E1cB reaction are a poor leaving group and rather easily lost proton.

Design a good E1cB reaction. *Hint*: What kind of leaving group do you need? What kind of carbon–hydrogen bond?

PROBLEM 7.32

The E2 reaction of alkyl fluorides does not go by this mechanism but its transition state *is* "E1cB-like." In the transition state for any E2 reaction, the bonds to hydrogen and the leaving group are both breaking (Fig. 7.98).

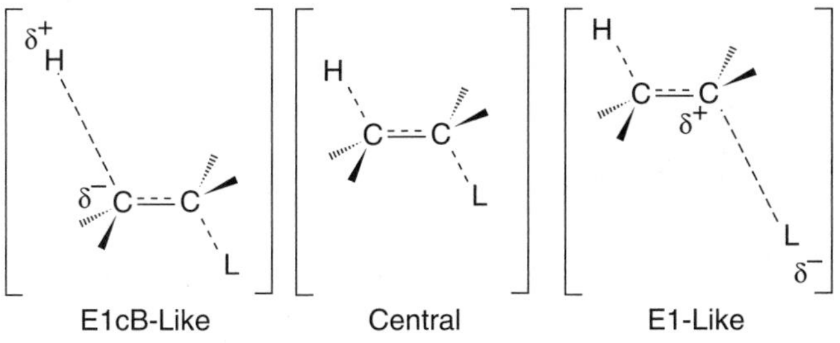

E1cB-Like   Central   E1-Like

FIGURE 7.98 In the E2 reaction, the transition state may or may not have the bonds to H and L breaking roughly simultaneously. If breaking of the C—H bond is further advanced in the transition state than is breaking of the C—L bond, the transition state will resemble the E1cB reaction. If the C—L bond is more broken in the transition state than is C—H, the transition state is E1-like.

In an E1cB-like transition state, the carbon adjacent to the carbon bearing the leaving group becomes partially negatively charged (Fig. 7.98). The highly polar carbon–fluorine bond will place a partial positive charge on the carbon to which the fluorine is attached. For an E1cB-like transition state, this is a favorable situation (Fig. 7.99).

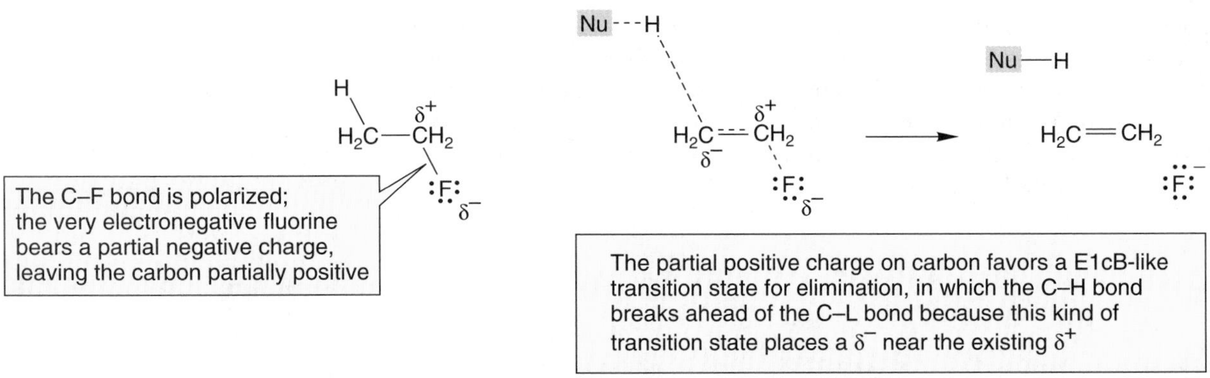

$Nu \cdots H$

H_2C—CH_2 δ^+

:F: δ^-

The C–F bond is polarized; the very electronegative fluorine bears a partial negative charge, leaving the carbon partially positive

H_2C δ^- === CH_2 δ^+ :F: δ^- ⟶ Nu—H $H_2C{=}CH_2$:F:$^-$

The partial positive charge on carbon favors a E1cB-like transition state for elimination, in which the C–H bond breaks ahead of the C–L bond because this kind of transition state places a δ^- near the existing δ^+

FIGURE 7.99 An E1cB-like transition state is favored by strongly electron-withdrawing groups such as fluorine.

In the example shown in Figure 7.100, the transition state for Hofmann elimination has a partial negative charge on a primary carbon, whereas the transition state for Saytzeff elimination has a partial negative charge on a secondary carbon (Fig. 7.100).

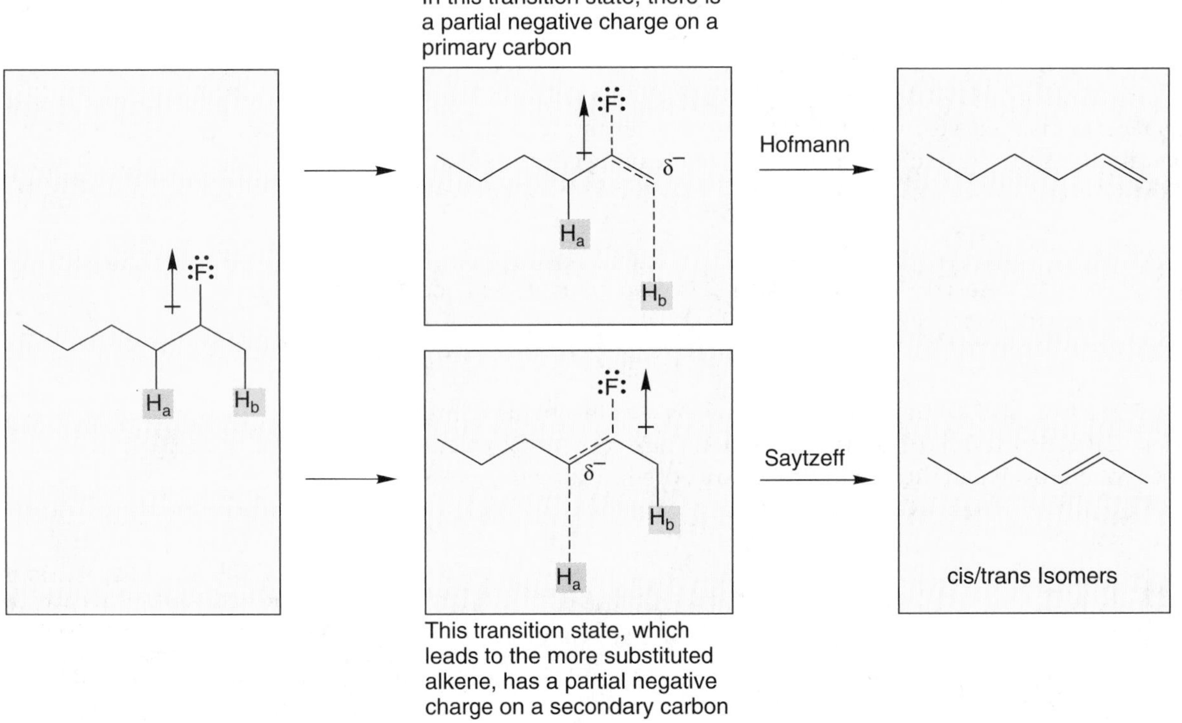
In this transition state, there is a partial negative charge on a primary carbon
:F:
H_a
H_b
δ^-
Hofmann
Saytzeff
cis/trans Isomers
This transition state, which leads to the more substituted alkene, has a partial negative charge on a secondary carbon

FIGURE 7.100 In the transition state for Hofmann elimination, a partial primary carbanion is formed. In the transition state for Saytzeff elimination, it is a partial secondary carbanion.

In contrast to carbocations and radicals, less substituted carbanions are more stable than more substituted carbanions. Methyl and primary carbanions are more stable than secondary carbanions, and secondary are more stable than tertiary (Fig. 7.101).

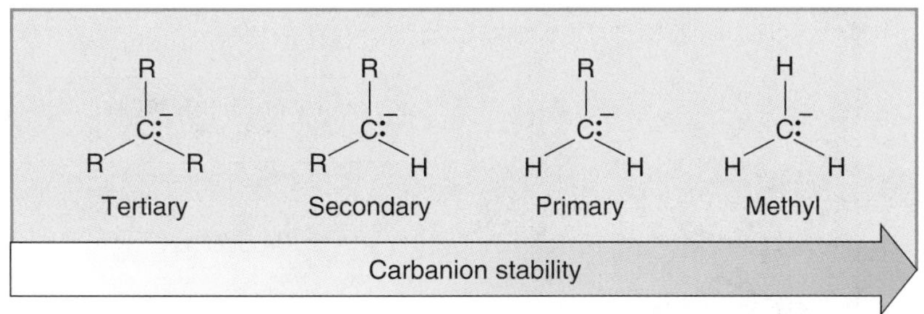

FIGURE **7.101** The order of carbanion stability is the opposite of that for carbocations and radicals partly because of the greater ease of solvation for the smaller carbanions. Thus methyl carbanions are more stable than primary carbanions, primary carbanions are more stable than secondary carbanions, and secondary carbanions are more stable than tertiary carbanions.

So, the more favorable transition state is the one with the developing negative charge on the primary carbon, the one leading to the less stable alkene (Fig. 7.100).

7.14 SUMMARY

NEW CONCEPTS

The most important new idea in this chapter is the expansion of the concept of Brønsted acidity and basicity to embrace the species known as Lewis acids and Lewis bases. The relatively narrow notion that acids and bases compete for a proton (Brønsted acids and bases) is extended to define all species with reactive pairs of electrons as Lewis bases, and all molecules that will react with Lewis bases as Lewis acids. The key word here is "reactive." All bonds are reactive under some circumstances, and therefore all molecules can act as Lewis bases at times.

Another way to describe the reaction of Lewis bases with Lewis acids is to use the graphic device showing the stabilizing interaction of a filled orbital (Lewis base) with an empty orbital (Lewis acid) (Fig. 7.12). This idea can be generalized if you recall that the strength of orbital interaction depends on the energy difference between the orbitals involved. The closer the orbitals are in energy the stronger their interaction and the greater the resulting stabilization. Therefore the strongest interactions between orbitals are likely to be between the highest occupied molecular orbital (HOMO) of one molecule (**A**, a Lewis base) and the lowest unoccupied molecular orbital (LUMO) of the other (**B**, a Lewis acid) (Fig. 7.14). This notion is used in this chapter to explain the overwhelming preference for backside attack in the S_N2 reaction.

Stereochemical experiments are used here for the first time to probe the details of reaction mechanisms. The observation of inversion of configuration in the S_N2 reaction shows that the incoming nucleophile must approach the leaving group from the rear. Similarly, if the stereochemistry of the E2 reaction is monitored, it can be shown that there is a strong preference for the anti (180°) E2 reaction in acyclic molecules.

The curved arrow formalism is used extensively in this chapter. This bookkeeping device is extremely useful in keeping track of the pairs of electrons involved in the making and breaking of bonds in a reaction. It is important not to confuse this formalism with a reaction mechanism.

The terms "transition state" (the structure representing the high-energy point separating starting material and products) and "activation energy" (the height of the transition state above the starting material) appear several times in this chapter. Chapter 8 will continue a detailed discussion of these terms (Fig. 7.102).

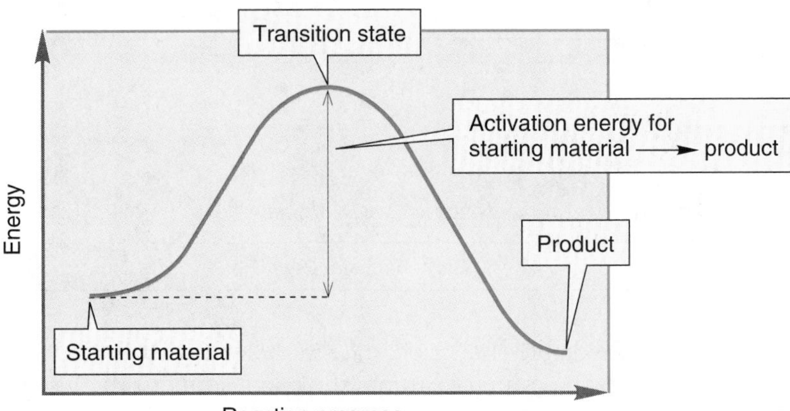

FIGURE 7.102 A typical Energy versus Reaction progress diagram.

REACTIONS, MECHANISMS, AND TOOLS

Four exceptionally important prototypal reactions are introduced in this chapter: the S_N1, S_N2, E1, and E2 reactions. The E1 and S_N1 reactions involve the same carbocationic intermediate. Addition of a nucleophile completes the S_N1 reaction; removal of a proton finishes the E1 (Fig. 7.73).

The S_N2 reaction and the E2 reaction are also competitive processes. In the S_N2 reaction, a nucleophile attacks the rear of a carbon-leaving group bond. In the E2 reaction, it is the adjacent hydrogen that is removed along with the leaving group (Fig. 7.77).

For primary substrates, it is the S_N2 and E2 reactions that compete. For tertiary compounds in polar solvents with weak bases, it is the S_N1 and E1 reactions that are important. If strong bases are employed in nonionizing solvents, it will be the E2 reactions that are most important. For secondary substrates all four reactions compete.

It is absolutely vital to have these four building block reaction mechanisms under complete control.

SYNTHESES

The E1 and E2 reactions allow the synthesis of alkenes from starting alkyl halides (Fig. 7.103).

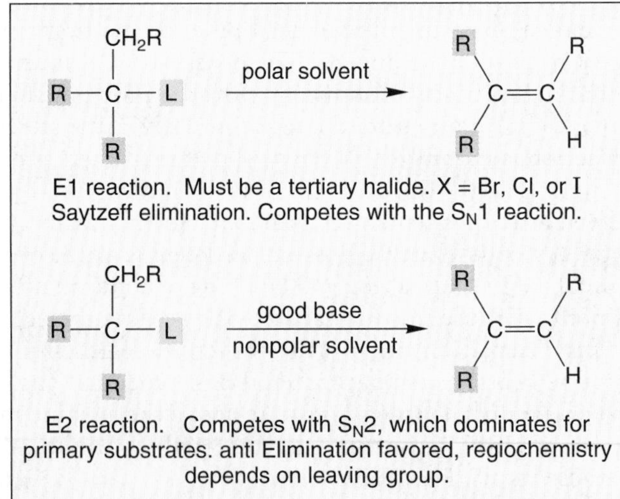

FIGURE 7.103 Synthetically useful reactions from chapter 7.

2. Substituted alkanes

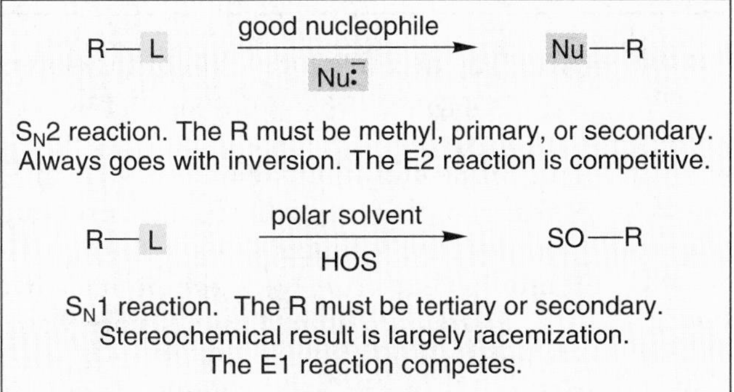

R—L
good nucleophile
Nu:
Nu—R
S_N2 reaction. The R must be methyl, primary, or secondary. Always goes with inversion. The E2 reaction is competitive.
R—L
polar solvent
HOS
SO—R
S_N1 reaction. The R must be tertiary or secondary. Stereochemical result is largely racemization. The E1 reaction competes.

3. Alkyl halides

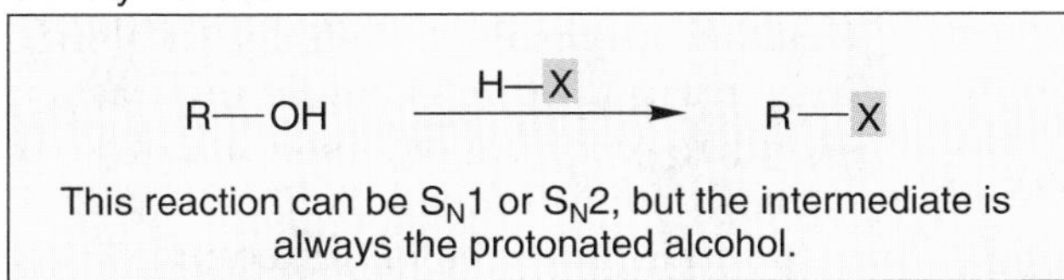

4. Alkyl sulfonates

$$R{-}OH \xrightarrow{Cl{-}SO_2{-}R} R{-}O{-}SO_2{-}R$$

These sulfonates are good substrates for further conversions using S_N1 and S_N2 reactions.

FIGURE 7.103 (CONTINUED)

The S_N1 and S_N2 reactions allow the conversion of alkyl halides into a host of new structures. These were summarized earlier in Figure 7.94. Remember that the S_N2 reaction works only for primary and secondary halides and the S_N1 reaction works only for tertiary and secondary molecules.

Alcohols can also be used as starting materials in these reactions, but the OH group is a very poor leaving group and must be transformed by protonation into water, or through other reactions into some other good leaving group.

COMMON ERRORS

Students of organic chemistry long for certainty. They have not yet learned to love the messiness of the subject, or at least to accommodate to it. That the answer to a question, What happens if we treat A with B? might be, All sorts of things, can be profoundly unsettling. In this chapter, we meet four basic, building blocks: the S_N1, S_N2, E1, and E2 reactions. It is a sad fact that in most instances they compete with each other. We can usually adjust reaction conditions to favor the reaction we desire, but it will be a rare occasion when we achieve perfect selectivity.

There is much detail in this chapter, but that is not usually a problem for most students. Good nucleophiles can be distinguished from good bases, good leaving groups can be identified, and so on. What is hard is to learn to adjust reaction conditions (reagents, temperature, and solvent) so as to achieve the desired selectivity; to favor the product of just one of these four reactions. That's hard in practice as well as in answering a question on an exam or problem set. To a certain extent we all have to learn to live with this problem!

7.15 KEY TERMS

Activation energy The height of the transition state above the starting material. It is this amount of energy that is required for a molecule of starting material to be transformed into product.

Alkyl halides Compounds of the formula $C_nH_{2n+1}X$, where X = F, Cl, Br, or I.

anti Elimination An elimination reaction in which the dihedral angle between the breaking bonds, usually C—H and C—L, is 180°.

Arrow formalism A mapping device for chemical reactions. The electron pairs (lone pairs or bond pairs) are "pushed" using curved arrows that show the bonds that are forming and breaking in the reaction.

Brønsted acid A donor of a proton.

Brønsted base An acceptor of a proton.

Conjugate acid Some molecule plus a proton. Conjugate acids and bases are related by the gain and loss of a proton.

Conjugate base Some molecule less a proton. Conjugate acids and bases are related by the gain and loss of a proton.

E1cB Reaction An elimination reaction in which the first step is loss of a proton to give an anion. The anion then internally displaces the leaving group in a second step.

E1 Reaction The unimolecular elimination reaction. The ionization of the starting material is followed by the loss of a proton to base.

E2 Reaction The bimolecular elimination reaction. The proton and leaving group are lost in a single, base-induced step.

Hofmann elimination The formation of the less substituted alkene in an elimination reaction.

HOMO The highest occupied molecular orbital.

Inversion of stereochemistry The change in the handedness of a molecule during certain reactions. Generally, inversion of stereochemistry means that an (*R*) starting material would be transformed into an (*S*) product.

Leaving group The departing group in a substitution or elimination reaction.

Lewis acid Any species that reacts with a Lewis base.

Lewis base Any species with a reactive pair of electrons.

LUMO The lowest unoccupied molecular orbital

Nucleophile Generally, a Lewis base. A strong nucleophile has a high affinity for a carbon 2*p* orbital.

pK_a The negative logarithm of the acidity constant. A strong acid has a low pK_a, a weak acid has a high pK_a.

Product-determining step The step in a multistep reaction that determines the structure or structures of the products.

Protic solvent A solvent containing a hydrogen easily lost as a proton. Examples are water and most alcohols.

Rate-determining step The step in a reaction mechanism with the highest activation energy.

Reaction mechanism Loosely speaking, How does the reaction occur? How do the reactants come together? Are there any intermediates? What do the transition states look like? More precisely, a determination in terms of structure and energy of the stable molecules, reaction intermediates, and transition states involved in the reaction, along with a consideration of how the energy changes as the reaction progresses.

Regiochemistry The orientation of a reaction taking place on an unsymmetrical substrate.

Retention of stereochemistry The preservation of the handedness of a molecule in a reaction. Generally, retention of stereochemistry means that an (*R*) starting material would be transformed into an (*R*) product.

Saytzeff elimination Formation of the more substituted alkene in an elimination reaction.

S_N1 Reaction Substitution, nucleophilic, unimolecular. An initial ionization is followed by attack of the nucleophilic solvent.

S_N2 Reaction Substitution, nucleophilic, bimolecular. In this reaction, the nucleophile dispaces the leaving group by attack from the rear. The stereochemistry of the starting material is inverted.

Solvolysis reaction In such reactions, the solvent acts as the nucleophile.

Substitution reaction The replacement of a leaving group in R—L by a nucleophile, $Nu:^-$. Typically,

$$Nu:^- + R{-}L \longrightarrow Nu{-}R + L:^-$$

syn Elimination An elimination reaction in which the dihedral angle between the breaking bonds, usually C—H and C—L, is 0°.

Transition state The high point in energy between starting material and product. The transition state is an energy maximum and not an isolable compound.

7.16 ADDITIONAL PROBLEMS

PROBLEM 7.33 The following molecules can form conjugate bases by loss of a proton from more than one position (arrows). Start by drawing a good Lewis structure (electron dots are deliberately left out) and then choose which proton will be lost more easily. Explain your choice.

(a) CH_3CH_2—CN

(b) HO—CR_2—H

(c) CH_3CH_2—C(=O)—OR

PROBLEM 7.34 The following molecules can accept a proton to form their conjugate acids by addition of a proton at more than one possible place (arrows). Chose the position at which protonation will be preferred and explain your choice. Once again, it makes sense to start by drawing a good Lewis structure for the molecule in question.

(a) $(CH_3)_3C$—C(=O)—OH

(b) $(H_3C)_2C$=O

(c) $(H_3C)_2C$=C(CH_3)H

PROBLEM 7.35 Write arrow formalisms for the following reactions. Charges and electron dots are deliberately left out of the reagents and products, so you must start by putting them in.

(a) H_3C—C(=O)—H_2C—H $\xrightarrow{^-OH}$ H_3C—C(=O)—H_2C + H—OH

(b) $(H_3C)_2C$=O $\xrightarrow{^-OH}$ HO—C(H_3C)(H_3C)—O

(c) $(CH_3)_3C$—C(=O)—OH $\xrightarrow{\overset{+}{H}OH_2}$ $(CH_3)_3C$—C(=O—H)—OH + H—OH

PROBLEM 7.36 Devise S_N2 reactions that would lead to the following products:

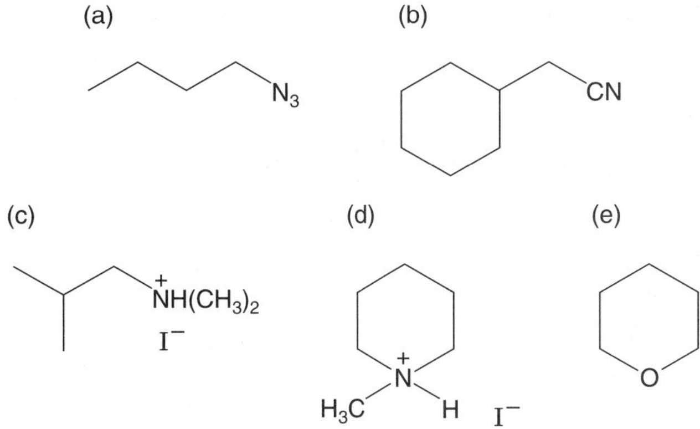

PROBLEM 7.37 Predict the products of the following S_N2 reactions:

(a) pyridine + CH_3CH_2—I: ⟶ ?

(b) CH_3CH_2—$\ddot{O}$:$^-$ + (propyl)—I: ⟶ ?

(c) :I:$^-$ + (cyclohexylmethyl)—Br: ⟶ ?

(d) (cyclopentane with :I: and CH_3) + $^-$:CN: ⟶ ?

(e) $^-$:CN: + (bicyclic compound with :I:) ⟶ ?

PROBLEM 7.38 The S_N2 reactions at the allyl position are especially facile. For example, allyl compounds are even more reactive than methyl compounds (Table 7.4, p. 251). Explain this rate difference. Why are allyl compounds especially fast? *Hint*: This, like all "rate" questions, can be answered by examining the structures of the transition states involved.

$$H_2C{=}CH{-}CH_2{-}X \xrightarrow[S_N2]{Nu:^-} H_2C{=}CH{-}CH_2{-}Nu + X:^-$$

PROBLEM 7.39 Predict the products of the following S_N1 reactions:

(a) H₃C / I: / H / CH₃ on cyclopentane $\xrightarrow{H_2\ddot{O}:}$?

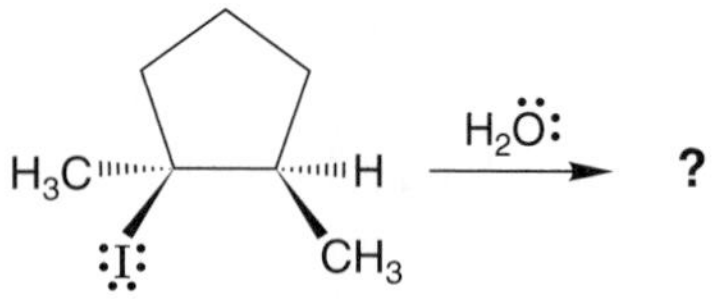

(b) $\ddot{I}$: $\xrightarrow{CH_3CH_2\ddot{O}H}$?

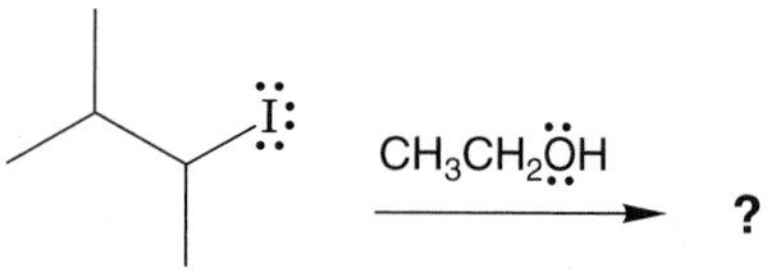

(c) $\ddot{Br}$: $\xrightarrow{H_2\ddot{O}:}$?

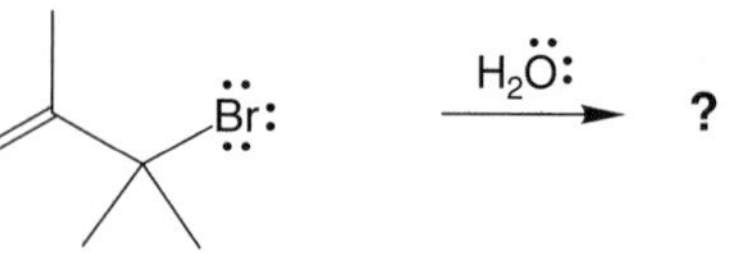

(d) $\ddot{Br}$: $\xrightarrow[CH_3CH_2\ddot{O}H,\ CH_3\ddot{O}H]{H_2\ddot{O}:}$?

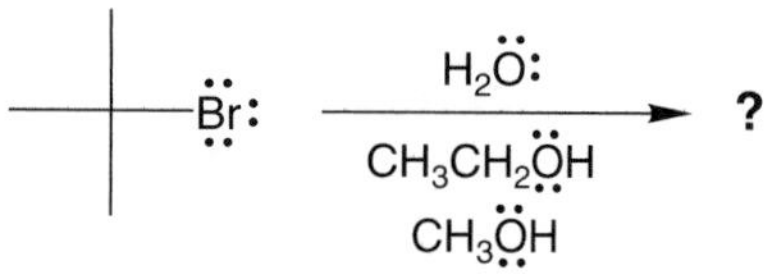

PROBLEM 7.40 In the following pairs of compounds, rank the molecules in order of rate of reaction in the S_N2 process. Explain your reasoning.

(a) Br Br

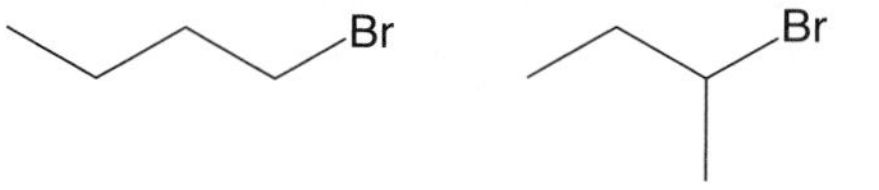

(b) Br I

(c) I I

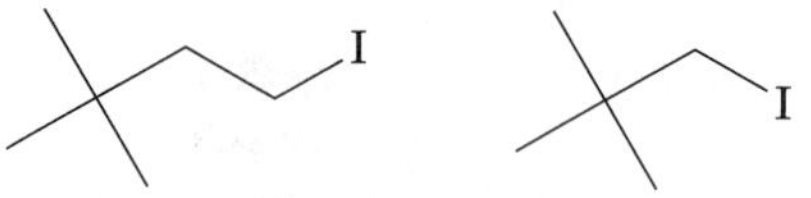

(d) Br Br

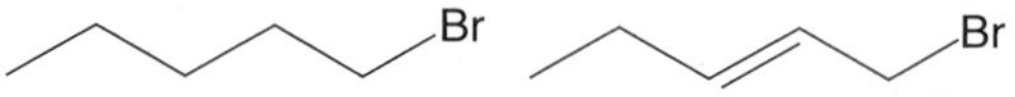

PROBLEM 7.41 In the following pairs of compounds, rank the molecules in order of rate of reaction in the S_N1 process. Explain your reasoning.

(a) I I

(b) I Br

(c) I I

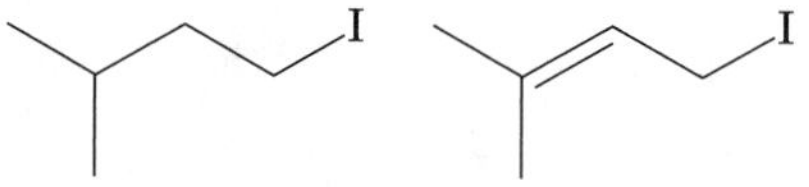

PROBLEM **7.42** You are given a supply of ethyl iodide, *tert*-butyl iodide, sodium ethoxide, and sodium *tert*-butoxide. Your task is to use the S_N2 reaction to make as many different ethers as you can. In principle, how many are possible? In practice, how many can you make?

PROBLEM **7.43** Each of the following molecules contains two halogens of different kinds. They are either in different positions in the molecule or are different halogens entirely.

(a) Br, Br

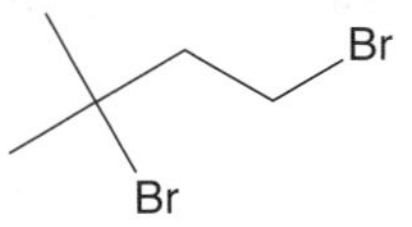

(b) I, I

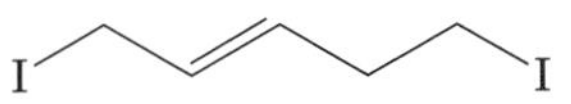

(c) Cl, Br

In each case, determine which of the two halogens will be the more reactive in the S_N2 reaction. Explain.

PROBLEM **7.44** Each of the following molecules contains two halogens at different positions in the molecule. In each case, determine which of the two halogens will be the more reactive in the S_N1 reaction. Explain.

(a) Br, Br

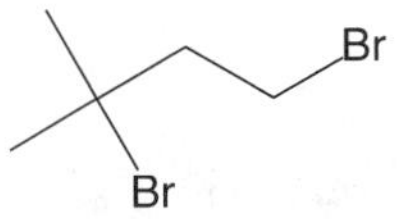

(b) I, I

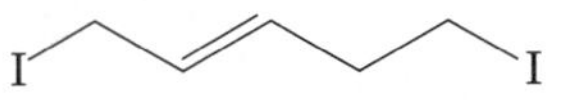

(c) Br, Br, CH_3

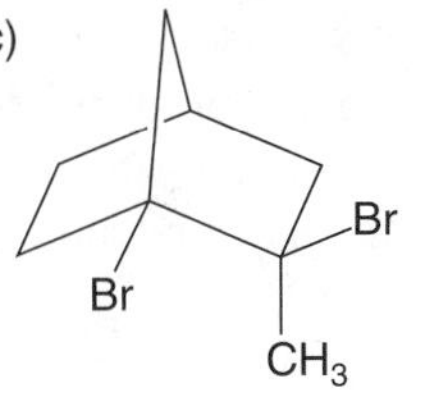

PROBLEM **7.45** You have two bottles containing the two possible diastereomers of the compound shown below ($C_8H_{11}IO_2$). When the salts of the carboxylic acids are made, only one of them is converted into a new compound, $C_8H_{10}O_2$. What is the new compound, and how does its formation allow you to determine the stereochemistry of the original two compounds? The squiggly bond indicates undefined stereochemistry.

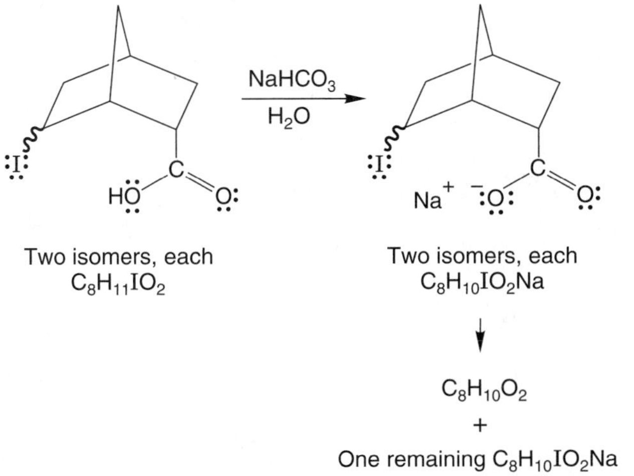

:I: HO C O: $\xrightarrow[H_2O]{NaHCO_3}$:I: Na^+ $^-$:O: C O:

Two isomers, each $C_8H_{11}IO_2$

Two isomers, each $C_8H_{10}IO_2Na$

↓

$C_8H_{10}O_2$

+

One remaining $C_8H_{10}IO_2Na$

PROBLEM **7.46** You have two bottles containing the two possible diastereomers of the compound shown below (C_5H_9IO). When the molecules are treated with a strong base, such as sodium hydride, only one of them is converted into a new compound, C_5H_8O. What is the new compound, and how does its formation allow you to determine the stereochemistry of the original two compounds?

:I: :OH $\xrightarrow{NaH}$ C_5H_8O + $C_5H_8IO^-$ Na^+

C_5H_9IO

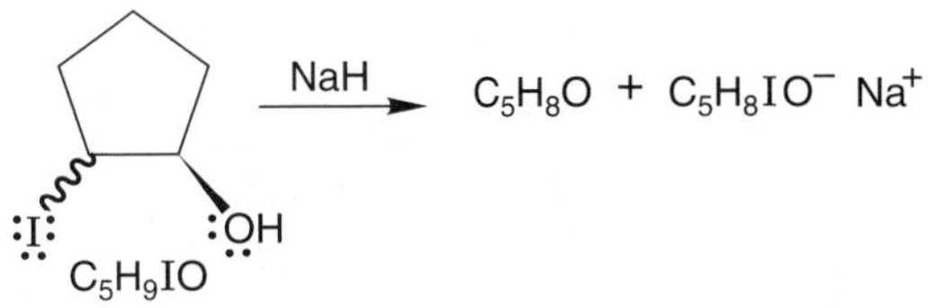

PROBLEM **7.47** Predict the products of the following E2 reactions. If more than one product is expected, indicate which will be the major compound formed.

(a) Br: $\xrightarrow[CH_3CH_2\ddot{O}H]{CH_3CH_2\ddot{O}:^-}$

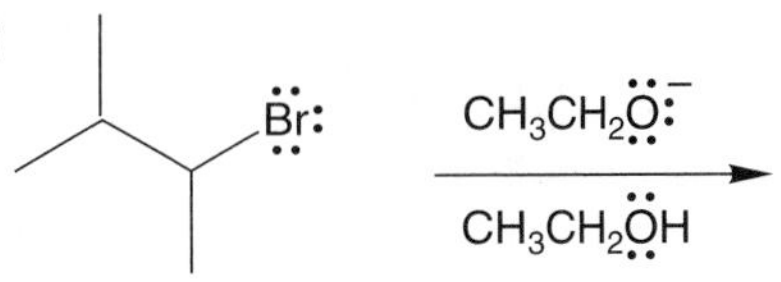

PROBLEM 7.47 (Continued)

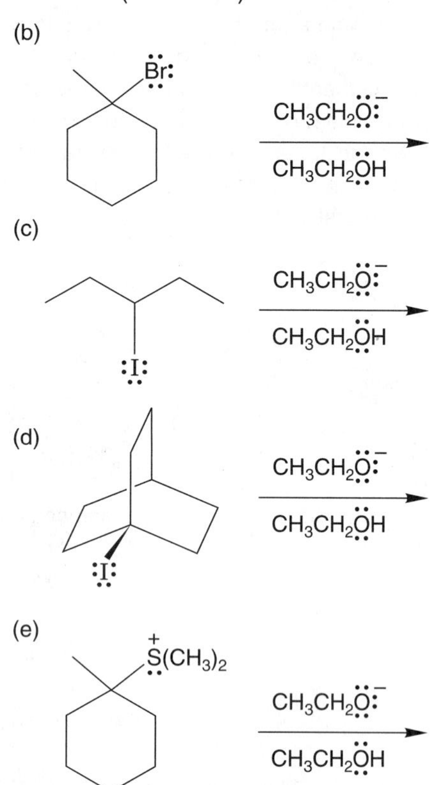

(d) $\overset{+}{S}(CH_3)_2$ $\xrightarrow{H_2\ddot{O}:}$

PROBLEM 7.48 Predict the products of the following E1 reactions. If more than one product is expected, indicate which will be the major compound formed.

(a) Br: $\xrightarrow[CH_3CH_2\ddot{O}H]{H_2\ddot{O}:}$

(b) Br: $\xrightarrow{H_2\ddot{O}:}$

(c) I: $\xrightarrow{H_2\ddot{O}:}$

PROBLEM 7.49 (a) Write the "obvious" arrow formalism for the following transformation. Of course it is going to be wrong; you are being set up here. But don't mind, it's all a learning experience.

SO_2, :O:, CH_3, $\ddot{C}H^-$, OTs $\longrightarrow$ SO_2, $:\ddot{O}:^-$, CH—CH_3, OTs

(b) OK, what you wrote is almost certainly wrong. To find the correct mechanism, measure the kinetics of this process and discover that the reaction is bimolecular. Now write another arrow formalism.

(c) What's wrong with the mechanism suggested by the original arrow formalism? *Hint*: Think about the geometry of the transition state for the S_N2 reaction.

PROBLEM 7.50 (a) Menthyl chloride reacts with sodium ethoxide in ethyl alcohol to give a single product as shown. Analyze this reaction mechanistically to explain why this is the only alkene produced. *Hint*: Think in three dimensions.

$CH(CH_3)_2$, Cl, CH_3 $\xrightarrow[CH_3CH_2\ddot{O}H]{CH_3CH_2\ddot{O}:^-\ Na^+}$ $CH(CH_3)_2$, CH_3

Menthyl chloride

By contrast, neomenthyl chloride treated in the same fashion gives an additional, major, product. Moreover, menthyl chloride reacts much more slowly than neomenthyl chloride. Explain these observations mechanistically.

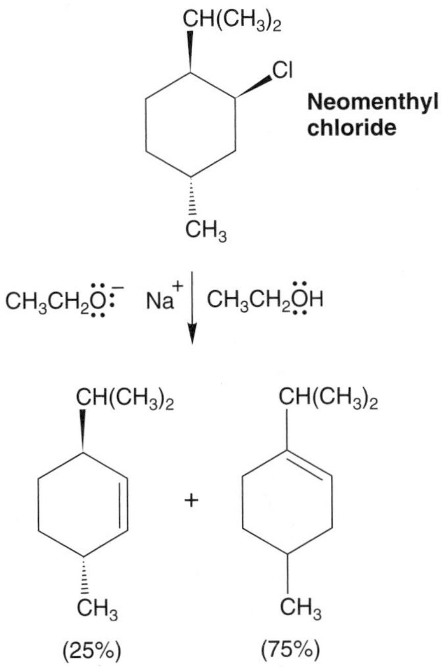

(b) When menthyl chloride is treated with 80% aqueous ethyl alcohol with no ethoxide present, once again, two products are formed in the ratio shown. Explain mechanistically.

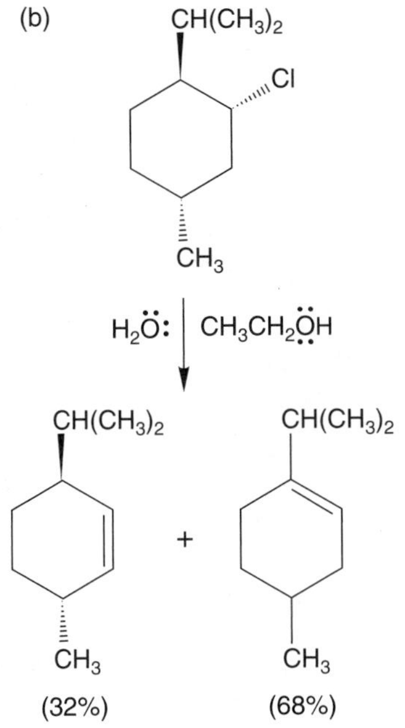

Steric effects are important in solvolysis reactions. Problems 7.51 and 7.52 illustrate this point.

PROBLEM **7.51** Explain the following rates for S_N1 solvolysis. Construct an Energy versus Reaction progress diagram showing the two reactions.

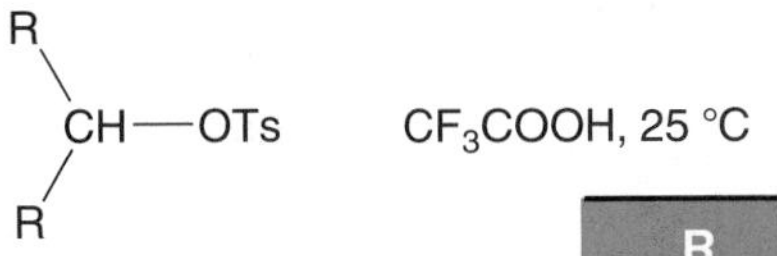

R	Rate
CH_3	1
$(CH_3)_3C$	10^5

PROBLEM **7.52** Explain the relative rates of the following compounds in the S_N1 reaction:

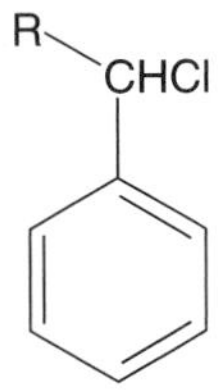

R	Relative Rate
CH_3	540
CH_3CH_2	125
$(CH_3)_2CH$	27
$(CH_3)_3C$	1

PROBLEM **7.53** You are given a supply of 2-bromo-1-methoxybicyclo[2.2.2]octane. Your task is to prepare 1-methoxybicyclo[2.2.2]oct-2-ene (**A**) from this starting material. In your first attempt, you treat the bromide with sodium ethoxide in ethyl alcohol. Although a small amount of the desired alkene is formed, most of the product is 2-ethoxy-1-methoxybicyclo[2.2.2]octane (**B**).

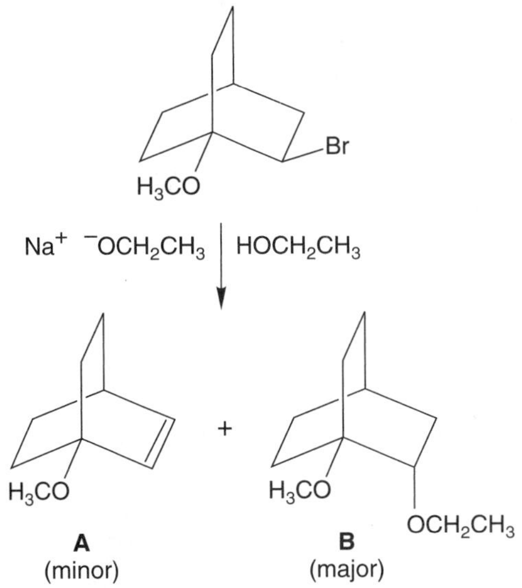

Suggest a simple experimental modification that should result in an increase in the amount of the desired alkene, and a decrease in the amount of the unwanted substitution product, **B**.

PROBLEM 7.54 A common procedure to increase the rate of an S_N2 displacement is to add a catalytic (very small) amount of sodium iodide. Notice that even though the rate increases, the structure of the product is not the iodide. Explain the function of the iodide. How does it act to increase the rate, and why is only a catalytic amount necessary?

R–CH₂CH₂CH₂CH₂–Cl + Na^+ Nu^-

(small amount) Na^+ I^- ↓ acetone (solvent)

R–CH₂CH₂CH₂CH₂–Nu

Equilibria

8

"I hope," the Golux said, "that this is true. I make things up, you know."

—James Thurber*
The 13 Clocks

In Chapter 7, we saw some examples of chemical equilibria. For example, the S_N2 reaction involves an equilibrium in which two nucleophiles (Nu:⁻ and L:⁻) compete for the R group (Fig. 8.1).

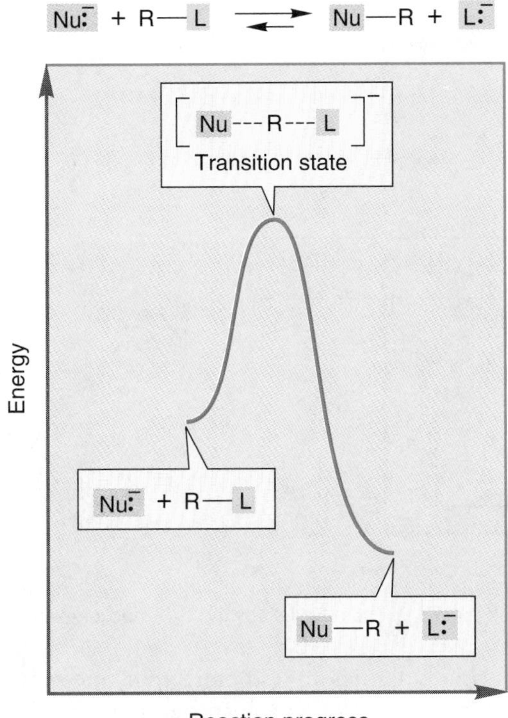

FIGURE **8.1** The S_N2 reaction is an equilibrium in which two nucleophiles (Lewis bases) compete for a Lewis acid.

*James Grove Thurber (1894–1961) was an American writer and cartoonist, much of whose work first appeared in the *New Yorker* magazine. *The 13 Clocks* is a fairy tale of surpassing charm and wit written in Bermuda.

All chemical reactions can be regarded as equilibria. Multistep reactions are simply series of equilibria in which species occupying energy minima are separated by barriers, the tops of which are called **transition states**. The S_N1 solvolysis reaction of *tert*-butyl iodide in water makes a good example. Here we have three equilibria. The first is between *tert*-butyl iodide and the *tert*-butyl cation and iodide ion. This equilibrium is highly unfavorable, as the cation–anion pair is far less stable than the covalent iodide. The second equilibration represents the exothermic capture of the cation by water to give the protonated alcohol, an **oxonium ion**. Finally, the oxonium ion is deprotonated to give the ultimate product, *tert*-butyl alcohol (Fig. 8.2).

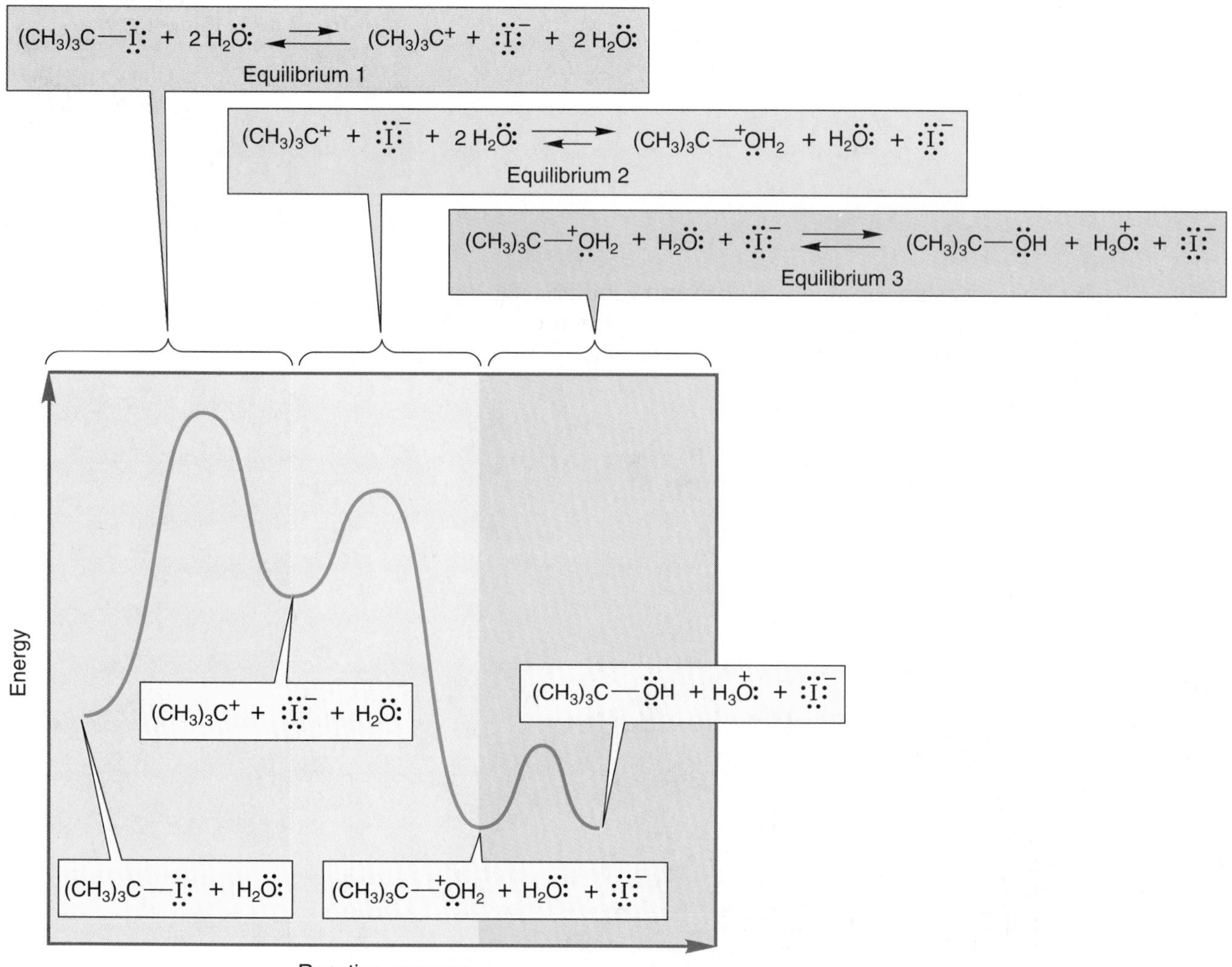

FIGURE 8.2 Multistep reactions—here typified by the S_N1 reaction—are series of equilibria.

In practice, thermodynamics may dictate that one side of an equilibrium be overwhelmingly favored over the other. Nevertheless, we can still apply techniques for studying equilibria to the reaction, deriving information about the position of the equilibrium (thermodynamics), and the rate of reaction (kinetics). The concepts we will develop in this short chapter are extremely general and very important.

8.1 EQUILIBRIUM

For a general chemical reaction,

$$aA + bB \rightleftarrows cC + dD \tag{8.1}$$

the concentrations of the species on the two sides of the arrows are related to each other by an **equilibrium constant, *K***:

$$[C]^c\,[D]^d = K[A]^a\,[B]^b \tag{8.2}$$

or, more familiarly:

$$K = \frac{[C]^c\,[D]^d}{[A]^a\,[B]^b} \tag{8.3}$$

The small letter superscripts are the coefficients in the balanced Eq. (8.1), and the capital letters represent the molar concentrations of the products and reactants at equilibrium. The concepts of starting materials and products are arbitrary ones, arising from practical considerations—we generally are interested in making the reaction run one way or the other, and by convention the desired reaction is written left to right. But it need not be; it is just as valid to look at the reaction in the other direction.

The reaction will be a successful one if the compounds we seek, the products, are strongly favored at equilibrium. So we are very interested in what determines the magnitude of this equilibrium constant (K): the relative free energies ($\Delta G°$) of the compounds involved in the chemical reaction. The exact relationship between K and the relative energies of the starting materials and products is shown in Eq. (8.4):

$$\Delta G° = -RT\ \ln K \tag{8.4}$$

or, if we want to use base 10 rather than natural logarithms, in Eq. (8.5):

$$\Delta G° = -2.3RT\ \log K \tag{8.5}$$

R is the gas constant, 1.99×10^{-3} kcal/deg·mol, T is the absolute temperature, and $\Delta G°$, the **Gibbs free energy change**, is the difference in energy between starting materials and products in their standard states.

Up to now, we have more or less equated energy with $\Delta H°$, the **enthalpy**. Enthalpy, or the heat of reaction, is a measure of the bond energy changes in a reaction. The Gibbs free energy ($\Delta G°$) is more complicated than this, as we will see in Section 8.2, and includes not only $\Delta H°$, but **entropy** ($\Delta S°$) as well. The entropy of a reaction is a measure of the change in amount of order in starting materials and products. The higher the degree of order, the greater the energy requirements for the reaction.

A number of times in Chapter 7 we drew diagrams that described the course of a reaction. We can now label those diagrams a bit more completely. The diagrams are plots of energy versus something called, most often, "reaction progress." Figures 8.3 and 8.4 show an exothermic reaction (products more stable than starting materials) and an endothermic reaction (starting material more stable than products). Actually, the term "exothermic" really refers to the enthalpy change, $\Delta H°$, which may or may not be in the same direction as the free energy change, $\Delta G°$, which involves entropy, $\Delta S°$, as well as enthalpy, $\Delta H°$. If we really mean free energy, ΔG, the proper terms are **exergonic** and **endergonic**.

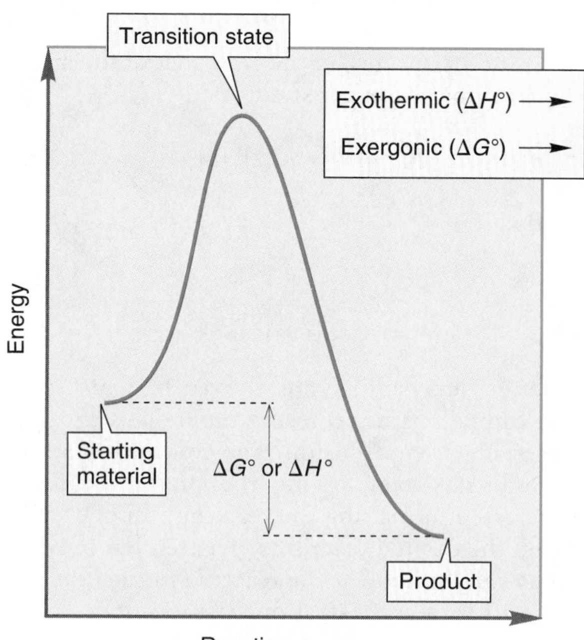

FIGURE 8.3 An exothermic (exergonic) reaction in which products are more stable than starting materials. In such a case, $\Delta H°$ ($\Delta G°$) is negative and the equilibrium constant K is greater than 1.

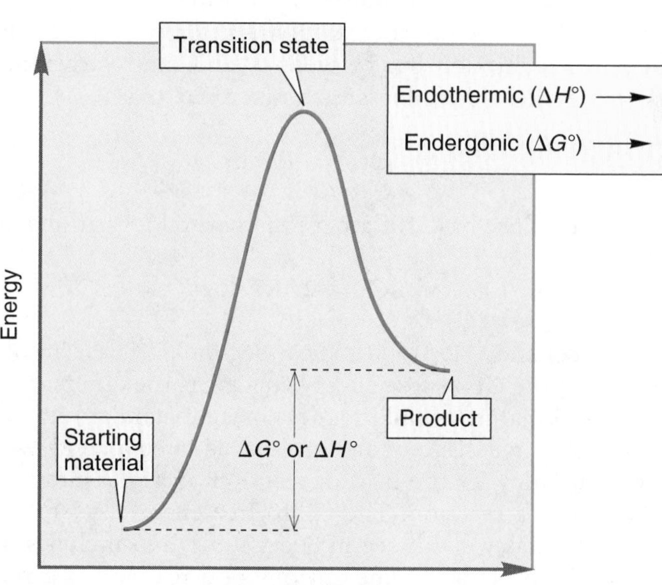

FIGURE 8.4 An endothermic (endergonic) reaction. The products are less stable than the starting materials. In this case, $\Delta H°$ ($\Delta G°$) is positive and K is less than 1.

For the exothermic reaction the difference in enthalpy between products and starting material is negative, and the equilibrium constant (K) is greater than 1. When we look at an endothermic reaction (or read the first reaction backward to produce an endothermic process), the enthalpy difference is positive and K is less than 1.

More stable products are favored at equilibrium. By how much? Note that the relationship between the difference in free energy between starting material and products (ΔG°) and the equilibrium constant K is logarithmic. We can rewrite Eqs. (8.4) and (8.5) to emphasize that point:

$$\Delta G^\circ = -RT \ln K \qquad \text{or} \qquad K = e^{-\Delta G^\circ/RT} \qquad (8.4)$$

or, in base 10:

$$\Delta G^\circ = -2.3RT \log K \qquad \text{or} \qquad K = 10^{-\Delta G^\circ/2.3RT} \qquad (8.5)$$

These equations show that a small difference in energy between starting materials and products results in a large excess of one over the other at equilibrium. Let's do a few calculations to show this. First, let's take a case in which products are more stable than starting materials by the seemingly tiny amount of 1 kcal/mol, $\Delta G^\circ = -1$ kcal/mol.

$$K = 10^{-\Delta G^\circ/2.3RT} \qquad \text{or} \qquad K = 10^{1/2.3RT}$$

at 25 °C, $2.3RT = 1.364$ kcal/mol,* so $K = 10^{1/(1.364)} = 10^{0.7331} = 5.4$.

An equilibrium constant of 5.4 translates into 84.4% product at equilibrium! One would very often be quite happy to find a reaction that would give about 85% product, and a mere 1 kcal/mol suffices to ensure this. Conversely, if we are "fighting" a 1-kcal/mol endothermic process it will be a futile fight indeed, because equilibrium will settle out at 84.4% starting material. *A little bit of energy goes a long way at equilibrium.* Table 8.1 summarizes a number of similar calculations.

TABLE **8.1** The Relationship between ΔG° and K

ΔG° (kcal/mol)	K	More Stable Isomer (%)
− 0.1	1.2	54.5
− 0.5	2.4	69.7
− 1	5.4	84.4
− 2	29.3	96.7
− 5	4631	99.98
−10	2.1×10^7	99.999996

PROBLEM **8.1**

Calculate the energy difference between starting material and products at 25 °C given equilibrium constants of 7 and 14.

In practice, there often is something we can do to drive a reaction in the direction we want. **Le Chatelier's principle**† states that a system at equilibrium responds to stress in such a way as to relieve that stress. This statement means that if we increase the concentration of one of the start-

*At 25 °C, the value of $2.3RT$ is about 1.4 kcal/mol. This number is useful to remember because with it one can do calculations easily at odd moments without having to look up the value of R and multiply it by T.

†Henri Louis Le Chatelier (1850–1936). The mathematics of this principle had been worked out earlier by the American Josiah Willard Gibbs (1839–1903), who donated the G in Gibbs to ΔG. Le Chatelier knew of Gibbs' contributions, and was a great popularizer of Gibbs' work in France.

ing materials, an equilibrium process will be driven toward products. In Eqs. (8.1–8.3), if we increase $[A]^a$, the quantity $[C]^c[D]^d/[B]^b$ must also increase to maintain K, and we will have more product. Similarly, if we can remove one product as it is formed, the equilibrium will be driven toward products. Perhaps either C or D is a solid that crystallizes out of solution or is a gas that can be driven off. As the concentration of this compound drops, the equilibrium will be disturbed and more C or D will be formed.

In a multistep reaction, it often does not matter that an early step involves an unfavorable equilibrium, as long as the overall reaction is exothermic. Figure 8.5 shows such a process for the conversion of A into C via an intermediate B.

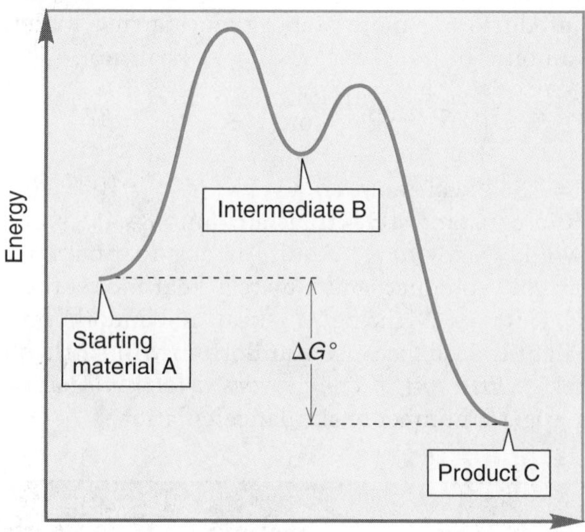

FIGURE **8.5** As long as equilibrium is reached, it does not matter that the formation of B from A is endothermic. As long as all barriers can be traversed, the ratio of C and A will be governed by the ΔG° between C and A. So, A → B is endothermic, but, overall, A → C is exothermic ($\Delta G^\circ < 0$).

As long as some B is present in equilibrium with starting material A, the reaction will be successful. As the equilibrium with B is established, B will be converted into the much more stable final product C. As the small amount of B present in equilibrium with A is depleted, the A $\rightleftarrows$ B equilibrium will be reestablished and more B will be formed from A. In the long run, A will be largely converted into C, and as long as there is enough energy to reach the intermediate B, the final mixture of A and C will depend on the ΔG° between these two compounds.

8.2 GIBBS STANDARD FREE ENERGY CHANGE

The parameter ΔG° is composed of enthalpy (ΔH°, the change in bond strengths in the reaction) and entropy (ΔS°, the change in freedom of motion in the reaction). The exact relationship is,

$$\Delta G^\circ = \Delta H^\circ - T\Delta S^\circ \tag{8.6}$$

where T is the absolute temperature. The enthalpy term (ΔH°) is most directly related to the bond strengths in starting materials and products. We can make rather accurate estimates of this term by knowing the bond dissociation energies for a variety of bonds. We know a few of these values already from our discussion of alkanes. Table 8.2 gives several more.

TABLE 8.2 Some Bond Dissociation Energies

Bond	Bond Dissociation Energy (kcal/mol)
I–I	36.1
Br–Br	46.1
Cl–Cl	59.0
F–F	38.0
HO–OH	51
$(CH_3)_3CO{-}OC(CH_3)_3$	38
H–H	104.2
H–I	71.3
H–Br	87.6
H–Cl	103.2
H–F	136.3
H–OH	119
$H{-}OCH_3$	104.4
$H{-}NH_2$	107.4
$H{-}CH_3$	104.8
$H{-}CH_2CH_3$	100.3
$H{-}CH(CH_3)_2$	96.0
$H{-}C(CH_3)_3$	93.3
$I{-}CH_3$	57
$Br{-}CH_3$	70.0
$Cl{-}CH_3$	85
$F{-}CH_3$	108
$HO{-}CH_3$	92
$H_2N{-}CH_3$	85
$H_3C{-}CH_3$	90
$H_3C{-}CH_2CH_3$	85
$H_2C{=}CH_2$ (π bond only)	66
$H_2C{=}CH_2$ (total)	172
$HC{\equiv}CH$ (total)	230

PROBLEM 8.2

Estimate the $\Delta H°$ for the following reactions:

(a) $(CH_3)_2CH{-}I + CH_3O^- \longrightarrow (CH_3)_2CH{-}OCH_3 + I^-$

(b) $H_2 + H_2C{=}CH_2 \longrightarrow H_3C{-}CH_3$

(c) $Br_2 + H_2C{=}CH_2 \longrightarrow BrCH_2{-}CH_2Br$

(d) $HCl + H_2C{=}CH_2 \longrightarrow ClCH_2{-}CH_3$

PROBLEM 8.3

The bond dissociation energies for the carbon–hydrogen bonds in Table 8.2 show a steady decline as the breaking carbon–hydrogen bond becomes more substituted. What can you infer from these data about the stability of the neutral species called "free radicals," R• ?

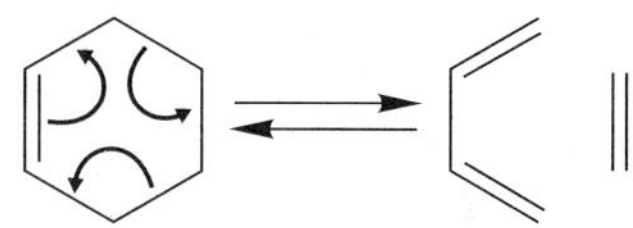

FIGURE 8.6 This reaction is a process in which one direction is favored by enthalpy and the other by entropy. Although ΔH favors cyclohexene, at very high temperature the formation of two compounds from one can drive this reaction to the right.

The entropy term ($\Delta S°$) is related to freedom of movement. The more restricted or "ordered" a molecule, the more negative the entropy. Since it is the negative of $\Delta S°$ that is related to $\Delta G°$ [Eq. (8.6)], the more negative the entropy, the higher the free energy ($\Delta G°$) of the reaction. Note also that the entropy term is temperature-dependent. Entropy becomes more important at high temperature. Imagine the reaction in Figure 8.6, in which strong carbon–carbon σ bonds are broken and weaker carbon–carbon π bonds are made (Table 8.2, p. 307). At the same time, one molecule is made into two other, smaller molecules.

*PROBLEM 8.4 Use $\Delta H°$ values from Table 8.2, as well as the value for the bond dissociation energy of $H_3C–CH_2CH_3 \rightarrow H_3C\bullet\ \bullet CH_2CH_3$ of 85 kcal/mol, to estimate $\Delta H°$ for the reaction in Figure 8.6.

ANSWER Compare the bonds that are breaking with those that are being made. Two carbon–carbon single bonds are being broken. As the problem says, each is worth 85 kcal/mol. In addition, a π bond (66 kcal/mol) is broken in the reaction. So, the bonds broken are worth a total of 236 kcal/mol. Three π bonds are made, each worth 66 kcal/mol, for a total of 198 kcal/mol. The left-to-right reaction is endothermic by about 38 kcal/mol, as the bonds made are weaker than those broken.

$$
\begin{aligned}
2 \times \text{C–C} &= +170 \text{ kcal/mol (bonds broken)} \\
\text{C=C } (\pi) &= +\ 66 \text{ kcal/mol (bond broken)} \\
3 \times \text{C=C } (\pi) &= \underline{-198 \text{ kcal/mol}} \text{ (bonds made)} \\
\Delta H° &= +\ 38 \text{ kcal/mol}
\end{aligned}
$$

At low temperature, this reaction will certainly be endothermic because of the unfavorable $\Delta H°$ term. Yet, at high temperature the $T\Delta S°$ term will become more important, and the reaction will become more favorable, largely because of the increase in the number of particles and the increased freedom of motion in the system. An unfavorable reaction can even become favorable at high temperature where the entropy term becomes more important in determining the value of $\Delta G°$.

So now we know something of the factors that affect equilibrium. We have said nothing of the factors influencing the *rate* of establishing that equilibrium, and this is quite another matter.

8.3 RATES OF CHEMICAL REACTIONS

Just because a reaction is exothermic does not mean that it will be a rapid reaction, or even that it will proceed at all under normal conditions. First of all, an exothermic reaction can still be endergonic if entropy plays an especially important role. Entropy does not raise its ugly head in this way often, but there is another much more frequently encountered reason why an *exergonic* reaction might not be a *fast* reaction. As we have seen, there are barriers to chemical reactions. In principle, a very stable product can be separated from a much less stable starting material by a barrier high enough to stop all reaction (Fig. 8.7).

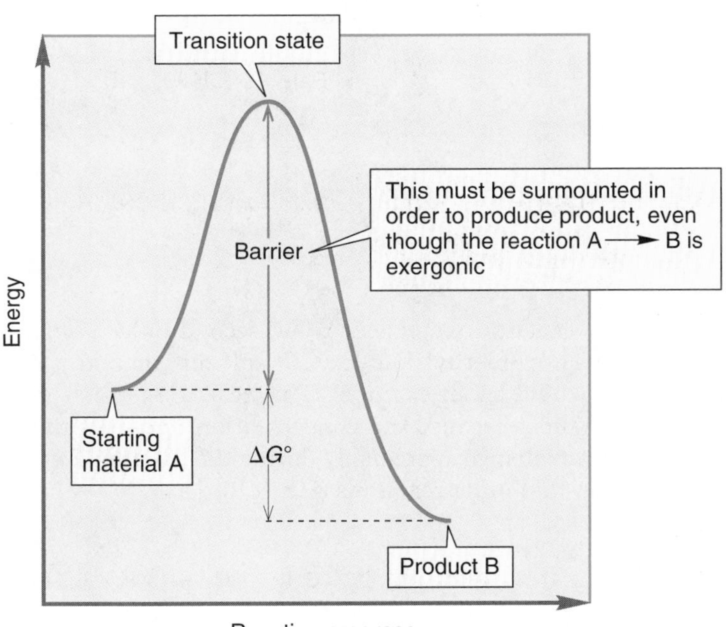

FIGURE 8.7 Rates of reactions are not determined by the $\Delta G°$ between starting material and product, but by the height of the barrier separating starting material and transition state. Some very unstable materials are protected by high barriers, and thus can be isolated.

Collections of molecules exist at a given temperature in ranges of energies called **Boltzmann distributions**. The energies of most molecules cluster about an average value. As energy is provided to the molecules, usually through the application of heat, the average energy increases. Although there is always a small number of highly energetic molecules present (the high-energy tail of the Boltzmann distribution), in practice there may not be enough molecules of sufficient energy to make the reaction proceed at an observable rate if the barrier is high enough. The application of energy in the form of heat produces more molecules with sufficient energy to traverse the barrier (Fig. 8.8). A reaction rate approximately doubles for every increase in temperature of 10 °C.

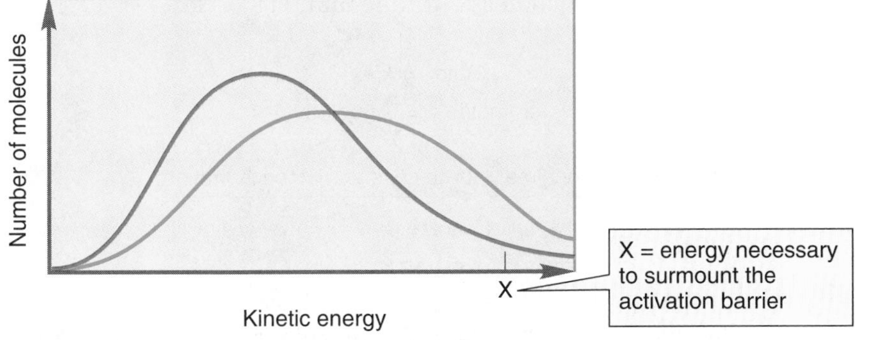

FIGURE 8.8 Boltzmann distributions of molecules at two temperatures. In this figure, the blue line is the distribution at higher temperature and the red line is the distribution at lower temperature.

As we already saw in our discussion of the S_N2 and S_N1 reactions, rates of chemical reactions also depend on the concentrations of reactants. The rates of most organic reactions either depend only on the concentration of a single reactant (S_N1, for example) and thus are **first-order reactions**, or upon the concentrations of two species (S_N2) and are **second-order reactions** (Fig. 8.9). There are examples of higher order reactions, but they are relatively rare.

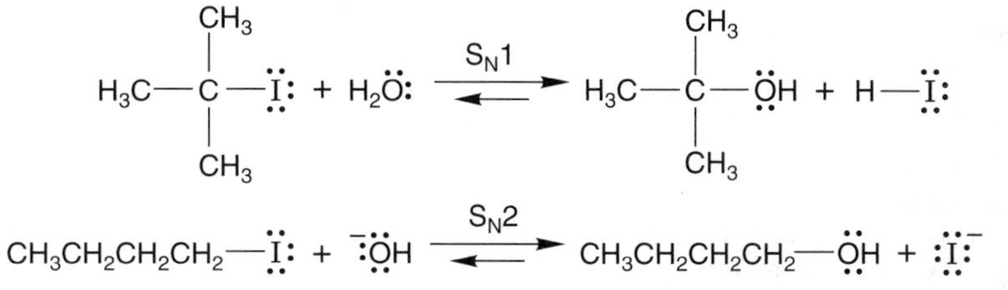

$$(CH_3)_3C{-}I + H_2O \underset{}{\overset{S_N1}{\rightleftharpoons}} (CH_3)_3C{-}OH + H{-}I \qquad \text{Rate} = k[(CH_3)_3C{-}I]$$

$$CH_3CH_2CH_2CH_2{-}I + {}^{-}OH \overset{S_N2}{\rightleftharpoons} CH_3CH_2CH_2CH_2{-}OH + I^- \qquad \text{Rate} = k[CH_3CH_2CH_2CH_2{-}I][^{-}OH]$$

FIGURE 8.9 Rate laws for the S_N1 and S_N2 reactions.

Sometimes a reaction we expect to be second order is not. The S_N2 solvolysis reaction of methyl iodide with solvent (therefore in great excess) ammonia would be an example. This reaction is first order in substrate, methyl iodide, because the concentration of one of the reactants, ammonia, does not change appreciably during the reaction. Such reactions are called **pseudo-first-order reactions** (Fig. 8.10).

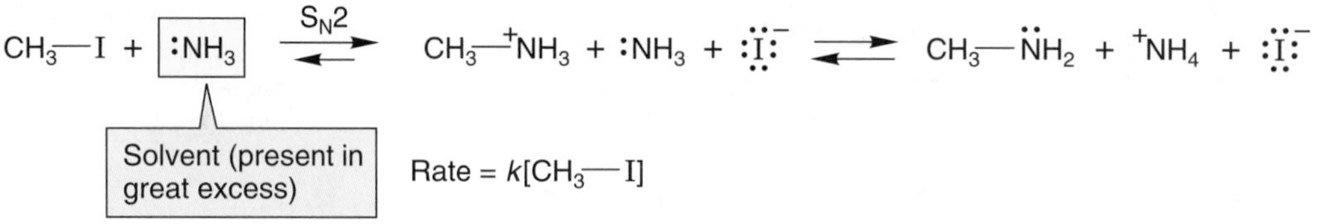

$$CH_3{-}I + :NH_3 \overset{S_N2}{\rightleftharpoons} CH_3{-}\overset{+}{N}H_3 + :NH_3 + :I:^- \rightleftharpoons CH_3{-}NH_2 + {}^{+}NH_4 + :I:^-$$

Solvent (present in great excess)

$$\text{Rate} = k[CH_3{-}I]$$

FIGURE 8.10 A pseudo-first-order reaction. The concentration of the nucleophile, ammonia, does not change during the reaction.

8.4 RATE CONSTANT

In the rate laws mentioned so far, there has always been a proportionality constant, the **rate constant (*k*)**, which is a fundamental property of any given chemical reaction. A rate must have the units of concentration per unit time, or moles per liter per unit time. For a first-order process such as the S_N1 reaction, the rate constant k must have the units of reciprocal seconds (s^{-1}) (if we use the second as our time unit, Fig. 8.11).

$$\text{Rate} = k[A]$$

$$\text{or, mol/L/s} = k\,(\text{mol/L})$$

concentration/time = (rate constant) (concentration)

$$\text{mol/L/s} \times \text{L/mol} = k$$

$$\text{and } 1/\text{s} = \text{s}^{-1} = k$$

FIGURE 8.11 For a first-order reaction, the rate constant (k) has the units of reciprocal time (time^{-1}).

For a second-order reaction, k must be in units of reciprocal moles per liter per second $(\text{mol/L})^{-1}\text{s}^{-1}$ (Fig. 8.12).

$$\text{Rate} = k[A][B]$$

$$\text{Rate} = \text{mol/L/s} = k \times \text{mol/L} \times \text{mol/L}$$

Rate = concentration/time = (rate constant) (concentration) (concentration)

$$\text{so, } k = \text{mol/L/s} \times \text{L}^2/\text{mol}^2 = \text{L/mol/s} = (\text{mol/L})^{-1}\text{s}^{-1}$$

FIGURE 8.12 For a second-order reaction, the rate constant (k) has the units of reciprocal moles per liter per second $(\text{mol/L})^{-1}\ \text{s}^{-1}$.

We can determine the value of k by measuring how the reaction rate varies with the concentration(s) of the reactants.

It is easy to confuse the *rate* of a reaction, which is simply how fast the concentrations of the reactants are changing, with the *rate constant*, which tells us how the rate will change as a function of reactant concentration. The rate of a reaction will vary with concentration. This variation is easy to see by imagining the limiting case in which one reactant is used up. When its concentration goes to zero, the reaction rate must also be zero. On the other hand, the rate constant is an intrinsic property of the reaction. It will vary with temperature, pressure, and solvent, but does not depend on the concentrations of reactants.

Only when the concentrations of all reactants are 1 M are the rate and the rate constant the same. The rate constant is thus a measure of the rate under standard conditions (Fig. 8.13).

$$\text{Rate} = k[\text{A}][\text{B}][\text{C}][\text{D}] \ldots [\text{X}] \qquad \text{if all } [\ ] = 1\ M$$

$$\text{Rate} = k[1][1][1][1] \ldots [1] \text{ and therefore, rate} = k$$

FIGURE 8.13 If the concentrations of the reactants are all 1 M, the rate constant (k) and the rate are the same.

8.5 ENERGY BARRIERS IN CHEMICAL REACTIONS: THE TRANSITION STATE AND ACTIVATION ENERGY

We have already mentioned that even exothermic reactions may be very slow, indeed immeasurably slow. And it's a good thing. Combustion is a very exothermic chemical reaction, and yet we and our surroundings exist more or less tranquilly in air without bursting into flame. Somehow we are protected by an energy barrier. Methane and oxygen do not react unless energy, called **activation energy**, is supplied. If we are considering free energy, the activation energy is $\Delta G^{\ddagger}$; if we are speaking of enthalpy, the activation energy is $\Delta H^{\ddagger}$. The double dagger (‡) is always used when we are referring to activation parameters involving the energy of a transition state. A match will suffice to provide the energy in this example. This highly exothermic process (see Chapter 3, p. 87, 88) then provides its own energy as highly stable molecules are produced from less stable ones (Fig. 8.14).

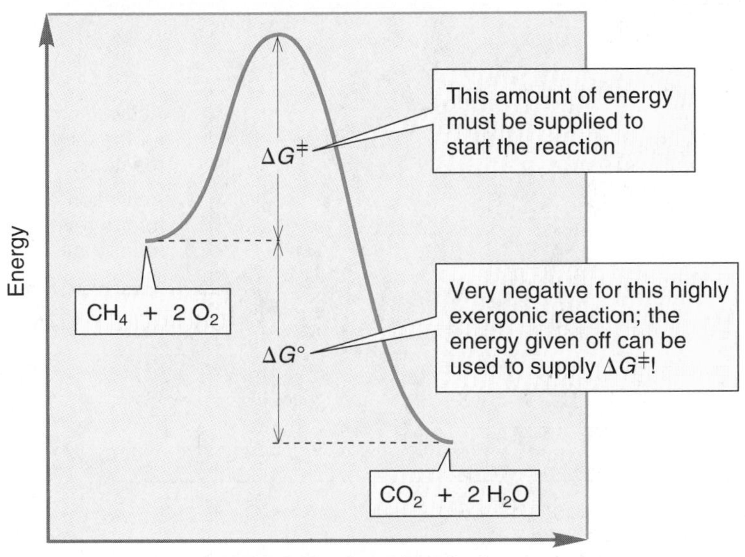

FIGURE 8.14 An exergonic reaction can provide the energy necessary for molecules to pass over an activation barrier. Once such reactions are started by the application of energy (a match), they continue until one reactant is used up.

We are protected from spontaneous combustion in the same way we are protected from hungry bears and tigers during a visit to the zoo. At the zoo the barrier is tangible and operates in a clear way to ensure our survival. Activation barriers are less visible but no less necessary for the continuance of our lives. In this important section, we will discuss these energy barriers to chemical reactions.

PROBLEM 8.5 Identify the activation energy for the reverse of the reaction in Figure 8.14, the formation of methane and oxygen from carbon dioxide and water.

In a process such as the S_N1 reaction, a bond must be broken. Consider the ionization of *tert*-butyl iodide. The first step in this reaction is the ionization of the substrate, and surely this will involve an energy cost in breaking the carbon leaving group bond (Fig. 8.15).

$$(CH_3)_3C—\ddot{\underset{..}{I}}: \underset{}{\overset{S_N1}{\rightleftharpoons}} (CH_3)_3C^+ + :\ddot{\underset{..}{I}}:^-$$

FIGURE 8.15 In the ionization of *tert*-butyl iodide, energy must be provided to break the carbon–iodine bond. The enthalpy change $\Delta H°$ (and $\Delta G°$) is positive for this reaction.

This cost is partially offset by an increase in entropy. We form two particles out of one and this will be favorable. But we have to be careful when we estimate entropy changes because there are invisible participants in most reactions, including this one. These are the solvent molecules. They will have to be disordered and reordered in any ionic reaction in solution and the net change in entropy may not be at all easy to estimate.

*PROBLEM 8.6 Construct an Energy versus Reaction progress diagram (see Fig. 8.9) for the reaction of *tert*-butyl iodide and water, showing clearly the activation energies and transition states for all steps. A discussion of this reaction is coming, but try to anticipate it here.

*PROBLEM 8.7 Show the activation energies for the reverse reactions involved in the formation of *tert*-butyl iodide from *tert*-butyl alcohol and hydrogen iodide (HI).

ANSWERS

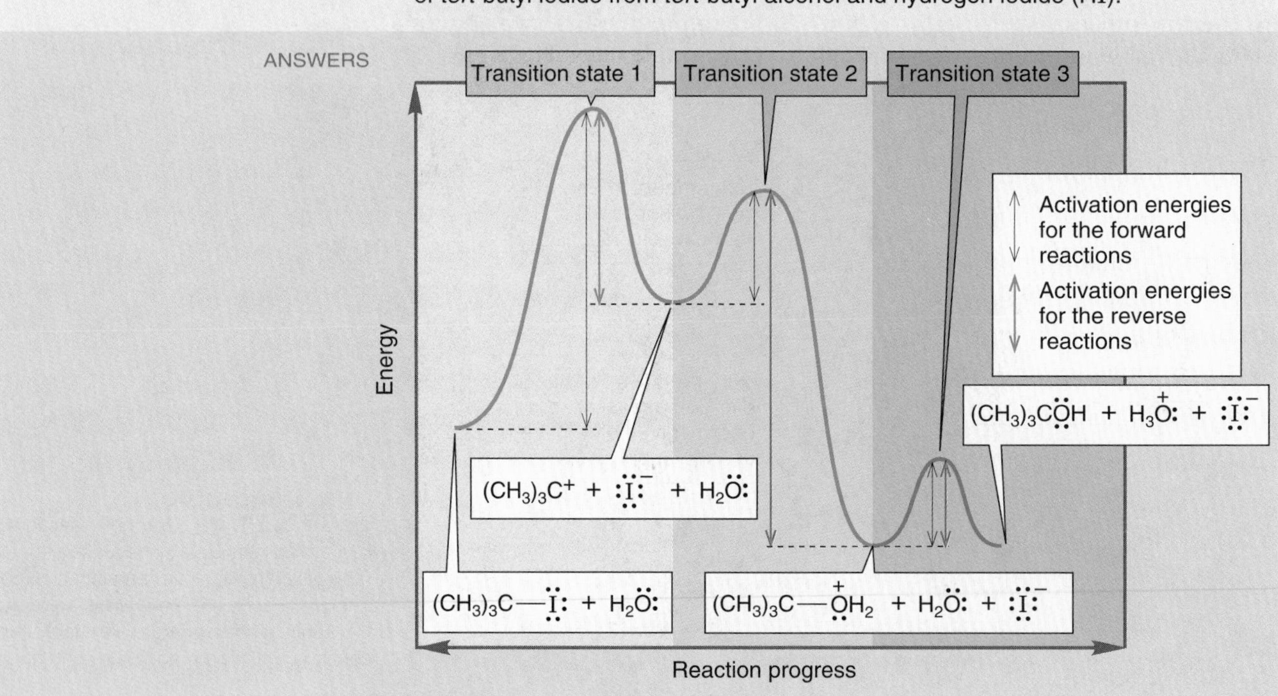

PROBLEM 8.8

Show that at equilibrium, the equilibrium constant (K) is given by the ratio of the rate constants for the forward and reverse reactions. *Hint*: At equilibrium the rates of the forward and reverse reactions are equal.

The situation is a little different for the S_N2 reaction. Here we must also break the carbon leaving group bond, but this energy cost will be partially offset by the formation of the bond from carbon to the entering nucleophile. We might imagine that there would also be a substantial entropy cost in this reaction as we must order the reactants in a special way for the reaction to proceed. The S_N2 reaction can occur only if the nucleophile attacks from the rear of the departing leaving group and this requires a specific arrangement of the molecules and ions involved in the reaction (Fig. 8.16).

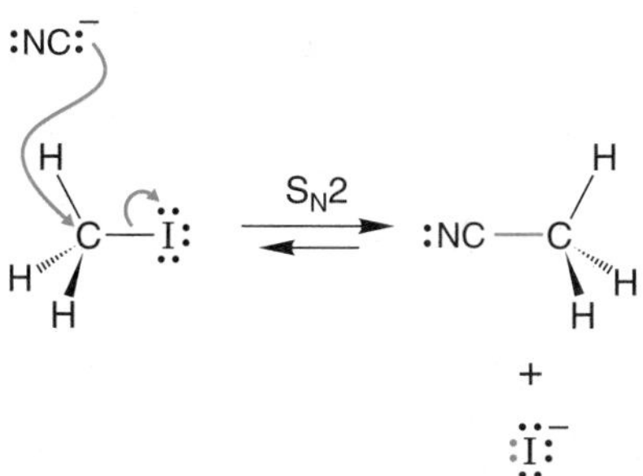

FIGURE 8.16 The transition state for the S_N2 reaction has severe ordering requirements. Attack of the nucleophile must be from the rear for the reaction to succeed.

We can follow the course of both these reactions in diagrams that plot Energy versus Reaction progress. One usually doesn't specify reaction progress in detail because there are many possible measures of this quantity, and for this kind of qualitative discussion it doesn't much matter what we pick. In the S_N2 reaction, for example, we might choose to plot energy versus the length of the carbon leaving group bond (which is increasing in the reaction), or the length of the carbon–nucleophile bond, which decreases through much of the process. The overall picture is the same.

The point of maximum energy in the reaction is called the transition state. If we have enough energy for molecules to cross the barrier in reasonable numbers, we will have a measurable rate of reaction and the final mixture of starting material and products will depend on the $\Delta G°$ between starting materials and products. Figure 8.17 shows a diagram for the

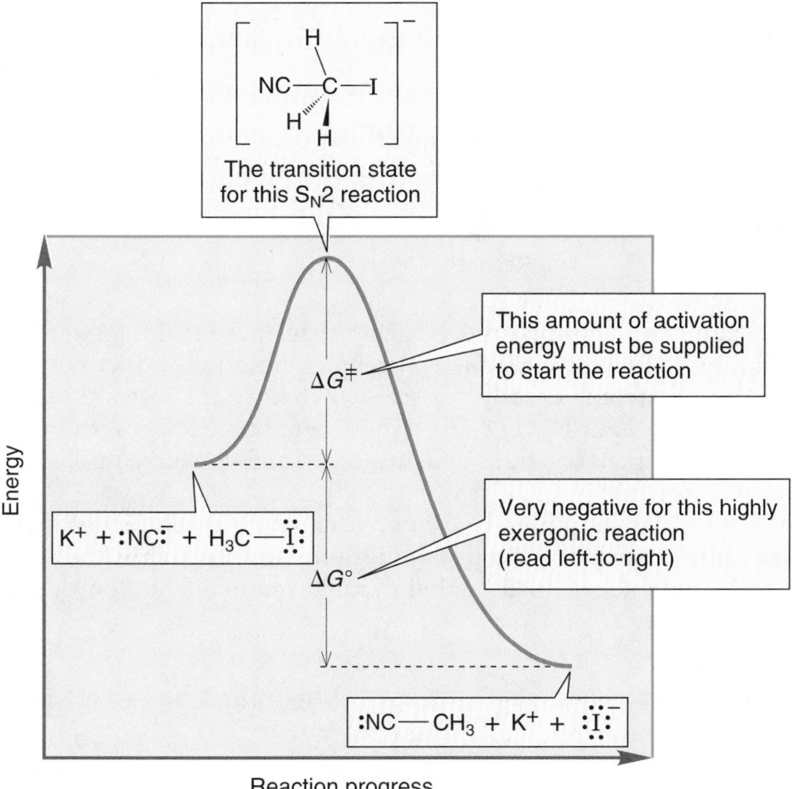

FIGURE 8.17 An energy diagram for the S_N2 displacement of iodide by cyanide (and its reverse reaction, the much less favorable displacement of the strong nucleophile cyanide by iodide). The activation energy for the exergonic displacement of iodide is shown.

exothermic formation of methyl cyanide (CH_3CN) and potassium iodide (KI) from methyl iodide (CH_3I) and potassium cyanide (KCN). The transition state is the high point in energy for the reaction. The activation energy is the height of the transition state above the starting material. At equilibrium, the ratio of product and starting material will depend on $\Delta G°$, the difference in free energy between starting material and product. The reaction shown has a negative $\Delta G°$, and will be successful in producing product.

In Figure 8.18, this same reaction is considered from the reverse point of view. This figure describes the endothermic formation of methyl iodide from methyl cyanide and iodide ion.

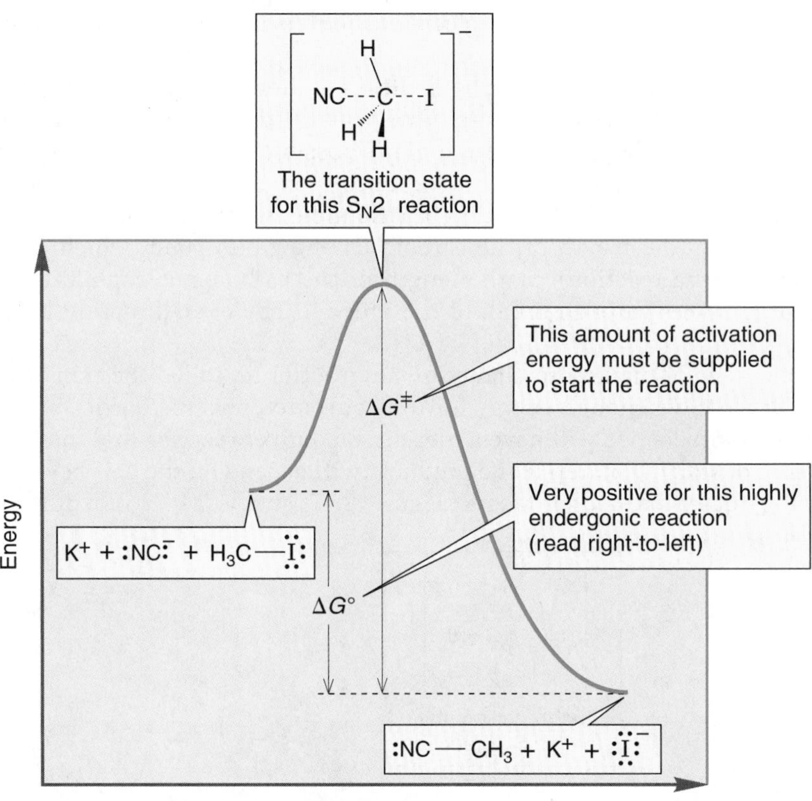

FIGURE **8.18** This energy diagram for the same reaction emphasizes the reversal. Here the activation energy for displacement of cyanide by iodide is shown.

The activation energy is much higher than that for the forward reaction, and $\Delta G°$ is positive for the endergonic formation of methyl iodide. The transition state is exactly the same as that for the forward reaction. This reaction will not be very successful. It not only requires overcoming a high activation energy, but is "fighting" an unfavorable equilibrium as $\Delta G°$ is positive.

Of course, this separation of the two reactions is absolutely artificial—we are really describing a single equilibrium between methyl iodide and cyanide ion on one side and methyl cyanide and iodide ion on the other (Fig. 8.19).

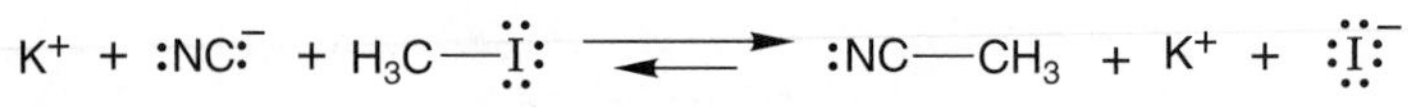

FIGURE **8.19** The reversible competition of cyanide and iodide for the Lewis acidic carbon.

$$K^+ + :NC:^- + H_3C—\ddot{I}: \rightleftharpoons :NC—CH_3 + K^+ + :\ddot{I}:^-$$

The exergonic formation of methyl cyanide and the endergonic formation of methyl iodide are one and the same reaction. The reactions share exactly the same transition state; only the ease of reaching it is different ($\Delta G^{\ddagger}_{forward} < \Delta G^{\ddagger}_{reverse}$). The energy difference between the starting material and the product ($\Delta G°$) is the difference between the two activation energies.

The rates of the forward and reverse reactions, determined by the magnitudes of the activation energies for the two halves of the reaction, are intimately related to the equilibrium constant for the reaction K, determined by $\Delta G°$ [Eq. (8.7)].

$$\Delta G^{\ddagger}_{reverse} - \Delta G^{\ddagger}_{forward} = \Delta G° = -RT \ln K \qquad (8.7)$$

The principle of **microscopic reversibility** says that if we know the mechanism of a forward reaction, we know the mechanism of the reverse. One diagram suffices to tell all about both forward and reverse reactions. Figures 8.17 and 8.18 are in reality a single diagram describing both reactions, in this case one "forward" and exergonic, the other "reverse" and endergonic. There is no lower energy route from methyl cyanide and iodide ion back to methyl iodide and cyanide ion. The lowest energy path back is exactly the reverse of the path forward.

The classic analogy for this idea pictures the molecules as crossing a high mountain pass. The best (lowest energy, or lowest altitude) path from village A to village B is the best (lowest energy, lowest altitude) path back from village B to village A. If there were a lower energy route back from B to A, it would have been the better route from A to B as well (Fig. 8.20).

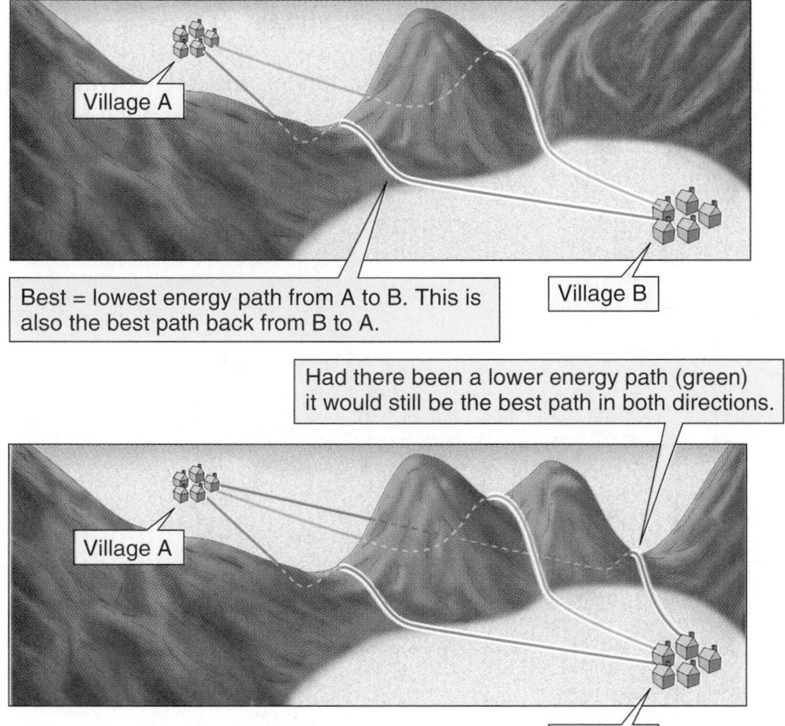

FIGURE **8.20** Molecules, unlike explorers, always take the best (lowest energy) path. You might take the red path from A to B and the orange path back, but a molecule wouldn't.

The Energy versus Reaction progress diagram for the S_N1 reaction is more complicated, because there are more steps. We dealt with the description of this reaction as a sequence of equilibria earlier (Fig. 8.2). Let's again take the reaction of *tert*-butyl iodide with solvent water as an example. The first step is the formation of an intermediate carbocation. We expect this first reaction to be quite endothermic as a carbon leaving group bond must be broken and ions must be formed. The high-energy point in this reaction, transition state 1, occurs as the carbon–iodine bond stretches (Fig. 8.21).

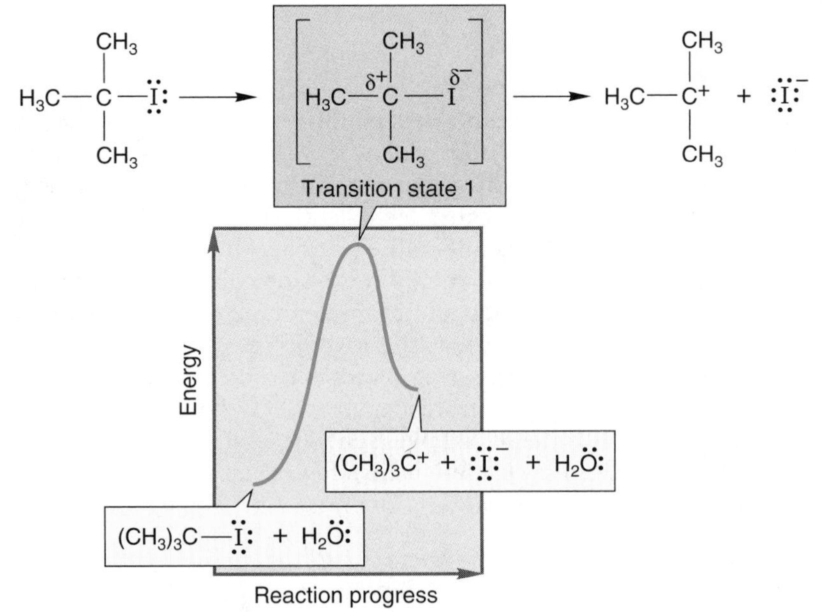

FIGURE 8.21 The first step in the S_N1 reaction of *tert*-butyl iodide in water. The formation of the cation is highly endothermic.

The second step involves the reaction of the nucleophilic solvent water with the carbocation, and is highly exothermic. A bond is made and none is broken. There is a second barrier, and therefore another transition state (2) for this second step of the reaction (Fig. 8.22).

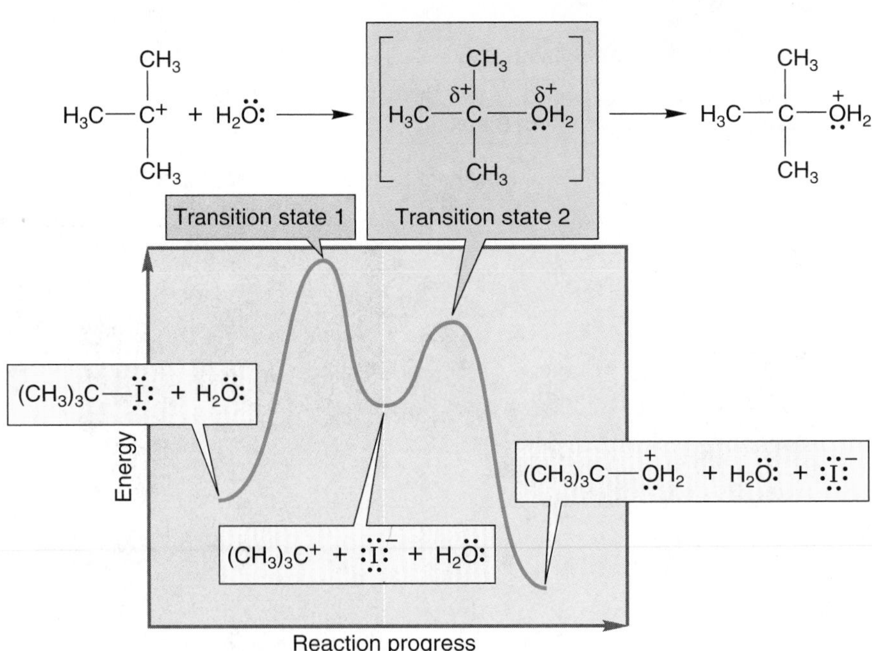

FIGURE 8.22 By contrast, the second step in this S_N1 reaction, the capture of the cation by water, is highly exothermic. Transition state 2 is lower than transition state 1.

There is a third and final step in this particular S_N1 reaction, as a proton must be removed by a base to give the ultimate product, *tert*-butyl alcohol. So there is still another barrier and a third transition state (Fig. 8.23).

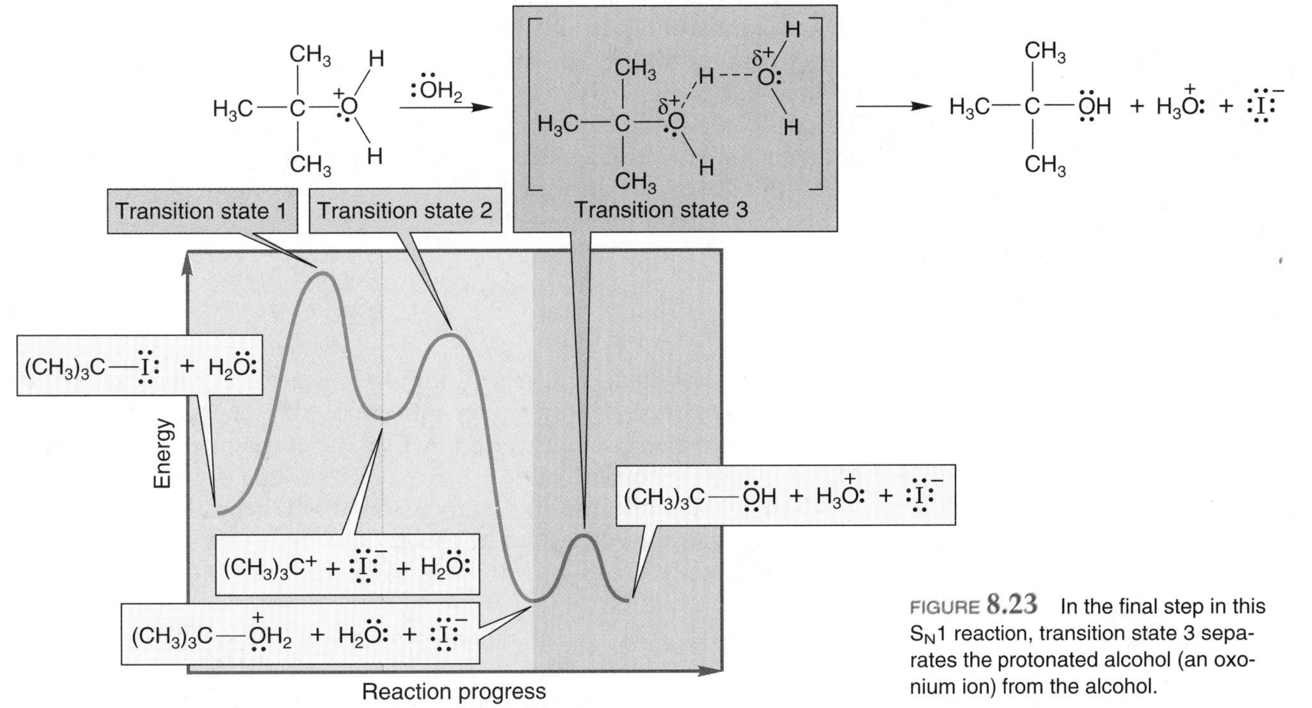

FIGURE 8.23 In the final step in this S_N1 reaction, transition state 3 separates the protonated alcohol (an oxonium ion) from the alcohol.

PROBLEM 8.9

Write an Energy versus Reaction progress profile for the S_N1 reaction of *tert*-butyl iodide in H_2O/KOH containing an additional nucleophile, Nu:$^-$. Note that the major product is not necessarily *tert*-butyl alcohol (Chapter 7, p. 269).

There are six activation energies in the completed diagram for this S_N1 reaction, three for the forward reactions and three for the reverse reactions (Fig. 8.24).

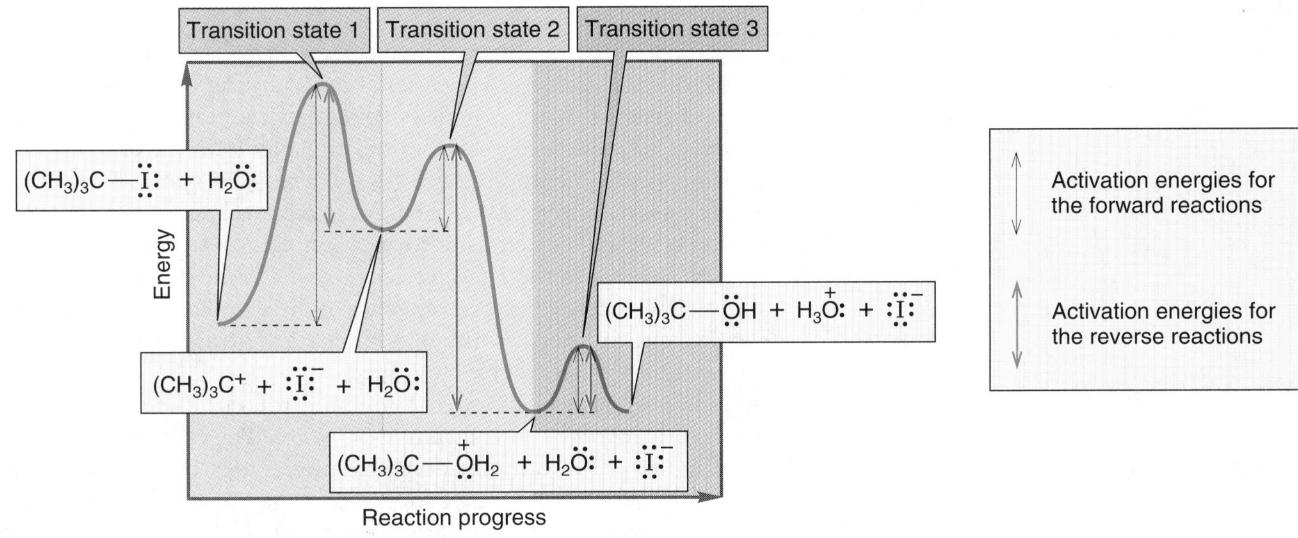

FIGURE 8.24 There are six activation energies involved in this S_N1 solvolysis. In this figure the black double-headed arrows represent activation energies for the forward reactions and the red double-headed arrows represent activation energies for the reverse reactions.

Now, concentrate on these barriers separating the various intermediates. The energy barriers are made up of the same quantities, enthalpy and entropy, that define the energy of any chemical species. Recall that $\Delta G° = \Delta H° - T\Delta S°$ [Eq. (8.6)]. Transition state theory assumes that we can treat the transition state as if it were a real molecule occupying a potential energy minimum. It is important to emphasize that it is *not,* however. It occupies an energy maximum and can never be bottled and examined as real molecules can. Nonetheless it has an energy, as indeed does every point in the energy diagrams on the preceding pages. The energy of the transition state, and thus the height of the energy barrier in the reaction, can be calculated using Eq. 8.8.

$$\Delta G^{\ddagger} = \Delta H^{\ddagger} - T\Delta S^{\ddagger} \tag{8.8}$$

Remember: The double daggers (‡) simply indicate that we are referring to a transition state, not to an energy minimum. The energy barrier, $\Delta G^{\ddagger}$, is the Gibbs free energy of activation, $\Delta H^{\ddagger}$ is the enthalpy of activation, and $\Delta S^{\ddagger}$ is the entropy of activation.

Clearly, the rate constant for any reaction will depend on the height of the barrier, and thus on $\Delta G^{\ddagger}$. The theoretical study of reaction rates gives us the exact equation relating the rate constant k and $\Delta G^{\ddagger}$ [Eq. (8.9)].

$$k = \nu^{\ddagger} e^{-\Delta G^{\ddagger}/RT} \quad \text{or} \quad k = \nu^{\ddagger} e^{-\Delta H^{\ddagger}/RT} e^{\Delta S^{\ddagger}/R} \tag{8.9}$$

It is worth dissecting this equation a bit further. The quantity $\nu^{\ddagger}$ is a frequency—its units are reciprocal seconds (s^{-1}) in this case. In qualitative terms, ν is the frequency with which molecules with enough energy to cross the barrier actually do so. So, the rate constant k is equal to $e^{-\Delta G^{\ddagger}/RT}$ (the fraction of molecules with enough energy to cross the barrier) times $\nu^{\ddagger}$ (the frequency with which such molecules actually do cross the barrier).

8.6 REACTION MECHANISM

Throughout this book we will frequently be concerned with the determination of reaction mechanism. A mechanism is nothing less than the description of the structures and energies of the starting materials and products of a reaction, as well as of any reaction intermediates. In addition, all of the transition states (energy maxima) separating the stable molecules lying in energy minima must be described. It is relatively easy to deal with energy minima—we can often isolate the molecules themselves, take their spectra, and measure their properties. As we can by definition never isolate a transition state, we can never make a direct measurement on it. Yet, if we are to have a good picture of a reaction we must arrive at good descriptions of the transition states involved.

We can usually measure the rate law, which tells us the number and kinds of molecules involved in the transition state. The rate law does not tell us anything about the orientation of the molecules in the transition state. Usually, stereochemical experiments do that. Recall the S_N2 reaction, in which inversion of configuration is observed, thus specifying the direction of approach of the entering nucleophile in the transition state (Chapter 7, p. 247; Fig. 8.25).

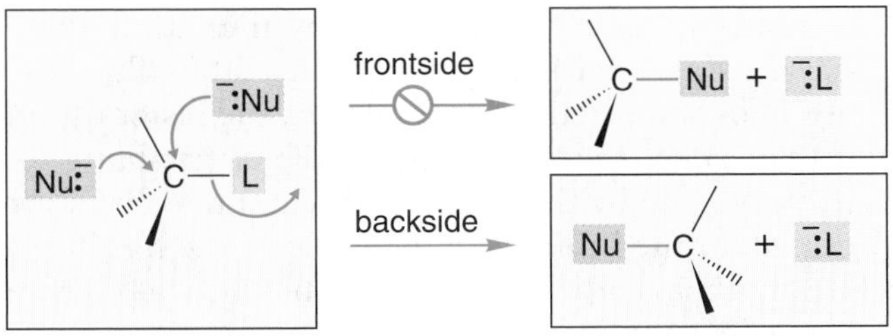

FIGURE 8.25 In the S_N2 reaction, kinetics cannot tell frontside from backside displacement. It takes a stereochemical experiment to do that.

This observation of inversion in the S_N2 reaction tells us that the incoming nucleophile must approach from the rear of the departing leaving group. The rate law, rate = $k[\text{Nu:}^-]$ [R–X] does not tell us that, only that both the nucleophile and the substrate are involved in the transition state. As we work through the important reactions of organic chemistry we will see many similar experiments designed to determine the stereochemistry of the reaction. The energy of the transition state can be measured if good quantitative kinetic experiments can be performed. If the rate constant for the reaction is determined at a variety of temperatures, the height of the transition state can be calculated. But we can do this only for the step with the highest energy transition state. In a multistep reaction, we can only measure the rate for the slowest step in the sequence.

This slow step is called the **rate-determining step**, or, sometimes, the rate-limiting step. It is always the step in the reaction with the highest energy transition state. We cannot measure the rates of faster steps in the reaction. In the S_N1 reaction, for example, the slow step is the ionization of the substrate to a carbocation. This ion may then be captured by a host of nucleophiles in following faster steps. There will be a different product for every nucleophile (Chapter 7, p. 269; Fig. 8.26).

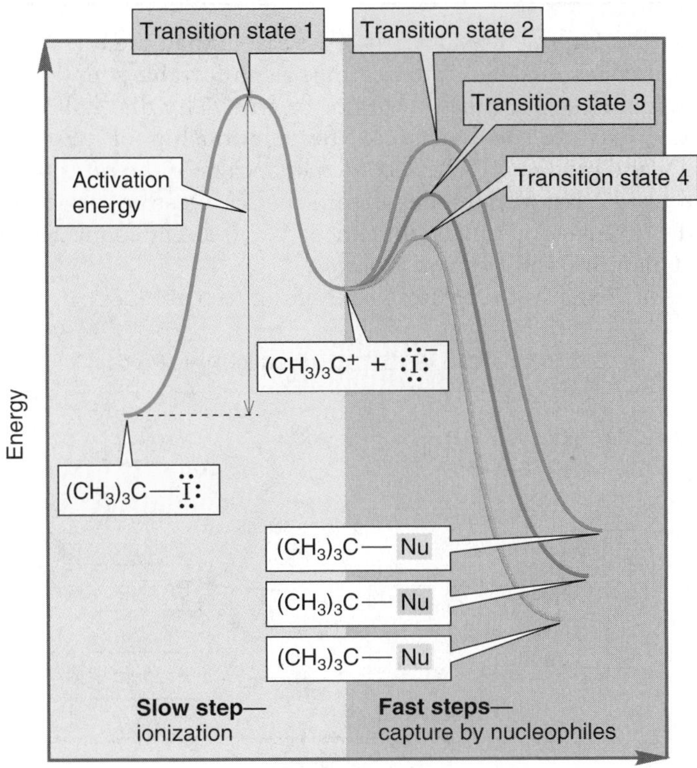

FIGURE 8.26 In this S_N1 reaction, the rate-determining step is the ionization of the starting iodide to give the *tert*-butyl cation. The faster, product-determining steps follow, but their rates cannot be measured kinetically (unless the intermediate carbocation can be isolated and investigated independently).

If we determine the rate of this process by measuring the disappearance of *tert*-butyl iodide over time, we can ultimately find the height of transition state 1 for the reaction. The rates of the faster captures of the carbocation *do not affect the rate of ionization of tert-butyl iodide*. The carbocation is gobbled up by the nucleophiles at a rate faster than that at which it is formed. In the limit, each molecule of the *tert*-butyl cation is captured before another is produced from *tert*-butyl iodide. The rate of formation of product depends on how fast the *tert*-butyl cation is produced in the slow, rate-determining step, but not on the rate of the fast capture by a nucleophile.

Imagine a cage of especially ravenous rats into which we are slowly dropping incredibly delicious food pellets. At first, the pellets are gobbled up by the hungry rats at a rate far faster than they are delivered to the cage. The slow step in this reaction, the step that determines the rate of disappearance of the food, is the rate at which the pellets are delivered to the cage. This process is analogous to a reaction in which the first step (food dropping into the cage) is slower than the second step (the rats eating the food). Later, however, a different situation obtains. Even these rats can become satiated and the food pellets accumulate in the cage as the no-longer starving rodents largely ignore them, only taking a nibble now and then. The second step is now slower than the first step. Now the step that determines the rate of disappearance of the food is the rate at which the rats eat the pellets. This second "reaction" has a different rate-determining step than the first.

8.7 THE HAMMOND POSTULATE: THERMODYNAMICS VERSUS KINETICS

We are used to saying, Product Y is more stable than product X, and therefore is formed faster. This statement has a comfortable sound to it, and intuitively makes sense. But it need not be true! The first half of the statement describes **thermodynamics**, the relationship of the energies of product X and product Y. The second half speaks of the *rates* of formation of X and Y, which is a statement about **kinetics**. Despite the good intuitive feel of the original statement, there need be no direct connection between thermodynamic stabilities and kinetics.

Figure 8.27 shows the relative energies of two products, X and Y, being

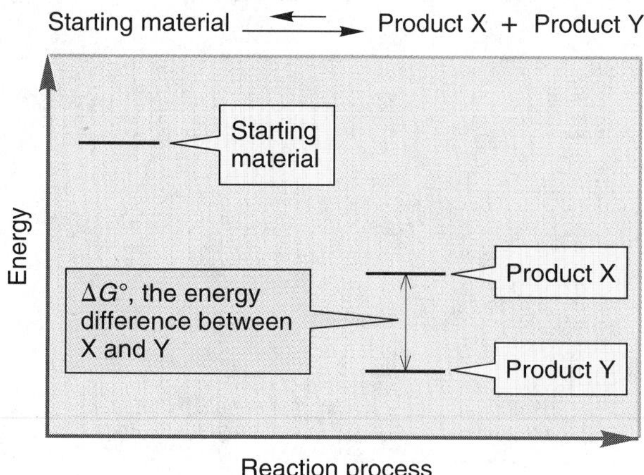

FIGURE 8.27 Two products, X (less stable) and Y (more stable), being formed from starting material.

formed from a starting material. One might expect that it would be the $\Delta G°$ between the two products X and Y that will determine the amount of the two compounds formed, but this is true only if X and Y are in equilibrium, via the starting material (Fig. 8.28).

Product X ⇌ Starting material ⇌ Product Y

FIGURE 8.28 If X and Y are in equilibrium with starting material (and thus with each other) it will be the $\Delta G°$ between X and Y that determines the ratio of products.

It is conceivable that the products are *not* in equilibrium. There might not be enough energy for the products to go back to the starting material. Under these conditions, the amounts of X and Y will be determined not by their relative energies but only by the heights of the transition states leading to them (Fig. 8.29).

Figure 8.29 shows a conventional picture in which the transition state leading to the more stable product, Y, is lower in energy than the transition state leading to the higher energy product, X. Under such conditions, the major product will be the more stable compound, Y. We have seen many examples of reactions like this.

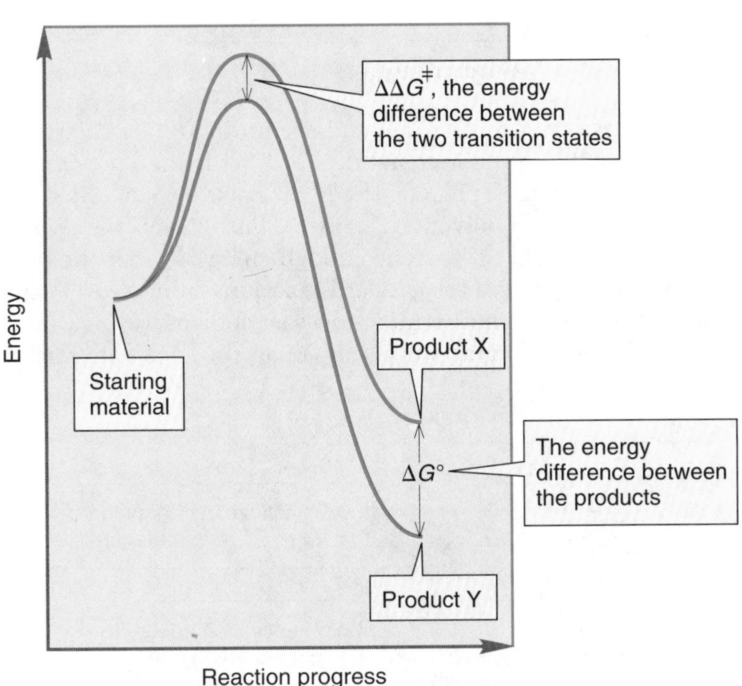

FIGURE 8.29 The usual picture is this: The higher transition state leads to the less stable product. The lower transition state leads to the more stable product.

PROBLEM 8.10

Identify the activation energies for the reactions of X and Y to give starting material (Fig. 8.29).

But it could be the case that the transition state leading to the more stable compound, Y, is actually higher in energy than the transition state leading to the less stable product, X (Fig. 8.30).

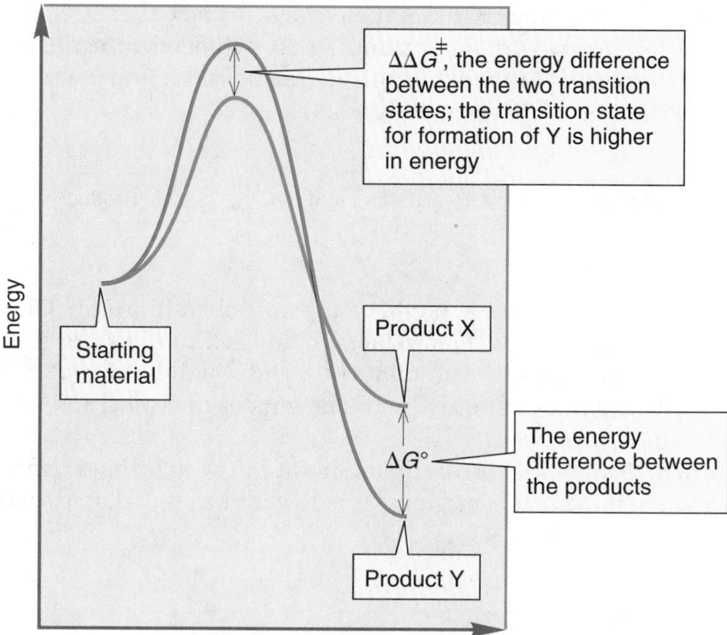

FIGURE 8.30 But the situation in Figure 8.29 could be otherwise; the higher transition state could lead to the lower energy product, and the lower energy transition state to the higher energy product.

This inversion will not matter if X and Y equilibrate via the starting material. In such a case, the greater stability of Y will win out eventually, and the ratio of X and Y will depend on the $\Delta G°$ between them. Such a reaction is said to be under **thermodynamic control**. *But what if the two products do not equilibrate*? What if there is enough energy so that the forward reactions are possible, but the backward reactions from X or Y to starting material are not? In the exothermic processes shown, the forward reaction is easier—has a lower activation barrier—than the backward reaction (Fig. 8.31).

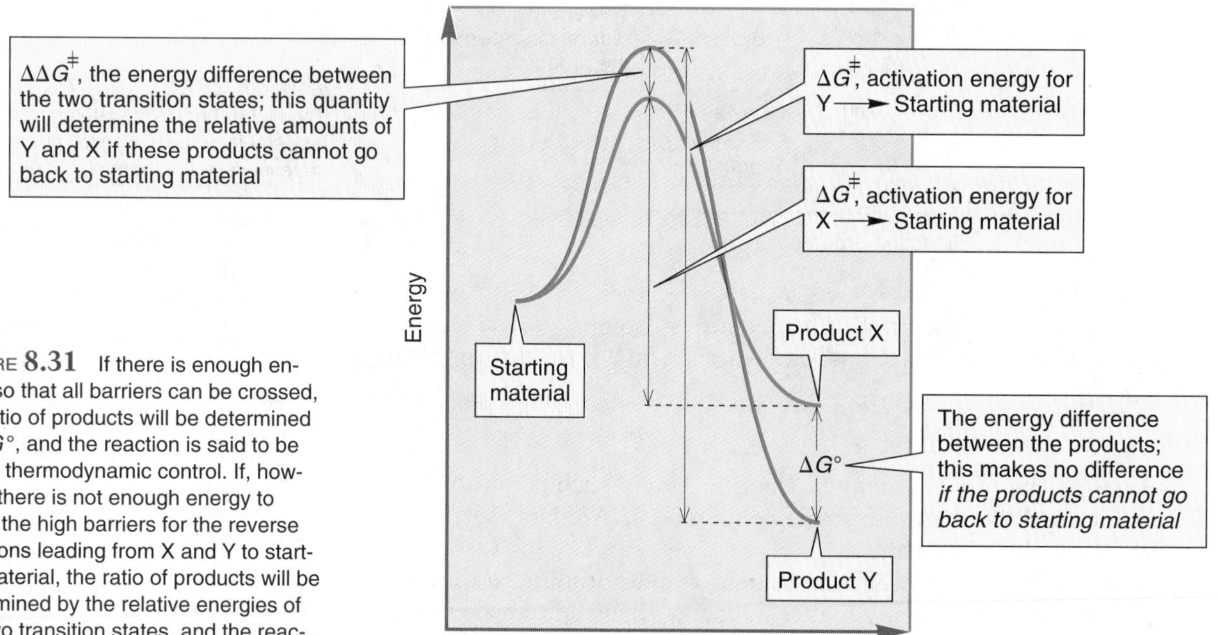

FIGURE 8.31 If there is enough energy so that all barriers can be crossed, the ratio of products will be determined by $\Delta G°$, and the reaction is said to be under thermodynamic control. If, however, there is not enough energy to cross the high barriers for the reverse reactions leading from X and Y to starting material, the ratio of products will be determined by the relative energies of the two transition states, and the reaction is said to be under kinetic control.

Under these circumstances (which are not all that uncommon), the relative amounts of X and Y depend only on the relative heights of the transition states leading to them. Such a reaction is said to be under **kinetic control**.

There is no *necessary* connection between thermodynamics and kinetics. In our original statement, Product Y is more stable than product X, and *therefore* is formed faster; we are equating thermodynamics and kinetics. We tacitly assume that it will always be the case that the product of greater stability (lower energy) will be reached at a faster rate by the lower energy (more stable) transition state (Fig. 8.29).

In principle, it could be that the transition state for the reaction leading to the more stable product is actually higher in energy than the transition state leading to the less stable product. Such a situation can lead to the preferred formation of the less stable product, if the reaction is under kinetic control (Fig. 8.31).

Fortunately, there are factors that operate to preserve a parallelism between the energies of products and the transition states leading to them. There are examples in which the higher energy transition state leads to the lower energy product, and *vice versa,* but they are relatively rare. We will be at pains to point them out as they occur. The rest of this section will try to show you why the parallelism between product energy and transition state energy exists for many reactions.

Consider, for a start, Problem 8.11.

*PROBLEM **8.11**

Alkenes can be protonated to give carbocations. Unsymmetrical alkenes, such as 2-methyl-2-butene, could be protonated at either end of the double bond to give two different carbocations. Predict the direction of protonation of 2-methyl-2-butene and explain your answer. (This reaction will be described in detail in Chapter 9) See Figure 8.32.

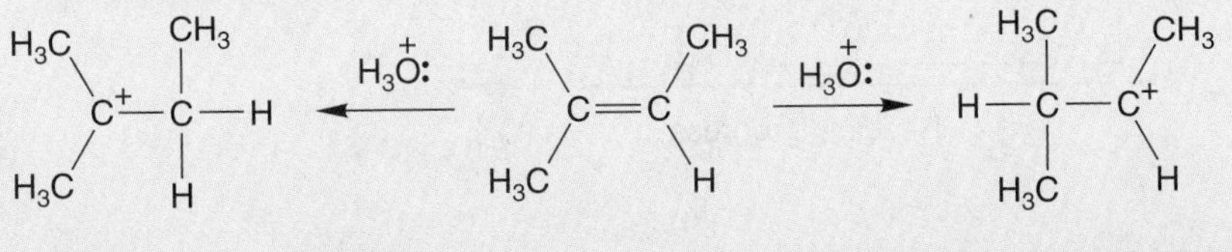

$(H_3C)_2C^+{-}CH(CH_3){-}H \xleftarrow{H_3\overset{+}{O}:} (H_3C)_2C{=}C(CH_3)H \xrightarrow{H_3\overset{+}{O}:} H{-}C(CH_3)_2{-}C^+(CH_3)H$

FIGURE **8.32**

ANSWER

The choice is between the formation of a secondary and a tertiary carbocation. It is surely tempting to predict that the more stable tertiary carbocation will be formed faster than the less stable secondary carbocation.

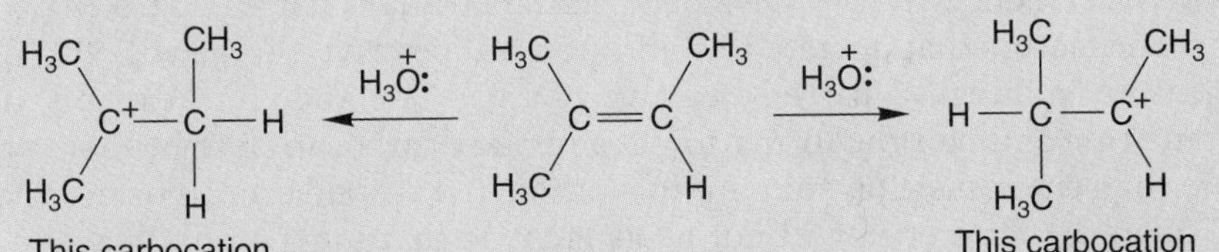

$(H_3C)_2C^+{-}CH(CH_3){-}H \xleftarrow{H_3\overset{+}{O}:} (H_3C)_2C{=}C(CH_3)H \xrightarrow{H_3\overset{+}{O}:} H{-}C(CH_3)_2{-}C^+(CH_3)H$

This carbocation is tertiary

This carbocation is secondary

However, this comfortable reasoning equates thermodynamics (more stable) with kinetics (will be formed faster), and this equation is not necessarily valid. Read on.

In one sense, the answer to this problem is obvious; the more stable tertiary carbocation will be formed faster than the less stable secondary carbocation. But this reasoning again equates *thermodynamics* "the *more stable* tertiary carbocation ... " with *kinetics* "will be formed *faster* than ... ." Before we accept it, we had better think it through. Look first at the Energy versus Reaction progress diagram (Fig. 8.33).

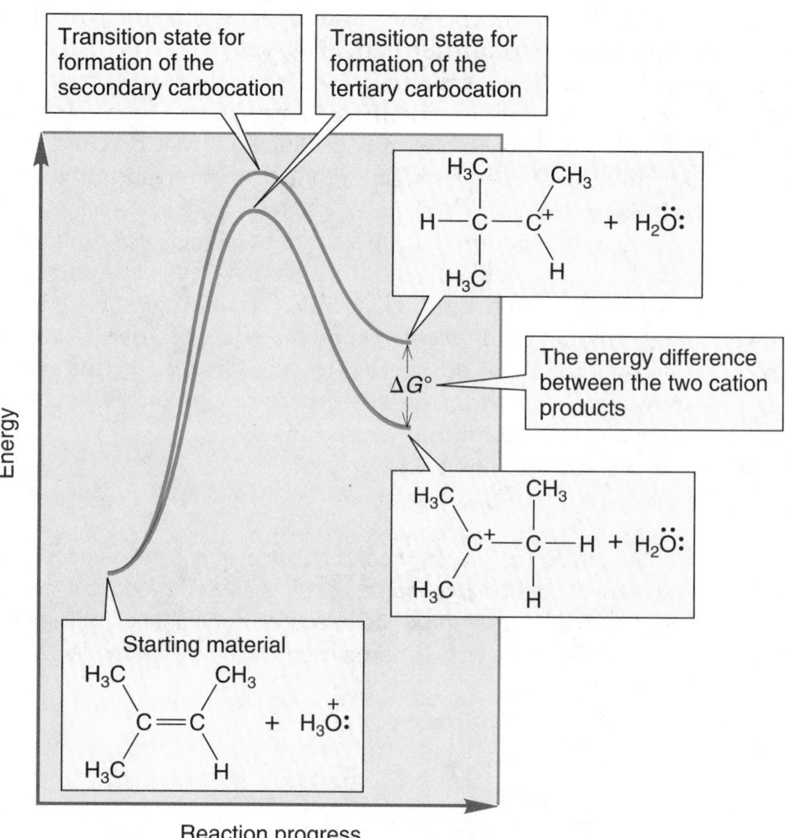

FIGURE 8.33 The more substituted cation will be at lower energy than the less substituted cation. Note also that the transition state for formation of the tertiary cation is lower than the transition state for formation of the secondary cation. Why?

The tertiary cation is more substituted than the secondary cation and thus is more stable (Chapter 7, p. 268). Substitution stabilizes the positive charge on carbon in these species. In the transition states for formation of the carbocations, the new bond from carbon to hydrogen will be partially formed and the carbons will have developed partial positive charges (Fig. 8.34).

The factors that operate to stabilize a full positive charge will also operate to stabilize a partial positive charge. One therefore expects the transition state for the formation of a tertiary carbocation to be more stable than the transition state for formation of a secondary carbocation, although the difference will not be as great as that for the fully developed cations (Fig. 8.35).

This specific situation can be generalized, in what is called the **Hammond postulate**, after its originator, G. S. Hammond (b. 1921). "For an endothermic reaction the transition state will resemble the product." Note immediately that this statement can be turned around. "For an

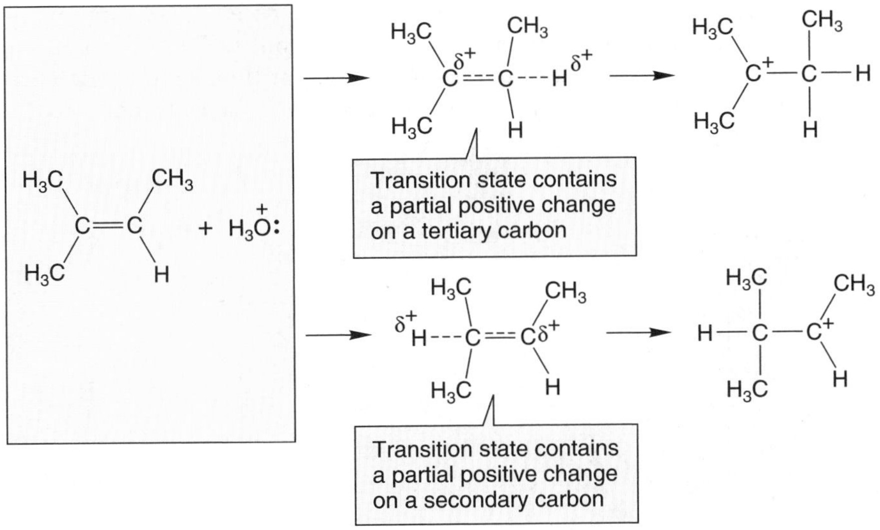

Transition state contains a partial positive change on a tertiary carbon

Transition state contains a partial positive change on a secondary carbon

FIGURE 8.34 In the two transition states, partial positive charge has built up on carbon. A partial positive charge on a tertiary carbon is more stable than a partial positive charge on a secondary carbon.

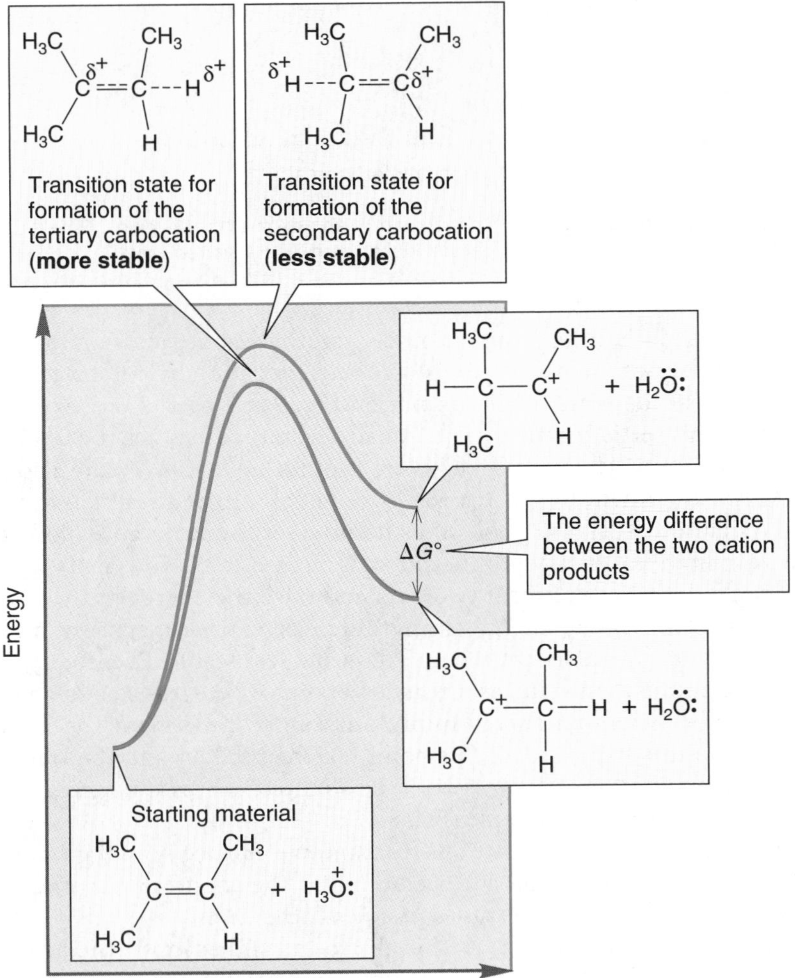

FIGURE 8.35 The energy difference between the tertiary and secondary cationic products will be reflected in the transition states for cation formation. There will be less difference in energy between the two transition states (bearing partial positive charges) than there is between the two products, in which there are full positive charges.

exothermic reaction the transition state will resemble the starting material." Figure 8.36 shows an endothermic reaction, but we know that all we have to do to see an exothermic reaction is look at the endothermic reaction in the other direction! So these two versions of the Hammond postulate are really the same.

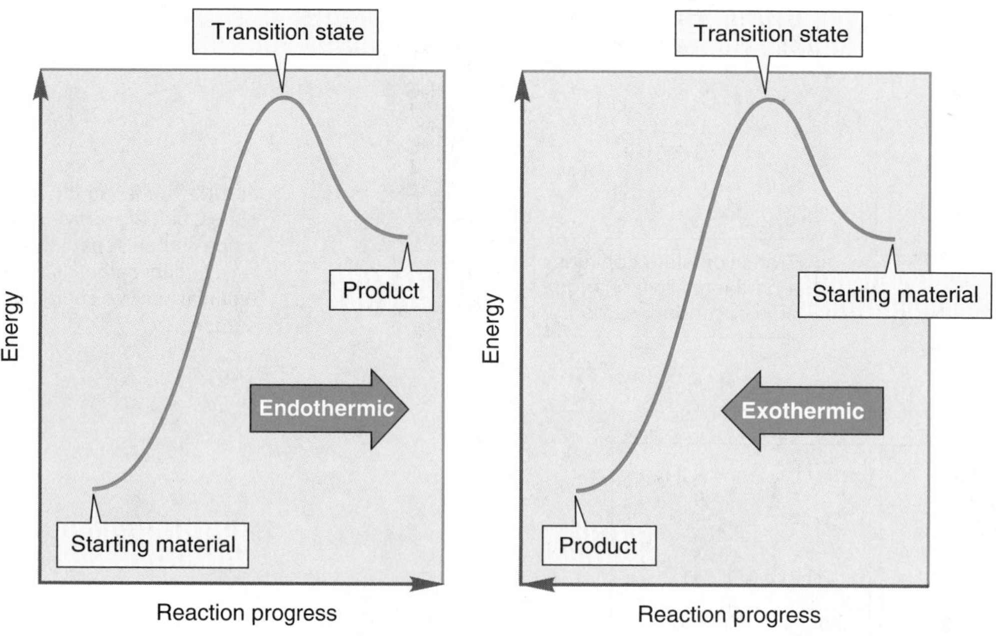

FIGURE 8.36 Energy diagrams for endothermic and exothermic reactions. Note that these are really exactly the same reaction looked at in different directions.

In the endothermic protonation of 2-methyl-2-butene, we rationalized the faster formation of the more stable tertiary cation by pointing out that in the transition state for its formation there was partial positive charge built up on a tertiary carbon. The transition state resembles the final product, just as the Hammond postulate would have it. We can also do a thought experiment to see that the more endothermic a reaction the more the transition state will look like the product. Imagine an endothermic reaction with the energy profile shown in Figure 8.36. Now apply a pair of molecule-grabbing chemical tweezers and raise the energy of the product without changing the energy of anything else. As we magically increase the energy of the product it becomes less and less stable. Eventually, it *becomes* the transition state, as it reaches a point where it is so unstable that it no longer lies in an energy minimum. Now think about the moment just before this happens. At this point (ε), the product and the transition state are so close to each other that the slightest increase in energy in the product makes them the same (Fig. 8.37).

By reflecting on this limiting situation, we can see how the transition state and the product approach each other as the energy of the product increases and the reaction becomes more endothermic.

It is a simple matter to turn this all around and examine the other side of the Hammond postulate, "The more exothermic a reaction the more the transition state resembles starting material." Figure 8.37 will do it for you if you simply read it backward, from right to left. See what happens to

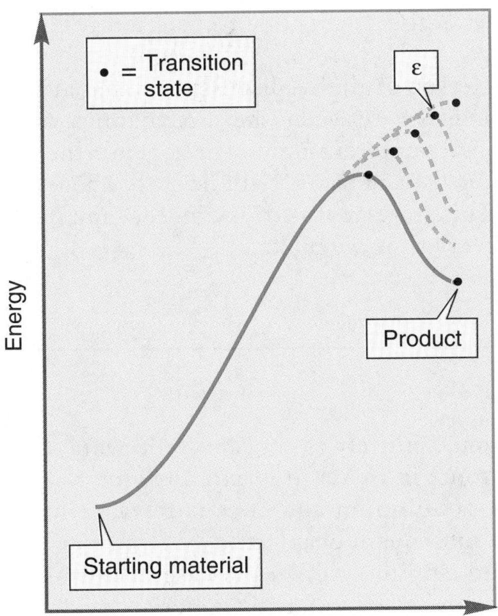

FIGURE 8.37 A thought experiment in which we raise the energy of the product without changing the energy of the starting material. As the product gets higher and higher in energy, the transition state for its formation resembles it more and more. The easiest way to see this is to look at point ε at which *any* further increase in energy makes the transition state and the product the same. At this point the product and the transition state are almost identical. Therefore, the more endothermic a reaction, the more the transition state will look like the product. It is easy to see the other half of the Hammond postulate by reading this diagram backward (the more exothermic the reaction, the more the transition state will resemble the starting material).

Aconitase

Aconitase

Citrate ⇌ ***cis*-Aconitate** ⇌ **Isocitrate**

In this chapter, we discuss the interconversion of compounds by passing from starting material to product over a high-energy point called the transition state. The higher the transition state, the slower the reaction. In the laboratory, we affect the rate of reaction by adding acid or base catalysts, changing solvents, or sometimes simply by supplying lots of energy. Our bodies run on the same chemistry found in the laboratory. For example, the substitution and elimination reactions described in Chapter 7 are all reasonably common in the transformations of biomolecules carried out in us.

Nature can't overcome activation barriers to reactions important to us by adding catalysts such as sulfuric acid, and turning up the heat in our bodies to provide energy becomes uncomfortable very quickly indeed. So, Nature does it another way, and it is most successful. Nature uses enzymes, polyamino acid chains with catalytic activity, to lower the barriers to chemical transformations. Aconitase, an enzyme of molecular weight 89,000 (!) catalyzes the conversion of citrate to *cis*-aconitate, an elimination reaction. Aconitate is then hydrated to give isocitrate. So, this enzyme makes possible one of the key reactions of Chapter 7, elimination, and then reverses the process, but in the other direction, to give the isocitrate.

The vast molecular architecture of this enormous molecule serves to bind the substrate precisely, holding it in perfect position for the side chains of the amino acids to act as catalysts, speeding the elimination and addition reactions. Spectacular rate accelerations can be achieved by enzymes, allowing all sorts of reactions to take place at 37 °C that could not occur otherwise.

the transition state as the "right-to-left" reaction becomes less and less exothermic.

The take-home lesson of all this discussion is that although one cannot take the correspondence between thermodynamics and kinetics as a given—there will be counterexamples—in general, the statement, "Y is more stable and thus formed faster" will be true. There is more to it than it appears at first. This statement is tricky indeed, and it is worth stopping and examining it every time we make it.

8.8 SUMMARY

NEW CONCEPTS

This chapter is devoted entirely to concepts. You won't find a new mechanism or synthetic route in it. Yet, it is extremely important because what we say here about equilibrium and rates is relevant to all the reactions that will appear in subsequent chapters.

All reactions are equilibria, or in the case of multistep reactions, sequences of equilibria between a starting material and a product. The equilibrium constant K is related to the difference in energy between starting material and products in a logarithmic fashion ($\Delta G° = -RT \ln K$ or $K = e^{-\Delta G°/RT}$). Therefore, a small amount of energy difference (ΔG) has a great influence on the equilibrium constant K.

The ratios of products depend on either the energy difference between the products $\Delta G°$ (thermodynamic control), or on the relative heights of the transition states leading to products (kinetic control).

Differences in G, the Gibbs free energy, whether they be $\Delta G°$, the difference in energy between starting material and product, or $\Delta G^{\ddagger}$, the difference in energy between starting material and the transition state (activation energy), are made up of entropy ($\Delta S°$) and enthalpy ($\Delta H°$) terms ($\Delta G° = \Delta H° - T\Delta S°$ or $\Delta G^{\ddagger} = \Delta H^{\ddagger} - T\Delta S^{\ddagger}$).

The rate of a reaction depends on temperature, the concentration of the reactant(s), and the rate constant for the reaction, k. The rate constant is a property of the reaction and depends on temperature, pressure, and solvent, but not upon the concentrations of the reactants.

The rate-determining step of a reaction is the step with the highest energy transition state. It may, or may not be the same as the product-determining step. For example, in the S_N2 reaction it is the same, but in the S_N1 reaction it is not.

The principle of microscopic reversiblility says that the lowest energy path in one direction of an equilibrium reaction is the lowest energy path for the reverse reaction.

The Hammond postulate says the transition state for an endothermic reaction resembles the product, or, equivalently, that the transition state for an exothermic reaction resembles the starting material.

COMMON ERRORS

The most common problem experienced with the material in this chapter is the confusion between thermodynamics and kinetics. Rates of reactions (kinetics) are determined by the relative heights of the available transition states leading to products. If there is insufficient energy available for the products to re-form starting material, it will be the relative heights of the transition states that determine the mixture of products, no matter what the energy differences among those products (Fig. 8.38).

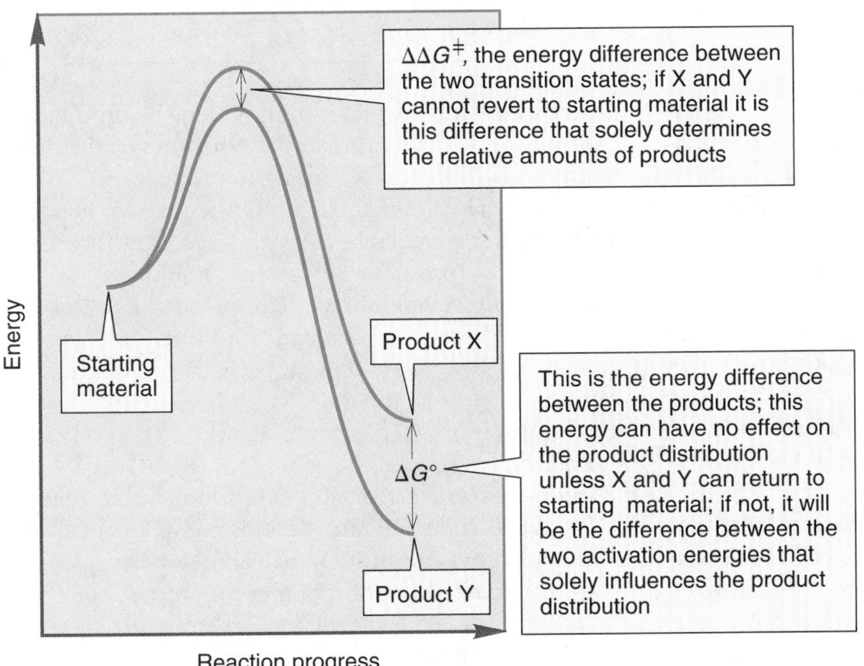

FIGURE 8.38 Kinetic control of product distribution.

On the other hand, if equilibrium is established; if the products can revert to starting material, it will be the relative energies of the product molecules that determine how much of each is formed (Fig. 8.39).

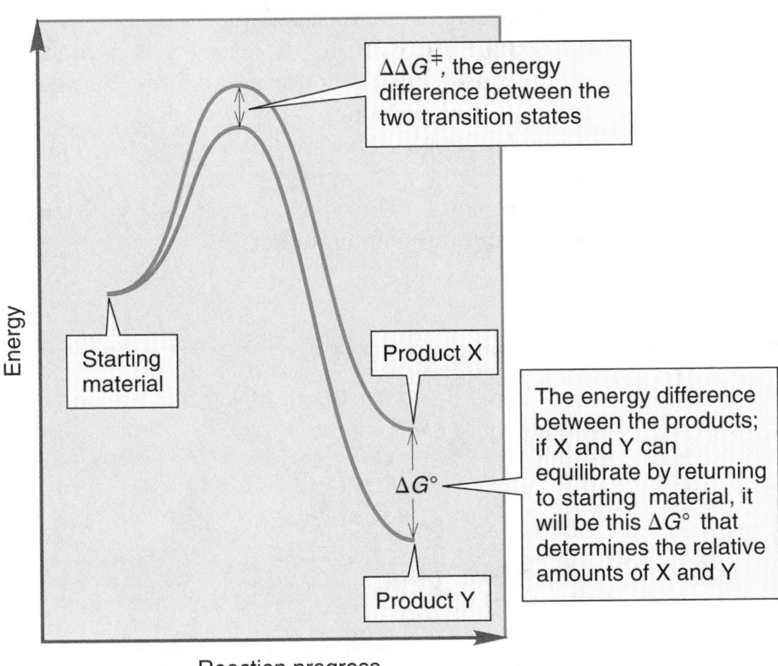

FIGURE 8.39 Thermodynamic control of product distribution.

Remember also to be careful to distinguish between a *rate* of a reaction and the *rate constant* for that reaction. The rate depends on the concentration(s) of the reacting molecules; the rate constant does not, and is a fundamental property of the reaction.

8.9 KEY TERMS

Activation energy ($\Delta G^{\ddagger}$) The difference in free energy between the starting material and the transition state in a reaction.

Boltzmann distribution The range of energies of a set of molecules at a given temperature.

Endergonic In an endergonic reaction the products are less stable than the starting materials.

Enthalpy change ($\Delta H°$) The difference in total bond energies between starting material and product in their standard states.

Entropy change ($\Delta S°$) The difference in disorder between the starting material and product in their standard states.

Equilibrium constant (K) The equilibrium constant is related to the difference in energy between starting material and products ($\Delta G°$) in the following way: $K = e^{-\Delta G°/RT}$.

Exergonic In an exergonic reaction, the products are more stable than the starting materials.

First-order reaction A reaction for which the rate depends on the product of a rate constant and the concentration of a single reagent.

Gibbs free energy change ($\Delta G°$) The difference in free energy during a reaction. The parameter $\Delta G°$ is composed of an enthalpy ($\Delta H°$) term and an entropy ($\Delta S°$) term. $\Delta G° = \Delta H° - T\Delta S°$.

Hammond postulate The transition state for an endothermic reaction will resemble the product. It can be equivalently stated as, The transition state for an exothermic reaction will resemble the starting material.

Kinetic control A reaction in which the product distribution is determined by the heights of the different transition states leading to products.

Kinetics The determination of the rates of reaction.

Le Chatelier's principle A system at equilibrium adjusts so as to relieve any stress upon it.

Microscopic reversibility The notion that the lowest energy path for a reaction in one direction will also be the lowest energy path in the other direction.

Oxonium ion A molecule containing a trivalent, positively charged, oxygen atom (R_3O^+).

Pseudo-first-order reaction A bimolecular reaction in which the concentration of one reagent (usually the solvent) does not change appreciably.

Rate constant (k) A fundamental property of a reaction that depends on the temperature, pressure, and solvent, but not on the concentrations of the reactants.

Rate-determining step The step in a reaction with the highest activation energy.

Second-order reaction A reaction for which the rate depends on the product of a rate constant and the concentrations of two reagents.

Thermodynamic control A reaction in which the product distribution is determined by the relative energies of the products.

Thermodynamics The study of energetic relationships.

Transition state The high point in energy between starting material and product.

8.10 ADDITIONAL PROBLEMS

PROBLEM 8.12 The Finkelstein reaction, shown below, is often used to prepare alkyl iodides from alkyl chlorides or bromides through an S_N2 reaction:

$$R{-}X + NaI \rightleftharpoons R{-}I + NaX$$

X = Cl or Br

(a) Suggest a simple way to drive this equilibrium toward the desired iodide.

(b) In practice, the Finkelstein reaction is usually run in acetone solvent, as sodium iodide, but not sodium chloride or sodium bromide, is soluble in acetone. Explain carefully how this difference in solubilities drives the equilibrium toward the desired iodide.

PROBLEM 8.13 Alcohol dehydration and alkene hydration is an equilibrium process.

OH $\underset{}{\overset{H_3\ddot{O}^+}{\rightleftharpoons}}$ + $H_2\ddot{O}$:

(a) Suggest experimental conditions that will favor cyclohexene.

(b) Suggest experimental conditions that will favor cyclohexanol.

PROBLEM **8.14** Calculate the energy differences between the carbonyl (C=O) compounds and their hydrates at 25 °C.

(a)

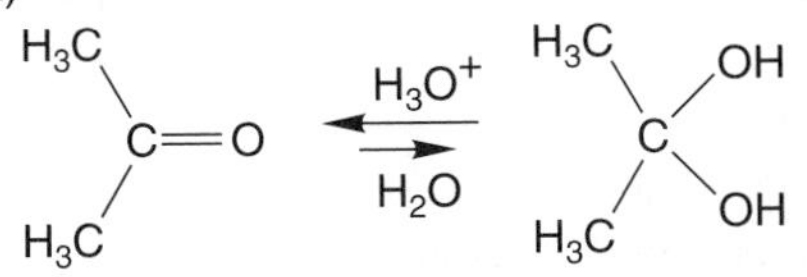

$(H_3C)_2C{=}O \underset{H_2O}{\overset{H_3O^+}{\rightleftharpoons}} (H_3C)_2C(OH)_2 \quad K = 1.4 \times 10^{-3}$

(b)

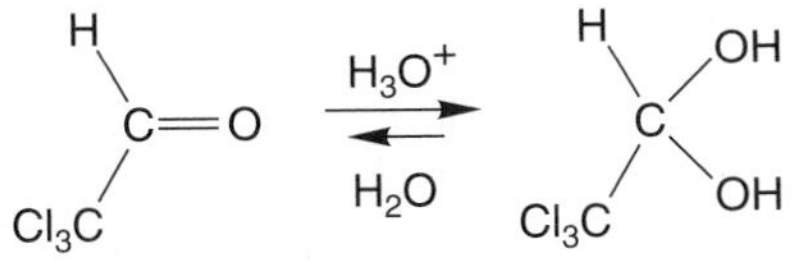

$(Cl_3C)HC{=}O \underset{H_2O}{\overset{H_3O^+}{\rightleftharpoons}} (Cl_3C)HC(OH)_2 \quad K = 2.8 \times 10^{4}$

PROBLEM **8.15** Under certain highly acidic conditions, the three bicyclooctanes shown below can be isomerized. Given the following ratios at equilibrium, calculate the relative energies of the three isomers at 25 °C.

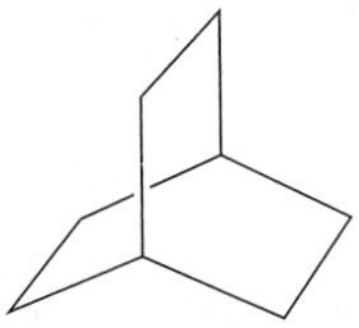

Bicyclo[2.2.2]octane
(3.66%)

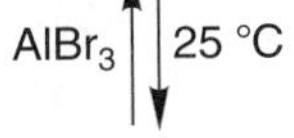

$AlBr_3$ | 25 °C

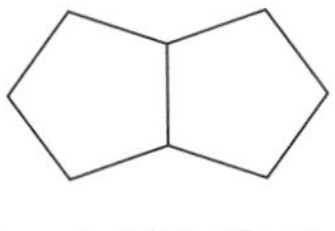

Bicyclo[3.3.0]octane
(32.95%)

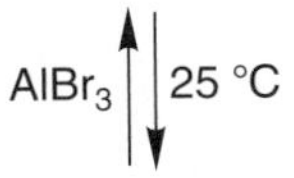

$AlBr_3$ | 25 °C

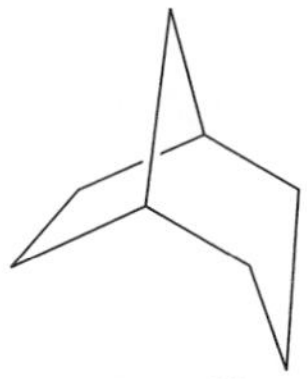

Bicyclo[3.2.1]octane
(63.35%)

PROBLEM **8.16** It is possible to equilibrate the following two ethers. It is found that **B** is favored over **A** by 1.51 kcal/mol. What is the equilibrium constant for this interconversion at 25 °C?

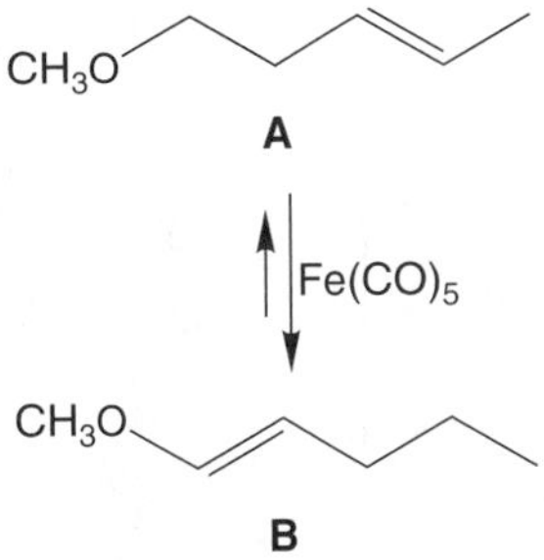

CH_3O (A)

$Fe(CO)_5$

CH_3O (B)

PROBLEM **8.17** Use the bond dissociation energies of Table 8.2 (p. 307) to estimate Δ*H*° for the following reactions. Reaction mechanisms are not necessary!

(a)

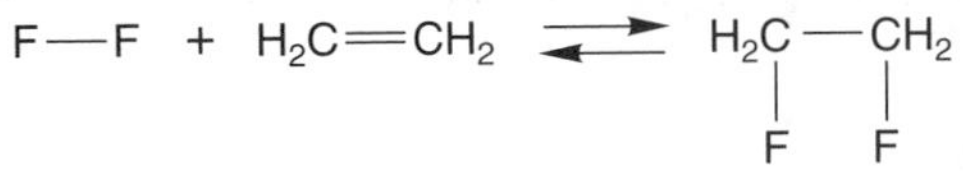

$F{-}F + H_2C{=}CH_2 \rightleftharpoons H_2C(F){-}CH_2(F)$

(b)

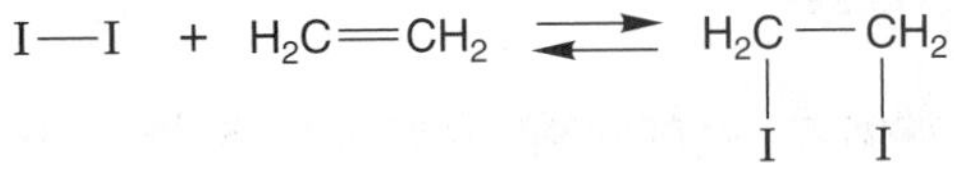

$I{-}I + H_2C{=}CH_2 \rightleftharpoons H_2C(I){-}CH_2(I)$

(c)

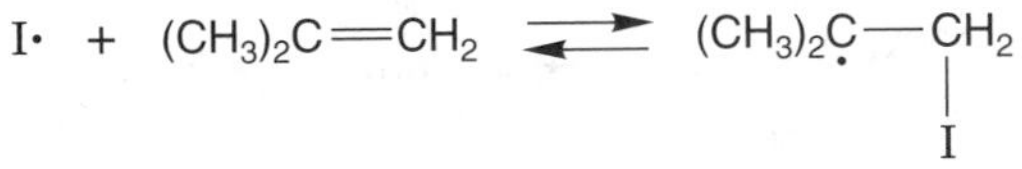

$I\cdot + (CH_3)_2C{=}CH_2 \rightleftharpoons (CH_3)_2\dot{C}{-}CH_2(I)$

(d)

$(CH_3)_2\dot{C}{-}CH_2(Cl) + H{-}Cl \rightleftharpoons (CH_3)_2C(H){-}CH_2(Cl) + Cl\cdot$

PROBLEM **8.18** Use the data in Table 8.2 (p. 307) to design two endothermic reactions. Don't worry about mechanism! This problem does not ask you to design a *reasonable* reaction—it merely asks you to find reactions that would be endothermic *if* they occurred.

PROBLEM **8.19** For the equilibrium A + B ⇄ C, calculate:
(a) Δ*G*° at 25 and 200 °C given Δ*H*° = –14.0 kcal/mol and Δ*S*° = –15.8 cal/deg·mol
(b) Δ*G*° at 25 and 200 °C given Δ*H*° = –6.0 kcal/mol and Δ*S*° = –15.8 cal/deg·mol.

PROBLEM **8.20** Construct Energy versus Reaction progress diagrams for E1 and E2 reactions that are overall (a) exothermic, (b) endothermic. Use a tertiary iodide as a sample substrate molecule.

PROBLEM 8.21 For the following two reactions, described by the two Energy versus Reaction progress diagrams, (1) and (2):

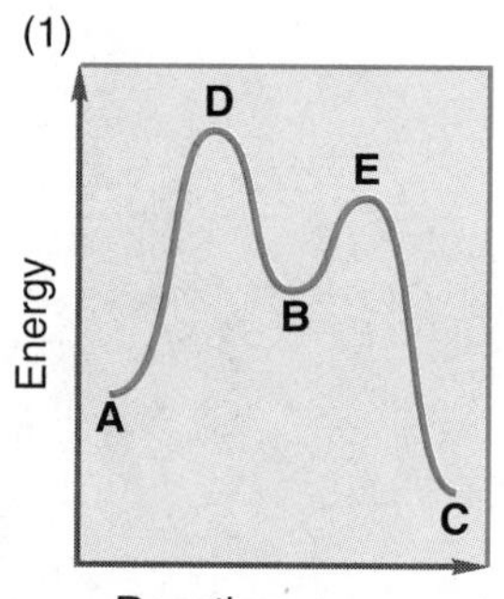

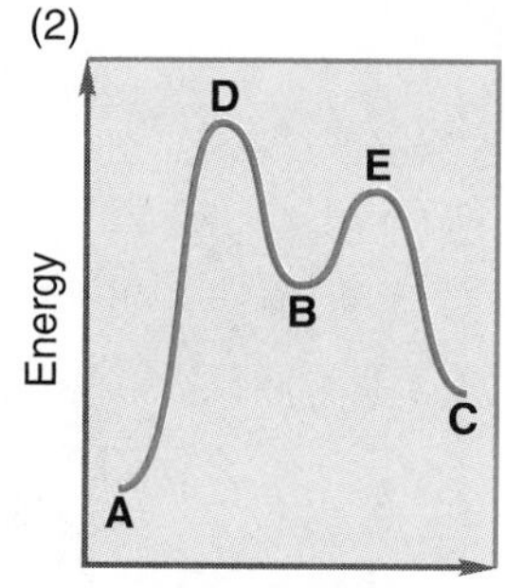

(a) In each case, what compounds will be present at the end of the reaction?

(b) In each case, which compound will be present in the largest amount? Which will be present in the smallest amount?

(c) What is the rate-determining step for the left-to-right reaction in each case?

(d) Use the diagrams to point out in each case the activation energy for the reactions:

A → **B**, **A** → **C**, **C** → **B**, and **C** → **A**

(e) In each case, what is the rate-determining step for the reaction **A** → **C**?

PROBLEM 8.22 Explain in painstaking detail why one is justified in saying, The reaction of 2-methyl-2-butene with HCl leads to the tertiary chloride because a tertiary carbocation is more stable than a secondary carbocation. You might start with the construction of Energy versus Reaction progress diagrams and good drawings for the transition states.

PROBLEM 8.23 Find the errors in the following Energy versus Reaction progress diagrams.

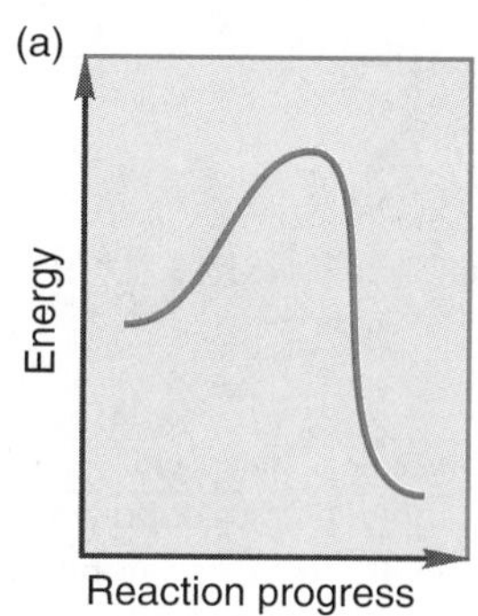

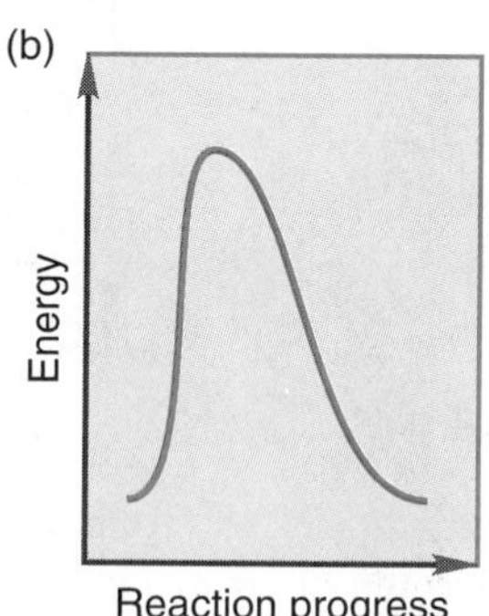

9

Additions to Alkenes 1

> In the last third of his life, there came over Laszlo Jamf—so it seemed to those who from out in the wood lecture halls watched his eyes slowly granulate, spots and wrinkles grow across his image, disintegrating it towards old age—a hostility, a strangely *personal* hatred, for the covalent bond.
>
> —Thomas Pynchon*
> *Gravity's Rainbow*

In this chapter, we continue a discussion of fundamental building block reactions of organic chemistry. In Chapter 7, we looked at the S_N1, S_N2, E1, and E2 reactions, and we continue here with an examination of additions to π systems. We start with reactions of H–Br and H–Cl. Then we move on to related H–X additions (Fig. 9.1).

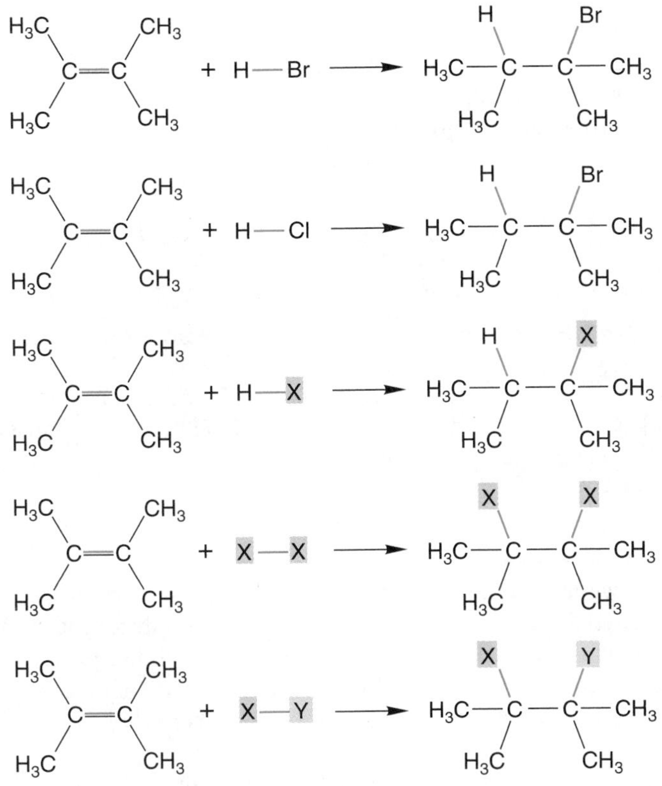

FIGURE 9.1 A variety of additions to alkenes. Many H–X, X_2, and X–Y reagents undergo the addition reaction.

*Thomas Pynchon is an American author born in 1937.

Polar reactions such as these are best understood in terms of filled–empty, HOMO–LUMO orbital interactions (Fig. 9.2). As we have seen many times before, when a filled, occupied orbital overlaps an empty orbital, the two electrons are stabilized in the new, lower energy molecular orbital. Many more examples of this kind of stabilization will be seen in this chapter.

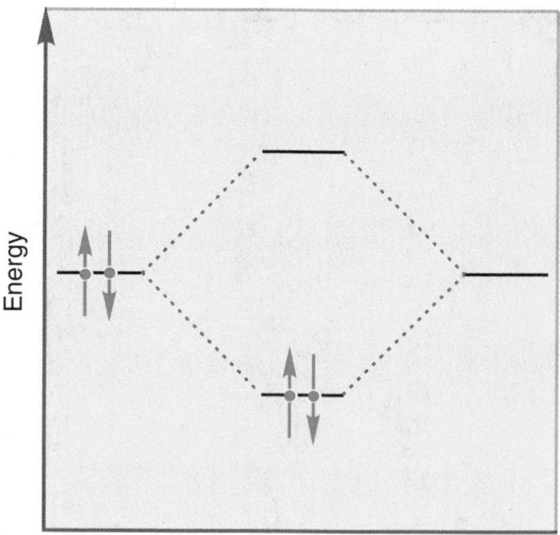

FIGURE 9.2 Stabilization results from overlap of a filled and empty orbital.

Our growing catalog of reactions will continue to add to your expertise in synthesis. The kinds of molecules we know how to make will increase sharply. Addition of hydrogen bromide and hydrogen chloride to alkenes provides access to alkyl halides. We already have the S_N1 and S_N2 reactions available for transforming these molecules into other compounds (Fig. 7.94, p. 287).

We are going to look first at the addition of hydrogen chloride to alkenes to yield alkyl chlorides (Fig. 9.3). We must lead into this with a look at the structure and properties of molecules such as hydrogen chloride.

$(H_3C)_2C{=}C(CH_3)_2$ + H—Cl ⟶ H_3C—C(H)(CH_3)—C(Cl)(CH_3)—CH_3

2,3-Dimethyl-2-butene **2-Chloro-2,3-dimethylbutane**

FIGURE 9.3 The addition of hydrogen chloride to 2,3-dimethyl-2-butene. The π bond of the alkene and the σ bond of H—Cl have been converted into a pair of σ bonds attaching H and Cl to carbon atoms.

PROBLEM 9.1 Use the bond dissociation energies of Table 8.2 (p. 307) to estimate the exothermicity or endothermicity of the reaction in Figure 9.3.

9.1 ELECTRONEGATIVITY

All bonds between different atoms are polar—the electrons in the bond cannot be shared equally in a covalent bond between different atoms. The limit of this phenomenon is an ionic bond in which two oppositely charged species are held together by the electrostatic attraction between them. Potassium chloride and sodium fluoride are examples (Fig. 9.4).

$_{19}K^+$	$_{17}Cl^-$	$_{11}Na^+$	$_{9}F^-$
$1s^22s^22p^63s^23p^6$	$1s^22s^22p^63s^23p^6$	$1s^22s^22p^6$	$1s^22s^22p^6$

FIGURE 9.4 Two examples of charge-separated, ionic bonding. The atoms are held together by the attraction between opposite electrical charges, not by electron sharing in a covalent bond.

Charges also exist in covalent bonds between different atoms. However, these are partial charges and are not fully developed. In such a bond, the electrons will be polarized toward the more electronegative atom, the atom with the greater attraction for the electrons in the shared orbital. As first mentioned in Chapter 1 (p. 14), **electronegativity** is the ability of an atom to attract electrons. Some electronegativities are collected again in Table 9.1.

TABLE 9.1 Some Electronegativities

Electronegativities

1 (IA)	2 (IIA)	3 (IIIA)	4 (IVA)	5 (VA)	6 (VIA)	7 (VIIA)
H 2.3						
Li 0.9	Be 1.6	B 2.1	C 2.5	N 3.1	O 3.6	F 4.2
Na 0.9	Mg 1.3	AL 1.6	Sc 1.9	P 2.3	S 2.6	Cl 2.9
K 0.7						Br 2.7
Rb 0.7						I 2.4

Notice that the most electronegative elements are in the upper right of the periodic table. The ability to attract electrons is related to the effective nuclear charge of an atom, which in turn depends on the effectiveness of the electrons surrounding the nucleus in shielding an additional negatively charged electron from the nucleus and its positive charge. As before (Chapter 7, p. 236), a dipole in a polarized bond is indicated with an arrow pointing toward the more electronegative element (Fig. 9.5).

Molecules such as hydrogen chloride and hydrogen bromide are very polar, even though they are covalently, not ionically, bonded molecules. The electrons in the bond joining the hydrogen and the halogen are not shared equally, and the dipole results from that unequal sharing. Hydrogen chloride has a dipole moment of 1.08 D. Although hydrogen chloride is not an ionic compound, it readily donates a proton, the positive end of its dipole. This familar property is what makes hydrogen chloride a good Brønsted acid. Now let's apply these ideas to the reaction of alkenes with acids such as hydrochloric acid and hydrobromic acid.

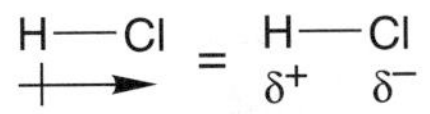

H—Cl (dipole arrow) = $\overset{\delta^+}{H}$—$\overset{\delta^-}{Cl}$

FIGURE 9.5 In any unsymmetrical covalent bond, there will be a dipole with the negative end on the more electronegative atom. This dipole is symbolized by an arrow, which represents the unequal sharing of the electrons in the bond. Partial charges are shown as δ^+ and δ^-.

9.2 MECHANISM OF THE ADDITION OF HYDROGEN HALIDES TO ALKENES

Consider the reaction of 2,3-dimethyl-2-butene with hydrogen chloride to give 2-chloro-2,3-dimethylbutane (Fig. 9.3).

The arrow formalism maps out the process for us and develops the picture of a two-step mechanism. In the first step, the alkene, with the filled π orbital acting as base, is protonated by hydrogen chloride to give a carbocation and a chloride ion. In the second step, chloride acts as nucleophile and adds to the strongly Lewis acidic cation to give the product (Fig. 9.6). Let's pause for a moment to look at that cation.

1 Protonation

2 Addition of Cl^-

FIGURE 9.6 The first step in this two-step reaction is the protonation of the alkene to give a carbocation. In the second step, a chloride ion adds to the cation to give the final product.

In Chapter 3 (p. 82), we described the methyl cation ($^+CH_3$). The cation formed by protonation of 2,3-dimethyl-2-butene is related to the methyl cation, but the three hydrogens of $^+CH_3$ have been replaced with three alkyl groups: two methyls and an isopropyl (Fig. 9.7).

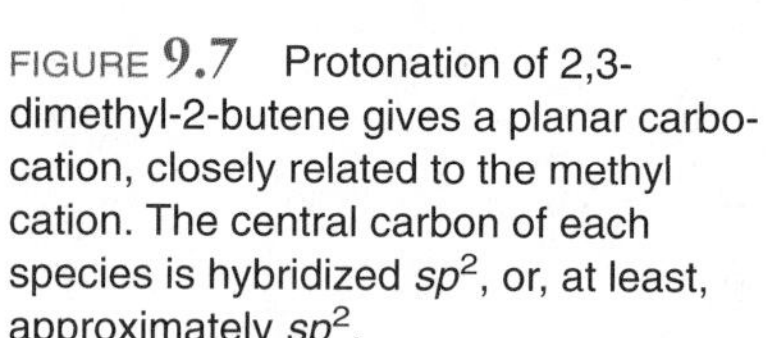

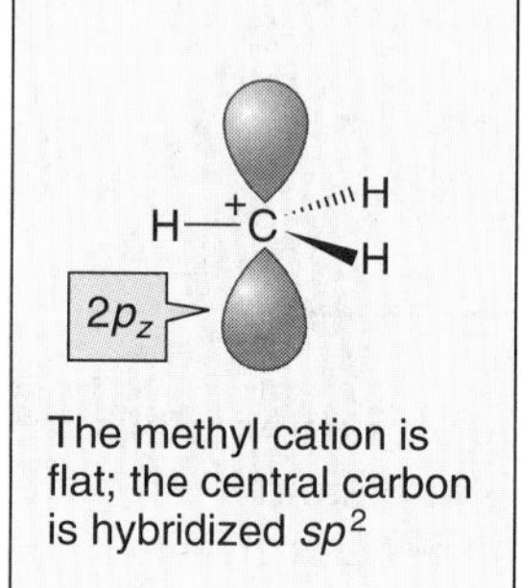

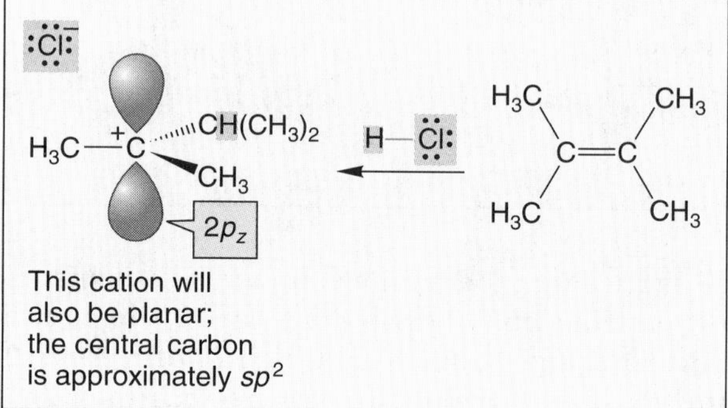

FIGURE 9.7 Protonation of 2,3-dimethyl-2-butene gives a planar carbocation, closely related to the methyl cation. The central carbon of each species is hybridized sp^2, or, at least, approximately sp^2.

$_6C$ $(1s^22s^22p^2)$
Neutral (no charge)

This $_6C$ has only five electrons surrounding it: $1s^2$, and three shared in the bonds to the three R groups; therefore, it has a single positive charge

FIGURE 9.8 The determination of the charge on carbon in a carbocation.

No great structural differences appear, however. The trigonal (attached to three groups) carbon is hybridized sp^2 and the C–C–C angles will be roughly 120°. The empty carbon $2p_z$ orbital extends above and below the plane of the central carbon and the three carbons attached to it (Fig. 9.7). Let's even verify that the carbon is positively charged (Fig. 9.8).

In a neutral carbon atom, the six positive nuclear charges are balanced by six electrons. The central carbon in this cation has a pair of 1*s* electrons and a half-share in the electrons in the three covalent bonds to the alkyl groups, for a total of five. The six positive charges in the nucleus are balanced by only five electrons, and so the carbon atom is positively charged.

Now look at both steps in the addition reaction from a HOMO–LUMO point of view. In the first step, the alkene is the base. More precisely, the base is the filled π orbital of the alkene. What empty orbital acts as the acid? The arrow formalism leads one to focus on the pair of electrons in the hydrogen–chlorine bond. But we know this is misleading; interactions of two filled orbitals are destabilizing. The π orbital must be interacting with σ^* of hydrogen chloride, the empty, antibonding counterpart of σ (Fig. 9.9).

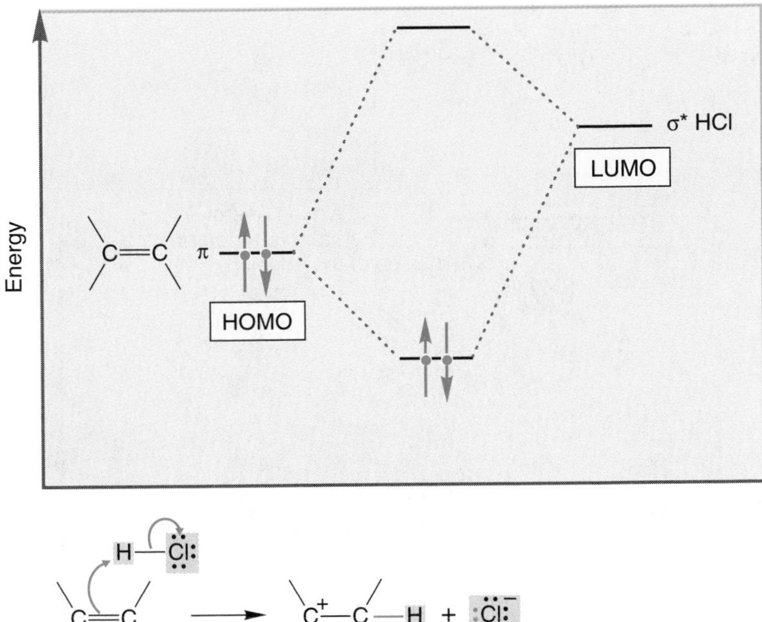

FIGURE 9.9 In the first step of the reaction, it is π of the alkene that is the HOMO and σ* of hydrogen chloride that is the LUMO.

In the second step of the sequence, the chloride ion is the nucleophile, and it reacts through one of its filled orbitals containing nonbonding pairs of electrons. The Lewis acid is also obvious—it is the carbocation reacting through the empty 2*p* orbital (Fig. 9.10).

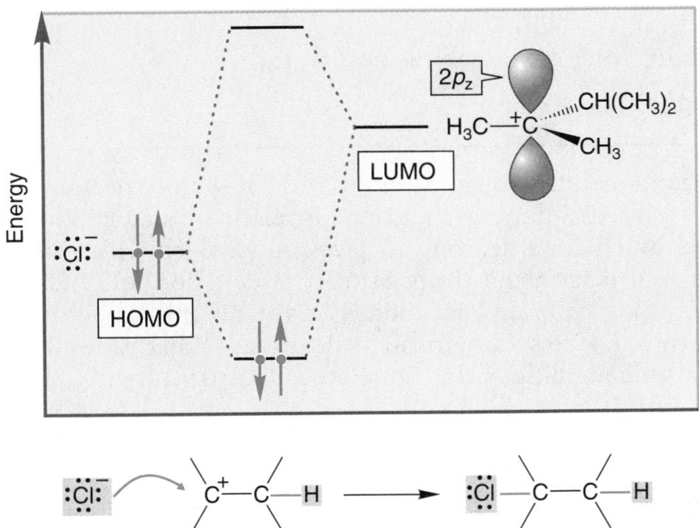

FIGURE 9.10 In the second step, the HOMO is one of the filled nonbonding orbitals on chloride and the LUMO is the empty 2*p* orbital on the central carbon of the cation.

The first step in this sequence, the protonation of the alkene to give a carbocation, is the slow step in the reaction. The rate-determining step, the one with the highest energy transition state, is this addition of a proton. As this step is endothermic, the transition state will resemble the product carbocation, and it will have substantial positive charge developed on carbon (Fig. 9.11).

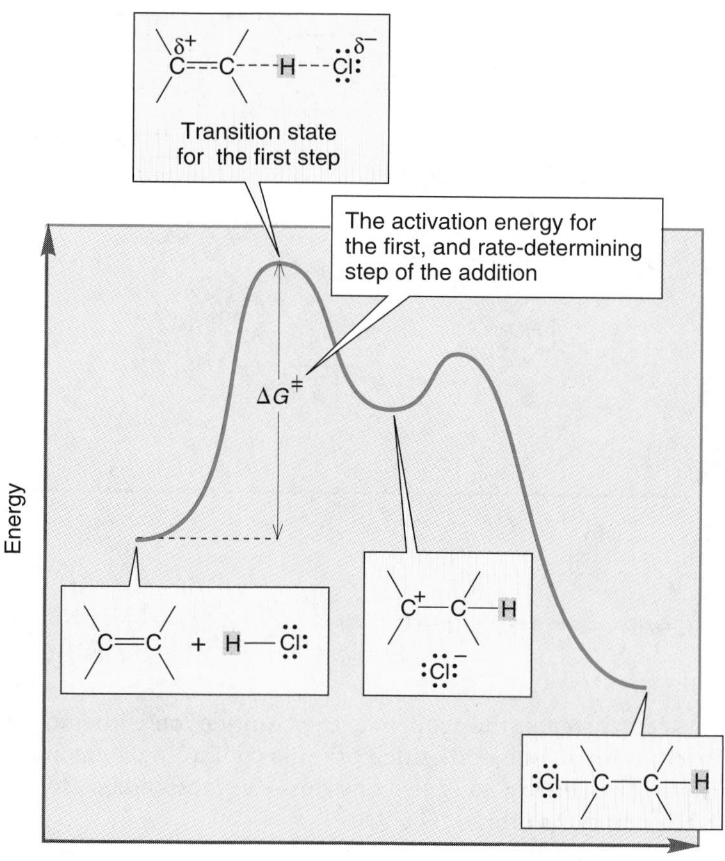

FIGURE 9.11 In this simple addition of hydrogen chloride to an alkene, the first step, the endothermic formation of a carbocation, is the slow, rate-determining step. In the transition state for this step, the positive charge accumulates on one of the carbons of the starting alkene.

That notion leads directly to the next topic.

9.3 REGIOCHEMISTRY

The reaction of hydrogen chloride with a symmetrically substituted alkene such as 2,3-dimethyl-2-butene is relatively simple. Protonation is followed by chloride addition to give the alkyl chloride. There are no choices to be made about the position of protonation, and once the carbocation is made, chloride formation is easy to rationalize. What happens if the alkene is not so symmetrically substituted? What happens if the ends of the alkene are differently substituted? Figure 9.12 poses the question

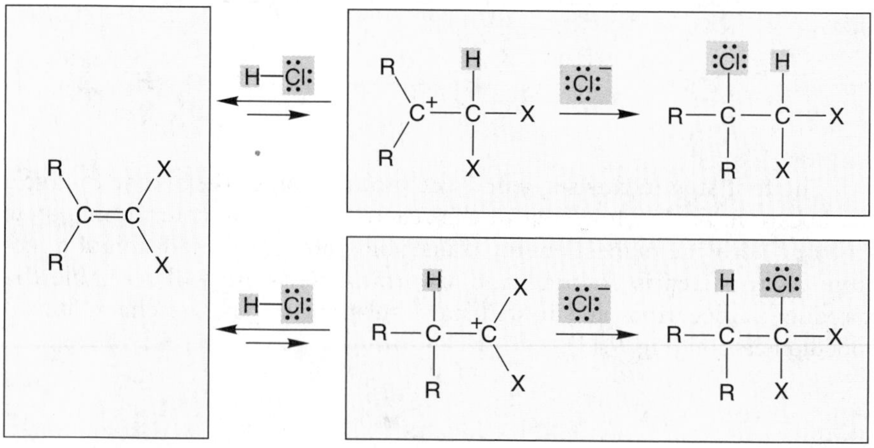

FIGURE 9.12 For an unsymmetrical alkene, there are two possible intermediate cations, and therefore two possible chloride products.

again. The answer is that it depends on *how* the ends of the alkene are different. The preferential formation of one isomer in those cases where a choice is possible is called regioselectivity, or if the choice is overwhelmingly in one direction, regiospecificity. The next few sections will discuss factors influencing **regiochemistry**.

*PROBLEM 9.2

Chloride formation isn't the only process possible, however. In this reaction, there is a small amount of an alkene isomeric with the 2,3-dimethyl-2-butene produced. What is this other alkene, and what is the mechanism of its formation?

ANSWER

Remember the E1 reaction! Just because we are starting a new subject does not mean that old material disappears. Cations will undergo the E1 and S_N1 reactions as well as the addition reactions encountered in this chapter. The cation formed on protonation of 2,3-dimethyl-2-butene can lose a proton in E1 fashion in two ways. Path (a) simply reverses to give starting compounds back, whereas (b) leads to 2,3-dimethyl-1-butene.

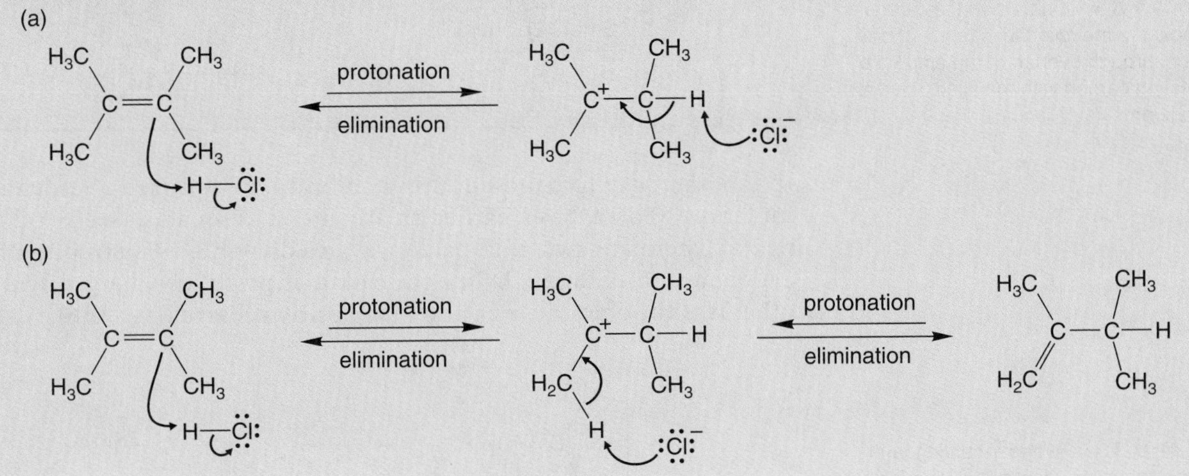

(a) $(H_3C)_2C{=}C(CH_3)_2$ + H—Cl: ⇌ (protonation / elimination) $(H_3C)_2C^+{-}C(CH_3)_2{-}H$ + :Cl:⁻

(b) $(H_3C)_2C{=}C(CH_3)_2$ + H—Cl: ⇌ (protonation / elimination) $(H_3C)(H_2C{-}H)C^+{-}C(CH_3)_2{-}H$ + :Cl:⁻ ⇌ (protonation / elimination) $(H_3C)(H_2C{=})C{-}C(CH_3)_2{-}H$

9.4 RESONANCE EFFECTS

When hydrogen chloride adds to vinyl chloride (chloroethylene), the major product is 1,1-dichloroethane, *not* 1,2-dichloroethane (Fig. 9.13).

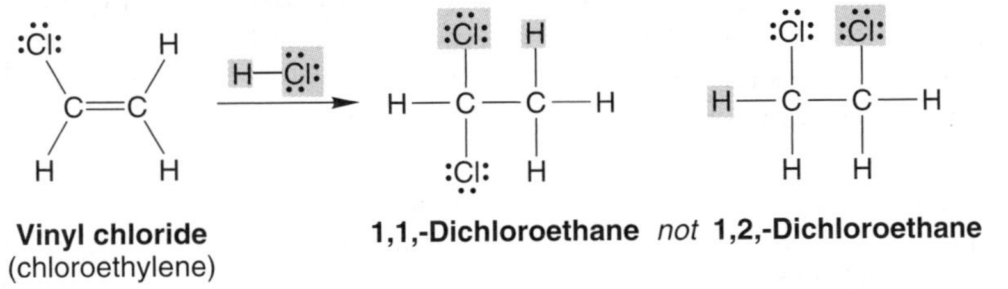

Vinyl chloride (chloroethylene) + H—Cl: → **1,1,-Dichloroethane** *not* **1,2,-Dichloroethane**

FIGURE 9.13 Addition of hydrogen chloride to vinyl chloride gives 1,1-dichloroethane, *not* 1,2-dichloroethane.

The best way to rationalize the regiospecificity of this reaction is to examine the two possible intermediate carbocations (Fig. 9.14) and see which of the two is more stable.

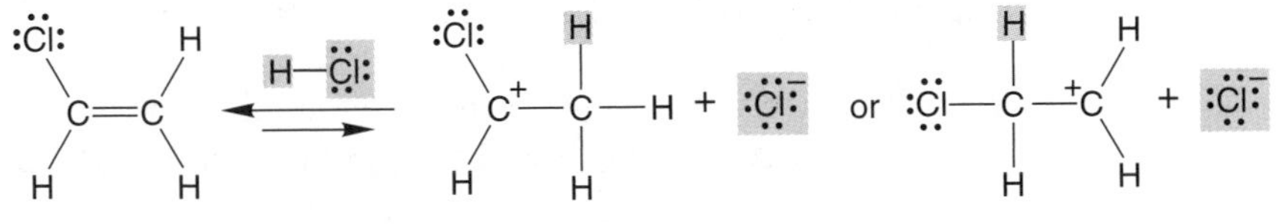

:Cl:(H)C=C(H)(H) + H—Cl: ⇌ :Cl:(H)C⁺—C(H)(H)—H + :Cl:⁻ or :Cl—C(H)(H)—C⁺(H)(H) + :Cl:⁻

FIGURE 9.14 The two cations that could be formed by protonation of vinyl chloride.

Remember: The carbocation has an empty *p* orbital on carbon. In one of the two possible cations this empty orbital overlaps a *p* orbital on chlorine containing two electrons. This overlap of filled and empty orbitals produces strong stabilization (Fig. 9.15). In this ion, the chlorine atom helps bear the positive charge. This stabilization is not available to the other cation, which, therefore, is much harder to form.

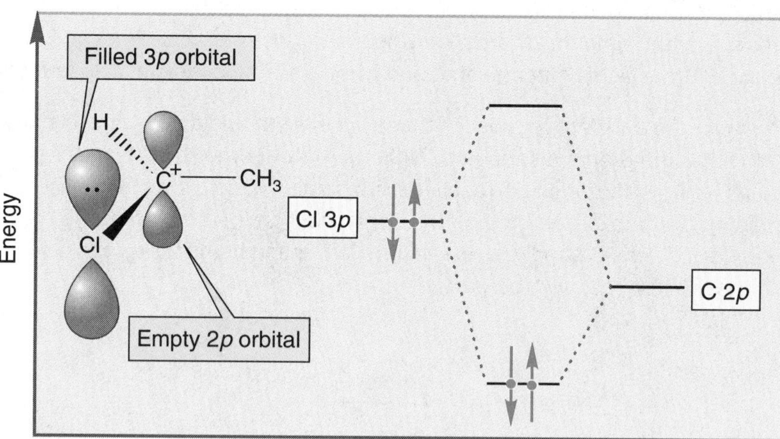

FIGURE 9.15 Only when the charge is adjacent to the chlorine is there stabilization through overlap of the empty 2*p* orbital on carbon with the filled 3*p* orbital on chlorine.

Another way to represent this stabilization is through a resonance formulation. (For a brief earlier treatment of resonance see Section 1.3, p. 24.) There are two reasonable ways to draw a Lewis structure for the cation next to chlorine. In one the carbon is positively charged, and in the other it is the chlorine that bears the positive charge (Fig. 9.16).

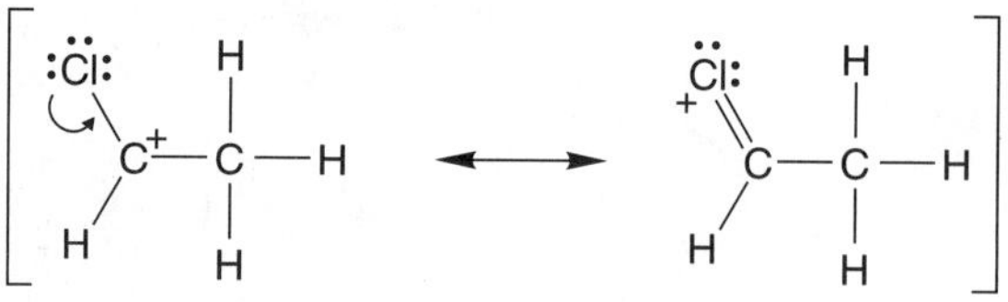

In this resonance form, the carbon bears the positive charge

In this resonance form, the + resides on the chlorine

FIGURE 9.16 Two resonance forms for the cation adjacent to the chlorine. The charge is borne by both the carbon and chlorine. Neither of these two resonance forms gives a complete picture of this cation. It is the combination of resonance forms that does this.

The two representations differ only in their distribution of electrons. They are two different electronic representations of the same structure. By itself neither is exactly right. Neither by itself shows that two atoms share the charge. Taken together they do. These two representations are resonance forms. The real structure of this cation is not well represented by either form alone, but the two combined do an excellent job.

The molecular orbital picture we first developed does a good job of showing the stabilization, but it is not particularly well suited for bookkeeping purposes. In practice, the ease of bookkeeping—of easily drawing the next chemical reaction—makes the resonance picture simpler to use. The price is that you have to draw two structures (or more in other cases), or adopt some other code that indicates the presence of more than one representation for the molecule. We will see many examples in the next pages and chapters.

Here is a reaction that we will return to in some detail later, but can examine quickly now. Hydrogen chloride adds to 1,3-butadiene very easily. One product you may be able to predict, but the other will probably not be immediately obvious (Fig. 9.17).

1,2-Addition (75–80%)

1,4-Addition (20–25%)

FIGURE 9.17 Addition of hydrogen chloride to 1,3-butadiene gives two products. Hydrogen chloride adds in both 1,2- and 1,4-fashion.

Look at the possibilities. There are two possible cations that can arise from protonation of 1,3-butadiene by hydrogen chloride (Fig. 9.18). In one, the allyl cation, the positive charge is on the carbon adjacent to the double bond; in the other, it is isolated on a terminal primary carbon.

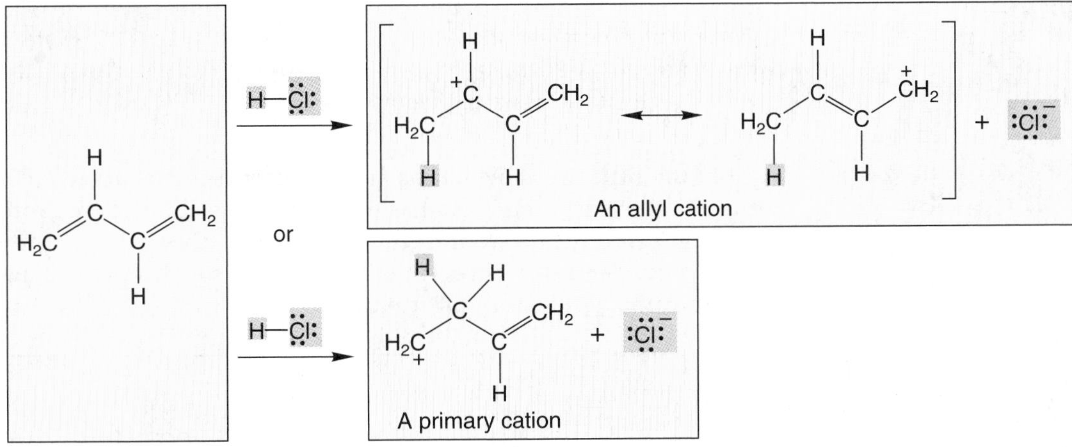

FIGURE 9.18 In principle, a proton can be added to 1,3-butadiene to give either an allyl cation or a primary cation.

The choice between these two ions becomes clear if we draw out orbital pictures of the two carbocations. In the allyl cation, the *p* orbital on the positively charged carbon atom overlaps the π orbitals of the double bond (Fig. 9.19). We have seen the system of three parallel, overlapping *p*

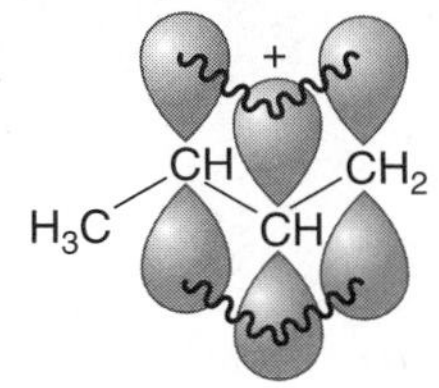

Protonation at the end carbon gives a delocalized allyl cation

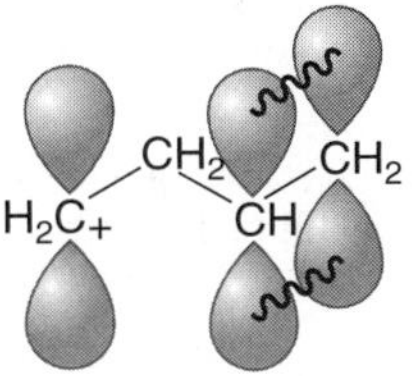

Protonation at an internal carbon gives a localized primary cation

FIGURE 9.19 In the allyl cation, the charge is shared by two carbons; in the primary cation it is localized on a single carbon.

orbitals before (Problem 1.16c; Problem 2.21). Here we have an allyl cation again, and the molecular orbital description is just the same as before. The three $2p$ atomic orbitals can be combined to yield three molecular orbitals. There are only two electrons in an allyl cation and they will occupy the lowest, most stable molecular orbital (Fig. 9.20). The presence of the double bond is highly stabilizing for this cation, as the charge is delocalized over the allyl system.

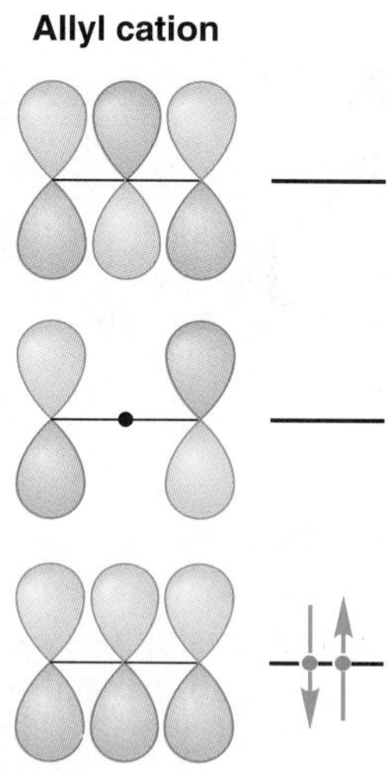

Allyl cation

FIGURE 9.20 In the allyl cation, it is only the lowest energy molecular orbital that is occupied.

PROBLEM 9.3 Derive the π orbitals of the allyl system shown in Figure 9.20. *Hint*: Do this by placing a carbon $2p$ orbital in between the two p orbitals of ethylene. *Another Hint*: Watch out for "net zero" interactions!

In the other possible cation in this reaction, the charge is localized at the end of the molecule and has no resonance stabilization. Protonation at the end carbon gives an ion (the allyl cation) that can be represented by more than one Lewis structure. The choice is easy once we have looked carefully at the structures (Fig. 9.21).

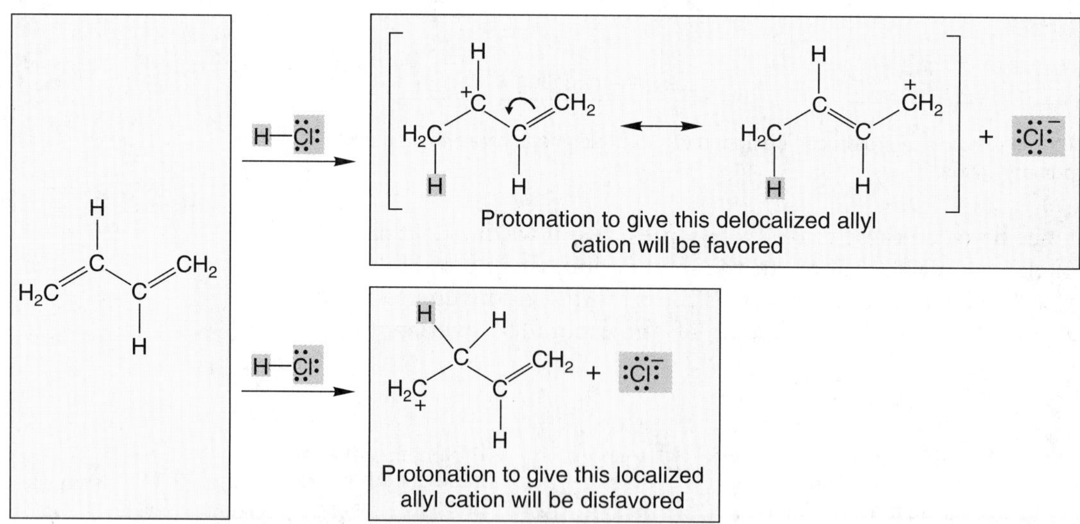
H—Cl:
Protonation to give this delocalized allyl cation will be favored
Protonation to give this localized allyl cation will be disfavored

FIGURE 9.21 Formation of the lower energy delocalized allyl cation will be favored over formation of the higher energy localized primary cation.

The real structure is best represented as a combination of the two Lewis (resonance) forms in the figure. This way of looking at the allyl cation is especially helpful as it points out simply and clearly which carbons share the positive charge. When the chloride ion approaches the allyl cation it can add at either of the two carbons that share the positive charge to give the two products. Be careful not to think of this resonance-stabilized structure as spending part of its time with the charge on one carbon and part of the time with the charge on the other! The cation has a *single* structure in which the two end carbons share the positive charge.

This single, summary structure is often shown using dashed partial bonds to represent the bonds that are double in one resonance form and single in the other. The charge is placed at the midpoint of the dashes. Alternatively, one resonance form is drawn out in full and the other atoms sharing the charge are shown with a (+) or (–) (Fig. 9.22).

Convention Alert!

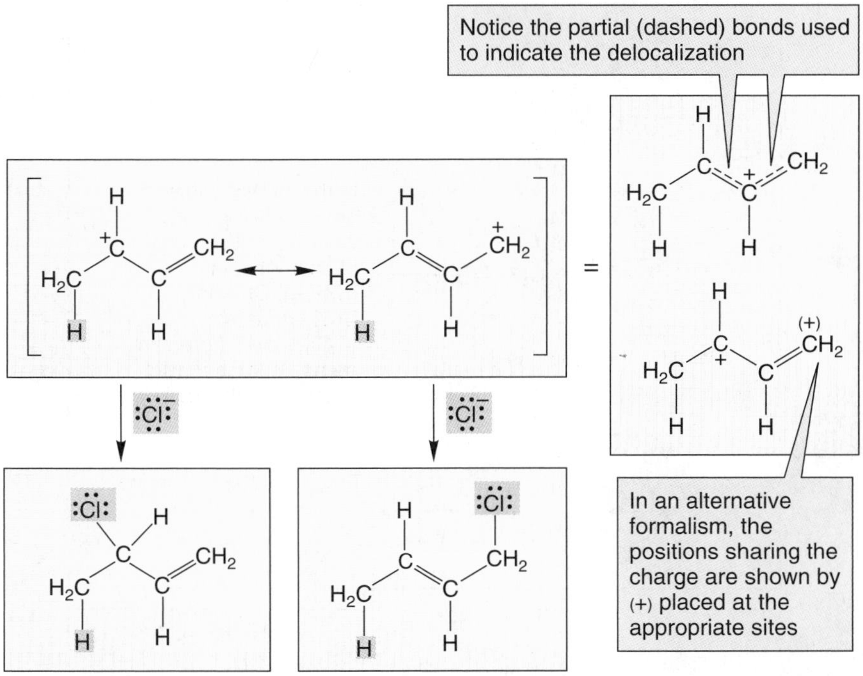

Notice the partial (dashed) bonds used to indicate the delocalization

In an alternative formalism, the positions sharing the charge are shown by (+) placed at the appropriate sites

FIGURE 9.22 In the allyl cation, the positive charge is shared by two carbons, and therefore chloride can add at two positions to give the two observed products. The resonance formulation shows this especially clearly. Note the summary structures in which dashed bonds or charges in parentheses are used to indicate delocalization without writing out all the resonance forms.

So, the resonance formulation is very useful and we are going to review it thoroughly in a moment. But first, let's ask if the molecular orbital description can predict both products. Of course it can, or it would hardly be very useful. What happens as the chloride ion, with its filled nonbonding orbitals, begins to interact with the allyl cation? Stabilizing interactions occur between filled orbitals and empty orbitals. Chloride bears the filled orbital, so we must look for the lowest *unoccupied* molecular orbital (LUMO), of allyl. Figure 9.23 shows it, Φ_2. There are two points at which chloride can add to Φ_2, and they lead to the two observed products. Note that the middle carbon, through which the node passes, cannot be attacked. So the molecular orbital description also explains the regiochemistry of the addition.

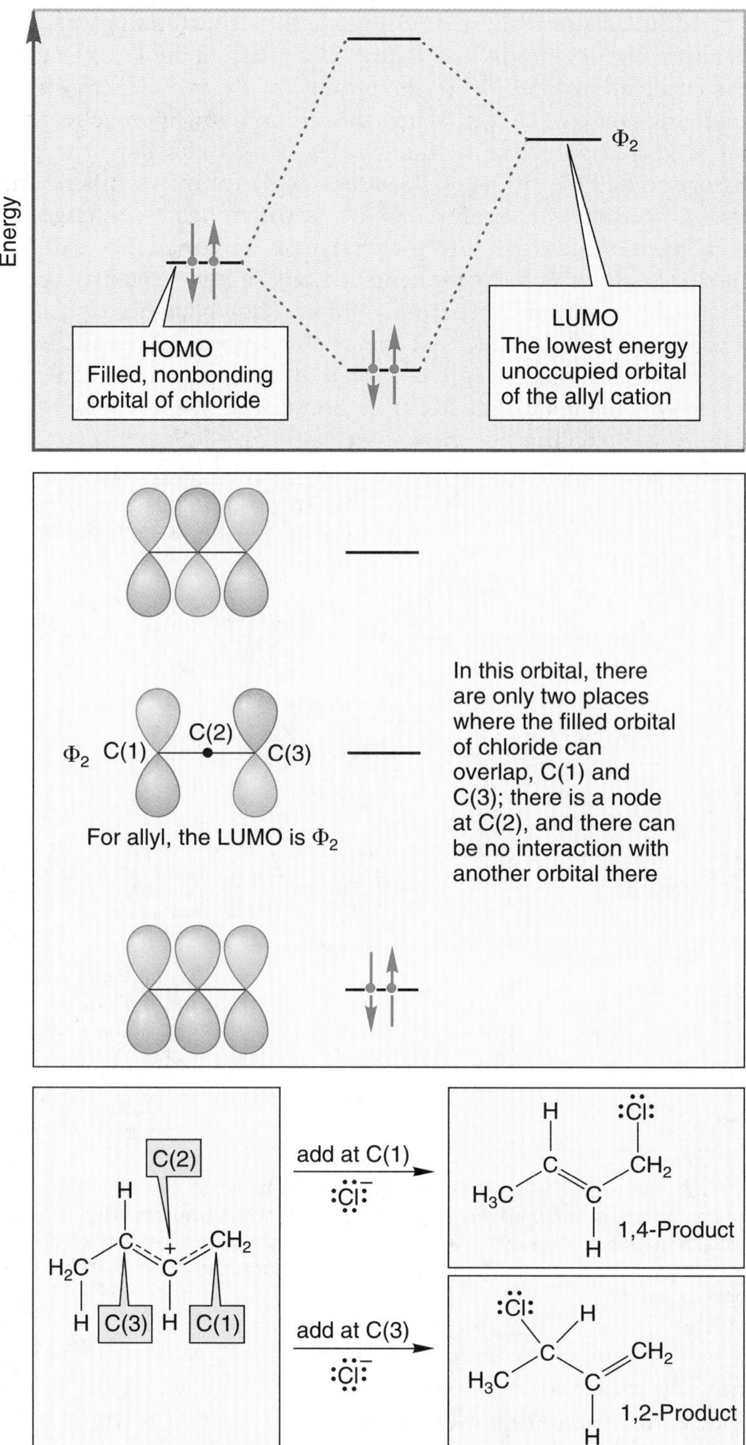

FIGURE 9.23 In this reaction, the HOMO–LUMO interaction is between a filled nonbonding orbital on chloride and the lowest energy unoccupied orbital of allyl, in this case, Φ_2.

Whether one uses resonance forms or molecular orbital descriptions is largely a matter of taste. Resonance forms are the traditional way, and they are extremely good at letting you see where the charge resides. We will use them extensively. It's worth taking time now to review the whole resonance system.

9.5 REVIEW OF RESONANCE: HOW TO WRITE RESONANCE FORMS

A molecule, best described as a resonance hybrid, *does not,* repeat *not,* spend part of its time as one structure and part as another. That is chemical equilibrium (Fig. 9.24).

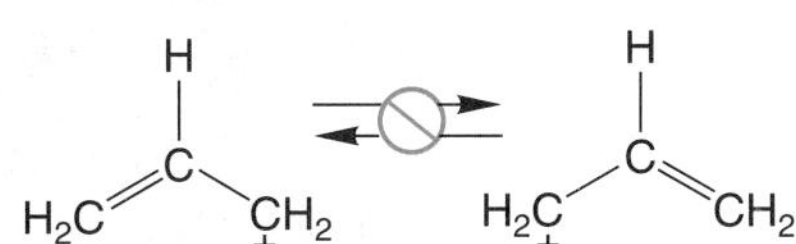

FIGURE **9.24** The allyl cation does not spend part of its time as one of these structures and part of its time as the other. That would be an example of equilibrium, not resonance. The two are very different phenomena.

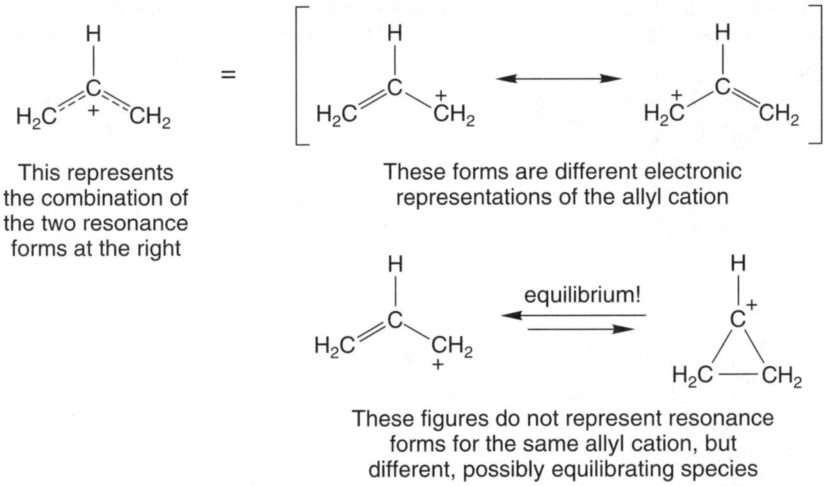

The real structure of a resonance-stabilized molecule is a hybrid, or combination, of the resonance forms contributing to the structure. These forms represent different electronic descriptions of the molecule. *Resonance forms differ only in the distributions of electrons and never in the positions of atoms.* If you have moved an atom you have written a chemical equilibrium (Fig. 9.25).

This represents the combination of the two resonance forms at the right

These forms are different electronic representations of the allyl cation

equilibrium!

These figures do not represent resonance forms for the same allyl cation, but different, possibly equilibrating species

FIGURE **9.25** Resonance is always indicated by the special double-headed (↔) resonance arrow. Equilibrium is shown by two arrows headed in different directions (⇄).

It is for this phenomenon of resonance that the special, double-headed arrow is strictly reserved. This special arrow is an important piece of the grammar of organic chemistry.

Convention Alert!

It is important to be able to estimate the relative importance of resonance forms in order to get an idea of the best way to represent a molecule. Some guidelines are listed below. Fortunately, this is an area in which common sense does rather well.

1. Identical resonance forms contribute equally. The allyl cation is a good example. Two equivalent forms are possible, and they contribute equally to the hybrid, real structure (Fig. 9.26).

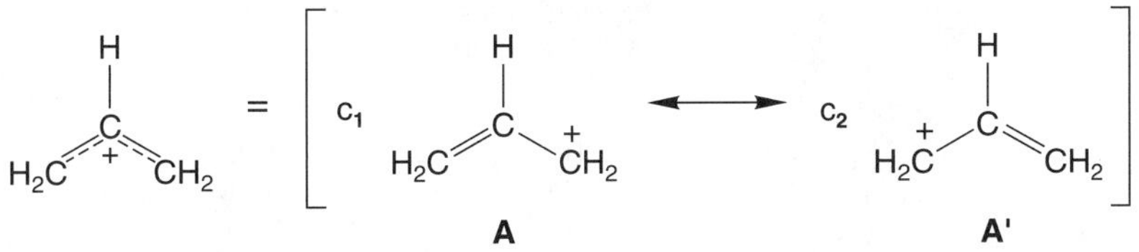

c_1 **A** c_2 **A'**

FIGURE **9.26** The allyl cation is an equal combination of the two equivalent resonance forms **A** and **A′**. In this case, the weighting factors, c_1 and c_2, are equal.

If we wrote an equation to represent the mathematics involved, it would be

Allyl = c_1 (form **A**) + c_2 (form **A′**), where $c_1 = c_2$. Here c represents the weighting factors, or coefficients. In this case, $c_1 = c_2$.

2. The more bonds in a resonance form the more stable it is. 1,3-Butadiene can be written as a resonance hybrid of several structures (Fig. 9.27).

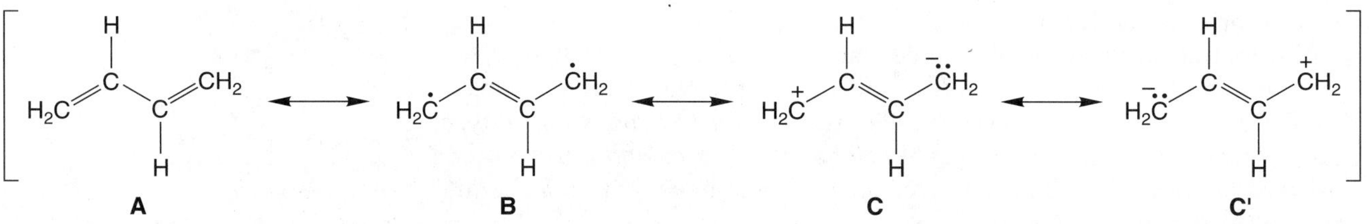

FIGURE 9.27 1,3-Butadiene can be described as a combination of four resonance forms, **A**, **B**, **C**, and **C′**. Structure **A**, however, contributes far more to the combination than the others because it has more bonds. 1,3-Butadiene is reasonably well described by the single form **A**.

The structures **B**, **C**, and **C′** contain fewer bonds than structure **A**, and will contribute far less than **A** to the real 1,3-butadiene. Forms **C** and **C′** are equivalent and will contribute equally. This does not mean that **B**, **C**, and **C′** do not contribute at all! It simply means that in our equation

$$\text{Butadiene} = c_1\,(\mathbf{A}) + c_2\,(\mathbf{B}) + c_3\,(\mathbf{C}) + c_3\,(\mathbf{C'})$$

The weighting factor c_1 will be much larger than the other coefficients. 1,3-Butadiene looks much more like form **A** than forms **B**, **C**, or **C′**.

3. Resonance forms must have the same number of paired and unpaired electrons. As we saw in Chapter 1 (p. 30), two electrons occupying the same orbital must have opposite, or paired spin quantum numbers. Most organic molecules, including our sample molecule, 1,3-butadiene, have all paired electrons. All resonance forms for 1,3-butadiene must also have all paired electrons. Consider form **B** in the resonance hybrid shown in Figure 9.27. It is often the convention to emphasize that all electrons are paired in this molecule by appending little arrows to the single electrons in this form (Fig. 9.28). Opposed arrows (↓↑) indicate paired spins, identical arrows (↑↑,↓↓) indicate parallel (same direction) spins.

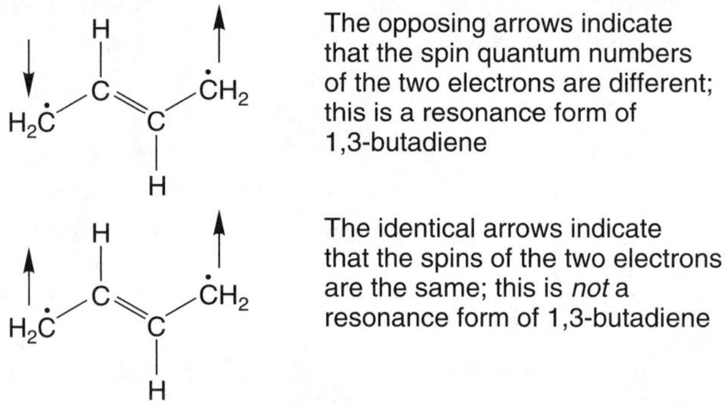

The opposing arrows indicate that the spin quantum numbers of the two electrons are different; this is a resonance form of 1,3-butadiene

The identical arrows indicate that the spins of the two electrons are the same; this is *not* a resonance form of 1,3-butadiene

FIGURE 9.28 Resonance forms must have the same number of paired and unpaired electrons. In 1,3-butadiene, all electrons are paired, and all resonance forms contributing to the structure must also have all-paired electrons.

4. Separation of charge is bad. In our equation for butadiene in Figure 9.27, the neutral forms **A** and **B** will contribute more than **C** or **C′**, both of which require a destabilizing separation of charge. (Form **B** will still contribute less than form **A** because it has fewer bonds.)

5. In charged species (ions), delocalization is especially important. As we have mentioned a number of times, small charged atoms are usually most unstable. Delocalization of electrons that allows more than one atom to share a charge is stabilizing in energy terms. The allyl cation is a good example—the two end carbons each bear one-half of the positive charge (Fig. 9.26).

   The protonation of vinyl chloride is another good example. Here the charge is shared by the carbon and the chlorine, and the resonance forms show it (Fig. 9.29).

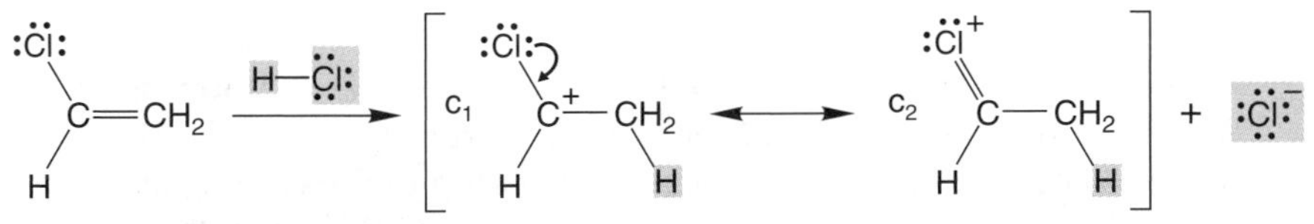

FIGURE 9.29 Protonation of vinyl chloride leads to a resonance-stabilized cation. The charge is shared by carbon and chlorine.

PROBLEM 9.4

Use the molecular orbitals constructed in Problem 9.3 to show the electronic occupancy of the allyl cation, radical, and anion.

PROBLEM 9.5

Consider the cyclic version of the allyl system, the cyclopropenyl molecule. Take as your starting point the molecular orbitals of allyl and construct the molecular orbitals of cyclopropenyl. Recall your construction of cyclic H_3 from linear H_3 in Problem 2.19 (p. 64).

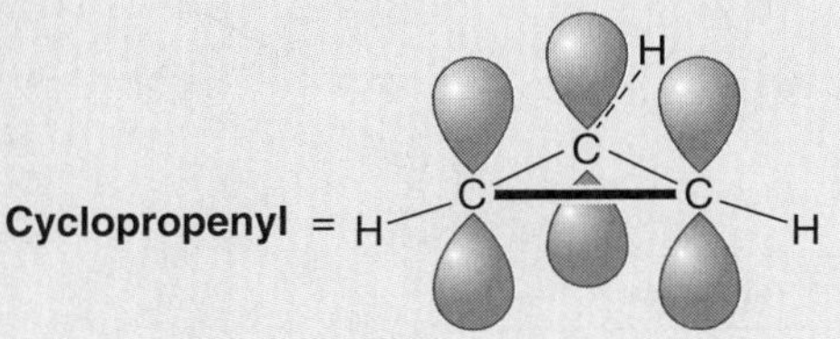

Cyclopropenyl =

6. Electronegativity is important. In the allyl anion (Fig. 9.30), there are two equivalent resonance forms. As in the allyl cation, the charge resides equally on the two end carbons.

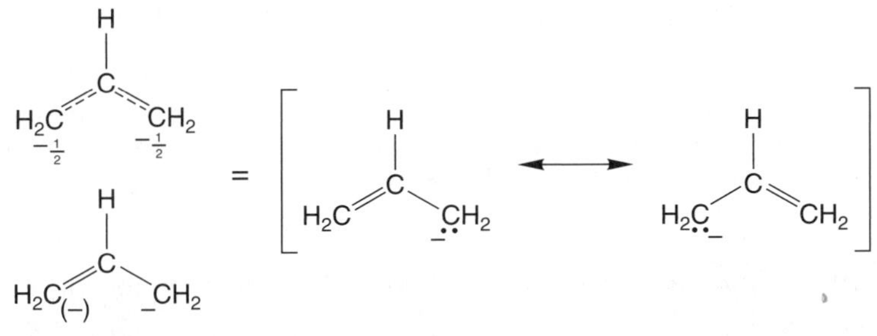

FIGURE 9.30 In the allyl anion, it is a negative charge that is shared between the two end carbons.

The related enolate anion, in which one CH_2 of allyl is replaced with an oxygen (Fig. 9.31), also has two forms, but they do not contribute equally because they are different. Each has the same number of bonds, so we cannot choose the better representation that way. However, one has the charge on the relatively electronegative oxygen while in the other form it is on carbon. Form **A** is the better representation for the enolate anion, although both contribute. Mathematically, we would say that the weighting factor for the first form, c_1, is larger than that for the second, c_2.

In this case, $c_1 > c_2$

FIGURE 9.31 The enolate anion is also described by two resonance forms, but form **A**, in which the electronegative oxygen atom bears the negative charge is weighted more heavily than form **B**, in which carbon is negatively charged ($c_1 > c_2$). The enolate anion is better represented by form **A** than form **B**, but best represented by a combination, or hybrid, of the two forms.

7. Watch out! Geometry is important. Is the "enolate" anion in Figure 9.32 resonance stabilized? On paper it certainly looks like it is, and there is at least a typographical relationship to the enolate anion we saw before in Figure 9.31. But we have been fooled again by the two-dimensional surface. The three-dimensional structure of the cage molecule prevents substantial overlap of the orbitals. Even though the paper doesn't protest when you draw the second resonance form, in reality, it is not there. The charge is borne on only one carbon—the bridgehead—and is not shared by another. Build a model and you will see it easily.

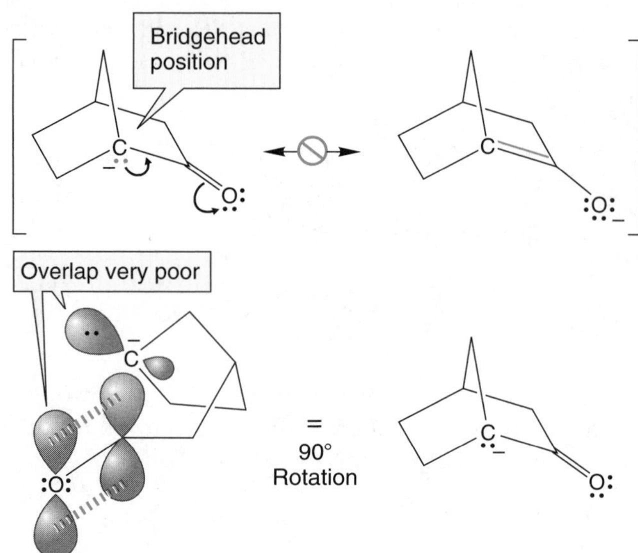

FIGURE 9.32 In this "enolate" anion, the orbital containing the pair of electrons does not overlap well with the *p* orbitals of the carbonyl group. There is no resonance, no matter what the deceptive two-dimensional surface says.

PROBLEM 9.6 Add dots for the electron pairs and write resonance forms for the following structures:

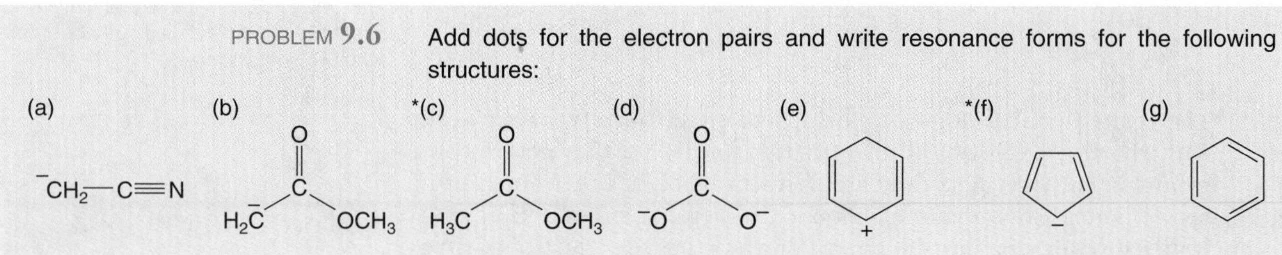

ANSWER (CONTINUED)

(c)

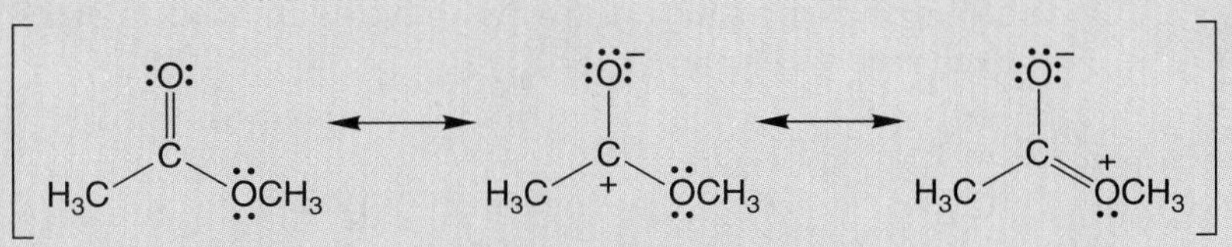

(f)

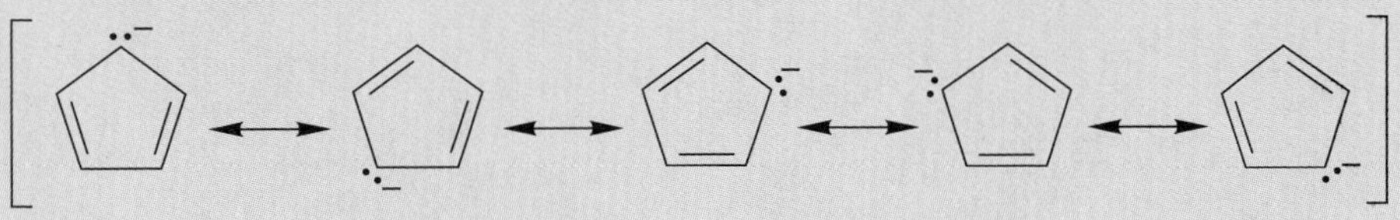

Write Lewis structures and resonance forms for the following compounds. If you have problems seeing the structures of some of these molecules, see the end-paper at the front of this book.

PROBLEM 9.7

(a) $HONO_2$ (b) $^{-}OSO_2OH$ (c) CH_3COO^{-}

Which of the following pairs of structures are not resonance forms? Why not? You may have to add dots to make good Lewis structures first.

PROBLEM 9.8

(a)

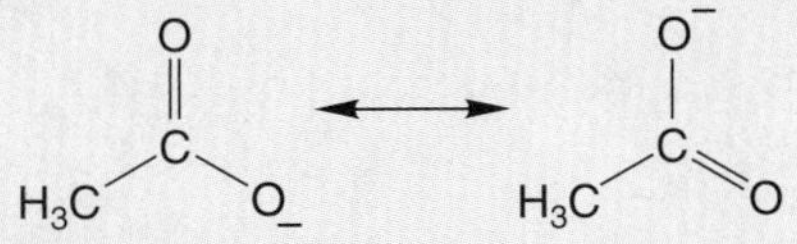

*(b)

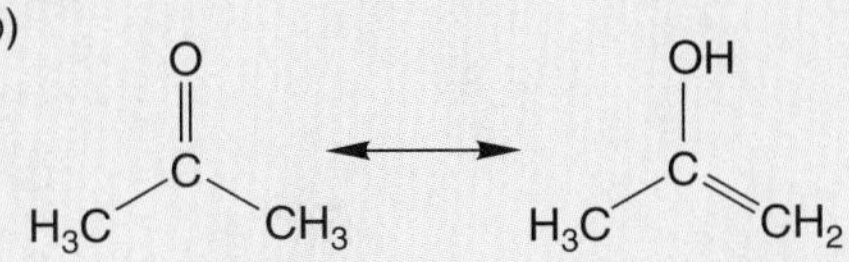

(c)

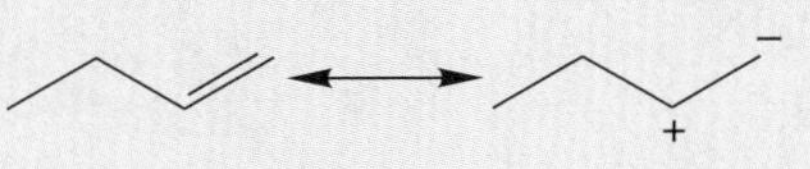

(d)

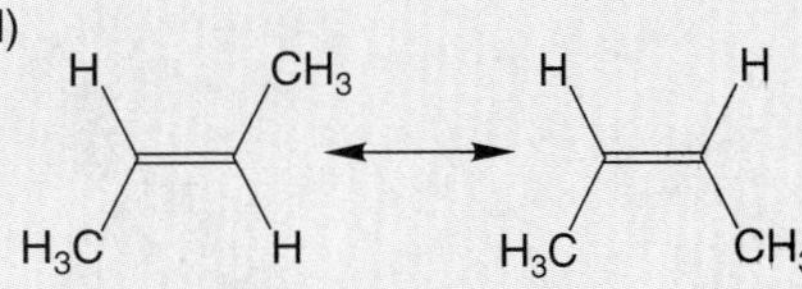

(e)

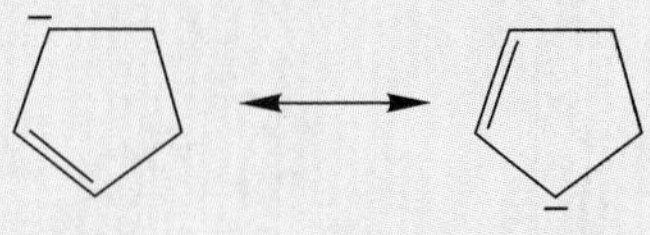

(f)

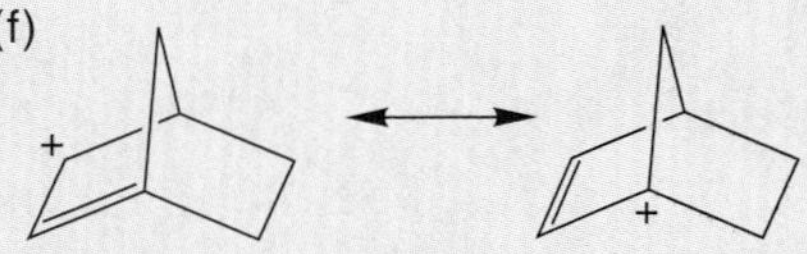

ANSWER

(b) These are not resonance forms. Atoms have been moved, not just electrons. These two molecules are in equilibrium. Molecules related through the change of position of a single hydrogen are called "tautomers."

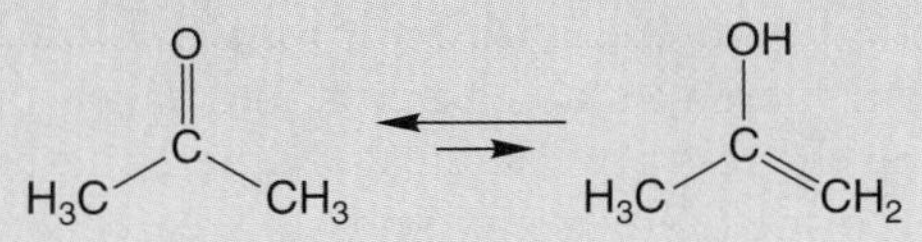

PROBLEM 9.9 In the following pairs of resonance forms, assign weighting factors. In each case, tell us which form you think is more important, and therefore a better representation of the structure. You will have to add dots to make good Lewis structures first.

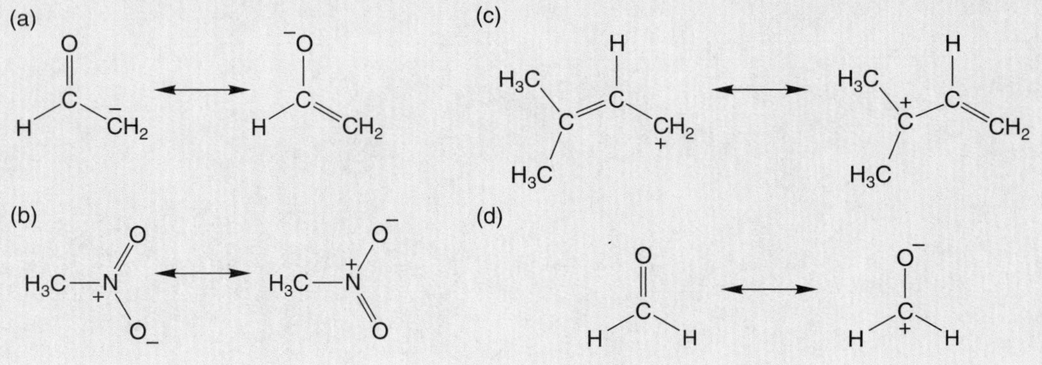

(a) (c)

(b) (d)

9.6 RESONANCE AND THE STABILITY OF CARBOCATIONS

We have already seen one example of how an atom capable of delocalizing electrons can help share a charge and influence the regiochemistry of an addition reaction. Addition of hydrogen chloride to vinyl chloride gives only 1,1-dichloroethane *not* 1,2-dichloroethane (p. 339). Protonation occurs so as to produce a carbocation adjacent to the chlorine, which is far more stable than the alternative in which the positive charge sits localized on a primary carbon (Fig. 9.33).

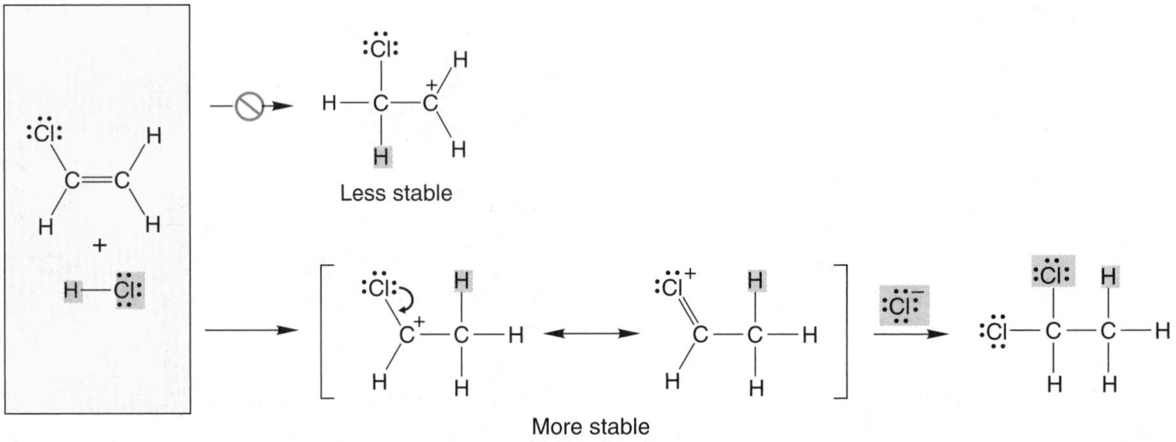

Less stable

More stable (delocalized)

FIGURE 9.33 In addition of hydrogen chloride to vinyl chloride, it is formation of the more stable, resonance-stabilized carbocation that determines the product.

In Section 9.2 (p. 335), we looked at the reaction of the symmetrically substituted 2,3-dimethyl-2-butene with hydrogen chloride. In the formation of the carbocation, there was no choice to be made—only one cation could be formed. When the less symmetrical alkene 2-methyl-1-butene reacts with hydrogen chloride there are two possible carbocations. The only chloride formed is the one produced from the more stable, tertiary carbocation. The product is the one in which the hydrogen has added to the less substituted end of the alkene and the chlorine to the more substituted end (Fig. 9.34).

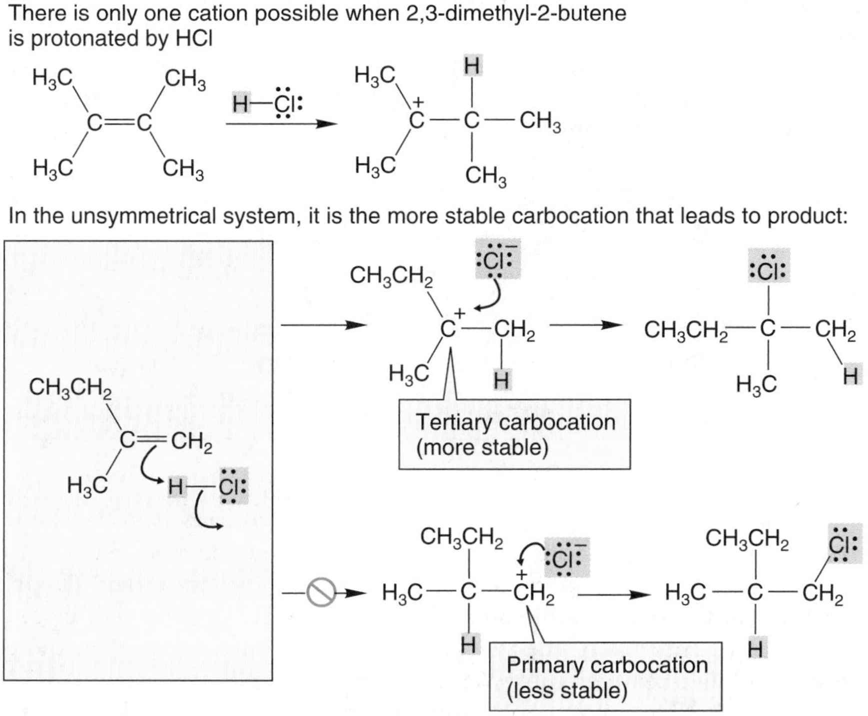

There is only one cation possible when 2,3-dimethyl-2-butene is protonated by HCl

$(H_3C)_2C{=}C(CH_3)_2 \xrightarrow{H-Cl} (H_3C)_2\overset{+}{C}-CH(CH_3)-CH_3$

In the unsymmetrical system, it is the more stable carbocation that leads to product:

$CH_3CH_2(H_3C)C{=}CH_2 + H-Cl \longrightarrow CH_3CH_2(H_3C)\overset{+}{C}-CH_2H \xrightarrow{Cl^-} CH_3CH_2-C(Cl)(H_3C)-CH_2H$

Tertiary carbocation (more stable)

$CH_3CH_2(H_3C)C{=}CH_2 + H-Cl \not\longrightarrow H_3C-C(CH_3CH_2)(H)-\overset{+}{C}H_2 \xrightarrow{Cl^-} H_3C-C(CH_3CH_2)(H)-CH_2Cl$

Primary carbocation (less stable)

FIGURE 9.34 In additions to unsymmetrical alkenes, formation of the more stable carbocation leads to product.

Unsymmetrical alkenes such as 2-pentene, from which two secondary carbocations of roughly equal stability can be formed, do give two products in comparable amounts (Fig. 9.35).

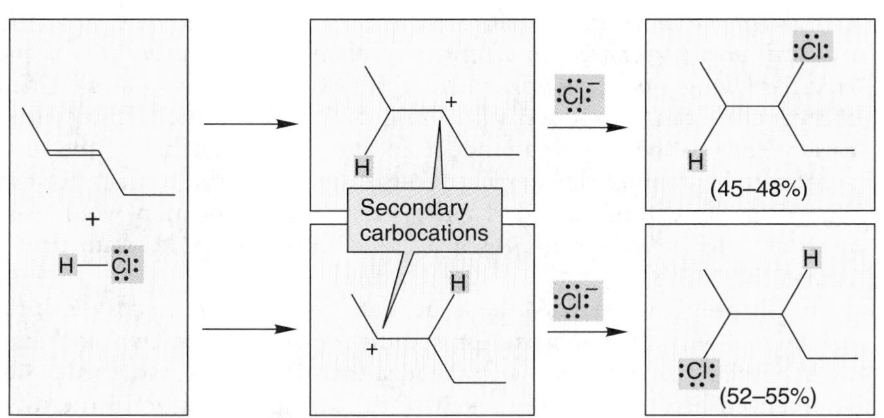

FIGURE 9.35 When protonation of an unsymmetrical alkene leads to two carbocations of similar stability, both are formed, and two products are produced.

These phenomena have been known a long time, and were first correlated by the Russian chemist Vladimir V. Markovnikov (1838–1904) in 1870. The observation that the more substituted halide is formed now bears his name in **Markovnikov's rule**. Rules are useless (or worse) unless we understand them, so let's see why Markovnikov's rule works.

The results of the previous reactions show that the mechanism of HX addition must involve formation of the more substituted cation (Fig. 9.34), as the less substituted cation would lead to a product that is not observed, or is only formed in minor amounts. In turn, a reasonable presumption is that the more substituted a carbocation is, the more stable it is. As first seen in Chapter 7 (p. 268) this is the case, and Figure 9.36 shows again the order of carbocation stability. *Remember*: These differences in stability are large. In practice, reactions in which tertiary and secondary carbocations are intermediates are common, but primary and methyl carbocations cannot be formed.

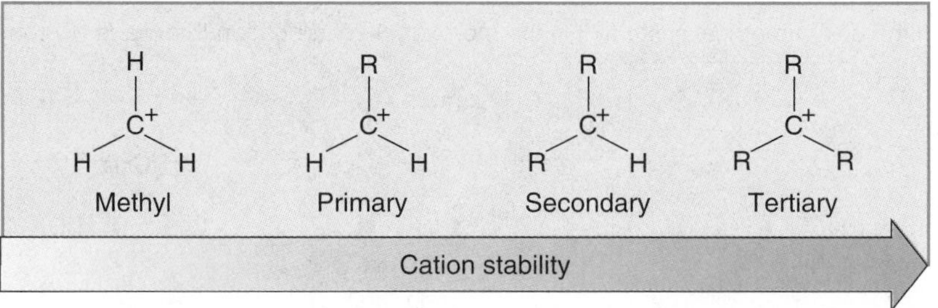

FIGURE 9.36 The more substituted a carbocation, the more stable it is (R = alkyl group).

How does the difference in energy of these two carbocations, the tertiary and secondary cationic intermediates in the addition reaction, become translated into a preference for one product over the other? The relative rate of the two possible protonation steps will not depend directly on the relative energies of the two carbocations themselves, but on the energy difference between the two transition states leading to them ($\Delta\Delta G^{\ddagger}$ in Fig. 9.37). But both protonation steps are endothermic and it is likely that the transition states will bear a strong resemblance to the cations. (Remember the Hammond postulate, "For an endothermic reaction the transition state will resemble the product," Chapter 8, p. 320) In these two transition states, positive charge is building up on carbon. There are partial positive charges in the transition states. *A tertiary partial positive charge will be more stable than a secondary partial positive charge, just as a full positive charge on a tertiary carbon is more stable than a full positive charge on a secondary carbon.* Once the carbocation is formed, the structure of the product is determined. Capture of the tertiary carbocation leads to the observed product, in which X is attached to the more substituted carbon of the starting alkene (the site of the tertiary positive charge in the cation). Similarly, the secondary carbocation would lead inevitably to a product in which X is attached to the less substituted carbon (Fig. 9.37).

In Chapter 7, we avoided the issue, but it is now time to deal with the question of why alkyl substitution stabilizes carbocations. Our mechanistic hypothesis is consistent with the idea that the more substituted halide is produced because it is the result of the initial formation of the more substituted (and hence more stable) carbocation (Fig. 9.38).

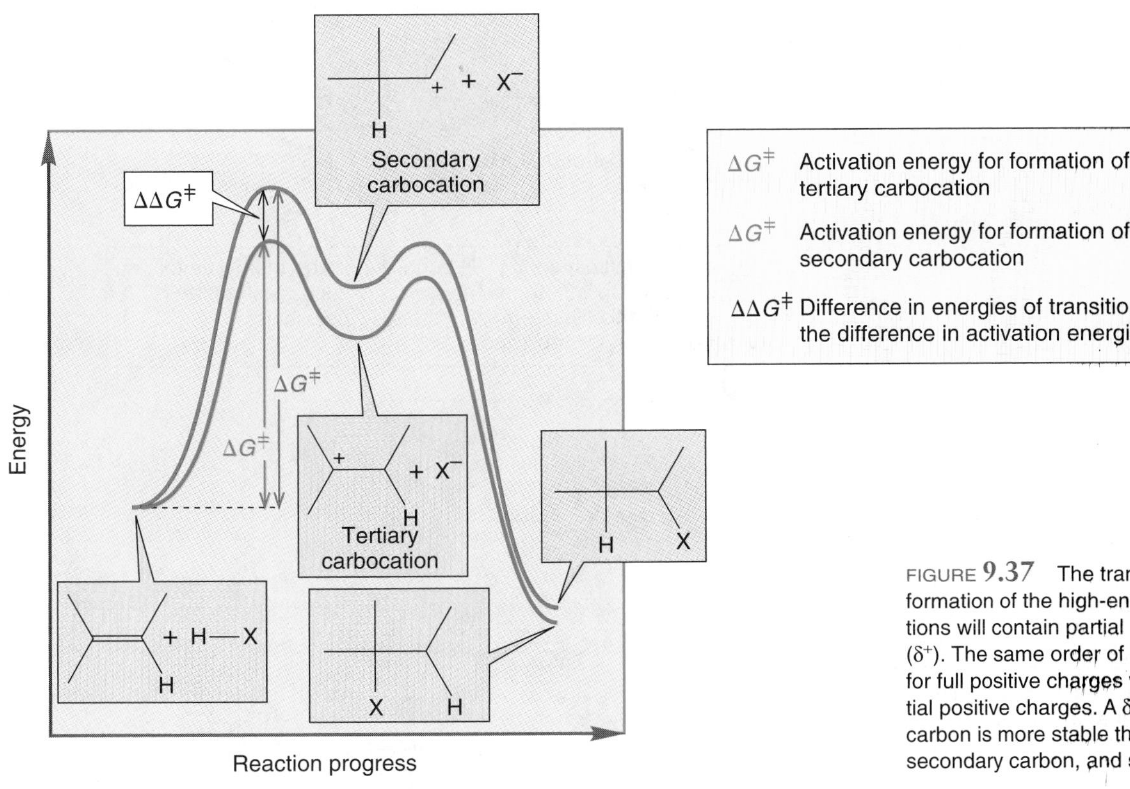

FIGURE 9.37 The transition states for formation of the high-energy carbocations will contain partial positive charges (δ^+). The same order of stability as holds for full positive charges will apply to partial positive charges. A δ^+ on a tertiary carbon is more stable than a δ^+ on a secondary carbon, and so on.

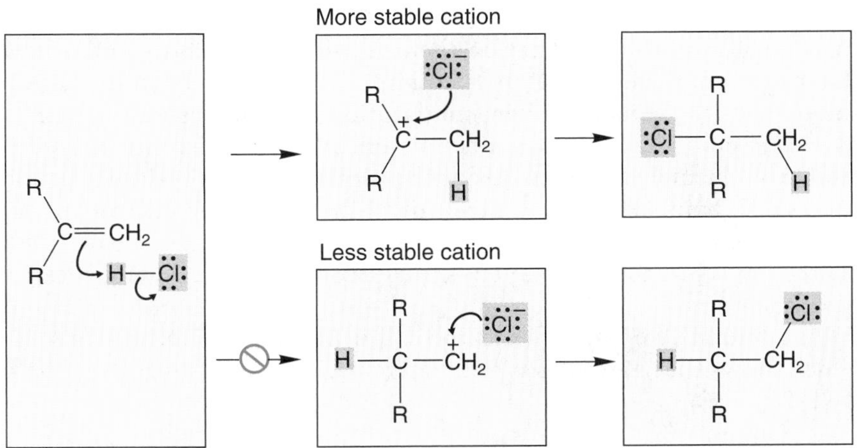

FIGURE 9.38 For an unsymmetrical alkene, protonation to give the more stable carbocation will always lead to the product in which the nucleophile, here chloride, is attached to the more substituted carbon.

But this consistency doesn't answer anything; we have just pushed the question further back. *Why* are more substituted carbocations more stable than less substituted ones? To answer such questions, we start with structure. Carbocations are flat, and the central trigonal carbon is sp^2 hybridized. Figure 9.39 contrasts the methyl cation ($^+CH_3$) with the ethyl cation ($^+CH_2CH_3$).

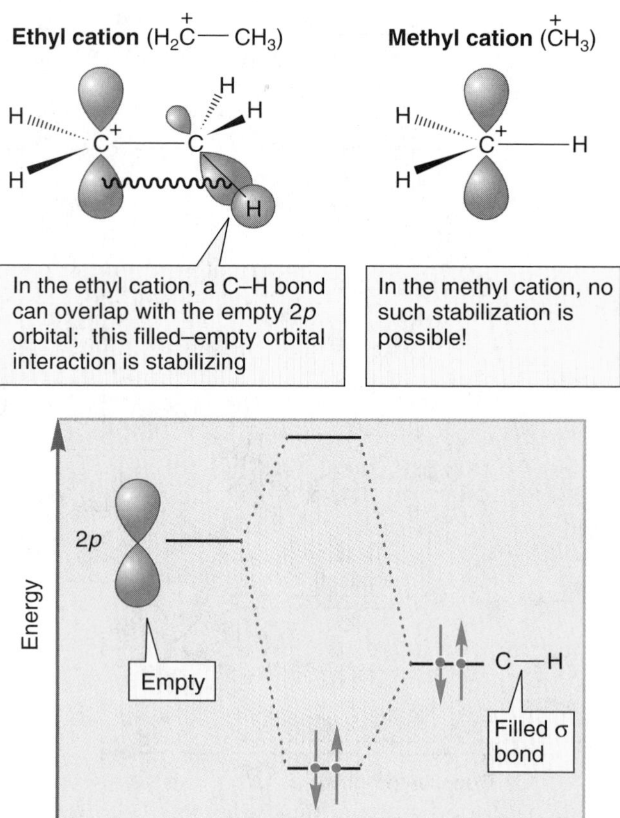
Ethyl cation ($H_2\overset{+}{C}$— CH_3)
Methyl cation ($\overset{+}{C}H_3$)
H
H
H
C+
C
H
H
H
H
C+
H
H
In the ethyl cation, a C–H bond can overlap with the empty 2p orbital; this filled–empty orbital interaction is stabilizing
In the methyl cation, no such stabilization is possible!
Energy
2p
Empty
C—H
Filled σ bond

FIGURE 9.39 The ethyl cation is stabilized by overlap of a filled carbon–hydrogen bond with the empty 2*p* orbital on carbon. No such stabilization is possible for the methyl cation.

Note that in the ethyl cation there is a filled carbon–hydrogen bond that can overlap with the empty *p* orbital. This phenomenon has the absolutely grotesque name of **hyperconjugation**, but there is nothing particularly "hyper" about it. It is a simple effect and can be rationalized in familar molecular orbital or resonance terms. The empty–filled interaction is stabilizing (Fig. 9.39) and nothing like it is available to the methyl cation. The more interactions of this kind that are possible, the more stable the ion. Thus the *tert*-butyl cation is most stable, the isopropyl cation next, and so on.

In addition, we could write the following kind of resonance form, which shows the charge delocalization from carbon to hydrogen especially well (Fig. 9.40):

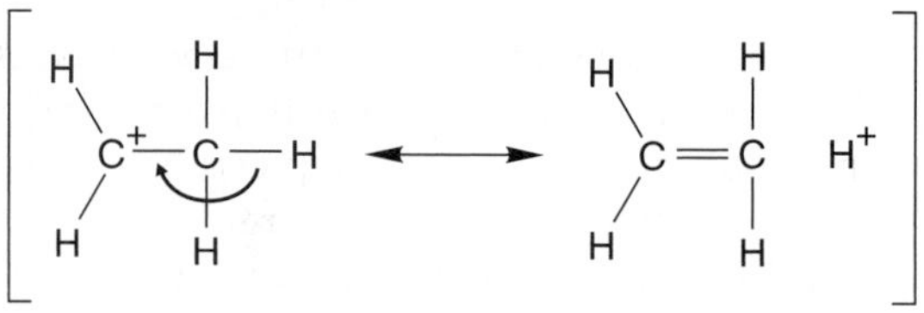

$$\left[H_2\overset{+}{C}-CH_2-H \longleftrightarrow H_2C{=}CH_2 \quad H^+ \right]$$

FIGURE 9.40 A resonance description of the ethyl cation ($^+CH_2CH_3$).

It's important to see that the second carbon–carbon bond in the resonance form of Figure 9.40 is *not* identical to the π bond of an alkene. It is not formed from 2*p*–2*p* overlap but rather from 2*p*–sp^3 overlap (Fig. 9.41). Be sure you see the difference between these two fundamentally different, but similar-looking structures.

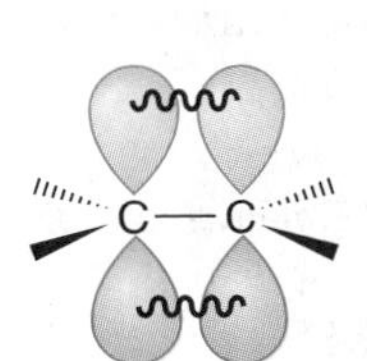

The π bond in ethylene is made up of 2*p*–2*p* overlap

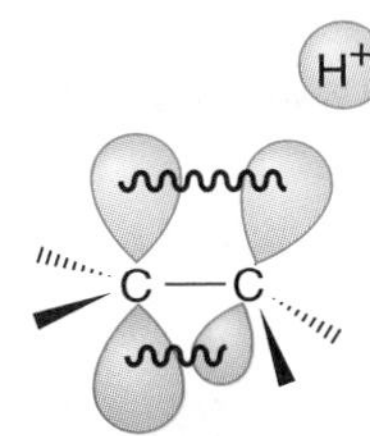

In the "hyperconjugative" resonance form for the ethyl cation, the "double" part of the double bond is not made up of 2*p*–2*p* overlap, but of 2*p*–sp^3 overlap

FIGURE **9.41** The hyperconjugative resonance form does not contain a 2*p*–2*p* π bond. The apparent "double" part of the double bond is in fact composed of 2*p*–sp^3 overlap. The two look the same in schematic form, but are very different.

PROBLEM **9.10**

It is often claimed that the *tert*-butyl cation has 10 resonance forms, 1 in which the carbon is positive and 9 others in which each hydrogen bears the positive charge. Is this true?

*PROBLEM **9.11**

In Section 4.5 (p. 133), we described the stabilization of alkenes by attached alkyl groups in different terms, focusing on the σ bond system. We could construct a similar treatment here. Provide a justification of the relative stability of differently substituted carbocations using an analysis of the σ system.

ANSWER

In explaining the increased stability of alkenes with increasing substitution, we focused on the strengths of the carbon–carbon σ bonds as substitution increased. The more *s* character in a bond the lower energy it is; an sp^2–sp^2 bond is lower energy (more stable) than an sp^3–sp^2 bond, for example. We can look at carbocations in exactly this way. Compare the *tert*-butyl cation with the isomeric ethylmethyl cation, for example:

$(CH_3)_3C^+$ — Three sp^2–sp^3 bonds (lower energy, more stable)

$H_3C{-}CH^+{-}CH_2{-}CH_3$ — Two sp^2–sp^3 bonds one sp^3–sp^3 bond (higher energy, less stable)

PROBLEM **9.12**

Do you think the π part of the double bond in the resonance form in Figure 9.41 is stronger or weaker than a normal π bond?

We have just seen that carbocations are stabilized by delocalization through the overlap of filled and empty orbitals. One example of this effect is fairly straightforward, as nonbonding, or lone-pair, electrons on adjacent atoms interact with an empty orbital. Less obvious is the last phenomenon, the overlap of filled σ orbitals of carbon–hydrogen bonds with empty orbitals, called hyperconjugation. Now we proceed to other factors influencing the stability of cations.

9.7 INDUCTIVE EFFECTS ON ADDITION REACTIONS

Polar bonds in a molecule can have a major effect on both the rate and the regiochemistry of an addition reaction. Let's look first at the series of ethylenes in Figure 9.42.

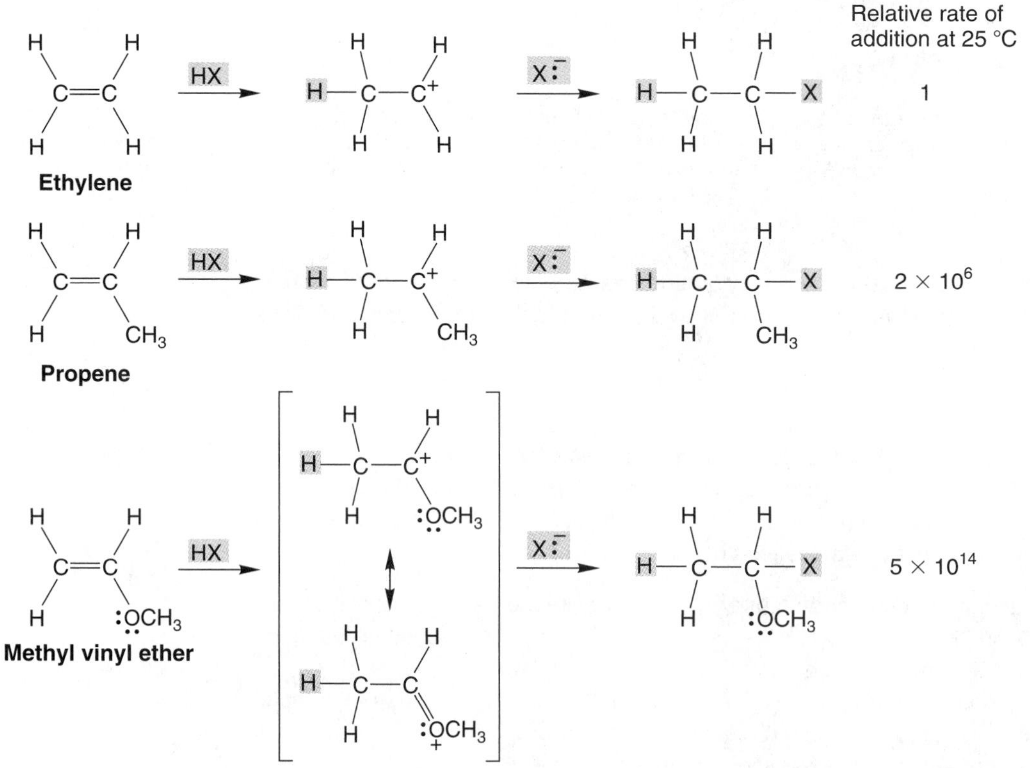

Alkene		Intermediate cation		Product	Relative rate of addition at 25 °C
Ethylene ($H_2C{=}CH_2$)	HX →	$H_3C{-}CH_2^+$	X:⁻ →	$H_3C{-}CH_2{-}X$	1
Propene ($H_2C{=}CH{-}CH_3$)	HX →	$H_3C{-}C^+H{-}CH_3$	X:⁻ →	$H_3C{-}CH(X){-}CH_3$	2×10^6
Methyl vinyl ether ($H_2C{=}CH{-}\ddot{O}CH_3$)	HX →	$[H_3C{-}C^+H{-}\ddot{O}CH_3 \leftrightarrow H_3C{-}CH{=}\overset{+}{O}CH_3]$	X:⁻ →	$H_3C{-}CH(X){-}\ddot{O}CH_3$	5×10^{14}

FIGURE 9.42 Typical rates of addition for a series of substituted ethylenes. The reaction is, in fact, a hydration reaction in which HOH adds across the double bond.

Ethylene itself will protonate only very slowly (see Section 9.13 for a further discussion of the protonation of ethylene). Propene will surely protonate so as to give the more stable secondary cation, and the methoxyl group of methyl vinyl ether will provide resonance stabilization to an adjacent positive charge. Resonance stabilization by the methyl and methoxyl groups of propene and methyl vinyl ether has a great accelerating effect on the rate of protonation, the rate-determining step in this reaction. Figure 9.42 shows the rates of protonation compared to that of unsubstituted ethylene itself.

PROBLEM 9.13 Why will the rate of protonation of ethylene be especially slow?

Substituents can slow reactions as well, especially if they are not in a position to provide resonance stabilization. Figure 9.43 shows typical rates for addition to some substituted alkenes. To see why bromine and chlorine so dramatically slow the rate of addition we must again look more closely at the bond from carbon to halogen (see p. 236 for an earlier look).

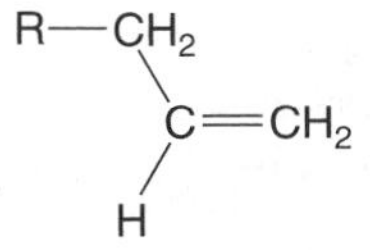

R = H	Br	Cl
Rate 1	10^{-4}	10^{-4}

FIGURE 9.43 Typical rates for an addition to two halo-substituted ethylenes. If the parent compound, propene, is given a relative rate of 1, the bromo- and chloro-substituted compounds are much slower. Why does substitution by halogen slow the reaction so dramatically?

The two atoms in the bond, carbon and halogen, are different, so the bond between them must be polarized. The electrons in the bond cannot be equally shared. Bromine and chlorine are far more electronegative than carbon (see Table 9.1, p. 335) and they will attract electrons strongly. The carbon–halogen bond will be polarized as shown in Figure 9.44. The carbon will bear a partial positive charge (δ^+) and the more electronegative halogen a partial negative charge (δ^-).

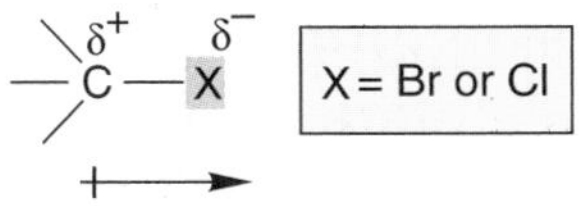

X = Br or Cl

FIGURE 9.44 The dipole in the bond between carbon and the relatively electronegative halogen atom bromine or chlorine will have its positive end on carbon and its negative end on the halogen.

For the molecules in Figure 9.43, protonation in the Markovnikov sense places a full positive charge adjacent to the already partially positive carbon. This will surely be destabilizing, and accounts for the decrease in rate for addition reactions to the bromine- and chlorine-substituted molecules (Fig. 9.45).

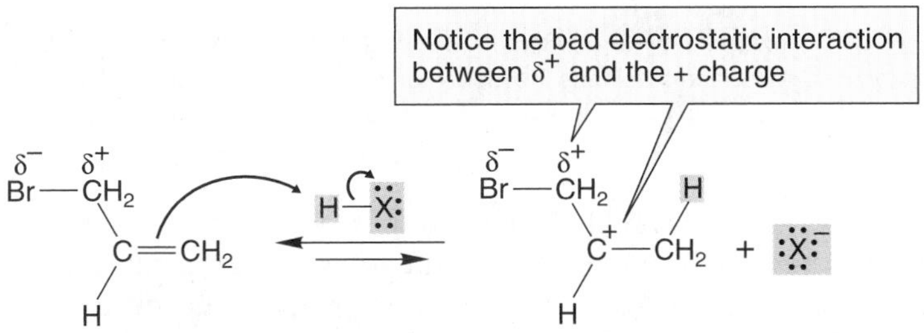

Notice the bad electrostatic interaction between δ^+ and the + charge

FIGURE 9.45 Protonation gives an intermediate destabilized by the electrostatic interaction between the newly formed full positive charge and the partial positive charge (δ^+).

Effects such as these, which are transmitted through the σ bonds, are called **inductive effects**.

But wouldn't the methoxyl group of methyl vinyl ether act in the same way? Oxygen is also more electronegative than carbon, and wouldn't this act to discourage any addition that puts more positive charge on the already partially positive carbon adjacent to oxygen? Why does addition to methyl vinyl ether occur so rapidly (Fig. 9.42) in the Markovnikov sense? Figure 9.46 frames the question, and answers it as well.

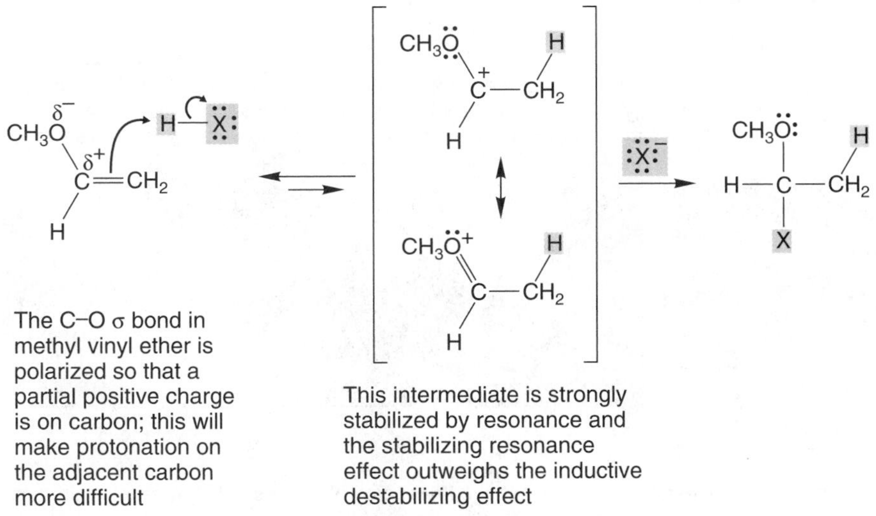

The C–O σ bond in methyl vinyl ether is polarized so that a partial positive charge is on carbon; this will make protonation on the adjacent carbon more difficult

This intermediate is strongly stabilized by resonance and the stabilizing resonance effect outweighs the inductive destabilizing effect

FIGURE 9.46 Stabilizing resonance effects and destabilizing inductive effects can "fight" each other. Generally, resonance stabilization is more important.

The answer is simply that the resonance effect of methoxy overwhelms the inductive effect. The two effects *are* in competition, and the stronger resonance effect wins out. Now let's look at a number of related addition reactions.

9.8 H–X ADDITION REACTIONS: HYDRATION

We have seen addition reactions of H–Br and H–Cl. It should not be surprising that other acidic H–X compounds undergo similar addition reactions. Hydrogen iodide is a simple example, and H–I addition rather closely resembles the reactions we have already studied in this chapter. Markovnikov's rule is followed, for example (Fig. 9.47).

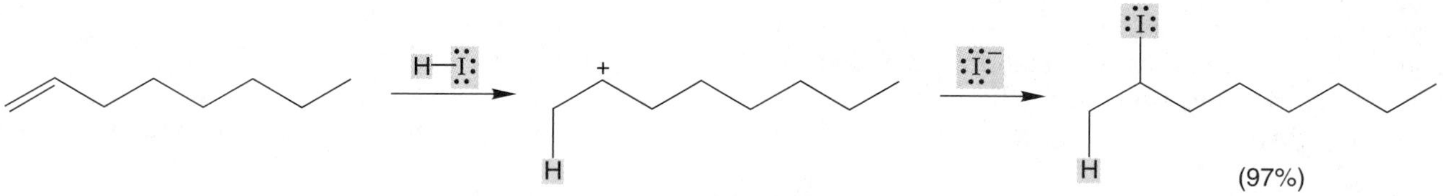

(97%)

FIGURE 9.47 Addition of hydrogen iodide follows the Markovnikov rule; the more stable carbocation is formed, and this leads inevitably to the observed iodide in which the iodine atom becomes attached to the more substituted carbon and the hydrogen to the less substituted carbon.

Water (H–O–H) is an H–X reagent, but not normally a strong enough acid to undergo addition to simple alkenes at a useful rate. If one mixes water and a typical alkene, no reaction occurs. The difference is that H–Cl and H–Br are strong enough acids to protonate the alkene and start things going. Water is not. The pK_a values of H–Br, H–Cl, and H–O–H are –9, –7, and +15.7, respectively. Water is a less powerful acid than H–Br and H–Cl by factors of about 10^{25} and 10^{23}! Acid must be added to protonate the alkene and thus generate a carbocation that can continue the reaction. The acid **catalyst** in this reaction sets things going. The catalyst has no effect on the position of the final equilibrium, but it does lower the energy requirement for generating product. A catalyst provides a new mechanistic pathway for product formation. It is able to carve a new pathway along a lower slope of the mountain we first encountered in Section 8.5 (Fig. 8.20, p. 315; Fig. 9.48).

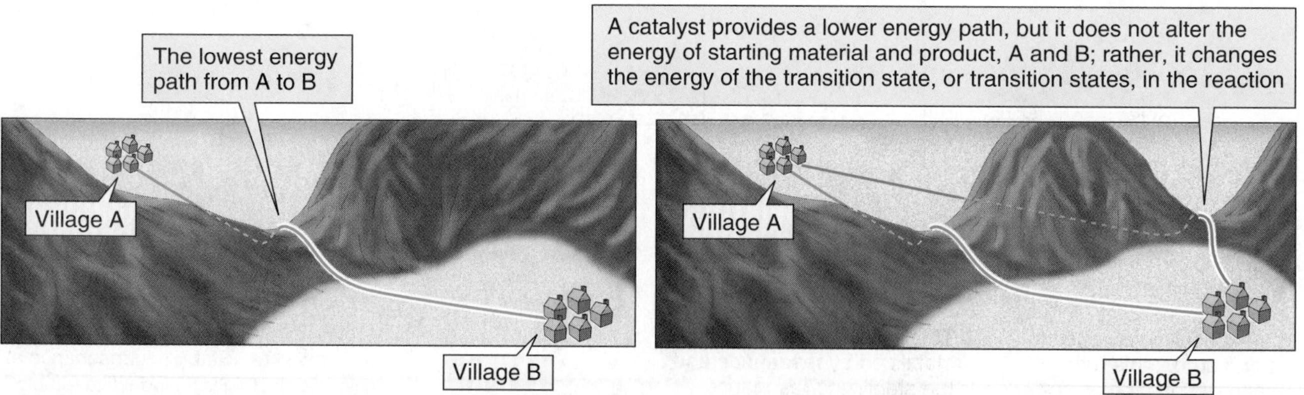
The lowest energy path from A to B
Village A
Village B
A catalyst provides a lower energy path, but it does not alter the energy of starting material and product, A and B; rather, it changes the energy of the transition state, or transition states, in the reaction
Village A
Village B

FIGURE 9.48 A catalyst does not change the energy of either starting material or product, but does provide a lower energy path between them.

As can be seen in Figure 9.49, the catalyst H_3O^+ is used up in the first step of the reaction, only to be regenerated in the last step. A number of acids can be used in the catalytic role; sulfuric (H_2SO_4 or $HO{-}SO_2{-}OH$) and nitric (HNO_3 or $HO{-}NO_2$) are commonly used to promote the addition reaction. In water containing a little sulfuric acid, 2,3-dimethyl-2-butene is protonated to generate the tertiary carbocation. This intermediate is then attacked by water (not hydroxide, ^-OH) to give the **oxonium ion**, which can be deprotonated by water to regenerate the catalyst, H_3O^+ and produce a molecule of an **alcohol**, a compound of the general formula R–OH (Fig. 9.49).

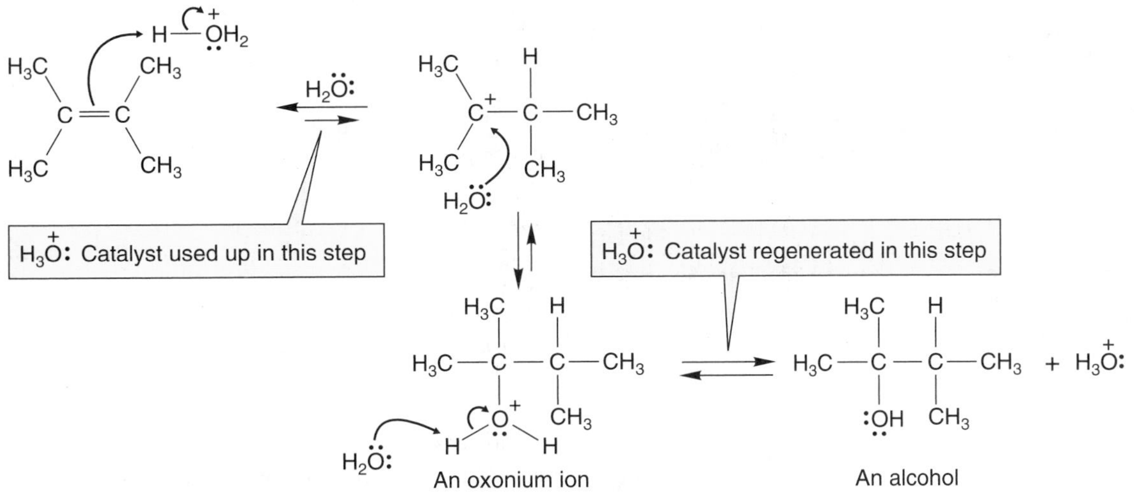

FIGURE 9.49 The mechanism for the acid-catalyzed hydration of 2,3-dimethyl-2-butene. Note that the catalyst H_3O^+, is consumed in the first step but regenerated in the last step.

This mechanism would predict, indeed require, that the Markovnikov rule be followed, and that sets up a mechanistic test. We will examine the hydration of isobutene to probe the regiochemistry of the reaction. Before we even start, it is worth thinking a bit about the results. What will we know after we have done the experiment? There are only two general results possible: Either the Markovnikov rule is followed or it is not (Fig. 9.50).

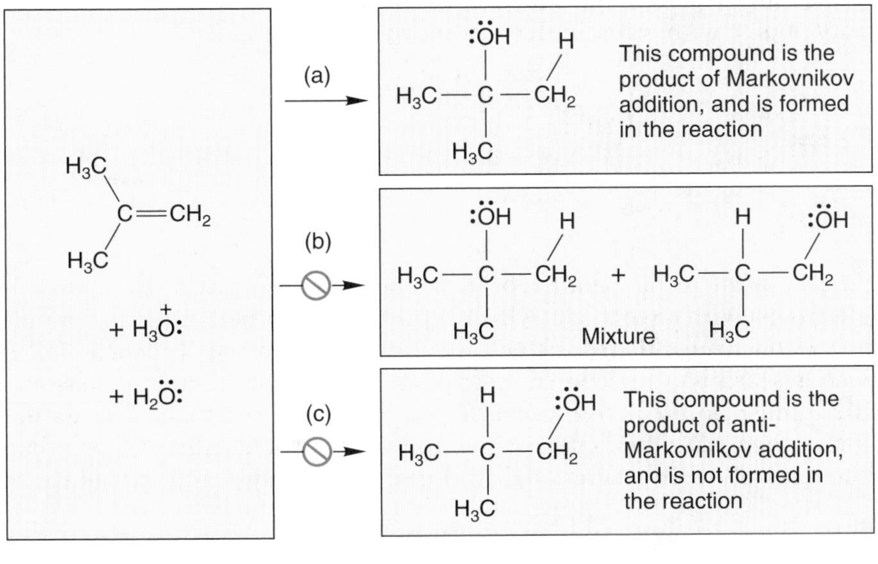

FIGURE 9.50 Possible results in the hydration of isobutene. The Markovnikov rule can be followed, as in (a), or not, as in (b) or (c).

If it is not, we can say with certainty that the mechanism we have suggested for this reaction is wrong. A crucial step in our mechanism involves the formation of a carbocation. When there is a choice, as there certainly is with isobutene, the more stable carbocation must be formed. Once formed, it must lead to the product in which the hydroxyl group is attached to the more substituted carbon (Fig. 9.51). If our postulated mechanism is correct, there is no other possible result. So if we get the "wrong" alcohol, or a mixture of alcohols [(b) and (c) in Fig. 9.50], we know the mechanism is wrong. It may need only a minor adjustment or it could require complete scrapping, but it cannot be correct as written.

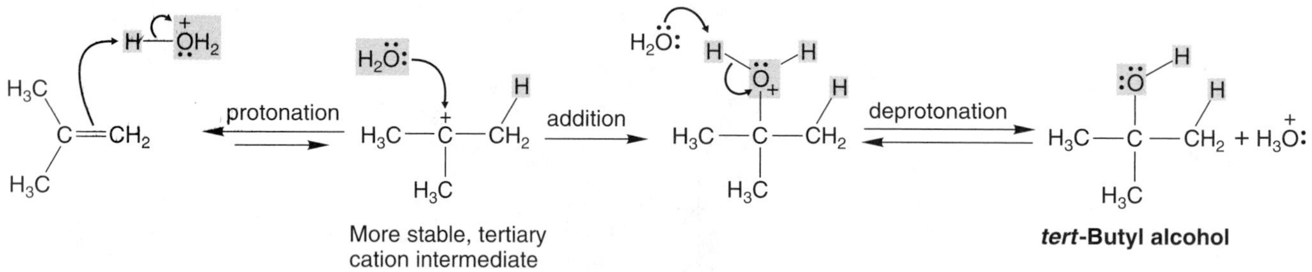

protonation

More stable, tertiary cation intermediate

addition

deprotonation

tert-Butyl alcohol

FIGURE 9.51 The mechanism for Markovnikov addition of water to isobutene. The product must be *tert*-butyl alcohol if this is indeed the mechanism.

More interesting, and less obvious, is what we know if our prediction turns out to be correct. Have we proved the mechanism? Absolutely not. Our mechanistic hypothesis remains just a hypothesis: a guess at how the reaction goes that is consistent with all experiments that have been done so far. There may be some other test experiment that would not work out as well. The sad fact is that no number of experiments we can do will suffice to prove our mechanism. We are doomed to be forever in doubt, and our mechanism to be forever suspect. There is no way out. Indeed all the mechanisms in this book are wrong, at least in the sense that they are broad-brush treatments that do not look closely enough at detail. Organic chemistry has come a long way since the days of the "rectangular hypothesis," which was little more than a bookkeeping device. In olden times it was common practice to write a reaction such as the S_N1 reaction of *tert*-butyl iodide with water in the way shown in Figure 9.52.

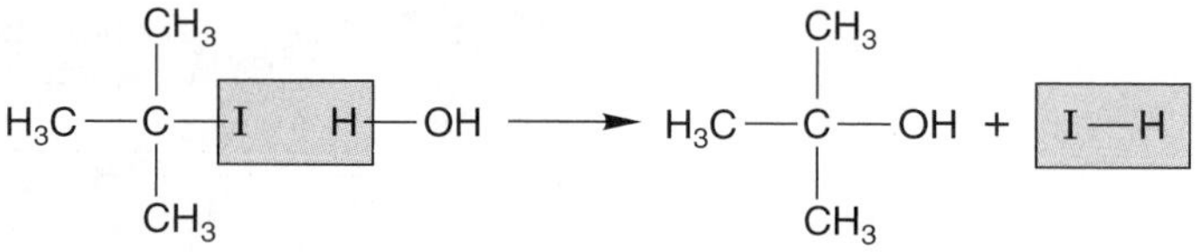

FIGURE 9.52 The "rectangular hypothesis" applied to the reaction of *tert*-butyl iodide with water.

To use the rectangular hypothesis, one simply surrounds the appropriate groups with a rectangle, which somehow removes the captured atoms and reconstitutes them, in this case as hydrogen iodide. This archaic device is scarcely a reasonable description of how the reaction occurs and has almost no predictive value. If you know one rectangle, you do not know them all! But let's not get too smug. We do know more than chemists of other generations. We know, for example, that carbocations are involved in the reactions we are currently studying, and molecular orbital theory gives us additional insight into the causes of these reactions. But we are still very ignorant, and I am certain that much of what we "know" is wrong, at least in detail. Chemists of the future will regard us as primitives, and they will be quite right. It doesn't matter, of course. Our job is to do the best we can; to add what we can to the fabric of knowledge and to provide some shoulders for those future chemists to stand on, as our earlier colleagues did for us.

In this case, things turn out well for our mechanistic hypothesis, and the Markovnikov rule is followed in hydration reactions.

There are variations on this reaction. Ethylene, for example, is hydrated only under more forcing conditions—concentrated sulfuric acid—and the initial product formed is not the alcohol, but ethyl hydrogen sulfate, which is converted in a second step into the alcohol (Fig. 9.53). It is not easy to be sure of the mechanism of this reaction. The concentrated sulfuric acid may be required to produce the very unstable primary carbocation, or a different, perhaps bridged cation may be involved (see Section 9.13).

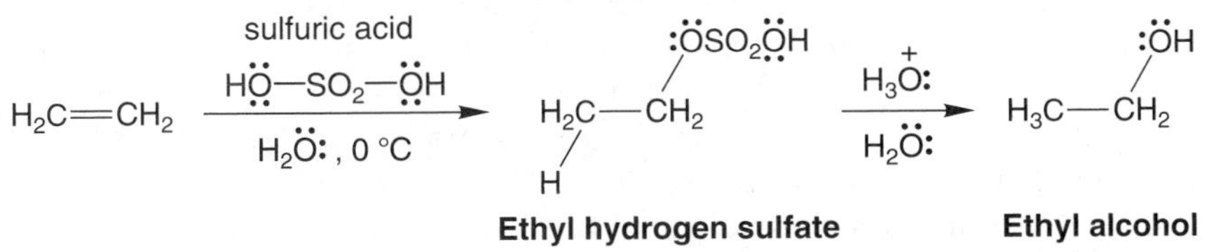

$$H_2C{=}CH_2 \xrightarrow[H_2\ddot{O}:,\ 0\ °C]{\text{sulfuric acid}\quad H\ddot{O}{-}SO_2{-}\ddot{O}H} \underset{\textbf{Ethyl hydrogen sulfate}}{H(H_2C{-}CH_2){:}\ddot{O}SO_2\ddot{O}H} \xrightarrow[H_2\ddot{O}:]{H_3\overset{+}{\ddot{O}}:} \underset{\textbf{Ethyl alcohol}}{H_3C{-}CH_2{:}\ddot{O}H}$$

FIGURE 9.53 Ethyl hydrogen sulfate is an intermediate in the hydration of ethylene.

9.9 DIMERIZATION AND POLYMERIZATION OF ALKENES

Let's look again at the hydration of isobutene. What if we change the conditions a bit? What if both the temperature and concentration of sulfuric acid are increased? Figure 9.54 shows that two new products containing eight carbons are formed. These are dimeric products, formed somehow by the combining of two molecules of isobutylene.

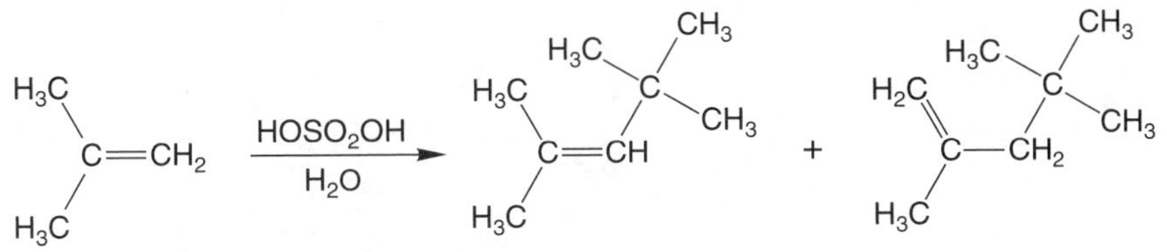

$$(H_3C)_2C{=}CH_2 \xrightarrow[H_2O]{HOSO_2OH} (H_3C)_2C{=}CH{-}C(CH_3)_3 \; + \; H_2C{=}C(CH_3){-}CH_2{-}C(CH_3)_3$$

FIGURE 9.54 If the hydration of isobutene (C_4H_8) is carried out at higher temperature and with a high concentration of sulfuric acid, two dimeric products (C_8H_{16}) are formed.

Surely, increasing the acid concentration can only make protonation of the alkene more probable, so the first step of the reaction is unchanged. The initially formed carbocation finds itself in the presence of less water than it did in the reaction we looked at before. Accordingly, formation of the alcohol through reaction with water is less likely. We might ask what other Lewis bases are available for the strongly Lewis acidic cation? The most obvious one is isobutene itself. The *tert*-butyl cation can react with isobutene, just as did the proton. Both the proton and the carbocation are Lewis acids and both can add to the double bond to give a new cation (Fig. 9.55). There is no essential difference in the two reactions, and once more it will be the more stable, tertiary carbocation that will be formed on addition.

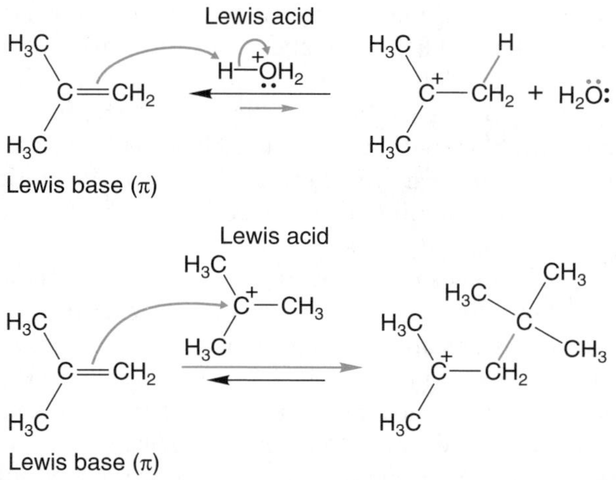

FIGURE 9.55 Many different Lewis acids add to isobutene (and other alkenes). Here we see H_3O^+ and the *tert*-butyl cation adding to isobutene to give new tertiary carbocations. In each case addition will be in the Markovnikov sense, as it will be the more stable carbocation that is preferentially formed.

Now we have a structure that contains eight carbons, as do the two new products. It looks as though we are on the right track. The only remaining trick is to see how to lose the "extra" proton and form the new alkenes. If a proton is lost from the two different positions adjacent to the positive charge, we get the two products. We have seen similar deprotonations before in our discussion of the E1 reaction (Chapter 7, p. 273), and we have also seen the reverse reaction. Loss of these protons is just the reverse of protonation of an alkene (Fig. 9.56).

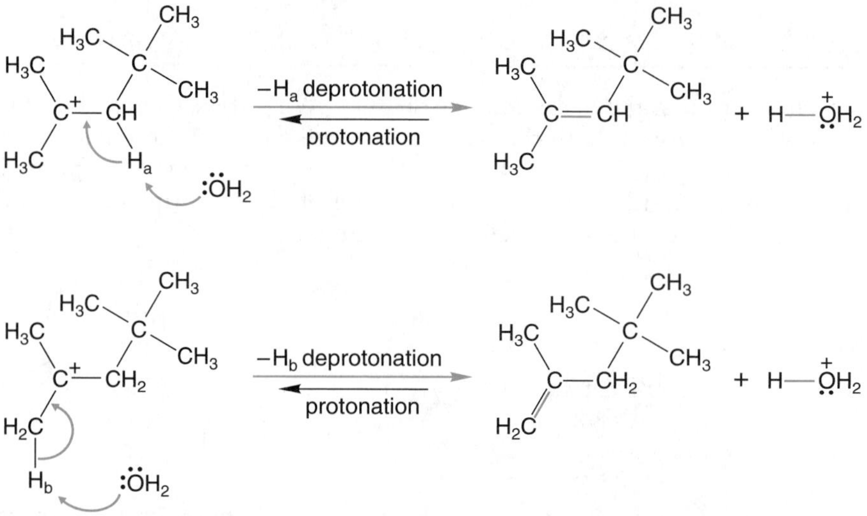

FIGURE 9.56 Loss of the two different protons (H_a and H_b) on carbons adjacent to the carbon bearing the positive charge leads to the two products.

The proton can't just leave, however. It must be removed by a base. In concentrated sulfuric acid, strong bases do not abound. There are some bases present, however, and the carbocation is a high-energy species, and thus rather easily deprotonated. Under these conditions both water and bisulfate ion are capable of assisting the deprotonation (Fig. 9.56).

Note the exact correspondence of the steps involved in alkene protonation and deprotonation. The two reactions are just the two sides of an equilibrium (Fig. 9.57).

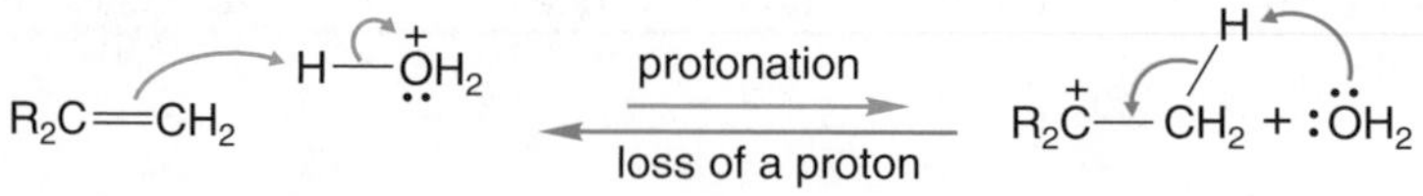

FIGURE 9.57 Protonation and deprotonation are just two sides of an equilibrium.

If the concentration of acid is even higher, the base concentration will be even lower and both alcohol formation and deprotonation will be slowed (both require water to act as Lewis base). There is no reason why the dimerization process cannot continue indefinitely under such conditions, leading first to a trimer (a molecule formed from three molecules of isobutylene) and, eventually, to a polymer (a molecule formed by combination of many starting alkene molecules). The cation-induced polymerization is called, logically enough, **cationic polymerization**, and ends only when the alkene concentration decreases sufficiently to slow the addition step (Fig. 9.58).

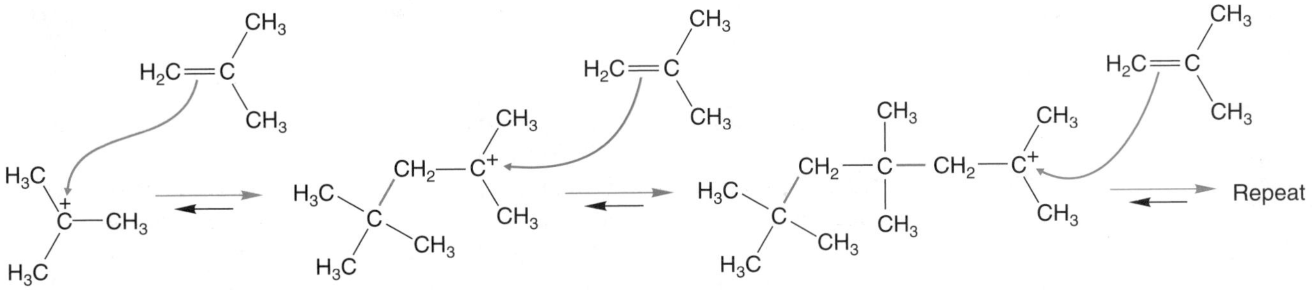

FIGURE 9.58 The beginning stages of the cationic polymerization of isobutene.

PROBLEM 9.14

Treatment of 2-ethyl-1-butene with H_3O^+/H_2O leads to three products of the formula $C_{12}H_{24}$. Give structures and a mechanism for their formation.

9.10 HYDROBORATION

In the previous example, we saw the addition of a carbocation to an alkene to generate a new carbocation. The familiar theme of overlap between filled (alkene π in this case) and empty (carbon $2p$) orbitals was recapitulated. "Lewis bases react with Lewis acids." Other reagents containing empty p orbitals, all good Lewis acids, might also be expected to add to alkenes, and so they do.

We know another kind of molecule that has an empty $2p$ orbital, the trigonal boranes, BF_3 and BH_3 (Chapter 3, p. 75; Fig. 9.59). There is no positive charge on these compounds, but they should still be strong Lewis acids.

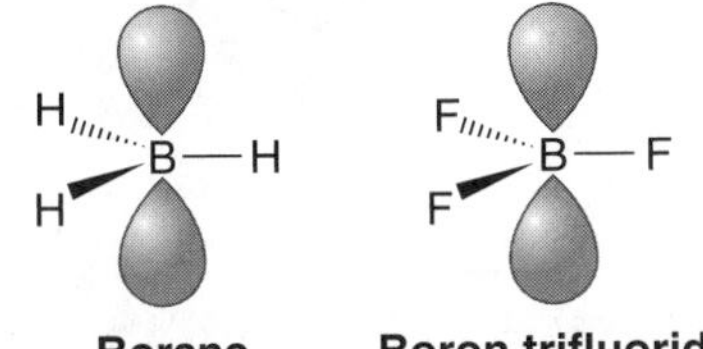

FIGURE 9.59 Borane and boron trifluoride are both Lewis acids and will react like cations.

PROBLEM 9.15

Verify that boron in BF_3 and BH_3 is neutral, and that each boron has an empty $2p$ orbital.

Although BF_3 is known and even commercially available, free BH_3 is unavailable as it spontaneously dimerizes to diborane (B_2H_6), an unpleasantly flammable and toxic gas. It is tempting to imagine a structure like ethane's ($H_3C{-}CH_3$ = C_2H_6) for diborane ($H_3B{-}BH_3$ = B_2H_6), but we would be wrong to do so. An ethane-like structure requires 14 electrons for 7 boron–hydrogen and boron–boron σ bonds. The 6 hydrogens contribute 6 electrons and the 2 borons another 6 to give a total of 12, so we are 2 electrons shy of the required number. The real structure is shown in

Figure 9.60, along with a resonance formulation. The molecule B_2H_6 is a hybrid of the two resonance forms **A** and **A′**.

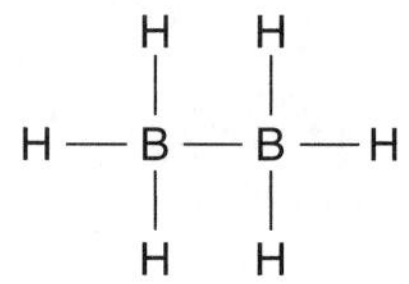

This structure cannot be B_2H_6 because there are only 12 electrons available for bonding in this molecule, and this formulation requires 14

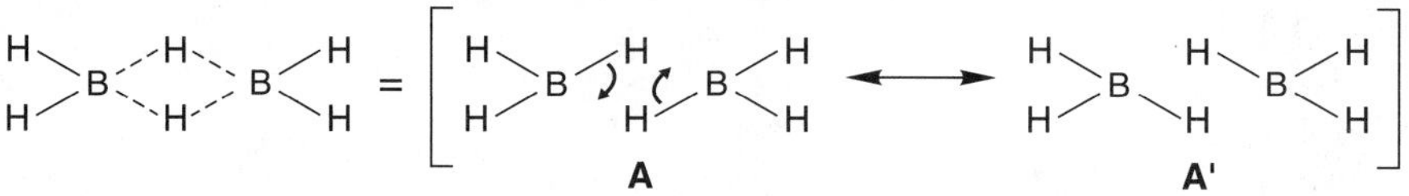

FIGURE 9.60 The real structure of B_2H_6 is a resonance hybrid of two forms, **A** and **A′**.

In this molecule, partial bonds (shown as dashed bonds) are used to connect the atoms. This kind of bonding is more common than once thought. It abounds in inorganic chemistry and in the organic chemistry of electron-deficient species such as carbocations as well (see Section 9.13).

When diborane dissolves in diethyl ether ($CH_3CH_2{-}O{-}CH_2CH_3$) a complex of ether and borane is formed. This complex of borane and an ether can be used as a source of BH_3. Often the cyclic ether tetrahydrofuran (THF) is used (Fig. 9.61). The boron trifluoride–ether complex is so stable that it can even be distilled.

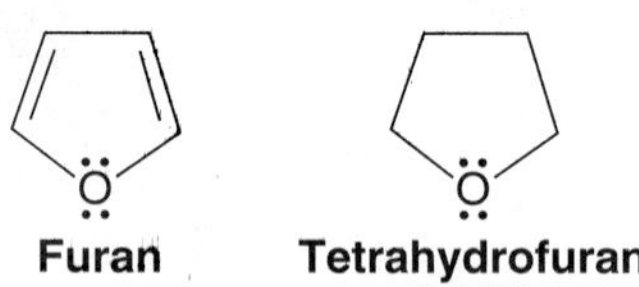

FIGURE 9.61 Furan and THF.

*PROBLEM 9.16 Propose a structure for the ether–borane complex, and suggest a mechanism for its formation.

ANSWER Borane is a Lewis acid, as the boron atom has an empty 2*p* orbital. An ether, here shown as THF, is a Lewis base. The reaction between the Lewis base and Lewis acid produces the borane–THF complex.

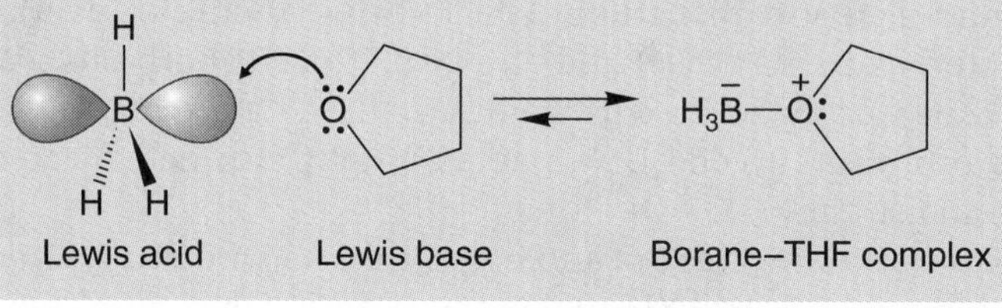

Lewis acid Lewis base Borane–THF complex

When the borane–ether complex is allowed to react with an alkene, there is a rapid addition of the borane across the double bond. The initial product is an alkylborane and the process is called **hydroboration** (Fig. 9.62).

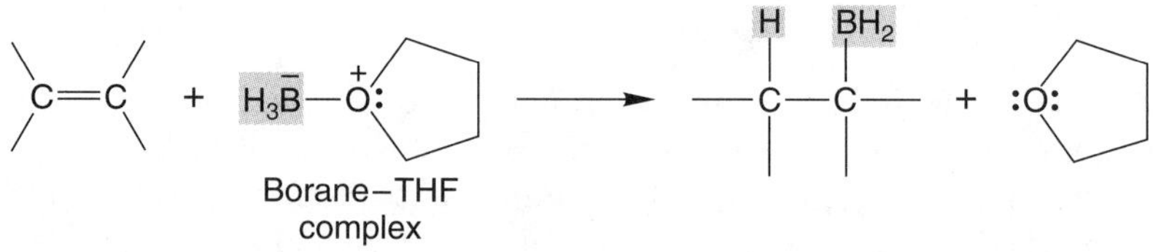

Borane–THF complex

FIGURE 9.62 A schematic hydroboration. The elements of BH_3 have been added across the double bond.

As in the addition of a carbocation to a double bond, the initial interaction is between the empty 2*p* orbital (in this case on neutral boron, not positive carbon) and the filled π orbital of the alkene (Fig. 9.63).

If the mechanism were directly parallel to that of cation addition, we would form the charge-separated molecule shown in Figure 9.63. A hydride ($H:^-$) could then be delivered from the negative boron to the positively charged carbon to complete the addition (Fig. 9.64).

We know this mechanism is not exactly correct for two reasons, one of which will have to wait until Section 9.12 to be explained. Hydroboration can be shown to proceed in syn fashion. The two pieces, H and BH_2, are delivered from the same side of the alkene. Figure 9.65 shows two examples, the hydroborations of 1-methylcyclopentene and *cis*-1,2-dideuterio-1-hexene. These reactions exclusively produce the products of syn addition shown in Figure 9.65.

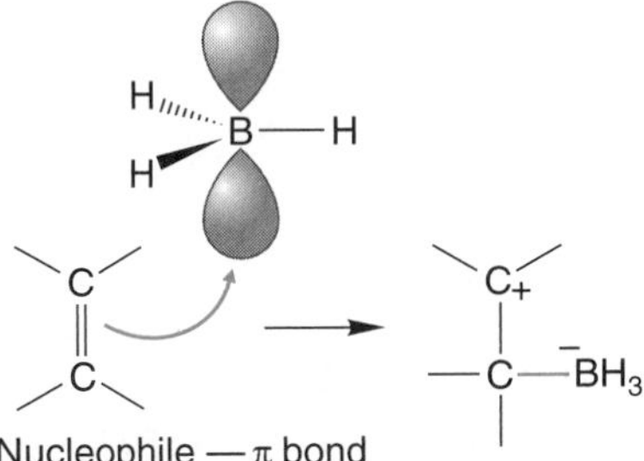

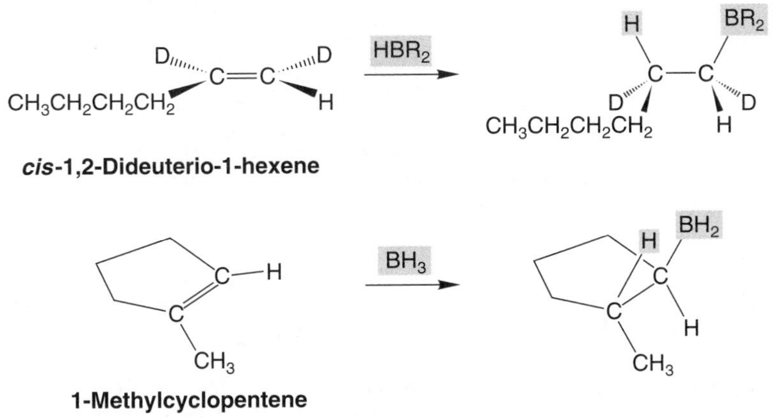

FIGURE 9.63 A possible first step in addition of BH_3 to an alkene. The π bond of the alkene acts as nucleophile and reacts with the strong Lewis acid, BH_3.

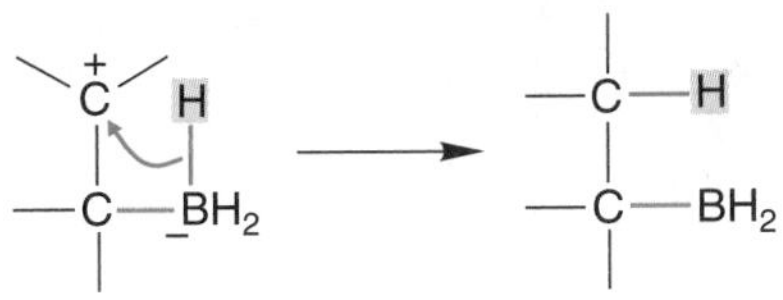

FIGURE 9.64 A possible second step in the hydroboration reaction is transfer of a hydride ion ($H:^-$).

cis-1,2-Dideuterio-1-hexene

1-Methylcyclopentene

FIGURE 9.65 These two stereochemical experiments show that addition of BH_3 proceeds in a syn fashion. The H and BH_2 groups are delivered to the same side of the alkene (R = cyclohexyl).

The mechanism outlined so far has great problems in accounting for this observation. If a planar cation were formed on addition to the deuterated 1-hexene, we would expect the rapid (~10^{11} s^{-1}) rotations about carbon–carbon single bonds to produce two products, the diastereomers shown in Figure 9.66.

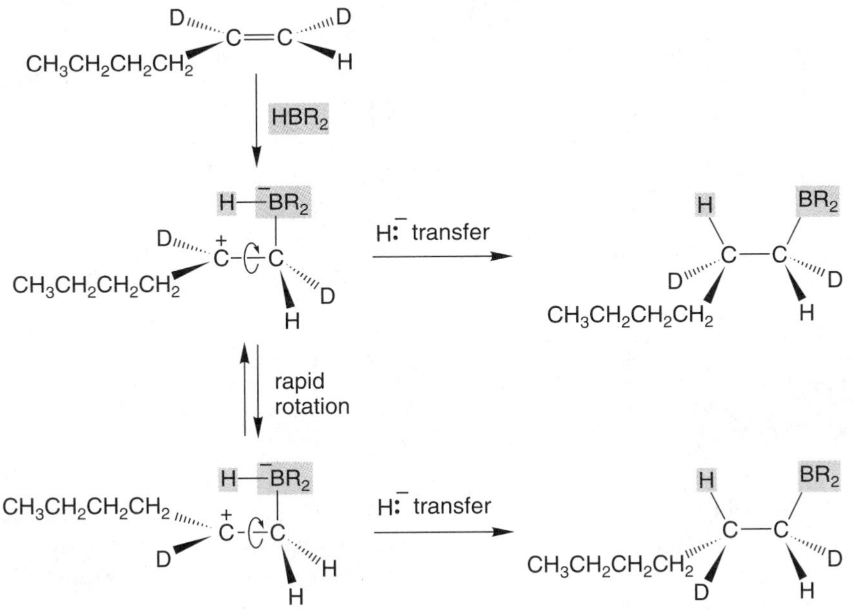

FIGURE 9.66 Rotation about carbon–carbon bonds is very fast and the two-step mechanism predicts that two products should be formed from the hydroboration of *cis*-1,2-dideuterio-1-hexene. Both syn and anti addition products should be formed. However, only the product of syn addition is found (R = cyclohexyl).

As our mechanism cannot easily accommodate the exclusive syn addition of BH_3 to alkenes, it must be modified. The current hypothesis suggests a concerted, or one-step, addition to alkenes in which both new bonds, the one to hydrogen and the one to boron, are formed at the same time (Fig. 9.67). There is no free ionic intermediate as both new bonds are formed in a single step. The observed syn addition is demanded by this modified mechanism.

Transition state

FIGURE 9.67 Syn addition is accommodated by a mechanism in which both new bonds are forming in the transition state. There are no intermediates, and therefore no chance for rotation and scrambling of stereochemistry.

*PROBLEM 9.17 Draw Energy versus Reaction progress diagrams for the concerted and unmodified, two-step mechanisms for the addition of BH_3 to alkenes.

ANSWER The crucial difference is the presence of an intermediate in the two-step process. The one-step (concerted) reaction simply passes from starting material to product over a single transition state.

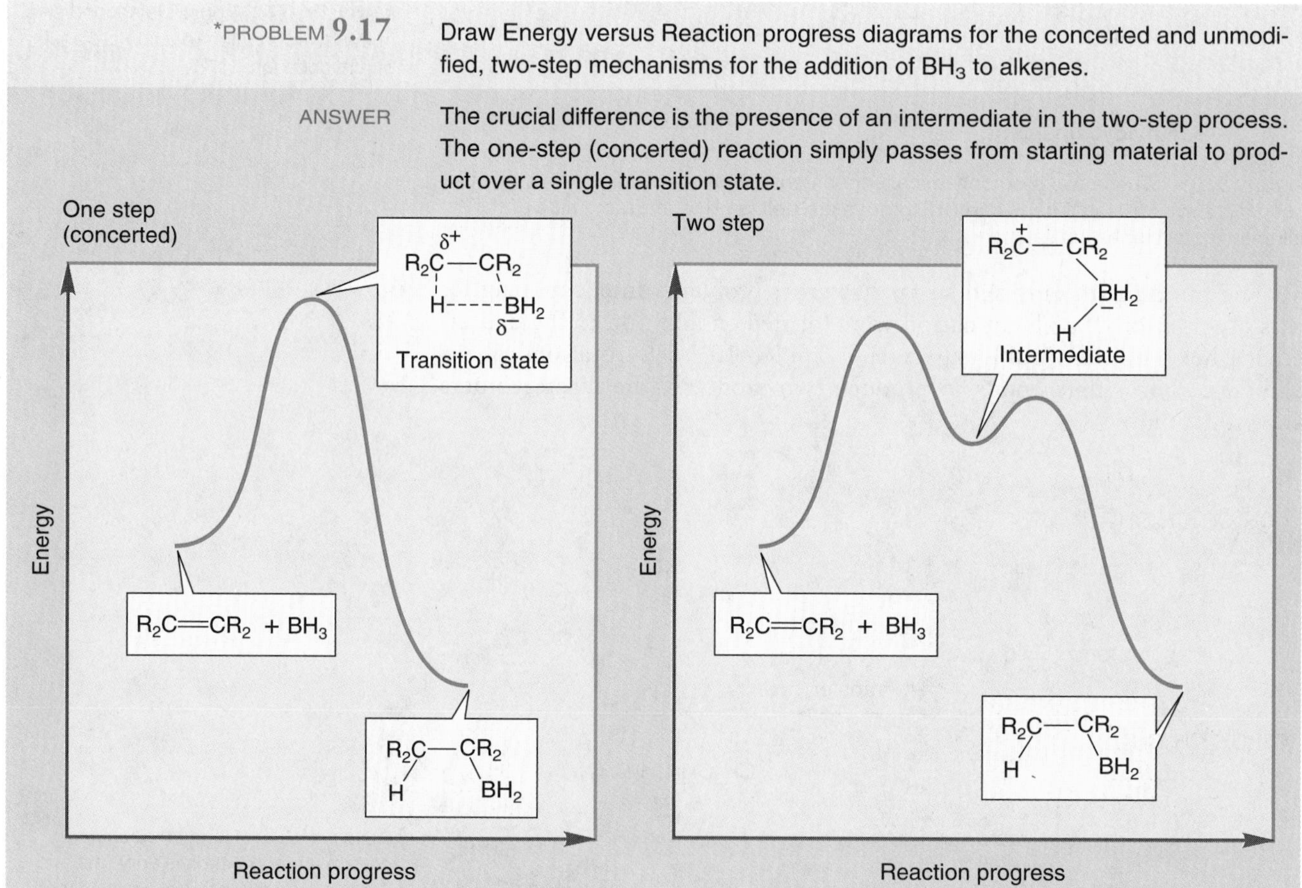

This example is typical of the use of stereochemistry in determining reaction mechanism. One reason we spent so much time on the stereochemical relationships of molecules of various shapes (diastereomers, enantiomers, and meso forms) was that we will use these relationships over and over again as we dig into questions of reaction mechanism. If it is not obvious what we are talking about in the last few paragraphs, and if Figures 9.65–9.67 are at all unclear, be sure to go back to Chapter 5 and work over the stereochemical arguments again. Otherwise, you run the risk of being lost in further discussion. The key points to see are the difference between the two hypothetical products shown in Figure 9.66 and their required formation from the rotating cation. A good test is to see if you can do Problems 9.18 and 9.19 easily. If you can, you are in good shape.

PROBLEM 9.18

Bromine adds to alkenes to give 1,2-dibromo compounds. From the following data on the reaction of bromine with the 2-butenes, determine if the addition is syn or anti. This problem does not pose a mechanism question—you do not have to draw arrow formalisms. You only need to determine from the structure of the product the direction from which the two bromines have added.

PROBLEM 9.19

Devise a way to determine the stereochemistry of the addition of hydrogen bromide to alkenes. Assume you can make any necessary starting material.

Now another detail raises its ugly head. If we follow the regiochemistry of the hydroboration of 1-hexene (Fig. 9.65), the boron winds up attached to the less substituted carbon. Similarly, 2-methyl-1-butene gives 99% addition of the boron to the less substituted position. So what? If the addition went through an ionic intermediate, all would be fine: Addition of the Lewis acidic boron to the less substituted end of the alkene produces the more substituted and thus more stable cation (Fig. 9.68).

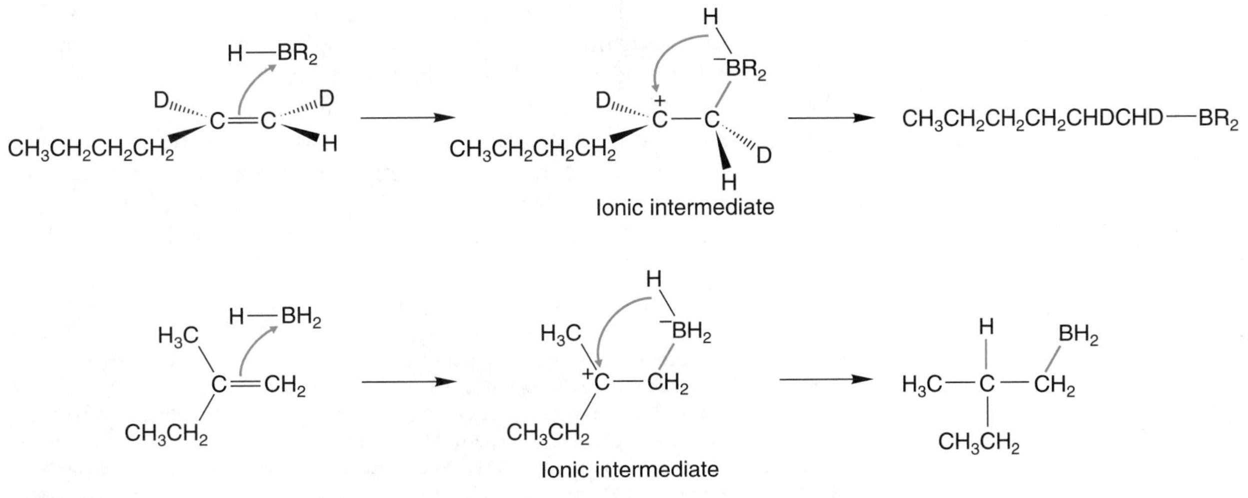

FIGURE 9.68 This two-step mechanism rationalizes Markovnikov addition, but is a total failure at explaining syn addition. How can we combine the two mechanisms to accommodate all the experimental data?

Unfortunately, we have just finished demolishing this simple mechanism, and we are stuck with explaining the regiospecific addition with our new, concerted (no intermediates) mechanism.*

There are at least two factors, one simple, the other more complicated, that influence the regiospecific addition in the hydroboration reaction. Consider again the reaction of BH_3 with 2-methyl-1-butene (Fig. 9.68). First of all, there are steric differences in the two possible concerted additions (Fig. 9.69). In the addition leading to the observed product, the larger end of the adding molecule, the BH_2 end, is nearer to the smaller hydrogens than to the larger alkyl groups. In the path leading to the product that is not observed, the larger end opposes the relatively large methyl and ethyl groups.

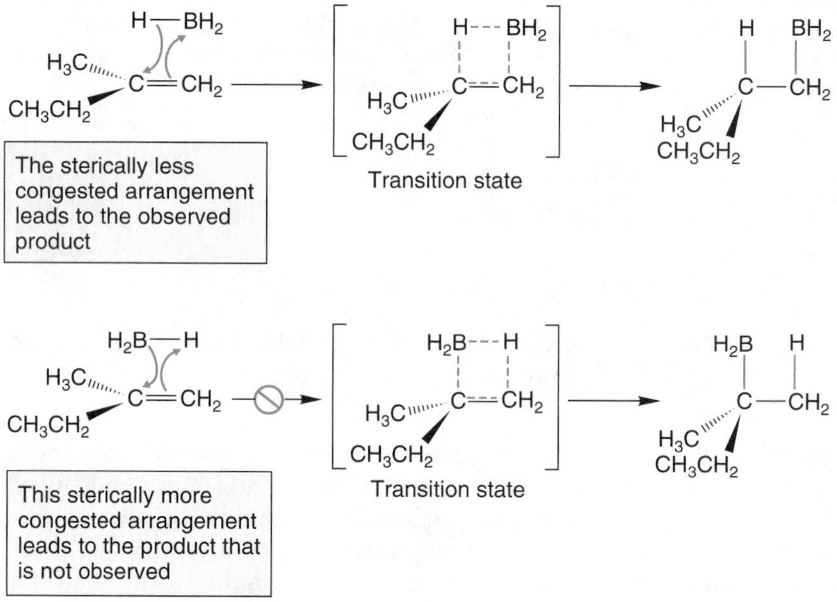

H—BH2
H3C
C═CH2
CH3CH2
The sterically less congested arrangement leads to the observed product
H---BH2
H3C
C═CH2
CH3CH2
Transition state
H BH2
C—CH2
H3C
CH3CH2

H2B—H
H3C
C═CH2
CH3CH2
This sterically more congested arrangement leads to the product that is not observed
H2B---H
H3C
C═CH2
CH3CH2
Transition state
H2B H
C—CH2
H3C
CH3CH2

FIGURE 9.69 Steric factors favor addition of the larger BH_2 group to the less congested end of the alkene.

That's the simple part; the more subtle effect has to do with the way in which the two new bonds are made in the hydroboration reaction. The two new bonds are formed in the same step, without a cationic intermediate, but there is no need for the new bonds to develop to the same extent as the reaction proceeds. Indeed, in a philosophical sense, they cannot, for they are different. In the transition state, one bond will be further along in its formation than the other. This differential progress can be symbolized by different length bonds as in Figure 9.70. In such a mechanism, a full positive charge is not developed on carbon, *but a partial one is.*

*If you feel a bit frustrated at this point, that's exactly what I want. I can well imagine you thinking, Why doesn't he just tell me what's going on in this reaction? Why all the going back and forth and looking at wrong (incomplete, really) mechanisms? The reason is that this book is not just about the "facts." It is also an attempt to show how science, or at least chemistry, works. Your frustration is an exact mirror of how a practicing chemist often feels. How many times have I done "one experiment too many?" One has a theory, it matches what's known about the reaction, and then some experiment raises doubts and all has to be rethought. The popular stereotype sees scientists as dispassionate seekers after Truth who would greet such events with undisguised joy—what a pleasure it is to find there is more to understand! Not a chance. One's initial reaction is much more likely to be to wish the experiment had never been done and that the uncomfortable new data would just vanish.

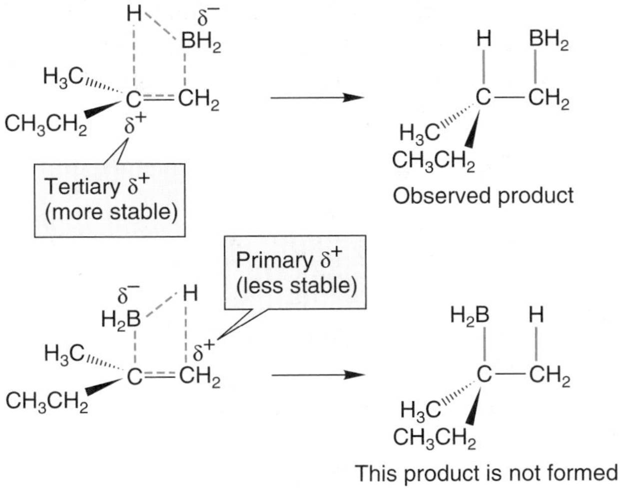

FIGURE 9.70 In the two possible transition states for the concerted (one-step) addition, there will be partial positive charge developed on carbon. This δ^+ will be more stable at the more substituted position. Polar factors contribute to the observed regiochemistry of the hydroboration reaction.

All the factors that operate to stabilize a full positive charge will operate to stabilize a partial positive charge. A more substituted partial positive charge is more stable than a less substituted partial positive charge, and this favors the observed product. So, the initial product of hydroboration of an alkene is the monoalkylborane formed by syn addition in which boron becomes attached to the less substituted end of the alkene partner (Fig. 9.70). Is this anti-Markovnikov addition? No!

Here is an instance where rules can be confusing. When we say "Markovnikov addition," we are used to seeing the hydrogen attached to the less substituted end of the alkene and the X group to the other end (Fig. 9.71).

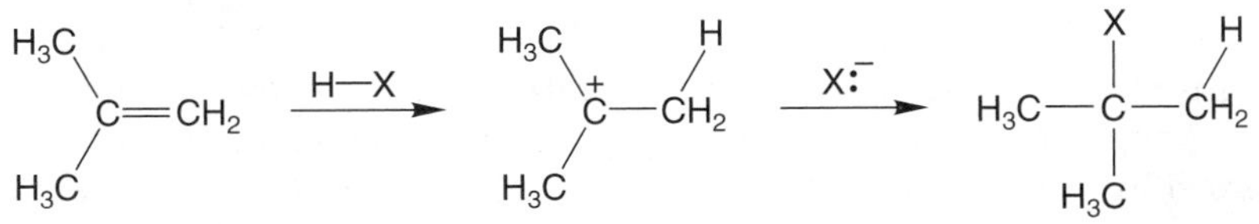

FIGURE 9.71 Typical Markovnikov addition in which the more stable cation is captured to give the product in which H is attached to the less substituted end of the original alkene, and X is attached to the more substituted end.

When $H{-}BH_2$ adds, it is the boron that is the positive end of the dipole, the Lewis acid in the addition. It is the boron, not the hydrogen, that becomes attached to the less substituted end of the alkene. As Figure 9.72 tries to show, it is the terminology that gets complicated, not the reaction mechanisms. There is danger in learning rules; it is much safer to understand the reaction mechanism.

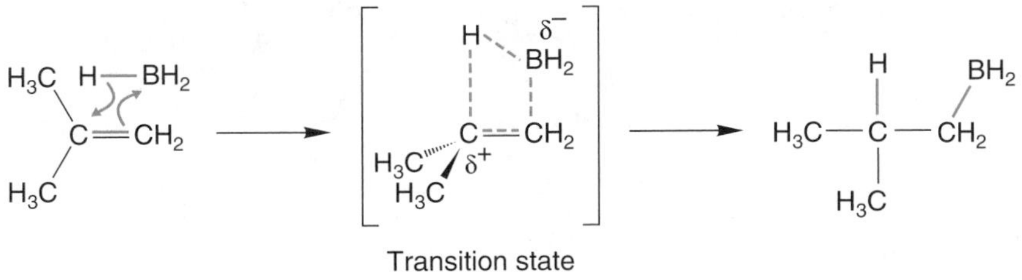

FIGURE 9.72 This reaction is still Markovnikov addition. The positive end of the dipole, boron in this case, adds to the less substituted end of the alkene. Positive charge develops on the more substituted carbon.

This isn't the end of it. The initially produced monoalkylborane still has a trigonal boron; therefore, it still has an empty $2p$ orbital (Fig. 9.73). It's still a Lewis acid, and it can participate in the hydroboration reaction to give the product of two hydroborations of the alkene, a dialkylborane.

Nor must the reaction stop here. The dialkylborane is also a Lewis acid, and it can do one more hydroboration to give a trialkylborane (Fig. 9.73). Now, however, all the original hydrogens are used up and there can be no further hydroborations.

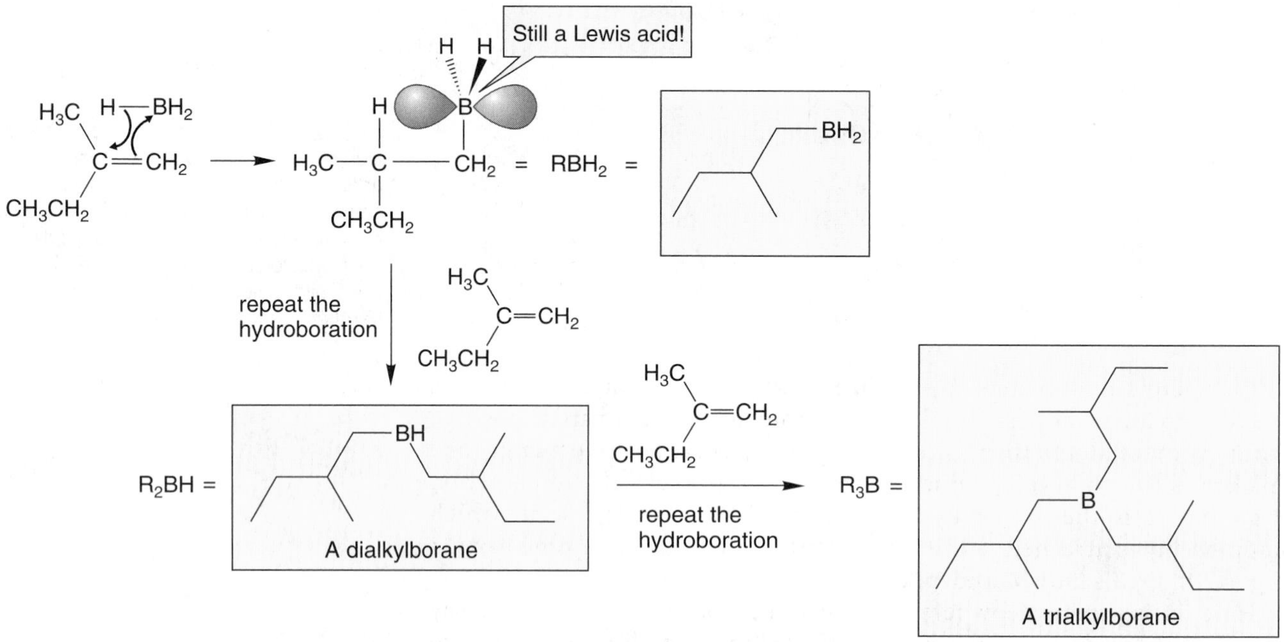

FIGURE 9.73 The alkylboranes formed in the hydroboration reaction of 2-methyl-1-butene are themselves Lewis acids and can undergo further reactions with the alkene. Be sure you can write out the details of these reactions.

In practice, the number of hydroborations depends on the size of the alkyl groups in the alkene. When these are rather large, further reaction is retarded by steric effects—the alkyl groups just get in the way. For example, 2-methyl-2-butene hydroborates only twice, and 2,3-dimethyl-2-butene only once (Fig. 9.74).

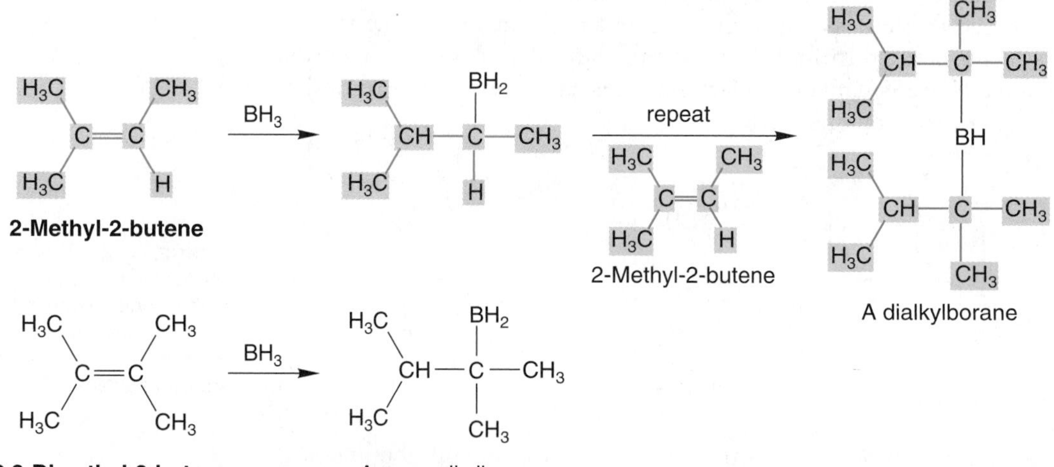

FIGURE 9.74 In practice, the number of hydroborations for a given alkene is strongly dependent on steric effects. The more hindered the alkene, the fewer the number of possible additions.

*PROBLEM 9.20

Provide a mechanism for the following reaction:

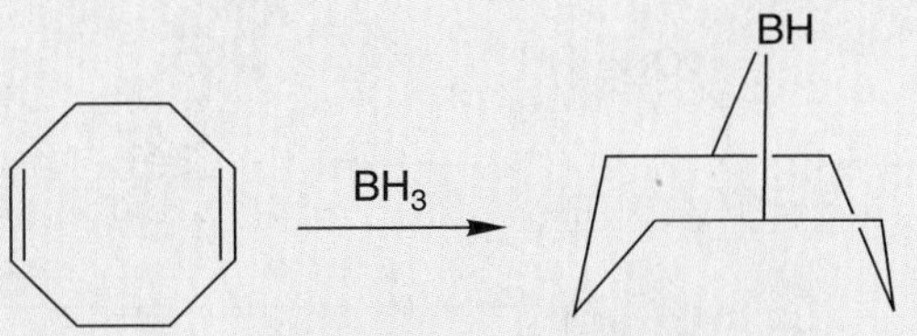

ANSWER

The first hydroboration goes in normal fashion.

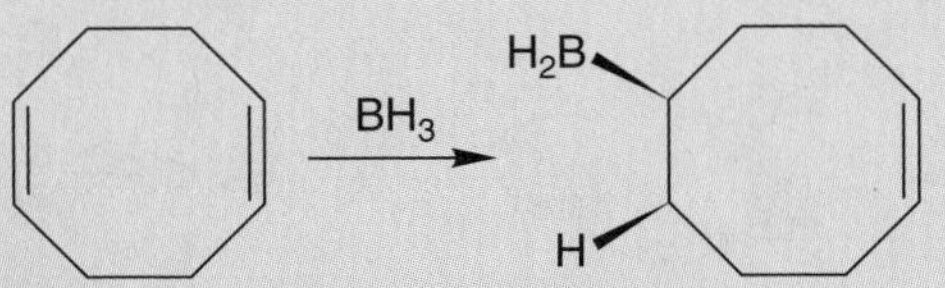

Now a second, intramolecular hydroboration takes place as the remaining double bond in the ring attacks the boron. It takes some unraveling to see the structure of the final product. It is important to begin to hone your skills at this kind of thing, and this problem provides practice.

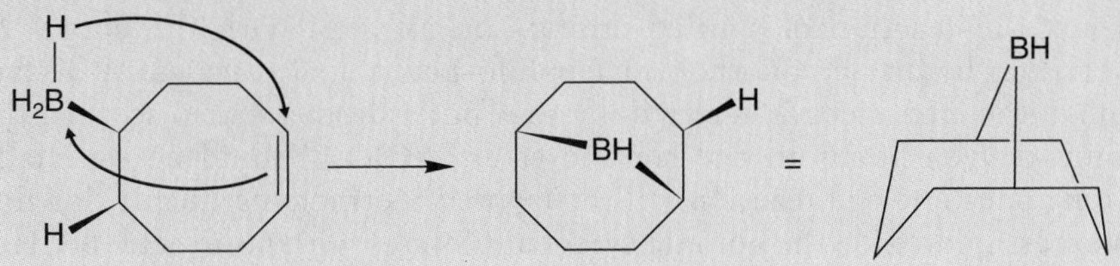

9.11 UTILITY OF HYDROBORATION: ALCOHOL FORMATION

It would be presumptuous to attempt to summarize the utility of the hydroboration reaction here. There are many variations of this reaction, each designed to accomplish a specific synthetic transformation. A book could be, and has been, written on the subject, and Professor H. C. Brown (b. 1912) of Purdue won the Nobel prize for chemistry in 1979 for his discovery and exploitation of hydroboration. I'll only mention one especially important reaction here. When one molecule of trialkylborane is treated with hydrogen peroxide (H_2O_2) and hydroxide ion (HO^-), three molecules of an alcohol are formed along with boric acid (Fig. 9.75).

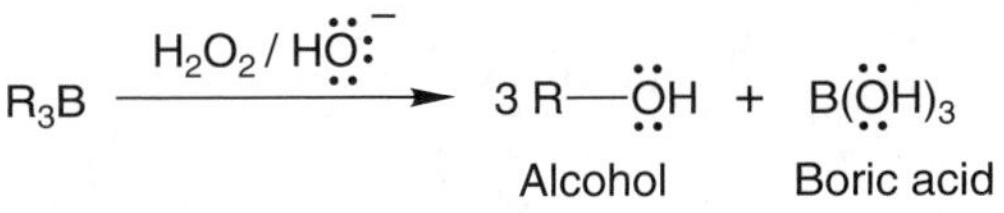

$$R_3B \xrightarrow{H_2O_2\,/\,HO^-} 3\,R{-}OH + B(OH)_3$$

Alcohol　Boric acid

FIGURE 9.75 Alcohols are formed when alkylboranes are treated with basic hydrogen peroxide.

We are not yet up to a full discussion of the mechanism of this reaction, which is outlined in Figure 9.76, but a number of important points are within our powers.

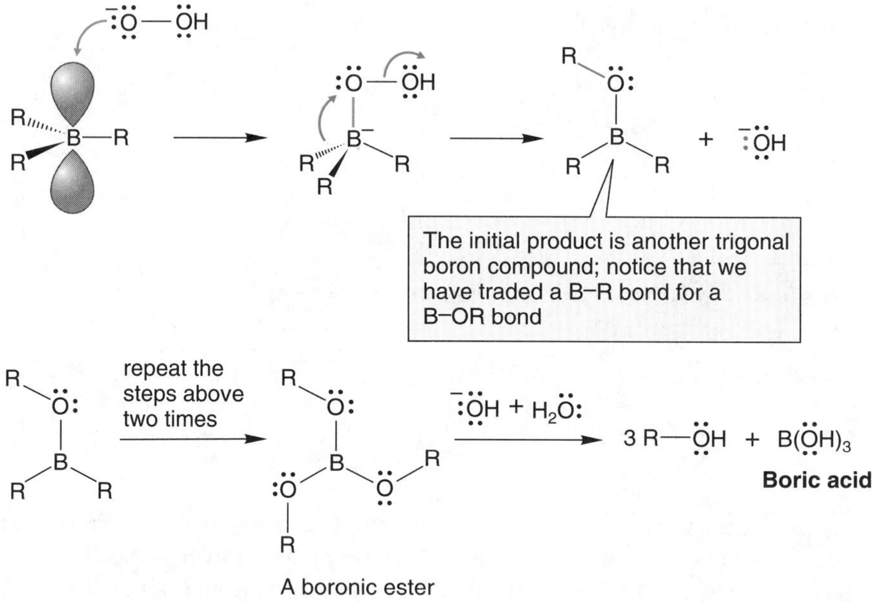

The initial product is another trigonal boron compound; notice that we have traded a B–R bond for a B–OR bond

repeat the steps above two times

A boronic ester

$3\ R{-}OH + B(OH)_3$

Boric acid

FIGURE 9.76 The mechanism of alcohol formation from alkyl boranes.

The first step, the addition of peroxidate ion (H–O–O⁻) to the borane, is just the reaction of a filled orbital—on oxygen—with the empty 2*p* orbital on boron. It's another nucleophile–Lewis acid reaction. A discussion of the later steps is deferred for now, but at least you can see that the boron–oxygen bonds in the boronic ester [$B(OR)_3$] should be especially strong ones and the reaction will be favored thermodynamically. Finally, in excess hydroxide the boronic ester equilibrates with boric acid [$B(OH)_3$] and the alcohol (ROH).

PROBLEM 9.21 Why are these boron–oxygen bonds especially strong?

*PROBLEM 9.22 Write a mechanism for the formation of boric acid, $B(OH)_3$, from hydroxide ion, HO^-, and a boronic ester, $B(OR)_3$.

ANSWER The boron atom in a boronic ester, $B(OR)_3$, remains a Lewis acid, as it still has an empty 2*p* orbital. Hydroxide is a strong nucleophile and adds to the Lewis acidic boron atom. Loss of ^-OR completes the first stage. Two repetitions lead to boric acid.

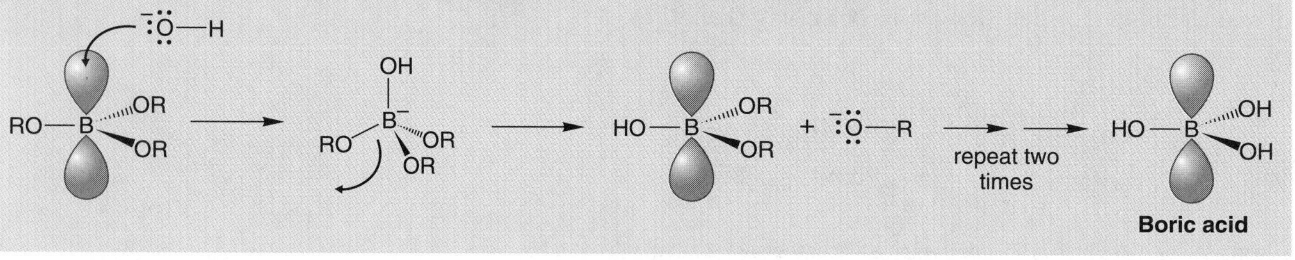

repeat two times

Boric acid

For once, questions of mechanism are secondary to the utility of this reaction. Note one very important thing: In the overall reaction we have produced the *less* substituted alcohol from the alkene. This reaction is an overall anti-Markovnikov addition (Fig. 9.77).

$$(H_3C)(CH_3CH_2)C{=}CH_2 \xrightarrow{BH_3} \text{A trialkylborane } [CH_3CH_2CH(CH_3)CH_2]_3B \xrightarrow{H_2O_2 / HO^-} (H_3C)(CH_3CH_2)CH{-}CH_2OH + B(OH)_3$$

This compound is the product of overall anti-Markovnikov addition of water to the alkene

Three molecules of alcohol

A trialkylborane

FIGURE 9.77 Hydroboration–oxidation accomplishes the addition of water to an alkene in an anti-Markovnikov sense. The hydroxyl group enters at the position to which the boron was attached: the less substituted end of the alkene.

Ethyl Alcohol

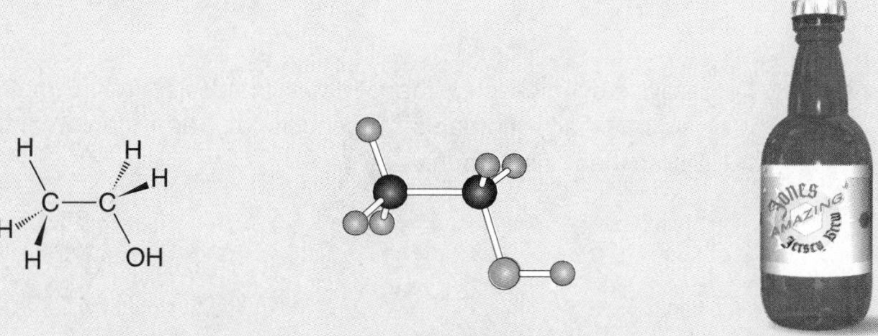

Ethyl alcohol, or ethanol, can be produced by several reactions described in this chapter. Although it has other uses, most ethyl alcohol is made for the production of alcoholic beverages, a use deeply rooted in history. Indeed, the rise of civilization has been attributed to the discovery of beer, and the necessity to give up a nomadic life to attend to the cultivation of the ingredients. In 1995, the average adult American consumed 32 gal of beer and more than 4 gal of wine and spirits.

Ethyl alcohol acts as a depressant in humans, interfering with neurotransmission. It works by binding to one side of the synapse, changing its shape slightly and making it able to bind γ-amino butyric acid (GABA) more efficiently. Bound GABA widens the synaptic channel, allowing chloride ions to migrate in, thus changing the voltage across the synapse. The nerve cell is less able to fire, and neurotransmission is inhibited. Your body cleanses itself of alcohol by converting it into acetaldehyde with the enzyme alcohol dehydrogenase. A second enzyme, aldehyde dehydrogenase, completes the conversion into acetic acid. There are individuals, indeed whole races, who are short of aldehyde dehydrogenase and thus overly susceptible to the deleterious effects of alcohol. By contrast, alcoholics have been found to contain elevated levels of this enzyme.

$$CH_3CH_2OH \xrightarrow[\text{dehydrogenase}]{\text{alcohol}} H_3C{-}CHO \xrightarrow[\text{dehydrogenase}]{\text{aldehyde}} H_3C{-}COOH$$

Acetaldehyde **Acetic acid**

Our previous method for alcohol synthesis, hydration of alkenes, necessarily produced the *more* substituted alcohol. So now we have complementary synthetic methods for producing both possible alcohols from a given alkene. Direct hydration gives the more substituted product (Markovnikov addition) and the indirect hydroboration method gives the less substituted alcohol (anti-Markovnikov addition, Fig. 9.78). Be sure to add these reactions to your file-card collection of synthetically useful reactions.

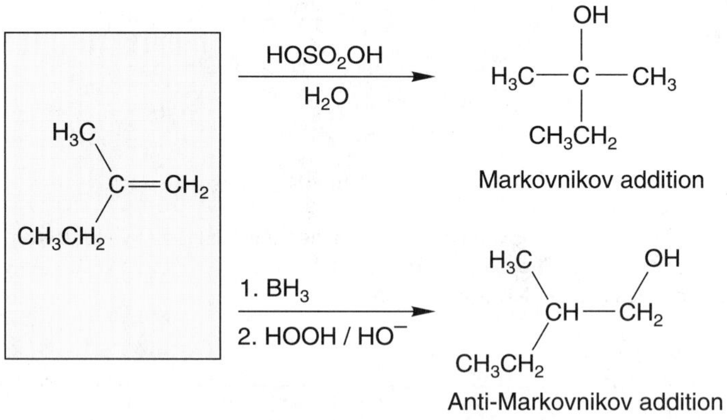

$HOSO_2OH$ / H_2O → Markovnikov addition

1. BH_3 / 2. HOOH / HO^- → Anti-Markovnikov addition

FIGURE 9.78 Two methods of alkene hydration with contrasting regiochemical results. Hydration in acid gives the more substituted (Markovnikov) alcohol, whereas hydroboration–oxidation gives the less substituted (anti-Markovnikov) alcohol. It is important to understand why the regiochemistry is different for the two reactions.

PROBLEM 9.23 Devise syntheses for the following molecules from 2-methyl-2-butene. You may also use any inorganic reagent as well as organic compounds containing no more than one carbon.

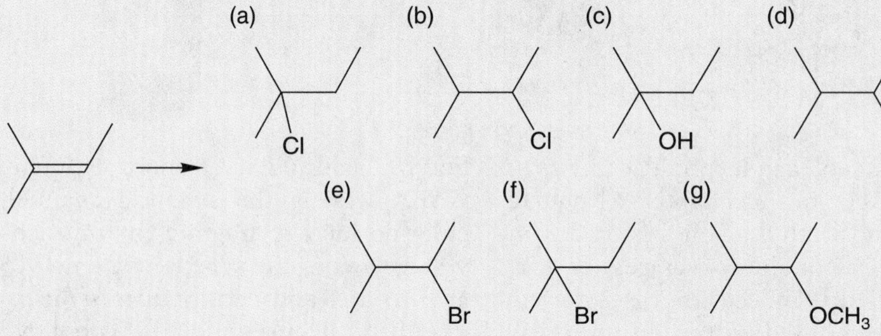

(a) (b) (c) (d) (e) (f) (g)

9.12 REARRANGEMENTS DURING H–X ADDITION TO ALKENES

When H–Cl adds to isopropylethylene (3-methyl-1-butene), a quite surprising thing happens. We do get some of the product we expect, 2-chloro-3-methylbutane, but it's not even the major compound formed. The major product is 2-chloro-2-methylbutane (Fig. 9.79).

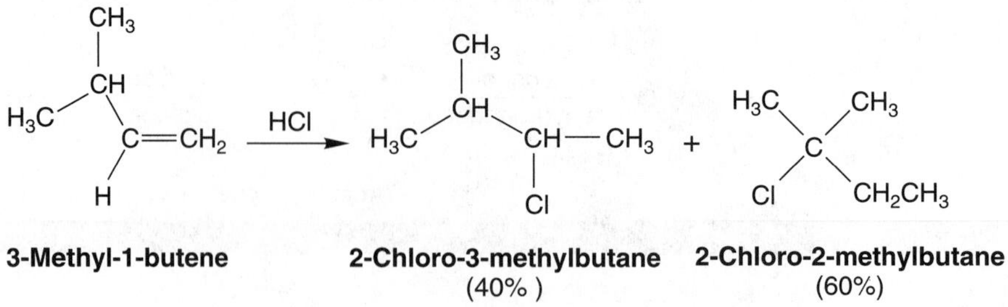

3-Methyl-1-butene —HCl→ **2-Chloro-3-methylbutane** (40%) + **2-Chloro-2-methylbutane** (60%)

FIGURE 9.79 Reaction of hydrogen chloride with 3-methyl-1-butene gives a new product in addition to the expected one.

The obvious question is, Why? The danger is that our mechanism for addition will have to be severely changed to accommodate these new data. In fact, we will be able to graft a small modification onto our general mechanism for addition reactions and avoid drastic changes. The way to attack a problem of this kind—the explanation of the formation of an unusual product—is to write out carefully the mechanism for the reaction as we have developed it so far, and see if we can see a way to rationalize the new product. In this case, protonation of 3-methyl-1-butene can give either a secondary or primary carbocation. There is no real choice here—the secondary carbocation is far more stable, and will surely be the one formed (Fig. 9.80).

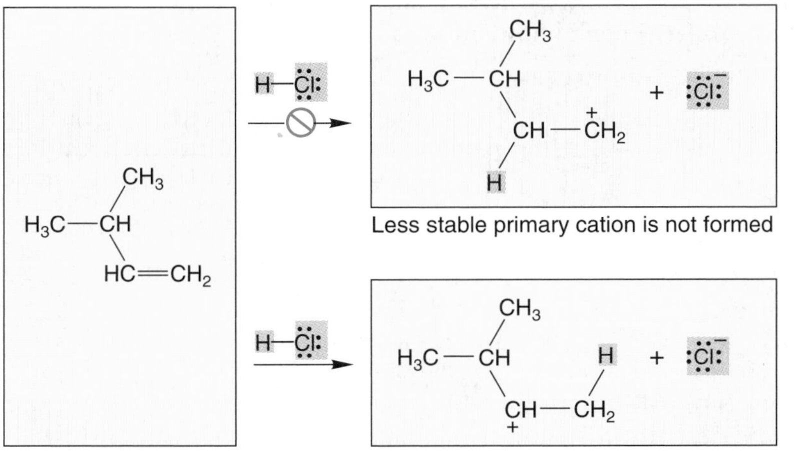

Less stable primary cation is not formed

More stable secondary cation is preferred

FIGURE 9.80 Protonation of 3-methyl-1-butene will surely give the relatively more stable secondary carbocation, not the hideously unstable primary one.

If the chloride ion simply adds to the positively charged carbon, we get the minor product of the reaction, and the one we surely expected. To get the major product, a **rearrangement** must take place. A hydrogen must move, with its pair of electrons, from the carbon adjacent to the cation to the positively charged carbon itself (Fig. 9.81). Such migrations of hydrogen with a pair of electrons are called **hydride shifts**.

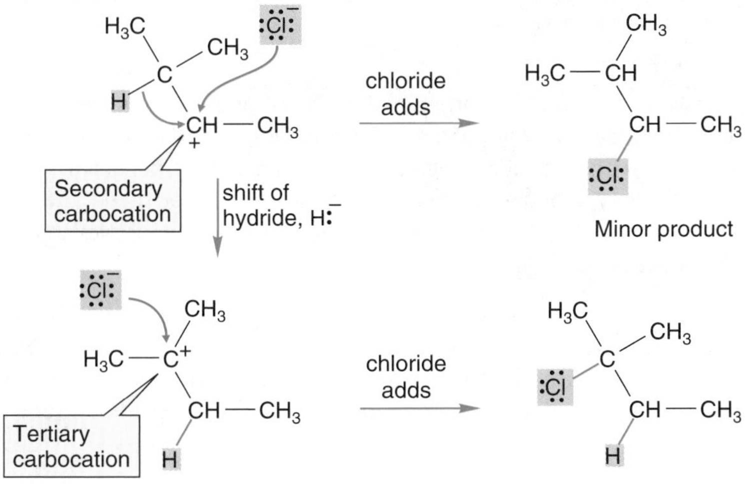

Secondary carbocation

chloride adds

shift of hydride, H:$^-$

Minor product

Tertiary carbocation

chloride adds

Major product

FIGURE 9.81 The major product is formed through a mechanism involving the migration of hydrogen with its pair of electrons (hydride, H:$^-$) to give the more stable tertiary carbocation from the less stable secondary ion.

Why should this happen? Note that in such a rearrangement we have generated a lower energy tertiary carbocation from a higher energy secondary carbocation. The cation is certainly "better off" (lower in energy) thermodynamically if the hydride moves. In practice, such rearrangements are quite common. Hydride shifts to produce more stable cations from less stable ones are easy. Indeed, not only hydride shifts but alkyl shifts, called **Wagner–Meerwein rearrangements** (Georg Wagner, 1849–1903; Hans Lebrecht Meerwein, 1879–1965) are also common. For example, 3,3-dimethyl-1-butene adds hydrogen chloride to give not only 2-chloro-3,3-dimethybutane but also 2-chloro-2,3-dimethylbutane. The new, rearranged compound results from an alkyl shift—in this case the migration of a methyl group with its pair of electrons (Fig. 9.82). Why is this reaction called an "alkyl shift," when by analogy to "hydride shift" it should be "alkide shift?" I don't know, but it is!

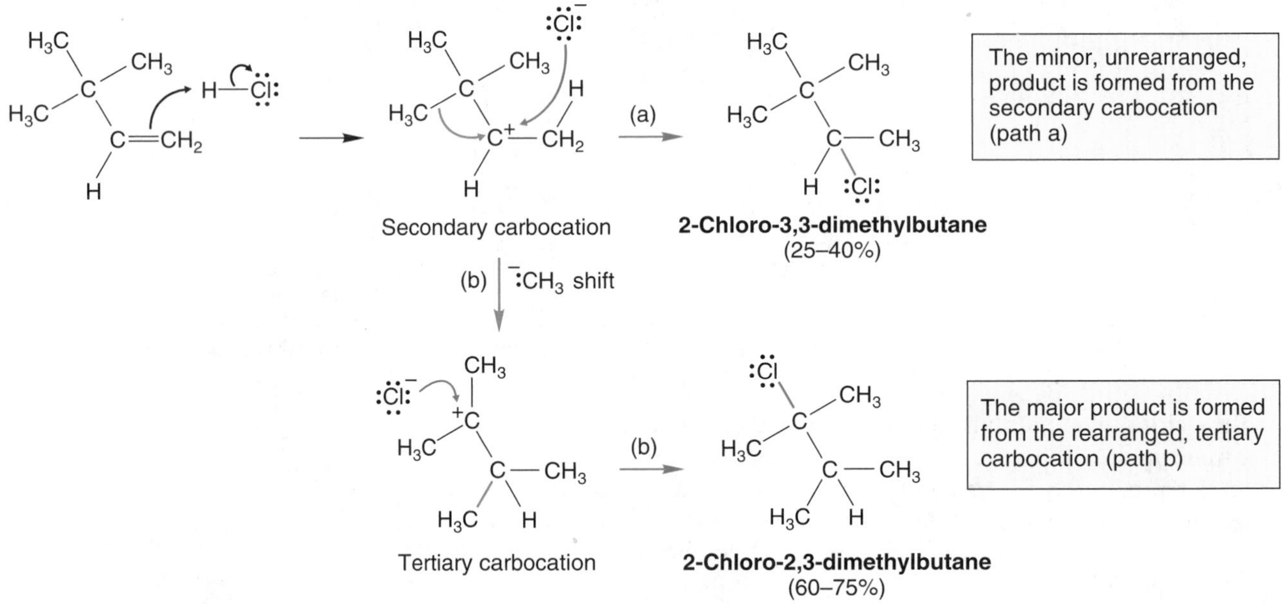

Secondary carbocation

2-Chloro-3,3-dimethylbutane (25–40%)

The minor, unrearranged, product is formed from the secondary carbocation (path a)

(b) $^{-}$:CH_3 shift

Tertiary carbocation

2-Chloro-2,3-dimethylbutane (60–75%)

The major product is formed from the rearranged, tertiary carbocation (path b)

FIGURE 9.82 Formation of the major product involves a mechanism in which a shift of a methyl group with its pair of electrons generates a more stable tertiary carbocation from a less stable secondary carbocation.

PROBLEM 9.24 A hydride shift to a positively charged carbon is just another example of a Lewis base–Lewis acid reaction. Although it looks very different from the more obvious Lewis acid–nucleophile reactions we have studied, it really is quite similar. In this case, there is a very strong and strategically located Lewis acid and a weak, but nonetheless reactive enough nucleophile. Identify the nucleophile and Lewis acid in this reaction.

Such rearrangements are so common that they have come to be diagnostic for reactions involving carbocations. Molecules prone to such rearrangements are used to test for carbocation involvement. Here is an example. Attempted S_N1 solvolysis of neopentyl iodide (1-iodo-2,2-dimethylpropane) in water leads not to neopentyl alcohol, as might have been anticipated, but instead to a rearranged alcohol, 2-methyl-2-butanol (Fig. 9.83).

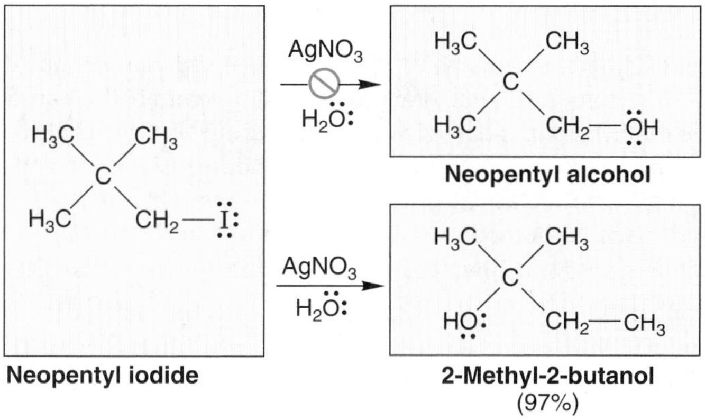

FIGURE 9.83 The solvolysis of neopentyl iodide in water does not take a normal course.

These results imply the scenario outlined in Figure 9.84, in which an unstable primary carbocation rearranges to a much more stable tertiary carbocation through the shift of a methyl group. In fact, the presence of the primary ion is so unsettling—they are most unstable and not likely to be formed—that a variant of this mechanism has been proposed in which the methyl group migrates as the iodide departs (Fig. 9.84).

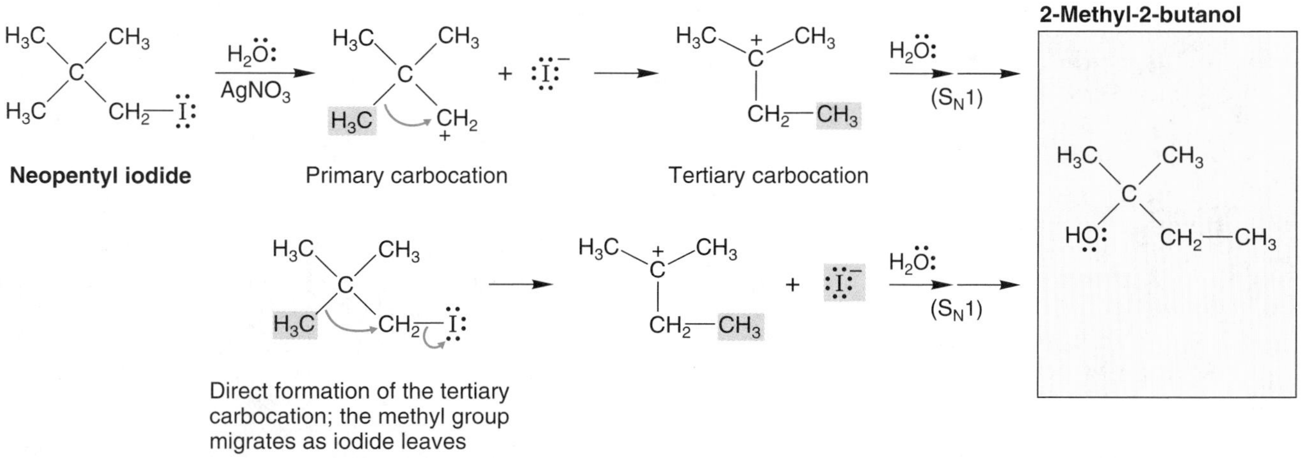

FIGURE 9.84 Rearrangements are common in many reactions in which carbocations are involved. The E1 and S_N1 reactions are good examples.

Under certain conditions of very low temperature and highly polar but nonnucleophilic solvents, some carbocations can be observed spectroscopically. Under such conditions it is possible to determine that many carbocations are rapidly rearranging as we observe them (Fig. 9.85).

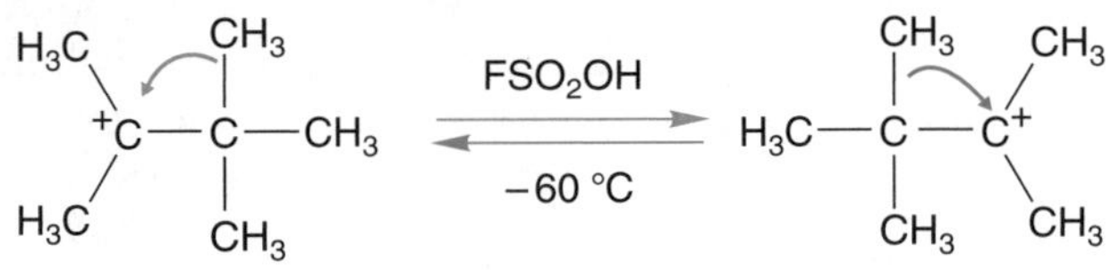

FIGURE 9.85 Under some conditions relatively stable carbocations can be detected spectroscopically and shown to be rearranging. The two tertiary carbocations in the figure are rapidly interconverting at −60 °C.

H3C
CH3
+C—C—CH3
H3C
CH3
FSO2OH
−60 °C
CH3
CH3
H3C—C—C+
CH3
CH3

PROBLEM 9.25 Use a molecular orbital argument to determine if the rearrangement shown in Figure 9.85 is favorable. First, draw a schematic picture of the transition state in which the methyl group is half-migrated. Next, sketch in the orbitals involved in bonding the migrating methyl to the carbon framework. Finally, combine these orbitals to make the molecular orbitals for the transition state, order them in energy, and place the proper number of electrons in them. *Hint*: Recall the construction of cyclic H_3 in Chapter 2 (p. 64) and the cyclopropenium ion in Problem 9.5 (p. 347).

Another use of rearrangement reactions in mechanism determination involves the hydroboration reaction. Hydroboration of 3,3-dimethyl-1-butene *leads to no products of rearrangement.* Thus it is extremely unlikely that this reaction can involve a carbocation as an intermediate. If it did, it seems quite certain that rearranged products would be seen (Fig. 9.86). The lack of rearrangements is the second piece of evidence that the hydroboration mechanism is different from that for simple H–X addition (for the first piece of evidence see p. 365).

BH3
H2O2/NaOH
B
A
This rearranged borane, **A**, should go on to form rearranged alcohol, **B**, which is *not* observed; therefore hydroboration cannot involve free ions

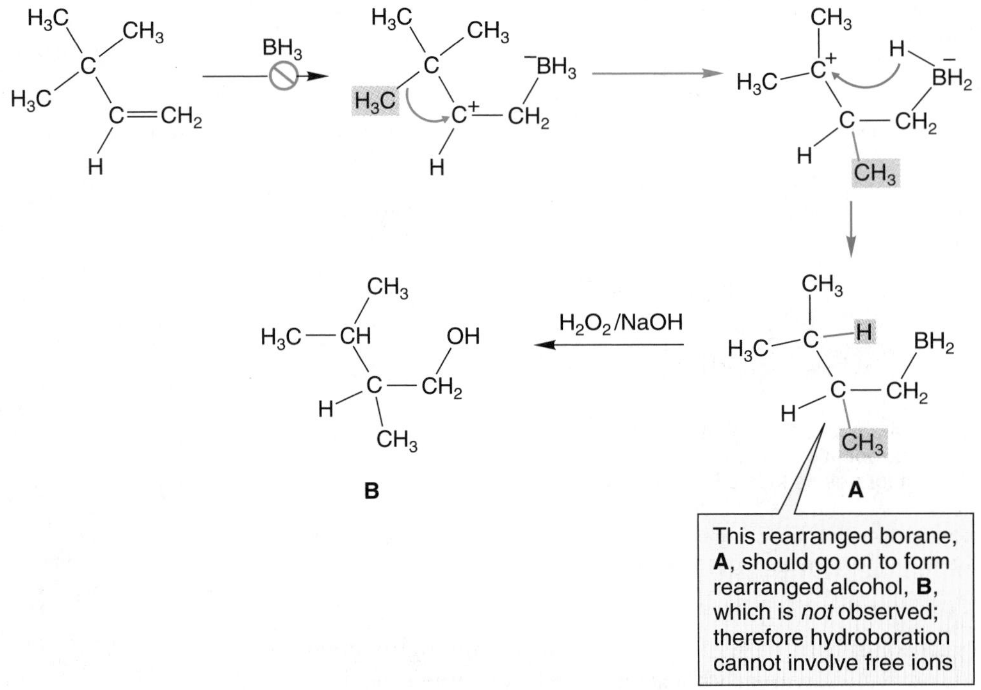

FIGURE 9.86 If the mechanism of hydroboration involved carbocations, rearrangements should be observed. They are not, and this constitutes strong evidence against the two-step mechanism for hydroboration.

Write a mechanism for the following reaction:

*PROBLEM 9.26

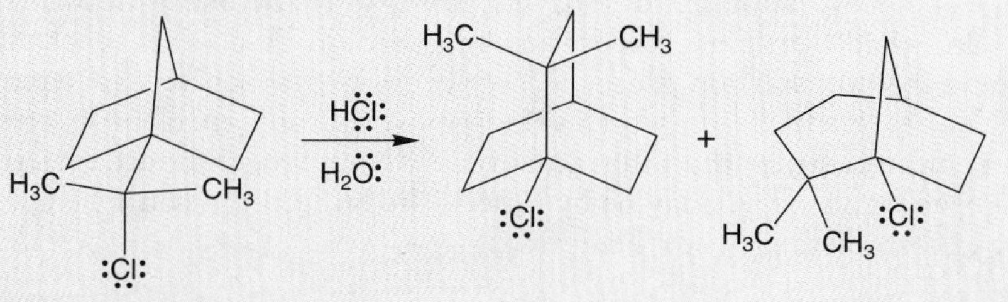

ANSWER

This reaction is just a Wagner–Meerwein rearrangement (p. 376) in a difficult context. The S_N1 solvolysis leads to a tertiary carbocation. Migration of either of the two different adjacent carbon atoms leads to a bridgehead cation that is captured by chloride to give the two products. The figure first shows the rearrangement of a methyl group in a geometry designed to resemble that of the more complicated molecule in the problem.

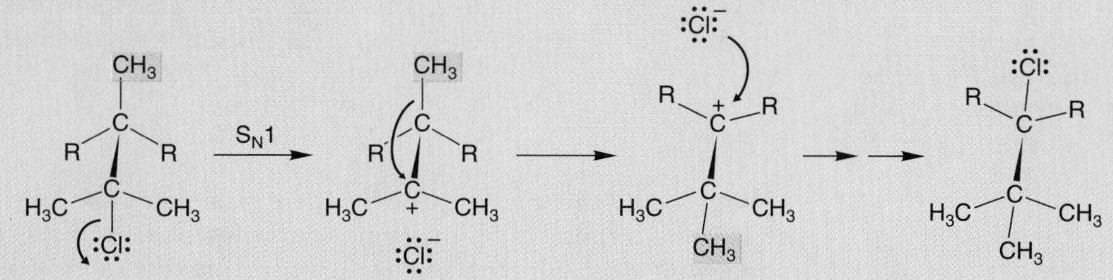

In the bicyclic molecule, S_N1 solvolysis also leads to a tertiary carbocation. Now, either of the two different adjacent carbons (shown in red) can migrate (as does the red methyl group above) to give a new carbocation. Capture by nucleophile leads to product.

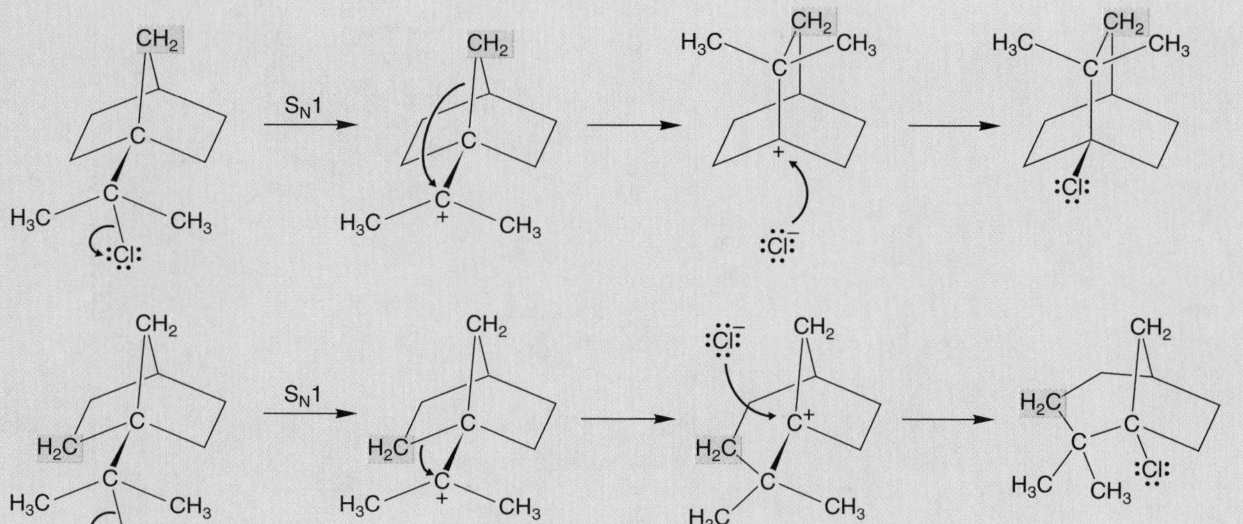

9.13 SOMETHING MORE: PROTONATION OF ETHYLENE

Ethylene adds hydrogen bromide and hydrogen chloride to give ethyl bromide and ethyl chloride, respectively. If we write the usual mechanisms, we find that a primary carbocation must be invoked. This mechanism should at least give you pause, because primary carbocations are very unstable. Yet, the reaction works—there must be *some* mechanism. It has been suggested that the intermediate may be hydrogen bridged and that the open ethyl cation may be bypassed. The bridged ion could be opened in S_N2 fashion to give ethyl chloride (Fig. 9.87).

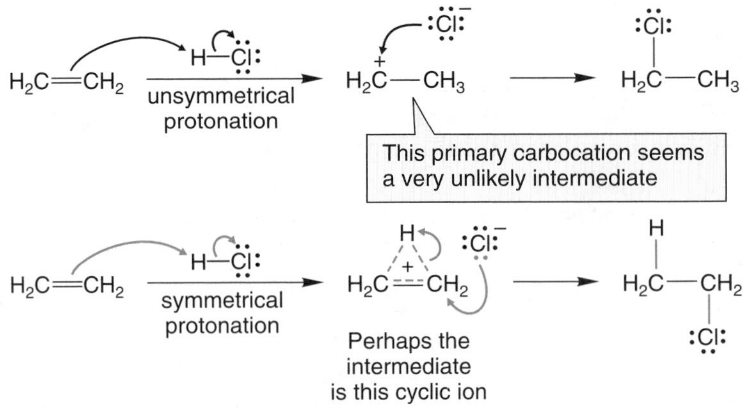

$H_2C{=}CH_2$ + H—Cl: → (unsymmetrical protonation) → $H_2\overset{+}{C}{-}CH_3$ + :Cl:⁻ → $H_2C(Cl){-}CH_3$

This primary carbocation seems a very unlikely intermediate

$H_2C{=}CH_2$ + H—Cl: → (symmetrical protonation) → $H_2C{=}CH_2$ bridged by H (+) + :Cl:⁻ → $H{-}H_2C{-}CH_2{-}Cl$

Perhaps the intermediate is this cyclic ion

FIGURE 9.87 Two possibilities for the structure of the ion formed by protonation of ethylene.

Why should this cyclic species be more stable than the open cation? If we imagine forming it by the symmetrical interaction of a hydrogen 1*s* orbital with the π orbital of ethylene, we can see that there is some stabilization in the cyclic intermediate (Fig. 9.88).

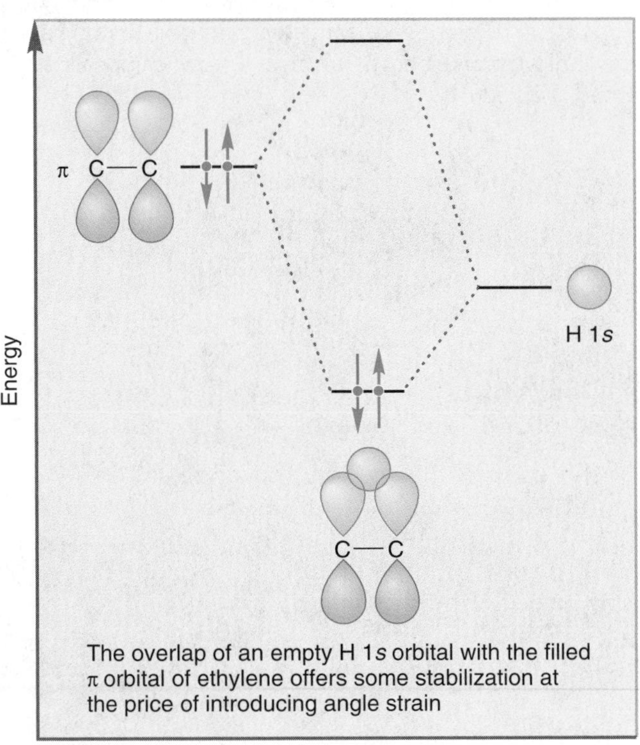

Energy
π C—C
H 1s
C—C
The overlap of an empty H 1s orbital with the filled π orbital of ethylene offers some stabilization at the price of introducing angle strain

FIGURE 9.88 A molecular orbital picture of the bridged version of the ethyl cation.

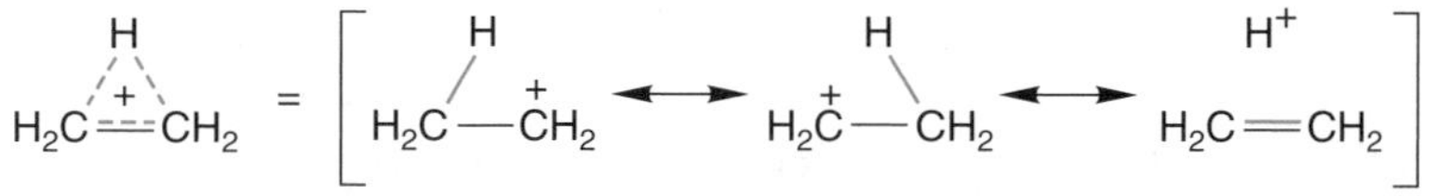

FIGURE 9.88 (CONTINUED) A resonance formulation shows the delocalization of charge in the cyclic ion.

The resonance forms for this species show how both of the carbons and the hydrogen share the positive charge. Might such stabilization be more effective than the hyperconjugation available to the open ethyl cation? There are prices to be paid in the cyclic intermediate. There is strain, for example, as the C–C–H internuclear angle is 60°, which is far from ideal.

Theory indicates that there may be something to the argument that the cyclic intermediate is formed in this reaction, as it is calculated to be about 3–4 kcal/mol more stable than the open ion. But remember, the calculations refer to the gas phase where solvation is absent, and the situation in solution is not guaranteed to resemble that in the gas phase. Moreover, the kinetics of the apparently simple addition reaction are very complicated indeed, and there are indications that termolecular (three-molecule) reactions are important in many addition reactions (Fig. 9.89). So the mechanistic picture for this simplest of alkene addition reactions is messy and complicated, which gives you an idea of how primitive our understanding of Nature is!

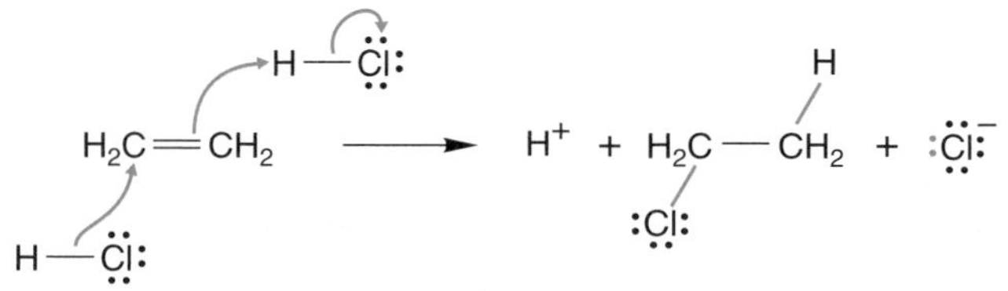

FIGURE 9.89 Kinetic measurements show that termolecular reactions may be quite important in the addition of H–X species to simple alkenes. This "simple" process is apparently quite complicated!

9.14 SUMMARY

NEW CONCEPTS

The central topic of this chapter is the addition of H–X molecules to alkenes. These reactions begin by the addition of the positive end of the dipole to the alkene in such a way as to form the more stable carbocation. The anion (X^-) then adds to the cation to form the final addition product. The direction of addition—the regiochemistry of the reaction—is determined by the relative stabilities of the possible intermediate carbocations. Resonance and inductive effects are important factors that influence stability.

Alkyl groups stabilize carbocations by a resonance effect in which a filled σ orbital of the alkyl group overlaps with the empty 2*p* orbital on carbon. This effect is known by the curious name of "hyperconjugation" (Figs. 9.39 and 9.40).

In this chapter, we continue to use stereochemical experiments to determine reaction mechanisms. Generally, one-step (concerted) reactions preserve the stereochemical relationships of groups in the reacting molecules. Multistep processes introduce the possibility of rotation around carbon–carbon bonds and loss of the initial stereochemistry.

Remember: No experiment can prove a mechanism. We are always at the mercy of the next experiment, the results of which might contradict our current mechanistic ideas. Experiments can certainly disprove a mechanism, but they can never prove one.

REACTIONS, MECHANISMS, AND TOOLS

Many acids (H–Br, H–Cl, H–I, $HOSO_2OH$, generally, H–X) will add directly to alkenes. The first step is addition of a proton to give the more stable carbocation. Inductive and resonance effects influence the stability of the intermediate carbocations. In the second step of the reaction, the negative end of the original H–X dipole adds to complete the addition process. The regiochemistry of the addition is determined by the formation of the more stable carbocation in the original addition (Fig. 9.34).

Other H–X molecules are not strong enough acids to protonate alkenes. Water (HOH) is an excellent example. However, such molecules will add if the reaction is acid catalyzed. Enough acid catalyst is added to give the protonated alkene, which is attacked by water. The catalyst is regenerated in the last step and recycles to carry the reaction further (Fig. 9.49).

If the concentration of alkene is high enough, it can compete with other Lewis bases in the system (for example, X^-), and dimerization or even polymerization of the alkene can take place. This reaction is not mysterious; the alkene is merely acting as Lewis base toward the carbocation as Lewis acid (Fig. 9.58).

In hydroboration, the boron atom of "BH_3" adds to an alkene to give an alkylborane. The mechanism involves a single step in which the new carbon–hydrogen and carbon–boron bonds are both made. Subsequent further hydroborations can give dialkyl- and trialkylboranes. Treatment of these boranes with basic hydrogen peroxide leads to alcohols (Figs. 9.75 and 9.76). Hydroboration is especially important because it gives access to the less substituted alcohol (anti-Markovnikov addition).

Additions that go through carbocationic intermediates can be complicated by rearrangements of hydride ($H:^-$) or alkyl groups ($R:^-$) to produce more stable carbocations from less stable ones (Fig. 9.82).

SYNTHESIS

In this chapter, we introduce the synthesis of alkyl halides and alcohols from alkenes through addition reactions. Hydroboration allows us to do addition reactions in the anti-Markovnikov sense. The S_N2 and S_N1 reactions from earlier chapters allow us to do further transformations of the alcohols and halides. The new synthetic methods are summarized in Figure 9.90.

1. Alkyl halides

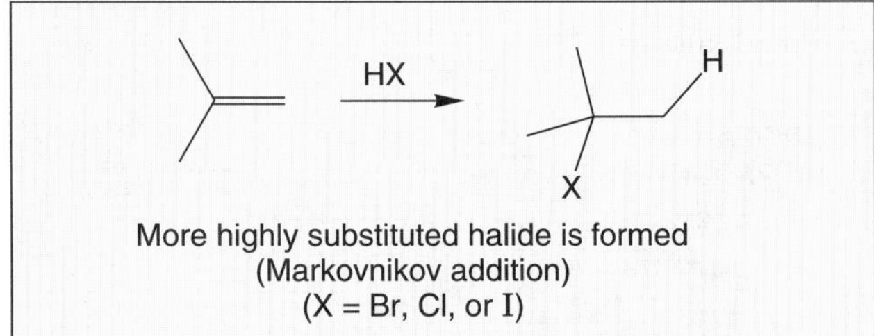

2. Allyl halides

HX

H

X

1,2- and 1,4-Addition compete

FIGURE 9.90 The new synthetic reactions of Chapter 9.

3. Alkylboranes

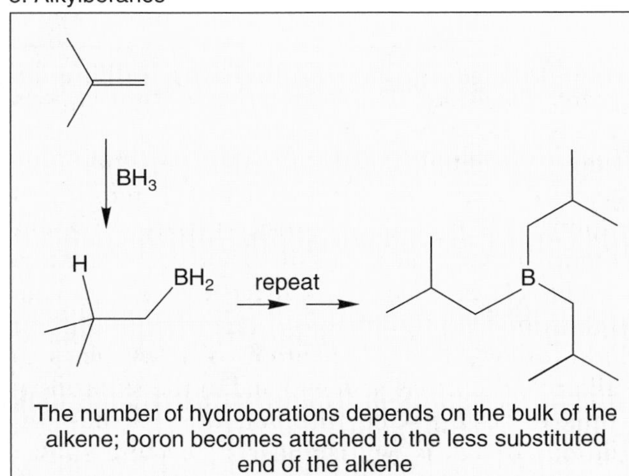
BH$_3$
H
BH$_2$
repeat
B
The number of hydroborations depends on the bulk of the alkene; boron becomes attached to the less substituted end of the alkene

4. Alcohols

H$_3$O$^+$

H

HO

More highly substituted alcohol is formed (Markovnikov addition)

BH$_2$

H$_2$O$_2$

HO$^-$

3

OH

Less highly substituted alcohol is formed (anti-Markovnikov addition)

FIGURE **9.90** (CONTINUED)

COMMON ERRORS

Keeping clear the difference between resonance and equilibrium is a constant difficulty for many students. Resonance forms are simply different electronic descriptions for a single molecule. The key word is "electronic." In resonance, only electrons are allowed to move to produce the various representations of the molecule. Atoms may not change their positions. If they do, we are no longer talking about resonance, but equilibrium. Be careful! This point is trickier than it sounds.

Remember also that the two-dimensional paper surface can fool us. Molecules are three dimensional. To have delocalization of electrons, orbitals must overlap. Sometimes it looks in two dimensions as though resonance forms exist when in fact they do not in the real world of three dimensions. Two excellent examples appear in Figure 9.32 (p. 348) and in Problem 9.8f (p. 349).

The grammar of chemistry—our use of arbitrary conventions—is important, because we need to be able to be precise in communicating with each other. The double-headed resonance arrow is reserved for resonance, and never used for anything else. There are several schemes for representing resonance forms, ranging from drawing them all out in full, to the summary structures of Figure 9.22 (p. 343).

Both mechanistic analysis and synthesis are gaining in complexity. There is much stereochemical detail to keep track of in many mechanistic analyses, for example. Perhaps the most complex mechanism considered so far is that for hydroboration. If you understand why it was necessary to modify the standard mechanism for the addition of Lewis acids to alkenes to accommodate the experimental observations, you are in fine shape so far.

There are no really complex synthetic procedures yet. However, even very simple steps taken in sequence can lead to difficulty. This area will rapidly proliferate and become more difficult, so do not be lulled by the deceptive simplicity of the reactions so far.

9.15 KEY TERMS

Alcohol (R–OH) A molecule containing a simple hydroxyl group.

Catalyst A catalyst functions to increase the rate of a chemical reaction. It is ultimately unchanged by the reaction and functions not by changing the energy of the starting material or product but by providing a lower energy pathway. Thus it operates to lower the energies of the transition states involved in the reaction.

Cationic polymerization A reaction in which an initially formed carbocation adds to an alkene that in turn adds to another alkene. Repeated additions can lead to polymer formation.

Electronegativity The ability of an atom to attract electrons.

Hydration The addition of water to an alkene. This reaction is generally acid catalyzed and leads to the most highly substituted alcohol possible.

Hydride shift The migration of hydrogen with a pair of electrons ($H:^-$).

Hydroboration The addition of BH_3 (in equilibrium with the dimer, B_2H_6) to alkenes to form alkylboranes. These alkylboranes can be further converted into alcohols. Addition to an unsymmetrical alkene proceeds so as to give the less substituted alcohol.

Hyperconjugation The stabilization of carbocations through the overlap of a filled σ orbital with the empty $2p$ orbital on carbon.

Inductive effects Electronic effects transmitted through the σ bonds. All bonds between different atoms are polar, and thus many molecules contain dipoles. These dipoles can affect reactions through induction.

Markovnikov's rule In additions of Lewis acids to alkenes, the Lewis acid will add to the less substituted end of the alkene. The rule works because additions of Lewis acids produce the more substituted, more stable carbocation.

Rearrangement The migration of an atom or group of atoms from one place to another in a molecule. Rearrangements are exceptionally common in reactions involving carbocations.

Regiochemistry (regioselectivity) (regiospecificity) If a reaction may produce one or more isomers and one predominates, the reaction is called regioselective. If one product strongly predominates, the reaction is regiospecific.

Wagner–Meerwein rearrangement The 1,2-migration of an alkyl group in a carbocation.

9.16 ADDITIONAL PROBLEMS

This chapter and Chapter 10 continue our cataloging of the basic reactions of organic chemistry. To the S_N1, S_N2, E1, and E2 reactions we now add a variety of addition reactions. Although there are several different mechanisms for additions, many take place through a three-step sequence of protonation, addition, and deprotonation. The following new problems allow you to practice the basics of addition reactions and to extend yourself to some more complex matters. Even simple additions become complicated when they occur in intramolecular fashion, for example. These problems also allow you to explore the influence of resonance and inductive effects, and to use the regiochemistry and stereochemistry of addition to help work out the probable mechanisms of reactions.

Your sophistication in synthesis is also growing, and the variety of addition reactions encountered in this chapter adds to the ways you have available to make differently substituted molecules. You are still not quite ready to undertake multistep syntheses, but you are now very close. In anticipation of these tougher problems, be sure to start working out synthetic questions backward. Always ask the question, What molecule is the *immediate* precursor of the target? Don't start, even with simple problems, thinking of how an ultimate starting material might be transformed into product. That approach will work in one- or two-step synthesis, but becomes almost impossible to do efficiently when we come to longer multistep syntheses.

PROBLEM **9.27** See if you can write the π molecular orbitals of allyl from memory. Show the electronic configuration (orbital occupancy) for the allyl cation, radical, and anion.

PROBLEM **9.28** The molecular orbitals for pentadienyl are

Pentadienyl

Views from the top of pentadienyl; beneath every (+) is a (–) and beneath every (–) is a (+); nodes are shown as (0)

= + − + − +

= + − 0 + −

= + 0 − 0 +

= + + 0 − −

= + + + + +

Compare these molecular orbitals with those of allyl. What rules can you develop to predict the molecular orbitals of any odd-carbon, fully conjugated (2*p* orbital on each carbon) chain? It is probably easiest to work from the schematic top views at the right of the figure.

PROBLEM **9.29** Apply your rules to the molecular orbitals of heptatrienyl.

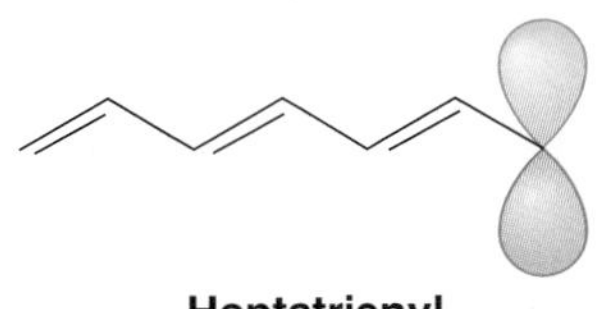

Heptatrienyl

PROBLEM **9.30** Draw a third resonance form for the allyl cation that involves a three-membered ring. Be careful not to move any atoms! What is the relative importance of this resonance form? Explain.

PROBLEM **9.31** Write resonance forms for the following species. You will first have to write good Lewis structures, as the electrons have been deliberately left out. You will also have to write a full Lewis structure for the nitro (NO_2) group.

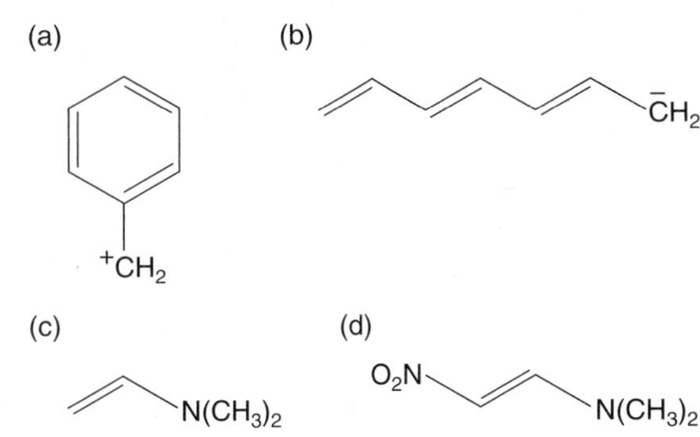

PROBLEM **9.32** Which of the following pairs represent resonance forms and which do not? Explain your choices.

(a) cyclopentenyl cation ⟷ cyclopentenyl cation

(b) $H_2\overset{+}{C}—\overset{-}{\ddot{C}}H_2$ ⟷ $H_2\overset{-}{\ddot{C}}—\overset{+}{C}H_2$

(c) $H_2C(I)—CH_2—\overset{-}{\ddot{C}}H_2$ ⟷ $H_2\overset{-}{\ddot{C}}—CH_2—CH_2(I)$

(d) $:C{=}\ddot{O}:$ ⟷ $\overset{-}{:}C{\equiv}\overset{+}{O}:$

PROBLEM 9.33 Analyze the relative importance of the following two resonance forms. What would you say to someone who claimed that form (b) must be unimportant because the positive charge is on the more electronegative atom? You will have to add electron dots to make good Lewis structures first.

CH_3O–$\overset{+}{C}$H–CH_3 ⟷ CH_3O^+=CH–CH_3

(a) (b)

PROBLEM 9.34 Predict the regiochemistry of the following addition reactions of a generic acid, H–X. Explain your predictions

(a) CH_2=CH–CH_3 (b) Br–CH_2–CH=CH–CH_3 (c) $(CH_3O)_2C$=$C(CH_3)_2$

(d) Ph–CH=CH–CH_3 (e) $(F_3C)_2C$=$C(CH_3)_2$

PROBLEM 9.35 What will be the structure of the polymer produced from the following two monomers through cationic polymerization?

PROBLEM 9.36 At –78 °C, hydrogen bromide adds to 1,3-butadiene to give the two products shown. Carefully write a mechanism for this reaction, and rationalize the formation of two products.

1,3-butadiene + HBr —(–78 °C)→ 3-bromo-1-butene (81%) + 1-bromo-2-butene (18%)

PROBLEM 9.37 Reaction of *cis*-2-butene with hydrogen bromide gives racemic 2-bromobutane.

(a) Show carefully the origin of the two enantiomers of 2-bromobutane.

(b) In addition, small amounts of two alkenes isomeric with *cis*-2-butene can be isolated. What are these two alkenes and how are they formed?

In the following four problems, reactions appear that are quite similar to those discussed in this chapter, but are not exactly the same. Each raises one or more slightly different mechanistic points. In this book, we aim to develop an ability to understand new observations in the light of old knowledge. In a sense, you are being put in the place of the research worker being presented with new experimental results. In each case, explain the products mechanistically.

PROBLEM 9.38

(a) H_2C=$C(CH_3)_2$ —(H_3O^+, CH_3OH)→ $(CH_3)_3C$—OCH_3

(b) dihydropyran —(H_3O^+, CH_3OH)→ 2-methoxytetrahydropyran (OCH_3)

PROBLEM 9.39

$(H_3C)_2C$=$C(CH_3)_2$ —(H_3O^+, H_2S)→ $(H_3C)_2HC$–$C(CH_3)_2$–SH + $(H_3C)_2HC$–$C(CH_3)_2$–S–$C(CH_3)_2$–$CH(CH_3)_2$

PROBLEM **9.40**

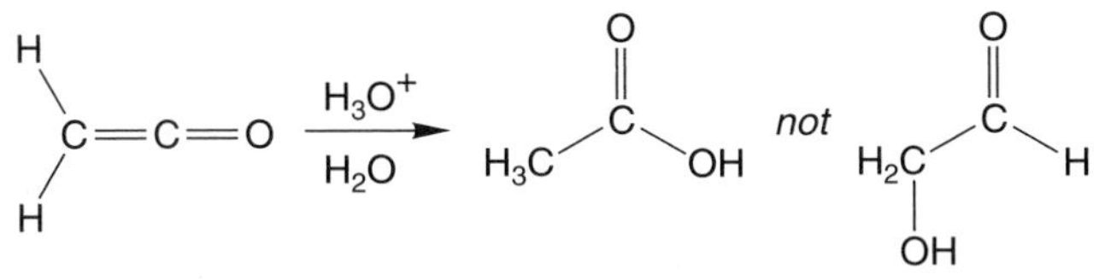

$H_2C{=}C{=}O \xrightarrow[H_2O]{H_3O^+} H_3C{-}C({=}O){-}OH$ *not* $H_2C(OH){-}C({=}O){-}H$

PROBLEM **9.41**

(a)

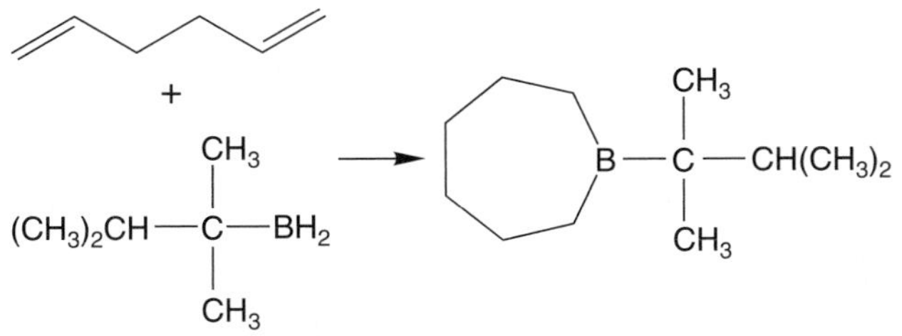

\+ $(CH_3)_2CH{-}C(CH_3)_2{-}BH_2$ → $B{-}C(CH_3)_2{-}CH(CH_3)_2$

(b)

BH_3

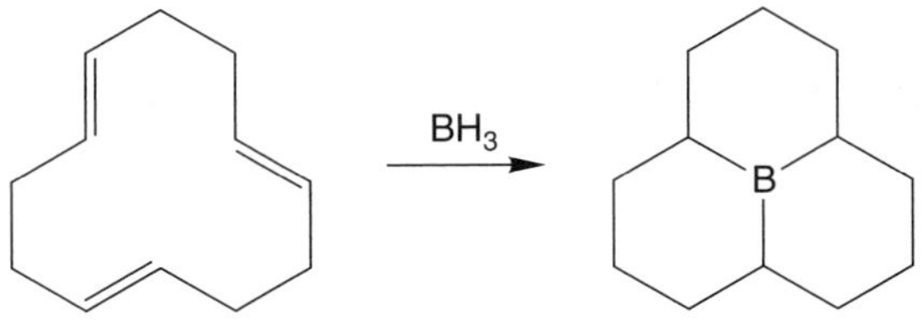

PROBLEM **9.42** Given the following experimental observation and the knowledge that the initial hydroboration step involves a syn addition, what can you say about the overall stereochemistry of the migration and oxidation steps of the organoborane (see Fig. 9.76, p. 372). Is there overall retention or inversion of configuration?

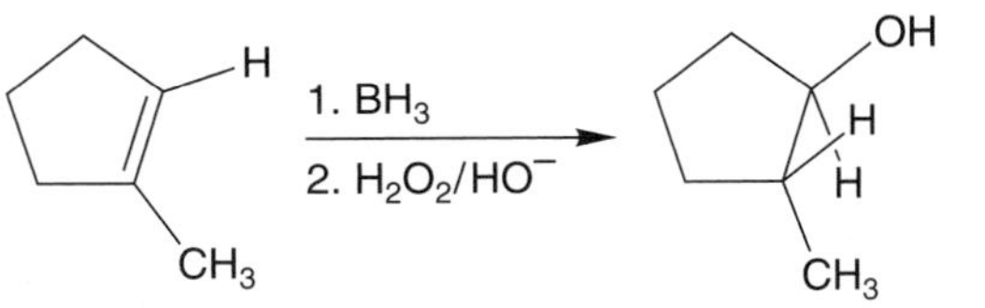

1. BH_3
2. H_2O_2/HO^-

PROBLEM **9.43** Consider the syntheses of 2-chloro-3-methylbutane and 2-bromo-3-methylbutane suggested in the answer to Problem 9.23 (b) and (e), given in the Study Guide. Carefully consider the last step in the proposed syntheses, the reaction of 3-methyl-2-butanol with hydrogen bromide or hydrogen chloride. Do you see any potential problems with this step? What else might happen?

PROBLEM **9.44** Propose syntheses of the following molecules starting from methylenecyclohexane, alcohols, and iodides containing no more than four carbon atoms, and any inorganic materials (no carbons) you want.

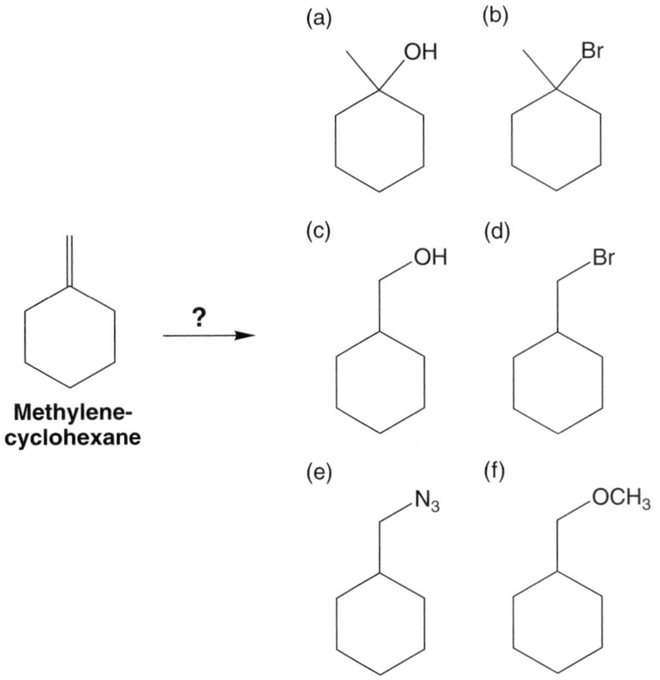

Methylene-cyclohexane →?

(a) OH (b) Br (c) OH (d) Br (e) N_3 (f) OCH_3

In each of the following reactions, more than a simple addition takes place. Keep your wits about you, think in simple stages, and work out mechanisms.

PROBLEM **9.45** Provide a mechanism for the following change:

$\xrightarrow[H_2O]{H_3O^+}$

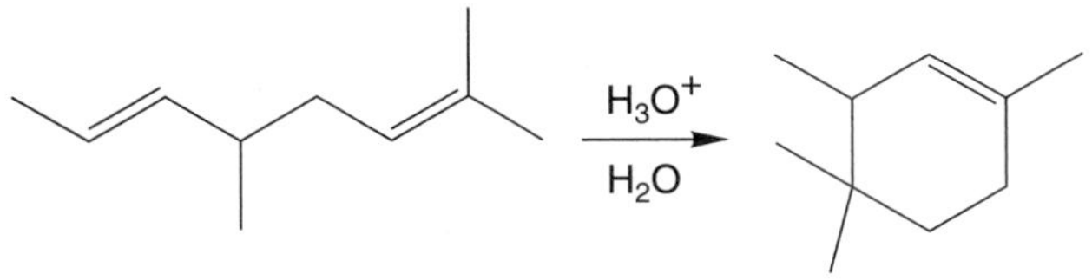

PROBLEM **9.46** Provide a mechanism for the following change:

$\xrightarrow[H_2O]{H_3O^+}$ +

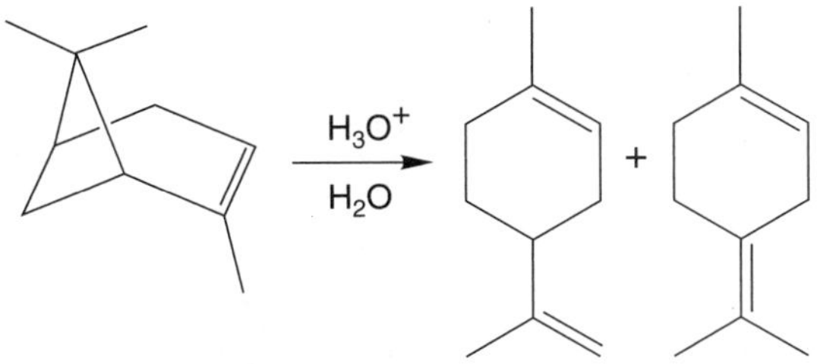

PROBLEM 9.47 Provide a mechanism for the following change:

Hint for the hard part: Think Wagner–Meerwein (p. 376).

Additions to Alkenes 2; Additions to Alkynes

Victor Serge said, "I followed his argument
With the blank uneasiness which one might feel
In the presence of a logical lunatic."
—Wallace Stevens*

We continue the discussion of additions to π systems in this chapter. Most reactions will follow the pattern seen in Chapter 9 in which a Lewis acid (usually a cation) is formed and subsequently is captured by a nucleophile, often an anion. Other reactions follow a different path. We begin with one of these, hydrogenation, the addition of H_2. We then move on to reactions of other X_2 reagents, which often proceed through the formation of an intermediate, but nonisolable, three-membered ring. Next, we take up additions in which the three-membered ring is stable, and finish with reactions forming different sized rings. We will also discuss the related additions of alkynes. Throughout this chapter, attention is paid to what all these reactions allow us to do in the way of synthesis.

10.1 ADDITION OF X_2 REAGENTS

In this section, we see two very different reactions of X_2 reagents. In the first, the one-step addition of hydrogen to alkenes, a catalyst is required to get things going. The second reaction, halogen addition, fits better into the pattern of Chapter 9, in which a two-step process is followed.

10.1a Hydrogenation

When an alkene is mixed with hydrogen, nothing happens (unless one is careless with oxygen and a match). If the two reagents are mixed in the presence of any of a large number of metallic catalysts, however, an efficient addition of hydrogen converts the alkene into an alkane. This reaction is called **hydrogenation**, and it can be achieved by two different catalytic processes. First, there is **heterogeneous catalysis**, in which an

*Wallace Stevens (1878–1958) was an American poet whose day job was being an insurance lawyer. It was only at the end of his life that he received any recognition as a poet.

insoluble catalyst, often palladium (Pd) on charcoal, is used. **Homogeneous catalysis**, in which a soluble catalyst, often Wilkinson's catalyst, after Geoffrey Wilkinson (b. 1921), can also be employed (Fig. 10.1).

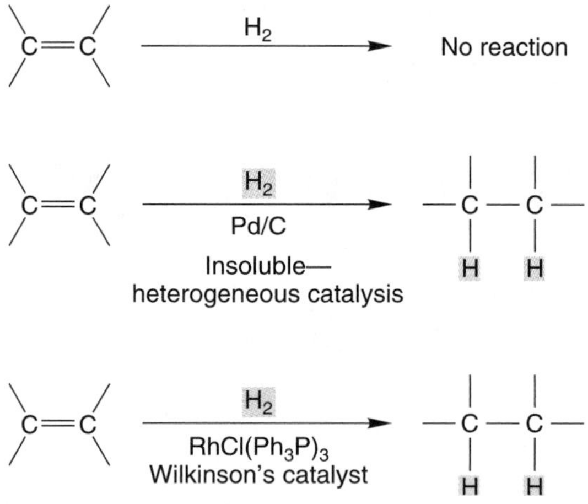

FIGURE 10.1 Although there is no reaction between hydrogen and an alkene, these species do react very quickly in the presence of a metal catalyst such as palladium adsorbed on charcoal (Pd/C). Many other insoluble metallic catalysts work as well in heterogeneous catalysis. Wilkinson's catalyst is typical of soluble molecules that also catalyze hydrogenation in homogeneous catalysis.

PROBLEM 10.1 Use bond dissociation energies (Table 8.2, p. 307) to calculate the exothermicity or endothermicity of the hydrogenation of ethylene to ethane.

Other, nonmetallic, organic reducing agents can be used in hydrogenation as well. **Diimide (HN=NH)** is a good example. Diimide is not stable under normal conditions and is made only as it is needed. Diimide exists in two forms, syn and anti, but only the less stable, syn form is active in hydrogenation (Fig. 10.2).

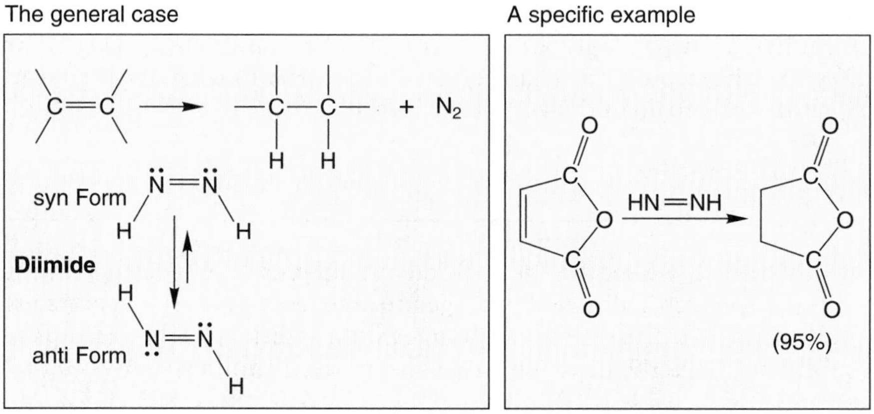

FIGURE 10.2 An effective nonmetallic reducing agent for alkenes is "diimide" (HNNH).

If hydrogenation is carried out with a ring compound, a cycloalkene, we can see that transfer of hydrogen takes place in predominately syn fashion. Both new hydrogens are delivered from the same side of the alkene (Fig. 10.3).

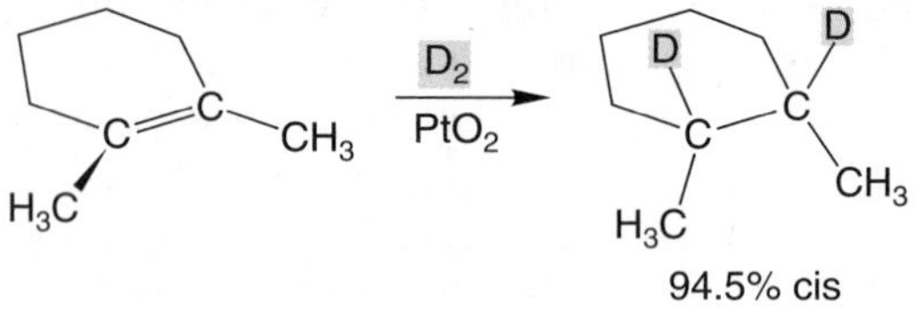

FIGURE 10.3 Metal-catalyzed addition to cycloalkenes shows that deuterium is added in syn fashion.

The mechanistic details of this reaction remain vague. Presumably, the metal surface binds both hydrogen and alkene, weakening the π bond of the alkene and the σ bond of hydrogen. The alkane is formed by addition of a weakened hydrogen molecule to the coordinated alkene. As the transfer is completed, the product alkane detaches from the metal surface (Fig. 10.4).

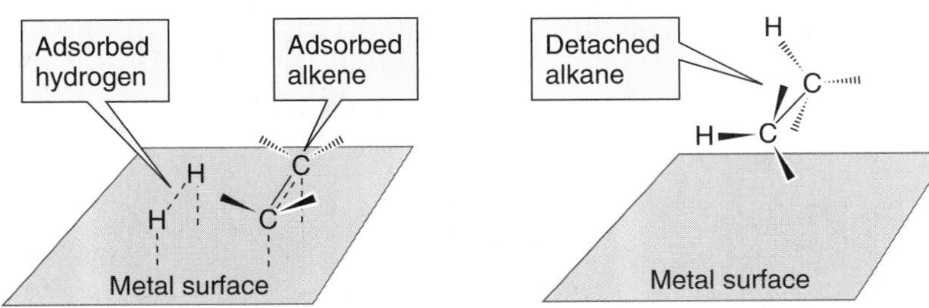

FIGURE **10.4** Both the alkene and hydrogen are adsorbed on the metal surface, weakening the π bond of the alkene and the σ bond of hydrogen. Hydrogens are transferred to generate the alkane that then detaches from the metal surface.

There are many different kinds of selectivity in the hydrogenation reaction. Not all carbon–carbon double bonds are hydrogenated at the same rate, and the differences in rate can often be used to hydrogenate one double bond in the presence of another. Generally, the more sterically hindered a double bond, the slower it is hydrogenated. This rate change is probably related to ease of adsorption on the catalyst surface. Large groups hinder adsorption, and therefore slow the rate of hydrogenation. One can often hydrogenate a less substituted double bond in the presence of more substituted double bonds, for example (Fig. 10.5).

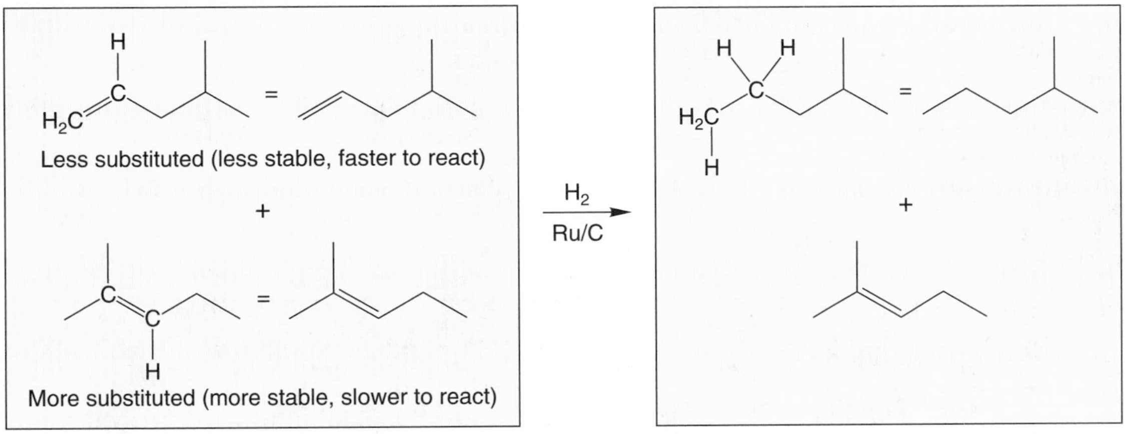

Less substituted (less stable, faster to react)

\+

More substituted (more stable, slower to react)

H_2, Ru/C

FIGURE **10.5** Steric factors are important in the attachment to the metal surface. Less hindered alkenes are absorbed more easily than compounds containing more encumbered double bonds. It is possible to hydrogenate less substituted double bonds in the presence of more hindered, more substituted double bonds.

Convention Alert!

We have to pay attention to the drawing conventions (codes) here. We do not always draw out all of the carbon and hydrogen atoms—we are beginning to adopt the convention most used by chemists—the "stick" drawings that show only the carbon framework of the molecule. This figure shows two levels of code, but we will often lapse into the most schematic representations of molecules.

PROBLEM **10.2** Hydrogenation of β-pinene gives only one of the possible diastereomeric products. Explain the direction of hydrogenation shown in Figure 10.6.

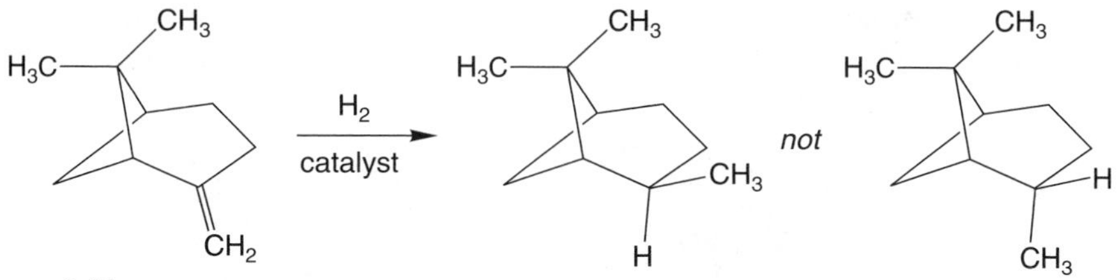

H₃C, CH₃, CH₂, H₂, catalyst, not, H, CH₃

FIGURE **10.6** **β-Pinene**

Now let's consider the energetics of this reaction. Is hydrogenation likely to be exothermic or endothermic? As shown earlier in Problem 10.1, a simple and common technique for making such estimates uses bond energies. In the hydrogenation reaction, two bonds are breaking: the hydrogen–hydrogen σ bond and the π bond of the alkene. Two bonds are also being made in this reaction: the two new σ bonds in the alkane product. If we know bond energies (Table 8.2, p. 307), we can estimate $\Delta H°$ for this reaction. We made this estimate in Problem 10.1, and Figure 10.7 repeats the calculation for this reaction. The process should be exothermic by about 30 kcal/mol.

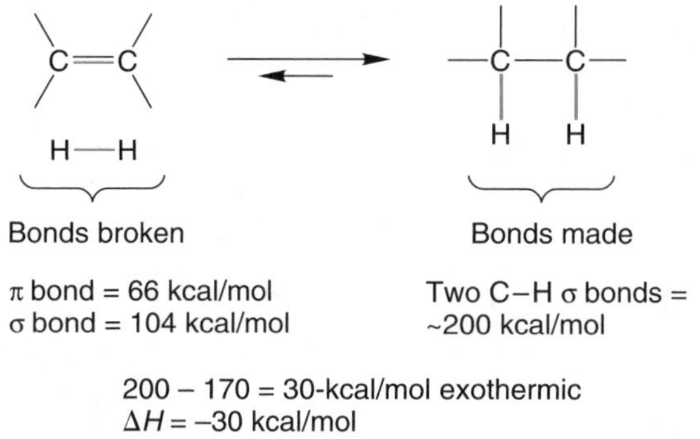

Bonds broken

π bond = 66 kcal/mol
σ bond = 104 kcal/mol

Bonds made

Two C–H σ bonds = ~200 kcal/mol

200 – 170 = 30-kcal/mol exothermic
$\Delta H = -30$ kcal/mol

FIGURE **10.7** In a typical hydrogenation, it is the π bond of the alkene and the σ bond of hydrogen that are broken. Two new carbon–hydrogen bonds are made. An analysis of the bond enthalpies involved allows a prediction that the hydrogenation of an alkene should be exothemic by approximately 30 kcal/mol.

TABLE **10.1** Some Heats of Hydrogenation

Isomer	$\Delta H^{\circ}_{H_2}$ (kcal/mol)
$H_2C{=}CH_2$	–32.7
$H_2C{=}CH{-}CH_3$	–30.0
$H_2C{=}CH{-}CH_2CH_3$	–30.3
cis $CH_3{-}CH{=}CH{-}CH_3$	–28.6
trans $CH_3{-}CH{=}CH{-}CH_3$	–27.6
$(CH_3)_2C{=}CH_2$	–28.4
$(CH_3)_2C{=}CHCH_3$	–26.9
$H_2C{=}CH{-}CH_2CH_2CH_2CH_3$	–29.6
cis $CH_3CH_2{-}CH{=}CH{-}CH_2CH_3$	–27.3
trans $CH_3CH_2{-}CH{=}CH{-}CH_2CH_3$	–26.4
$H_2C{=}C(CH_3){-}CH_2CH_2CH_3$	–28.0
$(CH_3)_2C{=}C(CH_3)_2$	–26.6

Table 10.1 collects some measured heats of hydrogenation. We can see that the crude bond energy calculation of Figure 10.7 has actually done quite well; the heats of hydrogenation of simple alkenes are quite close to our guess of 30 kcal/mol.

PROBLEM **10.3**

Use the values in Table 10.1 to estimate the relative energies of the hexenes. Compare your estimates with those generated from heats of formation in Chapter 4, (p. 134).

PROBLEM **10.4**

As a mixture of an alkene and hydrogen lies some 30 kcal/mol above the alkane product, why doesn't the reaction occur spontaneously? Use an Energy versus Reaction progress diagram to help explain *in general* the role of the metal catalyst (Chapter 8, Box, p. 327).

10.1b Additions of Halogens

Molecular bromine and chlorine (Br_2 and Cl_2) do not need a catalyst to add to alkenes to give **vicinal** dihalides (Fig. 10.8). "Vicinal" means "disubstituted on adjacent (1,2) positions."

If we take the addition of hydrogen bromide to an alkene as a model for this reaction, we can build a first-guess mechanism quite quickly. Just as the alkene reacts with hydrogen bromide to produce a carbocation and bromide (Br^-), so reaction with bromine–bromine might give a brominated cation and bromide (Fig. 10.9). The π bond of the alkene acts as a nucleophile, displacing Br^- from bromine. Readdition of bromide to the carbocation would give the dibromide product.

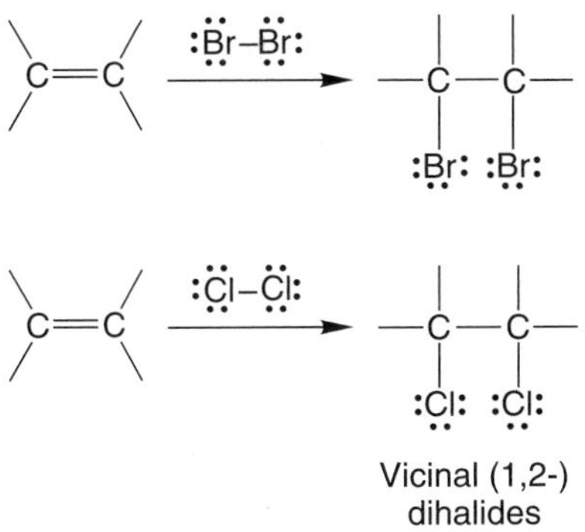

Vicinal (1,2-) dihalides

FIGURE **10.8** Bromine and chlorine add to alkenes to produce vicinal dihalides.

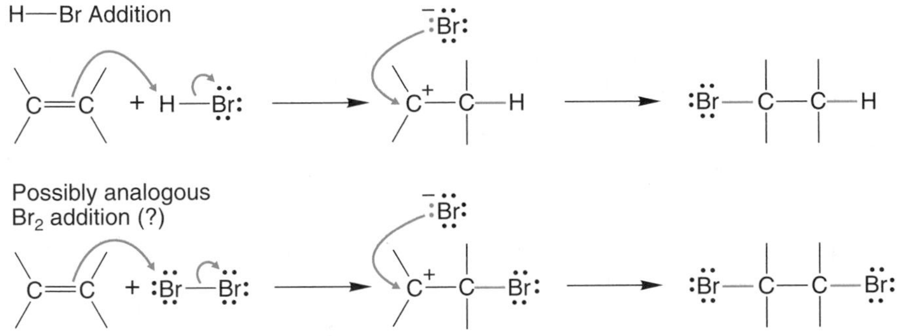

H—Br Addition

Possibly analogous Br_2 addition (?)

FIGURE **10.9** A first-guess mechanism for the reaction of an alkene and bromine. For analogy it draws on the mechanism of hydrogen bromide addition to an alkene.

There are cases of bromination in which this *is* the correct mechanism, but for most alkenes it is not quite right. First, carbocation rearrangements (Chapter 9, p. 374) are not observed in most bromination and chlorination reactions. For example, addition of bromine to 3,3-dimethyl-1-butene gives only 1,2-dibromo-3,3-dimethylbutane, *not* 1,3-dibromo-2,3-dimethylbutane, as would be expected if an open cation were involved (Fig. 10.10).

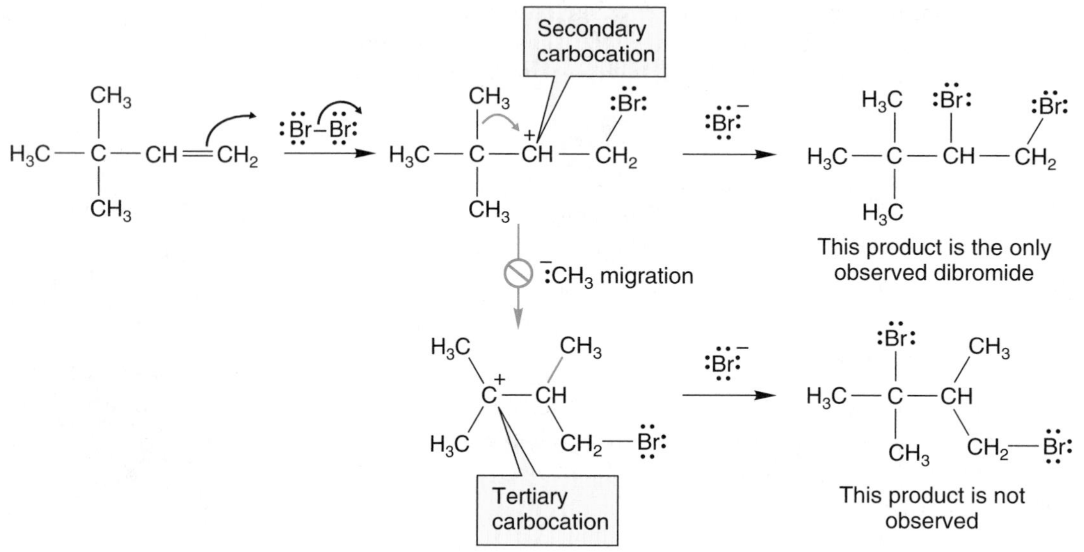

FIGURE **10.10**

Second, if a cycloalkene is used, we can easily see that the product is exclusively the *trans*-1,2-dibromide. The cis compound cannot be detected (Fig. 10.11).

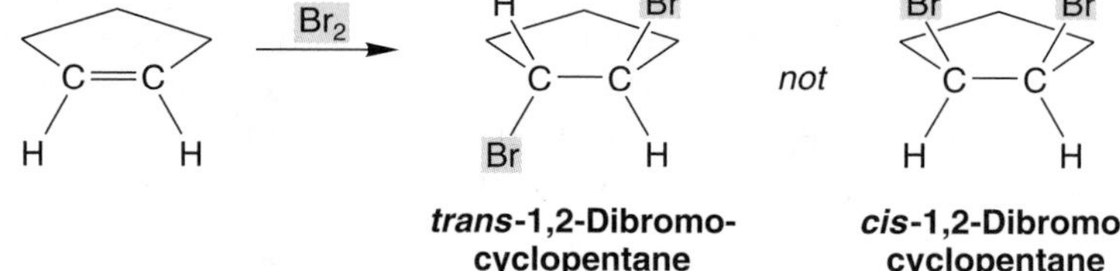

FIGURE **10.11** Addition of Br_2 (or Cl_2) proceeds stereospecifically anti to give the trans dihalide.

If the addition were to involve an open cation, we would expect the bromide ion to add from both sides to produce a mixture of products. The yields of cis and trans compounds would not be equal, but both compounds should be formed (Fig. 10.12).

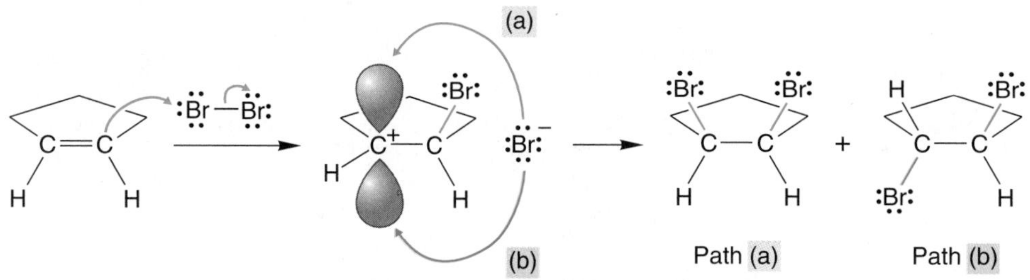

FIGURE **10.12** The first-guess mechanism predicts both anti (b) and syn (a) addition of the halogen. As only anti addition is observed, our initial mechanistic ideas cannot be completely correct. We need a new mechanism.

But only the trans dibromide is seen. So our first-guess mechanism needs modification. How do we explain the simultaneous preference for anti addition, and absence of rearrangement? It seems that an open carbocation, which would lead to cationic rearrangements and permit both syn and anti addition, must be avoided in some way.

What if the initial displacement of Br^- from Br_2 were to take place in a symmetrical fashion? Once again, the filled π orbital of the alkene acts as nucleophile in an S_N2 reaction in which Br^- is displaced from Br_2 (Fig. 10.13). As the figure shows, a bromine-containing three-membered ring, a **bromonium ion** (brom = bromine; onium = positive), is formed. An open carbocation is avoided and rearrangements cannot occur.

Symmetrical displacement

A bromonium ion

FIGURE 10.13 A symmetrical attack of the alkene on bromine yields a cyclic bromonium ion along with bromide, Br^-.

But what of the preference for anti addition? Look closely at the second step of the reaction, the opening of the bromonium ion by the nucleophile Br^-. This reaction is an S_N2 displacement, and therefore must take place with inversion, through addition from the rear as we saw so often in Chapter 7 (p. 244) (Fig. 10.14). Overall anti addition of the two bromine atoms is assured.

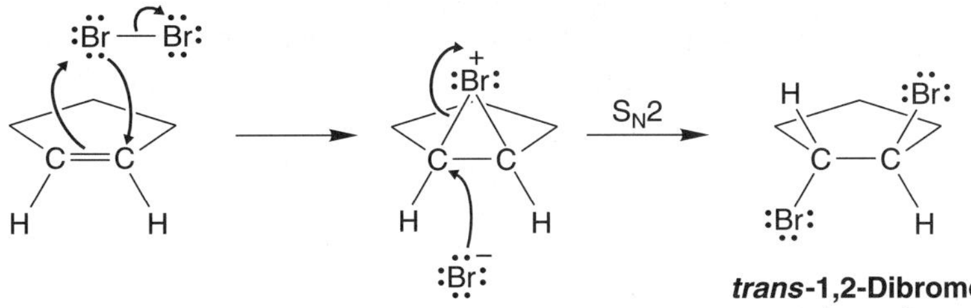

S_N2

***trans*-1,2-Dibromo–cyclopentane**

FIGURE 10.14 Addition of bromide to the cyclic bromonium ion is an S_N2 reaction and must occur from the rear, with inversion. This anti addition of the two bromine atoms requires the formation of *trans*-1,2-dibromocyclopentane.

*PROBLEM 10.5

In Figure 10.14, the reaction starts with achiral molecules. Yet the product, as drawn in the figure, is a single enantiomer. Clearly something is awry. What is it? *Hint*: Convention Alert! (Chapter 5, p. 188).

ANSWER

You were warned about this! Of course there can be no optical activity produced from a reaction of two achiral reagents. In this reaction, opening must take place in two equivalent ways to produce a racemic pair of enantiomers. In situations such as this one, it is the custom not to draw both enantiomers.

Convention Alert!

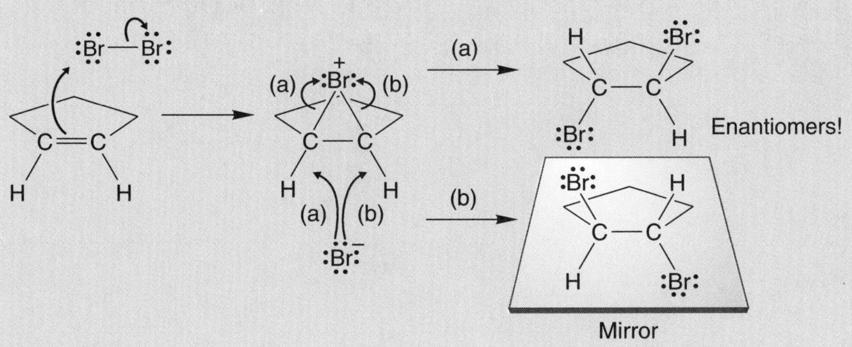

So, a modification of our earlier addition mechanism suffices to explain the reaction. In contrast to the additions to alkenes we have seen earlier (H—X, carbocations, and boranes), in bromination and chlorination a three-membered ring is produced as an intermediate. In some, very specialized cases the bromonium ion can even be isolated, and it seems its intermediacy in most brominations is assured.

PROBLEM 10.6 One case in which the bromonium ion is stable to further addition of nucleophiles is the reaction of adamantylideneadamantane with bromine (Fig. 10.15). Can you see why this cyclic ion might be stable?

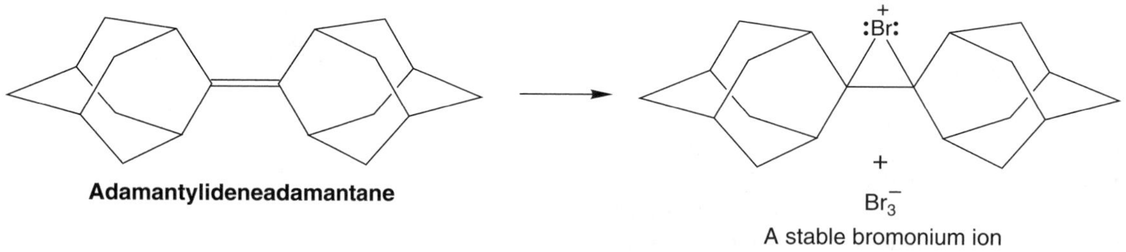

:Br:+

+

Br_3^-

Adamantylideneadamantane

A stable bromonium ion

FIGURE 10.15

*PROBLEM 10.7 Addition of Br_2 to acenaphthylene gives large amounts of the product of cis addition along with the usual trans dibromide. Give a mechanism for this bromination and, more important, explain why the mechanism of the reaction shown in Figure 10.16 is different from the one we just outlined.

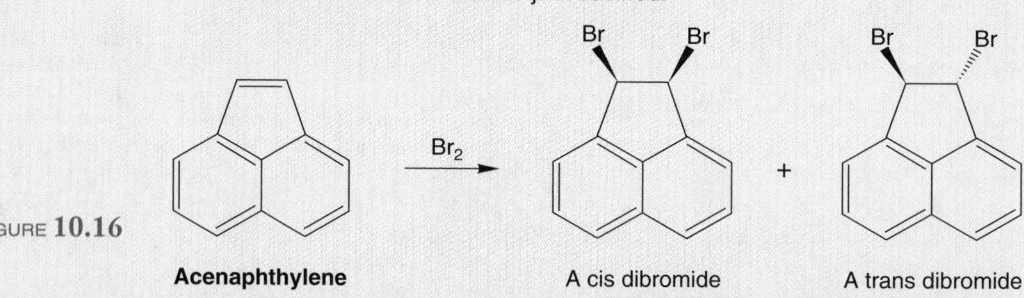

FIGURE 10.16

ANSWER In this case, formation of the open carbocation competes favorably with formation of the bromonium ion. The reason is that the open cation is exceptionally well stabilized by resonance. The open cation can lead to both cis and trans products, whereas the bromonium ion is destined to give only trans dibromide. Note again the convention we use in which a charge in parentheses is used to show the various atoms sharing the charge. If it is not obvious that the charge is delocalized to these positions, draw out the resonance forms.

Convention Alert!

ANSWER (CONTINUED)

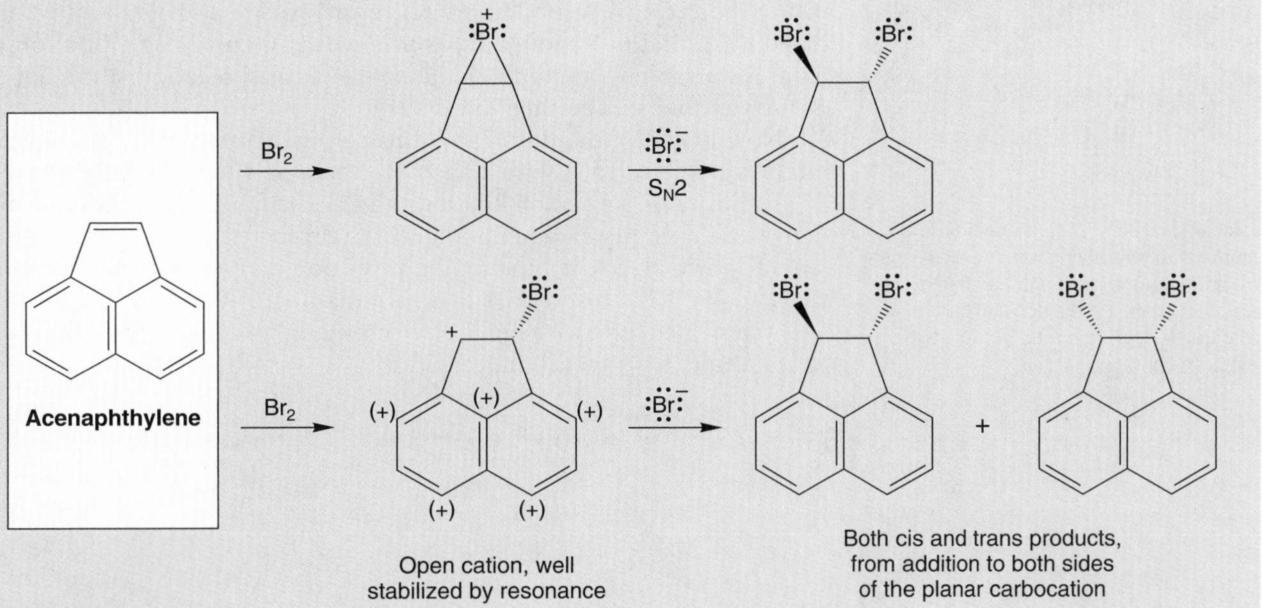

Acenaphthylene

Br_2

$:\ddot{Br}:^+$

$:\ddot{Br}:^-$ S_N2

Br_2

(+) (+) (+) (+) (+)

$:\ddot{Br}:^-$

Open cation, well stabilized by resonance

Both cis and trans products, from addition to both sides of the planar carbocation

The bromonium ion is not completely new; we have seen a similar-looking intermediate before. Recall the earlier suggestion that the structure of the ethyl cation might be bridged (Section 9.13, p. 380). One might call such a species a **protonium ion** (prot = H; onium = positive). But the two species are really comparable only in the way they look on paper. The protonium ion is an example of three-center, two-electron bonding—the six electrons required to make up the three full bonds in the three-membered ring are not available. Only the electrons in the carbon-hydrogen bond can be used to hold the three atoms (C, C, H) together. In the bromonium ion, there is the full complement of six electrons for the three σ bonds binding the two carbons and the bromine. The bromonium ion is a normal compound; the protonium ion is not. To see this more clearly, imagine forming the two species from open cations (Fig. 10.17).

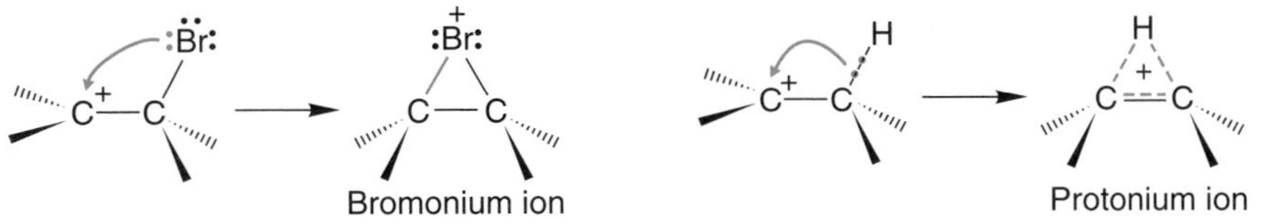

Bromonium ion

Protonium ion

FIGURE 10.17 The bromonium ion is a normal compound. There is nothing strange about it. A cyclic protonium ion, which looks the same on paper is, in fact, very different. This figure shows the formation of bromonium and protonium ions from open carbocations.

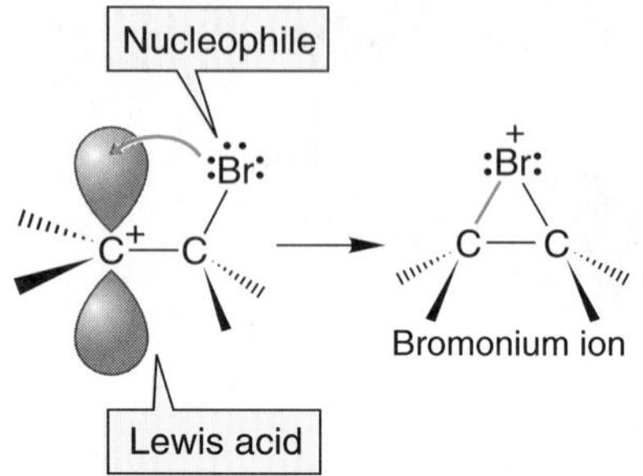

FIGURE 10.18 To form the bromonium ion, a pair of nonbonding electrons on bromine (the nucleophile) attacks the empty *p* orbital on the positive carbon (the Lewis acid) to form a normal, two-electron bond.

This point is important, so let's look at it slowly. The bromonium ion is simply made by using one of bromine's electron pairs to create the new bond to carbon. Bromine gains a positive charge, and there is nothing unusual about this reaction. In orbital terms, we are allowing a filled, nonbonding orbital on bromine to overlap with an empty 2*p* orbital on carbon. It's a garden-variety nucleophile–Lewis acid reaction, although in a somewhat unusual setting (Fig. 10.18).

Now look at protonium ion formation in a similar way. There are *no* nonbonding electrons with which to close the ring! The only electrons available are in the carbon–hydrogen bond itself. If these are used to overlap with the empty carbon 2*p* orbital, we get the protonium ion, but there are only two electrons binding the three atoms in the new dashed bonds (Fig. 10.19). This reaction still involves a nucleophile and a Lewis acid, but the nucleophile is the pair of electrons in the filled carbon–hydrogen σ bond, and is very weak indeed.

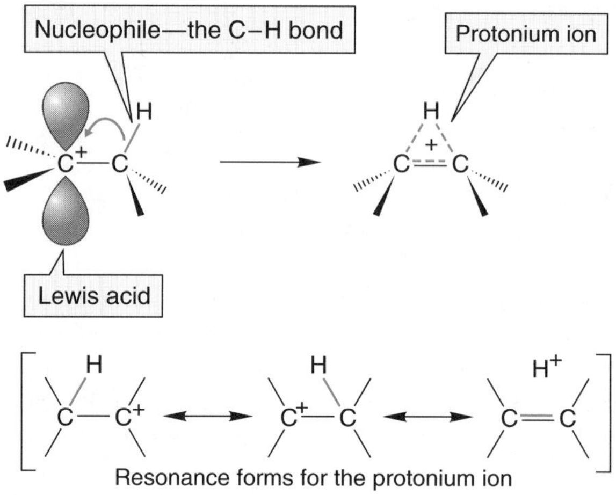

Nucleophile—the C–H bond

Protonium ion

Lewis acid

Resonance forms for the protonium ion

FIGURE 10.19 There are no nonbonding electrons on hydrogen. If a cyclic ion is to be formed, it must be by using the only electrons available, the pair in the carbon–hydrogen bond, as the nucleophile. A species is formed in which only two electrons bind three atoms: the two carbons and the hydrogen. Resonance forms for the protonium ion are shown.

The bromonium ion and protonium ion are *very* different.

Normally, the nonnucleophilic solvent carbon tetrachloride (CCl_4) is used for the bromination or chlorination of alkenes. But the reaction will also work in protonic solvents such as water and simple alcohols. In this kind of reaction, new products appear that incorporate molecules of the solvent (Fig. 10.20).

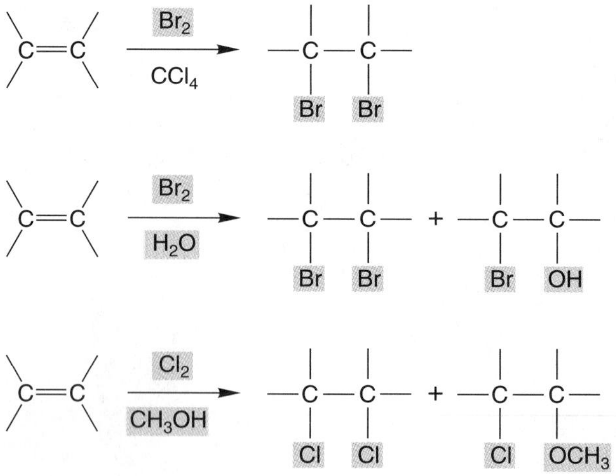

Br_2 / CCl_4 → Br, Br

Br_2 / H_2O → Br, Br + Br, OH

Cl_2 / CH_3OH → Cl, Cl + Cl, OCH_3

FIGURE 10.20 In the presence of nucleophilic solvents, addition reactions of bromine and chlorine give products incorporating solvent molecules as well as dihalides.

How do the OH and OR groups get into the product molecule? To see the answer write out the mechanism and look for an opportunity to make the new products. The first step is just the same, in this case formation of the cyclic chloronium ion (Fig. 10.21). Normally, the chloronium is opened in S_N2 fashion by the nucleophilic Cl^- to give the dichloride. This opening must be part of the mechanism of dichloride formation in these examples. But in these reactions there are other nucleophiles present: molecules of water or alcohol. Of course, they are weaker nucleophiles than the negatively charged Cl^-, but they are present as solvent, and have an enormous numerical advantage over the more powerfully nucleophilic chloride, which is present only in relatively low concentration (Fig. 10.21).

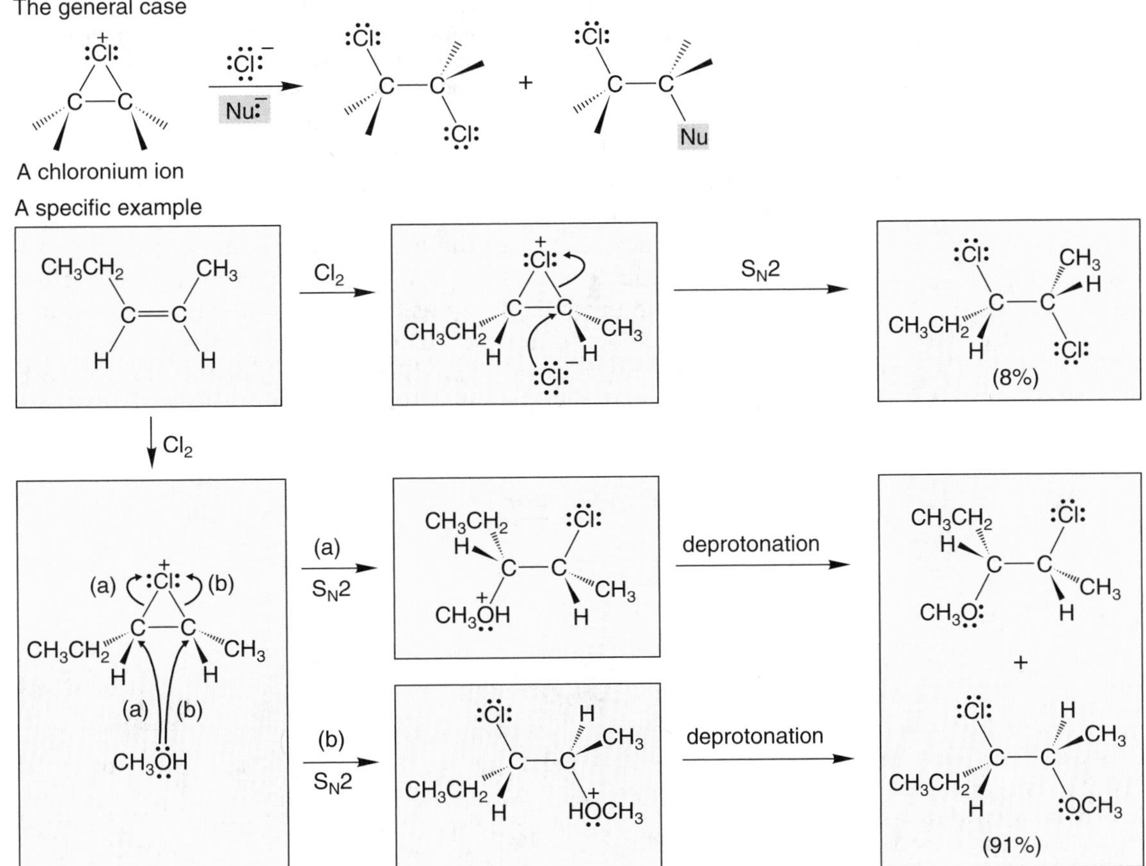

FIGURE **10.21** Once the chloronium ion is formed, it is vulnerable to attack by *any* nucleophile present. When a nucleophilic solvent is used in the reaction, it can attack the cyclic ion to give the solvent-containing product. In this case, addition of chloride accounts for 8% of the product, whereas the solvent methyl alcohol produces 91%. Notice that methyl alcohol can produce two products [(a) and (b)].

If this mechanism is correct, we can predict that the required inversion in the opening of the cyclic ion must produce a trans stereochemistry in the products. So, we have the opportunity to test the mechanism, and it turns out that addition of bromine to cyclohexene in the solvent water exclusively produces the compound resulting from anti addition in which the bromine and hydroxyl groups have a trans relationship. The bromonium ion mechanism is correct (Fig. 10.22).

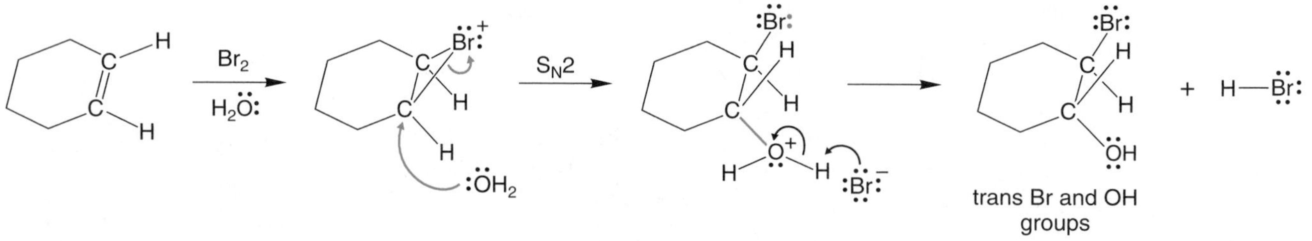

trans Br and OH groups

FIGURE 10.22 This mechanism demands that the new products be formed by anti addition, as they are. This stereochemical experiment uses cyclohexene to show that the bromine and solvent molecule become attached from different sides of the ring through the anti addition enforced by the presence of the bromonium ion.

Addition of bromine in nucleophilic solvents presents us with another opportunity. We can measure the regiospecificity of the reaction. When two bromines add, as in the bromination of 2,3,3-trimethyl-1-butene in carbon tetrachloride, there is no way to determine the regiochemistry. Bromination in water allows us to see where the hydroxyl group goes as it opens the bromonium ion. There are two possibilities. The hydroxyl group can attach to the more [Fig. 10.23(a)] or less [Fig. 10.23(b)] substituted carbon of the original alkene.

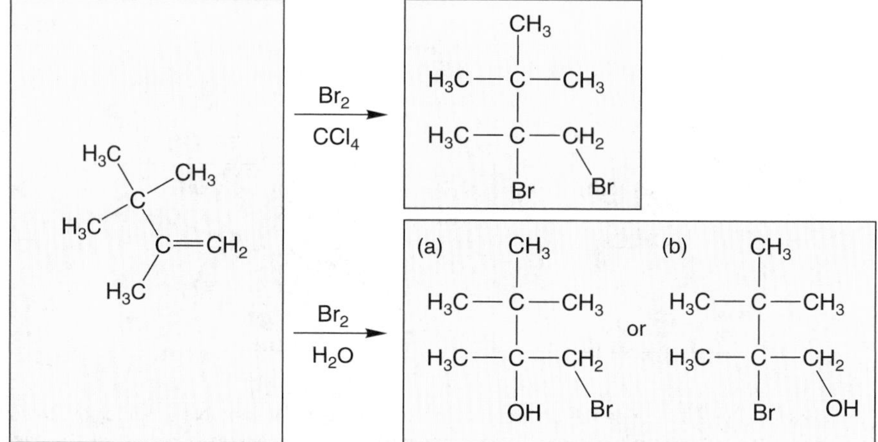

FIGURE 10.23 When the alkene is unsymmetrical, there are two possibilities for the regiochemistry of the bromination reaction in water. The hydroxyl group can become attached to the more substituted (a) or less substituted (b) carbon in the product.

In practice, the hydroxyl group becomes attached to the more substituted carbon. It is as if the reaction were proceeding through a carbocation (Fig. 10.24). But we are quite certain this is not the case. The problem now is to reconcile the observed regiospecificity with the intermediacy of a bromonium ion (recall trans addition and the lack of carbocation rearrangements).

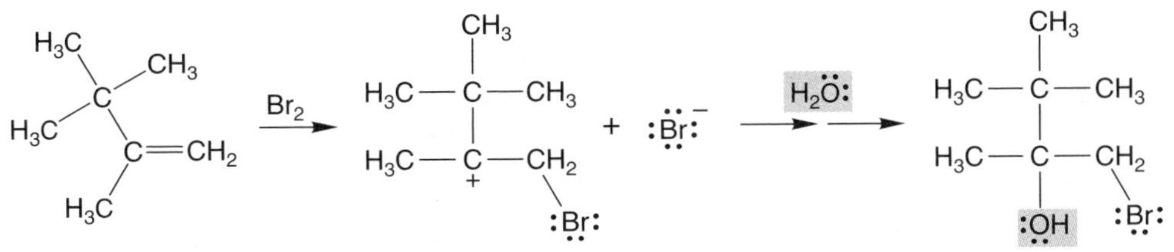

FIGURE 10.24 The bromine becomes attached to the less substituted carbon and the solvent to the more substituted carbon of the alkene. It is *as if* the reaction went through a carbocationic intermediate. Yet, we are quite certain it does not! The problem now is to resolve this seeming contradiction.

The bromonium ion is only *symmetrical* when formed from an alkene in which the two ends are *the same.* In all other cases, the two carbon–bromine bonds must be of different strength. In resonance terms, we would say that the weighting factor (coefficient) for the more stable form was greater than that for the less stable form ($c_1 > c_2$, Fig. 10.25). The more substituted carbon–bromine bond is longer and weaker than the less substituted bond in this positively charged intermediate.

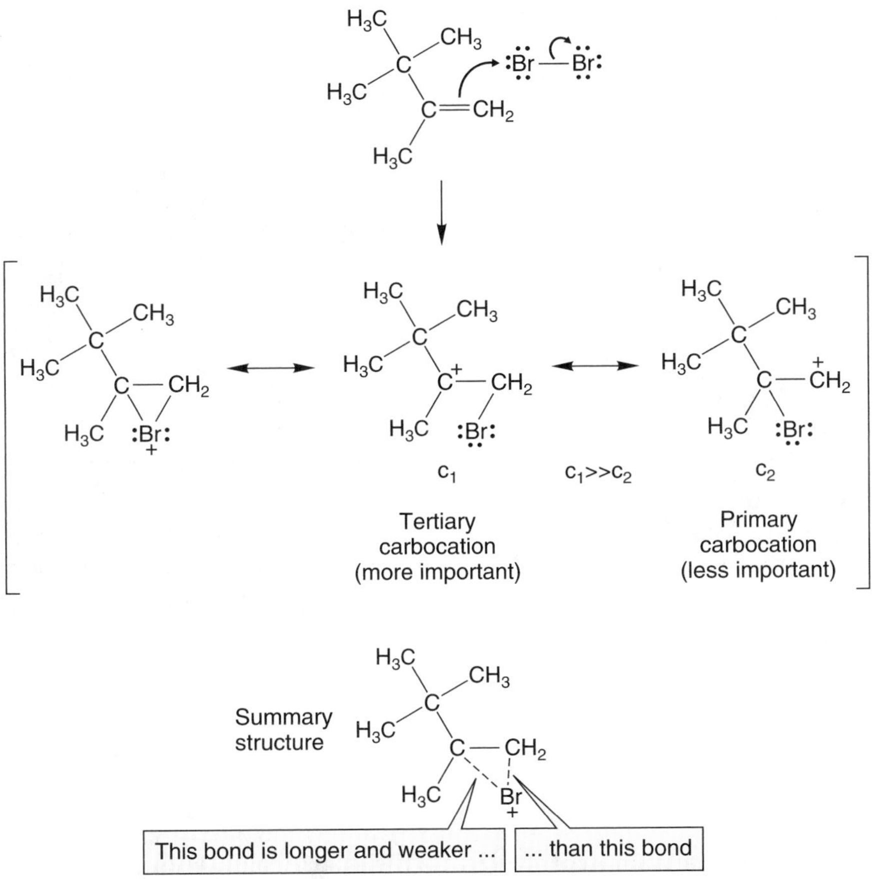

FIGURE 10.25 A resonance description of the bromonium ion. The two carbons share the positive charge, but not equally. The charge will be more stable on the more substituted carbon. Accordingly, the bond from the more substituted carbon to bromine will be weaker and longer than the bond from the less substituted carbon to bromine, and thus more easily broken.

Accordingly, the transition state for breaking the more substituted bond will be lower in energy than that for breaking the less substituted bond. Therefore, it is at the more substituted position that nucleophilic solvents such as methyl alcohol and water will add to the bromonium ion (Fig. 10.26).

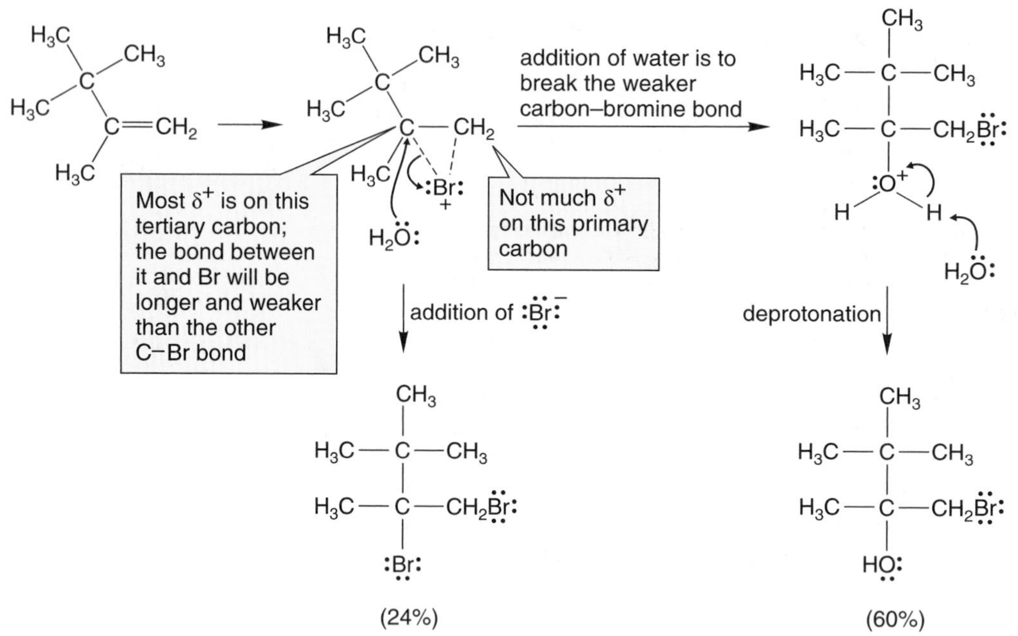

FIGURE 10.26 The weaker carbon–bromine bond is broken more easily than the stronger carbon–bromine bond.

The critical difference between addition of halogen (X_2) and the many HX additions we saw in Chapter 9 is the presence of the intermediate cyclic halonium ion. This intermediate is demanded by the observed overwhelming preference for anti addition. Now we move on to variations on this theme; to other addition reactions involving cyclic intermediates of various kinds.

10.2 HYDRATION THROUGH MERCURY COMPOUNDS: OXYMERCURATION

For small scale hydrations, it is convenient to use mercury salts such as mercuric acetate, $Hg(OAc)_2$, to carry out the reaction. In this two-step process, an alkene is first treated with mercuric acetate and then reduced with the reagent sodium borohydride ($Na^+ \, ^-BH_4$) (Fig. 10.27).

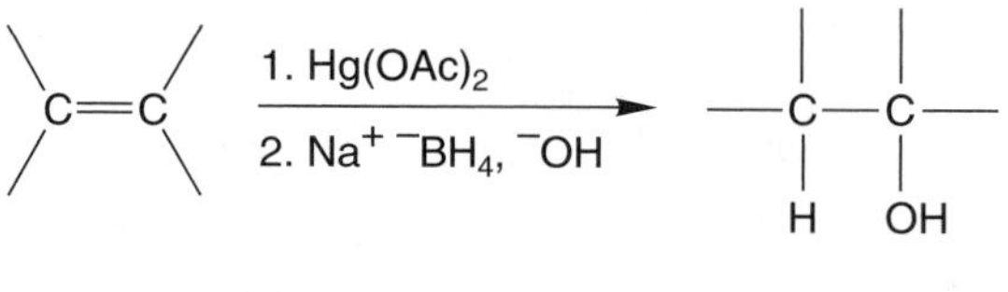

FIGURE 10.27 Alcohols can be produced from alkenes by oxymercuration followed by reduction with sodium borohydride.

Convention Alert!

The abbreviation AcO^- represents the acetate ion (CH_3COO^-), the conjugate base of acetic acid, CH_3COOH (often abbreviated as AcOH). Lewis structures are given in Figure 10.28, which shows the relationship between the acetate ion and its conjugate acid (AcOH). Note the resonance stabilization of the acetate ion—the two oxygen atoms are equivalent.

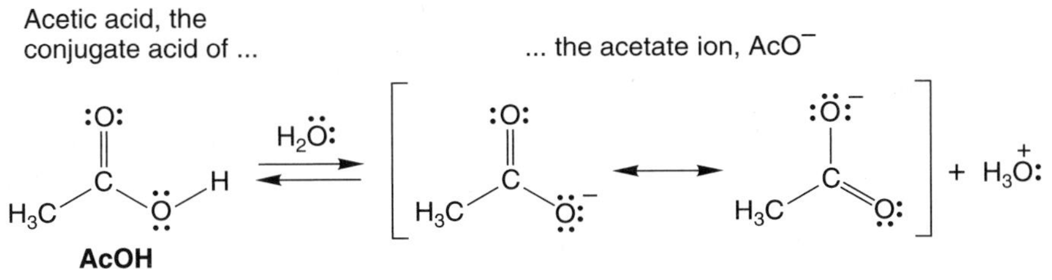

Acetic acid, the conjugate acid of ... ... the acetate ion, AcO^-

AcOH

FIGURE 10.28 Acetic acid (AcOH) and its conjugate base, the resonance-stabilized acetate ion AcO^-.

PROBLEM 10.8

The pK_a of acetic acid is 4.5. The corresponding two-carbon alcohol, ethyl alcohol (CH_3CH_2OH), is a much weaker acid, as its pK_a of 17 demonstrates. Explain why acetic acid is much more acidic than ethyl alcohol.

The first step in the reaction is the formation of a cyclic, mercurinium ion (Fig. 10.29). We can be quite certain of this because addition involves no carbocationic rearrangements. Open carbocations are unlikely intermediates, as they would surely undergo carbocationic rearrangements, and these are not seen even in systems normally prone to rearrangement.

*PROBLEM 10.9

Design an experiment to test for carbocationic rearrangements in the oxymercuration reaction.

ANSWER

It's simple. Just use any alkene in which migration of hydride ($H:^-$) or $R:^-$ would give a more thermodynamically stable cation. 3-Methyl-1-butene would work well. If addition were to give an open carbocation rather than the cyclic mercurinium ion, rearrangements would surely occur. That they are *not* observed provides strong evidence that open cations are not involved in oxymercuration.

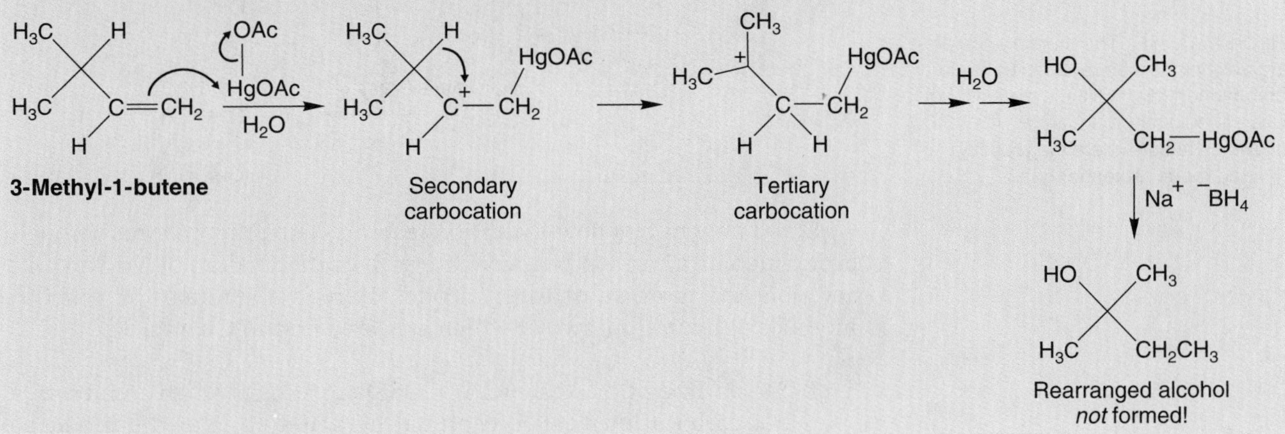

3-Methyl-1-butene Secondary carbocation Tertiary carbocation

Rearranged alcohol *not* formed!

With nonsymmetrical alkenes, the two bonds to mercury will not be equally strong. The bond between mercury and the more substituted carbon will be longer and weaker than the bond to the less substituted carbon. It is at this end that water opens the ring (Fig. 10.29). So, the first product of this reaction is the mercury-containing alcohol shown in Figure 10.29.

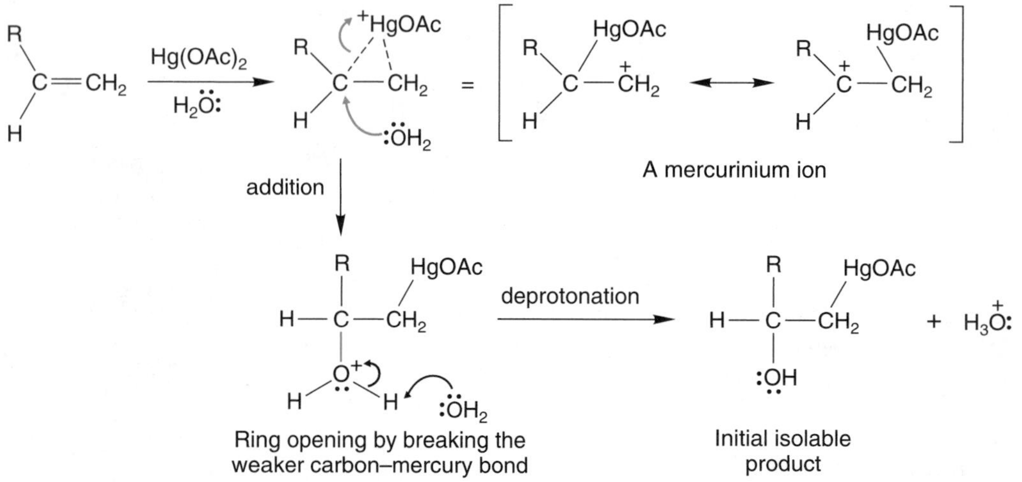

FIGURE 10.29 Oxymercuration begins by attack of the alkene on the Lewis acidic mercuric acetate to form a cyclic mercurinium ion. The nucleophilic solvent, water, then adds in S_N2 fashion to give the first product, a mercury-containing alcohol, where R = simple alkyl group.

Oxymercuration is useful because the reagent sodium borohydride ($Na^+ \ ^-BH_4$) efficiently replaces the mercury with hydrogen. Figure 10.30 shows the reaction with sodium borohydride and sodium borodeuteride. The use of the deuterated molecule allows us to see with certainty the position of the entering deuterium.

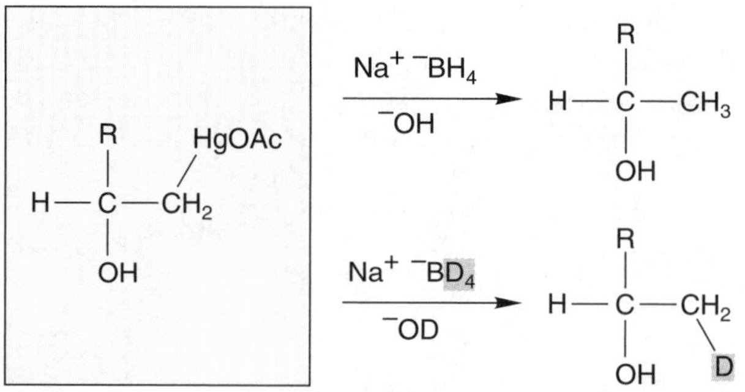

FIGURE 10.30 The mercury-containing alcohols can be reduced by sodium borohydride ($Na^+ \ ^-BH_4$) to give alcohols. Use of sodium borodeuteride ($Na^+ \ ^-BD_4$) allows us to see where the reducing hydride (here deuteride) goes.

So we now have a new hydration reaction; one that proceeds through a cyclic mercurinium ion but ultimately gives the product of Markovnikov addition, the more substituted alcohol. Here is a summary of our three methods of hydration, two of which appeared first in Chapter 9:

1. *Direct hydration* (Chapter 9, p. 358) proceeds through an intermediate carbocation that is captured by water to give the product of Markovnikov addition. The reaction is limited in utility because rearrangements of the initially formed cation to more stable species can lead to undesired, rearranged products.
2. *Hydroboration* (Chapter 9, p. 363) first generates an alkylborane that subsequently reacts with peroxide in base to give the product of anti-Markovnikov addition.
3. *Oxymercuration* proceeds through a cyclic mercury-containing ion and also gives the product of Markovnikov addition. No rearrangements take place, which is sometimes a great advantage.

Work out the consequences of applying the three procedures outlined above to 3-methyl-1-butene.

PROBLEM 10.10

10.3 OTHER ADDITION REACTIONS INVOLVING THREE-MEMBERED RINGS: OXIRANES AND CYCLOPROPANES

10.3a Oxiranes

We have seen that many addition reactions begin with formation of a three-membered ring. These intermediates have almost always been rapidly converted into the final products through addition of a nucleophile. This state of affairs will now change dramatically, as addition reactions are the best routes to certain stable three-membered rings: the cyclopropanes and **oxiranes (epoxides)**.

When an alkene reacts with a perbenzoic acid or trifluoroperacetic acid, three-membered rings containing one oxygen atom are isolated. An organic acid has the formula R—COOH. A *per*acid has the formula R—COO$\underline{O}$H. Figure 10.31 shows the difference in more detail. The "per" indicates an "extra" oxygen is present. Thus hydrogen *per*oxide is HO$\underline{O}$H, whereas the normal oxide of hydrogen is water, HOH.

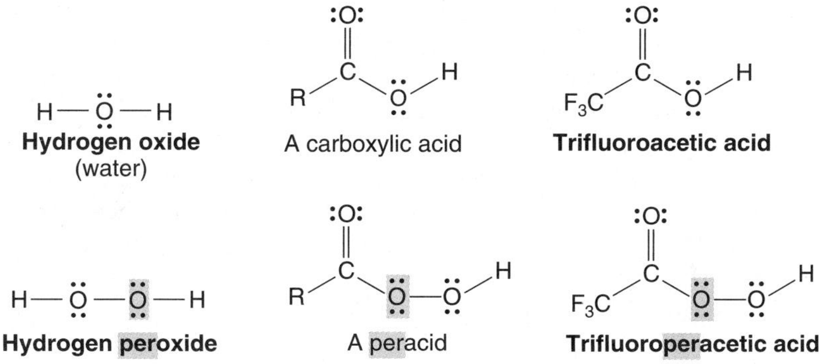

FIGURE 10.31 The prefix per- means extra. So a "peroxide" and a "peracid" each contain an extra oxygen atom, one more than the oxide or acid.

The epoxidation reaction resembles the formation of a bromonium ion; a leaving group is displaced by the alkene acting as nucleophile in an S_N2 reaction (Fig. 10.32). At the same time, a hydrogen is transferred to the carbonyl (C=O) oxygen. The result is a stable three-membered ring, an epoxide.

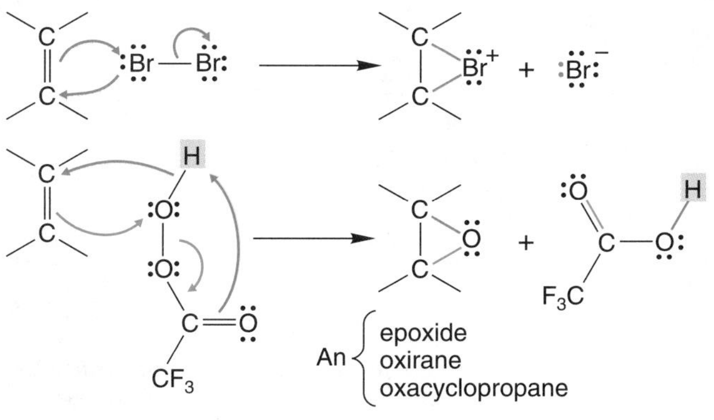

FIGURE 10.32 Epoxidation is closely related to bromination. In epoxidation, there is an additional transfer of hydrogen to give the isolable, stable epoxide, or oxirane.

It is important to be clear why we are so sure that the ring is formed in a single step. Why not just form the carbocation and then close up to the oxirane (Fig. 10.33)?

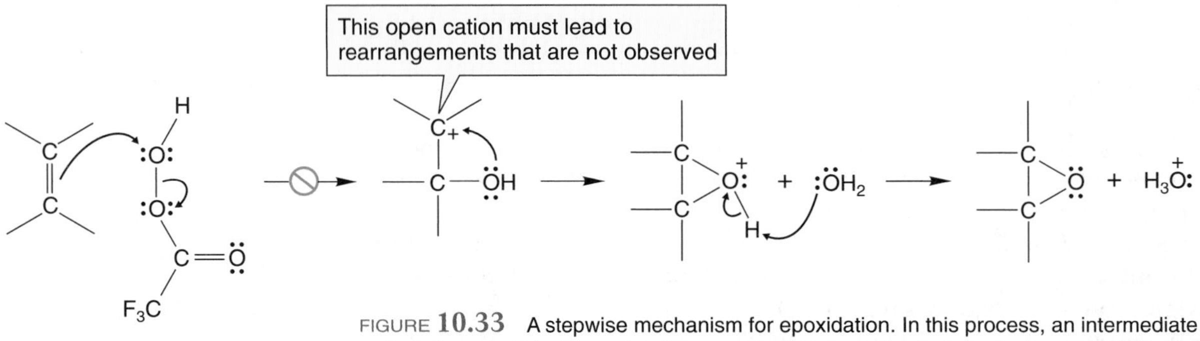

This open cation must lead to rearrangements that are not observed

FIGURE 10.33 A stepwise mechanism for epoxidation. In this process, an intermediate carbocation would be formed and then captured intramolecularly by the oxygen. If this mechanism were correct, carbocationic rearrangements should be observed, but they are not.

If a free cation were involved, carbocation rearrangements would be expected, and they are not found. Moreover it is easy to monitor the stereochemistry of this reaction. Concerted (no intermediates) formation of the oxirane explains the observed syn addition very well. A reaction in which the two new bonds are formed at the same time cannot change the stereochemical relationships of the groups in the original alkene. On the other hand, a mechanism involving an open carbocation would very likely lead to a mixture of stereoisomeric oxiranes (Fig. 10.34). If only one bond were made, there would be a carbocationic intermediate in which rapid rotations about carbon–carbon single bonds would take place. These rotations would scramble the original stereochemical relationship present in the alkene, and this is not observed. Oxirane formation retains the stereochemical relationships present in the starting alkene.

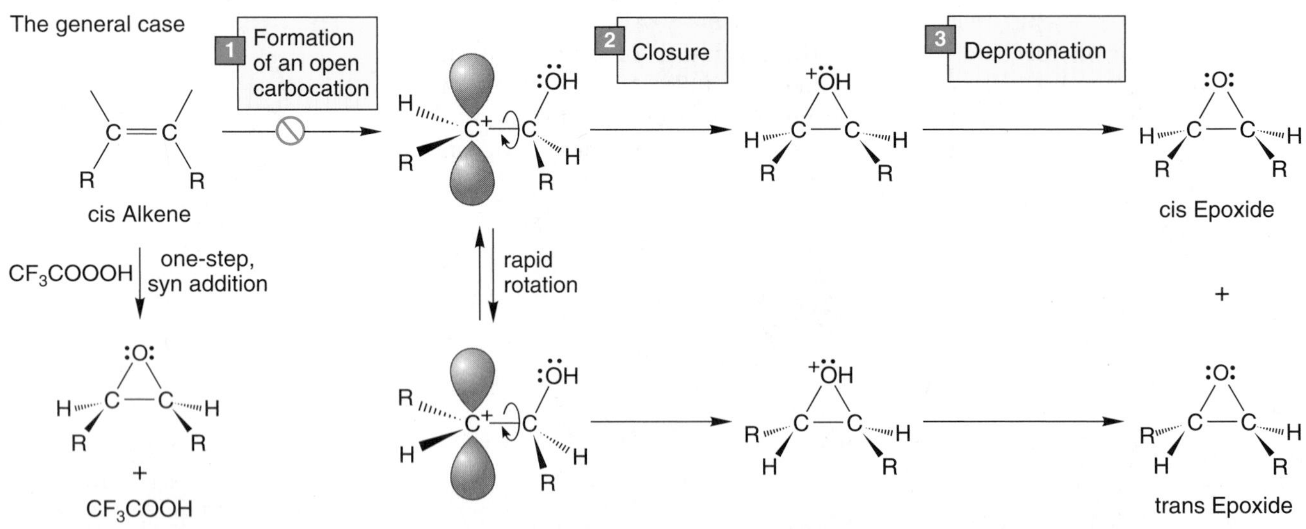

FIGURE 10.34 A test of the mechanism of epoxidation. A stepwise mechanism predicts scrambling of stereochemistry, whereas a concerted, one-step, mechanism must retain the stereochemistry of the starting alkene. As cis alkenes give only cis epoxides, the reaction must be concerted.

Specific examples

FIGURE **10.34** (CONTINUED)

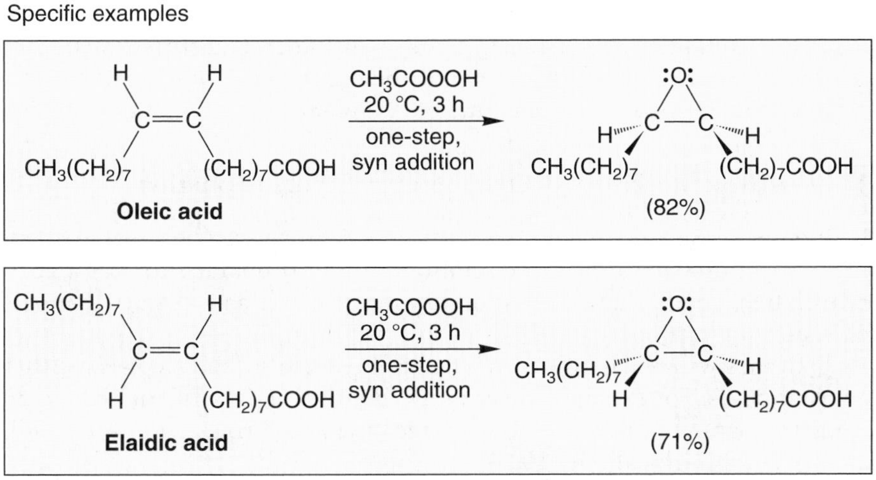

10.3b Asymmetric (Sharpless) Epoxidation

In recent years, an enormously useful synthetic technique for producing enantiomeric oxiranes (epoxides) has appeared. Discovered by Barry Sharpless (b. 1941), now at the Scripp's Research Institute, it uses a witches' brew of titanium isopropoxide, *tert*-butyl peroxide, and one enantiomer of a tartaric ester* to react with an allylic alcohol ($R{-}CH{=}CH{-}CH_2OH$). Part of the great utility of this procedure comes from the observation that the two enantiomers of tartaric ester lead to two different stereochemistries of product oxirane (Fig. 10.35).

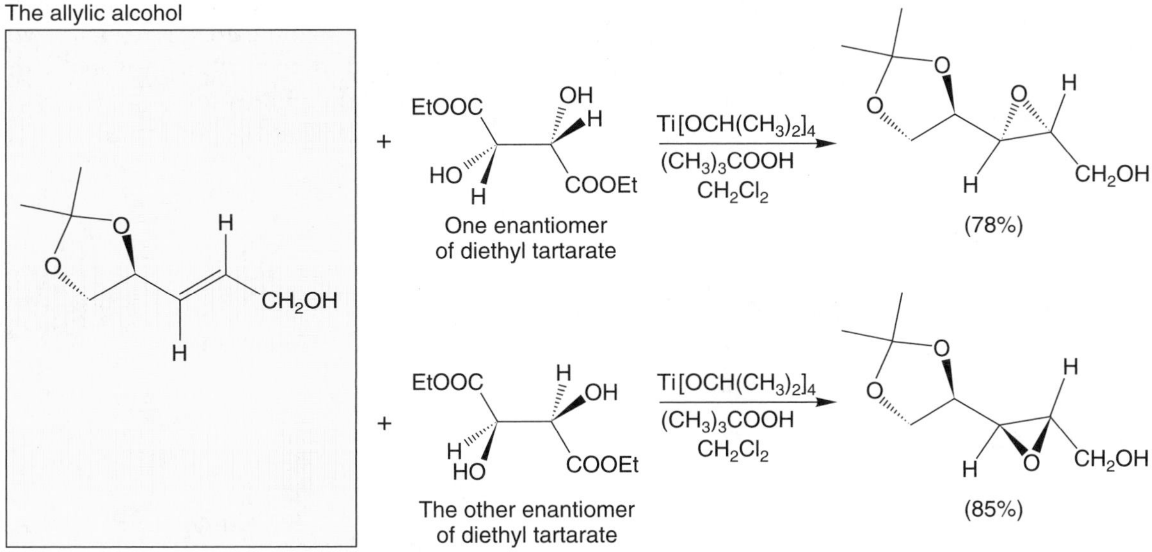

FIGURE **10.35** Assymetric (Sharpless) oxidation of alkenes ($Et = CH_3CH_2$).

The mechanism of this complex reaction involves the titanium compound acting as a clamp, holding the alkene, peroxide, and tartaric ester together. Because the tartaric ester is asymmetric, the clamped combination

*An ester ($R{-}COOR'$) is a derivative of a carboxylic acid ($R{-}COOH$).

of molecules is also asymmetric. In one cluster, the oxirane oxygen is delivered from one side, whereas in the enantiomeric cluster formed from the other tartaric ester, it comes from the other side. These epoxides can then be transformed into all manner of compounds, as we will now see.

10.3c Further Reactions of Oxiranes: 1,2-Glycol Formation

Unlike the closely related bromonium ions, oxiranes can be isolated under many reaction conditions. The bromonium ion is doomed to bear a positive charge, and is therefore a more powerful Lewis acid than the neutral oxirane.

Oxiranes will react, however, when treated in a second step with either acids or bases. For example, reaction with either H_3O^+/H_2O or HO^-/H_2O leads to opening of the three-membered ring and formation of a 1,2-diol, known as a **1,2-glycol** (Fig. 10.36).

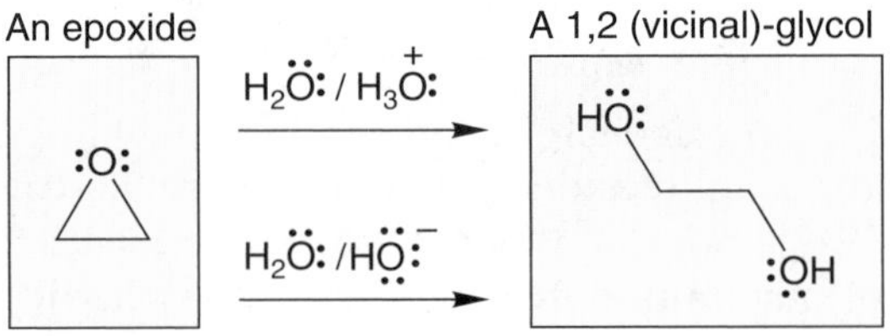

FIGURE 10.36 Oxiranes will open in either acid or base to give 1,2-diols, called 1,2-glycols.

The mechanisms of these ring openings are straightforward extensions of reactions you already know. In base, the strongly nucleophilic hydroxide ion attacks the oxirane and displaces the oxygen atom from one carbon in an S_N2 reaction (Fig. 10.37). In acid, there is no strong nucleophile, but protonation of the oxygen atom of the oxirane creates a good Lewis acid. Water is a strong enough nucleophile to open the protonated oxirane to give, after deprotonation, the 1,2-diol (Fig. 10.37).

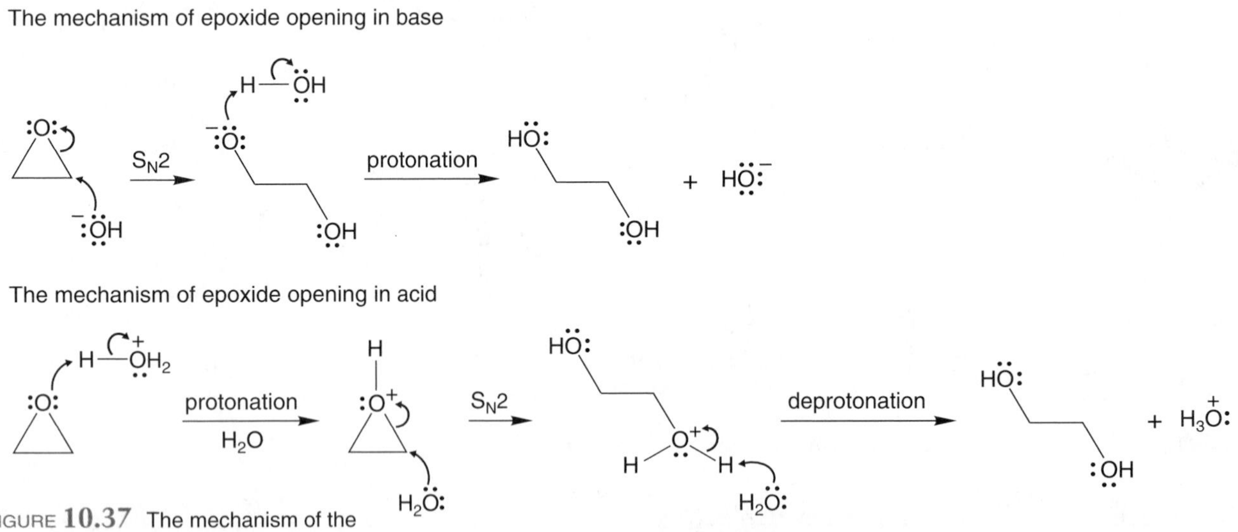

FIGURE 10.37 The mechanism of the reaction in base is a simple S_N2 opening of the ring. Hydroxide is the nucleophile and R–O⁻ the leaving group. In acid, the oxirane is first protonated, then opened in S_N2 fashion. The nucleophile is water and the leaving group is R–OH.

There is a strong parallel here to reactions we have already seen. Hydroxide is a poor leaving group. Treatment with acid or formation of sulfonate esters converts hydroxide into water or another good leaving group, and a number of reactions require this transformation (see Chapter 7, p. 260 for examples). The opening of the oxirane in acid involves exactly the same kind of transformation; protonation facilitates the opening by making the leaving group ROH not RO^- (Fig. 10.38).

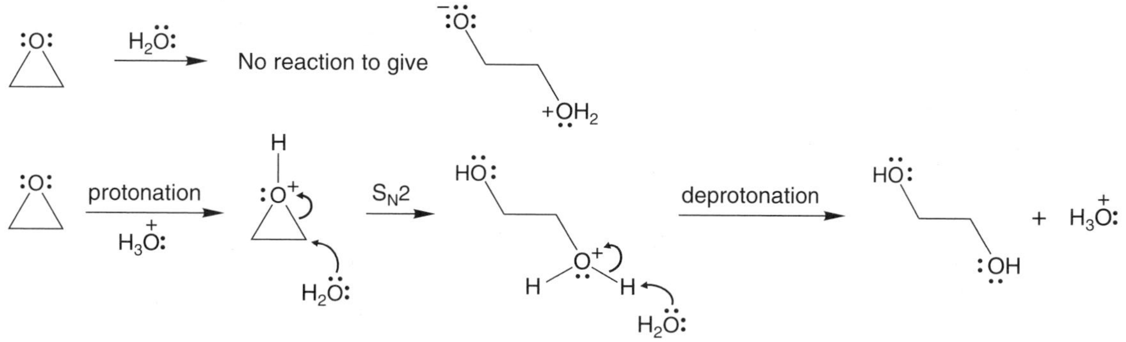

FIGURE 10.38 Water is not a strong enough nucleophile to open the unprotonated oxirane, as R—O⁻ is too poor a leaving group. Acid catalysis changes the leaving group to R—OH, which is much easier to displace.

*PROBLEM 10.11

If hydroxide is such a poor leaving group, one would expect that RO^- would also be a bad leaving group, and it is. In fact, simple ethers (R—O—R) are not generally cleaved in strong base. Why, then, do oxiranes open so easily in base (Fig. 10.36)? What is special about the three-membered ring that facilitates the displacement?

ANSWER

The difference appears because of the strain energy of the three-membered ring. The energy of the starting material is raised, and therefore the activation energy for reaction declines. Cyclopropane is strained by some 27 kcal/mol, and an oxirane cannot be very different. Both of these species will be more reactive than their unstrained counterparts, and ring openings of epoxides by nucleophiles are common.

A very rare reaction; alkoxide is a poor leaving group

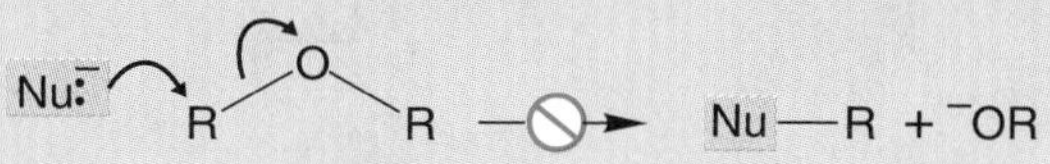

The strain in the three-membered ring raises the energy of the starting material and makes this ring opening possible

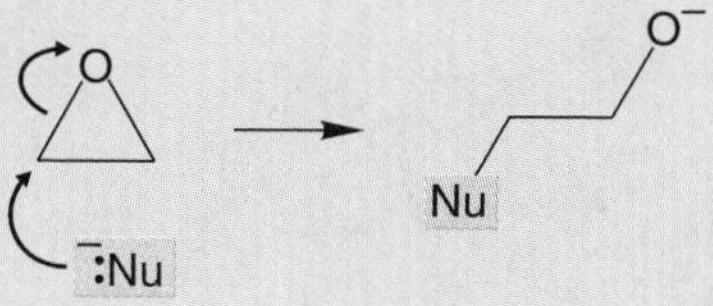

Unsymmetrical oxiranes can open in two ways and the regiochemical result is generally different in acid and base (Fig. 10.39).

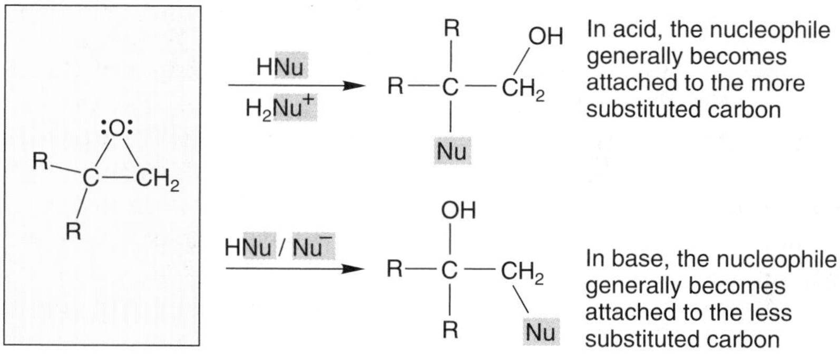

FIGURE 10.39 An unsymmetrical oxirane generally opens with different regiochemistry in acid and base.

How does the difference in regiochemistry arise? Why should the two openings be different? In base, the ring opening is an S_N2 displacement by hydroxide on the unprotonated oxirane. As in any S_N2 reaction, steric matters will be important, and opening will occur at the sterically more accessible, less substituted carbon (Fig. 10.40).

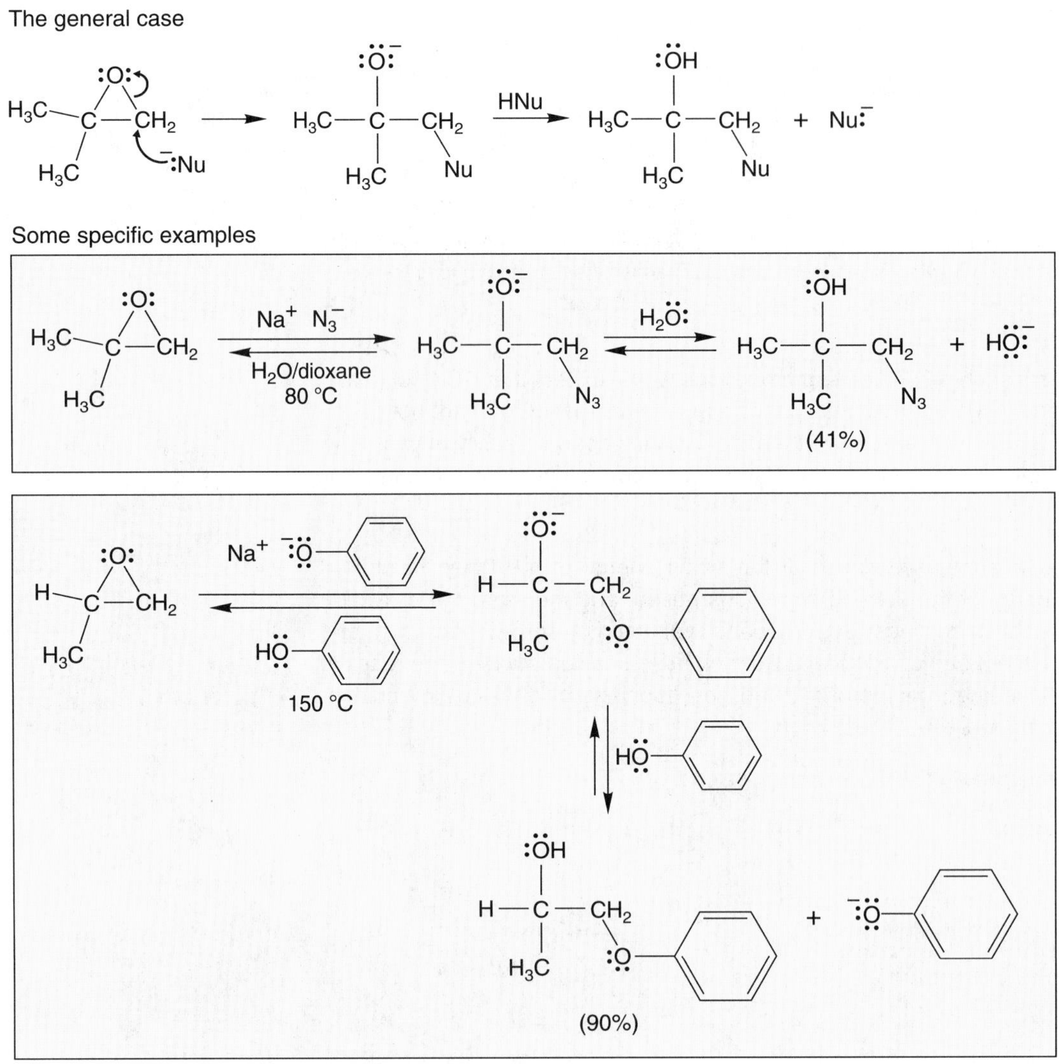
The general case
HNu
Some specific examples
Na+ N3−
H2O/dioxane
80 °C
H2O:
(41%)
Na+
150 °C
(90%)

FIGURE 10.40 In base, the nucleophile will attack at the sterically less encumbered carbon.

In acid, the first step is the protonation of the oxygen of the three-membered ring. Recall the discussion of the regiochemistry of the opening of unsymmetrical bromonium ions (p. 400). The positive charge is borne not only by the oxygen but also by the two carbons of the oxirane. Most of the positive charge on carbon will reside at the more substituted position. The more substituted carbon–oxygen bond will be longer and weaker than the other, less substituted carbon–oxygen bond (Fig. 10.41). As in unsymmetrical bromonium ions, the transition state for addition at the more substituted position will be lower in energy than that for addition at the less substituted position.

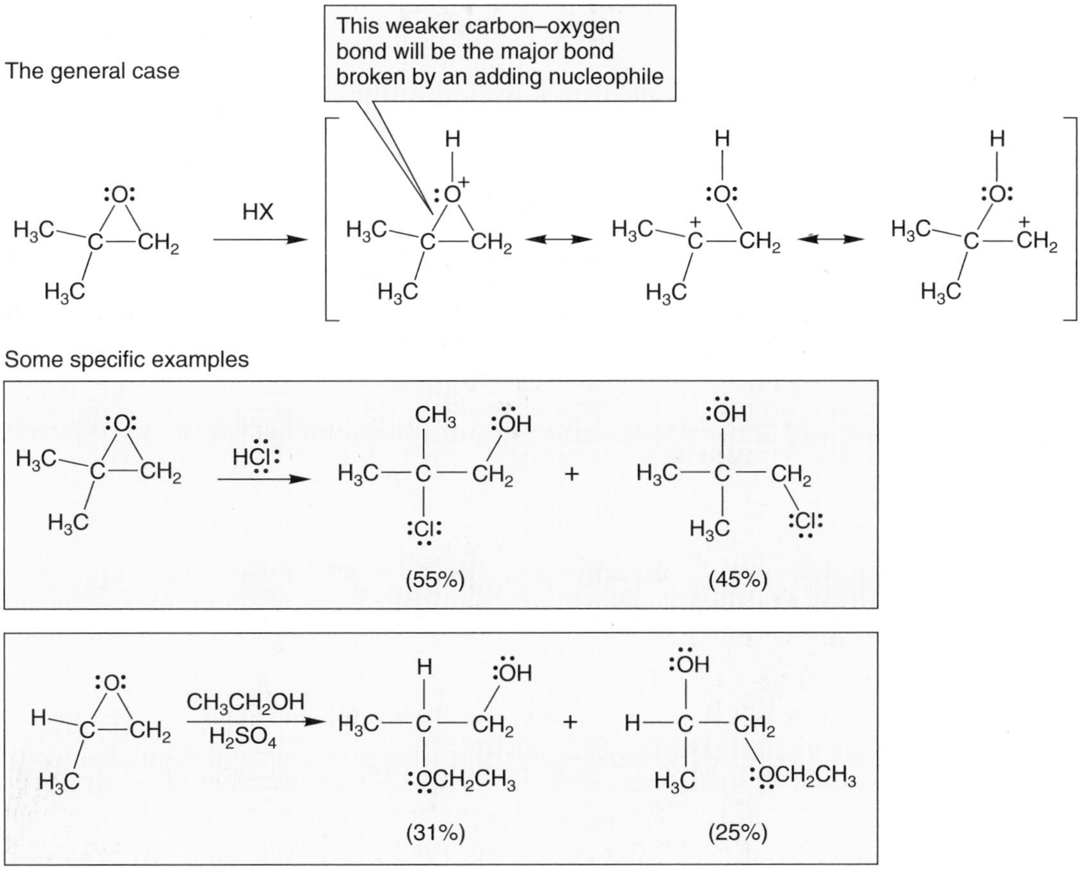

The general case

This weaker carbon–oxygen bond will be the major bond broken by an adding nucleophile

Some specific examples

(55%) (45%)

(31%) (25%)

FIGURE **10.41** Both carbons help bear the positive charge in the protonated oxirane. Most of the positive charge on carbon will be on the more substituted position, and that is where the nucleophile will add. This reaction is much like the opening of an unsymmetrical bromonium ion by water or alcohol.

So, the apparently mysterious change in regiochemistry of the opening in acid and base can be reconciled easily. The trick is to learn to see new reactions in terms of what you already know. Books will always be organized in this way. Questions are set up by recently discussed phenomena. Nature is not always so cooperative and organized, however.

10.3d Carbenes

Additions to alkenes lead to cyclopropanes through intermediates called **carbenes**. Carbenes are neutral compounds containing divalent carbon, and are often formed from nitrogen-containing molecules called diazo compounds. Diazo compounds have the attractive feature (aside from being sources of carbenes) of often being quite handsomely colored. Diazomethane (CH_2N_2) is a bright yellow gas, diazocyclopentadiene an iridescent orange liquid, and 4,4-dimethyldiazocyclohexa-2,5-diene an almost blue liquid (Fig. 10.42). However, diazo compounds have a downside: They are certainly poisonous, demonstrably explosive, and probably

$$H_2\bar{\ddot{C}}—\overset{+}{N}≡N: \longleftrightarrow H_2C=\overset{+}{N}=\ddot{N}:^- \longleftrightarrow H_2\overset{+}{C}—\ddot{N}=\ddot{N}:^-$$

Diazomethane (a yellow gas)

N_2

Diazocyclopentadiene (an orange liquid)

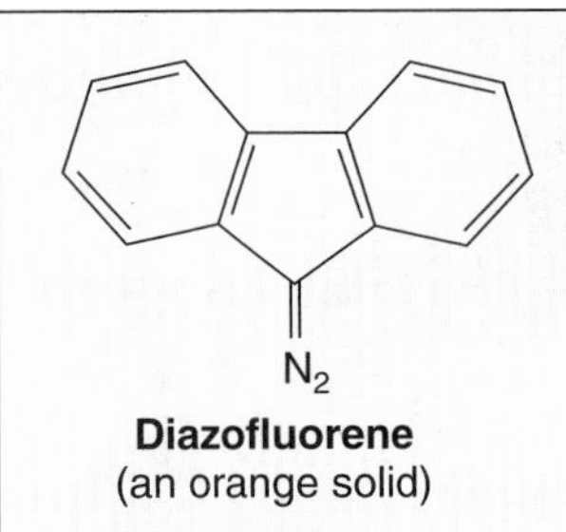

Diazofluorene (an orange solid)

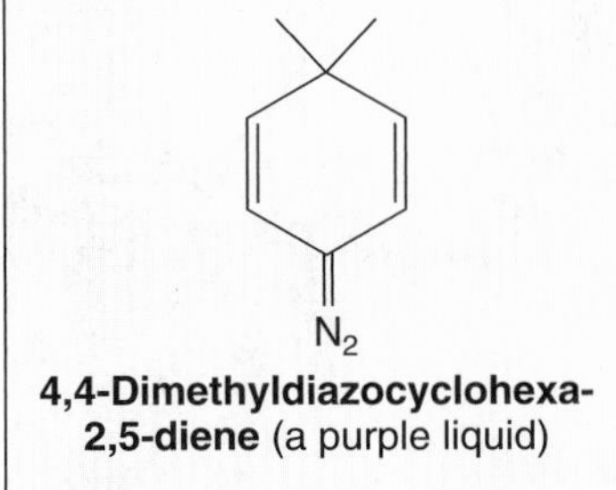

4,4-Dimethyldiazocyclohexa-2,5-diene (a purple liquid)

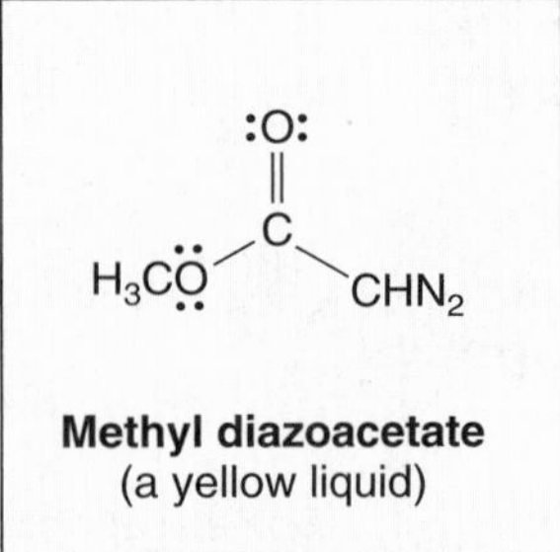

Methyl diazoacetate (a yellow liquid)

FIGURE 10.42 Some diazo compounds. Resonance forms are drawn only for diazomethane.

PROBLEM 10.12 Write resonance forms for 4,4-dimethyldiazocyclohexa-2,5-diene in Figure 10.42.

carcinogenic. So one must be careful (and brave, and perhaps a little crazy) to do carbene chemistry!

But this care is worth it because carbenes are fascinating species. Heat (Δ) or light ($h\nu$) liberates them from their parent diazo compounds through loss of nitrogen, and they react rapidly with alkenes, generally used as solvent for the reaction, to give cyclopropanes. Another common source of carbenes is the hydrolysis of halocarbons. Treatment of chloroform ($CHCl_3$) or bromoform ($CHBr_3$) with strong bases, such as hydroxide or potassium *tert*-butoxide gives dichlorocarbene or dibromocarbene (Fig. 10.43).

$$H_2\bar{\ddot{C}}—\overset{+}{N}≡N: \xrightarrow[\text{(or } \Delta)]{h\nu} :CH_2 + N_2 \xrightarrow{\text{cyclohexene}} \text{bicyclo product } CH_2$$

Diazomethane

$$CHCl_3 \xrightarrow{K^+ \ ^-:\ddot{O}C(CH_3)_3} :CCl_2 \xrightarrow{\text{cyclohexene}} \text{bicyclo product } CCl_2$$

Chloroform

$$CHBr_3 \xrightarrow{K^+ \ ^-:\ddot{O}C(CH_3)_3} :CBr_2 \xrightarrow{\text{cyclohexene}} \text{bicyclo product } CBr_2$$

Bromoform

FIGURE 10.43 Loss of nitrogen from a diazo compound liberates a carbene that can add to alkenes to give a cyclopropane. Dihalocarbenes can be made by treatment of many halogenated alkanes with strong base.

PROBLEM **10.13**

Propose a mechanism for the hydrolysis of chloroform ($CHCl_3$, $pK_a = 29$) to dichlorocarbene. *Hint*: When the reaction is carried out in D_2O/DO^-, recovered chloroform is partially deuterated ($CDCl_3$).

The additions of most carbenes occur in stereospecifically syn fashion, which means that cis alkenes react with carbenes to give only cis cyclopropanes and trans alkenes give the corresponding trans cyclopropanes (Fig. 10.44).

The general case

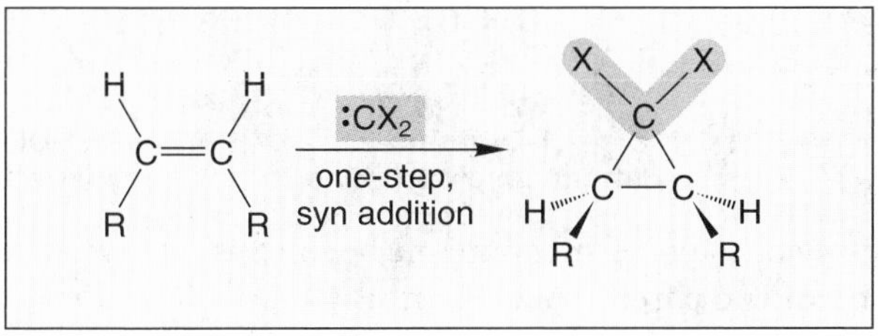

:CX_2 one-step, syn addition

Some specific examples

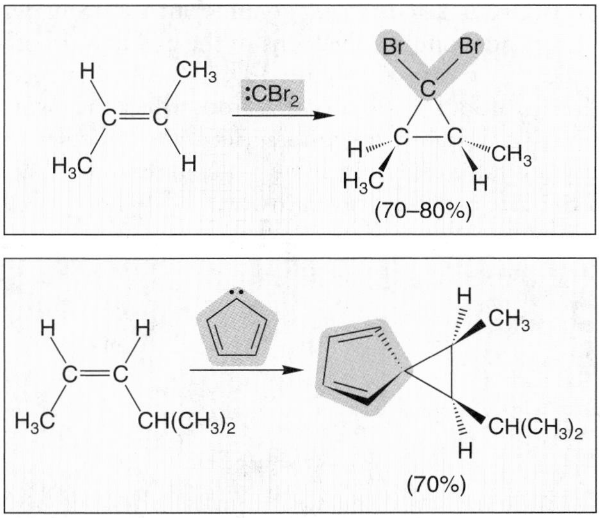

:CBr_2

(70–80%)

(70%)

FIGURE **10.44** Addition of carbenes to alkenes is a syn process. cis Alkenes give cis cyclopropanes and trans alkenes give trans cyclopropanes.

The stereospecificity of the addition reactions of carbenes tells us that there can be no intermediates capable of rotation about carbon–carbon bonds in the ring-forming reaction. The mechanism involves a single step; in the addition both new ring bonds are formed at the same time. A two-step addition would lead to formation of both cis and trans cyclopropanes (Fig. 10.45).

One-step reaction

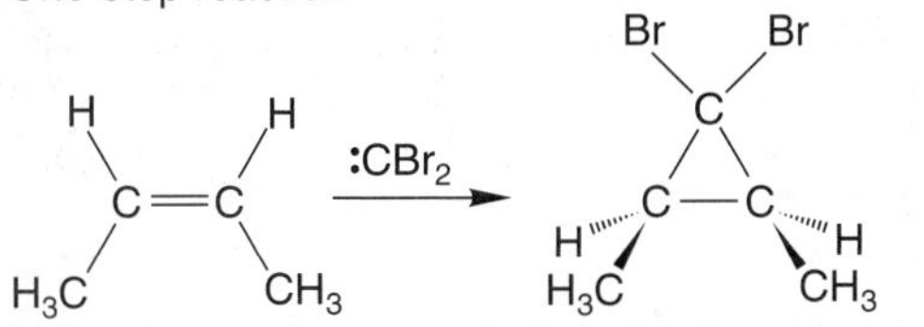

:CBr_2

***cis*-1,1-Dibromo-2,3-dimethylcyclopropane**

FIGURE **10.45** A two-step mechanism for addition of carbenes to alkenes would produce both cis and trans cyclopropanes from a cis alkene.

FIGURE 10.45 (CONTINUED)

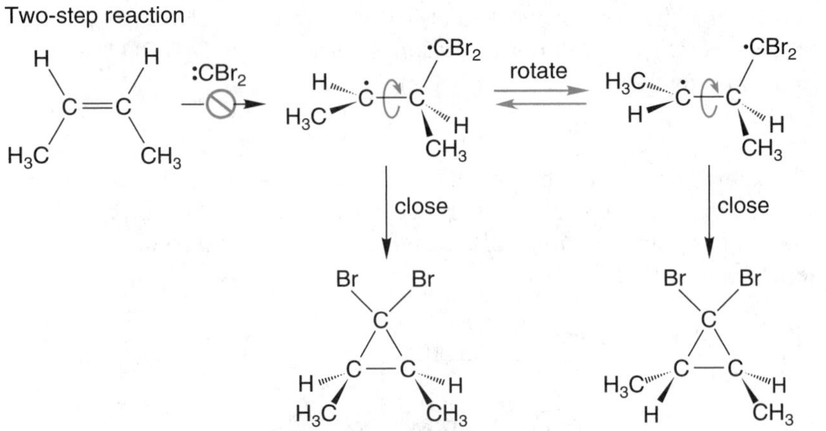

Two-step reaction

close

rotate

close

***cis*-1,1-Dibromo-2,3-dimethylcyclopropane**

***trans*-1,1-Dibromo-2,3-dimethylcyclopropane**

Empty $2p_z$ orbital

103°

Filled hybrid orbital, ~sp^2

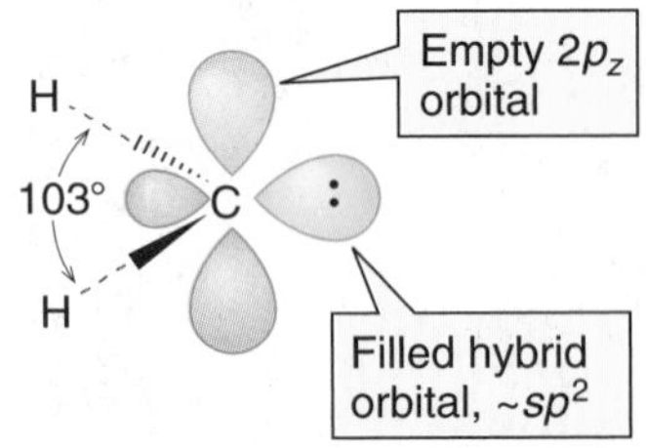

FIGURE 10.46 The structure of methylene.

Our brief discussion of mechanism begins, as always, with structure. The structure of the parent species, methylene (:CH_2), formed by irradiation of diazomethane, is shown in Figure 10.46.

At first glance, the mechanism of addition to alkenes may seem obvious; we have two new bonds to make and four electrons with which to make them—two nonbonding electrons in the carbene sp^2 orbital, and the pair in the alkene's π orbital. Figure 10.47 shows this mechanism. But there is one big difficulty: Figure 10.47 describes the overlap of a pair of filled orbitals, a destabilizing interaction. We must search for a filled orbital–empty orbital interaction if we want to describe this reaction reasonably and to write a low-energy pathway.

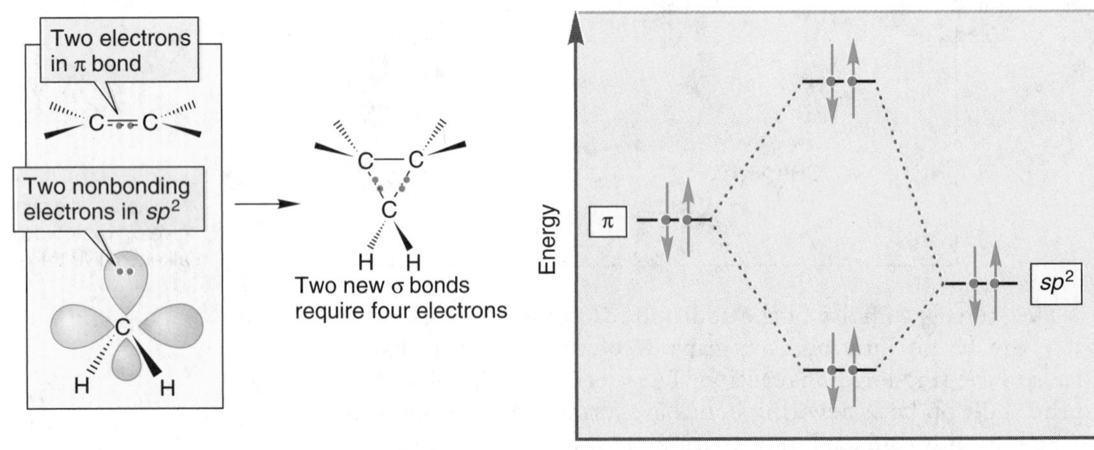

FIGURE 10.47 The simplest "just push 'em together" mechanism for cyclopropane formation. The four electrons in the π orbital of the alkene and the hybrid orbital of methylene are used to make the two new σ bonds in the cyclopropane. However, this requires the destabilizing interaction of these two filled orbitals.

The symmetrical approach of Figure 10.47 provides us with two possibilities: HOMO (carbene)–LUMO (alkene) or HOMO (alkene)–LUMO (carbene). As Figure 10.48 shows, neither of these interactions works—both are "net zero," orthogonal interactions in which the bonding overlap is exactly canceled by an antibonding overlap (Fig. 10.48).

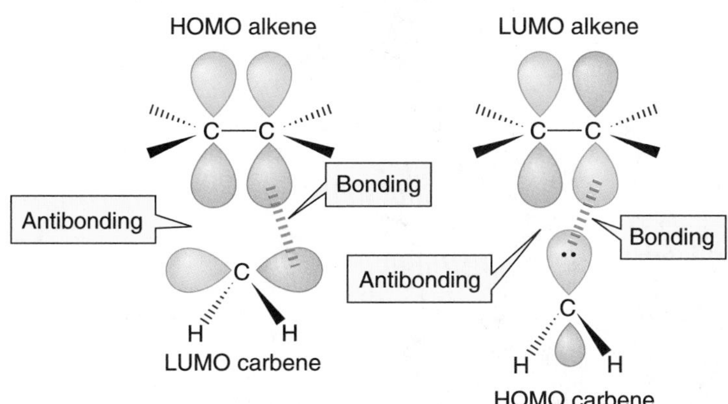

FIGURE 10.48 Neither HOMO–LUMO interaction is stabilizing in the symmetrical approach of methylene to the alkene. In each case, the bonding interactions are exactly canceled by the antibonding interactions.

There is, in fact, nothing favorable about the seemingly obvious symmetrical approach. That much we can see for ourselves—finding the real reaction path is harder. We do know that overlap between a filled and empty orbital will be important in any low-energy pathway. The best current calculations show an approach in which the carbene's empty *p* orbital (LUMO) first overlaps with the alkene's filled π orbital (HOMO). As the reaction continues, the carbene tips over to achieve the geometry of the final cyclopropane, and the filled sp^2 HOMO of the carbene overlaps with π^* (LUMO) of the alkene (Fig. 10.49). Neither theory nor experiment gives any indication of an intermediate. The reaction apparently is concerted, but unsymmetrical.

Pyrethrins

Pyrethrin I

Cyclopropanes are occasionally found in Nature, and one class of naturally occurring cyclopropanes, the pyrethroids, are useful insecticides. These compounds are highly toxic to insects but not to mammals, and are rapidly biodegradable. These cyclopropanes do not persist in the environment. The naturally occurring pyrethroids are found in the members of the chrysanthemum family and are formally derived from chrysanthemic acid. Many modified pryrethrins not found in Nature have been made in the laboratory and several are widely used. The molecule shown is pyrethrin I, or, more systematically, 2,2-dimethyl-3-(2-methyl-1-propenyl)cyclopropanecarboxylic acid 2-methyl-4-oxo-3-(2,4-dipentadienyl)-2-cyclopenten-1-yl ester. See why the shorthand is used?

Regardless of the mechanism, the reactions of carbenes with alkenes provide the most widely used synthetic route to cyclopropanes.

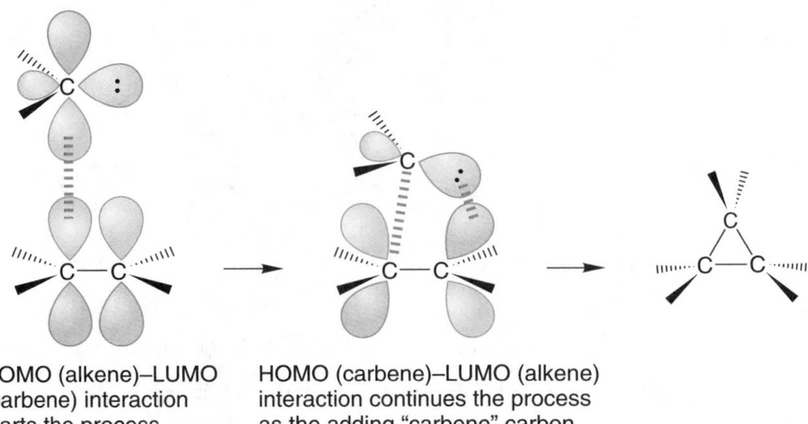

HOMO (alkene)–LUMO (carbene) interaction starts the process
HOMO (carbene)–LUMO (alkene) interaction continues the process as the adding "carbene" carbon tips over to its final position

FIGURE 10.49 This figure shows the calculated unsymmetrical approach of the carbene to an alkene. In the first stage, the LUMO of the carbene overlaps with π, the HOMO of the alkene. This stage is followed by one in which the carbene tips so that the HOMO of the carbene can overlap with the LUMO (π^*) of the alkene.

10.4 DIPOLAR ADDITION REACTIONS

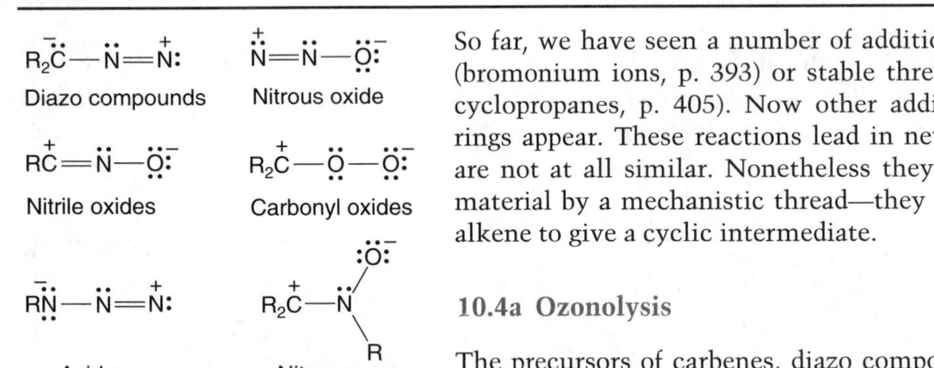

FIGURE 10.50 Some 1,3-dipolar reagents. Good, neutral resonance forms cannot be written. For each molecule, only the resonance form emphasizing the 1,3-dipole is shown (see Problem 10.14).

So far, we have seen a number of addition reactions leading to transient (bromonium ions, p. 393) or stable three-membered rings (oxiranes and cyclopropanes, p. 405). Now other additions leading to different sized rings appear. These reactions lead in new directions, and their products are not at all similar. Nonetheless they are tied together and to earlier material by a mechanistic thread—they involve an initial addition to an alkene to give a cyclic intermediate.

10.4a Ozonolysis

The precursors of carbenes, diazo compounds, are members of a class of molecules called **1,3-dipoles**, or **1,3-dipolar reagents**. They are so-called because good neutral resonance forms cannot be written for them. Figure 10.50 shows a number of typical 1,3-dipoles.

PROBLEM 10.14 Draw more resonance forms for the 1,3-dipoles in Figure 10.50.

An important characteristic of 1,3-dipoles is that they add to alkenes and alkynes to give five-membered rings. Figure 10.51 shows the addition of a substituted diazomethane to an alkyne.

A pyrazole

FIGURE 10.51 Addition of a substituted diazomethane to an alkyne makes a five-membered ring called a pyrazole.

PROBLEM 10.15

Use the arrow formalism to show the compounds formed from reaction of the azides, nitrile oxides, and carbonyl oxides shown in Figure 10.50 with a typical alkene and alkyne.

Ozone (O_3) is a particularly important 1,3-dipole (Fig. 10.52). The ozone molecule is often in the news these days. There is too little of it in the upper atmosphere, where it functions to screen the surface of the earth from much dangerous ultraviolet (UV) radiation. Yet, in all-too-many places near the surface, there is too much of it. Ozone is a principal component of photochemical smog.

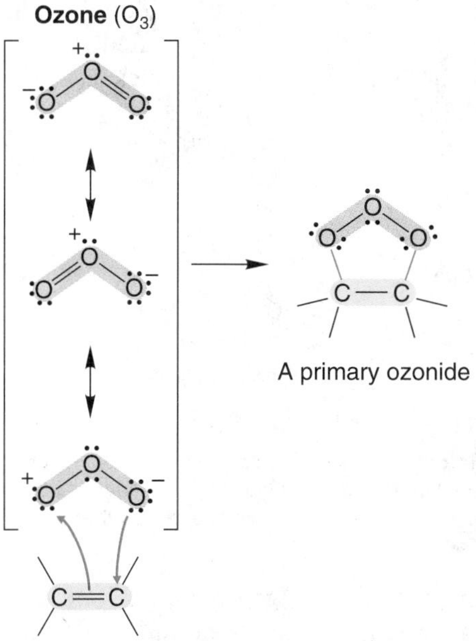

Ozone (O_3)

A primary ozonide

FIGURE 10.52 A 1,3-dipolar addition of ozone to an alkene gives a primary ozonide, a five-membered ring containing three oxygen atoms in a row.

The first step in the process called **ozonolysis** involves the reaction of ozone with most alkenes to give the ozonide shown in Figure 10.52. The arrow formalism lets us map out the formation of the ozonide. The π bond of the alkene reacts with ozone to produce a five-membered ring containing three oxygens, the **primary ozonide** or **molozonide**. But these primary ozonides are extremely difficult to handle, and it took great experimental skill on the part of Rudolf Criegee (1902–1975) and his co-workers at the University of Karlsruhe in Germany to isolate the ozonide from addition to *trans*-di-*tert*-butylethylene (Fig. 10.53).

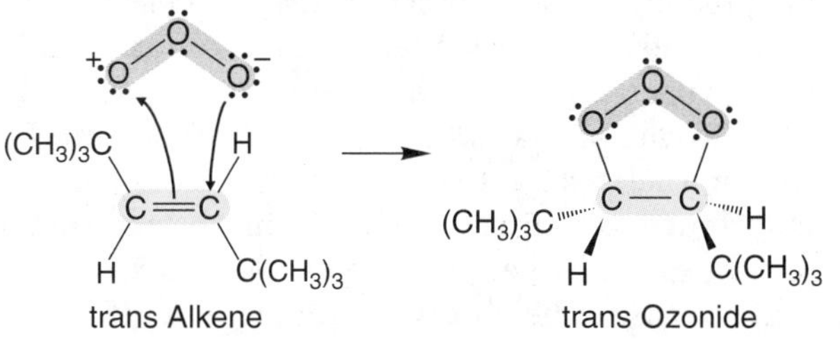

$(CH_3)_3C$ H H $C(CH_3)_3$

trans Alkene

$(CH_3)_3C$ H H $C(CH_3)_3$

trans Ozonide

FIGURE 10.53 A concerted mechanism predicts that the stereochemical relationship of the alkyl groups in the original alkene will be preserved in the ozonide, and this is what happens.

As implied previously, unless care is taken these initially formed ozonides are not stable, but rearrange to other compounds, known simply as ozonides (Fig. 10.54).

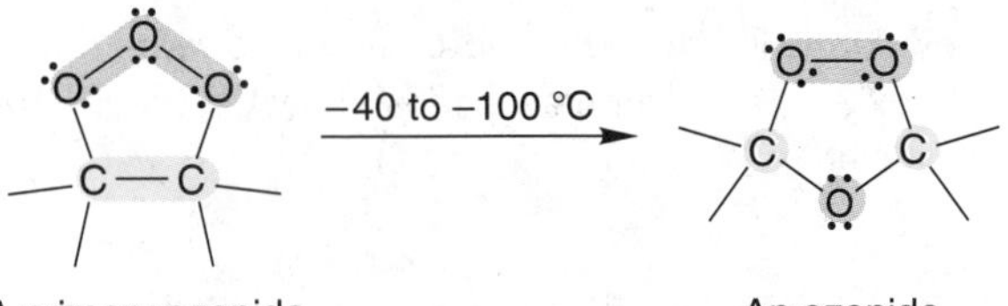

FIGURE 10.54 Unless great care is taken, primary ozonides rearrange to more stable ozonides.

The rearrangement of primary ozonides into these other ozonides involves a pair of reverse and forward 1,3-dipolar additions. The first step in the rearrangement is the opening of the primary ozonide to give a molecule containing a carbon–oxygen double bond called a **carbonyl compound**, along with a molecule called a carbonyl oxide, a new 1,3-dipole (Fig. 10.50). These two species can recombine to give the ozonide (Fig. 10.55).

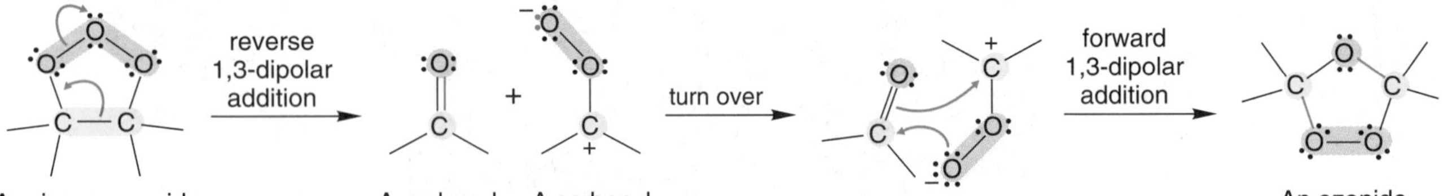

FIGURE 10.55 The mechanism of rearrangement involves two steps. In the first, a reverse 1,3-dipolar addition generates a carbonyl compound and a 1,3-dipole, a carbonyl oxide. The carbonyl oxide readds to the alkene in the other sense to give the new ozonide.

The mechanism of rearrangement from one ozonide to another may look a bit bizarre, and there can be no hiding the fact that it's not simple. It is not really far from things we already know, however. More frustrating is our distance from being able to predict the course of complicated reactions. At this point, you have little choice but to learn the course of the ozonolysis reaction, and that is not intellectually satisfying. It would be a far better situation if you were able to predict what must happen in this multistep, multi-intermediate process. Perhaps it is some comfort to know that you are not at all far behind the current frontier. Organic chemists have only recently emerged from the "gee whiz" phase to the point where advances in experimental skill (mostly in analytical techniques, if one is to be honest) and theoretical methods have begun to allow reliable predictions to be made in risky cases. All too often we are reduced to making explanations after the fact; to rationalizing, as we are just about to do with the rearrangement from one ozonide to the other.

We will take it one step at a time, working backward. First of all, the final addition of the carbonyl oxide, another 1,3-dipole, to a carbon–oxygen double bond is very little removed from the addition of ozone to an alkene. Like ozone, the carbonyl oxide can add to π bonds. In this case, such an addition leads to the observed ozonide (Fig. 10.56).

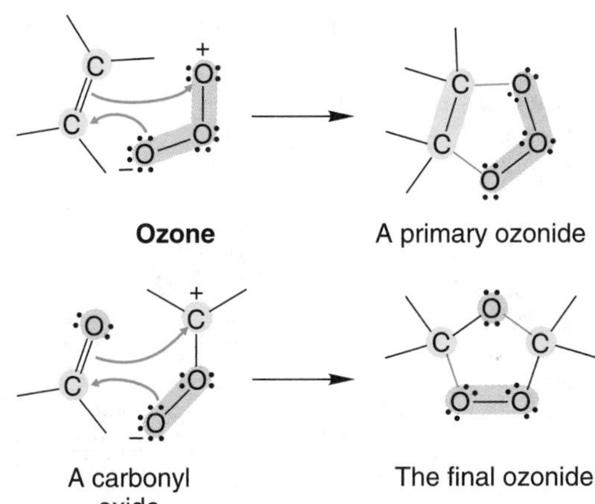

FIGURE 10.56 Like diazo compounds and ozone, a carbonyl oxide is a 1,3-dipole and adds to π systems to give five-membered ring compounds.

The first step in the conversion of one ozonide into the other is another 1,3-dipolar addition, but this time in reverse, and it is therefore a bit harder to see (Fig. 10.57). The reaction is completed when the carbonyl oxide turns over and readds to the carbonyl compound (Fig. 10.56).

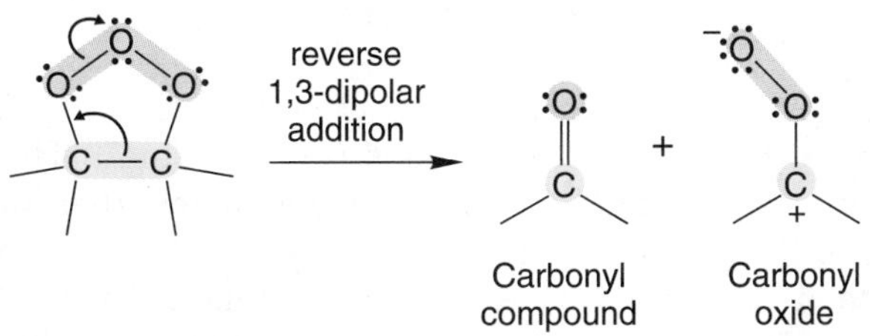

FIGURE 10.57 The first step in the formation of the ozonide from the primary ozonide is a reverse 1,3-dipolar addition to generate a carbonyl compound and a carbonyl oxide.

The overall reaction is a series of three 1,3-dipolar additions: First, ozone adds to the alkene to give the primary ozonide. The primary ozonide undergoes a reverse 1,3-dipolar addition to give a carbonyl compound and a new 1,3-dipole, the carbonyl oxide. Finally, the carbonyl oxide readds to the carbonyl compound in the opposite sense to give the new ozonide (Fig. 10.58).

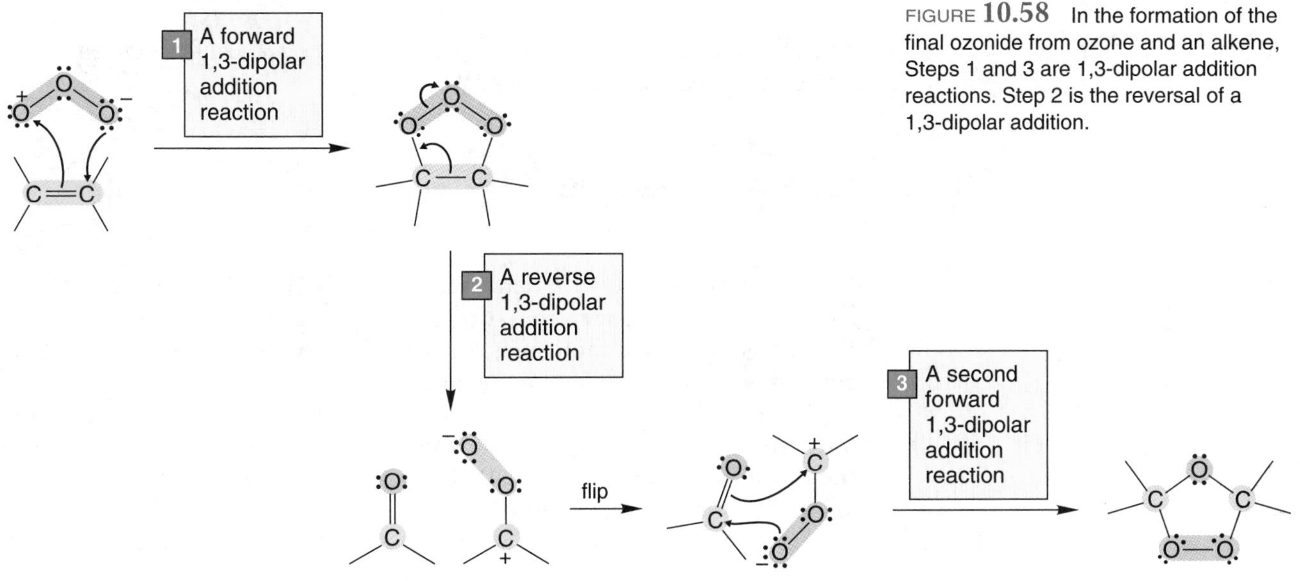

FIGURE 10.58 In the formation of the final ozonide from ozone and an alkene, Steps 1 and 3 are 1,3-dipolar addition reactions. Step 2 is the reversal of a 1,3-dipolar addition.

The reaction is a sequence of three reversible 1,3-dipolar additions driven by thermodynamics toward the relatively stable ozonide. We have now described the reaction; the tougher part is to explain why it runs the way it does. Why does thermodynamics favor the final, more stable ozonide over the primary ozonide (Fig. 10.59)?

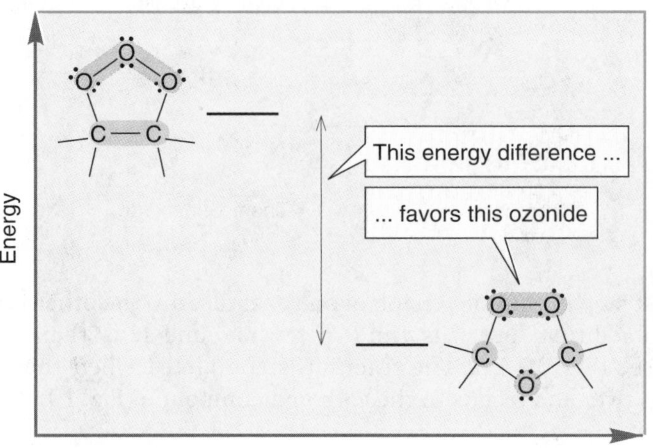

FIGURE **10.59** The ozonide contains fewer weak oxygen–oxygen bonds than the primary ozonide and is more stable.

The oxygen–oxygen bond is quite weak (~40 kcal/mol). Much more stable, though, is the carbon–oxygen bond (~92 kcal/mol). The most important factor in making the final ozonide more stable than the primary ozonide is the difference in the numbers of weak oxygen–oxygen bonds and strong carbon–oxygen bonds. As long as the kinetic barriers to the forward and reverse 1,3-dipolar additions are not too high, thermodynamics will favor formation of the stronger carbon–oxygen bonds and the ultimate result will be the final, more stable, ozonide.

PROBLEM **10.16** Use the table of bond dissociation energies (Table 8.2, p. 307) to do a full calculation of the $\Delta H°$ difference between the primary and final ozonide.

PROBLEM **10.17** Why are oxygen–oxygen bonds so weak? *Hint*: It is not just electron–electron repulsion. Do a quick molecular orbital analysis.

PROBLEM **10.18** Complete the Energy versus Reaction progress diagram of Figure 10.59.

Why do we spend so much time with this complicated reaction? We have already mentioned generality. A vast number of 1,3-dipolar agents—the charge-separated molecules known as **zwitterions**—have been made, and many will add to alkenes and alkynes to give five-membered rings.

Second, these ozonides are very useful compounds. Their further transformations give us entry into new classes of compounds. Depending on the structure of the starting alkene, reactions of the ozonide can yield aldehydes, ketones, or **carboxylic acids**. If the ozonide is reduced, the products are aldehydes or ketones, depending on the structure of the ozonide.

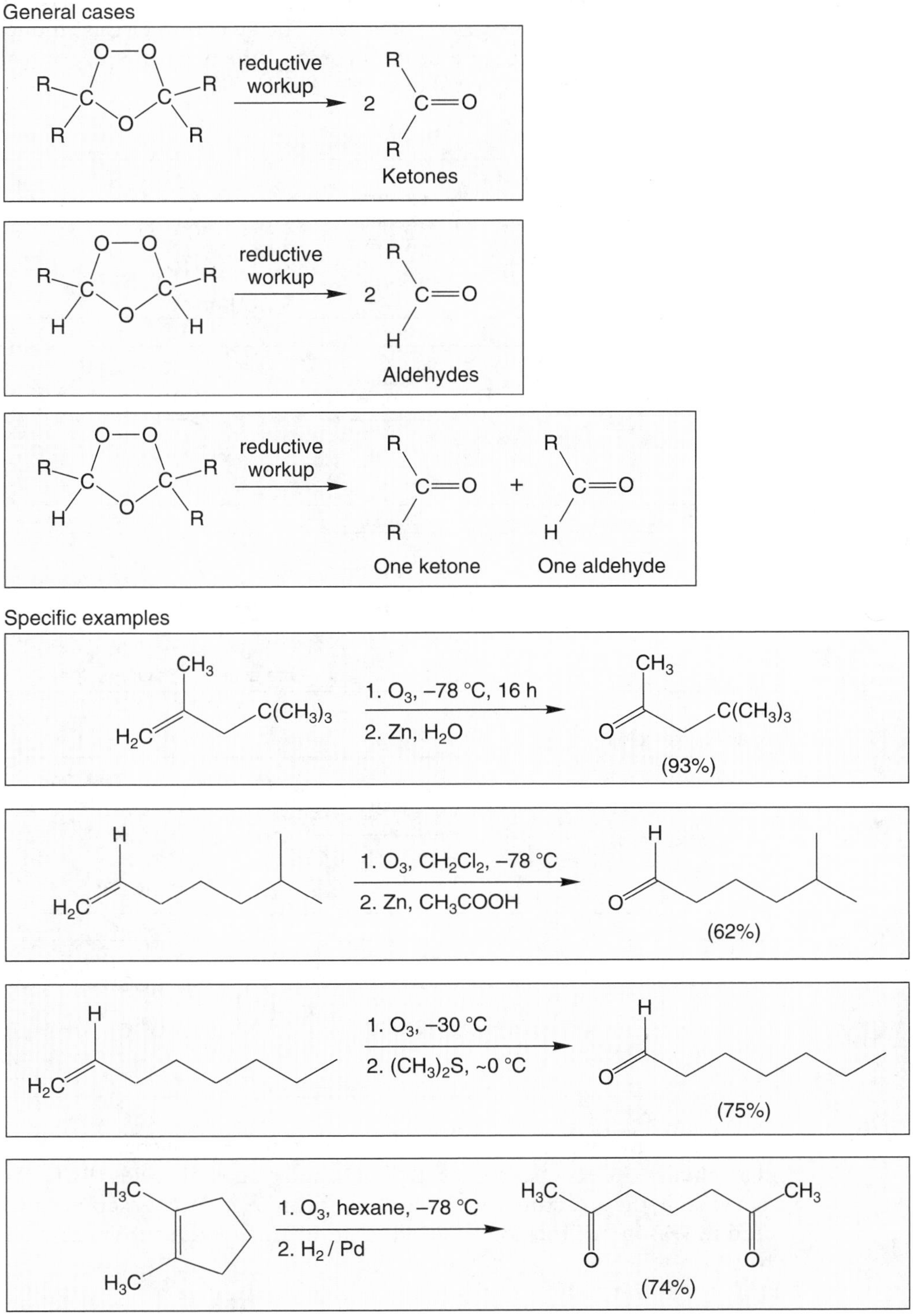

General cases

Specific examples

FIGURE **10.60** Reduction of the final ozonide with dimethyl sulfide or H_2/Pd yields aldehydes or ketones depending on the number of R groups on the ozonide.

Many reducing agents are known, among them H_2/Pd, Zn, and $(CH_3)_2S$ (dimethyl sulfide) (Fig. 10.60).

Oxidation of the ozonide, usually with hydrogen peroxide (H_2O_2), leads either to ketones or carboxylic acids. The structure of the product again depends on the structure of the ozonide, which itself is dependent on the structure of the original alkene (Fig. 10.61).

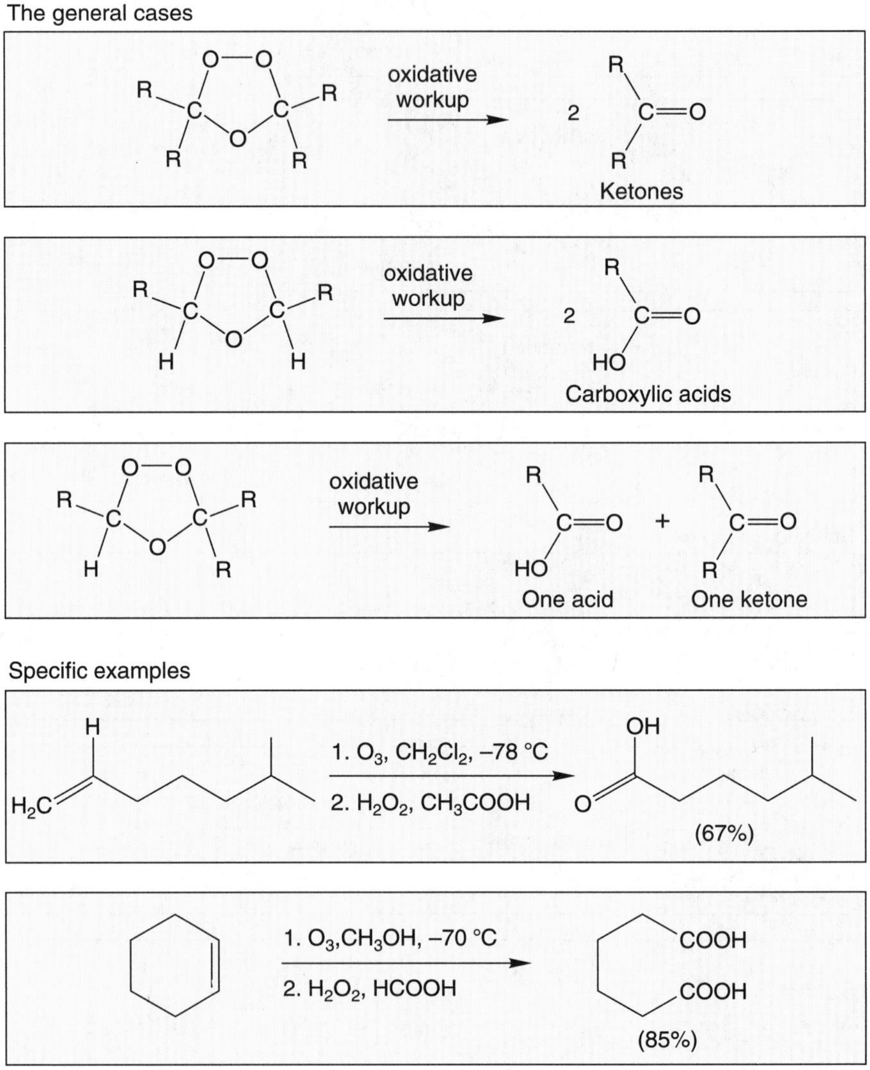

FIGURE 10.61 Oxidation of the final ozonide gives ketones or carboxylic acids.

This reaction provides a synthetic opportunity. We now have syntheses of aldehydes (RCH=O), ketones (R_2C=O), and carboxylic acids (RCOOH). In addition, the structures of the products of an ozonolysis reaction can be used to reason out the composition of an unknown alkene. Try Problem 10.19.

PROBLEM 10.19

Ozonolysis of alkenes can give the products shown below. Supply structures for the starting alkenes. Specify the nature of the workup step in the process (oxidation or reduction?).

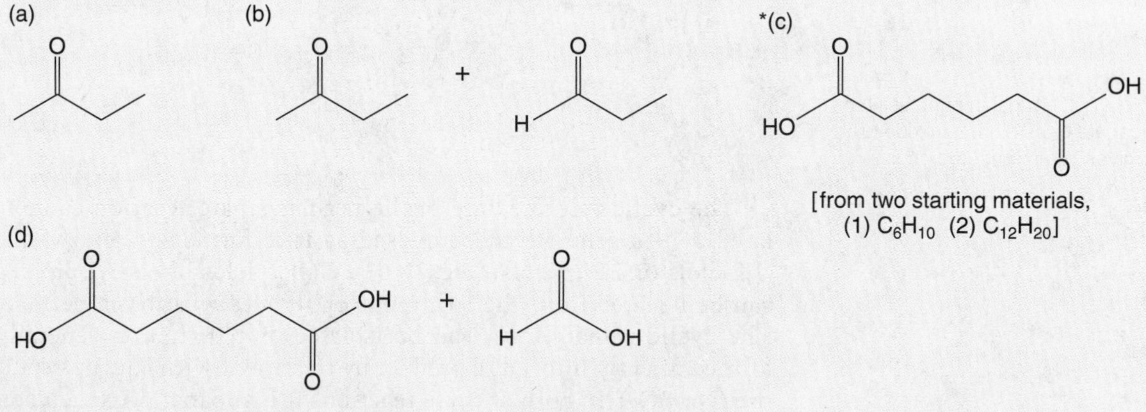

(a) (b) + *(c)

HO OH

[from two starting materials, (1) C_6H_{10} (2) $C_{12}H_{20}$]

(d)

HO OH + H OH

ANSWER (c)

Ozonolysis converts carbon–carbon double bonds into two carbon–oxygen double bonds.

C=C $\xrightarrow{O_3}$ C=O + O=C

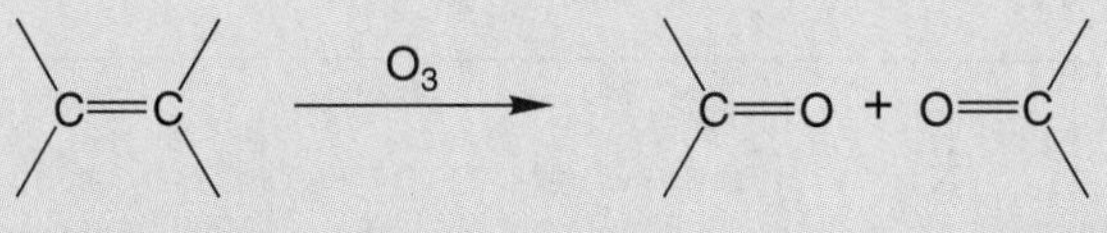

In this case, both new carbon–oxygen double bonds are within the same product molecule, which means that there must be a cyclic starting material. The product contains carboxylic acids, so we know that an oxidative workup is required. A reductive workup would give an aldehyde.

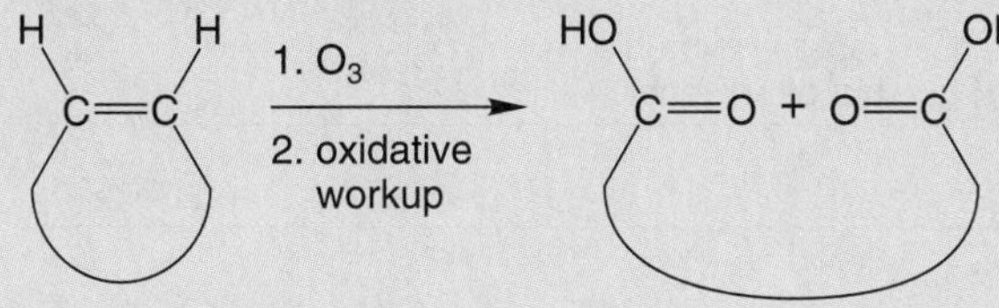

H H C=C — 1. O_3 2. oxidative workup → HO C=O + O=C OH

There are several ways to construct an appropriate starting material. Two are shown below.

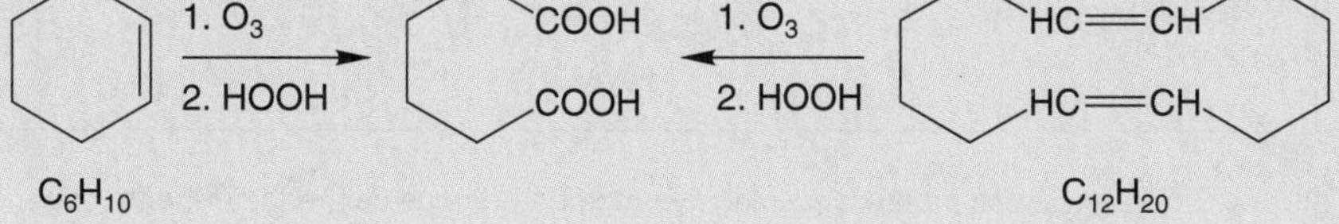

C_6H_{10} — 1. O_3 2. HOOH → COOH COOH ← 1. O_3 2. HOOH — HC=CH HC=CH $C_{12}H_{20}$

10.4b Oxidation with Permanganate or Osmium Tetroxide

Although they are not 1,3-dipolar reagents, both potassium permanganate ($KMnO_4$) and osmium tetroxide (OsO_4) add to alkenes to form five-membered rings in reactions that are related to the additions of ozone and other 1,3-dipoles (Fig. 10.62).

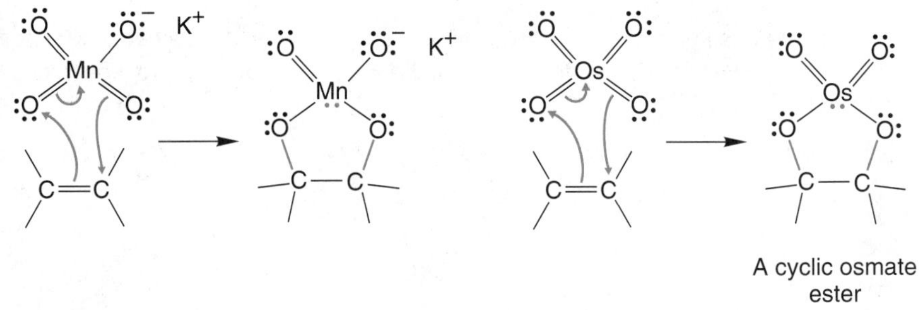

FIGURE 10.62 Addition of potassium permanganate or osmium tetroxide to alkenes gives five-membered rings.

A cyclic osmate ester

The cyclic intermediate in the permanganate reaction cannot be isolated and is generally decomposed as it is formed to give vicinal diols (1,2-diols or 1,2-glycols) (Fig. 10.63). Good yields of cis-1,2-diols (glycols) can be isolated from the treatment of alkenes with basic permanganate. The cyclic osmate ester can be isolated, but it, too, is generally transformed directly into a diol product by treatment with aqueous sodium sulfite (Na_2SO_3). In both of these reactions it is the metal–oxygen bonds, not the carbon–oxygen bonds, that are broken. There are many variations of these reactions, which are among the best ways of synthesizing 1,2-diols.

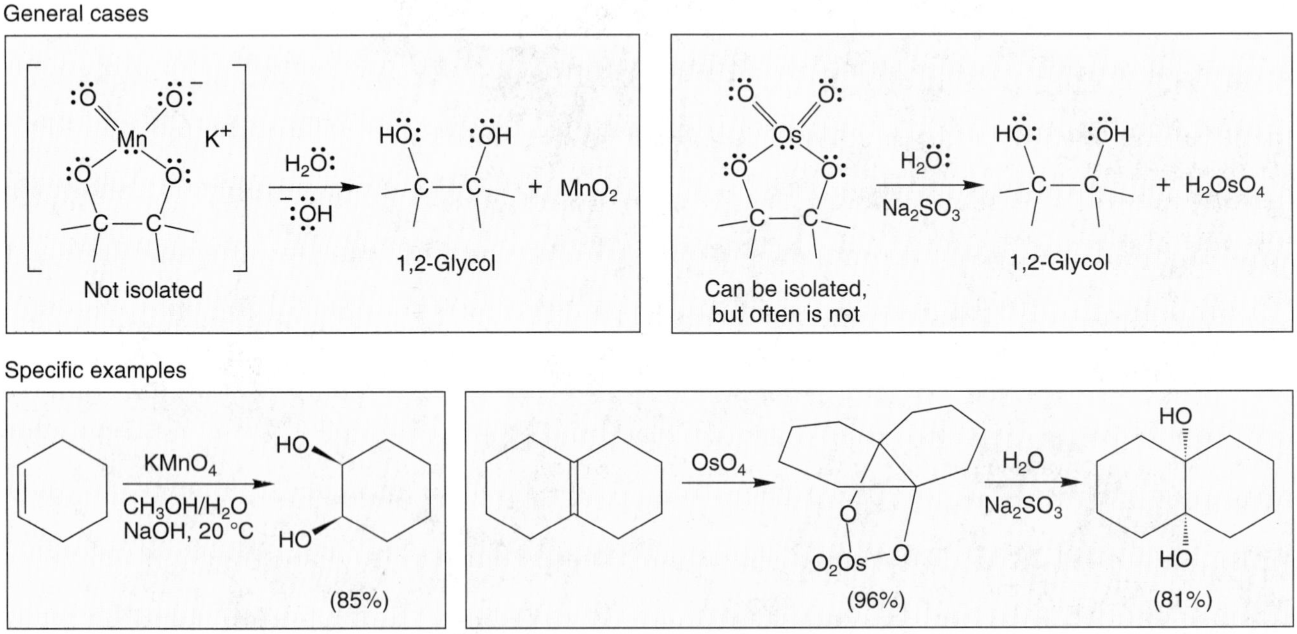

General cases

Not isolated

H_2O, ^-OH

1,2-Glycol + MnO_2

Can be isolated, but often is not

H_2O, Na_2SO_3

1,2-Glycol + H_2OsO_4

Specific examples

$KMnO_4$, CH_3OH/H_2O, NaOH, 20 °C (85%)

OsO_4 (96%) H_2O, Na_2SO_3 (81%)

FIGURE 10.63 These metal-containing five-membered rings can react further to generate 1,2-diols, which are also called 1,2-glycols.

PROBLEM 10.20 What mechanistic conclusions can be drawn from the observation that reaction of OsO_4 with *cis*-2-butene gives the single product shown in Figure 10.64?

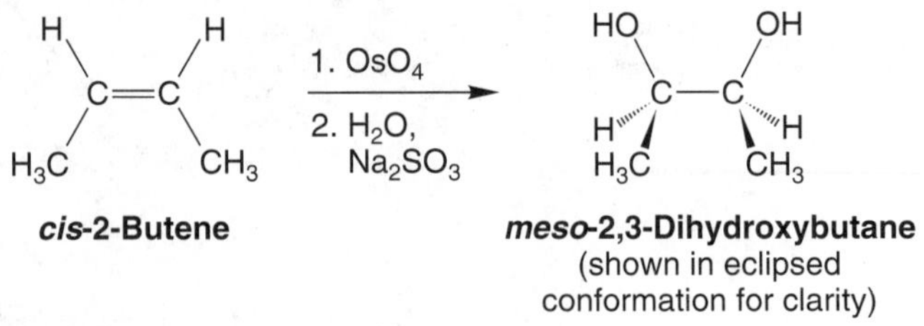

cis-2-Butene 1. OsO_4 2. H_2O, Na_2SO_3 **meso-2,3-Dihydroxybutane** (shown in eclipsed conformation for clarity)

FIGURE 10.64

10.5 ADDITION REACTIONS OF ALKYNES: H–X ADDITION

Much of Chapters 9 and 10 has been devoted to the addition of various reagents to carbon–carbon double bonds. Alkynes contain two π bonds and it would be quite astonishing if they did not undergo similar addition reactions. If we keep in mind what we have learned about alkene additions, it is reasonably easy to work though alkyne additions. The presence of the second double bond will add mechanistic complications, however, and some synthetic opportunities as well.

Addition of H–Br or H–Cl to 3-hexyne gives the corresponding vinyl halides. It is tempting to begin with a protonation of the π system to give a carbocation and a halide ion. In this case, the positive ion would be a vinyl cation. Addition of the halide to the cation would give the vinyl halide. The stereochemistry of this addition is mixed, as would be expected from an open cation, although the trans diethyl compound is favored (Fig. 10.65).

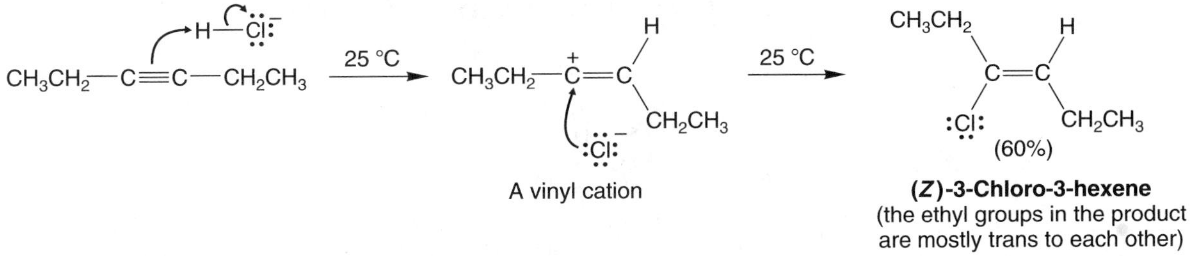

$CH_3CH_2-C\equiv C-CH_2CH_3$ + H–Cl: $\xrightarrow{25\,°C}$ $CH_3CH_2-\overset{+}{C}=C(H)(CH_2CH_3)$ + :Cl:⁻ $\xrightarrow{25\,°C}$ (CH₃CH₂)(Cl)C=C(H)(CH₂CH₃) (60%)

A vinyl cation

(Z)-3-Chloro-3-hexene
(the ethyl groups in the product are mostly trans to each other)

FIGURE 10.65 Hydrogen halides add to alkynes as well as to alkenes.

However, there are problems with this simple extension to alkynes of the mechanism for additions to alkenes. Vinyl cations are unusual species and we must stop for a moment to consider their structures. The positive carbon in a vinyl cation is attached to two groups: the other carbon in the original triple bond and an R group. Accordingly, we would expect *sp* hybridization, and that leads to the structure in Figure 10.66.

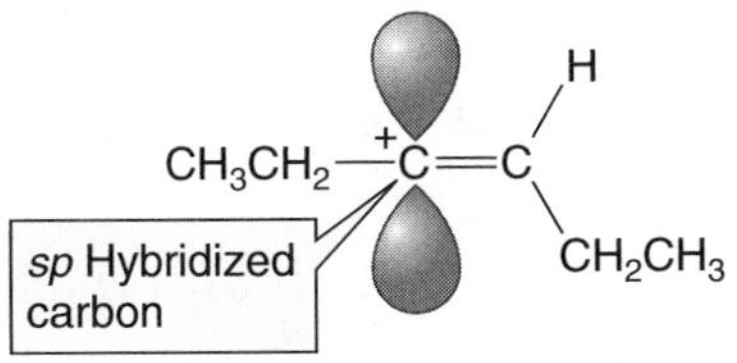

Measurements in the gas phase show that vinyl cations are very unstable. Remember that these gas-phase values do not include the solvent, which will be very important to any cationic species. The absolute values of the energy differences are not important, but the relative stability order of the various carbocations is (Table 10.2).

$CH_3CH_2-\overset{+}{C}=C(H)(CH_2CH_3)$

sp Hybridized carbon

FIGURE 10.66 A vinyl cation contains an *sp* hybridized carbon.

TABLE 10.2 Heats of Formation for Some Carbocations

Cation	Substitution	ΔH_f° (kcal/mol)
$^+CH{=}CH_2$	Primary vinyl	285
$^+CH_3$	Methyl	261.3
$^+CH{=}CH{-}CH_3$	Primary vinyl	252 (calculation)
$CH_3{-}\overset{+}{C}{=}CH_2$	Secondary vinyl	231
$^+CH_2CH_3$	Primary	215.6
$^+CH_2CH_2CH_3$	Primary	211
$^+CH_2CH_2CH_2CH_3$	Primary	203
$H_3C\overset{+}{C}HCH_2CH_3$	Secondary	183
$(CH_3)_2\overset{+}{C}H$	Secondary	190.9
$CH_2{=}CH{-}\overset{+}{C}H_2$	Allyl	226
$(CH_3)_3C^+$	Tertiary	165.8

Table 10.2 shows that, just as with alkyl carbocations, secondary vinyl cations are more stable than the primary species. However, it also shows that even a secondary vinyl cation is only about as stable as a primary alkyl carbocation. Primary carbocations serve as a kind of mechanistic stop sign: They are generally considered too unstable to be viable intermediates in most reactions. Presumably, we should treat unstabilized vinyl cations the same way. What intermediate could we use to replace the highly unstable vinyl cations in additions to acetylenes? Perhaps cyclic intermediates are involved. In Section 9.13 (p. 380), we discussed a cyclic protonium ion formed by protonation of ethylene. Moreover, acetylenes are known to form complexes with hydrohalic acids (HX). A cyclic protonium ion or complex could accommodate the complicated kinetics, which show that more than one molecule of halide is involved, and also account for the generally observed predominance of trans addition (Fig. 10.67). We will write cyclic ions in the answers to problems, but you should know that the intermediacy of vinyl cations in these reactions is not a fully resolved issue. There is still lots to do in organic chemistry.

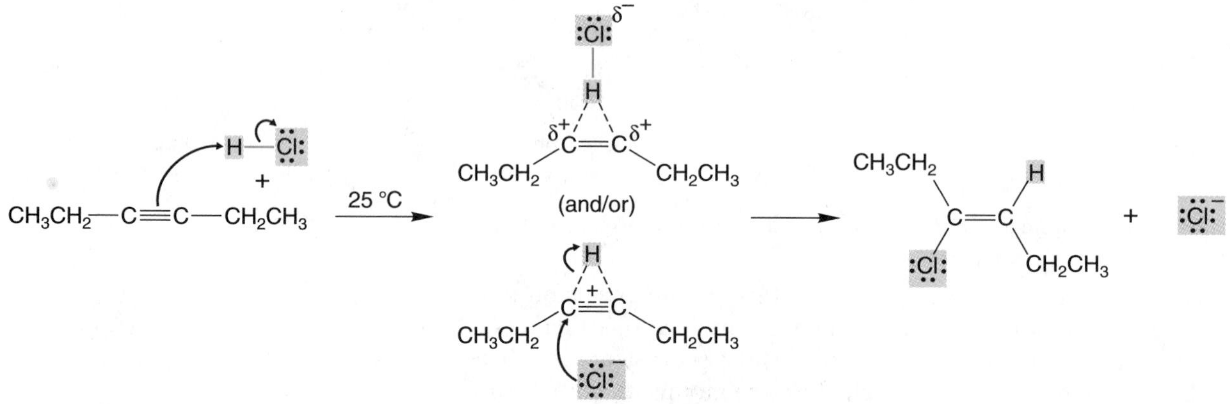

FIGURE 10.67 Possible cyclic intermediates in the reaction of an alkyne and hydrogen chloride.

This modified mechanism for addition to alkynes also predicts that Markovnikov addition should be observed: that the more substituted secondary vinyl halide should be formed whenever there is a choice. In the cyclic intermediate, there will be partial positive charge on the carbons of the original acetylene. The partial bond to hydrogen from the *more* substituted carbon of an unsymmetrical acetylene will be weaker than the bond from the *less* substituted carbon because the partial positive charge will be more stable at the more substituted position. Accordingly, addition of halide would be expected to take place more easily at the more substituted position leading to Markovnikov addition (Fig. 10.68).

FIGURE 10.68 Addition of a nucleophile to the cyclic ion will take place at the more substituted position.

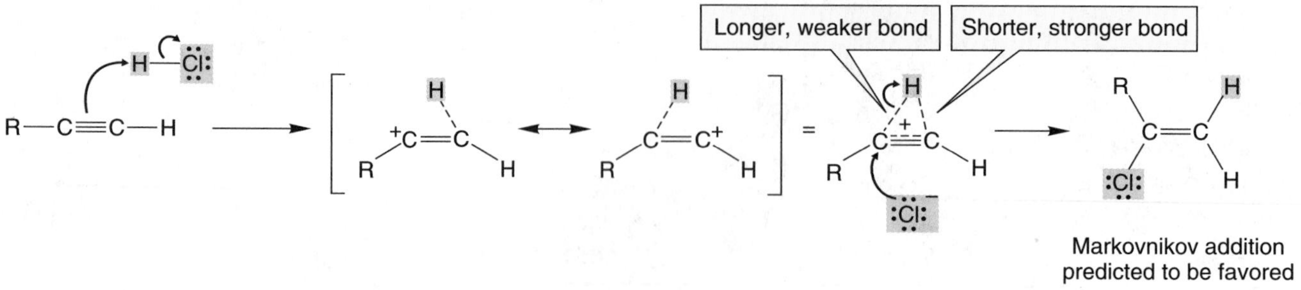

Let's look at the experimental data. There is no difference in the two ends of the molecule such as 3-hexyne, so let's examine the addition to a terminal acetylene—a 1-alkyne—to see if Markovnikov addition occurs. Our mechanism predicts that the more substituted halide should be the product, and it is (Fig. 10.69).

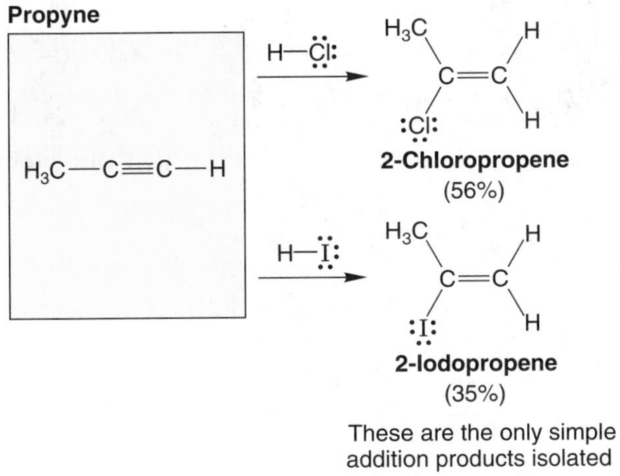

Propyne

$H_3C—C≡C—H$

H—Cl: → **2-Chloropropene** (56%)

H—I: → **2-Iodopropene** (35%)

These are the only simple addition products isolated

FIGURE 10.69 Markovnikov addition generally predominates in reactions of HX with acetylenes.

In addition to the mechanistic complexity, additions of HX to acetylenes are usually not practical sources of vinyl halides. The vinyl halide products also contain π bonds and often compete favorably with the starting alkyne in the addition reaction. The vinyl halide products are not symmetrical and two products of further reaction are possible. It makes a nice problem to figure out which way the second addition of H–X should go.

As always in a mechanistic problem of this sort, the answer only appears through an analysis of the possible pathways for further reaction. Draw out both mechanisms, compare them at every point, and search each step for differences. We will use the reaction of propyne with hydrogen chloride as a prototypal example. The initial product is 2-chloropropene, as shown in Figure 10.69. There are two possible ions from the further reaction of the vinyl halide with hydrogen chloride. In (a), the positive charge is adjacent to a chlorine, but in (b) it is not (Fig. 10.70).

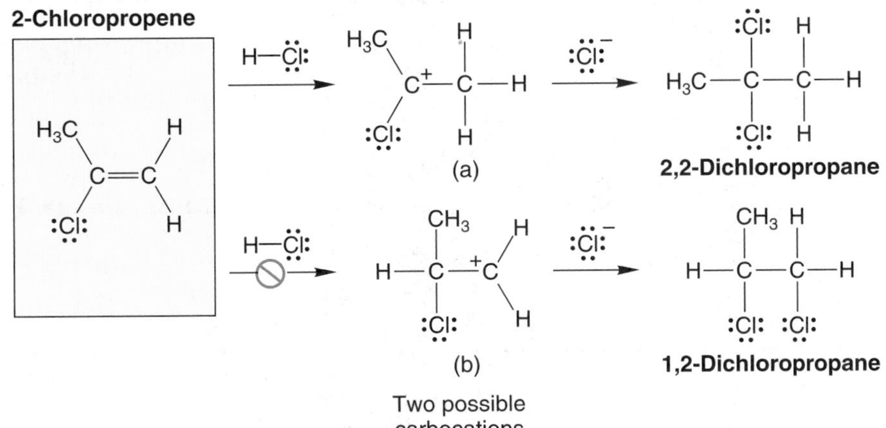

FIGURE 10.70 The initial products of addition of HX to alkynes are vinyl halides and can themselves react with HX to give addition products. Protonation of 2-chloropropene can give two possible cations. In (a), the positive charge is adjacent to the chlorine, but in (b) it is not.

In (a), the cation is adjacent to chlorine, and there is a way for the chlorine to share the positive charge. The charge is shared by the two atoms through $2p$–$3p$ overlap. In molecular orbital terms, we see that a filled $3p$ orbital on chlorine overlaps with the empty $2p$ orbital on carbon, and this filled–empty overlap is stabilizing. The simple primary carbocation (b) is much less stable, and products from it are not observed (Fig. 10.71).

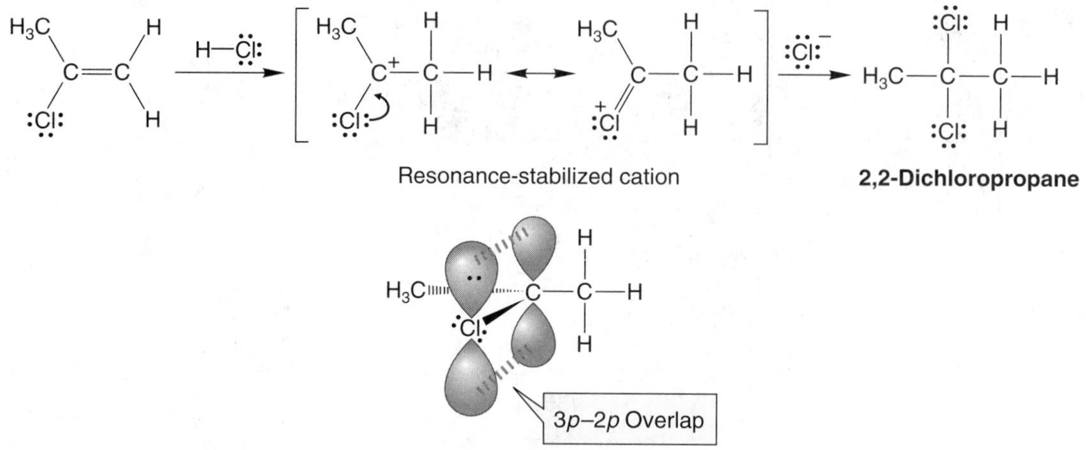

Resonance-stabilized cation

2,2-Dichloropropane

$3p$–$2p$ Overlap

FIGURE 10.71 The cation adjacent to chlorine is stabilized by resonance; the other is not.

So, addition of a second halogen will preferentially give the **geminal** dihalide (geminal means 1,1-disubstituted) not the vicinal (1,2-disubstituted) compound because the cation adjacent to chlorine is lower energy, and leads inevitably to the observed geminal product. The reactions of Figure 10.69 are complicated by the formation of compounds arising through double addition of hydrogen halide. The double addition always occurs so as to give the geminal dihalide (Fig. 10.72).

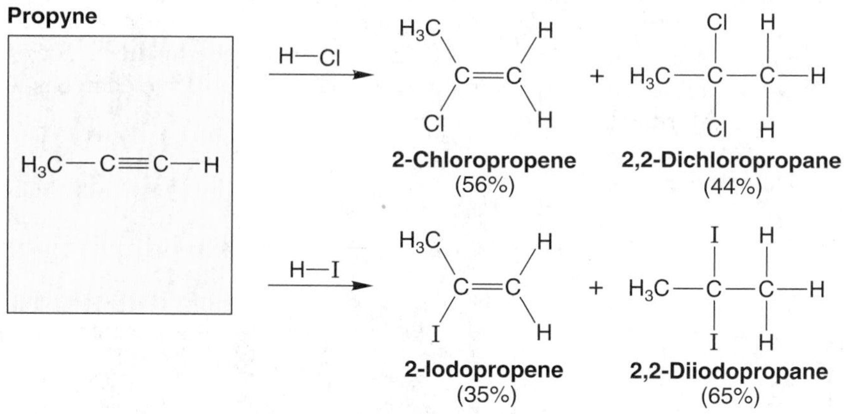

FIGURE 10.72 Addition of hydrogen halides to acetylenes gives mixtures of products of mono- and diaddition.

10.6 ADDITION OF X_2 REAGENTS TO ALKYNES

The pattern of alkene addition is also followed when X_2 reagents add to alkynes, although the mechanisms of these reactions are not well worked out. Both ionic and neutral intermediates (i.e., radicals, see Chapter 11) are involved. Alkynes add Br_2 or Cl_2 to give vicinal,1,2–dihalides. A second addition can follow to give the tetrahalides (Fig. 10.73).

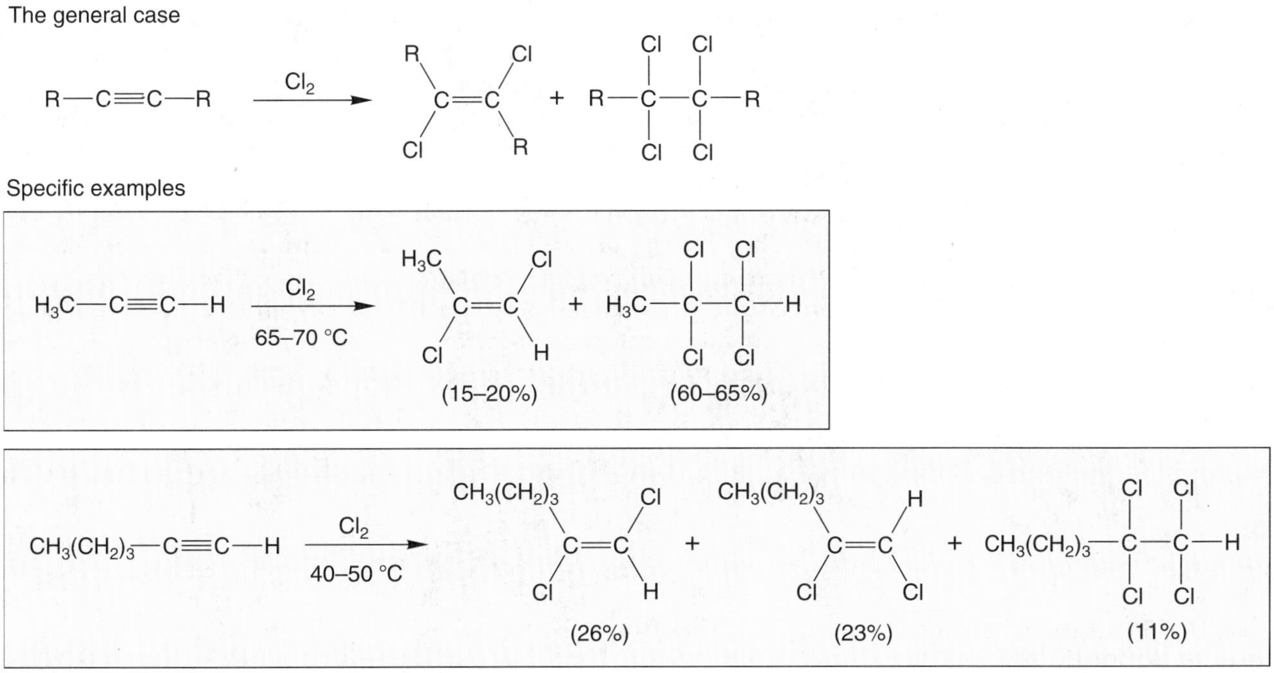

The general case

$$R{-}C{\equiv}C{-}R \xrightarrow{Cl_2} (Cl)(R)C{=}C(Cl)(R) + R{-}CCl_2{-}CCl_2{-}R$$

Specific examples

$$H_3C{-}C{\equiv}C{-}H \xrightarrow[65\text{–}70\ °C]{Cl_2} (H_3C)(Cl)C{=}C(Cl)(H)\ (15\text{–}20\%) + H_3C{-}CCl_2{-}CCl_2{-}H\ (60\text{–}65\%)$$

$$CH_3(CH_2)_3{-}C{\equiv}C{-}H \xrightarrow[40\text{–}50\ °C]{Cl_2} (CH_3(CH_2)_3)(Cl)C{=}C(Cl)(H)\ (26\%) + (CH_3(CH_2)_3)(Cl)C{=}C(H)(Cl)\ (23\%) + CH_3(CH_2)_3{-}CCl_2{-}CCl_2{-}H\ (11\%)$$

FIGURE **10.73** Addition of Cl_2 (or Br_2) to alkynes gives both di- and tetrahalides.

10.7 HYDRATION OF ALKYNES

Like alkenes, acetylenes can be hydrated. The reaction is generally catalyzed by mercuric ions in an oxymercuration process (see Section 10.2, p. 402 for the related alkene reaction), although simple acid catalysis is also known (Fig. 10.74). In contrast to the oxymercuration of alkenes, no second, reduction step is required.

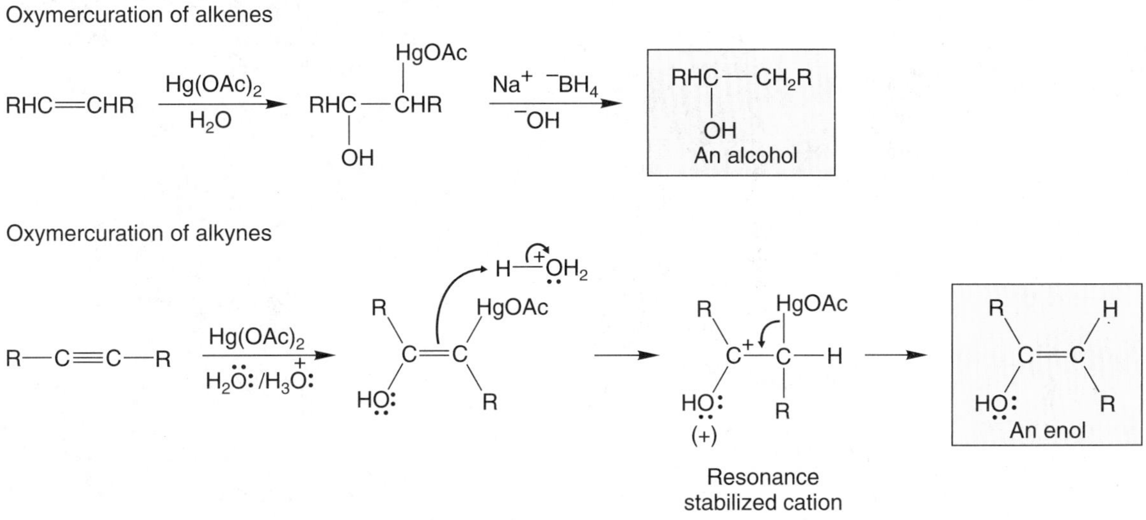

Oxymercuration of alkenes

$$RHC{=}CHR \xrightarrow[H_2O]{Hg(OAc)_2} RHC(OH){-}CHR(HgOAc) \xrightarrow[^-OH]{Na^+\ ^-BH_4} RHC(OH){-}CH_2R$$

An alcohol

Oxymercuration of alkynes

$$R{-}C{\equiv}C{-}R \xrightarrow[H_2\ddot{O}:/H_3\overset{+}{O}:]{Hg(OAc)_2} (R)(H\ddot{O}:)C{=}C(HgOAc)(R) \xrightarrow{H{-}\overset{+}{O}H_2} (R)(H\ddot{O}:)\overset{+}{C}{-}C(HgOAc)(H)(R) \longrightarrow (R)(H\ddot{O}:)C{=}C(H)(R)$$

(+)

Resonance stabilized cation

An enol

FIGURE **10.74** The oxymercuration of alkynes resembles the oxymercuration of alkenes. In the alkyne case, the product is an enol, which can react further.

By strict analogy to the oxymercuration of alkenes, the product should be a hydroxy mercury compound, but the second double bond exerts its influence and further reaction takes place. The double bond is protonated and mercury is lost to generate a species called an **enol**. This molecule is part alk*ene* and part alcoh*ol*, hence the name.

Enols are extraordinarily important compounds, and more than one chapter will be devoted to their chemistry. Here their versatility is exemplified by their conversion into **ketones**. Ketones are compounds in which the carbon of a carbon–oxygen double bond is attached to two other carbons. They are the final products of the hydration reactions of alkynes. The mechanism of their formation from enols is simple; the alkene part is protonated on carbon and deprotonated on oxygen to generate the compound containing the carbon–oxygen double bond. Note that this gives you another synthesis of carbonyl compounds (Fig. 10.75).

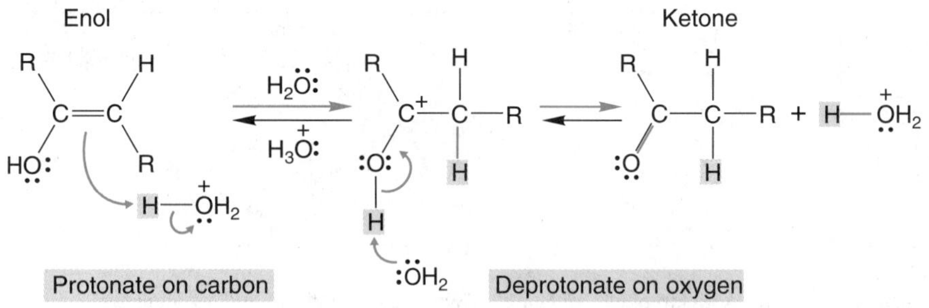

Enol

Ketone

Protonate on carbon

Deprotonate on oxygen

FIGURE 10.75 Protonation of carbon, followed by deprotonation at oxygen, generates the carbonyl compound. This sequence is a general reaction of enols.

PROBLEM 10.21 Draw a mechanism for the acid-catalyzed hydration of an acetylene.

*PROBLEM 10.22 Why was I so sure the enol would protonate at the end to give the cation shown in Figure 10.75? There is another cation possible. Why is it not formed?

ANSWER Protonation in the sense shown leads to a resonance-stabilized carbocation. The adjacent oxygen shares the positive charge. Deprotonation gives the ketone. Protonation in the other sense leads to a carbocation that is not resonance stabilized, and which is therefore of much higher energy.

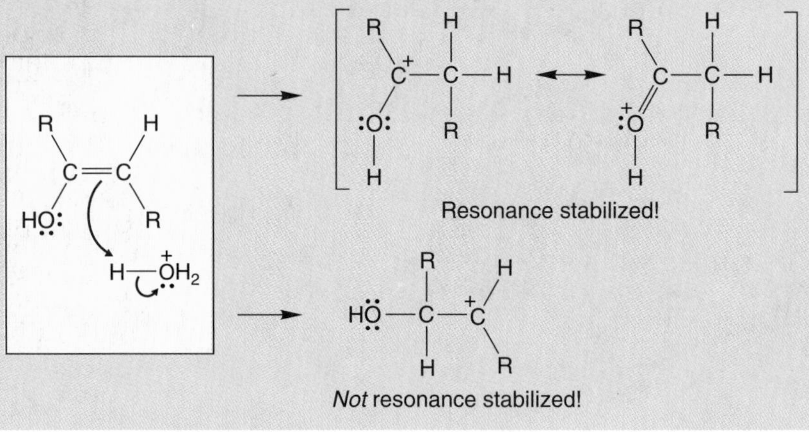

Resonance stabilized!

Not resonance stabilized!

PROBLEM 10.23 In general, hydration of unsymmetrical acetylenes such as 2-pentyne is not a practical source of ketones. Why not?

*PROBLEM **10.24**

Terminal acetylenes are certainly unsymmetrical, yet in these cases hydration is a useful process. In principle, either ketones or aldehydes could be formed, but in practice only the ketones are produced. Write mechanisms for both ketone and aldehyde formation and explain why the product is always the ketone. You may write open vinyl cations in your answer for simplicity, but be aware that the real intermediate(s) may be more complex cyclic species.

ANSWER

Follow through the two mechanisms, look for the point at which the pathways diverge, and try to see why the pathway to the ketone is favored. In this case, it is the initial protonation of the alkyne that determines the final product. Protonation gives the more substituted, more stable vinyl cation (or a cyclic intermediate in which the more substituted position bears most of the positive charge). Addition of water to give the enol is followed by conversion of the enol to the final product, the ketone (see Fig. 10.75 for mechanistic details of the formation of the ketone from the enol).

R—C≡C—H

$H_3\ddot{O}^+$ → R—C⁺=CH₂ → $H_2\ddot{O}$: addition–deprotonation → (HO)(R)C=CH₂

More stable (the product of initial protonation)

$H_3\ddot{O}^+$ (blocked) → (R)(H)C=C⁺—H → $H_2\ddot{O}$: addition–deprotonation → (R)(H)C=C(H)(:ÖH)

Less stable (not formed)

(HO)(R)C=CH₂ → $H_3\ddot{O}^+$, $H_2\ddot{O}$: → R—C(=O)—CH₃

The ketone is formed by hydrolysis of the product enol

[(R)(H)C=C(H)(:ÖH) → $H_3\ddot{O}^+$, $H_2\ddot{O}$: → H—C(R)(H)—C(=O)H

The aldehyde would have been formed if the "wrong" enol had been produced]

Nearly all ketones and aldehydes containing a hydrogen on the carbon adjacent to the carbon–oxygen double bond are in equilibrium with related enol forms. How much enol is present at equilibrium is a function of the detailed structure of the molecule, but there is almost always some enol present. In simple compounds, the ketone form is greatly favored (Chapter 16).

PROBLEM **10.25**

Write a mechanism for the acid-catalyzed conversion of diethyl ketone into its enol form.

10.8 HYDROBORATION OF ALKYNES

Like alkenes, alkynes can be hydroborated by boranes. The mechanism is very similar to that for alkenes; the stereochemistry of addition of HBH_2 is cis, and it is the boron that becomes attached to the less substituted end of the starting alkyne (Fig.10.76).

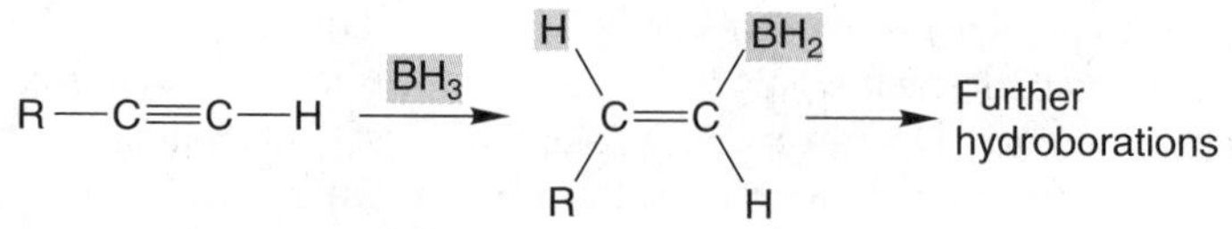

FIGURE 10.76 Hydroboration of alkynes gives vinylboranes that can react further (R = alkyl).

The initial products of hydroboration are prone to further hydroboration reactions. This complication can be avoided by using a sterically hindered borane so that further reaction is slowed. A favorite reagent is the borane formed by reaction of two equivalents of 2-methyl-2-butene with BH_3 (Fig. 10.77).

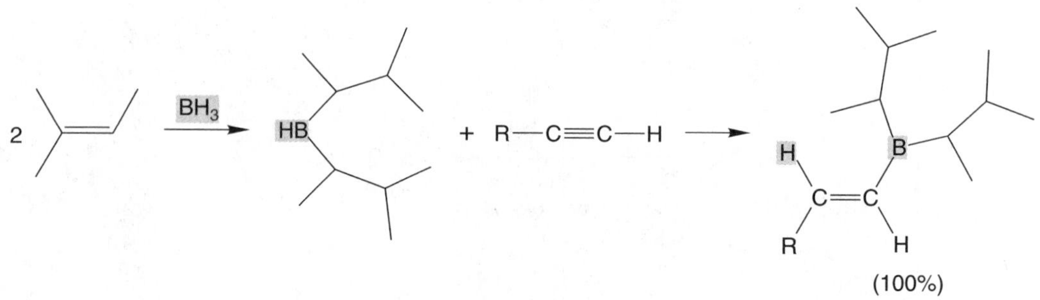

FIGURE 10.77 This sterically hindered dialkylborane reacts only once with alkynes to give an isolable vinylborane (R = alkyl).

Like alkylboranes, these alkenylboranes can be converted into alcohols by treatment with basic peroxide (Fig. 10.78).

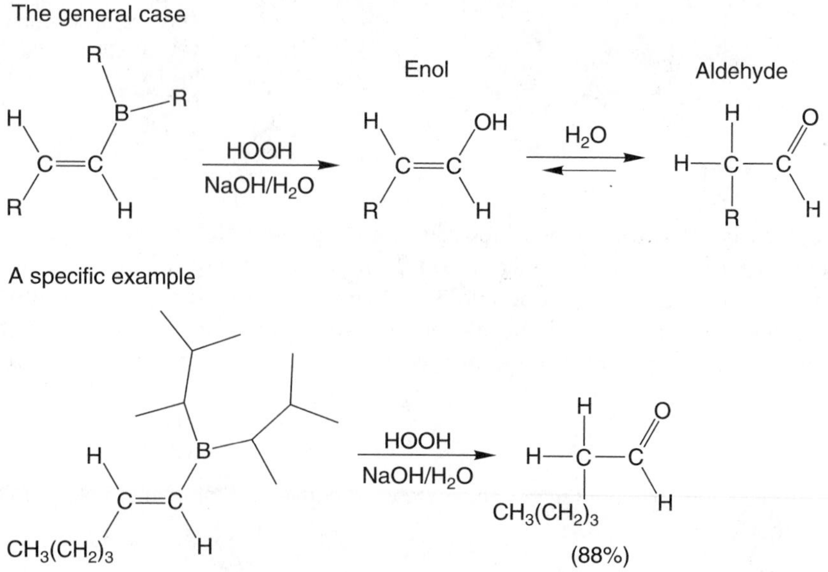

FIGURE 10.78 Oxidation of the vinylborane generates an enol that can be hydrolyzed to an aldehyde.

The products are enols, and they are in equilibrium with the corresponding carbonyl compounds (Fig. 10.78). Here's an important point: This reaction has the opposite regiochemistry from the hydration reaction of alkynes. This situation is exactly the same as the one that exists with the alkenes: Hydration of an alkene gives overall Markovnikov addition, whereas the hydroboration–oxidation sequence gives the anti-Markovnikov product. The ultimate formation of carbonyl compounds disguises the matter, but if we look back one step to the enols, we can see that in alkyne reactions the same regioselectivity holds. Hydration gives Markovnikov addition and, ultimately, the ketone. The hydroboration–oxidation reaction gives anti-Markovnikov addition, and, ultimately, the aldehyde (Fig. 10.79).

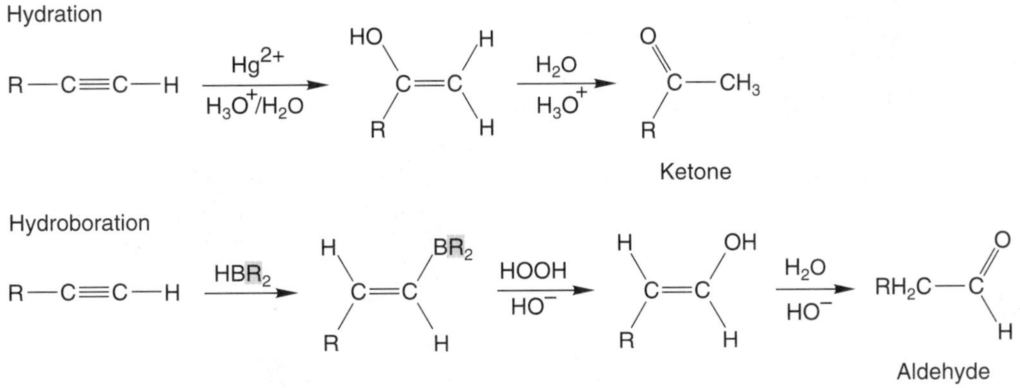

Hydration

R—C≡C—H $\xrightarrow[H_3O^+/H_2O]{Hg^{2+}}$ (HO)(R)C=C(H)(H) $\xrightarrow[H_3O^+]{H_2O}$ R—C(=O)—CH_3

Ketone

Hydroboration

R—C≡C—H $\xrightarrow{HBR_2}$ (H)(R)C=C(BR_2)(H) $\xrightarrow[HO^-]{HOOH}$ (H)(R)C=C(OH)(H) $\xrightarrow[HO^-]{H_2O}$ RH_2C—C(=O)H

Aldehyde

FIGURE 10.79 It is possible to make two carbonyl compounds from a terminal alkyne. Mercuric ion-catalyzed hydration gives the ketone, and hydroboration–oxidation the aldehyde. Be sure you see the mechanistic differences that require the different products.

Be sure you see the similarities here as well as the mechanistic underpinnings of the different paths followed. There is too much to memorize. Be sure to annotate your file cards as you update your catalog of synthetic methods.

10.9 HYDROGENATION OF ALKYNES: SYN HYDROGENATION

Like alkenes, alkynes can be hydrogenated in the presence of many catalysts. Alkenes are the first products, but they are not stable under the reaction conditions and are themselves hydrogenated further to give alkanes (Fig. 10.80). This reaction is your second synthesis of alkanes.

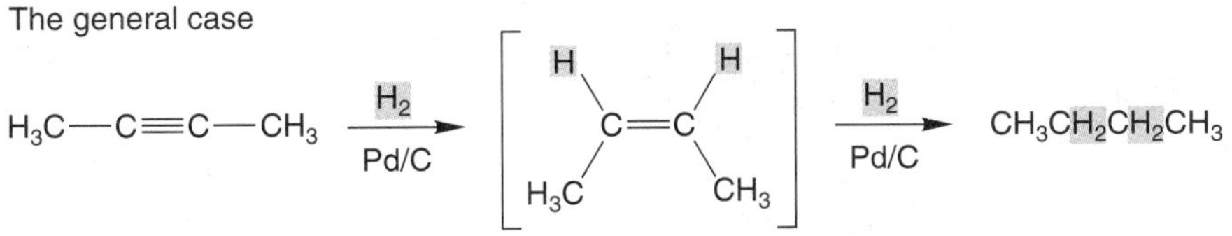

The general case

H_3C—C≡C—CH_3 $\xrightarrow[Pd/C]{H_2}$ [(H)(H_3C)C=C(H)(CH_3)] $\xrightarrow[Pd/C]{H_2}$ $CH_3CH_2CH_2CH_3$

FIGURE 10.80 Hydrogenation of an alkyne proceeds all the way to the alkane stage. The intermediate alkene cannot usually be isolated.

It would be extremely convenient to be able to stop at the alkene stage, and special, "poisoned" catalysts have been developed to do just this. A favorite is the Lindlar catalyst—Pd on calcium carbonate that has been treated with lead acetate in the presence of certain amines (H. H. M. Lindlar, b. 1909). For many years it was thought that the lead treatment poisoned the palladium catalyst through some sort of alloy formation, rendering it less active and the reaction more selective. It now appears that the treatment forms no alloys, but does modify the surface of the palladium. Regardless, the reaction can be stopped at the alkene stage, and the alkenes formed by hydrogenation with Lindlar catalyst have cis stereochemistry, as the usual syn addition of hydrogen has occurred (Fig. 10.81).

A specific example

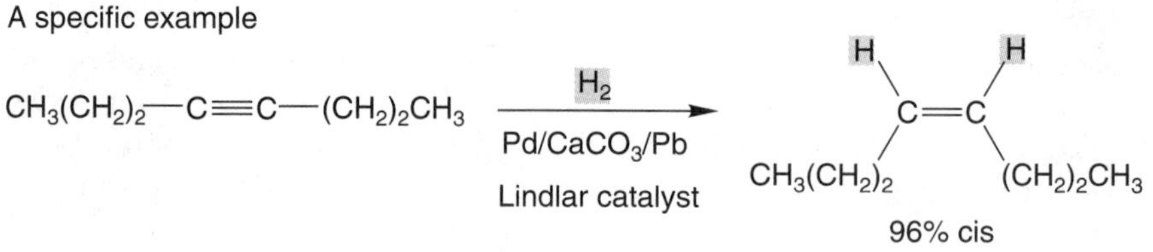

$$CH_3(CH_2)_2-C\equiv C-(CH_2)_2CH_3 \xrightarrow[\text{Pd/CaCO}_3\text{/Pb, Lindlar catalyst}]{H_2} \text{cis-}CH_3(CH_2)_2(H)C=C(H)(CH_2)_2CH_3 \text{ (96\% cis)}$$

FIGURE 10.81 If the Lindlar catalyst is used, the cis alkene can be obtained.

10.10 REDUCTION BY SODIUM IN AMMONIA: ANTI HYDROGENATION

In a reaction that must at first seem mysterious, an alkyne dissolved in a solution of sodium in liquid ammonia forms the trans alkene (Fig. 10.82). The practical consequences of this are that you now have a way of synthesizing either a cis or trans alkene from a given alkyne. So, update your file cards.

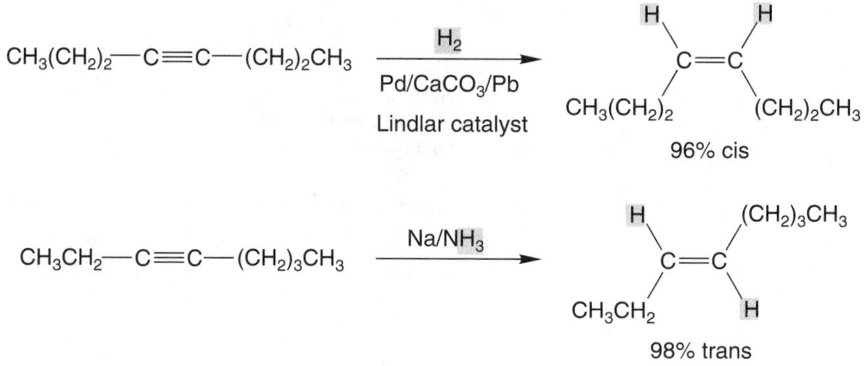

$$CH_3(CH_2)_2-C\equiv C-(CH_2)_2CH_3 \xrightarrow[\text{Pd/CaCO}_3\text{/Pb, Lindlar catalyst}]{H_2} \text{cis-}CH_3(CH_2)_2(H)C=C(H)(CH_2)_2CH_3 \text{ (96\% cis)}$$

$$CH_3CH_2-C\equiv C-(CH_2)_3CH_3 \xrightarrow{Na/NH_3} \text{trans-}CH_3CH_2(H)C=C(H)(CH_2)_3CH_3 \text{ (98\% trans)}$$

FIGURE 10.82 An alkyne can give either the cis or trans alkene. Hydrogenation with Lindlar catalyst gives the cis alkene, whereas reduction with sodium in ammonia gives the trans.

When sodium is dissolved in ammonia, it loses its odd electron to give a sodium ion and a **solvated electron** (Fig. 10.83).

$$Na\cdot + NH_3 \longrightarrow Na^+ + e^-[NH_3]_n$$

Solvated electron

FIGURE 10.83 The first step in the reaction is the generation of a solvated electron and a sodium ion.

This complexed electron can add to an acetylene to give a species called a **radical anion** (Fig. 10.84).

$$H_3C-C\equiv C-CH_3 + e^-[NH_3]_n \longrightarrow NH_3 + [H_3C-C\equiv C-CH_3]^{\cdot -}$$

A radical anion

FIGURE 10.84 The solvated electron can add to the alkyne to generate an acetylene radical anion.

I vividly recall being quite uncertain about the structure of this species when I was first studying organic chemistry. I suspect I solved the problem of dealing with it by memorizing this reaction and hoping that no one would ever ask me to explain what was happening. But it's not so hard and you should not repeat my error. First of all, where is the "extra" electron in the radical anion? It is most certainly not in one of the bonding π orbitals of the acetylene, as they are each filled with two electrons already. If it is not there, it must be in one of the antibonding, π^*, orbitals (Fig. 10.85).

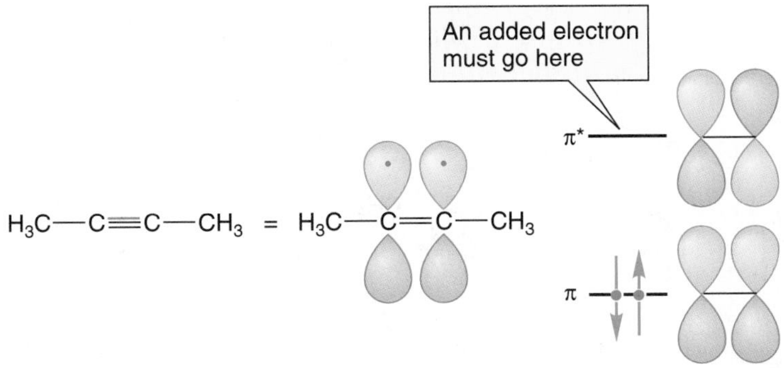

An added electron must go here

$H_3C—C≡C—CH_3$ = $H_3C—C=C—CH_3$

π^*

π

FIGURE 10.85 The additional electron must go into an antibonding, π^*, orbital of the alkyne.

Why should the molecule put up with this electron in an antibonding orbital? The answer is twofold. First of all, the molecule is held together by many electrons occupying bonding orbitals, and a single electron in a high-energy orbital is not destabilizing enough to overcome all the bonding interactions. Figure 10.86 gives both molecular orbital and resonance pictures of the radical anion.

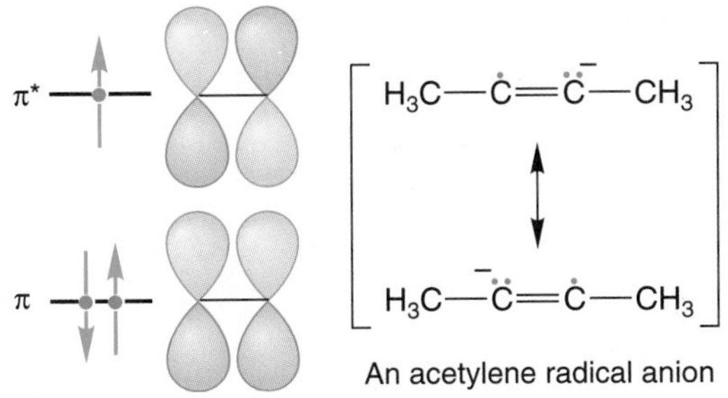

π^*

π

$H_3C—\dot{C}=\ddot{C}^-—CH_3$ ↔ $H_3C—\ddot{C}^-=\dot{C}—CH_3$

An acetylene radical anion

FIGURE 10.86 Molecular orbital and resonance descriptions of the radical anion.

Second, the radical anion *is* quite unstable and not much of it will be formed at equilibrium. Its very instability contributes to its great base strength, however, and the small amount present at equilibrium is rapidly protonated by ammonia to give the amide ion ($^-NH_2$) and the vinyl radical (Fig. 10.87).

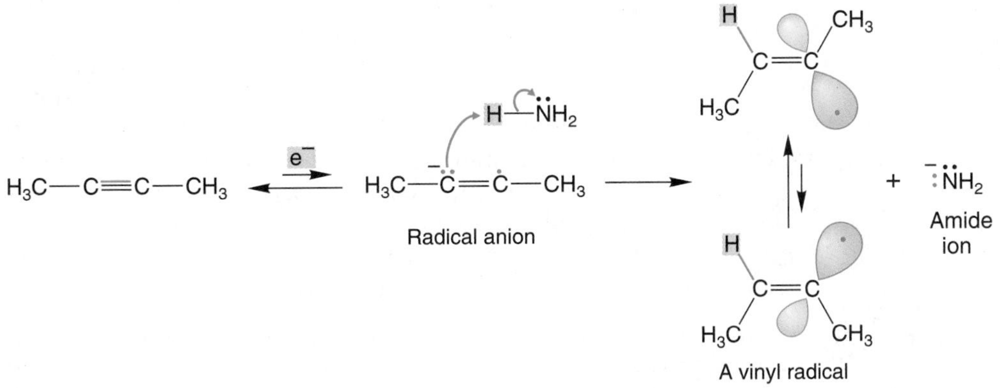

$H_3C—C≡C—CH_3$ ⇌ (e^-) $H_3C—\ddot{C}^-=\dot{C}—CH_3$ + $H—\ddot{N}H_2$ → vinyl radical + $:\ddot{N}H_2^-$

Radical anion

A vinyl radical

Amide ion

FIGURE 10.87 The radical anion can abstract a proton from ammonia to generate a vinyl radical and the amide ion ($^-NH_2$). The trans form, in which the alkyl groups are as far from each other as possible, will be favored.

There are two forms of the vinyl radical, and it will be the trans form, with the alkyl groups as far from each other as possible, that is the more stable. A solvated electron can be donated to this radical forming the vinyl anion, which is rapidly protonated by ammonia to give the final product, the trans alkene (Fig. 10.88).

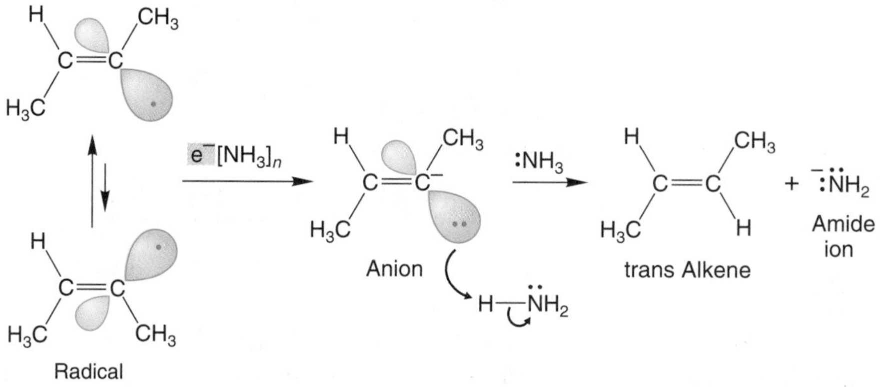

FIGURE 10.88 The trans vinyl radical adds a solvated electron to give the trans vinyl anion. Protonation by ammonia gives the trans alkene.

10.11 SOMETHING MORE: TRIPLET CARBENE REACTIONS

We noted earlier (Section 10.3d) that divalent carbon reacted with alkenes to form cyclopropanes. The addition was syn, and we took this as evidence of a concerted, one-step addition (Fig. 10.89).

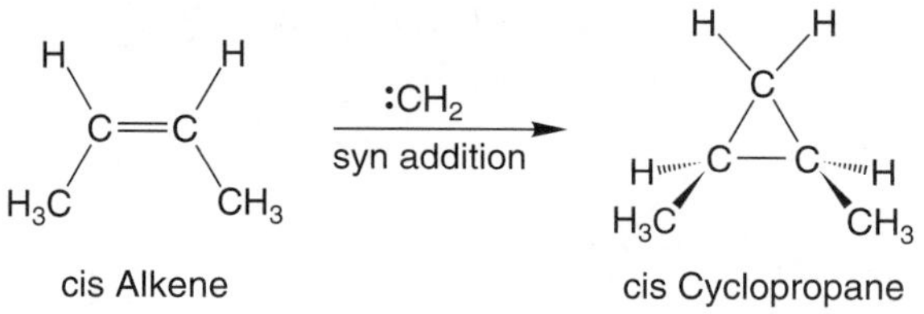

FIGURE 10.89 The stereospecific addition of the simplest carbene, methylene ($:CH_2$), to an alkene.

One of the reasons carbenes are so interesting is that unlike almost all other reactive intermediates, two reactive states exist, quite close in energy to each other. In Section 10.3d, we looked at reactions of a species in which two spin-paired electrons occupied the relatively low-energy hybrid orbital and the higher energy $2p$ orbital was empty. In fact, this species, called a **singlet carbene**, generally lies a few kilocalories per mole *higher* in energy than another form of divalent carbon called a **triplet carbene**. In the triplet species, one electron occupies the hybrid orbital and the other has been promoted to the previously empty $2p$ orbital. Of course, this incurs an energy cost—the approximately sp^2 hybrid orbital has more s character than the $2p$ orbital, and thus an electron in it is more stable than an electron in a pure $2p$ orbital. But there are energetic compensations. For example, the two nonbonding electrons are further apart in the triplet and thus electron–electron repulsion is lowered. With some exceptions (R = halogen, for example), the triplets are somewhat more stable than the corresponding singlets. For many carbenes the two species are in equilibrium, and sorting out the chemistry can be complicated (Fig. 10.90).

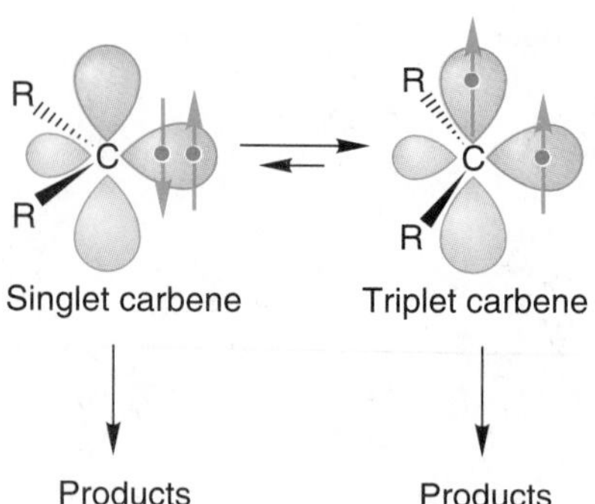

FIGURE 10.90 There are two low-energy "spin states" of a carbene. In one, the spin quantum numbers of the nonbonding electrons are opposite (paired). In this singlet state, the nonbonding electrons will both occupy the relatively low-energy hybrid orbital. In the other spin state, the spin quantum numbers of the nonbonding electrons are the same and they *may not* occupy the same orbital. In this triplet state, both the hybrid and $2p$ orbitals are half-occupied. Either spin state may undergo reactions.

*PROBLEM **10.26** Singlet $:CF_2$ is much more stable than the corresponding triplet. Look at the structure of this intermediate and explain why the singlet is especially favored in this case.

ANSWER In the singlet state, the empty $2p$ orbital on carbon overlaps with a filled $2p$ orbital on fluorine. This highly stabilizing filled–empty interaction is not present in the triplet, in which the carbon $2p$ orbital is singly occupied.

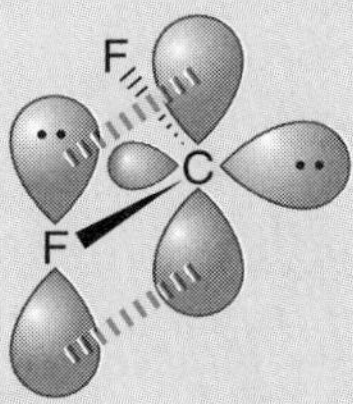

Singlet: Only one filled orbital on F is shown; the key point is the stabilizing overlap of this filled orbital with the empty $2p$ orbital on carbon

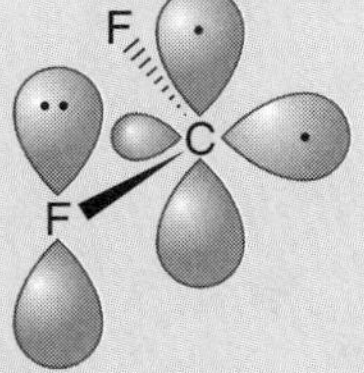

Triplet: The $2p$ orbital on carbon is half-filled; overlap with the filled orbital on the adjacent fluorine is not as stabilizing as in the case of the singlet

Normally, even though the singlet is disfavored at equilibrium, its reactions are observed. The singlet is usually *much* more reactive than the triplet. Even though there is more triplet present at equilibrium, its reactions cannot compete with those of the small amount of the much more reactive singlet. This point is important. A species does not have to be present in great quantity for its chemical reactions to dominate those of a more stable, but relatively unreactive equilibrium partner. The activation barriers surrounding the more stable partner may be higher than those surrounding the less stable partner. Figure 10.91 shows this situation in an energy diagram.

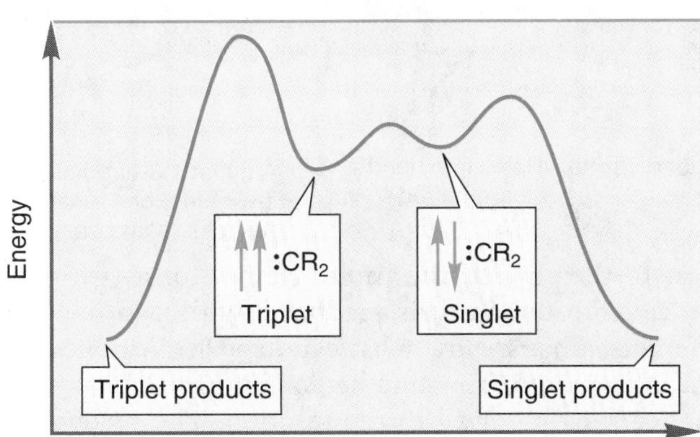

FIGURE **10.91** Even though the triplet is usually lower in energy, and thus favored at equilibrium, most of the chemistry comes from the singlet. The triplet is protected by quite high activation barriers. As long as some of the singlet is formed in equilibrium with the triplet, it will account for most or all of the chemistry.

However, triplet carbenes can sometimes be generated directly and their reactions observed. Like their singlet counterparts, they add to alkenes to give cyclopropanes (Fig. 10.92).

FIGURE **10.92** Like singlets, triplets add to alkenes to give cyclopropanes.

Now let's look at electron spin. For a singlet, the two nonbonding electrons on the carbene have paired (opposite) spins and can produce the two new cyclopropane carbon–carbon bonds without any problem by combining with the two paired electrons in the π bond. As we saw earlier (Fig. 10.44, p. 413) the reaction of a singlet with an alkene takes place in a single step, preserving the original stereochemical relationships of the groups on the alkene (Fig. 10.93).

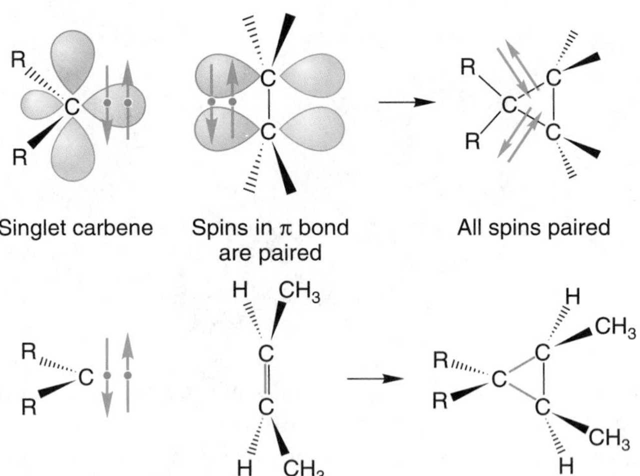

FIGURE 10.93 Addition of the singlet carbene can, and does, take place in a single step. Notice that the electrons involved have paired spins and can form two new cyclopropane bonds in a single step.

What do we expect of the triplet? In particular, what will be the stereochemistry of the addition reaction? The first step in the addition of triplets to alkenes is shown in Figure 10.94.

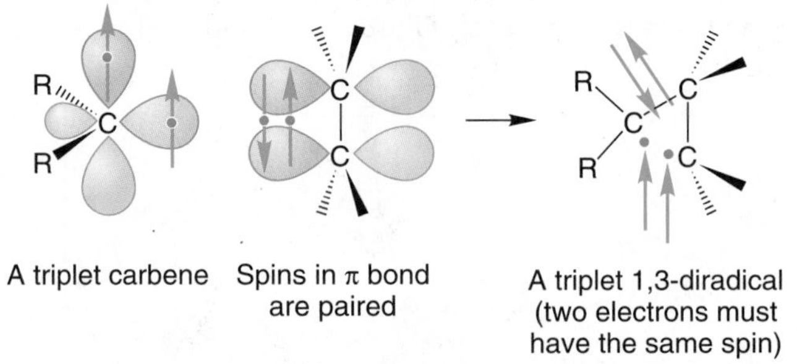

A triplet carbene

Spins in π bond are paired

A triplet 1,3-diradical (two electrons must have the same spin)

FIGURE 10.94 Only one bond can be formed in the reaction of the triplet with an alkene. The first step in triplet addition is reaction with the π bond of the alkene to give a 1,3-diradical.

In contrast to the singlet, the triplet cannot form two new bonds. A new species, called a **diradical**, is produced, with two nonbonding electrons on different carbons. Now what can happen? Why not simply close up to the three-membered ring and be done with it? Look closely at the spins of the electrons involved in the reaction. The second bond cannot close! The two electrons that would make up the new bond have the same spin and cannot occupy the same orbital (Fig. 10.95).

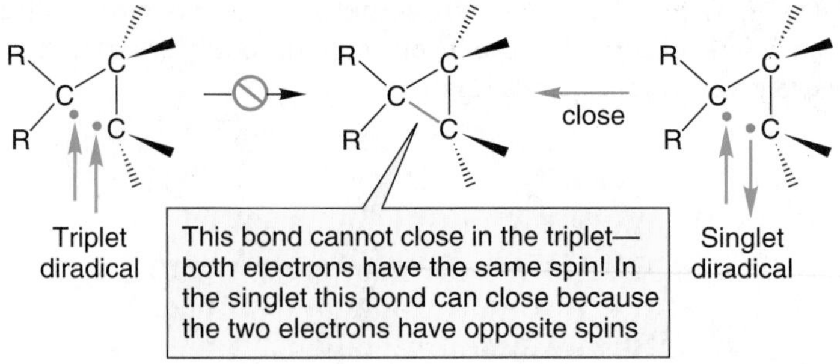

Triplet diradical

close

Singlet diradical

This bond cannot close in the triplet—both electrons have the same spin! In the singlet this bond can close because the two electrons have opposite spins

FIGURE 10.95 Two electrons with the same spin may not occupy the same orbital, and so there is no chance to form the second cyclopropane bond until a spin changes (flips). Spin flip is a quantum mechanical process and is not induced by a mere bond rotation!

The triplet, with two nonbonding electrons of the same spin, must produce an intermediate that cannot close to a cyclopropane until a spin is somehow changed. Such spin "flips" are not impossible, but they are usu-

The general case

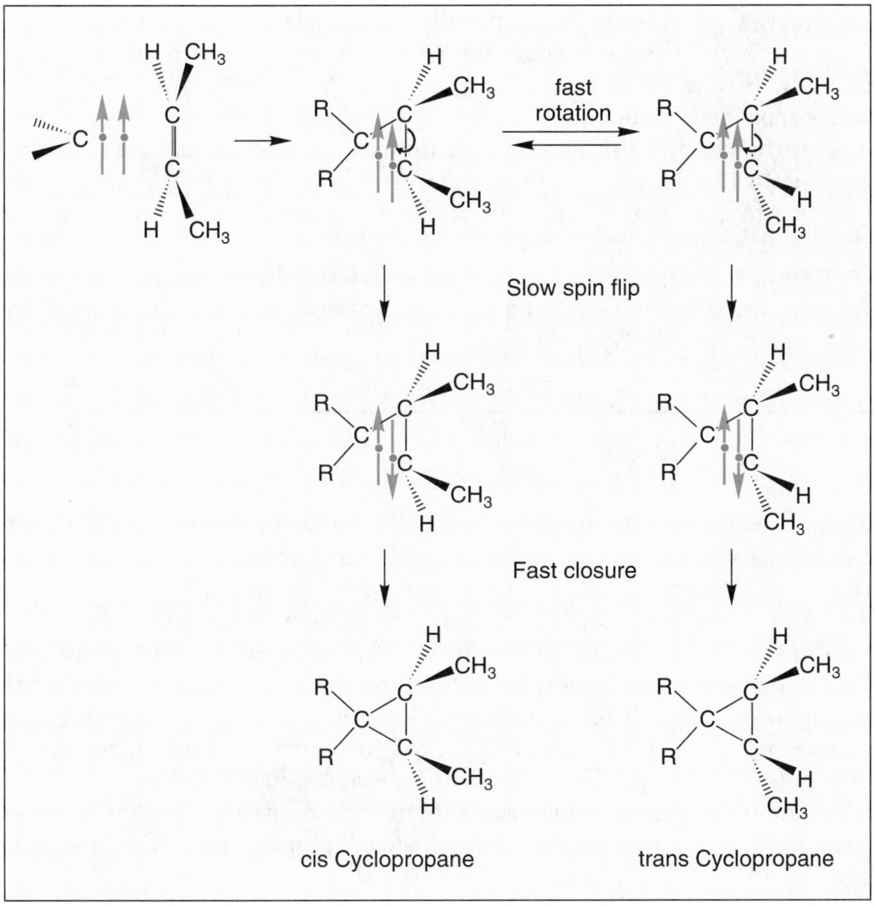

A specific example

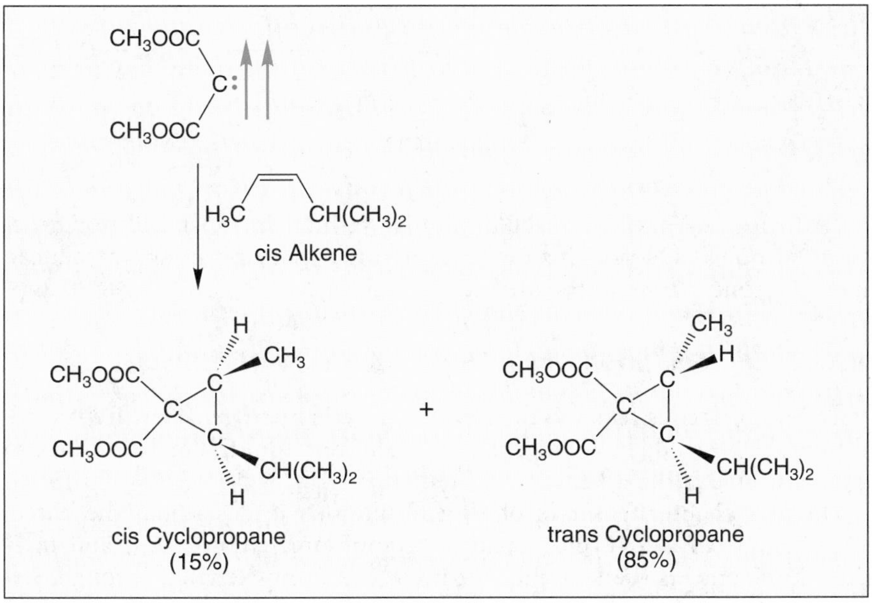

FIGURE 10.96 Triplet carbene addition to alkenes. The formation of an intermediate 1,3-diradical induces scrambling of stereochemistry. Unlike singlets, which add in a single step to generate cyclopropanes stereospecifically, triplets give both cis and trans cyclopropanes.

ally slow on the time scale of molecular processes. In particular, they are slow compared to the very rapid rotations about carbon–carbon single bonds. Therefore the stereochemistry of the original alkene need not be maintained in the eventual product of the reaction, the cyclopropane (Fig. 10.96). If rotation about the carbon–carbon bond is fast compared to changing of the spin quantum number of one electron (spin flip), it will not matter whether cis or trans alkene is used in the reaction with triplet methylene. The product will be a mixture of cis and trans 1,2-dialkylcyclopropanes in either case.

It turns out that triplets do give mixtures of the cis and trans cyclopropanes, and the examination of the stereochemistry of the addition reaction of carbenes has become the method of choice for determining the spin state of the reacting carbene.

This discussion of triplet carbenes and diradicals leads directly to Chapter 11, which will concentrate on the chemistry of radicals: species with a single nonbonding electron.

10.12 SUMMARY

NEW CONCEPTS

This chapter continues our discussion of addition reactions, and many of the mechanistic ideas of Chapter 9 apply. For example, many polar addition reactions start with the formation of the more stable carbocation. The most important new concept in this chapter is the requirement for overall trans addition introduced by three-membered ring intermediates. This concept appears most obviously in the trans bromination and chlorination of alkenes. The openings of the intermediate bromonium and chloronium ions by nucleophiles are S_N2 reactions in which inversion is always required. This process insures trans addition (Fig. 10.14).

Remember also the formation of oxiranes, stable three-membered rings containing an oxygen atom, and cyclopropane formation through carbene additions to alkenes (Fig. 10.97).

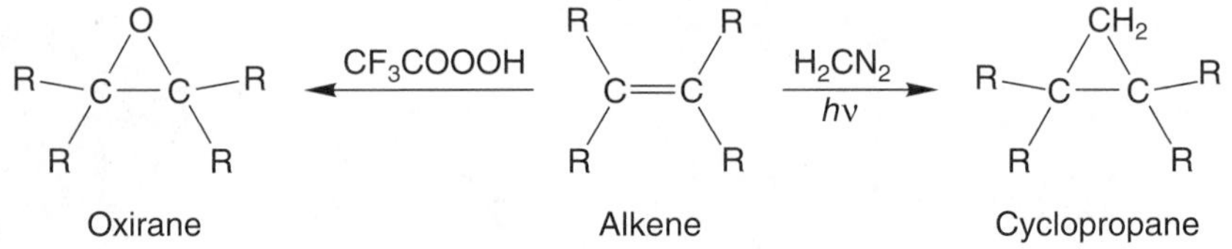

FIGURE 10.97 Stable three-membered rings can be formed in some cases. Two examples are epoxidation and carbene addition.

Alkenes and alkynes react in similar fashion in most addition reactions, although additional complications and opportunities are introduced by the second π bond in the alkynes.

REACTIONS, MECHANISMS, AND TOOLS

Both homogeneous and heterogeneous catalytic hydrogenation are syn additions, as both added hydrogens are transferred from the metal surface to the same side of the alkene.

In this chapter, a series of ring formations is described: the three-membered oxiranes, cyclopropanes, bromonium, chloronium, and mercurinium ions, as well as the five-membered rings formed through 1,3-

dipolar additions, ozonolysis, and the additions of osmium tetroxide and potassium permanganate (Fig. 10.98).

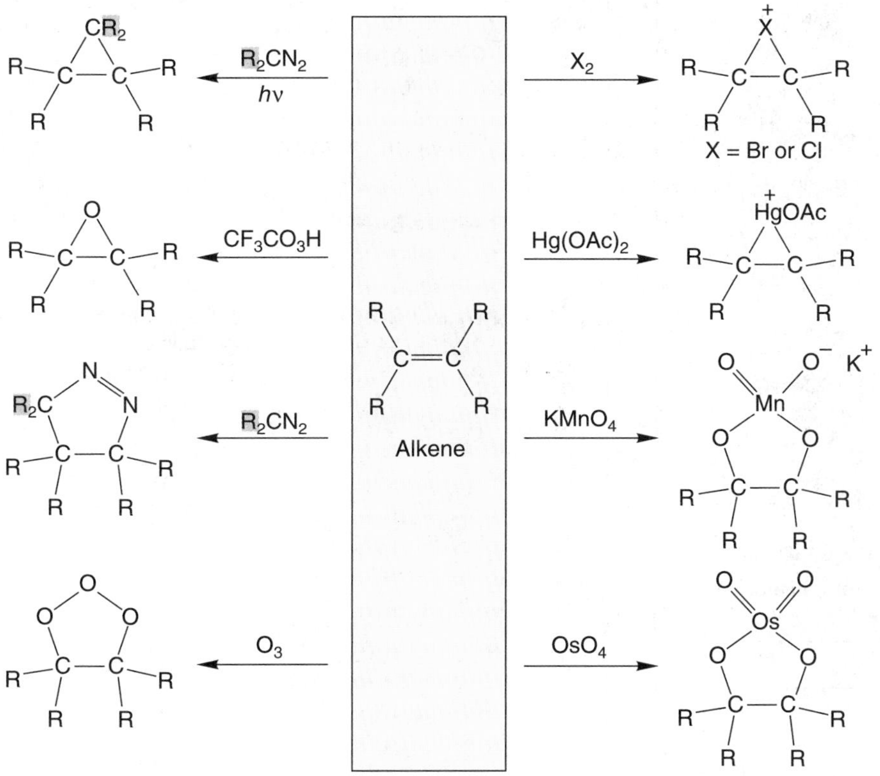

FIGURE **10.98** A summary of the ring-forming reactions in Chapter 10. Many of these compounds will react further to give the final products of the reaction.

Cyclopropanes and oxiranes can be isolated, but the other rings react further under the reaction conditions to give the final products. Oxiranes can also be opened in either acid or base to give new products. Additions to alkynes resemble those of alkenes, but the second π bond adds complications, and further addition reactions or rearrangements occur. Alkynes can be reduced in either cis or trans fashion (Fig. 10.82).

SYNTHESIS

This chapter provides many new synthetic methods (Fig. 10.99).

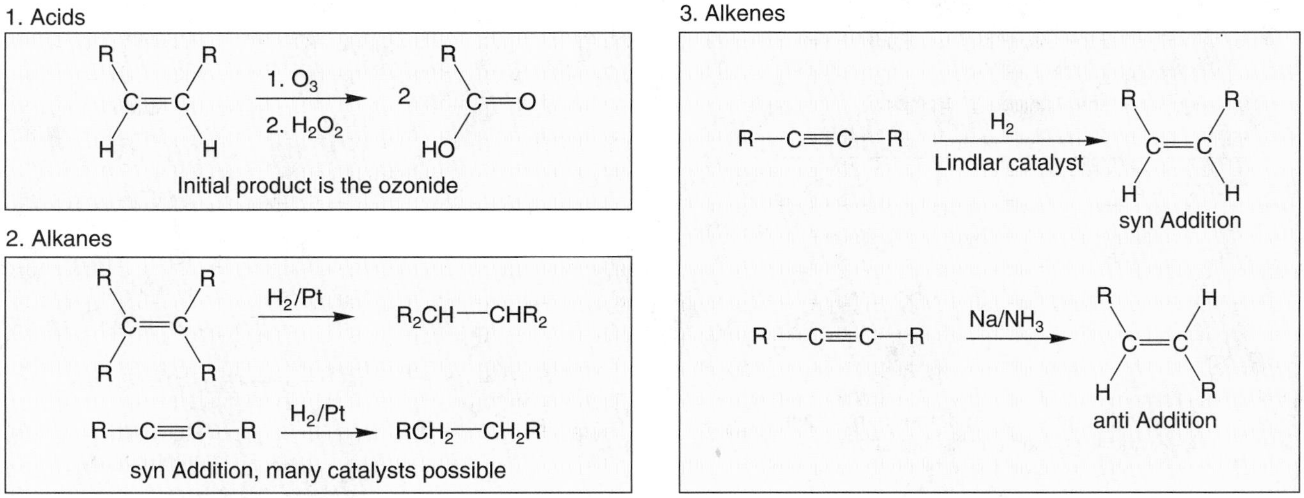

FIGURE **10.99** A summary of the new synthetically useful reactions of Chapter 10.

4. Alcohols

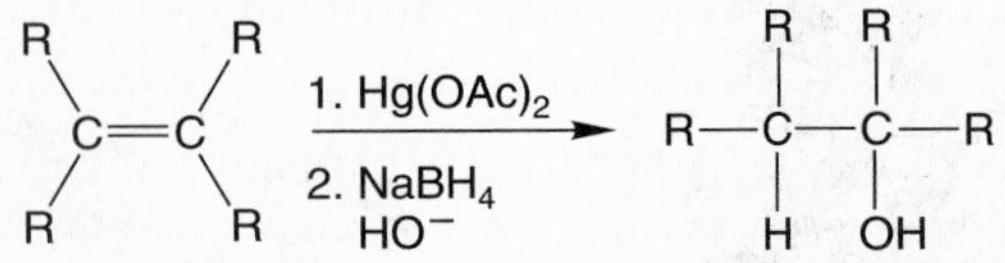

$R_2C{=}CR_2 \xrightarrow[\text{2. } NaBH_4,\ HO^-]{\text{1. } Hg(OAc)_2} R{-}CR(H){-}CR(OH){-}R$

Markovnikov addition, no rearrangements

5. Aldehydes

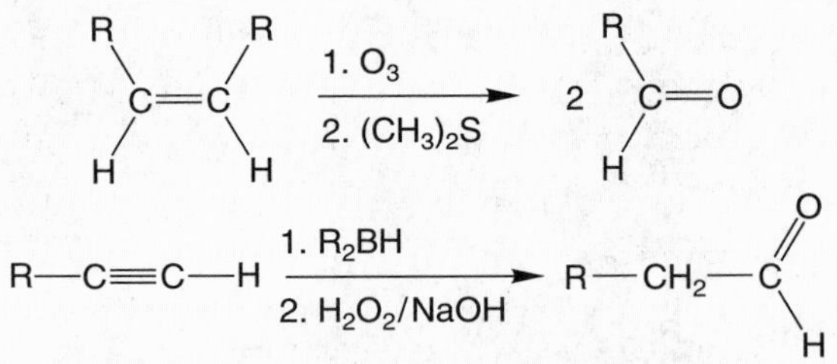

$RHC{=}CHR \xrightarrow[\text{2. } (CH_3)_2S]{\text{1. } O_3} 2\ RHC{=}O$

$R{-}C{\equiv}C{-}H \xrightarrow[\text{2. } H_2O_2/NaOH]{\text{1. } R_2BH} R{-}CH_2{-}CHO$

A vinyl borane and an enol are intermediates

6. Bromo alcohols and chloro alcohols

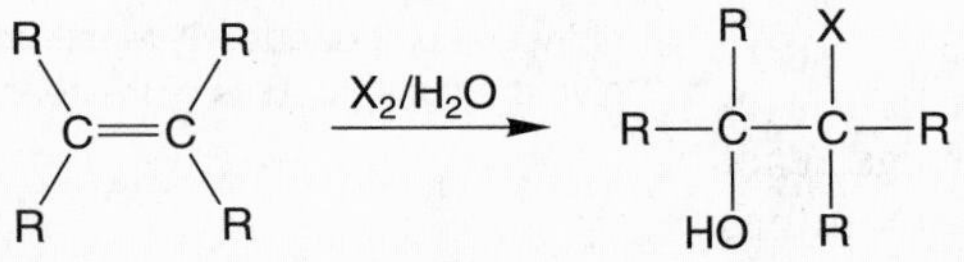

$R_2C{=}CR_2 \xrightarrow{X_2/H_2O} R{-}CR(OH){-}CR(X){-}R$

anti Addition, X = Br or Cl

7. Bromo ethers and chloro ethers

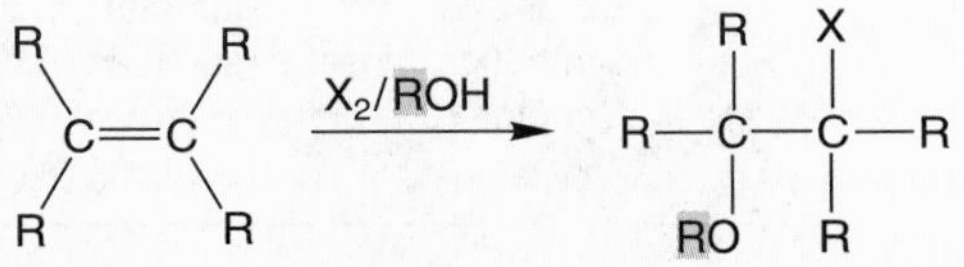

$R_2C{=}CR_2 \xrightarrow{X_2/ROH} R{-}CR(OR){-}CR(X){-}R$

anti Addition, X = Br or Cl

8. Cyclopropanes

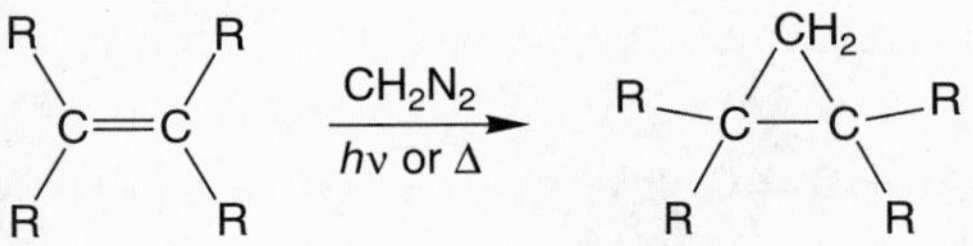

$R_2C{=}CR_2 \xrightarrow[h\nu \text{ or } \Delta]{CH_2N_2}$ cyclopropane ($R_2C{-}CR_2$ bridged by CH_2)

cis Addition for singlet carbene, but both cis and trans addition for triplet carbene

9. Dihalides

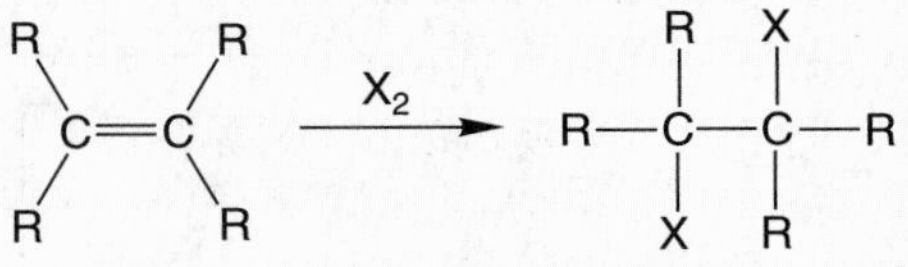

$R_2C{=}CR_2 \xrightarrow{X_2} R{-}CR(X){-}CR(X){-}R$

Vicinal dihalide, anti addition, X = Br or Cl

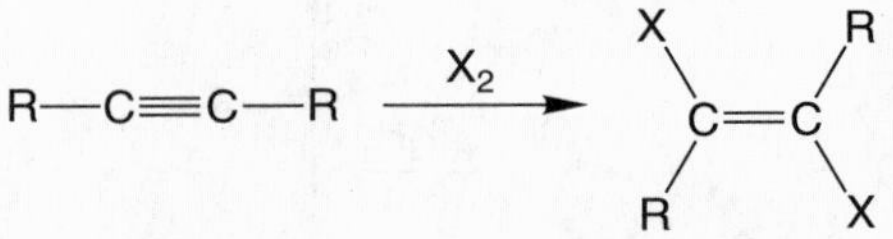

$R{-}C{\equiv}C{-}R \xrightarrow{X_2} RXC{=}CXR$

Vicinal dihalide, mostly anti addition, X = Br or Cl, further reaction likely

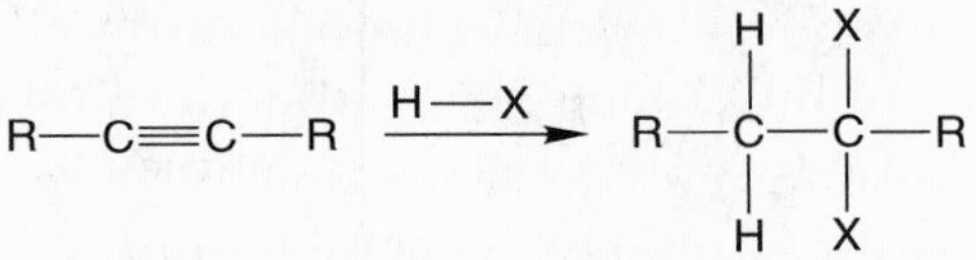

$R{-}C{\equiv}C{-}R \xrightarrow{H{-}X} R{-}CH_2{-}CX_2{-}R$

Geminal dihalide, X = Br, Cl, or I, double HX addition

10. Heterocyclic five-membered rings

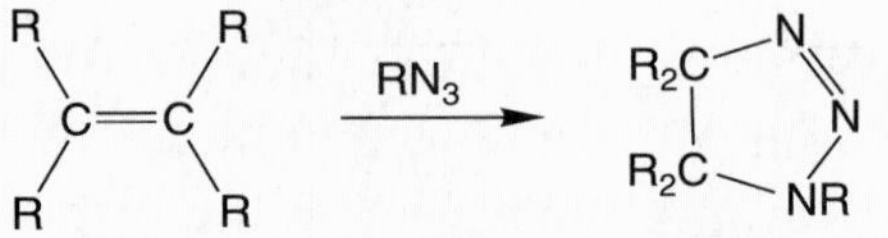

$R_2C{=}CR_2 \xrightarrow{RN_3}$ five-membered ring ($R_2C{-}N{=}N{-}NR{-}CR_2$)

This is only one example; there is a different product for every 1,3-dipole; alkynes work too

11. 1,2-Glycols

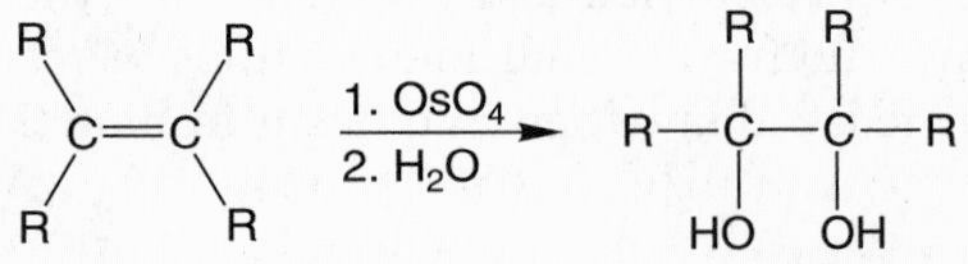

$R_2C{=}CR_2 \xrightarrow[\text{2. } H_2O]{\text{1. } OsO_4} R{-}CR(OH){-}CR(OH){-}R$

syn Addition

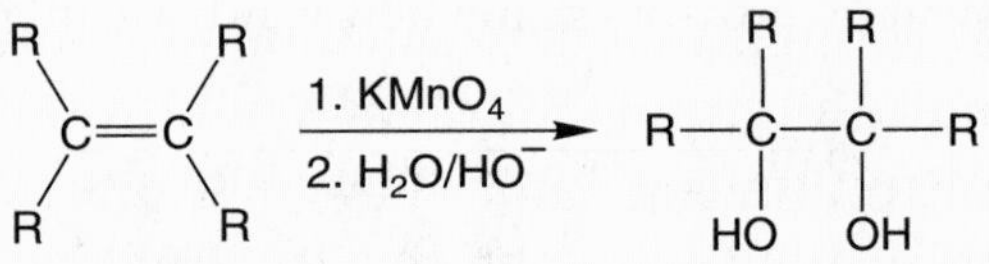

$R_2C{=}CR_2 \xrightarrow[\text{2. } H_2O/HO^-]{\text{1. } KMnO_4} R{-}CR(OH){-}CR(OH){-}R$

syn Addition

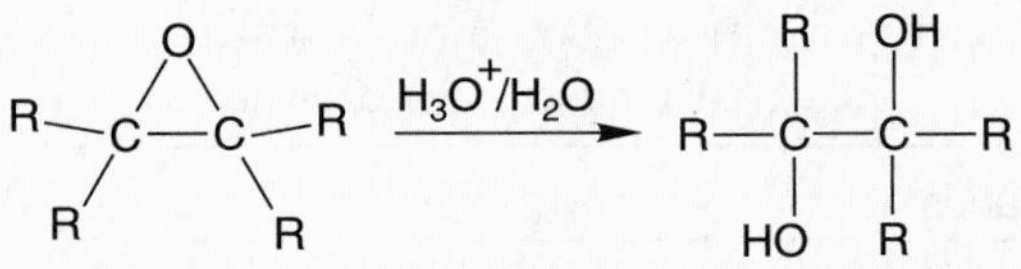

epoxide $\xrightarrow{H_3O^+/H_2O} R{-}CR(OH){-}CR(OH){-}R$

anti Addition

epoxide $\xrightarrow{H_2O/HO^-} R{-}CR(OH){-}CR(OH){-}R$

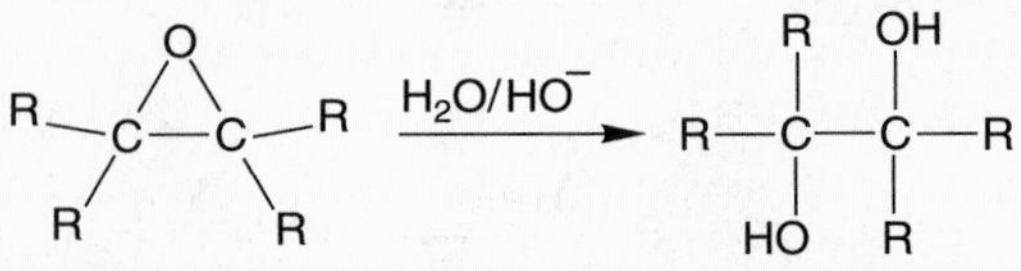

anti Addition

FIGURE 10.99 (CONTINUED)

12. Ketones

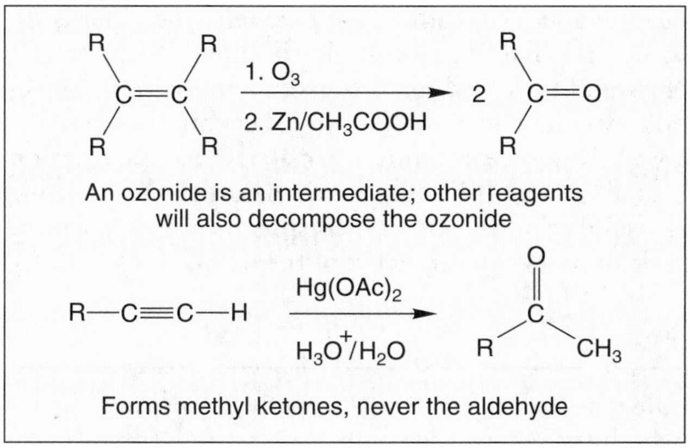

$$R_2C{=}CR_2 \xrightarrow[\text{2. Zn/CH}_3\text{COOH}]{\text{1. O}_3} 2\ R_2C{=}O$$

An ozonide is an intermediate; other reagents will also decompose the ozonide

$$R{-}C{\equiv}C{-}H \xrightarrow[\text{H}_3\text{O}^+/\text{H}_2\text{O}]{\text{Hg(OAc)}_2} R{-}C({=}O){-}CH_3$$

Forms methyl ketones, never the aldehyde

13. Oxiranes (epoxides)

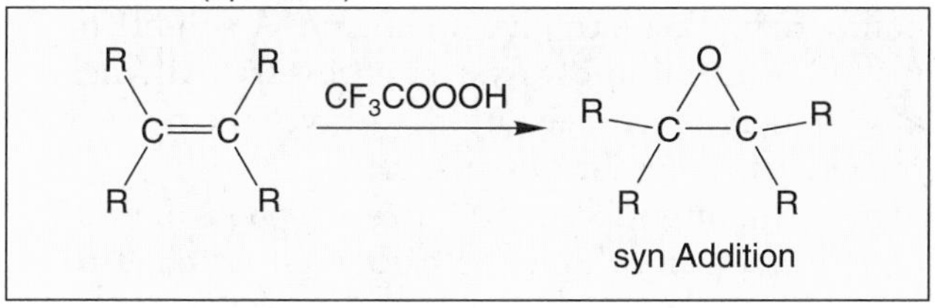

$$R_2C{=}CR_2 \xrightarrow{\text{CF}_3\text{COOOH}} R_2C{-}CR_2 \text{ (oxirane, O bridging both C)}$$

syn Addition

14. Tetrahalides

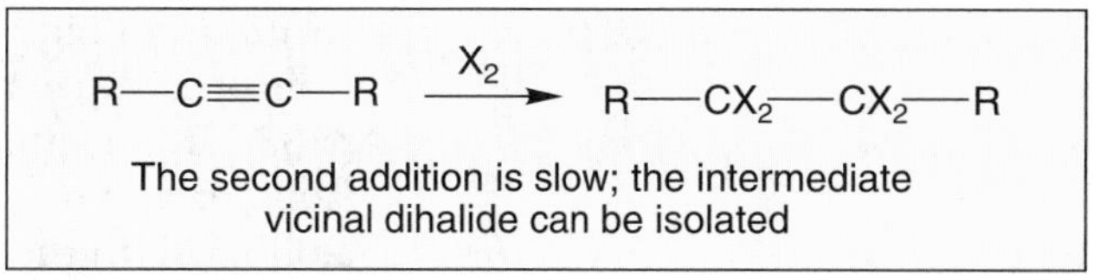

$$R{-}C{\equiv}C{-}R \xrightarrow{X_2} R{-}CX_2{-}CX_2{-}R$$

The second addition is slow; the intermediate vicinal dihalide can be isolated

15. Vinyl halides

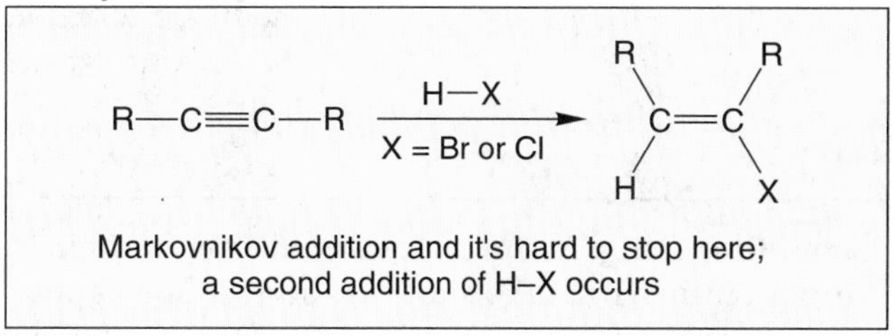

$$R{-}C{\equiv}C{-}R \xrightarrow[\text{X = Br or Cl}]{\text{H}{-}\text{X}} R(H)C{=}C(R)X$$

Markovnikov addition and it's hard to stop here; a second addition of H–X occurs

FIGURE 10.99 (CONTINUED)

COMMON ERRORS

What psychopathology exists in this chapter comes not so much from "something everyone always gets wrong" or from an overwhelmingly difficult concept, but from the mass of detail. There have been those who have succumbed to despair at sorting out all the stereochemical nuances, at dealing with the many new mechanisms and synthetic methods—at just making rational sense of all the material in this and Chapter 9. There certainly is beginning to be a lot of information, and you must be very careful to keep your new synthetic methods properly cataloged. There are unifying principles that may help you to keep from getting lost in the mechanism jungle. "Lewis acids react with Lewis bases." The Lewis base or nucleophile in most of the reactions in this chapter and Chapter 9 is the π bond of an alkene or an alkyne. The Lewis acids are numerous, but most add to give intermediate, or even stable three-membered rings (not all: hydrogenation does not, and 1,3-dipolar addition leads to a host of five-membered rings). The three-membered rings are themselves often prone to further reaction with nucleophiles, and the final products of reaction may be quite far removed in structure from the starting material!

Moreover, not all the mechanistic details are known about all these processes. There are reactions about which we need much more information. The reaction of acetylenes with halogens and HX are examples, and even hydrogenation has a complicated mechanism, still somewhat obscure to organic chemists.

A good technique that helps one not to get too overwhelmed or lost is to anchor oneself in one or two basic reactions and then to generalize; to relate other reactions to the anchor reaction. For example, the polar addition of hydrogen chloride and hydrogen bromide to alkenes (Chapter 9, p. 335) is within anyone's ability to master. Extensions to hydration reactions of alkenes and alkynes, and to hydroboration become easier if the analogy with the anchor is always kept in mind. Similarly, use the reaction of alkenes with Br_2 as an anchor on which to hang the other ring-forming addition reactions. It is useful to start a set of mechanism cards to go along with your synthesis cards in order to keep track of the detail.

10.13 KEY TERMS

Bromonium ion A three-membered ring containing bromine that is formed by the reaction of an alkene with Br_2. The bromine atom in the ring is positively charged.

Carbene A short-lived neutral intermediate containing a divalent carbon atom. See also **singlet** and **triplet carbene**.

Carbonyl compound A compound containing a carbon–oxygen double bond.

Carboxylic acid A compound of the structure

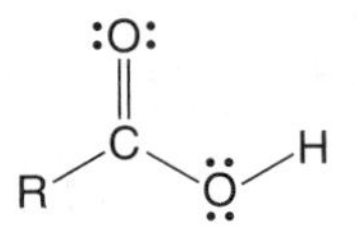

:O:
‖
R–C–Ö–H

Diimide HN=NH A nonmetallic hydrogenating agent.

1,3-Dipoles A class of reactive molecule, containing both positive and negative charges. These species undergo addition to π systems to give five-membered rings.

Diradical A diradical has a single nonbonding electron on each of two atoms.

Enol A vinyl alcohol. Enols are in equilibrium with carbonyl compounds.

Epoxide A three-membered ring containing one oxygen atom. See **oxirane**.

Geminal 1,1-Disubstituted.

1,2-Glycol A vicinal dialcohol. A molecule bearing hydroxyl groups on adjacent carbons. 1,2-Glycols come from the treatment of alkenes with osmium tetroxide or potassium permanganate followed by hydrolysis or by the acid- or base-catalyzed opening of epoxides.

Heterogeneous catalysis A catalytic process in which the catalyst is insoluble.

Homogeneous catalysis A catalytic process in which the catalyst is soluble.

Hydrogenation Addition of hydrogen (H_2) to the π bond of an alkene to give an alkane. A soluble or insoluble metallic catalyst is necessary. Alkynes also undergo hydrogenation to give alkanes, or under special conditions, alkenes.

Ketone A compound containing a carbon–oxygen double bond in which the carbon is attached to two other carbons (not hydrogen).

Molozonide The initial product of the reaction of ozone and an alkene. This molecule contains a five-membered ring with three oxygen atoms in a row. See **primary ozonide**.

Oxirane A three-membered ring containing one oxygen atom. See **epoxide**.

Oxymercuration The mercury-catalyzed conversion of alkenes into alcohols. Addition is in the Markovnikov sense, and there are no rearrangements. A three-membered ring containing mercury is an intermediate in the reaction. Alkynes also undergo oxymercuration to give enols that are rapidly converted into carbonyl compounds under the reaction conditions.

Ozonide The product of rearrangement of the primary ozonide formed on reaction of ozone with an alkene.

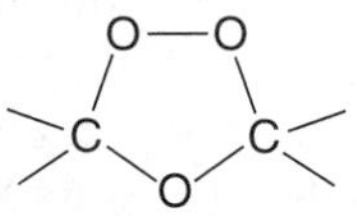

O—O
C   C
  O

Ozonolysis The reaction of ozone with π systems. The intermediate products are ozonides that can be transformed into carbonyl-containing compounds of various kinds.

Primary ozonide The initial product of the reaction of ozone and an alkene. This molecule contains a five-membered ring with three oxygen atoms in a row. See **molozonide**.

Protonium ion A three-membered ring in which a hydrogen atom bridges two carbons. In this ion two electrons bind three atoms, the two carbons and the hydrogen. It is very different from a bromonium ion in which the three atoms in the ring are connected by normal two-electron bonds.

Radical anion The species formed when an electron is added to a π system. In this case, the "extra" electron must be in an antibonding, π*, orbital.

Singlet carbene A singlet carbene contains only paired electrons. In a singlet carbene, the two nonbonding electrons have opposite spins and occupy the same orbital.

Solvated electron The species formed when sodium is dissolved in ammonia. The product is a sodium ion and an electron surrounded by ammonia molecules. This electron can add to alkynes (and some other π systems) in a reduction step.

Triplet carbene In a triplet carbene, the two nonbonding electrons have the same spin and must occupy different orbitals.

Vicinal 1,2-Disubstituted.

Zwitterion An internal salt. A molecule containing both plus and minus charges.

10.14 ADDITIONAL PROBLEMS

PROBLEM **10.27** Show the major organic product(s) expected when 1-methylcyclohexene reacts with the following reagents. Pay close attention to stereochemistry and regiochemistry where appropriate.

(a) D_2/Pd/C
(b) Br_2/CCl_4
(c) Br_2/CH_3OH
(d) $Hg(OAc)_2$, H_2O, then $NaBD_4$ in base (no stereochemical preference in this reaction)
(e) B_2H_6 (BH_3) in THF, then H_2O_2/HO^-
(f) CF_3COOOH
(g) CH_2N_2, $h\nu$
(h) O_3, then $(CH_3)_2S$
(i) OsO_4, then $NaHSO_3$, H_2O
(j) HN=NH

PROBLEM **10.28** Show in detail how both enantiomers of product are formed in Problem 10.27(a).

PROBLEM **10.29** Show the major organic products expected when the following acetylenes react with the reagents shown. Pay attention to stereochemistry and regiochemistry where appropriate.

(a) cyclohexyl–C≡CH $\xrightarrow{\text{HBr}}$

(b) cyclohexyl–C≡CH $\xrightarrow{Cl_2}$

(c) cyclohexyl–C≡CH $\xrightarrow[H_2O/H_3O^+]{Hg^{2+}}$

(d) cyclohexyl–C≡CH $\xrightarrow[\text{2. } H_2O_2/NaOH]{\text{1. (disiamyl)}_2\text{BH}}$

(e) cyclohexyl–C≡C–cyclohexyl $\xrightarrow[\text{Lindlar catalyst}]{H_2}$

(f) cyclohexyl–C≡C–cyclohexyl $\xrightarrow[NH_3]{Na}$

PROBLEM **10.30** In Section 10.1b (p. 393), we considered two possible mechanisms for the addition of Br_2 to alkenes. Ultimately, a stereochemical experiment that used a ring compound was used to decide the issue in favor of a mechanism in which addition proceeded through a bromonium ion rather than an open carbocation. Use a detailed stereochemical analysis to show how the experimental results shown below are accommodated by an intermediate bromonium ion but not by an open carbocation.

cis-2-Butene $\xrightarrow[CCl_4]{Br_2}$ Racemic dibromide

trans-2-Butene $\xrightarrow[CCl_4]{Br_2}$ Meso dibromide

PROBLEM **10.31** Explain the following regiochemical results mechanistically.

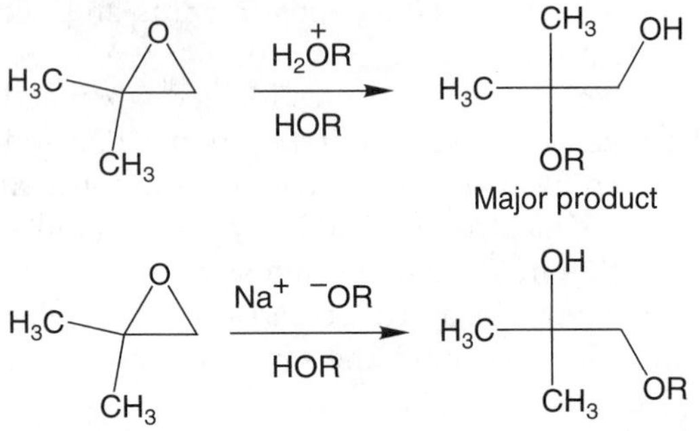

$H_2\overset{+}{O}R$ / HOR — Major product

$Na^+ \ ^-OR$ / HOR — Major product

PROBLEM **10.32** In Section 10.3a (p. 405), you saw that epoxidation of an alkene with a peracid such as trifluoroperacetic acid results in addition of an oxygen atom to the alkene in such as way as to preserve the stereochemical relationships originally present in the alkene in the product epoxide. That is, cis alkene leads to cis epoxide, and trans alkene leads to trans epoxide. There is another route to epoxides that involves cyclization of halohydrins. Use *cis*-2-butene as a substrate to analyze carefully the stereochemical outcome of this process. What happens to the alkene stereochemistry?

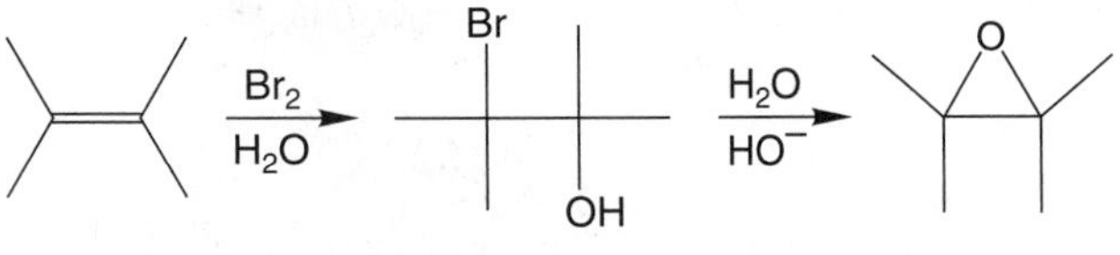

Br_2 / H_2O → A bromohydrin → H_2O / HO^-

PROBLEM **10.33** Show the final products of the reactions below. Pay attention to stereochemistry.

(a) CH_3 — CF_3COOOH → H_2O / NaOH →

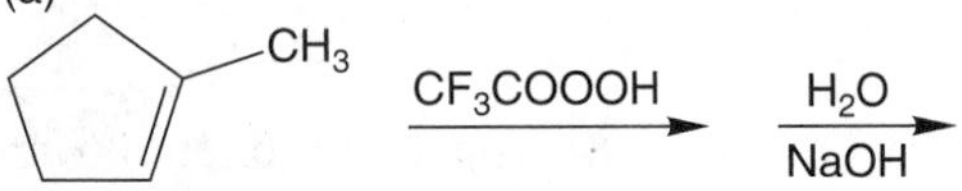

(b) CH_3 — $KMnO_4$ / $NaOH/H_2O$ →

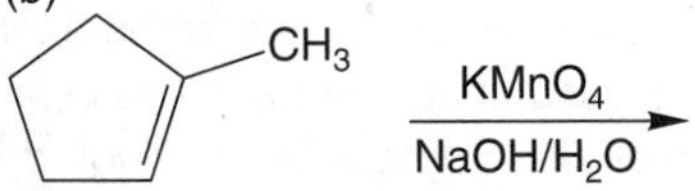

PROBLEM **10.34** Here is a new reaction that you will see again in Chapter 17. Glycols react with periodic acid to form a cyclic periodate intermediate. The periodate then decomposes to a pair of carbonyl compounds.

If the product from (b) in Problem 10.33 is cleaved with periodic acid, what would be the final product of the reaction sequence?

From a synthetic point of view, this two-step sequence—the treatment of an alkene with basic permanganate followed by periodate cleavage—is equivalent to what other direct method of cleaving carbon–carbon double bonds? What other way do you know to convert an alkene into a pair of aldehydes?

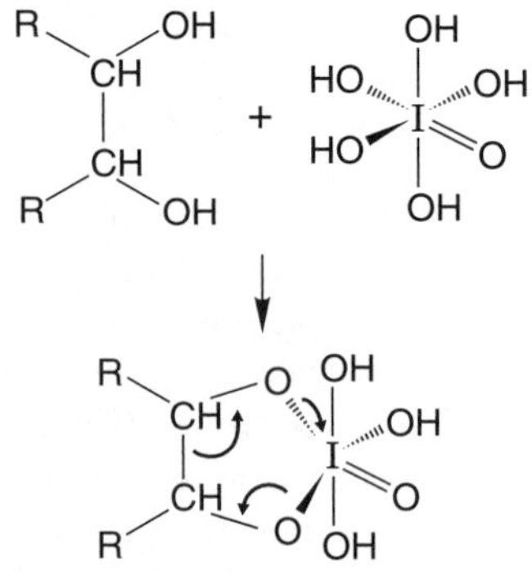

Periodate intermediate

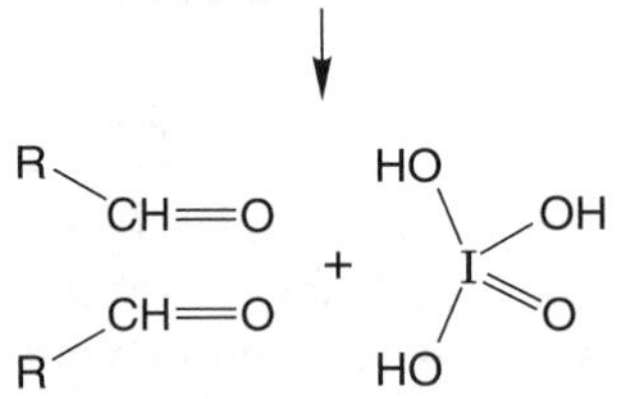

PROBLEM **10.35** Why doesn't the product of Problem 10.33a react with periodate?

PROBLEM **10.36** Incomplete catalytic hydrogenation of a hydrocarbon, **1** (C_5H_8), gives a mixture of three hydrocarbons, **2**, **3**, and **4**. Ozonolysis of **3**, followed by reductive workup, gives formaldehyde and 2-butanone. When treated in the same way, **4** gives formaldehyde and isobutyraldehyde. Provide structures for compounds **1**–**4** and explain your reasoning.

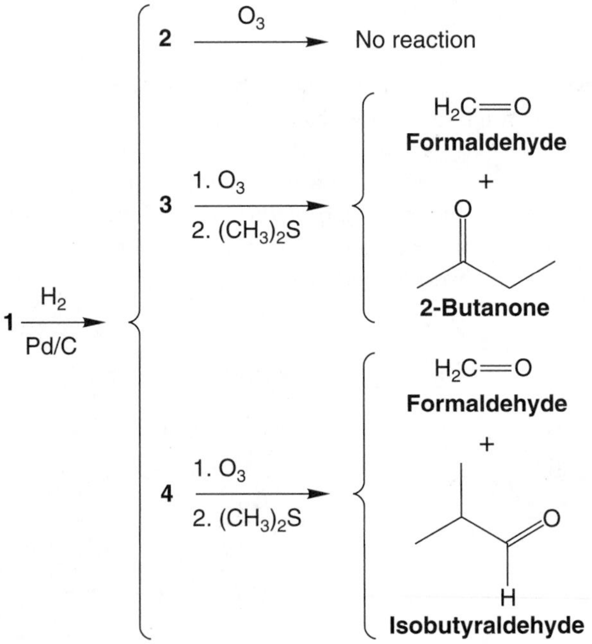

1 —H_2, Pd/C→

2 —O_3→ No reaction

3 —1. O_3; 2. $(CH_3)_2S$→ $H_2C{=}O$ **Formaldehyde** + **2-Butanone**

4 —1. O_3; 2. $(CH_3)_2S$→ $H_2C{=}O$ **Formaldehyde** + **Isobutyraldehyde**

PROBLEM **10.37** α-Terpinene (**1**) and γ-terpinene (**2**) are isomeric compounds ($C_{10}H_{16}$) that are constituents of many plants. Upon catalytic hydrogenation, they both afford 1-isopropyl-4-methylcyclohexane. However, on ozonolysis followed by oxidative workup, each compound yields different products. Provide structures for **1** and **2** and explain your reasoning.

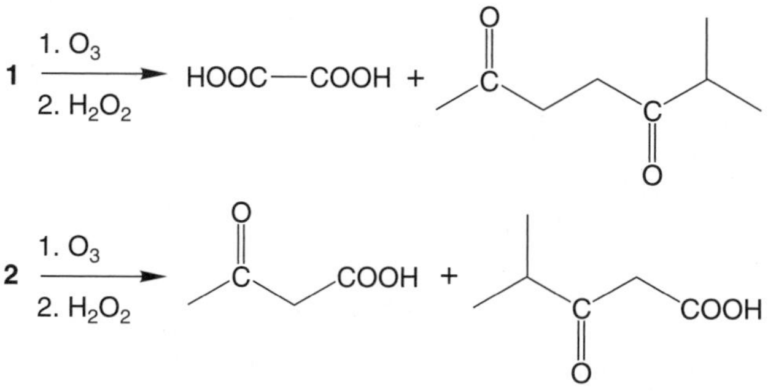

1 —(1. O_3; 2. H_2O_2)→ HOOC—COOH + [diketone]

2 —(1. O_3; 2. H_2O_2)→ [keto acid] COOH + [keto acid] COOH

PROBLEM **10.38** In practice, it is often very difficult to isolate ozonides from ozonolysis of tetrasubstituted double bonds. The product mixture depends strongly on reaction conditions, but the products shown below can all be isolated. Construct reasonable arrow formalisms for all of them.

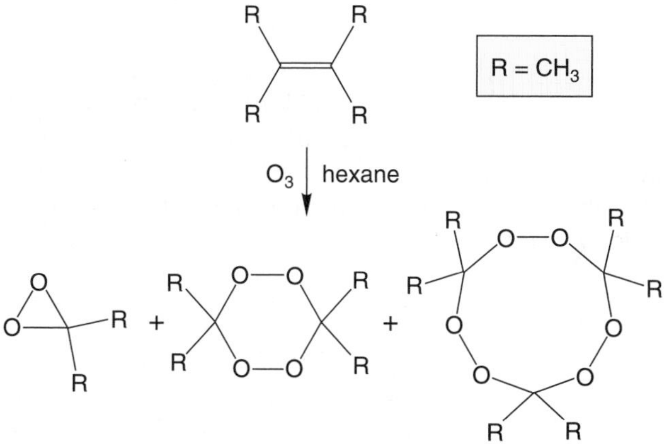

R = CH_3

O_3 | hexane

PROBLEM **10.39** Ozonolysis of **1** in solvent acetone leads to **2** as the major product. Explain.

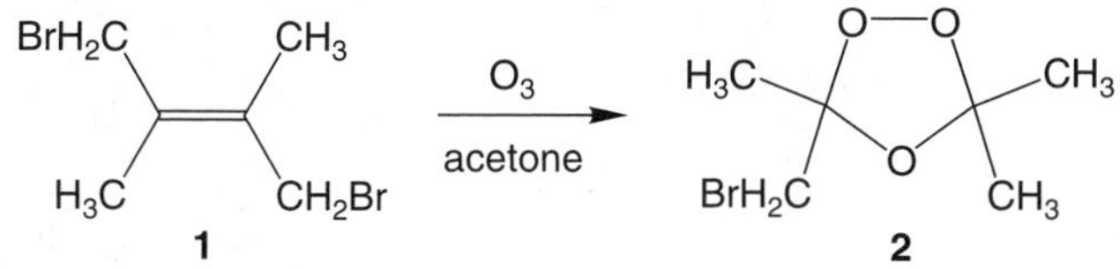

BrH_2C, CH_3, H_3C, CH_2Br **1** —(O_3, acetone)→ H_3C, CH_3, BrH_2C, CH_3 **2**

PROBLEM **10.40** *trans*-β-Methylstyrene reacts with phenyl azide to give a single product, triazoline (**1**). What other stereoisomeric products might have been produced? Draw an arrow formalism for this reaction and explain what mechanistic conclusions can be drawn from the formation of a *single* isomer of **1**.

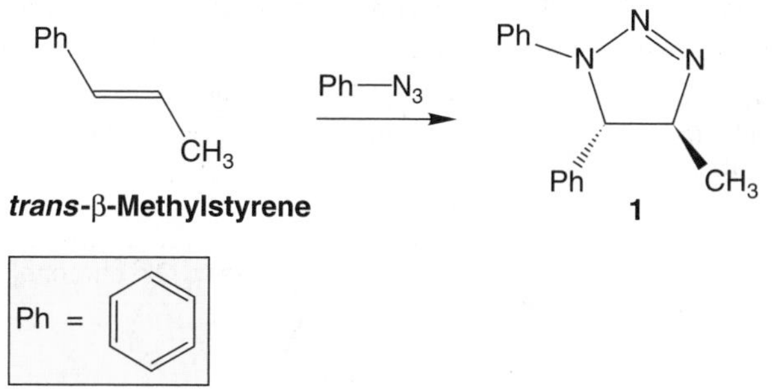

***trans*-β-Methylstyrene** —(Ph—N_3)→ **1**

Ph =

PROBLEM **10.41** Triazoline (**1**) (formed in Problem 10.40) decomposes upon photolysis or heating to give aziridine (**2**). Write two mechanisms for the conversion of **1** to **2**; one a concerted, one-step reaction, and the other is a nonconcerted, two-step process. How would you use the stereochemically labeled triazoline (**1**) to tell which mechanism is correct?

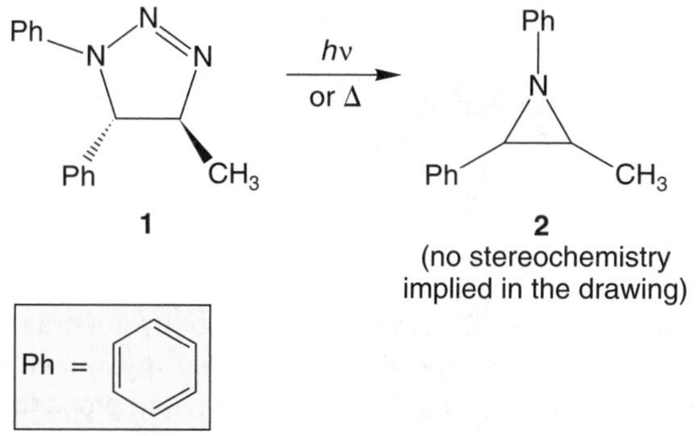

1 —(hν or Δ)→ **2**

(no stereochemistry implied in the drawing)

Ph =

PROBLEM **10.42** Explain the formation of aziridine (**2**) in the following reaction of azide **1**. *Hints*: Draw a full Lewis structure for the azide, and see Section 10.3d (p. 411).

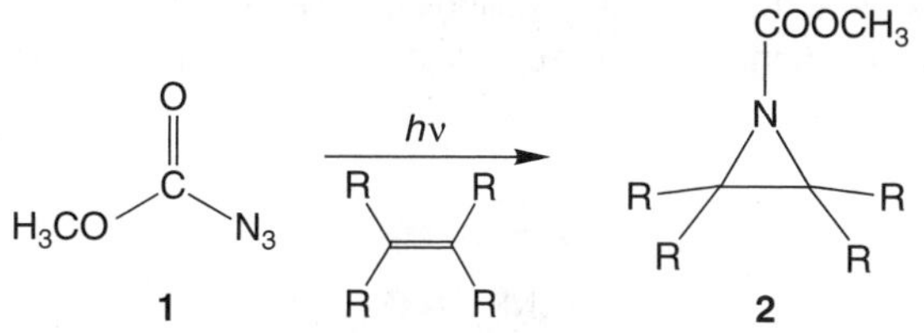

H_3CO–C(=O)–N_3 **1** —(hν, $R_2C{=}CR_2$)→ **2** ($COOCH_3$)

PROBLEM **10.43** What does the formation of only cis aziridine (**2**) from irradiation of azide (**1**) in *cis*-2-butene tell you about the nature of the reacting species in Problem 10.42?

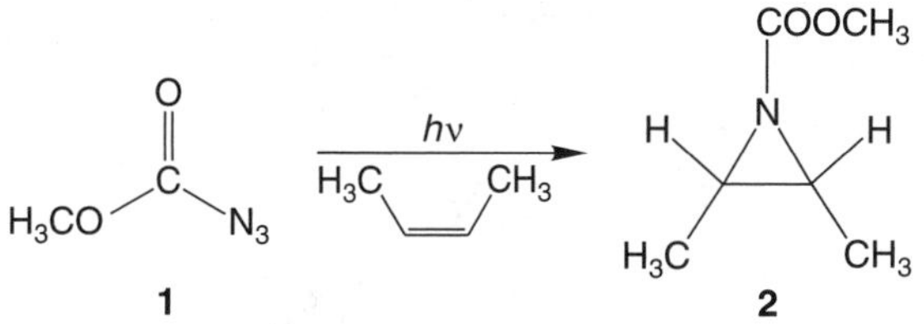

H_3CO–C(=O)–N_3 **1** —(hν, H_3C CH_3)→ **2** ($COOCH_3$, H, H, H_3C, CH_3)

PROBLEM **10.44** A reaction of carbenes not mentioned in the text is called "carbon–hydrogen insertion." In this startling reaction, a reactive carbene ultimately places itself in between the carbon and hydrogen of a carbon–hydrogen bond.

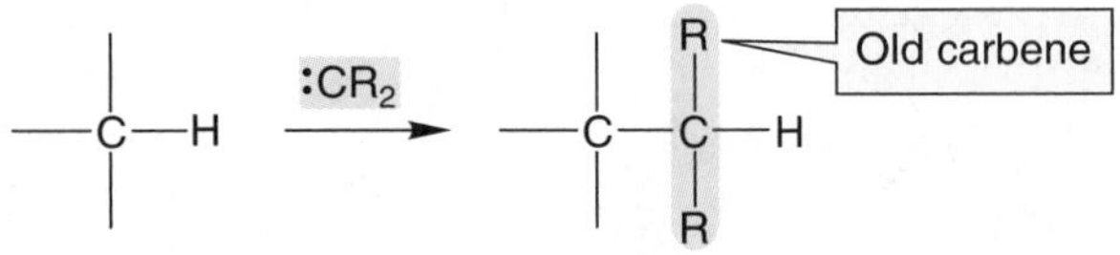

$$\text{—C—H} \xrightarrow{:CR_2} \text{—C—CR}_2\text{—H}$$

Old carbene

(a) Write two mechanisms for this reaction, one single step and the other two steps.
(b) In 1959, in a classic experiment in carbene chemistry, Doering and Prinzbach used ^{14}C-labeled isobutene to distinguish the two mechanisms. Explain what their results mean. The figure shows only one product of the many formed in the reaction. Focus on this compound only.

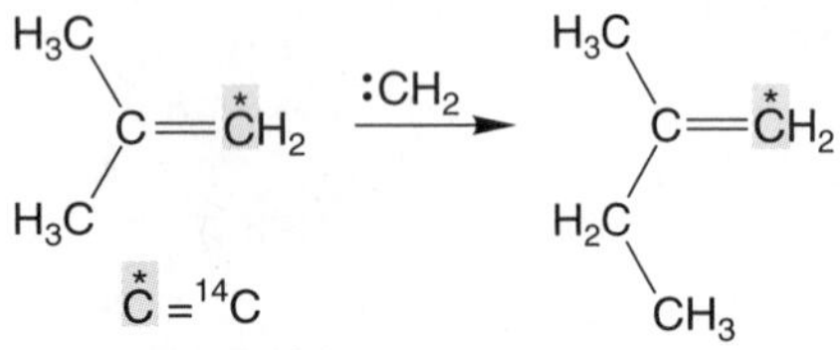

$$(H_3C)_2C{=}\overset{*}{C}H_2 \xrightarrow{:CH_2} (H_3C)(CH_3CH_2)C{=}\overset{*}{C}H_2$$

$\overset{*}{C} = {}^{14}C$

PROBLEM **10.45** Contrast the results of hydroboration–oxidation and mercury (Hg^{2+})-catalyzed hydration for 2-pentyne and 3-hexyne. Would any of these procedures be a practical preparative method? Explain.

PROBLEM **10.46** Recall that terminal alkynes are among the most acidic of the hydrocarbons (Chapter 4, p. 149), and that the acetylide ions can be used in S_N2 alkylation reactions with an appropriate alkyl halide. For example,

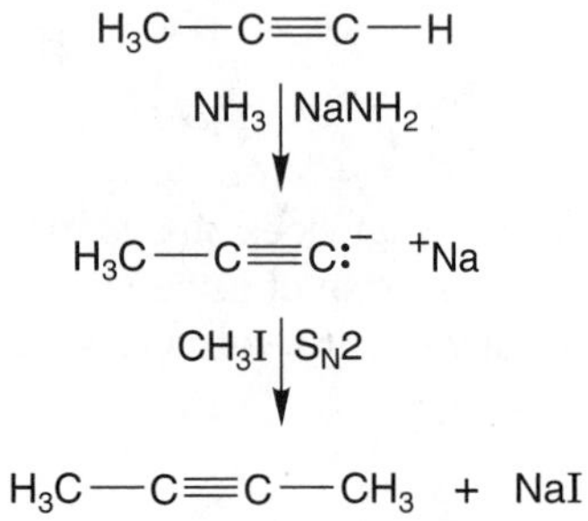

$$H_3C{-}C{\equiv}C{-}H \xrightarrow{NH_3 \;|\; NaNH_2} H_3C{-}C{\equiv}C{:}^- \;{}^+Na \xrightarrow{CH_3I \;|\; S_N2} H_3C{-}C{\equiv}C{-}CH_3 + NaI$$

Provide syntheses for the following molecules, free of other isomers. You must use alkynes containing no more than four carbon atoms as starting materials. You may use inorganic reagents of your choice, and other organic reagents containing no more than two carbon atoms. Mechanisms are not required.

(a) $H_3C{-}C({=}O){-}CH_2CH_2CH_2CH_3$

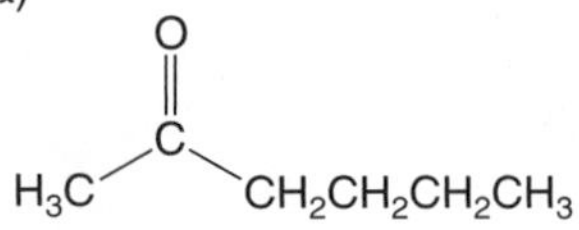

(b) $H{-}C({=}O){-}CH_2CH_2CH_2CH_3$

(c) epoxide: O bridging two carbons; H and CH_3CH_2 on one carbon, H and $CH_2CH_2CH_2CH_3$ on the other

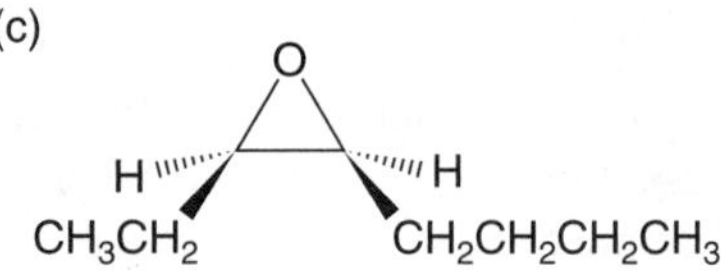

(d) epoxide: O bridging two carbons; H and $CH_3CH_2CH_2$ on one carbon, $CH_2CH_2CH_3$ and H on the other

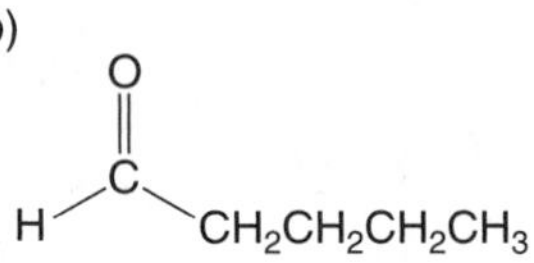

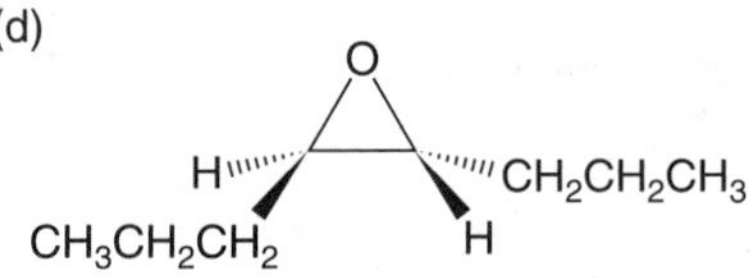

PROBLEM **10.47** Many additions of bromine to acetylenes give only trans dibromide intermediates. However, there are exceptions. Here is one. Explain why phenylacetylene gives both cis and trans dibromides on reaction with bromine.

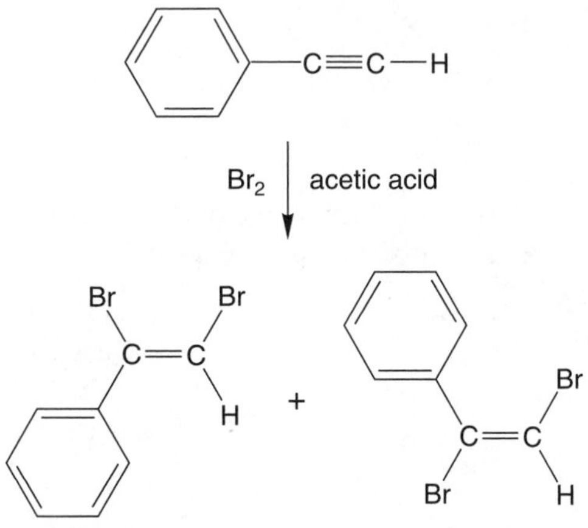

$$C_6H_5{-}C{\equiv}C{-}H \xrightarrow{Br_2 \;|\; \text{acetic acid}} (Br)(C_6H_5)C{=}C(Br)(H) \;(31\%) + (C_6H_5)(Br)C{=}C(Br)(H) \;(69\%)$$

Radical Reactions

Knowledge may have its purposes,
But guessing is always
More fun than knowing.
—W. H. Auden*

Although there have been many variations in the details, with few exceptions all of the reactions we have studied so far have been polar ones. The S_N1 and S_N2 substitutions, the addition reactions of HX and X_2 reagents, and the E1 and E2 reactions all involve cationic and anionic intermediates, or at least have obviously polar transition states.

Now we come to a series of quite different processes involving the decidedly nonpolar, neutral intermediates called **radicals** or, sometimes, **free radicals**. We will also encounter **chain reactions**, in which a small number of radicals can set a cycling reaction in motion. In such a case, a few starter molecules can determine the course of an entire chemical process.

The question of selectivity is prominent in this chapter. What factors influence the ability of a reactive species to pick and choose among various reaction pathways? In particular, what is the connection between the energy of a reactive molecule and its reactivity?

Don't forget—throughout our discussion of these radical reactions, the motion of single electrons is shown with single-barbed, "fishhook" arrows.

Convention Alert!

We begin with a section on the formation and simple reactions of radicals, and then move on to structure before considering the more complicated chain reactions.

*PROBLEM 11.1

Find an exception to the generalization in the first paragraph. What reaction have we studied that involves no polar intermediates, and doesn't have a highly polar transition state? *Hint*: Think "recent."

ANSWER

One example is the reaction most recently studied in Section 10.11 (p. 436), the addition of a triplet carbene to an alkene. Only neutral intermediates are involved in this reaction. Another example is hydrogenation, the conversion of an alkene

*Wystan Hugh Auden (1907–1975) was one of the best known and most influential British poets of his generation.

ANSWER (CONTINUED) and hydrogen into an alkane (Chapter 10, p. 389). Remember the use of the vertical arrows to indicate election spin.

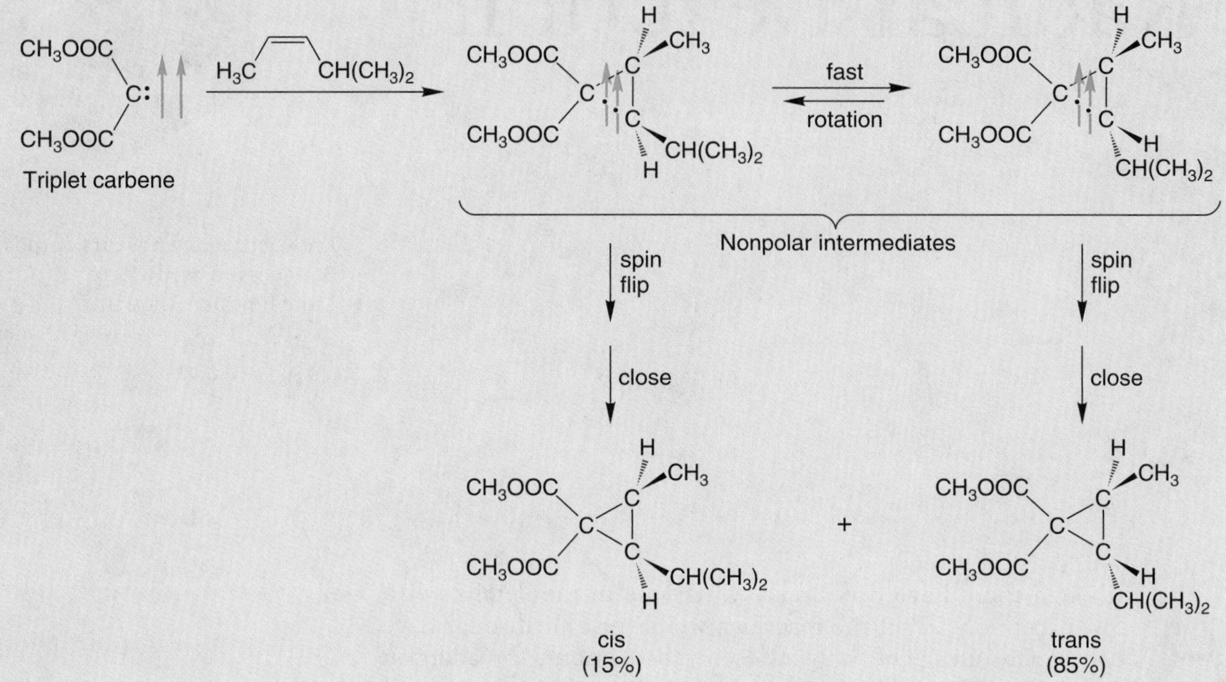

Triplet carbene

fast

rotation

Nonpolar intermediates

spin flip

close

cis (15%)

+

trans (85%)

PROBLEM 11.2 Find a reaction that involves no intermediates but certainly has a highly polar transition state.

11.1 FORMATION AND SIMPLE REACTIONS OF RADICALS

We have actually seen some radicals before (in Chapters 1 and 2) when we considered the formation of molecules such as hydrogen, methane, and ethane from their constituent parts: hydrogen atoms and methyl radicals (see Sections 1.3 and 2.1; Fig. 11.1). We also encountered them briefly in Section 3.4 (p. 81) in which we speculated a bit about the structure of radicals.

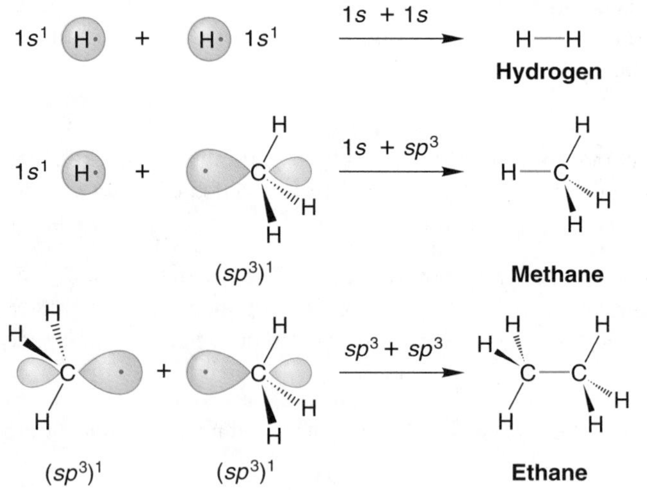

$1s^1$ H· + H· $1s^1$ → ($1s + 1s$) H—H **Hydrogen**

$1s^1$ H· + $(sp^3)^1$ → ($1s + sp^3$) **Methane**

$(sp^3)^1$ + $(sp^3)^1$ → ($sp^3 + sp^3$) **Ethane**

FIGURE 11.1 The formation of hydrogen, methane, and ethane through reactions of the radicals, H· and ·CH_3.

Why can't we just reverse the bond-forming process to produce radicals from these molecules? Energy will surely have to be added in the form of heat, because bond forming is highly exothermic and bond breaking is highly endothermic. Figure 11.2 shows schematically the stabilization gained by bond formation from the overlap of two singly occupied sp^3 orbitals. In order to reverse the bond-making process and form two radicals, energy must be provided up to the amount of the stabilization achieved in the bond making. For example, we can apply heat to ethane in a process called **pyrolysis** or **thermolysis**, eventually providing 90 kcal/mol and breaking the carbon–carbon bond (Fig. 11.2). Carbon–carbon bond cleavage is the lowest energy process easily detectable in ethane.

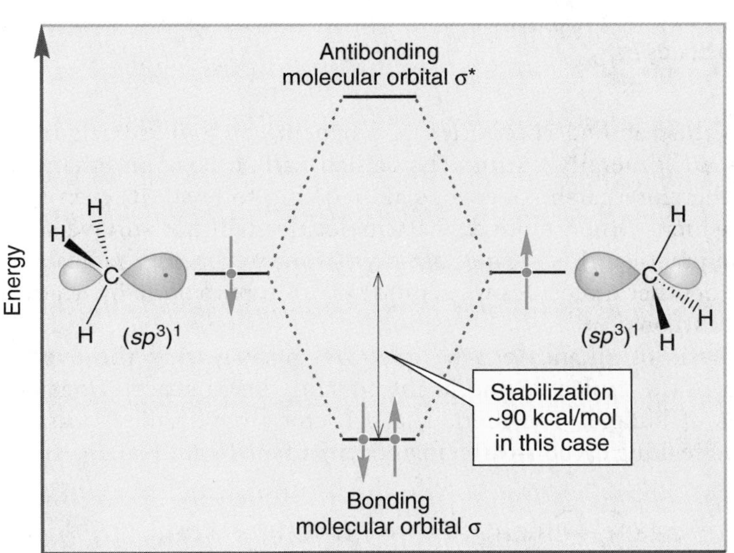

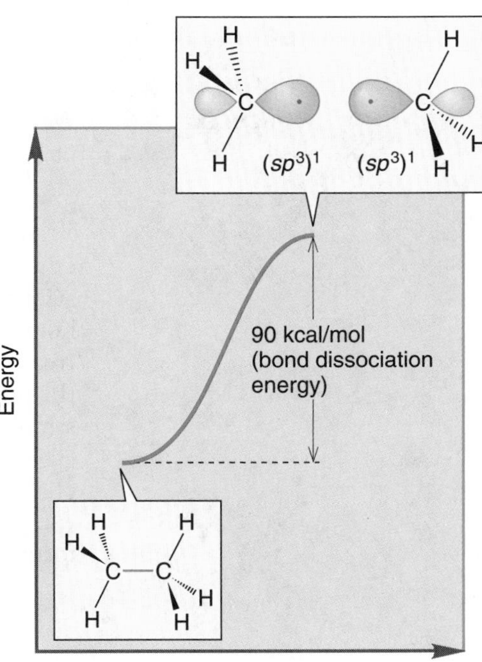

FIGURE **11.2** A schematic picture of the formation of a covalent bond through overlap of two half-filled sp^3 orbitals. To reverse the process, energy must be supplied. For example, methyl radicals can be formed by supplying enough energy (90 kcal/mol) to break the carbon–carbon bond in ethane.

*PROBLEM **11.3**

What unseen process is going on in ethane as we heat it?

ANSWER

Don't forget rotation about the carbon–carbon bond! No new product is formed, but this invisible reaction is still occurring.

The energy required to break the bond is called the **bond dissociation energy**. A series of these values was collected in Table 8.2 (p. 307). Note that the bond breaks to give two neutral species, the two methyl radicals, and not to give two polar species, the methyl cation and the methyl anion (Fig. 11.3).

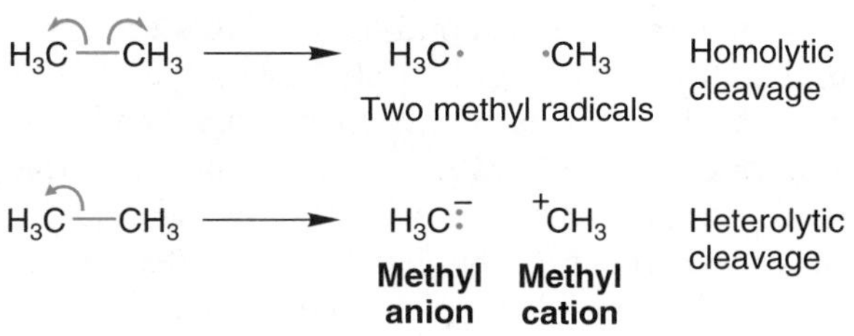

$H_3C{-}CH_3 \longrightarrow H_3C\cdot \quad \cdot CH_3$ Two methyl radicals — Homolytic cleavage

$H_3C{-}CH_3 \longrightarrow H_3C{:}^- \quad ^+CH_3$ **Methyl anion** **Methyl cation** — Heterolytic cleavage

FIGURE 11.3 Two possible bond breakings in ethane. The homolytic cleavage requires much less energy than the heterolytic formation of two ions.

Bond breaking to give a pair of neutral radicals is called **homolytic bond breaking**, whereas bond breaking to make the ions is called **heterolytic bond breaking**. Recall the use of single-barbed fishhook arrows in writing arrow formalisms for reactions involving single electron species (radicals).

PROBLEM 11.4 Why does the carbon–carbon bond in ethane break rather than one of the carbon–hydrogen bonds?

Although this method is effective in producing methyl radicals from ethane, it is not generally useful. The carbon–carbon bond in ethane is strong, and therefore harsh conditions are required to break it, even in a homolytic fashion. Other, more delicate molecules will not survive such treatment. In practice, this means that if we want to generate radicals in the presence of other molecules we should not attempt to do so by heating simple hydrocarbons.

The pyrolysis of alkanes can be effective in decreasing the overall chain length of the molecule and in introducing unsaturation. Consider the pyrolysis of butane. The methyl, ethyl, and propyl radicals can be formed from breaking carbon–carbon bonds in butane (Fig. 11.4).

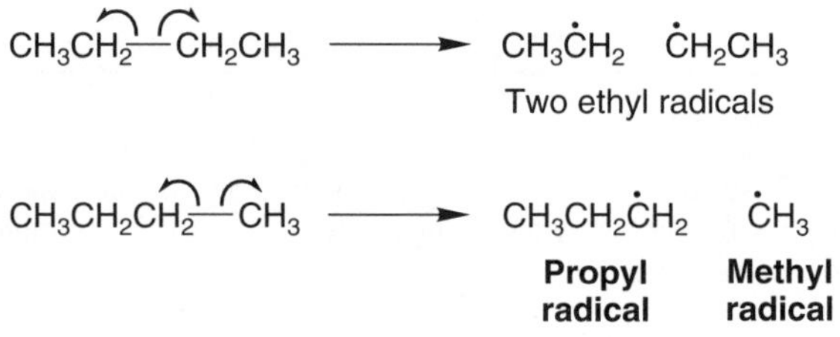

$CH_3CH_2{-}CH_2CH_3 \longrightarrow CH_3\dot{C}H_2 \quad \dot{C}H_2CH_3$ Two ethyl radicals

$CH_3CH_2CH_2{-}CH_3 \longrightarrow CH_3CH_2\dot{C}H_2 \quad \dot{C}H_3$ **Propyl radical** **Methyl radical**

FIGURE 11.4 There are two possible homolytic cleavages of carbon–carbon bonds in butane. Two ethyl radicals or a methyl radical and a propyl radical can be produced.

One reaction possible for these new radicals is abstraction of hydrogen from butane to make the smaller hydrocarbons methane, ethane, and propane, along with the butyl and *sec*-butyl radicals. There are two positions from which hydrogen abstraction can take place in butane, the primary C–H of the methyl group or the secondary C–H of the methylene group. We will consider the question of selectivity, of choosing between the two different positions, a little later in this chapter (Fig. 11.5).

PROBLEM 11.5 Draw the transition states for the two possible abstractions of hydrogen from butane by a methyl radical.

It is also possible for one radical to abstract a hydrogen from the β-position of another radical in a process called **disproportionation**, which

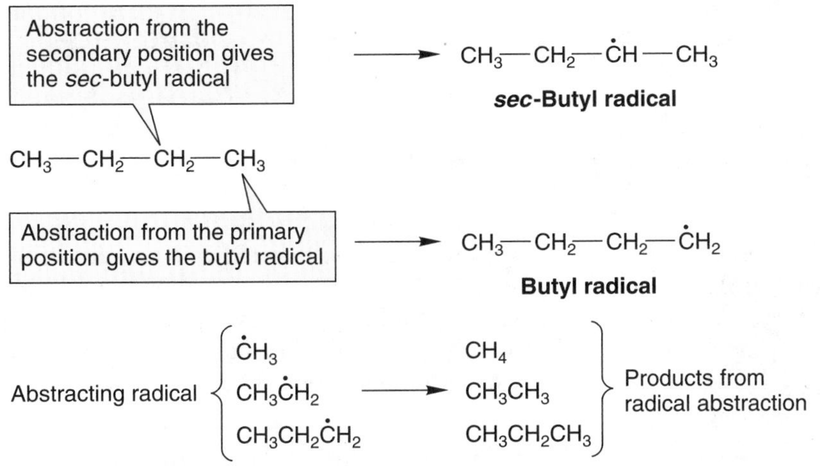

Abstraction from the secondary position gives the *sec*-butyl radical

$CH_3—CH_2—CH_2—CH_3$

Abstraction from the primary position gives the butyl radical

→ $CH_3—CH_2—\dot{C}H—CH_3$ **sec-Butyl radical**

→ $CH_3—CH_2—CH_2—\dot{C}H_2$ **Butyl radical**

Abstracting radical {$\dot{C}H_3$, $CH_3\dot{C}H_2$, $CH_3CH_2\dot{C}H_2$} → {CH_4, CH_3CH_3, $CH_3CH_2CH_3$} Products from radical abstraction

FIGURE 11.5 Hydrogen abstraction from butane by radicals leads to the formation of lower molecular weight hydrocarbons and two different butyl radicals.

produces an alkane and an alkene. In the propyl radical, for example, a carbon–hydrogen bond could be broken by a methyl radical to give propene and methane. In the terminology of radical chemistry, it is the α–β bond that is broken. The point of reference is the radical itself; the adjacent atom, the carbon, is α, and the next atom, the hydrogen, is β. There are many radicals that can do the abstracting (Fig. 11.6).

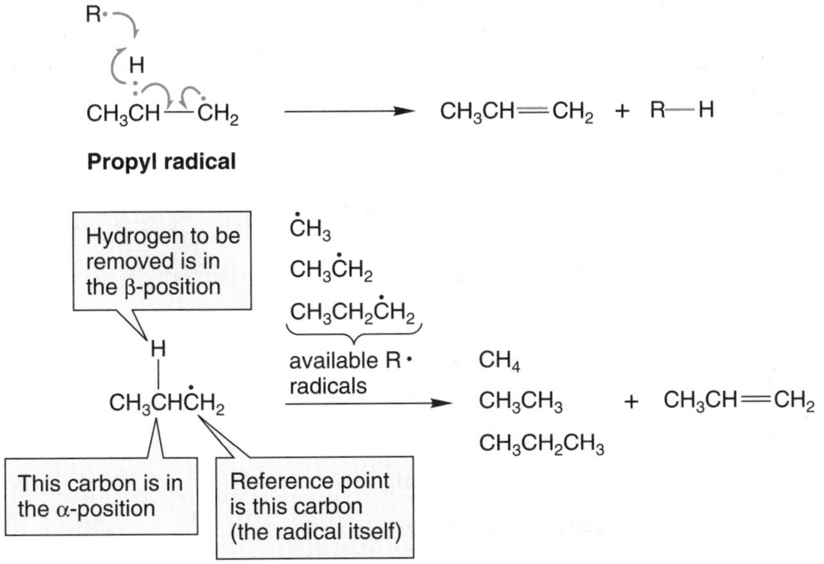

R· + $CH_3CH(H)—\dot{C}H_2$ → $CH_3CH{=}CH_2$ + R—H

Propyl radical

Hydrogen to be removed is in the β-position

$CH_3CH(H)\dot{C}H_2$ —(available R· radicals: $\dot{C}H_3$, $CH_3\dot{C}H_2$, $CH_3CH_2\dot{C}H_2$)→ CH_4, CH_3CH_3, $CH_3CH_2CH_3$ + $CH_3CH{=}CH_2$

This carbon is in the α-position

Reference point is this carbon (the radical itself)

FIGURE 11.6 The disproportionation reaction produces an alkane and an alkene from a pair of radicals. A hydrogen atom of one radical is abstracted by another radical.

PROBLEM 11.6

Draw the transition state for hydrogen abstraction in a disproportionation reaction.

Another reaction possible for many radicals is β-cleavage (Fig. 11.7). If an α–β carbon–carbon bond is present, it can break to give a new radical and an alkene. This cleavage is simply the reverse of an addition of a radical to an alkene (see Section 11.4, p. 462 for a fuller discussion of this process). In the *sec*-butyl radical, for example, cleavage between the methylene and methyl groups produces a methyl radical and propene. The β-cleavage reaction is favored at high temperature, and can be quite important in high-temperature reactions of alkanes.

FIGURE 11.7 β-Cleavage of the *sec*-butyl radical.

$$H_3C \cdot\!\cdot\, CH_2 - \dot{C}H - CH_3 \longrightarrow H_3C\cdot + CH_2{=}CH{-}CH_3$$

PROBLEM 11.7 Why will β-cleavage be favored at high temperature?

To make the issue even more complex, two radicals can occasionally react with each other (dimerize) to give new hydrocarbons (dimers) that can reenter the reaction sequence and provide new species. For example, two propyl radicals might combine to form hexane. Further pyrolysis of hexane would lead to *many* other products (Fig. 11.8).

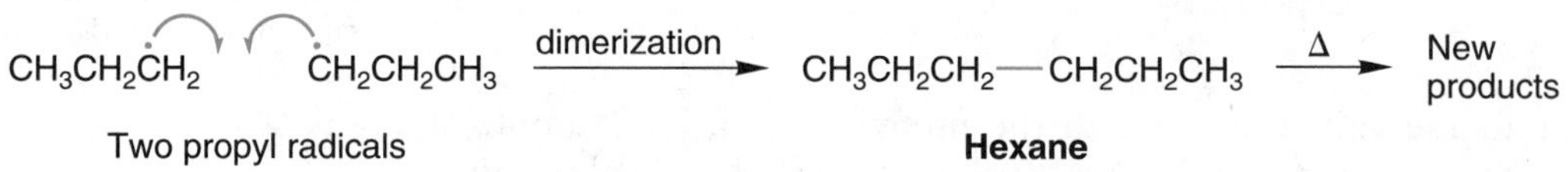

$$CH_3CH_2\dot{C}H_2 \quad \dot{C}H_2CH_2CH_3 \xrightarrow{\text{dimerization}} CH_3CH_2CH_2{-}CH_2CH_2CH_3 \xrightarrow{\Delta} \text{New products}$$

Two propyl radicals — **Hexane**

FIGURE 11.8 Radicals can combine in a very exothermic formation of new, larger hydrocarbons. These new products now enter the pyrolysis process and lead to other products.

To summarize: First, the reactions available to an alkyl radical include dimerization, hydrogen abstraction to make a saturated compound and a new radical, as well as disproportionation (hydrogen abstraction from another radical to give an alkene and an alkane) and β-cleavage (the fragmentation of a radical into a new radical and an alkene) (Fig. 11.9).

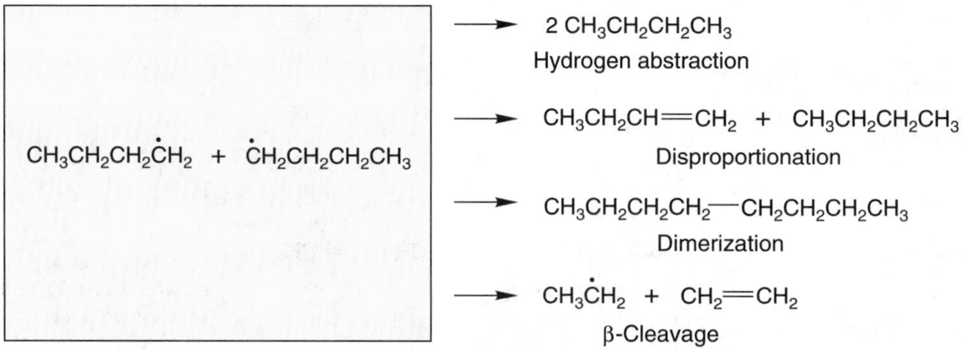

$$CH_3CH_2CH_2\dot{C}H_2 + \dot{C}H_2CH_2CH_2CH_3$$

$\longrightarrow$ 2 $CH_3CH_2CH_2CH_3$
Hydrogen abstraction

$\longrightarrow$ $CH_3CH_2CH{=}CH_2 + CH_3CH_2CH_2CH_3$
Disproportionation

$\longrightarrow$ $CH_3CH_2CH_2CH_2{-}CH_2CH_2CH_2CH_3$
Dimerization

$\longrightarrow$ $CH_3\dot{C}H_2 + CH_2{=}CH_2$
β-Cleavage

FIGURE 11.9 Simple reactions of radicals.

Clearly, little specificity is possible in this pyrolytic process, and one might at first expect that there would be little utility in such a series of reactions. That's not quite right. The petroleum industry relies on this kind of **hydrocarbon cracking** to convert the high molecular weight hydrocarbons making up much of crude oil into the much more valuable lower molecular weight gasoline fractions. It is also possible to do the pyrolysis in the presence of hydrogen to help produce smaller alkanes from the radicals that are produced initially.

PROBLEM 11.8 What function does the hydrogen perform in such a reaction?

*PROBLEM **11.9**

What products do you expect from the thermal cracking of hexane?

ANSWER

The point of this problem is to show you how incredibly complicated this "simple" process becomes even in a relatively small alkane such as hexane. First of all, carbon–carbon bond breakings give the methyl, ethyl, propyl, butyl, and pentyl radicals, which can abstract hydrogen to give methane, ethane, propane, butane, and pentane.

$\xrightarrow{\Delta}$ $\cdot CH_3$ → CH_4

→ $\cdot CH_2CH_3$ → CH_3CH_3

Disproportionation gives the alkanes already mentioned, along with unsaturated compounds, ethylene, propene, 1-butene, and 1-pentene.

$\cdot CH_2CH_3$ → $CH_2{=}CH_2$

β-Cleavage of these radicals gives more ethylene and the methyl, ethyl, and propyl radicals.

$\xrightarrow{\beta\text{-cleavage}}$ $H_2C{=}CH_2 + \cdot CH_3$

→ $H_2C{=}CH_2 + \cdot CH_2CH_3$

→ $H_2C{=}CH_2 + \cdot$

Of course, the alkanes and alkenes formed in these reactions can react further. Propane, butane, and pentane as well as the alkenes formed in the disproportionation and β-cleavage steps can participate in similar radical reactions. Moreover, dimerization reactions have the potential of introducing larger hydrocarbons, only two of which are shown below. These dimers can go on to produce other molecules as they too begin to participate in the radical reactions. It's an incredible mess.

→ Decane →

\+ → Nonane →

Many more products

Now comes the big question, How can ethane be modified in order to make the carbon–carbon bond weaker, and thus more easily broken?

PROBLEM **11.10**

Construct (on paper) a molecule with a carbon–carbon bond weaker than the one in ethane.

There are many other possible answers to Problem 11.10. For example, we could induce some strain in the molecule, raising its energy and making bond breaking easier. Cyclopropane is a nice example. This small ring compound contains both severe angle strain and substantial torsional strain (Section 6.2, p. 200). The strain energy of cyclopropane results in a low-bond dissociation energy of only 65 kcal/mol (Fig. 11.10).

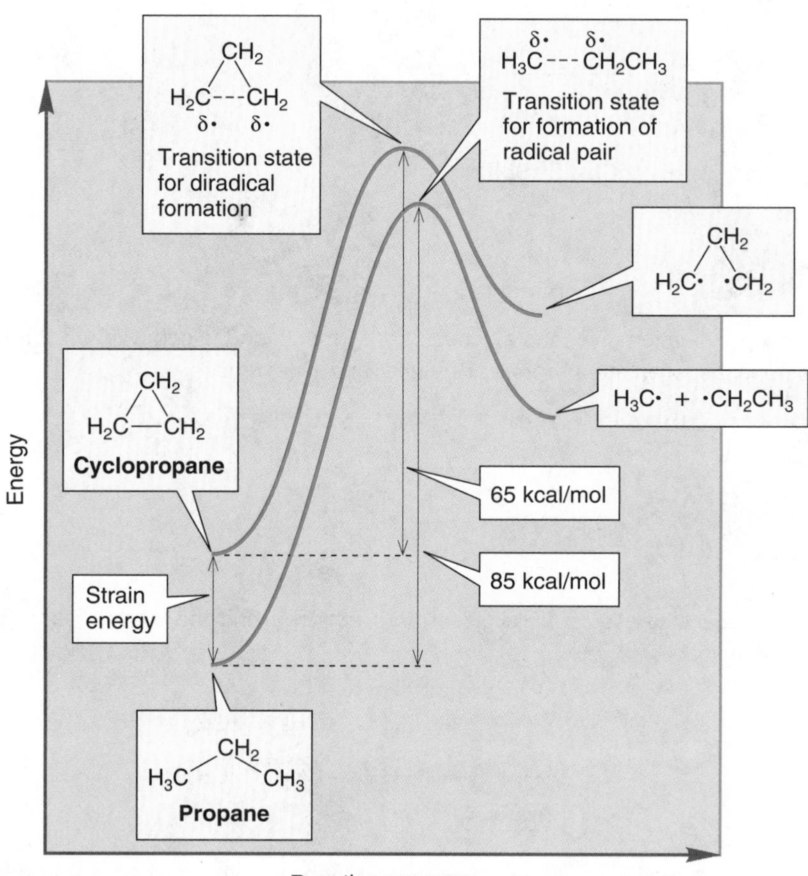

FIGURE **11.10** Strain can raise the energy of a hydrocarbon making it easier to break a carbon–carbon bond

Breaking a carbon–carbon bond in cyclopropane doesn't yield two new separated radicals, but a 1,3-**diradical**, in which two nonbonding electrons are present in one molecule. We haven't really done exactly what we wanted—produce two new radicals from a single molecule.

So let's try another way to modify ethane to make it easier to cleave. We just tried making the starting material *less stable* by introducing strain. How about taking the opposite approach by making the products of the reaction, the two radicals, *more stable*?

PROBLEM **11.11** Construct an Energy versus Reaction progress diagram for this second general approach to making the product of the reaction more stable.

The question now becomes, How can we make a radical more stable than the methyl radical? We know that delocalization of electrons is stabilizing, so one approach would be to construct a molecule that will yield resonance-stabilized radicals. That should lower the bond strength of the crucial carbon–carbon bond. For example, introduction of a single vinyl group (as in 1-butene) lowers the bond dissociation energy by about 13 kcal/mol to 72 kcal/mol. To calculate this number, the bond dissociation energy of propane and 1-butene are compared. Propane breaks a carbon–carbon bond to give a methyl radical and an ethyl radical. 1-Butene breaks a carbon–carbon bond to give a methyl radical and an allyl radical. The difference in the energy required for the two bond breakings is reasonably attributed to the stabilization of the allyl radical, and, of course, of the transition state leading to it. Introduction of a second vinyl group, as in 1,5-hexadiene, reduces the bond dissociation energy by approximately an additional 10 kcal/mol (Fig. 11.11). The key to lowering the bond dissociation energy is the formation of allyl radicals with their stabilizing delocalization of the nonbonding electrons.

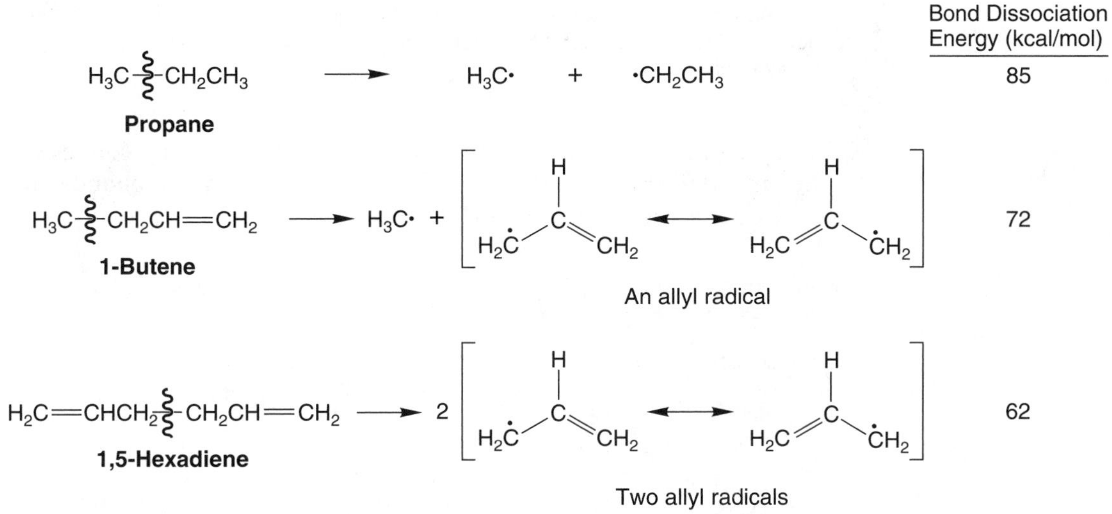

Bond Dissociation Energy (kcal/mol)

H_3C–CH_2CH_3 → $H_3C\cdot$ + $\cdot CH_2CH_3$ 85

Propane

H_3C–$CH_2CH{=}CH_2$ → $H_3C\cdot$ + [$H_2\dot{C}$–CH=CH_2 ⟷ H_2C=CH–$\dot{C}H_2$] 72

1-Butene

An allyl radical

$H_2C{=}CHCH_2$–$CH_2CH{=}CH_2$ → 2 [$H_2\dot{C}$–CH=CH_2 ⟷ H_2C=CH–$\dot{C}H_2$] 62

1,5-Hexadiene

Two allyl radicals

FIGURE **11.11** Stabilization of the products of bond breaking through delocalization can make the bond cleavage process easier.

PROBLEM **11.12**

There is a lower energy process occurring in 1,5-hexadiene, although no *new* molecule appears! Can you draw an arrow formalism for the conversion of one molecule of 1,5-hexadiene into another?
Hint: 1,1,6,6-Tetradeuterio-1,5-hexadiene equilibrates on heating with 3,3,4,4-tetradeuterio-1,5-hexadiene.

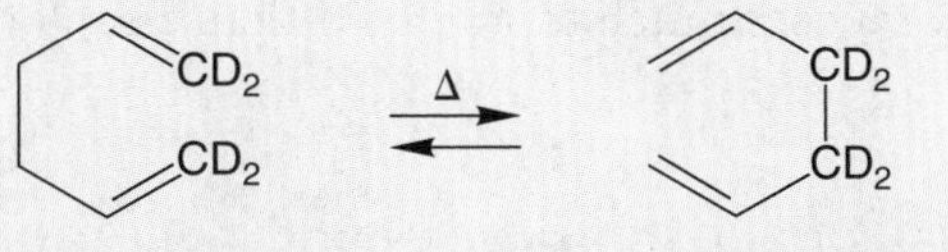

CD_2 CD_2 ⇌ (Δ) CD_2 CD_2

1,1,6,6-Tetradeuterio-1,5-hexadiene **3,3,4,4-Tetradeuterio-1,5-hexadiene**

Another technique for producing radicals is to use molecules containing especially weak, and therefore easily broken bonds. Examples are acyl **peroxides** (Fig. 11.12). The oxygen–oxygen bond in simple peroxides is already quite weak (~38 kcal/mol; Table 8.2, p. 307). The addition of the carbon–oxygen double bond of the **acyl group (R–C=O)**, helps delocalize the nonbonding electron in the carboxyl radical, and further lowers the bond dissociation energy to about 29 kcal/mol (Fig. 11.12).

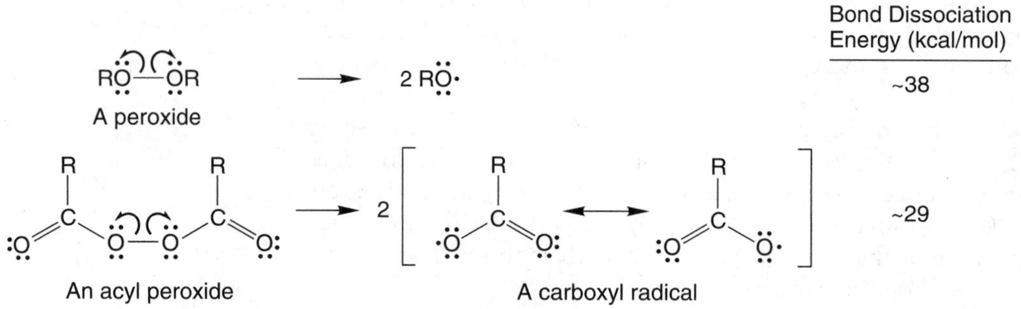

Reaction	Bond Dissociation Energy (kcal/mol)
A peroxide: RÖ—ÖR ⟶ 2 RÖ·	~38
An acyl peroxide ⟶ 2 [A carboxyl radical]	~29

FIGURE 11.12 In alkyl peroxides, the weak oxygen–oxygen bond is easily cleaved. In acyl peroxides, the product radicals are resonance stabilized and the oxygen–oxygen bond is even weaker.

Some molecules give stable gases (often nitrogen or carbon dioxide) on heating, and thus produce radicals irreversibly. **Azo compounds** and acyl peroxides are good examples (Fig. 11.13).

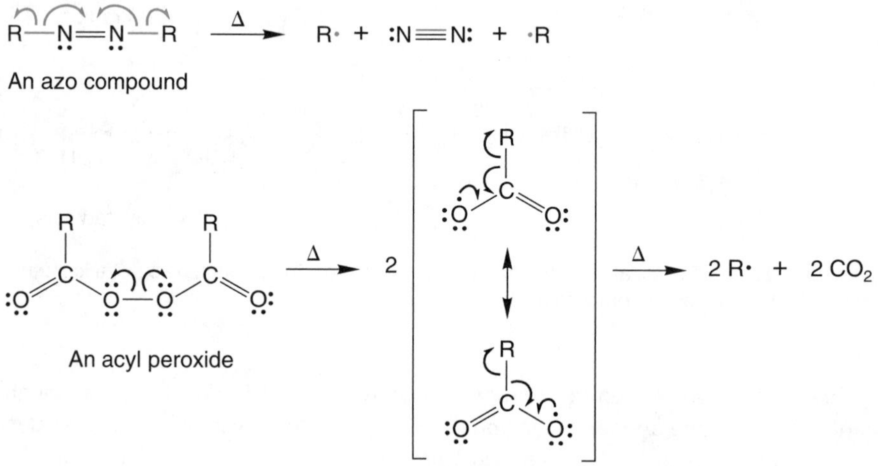

R—N̈=N̈—R $\xrightarrow{\Delta}$ R· + :N≡N: + ·R

An azo compound

An acyl peroxide $\xrightarrow{\Delta}$ 2 [carboxyl radical] $\xrightarrow{\Delta}$ 2 R· + 2 CO_2

FIGURE 11.13 Some molecules give free radicals irreversibly through the formation of gases.

A particularly useful source of radicals is azoisobutyronitrile (AIBN), a molecule that decomposes much more easily than simpler azo compounds (Fig. 11.14).

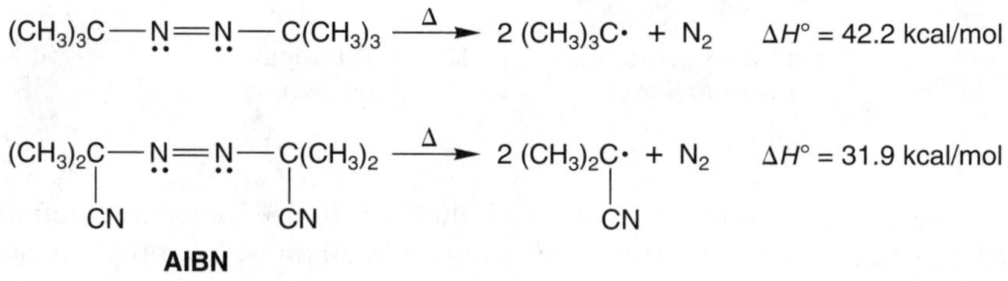

$(CH_3)_3C—N=N—C(CH_3)_3 \xrightarrow{\Delta} 2\ (CH_3)_3C\cdot + N_2 \quad \Delta H^\circ = 42.2$ kcal/mol

$(CH_3)_2C(CN)—N=N—C(CH_3)_2(CN) \xrightarrow{\Delta} 2\ (CH_3)_2C(CN)\cdot + N_2 \quad \Delta H^\circ = 31.9$ kcal/mol

AIBN

FIGURE 11.14 Azoisobutyronitrile is an especially easily cleaved azo compound.

PROBLEM 11.13 Why does AIBN decompose much more easily than a simple azo compound?

11.2 STRUCTURE OF RADICALS

We have already encountered both carbocations and carbanions, and spent some time discussing their structures (Section 3.4, p. 81). The methyl cation, and carbocations in general, are flat, sp^2 hybridized molecules with an empty $2p$ orbital extending above and below the plane of the three substituents on carbon. By contrast, the methyl anion is pyramidal. One might well guess that the methyl radical, with a single nonbonding electron, would have a structure intermediate between that of the two ions (Fig. 11.15).

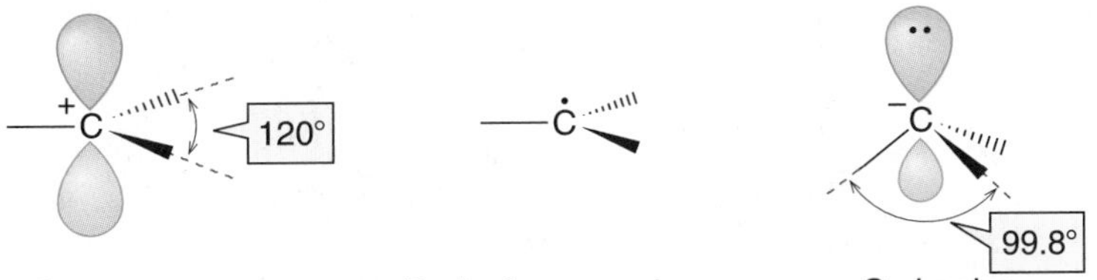

Carbocation, planar

Radical—somewhere in between the cation and anion

Carbanion, pyramidal

FIGURE 11.15 Radicals are intermediate in structure between the planar cations and the pyramidal carbanions.

Both theory and experiment agree that the structure of simple radicals is hard to determine! However, it is clear that these molecules are not very far from planar. If the molecules are not planar, the pyramid is *very* shallow and the two possible forms are rapidly inverting (Fig. 11.16).

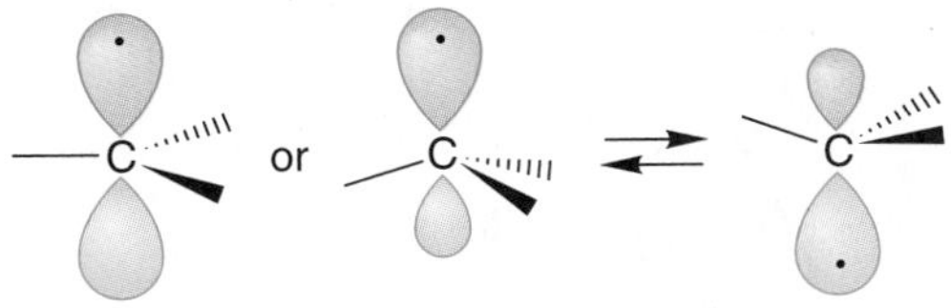

Planar ... or ... very shallow pyramid

FIGURE 11.16 Alkyl radicals are either flat or very shallow, rapidly interconverting pyramids.

PROBLEM 11.14

Draw the transition state for the inversion process in Figure 11.16.

11.3 STABILITY OF RADICALS

Like carbocations, more substituted radicals are more stable than less substituted ones. Table 11.1 shows the ΔH_f° of the radicals in the simple series: methyl, ethyl, isopropyl, allyl, and *tert*-butyl. *Remember*: The more negative (or less positive) the heat of formation, the more stable the species.

TABLE 11.1 Heats of Formation of Simple Radicals

Radical		
Name	Formula	Heat of Formation (ΔH_f°, kcal/mol)
Methyl	$\cdot CH_3$	34.9
Ethyl	$\cdot CH_2CH_3$	28.4
Isopropyl	$\cdot CH(CH_3)_2$	21.3
Allyl	$\cdot CH_2{-}CH{=}CH_2$	17.3
tert-Butyl	$\cdot C(CH_3)_3$	11.6

In explaining the order of stability for carbocations (tertiary > secondary > primary > methyl), we drew resonance forms in which a filled carbon–hydrogen bond overlapped with the empty 2*p* orbital (Fig. 11.17).

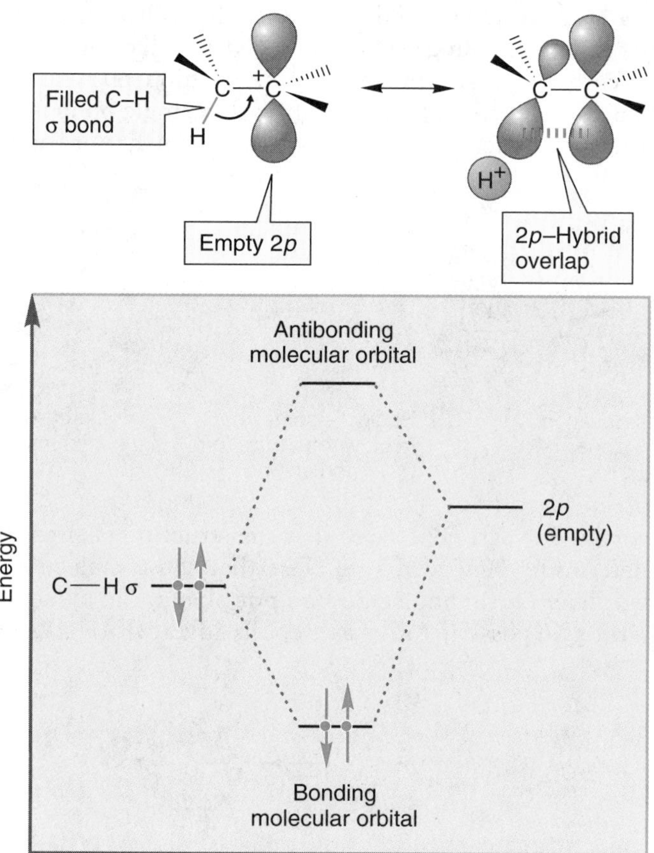

FIGURE **11.17** Resonance stabilization of a carbocation through "hyperconjugation," shown both by orbital overlap pictures and an interaction diagram.

We can draw similar resonance forms for radicals. As before, the more alkyl groups attached to the central carbon bearing the nonbonding electron, the more resonance forms are possible. But there is an important difference. Stabilization of the carbocation involves overlap of a filled carbon–hydrogen bond with an empty 2*p* orbital, whereas in the radical, overlap is between a filled carbon–hydrogen bond and a *half-filled* 2*p* orbital (Fig. 11.18).

Why should this interaction be stabilizing? The evident problem is that this is a *three*-electron system. One electron must occupy the antibonding orbital, and this is certainly destabilizing. Despite this, the *overall* system is still stabilized because two electrons are able to occupy the low-energy bonding orbital. Recall He_2^+ in which a similar orbital system was encountered. The He_2^+ molecule is held together (bound) by over 60 kcal/mol (Chapter 2, p. 54; Fig. 11.19).

As we showed earlier (p. 457), radicals substituted with electron delocalizing groups such as vinyl are even more stable. If they are immobilized, even quite simple radicals can be made and studied spectroscopically at low temperature. A few highly delocalized radicals are stable under normal, room temperature conditions. For example, the triphenylmethyl

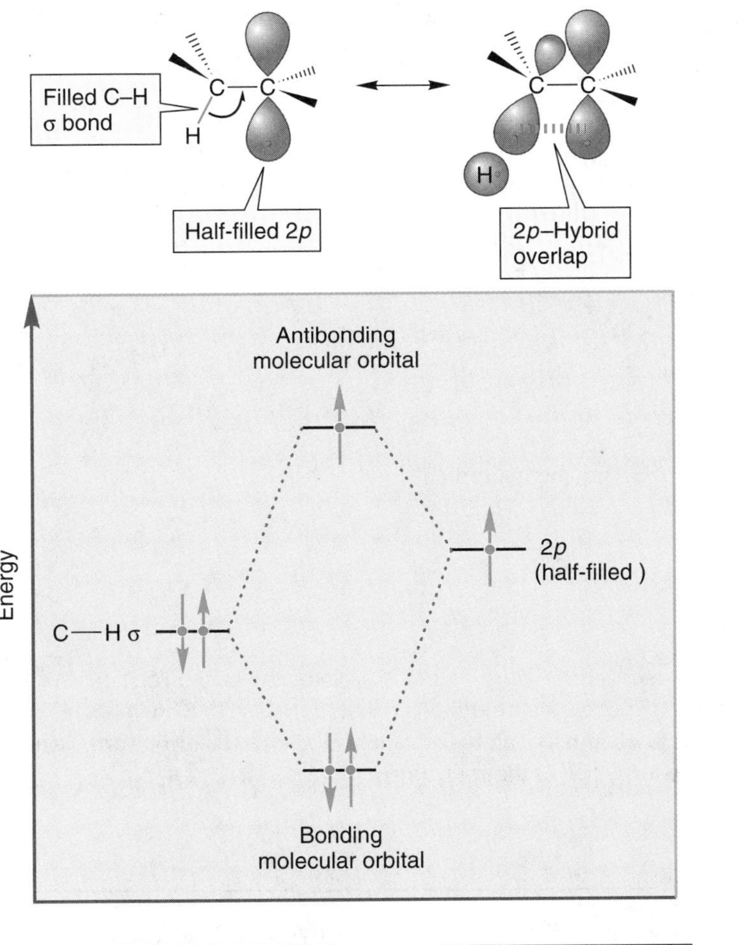

FIGURE 11.18 Resonance stabilization of a free radical, again shown both through orbital pictures and an interaction diagram.

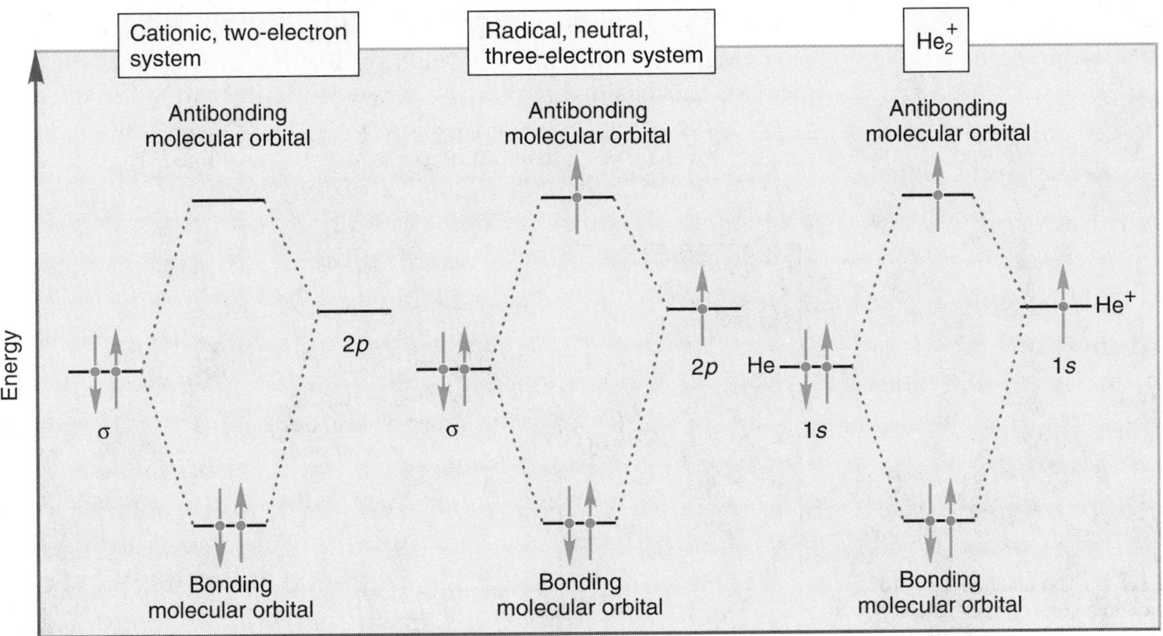

FIGURE 11.19 A comparison of the stabilizations of carbocations, radicals, and He_2^+. Notice the electron in the antibonding orbital of He_2^+ and the radical.

radical happily exists in solution as an equilibrium mixture with a dimer (Fig. 11.20).

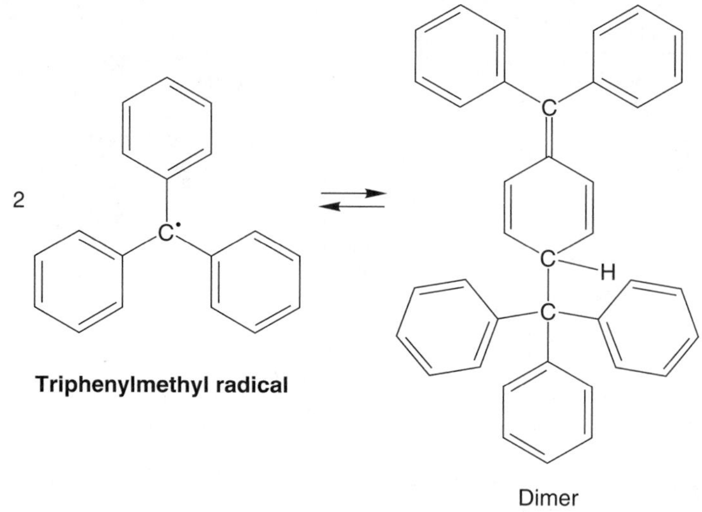

2 Triphenylmethyl radical ⇌ Dimer

FIGURE 11.20 The triphenylmethyl radical exists in equilibrium with a dimer.

PROBLEM 11.15 As you can see from Figure 11.20, the dimer is not the "obvious" one, hexaphenylethane, but a much more complicated molecule. Can you see why hexaphenylethane is a most unstable molecule, and draw an arrow formalism mechanism for the formation of the real dimer?

11.4 RADICAL ADDITION TO ALKENES

In Section 9.3 (p. 338), the addition of hydrogen bromide to alkenes was presented as a straightforward proposition, with the reaction always leading to Markovnikov addition. The historical reality was far from that simple. Our analysis in Chapter 9 stressed the initial formation of the more stable carbocation followed by addition of the residual bromide ion. The product of Markovnikov addition must be produced in this reaction (Fig. 11.21).

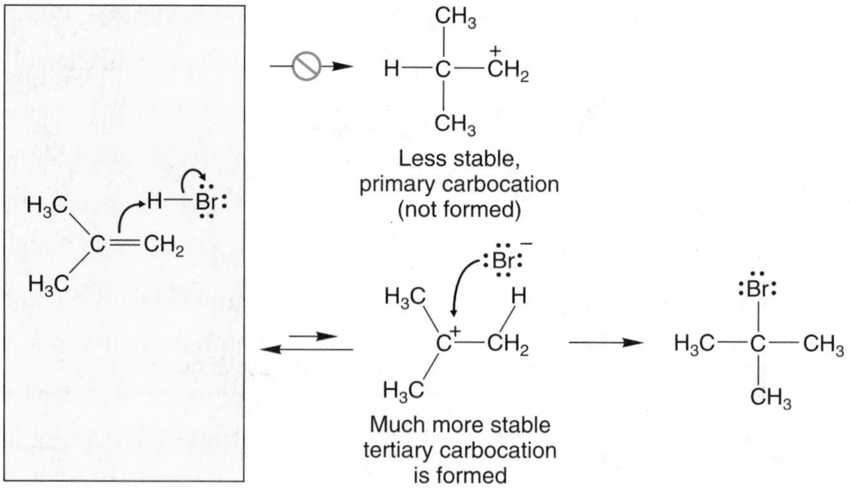

Less stable, primary carbocation (not formed)

Much more stable tertiary carbocation is formed

FIGURE 11.21 The addition of hydrogen bromide to an unsymmetrical alkene follows the Markovnikov rule because protonation of the alkene must give the more substituted (more stable) carbocation.

In fact, the regiochemistry of the reaction appeared strangely dependent on the source of the starting materials and, it seemed, even on the place in which the reaction was carried out. In some worker's hands, the reaction did follow the Markovnikov path; in other places, with other investigators, anti-Markovnikov addition was dominant. Confusion and argument was the rule of the day in the late 1800s. Oddly enough, the problem was restricted to hydrogen bromide addition. Both hydrogen chloride and hydrogen iodide addition remained securely Markovnikov in everyone's hands (Fig. 11.22).

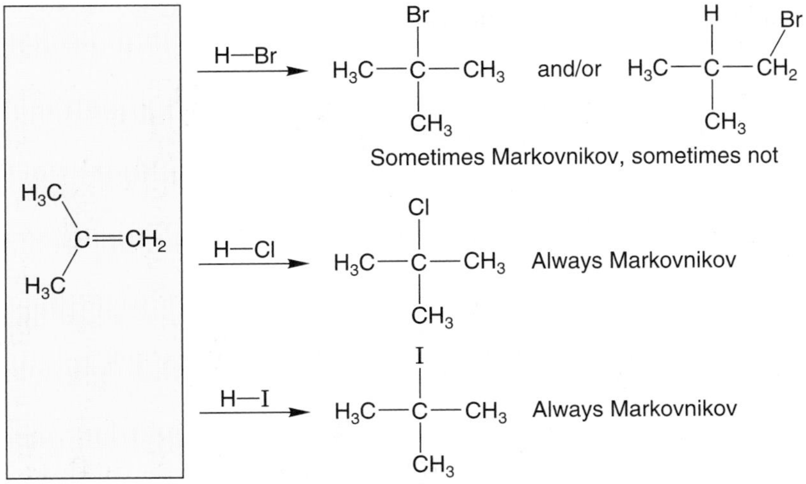

H—Br → H_3C—C(Br)(CH_3)—CH_3 and/or H_3C—C(H)(CH_3)—CH_2Br

Sometimes Markovnikov, sometimes not

$(H_3C)_2C{=}CH_2$

H—Cl → H_3C—C(Cl)(CH_3)—CH_3 Always Markovnikov

H—I → H_3C—C(I)(CH_3)—CH_3 Always Markovnikov

FIGURE 11.22 Both hydrogen chloride and hydrogen iodide addition always follow the Markovnikov rule, but hydrogen bromide addition is strangely capricious.

In the 1930s, largely through the efforts of M. S. Kharasch [Kharasch, 1895–1957, is reported by one of his students, F. R. Mayo (b. 1908), to have suggested that the direction of addition may have been affected by the phases of the moon], it was discovered that the purity of the compounds used was vital. Pure materials led to Markovnikov addition, and we can remain content with the mechanism for the polar reaction. Impure reagents led to predominately anti-Markovnikov addition. A special culprit was found to be peroxides, R–O–O–R (Fig. 11.23). Breaking the weak oxygen–oxygen bond in peroxides provides a relatively easy route to free radicals. The mechanism of addition in the presence of peroxides is different from the familiar polar one, and can be triggered by these small amounts of radicals.

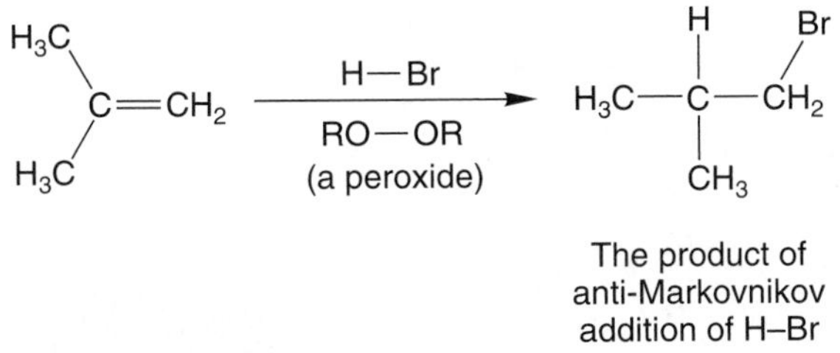

$(H_3C)_2C{=}CH_2$ —[H—Br; RO—OR (a peroxide)]→ H_3C—C(H)(CH_3)—CH_2Br

The product of anti-Markovnikov addition of H–Br

FIGURE 11.23 Addition of hydrogen bromide to alkenes in the presence of peroxides is anti-Markovnikov.

11.4a The Radical Chain Mechanism of HBr Addition

Peroxides contain a very weak oxygen–oxygen bond and are easily cleaved by heat or light to give two R–O· radicals (Fig. 11.24).

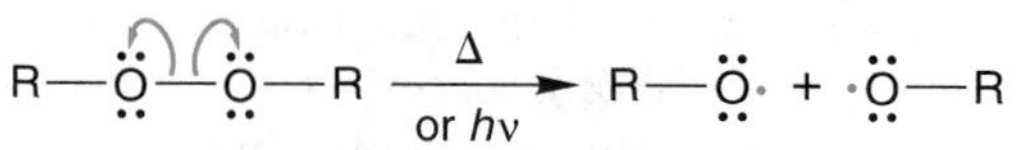

FIGURE 11.24 The oxygen–oxygen bond is weak and can be easily cleaved to give a pair of RO· radicals.

A free radical (for example, R–O·) can react with H–Br by hydrogen abstraction to form R–OH and a new radical, a bromine atom (Fig. 11.25, Step 2). Addition of the bromine atom to an alkene gives still another free radical (Step 3), which can play the role of R–O· and produce an alkyl bromide and another bromine atom (Step 4). The bromine atom can recycle to Step 3 and start the reaction over again. The overall result is the addition of hydrogen bromide to the alkene.

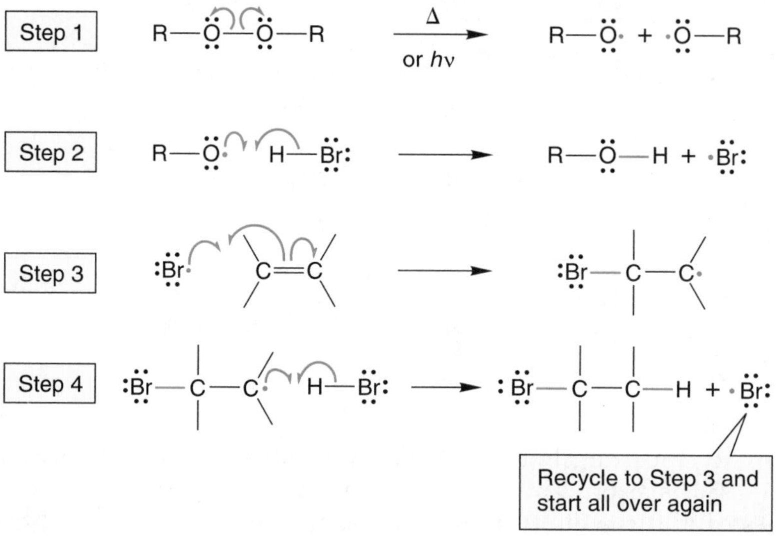

Step 1

Step 2

Step 3

Step 4

Recycle to Step 3 and start all over again

FIGURE 11.25 A radical-induced mechanism for addition of hydrogen bromide to an alkene. Alkoxy radicals are first formed through the dissociation of a peroxide. In Step 2, an alkoxy radical abstracts a hydrogen atom from hydrogen bromide to give an alcohol and a bromine atom. In Step 3, the bromine atom adds to the alkene to generate an alkyl radical that can continue the reaction by abstracting hydrogen from hydrogen bromide. A bromine atom and the addition product are produced. The newly formed bromine atom can start the reaction again.

PROBLEM 11.16 Why is abstraction of H from H–Br preferred to abstraction of Br? *Hint*: The bond dissociation energy of HO–Br is 56 kcal/mol.

Of course, when two radicals meet they can react in a very exothermic fashion to form a new bond. There are many possibilities for this kind of reaction using the radicals in Figure 11.25. A few ways are shown in Figure 11.26.

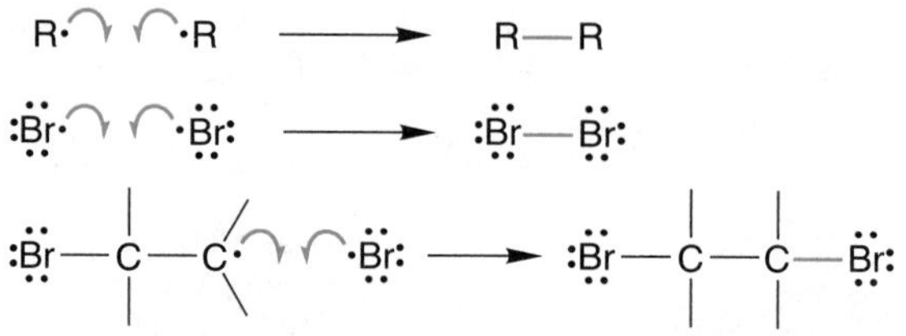

FIGURE 11.26 Whenever two radicals come together they can react to form a bond. This reaction removes two radicals and terminates this chain reaction.

We have just outlined the radical-induced addition of hydrogen bromide to an alkene. The steps of this chain reaction naturally divide into three kinds:

Step 1. **Initiation**. The reaction cannot start without the presence of a radical. Although there are now known examples of stable radicals, usually a radical must be formed to start the process, in a reaction known as an **initiation step**. Peroxides are good generators of radicals because of the weakness of their oxygen–oxygen bond. Once a radical is created in the presence of hydrogen bromide, a second initiation step, abstraction of hydrogen, takes place to form the chain-carrying bromine atom (Fig. 11.27).

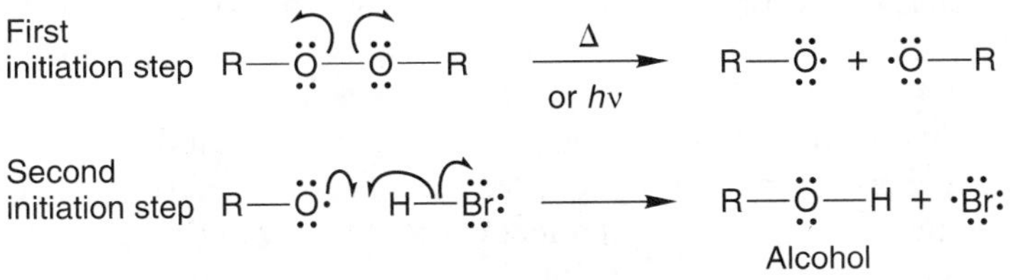

First initiation step R—Ö—Ö—R —Δ or *hν*→ R—Ö· + ·Ö—R

Second initiation step R—Ö· H—Br: ——→ R—Ö—H + ·Br:
Alcohol

FIGURE 11.27 The two initiation steps are the formation of two alkoxy radicals by the cleavage of the oxygen–oxygen bond in the peroxide, and the abstraction of hydrogen by RO· to give an alcohol and a bromine atom.

Step 2. **Propagation**. Once the reaction is started by the formation of a bromine atom in the initiation steps, the chain can begin to grow. The steps making up the chain are called **propagation steps**. In this example, the chain-carrying propagation steps are the addition of the bromine atom to the alkene to give a new, carbon-centered alkyl radical and the subsequent abstraction of hydrogen from hydrogen bromide by this radical (Fig. 11.28).

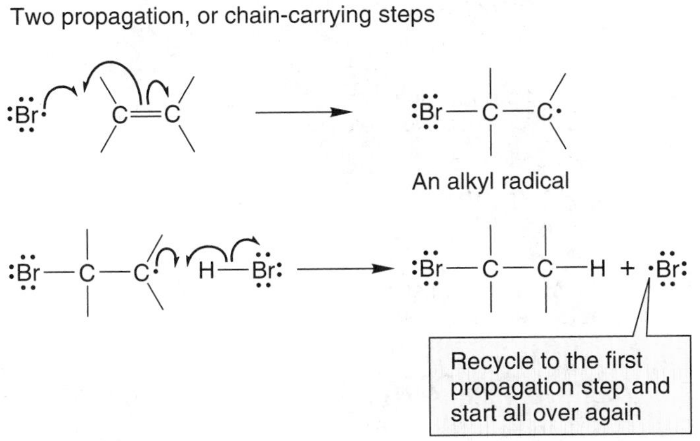

Two propagation, or chain-carrying steps

:Br· C=C ——→ :Br—C—C·
An alkyl radical

:Br—C—C· H—Br: ——→ :Br—C—C—H + ·Br:

Recycle to the first propagation step and start all over again

FIGURE 11.28 In the first propagation step, a bromine atom adds to the alkene to generate an alkyl radical. In the second propagation step, this alkyl radical abstracts hydrogen from hydrogen bromide and generates a new bromine atom. This bromine atom can recycle to propagation Step 1 and carry the chain further.

The two radicals involved in the chain-carrying steps, the bromine atom and the alkyl radical, are called "chain-carrying radicals." The chain continues as the newly formed bromine atom adds to another alkene to start the process over again.

Step 3. **Termination**. Any time one of the chain-carrying radicals (in this case a bromine atom or one of the alkyl radicals) is annihilated by combination with another radical, the chain reaction is stopped. Without such steps a single radical could initiate a chain reaction that would continue until starting material was used up. In practice, this is not the case, and a competition is set up between the propagation steps carrying the chain reaction and the steps ending it. Such reactions are called **termination steps** and there are many possibilities (Fig. 11.29).

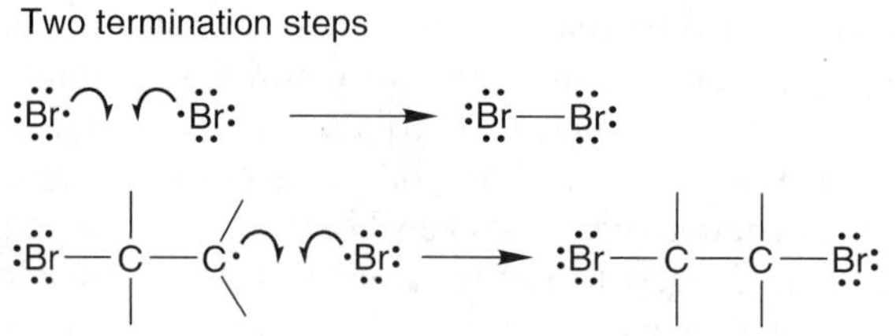

Two termination steps

FIGURE 11.29 Two possible termination steps for the chain reaction.

For a chain reaction to be effective in producing large amounts of products, the propagation steps must be much more successful than the termination steps. This is often the case, as for a termination step to occur two radicals must wander through the solution until they find each other. Typically, the chain-carrying propagation steps require reaction with a molecule present in relatively high concentration. So, even though termination steps often can take place very easily once the reactants find each other, propagation steps can compete successfully, and chain lengths of many thousands are common. Under such circumstances, the small amounts of product formed by the occasional termination step are inconsequential.

11.4b Inhibitors

Small amounts of substances that react with the chain-carrying radicals can stop chain reactions by removing the chain-carrying molecules from the process. These substances are called **inhibitors**. They act as artificial terminators of the chains (Fig. 11.30).

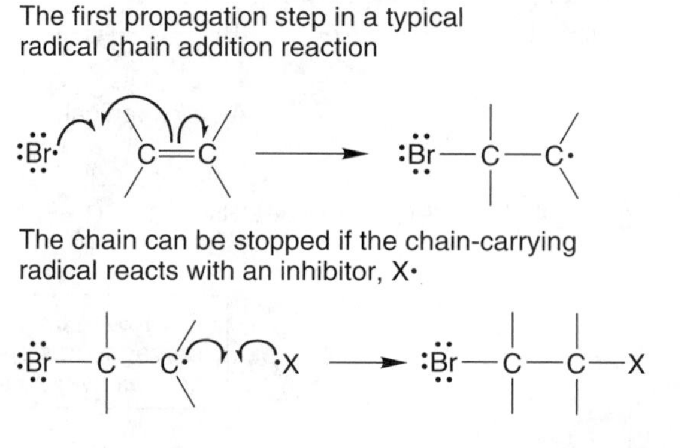

The first propagation step in a typical radical chain addition reaction

The chain can be stopped if the chain-carrying radical reacts with an inhibitor, X·

FIGURE 11.30 An inhibitor of radical chain reactions can act by intercepting one of the chain-carrying radicals.

Once again, a few molecules can have an impact out of proportion to their numbers, as stopping one chain can mean stopping the formation of thousands of product molecules. Oxygen (O_2), itself a diradical, can act in this way, and many radical reactions will not proceed until the last oxygen molecules are scavenged from the system, thus allowing the radical chains to carry on. Radical inhibitors are often added to foods to retard radical reactions causing spoilage. Figure 11.31 shows a few typical radical inhibitors. It is not important that you memorize these structures. What is worth remembering is the general structural features of these materials; also, you should think a bit about how they might function in intercepting radicals.

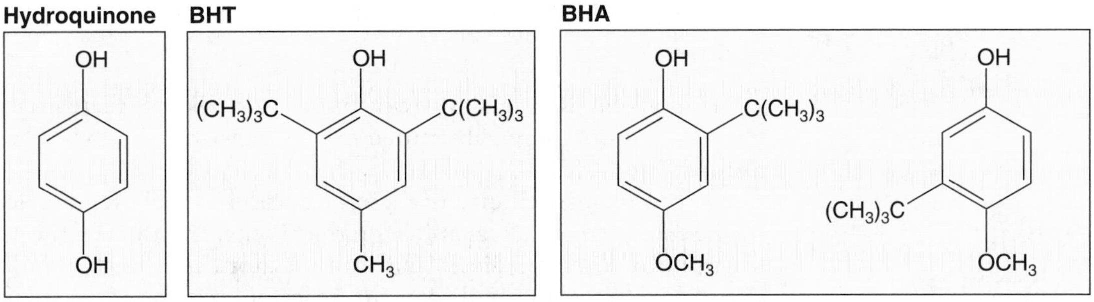

FIGURE **11.31** Some radical inhibitors. Both butylated hydroxy toluene (BHT) and butylated hydroxy anisole (BHA) are widely used in the food industry.

Vitamin E

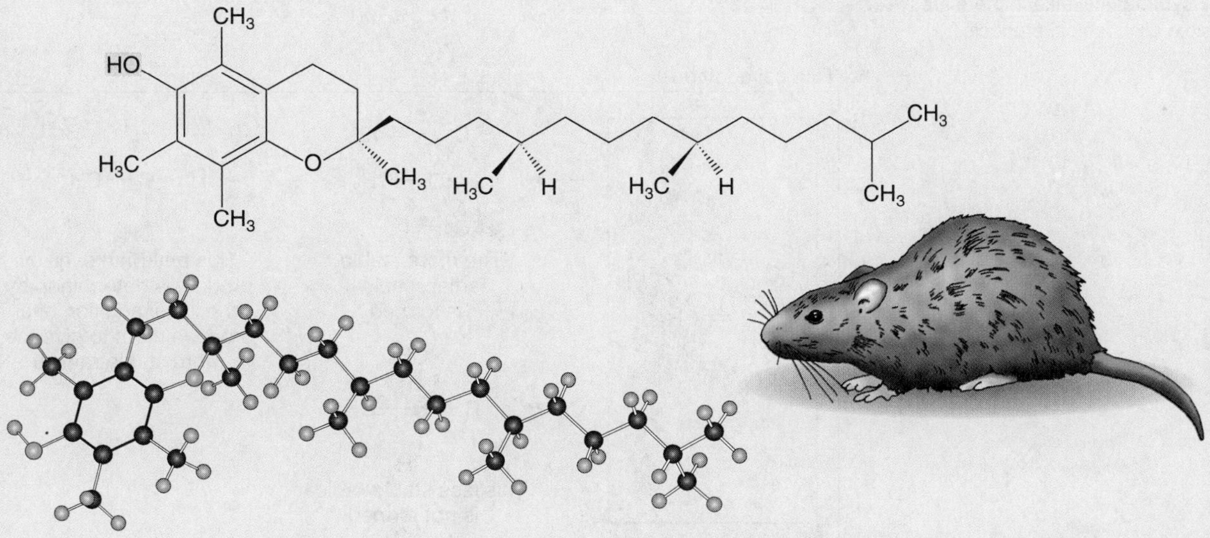

Vitamin E is essential to rats for proper liver function and nerve and muscle condition. At least one observer has pointed out that as rats and humans are so alike, vitamin E might be vital for humans as well. Despite the obvious truth of the observation, there is no proven need for vitamin E in humans.

However, there is one area in which vitamin E, and molecules like it, are drawing attention as possible effective therapeutic agents, and this involves the ability of vitamin E to act as an antioxidant. Free radicals are strongly implicated in the aging process. Cell membranes are made up of fatty acids, long-chain hydrocarbon carboxylic acids, and these molecules are quite vulnerable to attack by reactive free radicals. It is speculated that vitamin E acts by intercepting free radicals, thus slowing the aging process. It does this through the red hydroxyl group on the six-membered ring. A radical (RO·) plucks a hydrogen atom from this hydroxyl to make RO—H, and destroys itself. Can you see why it is especially easy for the radical to remove this OH hydrogen?

The long hydrocarbon chain of vitamin E also has its function. It aids in delivery of vitamin E to the cell membrane by making the molecule soluble in the fatty acids, which also contain long greasy hydrocarbon chains.

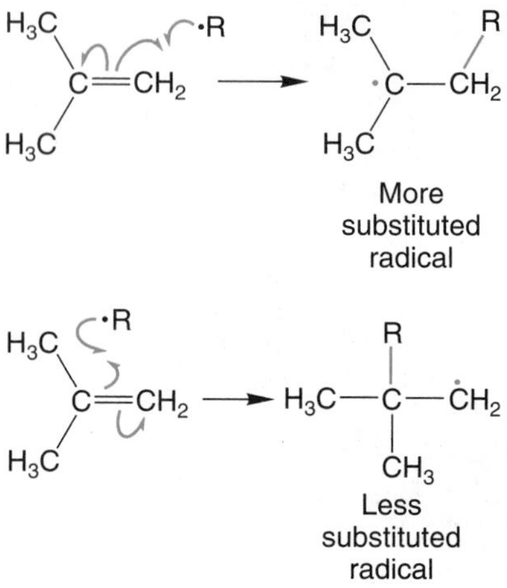

11.4c Regiospecificity of Radical Additions: Anti-Markovnikov Addition

When a radical (R·) adds to an unsymmetrical alkene, there are two possible outcomes: a more substituted or less substituted alkyl radical can be formed (Fig. 11.32).

So, the regiospecificity of radical addition to isobutene is determined by the direction of radical attack on the alkene, and thus by the stability of the initially formed radical. The bromine atom adds to give the more substituted, more stable radical. The reaction continues when the newly formed radical abstracts a hydrogen atom from hydrogen bromide to give, ultimately, the compound in which the bromine is attached to the *less* substituted end of the original alkene. This reaction is in contrast to polar additions in which the first step is protonation of the alkene to give the more substituted carbocation, which is attacked by bromide ion to give the *more* substituted bromide (Fig. 11.33).

More substituted radical

Less substituted radical

FIGURE 11.32 When a radical adds to an unsymmetrical alkene there are two possible alkyl radical products.

Radical addition

The more stable, tertiary radical is formed

This anti-Markovnikov product is determined by the initial addition of a bromine atom to give the more stable radical

This less stable radical is not formed

Contrast radical addition with polar addition of H–Br

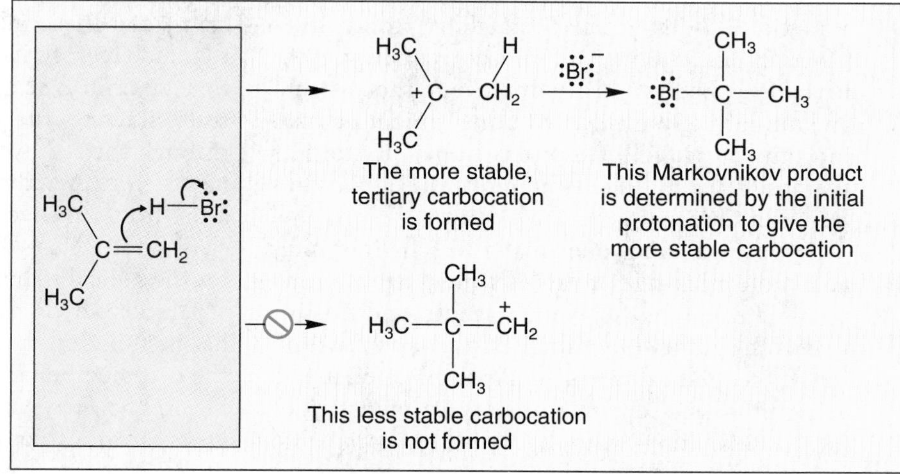

FIGURE 11.33 As in the cationic reaction, the first step in a radical addition to an alkene is the formation of the more substituted, and therefore more stable species. Notice, however, that in the cationic reaction it is the hydrogen atom that adds first, and in the radical reaction it is the bromine atom.

The historic difficulty in determining the regiospecificity of hydrogen bromide addition had only to do with recognizing that radical initiators such as peroxides had to be kept out of the reaction. If even a little peroxide was present, the chain nature of the radical addition of hydrogen bromide could overwhelm the nonchain polar reaction, and produce large amounts of anti-Markovnikov addition.

This understanding of the causes of the regiochemical outcome of these reactions presents an opportunity for synthetic control. We now have ways to do Markovnikov addition of hydrogen chloride and hydrogen iodide, and either Markovnikov or anti-Markovnikov addition reactions for hydrogen bromide. Of course, the halides formed as products in these reactions can be used for further synthetic work.

PROBLEM **11.17**

Show how to carry out the following conversions. You need not write mechanisms in such problems unless they help you to see the correct synthetic pathways.

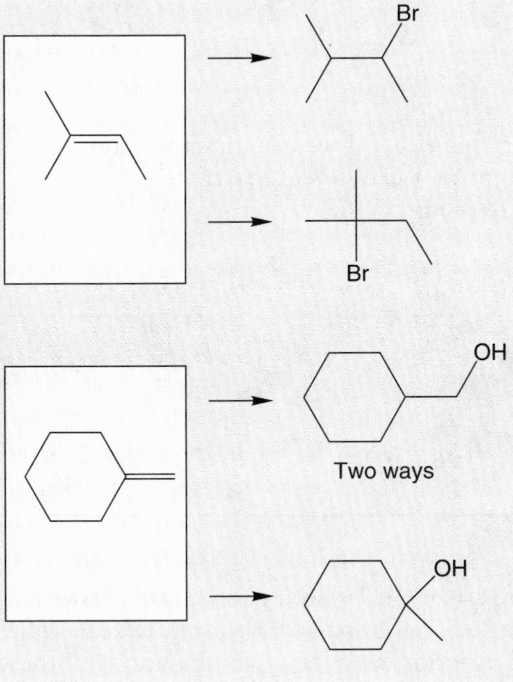

Br

Br

OH

Two ways

OH

Neither hydrogen chloride nor hydrogen iodide will add in anti-Markovnikov fashion, even if radical initiators are present. Now we have to see why this is so.

11.4d Thermochemical Analysis of HX Additions to Alkenes

Why does this chain reaction succeed with H—Br but not for other hydrogen halides? We can *write* identical radical chain mechanisms for H—Cl and H—I, but these must fail for some reason (Fig. 11.34).

Why does the following mechanism not operate for H–Cl or H–I?

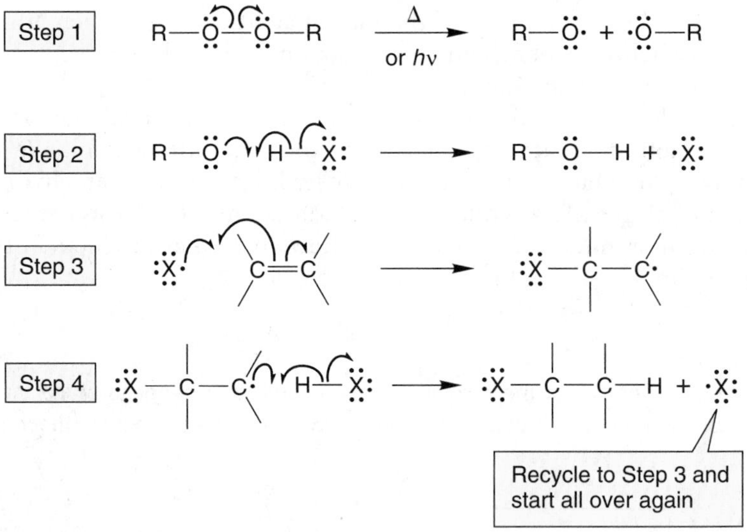

Step 1: $R\text{—}\ddot{O}\text{—}\ddot{O}\text{—}R \xrightarrow[\text{or } h\nu]{\Delta} R\text{—}\ddot{O}\cdot + \cdot\ddot{O}\text{—}R$

Step 2: $R\text{—}\ddot{O}\cdot + H\text{—}\ddot{X}: \longrightarrow R\text{—}\ddot{O}\text{—}H + \cdot\ddot{X}:$

Step 3: $:\ddot{X}\cdot + C{=}C \longrightarrow :\ddot{X}\text{—}C\text{—}C\cdot$

Step 4: $:\ddot{X}\text{—}C\text{—}C\cdot + H\text{—}\ddot{X}: \longrightarrow :\ddot{X}\text{—}C\text{—}C\text{—}H + \cdot\ddot{X}:$

Recycle to Step 3 and start all over again

FIGURE 11.34 It is possible to write analogous mechanisms for radical-induced additions of H–Cl and H–I to alkenes. Why don't these mechanisms operate? (X = Cl or I.)

We can understand this effect if we examine the thermochemistry of the two propagation steps for the different reactions.

The energetics of the first propagation step are shown in Figure 11.35:

$X\cdot + H_2C{=}CH_2 \longrightarrow X\text{—}CH_2CH_2\cdot$

X	Bond Strength (π bond)	Bond Strength (X–C in $X\text{–}CH_2CH_2\cdot$)	ΔH (kcal/mol)
Cl	66	82	–16
Br	66	69	– 3
I	66	54	+12

FIGURE 11.35 The thermochemistry of the first propagation step; the addition of halogen radicals to ethylene.

These values are estimated by noting that in this first reaction a π bond is broken and a X–C bond is made. The overall exothermicity or endothermicity of the reaction will be a combination of these two bond strengths. A π bond is worth 66 kcal/mol, and we approximate the X–C bond strength in the product by taking the values for ethyl halides. We are making all sorts of assumptions, but the errors involved turn out to be relatively small. For both Br and Cl this first step is exothermic—the product is more stable than starting material and will be favored at equilibrium. For I, the reaction is substantially *endothermic.* The product will be strongly disfavored and the reaction will be very slow. So now we know why the addition of H–I fails; the first propagation step of the reaction is highly endothermic and does not compete with the termination steps. But both the bromine case, which we know succeeds, and the chlorine case, which we know fails, have exothermic first steps. Let's look at the second propagation step in Figure 11.36.

$$H{-}X + \cdot CH_2{-}CH_2X \longrightarrow \cdot X + H{-}CH_2{-}CH_2{-}X$$

X	Bond Strength (H–X)	Bond Strength (H–C in H–CH_2–CH_2–X)	ΔH (kcal/mol)
Cl	103	98	+ 5
Br	87	98	–11
I	71	98	–27

FIGURE 11.36 The thermochemistry of the second propagation step, the abstraction of a hydrogen atom from H–X.

The second step is exothermic for Br and I, but *endothermic for Cl.* So, we can expect this reaction to succeed for Br and I, but fail for Cl. If we sum up the analysis for the two propagation steps we see that it is only for H–Br addition that *both* steps are exothermic. For H–Cl and H–I, one of the two propagation steps is endothermic, and the radical addition reaction cannot successfully compete with the termination steps. The polar addition of H–X wins out.

11.5 OTHER RADICAL ADDITION REACTIONS

Hydrogen bromide addition is only one example of a vast number of radical chain reactions. For example, in the presence of initiators carbon tetrahalides, such as CCl_4 and CBr_4, will add to alkenes. The steps in the peroxide-initiated reaction are shown in Figure 11.37.

1. ROOR ⟶ RO· + ·OR (initiation)
2. RO· + CCl_4 ⟶ ROCl + $\cdot CCl_3$ (initiation)
3. $\cdot CCl_3$ + $CH_2{=}CH_2$ ⟶ $\cdot CH_2CH_2CCl_3$ (propagation)
4. $\cdot CH_2CH_2CCl_3$ + CCl_4 ⟶ $ClCH_2CH_2CCl_3$ + $\cdot CCl_3$ (propagation) (recycle to Step 3)
5. Many possible termination steps

FIGURE 11.37 The mechanism for the radical-induced addition of carbon tetrachloride to alkenes.

PROBLEM 11.18

Write a number of termination steps for this reaction.

Under some conditions, alkenes can be polymerized by free radicals. In the examples we have already discussed, radical addition to an alkene produced a new radical that abstracted hydrogen (or halogen in the case of CX_4) in a second propagation step to give the product of overall addition to the alkene. What if there were no easy second step? If the concentration of alkene were very high, for example, or if the addition step were especially favorable, the abstraction reaction might not compete effectively with addition. Under such conditions the newly formed alkyl radical can undergo repeated addition reactions until either a termination step intercedes or the supply of alkene is exhausted (Fig. 11.38).

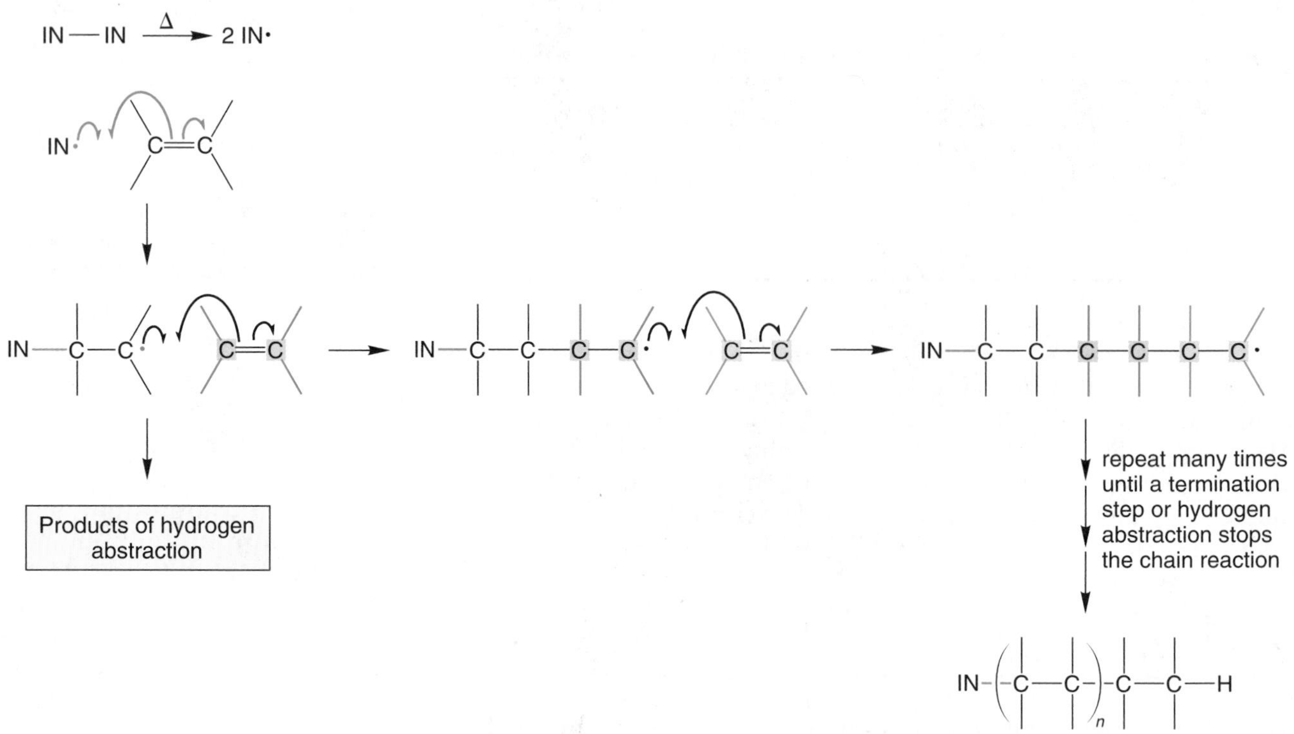

IN—IN →(Δ) 2 IN·

Products of hydrogen abstraction

repeat many times until a termination step or hydrogen abstraction stops the chain reaction

FIGURE 11.38 In the radical-induced polymerization of alkenes, repeated additions grow a long chain. A high concentration of alkene favors this process (IN· = initiator radical).

PROBLEM 11.19 Styrene (vinylbenzene, Fig. 11.39) is an example of a molecule in which the addition of a radical is especially favorable. Can you see why?

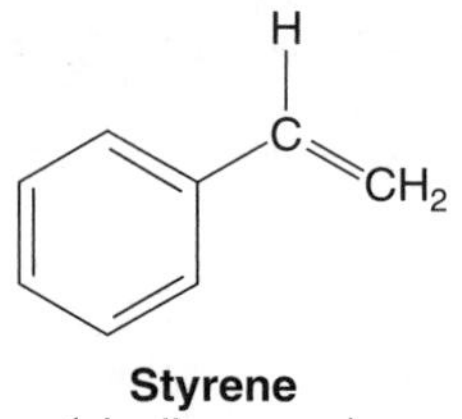

Styrene
(vinylbenzene)

FIGURE 11.39 Styrene (vinylbenzene).

*PROBLEM 11.20 Write the structure of polystyrene, the product of the polymerization of styrene (vinylbenzene).

ANSWER An initiator radical (IN·) can add to styrene to give two possible intermediate radicals. Resonance stabilization makes the radical formed from addition at the β-position much more stable than that produced from addition at the α-position.

ANSWER (CONTINUED)

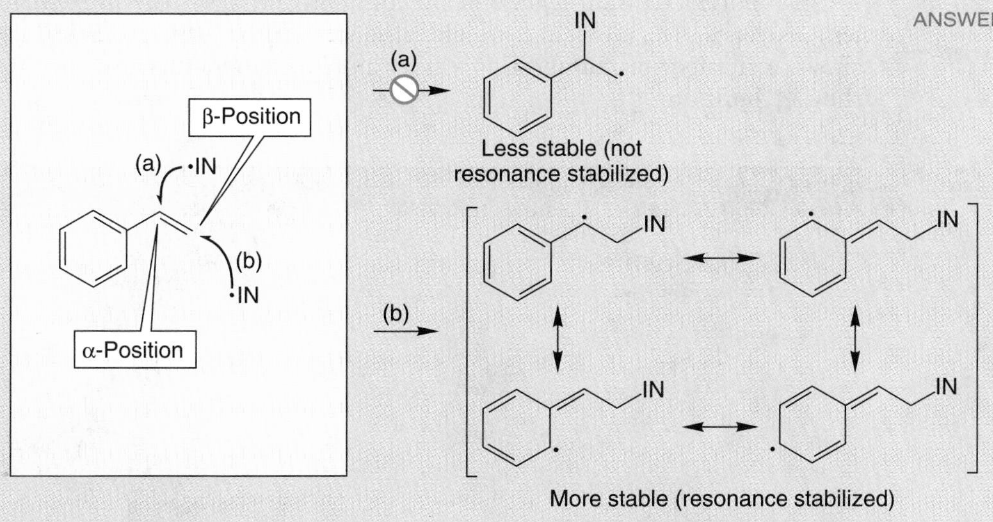

Now, this new radical can add to styrene as the initiator radical did. The same two choices exist; addition along (a) or (b). Again, and for the same reasons, (b) is favored since it gives the resonance-stabilized intermediate.

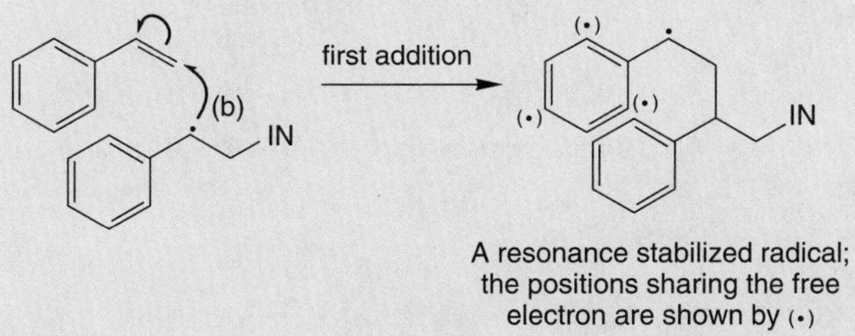

A resonance stabilized radical; the positions sharing the free electron are shown by (•)

Repeated additions in the (b) sense lead to the polymer

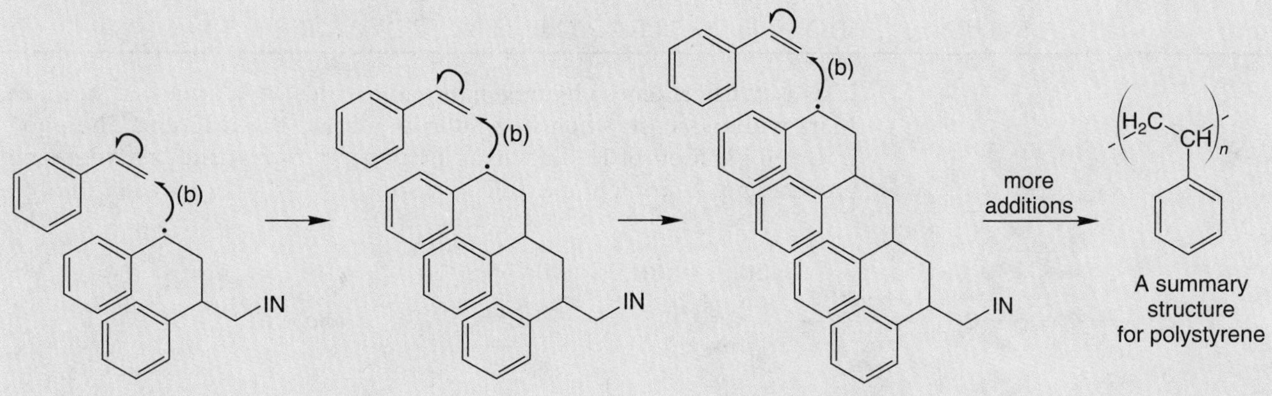

A summary structure for polystyrene

Such polymerization reactions are common and have very long chain lengths. Very high molecular weight products can be formed. Table 11.2 shows a number of common polymers and the monomers from which they are built up.

TABLE 11.2 Some Monomers and Polymers

Monomer	Polymer Formula	Name
$H_2C{=}CH_2$ Ethylene	$(CH_2)_n$	Polyethylene, polymethylene
$H_2C{=}CHCH_3$ Propylene	$-(CH(CH_3)-CH_2)_n-$	Polypropylene
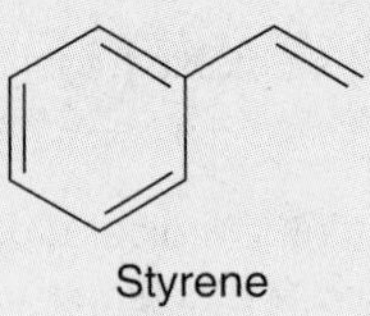 Styrene	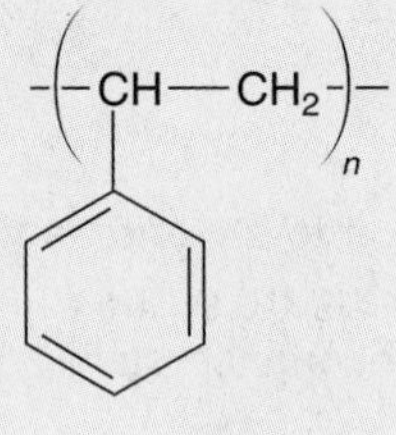 $-(CH(C_6H_5)-CH_2)_n-$	Polystyrene
$F_2C{=}CF_2$ Tetrafluoroethylene	$(CF_2)_n$	Poly(tetrafluoro-ethylene), Teflon
$H_2C{=}CHCl$ Vinyl chloride	$-(CH(Cl)-CH_2)_n-$	Poly(vinyl chloride)
$H_2C{=}CHCOOCH_3$ Methyl acrylate	$-(CH(COOCH_3)-CH_2)_n-$	Polyacrylate

11.6 RADICAL-INITIATED ADDITION OF HBr TO ALKYNES

Just as with alkenes, hydrogen bromide adds to alkynes in the anti-Markovnikov sense when free radicals initiate the reaction. The photo-initiated reaction of H—Br with 1-propyne gives *cis*- and *trans*-1-bromopropene, *not* 2-bromopropene (Fig. 11.40).

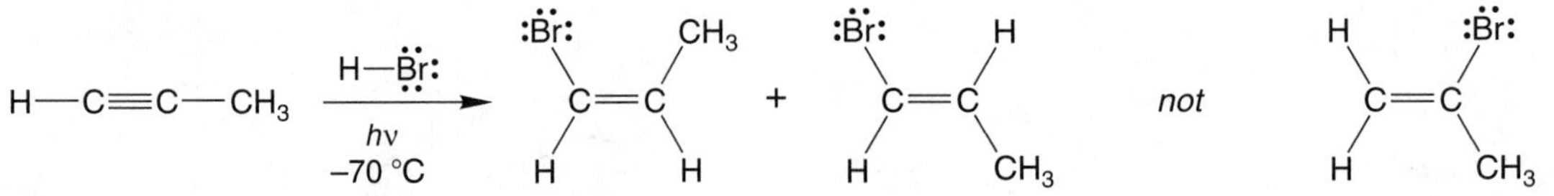

$$H{-}C{\equiv}C{-}CH_3 \xrightarrow[h\nu,\ -70\,^\circ C]{H-Br} \text{cis-}BrHC{=}CHCH_3 + \text{trans-}BrHC{=}CHCH_3 \quad \textit{not} \quad H_2C{=}CBrCH_3$$

***cis*- and *trans*-1-Bromopropene** **2-Bromopropene**

FIGURE 11.40 Just as with alkenes, the radical-induced addition of hydrogen bromide to an alkyne gives anti-Markovnikov addition.

The free radical chain mechanism for addition to alkynes is quite analogous to that operating in additions to alkenes. The crucial step is the formation of the more stable vinyl radical rather than the less stable vinyl radical (Fig. 11.41).

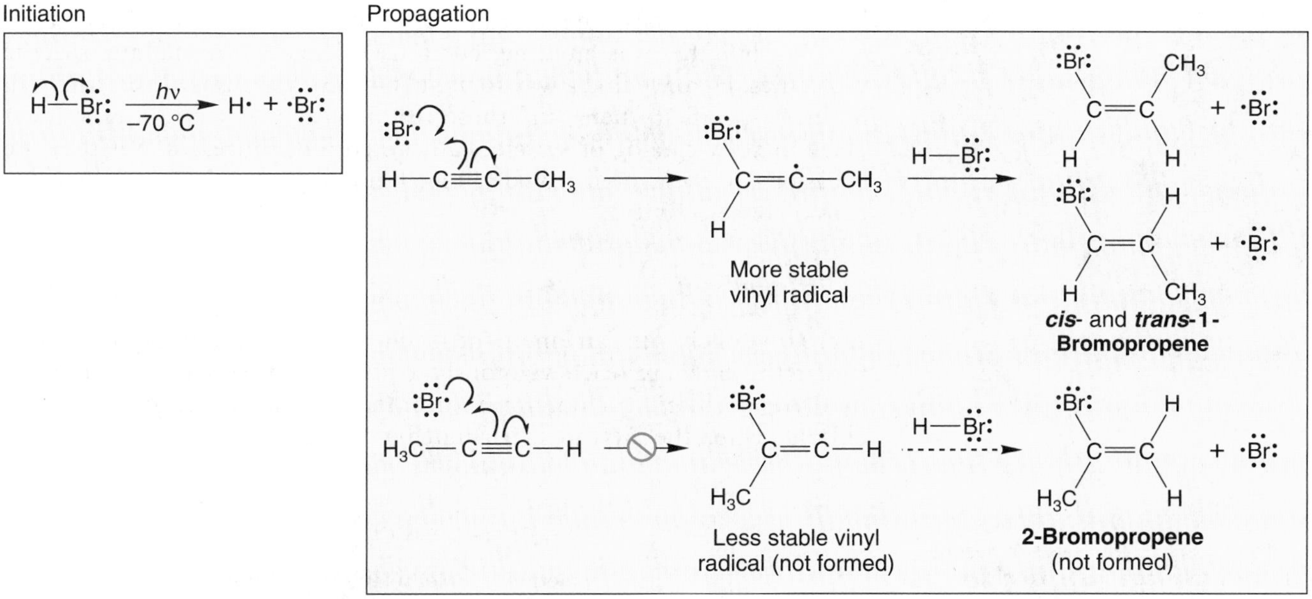

FIGURE 11.41 Addition of a bromine atom to 1-propyne gives the more stable vinyl radical, which then abstracts the hydrogen atom from hydrogen bromide to give the anti-Markovnikov product.

In principle, a second radical addition of hydrogen bromide could form either a **geminal** (1,1) or **vicinal** (1,2) dihalide. In contrast to ionic reactions it is the vicinal dibromide that is preferred (Fig. 11.42).

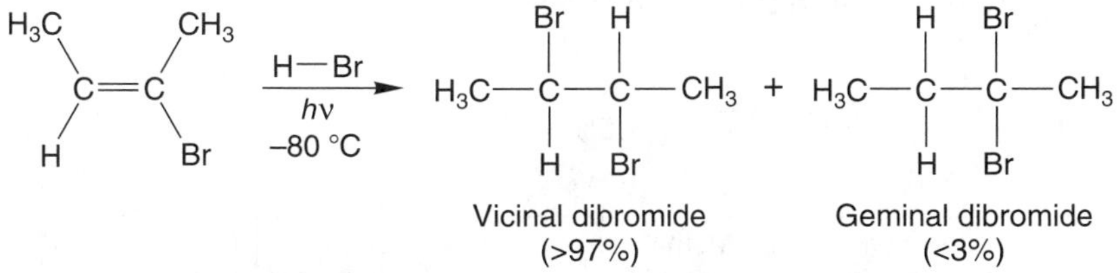

FIGURE 11.42 A second, radical addition of hydrogen bromide to a vinyl bromide generates the 1,2-dibromo compound, *not* the 1,1-dibromo compound. Problem 11.21 asks you to explain why.

PROBLEM 11.21

Explain in resonance and molecular orbital terms why it is the vicinal dihalide that is formed. Be careful! This question is more subtle than it looks at first. Consider the stabilization of an adjacent radical (a half-filled orbital) by an adjacent bromine atom. Why is this a stabilizing interaction? *Hint*: Recall the molecule He_2^+ from Chapter 2 (p. 54) and see page 460 in this chapter.

PROBLEM 11.22

Now predict whether R· will abstract hydrogen from the methylene or methyl position of diethyl ether ($CH_3CH_2OCH_2CH_3$). Explain.

11.7 PHOTOHALOGENATION

So far, we have seen a variety of reactions of radicals with alkenes. Alkanes are much less reactive than alkenes, and we have not seen ways to induce these notoriously unreactive species into reaction in a specific way. We have seen pyrolysis, the thermal cracking of saturated hydrocarbons (p. 454), but this was anything but a useful method of making specific molecules. Here we find a way to use alkanes in synthesis. There still will be little specificity here, and these reactions will not find a prominent place in your catalog of synthetically important processes. Yet they are occasionally useful, and provide a nice framework on which to base a discussion of selectivity.

11.7a Methane

A mixture of chlorine and methane is indefinitely stable at room temperature in the dark, but reacts vigorously to give a mixture of methyl chloride, methylene chloride, chloroform (trichloromethane), and carbon tetrachloride when the mixture is irradiated or heated (Fig. 11.43).

$$CH_4 + Cl_2 \xrightarrow[\text{or } \Delta]{h\nu} \underset{\textbf{Methyl chloride}}{CH_3Cl} + \underset{\textbf{Methylene chloride}}{CH_2Cl_2} + \underset{\textbf{Chloroform}}{CHCl_3} + \underset{\textbf{Carbon tetrachloride}}{CCl_4}$$

FIGURE 11.43 Photolysis of chlorine in methane gives a mixture of chlorinated methanes.

The product composition depends on the amount of chlorine available and the time the reaction is allowed to run. If we use a sufficient amount of chlorine and allow the reaction to run long enough, all the methane can be converted into carbon tetrachloride. By contrast, measuring the product composition as a function of time shows that the initial product of the reaction is methyl chloride. Methylene chloride and chloroform are formed on the way to carbon tetrachloride (Fig. 11.44).

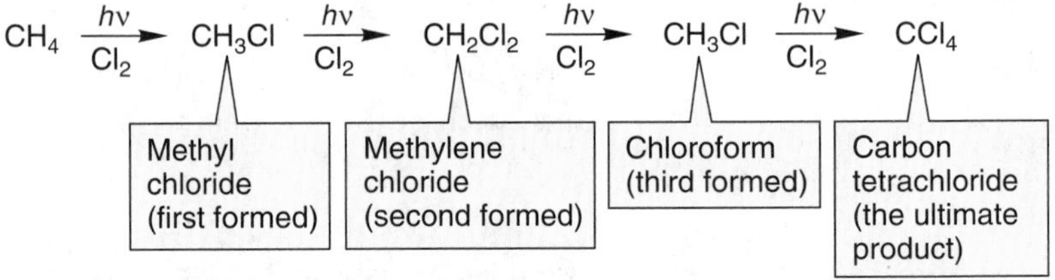

FIGURE 11.44 The chlorinated compounds are formed by sequential chlorination of methane. Methyl chloride is the initial product.

Let's first examine the formation of the initial product, methyl chloride. The overall reaction is exothermic by about 24 kcal/mol (Fig. 11.45).

$$H_3C\text{—}H + Cl\text{—}Cl \longrightarrow H_3C\text{—}Cl + H\text{—}Cl$$

105 kcal/mol (H₃C—H) | 59 kcal/mol (Cl—Cl) | 85 kcal/mol (H₃C—Cl) | 103 kcal/mol (H—Cl)

Bonds made 188 kcal/mol
Bonds broken 164 kcal/mol
24 kcal/mol exothermic

FIGURE 11.45 The overall formation of methyl chloride is exothermic by about 24 kcal/mol (103 + 85) − (105 + 59) = 24.

The first step is the thermal or photochemical breaking of the chlorine–chlorine bond. This initiation reaction is followed by the first propagation step, the abstraction of a hydrogen atom from methane by a chlorine atom to produce a methyl radical and hydrogen chloride. In the second propagation step, the newly produced methyl radical abstracts a chlorine atom from a chlorine molecule to give methyl chloride and another chlorine atom that can carry the chain. There are many possible termination steps; Figure 11.46, which shows the overall mechanism, includes only one.

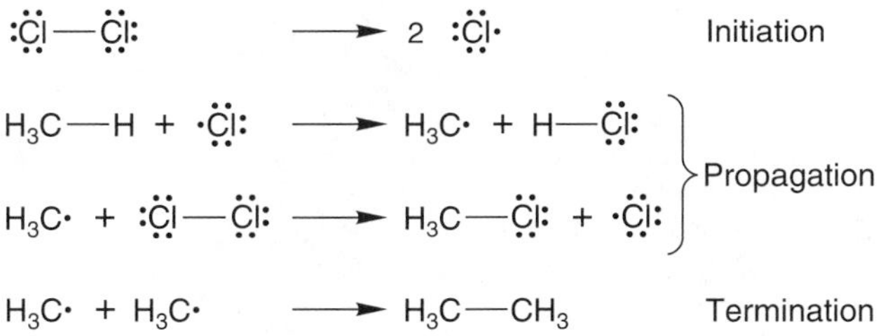

$$:\ddot{\underset{..}{Cl}}-\ddot{\underset{..}{Cl}}: \longrightarrow 2\ :\ddot{\underset{..}{Cl}}\cdot \qquad \text{Initiation}$$

$$H_3C-H + \cdot\ddot{\underset{..}{Cl}}: \longrightarrow H_3C\cdot + H-\ddot{\underset{..}{Cl}}: \qquad \text{Propagation}$$

$$H_3C\cdot + :\ddot{\underset{..}{Cl}}-\ddot{\underset{..}{Cl}}: \longrightarrow H_3C-\ddot{\underset{..}{Cl}}: + \cdot\ddot{\underset{..}{Cl}}: \qquad \text{Propagation}$$

$$H_3C\cdot + H_3C\cdot \longrightarrow H_3C-CH_3 \qquad \text{Termination}$$

FIGURE 11.46 The mechanism for the formation of methyl chloride from the photolysis of chlorine in methane.

Now let's look at the thermochemistry of the propagation steps. Although the second step is exothermic by 26 kcal/mol and thus is very favorable, the first propagation step is endothermic by about 2 kcal/mol (Fig. 11.47).

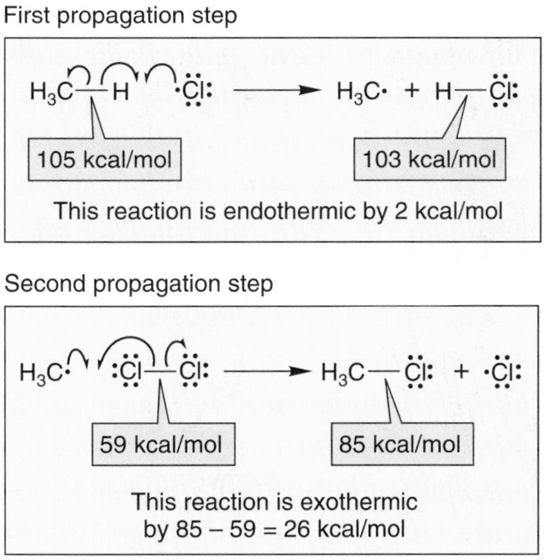

First propagation step

$$H_3C-H + \cdot\ddot{\underset{..}{Cl}}: \longrightarrow H_3C\cdot + H-\ddot{\underset{..}{Cl}}:$$

105 kcal/mol (H₃C–H) | 103 kcal/mol (H–Cl)

This reaction is endothermic by 2 kcal/mol

Second propagation step

$$H_3C\cdot + :\ddot{\underset{..}{Cl}}-\ddot{\underset{..}{Cl}}: \longrightarrow H_3C-\ddot{\underset{..}{Cl}}: + \cdot\ddot{\underset{..}{Cl}}:$$

59 kcal/mol (Cl–Cl) | 85 kcal/mol (H₃C–Cl)

This reaction is exothermic by 85 – 59 = 26 kcal/mol

FIGURE 11.47 The thermochemistry of the two propagation steps in the formation of methyl chloride. The second step is highly exothermic but the first step is slightly endothermic.

Will this endothermic step be enough to stop the reaction? No. Although a 2-kcal/mol energy difference produces an unfavorable equilibrium mixture, the second highly exothermic step results in the overall reaction being favored. The small number of methyl radicals formed reacts rapidly to give methyl chloride, thus disrupting the slightly unfavorable equilibrium.

$$Cl\cdot + CH_4 \rightleftarrows Cl{-}H + \cdot CH_3$$

Reestablishment of this equilibrium generates more methyl radicals, which go on to make more methyl chloride and chain-carrying chlorine atoms (Fig. 11.48).

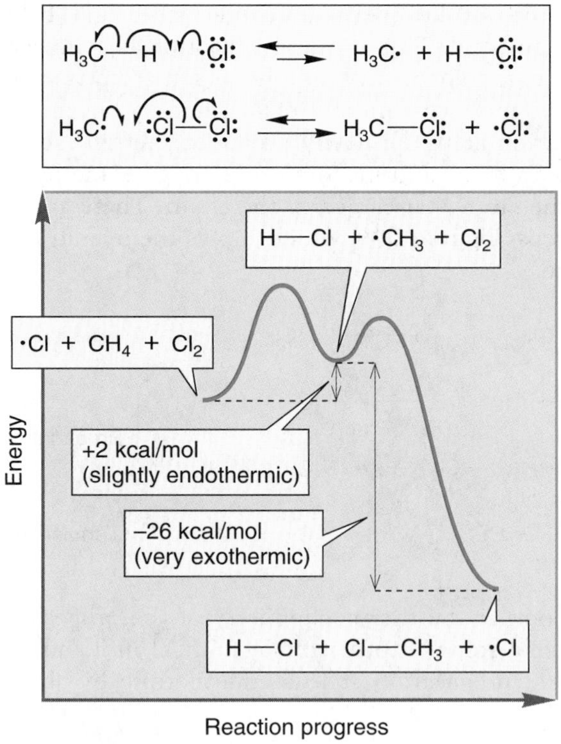

FIGURE 11.48 The high exothermicity of the second step overcomes the slight endothermicity of the first step in the formation of methyl chloride from methane and a chlorine atom.

As methyl chloride builds up in the reaction, it, too, begins to react with chlorine atoms in another chain reaction (Fig. 11.49).

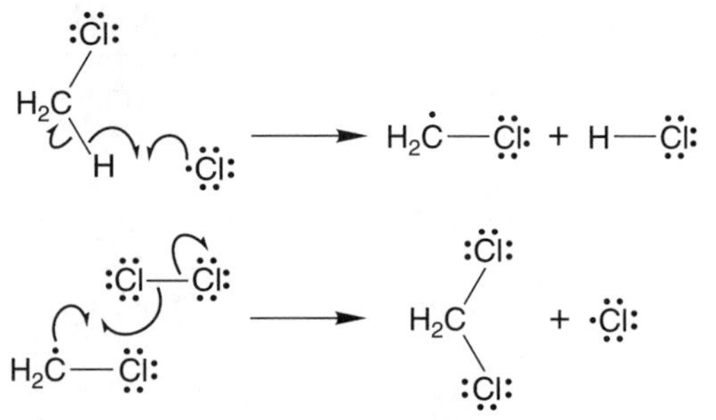

FIGURE 11.49 The formation of methylene chloride (H_2CCl_2) from methyl chloride and a chlorine atom.

Moreover, the carbon–hydrogen bond in methyl chloride is weaker than the one in methane, and thus easier to break. So methyl chloride competes favorably with methane and cannot build up in the reaction unless there is a vast excess of methane.

PROBLEM 11.23 Explain why the carbon–hydrogen bond in methyl chloride is easier to break than the carbon–hydrogen bond in methane itself.

Methylene chloride is more reactive than either methyl chloride or methane, and so it reacts to form chloroform as it builds up. Chloroform is more reactive still, and rapidly produces carbon tetrachloride through abstraction of its single hydrogen (Fig. 11.50).

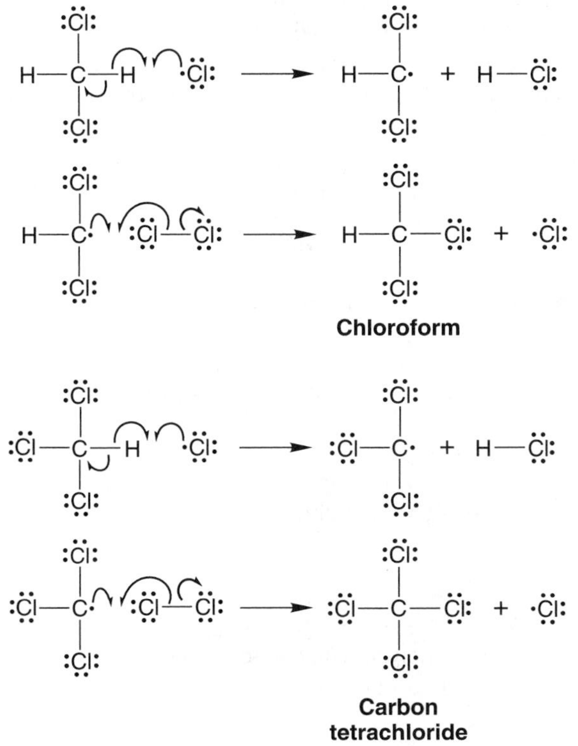

Chloroform

Carbon tetrachloride

FIGURE **11.50** The formation of chloroform ($HCCl_3$) from methylene chloride and of carbon tetrachloride (CCl_4) from chloroform.

Methane can also be photohalogenated with bromine, but the overall reaction is much less exothermic than is chlorination. Moreover, the first propagation step is endothermic by the substantial amount of 17 kcal/mol (Fig. 11.51). As we will see, this strongly endothermic step will have important consequences for the selectivity of the reaction when a choice of carbon–hydrogen bonds exists.

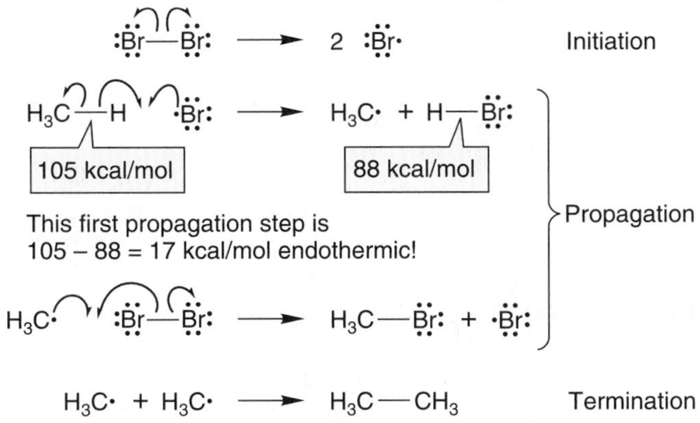

Initiation

105 kcal/mol

88 kcal/mol

This first propagation step is
105 – 88 = 17 kcal/mol endothermic!

Propagation

Termination

FIGURE **11.51** The formation of methyl bromide from bromine and methane.

11.7b Other Alkanes

If we irradiate a mixture of chlorine and a hydrocarbon more complicated than methane or ethane, there is usually more than one possible initial product. For example, butane can give either butyl chloride or *sec*-butyl chloride. Figure 11.52 shows that the secondary carbon–hydrogen bonds of butane are more reactive than the primary bonds. Abstraction by a chlorine atom occurs more easily at the secondary position than at the primary position. The percentages of products would indicate that the preference is 72.2:27.8, or a factor of 2.6. However, this simple analysis ignores the fact that there are more primary carbon–hydrogen bonds than secondary. In order to correct this value for the statistical advantage of the primary bonds, we must multiply it by $\frac{6}{4}$ to produce the true factor of 3.9. On a per hydrogen basis, secondary hydrogens are abstracted more easily than primary hydrogens by a factor of 3.9. Here is an example of selectivity (Fig. 11.52).

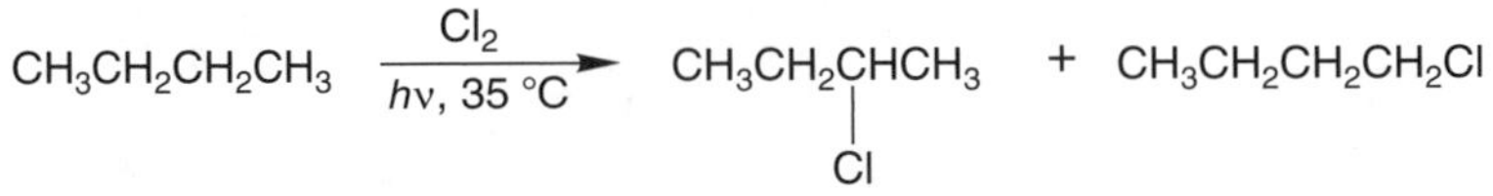

FIGURE **11.52** The photochlorination of butane can give two products, butyl chloride and *sec*-butyl chloride. The secondary (methylene) positions are favored by a factor of 72.2:27.8 = 2.6. However, there are only four methylene hydrogens and six methyl hydrogens, so this factor must be corrected for the statistical advantage in favor of methyl hydrogens. $2.6 \times \frac{3}{2} = 3.9$. On a per hydrogen basis, abstraction of a secondary hydrogen is favored over abstraction of a primary hydrogen by this amount.

$$CH_3CH_2CH_2CH_3 \xrightarrow[h\nu,\ 35\ ^\circ C]{Cl_2} CH_3CH_2CH(Cl)CH_3 + CH_3CH_2CH_2CH_2Cl$$

sec-Butyl chloride (72.2%) **Butyl chloride** (27.8%)

PROBLEM **11.24** Write a reaction mechanism for the photochlorination of ethane.

PROBLEM **11.25** Analyze the thermochemistry of the two propagation steps for this reaction. The bond dissociation energies of the chlorine–chlorine and carbon–chlorine bonds of ethyl chloride, and the carbon–hydrogen bond of ethane are 59, 81.5, and 100.3 kcal/mol, respectively.

Selectivity is increased when the possibilities include the formation of a tertiary radical. Photochlorination of isobutane gives a 64:36 mixture of isobutyl chloride and *tert*-butyl chloride. The primary/tertiary ratio is 64:36 = 1.78. However, there are nine primary hydrogens and only a single tertiary hydrogen in isobutane, so on a per hydrogen basis the primary/tertiary reactivity ratio is $1.78 \times \frac{1}{9} = 0.198$. Put another way, the tertiary hydrogen is favored by 1:0.198 = 5.1 (Fig. 11.53).

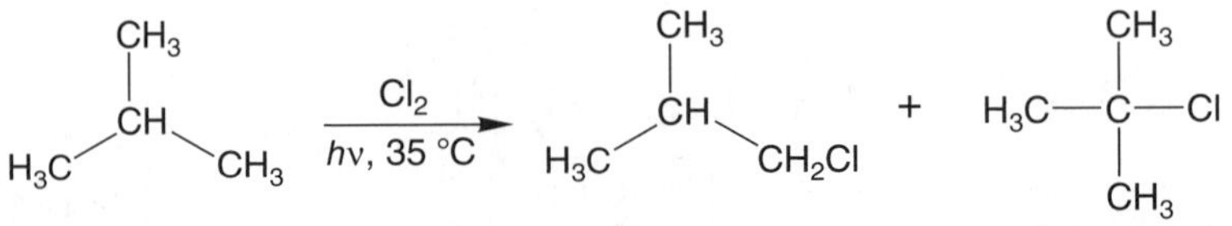

FIGURE **11.53** The photochlorination of isobutane gives *tert*-butyl chloride and isobutyl chloride. For a chlorine atom, abstraction of a tertiary hydrogen is favored over abstraction of a primary hydrogen by a factor of about 5 on a per hydrogen basis.

$$(CH_3)_2CHCH_3 \xrightarrow[h\nu,\ 35\ ^\circ C]{Cl_2} (CH_3)_2CHCH_2Cl + (CH_3)_3C{-}Cl$$

Isobutyl chloride (64%) ***tert*-Butyl chloride** (36%)

Bromine is much more selective in the photohalogenation reaction than is chlorine. That is, bromine is able to choose among the various reaction possibilities with more discrimination; to pick out the most favorable abstraction pathway by a wider margin than chlorine. For example, a bromine atom prefers secondary to primary hydrogens by a factor of 82, and tertiary hydrogens to primary hydrogens by a factor of about 1600 (Fig. 11.54; Table 11.3).

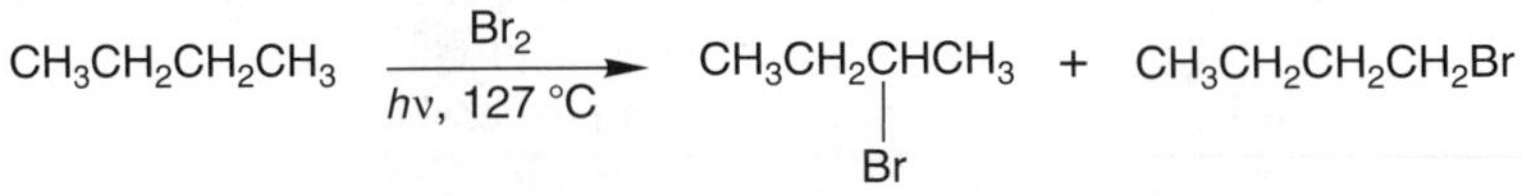

FIGURE **11.54** Photobromination is far more selective than photochlorination.

$$CH_3CH_2CH_2CH_3 \xrightarrow[h\nu,\ 127\ ^\circ C]{Br_2} CH_3CH_2CH(Br)CH_3 + CH_3CH_2CH_2CH_2Br$$

(98.2%) (1.8%)

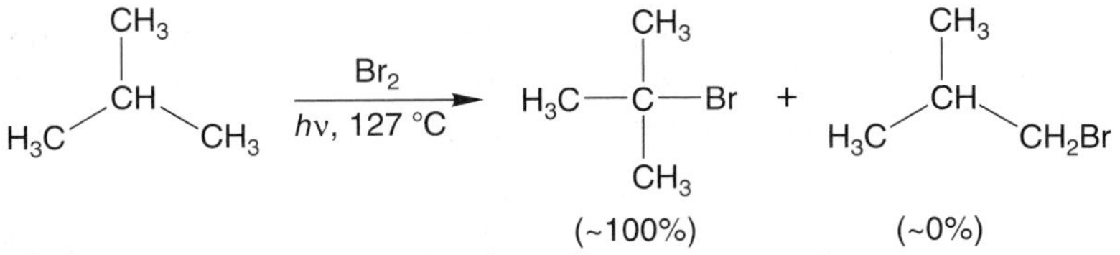

$$(CH_3)_2CH{-}CH_3 \xrightarrow[h\nu,\ 127\ ^\circ C]{Br_2} (CH_3)_3C{-}Br\ (\sim 100\%) + (CH_3)_2CH{-}CH_2Br\ (\sim 0\%)$$

FIGURE 11.54 (CONTINUED)

TABLE 11.3 Selectivities of Halogen Atoms in Carbon–Hydrogen Bond Abstraction

Halogen	Primary	Secondary	Tertiary
F	1	1.2	1.4
Cl	1	3.9	5.1
Br	1	82	1600

Neither fluorine nor iodine undergoes a useful photohalogenation reaction. The reaction of fluorine with carbon–hydrogen bonds is overwhelmingly exothermic and can even occur explosively. Almost no selectivity is possible (Table 11.3). In direct contrast to the reaction of fluorine atoms, reaction of iodine atoms with hydrocarbons is highly endothermic. It is so slow as to be of no synthetic utility.

*PROBLEM 11.26

Use the bond dissociation energies of Table 8.2 (p. 307) to estimate the thermochemistry of the propagation steps for photochemical fluorination and iodination of methane.

ANSWER

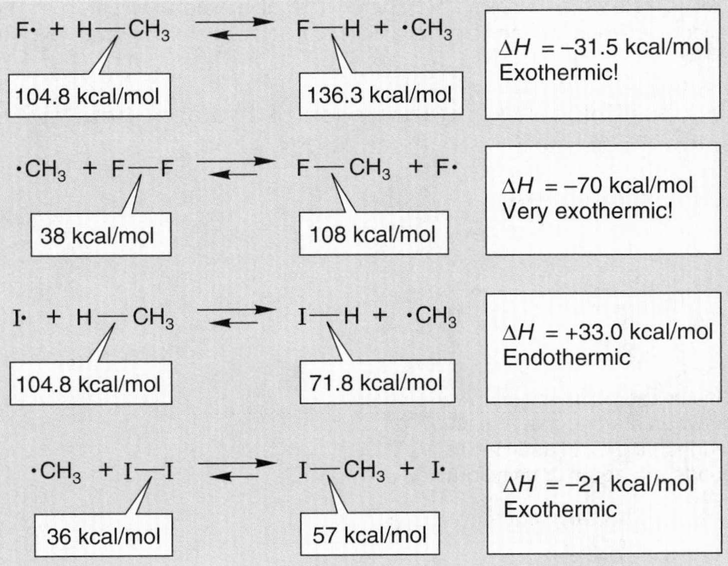

$F\cdot + H{-}CH_3 \rightleftharpoons F{-}H + \cdot CH_3$ (H–CH₃: 104.8 kcal/mol; F–H: 136.3 kcal/mol) $\Delta H = -31.5$ kcal/mol Exothermic!

$\cdot CH_3 + F{-}F \rightleftharpoons F{-}CH_3 + F\cdot$ (F–F: 38 kcal/mol; F–CH₃: 108 kcal/mol) $\Delta H = -70$ kcal/mol Very exothermic!

$I\cdot + H{-}CH_3 \rightleftharpoons I{-}H + \cdot CH_3$ (H–CH₃: 104.8 kcal/mol; I–H: 71.8 kcal/mol) $\Delta H = +33.0$ kcal/mol Endothermic

$\cdot CH_3 + I{-}I \rightleftharpoons I{-}CH_3 + I\cdot$ (I–I: 36 kcal/mol; I–CH₃: 57 kcal/mol) $\Delta H = -21$ kcal/mol Exothermic

What is the reason for the high selectivity of bromine as compared to chlorine? Recall that the abstraction reaction is close to thermoneutral* for chlorine, but very endothermic for bromine (p. 479). The Hammond postulate (the transition state for an endothermic process resembles the product) tells us that the transition state for the abstraction of hydrogen by chlorine will be nearly halfway in between starting material and product, but very product-like for bromine. Because the reaction is more en-

*Watch out! One of the things chemistry professors laugh about is the number of people who call "thermoneutral" reactions "thermonuclear" reactions. Believe it or not, this happens a lot!

dothermic, the transition state for abstraction of hydrogen by bromine will be product-like. There is a strongly developed hydrogen–bromine bond and a large degree of radical character on carbon in the transition state (Fig. 11.55).

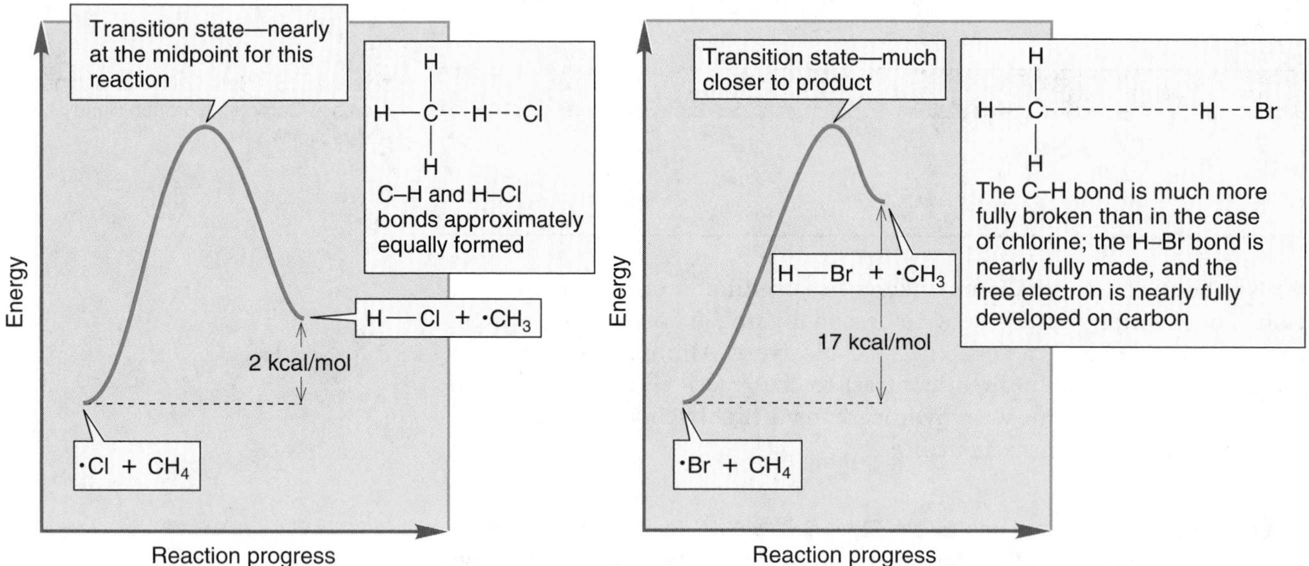

FIGURE 11.55 For the nearly thermoneutral abstraction of hydrogen by a chlorine atom, the transition state will be almost halfway between starting material and products. For the highly endothermic abstraction of hydrogen by bromine, the transition state will be quite product-like.

The transition states for abstraction of hydrogen from methane by chlorine and bromine are very different (Fig. 11.56).

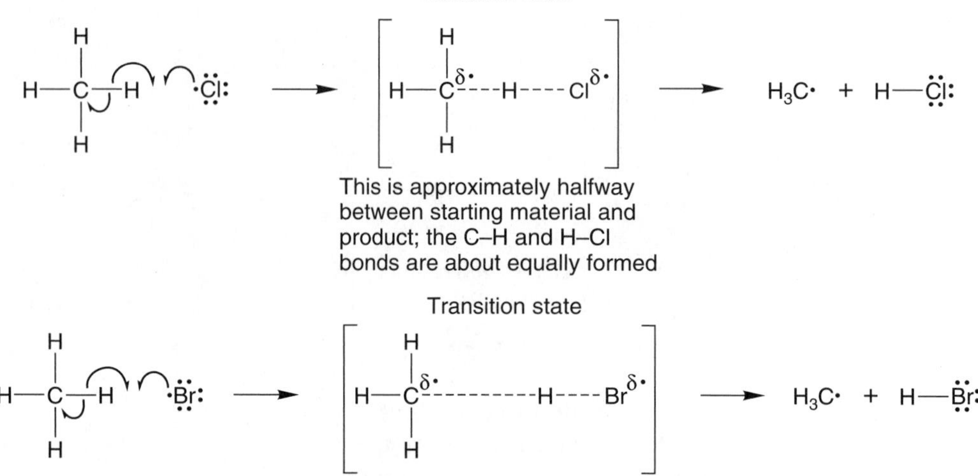

FIGURE 11.56 In the transition state for abstraction by chlorine, the carbon–hydrogen and hydrogen–chlorine bonds will be approximately equally formed. For bromine, with a very product-like transition state, the hydrogen–bromine bond will be nearly completely formed and the radical on carbon will be well developed.

In the transition states for abstractions by bromine, the radical on carbon is far more developed than it is in the abstractions by chlorine, which means that differences in the stabilities of different radicals will be far more important in the bromine abstractions than in the chlorine abstractions (Fig. 11.57).

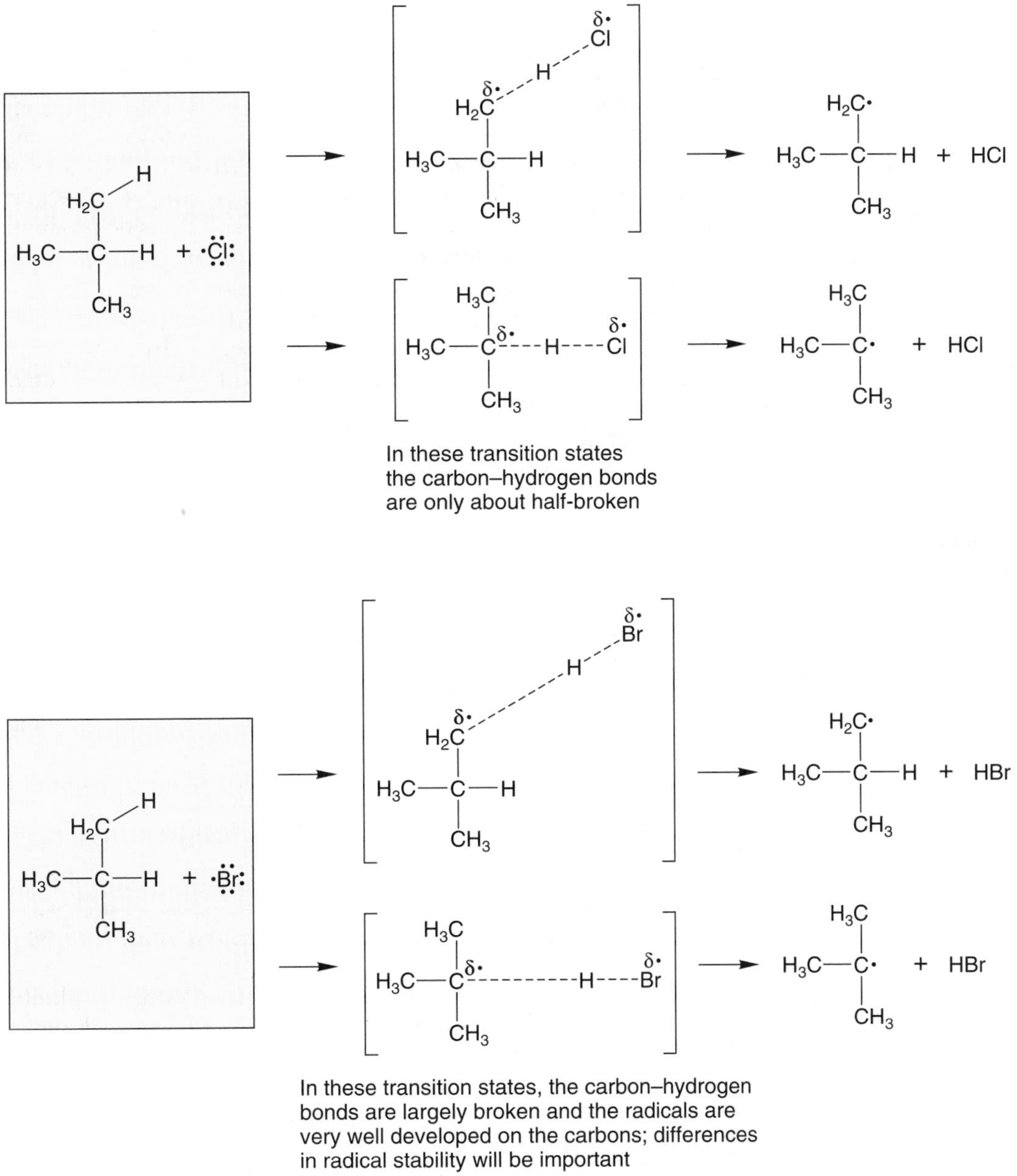

In these transition states the carbon–hydrogen bonds are only about half-broken

In these transition states, the carbon–hydrogen bonds are largely broken and the radicals are very well developed on the carbons; differences in radical stability will be important

FIGURE **11.57** The "late" (product-like) transition state for bromine abstraction means that the radical on carbon will be well developed and the substitution pattern of the carbon atom (tertiary, secondary, primary, or methyl) will be relatively important.

When questions of selectivity arise, it is always important to consider what the transition state looks like.

11.8 ALLYLIC HALOGENATION

One of the goals of synthetic organic chemists is to increase selectivity—to be able to do chemical transformations at specific sites in a molecule without affecting the rest of the molecule. Selectivity is useful—even necessary—when one is dealing with a complicated molecule. We must have ways of inducing change at one place in a molecule without affecting others if we are to be able to manipulate molecules rationally. It's quite clear that the photochlorination reaction of alkanes is not a good way of doing this; it is far too unselective. Chlorination takes place all too easily at the various positions in the hydrocarbon.

Photobromination more closely approaches the goal of perfect selectivity because the bromine atom abstracts tertiary hydrogens much more readily than secondary or primary hydrogens. But even this reaction is not without its problems. It is a slow process and the side products of di- and polybromination are formed.

There are some instances where the chlorination or bromination reaction can be used to good effect, however. Alkenes containing *allylic* hydrogens can sometimes be halogenated specifically at the allylic position in a process called **allylic halogenation**, another free radical chain reaction. For example, when low concentrations of bromine are photolyzed in a complicated cyclohexene, the benzoate ester of cholesterol, the product is exclusively brominated in the allylic position (Fig. 11.58).

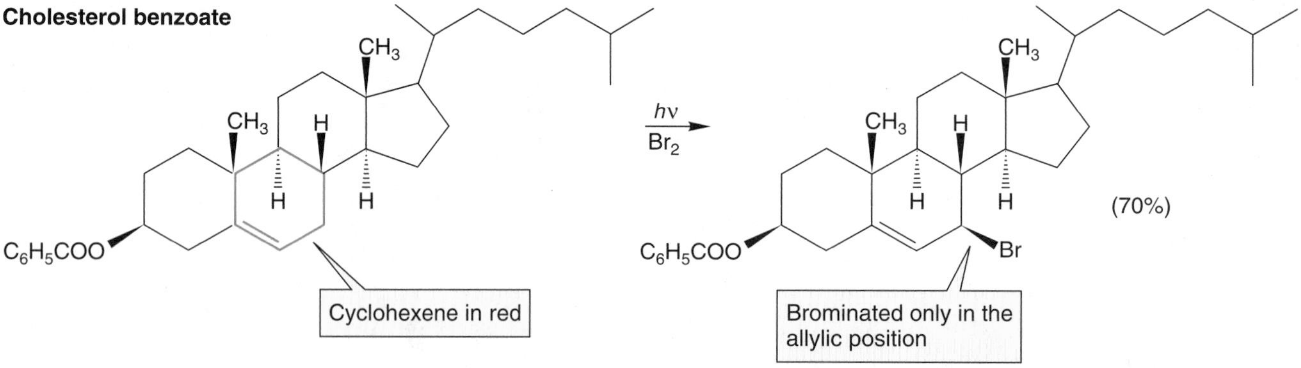

Cholesterol benzoate

CH$_3$ CH$_3$ H H H C_6H_5COO

Cyclohexene in red

$\xrightarrow[Br_2]{h\nu}$

CH$_3$ CH$_3$ H H H C_6H_5COO Br (70%)

Brominated only in the allylic position

FIGURE 11.58 When low concentrations of bromine are photolyzed in the benzoate ester of cholesterol, the product is exclusively the allylic bromide.

The mechanism is a typical radical chain process in which a low concentration of bromine atoms is first produced by breaking of the weak bromine–bromine bond (initiation step). Abstraction of hydrogen is the first propagation step, but there are three possible kinds of hydrogen that could be removed from a cyclohexene, H_v (vinyl), H_a (allylic), and H_m (methylene) (Fig. 11.59).

As shown in Figure 11.60, each of these radicals could lead to a bromocyclohexene through abstraction of a bromine atom in the second propagation step. The chain is then carried by the new bromine atom formed in this step.

As there is only one product formed, the allylic bromide, the allylic radical must be formed with high selectivity. Why? *It is only this radical that is resonance stabilized.* There are two equivalent resonance forms for the allyl radical and this delocalization makes it far more stable than the other two possibilities (Fig. 11.61).

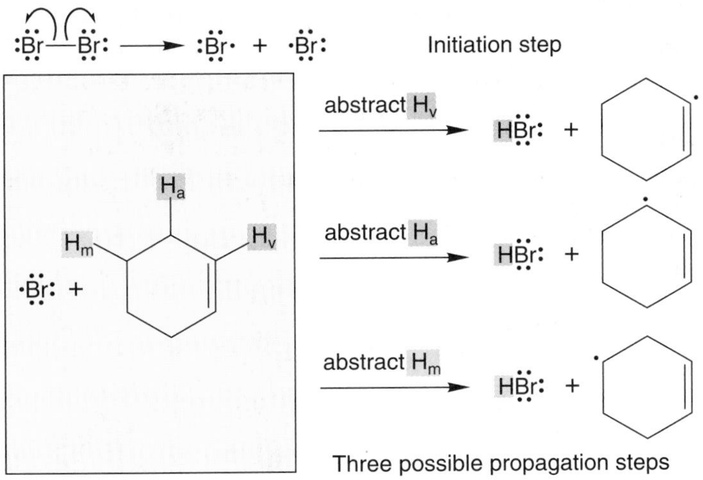

FIGURE **11.59** Three possible propagation steps in the allylic bromination of a cyclohexene. In principle, a bromine atom could abstract a vinyl (H_v), allylic (H_a), or methylene (H_m) hydrogen.

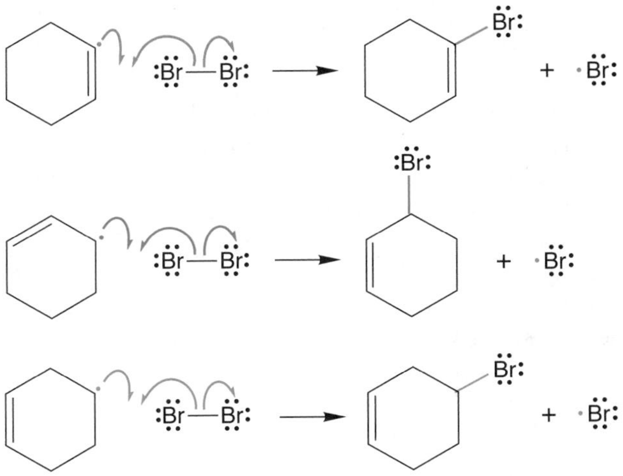

FIGURE **11.60** In a propagation step, any of the three carbon radicals could abstract bromine to give bromides and a chain-carrying bromine atom. The formation of only the allylic bromide tells us that only the allylic radical is produced originally.

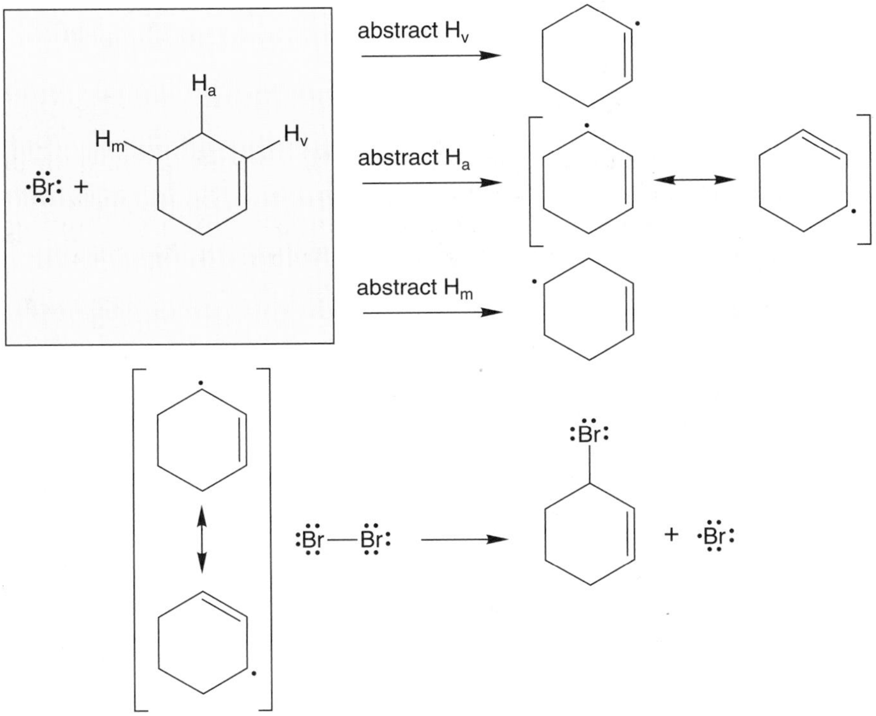

FIGURE **11.61** The allylic radical is resonance stabilized and will be formed much more easily than the others. Further reaction leads to the product of allylic bromination.

One might hope that the photohalogenation reaction would be general; that photochlorination would be as useful as photobromination. Alas it is not, and Table 11.3 shows why. A chlorine atom is far less selective than a bromine atom. Utility in synthesis comes from selectivity—from doing one reaction, not several. Photobromination is useful because the bromine atom picks and chooses among the various possible carbon–hydrogen bonds with fairly high discrimination. A chlorine atom does not. For example, the allylic position in cyclohexene is only slightly more reactive toward a chlorine atom than the other methylene position (Fig. 11.62).

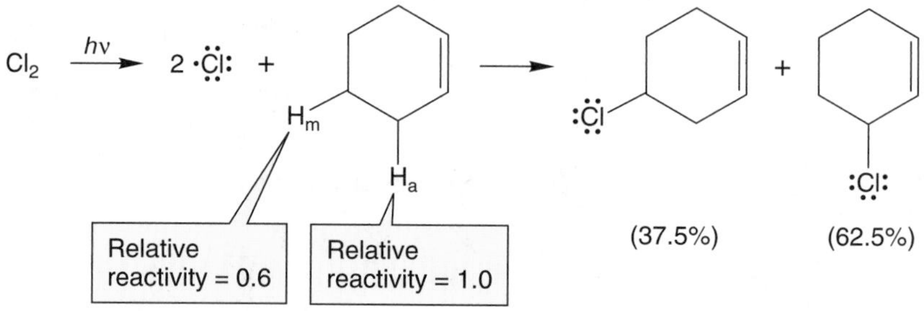

Cl_2 —hν→ 2 ·Cl: + cyclohexene → chlorocyclohexene (37.5%) + 3-chlorocyclohexene (62.5%)

H_m: Relative reactivity = 0.6

H_a: Relative reactivity = 1.0

FIGURE 11.62 Chlorine is much less selective than bromine. The allylic chloride is not the only major product formed in this reaction.

In photohalogenation reactions, keeping the concentration of bromine low is essential to success. If there is too much halogen, polar addition of bromine or chlorine to give vicinal dihalides competes with formation of the allylic halide, and the desired selectivity is lost. Methods have been developed to ensure that this low concentration is maintained, and to lower the temperature necessary for the initiation step. One of these methods involves the molecule **N-bromosuccinimide**, generally known as NBS, in the solvent carbon tetrachloride (CCl_4) (Fig. 11.63).

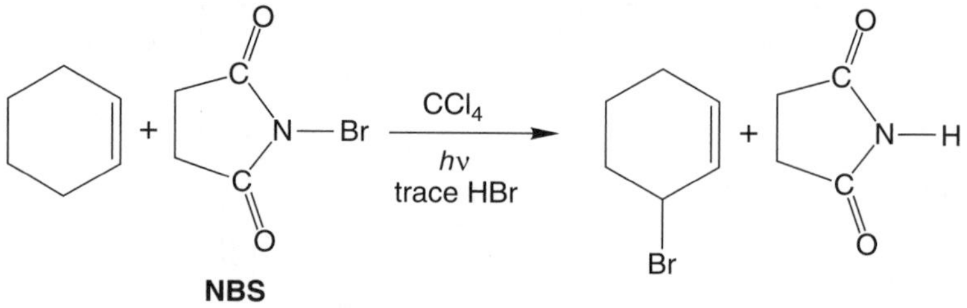

cyclohexene + NBS —CCl_4, hν, trace HBr→ 3-bromocyclohexene + succinimide (N—H)

NBS

FIGURE 11.63 A common and very effective method of allylic bromination uses NBS.

The first step in this reaction is formation of a bromine atom through dissociation of the nitrogen–bromine bond of NBS. The bromine atom then abstracts a hydrogen from the allylic position of the alkene to give hydrogen bromide and a resonance-stabilized allyl radical (Fig. 11.64).

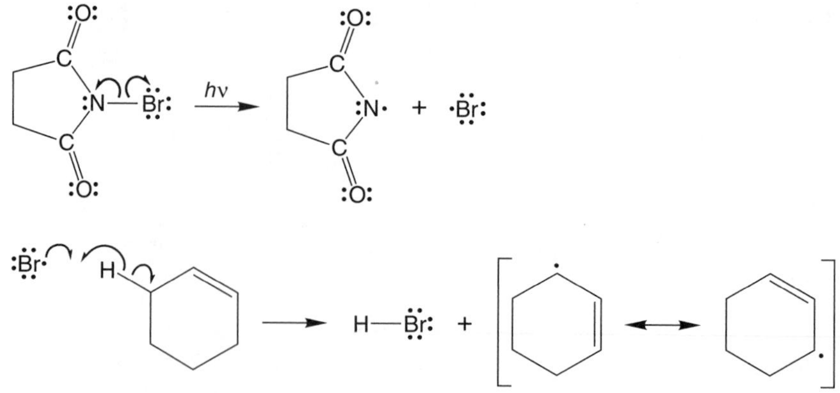

NBS —hν→ succinimidyl radical (:N·) + ·Br:

:Br· + cyclohexene (allylic H) → H—Br: + [cyclohexenyl radical ←→ cyclohexenyl radical]

FIGURE 11.64 The first two steps in the bromination of cyclohexene using NBS. A bromine atom is formed in an initiation step and then abstracts an allylic hydrogen from cyclohexene to form the resonance-stabilized cyclohexenyl radical.

The hydrogen bromide then reacts with NBS to form bromine, which in turn reacts with the allyl radical to give the allyl bromide and a new, chain-carrying bromine atom. The NBS is only slightly soluble in carbon tetrachloride, thus ensuring the low concentration of bromine necessary for the success of this reaction (Fig. 11.65).

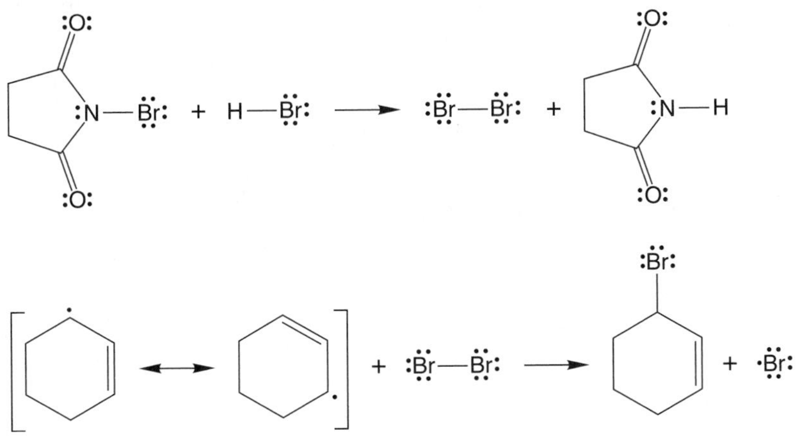

FIGURE 11.65 A molecule of bromine is formed, which then reacts with the cyclohexenyl radical to give the product and a new bromine atom.

Suggest a mechanism for the formation of bromine from NBS and hydrogen bromide. *Hint*: The oxygen of a carbon–oxygen double bond is a base.

PROBLEM 11.27

Selectivity is not possible in a disubstituted alkene such as *trans*-2-pentene. Explain.

*PROBLEM 11.28

ANSWER

In *trans*-2-pentene there are two allylic positions, and two resonance-stabilized radicals can be formed through hydrogen abstraction by a bromine atom.

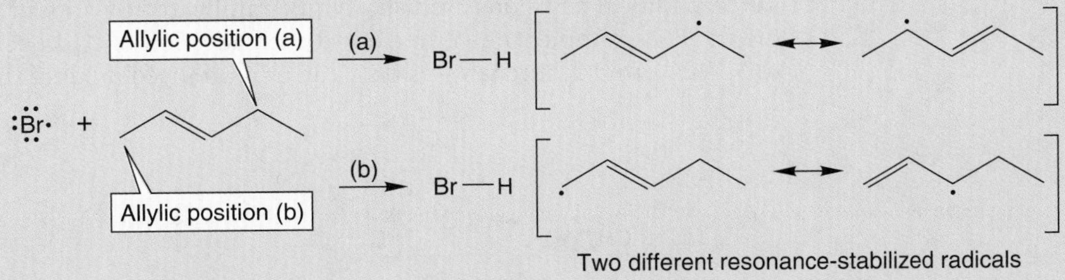

Allylic position (a)

Allylic position (b)

Two different resonance-stabilized radicals

Once the two allyl radicals are formed, the chain can continue as abstraction from Br_2 takes place. Notice that only three different products are formed because the two carbons sharing the free electron in the intermediate from abstraction (a) are equivalent.

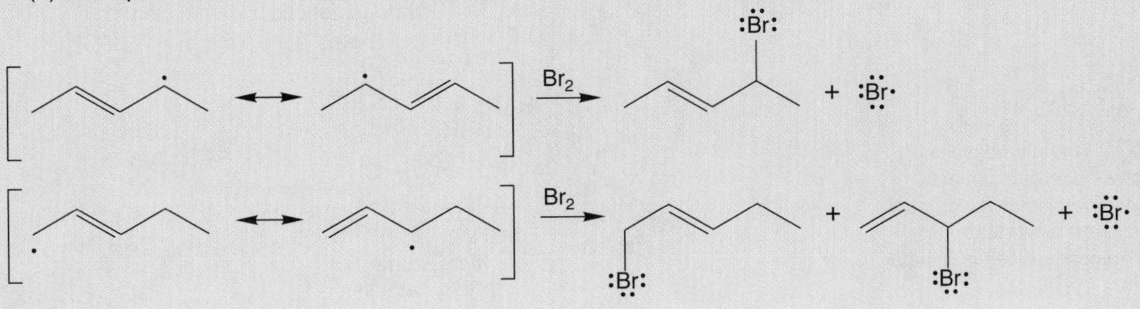

11.9 SOMETHING MORE: REARRANGEMENTS (AND NONREARRANGEMENTS) OF RADICALS

We have seen in Chapters 9 and 10 that polar additions to alkenes are bedeviled by shifts of hydride or alkyl groups when there is a possibility of forming a more stable cation from a less stable one. A typical example, the polar addition of hydrogen chloride to 3,3-dimethyl-1-butene, is shown again in Figure 11.66.

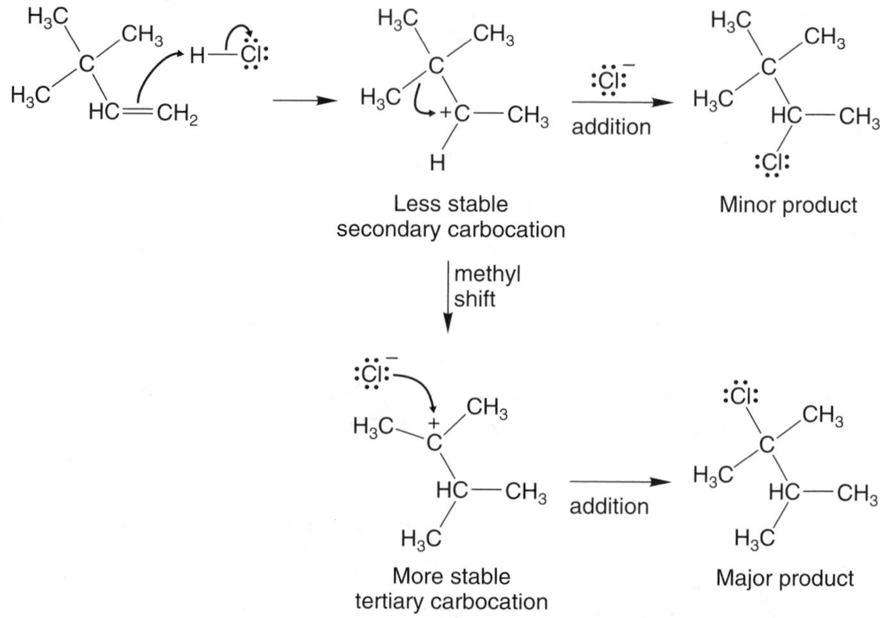

Less stable secondary carbocation

addition

Minor product

methyl shift

More stable tertiary carbocation

addition

Major product

FIGURE 11.66 Rearrangements occur in the polar addition of hydrogen chloride to 3,3-dimethyl-1-butene.

Yet free radical additions are usually uncomplicated by similar rearrangements. For example, the peroxide-initiated addition of hydrogen bromide to 3,3-dimethyl-1-butene gives only the unrearranged product (Fig. 11.67).

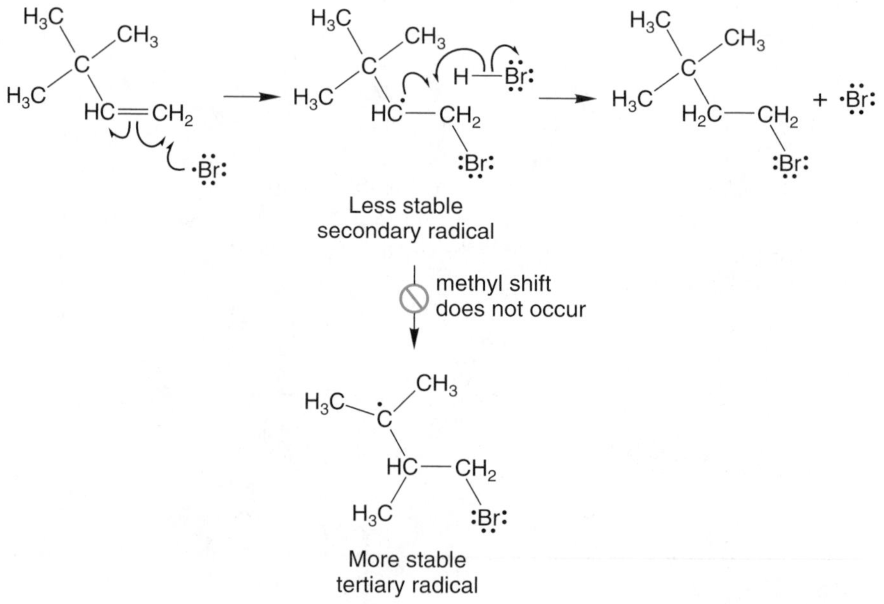

Less stable secondary radical

methyl shift does not occur

More stable tertiary radical

FIGURE 11.67 The related radical additions involve no rearrangements. For example, the peroxide-initiated addition of hydrogen bromide to 3,3-dimethyl-1-butene proceeds without rearrangement.

We might have expected rearrangement through the shift of a methyl group to give the more stable tertiary radical from the less stable secondary one, yet this doesn't happen. In fact, the situation is quite dramatic: 1,2-hydride shifts (migration of H:$^-$) are among the fastest known organic reactions and the simple 1,2-shift of a hydrogen atom (or an alkyl radical) is unknown (Fig. 11.68).

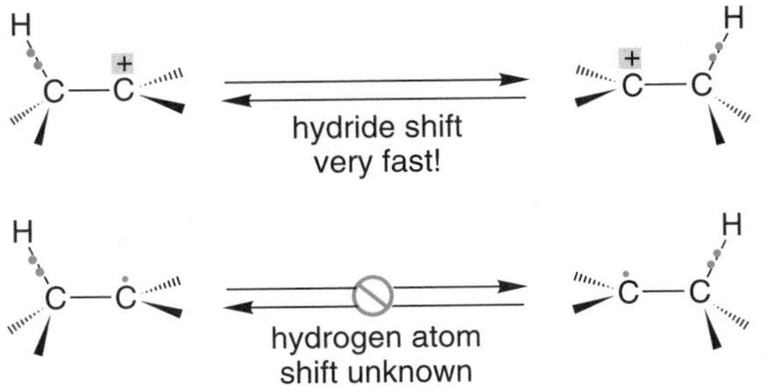

hydride shift very fast!

hydrogen atom shift unknown

FIGURE **11.68** The simple 1,2-shift of a hydrogen atom (H·) is not known. Hydride (H:$^-$) shifts are very common.

This situation seems mysterious, but a look at the molecular orbitals involved in the transition state for the reaction shows the reason. There is more than one way to approach this problem, but I think an analysis of the transition state for the rearrangement is the easiest. In the transition state for any symmetrical 1,2-shift of hydrogen, the hydrogen atom is half-transferred from one carbon to the other (Fig. 11.69).

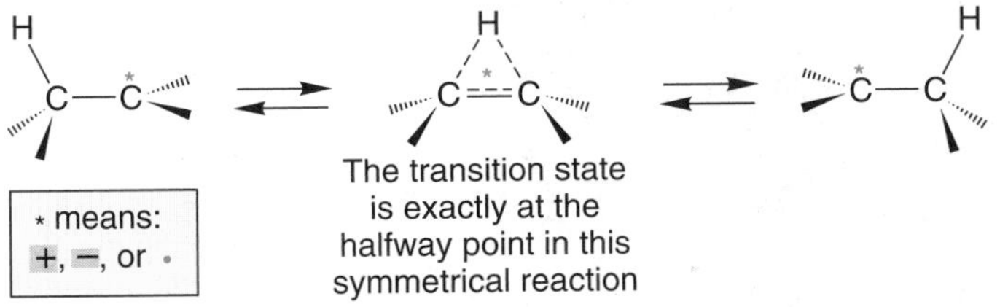

* means: +, −, or ·

The transition state is exactly at the halfway point in this symmetrical reaction

FIGURE **11.69** In the transition state for any symmetrical 1,2-shift, the hydrogen is halfway between the two carbons. Here the symbol * = +, ·, or –.

What will the molecular orbitals of the transition state look like? To answer this question, we need only construct the molecular orbitals for the system of a hydrogen 1*s* orbital interacting with π and π* orbitals of an alkene. For a model of how to do a closely related system, see the discussion of cyclic H_3 in Problem 2.19 (p. 64). The combination of 1*s*, π, and π* will yield three new molecular orbitals **A**, **B**, and **C**. They are constructed in Figure 11.70.

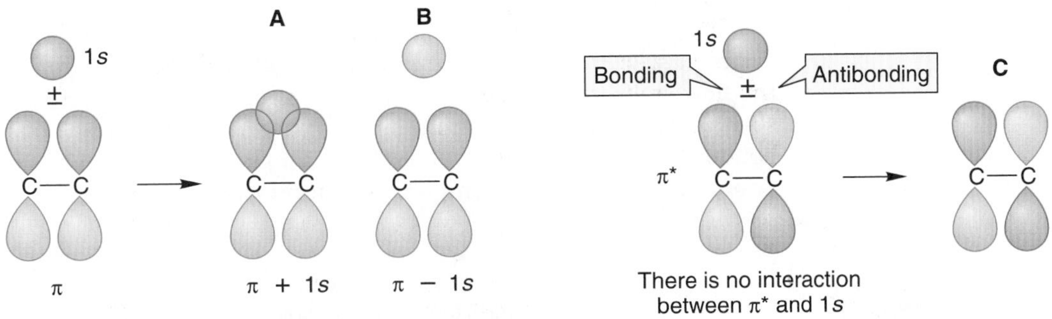

FIGURE **11.70** The three molecular orbitals for the transition state in this reaction can be easily constructed from H (1*s*), π, and π*.

The 1*s* orbital interacts with π to form two new molecular orbitals (**A** and **B**), but there is no interaction between 1*s* and π* in this geometry. Accordingly, π* becomes **C**, the third molecular orbital for the transition state. To order the new molecular orbitals (**A**, **B**, **C**) in energy, we need only count nodes. Molecular orbital **A** has no new nodes, but **B** and **C** each has one additional node (Fig. 11.71).

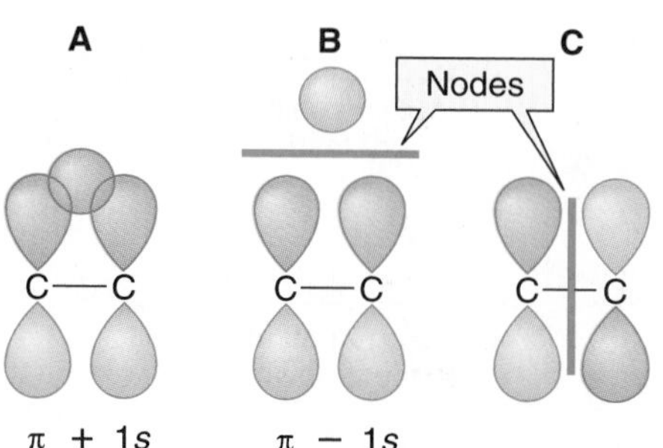

FIGURE 11.71 The new molecular orbitals (**A–C**) can be ordered in energy by counting nodes.

The ordering of the new molecular orbitals is shown in Figure 11.72.

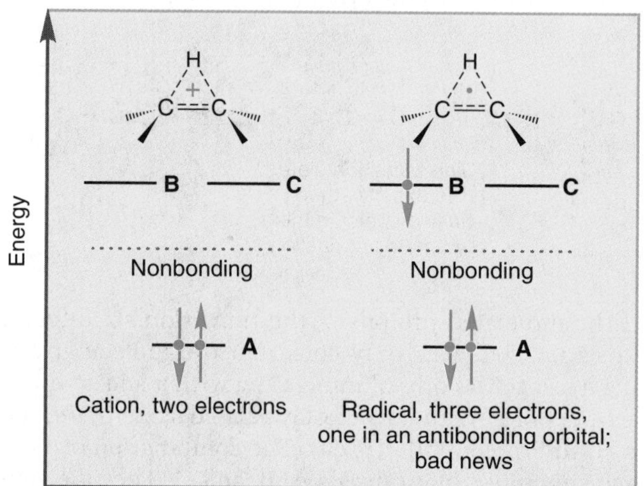

FIGURE 11.72 The transition state will have a single bonding molecular orbital (**A**) and two, equi-energetic antibonding molecular orbitals (**B** and **C**). For the cationic rearrangement, the two electrons involved nicely fill the bonding molecular orbital (**A**). In the radical case, there is another electron and it must occupy an antibonding molecular orbital (**B** or **C**) in the transition state. Occupancy of this antibonding orbital is unfavorable.

Now, how many electrons do we need to put into the orbital system? For the cationic reaction, there are only two and they will go nicely into the new stabilized bonding molecular orbital, **A**. However, for the radical shift there is another electron, and it must be placed in one of the antibonding molecular orbitals, **B** or **C**. This high-energy electron is evidently destabilizing enough to completely shut off the reaction (Fig. 11.72).

PROBLEM 11.29 Analyze the shift of hydrogen in the related anionic system.

*PROBLEM **11.30**

1,2-Shifts of vinyl groups are known in radicals. Can you see how this reaction might differ from the 1,2-shift of a hydrogen atom or a methyl radical, both of which are unknown?

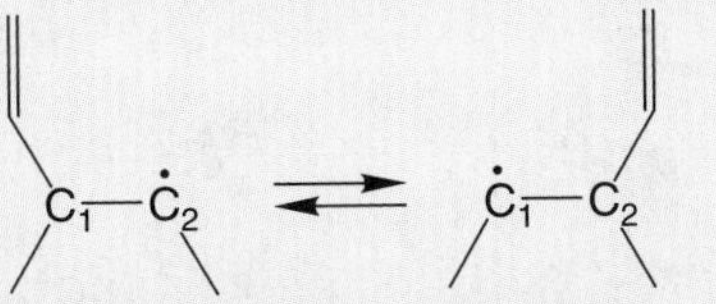

ANSWER

A vinyl group, or any group with *p* orbitals, is not restricted to using the σ bond in the migration. In this case, the radical on C(2) could add to the vinyl group to give the symmetrical intermediate **A**.

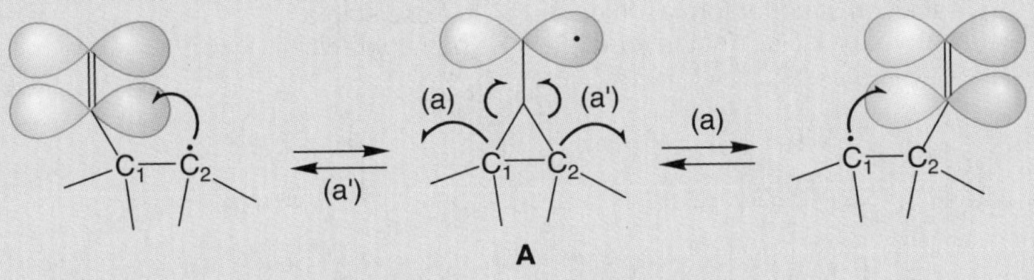

Opening of intermediate **A** can occur in two directions (a and a') to regenerate the starting material (a') or complete the vinyl migration (a).

*PROBLEM **11.31**

We have already seen a reaction much like the opening of the cyclic intermediate shown in the answer to Problem 11.30. What is the name of this process?

ANSWER

It is just a β-cleavage (p. 453).

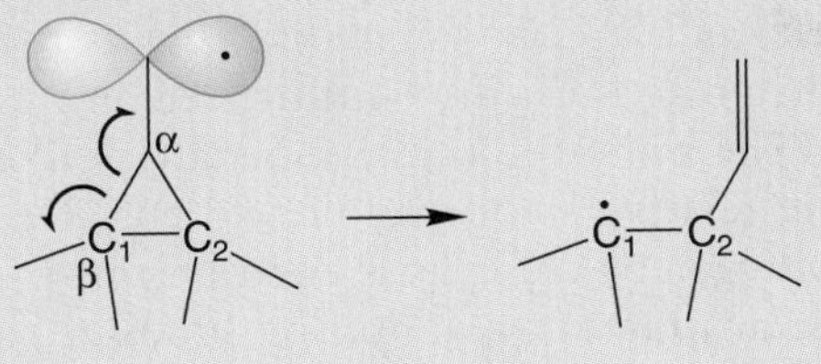

11.10 SUMMARY

NEW CONCEPTS

One of the most important concepts of this chapter is that of selectivity. It is not really new as you have seen many selective reactions before. A simple example is the protonation of isobutene to give the more stable tertiary carbocation rather than the less stable primary carbocation (Fig. 11.21).

In this chapter, you also see selectivity when a radical abstracts one of several possible hydrogens from an alkane. The ease of hydrogen abstraction is naturally determined by the energies of the transition states leading to the products. A tertiary radical is more likely to be formed than a primary radical. In the transition state for radical formation, the factors that make a tertiary radical more stable than a primary radical will operate to stabilize the partially formed tertiary radical over the partially formed primary radical (Fig. 11.73).

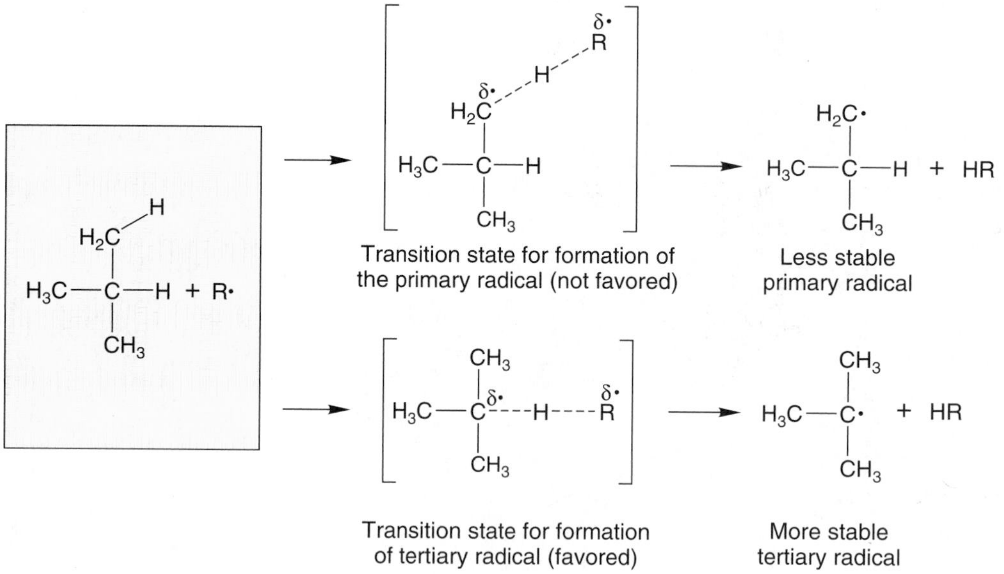

FIGURE 11.73 In the transition state for formation of the tertiary radical there is a partially formed tertiary radical. Similarly, in the transition state for formation of the primary radical there is a partially formed primary radical. The partially formed tertiary radical is more stable and will be preferred. The reaction will selectively form the more stable tertiary radical.

What is more subtle, and at least somewhat new, is the idea that the degree of selectivity in a reaction will be influenced by *how product-like* the transition state is. The more the transition state is like the product, the more important the relative stabilities of two possible products will be. Abstraction of a hydrogen by chlorine is endothermic by about 2 kcal/mol (p. 477). The same abstraction reaction by bromine is endothermic by about 17 kcal/mol. The transition state for the much more endothermic abstraction reaction by bromine will be more product-like than that for abstraction by chlorine. So we expect bromine to be more selective than chlorine, as the carbon–hydrogen bond will be more broken in the transition state. The stability of the product radical will be more important in the transition state for hydrogen abstraction by a bromine than it will in the transition state for abstraction by a chlorine (Fig. 11.57).

There are a few smaller concepts as well. Radical stability parallels carbocation stability. Tertiary radicals are more stable than secondary, secondary more stable than primary, and primary more stable than methyl. The differences aren't as great for radicals as for cations. Delocalization helps stabilize radicals by spreading the odd electron over two or more atoms.

Some reaction mechanisms are cyclic. Radical chain reactions produce a chain-carrying radical in their last, product-making step. The new radical can begin another series of propagation steps leading to another molecule of product. Chain reactions only stop when starting material is exhausted or a chain is terminated by bond formation by a pair of radicals.

Chains can be set going by small amounts of radicals released by radical initiators. In this way, a small number of chain-initiating radicals can have an impact far out of proportion to their numbers.

REACTIONS, MECHANISMS, AND TOOLS

The reactions in this chapter involve neutral free radicals, in contrast to the ionic species of earlier chapters. Chain reactions are characteristic of radical chemistry, and consist of several different kinds of steps. (1) Initiation: A small number of radicals serves to start the reaction. (2) Propagation: Radical reactions are carried out that generate a product molecule and a new, chain-carrying radical, which recycles to the beginning of the chain and starts a new series of product-forming steps. (3) Termination: Radical recombinations destroy chain-carrying species and end chains. The success of a chain reaction depends on the relative success of the chain-carrying and termination steps. Figure 11.25 gives an example of a typical chain reaction.

Important reactions of radicals include abstraction, additions to alkenes and alkynes, β-cleavage, and disproportionation, but not simple 1,2-rearrangements (Fig. 11.74).

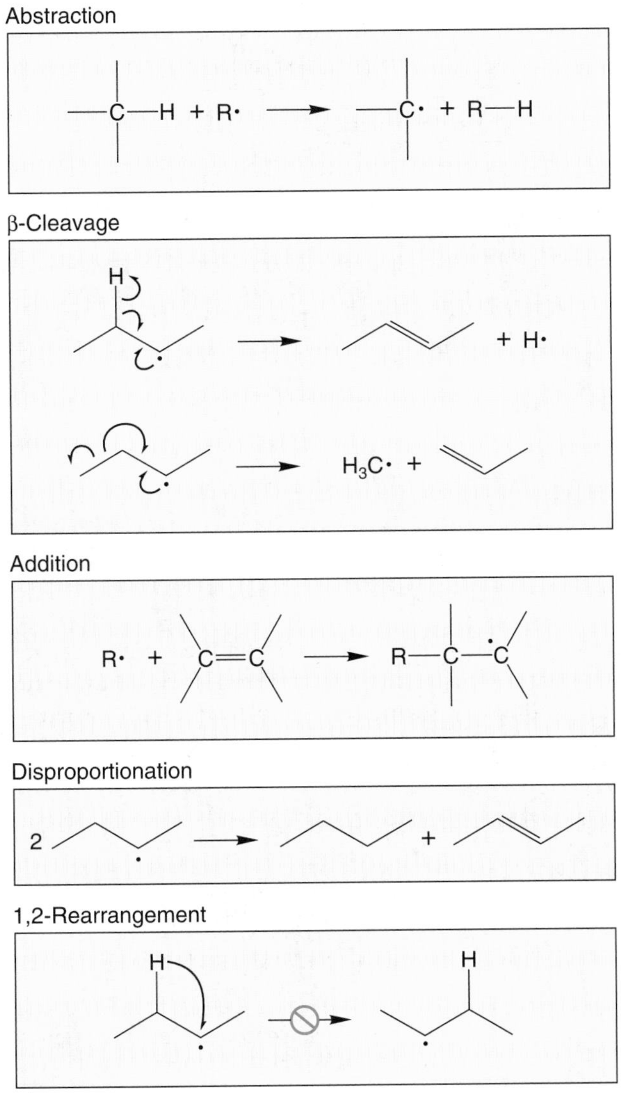

Abstraction

—C—H + R· ⟶ —C· + R—H

β-Cleavage

⟶ + H·

⟶ H_3C· +

Addition

R· + C=C ⟶ R—C—Ċ

Disproportionation

2 ⟶ +

1,2-Rearrangement

H ⟶ H

FIGURE 11.74 Some typical radical reactions.

SYNTHESES

The new synthetic reactions of this chapter are shown in Figure 11.75.

1. Alkanes

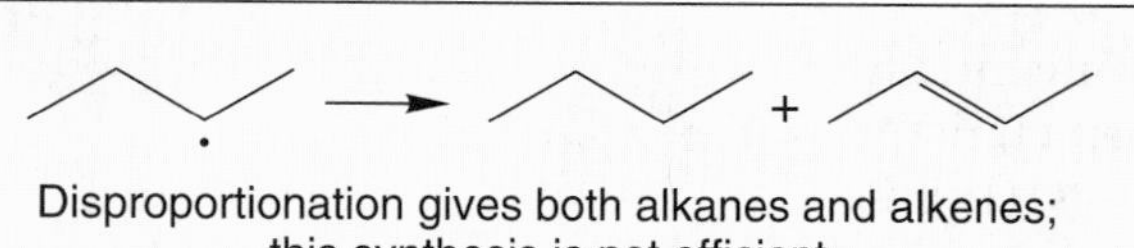

Disproportionation gives both alkanes and alkenes; this synthesis is not efficient

2. Alkenes

Disproportionation gives both alkanes and alkenes; this synthesis is not efficient

3. Alkyl bromides

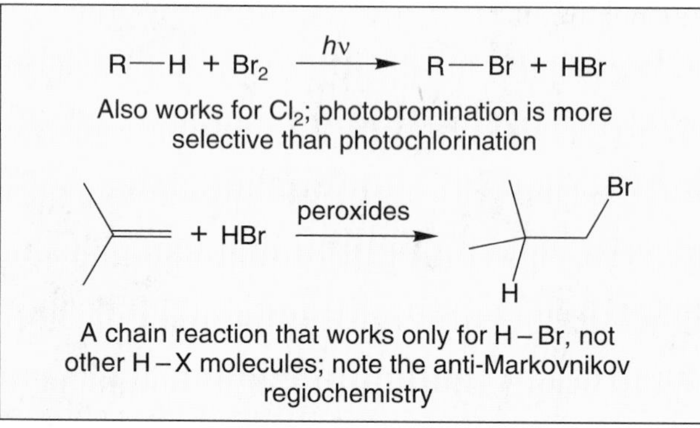

R—H + Br₂ —hν→ R—Br + HBr
Also works for Cl₂; photobromination is more selective than photochlorination
+ HBr
peroxides
Br
H
A chain reaction that works only for H–Br, not other H–X molecules; note the anti-Markovnikov regiochemistry

4. Allylic halides

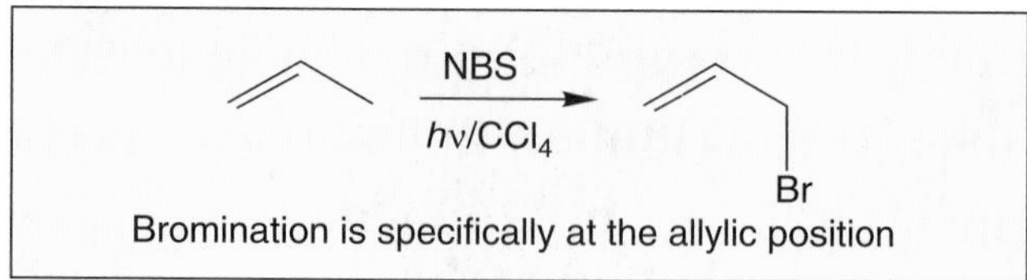

5. Radicals

$$\text{R—R} \xrightarrow{\Delta} 2\ \text{R}\cdot$$

Pyrolysis of hydrocarbons is an unselective method for making radicals; many different radicals are typically produced

$$\text{R—N=N—R} \xrightarrow{\Delta} \text{R}\cdot + N_2 + \text{R}\cdot$$

Pyrolysis of azo compounds

6. Vicinal dihalides

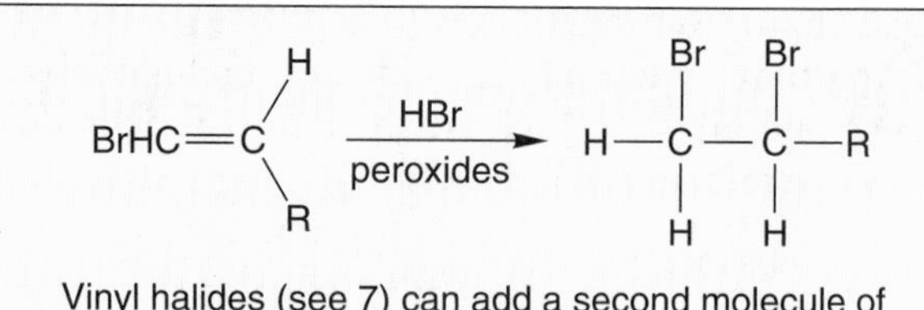

7. Vinyl bromides

$$\text{HC}\equiv\text{C—R} \xrightarrow[\text{peroxides}]{\text{HBr}} \text{BrHC=CHR}$$

Radical-induced addition to alkynes gives anti-Markovnikov addition; cis and trans products are formed

FIGURE 11.75 Synthetic reactions of Chapter 11.

COMMON ERRORS

A very common mistake is to confuse chain-carrying and termination steps in radical reactions. In the photobromination of an alkane such as methane, the product is formed in a radical chain process in which bromine and methyl radicals carry the chain (Fig. 11.76).

$$:\ddot{\text{Br}}—\ddot{\text{Br}}: \longrightarrow 2\ :\ddot{\text{Br}}\cdot$$

$$H_3C—H\ \ \cdot\ddot{\text{Br}}: \longrightarrow H_3C\cdot + H—\ddot{\text{Br}}:$$

$$H_3C\cdot\ \ :\ddot{\text{Br}}—\ddot{\text{Br}}: \longrightarrow H_3C—\ddot{\text{Br}}: + \cdot\ddot{\text{Br}}:$$

Product, along with a bromine atom that can recycle and start product formation again

FIGURE 11.76 The radical chain mechanism for the formation of methyl bromide from methane and bromine.

However, it is all too tempting to imagine that the product, methyl bromide, is formed by the combination of methyl and bromine radicals! In fact, although this step does produce a molecule of product, it destroys two chain-carrying radicals. This step terminates chains and does not contribute significantly to product formation (Fig. 11.77).

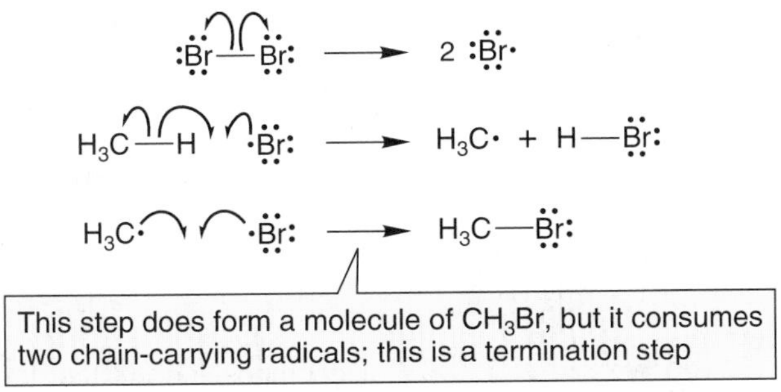

$$:\ddot{Br}-\ddot{Br}: \longrightarrow 2\ :\ddot{Br}\cdot$$

$$H_3C-H \quad \cdot\ddot{Br}: \longrightarrow H_3C\cdot + H-\ddot{Br}:$$

$$H_3C\cdot \quad \cdot\ddot{Br}: \longrightarrow H_3C-\ddot{Br}:$$

This step does form a molecule of CH_3Br, but it consumes two chain-carrying radicals; this is a termination step

FIGURE 11.77 Although the combination of a methyl radical and a bromine atom does form one molecule of methyl bromide, it terminates chains and does not contribute significantly to overall product formation.

11.11 KEY TERMS

Acyl group (R–C=O) The group

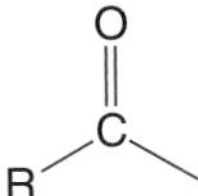

Allylic halogenation Specific formation of a carbon–halogen bond at the position adjacent to a carbon–carbon double bond.

Azo compounds Compounds of the structure, R–N=N–R.

Bond dissociation energy The energy required to break a bond homolytically to give a pair of radicals.

β-Cleavage The fragmentation of a radical into a new radical and an alkene through breaking of the α–β bond.

Chain reaction A cycling reaction in which the species necessary for the first step of the reaction is produced in the last step. This intermediate then recycles and starts the process over again.

Diradical A species containing two unpaired electrons, usually on different atoms.

Disproportionation The reaction of a pair of radicals to give a saturated and unsaturated molecule by abstraction of a hydrogen by one radical from the position adjacent to the free electron of the other radical.

Free radical A neutral molecule containing an odd, unpaired electron. Also simply called "radical."

Geminal 1,1-Disubstituted.

Heterolytic bond breaking The dissociation of two bonded species into a pair of ions.

$$X-Y \rightarrow X^+ + Y^-$$

Homolytic bond breaking The dissociation of two bonded species into a pair of neutral radicals.

$$X-Y \rightarrow X\cdot + Y\cdot$$

Hydrocarbon cracking The thermal treatment of high molecular weight hydrocarbons to give lower molecular weight fragments. Bonds are broken to give radicals that abstract hydrogen to give alkanes, undergo β-cleavage, and disproportionate to give alkenes and alkanes.

Inhibitors A species that can react with a radical, thus destroying it. Inhibitors interrupt chain reactions.

Initiation step The first step in a radical chain reaction in which a free radical is produced that can start the chain-carrying or propagation steps.

***N*-Bromosuccinimide (NBS)** An effective agent for allylic bromination.

Peroxides Compounds of the structure R–O–O–R.

Photohalogenation The formation of alkyl halides through the irradiation of molecular X_2 (X = Br or Cl) in an alkane.

Propagation steps The product-producing steps of a chain reaction. In the last propagation step, a molecule of product is formed along with a new radical that can carry the chain.

Pyrolysis The process of inducing chemical change by supplying heat energy. See **Thermolysis**.

Radical A neutral molecule containing an odd, unpaired electron. Also called a "free radical."

Termination step The mutual annihilation of two radicals. Termination steps destroy chain-carrying radicals and end chains through bond formation.

Thermolysis The process of inducing chemical change by supplying heat energy. See **Pyrolysis**.

Vicinal 1,2-Disubstituted.

11.12 ADDITIONAL PROBLEMS

PROBLEM **11.32** Give the major product(s) expected for each of the following reactions. Pay attention to regiochemistry and stereochemistry where appropriate. Mechanisms are not required, although they may be useful in finding the answers.

(a) 1-methylcyclohexene (CH_3) $\xrightarrow[\text{ROOR}]{\text{HBr}}$

(b) 1-methylcyclohexene (CH_3) $\xrightarrow{\text{HBr}}$

(c) 1-methylcyclohexene (CH_3) $\xrightarrow[\text{ROOR}]{\text{HCl}}$

(d) $(H_3C)_2C{=}CH_2$ + CCl_4 $\xrightarrow{h\nu}$

(e) $H_3C(H)C{=}C(H)COOCH_3$ $\xrightarrow[h\nu]{\text{NBS, } CCl_4}$

(f) cyclohexyl$-C{\equiv}CH$ $\xrightarrow{\text{HBr}}$

(g) cyclohexyl$-C{\equiv}CH$ $\xrightarrow[\text{ROOR}]{\text{HBr}}$

PROBLEM **11.33** Predict the products of hydrogen bromide addition to 1-butyne. Consider both ionic and radical-initiated reactions. Rationalize your predictions with arrow formalism descriptions. Remember that a second molecule of hydrogen bromide can add to the initial product(s) in both cases.

(a) $CH_3CH_2{-}C{\equiv}CH$ $\xrightarrow{\text{HBr}}$

(b) $CH_3CH_2{-}C{\equiv}CH$ $\xrightarrow[\text{ROOR}]{\text{HBr}}$

PROBLEM **11.34** The relative rates of recombination of methyl and isopropyl radicals are quite different. Give two possible reasons for this difference.

Reaction	Relative Rate
$2\ H_3C\cdot \longrightarrow H_3C{-}CH_3$	22
$2\ (H_3C)_2\dot{C}H \longrightarrow (H_3C)_2CH{-}CH(CH_3)_2$	1

PROBLEM **11.35** Decomposition of the following azo compounds at 300 °C occurs at quite different rates. Give two possible explanations for these differences.

$$R_3C{-}N{=}N{-}CR_3 \xrightarrow{300\ ^\circ C} N_2 + 2\ R_3C\cdot$$

R	Relative Rate
CH_3, CH_3, CH_3	13,000
CH_3, CH_3, H	111
CH_3, H, H	11
H, H, H	1

PROBLEM **11.36** Predict the stereochemical results of the photochlorination of the optically active compound **1**. The (*) indicates a single enantiomer, here (*R*). It is the tertiary carbon–hydrogen bond that is chlorinated.

$$CH_3CH_2{-}C^*H(CH_3){-}CH_2Cl\ (\mathbf{1}) \xrightarrow[h\nu]{Cl_2} CH_3CH_2{-}CCl(CH_3){-}CH_2Cl$$

No stereochemistry implied in this drawing

PROBLEM **11.37** In Problem 11.24, we examine the photochlorination of ethane to give ethyl chloride. As was the case in the photochlorination of methane (p. 476), products containing more than one chlorine are also isolated in the photochlorination of ethane. What is the structure of the dichloroethane produced in this reaction? Rationalize your prediction with an analysis of the mechanism of the reaction.

PROBLEM **11.38** Write a detailed mechanism for the bromination of ethyl phenylacetate (**1**) with NBS.

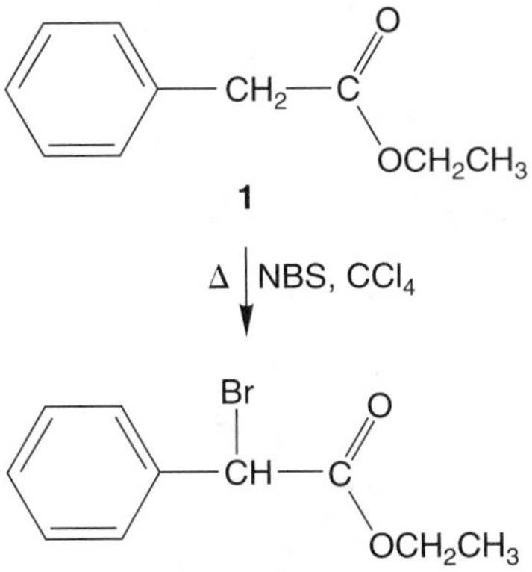

PROBLEM **11.39** Write a detailed mechanism for the photochlorination of **1** to give **2**. Why is compound **1** reactive even though it contains no π bond?

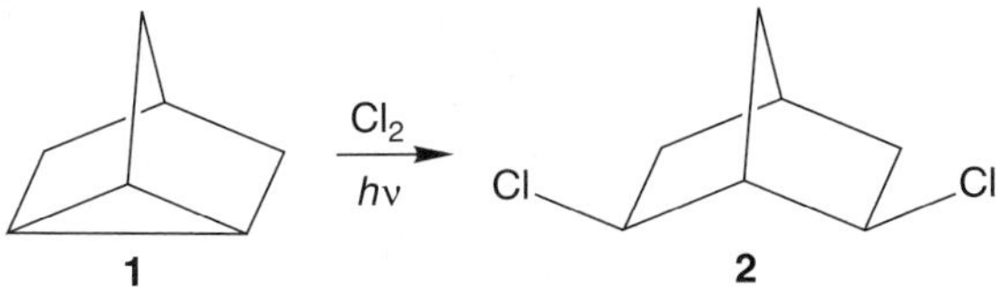

PROBLEM **11.40** Here is a more complicated photochlorination of a compound containing a three-membered ring. Explain the formation of the following products from spiropentane (**1**).

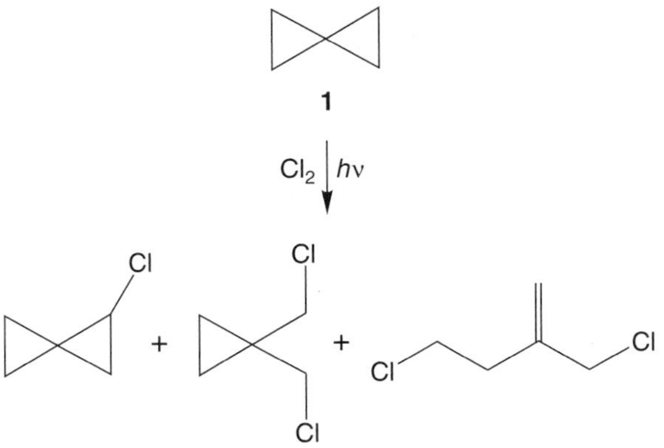

PROBLEM **11.41** Photolysis of iodine in the presence of *cis*-2-butene leads to isomerization of the alkene to a mixture of *cis*- and *trans*-2-butene. Explain.

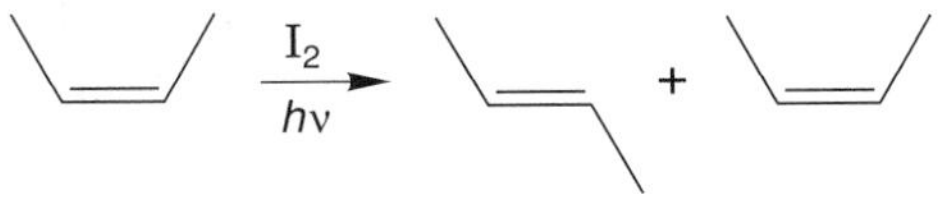

PROBLEM **11.42** Because of their high-potency, low-mammalian toxicity, and photostability, certain esters of halovinylcyclopropanecarboxylic acids (pyrethroids) are promising insecticides (see Chapter 10 Box, p. 415). An important precursor to this class of insecticides is the carboxylic acid (**1**); prepared as follows:

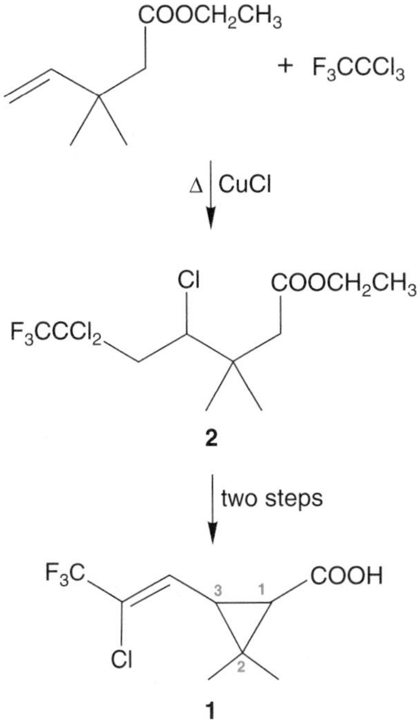

(a) Write a mechanism for the formation of **2**. Be sure to account for the observed regiochemistry. *Hint*: the CuCl can abstract a chlorine atom.

(b) What transformations must be accomplished in the **2** → **1** conversion? Don't worry about mechanisms.

(c) Given the stereochemistry about the carbon–carbon double bond in **1**, how many stereoisomers of **1** are possible? The most potent insecticides are derived from the cis (1*R*, 3*S*) isomer of **1**. Draw this isomer.

PROBLEM **11.43** Write a mechanism for the following reaction of 1-octene.

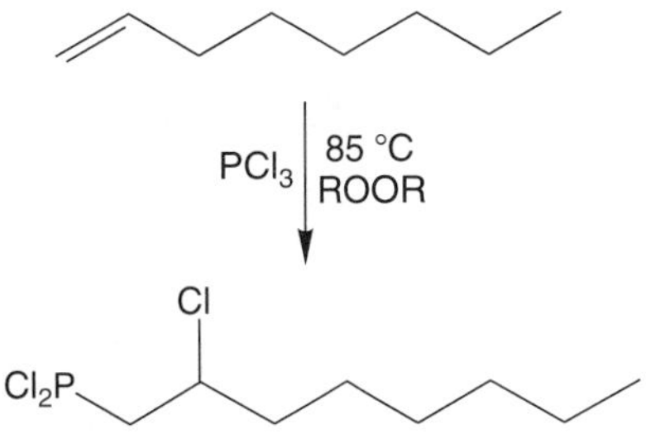

PROBLEM **11.44** Provide mechanisms for the formation of the following products:

77 °C | Ph—CH_3

$(CH_2)_4$ $(CH_2)_4$ $(CH_2)_8$ $(CH_2)_4$ OH Ph Ph Ph =

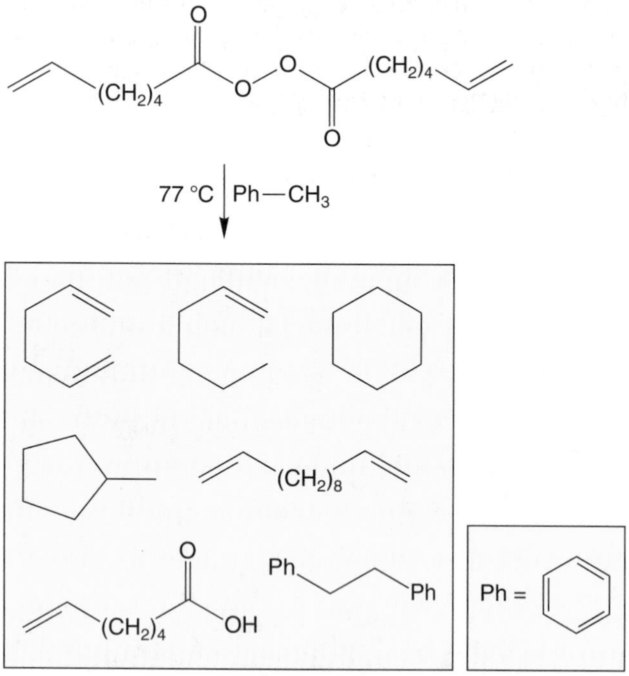

PROBLEM **11.45** Provide mechanisms for the formation of the following products:

$CH_3(CH_2)_4$—C≡CH

CCl_4 | ROOR, 77 °C

Cl $CH_3(CH_2)_4$ =$CHCCl_3$ + H_3C =CCl_2

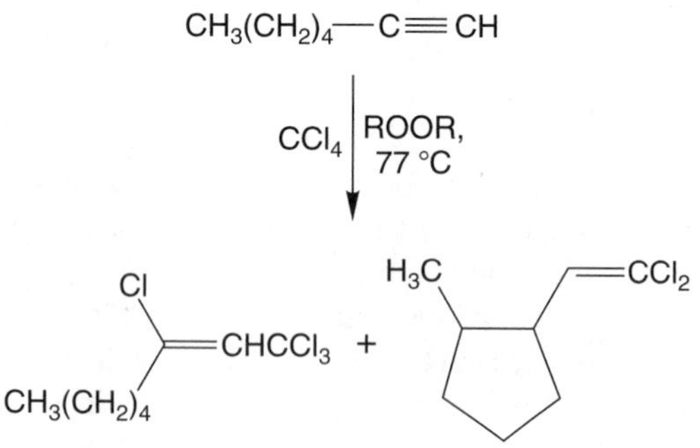

PROBLEM **11.46** Propose a mechanism for the following reaction. Be sure to rationalize the preferred formation of diene **1**.

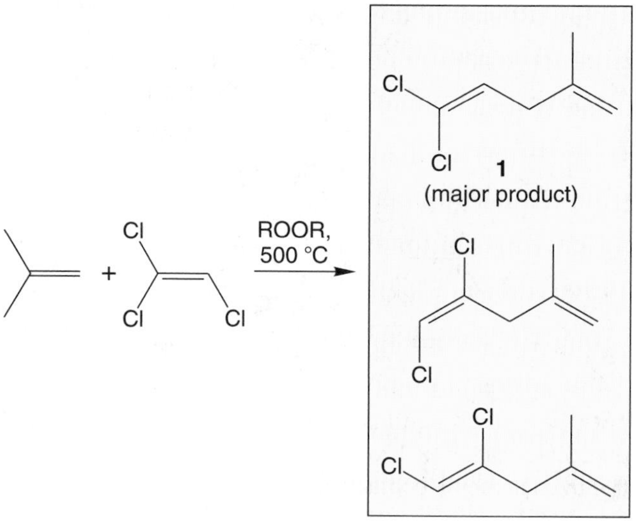

+ Cl Cl Cl ROOR, 500 °C

Cl Cl **1** (major product) Cl Cl Cl Cl

PROBLEM **11.47** Write a mechanism for the formation of compound **1**.

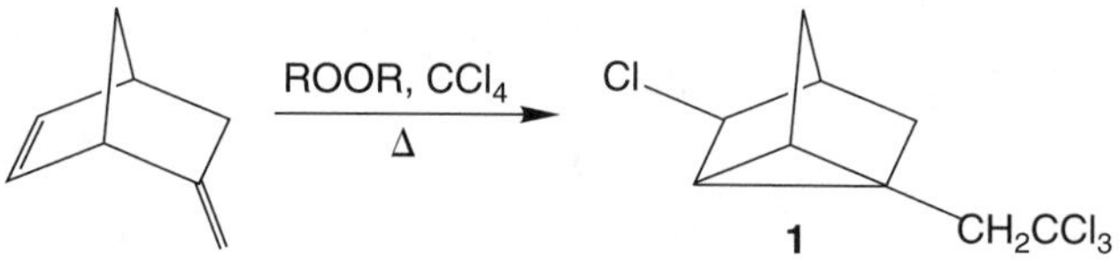

ROOR, CCl_4 Δ

Cl **1** CH_2CCl_3

Dienes and the Allyl System: 2*p* Orbitals in Conjugation

It's *deja vu* all over again.
—attributed to Yogi Berra,*
perhaps apocryphally

Sometimes the whole really is greater, or at least different, from the sum of the parts. In this chapter, we see just such a situation. One cannot always generalize from the properties of alkenes to those of polyalkenes. **Dienes**, compounds containing two double bonds (Chapter 4, p. 128), often have properties that are quite different from those of separated, individual alkenes. This difference depends on a phenomenon called conjugation. **Conjugated double bonds** are connected in 1,3-fashion. A molecule containing widely separated double bonds will generally react as would the individual alkenes. Double bonds in a 1,3-relationship react quite differently. A conjugated diene constitutes a separate functional group (Fig. 12.1).

Unconjugated molecules

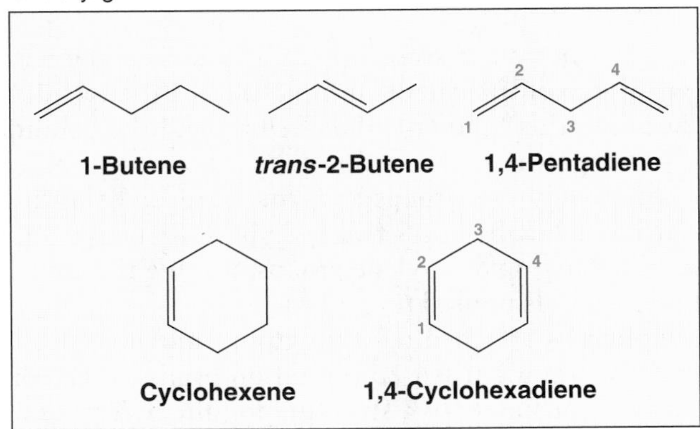

Conjugated dienes

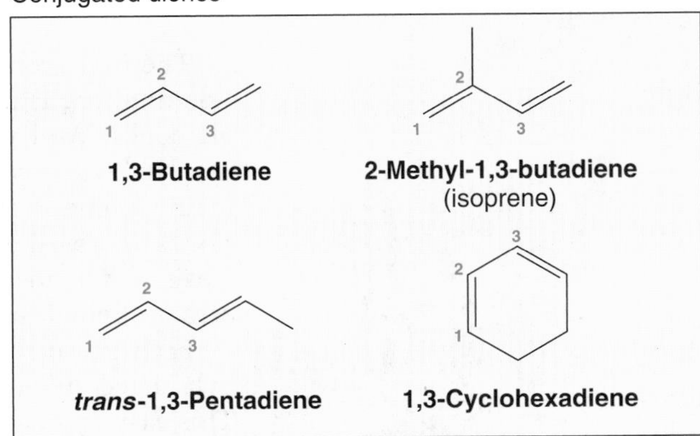

FIGURE **12.1** Some conjugated and unconjugated molecules.

Now we will generalize our discussion of the phenomenon of conjugation and its effect on chemical reactions, and we will develop further the

*Lawrence "Yogi" Berra (b. 1925) was a superb long-time catcher (and sometime outfielder and first baseman) for the New York Yankees.

chemistry of a conjugated molecule we already encountered, allyl. Figure 12.2 shows the neutral allyl radical, a system of three parallel 2*p* orbitals.

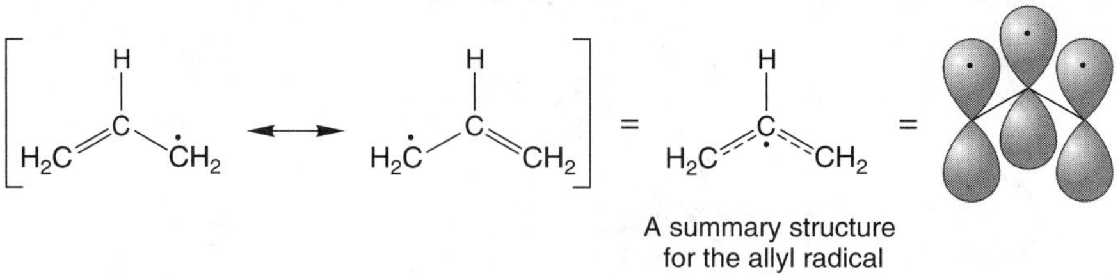

FIGURE 12.2 The allyl radical. Allyl has a 2*p* orbital on each carbon atom.

Finally, we will encounter spectroscopy for the first time. We will see the outlines of how large molecules are constructed in Nature from a rather simple building block, the diene isoprene (2-methyl-1,3-butadiene), shown in Figure 12.1.

12.1 ALLENES: 1,2-DIENES

The simplest diene possible is propadiene, almost always known as **allene*** (Fig. 12.3). Allene is the general term for molecules containing two double bonds that share a carbon atom. Allenes are commonly named as dienes or derivatives of allene (Fig. 12.3).

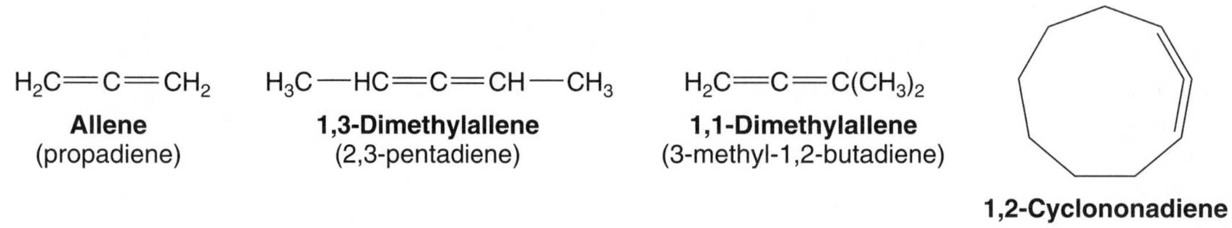

FIGURE 12.3 Some allenes or 1,2-dienes. These compounds can be named as either allenes or dienes.

The first problem is to work out a structure for allene. The key position in allene is certainly the central carbon atom shared by the two double bonds. We have not seen this kind of carbon before, and the bonding involved will have important structural consequences. The central carbon in any allene is attached to only two groups and, therefore, *sp* hybridization is appropriate. The central carbon uses two *sp* hybrid orbitals to form a pair of σ bonds with the flanking methylene groups, leaving the central carbon's remaining two *p* orbitals unused (Fig. 12.4).

Each end carbon, though, is attached to three groups and is therefore hybridized sp^2, which leaves one unused 2*p* orbital on each end carbon. This easy approach leads immediately to a structure for allene (Fig. 12.5). Notice how each 2*p* orbital of the sp^2 hybridized end carbons overlaps perfectly with one of the perpendicular 2*p* orbitals of the central, *sp* hybridized carbon. The important thing is not to be panicked by the odd

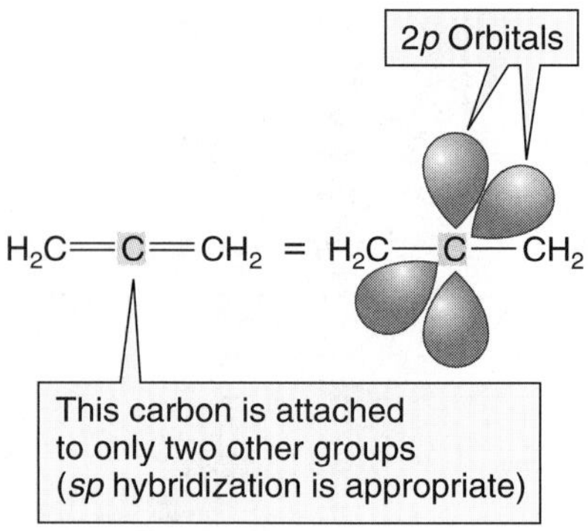

FIGURE 12.4 The central carbon of an allene is *sp* hybridized.

*There is a chance for confusion here between the system of three parallel 2*p* orbitals called "allyl" and the allenes in which two double bonds are attached to one carbon. Be careful; these species are very different!

central carbon and to keep in mind how we dealt with new structures before. Assign an appropriate hybridization, be sure to pay attention to the geometrical requirements of the hybrid, and the structure will always work out.

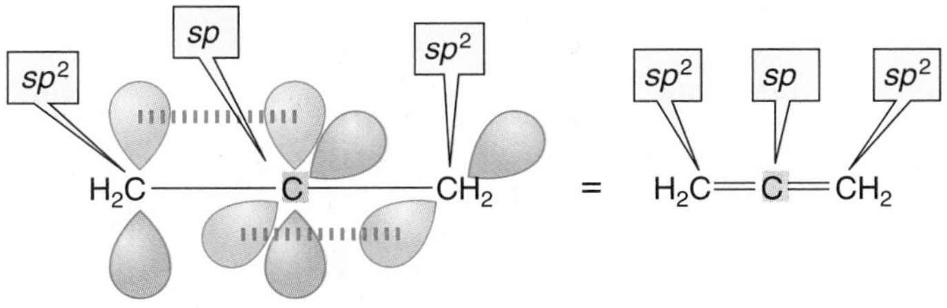

Notice that the two pairs of overlapping *2p* orbitals are in perpendicular planes

FIGURE 12.5 An orbital picture of an allene. Notice the perpendicular π systems.

But we do have to be careful and pay attention to detail. Finding a reasonable structure for allene is a classic example of a problem asked at this point in every course in organic chemistry (see Problems 12.1 and 12.2 and the following answers).

*PROBLEM 12.1

It would appear that there could be cis and trans isomers of 1,3-dimethylallene, but there aren't. Explain (Fig. 12.6).

FIGURE 12.6

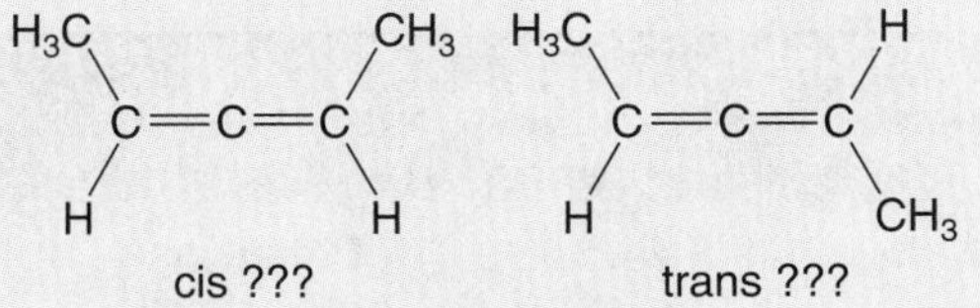

1,3-Dimethylallene
(2,3-pentadiene)

ANSWER

The two-dimensional paper will fool us all the time if we do not think in three dimensions. The end groups of an allene are not in the same plane, but lie in perpendicular planes. The words cis (same side) and trans (opposite sides) have absolutely no meaning for an allene.

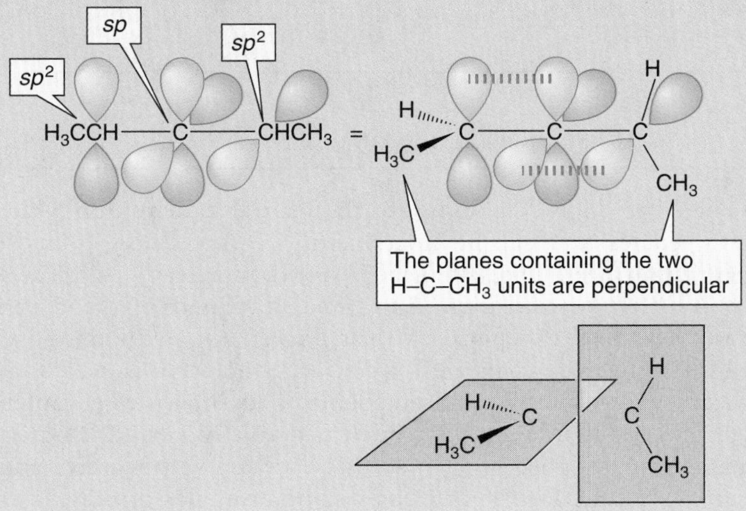

*PROBLEM 12.2 Yet, two isomers of 1,3-dimethylallene do exist. What are they?

ANSWER There are two isomers of 1,3-dimethylallene, a chiral molecule. The mirror image of 1,3-dimethylallene is not superimposable on the original; therefore a pair of enantiomers exists.

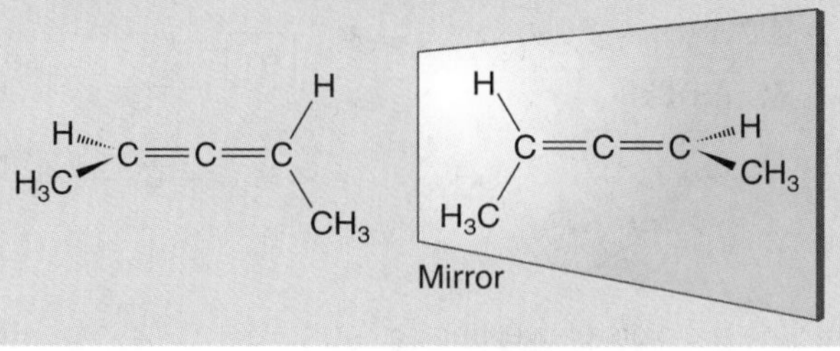

Mirror

Allenes are high-energy molecules, just as are alkynes. The *sp* hybridization of the central carbon is not an efficient way of using orbitals to make bonds. This inefficiency is reflected in relatively positive heats of formation (ΔH_f° values). Table 12.1 gives a few heats of formation and compares them to the heats of formation for other small hydrocarbons. As you can see, allenes are roughly of the same energy as alkynes. This similarity is not surprising; the two molecules contain the same number of π bonds.

TABLE 12.1 Heats of Formation for Some Small Hydrocarbons

Alkanes	ΔH_f° (kcal/mol)	Alkenes	ΔH_f° (kcal/mol)	Alkynes	ΔH_f° (kcal/mol)	Allenes	ΔH_f° (kcal/mol)
Propane	−25.0	Propene	4.8	Propyne	44.6	Propadiene $H_2C{=}C{=}CH_2$	47.7
Butane	−30.2	1-Butene	−0.1	1-Butyne	39.5	1,2-Butadiene $H_2C{=}C{=}CH{-}CH_3$	38.8
		2-Butene	−1.9 (cis) −2.9 (trans)	2-Butyne	34.7		
						1,2-Pentadiene $H_2C{=}C{=}CH{-}CH_2CH_3$	33.6
						2,3-Pentadiene $H_3C{-}CH{=}C{=}CH{-}CH_3$	31.8

12.2 RELATED SYSTEMS: KETENES AND CUMULENES

In allenes, the double bonds share the central carbon atom. There are other molecules in which heteroatoms (non-carbon heavy atoms such as oxygen and nitrogen) take the place of one or more carbons of the allenes and/or in which there are more than two double bonds in a row. When one end carbon of allene is replaced with an oxygen, the compounds are called **ketenes**. If both end atoms are oxygens, the molecule is carbon dioxide. Hydrocarbons with three attached double bonds are called **cumulenes**, although they are more properly named as derivatives of 1,2,3-butatrienes. Cumulenes are strained enough so that they are very reactive and most difficult to handle. Figure 12.7 gives some examples of these and other molecules:

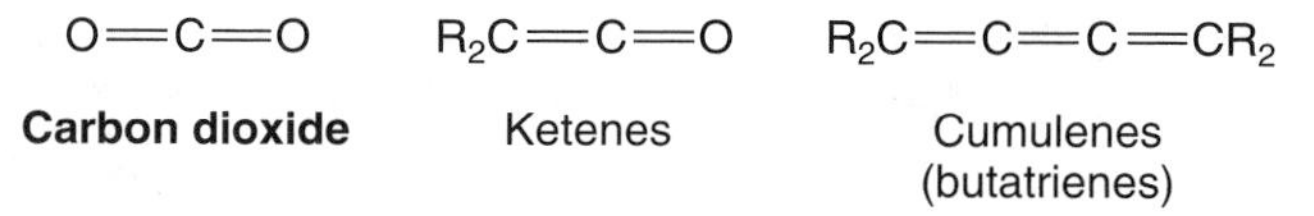

$O{=}C{=}O$ — **Carbon dioxide**

$R_2C{=}C{=}O$ — Ketenes

$R_2C{=}C{=}C{=}CR_2$ — Cumulenes (butatrienes)

FIGURE 12.7 Some oxygen-substituted allenes and some cumulenes.

PROBLEM 12.3

Draw Lewis dot structures for all the molecules in Figure 12.7 as well as for the related molecules, carbon disulfide (S=C=S), isocyanates (R–N=C=O), carbodiimides (RN=C=NR), and carbon suboxide (O=C=C=C=O).

PROBLEM 12.4

Unlike 1,3-dimethylallene, 2,3,4-hexatriene (1,4-dimethylcumulene) exists as a pair of cis and trans isomers. Explain.

PROBLEM 12.5

Demonstrate that 2,3,4-hexatriene is achiral.

12.3 ALLENES AS INTERMEDIATES IN THE ISOMERIZATION OF ACETYLENES

Now that we have the structures of allenes and cumulenes under control, it is time to look at reactivity. Our brief look at allenes permits us to understand one of the most astonishing reactions of alkynes; the base-catalyzed isomerization of internal triple bonds to the terminal position (Fig. 12.8).

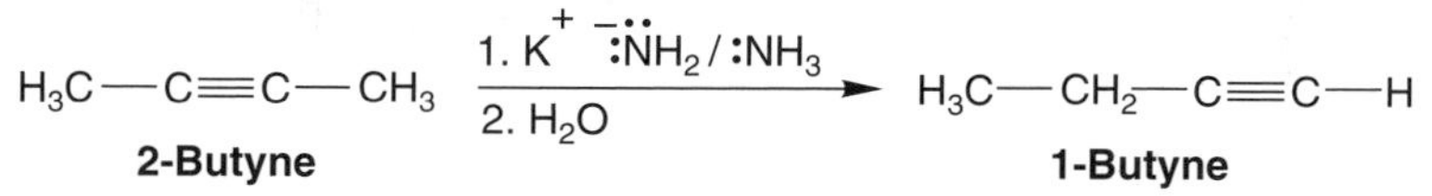

$$H_3C{-}C{\equiv}C{-}CH_3 \xrightarrow[2.\ H_2O]{1.\ K^+\ ^-{:}\ddot{N}H_2/{:}NH_3} H_3C{-}CH_2{-}C{\equiv}C{-}H$$

2-Butyne → **1-Butyne**

FIGURE 12.8 The isomerization of 2-butyne to 1-butyne can be carried out in strong base.

Let's look at the isomerization of 2-butyne to 1-butyne in detail. We do not begin with allenes, rather we encounter them as intermediates in this reaction. When 2-butyne is treated with a very strong base such as an amide ion ($^-NH_2$), a proton can be removed from one of the methyl groups to give a resonance-stabilized carbanion (Fig. 12.9)

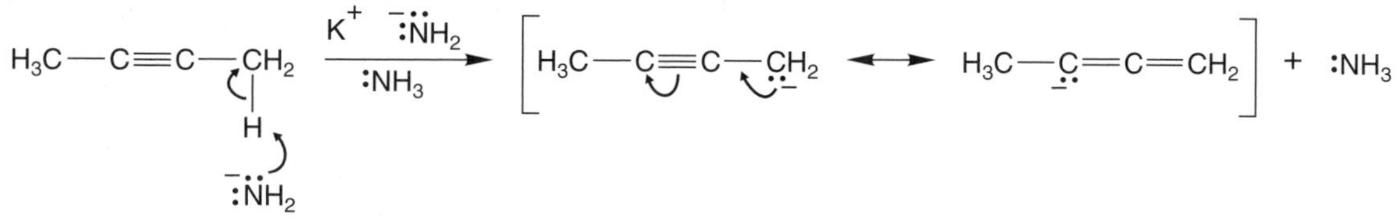

$$H_3C{-}C{\equiv}C{-}CH_2{-}H \ \ ^-{:}\ddot{N}H_2 \xrightarrow[{:}NH_3]{K^+\ ^-{:}\ddot{N}H_2} \left[H_3C{-}C{\equiv}C{-}\ddot{C}H_2^- \longleftrightarrow H_3C{-}\ddot{C}^-{=}C{=}CH_2\right] + {:}NH_3$$

FIGURE 12.9 Removal of a proton from 2-butyne gives a resonance-stabilized carbanion.

Ammonia and the carbon–hydrogen bonds of the methyl groups of 2-butyne both have pK_a values of about 38. Therefore, we can expect reasonable amounts of the carbanion to be produced on reaction with amide ion ($^-NH_2$). 2-Butyne is much more acidic than the parent hydrocarbon butane because the anion formed from 2-butyne is resonance stabilized, and that from butane is not (Fig. 12.9).

What can happen to this carbanion? It can be reprotonated by ammonia to regenerate 2-butyne, but this surely gets us nowhere, even though it is a reasonable reaction, and must happen frequently. Suppose, however, that the anion reprotonates at the *other* position sharing the negative charge? The product is 1,2-butadiene, an allene (Fig. 12.10).

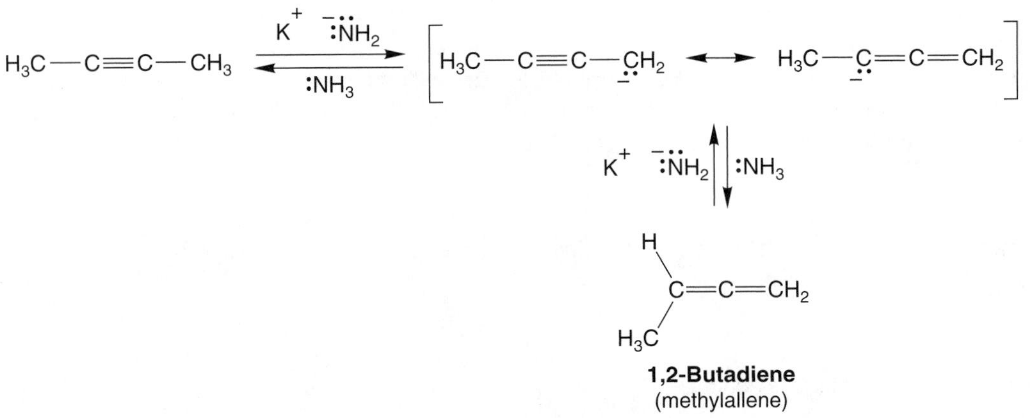

FIGURE 12.10 This anion can reprotonate in two ways. Reattachment of a proton at the end re-forms 2-butyne, but protonation at the other carbon sharing the negative charge produces an allene.

The pK_a of the allene at the terminal methylene position is also about 38, so the amide ion is strong enough to remove the terminal proton to generate a new, resonance- stabilized anion (Fig. 12.11).

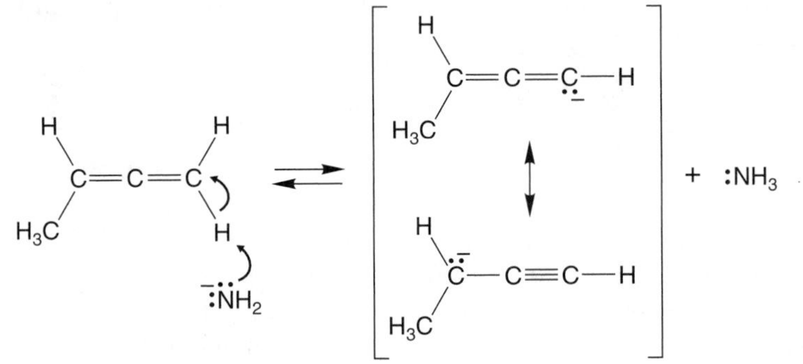

FIGURE 12.11 The allene is also acidic, as a resonance-stabilized anion can be formed through removal of a proton from the end carbon.

Reprotonation at one carbon sharing the negative charge regenerates the allene, but reprotonation at the other carbon yields a new, terminal acetylene, 1-butyne (Fig. 12.12).

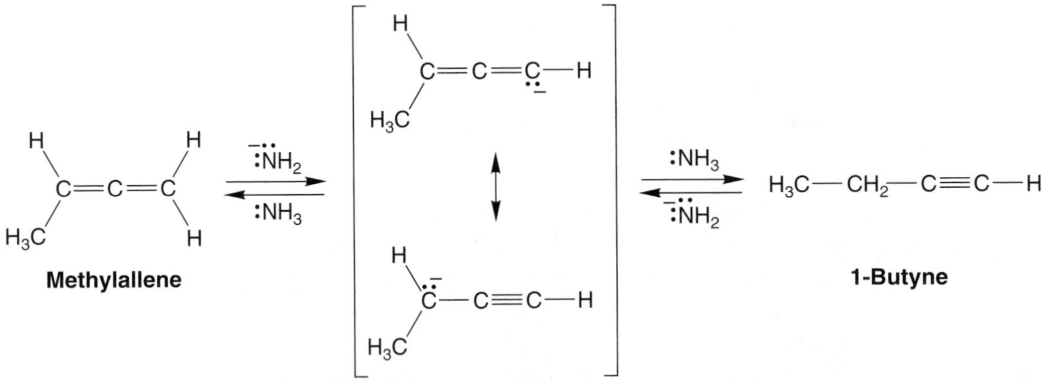

FIGURE 12.12 Reprotonation either re-forms the allene or gives the isomerized alkyne, 1-butyne.

1-Butyne (pK_a = 25, Chapter 4, p. 149) is a *much* stronger acid than any other molecule present in the mixture and will be rapidly, and in a practical sense, irreversibly deprotonated to give the acetylide (Section 4.14). This deprotonation is the key step in this reaction. No matter where the equilibria in the earlier steps of this reaction settle out, the thermodynamic stability of the acetylide anion will drag the *overall* equilibrium far to the right. When water is added to neutralize the solution and end the reaction, the acetylide will be protonated and 1-butyne can be isolated (Fig. 12.13).

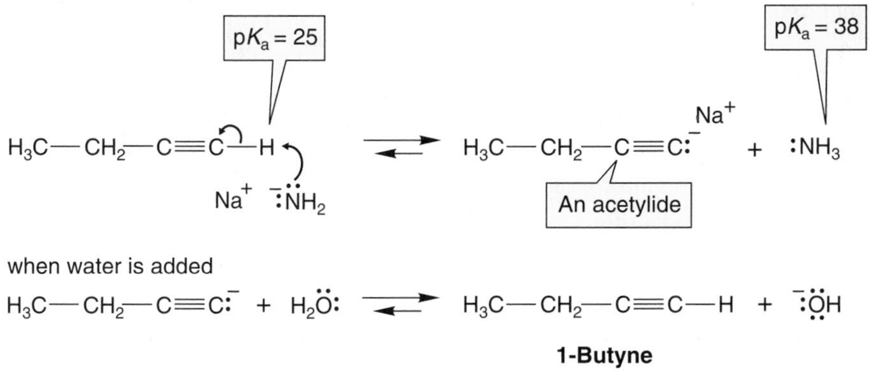

pK_a = 25

pK_a = 38

$H_3C—CH_2—C≡C—H$ + Na^+ $^-:NH_2$ ⇌ $H_3C—CH_2—C≡C:^-$ Na^+ + $:NH_3$

An acetylide

when water is added

$H_3C—CH_2—C≡C:^-$ + $H_2O:$ ⇌ $H_3C—CH_2—C≡C—H$ + $^-:OH$

1-Butyne

FIGURE **12.13** 1-Butyne is by far the strongest acid in this equilibrating system. Removal of the acetylenic hydrogen gives an acetylide, the thermodynamic energy minimum in this reaction. When the reaction mixture is quenched by the addition of water, the acetylide protonates to give 1-butyne. It is critical to understand why it is not possible to re-enter the equilibrium at this point. After acidification there is no base present strong enough to remove the acetylenic hydrogen!

The overall reaction is summarized in Figure 12.14.

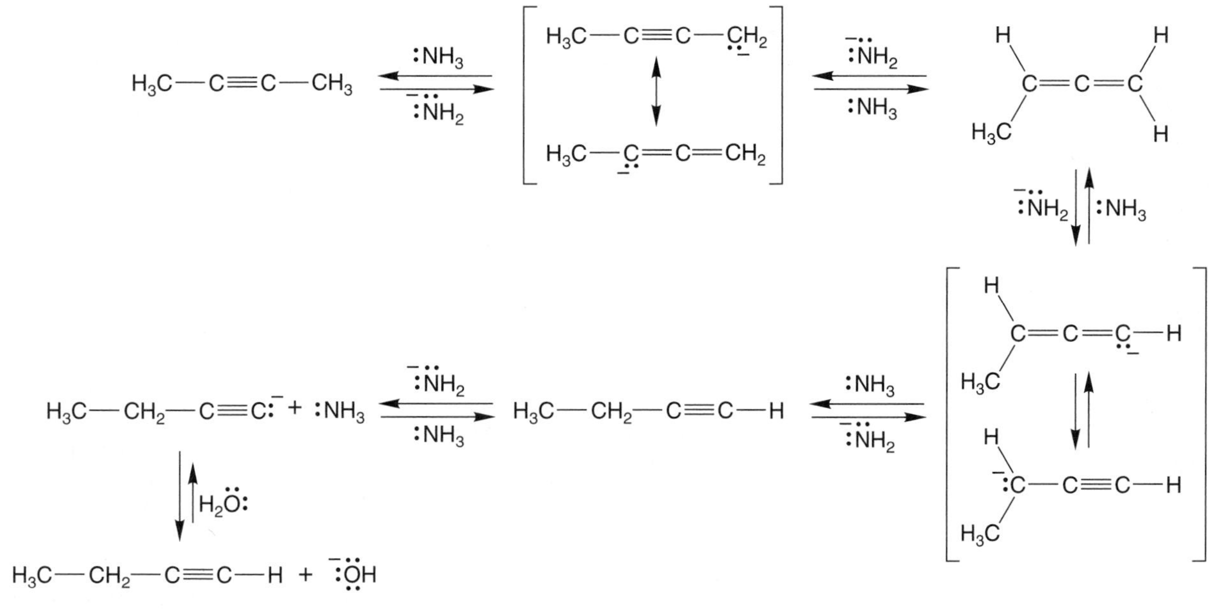

$H_3C—C≡C—CH_3$ ⇌ ($^-:NH_2$ / $:NH_3$) [$H_3C—C≡C—CH_2^-$ ↔ $H_3C—C^-=C=CH_2$] ⇌ ($:NH_3$ / $^-:NH_2$) $(H)(H_3C)C=C=C(H)(H)$

⇌ ($^-:NH_2$ / $:NH_3$) [$(H)(H_3C)C=C=C^-—H$ ↔ $(H)(H_3C)^-C—C≡C—H$] ⇌ ($:NH_3$ / $^-:NH_2$) $H_3C—CH_2—C≡C—H$ ⇌ ($^-:NH_2$ / $:NH_3$) $H_3C—CH_2—C≡C:^-$ + $:NH_3$

↓ $H_2O:$

$H_3C—CH_2—C≡C—H$ + $^-:OH$

FIGURE **12.14** The full mechanism for the isomerization of 2-butyne to 1-butyne in strong base.

PROBLEM **12.6**

Write a mechanism for the isomerization of 3-hexyne to 1-hexyne in the presence of sodium amide ($NaNH_2$).

*PROBLEM 12.7 Di-*sec*-butylacetylene (3,6-dimethyl-4-octyne) fails to isomerize to another alkyne when treated with $NaNH_2/NH_3$. Explain.

ANSWER Isomerization is blocked by the alkyl groups. An allene can be formed, but the reaction can go no further as there is no allenic hydrogen that can be removed in base to produce a new resonance-stabilized anion.

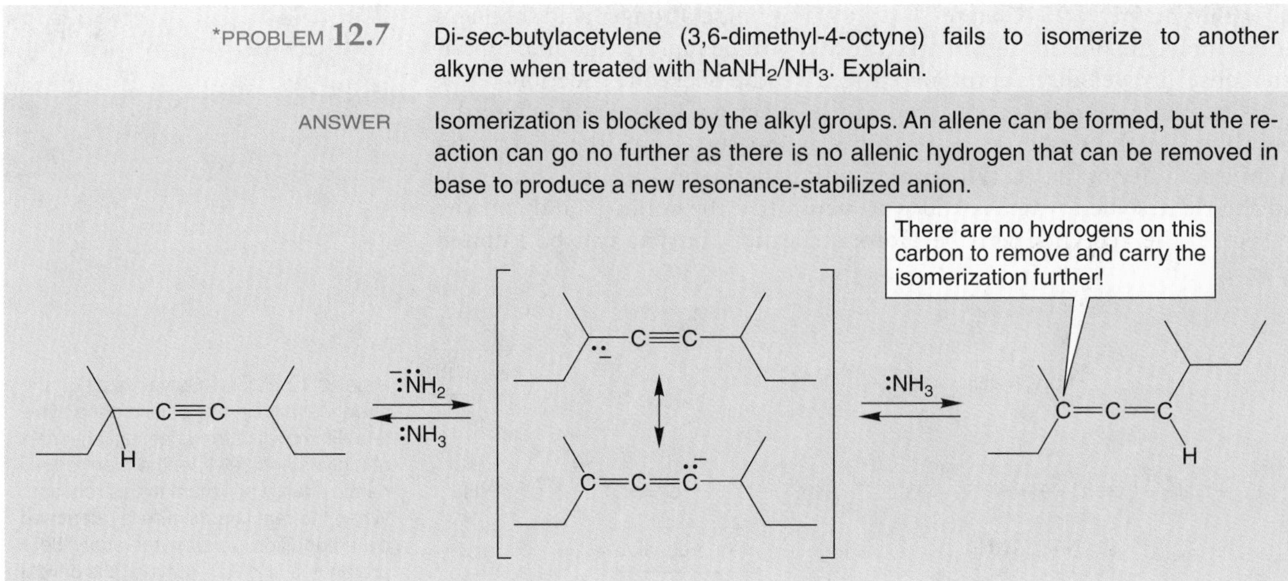
There are no hydrogens on this carbon to remove and carry the isomerization further!
C≡C
H
:NH2
:NH3
C≡C
C=C=C
:NH3
C=C=C
H

12.4 1,3-DIENES

The bonding in 1,3-dienes is not so obviously exotic as it is in 1,2-dienes, the allenes. Nevertheless, there are some subtle effects with far-reaching implications. The source can be seen from simple, schematic orbital drawings of 1,3-butadiene and 1,4-pentadiene (Fig. 12.15).

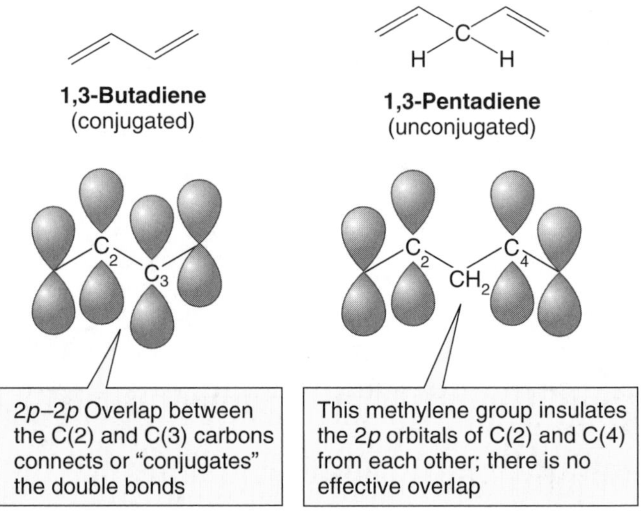
1,3-Butadiene
(conjugated)
1,3-Pentadiene
(unconjugated)
C
H
H
C2
C3
C2
CH2
C4
2p–2p Overlap between the C(2) and C(3) carbons connects or "conjugates" the double bonds
This methylene group insulates the 2p orbitals of C(2) and C(4) from each other; there is no effective overlap

FIGURE 12.15 In a conjugated diene, the orbitals of the two double bonds overlap; in an unconjugated diene, they do not.

In the 1,3-diene, the two double bonds are connected through 2*p*–2*p* overlap. These double bonds are called "conjugated." By contrast, in the 1,4-diene the two double bonds are insulated by a methylene group. These bonds are "not conjugated," or "unconjugated," and will behave much like two separate alkenes. The question we now have to examine is how the chemical behavior of a 1,3-diene will be influenced by the overlap between the 2*p* orbitals on C(2) and C(3). To start, let's take a closer look at the structure of 1,3-butadiene, giving special attention to the π system.

Sketch out the σ system of 1,3-butadiene. What orbitals overlap to form the carbon–carbon and carbon–hydrogen bonds?

PROBLEM 12.8

First, let's look at 1,3-butadiene from a resonance point of view (see Section 9.5, p. 346). 1,3-Butadiene does have resonance forms that will contribute to the overall electronic structure of the molecule (Fig. 12.16).

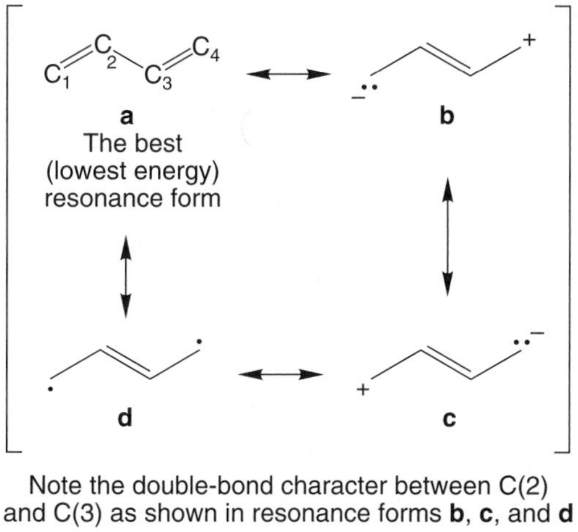

a
The best
(lowest energy)
resonance form

b

c

d

Note the double-bond character between C(2) and C(3) as shown in resonance forms **b**, **c**, and **d**

FIGURE 12.16 Resonance forms contributing to the structure of 1,3-butadiene.

The best resonance form, and therefore the best single description of the molecule, is the diene structure we conventionally write. This structure has one more bond than the others and involves no charge separation. Thus, it wins by almost all of our criteria for estimating the relative energies of resonance forms (Chapter 9, p. 345). Nevertheless, the other resonance forms show that there is some "double-bond character" to the C(2)–C(3) bond (Fig. 12.16).

For each of the other resonance forms in Figure 12.16 indicate what makes it less stable than the conventional structure, **a**.

PROBLEM 12.9

Conjugation can be seen, perhaps even more clearly, if we examine the π molecular orbitals of 1,3-butadiene. In butadiene, we have a system of four 2*p* orbitals on four adjacent carbons. These four atomic 2*p* orbitals will overlap to produce four π molecular orbitals. We can get the symmetries of the four new molecular orbitals in a number of equivalent ways. It is always easiest to build up something new out of things we know already, so it's productive to construct the molecular orbitals of butadiene out of two sets of the π molecular orbitals of ethylene, π and π*. One way to accomplish this is to allow the π molecular orbitals of ethylene to overlap end to end. In producing these π molecular orbitals in this way, we have made an important approximation. As often mentioned before, the closer in energy two orbitals are, the stronger the stabilization resulting from their overlap. In practice, all orbital overlap except for the strongest interactions can be ignored, at least at this level of theory. This observation means that we need to look only at the result of the $\pi \pm \pi$ and $\pi^* \pm \pi^*$ interactions and do not have to consider $\pi \pm \pi^*$.

The symmetries of the new butadiene orbitals can be easily derived from the symmetries of the starting ethylene orbitals (Fig. 12.17).

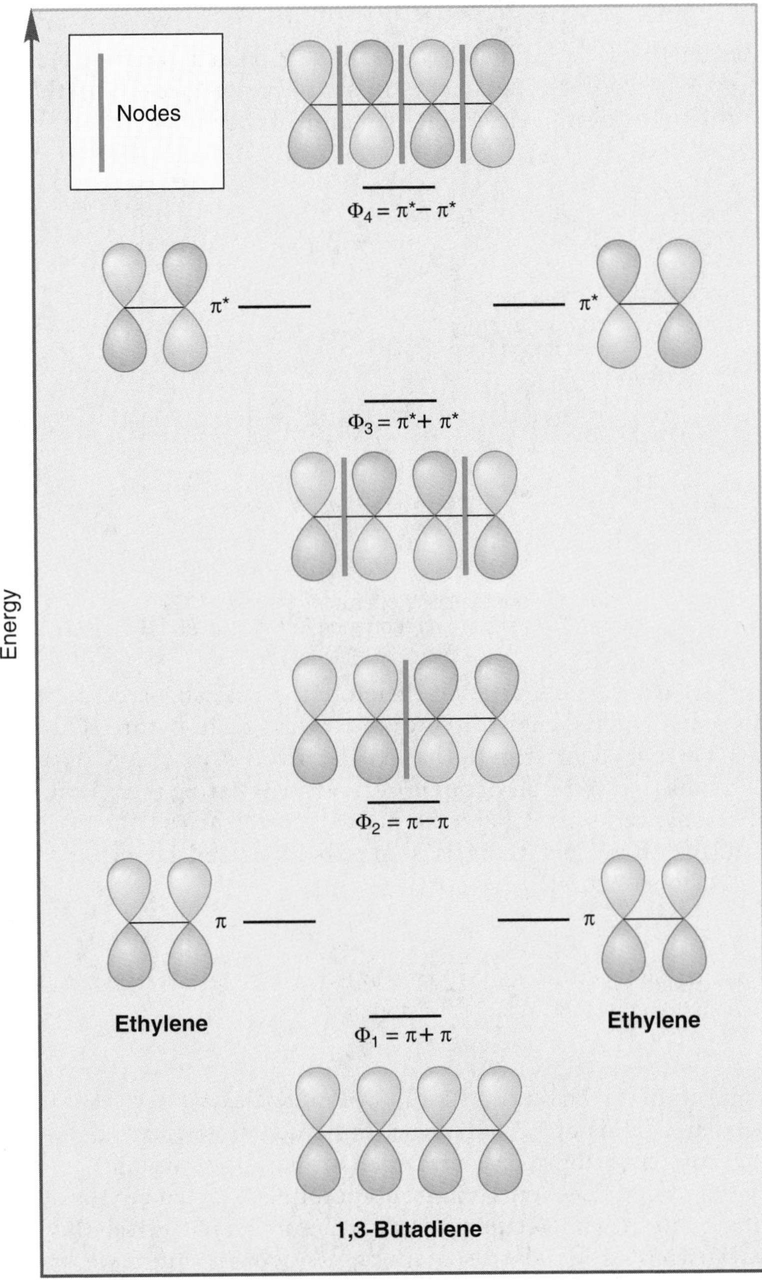

FIGURE **12.17** The schematic formation of the π molecular orbitals of 1,3-butadiene from the π molecular orbitals of ethylene.

To order the new molecular orbitals in energy, count nodes. Orbital Φ_1 has no new nodes, Φ_2 has one, Φ_3 has two, and Φ_4 has three. Not all methods of construction will give the correct molecular orbitals *in the correct order,* so you must always be careful to do an energy ordering. If you have a sense of *deja vu* (if this construction seems familiar to you), you are right. We made HHHH in exactly the same way in Chapter 2 (p. 55). The only difference is the use of 2*p* orbitals here and 1*s* orbitals in Chapter 2. Go back and see that the form of the molecular orbitals is exactly the same in the two cases.

Construct the π molecular orbitals of 1,3-butadiene from those of ethylene by putting one ethylene "inside" the other (Fig. 12.18).

PROBLEM 12.10

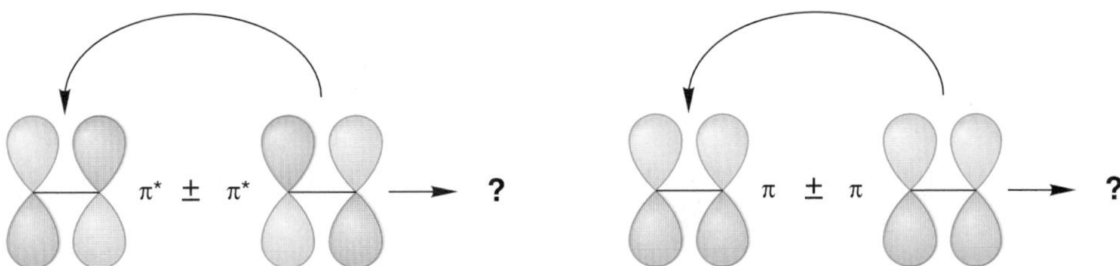
π* ± π*
?
π ± π
?

FIGURE 12.18

12.5 THE PHYSICAL CONSEQUENCES OF CONJUGATION

In 1,3-butadiene, or any conjugated diene, there are four electrons to accommodate in the π molecular orbital system. The lowest two bonding molecular orbitals are occupied and the two higher energy, antibonding molecular orbitals are empty. The HOMO for butadiene is Φ_2, and to a first approximation, Φ_2 will control the reactivity of the molecule. It is Φ_2 that looks most like the traditional representation of 1,3-butadiene. In the resonance formulation it is the traditional picture of 1,3-butadiene that is surely the lowest energy resonance form, and thus the best representation of the molecule. So we see the usual (and necessary) convergence between the resonance and molecular orbital way of looking at structures. Each method generates a picture of 1,3-butadiene that looks mostly like a molecule containing a simple pair of double bonds. Both methods warn us, however, that there are likely to be some consequences of conjugation in this molecule. This is shown by the C(1)–C(2)–C(3)–C(4) connection of Φ_1 in the molecular orbital picture, and by the existence of resonance forms in which there is double-bond character between C(2) and C(3) (Fig. 12.19).

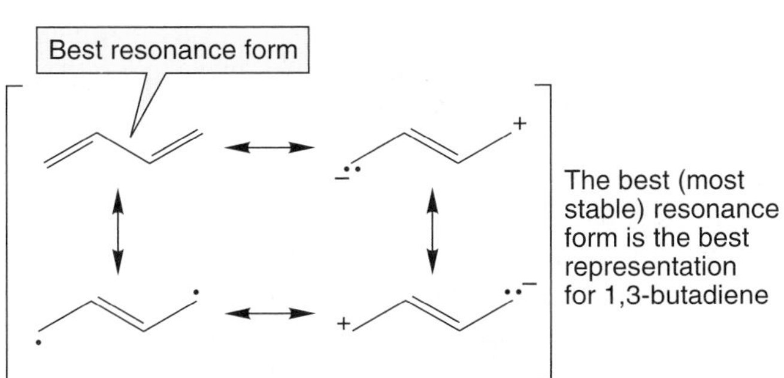
Best resonance form
The best (most stable) resonance form is the best representation for 1,3-butadiene

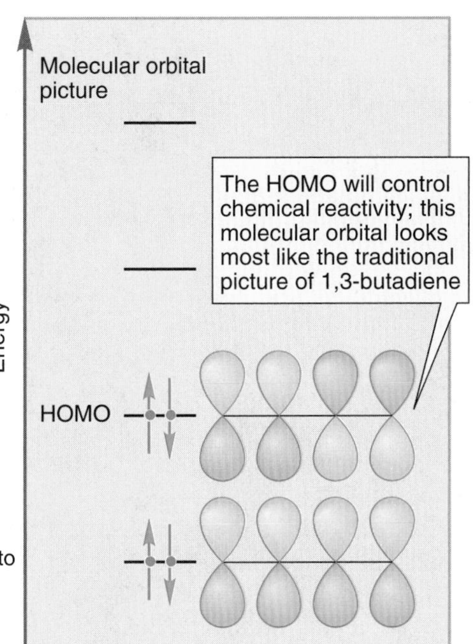
Molecular orbital picture
The HOMO will control chemical reactivity; this molecular orbital looks most like the traditional picture of 1,3-butadiene
Energy
HOMO

FIGURE 12.19 Both the resonance and molecular orbital views of 1,3-butadiene lead to the conclusion that the best view of 1,3-butadiene is of a molecule containing two traditional double bonds. However, both pictures make it clear that there will be some consequences of conjugation between the two double bonds.

What will be the effects of this C(2)—C(3) overlap on the structure of 1,3-butadiene? Carbon–carbon single bonds are relatively long; for example, in butane the carbon–carbon bond is 1.53 Å. The carbon–carbon double bond in 2-butene is only 1.32 Å, reflecting the presence of two bonds between the carbons, and the carbon–carbon triple bond in acetylene, 1.20 Å, is even shorter. To the extent that there is double-bond character between C(2) and C(3), the bond in 1,3-butadiene should be shortened. And so it is, the C(2)—C(3) bond distance is only 1.47 Å. It has been argued that the shortening is indicative of conjugation between the two double bonds (Fig. 12.20).

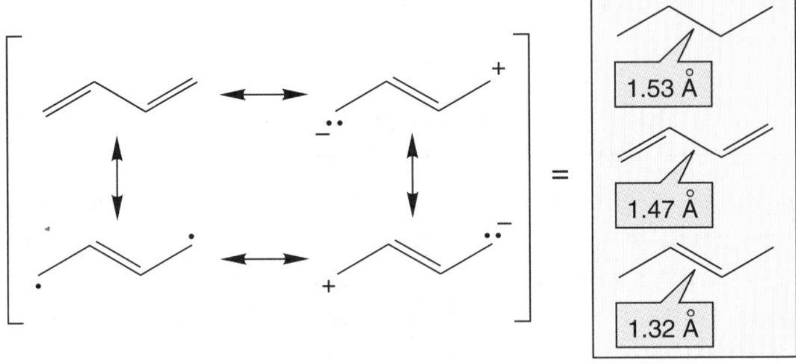

FIGURE 12.20 The central carbon–carbon bond length in 1,3-butadiene is in between that for an alkane and an alkene.

*PROBLEM 12.11 Life is never simple. There is a persuasive counterargument that attributes the shortening of the C(2)—C(3) bond to other factors. Can you come up with another explanation? *Hint for this super-tough question*: We have ignored the σ system in this argument. What kind of a σ bond joins C(2) and C(3)? So what? See Problem 12.8 (p. 507).

ANSWER The two inner carbons of 1,3-butadiene are joined by an sp^2–sp^2 σ bond, in contrast to the sp^3–sp^3 σ bonds of butane. An sp^2 orbital has more *s* character than an sp^3 orbital, and bonds made from sp^2–sp^2 overlap will be stronger and shorter than those made from sp^3–sp^3 overlap. Accordingly, some of the shortening of the central σ bond in 1,3-butadiene may be attributed to the σ system and not to π orbital overlap.

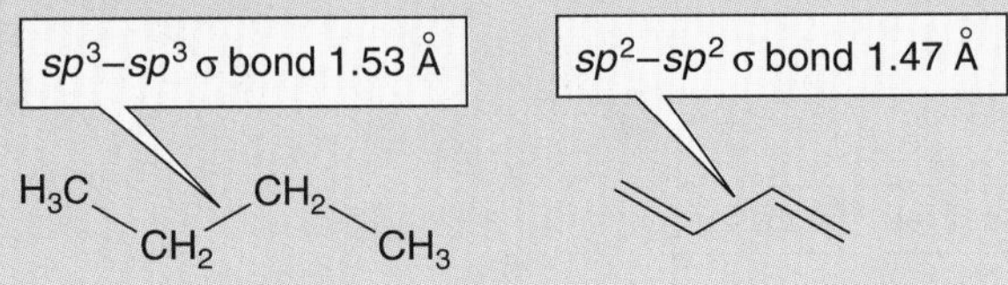

Now the question becomes, How strong is the C(2)—C(3) overlap? What consequences will this overlap have for the structure of conjugated 1,3-dienes? Overlap between 2*p* orbitals on adjacent positions has profound consequences for the structure of alkenes, for example, and the same might be true for dienes. It is the 1,2-overlap between 2*p* orbitals that makes cis and trans isomers of alkenes possible (Fig. 12.21).

Might the same be true for dienes? Will overlap between the 2*p* orbitals at the C(2) and C(3) positions create a large enough barrier so that isomers, called the s-cis and s-trans conformations, could be isolated? The "s" nomenclature in dienes stands for "single" as these are two conformations involving stereochemistry about the single bond of the diene (Fig. 12.22).

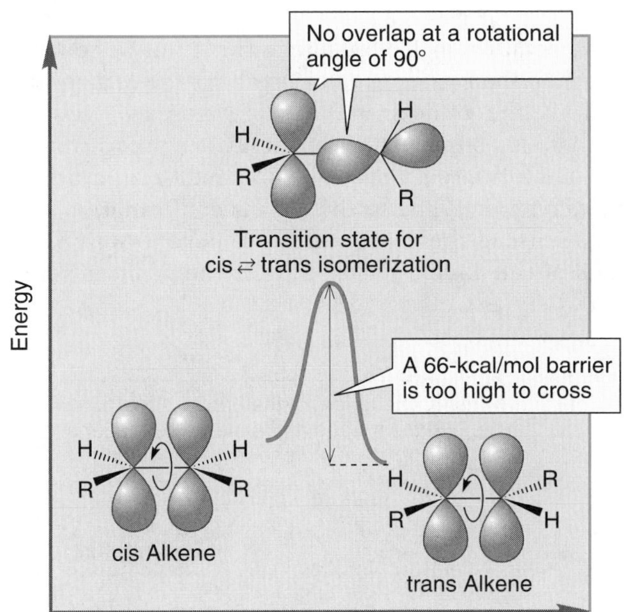

FIGURE 12.21 In alkenes, rotation about the carbon–carbon double bond requires overcoming a barrier of 66 kcal/mol. This barrier is too high to be surmounted under most conditions.

FIGURE 12.22 Interconversion of the s-trans and s-cis conformations of 1,3-butadiene.

As we might guess, there is a barrier to rotation, but it's not very large. Overlap between C(2) and C(3) is not nearly as important as that between the orbitals at the two adjacent carbons in ethylene. At the transition state for the rotation of 1,3-butadiene about its central, C(2)–C(3) bond, overlap between the 2*p* orbitals on C(2) and C(3) vanishes, but C(1)–C(2) and C(3)–C(4) overlap is undisturbed (Fig. 12.23).

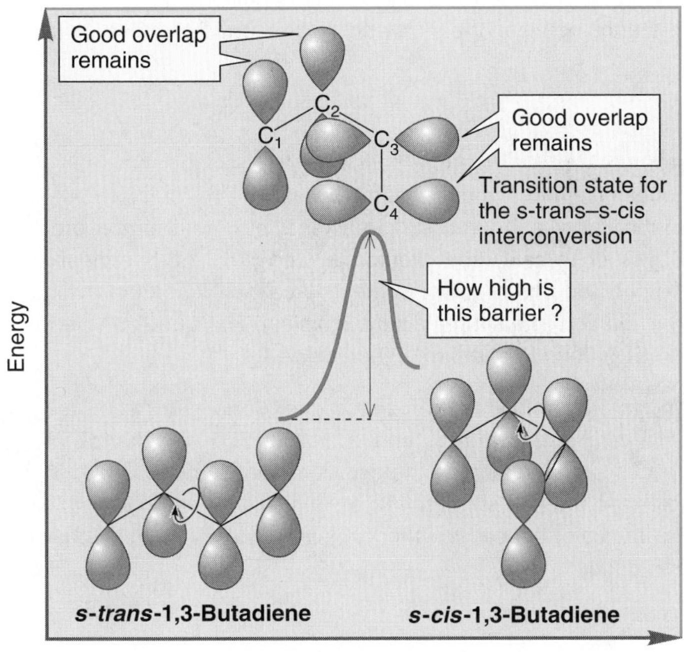

FIGURE 12.23 The barrier for the interconversion of the s-trans and s-cis conformations of 1,3-butadiene.

The barrier to rotation in 1,3-butadiene turns out to be 4–5 kcal/mol, only a bit larger than the barrier to rotation about the carbon–carbon bond in an alkane. It is the trans form that is favored at equilibrium, by 2–3 kcal/mol. There is a simple steric effect favoring the s-trans conformation. In the s-cis conformation the two inside hydrogens are quite close to each other (make a model!) and in the s-trans conformation they are not. In order to make sense of these structures, we have to watch out not only for the locations of the double bonds, but for the positions of the hydrogens attached to them as well (Fig. 12.24).

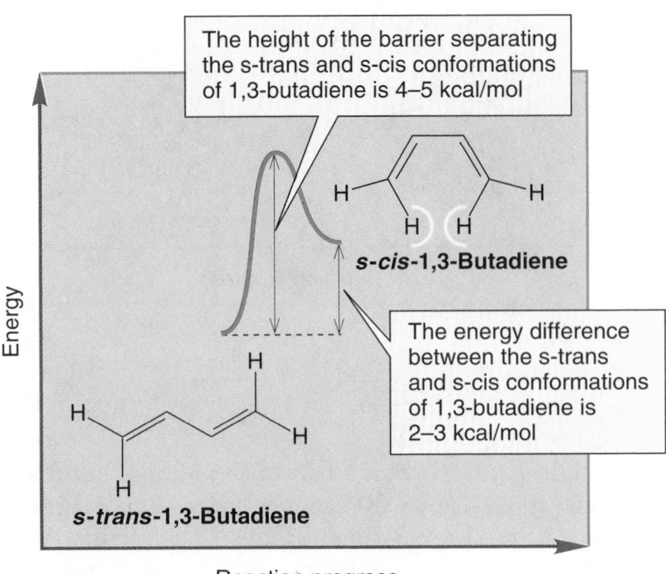

FIGURE 12.24 An energy diagram for the interconversion of the s-trans and s-cis conformations of 1,3-butadiene.

PROBLEM 12.12 Calculate the amount of the s-cis form present at equilibrium at 25 °C, assuming an energy difference between the s-cis and s-trans forms of 2.5 kcal/mol.

***PROBLEM 12.13** There is another kind of rotation possible, that about the C(1)—C(2) or C(3)—C(4) bonds. We might guess that the barrier to this rotation would be little different from that for rotation in a simple alkene, but, as a former president of the United States once said, "that would be wrong." In 1,3-butadiene, it takes only 52 kcal/mol to do this rotation, some 14 kcal/mol less than the barrier of 66 kcal/mol for rotation in a simple alkene. Explain. *Hint*: Look at the transition state for C(1)—C(2) bond rotation in 1,3-butadiene.

ANSWER The transition state for rotation about one of the "double" bonds in a 1,3-butadiene contains a localized radical and a delocalized allyl radical. It is this delocalization of the allyl radical in the transition state that makes this rotation more favorable than the related motion in a simple ethylene. In a simple alkene, two localized radicals appear in the transition state for rotation about a carbon–carbon bond.

ANSWER (CONTINUED)

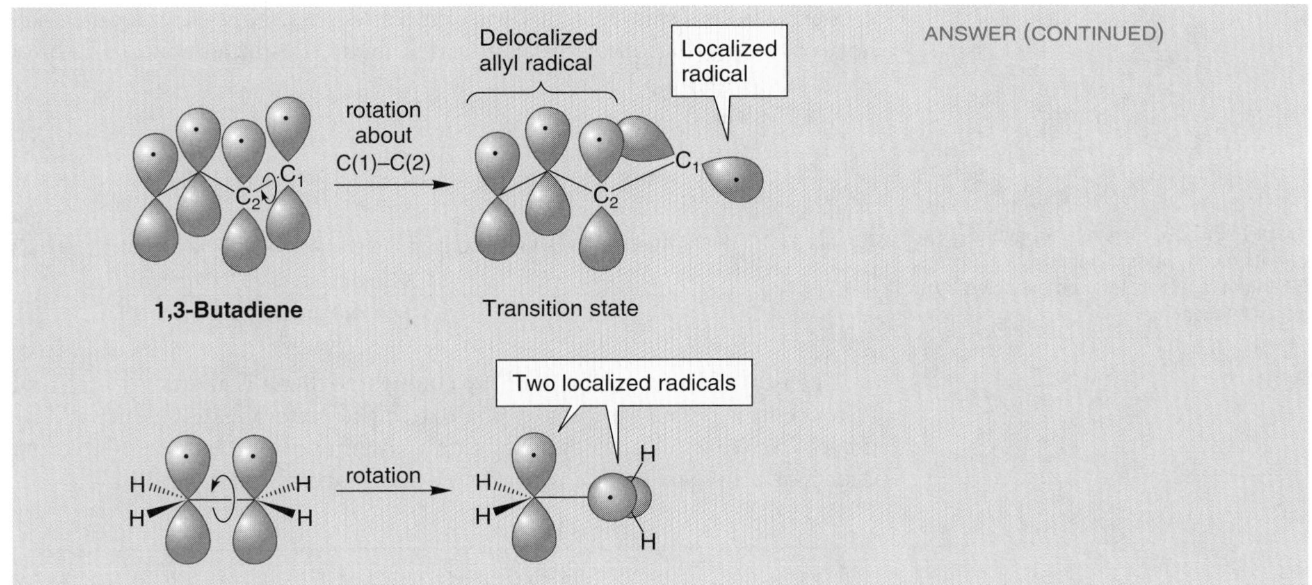

PROBLEM **12.14**

Design an experiment to determine whether rotation about the C(1)–C(2) bond is possible. Assume you have the ability to make any labeled molecules you need.

As both the resonance and molecular orbital pictures indicate, there is 2*p*–2*p* overlap between C(2) and C(3), but it does not dominate the structure of 1,3-butadienes the way 2*p*–2*p* overlap does for alkenes. We can get another idea of the magnitude of this interaction from measuring the heats of hydrogenation of a series of dienes (see Chapter 10, p. 392).

We begin here with one of the butene isomers, 1-butene. When 1-butene is hydrogenated to butane, the reaction is exothermic by 30.3 kcal/mol (Fig. 12.25). A simple-minded first guess would be that a molecule containing two double bonds would be hydrogenated in a reaction twice as exothermic as the hydrogenation of 1-butene. And this is just right; the heat of hydrogenation of 1,4-pentadiene is –60.8 kcal/mol, and that of 1,5-hexadiene is –60.5 kcal/mol (Fig. 12.25).

H_2 / PtO_2 — Find ΔH = –30.3 kcal/mol

1,4-Pentadiene — H_2 / PtO_2 — Predict 2×-30.3 kcal/mol = –60.6 kcal/mol; Find ΔH = –60.8 kcal/mol

1,5-Hexadiene — H_2 / PtO_2 — Find ΔH = –60.5 kcal/mol

Average = ΔH = –60.65 kcal/mol

FIGURE **12.25** Hydrogenation of 1-butene is exothermic by 30.3 kcal/mol. Hydrogenation of unconjugated dienes is exothermic by almost exactly twice the heat of hydrogenation of 1-butene.

Both of our sample dienes were unconjugated. Now let's look at the heat of hydrogenation of the conjugated diene, 1,3-butadiene: ΔH is only –57.1 kcal/mol (Fig. 12.26).

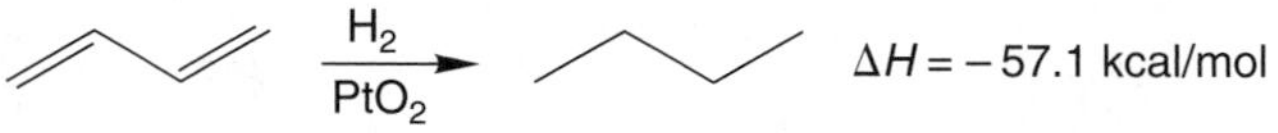

$\xrightarrow[PtO_2]{H_2}$ $\Delta H = -57.1$ kcal/mol

Average ΔH for unconjugated dienes = –60.7 kcal/mol
1,3-Butadiene = –57.1 kcal/mol
Difference = 3.6 kcal/mol

FIGURE **12.26** The heat of hydrogenation of 1,3-butadiene is 57.1 kcal/mol, smaller than expected by about 3.6 kcal/mol.

The heat of hydrogenation of the conjugated diene is about 3.6 kcal/mol lower than we would expect by analogy to the unconjugated dienes in Figure 12.25. This amount of energy can be attributed to the energy-lowering effects of conjugation. The situation is summarized in Figure 12.27.

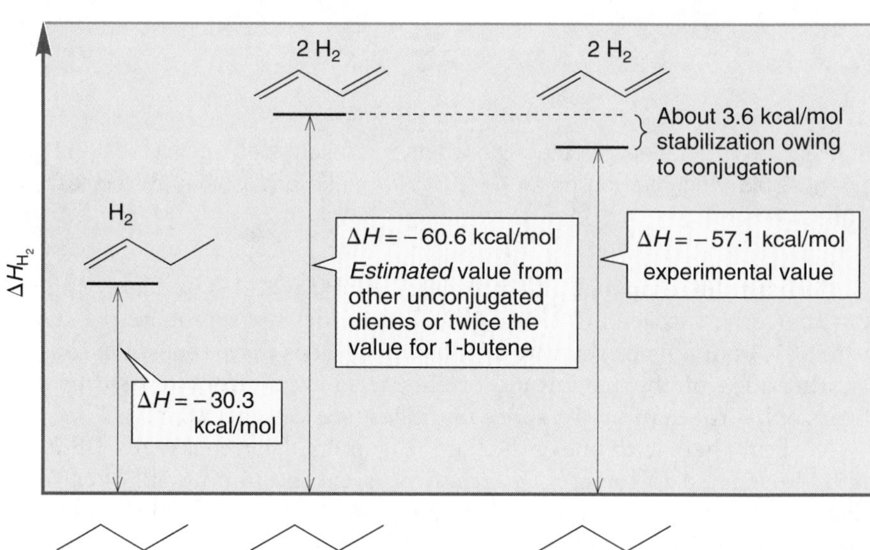

FIGURE **12.27** A summary of the heat of hydrogenation data.

So now we have two measures of the energetic value of conjugation in 1,3-butadiene, and they are remarkably close to each other. The first is the height of the barrier to rotation about the C(2)—C(3) bond (4–5 kcal/mol) and the other is the lowering of the heat of hydrogenation of a conjugated diene relative to that of an unconjugated diene (~3.6 kcal/mol).

Two conjugated double bonds are not utterly unconnected; the whole is somewhat different from the sum of the parts. Conjugation, the orbital overlap between C(2) and C(3) in a 1,3-diene, results in a shortening of the C(2)—C(3) single bond and in lower heats of hydrogenation for 1,3-dienes than for unconjugated systems. Now we move on to the consequences of further conjugation, and to an examination of the reactivity of these molecules. Here, too, we will see that conjugation influences the physical properties and chemical reactivity of these molecules.

12.6 MOLECULAR ORBITALS AND ULTRAVIOLET SPECTROSCOPY

In the early days of organic chemistry, there were two extraordinarily difficult general problems. One was to separate mixtures into pure compounds, and the other was to determine the structures of the compounds, once pure samples were obtained. How simple these problems seem! Yet, much of the mechanistic and synthetic organic chemistry of the last decades would have been quite impossible without the progress in solving these two old difficulties. We will have little to do with separation techniques in this book (your laboratory will take another approach), but we must certainly spend some time here on structure determination.

About 50 years ago, structure determination was an entire branch of chemistry. People spent whole careers working out the structures of complicated molecules by running chemical reactions. The products of many reactions of the unknown compound were analyzed and inferences were made regarding the structure of the unknown. This task was time-consuming and extraordinarily demanding of experimental technique, intellectual rigor, and the development, over years of work, of an intuitive sense of what was right, of what a given experiment might mean. In a way, this work had all the satisfactions and charms of handwork, of a craft. Nowadays, the work of years can usually be done in hours, and structural problems can be attacked that would have been beyond any chemical determination. **Spectroscopy**, the study of the interactions between electromagnetic radiation and atoms and molecules, has made the difference. First came the development of ultraviolet/visible (UV/vis) spectroscopy. We will deal with that subject in this chapter. Then followed infrared (IR) spectroscopy and finally nuclear magnetic resonance (NMR). The power of NMR is so great that this technique has largely eclipsed all other forms of structure determination. We will devote much of Chapter 15 to this topic.

The new spectrometers certainly brought progress, and no reasonable person would want to turn back the clock to the days of handcrafted structure determination. Yet, one can sympathize with the feeling that there has been some loss with the progress. We are somehow not as close to the molecules as we once were, even though we know much more about them. There is psychological distance involved here. One tends to believe what spectrometers say and to read too much into the data. In chemistry, nothing is more important than honing one's critical sense. It is somehow easier to ignore some small, but ultimately vital glitch on a piece of chart paper than to ignore the results of an experiment gone in an unexpected direction. The former somehow melts into the noise of the base line, while the latter teases the imagination.

One can also sympathize with those practitioners of the art of structure determination who saw the future when the first NMR machine was delivered, and realized that for them, "NMR" meant "No More Research." They were right.

Ultraviolet/visible spectroscopy is called **electronic spectroscopy** because it deals with the absorption of energy by electrons in molecules. The energy absorbed (ΔE) is used to promote an electron from one orbital to a higher energy one and the magnitude of ΔE is proportional to the frequency of the light, ν [Eq. (12.1); Fig. 12.28].

$$\Delta E = h\nu = hc/\lambda \qquad (12.1)$$

(h is Planck's constant, λ is the wavelength of the light, and c is the speed of light)

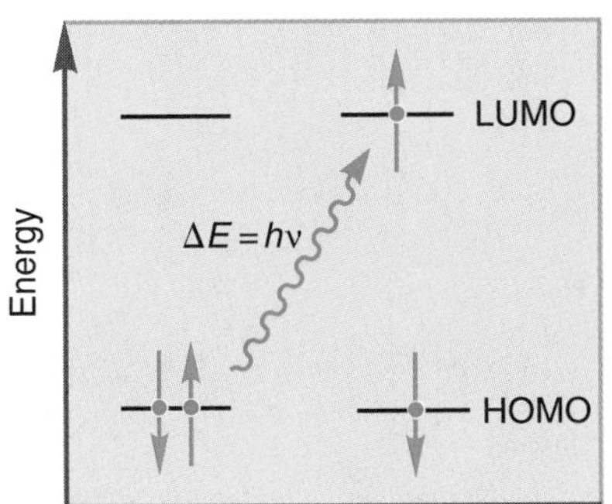

FIGURE **12.28** Absorption of UV light promotes an electron from the HOMO to the LUMO.

As electrons are restricted to orbitals of specific energy, ΔE is also quantized and may have only certain values. The distinction between **ultraviolet** and **visible spectroscopy** is an artificial one and relates only to the fact that the detectors built into our bodies, our eyes, are sensitive only to radiation of wavelength 400–800 nm (nm = nanometers; 1 nm = 10^{-9} m), which we call the visible spectrum. These machines are less limited in this respect than we are, and UV spectrometers easily detect radiation from 200–400 nm, as well as the entire visible range and beyond. At higher energy, below 200 nm, the atmosphere begins to absorb strongly, and spectroscopy in this range is more difficult, although by no means impossible.

A schematic spectrometer is shown in Figure 12.29. It consists of a source of radiation, a monochromator that is capable of scanning the range emitted by the source, a photomultiplier, and a detector.

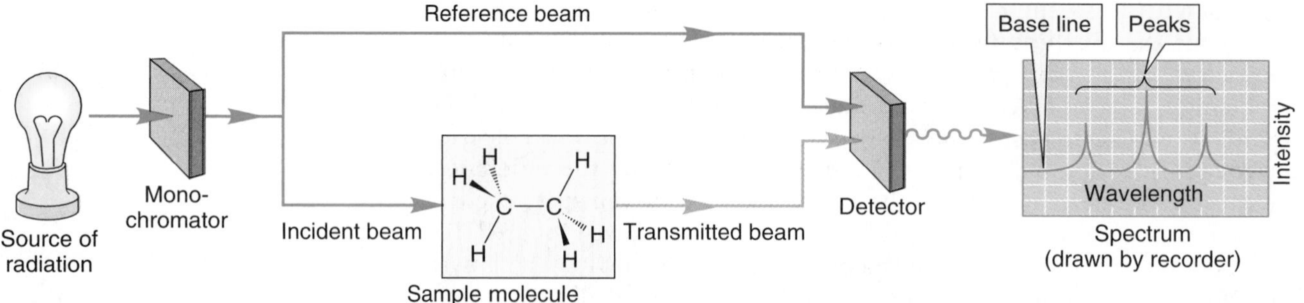

FIGURE **12.29** A schematic picture of a UV/vis spectrometer.

Generally, the spectrometer determines A, the absorbance, and plots A versus wavelength (λ in nm). The absorbance is calculated according to the Beer–Lambert law, Eq. (12.2).

$$A = \log I_0/I = \varepsilon lc \tag{12.2}$$

The intensity of the light entering the sample is I_0, the intensity of the light exiting the sample is I, the concentration in moles per liter is c, the length of the solvent cell is l, and the proportionality constant ε is called the **extinction coefficient**. Every compound has a characteristic ε, which depends not only on A, l and c [Eq. (12.2)], but also on wavelength, solvent, and temperature.

It is important to have a feel for the magnitudes of the energies involved in spectroscopy. Equation (12.1) repeats the familiar relationship between energy, Planck's constant h, and the frequency of light:

$$E = h\nu \tag{12.1}$$

As $\nu = c/\lambda$, and as we are interested in the energy per mole, Eq.(12.3) is the important one, where N is Avogadro's number:

$$E = Nhc/\lambda \tag{12.3}$$

If $h = 1.583 \times 10^{-37}$ kcal s/molecule, then the quantity Nhc is equal to 28.6×10^3 (nm)(kcal/mol).

PROBLEM 12.15

Calculate the energies involved in the range 200–800 nm, the region of UV/vis spectroscopy.

For a given molecule there will be many possible UV (or vis) transitions, as there are many different kinds of electrons even in a simple molecule. For ethylene there are several possible absorptions, as σ or π electrons are promoted to antibonding σ* or π* orbitals. Figure 12.30 ignores the carbon–hydrogen bonds and shows only σ → σ* and π → π* transitions. There are other possible absorptions, even without including the carbon–hydrogen σ and σ* orbitals: σ → π* and π → σ*, are two examples. Such transitions between orbitals of different symmetry are usually relatively weak, however.

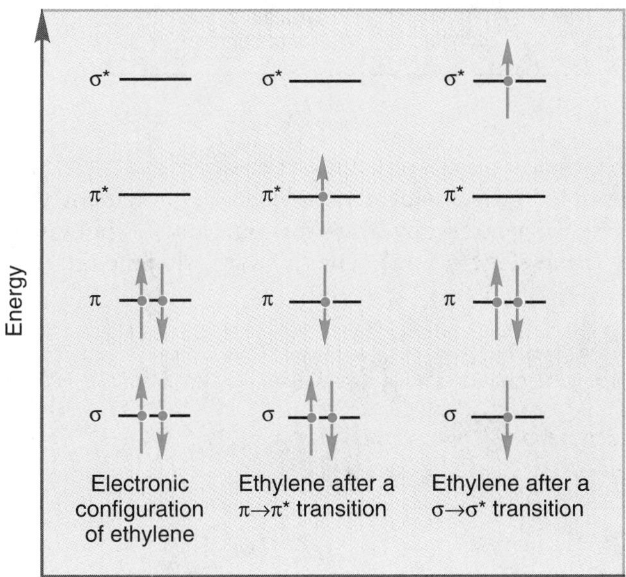

FIGURE 12.30 A π→π* and σ→σ* transition for ethylene.

The starting point and final destination of the electron are used to denote the transition; hence σ → σ*, σ → π*, π → π*, and π → σ* in this case. The energies, and therefore wavelengths of light involved, depend on the energy separating the orbitals. For ethylene, even the π → π* separation is great enough so that detection is difficult with common spectrometers. The π → π* absorption for ethylene occurs in the vacuum UV at about 165 nm, which corresponds to 173 kcal/mol, a high energy indeed. Little change is induced by alkyl substitution, and neither simple alkenes nor unconjugated dienes absorb strongly above 200 nm.

When a diene is conjugated, however, substantial changes occur in orbital energies, and therefore in the possible UV transitions. In general, the lowest energy (longest wavelength λ) transition possible for a polyene will be from the HOMO to the LUMO. In ethylene or butadiene, this is a π → π* transition. Figure 12.31 shows the relative energies of the π molecular orbitals of ethylene and 1,3-butadiene.

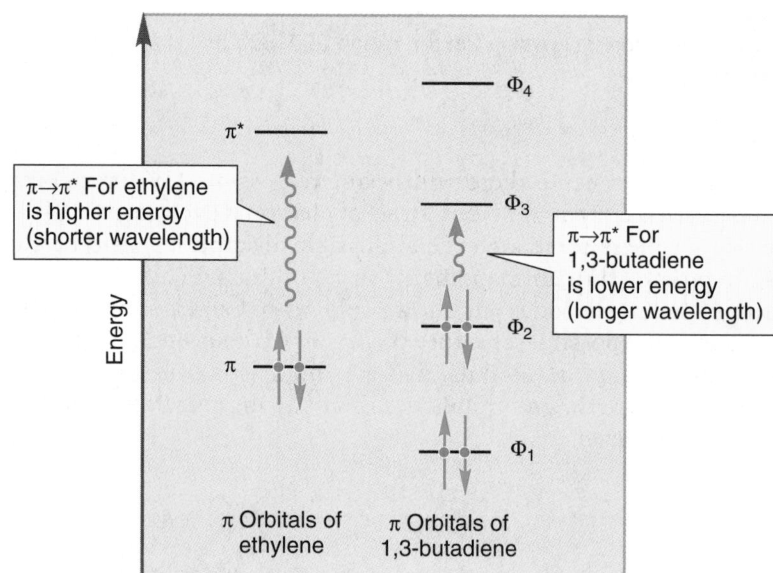

FIGURE **12.31** The HOMO–LUMO transition in ethylene is higher in energy (at shorter wavelength) than the HOMO–LUMO transition in 1,3-butadiene.

You can see that there is a much lower energy $\pi \rightarrow \pi^*$ (HOMO–LUMO) transition possible in the conjugated molecule 1,3-butadiene than there is in ethylene itself. Spectroscopy bears this out, and 1,3-butadiene absorbs at much longer wavelength (lower energy) than ethylene, 217 nm.

*PROBLEM **12.16** How much energy is contained in light of wavelength 217 nm?

ANSWER $E = Nhc/\lambda$ and the quantity $Nhc = 28.6 \times 10^3$ (p. 516).
So, $E = (28.6 \times 10^3)/217 = 132$ kcal/mol at 217 nm.

Further conjugation does two things. First, it reduces the energy gap between the HOMO and LUMO, and this pushes the position of the lowest energy absorption to longer wavelength (lower energy). Eventually, the HOMO–LUMO gap becomes small enough so that absorption enters the visible range, and we perceive highly conjugated polyenes as colored. The first colored polyenes are yellow, as it is violet light (400–420 nm) that is first absorbed as the HOMO–LUMO gap shrinks. β-Carotene, an important precursor to vitamin A, absorbs at 453 and 483 nm, and is orange (Fig. 12.32).

Second, as molecules become more complicated, and especially as conjugation is extended, the number of possible UV transitions increases and the spectra become complex. Figure 12.33 shows the spectrum of one moderately complicated molecule, which surely will resist a simple analysis, and which would be most difficult to predict in detail.

However, many important molecules contain simple diene or triene systems, either acyclic or contained in rings, and UV spectroscopy remains a useful tool for structure determination in such molecules.

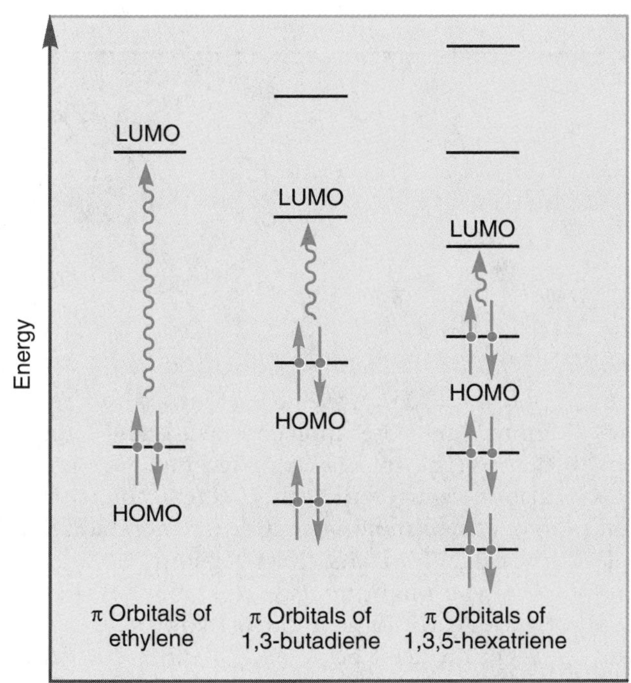

β-Carotene
λ_{max} = 453 and 483 nm

FIGURE 12.32 As conjugation increases, the HOMO–LUMO gap decreases and the position of the π→π* absorption shifts to longer wavelength (lower energy).

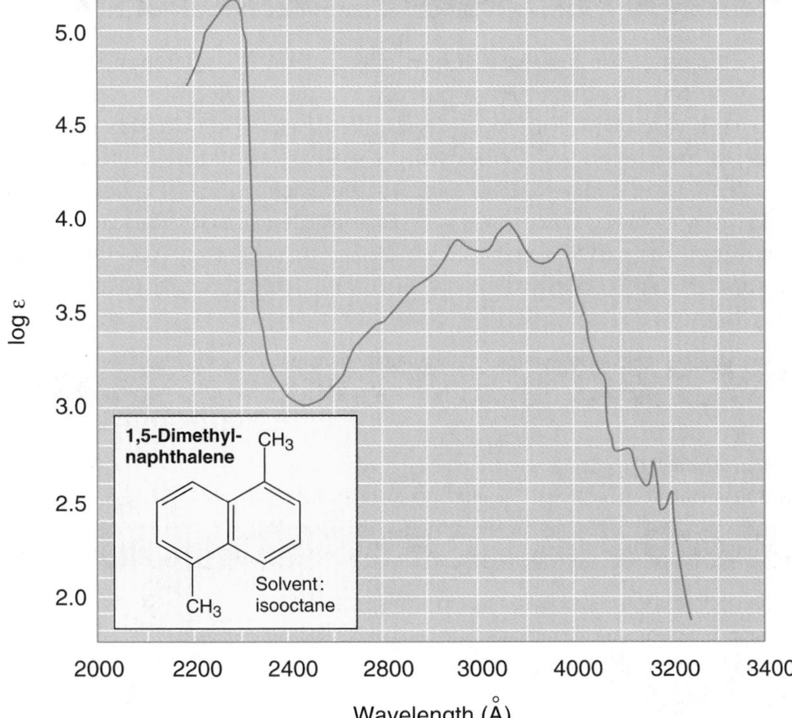

FIGURE 12.33 A complicated UV spectrum.

Bombykol

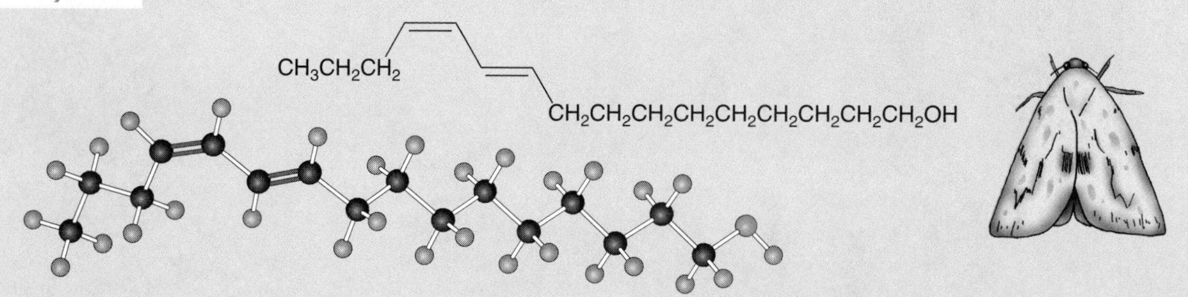

Most insects (and other animals, including humans) "talk" to each other by releasing chemicals that carry meaning. Such molecules are called "pheromones." Some are sexual attractants, others are trail markers, and still others carry an alarm signal. Here is bombykol, a simple cis, trans dienol that serves as the female silkworm moth's attractant for the male. The female of the species, *Bombyx mori,* releases bombykol from two glands located at her abdomen. The structure of bombykol was worked out by the German chemist Adolph Butenandt over a long period, ending in 1956. He collected 6.4 mg of the pure pheromone from 500,000 moth abdomens. Such work was extraordinarily difficult in those days, as he did not have the advantage of modern spectroscopy to help with structure determination. Eventually, the correct structure was determined and the molecule was made independently. The male moth's receptor for bombykol also recognizes the other stereoisomers, but the proper cis, trans isomer is far more active than the cis, cis or trans, trans versions.

As you might imagine, if bombykol is to function effectively as an attractant in the open air, reception must be very efficient; a little bombykol must go a long way if the silkworm moths are to have a satisfactory love life. In fact this is so; the male is staggeringly efficient at detecting this molecule. It is effective in doses as small as as 1×10^{-12} µg/mL.

For a nice introduction to pheromones, see William C. Agosta, *Chemical Communication,* Freeman, New York, 1992.

The position of longest wavelength absorption (λ_{max}) depends in a predictable way on the substitution pattern of such molecules. A set of empirical correlations was collated by R. B. Woodward (see the Box on p. 521) and was known then as "Woodward's rules" and now sometimes as "Woodward's first rules." These rules were extended by Louis Fieser (1899–1977) and Mary Fieser (b. 1909) to include cyclic dienes in the polycyclic compounds called steroids. Table 12.2 summarizes Woodward's and the Fiesers' rules. Woodward's very different second set of rules will appear in Chapter 22.

TABLE **12.2** Woodward's and Fiesers' Rules for Ultraviolet Spectra of Dienes

Woodward's Rules for Acyclic Dienes	(nm)
Base diene value	217
Each added alkyl group	+5
Each double-bond exocyclic in a six-membered ring	+5
Fiesers' Rules for Steroid Dienes	**(nm)**
Base diene with both double bonds in the same ring	253
Base diene with double bonds in different rings	214
Each additional conjugated double bond	+30
Each alkyl substituent	+ 5
Each exocyclic double bond	+ 5

Robert Burns Woodward

Robert Burns Woodward (1917–1979) was acknowledged as the greatest American organic chemist, and was certainly one of the greatest chemists of all time. There is remarkable agreement on the subjective evaluation of Woodward's preeminence among organic chemists, and it's worth taking a little time to try to see what set this man apart from his merely excellent colleagues. First of all, he was devoted to the subject and worked unbelievably hard at it. Although the main themes of his professional life were the elucidation of structure and the synthesis of molecules of ever more breathtaking complexity, his work is also notable for the attention he paid to mechanistic detail, and to the broad implications of the work. Working hard is not a matter of mere hours, but of maintaining focus as well. Woodward was able to *concentrate,* and had or developed a prodigious memory for minutiae. In another person that might amount to mere obsession, but Woodward was able to extract from the sea of detail the important facts that allowed him time and again to turn his work from the specific to the general. Early on he was able to propose correct structures for such disparate molecules as penicillin, tetracycline, and ferrocene (Fig. 12.34).

Woodward was able to pull obscure facts out of remote chemical hats and apply them to the matter at hand in a way that often transformed the very nature of the problem. The molecules he considered worthy of synthesis grew in complexity throughout his career, but it was more than the difficulty of the target that made his work so universally impressive. Woodward's work led places—it had "legs." For example, Woodward, primarily a synthetic chemist, made the observations [with Roald Hoffmann (b. 1937)] that led to the development and exploitation of what is arguably the most important *theoretical* construct of recent times. Hoffmann and Kenichi Fukui (b. 1918) of Kyoto University received the Nobel prize for this work in 1981. Many believe Woodward would have shared this Nobel prize (it would have been his second; his first was awarded in 1965 for "his outstanding achievements in the art of organic synthesis"), had he lived. Some of the molecules synthesized by Woodward and his co-workers are shown in Figure 12.35. Woodward said that the synthesis of quinine was conceived when he was 12 years old.

Penicillin V

Ferrocene

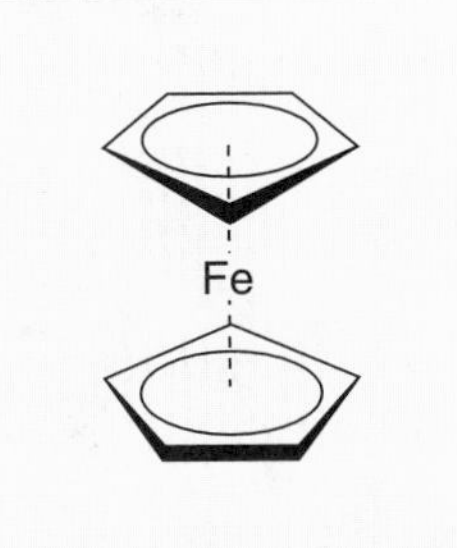

Tetracycline

FIGURE 12.34 Some complicated molecules whose structures were proposed by Woodward and his co-workers.

Should you memorize Woodward's rules? Frankly, I think not, although there are those who strongly disagree, and some of these people write examination questions. These rules were an important part of structure determination but are restricted in utility to relatively simple polyenes. You should know they exist, know what systems they apply to, and look them up when you need them. If you wind up in steroid research, you'll learn them.

Quinine, with W. von E. Doering, 1944

Patulin, 1950

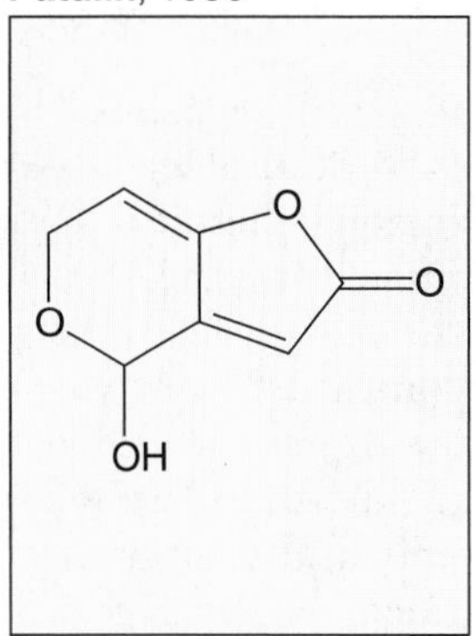

Cholesterol, 1951

Strychnine, 1954

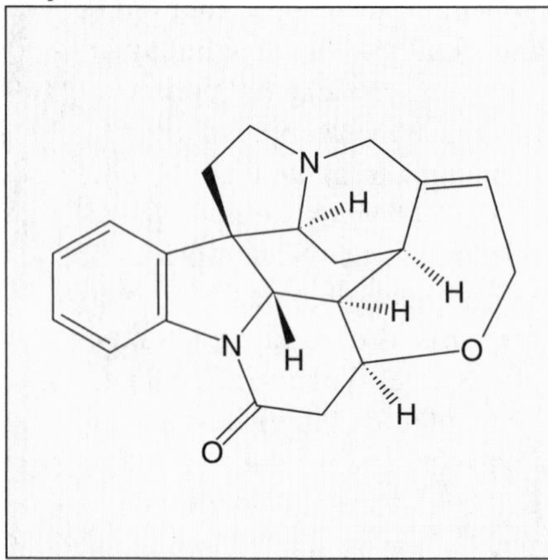

Reserpine, 1956

Triquinacene, 1964

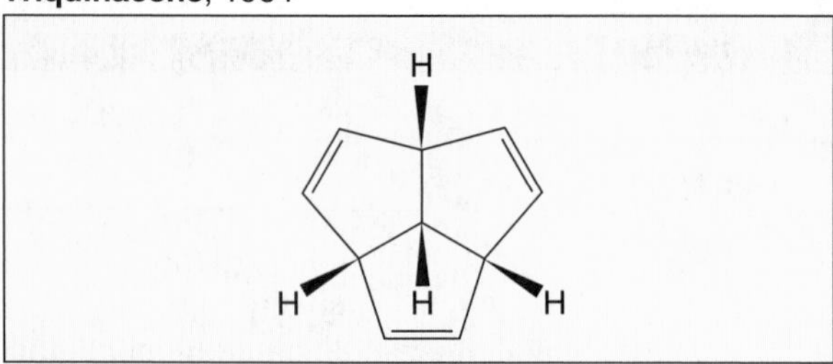

Cephalosporin C, 1966

Vitamin B$_{12}$ with A. Eschenmoser and a cast of thousands, 1972

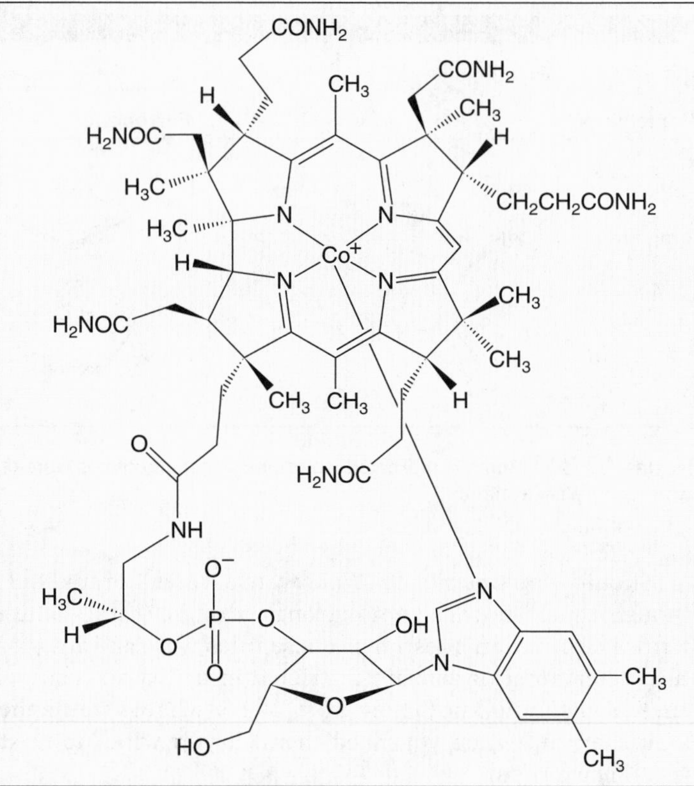

FIGURE 12.35 Some molecules synthesized by the Woodward group.

12.7 POLYENES AND VISION

When we hear or read the word "spectroscopy" we tend to imagine rooms full of expensive machines measuring away. It's quite an inhuman vision. Yet we are spectrometers ourselves! All of our sensory experiences, most obviously vision, are examples of spectroscopic measurements. Those machines in our imaginations are mere extensions of our biological, in-house, equipment. When you see a red apple among the green ones, and decide that it is the ripe red one you will eat, you are acting on a spectroscopic measurement. Let's examine vision a little bit here, because it depends on polyene chemistry.

Many polyenes are important in biological processes, and as we might imagine, polyenes with visible spectra are important in human vision. Vitamin A is *trans*-retinol, for example (Fig. 12.36). We have two enzymes, one of which, retinol dehydrogenase, oxidizes retinol to retinal and another, retinal isomerase, which isomerizes one double bond of *trans*-retinal to produce *cis*-retinal (Fig. 12.36).

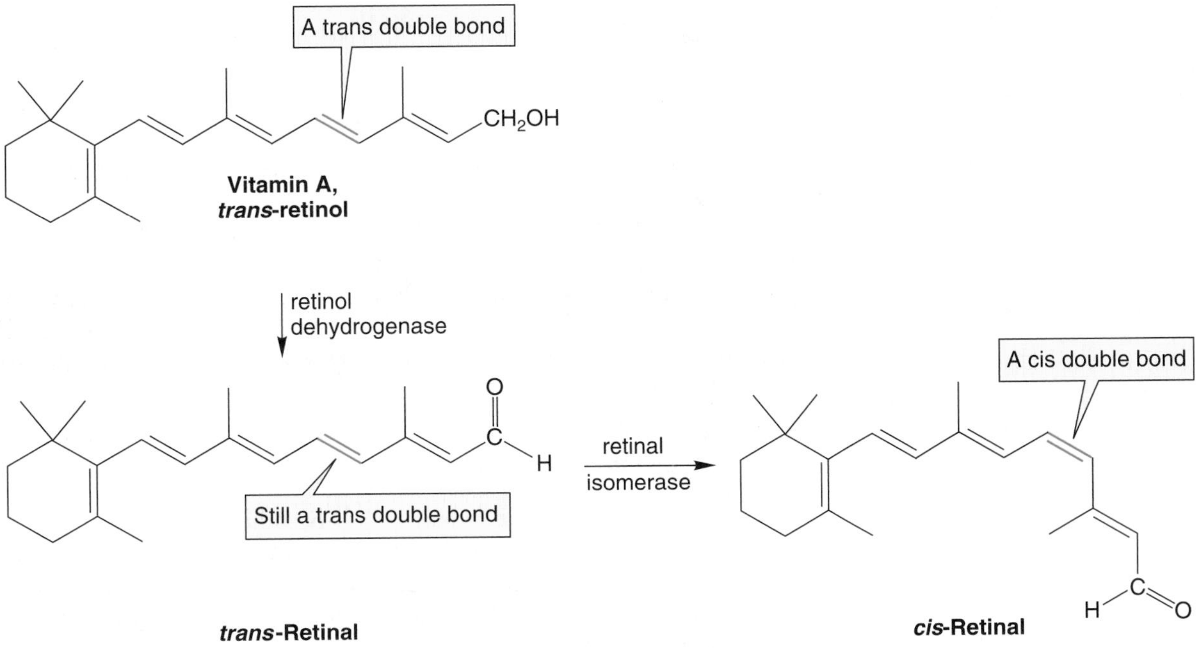

A trans double bond

CH_2OH

Vitamin A,
***trans*-retinol**

retinol
dehydrogenase

A cis double bond

O
C
H

retinal
isomerase

Still a trans double bond

H C O

***trans*-Retinal**

***cis*-Retinal**

FIGURE 12.36 Vitamin A (*trans*-retinol) is oxidized to *trans*-retinal and then enzymatically isomerized to *cis*-retinal.

The isomerized *cis*-retinal (but *not trans*-retinal) has the proper shape to react with a protein called opsin to form rhodopsin (Fig. 12.37). Rhodopsin has a strong absorption in the visible spectrum, which isomerizes the cis double bond back to trans. This absorption of visible light changes the shape of the molecule severely, destroying the crucial fit of molecule and protein. The attachment point is exposed, and the retinal is detached from the protein, opsin. A nerve impulse is generated and perceived by our brains as light (Fig. 12.37). The chemistry involved in the attachment and detachment of retinal and opsin is simple and will be covered in Chapter 16.

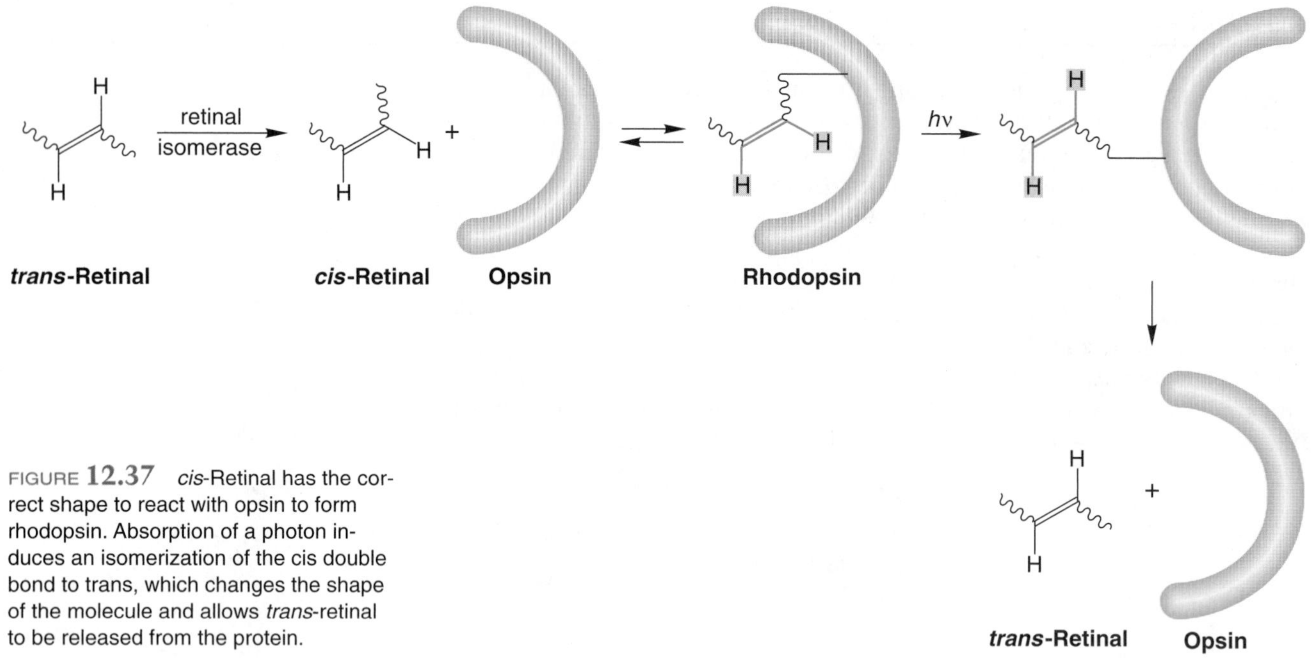

trans-Retinal — retinal isomerase → *cis*-Retinal + Opsin ⇌ Rhodopsin — $h\nu$ → ↓ *trans*-Retinal + Opsin

FIGURE 12.37 *cis*-Retinal has the correct shape to react with opsin to form rhodopsin. Absorption of a photon induces an isomerization of the cis double bond to trans, which changes the shape of the molecule and allows *trans*-retinal to be released from the protein.

12.8 THE CHEMICAL CONSEQUENCES OF CONJUGATION: ADDITION REACTIONS OF DIENES

Now we pass on to reactions (at last!). Conjugation affects not only the physical properties of molecules but their reactions as well. Although the basic chemistry of dienes does resemble that of alkenes, there are detailed, but important differences introduced by the presence of the *pair* of connected (conjugated) double bonds.

12.8a Addition of HBr and HCl

When hydrogen chloride adds to 1,3-butadiene, two major products, 3-chloro-1-butene (75.5%) and *trans*-1-chloro-2-butene (24%), are formed. There is a very small amount of *cis*-1-chloro-2-butene formed (0.5%). The main product is formed through 1,2-addition and the minor products by 1,4-addition (Fig. 12.38).

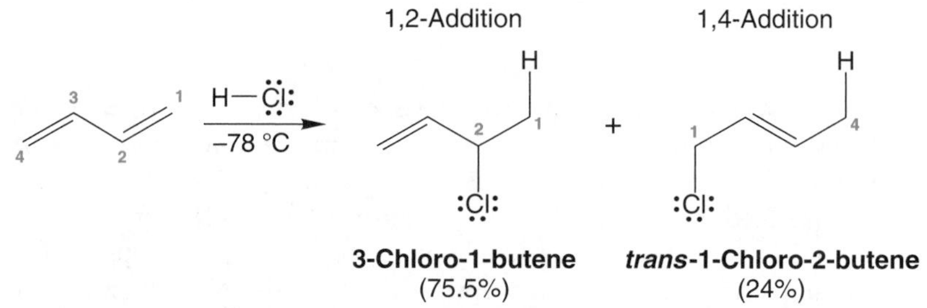

H—Cl: , −78 °C → 1,2-Addition: **3-Chloro-1-butene** (75.5%) + 1,4-Addition: ***trans*-1-Chloro-2-butene** (24%)

FIGURE 12.38 Addition of hydrogen chloride to 1,3-butadiene produces the products from both 1,2- and 1,4-addition.

We have seen this reaction before briefly in the discussion of resonance in Chapter 9 (p. 341). The source of the two products becomes clear if we work through the mechanism step by step. The first reaction must be protonation of butadiene by hydrogen chloride. There is a choice of two cations, one of which is a resonance-stabilized allylic cation, and the other an unstabilized primary carbocation. The first of these will be much lower in energy and its formation will be greatly preferred (Fig. 12.39).

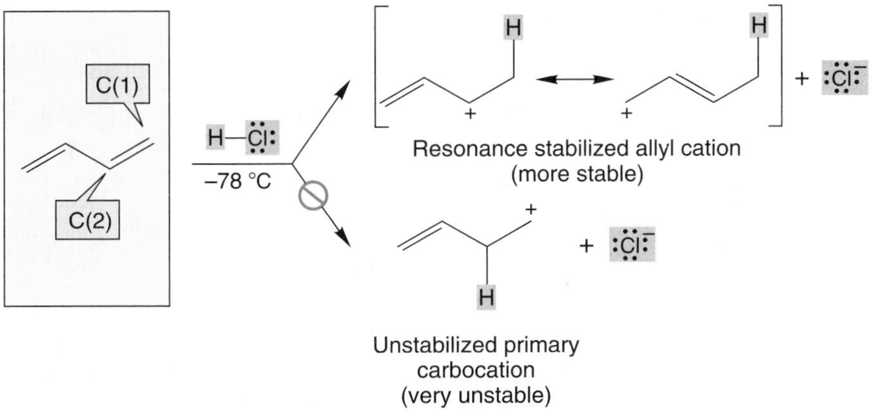

FIGURE **12.39** Protonation at C(1) of 1,3-butadiene gives a resonance-stabilized cation, but protonation at C(2) does not.

The resonance description of the allyl cation shows the source of the two products, as chloride can add to each of the two carbons sharing the positive change to give the two observed alkyl chloride products (Fig. 12.40).

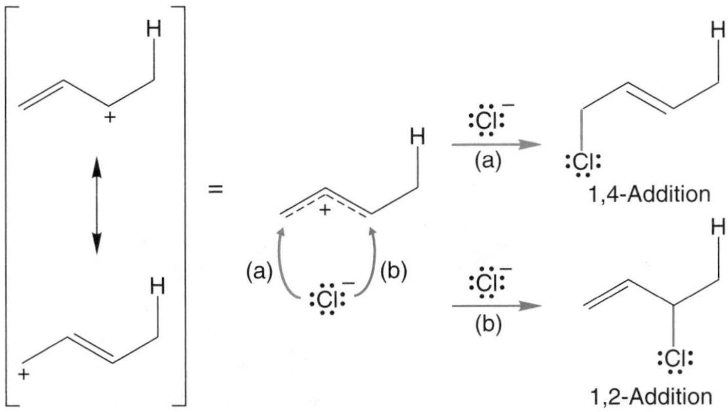

FIGURE **12.40** Addition of the nucleophile at the two positions (a and b) sharing the positive charge gives the products of 1,4- and 1,2-addition.

PROBLEM **12.17**

When treated under S_N1 conditions, the two chlorides in Figure 12.41 give the same pair of alcohols in the same ratio. Provide structures for the two alcohols and explain mechanistically.

FIGURE **12.41**

12.8b Addition of Br_2 and Cl_2

Other addition reactions of 1,3-dienes also produce two products. Like simple alkenes, dienes react with bromine or chlorine to give dihalides. As in hydrogen halide addition, products of both 1,2- and 1,4-addition are formed (Fig. 12.42).

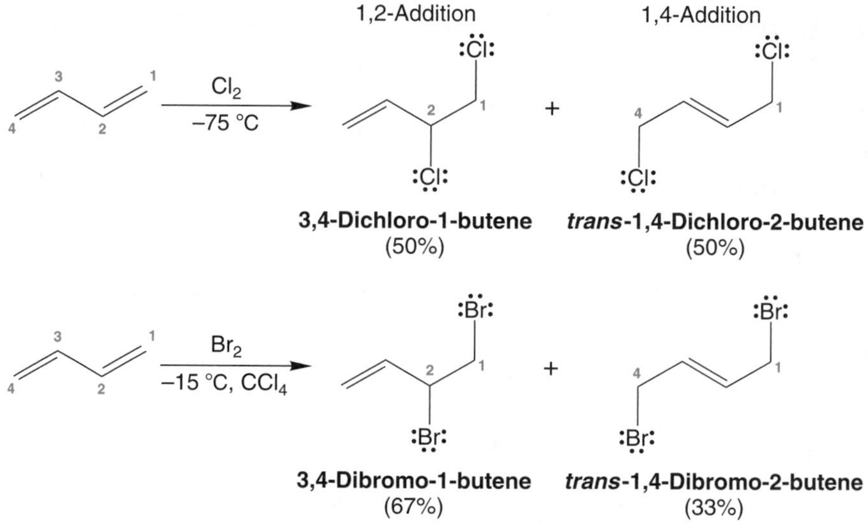

FIGURE 12.42 Addition of bromine or chlorine to conjugated dienes also gives two products. The red numbers show the origins of the terms 1,2- and 1,4- addition.

*PROBLEM 12.18 One potential source of the 1,4-addition product is the bromonium ion shown in Figure 12.43. Look carefully at the products of the reaction of 1,3-butadiene and bromine (Fig. 12.42). Explain why the mechanism of Figure 12.43 is certainly wrong.

FIGURE 12.43

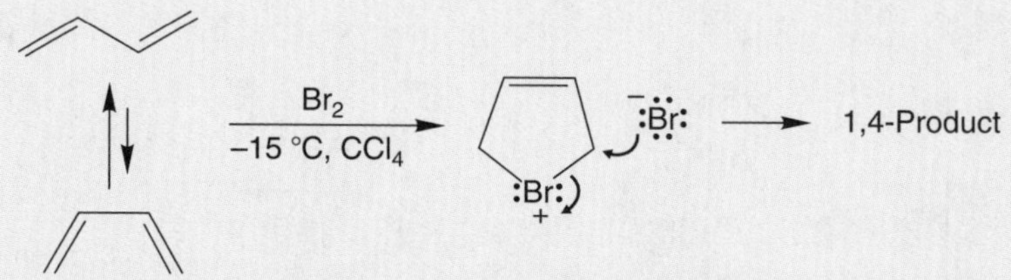

ANSWER Although there is nothing wrong structurally or electronically with the five-membered ring bromonium ion, it cannot be involved in this reaction. Opening of this ion must lead to an alkene containing a cis double bond, and the 1,4-product is, in fact, overwhelmingly trans (Fig. 12.42). The technique for solving this problem is to follow through using the five-membered bromonium ion as the intermediate, see what the structure of the product must be, and compare that to what is actually observed. If we pay careful attention to stereochemistry, we can see that the five-membered ring cannot be involved.

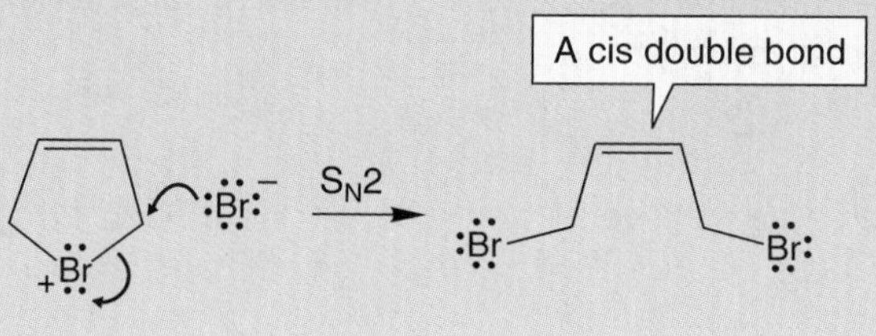

12.9 THERMODYNAMIC AND KINETIC CONTROL OF ADDITION REACTIONS

The addition reactions in the previous sections allow us to look at general properties of chemical reactions and to address fundamental issues of reactivity. Just what are the factors that influence the direction a reaction takes? Why is one product formed more than another? Is it the energies of the products themselves, or something else?

There are some curious effects of temperature and time on the addition reactions of both H–X and X_2. If the reactions are run at low temperature, it is the product of 1,2-addition that is generally favored. The more severe the reaction conditions, the more product of 1,4-addition is formed (Fig. 12.44).

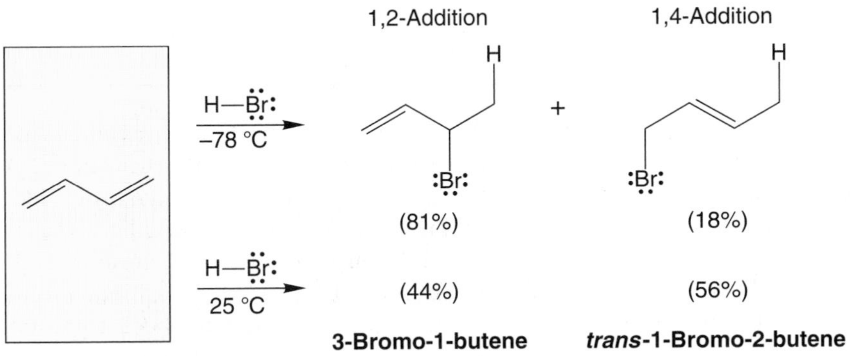

1,2-Addition

1,4-Addition

H–Br: −78 °C → (81%) + (18%)

H–Br: 25 °C → (44%) (56%)

3-Bromo-1-butene **_trans_-1-Bromo-2-butene**

FIGURE 12.44 The product of 1,4-addition is favored at high temperature.

Indeed, the products themselves will equilibrate if allowed to stand in solution. The same mixture of 1,2- and 1,4-addition products is formed from each compound, and the product of 1,4-addition is favored. Here is an example using the products of addition of Cl_2 to 1,3-butadiene (Fig. 12.45).

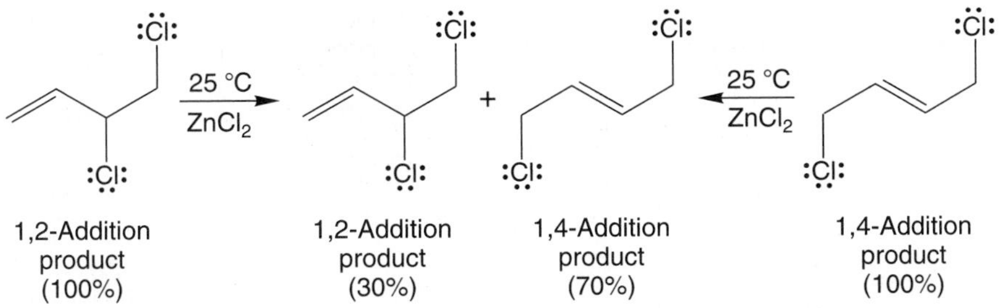

1,2-Addition product (100%) → 25 °C, $ZnCl_2$ → 1,2-Addition product (30%) + 1,4-Addition product (70%) ← 25 °C, $ZnCl_2$ ← 1,4-Addition product (100%)

FIGURE 12.45 The products of 1,2- and 1,4-addition equilibrate under the reaction conditions. The same mixture is formed from either isomer.

Several questions now confront us: (1) Why is the product of 1,4-addition favored? (2) How do the products of 1,2- and 1,4-additions equilibrate? (3) What are the reasons behind the curious temperature effect?

The first question is easy. The product of 1,4-addition contains a disubstituted double bond, whereas the 1,2-product is only monosubstituted, and therefore less stable *Remember*: The more substituted an alkene, the more stable it is (see Chapter 4, p. 133; Fig. 12.46).

The second question isn't hard either. For example, S_N1 solvolysis of either of the two products in Figure 12.45 leads to the same, resonance-stabilized allylic cation. The chloride ion can add at the two positions sharing the positive change and at equilibrium it will be the more stable product of 1,4-addition that dominates (Fig. 12.47).

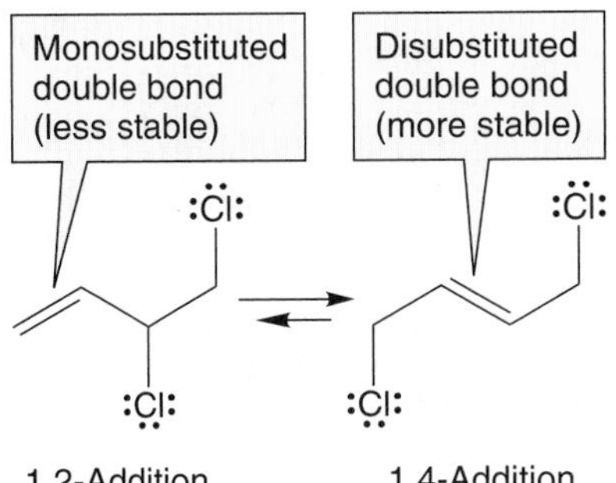

FIGURE 12.46 The product of 1,4-addition contains a disubstituted double bond. It is more stable than the product of 1,2-addition, which is a monosubstituted alkene.

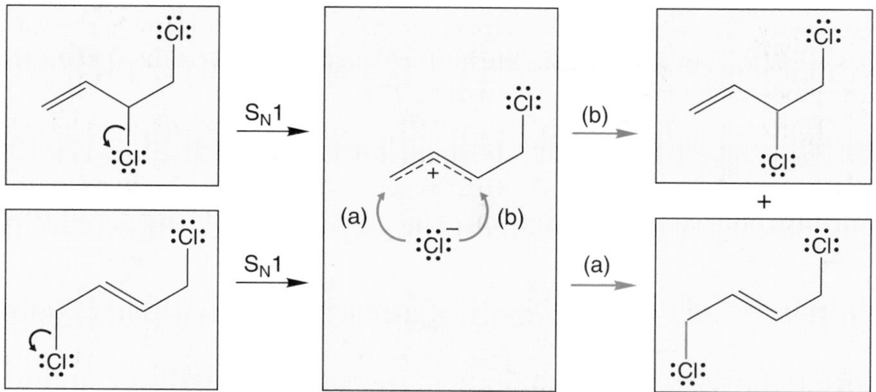

FIGURE 12.47 The 1,2- and 1,4-addition products equilibrate through the formation of a resonance-stabilized allyl cation.

The last question is tougher. Why should one product be favored under mild conditions and another at long time or under harsher conditions? At long time and relatively harsh reaction conditions, thermodynamics takes over and the more stable product prevails. At short time, or under mild conditions, it is not the *more stable* 1,4-compound that is favored, but the *less stable* product of 1,2-addition. But if the 1,2-product is formed *first*, the rate of its formation is by definition faster than that of the 1,4-product. Therefore, the transition state for formation of the less stable 1,2-product must be lower in energy than the transition state for formation of the more stable 1,4-product. The *lower energy* product is formed by passing over the *higher energy* transition state and the *higher energy* product by passing over the *lower energy* transition state. In chapter 8, you were warned that such cases would occur; here is one. An energy diagram for the reaction of chlorine with 1,3-butadiene should make everything clear (Fig. 12.48).

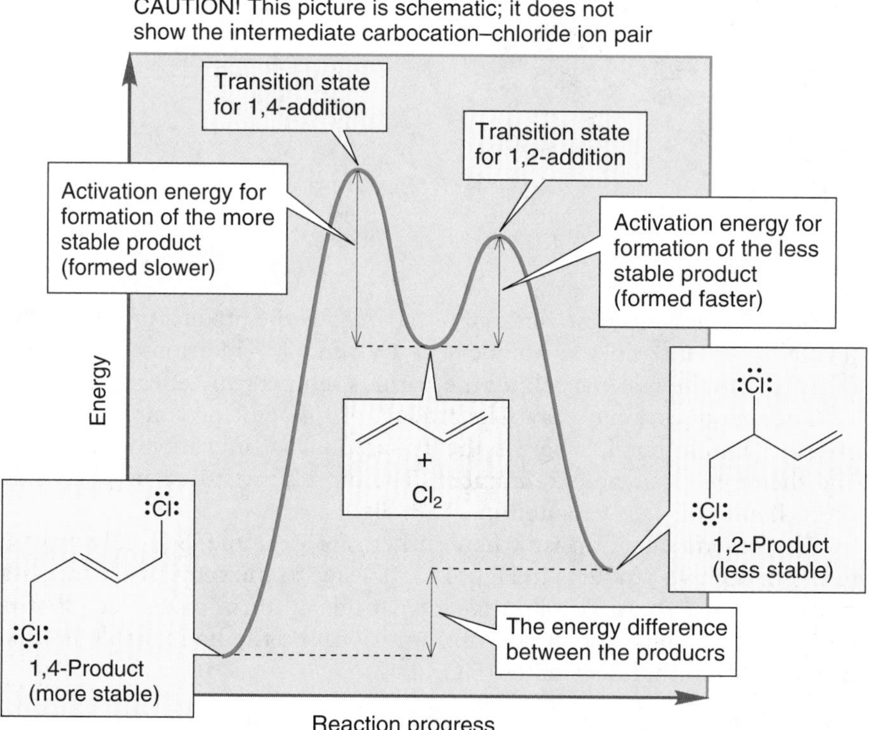

FIGURE 12.48 An energy diagram that summarizes 1,2- and 1,4-addition. The 1,2-product is the major compound formed under conditions of kinetic control. The 1,4-product is the more stable compound and is favored under conditions of thermodynamic control.

The 1,2-addition product is formed under conditions of **kinetic control** (short time, mild conditions), and the 1,4-product is favored under conditions of **thermodynamic control** (long time, severe conditions). To form the 1,2-product only the lower transition state must be passed, but to produce the more stable 1,4-product it is the higher energy transition state that must be crossed (Fig. 12.48).

Now we have generated a fourth question, and it's a nice one. We know why thermodynamics favors the 1,4-product: a disubstituted alkene is more stable than a monosubstituted alkene. But why does kinetic control favor the product of 1,2-addition?

Two possibilities suggest themselves. Let's examine the addition of hydrogen chloride to 1,3-butadiene. Look first at the resonance forms for the allyl cation. The two resonance forms are not equivalent; one is a secondary carbocation, the other a primary carbocation. The first will be weighted more heavily than the second, and most of the positive charge in the ion will be at the secondary position. Perhaps it is this more positive position that is attacked by the anion under kinetic control to give the 1,2-addition product (Fig. 12.49).

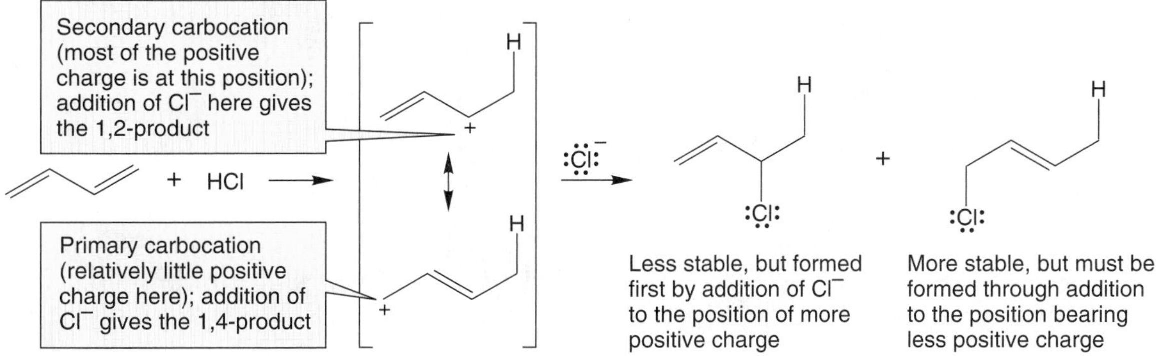

FIGURE 12.49 The resonance forms for the allyl cation formed from the protonation of 1,3-butadiene. The secondary carbocation will contribute more to the hybrid than will the primary carbocation. Addition of chloride at the two positions bearing positive charge gives the two products. The site of more positive charge gives the major product.

A second possiblity is even more straightforward. Where is the chloride ion after the initial protonation at C(1)? At the moment of its birth, the chloride ion is much closer to C(2) than to C(4). Perhaps addition of chloride is faster at the 2-position than at the 4-position by virtue of simple proximity (Fig. 12.50).

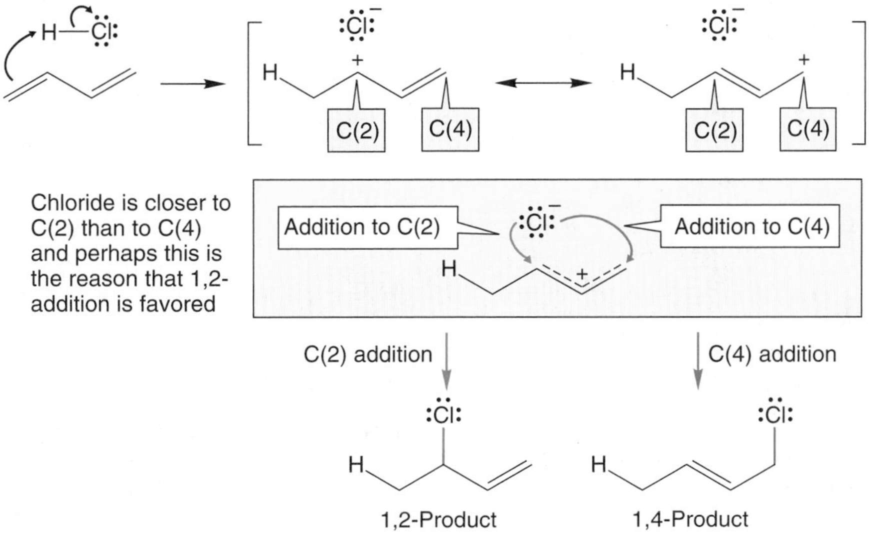

C(2) C(4) C(2) C(4)

Chloride is closer to C(2) than to C(4) and perhaps this is the reason that 1,2-addition is favored

Addition to C(2) Addition to C(4)

C(2) addition C(4) addition

1,2-Product 1,4-Product

FIGURE 12.50 After protonation, the chloride ion is closer to C(2) than to C(4). Perhaps this accounts for the preference for 1,2-addition.

There is a brilliant experiment that settles the issue, done by Eric Nordlander (1934–1984) at Case Western Reserve University. He studied the addition of D–Cl, not to 1,3-butadiene, but to 1,3-pentadiene. In this molecule, addition of D^+ yields a resonance-stabilized diene in which both resonance forms are secondary cations. Indeed these resonance forms are identical except for the remote isotopic label (Fig. 12.51).

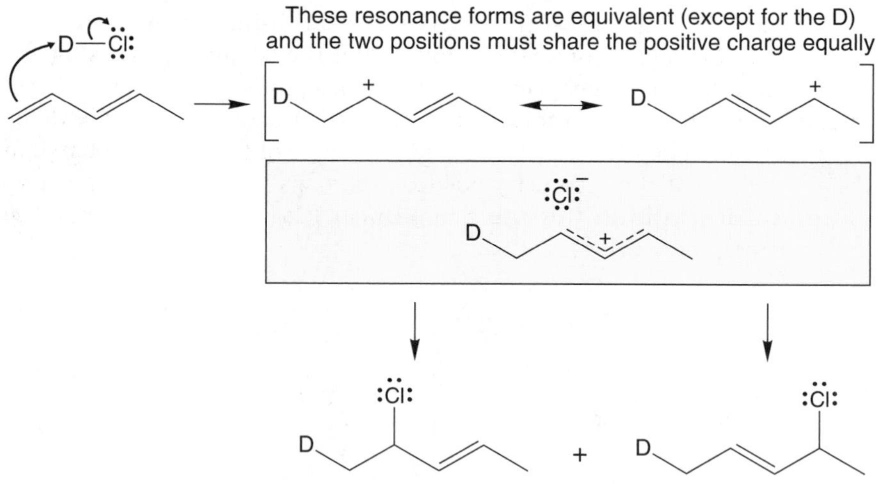

These resonance forms are equivalent (except for the D) and the two positions must share the positive charge equally

FIGURE 12.51 Protonation (D^+ in this case) of 1,3-pentadiene gives a resonance-stabilized allylic carbocation in which both contributing resonance forms are secondary. Addition of chloride at the two positions must take place to give equal amounts of 1,2- and 1,4-adduct.

PROBLEM 12.19 In Figure 12.51, why is protonation preferred at the end carbon [C(1)] rather than at C(4)?

If the source of preference for 1,2-addition depends on the difference between a secondary and primary carbocation, that preference should disappear in the reaction with 1,3-pentadiene. If proximity is at the root of the effect, 1,2-addition should still be favored. For once simplicity wins out, and 1,2-addition is still preferred (Fig. 12.52).

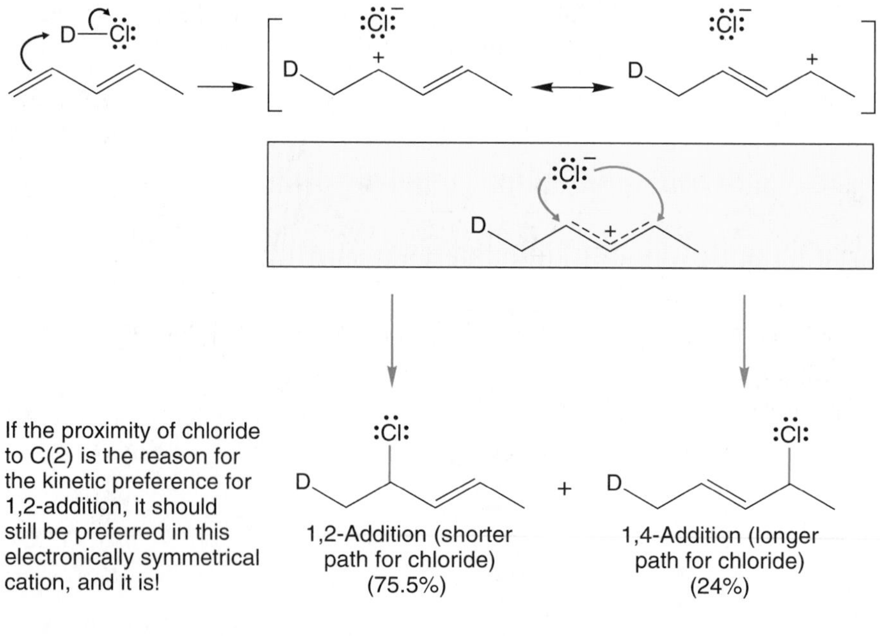

If the proximity of chloride to C(2) is the reason for the kinetic preference for 1,2-addition, it should still be preferred in this electronically symmetrical cation, and it is!

1,2-Addition (shorter path for chloride) (75.5%)

1,4-Addition (longer path for chloride) (24%)

FIGURE 12.52 1,2-Addition is still preferred in 1,3-pentadiene. The kinetic preference for 1,2-addition is apparently a simple proximity effect.

In the addition reactions of conjugated butadienes, the formation of a resonance-stabilized allylic cation is crucial. We will now continue with an examination of a few more reactions in which allylic stabilization is important.

12.10 THE ALLYL SYSTEM: THREE OVERLAPPING 2*p* ORBITALS

We have met the structure of the allyl system several times already (Problem 1.16; Chapter 2, p. 65; Problem 9.3), but Figure 12.53 again summarizes both the molecular orbital and resonance treatment of this system of three contiguous 2*p* orbitals.

Allyl

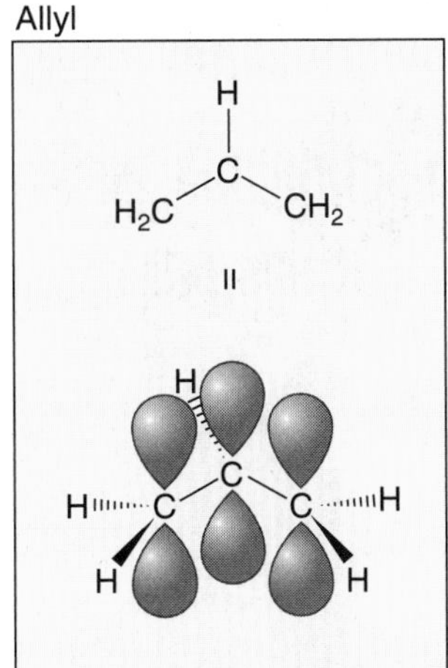

Molecular orbitals for allyl

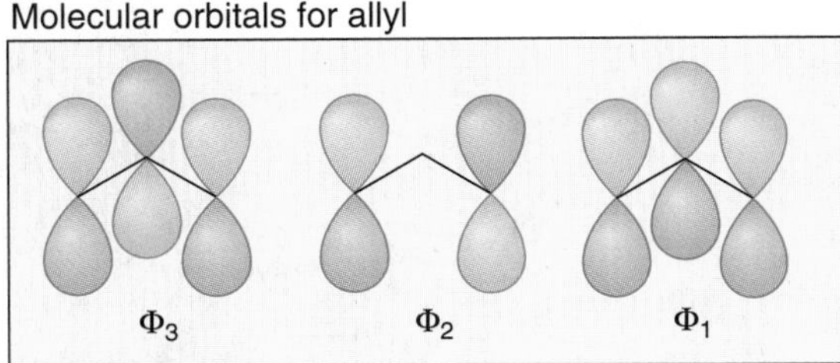

Resonance forms for allyl

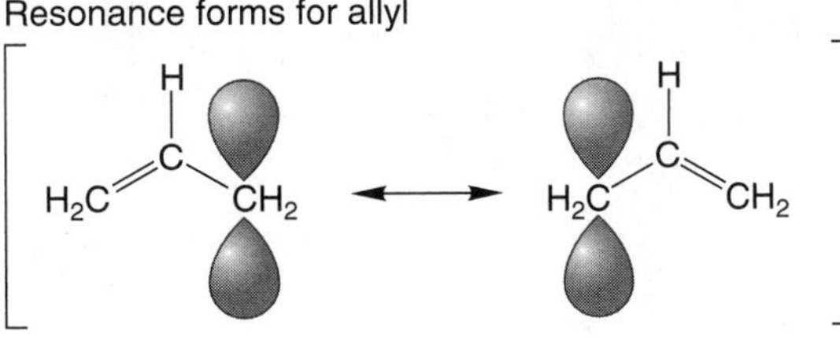

FIGURE 12.53 The resonance and molecular orbital treatment of the allyl system.

PROBLEM 12.20

Add electrons to both the resonance and molecular orbital descriptions in Figure 12.53 to form the allyl anion, radical, and cation.

12.11 THE ALLYL CATION: S_N1 SOLVOLYSIS OF ALLYLIC HALIDES

The resonance stabilization of the allyl cation makes allyl halides relatively reactive under S_N1 conditions (Fig. 12.54).

Allyl iodide

H_2O / S_N1

capture by water

deprotonation

FIGURE 12.54 The S_N1 solvolysis of allyl iodide in water. This reaction proceeds much faster than the solvolysis of propyl iodide.

As mentioned earlier, the allyl cation can be captured at either carbon sharing the positive charge. Table 12.3 gives some average rates of S_N1 solvolysis for a few common structural types. Primary allyl halides react much faster than simple primary halides. Secondary and tertiary allyl halides also ionize faster than their non-allylic counterparts. Resonance stabilization does have an accelerating effect on the ionizations (Table 12.3).

TABLE 12.3 Relative Rates of S_N1 Solvolysis of RCl in 50% Ethyl Alcohol at 45 °C[a]

Molecule	Ion	Relative Rate
Cl	Primary[a]	1.0[a]
Cl	Secondary	1.7
Cl	Allyl	14.3
Cl	Allyl	21.4
Cl	Methallyl	1300
Cl	Methallyl	1157
Cl	Tertiary	3×10^4
Cl	Dimethylallyl	1.9×10^6
Cl	Dimethylallyl	7.9×10^6

[a]The value for the primary substrate is suspect, and may include some contribution from S_N2 reactions.

*PROBLEM 12.21 Explain why the last two entries in Table 12.3 react at nearly the same rate and are so much faster than the others.

ANSWER

The two compounds ionize to the same carbocation intermediate. Although the two transition states are different, the ion will be partially developed in the transition state, and the rates of chloride loss will be quite similar.

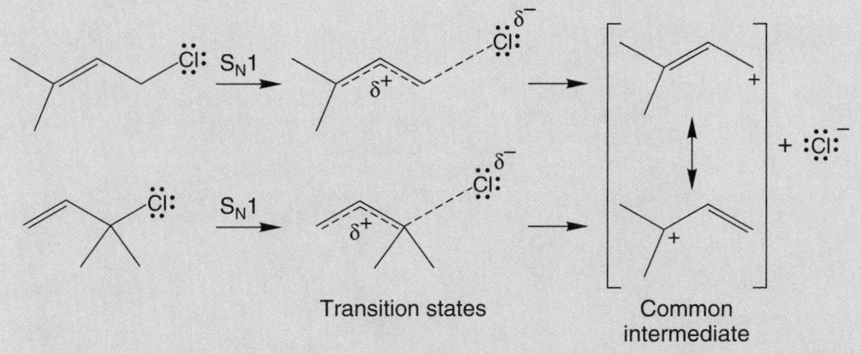

S_N1 S_N1 δ^- δ^+ + :Cl:$^-$

Transition states

Common intermediate

Both compounds ionize faster than the others in Table 12.3 because of the tertiary δ^+ in the transition state. A tertiary partial positive charge is more stable than a secondary or primary partial positive charge. For example,

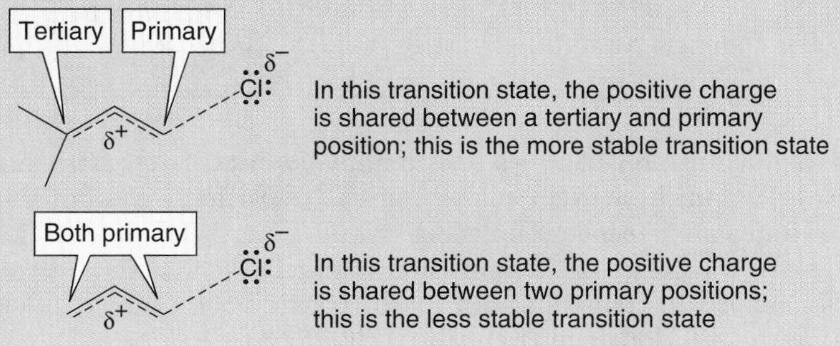

Tertiary Primary

In this transition state, the positive charge is shared between a tertiary and primary position; this is the more stable transition state

Both primary

In this transition state, the positive charge is shared between two primary positions; this is the less stable transition state

Remember: We are doing a dangerous thing here. We are again equating thermodynamic stability (a secondary allyl cation is more stable than a simple secondary cation) with kinetics (therefore secondary allylic halides form cations faster than do simple secondary halides). Although justified here, we must be clear on the reason why thermodynamics and kinetics are closely related in this example. Work Problem 12.22.

PROBLEM **12.22**

Explain carefully, with the use of an energy diagram, why the thermodynamic stability of the product cation is related to the ease of ionization of the corresponding halide. Remember the Hammond postulate (Chapter 8, p. 320), and consider the transition state for ionization.

12.12 S_N2 REACTIONS OF ALLYLIC HALIDES

Not only do allylic halides react faster than simple halides in the S_N1 reaction, but they are accelerated in S_N2 reactions as well. Table 12.4 gives some average rates of S_N2 reactions for a variety of halides.

TABLE 12.4 Relative Rates of S_N2 Reaction with Ethoxide in Ethyl Alcohol at 45 °C

Molecule	Substitution	Relative Rate
 1-chloropropane	Primary	1.0
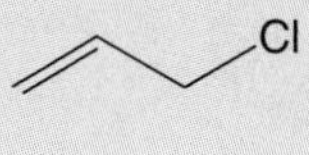 allyl chloride	Allyl	37.0
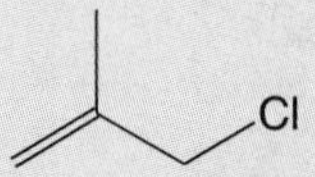 3-chloro-2-methylpropene	Allyl	33.0
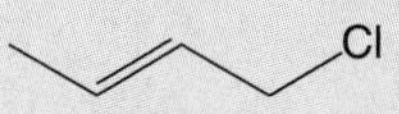 1-chloro-2-butene	Methallyl	97
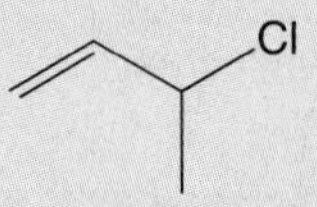 3-chloro-1-butene	Methallyl	1.85
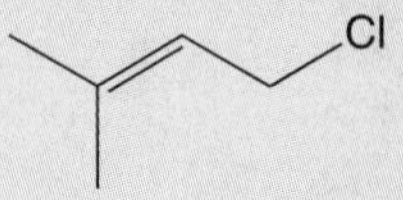 1-chloro-3-methyl-2-butene	Dimethylallyl	556

When we are considering a question involving rates, the answer can always be found through an analysis of the transition states for the reactions. Rates are dependent on the activation energy for the reaction, and this is just the height of the transition state relative to the starting halide. Let's compare the transition state for the reaction of a 1-propyl halide with that for the reaction of an allyl halide (Fig. 12.55).

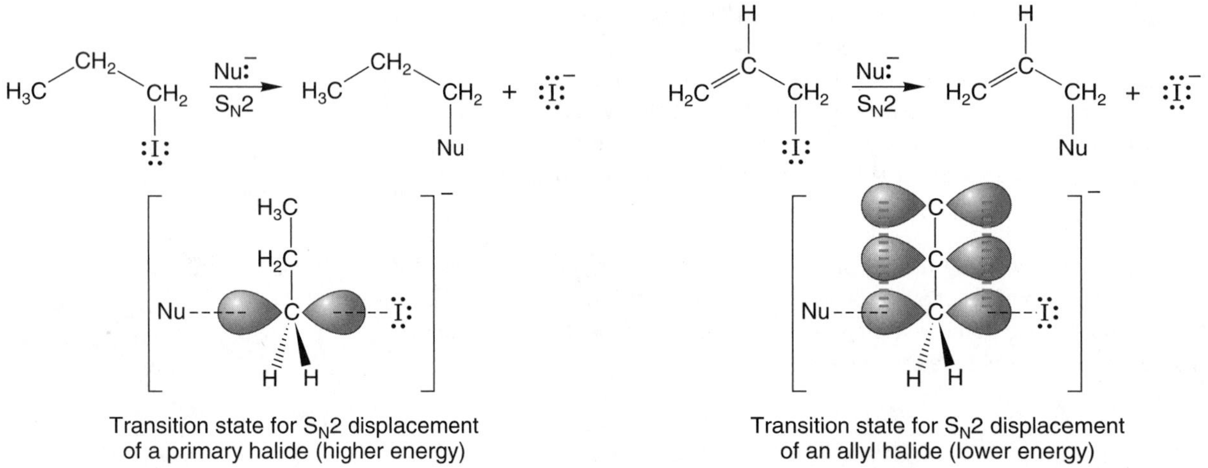

FIGURE 12.55 The transition states for S_N2 displacement of a primary and allyl halide. In the transition state for the allylic system there is resonance stabilization; in that for the primary system there is not. The delocalized transition state is more stable, and the rates of S_N2 displacements in allyl systems are especially fast.

In the transition state for the reaction of the allyl halide, there is overlap between the 2*p* orbital developed on the carbon at which displacement occurs and the $2p$ orbitals on the attached double bond. This transition state contains an allyl system. It will benefit energetically from the delocalization of electrons, and will lie at lower energy than the transition state for S_N2 displacement in the 1-propyl system, in which there is no delocalization.

12.13 THE ALLYL RADICAL

It is much easier to form a resonance-stabilized allylic radical than an undelocalized one. We have already seen the effects of delocalization in reactions involving allylic radicals in Chapter 11 (p. 484). For example, the regiospecific bromination of the allylic position of alkenes with NBS is possible because of the relative ease of forming the delocalized allyl radical (Fig. 12.56).

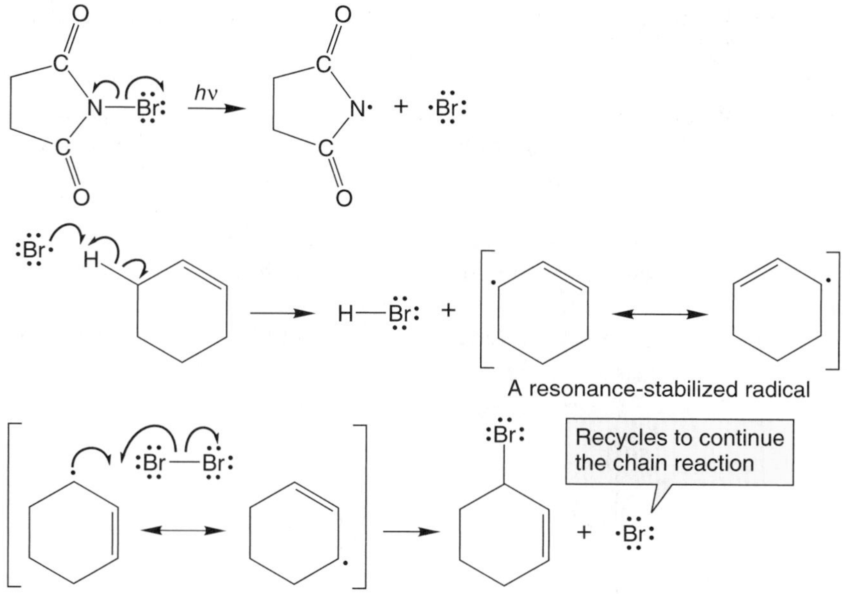

A resonance-stabilized radical

Recycles to continue the chain reaction

FIGURE 12.56 Allylic bromination of cyclohexene using NBS. Only the allylic position is substituted in this reaction.

12.14 THE ALLYL ANION

Now let's look at the acidity of allylic hydrogens to see if the formation of a resonance-stabilized allyl anion has any effect. We would surely expect it to. If delocalization of electrons is important, then removal of a proton at the allylic position should be easier than at a position not leading to a delocalized anion. A comparison of the acidities of propane and propene should test this idea (Fig. 12.57).

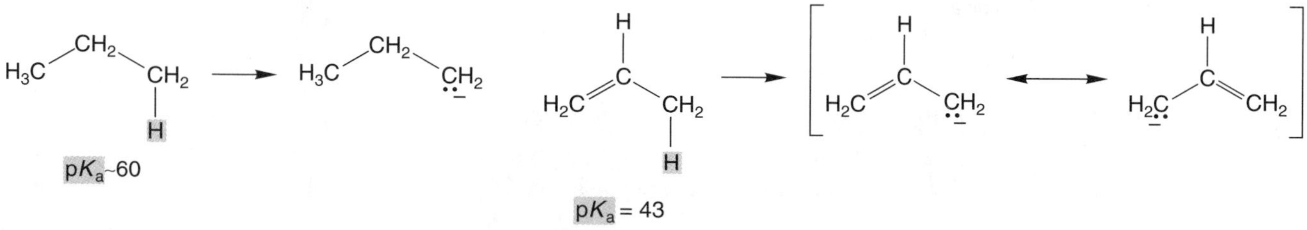

pK_a~60

$pK_a = 43$

FIGURE 12.57 If delocalization is important, it should be easier to remove a proton to give the resonance-stabilized allyl anion than to form the undelocalized propyl anion. It certainly is.

In fact, the pK_a of propene is 43, some 15–20 pK_a units lower than that for propane (pK_a ~ 60). The uncertainty arises from the difficulty of measuring the pK_a of an acid as weak as a saturated hydrocarbon. In any event, propene is a much stronger acid than propane, and it is reasonable to suppose that the formation of the delocalized allyl anion is responsible for the increased acidity of the alkene.

12.15 THE DIELS–ALDER REACTION OF DIENES

In 1950, Otto Diels (1876–1954) and his student Kurt Alder (1902–1958) received the Nobel prize for their work on what has appropriately come to be known as the **Diels–Alder reaction**, which they discovered in 1928. (So much for timeliness; at least they lived long enough to outlast the Nobel committee's renowned conservatism.) We will have more to say about its mechanism in Chapter 22, but the Diels–Alder reaction is characteristic of 1,3-dienes and is of enormous synthetic utility. Therefore we must outline it here.

When energy is supplied to a mixture of a 1,3-diene and a simple alkene (a dienophile), a ring-forming reaction takes place to produce a cyclohexene. Our discussion of mechanism begins with the arrow formalism, which points out which bonds are broken and shows where the new bonds are made (Fig. 12.58).

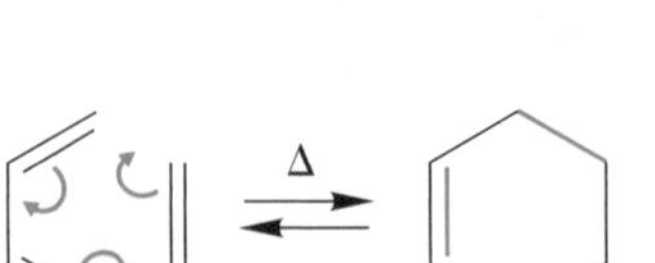

FIGURE 12.58 The Diels–Alder reaction between a 1,3-diene and an alkene produces a cyclohexene. The arrow formalism description shows the bonds made and broken in the forward reaction.

What orbitals are involved? The Diels–Alder reaction begins with the overlap of the π systems of the alkene (the dienophile) and the diene. The two reaction partners approach in parallel planes and as the bonds form, the end carbons of both the diene and alkene rehybridize from sp^2 to sp^3 (Fig. 12.59).

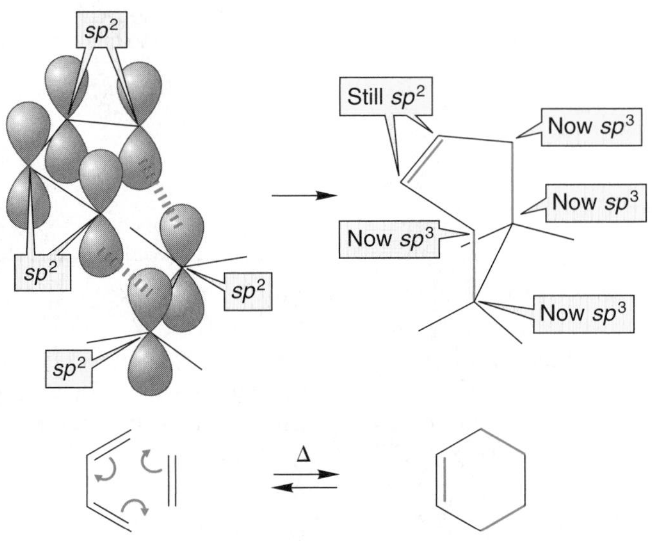

FIGURE 12.59 In the Diels–Alder reaction, the two participants approach each other in parallel planes. As the reaction occurs, the end atoms of the diene and the dienophile rehybridize from sp^2 to sp^3.

PROBLEM 12.23 Curiously, it matters just how energy is supplied to these molecules. If we heat the reactants up, the reaction occurs, but if we attempt to supply energy photochemically, it generally fails. We will examine such matters in detail in Chapter

PROBLEM 12.23 (CONTINUED)

22, but we might start to think about it now. When we think about reactions of atoms we do not consider every electron in the atom, just those most loosely held, the valence electrons. Similarly, in molecular reactions it is the most weakly held electrons, those in the HOMO, that control reactivity. The electrons in the HOMO might well be thought of as the valence electrons of molecules. Remember what we said about the interaction of light and molecules in Section 12.6 on UV–vis spectroscopy (Fig. 12.28, p. 515). First, consider the phase relationships in the possible HOMO–LUMO interactions in the transition state for a simple Diels–Alder reaction. Then, think about how the absorption of a *single* light photon might change those interactions.

PROBLEM 12.24

Write an arrow formalism for the reverse Diels–Alder reaction.

It is possible to estimate the thermochemistry in the Diels–Alder reaction by comparing the bonds made and broken in the reaction. Three π bonds in the starting material are converted into one π bond and two σ bonds in the product. Accordingly, the reaction is calculated to be exothermic by approximately 18 kcal/mol (Fig. 12.60).

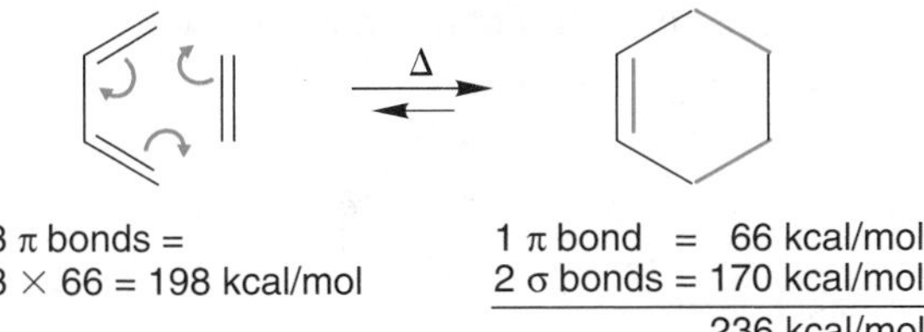

3 π bonds =
3 × 66 = 198 kcal/mol

1 π bond = 66 kcal/mol
2 σ bonds = 170 kcal/mol
236 kcal/mol

The reaction is 236 − 198 = 38 kcal/mol exothermic;
$\Delta H = -38$ kcal/mol

FIGURE 12.60 The simplest Diels–Alder reaction is exothermic by about 38 kcal/mol.

PROBLEM 12.25

Why is the bond strength of the new σ bonds in cyclohexene (85 kcal/mol) used in Figure 12.60 lower than that in ethane (90 kcal/mol)?

As shown in the previous figures, the Diels–Alder reaction requires the s-cis form of the diene component. However, it is the s-trans arrangement that is the more stable one, by approximately 3 kcal/mol (p. 512). If equilibrium favors the s-trans form strongly, why is the Diels–Alder reaction observed at all? Even though there is little s-cis form at equilibrium, reaction with the dienophile disturbs the equilibrium between the s-trans and s-cis form by depleting the s-cis partner. Reestablishment of equilibrium generates more s-cis molecules, which can react in turn. Eventually, all the diene can be converted into Diels–Alder product, even though at any moment there is very little of the active s-cis form present (Fig. 12.61).

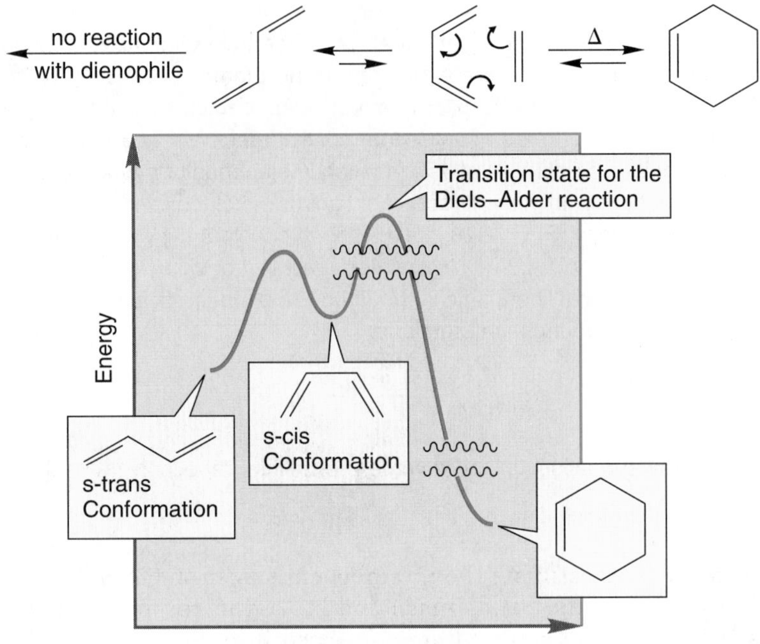

FIGURE 12.61 Even though the unreactive s-trans form of 1,3-butadiene is favored at equilibrium, the small amount of the s-cis form present can lead to product. The vertical axis of the figure is not to scale. The squiggly lines indicate gaps.

When there is no possibility of any s-cis form there can be no Diels–Alder reaction. For example, although 1,2-dimethylenecyclohexane reacts normally with dieneophiles, 3-methylenecyclohexene does not. The product of this reaction would contain a double bond far too strained to be formed (Fig. 12.62).

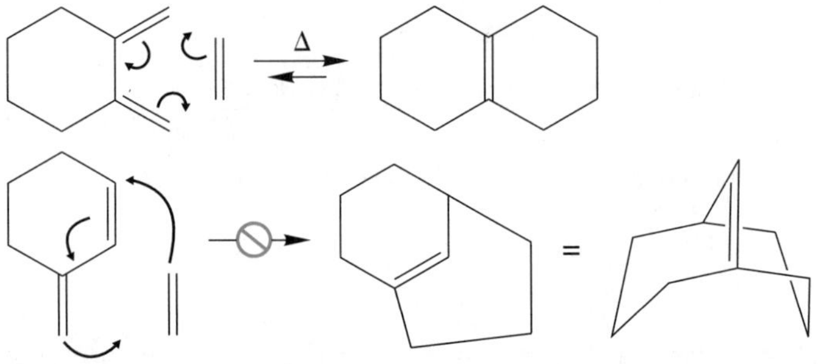

FIGURE 12.62 When there can be no s-cis form of the diene, there can be no Diels–Alder reaction.

*PROBLEM 12.26 Identify the source of strain in the putative product of Diels–Alder addition to 3-methylenecyclohexene in Figure 12.62. Use your models.

ANSWER Although the paper puts up with the double bond drawn in Figure 12.62, Nature won't. The orbitals making up this bridgehead double bond do not, in reality, overlap very well.

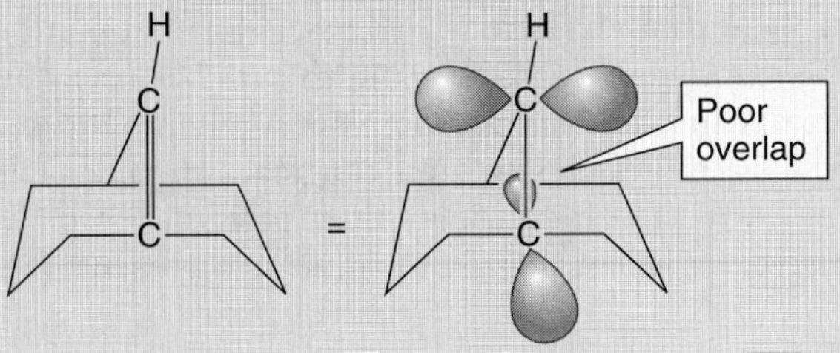

Another serious mechanistic question involves the timing of the formation of the two new σ bonds in the product cyclohexene. Are they made simultaneously in a one-step or "concerted" reaction, or are they formed in two separate steps (Fig. 12.63)?

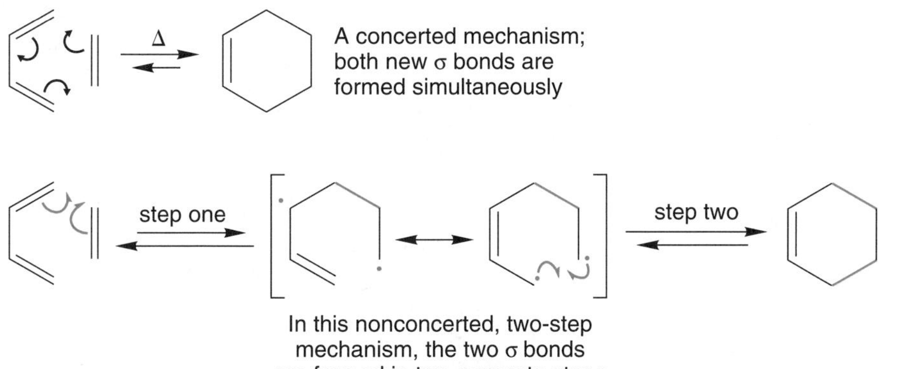

FIGURE 12.63 The mechanism of the Diels–Alder reaction could be concerted or involve two steps.

This question asks if there is an intermediate in the reaction or not. Figure 12.63 shows the two arrow formalisms and Figure 12.64 gives Energy versus Reaction progress diagrams for both the one- and two-step processes.

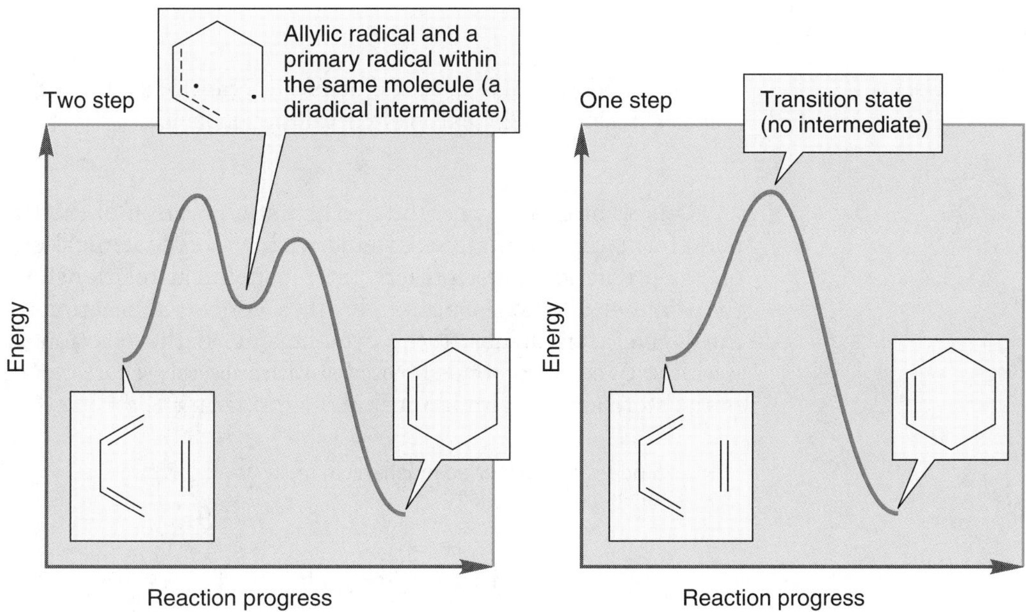

FIGURE 12.64 Energy versus Reaction progress diagrams for the concerted and two-step mechanisms.

We might immediately be suspicious of the two-step mechanism. It involves an intermediate allylic diradical (Figs. 12.63 and 12.64). Why don't we observe closure of the diradical at both possible positions? Should there not be a vinylcyclobutane produced as well as the cyclohexene (Fig. 12.65)?

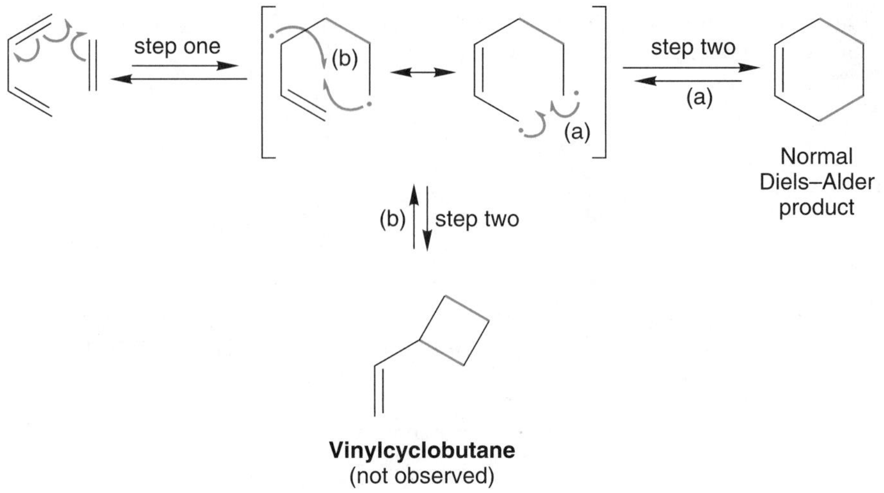

step one
(b)
(a)
step two
(a)
Normal Diels–Alder product
(b) step two
Vinylcyclobutane (not observed)

FIGURE 12.65 The two-step reaction could lead to vinylcyclobutane. This product is not observed.

In fact, simple Diels–Alder reactions do not lead to vinylcyclobutanes, but give cyclohexenes exclusively. Although this would seem at first a powerful argument favoring the concerted, one-step mechanism, which can only give cyclohexenes, it could be countered that the transition state for closing the four-membered ring will contain much of the strain present in all cyclobutanes, and will therefore be high in energy.

PROBLEM 12.27 Write an Energy versus Reaction progress diagram for the two-step Diels–Alder reaction that includes formation of the vinylcyclobutane.

As is so often the case, the mechanistic question of the timing of new bond formation is finally answered with a stereochemical experiment. A one-step reaction must preserve stereochemical relationships present in the original alkene. For example, in a concerted reaction, a cis alkene must give a cis disubstituted cyclohexene. If the reaction is concerted, with the two new σ bonds formed simultaneously, there can be no change in the stereochemical relationship of groups on the alkene (Fig. 12.66).

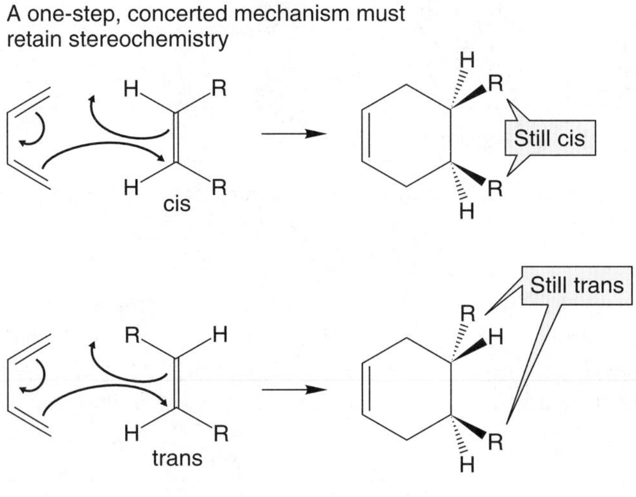

A one-step, concerted mechanism must retain stereochemistry
H R H R cis
H R R H
Still cis
R H H R trans
R H R H
Still trans

FIGURE 12.66 A one-step, concerted mechanism must preserve the stereochemical relationships of the groups on the alkene.

Diene
H
R
cis
H
R
Dieneophile
H
R
Still cis
H
R
A cyclohexene

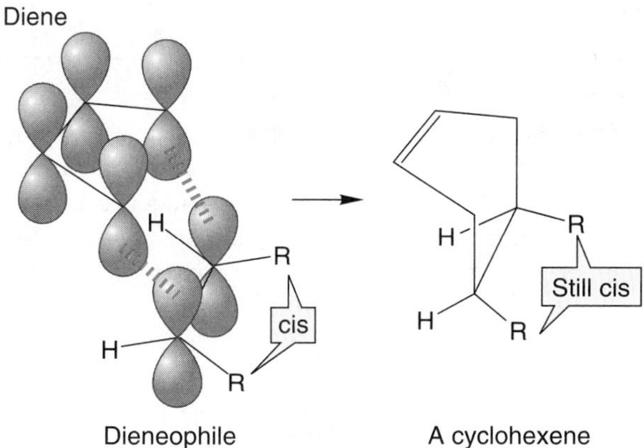

FIGURE 12.66 (CONTINUED)

If the bonds are formed in two steps, there should be time for rotation about the carbon–carbon σ bond in the original alkene. If the mechanism is stepwise, it should lead to stereochemical scrambling of the groups originally attached to the alkene (Fig. 12.67).

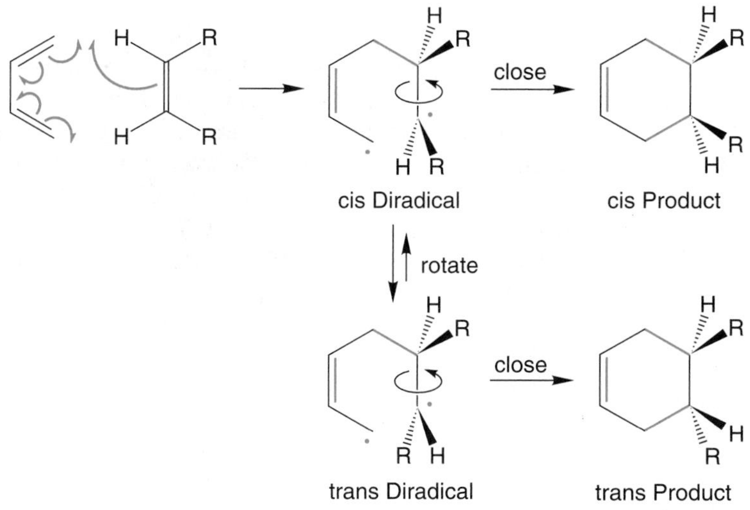

H
R
H
R
close
cis Diradical
cis Product
rotate
close
trans Diradical
trans Product

FIGURE 12.67 A two-step, nonconcerted mechanism predicts a mixture of stereoisomers.

In fact, retention of stereochemistry is observed. The reaction is a concerted, one-step process (Fig. 12.68).

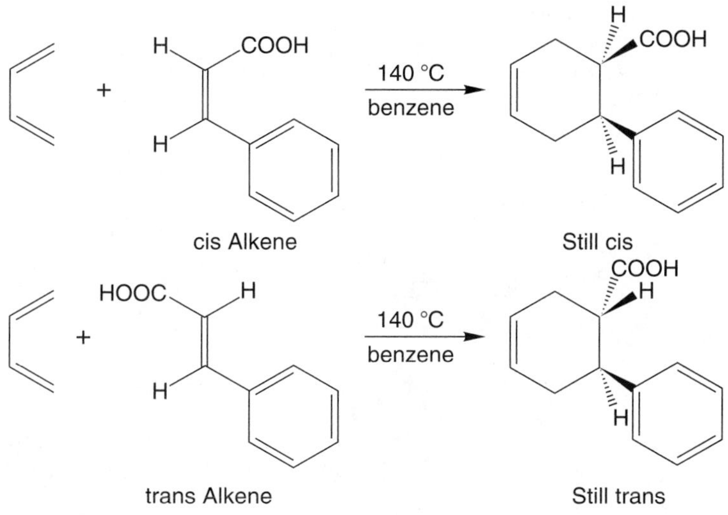

H
COOH
H
140 °C
benzene
cis Alkene
Still cis
HOOC
H
H
140 °C
benzene
trans Alkene
Still trans

FIGURE 12.68 The preservation of stereochemistry in this reaction favors the concerted mechanism.

PROBLEM 12.28 The experiment described in Figures 12.66–12.68 is usually easy to understand, but there is another stereochemical experiment possible. It tests the same thing, but is harder to see without models. So, if this problem is difficult to follow, by all means work it through with models. Let's follow the stereochemical fate of groups attached to the ends of the diene, not the alkene. Analyze the reaction of a *trans,trans*-1,3-diene and ethylene. There is only a single cyclohexene product, with the R groups cis (Fig. 12.69). Explain how this result shows that the mechanism of cycloaddition cannot be a two-step process.

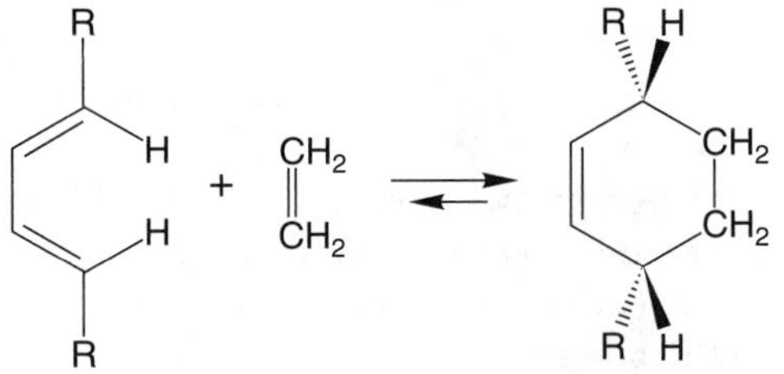
R
H
H
R
+
CH_2
CH_2
R H
CH_2
CH_2
R H
trans,trans Diene
cis Cycloalkene

FIGURE 12.69

In the preceding figures, mechanistic points have been illustrated using very simple examples. In fact, the simplest Diels–Alder reaction, ethylene and 1,3-butadiene, has a rather high activation enthalpy of about 27 kcal/mol and requires high pressure and temperature for success. Even at 200 °C, a reaction at 4500 psi yielded only 18% cyclohexene after 17 h. Other Diels–Alder reactions proceed under milder conditions. Reactions are more facile when the dieneophile bears one or more electron-withdrawing groups, and if the diene can easily achieve the required s-cis conformation. Some examples are shown in Figure 12.70.

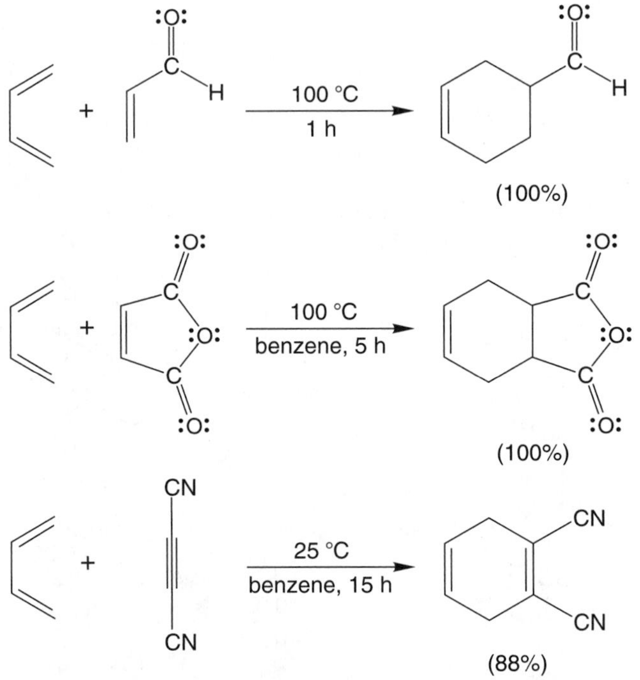
:O:
C
H
+
100 °C
1 h
:O:
C
H
(100%)
:O:
C
+
:O:
C
:O:
100 °C
benzene, 5 h
:O:
C
:O:
C
:O:
(100%)
CN
+
CN
25 °C
benzene, 15 h
CN
CN
(88%)

FIGURE 12.70 Diels–Alder reactions between a diene and alkenes or alkynes bearing electron-withdrawing groups are relatively fast.

Why is this reaction so important and so useful? Six-membered rings are very common, and the Diels–Alder reaction builds six-membered rings from relatively simple building blocks. Moreover, the reaction is very general. Most dienes react with most dienophiles to give Diels–Alder products. From now on, every cyclohexene you see must trigger the thought "can be made by a Diels–Alder reaction" in your mind. That's an important concept for both synthesis and problem solving.

Like most reactions in organic chemistry, the Diels–Alder reaction is not a one-way street. Reverse Diels–Alder reactions are common, especially at high temperature. *Remember*: $\Delta G = \Delta H - T\Delta S$ (Chapter 8, p. 306), and the temperature-dependent ΔS term can overwhelm the temperature-independent ΔH term at high temperature. The reverse Diels–Alder reaction makes two molecules from one and will be favored by entropy, ΔS. Cyclohexenes are not only formed in a reaction of a diene and a dienophile, but can serve as sources of dienes through the reverse Diels–Alder reaction (Fig.12.71).

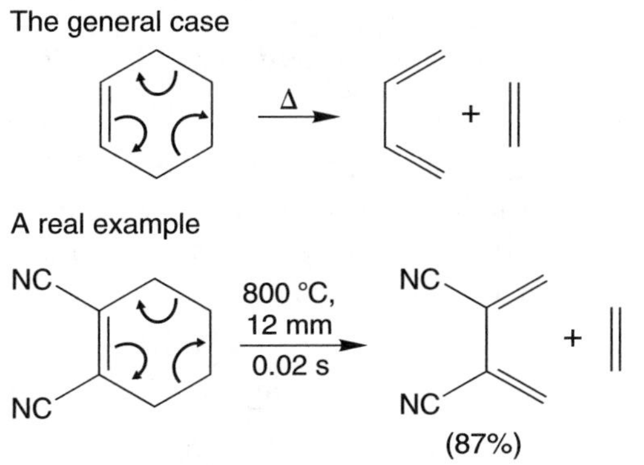

The general case

Δ

A real example

NC NC 800 °C, 12 mm 0.02 s NC NC (87%)

FIGURE 12.71 The reverse Diels–Alder reaction.

Now let's press on to look at other uses of this reaction, surely one of the most widely used in synthesis of all the reactions of organic chemistry. Chemists often use bicyclic molecules to investigate effects requiring a fixed geometrical arrangement of orbitals or groups. Consider the simple Diels–Alder reaction of an alkene with a common cyclic diene, cyclopentadiene (Fig. 12.72).

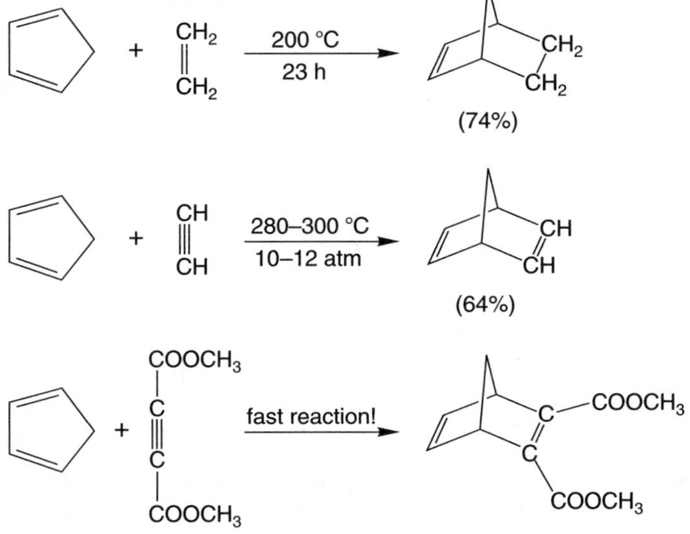

CH_2 CH_2 200 °C 23 h CH_2 CH_2 (74%)

CH CH 280–300 °C 10–12 atm CH CH (64%)

$COOCH_3$ C C $COOCH_3$ fast reaction! C $COOCH_3$ C $COOCH_3$

FIGURE 12.72 The Diels–Alder reaction between cyclic dienes and alkenes (or alkynes) gives bicyclic molecules.

Presto! Bicyclic compounds! But this instant synthetic expertise is not without its price. There is a new stereochemical effect to be considered. When a substituted alkene reacts with a cyclic diene, there are two possible stereoisomeric products even though both new σ bonds are formed simultaneously. The substituents on the alkene must retain their stereochemical relationship during the reaction, but there are still two possible products. The substituents on the alkene can point toward the newly formed double bond or away from it (Fig. 12.73).

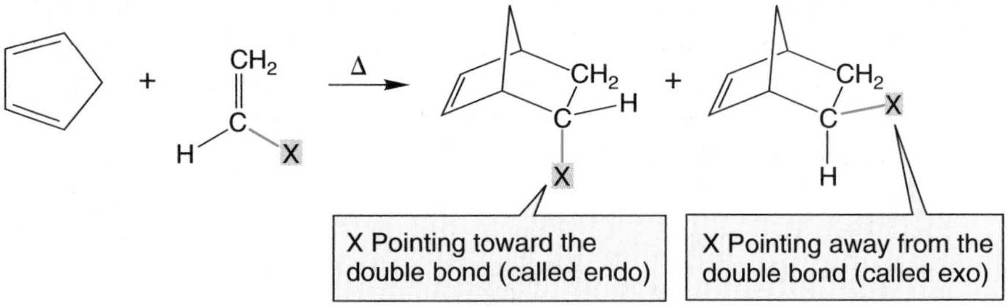

X Pointing toward the double bond (called endo)

X Pointing away from the double bond (called exo)

FIGURE 12.73 In principle, this kind of Diels–Alder reaction can give either endo or exo product.

The compound with the substituents aimed toward the double bond is called **endo** and that with the groups aimed away from the double bond, **exo**. The exo isomers are usually more stable, but the endo compounds are kinetically favored in the Diels–Alder reaction.

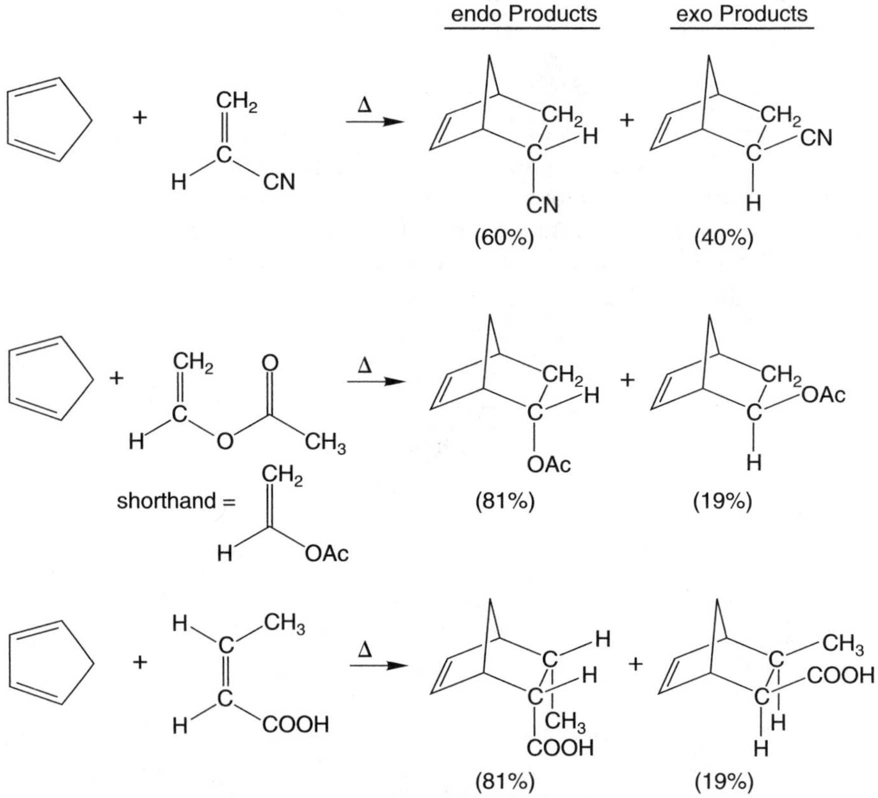

endo Products (60%) exo Products (40%)

shorthand =

endo (81%) exo (19%)

endo (81%) exo (19%)

FIGURE 12.74 It is the endo product that is favored in most Diels–Alder reactions. In each case, the percentage of endo product is greater than the percentage of exo product.

The exo compounds are more stable for steric reasons: the substituent points away from the more congested part of the molecule. Indeed, the transition state for exo product formation must also be less crowded than the transition state for endo product formation. But the transition state for endo product benefits from another factor and this overwhelms steric considerations. In the endo transition state, but not in the exo transition state, there can be "secondary orbital overlap." This fancy-sounding phrase simply means that there can be interaction between the "back" diene orbitals and orbitals on the substituent X only in the endo transition state. Figure 12.75 shows this effect for the first example of Figure 12.74.

CH_2 H CN
endo Product
exo Product

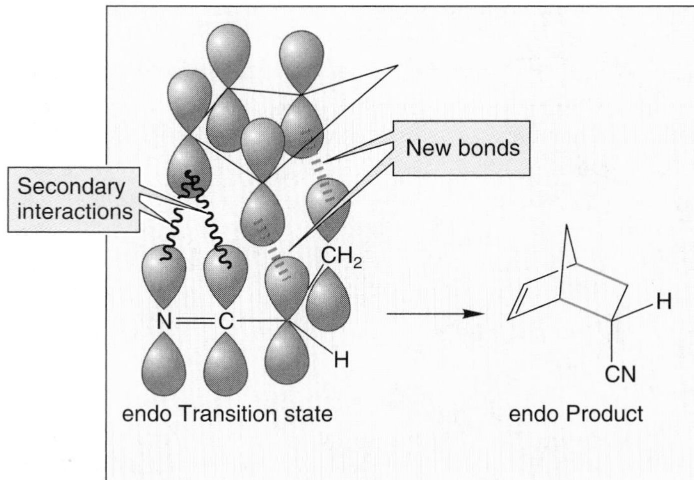

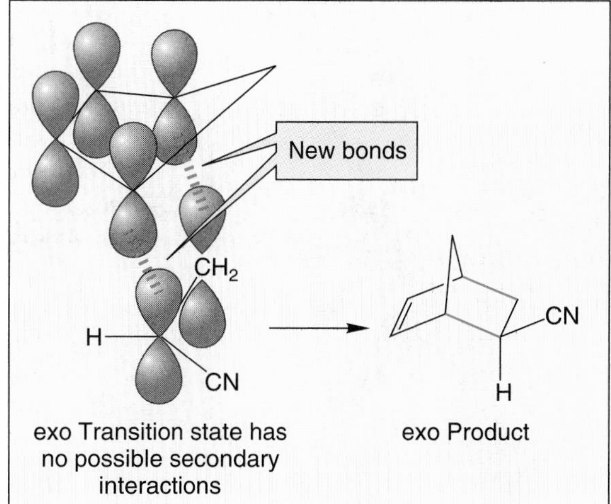

FIGURE 12.75 Secondary orbital interactions stabilize the endo transition states.

Even more complicated polycyclic structures can be made from the Diels–Alder reaction between two ring compounds. In the simplest case, the reaction of a cyclic 1,3-diene with a cycloalkene, it is again endo addition that is preferred (Fig. 12.76).

Δ
endo Product (98%)
exo Product (2%)

FIGURE 12.76 The Diels–Alder reaction between a cyclic diene and a cyclic dienophile gives a polycyclic product. Once again, it is the endo form that is favored.

Two cyclic dienes can also react in Diels–Alder fashion. The self-condensation of cyclopentadiene is the classic example. We'll illustrate this reaction with a pair of problems.

*PROBLEM 12.29 You attempt to prepare 1,3-cyclopentadiene from the base-catalyzed loss of hydrogen bromide (E2 reaction) from 3-bromocyclopentene (Fig. 12.77). Instead of isolating a compound of the formula C_5H_6, you get a high yield of a compound of the formula $C_{10}H_{12}$. Fame, fortune, and the Nobel prize await you if you can explain this result (at least if you were explaining it in 1928). Be careful—although this question is obviously set up in the text, it is not so easy to write a structure correct in every detail.

FIGURE 12.77

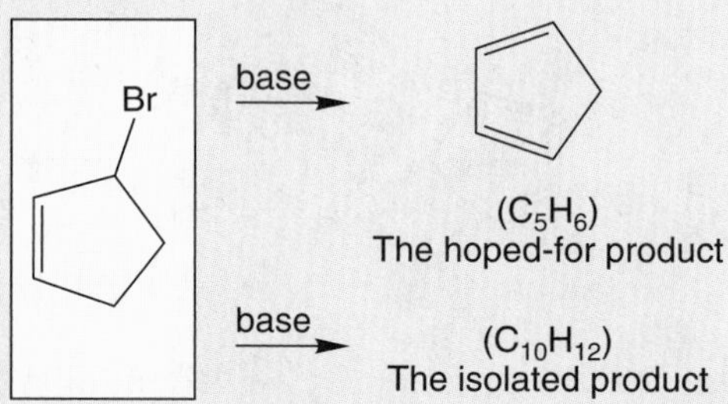

ANSWER The formula of the product $C_{10}H_{12}$ is a clue. Clearly, two C_5H_6 molecules have joined to make the $C_{10}H_{12}$ product. The reaction is, in fact, successful. The problem is that cyclopentadiene dimerizes in a Diels–Alder reaction very quickly. The molecule you isolate is the Diels–Alder dimer. Note formation (mainly) of the endo adduct. Anytime you see a 1,3-diene, especially a cyclic one, think Diels–Alder.

(C_5H_6) + (C_5H_6) $\xrightarrow{\Delta}$ ($C_{10}H_{12}$)

*PROBLEM 12.30 In a fit of frustration or inspiration you decide to boil (reflux) your unanticipated $C_{10}H_{12}$, and you do so by attaching a long condenser to a receiving vessel kept very cold. Your $C_{10}H_{12}$ boils at about 165 °C, but you collect a product of the formula C_5H_6 boiling at 40 °C. Moreover, although this product is stable at low temperature for a long time, at room temperature it rapidly reverts to your starting $C_{10}H_{12}$. Explain.

ANSWER At 165 °C, the Diels–Alder reaction reverses, and the low-boiling cyclopentadiene can be collected and kept at low temperature. If it is allowed to warm up, it will again form the Diels–Alder dimer.

165 °C → + (Cold) warm →

12.16 ISOPRENE: BIOSYNTHESIS OF TERPENES

When we discussed cationic addition reactions (Chapters 9 and 10), we included a section on the polymerization reactions of alkenes (Section 9.9, p. 361). Dienes form quite stable intermediates in addition reactions and are especially prone to polymerization.

Write a mechanism for the radical-induced polymerization of 1,3-butadiene. PROBLEM 12.31

Nature uses related reactions (the cart is way before the horse here; we make approximations of Nature, not the other way round!). Natural rubber is polymerized **isoprene** (2-methyl-1,3-butadiene), for example (Fig. 12.78).

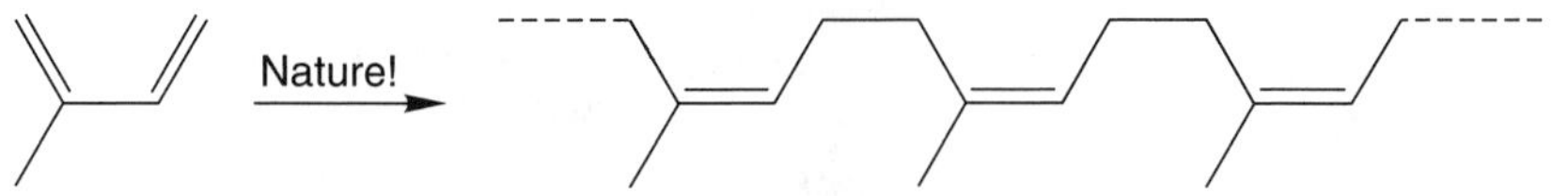

Nature!

Polyisoprene
one form of natural rubber
with all double bonds (Z)

FIGURE 12.78 Natural rubber is a polyisoprene.

There is another, stereoisomeric form of natural rubber, also a polymer of isoprene, called by the wonderful name gutta percha.* What is it? PROBLEM 12.32

Polyisoprene is a limiting case—a molecule composed of a vast number of isoprene units. Many important natural products are constructed of smaller numbers of isoprene units put together in various ways, and the reactions used to combine the diene units can be largely understood in terms of reaction mechanisms we know well. 3-Methyl-3-buten-1-ol is converted first into the corresponding 3-methyl-3-buten-1-ol **pyrophosphate** (isopentenyl pyrophosphate) by reaction with pyrophosphoric acid (Fig. 12.79).

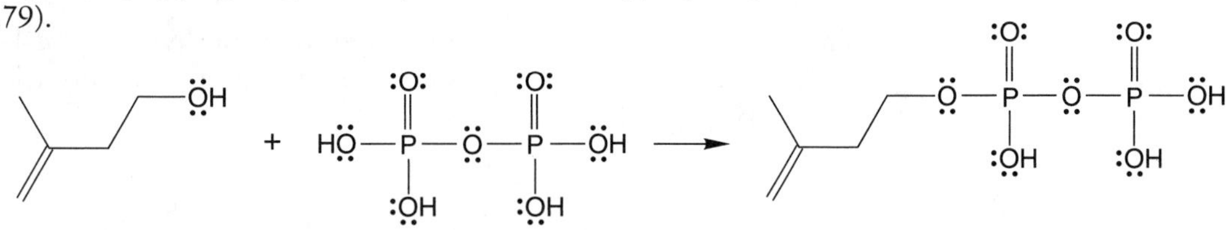

3-Methyl-3-buten-1-ol + **Pyrophosphoric acid** → **3-Methyl-3-buten-1-ol pyrophosphate** (isopentenyl pyrophosphate)

FIGURE 12.79 In the biosynthesis of terpenes, 3-methyl-3-buten-1-ol is first converted into the corresponding pyrophosphate.

Recall the discussion of various ways to change OH, a very poor leaving group, into much better leaving groups (Chapter 7, p. 259). Earlier, we saw examples of protonation to make the leaving group water, or of

*Gutta percha comes from the Malay, *getah percha* (gum of the percha tree).

formation of a sulfonate ester (Fig. 12.80). Formation of the pyrophosphate ester serves exactly the same purpose and is widely used in Nature to generate a good leaving group from a poor one.

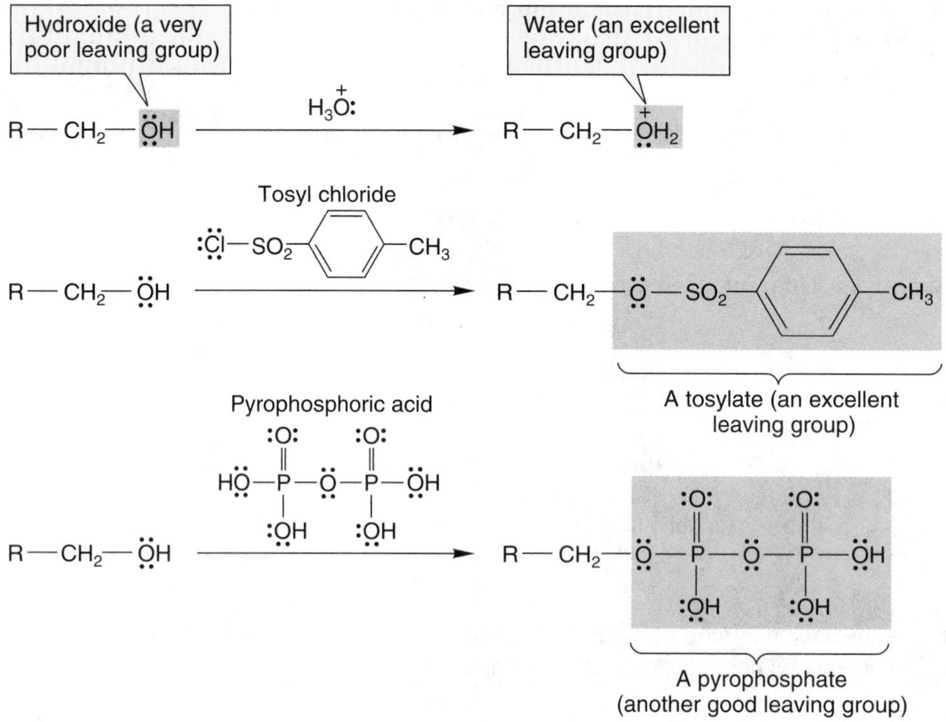

FIGURE 12.80 Three ways of converting OH, a poor leaving group, into a good leaving group.

PROBLEM 12.33 Explain why the pyrophosphate ester in Figure 12.80 is an excellent leaving group.

There is an enzyme that will equilibrate 3-methyl-3-buten-1-ol pyrophosphate with an isomeric pyrophosphate, 3-methyl-2-buten-1-ol pyrophosphate (Fig. 12.81).

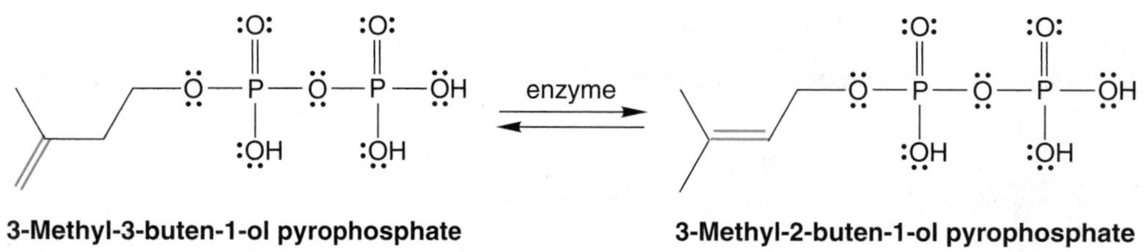

FIGURE 12.81 An enzyme converts 3-methyl-3-buten-1-ol pyrophosphate into a new, allylic pyrophosphate.

This new pyrophosphate now has a good leaving group *in an allylic position.* The combination contributes to an easy ionization to produce an allyl cation that can be attacked by 3-methyl-3-buten-1-ol pyrophosphate to produce the C_{10} compound, geranyl pyrophosphate (Fig. 12.82).

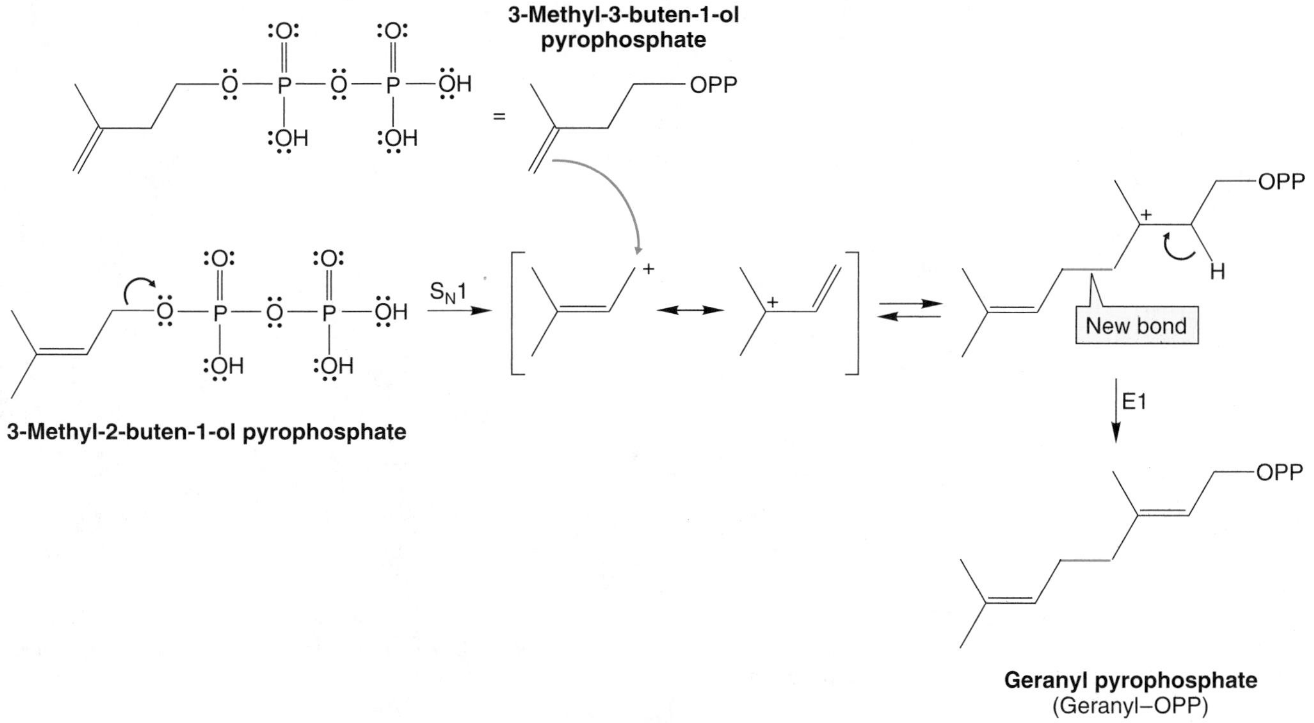

FIGURE 12.82 Loss of the pyrophosphate gives an allylic cation that can be attacked by a molecule of 3-methyl-3-buten-1-ol to give a new cation. Deprotonation gives geranyl pyrophosphate.

Geranyl pyrophosphate itself has a good leaving group in an allylic position and this process can be repeated to produce the C_{15} compound, farnesyl pyrophosphate (Fig. 12.83).

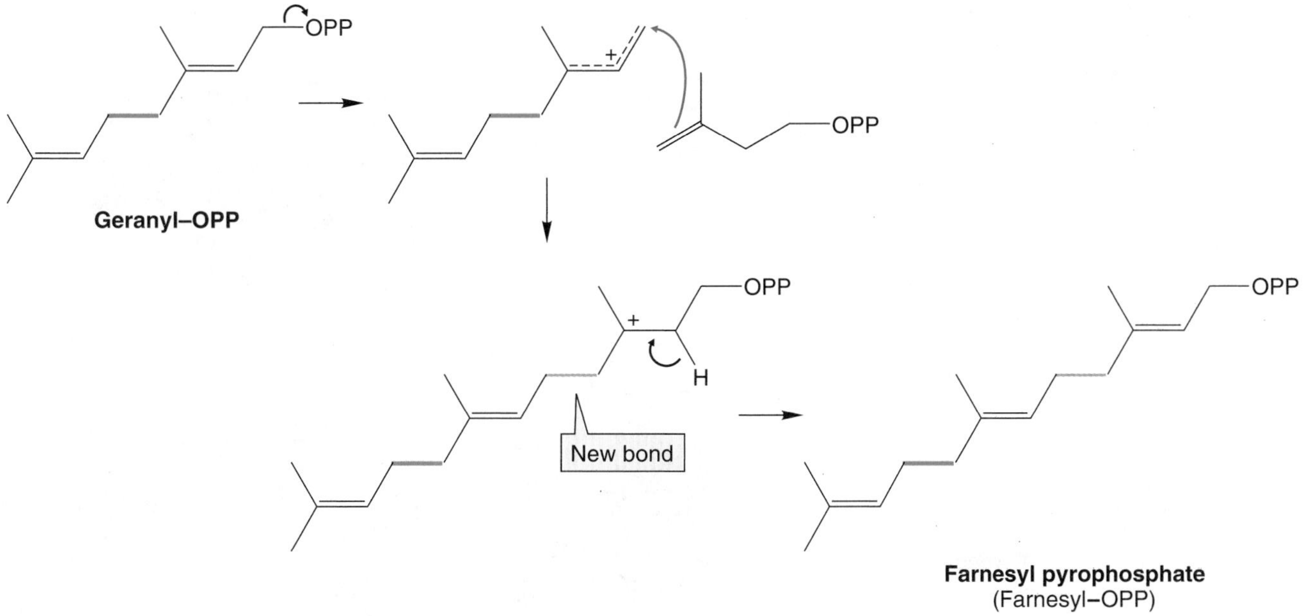

FIGURE 12.83 This process can be repeated to give farnesyl pyrophosphate.

Hydrolysis of geranyl pyrophosphate and farnesyl pyrophosphate produces geraniol and farnesol (Fig. 12.84).

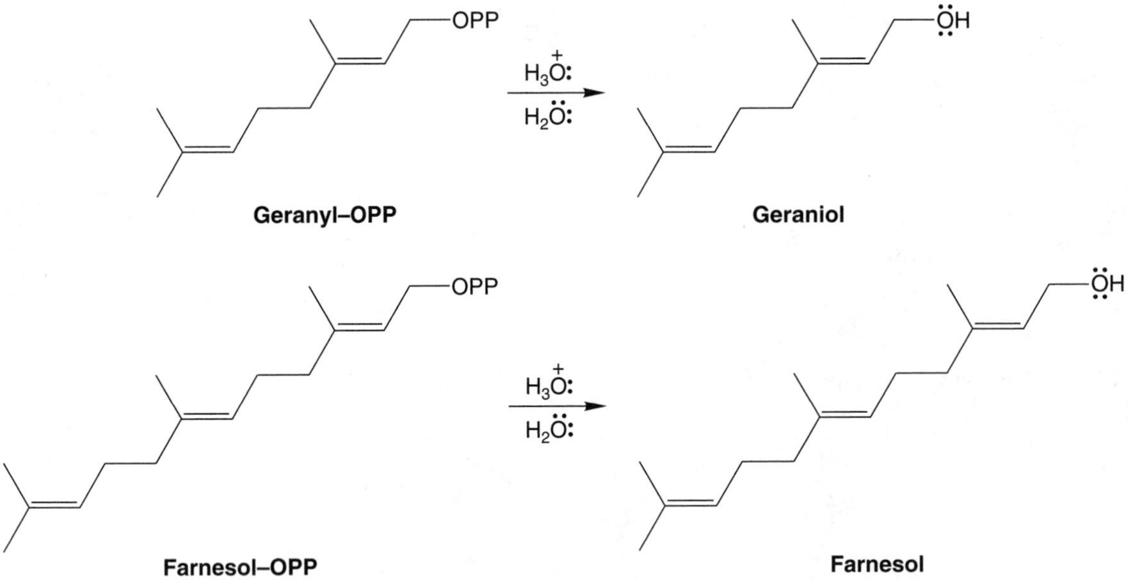

OPP
$H_3\overset{+}{O}$:
$H_2\ddot{O}$:
$\ddot{O}H$
Geranyl–OPP
Geraniol
OPP
$H_3\overset{+}{O}$:
$H_2\ddot{O}$:
$\ddot{O}H$
Farnesol–OPP
Farnesol

FIGURE 12.84 The alcohols geraniol and farnesol are formed by hydrolysis of the related pyrophosphates.

Notice that both geranyl and farnesyl pyrophosphate are formed by head-to-tail combinations of isoprene (Fig. 12.85).

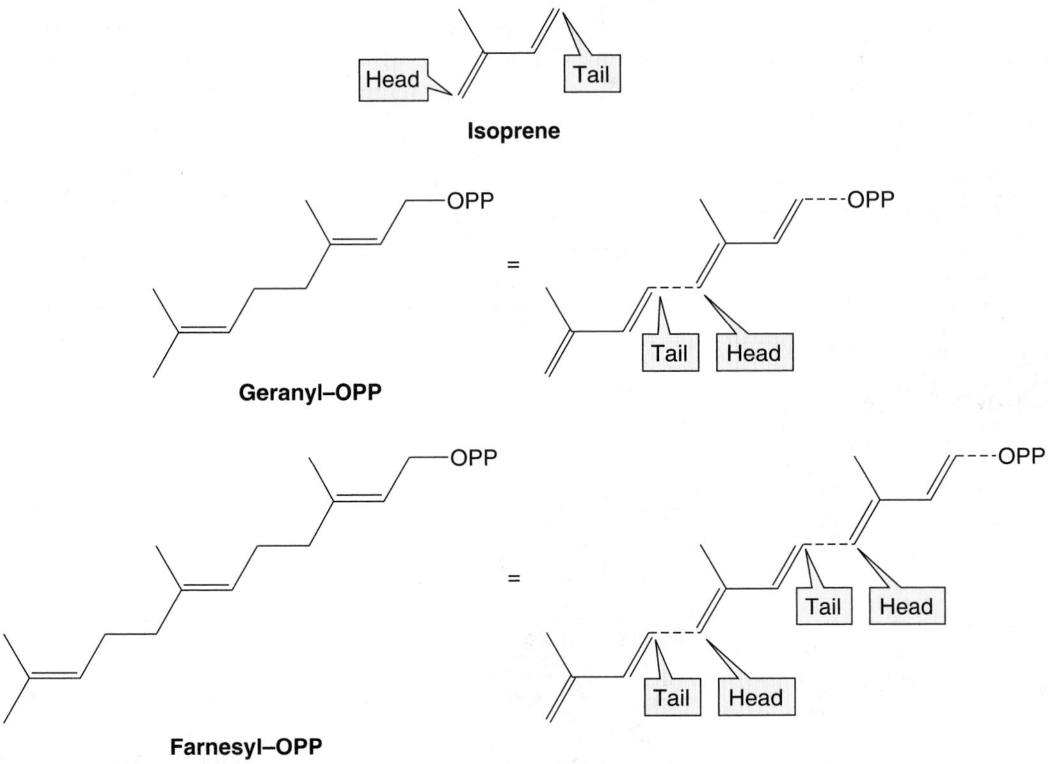

Head
Tail
Isoprene
OPP
=
OPP
Tail
Head
Geranyl–OPP
OPP
=
OPP
Tail
Head
Tail
Head
Farnesyl–OPP

FIGURE 12.85 Geranyl and farnesyl pyrophosphates are formed by head-to-tail attachment of isoprene units.

The limit of this head-to-tail combining is polyisoprene, natural rubber. Figure 12.86 shows the schematic formation of the polymer of isoprene called gutta percha (p. 547).

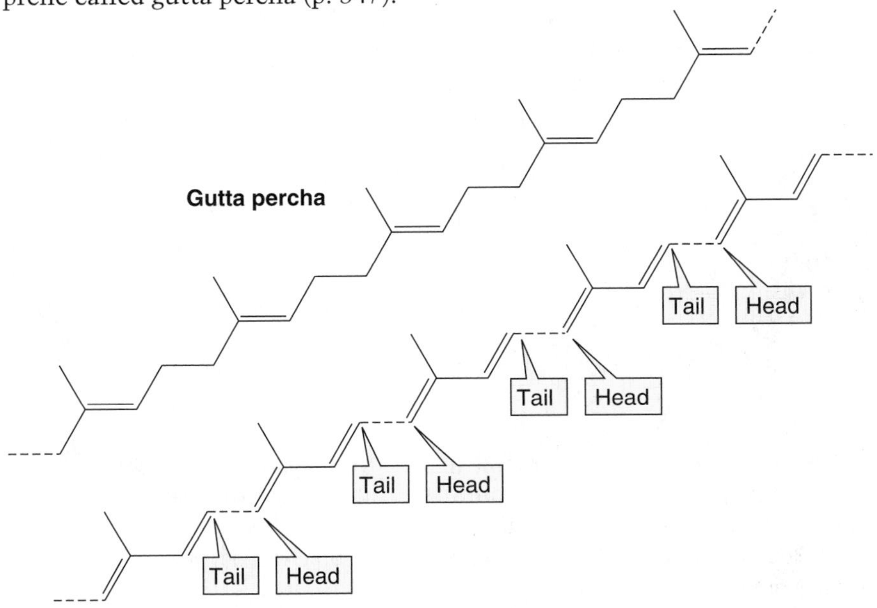

Gutta percha

Tail Head

Tail Head

Tail Head

Tail Head

FIGURE 12.86 Gutta percha, one form of natural rubber, is a head-to-tail polymer of isoprene.

Molecules formed from isoprene units are called **isoprenoids**, or **terpenes**. Two units of isoprene combine to form terpenes, C_{10} compounds such as geraniol. Three isoprene units form sesquiterpenes, C_{15} compounds such as farnesol. Diterpenes are C_{20} compounds, triterpenes are C_{30} compounds, and so on. Myriad terpenoid compounds are known, and most can be traced to isoprene as starting material. Exceptions are known, but most terpenes, including those in Figure 12.87, follow the **isoprene rule**, which dictates the head-to-tail formation described above.

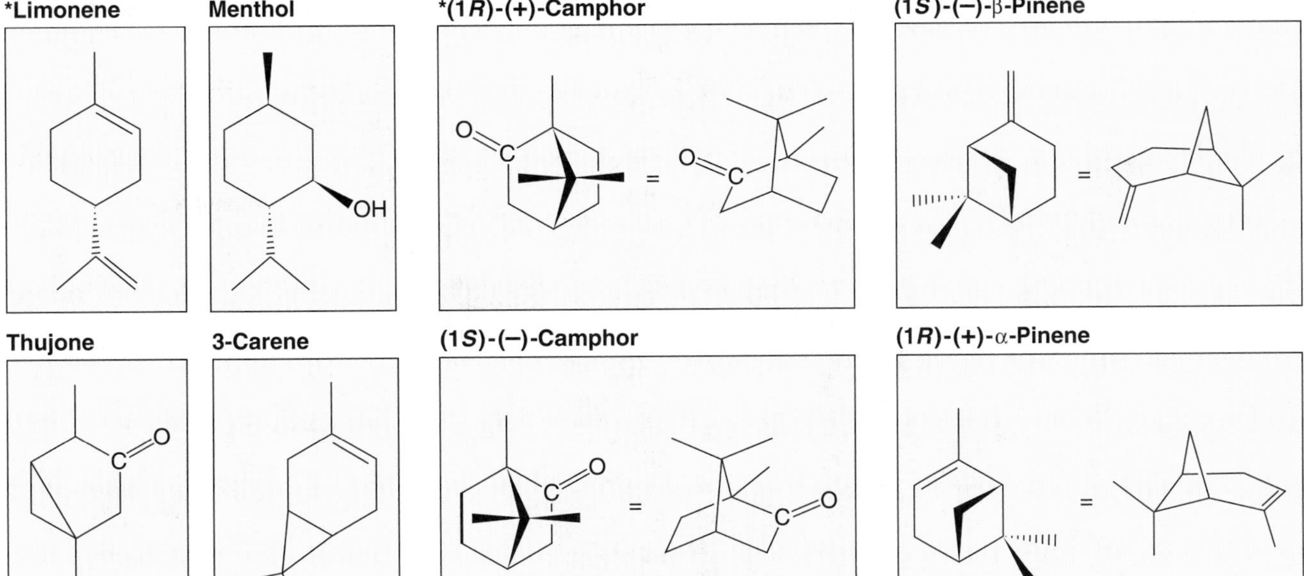

FIGURE 12.87 Some terpenes.

*PROBLEM 12.34 Dissect the molecules marked with an asterisk in Figure 12.87 into isoprene units.

ANSWER The isoprene units can usually be found by searching for the methyl groups first. One of these problems is easy, the other is more complicated.

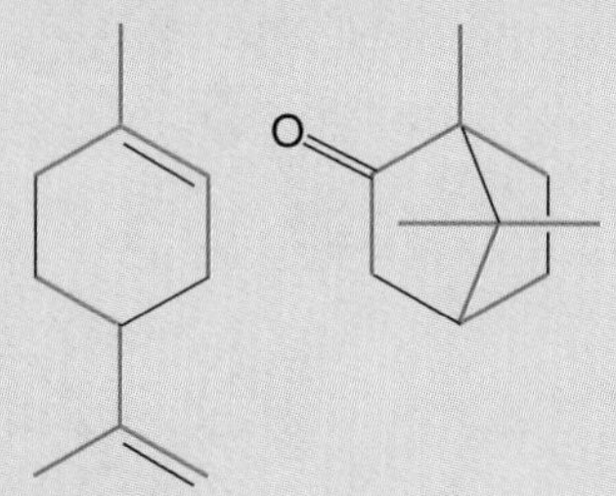

PROBLEM 12.35 Devise a mechanism for converting geranyl pyrophosphate into racemic limonene (Fig. 12.87). Such processes are usually enzyme mediated in Nature, and for good reason. When you try this problem you should encounter a road-block. It can't really be done without a cis/trans isomerization, very likely carried out in Nature by the enzyme.

12.17 STEROID BIOSYNTHESIS

There is an extraordinarily important class of molecules called steroids, which are formed from isoprene units. Two units of farnesyl pyrophosphate can be reductively coupled to give the symmetrical triterpene called squalene (Fig. 12.88).

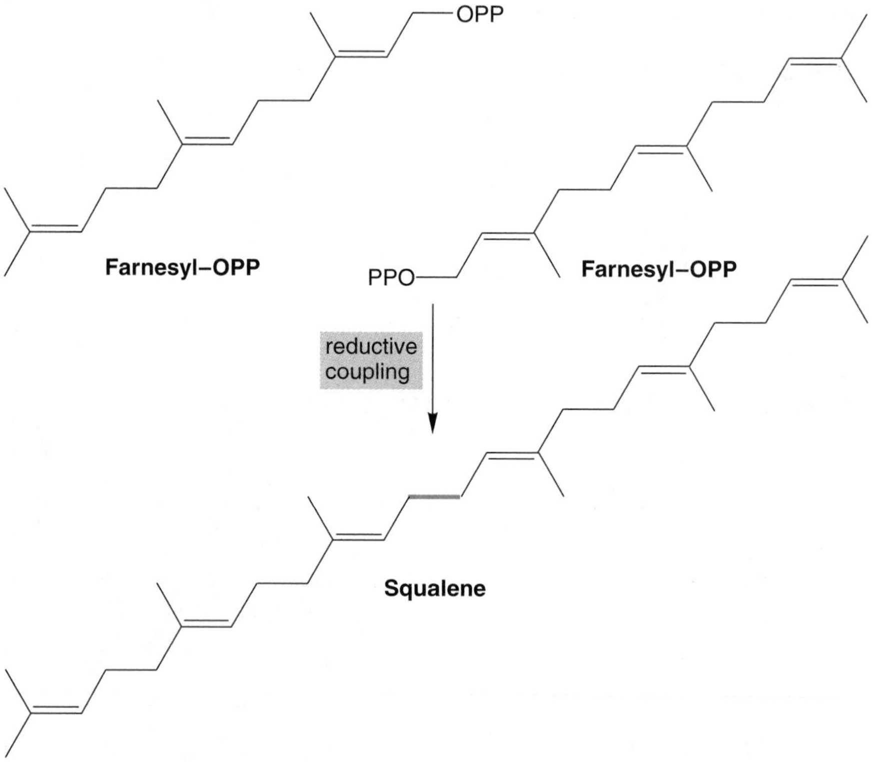

OPP

Farnesyl–OPP

PPO

Farnesyl–OPP

reductive coupling

Squalene

FIGURE 12.88 Reductive coupling of two molecules of farnesyl pyrophosphate gives squalene, a triterpene.

Squalene can be epoxidized enzymatically to give squalene oxide, which is written in a suggestive way in Figure 12.89. Protonation of

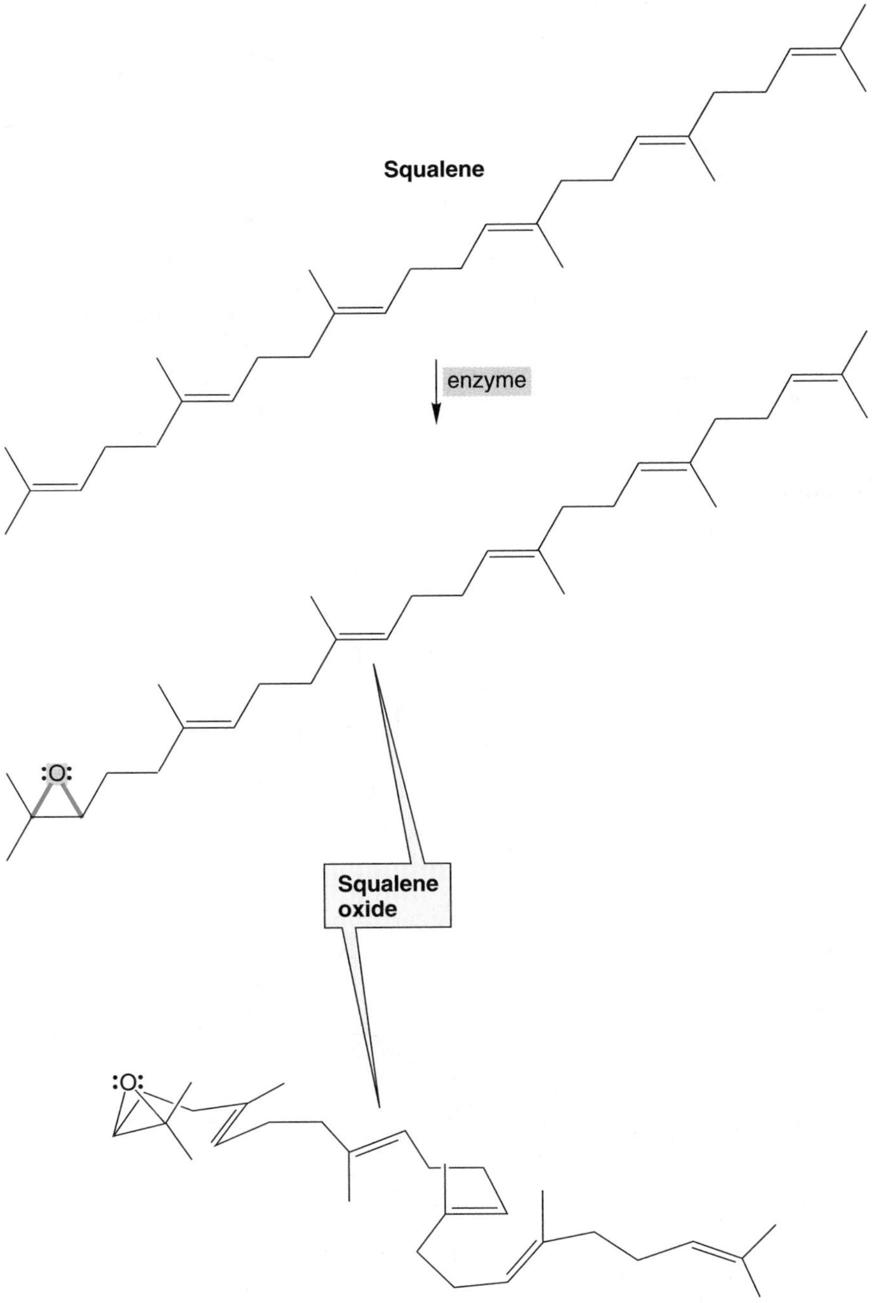

Squalene

enzyme

:O:

Squalene oxide

:O:

FIGURE 12.89 Enzymatic epoxidation of squalene gives squalene oxide.

squalene epoxide leads to a relatively stable tertiary carbocation that is attacked by the proximate nucleophilic double bond to give a new cation. This cation is attacked in turn by another nearby alkene. The process continues until the molecule is sewn up in the four-ring **steroid** skeleton as shown in Figure 12.90.

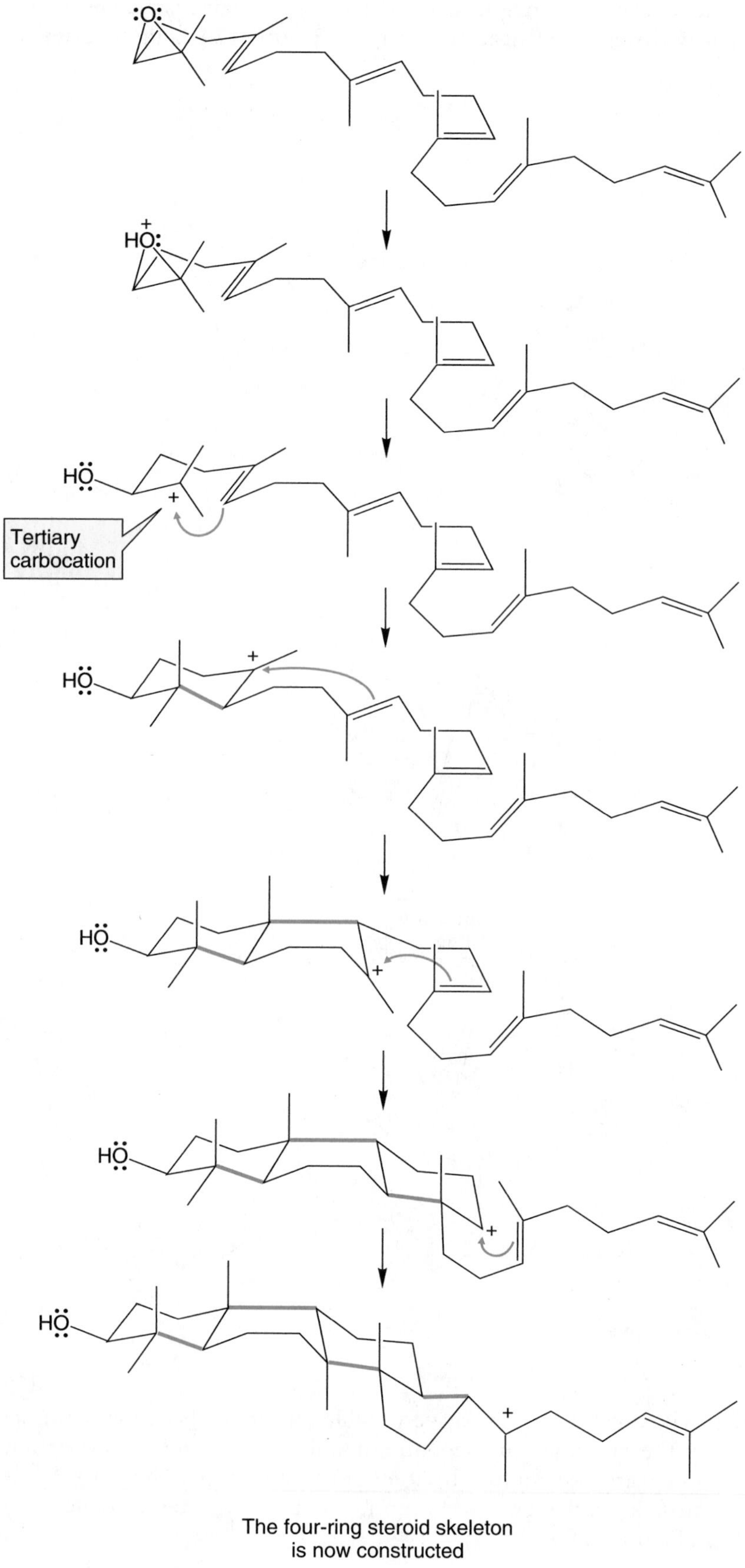

:O:

HȮ:

HÖ

Tertiary carbocation

HÖ

HÖ

HÖ

HÖ

The four-ring steroid skeleton is now constructed

FIGURE 12.90 Acid-catalyzed opening of the oxirane gives a tertiary cation that can undergo a series of ring closings to give a new, tertiary carbocation in which the four-ring steroid skeleton has been constructed.

The series of rearrangements summarized in Figure 12.91 ensues and a proton is finally lost to give the steroid known as lanosterol.

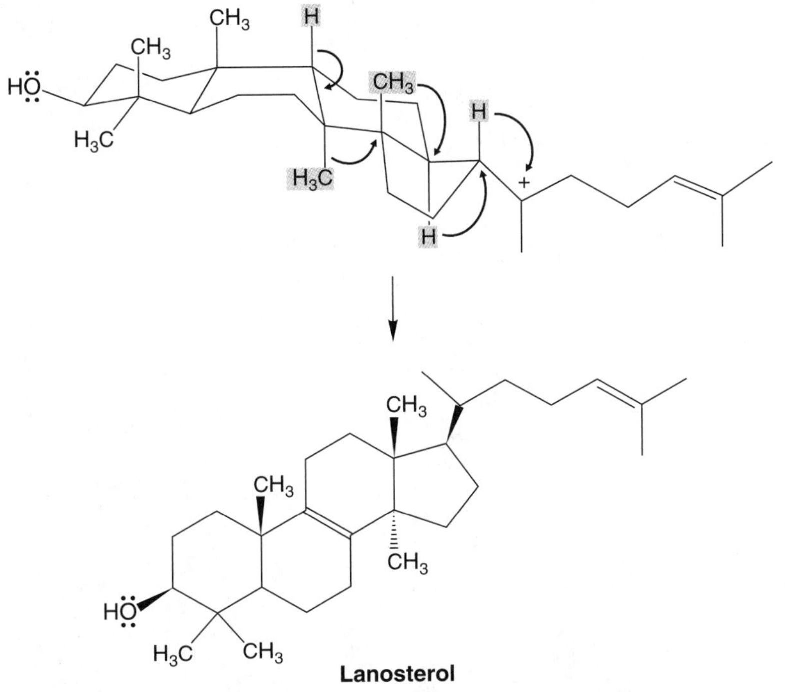

FIGURE 12.91 A series of skeletal rearrangements, followed by removal of a proton, gives the steroid lanosterol.

*PROBLEM 12.36

Starting at the right-hand end of the molecule write a stepwise mechanism for the conversion of the ion in Figure 12.91 to lanosterol. Take the several rearrangements one at a time and draw out each reaction intermediate.

ANSWER

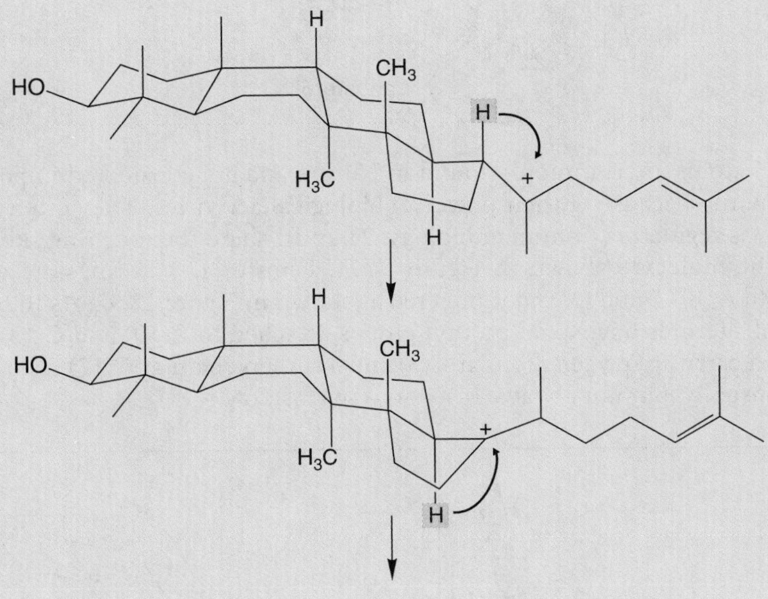

ANSWER (CONTINUED)

Steroids are commonly found in Nature and are considered important primarily because of their powerful biological activities. They often function as regulators of human biology. They all share the four-ring skeleton of the molecule shown in Figure 12.91, lanosterol. The rings are designated A, B, C, and D, and numbered as shown in Figure 12.92. As in lanosterol, steroids have axial methyl groups attached to C(10) and C(13), and often carry an oxygen atom at C(3) and a carbon chain at C(17). The rings are usually, but not always, fused in trans fashion.

FIGURE 12.92 The steroid ring system.

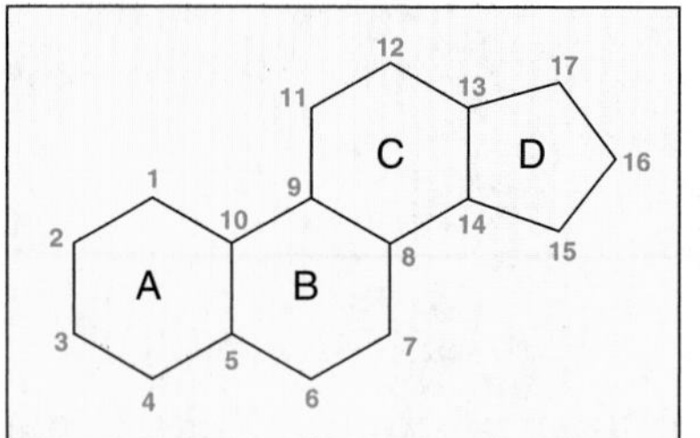

Lanosterol

HO

FIGURE 12.92 (CONTINUED)

Steroids were, and still can be, isolated from natural sources. Such important biomolecules quickly became the subject of attempts at chemical modification and total synthesis in the hopes of uncovering routes to new molecules of greater and different bioactivities. Many naturally occurring steroids have been constructed in the laboratory and numerous "unnatural" steroids have been created as well. Figure 12.93 shows a few important steroids.

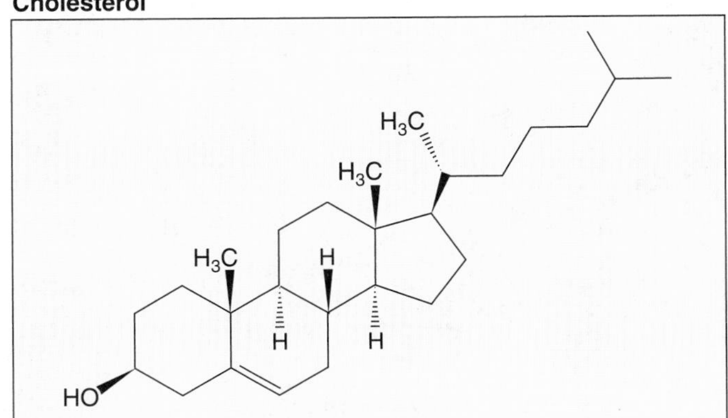

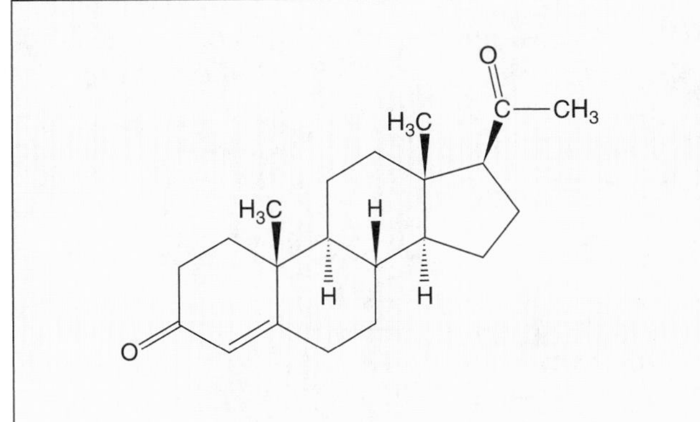

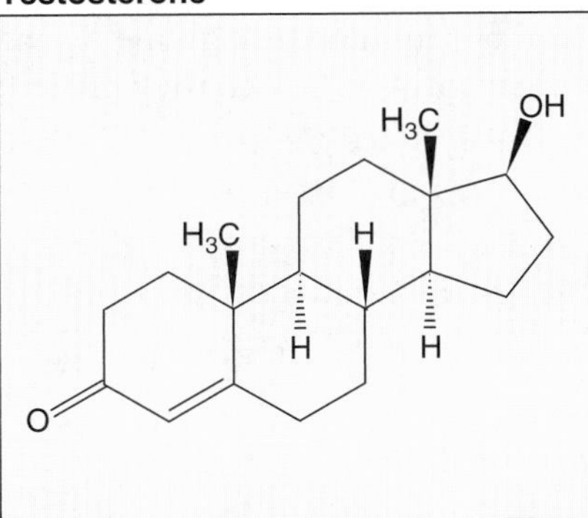

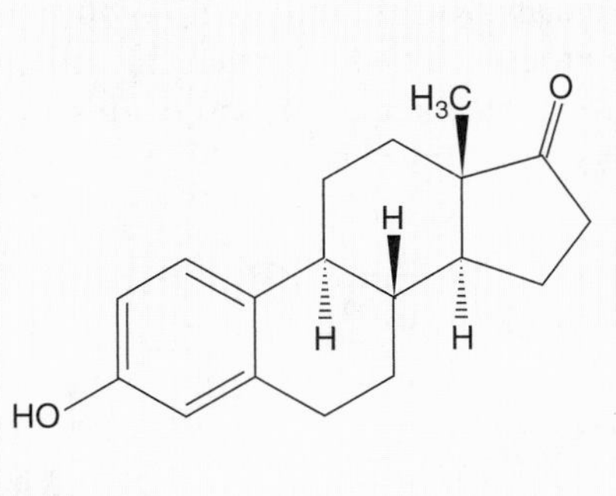

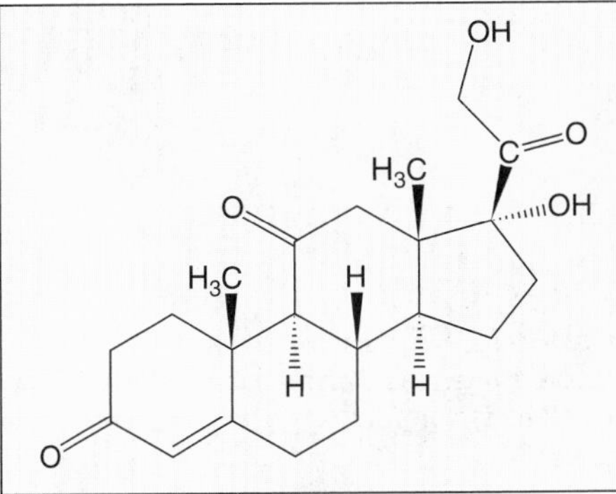

FIGURE 12.93 Some naturally occurring steroids.

Whole industries have been derived from this work. Most famous are the steroids used in our attempts to control human fertility. "The pill" contains synthetic steroids that mimic the action of the body's natural steroids, the estrogens and progesterone, which are involved in regulation of egg production and ovulation. The body is essentially tricked into behaving as if it were pregnant, so that neither egg production nor development can take place. It should be no surprise that it has been difficult to use such powerfully biologically active molecules in a completely specific way. The goal of this work is to interfere specifically with fertility without doing *anything* else. Considering the activity of the molecules involved and the complexity of the mechanisms we are trying to regulate, it is remarkable that we have been so successful. This control is not totally without cost, however. In fertility, those costs appear as the side effects of the pill, which occur in a few women. Perhaps the most difficult of human problems is to find the balance between benefits and risks in cases such as this. Although such problems have been with us throughout our history, questions of biology are particularly vexing, it seems. It is easier to judge that the benefits of the automobile outweigh the societal costs than to make a decision regarding the steroids used in the pill. This problem is certain to become much worse as our ability to manipulate complicated molecules continues to grow.

12.18 SOMETHING MORE: THE S_N2' REACTION

Earlier (Section 12.12, p. 533), we saw that allyl halides underwent the S_N2 reaction faster than saturated alkyl halides. There is another substitution pathway open to allylic halides called the **S_N2' reaction** (Fig. 12.94).

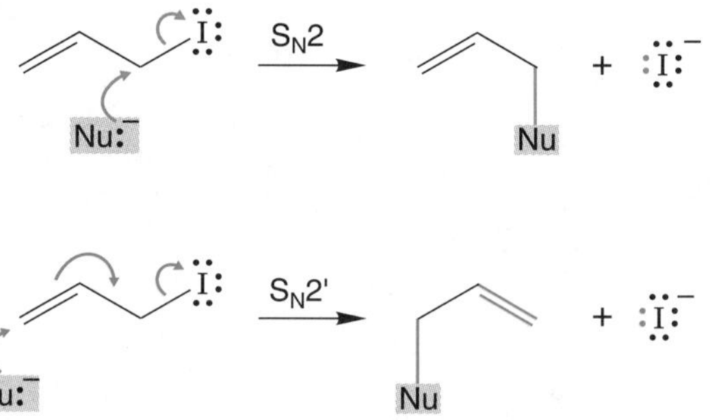

Nu:⁻ + allyl iodide —S_N2→ allyl–Nu + :I:⁻

Nu:⁻ + allyl iodide —S_N2'→ Nu–CH₂–CH=CH₂ + :I:⁻

FIGURE 12.94 The S_N2 reaction compared with the S_N2' reaction (Nu = nucleophile).

The S_N2' reaction is indistinguishable from the S_N2 reaction in the example shown in Figure 12.94, but a more substituted allylic halide reveals the process. The S_N2' reaction competes especially well with the normally faster S_N2 reaction in molecules in which the S_N2 process is retarded by steric blocking (Fig. 12.95).

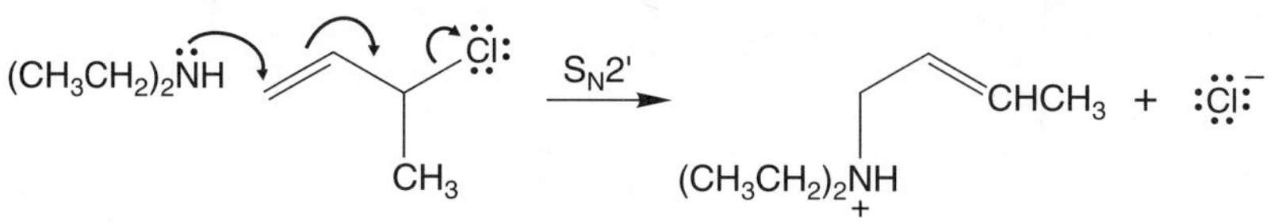

$(CH_3CH_2)_2\ddot{N}H$ + allylic chloride (CH_3) —S_N2'→ $(CH_3CH_2)_2\overset{+}{N}H$–CH₂–CH=CHCH₃ + :Cl:⁻

FIGURE 12.95 The S_N2' reaction is favored when the S_N2 reaction is slowed by steric factors.

PROBLEM 12.37 Devise an experiment using isotopic labeling that would determine if the S_N2' reaction is occurring in the reaction in Figure 12.94.

The stereochemistry of the S_N2 reaction has been well worked out and has been described in detail in Chapter 7 (p. 244). The S_N2' reaction has also been investigated and Problem 12.38 allows you to put yourself in the place of R. M. Magid (b. 1938) and his co-workers. They used the following experiment to determine the stereochemistry of the S_N2' reaction in the simple acyclic molecule shown (Problem 12.38).

PROBLEM 12.38

Use the following data to determine whether the nucleophile approaches the double bond from the same side as the leaving group or from the opposite side (Fig. 12.96):

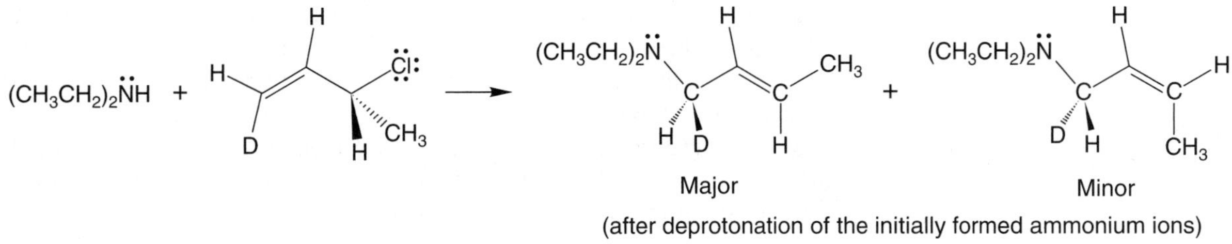

$(CH_3CH_2)_2\ddot{N}H$ + [allylic chloride] → Major + Minor

Major

Minor

(after deprotonation of the initially formed ammonium ions)

FIGURE 12.96

12.19 SUMMARY

NEW CONCEPTS

In this chapter, the chemical and physical consequences of the overlap of 2*p* orbitals (conjugation) are explored. In structural terms, conjugation appears as a short C(2)–C(3) bond and in a small barrier to rotation around this bond (Fig. 12.20).

Conjugation leads to the formation of allyl systems in addition reactions to dienes, and is a requirement for the Diels–Alder reaction (Fig. 12.97).

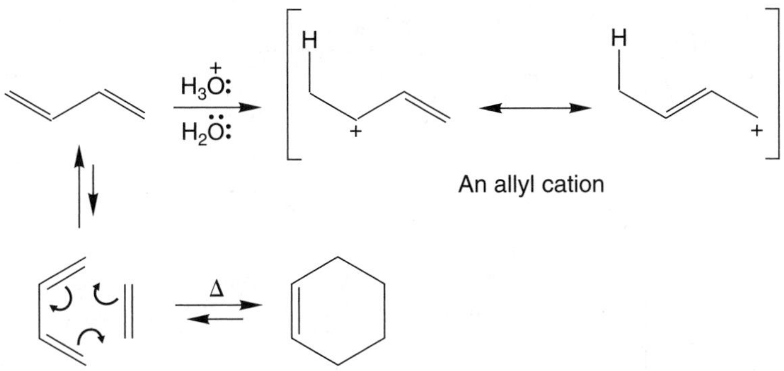

$H_3\ddot{O}^+$: / $H_2\ddot{O}$:

An allyl cation

Δ

FIGURE 12.97 Two typical reactions of conjugated dienes: addition to give an allyl cation and the Diels–Alder reaction.

An important fundamental concept in this chapter involves the difference between thermodynamic and kinetic control of a reaction. When the product distribution depends on the relative energies of the products, the reaction is under thermodynamic control. Thermodynamic control occurs when there is sufficient energy available so that the activation barriers between starting materials and products can be crossed in both directions. By contrast, if products are not freely equilibrating with starting material;

if there is not enough energy to carry out the reverse reactions, then it will be the energies of the transition states that control the distribution of products. This results in kinetic control.

In UV/vis spectroscopy, absorption of light results in the promotion of an electron from the HOMO to the LUMO.

REACTIONS, MECHANISMS, AND TOOLS

The base-catalyzed isomerization of disubstituted to terminal acetylenes takes place through allene intermediates (Fig. 12.14). Addition reactions to 1,3-dienes can give products of both 1,2- and 1,4-additions (Fig. 12.42). The Diels–Alder reaction forms cyclohexenes and cyclohexadienes from the thermal reaction of 1,3-dienes and alkenes or alkynes. Stereochemical labeling experiments show that both new σ bonds are formed simultaneously (Figs. 12.68 and 12.69). In the S_N2' reaction, a nucleophile displaces a leaving group through attack at an adjacent double bond (Fig. 12.95). Many natural products (terpenes) are produced through the combination of various numbers of isoprene units.

SYNTHESIS

The most important synthetic methods described here are a route to allylic halides through additions to conjugated dienes and the Diels–Alder synthesis of cyclohexenes and cyclohexadienes. Be especially careful about the Diels–Alder reaction: There are many structural possibilities for the products depending on the complexity of the diene and dieneophile used as starting materials. The S_N2' reaction can be another source of

1. Allylic halides

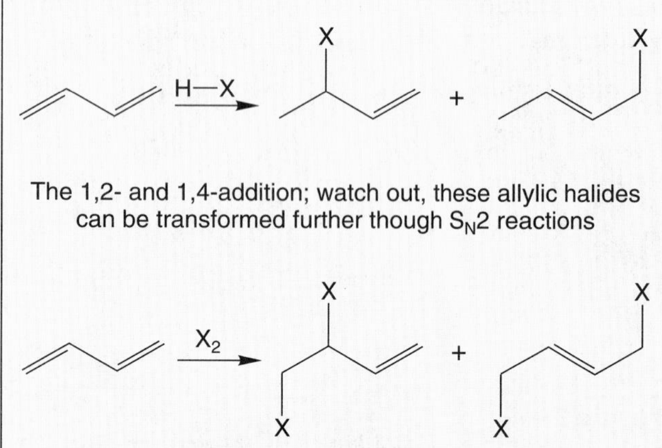

The 1,2- and 1,4-addition; watch out, these allylic halides can be transformed further though S_N2 reactions

Both 1,2- and 1,4-addition takes place

2. Other allylic compounds

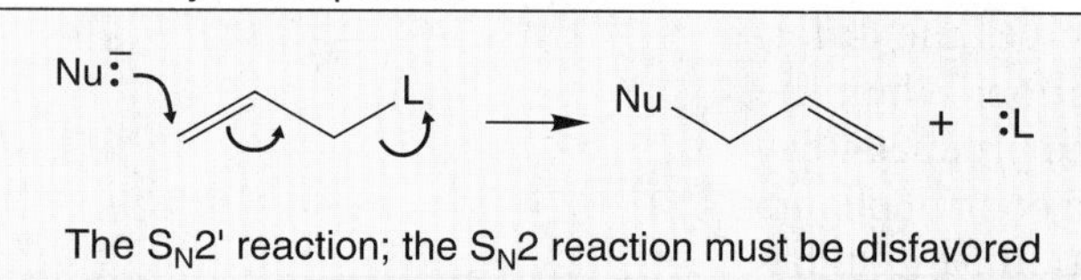

The S_N2' reaction; the S_N2 reaction must be disfavored

3. Cyclohexenes

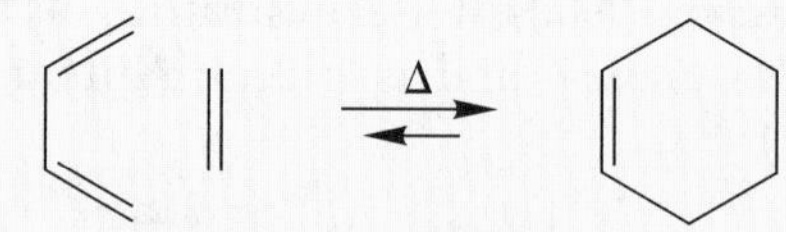

Diels–Alder reaction; watch out for structural variation; the reverse reaction is a synthesis of 1,3-dienes and alkenes from cyclohexenes

4. Cyclohexadienes

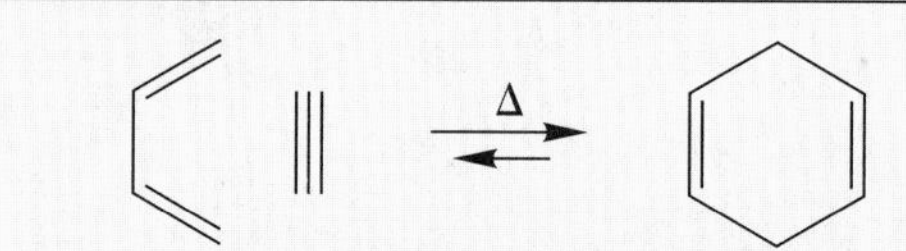

The dienophile is an alkyne; watch out for structural variation and the reverse reaction

5. Dienes

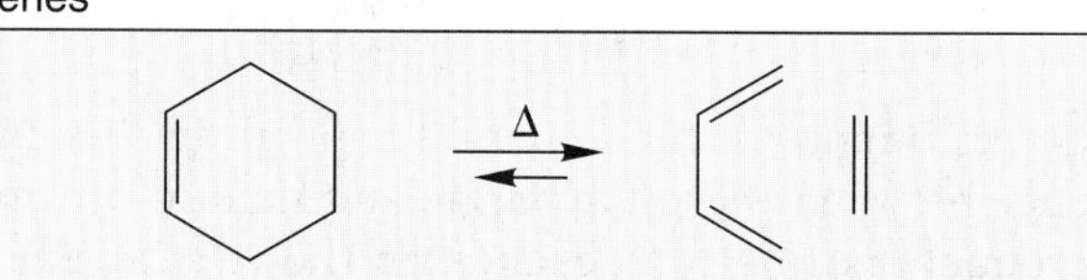

The reverse Diels–Alder reaction; watch out for structural variation; this picture gives only the simplest version

FIGURE 12.98 A summary of the new synthetic reactions of Chapter 12.

allylic compounds, but its synthetic utility is limited to those cases in which the generally more favorable S_N2 reaction is unfavorable (Fig. 12.98).

COMMON ERRORS

Be sure that it is clear how the structures of allenes and cumulenes can be derived from an analysis of the hybridizations of the atoms involved. Although the structures all *look* flat on a two-dimensional page, they are not all planar by any means.

Although the Diels–Alder reaction can be quickly drawn using only three little arrows, and although it is not difficult to see that all cyclohexenes can be constructed on paper from the reaction of a diene and a dienophile, very complex structures can be built very quickly using Diels–Alder reactions. The simplicity can quickly disappear in the wealth of detail! Try to see to the essence of the problem; locate the cyclohexene, and then take it apart to find the building block diene and dienophile.

The reverse Diels–Alder reaction is even harder to find. Not only can cyclohexenes be constructed through Diels–Alder reactions, but they can serve as sources of dienes and dienophiles as well. It is not simple to find reverse Diels–Alder reactions!

It is time to mention again another very common conceptual error that often traps students. Look, for example, at Figure 12.40. Protonation of 1,3-butadiene leads to an allyl cation in which two different carbons share the positive charge. Addition of chloride at these two positions leads to the two products. It is very easy to think of the resonance forms of the allyl cation as each having a separate existence; of chloride adding to one resonance form to give one product and to the other resonance form to give the other product. This notion is absolutely wrong! The allyl cation is a single species with two partially positive carbon atoms. Chloride adds to the allyl cation, not to one resonance form or the other. The figure tries to help out by showing a summary structure, but this is not always done, and you must be careful (Fig. 12.40).

12.20 KEY TERMS

Allene A 1,2-diene.

Cumulene Any molecule containing at least three consecutive double bonds, $R_2C{=}C{=}C{=}CR_2$.

Conjugated (double bonds) Double bonds in a 1,3-relationship are conjugated.

s-cis The less stable, coiled form of a 1,3-diene.

Diels–Alder reaction The concerted reaction of an alkene or alkyne with a 1,3-diene to form a six-membered ring.

Diene Any molecule containing two double bonds.

Electronic spectroscopy The measurement of the absorption of energy when electromagnetic radiation of the proper energy is provided. An electron is promoted from the HOMO to the LUMO.

endo Aimed "inside" the cage in a bicyclic molecule. In a Diels–Alder reaction, the endo product generally has the substituents aimed toward the newly produced double bond.

exo Aimed "outside" the cage in a bicyclic molecule. In a Diels–Alder reaction, the exo product generally has the substituents aimed away from the newly produced double bond.

Extinction coefficient The proportionality constant ε in Beer's law, $A = \log I_0/I = \varepsilon lc$.

Isoprene 2-Methyl-1,3-butadiene.

Isoprene rule Most terpenes are composed of isoprene units combined in "head-to-tail" fashion.

Isoprenoids Compounds whose carbon skeleton is composed of isoprene units. Terpenes.

Ketenes Compounds of the structure: $R_2C{=}C{=}O$.

Kinetic control In a reaction under kinetic control, the product distribution depends on the relative energies of the transition states leading to products, not on the energies of the products themselves.

Pyrophosphate The group

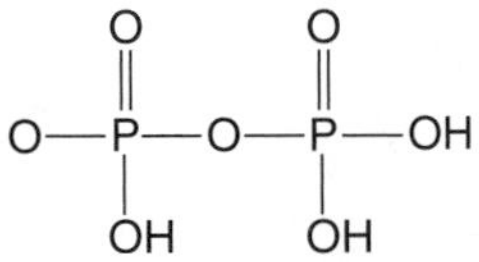

O—P(=O)(OH)—O—P(=O)(OH)—OH

Nature uses pyrophosphates as leaving groups.

Spectroscopy The study of the interactions between electromagnetic radiation and atoms and molecules.

S_N2' Reaction Nucleophilic displacement of a leaving group through attack at an allylic position.

Nu:⁻ + CH₂=CH–CH₂–L ⟶ Nu–CH₂–CH=CH₂ + L:⁻

Steroid A class of four-ring compounds, always containing three six-membered rings and one five-membered ring. The ring system can be substituted in many ways, but the rings are always connected in the following fashion:

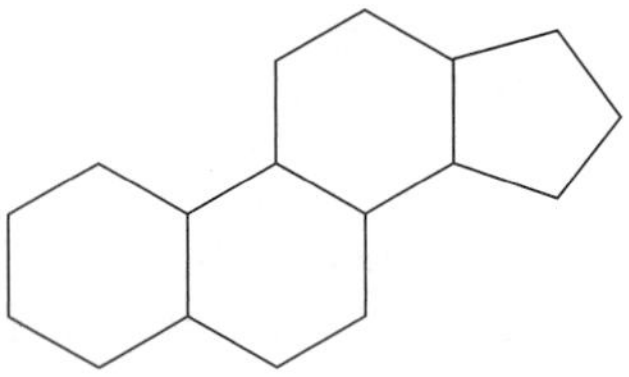

Terpenes Compounds whose carbon skeleton is composed of isoprene units.

Thermodynamic control In a reaction under thermodynamic control, the product distribution depends on the energy differences between the products.

s-trans The more stable, extended form of a 1,3-diene,

Ultraviolet (UV) spectroscopy Electronic spectroscopy using light of wavelength 200–400 nm.

Visible (vis) spectroscopy Electronic spectroscopy using light of wavelength 400–800 nm.

12.21 ADDITIONAL PROBLEMS

PROBLEM **12.39** Use the rules given in Table 12.2 to calculate λ_{max} for the following steroids:

(a) H₃C, H₃C, C_8H_{17}, HO

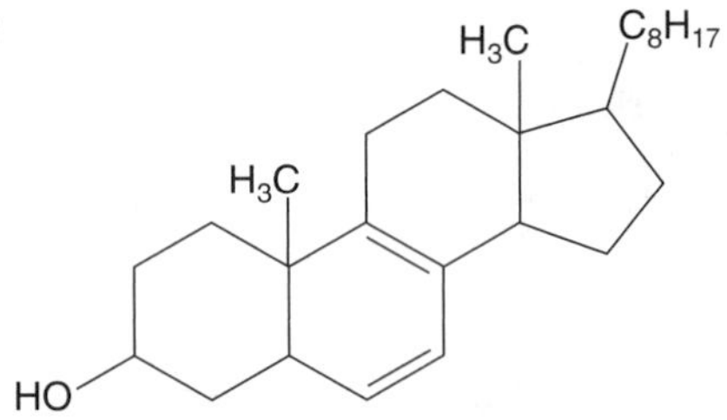

(b) H₃C, H₃C, C_9H_{19}, AcO

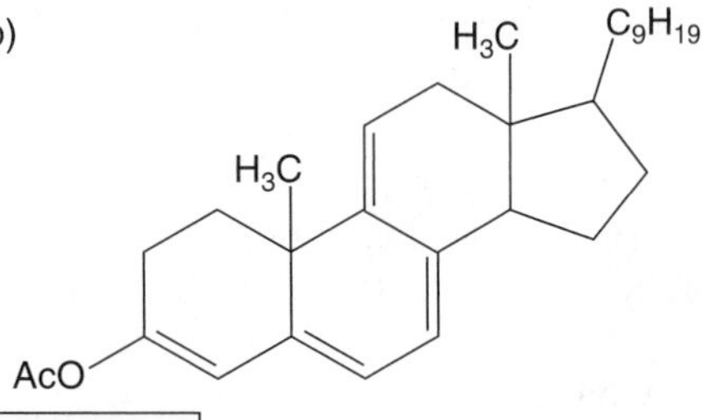

Ac = CH_3CO

PROBLEM **12.40** Write resonance forms for the following ions:

(a)

(b)

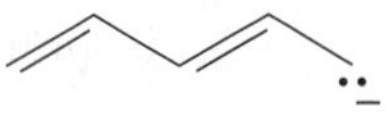

(c)

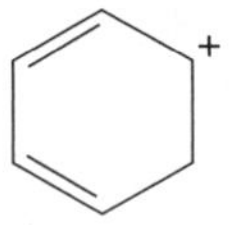

(d)

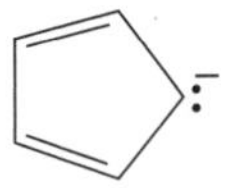

PROBLEM **12.41** Are any of the following compounds chiral? Explain with good drawings.

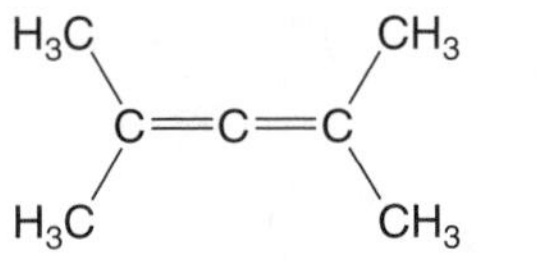

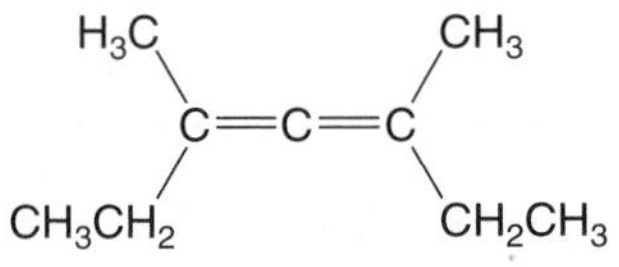

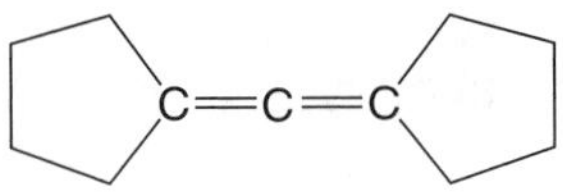

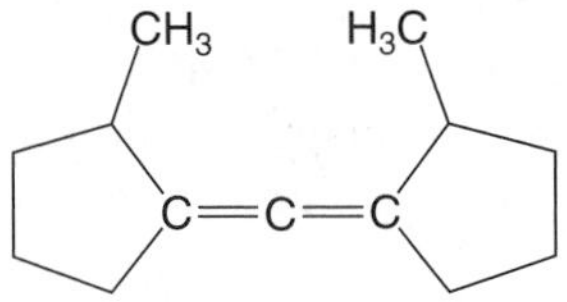

$(H_3C)_2C{=}C{=}C(CH_3)_2$

$(H_3C)(CH_3CH_2)C{=}C{=}C(CH_3)(CH_2CH_3)$

C=C=C (cyclopentylidene)

CH₃, H₃C, C=C=C

PROBLEM **12.42** Show the main products of the reactions of *trans,trans*-2,4-hexadiene under the following conditions. There may be more than one product in some cases.

(a) HCl (b) H_2/Pd (c) Cl_2/CCl_4

(d) Δ, $H_3COOC{-}C{\equiv}C{-}COOCH_3$

PROBLEM **12.43** Suggest a plausible arrow formalism mechanism to account for the products in the following reaction:

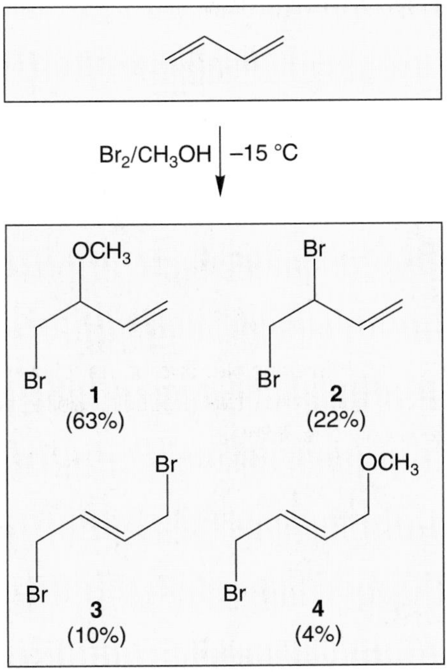

Br_2/CH_3OH | −15 °C

OCH_3 Br **1** (63%)

Br Br **2** (22%)

Br Br **3** (10%)

OCH_3 Br **4** (4%)

PROBLEM **12.44** Predict the products of the reaction of hydrogen chloride with isoprene under kinetic and thermodynamic control.

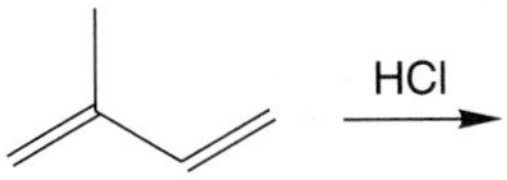

HCl

Isoprene

PROBLEM **12.45** Devise syntheses for the following molecules. You may use 1-butyne and ethyl iodide as your only sources of carbon, as well as any inorganic reagents you need.

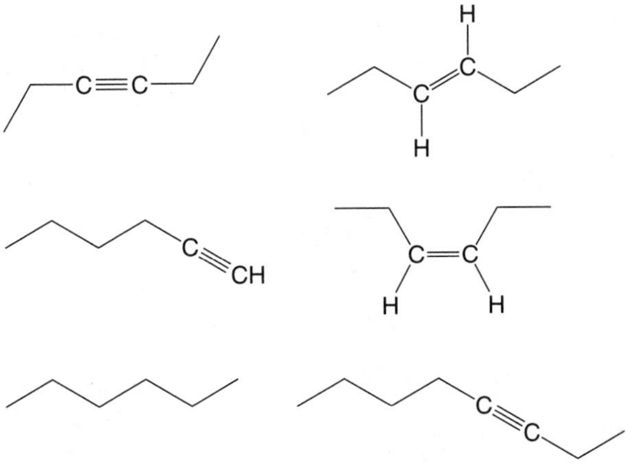

PROBLEM **12.46** Conversion of 2-butyne into 1-butyne is easy. The reverse reaction is more difficult. Suggest routes for doing both these conversions. Mechanisms are not required.

$$H_3C—C{\equiv}C—CH_3 \xrightarrow{?} CH_3CH_2—C{\equiv}C—H$$

$$CH_3CH_2—C{\equiv}C—H \xrightarrow{?} H_3C—C{\equiv}C—CH_3$$

PROBLEM **12.47** Predict the products of the following Diels–Alder reactions:

(a) + $COOCH_3$ $\xrightarrow[0–25\ °C]{ether}$

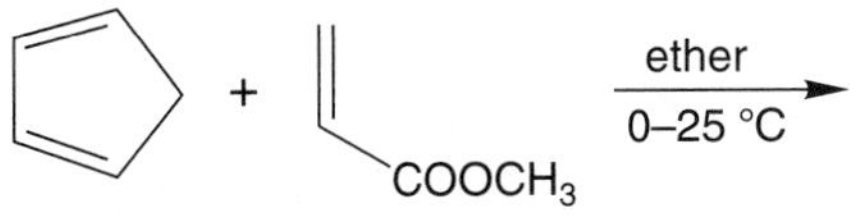

(b) + $COOCH_3$ $COOCH_3$ $\xrightarrow[150\ °C]{toluene}$

(c) + O O $\xrightarrow[0\ °C]{alcohol}$

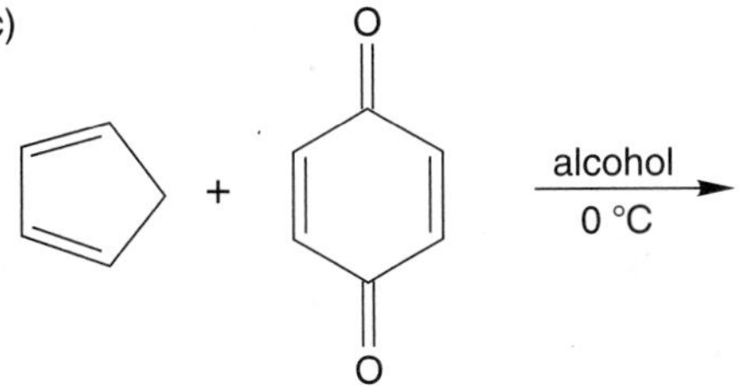

(d) Ph Ph Ph Ph =O + $COOCH_3$ $COOCH_3$ Ph =

dichlorobenzene | Δ

$C_{10}H_6O_4Ph_4$

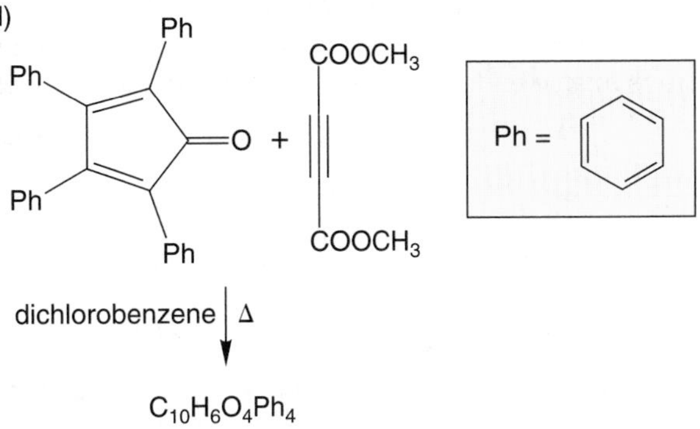

(e) CH_3 C_8H_{17} H_3C + O O O $\xrightarrow[Δ,10\ h]{toluene}$

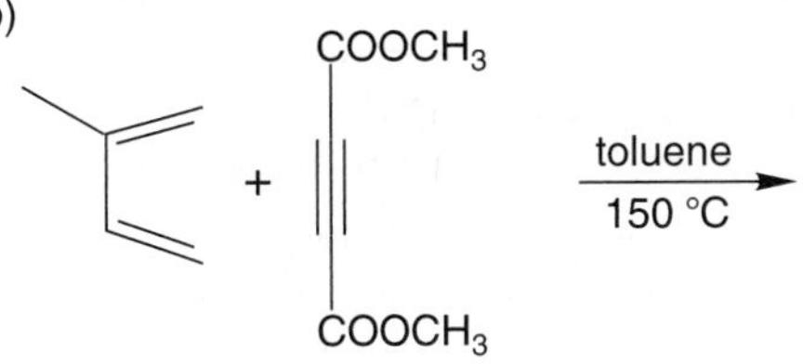

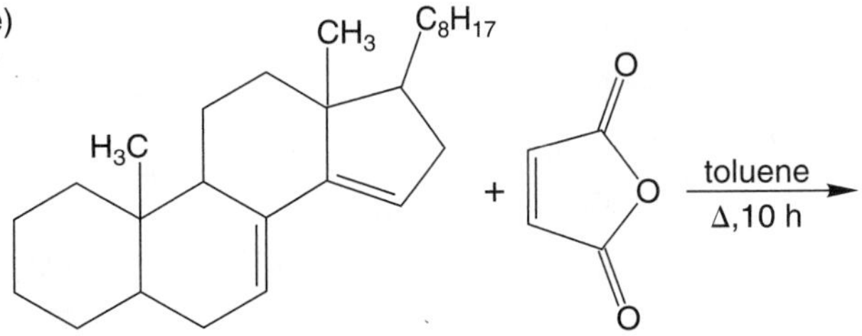

(f)

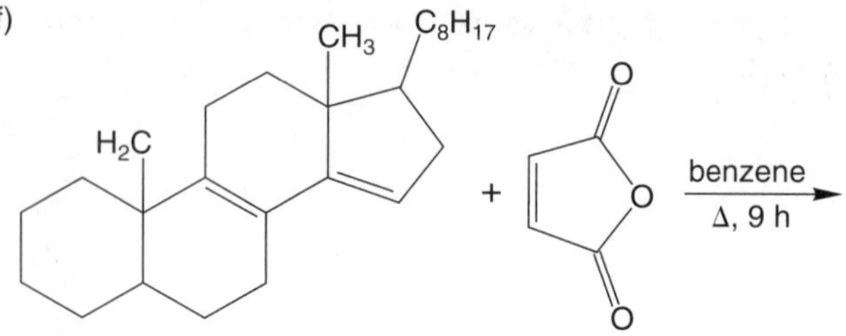

CH_3 C_8H_{17} H_2C + benzene, Δ, 9 h →

(g) $COOCH_3$ ... + OCH_2CH_3 —dioxane, 25 °C→ $C_8H_8N_2O_4$

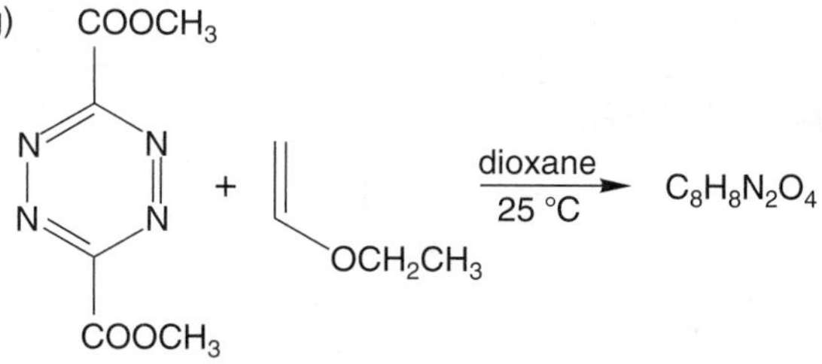

PROBLEM 12.48 In Problem 12.47(c), the initially formed adduct can be closed in a photochemical reaction to give a complex cage structure (shown below). Write an arrow formalism for this process and explain how the observation of this reaction was helpful in determining the stereochemistry (exo or endo) of the original Diels–Alder adduct.

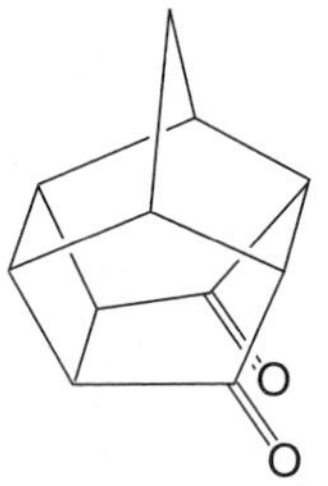

PROBLEM 12.49 Use a carefully constructed Energy versus Reaction progress diagram to rationalize the following data. Be sure to explain why the product at low temperature is mainly endo and why the product at high temperature is mainly exo.

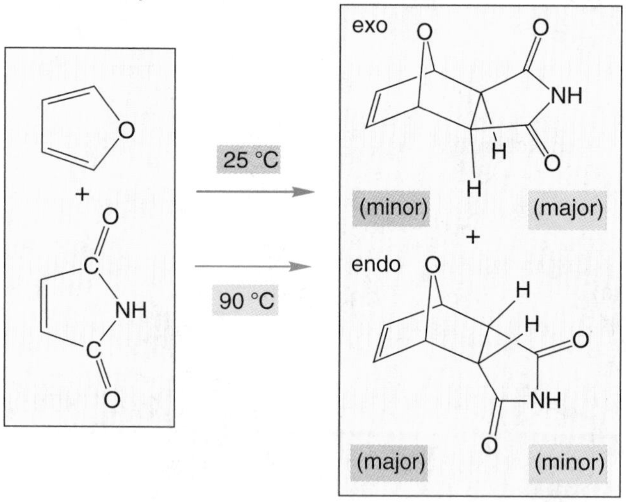

25 °C / 90 °C

exo: (minor) (major)

endo: (major) (minor)

PROBLEM 12.50 Here are two reactions of allyl systems related to the Diels–Alder reaction. Use a HOMO–LUMO molecular orbital analysis to show why reaction (a) succeeds whereas reaction (b) fails.

(a)

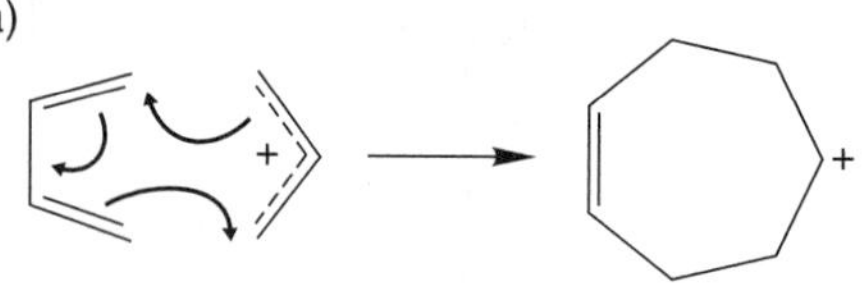

(b)

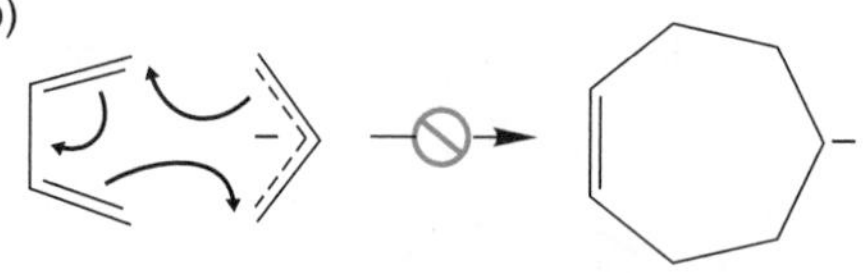

PROBLEM 12.51 In principle, one of the following reaction sequences could give two products, while the other must produce only one. Explain.

(a) Δ, then H_2/Pd

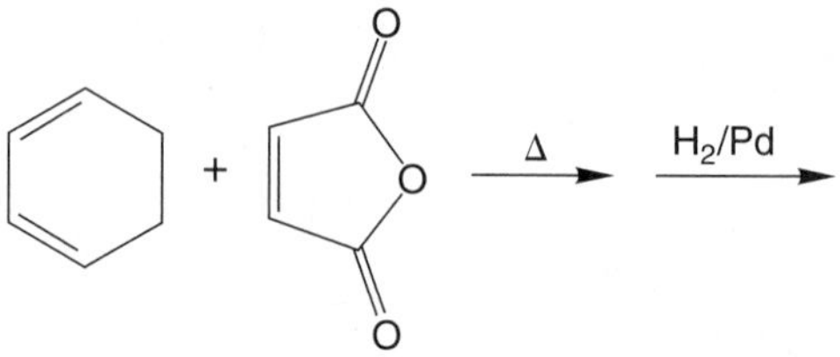

(b) Δ, then H_2/Pd

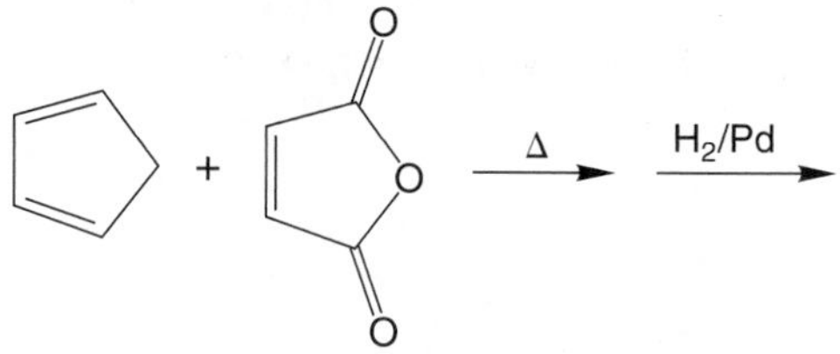

PROBLEM 12.52 In 1951, the great Kurt Alder himself examined the reactions of a mixture of dienes **1** and **2** with maleic anhydride. He discovered that diene **1** reacted at

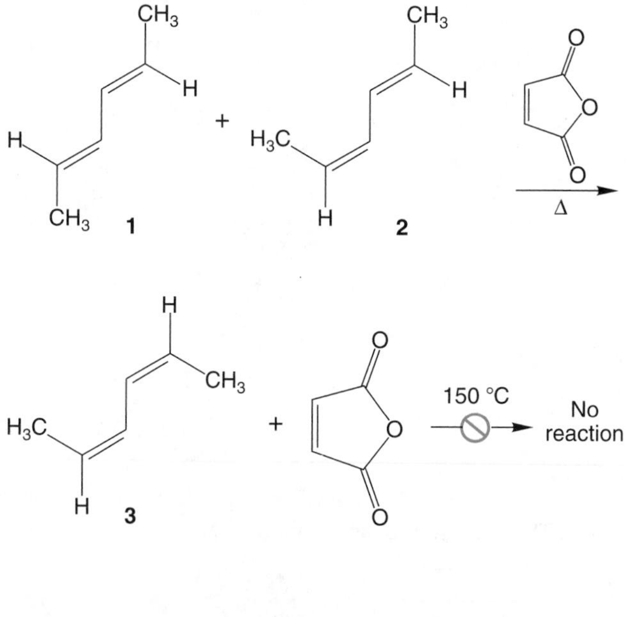

1 + **2** + maleic anhydride —Δ→

3 + maleic anhydride —150 °C⊘→ No reaction

35 °C to give a single adduct. Diene **2** reacted only at 150 °C to give a single, different, adduct. Write structures for the two products and explain the difference in the ease of reaction. Why is reaction of **1** so much easier than reaction with **2**? Alder never actually looked at the reaction with diene **3**, but if he had, I predict that no reaction would have occurred at 150 °C. Explain.

PROBLEM **12.53** Propose a synthesis of the powerful vesicant (blister inducer) cantharidin, the active ingredient in the putative aphrodisiac "Spanish fly." You may assume that organic starting materials containing no more than six carbon atoms are available, along with any inorganic materials you may need. *Hint*: The thermodynamically more stable exo Diels–Alder adducts of furan are generally isolable.

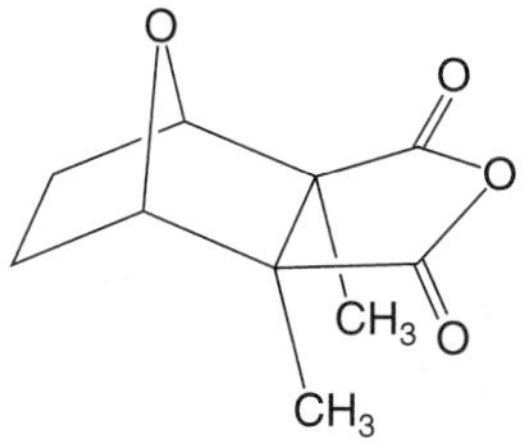

O O O CH_3 O CH_3

Cantharidin

PROBLEM **12.54** Write arrow formalism mechanisms for the following transformations:

(a)

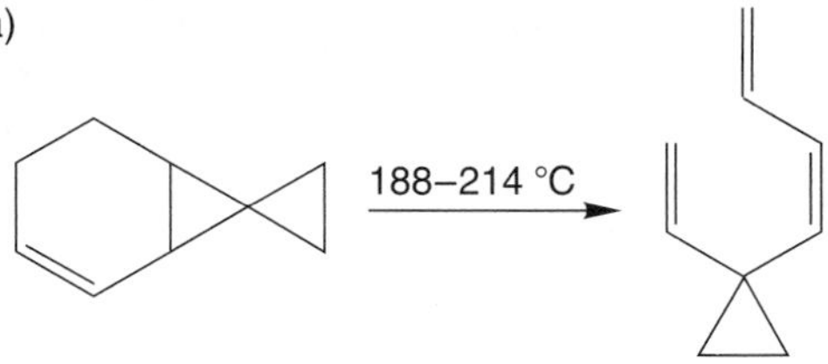

188–214 °C

(b) H_3COOC $COOCH_3$

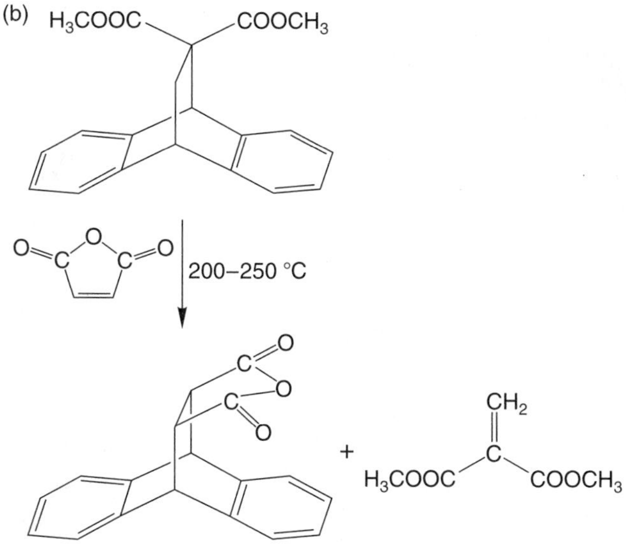

O O O

200–250 °C

O O O

CH_2

+ H_3COOC C $COOCH_3$

(c)

O C Ph O

O C O $COOCH_2CH_3$

140 °C

CH_3CH_2OOC PhCOO O C O

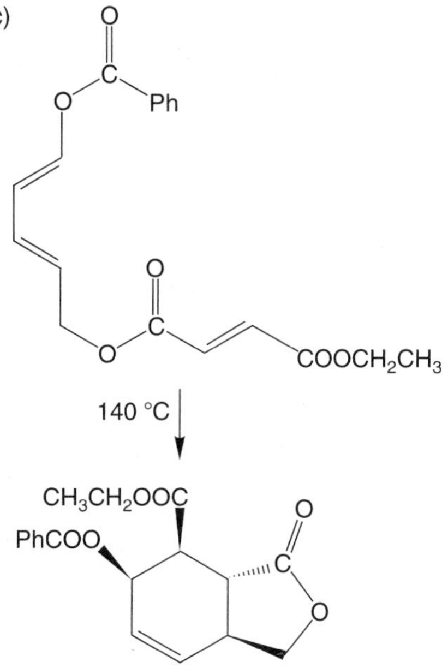

(d)

O C O O H_3COOC

25 °C

O O C O H $COOCH_3$

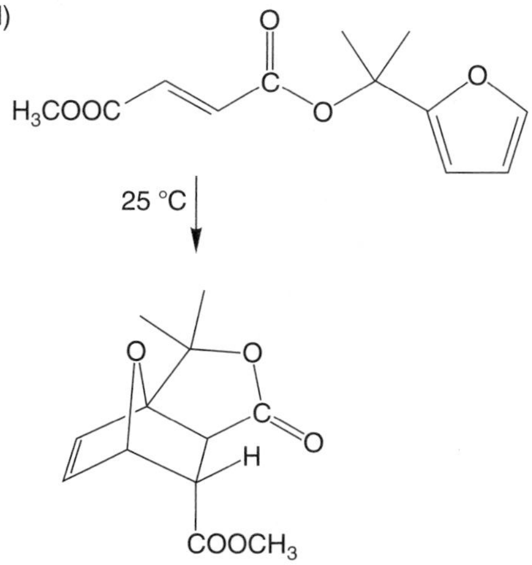

PROBLEM **12.55** Write an arrow formalism mechanism for the following bimolecular process. Be sure to account for the stereochemistry of the product.

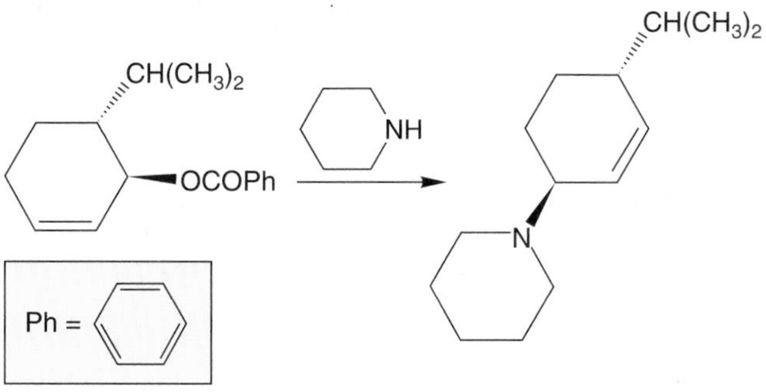

$CH(CH_3)_2$ OCOPh

NH

$CH(CH_3)_2$ N

Ph = (benzene ring)

Conjugation and Aromaticity

13

Curious, though, isn't it, um, Patwardhan, that the number, er, six should be, um, embodied in one of the most, er, er, beautiful, er, shapes in all nature: I refer, um, needless to say, to the, er, benzene ring with its single, and, er, double carbon bonds. But is it, er, truly symmetrical, Patwardhan, or, um, asymmetrical? Or asymmetrically symmetrical, perhaps … .

—Vikram Seth*
A Suitable Boy

Chapter 12 was devoted to the consequences of conjugation—the overlap of $2p$ orbitals—in acyclic systems. Now we will see what happens when conjugated systems are cyclized, turned into rings. Nothing much changes if the atoms of the rings do not maintain orbital overlap throughout the ring—if they are not fully conjugated. 1,3-Cyclohexadiene is not especially different from 1,3-pentadiene, for example. There are small differences in chemical and physical properties, but the two molecules clearly belong to the same chemical family. Hydrogenation is easy, addition reactions abound, the Diels–Alder reaction is common, and so on (Fig. 13.1).

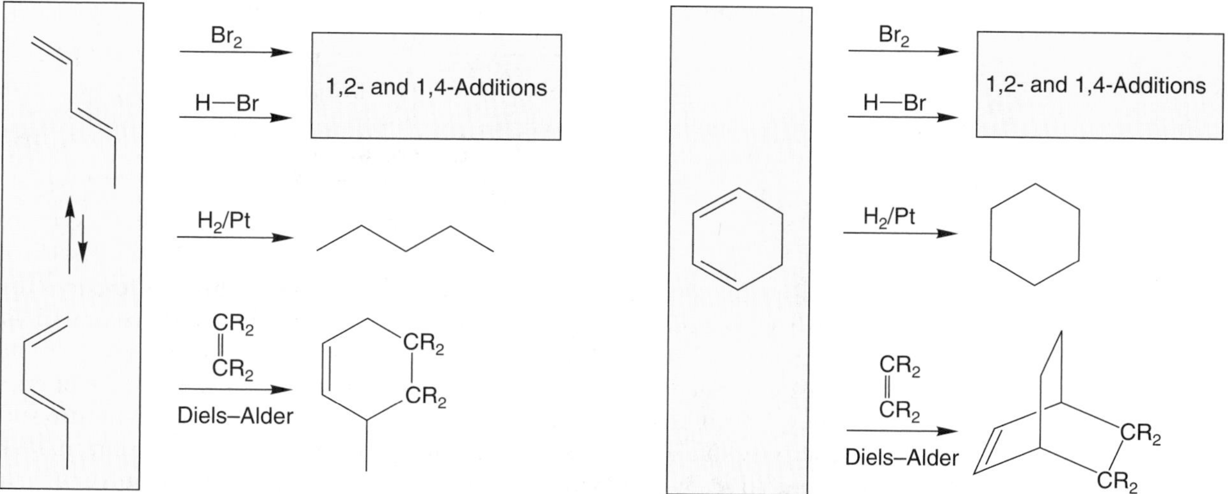

FIGURE 13.1 Cyclic and acyclic dienes have similar chemical properties.

*Vikram Seth, a poet and virtuoso tabla player, was born in Calcutta in 1952 and educated in India, Britain, and the United States. This brilliant first novel gets the MJ 10-star recommendation.

When the conjugation is continued around the ring to give what we might call "1,3,5-cyclohexatriene," matters change. None of the reactions shown in Figure 13.1 is easy, and this molecule appears to be remarkably unreactive. It is clearly in a different class from 1,3-cyclohexadiene and the acyclic dienes (Fig. 13.2).

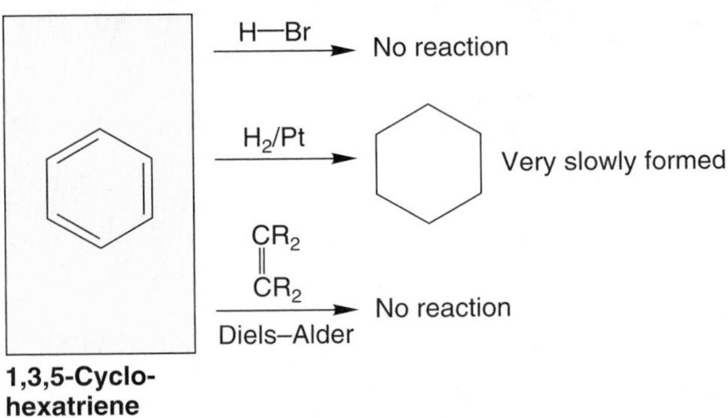

FIGURE 13.2 Benzene (1,3,5-cyclohexatriene) does not undergo most of the reactions of normal alkenes.

But not all ring compounds containing double bonds (cyclic polyenes), behave in this way. Some are normal cycloalkenes; cycloheptatriene and cyclooctatetraene are examples of this kind of molecule. Others are exceptionally unstable; cyclobutadiene is the archetype of this kind of cyclic polyene (Fig. 13.3).

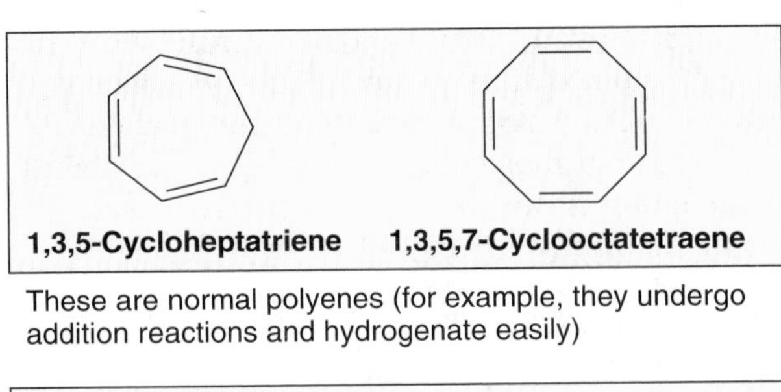

These are normal polyenes (for example, they undergo addition reactions and hydrogenate easily)

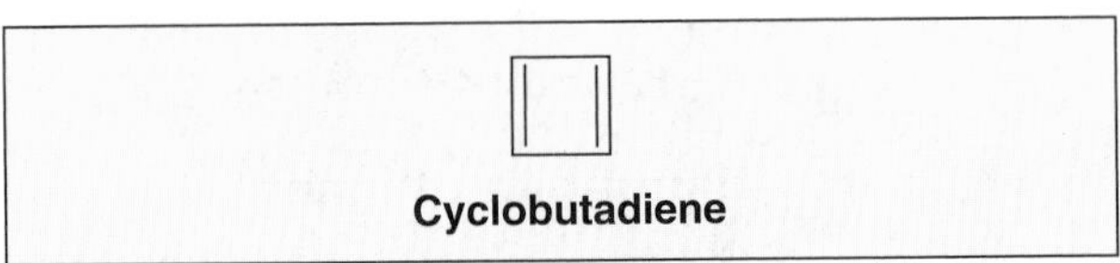

This molecule is extraordinarily unstable—it can only be isolated at very low temperature

FIGURE 13.3 Other polyenes are either normal in their chemical properties or exceptionally unstable.

In this chapter, we will uncover the sources of these differences in stability between cyclic polyenes and begin to see the chemical and physical consequences of this special kind of conjugative interaction of 2*p* orbitals. We will encounter the molecule **benzene** in which the overlap of six carbon 2*p* orbitals in a ring has great consequences for both structure and reactivity. We will also see a generalization of the properties that make benzene so stable, and will learn how to predict which cyclic polyenes should share benzene's stability and which should not.

13.1 THE STRUCTURE OF BENZENE

The molecule in Figure 13.2 that we called 1,3,5-cyclohexatriene was isolated from the thermal rendering of whale blubber in the last century. Although the formula was established as C_6H_6 [or better, $(CH)_6$], there remained a serious problem: What was its structure? There are many possibilities, among them the four shown in Figure 13.4. Two of these, Dewar benzene and Ladenburg benzene, are named for the people who suggested them, Sir James Dewar (1842–1923) and Albert Ladenburg (1842–1911). The third, benzvalene, is more cryptically named. The last compound, 3,3′-bicyclopropenyl, was first made only in 1989 by a group headed by Professor W. E. Billups (b. 1939) at Rice University.

The German chemist Friedrich August Kekulé von Stradonitz (1829–1896), suggested the ring structure.* Legend has it that Kekulé was provoked by a dream of a snake biting its tail. Kekulé describes dozing and seeing long chains of carbon atoms twisting and turning until one "gripped its own tail and the picture whirled scornfully before my eyes." Kekulé's suggestion for the structure of benzene is shown in cyclohexatriene form in Figure 13.5.

"Dewar" benzene
(bicyclo[2.2.0]hexa-2,5-diene)

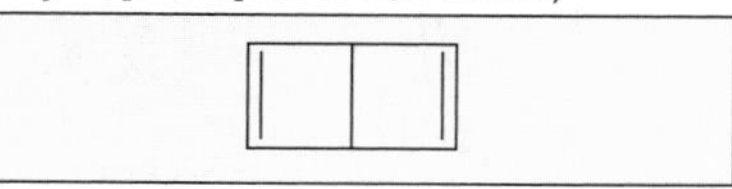

Benzvalene
(tricyclo[3.1.0.0^{2,6}]hex-3-ene)

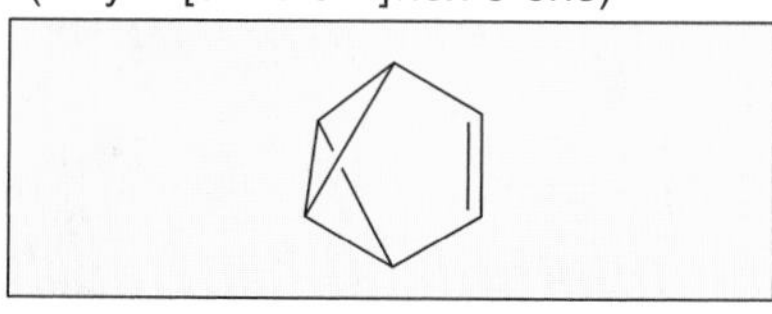

"Ladenburg" benzene
or prismane
(tetracyclo[2.2.0.0^{2,6}.0^{3,5}]hexane)

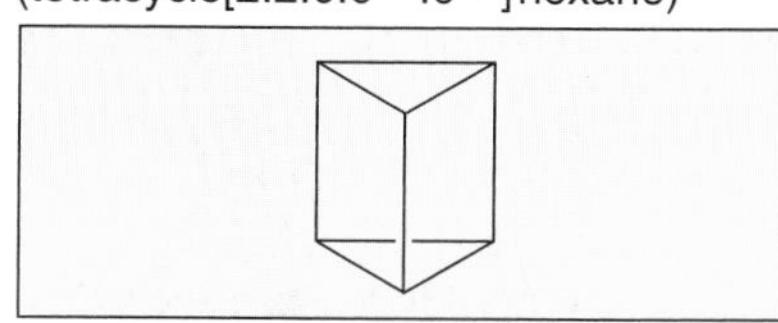

3,3′-Bicyclopropenyl

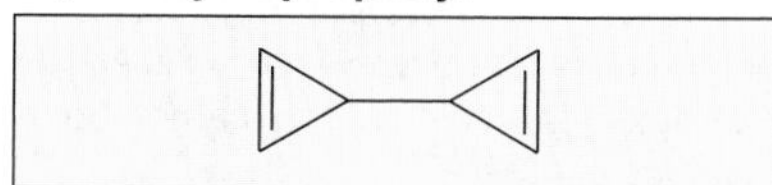

FIGURE 13.4 Some possible structures for benzene, $(CH)_6$.

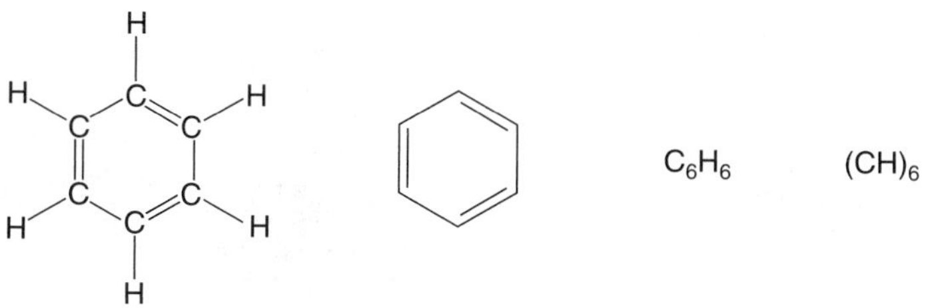

FIGURE 13.5 The structure of benzene, drawn as 1,3,5-cyclohexatriene.

Traditional, or Kekulé, structures

The same structures written by J. J. Loschmidt in 1861

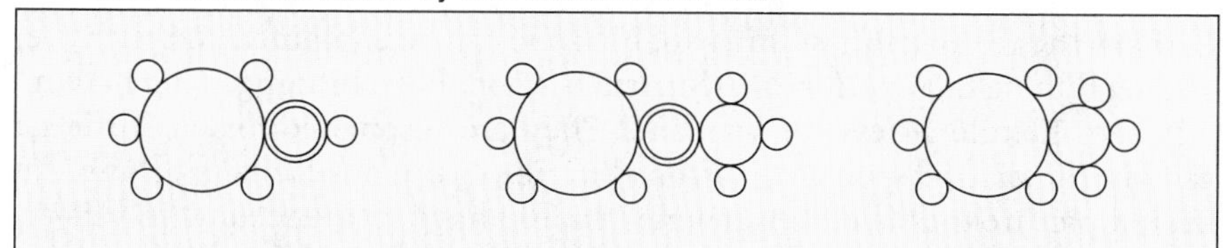

FIGURE 13.6 Josef Loschmidt's formulations of benzene structures.

But there were problems with the Kekulé formulation as well as with the others. Benzene didn't react with halogens or hydrogen halides like

*Kekulé gets the lion's share of the credit, but in the mid-1800s there were others hot on the trail of structural theory in general and a decent representation of benzene in particular. One such chemist was Alexandr M. Butlerov (1828–1886) from Kazan, who may well have independently conceived of the ring structure for benzene, and who surely contributed greatly, if quite invisibly in the West, to the development of much of modern chemistry. Another, even more obscure contributor, was the Austrian chemist–physicist Johann Josef Loschmidt (1821–1895) who, four years before Kekulé's proposal in 1865, clearly wrote a ring structure for benzene. Figure 13.6 shows how forward-looking were the structures proposed by this too-little-known Austrian schoolteacher.

any self-respecting polyene did, for example (Fig. 13.2). Hydrogenation was much slower than for other alkenes and required severe conditions. Benzene was clearly a remarkably stable compound, and the cyclohexatriene structure couldn't account for this.

Moreover, it eventually became known that the structure of benzene was that of a *regular* hexagon. Cyclohexatriene is immediately ruled out as a structural possibility, because it must contain alternating long single bonds and short double bonds. We might expect normal double bonds of about 1.33 Å and single bonds of about the length of the C(2)–C(3) bond in 1,3-butadiene: 1.47 Å. Instead, benzene has a uniform carbon–carbon bond distance of 1.39 Å. Note that the real bond distance, 1.39 Å, is just about the average of the double-bond distance, 1.33 Å, and the single-bond distance, 1.47 Å (Fig. 13.7).

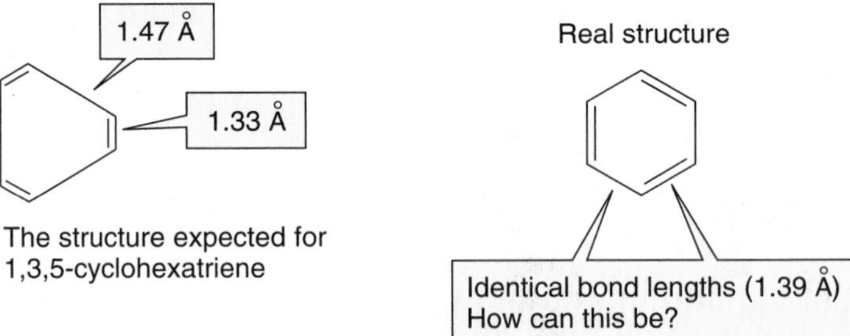

13.7 Any 1,3,5-cyclohexatriene structure must contain alternating long single bonds and short double bonds. This is not the case for benzene, which is a regular hexagon with a carbon–carbon bond distance of 1.39 Å.

At this point, we can probably see the beginnings of the answer to these structural problems. Kekulé's static cyclohexatriene formulation doesn't take account of resonance stabilization.

PROBLEM 13.1 One early problem with the Kekulé structure was that only one 1,2-disubstituted isomer of benzene could be found for any given substituent. That is, there is only one 1,2-dimethylbenzene (*o*-xylene). Explain why this observation would have been hard for Kekulé to explain.

13.2 A RESONANCE PICTURE OF BENZENE

Perhaps the remarkable stability of benzene can be explained simply by realizing that delocalization of electrons is strongly stabilizing. There are really *two* **Kekulé forms** for benzene. They are resonance forms, differing only in the distribution of electrons but not in the positions of atoms. An orbital picture for "cyclohexatriene" should have influenced us in this direction. Each carbon is hybridized sp^2 and there is, therefore, a $2p$ orbital on every carbon. The structure in Figure 13.8(a) stresses overlap between the $2p$ orbitals on C(1) and C(2), C(3) and C(4), and C(5) and C(6), but not between C(2) and C(3), C(4) and C(5), or C(6) and C(1). This is suspicious, as there is a $2p$ orbital on every carbon atom and the drawing of Figure 13.8(a) takes no account of the symmetry of the situation. If there is overlap between the orbitals on C(1) and C(2), then there must be equivalent overlap between the $2p$ orbitals on C(2) and C(3) as well, as in the structure in Figure 13.8(b). Drawing both Kekulé forms emphasizes this, and produces a regular hexagon as the combination of the two Kekulé resonance forms.

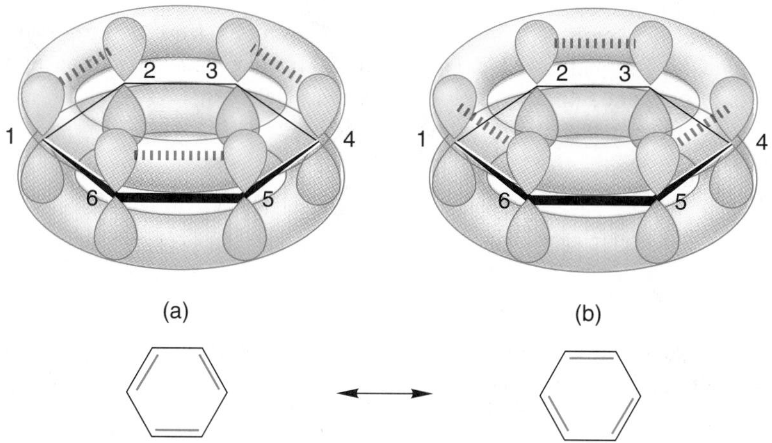

FIGURE 13.8 In the Kekulé form (a) the overlap between C(1) and C(2), C(3) and C(4), and C(5) and C(6) is stressed, but the other, equivalent overlaps between C(2) and C(3), C(4) and C(5), and C(6) and C(1) are ignored. Form (b) remedies this omission. The two Kekulé forms together provide a real structure for benzene.

So, our picture of benzene has now been elaborated somewhat to show the orbitals. The 2*p* orbitals on each carbon, extending above and below the plane containing the six carbons and their attached hydrogens, overlap to form a circular cloud of electron density above and below the plane of the ring.

A convention has been adopted that attempts to show the cyclic overlap of the six 2*p* orbitals. Instead of drawing out formal double bonds as in the Kekulé forms, a circle is inscribed inside the ring (Fig. 13.9). The cyclic overlap of 2*p* orbitals is nicely evoked by this formulation, but a price is paid. The Kekulé forms are easier for bookkeeping purposes: for keeping track of bonds making and breaking in chemical reactions. We will use both formulations in this book.

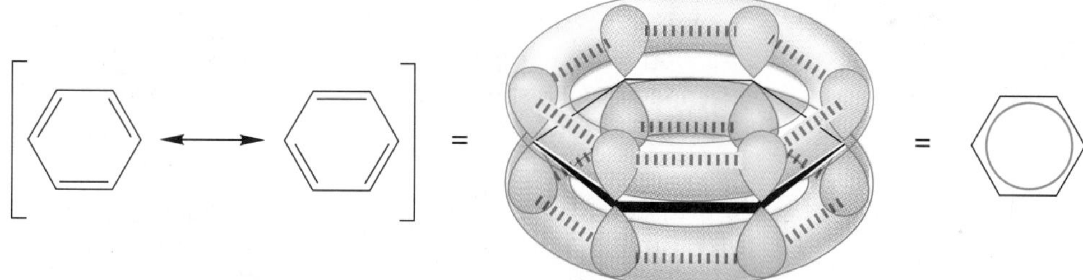

FIGURE 13.9 There is a ring of electron density above and below the six-membered benzene ring. The circle in the center of the benzene ring is evocative of the cyclic overlap of six 2*p* orbitals.

*PROBLEM 13.2

There are other resonance forms for benzene, which can be included to get a slightly better electronic description of the molecule than is provided by the pair of Kekulé forms alone. In these other resonance forms, different sets of overlapping 2*p* orbitals from those shown in the Kekulé forms are emphasized. In these electronic structures, overlap between two *p* orbitals on "across the ring" or "para" carbons is taken into account. Draw these resonance forms, called **Dewar forms** after Sir James Dewar, including pictures of the orbital lobes involved in the "cross-ring" bond. Be careful! This problem is harder than it looks.

ANSWER

Any resonance form represents one electronic description for a molecule and ignores all others. For example, in one resonance form for the allyl cation, over-

ANSWER (CONTINUED) lap between the $2p$ orbitals on C(1) and C(2) is emphasized, and the third $2p$ orbital on C(3) is written as if it did not overlap. This error is taken into account in the other resonance form that shows overlap between C(3) and C(2), and ignores any overlap with the $2p$ orbital on C(1).

Summary structure

The two Kekulé forms of benzene make similar approximations and only do a good job of giving the structure of benzene when taken in combination (Fig. 13.8).

Other choices might have been made as to which orbital overlaps to emphasize. For example, in the Dewar forms, 1,4-overlap is explicitly emphasized. There are three Dewar forms that differ in which 1,4-overlap (1,4; 3,6; or 2,5) is counted.

But this business is tricky. By now, your eye is trained to see this central bond as a σ bond, *and it isn't*! As in the Kekulé forms, it is $2p$–$2p$, π overlap that is being considered here. For the central bond to be a σ bond, atoms would have to move, and that cannot be done in resonance forms. Only electrons can move. A drawing should make it clear.

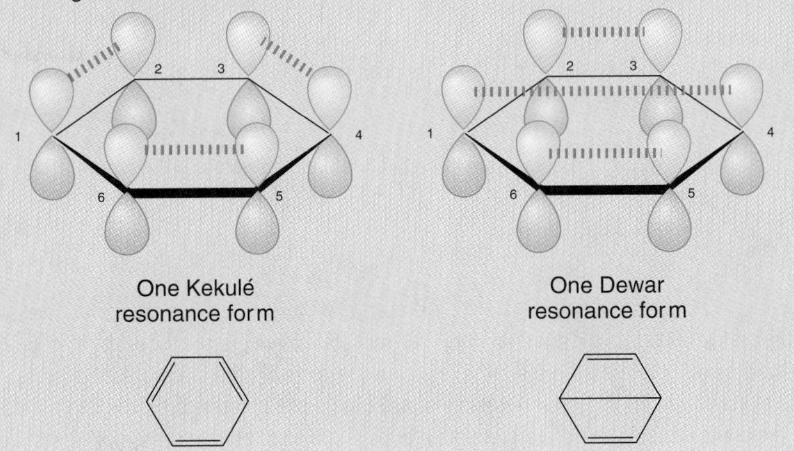

*PROBLEM **13.3** Do you think these Dewar forms will be important, contributing strongly to the benzene structure? Why or why not?

ANSWER Certainly not! The Dewar forms each have a long π bond that cannot be very strong. In the last figure for Problem 13.2, it is the C(1)–C(4) bond that is the

ANSWER (CONTINUED)

problem. Overlap will be small, bond strength low, and the contribution of Dewar forms to the resonance hybrid of benzene, though not zero, cannot be large.

*PROBLEM 13.4

There is a real compound $(CH)_6$ called Dewar benzene containing a pair of fused (they share an edge) cyclobutene rings (Dewar benzene is correctly named bicyclo[2.2.0]hexa-2,5-diene). This molecule is not flat, and therefore *not* a resonance form of benzene, but a separate molecule (Fig. 13.10). Make a good three-dimensional drawing of Dewar benzene.

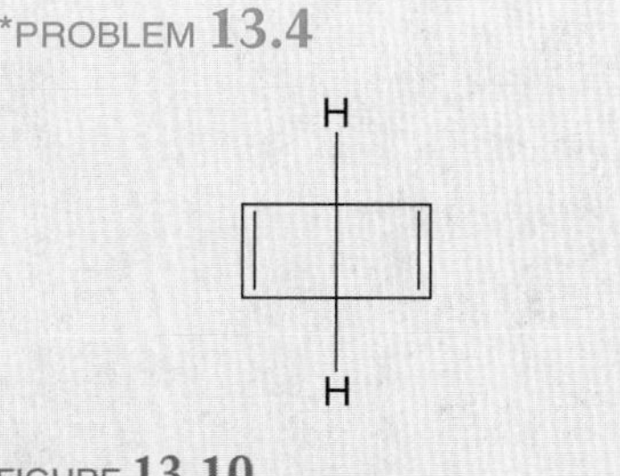

FIGURE 13.10

ANSWER

In this case, in the real Dewar benzene, the central bond is a σ bond. The two "hinge" carbons are hybridized approximately sp^3, and the molecule is shaped like an open book.

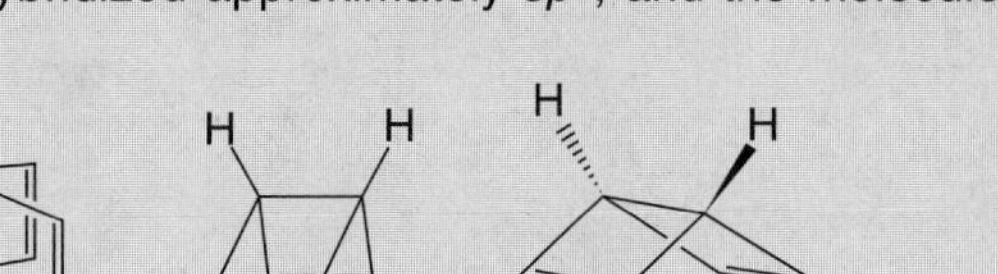

Three views of Dewar benzene

13.3 THE MOLECULAR ORBITAL PICTURE OF BENZENE

The raw material for a calculation of the molecular orbitals of benzene is six $2p$ orbitals in a ring. Linear combination of these atomic orbitals leads to a set of six molecular orbitals: three bonding and three antibonding. Their symmetries and ordering in energy are shown in Figure 13.11.

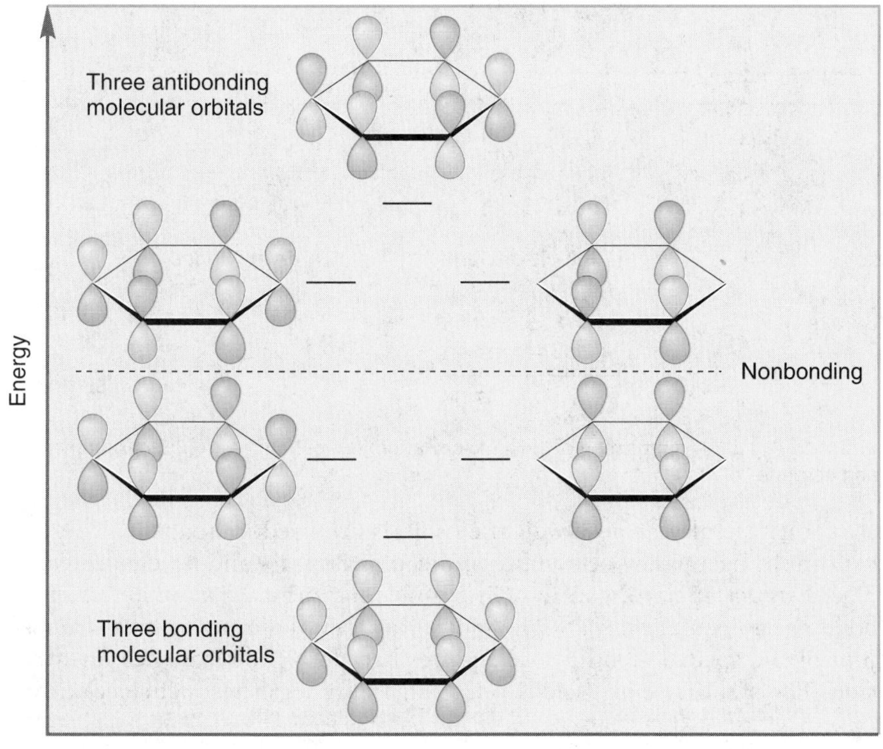

FIGURE 13.11 The π molecular orbitals of benzene.

There are six electrons to go into the system, one from each carbon, and they naturally occupy the three lowest energy, bonding molecular orbitals. The antibonding molecular orbitals are empty. If we compare benzene to a system of three double bonds not tied back into a ring (1,3,5-hexatriene is a good model) we can see that there are substantial energetic savings involved in the ring structure. One can calculate the energy difference in terms of a quantity (β) whose value is approximately –18 kcal/mol (Fig. 13.12). If we sum all the energies of the electrons in Figure 13.12, we find an advantage for benzene over 1,3,5-hexatriene of 1β. The energetic advantage over three separated ethylenes is even greater: 2β (Fig. 13.12).

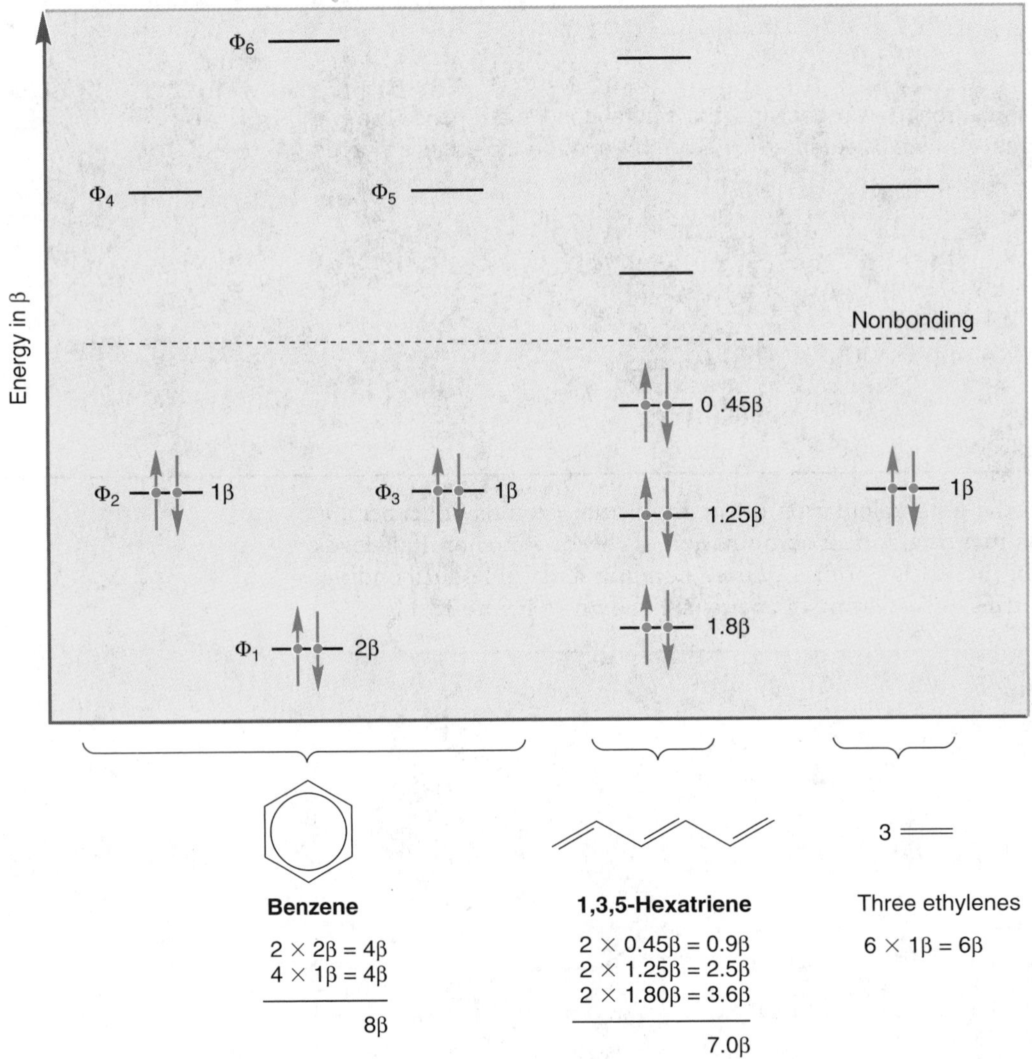

FIGURE **13.12** A comparison of the π molecular orbitals of benzene, 1,3,5-hexatriene, and ethylene.

Look carefully at how well the orbitals are used in benzene. There is maximum occupancy of bonding molecular orbitals and no electrons are placed in destructively antibonding molecular orbitals or wasted in nonbonding molecular orbitals. The molecular orbital picture gives a stronger hint of the special stability of benzene than does the resonance formulation. The stability comes not from a large number of resonance contributors, but from an especially good, especially low-energy, set of orbitals. We will see other examples. Now the question is, Just how good is this electronic arrangement? *How much* more stable is benzene than we might have expected from a cyclohexatriene model?

13.4 QUANTITATIVE EVALUATIONS OF RESONANCE STABILIZATION (DELOCALIZATION ENERGY) IN BENZENE

Although we can appreciate that delocalization is an energy-lowering phenomenon, it is not easy to see intuitively whether this effect is big or small. In this section, we see two experimental procedures for determining the magnitude of resonance stabilization.

13.4a Heats of Hydrogenation

The heat of hydrogenation of *cis*-2-butene is –28.6 kcal/mol (Chapter 10, p. 392). There is no change induced by incorporation of a double bond into a six-membered ring, as the heat of hydrogenation of cyclohexene is also –28.6 kcal/mol. A reasonable first guess about the heat of hydrogenation of a compound containing two double bonds in a ring would be just twice the value for cyclohexene, or $2 \times -28.6 = -57.2$ kcal/mol. This guess is very close to correct; the heat of hydrogenation of 1,3-cyclohexadiene is –55.4 kcal/mol. There is no special effect apparent from confining two double bonds in a six-membered ring. The difference between the observed heat of hydrogenation for 1,3-cyclohexadiene (–55.4) and the predicted value (–57.2) is approximately what one would expect, as our prediction takes no account of the stabilizing effects of conjugation in 1,3-cyclohexadiene (Chapter 12, p. 514; Fig. 13.13).

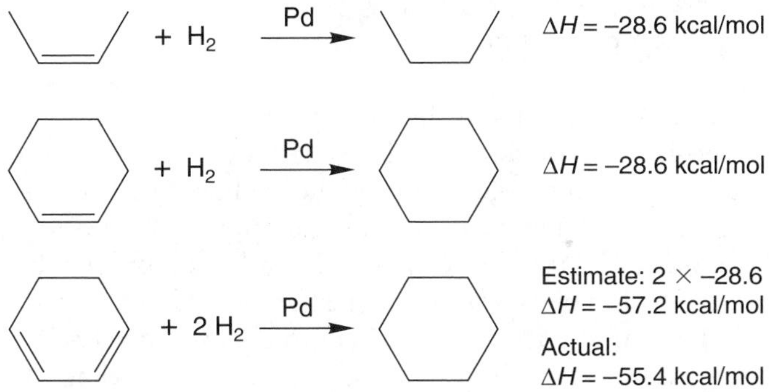

+ H_2 —Pd→ $\Delta H = -28.6$ kcal/mol

+ H_2 —Pd→ $\Delta H = -28.6$ kcal/mol

+ 2 H_2 —Pd→ Estimate: 2×-28.6 $\Delta H = -57.2$ kcal/mol; Actual: $\Delta H = -55.4$ kcal/mol

FIGURE 13.13 Heats of hydrogenation for some alkenes and cycloalkenes.

Now let's guess (calculate) the heat of hydrogenation for 1,3,5-cyclohexatriene, the hypothetical molecule with three double bonds in a six-membered ring. We might make this estimate by assuming that the third double bond will increase the heat of hydrogenation of 1,3-cyclohexadiene by as much as the addition of a second double bond did for cyclohexene [$-55.4 - (-28.6) = -26.8$ kcal/mol]. This increment would give a predicted value of $-55.4 + (-26.8) = -82.2$ kcal/mol for the heat of hydrogenation of the hypothetical 1,3,5-cyclohexatriene. Now we can measure the heat of hydrogenation of the real molecule, benzene. Hydrogenation is extremely slow under normal circumstances but there are special catalysts that can speed the reaction up a bit, and eventually even the stable benzene will hydrogenate to give cyclohexane. The measured heat of hydrogenation for benzene, –49.3 kcal/mol, is much lower than the calculated value for 1,3,5-cyclohexatriene. Figure 13.14 combines all these numbers and shows graphically how this procedure yields the amount by which benzene is more stable than the hypothetical 1,3,5-cyclohexatriene, –32.9 kcal/mol (Fig. 13.14).

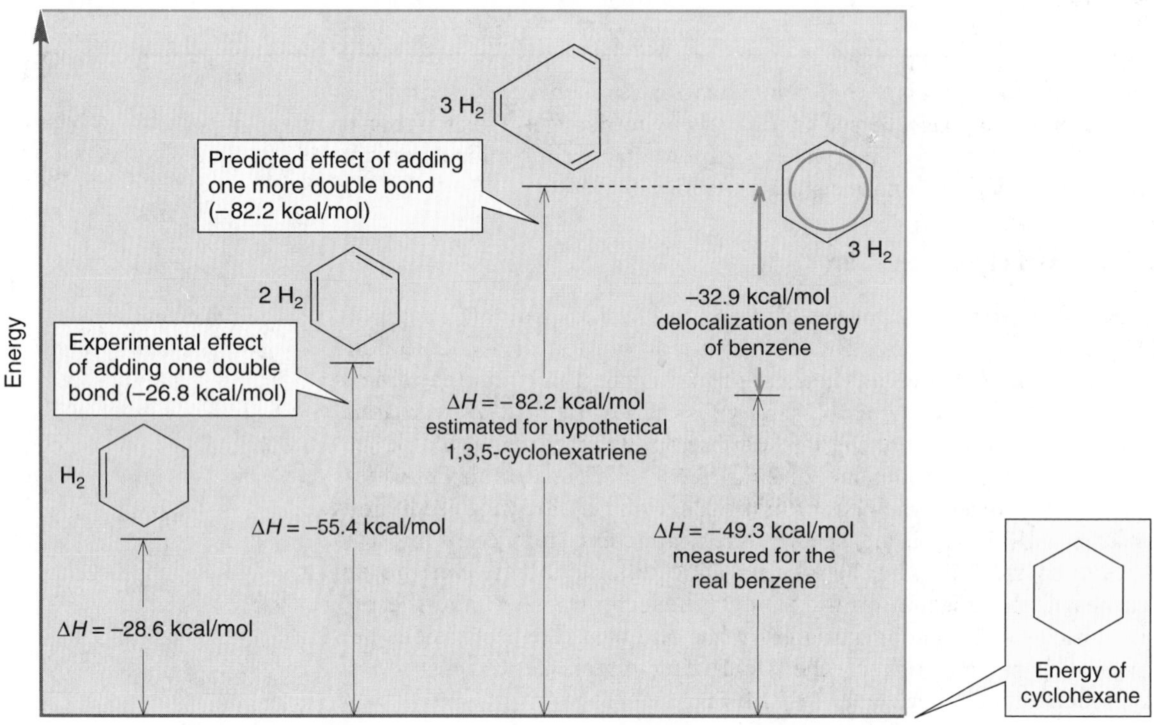

FIGURE **13.14** The value for the special stability of benzene as derived from measured and calculated heats of hydrogenation.

To summarize: We calculated the heat of hydrogenation of the unreal, unknown, hypothetical model molecule 1,3,5-cyclohexatriene by using values for the heats of hydrogenation of the real molecules cyclohexene and 1,3-cyclohexadiene. The heat of hydrogenation of 1,3-cyclohexadiene was calculated quite accurately, so we can be rather confident that our estimate for 1,3,5-cyclohexatriene won't be far off. That's important, because we can never make the cyclic triene to check the calculation. Any synthetic route will inevitably produce the much more stable real molecule, benzene. Next, measure the heat of hydrogenation for the real molecule, benzene, which turns out to be far lower than our estimate. This difference (Fig. 13.14) represents the amount of energy the special stabilization of benzene is worth. It is more than 30 kcal/mol and is called the **resonance energy** or **delocalization energy** of benzene.

13.4b Heats of Formation

The delocalization energy of benzene can be estimated in another way, using heats of formation (ΔH_f°). The chemistry of 1,3,5,7-cyclooctatetraene tells us there is no special stabilization in this molecule. Hydrogenation is fast and addition reactions occur with ease. The ΔH_f° for cyclooctatetraene is +71.23 kcal/mol, or +8.9 kcal/mol per CH unit. The ΔH_f° for benzene is +19.82 kcal/mol, or 3.3 kcal/mol per CH unit. Therefore, benzene is about 5.6 kcal/mol more stable per CH unit than a cyclic polyene with no special stability (8.9 – 3.3 = 5.6 kcal/mol). As there are six CH units in benzene we can estimate the special stability (resonance or delocalization energy) to be 6 × 5.6 = 33.6 kcal/mol. The two methods, which use heats of hydrogenation and heats of formation, are in rather good agreement (Fig. 13.15).

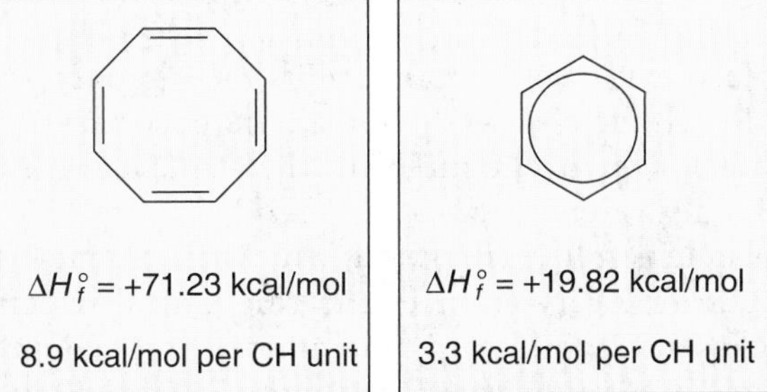

6 x (8.9 – 3.3) = 33.6 kcal/mol delocalization energy — this value compares well with that calculated from heat of hydrogenation data (32.9 kcal/mol)

FIGURE **13.15** The delocalization energy of benzene as calculated from heats of formation.

In fact, there was an earlier estimate of the delocalization energy. In Figure 13.12 (p. 574) the difference between benzene and three ethylenes (three separate double bonds) was calculated in terms of β. Benzene was found to be more stable by a factor of 2β. As β is approximately –18 kcal/mol, the special stability of benzene can be calculated to be about 36 kcal/mol, very close to the two experimental measurments (32.9 and 33.6 kcal/mol).

This special stability has come to be known as benzene's **aromatic character** or **aromaticity**, a name derived in an obvious way from the pleasant (?) odor of many benzene derivatives. In the next few pages, we will expand on the theme of aromaticity, eventually ending up with a general method for finding aromatic compounds.

Vanillin

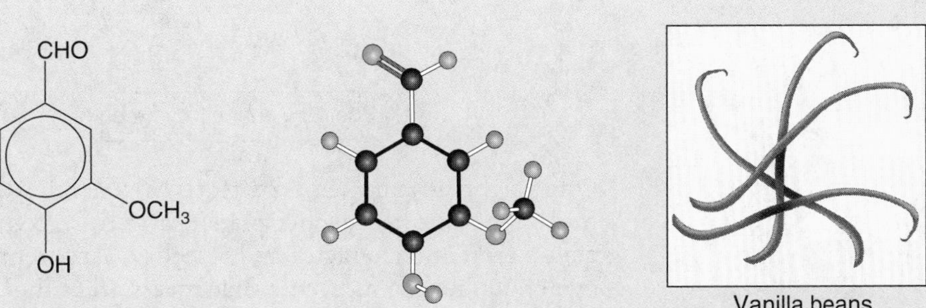

Vanilla beans

The term "aromatic" was coined because many such compounds were odoriferous. Vanillin, is a good example, although there are a great many more. Vanillin is found in nature attached to a sugar molecule, glucose, in the seed pods of the vanilla orchid, *Vanilla fragrens.* It appears as well in many other natural sources found in the tropical world.

It is now made synthetically from eugenol. Can you devise a synthesis of vanillin? *Hint*: Work backward from vanillin to eugenol, and you will see the tricky step necessary.

HO

H_3CO

Eugenol

13.5 A GENERALIZATION OF AROMATICITY: HÜCKEL'S $4n + 2$ RULE

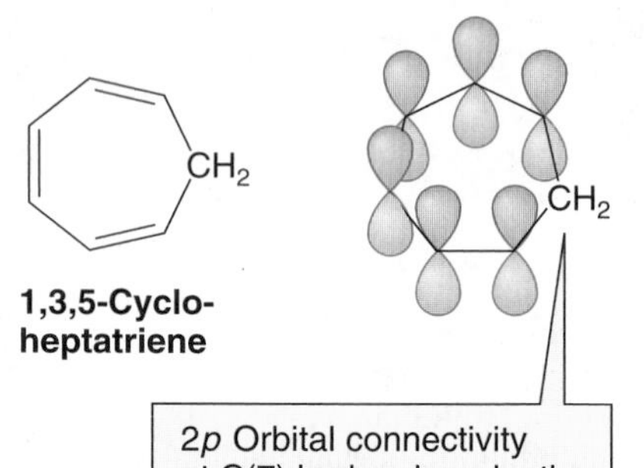
CH2
CH2
1,3,5-Cyclo-
heptatriene
2p Orbital connectivity
at C(7) broken here by the
CH2 group (no 2p orbital)

FIGURE 13.16 In 1,3,5-cycloheptatriene, orbital connectivity is broken by the CH_2 group. This molecule is not aromatic, and behaves as a normal triene.

It would certainly be surprising if there were not other compounds sharing the special stability we have just called aromaticity. To begin our search for other aromatic compounds, and, ultimately, a general way of finding such compounds, let's review the structural features of the prototype, benzene, that lead to special stability.

1. The molecule is cyclic. Aromaticity is a property of ring compounds. Recall the discussion of the difference between the molecular orbitals for the cyclic molecule, benzene, and the acyclic molecule, 1,3,5-hexatriene (Fig. 13.12, p. 574).
2. The molecule is **fully conjugated**, which simply means that there is a $2p$ orbital on every atom in the ring, and that connectivity between adjacent $2p$ orbitals is maintained throughout. 1,3,5-Cycloheptatriene is not aromatic, because the $2p$ orbitals at the end of the cycle do not overlap. Its properties resemble those of 1,3,5-hexatriene (Fig. 13.16).
3. The molecule is planar. Why should this make a difference? It insures optimal overlap of the $2p$ orbitals. If the ring is not planar, or at least quite close to planar, the $2p$ orbitals do not overlap well and some connectivity between orbitals is lost. Large deviations from planarity can nearly or completely uncouple the cycle of orbitals. An example is the tub-shaped molecule cyclooctatetraene (Fig. 13.17).

FIGURE 13.17 Nonplanarity can uncouple the $2p$ orbitals and interrupt orbital connectivity. An example is the nonplanar cyclooctatetraene.

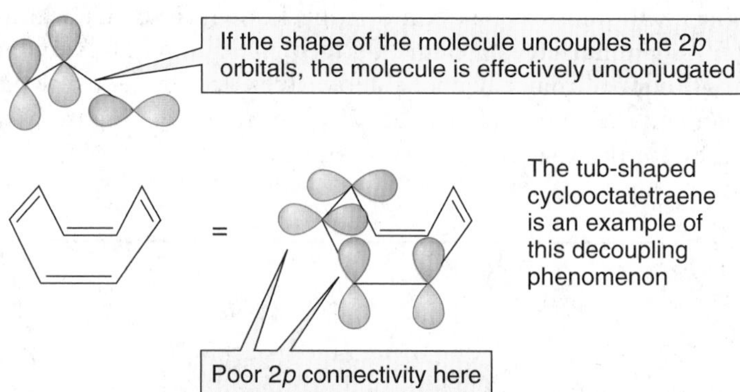
If the shape of the molecule uncouples the 2p orbitals, the molecule is effectively unconjugated
=
The tub-shaped cyclooctatetraene is an example of this decoupling phenomenon
Poor 2p connectivity here

One might now be tempted to try a generalization. Might not all molecules that are cyclic, planar, and fully conjugated share benzene's aromatic character and stability? Apparently not. Recall that cyclobutadiene, a molecule that meets all of the three criteria mentioned above, is most *unstable* (Fig. 13.18).

There must be more to it, and in our search for further criteria, we must take into account the difference between benzene and cyclobutadiene. The crucial generalization was made by the German Erich Hückel (1896–1984) in the early 1930s. **Hückel's rule** can be added to our list of three criteria (also recognized by Hückel) developed earlier.

4. Aromatic systems will contain $4n + 2$ π electrons, where n is an integer, 0, 1, 2, 3 ... . *Please note carefully that "n" has nothing to do with the number of atoms in the ring. We are considering only the number of π electrons here.*

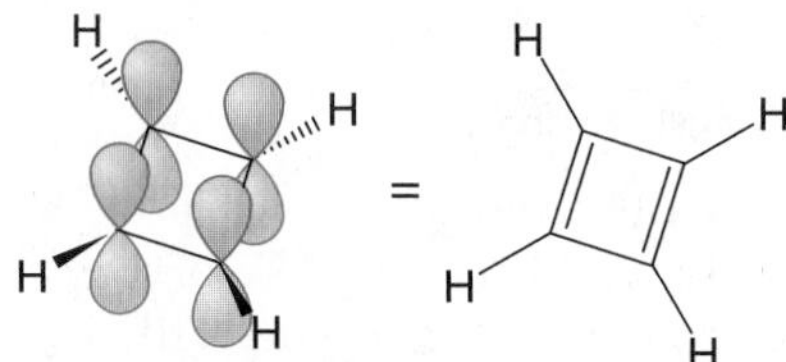
H
H
H
H
=
H
H
H
H
A most unstable molecule even though it is planar, cyclic, and fully conjugated

FIGURE 13.18 Not all cyclic, planar, and fully conjugated molecules are aromatic. Cyclobutadiene meets these three criteria but is *very* unstable.

We are now faced with the most important question of this section: Why should the number of π electrons make a difference? Why does Hückel's rule work?

Briefly stated, it works because it is *only* those molecules containing $4n + 2$ π electrons (and satisfying the other three criteria for aromaticity) that will have molecular orbitals arranged like those of benzene. All bonding molecular orbitals must be fully occupied and there may be no electrons in destabilizing antibonding molecular orbitals or wasted in nonbonding molecular orbitals. It's best to look at some examples now.

cis-1,3,5-Hexatriene can serve as an example for all acyclic molecules. Although it is fully conjugated (there is a $2p$ orbital on every carbon), and could be planar without inducing great strain, it most certainly is not cyclic. The lack of a ring means that there is no chance of strong overlap between the $2p$ orbitals on C(1) and C(6) and, therefore, no chance of aromatic character (Fig.13.19).

1,3,5-Cycloheptatriene is a cyclic molecule that is not fully conjugated. The connectivity of the $2p$ orbitals is broken by the methylene group at C(7) (Fig. 13.16). Cyclobutadiene is a planar, cyclic molecule that is fully conjugated. It fails only the fourth criterion as it has $4n$ ($n = 1$) π electrons, not a "Hückel number" of $4n + 2$.

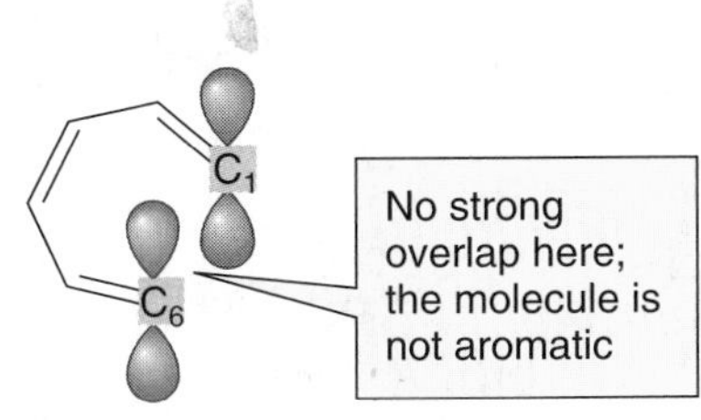

C_1
C_6
No strong overlap here; the molecule is not aromatic

1,3,5-Hexatriene

FIGURE 13.19 *cis*-1,3,5-Hexatriene is fully conjugated and can be planar, but the lack of a ring structure means that overlap between the $2p$ orbitals on C(1) and C(6) is essentially zero.

Convention Alert!

Before we can go on, we need a reliable way of finding the molecular orbital patterns for these molecules. Consider a device called a **Frost circle**, or sometimes, a Frost device. It is named for the American Arthur A. Frost (b. 1909) who devised it in 1953. It is a method of finding, mercifully without the use of mathematics, the relative energies of the molecular orbitals for planar, cyclic, fully conjugated molecules. One merely inscribes the appropriate (same size as the ring) polygon in a circle (radius 2β), *vertex down,* and the intersections of the ring with the circle will mark the positions of the molecular orbitals.

For example, to locate the molecular orbitals of benzene, inscribe a hexagon in a circle, vertex down, with the nonbonding line dividing the hexagon exactly in half (Fig. 13.20).

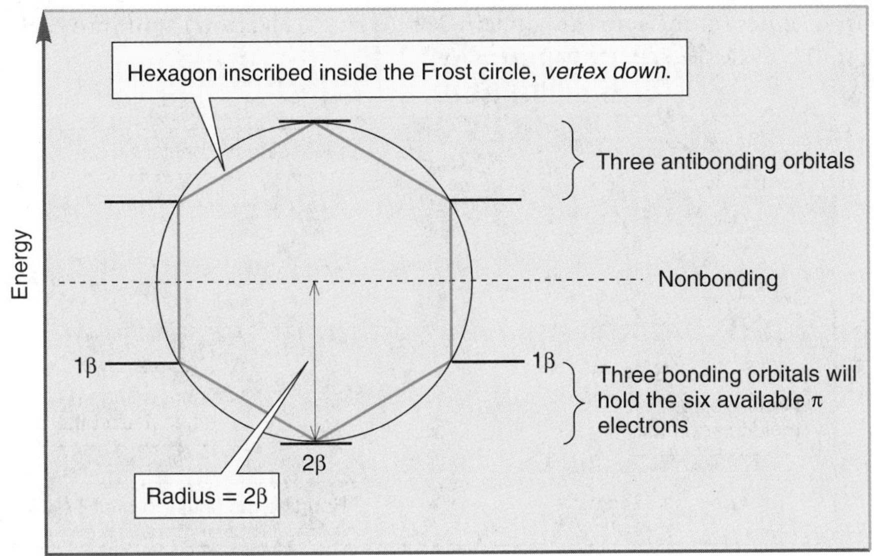

Hexagon inscribed inside the Frost circle, *vertex down.*
Three antibonding orbitals
Energy
Nonbonding
1β
1β
Three bonding orbitals will hold the six available π electrons
2β
Radius = 2β

FIGURE 13.20 The relative energies of the molecular orbitals of benzene derived from a Frost circle.

The familiar (p. 573) arrangement of molecular orbitals appears, and with some plane geometry we can even get the relative energies of the molecular orbitals exactly in terms of β (the radius of the Frost circle is 2β). For benzene, the available six π electrons completely fill the three available bonding molecular orbitals with no offending electrons remaining to be placed in the antibonding set. There are no nonbonding molecular orbitals.

If we follow this procedure for cyclobutadiene, a set of four molecular orbitals appears. One is bonding, one antibonding, and two are equienergetic (degenerate), nonbonding molecular orbitals (Fig. 13.21). The related set of molecular orbitals for square H_4 was constructed in Section 2.3 (p. 55).

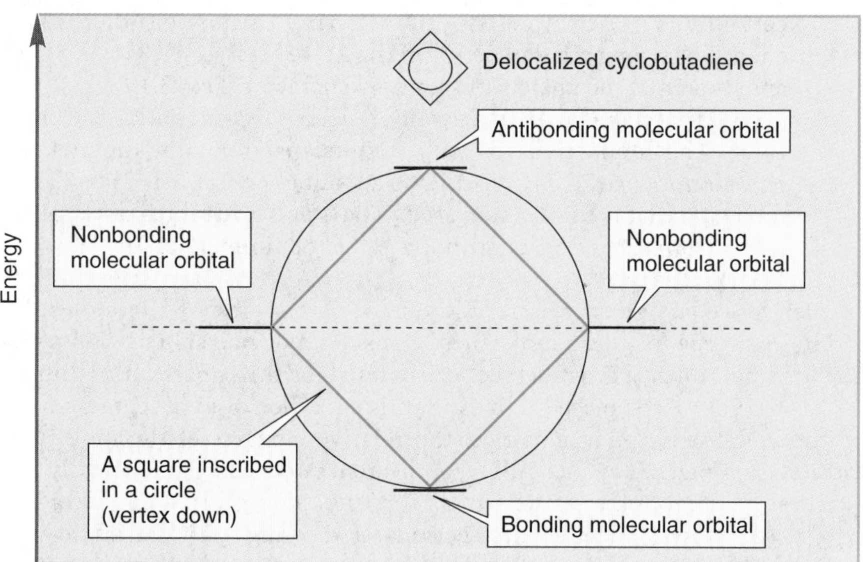

FIGURE 13.21 The relative energies of the molecular orbitals of cyclobutadiene derived from a Frost circle.

Four electrons are available and two of them will surely go into the bonding molecular orbital. There is no other bonding molecular orbital to accommodate the remaining two electrons. They will have to occupy the nonbonding degenerate pair, and will be most stable in this square arrangement if the spins are unpaired and a single electron is put into each orbital of the degenerate pair (Fig. 13.22).

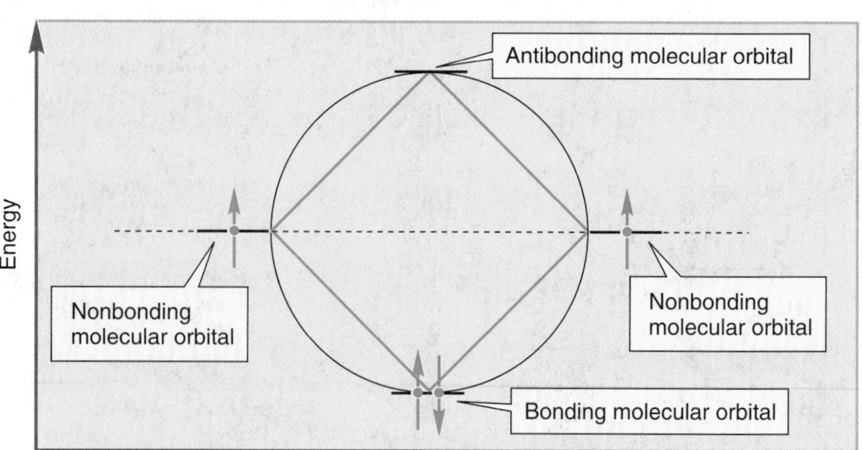

FIGURE 13.22 The electronic occupancy of the four molecular orbitals for square cyclobutadiene.

This arrangement is surely not particularly stable, and we would not expect the properties of cyclobutadiene to resemble those of benzene, as they most emphatically do not. Moreover, cyclobutadiene must pay a severe price in strain as the internal angles must be 90° (Fig. 13.23).

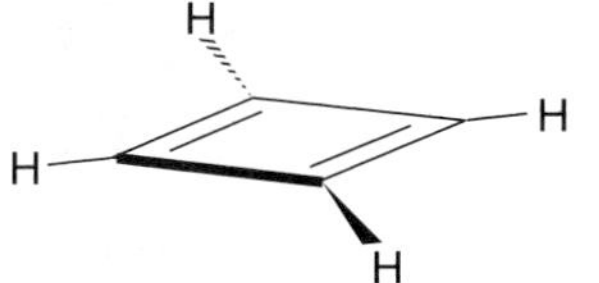

FIGURE 13.23 Cyclobutadiene has severe angle strain.

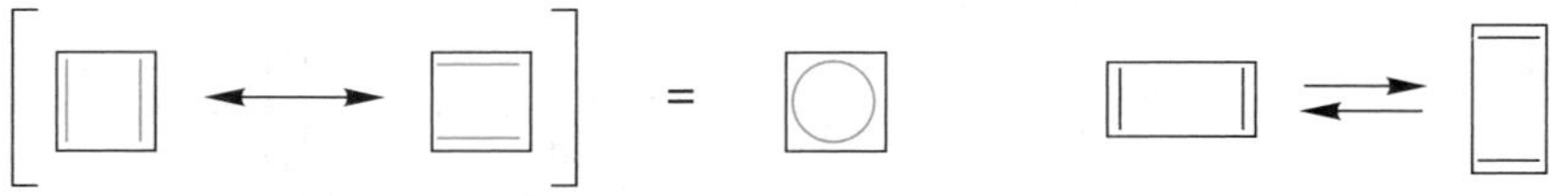

Cyclobutadiene can be isolated, but only if it is kept very cold (its IR spectrum was measured at 8 K = – 265 °C), and segregated from other reactive molecules, including itself. Theory and experiment agree that cyclobutadiene distorts from the square arrangement to form a rectangle with short carbon–carbon double bonds and longer carbon–carbon single bonds. This distortion minimizes the connection (conjugation) between the *p* orbitals (Fig. 13.24).

Square, delocalized cyclobutadiene

Rectangular cyclobutadiene

FIGURE 13.24 In rectangular cyclobutadiene, orbital connectivity is minimized.

1,3,5,7-Cyclooctatetraene is another example of a $4n$ molecule. Here we find neither the special stability of benzene, nor the special instability of cyclobutadiene. The molecule behaves like four separated, normal alkenes. Let's first examine planar cyclooctatetraene. The Frost circle device again determines the relative positions of the molecular orbitals (Fig. 13.25).

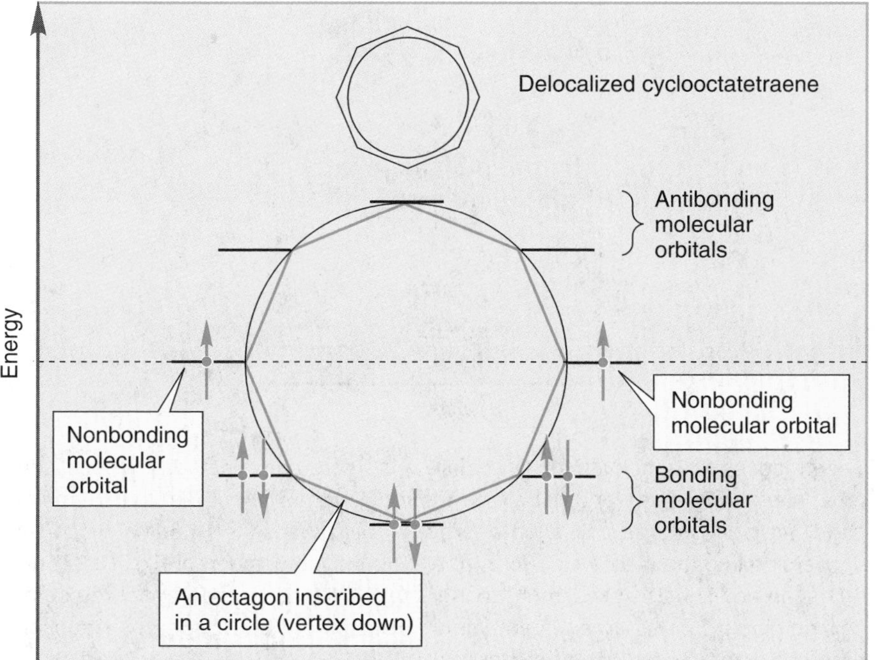

FIGURE 13.25 The molecular orbitals of planar cyclooctatetraene. Two electrons must occupy nonbonding orbitals.

Like cyclobutadiene, this molecule must have two electrons in nonbonding molecular orbitals, and there will be no special stability. Like planar cyclobutadiene, planar cyclooctatetraene has both angle and torsional strain. The angle in a regular octagon is 135°, quite far from the optimum sp^2 value of 120°. Unlike cyclobutadiene, there is a way for cyclooctatetraene to get out of its misery. Cyclobutadiene must be planar, or very nearly so. By contrast, cyclooctatetraene can distort easily into a tublike form in which most of the angle and some of the torsional strain of the planar form is relieved (Fig. 13.26).

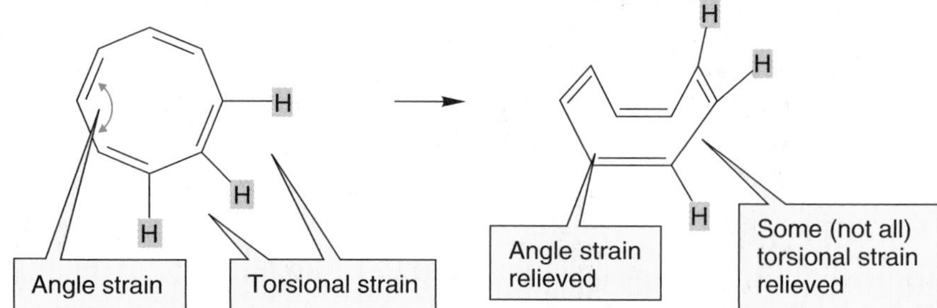

FIGURE 13.26 In planar cyclooctatetraene there is both angle and torsional strain. Both can be minimized by distorting into a tub.

Kinetic measurements reveal that two tubs interconvert rather easily. The energy difference between the tub and the planar, delocalized species can be extracted from the kinetic data. It turns out that the tub is 17 kcal/mol more stable than the planar, symmetrical, fully conjugated octagon (Fig. 13.27).

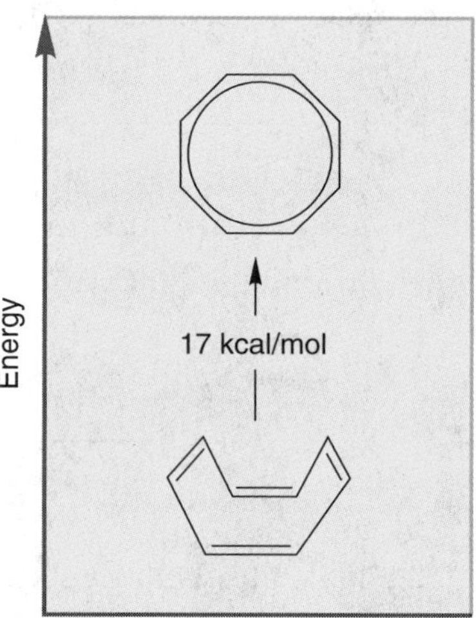

FIGURE 13.27 The tub-shaped cyclooctatetraene is 17 kcal/mol more stable than the planar, delocalized structure.

PROBLEM 13.5 The process of interconversion is actually a bit more complicated. Two localized tubs (alternating single and double bonds) equilibrate without the double bonds changing position. The activation energy for this process is 14.6 kcal/mol. The conversion into the delocalized form (the symmetrical octagon of Fig. 13.27) requires an additional 2.4 kcal/mol. Neither the localized planar form nor the delocalized planar form is an energy minimum. Sketch out Energy versus Reaction progress diagrams for these processes.

PROBLEM 13.6 Which of the molecules in Figure 13.28 are properly termed "aromatic" and which are not?

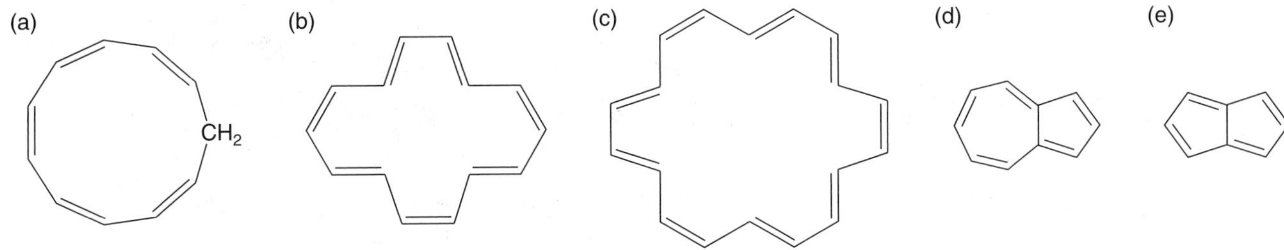

FIGURE **13.28**

So we have now seen two examples of the lack of aromaticity in 4*n* molecules; it would seem to be time to look at some potentially aromatic 4*n* + 2 molecules. It is in the stability of certain ions that the most spectacular examples have appeared. As we have stressed over and over, small carbocations are most unstable. But there are a few exceptions to this generality. One of them is the **cycloheptatrienylium ion**, or **tropylium ion** ($C_7H_7^+$), the ion derived from the loss of hydride from 1,3,5-cycloheptatriene, also called tropilidene (Fig. 13.29).

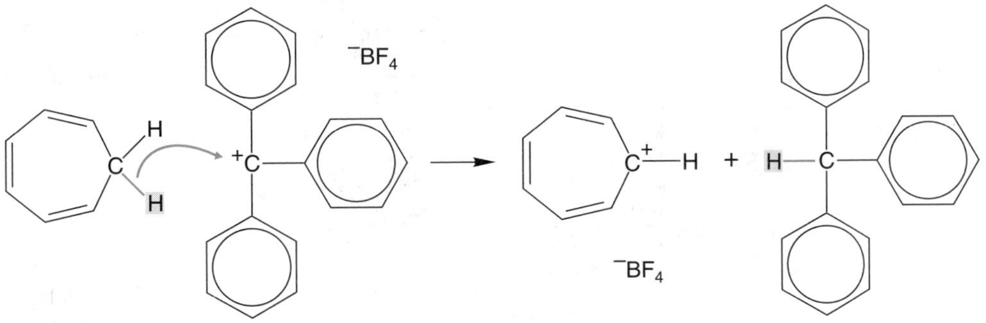

$^-BF_4$

H

C

H

$^+$C

C$^+$—H + H—C

$^-BF_4$

1,3,5-Cyclo-heptatriene (tropilidine)

Trityl cation

Tropylium fluoborate

Triphenylmethane

FIGURE **13.29** The tropylium ion ($C_7H_7^+$) can be made by transfer of hydride (H:$^-$) from 1,3,5-cycloheptatriene to the trityl cation.

Hydride cannot be simply lost from cycloheptatriene, but it can be *transferred* to the **trityl cation**, which is itself a quite stable carbocation.

Why is the trityl cation relatively stable? PROBLEM **13.7**

Tropylium fluoborate appears to be stable indefinitely under normal conditions. It sits uncomplaining on the desk top, for all the world resembling table salt. This stuff is no ordinary carbocation!

Note first that the tropylium ion has one important difference from its parent, 1,3,5-cycloheptatriene. It is fully conjugated. The removal of a hydride from the cyclic triene generates a 2*p* orbital at the 7-position, which once insulated the ends of the π system from each other. In the cation they are connected; this molecule is fully conjugated (Fig. 13.30).

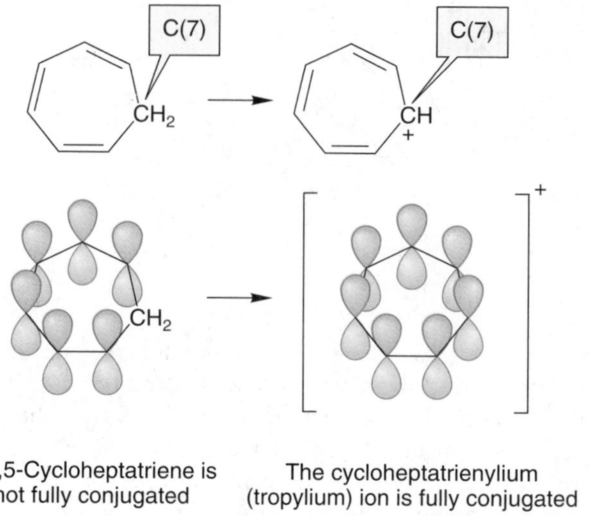

C(7) CH_2 C(7) CH +

CH_2

1,3,5-Cycloheptatriene is not fully conjugated

The cycloheptatrienylium (tropylium) ion is fully conjugated

FIGURE 13.30 In the tropylium ion (the cycloheptatrienylium ion) orbital connectivity is continuous around the ring.

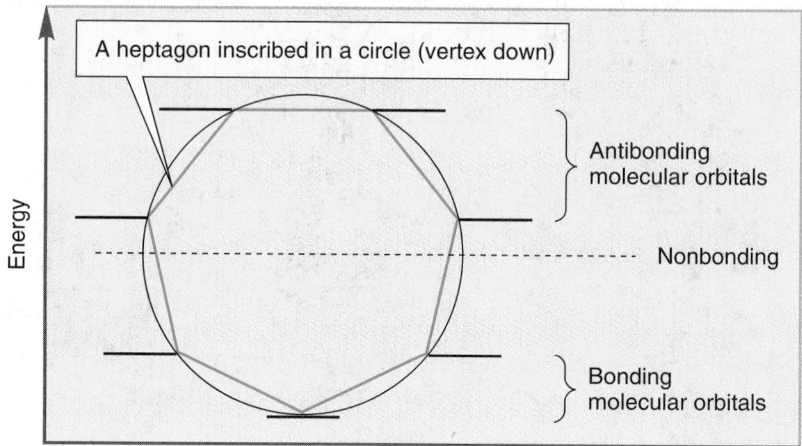

FIGURE 13.31 The relative energies of the molecular orbitals of the tropylium ion as derived from a Frost circle.

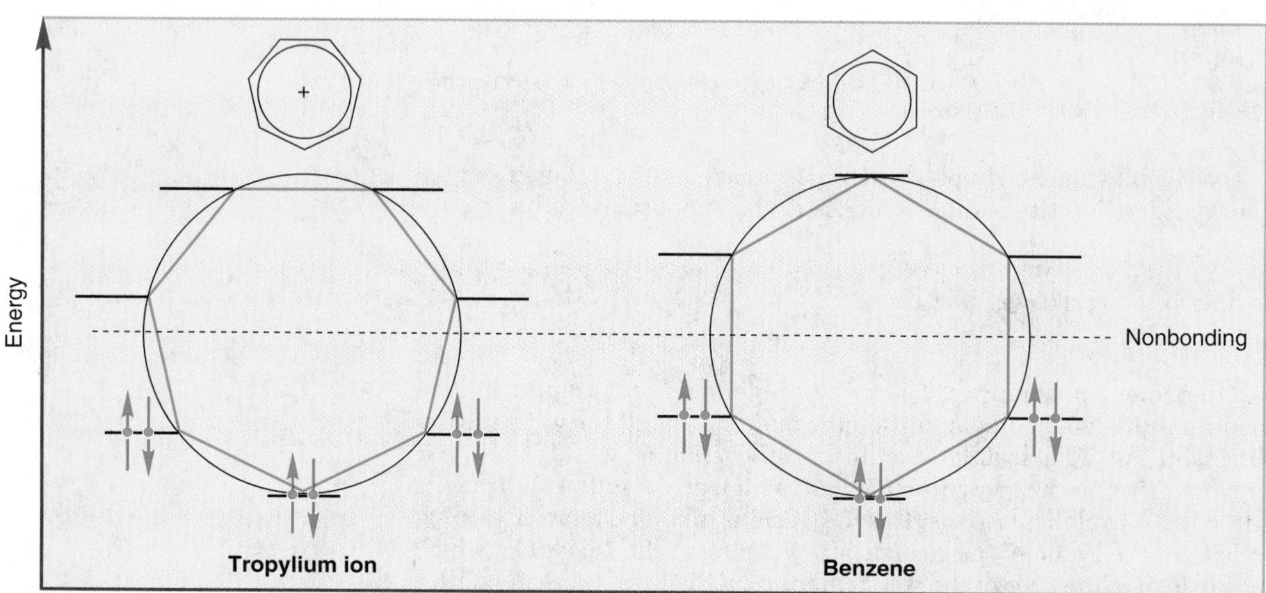

FIGURE 13.32 In the tropylium ion, the three bonding molecular orbitals are fully occupied and the antibonding molecular orbitals are unoccupied. The molecular orbital picture of the tropylium ion closely resembles that of benzene.

The Frost circle device lets us see the relative energies of the molecular orbitals for this system (Fig. 13.31).

How many electrons must go into the π system? In the cation, there is an empty 2*p* orbital on C(7) and the only π electrons are the six from the three original double bonds. These electrons can fit nicely into the three bonding molecular orbitals of the π system. Note how this system resembles the arrangement of benzene—all the bonding molecular orbitals are full, and no antibonding or nonbonding molecular orbitals are occupied (Fig. 13.32). This ion qualifies as aromatic, and the increased stability *relative to other carbocations* is dramatic indeed. Notice again the difference between the number of atoms involved in the ring—here seven—and the number of π electrons—here six = $(4n + 2)$, $n = 1$.

There is an anionic counterpart to this stable cation. The pK_a of propene is 43, and the pK_a of propane is greater than 60. The added delocalization in the allyl anion makes propene a stronger acid than propane by a factor of about 10^{20} (Fig. 13.33).

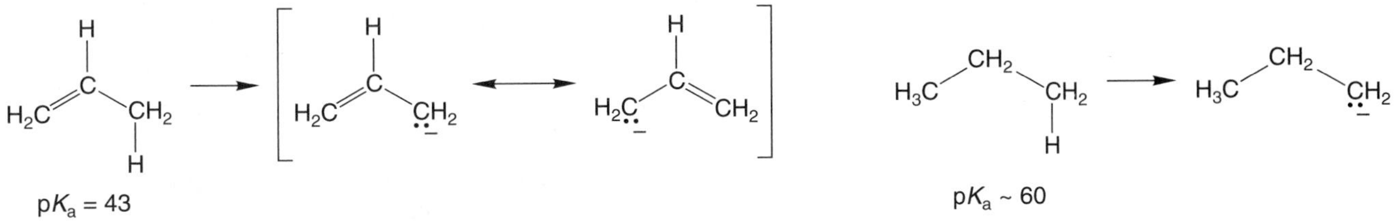

pK_a = 43

pK_a ~ 60

FIGURE **13.33** Resonance stabilization of its conjugate base makes propene much more acidic than propane.

But the pK_a of 1,3-cyclopentadiene is 16. Cyclopentadiene is a *much* stronger acid than propene. Look carefully at the structure of the **cyclopentadienyl anion**. Here too, we have a planar, cyclic, and fully conjugated system. The molecular orbitals can be derived from a Frost circle (Fig. 13.34).

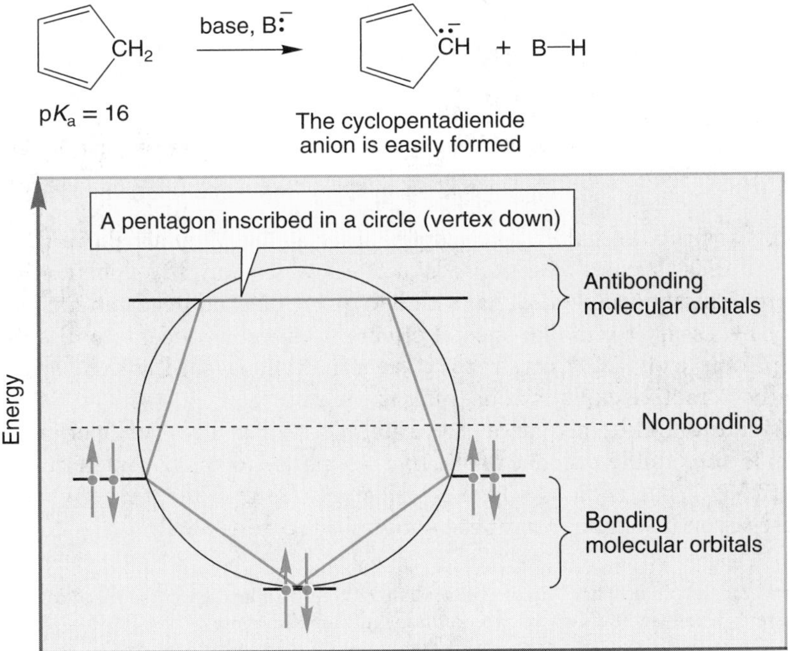

FIGURE **13.34** Cyclopentadiene is a quite strong acid. The cyclopentadienyl anion is an aromatic species. For an anion, it is extremely stable.

There are six electrons to put into the molecular orbitals, and, as in the tropylium ion or benzene, they fully occupy the set of three bonding molecular orbitals without requiring antibonding or nonbonding molecular orbitals to be occupied (Fig. 13.34). The cyclopentadienyl anion can be described as aromatic, and *for an anion,* this species is remarkably stable. Do not fall into the trap of expecting this anion to be as stable as benzene. Benzene is a neutral molecule in which all of carbon's valences are satisfied. The anion is charged and contains unsatisfied valence. It is very stable, but only in comparison with the rest of the family of anions.

But wait. We have been at pains to equate stability with delocalization of electrons, which results in a distribution of charge to several atoms in the molecule (Fig. 13.33). Both the tropylium ion and the cyclopentadienyl anion can be described by a most impressive array of resonance forms (Fig. 13.35).

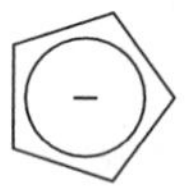

Cyclopentadienide ion

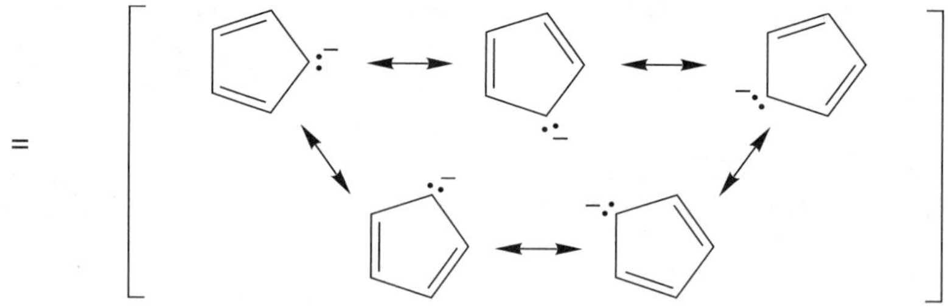

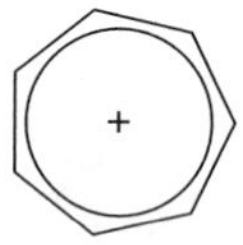

Cycloheptatrienylium ion (tropylium ion)

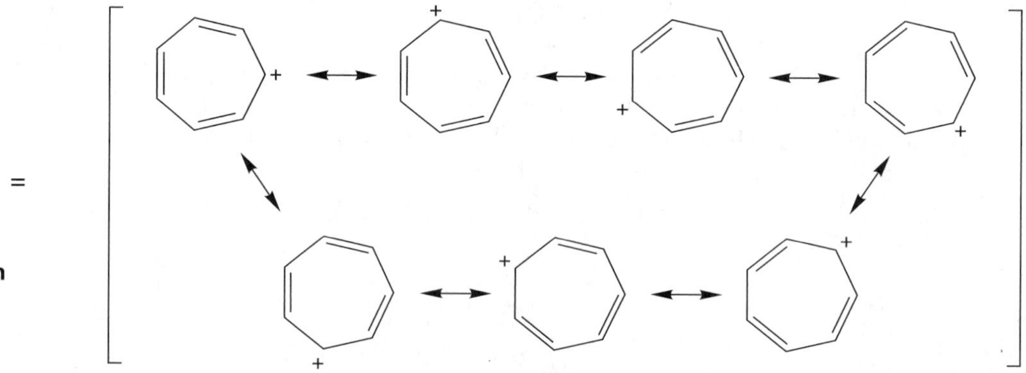

FIGURE 13.35 Resonance descriptions of the tropylium (cycloheptatrienylium) ion and the cyclopentadienyl anion.

Perhaps the aromaticity concept is unecessary. Why attribute the stability of these species to some special effect when simple delocalization of electrons may do as well? That's a good question, and needs an answer. It is not necessary to invoke special effects if simpler concepts will suffice. The principle of "Ockham's razor"* says that when faced with a variety of equally attractive explanations, pick the simplest.

Wielding Ockham's razor, we might say that the cyclopentadienyl anion is very stable because of the five resonance forms so apparent in Figure 13.35. What then would we expect of the cycloheptatrienyl anion, whose seven resonance forms are shown in Figure 13.36?

*Named for the philosopher William of Ockham (1285–1349; he died in the plague), who put it better: "What can be done with fewer is done in vain with more."

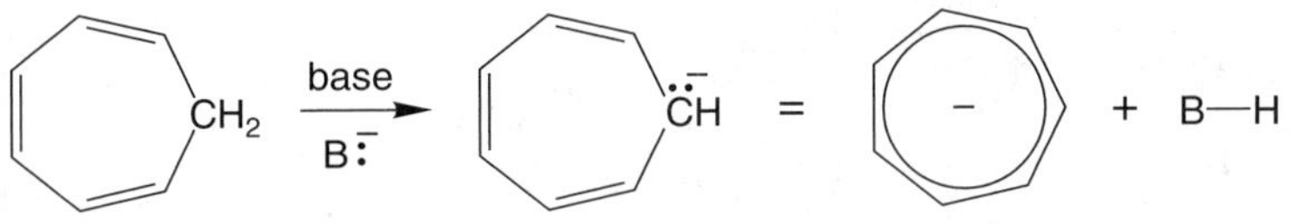

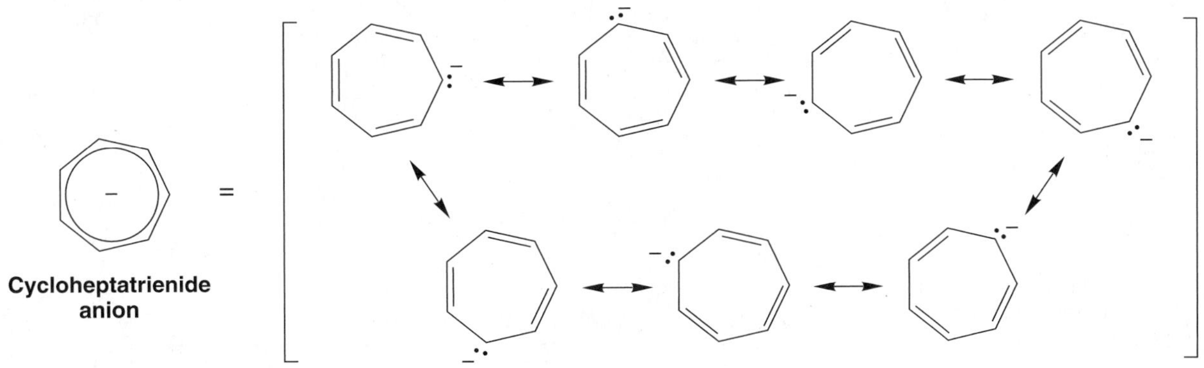

FIGURE 13.36 The seven resonance forms for the cycloheptatrienyl anion.

If an ion with five resonance forms is good, surely one with seven forms will be better. Our explanation predicts a formidable acidity for 1,3,5-cycloheptatriene; in particular, it must be more acidic than cyclopentadiene. Its pK_a must be < 16. But we are wrong; the pK_a of cycloheptatriene is about 39. Cycloheptatriene is a *weaker* acid than cyclopentadiene by a factor of 10^{23} (Fig. 13.37)!

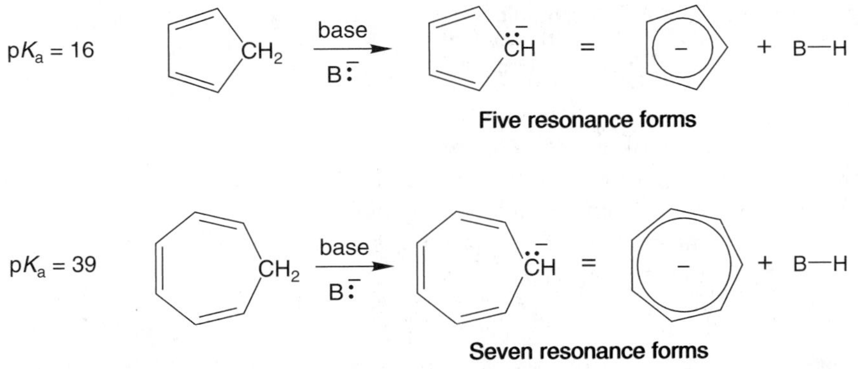

FIGURE 13.37 Cycloheptatriene is a weaker acid than cyclopentadiene by a factor of 10^{23}.

The number of resonance forms is *not* always a predictor of stability. No longer are we faced with equally attractive explanations. The aromaticity concept is necessary to make sense of the data. A look at the molecular orbitals of cycloheptatriene shows the source of the instability. The cycloheptatrienyl anion must contain electrons in antibonding orbitals (Fig. 13.38)!

Now let's search for other neutral compounds exhibiting aromaticity. Of course, all manner of substituted benzenes exist, but these hardly constitute exciting new examples. It is possible to combine benzene rings so that they share an edge in what is called a fused relationship. These molecules are also aromatic. Atoms other than carbon can be introduced into the ring while maintaining the sextet of electrons to produce **heterobenzenes**, fully conjugated, six-membered rings containing one or more noncarbon atoms (N, O, S, etc.). Some smaller rings also contain a sextet of π

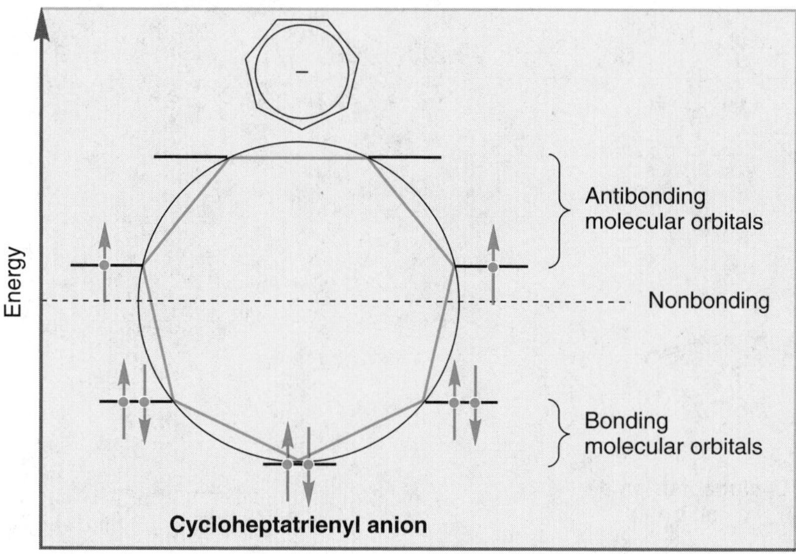

FIGURE **13.38** In the cycloheptatrienyl anion, two electrons must occupy antibonding orbitals.

electrons; pyrrole is an example. Molecules of these types are shown in Figure 13.39, and will be considered further in later sections.

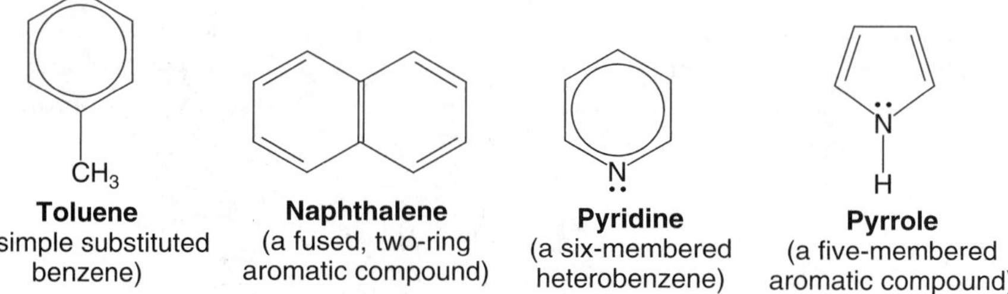

FIGURE **13.39** Some simple extensions of aromaticity: substituted, fused, and heterobenzene compounds, as well as a heteroaromatic compound, pyrrole.

What we are looking for here are not examples of relatively straightforward extensions, but new $4n + 2$ molecules. For benzene, and even the stable ions we have been considering, n has always been equal to 1, and $4n + 2 = 6$. If the concept of aromaticity is valid, there should be examples where $n = 0$, 2, 3, and so on.

PROBLEM 13.8 It is startlingly easy to form the cyclopropenyl cation (Fig. 13.40). Explain and classify the cyclopropenyl cation in terms of *n*.

PROBLEM 13.9 Do you expect the corresponding anion to be stable? Explain.

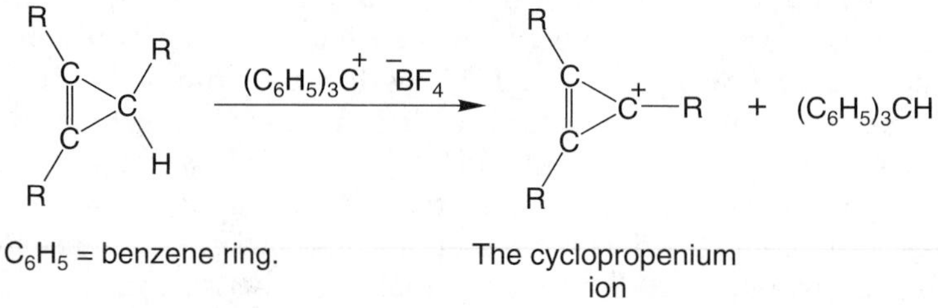

R, R, R, H (cyclopropene) $\xrightarrow{(C_6H_5)_3\overset{+}{C}\ \overset{-}{B}F_4}$ R, R, R (cyclopropenium cation) + $(C_6H_5)_3CH$

C_6H_5 = benzene ring.

The cyclopropenium ion

FIGURE **13.40**

Let's test this notion. Consider the cyclodecapentaenes, cyclic molecules with 10 π electrons ($4n + 2$, $n = 2$) (Fig. 13.41). Are they as stable as the six-electron molecules we have been examining?

***cis,cis,cis,cis,cis*-**
Cyclodecapentaene

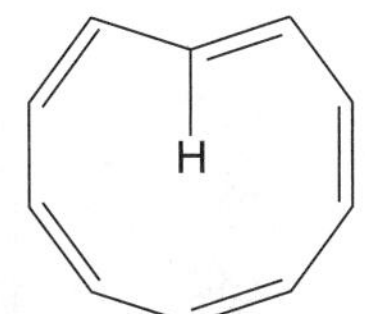

***trans,cis,cis,cis,cis*-**
Cyclodecapentaene

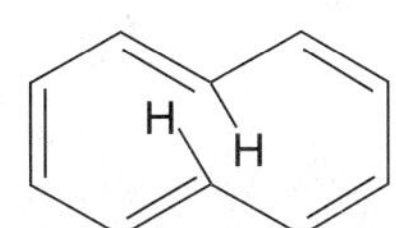

***trans,cis,trans,cis,cis*-**
Cyclodecapentaene

FIGURE **13.41** Some cyclodecapentaenes. These molecules have 10 π electrons: $4n + 2 = 10$, $n = 2$.

PROBLEM **13.10**

The several cyclodecapentaenes of Figure 13.41 differ in the stereochemistries of the double bonds. Why isn't there more than one isomer of benzene: a six-membered ring containing three double bonds?

A remarkable amount of effort was expended on the synthesis of the cyclodecapentaenes before they were successfully made. The magnitude of this effort should make us suspicious; benzenes are everywhere and extraordinarily easy to make. Moreover, the known cyclodecapentaenes are not only not especially stable, they are instead strikingly <u>un</u>stable. So, we have some explaining to do. We have been relying on the theory of special stability conferred by the aromatic $4n + 2$ number of π electrons, and the cyclodecapentaenes are members of the $4n + 2$ club. If we cannot explain why these molecules are unstable, our theory must be abandoned. It turns out not to be too difficult to resolve the problem. Aromatic stability is not some magical force. It is just one kind of stabilizing effect, which in the face of stronger destabilizing effects will not prevail. Put simply, the planar cyclodecapentaenes are destabilized by an amount greater than aromaticity can overcome. What might those destabilizing effects be? Look first at the molecule in Figure 13.42, the trans,cis,trans,cis,cis isomer. The two hydrogens inside the ring occupy the same space. This crowding induces extraordinary strain that is greater than any resonance stabilization (Fig. 13.42).

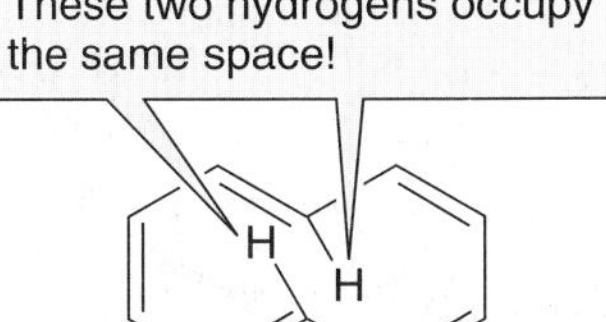

***trans,cis,trans,cis,cis*-**
Cyclodecapentaene

FIGURE **13.42** In *trans,cis,trans,cis,-cis*-cyclodecapentaene two inside hydrogens occupy the same space. Aromaticity cannot compensate for this strong destabilizing effect.

The trans,cis,cis,cis,cis isomer has some of the same problems, although here only one hydrogen is inside the ring. In this molecule, however, there is severe angle strain, and the planar form, required for aromaticity, is badly destabilized. The all-cis isomer has even more angle strain, and it too can be planar, and therefore aromatic, only at a prohibitively high energy cost. The benefits of delocalization are overwhelmed by the angle strain (Fig. 13.43).

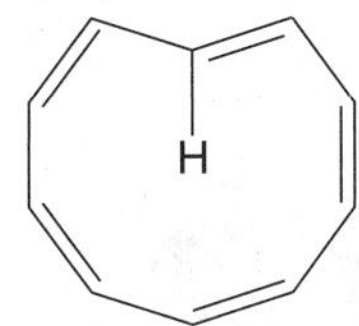

***trans,cis,cis,cis,cis*-**
Cyclodecapentaene
(angle and other strain caused by the inside hydrogen)

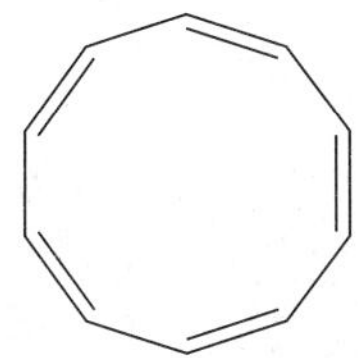

***cis,cis,cis,cis,cis*-**
Cyclodecapentaene
(angle strain)

FIGURE **13.43** *trans,cis,cis,cis,cis*-Cyclodecapentaene and the all-cis isomer have severe angle strain and cannot be planar. They are polyenes, not aromatic molecules.

PROBLEM **13.11** The cyclodecapentaenes actually escape their misery by forming a new, cross-ring bond (Fig. 13.44). As they warm up, a bicyclo[4.4.0]decatetraene is formed. Write an arrow formalism for this reaction.

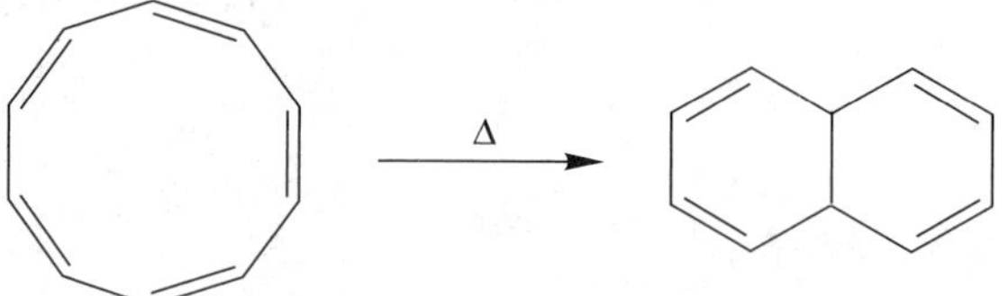
Δ
cis,cis,cis,cis,cis-Cyclodecapentaene
Bicyclo[4.4.0]deca-2,4,7,9-tetraene

FIGURE **13.44**

A good set of models will show this easily. Construct the planar, all-cis cyclodecapentaene and the strain will be apparent. If you are lucky, the plastic will mimic the molecule and the model will snap into a pretzel shape, which approximates the energy minimum form of the molecule. These molecules are best described as strained polyenes. They fail the test of planarity for aromatic molecules because the planar structures are much more strained than the nonplanar forms.

So we haven't really judged the limits of the theory of aromaticity by using the cyclodecapentaenes. We might be able to if the offending strain could somehow be removed. On paper this is simple. For example, we might just erase the two inside hydrogens in one of the molecules (Fig. 13.45).

A particularly clever German chemist, Emanuel Vogel (b. 1927), found a way to do this erasure chemically. He used a remarkable synthesis to replace the offending inside hydrogens with a bridging methylene group (Fig. 13.46).

H
H
***trans,cis,trans,cis,cis*-Cyclodecapentaene**
(strain caused by the inside red hydrogens)
magic hydrogen eraser

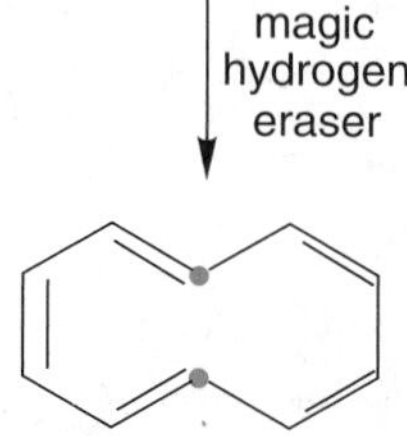
(much less strain! But ... what is going on at the red dotted positions?)

FIGURE **13.45** The strain caused by the two inside hydrogens in one cyclodecapentaene can be removed by removing the offending hydrogens. Unfortunately, this doesn't give a real molecule—there are two trivalent carbons at the "dotted" positions.

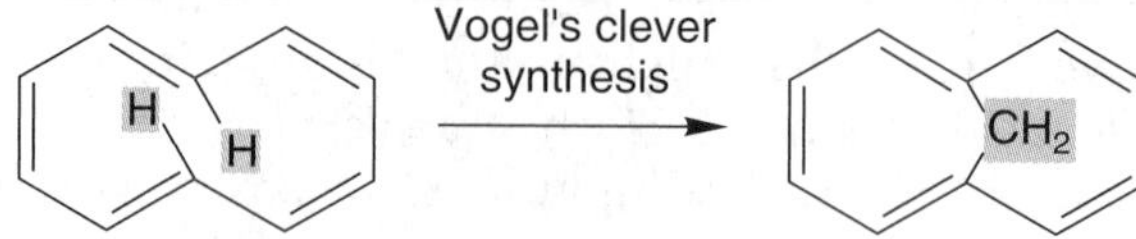
H
H
Vogel's clever synthesis
CH_2

FIGURE **13.46** Vogel's bridged cyclodecapentaene. The two inside hydrogens are gone, replaced with the red methylene group. The removal of the offending hydrogens eliminates the strain that comes from their "bumping."

The resulting molecule, a bridged 10 π electron system, isn't a perfect cyclodecapentaene. The overlap between the 2*p* orbitals at the bridgehead and adjacent positions isn't optimal, for example (Fig. 13.47). Still, the molecule maintains enough orbital overlap to remain fully conjugated and satisfy the criteria for aromaticity. Both its chemical and physical properties allow it to be classified as aromatic.

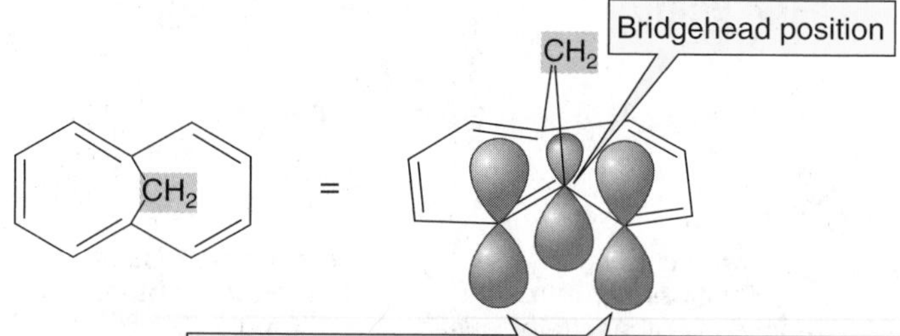
CH_2
=
CH_2
Bridgehead position
Overlap here is not ideal (make a model!), but it is good enough to allow substantial electron delocalization

FIGURE **13.47** Vogel's molecule is fully conjugated, but orbital overlap is not optimal, as it is, for example, in benzene.

PROBLEM 13.12

There are ions containing 10 π electrons. One of them, the cyclooctatetraene dianion, has long been described as a triumph of the theory of aromaticity. Here we have a relatively stable *dianion.* Aromaticity must compensate for strain effects as well as the repulsive destabilization of the two negative charges (Fig. 13.48). Yet this is not a comforting example, at least to me. It certainly represents a synthetic triumph [it was made in 1960 by Professor Thomas J. Katz (b. 1936) of Columbia University] and is a fascinating molecule, but I am not sure that it really supports the theory of aromaticity. From a careful examination of the molecular orbital picture, explain why I am discomforted by this example. *Hint*: Look carefully at our definition of aromaticity on page 579. Does this dianion really fit?

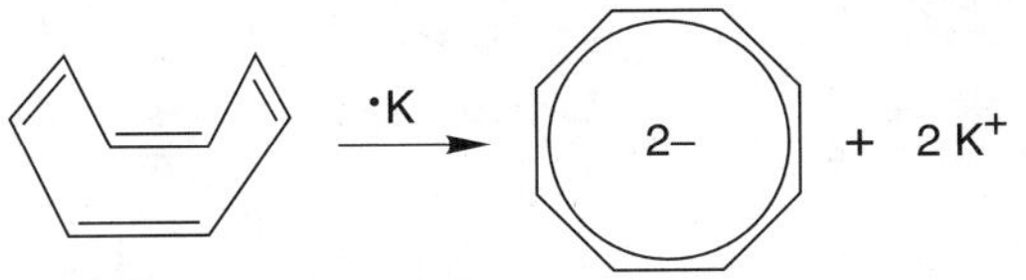

•K → 2– + 2 K^+

FIGURE 13.48

13.6 ANNULENES

The generic term for monocyclic, fully conjugated molecules is **annulene.** Benzene could be called [6]annulene, square cyclobutadiene is [4]annulene, and planar cyclooctatetraene is [8]annulene (Fig. 13.49).

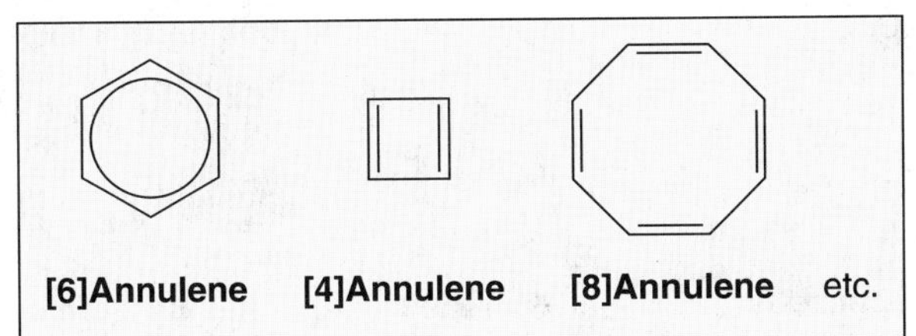

FIGURE 13.49 Some simple annulenes.

Other annulenes appear to be aromatic. These annulenes are large enough so that the destabilizing steric effects present in the cyclodecapentaenes ([10]annulenes), are diminished, and severe angle strain is avoided. A nice example is [18]annulene, a planar molecule in which all carbon–carbon bond lengths are very similar. It contains $4n + 2$ π electrons, where $n = 4$ (Fig. 13.50).

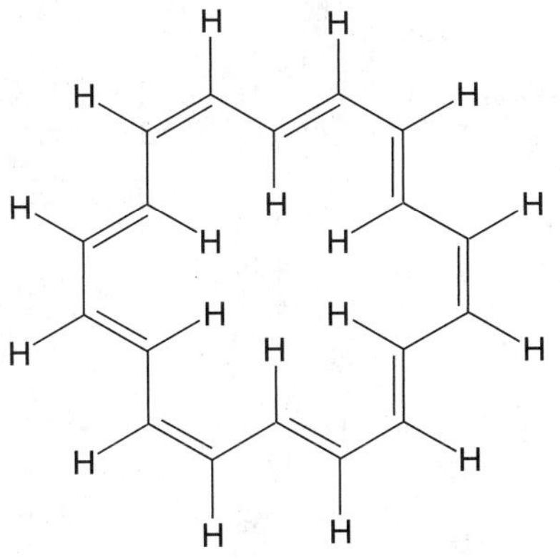

[18]Annulene

FIGURE 13.50 [18]Annulene, a planar, aromatic molecule. This molecule has $4n + 2$ π electrons ($n = 4$).

X

Generic name
Phenyl X

CH_3

Toluene
(methylbenzene)

CH_2CH_3

Ethylbenzene

$CH(CH_3)_2$

Cumene
(isopropylbenzene)

NH_2

Aniline
(benzenamine or aminobenzene)

OH

Phenol
(benzenol or hydroxybenzene)

OCH_3

Anisole
(methyl phenyl ether or methoxybenzene)

Br

Bromobenzene

$HC{=}CH_2$

Styrene
(vinylbenzene or ethenylbenzene)

COOH

Benzoic acid
(benzenecarboxylic acid)

CHO

Benzaldehyde
(benzenecarboxaldehyde)

Biphenyl

CH_2X

Benzyl X
Generic name

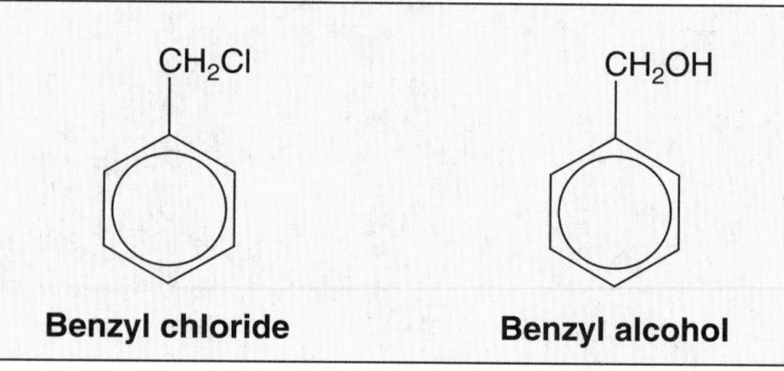

FIGURE 13.51 Some simple monosubstituted benzenes. Some common names are widely used.

The description of this molecule as aromatic derives largely from an examination of its nuclear magnetic resonance (NMR) spectrum and that must wait for Chapter 15, but [18]annulene does appear to be aromatic.

So far, we have concentrated on the phenomenon of aromaticity as applied to benzene itself and larger and smaller polyenes. We saw two experimental methods of evaluating the stabilization of aromaticity quantitatively, and with the aid of Hückel's insight, we found a generalization of the phenomenon. Now we move on to ways of varying the structure of our prototypal example, benzene, and then take a quick look at reactivity.

13.7 SUBSTITUTED BENZENES

13.7a Monosubstitution

Monosubstituted benzenes are usually named as derivatives of benzene, and the names are written as single words. There are numerous common names that have survived a number of attempts to systemize them out of existence. You really do have to know a number of these names in order to be literate in organic chemistry. I don't think the systematic "benzenol" will replace "phenol" in your lifetime, for example. At least I hope it doesn't. Figure 13.51 shows some simple benzenes with both the systematic and common names.

Two of the most often used common names are **phenyl** (sometimes even **Ph**), which stands for a benzene ring, and **benzyl**, which stands for the Ph–CH_2 group. Now is a good time to summarize the various abbreviations for benzene rings, and Figure 13.52 does just that.

Convention Alert!

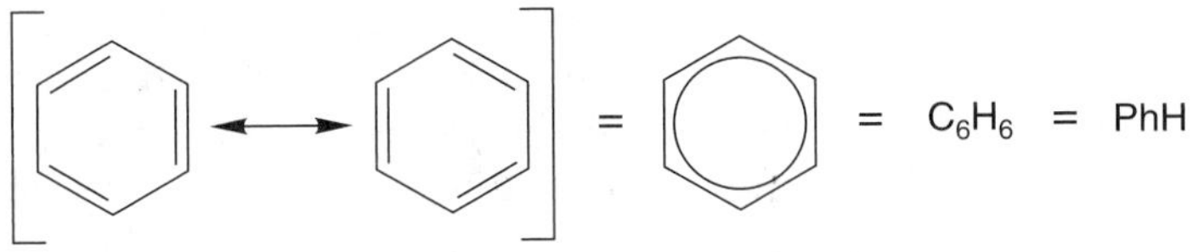

= C_6H_6 = PhH

A phenyl group
Ph = C_6H_5 = ϕ
A benzyl group
$PhCH_2$ = $C_6H_5CH_2$ = Bz

FIGURE 13.52 Abbreviations for benzene groups and benzyl groups.

There are three possible isomers of disubstituted benzenes. 1,2-Substitution is known as **ortho** (= o-), 1,3-substitution as **meta** (= m-), and 1,4-substitution as **para** (= p-).

*PROBLEM 13.13

Indeed, the isolation of three, and only three isomers of disubstituted benzenes was a crucial factor in the assignment of the Kekulé form as the correct structure of benzene. Use the fact that there are three and only three achiral isomers of a disubstituted benzene ($C_6H_4R_2$) to criticize the suggestions that benzene might have the following structures (Fig. 13.53).

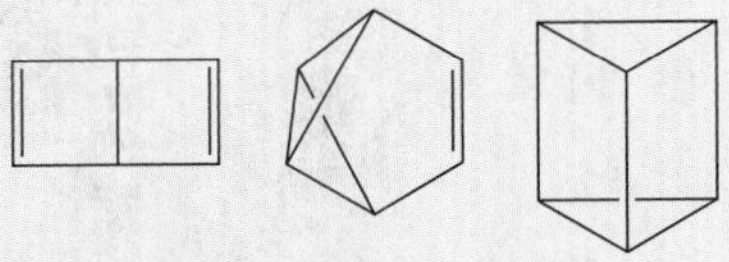

FIGURE 13.53

ANSWER

The first two are easy, and it is relatively simple to find that there are more than three possible disubstituted versions. For Ladenburg benzene (prismane) there

ANSWER (CONTINUED) are only three positional possibilites, but one is chiral. So none of these molecules can be the real benzene.

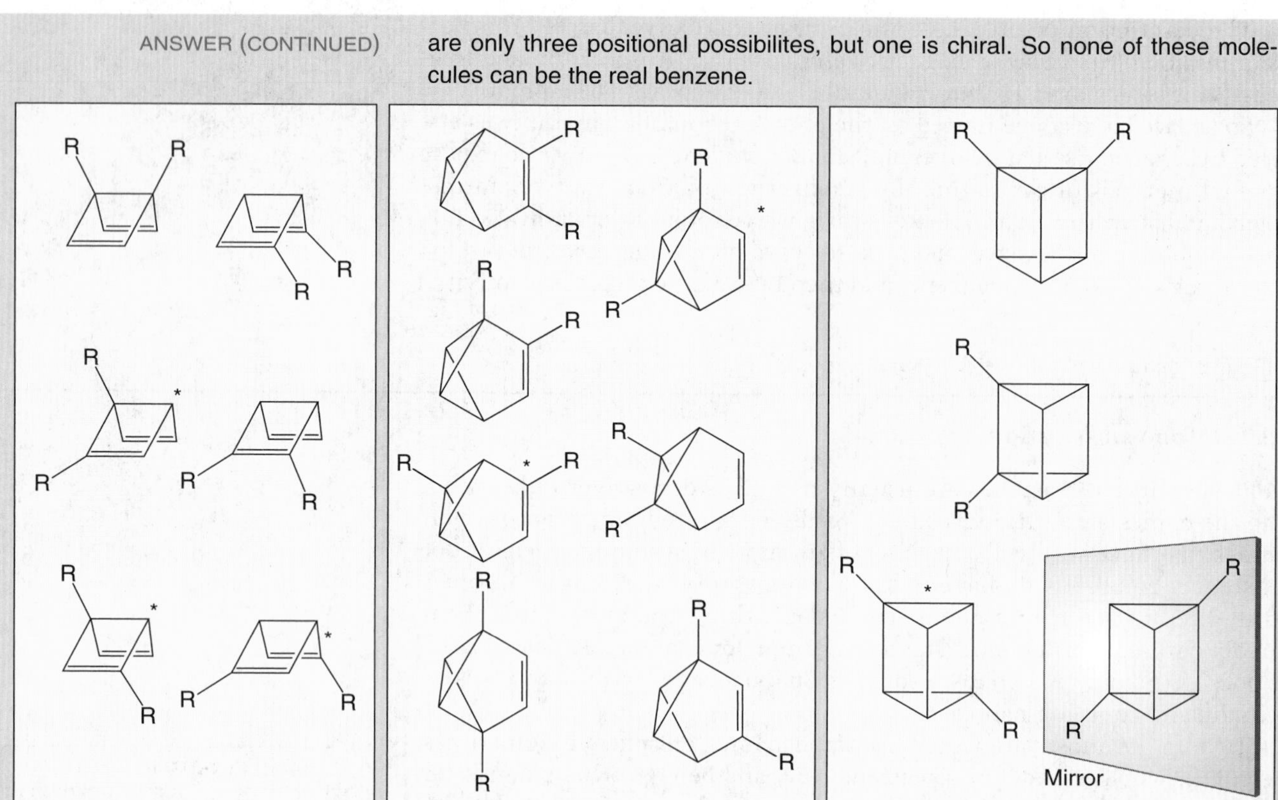

* = chiral isomers

In practice, either the numbers or the "ortho, meta, para" designations are used. If there is a widely used common name, it is used as a base for the full name. For example, two of the molecules of Figure 13.54 are quite properly named as *o*-bromotoluene and *p*-nitrophenol, respectively.

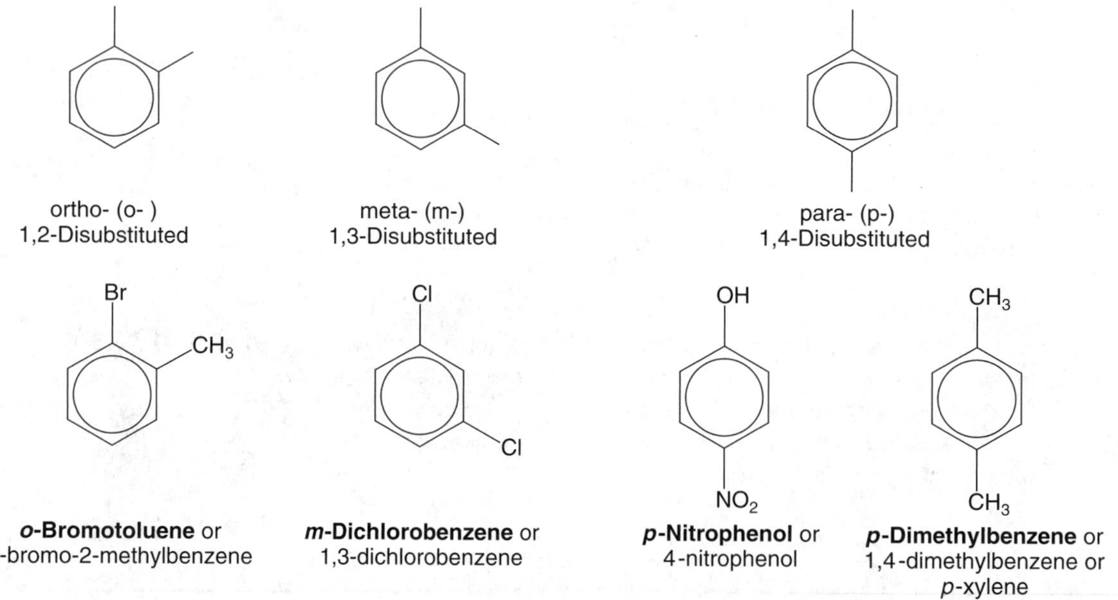

FIGURE **13.54** The three possible substitution patterns for disubstituted benzenes and some disubstituted benzenes.

13.7b Polysubstitution

There are polysubstituted benzenes that are also known by common names. Figure 13.55 shows a sampling of these molecules.

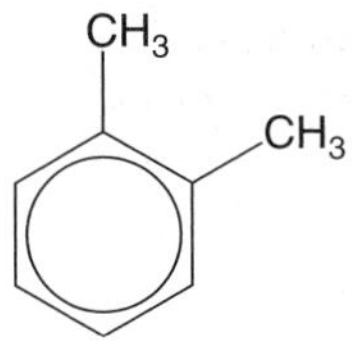

***o*-Xylene**
(1,2-dimethylbenzene)

***m*-Xylene**
(1,3-dimethylbenzene)

***p*-Xylene**
(1,4-dimethylbenzene)

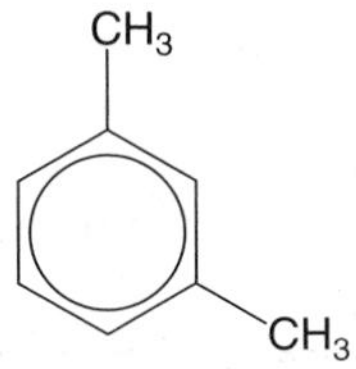

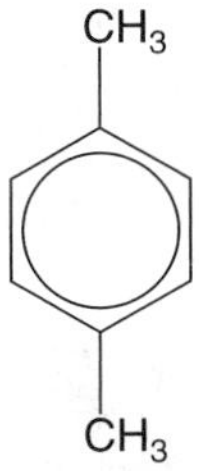

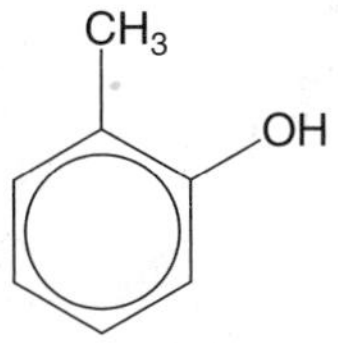

***o*-Cresol**
(2-methylphenol)

***m*-Cresol**
(3-methylphenol)

***p*-Cresol**
(4-methylphenol)

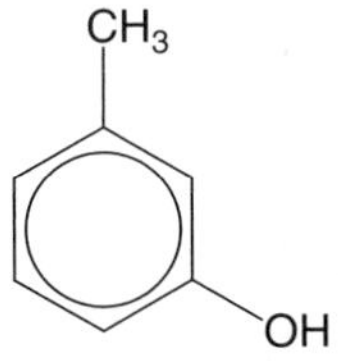

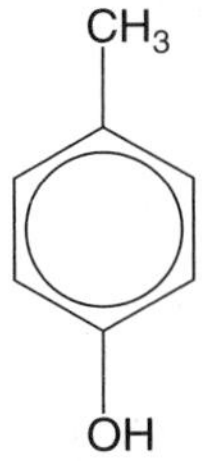

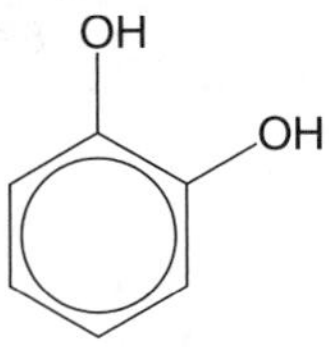

Catechol
(*o*-dihydroxy-benzene)

Resorcinol
(*m*-dihydroxy-benzene)

Hydroquinone
(*p*-dihydroxy-benzene)

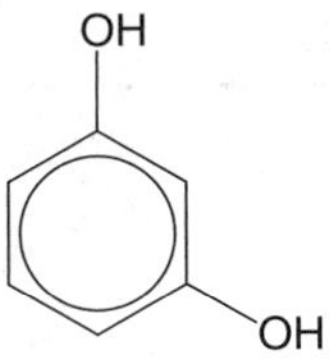

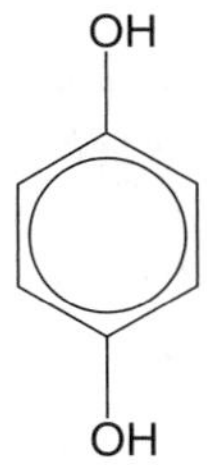

FIGURE 13.55 Some common names for disubstituted benzenes.

The substituents on polysubstituted benzenes are numbered and named alphabetically. Figure 13.56 gives a few examples.

Mesitylene
(1,3,5-trimethylbenzene)

1,3.5-Trichlorobenzene

2-Bromo-4-chloro-6-fluorophenol

4-Bromo-2-chlorotoluene

2,4-Dibromobenzoic acid

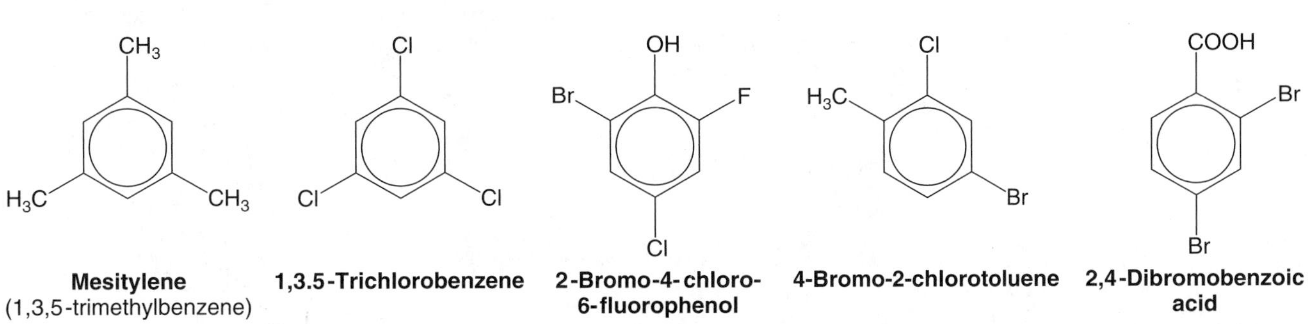

FIGURE 13.56 Some named trisubstituted benzenes.

13.8 PHYSICAL PROPERTIES OF SUBSTITUTED BENZENES

The physical properties of substituted benzenes resemble those of alkanes and alkenes of similar shape and molecular weight. Table 13.1 collects some physical properties for a number of common substituted benzenes. Notice the effects of symmetry. *p*-Xylene melts at a much higher temperature than the ortho or meta isomer, for example. Many para isomers have high melting points, and crystallization can sometimes be used as a means of separating the para regioisomer from the others.

TABLE 13.1 Physical Properties of Some Substituted Benzenes

Name	bp (°C)	mp (°C)	Density (g/mL)
Cyclohexane	80.7	6.5	0.78
Benzene	80.1	5.5	0.88
Methylcyclohexane	100.9	−126.6	0.77
Toluene	110.6	− 95	0.87
o-Xylene	144.4	− 25.2	0.88
m-Xylene	139.1	− 47.9	0.86
p-Xylene	138.3	13.3	0.96
Aniline	184.7	− 6.3	1.02
Phenol	181.7	43	1.06
Anisole	155	− 37.5	1.0
Bromobenzene	156.4	− 30.8	1.5
Styrene	145.2	− 30.6	0.91
Benzoic acid	249.1[a]	122.1	1.1
Benzaldehyde	178.6	− 26	1.0
Nitrobenzene	210.8	5.7	1.2
Biphenyl	255.9	71	0.87

[a] At 10-mm pressure.

Convention Alert!

Aromatic compounds are called **arenes** and the general abbreviation "Ar = aryl" is often used, in analogy to the R used for a general alkyl species.

13.9 HETEROBENZENES AND OTHER HETEROCYCLIC AROMATIC COMPOUNDS

If a CH unit of benzene is replaced with a nitrogen, the result is **pyridine**, another aromatic six-membered ring (Fig. 13.57). From the provincial view of organic chemistry all heavy atoms except carbon are called "heteroatoms," and pyridine is a heterobenzene, a member of the general class of compounds called **heteroaromatic compounds**, aromatic molecules con-

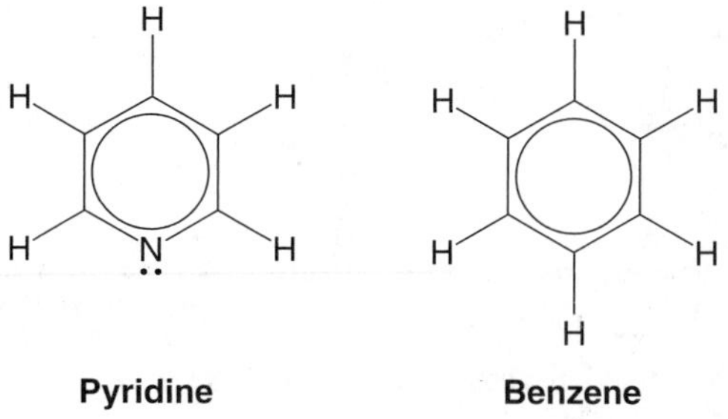

Pyridine **Benzene**

FIGURE 13.57 Pyridine, a heterobenzene.

taining at least one non-carbon heavy atom.

The electron counting for pyridine can be confusing at first because of the extra pair of electrons on nitrogen. Doesn't pyridine have eight π electrons and thus violate the $4n + 2$ rule? Put a better way, we have to worry whether antibonding orbitals are occupied in pyridine. A good orbital drawing shows that the "extra" two electrons are not in the π system, and pyridine is not in violation of the Hückel rule (Fig. 13.58).

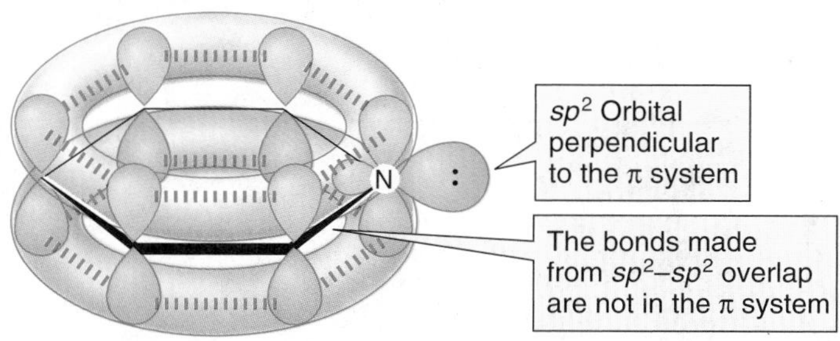

FIGURE 13.58 Pyridine has only 6 π electrons. The lone-pair electrons on nitrogen are in an sp^2 orbital perpendicular to the π system, not in the π system itself.

Other six-membered rings containing one heteroatom are known, but none is very stable. Oxabenzene must be cationic and the charge will contibute to making it a reactive molecule. Aromaticity must overcome the destabilizing influences of the charged atom and doesn't succeed in this case (Fig. 13.59).

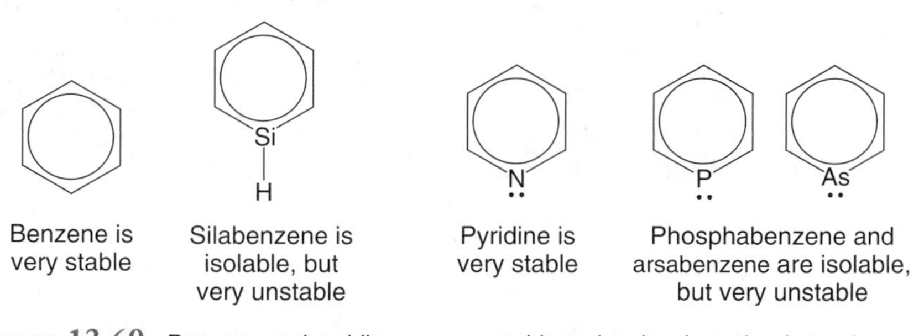

FIGURE 13.59 Oxabenzene must be charged and is unlikely to be very stable.

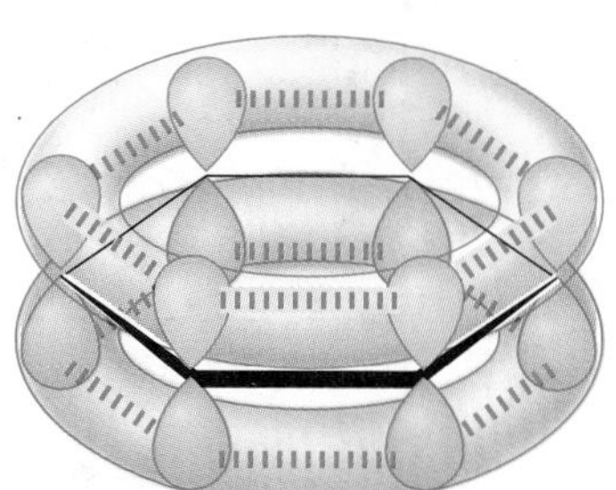

More curious is the relative instability of silabenzene, and the second- and third-row analogues of pyridine, phosphabenzene, and arsabenzene (Fig. 13.60).

Silabenzene is isolable, but very unstable

FIGURE 13.60 Benzene and pyridine are very stable molecules, but other heterobenzenes are unstable.

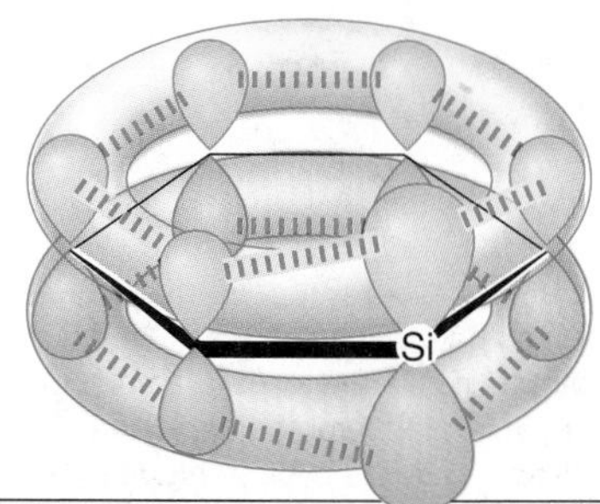

Si
In silabenzene, the silicon atom provides a $3p$ orbital, not a $2p$ orbital, to the π system; overlap is hinderedby size differences in the orbitals as well as by the long C–Si σ bonds; the extra nodes in the $3p$ orbital are not shown

FIGURE 13.61 The C ($2p$)–Si ($3p$) overlap is not optimal, and the carbon–silicon σ bonds are quite long. This poor overlap destabilizes silabenzene and the other heteroaromatics of Figure 13.60.

These molecules are all known but each is extremely reactive. In fact, it is quite a challenge just to detect these compounds. Unlike oxabenzene, there is no problem of charge here, but there is one of poor orbital overlap. Silicon is in the third row of the periodic table, and so in silabenzene there is no sextet of overlapping $2p$ orbitals but instead there is a ring of five $2p$ orbitals and one $3p$ orbital (Fig. 13.61).

Delocalization is hindered by the discrepancy in size between the $2p$ and $3p$ orbitals and by the long carbon–silicon σ bonds. Put another way, there is less stabilization to be gained from the overlap of a $2p$ orbital with a $3p$ orbital than there is from overlap of two $2p$ orbitals. Stabilization is maximized when the energies of the overlapping orbitals are equal, as they are for a pair of $2p$ orbitals. The principal quantum number (n) is higher for silicon's $3p$ orbital and this orbital will not match a $2p$ orbital well in energy (Fig. 13.62). Similar problems exist for phosphabenzene and arsabenzene.

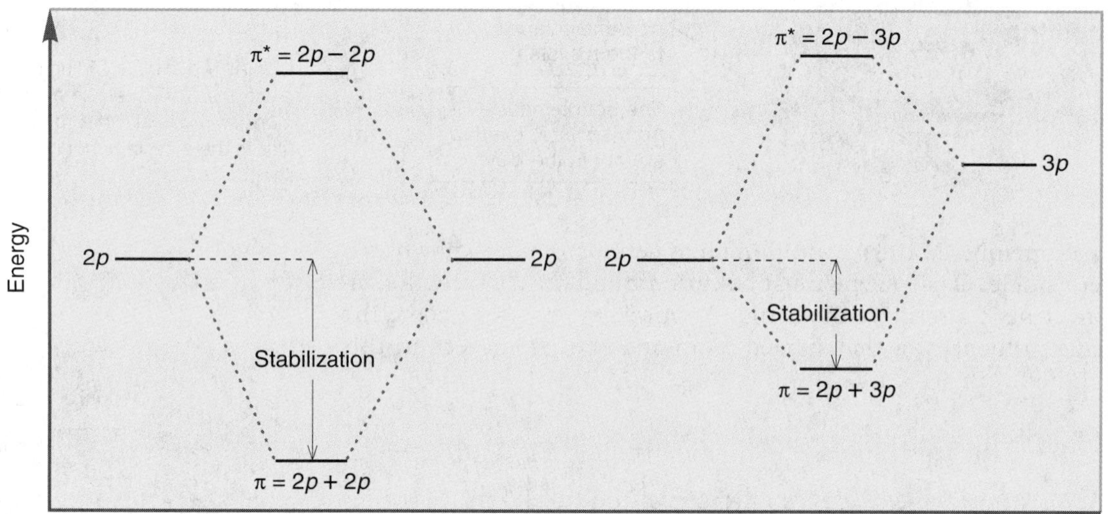

FIGURE 13.62 The stabilization afforded by overlap of the two degenerate (equi-energetic) $2p$ orbitals is greater than that from $2p$–$3p$ overlap.

The neutral, five-membered heterocyclic ring compounds, **pyrrole** and **furan**, also show aromatic character. Even **thiophene** is apparently aromatic (Fig. 13.63).

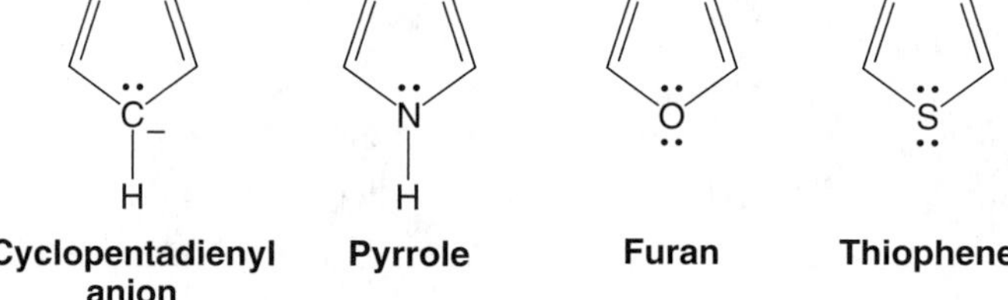

FIGURE 13.63 Four five-membered aromatic rings. There are still $4n + 2$ π electrons in the three neutral molecules, which are isoelectronic (same number of electrons) with the cyclopentadienyl anion.

Furan and pyrrole can complete an aromatic sextet of π electrons by using a pair of nonbonding electrons in a $2p$ orbital. These molecules have the same number of π electrons as the cyclopentadienyl anion, but none of the problems induced by the charge on the all-carbon molecule. In thiophene and furan, there is a second pair of electrons remaining in an sp^2 orbital perpendicular to the π orbital system (Fig. 13.64).

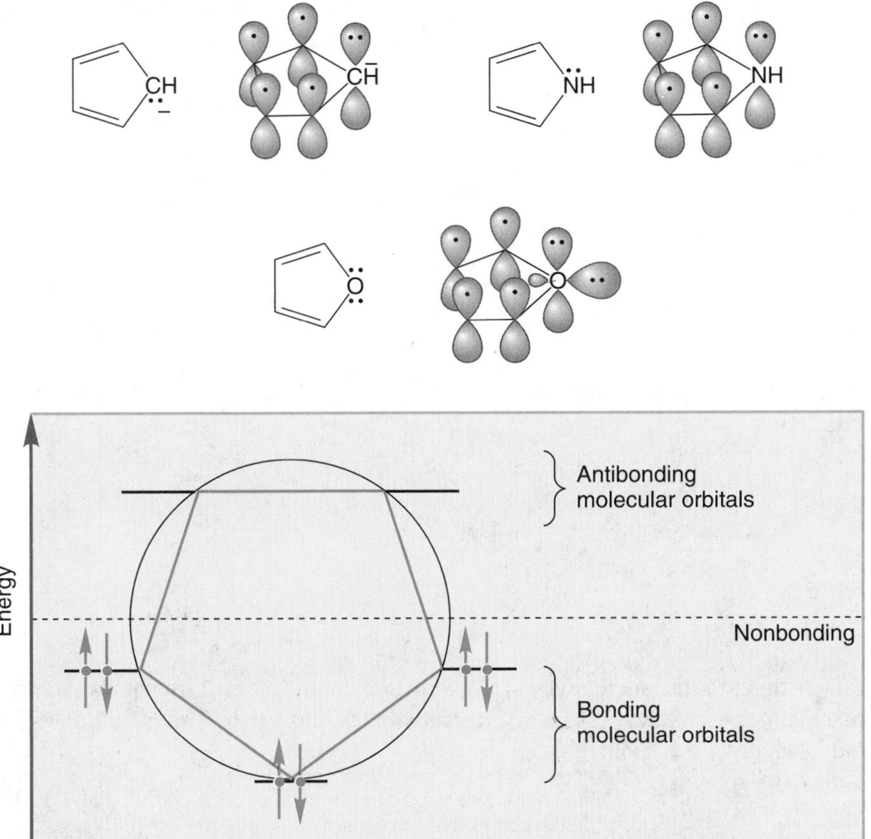

FIGURE 13.64 Orbital pictures of the isoelectronic five-membered aromatic compounds.

13.10 POLYNUCLEAR AROMATIC COMPOUNDS

At this point, we can see benzene as a building block from which more complicated, polycyclic molecules can be constructed. For example, benzene and other aromatic rings can share edges to form fused, **polynuclear aromatic compounds**. The simplest of these is naphthalene, in which two benzene rings share an edge. In quinoline and isoquinoline, it is a benzene ring and a pyridine that share an edge (Fig. 13.65).

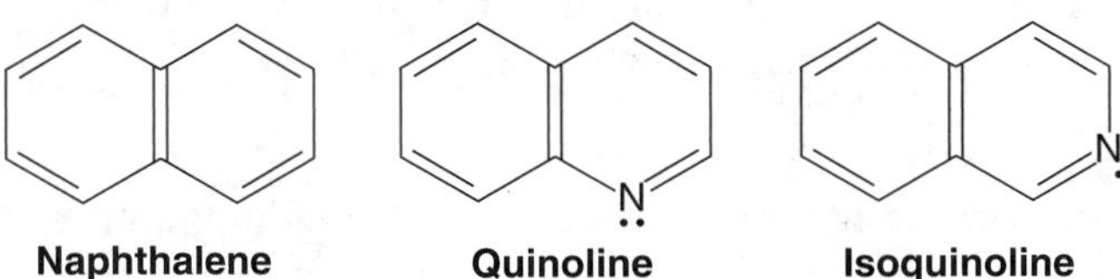

FIGURE 13.65 Three two-ring fused aromatic compounds.

PROBLEM 13.14

Draw schematic orbital pictures of naphthalene and isoquinoline. Put in the π electrons as dots. It is not necessary to draw the carbon–hydrogen σ bonds.

PROBLEM 13.15

Can two benzene rings share a single carbon to form a spiro substituted compound?

Clearly we can continue this building process. Three fused benzene rings produces either anthracene or phenanthrene (Fig. 13.66).

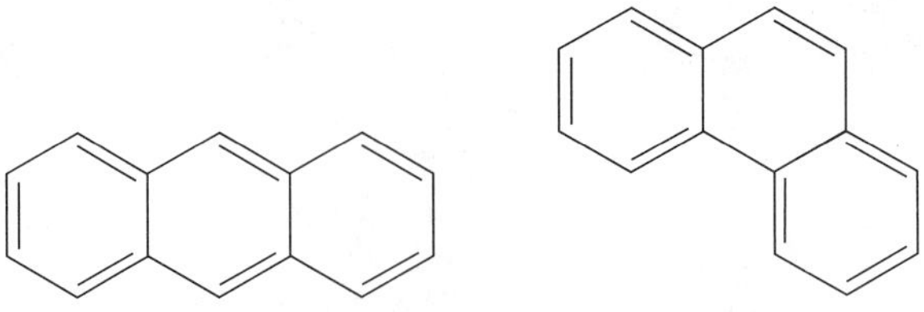

Anthracene Phenanthrene

FIGURE 13.66 Aromatic hydrocarbons containing three fused benzene rings.

*PROBLEM 13.16 Why is the following molecule not stable (Fig. 13.67)? *Hint*: Kekulé forms are often useful.

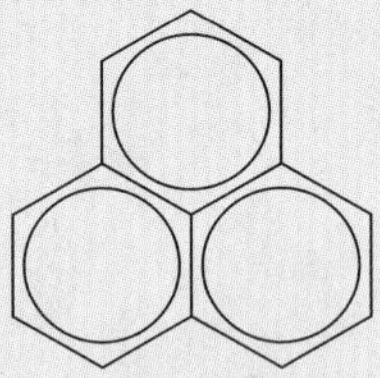

FIGURE 13.67

ANSWER The circles are misleading. Drawing the molecule in Kekulé form allows you to see that there is no way for each carbon to be hybridized sp^2 if each carbon has four valences.

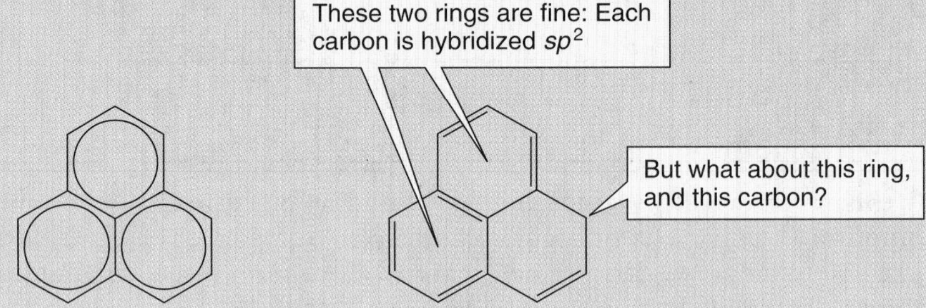
These two rings are fine: Each carbon is hybridized sp^2
But what about this ring, and this carbon?

PROBLEM 13.17 Draw all the fused polynuclear aromatic compounds made from four benzene rings. Be careful. This question is much harder than it seems.

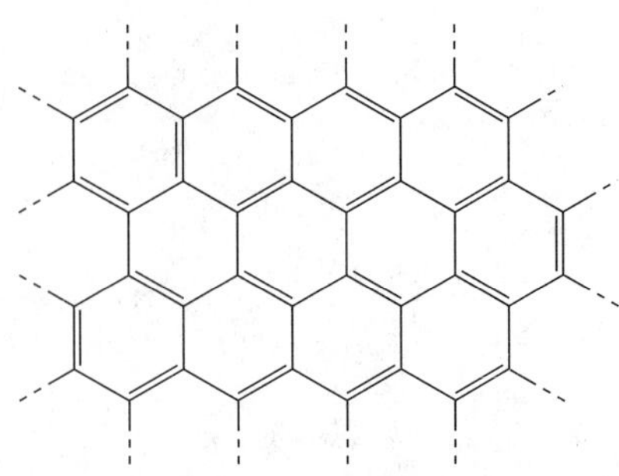

FIGURE 13.68 Graphite, an infinitely extended planar array of fused benzene rings.

The limit of the ring fusing business is reached in graphite, a polymer composed of fused benzene rings extended in what amounts to an infinite plane. Of course, the plane is not really infinite, but it is large enough so that the edges do not affect the physical or chemical properties of graphite. Graphite's lubricating properties are related to the ease of sliding the planar layers of graphite over each other (Fig. 13.68).

In the mid-1980s, R.F. Curl (b. 1933), H. W. Kroto (b. 1939), R. E. Smalley (b. 1943), and their co-workers were attempting to explain the striking abundance of certain fragments observed when a carbon rod was vaporized. In the face of much skepticism in the chemical community, they proposed that the major fragment, C_{60}, had the structure of a soccer ball of pure carbon. They turned out to be absolutely right, and named this molecule "buckminsterfullerene" after R. Buckminster Fuller (1895–1983), the inventor of the geodesic dome. This compound, which can now be made easily in multigram lots through the vaporization of carbon rods, is a new form of carbon, and a testimony to the proposition that organic chemistry is hardly played out as a provider of exciting new molecules (Fig. 13.69). Carl, Kroto, and Smalley shared the 1996 Nobel prize in chemistry for this discovery.

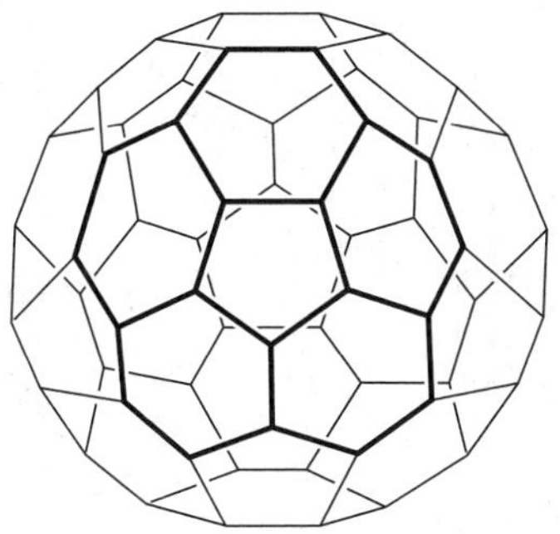

FIGURE **13.69** Buckminsterfullerene; C_{60}, a soccer ball shaped form of carbon. The double bonds (or circles) are left out for clarity. The molecule is fully aromatic.

PROBLEM **13.18** The ring-fusing process produces some interesting molecules along the way to graphite (Fig. 13.70). "Hexahelicene" and "twistoflex" are chiral, for example. Why?

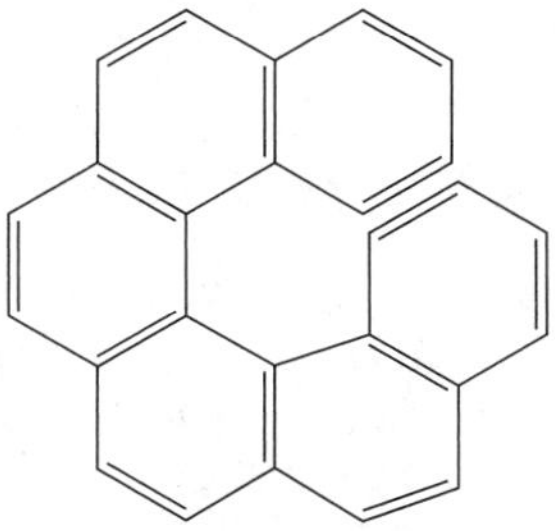

Hexihelicene

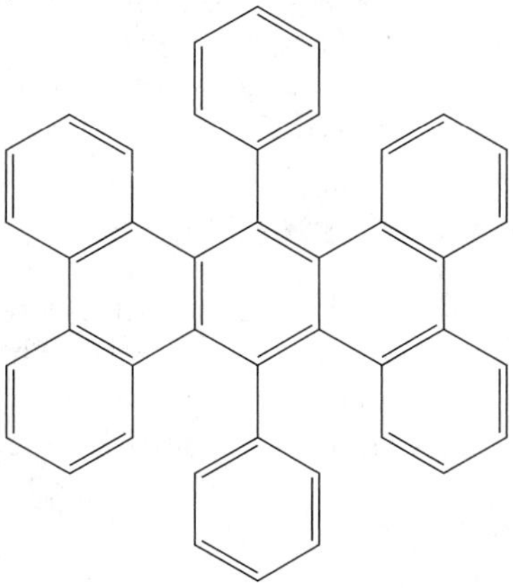

Twistoflex

FIGURE **13.70**

Expanding this building block process to generate ever larger and varied polycyclic aromatic compounds, may be fun, but there is a downside to the process. Many of the polycyclic aromatic compounds (and some of the simple ones as well), are carcinogenic. Some of the worst are shown in Figure 13.71. A prudent person avoids exposure to the known strong carcinogens, and minimizes contact with related aromatic compounds. It is less easy to avoid such compounds than it sounds. It is clearly simple to avoid bathing in benzene, and usually (but not always) possible to substitute other molecules when benzene is called for in a chemical recipe. However, one of the bad carcinogens, **benz[*a*]pyrene**, is a product of the internal combustion engine as well as combustion (cigarettes, charcoal-broiled meat) in general. The control of such molecules is not only physically difficult, but intersects with politics and economics as well. This situation creates serious problems of social policy for, beyond the questions of public health, there are fortunes to be made and elections to be won. Such matters seem often to complicate the purely scientific questions!

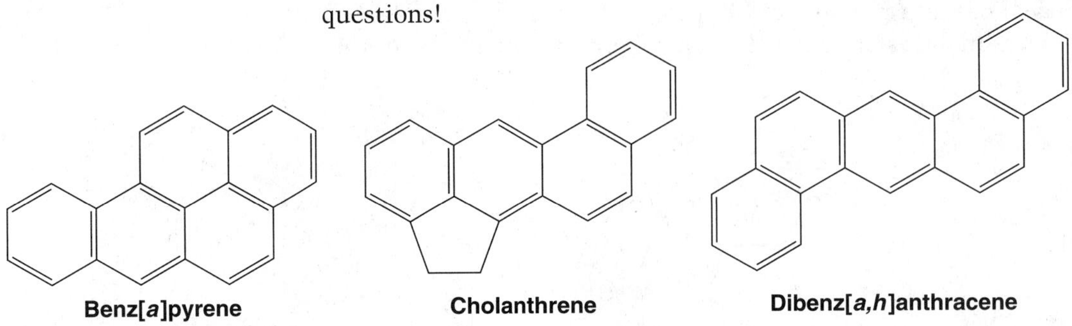

Benz[*a*]pyrene　Cholanthrene　Dibenz[*a,h*]anthracene

FIGURE 13.71 Some carcinogenic aromatic hydrocarbons.

13.11 INTRODUCTION TO THE CHEMISTRY OF BENZENE

We will take a quick look here at reactions. We will first introduce a general reaction that preserves the aromatic character of benzene. It is risky even to suggest that molecules have desires, but the imperative in benzene chemistry definitely is, Preserve the Aromatic Sextet! A less flamboyant way of saying this is to point out that thermodynamics will surely favor very stable molecules, and therefore benzene chemistry is likely to lead to more very low energy aromatic compounds. However, occasionally there are useful reactions in which the aromatic stabilization is lost, and we will see one of them soon in Section 13.11b.

13.11a Aromatic Substitution

We have already mentioned that benzene does not react with Br_2 or HBr, as do almost all alkenes (p. 568). This lack of reactivity results from the stabilization of benzene by aromaticity, which produces activation barriers that are too high for the addition reactions (Fig. 13.72).

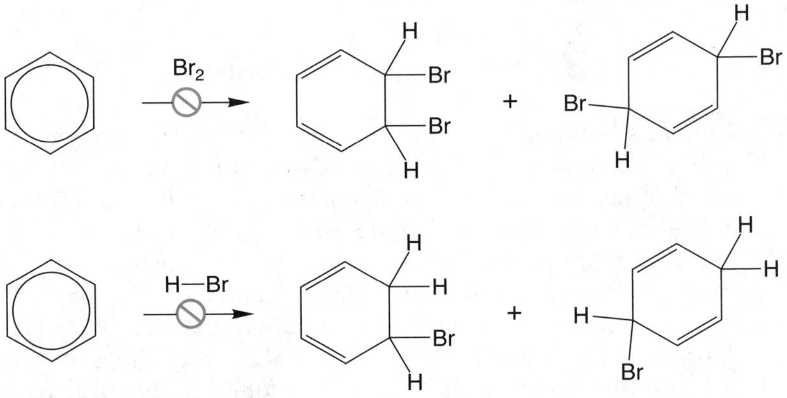

Br_2 H Br Br H + H Br Br H

H—Br H H Br H + H H H Br

FIGURE 13.72 Benzene does not react with Br_2 or HBr. Simple addition reactions do not occur.

If we use a deuterated acid we can see that there is a reaction taking place, at least in strong acid, but the product is merely the exchanged, deuterated benzene, not an addition product (Fig. 13.73).

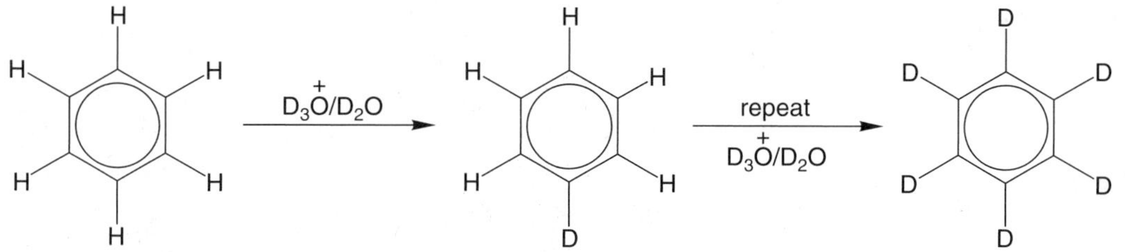

FIGURE 13.73 The hydrogens of benzene can be exchanged for deuterium in deuterated acid.

What is happening? Why is the reaction diverted from ordinary addition? The answers to these questions will serve to summarize much of the chemistry of aromatic compounds, which will be encountered in detail in Chapter 14. In deuterio acid, addition of a deuteron gives a resonance-stabilized carbocation, but aromaticity is lost (Fig. 13.74).

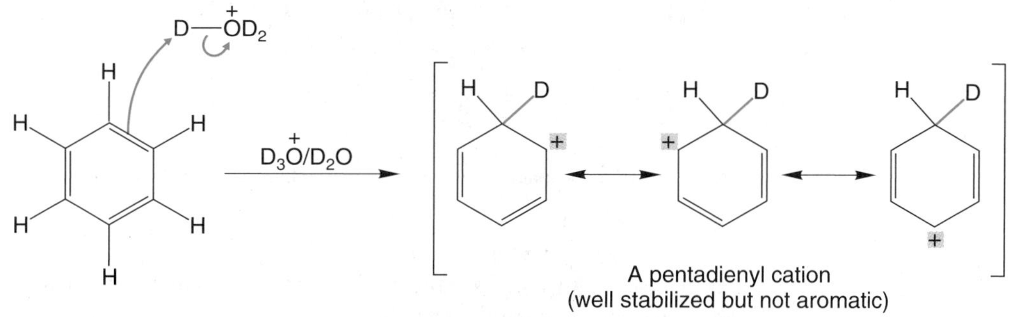

FIGURE 13.74 Addition of a deuteron to benzene gives a resonance-stabilized, but nonaromatic, carbocation.

*PROBLEM 13.19

Be sure you see that the ion in Figure 13.74 is not aromatic. Why is aromaticity lost?

ANSWER

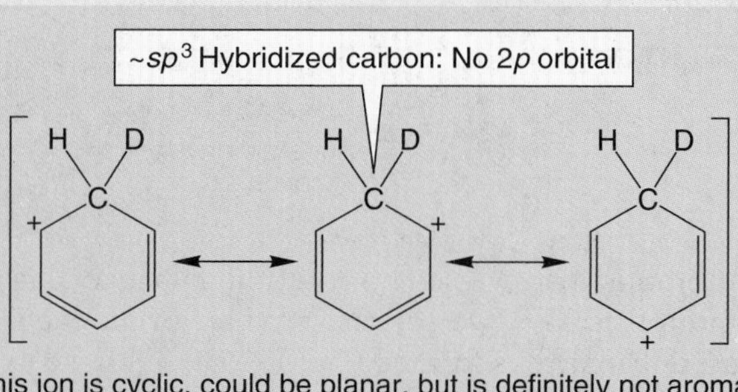

This ion is cyclic, could be planar, but is definitely not aromatic, as the CHD group interrupts the connectivity of 2*p* orbitals around the ring; this molecule is not fully conjugated

In the normal addition reaction of alkenes or dienes, a nucleophile such as bromide would attack this carbocation to give the final product. In this case, we would anticipate the two products of 1,2- and 1,4-addition (Fig. 13.75).

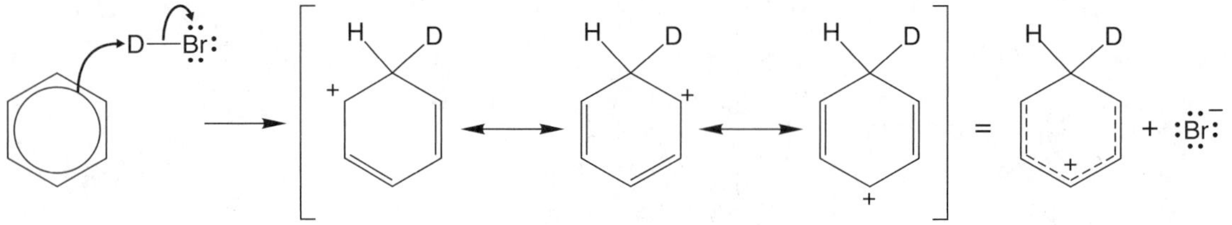

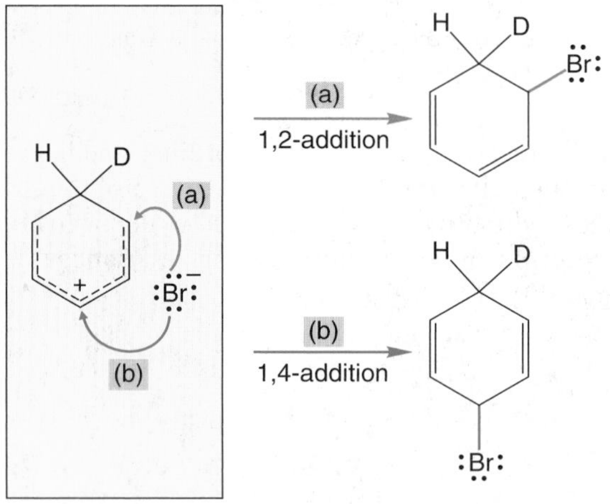

FIGURE 13.75 Addition of Br^- to this ion would give two nonaromatic products.

But aromaticity, the more than 30 kcal/mol of delocalization energy, has been lost in each of these products. Their formation from benzene is generally endothermic. A far better reaction that regenerates the benzene ring is possible. Loss of a proton (or deuteron) regenerates the aromatic sextet of π electrons (Fig. 13.76).

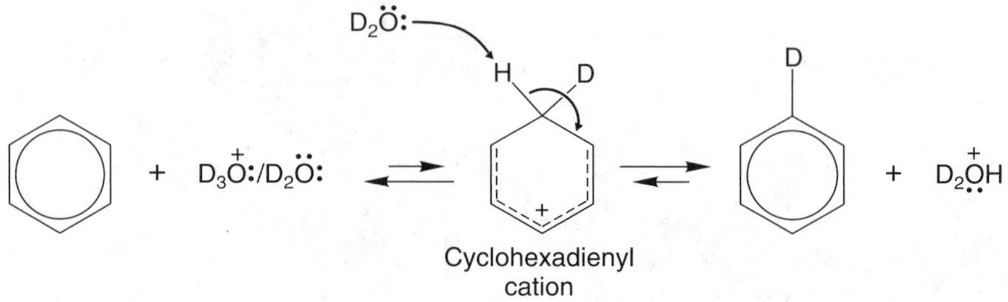

FIGURE 13.76 Loss of a proton or deuteron from the carbocation regenerates an aromatic compound.

If a deuteron is lost, we see no reaction as benzene is simply regenerated. If a proton is lost, we see the product of an exchange reaction in which deuteriobenzene is formed (Fig. 13.76). If a large excess of deuterio acid is used, all of the hydrogens can be exchanged to form hexadeuteriobenzene. We will elaborate on this reaction in Chapter 14.

*PROBLEM 13.20 Construct an Energy versus Reaction progress diagram for the conversion of benzene into deuteriobenzene.

ANSWER The first step is protonation (using a deuteron, not a proton) of the benzene ring to give the resonance-stabilized cyclohexadienyl cation. This process is surely very endothermic, as aromaticity is lost. The second half of this symmetrical

ANSWER (CONTINUED)

reaction (except for the difference between the isotopes H and D) involves the deprotonation of the intermediate to produce monodeuterated benzene. If there is a sufficient source of deuterium, the reaction can continue until all H atoms are replaced with D atoms.

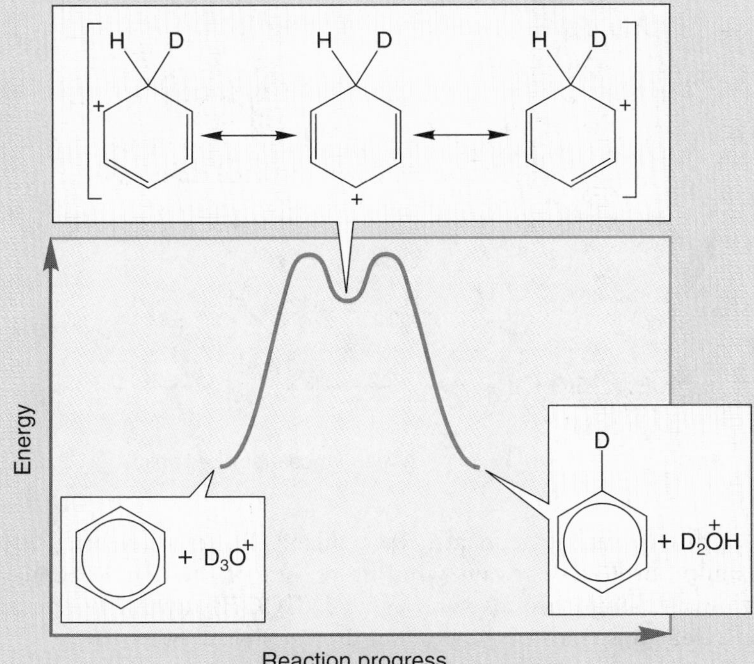

The second step is the reverse of the first, and simply is the exothermic deprotonation of the intermediate pentadienyl cation to regenerate the aromatic ring.

13.11b Birch Reduction of Aromatic Compounds

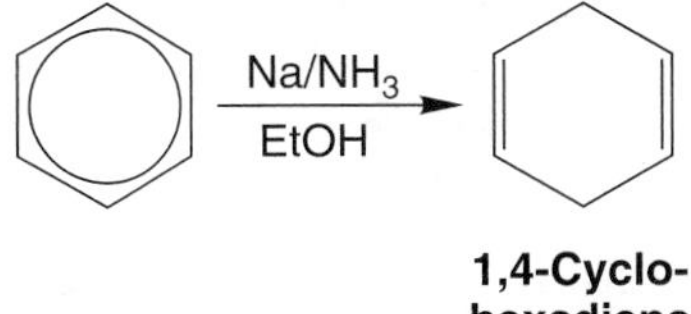

FIGURE 13.77 Birch reduction of aromatic compounds leads to 1,4-cyclohexadienes.

Now we come to a reaction in which aromaticity *is* lost, the Birch reduction. Alkynes can be treated with sodium in liquid ammonia to give, ultimately, trans alkenes (see Chapter 10, p. 434). Although catalytic hydrogenation of benzene is difficult (p. 568), treatment of benzene with sodium in liquid ammonia and a little alcohol does lead to a reduced product, in this case, 1,4-cyclohexadiene (Fig. 13.77). The reaction is named after the Australian chemist Arthur J. Birch (b. 1915).

Immediately note the important difference between this method of reducing the benzene ring and a typical catalytic reduction in which all carbon–carbon π bonds are reduced. In this case, the ring is not totally reduced; the reaction stops at the diene stage.

The mechanism of this **Birch reduction** is similar to that for dissolving metal reduction of alkynes (Chapter 10, p. 434), and starts in the same way, with a transfer of an electron from the metal to one of the antibonding orbitals of benzene. The product is a resonance-stabilized **radical anion** (Fig. 13.78).

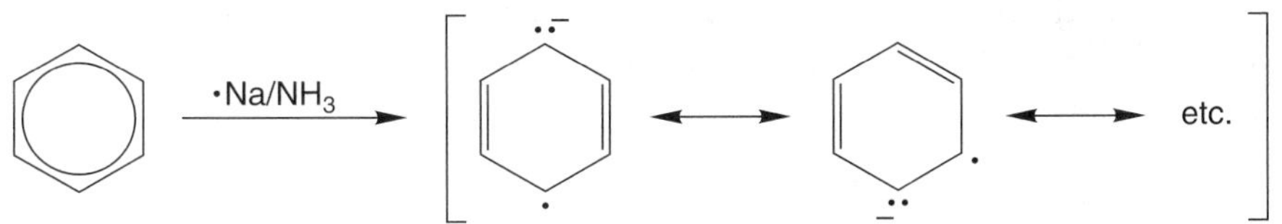

FIGURE 13.78 Transfer of an electron from sodium to benzene leads to a resonance-stabilized radical anion.

Protonation by alcohol gives a radical that can undergo a second electron transfer–protonation sequence to complete the reduction (Fig. 13.79).

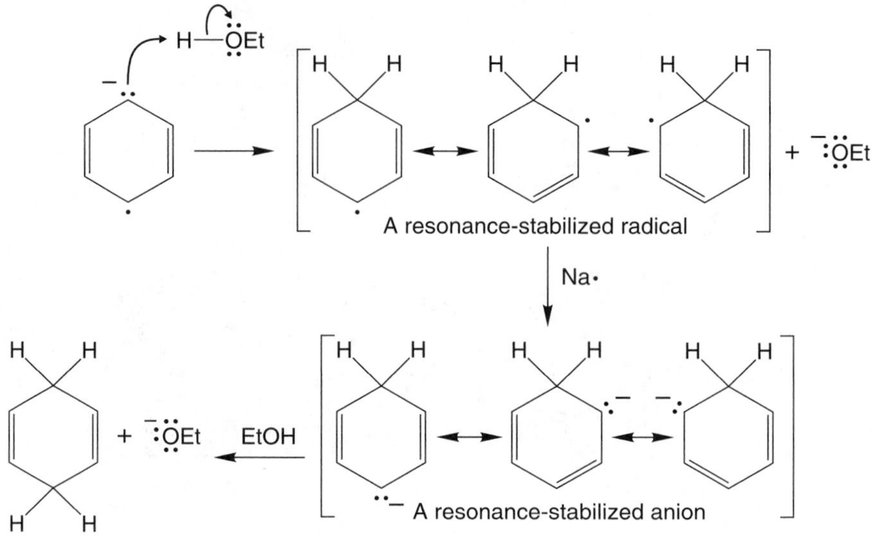

FIGURE 13.79 The radical anion is protonated to give a resonance-stabilized cyclohexadienyl radical that goes on to produce the 1,4-cyclohexadiene through another reduction–protonation sequence (Et = CH_2CH_3).

Substituted benzenes can also be reduced. Electron-releasing groups such as alkyl or alkoxy always wind up on one of the double bonds, but electron-withdrawing groups such as acid (COOH) invariably appear on the methylene positions of the product dienes (Fig. 13.80).

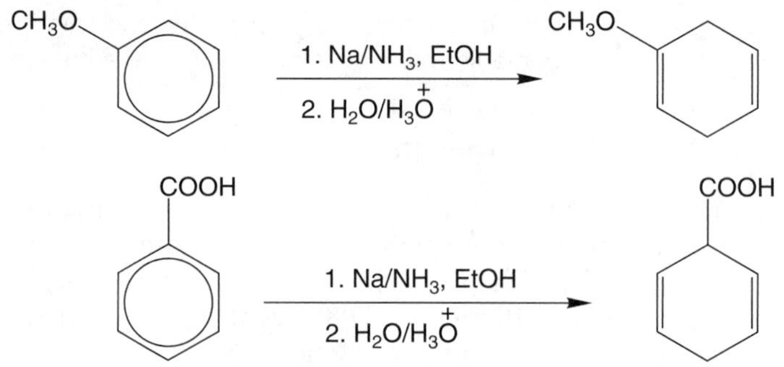

FIGURE 13.80 In the Birch reduction of substituted benzenes, electron-releasing groups appear on one of the double bonds, but electron-withdrawing groups are located at the methylene position (Et = CH_2CH_3).

13.12 THE BENZYL GROUP AND ITS REACTIVITY: ACTIVATION OF THE ADJACENT (BENZYL) POSITION BY BENZENE RINGS

Thus far we have been looking at processes directly involving the benzene ring itself. Here we sketch out some reactions in which benzene plays a major role in the progress of the reaction, but in which the action takes place more on the outskirts of the ring. The carbon adjacent to a benzene ring is called the benzylic carbon, or the benzylic position. The **benzyl group** is Ph—CH_2. In many ways, the reactivity of the benzylic position resembles that of the allylic position. Indeed, an orbital at the benzylic position is stabilized by overlap with adjacent orbitals in much the same way as is an orbital at the allyl position (Fig. 13.81).

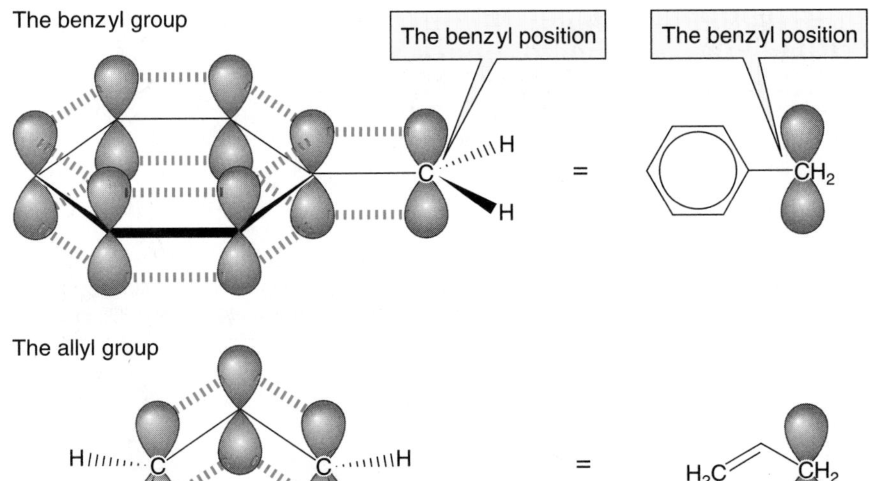

The allyl group

H C C H

H H

= H_2C CH_2

FIGURE **13.81** The position adjacent to a benzene ring is quite analogous to the allylic position.

13.12a Substitution Reactions of Benzyl Compounds

Benzyl compounds are unusually active in both the S_N1 and S_N2 reactions. Clearly, the benzene ring helps enormously to stabilize the carbocation formed in the rate-determining ionization of an S_N1 reaction. For example, benzyl chloride reacts 145 times faster than isopropyl chloride in a typical S_N1 process (Fig. 13.82).

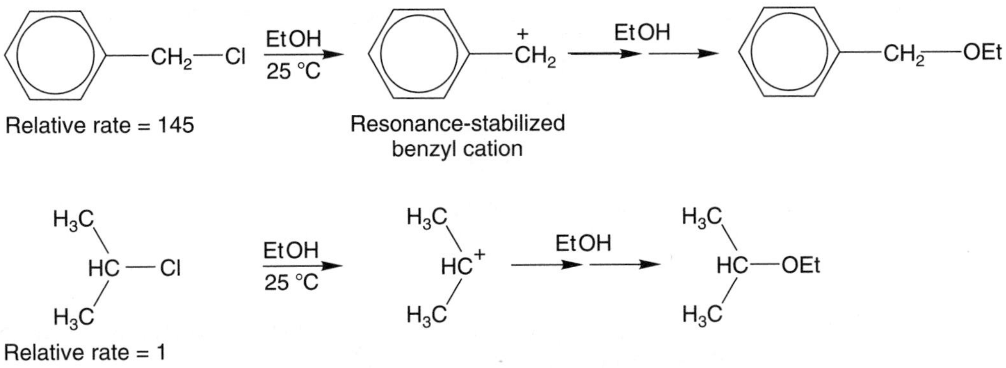

C_6H_5—CH_2—Cl —(EtOH, 25 °C)→ C_6H_5—$\overset{+}{C}H_2$ —EtOH→ C_6H_5—CH_2—OEt

Relative rate = 145

Resonance-stabilized benzyl cation

$(H_3C)_2$HC—Cl —(EtOH, 25 °C)→ $(H_3C)_2HC^+$ —EtOH→ $(H_3C)_2$HC—OEt

Relative rate = 1

FIGURE **13.82** The resonance stabilization of the benzyl cation makes its formation relatively easy. Accordingly, S_N1 reactions of benzyl compounds containing good leaving groups are quite fast.

The reason is simple: The benzene ring helps to stabilize the positive charge on the intermediate carbocation and, of course, in the transition state leading to it. In the benzyl cation itself, the positive charge is borne by 4 carbons, in the benzhydryl cation ($Ph_2\overset{+}{C}H$); the **benzhydryl group** is Ph_2CH) by 7 carbons, and in the trityl cation (Ph_3C^+) by 10 carbons (Fig. 13.83).

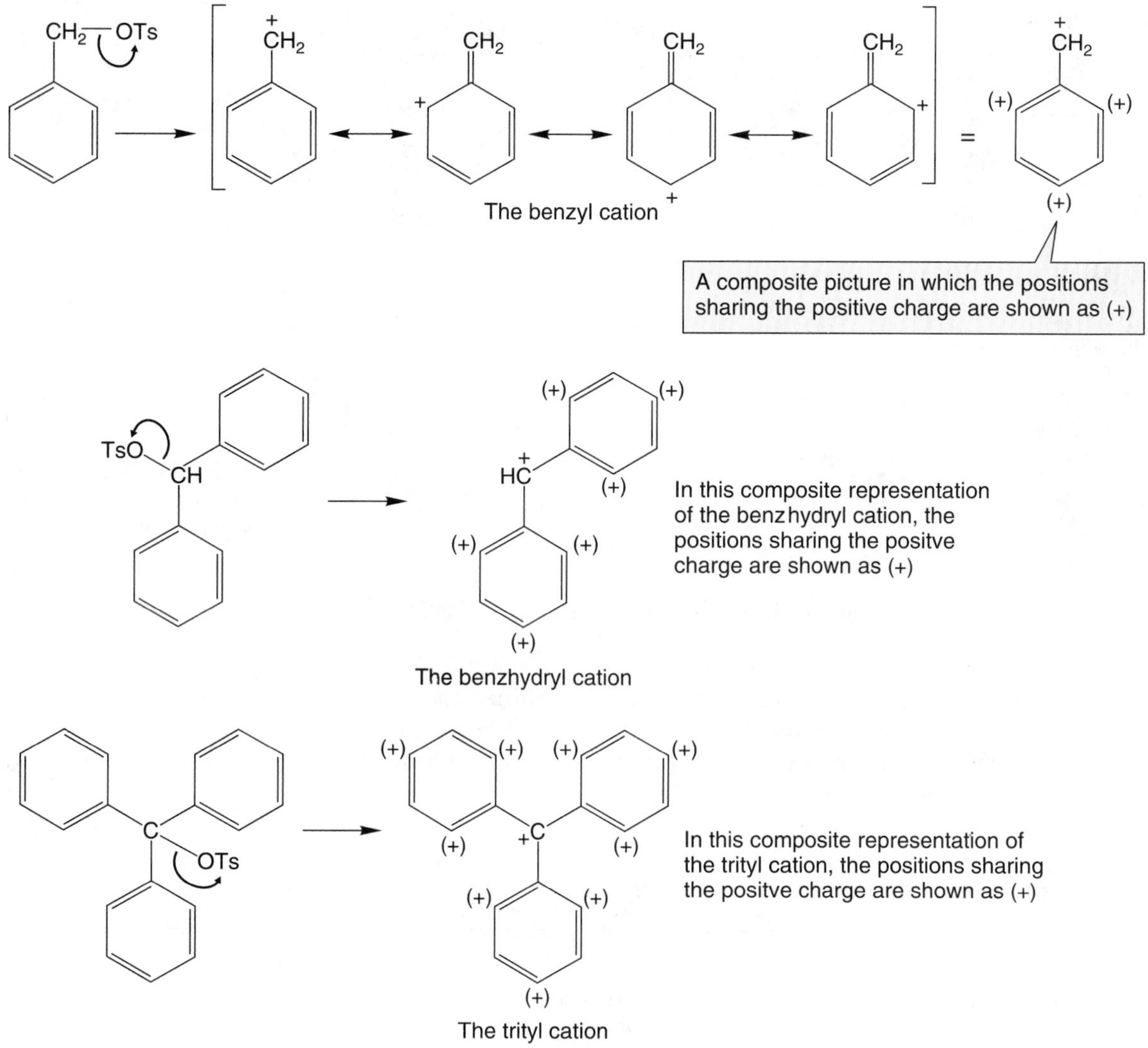

The benzyl cation

A composite picture in which the positions sharing the positive charge are shown as (+)

In this composite representation of the benzhydryl cation, the positions sharing the positve charge are shown as (+)

The benzhydryl cation

In this composite representation of the trityl cation, the positions sharing the positve charge are shown as (+)

The trityl cation

FIGURE **13.83** The benzyl cation is stabilized by 4 resonance forms, the benzhydryl cation by 7, and the trityl cation by 10. Remember tosylate (OTs) is a good leaving group.

PROBLEM **13.21** Why is trypticenyl chloride not nearly as reactive in the S_N1 reaction as trityl chloride (Fig. 13.84)?

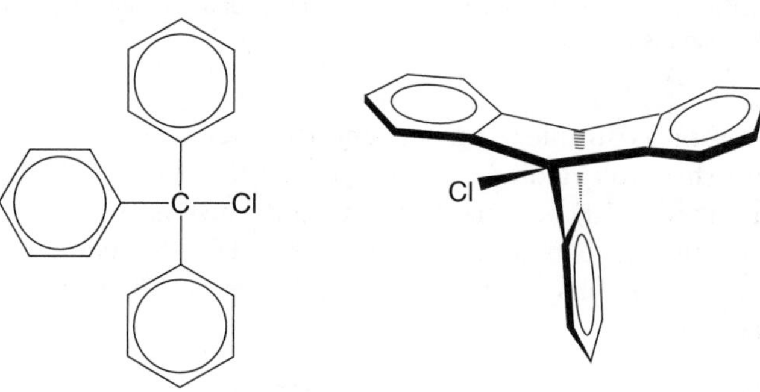

FIGURE **13.84**

Benzyl halides are also especially reactive in the S_N2 reaction. The reason is the same as that for the enhanced reactivity of allyl chloride (Chapter 12, p. 533). The transition state for S_N2 displacement is delocalized, and therefore especially stable. If the transition state for a reaction is at relatively low energy, the activation energy for the reaction will also be relatively low and the reaction itself will be fast (Fig. 13.85).

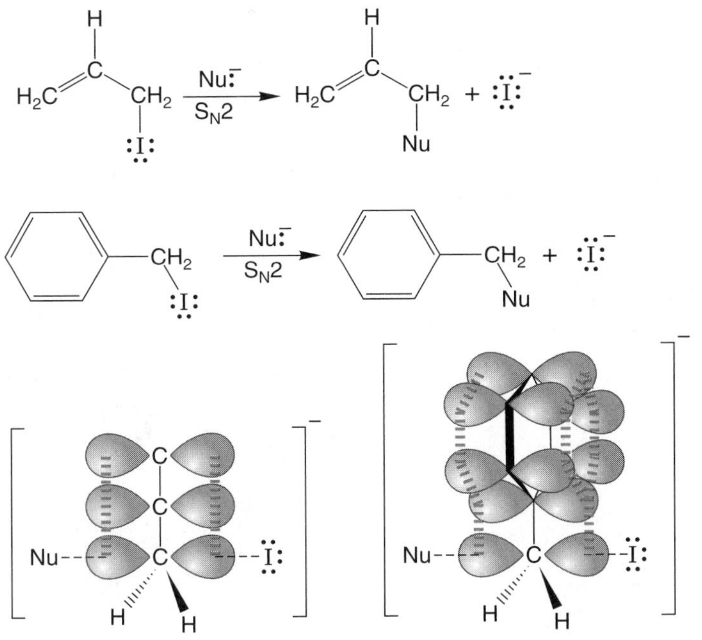

Transition state for S_N2 displacement of an allyl halide

Transition state for S_N2 displacement of a benzyl halide

FIGURE 13.85 The transition state for S_N2 displacement at the benzyl position is stabilized through orbital overlap with the ring.

Table 13.2 compares the rates of reaction of benzyl iodide with other alkyl iodides in an S_N2 reaction.

TABLE 13.2 Some Relative Rates of S_N2 Displacements in the Reaction of R—I with Iodide Ion at 50 °C in Ethyl Alcohol Solvent

R group	Relative Rate
Ethyl	1.0
Propyl	0.6
Butyl	0.4
Isopropyl	0.04
Allyl	33
Benzyl	78

13.12b Radical Reactions at the Benzyl Position

A benzene ring will also stabilize an adjacent half-filled orbital (a free radical) through resonance (Fig. 13.86).

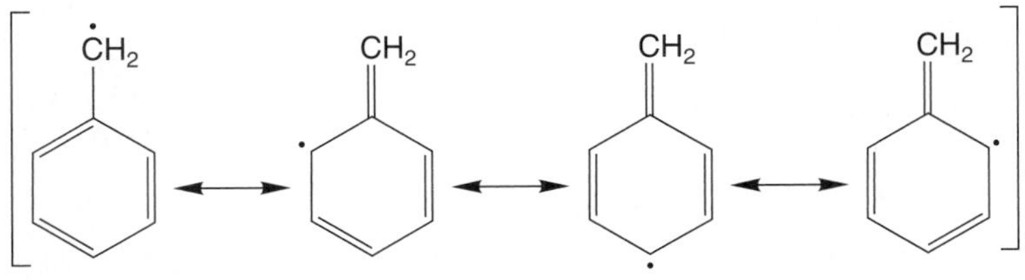

FIGURE 13.86 The resonance forms for the benzyl radical.

In Chapter 11 (p. 484), we discussed radical reactivity at the allylic position, and very little need be added to get a description of benzylic reactivity. Toluene, for example, can be brominated by *N*-bromosuccinimide (NBS) to give benzyl bromide (Fig. 13.87).

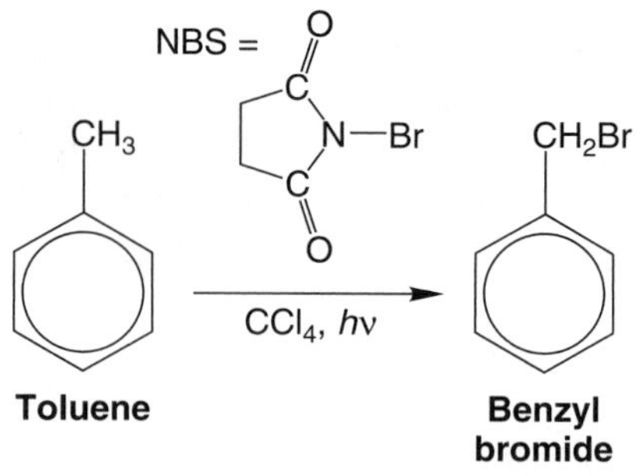

FIGURE 13.87 The photobromination of toluene with NBS gives benzyl bromide.

PROBLEM 13.22 Write a mechanism for this reaction. *Hint*: See Chapter 11 (p. 486) for a description of the reaction of NBS with allyl compounds.

*PROBLEM 13.23 Ethylbenzene gives only the product of bromination at the α-position (the position adjacent to the ring) when treated with NBS in CCl_4. Explain.

ANSWER Abstraction from the α-position gives a resonance-stabilized benzyl radical. Abstraction from the β-position does not! This radical is not resonance stabilized, and will be formed much more slowly than the α-radical. Only product from the α-radical is formed.

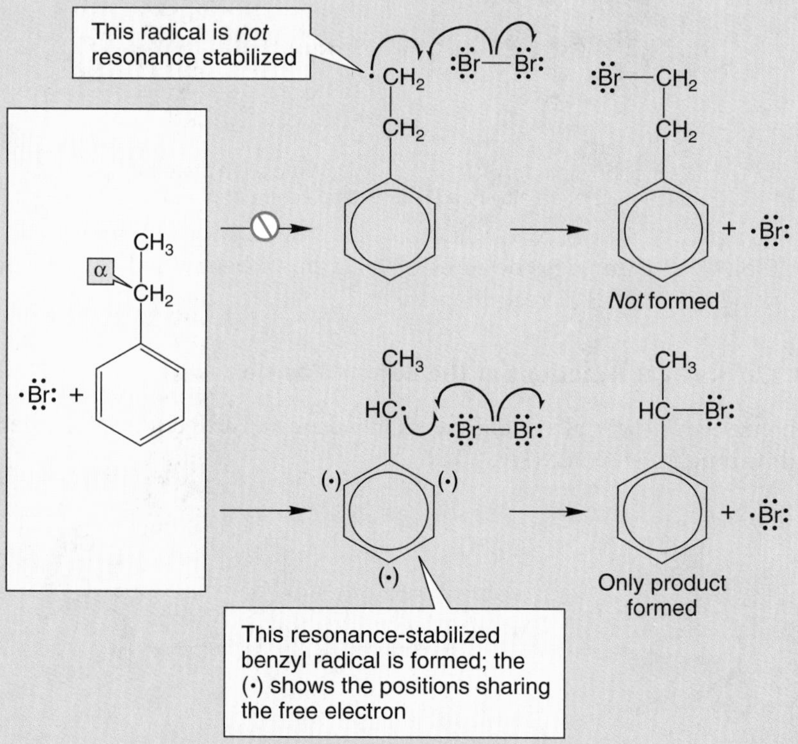

As your mechanism for Problem 13.23 would indicate, abstraction of hydrogen is especially easy at the α-position of ethylbenzene. This abstraction occurs easily because of the resonance stabilization of the benzyl radical (Fig. 13.86). Because bromine will not attack the benzene ring without a catalyst, benzyl compounds can often be brominated by simply heating or photolyzing bromine in their presence (Fig. 13.88).

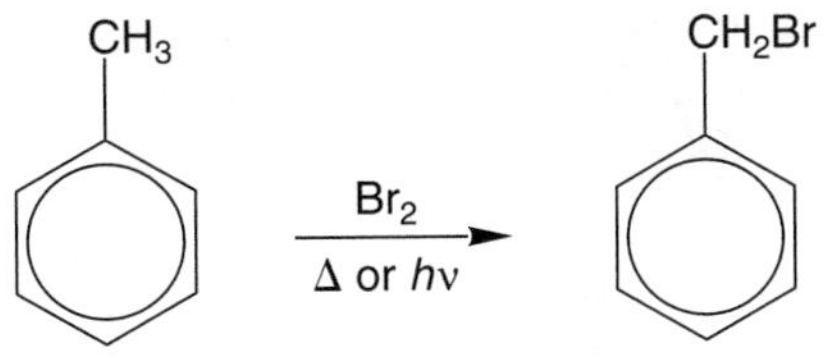

FIGURE 13.88 The thermal or photochemical decomposition of bromine in toluene leads to benzyl bromide through radical bromination at the benzyl position.

13.12c Oxidation at the Benzyl Position

Both free radicals and the special activity of the benzyl position play a role in the reaction of alkylbenzenes with $KMnO_4$ or H_2CrO_7 to give **benzoic acids**. Side chains are "chewed down" to carboxylic acids no matter what their length. Although mechanistic details are vague, the key observation is that tertiary side chains are untouched by oxidizing agents. There must be at least one hydrogen in the benzylic position for the reaction to succeed (Fig. 13.89).

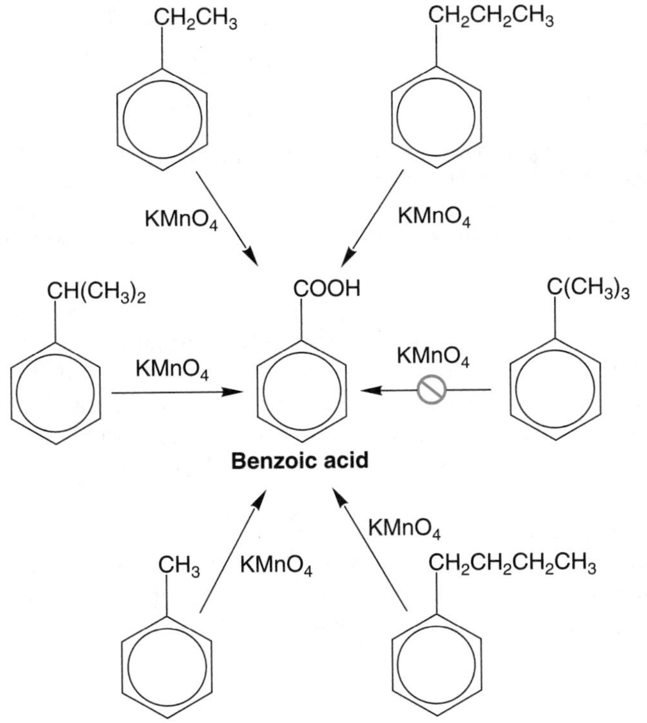

FIGURE 13.89 Oxidizing agents convert most alkyl side chains on benzene rings into the acid group, COOH. However, for the oxidation to succeed there must be at least one benzyl hydrogen.

13.13 SOMETHING MORE. THE MECHANISM OF CARCINOGENESIS BY POLYCYCLIC AROMATIC COMPOUNDS

We mentioned earlier that some polycyclic aromatic hydrocarbons were strongly carcinogenic. Although all the details of this carcinogenesis are not worked out, the broad outlines are known for some molecules. Deoxyribonucleic acid (DNA) and ribonucleic acid (RNA) are huge biomolecules that contain and help transcribe genetic information (Chapter 26). Both DNA and RNA are phosphate-linked sequences of nucleotides, which are sugar molecules attached to heterocyclic bases. These molecules bear large numbers of nucleophilic groups that can be alkylated by other molecules in what seem to be simple S_N2 reactions (Fig. 13.90).

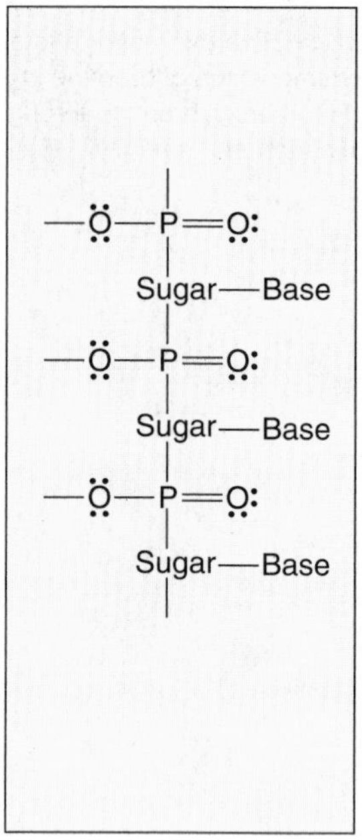

The structures of the four bases attached to sugars

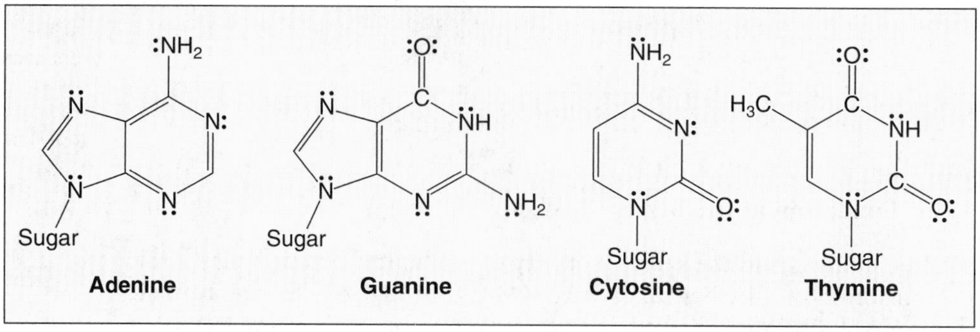

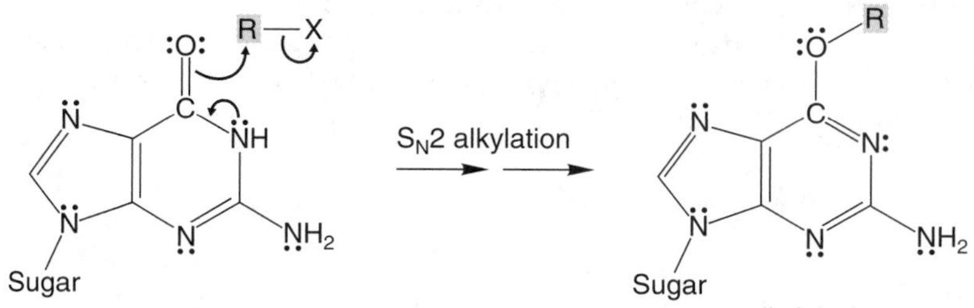

FIGURE 13.90 The DNA polymer is composed of phosphate–linked sugar-base units. Two strands of DNA are held together by hydrogen bonding between pairs of bases. Alkylation of the bases changes the size and shape of one of the units. This change can interfere with hydrogen bonding and base pairing. In turn, this can lead to mutations and cancer.

Alkylation changes the size and shape of the DNA molecule and makes mismatched base pairing through hydrogen bonding more likely (see Chapter 26 for more details of base pairing). This mismatch can lead to errors in replication and to mutations. Most mutations are harmless to the whole organism, as they lead only to the death of the individual cell. On rare occasions, however, a mutation can lead to uncontrolled cell division and cancer.

The hydrocarbons themselves are only indirectly responsible for the chemical processes that lead to cancer. Aromatic hydrocarbons are not very reactive and there is no obvious way for them to interact chemically with the nucleophilic bases in DNA or RNA (Fig. 13.91).

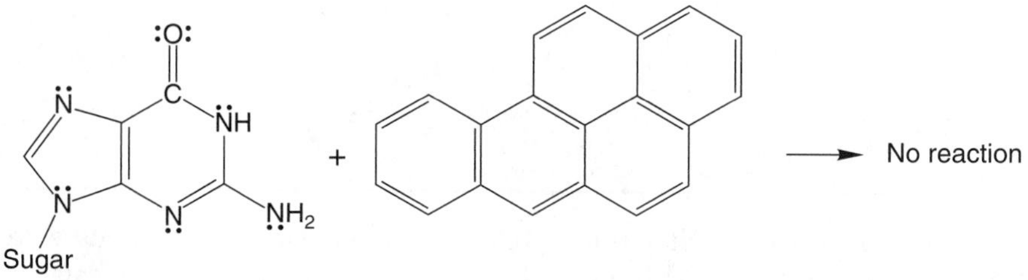

FIGURE 13.91 Aromatic hydrocarbons do not react with nucleophiles, and there is no way for a molecule like benz[*a*]pyrene to alkylate one of the bases directly.

However, there are enzymes that can modify benzene rings. Nonpolar molecules such as aromatic hydrocarbons accumulate in fat cells and methods have been evolved to purge our bodies of such molecules. These methods generally involve making the hydrocarbons more water soluble by adding polar groups. For example, benz[*a*]pyrene is epoxidized by the enzyme P-450-mono-oxygenase and the product epoxide is then opened enzymatically to give a trans diol (Fig. 13.92).

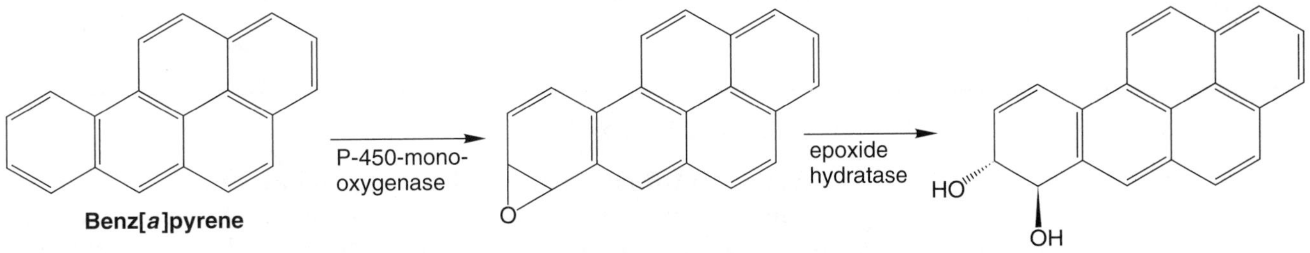

FIGURE 13.92 Benz[*a*]pyrene is enzymatically epoxidized and ring opened to give a polar trans 1,2-diol.

However, a second enzymatic epoxidation, again by P-450-mono-oxygenase, gives a molecule that reacts with guanine, one of the DNA bases, to give a molecule of DNA that is severely encumbered by the large, alkylated guanine (Fig. 13.93). This modified base has difficulty in maintaining normal hydrogen bonding with another complementary base, and this mismatch can lead to mutations.

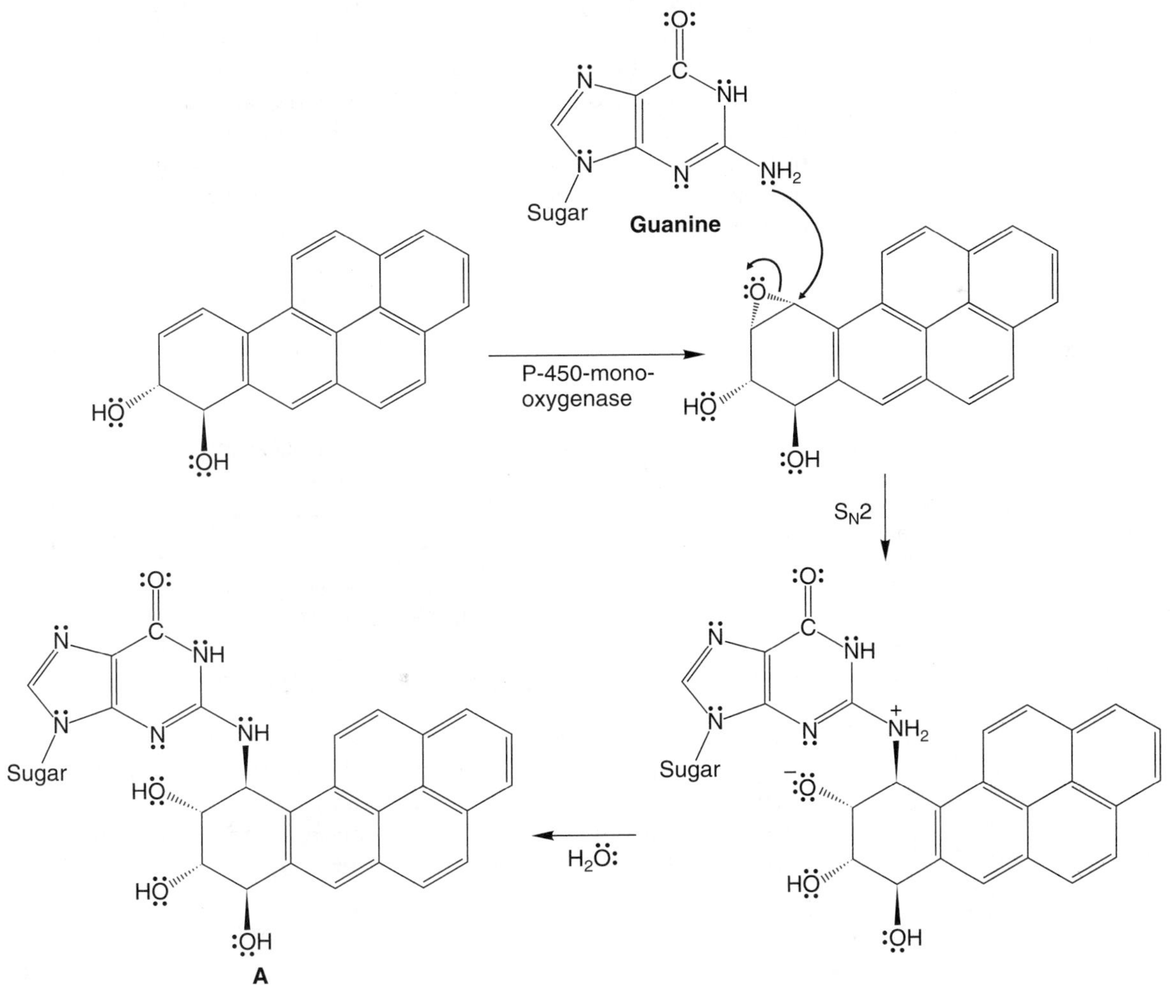

FIGURE 13.93 A second epoxide is formed. This epoxide is opened by guanine to give a highly encumbered molecule, **A**, that has problems maintaining proper hydrogen bonding.

This alkylation is not a new reaction. You already know of the opening of epoxides in both base and acid (Chapter 10, p. 408), and this is nothing more than a complicated version of that simple reaction.

13.14 SUMMARY

NEW CONCEPTS

The material in this chapter is composed almost entirely of concepts. There are few new reactions or synthetic procedures. We concentrate here on the special stability of some planar, cyclic, and fully conjugated polyenes. The special stability called *aromaticity* is encountered when the polyene has a molecular orbital system in which all bonding molecular orbitals are completely filled and there are no electrons in antibonding or nonbonding molecular orbitals.

These especially stable molecular orbital systems will be found in planar, cyclic, fully conjugated polyenes that contain $4n + 2$ π electrons (Hückel's rule).

Heats of hydrogenation or heats of formation can be used to calculate the magnitude of the stabilization. For benzene, the delocalization energy or resonance energy amounts to more than 30 kcal/mol.

It's vital to keep clear the difference between resonance forms and molecules in equilibrium. In this chapter, that difference is exemplified by Dewar benzene, bicyclo[2.2.0]hexa-2,5-diene, and the Dewar resonance forms contributing slightly to the structure of benzene. These are shown in Problems 13.2–13.4. As always, resonance forms are related only by the movement of electrons, whereas equilibrating molecules have different geometries—different arrangements of atoms in space (Fig. 13.94).

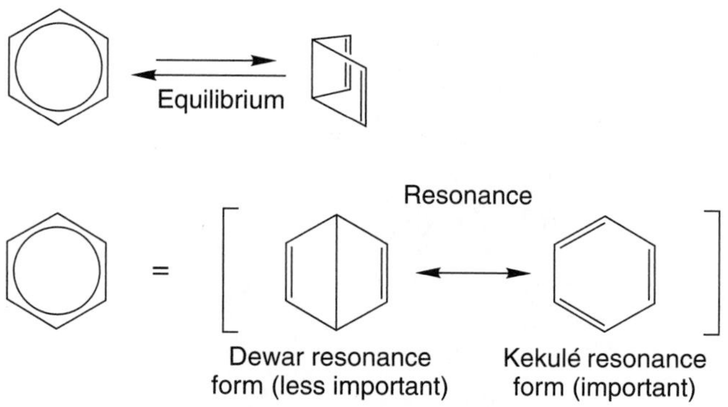

Equilibrium

Resonance

Dewar resonance form (less important)

Kekulé resonance form (important)

FIGURE 13.94 Benzene and Dewar benzene are different molecules; the Kekulé and Dewar resonance forms contribute to the real structure of benzene.

REACTIONS, MECHANISMS, AND TOOLS

In this chapter, the first reaction mechanism encountered is the important and general mechanism for electrophilic substitution of benzene. A host of substitution reactions will be studied in Chapter 14 and are exemplified here by deuterium exchange. The aromatic ring is destroyed by an endothermic addition of D^+, but reconstituted by an exothermic loss of H^+ (Fig. 13.76).

SYNTHESIS

There is rather little in the way of new synthetic procedures in this chapter. You might remember the formation of the tropylium ion by hydride abstraction and the Birch reduction of benzenes to 1,4-cyclohexadienes (Fig. 13.95). You also have a method of synthesizing deuteriobenzene through the acid-catalyzed exchange reaction of benzene. This reaction will serve as the prototype of many similar substitution reactions to be found in Chapter 14.

A benzene ring activates the adjacent, or benzyl position. There are a number of synthetically useful reactions that take advantage of this activation.

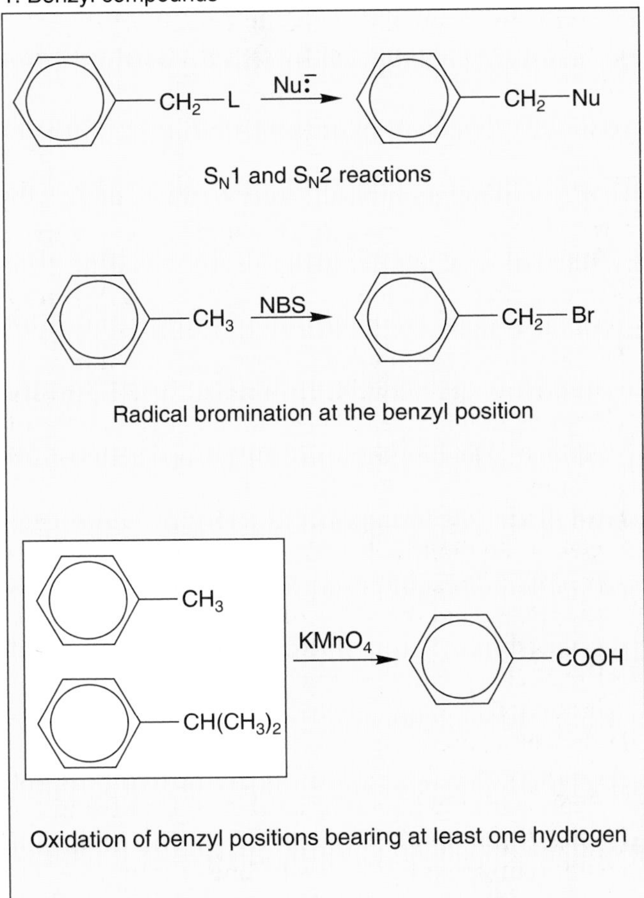

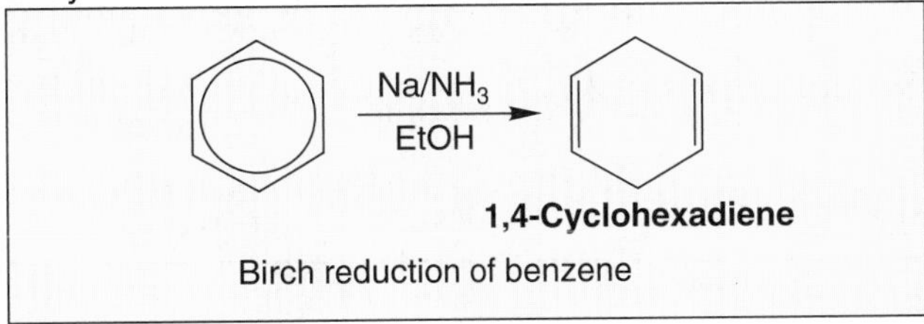

3. Deuteriobenzenes

$D_3\overset{+}{O}/D_2O$ → (D-benzene) + $D_2\overset{+}{O}H$

Acid-catalyzed exchange of hydrogens on a benzene ring; all hydrogens can eventually be exchanged

4. Tropylium ions

(cycloheptatriene C(H)(H)) $\overset{-}{B}F_4$ $\overset{+}{C}Ph_3$ → $\overset{+}{C}$—H + $HCPh_3$ $\overset{-}{B}F_4$

Tropylium fluoborate

Hydride transfer from cycloheptatriene to the trityl cation ($^{+}CPh_3$)

FIGURE 13.95 The synthetically useful reactions of Chapter 13.

COMMON ERRORS

Certainly the most common error is to overgeneralize Hückel's rule. We often forget to check each potentially aromatic molecule to be certain it satisfies all the criteria. The molecule must be cyclic or there cannot be the fully connected set of *p* orbitals necessary (Fig. 13.19). The molecule must be planar so that the *p* orbitals can overlap effectively (Fig. 13.17). There may be no positions in the ring at which *p–p* overlap is interrupted (Fig. 13.16). Finally, there must be the proper number of $(4n + 2)$ π electrons. Be certain you are counting π electrons, and not including in your count electrons not in the π system. The classic example is pyridine, which contains two paired electrons in an sp^2 orbital that are often mistakenly counted as π electrons (Fig. 13.58). If, and only if, *all* of these criteria are satisfied do we find aromaticity.

Do not misapply the Frost circle device. The most common error is to forget to inscribe the circle vertex down. This requirement is hard to remember because at this level of discussion this point is arbitrary—there is no easy way to figure it out if you don't remember the proper convention.

Do not confuse absolute and relative stability. For example, the cyclopentadienide anion is aromatic, and therefore exceptionally stable, *but only within the family of anions.* The negatively charged cyclopentadienide anion is not as stable as benzene.

As always, you must be clear as to the difference between resonance and equilibrium. Be certain that you know the difference between the two Kekulé resonance forms for benzene and the hypothetical molecule 1,3,5-cyclohexatriene.

13.15 KEY TERMS

Annulene A cyclic polyene that is at least formally fully conjugated.

Arene An aromatic compound containing a benzene ring or rings.

Aromatic character See **aromaticity**.

Aromaticity The special stability of planar, cyclic, fully conjugated molecules with $4n + 2$ π electrons. Such molecules will have molecular orbital systems with all bonding molecular orbitals completely filled and all antibonding and nonbonding molecular orbitals empty.

Benzene The archetypal aromatic compound; a planar, regular hexagon of sp^2 hybridized carbons. The six $2p$ orbitals overlap to form a six-electron cycle above and below the plane of the ring. The molecular orbital system has three fully occupied bonding molecular orbitals and three unoccupied antibonding orbitals.

Benz[*a*]pyrene A powerfully carcinogenic polynuclear aromatic compound composed of five fused benzene rings.

Benzhydryl group The group Ph_2CH.

Benzoic acid Ph–COOH.

Benzyl group The group Ph–CH_2.

Birch reduction The conversion of aromatic compounds into 1,4-cyclohexadienes through treatment with sodium in liquid ammonia–ethyl alcohol. Radical anions are the first formed intermediates.

Cycloheptatrienylium ion See **tropylium ion**.

Cyclopentadienyl anion A five-carbon aromatic anion containing 6 π electrons ($4n + 2$, $n = 1$).

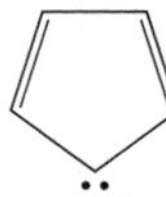

Delocalization energy The energy lowering conferred by the delocalization of electrons. In benzene, this is the amount by which benzene is more stable than the hypothetical 1,3,5-cyclohexatriene containing three localized double bonds. See **resonance energy**.

Dewar forms Resonance forms for benzene in which overlap between $2p$ orbitals on two para carbons is emphasized. These forms superficially resemble Dewar benzene (bicyclo[2.2.0]hexa-2,5-diene).

Frost circle A device used to find the relative energies of the molecular orbitals of planar, cyclic, fully conjugated molecules. A polygon corresponding to the ring size of the molecule is inscribed in a circle, vertex down. The intersections of the polygon with the circle give the relative positions of the molecular orbitals.

Fully conjugated In a fully conjugated molecule, every carbon has a p orbital that overlaps effectively with the p orbitals on the adjacent atoms.

Furan An aromatic five-membered ring compound containing four CH units and one O atom.

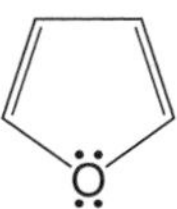

Heteroaromatic compound See **heterobenzene**.

Heterobenzene A benzene ring in which one (or more) ring carbons is replaced with another heavy (non-hydrogen) atom.

Hückel's rule All planar, cyclic, fully conjugated molecules with $4n + 2$ π electrons will be aromatic (especially stable). The rule works because such molecules will have molecular orbital systems in which all bonding molecular orbitals are completely full and in which no antibonding or nonbonding molecular orbitals are occupied.

Kekulé forms Resonance forms for benzene in which

overlap between adjacent carbons is emphasized. These forms superficially resemble 1,3,5-cyclohexatriene.

Meta 1,3-Substitution on a benzene ring.

Ortho 1,2-Substitution on a benzene ring.

Para 1,4-Substitution on a benzene ring.

Phenyl group The group C_6H_5. A substituted benzene ring, abbreviated, Ph.

Polynuclear aromatic compound An aromatic molecule composed of two or more fused aromatic rings.

Pyridine Azabenzene. A benzene in which one CH unit has been replaced by a nitrogen atom.

Pyrrole An aromatic five-membered ring compound containing four CH units and one NH unit.

Radical anion A negatively charged molecule containing both a pair of electrons and an odd, unpaired electron.

Resonance energy The energy lowering conferred by the delocalization of electrons. In benzene, this is the amount by which benzene is more stable than the hypothetical 1,3,5-cyclohexatriene containing three localized double bonds. See **delocalization energy**.

Thiophene An aromatic five-membered ring compound containing four CH units and one S atom.

Trityl cation The triphenylmethyl cation.

Tropylium ion The 1,3,5-cycloheptatrienylium ion. This ion has $4n + 2$ π electrons ($n = 1$) and is aromatic.

13.16 ADDITIONAL PROBLEMS

PROBLEM **13.24** Give names for the following compounds:

(a) F, Br

(b)

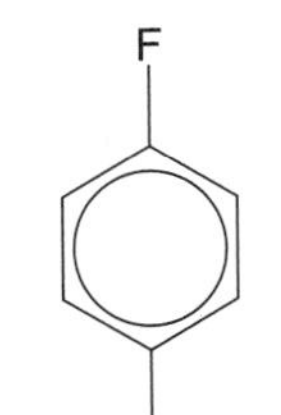

(c) F, Br

(d)

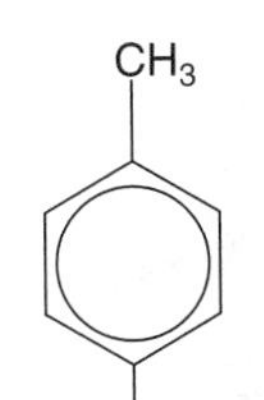

(e)

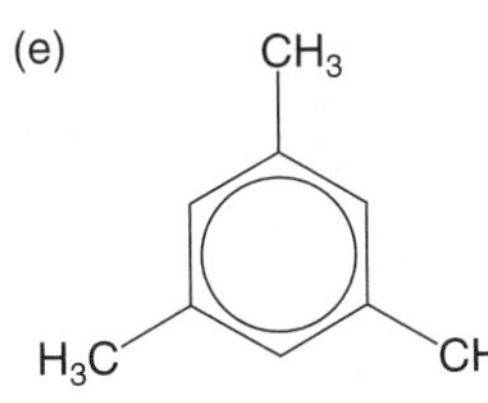

(f)

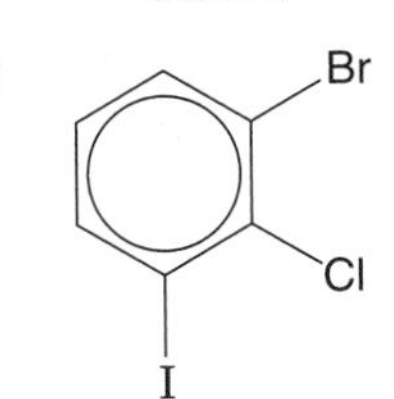

PROBLEM **13.25** Write structures for the following compounds:

(a) 2,4,6-Trinitrotoluene (TNT)
(b) 4-Bromotoluene
(c) *p*-Bromotoluene
(d) *m*-Dihydroxybenzene
(e) 3-Hydroxyphenol
(f) *m*-Hydroxyphenol
(g) 1,2,4,5-Tetramethylbenzene (durene)
(h) *o*-Deuteriobenzenesulfonic acid (Note the "i" in "deuter*i*o". It is not "deutero".)
(i) *p*-Aminoaniline
(j) *m*-Chlorobenzoic acid

PROBLEM **13.26** Write arrow formalisms for the conversions of Dewar benzene, Ladenberg benzene (prismane), benzvalene, and 3,3′-bicyclopropenyl (Fig. 13.4) into benzene. Some of these may be quite challenging tasks.

PROBLEM **13.27** Which of the following heterocyclic compounds might be aromatic? Explain. It will help if you start by drawing good Lewis structures for all the compounds.

(a)

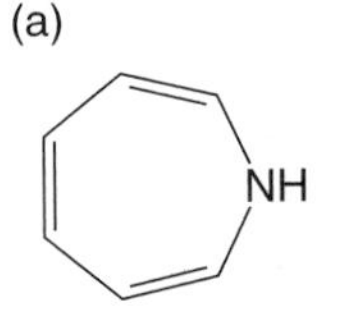

(b)

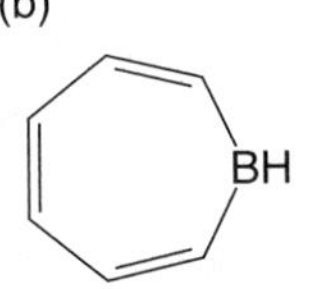

(c)

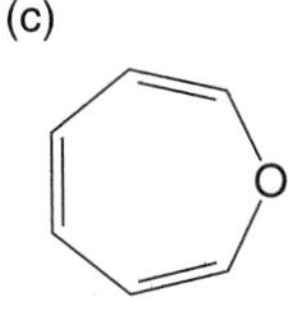

(d)

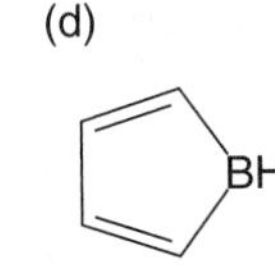

(e)

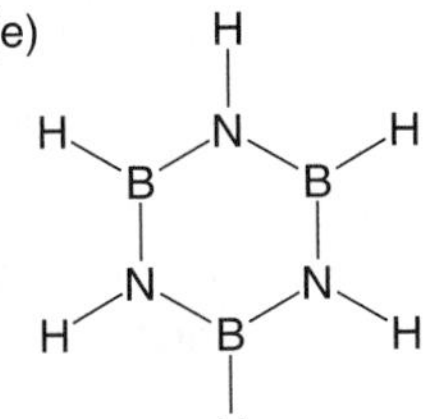

PROBLEM 13.28 Which of the following hydrocarbons might be aromatic? Explain.

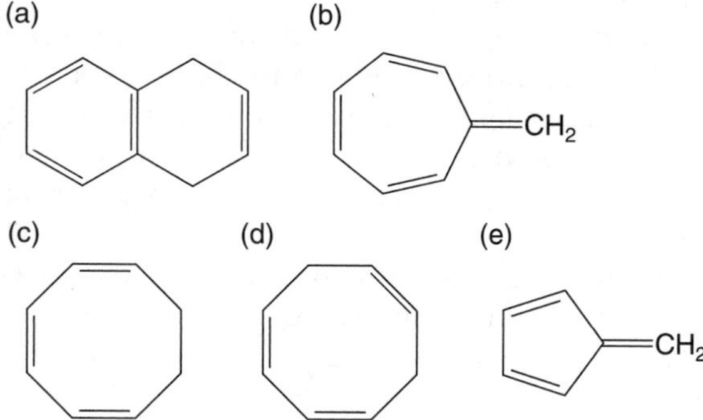

PROBLEM 13.29 Which of the following ions might be aromatic? Explain.

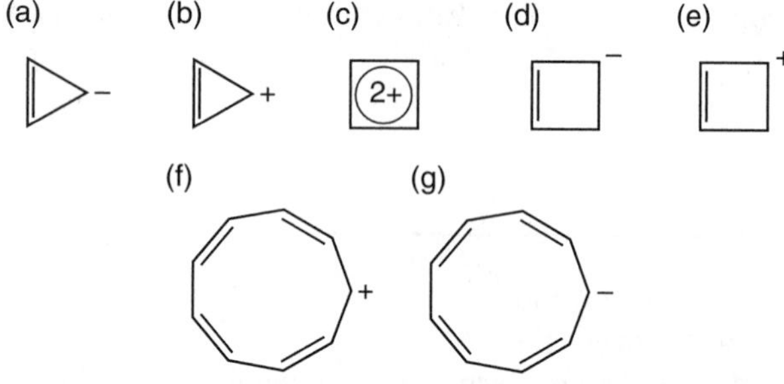

PROBLEM 13.30 The heat of hydrogenation of 1,3,5,7-cyclooctatetraene (COT) is about 101 kcal/mol. The heat of hydrogenation of cyclooctene is about 23 kcal/mol. Is COT aromatic? Explain.

PROBLEM 13.31 As early as 1891 the German chemist G. Merling showed that the bromination of 1,3,5-cycloheptatriene leads to a liquid dibromide. When the dibromide is heated, HBr is evolved and a yellow solid of the formula C_7H_7Br (mp 203 °C) can be isolated. This compound is soluble in water, but cannot be reisolated from water solution. Instead, ditropyl ether is obtained. Merling could not explain what was happening, but perhaps you can.

Br_2 → $C_7H_8Br_2$ →(Δ) HBr + C_7H_7Br →(H_2O)

Ditropyl ether

PROBLEM 13.32 Most diazo compounds are exceedingly, and sometimes disastrously unstable. They are notorious explosives. By contrast, the brilliant, iridescent orange compound diazocyclopentadiene is quite stable. Explain.

=N_2

Diazocyclopentadiene

PROBLEM 13.33 Construct the π molecular orbitals for benzene (p. 573) from the π molecular orbitals of allyl. *Remember*: Only combinations of orbitals of comparable energy need be considered.

PROBLEM 13.34 Vogel and Roth's synthesis of the bridged cyclodecapentaene (**1**) is outlined below. Supply the missing reagents (a, b, c, and d) and an arrow formalism for the last step.

a, b, Na / NH_3, c, d

1

PROBLEM **13.35** Pyridine and imidazole are modest Brønsted bases at nitrogen, whereas pyrrole is not.

Pyridine **Pyrrole** **Imidazole**

In fact, pyrrole is protonated only in strong acid, and protonation occurs at C(2), not on nitrogen. First, explain why pyridine and imidazole are basic and pyrrole is not. Then, explain why eventual protonation of pyrrole is at carbon, not nitrogen. **Caution!** You will have to write full Lewis structures to answer this question.

PROBLEM **13.36** Explain why the 1,2-bond of anthracene is shorter than the 2,3-bond.

1.37 Å

1.42 Å

Anthracene

PROBLEM **13.37** The nitronium ion ($^+NO_2$) is known and can react with simple benzenes to give nitrobenzenes. Write an arrow formalism mechanism for this reaction.

$^+NO_2$ $^-BF_4$ → NO_2

PROBLEM **13.38** The molecule shown below undergoes reaction with cupric nitrate in acetic anhydride (a source of $^+NO_2$) to give a high yield of the substituted product shown. Write a mechanism for this reaction.

$Cu(NO_3)_2$, acetic anhydride, 0 °C

PROBLEM **13.39** The compound shown below was synthesized by Professor I. M. Dimm who was astonished to find that it was exceedingly stable. Prof. Dimm had expected world renown for making what he thought would be a most unstable compound. Dimm reasoned that with six double bonds and a triple bond in the molecule, there would be 16 π electrons and thus, no aromatic character ($4n$, $n = 4$, not $4n + 2$). Where was his reasoning wrong?

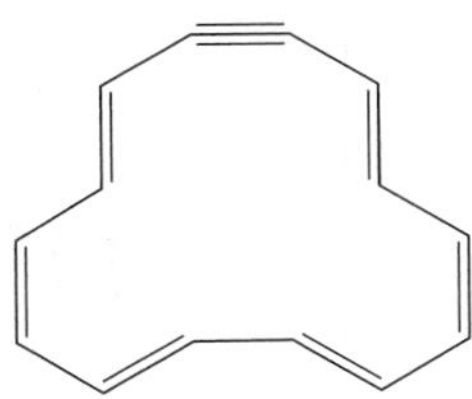

PROBLEM **13.40** Even though naphthalene and anthracene are polynuclear aromatic compounds, they undergo the Diels–Alder reaction. For example, anthracene readily reacts with maleic anhydride. However, the reaction of naphthalene with maleic anhydride proceeds efficiently only under high pressure.

xylene, 140 °C, 10 min → ?

100 °C, 9.6 kbar, 17 h → ?

(a) Draw the structures of the Diels–Alder adducts formed in these reactions.

(b) Explain why anthracene is so much more reactive than naphthalene in the Diels–Alder reaction. *Hint*: The calculated resonance energies of naphthalene and anthracene are 61 and 84 kcal/mol, respectively.

PROBLEM **13.41** Provide a mechanism for the following reaction sequence:

1. Na/NH_3, ether, −78 °C
2. CH_3I
3. H_2O, H_3O^+

PROBLEM **13.42** A certain alkylbenzene was known to have one of the isomeric structures shown below.

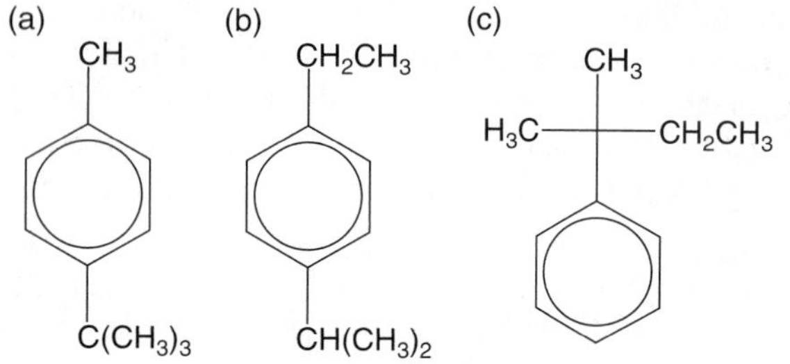

(a) CH_3 / $C(CH_3)_3$ (b) CH_2CH_3 / $CH(CH_3)_2$ (c) CH_3, H_3C—C—CH_2CH_3

Explain how the structure of the unknown compound could be determined by oxidation with $KMnO_4$.

Substitution Reactions of Aromatic Compounds

Even electrons, supposedly the paragons of unpredictability, are tame and obsequious little creatures that rush around at the speed of light, going precisely where they are supposed to go. They make faint whistling sounds that when apprehended in varying combinations are as pleasant as the wind flying through a forest, and they do exactly as they are told. Of this, one can be certain.

—Mark Helprin*
Winter's Tale

In Chapters 12 and 13, we examined the structural consequences of conjugation, the overlap of 2*p* orbitals. This story culminates in the enormously stable aromatic molecules, with the parent compound, benzene, being the prototype for all these "Hückel" compounds. In this chapter, we describe substitution reactions of aromatic compounds, an idea already introduced in Chapter 13 (p. 602). The chemistry of aromatic compounds is dominated by a tendency to retain the stable aromatic sextet of electrons. It takes a great deal of energy to destroy this arrangement, and it is very easily regained.

Because aromatic compounds are so stable, so low in energy, they are often surrounded by especially high energy barriers, and the activation energies for their reactions are usually formidably large. The transformation of an aromatic compound into a nonaromatic compound is most often highly endothermic (Fig. 14.1), so the transition state will lie close to the relatively high energy nonaromatic product (recall the Hammond postulate, Chapter 8, p. 320). By contrast, the conversion of a nonaromatic compound into an aromatic molecule is likely to be very exothermic (Fig. 14.2). In such a case, the transition state will resemble the starting material. So, if we again recall the Hammond postulate (and why it works), we will expect that only small activation barriers will need to be surmounted in such reactions. A complete energy diagram for a typical conversion of one aromatic compound into another is shown in Figure 14.3.

Transition state
Activation energy
Energy
Aromatic compound
Nonaromatic compound
Reaction progress

FIGURE **14.1** A typical Energy versus Reaction progress diagram for the conversion of an aromatic compound into a nonaromatic compound. Aromaticity is lost in this reaction, and there is a high activation energy.

*Mark Helprin (b. 1947) is an American author. His wonderfully imaginative novels are *Winter's Tale, A Soldier of the Great War,* and *Memoire from Antproof Case.*

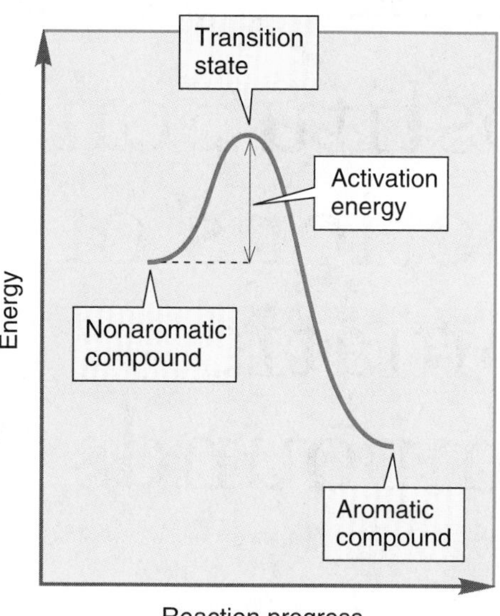

FIGURE **14.2** A typical Energy versus Reaction progress diagram for the conversion of a nonaromatic compound into an aromatic compound. Aromaticity is regained in this reaction, and there is a correspondingly low activation energy.

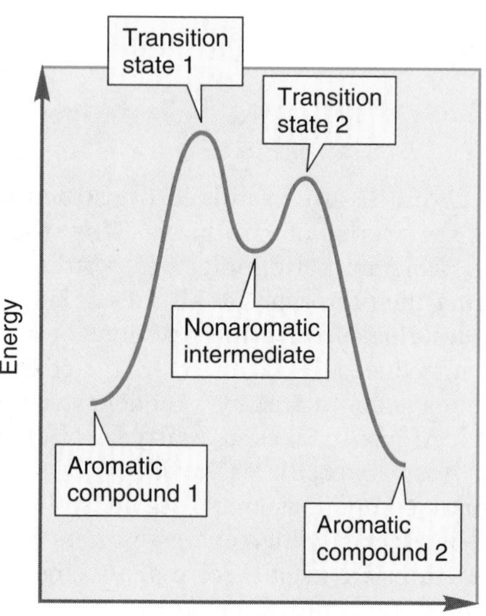

FIGURE **14.3** A typical Energy versus Reaction progress diagram for the conversion of one aromatic compound into another.

In this chapter, we will see many examples of a general reaction of aromatic compounds in which E^+, an **electrophile**, a "lover of electrons" (in other words, a Lewis acid), substitutes for one of the hydrogens on a benzene ring. Note that the aromatic sextet, represented by the circle in the hexagon, is retained in the overall reaction (Fig. 14.4).

Later in this chapter, we take up the question of how an initial substitution on the ring affects further reaction. Some groups make a second substitution reaction easier, some make it harder. We begin with reactions that are rather unusual for benzene rings, simple additions in which aromaticity is lost.

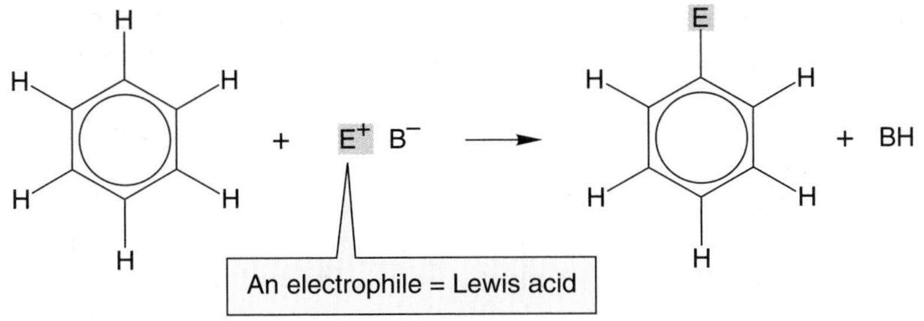

FIGURE **14.4** A generic aromatic substitution reaction. Note that the aromatic sextet is retained.

14.1 ADDITION REACTIONS OF BENZENES TO GIVE NONAROMATIC COMPOUNDS: HYDROGENATION

This section will be short simply because there are so few of these reactions! If one were to try to predict the reactions of benzene from first principles, one would probably start by examining the reactions of other π systems and applying them to aromatic compounds. For example, we know that hydrogenation of cyclohexene is exothermic by 28.6 kcal/mol (Chapter 13, p. 575). It is no great surprise to find that absorption of 3 mol of hydrogen by benzene to give cyclohexane is exothermic as well; indeed we examined this reaction in some detail in Chapter 13 (see p. 576). Although the reaction is still very exothermic, it is far less exothermic than a guess based on the value of –28.6 kcal/mol for a simple alkene would suggest. The difference between the estimated value and the measured value was taken as a measure of the delocalization energy of benzene, approximately 33 kcal/mol (Fig. 14.5).

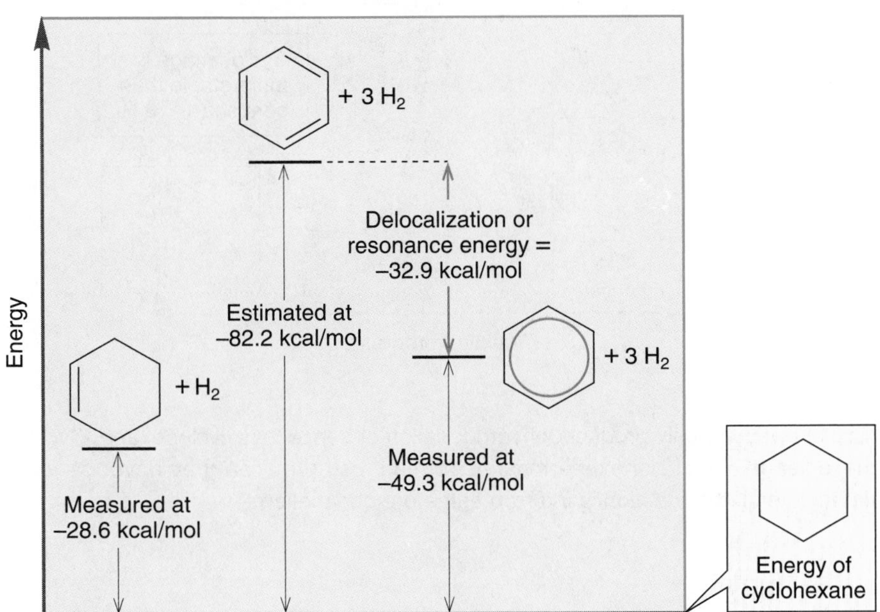

FIGURE **14.5** Measurements of the heats of hydrogenation of cyclohexene and benzene lead to a value for the delocalization, or resonance energy, of benzene.

Often, conditions under which alkenes are swiftly transformed into alkanes are quite without effect on benzene and other simple aromatic compounds. Indeed, benzene can be used as solvent in many hydrogenation reactions. We can eventually make the reaction occur: high temperature and pressure will help, as does the use of especially active catalysts, such as rhodium or ruthenium (Fig. 14.6).

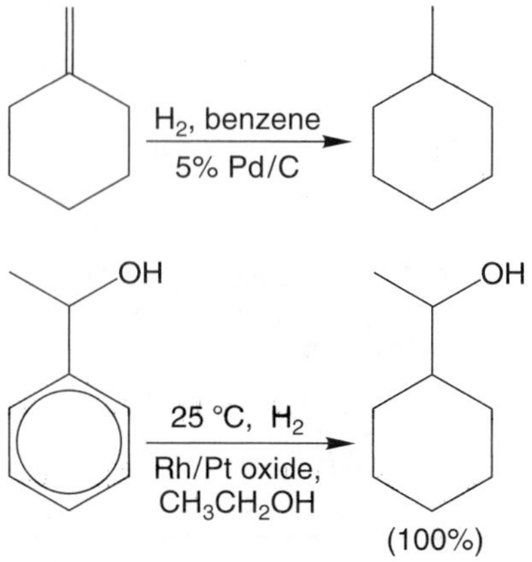

The reaction is highly exothermic, by 49.3 kcal/mol in the case of the parent benzene. The combination of benzene and three hydrogen molecules does lie above cyclohexane, the product of hydrogenation, by a substantial amount, but benzene is protected by high activation barriers. The heat, pressure, and special catalysts are needed to help overcome these barriers. Benzene rests in a valley, separated from other valleys—even lower energy valleys—by especially high mountains.

In the language of chemistry, we say that benzene is both thermodynamically and kinetically very stable. The valley containing benzene is low (thermodynamics), and the mountains surrounding it (the activation barriers, kinetics) are very high. Benzene is even protected by these barriers from undergoing reactions that are, ultimately, exothermic. The example of benzene plus hydrogen nicely illustrates this point, as the combination of these molecules lies higher in energy than the product of hydrogenation, cyclohexane (Fig. 14.7).

H_2, benzene / 5% Pd/C

OH — 25 °C, H_2 / Rh/Pt oxide, CH_3CH_2OH — OH (100%)

FIGURE 14.6 Benzene does not hydrogenate under the same conditions as alkenes.

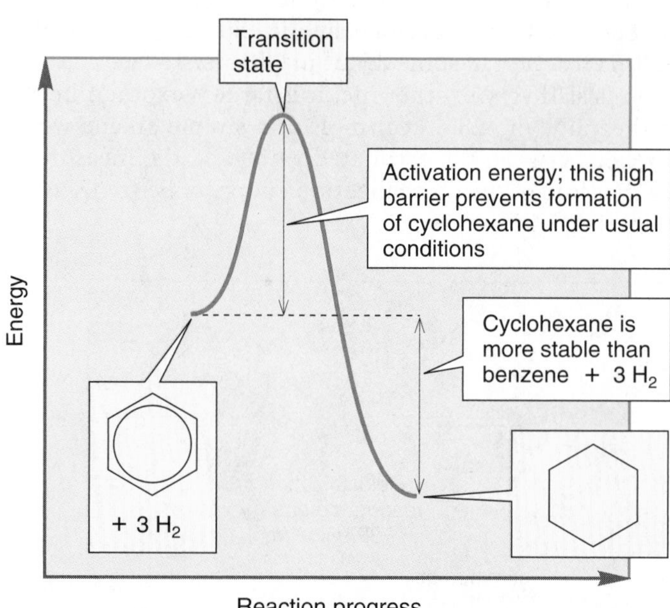
Transition state
Activation energy; this high barrier prevents formation of cyclohexane under usual conditions
Cyclohexane is more stable than benzene + 3 H_2
+ 3 H_2
Energy
Reaction progress

FIGURE 14.7 The activation barrier for cyclohexane formation is high, and makes this reaction difficult under normal conditions.

PROBLEM 14.1 Explain why the only product of hydrogenation of benzene is cyclohexane. Cyclohexadienes and cyclohexene must be formed first. Why can they never be isolated, even if the reaction is interrupted before completion?

14.2 DIELS–ALDER REACTIONS

Benzene contains a conjugated π system. Such molecules are known to undergo the Diels–Alder reaction (Chapter 12, p. 536). Yet, it is extremely difficult to induce the Diels–Alder reaction for simple benzenes, and yields are often poor. Harsh conditions (temperature as high as 200 °C, for example) and very reactive dienophiles such as hexafluoro-2-butyne are necessary (Fig. 14.8).

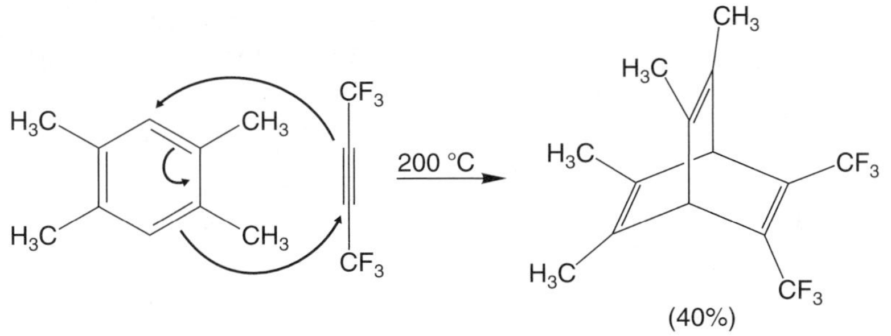

200 °C

(40%)

FIGURE **14.8** Benzenes will undergo the Diels–Alder reaction, but only with reactive dienophiles under forcing conditions.

The reason for the lack of reactivity is simple: Benzene is too stable (lies in a deep energy well) and the activation energies for this reaction are very high.

To see if all this makes sense, let's try a thought experiment. Suppose we were set the task of finding a way to make the Diels–Alder reaction possible for benzenes. We might first try to use as reactive (high energy) a dienophile as possible, as we just saw in Figure 14.8. But there is another way. We might try instead to destabilize the benzene so that surmounting the activation barriers for Diels–Alder reaction becomes easier. Figure 14.9 shows this idea in graphic form.

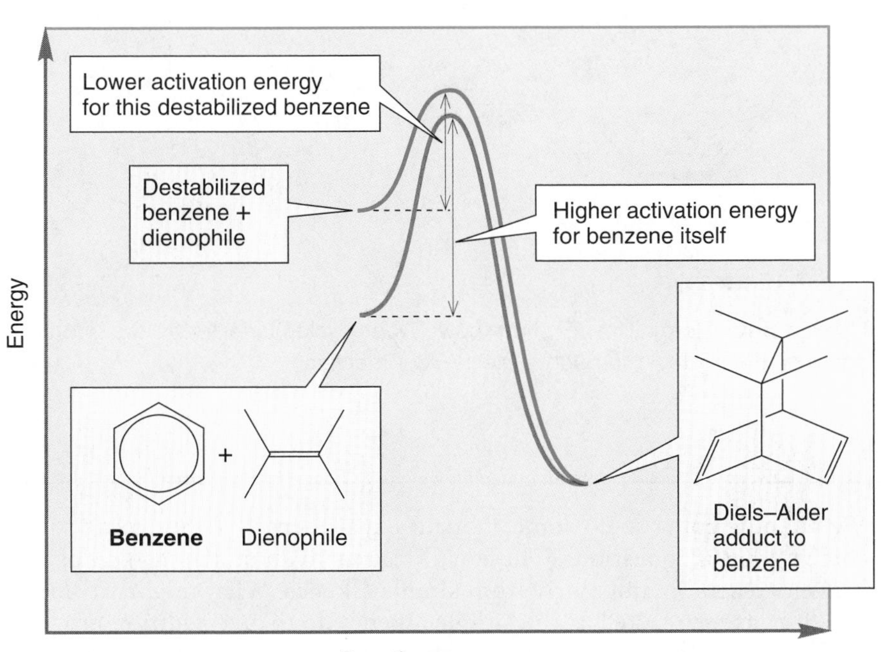

FIGURE **14.9** If benzene is destabilized, the activation barrier for reaction should be reduced.

There is a nice practical example of this idea. Imagine a benzene ring bridged by a chain of methylene groups between the 1- and 4-positions (Fig. 14.10). Such a molecule is called an **[*n*]paracyclophane**, where *n* gives the number of methylene groups in the bridge. As long as there are enough methylene groups (at least 7) there is no problem, and the molecule behaves like a simple benzene. But as [*n*] grows small (molecules with $n = 5$ and 6 are known), the ring becomes warped, and must assume a boat shape (Fig. 14.10), which severely destabilizes the molecule. In this case, the models tell all. Make a model of a paracyclophane and reduce the number of bridging methylenes one by one. The benzene ring will increasingly deform as this happens. The source of the destabilization should also become apparent. Try Problem 14.2.

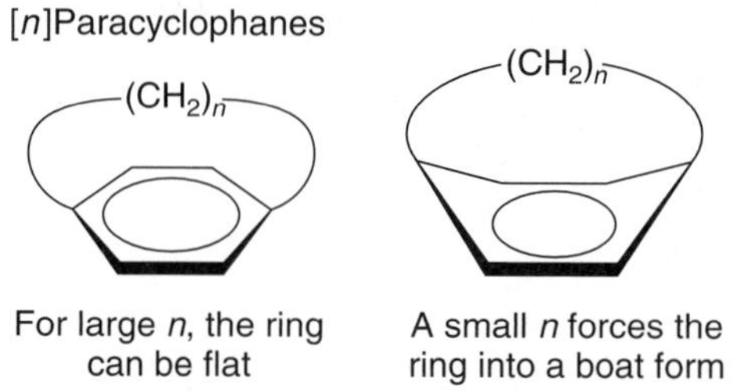

[*n*]Paracyclophanes

$(CH_2)_n$

$(CH_2)_n$

For large *n*, the ring can be flat

A small *n* forces the ring into a boat form

FIGURE **14.10** A benzene bridged from the 1- to 4-positions by *n* methylene groups is called an [*n*]paracyclophane. If *n* is small, the 1,4-bridge will force the benzene ring into a boat shape. This deformation accomplishes the task of destabilizing the molecule.

PROBLEM **14.2** Why are these boat-shaped benzenes less stable than flat ones?

As anticipated, these destabilized benzenes are much more reactive in the Diels–Alder reaction than their undistorted relatives (Fig. 14.11).

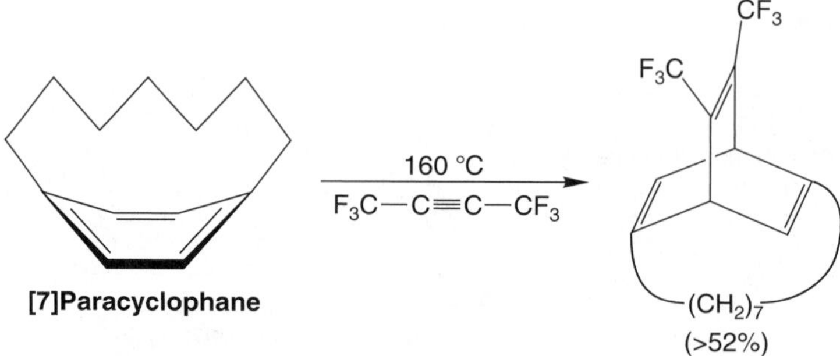

FIGURE **14.11** [7]Paracyclophane undergoes a Diels–Alder reaction at 160 °C.

PROBLEM **14.3** Use an arrow formalism to show how [7]paracyclophane forms the product shown in Figure 14.11 through a Diels–Alder reaction.

14.3 OTHER ADDITION REACTIONS

We are now going to do some thought experiments. We are going to ask some "What if" questions. These will lead us to an understanding of why benzene reacts so differently from simple alkenes. What might we anticipate if benzene reacted as the simple alkenes do to give addition products with X_2 and HX reagents? The answer to this hypothetical question is more complicated than it appears at first. Let's work through a prediction for the addition of HBr to benzene, a reaction that in fact does not occur. By analogy to alkene reactions, the first step should be the addition of a proton to the π system of benzene to give an intermediate carbocation (Fig. 14.12).

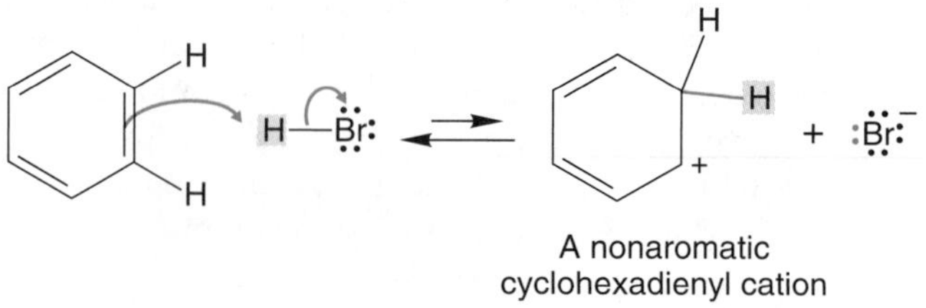

A nonaromatic cyclohexadienyl cation

FIGURE **14.12** Protonation of the benzene ring gives a cyclohexadienyl cation. Note that aromaticity is lost.

Simple addition of the bromide ion to the cation, again by analogy to the addition of HBr to an alkene, would generate a cyclohexadienyl bromide (Fig. 14.13).

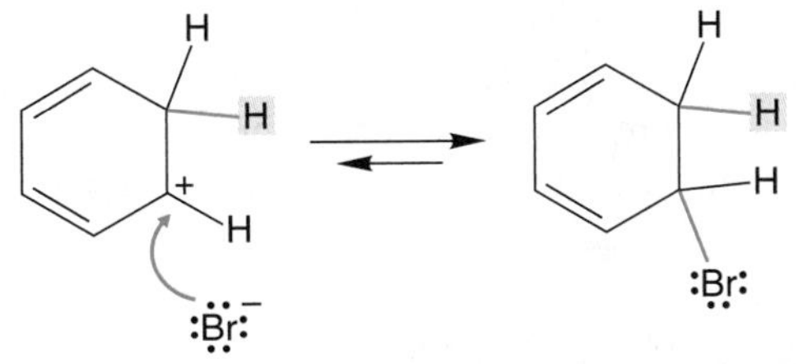

FIGURE **14.13** Addition of bromide to the cation would give a cyclohexadienyl bromide.

But we have forgotten to look closely at the structure of the carbocation formed by the addition of a proton. Just as is the case with simple dienes, the positive charge is not localized on one atom, but instead is shared by two other carbons. The resonance picture of Figure 14.14 shows a better representation of the cyclohexadienyl cation.

Summary structures

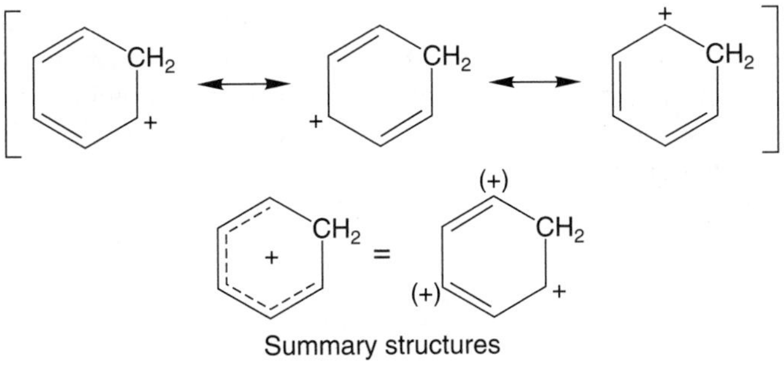

FIGURE **14.14** The cyclohexadienyl cation is resonance stabilized. The positive charge is shared by three carbons.

Convention Alert!

Look carefully again at the shorthand used to indicate the delocalized species. The dashed loop of Figure 14.14 indicates delocalization quite clearly, but it does not do a good job of showing which atoms share the positive charge. To get this information, you must draw out the resonance forms, or use a single resonance form with the other positive positions labeled with (+). Keep in mind that the dashed loop does not mean that *all* the atoms surrounding the loop are partially positive.

So, addition of bromide would take place at the three carbons sharing the positive charge to produce two different products (Fig. 14.15).

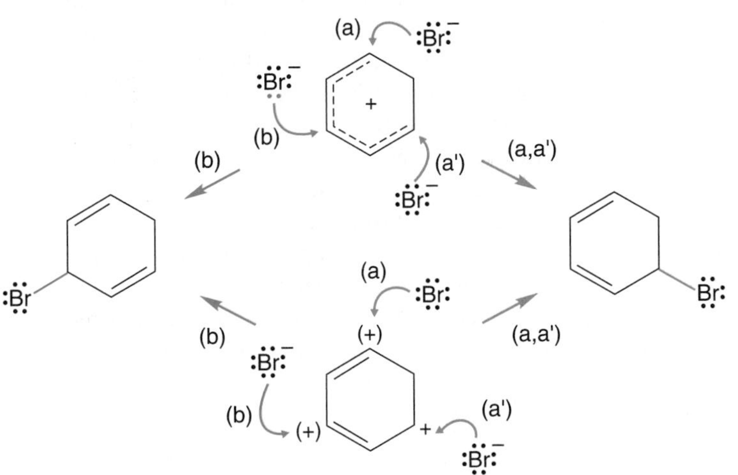

FIGURE **14.15** Capture of the cyclohexadienyl cation at the two different positions bearing positive charge must give two cyclohexadienyl bromides. These additions are shown using both of the summary structures of Figure 14.14.

Yet, neither of these compounds is the product of the reaction! In fact, there is no apparent reaction at all, and benzene is recovered unchanged from the attempted addition reaction. One searches in vain for the predicted cyclohexadienyl bromides.

Benzene fails to react with most HX reagents under conditions that easily induce addition to alkenes and dienes. The X_2 reagents, such as bromine and chlorine, are similarly unreactive. Benzene lies in too deep a well, and is protected by activation barriers that are too high for these reactions to occur.

14.4 SUBSTITUTION REACTIONS OF BENZENE

In this section, we see how benzene *does* react. What we will see is the result of an energetic imperative to preserve the aromatic sextet. Benzene reacts not by addition, which would destroy aromatic stabilization, but by substitution reactions that preserve the aromatic sextet and stabilization. The prototypal example of the substitution process is H/D exchange. If sufficiently acidic conditions are used, we can see that the apparent non-reaction of HX with benzene is an illusion. When benzene is treated with D_2SO_4, the recovered benzene has incorporated deuterium, with the amount of incorporation depending on the strength of acid used and the time of the reaction. Ultimately, excess deuteriosulfuric acid will produce fully exchanged benzene (Fig. 14.16).

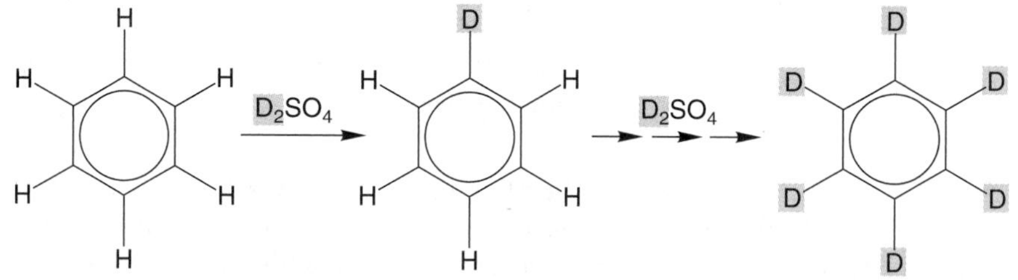

$$C_6H_6 \xrightarrow{D_2SO_4} C_6H_5D \xrightarrow{D_2SO_4} \cdots \rightarrow C_6D_6$$

FIGURE 14.16 The hydrogens in benzene can be exchanged for deuteriums in deuteriosulfuric acid.

As deuterium must become incorporated onto the ring, perhaps the reaction shown in Figure 14.12 is not so wrong after all. If benzene is protonated to produce the cyclohexadienyl cation shown in Figure 14.17, there are two possible further reactions.

As shown in Figure 14.17, the anion produced in the reaction, bisulfate, could add to give two cyclohexadienyl compounds. We imagined a related reaction in Section 14.3, where the Lewis base was bromide, Br^-. In this reaction, aromaticity is lost, and we would not expect this process to be favorable energetically. Alternatively, a proton might be lost to reverse the initial addition reaction. The aromatic sextet is regenerated. In the absence of an isotope such as deuterium, this reaction is invisible as it simply regenerates starting material (Fig. 14.18).

If it is a deuteron that is added to produce an isotopically labeled cyclohexadienyl cation, either a deuteron or a proton could be lost to regenerate benzene. If the proton is lost, deuteriobenzene will appear as product, as one hydrogen will have been exchanged for a deuterium. Repetition of the reaction can serve to exchange all the hydrogens for deuterons (Fig. 14.19).

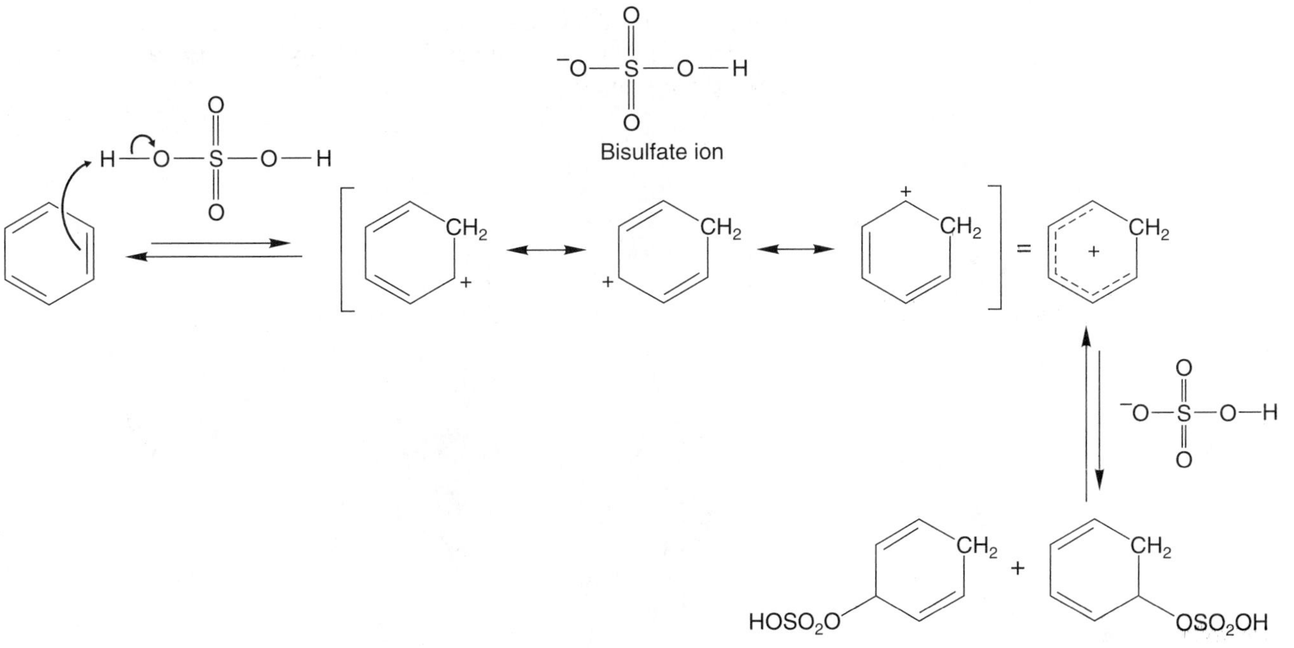

FIGURE **14.17** Protonation of benzene by sulfuric acid, followed by addition of the bisulfate ion, would produce a pair of addition products.

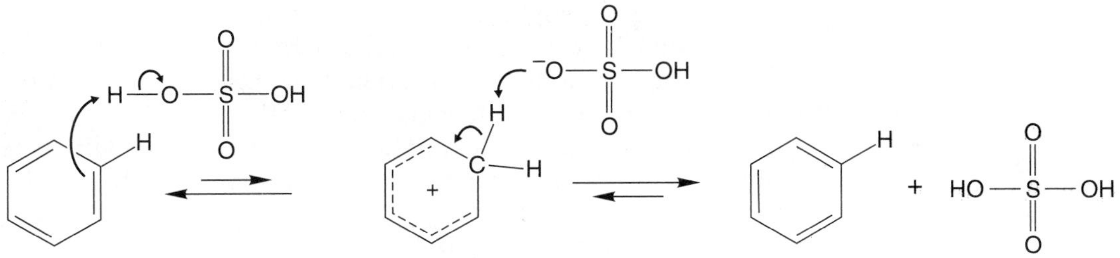

FIGURE **14.18** The protonation reaction is reversible. Bisulfate could remove a proton to regenerate benzene.

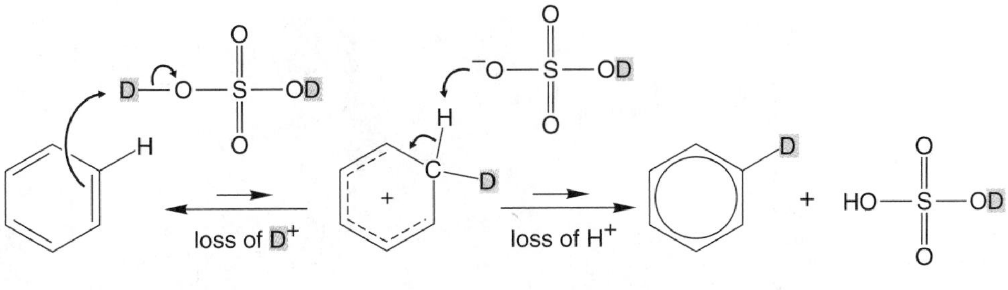

FIGURE **14.19** If deuteriosulfuric acid is used, the inital addition must generate a deuterated cyclohexadienyl cation. Reversal of the reaction will regenerate benzene, but loss of a proton will give the exchanged benzene. In an excess of D_2SO_4, all the protons of benzene can be exchanged for deuterons.

An Energy versus Reaction progress diagram for this reaction shows the endothermic formation of the cyclohexadienyl cation and its exothermic regeneration of an aromatic system (Fig. 14.20).

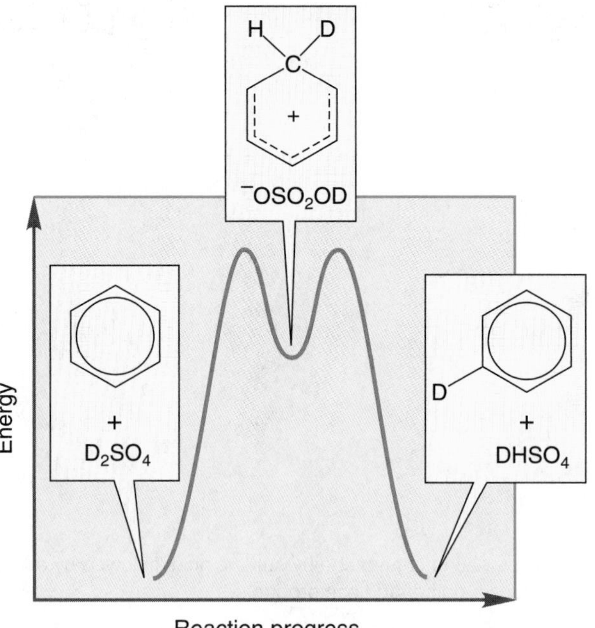

FIGURE **14.20** An Energy versus Reaction progress diagram for the exchange reaction.

If the cyclohexadienyl cation is formed, why aren't normal addition reactions observed? Loss of a proton (or deuteron) regenerates an aromatic system in a highly exothermic reaction. Were addition to occur, the destruction of the relatively high-energy carbocation would still be exothermic, but no aromatic molecule would be generated. The addition reaction is much less exothermic than the rearomatization reaction (Fig. 14.21).

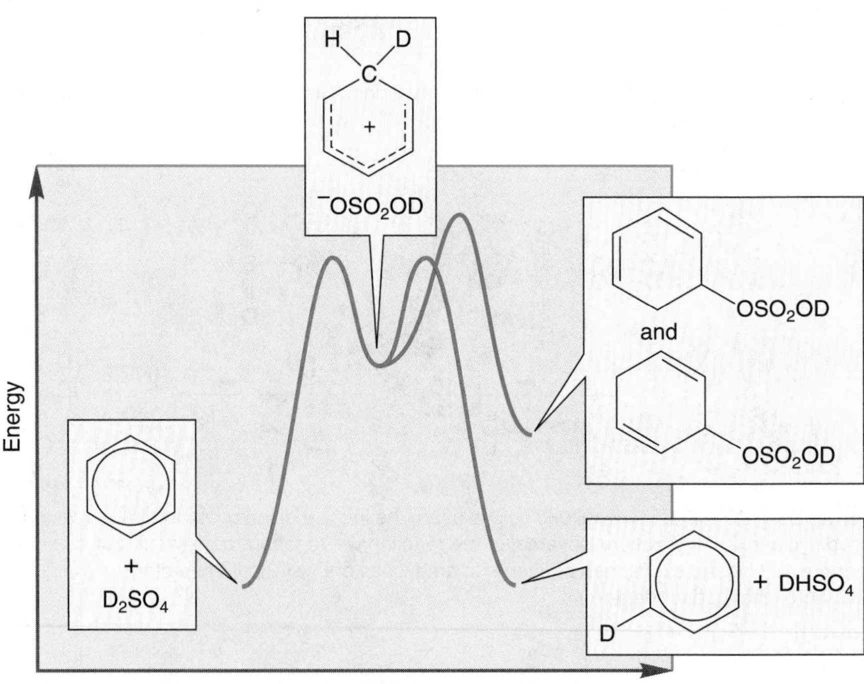

FIGURE **14.21** Both proton removal and addition are exothermic reactions for the cyclohexadienyl cation, but the regaining of aromaticity is much more so.

14.5 ELECTROPHILIC AROMATIC SUBSTITUTION

We have just seen one example of substitution of a benzene ring. Now we generalize, discovering many variations on this theme. The main points to remember are, first, that aromaticity is not easily disrupted, and second, that once the aromatic stabilization is lost, it is easily regained.

14.5a The General Reaction

The exchange reaction of benzene in deuterated acid serves as a prototype for a host of similar reactions in which an electrophile, E^+, is substituted for a hydrogen on the ring (Fig. 14.22).

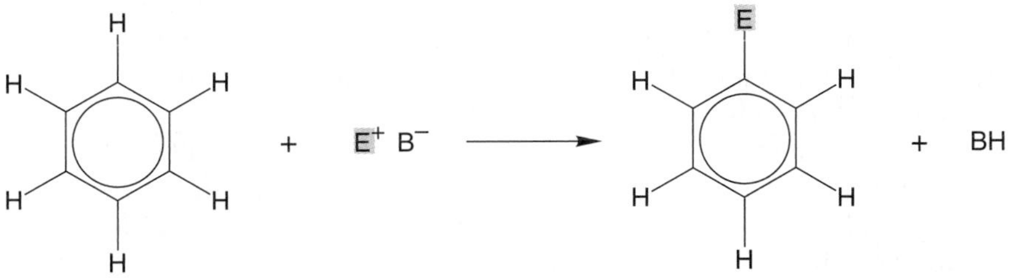

FIGURE 14.22 A generic aromatic substitution reaction.

The reaction is always referred to as **electrophilic aromatic substitution**, since the emphasis is upon the agent doing the substituting, the electrophile. Fair enough, but in every electrophilic reaction there must be a nucleophile as well. In this case, the π system of the benzene ring acts as the nucleophile. In orbital terms, it is one of the degenerate (equi-energetic) filled molecular orbitals of benzene that overlaps with the empty orbital on the electrophile (Fig. 14.23).

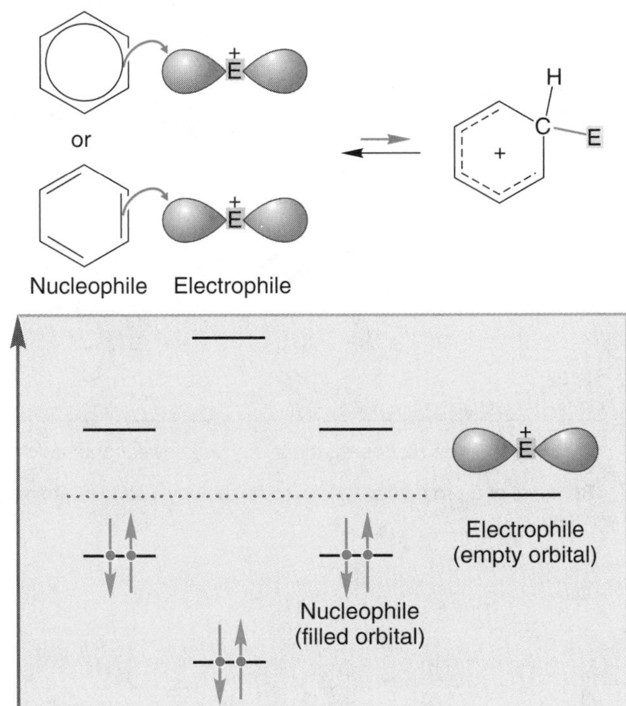

FIGURE 14.23 One of benzene's equi-energetic (degenerate) HOMOs overlaps with the empty orbital on E^+ in a Lewis base–Lewis acid reaction.

As in the exchange reaction of D for H, the general reaction is completed by the removal of a proton from the intermediate cyclohexadienyl cation by a base, regenerating the aromatic benzene ring (Fig. 14.24).

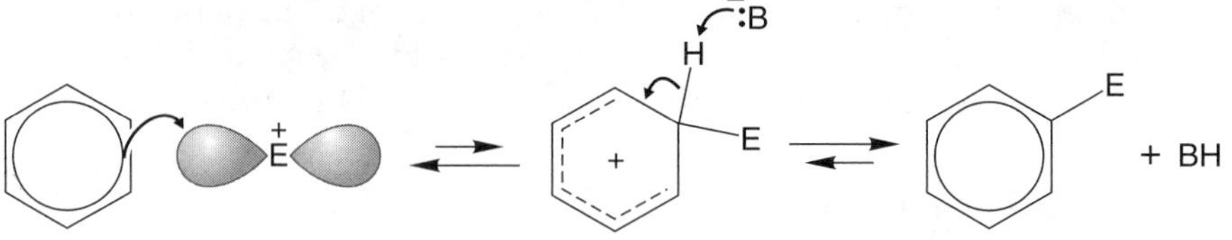

FIGURE 14.24 The general aromatic substitution mechanism involves formation of a cyclohexadienyl cation, followed by proton removal, regenerating the aromatic ring.

14.5b Sulfonation

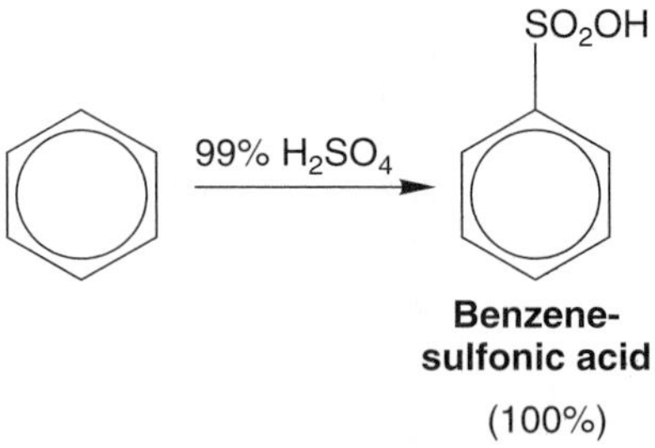

When benzene is treated with very concentrated sulfuric acid, or a solution of sulfur trioxide in sulfuric acid, called "oleum," benzenesulfonic acid (Fig. 14.25) is the product. The electrophile in this reaction can be SO_3 itself, or protonated SO_3 ($HO\overset{+}{S}O_2$), either of which can add to benzene to give a cyclohexadienyl cation (Fig. 14.26).

As in the general mechanism for aromatic substitution, a base now removes a proton to give the product. What base can exist in concentrated sulfuric acid? Lewis bases abound: The SO_3 itself is one, as is sulfuric acid. Proton loss might even be intramolecular as shown in the second part of Figure 14.27. The cyclohexadienyl cation can recover aromaticity by donating a proton to one of the bases present. The structure of the product of this reaction, benzenesulfonic acid, gives students problems because the coded structures, SO_3H or SO_2OH hide the detail. The full structure is drawn out in Figure 14.27.

Sulfonation of benzene is reversible, especially at high temperature. Treatment of benzenesulfonic acid in acidic steam regenerates the parent benzene (Fig. 14.28).

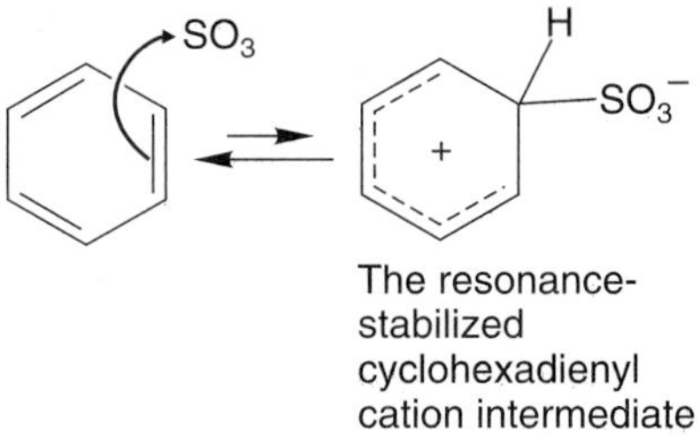

Benzenesulfonic acid (100%)

FIGURE 14.25 Sulfonation of benzene to give benzenesulfonic acid.

The resonance-stabilized cyclohexadienyl cation intermediate

FIGURE 14.26 The SO_3 molecule adds to the benzene ring to form a cyclohexadienyl cation.

Three representations of benzenesulfonic acid

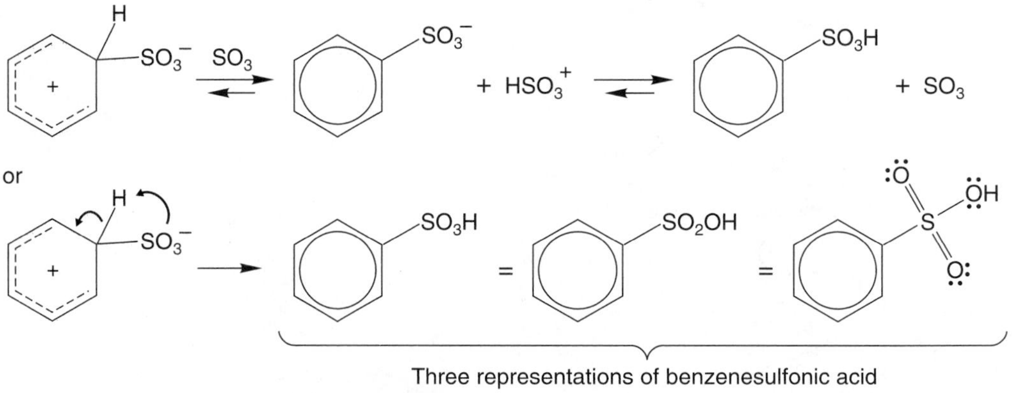

FIGURE 14.27 Removal of a proton from the cyclohexadienyl cation intermediate regenerates the aromatic ring.

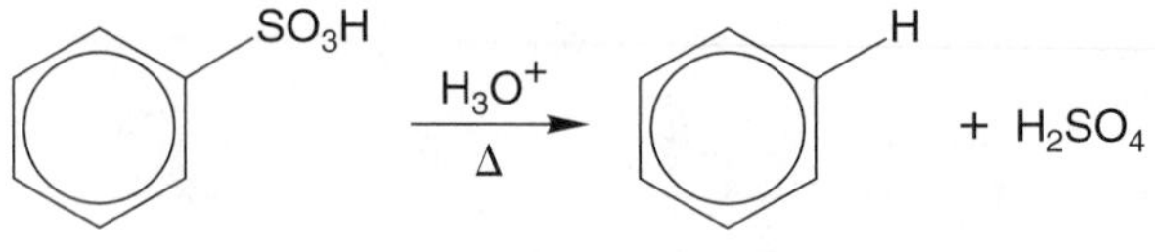

FIGURE 14.28 Sulfonation is reversible.

*PROBLEM **14.4**

Write a detailed mechanism for the reaction in Figure 14.28.

ANSWER

The mechanism for the transformation of benzenesulfonic acid into benzene is just the reverse of the formation of benzenesulfonic acid from benzene. We have already written something very much like this mechanism in the earlier figures when we drew out the mechanism for the forward version of this reaction. Protonation gives a cyclohexadienyl cation that can either lose a proton to revert to benzenesulfonic acid, or lose the sulfonic acid group to give benzene.

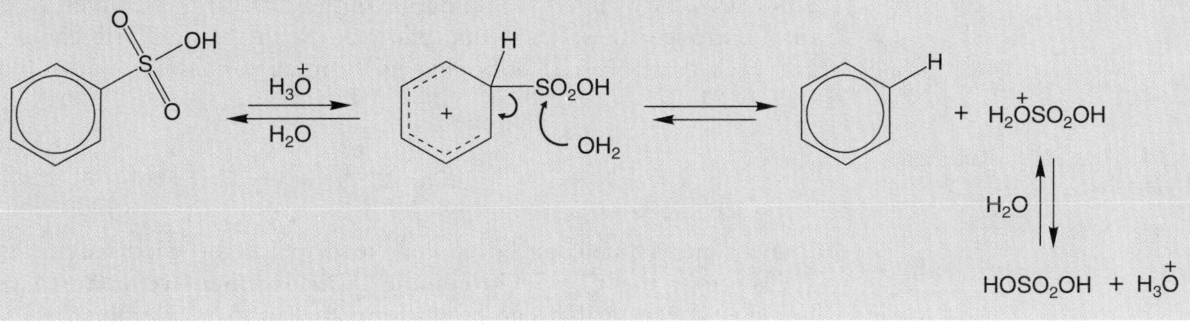

14.5c Nitration

In concentrated nitric acid ($HONO_2$), often in the presence of sulfuric acid (H_2SO_4), benzene is converted into nitrobenzene (Fig. 14.29).

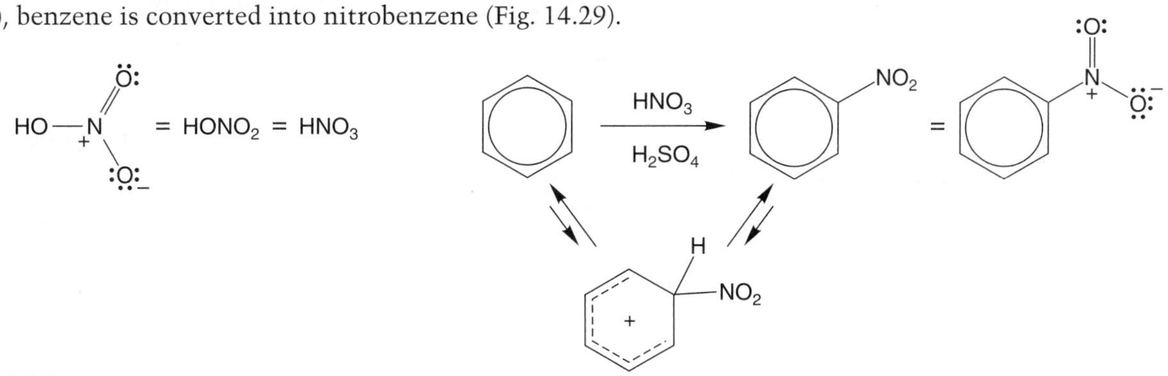

FIGURE **14.29** Nitration of benzene.

In strong acid, nitric acid is protonated to give $H_2\overset{+}{O}NO_2$. Loss of water can generate a powerful Lewis acid, the nitronium ion, $^+NO_2$, that can act as the electrophile in an aromatic substitution reaction. Indeed, stable nitronium salts are known and can also effect nitration (Fig. 14.30).

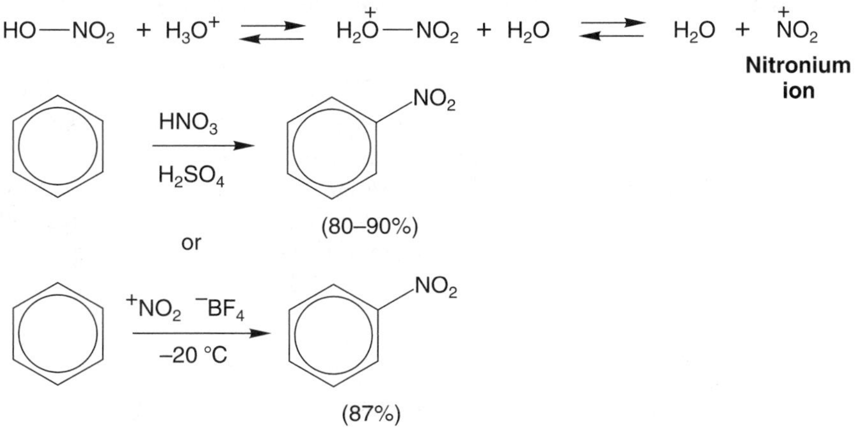

FIGURE **14.30** The powerful electrophile $^+NO_2$ is formed by loss of water from protonated nitric acid. Aromatic substitution follows. Alternatively, stable nitronium salts can be used to nitrate benzenes.

As usual, the reaction is completed by deprotonation of the cyclohexadienyl cation by one of the Lewis bases present in the mixture.

14.5d Halogenation

Unlike simple alkenes, benzene fails to react with Br_2 or Cl_2. It takes a catalyst, usually $FeBr_3$ or $FeCl_3$, to initiate the reaction. Chlorine is not a strong enough electrophile (Lewis acid) to react with the weakly nucleophilic benzene by itself. The function of the catalyst is to convert chlorine into a stronger Lewis acid that can react with the aromatic compound. Ferric chloride ($FeCl_3$) does this by forming a complex with chlorine (Fig. 14.31).

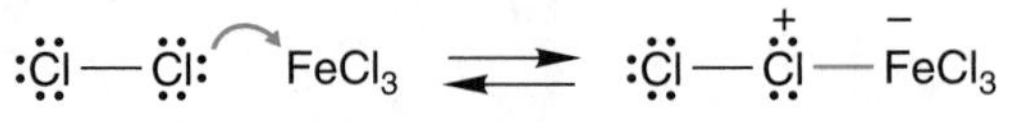

FIGURE 14.31 Complex formation from Cl_2 and $FeCl_3$.

Benzene is nucleophilic enough to displace the good leaving group $FeCl_4^-$ from the Cl_2–$FeCl_3$ complex. Displacement generates a cyclohexadienyl cation that can be deprotonated to give chlorobenzene. This reaction may look strange, but it has a most familiar analogy in the protonation of an alcohol to convert OH into the good leaving group $^+OH_2$ (Fig. 14.32). The $FeCl_3$ plays the role of the proton in the alcohol reaction. The mechanism of bromination using Br_2 and $FeBr_3$ is similar.

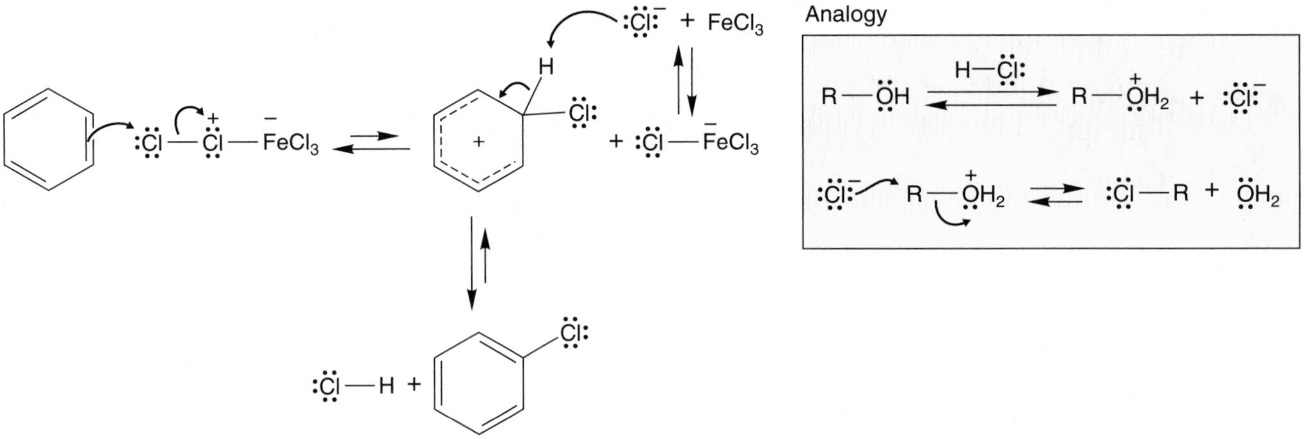

FIGURE 14.32 Chlorination of benzene.

PROBLEM 14.5 Write a mechanism for the bromination of benzene using $Br_2/FeBr_3$.

These reactions (sulfonation, nitration, chlorination, and bromination) serve to underscore the general principles of electrophilic aromatic substitution. In each, an electrophile powerful enough to react with the weakly nucleophilic benzene ring must be generated. The addition reaction first generates a resonance-stabilized (but not aromatic) cyclohexadienyl cation that can be deprotonated to regenerate an aromatic system. The general mechanism is shown again in Figure 14.33.

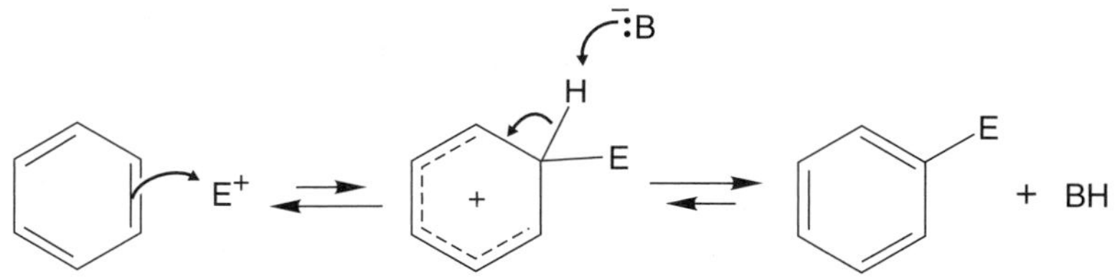

FIGURE **14.33** Once again, here is the general mechanism for aromatic substitution, where $E^+ = D^+$, $^+SO_2OH$, $^+NO_2$, ^+Cl, or ^+Br.

We are now able to make a few substituted aromatic compounds, and this capability will shortly prove quite useful. But much of the interest in synthetic chemistry involves the construction of carbon–carbon bonds. We have made no progress at all in finding ways to make bonds from benzene to carbon. In Sections 14.6 and 14.7, we do just that by making use of another electrophilic aromatic substitution reaction quite closely related to the bromination and chlorination reactions we have just seen.

14.6 THE FRIEDEL–CRAFTS ALKYLATION REACTION

The Friedel–Crafts alkylation reaction is named for Charles Friedel (1832–1899) and James M. Crafts (1839–1917), who discovered it by accident when they tried to synthesize pentyl chloride from pentyl iodide through reaction with aluminum chloride in an aromatic solvent. Instead of the hoped-for chloride, substituted aromatic hydrocarbons appeared. The reaction is closely related to bromination and chlorination of benzene. If we treat benzene with isopropyl bromide in the hope that a nucleophilic attack of benzene on isopropyl bromide will produce isopropylbenzene (cumene), we are sure to be disappointed. Benzene is by no means a strong nucleophile and isopropyl bromide can scarcely be described as a powerful electrophile. The proposed reaction is hopeless, and utterly fails to give product (Fig. 14.34).

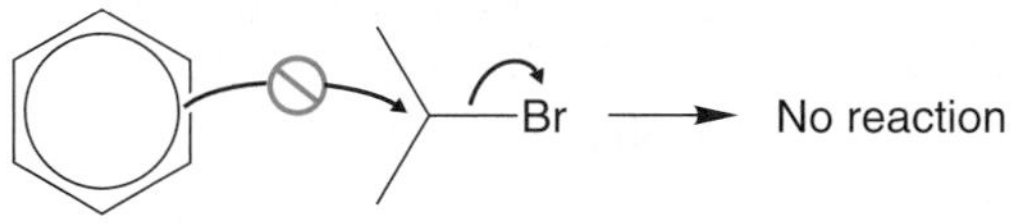

FIGURE **14.34** Benzene fails to react with isopropyl bromide (and other simple halides).

So why even bring the subject up? It does look a bit like another reaction that fails to give product—the attempt to form chlorobenzene or bromobenzene from chlorine or bromine and benzene. Those halogenations are transformed into a success by the addition of a catalyst, $FeCl_3$ or $FeBr_3$, which converts the halogen, a weak electrophile, into a much stronger Lewis acid (Figs. 14.31 and 14.32). Perhaps a related technique will work here. Indeed it does, as the addition of a small amount of the catalyst $AlBr_3$ or $AlCl_3$ to a mixture of benzene and isopropyl bromide or chloride results in the formation of isopropylbenzene. The mechanism should be easy to write now. A complex is first formed between isopropyl bromide and $AlBr_3$. This complex can be attacked by benzene to give a resonance-stabilized cyclohexadienyl cation. Deprotonation of the intermediate cyclohexadienyl cation gives isopropylbenzene (Fig. 14.35).

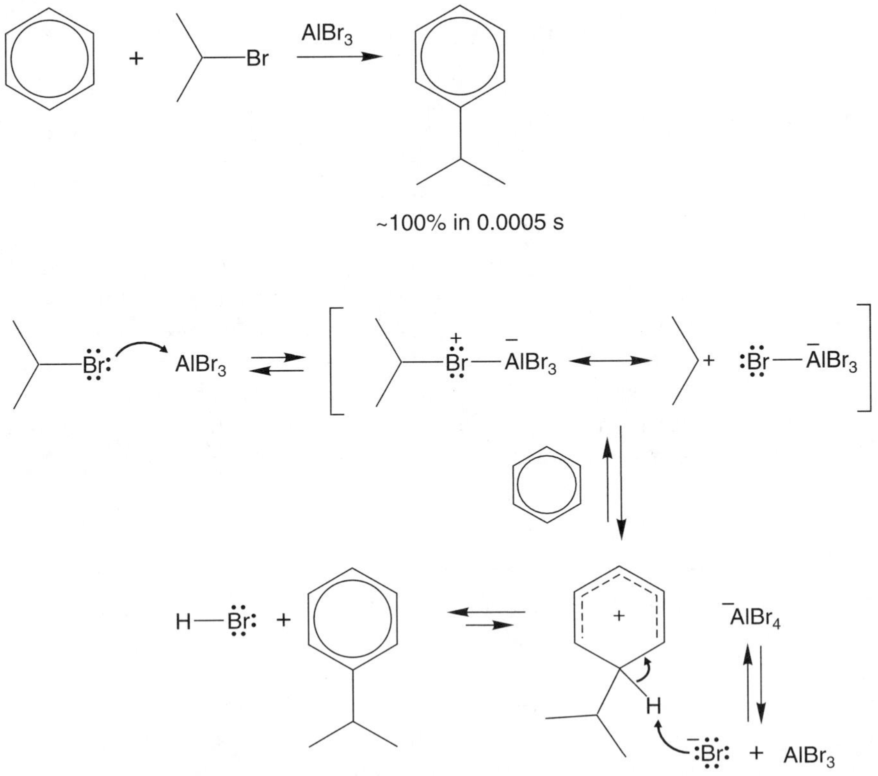

FIGURE 14.35 The mechanism for Friedel–Crafts alkylation of benzene using isopropyl bromide and aluminum bromide ($AlBr_3$).

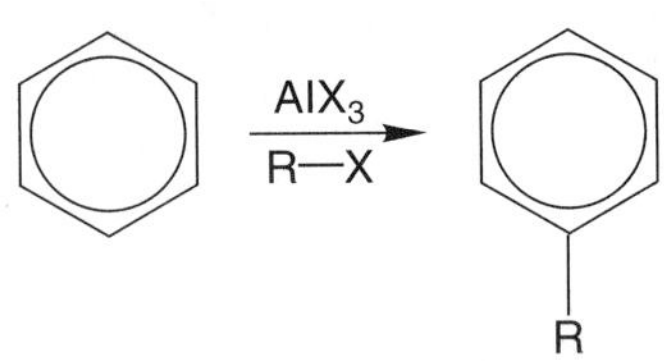

FIGURE 14.36 Friedel–Crafts alkylation succeeds for a wide variety of alkyl halides, where X = Cl or Br and R = CH_3, CH_3CH_2, $(CH_3)_2CH$, $(CH_3)_3C$.

Whether or not the complex in Figure 14.35 can actually form a free carbocation depends on the alkyl halide used in the reaction. Friedel–Crafts alkylation succeeds with a variety of alkyl halides, including methyl-, ethyl-, isopropyl-, and *tert*-butyl chlorides and bromides (Fig. 14.36).

We would not expect the methyl or ethyl cation to be formed in these reactions, and surely these alkylations take place by attack of benzene on the complexed halide. *tert*-Butyl chloride, on the other hand, can produce the relatively stable *tert*-butyl cation, not an unlikely intermediate at all. So, the detailed mechanism of the Friedel–Crafts alkylation depends on the nature of the R group.

*PROBLEM 14.6 Why is attack of benzene on complexed *tert*-butyl bromide to give a cyclohexadienyl cation an unlikely mechanistic step (Fig. 14.37)?

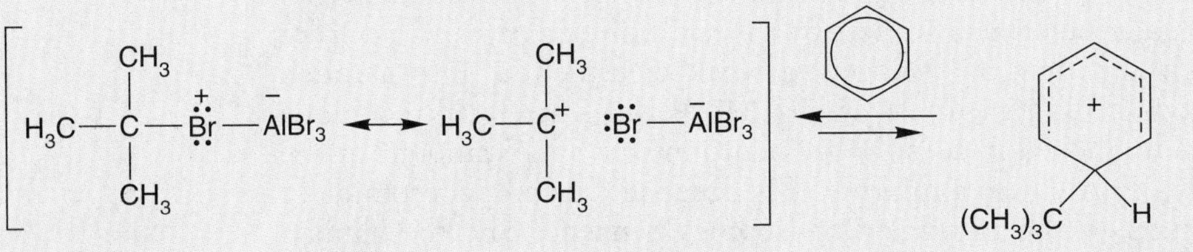

FIGURE 14.37

ANSWER The S_N2 reaction lives! Old material doesn't go away; this fundamental reaction is as important now as it was earlier. Nucleophilic attack of benzene on the

ANSWER (CONTINUED)

tert-butyl bromide–$AlBr_3$ complex is an S_N2 reaction, and these do not occur at tertiary carbons.

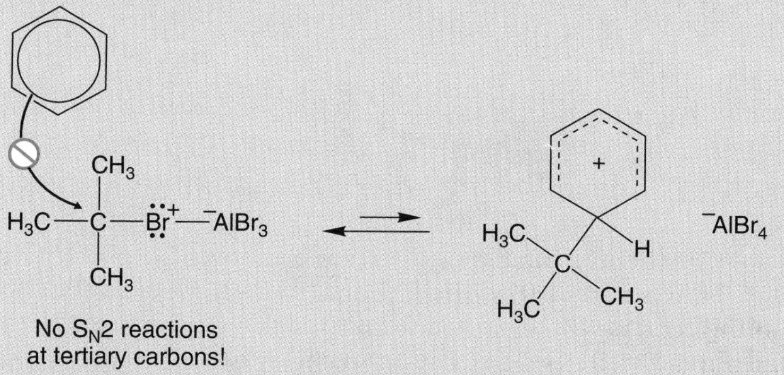

No S_N2 reactions at tertiary carbons!

This reaction must go through formation of the free *tert*-butyl cation.

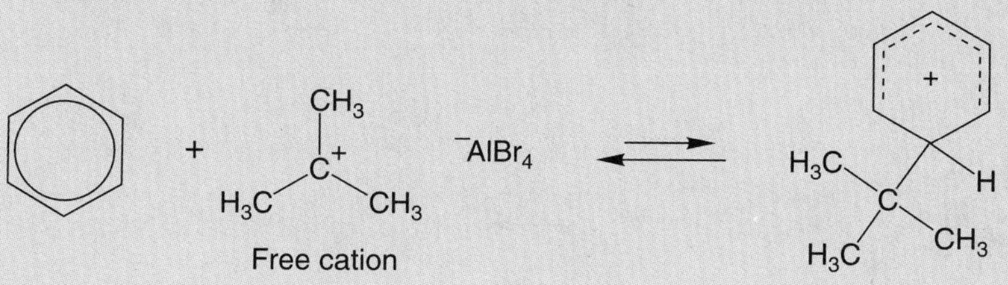

Free cation

PROBLEM 14.7

The synthesis of toluene by the aluminum chloride catalyzed Friedel–Crafts alkylation of benzene with methyl chloride is badly complicated by the formation of di-, tri-, and polymethylated benzenes. It appears that the initial product of the reaction, methylbenzene (toluene), is more reactive in the Friedel–Crafts reaction than is benzene. Analyze the mechanism of electrophilic aromatic substitution to see why toluene is more reactive than benzene. *Hint*: Look carefully at substitution in the position directly across the ring from the methyl group (the para position) for both benzene and toluene, and look for differences.

PROBLEM 14.8

How would you minimize the overalkylation mentioned in Problem 14.7 and achieve a reasonable synthesis of toluene from benzene?

Not only do the mechanistic details of the reaction depend on the alkyl group (R), but even the gross outcome of the reaction depends on the structure of R. Although the alkyl halides we have mentioned so far all succeed in making the related alkylated benzenes, there are many that do not. Suppose we set out to make propylbenzene by the Friedel–Crafts alkylation of benzene using propyl chloride, benzene, and aluminum chloride. If we run this reaction, all looks well at first. The products formed are of the correct molecular weight, but a closer look uncovers a problem. Some propylbenzene is formed, but the major product is isopropylbenzene (Fig. 14.38).

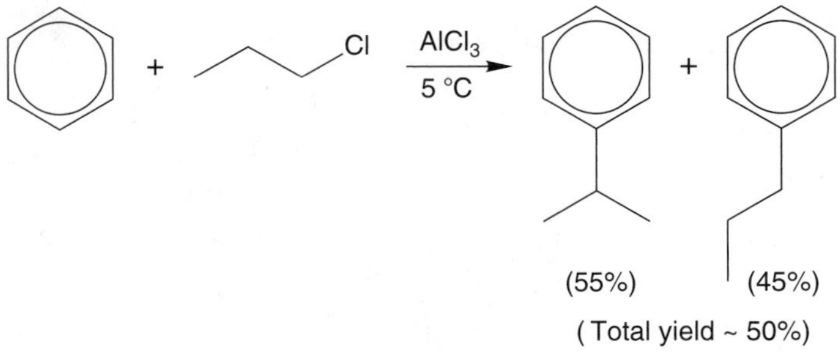

FIGURE 14.38 An attempt to make propylbenzene by Friedel–Crafts alkylation gives isopropylbenzene as the major product.

A close look at the mechanism of the reaction shows the problem. In the complex between the alkyl halide and aluminum chloride, the carbon–chlorine bond is weakened and the positive charge is shared by the chlorine and the adjacent carbon. The resonance form shown in Figure 14.39 shows this clearly.

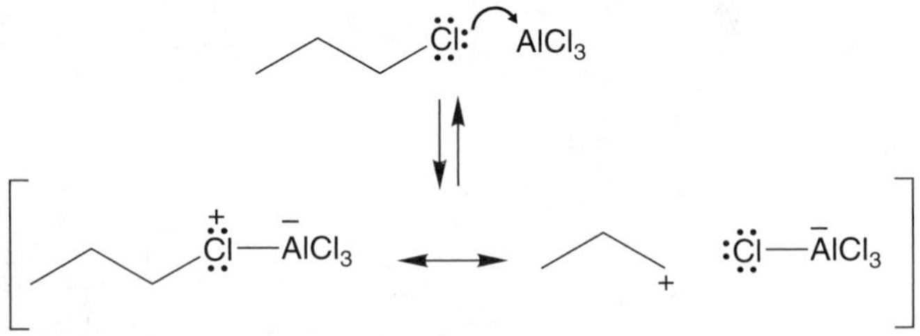

FIGURE 14.39 In the complex formed between propyl chloride and aluminum chloride ($AlCl_3$), the positive charge is partially on the primary carbon adjacent to the chlorine.

The complex formed from propyl chloride and aluminum chloride contains a partial positive charge on a primary carbon, a most unhappy state of affairs. When this happens in reactions of methyl chloride or ethyl chloride, there is nothing that can be done about it, but propyl chloride has a way out. A rearrangement of hydride generates the much more stable complex, or even a free cation in which a secondary carbon bears the positive charge. Straightforward alkylation gives the observed major product, isopropylbenzene (Fig. 14.40).

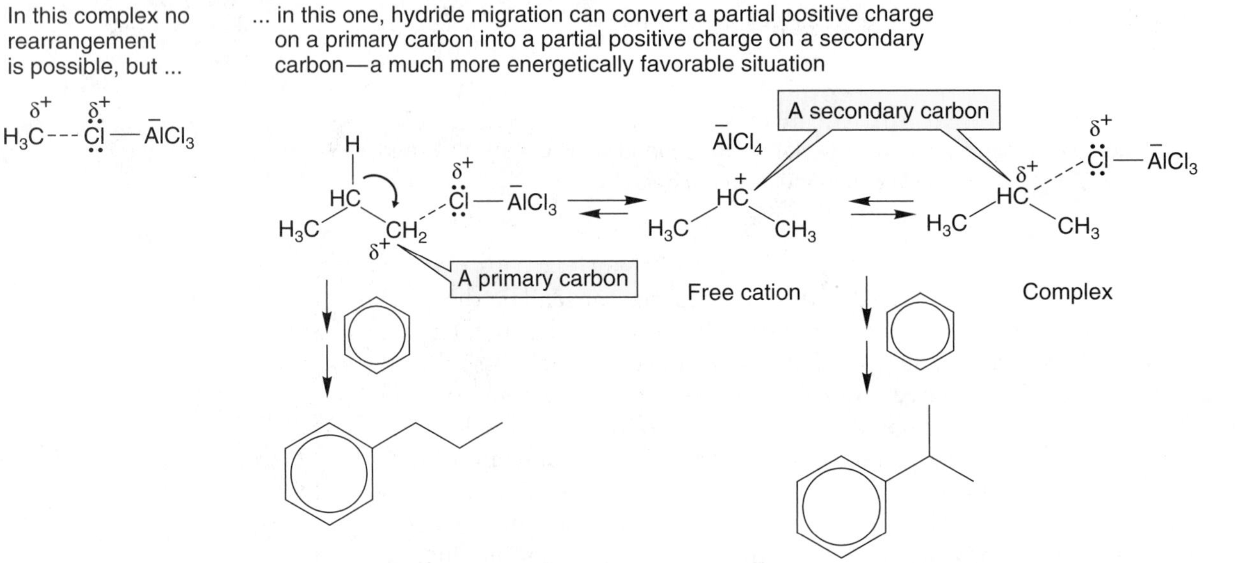

FIGURE 14.40 A hydride shift can convert a less stable complex with a partial positive charge on a primary carbon into a more stable complex in which the partial positive charge is on a secondary carbon. The substitution reaction follows.

These rearrangements are very common, and are encountered whenever a hydride or alkyl shift will generate a more stable cation, or complex (Fig. 14.40). This phenomenon is not new. Recall, for example, the rearrangements of hydride and alkyl groups encountered in reactions involving carbocations (Fig. 14.41; also Chapter 9, p. 374).

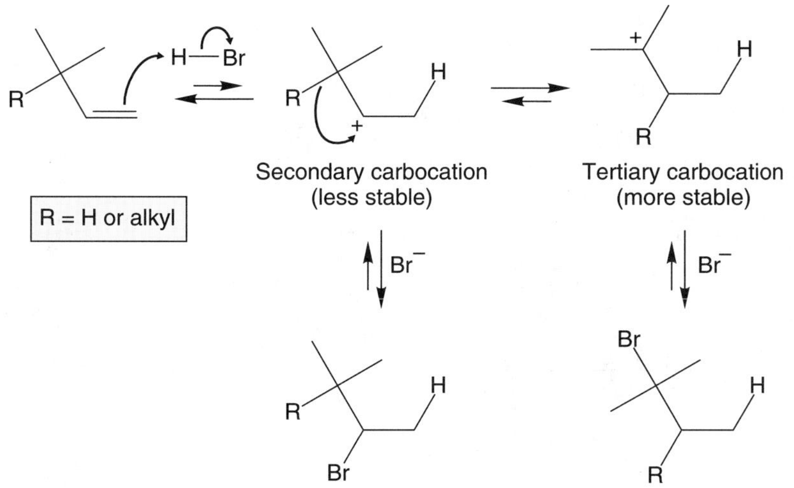

R = H or alkyl

Secondary carbocation (less stable)

Tertiary carbocation (more stable)

FIGURE **14.41** Hydride shifts are commonly encountered in addition reactions whenever a more stable carbocation can be generated from a less stable one.

So, Friedel–Crafts alkylation is limited to those alkyl halides for which rearrangement is not a problem. In all other cases, rearrangements can be expected. You just cannot make pure propylbenzene by a Friedel–Crafts alkylation. But there is a way out; you can use another reaction discovered by Friedel and Crafts.

PROBLEM **14.9**

tert-Butylbenzene is converted into benzene on treatment with $AlCl_3$ and a trace of HCl. Write a mechanism for this reaction.

14.7 FRIEDEL–CRAFTS ACYLATION

A variation of the Friedel–Crafts alkylation (**Friedel–Crafts acylation**) introduces a new group, called an **acyl group** (RC=O) (Fig. 14.42). An acyl chloride (RCOCl) is also known as an **acid chloride**. We will take up the chemistry of all manner of acyl derivatives in Chapter 20, but we need to use them here. In practice, they are rather easy to make from the relatively available carboxylic acids, RCOOH, by treatment with thionyl chloride ($SOCl_2$) (Fig. 14.43).

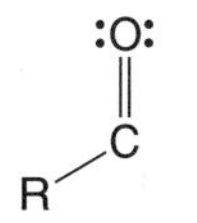

The acyl group

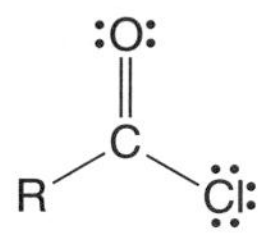

An acyl chloride or an acid chloride

FIGURE **14.42** The acyl group.

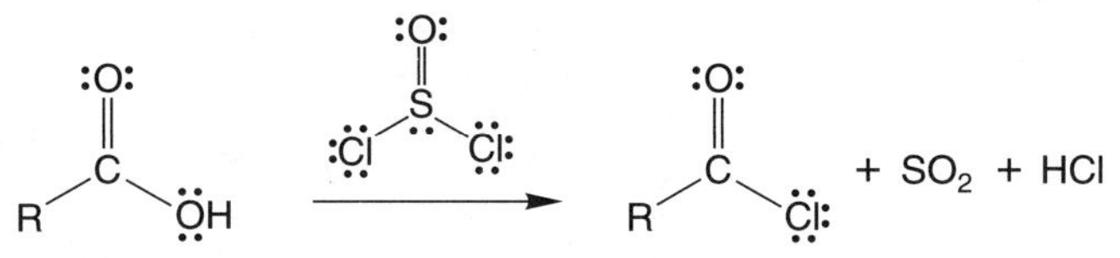

A carboxylic acid → An acid chloride + SO_2 + HCl

FIGURE **14.43** Acid chlorides can be made by the reaction of carboxylic acids with thionyl chloride.

Acyl chlorides react with benzene under the influence of an equivalent (not a catalytic amount) of $AlCl_3$. A molecule of $AlCl_3$ is needed for each molecule of product. By now it should be easy to write a general mechanism. A complex is first formed with $AlCl_3$. But which complex? Unlike alkyl chlorides, acyl chlorides contain two nucleophilic sites, the oxygen and chlorine atoms. The evidence is that both complexes are formed, and that the two are in equilibrium. The resonance-stabilized **acylium ion** can be formed by dissociation of either complex (Fig. 14.44).

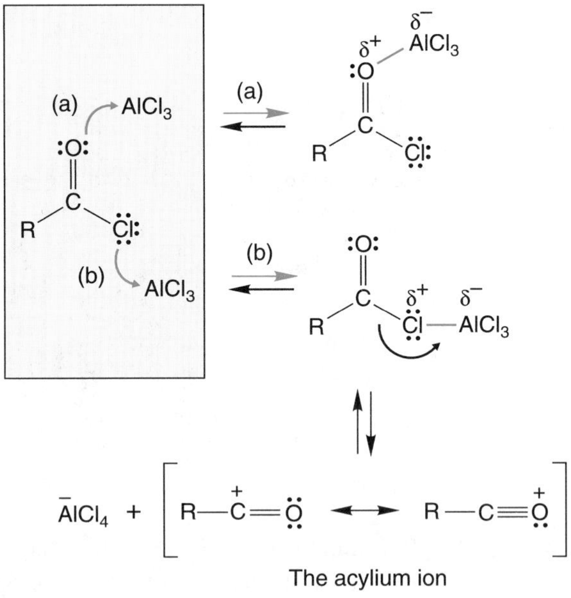

FIGURE 14.44 The acid chloride–$AlCl_3$ complex can dissociate to give the acylium ion.

The acylium ion can then add to the benzene ring to generate an intermediate cyclohexadienyl cation. Deprotonation by a Lewis base completes the reaction and produces acyl benzenes (Fig. 14.45). As mentioned above, unlike Friedel–Crafts alkylation, Friedel–Crafts acylation is not a catalytic reaction. The product, an acyl benzene, is also reactive toward $AlCl_3$ and forms its own 1:1 complex as it is produced. So, for every molecule of acyl benzene formed in the reaction, one molecule of $AlCl_3$ is used up in complex formation. A stoichiometric amount of $AlCl_3$ is required in this reaction. At the end of the reaction, addition of water destroys the complex and generates the free, acylated benzene (Fig. 14.45).

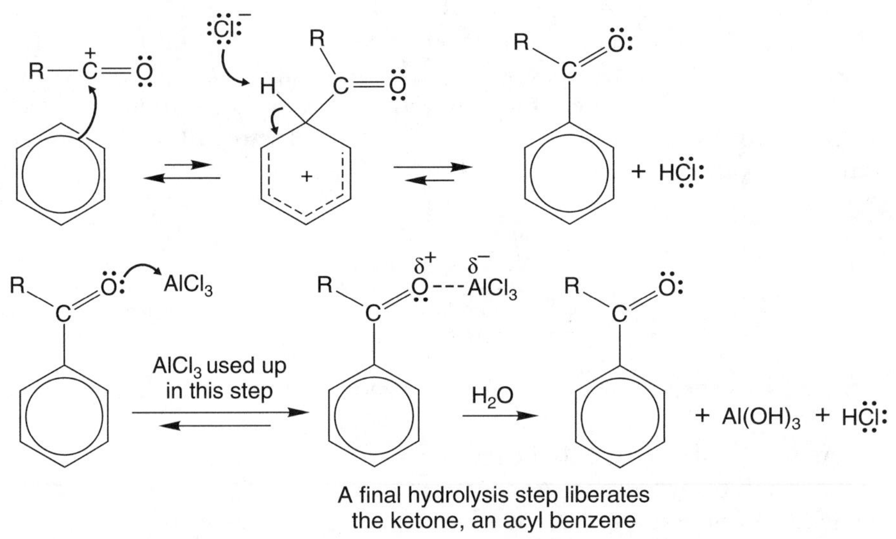

FIGURE 14.45 The mechanism of Friedel–Crafts acylation. Aromatic ketones can form complexes with $AlCl_3$, which is the reason why a full equivalent of $AlCl_3$ is necessary for Friedel–Crafts acylation.

Despite the mechanistic complexity, in practice the Friedel–Crafts acylation reaction works very well, and a variety of phenyl-substituted ketones can be produced in good yield. Be certain to add this reaction to your collection of "synthesis" file cards (Fig. 14.46).

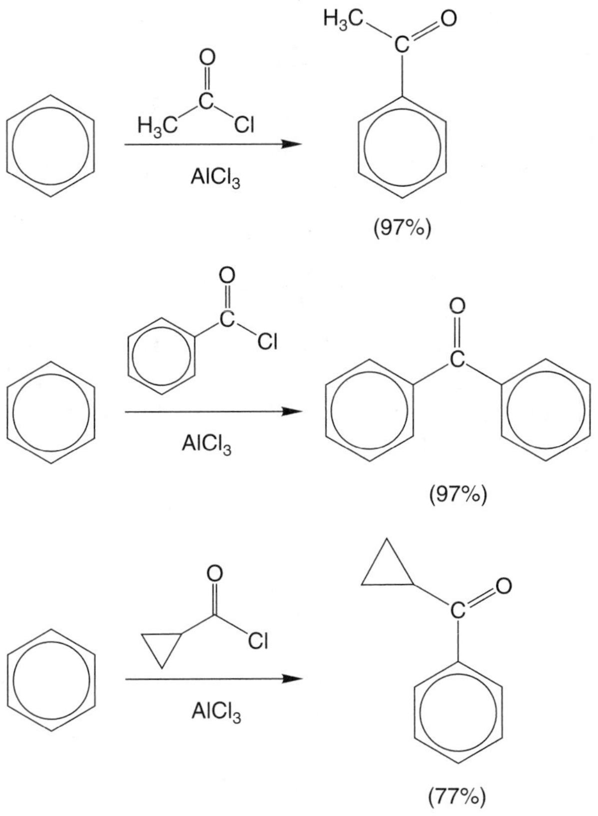

H_3C / C=O / Cl; $AlCl_3$ (97%)

$AlCl_3$ (97%)

$AlCl_3$ (77%)

FIGURE 14.46 Some typical Friedel–Crafts acylations.

Figure 14.47 shows another successful acylation, but this reaction contains a surprise. The reaction of butyryl chloride with benzene yields phenyl propyl ketone (butyrophenone) This result implies that acyl complexes are not prone to rearrangement (Fig. 14.47).

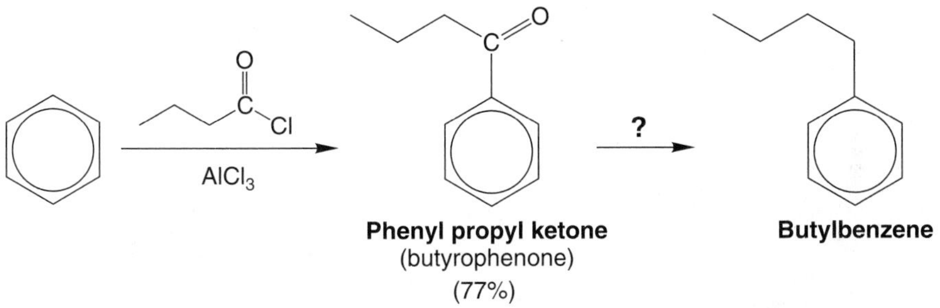

$AlCl_3$?

Phenyl propyl ketone
(butyrophenone)
(77%)

Butylbenzene

FIGURE 14.47 Friedel–Crafts acylation is not complicated by hydride rearrangements.

*PROBLEM 14.10

Explain why these acylium ions (or acyl complexes) are not likely to rearrange. Two reasons should be apparent. First, what differentiates the acylium ion from simple alkyl carbocations? Next, look closely at the product of rearrangement and consider the polarity of the carbon–oxygen double bond.

ANSWER The first reason has to do with the stability of the acylium ion itself. The acylium ion is well stabilized by resonance, and is therefore less likely to rearrange than simple alkyl carbocations.

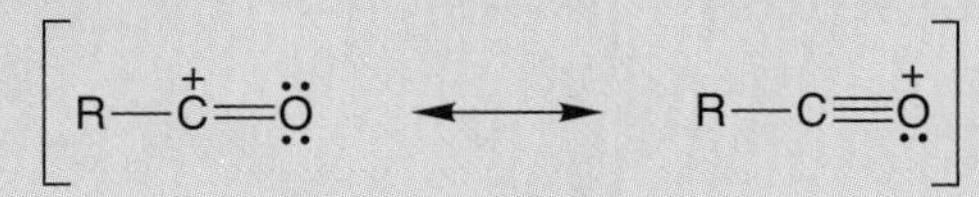

$$\left[R-\overset{+}{C}=\ddot{O} \longleftrightarrow R-C\equiv\overset{+}{O} \right]$$

An acylium ion

Second, consider the structure of the putative rearranged cation.

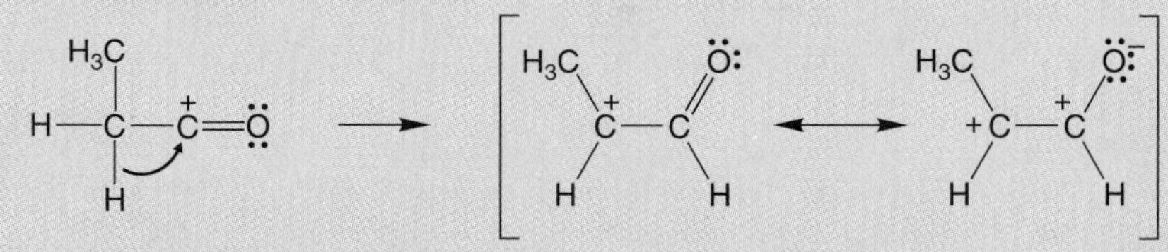

The carbonyl group is strongly polarized in the sense shown in the drawing. The carbon is δ^+ and the oxygen is δ^-. Rearrangement opposes a full and partial positive charge. This situation is very bad energetically and will act to retard migration of hydrogen or an alkyl group.

The four-carbon chain is intact in the product, and if we had a way to transform the carbon–oxygen double bond into a methylene group, we would have a way to make butylbenzene without rearrangement.

In fact, there are many such ways, two of which are shown in Figure 14.48. The Clemmensen reduction, named for its discoverer E. C. Clemmensen (1876–1941), takes place under strongly acidic conditions, whereas the Wolff–Kishner reduction (Ludwig Wolff, 1857–1919; N. Kishner, 1867–1935) requires strong base.

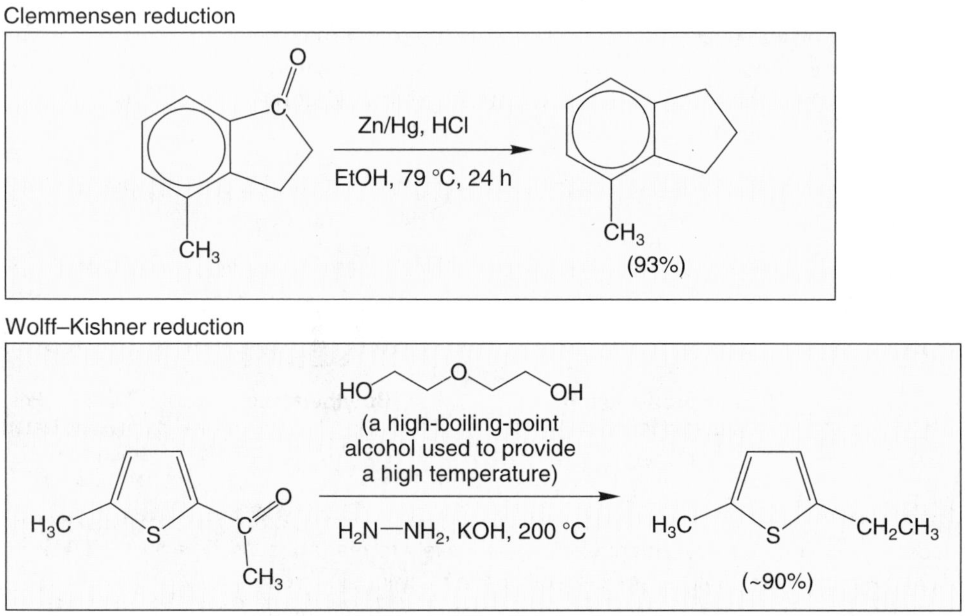

Clemmensen reduction

Zn/Hg, HCl

EtOH, 79 °C, 24 h

(93%)

Wolff–Kishner reduction

(a high-boiling-point alcohol used to provide a high temperature)

H_2N-NH_2, KOH, 200 °C

(~90%)

FIGURE **14.48** Two ways of reducing ketones to hydrocarbons.

PROBLEM **14.11**

Other reagents besides acid chlorides can lead to acylation reactions. Write a mechanism for the reaction in Figure 14.49.

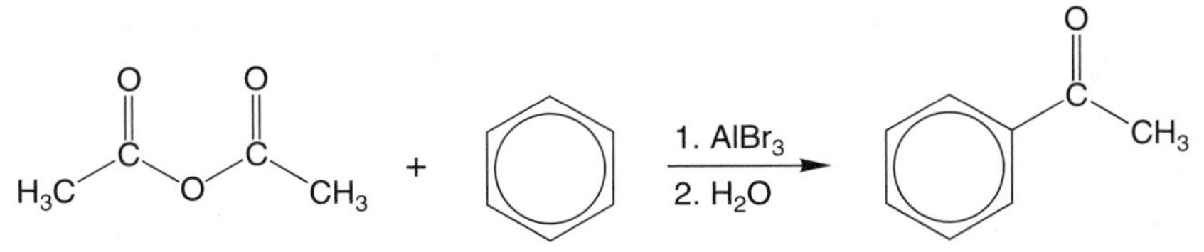

$$CH_3COOCOCH_3 + C_6H_6 \xrightarrow[2.\ H_2O]{1.\ AlBr_3} C_6H_5COCH_3$$

Acetic anhydride **Acetophenone** (methyl phenyl ketone)

FIGURE **14.49** Acylation using acetic anhydride and $AlBr_3$.

14.8 STABLE CARBOCATIONS IN "SUPERACID"

The Friedel–Crafts reactions succeed because relatively stable carbocations can be formed by the reaction of aluminum chloride (an exceptionally strong Lewis acid with respect to halide as Lewis base) in solvents that are nonnucleophilic. The strongest nucleophile available is benzene, which attacks the cation to start the Friedel–Crafts reaction. If stable carbocations cannot be formed, as with methyl chloride or ethyl chloride, the $AlCl_3$ complexes can also serve to alkylate benzenes.

How would we modify the reaction conditions so as to maximize the possibility of generating the carbocation and minimize the chances for its further reaction? Antimony pentafluoride (SbF_5) is an even stronger Lewis acid than $AlCl_3$ toward halogen, particularly fluorine. When SbF_5 removes a fluoride ion from an alkyl compound the anion antimony hexafluoride (SbF_6^-) is produced (Fig. 14.50). This anion is nonnucleophilic at antimony, and a poor nucleophile at fluorine as well.

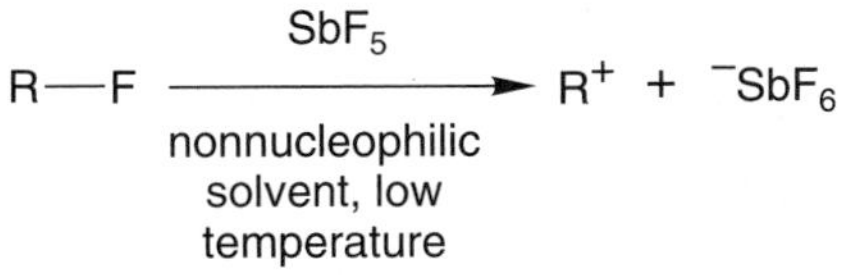

$$R{-}F \xrightarrow[\text{nonnucleophilic solvent, low temperature}]{SbF_5} R^+ + {}^-SbF_6$$

FIGURE **14.50** Cation formation through reaction of an alkyl fluoride with SbF_5.

PROBLEM **14.12**

Explain why SbF_6^- is nonnucleophilic at antimony and a very poor nucleophile at fluorine.

Treatment of alkyl fluorides with SbF_5 at very low temperatures in highly polar, nonnucleophilic solvents such as SO_2, SO_2ClF, or FSO_3H (so-called **superacid** conditions) can generate many stable carbocations. This process is Friedel–Crafts chemistry carried to extremes. Methyl and primary carbocations cannot be generated this way, but tertiary and secondary cations not prone to rearrangement to more stable cations can be (Fig. 14.51).

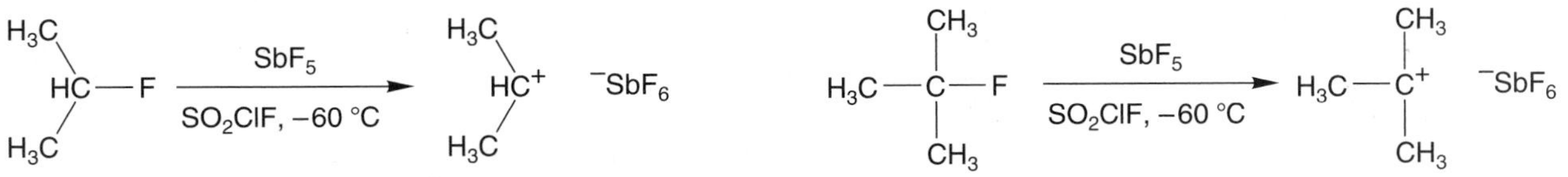

$$(H_3C)_2HC{-}F \xrightarrow[SO_2ClF,\ -60\ °C]{SbF_5} (H_3C)_2HC^+ \quad {}^-SbF_6$$

$$(H_3C)_3C{-}F \xrightarrow[SO_2ClF,\ -60\ °C]{SbF_5} (H_3C)_3C^+ \quad {}^-SbF_6$$

FIGURE **14.51** Stable secondary and tertiary carbocations can be formed in superacid.

Many carbocations, once thought impossible species even as reaction intermediates, can be generated and studied under superacid conditions. This work, pioneered by George A. Olah (b. 1927) now at the University of Southern California, has been one of the most exciting research areas of recent times and earned Olah the Nobel prize in 1994. We will see a number of examples of the utility of looking at cations in superacid in future chapters.

14.9 SUMMARY OF SIMPLE AROMATIC SUBSTITUTION: WHAT WE CAN DO SO FAR

We have uncovered six productive reactions so far. We can brominate or chlorinate benzene. We can nitrate and reversibly sulfonate benzene. The two Friedel–Crafts reactions allow the construction of carbon–carbon bonds to give some alkyl benzenes and give you another synthetic route to compounds containing carbon–oxygen double bonds. Figure 14.52 summarizes the situation.

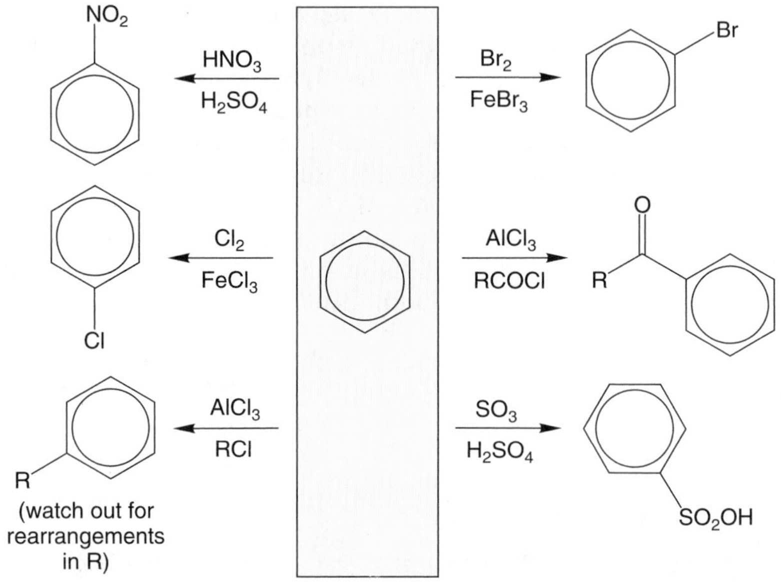

FIGURE 14.52 Aromatic substitution reactions encountered so far.

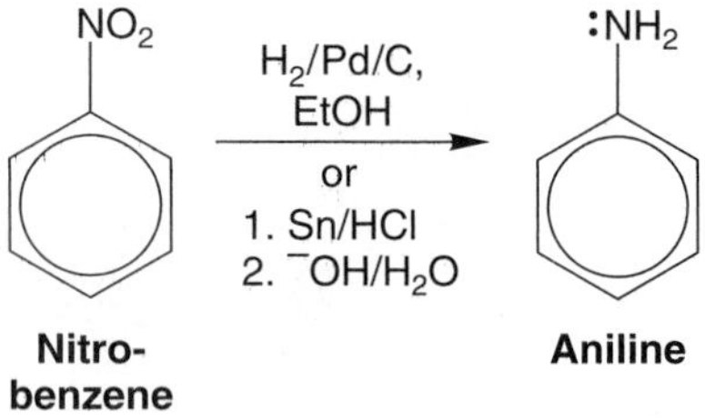

NO_2
HNO_3
H_2SO_4
Br_2
$FeBr_3$
Br
Cl_2
$FeCl_3$
$AlCl_3$
RCOCl
O
R
Cl
$AlCl_3$
RCl
SO_3
H_2SO_4
R
(watch out for rearrangements in R)
SO_2OH

NO_2
H_2/Pd/C, EtOH
or
1. Sn/HCl
2. $^-OH/H_2O$
:NH_2
Nitrobenzene
Aniline

FIGURE 14.53 The reduction of nitrobenzene to aniline.

Although there are relatively few groups we can attach to a benzene ring directly, we can often transform one group into another through chemical reactions. It is nitrobenzene that is the key to many of these transformations. Nitrobenzene itself is relatively unreactive, but the nitro group can be reduced in many ways to the amino group. This transformation opens the way for the synthesis of a host of new molecules (Fig. 14.53).

Aminobenzene, generally called by its common name **aniline**, can be treated with nitrous acid (HONO) or its sodium salt ($Na^+ \, ^-ONO$) and HCl, to produce benzenediazonium chloride, which is explosive when dry, but a relatively safe material when wet (Fig. 14.54).

The mechanism for this reaction contains a lot of steps, but is really within our range now. A critical intermediate is dinitrogen trioxide, formed in equilibrium with nitrous acid by loss of water. When nitrous

:NH_2 NaONO/HCl :Cl:$^-$ $^+N_2$

Aniline **Benzene-diazonium chloride**

FIGURE **14.54** Reaction of aniline with nitrous acid to give benzenediazonium chloride.

Draw resonance forms for benzenediazonium chloride. PROBLEM **14.13**

Aniline

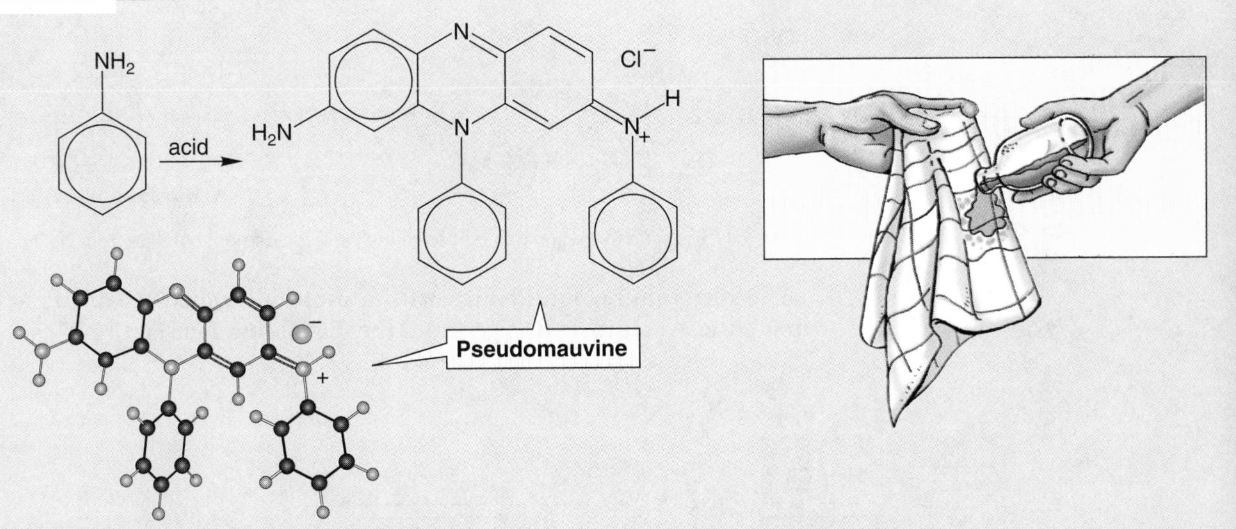

Aniline featured prominantly in the development of the synthetic dye industry. In 1856, the 17-year-old William Henry Perkin, on holiday from the Royal College of Chemistry where he was studying with August Wilhelm Hofmann, was working in his home lab (isn't this what you do on your vacations?). As an unexpected outcome of a failed attempt to make quinine, Perkin obtained a black goop that stained his laboratory rag a beautiful mauve. This dye is the molecule shown above, pseudomauvine. Over Hofmann's objections, Perkin and his family set out to commercialize this dye. It was not a vast financial success, as colorfastness became a problem, and mauve was very expensive, being roughly the same price per pound as platinum at the time. It was overtaken by other, related dyes made in similar ways from substituted anilines. Nonetheless, the serendipitous synthesis of mauve marks the beginnings of a great chemical industry.

It shouldn't be difficult to find the four molecules of aniline within the structure of pseudomauvine. It is harder to map out a general mechanism through which the four molecules come together; keep in mind that the necessary oxidations would be easy to accomplish under Perkin's crude reaction conditions.

acid is protonated, a good leaving group, water, is created and can be displaced by nitrite ion (Fig. 14.55).*

*This reaction is actually a bit more complicated than is shown in the figure. It involves addition of nitrite to the nitrogen–oxygen double bond followed by loss of water. We will explore this kind of mechanism in detail beginning in Chapter 16.

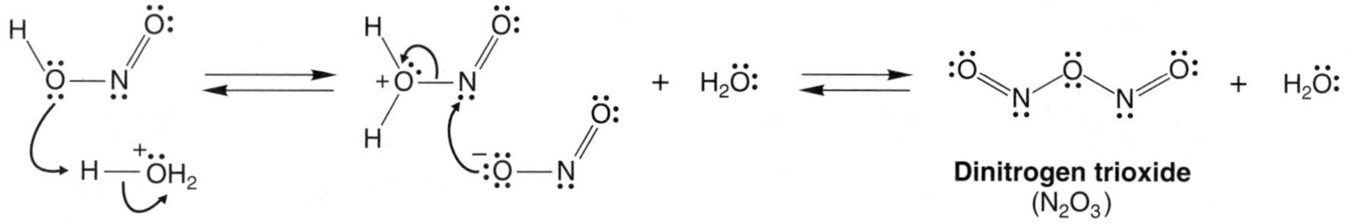

Dinitrogen trioxide (N_2O_3)

FIGURE 14.55 The formation of dinitrogen trioxide, N_2O_3.

Displacement of the good leaving group nitrite ($^-$ONO) from N_2O_3 by the Lewis basic nitrogen of aniline leads, after removal of a proton, to *N*-nitrosoaniline (Fig. 14.56).

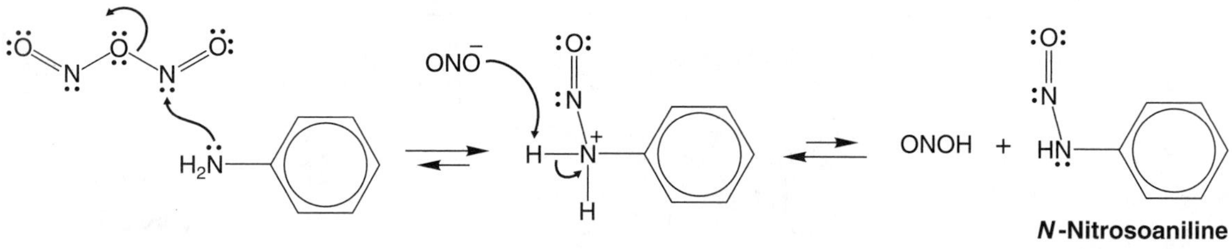

***N*-Nitrosoaniline**

FIGURE 14.56 The formation of a nitrosoamine by displacement of nitrite from N_2O_3.

The nitrosamine equilibrates with a molecule called a diazotic acid, which loses water in acid to generate the **diazonium ion** (Fig. 14.57).

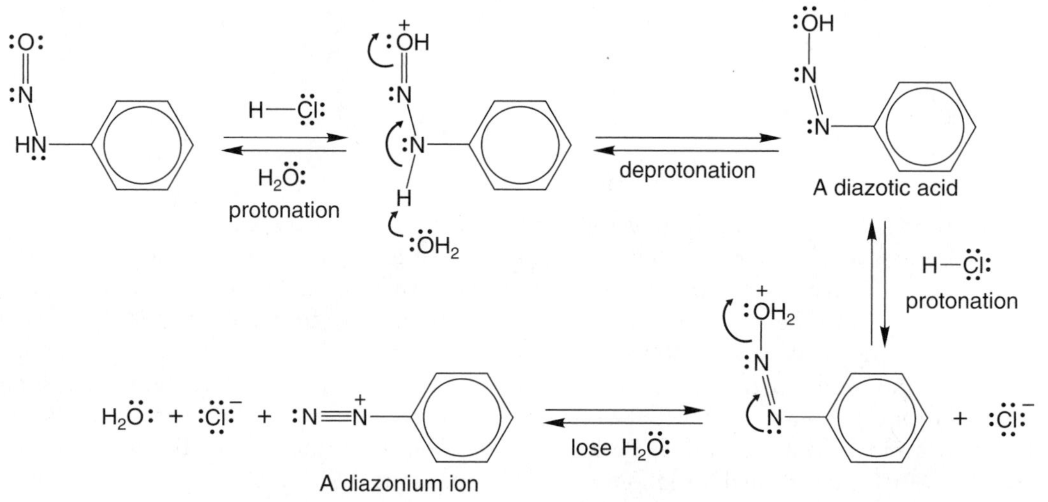

H—Cl: / H_2O: protonation

deprotonation

A diazotic acid

H—Cl: protonation

lose H_2O:

A diazonium ion

FIGURE 14.57 Loss of water from the protonated diazotic acid gives the diazonium ion, benzenediazonium chloride.

These transformations are probably bewilderingly complex right now, even though they are made up of very simple reactions in sequence. It certainly takes some practice to get them right. However, many reaction mechanisms involve similar sequences of protonations, deprotonations, and additions and losses of water molecules. Now is the time to start practicing them. We will encounter many more when we discuss carbonyl chemistry in Chapters 16–20.

The diazonium ion is important because it is the critical intermediate in a number of transformations lumped under the heading of the **Sandmeyer reaction**, named for Traugett Sandmeyer (1854–1922). Cuprous salts convert the diazonium ion into the related halide or cyanide. In recent times, numerous variations on the basic theme of the Sandmeyer reaction have improved yields and reduced side products. Figure 14.58 shows three typical Sandmeyer reactions.

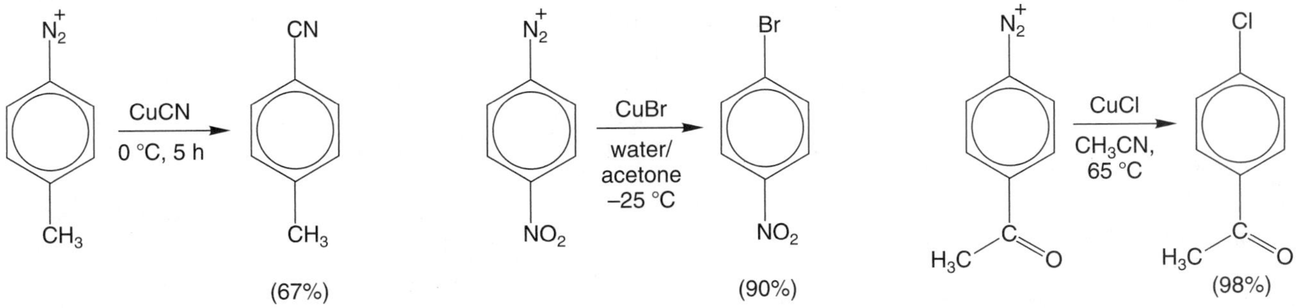

FIGURE **14.58** Three Sandmeyer reactions. Decomposition of the diazonium ion with cuprous salts leads to substitution.

Reagents other than cuprous salts are also effective. Treatment with potassium iodide (KI) or fluoboric acid (HBF_4) gives the iodide or fluoride (the Schiemann reaction; G. Schiemann 1899–1969), respectively, and hydrolysis gives **phenol**. Heating with hypophosphorus acid (H_3PO_2) removes the diazonium group and replaces it with hydrogen (Fig. 14.59).

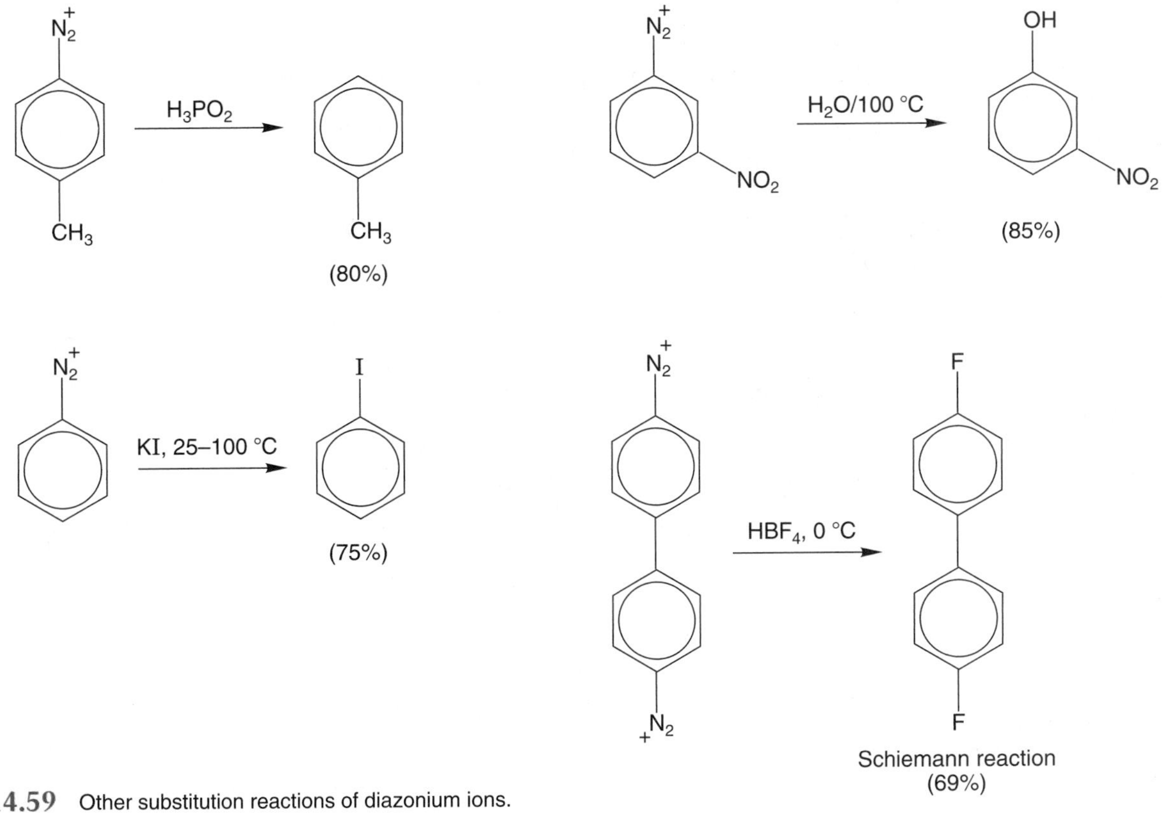

FIGURE **14.59** Other substitution reactions of diazonium ions.

Of course, you are now clamoring for the mechanisms for all these changes.

PROBLEM **14.14**

We reject out of hand the notion that any of these net displacement reactions could go by something as simple as an S_N2 reaction. Why?

Sadly, only the outlines of the mechanisms are known, and much remains to be filled in. The details of many reactions are imperfectly understood, and there remains a delightful (to some) atmosphere of magic about parts of organic chemistry. What are we to make of the admonition in one procedure for the Sandmeyer reaction to "stir the mixture with a lead rod." What happens if we use a glass rod? Although one would think that the reaction might still succeed, there are examples to the contrary, and perhaps microscopic pieces of lead do affect the success of the reaction. I'd use lead.*

In the reactions involving copper, electron-transfer-mediated radical reactions are certainly involved. In HCl, cuprous chloride (CuCl) is in equilibrium with the dichlorocuprate ion $^{-}CuCl_2$ (Fig. 14.60). The dichlorocuprate ion can transfer an electron to the diazonium ion, which can then lose molecular nitrogen to give a phenyl radical. In turn, the phenyl radical can pluck a chlorine atom from the cuprate to give the chlorobenzene and a molecule of cuprous chloride (Fig. 14.61).

$CuCl + HCl \rightleftharpoons H^+ \; ^{-}CuCl_2$

The dichlorocuprate ion

FIGURE 14.60 Formation of the dichlorocuprate ion from CuCl and HCl.

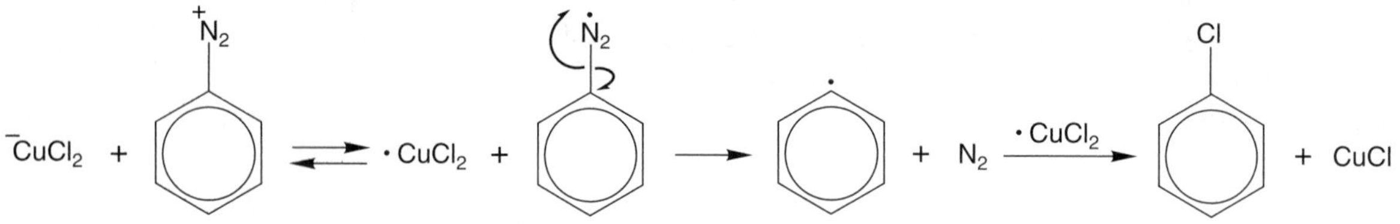

FIGURE 14.61 The mechanism for the formation of chlorobenzene in the Sandmeyer reaction.

Other Sandmeyer reactions apparently involve a phenyl cation formed by the irreversible S_N1 ionization of benzenediazonium chloride (Fig. 14.62).

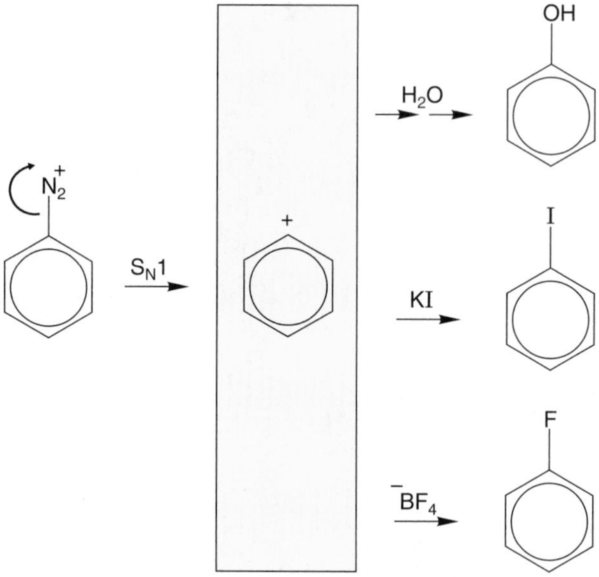

FIGURE 14.62 Other conversions of benzenediazonium chloride involve the phenyl cation.

*Speaking of lead and magic, two of my chemical relatives, Milton Farber and Adnan Abdulrida Sayigh (~1922–1980) once discovered a useful reaction of lead dioxide, and the work was duly published as an effective way to make alkenes from 1,2-diacids. However, the reaction only worked when one particular bottle of lead oxide was used, and once that bottle was used up no one could reproduce the yields achieved earlier. All other sources of lead dioxide failed! Apparently, the particle size of the lead dioxide was critical. (What's a "chemical relation"? That's someone who did his or her graduate work with the same professor as you. Farber and Sayigh both worked with the same person I did and hence are my chemical brothers.)

PROBLEM 14.15

Draw resonance forms for the phenyl radical Ph· (be careful).

*PROBLEM 14.16

In the formation of diazonium ions from aromatic amines, side products are occasionally encountered, as shown in Figure 14.63. Give a mechanism for the formation of these compounds.

NH$_2$ — NaNO$_2$/HCl → N$_2^+$ + N=N, NH$_2$

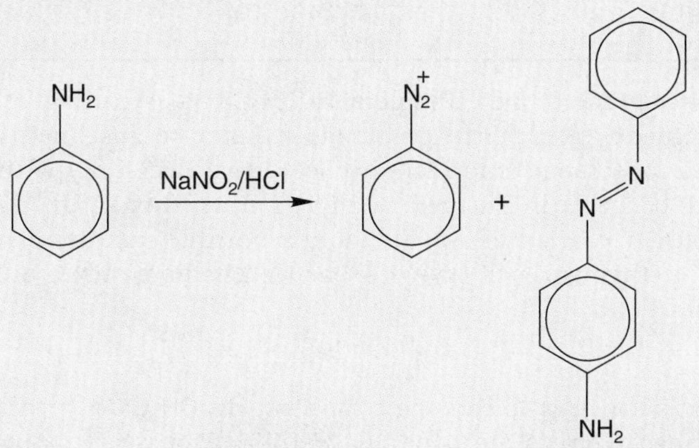

An azo compound

FIGURE 14.63

ANSWER

A resonance formulation of the benzenediazonium ion shows that the terminal nitrogen is a Lewis acid. It can act as an E$^+$ reagent and add to the highly activated ring of aniline. This reaction gives a cyclohexadienyl cation intermediate that can be deprotonated by any base in the system to give the observed side product, an azo compound.

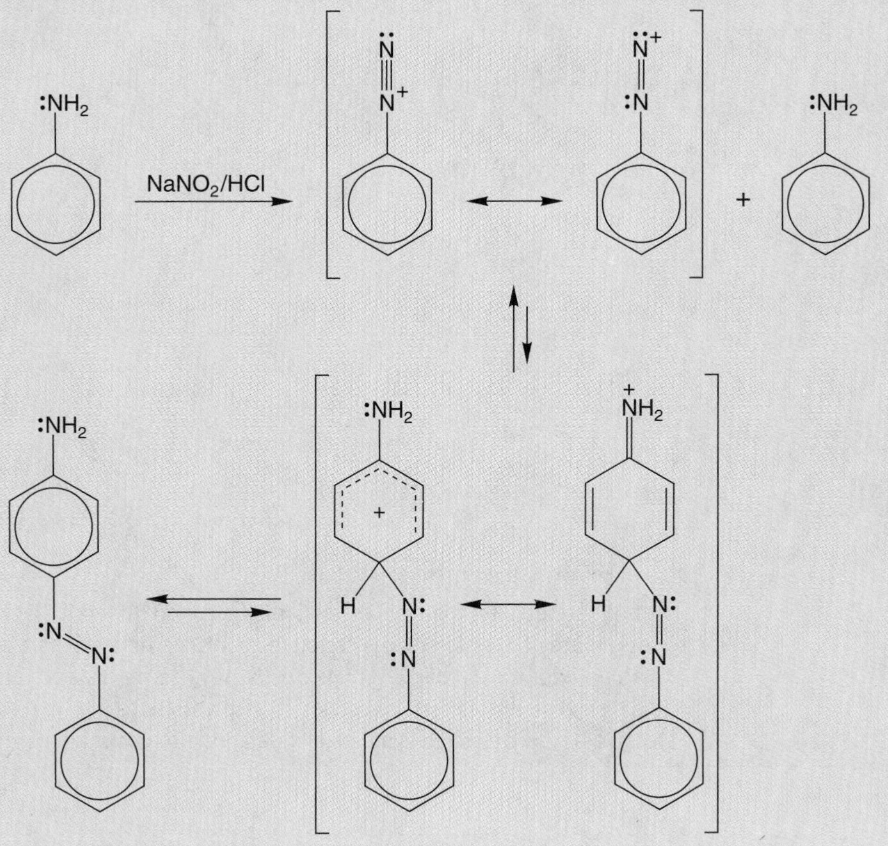

:NH$_2$ — NaNO$_2$/HCl → [N$^+$ ⟷ :N, N$^+$] + :NH$_2$

:NH$_2$, :N, N: ⇌ [:NH$_2$, H, N:, :N ⟷ NH$_2^+$, H, N:, :N]

It doesn't matter that some mechanistic details are sketchy; the formation of the diazonium ion from aniline (and therefore ultimately from the easily made nitrobenzene) and its further transformations gives you the ability to make a great number of substituted benzenes from very simple precursors.

14.10 DISUBSTITUTED BENZENES: ORTHO, META, AND PARA SUBSTITUTION

We have already mentioned (Problem 14.7) that the synthesis of methylbenzene (toluene) from benzene is made difficult because once a significant amount of toluene is formed, it competes successfully with the unsubstituted benzene in further methylation reactions. Unless one is careful to use a vast excess of benzene, a horrible mixture of benzene, toluene, dimethylbenzenes (xylenes), and even more substituted molecules results.

*PROBLEM **14.17** If one attempts to methylate benzene completely with $CH_3Cl/AlCl_3$, the product contains a greenish solid of the formula $C_{13}H_{21}AlCl_4$. Propose a structure for the compound and a mechanism for its formation.

ANSWER This problem involves a sequence of Friedel–Crafts alkylations. Toluene is made first, then the xylenes, then trimethylbenzenes, and so on, until hexamethylbenzene is formed.

Xylenes

CH_3 CH_3 CH_3 CH_3

CH_3Cl $AlCl_3$ CH_3Cl $AlCl_3$ +

Toluene

CH_3 CH_3Cl $AlCl_3$

CH_3 CH_3 H_3C CH_3 CH_3

repeat three times

H_3C CH_3

CH_3 CH_3

Hexamethylbenzene

What happens if another Friedel–Crafts reaction occurs? The intermediate is the heptamethylcyclohexadienyl cation tetrachloroaluminate salt, and this material is stable under the reaction conditions. Why? The resonance forms contributing to the structure are all tertiary carbocations, and there is no proton that can be lost to give a new aromatic system. All positions are occupied by methyl groups.

ANSWER (CONTINUED)

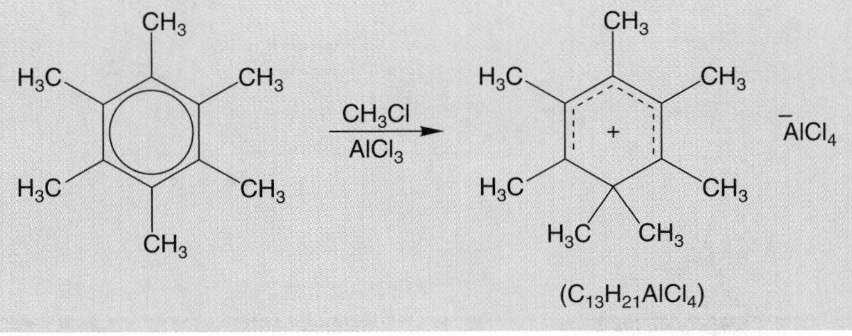

The general question now arises of how the substituted benzenes we have learned to make will react with electrophilic reagents. What are the possibilities? There are three kinds of disubstituted benzene, called **ortho** (1,2-disubstituted), **meta** (1,3-disubstituted), and **para** (1,4-disubstituted), as shown in Figure 14.64.

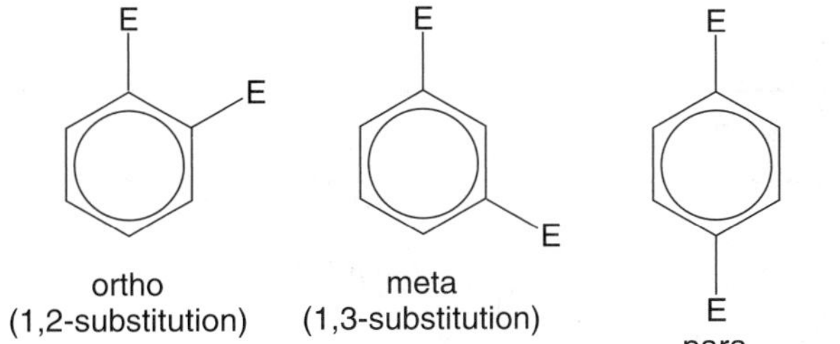

FIGURE **14.64** The three possible disubstituted benzenes.

Substituents influence the distribution of the three isomeric disubstituted products dramatically. Further substitution of monosubstituted benzenes by no means proceeds to give a statistical distribution of two parts ortho product, two parts meta product, and one part para product. In fact, if one looks at the reactions of a number of monosubstituted benzenes, a pattern emerges. Some groups produce predominately ortho and para disubstituted compounds while for others meta is dominant. There is also a correlation of the position of further substitution with the rate of the reaction. Substituted benzenes that give mainly ortho and para products usually react faster than benzene itself. Substituted benzenes that give mainly meta product react more slowly than benzene itself (Table 14.1).

A look at the general mechanism of substitution shows how different groups influence the position of subsequent reaction. We can also see how these orientation (regiochemical) effects are connected to the rates of the reactions. Let's look first at methoxybenzene **(anisole)**. Methoxy is one of the groups most effective at directing further substitution to the ortho and para positions and at increasing the rate of reaction. We will work through the mechanism for both meta and para substitution and look for a point of difference. Figure 14.65 shows the mechanism for meta substitution of anisole by a generic E^+ reagent, and compares the reaction to that of benzene itself. As usual, a resonance-stabilized cyclohexadienyl cation is formed first, and then deprotonated to give the substituted product.

TABLE 14.1 Patterns of Further Reactions of Substituted Benzenes

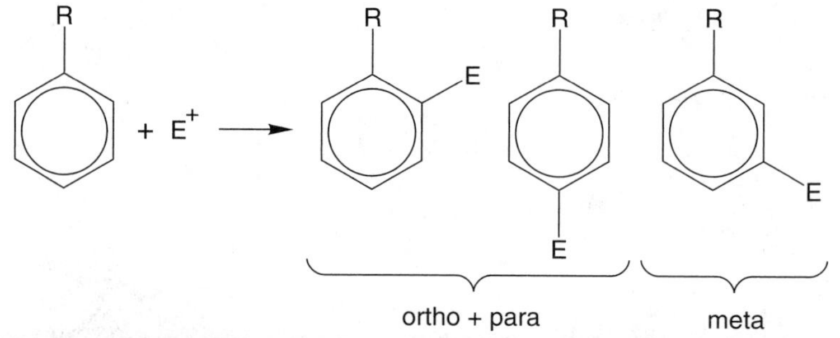

R	Position[a]	Relative Rate
H (benzene)		1
NH_2, NHR, NR_2	o/p	Very fast
OH	o/p	Very fast
OR	o/p	Very fast
Alkyl	o/p	Fast
NH—C(=O)R'	o/p	Fast
I, Cl, Br, F	o/p	Slow
—C(=O)R' (R' = OH, OR, NH_2, Cl, alkyl, aryl)	m	Slow
SO_2OH	m	Slow
NO_2	m	Slow
CN	m	Slow
$^+NR_3$	m	Slow

[a] Ortho = o, para = p, meta = m.

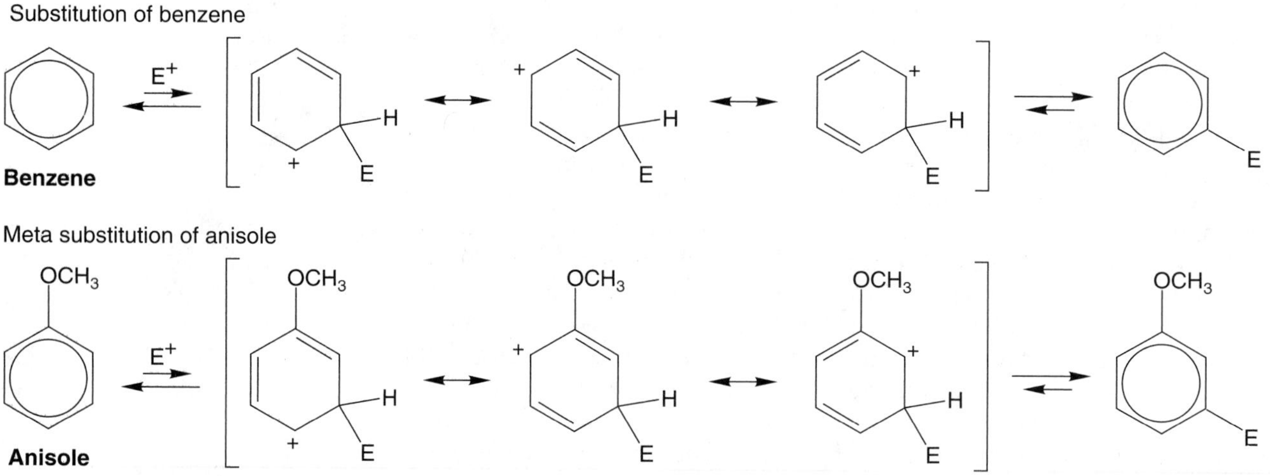

FIGURE 14.65 The intermediate cyclohexadienyl cations involved in the substitution of benzene and meta substitution of anisole.

At first glance, substitution at the para position seems similar (Fig. 14.66).

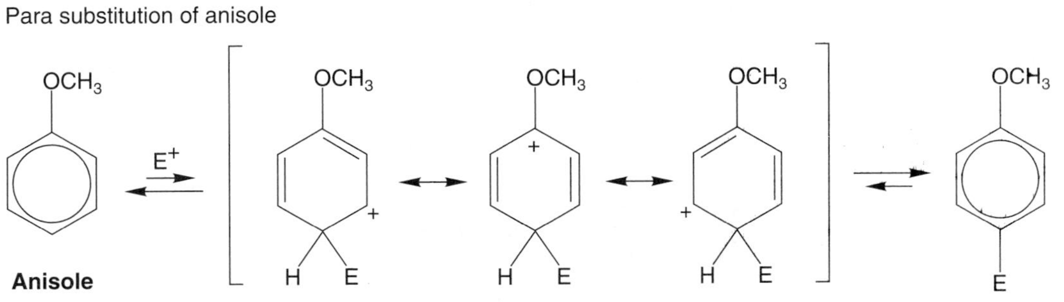

Para substitution of anisole

OCH_3 Anisole $\underset{}{\overset{E^+}{\rightleftharpoons}}$ [OCH_3 ... H E ↔ OCH_3 ... H E ↔ OCH_3 ... H E] ⇌ OCH_3 ... E

FIGURE **14.66** The intermediate cyclohexadienyl cation involved in para substitution of anisole.

But look carefully at the intermediate in para substitution, the resonance-stabilized cyclohexadienyl cation. Remember our discussion of the stabilization of adjacent cations by oxygen (Chapter 9, p. 356). In this intermediate, there is a fourth resonance form in which oxygen shares the positive charge. This fourth form stabilizes the cation relative to that formed in the meta substitution or in substitution of benzene itself (Fig. 14.67).

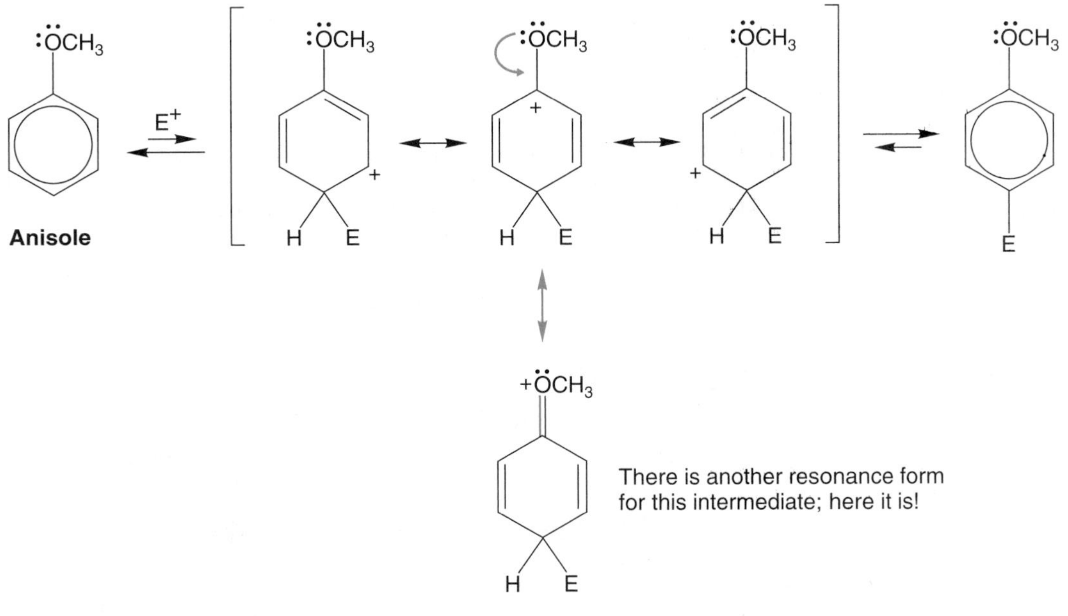

$:\ddot{O}CH_3$ Anisole $\overset{E^+}{\rightleftharpoons}$ [$:\ddot{O}CH_3$ H E ↔ $:\ddot{O}CH_3$ H E ↔ $:\ddot{O}CH_3$ H E] ⇌ $:\ddot{O}CH_3$ E

↕

$+\ddot{O}CH_3$ H E

There is another resonance form for this intermediate; here it is!

FIGURE **14.67** In the intermediate for para substitution, there is a fourth resonance form in which oxygen bears the positive charge.

The intermediate formed in para substitution of anisole is more stable than that formed in meta substitution of anisole or in substitution of benzene itself. Accordingly, the transition state leading to the para intermediate will be more stable than the transition states for meta substitution or substitution of benzene itself. We can expect para substitution to occur more rapidly than the other two reactions (Fig. 14.68).

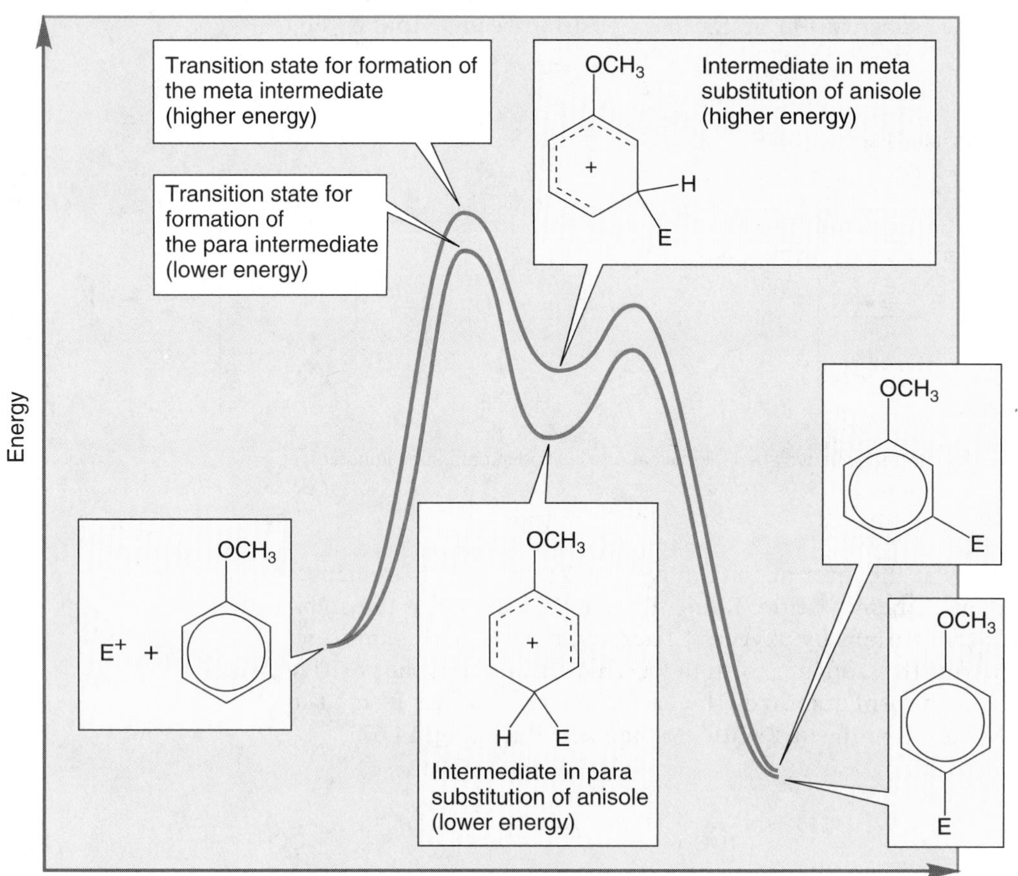

FIGURE **14.68** The intermediate for para substitution, with four resonance forms, is lower in energy than the intermediate for meta substitution (three forms) or that for substitution of benzene (also three forms). The transition state for para substitution of anisole will also be lower in energy than the transition state for meta substitution.

In fact, the chlorination of anisole gives only ortho and para substituted product (Fig. 14.69).

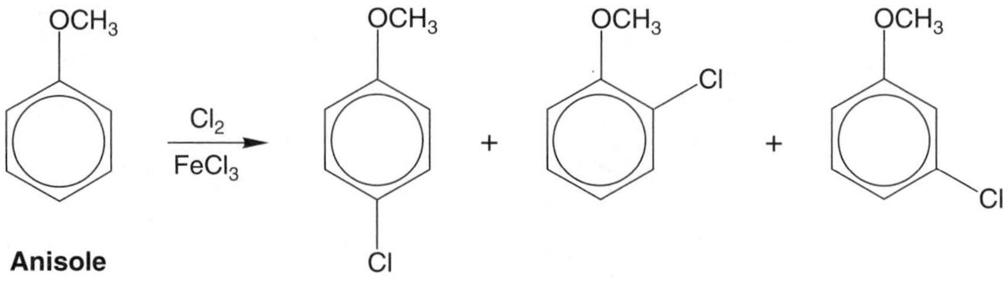

Anisole $\xrightarrow[FeCl_3]{Cl_2}$ (65.1%) + (34.9%) + (0%)

FIGURE **14.69** The chlorination of anisole.

*PROBLEM **14.18** Work out the structure of the intermediate formed by substitution of anisole at the ortho position.

ANSWER The figure shows the four resonance forms of the cyclohexadienyl cation intermediate. As in the para case, the methoxyl group is in a position to help share the positive charge. The resulting delocalization contributes to the stability of the

ANSWER (CONTINUED)

intermediate and makes it easy to form, relative to the intermediate formed in meta substitution, or from benzene itself (Fig. 14.65).

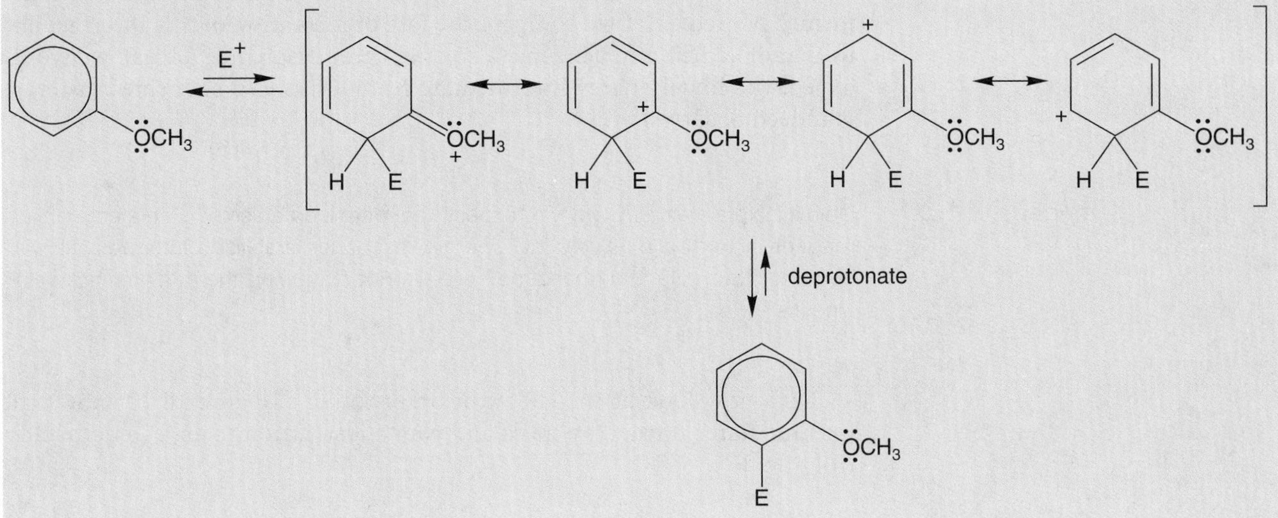

*PROBLEM **14.19**

Why is ortho substitution often less favorable than para substitution, even though ortho substitution is favored statistically by a 2:1 factor?

ANSWER

The obvious answer is that substitution in the ortho position encounters steric problems that are absent in the para position.

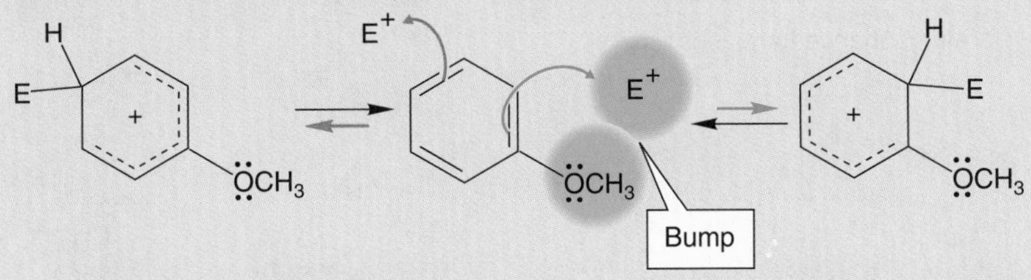

In answering Problem 14.19, you almost certainly hit upon this sensible reason for the dominance of para substitution over ortho, even in the face of a 2:1 statistical factor favoring ortho. Steric considerations will surely favor the para position. The problem is that sometimes the ortho position is *favored* over the para, and the statistical advantage doesn't seem adequate to explain this preference. The answer almost certainly lies in a molecular orbital treatment of the reaction. We can write resonance forms for the intermediate cations in para and ortho substitution, as in Figure 14.67 and the answer to Problem 14.18. When we do this, no distinction between the intermediates for ortho and para substitution is obvious, as both intermediates have similar resonance structures. In reality, this is too simple a treatment. The orbital lobes at the ortho and para positions are not all of equal size. There are weighting factors (coefficients) that give the relative magnitude of the wave function at the positions sharing the charge, and it is sometimes the case that the sum of the coefficients at the ortho position (remember the twofold statistical advantage) is greater than that at the para position, thus favoring ortho substitution. The ortho/para ratio is a very complicated area, and there is more research to be done.

The relatively low energy of the intermediate formed by addition at the ortho or para substitution (four resonance forms) means that this reaction will prevail over meta attack (the intermediate has only three resonance forms). Moreover, if we compare the rate of formation of this intermediate to that of addition to benzene (also only three resonance forms), we would expect the lower energy intermediate from anisole to be formed faster, as is indeed observed.

PROBLEM 14.20 Hold it! Explain *exactly why* we "expect" that attack on anisole by an electrophilic reagent to form a relatively low energy, resonance-stabilized intermediate will be faster than attack on benzene? Are we not confusing thermodynamics with kinetics?

Let's now look at the substitution of toluene. Figure 14.70 shows the intermediates formed by meta and para substitution using a generic electrophile, E^+.

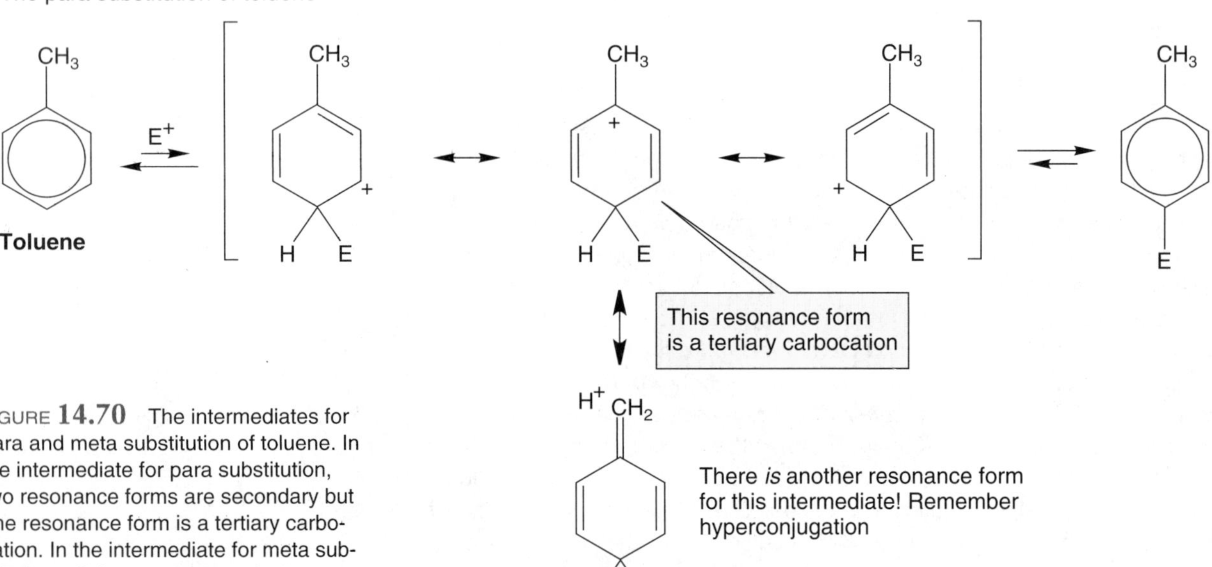

FIGURE 14.70 The intermediates for para and meta substitution of toluene. In the intermediate for para substitution, two resonance forms are secondary but one resonance form is a tertiary carbocation. In the intermediate for meta substitution, all three resonance forms are secondary.

This time there is no obvious fourth resonance form, as there was in the para (and ortho) substitution of anisole. Why is ortho/para substitution still preferred, and why does toluene react faster than unsubstituted benzene (Table 14.1, p. 652)? Look at the positions sharing the positive charge. In the intermediate formed from meta substitution, the three positions sharing the positive charge are all secondary. In the intermediate formed by para substitution, one of the three positions is a tertiary carbon and will be especially effective in stabilizing a positive charge. So there really is a fourth resonance form! Recall the description of the stabilization of adjacent positive charge by alkyl groups (Chapter 9, p. 350). An alkyl group can stabilize an adjacent positive charge in a hyperconjugative way (Fig. 14.70).

The intermediate for para substitution, *and the transition state leading to it,* will be lower in energy than those involved in meta substitution (Fig. 14.71).

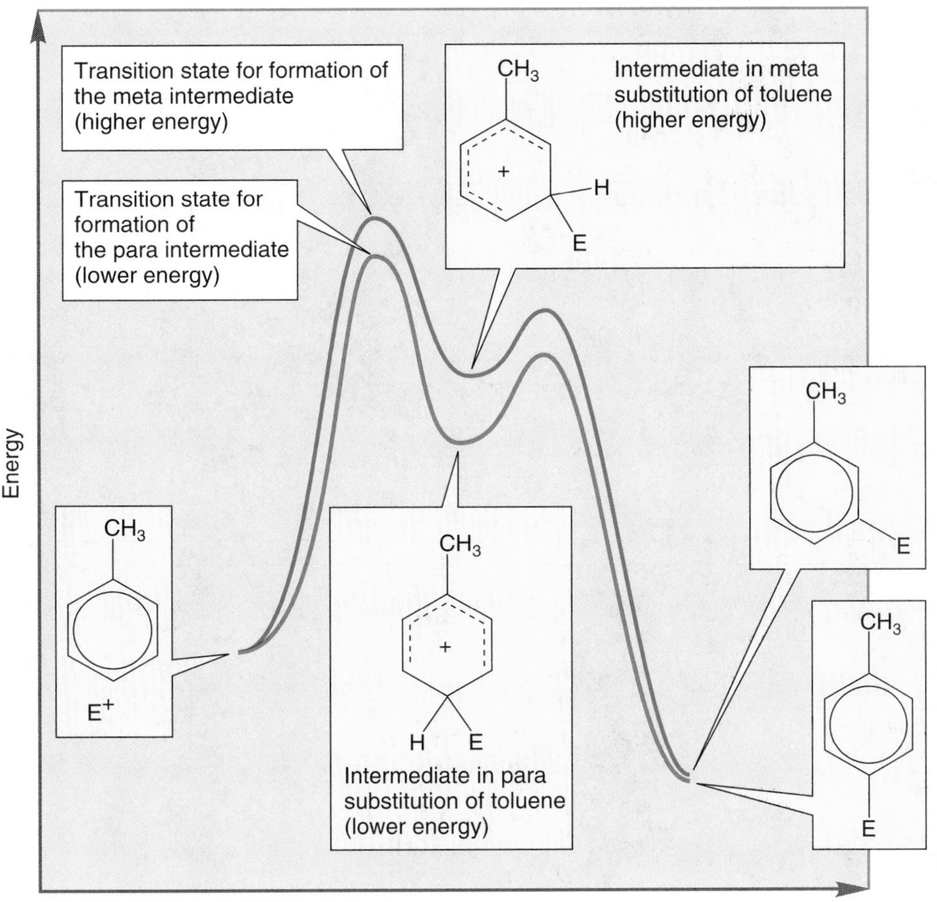

FIGURE **14.71** The intermediate and transition state for para substitution of toluene are lower in energy than those for meta substitution.

PROBLEM **14.21** Write a mechanism for the ortho substitution of toluene. Compare this reaction to meta substitution.

PROBLEM **14.22** Explain why bromobenzene and aniline direct subsequent substitution reactions ortho/para.

This theoretical analysis is borne out in practice. Figure 14.72 shows some typical substitution reactions of toluene.

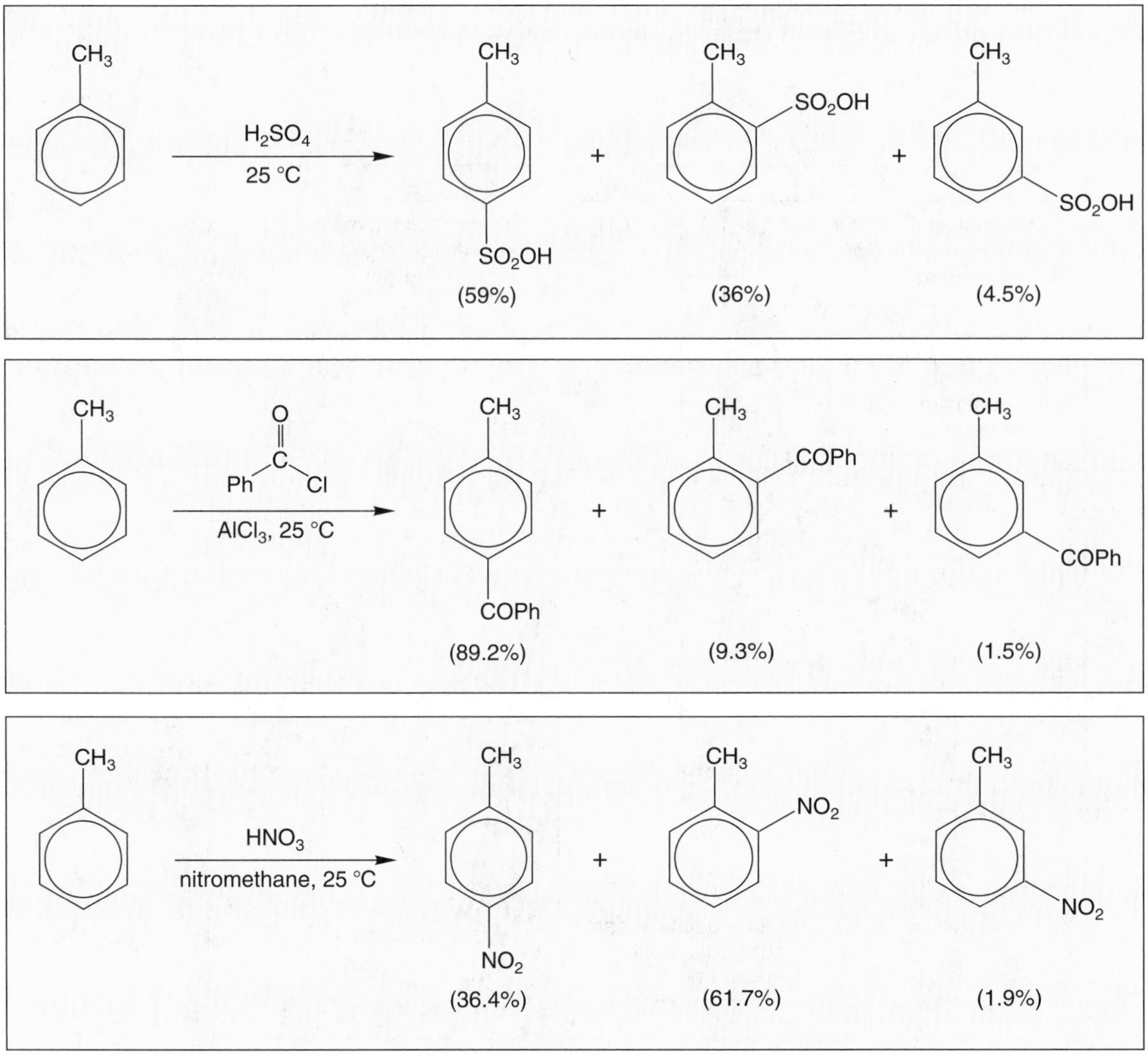

FIGURE **14.72** Some substitution reactions of toluene.

So, for many groups, meta substitution does not present a problem. However, separation of ortho and para products *is* a problem, especially as the two are often formed in comparable amounts. Happily, physical methods such as crystallization and distillation can generally be used successfully to separate the products. Electrophilic aromatic substitution is a practical source of many ortho and para disubstituted benzenes.

What about those groups that direct further substitution to the meta position and slow the rate? Why are the effects we have seen so far not general? Once more, the way to attack this problem is to write out the various possibilites and look for points of difference. Look first at the intermediates formed by reaction of the trimethylanilinium ion in Figure 14.73.

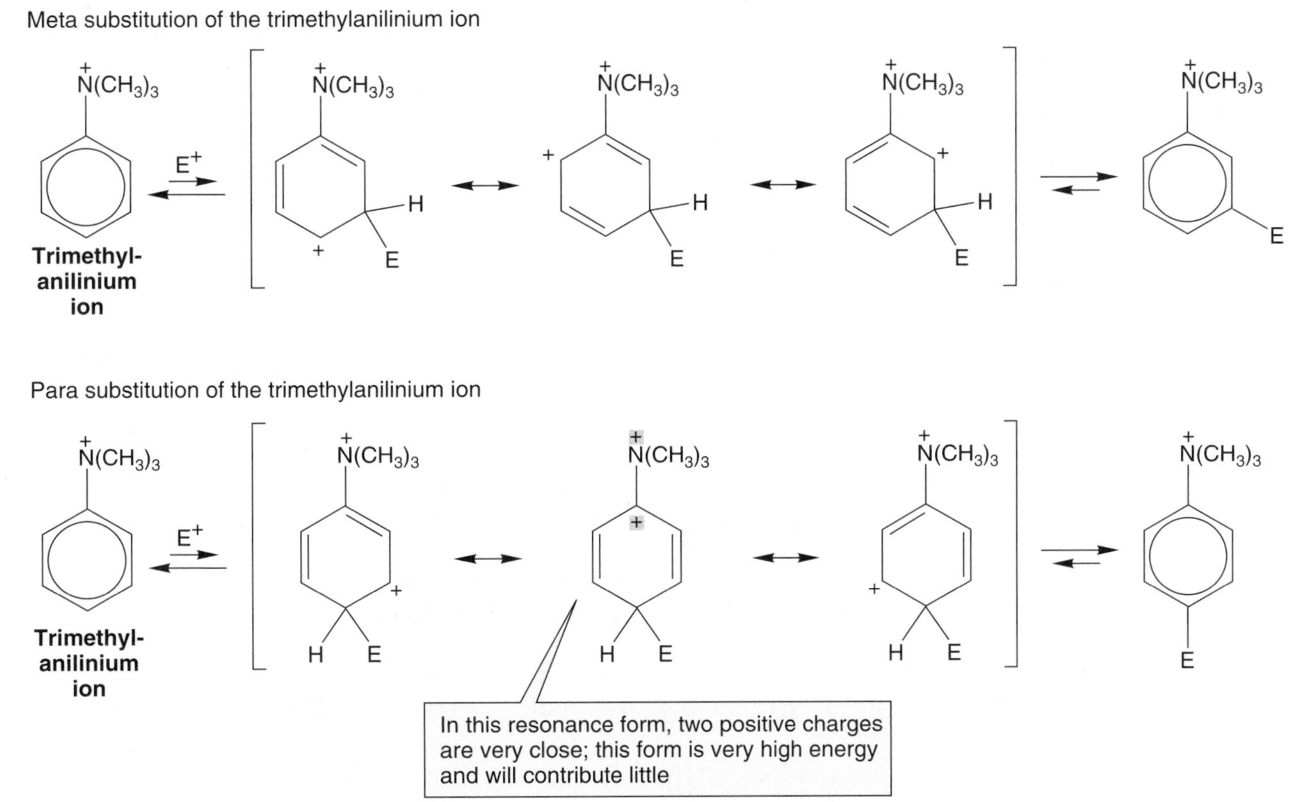

Meta substitution of the trimethylanilinium ion

Trimethyl-anilinium ion

Para substitution of the trimethylanilinium ion

Trimethyl-anilinium ion

In this resonance form, two positive charges are very close; this form is very high energy and will contribute little

FIGURE 14.73 Meta and para substitution of the trimethylanilinium ion. Both intermediates have three resonance forms, but the adacent positive charges in the intermediate for para substitution are especially destabilizing.

Both meta and para substitution give an intermediate with three resonance forms, so there is no gross difference here. Note that we are introducing a second positive charge into the molecule when we do *any* electrophilic aromatic substitution reaction of this molecule. This second charge is bound to slow the reaction relative to substitution of benzene itself. Now we know why this substitution reaction is especially slow, but we can't yet explain why the meta position is preferred. The difficulty in para substitution is that the reaction places the two positive charges especially close together. In para substitution, the new positive charge resides partly on the carbon bearing the original substituent. When this substituent can stabilize the charge through resonance, as in anisole or toluene, para substitution will be favored. When it brings two positive charges together, as in the trimethylanilinium ion, para substitution is decidedly disfavored for electrostatic reasons.

PROBLEM **14.23** Work out the structure of the intermediate formed by ortho substitution of the trimethylanilinium ion.

PROBLEM **14.24** Use an Energy versus Reaction progress diagram to compare ortho and meta substitution of the trimethylanilinium ion.

So the situation is clear for the trimethylanilinium ion. Meta substitution, which does not place two like charges on adjacent carbons, is favored over para (or ortho) substitution, which does. Figure 14.74 shows one example.

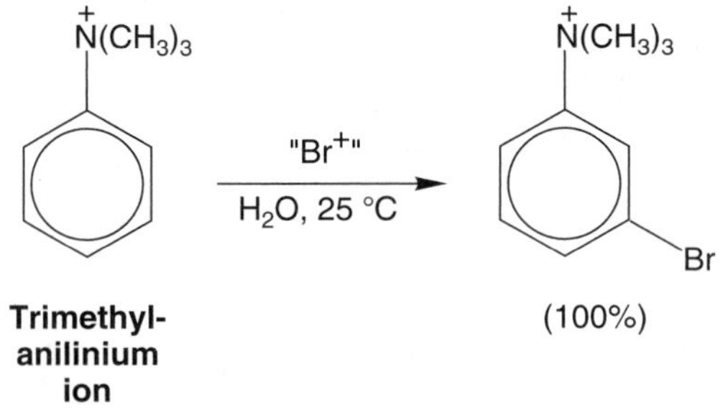

$\overset{+}{N}(CH_3)_3$ — "Br^+", H_2O, 25 °C → $\overset{+}{N}(CH_3)_3$, Br

Trimethyl-anilinium ion (100%)

FIGURE **14.74** The bromination of the trimethylanilinium ion.

What about the other meta directing groups? They do not bear obvious positive charges. Nitrobenzene is a good example. Nitrobenzene is so deactivated that it can be used as a solvent for Friedel–Crafts reactions of other aromatic compounds and not suffer attack itself. Further substitution on nitrobenzene is invariably largely in the meta position (Table 14.1; Fig. 14.75).

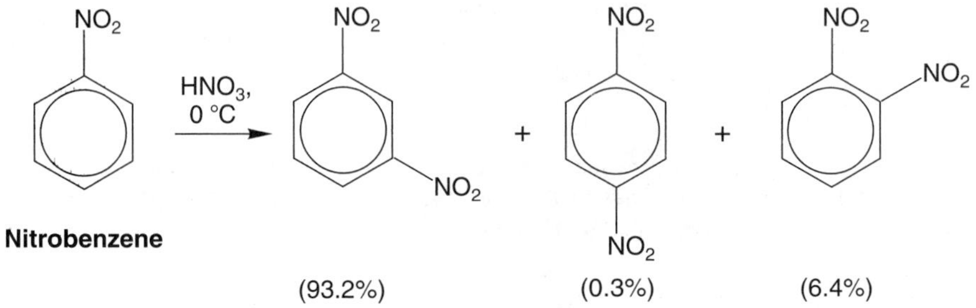

NO_2 — HNO_3, 0 °C → NO_2, NO_2 + NO_2, NO_2 + NO_2, NO_2

Nitrobenzene (93.2%) (0.3%) (6.4%)

FIGURE **14.75** Nitrobenzene nitrates very slowly at 0 °C, and almost entirely in the meta position.

Yet, there is no obvious positive charge as there was in the trimethylanilinium ion we analyzed before. A close look at the structure of the nitro group reveals the problem, which has been hidden by the too-schematic rendering of nitro as NO_2. If you draw out a nitro group carefully, you will see that there is no good uncharged structure, and that the nitrogen always bears a positive charge (Fig. 14.76). So there really is a pos-

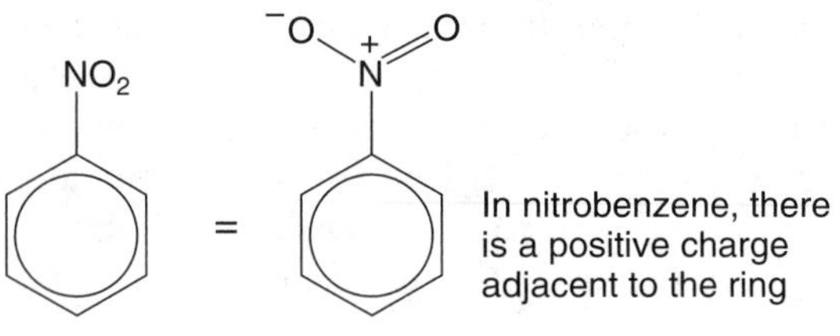

NO_2 = ^-O $\overset{+}{N}$ O

In nitrobenzene, there is a positive charge adjacent to the ring

FIGURE **14.76** The resonance description for the nitro group reveals that nitrogen bears a full positive charge. Further substitution will be directed to the meta position.

So,

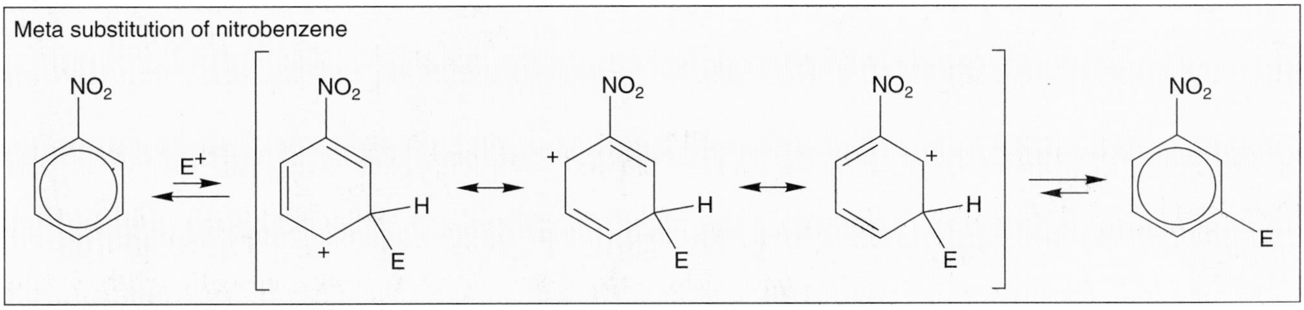

Meta substitution of nitrobenzene

is favored over

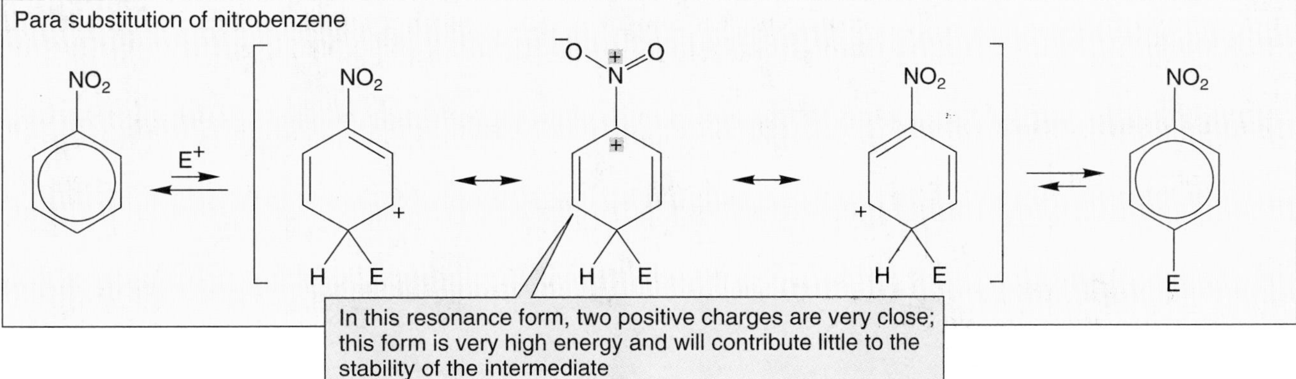
Para substitution of nitrobenzene
NO_2
E^+
NO_2
H E
H E
NO_2
H E
NO_2
E
In this resonance form, two positive charges are very close; this form is very high energy and will contribute little to the stability of the intermediate

FIGURE **14.76** (CONTINUED)

itive charge adjacent to the ring in nitrobenzene and the same factors that lead to meta substitution and a slow rate in the trimethylanilinium ion will operate for nitrobenzene as well .

But other meta directors do not have hidden positive charges. Ethyl benzoate is a good example. It too substitutes slowly, and predominantly in the meta position (Fig. 14.77).

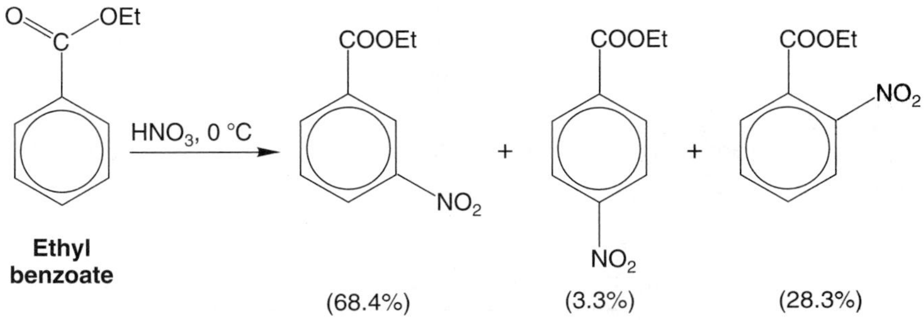

Ethyl benzoate $\xrightarrow{HNO_3,\ 0\ °C}$ (68.4%) + (3.3%) + (28.3%)

FIGURE **14.77** The nitration of ethyl benzoate gives mainly meta product.

The answer again comes from a close look at the structure of the group substituting the benzene ring, in this case a carbon–oxygen double bond, a carbonyl group. Although there are no hidden full positive charges as in nitro, the carbon–oxygen double bond is highly polar. Because oxygen is much more electronegative than carbon, the dipole in the bond will place a partial positive charge on the carbon next to the ring (Fig. 14.78).

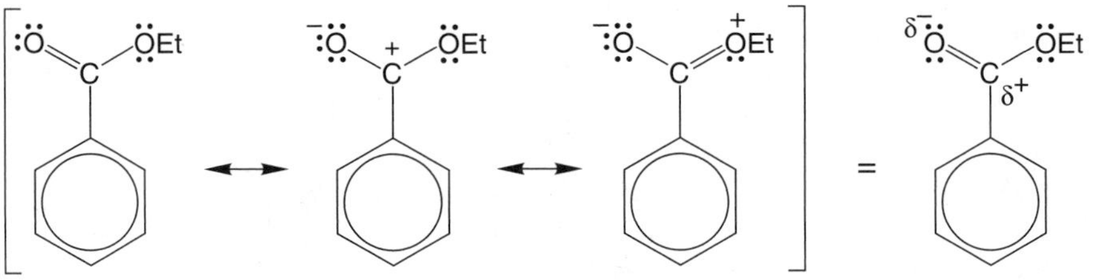

FIGURE 14.78 The carbonyl group is polar, with the carbon bearing a partial positive charge.

So, once again, ortho or para substitution will oppose like charges. This time it is not two full positive charges that must be adjacent, but a full and partial positive charge. No matter, meta substitution, which keeps the charges further apart, will still be favored, and the rate of the reaction will still be slow relative to that for benzene (Fig. 14.79).

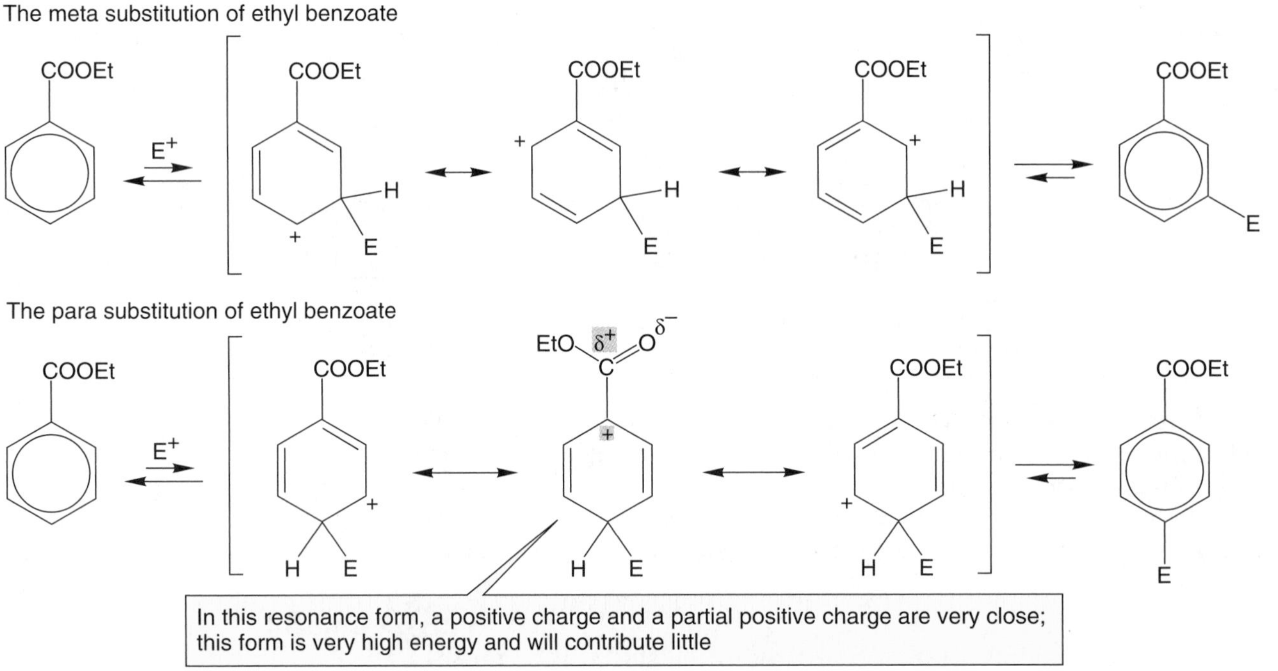

The meta substitution of ethyl benzoate

The para substitution of ethyl benzoate

In this resonance form, a positive charge and a partial positive charge are very close; this form is very high energy and will contribute little

FIGURE 14.79 Meta substitution will be preferred to para substitution. In the intermediate for para substitution, a full and partial positive charge are opposed on adjacent carbons.

PROBLEM 14.25 Explain why cyanobenzene (benzonitrile) substitutes preferentially in the meta position.

14.11 INDUCTIVE EFFECTS IN AROMATIC SUBSTITUTION

So far, we have mainly considered the way in which groups attached to a benzene ring can affect further reaction through resonance stabilization or electrostatic destabilization of an intermediate cyclohexadienyl cation. We should also consider the inductive effects of groups, which act by withdrawing or donating electrons through the σ bonds. For example, both the trimethylammonium group and the nitro group will be strongly electron withdrawing (Fig. 14.80).

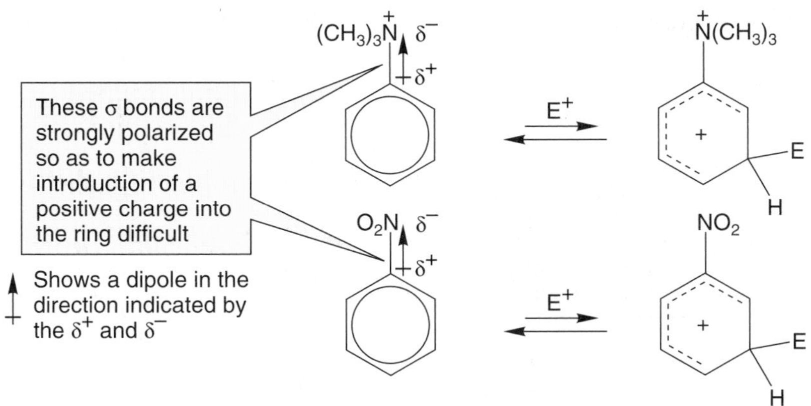

FIGURE **14.80** In the trimethylanilinium ion and nitrobenzene, the σ bond from the ring to the substituent will be polarized by the electron-withdrawing group. Further substitution is difficult.

These subsituents will reduce the electron density in the ring making the aromatic compound a weaker nucleophile, and thus a less active participant in any reaction with an electrophilic reagent. More important will be the destabilizing effect of electron withdrawal on the charged intermediate and the partially charged transition state leading to the intermediate. So, it is no surprise to find that both trimethylanilinium ion and nitrobenzene are very slow to react in electrophilic aromatic substitution. Similar rate-retarding effects might be expected to operate whenever an atom more electronegative than carbon is attached to the ring.

If that is the case, what is one to make of the rate *accelerating* effect of the methoxyl group? Anisole (methoxybenzene) reacts much faster than benzene itself, and yet the methoxyl group is certainly withdrawing electrons from the ring through induction. The dipole in the carbon–oxygen σ bond surely points toward the electronegative oxygen atom (Fig. 14.81).

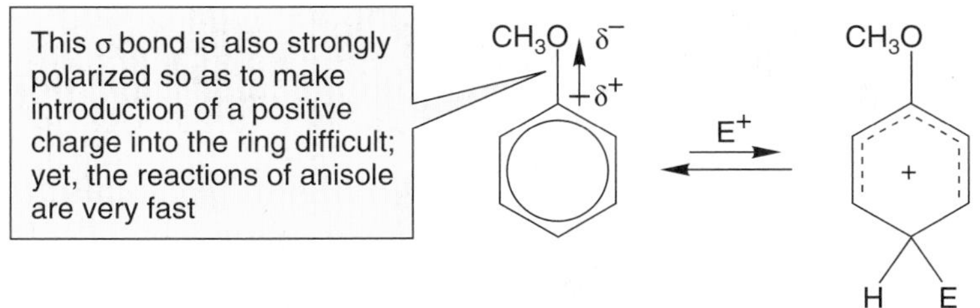

FIGURE **14.81** The carbon–oxygen bond in anisole is also electron withdrawing, yet substitution is directed ortho/para and the reactions are fast. The resonance effect outweighs this inductive effect.

The answer to this conundrum is simply that the resonance effect must substantially outweigh the inductive effect. The fourth resonance form in Figure 14.67 helps more in energy terms than the inductive withdrawal of σ electrons by methoxy hurts.

Thus, often we have two effects operating in opposition. In anisole, the resonance effect stabilizes the intermediate for ortho or para substitution and the transition states leading to them, accelerating the reaction and directing ortho/para. The inductive effect is decelerating for any substitution reaction. For anisole and most groups, the resonance effect wins out, but we can surely find examples where the two effects are largely in balance; where substitution is still ortho/para because of a resonance effect, but where a strong inductive effect slows the reaction relative to unsubstituted benzene. A look at Table 14.1 shows that there are such cases. For example, halobenzenes direct substitution to the ortho/para positions but substitution proceeds at a slower rate than that for reaction of benzene itself. In halobenzenes, *all* electrophilic aromatic substitution is retarded by the strong electron-withdrawing effect of the electronegative halogens (Fig. 14.82).

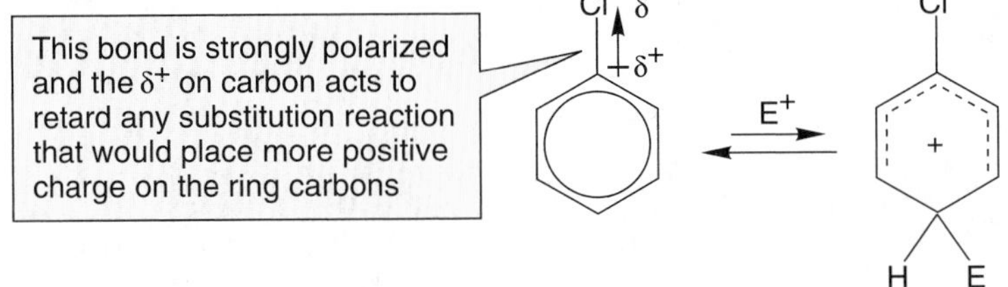
This bond is strongly polarized and the δ+ on carbon acts to retard any substitution reaction that would place more positive charge on the ring carbons
Cl
δ−
δ+
E+
Cl
+
H
E

FIGURE 14.82 Chlorine and the other halogens are strongly electron withdrawing and will slow the rate of further substitution.

However, ortho/para substitution is still favored over meta because the halogens can stabilize an adjacent positive charge by resonance (Fig. 14.83).

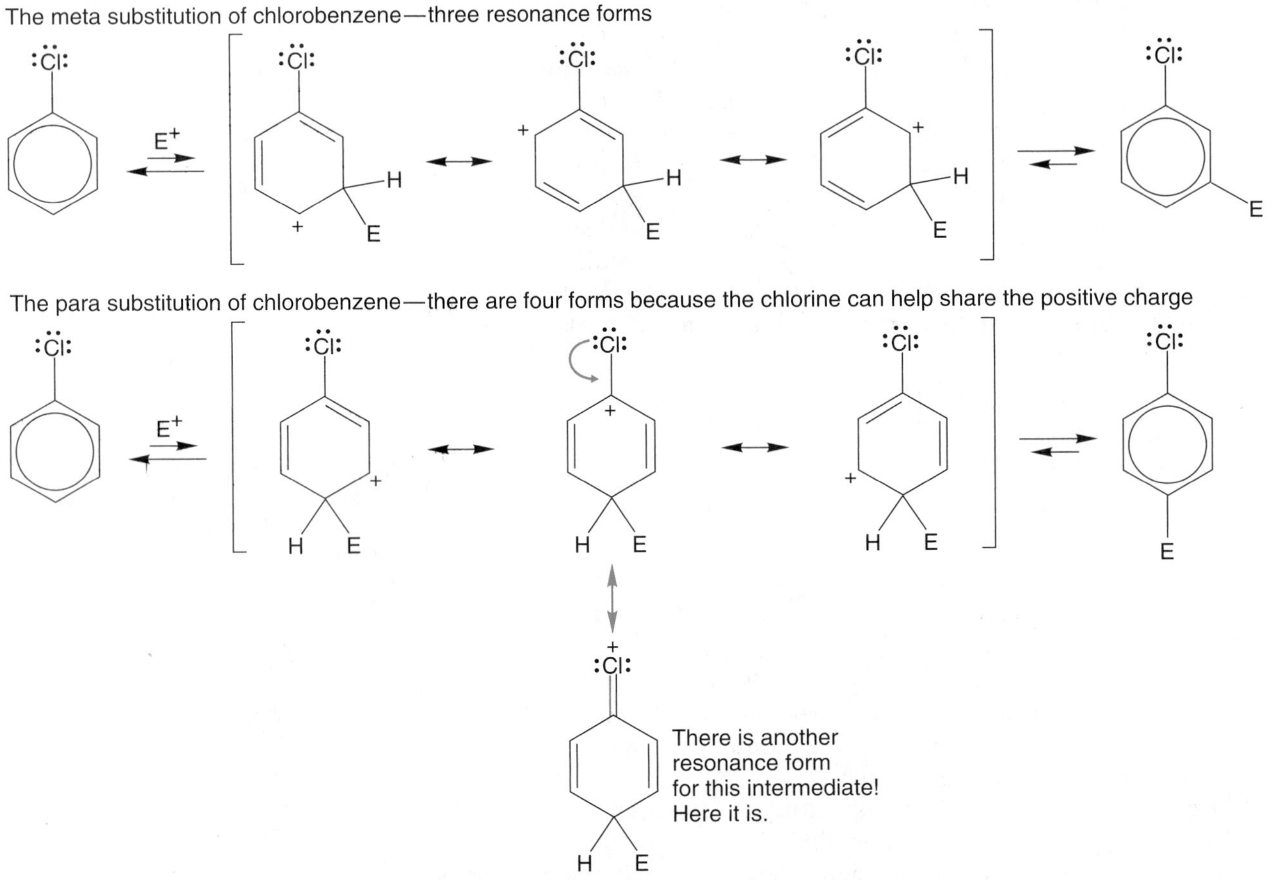

The meta substitution of chlorobenzene—three resonance forms

The para substitution of chlorobenzene—there are four forms because the chlorine can help share the positive charge

There is another resonance form for this intermediate! Here it is.

FIGURE 14.83 Chlorine and other halogens can help stabilize the intermediate for para substitution through a fourth resonance form. The halogens strongly withdraw electrons from the ring through an inductive effect, however, and substitution is slow relative to benzene.

The net result is an overall slowing of the rate. The rate of meta substitution, where the stabilizing resonance effect cannot operate, is slowest of all, and so meta product is greatly disfavored (Fig. 14.84).

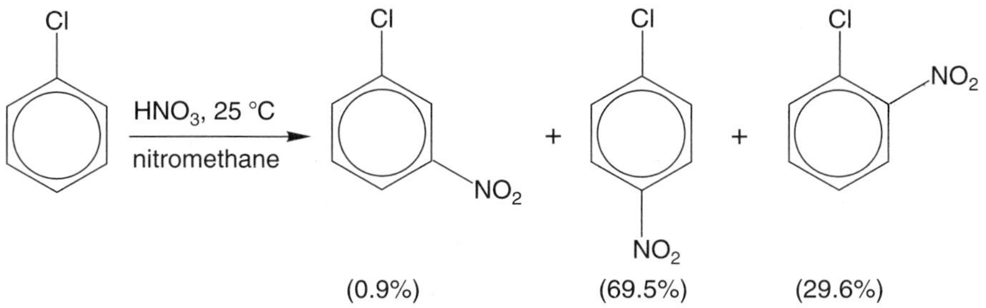

Cl

HNO_3, 25 °C

nitromethane

Cl, NO_2 (0.9%) + Cl, NO_2 (69.5%) + Cl, NO_2 (29.6%)

FIGURE **14.84** The nitration of chlorobenzene gives almost entirely ortho and para products.

14.12 POLYSUBSTITUTION OF AROMATIC COMPOUNDS AND SYNTHESIS OF MULTIPLY SUBSTITUTED BENZENES

Sometimes it is easy to predict the position of attack on a multiply substituted benzene ring. For example, if we have a ring substituted 1,3 with meta directing groups, further substitution will surely occur at the position meta to both groups, although the rate will be severely retarded by the presence of a pair of deactivating groups (Fig. 14.85).

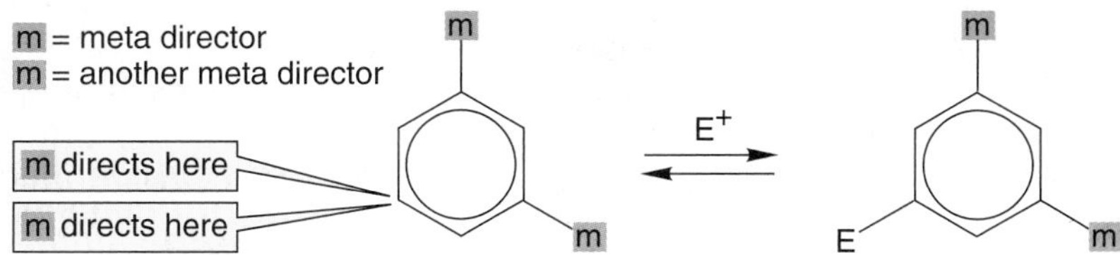

FIGURE **14.85** On a benzene ring substituted 1,3 with a pair of meta directing groups, both substituents direct further substitution to the same position.

For example, nitration of 1,3-dinitrobenzene gives only meta substitution (Fig. 14.86).

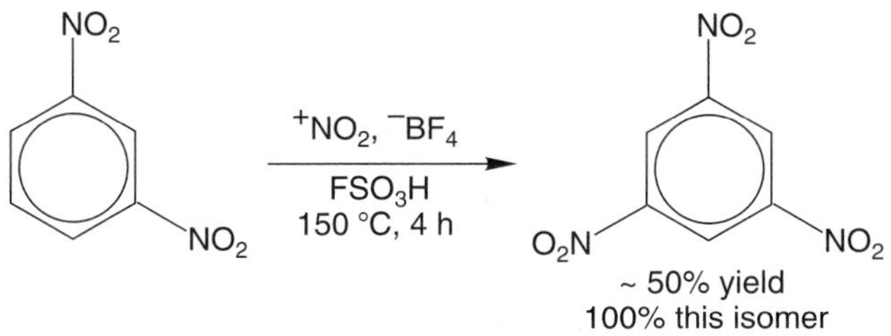

NO_2, NO_2

$^+NO_2$, $^-BF_4$

FSO_3H

150 °C, 4 h

NO_2, O_2N, NO_2

~ 50% yield

100% this isomer

FIGURE **14.86** Nitration of 1,3-dinitrobenzene gives a single product, 1,3,5-trinitrobenzene.

Further substitution of a molecule containing an ortho/para director and a meta director at the 1- and 4-positions is similarly easy to predict. Both groups direct the next substitution to the position ortho to the ortho/para directing group. An example of this kind of reaction is the nitration of 4-nitrotoluene (Fig. 14.87).

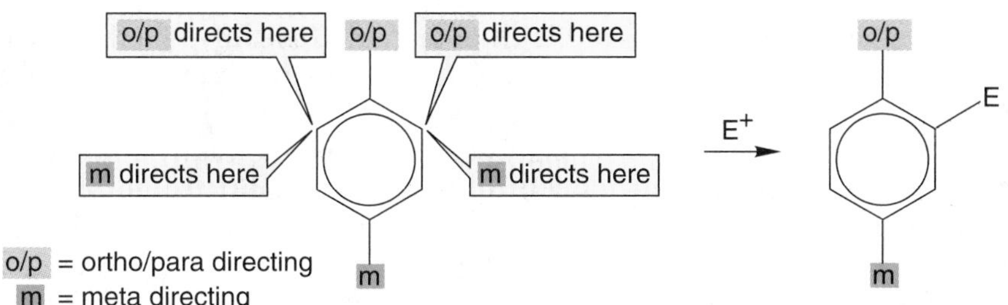

FIGURE **14.87** Substitution is also directed to a single position if a benzene ring is substituted 1,4 with one ortho/para directing group and one meta directing group. An example is the nitration of 4-nitrotoluene.

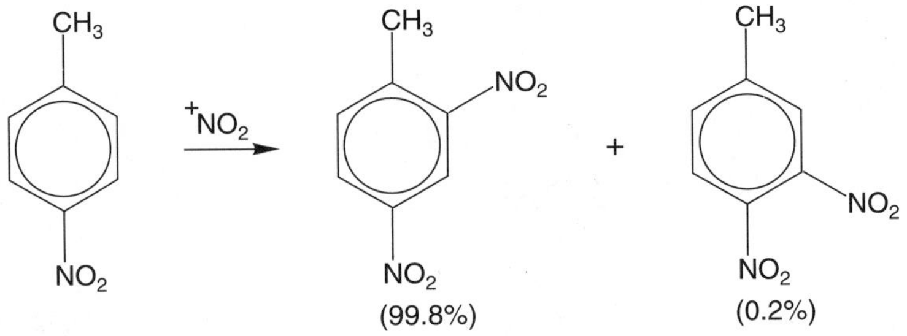

FIGURE 14.87 (CONTINUED)

In the example in Figure 14.87, we really didn't have to worry much about the directing effect of the nitro group, a rate-retarding substituent, which will be overwhelmed by the activating and ortho/para directing methyl group. However, there are less obvious situations. If the benzene ring bears different ortho/para directing groups at the 1- and 4-positions, the position of further substitution is not immediately clear. Sometimes steric effects determine the outcome. For example, Friedel–Crafts acylation of 4-isopropyltoluene gives reaction only at the less encumbered position adjacent to the methyl group (Fig. 14.88). In other cases, electronic factors determine the outcome, and further reaction will be at the position activated by the more strongly activating group. For example, Friedel–Crafts alkylation of 4-methylphenol (*p*-cresol) gives almost entirely substitution ortho to the strongly ortho/para directing hydroxyl group, and only traces of substitution ortho to the less activating methyl group (Fig. 14.88).

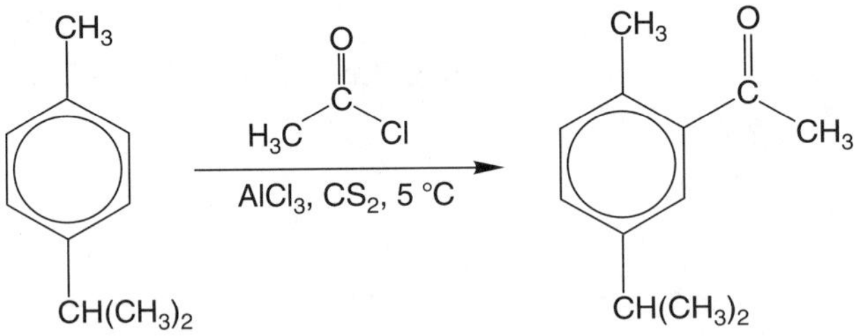

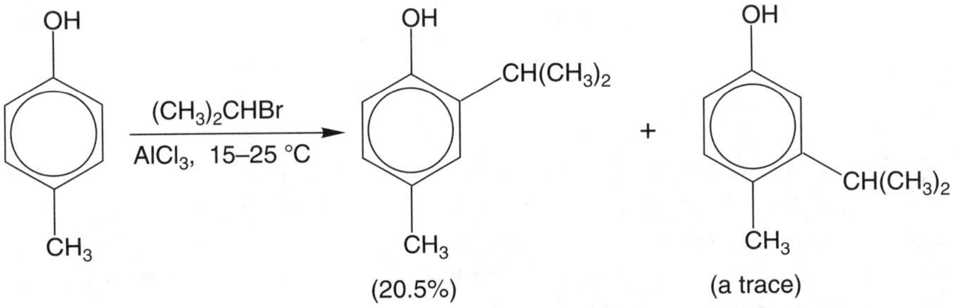

FIGURE 14.88 Two examples of reactions of benzenes substituted 1,4 with two different ortho/para directing groups. Steric effects decide the first case, electronic effects the second.

Other situations are more difficult to predict. Generally, the result will be determined by the sum of the steric and electronic effects of groups already present on the ring.

Some substituents are so strongly activating that no catalyst is needed, and it is often difficult to stop substitution before more than one new substituent has been added. For example, anilines are brominated with Br_2 alone to give 2,4,6-tribromoanilines. Phenols are also highly activated toward further substitution. Mild conditions are needed to restrict the reaction to monosubstitution (Fig. 14.89).

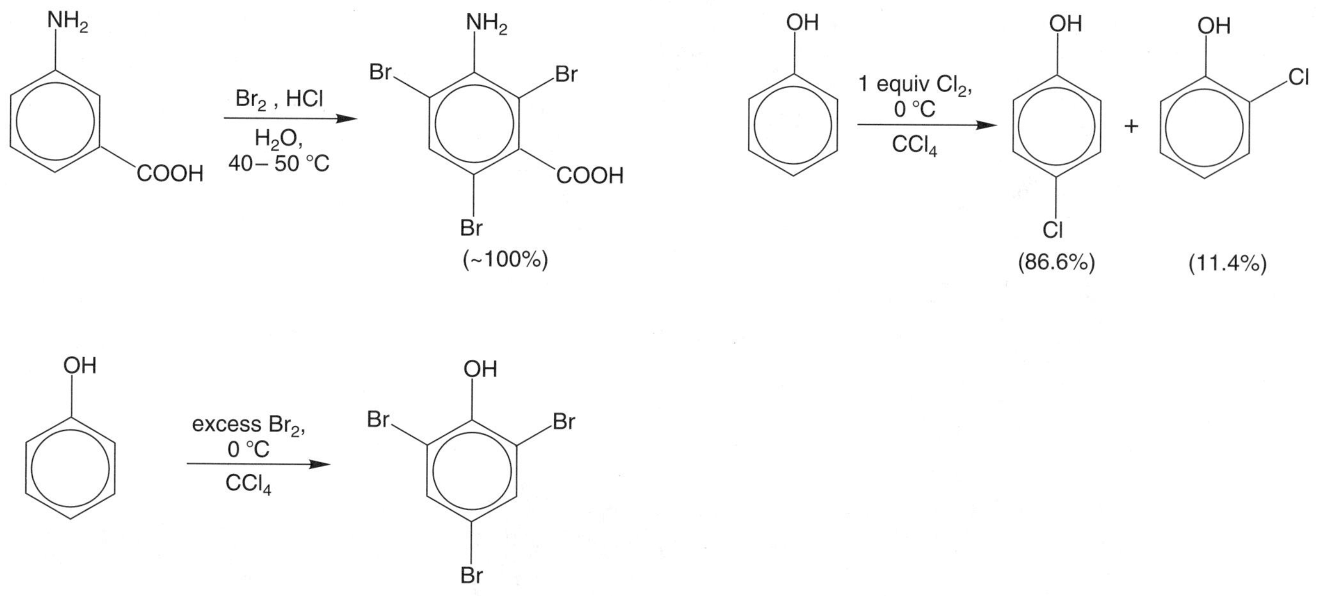

FIGURE **14.89** The bromination of anilines and phenols often leads to multiple additions of several bromines. The strongly activating groups, amino and hydroxyl, direct substitution to the ortho and para positions.

The hydroxyl group and the amino group are so strongly activating that neither monobromo nor dibromo compounds can be isolated unless great care is taken. Note that when the first bromine is added in the para position we have the general situation of Figure 14.88 in which the OH and the Br direct further substitution to different positions. The more powerfully activating OH and NH_2 groups win out (bromine is actually *de*activating), and the second and third bromines add ortho to the amino or hydroxyl group.

It is possible to deactivate the amino group by acylating it through reaction with acetyl chloride to give acetamidobenzene, also known as acetanilide (Fig. 14.90). *N*-Acylated aromatic amines still direct ortho/para, but substitution is slower than for the unsubstituted amine, and the reaction can be stopped after a single new substitution. The protecting acyl group can be removed later by treatment with acid or base. We will see the mechanisms for these reactions in Chapter 20 (Fig. 14.90).

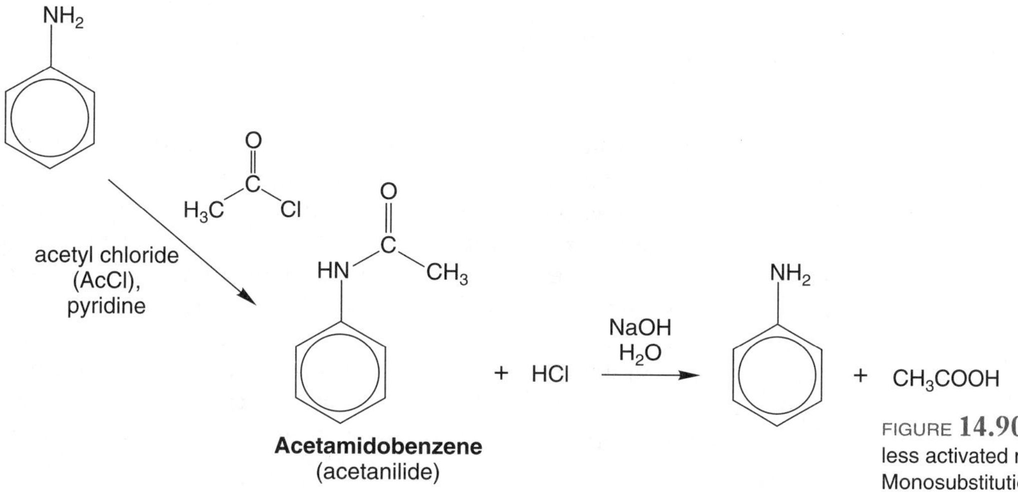

FIGURE **14.90** Acylated aniline is a less activated molecule than aniline. Monosubstitution can be achieved.

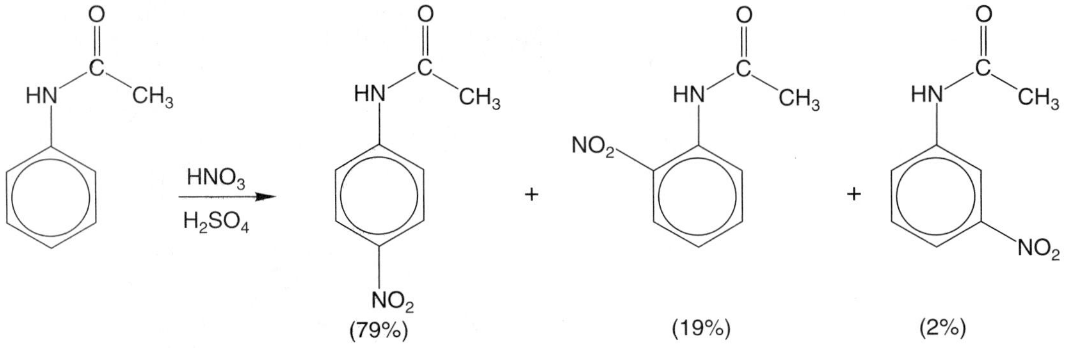

$$\text{acetanilide} \xrightarrow[H_2SO_4]{HNO_3} \text{para } (79\%) + \text{ortho } (19\%) + \text{meta } (2\%)$$

FIGURE 14.90 (CONTINUED)

PROBLEM 14.26 Through a mechanistic analysis of the substitution of acetamidobenzene (acetanilide), explain why this compound substitutes mainly para, but at a rate slower than aniline itself.

Effective use can sometimes be made of removable blocking groups on the ring. Suppose you are set the task of making pure *o*-bromophenol. Simple bromination of phenol will fail because of the dominant formation of the para isomer and polybrominated compounds (Fig. 14.91).

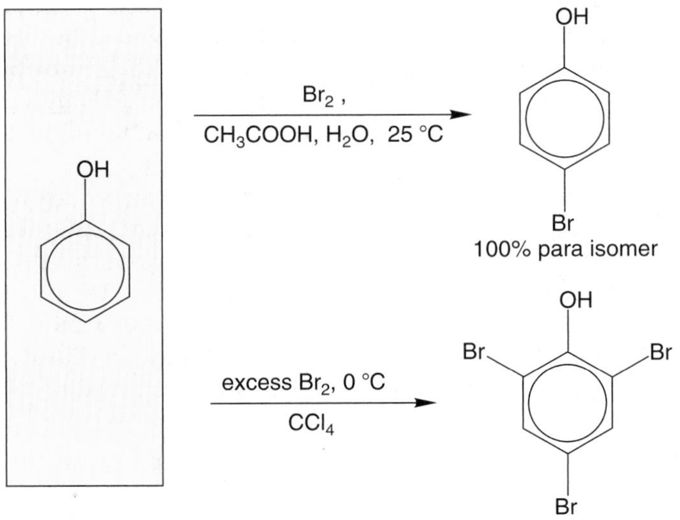

$$\text{phenol} \xrightarrow[CH_3COOH,\ H_2O,\ 25\ °C]{Br_2\,,} \text{p-bromophenol (100\% para isomer)}$$

$$\text{phenol} \xrightarrow[CCl_4]{\text{excess } Br_2,\ 0\ °C} \text{2,4,6-tribromophenol}$$

FIGURE 14.91 Direct bromination of phenol cannot give *o*-bromophenol.

The solution is first to sulfonate to give the 2,4-disulfonic acid of phenol. The deactivating effect of the sulfonic acid groups prevents further sulfonation. This compound can now be brominated at the position ortho to the activating hydroxyl group. The sulfonic acids have acted to block substitution at the other ortho and para positions. Removal of the blocking sulfonate groups with acidic steam completes the reaction (Fig. 14.92).

The ability to interconvert nitro and amino groups is also useful. The synthesis of 4-bromonitrobenzene from nitrobenzene is an example. Nitrobenzene cannot be brominated in the para position because the meta-directing nitro group will force formation of the meta isomer (Fig. 14.93).

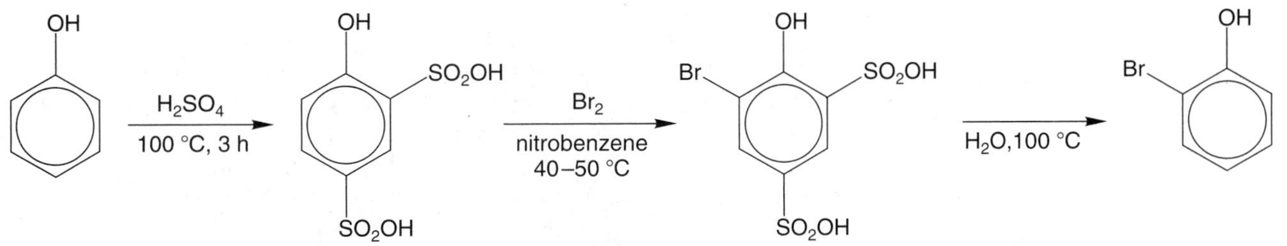

FIGURE 14.92 The para and one ortho position can be reversibly blocked by use of the sulfonic acid group. Once the sulfonic acid is attached, only one ortho position remains for further substitution. After bromination, the sulfonic acid groups can easily be removed.

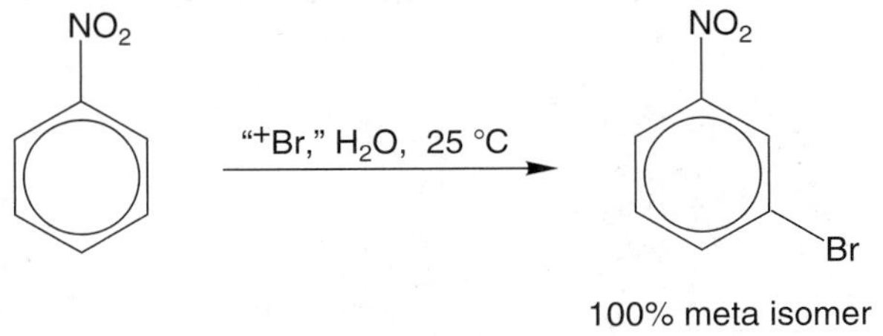

FIGURE 14.93 Nitrobenzene cannot be brominated in the ortho or para positions. Substitution is entirely meta.

The solution is to convert nitrobenzene into aniline by a reduction step, acetylate, and then brominate. Bromination now goes 100% in the para position. Now the amino group can be regenerated by removal of the acyl group and then oxidized to a nitro group (Fig. 14.94). The initial reduction of nitro to amino can be carried out catalytically, or by a variety of other reducing agents. The oxidation step is usually accomplished with trifluoroperacetic acid or *m*-chloroperbenzoic acid.

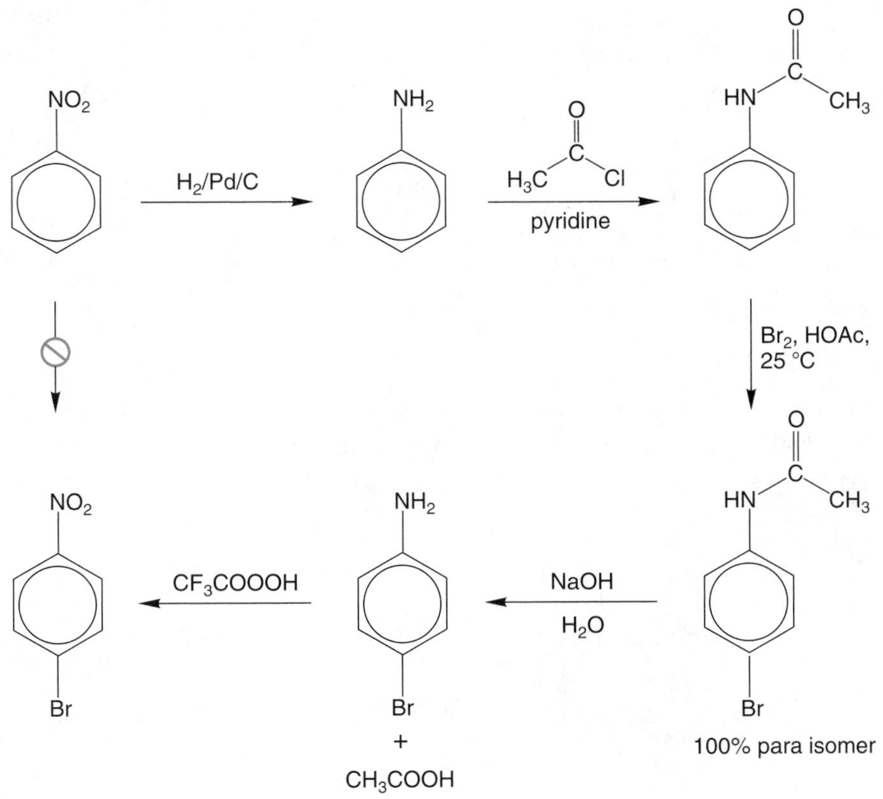

FIGURE 14.94 Conversion of the nitro group into the ortho/para-directing amino group followed by acylation allows the introduction of a single para bromine. Regeneration of the amine, followed by oxidation back to nitro, completes the reaction sequence.

The synthesis of a disubstituted benzene in which two meta-directing groups are substituted ortho or para to each other constitutes a general synthetic problem. To solve such a problem, the ability to convert a meta-directing group into an ortho/para director *and back again* is crucial. A similar difficulty is encountered in the construction of disubstituted benzenes bearing ortho/para-directing groups in meta positions.

*PROBLEM **14.27**

The nitration of aniline takes place in acid. Under acidic conditions the amino group must be largely protonated to give an anilinium ion, a meta-directing group. In fact, if the amino group is not acetylated, the product does contain substantial amounts of *m*-nitroaniline. Why is the product not entirely *m*-nitroaniline? *Hint*: Think about the relative rates of the reactions involved.

ANSWER

It is certainly the case that aniline will protonate in acid, and that the anilinium ion will substitute meta.

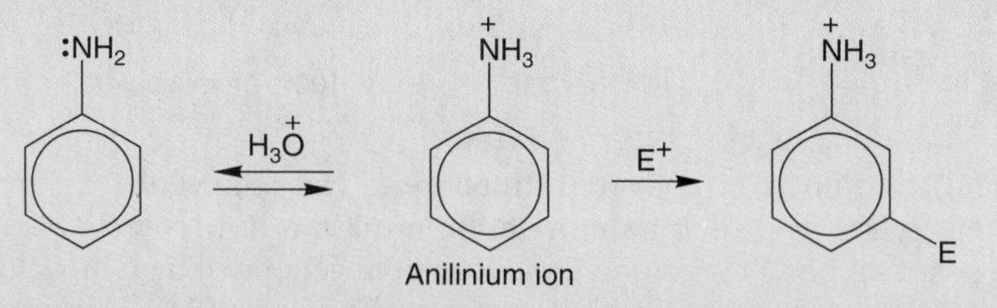

:NH_2 — $H_3\overset{+}{O}$ ⇌ $\overset{+}{N}H_3$ (Anilinium ion) — E^+ → $\overset{+}{N}H_3$ / E (meta)

Anilinium ion

However, as long as there is some unprotonated aniline present at equilibrium, it will continue to substitute ortho and para. It is estimated that some anilines react at 10^{19} times the rate of the deactivated anilinium ions! So, even if there is only a little free amine present there can still be a great deal of product formed from it.

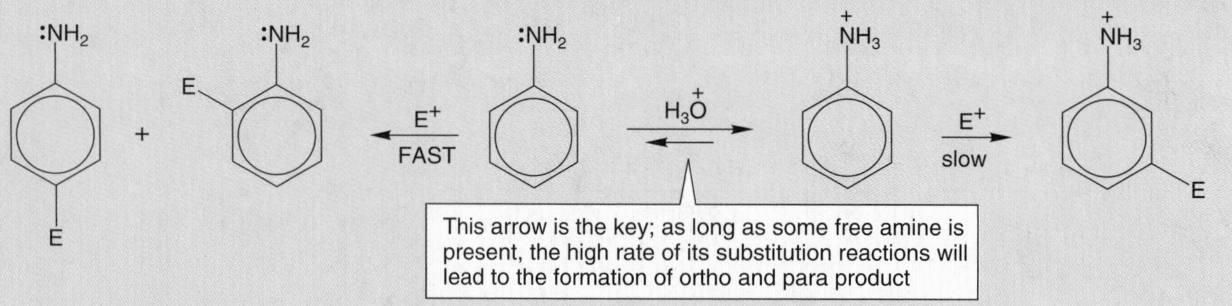

:NH_2 (para E) + :NH_2 (ortho E) ← E^+ FAST — :NH_2 — $H_3\overset{+}{O}$ ⇌ $\overset{+}{N}H_3$ — E^+ slow → $\overset{+}{N}H_3$ (meta E)

This arrow is the key; as long as some free amine is present, the high rate of its substitution reactions will lead to the formation of ortho and para product

PROBLEM **14.28**

Devise syntheses of the following substituted benzenes. You may start from benzene, organic reagents containing no more than two carbons, and any inorganic reagents you may need (no carbons).

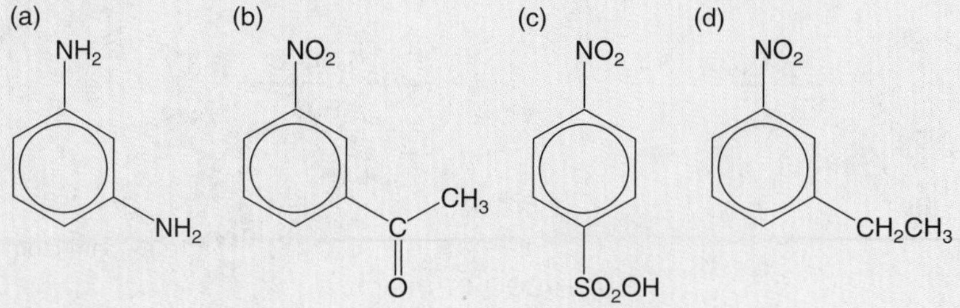

(a) NH_2, NH_2 (meta)

(b) NO_2, $C(=O)CH_3$ (meta)

(c) NO_2, SO_2OH (para)

(d) NO_2, CH_2CH_3 (meta)

14.13 NUCLEOPHILIC AROMATIC SUBSTITUTION

Normally, alkenes are unaffected by treatment with nucleophiles, and it is no surprise to find that benzene also survives treatment with strong nucleophiles. In particular, no reaction occurs when simple substituted benzenes are treated with base (Fig. 14.95).

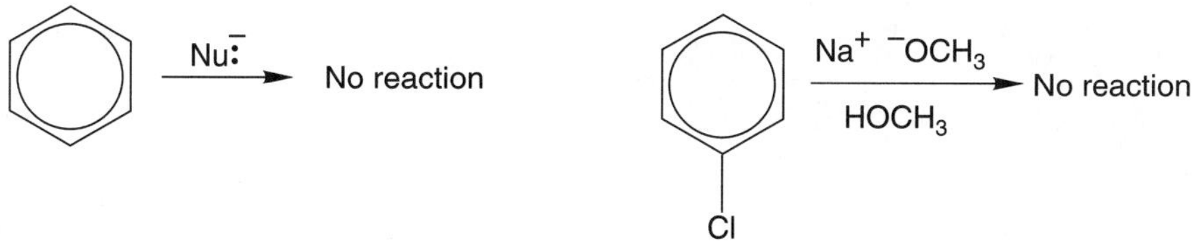

Nu: → No reaction

$Na^+ \ ^-OCH_3$ / $HOCH_3$ → No reaction

Cl

FIGURE 14.95 Benzenes do not normally react with nucleophiles to give substituted benzenes.

But a few specially substituted benzenes do undergo reaction in base. For example, 2,4-dinitrochlorobenzene is converted into 2,4-dinitrophenol by treatment with sodium hydroxide in water (Fig. 14.96).

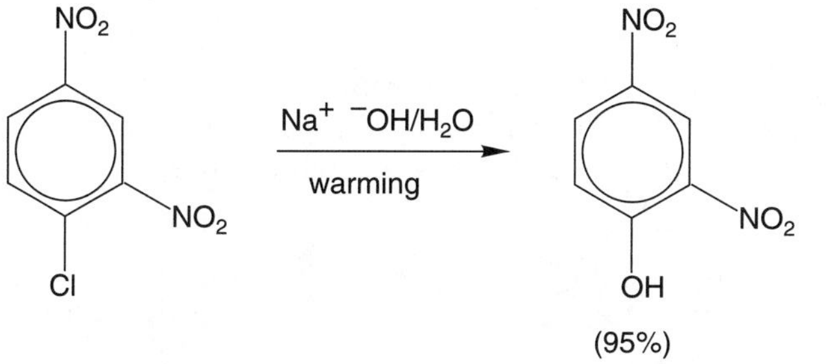

NO_2 NO_2 Cl — $Na^+ \ ^-OH/H_2O$, warming → NO_2 NO_2 OH (95%)

FIGURE 14.96 Some substituted benzenes are reactive toward nucleophiles. Electron-withdrawing groups are particularly effective, and 2,4-dinitrochlorobenzene can be swiftly transformed into the corresponding phenol by hydroxide ion in water.

The nitro group is essential, and it must be substituted in the right place relative to the chlorine. Chlorobenzene and *m*-nitrochlorobenzene do not react under these conditions, but *p*-nitrochlorobenzene does, although it substitutes much more slowly than the 2,4-dinitro compound (Fig. 14.97).

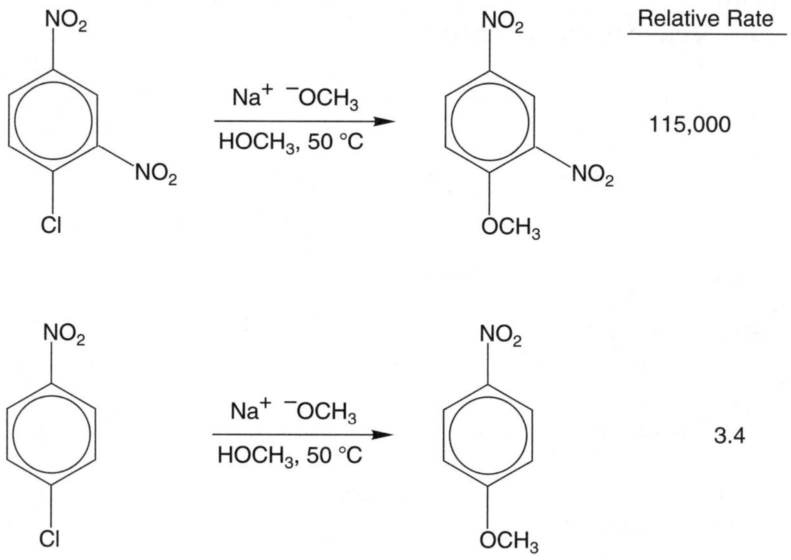

Relative Rate

NO_2 NO_2 Cl — $Na^+ \ ^-OCH_3$ / $HOCH_3$, 50 °C → NO_2 NO_2 OCH_3 — 115,000

NO_2 Cl — $Na^+ \ ^-OCH_3$ / $HOCH_3$, 50 °C → NO_2 OCH_3 — 3.4

FIGURE 14.97 For the reaction to succeed, at least one nitro group must be present. 2,4-Dinitro compounds (represented by the first molecule in this figure) react much faster than mononitro compounds (seen following both here and at the top of the next page).

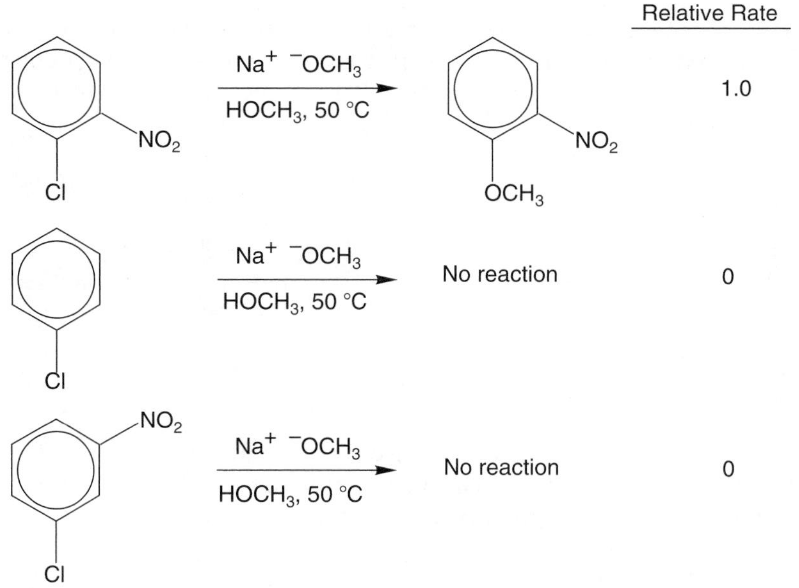
Relative Rate
NO2
Cl
Na+ −OCH3
HOCH3, 50 °C
NO2
OCH3
1.0
Cl
Na+ −OCH3
HOCH3, 50 °C
No reaction
0
NO2
Cl
Na+ −OCH3
HOCH3, 50 °C
No reaction
0

FIGURE **14.97** (CONTINUED)

Neither an S_N2 nor an S_N1 reaction seems likely. Attack from the rear is impossible, as the ring blocks the path of any entering nucleophile, and ionization would produce a most unstable carbocation (Fig. 14.98).

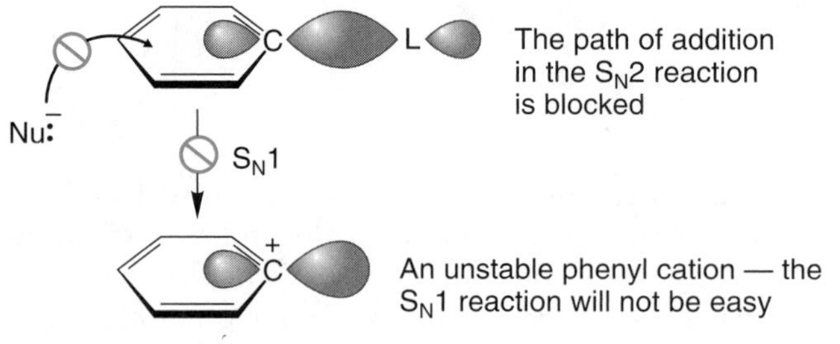
Nu:−
C
L
The path of addition
in the SN2 reaction
is blocked
SN1
C+
An unstable phenyl cation — the
SN1 reaction will not be easy

FIGURE **14.98** Neither an S_N2 nor S_N1 reaction is likely in this system.

The mechanism of this reaction involves addition to the benzene ring by methoxide to generate a resonance-stabilized anionic intermediate, called a **Meisenheimer complex**, after Jakob Meisenheimer (1876–1934) (Fig. 14.99).

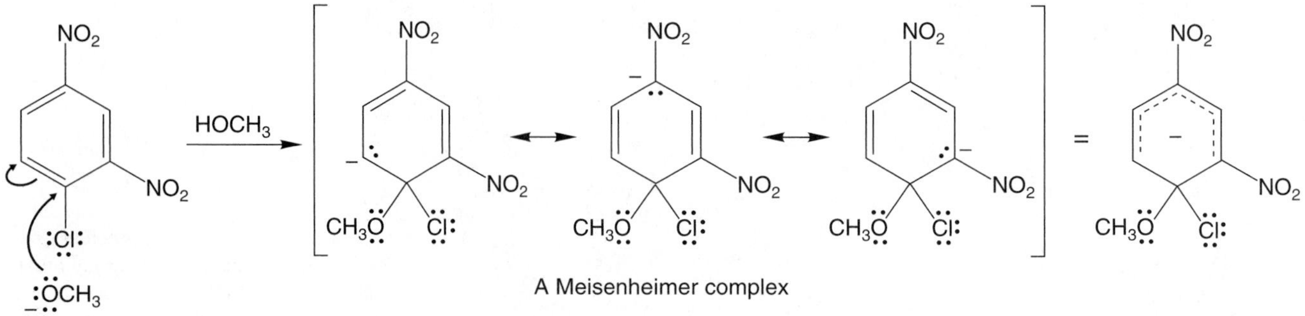
NO2
NO2
:Cl:
:OCH3
HOCH3
CH3O
Cl
A Meisenheimer complex

FIGURE **14.99** Addition to the ring by methoxide leads to a highly resonance-stabilized anion. Both nitro groups help bear the negative charge.

The nitro groups stabilize the negative charge through resonance, and therefore will exert their effect *only* when they are substituted at a carbon that helps to bear the negative charge. This stabilization by nitro is the reason that *o*- or *p*-nitrochlorobenzene undergoes the reaction, but the meta isomer does not (Fig. 14.100).

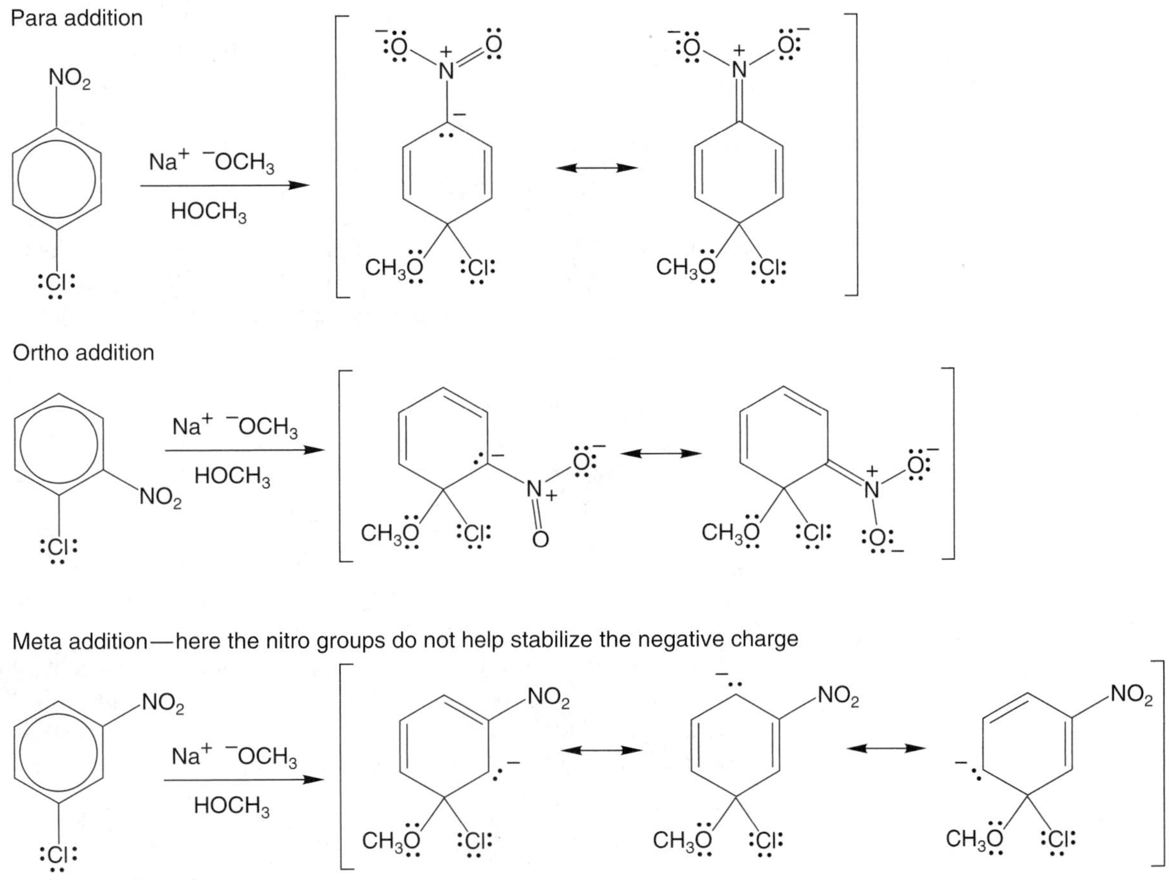

Para addition

Ortho addition

Meta addition—here the nitro groups do not help stabilize the negative charge

FIGURE **14.100** The stabilizing nitro groups must be in the proper position. Nitro groups meta to the point of attack cannot stabilize the negative charge by resonance.

The aromatic system can be regenerated by reversal of the attack of methoxide or by loss of chloride to give 2,4-dinitroanisole (Fig. 14.101).

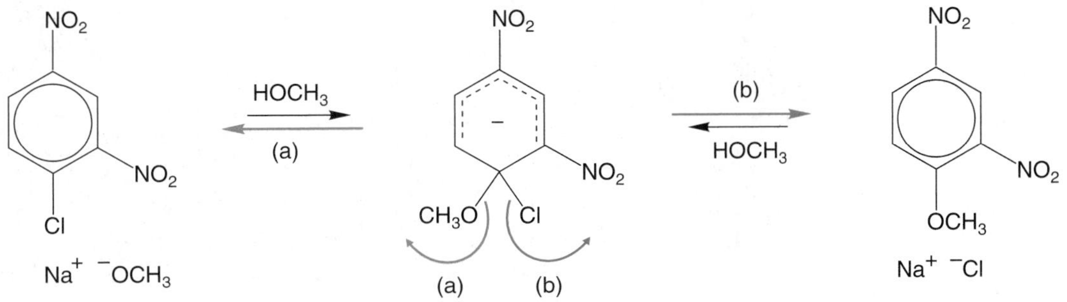

FIGURE **14.101** After addition, the reaction is completed by loss of chloride and regeneration of the aromatic ring. Loss of methoxide only leads back to starting material.

There are similarities between nucleophilic aromatic substitution and its more usual electrophilic counterpart. Each involves the formation of a resonance-stabilized intermediate, and each involves a temporary loss of aromaticity that is regained in the final step of the reaction. But the similarities are only so deep. One reaction is cationic; the other involves anions. Use the differing effects of a nitro group, strongly decelerating in the electrophilic substitution and strongly accelerating in the nucleophilic reaction, to keep the two mechanisms distinct in your mind.

14.14 SOMETHING MORE: BENZYNE

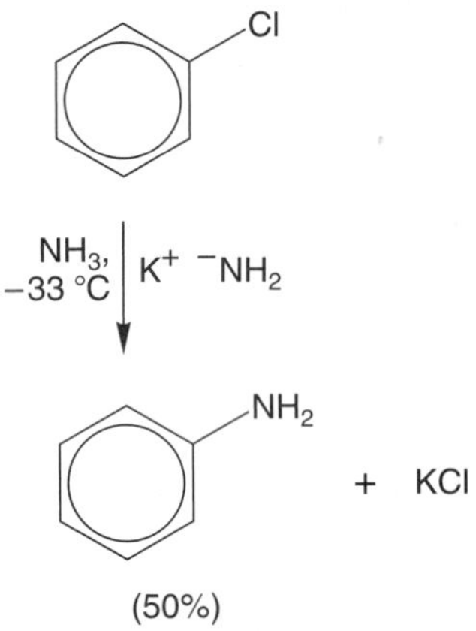

In Section 14.13, we mentioned the lack of reactivity of halobenzenes with bases. It took the activation of a nitro group for substitution to occur, for example (Figs. 14.95–14.97). In truth, if the base is strong enough, even chlorobenzene will react. In the very strongly basic medium potassium amide in liquid ammonia, chlorobenzene is converted into aniline (Fig. 14.102).

The mechanism is surely not an S_N2 displacement, and there is no reason to think nucleophilic addition would be possible to an unstabilized benzene ring. This difficult mechanistic problem begins to unravel with the observation by John D. Roberts (b. 1918) and his co-workers that labeled chlorobenzene produces aniline *labeled equally in two positions* (Fig. 14.103).

What reactions of halides do you know besides displacement? The E1 and E2 reactions compete with displacement reactions (Chapter 7, p. 273). If HCl were lost from chlorobenzene, a cyclic alkyne, called **dehydrobenzene** or **benzyne**, would be formed. This bent acetylene might be reactive enough to undergo an addition reaction with the amide ion. If this were the case, the labeled material must produce two differently labeled products of addition (Fig. 14.104).

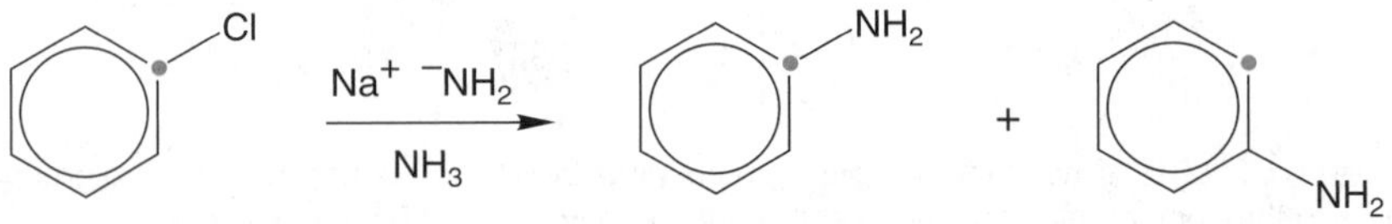

Cl
NH_3, −33 °C | K^+ $^-NH_2$
NH_2 + KCl
(50%)

FIGURE 14.102 Aniline can be formed from chlorobenzene by reaction with the very strong base, potassium amide (K^+ $^-NH_2$) in ammonia (NH_3).

Cl
Na^+ $^-NH_2$
NH_3
NH_2 + NH_2

FIGURE 14.103 A clue to the mechanism is provided by Roberts' observation of isotopic scrambling when labeled chlorobenzene is used in the reaction. The red dot = ^{14}C, an isotopic label.

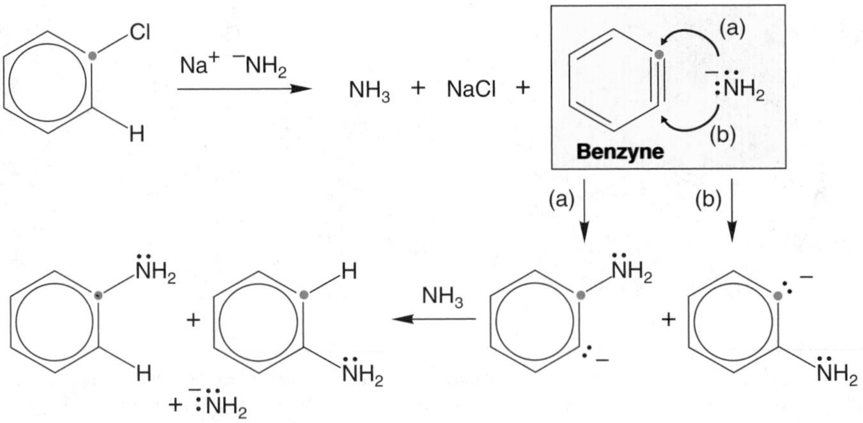

Cl, H
Na^+ $^-NH_2$
NH_3 + NaCl +
Benzyne
(a) (b) $:\ddot{N}H_2^-$
$\ddot{N}H_2$ + $\ddot{N}H_2$ ← NH_3 ← $\ddot{N}H_2$ + $\ddot{N}H_2$
H, H
+ $:\ddot{N}H_2^-$

FIGURE 14.104 If the symmetrical intermediate benzyne were formed, re-addition of the amide ion must produce the observed labeling results.

Benzyne is surely an unusual species and deserves a close look. Although this intermediate retains the aromatic sextet, the triple bond is badly bent, and there must be very severe angle strain indeed. Remember, the optimal angle in triple bonds is 180° (Chapter 4, p. 142). Moreover, the "third" bond is not composed of $2p$–$2p$ overlap like a normal alkyne, but of overlap of two hybrid orbitals (Fig. 14.105).

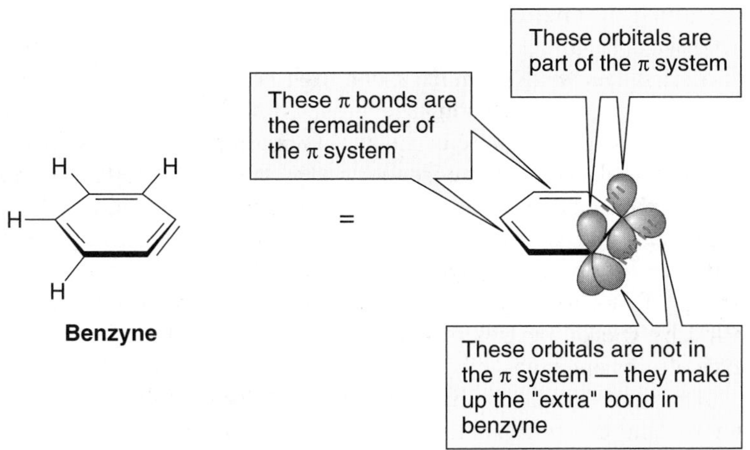
These orbitals are part of the π system
These π bonds are the remainder of the π system
H
H
H
H
=
Benzyne
These orbitals are not in the π system — they make up the "extra" bond in benzyne

FIGURE 14.105 The structure of benzyne.

One would therefore not be too surprised to find that this molecule is able to add strong bases such as the amide ion. In fact, one would expect very high reactivity for this strained acetylene. Problem 14.29 gives you a chance to see some of the reactions of benzyne.

PROBLEM 14.29 Provide arrow formalisms for the formation of the following products (Fig. 14.106):

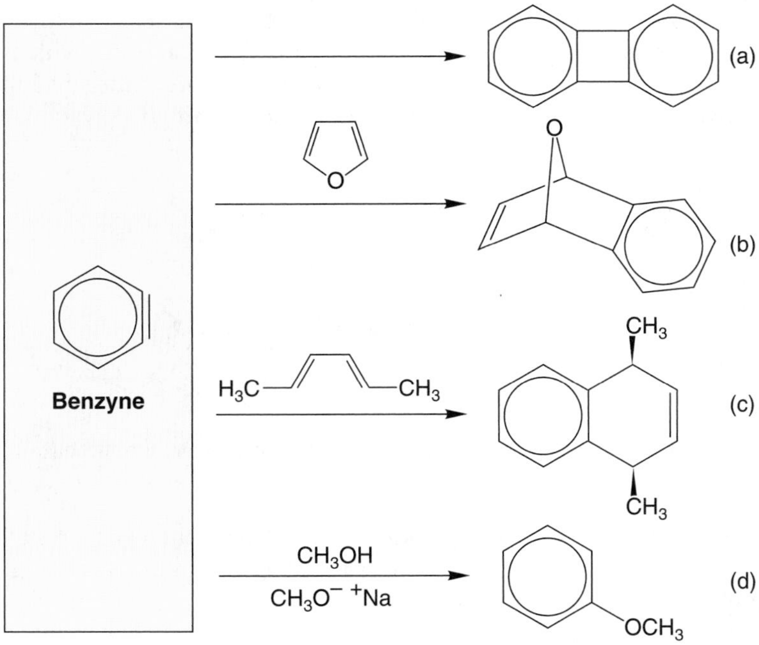

Benzyne
(a)
(b)
H_3C CH_3
CH_3
CH_3
(c)
CH_3OH
CH_3O^- ^+Na
OCH_3
(d)

FIGURE 14.106 Some reactions of benzyne.

14.15 SUMMARY

NEW CONCEPTS

This chapter involves the chemistry of aromatic compounds and is dominated by two basic interrelated notions. First, it is simply difficult to disturb the aromatic sextet, and second, once aromaticity is disrupted, it is easily regained. In chemical terms, this means that reactions of benzene and other simple aromatic compounds generally involve a highly endothermic first step with formation of a high-energy intermediate. In electrophilic aromatic substitution, the most common reaction of aromatic compounds, this intermediate is the cyclohexadienyl cation. As this intermediate lies far above benzene in energy, the transition state leading to it is also high in energy, and the reaction is difficult. Reactions of the cyclohexadienyl cation that regenerate aromatic systems are highly exothermic, and therefore have low activation barriers.

Further substitution reactions of an already substituted benzene are controlled by the properties of the first substituent. Both resonance and inductive effects are involved, but the resonance effect usually dominates. Many substituents can stabilize an intermediate cyclohexadienyl cation through resonance if substitution is in the ortho or para position. The extra stabilization afforded by the substituent leads to relatively high rates of reaction. Positively charged substituents generally substitute meta, as ortho or para substitution places two positive charges on adjacent positions. The introduction of a second positive charge means that these substitutions are relatively slow.

At low temperature, in sufficiently polar, but nonnucleophilic solvents, some carbocations are stable. Even under these superacid conditions, rearrangements through hydride or alkyl shifts to the most stable possible carbocation are common.

REACTIONS, MECHANISMS, AND TOOLS

By far the most important reaction in this chapter is electrophilic aromatic substitution by a variety of E^+ reagents. This reaction involves the initial formation of a resonance-stabilized, but not aromatic, cyclohexadienyl cation. Deprotonation regenerates an aromatic system (Fig. 14.107).

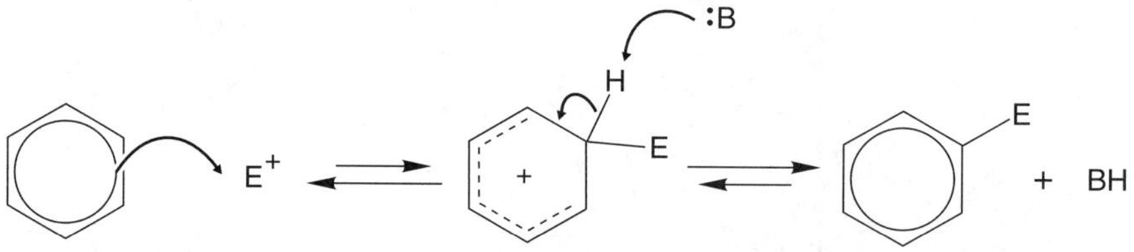

FIGURE 14.107 The general mechanism for electrophilic aromatic substitution.

Nucleophilic aromatic substitution occurs only when the ring carries substituents that can stabilize the negative charge introduced through addition of a nucleophile. Departure of a leaving group regenerates the aromatic system (Fig. 14.101).

SYNTHESES

This chapter provides many new syntheses of aromatic compounds (Fig. 14.109). Although we can now make all sorts of differently substituted benzenes, there is still a unifying theme: each product ultimately derives from a reaction of benzene with an E^+ reagent.

1. Alkanes

NH_2NH_2/base, ethylene glycol, Δ

Wolff–Kishner reduction

Zn/Hg/HCl, Δ

Clemmensen reduction

2. Cyanobenzenes

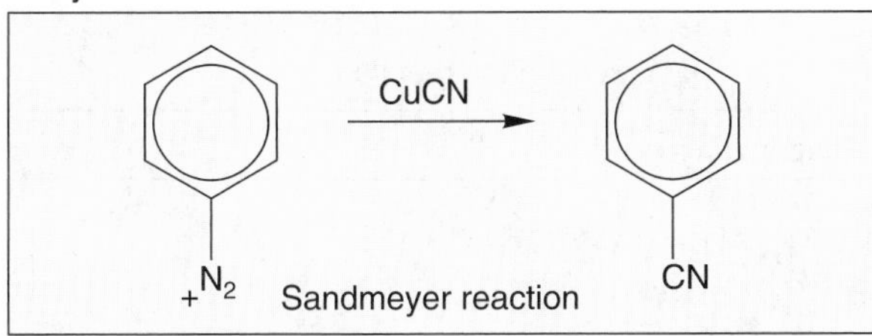

3. Cyclohexanes

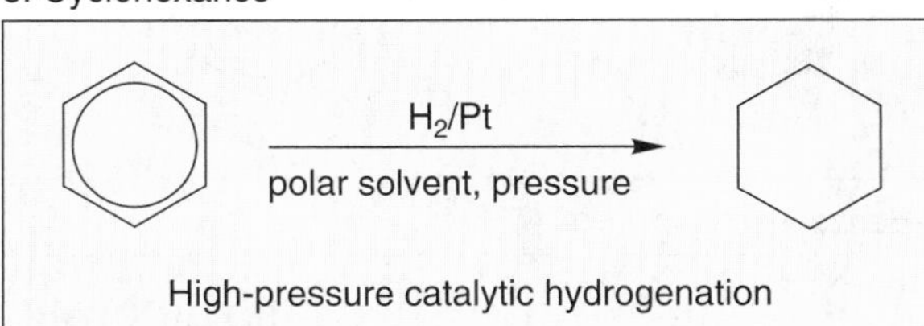

4. Benzenes

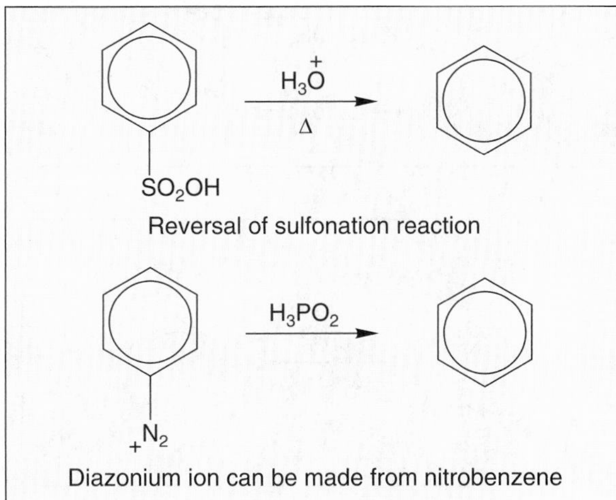

5. Deuterated benzenes

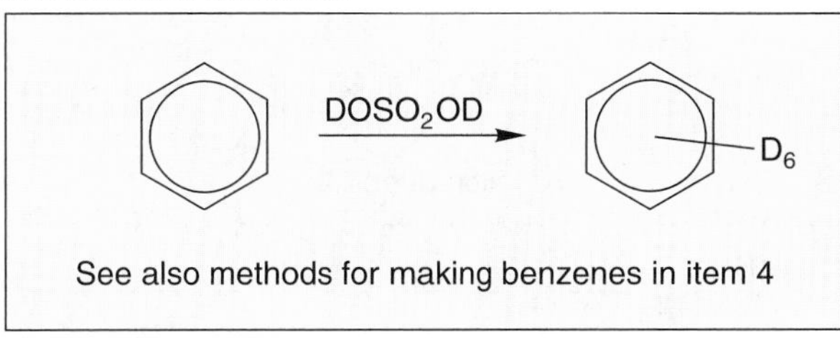

6. Alkylbenzenes

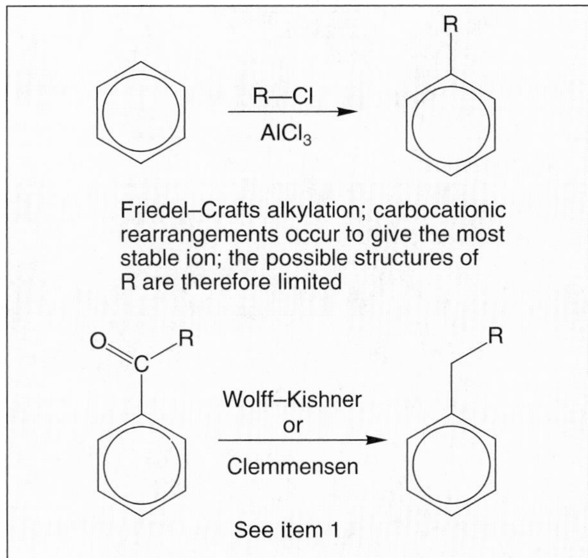

7. Phenyl ketones

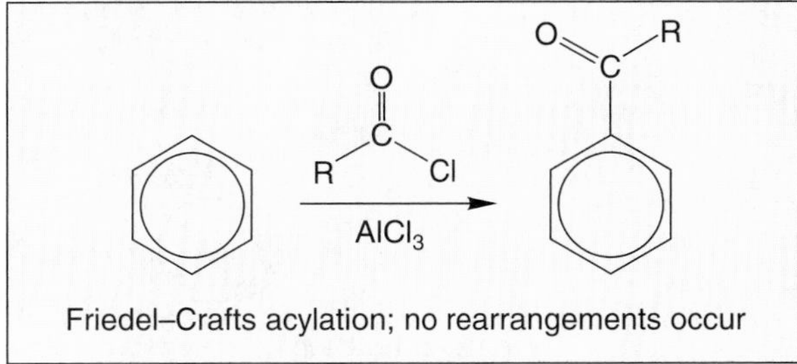

FIGURE **14.108** Synthetic reactions encountered in Chapter 14.

8. Nitrobenzenes

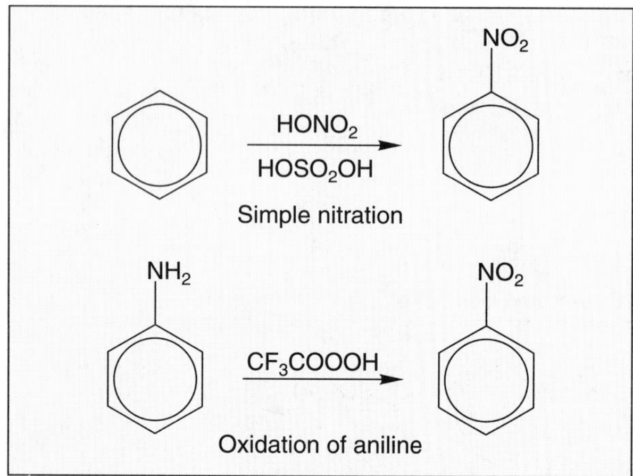

9. Anilines

NO_2 → NH_2

Sn/HCl or catalytic hydrogenation

Reduction of nitro groups

10. Halobenzenes

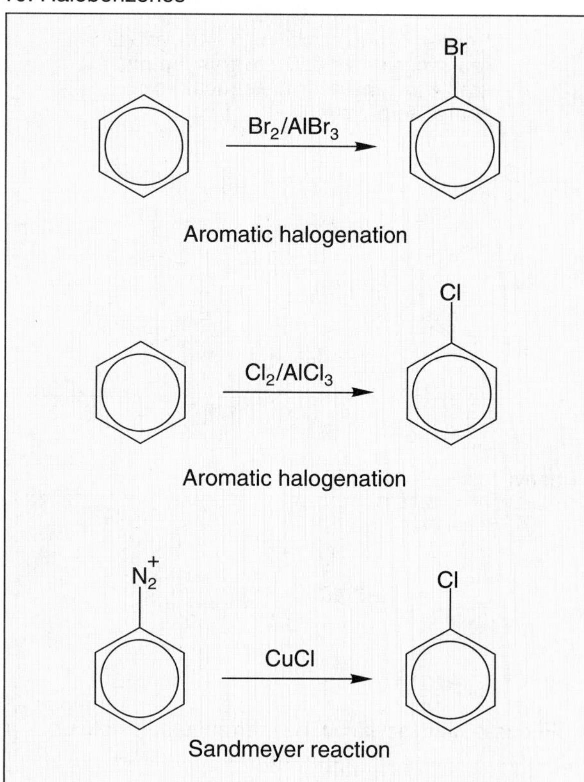

10. Halobenzenes (continued)

N_2^+ → Br

CuBr

Sandmeyer reaction

N_2^+ → I

KI

N_2^+ → F

HBF_4

Schiemann reaction

11. Diazonium ions

NH_2 → $N_2^+ \, ^-Cl$

HONO/HCl

12. Benzenesufonic acids

→ SO_2OH

SO_3

$HOSO_2OH$

Simple sulfonation

13. Phenols

N_2^+ → OH

H_2O

Δ

FIGURE **14.108** (CONTINUED)

COMMON ERRORS

The most common conceptual error is to attempt to analyze the course of aromatic substitution reactions through looking at ground-state molecules (the starting materials and products). The various resonance and inductive effects discussed in this chapter will be far more important in the charged intermediates and in the transition states leading to those intermediates. When analyzing one of these reactions, always look first at the cyclohexadienyl intermediate (a cation in electrophilic aromatic substitution and an anion in the less common nucleophilic reactions), and then consider the transition state leading to it.

Another potential error is failing to keep up with the proliferating synthetic detail. Be sure to look at each new reaction in this and subsequent chapters with an eye to using it in synthesis. Keep up your file cards!

14.16 KEY TERMS

Acid chloride Any compound containing the structure

Acyl group The group

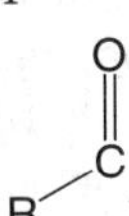

Acylium ion $(RC{=}O)^+$.

Aniline Aminobenzene.

Anisole Methoxybenzene, $PhOCH_3$.

Benzyne 1,2-Dehydrobenzene, C_6H_4.

Dehydrobenzene See **benzyne**.

Diazonium ion The group N_2^+ as in RN_2^+.

Electrophile A lover of electrons, a Lewis acid.

Electrophilic aromatic substitution The classic substitution reaction of aromatic compounds with Lewis acids. A hydrogen attached to the benzene ring is replaced by the Lewis acid and the aromatic ring is retained in the overall reaction.

Friedel–Crafts acylation The electrophilic substitution of aromatic molecules with acyl chlorides facilitated by strong Lewis acids, usually $AlCl_3$.

Friedel–Crafts alkylation The electrophilic substitution of aromatic molecules with alkyl chlorides catalyzed by strong Lewis acids, usually $AlCl_3$.

Halogenation The electrophilic substitution of aromatic systems using chlorine or bromine and Lewis acid catalysts, usually $FeCl_3$ or $FeBr_3$.

Meisenheimer complex The intermediate in nucleophilic aromatic substitution formed by the addition of a nucleophile to an aromatic compound activated by electron-withdrawing groups, often NO_2.

Meta substitution 1,3-Substitution on a benzene ring.

Nitration The electrophilic substitution of aromatic compounds by nitro groups using a source of the nitronium ion, $^+NO_2$.

Ortho substitution 1,2-Substitution on a benzene ring.

Para substitution 1,4-Substitution on a benzene ring.

Paracyclophane An aromatic molecule bridged across the ring with one or more chains.

Phenol

Sandmeyer reaction The reaction of an aromatic diazonium ion with cuprous salts to form substituted aromatic compounds.

Sulfonation The electrophilic substitution of aromatic compounds by $SO_3/HOSO_2OH$ to give benzenesulfonic acids.

Superacid One of many nonnucleophilic solvent systems so polar that they are able to stabilize carbocations, usually at low temperature.

14.17 ADDITIONAL PROBLEMS

PROBLEM **14.30** Explain why the trifluoromethyl group is meta directing in electrophilic aromatic substitution. Would you expect CF_3 to be activating or deactivating? Why?

PROBLEM **14.31** Show where the following compounds would undergo further substitution in electrophilic aromatic substitution. There may be more than one position for further substitution in some cases. Explain your reasoning.

Ac = $\overset{O}{\overset{\|}{C}}$—$CH_3$

PROBLEM **14.32** Predict the position of further substitution in biphenyl. Explain your reasoning.

Biphenyl

PROBLEM **14.33** Provide syntheses for the following compounds, as free of other isomers as possible. You do not need to write mechanisms, and you may start from benzene, inorganic reagents of your choice, organic reagents containing no more than four carbons, and pyridine.

(free of para)

PROBLEM **14.34** Provide syntheses for the following compounds, free of other isomers. You do not need to write mechanisms, and you may start from benzene, inorganic reagents of your choice, organic reagents containing no more than two carbons, and pyridine.

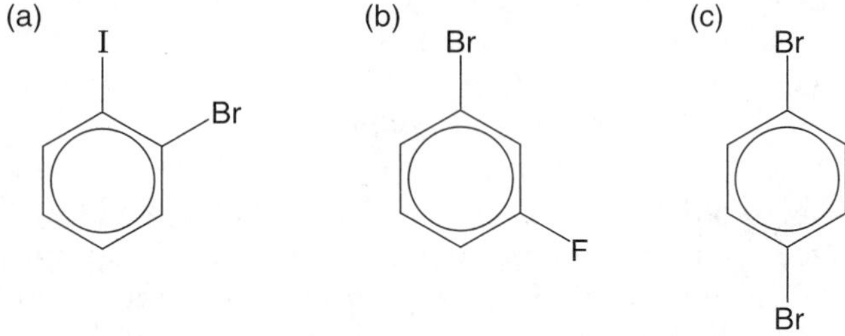

PROBLEM **14.35** Provide syntheses for the following compounds, as free of other isomers as possible. You do not need to write mechanisms, and you may start from benzene, inorganic reagents of your choice, organic reagents containing no more than three carbons, and pyridine.

PROBLEM **14.36** Provide structures for compounds **A–E**. Mechanisms are not required, but a mechanistic analysis of the various reactions will probably prove helpful in deducing the structures. *Hint*: You might look at Figure 14.49 (p. 643).

OCH_3 + (anhydride) —1. $AlCl_3$; 2. H_2O, H_3O^+→ **A** $C_{11}H_{12}O_4$

A —1. Zn/Hg/HCl; 2. H_2O→ **B** $C_{11}H_{14}O_3$

B —$SOCl_2$→ **C** $C_{11}H_{13}O_2Cl$

C —1. $AlCl_3$; 2. H_2O, H_3O^+→ **D** $C_{11}H_{12}O_2$

D —Zn/Hg, HCl→ **E** $C_{11}H_{14}O$

PROBLEM **14.37** One substituent you do not know how to attach to a benzene ring is the aldehyde group, HC=O. Here is a way. The price of this knowledge is that you must write a mechanism for the following aromatic substitution reaction.

CH_3 —CO, HCl, $AlCl_3$→ CH_3, CHO

PROBLEM **14.38** Lies! lies! lies! I can think of at least one way you *can* put an aldehyde group on a benzene ring with your current knowledge. It is a roundabout process, and so makes a good "roadmap" problem. Try this one. Provide structures for compounds **A–D**.

Benzene —Cl–C(=O)–CH_3, $AlCl_3$→ **A** —Zn/Hg, HCl→ **B**

Benzene —$AlCl_3$, CH_3CH_2Cl→ **B**

B —hν, Cl_2→ **C** —$(CH_3)_3CO^-$ Na^+→ **D** —1. O_3; 2. $(CH_3)_2S$→ O=C–H (benzaldehyde)

14.39 Write mechanisms for the following two related reactions:

CH_3, O=C, CH_3 —$HClO_4$→ CH_3, CH_3 (43%)

NH, N, O=C —75–80% H_2SO_4→ N, N (75%)

PROBLEM **14.40** Rationalize the following data. An Energy versus Reaction progress diagram will be useful. Be sure to explain why the low-temperature product is mainly naphthalene-1-sulfonic acid, and why the high-temperature product is mainly naphthalene-2-sulfonic acid. Note that naphthalene-2-sulfonic acid is more stable than naphthalene-1-sulfonic acid. You do not have to explain the relative energies of the two naphthalene sulfonic acids, although you might guess, and the answer will tell you.

concd H_2SO_4, 40 °C: SO_2OH (1-position) (major) + SO_2OH (2-position) (minor)

concd H_2SO_4, 160 °C: SO_2OH (1-position) (minor) + SO_2OH (2-position) (major)

PROBLEM **14.41** Write a mechanism for the following chloromethylation reaction:

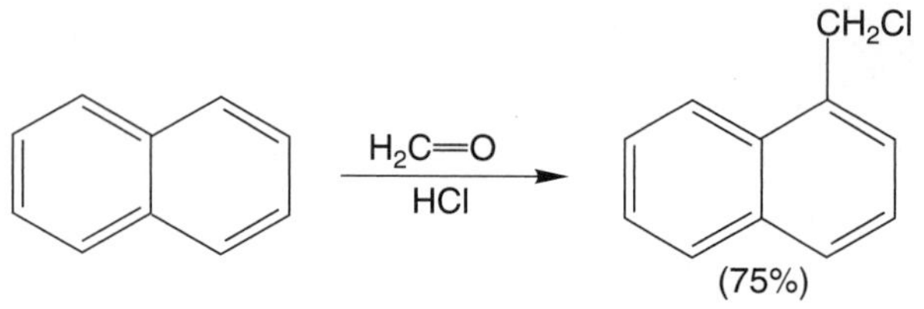

$H_2C{=}O$ / HCl; CH_2Cl (75%)

PROBLEM **14.42** The following wonderful question appeared on a graduate cumulative examination at Princeton in January, 1993. With a little care you can do it easily. Write a mechanism for this seemingly bizarre change. *Hint*: It is not bizarre at all. Think simple! First consider what HCl might do if it reacts with the starting material.

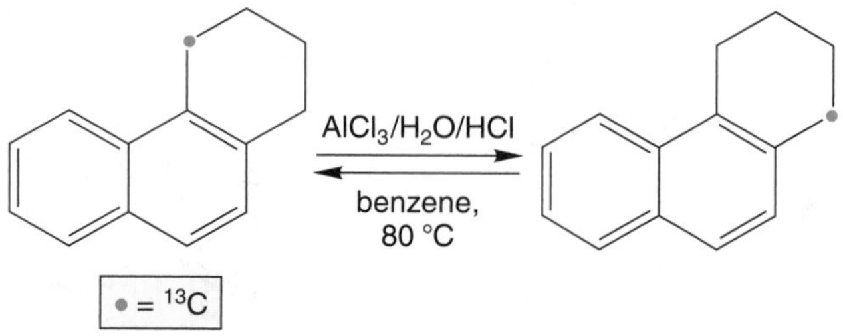

$AlCl_3/H_2O/HCl$ / benzene, 80 °C

• = ^{13}C

PROBLEM **14.43** At low temperature, only a single product (**A**) is formed from the treatment of *p*-xylene with isopropyl chloride and $AlCl_3$, which is surely no surprise, as there is only one position open on the ring. However, at higher temperature some sort of rearrangement occurs, as two products are formed in roughly equal amounts. First, write a mechanism for formation of both products. *Hint*: Note that HCl is a product of the first reaction to give **A**. Second, explain why product **A** rearranges to product **B**. What is the thermodynamic driving force for this reaction?

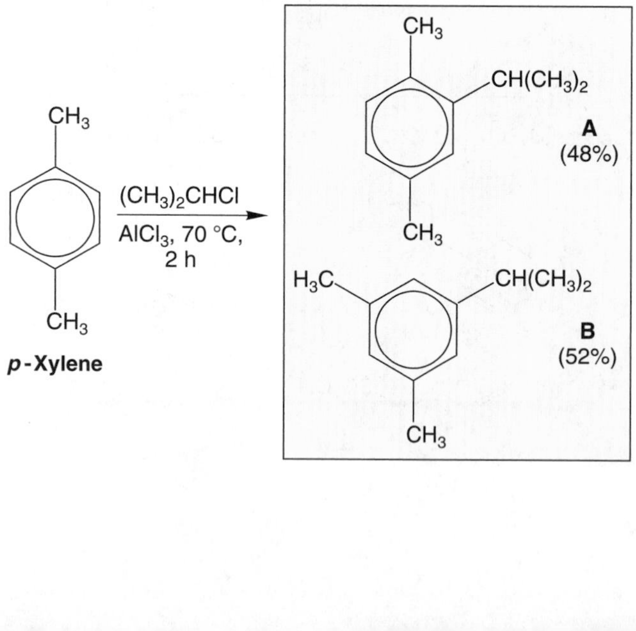

CH_3, CH_3, **p-Xylene**; $(CH_3)_2CHCl$ / $AlCl_3$, 70 °C, 2 h

CH_3, $CH(CH_3)_2$, CH_3 **A** (48%)

H_3C, $CH(CH_3)_2$, CH_3 **B** (52%)

PROBLEM **14.44** Provide mechanisms for the following reactions:

(a)

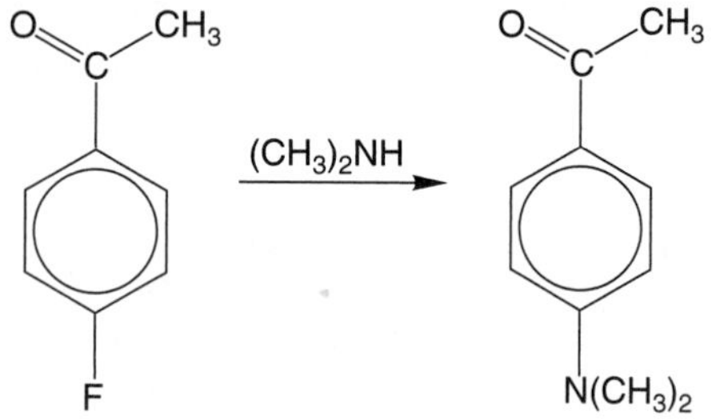

$O{=}C{-}CH_3$, F; $(CH_3)_2NH$; $O{=}C{-}CH_3$, $N(CH_3)_2$

(b)

O, C, O, O, C, O; *hν*; + CO_2

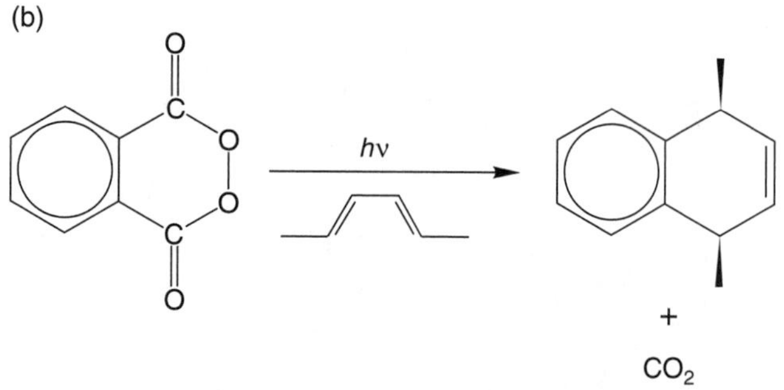

PROBLEM **14.45** Construct an Energy versus Reaction progress diagram for the reaction of 1-chloro-2,4-dinitrobenzene with sodium methoxide.

PROBLEM **14.46** When 1,2-dichloro-4-nitrobenzene is treated with an excess of sodium methoxide, three different substitution products (two monosubstituted, one disubstituted) could, in principle, be formed. In fact, only a single product is obtained.

(a) Draw the three possible products.

(b) Carefully analyze the mechanism for this nucleophilic aromatic substitution reaction. What single product will be formed and why?

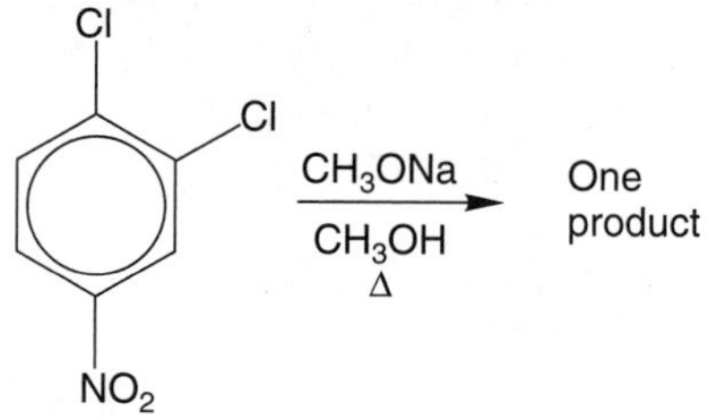

Cl, Cl, NO_2; CH_3ONa / CH_3OH / Δ; One product

PROBLEM **14.47** A useful benzyne precursor is benzenediazonium-2-carboxylate (**1**), a molecule readily formed from anthranilic acid. Upon heating, **1** affords benzyne, which can be trapped by anthracene to give the architectural marvel, triptycene (**2**).

(a) Provide an arrow formalism for the formation of benzyne on decomposition of benzenediazonium-2-carboxylate.

(b) Similarly, provide an arrow formalism for the formation of triptycene from benzyne and anthracene.

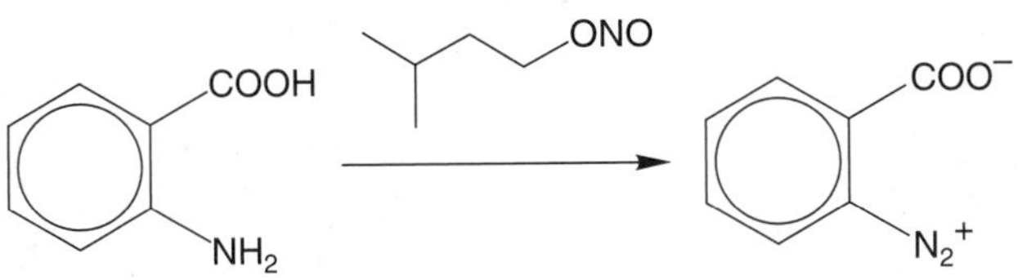

COOH, NH_2 — **Anthranilic acid**; ONO; COO^-, N_2^+ — **1**; Δ

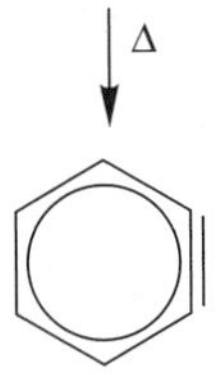

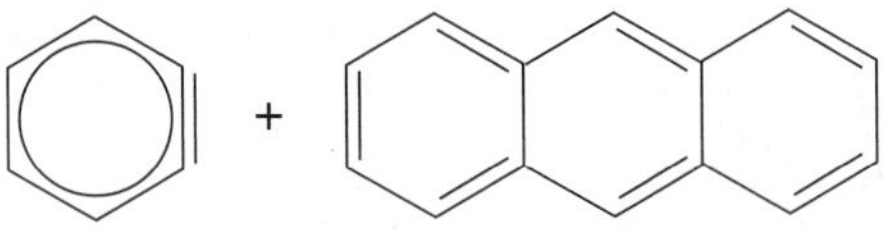

+ **Anthracene**

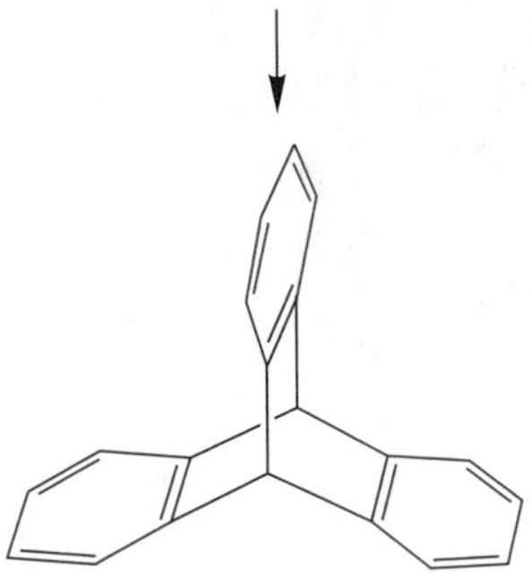

Triptycene
2

PROBLEM **14.48** Years before J. D. Roberts' classic labeling experiment (see Section 14.14, p. 674), it was known that nucleophilic aromatic substitution occasionally resulted in what was termed *cine* (from the Greek *kinema* meaning "movement") substitution. In this type of substitution reaction, the nucleophile becomes attached to the carbon adjacent to the carbon bearing the leaving group. Here are two examples of this kind of substitution:

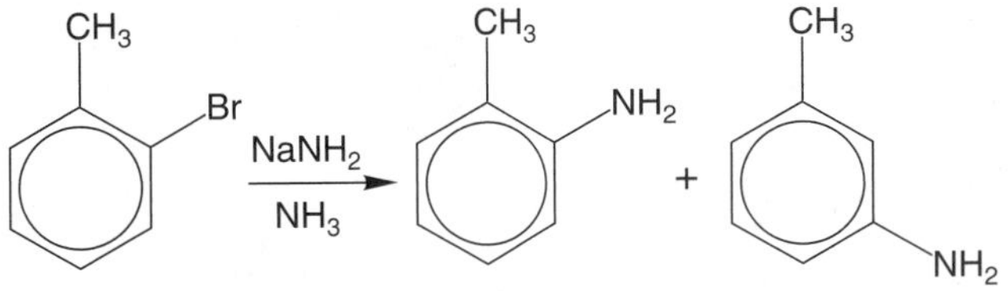

CH_3, Br —($NaNH_2$ / NH_3)→ CH_3, NH_2 + CH_3, NH_2

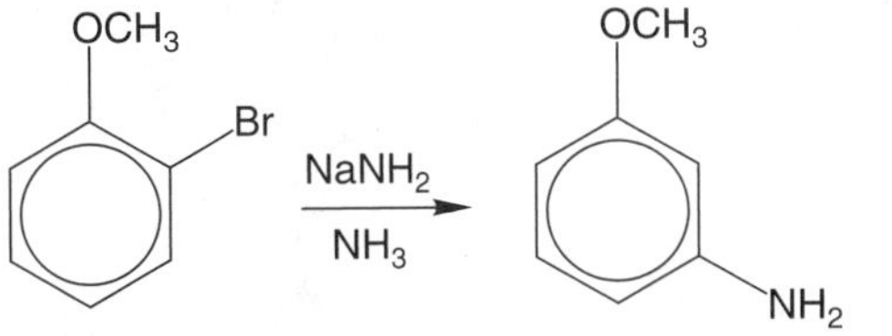

OCH_3, Br —($NaNH_2$ / NH_3)→ OCH_3, NH_2

Only isomer formed

Rationalize these *cine* substitutions through use of a substituted benzyne as an intermediate. Explain why only a single isomer is formed in the second experiment.

Analytical Chemistry

After a great expenditure of life and treasure a Daring Explorer had succeeded in reaching the North Pole, when he was approached by a Native Galeut who lived there.

"Good morning," said the Native Galeut. "I'm very glad to see you, but why did you come here?"

"Glory," said the Daring Explorer, curtly.

"Yes, yes, I know," the other persisted; "but of what benefit to man is your discovery? To what truths does it give access which were inaccessible before?—facts, I mean, having scientific value?"

"I'll be Tom scatted if I know," the great man replied frankly; "you will have to ask the Scientist of the Expedition."

But the Scientist of the Expedition explained that he had been so engrossed with the care of his instruments and the study of his tables that he had found no time to think of it.

—Ambrose Bierce*
"At the Pole," from *Fantastic Fables*

Years ago, all universities gave one or several courses in the subject of analytical chemistry. Sadly, in recent times this material has sometimes been folded into courses in physical and organic chemistry, and the subject "lost." Sometimes also lost, I fear, is the sense of how important the subject is. Nothing has influenced the development of organic chemistry more strongly over the last 30 years than progress in analytical chemistry. After a chemical reaction is run, the experimentalist is faced with two general problems before he or she can begin to think about what the results mean. The products of the reaction must first be separated, then identified. Separation was a severe problem in the early days of organic chemistry. For many compounds, the only available methods were distillation or, for solids, fractional crystallization. Fractional crystallization works well (if done properly, which it rarely is), but this method requires reasonably large amounts of material and, of course, works only for compounds that are solids at or near room temperature. By now, your laboratory experience has likely revealed that distillation is not often the final answer to securing large amounts of very pure materials. Distillation works well if there is a large difference between the boiling points of all the desired components of a mixture, and if substantial amounts of material are available. It seems that reactions in the research laboratory are all too prone to give several isomeric products, and distillation and fractional crystallization are often impractical methods. The vexing problem of separation was relieved somewhat by the advent of column chromatography, then more or less solved, at least in the research lab, by the discovery of gas and liquid chromatography. Other chromatographic techniques include thin-layer, thick-layer, and flash chromatography. Nowadays, it is the chemist's realistic expectation that any mixture can be separated. Even enantiomers are now often separable by chromatography.

*Ambrose Bierce (1842–1914?) was an American journalist and writer of bitter, sardonic tales who vanished during the Mexican civil war at the start of this century.

PROBLEM **15.1** How would a separation of enantiomers be possible? Are not all physical properties (except, of course, for the direction of rotation of the plane of plane-polarized light) identical for enantiomers? Can you guess the principle upon which the chromatographic separation of enantiomers must rest? *Hint*: What is true for enantiomers is not necessarily true for diastereomers (Chapter 5, p. 177).

Even given a successful separation, the chemist is now faced with the daunting task of working out the structures of the various components. Just how do we find out what's in that bottle? In the old days, structure determination was a respected profession, and many careers were built through the chemical determination of complex molecular structures. Diagnostic reactions were run, and the presence or absence of functional groups inferred from the results. Gradually, the results from many chemical experiments allowed the formulation of a reasonable structural hypothesis that could be tested through further chemical reactions. The intellectual content of such work was substantial, and the experimental skill required was prodigious. These days, the correlation of the results of chemical reactions with chemical structure has been largely replaced with the spectroscopic identification of compounds, and *every* chemist is an analytical chemist. The profession of analytical chemist has by no means disappeared, but rather has expanded to include us all. Ultraviolet/visible (UV/vis) spectroscopy (see Chapter 12, p. 514) was an early arrival, followed by mass spectrometry (MS), which is capable of determining molecular weights, and infrared (IR) spectroscopy, in which the vibrations of atoms in bonds are detected. At the end of the 1950s, nuclear magnetic resonance (NMR) spectroscopy appeared,* and the real revolution in the determination of the structures of organic molecules began. In this chapter, we will briefly discuss chromatography, MS, and IR spectroscopy, and then proceed to a more detailed investigation of NMR.

15.1 CHROMATOGRAPHY

In Chapter 5 (p. 179), we briefly encountered column chromatography. All chromatography works in much the same way. The components of a mixture are forced to equilibrate between an immobile material, the stationary phase, and a moving phase. The more time one component spends adsorbed on the stationary phase, the more slowly it moves. If it is possible to monitor the moving phase as it emerges from the chromatograph, each

*Professor Martin Saunders (b. 1931) told me of the arrival of the long and eagerly awaited first NMR spectrometer at Yale in the late 1950s. The magnet arrived in a plywood carton labeled in bold letters, "2.5 tons." Unfortunately, the workman who was to bring the machine into the laboratory didn't believe that such a heavy object could be in such a flimsy box and attempted to use his forklift, rated at 1 ton, to bring it in. The forklift did its best and labored mightily to lift the object into the air. It succeeded at this task, but then, apparently exhausted, gave out and dropped the magnet to the stone floor, where it burst its packing and lay blocking the entrance. The workman then attached a chain to the magnet and dragged it about 30 feet across the cobblestones to the side. Naturally Professor Saunders, whose professional fate depended to a fair extent on the performance of this instrument, was less than happy about this state of affairs. However, the story has a happy ending. The machine functioned brilliantly after minor repairs, and Professor Saunders has had a long and outstanding career at Yale. He claims that he advocated to Varian Associates, the manufacturer of the instrument, that they add a 5-foot drop to a cobblestone floor to the manufacturing process for their magnets, in order to bring all their magnets up to the standard of this one. Apparently, they ignored his advice.

component of the original mixture can be detected and collected as it appears. The efficiency of column chromatography can be greatly increased by a technique called **high-performance liquid chromatography** (**HPLC**). In this variation, high-pressure pumps force the liquid phase through a column of densely packed uniform microspheres, which provide an immense surface for adsorption. As the liquid elutes, a detector, often a small UV/vis spectrometer, monitors the composition. Figure 15.1 shows a schematic instrument as well as a nice separation of a series of organic compounds.

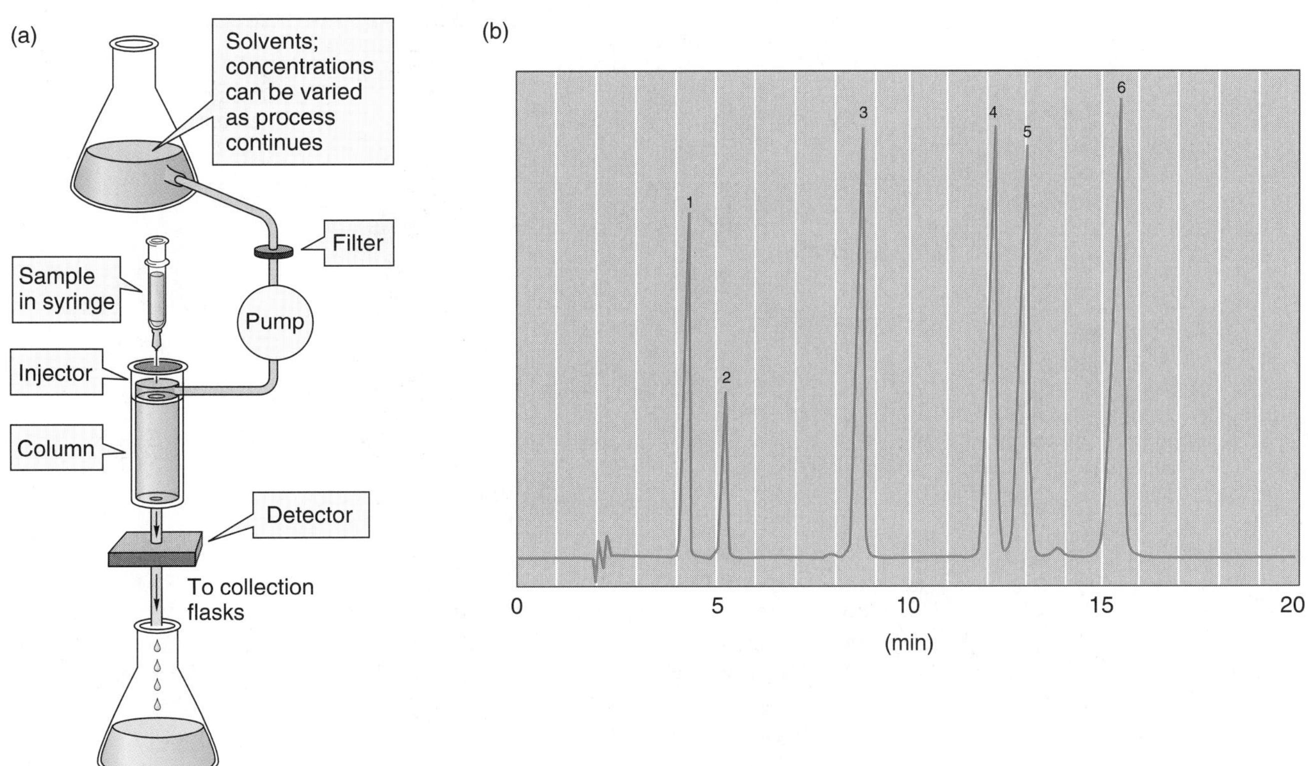

FIGURE 15.1 (a) A schematic view of a liquid chromatograph. (b) A chromatogram showing the separation of a mixture of organic compounds into its components.

For separating many organic molecules, **gas chromatography** (**GC**) is the technique of choice. Gas chromatographs are remarkably simple and it is still possible to build one on your own that will function well in a research setting. The stationary phase consists of a high-boiling material adsorbed on a solid support. High molecular weight silicone oil polymers are often used, as well as many other organic chemicals. The moving phase is a gas, usually helium or hydrogen. Typically, the solid support is packed in a coiled column through which the carrier gas flows. The mixture to be separated is injected at one end. As the carrier gas moves the mixture through the column, the various components are differentially adsorbed by the stationary phase and exist in an equilibrium between the gas phase and solution. The more easily a component is adsorbed, the more slowly it moves through the column. At the end of the device, there is a detector that plots the amount of material emerging against time. A schematic instrument and a typical chromatogram are shown in Figure 15.2.

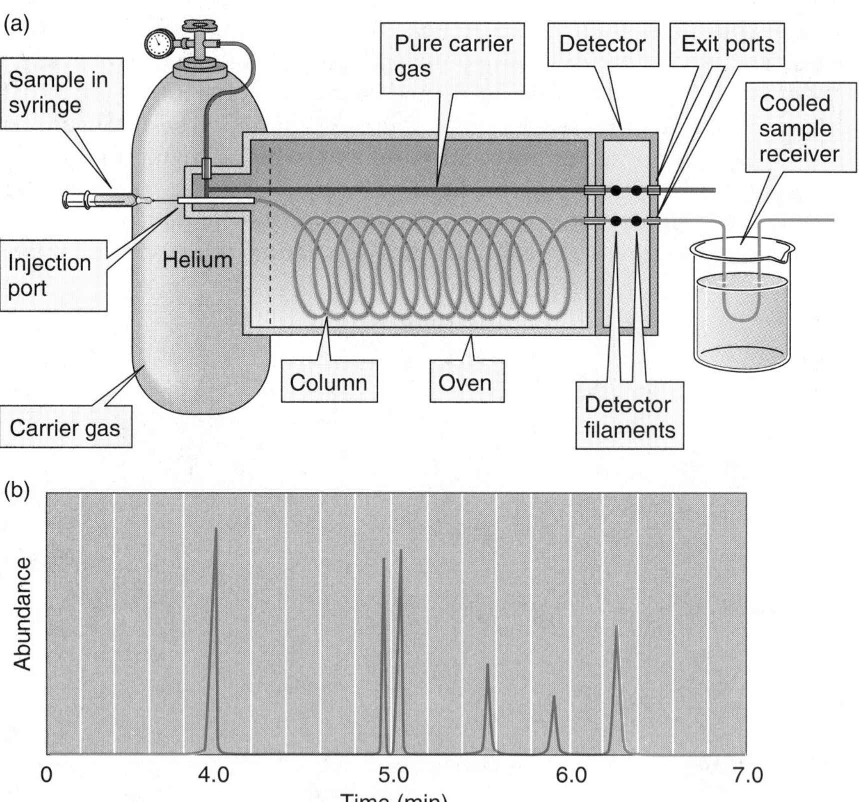

FIGURE 15.2 (a) A schematic view of a gas chromatograph. (b) A chromatogram showing the separation of a mixture of complex alkaloids into its components.

There are many kinds of detectors, but two common ones use thermal conductivity or flame ionization. In a thermal conductivity detector, the carrier gas passes over two sets of hot filaments whose temperature depends on the composition of the gas flowing over them. The detector can be set to zero for pure carrier gas passing through the instrument over both filaments. As the components of the mixture elute over one set of filaments, their temperature will change and the current needed to bring them back to the zero point can be measured and plotted. A flame ionization detector measures the current induced by ions created as the effluent is burned. Again, a zero point for pure carrier gas is determined and deviations from zero detected and plotted as the components of the mixture elute. A typical gas chromatogram is shown in Figure 15.2(b).

All sorts of fancy bells and whistles are possible, and the art of creating efficient columns has progressed immensely from the days when one packed large-bore columns by hanging them down a stairwell, and pouring in the stationary phase. The resulting columns were then tested for efficiency in the sometimes vain hope that one would do a good enough job of separating the mixture of products. One especially effective modification is the construction of capillary columns. Such columns can be several hundred feet long and contain only a tiny amount of stationary phase coated on the walls. They often have immense separating powers. Yet, for the practicing chemist, the old style simple packed columns are still very important, as they allow for the collection of the components of the mixture as they pass out of the column. Not only can the mixtures be separated, but the pure components can be isolated for identification by the simple means of attaching a cooled receiver to the end of the chromatograph (Fig. 15.2).

A particularly useful modification of a gas chromatograph is to attach it to a mass spectrometer (**GC/MS**) or infrared spectrometer (**GC/IR**) so that the various components can be analyzed directly as they emerge from the column. In the following sections, we will see what kinds of information MS and IR spectroscopy can give about molecules.

15.2 MASS SPECTROMETRY

First of all, never make the mistake of calling it "mass spectroscopy." Spectroscopy involves the absorption of electromagnetic radiation, and **mass spectrometry** (**MS**) is different, as we will see. The mass spectrometrists sometimes get upset if you confuse the issue. In MS, a compound is bombarded with very high energy electrons. Typically, electrons of 70 eV are used.

*PROBLEM 15.2

How many kilocalories per mole does 70 eV represent? See Chapter 1, page 11.

ANSWER

One electron volt (eV) is 23.06 kcal/mol, so 70 eV is 1614 kcal/mol. The point of this problem is to show that the incoming electron is carrying a *lot* of energy!

One thing these very high energy electrons can do is eject an electron from a molecule, M, to give the corresponding **radical cation**, $M^{\cdot+}$ (Fig. 15.3). Now what can happen to this radical cation? In a typical mass spectrometer, it is first accelerated toward a negatively charged plate (Fig. 15.4). There is a slit in the plate, and some of the accelerated ions do not hit the plate but pass through the slit to form a beam. The beam of ions then follows a curved path between the poles of a magnet. In a magnetic field, a moving charged particle is deflected by an amount proportional to the strength of the magnetic field. In this case, too little deflection (too weak a magnetic field) results in the ion hitting the far wall of the tunnel and in destruction of the ion. If there is too much deflection (too strong a magnetic field), the ion impinges on the near wall and is also lost. If, however, the strength of the magnetic field is just right, the ion will follow a curved path to the detector and be "seen." The magnetic field can be

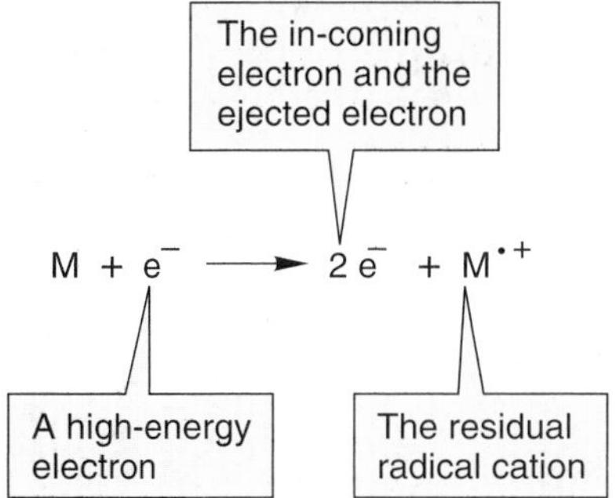

FIGURE 15.3 When bombarded with high-energy electrons, molecules can lose an electron to give a radical cation.

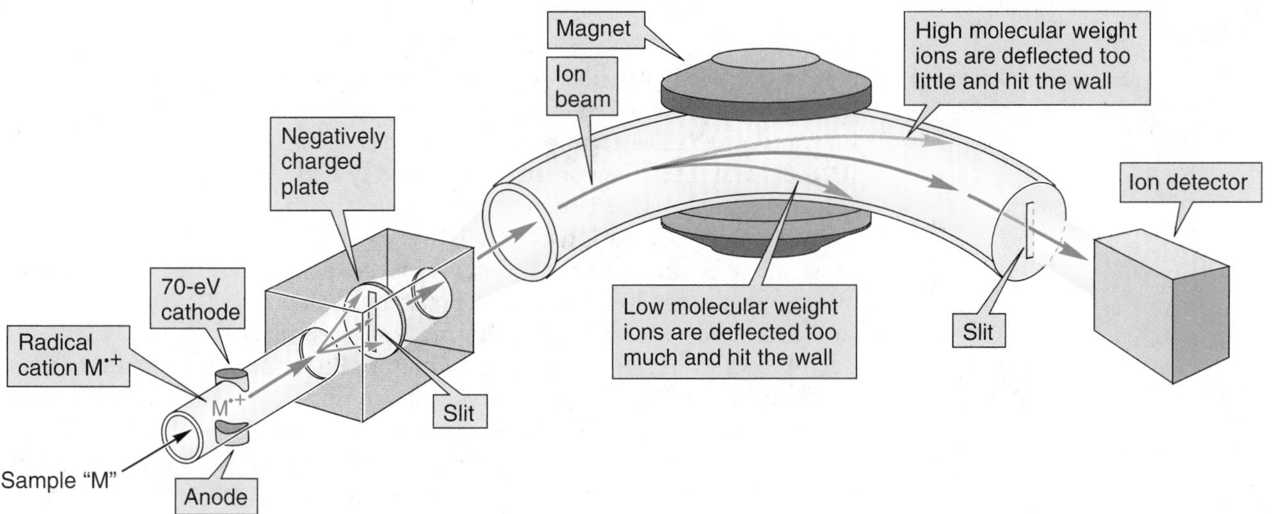

FIGURE 15.4 A schematic view of a mass spectrometer.

tuned so that ions of different mass are curved just the right amount to be detected. For ions of low mass, a relatively low magnetic field will be necessary; for ions of higher mass, the field will have to be stronger to do the required bending of the path.

Mass spectra are reported as graphs of the mass-to-charge, m/z, ratio versus intensity. As long as we are dealing with singly charged ions, the positions of the signals will correspond to the molecular weights of the ions detected.

So far, we have a device that can detect the ions, $M^{\cdot +}$, formed by bombardment of a molecule M. In our detector, we should see an ion p, the **parent ion** of mass m of M. For cyclohexane, we should see a parent ion of 84 mass units, as the molecular weight of cyclohexane (C_6H_{12}) is 84. In the mass spectrum of cyclohexane, there is a small peak above the parent ion, at $m/z = 85$. It is only about 7% of the parent ion, and therefore one must look closely in order to see it. This **p + 1 peak** is not the result of an impurity, however, as it is always there no matter how carefully the cyclohexane is purified. The reason is that carbon and hydrogen have naturally occurring isotopes, and the mass spectrometer can detect molecules containing these heavy isotopes. Table 15.1 gives the relative abundance of some common isotopes.

TABLE **15.1** Some Common Isotopes and Their Abundances

Isotope	Abundance	Isotope	Abundance
^{1}H	99.985	^{31}P	100
^{2}H (D)	0.015	^{32}S	95.02
^{12}C	98.90	^{33}S	0.75
^{13}C	1.10	^{34}S	4.21
^{14}N	99.63	^{36}S	0.02
^{15}N	0.37	^{35}Cl	75.77
^{16}O	99.762	^{37}Cl	24.23
^{17}O	0.038	^{79}Br	50.69
^{18}O	0.200	^{81}Br	49.31
^{28}Si	92.23	^{127}I	100
^{29}Si	4.67		
^{30}Si	3.10		

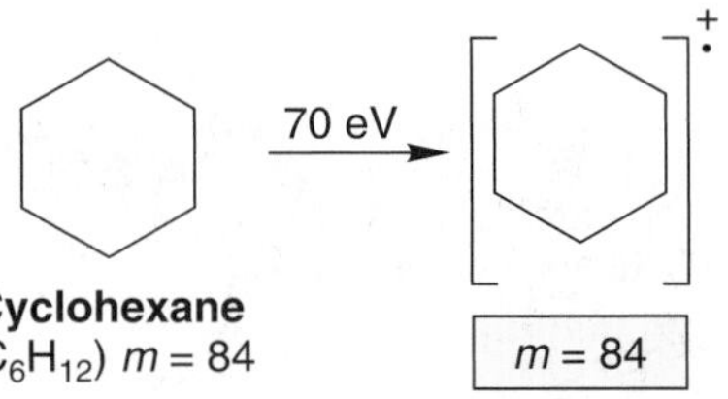

70 eV

Cyclohexane (C_6H_{12}) $m = 84$

$m = 84$

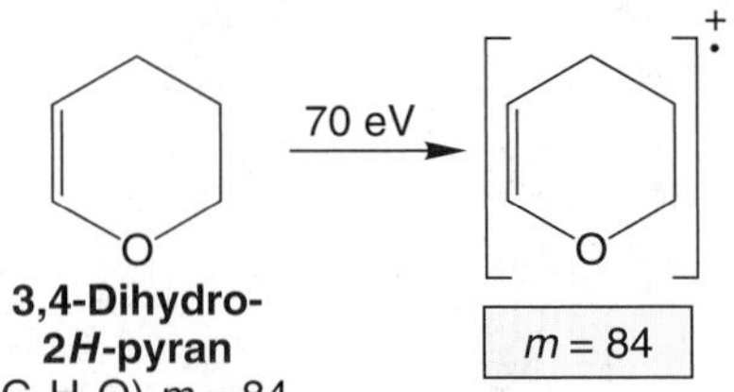

70 eV

3,4-Dihydro-2*H*-pyran (C_5H_8O) $m = 84$

$m = 84$

FIGURE **15.5** Many different ions can have a molecular weight of 84. Two possibilities are cyclohexane and 3,4-dihydro-2*H*-pyran.

For cyclohexane, there is a 1.1% probability that *each* carbon will be a ^{13}C, and therefore a 6.6% probability that the mass of cyclohexane will be one unit high. Deuterium (^{2}H or D) has a natural abundance of only 0.015%, and therefore contributes relatively little to the p + 1 peak. Sometimes the presence of isotopes is useful in the diagnosis of structure. For example, the observation of a roughly 1:1 doublet of peaks indicates the presence of bromine in the molecule, as the pool of natural bromine consists of an almost 50:50 mixture of ^{79}Br and ^{81}Br.

But how do we know that the mass spectrum with $m/z = 84$ is really cyclohexane, C_6H_{12}? If the starting material is of unknown structure, we probably do not know the formula of the molecule. The detection of a parent ion of $m/z = 84$ only tells us that the molecular weight of the unknown is 84, nothing more. Any molecule of $m = 84$ could give the parent ion. In this case, we might be looking at 3,4-dihydro-2*H*-pyran, for example ($C_5H_8O = 84$, Fig. 15.5).

Find other possible structures that might fit the observed *m/z* of 84.

PROBLEM 15.3

These days, mass spectrometers are so accurate that we can actually tell C_6H_{12} from C_5H_8O. The two species do not have the same molecular weight *if we measure molecular weights accurately enough.* Table 15.2 gives the accurate atomic weights for a variety of isotopes as well as the average values.

TABLE 15.2 Some Atomic Weights

Element	Atomic Weight	Element	Atomic Weight
^{1}H	1.008	^{19}F	19.998
^{2}H	2.010	^{28}Si	27.977
H (av)	1.008	^{29}Si	28.975
^{10}B	10.013	^{30}Si	29.974
^{11}B	11.009	Si (av)	28.086
B (av)	10.810	^{31}P	30.974
^{12}C	12.000	^{32}S	31.972
^{13}C	13.003	^{33}S	32.972
C (av)	12.011	^{34}S	33.968
^{14}N	14.003	^{36}S	35.967
^{15}N	15.000	S (av)	32.066
N (av)	14.007	^{35}Cl	34.969
^{16}O	15.995	^{37}Cl	36.966
^{17}O	16.999	Cl (av)	35.453
^{18}O	17.999	^{79}Br	78.918
O (av)	15.999	^{81}Br	80.916
		Br (av)	79.904
		^{127}I	126.905

When a molecular formula is calculated from combustion data, it is the average values in Table 15.2 that must be used, because unless there has been a deliberate introduction of an isotope during the synthesis of the material, it will be the natural isotopic mixture that appears in the compound. The mass spectrometer, on the other hand, is able to separate ions containing different isotopes. Each peak represents a particular set of isotopes, for example, $^{12}C_6{}^{1}H_{12}$ or $^{12}C_5{}^{1}H_8{}^{16}O$. So, a good mass spectrometer can give us the molecular formula of an unknown, as long as we can detect the parent ion (Fig. 15.6).

C_6H_{12}

$6\ ^{12}C = 6 \times 12.000 = 72.000$
$12\ ^{1}H = 12 \times 1.008 = 12.096$
84.096 — *m* of C_6H_{12}

C_5H_8O

$5\ ^{12}C = 5 \times 12.000 = 60.000$
$8\ ^{1}H = 8 \times 1.008 = 8.064$
$1\ ^{16}O = 15.995 = 15.995$
84.059 — *m* of C_5H_8O

FIGURE 15.6 If molecular weights are measured carefully enough, mass spectrometry can differentiate between different molecules of the nominal molecular weight of 84.

As the full mass spectra of the two *m* = 84 molecules show (Fig. 15.7), mass spectra are not simple composites of parent and p + 1 peaks. There are many lower molecular weight ions formed besides the parent and p + 1 peaks, to give a characteristic array of ions known as the **fragmentation pattern** for each molecule.

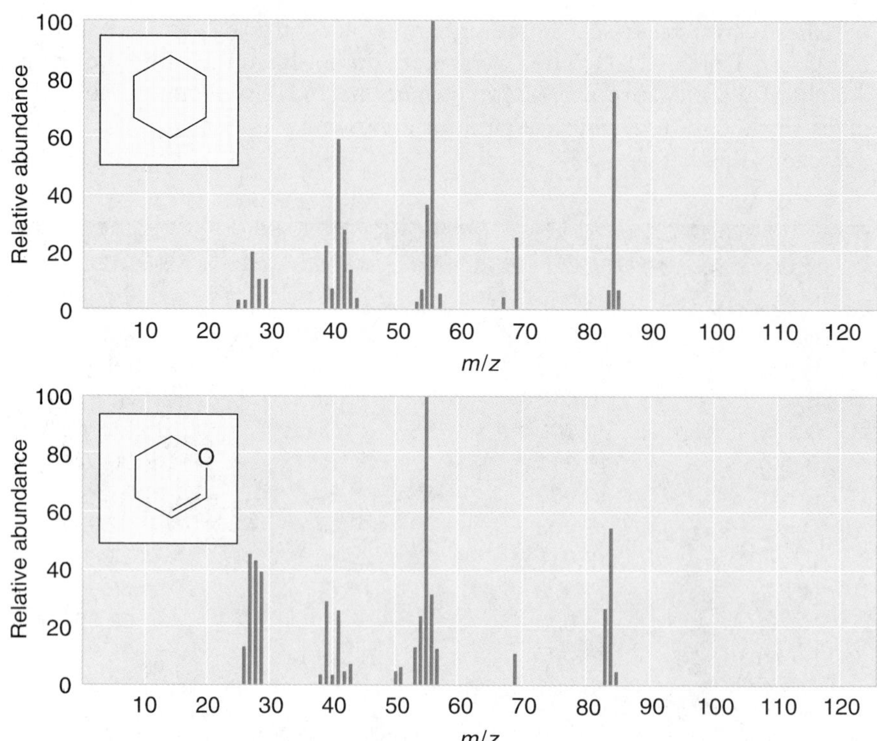

FIGURE 15.7 The mass spectra of cyclohexane and 3,4-dihydro-2*H*-pyran.

The original radical cation, $M^{\bullet+}$, can undergo many fragmentation reactions before it is accelerated into the magnetic field and passed on to the detector. The parent ion can react to give many **daughter ions**, and these appear as peaks in the final mass spectrum; the fragmentation pattern. Often these peaks are more intense than the parent ion. In all cases, the intensities of mass spectral peaks are given as percentages of the largest peak in the spectrum, which may be the parent ion, but usually is not. The largest peak is called the **base peak** of the spectrum.

Is there any way to tell how a given molecular ion will fragment? Yes and no. There is no way to predict the detailed pattern of daughter ions. Yet we can make general predictions by applying what we know about reaction mechanisms and the stabilities of ions. Fragmentations usually take place to give the most stable possible cations, and we are now able to make some reasonable guesses as to what these might be. First of all, note that there are two ways for a typical parent radical cation to produce two fragments (Fig. 15.8). When the radical cation $A{-}B^{\bullet+}$ produces A and B by fragmentation, the charge may go with either A or B. The detector can see only positively charged ions, and so any neutral species formed are invisible. In the fragmentation process, the charge will generally wind up where it is more stable, and we are often able to predict where it will go. For example, *tert*-butyl chloride gives large amounts of the *tert*-butyl cation, $m/z = 57$, in its mass spectrum, not Cl^+ $m/z = 35$ (Fig. 15.8). The charge is more easily borne on the tertiary cation than on the relatively electronegative chlorine. Nevertheless, it must be admitted that predicting is not always a straightforward process. High energies are involved, and there is plenty of energy available to do a lot of bond breaking. Rearrangements are also common, especially if a particularly stable ion can be reached.

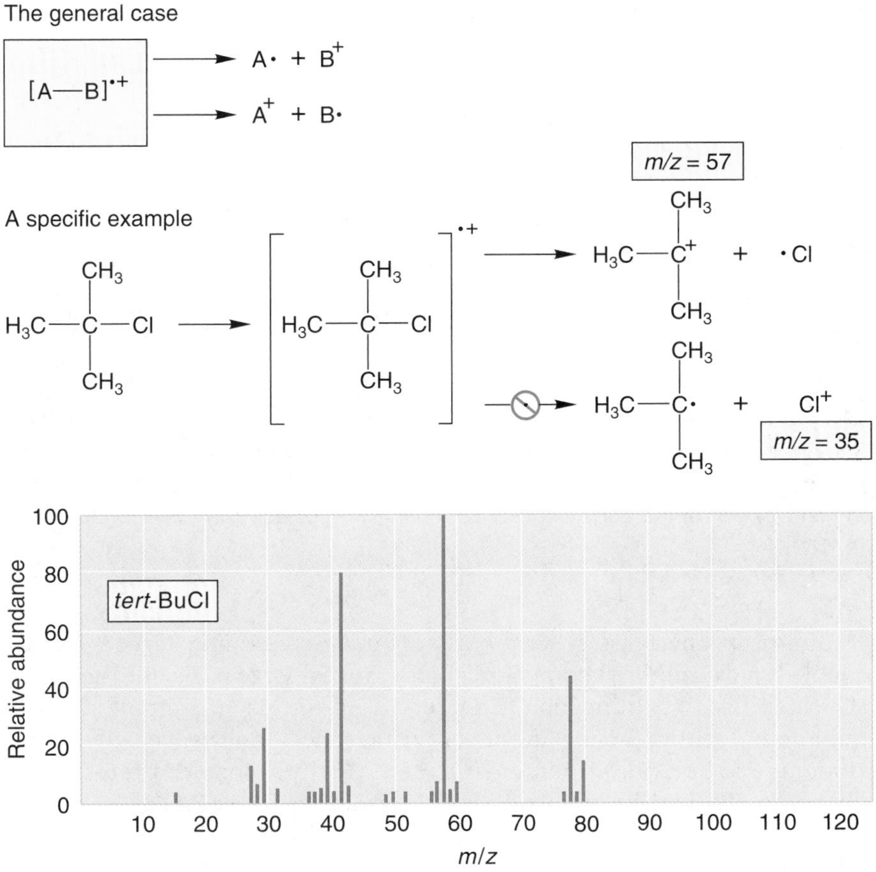

FIGURE 15.8 Two possible fragmentations for a general molecule A–B and the mass spectrum of *tert*-butyl chloride.

Some other examples will probably be useful. Figure 15.9 shows the mass spectrum of toluene.

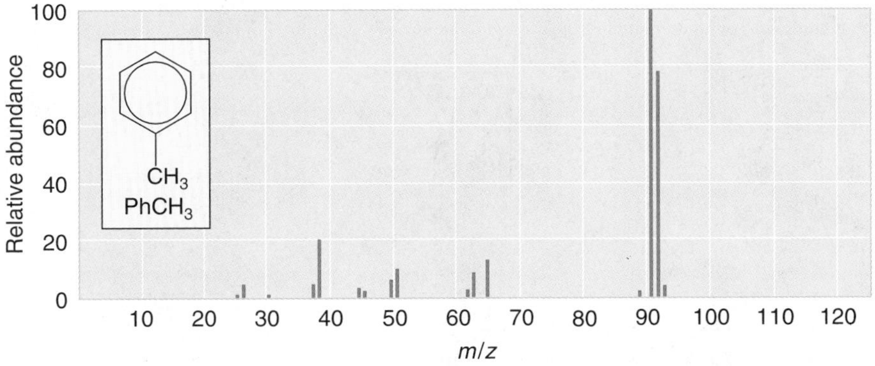

FIGURE 15.9 The mass spectrum of toluene.

The parent ion is visible as a strong peak at $m/z = 92$, but it is not the base peak. The largest peak in the spectrum is the p – 1 peak at $m/z = 91$. What can this be? A little thought brings us quickly to a reasonable answer. If a hydrogen atom (not a proton, which would be positively charged and leave a neutral fragment behind) is lost from the parent ion, it leaves behind a benzyl cation that is well stabilized by resonance (Fig. 15.10). Accordingly, loss of the hydrogen atom from the first-formed radical cation is relatively easy, and the largest peak in the spectrum belongs to $C_7H_7^+$, m/z = 91, the benzyl cation.

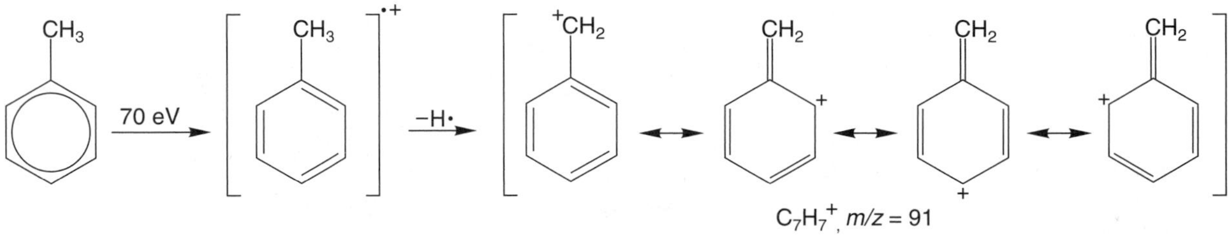

$C_7H_7^+$, m/z = 91

FIGURE 15.10 The molecular ion of toluene loses a hydrogen atom to give the benzyl cation.

PROBLEM 15.4 There is some evidence that the m/z = 91 peak may not be the benzyl cation but another species. Can you (a) think of a $C_7H_7^+$ species that might be even more stable than the benzyl cation, and (b; much harder) write an arrow formalism for the conversion of the benzyl cation into this new species? *Hint*: See Chapter 13 (p. 583).

Similarly, the mass spectra of compounds containing carbon–oxygen double bonds (carbonyl compounds) often reveal large peaks for the resonance-stabilized acylium ions (Chapter 14, p. 639). For example, 2,2,4,4-tetramethylpentan-3-one (di-*tert*-butyl ketone) ionizes to give large amounts of the acylium ion of m/z = 85. Acylium ions can break down further to give carbon monoxide and carbocations (Fig. 15.11).

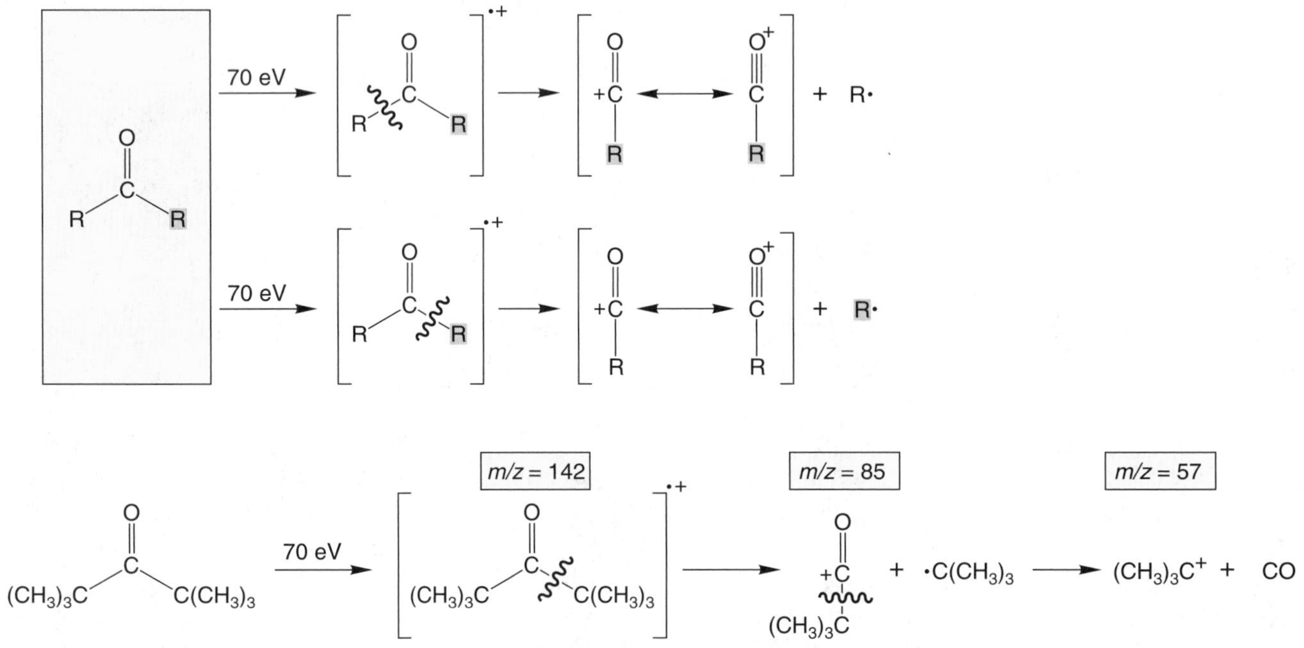

m/z = 142 m/z = 85 m/z = 57

$(CH_3)_3C$–CO–$C(CH_3)_3$ →(70 eV) $[(CH_3)_3C$–CO–$C(CH_3)_3]^{\bullet+}$ → $(CH_3)_3C$–C^+=O + $\bullet C(CH_3)_3$ → $(CH_3)_3C^+$ + CO

FIGURE 15.11 Carbonyl compounds usually give large amounts of a resonance-stabilized acylium ion.

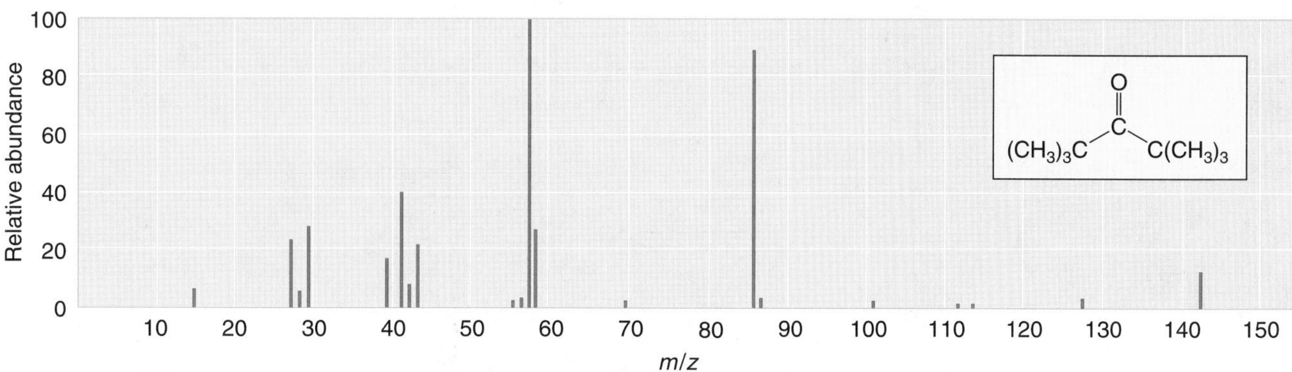

FIGURE **15.11** (CONTINUED)

Alkanes cleave so as to give the most stable possible carbocation (Fig. 15.12). For example, 2,2-dimethylpropane (neopentane) fragments to give an ion of $m/z = 57$, the *tert*-butyl cation, much more easily than it gives the far less stable methyl cation, $m/z = 15$.

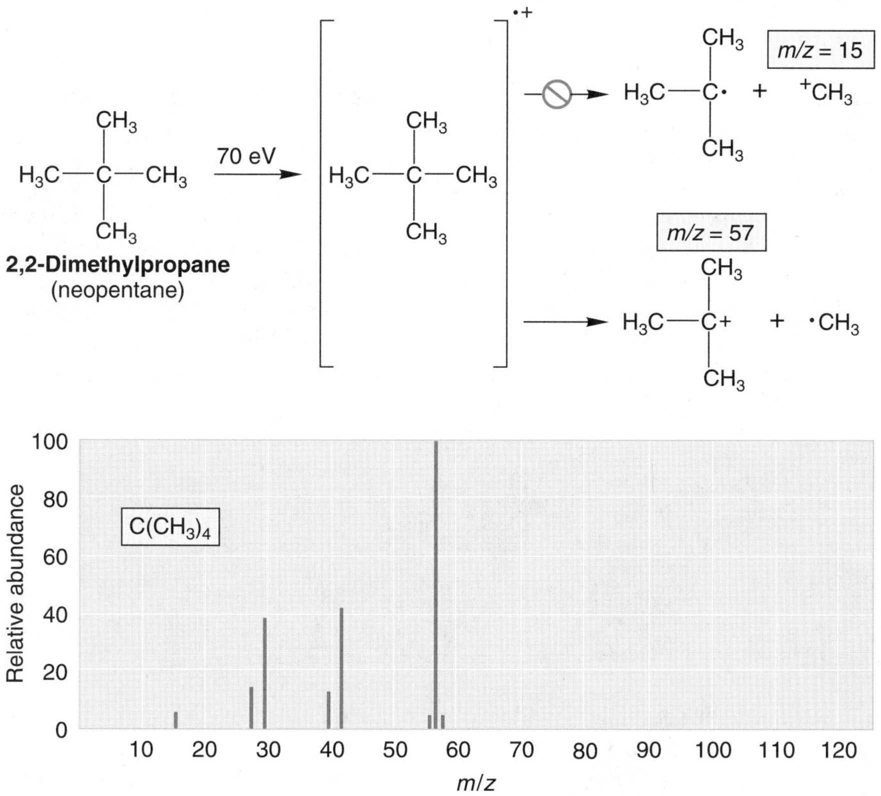

2,2-Dimethylpropane
(neopentane)

70 eV

$m/z = 15$

$m/z = 57$

$C(CH_3)_4$

FIGURE **15.12** As illustrated by the mass spectrum of 2,2-dimethylpropane, alkanes fragment so as to give the most stable carbocation possible.

There are other fragmentation reactions that resemble solution chemistry. For example, an important peak in the mass spectrum of 1-butanol is at $m/z = 56$. High-resolution analysis of this peak shows that its formula is C_4H_8. Apparently, the parent ion eliminates water to give a butene radical cation (Fig. 15.13). This behavior is typical for alcohols and has clear roots in the solution chemistry described in our sections on elimination reactions (Chapter 7, p. 273).

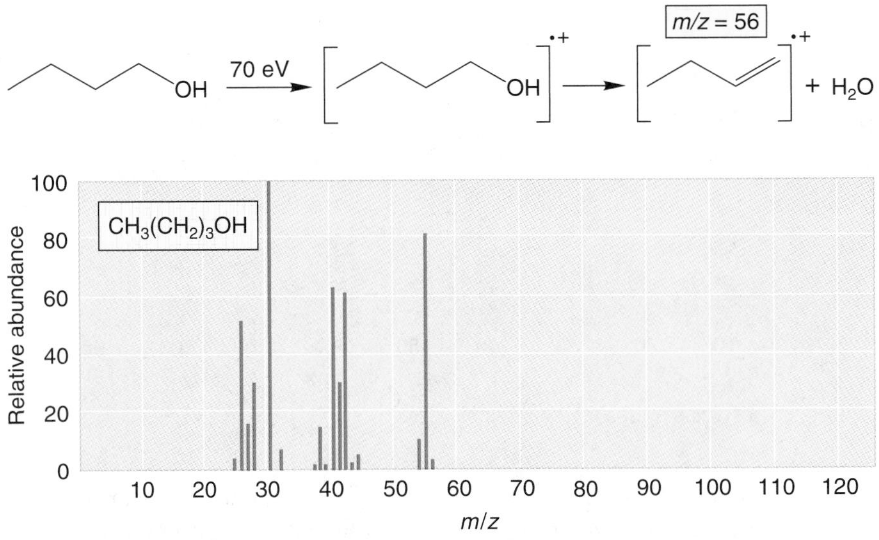

FIGURE 15.13 In the mass spectrometer, alcohols such as 1-butanol can eliminate water to give alkene radical cations.

PROBLEM 15.5 Draw a schematic molecular orbital diagram for the π orbitals of the butene radical cation, $C_4H_8^{\bullet+}$.

*PROBLEM 15.6 What do you suggest as a structure for the base peak ($m/z = 31$) in the mass spectrum of 1-butanol?

ANSWER The base peak is $m/z = 31$. The mass of 1-butanol ($C_4H_{10}O$) is 74, so some quite serious fragmentation must have taken place. A mass of 31 can only have two heavy (non-hydrogen) atoms, so it must be either a C_2 ($m = 24$) or CO ($m = 28$) compound. If the ion contained two carbons, it would have to be C_2H_7, an impossible formula, so it can only be CH_3O. The problem now is to find a reasonably stable ion of this formula that can be made from a fragmentation of 1-butanol. A simple carbon–carbon bond breaking leads to a resonance-stabilized cation that certainly is a good candidate.

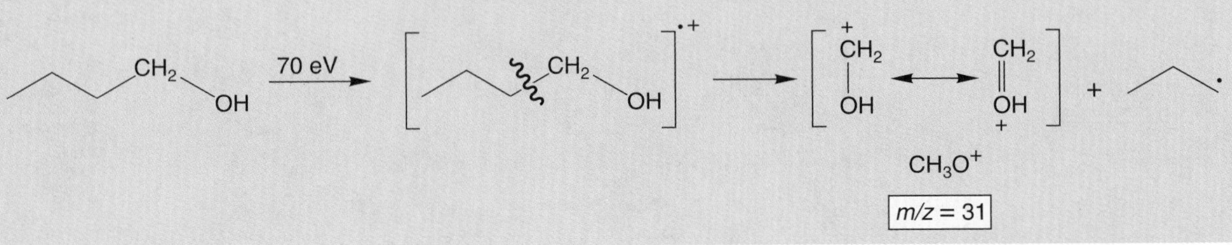

As mentioned earlier, rearrangements are common. Perhaps the most famous is named for Professor F. W. McLafferty (b. 1923) of Cornell University. He noticed that when a hydrogen was available at the γ-position, carbonyl compounds underwent what has come to be known as the **McLafferty rearrangement**. An alkene is produced as the neutral partner and the positive fragment is the radical cation of an alkenol called an enol. Note that when there is a hydrogen in the γ-position, it can be transferred to the Lewis basic oxygen atom through a relatively strain-free six-membered transition state. For example, the base (most prominent) peak in the mass spectrum of butyraldehyde is an ion of $m/z = 44$ produced by the McLafferty rearrangement (Fig. 15.14).

The general case

70 eV

An alkenol
(an enol)

A specific example

70 eV

Butyraldehyde
(butanal)

$m/z = 44$

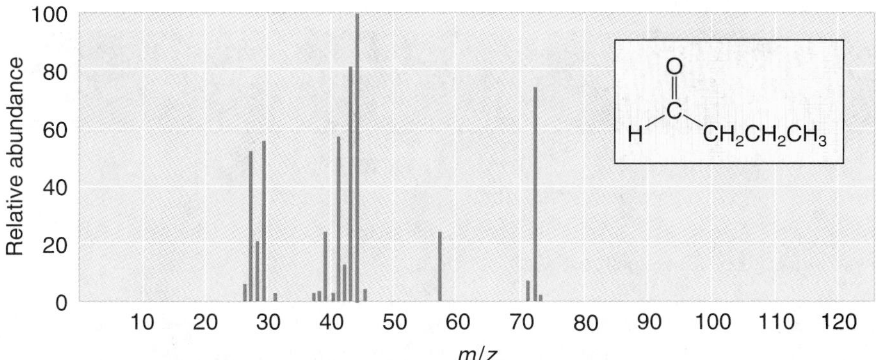

FIGURE 15.14 If there is a carbon–hydrogen bond in the γ-position, carbonyl compounds can fragment through the McLafferty rearrangement.

There are still plenty of mysteries in mass spectrometry, however. For example, my research group first made the molecule bicyclo[4.2.2]deca-2,4,7,9-tetraene ($m = 130$; Fig. 15.15) over 20 years ago, and were startled then to find that the base peak in the mass spectrum was at $m/z = 115$. It seemed likely that this ion (p – 15) represented the loss of a methyl group from the parent ion. We have been thinking about this process (not constantly!) over the ensuing two decades and still have no reasonable explanation for the loss of a methyl group.

70 eV

$m/z = 115$???

Bicyclo[4.2.2]deca-2,4,7,9-tetraene

$m/z = 130$

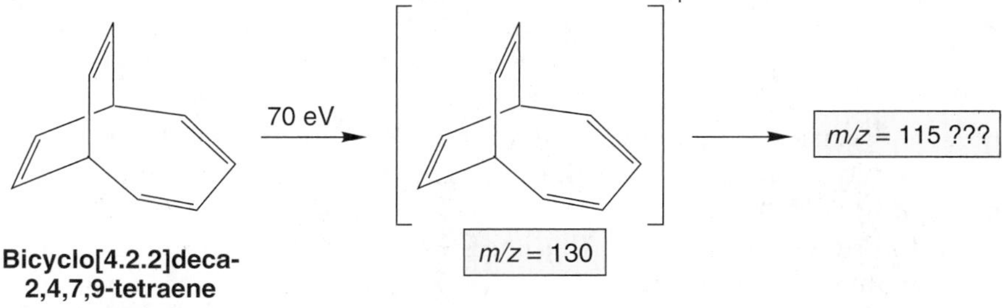

FIGURE 15.15 Bicyclo[4.2.2]deca-2,4,7,9-tetraene apparently loses a methyl group in the mass spectrometer. How?

Despite occasional mysteries, mass spectrometry is still highly useful, even in the bare bones form described here. It gives us a way to determine molecular composition through detection of a parent ion, and can give hints as to structure from the fragment ions produced.

15.3 INFRARED SPECTROSCOPY

Infrared (IR) spectroscopy is true spectroscopy—we are studying what happens when electromagnetic radiation is absorbed by a molecule. In Chapter 12, we discussed UV/vis spectroscopy, and IR spectroscopy is qualitatively similar. Figure 15.16 gives a schematic picture of the electromagnetic spectrum and identifies the regions used for UV/vis, IR, and the spectroscopy we will study in detail in the second half of this chapter, nuclear magnetic resonance (NMR).

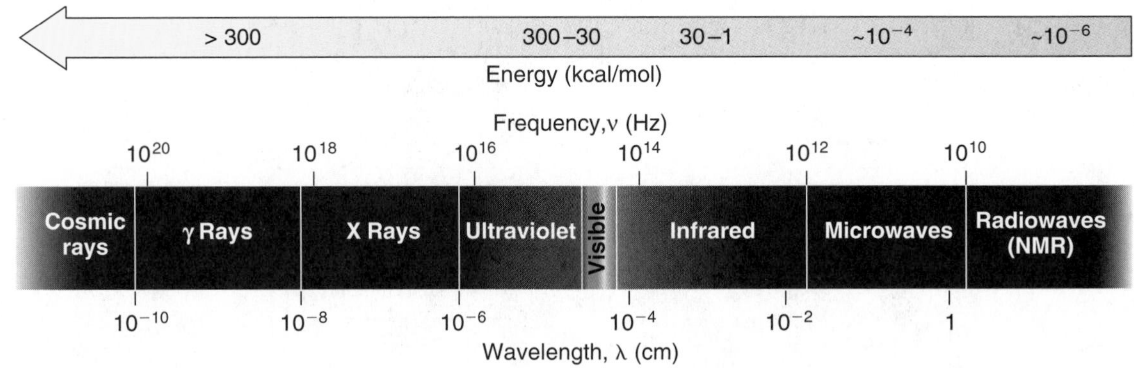

FIGURE **15.16** The electromagnetic spectrum.

The wavelength, λ (cm), is the length of a wave as measured from peak to peak or trough to trough (Fig. 15.16), and the frequency (ν) is the number of cycles passing a given point per unit time, usually seconds. So, the units of frequency (ν) are cycles per second (usually written as reciprocal seconds, or s^{-1}), or more simply, hertz (Hz). Wavelength and frequency are related by Eq. (15.1).

$$\nu = \frac{c}{\lambda} \qquad c = \text{speed of light in centimeters per second (cm/s)} \quad (15.1)$$

The quantity $1/\lambda = \tilde{\nu}$ is called the **wavenumber** and its units are expressed in terms of reciprocal centimeters (cm^{-1}). Physically, $\tilde{\nu}$ is the number of wavelengths contained in a centimeter (cm^{-1}) and is related to the more familiar frequency, ν, by the relationship in Eq. 15.2.

$$\nu = \frac{c}{\lambda} \quad \text{or} \quad \nu = c\tilde{\nu} \qquad (15.2)$$

Infrared spectra are quoted these days in wavenumbers (cm^{-1}), but in the older literature one also sees values in microns (μ). One micron (a micrometer, μm) is 10^{-4} cm.

Absorption of IR light involves energies of roughly 1–10 kcal/mol, and is associated with molecular vibrations. A chemical bond can be approximated by a picture of two weights connected by a spring. Hooke's law [Eq. (15.3)] covers weights connected by real springs.

$$\nu = k\sqrt{f\,(m_1 + m_2)/m_1 m_2} \qquad (15.3)$$

The frequency ν is related by a constant k to the masses of the weights (m_1 and m_2) and another constant, f, which is the **force constant** of the spring. Chemical bonds do behave rather like classical springs. So, like a real spring, the frequency of vibration depends on the masses of the atoms and the strength of the bond between them. Bond strength turns out to be the dominant feature, although mass does make a difference. We would expect strong bonds, with higher force constants, to absorb at higher wavenumbers. The family of carbon–carbon single bonds (~90 kcal/mol, ~1450 cm^{-1}), double bonds (~168 kcal/mol, ~1650 cm^{-1}), and triple bonds (~230 kcal/mol, ~2180 cm^{-1}) illustrates this point well.

However, not all vibrations can be detected. For a vibration to be infrared active, it must produce a net change in the dipole moment of the molecule, which means that symmetrical vibrations are weak or invisible in the IR. For example, ethylene itself shows no signal for the carbon–carbon double-bond stretch in the IR. However, Raman spectroscopy, which is related to IR spectroscopy, does detect symmetrical vibrations, and shows a band at 1623 cm^{-1} for ethylene.

15.3a The Infrared Spectrometer

A bare bones IR spectrometer simply compares the intensity of two beams of IR radiation, one of which passes through a reference cell containing only solvent and the other through a cell containing a sample dissolved in the same solvent. The signals resulting from absorptions by solvent are largely canceled, and the recorder plots the difference between the two signals as a function of frequency. Other arrangements can be made to determine the IR spectra of solid samples or liquid films. In descriptions of IR spectra, the solvent used is shown in parentheses. If no solvent is used, the sample is called neat. Spectra of solids are sometimes measured in pellets of KBr. Wherever the sample absorbs IR light, a peak will appear on the chart (Fig. 15.17).

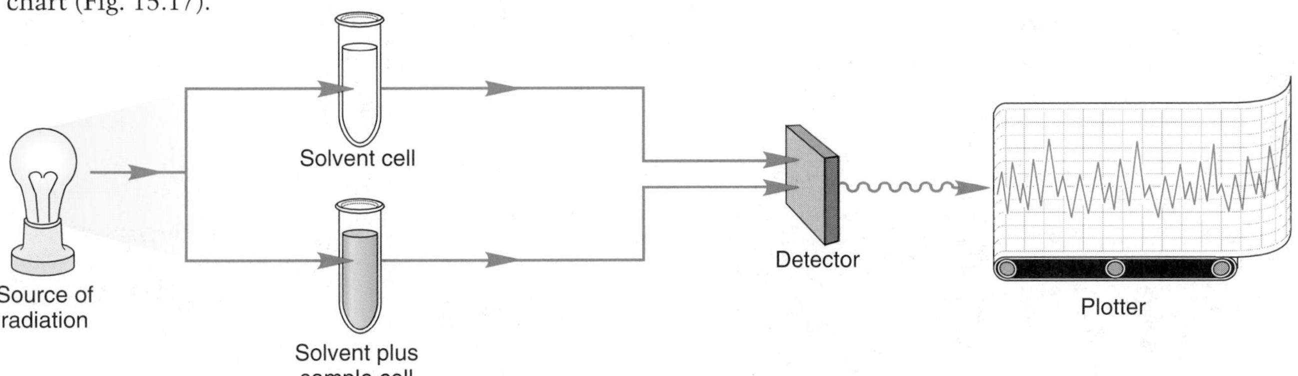

FIGURE **15.17** A schematic view of an IR spectrometer.

More sophisticated instruments are common today. Sensitivity is enhanced by pulse techniques and spectrometers are routinely attached to computers that can search for matches between the spectrum of an unknown and a library of known spectra. As with mass spectrometry, gas chromatographs can be attached to IR spectrometers and spectra can be determined as the individual components of a mixture elute from a column. This technique is called gas chromatography/infrared spectroscopy, or GC/IR.

15.3b Characteristic Infrared Absorptions

From the foregoing discussion, one might expect that IR spectroscopy would be a simple matter of noting what characteristic absorptions appear in the spectrum. An IR spectrum might be a very simple thing, composed only of signals for each kind of bond present in the sample molecule. There certainly are characteristic regions of absorption for certain bonds, but the practical situation is not quite so simple. Connected vibrating springs interact. Two springs attached to the same weight (or atom) do not vibrate independently—try it. If you set one spring in motion, the other moves as well. In fact, this is true even if the two springs (bonds) are attached to different atoms within the same molecule. The closer in energy the two vibrations are, the more strongly they affect each other. A molecule is like a collection of interacting springs, and it is often not possible to make a simple assignment of all the bands in an IR spectrum, although computational methods are becoming more proficient at this task. Most IR spectra are complicated series of bands from which we can extract information, but which we cannot completely rationalize in more than a general way. Figure 15.18 shows the IR spectra of two typical organic molecules.

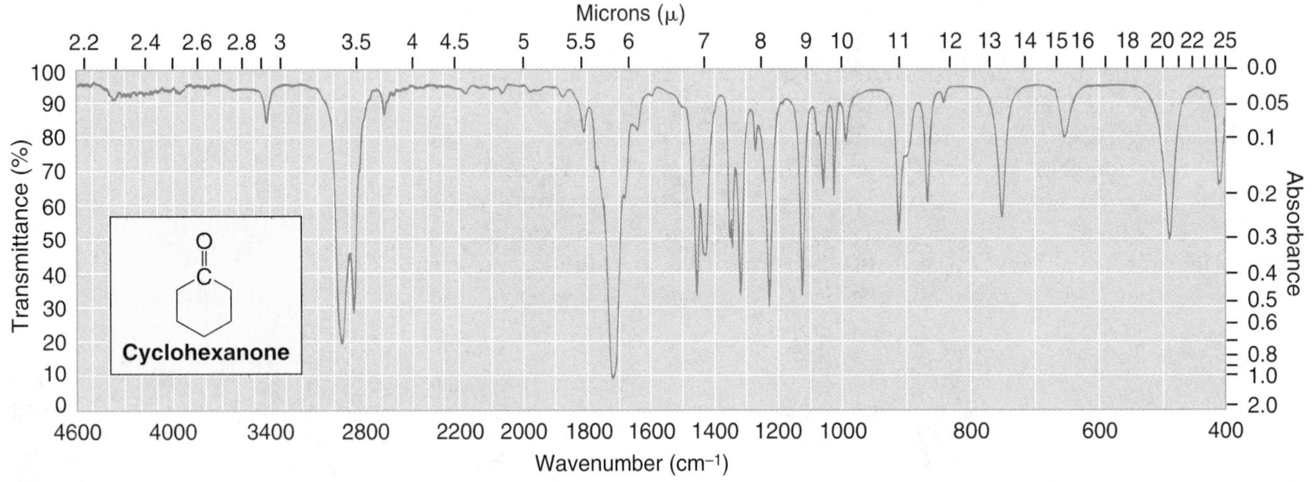

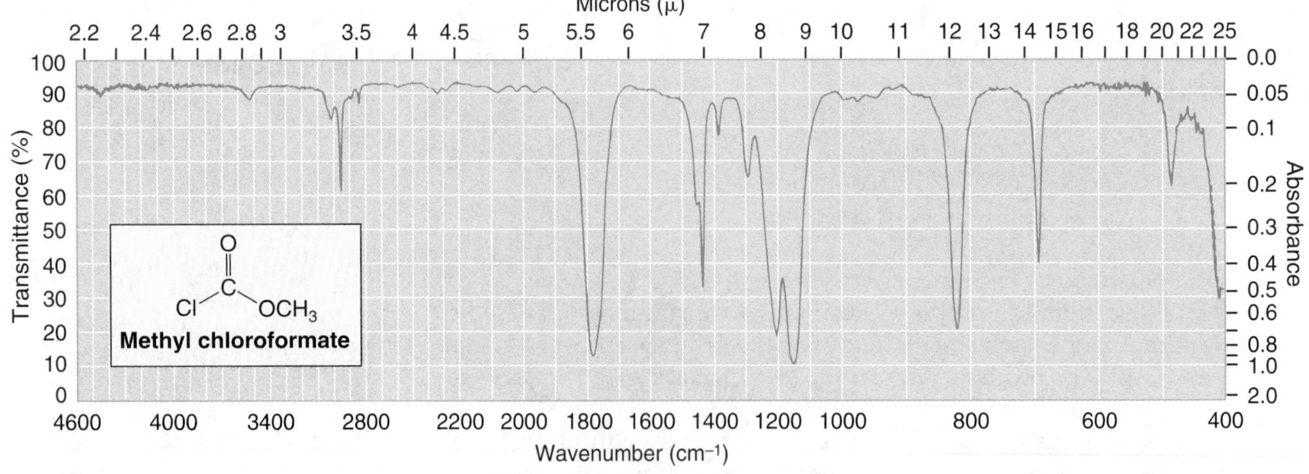

FIGURE 15.18 Two typical IR spectra.

Although this complexity may seem confusing at first, it is usually possible to gain a lot of information about the groups present in a molecule, even if we cannot assign all the bands, or draw a complete structure of the sample molecule. It is a great help in structure assignment to know what types of bonds are present in a molecule. There is also an important side benefit to these complicated IR spectra. The very complexity of the spectrum means that IR spectra of quite similar molecules are different. Each spectrum serves as a fingerprint of the molecule. If two IR spectra are identical (not similar—my old boss insisted on there being no difference greater than the width of the pen line drawn by the recorder) the compounds must be the same.

Table 15.3 gives the general positions of absorptions of a variety of functional groups, and once again we come to the question of memorization. Should one learn this chart by heart? Personally, I think not. It is important to know that this kind of general correlation exists, and you should have a rough idea of where some important functional groups absorb. If you come to use IR often you will automatically learn the relevant details of the chart, as you work out what the signals in your IR spectra tell you.

TABLE **15.3** Typical Infrared Absorptions of Functional Groups[a]

Functional Group	Position (cm^{-1})	Intensity[b]
Alkanes		
C–H	2980–2850	m–s (stretch)
C–C	1480–1420	m (bend)
Alkenes		
=C–H	3150–3000	m (stretch)
C=C	1680–1620	m–w (stretch)
(conj) C=C	1630–1600	m–w (stretch)
$R(H)C=CH_2$	995–985 915–905	s (out-of-plane bend)
$R(H)C=C(H)R$ (trans)	980–960	s (out-of-plane bend)
$R(H)C=C(H)R$ (cis)	730–665	s (out-of-plane bend) (br, variable)
$R_2C=CH_2$	895–885	s (out-of-plane bend)
$R_2C=C(H)R$	840–790	m (out-of-plane bend)

TABLE 15.3 (CONTINUED)

Functional Group	Position (cm^{-1})	Intensity[b]	
Alkynes			
≡C–H	3350–3300	s	(stretch)
C≡C	2260–2100	m–w	(stretch)
Alcohols			
O–H			
free	3650–3580	m	(stretch)
hydrogen bonded	3550–3300	br, s	(stretch)
C–O	1350–1250 1150–1050	s	(stretch)
Amines			
N–H	3500–3100 (two bands for primary amines one band for secondary amines)	br, m	(stretch)
C–N	~1200	m	(stretch)
Aromatic compounds			
=C–H	3080–3020	m–w	(stretch)
C=C	1650–1580	m–w	(stretch)
C–H			
mono	770–730 710–690	s	(out of plane bend)
ortho	770–735	s	(out-of-plane bend)
meta	900–860	m	(out of plane bend)
	810–750	s	(out-of-plane bend)
	725–680	m	(out-of-plane bend)
para	860–800	s	(out-of-plane bend)
Carbonyl compounds aldehydes, ketones			
(structure: ketone, O)			
C=O	1730–1700 (higher in strained cyclic molecules)	s	(stretch)
(structure: α,β-unsaturated ketone, O)	1680–1660	s	(stretch)
C–H (aldehydes)	2900–2700 (two bands)	w	(stretch)

TABLE 15.3 (CONTINUED)

Functional Group	Position (cm^{-1})	Intensity[b]	
Esters			
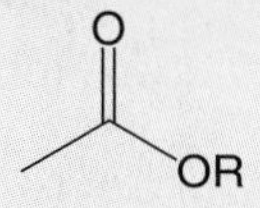	1750–1735	s (C=O)	(stretch)
	1300–1000	s (C–O)	(stretch)
Acids			
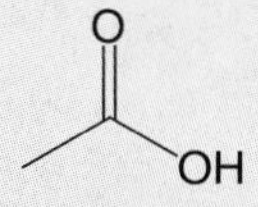	1730–1700	s (C=O)	(stretch)
	3200–2800	s, br (O–H)	(stretch)
Acid chlorides			
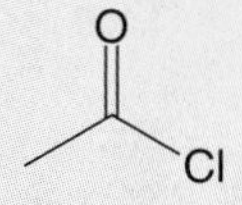	1820–1770	s (C=O)	(stretch)
Anhydrides			
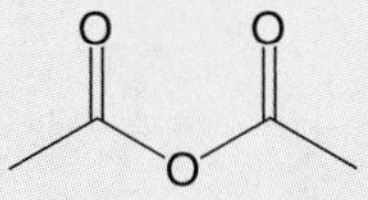	1820–1750 (two bands)	s (C=O)	(stretch)
	1150–1000	s (C–O)	(stretch)
Imines C=N	1680–1650	m	(stretch)
Cyanides (nitriles) C≡N	~2250	s	(stretch)

[a] There certainly is some subjectivity in this table, and the values represent average positions for "normal" compounds. Conjugation generally lowers double-bond stretching vibrations by about 20 cm^{-1}.

[b] Medium = m, strong = s, weak = w, broad = br.

PROBLEM 15.7

Use the data in Table 15.3 to identify the functional groups in the IR spectra of Figure 15.19. Perhaps, given the molecular formulas, you can make some guesses as to structure.

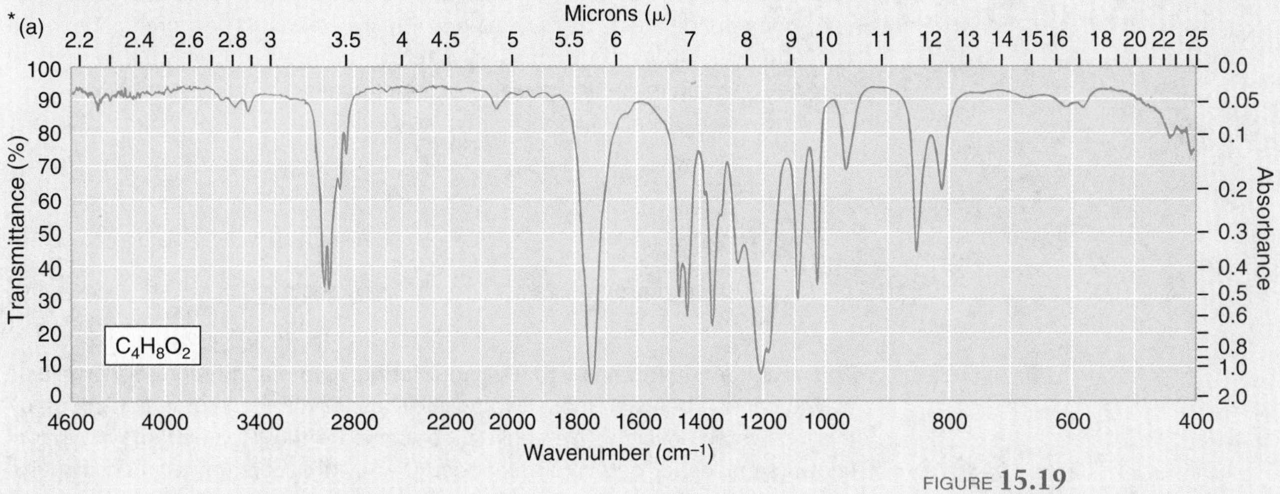

*(a) $C_4H_8O_2$

FIGURE 15.19

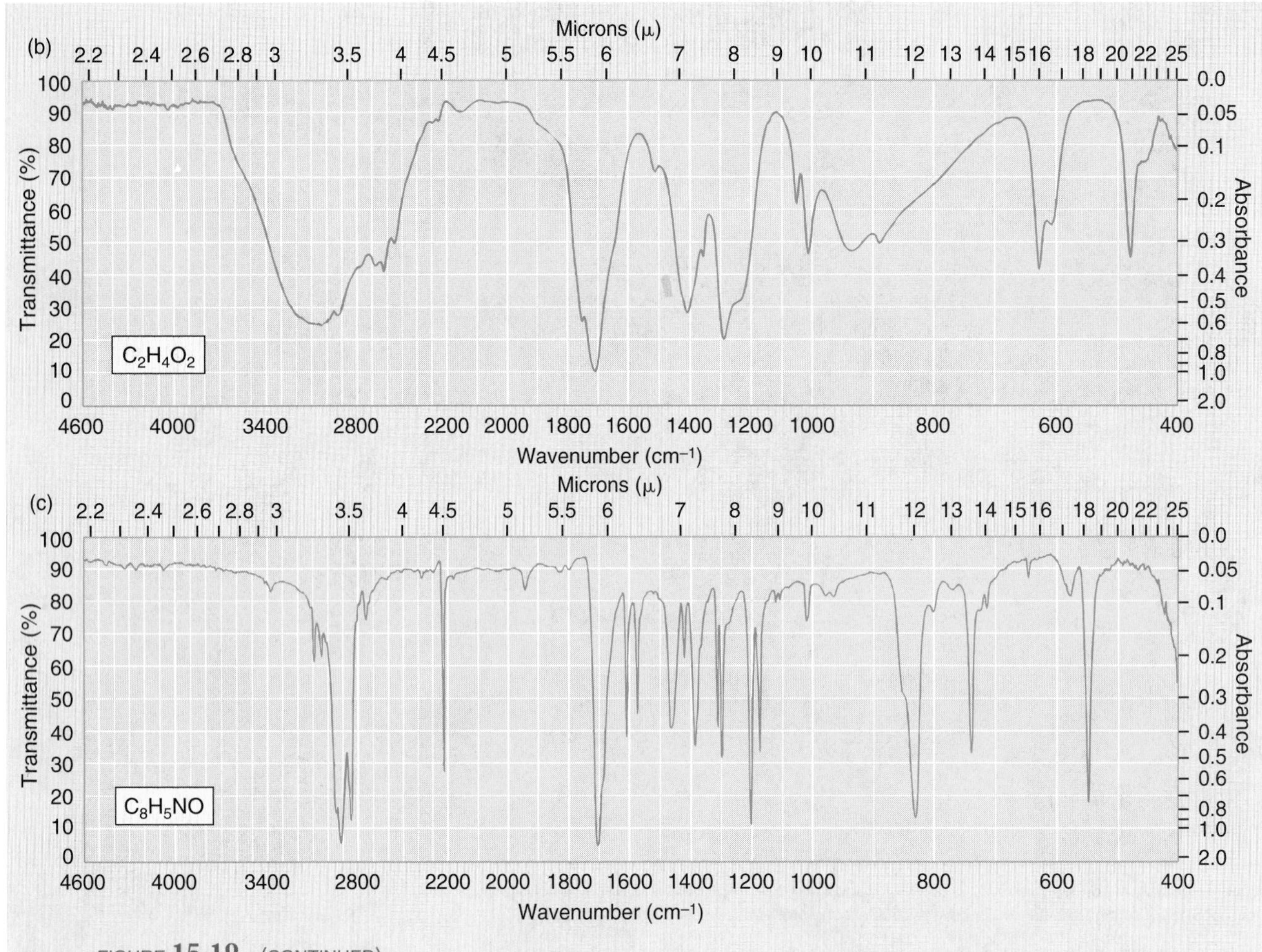

FIGURE **15.19** (CONTINUED)

ANSWER

(a) In the C—H region almost all the bands are below 3000 cm^{-1}, indicating that there are only alkane-like carbon–hydrogen bonds. No hydrogens are attached to double bonds, as they would absorb above 3000 cm^{-1}. There is a strong absorption at about 1740 cm^{-1}, too high to be an aldehyde, acyclic ketone, or acid. The compound must be an ester. Finally, there is a strong band at 1200 cm^{-1}, appropriate for the C—O stretch for an ester. Given the formula of $C_4H_8O_2$, one can make two guesses at the structure.

$H_3C-CH_2-C(=O)-OCH_3$ or $H_3C-C(=O)-OCH_2CH_3$

Methyl propionate **Ethyl acetate**

This problem tries to show that without further information there is no reliable way to choose between these two possibilities. Given the information so far, either answer is possible. We could either compare the IR spectrum with that of known samples, or obtain further spectra. An NMR spectrum would distinguish the two possibilities easily, as you will soon see. In fact, this compound is methyl propionate.

Maitotoxin

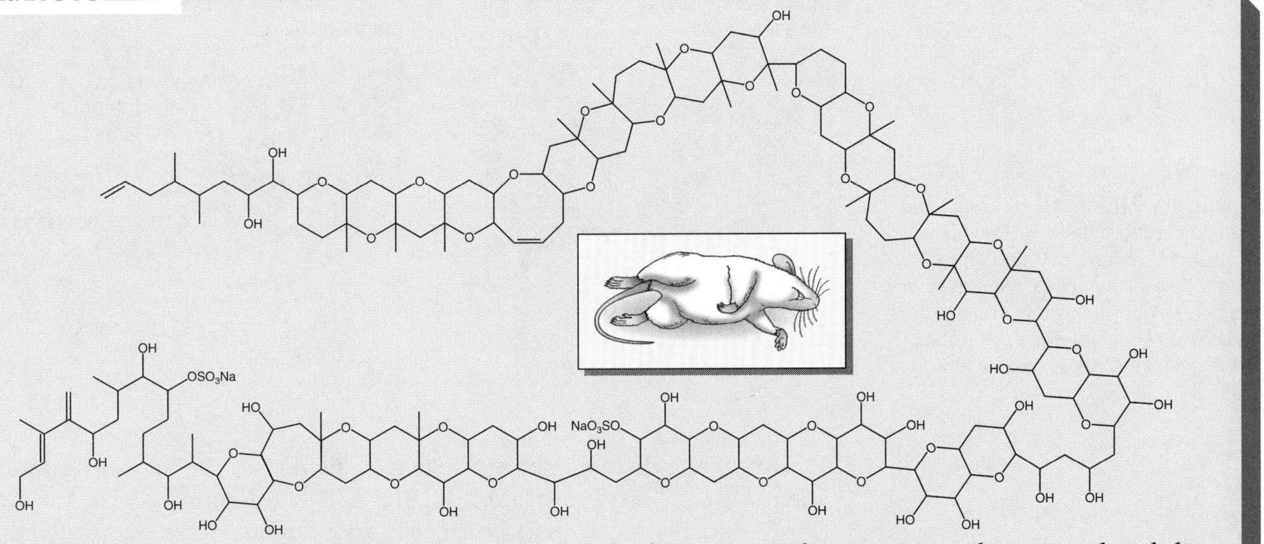

Maitotoxin, a molecule for which I maintain a certain affection, is extraordinarily lethal [for half the mice treated the lethal dose is only 50 ng/kg (LD_{50})] and is one of the largest natural products known. It is an excellent example of the frontiers of structure determination being explored by NMR spectroscopy. In 1993, a group of Japanese chemists reported a variety of sophisticated NMR studies that, together with MS, allowed them to work out the structure of maitotoxin.

15.4 NUCLEAR MAGNETIC RESONANCE SPECTROSCOPY

In contrast to IR spectroscopy, which gives an overall view of the structure of a molecule, **nuclear magnetic resonance** (**NMR**) can give a wealth of specific detail. It is the general feeling among organic chemists that given good NMR spectra of a molecule, the structure must follow. Today, the frontier of structure determination by NMR lies in immense biomolecules, and even these structures are yielding to an increasing variety of sophisticated NMR techniques.

As you can see from Figure 15.16, NMR spectroscopy involves energies much smaller than those in IR, UV, or vis spectroscopy. What happens when radiation of the wavelength 1–100 m (~10^{-6} kcal/mol) is absorbed? What change can such a tiny amount of energy induce? Even molecular vibrations require much higher energies (1–10 kcal/mol). Molecular rotations demand less energy (~10^{-4} kcal/mol), but even these motions require about 100 times more energy than that of the radio waves used for NMR. To see what happens when radio waves are absorbed by molecules, we must first go back to the beginning of our discussion of atomic structure, and learn a bit more about the structure of the atomic nucleus.

Like the electron (Chapter 1, p. 28), the nucleus has spin. For some nuclei (^{1}H, ^{13}C, ^{15}N, ^{19}F, ^{29}Si), the value of the nuclear spin (I) is ½. For others, the spin can take different values, such as 0 (^{12}C, ^{16}O), 1 (^{2}H, ^{14}N), 3⁄2 (^{11}B, ^{35}Cl), 5⁄2 (^{17}O) ... . A nonzero spin is a requirement for the NMR phenomenon. It is important that some very common isotopes have zero spin (^{12}C, ^{16}O) and will be NMR inactive. However, the most common nucleus in organic chemistry, hydrogen, is one of the nuclei with a spin of ½ and will be NMR active. The NMR effect arises as follows: Any spinning charged particle generates a magnetic field, and therefore we can think of the proton as a bar magnet. In the absence of a magnetic field, these bar magnets will be oriented randomly (Fig. 15.20).

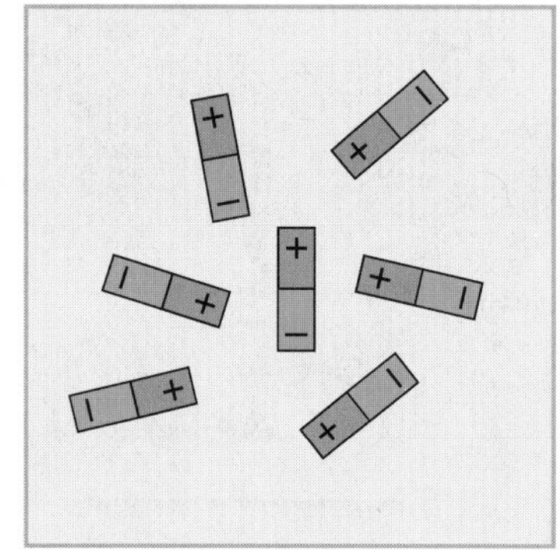

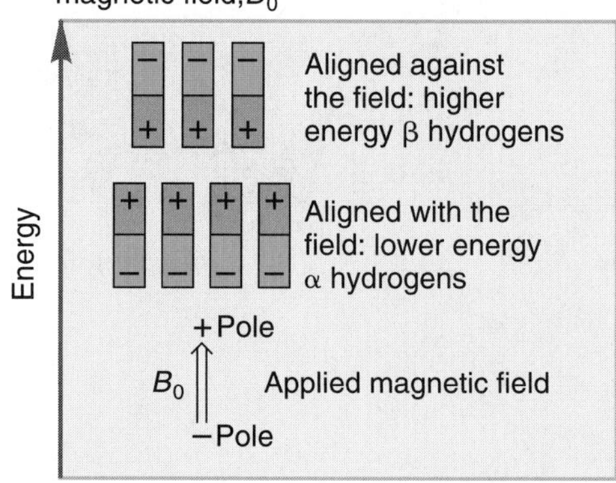

FIGURE 15.20 In the absence of an applied magnetic field, nuclear spins will be randomly oriented. When an external magnetic field is applied, alignment with the field will be slightly favored energetically over alignment against the field; there will be a small excess of molecules aligned with the field.

However, when we apply a magnetic field (B_0), the proton can either align with the field or against it (Fig. 15.20). Alignment with the field ($I = +\frac{1}{2}$, α hydrogen) will be slightly more favorable energetically than alignment against the field ($I = -\frac{1}{2}$, β hydrogen), and there will be an excess of nuclei in the lower energy, more favorable orientation. As the energy difference between the two orientations is very small, there will only be a slight excess, but it will exist, and there is the possibility of inducing transitions between the two orientations if the proper amount of energy is supplied. The function of the radio waves is to supply the energy necessary to change the orientation of the nuclear spin (often called "flipping" the spin). The disturbed equilibrium is reestablished when this absorbed energy is returned to the environment as heat (Fig. 15.21).

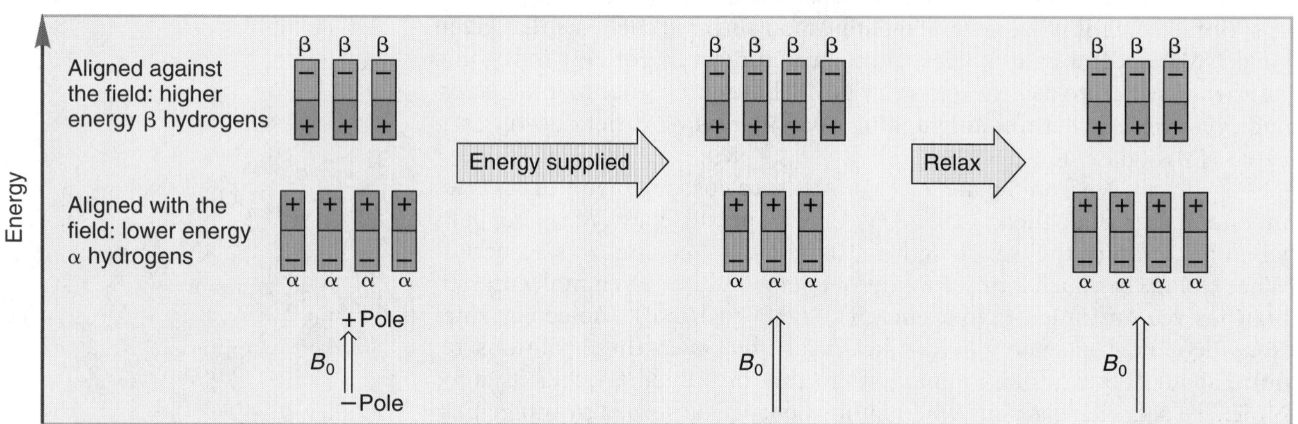

FIGURE 15.21 Absorption of energy can "flip" a nuclear spin, converting the low energy orientation into the higher one.

The magnitude of the separation of the two states—alignment with and against the applied field—depends on the strength of the applied magnetic field, B_0, as shown in Eq. (15.4) and Figure 15.22. For values of B_0 in the range commonly used, the frequencies ν for most common isotopes are in the 60–750-MHz range. The gyromagnetic ratio, γ, is characteristic of the nucleus; there is a different γ for every NMR active nucleus.

$$\nu = \gamma B_0 / 2\pi \tag{15.4}$$

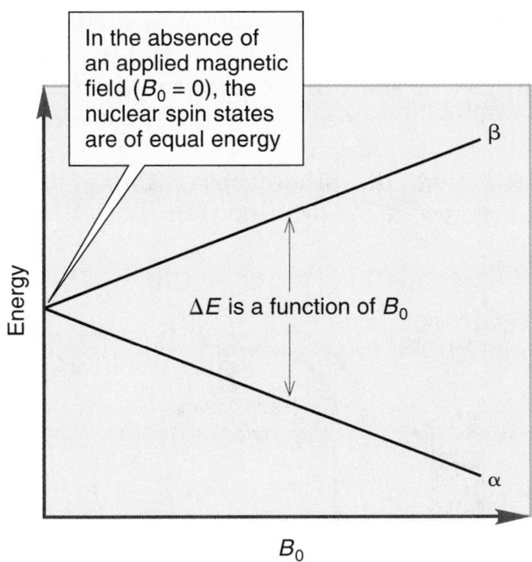

In the absence of an applied magnetic field (B_0 = 0), the nuclear spin states are of equal energy
β
Energy
ΔE is a function of B_0
α
B_0

FIGURE 15.22 The energy difference between the two spin states depends on the strength of the applied magnetic field B_0. The higher the field strength, the greater the energy difference.

If we can build an instrument capable of delivering the required radio waves and of detecting the absorption of tiny amounts of energy, we should be able to find a signal from any hydrogen-containing material. But won't there be problems from the other nuclei? Many common nuclei are NMR active, and there might be too many overlapping signals for us to make sense of the spectrum. We are saved from this fate by the gyromagnetic ratios, γ [Eq. (15.4)]. Gyromagnetic ratios are sufficiently different so that for the usual field strengths, there is a wide separation of signals from different nuclei. For example, signals from ^{13}C or ^{19}F nuclei appear far from those for ^1H.

*PROBLEM 15.8

Calculate the energy involved in an NMR transition at a field strength of 4.7 T. For H, γ is 2.7×10^8 rad T^{-1} s^{-1}, and Planck's constant is 9.5×10^{-14} kcal/mol·s.

ANSWER

Energy equals $h\nu$, Planck's constant $(h) \times$ frequency (ν): ($E = h\nu$). The frequency is given by Eq. (15.4), $\nu = \gamma B_0/2\pi$, so the useful relationship for this problem is

$$E = h\gamma B_0/2\pi$$
$$E = (9.5 \times 10^{-14})(2.7 \times 10^8)(4.7)/2(3.14) = 1.9 \times 10^{-5}$$

1.9×10^{-5} whats? Figure out the units using the same equation.

$$E = (\text{kcal/mol})(\text{s})(\text{rad T}^{-1}\text{ s}^{-1})(\text{T})/\text{rad} = \text{kcal/mol}$$

The answer is 1.9×10^{-5} kcal/mol or 0.019 cal/mol. Notice that this is *calories* per mole! These NMR transitions involve *tiny* energies.

15.4a The Nuclear Magnetic Resonance Spectrometer

A few milligrams of the sample, usually dissolved in a solvent, is placed in a tube suspended in a magnetic field. A typical solvent is deuteriochloroform ($CDCl_3$), although any material that will dissolve the sample molecule and not interfere with the NMR spectrum will do. Although the isotope deuterium (^2H) is NMR active, it will appear in a range far removed from that of hydrogen (^1H) and will not interfere with the spectrum. Normal chloroform, with its hydrogen, would interfere and cannot be used as a solvent. Two coils of wires surround the sample tube. One coil delivers radio-frequency radiation, the other detects absorption. Figure 15.23 shows a schematic picture of an NMR spectrometer.

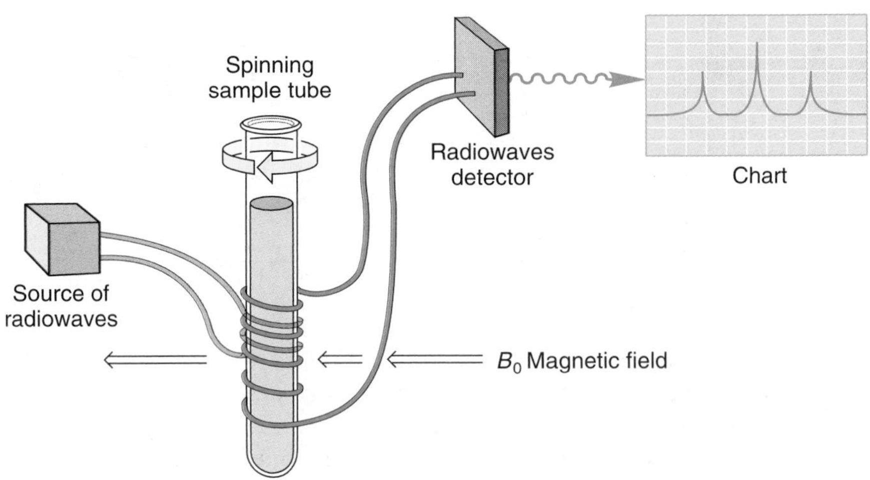

FIGURE 15.23 A schematic picture of an NMR spectrometer.

The NMR machines actually can operate in two modes. The magnetic field may be held constant and the frequency of the radiation supplied to the sample varied, or, as is more usual, the frequency may be kept constant and the magnetic field varied until resonance, absorption of energy, is achieved. In either case, the spectrum is plotted as shown in Figure 15.24.

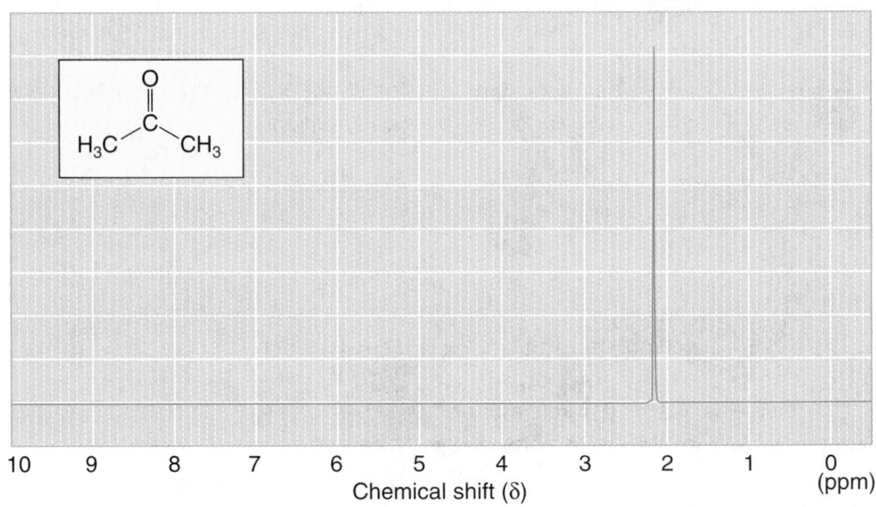

FIGURE 15.24 The 90-MHz ^1H NMR spectrum of acetone.

Resonance positions are always referenced to a standard signal, and given in parts per million (ppm) of applied magnetic field. The standard chosen is **tetramethylsilane (TMS)**, which shows a sharp signal in a position somewhat removed from most other resonance positions of hydrogens in organic molecules. The position of TMS is defined as the zero point of the **ppm scale**. Why is the somewhat odd ppm scale used? Let's assume that we are using a 60-MHz spectrometer and that the resonance frequency of the sample molecule appears at 30 Hz from TMS. If we measured the NMR spectrum of the same sample on one of the higher field instruments common today, say, 300 MHz, the resonance position would be different because the splitting between the two nuclear spin states depends on the strength of the applied magnetic field [Eq. (15.4); Fig. 15.22]. At 300 MHz, our sample hydrogen signal would appear at 150 Hz from TMS. Although the measured resonance position is different for every applied magnetic field, *the position is always the same in ppm.* In the example above, it appears at 0.5 ppm each time; $30/(60 \times 10^6)$, or $150/(300 \times 10^6)$ (Fig. 15.25). The ppm scale was devised so that chemists could report the *same* values for resonance positions regardless of the magnetic field strength of their spectrometer.

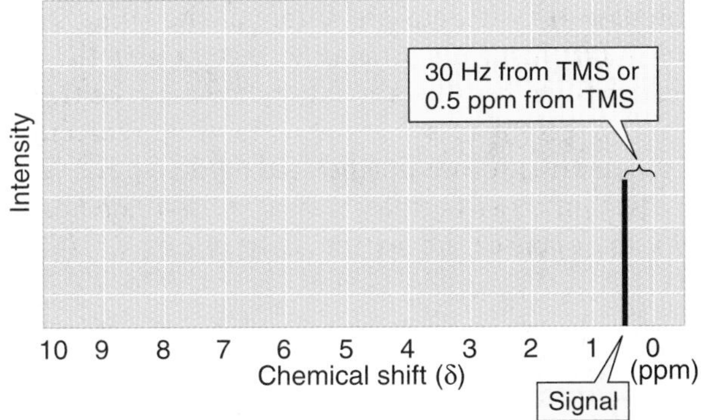

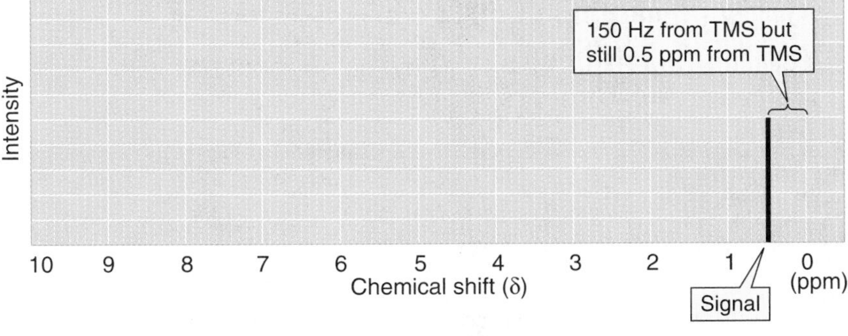

FIGURE 15.25 The ppm scale ensures that peak positions will be the same regardless of field strength.

15.4b The Chemical Shift

Up to this point, NMR spectroscopy might seem to be merely an expensive way of seeing if a compound contained hydrogen. We pay some hundreds of thousands of dollars for a 300 MHz machine, put a sample of an organic compound in it, and if we see absorption of radio waves at an appropriate frequency, we know the sample contains hydrogen. That's usually not very surprising if we are dealing with organic molecules. So what's the big deal?

It turns out that we don't see just one big signal from a hydrogen-containing molecule. Instead, there is a signal at a different position for every different hydrogen in the molecule. The resonance frequency for each hydrogen in the molecule is strongly dependent on its local environment.

*PROBLEM 15.9 The higher the applied field, the greater the dispersion between the different signals, and therefore the greater the resolving power of the instrument. This resolution is one of the reasons that chemists are willing to spend extra thousands of dollars to get extra megahertz. For example, two signals appearing at 30 and 36 Hz in a spectrum taken on a 60 MHz machine will be much better separated if the spectrum is obtained at 300 MHz. Show that this is true.

ANSWER At 60 MHz, the signals at 30 and 36 Hz are at 0.5 and 0.6 ppm, respectively [$30/(60 \times 10^6)$ and $36/(60 \times 10^6)$]. At 300 MHz, 0.5 and 0.6 ppm are 150 and 180 Hz, respectively. So, the two signals are separated by 6 Hz at 60 MHz, but by 30 Hz at 300 MHz. That's quite an improvement in separation.

We now have two questions to answer: First, we need to know how the dependence of the **chemical shift (δ)** on the local environment arises. In a sense, this is a nonquestion. If two hydrogens are different, they *must* give different NMR signals. It is in fact just a question of whether our ability to detect the difference is great enough, because the difference must be there. Electrons in molecules occupy regions of space defined by the molecular orbitals of the molecule. When an external magnetic field is applied (B_0), the electrons will circulate within those regions of space so as to create an induced magnetic field (B_i) that opposes the applied field B_0 (Fig. 15.26).

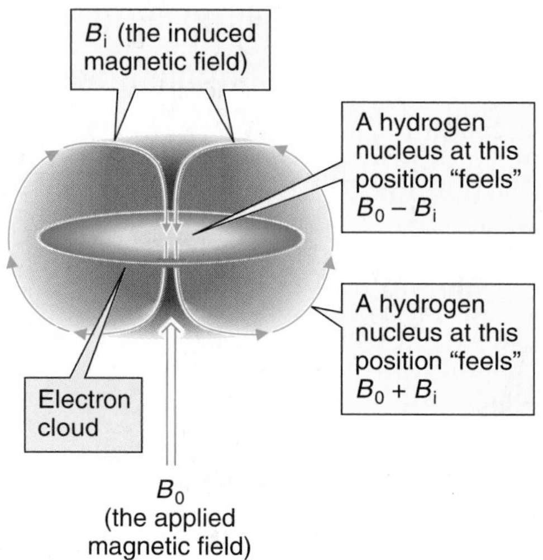
B_i (the induced magnetic field)
A hydrogen nucleus at this position "feels" $B_0 - B_i$
A hydrogen nucleus at this position "feels" $B_0 + B_i$
Electron cloud
B_0 (the applied magnetic field)

FIGURE 15.26 In an applied magnetic field (B_0), electrons will circulate so as to generate an induced magnetic field (B_i) that will oppose the applied field.

The net field ($B_0 \pm B_i$) will be slightly different at every different hydrogen in the molecule, and therefore the resonance frequency will also be different. Relative to the hydrogens in TMS, for most hydrogens it is necessary to decrease the natural B_0 for hydrogen because of the field induced by the electrons in the molecule. Such hydrogens are said to be "deshielded" by the induced field and will appear at a lower field than TMS. In a few instances, we will find that hydrogens are "shielded" and appear upfield of TMS. In these cases, the applied field, B_0, is opposed by the induced field, B_i, and it takes more applied field than usual to flip the nuclear spin.

So we can expect to see a signal for each different hydrogen in a molecule, and this leads to the second, and sometimes thorny question, What constitutes different hydrogens? There is a simple scheme for finding identical (chemically and magnetically equivalent) hydrogens. First, carry out mental substitutions of the hydrogens in question with a phantom group X. If the resulting molecules are identical, the hydrogens are also identical, or **homotopic**. Figure 15.27 gives an example using methylene chloride. Replacement of each hydrogen of methylene chloride with X (red and green in the figure) gives the same $CHCl_2X$. The two hydrogens of methylene chloride are homotopic and equivalent in all chemical and spectroscopic operations.

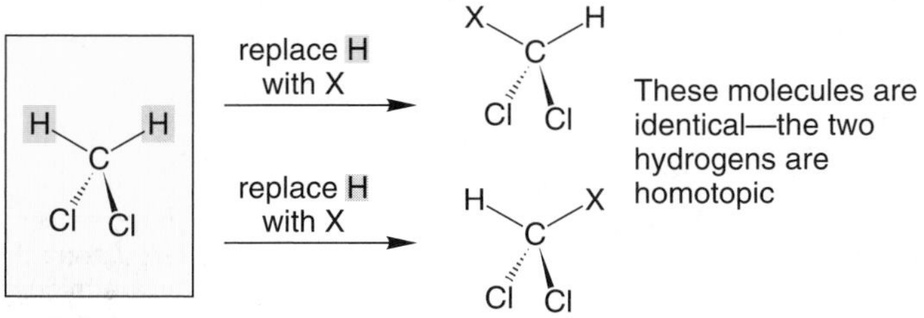

replace H with X

replace H with X

These molecules are identical—the two hydrogens are homotopic

FIGURE 15.27 Replacement of the indicated hydrogens of methylene chloride with X gives identical molecules. These methylene hydrogens are homotopic and can give only one signal.

Hydrogens in more complicated molecules are often not homotopic. Chlorofluoromethane is a good example. Replacement of the two hydrogens by the phantom X does not yield identical molecules, but enantiomers (Fig. 15.28).

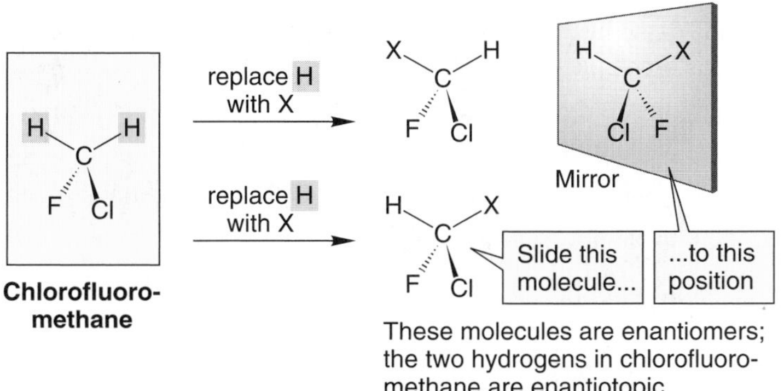

FIGURE 15.28 Replacement of the indicated hydrogens in chlorofluoromethane with X gives molecules that are enantiomers. These methylene hydrogens are called enantiotopic.

These methylene hydrogens are enantiotopic. **Enantiotopic hydrogens** are still equivalent in chemical reactions and in the NMR experiment.

*PROBLEM 15.10

Explain clearly why the enantiotopic hydrogens of chlorofluoromethane will show different signals if the NMR spectrum is obtained in a solvent consisting of a single enantiomer.

ANSWER

We will use a generic enantiomer C(ABDE) as solvent. Let's look at how this enantiomer approaches the two hydrogens in question. To make the comparison easier, we will stipulate that the enantiomer must approach A-first.

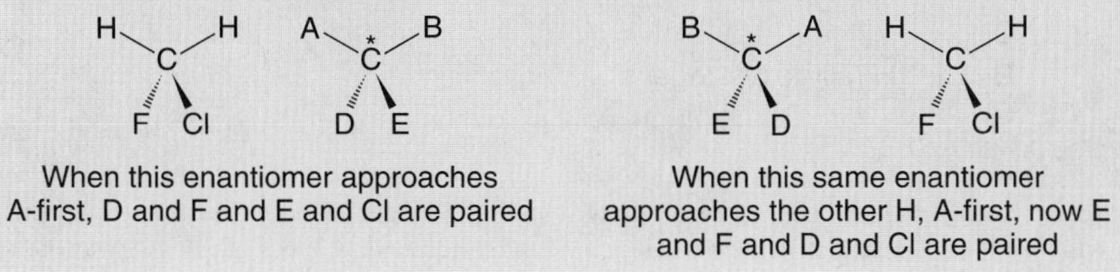

When this enantiomer approaches A-first, D and F and E and Cl are paired

When this same enantiomer approaches the other H, A-first, now E and F and D and Cl are paired

ANSWER (CONTINUED) The two hydrogens interact differently with the single enantiomer in this and all other approaches. In this solvent, the enantiotopic hydrogens will give different signals. Try another approach and show that the differences persist.

Finally, consider 1,2-dichlorofluoroethane. Now replacement of the two methylene hydrogens gives neither identical molecules nor enantiomers, but diastereomers (stereoisomers, but not mirror images, Chapter 5, p. 170; Fig. 15.29). These hydrogens are **diastereotopic**, and will react differently in all circumstances and will give different signals in the NMR spectrum of this molecule.

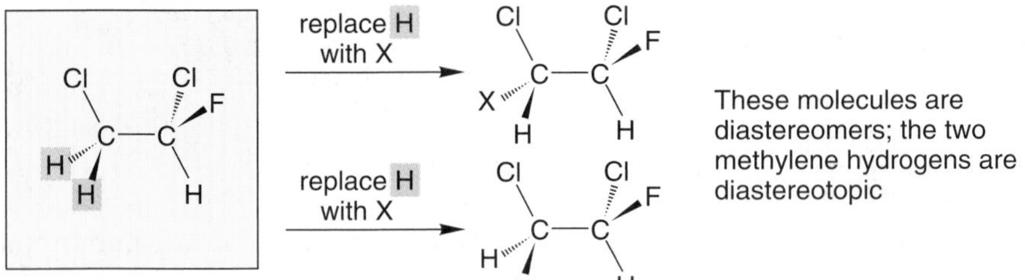

1,2-Dichloro-fluoroethane

replace H with X

replace H with X

These molecules are diastereomers; the two methylene hydrogens are diastereotopic

FIGURE 15.29 Replacement of the indicated hydrogens with X gives diastereomers. The methylene hydrogens are diastereotopic.

PROBLEM 15.11 Identify the indicated hydrogens in the following molecules as homotopic, enantiotopic, or diastereotopic (Fig. 15.30).

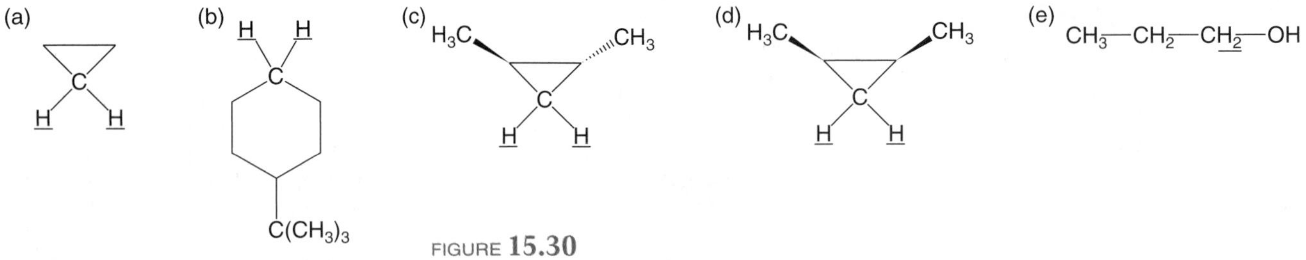

(a) (b) (c) (d) (e) CH_3—CH_2—CH_2—OH

FIGURE 15.30

There is a general correlation between the resonance frequency of hydrogens in similar positions in different molecules. All methyl groups attached to double bonds appear at roughly the same position, all aromatic hydrogens at another, and so on.* Table 15.4 gives a correlation chart for the peak positions of many different kinds of hydrogen.

TABLE 15.4 Approximate Chemical Shifts of Various Hydrogens[a,b]

Hydrogen	δ (ppm)
CH_3	0.8–1.0
CH_2	1.2–1.5
CH	1.4–1.7
C=C–CH (allylic hydrogens)	1.8–2.3

(Continued on next page)

Convention Alert!

*To say "aromatic hydrogen" is certainly wrong; the hydrogen isn't aromatic, it is the ring to which it is attached that is. Nonetheless, one must admit that it is slightly awkward to say "hydrogens attached to aromatic systems", and this kind of loose talk is common. One sees vinyl or olefinic hydrogen, cyclopropyl hydrogen, and so on.

TABLE 15.4 (CONTINUED)

Hydrogen	δ (ppm)
O=C–CH	2.0–2.5
Ph–CH (benzylic hydrogens)	2.3–2.8
≡C–H	2.5
R_2N–CH	2.0–3.0
I–CH	2.8–3.3
Br–CH	2.8–3.5
Cl–CH	3.1–3.8
F–CH	4.1–4.7
O–CH	3.1–3.8
=CH_2 (terminal alkene)	5.0
C=CH (internal alkene)	4.5–5.5
Ph–H (aromatic hydrogens)	7.0–7.5
O=CH (aldehyde hydrogens)	9.0–10.0
RCOOH	10–13

[a]These values are approximate. There will surely be examples that lie outside the ranges indicated. Use them as guidelines, not "etched in stone" inviolable numbers.

[b]Watch out for loose talk. For example, "aromatic hydrogen" means a hydrogen attached to a benzene ring.

It is also possible to estimate the number of hydrogens giving rise to a particular peak by measuring the relative intensities of the signals in a spectrum. This process is called **integration** of a signal. As an example, look at the ^{1}H NMR spectrum of 4,4-dimethyl-2-pentanone in Figure 15.31.

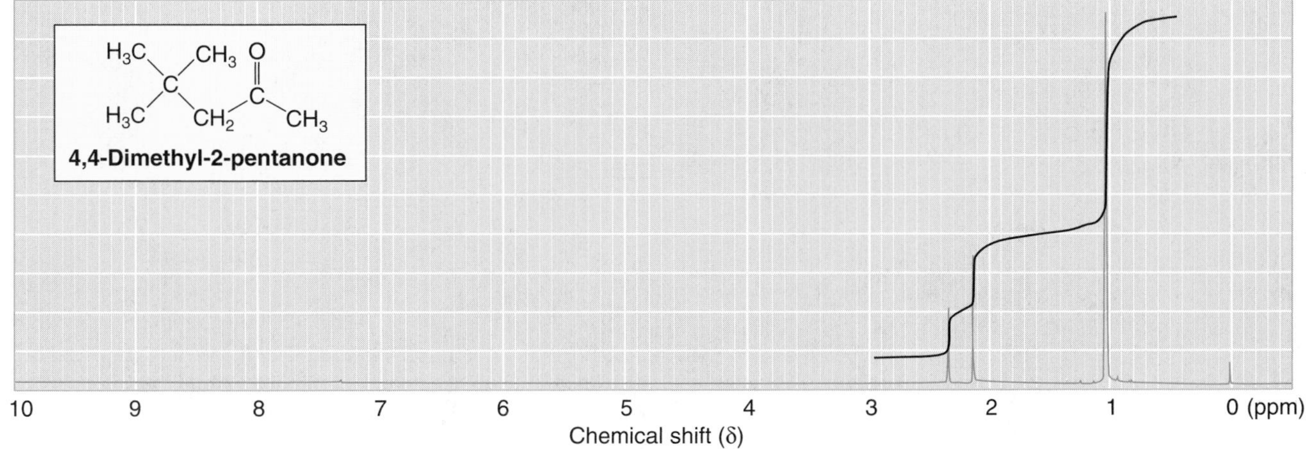

FIGURE 15.31 The ^{1}H NMR spectrum of 4,4-dimethyl-2-pentanone.

Three signals appear, and the integral shows that they are in the ratio 9:3:2. These days, integrals are calculated electronically and printed out as numbers on the chart paper. In older spectra, one sees a line traced by a pen, as in Figure 15.31. The integral is determined by measuring the vertical (not horizontal or diagonal) distance the pen moves as the signal appears. In this spectrum, the assignment is easy: The nine hydrogens of the *tert*-butyl group give rise to the highest field signal, the methyl group produces the signal for three hydrogens and the methylene group gives the peak for two hydrogens. There is no difficulty in making assignments for this simple example, but you should check the positions of the peaks against those predicted by Table 15.4 to be certain there is no error.

*PROBLEM **15.12** Which of the molecules in Figure 15.32 give rise to which spectra?

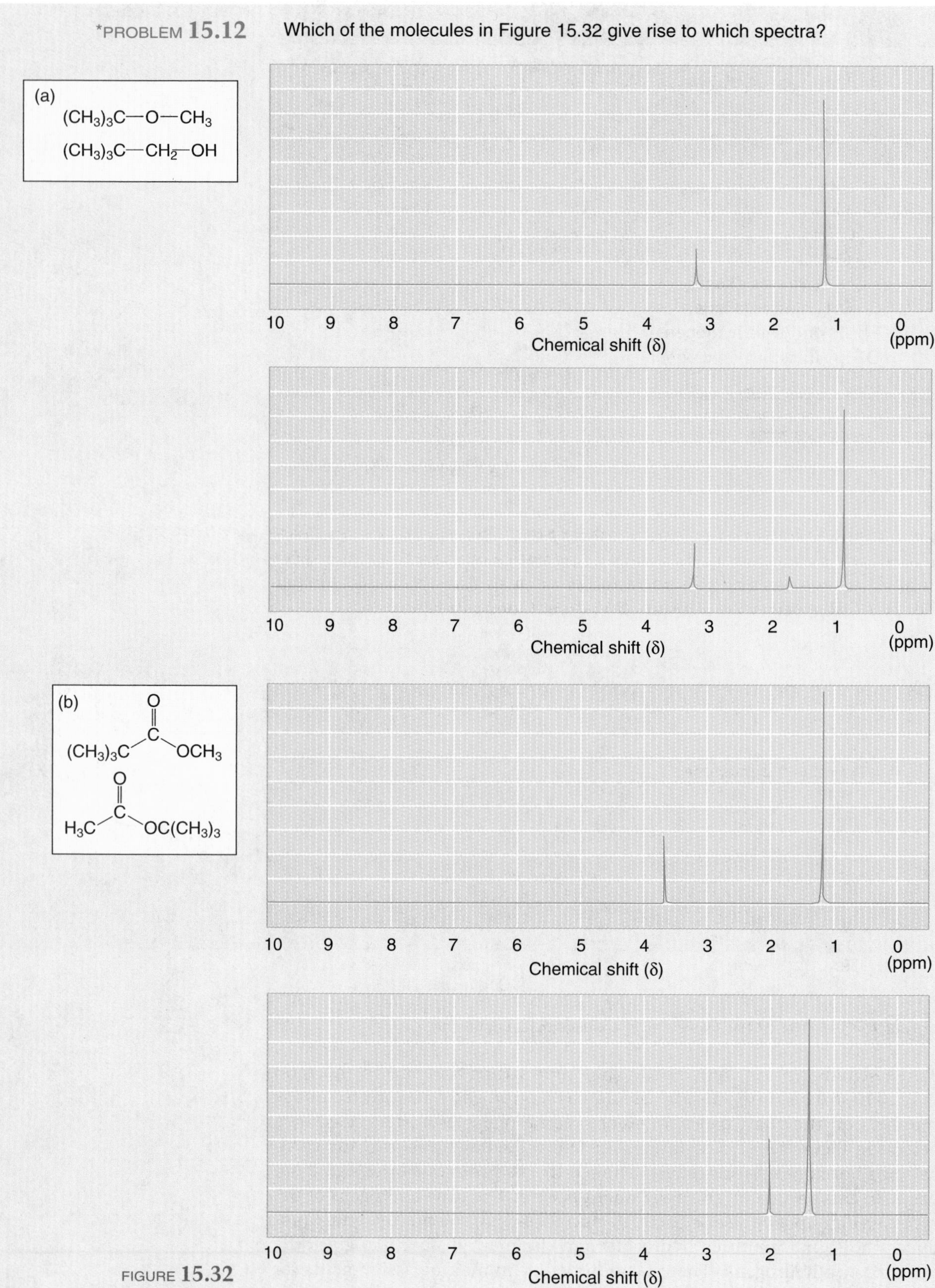

FIGURE **15.32**

*PROBLEM **15.12** (CONTINUED)

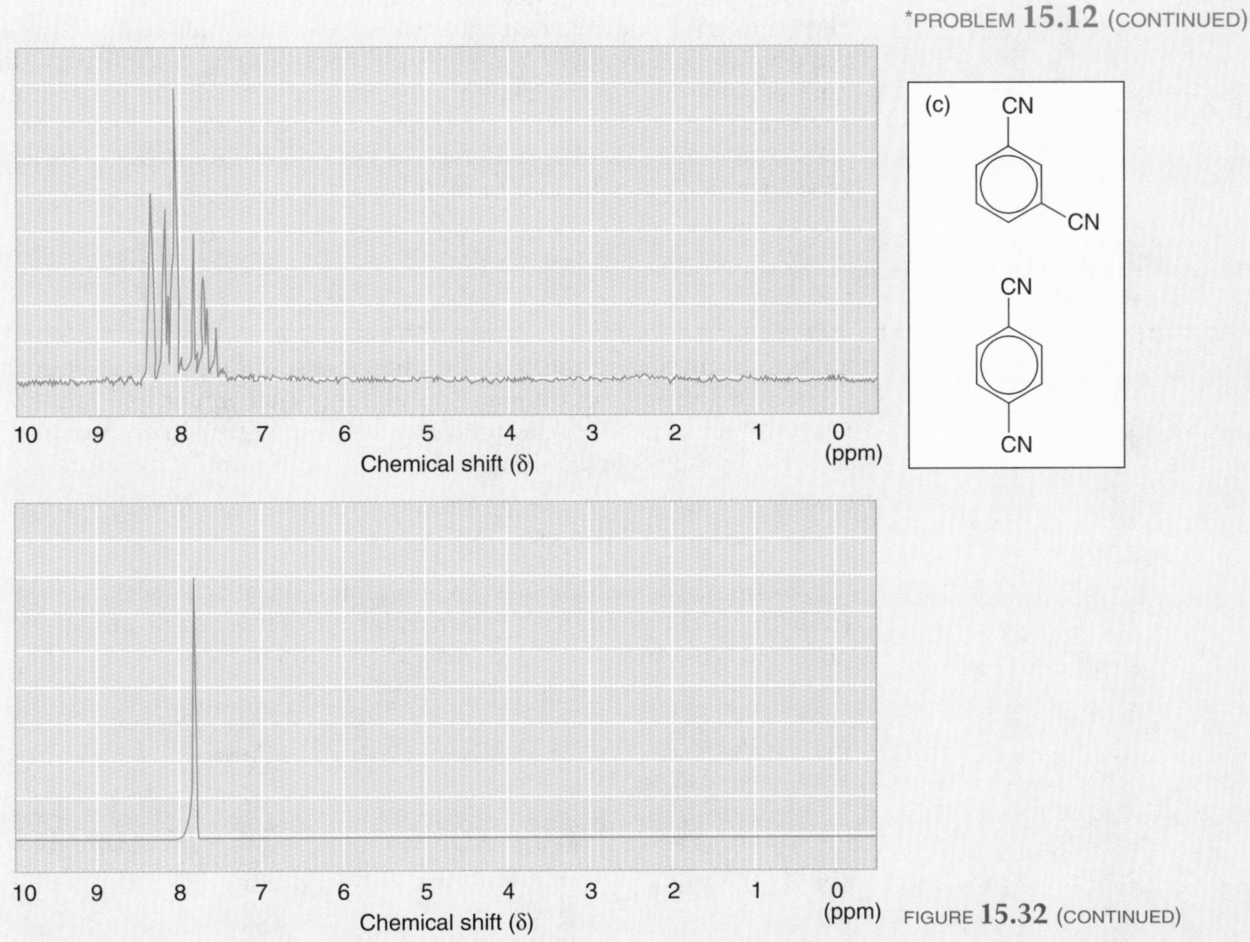

FIGURE **15.32** (CONTINUED)

ANSWER

(a) A signal integrating for nine hydrogens at about δ 1 ppm is a sure sign of the presence of a *tert*-butyl group. In the first spectrum, there is only one other signal, at δ 3.2 ppm. This signal is in an appropriate position for a methyl group next to an oxygen, and the compound is *tert*-butyl methyl ether. In the other compound, 2,2-dimethyl-1-propanol, we see two other signals; two hydrogens at δ 3.3 ppm (the methylene next to the oxygen) and one hydrogen at δ 1.7 ppm, the O<u>H</u>.

(b) The key to this problem is the position of the methyl signal. In the first spectrum, it is far downfield of the position in the second spectrum. The two *tert*-butyl groups are in about the same place. Hydrogens adjacent to oxygen appear well downfield of hydrogens next to a carbon–oxygen double bond (see Table 15.4), so it is easy to make the assignment of methyl trimethylacetate, $(CH_3)_3CCOOCH_3$, to the first spectrum and *tert*-butyl acetate, $CH_3COOC(CH_3)_3$, to the second.

(c) Here symmetry allows a quick assignment. 1,4-Dicyanobenzene has only one kind of hydrogen and must give rise to the second spectrum. 1,3-Dicyanobenzene has three different hydrogens and has the far less symmetrical first spectrum.

These problems are simple, and are designed to ease you into analyzing NMR spectra. In no area of organic chemistry is practice so necessary in the development of skill. You simply cannot become proficient at this kind of structure determination by reading about how it is done and studying Table 15.4. Working problems is an absolute must. Do the problems at the end of this chapter and seek out others. Other textbooks have good ones; don't be reluctant to dig them out.

15.5 SURVEY OF NMR SPECTRA OF ORGANIC MOLECULES

Table 15.4 gives a summary of the chemical shifts for hydrogens in the vicinity of a variety of common organic functional groups (pp. 712, 713). Use this table when working problems and you will gradually become familiar with the chemical shifts of commonly encountered hydrogens.

15.5a Saturated Alkanes

Signals from saturated alkanes (and the simplest cycloalkanes) are at relatively high field. Note that the signals for cyclopropanes are at unusually high field, sometimes even appearing above TMS (Fig. 15.33).

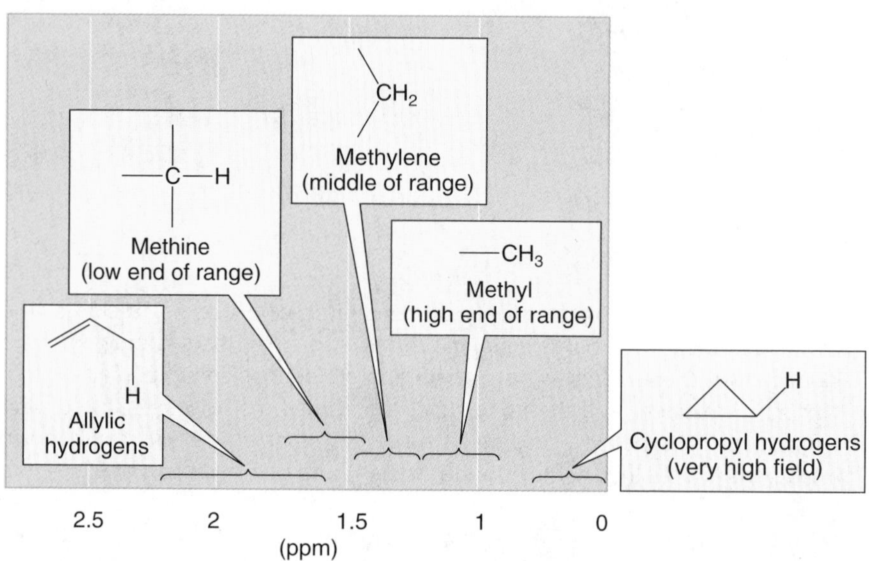

FIGURE 15.33 The ^{1}H NMR chemical shifts of alkanes.

15.5b Substitution by Electronegative Groups

Electron-withdrawing groups will reduce the electron density at neighboring hydrogens, which decreases the ability of nearby electrons to shield the nucleus, causing the chemical shifts to appear at relatively low field. Notice, for example, that as fluorine is the most electronegative halogen, alkyl fluorides appear at the lowest field of all the halogen-substituted molecules. Of course, the more electronegative groups in the neighborhood of a hydrogen, the greater the electron-withdrawing effect, and the lower field the resonance position. The chemical shifts of a variety of halogenated and oxygenated compounds are shown in Figure 15.34.

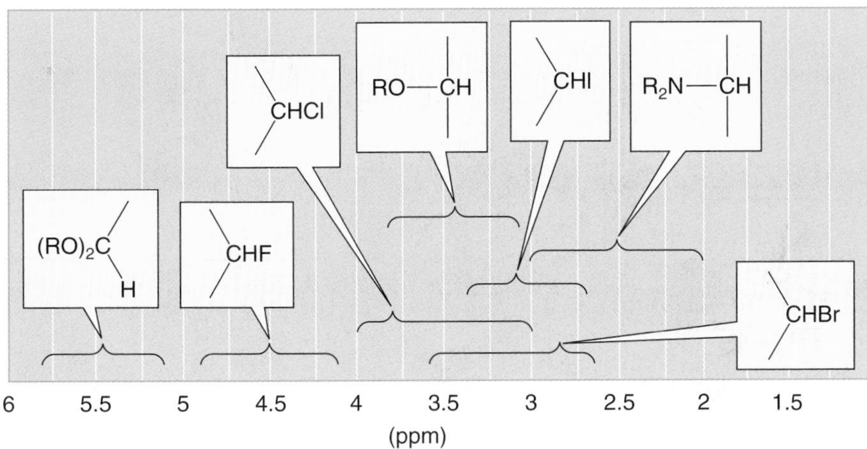

	X=F	X=Cl	X=Br	X=I
CH_3X	4.26	3.05	2.68	2.16
CH_2X_2		5.31	4.96	3.88
CHX_3		7.28	6.86	5.36

FIGURE 15.34 The ^{1}H NMR chemical shifts of alkanes substituted with electron-withdrawing groups.

15.5c Allylic Hydrogens

A carbon–carbon double or triple bond is electron withdrawing, and therefore will exert a deshielding effect, producing relatively low-field chemical shifts for hydrogens in the allylic position. The carbon–oxygen double bond exerts a similar, and greater effect (Fig. 15.35).

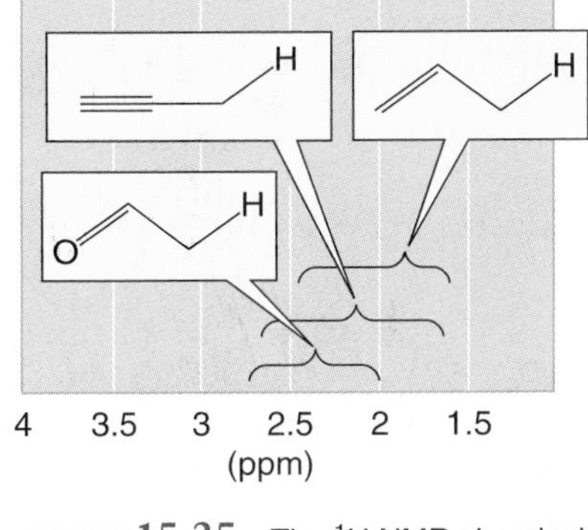

FIGURE 15.35 The ^{1}H NMR chemical shifts of hydrogens adjacent to double and triple bonds.

15.5d Alkenes

Hydrogens attached directly to a carbon–carbon double bond, "olefinic hydrogens," appear at fairly low field. Two effects are operating to deshield such hydrogens. In an applied magnetic field, B_0, the electrons in the π bond will circulate so as to create a magnetic field B_i in opposition to B_0. As shown in Figure 15.36, this opposition results in an *augmented* applied field in the region of the hydrogens attached to the double bond. In addition, the sp^2 carbon of the double bond has high *s* character and attracts electrons, thereby removing electrons from the vicinity of the hydrogen and deshielding it (Fig. 15.36).

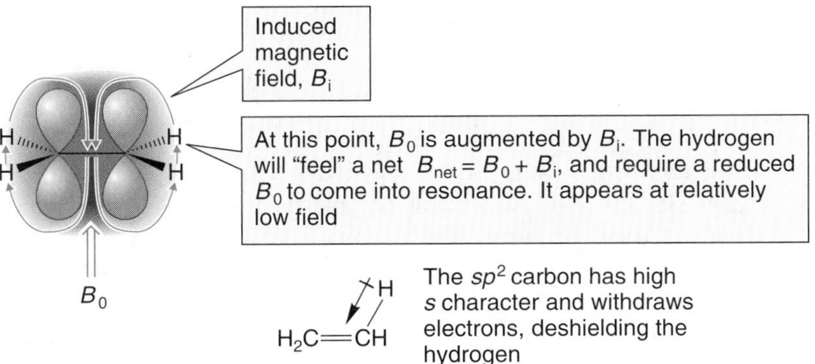

FIGURE 15.36 A carbon–carbon double bond acts in two ways to deshield attached hydrogens. Such deshielded hydrogens appear at relatively low field.

Accordingly, olefinic hydrogens are deshielded and appear at relatively low field (Fig. 15.37). There are important differences between the various kinds of hydrogens attached to double bonds. For example, terminal methylene hydrogens resonate at unusually high field for olefinic hydrogens and

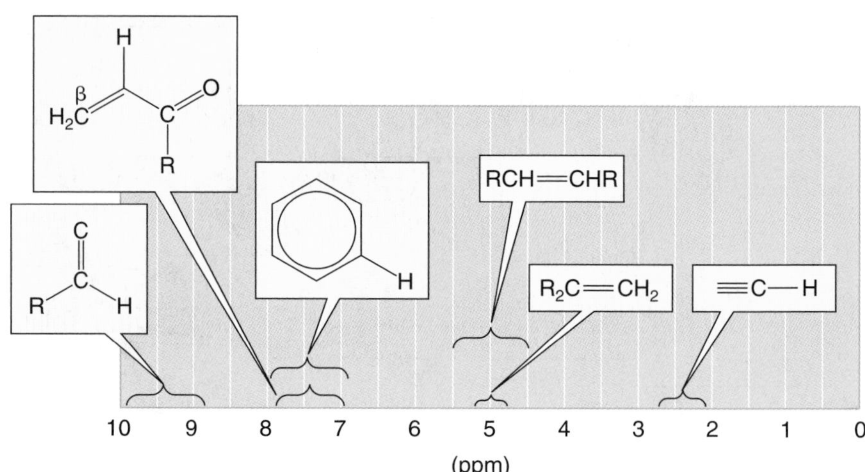

FIGURE 15.37 The ^{1}H NMR chemical shifts of alkenes and alkynes.

in a region of few other absorptions. Hydrogens attached to a double bond, and β to a carbonyl group, absorb at exceptionally low field, and the hydrogens of aldehydes are even lower (Fig. 15.37).

PROBLEM 15.13 Can you explain why vinyl hydrogens β to a carbonyl group are at very low fields?

*PROBLEM 15.14 Hydrogens attached directly to the carbon of a carbonyl group (aldehyde hydrogens) appear at especially low field, and must be strongly deshielded. Explain why.

ANSWER In the absence of other effects, we would expect to see the hydrogen of an aldehyde group at about the positon of hydrogens attached to carbon–carbon double bonds. However, the carbonyl group is very polar. The oxygen atom is much more electronegative than the carbon and will attract electrons strongly, becoming partially negative (δ^-) and leaving the carbon partially positive (δ^+). The same effect can be seen from a resonance formulation.

The electron density around carbon is depleted and the attached hydrogen less effectively shielded than usual. Accordingly, a less strong applied magnetic field is necessary to reach the resonance frequency, and the hydrogen appears at very low field.

15.5e Aromatic Compounds

Hydrogens on aromatic rings are found at even lower field than hydrogens attached to double bonds. In fact, they can quite reliably be diagnosed by the existence of signals in the extended range δ 6.5–8.0. The reason for the low-field resonance is similar to that for hydrogens attached to double bonds. The applied magnetic field induces a circulation of electrons in the ring (a "ring current") that creates its own magnetic field B_i opposing B_0. At the edge of the ring where the hydrogens are, the induced magnetic field augments B_0. Therefore, such hydrogens require a lower B_0 to bring them into resonance (Figs. 15.37 and 15.38).

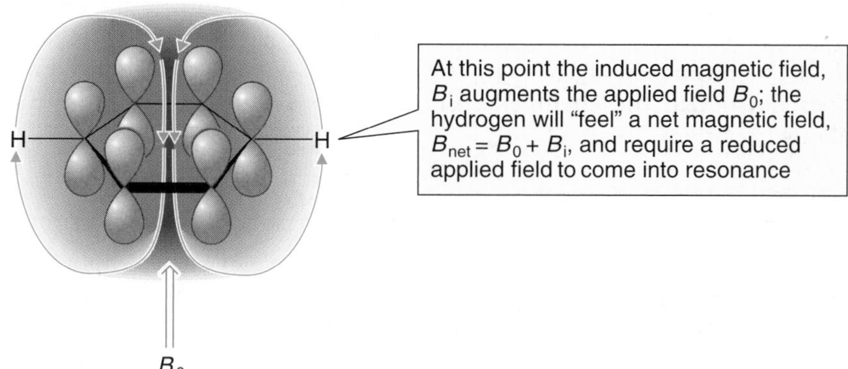

H
H
At this point the induced magnetic field, B_i augments the applied field B_0; the hydrogen will "feel" a net magnetic field, $B_{net} = B_0 + B_i$, and require a reduced applied field to come into resonance
B_0

FIGURE **15.38** The field induced (B_i) by the ring current of an aromatic ring in an applied magnetic field (B_0).

PROBLEM **15.15**

Explain the unusually high chemical shifts (above TMS) of the hydrogens indicated in the molecules in Figure 15.39.

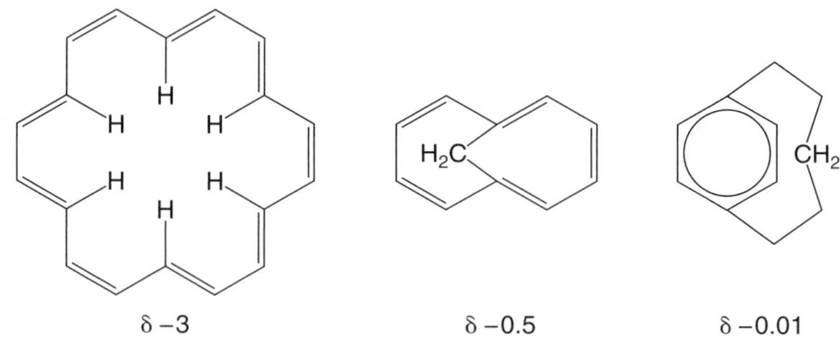

FIGURE **15.39**

15.5f Alkynes

Although the position of hydrogens α to a carbon–carbon triple bond (the propargyl position) is similar to that of allylic hydrogens, those hydrogens directly attached to the triply bonded carbon appear at substantially higher field than hydrogens attached to double bonds, δ 2.2–2.8 (Fig. 15.37). The upfield shift results from a strong shielding effect induced by circulation of electrons in the triple bond. Compare Figures 15.40 and 15.36 to be certain you see the different shielding effects of π bonds at work.

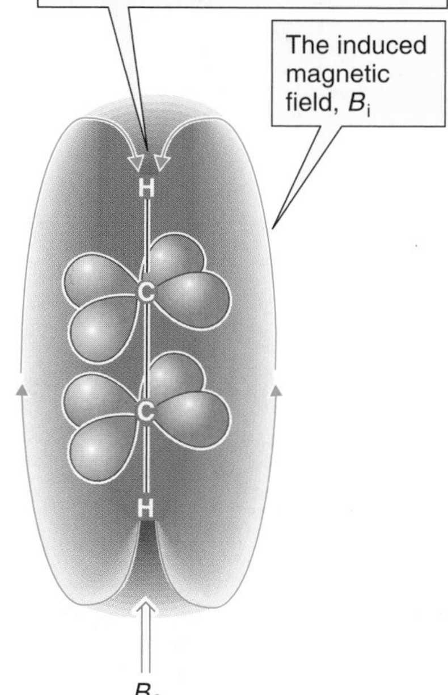

FIGURE **15.40** Acetylenic hydrogens are strongly shielded by the induced magnetic field.

15.5g Alcohols and Amines

Hydrogens attached to a carbon bearing the electronegative oxygen or nitrogen atom of an alcohol or primary or secondary amine absorb, as we might expect, in the same region as the related hydrogens in ethers and tertiary amines (Fig. 15.34). But what about the OH and NH hydrogens themselves? Here we do not see what a simple analysis would lead us to expect. First of all, the chemical shifts of OH and NH vary greatly from sample to sample. The problem is that these molecules are extensively hydrogen bonded, and the extent of hydrogen bonding depends on the concentration of the alcohol or amine. The chemical shift depends strongly on the chemical environment and this changes depending on the extent of hydrogen bonding. Generally, the less hydrogen bonding, the further upfield the resonance appears. In the gas phase, where hydrogen bonding does not occur, the chemical shifts of these hydrogens are quite high, about 1 ppm.

15.6 SPIN–SPIN COUPLING: THE COUPLING CONSTANT, *J*

The utility of NMR spectroscopy goes far beyond that of a simple hydrogen-detecting machine. We can determine the number of different kinds of hydrogen in a molecule, and gain a general idea of what chemical environments those hydrogens might occupy. But there is much more information available. Most NMR spectra are not as simple as those of Figure 15.31 or Problem 15.12. The signals in typical NMR spectra are not all single peaks, but are often multiplets containing much fine structure. Ethyl iodide makes a nice example. Our initial expectation might be that the spectrum would contain only two signals, one for the three equivalent hydrogens of the methyl (CH_3) group and another for the two hydrogens of the methylene (CH_2) group. A look at Table 15.4 would allow us to estimate the general position for each peak. What we see is somewhat different. There are signals at the positions we estimated, but they are composed of several lines, and are not singlets as were most of the signals of the molecules in Problem 15.12. The signal for the methyl hydrogens shows three lines (a triplet) in a 1:2:1 ratio centered at δ 1.85, and the methylene appears as four lines (a quartet) in the ratio 1:3:3:1, centered at δ 3.2. The spacings between the lines are all the same, 7.6 Hz (Fig. 15.41).

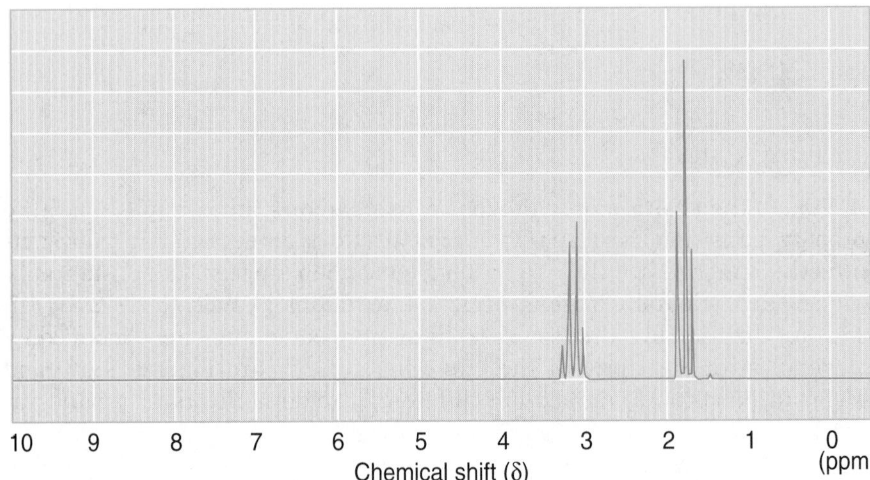

FIGURE 15.41 The ^{1}H NMR spectrum of ethyl iodide.

Multiple line signals are called "doublets" (d; two lines), "triplets" (t; three lines), "quartets" (q; four lines), and so on. Our job is now to see how these multiple signals arise, and then to see how we can use that information.

Let's examine the three-line signal (triplet) for the methyl hydrogens first. To do this, we look first at the neighboring methylene hydrogens. Each of the two equivalent methylene hydrogens has a spin of either +½ or −½. Therefore, there are four different combinations of these spins, (+½ +½), (+ ½ −½), (− ½ +½), and (−½ −½) (Fig. 15.42).

To explain the three-line signal for the methyl hydrogens of ethyl iodide, look first at the neighboring methylene hydrogens

$\pm\frac{1}{2}$H H$\pm\frac{1}{2}$

C

H_3C I

The possible combinations for the methylene groups are

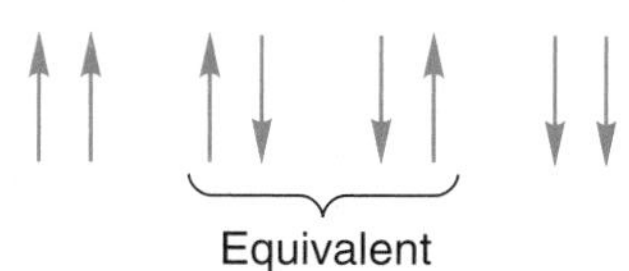

Equivalent

↓ Means spin = $-\frac{1}{2}$

↑ Means spin = $+\frac{1}{2}$

FIGURE **15.42** For two equivalent hydrogens there are four different combinations of the nuclear spins of the two hydrogens. Two of these combinations are equivalent.

The net magnetic field experienced by a methyl hydrogen will be modified by each of the different combinations. Each methyl hydrogen will experience an applied magnetic field (B_0) modified by each of the different spin combinations of the neighboring methylene hydrogens. So we now expect to see four lines, one for each possible spin combination. But two of these combinations, (+½ −½) and (−½ +½), are equivalent and will give rise to the same net magnetic field, which is the reason we see only a three-line signal in a 1:2:1 ratio for the methyl hydrogens (Fig. 15.43).

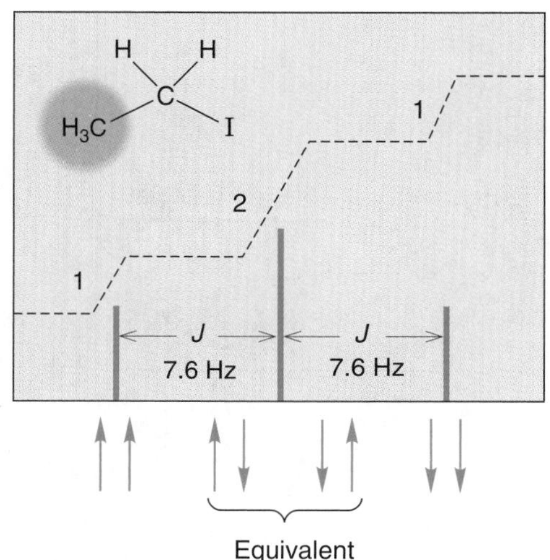

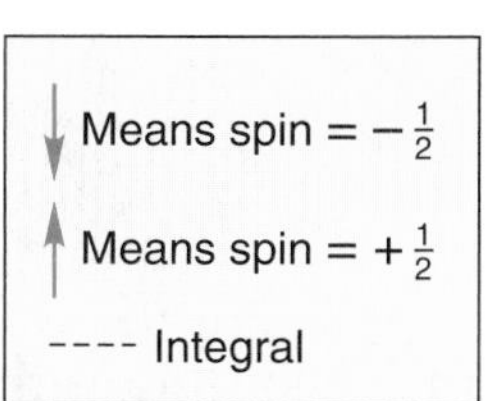

Equivalent

FIGURE **15.43** The methyl group adjacent to the two methylene hydrogens will be split into three lines in the ratio 1:2:1. The splitting between the lines is called the coupling constant, *J*.

The methyl hydrogens are said to be coupled to the adjacent methylene hydrogens. The splitting between the lines is the **coupling constant (*J*)**, and is measured in hertz. In ethyl iodide, this splitting is 7.6 Hz (Fig. 15.43).

Now let's rationalize the four-line signal for the methylene hydrogens of ethyl iodide. For the neighboring methyl group there are eight possible combinations of the nuclear spins of the three equivalent hydrogens (Fig. 15.44). Note that there are now two sets of three equivalent spin combinations (two up and one down or two down and one up). So the hydrogens of the methylene group feel an applied magnetic field modified by the four different combinations of spins of the adjacent methyl hydrogens. Accord-

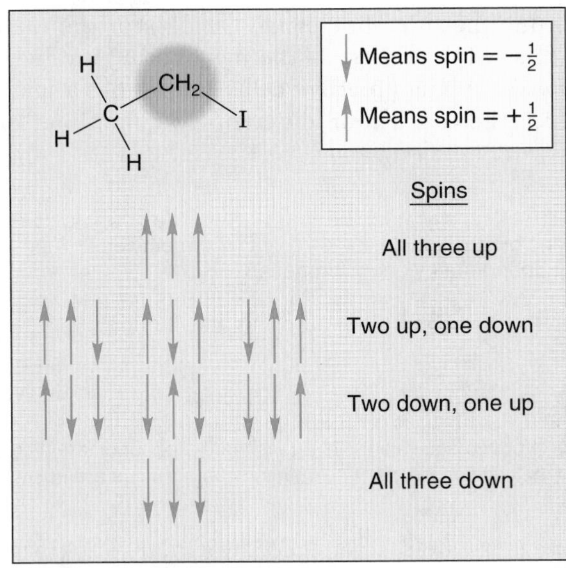

FIGURE 15.44 For the three methyl hydrogens there are eight possible spin combinations, but only four different energy combinations.

ingly, the spectrum consists of four lines in a 1:3:3:1 ratio, a quartet. The methylene hydrogens are coupled to the methyl hydrogens by the same coupling constant, $J = 7.6$ Hz, by which the methyl hydrogens are coupled to the methylene hydrogens (Fig. 15.45).

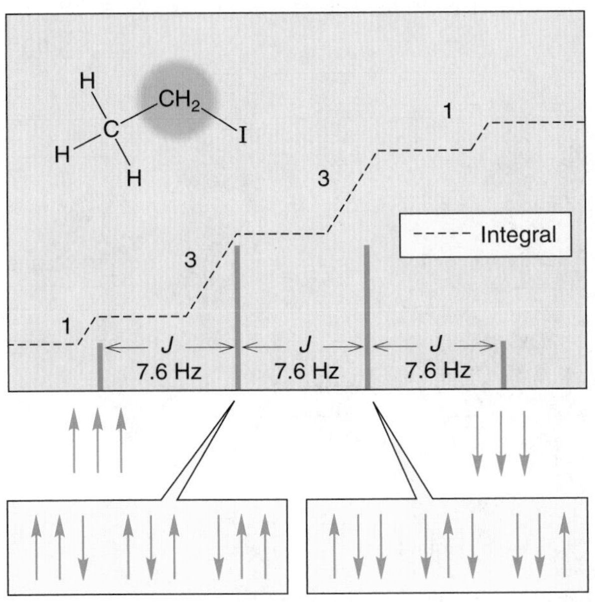

FIGURE 15.45 The methylene group of ethyl iodide will be split into four lines in the ratio 1:3:3:1.

In the general situation, for a given hydrogen, H_a, the NMR spectrum will consist of $n + 1$ lines where n is the number of equivalent adjacent hydrogens, H_b. The lines are separated by J Hz. The intensity of the lines can be determined by analyzing the various possible spins of the coupled hydrogens as we have been doing, or by using a device called "Pascal's triangle." In this device, the value (intensity) of a number is the sum of the two numbers above it. Figure 15.46 explains it better than these words. Pascal's triangle allows the rapid determination of the intensities of lines in NMR signals. Note that we are speaking here of the relative intensities of

	Signal for H_a	Number of equivalent adjacent H	Number of lines
H_a, H	1 1	1	2
H_a, H, H	1 2 1	2	3
H_a, CH_3	1 3 3 1	3	4
H_a, H_3C, CH_3	1 6 15 20 15 6 1	6	7

General case

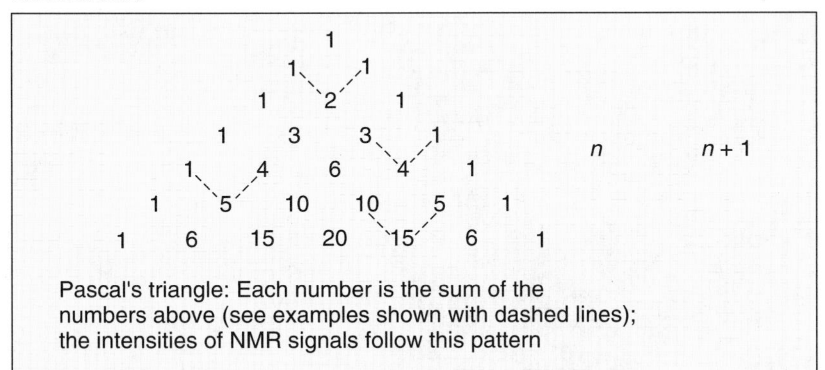

FIGURE **15.46** In the general case, there will be $n + 1$ lines for a hydrogen adjacent to n equivalent hydrogens.

the lines *within a given signal,* not of the intensities of signals for different hydrogens in the molecule.

There is no coupling between equivalent nuclei. But why is coupling observed only to hydrogens on the same or adjacent atoms and not to all other hydrogens in a molecule? Usually, only "two-bond" or "three-bond coupling" is large enough to be detectable (Fig. 15.47). In some cases, **long-range coupling** can be observed, especially if the hydrogens are connected in allylic fashion. The coupling constant between hydrogens on adjacent carbons is most sensitive to the dihedral angle between those hydrogens, and this can often be used in assigning structure. Figure 15.48 shows the **Karplus curve**, named for Professor Martin Karplus (b. 1930) of Harvard

		Magnitude of J (Hz)
H_a–C–H_a	Identical hydrogens do not couple	0
H_a–C–H_b	H_a is two bonds away from H_b; coupling over this distance is usually observable	2–30
H_a–C–C–H_c	H_a is three bonds away from H_c; coupling over this distance is usually observable	6–8
H_a–C–C–C–H_d	H_a is four bonds away from H_d; coupling is not usually observable	0–1
H_a–C–C=C–H_e	Ha is in the allylic position relative to H_e; coupling is usually observable	2–3

FIGURE **15.47** Coupling can be measured over a three-bond distance, but is not usually easily detectable over longer distances.

University, which shows the dependence of J on the dihedral angle for two hydrogens on adjacent sp^3 carbons. Figure 15.48 also shows the dependence of the coupling constant on the angle between hydrogens attached to sp^2 hydridized carbons.

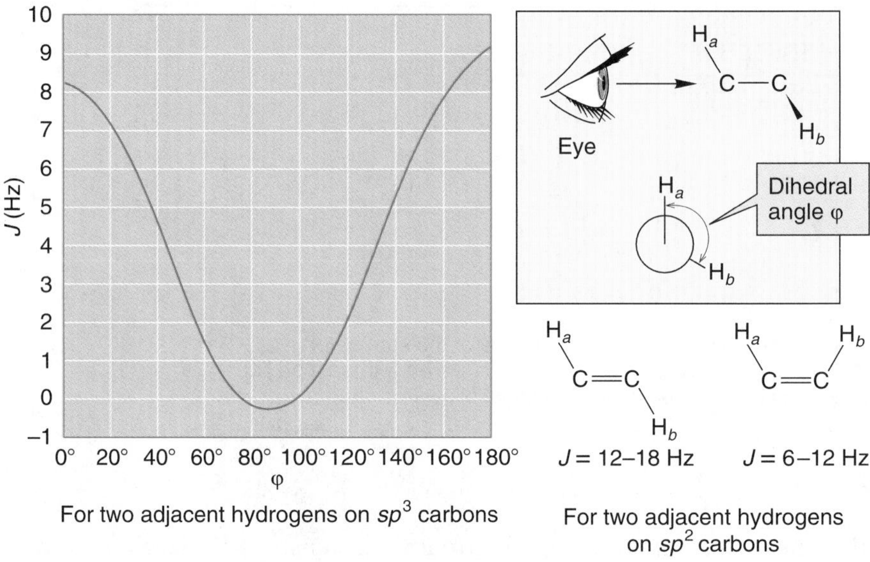

FIGURE **15.48** Coupling between two adjacent hydrogens is sensitive to the dihedral angle between them. The Karplus curve shows the calculated relationship between J and the dihedral angle.

PROBLEM **15.16** How many lines do you expect to see for the central hydrogen in 1,3-dichloro-2-methoxypropane (Fig. 15.49)?

OCH_3

Cl—CH_2—C(H)—CH_2—Cl

1,3-Dichloro-2-methoxypropane

FIGURE **15.49**

PROBLEM **15.17**

The two molecules in Figure 15.50 have coupling constants for the hydrogens attached to the three-membered ring of 8.4 and 5.3 Hz. Which compound has which coupling constant?

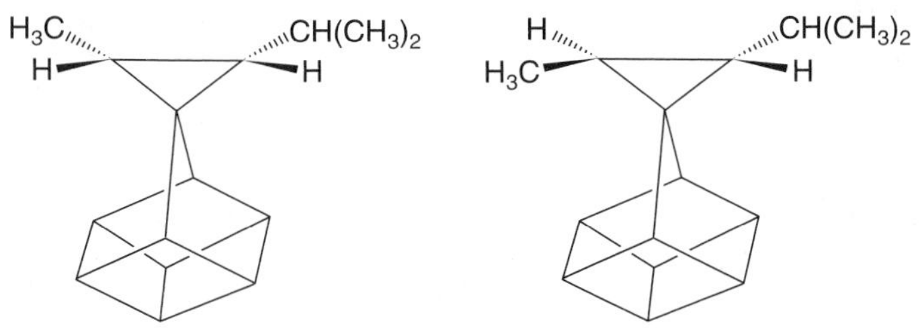

FIGURE **15.50**

The cis coupling constant, J_{cis}, is smaller than the trans coupling constant, J_{trans}. In alkenes, J_{cis} is in the range 6–12 Hz and J_{trans} is in the range 12–18 Hz. Notice that the ranges slightly overlap. One cannot always assign stereochemistry from just one stereoisomer. What would you make of a coupling constant of 12 Hz, for example? But if both isomers are available, the cis stereochemistry can be reliably assigned to the one with the smaller J value. Figure 15.51 gives some typical coupling constants in alkenes and aromatic compounds.

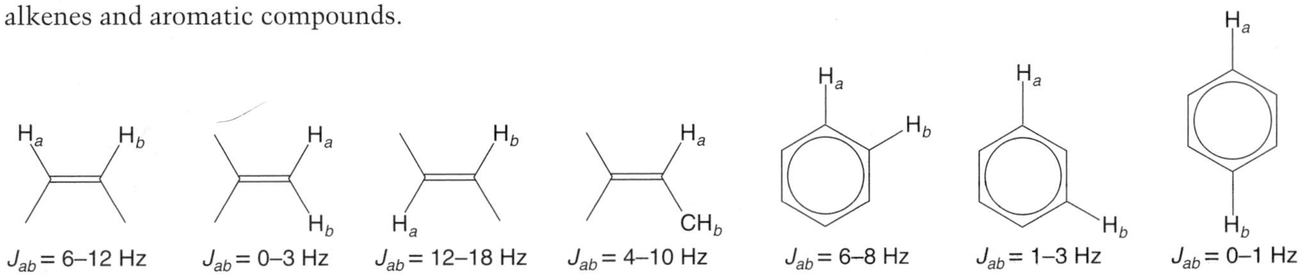

FIGURE **15.51** Some coupling constants for hydrogens attached to double bonds.

The splitting patterns of alcohols and amines are often not what we would expect. Consider the common alcohol, ethyl alcohol, whose ^{1}H NMR spectrum is shown twice in Figure 15.52. Notice also that these two spectra, taken from different catalogs in the chemical literature, show the concentration dependence of the O<u>H</u> signal. The methyl and methylene hydrogens are in the same place in the two spectra, but the two different samples reveal different chemical shifts for the O<u>H</u>. The spectrum of ethyl alcohol does show the expected triplet for the methyl group, but the methylene and hydroxyl hydrogens do not appear as a first-order analysis would predict. The most accurate description of them might be "blobs." What is going on?

Here the problem is one of chemical exchange. In the presence of either an acidic or basic catalyst, alcohols and amines exchange the OH or NH hydrogens. Figure 15.53 gives the mechanism for base-catalyzed exchange of an alcohol and acid-catalyzed exchange of an amine, using deuterated materials in order to be able to detect the exchange reaction. This exchange reaction can be used to locate the signals for OH and NH groups. The spectrum is taken, then a drop of D_2O is added to the sample tube. The tube is shaken and the spectrum taken again. All OH and NH peaks will disappear, as their protons are exchanged for deuterons.

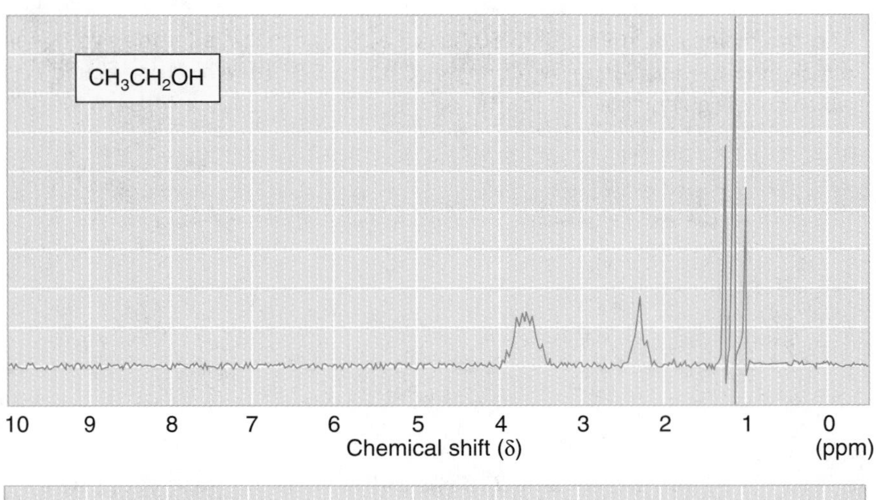

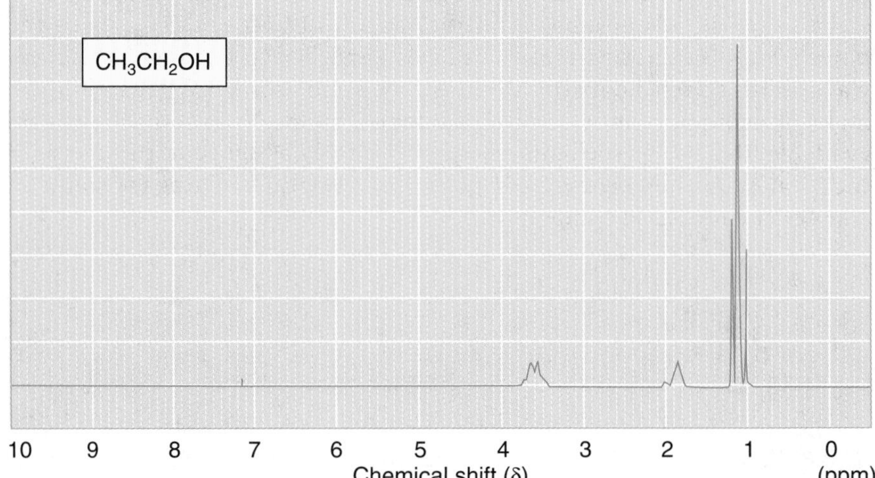

FIGURE **15.52** Two ^{1}H NMR spectra of ethyl alcohol.

$$CH_3CH_2\text{—}\ddot{\underset{..}{O}}\text{—}H + {}^{-}\!:\!B \rightleftharpoons CH_3CH_2\text{—}\ddot{\underset{..}{O}}\!:^{-} + B\text{—}H$$

$$CH_3CH_2\text{—}\ddot{\underset{..}{O}}\!:^{-} + D\text{—}B \rightleftharpoons CH_3CH_2\text{—}\ddot{\underset{..}{O}}\text{—}D + {}^{-}\!:\!B$$

$$(CH_3CH_2)_2\ddot{N}\text{—}H + D\text{—}\overset{+}{\underset{..}{O}}D_2 \rightleftharpoons (CH_3CH_2)_2\overset{+}{N}(D)\text{—}H + \ddot{\underset{..}{O}}D_2 \rightleftharpoons (CH_3CH_2)_2\ddot{N}\text{—}D + H\overset{+}{\underset{..}{O}}D_2$$

FIGURE **15.53** The base-catalyzed exchange reaction of alcohols, and the acid-catalyzed exchange reaction of amines.

PROBLEM **15.18** Write a mechanism for acid-catalyzed exchange of ethyl alcohol and the base-catalyzed exchange of diethylamine.

In IR spectroscopy, the absorption of energy is essentially instantaneous. The IR spectrometer takes a stop-action picture of the molecule. Nuclear magnetic resonance spectroscopy is different. Acquisition of the signal is much slower, about 10^{-3} s, and the exchange reaction is fast compared to the time it takes to make the NMR measurement. During this time the oxygen of a given molecule will be bonded to many different hydrogens. The two methylene hydrogens of ethyl alcohol "see" only an average of all spins of the hydroxyl hydrogen, not the two different $+\frac{1}{2}$ and $-\frac{1}{2}$ states. The same phenomenon serves to remove coupling between alkyl hydrogens and the NH hydrogens of primary and secondary amines.

If this explanation is correct, the coupling should be detectable in scrupulously purified ethyl alcohol because the exchange reaction will be stopped or, at least, greatly slowed. It is very hard to remove the last vestiges of acidic or basic catalysts from these polar molecules, but if the effort is made, the coupling returns, as it should. Absolutely pure ethyl alcohol does give the anticipated first-order triplet for the OH hydrogen (Fig. 15.54).

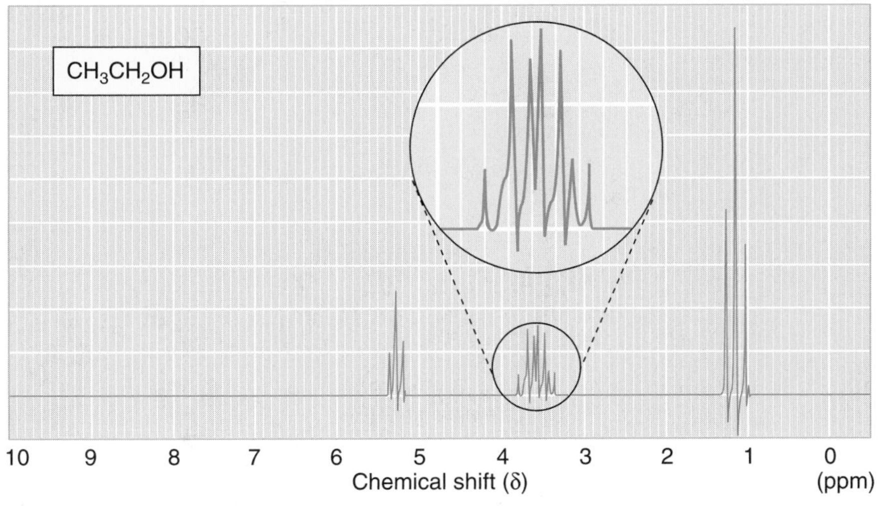

FIGURE 15.54 The ^{1}H NMR spectrum of absolutely pure ethyl alcohol. See Problem 15.39.

This section gives the first hint that NMR spectra are dependent on the rates of reactions. If the rate of the exchange reaction had not been fast, if it were slow on the NMR time scale, we would not see the averaged spectra that we do for alcohols and amines. We will return to this notion in Section 15.10.

To summarize: In the best of all worlds (which means in this case that your budget includes enough money to buy the highest field spectrometer available), we will see a different signal for every different hydrogen in our molecule, and we will be able to determine the number of equivalent adjacent hydrogens from the $n + 1$ rule. In practice, things may not be quite so simple. We can see some possible complications right away. Perhaps the signals will not be sufficiently separated, and the spectrum may consist of complicated overlapping multiplets. Here is an example that shows why chemists are so anxious to spend taxpayer's money on ever higher field spectrometers. The higher the field, the better the dispersion of the signals (recall Problem 15.9, p. 710). Remember that chemical shifts are dependent on the strength of the applied field and the higher the field, the more separated the signals. But there are more complex problems as well, and Section 15.7 touches upon one of them. Once again, we'll find that the solution to the difficulty involves using the highest possible magnetic field.

15.7 MORE COMPLICATED SPECTRA

Even at high field, NMR spectra are often not the simple "first-order" collections of lines we might derive from an analysis of the molecule using the $n + 1$ rule. A classic example of this phenomenon occurs in *AB* signals. For the two-spin system in Figure 15.55, we would expect to see a pair of doublets. Each hydrogen is different, and coupled to only one hydrogen on the adjacent carbon (Fig. 15.55).

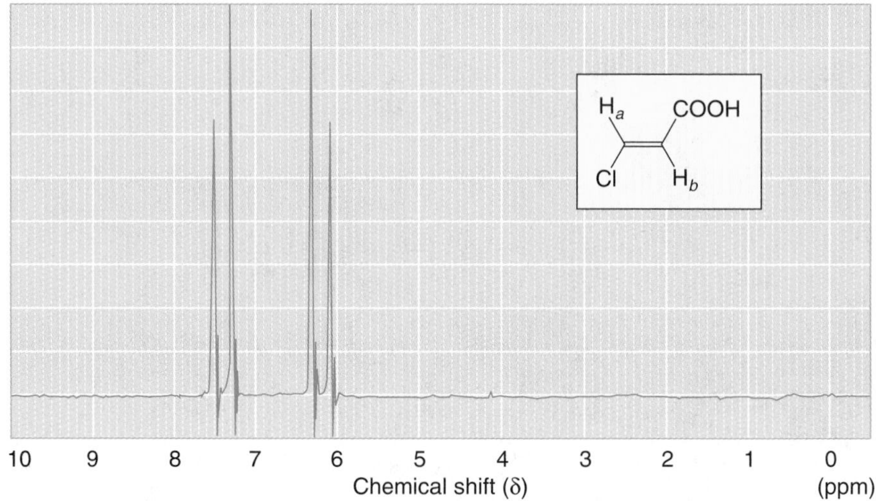

FIGURE 15.55 Two different hydrogens adjacent to each other will give rise to a pair of doublets as long as the hydrogens are very different.

In fact, we see this simple spectrum only when the two hydrogens, H_a and H_b, are *very* different: when they have very different chemical shifts. The more alike the two hydrogens, the more the spectrum deviates from first order, the spectrum predicted by application of the $n + 1$ rule. As the chemical environments of the two hydrogens become more similar, the chemical shifts for the corresponding NMR signals of course also become more alike. The more alike they are, the closer the signals are and the more pronounced the leaning of the peaks toward the center (Fig. 15.56).

In the limit, H_a and H_b become the same, and have the same chemical shift. As there is no coupling between equivalent hydrogens, the signal for them collapses to a sharp line. Figure 15.56 shows the gradual transition between the case in which the two hydrogens are very different and that in which they are identical. In the language of NMR, the very different system is called *AX*, the identical system A_2, and an intermediate case *AB* or *AM*.

So far, in the spectra we have seen, a hydrogen has only been split by a single kind of adjacent hydrogen, and it is simple to derive the expected multiplicity of the NMR peak from the $n + 1$ rule. In a molecule such as

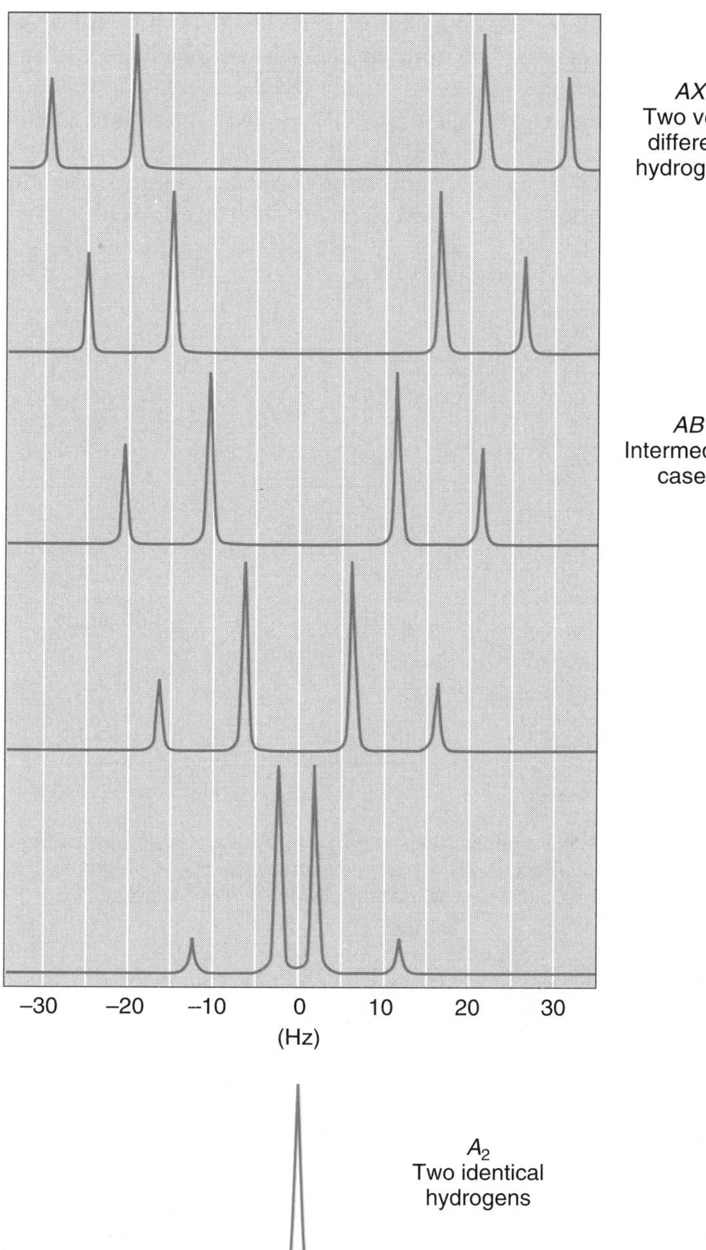

FIGURE 15.56 As the two hydrogens become more similar, the inner lines grow at the expense of the outer lines. The signals increasingly lean toward each other. In the limit, the two hydrogens become identical, and there is only a single line.

An *AMX* spin system

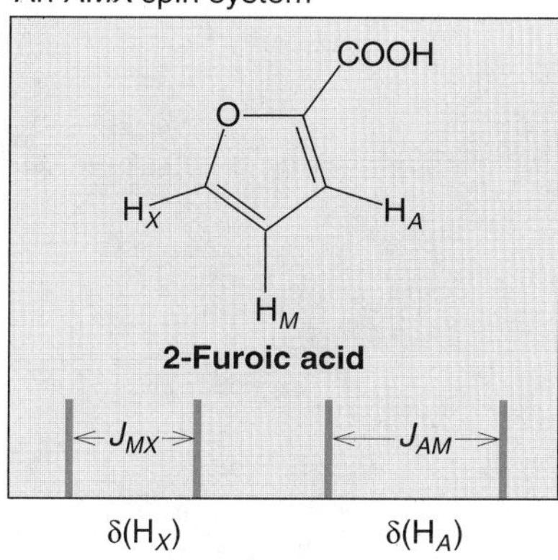

$\delta(H_X)$ $\delta(H_A)$

Predicted spectrum for H_A and H_X

FIGURE 15.57 In 2-furoic acid, H_A and H_X will appear as doublets. Each is strongly coupled to a single adjacent hydrogen, H_M. Each line will be further split by allylic coupling (J_{AX}) into narrow doublets.

2-furoic acid, the two different hydrogens on C(3) and C(5) (H_A and H_X) are each coupled to the single hydrogen on C(4) (H_M) to give a pair of doublets. This system of three very different hydrogens is called *AMX* (Fig. 15.57).

Hydrogen H_M itself is coupled differently to H_A and H_X by the coupling constants J_{AM} and J_{MX}. How can we figure out the pattern for H_M that should be observed in the spectrum? The first-order pattern can be determined by working out the number of lines from each coupling one at a time. We first determine the number of lines produced by coupling to H_A and then apply the coupling to H_X to each of the lines in a "tree diagram." For example, coupling of H_M to H_A will produce a doublet, separated by J_{AM}. Each of these two lines will be further split into two by coupling to H_X. The result is four lines, a doublet of doublets. Figure 15.58 shows the theoretical prediction and the real spectrum of 2-furoic acid.

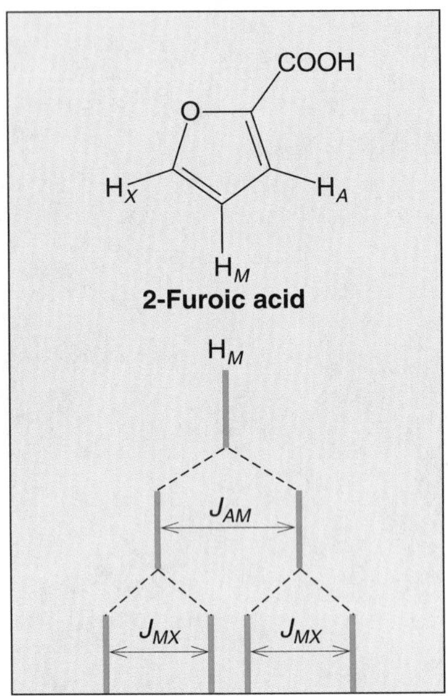

Predicted spectrum for H_M, four lines

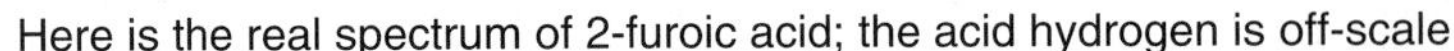
Here is the real spectrum of 2-furoic acid; the acid hydrogen is off-scale

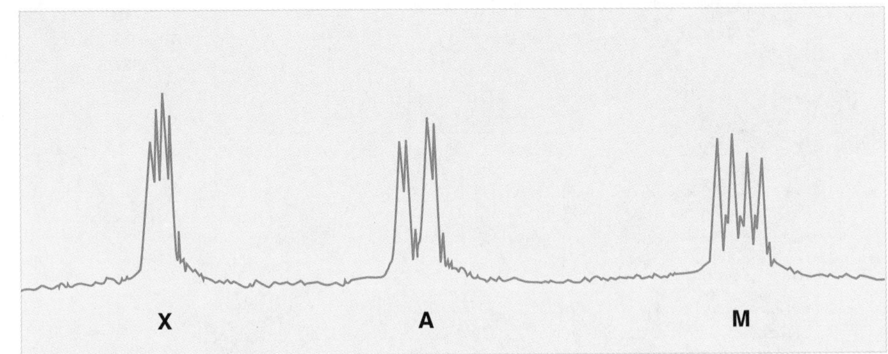

FIGURE 15.58 The first-order splitting pattern can be determined by calculating the couplings one at a time. In this case, H_M is split into a doublet by H_A and each line of the doublet is split again into doublets by coupling to H_X. Note the allylic coupling, J_{AX}.

First-order spectra are quite easy to predict. For example, according to the $n + 1$ rule, the two different methylene groups of a generic molecule X–CH_2–CH_2–Y should appear as a pair of 1:2:1 triplets. 2-Chloropropionic acid fits the pattern, and, as Figure 15.59 shows, satisfactorily produces the expected pair of methylene triplets in its ^{1}H NMR spectrum at δ 2.89 and 3.78.

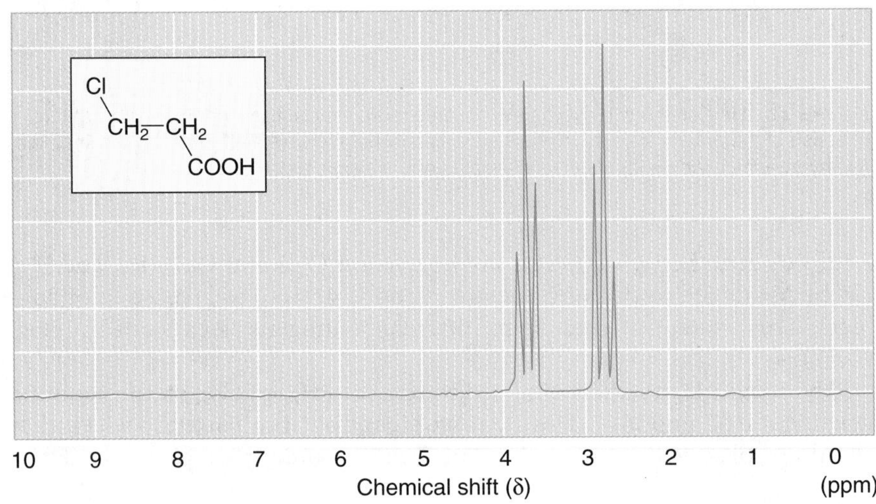

FIGURE 15.59 In 2-chloropropionic acid, each methylene group appears as a simple triplet.

However, when the molecules become more complex we can anticipate complicated spectra, even for first-order systems. Methyl butyrate is a simple molecule, and we can predict its first-order spectrum quite easily. The methyl group attached to oxygen has no neighboring hydrogens and appears as a singlet at δ 3.68. The other methyl group is adjacent to a pair of methylene hydrogens and appears as a triplet at δ 0.94. Similarly, the methylene group adjacent to the carbon–oxygen double bond is flanked by the same methylene group. It, too, is a triplet, this time at δ 2.30. But what about the other methylene group? It is split by three methyl hydrogens, and then split again by two methylene hydrogens. The result should be a quartet of triplets, or, equivalently, a triplet of quartets. Either way you work it out, the result is a 12-line pattern, a dodectuplet! It would be quite amazing if these lines did not overlap to give a pattern difficult to analyze. So they do, as Figure 15.60 shows.

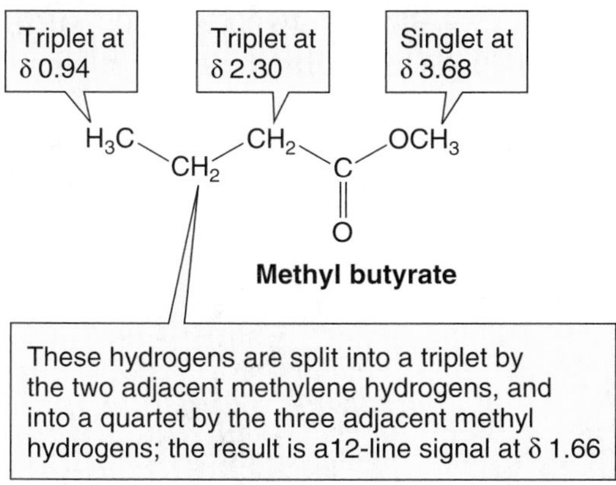

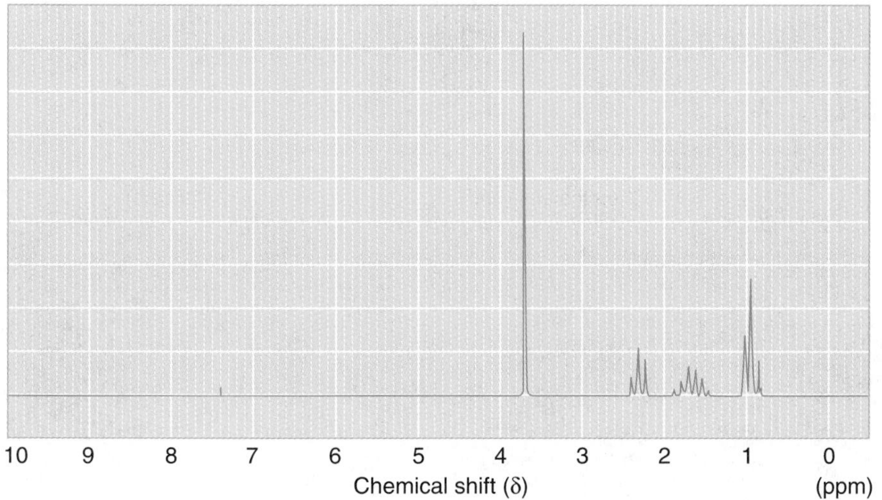

FIGURE 15.60 The multiplet at δ 1.66 results from the splitting of one methylene group by the adjacent methyl and methylene groups. The result is 12 lines that overlap in a complex pattern.

Clearly, even if the spectrum is first order we are likely to be faced with a seriously overlapping set of lines in a case such as this one, and interpretation may not be simple. Moreover, as we have already seen, NMR spectra are often not first order. Even in the simple system of two adjacent different hydrogens, the observed spectrum is often more complicated than what a first-order analysis would predict. First-order spectra are only observed when all the hydrogens in a molecule are very different. As they become less different, more complicated spectra appear. Finally, if we imagine carrying the process of making the hydrogens more and more alike to its logical conclusion, we get a single signal for equivalent hydrogens. But when are hydrogens "very different" and when are they not? In a practical sense, for hydrogens to give rise to first-order spectra, the difference in hertz in their chemical shifts ($\Delta\delta$) must be at least 10 times greater than their coupling constant, J_{ab} [Eq. (15.5)].

$$\Delta\delta > 10\ J_{ab} \tag{15.5}$$

That requirement is another reason why high-field NMR spectrometers are so much more useful than lower field instruments. Although the coupling constant J does not depend on the field strength, the value (in hertz) of the chemical shifts does (p. 707). The greater the field strength, the greater the chemical shift difference ($\Delta\delta$), and therefore the greater the likelihood that the condition of Eq. (15.5) will be satisfied. Figure 15.61 shows the ^{1}H NMR spectrum of butyl bromide at 60, 90, and 300 MHz. Even the modest change from 60 to 90 MHz results in substantial improvement in the spectrum.

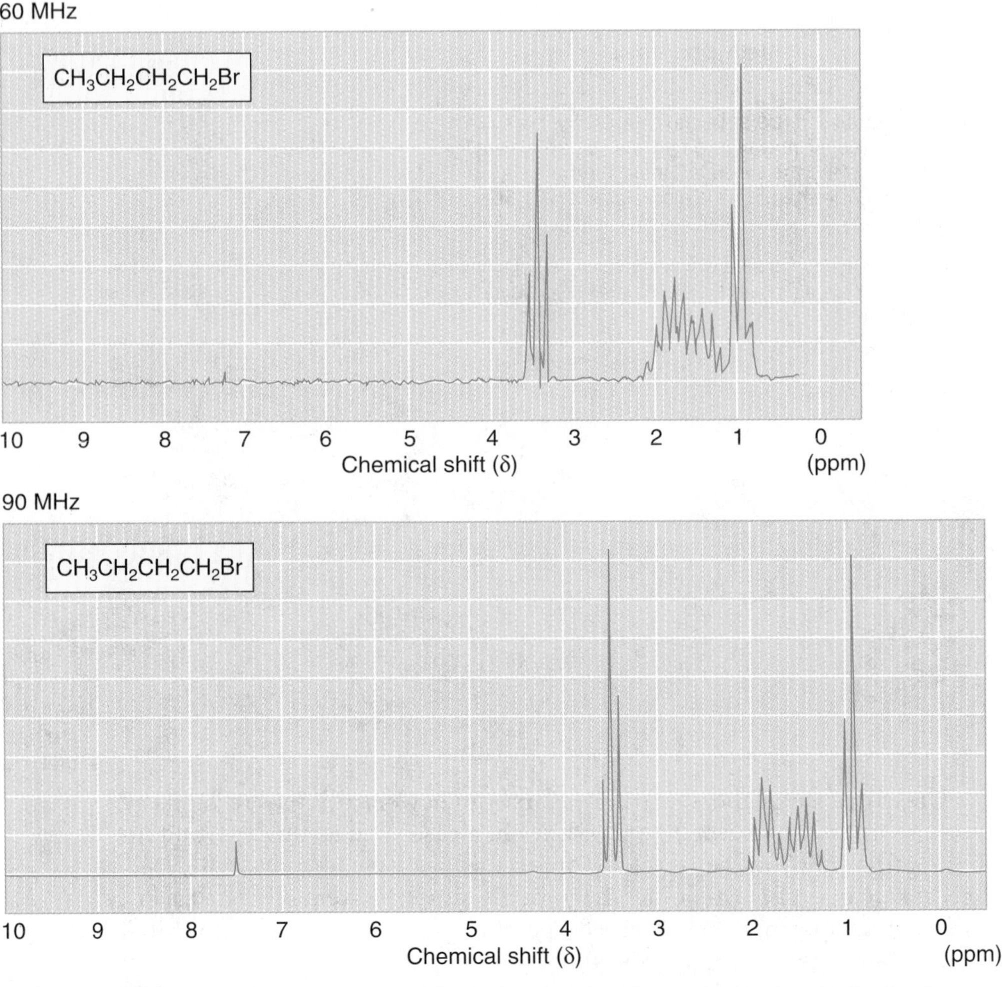

FIGURE 15.61 The ^{1}H NMR spectra of butyl bromide taken at 60, 90, and 300 MHz.

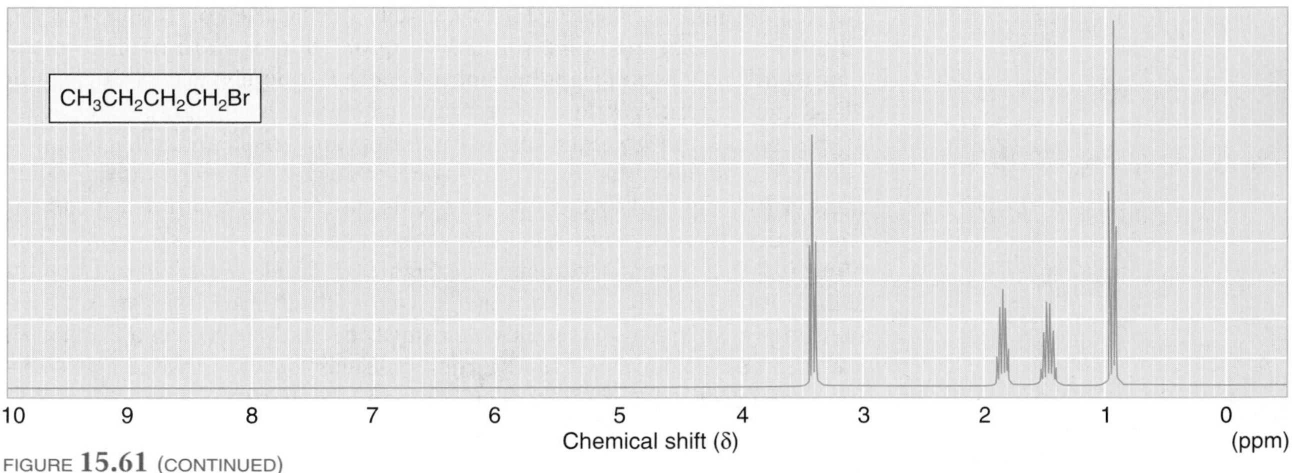

FIGURE **15.61** (CONTINUED)

15.8 DECOUPLED SPECTRA

When we discussed the NMR spectra of alcohols and amines, we encountered the phenomenon of chemical exchange of the OH and NH hydrogens. Rapid exchange of these hydrogens resulted in an averaging of the coupling to adjacent hydrogens. Exchange effectively **decoupled** the OH and NH hydrogens from hydrogens at adjacent positions. This same trick can be accomplished electronically, and it can be very useful, or even essential in the interpretation of NMR spectra. Suppose we are set the task of determining the stereochemistry of ethyl crotonate from its ^1H NMR spectrum (Fig. 15.62)? We might hope to do this by measuring the coupling

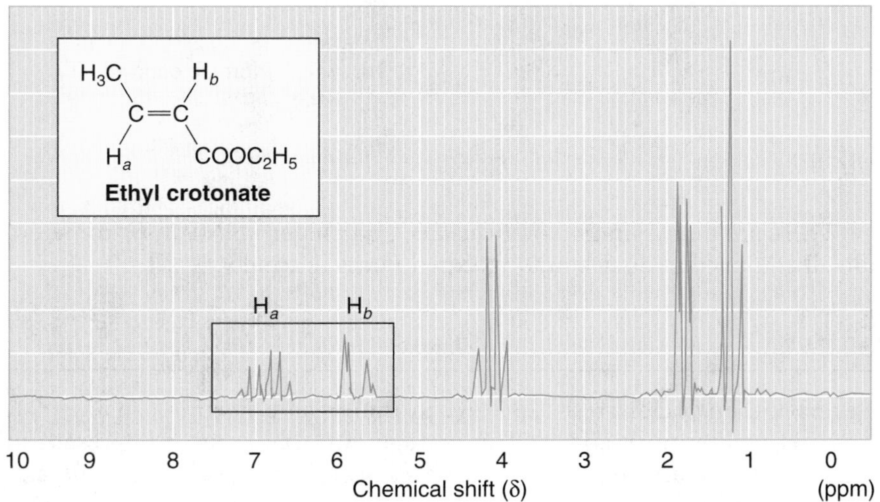

FIGURE **15.62** The ^1H NMR spectrum of ethyl crotonate. The coupling between H_a and H_b is obscured by the coupling to the adjacent methyl groups.

constant between the two vinyl hydrogens. If the compound is trans, we would expect *J* to be 12–18 Hz; if it is cis, *J* would be smaller, 6–12 Hz (Fig. 15.54). However, the rather complex spectrum makes the determination of *J* difficult. The problem is that both vinyl hydrogens are coupled to the methyl group, and this coupling leads to the complex signals we see. If we could somehow get rid of the coupling to the adjacent methyl group, the spectrum would surely simplify, and it might become possible to pick out the couplings between H_a and H_b. It is possible to accomplish this simplification through electronic decoupling. The spectrometer is modified to allow a second application of radio-frequency waves during the NMR experiment. While the spectrum is being measured, radio frequency at the resonance frequency of the methyl hydrogens is applied. The result is to induce rapid transitions between the two possible spin states for each methyl hydrogen. As in chemical exchange of ethyl alcohol, the spin of the methyl hydrogens is averaged to zero and there is no observable coupling to the adjacent hydrogen H_a. Figure 15.63 shows a blow-up of the signals for the two vinyl hydrogens with and without irradiation at the methyl group. It is now easy to pick out the coupling between H_a and H_b and the stereochemistry of the molecule can be determined with ease from the observed J of 17 Hz. The compound is trans.

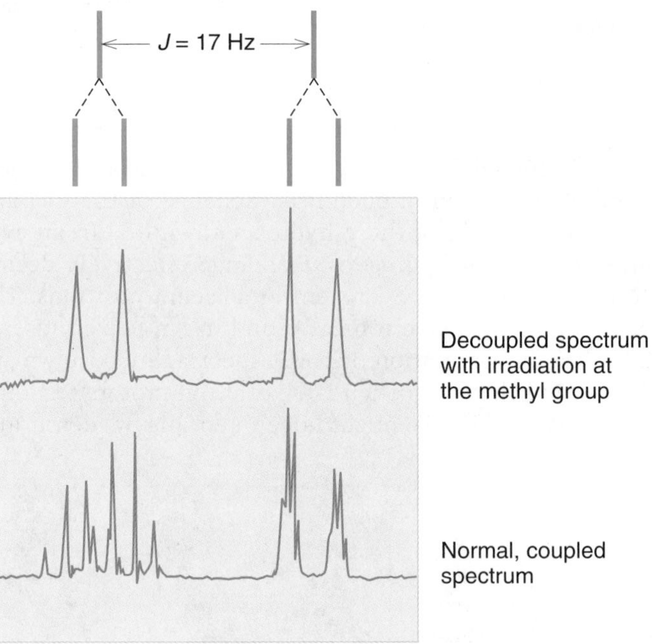

FIGURE 15.63 Irradiation at the resonance frequency of the allylic methyl group averages the spins of the methyl hydrogens and decouples the allylic methyl hydrogens. The coupling constant J between H_a and H_b is now easier to determine.

There are many similar applications, and modern NMR spectrometers are all equipped to do decoupling experiments routinely.

15.9 NUCLEAR MAGNETIC RESONANCE SPECTRA OF OTHER NUCLEI

Several other nuclei are of interest to the organic chemist, mainly ^{19}F, ^{15}N, ^{2}H (deuterium), and ^{13}C, an isotope of carbon, which is present in 1.1% natural abundance. Even though all organic compounds contain carbon, the natural abundance of ^{13}C is low enough so that coupling to ^{13}C is not observed in hydrogen spectra. These other nuclei can be used either in natural abundance or in enriched samples to provide their own NMR spectra. Recall that the constant γ in Eq. (15.4) will be different for ^{19}F, ^{2}H, ^{15}N, or ^{13}C, and their NMR signals will not overlap with the hydrogen spectra.

One other nucleus, ^{13}C, is comparable in importance to ^{1}H. Like hydrogen, ^{13}C has a spin of ½, and coupling between ^{13}C and ^{1}H will complicate ^{13}C spectra. Normally, ^{13}C spectra are run using a broad-band decoupling so that all hydrogens in the molecule are decoupled. All $^{13}C-^{1}H$ coupling is removed in this way. What about coupling between two ^{13}C atoms? Won't this complicate the spectrum? The chance of finding a ^{13}C at any given position is only 1.1% or 0.011. The chance of finding two ^{13}C atoms on adjacent positions is therefore $0.011 \times 0.011 = 0.00012$! At least for natural abundance spectra we can ignore $^{13}C-^{13}C$ coupling. Accordingly, ^{13}C spectra are collections of sharp lines, singlets. If the broad-band decoupler is turned off, a spectrum can be obtained through a technique called **off-resonance decoupling**, which allows the ^{13}C signals to be split according to the $n + 1$ rule, only by the *directly attached* (not adjacent) hydrogens. As there are no attached hydrogens, quaternary carbons will still appear as singlets, but methine carbons (one attached hydrogen) will be doublets, methylene carbons (two attached hydrogens) triplets, and methyl carbons (three attached hydrogens) will appear as quartets. Figure 15.64 summarizes this notion and shows the decoupled and off-resonance decoupled spectra of ethyl maleate.

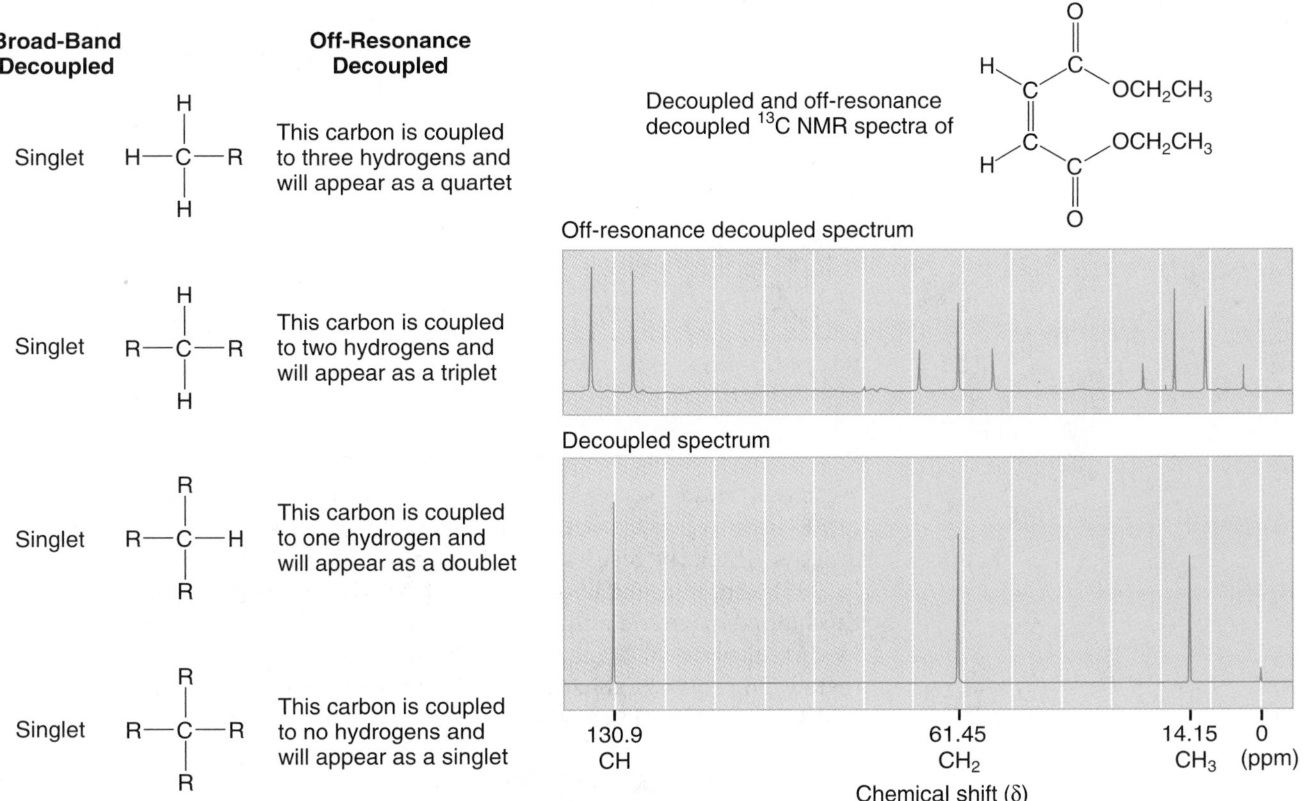

FIGURE **15.64** The ^{13}C NMR spectra are generally taken with coupling to hydrogen removed through a broad-band decoupling of all hydrogens. If that coupling is left in, as in an off-resonance decoupling experiment, the number of hydrogens attached to the carbon atoms can be determined.

Table 15.5 gives a series of chemical shift correlations for ^{13}C NMR. Notice the greatly expanded range of chemical shifts extending over 200 ppm, as opposed to the mere 10 ppm for ^{1}H. Dispersion is much greater for ^{13}C NMR than for ^{1}H NMR. Once again, the question of memorization arises. Use Table 15.5 to work problems, and you will automatically memorize the information you really need (if you do enough problems!).

TABLE 15.5 Some ^{13}C Chemical Shifts

Type of Carbon	Chemical Shift (δ)[a]
Alkanes	
Methyl	0–30
Methylene	15–55
Methine	25–55
Quaternary	30–40
Alkenes	
C=C	80–145
Alkynes	
C≡C	70–90
Aromatics	110–170
Benzene	128.7
Alcohols, ethers	
C–O	50–90
Amines	
C–N	40–60
Halogens	
C–F	70–80
C–Cl	25–50
C–Br	10–40
C–I	−20–10
Carbonyls, C=O	
R_2C=O	190–220
RXC=O (X = O or N)	150–180

[a]The chemical shift δ is in parts per million (ppm) from TMS.

15.10 SOMETHING MORE: DYNAMIC NMR

We have already seen that rates of hydrogen exchange are important in NMR spectroscopy. Recall that fast exchange of hydroxyl and amine hydrogens generally results in a decoupling of the OH and NH hydrogens (p. 725). An adjacent hydrogen "feels" only an averaged perturbation of the applied magnetic field if exchange is fast enough. If exchange is slowed by careful removal of all acidic or basic catalysts, the coupling can be detected. The *rate* of exchange is important to the spectrum. If it is fast, the NMR experiment detects an averaged spectrum; if it is slow, it sees a static picture.

The classic example of this phenomenon involves the spectrum of cyclohexane. As you know, there are two different kinds of hydrogen in cyclohexane, the axial and equatorial hydrogens, and these are interconverted by a ring flip, with an activation energy of about 11 kcal/mol (Chapter 6, p. 204). Despite the existence of two kinds of hydrogen, the room-temperature spectrum of cyclohexane shows only a single line at δ 1.4 ppm (Fig. 15.65). This simple spectrum appears because the rate of interconversion of axial and equatorial hydrogens is so fast, with respect to the time necessary for the NMR measurement, that an averaged spectrum appears. Suppose we decrease the energy available to the system by lowering the temperature. If we can decrease the rate of the ring flip suf-

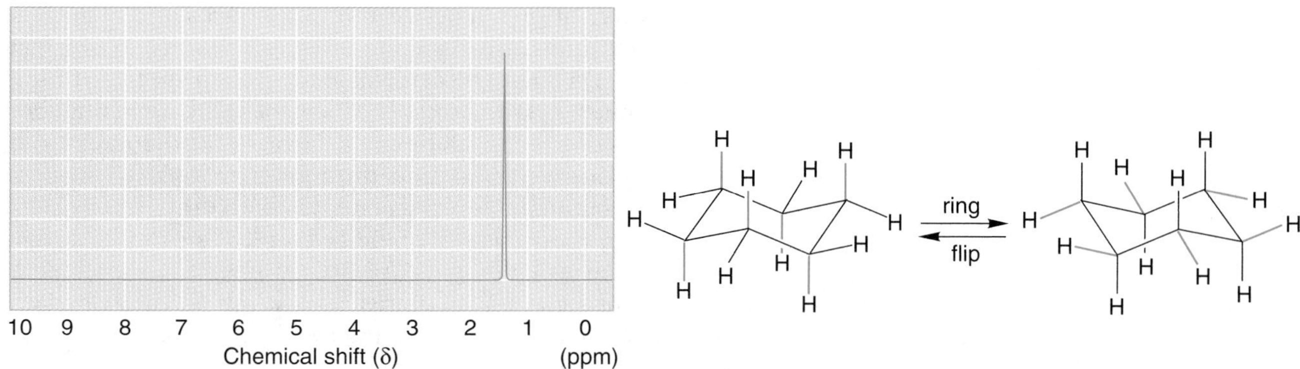

FIGURE 15.65 The two kinds of hydrogen in cyclohexane are averaged by rapid ring flipping. At room temperature there is only one sharp signal.

ficiently, the spectrometer will no longer detect an average of the two possible positions but will be able to see both axial and equatorial hydrogens. Figure 15.66 shows the ^{1}H NMR spectrum of undecadeuteriocyclohexane (cyclohexane-d_{11}) at –90 °C. Two equal signals appear, centered upon the position of the room temperature averaged signal.

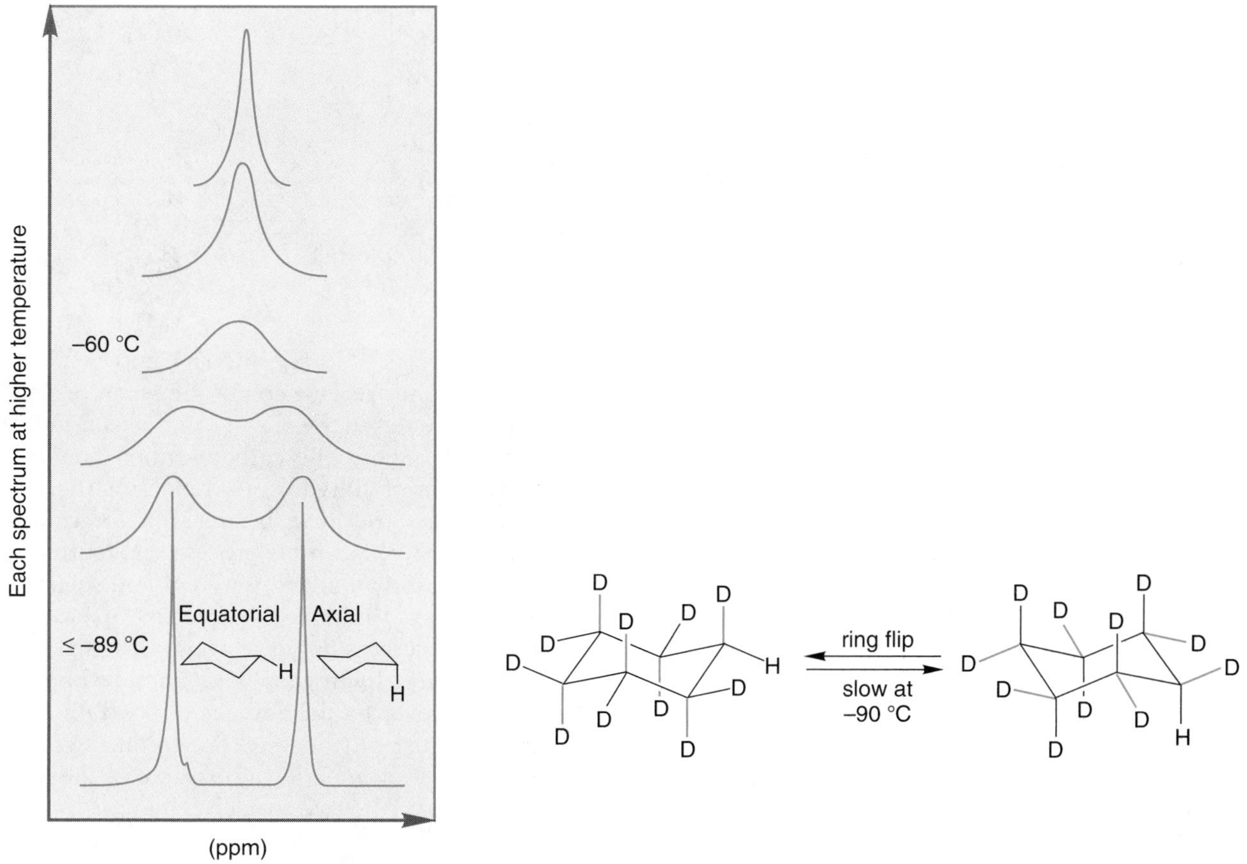

Each spectrum at higher temperature

–60 °C

≤ –89 °C

Equatorial Axial

(ppm)

ring flip

slow at –90 °C

FIGURE 15.66 At low temperature, ring flipping is slowed and the axial and equatorial hydrogens of cyclohexane can be resolved. As the temperature is raised, the –90 °C spectrum approaches the 25 °C spectrum.

At –90 °C, the interconversion between the two forms of cyclohexane is slow enough so that the NMR machine no longer detects an averaged signal. In practice, for a signal to be detectable for a given species, it must have a lifetime of about 1 s. As the temperature is raised to room temperature, the lifetime of individual cyclohexanes decreases as the rate of ring flipping increases, and an averaged spectrum appears.

Now, what happens at an intermediate temperature? Figure 15.66 shows you. A common sense answer would be correct here. If two signals are found at –90 °C, and a single signal at 25 °C, at intermediate temperatures the two kinds of signal must approach each other. The two sharp peaks of the –90 °C spectrum will slowly merge into the single peak of the 25 °C spectrum. In fact, it is possible to derive kinetic information from the temperature dependance of NMR spectra, and this is still another use of NMR spectroscopy.

More molecules than you might at first expect give this kind of averaged spectrum. It takes a detailed look at structure to see why. Consider ethyl alcohol. If we ignore the OH signal, which we know to be concentration and catalyst dependent, we expect to see only one peak for the three equivalent methyl hydrogens and another for the two equivalent methylene hydrogens. But wait—look at the Newman projection in Figure 15.67, which shows the most stable staggered conformation of ethyl alcohol.

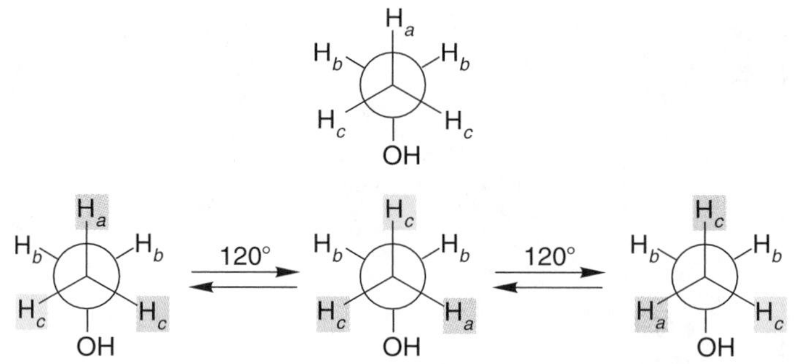

FIGURE 15.67 In static ethyl alcohol, there are three different kinds of hydrogen. However, rapid rotation around the carbon–carbon bond interconverts the methyl hydrogens and makes them equivalent. The two methylene hydrogens are enantiotopic.

There are three kinds of hydrogen (H_a, H_b, and H_c), not two. The three methyl hydrogens are *not* the same. Yet, we do not see separate signals for them. By now you should be able to see the answer to this conundrum. As shown in Figure 15.67, rotation about the carbon–carbon bond of ethyl alcohol interconverts the three methyl hydrogens, rendering them equivalent. Moreover, we know that the low barrier for rotation, about 3 kcal/mol, ensures a rapid rate for this simple process. In Figure 15.67, H_a and the two H_c hydrogens are individualized with colored squares. Note how each hydrogen occupies one of the three possible positions as rotation occurs. The NMR spectrometer detects only the averaged signal.

What about the two methylene hydrogens, H_b? The Newman projection of Figure 15.67 shows that even in the absence of rapid rotation they are equivalent, and therefore must *always* give rise to one signal. These hydrogens are enantiotopic. As we saw earlier, the origin of the name lies in the mirror-image relationship of the two to each other.

PROBLEM 15.19 In a solvent consisting of a single enantiomer, the two enantiotopic hydrogens of ethyl alcohol (H_b in Fig. 15.67) appear as separate signals. Explain.

*PROBLEM **15.20**

Use the device of replacing the methylene hydrogens (H_b in Fig. 15.67) with an X group to show that they are enantiotopic (p. 711).

ANSWER

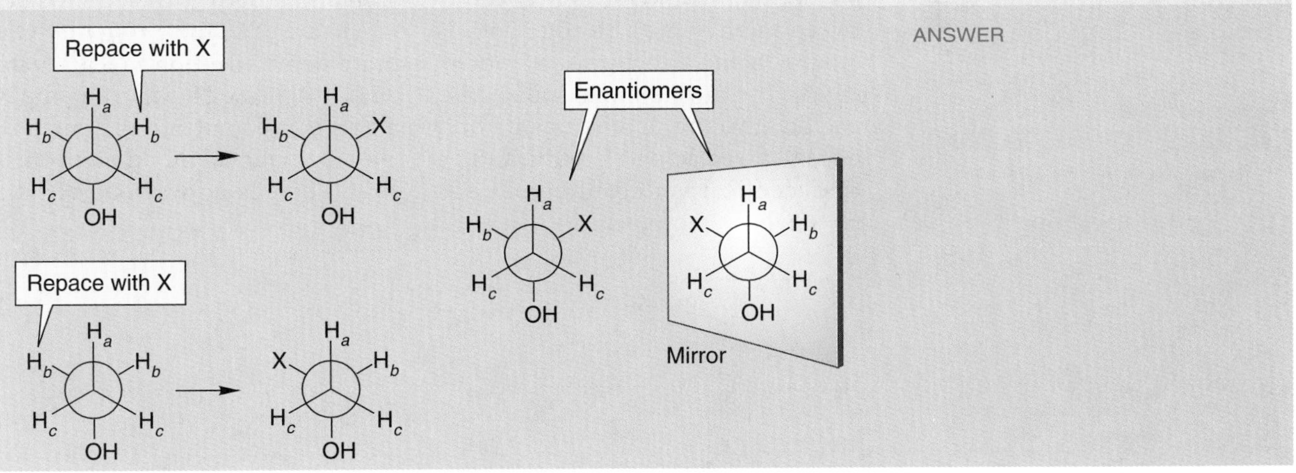

So, as long as rotation is fast, neither the individual methyl hydrogens nor the individual methylene hydrogens of ethyl alcohol are distinguishable by NMR spectroscopy.

Now let's change some things about ethyl alcohol. Let's first convert it into propyl alcohol by replacing one methyl hydrogen with a methyl group. There are now three sets of hydrogens, the methyl hydrogens and two sets of different methylene hydrogens. Rapid rotation ensures that the three methyl hydrogens will be equivalent, and each pair of methylene hydrogens consists of two magnetically equivalent, enantiotopic hydrogens (Fig. 15.68).

Now let's replace a hydrogen of propyl alcohol with a chlorine to make 2-chloro-1-propanol. Look at the Newman projection of this molecule in Figure 15.69. Are the two methylene hydrogens equivalent? No. The Newman projections of Figure 15.69 show the three rotational isomers and in *all* of them the two hydrogens are different. *Never* does H_x become equivalent to H_y. They are forever different no matter how fast rotation about the carbon–carbon bond is, and so must give rise to separate signals. These hydrogens are diastereotopic, and will give rise to distinct signals. This point is particularly difficult, so let's anticipate the Common Errors section a little bit and deal with the problem right here. Is not H_y in drawing **2** of Figure 15.69 in the same position as H_x in drawing **1**? Are the two not equivalent? Both are 120° from the hydroxyl group, and both lie in between a methyl group and a hydrogen (Fig. 15.69).

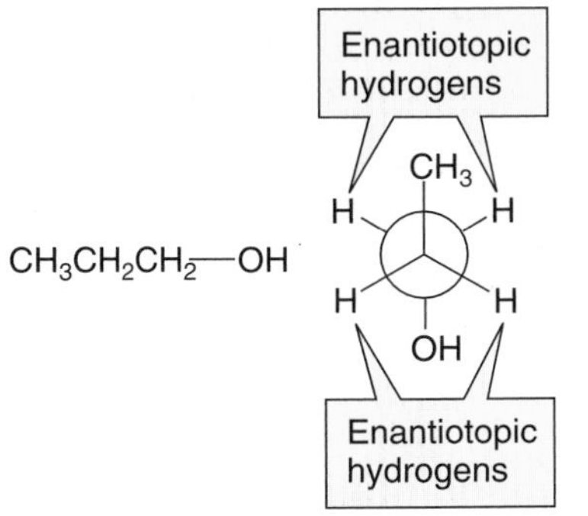

FIGURE **15.68** In propyl alcohol, the methyl hydrogens are homotopic and there are two pairs of enantiotopic methylene hydrogens.

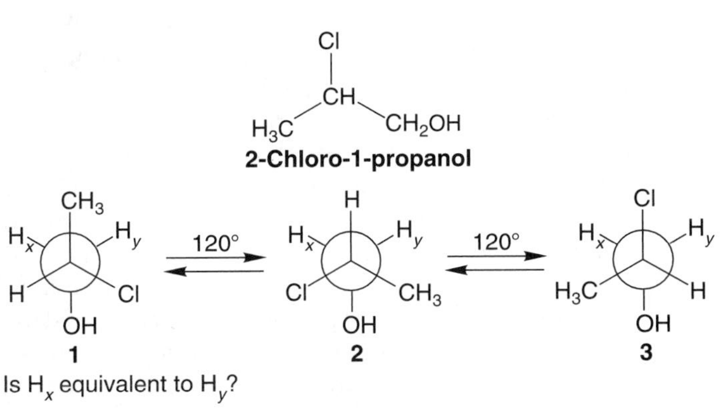

FIGURE **15.69** In 2-chloro-1-propanol, the two methylene hydrogens, H_x and H_y, are neither equivalent (homotopic) nor enantiotopic.

No, H_x in drawing **1** of Figure 15.69 is different from H_y in drawing **2.** A short demonstration shows that the two OH groups in drawings **1** and **2** are not identical. One is between hydrogen and chlorine, the other is between methyl and chlorine. Both H_x and H_y are 120° from *different* OH groups in the two forms. A more elaborate demonstration takes advantage of the device mentioned earlier (p. 711). Replace H_x with a group X and determine the relationship between this compound and the one produced by replacing H_y with X. These two compounds are different, neither identical nor enantiotopic (Fig. 15.70). There is no way out, H_x and H_y are diastereotopic.

replace H_x with X

replace H_y with X

These two structures are neither identical nor enantiomers; they are diastereomers, and H_x and H_y are diastereotopic hydrogens

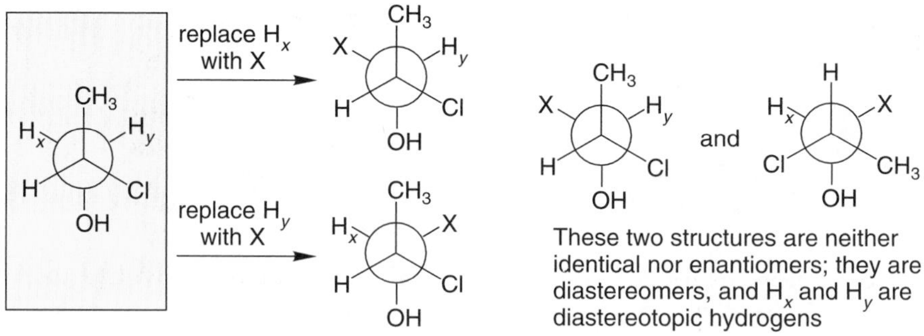

FIGURE **15.70** The two hydrogens are diastereotopic and will give different signals in the NMR spectrum.

PROBLEM **15.21** The methyl group of *trans*-1-*tert*-butyl-4-methylcyclohexane appears as a single doublet, but in *cis*-1-isopropyl-3-phenylcyclohexane there are two methyl doublets (Fig. 15.72). Explain.

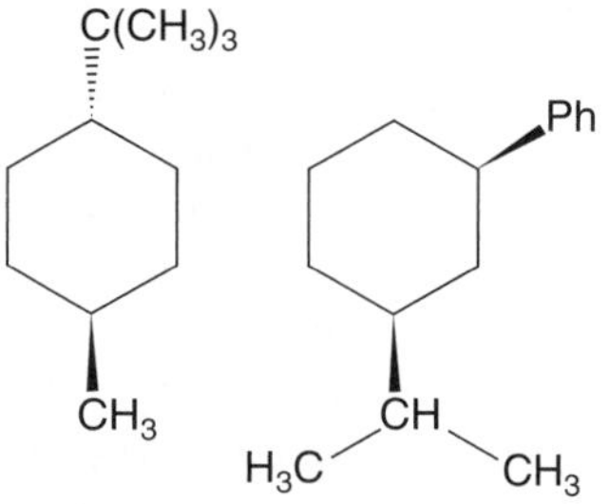

FIGURE **15.71**

15.11 SUMMARY

NEW CONCEPTS

This chapter recapitulates the ideas first introduced in Chapter 5 in the discussion of column chromatography, and introduces the much more versatile separation methods of gas chromatography and high-performance liquid chromatography. All chromatography works in much the same way: The components of a mixture are forced to partition themselves between a moving phase and a stationary phase. The more a given component exists in the moving phase, the faster it moves through the column; the more strongly it is absorbed, the longer it remains on the column.

In mass spectrometry, a molecule is first ionized by bombardment with high-energy electrons. The parent ion so formed can undergo numerous fragmentation reactions before it passes, along with the daughter ions formed through fragmentation, into a region between the poles of a magnet. Ions are separated by mass, and focused on a detector. The mass spectrometer gives the molecular weight of a molecule, provided that the parent ion can be detected, and high-resolution mass spectrometry can determine the molecular formula of an ion. Structural information can often be gained through an analysis of the important daughter ions in the fragmentation pattern.

Infrared spectroscopy detects vibrational excitation of chemical bonds. Functional groups have characteristic vibrational frequencies in the IR that depend on the masses of the atoms in the bond and, importantly, on the force constant of the bond. Analysis of an IR spectrum gives information on the kinds of functional groups present.

Nuclear magnetic resonance detects the energy absorbed in the transition from a low-energy state in which the nuclear spin is aligned with an external magnetic field to a slightly higher energy state in which it is aligned against the applied field. Hydrogens, ^{13}C, and some other nuclei have characteristic absorptions (the chemical shift, δ), which depend critically on the surrounding chemical environment. Hydrogen nuclei are coupled to adjacent hydrogens through a coupling constant, J. The number of lines observed for a particular hydrogen depends on the number of other nuclei to which it is coupled. For first-order spectra the number of lines equals $n + 1$, where n is the number of equivalent coupled nuclei. The NMR spectra can be integrated to reveal the relative number of hydrogens in each peak.

COMMON ERRORS

Practice, practice, practice! The successful interpretation of spectra requires it. The most common error is not to realize this. The memorization of tables of data will not make you into a spectroscopist, but working lots and lots of problems will.

A central question in this chapter, and in all of chemistry, is, What's different? When are two hydrogens equivalent and when are they not? This chapter offers a device that is useful in making such distinctions.

It is often difficult to keep track of the interplay of the applied magnetic field, B_0, the chemical shift, δ, and the coupling constant, J. Even at very high fields, first-order spectra are not always seen unless the coupled hydrogens have very different chemical shifts.

15.12 KEY TERMS

Base peak The largest peak in a mass spectrum, to which all other peaks are referred.

Chemical shift (δ) The position, on the ppm scale, of a peak in an NMR spectrum. The chemical shift for ^{1}H and ^{13}C is given relative to a standard, TMS, and is determined by the chemical environment surrounding the nucleus.

Coupling constant (J) The magnitude (in hertz) of J, the measure of the spin–spin interaction between two nuclei.

Daughter ion In MS, an ion formed by the fragmentation of the first-formed parent ion.

Decoupling The removal of coupling between hydrogens or other nuclei (see **coupling constant**) through either chemical exchange or electronic means.

Diastereotopic Diastereotopic hydrogens are different both chemically and spectroscopically under all circumstances.

Enantiotopic Enantiotopic hydrogens are chemically and spectroscopically equivalent except in the presence of optically active (single enantiomer) reagents.

Force constant A property of a bond related to the bond strength: to the stiffness of the bond. Bonds with high force constants absorb at high frequency in the IR.

Fragmentation pattern The characteristic spectrum

of ions formed by decomposition of a parent ion produced in a mass spectrometer when a molecule is bombarded by high-energy electrons.

Gas chromatogaphy (GC) A method of separation in which molecules are forced to equilibrate between the moving gas phase and a stationary phase packed in a column. The less easily a molecule is adsorbed in the stationary phase, the faster it moves through the column.

GC/IR The combination of gas chromatography and infrared spectroscopy in which the molecules separated by a gas chromatograph are led directly into an infrared spectrometer for analysis.

GC/MS The combination of gas chromatography and mass spectrometry in which the molecules separated by a gas chromatograph are led directly into a mass spectrometer for analysis.

High-performance liquid chromatography (HPLC) An especially effective version of column chromatography in which the stationary phase consists of many tiny spheres that provide an immense surface area for absorption.

Homotopic Homotopic hydrogens are identical, both chemically and spectroscopically, under all circumstances.

Infrared spectroscopy (IR) In IR, absorption of IR energy causes bonds to vibrate. The characteristic vibrational frequencies can be used to determine the functional groups present in a molecule.

Integration The determination of the relative numbers of hydrogens corresponding to the signals in an NMR spectrum.

Karplus curve The dependence between the coupling constant of two hydrogens on adjacent atoms and the dihedral angle between the carbon–hydrogen bonds.

Long-range coupling Any coupling between nuclei separated by more than three bonds. It is not usually observed in ^{1}H NMR spectroscopy.

Mass spectrometry (MS) The analysis of the ions formed by bombardment of a molecule with high-energy electrons. High-resolution MS can give the molecular formula of an ion, and an analysis of the fragmentation pattern can give information about the structure.

McLafferty rearrangement A common process in MS through which a hydrogen in the γ-position of a carbonyl compound is transferred to oxygen to give an enol radical cation and an alkene.

Nuclear magnetic resonance spectroscopy (NMR) Spectroscopy that detects the absorption of energy as the lower energy nuclear spin state in which the nuclear spin is aligned with an applied magnetic field flips to the higher energy spin state in which it is aligned against the field. See **chemical shift** and **coupling constant**.

Off-resonance decoupling A technique used in ^{13}C NMR in which coupling between ^{13}C and ^{1}H is restricted to the hydrogens directly attached to the carbon. The number of directly attached hydrogens can be determined from the multiplicity of the observed signal for the carbon.

Parent ion The ion directly formed in the mass spectrometer by ejection of an electron from the sample molecule. The parent ion commonly undergoes many fragmentation reactions before detection.

p + 1 peak An ion with a mass one greater than that of the parent ion. It appears because molecules contain atoms with isotopes of mass one greater than that of the most commonly observed isotope.

ppm scale The chemical shift scale commonly used in NMR spectroscopy in which the positions of absorptions are quoted as parts per million (ppm) of applied magnetic field. The chemical shift is independent of the magnetic field on the ppm scale.

Radical cation A species that is positively charged yet contains a single unpaired electron. These are commonly formed by ejection of an electron when a molecule is bombarded with high-energy electrons in a mass spectrometer.

Tetramethylsilane (TMS) The standard "zero point" on the ppm scale in NMR spectroscopy. The chemical shift of a nucleus is quoted in parts per million of applied magnetic field relative to the position of TMS.

Wavenumber The wavenumber ($\tilde{\nu}$) equals $1/\lambda$ or ν/c.

15.13 ADDITIONAL PROBLEMS

General advice: Although this section is longer than most problem sections, it cannot do more than illustrate some of the difficulties involved in going from a set of lines on chart paper, or a table of numbers, to a structure. Moreover, in the real world of chemistry there is close interaction between the reactions involved and the interpretation of spectra. The chemist has some idea (usually!) of what is likely to be present. Some of the later problems must be solved in this interactive way. The chemistry involved will suggest structures, and you will have to see if they are consistent with the data available. All of the following problems involve manipulating spectral data. You should start well equipped with a table of IR frequencies and ^{1}H NMR and ^{13}C NMR correlation charts for chemical shifts. A table of coupling constants will also be essential, and a copy of the Karplus curve is often useful. Do not be shy about looking values up. Most chemists analyze their "real world" spectra that way, and *everyone* starts that way.

Specific advice: In Chapter 4 we first encountered Ω, the number of "degrees of unsaturation." In this chapter, we see nitrogen-containing compounds and will have to modify our formula a bit. Here is the protocol for nitrogen and some other heteroatoms. Be sure to review Ω in Chapter 4 (p. 148).

1. Oxygen and divalent sulfur can be ignored.
2. Halogens (F, Cl, Br, and I) are counted as if they were hydrogens.
3. Treat nitrogen as if it were a carbon. However, as nitrogen is trivalent, not tetravalent, subtract one hydrogen per nitrogen from the total number of hydrogens computed for the saturated hydrocarbon.

We begin with some relatively straightforward problems largely involving infrared spectroscopy and mass spectrometry, and then move on.

PROBLEM **15.22** Here is a practical problem of no intellectual content whatsoever. In doing problems involving IR, it is often much easier to read the values on the chart paper in microns (μ), rather than wavenumber ($\tilde{\nu}$). Unfortunately, most of us have learned to think in wavenumbers. There is a simple conversion. Convert 3.00, 5.00, 7.00, 9.00, and 11.00 μ into wavenumbers.

PROBLEM **15.23** Calculate the degrees of unsaturation in the following molecules:

(a) (b) (c) (d)

(e) (f) (g)

PROBLEM **15.24** The IR spectrum of a molecule known to have structure **A**, **B**, **C**, or **D** is shown below. Which structure is most consistent with the spectrum? Explain your reasoning.

A **B**

C **D**

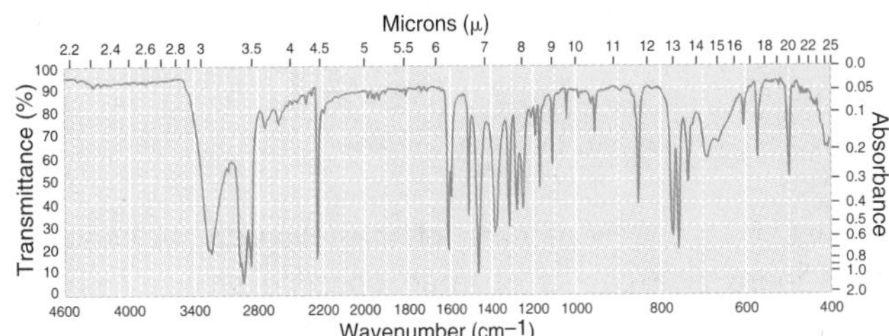

PROBLEM **15.25** The IR spectra of compounds **1–5** are shown below. Each compound is either an aldehyde, an anhydride, a carboxylic acid, an ester, or a ketone. Identify to which class each compound belongs based on its IR spectrum. Explain your reasoning by assigning characteristic and/or distinguishing absorptions. Each class is represented by one compound.

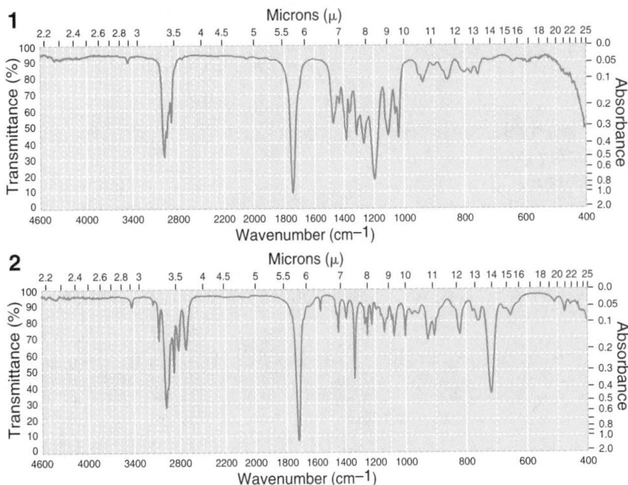

PROBLEM **15.25** (CONTINUED)

3

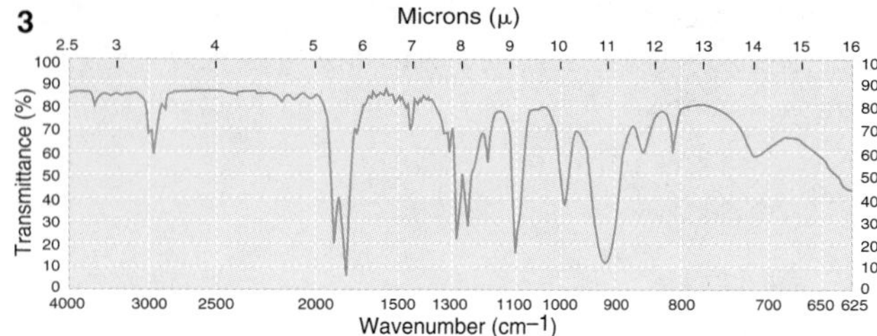

4

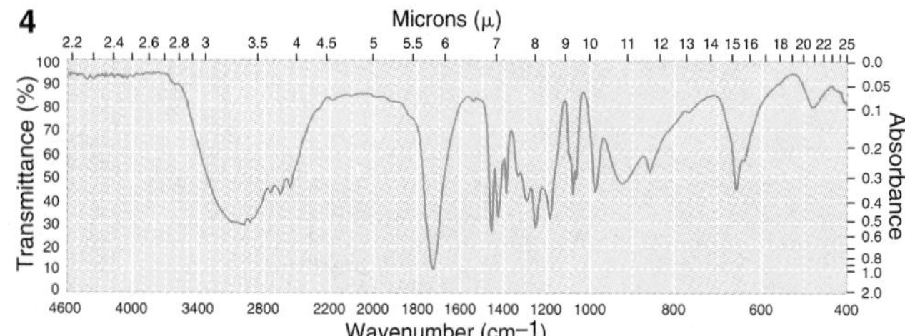

5

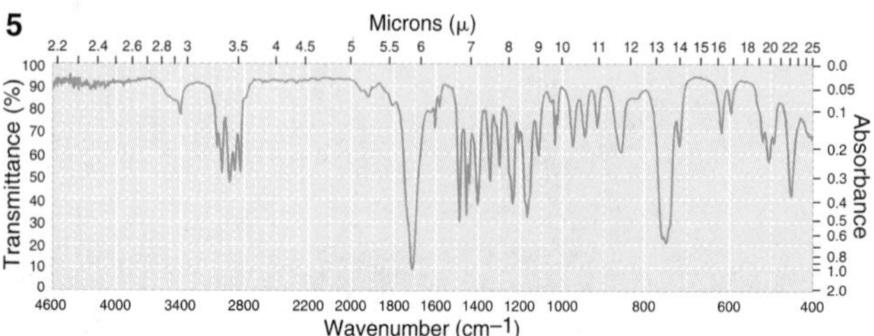

PROBLEM **15.26** For each of the following IR spectra (**1–5**) there is a choice of three possible types of compounds. For each spectrum, choose the most appropriate class of compound. Explain your reasoning by noting the presence or absence of characteristic bands in the spectrum.

1 Alcohol
Carboxylic acid
Phenol

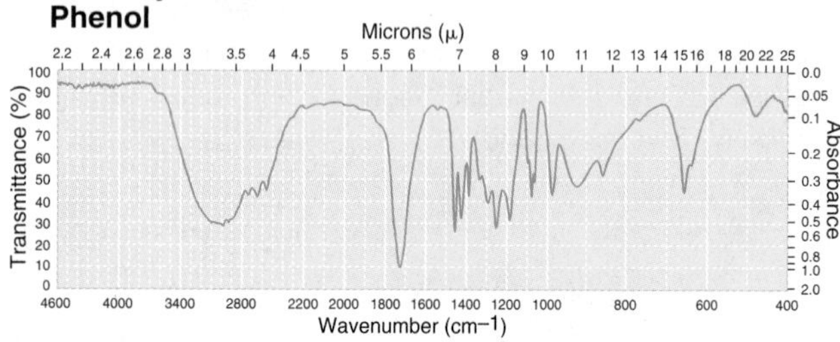

2 Aldehyde
Ester
Ketone

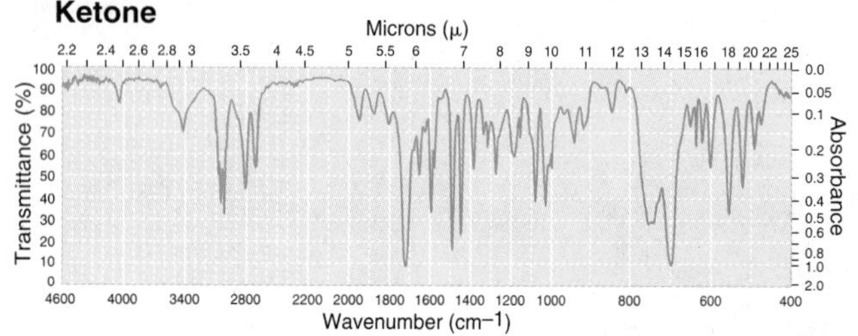

3 1-Alkyne
Symmetrical disubstituted alkyne
Cyanide

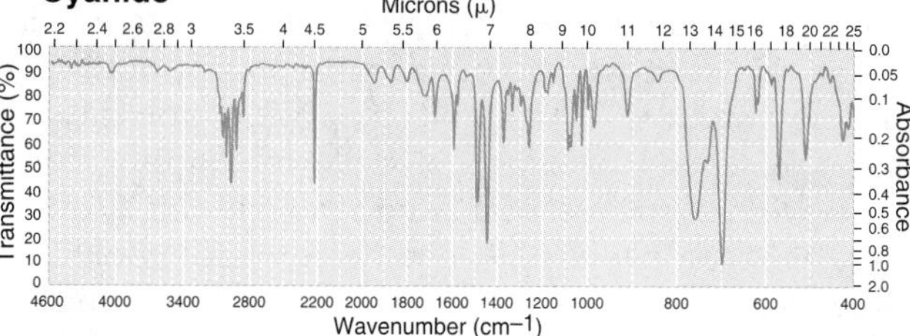

4 Primary amide
Primary amine
Nitro compound

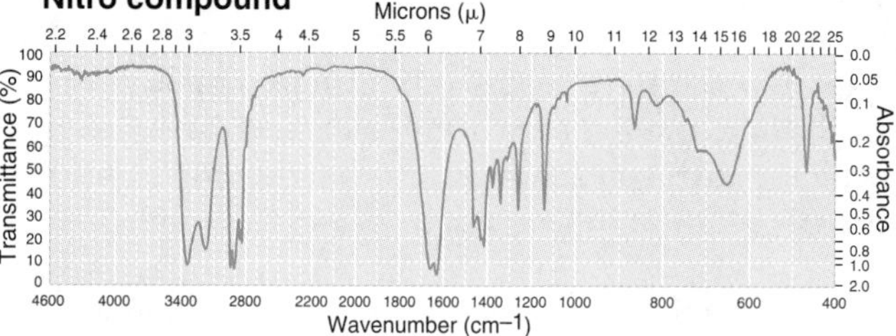

5 Anhydride
Carboxylic acid
Ester

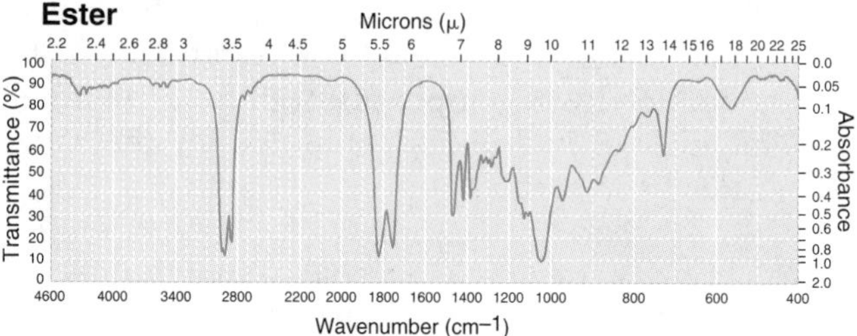

PROBLEM **15.27** Assign the IR spectra below to the correct octene isomer. Explain your reasoning.

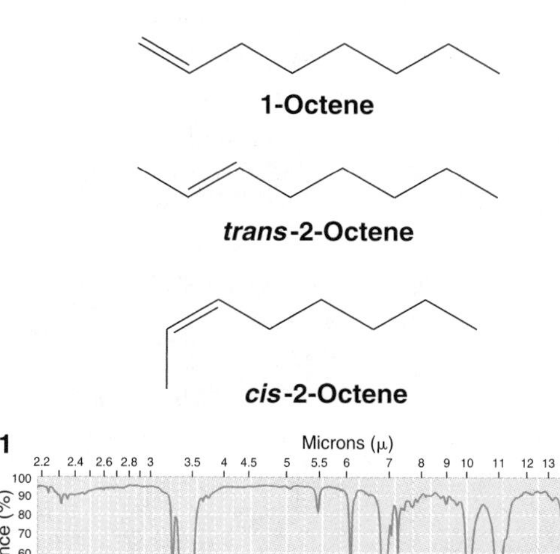

1

Microns (μ)

Transmittance (%)

Absorbance

Wavenumber (cm−1)

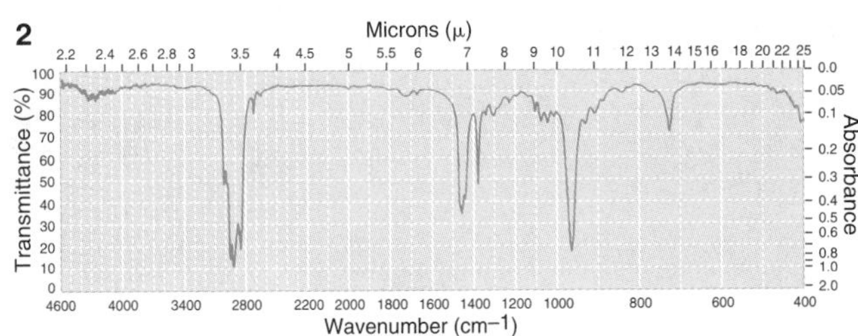

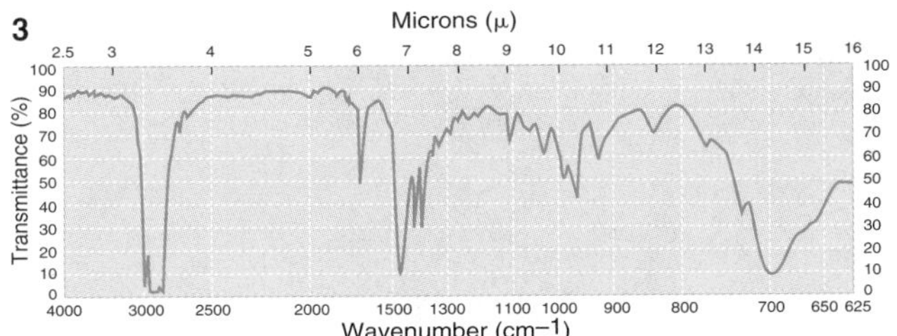

PROBLEM **15.28** The IR spectra below belong to 5-phenyl-1-pentyne and 6-phenyl-2-hexyne. Which is which? Explain your reasoning.

$PhCH_2CH_2CH_2C{\equiv}CH$ $\quad$ $PhCH_2CH_2CH_2C{\equiv}CCH_3$

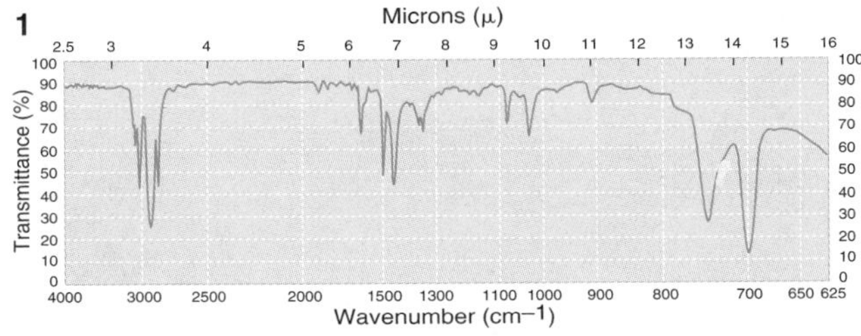

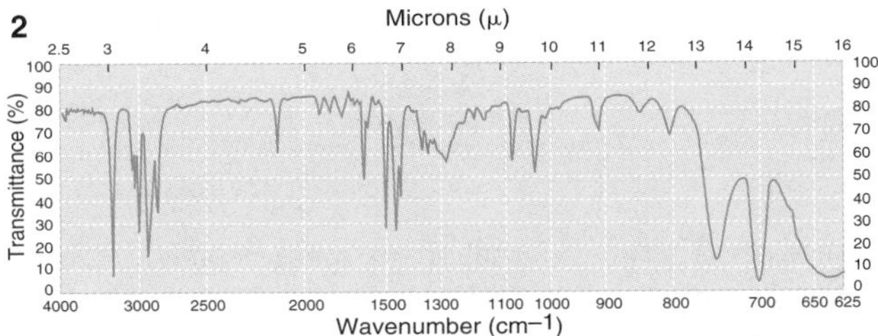

PROBLEM **15.29** Assign the IR spectra below to the correct chlorotoluene isomer.

CH_3 / Cl $\quad$ CH_3 / Cl $\quad$ CH_3 / Cl

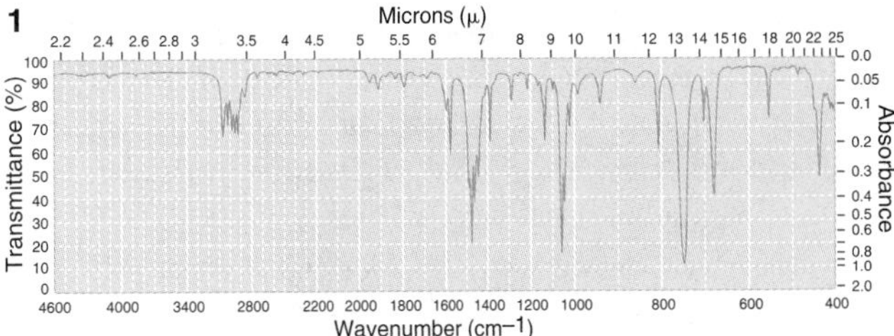

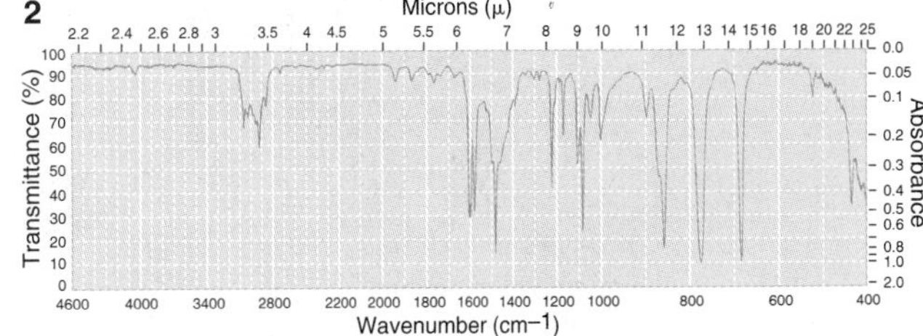

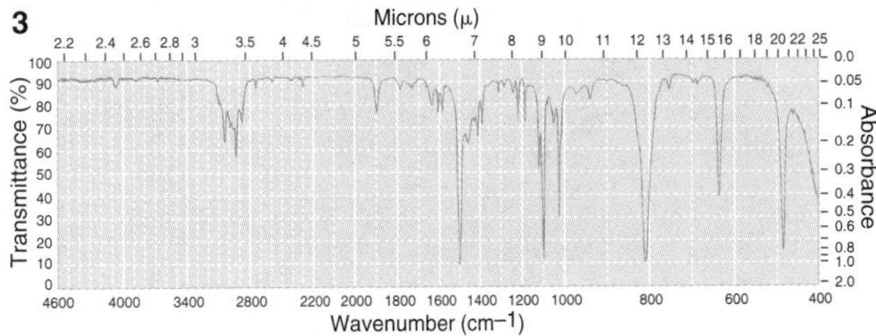

PROBLEM **15.30** Compound **A** has the formula $C_8H_7BrO_2$. The two most intense peaks in the mass spectrum are at $m/z = 216$ and 214, and they appear in almost equal amounts. Why? Compound **A** is an aldehyde. Explain why the second-most intense peaks in the mass spectrum come at m/z values of 215 and 213.

PROBLEM **15.31** Compound **A**, containing only C, H, and O, gave 80.00% C and 6.70% H upon elemental analysis. Summaries of the mass spectrum and the IR spectrum of compound **A** are shown below. Deduce the structure of **A** and explain your reasoning.

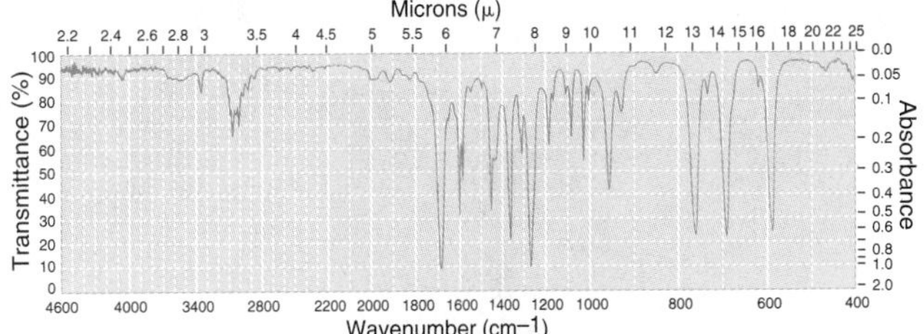

m/z	Relative Abundance
105	100
77	88
51	40
120	29
50	21
43	17
78	10
39	9

PROBLEM **15.32** Compound **A**, containing only C, H, and O, gave 70.60% C and 5.90% H upon elemental analysis. Summaries of the mass spectrum and the IR spectrum of compound **A** are shown below. Deduce the structure of **A** and explain your reasoning.

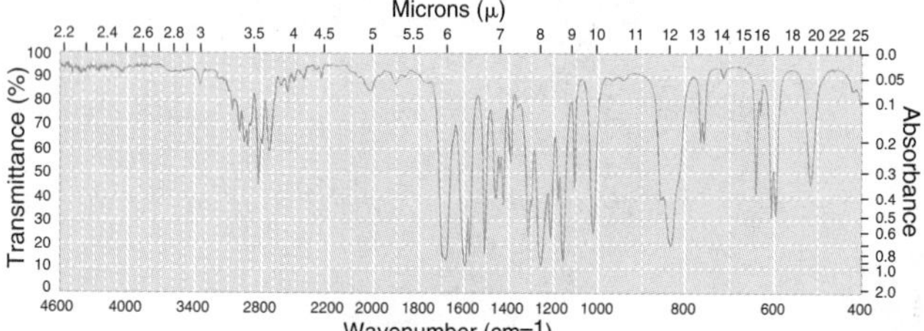

m/z	Relative Abundance
135	100
136	71
77	28
92	15
107	15
39	8
65	8
63	8

PROBLEM **15.33** Compounds **A**, **B**, and **C** all gave 71.17% C, 5.12 % H, and 23.71% N upon elemental analysis. Summaries of the mass spectra and the IR spectra of these compounds are shown below. Deduce the structures of **A**–**C** and explain your reasoning.

A

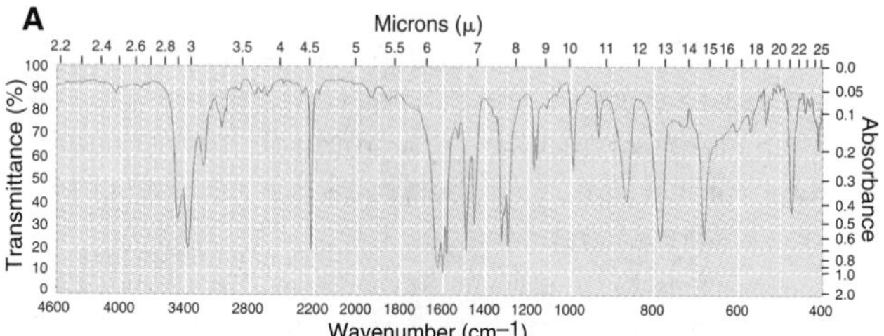

m/z	Relative Abundance
118	100
96	20
119	9
64	8
90	7
63	6
39	5
117	5

B

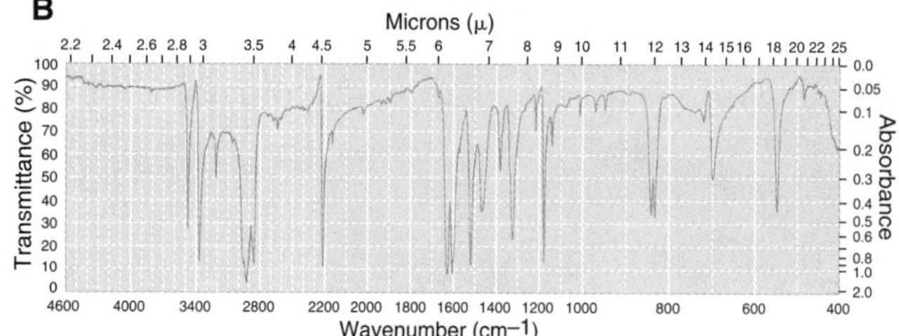

m/z	Relative Abundance
118	100
96	40
119	22
64	22
90	19
63	14
39	14
117	12

C

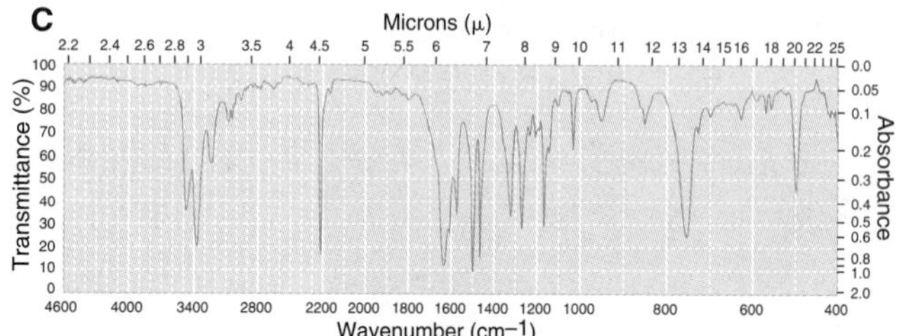

m/z	Relative Abundance
118	100
96	35
119	12
64	7
90	7
63	7
39	6
117	4

PROBLEM **15.34** Extraction of black Hemiptera bugs affords a colorless oil **A** of the formula $C_7H_{12}O$ possessing a sickening odor. The IR spectrum of **A** displays characteristic peaks at 2833 (w), 2725 (w), 1725 (s), 1380 (m), and 975 (m) cm^{-1}. Upon catalytic hydrogenation, the peak at 975 cm^{-1} disappears. Ozonolysis of compound **A**, followed by an oxidative workup, gives succinic acid ($HOOC-CH_2CH_2-COOH$) and propionic acid ($HOOC-CH_2CH_3$). Deduce a structure for **A** and explain your reasoning.

Now we move on to NMR spectroscopy. Again, we start with simple exercises:

PROBLEM **15.35** Explain carefully why the following two spectra of 1-bromobutane look so different:

(a) Spectrum taken at 60 MHz.

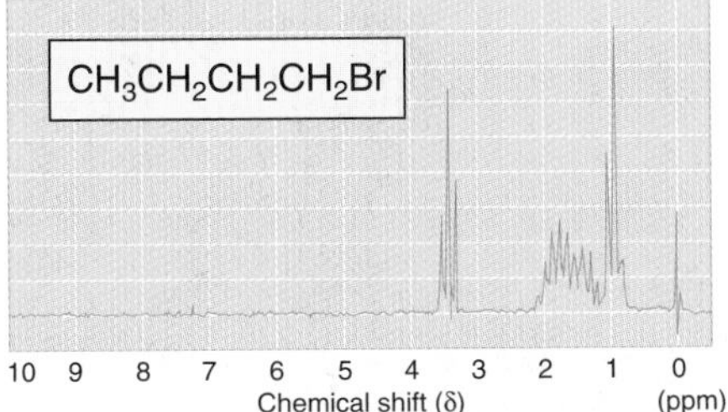

(b) Spectrum taken at 300 MHz.

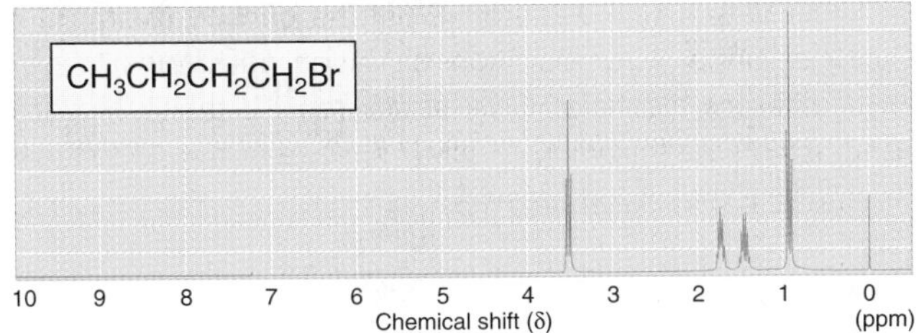

PROBLEM **15.36** Are the underlined hydrogens in the following molecules homotopic, enantiotopic, or diastereotopic? Justify your answers.

(a) H_3C, CH_3, H H

(b) H_3C, CH_2CH_3, H H

(c) H_3C, $CH(CH_3)_2$, H H

(d) H_3C, H, C, CH_3, CH_2CH_3, H H

PROBLEM **15.37** Assign the following chemical shifts to the hydrogens in compounds in (a–d). Use the $(n + 1)$ rule to predict the multiplicities of the peaks.

(a) $Cl_2CH—CHCl—CH_3$

δ
5.88
4.37
1.70

(b) $ClCH_2—CH(OCH_3)_2$

δ
4.53
3.53
3.43

(c) $(CH_3)_2CH—CH_2I$

δ
3.16
1.72
1.03

(d) $BrCH_2CH_2CH_2Br$

δ
3.59
2.36

PROBLEM **15.38** Predict the first-order spectra of the indicated hydrogens in the molecules in (a–c).

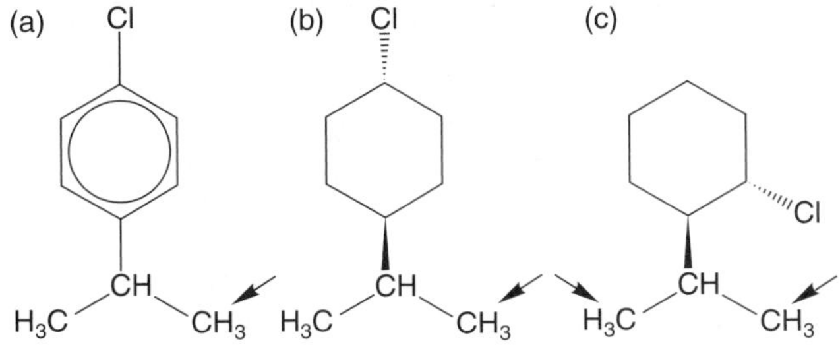

PROBLEM **15.39** Figure 15.54 shows an enlarged view of the multiplet centered at roughly δ 3.6 in the ^1H NMR spectrum of pure ethyl alcohol. Use a tree diagram to analyze this complex multiplet.

PROBLEM **15.40** The ^1H NMR spectra of the two scrupulously dry alcohols shown below have signals (among others) at δ 2.60 (d, $J = 3.8$ Hz, 1H) [**A**] and δ 2.26 (t, $J = 6.1$ Hz, 1H) [**B**]. Assign the spectra, and explain your reasoning.

OH OH

PROBLEM **15.41** When a drop of aqueous acid is added to the samples described in Problem 15.40, the signals described collapse to broad singlets. Explain.

PROBLEM **15.42** In Sections 15.4 and 15.9, it was claimed that the NMR signals for ^2H (deuterium) and ^{13}C will not overlap ^1H spectra. Verify that this statement is true. Show that the frequency (ν) will be different for ^1H, ^2H, and ^{13}C at a field strength (B_0) of 4.7 T. The gyromagnetic ratios for ^1H, ^2H, and ^{13}C are 2.7×10^8, 0.41×10^8, and 0.67×10^8 rad T^{-1} s^{-1}, respectively.

PROBLEM **15.43** Match the following ^1H NMR absorptions to the following compounds, each of which has a one-signal spectrum: δ 2.25, 0.04, 1.42, 0.90, 7.37 ppm.

$(H_3C)_3Si—Si(CH_3)_3$ $C(CH_3)_4$

CH_3, H_3C, CH_3, H_3C, CH_3, CH_3

PROBLEM 15.44 Match the following ^{13}C NMR absorptions to the following compounds, each of which has a one-signal spectrum: δ 26.9, −2.9, 59.7, 128.5.

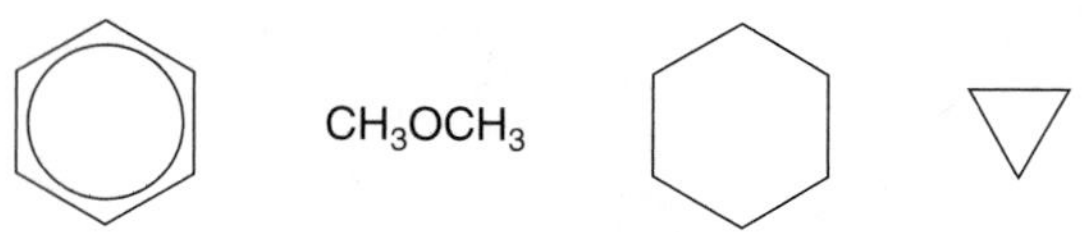

PROBLEM 15.45 The following spectrum appears in an old catalog of ^1H NMR spectra of organic compounds. Is the assigned structure correct? What's wrong?

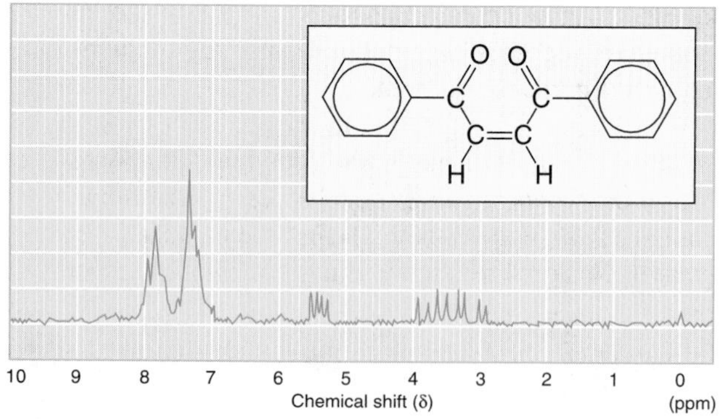

PROBLEM 15.46 In the synthesis of the compound described in Problem 15.45, HCl is produced. Use this information to postulate a reasonable structure for the compound whose ^1H NMR spectrum is shown above.

PROBLEM 15.47 Isobutene can be hydrated in acid, or under hydroboration–oxidation conditions to give two different alcohols. Give structures for these alcohols and assign the peaks in the following ^1H NMR spectra (**1** and **2**). Which peaks would vanish if a drop of D_2O were added?

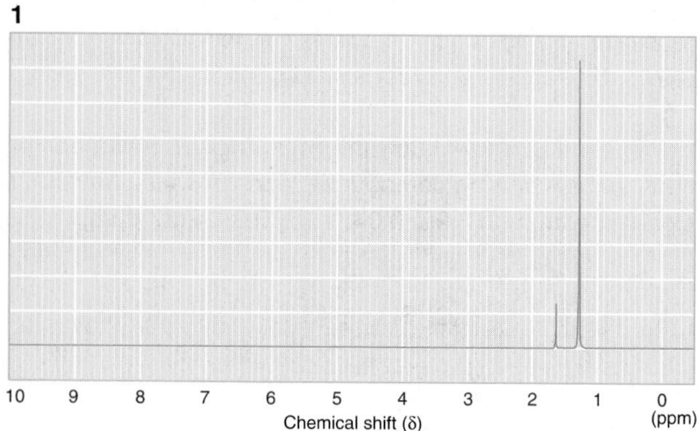

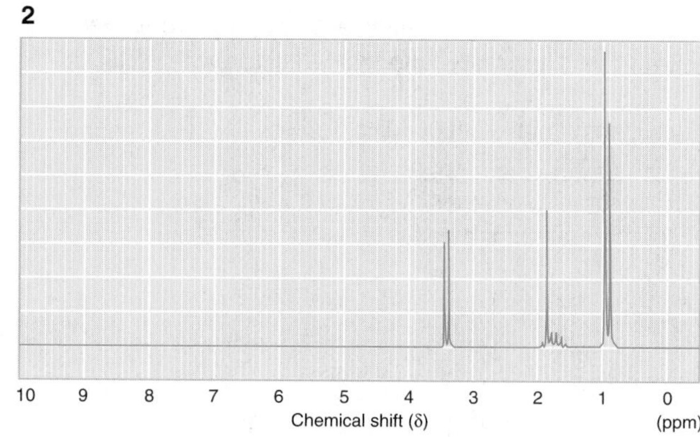

PROBLEM 15.48 2-Methyl-2-butene can be hydrated in acid, or under hydroboration–oxidation conditions to give two different alcohols. Give structures for these alcohols and assign the following ^1H NMR spectra (**1** and **2**) to these isomers. Explain your answers.

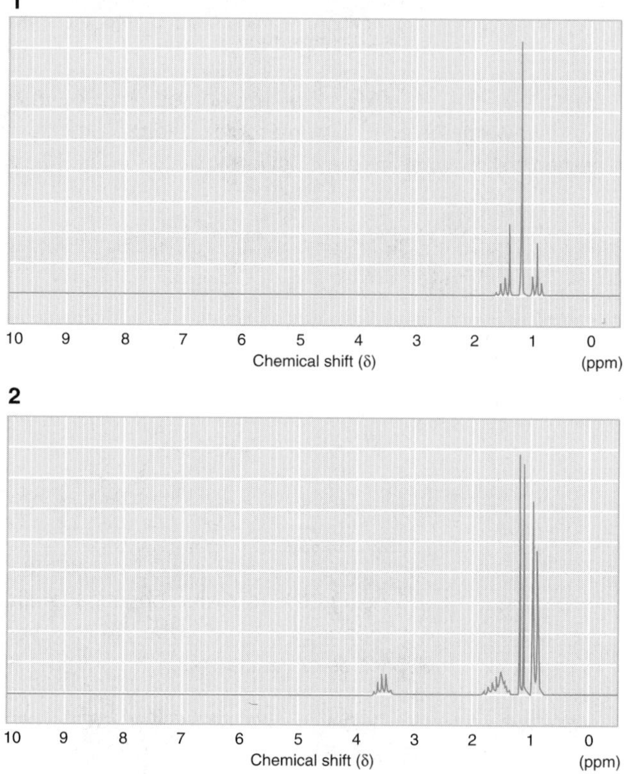

PROBLEM 15.49 On careful examination, it can be seen that the two upfield signals in spectrum 2 of Problem 15.48 (at δ 0.90 and 0.93) are both doublets. Explain.

PROBLEM **15.50** The ^{1}H NMR spectra for the following eight-membered ring compounds are summarized below. Assign structures, and explain your reasoning.

A: δ 1.54 (s), **B**: δ 5.74 (s), **C**: δ 2.39 (m, 4H), 5.60 (m, 4H), **D**: δ 1.20–1.70 (m, 4H), 1.85–2.50 (m, 4H), 5.35–5.94 (m, 4H).

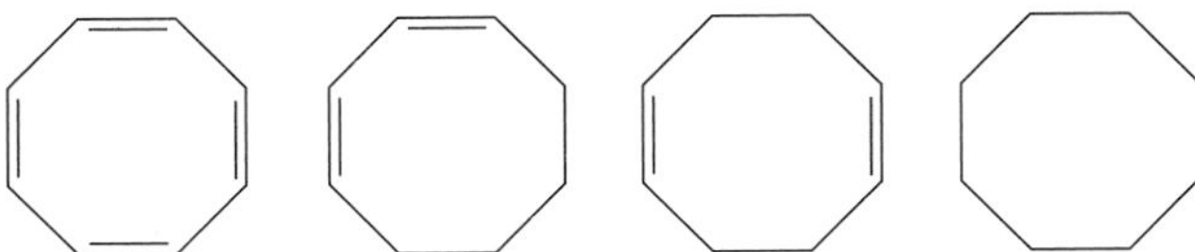

PROBLEM **15.51** You are given three bottles containing *o*-dichlorobenzene, *m*-dichlorobenzene, and *p*-difluorobenzene, along with broad-band decoupled ^{13}C NMR spectra of the three isomers. Assign the three spectra to the three compounds and explain your reasoning.

A: δ 127.0, 128.9, 130.6, 135.1 ppm. **B**: δ 127.7, 130.5, 132.6 ppm. **C**: δ 116.5, 159.1 ppm

PROBLEM **15.52** Could this same technique (looking at symmetry) allow you to assign the spectra of 4-methylcyclohexanone, 2-methylcyclohexanone, and 2,3-dimethylcyclopentanone, **A**, **B**, and **C**? If not, suggest an experiment, involving only ^{13}C NMR spectroscopy, that would let you finish the job. (*Hint*: See p. 735.)

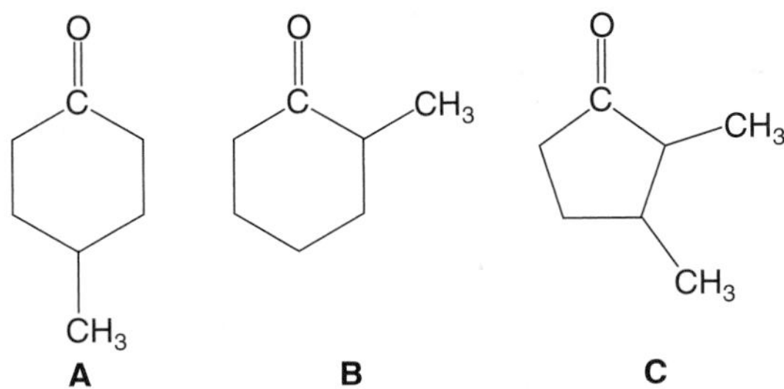

PROBLEM **15.53** The three isomers of pentane, C_5H_{12}, have the following ^{13}C NMR spectra. Assign the spectra of these three compounds.

A: δ 13.9, 22.8, 34.7, **B**: δ 22.2, 31.1, 32.0, 11.7, **C**: δ 31.7, 28.1

PROBLEM **15.54** Assign the following ^{1}H NMR spectra to the isomers of trichlorophenol shown below and explain your reasoning.

(a) OH, Cl, Cl, Cl (b) OH, Cl, Cl, Cl

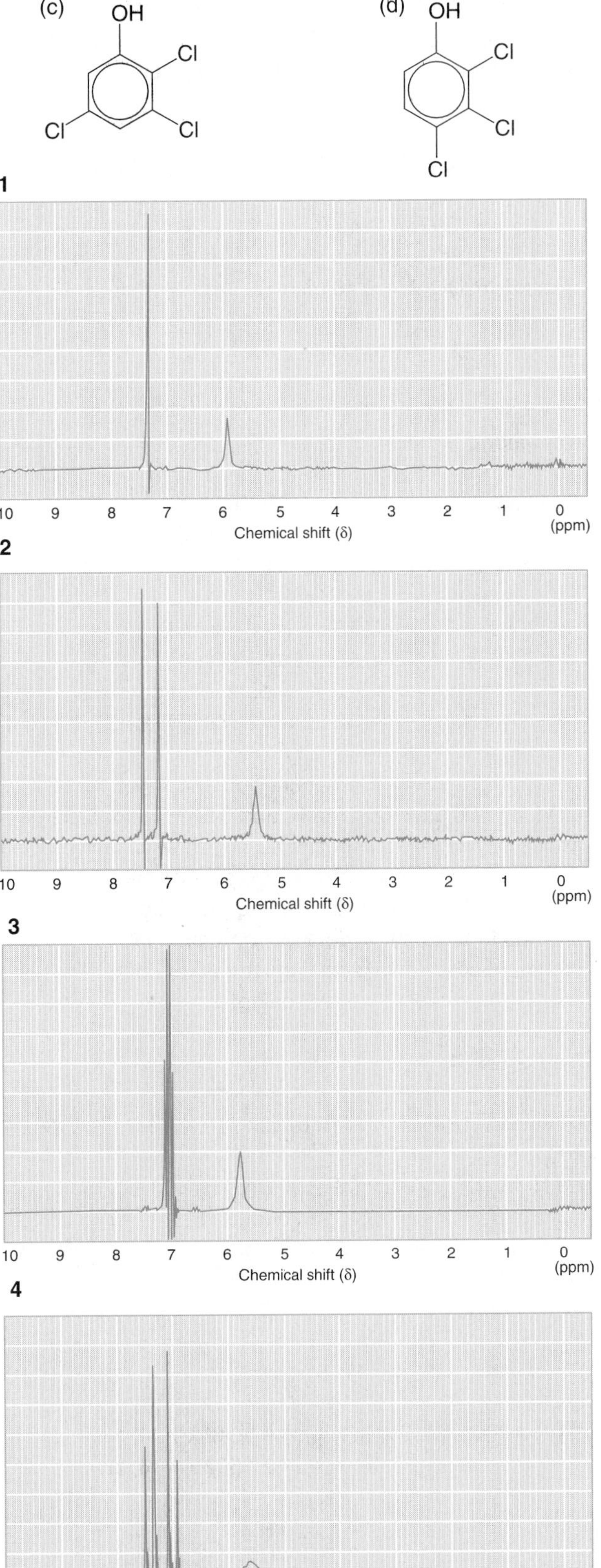

PROBLEM **15.55** Isomeric compounds **A** and **B** have the composition $C_{11}H_{12}O_4$. Spectral data are summarized below. Deduce structures for **A** and **B** and explain your reasoning.

Compound **A**

IR (KBr): 1720 (s), 1240 (s) cm^{-1}

^{1}H NMR ($CDCl_3$): δ 2.46 (s, 3H)
3.94 (s, 6H)
8.05 (d, J = 2 Hz, 2H)
8.49 (t, J = 2 Hz, 1H)

Compound **B**

IR (Nujol®): 1720 (s), 1245 (s) cm^{-1}

^{1}H NMR ($CDCl_3$): δ 2.63 (s, 3H)
3.91 (s, 6H)
7.28 (d, J = 8 Hz, 1H)
8.00 (dd, J = 8 Hz, 2Hz, 1H),
dd = doublet of doublets
8.52 (d, J = 2 Hz, 1H)

PROBLEM **15.56** Compounds **A** and **B**, containing only C, H, and O, gave 63.15% C, and 5.30% H upon elemental analysis. Their mass spectra do not show a parent ion, but data from freezing point depressions indicate a molecular weight of 150 g/mol. Catalytic hydrogenation of both **A** and **B** leads to **C**. Spectral data for compounds **A**, **B**, and **C** are summarized below. Assign structures for compounds **A**, **B**, and **C** and explain your reasoning.

Compound **A**

IR (Nujol): 1845 (m), 1770 (s) cm^{-1}

^{1}H NMR ($CDCl_3$): δ 2.03–2.90 (m, 2H)
3.15–3.70 (m, 1H)
5.65–6.25 (m, 1H)

Compound **B**

IR (Nujol): 1825 (m), 1755 (s) cm^{-1}

^{1}H NMR ($CDCl_3$/DMSO-d_6): δ 1.50–2.10 (m, 1H)
2.15–2.70 (m, 1H)

Compound **C**

IR (neat): 1852 (m), 1786 (s) cm^{-1}

^{1}H NMR ($CDCl_3$/DMSO-d_6): δ 1.10–2.30 (m, 4H)
3.05–3.55 (m, 1H)

PROBLEM **15.57** Spectral data for isomeric compounds **A** and **B** are summarized below. Assign structures for compounds **A** and **B**, and explain your reasoning.

Compound **A**

Mass spectrum: m/z = 148 (p, 7%), 106 (8%), 105 (100%), 77(29%), 51(8%)

IR (neat): 1675 (s), 1220 (s), 980 (s), and 702 (s) cm^{-1}

^{1}H NMR ($CDCl_3$): δ 1.20 (d, J = 7 Hz, 6H)
3.53 (septet, J = 7 Hz, 1H)
7.20–7.60 (m, 3H)
7.80–8.08 (m, 2H)

Compound **B**

IR (neat): 1705 (s), 738 (m), and 698 (s) cm^{-1}

^{1}H NMR ($CDCl_3$): δ 0.95 (t, J = 7 Hz, 3H)
2.35 (q, J = 7 Hz, 2H)
3.60 (s, 2H)
7.20 (s, 5H)

PROBLEM **15.58** Upon direct photolysis or heating at 220 °C, the dimer of 2,5-dimethyl-3,4-diphenyl-2,4-cyclopentadien-1-one (**A**) yields compound **B**, mp 166–168 °C.

O
H_3C C CH_3
H_5C_6 C_6H_5
A
(dimer)
$h\nu$, CH_3CN or 220 °C → **B**

Spectral data for compound **B** are summarized below. Deduce the structure of compound **B** and explain your reasoning. (You may have to make an educated guess about the actual stereochemistry of **B**.)

Compound **B**

Mass spectrum: m/z = 492 (p)

IR (CCl_4): 1704 cm^{-1}

^1H NMR ($CDCl_3$): δ 0.73 (s, 3H)
0.92 (s, 3H)
1.51 (s, 3H)
1.88 (s, 3H)
6.6–7.5 (m, 20H)

PROBLEM **15.59** An appreciation of the concepts of electronegativity (inductive effects) and of electron delocalization (resonance effects), combined with an understanding of ring current effects, permits both rationalization and prediction of many chemical shifts.

(a) The carbons of ethylene have a ^{13}C NMR chemical shift of δ 123.3 ppm. Rationalize the ^{13}C NMR chemical shifts of the β carbons of 2-cyclopenten-1-one (**A**) and methyl vinyl ether (**B**). (*Hint*: Think resonance!)

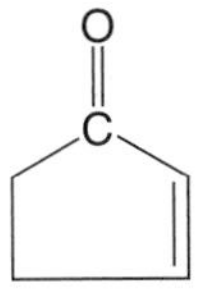

$CH_3O—CH{=}CH_2$

165.1 **A** 84.2 **B**

(b) The ^1H NMR chemical shifts of hydrogens ortho, meta, and para to a substituent on a benzene ring can also be correlated in this way. For reference sake, note that the hydrogens in benzene appear at about δ 7.37 ppm. Aniline (**C**) exhibits three shielded aromatic hydrogens (i.e., δ < 7.0), whereas the other two aromatic hydrogens appear in the normal range. On the other hand, two of the aromatic hydrogens in acetophenone (**D**) are deshielded (δ 7.9–8.1 ppm), whereas the other three aromatic hydrogens have a normal chemical shift. Finally, all five aromatic hydrogens in chlorobenzene (**E**) appear at δ 7.3 ppm. Assign the ^1H NMR chemical shifts of the aromatic hydrogens for **C**, **D**, and **E** by considering inductive, resonance, and ring current effects. Additionally, the ^{13}C NMR chemical shifts for the aromatic carbons of **C** and **E** are shown in color below. (The ^{13}C NMR chemical shift for benzene is δ 128.5 ppm.) Are these ^{13}C NMR chemical shifts consistent with the ^1H NMR chemical shifts for **C** and **E**?

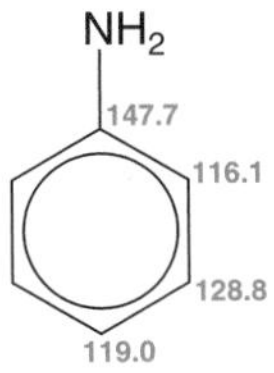

NH_2: 147.7, 116.1, 128.8, 119.0 — **C**

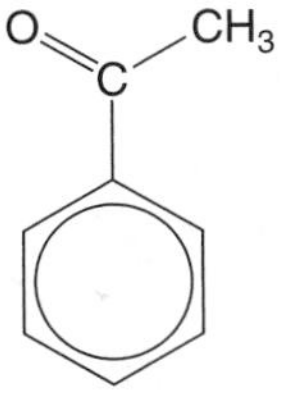

D

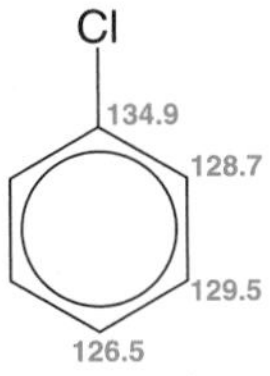

Cl: 134.9, 128.7, 129.5, 126.5 — **E**

PROBLEM **15.60** Compound **A** was isolated from the bark of the sweet birch *(Betula lenta)*. Compound **A** is soluble in 5% aqueous NaOH solution but not in 5% aqueous $NaHCO_3$ solution. The spectral data for compound **A** are summarized below. Deduce the structure of compound **A**. (Be sure to assign the aromatic resonances in the ^1H NMR spectrum. Note that J_{meta} and J_{para} were not observed.) *Hint*: The pK_a of phenol is about 10, and that of benzoic acid is about 4.5.

Compound **A**

Mass spectrum: m/z = 152 (p, 49%), 121 (29%), 120 (100%), 92 (54%)

IR (neat): 3205 (br), 1675 (s), 1307 (s), 1253 (s), 1220 (s), and 757 (s) cm^{-1}

^1H NMR ($CDCl_3$, 300 MHz): δ 3.92 (s, 3H)
6.85 (t, J = 8 Hz, 1H)
7.00 (d, J = 8 Hz, 1H)
7.44 (t, J = 8 Hz, 1H)
7.83 (d, J = 8 Hz, 1H)
10.8 (s, 1H)

^{13}C NMR ($CDCl_3$): δ 52.1 (q), 112.7 (s), 117.7 (d), 119.2 (d), 130.1 (d), 135.7 (d), 162.0 (s), 170.7 (s)

PROBLEM **15.61** Dehydration of α-terpineol (**A**) affords a mixture of at least two isomers, one of which, **B**, reacts with maleic anhydride to give **C**.

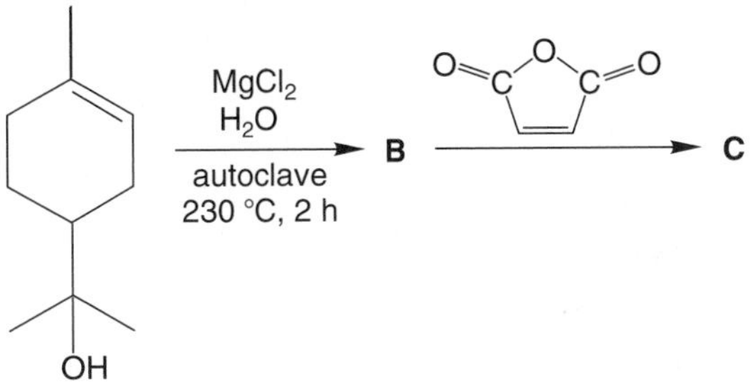

A (OH) —[$MgCl_2$, H_2O, autoclave, 230 °C, 2 h]→ **B** —[maleic anhydride]→ **C**

Isomer **B**, which exhibits the pleasant odor of lemons and can be isolated from the essential oils of cardamom, marjoram, and coriander, reacts with maleic anhydride to yield compound **C**, mp 64–65 °C. Spectral data for compound **C** are summarized below. Deduce the structures of compounds **B** and **C**. (You may have to make an educated guess about the stereochemistry of compound **C**.)

PROBLEM **15.61** (CONTINUED)

Compound **C**

Mass spectrum: $m/z = 234$ (p)

IR (Nujol): 1870 (sh), 1840 (m), and 1780 (s) cm^{-1}

^{1}H NMR ($CDCl_3$): δ 0.98 (d, $J = 6.5$ Hz, 3H)
1.08 (d, $J = 6.5$ Hz, 3H)
1.36 (m, 4H)
1.46 (s, 3H)
2.60 (m, 1H)
2.80 (d, $J = 9$ Hz, 1H)
3.20 (d, $J = 9$ Hz, 1H)
5.96 (m, 2H)

PROBLEM **15.62** The ^{1}H NMR spectra of 2,2,2-trichloroethanol and 2,2,2-trifluoroethanol are shown below. Assign the spectra and explain your reasoning.

CCl_3CH_2OH CF_3CH_2OH

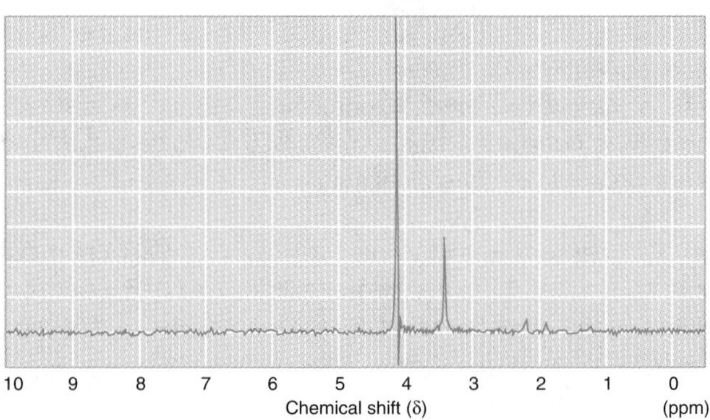

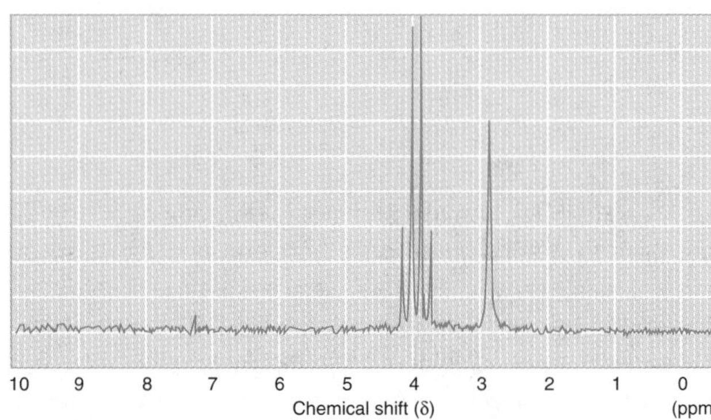

Carbonyl Chemistry 1: Addition Reactions

"You have two choices," Jamf cried, his last lecture of the year: outside were the flowery strokings of wind, girls in pale colored dresses, oceans of beer, male choruses intensely, movingly lifted as they sang *Semper sit in flores!* Semper sit in flo-ho-res ... "stay behind with carbon and hydrogen, take your lunch-bucket in to the works every morning with the faceless droves who can't wait to get in out of the sunlight—or move *beyond*. Silicon, boron, phosphorus—these can replace carbon, and can bond to nitrogen instead of hydrogen.—" a few snickers here, not unanticipated by the playful old pedagogue, be he always in flower: his involvement in getting Weimar to subsidize IG's Stickstoff Syndikat was well known—"move beyond life, toward the inorganic. Here is no frailty, no mortality—here is Strength, and the Timeless." Then his well-known finale, as he wiped away the scrawled C—H on his chalkboard and wrote, in enormous letters, Si—N.

—Thomas Pynchon*
Gravity's Rainbow

In Chapters 4 and 9, we examined the structure of alkenes and the addition reactions of their carbon–carbon double bonds. Indeed, we have been exploring various aspects of carbon–carbon double-bond chemistry since Chapter 9. Now it is time to extend what we know to the chemistry of carbon–oxygen double bonds, the **carbonyl group**. Although it bears many similarities to carbon–carbon double bonds, the carbon–oxygen double bond is not "just another double bond," as there are also many important differences. The chemistry of the carbonyl group is the cornerstone of synthetic chemistry. Much of our ability to make molecules depends on manipulation of carbonyl groups because they allow us to make carbon–carbon bonds. There are several centers of reactivity in a molecule containing a carbon–oxygen double bond, and this versatility contributes to the widespread use of carbonyl compounds in synthesis. There is the π bond itself (don't forget the antibonding π* orbital), and the lone-pair electrons on oxygen. In addition, there is special reactivity associated with the carbon–hydrogen bonds in the position adjacent (α) to the carbon–oxygen double bond (Fig. 16.1). The centerpiece reaction of this chapter is the acid- or base-induced addition to the carbon–oxygen double bond. The chemistry of the α-position will wait until Chapter 18.

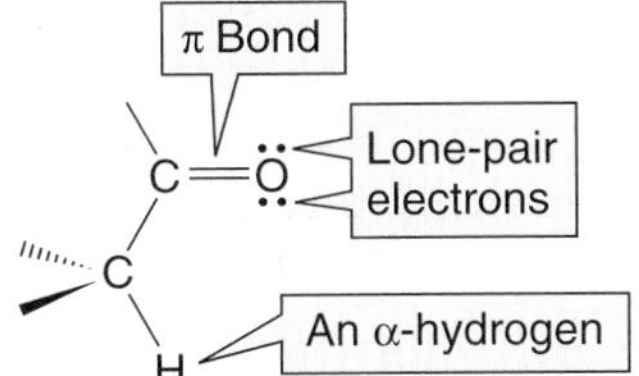

FIGURE 16.1 A generic carbonyl group has three loci of reactivity: the nonbonding, or lone-pair electrons; the π bond; and the α carbon–hydrogen bonds.

16.1 STRUCTURE OF THE CARBON–OXYGEN DOUBLE BOND

By analogy to the carbon–carbon double bond, we might expect sp^2 hybridization for the carbon atom. The $2p$ orbitals on carbon and oxygen

*Our first Pynchon quote was in Chapter 9. Same man, same book.

form the π bond, and two lone pairs of electrons remain on oxygen. Figure 16.2 gives a schematic orbital picture of both the σ and π systems for the simplest carbonyl compound, formaldehyde.

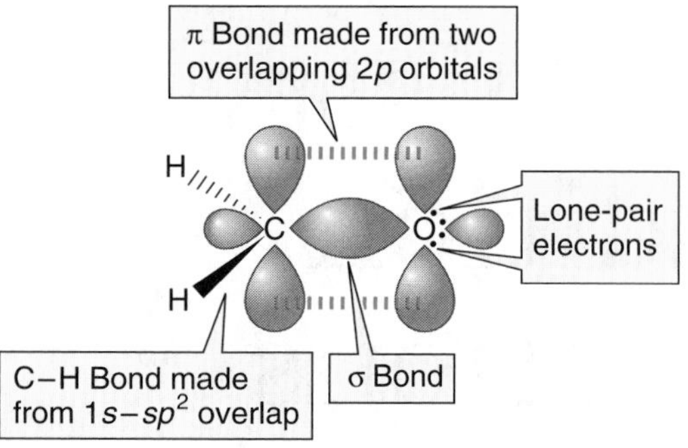

FIGURE **16.2** One orbital picture of the simplest carbonyl compound, formaldehyde.

PROBLEM **16.1** Make schematic, graphical orbital interaction diagrams for each different kind of bond in the molecule formaldehyde, shown in Figure 16.2.

The structures of some known carbonyl compounds show that this picture is a reasonable one. Bond angles are close to the 120° required by pure sp^2 hybridization, and of course we would not expect a molecule of less than perfect trigonal symmetry to have exactly sp^2 hybridization. Bond lengths are similar to those in simple alkenes, although the carbon–oxygen double bond is a bit shorter and stronger than its carbon–carbon counterpart (Fig. 16.3).

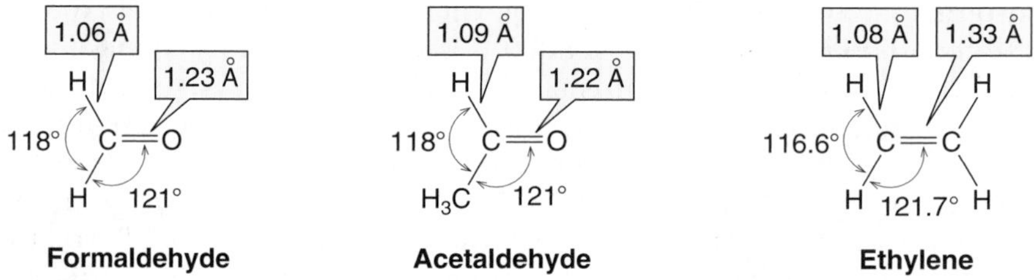

1.06 Å, 1.23 Å, 118°, 121° — **Formaldehyde**

1.09 Å, 1.22 Å, 118°, 121° — **Acetaldehyde**

1.08 Å, 1.33 Å, 116.6°, 121.7° — **Ethylene**

FIGURE **16.3** A comparison of the structures of formaldehyde, acetaldehyde, and ethylene.

The π bond is constructed in the same way as the related carbon–carbon π bond. The 2*p* orbitals on the adjacent carbon and oxygen atoms are allowed to interact in a constructive way to form a bonding π molecular orbital, and in a destructive, antibonding way to form π*. But there is a difference. In the alkenes, the constituent 2*p* orbitals are of equal energy, and therefore contribute equally to the formation of π and π*. In a carbonyl group, this is not the case. The 2*p* orbital on the more electronegative oxygen atom is lower in energy than the 2*p* orbital on a carbon atom, so the oxygen atom contributes more to π than the more energetically remote carbon 2*p* orbital. The bonding π molecular orbital will be constructed from more than 50% of the oxygen 2*p* atomic orbital and less than 50% of the carbon 2*p* atomic orbital. In a similar way, we can see that π* will be made up of more than 50% of the energetically close carbon 2*p* orbital and less than 50% of the energetically more remote 2*p* orbital on oxygen (Fig. 16.4).

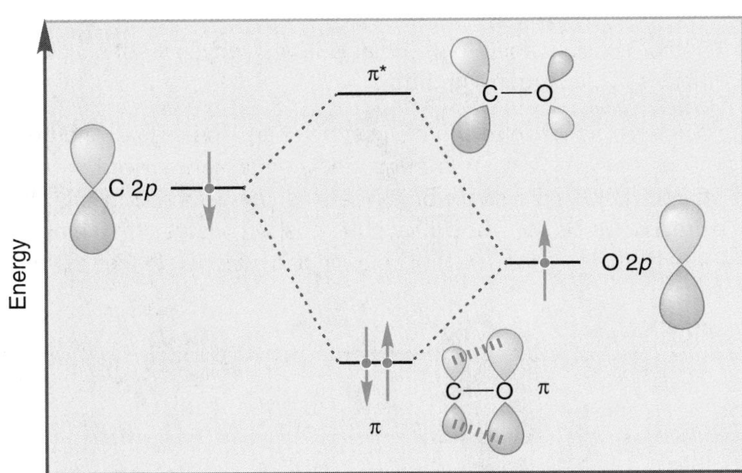

FIGURE 16.4 A graphical construction of π and π* for a carbonyl compound. The π orbital is constructed from more than 50% of the oxygen 2*p* orbital, and π* from more than 50% of the carbon 2*p* orbital.

So π and π* bear a general resemblance to π and π* of the alkenes, but are less symmetrical. The two electrons naturally occupy the lower energy bonding molecular orbital. In the π molecular orbital, there is a greater probability of finding an electron in the neighborhood of the relatively electronegative oxygen atom than near the more electropositive carbon, and this unequal sharing of electrons contributes to the dipole moment in molecules containing carbonyl groups. As can be seen from Figure 16.5, carbonyl compounds do have substantial dipole moments and are quite polar molecules.

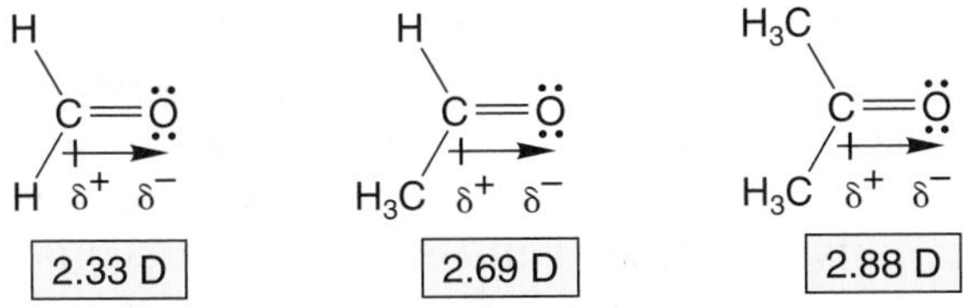

2.33 D   2.69 D   2.88 D

FIGURE 16.5 Carbonyl compounds are polar molecules with substantial dipole moments [measured in debye (D) units]. Notice the small bond dipole arrow that points from the positive end toward the negative end of the dipole.

PROBLEM 16.2

It is not just the π bonds that are unsymmetrical. Use the hybrid orbitals of oxygen and carbon to construct the σ and σ* orbitals.

PROBLEM 16.3

One way to construct a carbon–oxygen double bond within the hybridization formalism is to use *sp* hybridized oxygen. Make a drawing of this model in the style of Figure 16.2.

In resonance terms, we can describe the carbonyl group as a combination of the two major contributing forms shown in Figure 16.6. This picture emphasizes the Lewis acid character of the carbon of the carbon–oxygen double bond.

FIGURE 16.6 A resonance formulation of a carbonyl group. Note the polar resonance form.

*PROBLEM 16.4 There is another possible polar form. What is it, and why is it not considered a major contributor to the carbonyl group?

ANSWER The other polar resonance form has the positive charge on oxygen and the negative charge on carbon. Oxygen is more electronegative than carbon and the negative charge will be more stable on oxygen than on carbon. Similarly, the relatively electropositive carbon should have the positive charge, not the negative. This form is of high energy and cannot be a major contributor to the structure.

*PROBLEM 16.5 Why is the dipole moment in acetone larger than that in acetaldehyde (Fig. 16.5)?

ANSWER As the negative end of the dipole is the same for each species, there can be no answer there. However, the two methyl groups in acetone stabilize the positive charge much better than the methyl and hydrogen of acetaldehyde. The polar form is more stable in acetone than it is in acetaldehyde, and the compound has a larger dipole.

16.2 NOMENCLATURE OF CARBONYL COMPOUNDS

In a generic carbonyl compound $R_2C{=}O$, there is a vast array of possible substitution patterns, and there is a separate name for each one of them. Simple compounds in which both R groups are hydrocarbons such as alkyl, alkenyl, or aryl are called **ketones**. If one R group is hydrogen, the compound is an **aldehyde**, and the molecule in which both R groups are hydrogen is called **formaldehyde** (Fig. 16.7).

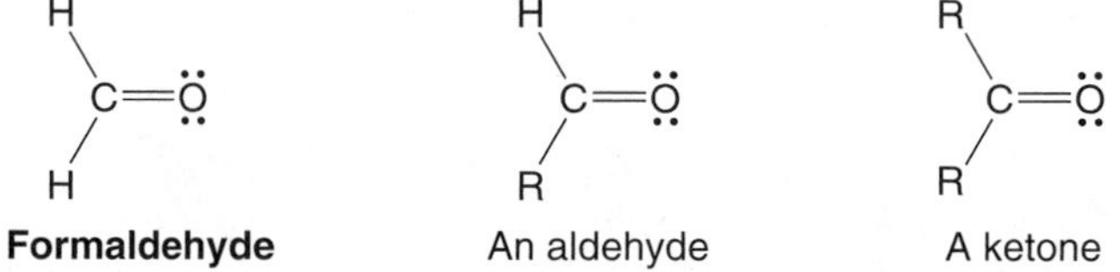

FIGURE 16.7 Different substitution patterns for simple carbonyl compounds.

Common names are normally used for the aldehydes containing four carbons or fewer, but thereafter the IUPAC system takes over, and the naming process becomes rational. The "e" of the parent alkane's name is dropped and the suffix "al" is added. In the priority system the aldehyde group is always number 1, and higher numbers are given to any substituents. When aldehydes are attached to rings, the term "carbaldehyde" or "carboxaldehyde" is used, as in "cyclopropanecarboxaldehyde." Some aldehydes and their common names are given in Figure 16.8.

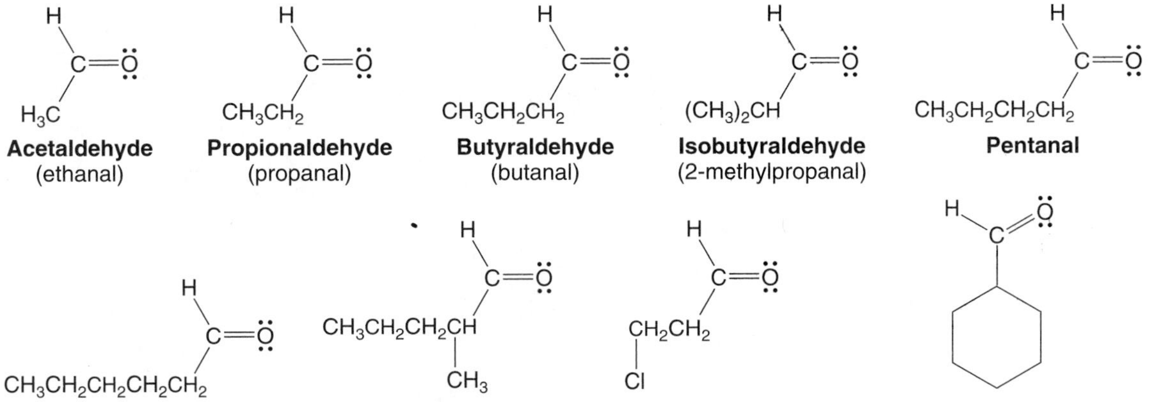

Acetaldehyde (ethanal)
Propionaldehyde (propanal)
Butyraldehyde (butanal)
Isobutyraldehyde (2-methylpropanal)
Pentanal
Hexanal
2-Methylpentanal
3-Chloropropanal
Cyclohexanecarboxaldehyde

FIGURE 16.8 Some simple aldehydes and their names.

Compounds containing two aldehydes are **dials**. In this case, the final "e" of the parent name is not dropped before adding "dial" (Fig. 16.9).

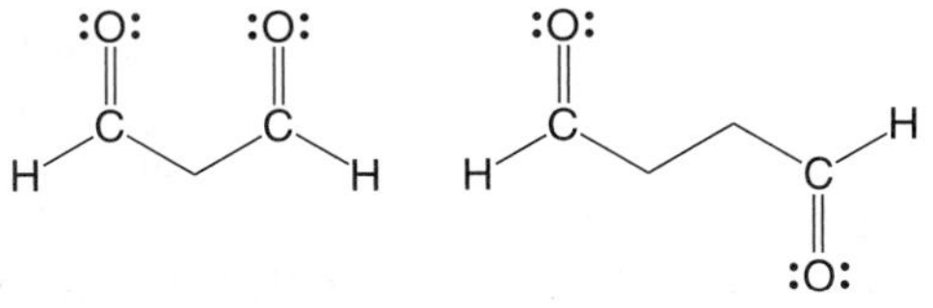

Propanedial
Butanedial

FIGURE 16.9 Dialdehydes are named as "dials."

Ketones, too, have common names, but most have been discarded. An exception is the name for the simplest member of the class, dimethyl ketone, which is still called **acetone**. Ketones are named in several ways. For example, acetone is also 2-oxopropane, propanone, or dimethyl ketone (Fig. 16.10). The rules are fairly simple: the "oxo" system treats the oxygen atom as a substituent that is given a number like any other attached group. In using the "one" system, the final "e" of the parent alkane is dropped, and the suffix "one" is appended. The number of the carbon atom bearing the carbon–oxygen double bond begins the name.

Acetone
dimethyl ketone
propanone
2-oxopropane

Ethyl methyl ketone
butanone
2-oxobutane

3-Methyl-2-pentanone
sec-butyl methyl ketone
3-methyl-2-oxopentane

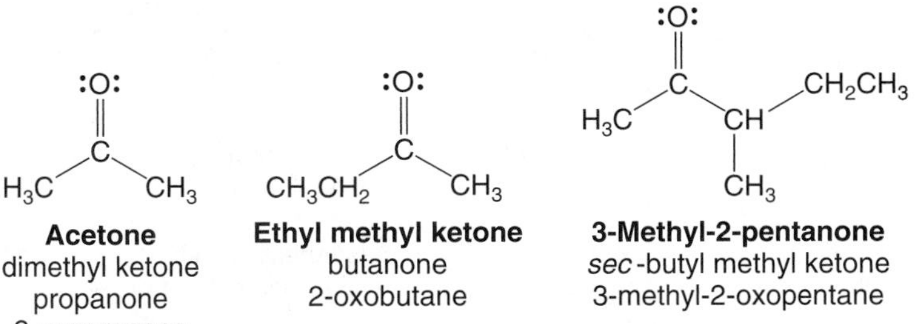

FIGURE 16.10 Simple ketones can be named in a variety of ways.

PROBLEM 16.6

Why is acetone properly named propanone and *not* 2-propanone?

Compounds containing two ketones are **diones**. As for dials, the final "e" of the parent framework is retained, and the name begins with the position numbers of the two carbonyl groups (Fig. 16.11). Ketoaldehydes

Civetone

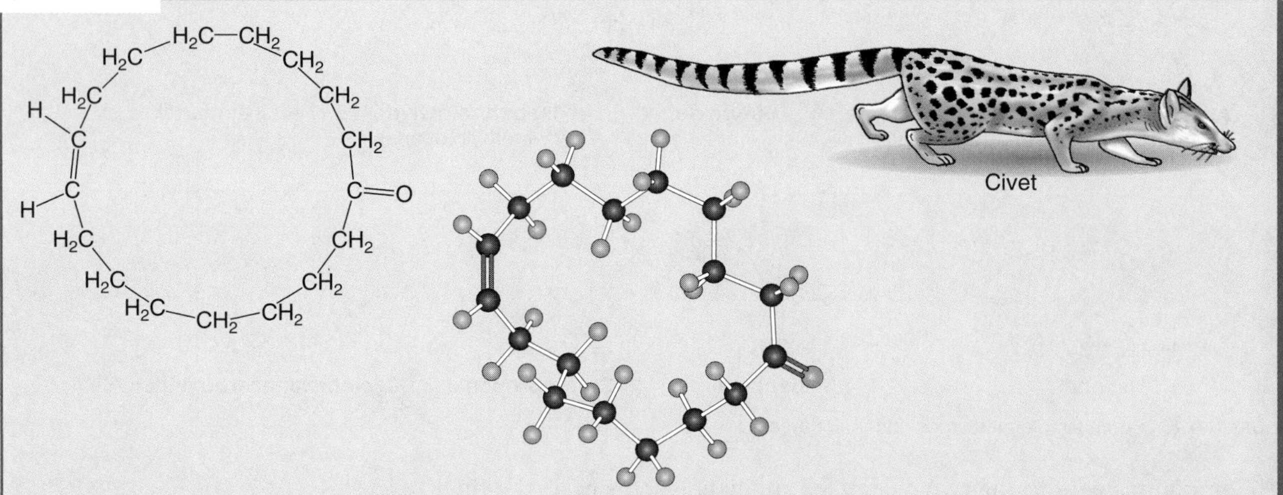

This simple, but unusual cycloalkenone is called civetone, as it is isolated from a gland present in both male and female civet cats, *Viverra civetta* and *Viverra zibetha.* The cats use it as a sexual attractant and marking agent. It has a strong musky odor, overwhelming when concentrated (to humans, that is, probably not to civet cats). In high dilution the odor becomes quite pleasant, and civetone and the related cyclic 15-membered ring ketone, muskone, are important starting materials in the perfumery industry. In the old days, you had to catch yourself a mess o' civet cats, described as "savage and unreliable," to make your perfumes. Nowadays, this compound can be made quite easily in the laboratory, a fact presumably reassuring to the civet cat population.

are named as aldehydes, not ketones, because the aldehyde group has a higher priority than a keto group (Fig. 16.11).

2,4-Pentanedione — H_3C–C(=O)–CH_2–C(=O)–CH_3

1,4-Cyclohexanedione

3-Oxohexanal — H–C(=O)–CH_2–C(=O)–$CH_2CH_2CH_3$

2-Oxopropanal (pyruvic aldehyde) — H–C(=O)–C(=O)–CH_3

FIGURE 16.11 Diketones are named as "diones." As for dials, the final "e" of the hydrocarbon backbone is retained. Ketoaldehydes are named as aldehydes, not ketones.

A great many common names are retained for aromatic aldehydes and ketones. For example, the simplest aromatic aldehyde is not generally called "benzenecarboxaldehyde" but **benzaldehyde**. That one's quite easy to figure out, but most are not. A few commonly used names are given in Figure 16.12.

Aromatic ketones are often named by treating the benzene ring as a **phenyl group**, as in isopropyl phenyl ketone (Fig. 16.13). A few can be named through a system in which the framework Ph—C=O is indicated by the term ophenone and the remainder of the molecule designated by one of

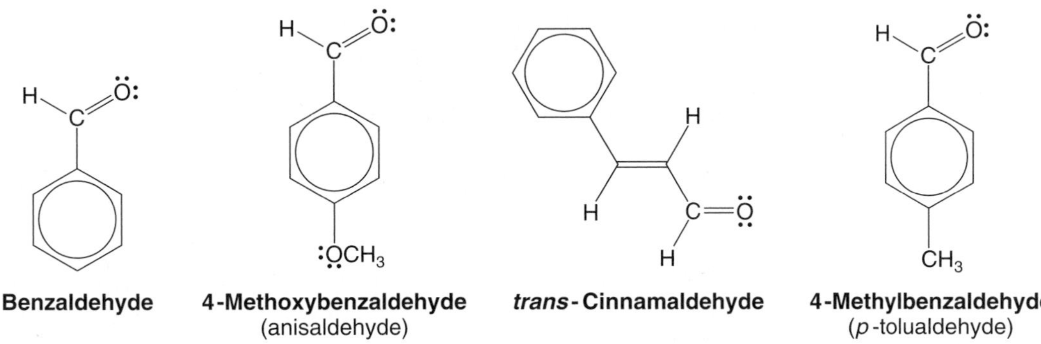

FIGURE **16.12** Some common names for aromatic aldehydes.

the alternative names in Figure 16.13. Thus, diphenyl ketone is also benzophenone, methyl phenyl ketone is acetophenone, ethyl phenyl ketone is propiophenone, and phenyl propyl ketone is butyrophenone (Fig. 16.13).

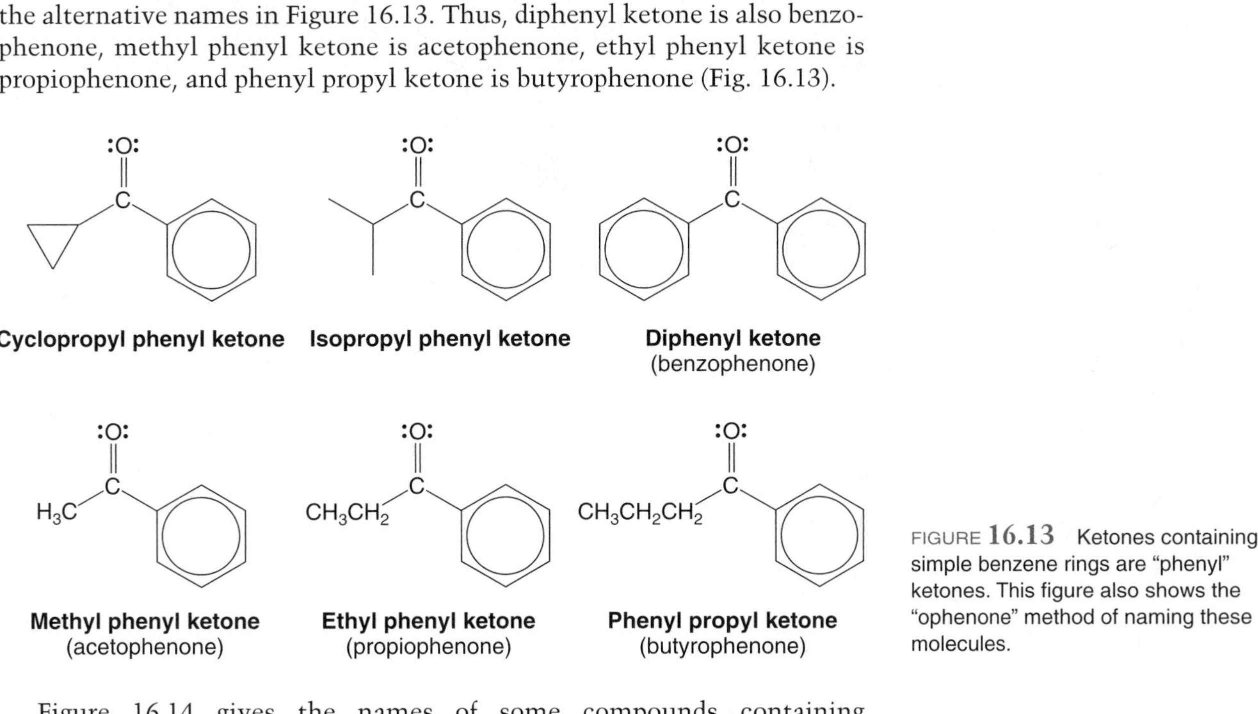

FIGURE **16.13** Ketones containing simple benzene rings are "phenyl" ketones. This figure also shows the "ophenone" method of naming these molecules.

Figure 16.14 gives the names of some compounds containing carbon–oxygen double bonds in which one of the R groups is neither hydrogen nor a hydrocarbon group. We will meet these compounds in later chapters and deal with nomenclature more extensively there.

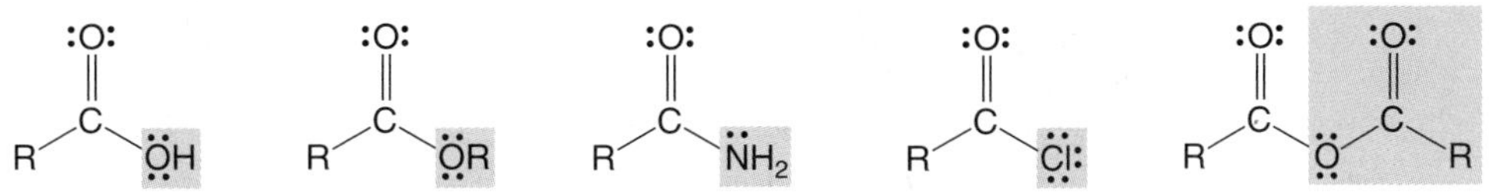

Carboxylic acid Ester Amide Acid chloride Anhydride

FIGURE **16.14** Some names for other carbonyl compounds in which one R group is neither hydrogen nor a hydrocarbon group.

16.3 PHYSICAL PROPERTIES OF CARBONYL COMPOUNDS

Carbonyl compounds are all somewhat polar, and the smaller, less hydrocarbon-like molecules are all at least somewhat water soluble. Acetone and acetaldehyde are miscible with water, for example. Their polarity means that they can profit energetically by associating in solution, and this means that their boiling points are considerably higher than those of hydrocarbons containing the same number of heavy atoms (carbons and oxygens). For example, propane and propylene are gases with boiling points of –42.1 and –47.4 °C, respectively. By contrast, acetaldehyde boils at +20.8 °C, about 63 °C higher. Acetone boils more than 60 °C higher than isobutene.

Table 16.1 gives some physical properties of a number of common carbonyl-containing compounds (and some related hydrocarbons in parentheses).

TABLE 16.1 Some Physical Properties of Carbonyl Compounds

Compound	bp (°C)	mp (°C)	Density (g/mL)	Dipole Moment (D)
Formaldehyde	– 21	– 92	0.815	2.33
Acetaldehyde	20.8	–121	0.783	2.69
Propionaldehyde	48.8	– 81	0.806	2.52
(Propane)[a]	– 42.1			
(Propene)[a]	– 47.4			
Benzaldehyde	178.6	– 26	1.04	
Acetone	56.2	– 95.4	0.789	2.88
(Isobutene)[a]	– 6.9			
Ethyl methyl ketone	79.6	– 86.3	0.805	
Cyclohexanone	155.6	– 16.4	0.948	
Methyl phenyl ketone	202.6	20.5	1.03	
Diphenyl ketone	305.9	48.1	1.15	2.98

[a] Hydrocarbons for comparison purposes.

16.4 SPECTROSCOPY OF CARBONYL COMPOUNDS

16.4a Infrared Spectroscopy

Carbonyl groups are common in Nature, and spectroscopy, especially IR spectroscopy, played an important role in identifying many medicinally and otherwise important compounds. The carbon–oxygen double bond has a diagnostically useful, and quite strong stretching frequency in the IR, generally near 1700 cm^{-1}. This absorption is at substantially higher frequency than that of carbon–carbon double bonds. This shift should be no surprise, as we already know that carbon–oxygen double bonds are stronger than carbon–carbon double bonds, and therefore require more energy for stretching (Chapter 15, p. 699). A useful diagnostic for aldehydes is the pair of bands observed in the C—H stretching region at about 2850 and 2750 cm^{-1}. Table 16.2 gives some IR stretching frequencies and some other data.

Note that conjugated carbonyl compounds are always found at lower frequency than their unconjugated relatives. For example, cyclohexanone appears at 1718 cm^{-1} and 2-cyclohexenone at 1691 cm^{-1}. Thus there is reduced double-bond character in the conjugated molecules. The resonance formulation of 2-cyclohexenone clearly shows the partial single-bond character of the carbon–oxygen bond in these compounds (Fig. 16.15).

TABLE 16.2 Some Spectral Properties of Carbonyl Compounds

Compound	IR (CCl_4, cm^{-1})	^{13}C NMR (δ C=O)	UV [nm (ε)]
Formaldehyde	1744 (gas phase)		270
Acetaldehyde	1733	199.6	293.4 (11.8)
Propionaldehyde	1738	201.8	289.5 (18.2)
Acetone	1719	205.1	279 (14.8)
Cyclohexanone	1718	207.9	285 (14)
Cyclopentanone	1751	213.9	299 (20)
Cyclobutanone	1788	208.2	280 (18)
Ethyl methyl ketone	1723	206.3	277 (19.4)
Methyl vinyl ketone	1684	197.2	219 (3.6)
2-Cyclohexenone	1691	197.1	224.5 (10,300) (π→π*)
Benzaldehyde	1710	191.0	244 (15,000) (π→π*)
			328 (20) (*n*→π*)
Diphenyl ketone	1666	195.2	252 (20,000) (π→π*)
			325 (180) (*n*→π*)
Methyl phenyl ketone	1692	196.8	240 (13,000) (π→π*)
			319 (50) (*n*→π*)

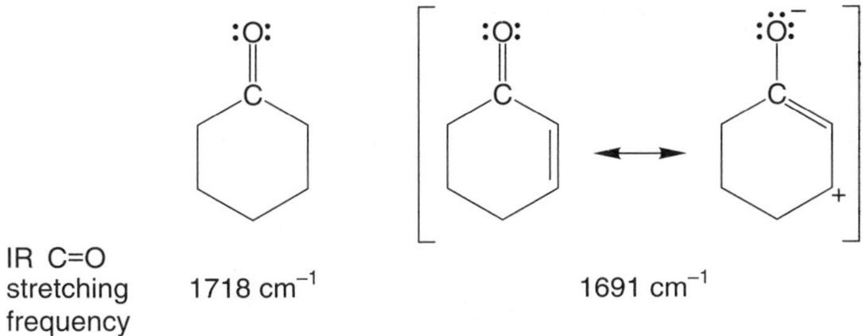

FIGURE 16.15 Conjugation reduces the double-bond character of a carbonyl group. The result is a shift of the IR stretching band to lower frequency.

16.4b Nuclear Magnetic Resonance Spectroscopy

The carbonyl group inductively withdraws electrons from the nearby positions, and this results in a deshielding effect. The further away from the carbonyl group the less effective the deshielding. This effect is well illustrated by methyl propyl ketone (Fig. 16.16).

The hydrogens directly attached to the carbonyl group resonate at very low field, δ ~ 8.5–10 ppm. Not only does the carbon bear a partial positive charge, but the applied magnetic field B_0 creates an induced field B_i in the carbonyl group that augments the applied field in the region of a directly attached hydrogen (Fig. 16.17). Accordingly, a relatively small applied B_0 is required to bring these hydrogens into resonance, and they appear at unusually low field (Chapter 15, p. 713).

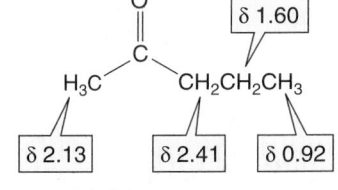

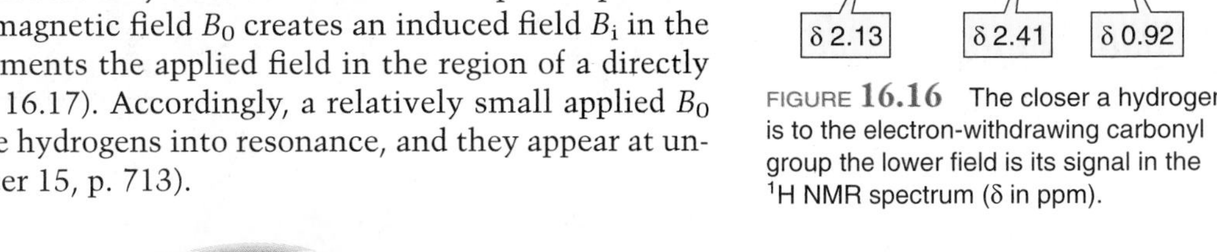

FIGURE 16.16 The closer a hydrogen is to the electron-withdrawing carbonyl group the lower field is its signal in the ^1H NMR spectrum (δ in ppm).

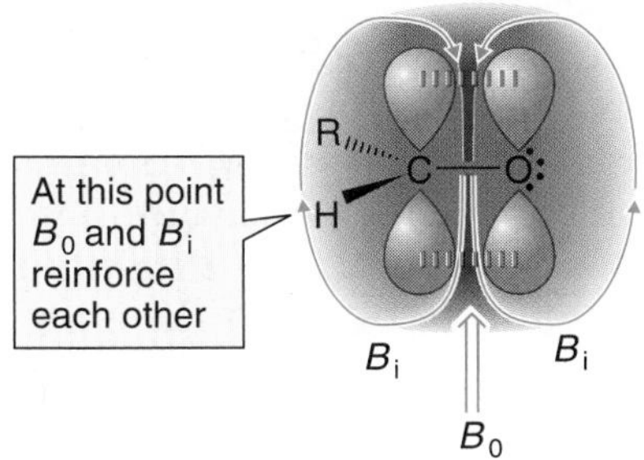

FIGURE 16.17 Aldehyde hydrogens are in a strongly deshielded region in which the induced magnetic field B_i augments the applied field B_0. These hydrogens resonate at exceptionally low field in ^1H NMR spectra.

The carbon atoms of carbonyl groups carry partial positive charges and are greatly deshielded by their relative lack of surrounding electrons. They, too, resonate at especially low field in the ^{13}C NMR. Table 16.2 shows typical resonance positions.

16.4c Ultraviolet Spectroscopy

Simple aldehydes and ketones, like simple alkenes, do not have a π–π* absorption in the region of the UV spectrum accessible to most spectrometers. However, unlike alkenes, there is another possible absorption which, though weak, is detectable. This absorption involves promotion of a nonbonding, or *n*, electron, to the antibonding π* orbital, the *n*→π* absorption. Table 16.2 gives a few absorptions.

As for alkenes (Chapter 12, p. 517), conjugation greatly reduces the energy necessary to promote an electron from the highest occupied π molecular orbital to the LUMO. So, the strong π →π* transition in unsaturated carbonyl compounds can be detected because of the conjugation-induced shift to the longer wavelengths (lower energy) observable by normal UV/vis spectrometers (Table 16.2; Fig. 16.18).

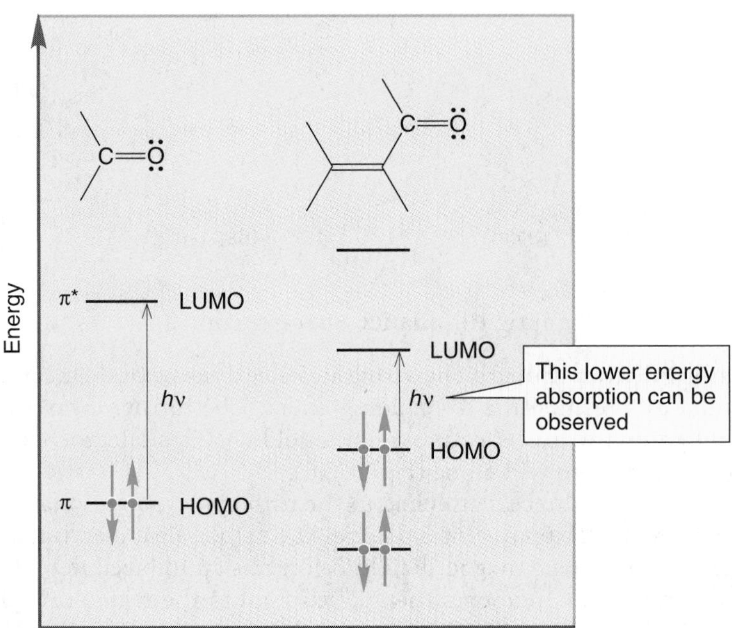

FIGURE **16.18** Conjugation leads to lower energy (longer wavelength) π→π* absorptions that are observable by UV/vis spectrometers.

16.4d Mass Spectrometry

There is a diagnostic cleavage reaction that occurs for many simple carbonyl-containing compounds. It involves breaking the bond attached directly to the carbonyl carbon, with the charge remaining on the carbonyl group. Cleavage of this bond generates a resonance-stabilized acylium ion. It is this resonance stabilization that accounts for the prominence of this particular fragmentation. Figure 16.19 shows the mass spectrum for acetone, the archetypal methyl ketone.

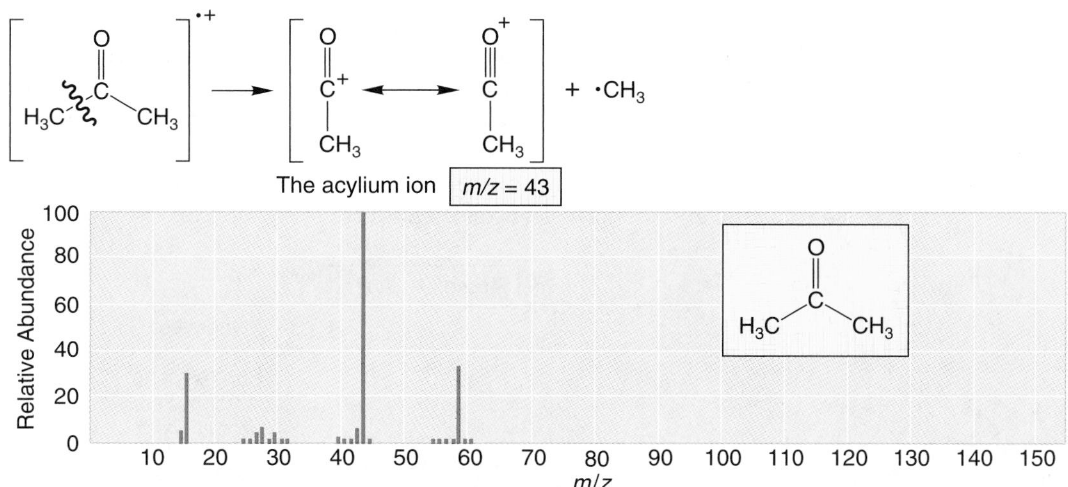

FIGURE 16.19 In the mass spectrum for acetone, the major peak results from cleavage of a methyl group and formation of the resonance-stabilized acylium ion.

16.5 REACTIONS OF CARBONYL COMPOUNDS: SIMPLE REVERSIBLE ADDITIONS

In simple addition reactions, carbonyl compounds behave as both Lewis acids and Lewis bases depending on the other reagents present. As we might expect, there is some correspondence between addition reactions of carbon–oxygen double bonds and those of the carbon–carbon double bonds we saw in Chapters 9 and 10. As we might also expect, what complications exist are introduced by the presence of two *different* atoms in the carbon–oxygen double bond. Figure 16.20 compares the reaction between water and a generic carbonyl compound (the formation of a **hydrate**) and the hydration of an alkene.

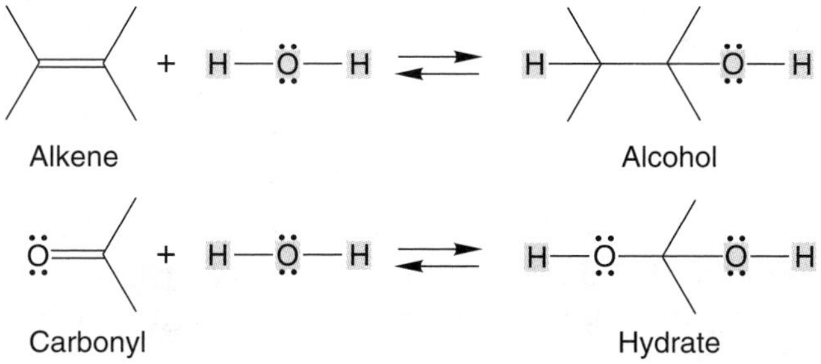

Alkene + H—Ö—H ⇌ Alcohol

Carbonyl + H—Ö—H ⇌ Hydrate

FIGURE 16.20 Addition of water to an alkene to give an alcohol and addition to a ketone to give a hydrate are analogous reactions.

Water acts as a nucleophile and adds to the Lewis acid carbonyl group. However, there is a choice to be made. In principle, water could add to either the carbon or oxygen end of the carbonyl group (Fig. 16.21).

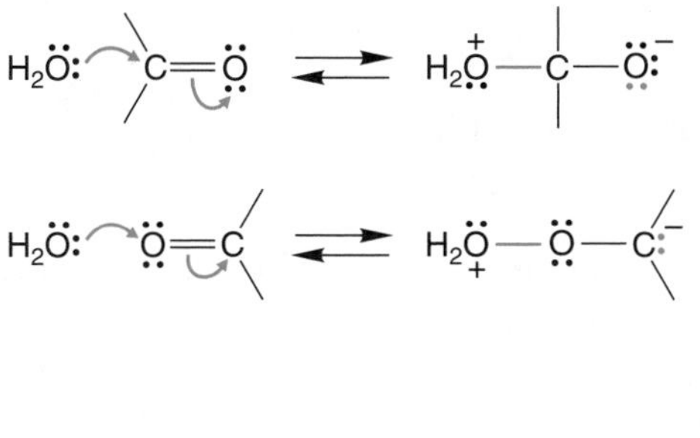

FIGURE 16.21 In principle, addition of water to a carbonyl group could occur in two ways.

Three factors must be considered in predicting which way the addition will go. First of all, addition to the carbon of the carbon–oxygen double bond places the negative charge on the highly electronegative oxygen atom, whereas addition at oxygen generates a much less stable carbanion. Second, addition to oxygen forms a weak oxygen–oxygen bond, whereas addition to carbon makes a stronger oxygen–carbon bond. Formation of the alkoxide must be vastly preferred (Fig. 16.22).

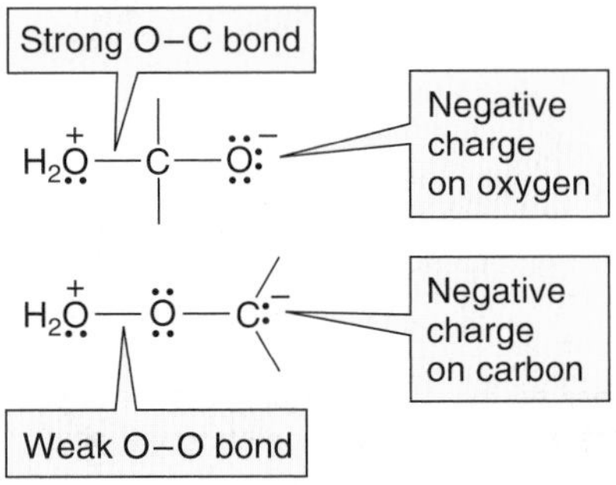

Strong O–C bond

Negative charge on oxygen

Negative charge on carbon

Weak O–O bond

FIGURE 16.22 Addition to the carbon end, which puts the negative charge on the relatively electronegative oxygen atom, will be greatly preferred.

*PROBLEM 16.7 Explain how this thermodynamic preference for placing a negative charge on oxygen, not carbon, can influence the *rates* of addition to the two different ends of the carbon–oxygen double bond.

ANSWER The rates of reaction depend, of course, on the heights of the transition states, not on the energies of the products in which a full negative charge resides on either oxygen or carbon. In the transition states for addition, there will be a partial negative charge on either oxygen or carbon. The factors that favor the presence of a full negative charge on oxygen rather than carbon (electronegativity) will

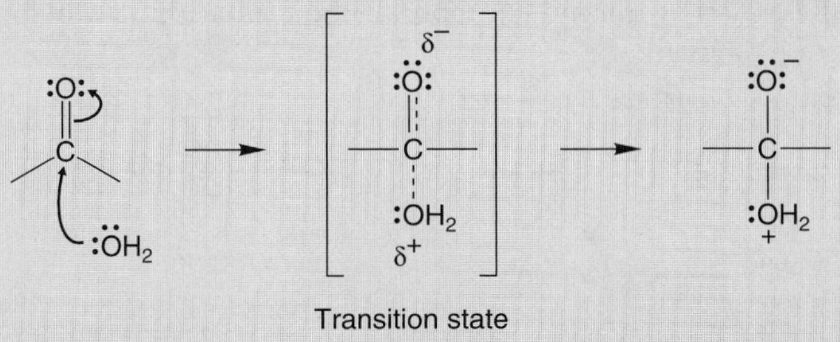

Transition state

also favor the presence of a partial negative charge (δ^-) on oxygen. In the transition state, a partial negative charge will be more stable on oxygen than on carbon.

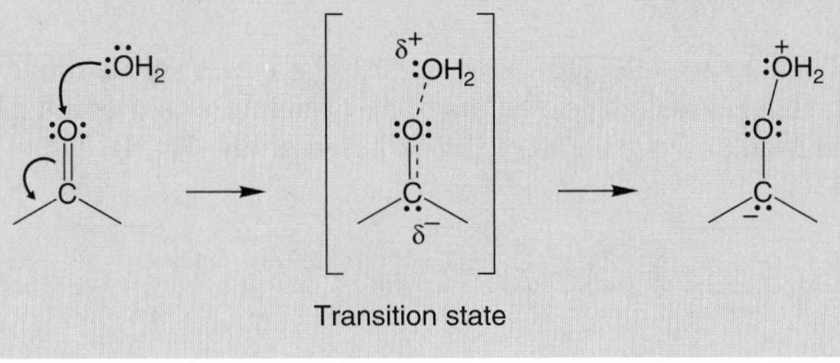

Transition state

But there is more to consider. If we look at the molecular orbitals involved, we can see how they influence the direction in which the reaction occurs. As in all polar reactions, we look for the combination of a nucleophile (here, water) and a Lewis acid. The Lewis acid must be the *empty* π^*

orbital of the carbonyl group. Figure 16.23 shows the orbital interactions graphically, and gives pictures of the orbital lobes involved.

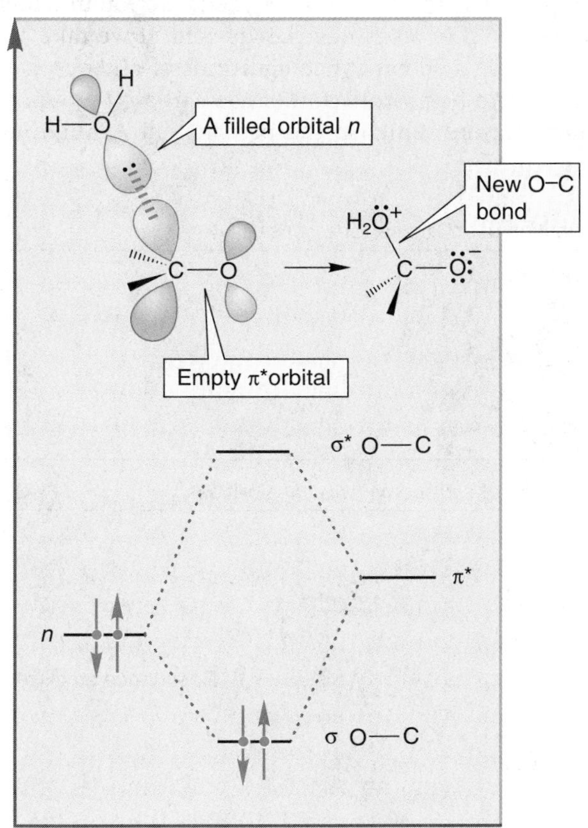

FIGURE **16.23** In the addition of water to a carbonyl, a filled *n* orbital of water overlaps with the empty π* orbital of the carbonyl group. This interaction between a filled and an empty orbital is stabilizing.

As we saw in our earlier section on structure (p. 754), neither π nor π* of the carbonyl group is symmetric with respect to the midpoint of the carbon–oxygen bond. In particular, π* is much larger in the region of the carbon atom than it is near the oxygen. As the magnitude of stabilization depends on the amount of orbital overlap, it is no surprise to see that addition at the carbon, where π* is largest, dominates the reaction. So, not only is the direction of the addition affected by the electronegativities of the two different atoms in the double bond, but it's affected by the shapes of the orbitals involved as well.

After the addition of water, protons are lost and added to give the hydrate. It is not necessary that the proton lost from one molecule be readded to the same molecule. The reaction is very likely to be intermolecular, as shown in Figure 16.24.

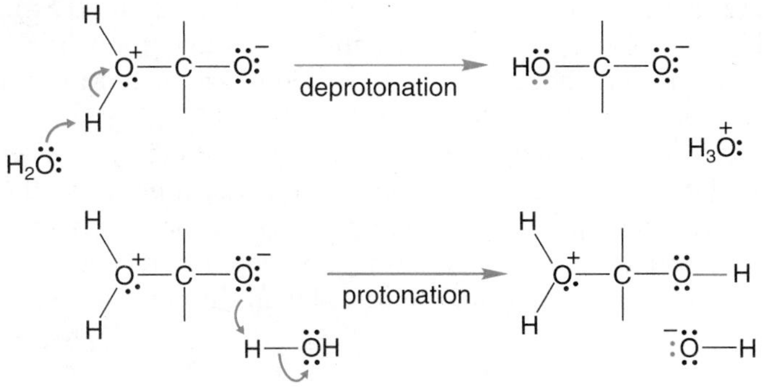

deprotonation

protonation

FIGURE **16.24** The hydration reaction is completed by a series of proton transfers. The order of the two steps is not certain.

Now consider how we might make the addition reaction easier. Two ways seem possible—we could make either the Lewis acid or nucleophile stronger. Let's start with the acid-catalyzed reaction in which the carbonyl group is converted into a stronger Lewis acid. If we take as our model for the mechanism the acid-catalyzed additions of alkenes, a reasonable first step would seem to be protonation of the carbonyl group. But here enters another complication. Which end is protonated? Again, there are two possiblities (Fig. 16.25).

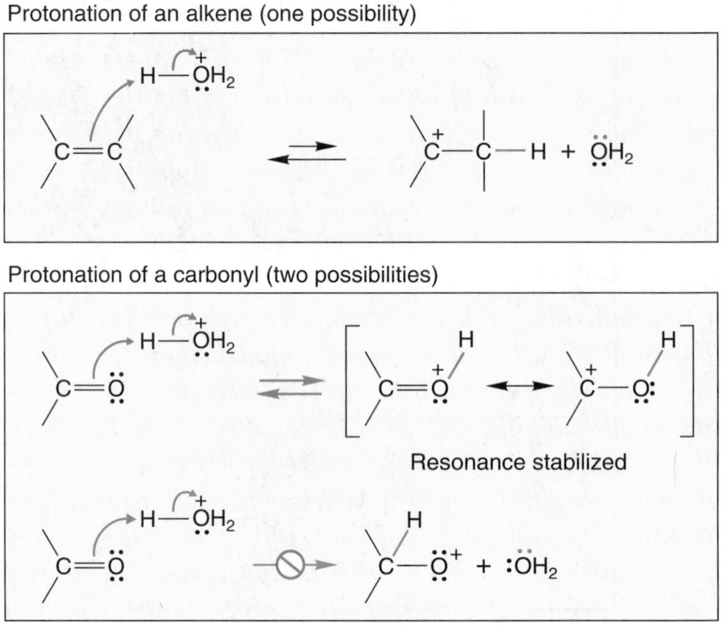

Protonation of an alkene (one possibility)

Protonation of a carbonyl (two possibilities)

Resonance stabilized

FIGURE 16.25 Protonation of the carbonyl group can occur in two ways. The placement of a proton on oxygen to give a resonance-stabilized cation will be greatly preferred.

The choice is easy this time. First, the oxygen atom is a strong Lewis base, as it has two lone pairs of nonbonding electrons. Moreover, it is surely better to protonate on oxygen, and thus place the positive charge on the relatively electropositive carbon rather than to protonate on carbon, which would place the positive charge upon the highly electronegative oxygen atom. Finally, protonation at oxygen gives a resonance-stabilized cation, whereas protonation at carbon does not.

So there is not really much choice in the direction of the protonation step, which creates a strongly Lewis acidic species, the protonated carbonyl compound. In the second step of the reaction, water acts as a nucleophile and adds to the Lewis acid (the protonated carbonyl) to generate a protonated 1,1-diol (a geminate diol). In the final step of the reaction, this species is deprotonated by a water molecule to generate the 1,1-diol itself (the hydrate) and regenerate the acid catalyst, H_3O^+. Note how closely the acid-catalyzed additions of water to an alkene and a carbonyl group are related. Each reaction is a sequence of three steps: protonation, addition of water, and deprotonation (Fig. 16.26).

What about making the nucleophile more reactive? There is another mechanism for hydration that operates under basic conditions. Here the active agent is the hydroxide ion, ^-OH. The first step in the base-catalyzed hydration reaction is the addition of the strong nucleophile hydroxide to the carbonyl group. Unlike protonation, the first step in the acid-catalyzed

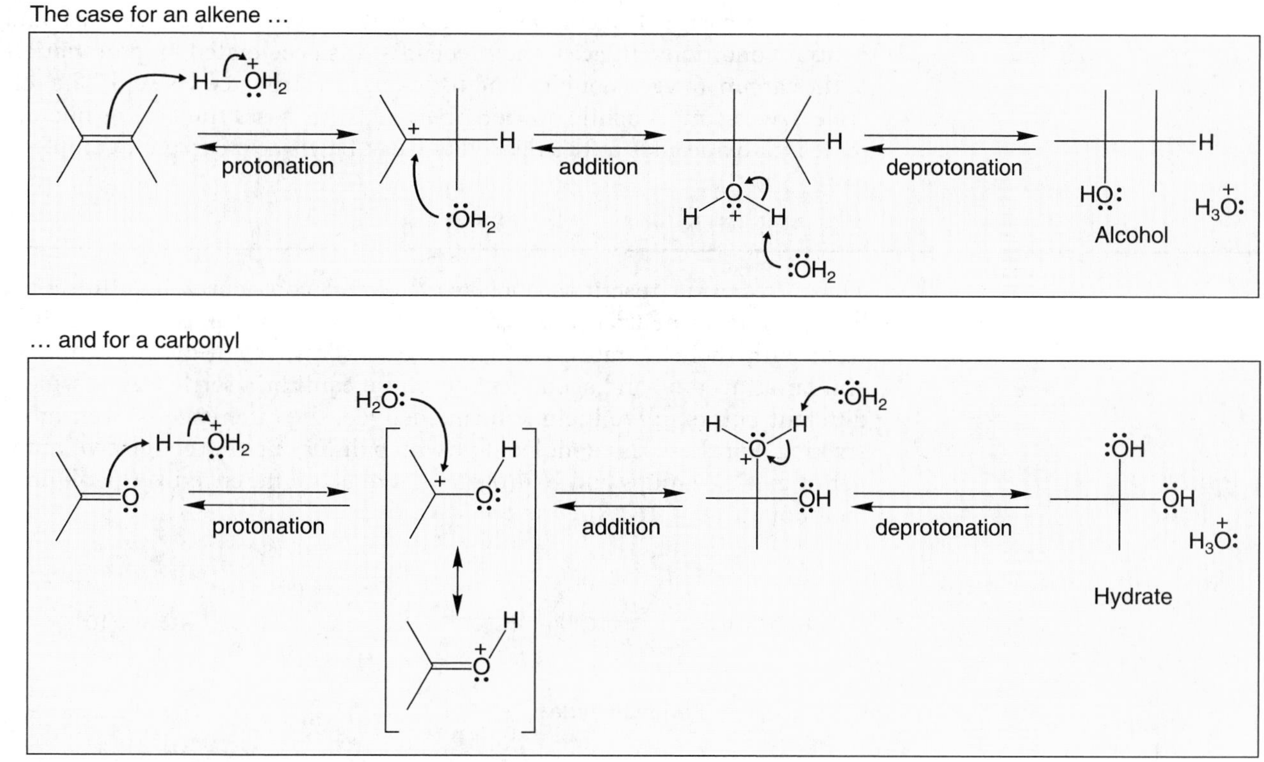

FIGURE 16.26 A comparison of the acid-catalyzed hydration reactions of an alkene and a carbonyl compound.

reaction, this reaction has no counterpart in the chemistry of simple alkenes. After addition, a proton is transferred from water to the newly formed alkoxide, generating the hydrate and regenerating the hydroxide ion (Fig. 16.27).

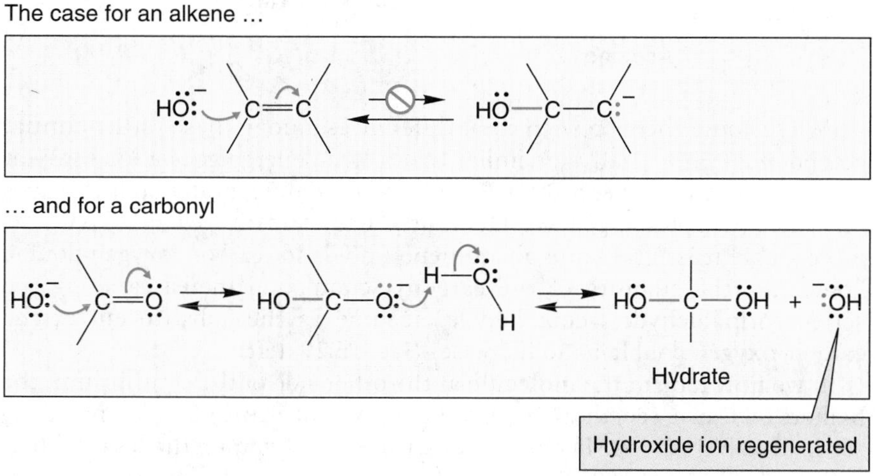

FIGURE 16.27 Although strong bases such as hydroxide ion do not add to alkenes, they do attack carbonyl groups. The final step in the base-catalyzed hydration reaction is protonation of the newly formed alkoxide ion by water. Hydroxide is regenerated.

PROBLEM 16.8

Explain why bases such as hydroxide ion do not add to the oxygen of the carbon–oxygen double bond, and do not add at all to alkenes.

To summarize: The carbonyl group is hydrated under acidic, basic, and neutral conditions. In acid, the mechanism is accelerated by protonation of the carbon–oxygen double bond to generate a strong Lewis acid. The relatively weak nucleophile, water, then adds. In base, the strong nucleophile hydroxide adds to the relatively weak Lewis acid carbonyl group.

16.6 EQUILIBRIUM IN ADDITION REACTIONS

These hydration reactions consist of several consecutive equilibria. In basic, acidic, or neutral water, simple carbonyl compounds are in equilibrium with their hydrates. As Figure 16.28 shows, even for molecules as similar as acetone and acetaldehyde, these equilibria settle out at widely different points. If we include formaldehyde, the difference is even more striking. For these three molecules, which differ only in the relative number of methyl groups and hydrogens, there is a difference in K of about a factor of 10^6 (Fig. 16.28).

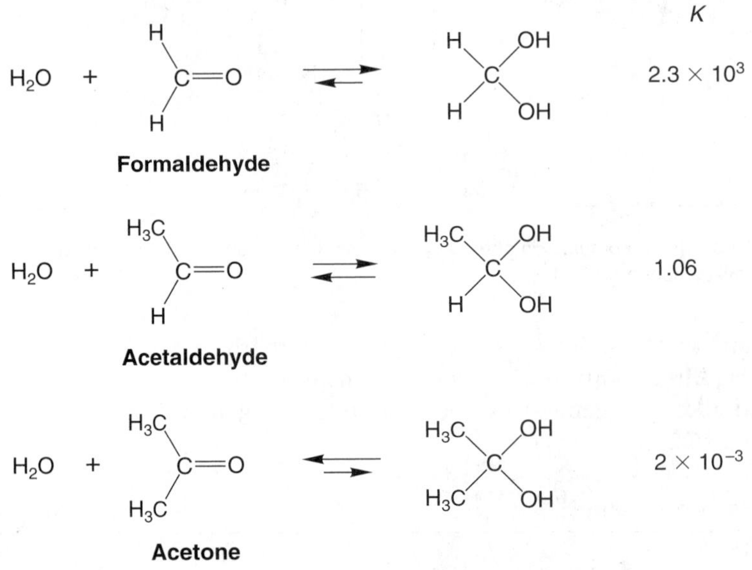

Reaction	K
H_2O + $H_2C{=}O$ (**Formaldehyde**) $\rightleftharpoons$ $H_2C(OH)_2$	2.3×10^3
H_2O + $H_3C(H)C{=}O$ (**Acetaldehyde**) $\rightleftharpoons$ $H_3C(H)C(OH)_2$	1.06
H_2O + $(H_3C)_2C{=}O$ (**Acetone**) $\rightleftharpoons$ $(H_3C)_2C(OH)_2$	2×10^{-3}

FIGURE 16.28 The equilibrium constant for hydration of a carbonyl-containing molecule can favor either the hydrate or the carbonyl compound.

Why should there be such great differences among these rather similar compounds? First of all, remember that a little energy goes a long way at equilibrium (Chapter 8, p. 305). Recall also that alkyl groups stabilize double bonds (Chapter 4, p. 133). The more highly substituted an alkene, the more stable it is. The same phenomenon holds for carbon–oxygen double bonds. So, the stability of the carbonyl compound increases along the series: formaldehyde, acetaldehyde, acetone as the substitution of the carbon–oxygen double bond increases (Fig. 16.29, left).

If we now look at the molecule at the other side of the equilibrium, the hydrate, we can see that increased methyl substitution will introduce steric destabilizations. The more substituted the hydrate, the less stable it is (Fig. 16.29, right).

We can immediately see the sources of the difference in K of 10^6. They are the increased stabilization of the starting carbonyl compounds and destabilization of the hydrates, both of which are induced by increased substitution by methyl groups (Fig. 16.30).

We can see other striking effects on this equilibrium process in Table 16.3. For example, benzene rings, which can interact with the carbon–

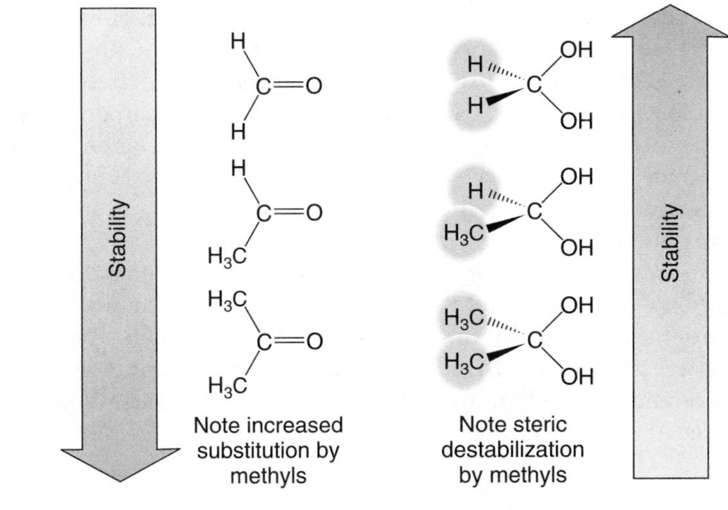

FIGURE 16.29 The stability of carbonyl groups depends greatly on the number of alkyl groups. Like alkenes, more substituted carbonyl groups are more stable than their less substituted counterparts. In the hydrates, increasing substitution results in increasing destabilization through steric interactions.

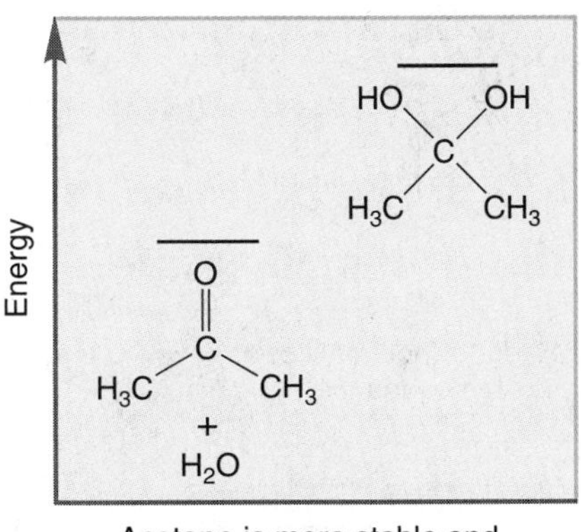

Acetone is more stable and the hydrate is less stable
$K < 1$

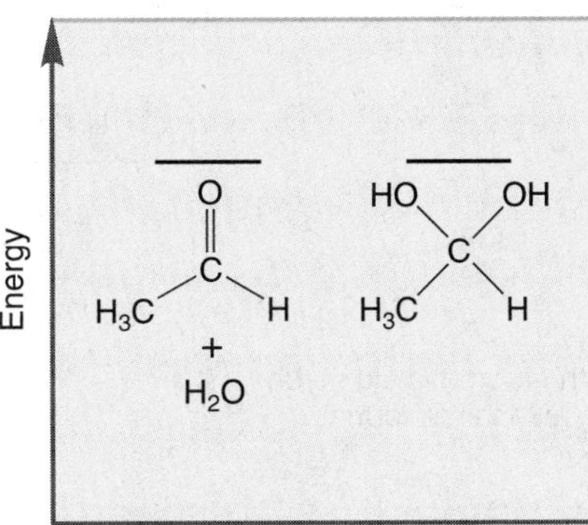

The aldehyde and hydrate are of comparable energy
$K \sim 1$

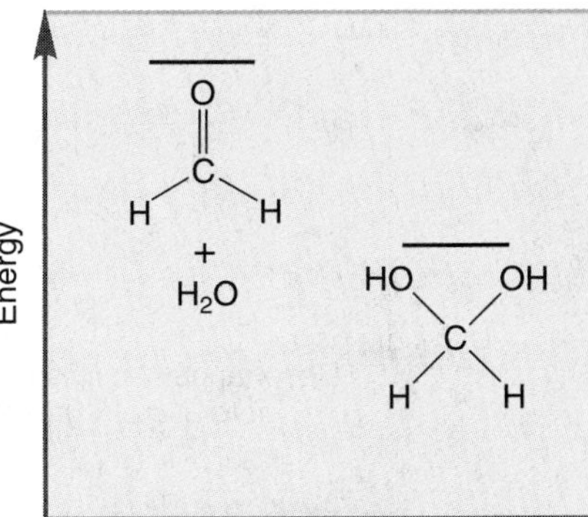

Formaldehyde is less stable and the hydrate is more stable
$K > 1$

FIGURE 16.30 The magnitude of the equilibrium constant (K) depends on the relative energies of the carbonyl compounds and their hydrates. We know that for acetaldehyde the energies of hydrate and aldehyde are comparable, as $K \sim 1$. Acetone is more stable than acetaldehyde and formaldehyde is less stable. Steric factors make the hydrate of acetone less stable than the acetaldehyde hydrate. By contrast, the hydrate of formaldehyde is more stable than the acetaldehyde hydrate. The equilibrium constants reflect these changes.

TABLE 16.3 Equilibrium Constants for Carbonyl Hydration

Compound		
Structure	Name	K
(benzene ring with CHO and Cl in meta positions)	*m*-Chlorobenzaldehyde	0.022
CH_3CH_2CHO	Propionaldehyde	0.85
$CH_3CH_2CH_2CHO$	Butyraldehyde	0.62
$(CH_3)_3CCHO$	*tert*-Butylcarboxaldehyde	0.23
CF_3CHO	Trifluoroacetaldehyde	2.9×10^4
CCl_3CHO	Trichloroacetaldehyde	2.8×10^4
$ClCH_2CHO$	Chloroacetaldehyde	37

oxygen double bond by resonance, decrease the amount of hydrate at equilibrium. However, electron-withdrawing groups in the position adjacent to the C=O bond, the α-position, strongly favor the hydrate. There also appears to be a steric effect. Large alkyl groups favor the carbonyl compound, as can be seen from the sequence in Table 16.3: propionaldehyde, butyraldehyde, and *tert*-butylcarboxaldehyde. Each of these phenomena can be understood by considering the relative stabilities of the starting materials and products.

Resonance stabilization of the carbonyl group is generally not accompanied by a resonance stabilization of the product hydrate. For example, the aromatic aldehyde, *m*-chlorobenzaldehyde (Table 16.3), is more strongly resonance-stabilized than acetaldehyde. There is no corresponding difference in the hydrates, and therefore the aromatic aldehyde is less hydrated at equilibrium than acetaldehyde (Fig. 16.31).

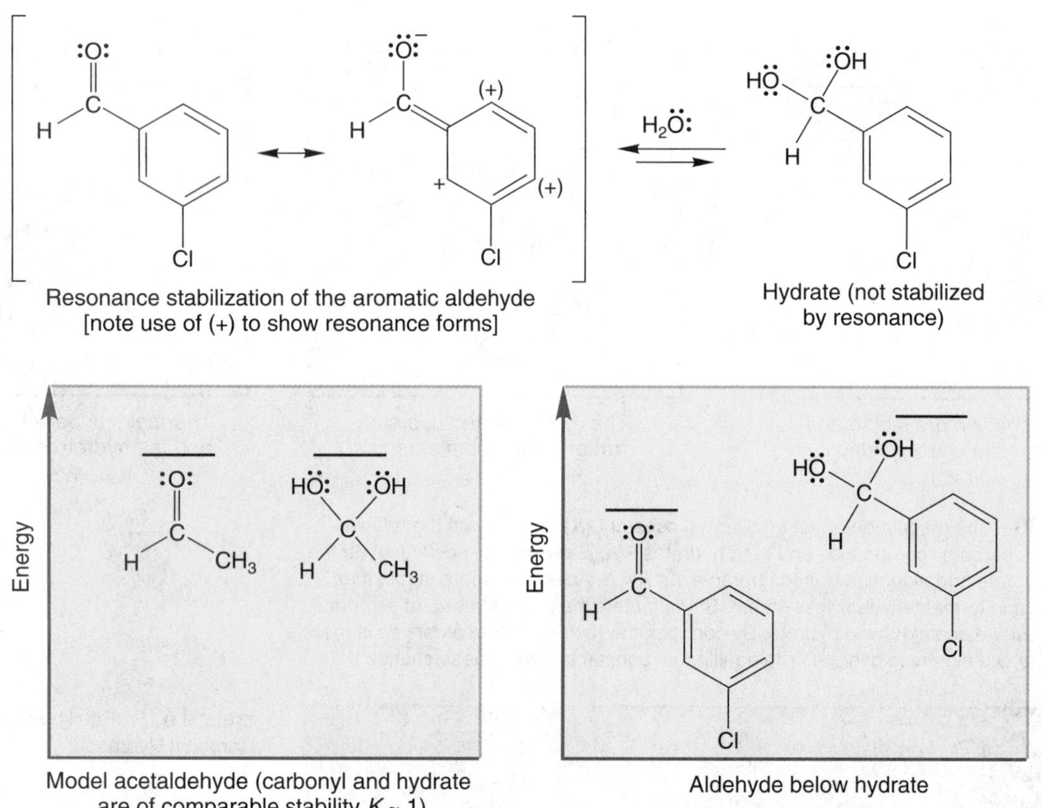

FIGURE 16.31 Resonance stabilization of the aromatic carbonyl compound is not accompanied by resonance stabilization of the product hydrate. Accordingly, the carbonyl compound is relatively favored at equilibrium.

Electron-withdrawing groups such as chlorine or fluorine sharply destabilize carbonyl groups when attached at the α-position. The carbon atom is already electron deficient, and further inductive removal of electrons is costly in energy terms. There is relatively little effect in the product hydrate, although there will be some destabilization in this molecule as well. So, as the diagram in Figure 16.32 shows, the strong destabilization of the carbonyl group favors the hydrate at equilibrium (Fig. 16.32).

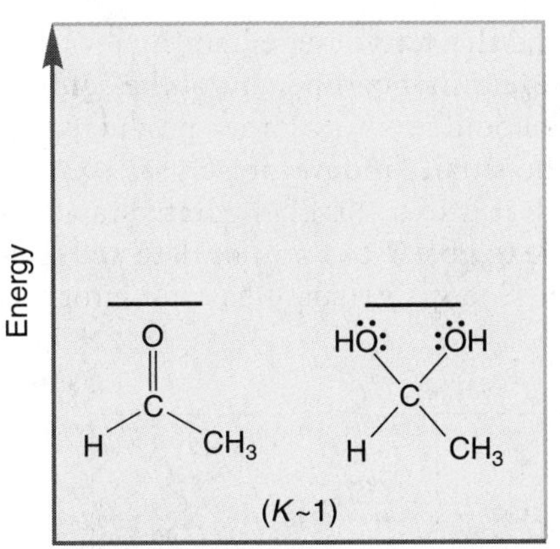

Model acetaldehyde (carbonyl and hydrate are of comparable stability)

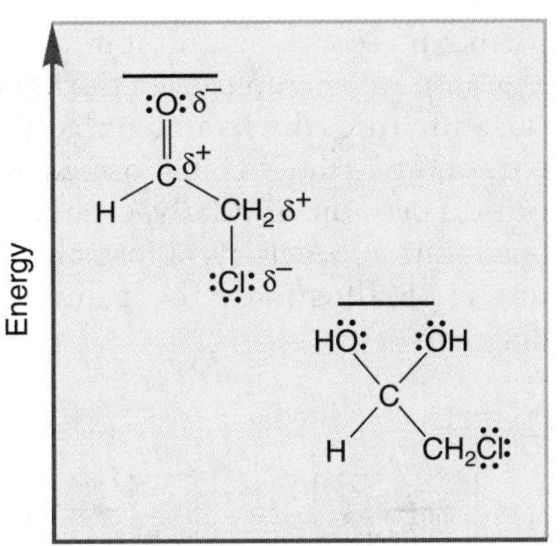

The inductive effect of the electronegative chlorine atom destabilizes the aldehyde relative to the hydrate

FIGURE 16.32 The energy of an α-halo carbonyl compound is raised by the opposed dipoles. Accordingly, the hydrate is favored at equilibrium.

The steric effect is slightly more subtle. In the carbonyl group, the carbon is hybridized sp^2 and the bond angles are therefore close to 120°. In the product hydrate, the carbon is nearly sp^3 hybridized, and the bond angles must be close to 109°. So, in forming the hydrate, groups are moved closer to each other, and for a large group that will be relatively destabilizing (Fig. 16.33).

:O:
C
R H
~120°
R and H are further apart

⇌

HÖ: :ÖH
C
R H
~109°
R and H are closer together (destabilizing for large R)

FIGURE 16.33 In the hydrate, groups are closer together than they are in the carbonyl compound. This proximity will destabilize the hydrate relative to the carbonyl compound.

It is less costly in energy terms to move a methyl group closer to a hydrogen than to move a *tert*-butyl group closer to a hydrogen. Accordingly, acetaldehyde is more strongly hydrated than is *tert*-butylcarboxaldehyde (pivaldehyde) (Fig. 16.34).

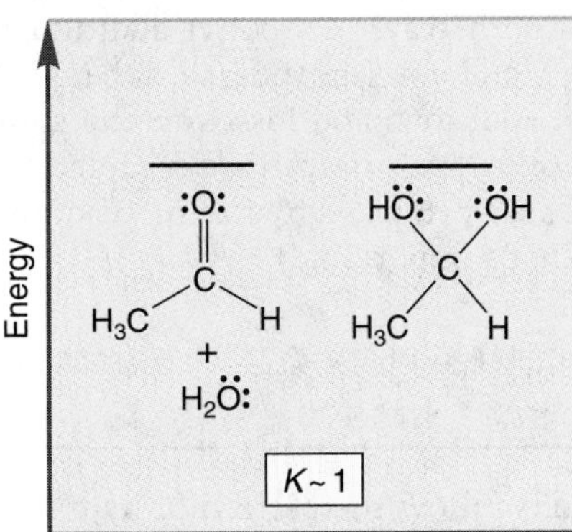

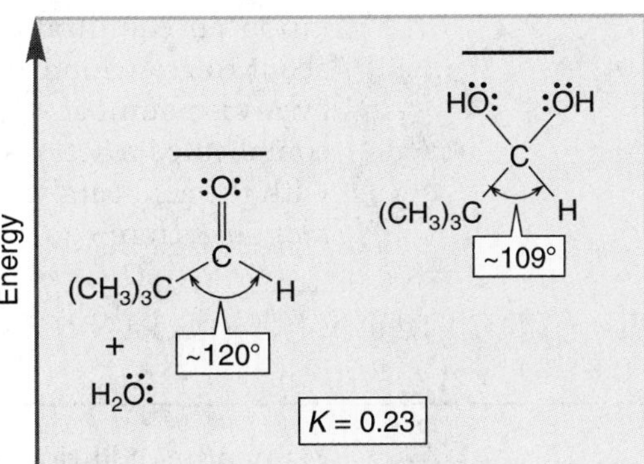

FIGURE 16.34 The larger the R group, the more important the steric effect. For example, *tert*-butylcarboxaldehyde is less hydrated at equilibrium than is acetaldehyde.

These equilibrium effects are important to keep in mind. Equally important is the notion that as long as there is *some* of the higher energy partner present at equilibrium, it can be the active ingredient in a further reaction. For example, even though it is a minor component of the equilibrium mixture, the hydrate of acetone could go on to other products (X), with more hydrate being formed as the small amount present at equilibrium is used up. Similarly, even though it is extensively hydrated in aqueous solution, reactions of formaldehyde to give Y can still be detected. Figure 16.35 illustrates the point with Energy versus Reaction progress diagrams.

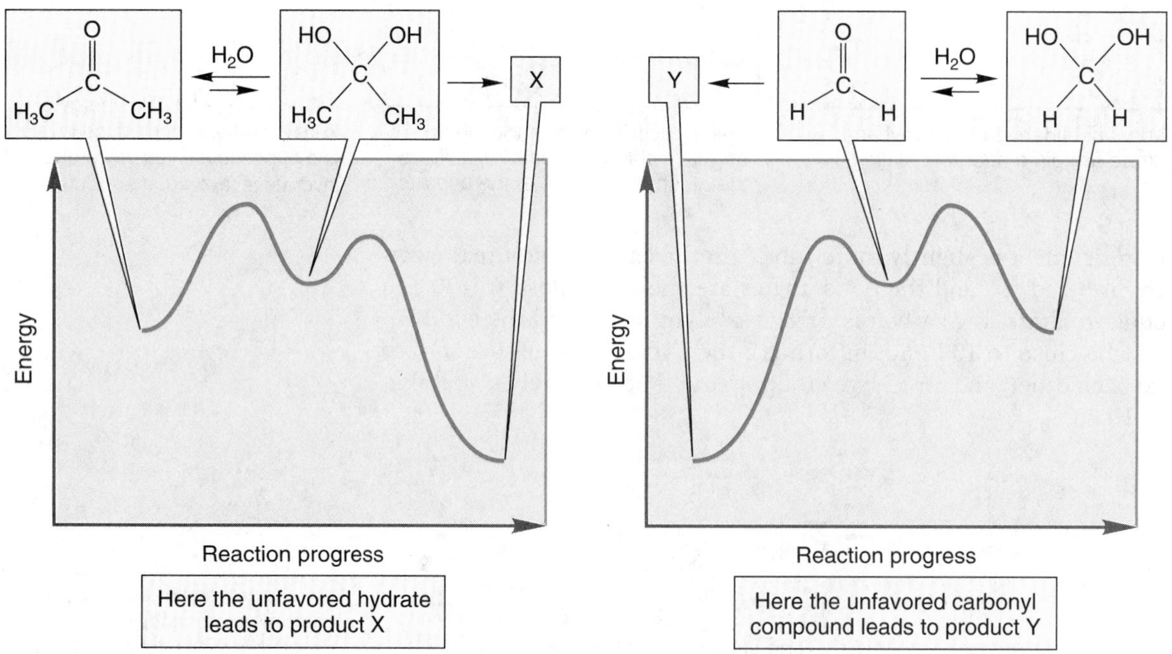

FIGURE 16.35 Molecules not favored at equilibrium may nonetheless lead to products. The two examples show a pair of carbonyl compound–hydrate equilibria in which it is the *unfavored* partner that leads to product. As the small amount of the higher energy molecule is used up, equilibrium will be reestablished and more reactive compound formed.

As usual, mechanisms stick in the mind most securely if we are able to write them backward. "Backward" is an arbitrary term anyway, as by definition an equilibrium runs both ways. Carbonyl addition reactions are about to grow more complex, and you run the risk of being overwhelmed by a vast number of proton additions and losses in the somewhat more complicated reactions that follow. Be sure you are completely comfortable with the acid-catalyzed and base-catalyzed hydration reactions *in both directions* before you go on. That's important.

16.7 OTHER ADDITION REACTIONS: ADDITIONS OF CYANIDE AND BISULFITE

Many other addition reactions follow the pattern of hydration. Two typical examples are cyanohydrin formation and bisulfite addition. Both of these reactions are generally base catalyzed. Because of an understandable and widespread aversion to working with the volatile and notorious HCN, cyanide additions are generally carried out with cyanide ion. Potassium cyanide ($K^+ \ ^-CN$) is certainly still poisonous, but it's not volatile, and one knows where it is at all times, at least as long as basic conditions are maintained. Cyanide ion is a good nucleophile and attacks the Lewis acidic carbonyl compound to produce an alkoxide. This alkoxide is protonated by solvent, often water, to give the final product, a **cyanohydrin** (Fig. 16.36).

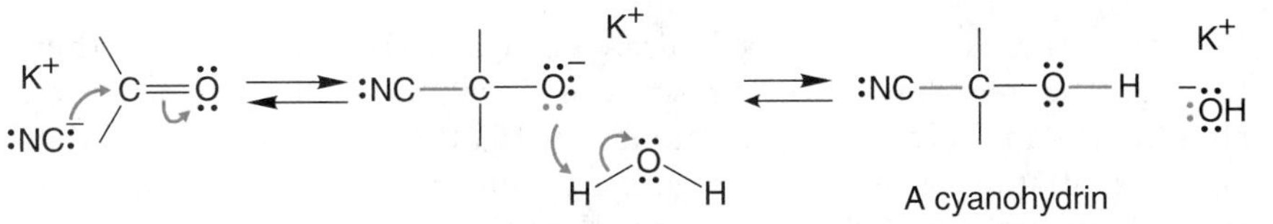

A cyanohydrin

FIGURE 16.36 Cyanohydrin formation. The carbonyl is first attacked by the Lewis base cyanide ion to give an alkoxide. The alkoxide is protonated in a second step to give the cyanohydrin.

PROBLEM 16.9

Why does acetaldehyde in an aqueous solution form an excellent yield of cyanohydrin when we know that acetaldehyde is substantially hydrated in water ($K \sim 1$)? Might it not be argued that, as about half the aldehyde is present in hydrated form at equilibrium, no more than about 50% cyanohydrin can be formed. Why is this argument wrong?

Bisulfite addition is similar, although the nucleophile in this case is sulfur. Sodium bisulfite adds to many aldehydes and ketones to give an addition product, often a nicely isolable solid (Fig. 16.37).

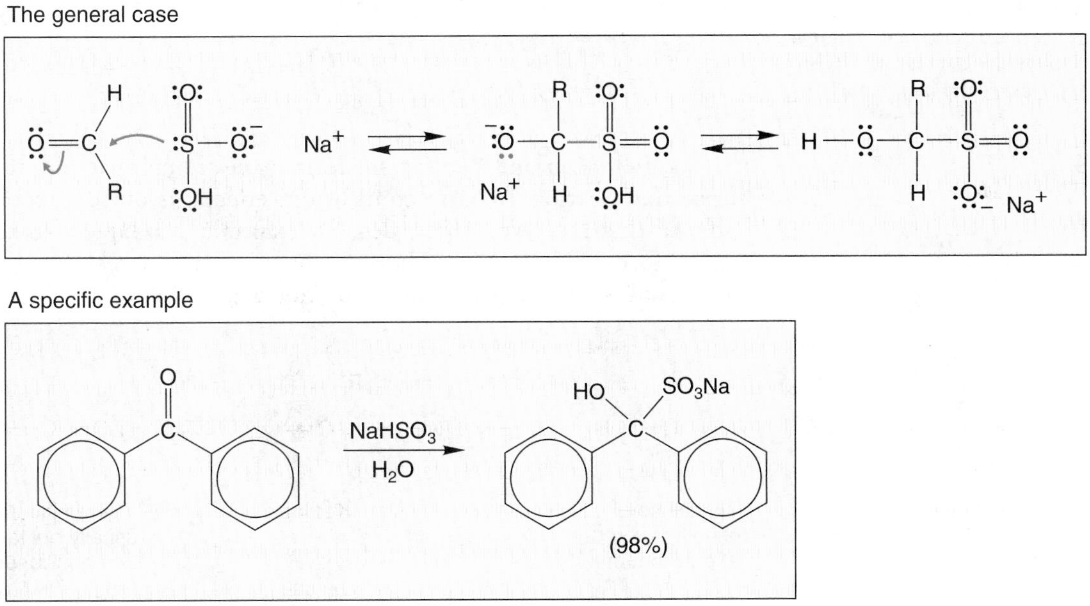

The general case

A specific example

NaHSO$_3$
H$_2$O
(98%)

FIGURE 16.37 Bisulfite addition is another base-catalyzed addition to carbonyls. In this case, it is a sulfur atom that acts as nucleophile.

PROBLEM 16.10

Bisulfite addition is generally far less successful with ketones than it is with aldehydes. Explain why.

The reactions we have seen so far have all been simple additions of a generic HX to the carbon–oxygen bond. Now we move on to related reactions in which there is one more dimension. In these new reactions, addition of HX is accompanied by the loss of some other group.

16.8 ADDITION REACTIONS FOLLOWED BY WATER LOSS: ACETAL AND KETAL* FORMATION

If aldehydes and ketones form hydrates in water (HOH), shouldn't there be a similar reaction in the presence of alcohols (ROH)? Indeed there should be, and there is. The reactions are slightly more complicated than simple hydration, at least under acidic conditions. In base, the two reactions are entirely analogous, so let's look at that reaction first.

Figure 16.38 contrasts base-catalyzed hydration with the related formation of a **hemiacetal** (from aldehydes) or hemiketal* (from ketones) in an alcohol and its corresponding alkoxide.

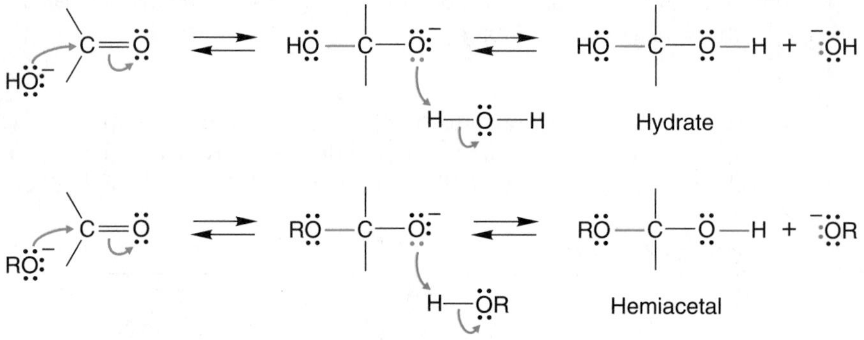

Hydrate

Hemiacetal

FIGURE 16.38 The reaction of a carbonyl group with hydroxide parallels the reaction with an alkoxide ion. In water, a hydrate is formed; in alcohol, the analogous product is called a hemiacetal.

As hydration and hemiacetal formation are comparable, it is no surprise that Section 16.6, on equilibrium effects, is quite directly transferable to this reaction. Aldehydes, for example, are largely in their hemiacetal forms in basic alcoholic solutions (in acid the reaction goes further, as we will see), whereas ketones are not (Fig. 16.39).

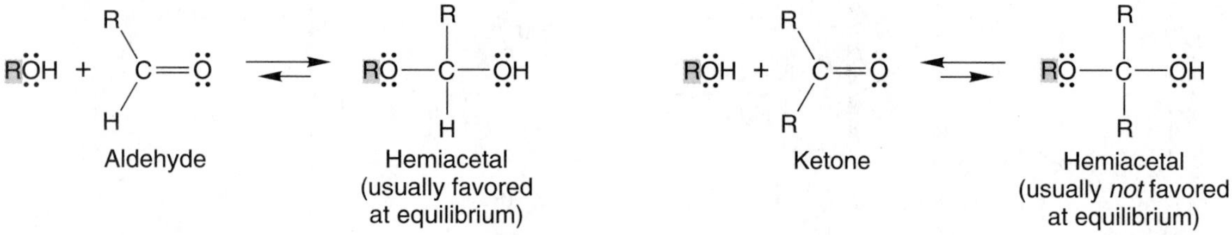

Aldehyde

Hemiacetal (usually favored at equilibrium)

Ketone

Hemiacetal (usually *not* favored at equilibrium)

FIGURE 16.39 Hemiacetals formed from aldehydes are generally favored at equilibrium, but hemiacetals formed from ketones are not.

Although most simple hemiacetals are not isolable, often the cyclic versions of these compounds are. For example, consider combining the

*Strictly speaking, neither ketal nor hemiketal is correct, as the IUPAC has dropped the names in favor of using acetal and hemiacetal for the species formed from both aldehydes and ketones. We will not use ketal or hemiketal, but be alert, because one still sees these terms occasionally.

carbonyl and hydroxyl components of the reaction within the same molecule. Now hemiacetal formation is an intramolecular process. If the ring size is neither too large nor too small, the hemiacetals are often more stable than the open-chain molecules (Fig. 16.40).

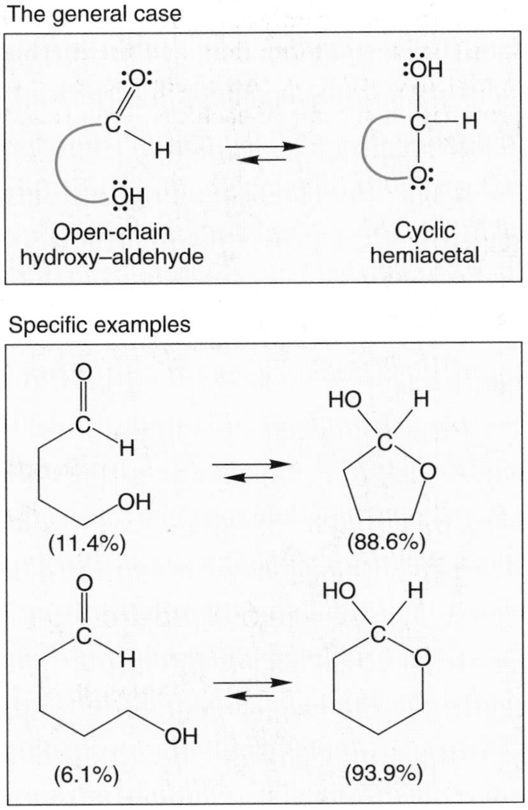

The general case

Open-chain hydroxy–aldehyde ⇌ Cyclic hemiacetal

Specific examples

(11.4%) ⇌ (88.6%)

(6.1%) ⇌ (93.9%)

FIGURE **16.40** Cyclic hemiacetals are often more stable than their open-chain counterparts.

PROBLEM **16.11**

Write a mechanism for the intramolecular formation of one of the hemiacetals of Figure 16.40.

The most spectacular examples of this are the sugars, polyhydroxylated compounds that exist largely, though not quite exclusively, in their cyclic hemiacetal forms (Chapter 23; Fig. 16.41).

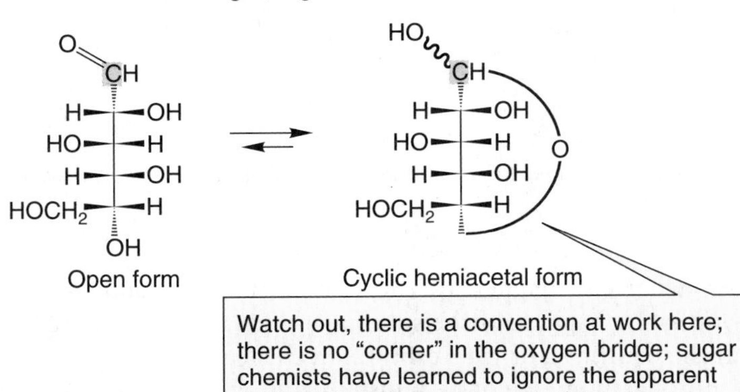
The sugar, D-glucose
Open form
Cyclic hemiacetal form
Watch out, there is a convention at work here; there is no "corner" in the oxygen bridge; sugar chemists have learned to ignore the apparent right angle in the bottom part of the structure

FIGURE **16.41** Sugars exist mainly in their cyclic hemiacetal forms.

PROBLEM 16.12 Construct a three-dimensional picture of D-glucose in its cyclic hemiacetal form. Note that the position of the OH on the carbon marked in red in Figure 16.41 can be either axial or equatorial. Place this OH in the energy-minimum position.

Now, what about the acid-catalyzed version of this reaction? The process begins with hemiacetal formation, and the mechanism parallels that of acid-catalyzed hydration. A three-step sequence of protonation, addition, and deprotonation appears in each case (Fig. 16.42).

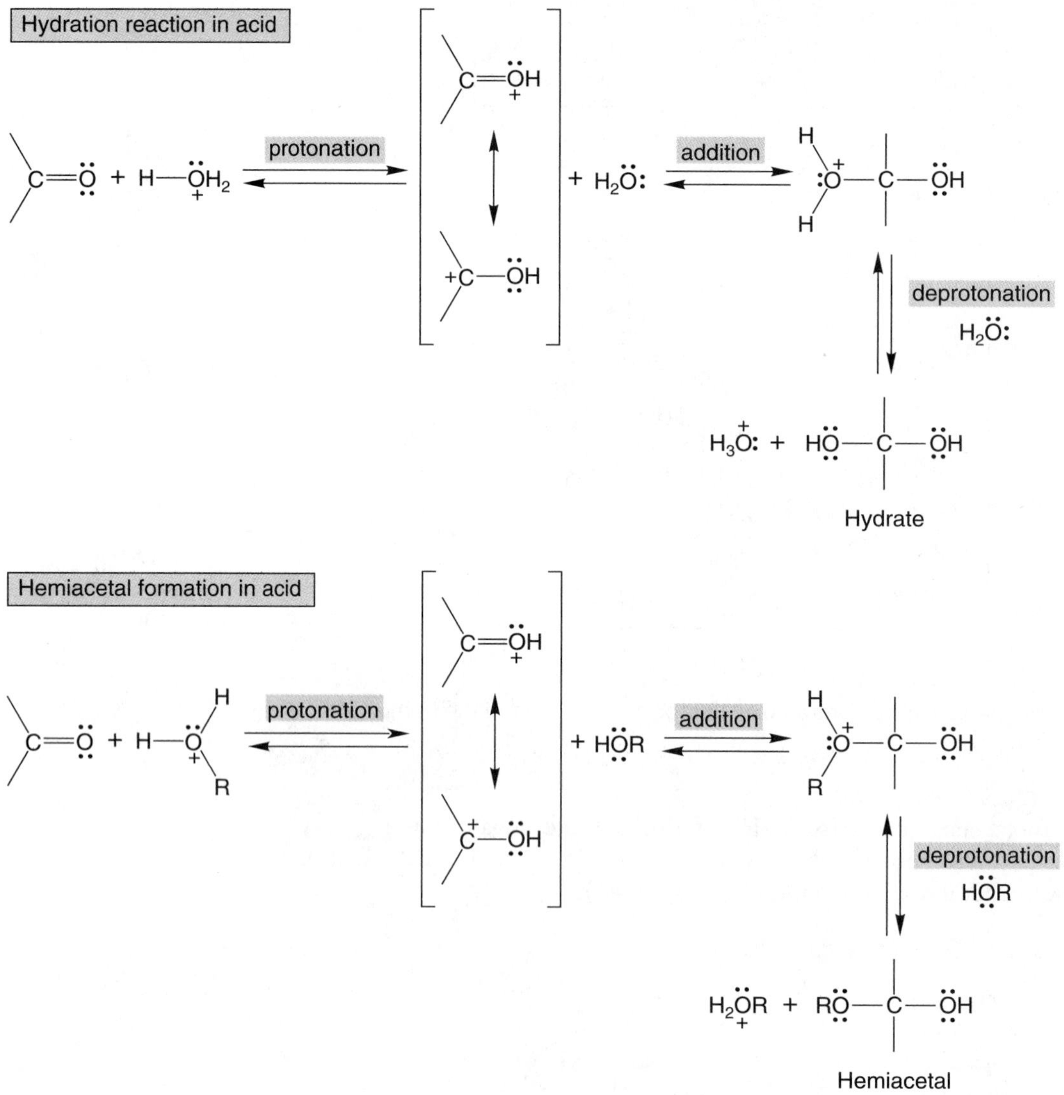

FIGURE 16.42 The acid-catalyzed formation of a hemiacetal parallels the acid-catalyzed hydration reaction.

Now, what can happen to the hemiacetals in acidic solution? One thing you know happens is the set of reactions indicated by the reverse arrows in the equilibria of Figure 16.42. In this case, the oxygen atom of the OR is protonated and alcohol is lost to give the resonance-stabilized protonated carbonyl compound. This intermediate is deprotonated to give the starting ketone or aldehyde. There is no need for us to write out the mechanism because it already appears exactly in the reverse steps of Figure 16.42. It is simply the forward reaction run directly backward.

But suppose it is not the OR that is protonated, but instead the new OH? Surely, if one of these reactions is possible, so is the other. If we keep strictly parallel to the reverse process, we can now lose water to give a resonance-stabilized species (Fig. 16.43), which is quite analogous to the resonance-stabilized protonated carbonyl compound of Figure 16.42.

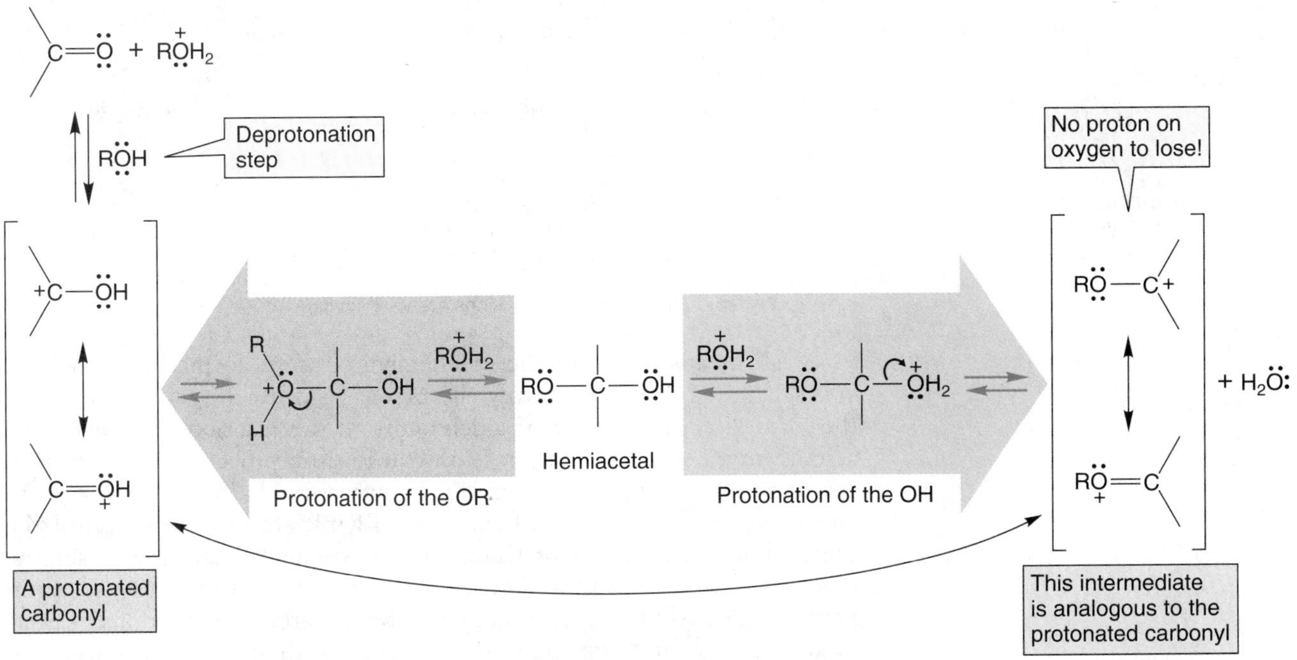

FIGURE 16.43 Protonation of the OH in the hemiacetal, followed by water loss, leads to a resonance-stabilized species that is analogous to the protonated carbonyl.

But now deprotonation is not possible—there is no proton on oxygen to lose, as its place is taken by an R group. Another reaction occurs instead, the addition of a second molecule of alcohol. Deprotonation gives a full **acetal** (Fig. 16.44).

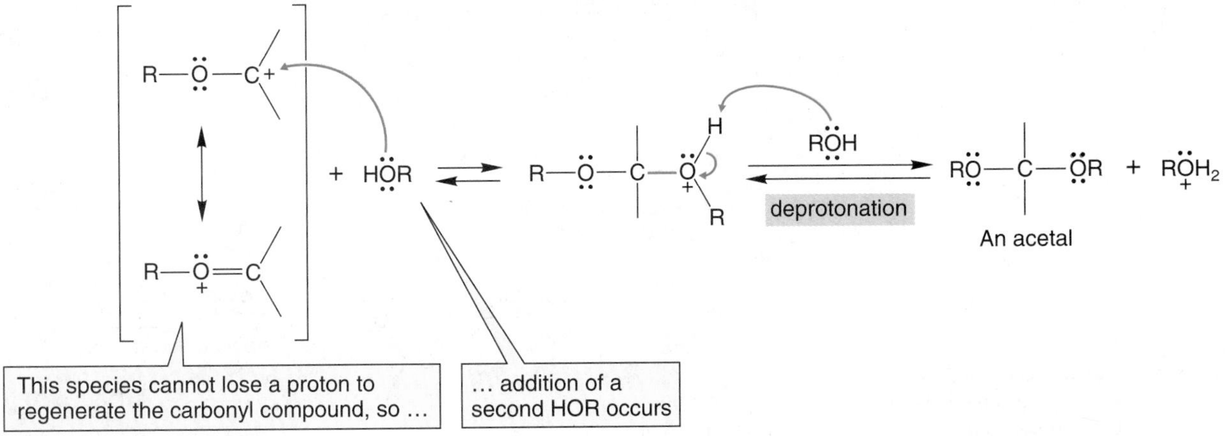

FIGURE 16.44 A full acetal is formed by addition of a second molecule of alcohol to the resonance-stabilized intermediate.

Notice that the second addition of alcohol in acetal formation is just like the first; the only difference is the presence of an H or R group in the species to be attacked (Fig. 16.45).

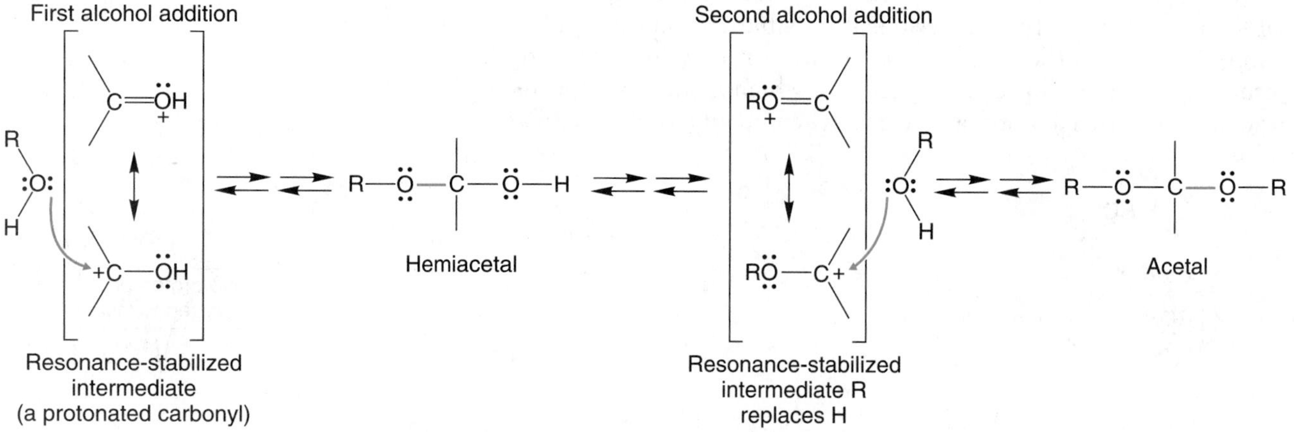

FIGURE 16.45 The two alcohol additions are very similar.

By all means work out the mechanism of acetal formation by setting yourself the problem of writing the mechanism for the *reverse* reaction, the formation of a ketone or aldehyde from the reaction of an acetal in acidic water. Acetal formation is an equilibrium process, and conditions can drive the overall equilibrium in the forward or backward direction. In an excess of alcohol, an acid catalyst will initiate the formation of the acetal, but in an excess of water, that same acid catalyst will set the reverse reaction in motion, and we will end up with the carbonyl compound. This point is important and practical. Carbonyl compounds can be hidden, stored or "protected," by converting them into their ketal or acetal forms, and then regenerating them as needed. The acetal constitutes your first **protecting group**. For example, it might be desirable to convert group X into group Y in a base-catalyzed reaction. But we know that carbonyl compounds are sensitive to base, and the starting material may be destroyed in the attempted conversion of X into Y. The solution is first to convert the carbonyl group into an acetal, then carry out the conversion of X to Y, and finally regenerate the carbonyl group (Fig. 16.46).

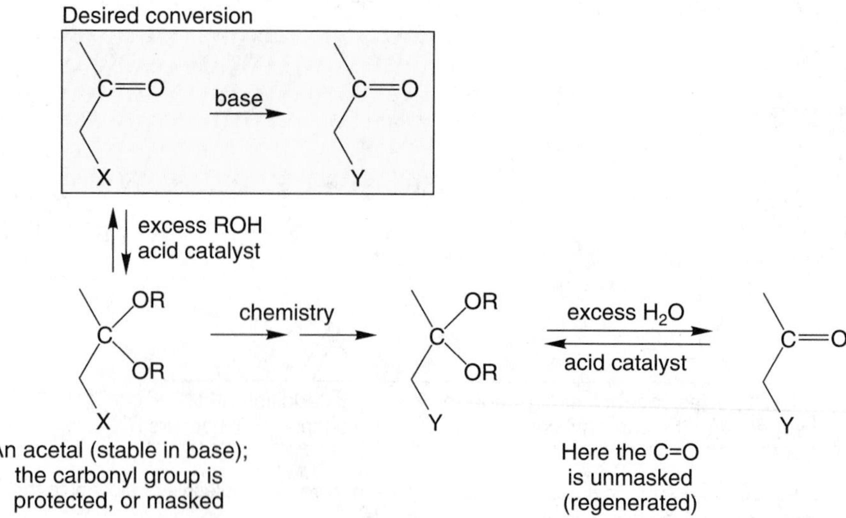
Desired conversion
base
excess ROH
acid catalyst
chemistry
excess H_2O
acid catalyst
An acetal (stable in base); the carbonyl group is protected, or masked
Here the C=O is unmasked (regenerated)

FIGURE 16.46 A protecting group can be used to mask a sensitive functional group while reactions are carried out on other parts of the molecule. When these are completed, the original functionality can be regenerated.

Write the mechanism for the formation of the acetal from treatment of cyclopentanone in ethylene glycol (1,2-dihydroxyethane) with an acid catalyst.

PROBLEM 16.13

Write a mechanism for the reverse reaction, the regeneration of the starting ketone on treatment of the acetal with aqueous acid.

PROBLEM 16.14

Explain clearly why hemiacetals, but not full acetals, are formed under basic conditions.

*PROBLEM 16.15

ANSWER

If we work through the mechanism we can see why. The first step is addition of an alkoxide to the carbonyl group. Protonation then gives the hemiacetal and regenerates the base.

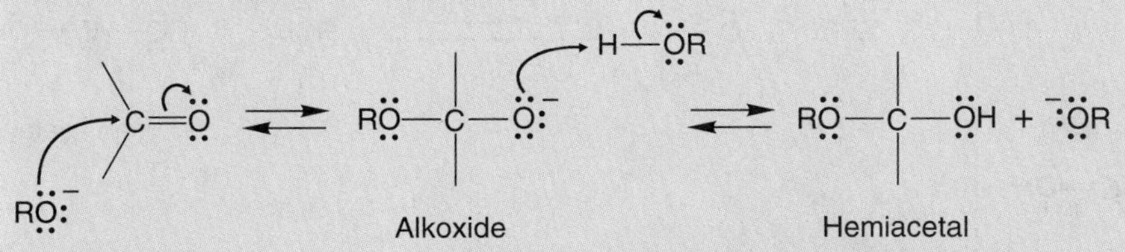

Alkoxide

Hemiacetal

Now what? In order to get a full acetal, RO^- must displace hydroxide! This reaction is most unfavorable. The more favorable reaction between RO^- and the hemiketal will be removal of a proton, *the reverse of the last step in the figure above.* In base, the reaction cannot go beyond hemiketal formation.

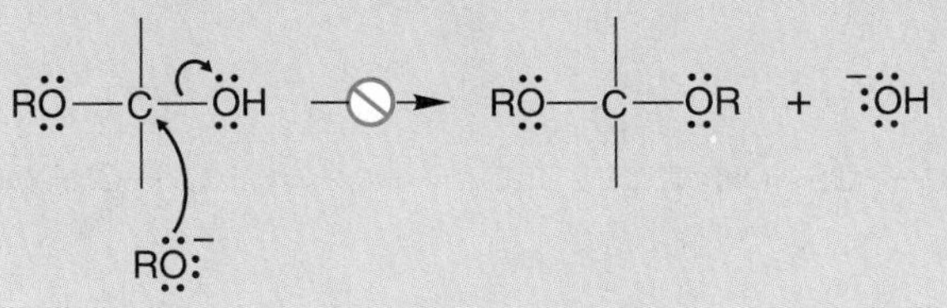

When acetone is treated with $H_2O^{18}/H_3\overset{+}{O}{}^{18}$, the acetone oxygen atom is found to be labeled with O^{18} (Fig. 16.47). Explain.

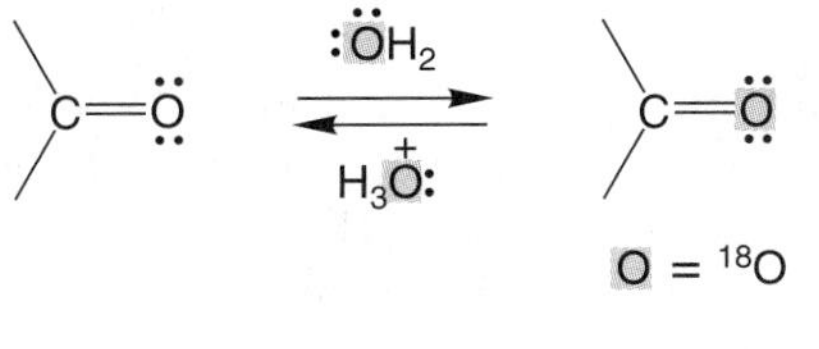

PROBLEM 16.16

O = ^{18}O

FIGURE 16.47

By now, you may be getting the impression that almost any nucleophile will add to carbonyl compounds. If so, you are quite right. So far, we have seen water, cyanide, bisulfite, and alcohols act as nucleophiles toward the Lewis acidic carbon of the carbon–oxygen double bond. All of the addition reactions are similar, with only the details being different. Those details are important though, as well as highly effective at camouflaging the essential similarity of the reactions. We'll now see just such an example.

16.9 ADDITION REACTION OF NITROGEN BASES: IMINE AND ENAMINE FORMATION

The nitrogen atom of an amine (RNH_2) is a nucleophile, and, like the related oxygen bases, it can add to carbonyl compounds. Indeed, the acid-catalyzed reaction between a carbonyl compound and an amine to give a **carbinolamine** is quite analogous to the reactions of water and alcohols we have just seen (Fig. 16.48).

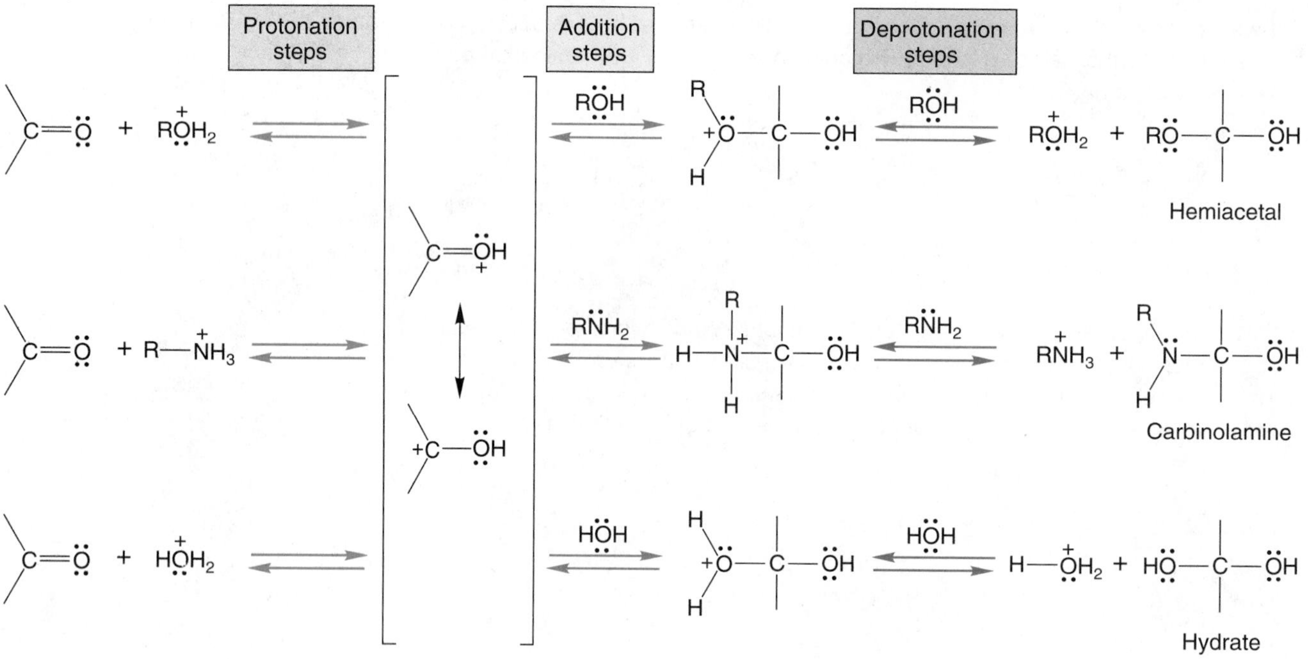

Protonation steps

Addition steps

Deprotonation steps

Hemiacetal

Carbinolamine

Hydrate

FIGURE 16.48 Reactions of carbonyl compounds to give hemiacetals and hydrates are completely analogous to carbinolamine formation. The usual sequence of protonation, addition, and deprotonation steps is followed.

PROBLEM 16.17 Write the base-catalyzed versions of the reactions in Figure 16.48.

Further steps in the reaction are all similar as well. An OH is protonated, and then lost as water (note that ^-OH is *not* the leaving group) to give a resonance-stabilized cation (Fig. 16.49).

In the first example in Figures 16.48 and 16.49, the adding nucleophile is an alcohol, ROH, and R^+ cannot usually be lost. Instead, the cation goes on to full acetal formation by a second addition of alcohol. For the second and third examples of Figures 16.48 and 16.49, however, there remains an

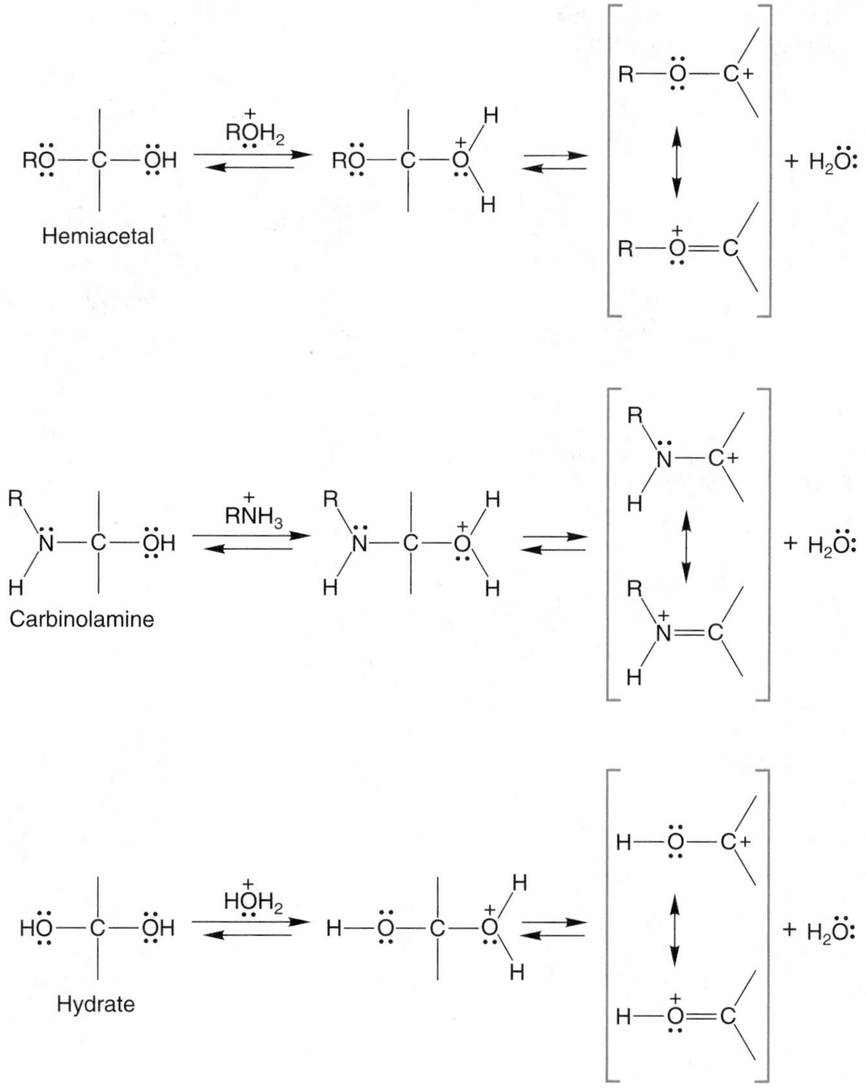

Hemiacetal

Carbinolamine

Hydrate

FIGURE 16.49 Loss of water leads to a resonance-stabilized intermediate in all three reactions.

available hydrogen that can be lost as a proton. In the third reaction (hydration), this simply completes the re-formation of the carbonyl group of the starting material, but in the second example it leads to the formation of a carbon–nitrogen double bond, called, in simple cases, a **Schiff base** or an **imine** (Fig. 16.50).

There is a vast variety of imines formed from specialized primary amines (RNH_2, monosubstituted derivatives of ammonia, NH_3), and each has its own common name. Many of them were once useful in analytical chemistry because they are generally solids, and could be characterized by their melting points. For example, formation of a 2,4-dinitrophenylhydrazone or a semicarbazone from an unknown was not only diagnostic for the presence of a carbonyl group, but yielded a specific derivative that could be compared to a similar sample made from a compound of known struc-

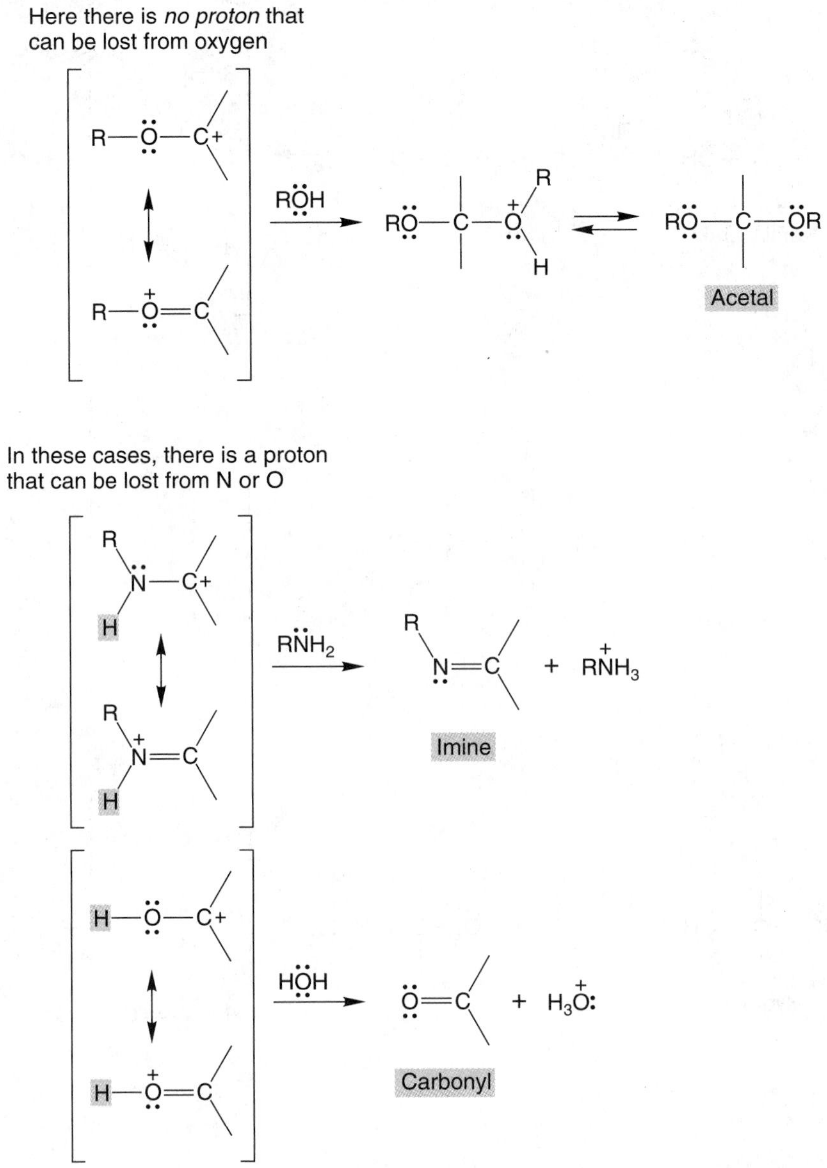

Here there is *no proton* that can be lost from oxygen

In these cases, there is a proton that can be lost from N or O

Acetal

Imine

Carbonyl

FIGURE 16.50 In the second two reactions, loss of a proton leads to an imine or carbonyl compound. In the first reaction, there is no proton to be lost, and a second molecule of alcohol adds to give an acetal.

ture. Figure 16.51 gives a few of these substituted imines. Be sure to note that all of these reactions are the same—the only difference is the identity of the R group.

PROBLEM 16.18 Write a mechanism for the general reaction of Figure 16.51, the acid-catalyzed reaction of a carbonyl compound with RNH_2.

Note also that the success of this reaction relies on the presence of at least two hydrogens in the starting amine. One is lost as a proton in the first step of the reaction, the formation of the carbinolamine. The second

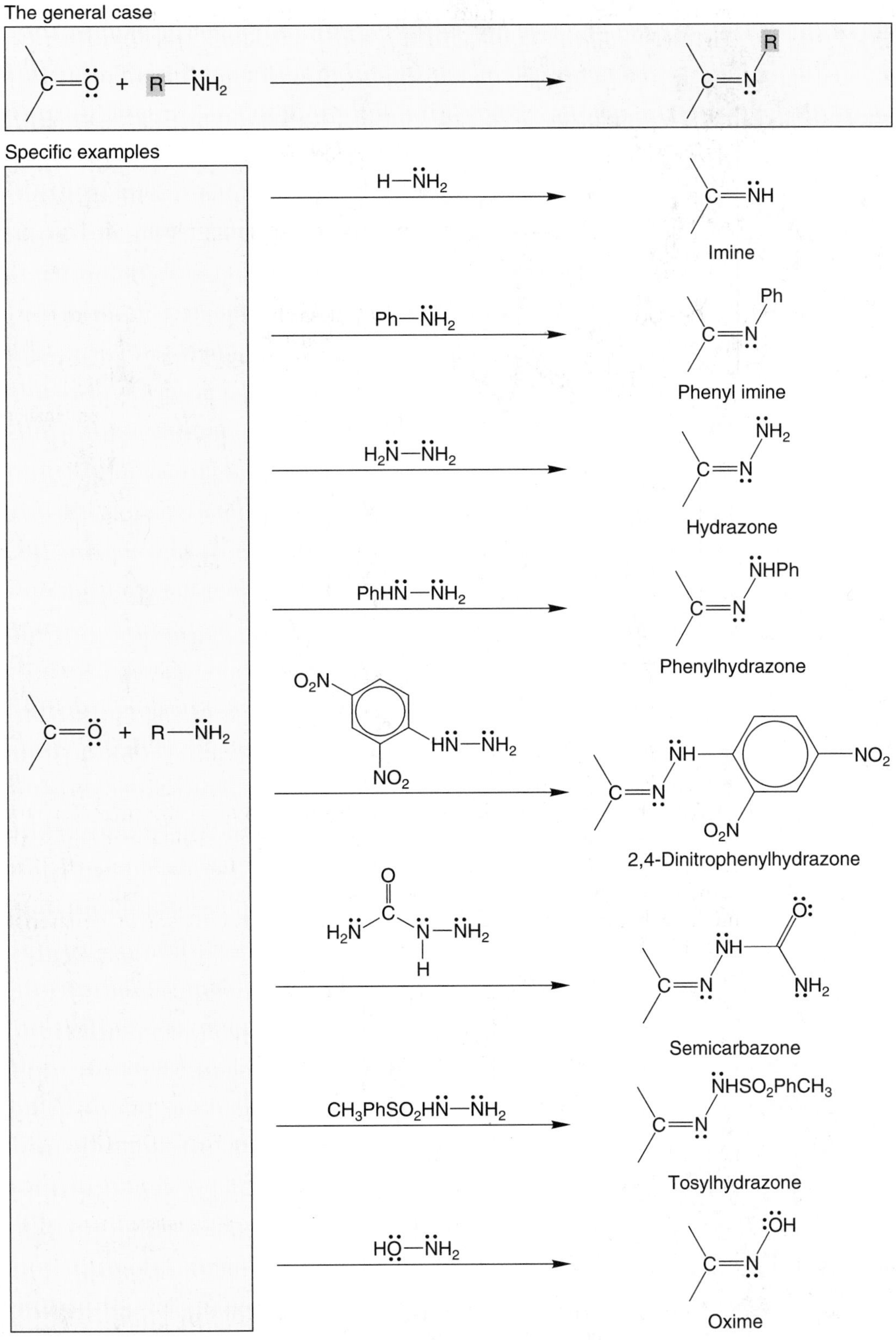

FIGURE 16.51 Some substituted imines formed from reactions of substituted amines with carbonyl compounds.

proton is lost in a deprotonation step as the final imine is produced (Fig. 16.52).

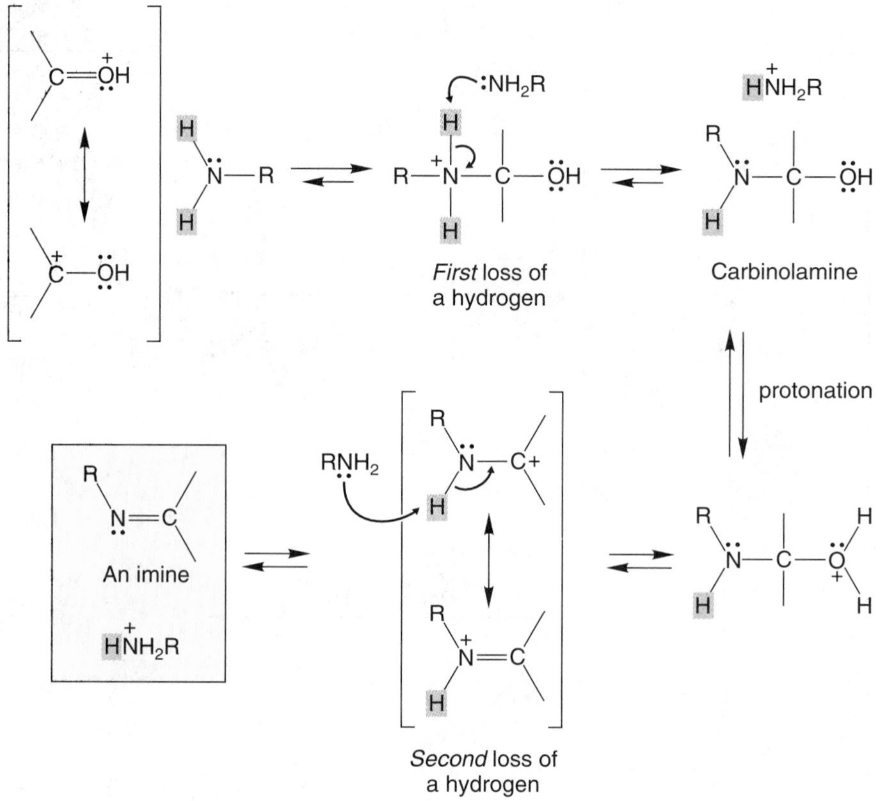

FIGURE 16.52 In order to form an imine, there must be *two* hydrogens on the starting amine. That is, the amine must be primary (RNH_2) or ammonia ($:NH_3$).

What happens if one or both of these hydrogens is missing? Tertiary amines (R_3N) carry no hydrogens and undergo no visible reaction with carbonyl groups. But this masks reality. In fact, addition to the carbon–oxygen double bond does take place, but the unstable intermediate has no available reaction except reversal to starting material (Fig. 16.53).

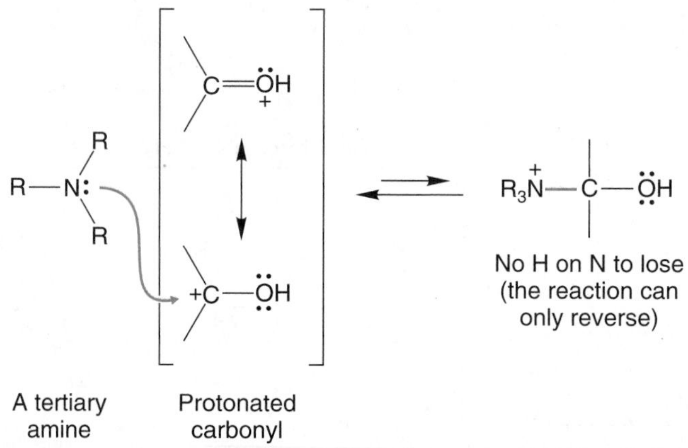

FIGURE 16.53 Tertiary amines (R_3N:) can add to carbonyl groups, but the only available further reaction is reversion to starting materials.

Secondary amines (R_2NH) have one hydrogen, and so carbinolamine formation is possible. This reaction can be followed by protonation and water loss to give a resonance-stabilized cation, called an **immonium ion**, or, sometimes, an imminium ion. But now imine formation is *not* possible as there remains no second proton on nitrogen to be lost. Figure 16.54 shows the reaction for secondary amines.

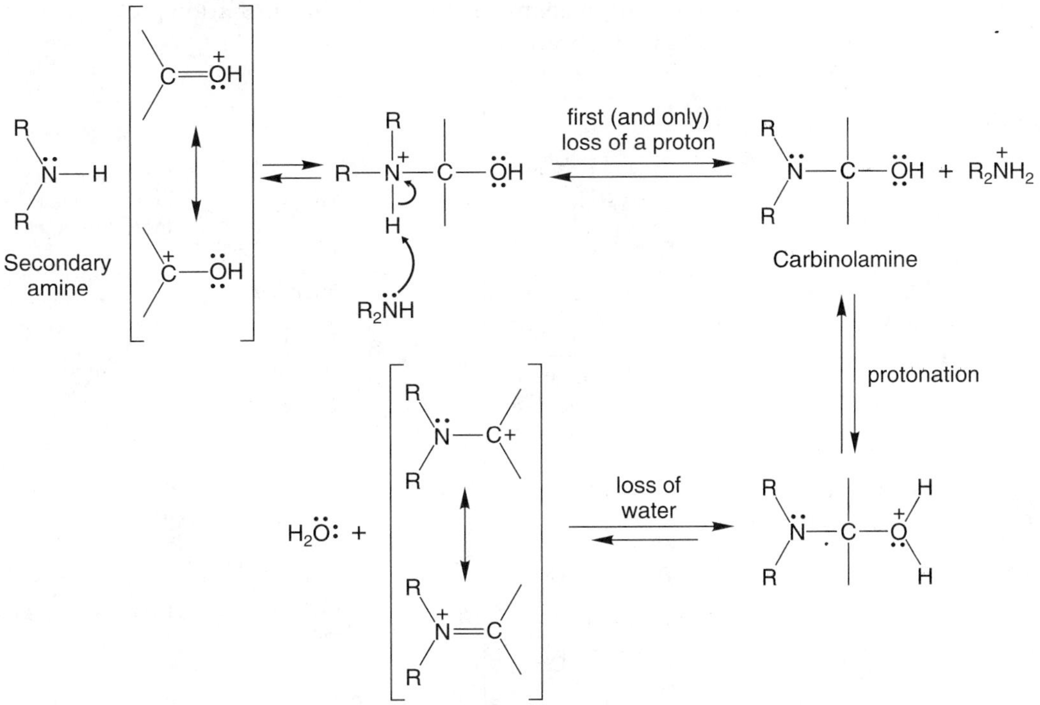

An immonium ion (no proton can be lost from nitrogen to give an imine because there is no H left attached to N)

FIGURE **16.54** For secondary amines (R_2NH) carbinolamine formation is possible, but there is no hydrogen remaining on nitrogen that can be lost to give an imine. Instead, water is lost to give a resonance-stabilized immonium ion.

Think now about what you know of the possibilities for proton loss by carbocations. The E1 reaction (Chapter 7, p. 273) is a perfect example. In the E1 reaction, a leaving group departs to give a carbocation, which reacts further by deprotonating at all possible adjacent positions. If an adjacent carbon-attached hydrogen is available, the immonium ion can do this same reaction. Figure 16.55 makes the analogy.

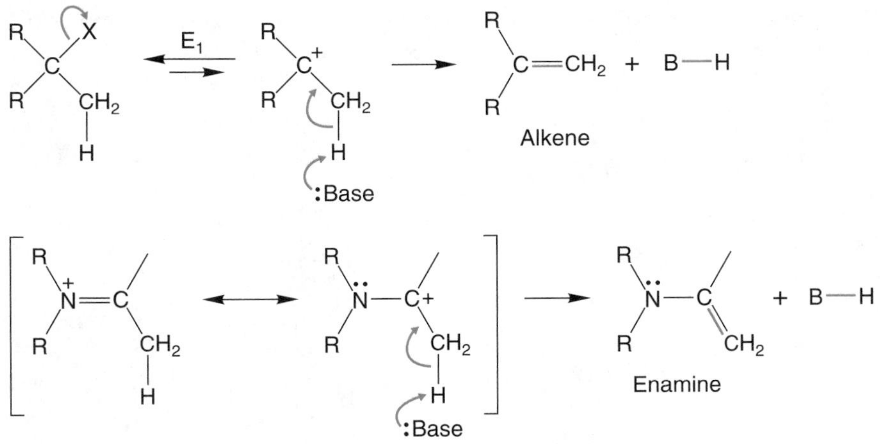

FIGURE **16.55** Deprotonation of an immonium ion to give an enamine is exactly like deprotonation of a carbocation to give an alkene.

This reaction gives a molecule called an **enamine**, which is the final product of the reaction of secondary amines and simple aldehydes and ketones. Now comes an important question: Why doesn't the immonium ion formed from primary amines or ammonia also lose a proton from carbon to give an enamine? Why is imine formation preferred when there is a choice (Fig. 16.56)?

Immonium ion from a secondary amine
Enamine
No choice here—base can only remove H^+ from C; there are no H atoms on N
Immonium ion from a primary amine
or
Enamine (less stable)
Imine (more stable)
Here there is a choice; base can remove H^+ from either C (red) or N (green)

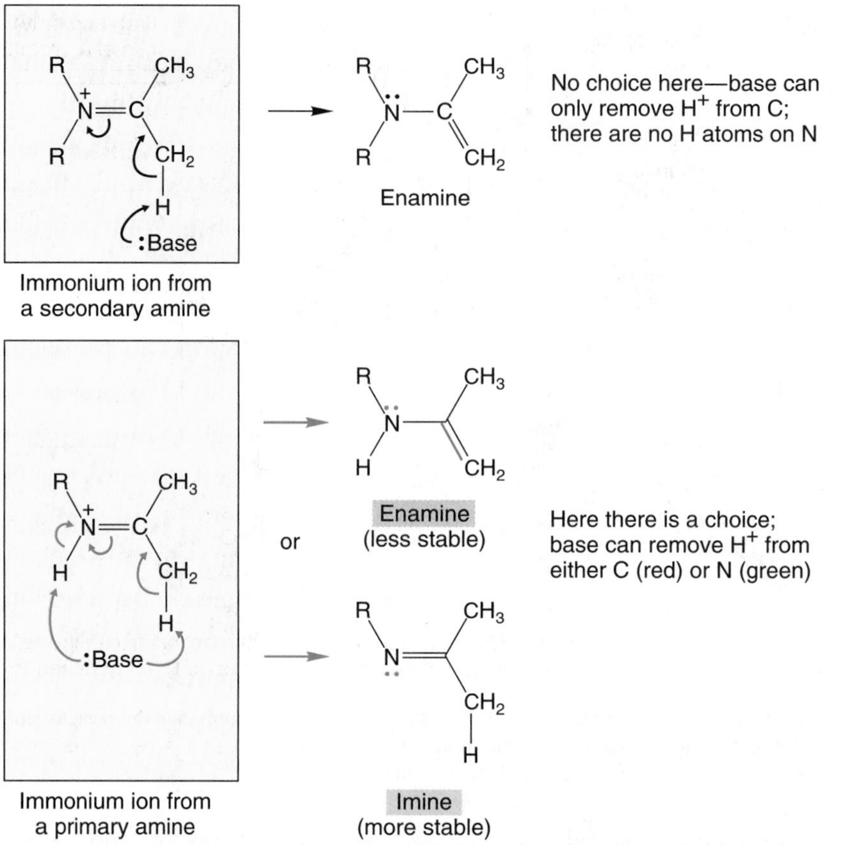

FIGURE 16.56 Why don't immonium ions formed from primary amines give enamines rather than imines?

The answer is extraordinarily simple: The imine is the more stable molecule and so is preferred when possible. It is only in the cases where imine formation is not possible, as when secondary amines are used, that enamines are formed.

*PROBLEM 16.19 Many enamines and imines are in equilibrium. Write a mechanism for the acid- and base-catalyzed reactions (both directions) in Figure 16.57.

$H_3\overset{+}{O}$:
$H_2\ddot{O}$:
Imine
Enamine
K^+ $^-$:ÖH
$H_2\ddot{O}$:

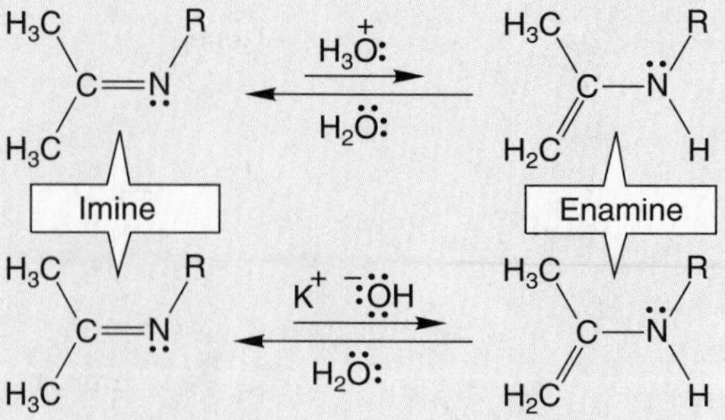

FIGURE 16.57

ANSWER

In the acid-catalyzed reaction, the imine is protonated on nitrogen and then deprotonated on carbon. Thermodynamics will determine where this equilibrium settles out.

Imine — protonation → [resonance forms] — deprotonation → Enamine + $H{-}\overset{+}{O}H_2$

In base, the protonation–deprotonation sequence of the first part of the problem is reversed. In base, a proton is first removed from carbon to give a resonance-stabilized anion. Reprotonation on nitrogen completes the equilibration. Once again, the relative thermodynamic stabilities of the imine and enamine will determine the position of the equilibration.

Imine — deprotonation → [resonance forms] — protonation → Enamine + $^{-}$:ÖH

16.10 ORGANOMETALLIC REAGENTS

Not all additions to carbonyl compounds are reversible. Some of the irreversible reactions are of great importance in synthetic chemistry, as they involve the creation of carbon–carbon bonds. In order to discuss these, we must first talk a bit about the formation of **organometallic reagents**.

When an alkyl halide (fluorides are generally exceptions) is added to a cold mixture of magnesium or lithium metal and an ether solvent [usually diethyl ether or tetrahydrofuran (THF)], the metal begins to disappear in an exothermic reaction. The end result is either a **Grignard reagent** or an **organolithium reagent** (Fig. 16.58).

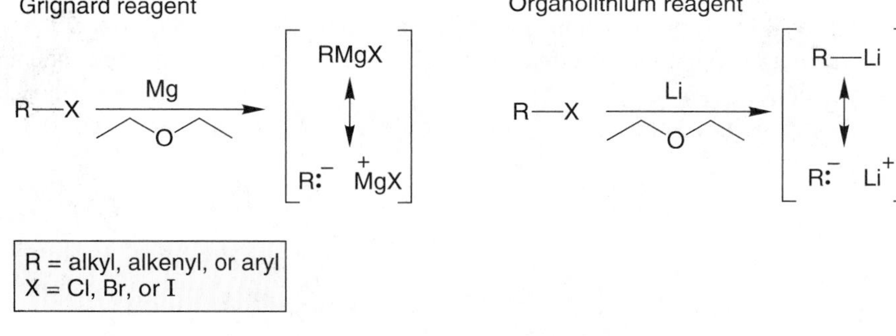

FIGURE **16.58** Treatment of alkyl chlorides, bromides, and iodides with Mg or Li leads to the formation of an organometallic reagent.

Grignard reagents were discovered by P. A. Barbier (1848–1922) and their chemistry was worked out extensively by his student, Victor Grignard (1871–1935), who received the 1912 Nobel prize for his work in this area. Neither the formation of the Grignard reagent nor its structure is an easy subject. The simplest formulation is "RMgX" and should be amplified by the addition of the ionic resonance form in Figure 16.58. The Grignard reagent is also in equilibrium with a mixture of the magnesium halide and dialkylmagnesium compound. Moreover, the ether solvent is essential for its formation. Grignard reagents incorporate two molecules of solvent, conveniently left out in the traditional formulation as RMgX. Whatever the detailed structure, the result is a highly polar reagent that is a very strong Lewis base.

R—X —(Mg, ether)→ R—Mg—X (with two coordinated ether molecules) ⇌ MgX_2 + R_2Mg

FIGURE **16.59** The Grignard reagent has a complex structure in which the ether solvent is important. The Grignard reagent (RMgX) is also in equilibrium with a dialkyl magnesium reagent (R_2Mg) and magnesium halide (MgX_2).

The mechanism of formation of the Grignard reagent involves radical-transfer reactions in which a transient alkyl radical is formed in the presence of a magnesium-centered radical (Fig. 16.60). Coupling generates RMgX, and further radical reactions can equilibrate this species with the dialkyl magnesium. Most alkyl, alkenyl, and aryl halides (except fluorides) can be made into Grignard reagents.

R—X :Mg ⟶ R· ·Mg—X ⟶ R—Mg—X

X = Cl, Br, or I

FIGURE **16.60** The mechanism of formation of the Grignard reagent involves a radical coupling reaction.

Organolithium reagents are even more complex. Here the ether solvent is not essential, but the reagents themselves are known not to be monomeric and to form aggregates the size of which depends on the nature of the solvent and the structure of the R group. As in RMgX, the representation of organolithium reagents as RLi, even with the addition of the polar resonance form, is a convenient simplification (Fig. 16.61).

R—X + Li ⟶ $(R—Li)_n$

FIGURE **16.61** Organolithium reagents are not monomeric, but oligomeric. The exact value of *n* depends on solvent and the structure of R.

Both of these organometallic reagents are extraordinarily strong bases, and they must be carefully protected from moisture and oxygen. Chemists form them under a strictly dry and inert atmosphere. Although these reagents do not contain free carbanions, *they act as if they did.* They are sources of $R:^-$. The very polar carbon–metal bond attacks all manner of

Lewis and Brønsted acids. Water is more than sufficiently strong enough to protonate a Grignard or organolithium reagent (Fig. 16.62).

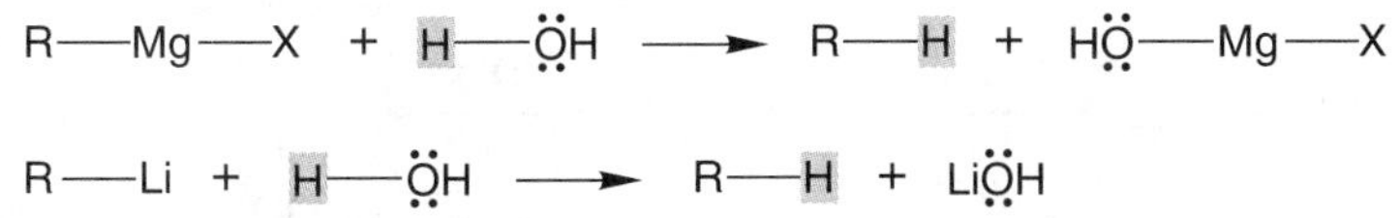

$$R{-}Mg{-}X + H{-}\ddot{\underset{..}{O}}H \longrightarrow R{-}H + H\ddot{\underset{..}{O}}{-}Mg{-}X$$

$$R{-}Li + H{-}\ddot{\underset{..}{O}}H \longrightarrow R{-}H + Li\ddot{\underset{..}{O}}H$$

FIGURE **16.62** The powerfully basic organometallic reagents are easily protonated by water to give hydrocarbons and metal hydroxides.

The end product is a hydrocarbon. This sequence constitutes a new synthesis of hydrocarbons from alkyl, alkenyl, or aryl halides, so update your file cards (Fig. 16.63). This reaction can often be put to good use in the construction of isotopically labeled reagents, because D_2O can be used in place of water to produce a specifically deuteriolabeled hydrocarbon.

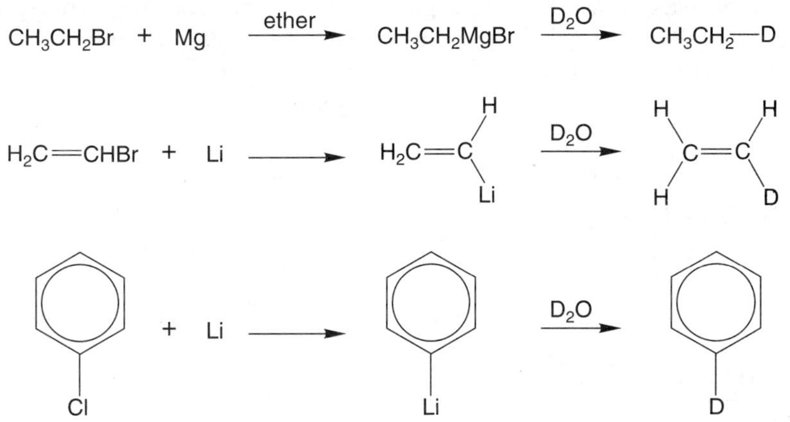

$$CH_3CH_2Br + Mg \xrightarrow{\text{ether}} CH_3CH_2MgBr \xrightarrow{D_2O} CH_3CH_2{-}D$$

$$H_2C{=}CHBr + Li \longrightarrow H_2C{=}CH(Li) \xrightarrow{D_2O} H_2C{=}CH(D)$$

$$C_6H_5Cl + Li \longrightarrow C_6H_5Li \xrightarrow{D_2O} C_6H_5D$$

FIGURE **16.63** Treatment of an organometallic reagent with D_2O gives deuteriolabeled molecules.

PROBLEM **16.20**

Given inorganic reagents of your choice (including D_2O), devise syntheses of the following molecules from the indicated starting materials (Fig. 16.64).

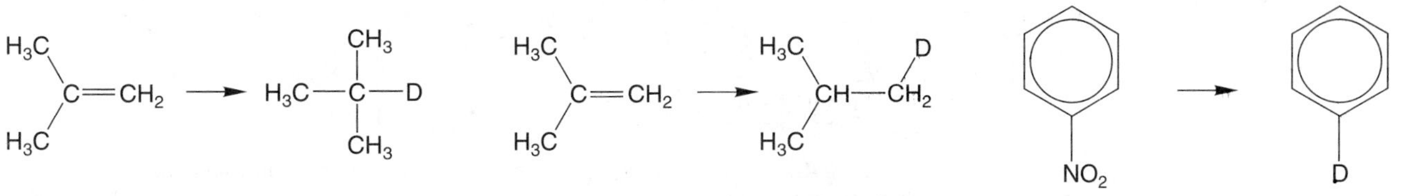

$$(H_3C)_2C{=}CH_2 \longrightarrow H_3C{-}C(CH_3)_2{-}D$$

$$(H_3C)_2C{=}CH_2 \longrightarrow (H_3C)_2CH{-}CH_2D$$

$$C_6H_5NO_2 \longrightarrow C_6H_5D$$

FIGURE **16.64**

Neither Grignard reagents nor organolithium reagents are good nucleophiles toward organic halides. If they were, the syntheses we use to generate the organometallic reagents in the presence of the starting organic halide would be quite ineffective, as hydrocarbons would be the major products. Small amounts of coupling are sometimes observed, especially with highly reactive halides such as allyl bromide (Fig. 16.65).

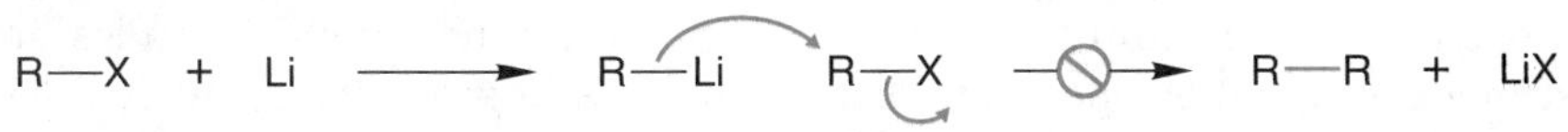

$$R{-}X + Li \longrightarrow R{-}Li \quad R{-}X \not\longrightarrow R{-}R + LiX$$

FIGURE **16.65** Organometallic reagents are strong bases, but not strong nucleophiles. If they were, displacement of halogen from R–X would give hydrocarbons in the organometallic-generating reaction of R–X and a metal.

There are other organometallic reagents that can be used to displace halides to good effect. Examples are the **organocuprates** formed from organocopper reagents and organolithium compounds. The organocopper reagents themselves can be made from organolithium compounds and cuprous halides (Fig. 16.66).

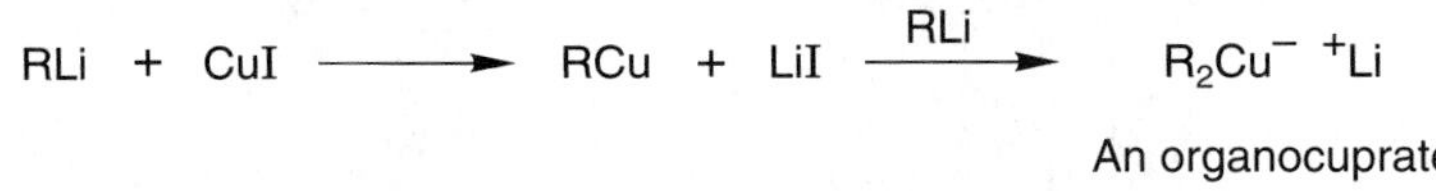

$$RLi + CuI \longrightarrow RCu + LiI \xrightarrow{RLi} R_2Cu^- \ {}^+Li$$

An organocuprate

FIGURE 16.66 The synthesis of lithium organocuprates from alkyllithiums and cuprous halides.

The organocuprates are quite versatile reagents, and undergo a variety of synthetically useful reactions. We will meet them again a number of times, but for now, note that they are able to displace chloride, bromide, and iodide ions, as well as other good leaving groups, from primary, and often secondary alkyl halides. This capability gives you another quite effective synthesis of hydrocarbons (Fig. 16.67).

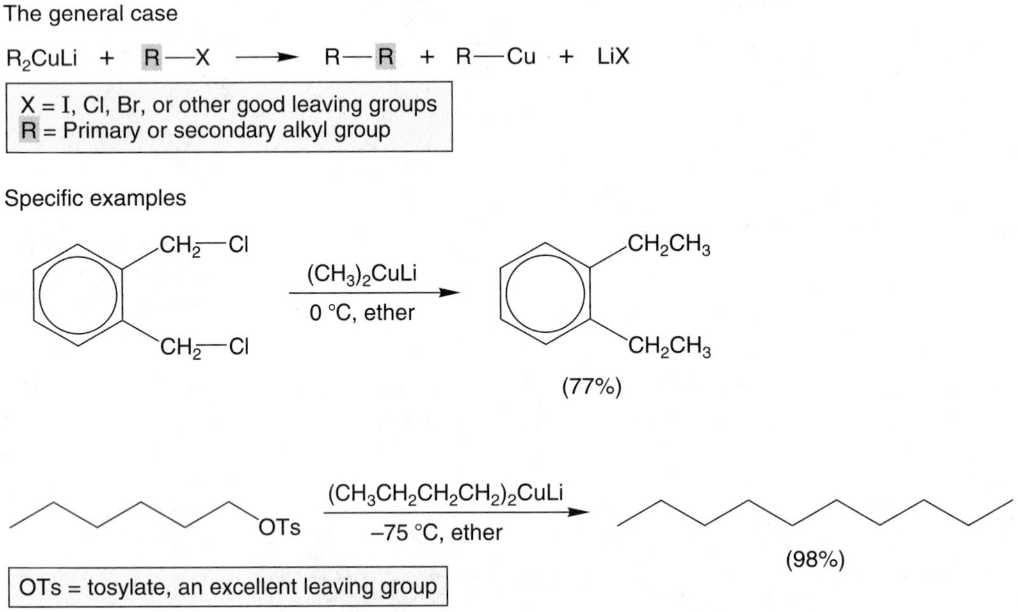

The general case

$$R_2CuLi + R{-}X \longrightarrow R{-}R + R{-}Cu + LiX$$

X = I, Cl, Br, or other good leaving groups
R = Primary or secondary alkyl group

Specific examples

$CH_2{-}Cl$, $CH_2{-}Cl$ (benzene ring) $\xrightarrow[0\ °C,\ ether]{(CH_3)_2CuLi}$ CH_2CH_3, CH_2CH_3 (benzene ring) (77%)

OTs $\xrightarrow[-75\ °C,\ ether]{(CH_3CH_2CH_2CH_2)_2CuLi}$ (98%)

OTs = tosylate, an excellent leaving group

FIGURE 16.67 Lithium organocuprates will react with primary and secondary alkyl halides to give hydrocarbons.

PROBLEM 16.21 One possible mechanism for hydrocarbon formation by organocuprates is an S_N2 displacement by one of the R groups on the alkyl halide. Can you devise an experiment to test this possibility?

Many metal hydrides such as lithium aluminum hydride undergo related reactions. Such reagents react violently with moisture to liberate hydrogen. Lithium aluminum hydride and some other metal hydrides are strong enough nucleophiles to displace halides from most organic halides. Lithium triethylborohydride, $LiHB(Et)_3$, is an especially effective displacing agent (Fig. 16.68).

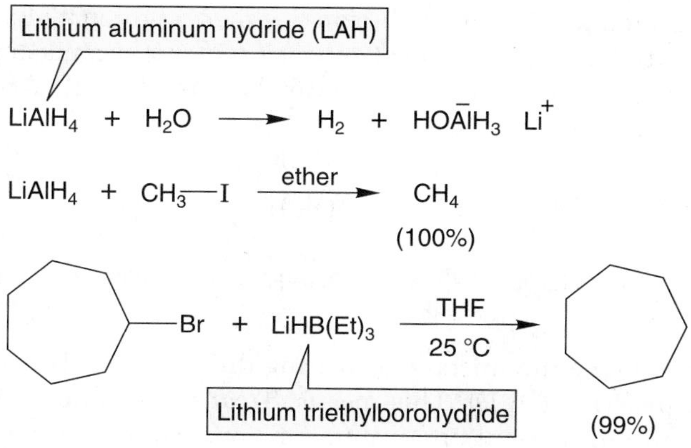

FIGURE 16.68 Metal hydrides react with water to generate hydrogen, and are often strong enough nucleophiles to displace halides and generate hydrocarbons.

Like other Lewis bases, Grignard reagents, organolithium reagents, and metal hydrides are also able to undergo addition reactions to carbonyl-containing compounds. These additions lead directly to the next subject.

6.11 IRREVERSIBLE ADDITION REACTIONS

Now that we know something of the structures of these organometallic reagents, it is time to look at their chemistry. This subject is important, because these reagents are able to form carbon–carbon bonds, a central problem in synthetic chemistry. Like oxygen- and nitrogen-centered nucleophiles, organometallic reagents (and metal hydrides) add to carbonyl compounds. Although there is no free R:$^-$ (or H:$^-$) in these reactions, the organometallic reagents are reactive enough to deliver an R:$^-$ group, acting as a nucleophile, to the Lewis acidic carbon of the carbonyl group (Fig. 16.69).

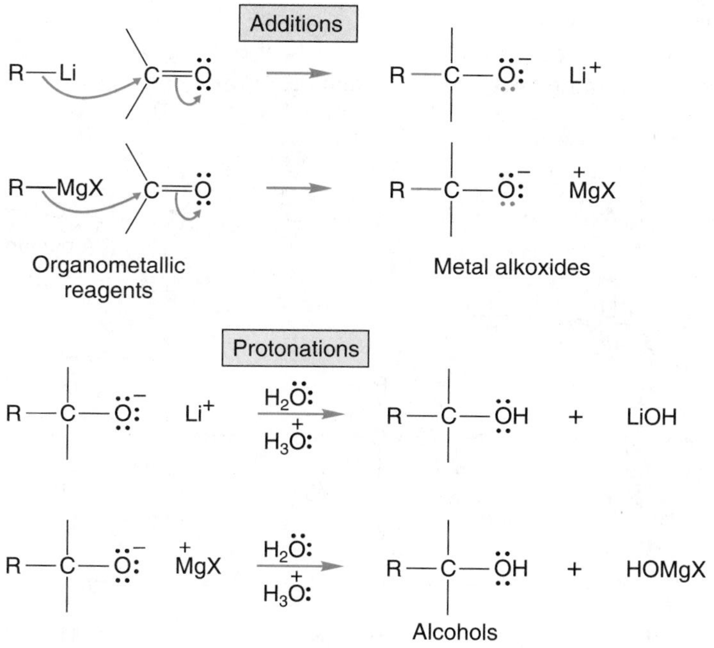

FIGURE 16.69 Organometallic reagents are strong enough nucleophiles to add to carbonyl compounds. When water is added in a second step, alcohols are produced.

The initial product is a metal alkoxide, which is protonated when the solution is neutralized by addition of aqueous acid in a second step. Water cannot be present at the start of the reaction because it would destroy the organometallic reagent (Fig. 16.70).

$$R{-}Li \xrightarrow{H_2\ddot{O}:} R{-}H + LiOH$$

$$R{-}MgX \xrightarrow[H_2\ddot{O}:]{} R{-}H + HOMgX$$

FIGURE 16.70 If water had been present at the beginning of the reaction, the organometallic reagent would have been destroyed through hydrocarbon formation.

It is easy to make the mistake of writing this reaction sequence as in Figure 16.71 [reaction (16.1)]. This means, Treat the ketone with a mixture of alkyllithium reagent and water. This procedure is impossible, as water would destroy the alkyllithium immediately (Figs. 16.62 and 16.70). The correct way to write the sequence is shown in Figure 16.71 [reaction (16.2)] and means, First allow the ketone to react with the alkyllithium reagent, and then, in a second step, add water.

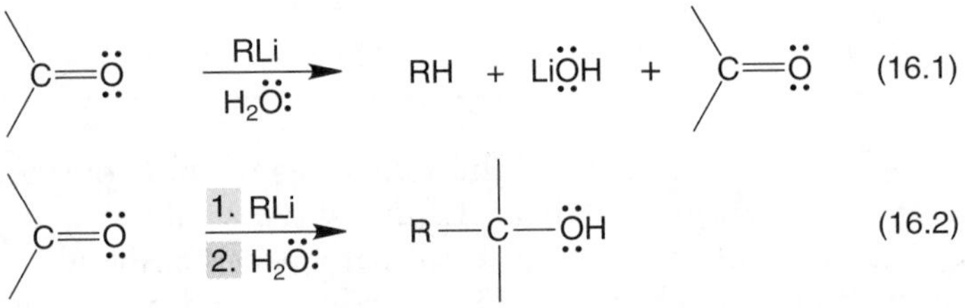

$$\text{>C=O} \xrightarrow[H_2\ddot{O}:]{RLi} RH + LiOH + \text{>C=O} \quad (16.1)$$

$$\text{>C=O} \xrightarrow[2.\ H_2\ddot{O}:]{1.\ RLi} R{-}C{-}OH \quad (16.2)$$

FIGURE 16.71 Reaction (16.1) means, Add a mixture of R–Li and water to the carbonyl compound. Under such conditions the organometallic reagent will be destroyed. Reaction (16.2) means, First add the organometallic reagent to the carbonyl compound and then in a second step, add water.

This procedure gives us a new and most versatile alcohol synthesis. New primary alcohols can be made through the reaction of an organometallic reagent with formaldehyde (Fig. 16.72). Figure 16.72 also shows the shorthand form of the reaction, which emphasizes the synthetic utility. Be careful to insert the missing details.

Similarly, reactions of organometallic reagents (RLi or RMgX) with aldehydes give secondary alcohols. Two kinds are possible, depending on the relationship between the two R groups. The R of the organometallic reagent can either be the same as that of the aldehyde or different (Fig. 16.72).

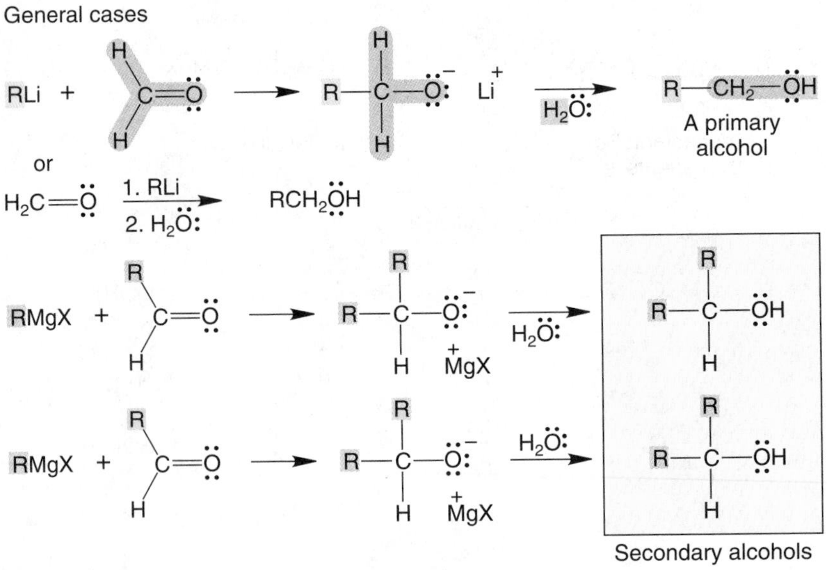

General cases

$$RLi + H_2C{=}\ddot{O} \longrightarrow R{-}CH_2{-}\ddot{O}:^-\ Li^+ \xrightarrow[H_2\ddot{O}:]{} R{-}CH_2{-}\ddot{O}H$$

A primary alcohol

or

$$H_2C{=}\ddot{O} \xrightarrow[2.\ H_2\ddot{O}:]{1.\ RLi} RCH_2\ddot{O}H$$

$$RMgX + RCH{=}\ddot{O} \longrightarrow R{-}CH(R){-}\ddot{O}:^-\ ^+MgX \xrightarrow[H_2\ddot{O}:]{} R{-}CH(R){-}\ddot{O}H$$

$$RMgX + RCH{=}\ddot{O} \longrightarrow R{-}CH(R){-}\ddot{O}:^-\ ^+MgX \xrightarrow{H_2\ddot{O}:} R{-}CH(R){-}\ddot{O}H$$

Secondary alcohols

FIGURE 16.72 Reaction of organometallic reagents with formaldehyde, followed by hydrolysis, leads to primary alcohols. The reaction of aldehydes with organometallic reagents, followed by hydrolysis, gives secondary alcohols.

A specific example

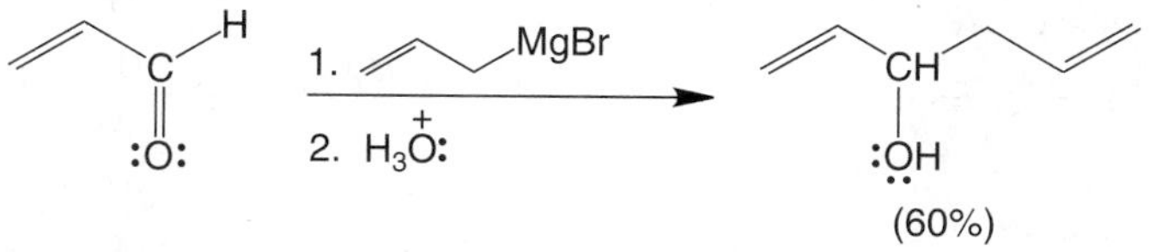

1. allyl-MgBr; 2. H_3O^+ → (60%)

FIGURE 16.72 (CONTINUED)

Ketones can react to give tertiary alcohols. Three different substitution patterns are possible depending on the number of different R groups in the reaction (Fig. 16.73).

The general case

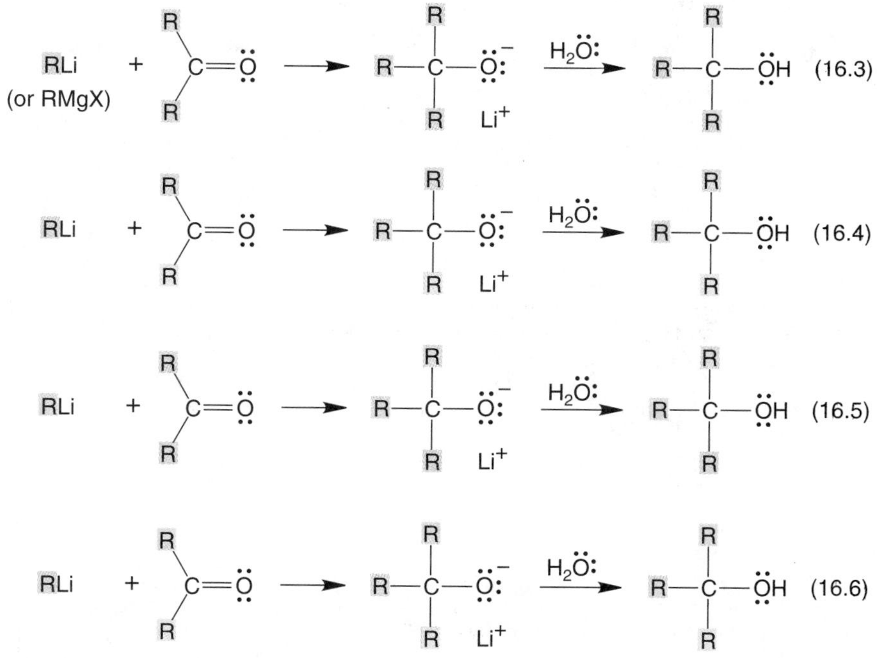

$$RLi \text{ (or RMgX)} + R_2C{=}O \longrightarrow R_3C{-}O^-\,Li^+ \xrightarrow{H_2O} R_3C{-}OH \quad (16.3)$$

$$RLi + R_2C{=}O \longrightarrow R_3C{-}O^-\,Li^+ \xrightarrow{H_2O} R_3C{-}OH \quad (16.4)$$

$$RLi + R_2C{=}O \longrightarrow R_3C{-}O^-\,Li^+ \xrightarrow{H_2O} R_3C{-}OH \quad (16.5)$$

$$RLi + R_2C{=}O \longrightarrow R_3C{-}O^-\,Li^+ \xrightarrow{H_2O} R_3C{-}OH \quad (16.6)$$

A specific example of reaction 16.5

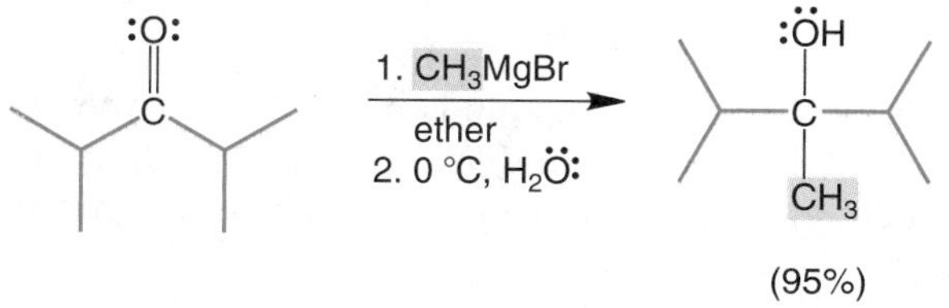

1. CH_3MgBr, ether; 2. 0 °C, H_2O → (95%)

FIGURE 16.73 The reaction of ketones with organometallic reagents, followed by hydrolysis, yields tertiary alcohols.

Metal hydrides ($LiAlH_4$, $NaBH_4$, and many others) react in a similar fashion to deliver hydride ($H:^-$) as the nucleophilic agent. Formaldehyde is reduced to methyl alcohol, aldehydes to primary alcohols, and ketones to secondary alcohols. The alcohols are formed when water is added in a second, quenching step. There is no way to make a tertiary alcohol through this kind of reduction (Fig. 16.74).

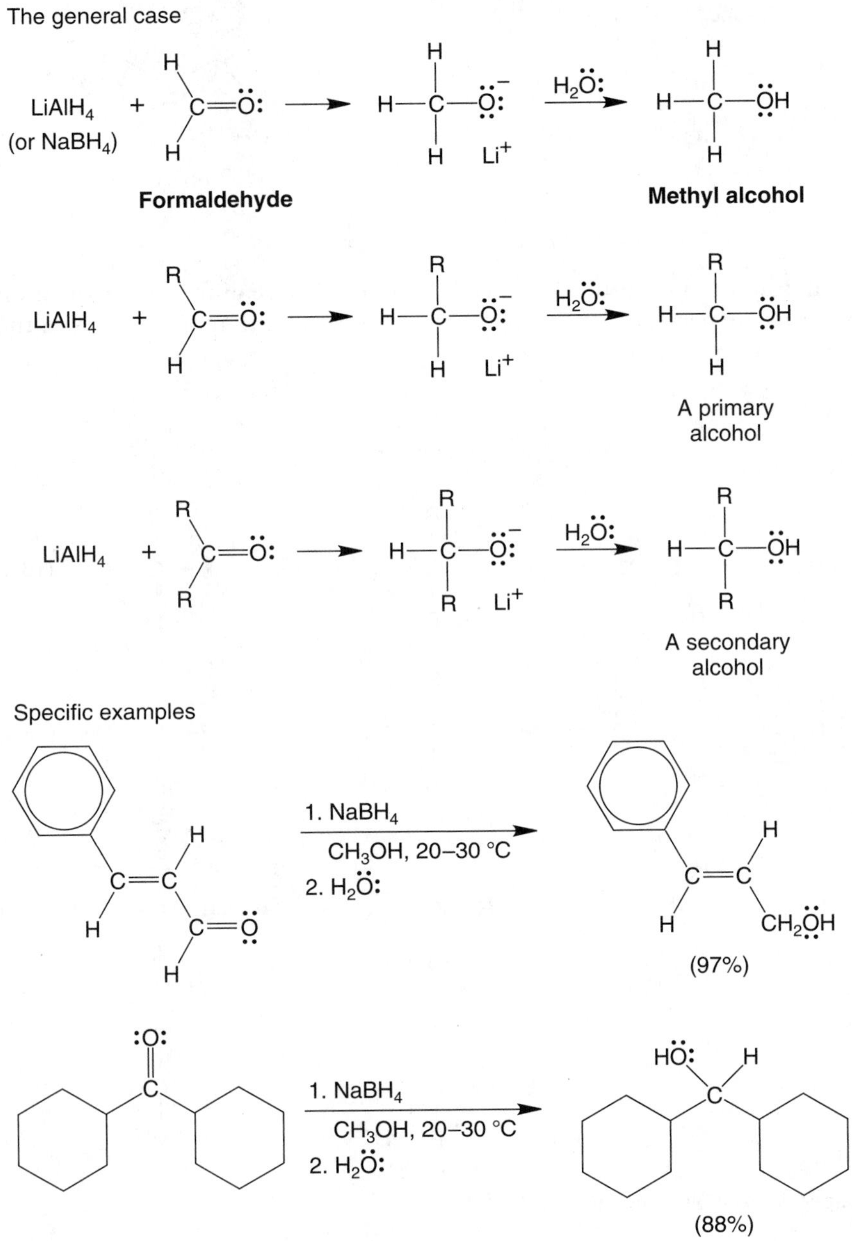

The general case

$LiAlH_4$ (or $NaBH_4$) + $H_2C{=}\ddot{O}{:}$ ⟶ $H_3C{-}\ddot{O}{:}^-\ Li^+$ $\xrightarrow{H_2\ddot{O}{:}}$ $H_3C{-}\ddot{O}H$

Formaldehyde **Methyl alcohol**

$LiAlH_4$ + $RHC{=}\ddot{O}{:}$ ⟶ $RCH_2{-}\ddot{O}{:}^-\ Li^+$ $\xrightarrow{H_2\ddot{O}{:}}$ $RCH_2{-}\ddot{O}H$

A primary alcohol

$LiAlH_4$ + $R_2C{=}\ddot{O}{:}$ ⟶ $R_2CH{-}\ddot{O}{:}^-\ Li^+$ $\xrightarrow{H_2\ddot{O}{:}}$ $R_2CH{-}\ddot{O}H$

A secondary alcohol

Specific examples

$C_6H_5CH{=}CH{-}CH{=}\ddot{O}$ $\xrightarrow[\text{2. } H_2\ddot{O}{:}]{\text{1. } NaBH_4,\ CH_3OH,\ 20\text{–}30\ °C}$ $C_6H_5CH{=}CH{-}CH_2\ddot{O}H$ (97%)

Dicyclohexyl ketone ($:\!O\!:$) $\xrightarrow[\text{2. } H_2\ddot{O}{:}]{\text{1. } NaBH_4,\ CH_3OH,\ 20\text{–}30\ °C}$ dicyclohexylmethanol ($H\ddot{O}{:}$, H) (88%)

FIGURE 16.74 Metal hydride reduction, followed by hydrolysis, gives methyl alcohol from formaldehyde, primary alcohols from aldehydes, and secondary alcohols from ketones. Tertiary alcohols cannot be made this way.

We have seen reduction of carbonyl compounds before. In Chapter 14 (p. 642), we encountered the Clemmensen and Wolff–Kishner reductions in which a carbon–oxygen double bond was reduced all the way to a methylene group (Fig. 16.75).

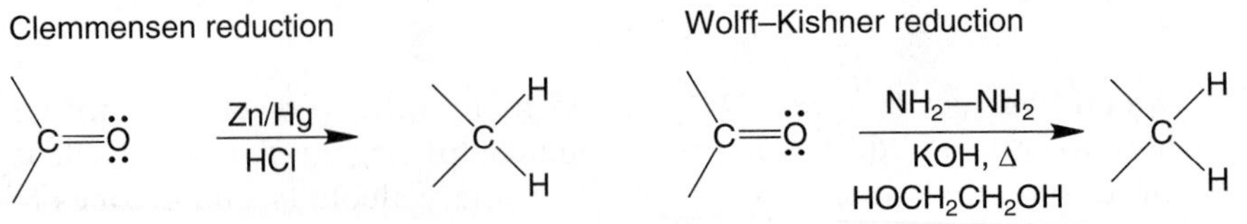

Clemmensen reduction

$C{=}\ddot{O}{:}$ $\xrightarrow[HCl]{Zn/Hg}$ CH_2

Wolff–Kishner reduction

$C{=}\ddot{O}$ $\xrightarrow[\text{KOH, }\Delta,\ HOCH_2CH_2OH]{NH_2{-}NH_2}$ CH_2

FIGURE 16.75 The Clemmensen and Wolff–Kishner reductions of the carbonyl group.

It's worth taking a moment to say a word again about strategy in doing synthesis problems, and to reinforce an old idea. It is almost never a good idea to try to see all the way back from the target molecule to the initial starting materials. Chemists have learned to be much less ambitious and to work back one step at a time. This process has been dignified with the title "retrosynthetic analysis" (encountered first in Chapter 10), which makes the idea seem far more complicated than it is. The idea behind this strategy is simply that we can almost always see the synthetic step immediately leading to the product. Indeed, the problem is more likely to be one of sorting among various possibilities than of finding a possible reaction. We have then created a new problem in synthesis for ourselves, and this one is treated in exactly the same way. We just ask ourselves for a set of reactions leading to the new target. Eventually, this technique will lead back to reagents that are readily available. Don't be too proud to work back step by step, even in simple problems. For example, the tertiary alcohol of Figure 16.76 could be made in three ways, starting from each of the three ketones shown in the figure. Now the problem is to devise a synthesis of the required ketone. Unfortunately, you don't yet have ways to make ketones, and so it is time to remedy that deficiency.

Remember also that retrosynthetic analysis has appropriated a special arrow for its own use. The double-lined arrow always points *from* the target molecule *to* the immediate precursor (Fig. 16.76).

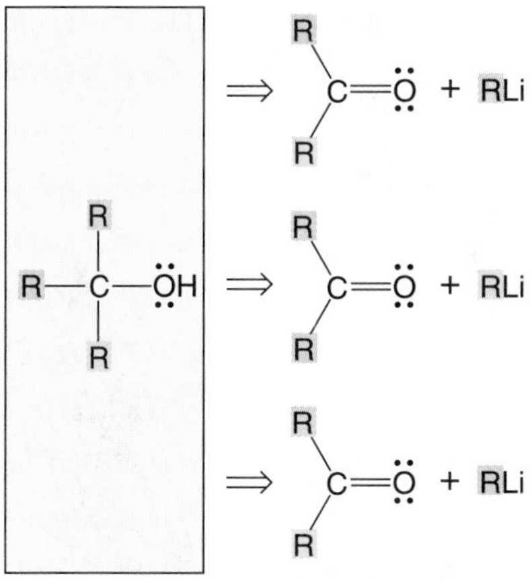

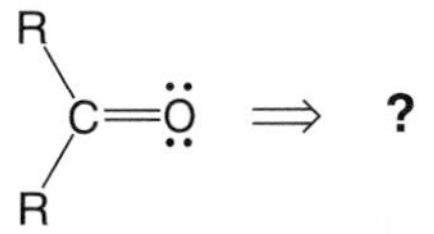

FIGURE 16.76 Retrosynthetic analysis generates three ketones as possible precursors of the tertiary alcohol. Notice the special retrosynthetic arrow (⇒), which always points from the target to the immediate precursor.

16.12 OXIDATION OF ALCOHOLS TO CARBONYL COMPOUNDS

So far, we have generated two new and very general alcohol syntheses, the reaction of aldehydes and ketones with organometallic reagents, and the reduction of carbonyl compounds by metal hydrides. There is a complement to the reduction reaction in the oxidation of alcohols to aldehydes and ketones (Fig. 16.77). These oxidations extend the utility of the alcohol syntheses we have just learned, and complicate your life by increasing the complexity of the molecules you are able to make.

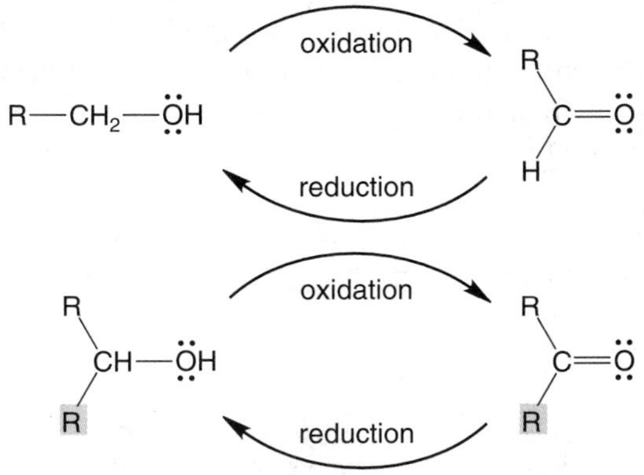

FIGURE 16.77 Alcohols can be oxidized to carbonyl compounds. This reaction is the reverse of what you have just learned—the reduction of carbonyl compounds to alcohols through reactions with metal hydride reagents.

Now your collection of syntheses can be extended by using the product alcohols as starting materials in the oxidation reaction. Tertiary alcohols cannot be simply oxidized further, but primary and secondary ones can be.

As shown in Figure 16.78, secondary alcohols can only give ketones, but primary alcohols can give either aldehydes or carboxylic acids (RCOOH) depending on the oxidizing agent used.

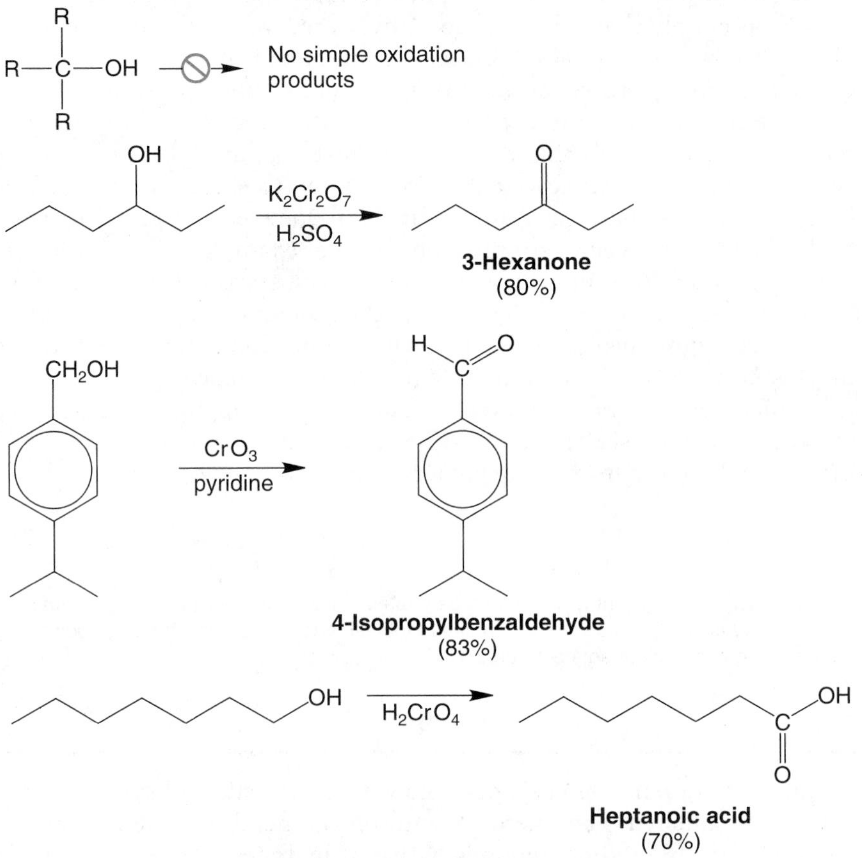

FIGURE 16.78 Tertiary alcohols cannot be easily oxidized, but secondary or primary alcohols can be. It is logical to infer that for the oxidation reaction to succeed there must be a carbon–hydrogen bond available.

How do these oxidation reactions work? A typical oxidizing agent for primary alcohols is chromium trioxide (CrO_3) in pyridine. As long as the reagent is kept dry, good yields of aldehydes can often be obtained (Fig. 16.78). Although a metal–oxygen bond is more complicated than a carbon–oxygen bond, the double bonds in CrO_3 and $R_2C{=}O$ react in similar ways. For example, the alcohol oxygen is a nucleophile and can attack the chromium–oxygen bond to give, ultimately, a chromate **ester**. This reaction of a Cr=O with an alcohol is analogous to the formation of a hemiacetal from the reaction of an alcohol and a C=O (Fig. 16.79).

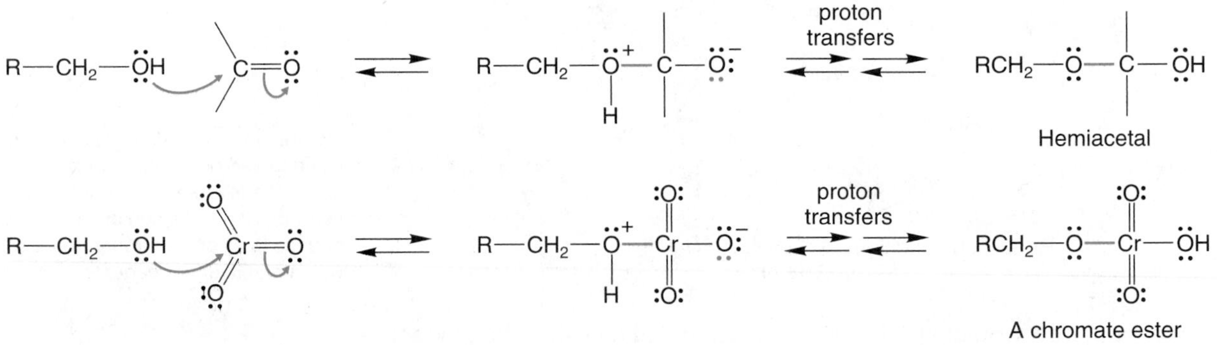

FIGURE 16.79 Nucleophilic alcohols can add to Cr=O bonds just as they do to C=O bonds. This reaction is the first step in the oxidation process, and in this case gives a chromate ester.

Pyridine (or any other base) now initiates an elimination reaction to give an aldehyde (an oxidized alcohol) and $HOCrO_2^-$ (a reduced chromium species) (Fig. 16.80).

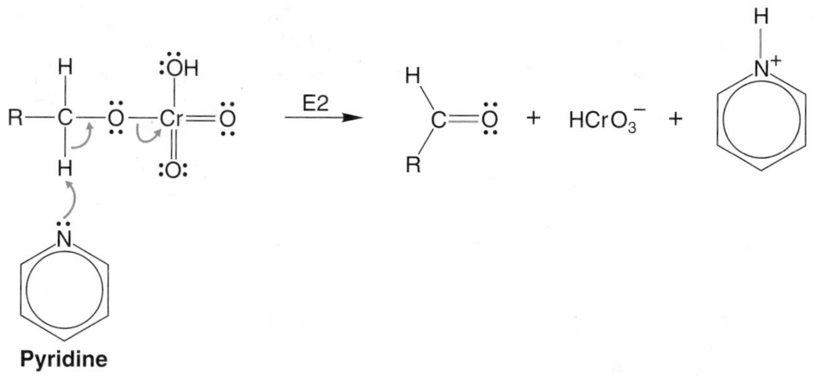

FIGURE 16.80 The second step in the oxidation reaction is a simple E2 reaction to generate the new C=O double bond.

It should now be clear why tertiary alcohols resist this reagent. The initial addition reaction can proceed, but there is no hydrogen to be removed in the elimination process (Fig. 16.81).

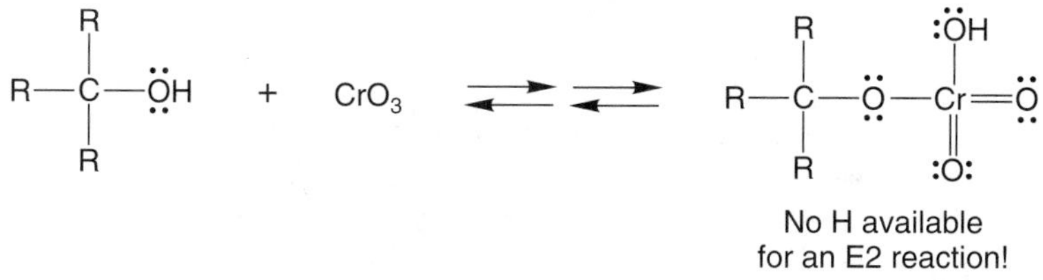

No H available for an E2 reaction!

FIGURE 16.81 A tertiary alcohol can add to the Cr=O double bond, but there is no hydrogen available to be lost in an E2 reaction.

Notice also how this seemingly strange reaction is really made up of two processes we have already encountered. The first step is addition of an oxygen base to a chromium–oxygen double bond. The second step is nothing more than a garden-variety elimination. The lesson here is not to be bothered by the identity of the atoms involved. When you see a reaction that looks strange, rely on what you already know. Make analogies, and more often than not you will come out all right. That is exactly what all chemists do when we see a new reaction. We think first about related reactions and try to find a way to generalize.

Chromium trioxide (CrO_3) will also oxidize secondary alcohols to ketones, but a more usual reagent is sodium chromate (Na_2CrO_4) or sodium dichromate ($Na_2Cr_2O_7$) in aqueous strong acid (Fig. 16.82).

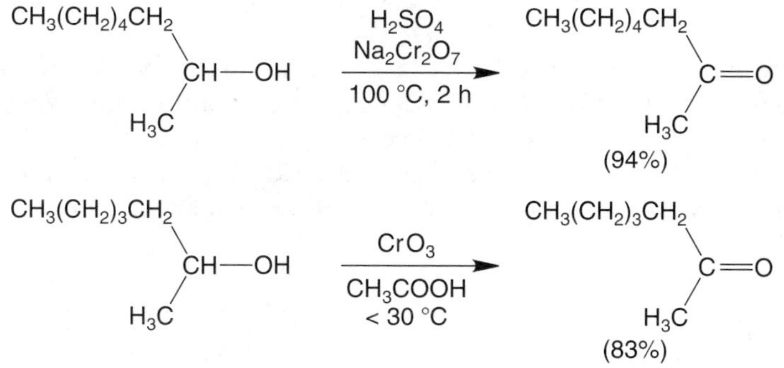

(94%)

(83%)

FIGURE 16.82 A typical reagent for oxidation of secondary alcohols to ketones is sodium dichromate in acid. Other oxidizing agents also work, however.

*PROBLEM 16.22 Write a mechanism for the oxidation of a secondary alcohol to a ketone by chromic oxide.

ANSWER The reaction mechanism exactly parallels that for primary alcohols. The first step is formation of a chromate ester through addition of the alcohol to CrO_3 followed by protonation and deprotonation steps. The reactions effect the change of leaving group necessary to the success of the following elimination reaction.

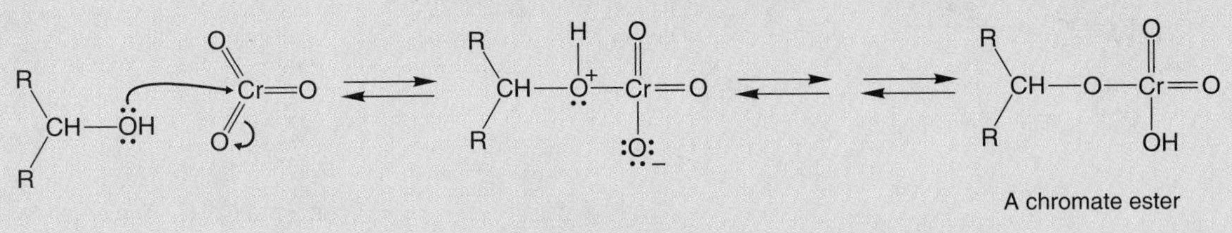

A chromate ester

An elimination using any base in the system (pyridine, or the alcohol itself are possibilities) generates the ketone.

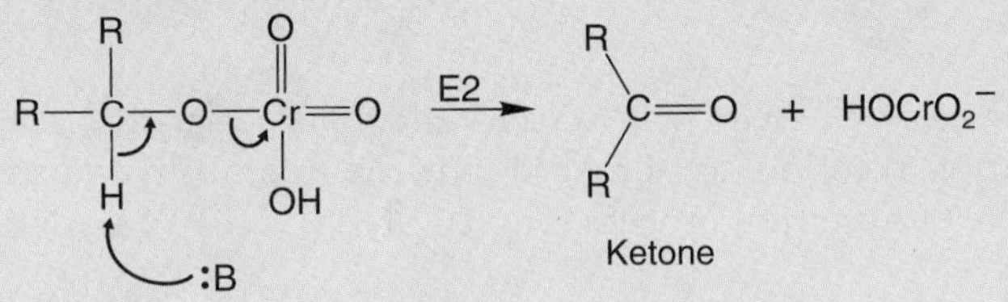

E2

$HOCrO_2^-$

Ketone

Under these conditions, or any aqueous conditions, it is difficult to stop oxidation of primary alcohols at the aldehyde stage. Overoxidation to the carboxylic acid is common. The difficulty is that aldehydes are hydrated in water (p. 763). These hydrates are themselves alcohols and can be oxidized further (Fig. 16.83).

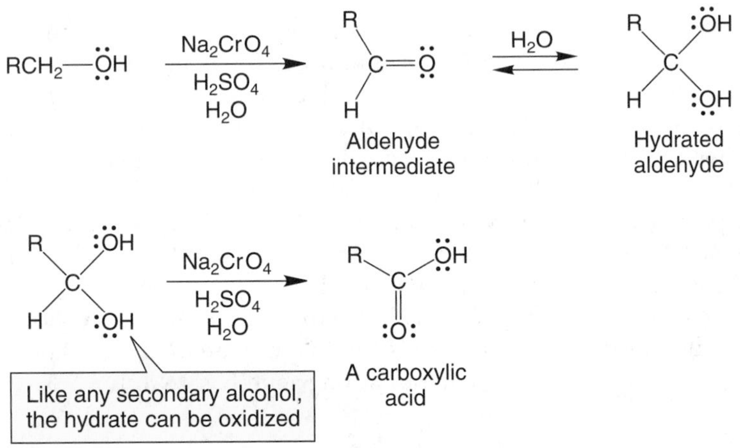

RCH_2—OH → (Na_2CrO_4, H_2SO_4, H_2O) Aldehyde intermediate ⇌ (H_2O) Hydrated aldehyde

Hydrated aldehyde → (Na_2CrO_4, H_2SO_4, H_2O) A carboxylic acid

Like any secondary alcohol, the hydrate can be oxidized

FIGURE 16.83 Oxidation of a primary alcohol leads to an aldehyde. Aldehydes are hydrated in the presence of water, and the hydrates are themselves alcohols. Further oxidation of the hydrate leads to the carboxylic acid.

There are many other oxidizing agents of varying strength and selectivity. Manganese dioxide (MnO_2) is effective at making aldehydes from allylic or benzylic alcohols, nitric acid (HNO_3) will oxidize primary alcohols to acids, and potassium permanganate ($KMnO_4$) is an excellent oxidizer of secondary alcohols to ketones and of primary alcohols to acids (Fig. 16.84).

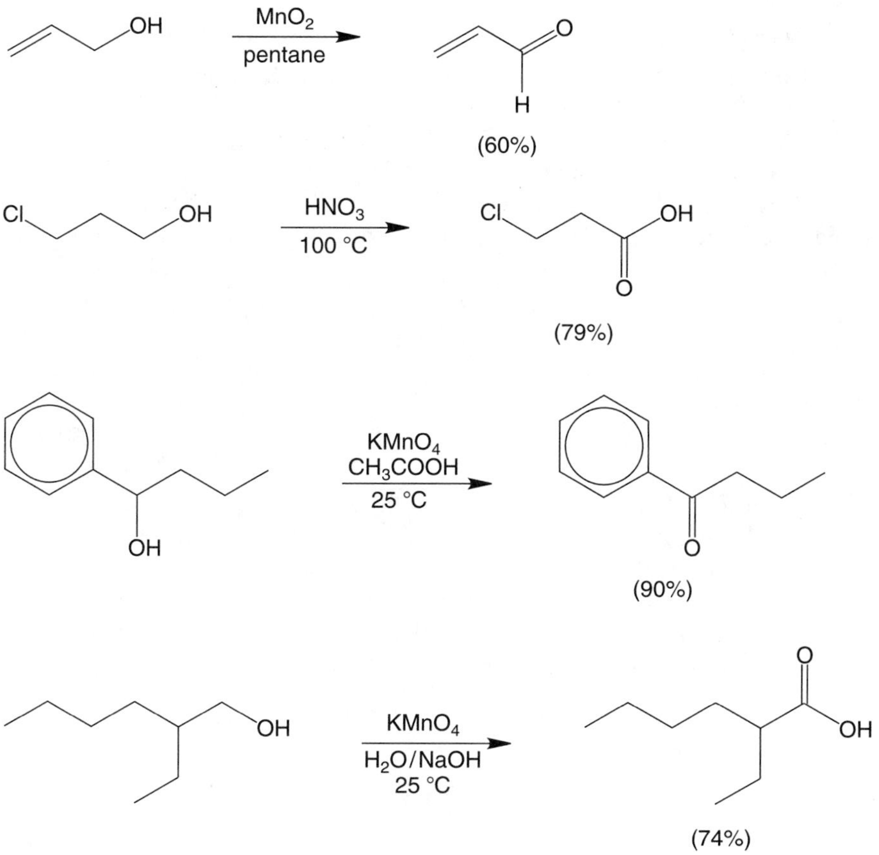

OH $\xrightarrow[\text{pentane}]{MnO_2}$ O H (60%)

Cl OH $\xrightarrow[100\ °C]{HNO_3}$ Cl OH O (79%)

OH $\xrightarrow[25\ °C]{KMnO_4,\ CH_3COOH}$ O (90%)

OH $\xrightarrow[H_2O/NaOH,\ 25\ °C]{KMnO_4}$ O OH (74%)

FIGURE **16.84** Some other oxidizing agents.

So, now we can reduce carbonyl compounds to the alcohol level with hydride reagents, and oxidize them back again using the various reagents of this section. Although one probably wouldn't want to spend one's life cycling a compound back and forth between the alcohol and carbonyl stage, these reactions are enormously useful in synthesis. Many alcohols are now the equivalent of aldehydes or ketones, because they can be oxidized, and carbonyl compounds are the equivalent of alcohols because they can be reduced with hydride or treated with an organometallic reagent.

16.13 THE WITTIG REACTION

As we have seen, all manner of nucleophiles add to carbon–oxygen double bonds, often leading to useful compounds such as alcohols. Here is an example of a synthetically useful carbonyl addition reaction that produces certain kinds of alkenes. This reaction was discovered in the laboratories of Georg Wittig (1897–1987) and is called the **Wittig reaction**. Wittig shared the 1979 Nobel prize largely for his work on this reaction. The process begins with the reaction between phosphines and alkyl halides to give phosphonium halides. Although we have not encountered reactions of phosphorus yet, it sits right below nitrogen in the periodic table and, like nitrogen, is a good nucleophile. Formation of the phosphonium halide takes place through an S_N2 displacement (Fig. 16.85).

FIGURE 16.85 Displacement of iodide by the nucleophilic phosphorus atom of triphenylphosphine leads to phosphonium ions.

(Ph)$_3$P: CH$_3$—I: →(S$_N$2) (Ph)$_3\overset{+}{P}$—CH$_3$ + :I:$^-$

Triphenylphosphine A phosphonium ion

The phosphonium halides contain acidic hydrogens that can be removed by strong bases such as alkyllithium reagents. The product is an **ylide**, a compound containing opposite charges on adjacent atoms (Fig. 16.86).

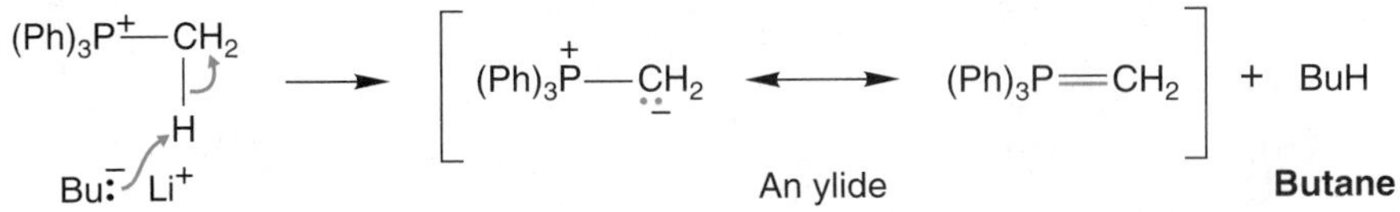

FIGURE 16.86 Protons adjacent to the positively charged phosphorus atom can be removed in strong base to give ylides.

(Ph)$_3$P$^+$—CH$_2$—H Bu:$^-$ Li$^+$ → [(Ph)$_3\overset{+}{P}$—$\overset{..}{C}$H$_2^-$ ⟷ (Ph)$_3$P=CH$_2$] + BuH

An ylide **Butane**

The carbon of the ylide is a nucleophile, and like other nucleophiles adds to carbon–oxygen double bonds. Intramolecular closure of the intermediate leads to a four-membered ring, an **oxaphosphetane**. Oxaphosphetanes are strained and unstable relative to their quite stable constituent parts, phosphine oxides and alkenes. Therefore, it is easy for these compounds to fragment (Fig. 16.87).

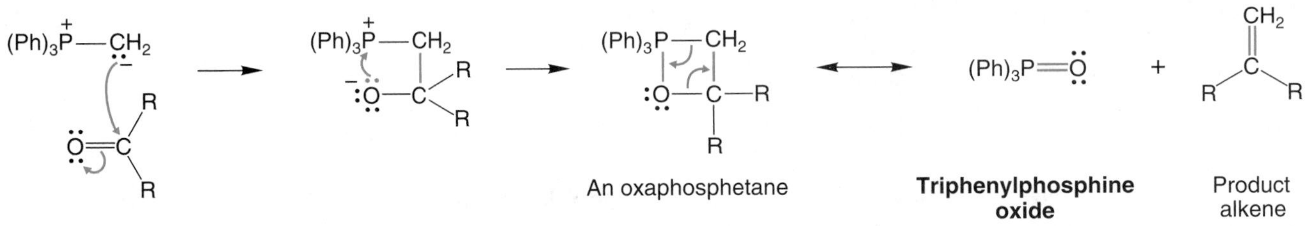

(Ph)$_3\overset{+}{P}$—CH$_2^-$ + O=CR$_2$ → (Ph)$_3\overset{+}{P}$—CH$_2$—CR$_2$—O$^-$ → (Ph)$_3$P—CH$_2$—C(R)$_2$—O (ring) ⟷ (Ph)$_3$P=O + R$_2$C=CH$_2$

An oxaphosphetane **Triphenylphosphine oxide** Product alkene

FIGURE 16.87 Ylides are nucleophiles and will add to carbonyl compounds to give intermediates that can close to oxaphosphetanes. These four-membered ring compounds can open to give triphenylphosphine oxide and the product alkenes.

The Wittig synthesis does contain some unfamiliar intermediates. You have not seen either an ylide or an oxaphosphetane before, for example. Yet the key reactions, the S$_N$2 displacement to form the phosphonium halide (Fig. 16.85) and the addition to the carbonyl group (Fig. 16.87), are simple variations on basic processes. In a practical sense, it is worth working through the new parts of the Wittig reaction because it is so useful synthetically. Note that the Wittig reaction is, in a formal sense, the reverse of ozonolysis (Chapter 10, p. 416). Figure 16.88 shows some specific examples.

Specific examples

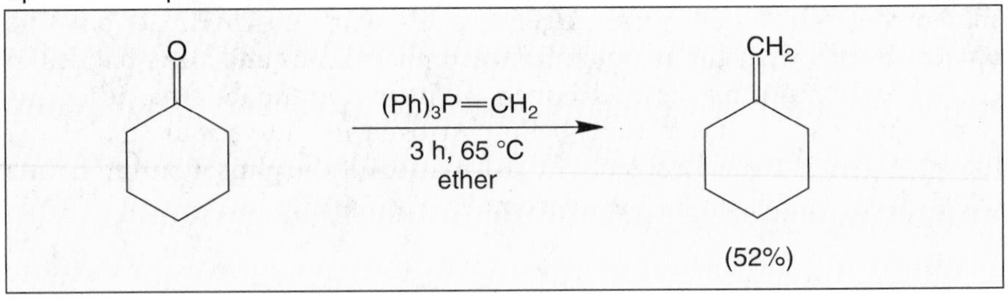

Cyclohexanone →((Ph)$_3$P=CH$_2$; 3 h, 65 °C; ether) Methylenecyclohexane (52%)

FIGURE 16.88 Some typical synthetic uses of the Wittig reaction.

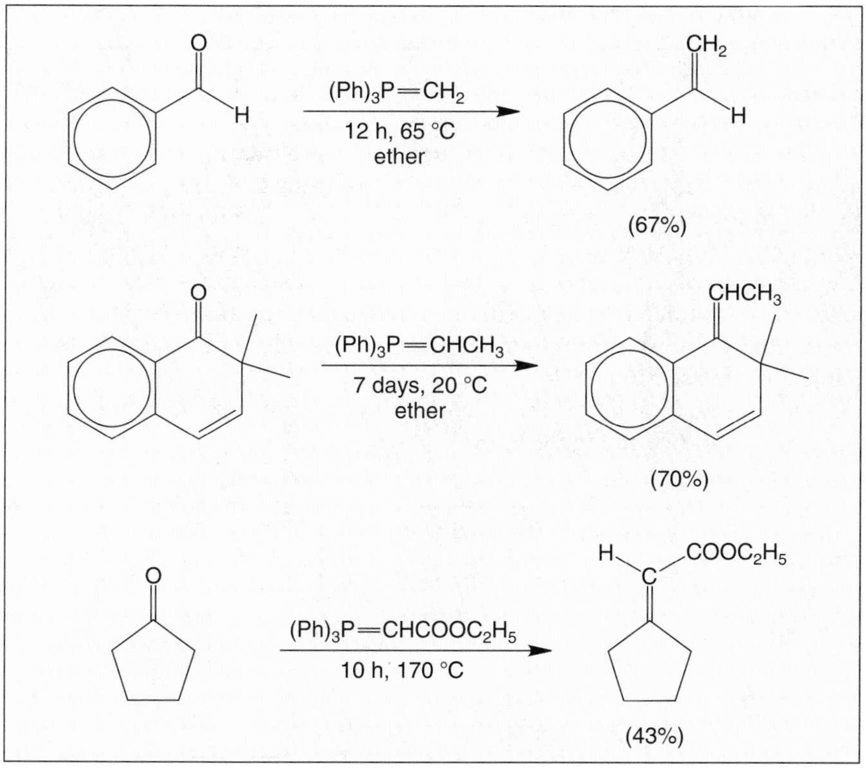

FIGURE 16.88 (CONTINUED)

$(Ph)_3P{=}CH_2$, 12 h, 65 °C, ether (67%)

$(Ph)_3P{=}CHCH_3$, 7 days, 20 °C, ether (70%)

$(Ph)_3P{=}CHCOOC_2H_5$, 10 h, 170 °C (43%)

General interconversions

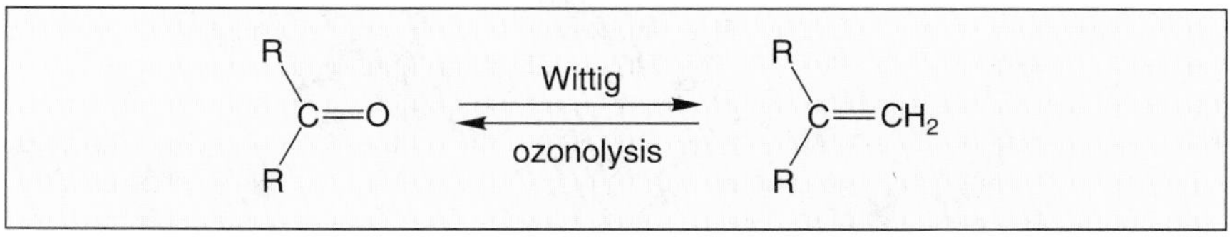

$R_2C{=}O$ ⇌ $R_2C{=}CH_2$ (Wittig; ozonolysis)

The Wittig reaction is very useful. How else could we convert cyclohexanone into pure methylenecyclohexane? We can use the reactions of this chapter to devise a different synthesis. Some thought might lead to a sequence in which cyclohexanone reacts with methyllithium to give, after hydrolysis, a tertiary alcohol. An acid-catalyzed elimination reaction would give some of the desired product, but there is no easy way to avoid the predominant formation of the undesired isomer, 1-methylcyclohexene (Fig. 16.89). The Wittig reaction solves this synthetic problem and is an important constituent of any synthetic chemist's bag of tricks.

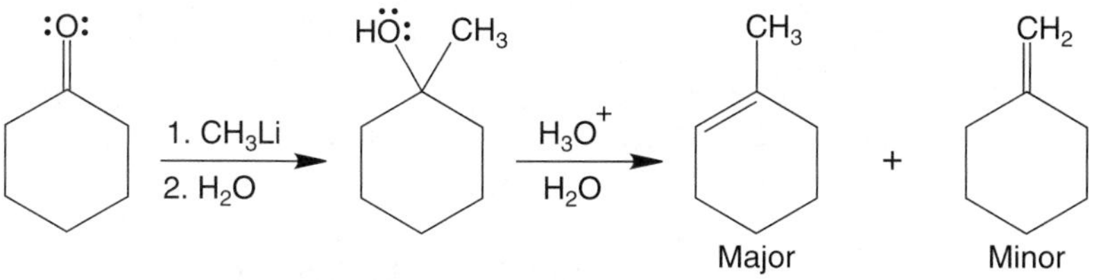

1. CH_3Li 2. H_2O; H_3O^+, H_2O; Major + Minor

FIGURE 16.89 This approach fails to give very much of the desired product. The Wittig reaction is the usual alternative.

PROBLEM 16.23

Explain why the acid-catalyzed dehydration reaction of Figure 16.89 will give mostly the undesired 1-methylcyclohexene.

16.14 SOMETHING MORE: BIOLOGICAL OXIDATION

We have seen several examples of redox reactions in the last few sections. Related processes are of vital importance in biological systems. We humans, for example, derive much of our energy from the oxidation of ingested sugars and fats. However, we do not use chromium reagents or nitric acid to accomplish these reactions, for these reagents are far too unselective and harsh. They would surely destroy most of our constituent molecules; a quite uncomfortable process one imagines. Instead, a series of highly selective biomolecules has evolved, each dedicated to a particular purpose. One of these, nicotinamide adenine dinucleotide (**NAD⁺**)* is the subject of this section.

Nicotinamide adenine dinucleotide is a pyridinium ion, and its full structure is shown in Figure 16.90.

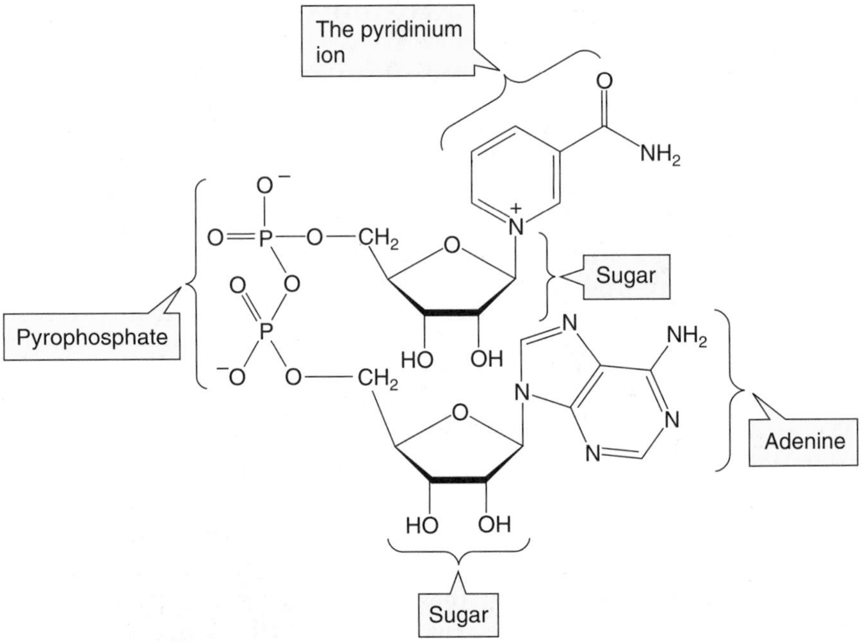

FIGURE 16.90 Nicotinamide adenine dinucleotide, NAD⁺.

The business end of the molecule, the nicotinamide, is connected through a sugar (ribose) and a pyrophosphate linkage to another ribose which is, in turn, attached to the base adenine. The molecule is a **coenzyme**, meaning that it can only operate in connection with an enzyme, in this case, alcohol dehydrogenase. The enzyme binds both ethyl alcohol, the molecule to be oxidized, and NAD⁺, which will be reduced.

The enzyme serves to bring the Lewis acid, NAD⁺, and the Lewis base, ethyl alcohol, together, but this bringing together is of immense importance to the reaction, as it allows the redox process to take place without the requirements that molecules in solution have for finding each other and orienting properly for reaction.

The NAD⁺ is a strong Lewis acid by virtue of the quaternary nitrogen with its positive charge. It can be reduced by transfer of a hydrogen, *with its pair of electrons* (a **hydride** reduction) from ethyl alcohol (Fig. 16.91). The reaction is completed by deprotonation to give the product acetalde-

*A glance at the structure of this molecule and some brief thoughts on how complicated its systematic name must be quickly reveal the obvious reasons why biochemists are addicted to acronyms.

hyde and removal of the products from the enzyme. This removal is easy because the enzyme has evolved to bind ethyl alcohol and NAD^+, not acetaldehyde and the reduced form of NAD^+, NADH.

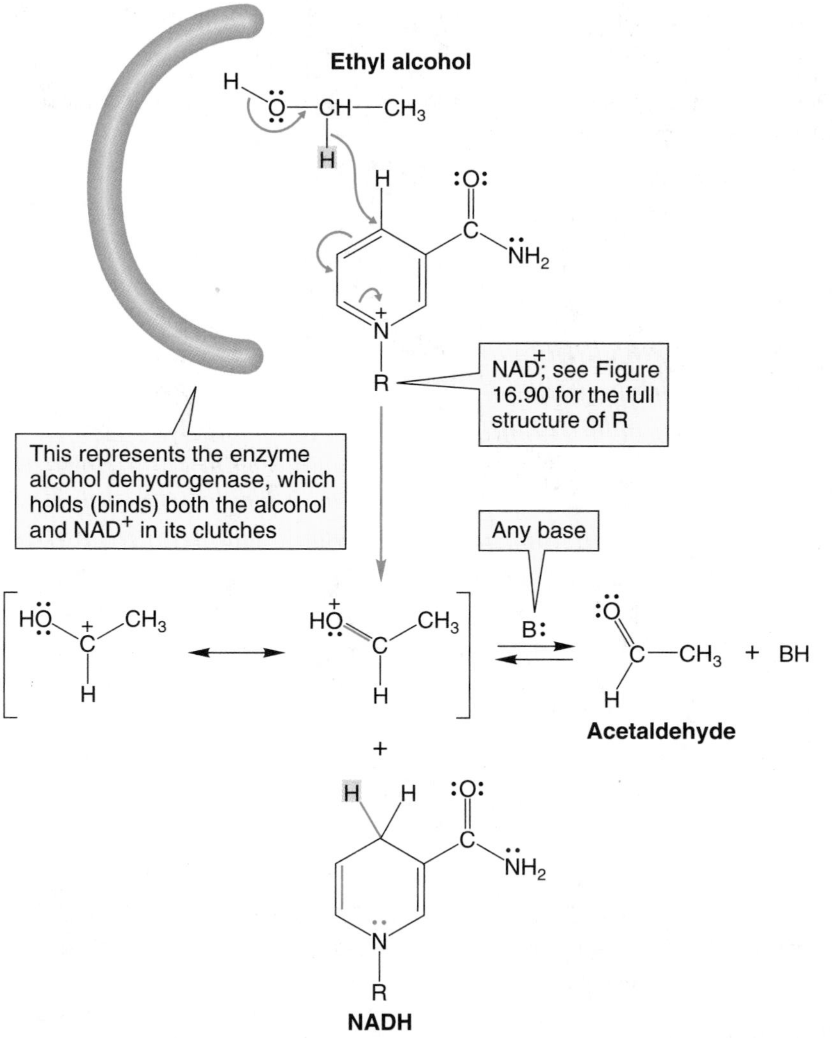

FIGURE 16.91 The NAD^+ is reduced by transfer of hydride ($H:^-$) from ethyl alcohol.

This hydride reduction may seem strange, but it is not. There is a close relative in the intramolecular hydride shifts in carbocation chemistry. These, too, are redox reactions. The originally positive carbon is reduced and the carbon from which the hydride moves is oxidized (Fig. 16.92).

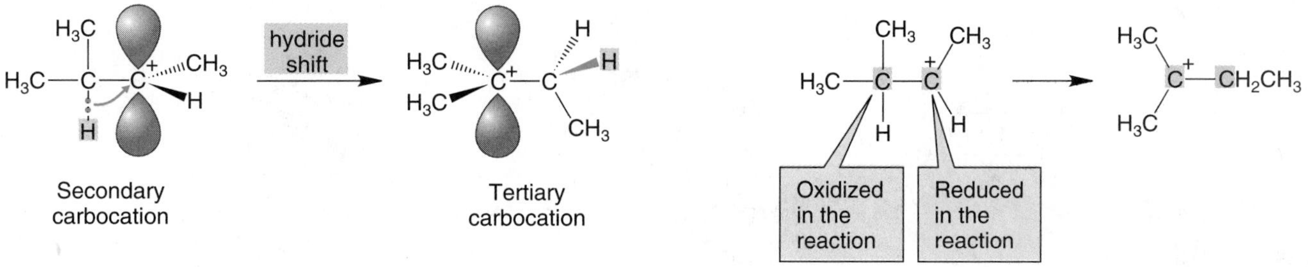

FIGURE 16.92 Hydride shifts in carbocations are also redox reactions.

The end products of the biological reaction are NADH, the reduced NAD^+, and the oxidized ethyl alcohol, acetaldehyde (Fig. 16.91).

*PROBLEM **16.24** Each of the two methylene hydrogens of ethyl alcohol can be replaced by deuterium to give a pair of chiral deuterioethanols. Show this process and indicate which of the new compounds is (*R*) and which is (*S*).

ANSWER Replacement of one methylene H with D gives the (*S*) enantiomer; replacement of the other gives the (*R*) enantiomer.

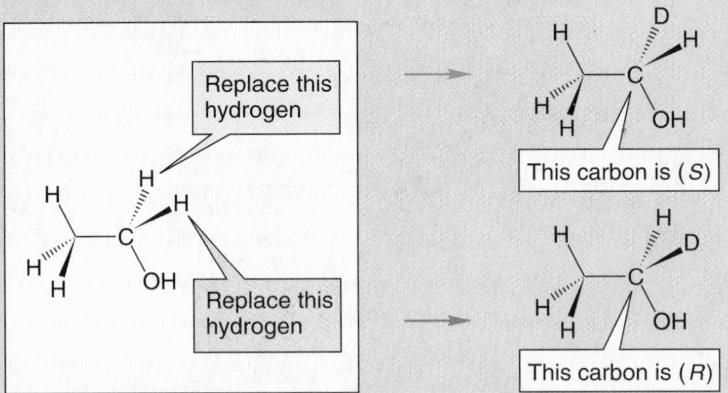

This structure shows the priorities for the (*R*) compound; the 1 ... 2 ... 3 arrow is clockwise

*PROBLEM **16.25** Oxidation of (*R*)-1-deuterioethanol produces deuterio NADH (NADD) and acetaldehyde. Predict the products of the oxidation of (*S*)-1-deuterioethanol with NAD^+. Explain carefully.

ANSWER The NAD^+ is chiral and will interact differently with the two enantiomers of deuterioethanol identified in Problem 16.24. The (*R*) enantiomer transfers a deuterium to give NADD. Although it is not easy to specify the details of why this occurs, the results of the experiment show that the sum of all steric and electronic interactions between the two molecules results in the deuterium being in the correct position for transfer in the (*R*) enantiomer.

This carbon is (*R*)

NAD^+ **NADD** **Acetaldehyde**

ANSWER (CONTINUED)

In the (*S*) enantiomer, the hydrogen occupies the same position as the deuterium in the (*R*) enantiomer. The (*S*) enantiomer must transfer this H in the reduction of NAD^+. This is exactly what happens.

This carbon is (*S*)

NAD^+ — **NADH** — **Deuterated acetaldehyde**

16.15 SUMMARY

NEW CONCEPTS

The centerpiece of this chapter, and its unifying theme, is the reaction in which nucleophiles of all kinds add to carbonyl groups. This process can be either acid or base catalyzed, and starts with the overlap of a filled orbital on the nucleophilic Lewis base with the empty π^* orbital of the carbonyl to give an alkoxide in which the highly electronegative oxygen atom bears the negative charge (Fig. 16.23).

Many addition reactions are equilibria, and in these cases the position of the equilibrium will depend on the relative stabilities of the starting materials and products. These stabilities depend, as always, on a variety of factors including electronic structure, substitution pattern, and steric effects.

The π system of the carbonyl group is constructed through taking combinations of carbon and oxygen $2p$ orbitals. This process leads to the π orbitals of Figure 16.4.

The concept of retrosynthetic analysis reappears in this chapter. Retrosynthetic analysis simply means that it is best to approach problems of synthesis by searching not for the *ultimate* starting material, but rather to undertake the much easier task of finding the *immediate* precursor of the product. This process can be repeated until an appropriate level of simplicity is reached (Fig. 16.93).

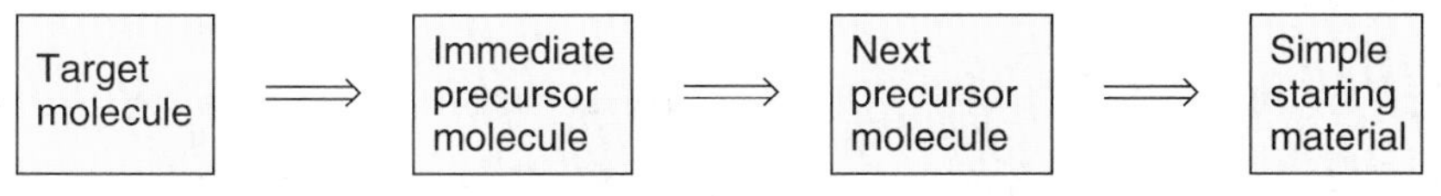

FIGURE 16.93 Retrosynthetic analysis.

REACTIONS, MECHANISMS, AND TOOLS

As mentioned in the previous section, there is a single, central reaction in this chapter, the addition of a nucleophile to the Lewis acid carbonyl group. The details of the mechanism vary, depending on whether the reaction is acid or base catalyzed, and on the substitution patterns of the nucleophile and the carbonyl-containing compound. Here are some general examples.

A simple case is the acid- or base-catalyzed reversible addition reaction (Fig. 16.36). Examples are hydration or cyanohydrin formation.

A slightly more complicated reaction involves an addition followed by loss of water. An example is the reaction of primary amines with carbonyl groups to give substituted imines (Fig. 16.52).

If there is no proton that can be removed after loss of water, a second addition reaction can occur, as in acetal formation (Fig. 16.45).

There are also irreversible additions, as in the reactions of organometallic reagents or metal hydrides with carbonyl compounds. Protonation of the initial products, alkoxides, gives alcohols (Fig. 16.69).

This chapter also introduces the oxidation reactions of alcohols. These reactions involve addition reactions in their first steps. For example, in the oxidation of primary alcohols by CrO_3, the first step is addition of the alcohol to the chromium–oxygen double bond (Figs. 16.79 and 16.80). The oxidation is completed by an elimination in which the new carbon–oxygen double bond is constructed.

SYNTHESIS

This chapter contains lots of new syntheses (Fig. 16.94).

1. Acetals

R(H)C=O $\xrightarrow[\text{ROH}]{\text{RO}^{+}\text{H}_2}$ RO—C(R)(H)—OR

R(R)C=O $\xrightarrow[\text{ROH}]{\text{RO}^{+}\text{H}_2}$ RO—C(R)(R)—OR

Acetals can be used as protecting groups for aldehydes and ketones; treatment with H_2O/H_3O^+ regenerates the aldehyde. The hemiacetal is an intermediate

2. Acids

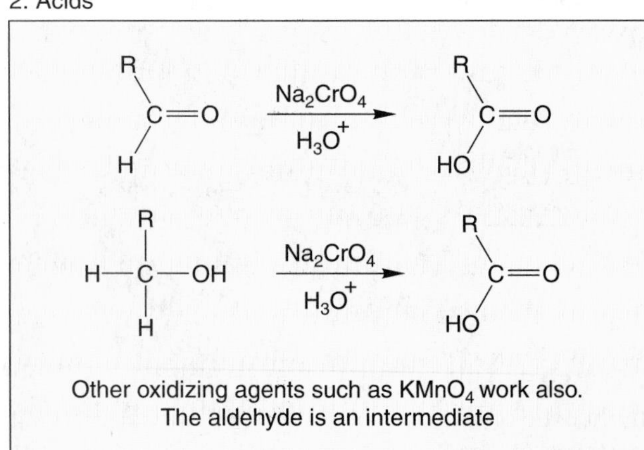

Other oxidizing agents such as $KMnO_4$ work also. The aldehyde is an intermediate

3. Alcohols

Primary alcohols

R(H)C=O $\xrightarrow[\text{2. } H_2O]{\text{1. } NaBH_4}$ H—C(R)(H)—OH

The alkoxide is an intermediate; $LiAlH_4$ can also be used

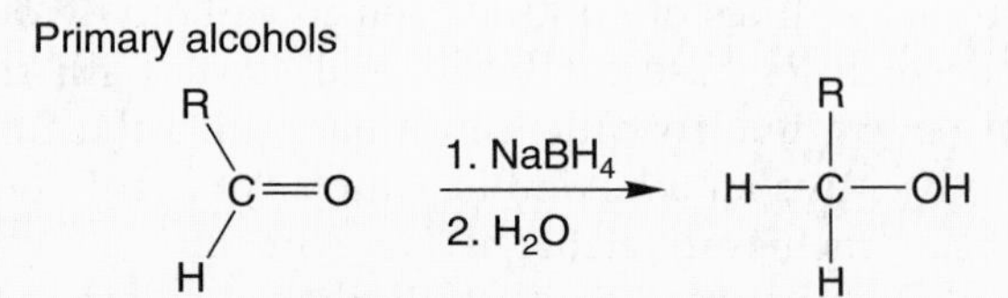

The alkoxide is an intermediate

Secondary alcohols

R(H)C=O $\xrightarrow[\text{2. } H_2O]{\text{1. RMgX or RLi}}$ R—C(R)(H)—OH

The alkoxide is an intermediate; the R groups can be the same or different

R(R)C=O $\xrightarrow[\text{2. } H_2O]{\text{1. } NaBH_4}$ R—C(R)(H)—OH

The alkoxide is an intermediate; the R groups can be the same or different, depending on the structure of the ketone; $LiAlH_4$ can also be used

FIGURE 16.94 The synthetic reactions of Chapter 16.

3. Alcohols (continued)

Tertiary alcohols

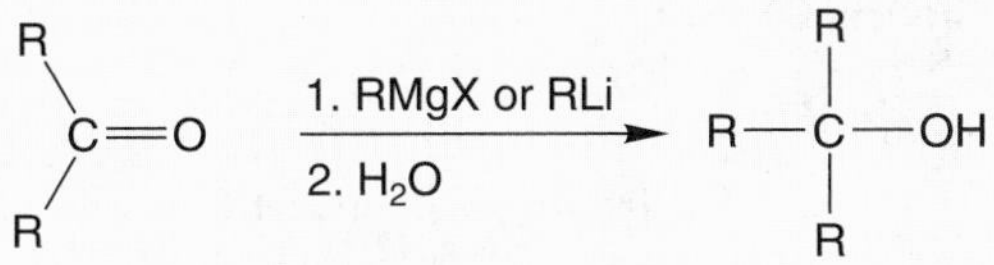

$$R_2C{=}O \xrightarrow[\text{2. } H_2O]{\text{1. RMgX or RLi}} R_3C{-}OH$$

The product can be $R_3C{-}OH$, $R_2RC{-}OH$, or RRRC–OH, depending on the structures of the starting ketone and organometallic reagent

4. Aldehydes

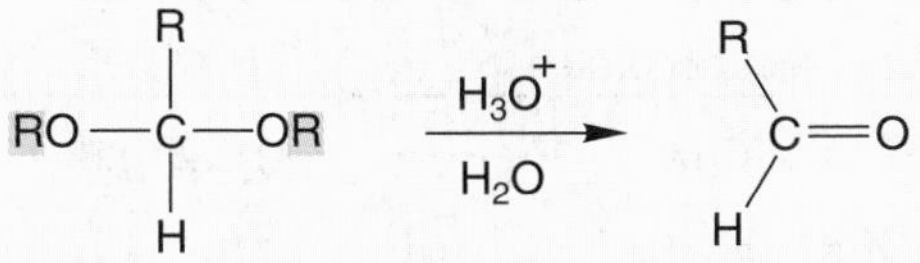

$$RO{-}CH(R){-}OR \xrightarrow[H_2O]{H_3O^+} RHC{=}O$$

Hydrolysis of an acetal

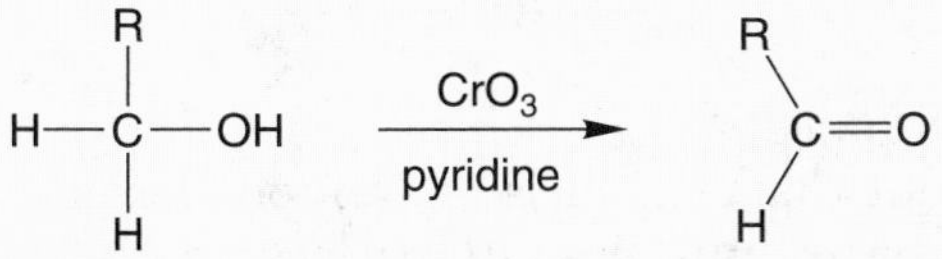

$$H{-}C(R)(H){-}OH \xrightarrow[\text{pyridine}]{CrO_3} RHC{=}O$$

Oxidation of a primary alcohol; water must be absent

5. Alkyllithium reagents

$$R{-}X \xrightarrow{Li} RLi$$

X = Cl, Br, or I

RLi is a simplistic picture of the reagent, which is not monomeric

6. Alkenes

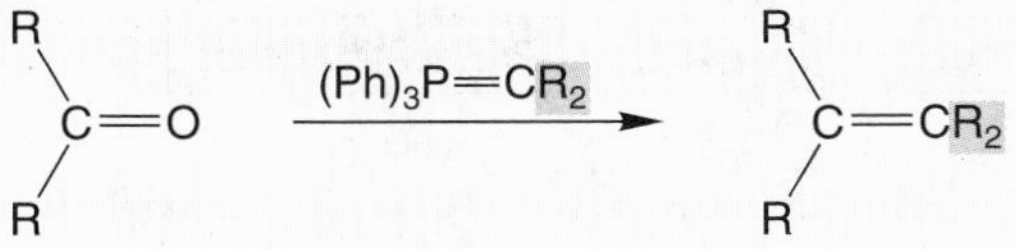

$$R_2C{=}O \xrightarrow{(Ph)_3P{=}CR_2} R_2C{=}CR_2$$

The Wittig reaction

7. Bisulfite addition products

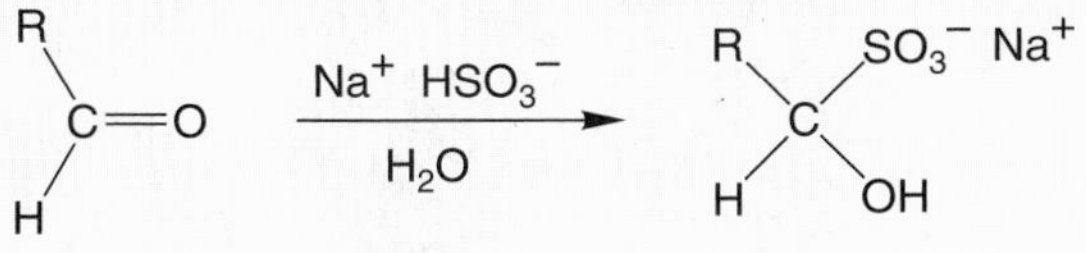

$$RHC{=}O \xrightarrow[H_2O]{Na^+\ HSO_3^-} RHC(OH)SO_3^-\ Na^+$$

Also works with some ketones

8. Cyanohydrins

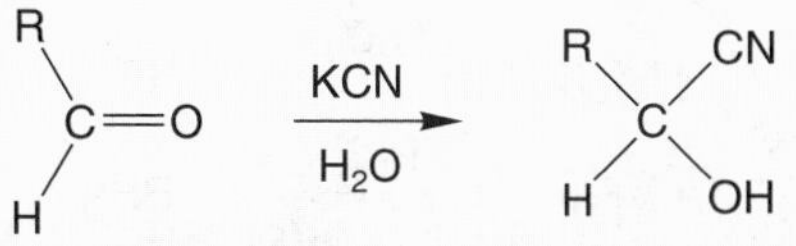

$$RHC{=}O \xrightarrow[H_2O]{KCN} RHC(OH)CN$$

Cyanohydrin formation is more favorable for aldehydes than for ketones

9. Enamines

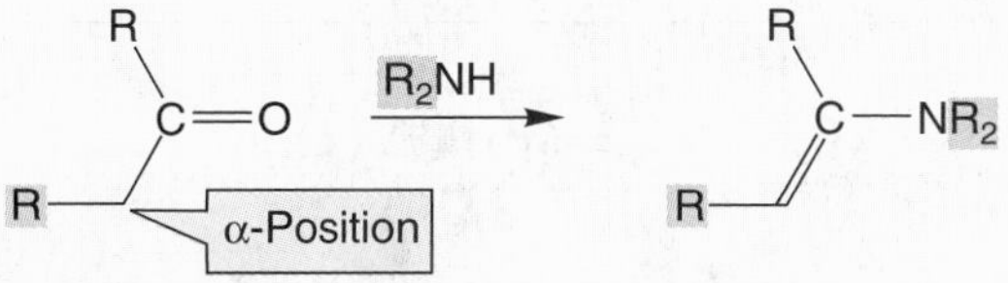

$$R{-}C(=O){-}CH_2{-}R \xrightarrow{R_2NH} R{-}C(NR_2){=}CH{-}R$$

α-Position

At least one α-hydrogen must be available; the amine must be secondary

10. Grignard reagents

$$R{-}X \xrightarrow[\text{ether}]{Mg} RMgX$$

The structure of the organometallic reagent is complicated, as RMgX is in equilibrium with other molecules; the ether solvent is also critical to the success of the reaction

11. Hemiacetals

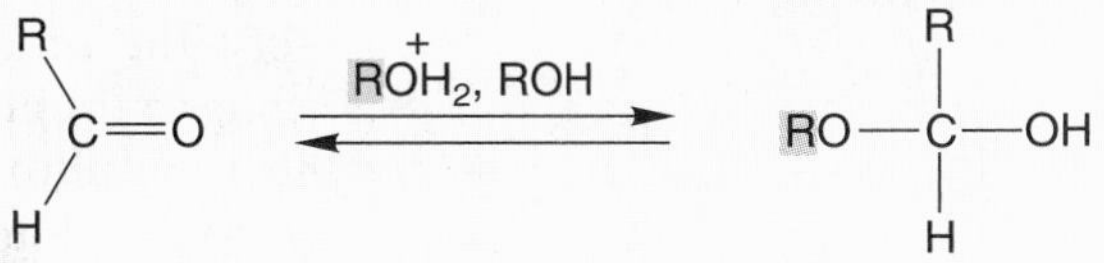

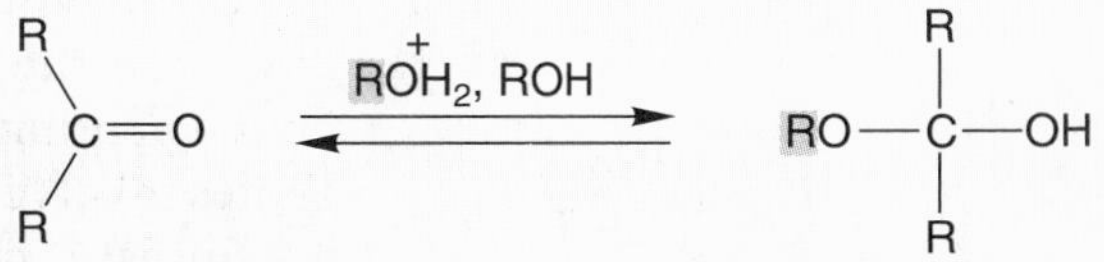

$$RHC{=}O \underset{}{\overset{R\overset{+}{O}H_2,\ ROH}{\rightleftharpoons}} RO{-}CH(R){-}OH$$

$$R_2C{=}O \underset{}{\overset{R\overset{+}{O}H_2,\ ROH}{\rightleftharpoons}} RO{-}CR_2{-}OH$$

Rarely isolable; exceptions are cyclic hemiacetals, especially sugars; the reaction usually goes further to give the full acetal

12. Hydrates

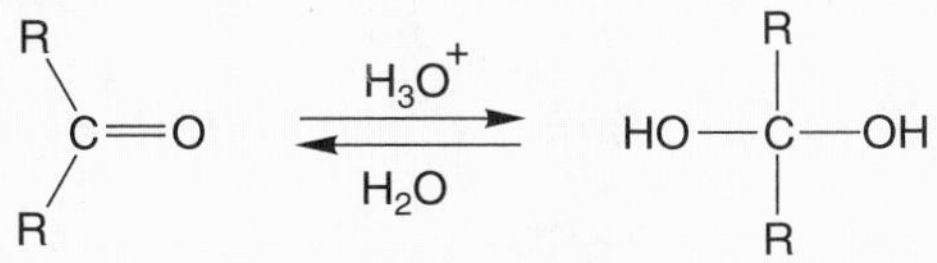

$$R_2C{=}O \underset{H_2O}{\overset{H_3O^+}{\rightleftharpoons}} HO{-}CR_2{-}OH$$

Unstable; for ketones the starting material is usually favored

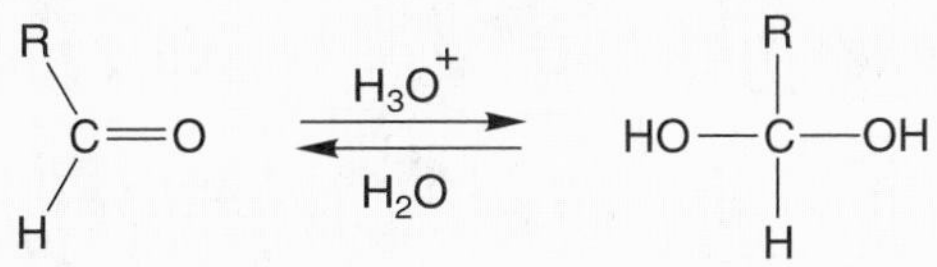

$$RHC{=}O \underset{H_2O}{\overset{H_3O^+}{\rightleftharpoons}} HO{-}CH(R){-}OH$$

More likely to be favored at equilibrium than the hydrated ketones; not usually isolable

FIGURE **16.94** (CONTINUED)

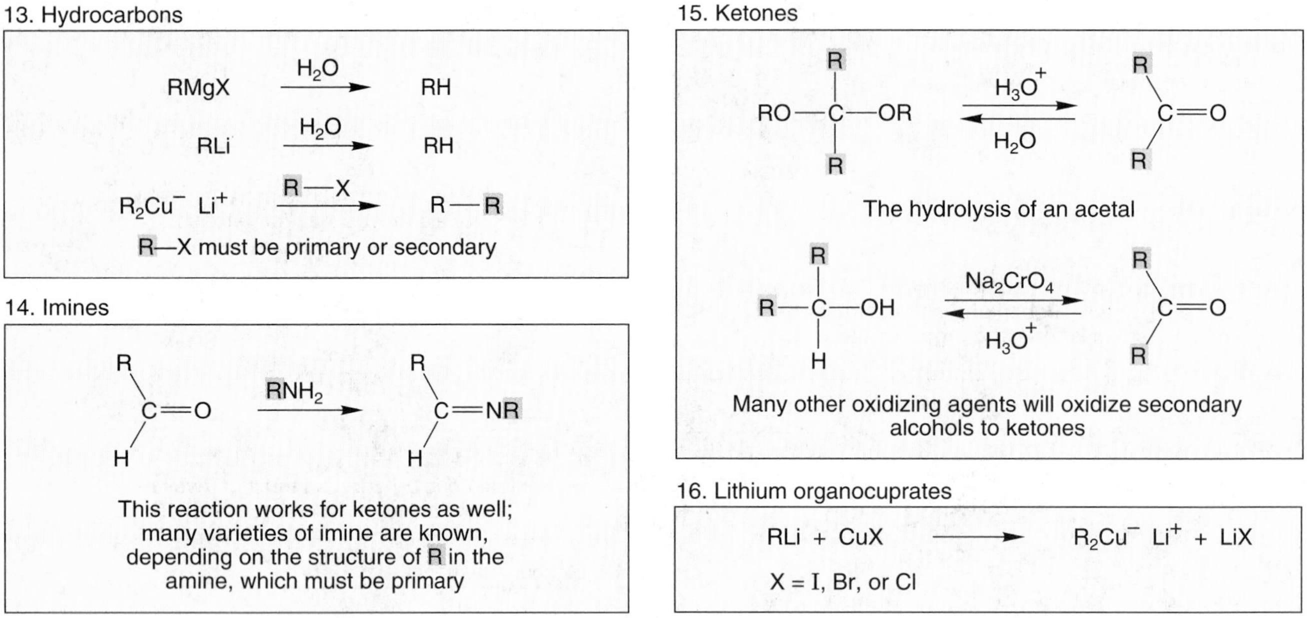

13. Hydrocarbons

$$RMgX \xrightarrow{H_2O} RH$$

$$RLi \xrightarrow{H_2O} RH$$

$$R_2Cu^- \; Li^+ \xrightarrow{R—X} R—R$$

R—X must be primary or secondary

14. Imines

$$RHC{=}O \xrightarrow{RNH_2} RHC{=}NR$$

This reaction works for ketones as well; many varieties of imine are known, depending on the structure of R in the amine, which must be primary

15. Ketones

$$RO—CR_2—OR \underset{H_2O}{\overset{H_3O^+}{\rightleftharpoons}} R_2C{=}O$$

The hydrolysis of an acetal

$$R—CHR—OH \underset{H_3O^+}{\overset{Na_2CrO_4}{\rightleftharpoons}} R_2C{=}O$$

Many other oxidizing agents will oxidize secondary alcohols to ketones

16. Lithium organocuprates

$$RLi + CuX \longrightarrow R_2Cu^- \; Li^+ + LiX$$

X = I, Br, or Cl

FIGURE 16.94 (CONTINUED)

COMMON ERRORS

By far the hardest thing about this chapter is the all-too-plentiful detail. To get this material under control it is absolutely necessary to be able to generalize. The easiest mistake to make is to memorize the details and lose the broad picture. Although there is one general principle in this chapter (nucleophiles add to carbon–oxygen double bonds), it is easy to get lost in the myriad protonation, addition, and deprotonation steps. The reversible addition reactions of this chapter are the hardest to keep straight. The following analysis is a model for the process of sorting necessary to focus on the fundamental similarities while being mindful of the small, but important differences introduced by structural changes in the molecules involved in these reactions.

In base, hydroxide, alkoxides, and amide ions all add to the carbonyl group to give alkoxides that can be protonated to produce hydrates, hemiacetals, and carbinolamines. The three reactions are very similar (Fig. 16.95).

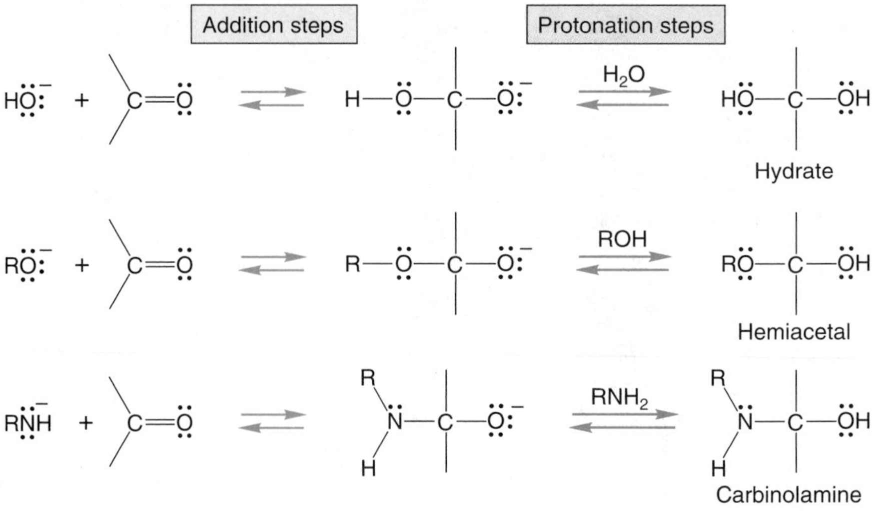

Addition steps | Protonation steps

$$HO^- + {>}C{=}O \rightleftharpoons H—O—C—O^- \xrightarrow{H_2O} HO—C—OH$$ Hydrate

$$RO^- + {>}C{=}O \rightleftharpoons R—O—C—O^- \xrightarrow{ROH} RO—C—OH$$ Hemiacetal

$$RNH^- + {>}C{=}O \rightleftharpoons R(H)N—C—O^- \xrightarrow{RNH_2} R(H)N—C—OH$$ Carbinolamine

FIGURE 16.95 Base-catalyzed additions to carbonyl groups.

In acid, similar reactions occur, and the three processes begin with closely related sequences of protonation, addition, and deprotonation (Fig. 16.96).

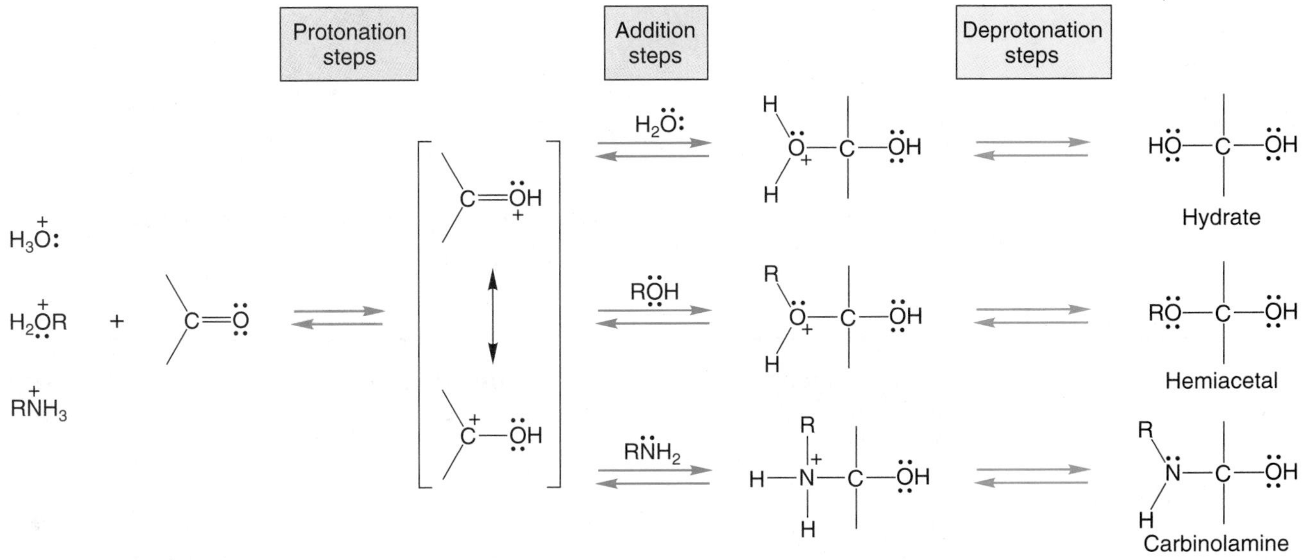

FIGURE 16.96 Acid-catalyzed additions to carbonyl groups.

In the acid-catalyzed reactions, water can be lost from the hydrate, hemiacetal, and carbinolamine to give new, resonance-stabilized intermediates. In the hydrate, this results only in re-formation of the original carbonyl group, but new compounds can be produced in the other two cases. The carbinolamines formed from primary amines follow a similar path that leads ultimately to imines. Hemiacetals can lose water but do not have a second proton that can be lost. Instead, they add a second molecule of alcohol to give full acetals (Fig. 16.97).

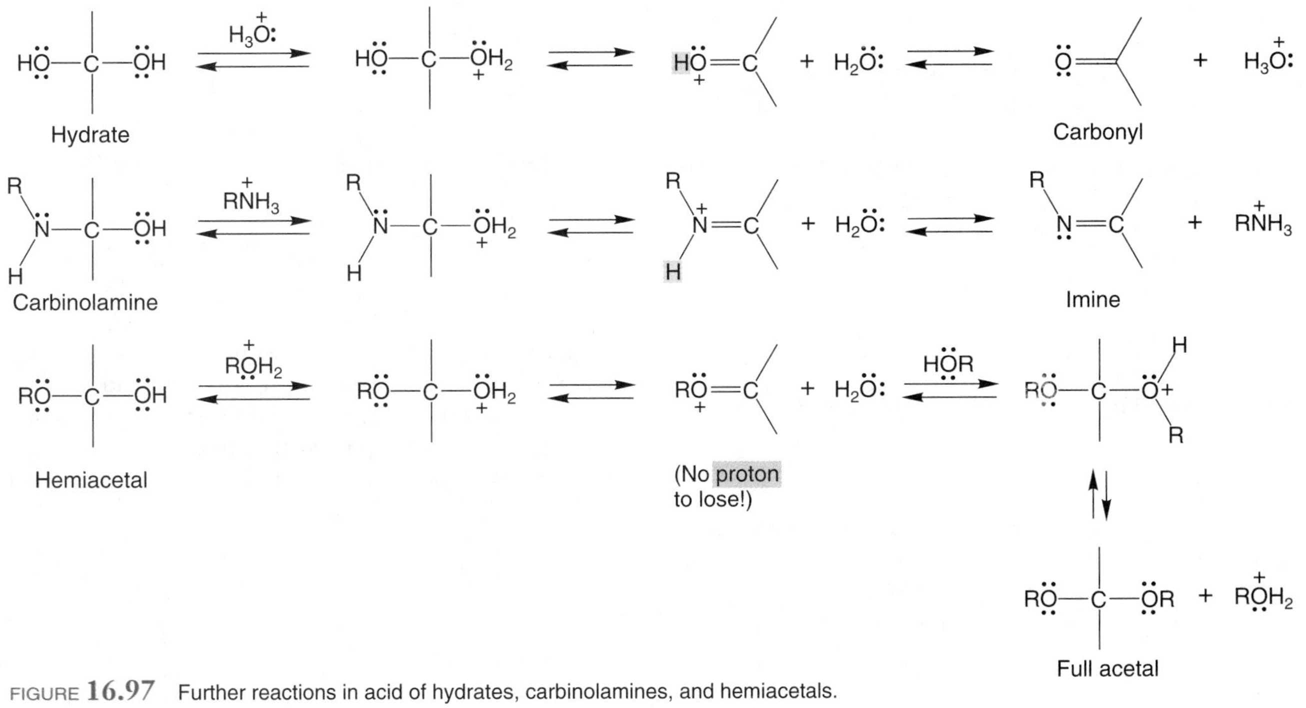

FIGURE 16.97 Further reactions in acid of hydrates, carbinolamines, and hemiacetals.

The carbinolamines formed from secondary amines and carbonyl compounds also have no second proton to lose and instead are deprotonated at carbon to give enamines (Fig. 16.98).

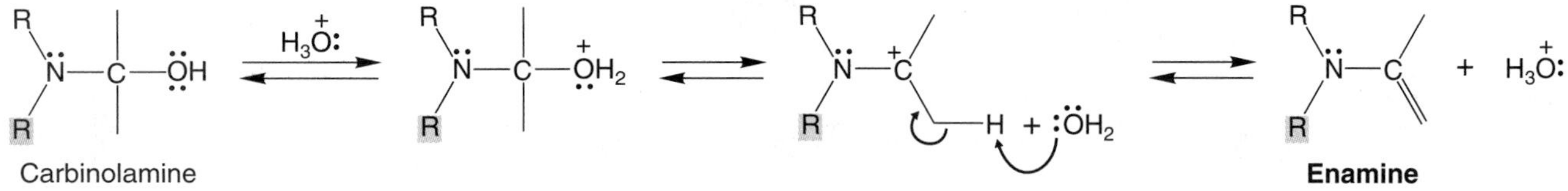

FIGURE 16.98 Further reactions of carbinolamines formed from secondary amines (RRNH).

16.16 KEY TERMS

Acetal The final product in the acid-catalyzed reaction of an aldehyde with an alcohol.

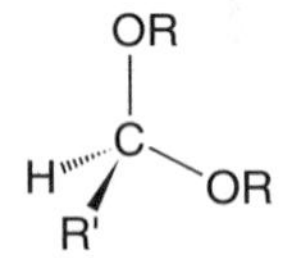

Acetone Dimethyl ketone, the simplest ketone.

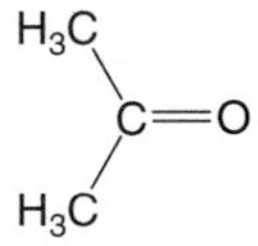

Aldehydes Compounds containing a monosubstituted carbon–oxygen double bond.

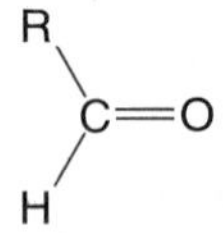

Benzaldehyde The common (and always used) name for the simplest aromatic aldehyde, "benzenecarboxaldehyde."

Carbinolamine The initial product in the reaction between a carbonyl-containing molecule $R_2C{=}O$, and an amine. It is analogous to a hemiacetal.

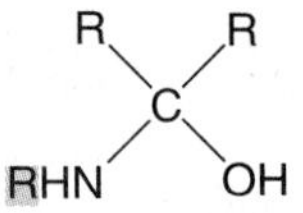

Carbonyl group The carbon–oxygen double bond.

Coenzyme A molecule able to carry out a chemical reaction with another molecule only in cooperation with an enzyme. The enzyme's function is often to bring the substrate and the coenzyme together.

Cyanohydrin The product of addition of hydrogen cyanide to a carbonyl compound.

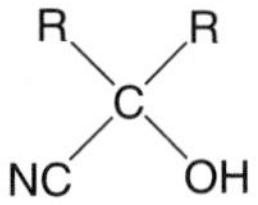

Dial A molecule containing two aldehyde groups, a dialdehyde.

Dione A compound containing two ketone groups, a diketone.

Enamine The nitrogen analogue of an enol, a vinyl amine.

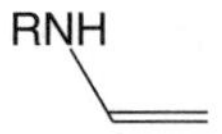

Ester A derivative of an acid in which an OR replaces the OH. For example:

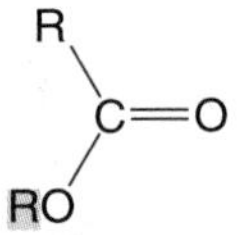

Formaldehyde The simplest possible aldehyde.

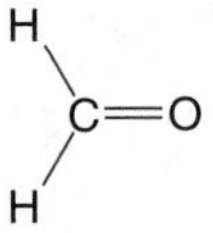

Grignard reagent (RMgX) A strongly basic organometallic reagent formed from a halide and magnesium in an ether solvent. An important and characteristic reaction is the addition to carbonyl groups.

Hemiacetal The initial product when an alcohol adds to an aldehyde or ketone.

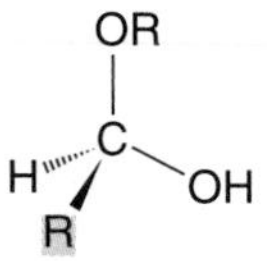

Further reaction gives the full acetal under acidic conditions.

Hydrate The product of the reaction of a carbonyl compound with water.

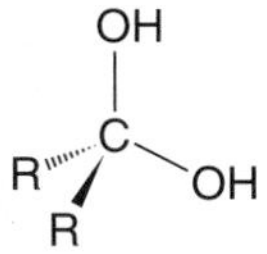

Both aldehydes and ketones can be hydrated.

Hydride A negatively charged hydrogen ion ($H:^-$) bearing a pair of electrons.

Imine The nitrogen analogue of a ketone or aldehyde.

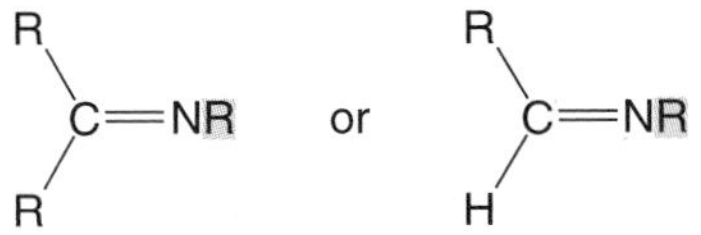

See **Schiff base**.

Immonium ion The nitrogen analogue of a protonated carbonyl group.

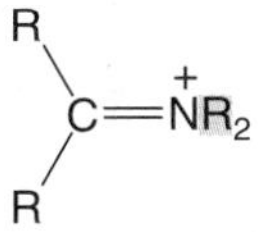

(R can also be H).

Ketones Compounds containing a disubstituted carbon–oxygen double bond.

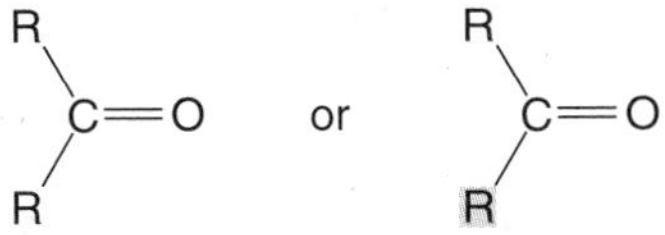

NAD⁺ Nicotinamide adenine dinucleotide, a biological oxidizing agent.

Organocuprates An organometallic reagent (R_2CuLi) notable for its ability to react with primary or secondary halides (R′–X) to generate hydrocarbons, R–R′.

Organolithium reagent (R–Li) A strongly basic organometallic reagent formed from a halide and lithium. A characteristic reaction is addition to carbonyl groups.

Organometallic reagents Molecules that contain both carbon and a metal. Usually carbon is at least partially covalently bonded to the metal. Examples are Grignard reagents, organolithium reagents, and lithium organocuprates.

Oxaphosphetane The four-membered intermediate in the Wittig reaction that contains two carbons, a phosphorus, and an oxygen.

Phenyl group A common name for the benzene ring as substituent, C_6H_5–X.

Protecting group Sensitive functional groups in a molecule can be protected by a reaction that converts them into a less reactive group, called a protecting group. A protecting group must be removable to regenerate the original functionality.

Schiff base The nitrogen analogue of a ketone or aldehyde.

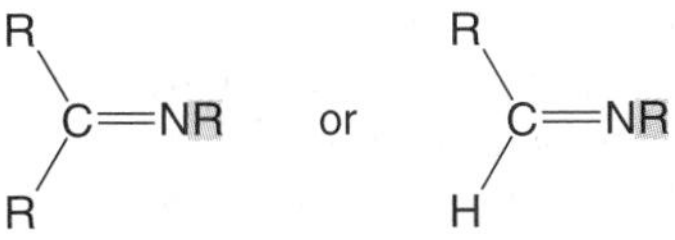

See **imine**.

Wittig reaction The reaction of an ylide with a carbonyl group to give, ultimately, an alkene.

Ylide A compound containing opposite charges on adjacent atoms.

16.17 ADDITIONAL PROBLEMS

PROBLEM **16.26** Name the following molecules:

(a)

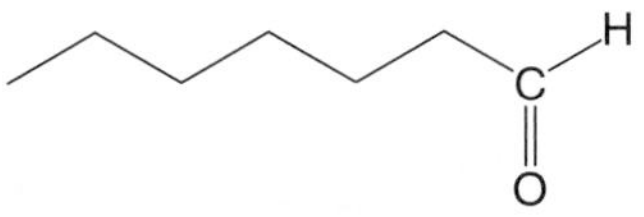

(b)

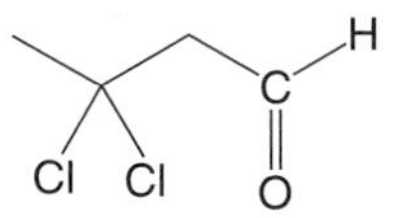

(c)

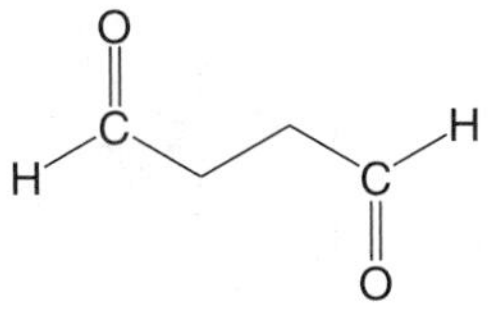

(d)

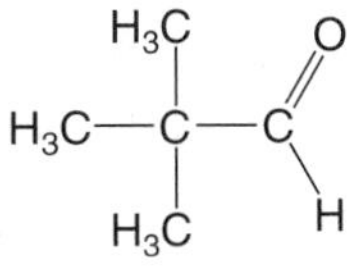

(e)

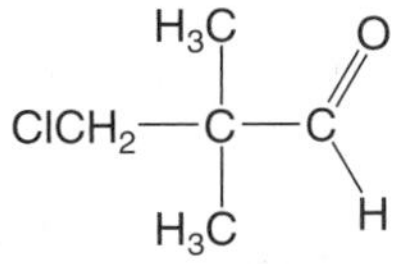

(f)

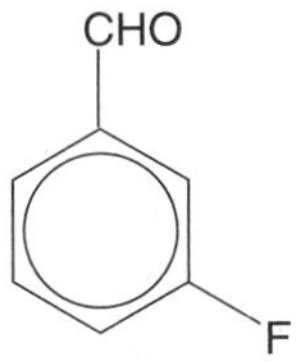

PROBLEM **16.27** Draw structures for the following compounds:

(a) *p*-Nitrobenzaldehyde

(b) 4-Methylhexanal

(c) 2-Methylpropiophenone (isobutyrophenone)

(d) *cis*-2-Bromocyclopropanecarboxaldehyde

PROBLEM 16.28 Name the following compounds:

(a) (b)

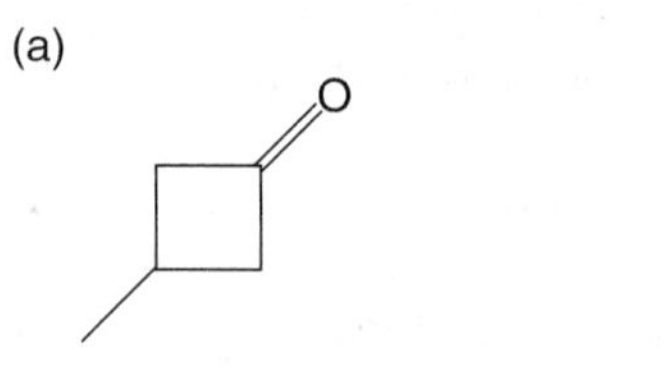

(c)

PROBLEM 16.29 Draw structures for the following compounds:

(a) Methyl propyl ketone
(b) 2-Pentanone
(c) 2,3-Hexanedione
(d) Butyrophenone
(e) 3-Bromo-4-chloro-5-iodo-2-octanone

PROBLEM 16.30 Draw structures for the following compounds:

(a) 3,5-Di-*tert*-butyl-4-hydroxybenzaldehyde
(b) 2-Acetylcyclopentanone
(c) 2,4'-Dichloro-4-nitrobenzophenone
(d) 6-Methyl-2-pyridinecarboxaldehyde
(e) *trans*-4-Phenyl-3-buten-2-one
(f) Aminoacetaldehyde dimethyl acetal

PROBLEM 16.31 Use the indicated spectroscopy to distinguish the following pairs of compounds:

(a) (IR)

(b) (IR)

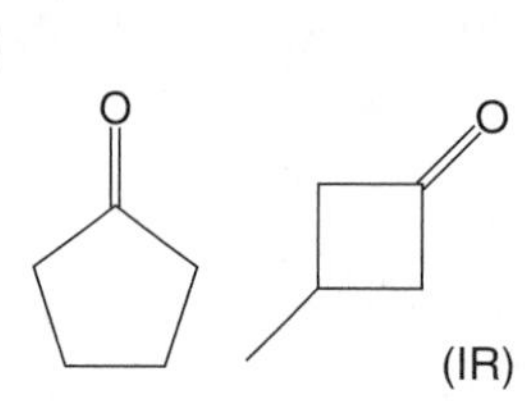

(c) (^{13}C NMR)

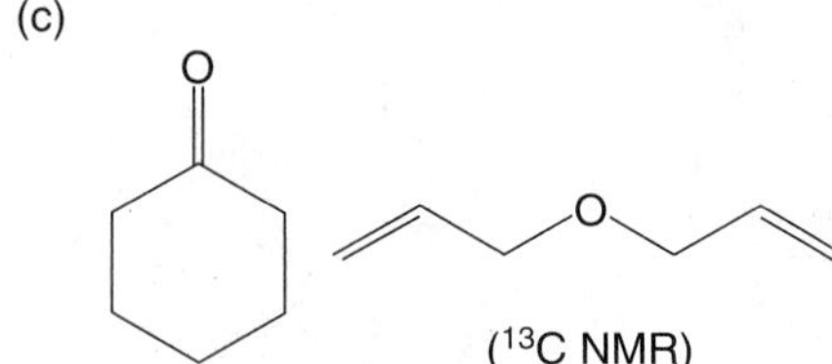

(d) (IR, ^{1}H NMR)

PROBLEM 16.32 Give the major organic products expected in each of the following reactions:

(a) HO–CH$_2$CH$_2$CH$_2$–OH; benzene, H_3O^+

(b) $PhNH_2$

(c) pyrrolidine (NH)

(d) 1. PhMgBr; 2. H_2O/H_3O^+

(e) $NaHSO_3$, H_2O; NaCN, H_2O

(f) 1. CH_3Li; 2. H_2O/H_3O^+

(g) $NaBH_4$, CH_3OH

(h) 1. $LiAlH_4$; 2. H_2O/H_3O^+

(i) $Na_2Cr_2O_7$, H_2SO_4

(j) $PhCH_2OH \xrightarrow{H_2CrO_4}$

(k) Ph–CH=CH–CH$_2$OH $\xrightarrow[\text{pyridine}]{CrO_3}$

(l) Ph–C(=O)–$(CH_2)_4CH_2I$ $\xrightarrow[\text{2. BuLi; 3. }\Delta]{\text{1. }Ph_3P}$

Hint: Part (l) involves an intramolecular reaction.

PROBLEM **16.33** Provide an argument not involving molecular orbitals to explain why a carbon–oxygen double bond protonates on O, not C.

PROBLEM **16.34** Write products and arrow formalism mechanisms for the following simple changes:

(a) cyclohexanone ethylene ketal $\xrightarrow{H_2O/H_3O^+}$

(b) N-cyclopentylidene pyrrolidinium ion $\xrightarrow{H_2O}$

(c) $(H_3CO)(HO)C(CH_3)(CH_3)$ $\xrightarrow[CH_3\overset{+}{O}H_2]{CH_3OH}$

(d) $(H_3CO)(HO)C(CH_3)(CH_3)$ $\xrightarrow{KOH/H_2O}$

(e) cyclohexanone ethylene ketal $\xrightarrow{KOH/H_2O}$

(f) cyclohexanone $\xrightarrow[NH_2OH]{^+NH_3OH}$

PROBLEM **16.35** The reduction of 4-*tert*-butylcyclohexanone with lithium aluminum hydride affords two isomeric products. What are the structures of these two products? Which isomer do you expect to predominate?

4-*tert*-butylcyclohexanone ($C(CH_3)_3$) $\xrightarrow[\text{2. }H_2O/H_3O^+]{\text{1. }LiAlH_4}$ Two isomeric products

PROBLEM **16.36** Write an arrow formalism for the following reaction:

Br–$CH_2CH_2CH_2$–C(=O)–CH_3 $\xrightarrow[\text{2. }H_2O/H_3O^+]{\text{1. Mg/THF}}$ 1-methylcyclobutanol (CH_3, OH)

PROBLEM **16.37** Starting from isopropyl alcohol as your primary source of carbon, incidental organic reagents such as bases and solvents, as well as inorganic reagents of your choice (including labeled materials as needed), devise syntheses of the following labeled compounds. It is not necessary to write mechanisms.

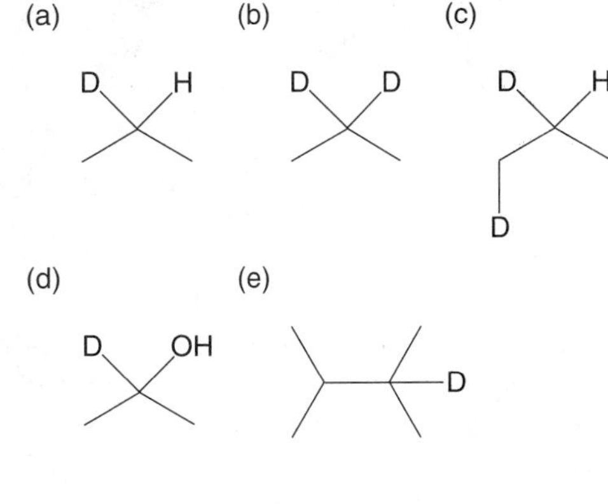

PROBLEM 16.38 Starting from propyl alcohol, organic reagents containing no more than one carbon, and incidental organic reagents such as bases and solvents, as well as inorganic materials, devise syntheses of the following compounds:

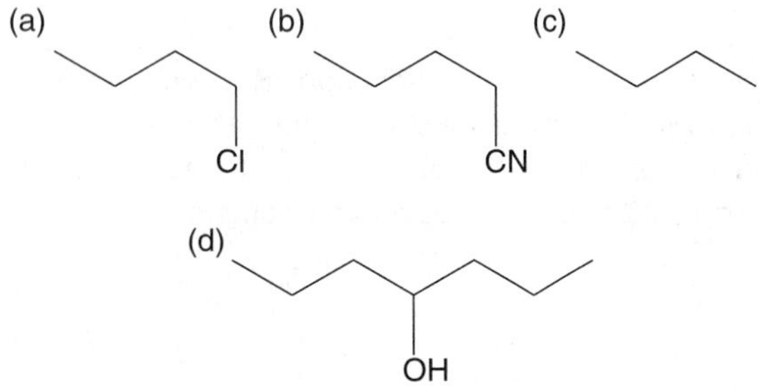

(a) Cl (b) CN (c)

(d) OH

PROBLEM 16.39 Starting from alcohols containing no more than four carbon atoms, inorganic reagents of your choice, and "special" organic reagents (such as ether and pyridine) provide syntheses of the following molecules:

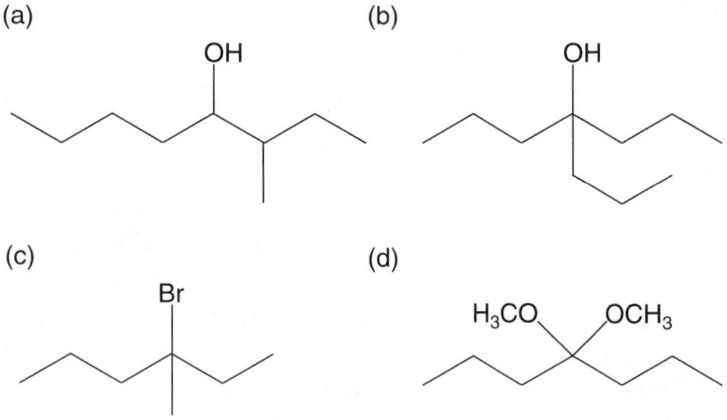

(a) OH (b) OH

(c) Br (d) H_3CO OCH_3

PROBLEM 16.40 Propose an arrow formalism mechanism for the following reaction:

H_2N–C(=O)–$NHCH_2CH(OCH_2CH_3)_2$ $\xrightarrow[H_2O]{H_3O^+}$ NH, C=O, NH

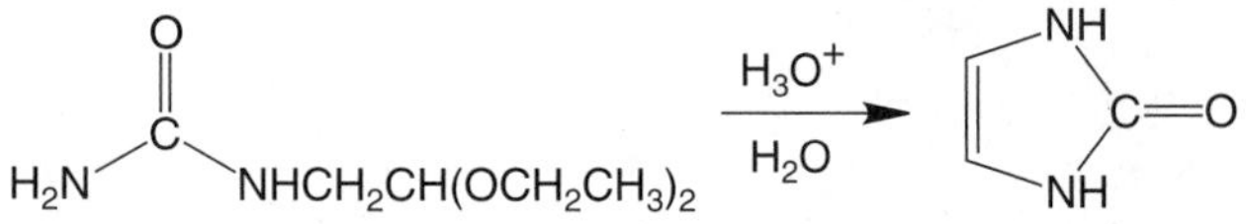

PROBLEM 16.41 Propose an arrow formalism mechanism for the following reaction. What is the driving force for this reaction? Why does it go from the seven-membered ring to the five-membered ring?

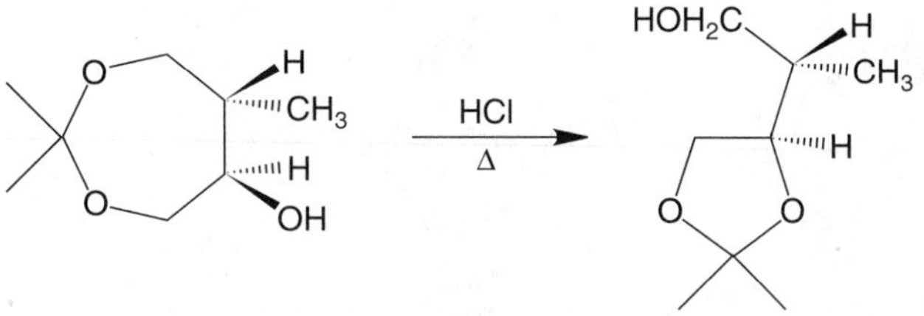

O, O, H, CH_3, H, OH $\xrightarrow[\Delta]{HCl}$ HOH_2C, H, CH_3, H, O, O

PROBLEM 16.42 Propose an arrow formalism mechanism for the following reaction:

O + HO, OH $\xrightarrow[H_2O]{HCl}$ O, O, HO

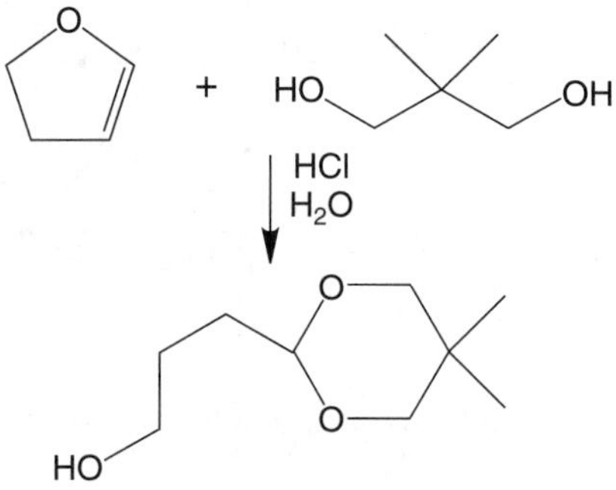

PROBLEM 16.43 Rationalize the following somewhat surprising result. *Hint*: This problem is not as difficult as it looks. First, consider what the reaction of the alcohol starting material with the chromium reagent will give. Second, what will happen when this compound reacts with more alcohol? Finally, use the chromium reagent again.

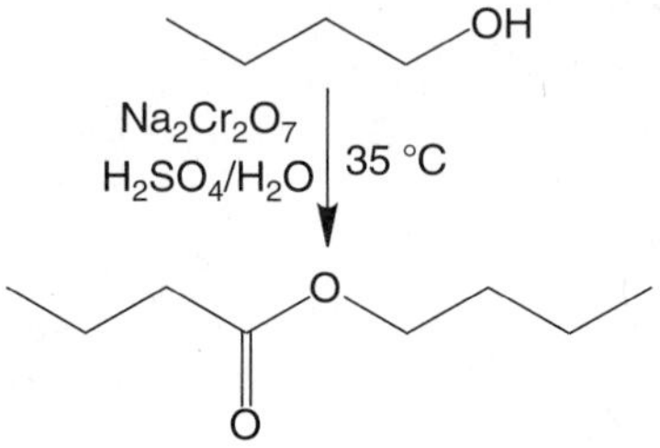

OH $\xrightarrow[H_2SO_4/H_2O,\ 35\ °C]{Na_2Cr_2O_7}$ O, O

PROBLEM 16.44 The reaction of acetone with methylamine reaches equilibrium with a carbinolamine and, eventually, an imine. There is another participant in the equilibrium called an aminal. Its formula is $C_5H_{14}N_2$. Provide a structure and a mechanism.

PROBLEM 16.45 Les Gometz, a professor from New York who is often disappointed in September, was interested in investigating the Diels–Alder reaction of methyl coumalate (**A**) and the enamine **B**. The Professor attempted to prepare **B** by allowing propionaldehyde and morpholine (**C**) to react.

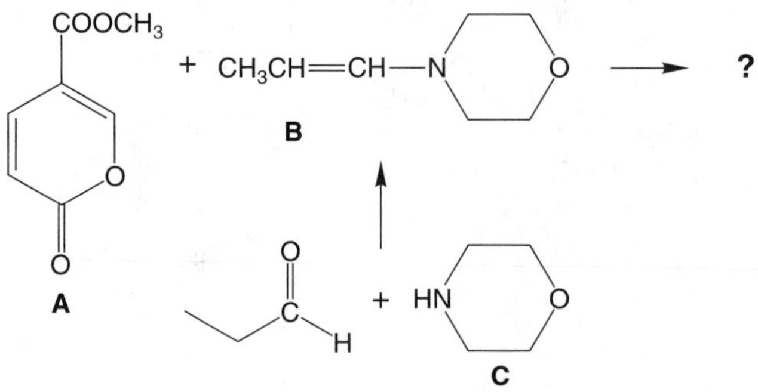

$COOCH_3$, O, O **A** + $CH_3CH{=}CH{-}N$ O **B** → ?

O, C, H + HN O **C**

After an appropriate time, Professor Gometz assayed the reaction mixture by ^{1}H NMR spectroscopy and was disappointed to find that only a small portion (<5–10%) of the desired enamine **B** had formed. Ever the optimist, he ran the Diels–Alder reaction with diene **A** anyway, using the reaction mixture containing what he knew to be only a small amount of **B**. He was delighted and astonished to obtain an 80% yield of cycloadducts.

Your task is to explain how Professor Gometz could get such a good yield of products when only a small amount of enamine **B** was present in the reaction mixture. *Hint*: See Problem 16.9.

PROBLEM **16.46** Provide structures for compounds **A–D**. Spectral data for compound **D** are shown below.
Compound **D**
Mass spectrum: m/z = 332 (p, 82%), 255 (85%), 213 (100%), 147 (37%), 135 (43%), 106 (48%), 77 (25%), 43 (25%).
IR (KBr): 3455 (s) and 1655 (s) cm^{-1}.
^{1}H NMR ($CDCl_3$): δ 2.58 (s, 3H), 2.85 (s, 1H, vanishes when a drop of D_2O is added), 3.74 (s, 3H), 6.77–7.98 (m, 13H).
^{13}C NMR (acetone-d_6): δ 27.0, 55.8, 82.0, 128.1–159.9 (12 lines), 197.8.

Br–C_6H_4–C(=O)CH_3 —[$HOCH_2CH_2OH$, H_3O^+]→ **A** —[Mg, THF, Δ]→ **B**

H_3CO–C_6H_4–C(=O)Ph —[1. **B**, Δ; 2. H_2O/H_3O^+]→ **C** —[H_2O/H_3O^+, THF, Δ]→ **D**

PROBLEM **16.47** Provide structures for compounds **A–D**. Spectral data for compound **D** are shown below. Although mechanisms are not required, a mechanistic analysis may prove helpful in deducing the structures.
Compound **D**
IR (neat): 2817 (w), 2717 (w), 1730 (s) cm^{-1}.
^{1}H NMR ($CDCl_3$): δ 0.92 (t, J = 7 Hz, 6H), 1.2–2.3 (m, 5H), 9.51 (d, J = 2.5 Hz, 1H).
^{13}C NMR ($CDCl_3$): δ 11.4, 21.5, 55.0, 205.0.

H_3C–C_6H_4–OCH_2Cl —[Ph_3P]→ **A** —[PhLi]→ **B** —[$(CH_3CH_2)_2C{=}O$]→ $Ph_3P{=}O$ + **C** —[H_2O/H_3O^+]→ **D**

The Chemistry of Alcohols Revisited and Extended: Glycols, Ethers, and Related Sulfur Compounds

Since things that here in order shall ensue,
Against all poysons have a secret power.
Peare, garlicke, reddish root, nuts, rape and rue—
But garlicke chiefe; for they that devoure
May drinke, and care not who their drinke do brewe.
May walk in aires infected, every houre
Sith garlicke them have power to save from death,
Bear with it though it maketh unsavory breath,
And scorne not garlicke, like some that thinke
It only maketh men winke, and drinke—and stinke.

—Sir John Harington*
The Englishman's Doctor

In Chapter 16, we developed a general alcohol synthesis through the reaction of organometallic and metal hydride reagents with carbonyl compounds. Earlier, we had seen the acid-catalyzed addition of water to alkenes (Chapter 9, p. 358). In fact, alcohol chemistry is scattered throughout the first 16 chapters of this book. Alcohols appear often as reactants, products, and as exemplars of various phenomena. In this summary chapter, the material first appearing in earlier chapters is collected and extended. Reactions already described will not be repeated at length, and if any of them seem unfamiliar, by all means go back to the original material. The page references are given. Some new chemistry of alcohols will appear, but most of the new material relates to relatives of alcohols: polyalcohols (especially 1,2-diols, or **1,2-glycols**), ethers, and the sulfur-containing **thiols** (**mercaptans**) and **thioethers** (**sulfides**).

Even after we complete this chapter we will not be finished with alcohol chemistry. New syntheses and reactions will continue to pop up. The idea of this summary chapter is not to exhaust the subject, but to provide an overview. The subject tends to get lost in the too-prolific detail, and reactions can seem to stand alone rather than as examples of general processes. One aim of this chapter is to tie together what we already know. Another is to introduce the relatives of alcohols and their chemistries. We will first review and collect what we know about alcohols (ROH), then venture into new, but related territory: the ethers (ROR), molecules with more than one OH group, and the sulfur-containing counterparts of alcohols and ethers.

*Harrington (1561–1612) is said to have translated this from verses from the Medical School of Salerno. I found it in a fine review on constituents of the genus *allium* by Professor Eric Block (b. 1944) of the State University of New York at Albany.

17.1 NOMENCLATURE

There are two ways to name alcohols, the systematic, IUPAC way, and the way chemists often do it, by using the splendidly vulgar common names. As usual, these common names are only used for the smaller members of the class, and the IUPAC system takes over as complexity increases.

Alcohols are named systematically by dropping the final "e" of the parent hydrocarbon and adding the suffix "ol." The functional group, OH, is given the lowest number possible, taking precedence over other groups in the molecule including the related thiol group, SH. The longest carbon chain is identified and the substituents numbered appropriately (Fig. 17.1).

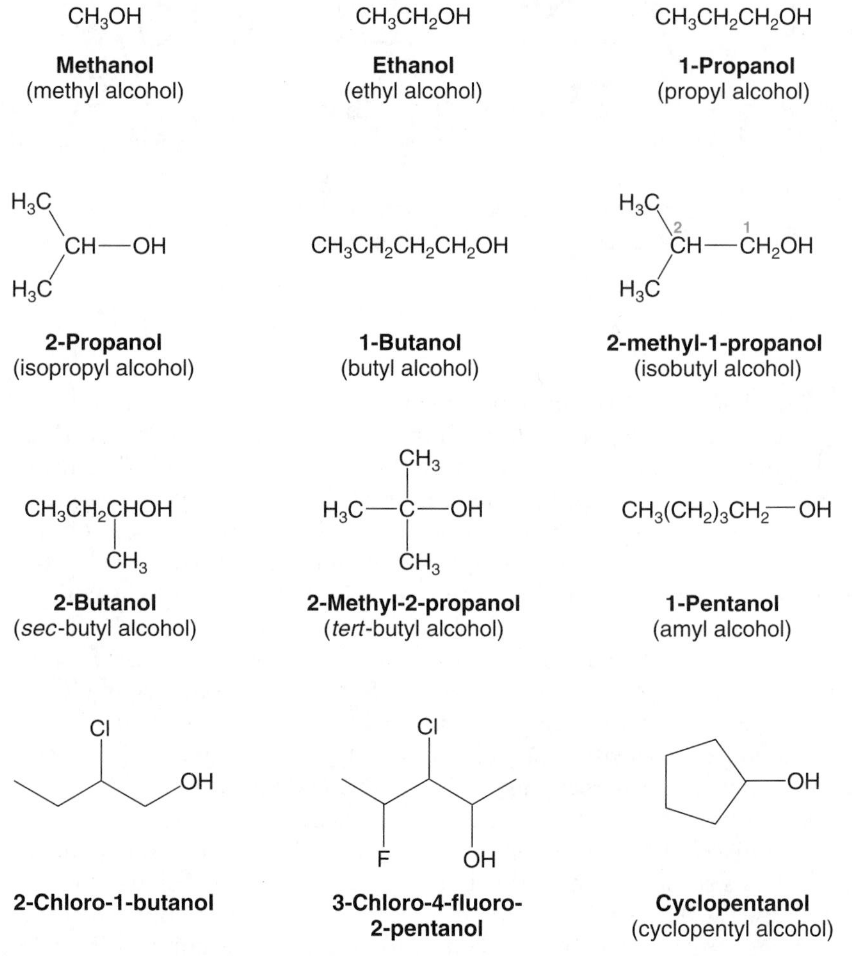

CH_3OH **Methanol** (methyl alcohol)

CH_3CH_2OH **Ethanol** (ethyl alcohol)

$CH_3CH_2CH_2OH$ **1-Propanol** (propyl alcohol)

$(H_3C)_2CH{-}OH$ **2-Propanol** (isopropyl alcohol)

$CH_3CH_2CH_2CH_2OH$ **1-Butanol** (butyl alcohol)

$(H_3C)_2\overset{2}{C}H{-}\overset{1}{C}H_2OH$ **2-methyl-1-propanol** (isobutyl alcohol)

$CH_3CH_2CH(CH_3)OH$ **2-Butanol** (*sec*-butyl alcohol)

$H_3C{-}C(CH_3)_2{-}OH$ **2-Methyl-2-propanol** (*tert*-butyl alcohol)

$CH_3(CH_2)_3CH_2{-}OH$ **1-Pentanol** (amyl alcohol)

2-Chloro-1-butanol

3-Chloro-4-fluoro-2-pentanol

Cyclopentanol (cyclopentyl alcohol)

FIGURE 17.1 Some systematic and commonly used names for alcohols.

The smaller alcohols, however, are commonly named by appending the word alcohol to the appropriate group name. Figure 17.1 gives some systematic and common names for the smaller alcohols. When we get past approximately five carbons, the systematic naming protocol takes over completely, and even the delightful name amyl for the five-carbon alcohols is disappearing.* However, some common names seem to be quite hardy. For example, the correct name for hydroxybenzene is benzenol,

*Amyl derives from the Latin word for starch, "amylum." The first amyl alcohol was isolated from the fermentation of potatoes.

but it appears that the traditional phenol will survive for a long time (Fig. 17.2)

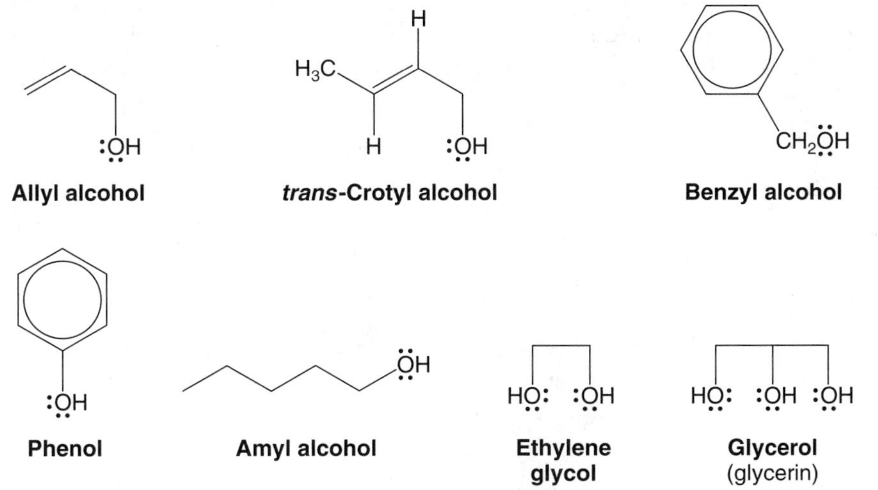

FIGURE 17.2 Some common names for alcohols.

PROBLEM 17.1

Provide IUPAC names for the following alcohols (Fig. 17.3):

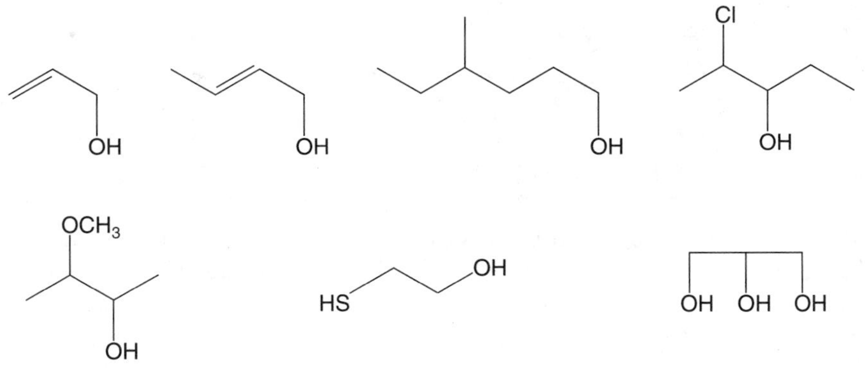

FIGURE 17.3

17.2 STRUCTURE OF ALCOHOLS

Alcohols are derivatives of water, and it will be no surprise to see that they rather closely resemble water structurally. The bond angle is expanded a bit from 104.5° in H—O—H to approximately 109° in simple alcohols R—O—H.

PROBLEM 17.2

What is the approximate hybridization of the oxygen atom in methyl alcohol (H_3C—O—H angle = 108.9°). How do you know?

The oxygen–hydrogen bond length is little changed from that of water, but the carbon–oxygen bond is a bit shorter than the carbon–carbon bond in ethane (Fig. 17.4).

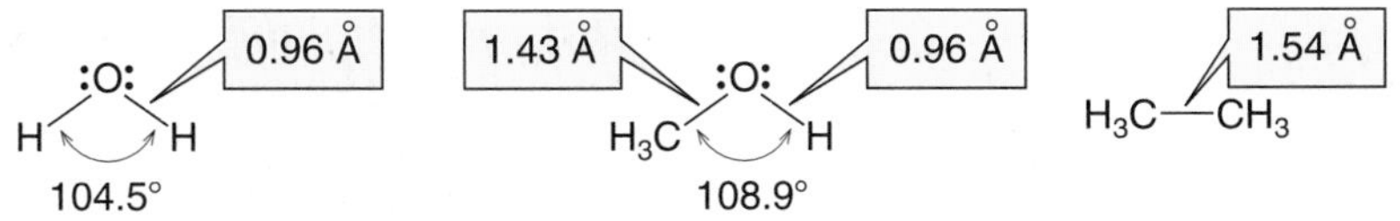

FIGURE 17.4 Bond lengths and angles for some hydroxy compounds.

The electronegative oxygen atom results in a carbon–oxygen bond that is stronger and shorter than the carbon–carbon bond in ethane. An electron in an orbital near oxygen is more stable than it would be in an orbital near carbon. The orbital interaction diagram of Figure 17.5 shows the construction of the carbon–oxygen bond as well as typical bond energies.

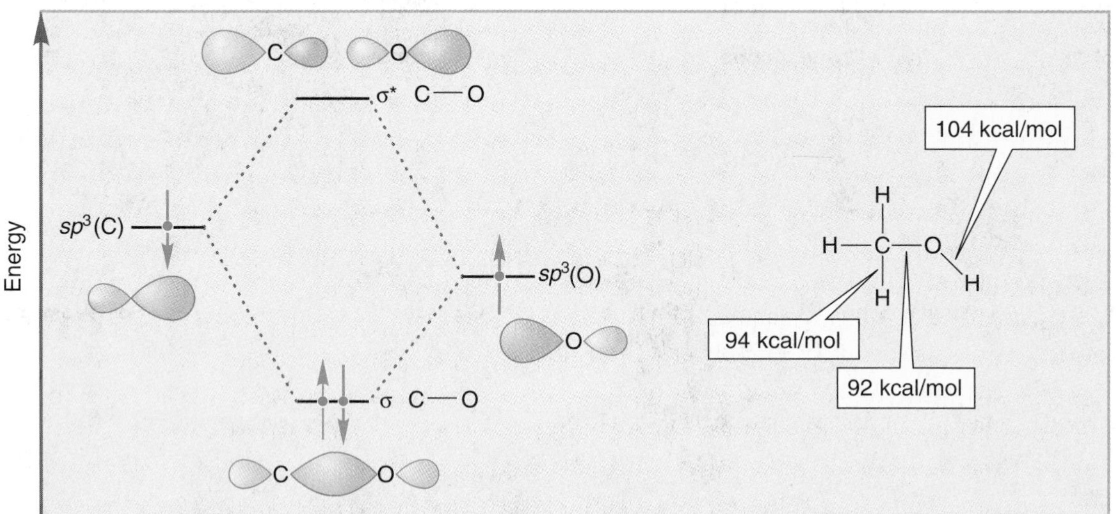

FIGURE 17.5 A graphical description of the formation of the carbon–oxygen σ bond.

17.3 PHYSICAL PROPERTIES OF ALCOHOLS

The presence of the electronegative oxygen atom ensures that bonds will be strongly polarized in alcohols, and that substantial dipole moments will exist. In solution, alcohols are strongly associated both because of dipole–dipole attractions and because of **hydrogen bonding**. The basic oxygen atoms form partial bonds to the acidic hydroxyl hydrogens. Such bonds can be quite strong, on the order of 5 kcal/mol, a value not high enough to make permanent dimeric or polymeric structures, but quite sufficient to raise the boiling points of alcohols far above those of the much less strongly associated alkanes. Figure 17.6 shows an ordered, hydrogen-bonded structure for water. In the absence of hydrogen bonding, water, with a molecular weight of only 18, would surely be a gas. The polarity of alcohols makes them quite water soluble, and most of the smaller molecules are miscible with, or at least highly soluble in water. Table 17.1 gives some physical properties of alcohols. For comparisons with the parent alkanes see Table 3.4 (p. 94).

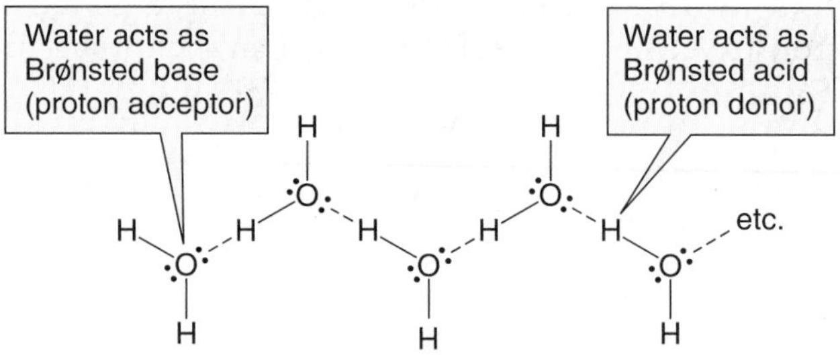

Water acts as Brønsted base (proton acceptor)

Water acts as Brønsted acid (proton donor)

FIGURE 17.6 An alcohol is both Brønsted acid and Brønsted base. This figure shows hydrogen bonding between molecules of water.

TABLE 17.1 Some Physical Properties of Alcohols

Compound	bp (°C)	mp (°C)	Density (g/cm^3)	Dipole Moment (D)
CH_3OH Methyl alcohol	65.15	− 93.9	0.79	1.7
CH_3CH_2OH Ethyl alcohol	78.5	−117.3	0.79	1.69
$(CH_3)_2CHOH$ Isopropyl alcohol	82.4	− 89.5	0.80	1.68
$CH_3CH_2CH_2CH_2OH$ Butyl alcohol	117.2	− 89.5	0.81	1.66
$(CH_3)_2CHCH_2OH$ Isobutyl alcohol	108	−108	0.80	1.64
$CH_3CH_2CH(CH_3)OH$ *sec*-Butyl alcohol	99.5		0.81	
$(CH_3)_3COH$ *tert*-Butyl alcohol	82.3	25.5	0.79	
OH Phenol	181.7	43	1.06	1.45

17.4 SPECTROSCOPIC PROPERTIES OF ALCOHOLS

17.4a Infrared Spectra

There is an IR stretching band at 3200–3600 cm^{-1} that is diagnostic for O—H (or N—H) bonds. It is difficult to be precise about the position of this band because it appears as a very broad signal. The reason is that we are really seeing the stretching frequencies for a large number of differently hydrogen-bonded molecules. The "O—H stretch" is really the result of a compendium of monomers, dimers, and oligomers (associations of several molecules). Figure 17.7 shows an IR spectrum for a typical simple alcohol, ethyl alcohol.

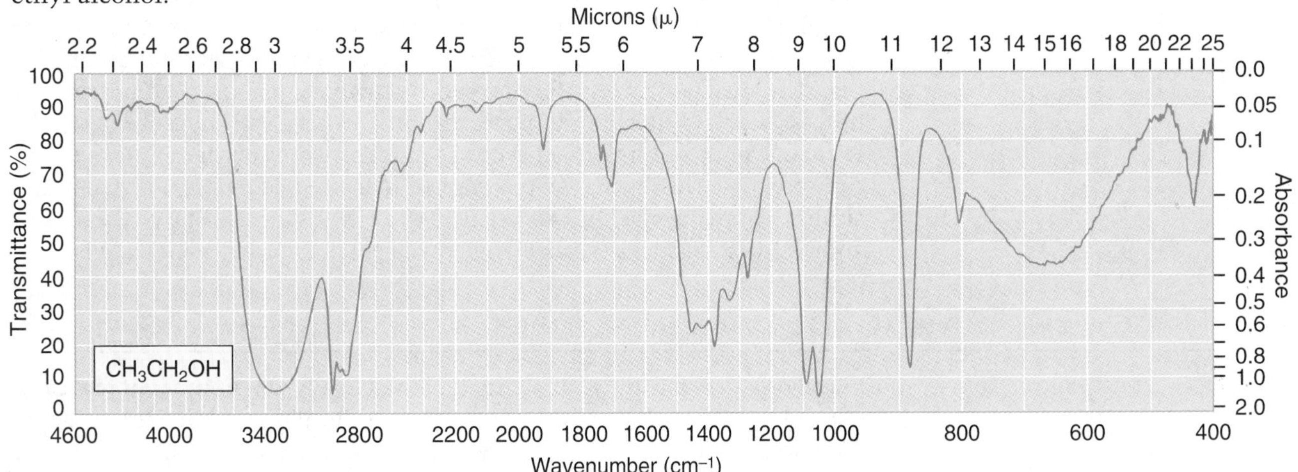

FIGURE 17.7 The IR spectrum of ethyl alcohol.

*PROBLEM 17.3 The O—H portion of the spectrum changes with concentration. As the solution becomes more dilute, the broad band centered at roughly 3300 cm^{-1} sharpens. Explain.

ANSWER As the concentration of alcohol drops, there will be fewer hydrogen-bonded dimers and oligomers (collections of several molecules). The broad band at 3200–3600 cm^{-1} is the result of the presence of these hydrogen-bonded molecules, not the monomer. At high dilution, the O—H stretching frequency of the isolated monomeric alcohol appears at approximately 3600 cm^{-1}.

PROBLEM 17.4 The O—H stretching IR band for *tert*-butyl alcohol is much sharper than that for methyl alcohol. Explain.

PROBLEM 17.5 The IR stretching frequency for the O—H group in the syn unsaturated molecule of Figure 17.8 is sharp at all concentrations and shows no change on dilution, whereas that for the related saturated molecule or the anti unsaturated molecule is broad and sharpens with dilution. Explain. *Hint*: Oxygen is not the only base that can participate in hydrogen bonding.

FIGURE 17.8 syn-Unsaturated anti-Unsaturated Saturated

A second important band in compounds containing C—O bonds is the stretching frequency found at about 1050–1250 cm^{-1}.

17.4b Nuclear Magnetic Resonance Spectra

The position of the OH hydrogen in the ^{1}H NMR spectrum depends strongly on the nature and concentration of the solvent, and cannot be simply tabulated. The electron-withdrawing oxygen atom results in a deshielding of hydrogens attached to adjacent carbons, and in rather low-field chemical shifts (δ 4–5 ppm, see Chapter 15, p. 716). Remember that unless great care is taken in purification, OH (and NH) hydrogens rapidly exchange and coupling to adjacent hydrogens is not observed. Indeed, one can often use this exchange reaction to identify OH and NH signals. One simply adds a drop of D_2O and shakes the NMR sample. The peaks for OH and NH will disappear as exchange takes place to produce OD and ND.

17.5 ACID AND BASE PROPERTIES OF ALCOHOLS

In the next few pages, we are going to look closely at a number of reactions of alcohols. These processes will depend on the ability of alcohols to act as both acids and bases. Accordingly, we will first make sure that we can see these roles clearly. Both Brønsted and Lewis basicity is centered upon the lone-pair electrons of the oxygen atom, which form bonds with a variety of acids, including the O–H hydrogens of other alcohols (hydrogen bonding). Protonation gives an oxonium ion, and serves to convert an alcohol (ROH) in which the potential leaving group is the very poor $^{-}$OH, into the protonated alcohol ($R\overset{+}{O}H_2$) in which the leaving group is the excellent OH_2 (Fig. 17.9).

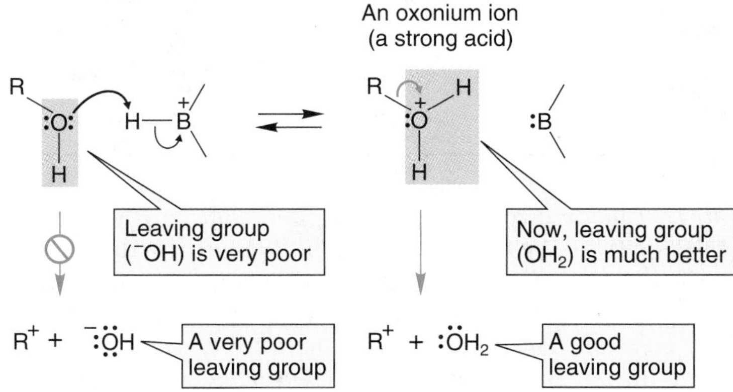

FIGURE 17.9 Protonation of an alcohol leads to an oxonium ion. In this process, the very poor leaving group OH is transformed into the good leaving group OH_2.

Protonated alcohols are very strong Brønsted acids. Therefore, in order to protonate an alcohol, it takes a strong acid, indeed. The conjugate base (B:$^{-}$) of the protonating acid (HB) must be a weaker base than the alcohol. The base must be an ineffective competitor in the equilibrium of Figure 17.9.

Oxonium ions are much better proton donors than are the alcohols themselves. Table 17.2 gives the pK_a values for a few protonated alcohols. Oxonium ions formed from alcohols have pK_a values in the –2 range, and those from ethers are even more acidic.

TABLE 17.2 Some pK_a Values for Simple Oxonium Ions

Compound	pK_a
$H_3\overset{+}{O}$	–1.74
$CH_3\overset{+}{O}H_2$	–2.2
$CH_3CH_2\overset{+}{O}H_2$	–1.9
$CH_3CH_2CH_2CH_2\overset{+}{O}H_2$	–2.3
$CH_3CH_2CH\overset{+}{O}H_2CH_3$	–2.2
$(CH_3)_3C\overset{+}{O}H_2$	–2.6
$R\overset{+}{O}HR'$	about –3.5

PROBLEM 17.6

You have encountered oxonium ions before. Write the mechanism for the solvolysis reaction of *tert*-butyl chloride in water and find the oxonium ion.

Brønsted acidity, of course, is centered upon the hydroxyl hydrogen. Loss of this hydrogen to a Brønsted base acceptor (it is not just ionized as H^+) leads to the **alkoxide ion** (RO^-) (Fig. 17.10).

Brønsted base

Alkoxide ion

FIGURE 17.10 Loss of a proton to some general Brønsted base leads to an alkoxide ion.

Alcohols are approximately as acidic as water. Table 17.3 gives the pK_a values for some simple alcohols and water.

TABLE 17.3 Some pK_a Values for Simple Alcohols

Compound		pK_a (loss of underlined H)
Water	H_2O	15.7
Methyl alcohol	$CH_3O\underline{H}$	15.2
Ethyl alcohol	$CH_3CH_2O\underline{H}$	15.9
Isopropyl alcohol	$(CH_3)_2CHO\underline{H}$	16.5
tert-Butyl alcohol	$(CH_3)_3O\underline{H}$	16.6

Note that the acidity order in aqueous solution is $CH_3OH > CH_3CH_2OH > (CH_3)_2CHOH > (CH_3)_3COH$. Apparently, the more alkyl groups on the alcohol, the weaker an acid it is. For many years, this was explained through an inductive effect. Alkyl groups were assumed to be intrinsically electron donating. If this were true, the product alkoxides would be destabilized by alkyl groups, and the acidity of the corresponding alcohol would be reduced (Fig. 17.11).

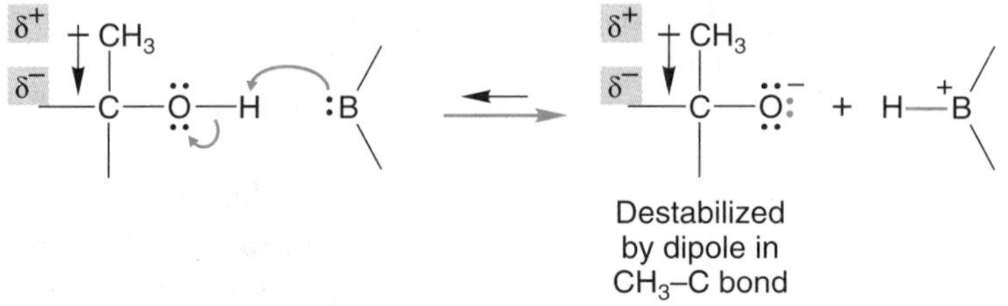

Destabilized by dipole in CH_3–C bond

FIGURE 17.11 If alkyl groups were electron releasing, formation of an alkoxide from an alcohol would be retarded by alkyl groups.

Inductive effects *are* important. For example, the pK_a of 2,2,2-trifluoroethanol ($CF_3CH_2O\underline{H}$) is 12.8, whereas the pK_a of ethyl alcohol is 15.9. The fluorinated alcohol is more than a thousand times as acidic as ethyl alcohol (remember the pK_a scale is logarithmic). Fluorinated alcohols are always stronger acids than their hydrogen-substituted counterparts.

*PROBLEM 17.7 Explain in detail why 2,2,2-trifluoroethanol is a stronger acid than ethyl alcohol.

ANSWER As fluorine is very electronegative, the fluorines are strongly electron withdrawing and will help stabilize the alkoxide ion. They will also stabilize the transition state leading to the alkoxide because partial negative charge has begun to develop on the oxygen atom. So the fluorines will have an effect on both the kinetic and thermodynamic properties of the alcohol.

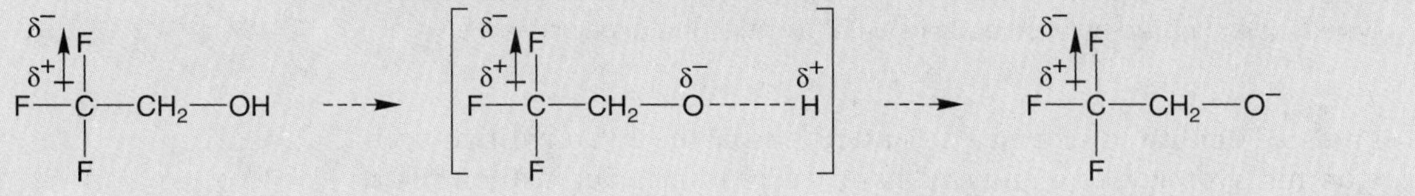

Yet this traditional, and simple, explanation that treats alkyl groups as intrinsically electron donating is not correct. In 1970, Professor John Brauman (b. 1937) and his co-workers at Stanford University showed that in the gas phase the *opposite* acidity order obtained. The intrinsic acidity of the four alcohols of Table 17.3 is exactly opposite to that found in solution. The acidity order measured in solution reflects a powerful effect of the solvent, not the natural acidities of the alcohols themselves. Organic ions are almost all unstable species, and the formation of the alkoxide anions depends critically on how easy it is to stabilize them through solvation. *tert*-Butyl alcohol is a weaker acid in solution than methyl alcohol because the large *tert*-butyl alkoxide ion is difficult to solvate. The more alkyl groups, the more difficult it is for the stabilizing solvent molecules to approach. Of course, in the gas phase where solvation is impossible, the natural acidity order is observed (Fig. 17.12).

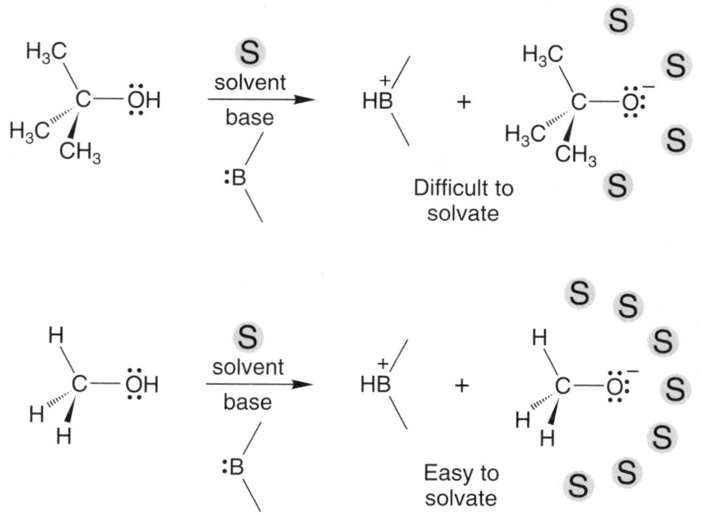

FIGURE 17.12 The smaller the alkoxide ion, the easier it is for solvent molecules to approach and stabilize it.

In the gas phase, the alkyl groups are actually operating so as to *stabilize* the charged alkoxide ions, presumably by withdrawing electrons, exactly in the opposite sense to what had been thought (Fig. 17.13)!

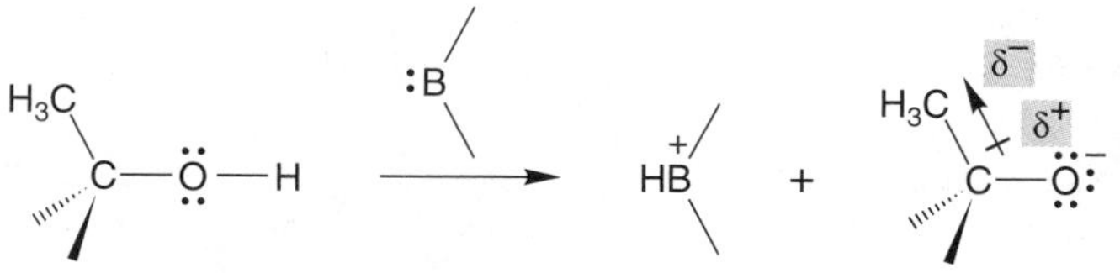

FIGURE 17.13 Alkyl groups can be electron withdrawing.

*PROBLEM 17.8

Can you describe this phenomenon—the stabilization of an adjacent pair of electrons by an adjacent alkyl group—in orbital terms? No orbital construction or complicated argument is necessary. A simple statement is all that is needed.

ANSWER

Alkyl groups have both filled and empty molecular orbitals (see the problems at the end of Chapter 2 for several examples). An adjacent pair of electrons can be stabilized through overlap with the LUMO of the alkyl group. (Similarly, an alkyl group stabilizes an adjacent empty orbital through overlap with the alkyl HOMO.)

What lessons are to be drawn from this? First, solvation is important and not completely understood. We are certain to find other phenomena best explained in terms of solvation in the future. Second, the gas phase is the ideal medium for revealing the intrinsic properties of molecules.

(Some would go further and say that calculation is the best way.) However, the practical world of solvated reactions is certainly real, and for us to understand reactivity we can no longer afford to ignore the solvent.

17.6 REACTIONS OF ALCOHOLS

Although much of this section is review, we will meet a few new reactions, and the old ones will appear in a new light. Here we focus on the alcohols themselves and their roles. Although we saw them previously, it was as participants in reactions, and it was the reactions on which we were focusing. First, let's see a new and simple transformation of the OH group.

17.6a Alkoxide Formation

When an alcohol is treated with a much stronger base, it can be converted entirely into the alkoxide. Note that the base used must be much stronger than the alkoxide or an equilibrium mixture of the base and the alkoxide will result. A favorite reagent is sodium hydride (Fig. 17.14). Sodium hydride has the advantage of producing hydrogen gas, which can easily be removed from the reaction mixture, thus driving the equilibrium toward the alkoxide.

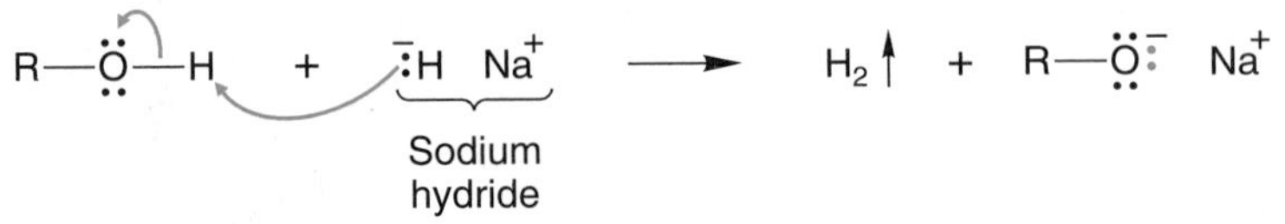

$$R{-}\ddot{O}{-}H \;+\; \underbrace{:\!H^- \; Na^+}_{\text{Sodium hydride}} \longrightarrow H_2\uparrow \;+\; R{-}\ddot{O}{:}^- \; Na^+$$

FIGURE 17.14 Sodium hydride irreversibly removes a proton to give the alkoxide, the conjugate base of the alcohol.

*PROBLEM 17.9 Is an equivalent amount of sodium hydroxide an effective reagent for forming sodium ethoxide from ethyl alcohol (Fig. 17.15)? Explain your answer using Table 17.3.

FIGURE 17.15

$$CH_3CH_2{-}\ddot{O}{-}H \;+\; Na^+ \; {}^-{:}\ddot{O}H \rightleftharpoons CH_3CH_2{-}\ddot{O}{:}^- \; Na^+ \;+\; H_2\ddot{O}{:}$$

ANSWER The pK_a values for water (15.7) and ethyl alcohol (15.9) are very close, and the equilibrium concentration of sodium ethoxide will be slightly less than that of hydroxide ion. Hydroxide ion is a slightly weaker base than ethoxide in solution. Thus, hydroxide will not be effective in making large amounts of ethoxide. At equilibrium both species will be present.

$$\underset{pK_a = 15.9}{CH_3CH_2{-}O{-}H} \;+\; {}^-OH \rightleftharpoons CH_3CH_2{-}O^- \;+\; \underset{pK_a = 15.7}{H{-}O{-}H}$$

Just as water reacts with metallic sodium or potassium to give the corresponding metal hydroxide, alkoxides can be made through a reaction in which sodium or potassium metal is dissolved directly in the alcohol. Hydrogen is liberated in what can be a most vigorous reaction indeed. Potassium is particularly active in all cases, and the reaction of sodium metal with smaller alcohols such as methyl and ethyl alcohol is also rapid and exothermic (Fig. 17.16).

$$2\,CH_3-\ddot{O}-H + 2\,Na\cdot \longrightarrow H_2\uparrow + 2\,[Na^+ \;^-\!:\ddot{O}-CH_3]$$

FIGURE 17.16 Alkoxides can be made through the reaction of metallic sodium or potassium with alcohols.

Alkoxides are named after the starting alcohol as metal alkoxides (Fig. 17.17).

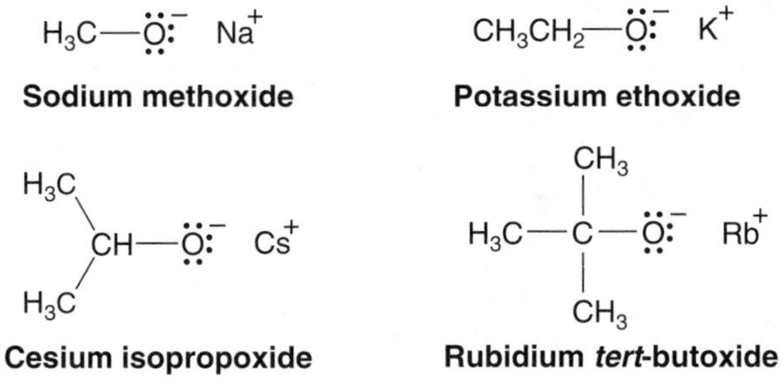

$H_3C-\ddot{O}:^-\ Na^+$ **Sodium methoxide**

$CH_3CH_2-\ddot{O}:^-\ K^+$ **Potassium ethoxide**

$(H_3C)_2CH-\ddot{O}:^-\ Cs^+$ **Cesium isopropoxide**

$(CH_3)_2(H_3C)C-\ddot{O}:^-\ Rb^+$ **Rubidium *tert*-butoxide**

FIGURE 17.17 Alkoxides are named as metal alkoxides.

17.6b Substitution Reactions

Alcohols are not generally active in the nucleophile-induced S_N2 displacement reaction (Chapter 7, p. 259). Hydroxide is a poor leaving group and the treatment of an alcohol with a nucleophile ($^-$:Nu) is likely to result in alkoxide formation (the nucleophile acts as a Brønsted base, removing the hydrogen) rather than displacement (Fig. 17.18).

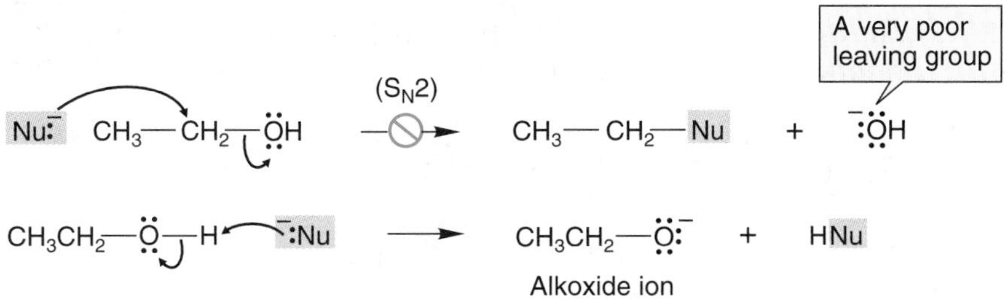

$$Nu:^- \ CH_3-CH_2-\ddot{O}H \xrightarrow{(S_N2)\ \oslash} CH_3-CH_2-Nu + \;^-\!:\ddot{O}H$$

A very poor leaving group

$$CH_3CH_2-\ddot{O}-H \;\; ^-\!:Nu \longrightarrow CH_3CH_2-\ddot{O}:^- + HNu$$

Alkoxide ion

FIGURE 17.18 Removal of a proton to give an alkoxide is a much easier reaction than displacement of the poor leaving group, hydroxide.

Reactions in acid are another matter, and there are many varieties of acid-catalyzed transformations of alcohols into new materials (Chapter 7, p. 260). A typical example is the formation of alkyl halides by treatment of an alcohol with the acid HX. Two different mechanisms are possible, depending on the structure of the R group of the alcohol (Fig. 17.19).

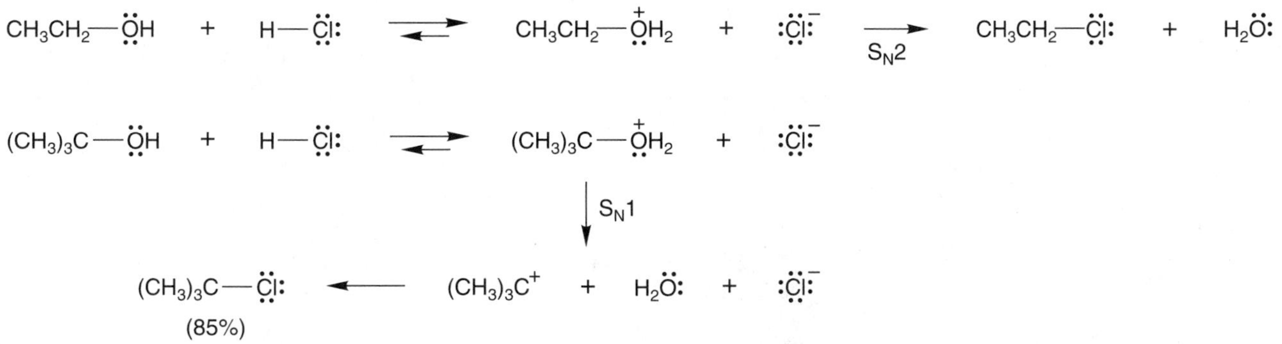

$$CH_3CH_2-\ddot{O}H + H-\ddot{C}l: \rightleftharpoons CH_3CH_2-\overset{+}{\ddot{O}}H_2 + :\ddot{C}l:^- \xrightarrow{S_N2} CH_3CH_2-\ddot{C}l: + H_2\ddot{O}:$$

$$(CH_3)_3C-\ddot{O}H + H-\ddot{C}l: \rightleftharpoons (CH_3)_3C-\overset{+}{\ddot{O}}H_2 + :\ddot{C}l:^-$$

$$\downarrow S_N1$$

$$(CH_3)_3C-\ddot{C}l: \longleftarrow (CH_3)_3C^+ + H_2\ddot{O}: + :\ddot{C}l:^-$$

(85%)

FIGURE 17.19 Two mechanisms for the conversion of an alcohol into a chloride in HCl.

In both cases, the key is the transformation of the leaving group from ^{-}OH, a very poor leaving group, into OH_2, an excellent one. If the R group is tertiary, the reaction proceeds through the cation in S_N1 fashion; if R is primary, water is displaced directly with inversion by the halide in an S_N2 reaction. Secondary R groups can react through both mechanisms (Fig. 17.19).

17.6c Elimination Reactions

Alcohols are also sources of alkenes through the elimination of water. Once more, the base-catalyzed reaction is generally ineffective, as hydroxide is not a good enough leaving group. Alkoxide formation is the likely result, as we saw in the last section (Fig. 17.20).

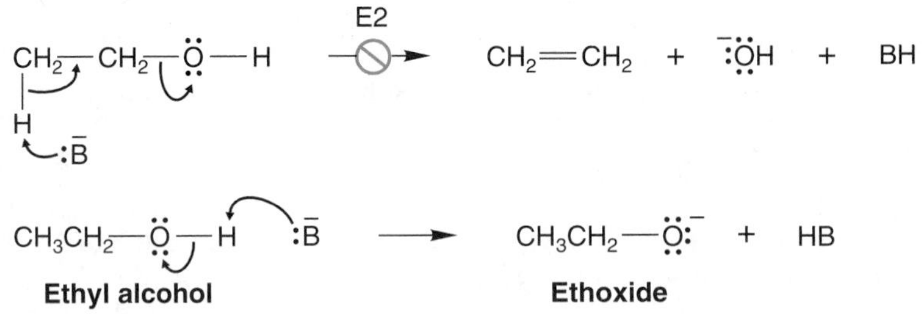

FIGURE 17.20 The base-catalyzed elimination of water from an alcohol is difficult because the leaving group must be hydroxide. Alkoxide formation is much easier.

As before, reaction in acid is more successful. The products of the reactions in Figure 17.19 are accompanied by the products of E1 and E2 reactions. The results of the competition between the E1 with the S_N1 and the E2 with the S_N2 reactions depend on the detailed structure of the alcohol, and the reaction conditions (Fig. 17.21).

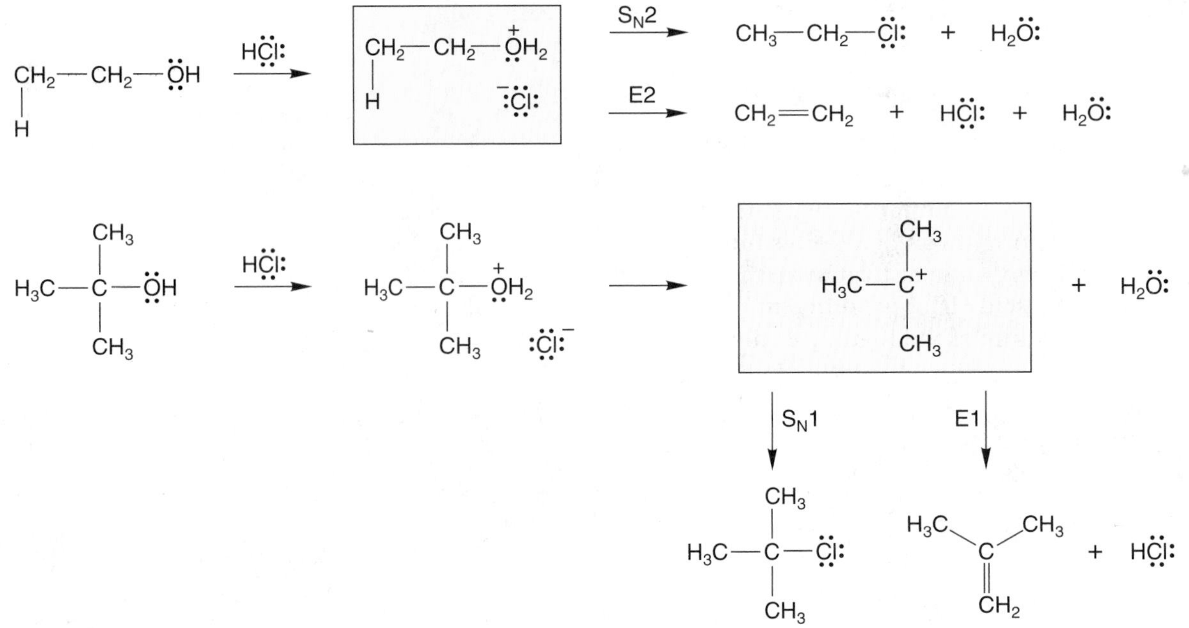

FIGURE 17.21 In the reactions of alcohols with acid, the S_N2 and E2 reactions compete, as do the S_N1 and E1 reactions.

17.6d Rearrangements in Alcohol Reactions

Any reaction in which a carbocation is an intermediate will be likely to involve a rearrangement if a more stable carbocation can be formed by hydride or alkyl shift. Be sure to be alert for rearrangements in the E1 elimination and S_N1 displacement reactions. Figure 17.22 gives just one example (Chapter 9, p. 374).

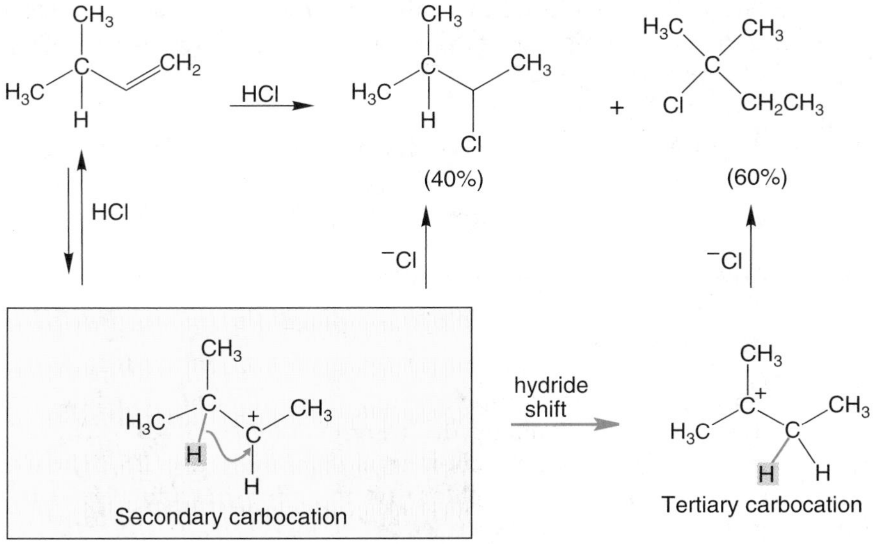

FIGURE 17.22 Rearrangements are common whenever a less stable carbocation can produce a more stable carbocation.

17.6e Ester Formation

In Chapter 7 (p. 260), we mentioned the formation of sulfonate esters on treatment of an alcohol with *p*-toluenesulfonyl chloride (tosyl chloride). At the time we weren't really ready for a mechanistic treatment, but now after the discussion in Chapter 16 of additions to carbon–oxygen and other double bonds, we are better equipped. The first step is the addition to the sulfur–oxygen double bond by the alcohol. This step is followed by proton loss and re-formation of the sulfur–oxygen double bond through loss of chloride ion (Fig. 17.23).

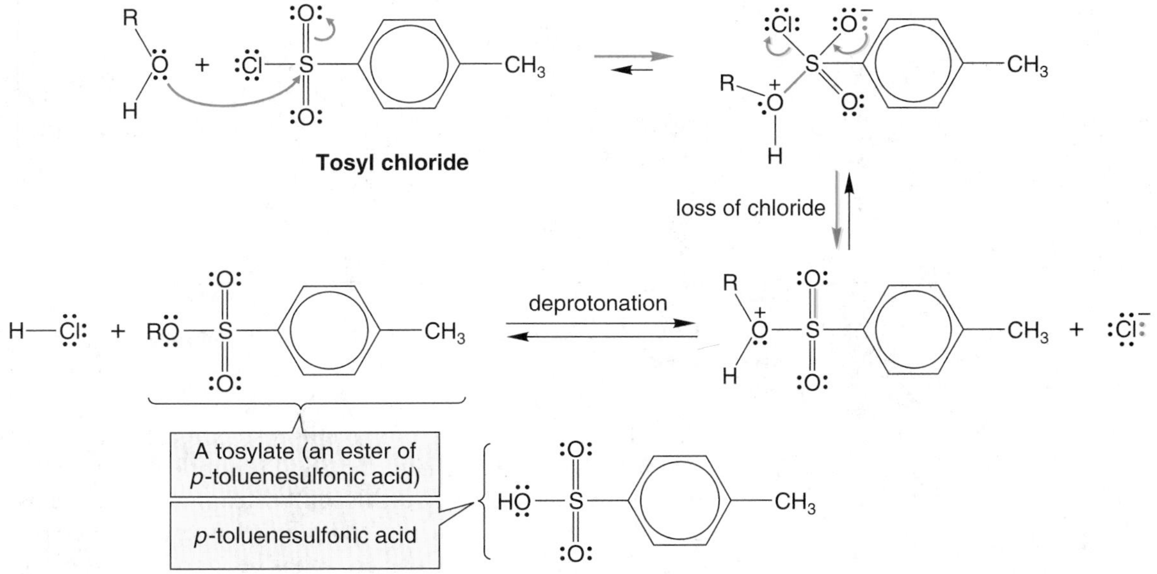

FIGURE 17.23 The mechanism of tosylate formation. A tosylate is an ester of *p*-toluenesulfonic acid.

In the product, the OH of *p*-toluenesulfonic acid has been formally replaced by an OR group from the alcohol. Such compounds are called **esters**. An ester is an acid in which one or more OH groups have been transformed into OR groups.

Sulfonate esters make excellent substrates for the S_N1 and S_N2 reactions, and their use complements the reactions of alcohols with HX described in Sections 17.6b and 17.6c (Fig. 17.24).

FIGURE 17.24 Treatment of tosylates in polar solvents leads to substituted alkanes and *p*-toluenesulfonic acid, HOTs.

$$CH_3OTs \xrightarrow{\text{EtOH, 75 °C}} CH_3OCH_2CH_3 + HOTs$$

$$CH_3CH_2OTs \xrightarrow{\text{EtOH, 75 °C}} CH_3CH_2OCH_2CH_3 + HOTs$$

$$(CH_3)_2CHOTs \xrightarrow{\text{EtOH, 75 °C}} (CH_3)_2CHOCH_2CH_3 + HOTs$$

17.6f Halide Formation

A variation on the sulfonate ester-forming reaction of Section 17.6e involves the treatment of alcohols with **thionyl chloride ($SOCl_2$)**. The end result is the conversion of an alcohol into a chloride (Fig. 17.25). The initial product in this reaction is a chlorosulfite ester formed in exactly the same way as are the tosylates described in Section 17.6e.

The general case

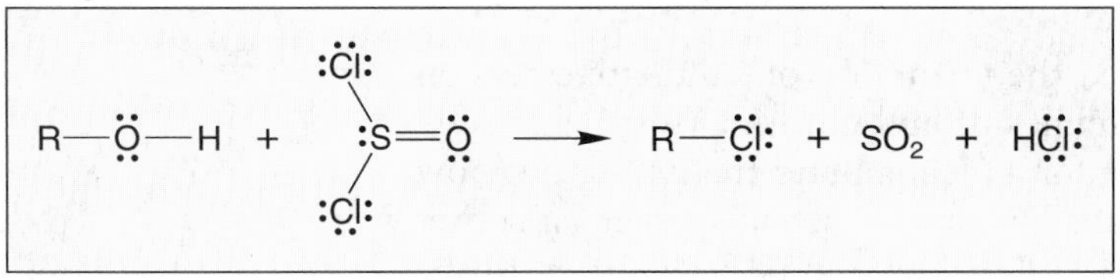

$$R{-}\ddot{O}{-}H + Cl_2S{=}O \longrightarrow R{-}Cl + SO_2 + HCl$$

Specific examples

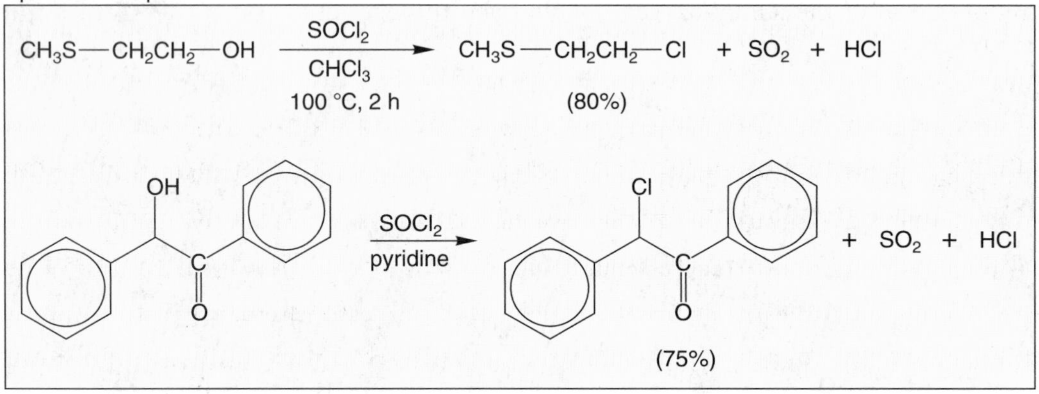

$$CH_3S{-}CH_2CH_2{-}OH \xrightarrow[\text{CHCl}_3,\ 100\ °C,\ 2\ h]{\text{SOCl}_2} CH_3S{-}CH_2CH_2{-}Cl + SO_2 + HCl \quad (80\%)$$

$$PhCH(OH)C(=O)Ph \xrightarrow[\text{pyridine}]{\text{SOCl}_2} PhCH(Cl)C(=O)Ph + SO_2 + HCl \quad (75\%)$$

FIGURE 17.25 Treatment of an alcohol with thionyl chloride results in the formation of an alkyl chloride.

Mechanisms of breakdown of the chlorosulfite ester involve both displacement by chloride ion to give the alkyl chloride, sulfur dioxide (SO_2), and a chloride ion, as well as direct decomposition to SO_2 and a pair of ions that recombine. In either case, the end product is the chloride (Fig. 17.26). Once again, the strategy in this reaction has been to convert a poor leaving group (^-OH) into a good one (SO_2 and Cl^-).

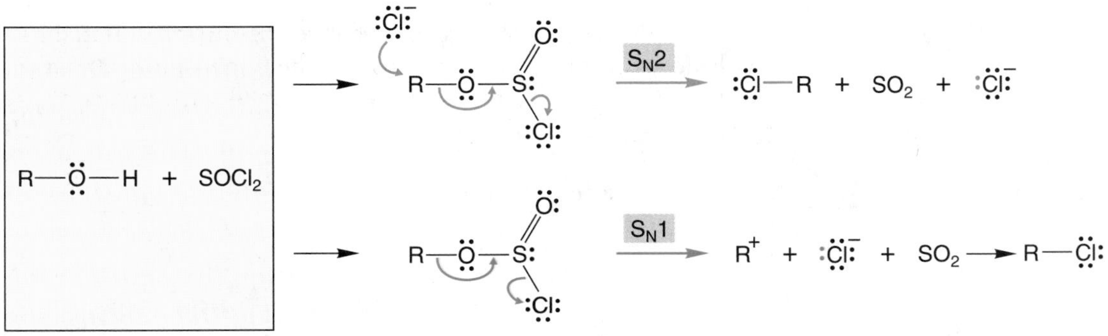

$$R{-}\ddot{O}{-}H + SOCl_2 \longrightarrow R{-}O{-}S(=O)Cl \xrightarrow{S_N2} :\ddot{Cl}{-}R + SO_2 + :\ddot{Cl}:^-$$

$$R{-}\ddot{O}{-}H + SOCl_2 \longrightarrow R{-}O{-}S(=O)Cl \xrightarrow{S_N1} R^+ + :\ddot{Cl}:^- + SO_2 \longrightarrow R{-}\ddot{Cl}:$$

FIGURE 17.26 Mechanisms of chloride formation from reaction of an alcohol with thionyl chloride.

PROBLEM 17.10

Why are sulfonate esters such good participants in the S_N1 and S_N2 reactions?

PROBLEM 17.11

Predict any side products in the three reactions of Figure 17.24 and in the reaction of *tert*-butyl chloride with ethyl alcohol.

PROBLEM 17.12

Treatment of an alkoxide with dimethyl sulfate leads to methyl ethers (Fig. 17.27). Write a mechanism for this reaction.

$$R{-}\ddot{O}:^- + CH_3{-}\ddot{O}{-}S(=O)_2{-}\ddot{O}{-}CH_3 \longrightarrow R{-}\ddot{O}{-}CH_3 + {}^-:\ddot{O}{-}SO_2OCH_3$$

Dimethyl sulfate

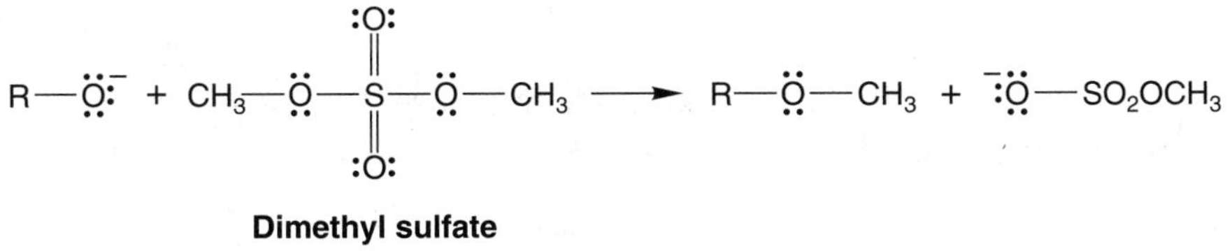

FIGURE 17.27

PROBLEM 17.13

Write a mechanism for the formation of the chlorosulfite ester from an alcohol and thionyl chloride. *Hint*: The sulfur–oxygen double bond undergoes addition reactions much like those of a carbon–oxygen double bond (Fig. 17.23).

Several related reactions involve the use of phosphorus reagents. Alkyl halides can be made from the treatment of an alcohol with phosphorus halides, PX_3 or PX_5, as shown in Figure 17.28. Many of these reagents cause rearrangement, especially in secondary systems, and the stereochemical outcome, retention or inversion, depends on solvent and other reaction conditions. Accordingly, modifications have been worked out in recent years in order to avoid such problems.

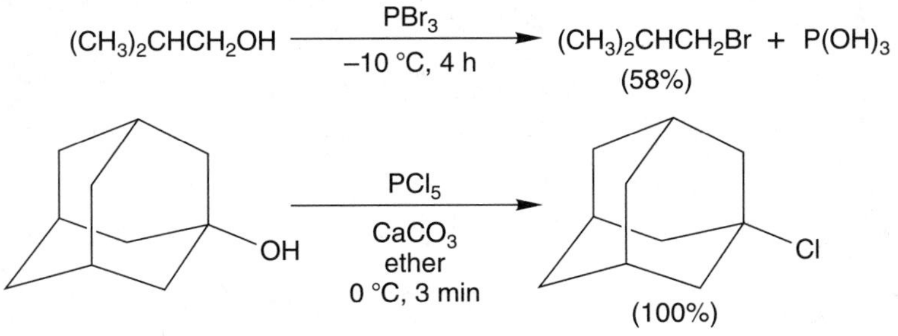

$$(CH_3)_2CHCH_2OH \xrightarrow[-10\ °C,\ 4\ h]{PBr_3} (CH_3)_2CHCH_2Br + P(OH)_3 \quad (58\%)$$

$$\text{1-adamantanol (OH)} \xrightarrow[\text{CaCO}_3,\ \text{ether},\ 0\ °C,\ 3\ min]{PCl_5} \text{1-chloroadamantane (Cl)} \quad (100\%)$$

FIGURE 17.28 Alkyl halides can be formed from the reaction of alcohols with phosphorus halides.

For example, alcohols can be converted into the related bromides and chlorides through treatment with triphenylphosphine and a carbon tetrahalide. The alkyl halide is generally formed with inversion, and a rough mechanism is sketched out in Figure 17.29.

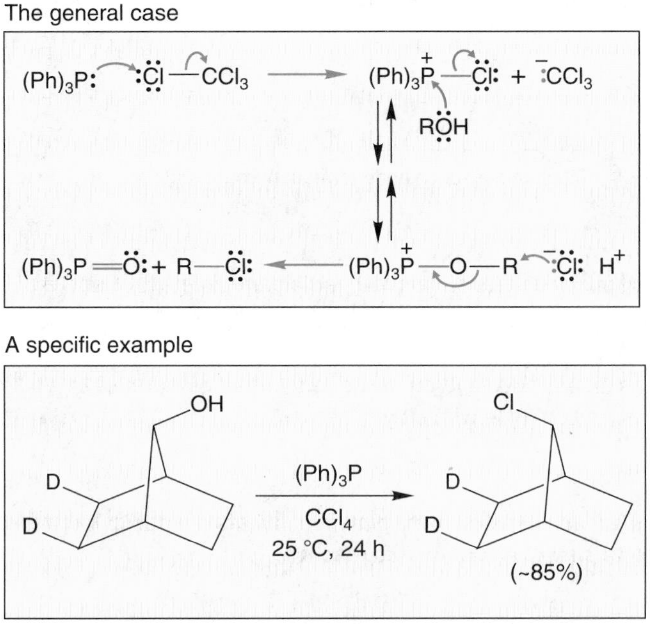

The general case

$$(Ph)_3P: + :\ddot{Cl}{-}CCl_3 \longrightarrow (Ph)_3\overset{+}{P}{-}\ddot{Cl}: + \overset{-}{:}CCl_3$$

$$\text{ROH} \rightleftharpoons (Ph)_3\overset{+}{P}{-}O{-}R \;\; \overset{-}{:}\ddot{Cl}: \;\; H^+ \longrightarrow (Ph)_3P{=}\ddot{O}: + R{-}\ddot{Cl}:$$

A specific example

(Ph)$_3$P, CCl$_4$, 25 °C, 24 h (~85%)

FIGURE 17.29 Halides are formed from the reaction of alcohols with the intermediate produced from triphenylphosphine and a carbon tetrahalide.

PROBLEM 17.14 Write a mechanism for the formation of alkyl bromides from the reaction of alcohols and PBr_3.

17.6g Oxidation of Alcohols

Reaction of primary alcohols with CrO_3/pyridine or allylic alcohols with MnO_2 results in the formation of an aldehyde. Organic acids are formed if water is present, or if $KMnO_4$ or Na_2CrO_4 in acid is used as the oxidizing agent (Fig. 17.30).

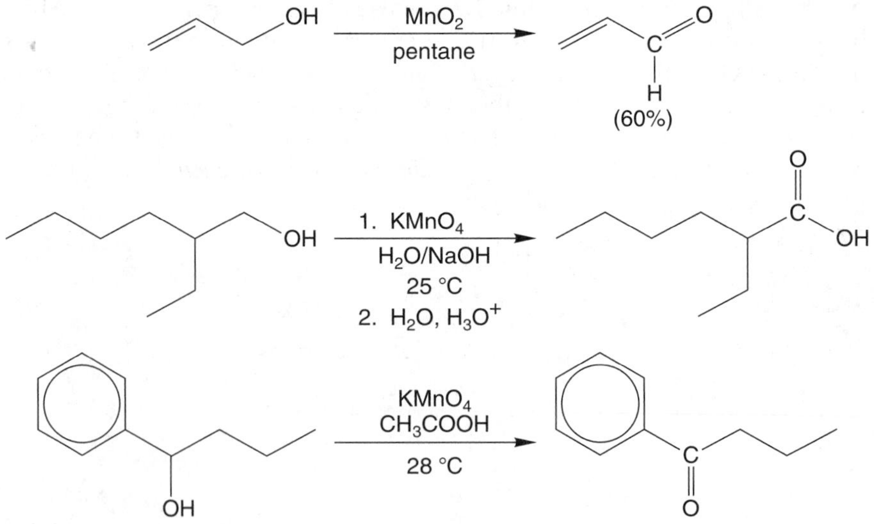

MnO_2, pentane (60%)

1. $KMnO_4$, H_2O/NaOH, 25 °C; 2. H_2O, H_3O^+

$KMnO_4$, CH_3COOH, 28 °C

FIGURE 17.30 Primary alcohols can be oxidized to either aldehydes or acids. Secondary alcohols are sources of ketones.

Secondary alcohols are oxidized in a similar fashion by many reagents to give ketones, but tertiary alcohols are unreactive. The mechanism involves formation of a chromate ester followed by an elimination reaction (Chapter 16, p. 795).

17.7 SYNTHESES OF ALCOHOLS

Now we pass from alcohols as participants in reactions to alcohols as targets of reactions. Here we see old and new ways to make these compounds. Once again, we see old material in new ways. Earlier we were looking primarily at reactions, studying their mechanisms and learning to generalize what we discovered about a particular process. Here we focus on many of the same reactions in a utilitarian way: How can we use them to make alcohols?

17.7a From Alkenes*

The elements of water can sometimes be added to alkenes, but side reactions often dominate (Fig. 17.31).

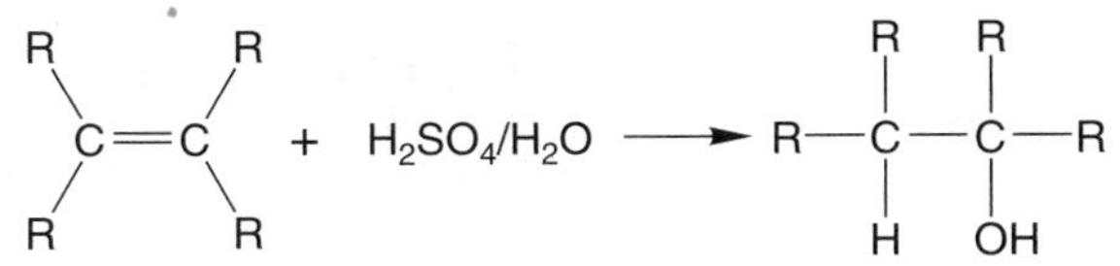

$$R_2C{=}CR_2 + H_2SO_4/H_2O \longrightarrow R{-}CR(H){-}CR(OH){-}R$$

FIGURE 17.31 The hydration of an alkene in acid.

More effective ways of synthesizing alcohols are the complementary processes of hydroboration–oxidation (Chapter 9, p. 363) and oxymercuration–reduction (Chapter 10, p. 402). Oxymercuration–reduction leads to Markovnikov addition (formation of the more substituted alcohol) but hydroboration–oxidation gives anti-Markovnikov addition (formation of the less substituted alcohol) (Fig. 17.32).

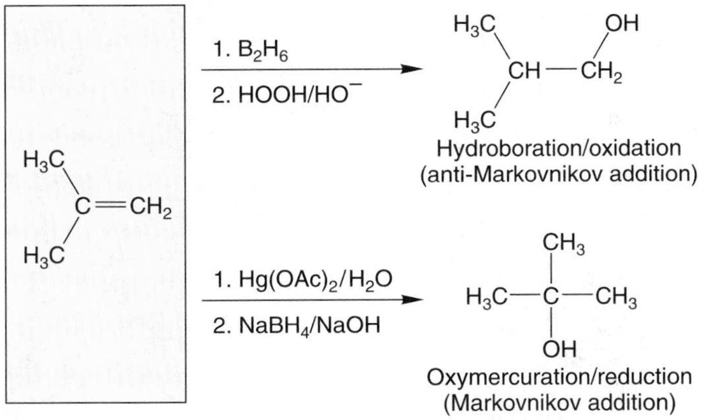

$(H_3C)_2C{=}CH_2$ $\xrightarrow[\text{2. HOOH/HO}^-]{\text{1. } B_2H_6}$ $(H_3C)_2CH{-}CH_2OH$

Hydroboration/oxidation (anti-Markovnikov addition)

$(H_3C)_2C{=}CH_2$ $\xrightarrow[\text{2. NaBH}_4\text{/NaOH}]{\text{1. Hg(OAc)}_2/H_2O}$ $H_3C{-}C(CH_3)(OH){-}CH_3$

Oxymercuration/reduction (Markovnikov addition)

FIGURE 17.32 Oxymercuration–reduction and hydroboration–oxidation.

PROBLEM 17.15

Explain mechanistically why these two reactions, hydroboration–oxidation and oxymercuration–reduction, lead to the opposite regiochemistries.

*See Chapter 9 (p. 358).

17.7b From Alkyl Halides

Displacement of primary or secondary halides by hydroxide leads to alcohols, although elimination reactions complicate the process (Chapter 7, p. 277). Most halides are formed from alcohols, so this reaction is of limited practical use (Fig. 17.33).

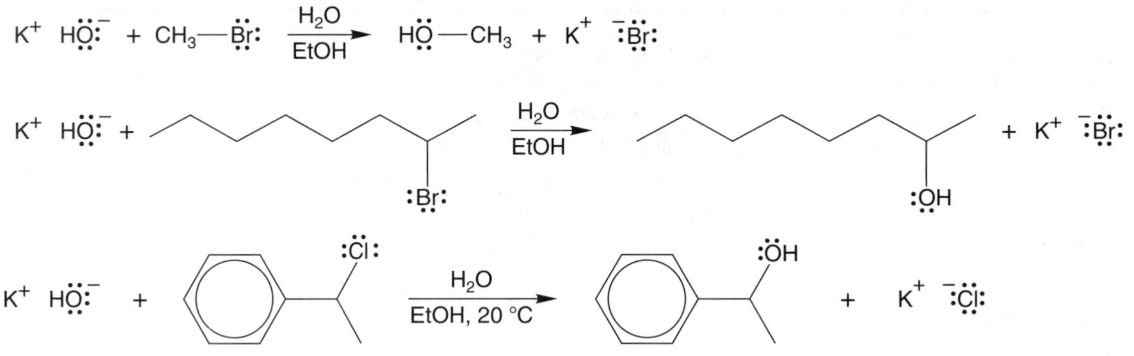

K^+ HO:⁻ + CH_3—Br: →(H_2O, EtOH) HO—CH_3 + K^+ ⁻:Br:

K^+ HO:⁻ + (secondary alkyl bromide) →(H_2O, EtOH) (secondary alcohol) + K^+ ⁻:Br:

K^+ HO:⁻ + (1-chloro-2-methylpropylbenzene) →(H_2O, EtOH, 20 °C) (alcohol) + K^+ ⁻:Cl:

FIGURE 17.33 The S_N2 displacement of bromides by hydroxide. Except for the methyl system, E2 elimination is a complicating side reaction.

PROBLEM 17.16 Predict the side products and stereochemical results for the reactions of Figure 17.33.

PROBLEM 17.17 How would you make ethyl iodide from ethyl alcohol? Show at least two ways.

17.7c From Carbonyl Compounds

Reduction of aldehydes or ketones by metal hydrides leads to primary and secondary alcohols (see Chapter 16, p. 793). Catalytic hydrogenation or dissolving metal reductions (for example, sodium in alcohol) can also be used in some cases (Fig. 17.34).

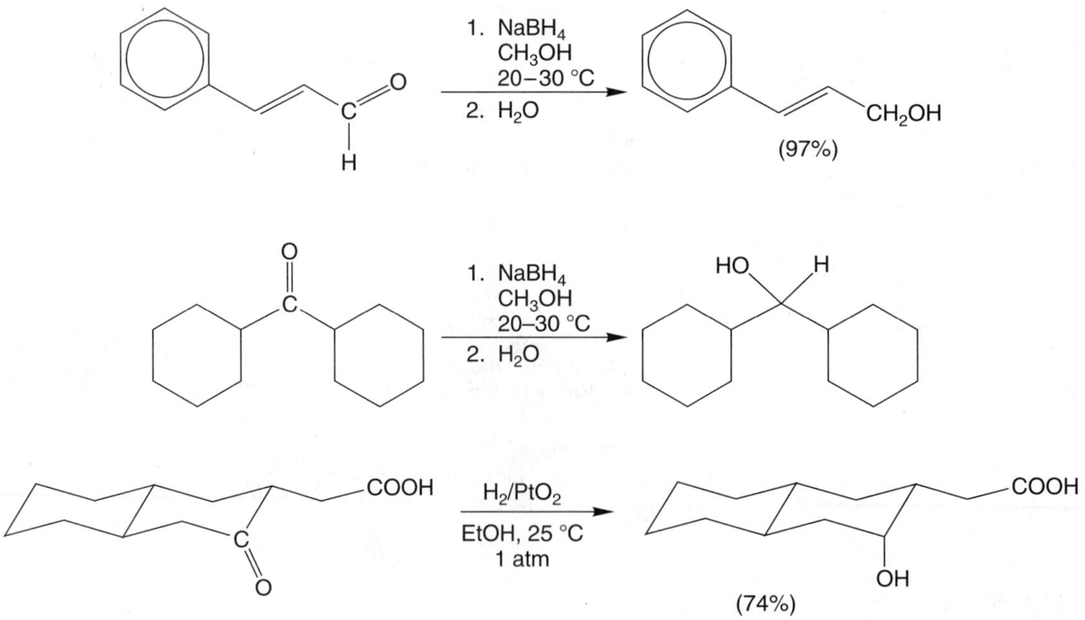

Cinnamaldehyde →(1. $NaBH_4$, CH_3OH, 20–30 °C; 2. H_2O) cinnamyl alcohol, CH_2OH (97%)

Dicyclohexyl ketone →(1. $NaBH_4$, CH_3OH, 20–30 °C; 2. H_2O) dicyclohexylmethanol (HO, H)

Keto acid (COOH) →(H_2/PtO_2, EtOH, 25 °C, 1 atm) hydroxy acid (OH, COOH) (74%)

FIGURE 17.34 A variety of reductions of carbonyl compounds to alcohols.

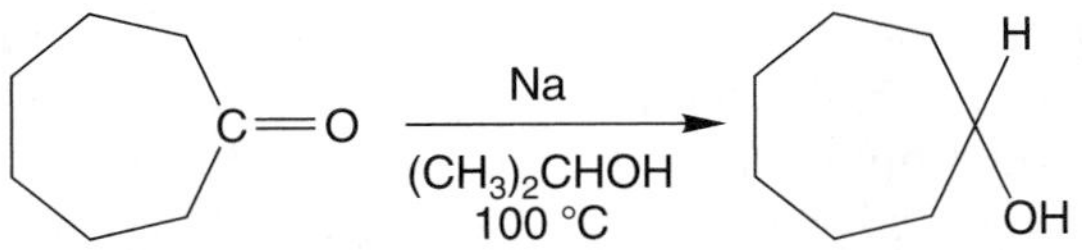

Na; $(CH_3)_2CHOH$; 100 °C

FIGURE 17.34 (CONTINUED)

More complex alcohols can be made through the related additions of organometallic reagents to aldehydes and ketones (Chapter 16, p. 791). Primary, secondary, and tertiary alcohols can all be made through a judicious use of this reaction sequence (Fig. 17.35).

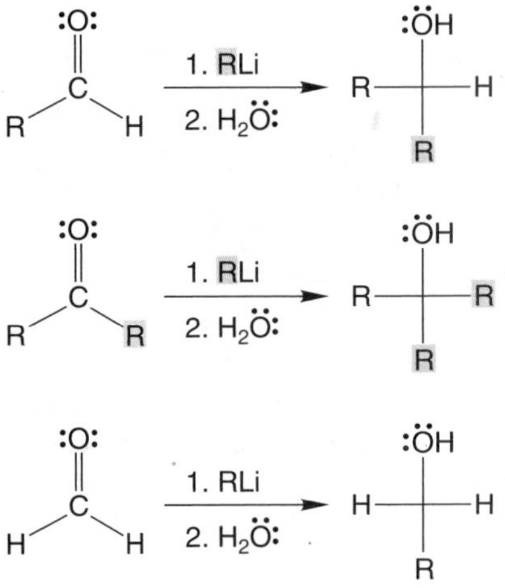

FIGURE 17.35 Reaction of carbonyl compounds with organometallic reagents leads to alcohols. Alkyllithium reagents are shown, but Grignards will do as well.

17.7d Special Reactions

Phenol can be made through the treatment of benzene diazonium chloride with water (Chapter 14, p. 647), and methyl alcohol is formed directly from a mixture of hydrogen and carbon monoxide passed over a complex metallic catalyst at high temperature and pressure (Fig. 17.36).

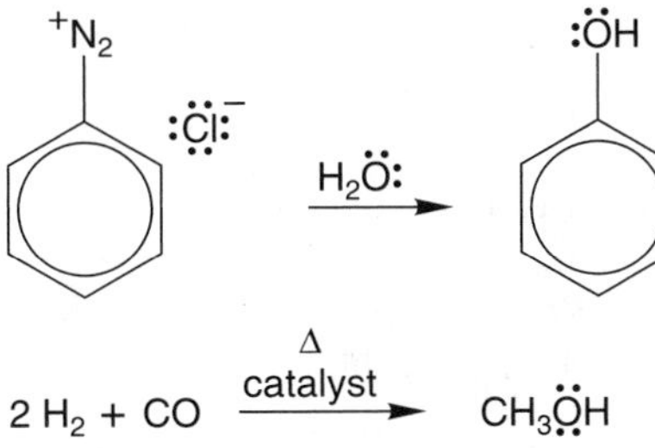

$$2\,H_2 + CO \xrightarrow[\text{catalyst}]{\Delta} CH_3OH$$

FIGURE 17.36 Special syntheses of methyl alcohol and phenol.

17.8 DIALCOHOLS: DIOLS OR GLYCOLS

Now we encounter molecules containing more than one OH group. Although we have met some of these molecules briefly before, it is now time to focus directly upon them.

17.8a Nomenclature

Although molecules containing two OH groups are logically enough called **diols**, there is an important common name as well, **glycol**. 1,1-Diols, or 1,1-glycols, are hydrates of aldehydes and ketones, and, as we have seen (Chapter 16, p. 768), are not often stable relative to their carbonyl-containing precursors and water (Fig. 17.37).

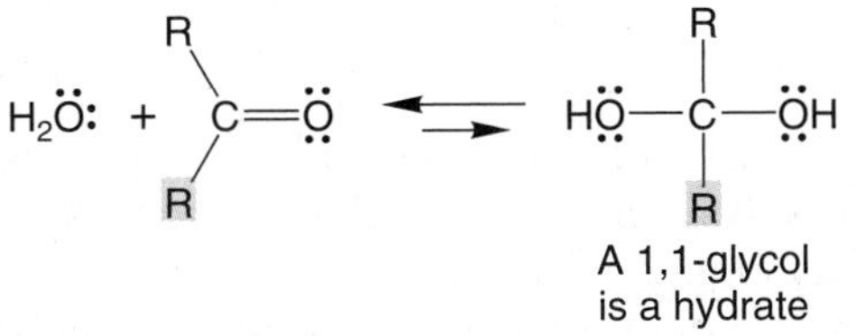

A 1,1-glycol is a hydrate

FIGURE 17.37 The hydration of carbonyl compounds gives 1,1-glycols called hydrates.

Some 1,2-diols or 1,2-glycols are familiar and often important molecules. 1,2-Ethanediol, or ethylene glycol, is the commonly used antifreeze, for example. Note that in naming these compounds, the final "e" of the alkane parent compound is not dropped, as it is in naming simple alcohols, and that the almost universally used common name can be very misleading. Ethylene glycol implies the presence of an alkene, and "propylene glycol" (1,2-propanediol) suggests a three-carbon unsaturated chain. In neither case is the implied unsaturation present (Fig. 17.38). Be aware that the real structures are based on *saturated* hydrocarbon chains.

Aromatic glycols are easily named systematically, but here the trivial names are holding on. Figure 17.39 shows the three dihydroxybenzenes and their names.

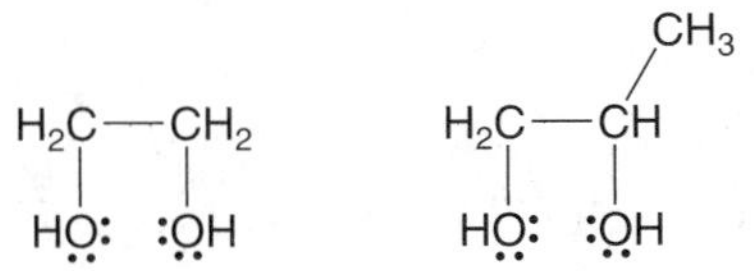

H_2C—CH_2
HO: :OH
CH_3
H_2C—CH
HO: :OH

Ethylene glycol or 1,2-ethanediol

Propylene glycol or 1,2-propanediol

FIGURE 17.38 1,2-Ethanediol and 1,2-propanediol are called ethylene glycol and propylene glycol, respectively.

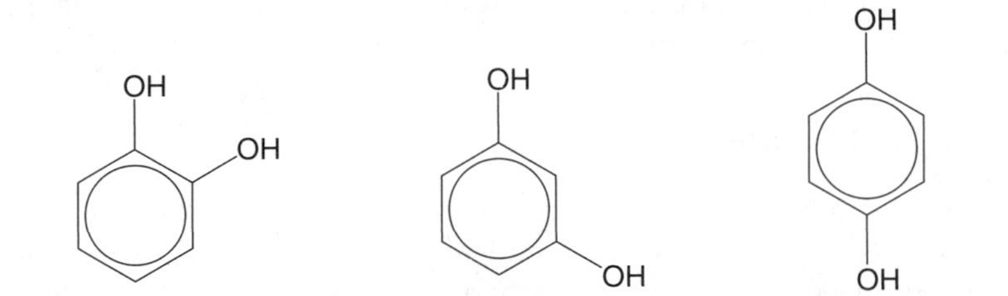

OH
OH
OH
OH
OH
OH

Catechol (1,2-benzenediol or *o*-hydroxyphenol)

Resorcinol (1,3-benzenediol or *m*-hydroxyphenol)

Hydroquinone (1,4-benzenediol or *p*-hydroxyphenol)

FIGURE 17.39 The three dihydroxybenzenes with their (systematic) common names.

17.8b Synthesis of Glycols

There are two important methods of generating 1,2-glycols. We have already studied both of them (Chapter 10, pp. 408, 423). One method involves the reaction of alkenes with osmium tetroxide (OsO_4) or basic potassium permanganate ($KMnO_4$) (Fig. 17.40).

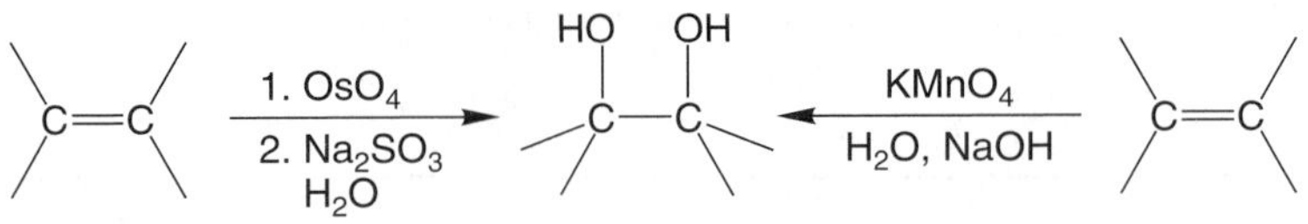

C=C
1. OsO_4
2. Na_2SO_3
H_2O
HO OH
C—C
$KMnO_4$
H_2O, NaOH
C=C

FIGURE 17.40 The cis 1,2-glycols can be made through oxidation with basic permanganate or osmium tetroxide.

The mechanism in each case involves the formation of a cyclic compound and subsequent hydroysis. Be sure you see why this cyclic intermediate demands the observed cis stereochemistry of the diol (Chapter 10, p. 423; Fig. 17.41).

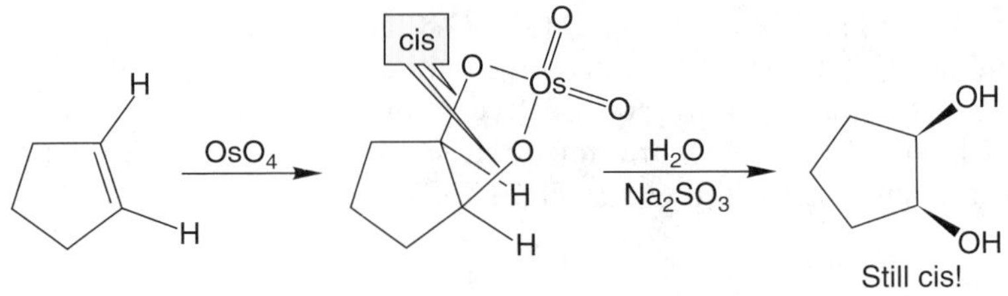

H
H
OsO_4
cis
O
O
Os
O
O
H
H
H_2O
Na_2SO_3
OH
OH
Still cis!

FIGURE 17.41 The presence of a cyclic intermediate ensures the observed cis stereochemistry of the product diol.

The second synthesis involves the **epoxides**, or **oxiranes**, more properly, if rarely, known as **oxacyclopropanes**. The synthetic route to them involves treatment of an alkene with a peracid, typically trifluoroperacetic acid, *m*-chloroperbenzoic acid, or perphthalic acid. This reaction is another process involving syn addition (Chapter 10, p. 405; Fig. 17.42). These oxygen-containing three-membered rings are the smallest examples of cyclic ethers. **Ethers**, as we will soon see, have the general composition ROR and in the epoxides the two R groups are tied into a ring. As the leaving group must be ^{-}OR, ethers are generally not very reactive in S_N2 displacement reactions. In contrast to most ethers, the epoxides are quite re-

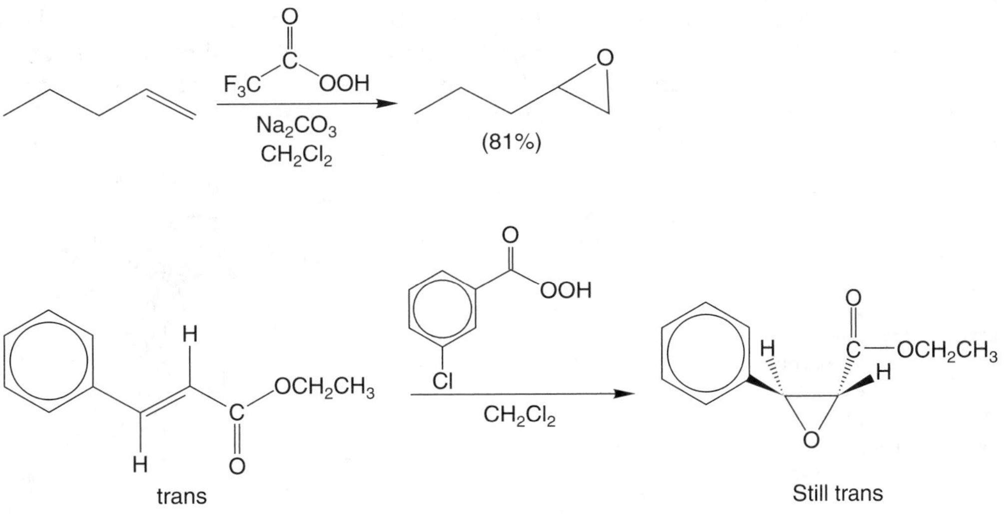

FIGURE 17.42 Epoxidation, the formation of oxiranes, involves a syn addition. The stereochemical relationships present in the alkene are maintained in the epoxide.

active in nucleophilic substitution reactions. In this case, these S_N2 reactions lead to ring-opened products (Fig. 17.43).

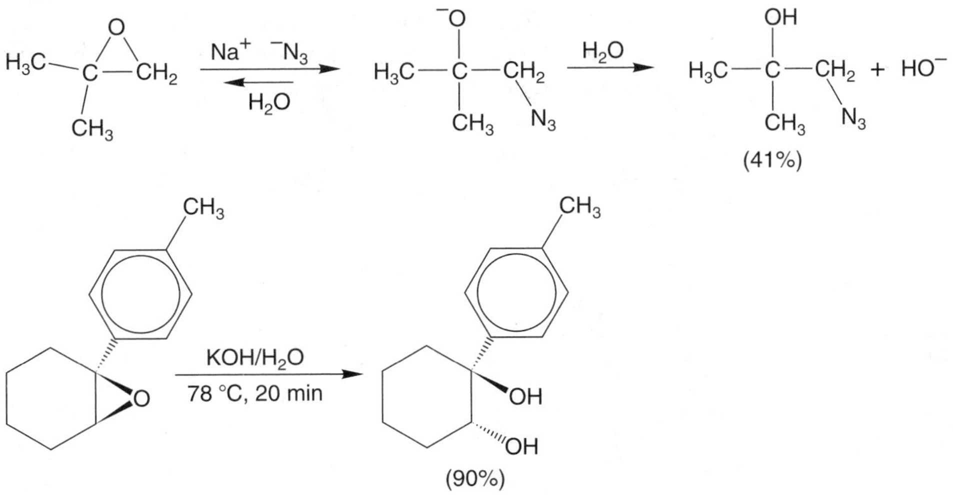

FIGURE 17.43 Two ring-opening substitution reactions of epoxides with nucleophiles.

PROBLEM 17.18

Suggest a reason for the increased reactivity of epoxides over acyclic and larger cyclic ethers.

We will return to this general reaction when epoxides become the main subject of discussion (Section 17.11), but look for a moment at the specific displacement reaction in which hydroxide plays the role of nucleophile (Fig. 17.43). The ring-opened product is a 1,2-glycol, and the reaction of epoxides with hydroxide is the second general synthetic route to such molecules. Note especially that the ring-opening S_N2 reaction must occur with displacement from the rear (Chapter 7, p. 244), and thus the diol is formed with overall trans stereochemistry. Figure 17.44 contrasts the stereochemical results of the two general routes to 1,2-diols. Reaction of an alkene with OsO_4 (or $KMnO_4$) followed by reduction with Na_2SO_3 (or hydrolysis), leads to the cis 1,2-diol, whereas formation of the epoxide, followed by ring opening, must give the trans 1,2-diol.

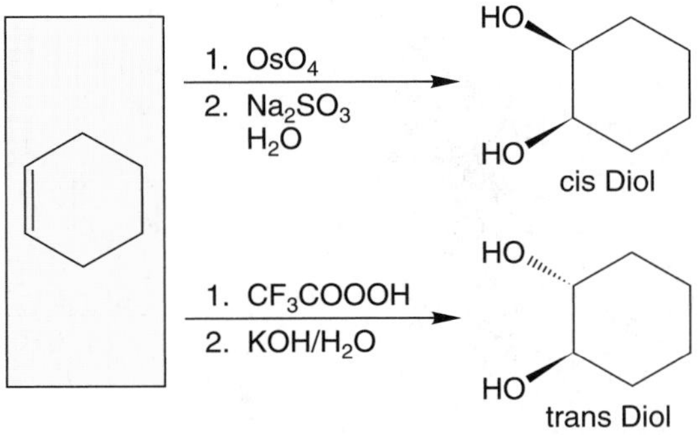

1. OsO_4
2. Na_2SO_3, H_2O

HO, HO — cis Diol

1. CF_3COOOH
2. KOH/H_2O

HO, HO — trans Diol

FIGURE 17.44 The two synthetic routes to 1,2-glycols lead to different stereochemistries. Oxidation by OsO_4 (or $KMnO_4$), followed by hydrolysis, gives the product of syn addition of two OH groups. Formation of the epoxide, followed by ring opening in S_N2 fashion by the nucleophile hydroxide, leads to the trans 1,2-diol through anti addition.

PROBLEM 17.19 Work out the structures of the products of treatment of *cis*- and *trans*-2-butene with aqueous basic potassium permanganate.

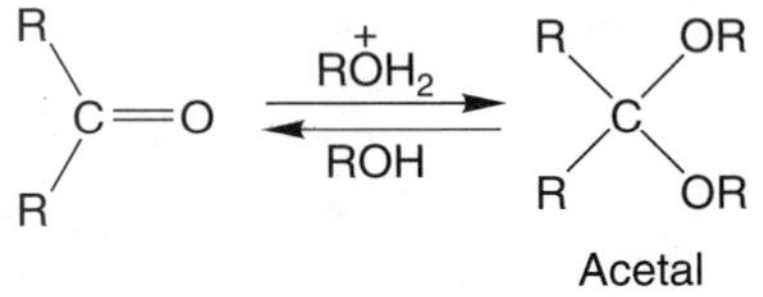

$\overset{+}{R}OH_2$ / ROH — Acetal

FIGURE 17.45 Formation of an acetal (R = aryl, alkyl, or H).

17.8c Reactions of 1,2-Glycols

As glycols are dialcohols, their chemistry generally resembles that of other, simpler alcohols. The presence of the second hydroxyl group adds some complications, however, and we will concentrate on these.

When aldehydes or ketones are treated with an excess of an alcohol and an acid catalyst, the carbonyl group is "protected" as the acetal (Chapter 16, p. 774; Fig. 17.45).

Ethylene glycol can undergo this reaction as well, but both of the alcohol groups are now within the same molecule. The result is a cyclic acetal (Fig 17.46).

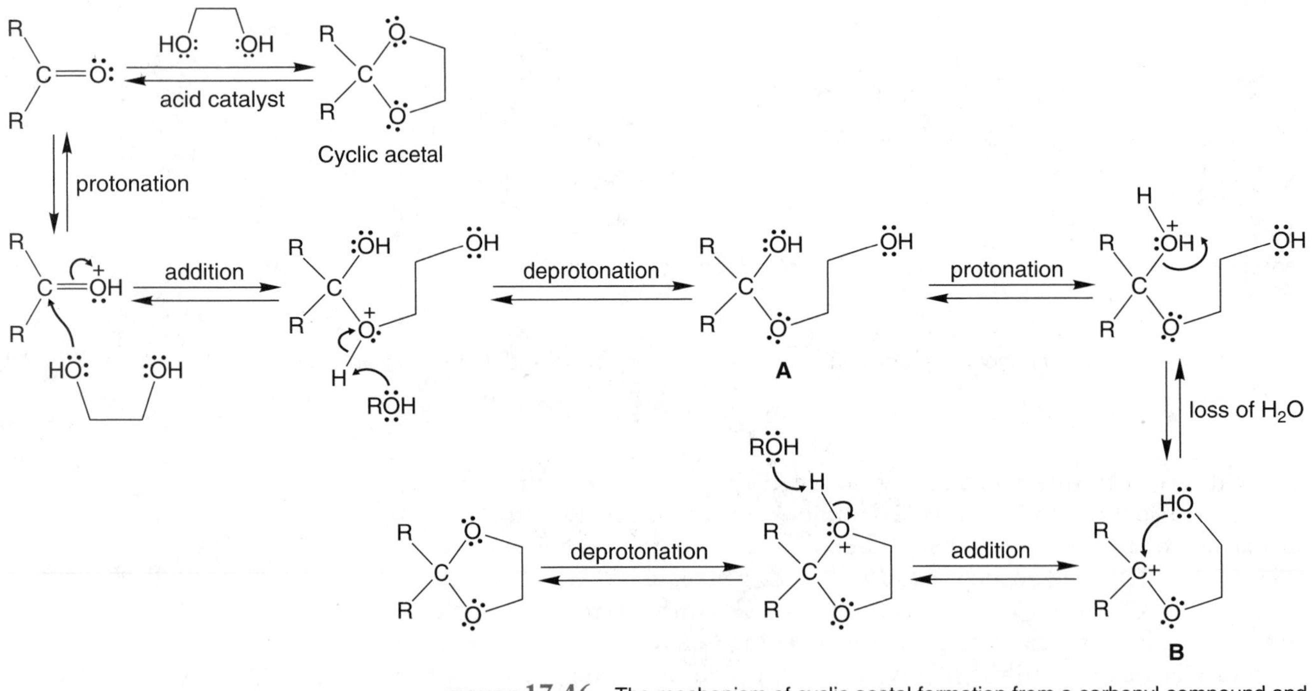

acid catalyst — Cyclic acetal

protonation — addition — deprotonation — A — protonation — loss of H_2O — B — addition — deprotonation

FIGURE 17.46 The mechanism of cyclic acetal formation from a carbonyl compound and ethylene glycol.

Like many reactions involving rings, cyclic acetal formation sometimes appears mysterious. Somehow, incorporating the constituent parts of a familiar intermolecular reaction within a single molecule leads to difficulties. So, let's pay close attention to the reaction. As in the intermolecular case, the first step is protonation of the carbonyl oxygen. Addition of one alcohol follows, and deprotonation gives hemiacetal **A**. In strict analogy to the intermolecular reaction, protonation of one hydroxyl group is followed by water loss to give the resonance-stabilized carbocation **B**. Now comes the difference. This cation is captured, not by a new molecule of alcohol, but by the hydroxyl group within the same molecule. A ring is constructed. Be sure to fix the analogy in your mind by working out the mechanism for the acyclic case.

17.8d Acid-Catalyzed Rearrangement of 1,2-Glycols: The Pinacol Rearrangement

A typical reaction of monoalcohols is loss of water to give an alkene, the E1 reaction (Chapter 7, p. 273; Fig. 17.47).

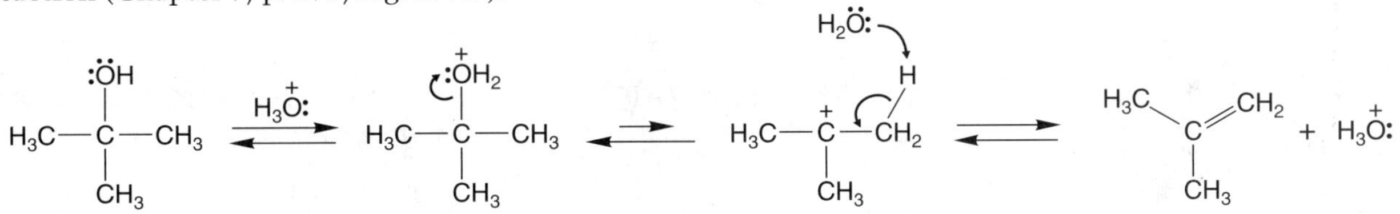

FIGURE 17.47 The E1 elimination of water from *tert*-butyl alcohol

If we attempt to dehydrate a typical 1,2-glycol, we are in for a surprise. Let's examine the possibilities and try to see why the reaction takes an unexpected course. Suppose we treat the glycol made from 2,3-dimethyl-2-butene with H_2SO_4 in an attempt to dehydrate the molecule. What do we expect? After protonation of one of the hydroxyl groups, dehydration will take place by an E1 process (Fig. 17.48).

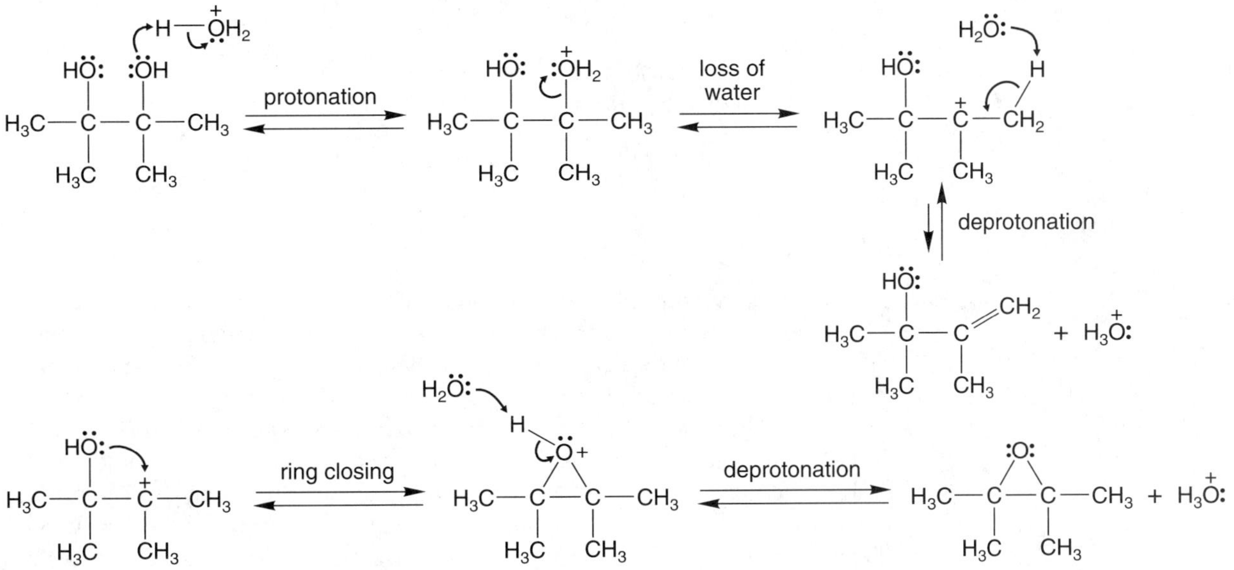

FIGURE 17.48 By analogy, one might expect an E1 reaction to take place in a 1,2-glycol to generate an alkene. Alternatively, an epoxide might be formed. In fact, neither of these processes is the major one when 2,3-dihydroxy-2,3-dimethylbutane is treated with acid.

Alternatively, the carbocation might be captured by the second hydroxyl group, which is, after all, lurking nearby, to generate the epoxide (Fig. 17.48). This alternative reaction is an example of a very common phenomenon, the "neighboring group effect," and we will devote an entire chapter (Chapter 23) to it later on. Although both of these reactions are reasonable, there is another, lower energy process that wins out. This reaction makes such a good problem that it would be cruel to avoid it. Before we go on to the details, see if you can determine the structure of the product, and then the mechanism of its formation in Problems 17.20 and 17.21 (Fig. 17.49). Be sure to work through the answers to these problems.

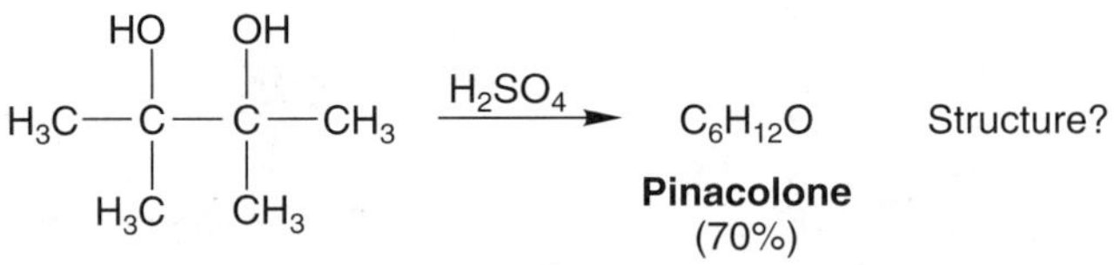

$$H_3C-C(OH)(CH_3)-C(OH)(CH_3)-CH_3 \xrightarrow{H_2SO_4} C_6H_{12}O$$

Pinacolone (70%)

Structure?

FIGURE 17.49 The real product, pinacolone, has the formula $C_6H_{12}O$.

*PROBLEM 17.20 Figure 17.50 shows the IR, ^{1}H NMR, and ^{13}C NMR spectra of the product of Figure 17.49. Provide a structure and explain your reasoning.

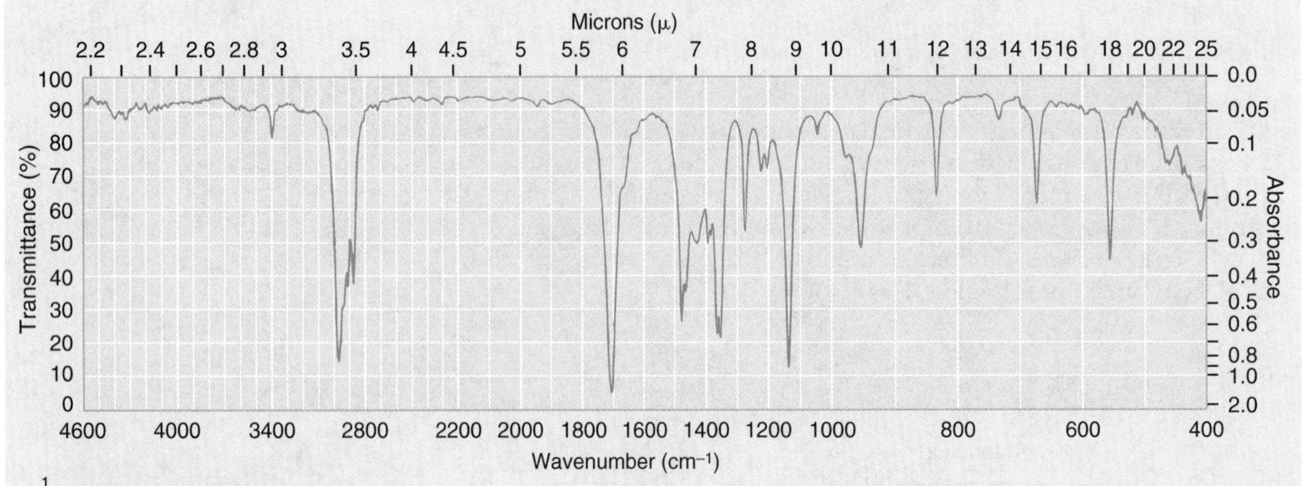

^{1}H NMR: δ 2.15 (1H), 1.16 (3H)

^{13}C NMR: δ 24.5, 26.4, 44.3, 213.9 ppm

FIGURE 17.50

ANSWER This problem is typical of a "real world" difficulty. These are exactly the kinds of data that the research chemist has when he or she faces the problem of determining the structure of an unknown compound. The chemist knows what molecules and reaction conditions produced the unknown, and can obtain the various spectra. The task is to fit all this together to produce a structure. The IR spectrum (1701 cm^{-1}) tells us that there is a saturated carbonyl group in the product. The low-field signal in the ^{13}C NMR spectrum confirms this analysis. There is no signal in the IR spectrum appropriate for an O—H stretch, so somehow both hydroxyl groups of the starting material have disappeared, and, of course, the formula shows that there is only one oxygen atom remaining in the molecule.

ANSWER (CONTINUED)

The ^1H NMR reveals that there are only two kinds of hydrogen, in the ratio 3:1. Because there are 12 hydrogens in the product, these signals correspond to 9 and 3 hydrogens, respectively. We know these hydrogens are not on adjacent carbons because there is no coupling between them. The larger signal is in the position appropriate for normal methyl groups. Nine equivalent methyl hydrogens come most easily from a *tert*-butyl group. That's a worthwhile practical hint—a singlet signal integrating for 9 hydrogens should always make you consider the presence of a *tert*-butyl group. Given the presence of a *tert*-butyl group, there is only one way to put the remaining CH_3CO atoms together in a reasonable way. The product is *tert*-butyl methyl ketone, otherwise known as pinacolone.

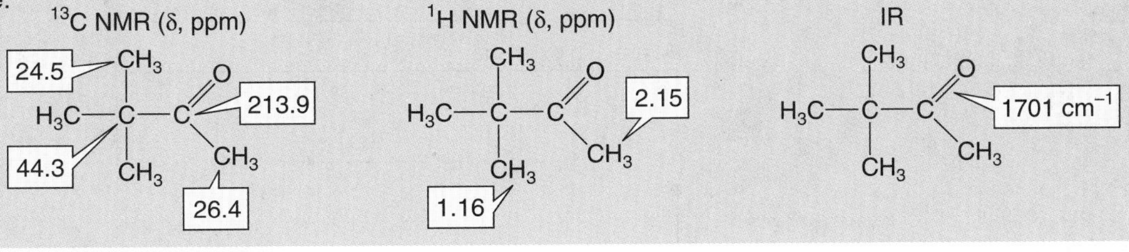

^{13}C NMR (δ, ppm): 24.5, 44.3, 213.9, 26.4

^1H NMR (δ, ppm): 1.16, 2.15

IR: 1701 cm^{-1}

*PROBLEM 17.21

What is the mechanism of pinacolone formation?

ANSWER

The first two steps in the mechanism seem inevitable—there just isn't much else possible other than protonation of an OH and loss of H_2O to give the tertiary carbocation, **A**. Whenever you see a carbocation, and especially when the unexpected has happened, you are well advised to consider the possibility of rearrangement. We have seen this kind of reaction many times in carbocation chemistry. In fact, it appears whenever it is possible to migrate a group to give a more stable cation. In this molecule, there is only one possible migrating group, one of the adjacent methyl groups. The thing to do now is to write the product of this rearrangement (cation **B**) and try to see if it looks more or less stable than the starting carbocation. We might suspect that we are on the right track because we have generated the appropriate carbon skeleton of pinacolone.

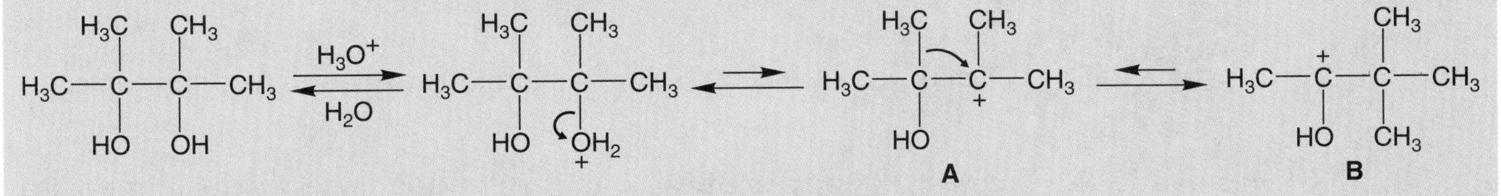

H_3O^+ / H_2O

A **B**

Cations flanked by heteroatoms bearing pairs of electrons are resonance stabilized, and it is this factor that makes the rearranged ion more stable than the starting cation. Deprotonation of the rearranged ion by water gives the final product. This reaction, called the **pinacol rearrangement**, takes its name from the trivial name of the diol used in this example, pinacol, and is common whenever a 1,2-diol is treated with acid.

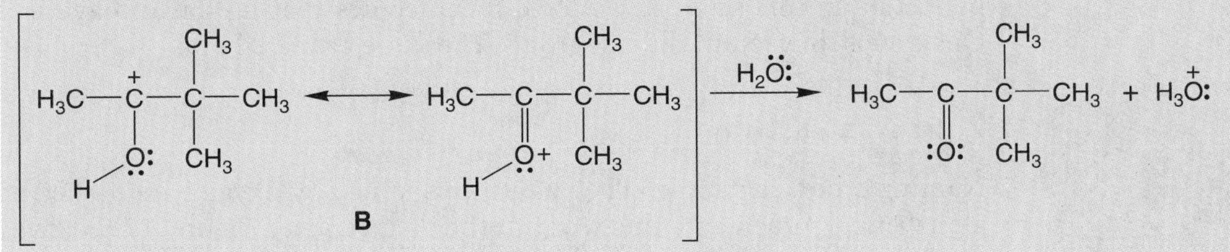

B

$H_2\ddot{O}$: → + $H_3\overset{+}{O}$:

17.8e Oxidative Cleavage of 1,2-Diols

When a 1,2-diol is oxidized by periodic acid (HIO_4) it is cleaved to a pair of carbonyl compounds. This reaction has long been used as a test for 1,2-diols (Fig. 17.51).

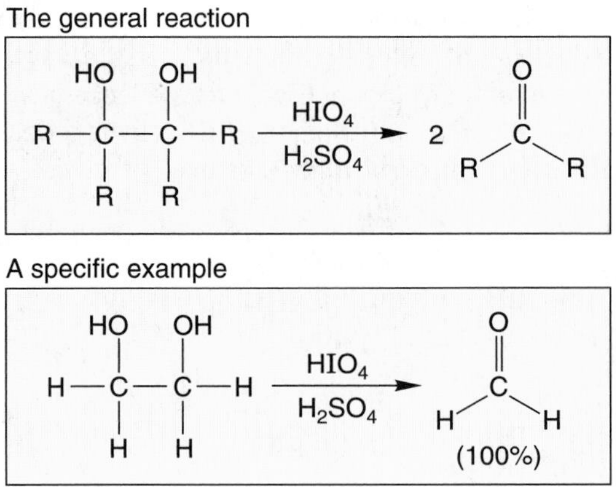

The general reaction

A specific example

(100%)

FIGURE 17.51 The cleavage of 1,2-diols by periodic acid.

The mechanism involves the formation of a cyclic periodate ester in a reaction that looks much like the formation of a cyclic acetal from a 1,2-diol and a ketone (Fig. 17.52).

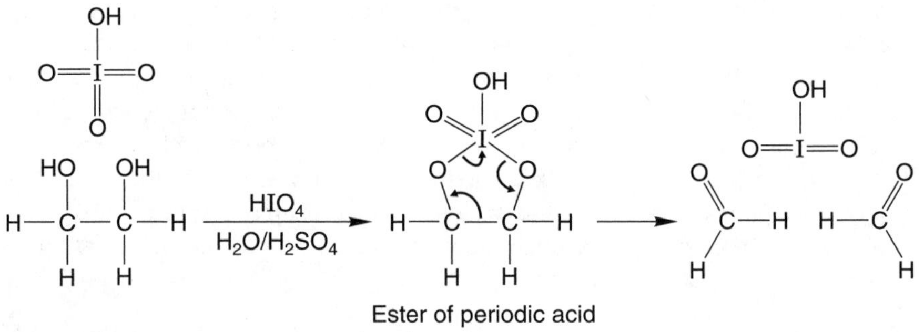

Ester of periodic acid

FIGURE 17.52 The cleavage reaction involves a cyclic ester of periodic acid that breaks down to give a pair of carbonyl compounds.

The intermediate breaks down to form the two carbonyl compounds.

PROBLEM 17.22 *trans*-1,2-Cyclohexanediol is much more reactive toward periodic acid than the cis isomer. Explain why.

17.9 ETHERS

Alcohol (ROH) chemistry depends on both the OH and the OR part of the molecule. In this section, we look at molecules that no longer have any OH groups, but retain the OR group. These are called ethers.

17.9a Nomenclature

Naming ethers is simple. The two groups joined by oxygen are named in alphabetical order and the word "ether" appended. Figure 17.53 gives some simple examples, as well as some common names.

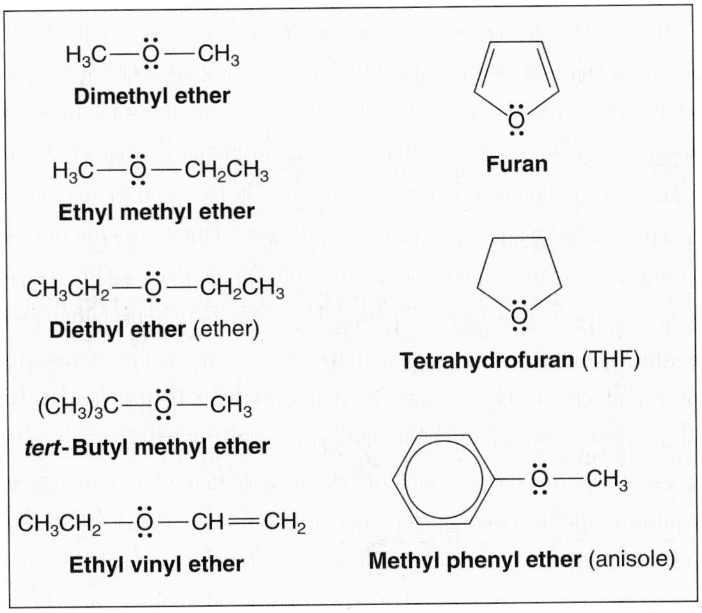

$H_3C-\ddot{O}-CH_3$
Dimethyl ether

$H_3C-\ddot{O}-CH_2CH_3$
Ethyl methyl ether

$CH_3CH_2-\ddot{O}-CH_2CH_3$
Diethyl ether (ether)

$(CH_3)_3C-\ddot{O}-CH_3$
***tert*-Butyl methyl ether**

$CH_3CH_2-\ddot{O}-CH=CH_2$
Ethyl vinyl ether

Furan

Tetrahydrofuran (THF)

$\ddot{O}-CH_3$
Methyl phenyl ether (anisole)

FIGURE 17.53 Some ethers and their common names.

17.9b Physical Properties

Ethers are polar, but only for the smallest members of the class does this polarity strongly affect physical properties. Others are sufficiently hydrocarbon-like so as to behave as their all-carbon relatives. For example, diethyl ether has nearly the same boiling point as pentane, and is only modestly soluble in water (Table 17.4).

TABLE 17.4 Some Physical Properties of Ethers

Compound	bp (°C)	mp (°C)	Density (g/mL)
CH_3-O-CH_3 Dimethyl ether	–25	–138.5	
$CH_3-O-CH_2CH_3$ Ethyl methyl ether	10.8		0.725
$CH_3CH_2-O-CH_2CH_3$ Diethyl ether	34.5	–116	0.714
$(CH_3)_3C-O-CH_3$ *tert*-Butyl methyl ether	55.2	–109	0.740
$CH_3CH_2-O-CH=CH_2$ Ethyl vinyl ether	35.5	–115.8	0.759
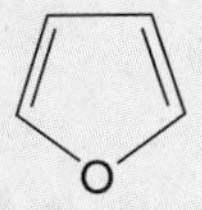 O Furan	31.4	– 85.6	0.951
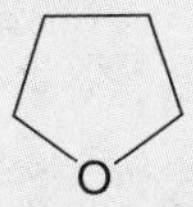 O Tetrahydrofuran (THF)	67	–108	0.889
$Ph-O-CH_3$ Anisole	155	– 37.5	0.996

17.9c Structure

The carbon–oxygen bond length in ethers is similar to that in alcohols, and the opening of the angle as we go from H–O–H (104.5°) to R–O–H (~109°), to R–O–R (~112°) continues. As the angle is approximately 112°, the oxygen is hybridized approximately sp^3.

17.9d Acidity and Basicity

Of course, ethers lack the OH group of their relatives water and the alcohols, and so are not Brønsted acids. The oxygen atom has two pairs of nonbonding electrons, however, and they make ethers weak Brønsted and Lewis bases. Ethers can be protonated, and the protonated form has a pK_a in the same range as those of the protonated alcohols, which means that ethers and alcohols are similarly strong bases (Fig. 17.54).

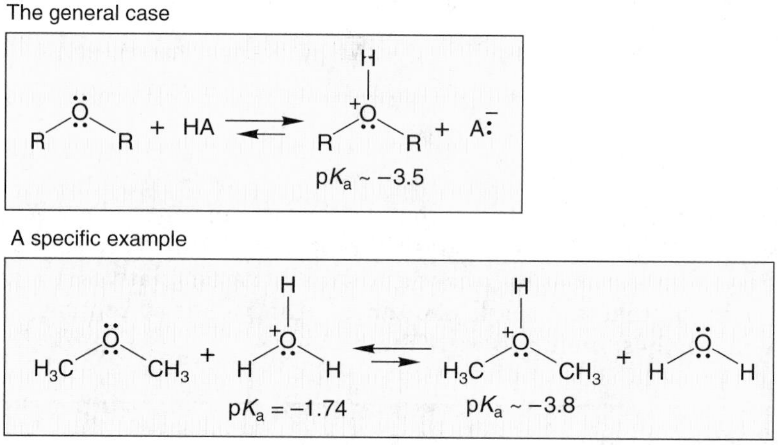

FIGURE 17.54 Ethers are Lewis bases and can be protonated. Protonated ethers (and protonated alcohols) are strong acids.

We have already seen one instance in which the nucleophilicity of ethers was a critical property. Recall that an ether solvent is vital to success in the formation of the Grignard reagent (Chapter 16, p. 787; Fig. 17.55). Ethers are stabilizing to Lewis acidic species. Boron trifluoride etherate is a commercially available complex of BF_3 and ether that can even be distilled. It is commonly used as a source of BF_3. For similar reasons, tetrahydrofuran (THF) is used to stabilize the highly reactive BH_3.

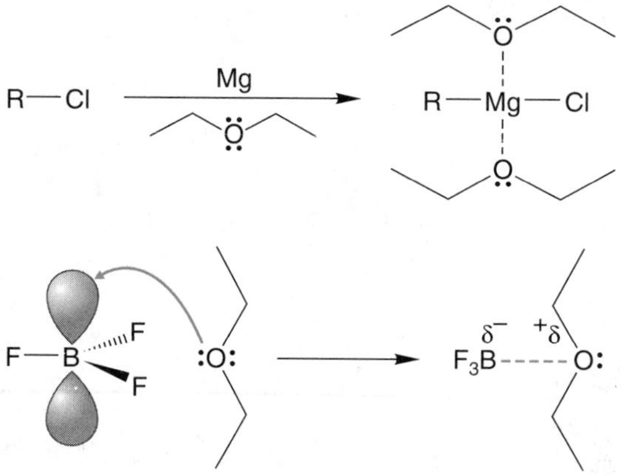

FIGURE 17.55 Ethers will react with Lewis acids as well as Brønsted acids. The BF_3–etherate complex is a stable source of diethyl ether and boron trifluoride.

17.9e Spectra

There is a characteristic C–O stretching frequency in the IR at 1050–1250 cm^{-1}. Like hydrogens α to alcohol oxygens, the α-hydrogens of ethers are deshielded by the electron-withdrawing effect of oxygen, and so they appear in roughly the same position in the ^1H NMR spectrum, about δ 3–4 ppm (Chapter 15, p. 716).

17.10 SYNTHESES OF ETHERS

As with the alcohols, we will encounter much familiar material in the following sections. However, now we see it from the perspective of the molecule, not the reaction.

17.10a From Alkoxides

A general ether synthesis involves an S_N2 displacement of a halide by an alkoxide (Chapter 7, p. 287; Fig. 17.56).

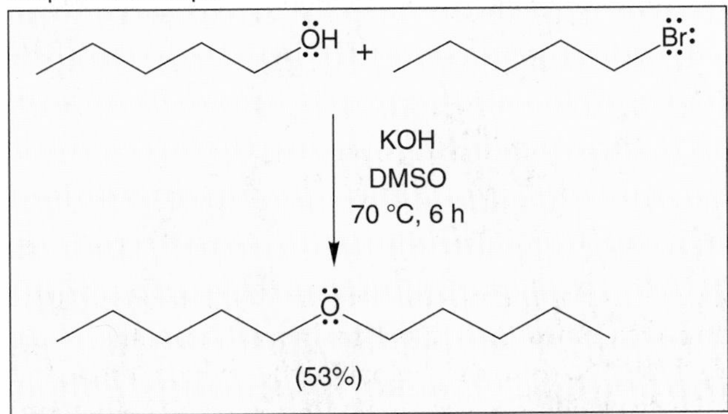

FIGURE 17.56 Alkoxides can displace halides in an S_N2 reaction to make ethers. This reaction is the Williamson ether synthesis.

This reaction is called the **Williamson ether synthesis** after A. W. Williamson (1824–1904). Several modifications of the original procedure have made this venerable method quite useful, although there are some restrictions. The reaction only works for alkyl halides that are active in the S_N2 reaction. Therefore, tertiary halides cannot be used. Sometimes there is an easy way around this problem, but sometimes there isn't. For example, *tert*-butyl methyl ether cannot be made from *tert*-butyl chloride and sodium methoxide, but it can be made from *tert*-butoxide and methyl iodide. However, there is no way to use the Williamson ether synthesis to make di-*tert*-butyl ether (Fig. 17.57).

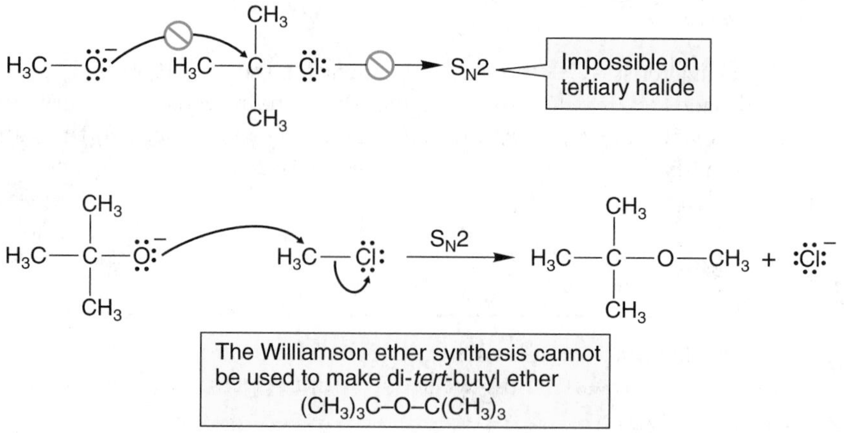

Impossible on tertiary halide

The Williamson ether synthesis cannot be used to make di-*tert*-butyl ether $(CH_3)_3C–O–C(CH_3)_3$

FIGURE 17.57 Tertiary halides cannot be used in the Williamson ether synthesis because they are too hindered to undergo the crucial S_N2 reaction.

Even secondary halides are of little use in this reaction. Alkoxides are strong bases (high affinity for a hydrogen 1*s* orbital) and the E2 reaction is a prominent side reaction when a secondary halide is used (Chapter 7, p. 277; Fig. 17.58).

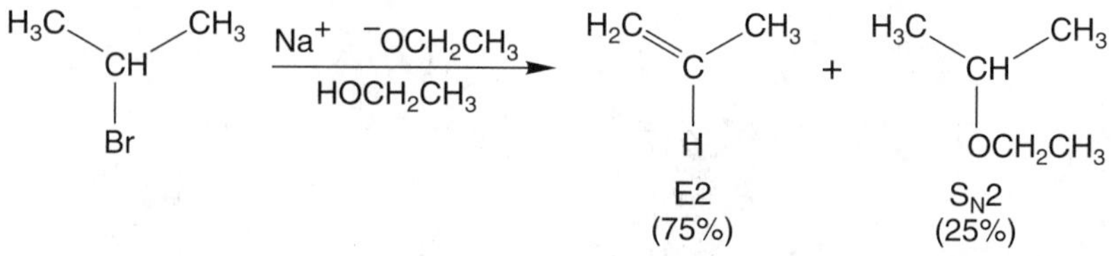

FIGURE 17.58 Secondary halides undergo substantial E2 elimination with alkoxides.

Be alert for intramolecular, ring-forming versions of the Williamson ether synthesis. There is no mystery in this reaction; it is simply the acyclic process applied in an intramolecular way (Fig. 17.59).

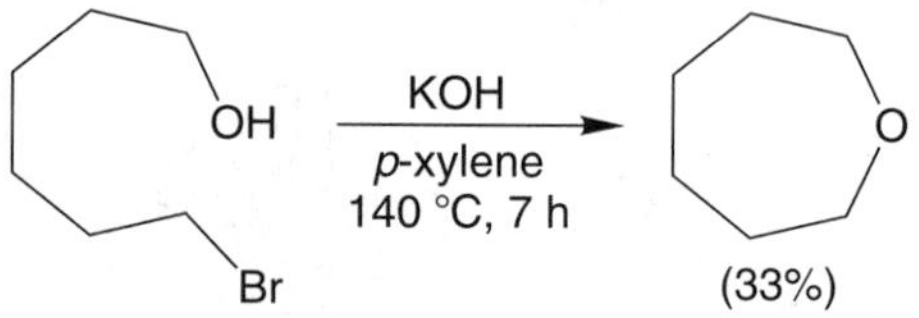

FIGURE 17.59 Cyclic ethers can be formed by intramolecular S_N2 reactions.

An important use of the intramolecular Williamson ether synthesis is in the making of oxiranes. A 2-halo-1-hydroxy compound (a halohydrin) is treated with base to form the haloalkoxide, which then undergoes an intramolecular backside displacement to give the three-membered ring (Fig. 17.60).

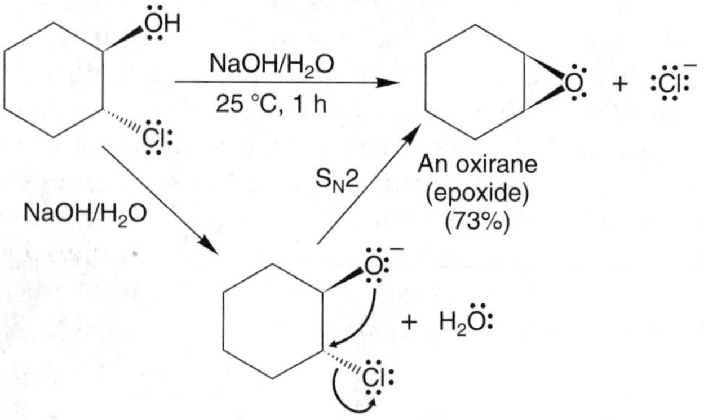

FIGURE 17.60 A common epoxide synthesis takes advantage of an intramolecular S_N2 reaction.

PROBLEM 17.23

How are 2-bromo-1-hydroxyalkanes made (Chapter 10, p. 398)?

*PROBLEM 17.24

Treatment of *cis*-2-butene with Br_2/H_2O gives a product (C_4H_9OBr), that reacts with sodium hydride to give a meso compound of the formula C_4H_8O. *trans*-2-Butene reacts in the same sequence to give a different compound (a racemic mixture) of the same formula, C_4H_8O. Explain mechanistically, paying close attention to stereochemical relationships.

ANSWER

Formation of the bromonium ion and opening by water leads to a racemic mixture of bromohydrins (3-bromo-2-butanol).

Treatment with NaH forms the alkoxides, and intramolecular S_N2 displacement gives the same meso epoxide in each case.

The keys to seeing this step of the problem are remembering that NaH is a useful base for forming alkoxides, and the recognition that a HBr is lost in going from C_4H_9OBr to C_4H_8O.

The *trans*-2-butene undergoes exactly the same reaction sequence, but the stereochemical relationships are different. The bromonium ion can be formed by addition of the bromine from either the top or bottom of the alkene to give a pair of enantiomeric ions.

ANSWER (CONTINUED) Opening of these bromonium ions with water leads to a pair of enantiomeric bromohydrins.

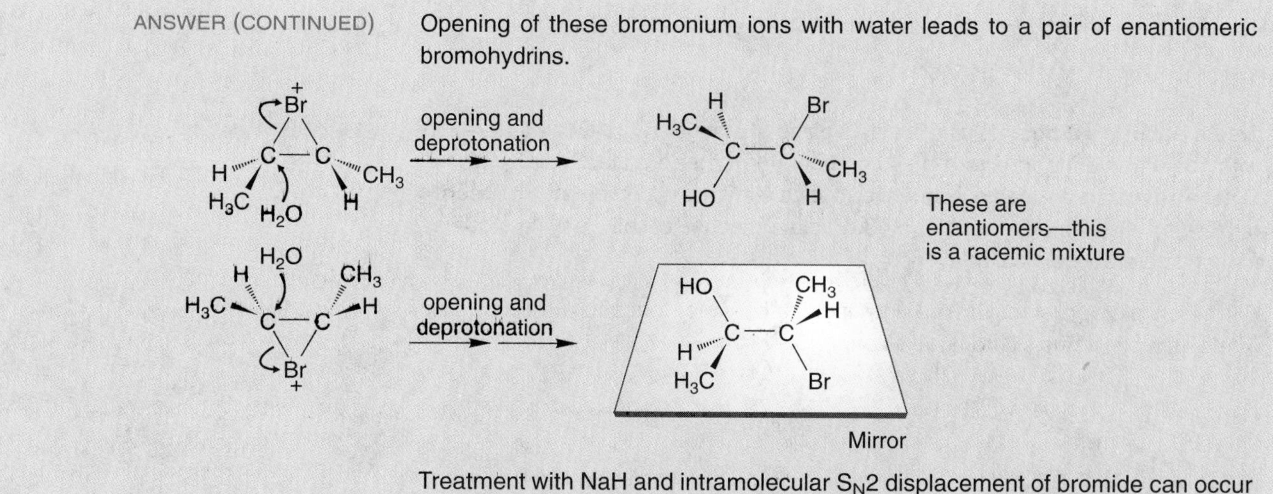

Treatment with NaH and intramolecular S_N2 displacement of bromide can occur in two ways to give the final racemic pair of enantiomeric epoxides.

NaH

NaH

(C_4H_8O)

(C_4H_8O)

A racemic mixture of enantiomeric epoxides

Mirror

17.10b From Alkenes

Treatment of an alkene with bromine in an alcohol solvent leads to the bromoalkyl ether through opening of the intermediate bromonium ion by alcohol. This reaction is usually complicated by the formation of the 1,2-dibromide (Fig. 17.61).

Br_2/CH_3OH, 0 °C

(17%) (83%)

Ph =

FIGURE 17.61 Bromoalkyl ethers can be made from alkenes by bromination in alcohol solvent. Chlorination in alcohol leads to related products.

Oxymercuration–reduction in alcohol can be used to make ethers in a process analogous to the synthesis of alcohols by oxymercuration–reduction in water (Chapter 10, p. 402). The overall result is attachment of the ether at the more substituted position of the starting alkene (Fig. 17.62).

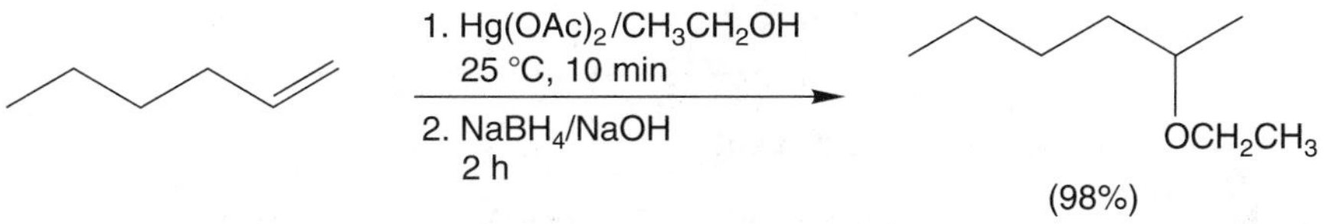

1. $Hg(OAc)_2/CH_3CH_2OH$ 25 °C, 10 min
2. $NaBH_4/NaOH$ 2 h

OCH_2CH_3

(98%)

FIGURE **17.62** Oxymercuration–reduction in alcohol also leads to ethers.

PROBLEM **17.25**

Write an arrow formalism mechanism for the reaction in Figure 17.62.

Three-membered ethers, the oxiranes, can be made directly from most alkenes by reaction with *m*-chloroperbenzoic acid or other peracids (Fig. 17.42).

17.10c Aromatic Ethers

Simple aromatic ethers can be made by using phenol as a source of the phenoxide ion in the Williamson synthesis (Fig. 17.63).

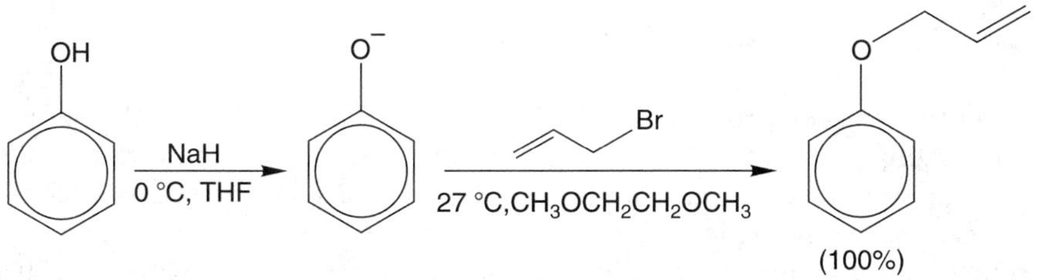

OH NaH 0 °C, THF O^- Br 27 °C,$CH_3OCH_2CH_2OCH_3$ O (100%)

FIGURE **17.63** Aryl ethers can be made by using phenoxides as S_N2 displacing agents in the Williamson ether synthesis.

PROBLEM **17.26**

Phenol (pK_a = 10) is a much stronger acid than cyclohexanol (pK_a = 16). Explain why.

PROBLEM **17.27**

Varying amounts of a side product are often formed in the reaction in Figure 17.63. Explain the formation of the compound shown in Figure 17.64 in the reaction of phenoxide with allyl bromide.

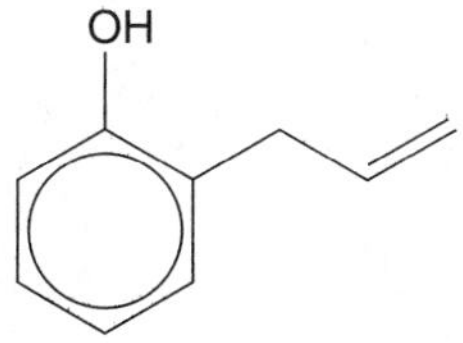

OH

FIGURE **17.64**

Specially activated aromatic rings will undergo nucleophilic aromatic substitution to give aromatic ethers (Chapter 14, p. 671). Note that the presence of the activating nitro groups is crucial; this synthesis is not general for aromatic ethers (Fig. 17.65).

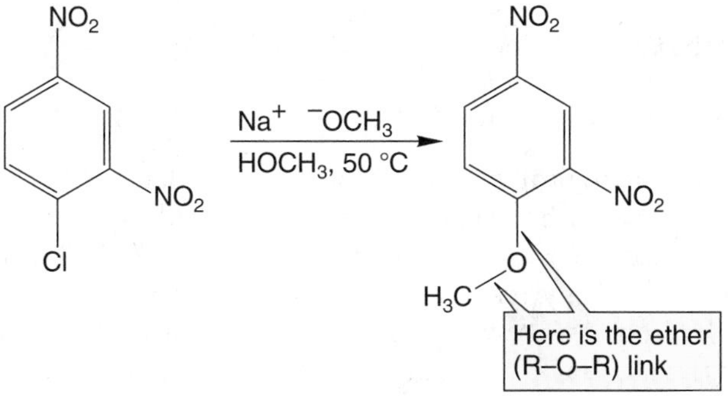

NO_2 NO_2 Cl $Na^+ \ ^-OCH_3$ $HOCH_3$, 50 °C NO_2 NO_2 O H_3C

Here is the ether (R–O–R) link

FIGURE **17.65** Nucleophilic aromatic substitution leads to aryl ethers.

17.11 REACTIONS OF ETHERS

17.11a Cleavage by Strong Acids

In strong halogenated acids such as HI or HBr, ethers can be cleaved to the corresponding alcohol and halide. In the presence of excess HBr, the alcohols formed in this reaction are often converted into the bromides (Fig. 17.66).

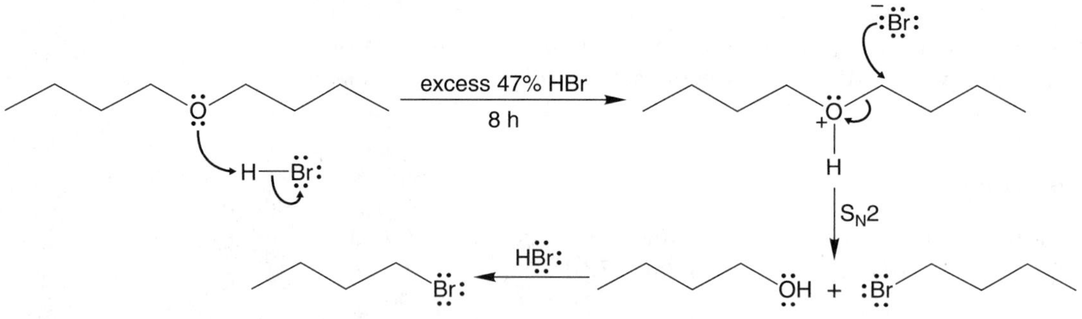

excess 47% HBr
8 h
S_N2
HBr:

FIGURE 17.66 The mechanism of the cleavage of ethers by strong haloacids.

In this reaction, the ether oxygen is first protonated, turning the potential leaving group in an S_N2 reaction from alkoxide, $^-$OR, into the much more easily displaced alcohol, HOR. The displacement itself is accomplished by the halide ion formed when the ether is protonated. This reaction is quite analogous to the reaction of alcohols with haloacids. The halide must be a nucleophile strong enough to do the displacement, and the reaction is subject to the limitations of any S_N2 reaction (Fig. 17.66).

PROBLEM 17.28 Clearly, an unsymmetrical ether can cleave in two different ways. Yet often only one path is followed. Explain the specificity shown in reaction (17.1) in Figure 17.67.

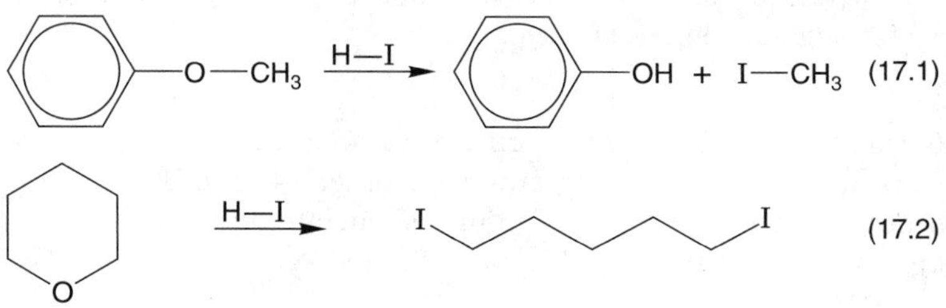

C_6H_5—O—CH_3 $\xrightarrow{\text{H—I}}$ C_6H_5—OH + I—CH_3 (17.1)

(tetrahydropyran) $\xrightarrow{\text{H—I}}$ I—$(CH_2)_5$—I (17.2)

FIGURE 17.67

PROBLEM 17.29 Provide a mechanism for reaction (17.2) in Figure 17.67.

17.11b Protecting Groups for Alcohols

It is often desirable to protect a hydroxyl group so that further reactions in base can take place without complication. One widely used method is to convert the hydroxyl group into a **tetrahydropyranyl (THP) ether** through reaction with dihydropyran (Fig. 17.68). The mechanism involves protonation of dihydropyran to give the resonance-stabilized cation, which is captured by the alcohol. The result is an acetal, a molecule stable to base, but not to acid (Chapter 16, p. 778). Regeneration of the alcohol can be accomplished through acid-catalyzed hydrolysis.

The general case

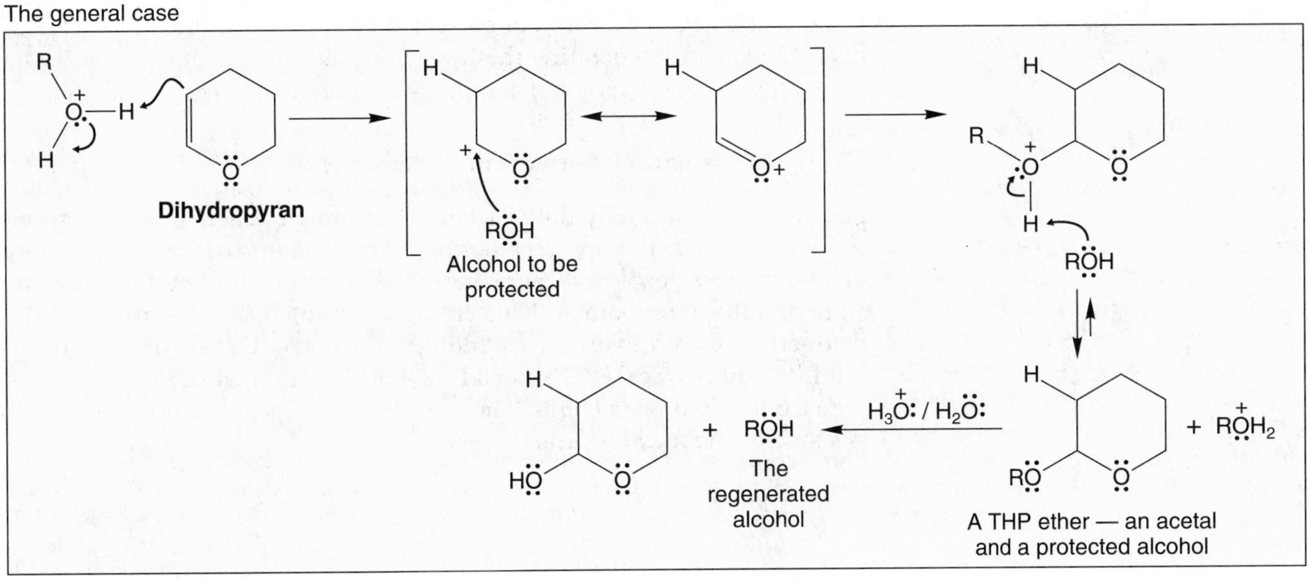

Dihydropyran

Alcohol to be protected

The regenerated alcohol

A THP ether — an acetal and a protected alcohol

A specific example

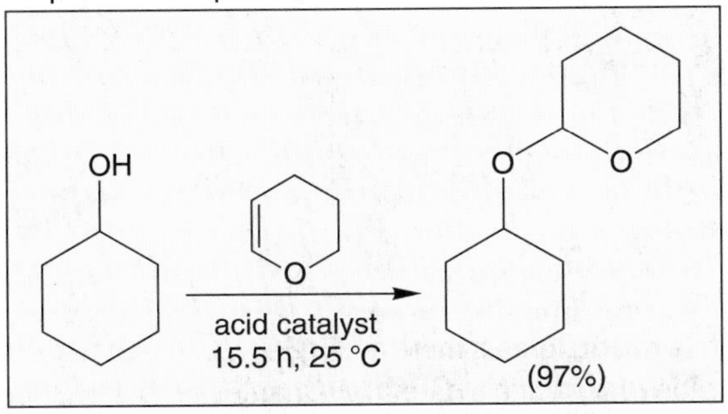

acid catalyst
15.5 h, 25 °C
(97%)

FIGURE 17.68 The use of THP ethers as protecting groups for alcohols.

The reaction of an alcohol with a trialkylsilyl halide leads to a trialkylsilyl ether in near-quantitative yield. The reaction depends for its success on the enormous strength of the silicon–oxygen bond (~ 110 kcal/mol). The silyl group can be removed through reaction with fluoride ion or hydrogen fluoride. The silicon–fluorine bond is even stronger than the silicon–oxygen bond (~140 kcal/mol), and the formation of the trialkylsilyl fluoride provides the thermodynamic driving force for the reaction (Fig. 17.69).

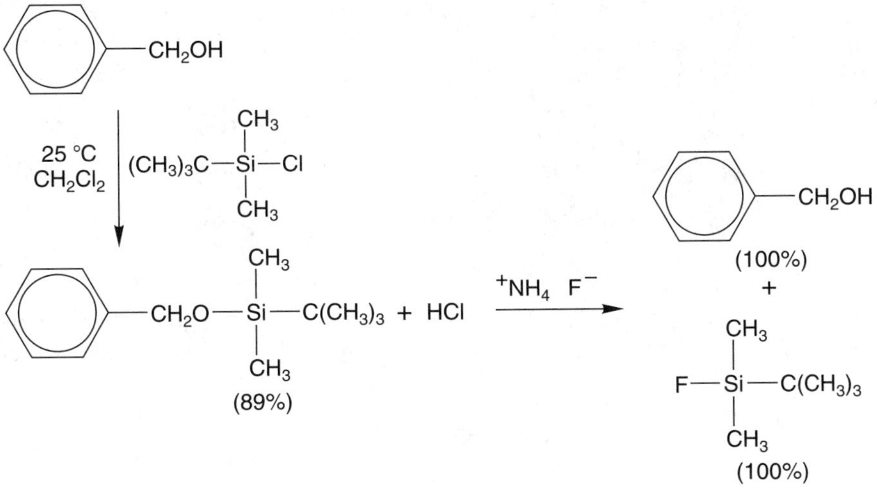

25 °C
CH_2Cl_2
$(CH_3)_3C{-}Si(CH_3)_2{-}Cl$

$C_6H_5CH_2OH \rightarrow C_6H_5CH_2O{-}Si(CH_3)_2{-}C(CH_3)_3$ + HCl (89%)

$\xrightarrow{^{+}NH_4\ F^{-}}$ $C_6H_5CH_2OH$ (100%) + $F{-}Si(CH_3)_2{-}C(CH_3)_3$ (100%)

FIGURE 17.69 The use of trialkylsilyl ethers as protecting groups for alcohols.

So, alcohols can be stored or protected by converting them into THP or silyl ethers, and regenerating them when needed. In Chapter 16, we saw a protecting group for carbonyl compounds—acetals (p. 778).

17.11c Autoxidation: Formation of Peroxides

The handling, especially the distillation, of simple ethers must be carried out with care. Left unprotected from oxygen, ethers form dangerously shock-sensitive, explosive peroxides. Old bottles containing ether are, quite literally, time bombs. The peroxide-forming reaction starts with the abstraction of an α-hydrogen by molecular oxygen. Oxygen is a diradical, and like other radicals (Chapter 11, p. 454) is able to abstract a hydrogen from carbon–hydrogen bonds. The α-position of ethers is especially active, as a stabilized radical results (Fig. 17.70).

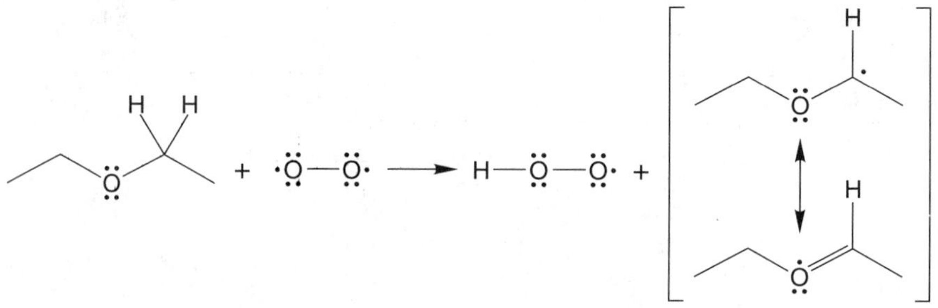

FIGURE 17.70 The diradical oxygen can abstract a hydrogen atom from the α-position of ethers.

*PROBLEM 17.30 As you may recall, the situation is not quite as simple as Figure 17.70 suggests. Explain the resonance stabilization of α-radicals carefully (Chapter 11, p. 461).

ANSWER The overlap symbolized by the resonance arrow in Figure 17.70 is between a filled orbital on oxygen and a half-filled orbital on carbon. There is stabilization of two electrons, but one electron must occupy a high-energy antibonding orbital. Nonetheless, this situation is still stabilizing.

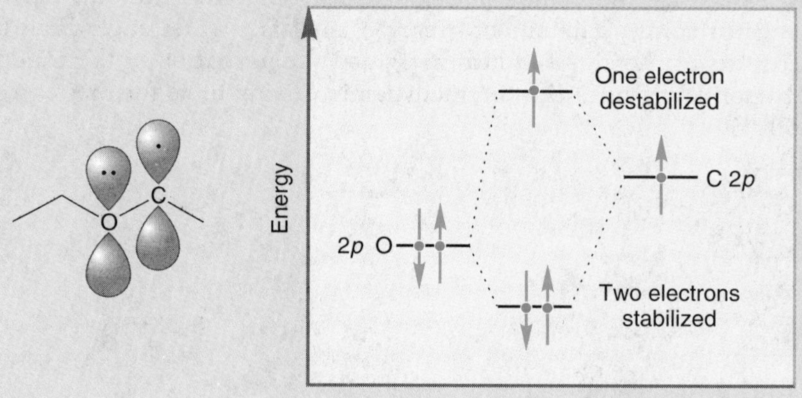

A new, peroxy radical results from combination of oxygen with the carbon radical, and the process can continue by further α-hydrogen abstraction to give the dangerous **hydroperoxides** or by radical coupling to produce **peroxides** (Fig. 17.71).

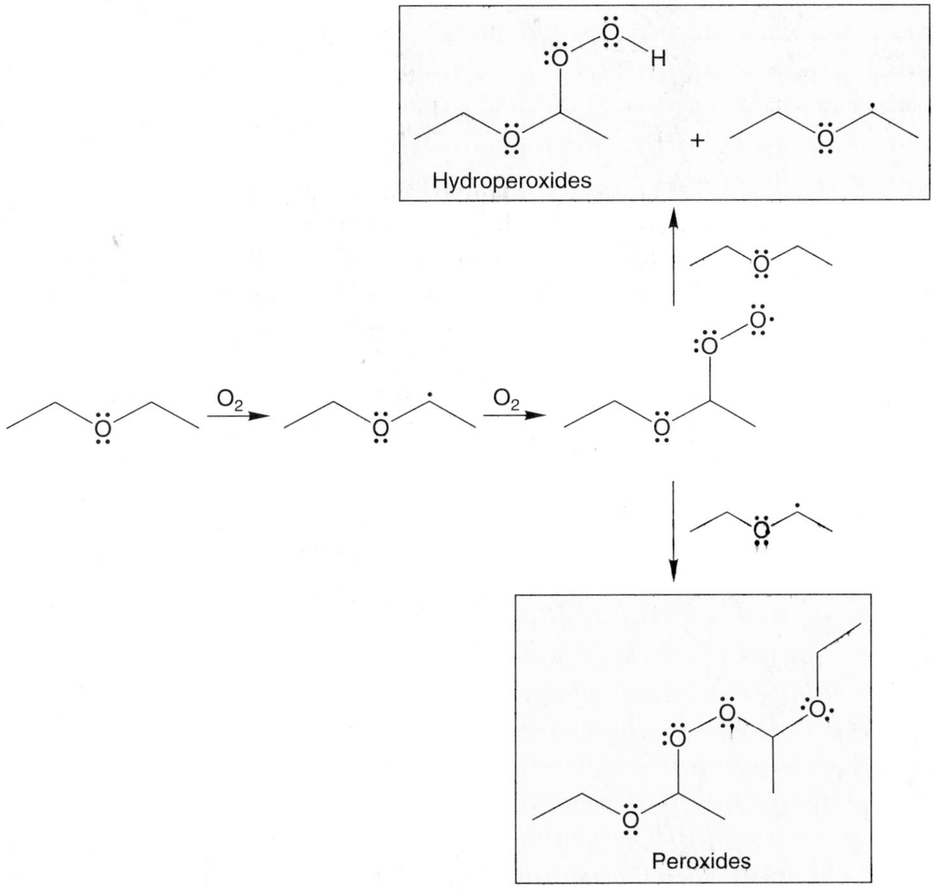

Hydroperoxides

O_2

O_2

Peroxides

FIGURE 17.71 Further reactions lead to hydroperoxides and peroxides.

Ethers are generally protected by the addition of a small amount of hydroquinone (Fig. 17.39), an inhibitor of radical reactions. Peroxides can be destroyed by boiling with lithium aluminum hydride.

How do you think lithium aluminum hydride reacts with peroxides? PROBLEM 17.31

17.11d Reactions of Oxiranes with Nucleophiles

Although most ethers are inert toward nucleophilic addition, and are excellent solvents for the strongly reactive organometallic reagents, we have already seen one example of an ether reacting with hydroxide by nucleophilic substitution. This process is the opening of oxiranes (epoxides) to give 1,2-glycols by the reaction with hydroxide (Chapter 10, p. 408; Fig. 17.43). Why should this three-membered ring be different from other acyclic and cyclic ethers (Fig. 17.72)?

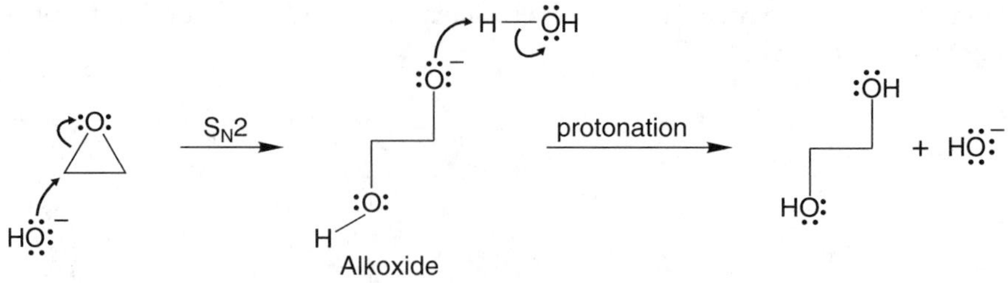

FIGURE 17.72 Hydroxide ion opens epoxides through an S$_N$2 reaction. Protonation of the initially produced alkoxide ion yields 1,2-glycols

The answer is simply that opening of the three-membered ring relieves strain and this relief is felt in the transition state for the reaction. In the transition state, the three-membered ring is partially opened and some of the strain inherent in the small ring is gone (Fig. 17.73).

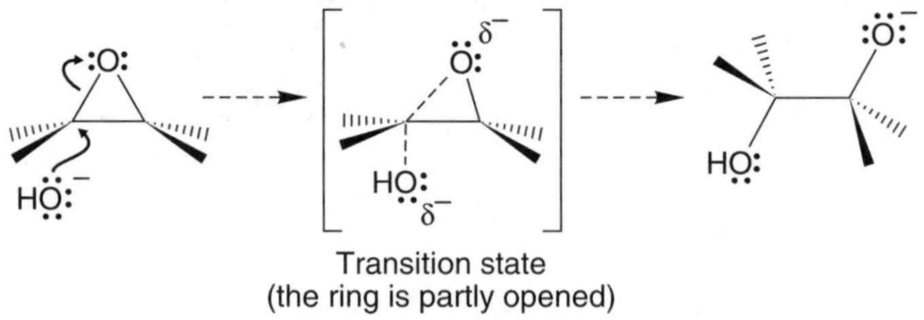

FIGURE 17.73 Epoxides are opened more easily than other cyclic ethers because ring strain is relieved in the transition state.

Oxiranes will also open to glycols under mildly acidic conditions (Chapter 10, p. 408; Fig. 17.74).

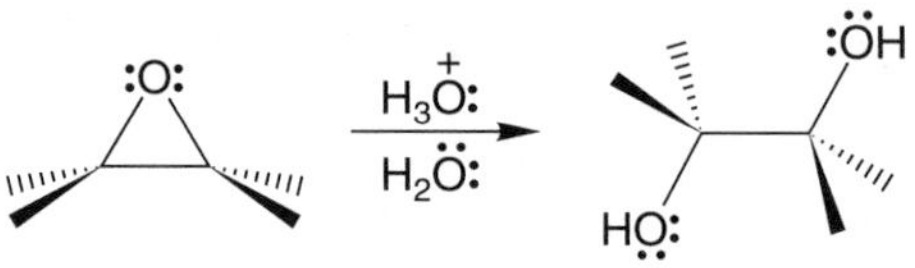

FIGURE 17.74 Epoxides are also cleaved in acid.

PROBLEM 17.32 Write a mechanism for the reaction in Figure 17.74.

All manner of nucleophiles reacts with oxiranes in S$_N$2 fashion to give ring-opened products. Addition is to the less hindered side of the ring (Fig. 17.75).

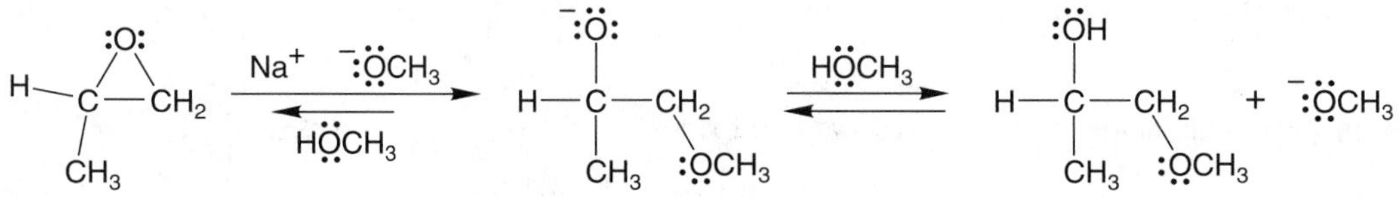

FIGURE 17.75 Opening of unsymmetrical epoxides in base occurs by attack at the sterically less hindered end of the epoxide.

Hydroxide is an obvious and useful example, but more important still is the reaction with organometallic reagents. Attack of the organometallic reagent initially gives an alkoxide. In a second step, water protonates the alkoxide and gives the final alcohol product (Fig. 17.76).

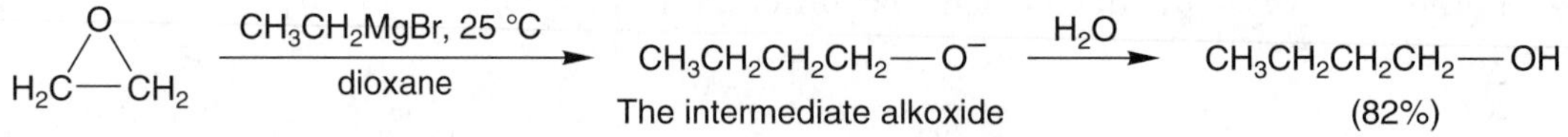

FIGURE 17.76 Opening of an epoxide by an organometallic reagent. The initial product is an alkoxide, which is protonated in a second step to give the alcohol.

The strongly basic organometallic reagents inevitably attack at the less substituted, less sterically hindered end of the oxirane, as did hydroxide ion (Fig. 17.77).

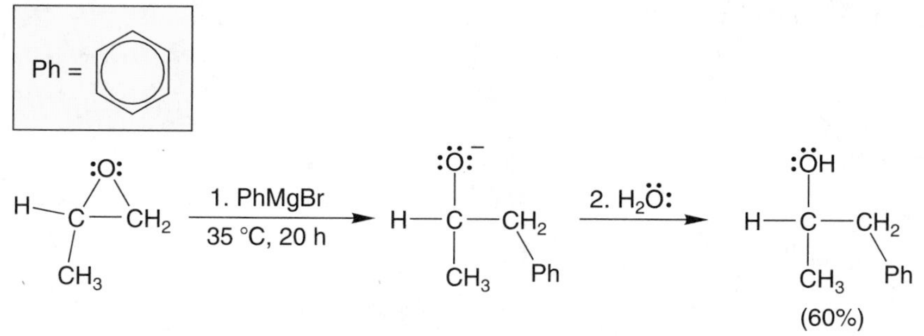

Ph =

H—C(CH$_3$)(—O—)CH$_2$ —(1. PhMgBr, 35 °C, 20 h)→ H—C(:Ö:$^-$)(CH$_3$)—CH$_2$—Ph —(2. H$_2$Ö:)→ H—C(:ÖH)(CH$_3$)—CH$_2$—Ph (60%)

FIGURE 17.77 The S_N2 ring opening occurs at the less sterically demanding end of the epoxide.

The reaction is a complement to the addition of organometallic reagents to carbonyl compounds. Reaction of a Grignard reagent (or organolithium reagent) with formaldehyde extends the chain of the organometallic reagent by one carbon, whereas reaction with ethylene oxide extends the chain by two carbon atoms (Fig. 17.78).

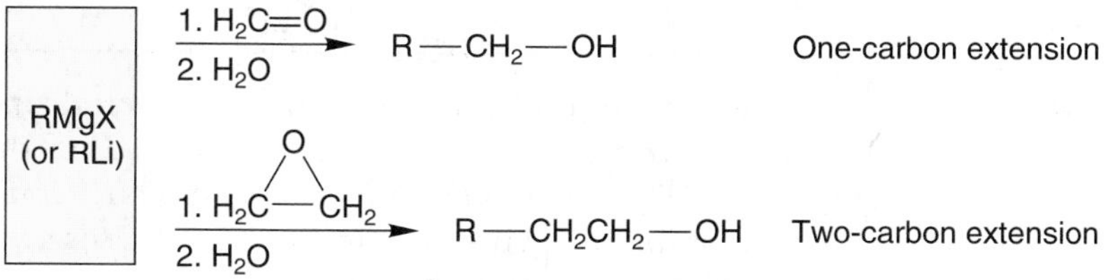

RMgX (or RLi)

1. $H_2C{=}O$, 2. H_2O → R—CH$_2$—OH One-carbon extension

1. H$_2$C(—O—)CH$_2$, 2. H_2O → R—CH$_2$CH$_2$—OH Two-carbon extension

FIGURE 17.78 Two chain-lengthening reactions. Addition to formaldehyde, followed by hydrolysis, extends the chain by one carbon; whereas reaction with ethylene oxide, followed by hydrolysis, lengthens the chain by two carbons.

Lithium aluminum hydride also opens oxiranes by nucleophilic addition, and this reaction provides you with yet another synthesis of alcohols. Addition of hydride is predominately from the less hindered side of the epoxide. Once again, an initially produced alkoxide is protonated by water in a second step (Fig. 17.79).

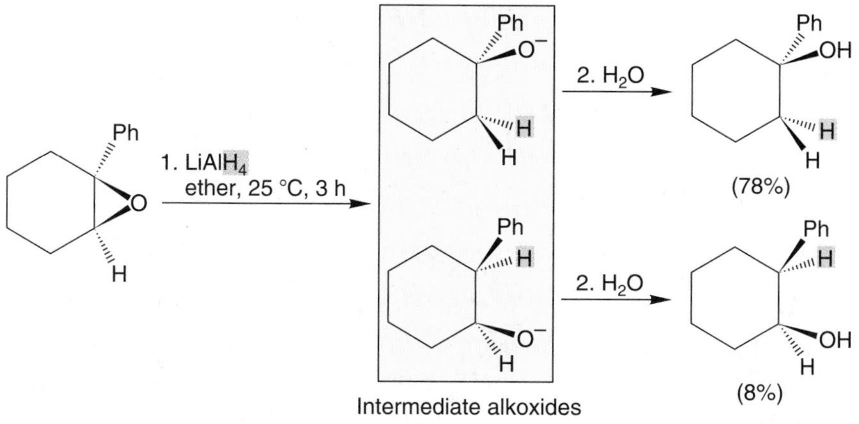

Ph, O, H —(1. LiAlH$_4$, ether, 25 °C, 3 h)→

Intermediate alkoxides: Ph, O$^-$, H, H —(2. H$_2$O)→ Ph, OH, H, H (78%)

Ph, H, O$^-$, H —(2. H$_2$O)→ Ph, H, OH, H (8%)

FIGURE 17.79 Lithium aluminium hydride reduction of epoxides, followed by hydrolysis, is yet another alcohol synthesis.

Organometallic reagents have drawbacks, however. Reduction of the epoxide by hydride transfer from the organometallic reagent sometimes occurs and rearrangements can be induced, especially by organolithium reagents. Moreover, Grignards and organolithium reagents will inevitably add to any carbonyl groups present in the molecule (Chapter 16, p. 791).

By contrast, the organocuprates formed from RLi and CuCN (Chapter 16, p. 789) open epoxides smoothly without rearrangement, and are not reactive enough to attack the carbon–oxygen double bond. The opening of the three-membered ring is from the rear, and occurs predominately at the less hindered position (Fig. 17.80).

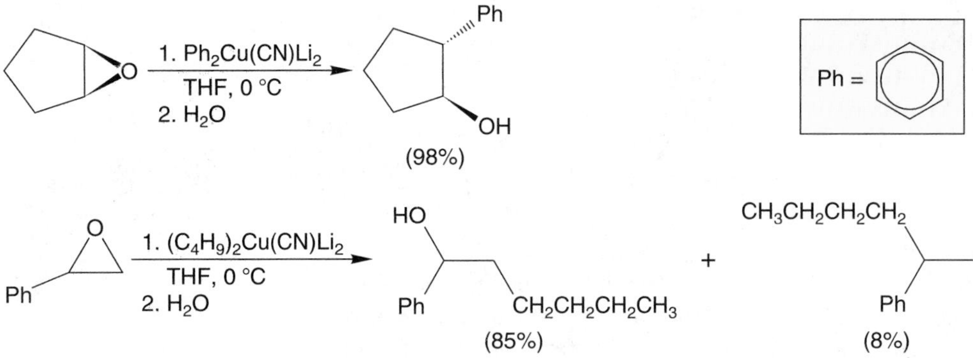
O
1. $Ph_2Cu(CN)Li_2$
THF, 0 °C
2. H_2O
Ph
OH
(98%)
Ph =
O
1. $(C_4H_9)_2Cu(CN)Li_2$
THF, 0 °C
2. H_2O
Ph
HO
Ph
$CH_2CH_2CH_2CH_3$
(85%)
+
$CH_3CH_2CH_2CH_2$
OH
Ph
(8%)

FIGURE 17.80 Organocuprates also open epoxides.

17.12 THIOLS (MERCAPTANS) AND THIOETHERS (SULFIDES)

Thiols and thioethers are the sulfur-containing counterparts of alcohols and ethers, and their chemistries are generally similar to those of their oxygen-containing relatives. These compounds have a poor reputation, as some thiols are quite extraordinarily evil-smelling. The skunk uses a variety of four-carbon thiols in its defense mechanism, for example. But this bad press is not entirely deserved; not all thiols are malodorous. Garlic derives its odor from small sulfur-containing compounds. You many never need to ward off a vampire with it, but garlic contains many sulfur compounds that have quite remarkably positive qualities. Figure 17.81 shows some of the sulfur-containing molecules that can be found in garlic.

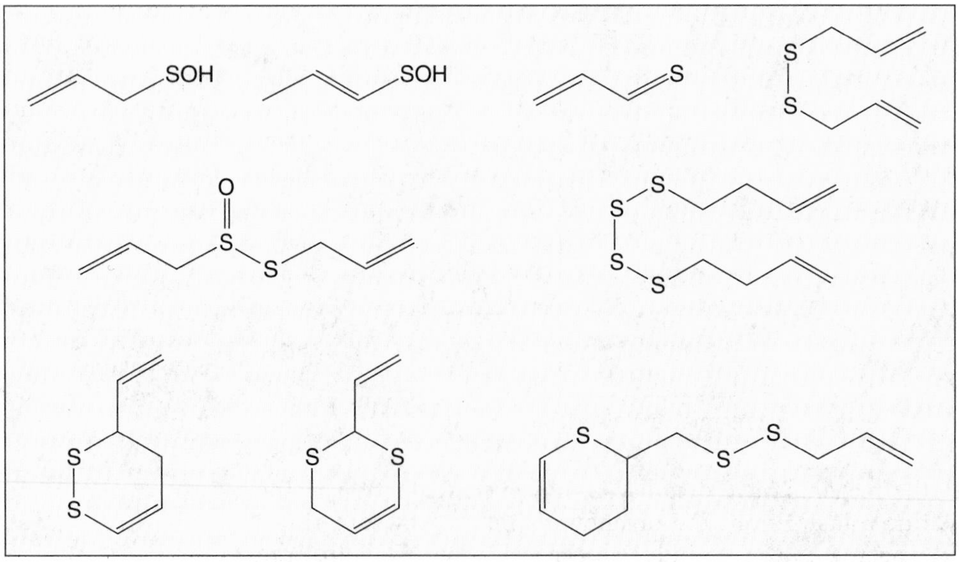
SOH
SOH
S
S
S
O
S
S
S
S
S
S
S
S
S
S
S
S
S
S

FIGURE 17.81 Some of the sulfur-containing molecules found in garlic.

Ajoene

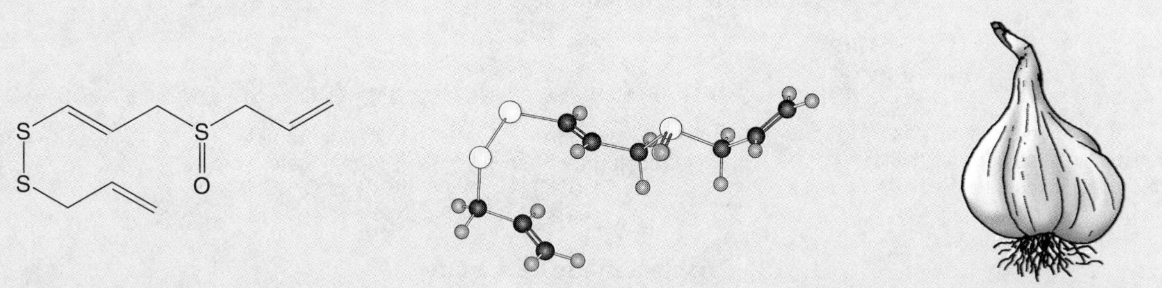

Ajoene is only one of the beneficial compounds present in garlic. Ajoene is lethal to certain tumor-prone cells, and promotes the antiaggretory action of some molecules on human blood platelets. It also seems effective in reducing repeat heart attacks among people who have already suffered an initial attack. Garlic contains many other compounds that also appear to have beneficial medicinal properties. And, of course, if a vampire is hot on your trail ...

17.12a Nomenclature

Thiols, or mercaptans, are named by adding the suffix *thiol* to the parent hydrocarbon name. Note that the final "e" is not dropped, as it is, for example, in alcohol nomenclature (Fig. 17.82).

$H_3C—\ddot{S}—H$ **Methanethiol** (methyl mercaptan)

$CH_3CH_2CH_2CH_2—\ddot{S}—H$ **Butanethiol** (butyl mercaptan)

Cyclohexanethiol

Phenyl mercaptan (benzenethiol) (thiophenol)

FIGURE 17.82 Some thiols, or mercaptans.

The sulfur counterparts of ethers are **sulfides**, which are named in the same way as ethers. The two groups attached to the sulfur atom are followed by the word sulfide. Disulfides are the counterparts of peroxides, but unlike the oxygen compounds, these molecules are not explosive. They are named in the same way as sulfides (Fig. 17.83).

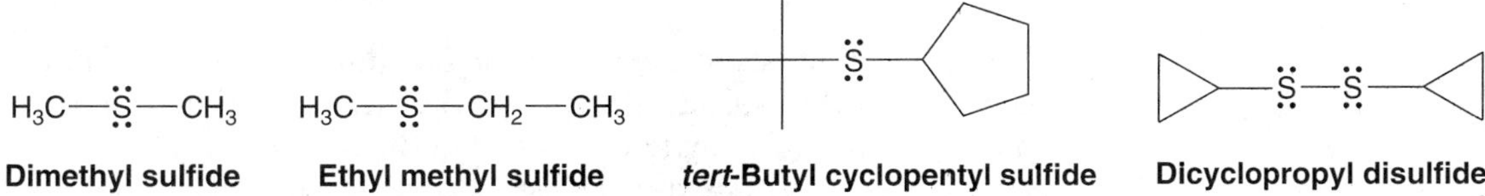

FIGURE 17.83 Some thioethers (sulfides) and disulfides.

17.13 REACTIONS AND SYNTHESES OF THIOLS AND SULFIDES

17.13a Acidity

Thiols are stronger acids than alcohols ($pK_a = 9–12$) and form **mercaptides**, the sulfur counterparts of alkoxides, when treated with base. The increased acidity of thiols makes formation of the conjugate base more favorable than from alcohols (Fig. 17.84).

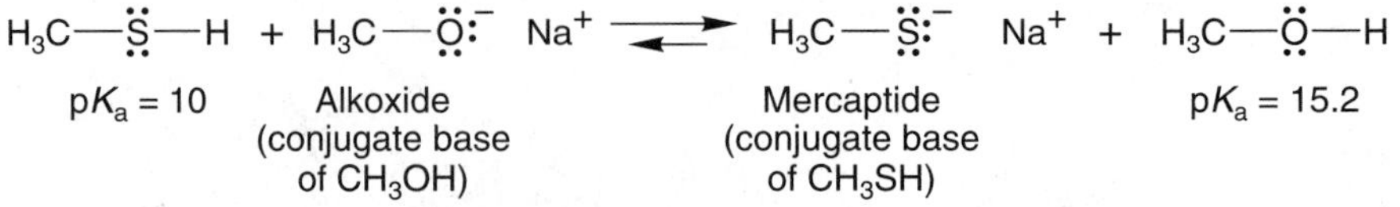

$$H_3C{-}\ddot{S}{-}H + H_3C{-}\ddot{O}{:}^- \; Na^+ \rightleftharpoons H_3C{-}\ddot{S}{:}^- \; Na^+ + H_3C{-}\ddot{O}{-}H$$

pK_a = 10; Alkoxide (conjugate base of CH_3OH); Mercaptide (conjugate base of CH_3SH); pK_a = 15.2

FIGURE 17.84 Mercaptides are easier to form than alkoxides because thiols are much more acidic than alcohols.

17.13b Displacement Reactions

The mercaptide ion itself ($^-$:SH), as well as alkyl mercaptides ($^-$:SR), are powerfully nucleophilic and can be used to great effect in the S_N2 reaction. This nucleophilicity leads to the most common synthesis of thiols and sulfides (Fig. 17.85).

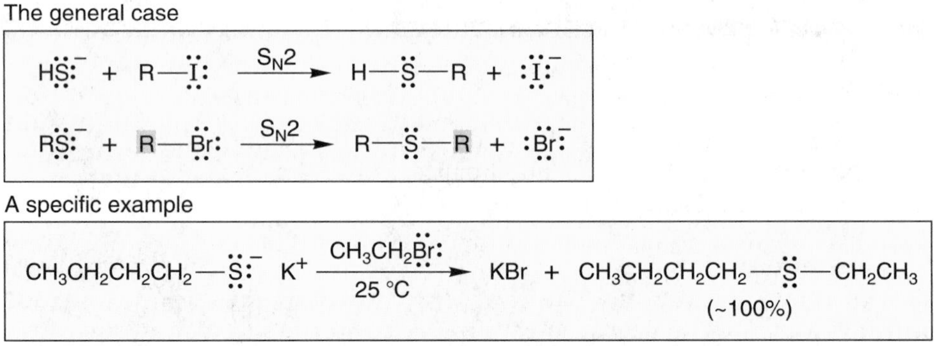

The general case

$$H\ddot{S}{:}^- + R{-}\ddot{I}{:} \xrightarrow{S_N2} H{-}\ddot{S}{-}R + {:}\ddot{I}{:}^-$$

$$R\ddot{S}{:}^- + R{-}\ddot{Br}{:} \xrightarrow{S_N2} R{-}\ddot{S}{-}R + {:}\ddot{Br}{:}^-$$

A specific example

$$CH_3CH_2CH_2CH_2{-}\ddot{S}{:}^- \; K^+ \xrightarrow[25\ °C]{CH_3CH_2\ddot{Br}{:}} KBr + CH_3CH_2CH_2CH_2{-}\ddot{S}{-}CH_2CH_3$$

(~100%)

FIGURE 17.85 The S_N2 reactions of mercaptide and substituted mercaptides lead to thiols and sulfides. One restriction is that the alkyl halide must be active in the S_N2 reaction.

Sulfides have two lone pairs of electrons and are also quite nucleophilic. Further S_N2 reactions give **sulfonium ions** (Fig. 17.86).

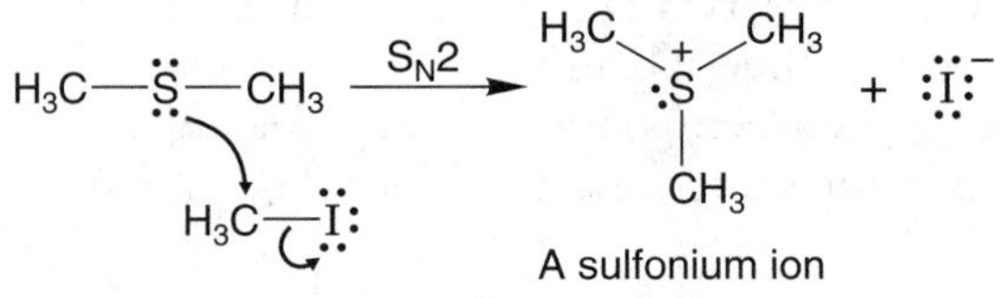

$$H_3C{-}\ddot{S}{-}CH_3 + H_3C{-}\ddot{I}{:} \xrightarrow{S_N2} (CH_3)_3S^+ + {:}\ddot{I}{:}^-$$

A sulfonium ion

FIGURE 17.86 Sulfides can also be alkylated to give sulfonium ions.

Sulfonium ions and the less stable oxonium ions are used as alkylating agents. Dimethyl sulfide and dimethyl ether are excellent leaving groups and can be displaced by nucleophiles in further S_N2 reactions (Fig. 17.87). Nature uses a variant of this reaction, displacement on the amino acid derivative (*S*)-adenosylmethionine, to do many methylation reactions.

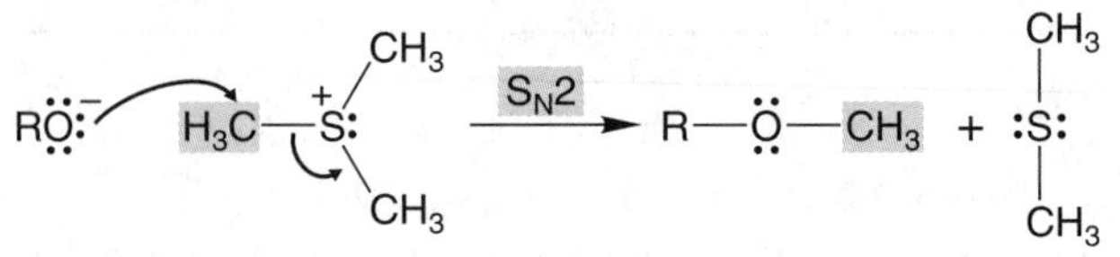

$$R\ddot{O}{:}^- + H_3C{-}\overset{+}{S}(CH_3)_2 \xrightarrow{S_N2} R{-}\ddot{O}{-}CH_3 + {:}S(CH_3)_2$$

FIGURE 17.87 Alkoxides can be alkylated by ions in an S_N2 reaction. Nature uses a variation of this process to do many alkylation reactions. The sulfonium ion active in many biological alkylations is (*S*)-adenosylmethionine.

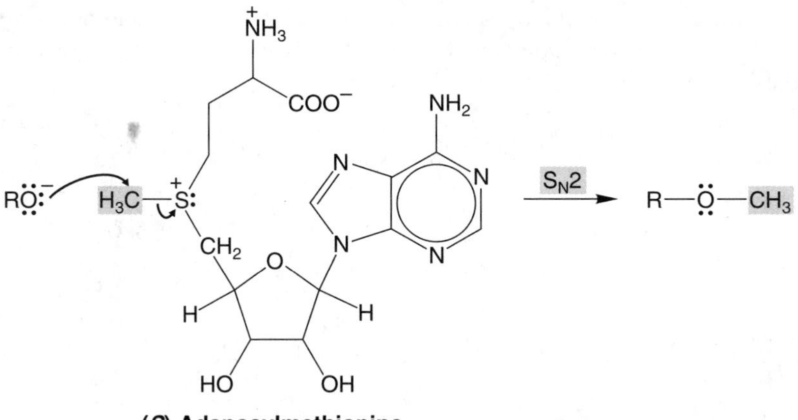

(*S*)-Adenosylmethionine

FIGURE 17.87 (CONTINUED)

PROBLEM 17.33

The reaction shown in Figure 17.88 was used to help clarify the S_N2 nature of the alkylation reactions of (*S*)-adenosylmethionine. Explain why the results are relevant to the mechanistic questions.

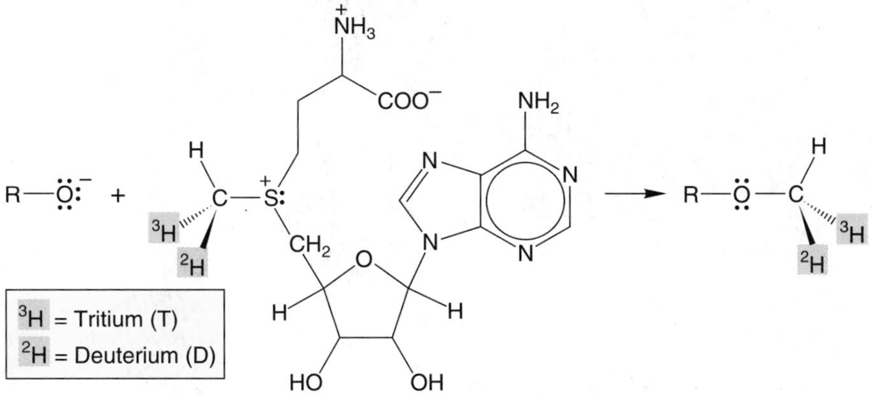

^3H = Tritium (T)
^2H = Deuterium (D)

FIGURE 17.88

The high nucleophilicity of sulfur makes intramolecular displacement reactions especially common. We will see this in Chapter 23 (neighboring group participation), but there is really nothing complicated that you can't figure out now. Try Problem 17.34.

*PROBLEM 17.34

Treatment of the following two compounds with HCl leads to the same chloride (Fig. 17.89). Explain, mechanistically. *Hint*: Watch the sulfur atom. What can sulfur do after an initial protonation of the oxygen atom?

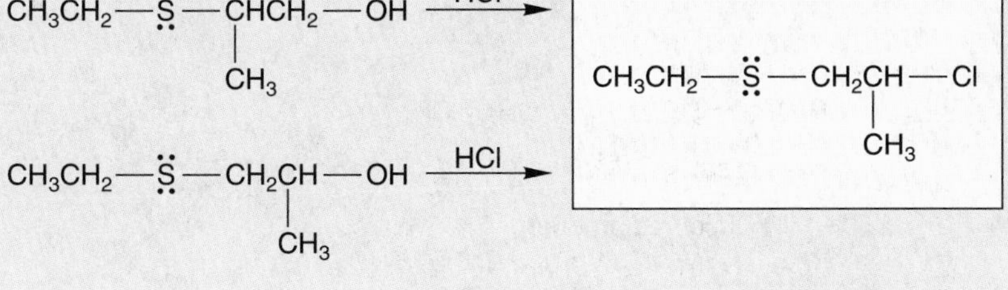

$CH_3CH_2—\ddot{S}—CH(CH_3)CH_2—OH \xrightarrow{HCl}$

$CH_3CH_2—\ddot{S}—CH_2CH(CH_3)—OH \xrightarrow{HCl}$

$CH_3CH_2—\ddot{S}—CH_2CH(CH_3)—Cl$

FIGURE 17.89

ANSWER Right now, this is a hard problem, at least without the rather broad hint. In HCl, the first step is surely protonation of the alcohol, converting the poor leaving group ^{-}OH into OH_2, a good leaving group.

$CH_3CH_2—\ddot{S}—CH(CH_3)—CH_2—\ddot{O}H + H—Cl \rightarrow CH_3CH_2—\ddot{S}—CH(CH_3)—CH_2—\overset{+}{O}H_2 + :Cl:^-$

$CH_3CH_2—\ddot{S}—CH_2—CH(CH_3)—\ddot{O}H + H—Cl \rightarrow CH_3CH_2—\ddot{S}—CH_2—CH(CH_3)—\overset{+}{O}H_2 + :Cl:^-$

Ordinarily, one would now expect the chloride ion to displace water and produce two different chlorides. But the hint tells you to focus on the sulfur. What can sulfur do? You know that sulfur is a powerful nucleophile, and it is lurking right there within the same molecule as the leaving group, water. Try a pair of intramolecular S_N2 displacements.

$CH_3CH_2—\ddot{S}—CH(CH_3)—CH_2—\overset{+}{O}H_2 \xrightarrow{S_N2}$

$CH_3CH_2—\ddot{S}—CH_2—CH(CH_3)—\overset{+}{O}H_2 \xrightarrow{S_N2}$

CH₂CH₃ / :S+ / HC—CH₂ / H₃C + H₂Ö:

A

(an episulfonium ion)

They give the same product, **A**! This molecule is called an episulfonium ion and it behaves much like a bromonium or oxonium ion. As with bromonium ions, chloride ion opens the episulfonium ion by adding to the more substituted carbon (Chapter 10, p. 400). Of course, as **A** is formed from *each* starting alcohol, the product must be the same in each case.

A + Cl^- —(a)⊘→ $CH_3CH_2—S—CH(CH_3)—CH_2—Cl$

A + Cl^- —(b)→ $CH_3CH_2—S—CH_2—CH(CH_3)—Cl$

Observed product

17.13c Reduction of Sulfur Compounds with Raney Nickel: Thioacetals

Like alcohols, thiols form adducts with aldehydes and ketones, called **thioacetals**. Unlike their oxygen counterparts, thioacetals are reduced by a catalyst called **Raney nickel** (essentially just hydrogen absorbed on finely divided nickel) to give hydrocarbons. This reduction gives you a new synthesis of hydrocarbons from carbonyl compounds (Fig. 17.90). Recall that we have already seen two other ways to do this transformation, the Clemmensen and Wolff–Kishner reductions (Chapter 14, p. 642). The reduction of thioacetals is really just a special case of the reduction of sulfides with Raney nickel.

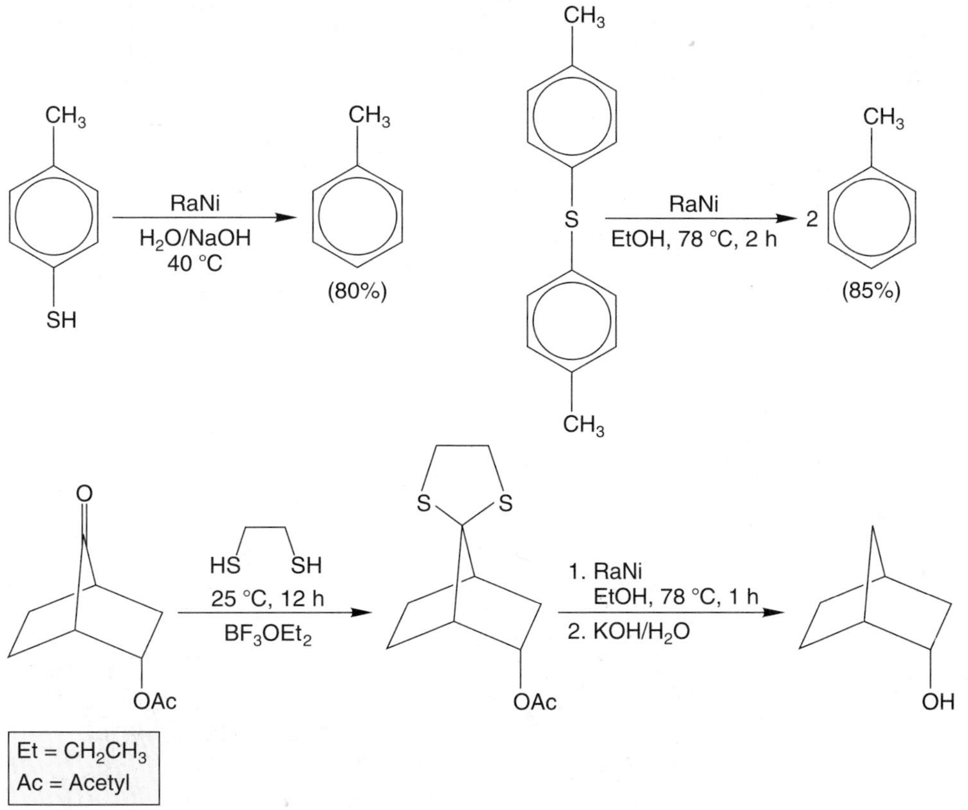

CH_3 SH RaNi $H_2O/NaOH$ 40 °C CH_3 (80%)

CH_3 S CH_3 RaNi EtOH, 78 °C, 2 h 2 CH_3 (85%)

O OAc HS SH 25 °C, 12 h BF_3OEt_2 S S OAc 1. RaNi EtOH, 78 °C, 1 h 2. KOH/H_2O OH

Et = CH_2CH_3
Ac = Acetyl

FIGURE 17.90 Thioacetals can be formed from aldehydes and ketones. Reduction of sulfides (and thioacetals) with Raney nickel gives hydrocarbons.

PROBLEM 17.35 What roles do the BF_3OEt_2 and KOH play in the last reaction of Figure 17.90?

17.13d Oxidation

Like alcohols, thiols can be oxidized, but oxidation takes place not at carbon, as with alcohols, but on sulfur to give **sulfonic acids** (Fig. 17.91).

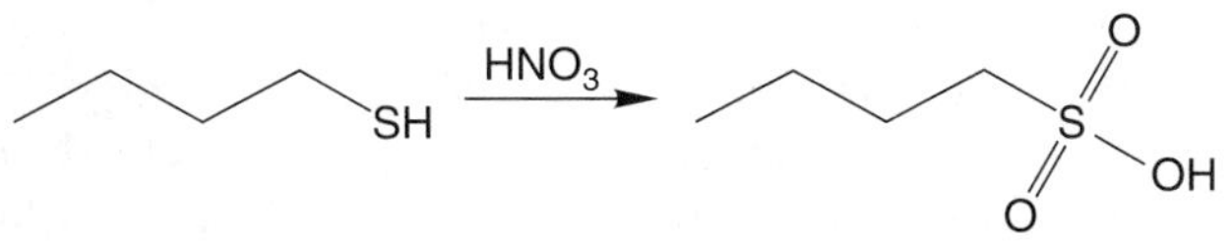

SH HNO_3 O S OH O

Butanethiol **Butanesulfonic acid** (85%)

FIGURE 17.91 Oxidation of mercaptans with nitric acid gives sulfonic acids.

Milder oxidation, using halogens (I_2 or Br_2) in base, is a general source of disulfides. This reaction probably involves a sequence of displacement reactions (Fig. 17.92).

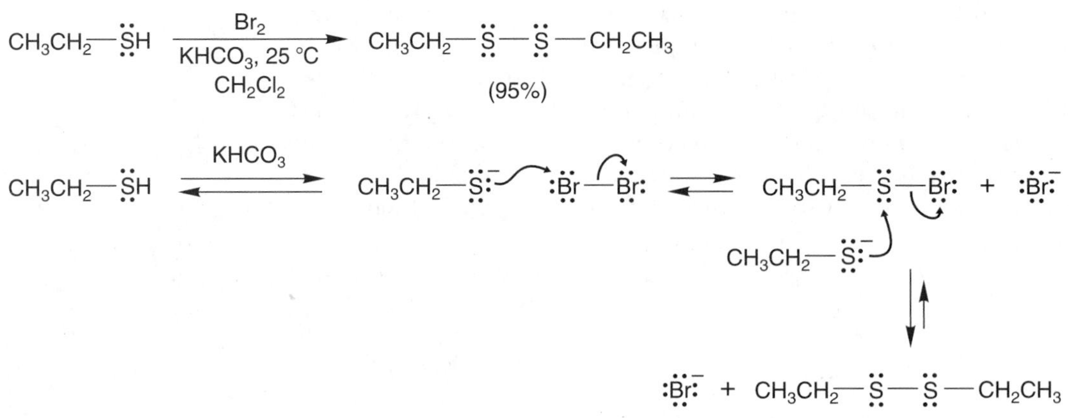

$$CH_3CH_2—SH \xrightarrow[KHCO_3,\ 25\ °C,\ CH_2Cl_2]{Br_2} CH_3CH_2—S—S—CH_2CH_3\ (95\%)$$

$$CH_3CH_2—SH \underset{}{\overset{KHCO_3}{\rightleftharpoons}} CH_3CH_2—S^- \ :Br—Br: \rightleftharpoons CH_3CH_2—S—Br: + :Br^- $$

$$CH_3CH_2—S^- + CH_3CH_2—S—Br \rightleftharpoons :Br^- + CH_3CH_2—S—S—CH_2CH_3$$

FIGURE 17.92 Milder oxidation with bromine or iodine in base gives disulfides.

Unlike ethers, sulfides can easily be oxidized to **sulfoxides**, usually with hydrogen peroxide, H_2O_2. Further or more vigorous oxidation gives **sulfones** (Fig. 17.93).

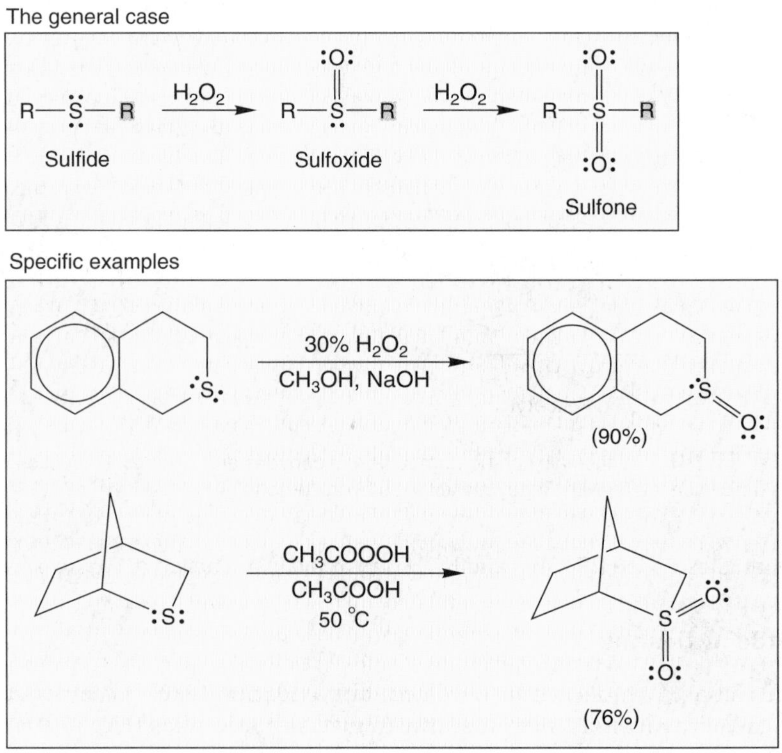

The general case

$$R—S—R \xrightarrow{H_2O_2} R—S(=O)—R \xrightarrow{H_2O_2} R—S(=O)_2—R$$

Sulfide; Sulfoxide; Sulfone

Specific examples

$$\xrightarrow[CH_3OH,\ NaOH]{30\%\ H_2O_2}\ (90\%)$$

$$\xrightarrow[CH_3COOH,\ 50\ °C]{CH_3COOOH}\ (76\%)$$

FIGURE 17.93 Sulfides can be oxidized to sulfoxides with hydrogen peroxide. Further or more vigorous oxidation yields sulfones.

Sulfoxides and sulfones are more complicated structurally than they appear at first. First, they are not planar, but approximately tetrahedral. Second, the uncharged resonance forms are not the major contributors to the structures. Better representations involve charge-separated resonance forms in which the sulfur octet is not expanded (Fig. 17.94).

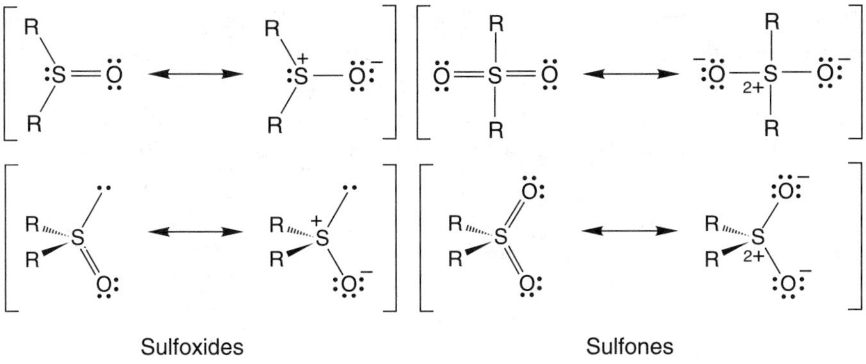

Sulfoxides

Sulfones

FIGURE 17.94 A resonance description of sulfoxides and sulfones.

Thiol and sulfide chemistry resembles alcohol and ether chemistry. There are differences, but many of the differences are quantitative, not fundamental. For example, mercaptides are better nucleophiles than alkoxides, but both species undergo the S_N2 reaction. There are areas in which substantial differences do appear; ethers cannot be reduced as can sulfides, and sulfur is far more prone to oxidation than oxygen.

17.14 SOMETHING MORE: CROWN ETHERS

Cyclic ethers are common solvents, and can often be prepared rather easily through intramolecular cyclization reactions (Williamson ether synthesis). It is possible to make rings in which there is more than one ether linkage. Figure 17.95 gives some examples of cyclic ethers and polyethers.

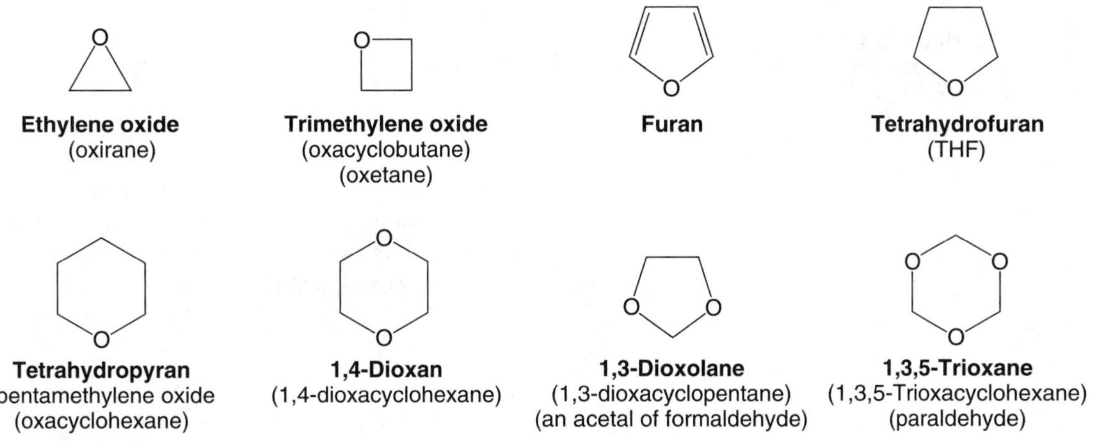

FIGURE 17.95 Some cyclic ethers and polyethers.

We have mentioned more than once the remarkable solvating powers of ethers. Ethers are nucleophiles and can donate electrons to Lewis acids, thus stabilizing them. One example of this is the requirement for an ether, usually diethyl ether or THF, in the formation of the Grignard reagent. Ethers are also polar compounds, and therefore ethers are able to stabilize other polar molecules through noncovalent dipole–dipole interactions.

This idea has been carried to extremes in work that ultimately led to the Nobel prize of 1987 for Charles J. Pedersen (b. 1904) of du Pont, Donald J. Cram (b. 1919) of UCLA, and Jean-Marie Lehn (b. 1939) of Université Louis Pasteur in Strasbourg for opening the field of "host–guest" chemistry. Pedersen discovered that certain cyclic polyethers had a remarkable affinity for metal cations. Molecules were constructed whose molecular shapes created different sized cavities into which different metal ions fit well. Because of their vaguely crown-shaped structures, these molecules came to be called **crown ethers**. Figure 17.96 shows two of them.

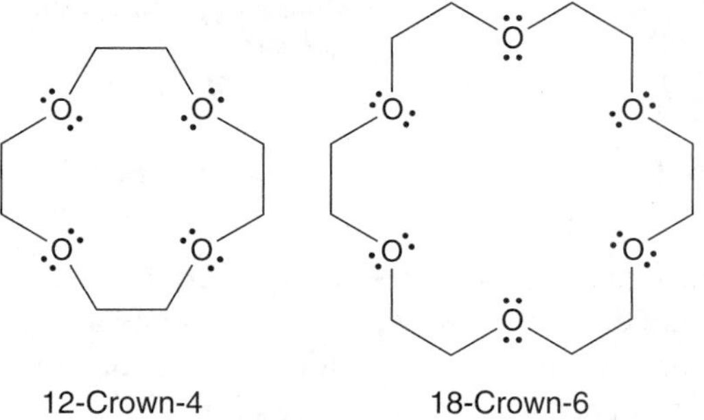

FIGURE 17.96 Two crown ethers

The molecule 12-crown-4 is the proper size to capture lithium ions, and 18-crown-6 has a remarkable affinity for potassium ions. The first number shows the number of atoms in the ring, and the final number shows the number of heteroatoms in the ring (Fig. 17.97).

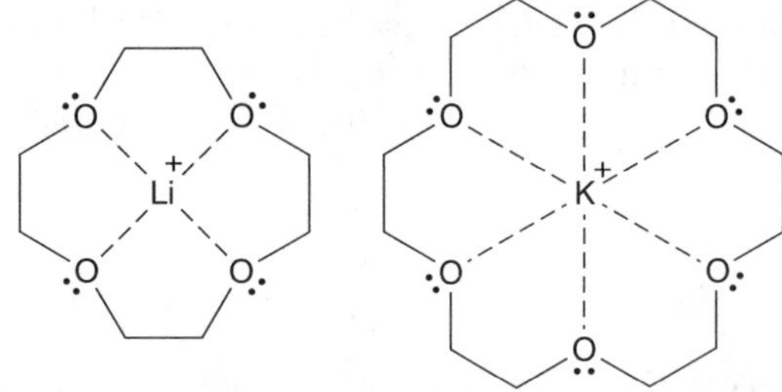

FIGURE 17.97 These crown ethers can stabilize positive ions of different sizes, depending on the size of the cavity.

Three-dimensional versions of these essentially two-dimensional molecules have now also been made, using both oxygen and other heteroatoms such as sulfur and nitrogen as the Lewis bases. These molecules are called **cryptands**, and can incorporate positive ions into the roughly spherical cavity within the cage (Fig. 17.98).

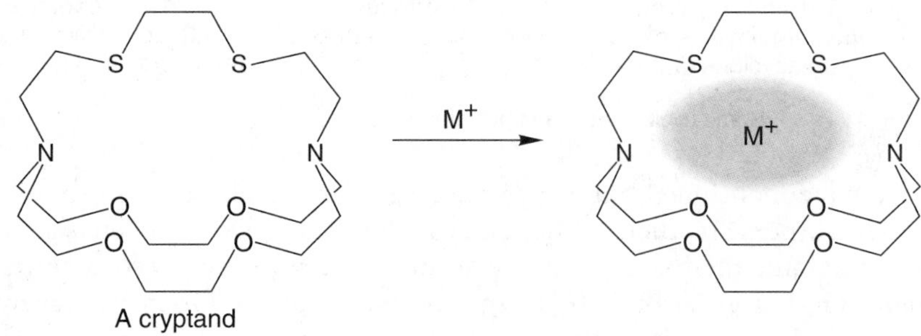

FIGURE 17.98 Three-dimensional cryptands can incorporate metal ions into a central cavity.

There are chemical and larger implications of this work. Potassium permanganate is an effective oxidizing agent. Its great insolubility in organic reagents limits its use, however. For example, the deep purple $KMnO_4$ is completely insoluble in benzene. It simply forms a solid suspension when added to the benzene, which remains a colorless liquid. The addition of a little 18-crown-6 results in the dissolution of the permanganate, and the benzene takes on the deep purple color of the inorganic ion. The potassium ion has been effectively solvated by the crown ether and now dissolves in the organic solvent. The permanganate counterion, though not solvated by the crown compound, accompanies the positive ion into solution, as it must to preserve electrical neutrality. The purple benzene solution can be used as an effective oxidizing agent.

Other uses are more exciting, though they still lie in the future. One can imagine schemes where specially designed crown compounds are used to scavenge poisonous heavy metal ions from water supplies, for example. There have been less humanitarian, though admittedly clever schemes, for extracting the minute amounts of gold ions present in seawater, by passing vast amounts of water over some yet-to-be-discovered crown compound.

17.15 SUMMARY

NEW CONCEPTS

In the bulk of this chapter we use concepts, reactions, and mechanisms we encountered earlier. We attempt to summarize what we learned, and to apply that knowledge to some new reactions. So, even though this is a summary chapter, there are several new reactions and even some new concepts worth keeping in mind.

The hydrogen bond is introduced. This kind of bond is nothing more than a partially completed Brønsted acid–base reaction in which a partial bond is formed between a pair of electrons on one atom and a hydrogen on another (Fig. 17.99).

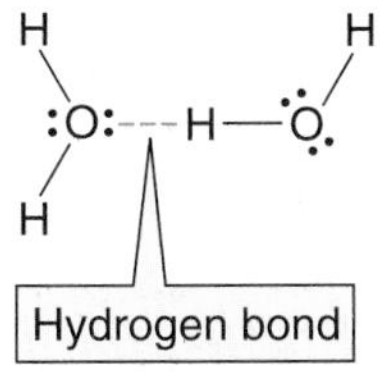

FIGURE 17.99 A hydrogen bond.

Throughout this chapter, the technique of changing a poor leaving group into a good one is used. In alcohols, ^{-}OH must play the part of the leaving group in either S_N2/S_N1 or E2/E1 reactions, and it is very poor at this role. One easy way to change OH into a good leaving group is to protonate it, converting the leaving group into water, OH_2. Alternatively, a sulfonate ester can be formed through treatment with a sulfonyl chloride (Fig. 17.23).

The notion of chain lengthening was introduced. Two protocols are mentioned, and both methods involve reactions of organometallic reagents. In the first reaction, formaldehyde is used to add one carbon; in the second, ethylene oxide is used to add two (Fig. 17.78).

A small point is the use of the intramolecular S_N2 reaction to form rings. A classic example is the Williamson ether synthesis, but one can imagine many variations (Fig. 17.59).

Sulfur chemistry appears in quantity for the first time. There is a general correspondence between the reactions of oxygen compounds and those of sulfur compounds. Important differences include the greater nucleophilicity of the sulfur compounds and the tendency of the sulfur-containing molecules to undergo oxidation reactions at sulfur, not carbon.

Cyclic polyethers, called crown ethers are introduced. These compounds have the ability to form complexes with metal ions depending on the match between the size of the ion and that of the cavity of the crown ether.

REACTIONS, MECHANISMS, AND TOOLS

In this chapter, many old friends reappear, including a number of S_N1, S_N2, E1, and E2 reactions, polar additions, and several reactions of carbonyl compounds. The old reactions have been described in detail elsewhere and are only collected here according to general mechanistic type. New variations of old reactions and genuinely new reactions will be described in more detail.

S_N1 and S_N2 Reactions The conversion of alcohols into halides, as well as the reverse reaction, the formation of alcohols from halides, both involve S_N1 and S_N2 reactions (Fig. 17.21).

E1 and E2 Reactions These eliminations include both the obvious conversion of alcohols into alkenes, but also the key step in the oxidation of alcohols to carbonyl compounds, in which an elimination occurs in a chromate ester (Chapter 16, p. 795; Fig. 17.100).

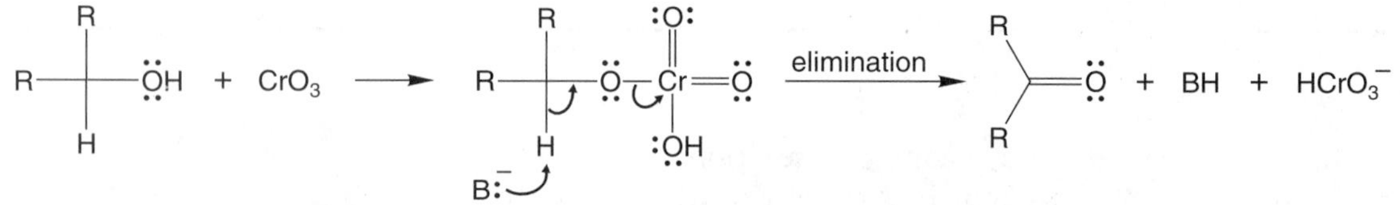

FIGURE 17.100 At the heart of the oxidation of alcohols to ketones is an elimination reaction.

Carbonyl Reactions Hydride and organometallic addition reactions lead, after a hydrolysis step, to alcohols. Acetals are formed through the acid-catalyzed reaction of carbonyl compounds with alcohols.

NEW AND SPECIALLY EMPHASIZED REACTIONS

Addition Reactions Additions of nucleophiles to epoxides, though straightforward examples of the S_N2 reaction, constitute an important and useful way of adding two carbons to a chain. Three base-catalyzed addition reactions were emphasized: those of $^-$OH, $^-$H, and $^-$R. Epoxides can also be opened under acidic conditions (Fig. 17.101).

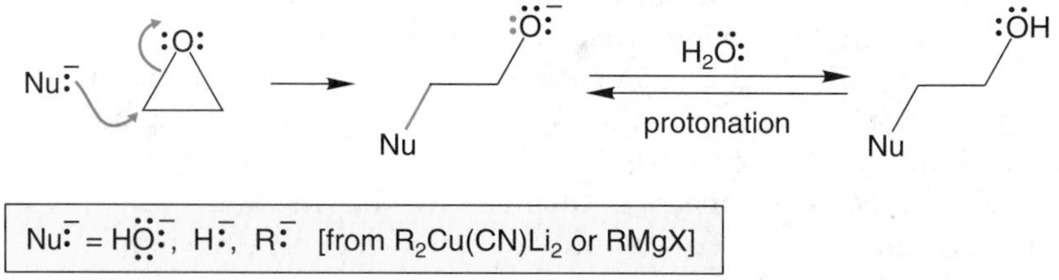

FIGURE 17.101 Epoxides can be opened by a variety of nucleophiles.

Alkoxide and Mercaptide Formation Alcohols are converted into alkoxides by treatment with certain metals or strong base. The analogous formation of mercaptides from thiols is easier because thiols are stronger acids than alcohols and proton removal from them is therefore easier (Fig. 17.102).

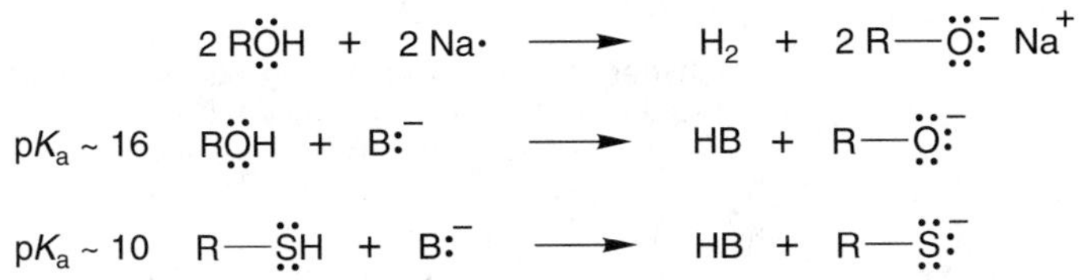

$$2\,RÖH + 2\,Na\cdot \longrightarrow H_2 + 2\,R{-}\ddot{O}{:}^-\ Na^+$$

$$pK_a \sim 16 \quad RÖH + B{:}^- \longrightarrow HB + R{-}\ddot{O}{:}^-$$

$$pK_a \sim 10 \quad R{-}\ddot{S}H + B{:}^- \longrightarrow HB + R{-}\ddot{S}{:}^-$$

FIGURE 17.102 Alkoxide and mercaptide formation.

Oxidation Reactions Several new oxidation reactions appear. We already know that alkenes are converted into cis 1,2-diols by potassium permanganate or osmium tetroxide (Chapter 10, p. 423). The cis diols can be cleaved to form carbonyl compounds by periodate oxidation. Ethers are easily oxidized at their α-positions by molecular oxygen to give the dangerous peroxides and hydroperoxides. Several oxidations at sulfur appear for the first time (Fig. 17.103).

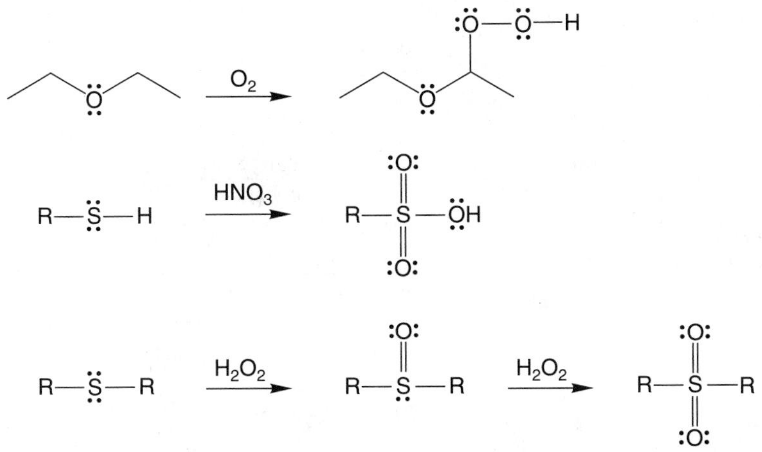

O_2 ; :Ö—Ö—H

HNO₃ → R—S(=O)₂—ÖH ; R—S̈—H

$$R{-}\ddot{S}{-}H \xrightarrow{HNO_3} R{-}S(=O)_2{-}\ddot{O}H$$

$$R{-}\ddot{S}{-}R \xrightarrow{H_2O_2} R{-}\ddot{S}(=O){-}R \xrightarrow{H_2O_2} R{-}S(=O)_2{-}R$$

FIGURE 17.103 Various new oxidation reactions.

Rearrangement Reactions The pinacol rearrangement converts a 1,2-diol into a carbonyl compound (Fig. 17.104).

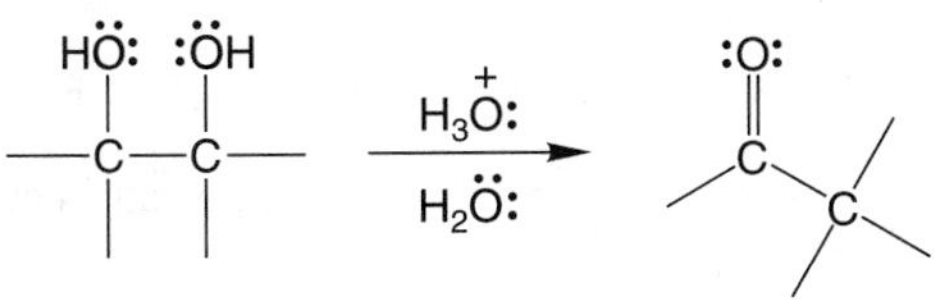

HÖ: :ÖH

$H_3\overset{+}{O}{:}$ / $H_2\ddot{O}{:}$

:O:

FIGURE 17.104 The pinacol rearrangement.

SYNTHESES

Many new synthetic reactions appear in this chapter. Both the new and the old syntheses are collected in this section (Fig. 17.105)

1. Acids

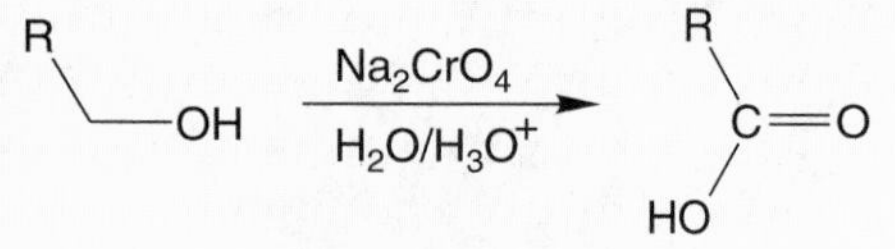

$RCH_2OH \xrightarrow[H_2O/H_3O^+]{Na_2CrO_4} RCOOH$

Other oxidizing agents also work

2. Alcohols

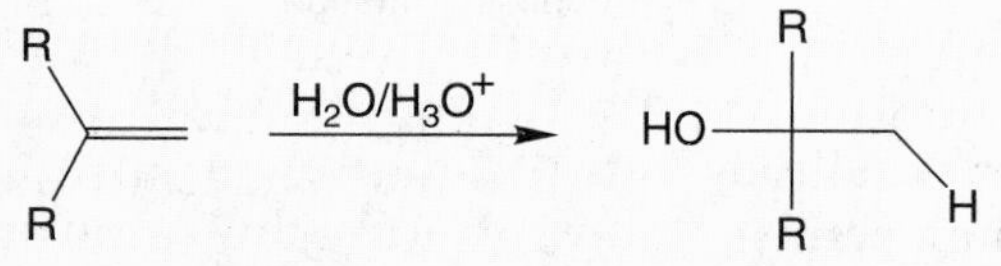

$R_2C{=}CR_2 \xrightarrow{H_2O/H_3O^+}$ HO—CR₂—CR₂H

Addition is in the Markovnikov sense; rearrangements are possible

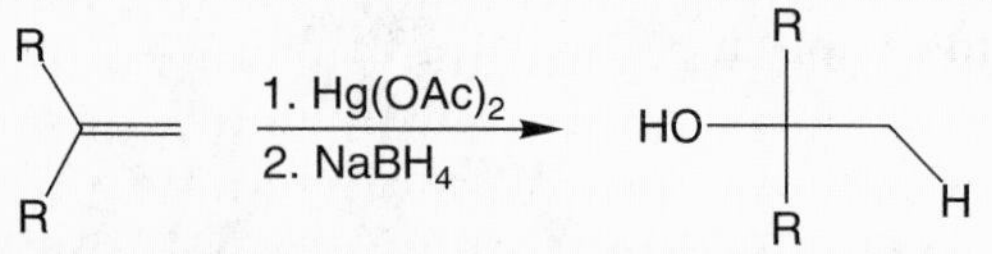

$R_2C{=}CR_2 \xrightarrow[2.\ NaBH_4]{1.\ Hg(OAc)_2}$ HO—CR₂—CR₂H

Oxymercuration–reduction also gives Markovnikov addition, but without rearrangement

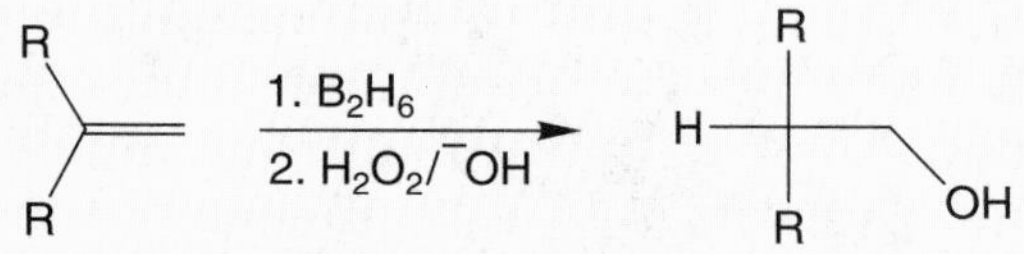

$R_2C{=}CR_2 \xrightarrow[2.\ H_2O_2/^-OH]{1.\ B_2H_6}$ H—CR₂—CR₂OH

Hydroboration–oxidation gives anti-Markovnikov addition; the stereochemistry of addition is cis

$R{-}X \xrightarrow{H_2O/H_3O^+} R{-}OH$

S_N1 reaction; R must be tertiary or otherwise stabilized; X = halide or other good leaving group

$R{-}X \xrightarrow{H_2O/HO^-} R{-}OH$

S_N2 reaction; R cannot be tertiary; X = halide or other good leaving group

Ethylene oxide $\xrightarrow[2.\ H_2O]{1.\ RMgX}$ RCH_2CH_2OH

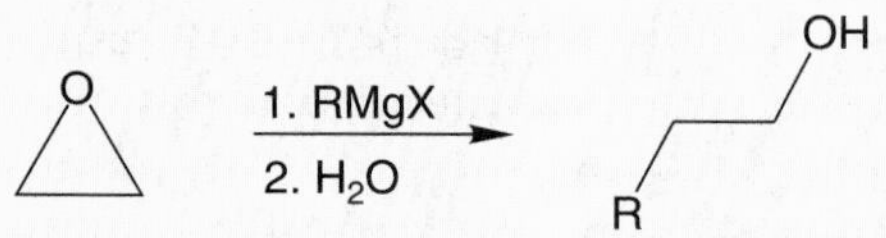

Organocopper reagents also work

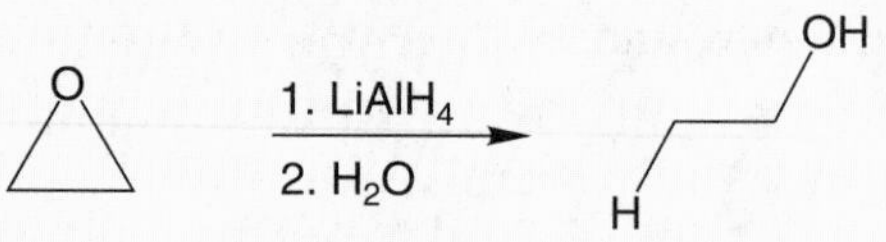

Ethylene oxide $\xrightarrow[2.\ H_2O]{1.\ LiAlH_4}$ HCH_2CH_2OH

2. Alcohols (continued)

$R_2C{=}O \xrightarrow[2.\ H_2O]{1.\ RMgX}$ R—CR₂—OH

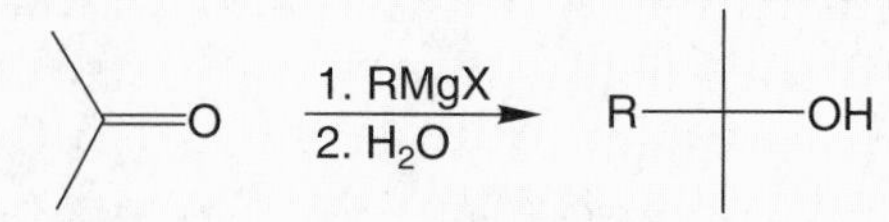

Aldehydes (except formaldehyde) give secondary alcohols; ketones give tertiary alcohols; organolithium reagents also work

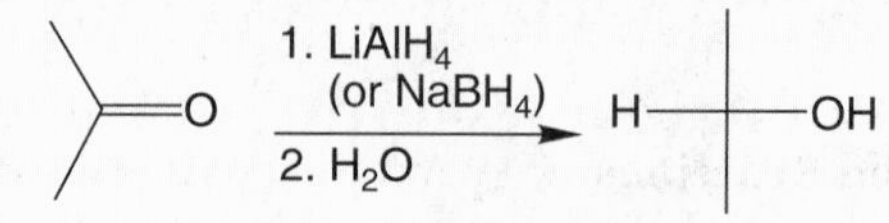

$R_2C{=}O \xrightarrow[2.\ H_2O]{1.\ LiAlH_4\ (or\ NaBH_4)}$ H—CR₂—OH

Aldehydes give primary alcohols; ketones give secondary alcohols

$C_6H_5N_2^+\ Cl^- \xrightarrow[\Delta]{H_2O} C_6H_5OH$

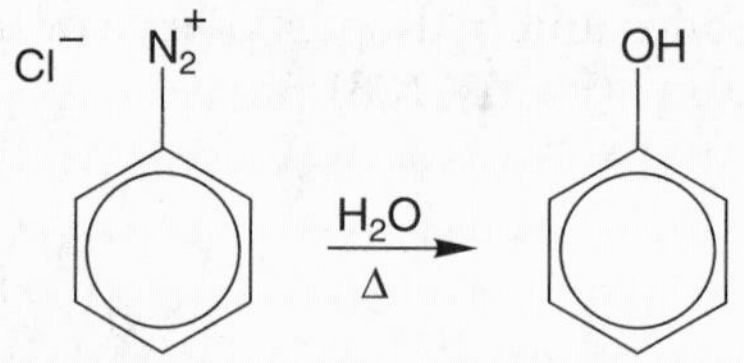

$CO + 2\,H_2 \xrightarrow[\Delta,\ pressure]{catalyst} CH_3OH$

3. Aldehydes

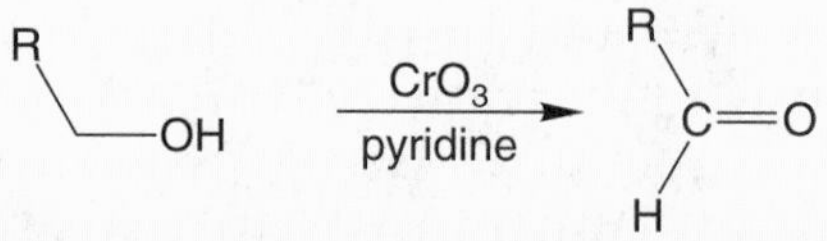

$RCH_2OH \xrightarrow[pyridine]{CrO_3} RCHO$

Oxidation of a primary alcohol; no water may be present; MnO_2 also works

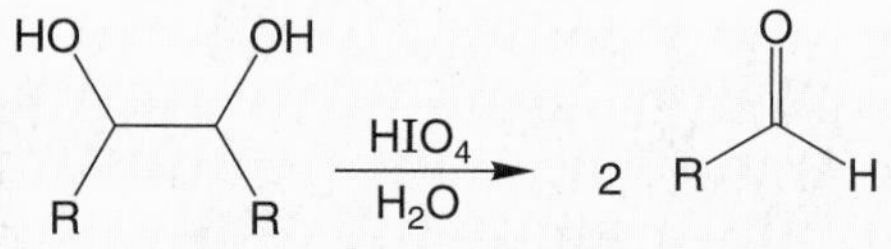

$RCH(OH){-}CH(OH)R \xrightarrow[H_2O]{HIO_4} 2\ RCHO$

Periodate cleavage of a 1,2-glycol; a cyclic ester is an intermediate

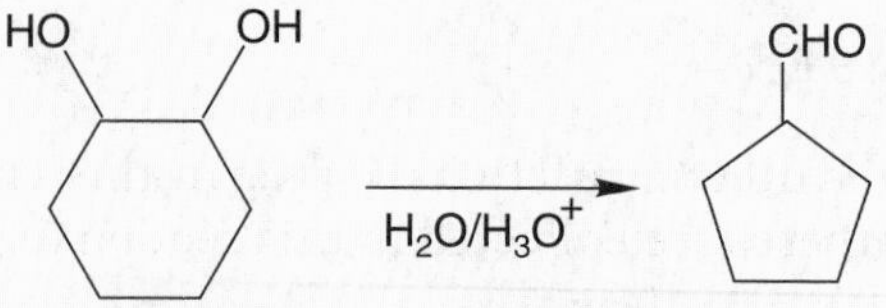

1,2-Cyclohexanediol $\xrightarrow{H_2O/H_3O^+}$ cyclopentanecarboxaldehyde (CHO)

In cyclic 1,2-diols, the pinacol rearrangement leads to ring-contracted aldehydes

FIGURE 17.105 Old and new syntheses mentioned in this chapter

4. Alkenes

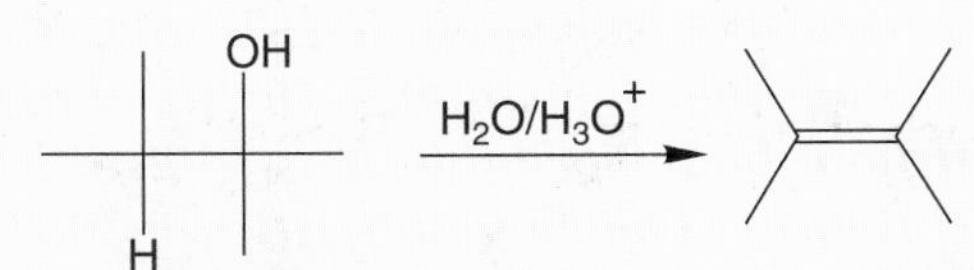

E1 reaction; E2 reactions also work provided the leaving group is changed from OH to something better

5. Alkoxides

R—OH $\xrightarrow{\text{Na or K}}$ H_2 + R—O^- M^+

R—OH $\xrightarrow{\text{strong base}}$ R—O^-

Sodium hydride is a favorite base for this reaction

RMgX or $R_2Cu(CN)Li_2$

The alkoxide is an intermediate in alcohol formation

$LiAlH_4$

The alkoxide is an intermediate in alcohol formation

6. Disulfides

RHS $\xrightarrow[\text{base}]{I_2}$ RSSR

Mild oxidation of thiols

7. Ethers

RO^- + R—X $\xrightarrow{S_N2}$ RO—R + X^-

Williamson ether synthesis; the alkyl halide, RX, must be able to undergo the S_N2 reaction

ROH + $CH_3OSO_2OCH_3$ ⟶ RO—CH_3

Reaction of alcohols with dimethyl sulfate gives methyl ethers

Br_2 / ROH

The ether oxygen always becomes attached to the more substituted end of the alkene

7. Ethers (continued)

1. $Hg(OAc)_2$/ROH 2. $NaBH_4$

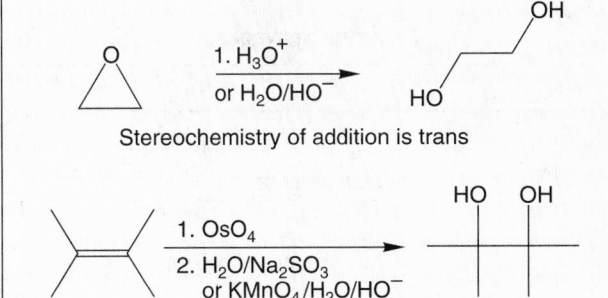

Oxymercuration–reduction in alcohol; note the Markovnikov addition

RO^- + $^+SR_3$ $\xrightarrow{S_N2}$ RO—R

Alkylation using sulfonium ions

ROH / RO^-

Nucleophilic aromatic substitution

8. Glycols

1. H_3O^+ or H_2O/HO^-

Stereochemistry of addition is trans

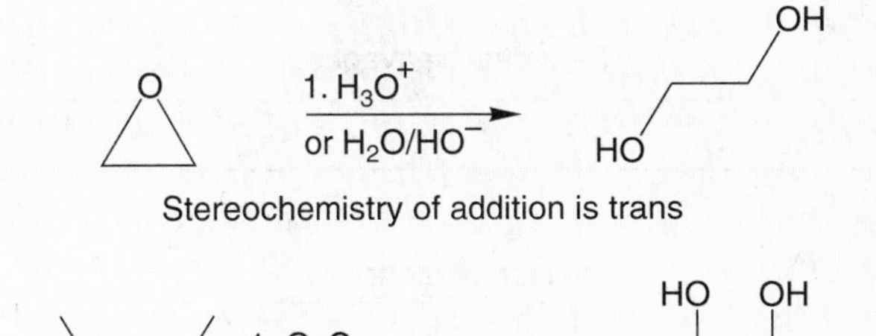

Stereochemistry of addition is cis

9. Halides

R—OH $\xrightarrow{HX}$ RX + H_2O

S_N1 or S_N2, depending on the structure of R

ROH + $SOCl_2$ ⟶ RCl + SO_2 + HCl

Other reagents such as PI_3, PBr_3, and PCl_5 also give halides

ROR $\xrightarrow{HX}$ RX + ROH

Ethers are cleaved by strong haloacids

FIGURE **17.105** (CONTINUED)

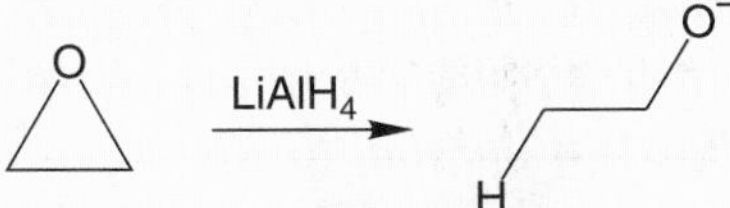

10. Hydrocarbons

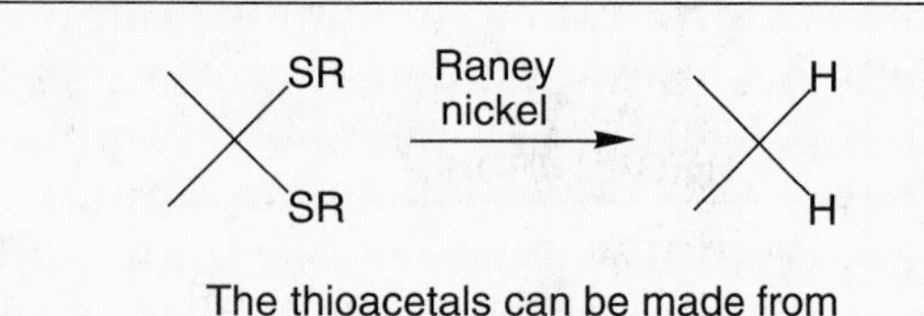

The thioacetals can be made from carbonyl compounds

11. Ketones

$R_2CHOH \xrightarrow[H_3O^+]{Na_2CrO_4} R_2C{=}O$

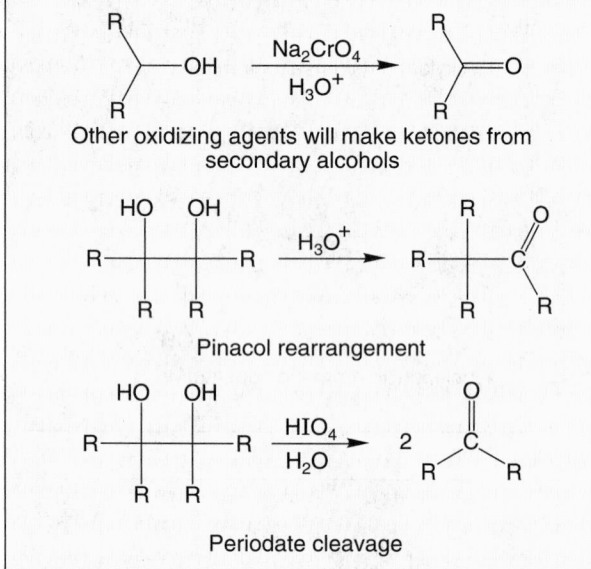

Other oxidizing agents will make ketones from secondary alcohols

$R_2C(OH)-C(OH)R_2 \xrightarrow{H_3O^+} R_3C-C(=O)R$

Pinacol rearrangement

$R_2C(OH)-C(OH)R_2 \xrightarrow[H_2O]{HIO_4} 2\ R_2C{=}O$

Periodate cleavage

12. Oxiranes

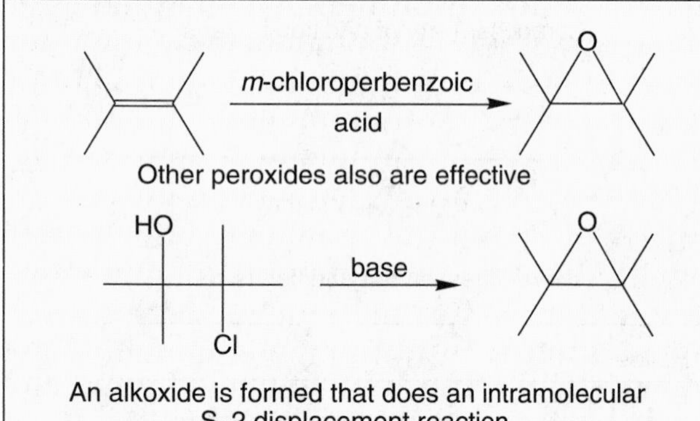

An alkoxide is formed that does an intramolecular S_N2 displacement reaction

13. Peroxides

$\text{Ethers} \xrightarrow{O_2} \text{Peroxides}$

A radical mechanism forms explosive peroxides if the ether has an α-hydrogen.

14. Sulfides

$RS^- + RX \longrightarrow RS{-}R + {}^-X$

This reaction is an S_N2 displacement so RX cannot be tertiary

15. Sulfones

$RSR + H_2O_2 \longrightarrow R{-}S(=O)_2{-}R$

The sulfoxide is an intermediate

16. Sulfonic acids

$RSH + HNO_3 \longrightarrow R{-}S(=O)_2{-}OH$

17. Sulfonium ions

$R_2S + RX \longrightarrow R_2\overset{+}{S}{-}R$

Another S_N2 reaction; R cannot be tertiary

18. Sulfoxides

$RSR + H_2O_2 \longrightarrow R{-}S(=O){-}R$

Further oxidation gives the sulfone

19. Thioacetals

$C{=}O \xrightarrow[\text{acid}]{RSH} C(SR)_2$

This reaction is completely analogous to the formation of acetals

20. Thiols

$HS^- + RX \longrightarrow HS{-}R$

Yet another S_N2 reaction; R cannot be tertiary

21. Tosylates

$ROH + TsCl \longrightarrow ROTs$

Tosylate is an ester of *p*-toluenesulfonic acid

FIGURE **17.105** (CONTINUED)

COMMON ERRORS

Once again, the easiest, and most serious mistake is to lose sight of generalities because of the prolific detail. There are so many different reactions and so many detailed differences in reactions that to memorize them all is a task beyond almost everyone. Try hard to see the relationships that tie reactions together.

17.16 KEY TERMS

Alkoxide ion The conjugate base of an alcohol, RO^-.

Crown ether A cyclic polyether often capable of forming complexes with metal ions. The ease of complexation depends on the size of the ring and the number of oxygen atoms in the ring.

Cryptand A three-dimensional, bicyclic, counterpart of a crown ether. Various heteroatoms (O, N, S) act to complex metal ions that fit into the cavity.

Diol A molecule containing two OH groups. Also called a glycol.

Epoxide A three-membered ring containing oxygen. Also called an oxirane or oxacyclopropane.

Ester An acid in which one or more OH groups has been converted into an OR group or groups.

Ether A compound of the general structure, ROR, or ROR′.

Hydrogen bonding A low-energy bond between a pair of electrons, usually on oxygen, and a hydrogen.

Glycol A dialcohol. See **diol**.

Hydroperoxide (ROOH) The prefix *per* means "extra oxygen."

Mercaptan (RSH) A thiol. The sulfur counterpart of an alcohol.

Mercaptide (RS^-) The sulfur counterpart of an alkoxide.

Oxacyclopropane A three-membered ring containing oxygen, also called an epoxide or oxirane.

Oxirane A three-membered ring containing oxygen, also called an epoxide or oxacyclopropane.

Peroxide (ROOR) The prefix *per* means "extra oxygen."

Pinacol rearrangement An acid-catalyzed rearrangement reaction leading from 1,2-diols to carbonyl compounds. The rearrangement is driven by the conversion of a localized carbocation into a resonance-stabilized carbocation.

Raney nickel Hydrogen absorbed on finely divided nickel.

Sulfide (RSR) The sulfur counterpart of an ether.

Sulfone ($R{-}SO_2{-}R$) The final product of oxidation of a sulfide with hydrogen peroxide.

Sulfonic acid A compound of the general structure RSO_2OH.

Sulfonium ion (R_3S^+) A positively charged sulfur ion.

Sulfoxide (R–SO–R) The initial product of oxidation of a sulfide by peroxide.

Tetrahydropyranyl (THP) ether A protecting group for an alcohol. These ethers are stable to base, but not to acid.

Thioacetal The sulfur counterpart of an acetal.

Thioether (RSR) The sulfur counterpart of an ether. A sulfide.

Thiol The sulfur counterpart of an alcohol, RSH.

Thionyl chloride ($SOCl_2$) An effective reagent for converting alcohols into chlorides.

Williamson ether synthesis The S_N2 reaction of an alkoxide with R–L to give an ether.

17.17 ADDITIONAL PROBLEMS

PROBLEM **17.36** Name the following compounds:

(a) OH

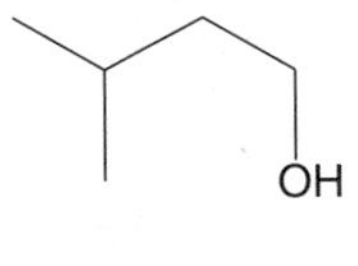

(b) OH

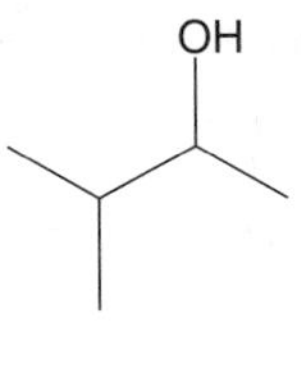

(c) OH

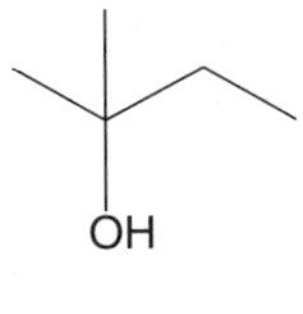

(d) OH

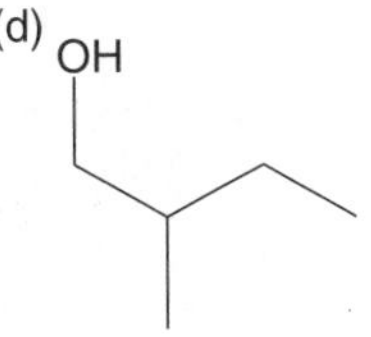

(e) OH F

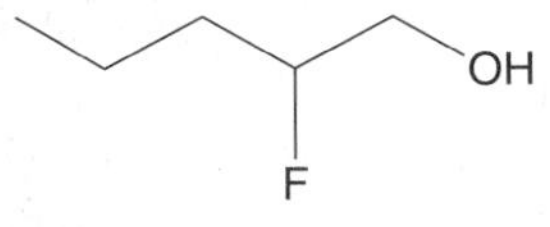

PROBLEM 17.36 (CONTINUED)

(f) (g)

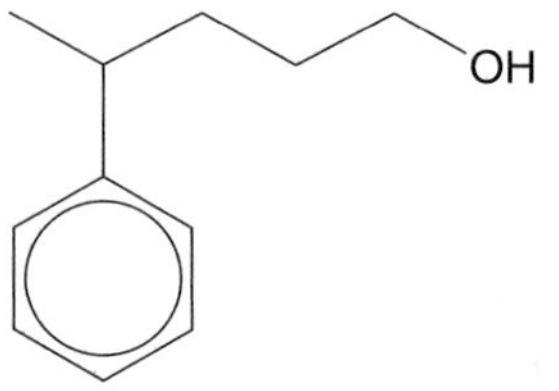

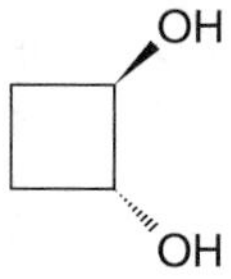

(h) 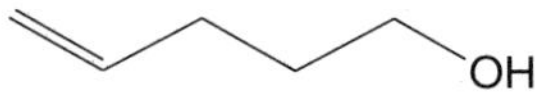 (i)

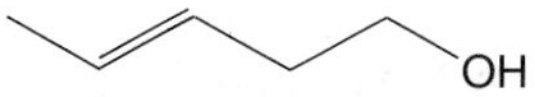

(j) 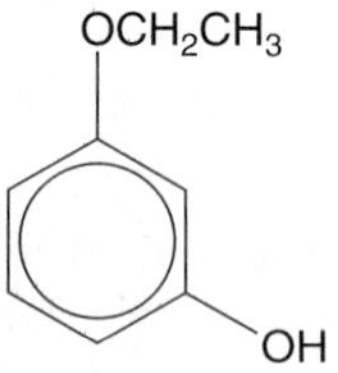 (k) 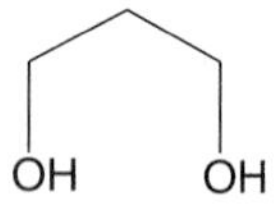 (l)

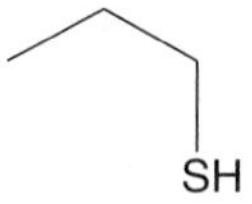

(m)

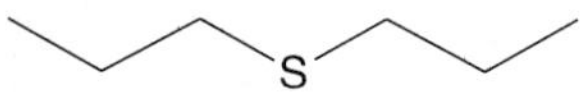

(n)

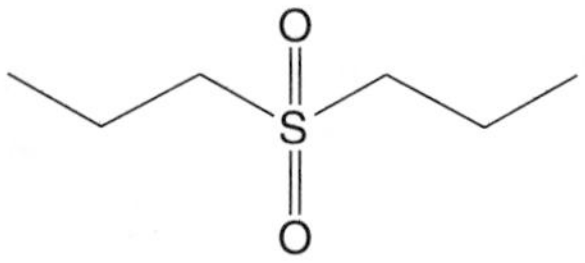

(o)

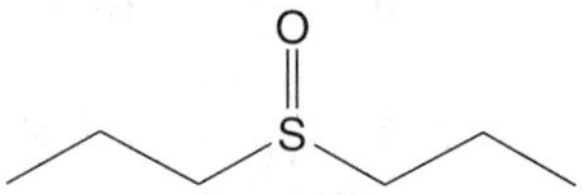

PROBLEM 17.37 Draw structures for the following compounds:

(a) 1,4-Butanediol
(b) 3-Amino-1-pentanol
(c) *cis*-4-*tert*-Butoxycyclohexanol
(d) 2,2,4,4-Tetramethyl-3-pentanol
(e) (*R*)-*sec*-Butyl alcohol
(f) *m*-Chlorobenzenethiol
(g) *tert*-Butyl cyclopentyl ether

PROBLEM 17.38 Draw structures for the following compounds:

(a) Neopentyl alcohol
(b) 18-Crown-6
(c) Tetrahydropyran
(d) THF
(e) Furan

PROBLEM 17.39 In the following molecules, there exists the possibility of forming more than one alkoxide or thiolate. In each case, explain which will be formed preferentially, and why.

(a) (solution)

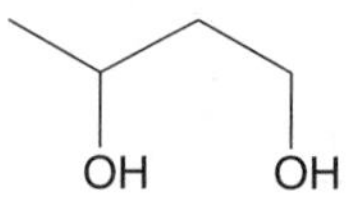

(b) (gas phase)

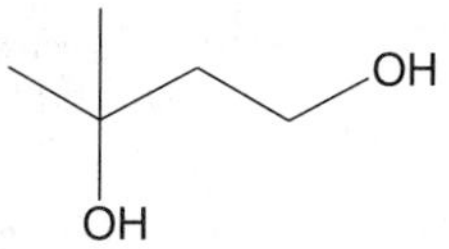

(c) (solution)

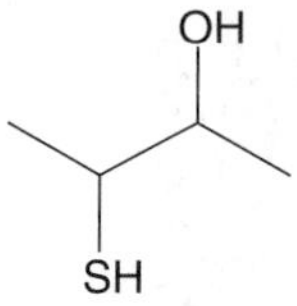

PROBLEM 17.40 Predict the relative acidities of the following three alcohols:

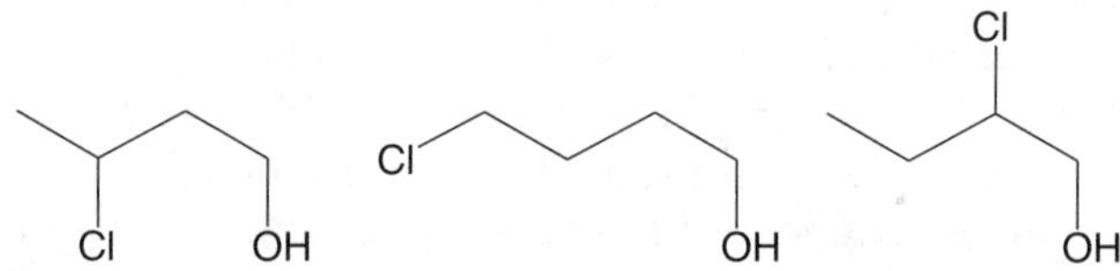

PROBLEM 17.41 Write all the isomers of the alcohols of the formula $C_6H_{14}O$.

PROBLEM 17.42 When reactions are run in THF, it is often possible to isolate traces of an impurity, compound **1**. Compound **1** gives 81.76% C and 10.98% H upon elemental analysis. Spectral data for **1** are summarized below. Deduce the structure of compound **1** and explain your reasoning.

Compound **1**
Mass spectrum: $m/z = 220$ (p, 23), 205 (100), 57 (27)
IR (Nujol): 3660 (m, sharp) cm^{-1}
^{1}H NMR ($CDCl_3$): δ 1.43 (s, 18H), 2.27 (s, 3H), 5.00 (s,1H), 6.98 (s, 2H)
^{13}C NMR ($CDCl_3$): δ 21.2 (q), 30.4 (q), 34.2 (s), 125.5 (d), 128.2 (s), 135.8 (s), 151.5 (s)

PROBLEM 17.43 Write all the possible products resulting from treatment of alcohol **1** with HCl/H_2O.

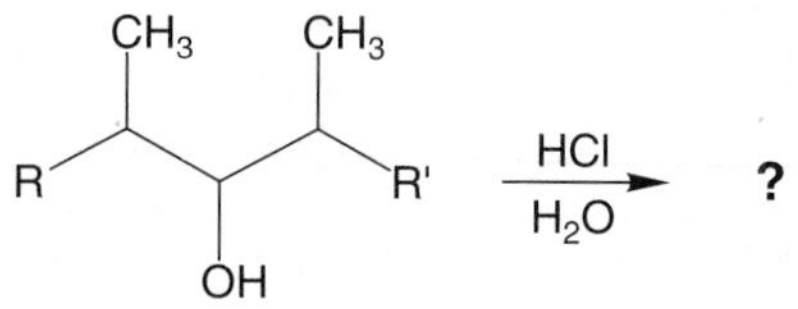

1

PROBLEM 17.44 How would you convert 1-butanol into the following compounds?

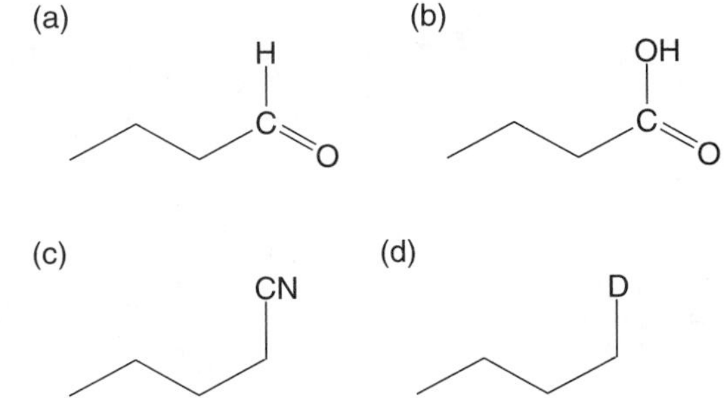

PROBLEM 17.45 There is one alcohol that can be synthesized in one or two easy steps from each of the following precursors. What is the alcohol, and how would you make it from each starting material?

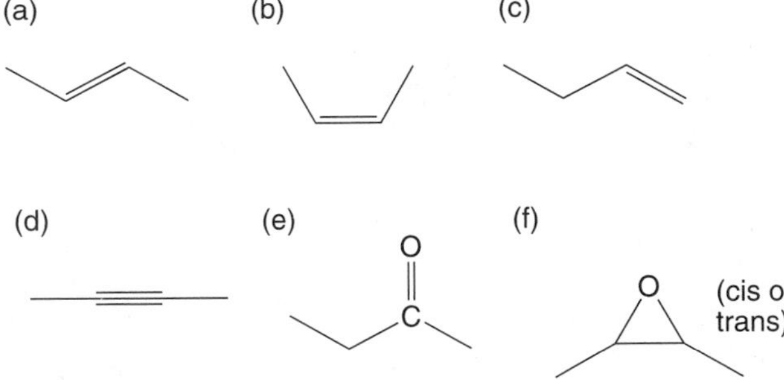

PROBLEM 17.46 Treatment of butyl ethyl ether with HI leads to both butyl iodide and ethyl iodide, along with the related alcohols. By contrast, when *tert*-butyl ethyl ether is treated the same way, only ethyl iodide and *tert*-butyl alcohol are formed. *tert*-Butyl iodide and ethyl alcohol are not produced. Explain.

$CH_3CH_2CH_2CH_2—O—CH_2CH_3$

↓ HI

$CH_3CH_2CH_2CH_2I + CH_3CH_2I$
$CH_3CH_2CH_2CH_2OH + CH_3CH_2OH$

$(CH_3)_3C—O—CH_2CH_3 \xrightarrow{HI} (CH_3)_3COH + CH_3CH_2I$

PROBLEM 17.47 Outline a method for converting (*R*)-2-hexanol into (*S*)-2-cyanohexane.

OH (*R*) —?→ CN (*S*)

PROBLEM 17.48 What are the principal organic products of the following reactions?

(a) $H_3C—O—CH_3 \xrightarrow{HI}$

(b) $H_3C—O—CH_3 \xrightarrow[CH_3CH_2OH]{CH_3CH_2O^-}$

(c) $(CH_3)_3C—O^- \xrightarrow{H_3C—I}$

(d) $H_3C—O^- \xrightarrow{(CH_3)_3C—I}$

(e) $CH_3OH \xrightarrow[pyridine]{TsCl}$

(f) Ph–CH(OH)–CH(OH)–CH_3 $\xrightarrow[H_2SO_4]{HIO_4}$

(g) Cl–CH_2CH_2–OH $\xrightarrow[H_2O]{NaOH}$

(h) $(H_3C)_2C{=}CH_2 \xrightarrow[CH_3OH]{Br_2}$

(i) $(H_3C)_2C{=}CH_2 \xrightarrow[H_2O]{H_3O^+}$

(j) $(H_3C)_2C{=}CH_2 \xrightarrow[2.\ NaBH_4/NaOH]{1.\ Hg(OAc)_2/CH_3CH_2OH}$

(k) $PhO^- \xrightarrow{H_3C—I}$

(l) $CH_3O^- \xrightarrow{Ph—I}$

(m) (dihydropyran, O) $\xrightarrow[CH_3\overset{+}{O}H_2]{CH_3OH}$

(n) $CH_3CH_2O—Si(CH_3)_3$ —KF / H_2O→

(o) —H_2O / H_3O^+→

(p) —1. $LiAlD_4$ 2. H_2O→

(q) PhO^- —$(CH_3)_3S^+$→

(r) —RaNi (H_2)→

(s) SH —HNO_3→

(t) CHO —1. $NaBH_4$ 2. H_2O→

PROBLEM **17.49** Predict a structure for the product and outline the steps in this change. *Hint*: See Problem 17.21 and Figure 17.49.

OH, OH —H_3O^+ / H_2O→ $C_7H_{12}O$ IR: (CCl_4) 1729 cm^{-1} ^1H NMR, δ 9.5 (1H)

PROBLEM **17.50** Provide structures for compounds **A–C**.

O —1. $LiAlH_4$ 2. H_2O→ $C_6H_{12}O$ **A** —PCl_3→ $C_6H_{11}Cl$ **B** —1. Li 2. $(CH_3)_2C{=}O$ 3. H_2O→ $C_9H_{18}O$ **C** —H_2O / H_3O^+→ H_3C, CH_3, C

PROBLEM **17.51** Provide structures for compounds **A–C**.

—Br_2 / H_2O→ C_8H_9BrO **A** —NaOH / H_2O→ C_8H_8O **B** —H_3O^+ / H_2O→ $C_8H_{10}O_2$ **C** —HIO_4 / H_2SO_4→ $H_2C{=}O$ + CHO

PROBLEM **17.52** Provide a mechanism for the following transformations:

OH, CH_3, CH_3 —H_3O^+ / H_2O→ CH_3, CH_3 + H_3C, CH_3, C

PROBLEM **17.53** This problem introduces another alcohol protecting group, methoxymethyl (MOM), particularly popular in the month of May. The MOM protecting group is stable to base, but can be cleaved upon treatment with mild acid. Provide an arrow formalism for the introduction of the MOM protecting group. Also explain why this protecting group is readily cleaved upon treatment with aqueous acid.

R—OH —1. NaH 2. CH_3OCH_2Cl→ $R—O—CH_2OCH_3$ —H_3O^+ / H_2O→ R—OH

PROBLEM **17.54** Write an arrow formalism mechanism for this reaction and explain why phenylacetophenone (**A**) is not formed in the reaction.

Ph—C(Ph)(OH)—C(H)(OH)—H —H_2O / H_3O^+→ Ph—C(Ph)(H)—C(=O)H

not

Ph—C(=O)—CH_2Ph **A**

Ph =

PROBLEM **17.55** Explain why it is possible to alkylate a phenol selectively in the presence of an alcohol.

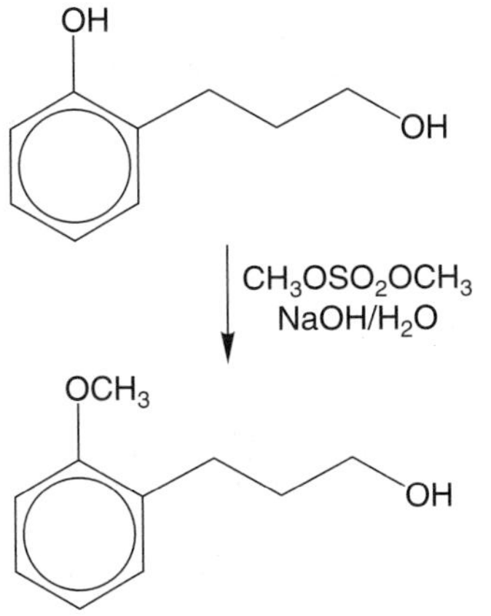

PROBLEM **17.56** Metaproterenol (orciprenaline, **1**) is a β-adrenoreceptor stimulant used therapeutically as a bronchodilator. In the synthesis of the hydrobromide salt of racemic **1**, outlined below, supply the appropriate reagents.

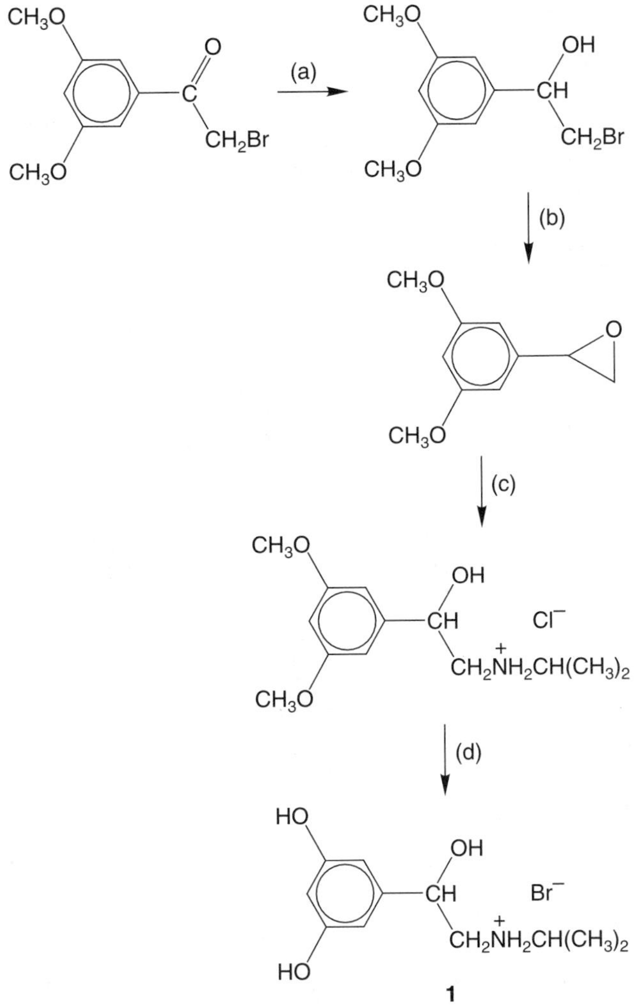

PROBLEM **17.57** Treatment of 2-thiabicyclo[2.2.1]-heptane (**1**) with 1 equiv of hydrogen peroxide affords a mixture of diastereomers **2** and **3** of molecular formula $C_6H_{10}OS$. Both diastereomers **2** and **3** yield the same product **4** ($C_6H_{10}O_2S$) upon treatment with peracetic acid. Compound **4** is also produced when **1** reacts with 2 equiv of peracetic acid. Propose structures for compounds **2**, **3**, and **4**. *Hint*: Start by drawing a good Lewis structure for the starting material **1** (see Fig. 17.94).

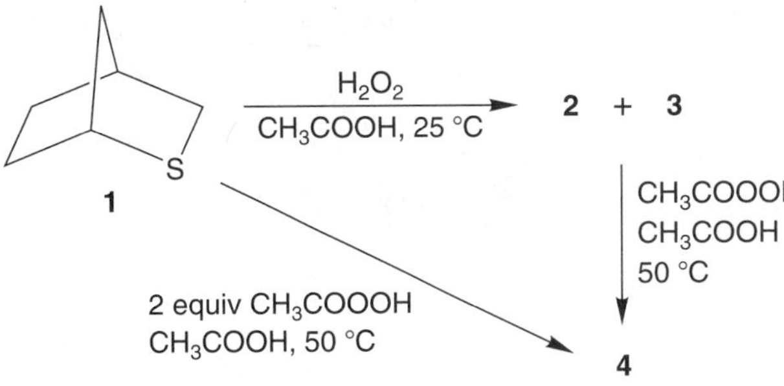

PROBLEM **17.58** When benzoin (**1**) is heated with thionyl chloride, only compound **2** is formed. However, when **1** and thionyl chloride are allowed to react at 0 °C, compounds **2** and **3** were isolated, as well as elemental sulfur.

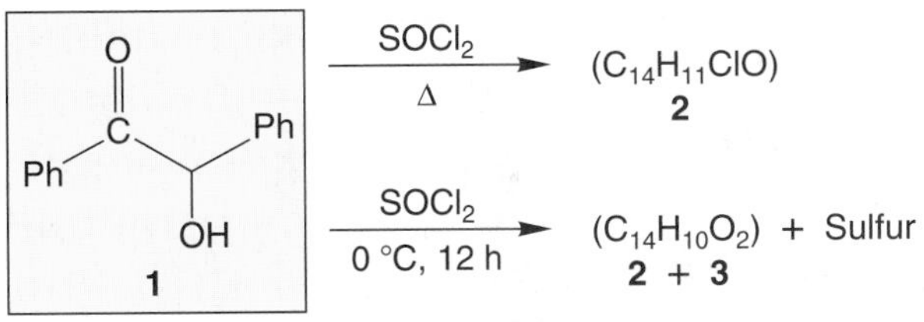

Spectral data for compounds **2** and **3** are summarized below. Propose structures for compounds **2** and **3** and provide an arrow formalism mechanism for the formation of **3**. *Hint*: It is probably useful to start with a determination of the structure of **3**, and then move on to questions of mechanism. *Another Hint*: Sulfur monoxide is known to give elemental sulfur through the following reaction:

$$4\,SO \rightleftharpoons 2\,SO_2 + S_2$$

Compound **2**
IR (Nujol): 1689 cm^{-1}
^1H NMR ($CDCl_3$): δ 6.33 (s, 1H), 7.05–7.60 (m, 8H), 7.75–8.05 (m, 2H)

Compound **3**
Mass spectrum: *m/z* = 210 (p, 4%), 106 (8%), 105 (100%), 77 (52%), 51 (24%), 50 (10%).
IR (Nujol): 1645 cm^{-1}
^1H NMR ($CDCl_3$): δ 7.10–7.60 (m, 6H), 7.73–8.00 (m, 4H)
^{13}C NMR ($CDCl_3$): δ 128.9 (d), 129.7 (d), 133.0 (s), 134.7 (d), 194.3 (s)

PROBLEM 17.59 Rationalize the following curious experimental results. When the optically active chlorosulfite **1** is heated in toluene, the chloride with inverted configuration is obtained. However, when **1** is decomposed in 1,4-dioxane (**2**), retention of configuration in the resulting chloride is observed. *Hint*: See Figure 17.26, and also think about the differences between toluene and 1,4-dioxane. A good Lewis structure for 1,4-dioxane may also be helpful.

Cl
S—O
O
H
*
Bu
CH_3
1

toluene
Δ

H
*
Bu
Cl
H_3C
Inverted

O O
2
Δ

H
*
Cl
Bu
CH_3
Retention

Carbonyl Chemistry 2: Reactions at the α-Position

I want to beg you, as much as I can,
dear sir, to be patient towards all
that is unsolved in your heart and
try to love the *questions themselves.*
—Rainer Maria Rilke*
Letters to a Young Poet

In this chapter, we begin an exploration of the chemistry of the α-position in carbonyl compounds that will continue for the better part of Chapters 18–20. Our ability to understand complex reactions and to construct complicated molecules will greatly increase. By this time, there really is little fundamentally new chemistry—there is some, but much of what we find from now on will be applications of reactions we already know, placed in more complicated settings. For example, in Chapter 16 we saw a vast array of additions of nucleophiles to carbon–oxygen double bonds. Here that same reaction appears, when a new nucleophile, an enolate, adds to carbonyl groups to yield products of remarkable complexity. The reaction is the same old addition, but the context makes it look complicated (Fig. 18.1).

FIGURE **18.1** Additions of nucleophiles to carbonyl compounds.

*Rilke (1875–1926) was a German lyric poet much influenced by the French sculptor, Rodin.

The chemistry in the next few chapters can be either daunting or exhilarating. Much depends on how well you have the basic reactions under control. If you do, you will be able to apply them in the new situations we are about to explore. If not, the next chapters will seem crammed with new and complicated reactions, each of which requires an effort to understand. Like anything else, learning chemistry can have a strong psychological component. When you are able to think, "That's not so bad, it's just an application of ... ," you gain confidence and what you see on the page is likely to stick in your mind. That's one reason we broke the chain of carbonyl chemistry with Chapter 17, which contained a fair amount of review and lots of problems that reached back into earlier chapters.

It is now possible to ask questions that stretch your knowledge and talent. It's hard to teach this kind of problem solving, as success has a lot to do with having worked lots of related problems in the past. Nevertheless, there are some techniques that I can try to pass on. Don't worry if you don't get every problem! My colleagues can write problems that I can't get. It's no disgrace to have trouble with a problem. There are several favorite problems that I've been working on (not continuously) for many years. You'll see some of them.

Success in the chemistry business has relatively little to do with how *fast* you can solve problems, at least not problems that require thought and the bringing together of information from different areas. People think in different ways and this shows up as answers appearing at different rates. There are those who are blindingly fast at this kind of thing and others who reach answers more slowly. There is nothing wrong with chewing over a problem several times, approaching it in several ways. Letting a particularly vexing problem rest for a while is often a good idea. Of course, this is the bane of exam writers (and even more so of exam answerers). How do you write an exam that allows for thought and doesn't simply select those who can answer quickly? It's hard, especially if you don't have the luxury of asking a small number of questions and giving lots of time.

What does make for success in the real world of organic chemistry? I think nothing correlates better than love of the subject, and by "subject" I do not necessarily mean what you are doing now, but rather really *doing* organic chemistry. Doing organic chemistry means not only liking the manipulative aspects of the subject, although that is a great help, but also, as Rilke said, of "loving the *questions themselves.*" At any rate, it is not the number of A's on your transcript that correlates with your later success in research!

We will start with an examination of the behavior of aldehydes and ketones as Brønsted acids, and then go on to see the many reactions that are possible once a proton has been lost.

18.1 ALDEHYDES AND KETONES ARE WEAK BRØNSTED ACIDS

One of the remarkable things about carbonyl compounds is that they are not only Lewis acids (see the addition reactions throughout Chapter 16), but are proton donors as well. The pK_a values of simple carbonyl compounds are generally in the high teens (Table 18.1).

It is the hydrogen at the α-position, the position adjacent to the carbon–oxygen double bond, that is lost to a base as a proton (Fig. 18.2). Our first task is to see why these compounds are weak acids at their

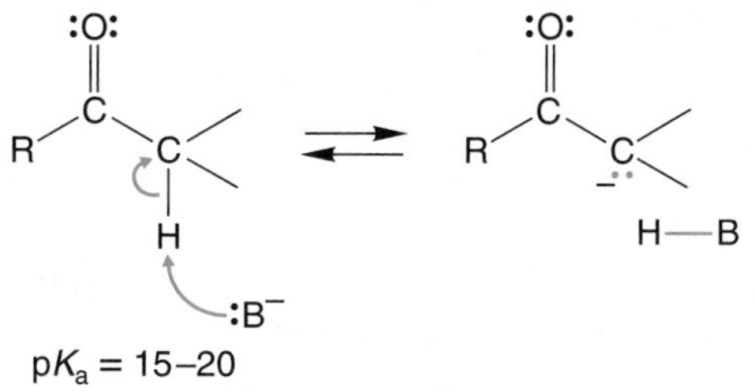

$pK_a = 15–20$

FIGURE 18.2 Carbonyl compounds bearing hydrogen at the α-position are weak acids, with pK_a values in the high teens.

TABLE 18.1 Some pK_a Values for Simple Ketones and Aldehydes

Compound[a]	pK_a
$CH_3CH_2COC\underline{H}_2CH_3$	19.9
CH_3COCH_3	18.9
$PhCOCH_3$	17.7
$PhCH_2COC\underline{H}_3$	18.3
$PhC\underline{H}_2COCH_3$	15.9
Cyclohexanone (α-hydrogen)	17.8
$C\underline{H}_3CHO$	16.5
$(CH_3)_2C\underline{H}CHO$	14.6

[a]The proton to be lost is underlined when there is a choice.

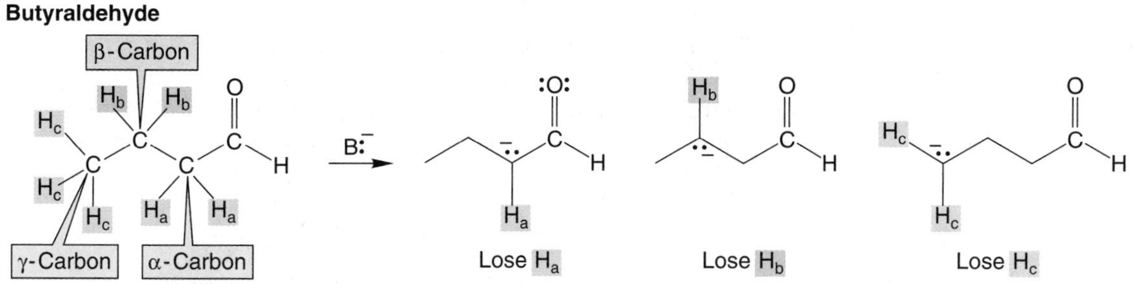

α-positions, and then to see what new chemistry results from this acidity.

We'll start with butyraldehyde ($pK_a = 16.7$) and look at all the possible anions we might form from it by breaking an sp^3–$1s$ carbon–hydrogen bond (Fig. 18.3).

FIGURE 18.3 There are three possible anions that can be formed from butyraldehyde through breaking an sp^3–$1s$ carbon–hydrogen bond.

First of all, the dipole in the carbon–oxygen bond will stabilize an adjacent anion more than a more remote anion. Loss of the α-hydrogen will lead to a more stable anion than loss of either the β- or γ-hydrogen (Fig. 18.4).

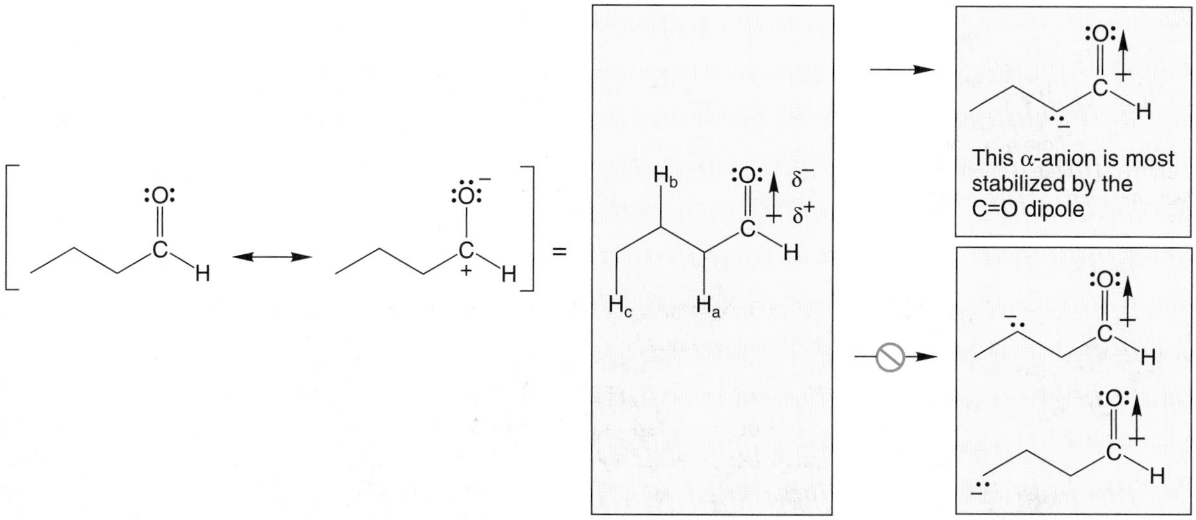

FIGURE 18.4 The dipole in the carbon–oxygen bond will stabilize an adjacent anion more than a more distant anion.

Moreover, one of these anions, but only one of them, is stabilized by resonance, as the oxygen atom can share the negative charge. It is this stabilization that makes the α-position, but only the α-position, a proton source. The intermediate is called an **enolate** anion, as it is part alkene (*ene*) and part alcoholate (*olate*). Please note again that the enolate anion is *not* part of the time an alkoxide and part of the time a carbanion, but all of the time a single species, the resonance-stabilized enolate (Fig. 18.5).

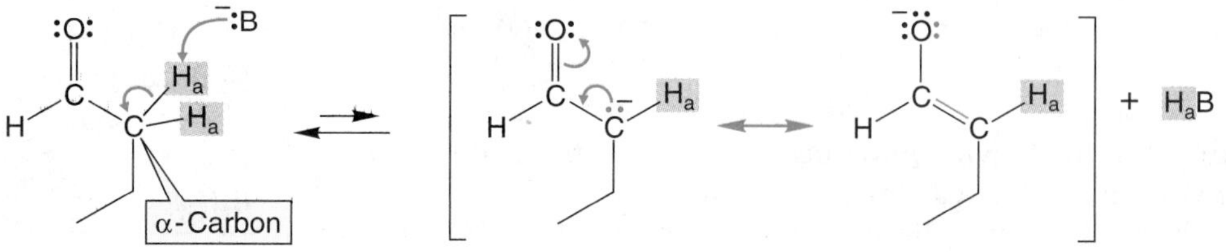

FIGURE 18.5 Loss of the α-hydrogen leads to a resonance-stabilized enolate anion.

The enolate anion can reprotonate in two ways. If it reacts with a proton source such as water at carbon, the original carbonyl compound is regenerated. If it protonates at oxygen, the neutral **enol** is formed. In base, carbonyl compounds are in equilibrium with their enol forms (Fig. 18.6).

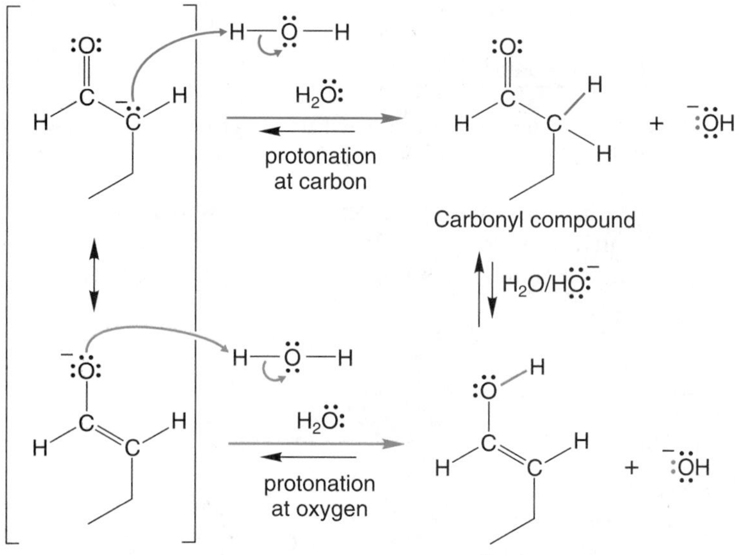

FIGURE 18.6 Reprotonation of the enolate can lead either to the original carbonyl compound or to the related enol.

*PROBLEM 18.1 Write a mechanism for the base-catalyzed equilibration of the carbonyl and enol forms of acetone.

ANSWER This problem is not hard, but it needs to be done right now. To be able to do the more complicated material that will appear later, these simple interconversions must become second nature. This problem is drill, but it is nevertheless very important.

Loss of either the α-hydrogen from acetone or the hydroxyl hydrogen of the enol form of acetone leads to the same enolate. Protonation at carbon gives the

ANSWER (CONTINUED)

ketone; protonation at oxygen gives the enol. It is formation of the enolate that achieves the equilibration.

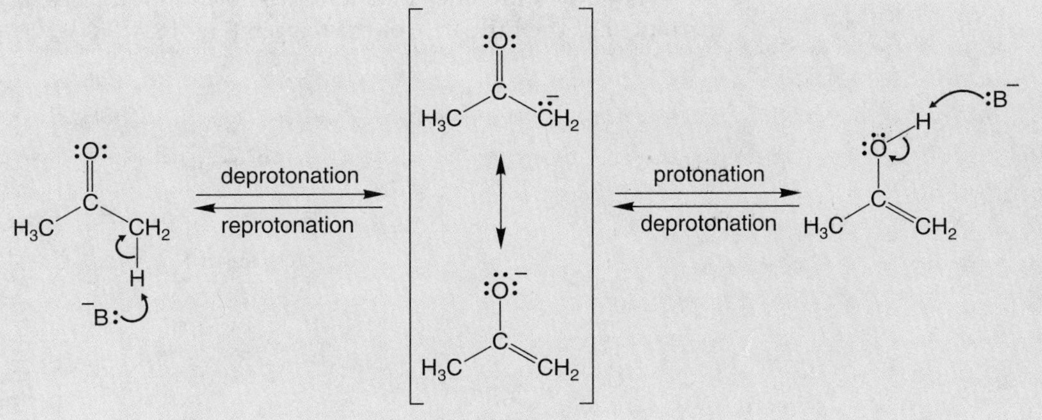

An orbital picture clearly shows the source of the stabilization of the enolate. The three 2*p* orbitals overlap to form an allyl-like system (Chapter 12, p. 531). Two of the four π electrons are accommodated in the lowest bonding molecular orbital, Φ_1, and two in the nonbonding orbital, Φ_2. The presence of the oxygen atom makes the enolate less symmetrical than an allyl system formed from three carbon 2*p* orbitals (Fig. 18.7).

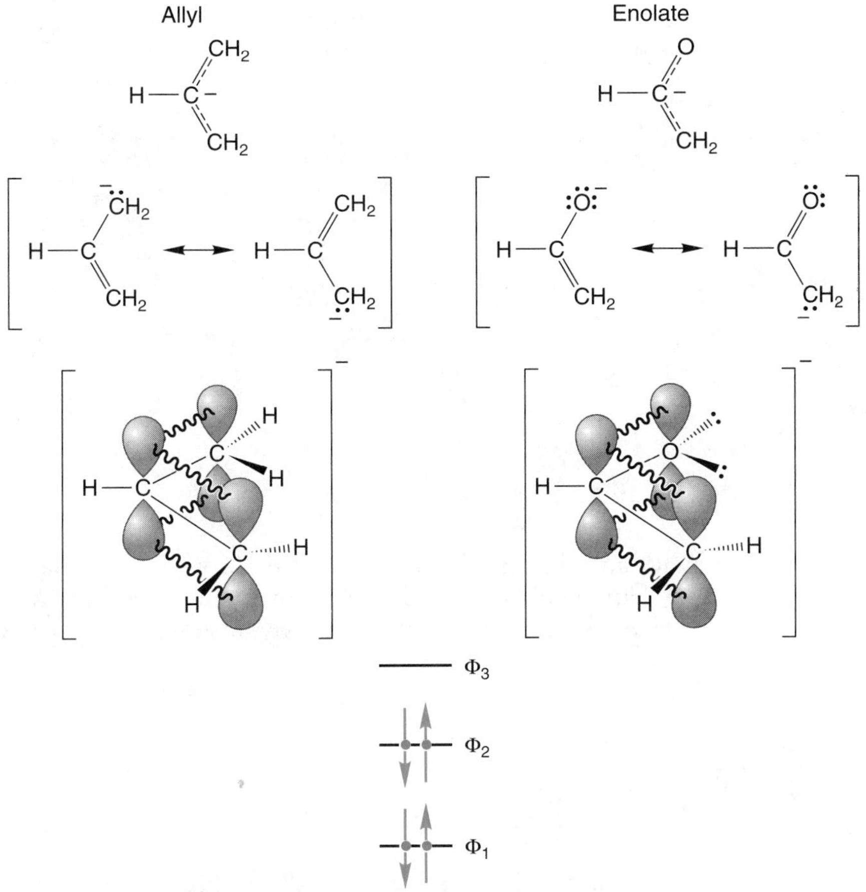

FIGURE **18.7** A comparison of the enolate and allyl anions.

How important is the electronegative oxygen atom to the stabilization of the enolate? The pK_a of propene is 42 and that of acetaldehyde is 16.5. There is a 10^{25} difference in acidity in the two molecules, and this can be attributed to the influence of the oxygen (Fig. 18.8).

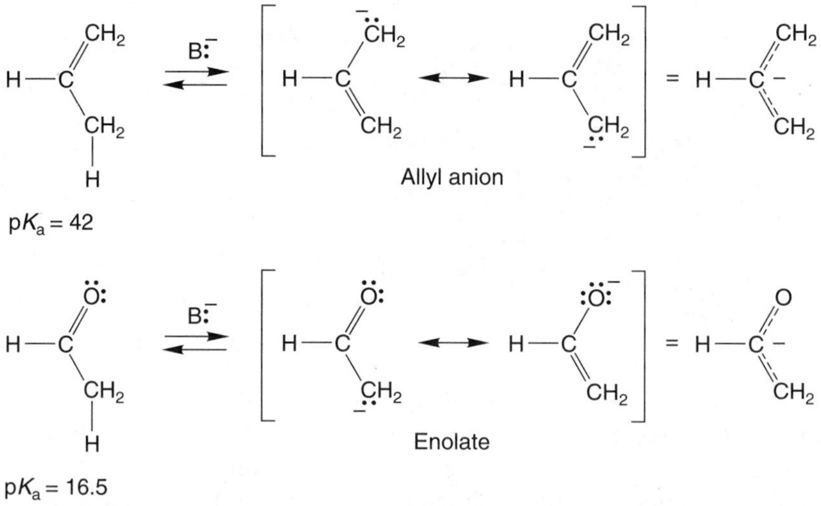

FIGURE 18.8 The oxygen atom of the enolate plays a crucial role in promoting the acidity at the α-position. Acetaldehyde is much more acidic than propene.

*PROBLEM 18.2 Propionaldehyde (propanal) can form two enols. What are they?

ANSWER There are stereochemical differences between the two enols. Like any unsymmetrical alkene, this enol can exist in (*Z*) or (*E*) forms.

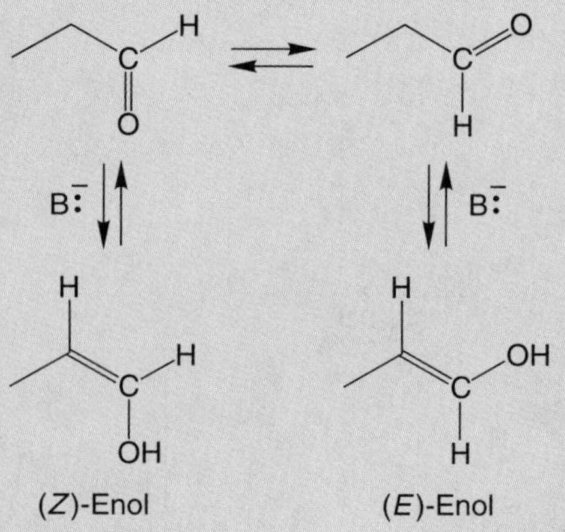

Enolate formation reveals itself most commonly through the exchange of α-hydrogens in deuterated base. For example, treatment of acetaldehyde with D_2O/DO^- results in exchange of all three α-hydrogens for deuterium (Fig. 18.9).

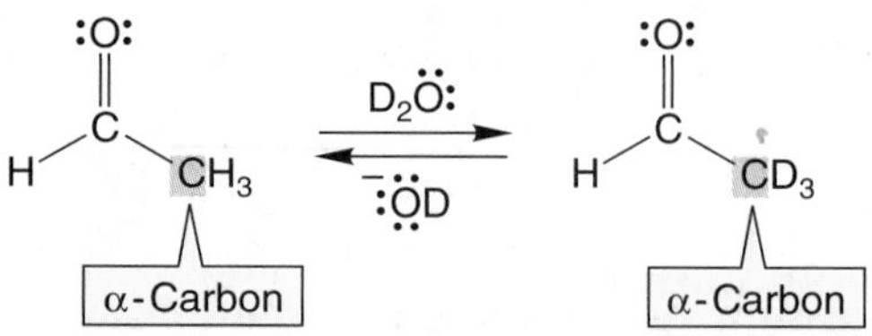

FIGURE 18.9 In D_2O/DO^-, the three α-hydrogens of acetaldehyde are exchanged for deuterium.

In hydroxide (here deuteroxide), not all of the acetaldehyde will be transformed into the enolate. The pK_a of acetaldehyde is 16.5 and that of water is 15.7. So, hydroxide is a *weaker* base than the enolate anion and at equilibrium only a small amount of the enolate anion will be present. Enolate formation is endothermic in this case (Fig. 18.10).

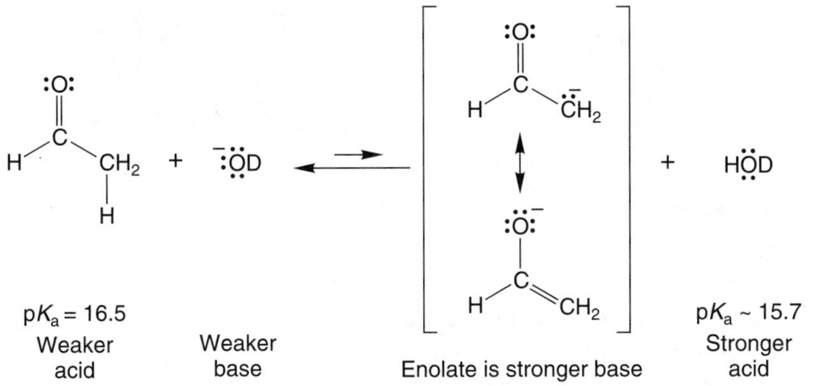

pK_a = 16.5 Weaker acid; Weaker base; Enolate is stronger base; pK_a ~ 15.7 Stronger acid

FIGURE 18.10 Enolate formation is an equilibrium reaction, and is endothermic in the case of acetaldehyde.

Moreover, these exchange reactions are typically run using catalytic amounts of base, DO^-, in a large excess of D_2O. Under such conditions only a small amount of enolate can be formed. However, any enolate present will react with the available water, here D_2O, to remove a deuteron and produce a molecule of exchanged acetaldehyde. A new molecule of DO^- that can carry the reaction further is also formed (Fig. 18.11).

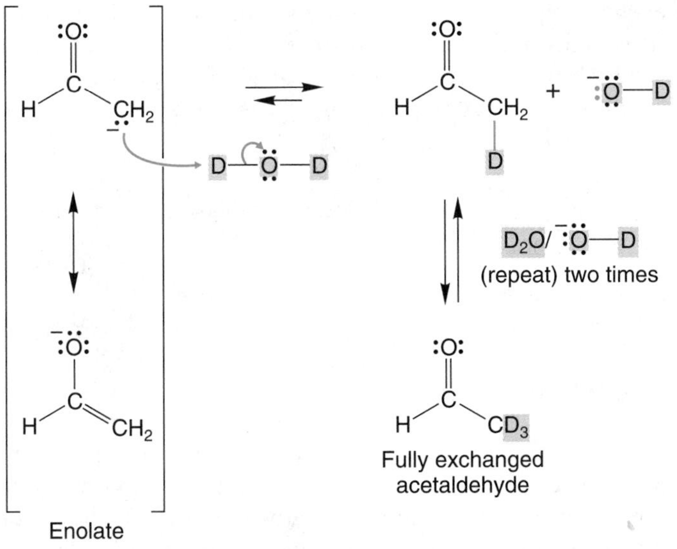

Enolate; D_2O/ $^-$O—D (repeat) two times; Fully exchanged acetaldehyde

FIGURE 18.11 The exchange reaction is a catalytic process, with deuteroxide ion (^-OD) acting as the catalyst.

Eventually all the available α-hydrogens will be exchanged for deuterons.

PROBLEM 18.3 Explain why the aldehydic hydrogen in acetaldehyde does not exchange in D_2O/DO^-.

PROBLEM 18.4 Explain why the bicyclic ketone in Figure 18.12 exchanges only the two α-hydrogens shown (H_α) and not the bridgehead hydrogen, which is also "α." *Hint*: Use models and examine the relationship between the bridgehead carbon–hydrogen bond and the π orbitals of the carbonyl group.

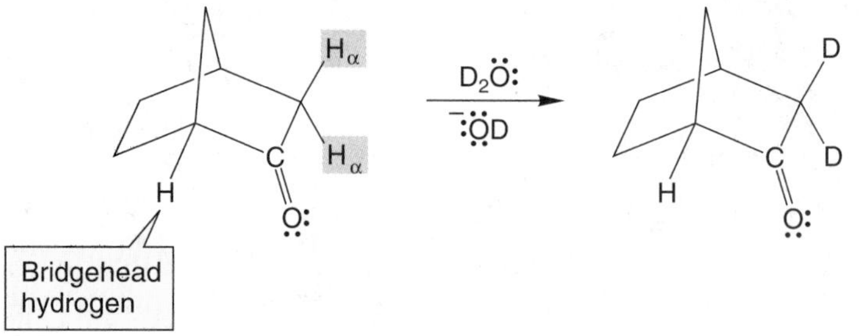

FIGURE 18.12

The α-hydrogens of acetaldehyde can also be exchanged under acidic conditions (Fig. 18.13).

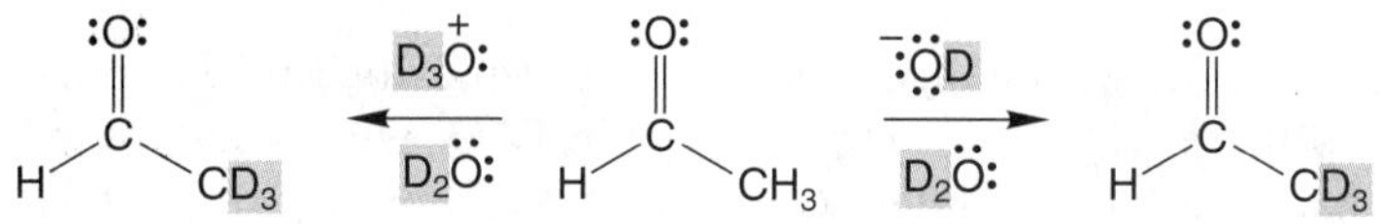

FIGURE 18.13 Exchange can also be carried out in deuterated acid, D_3O^+/D_2O.

This result illustrates a general principle: In carbonyl chemistry, there will usually (not quite always) be an acid-catalyzed version for every base-catalyzed reaction. The mechanisms of the two reactions will be different, however, even though the end results are similar or even identical. For example, in this acid-catalyzed reaction there is no base strong enough to remove an α-hydrogen from the starting acetaldehyde. The only reasonable reaction is protonation of the Brønsted basic oxygen by D_3O^+ to give a resonance-stabilized cation (Fig. 18.14).

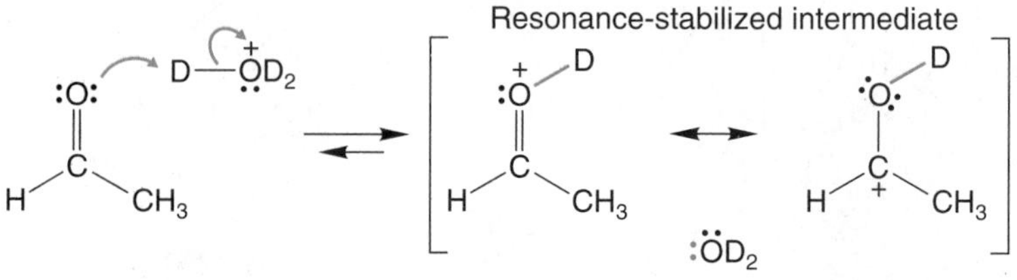

FIGURE 18.14 The first step in the acid-catalyzed exchange is addition of a D^+ to the carbonyl oxygen. A resonance-stabilized cation results.

Now we have created a powerfully acidic species, the protonated carbonyl compound. Removal of a deuterium from the oxygen simply reverses the reaction to regenerate acetaldehyde and D_3O^+, but deprotonation at carbon generates the enol. Note that the product is not an anionic *enolate,* but the neutral *enol.* Note also that in acid, as in base, carbonyl compounds that have α-hydrogens are in equilibrium with their enol forms (Fig. 18.15).

We will soon see that although enols are in equilibrium with the related keto forms, it is usually the keto forms that are favored. If this enol regenerates the carbonyl compound in D_3O^+, exchanged acetaldehyde will result. Notice that the steps in the enol → acetaldehyde reaction are simply the reverse of the acetaldehyde → enol reaction (Fig. 18.16).

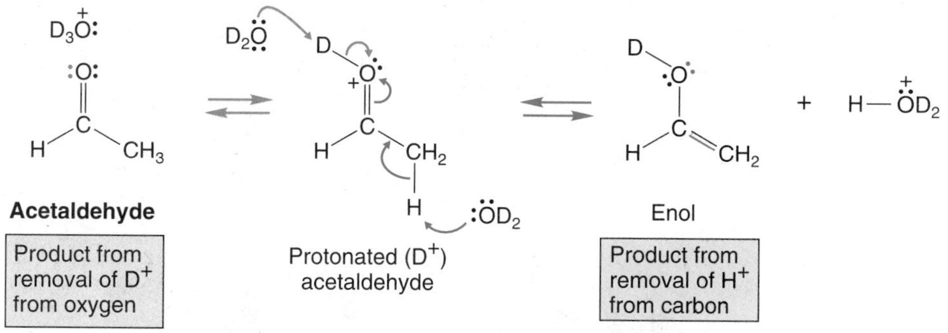

Acetaldehyde

Product from removal of D^+ from oxygen

Protonated (D^+) acetaldehyde

Enol

Product from removal of H^+ from carbon

FIGURE 18.15 Removal of a proton from carbon generates the neutral enol form. Removal of a deuteron from oxygen regenerates starting material.

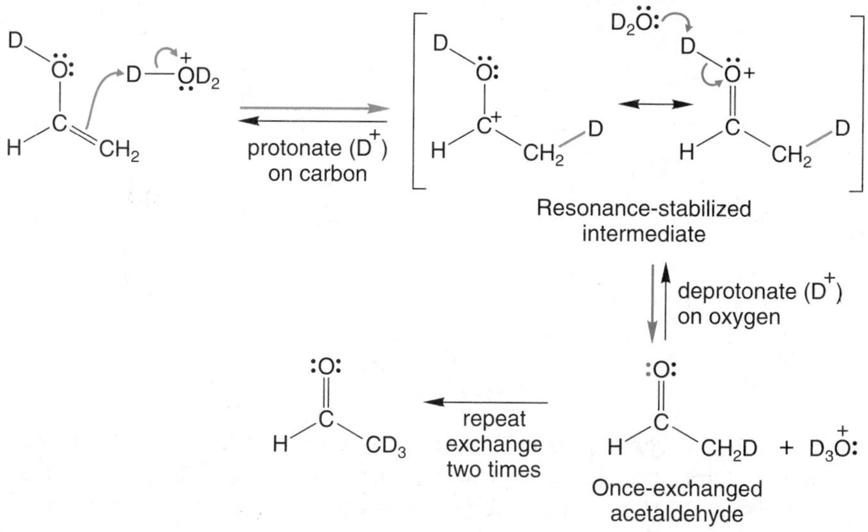

protonate (D^+) on carbon

Resonance-stabilized intermediate

deprotonate (D^+) on oxygen

repeat exchange two times

Once-exchanged acetaldehyde

FIGURE 18.16 Reaction of the enol with D_3O^+ will generate exchanged acetaldehyde.

*PROBLEM 18.5

Write a mechanism for the equilibration of acetaldehyde and its enol form in acid. This problem may seem mindless, and it certainly has little intellectual content at this point, but it is an important drill problem. Doing this problem is part of "reading organic chemistry with a pencil."

ANSWER

This process must become part of your chemical vocabulary from now on. The carbonyl oxygen is protonated and the resulting cation deprotonated at carbon to give the enol.

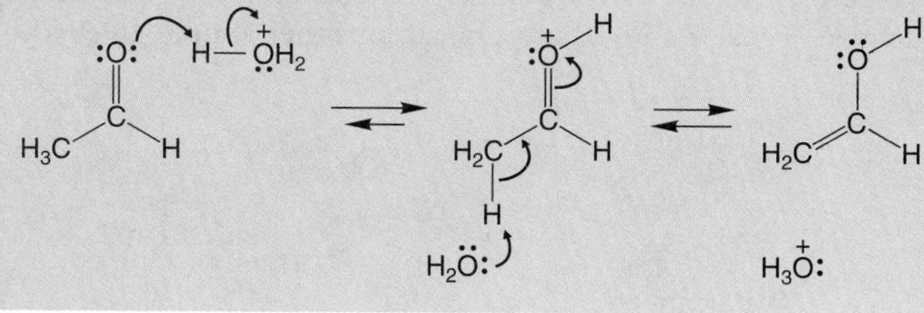

Recently, it has become possible to generate the simplest enol, vinyl alcohol (hydroxyethylene), in such a fashion that the acid or base catalysts of the previous reactions are absent. As long as it is protected from acids or bases, the enol is stable and can be isolated and studied (Fig. 18.17).

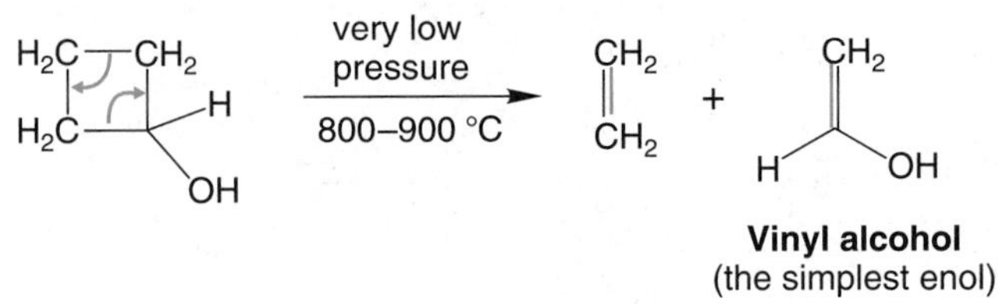

very low pressure
800–900 °C

Vinyl alcohol
(the simplest enol)

FIGURE 18.17 The parent of all enols, vinyl alcohol, is stable as long as it is protected from acid or base catalysts.

PROBLEM 18.6 It is impossible to protect vinyl alcohol from *all* acidic and basic materials, and so it eventually reverts to the carbonyl form, acetaldehyde. Explain what the unavoidable acid and base catalysts are.

The detailed structure of vinyl alcohol provides no surprises. Bond lengths and angles are not unusual (Fig. 18.18).

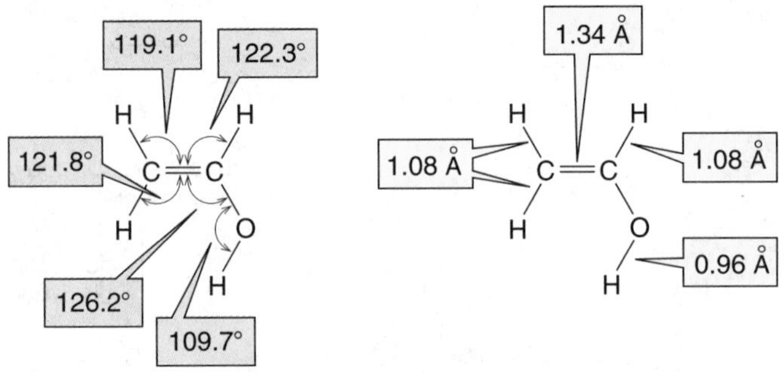

119.1°
122.3°
121.8°
126.2°
109.7°
1.34 Å
1.08 Å
1.08 Å
0.96 Å

FIGURE 18.18 The structure of vinyl alcohol.

So far we have seen that α-hydrogens can be removed, and that carbonyl compounds are often in equilibrium with enol forms. The next question to address is the position of the equilibrium between the carbonyl compound and its enol form. Where does the equilibrium lie, and how does it depend on structure (Fig. 18.19)?

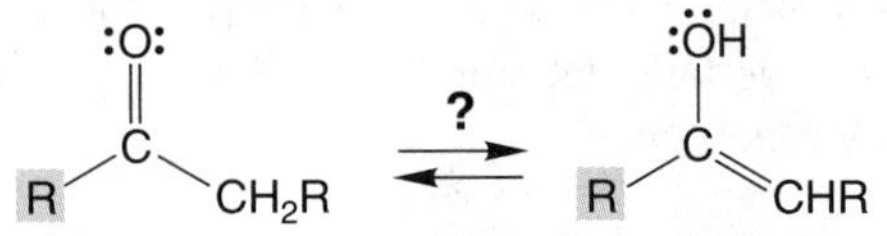

FIGURE 18.19 Where does the equilibrium between carbonyl compound and enol lie?

Simple aldehydes and ketones are predominantly in their carbonyl forms. Figure 18.20 shows the equilibrium constants, *K*, for acetone and acetaldehyde. These numbers are typical for other simple aldehydes and ketones.

Acetaldehyde $K \sim 5 \times 10^{-6}$

Acetone $K \sim 6 \times 10^{-8}$

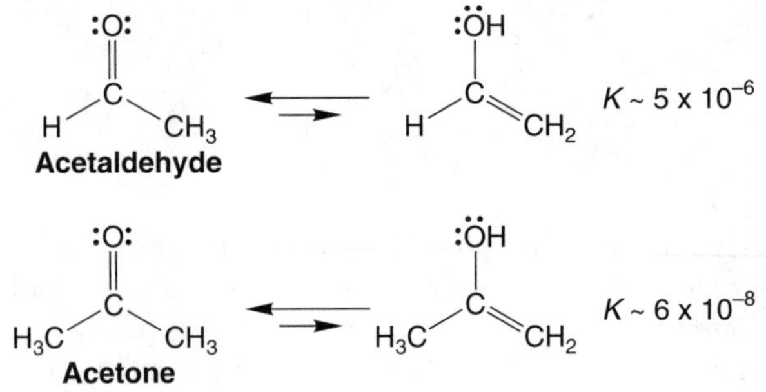

FIGURE 18.20 For simple aldehydes and ketones, it is the carbonyl form that is favored at equilibrium.

Neither simple ketones nor simple aldehydes are very highly enolized, although aldehydes generally exist somewhat more in their enol forms than ketones. The second methyl group stabilizes the carbon–oxygen double bond more than the carbon–carbon double bond (Fig. 18.21).

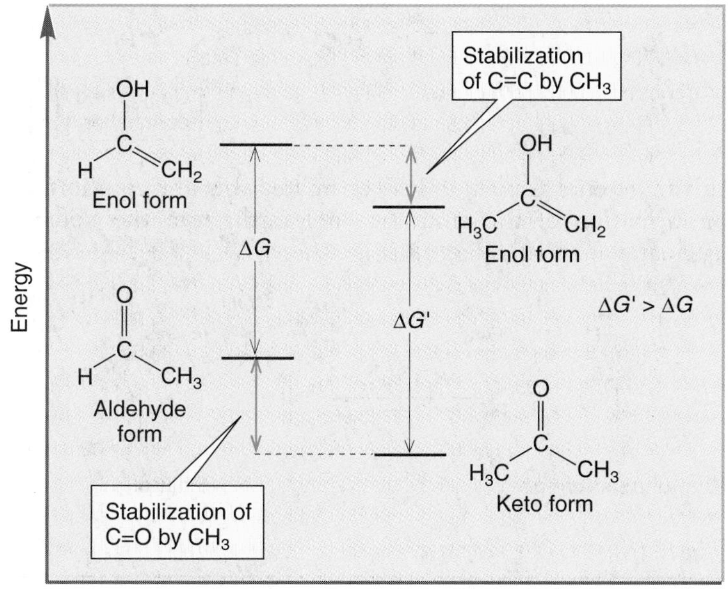

Energy

Stabilization of C=C by CH$_3$

Enol form

ΔG

$\Delta G'$

$\Delta G' > \Delta G$

Aldehyde form

Keto form

Stabilization of C=O by CH$_3$

FIGURE **18.21** Acetone is less enolized than acetaldehyde because of the greater stabilization provided to the keto form by the second methyl group.

PROBLEM **18.7**

From a bond strength analysis (ΔH) decide whether the keto or enol form of acetone is more stable. Use the data from Table 8.2 (p. 307).

More complex carbonyl compounds can be much more strongly enolized. β-Dicarbonyl compounds (1,3-dicarbonyl compounds) exist largely in their enol forms, for example (Fig. 18.22).

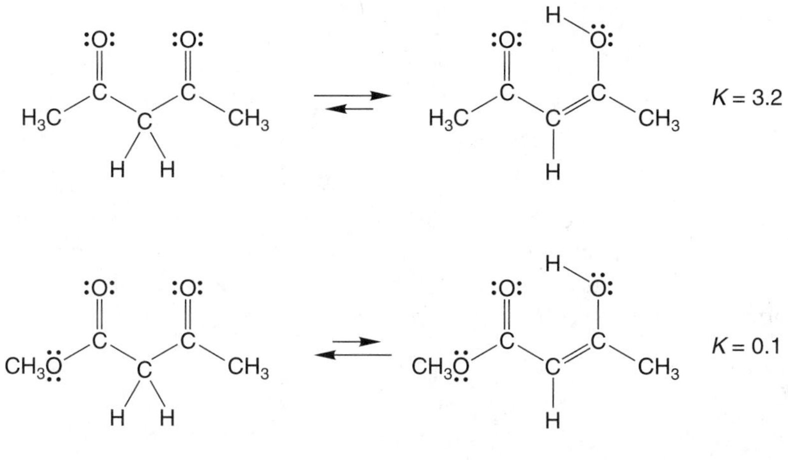

$K = 3.2$

$K = 0.1$

FIGURE **18.22** β-Dicarbonyl compounds are more enolized than are monoketones.

In the enol forms of these compounds, it is possible to form an intramolecular hydrogen bond, and that contributes to forcing the equilibrium position away from the diketo forms. There is also conjugation between the carbon–carbon double bond and the remaining carbonyl group in these enols (Fig. 18.23).

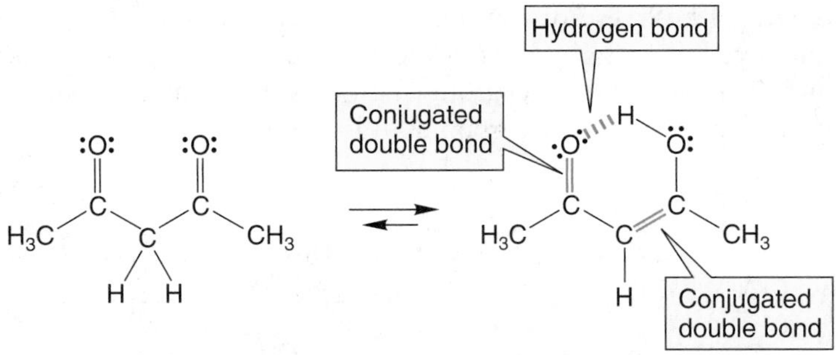

FIGURE 18.23 The formation of an intramolecular hydrogen bond, as well as conjugation between the carbon–carbon double bond and the carbonyl group, contributes to the increased stability of the enols in β-dicarbonyl compounds.

The ultimate enol is phenol. It is estimated that the equilibrium constant for formation of the aromatic enol form from the nonaromatic ketone is greater than 10^{13} (Fig. 18.24).

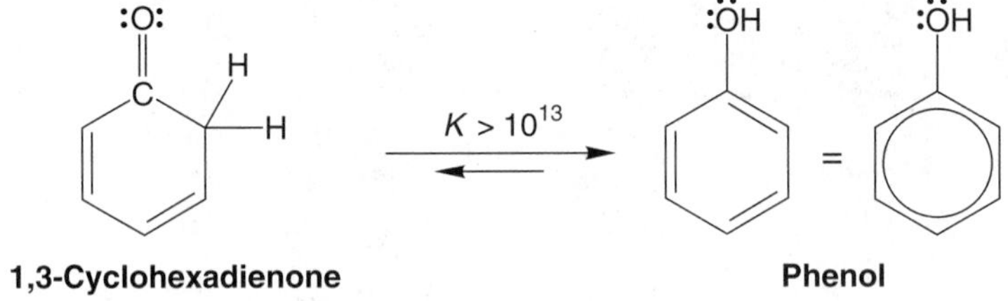

FIGURE 18.24 The equilibrium constant for enolization of cyclohexadienone is estimated to be greater than 10^{13}.

18.2 REACTIONS OF ENOLS AND ENOLATES

Now we know something of the origins of enols and enolates, and we have seen that enols are in equilibrium with carbonyl compounds. Now it is time to move on to a look at reactions of these species.

18.2a Exchange Reactions

We saw in the previous sections that as long as an enol or enolate can be formed, hydrogens in the α-position can be exchanged for deuterium through treatment with deuterated acid or base (Fig. 18.25).

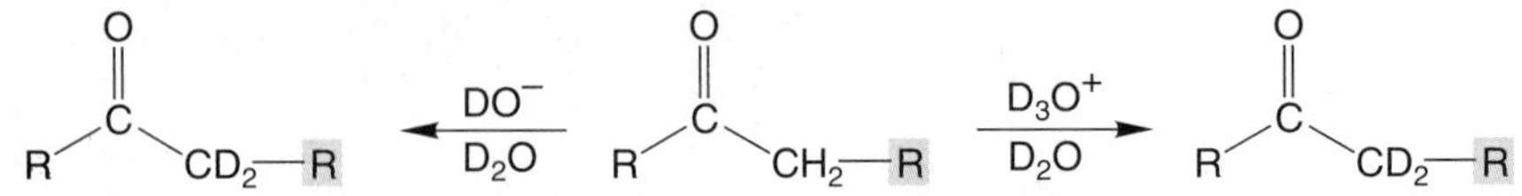

FIGURE 18.25 Exchange reactions of carbonyl compounds bearing α-hydrogens can be either acid or base catalyzed.

18.2b Racemization

Treatment of a simple optically active aldehyde or ketone with acid or base results in loss of optical activity, racemization, *as long as the stereogenic carbon is in the α-position to the carbon–oxygen double bond, and as long as there is an α-hydrogen* (Fig. 18.26).

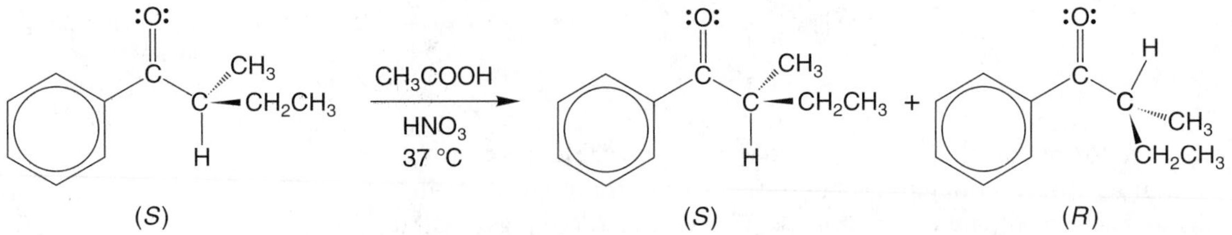

FIGURE 18.26 Optically active carbonyl compounds are racemized in acid or base, as long as an α-hydrogen is present on the stereogenic carbon.

Once more, the mechanism depends on enol formation in acid, or enolate formation in base. In acid, an achiral, planar enol is formed. Reformation of the keto form occurs by equal protonation from either side of the enol (Fig. 18.27).

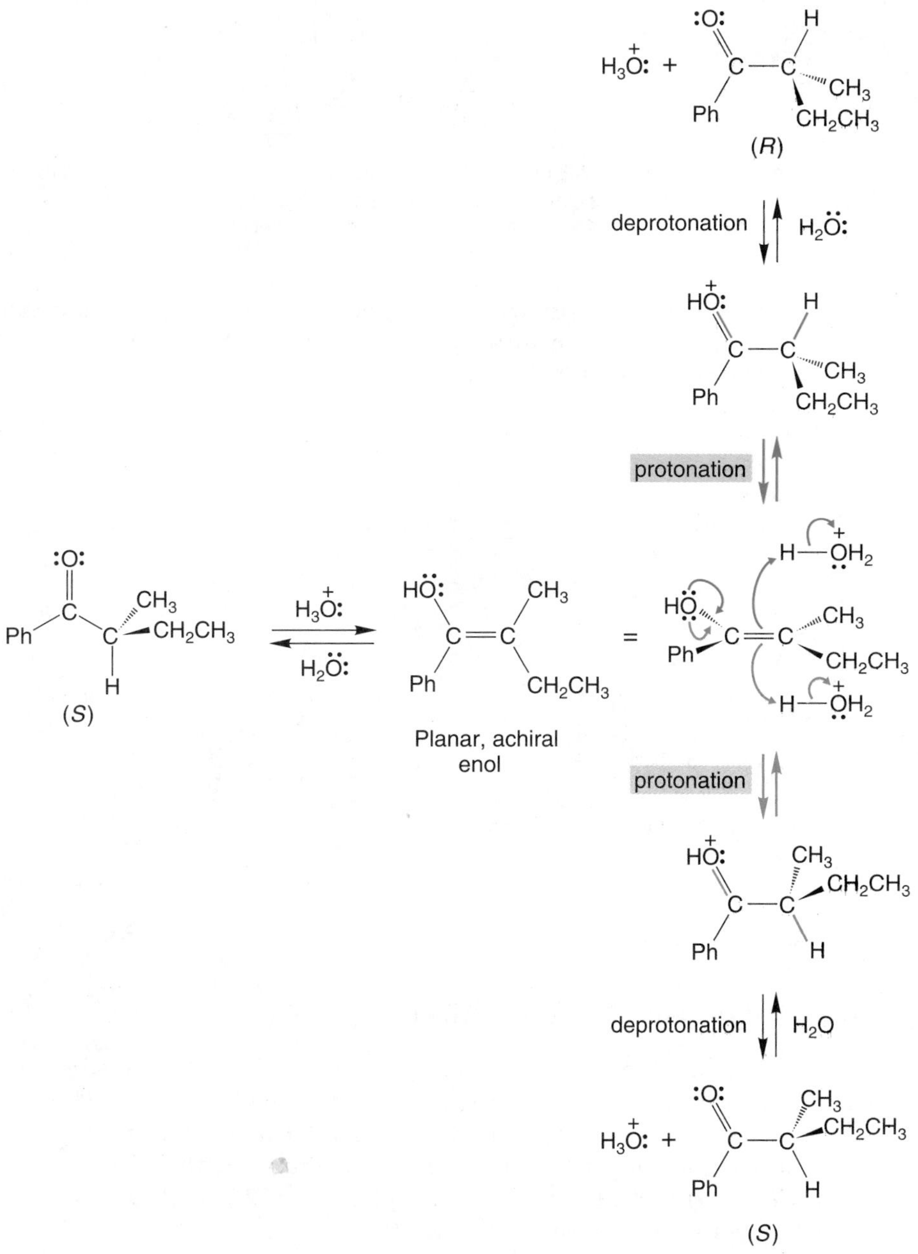

FIGURE 18.27 Protonation of the planar enol must result in formation of equal amounts of two enantiomers. An optically active carbonyl compound is racemized on enol formation.

In base, removal of the α-hydrogen generates an enolate anion that is itself achiral. For overlap between the 2*p* orbitals on carbon and oxygen to be maximized, the enolate must be planar, and the planar structure is not chiral (Fig. 18.28).

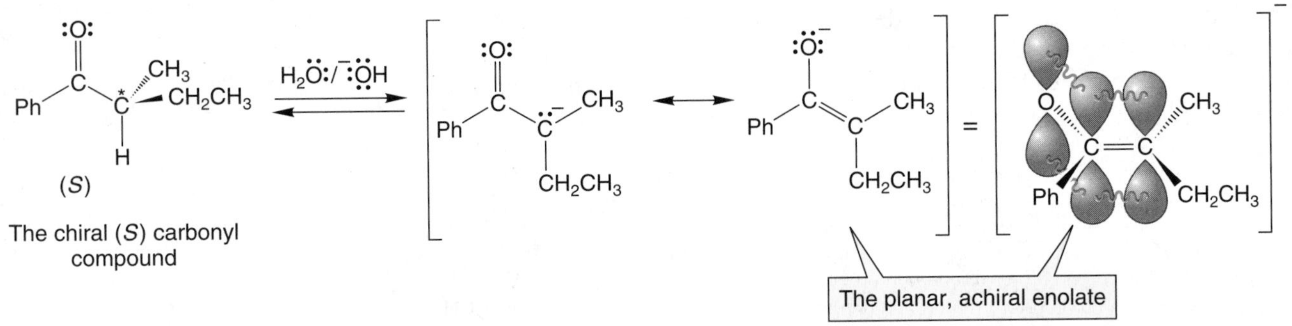

FIGURE 18.28 Resonance-stabilization of the enolate depends on overlap of the 2*p* orbitals on carbon and oxygen. Maximum overlap requires planarity, and the planar enolate is necessarily achiral. Racemization has been accomplished at the enolate stage.

A racemic mixture is formed when the enolate reprotonates. Addition of a proton *must* take place with equal probability at the two equivalent faces of the planar molecule (Fig. 18.29).

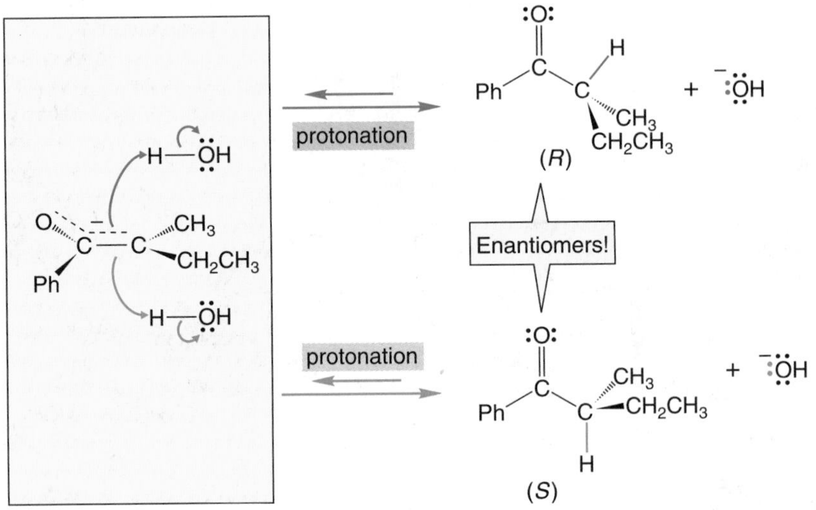

FIGURE 18.29 Reprotonation of the enolate from the two equivalent faces must give a pair of enantiomers.

18.2c Halogenation in the α-Position

Halogenation at the α-position is one of the relatively rare examples of a reaction that takes a different course in its acid- and base-catalyzed versions. As you will see, though, the two reactions are closely related.

Treatment of a ketone containing an α-hydrogen with halogen (X_2, X = I, Cl, or Br) in acid results in the replacement of one of the methyl hydrogens with a halogen atom (Fig. 18.30).

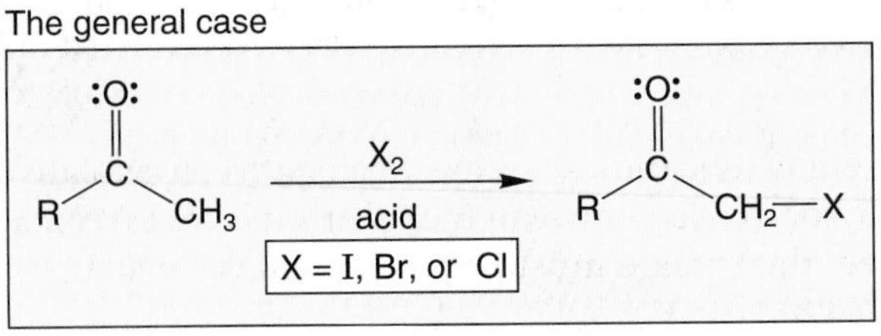

FIGURE 18.30 Treatment of a ketone containing an α-hydrogen with iodine, bromine, or chlorine in acid leads to α-halogenation.

A specific example

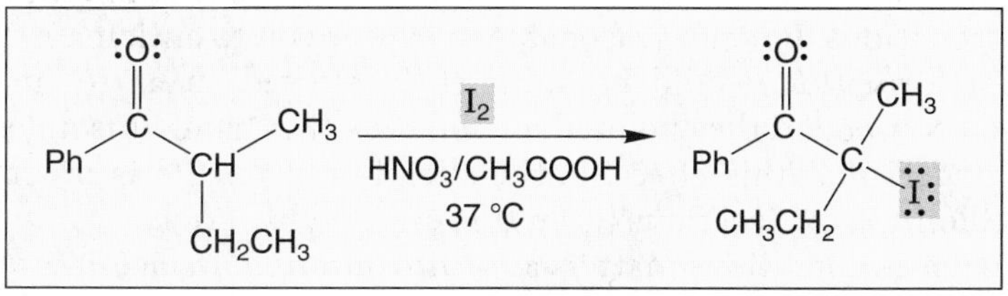

FIGURE 18.30 (CONTINUED)

The enol is formed first, and like most alkenes, reacts with iodine to displace iodide and form a resonance-stabilized cation (Fig. 18.31). The iodide ion deprotonates the intermediate cation generating the α-iodo carbonyl compound, and ending the reaction.

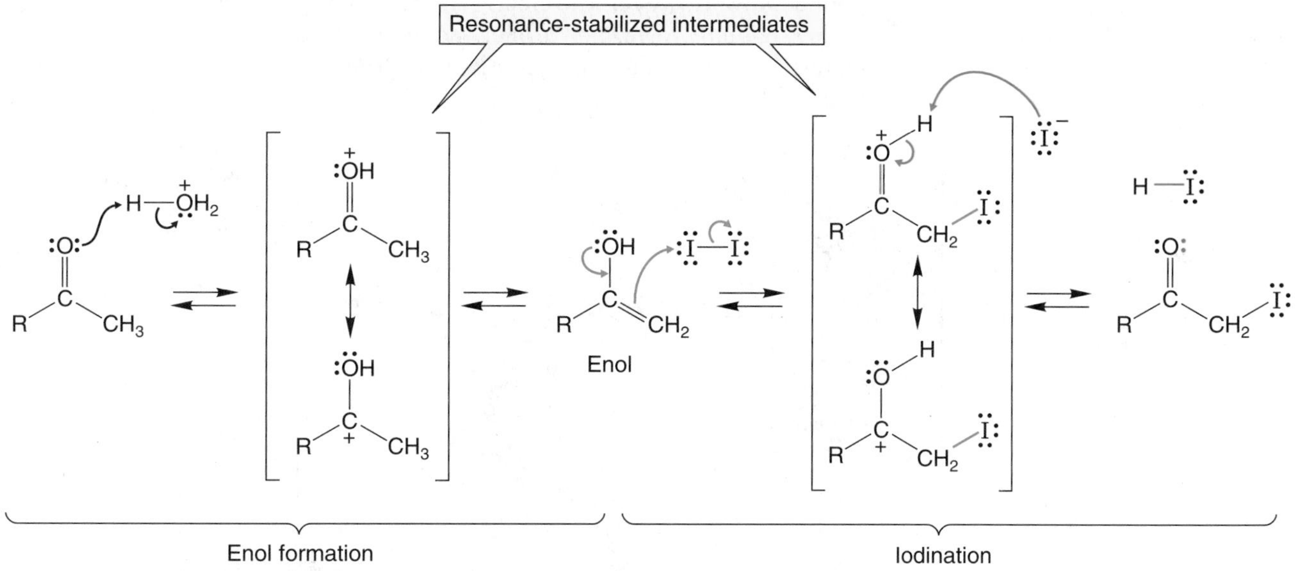

Resonance-stabilized intermediates

Enol formation

Iodination

FIGURE 18.31 Under acidic conditions the enol is formed, and then reacts with iodine to give the open, resonance-stabilized cation. Deprotonation leads to the α-iodide.

Earlier, we saw how deuterium replaces *all* α-hydrogens. Why doesn't the α-iodo compound react further to add more iodines? The way to answer a question such as this is to work through the mechanism of the hypothetical reaction and see if you can find a point at which it is disfavored. In this case, that point appears right away. It is enol formation itself that is slowed by the introduction of the first halogen. The introduced iodine inductively withdraws electrons and this disfavors the first step in enol formation, protonation of the carbonyl group (Fig. 18.32).

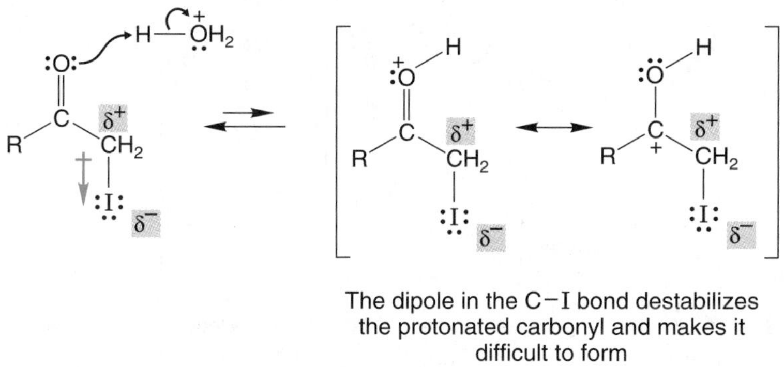

The dipole in the C−I bond destabilizes the protonated carbonyl and makes it difficult to form

FIGURE 18.32 Protonation of the α-iodoketone is disfavored by the electron-withdrawing character of the halogen.

It is not easy to *predict* the extent of this sequence of enol formations and halogenations. It is not reasonable to expect you to anticipate that further enol formation would be sufficiently slowed so as to make the monoiodo compound the end product. On the other hand, it is not so difficult to *understand* this effect. Rationalization of results is much easier than predicting new reactions (or in this case, nonreactions).

Under basic conditions matters are quite different. In fact, it is very difficult to stop the reaction at the monoiodo stage. The general result is that all α-hydrogens are replaced with halogen. For molecules with three α-hydrogens, methyl compounds, the reaction goes even further in what is called the **haloform reaction**.

In base, it is not the enol that is formed, but the enolate. This anion reacts in S_N2 fashion with iodine to generate the α-iodo carbonyl compound. Now, however, the electron-withdrawing inductive effect of iodine makes enolate formation easier, not harder, as for enols. The negative charge of the enolate is *stabilized* by the electron-withdrawing effect (Fig. 18.33).

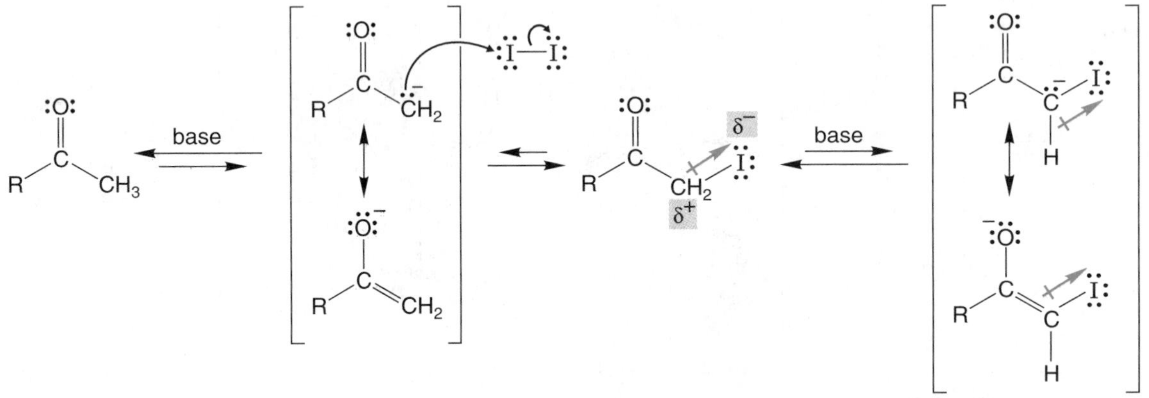

FIGURE 18.33 The initially formed α-iodo carbonyl compound is a stronger acid than the carbonyl compound itself. The introduced iodine makes enolate formation easier.

A second, and then a third displacement reaction on iodine generates the α,α,α-triiodo carbonyl compound. Similar reactions are known for bromine and chlorine (Fig. 18.34).

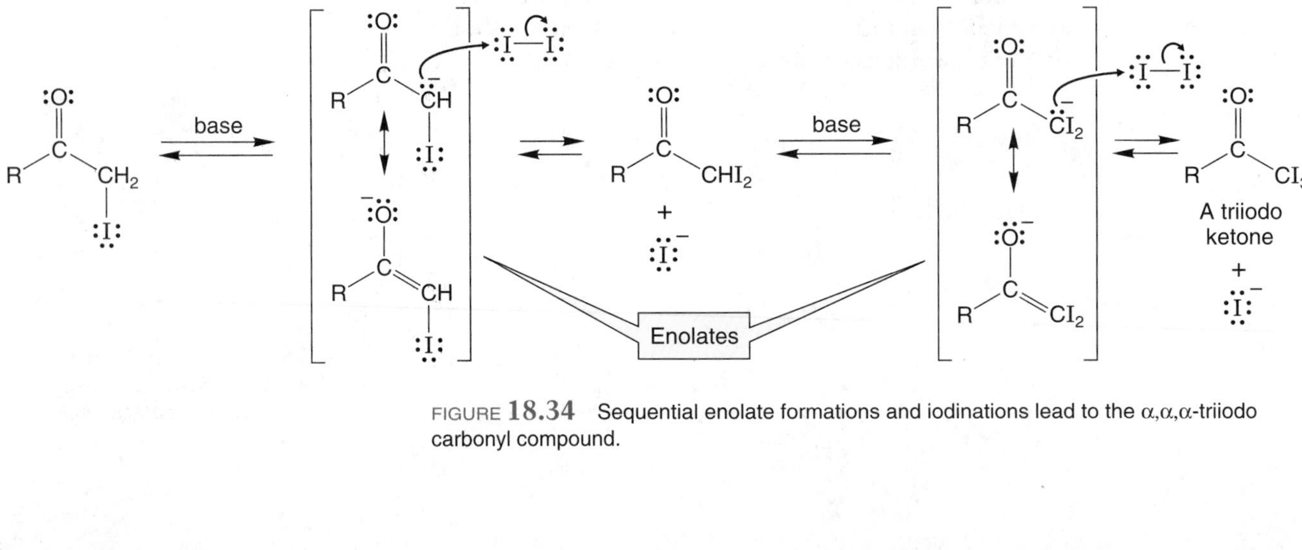

FIGURE 18.34 Sequential enolate formations and iodinations lead to the α,α,α-triiodo carbonyl compound.

In $NaOH/H_2O$, these trihalocarbonyl compounds react further to give, after a final acidification, a molecule of a carboxylic acid and a molecule of a **haloform** (trihalomethane) (Fig. 18.35).

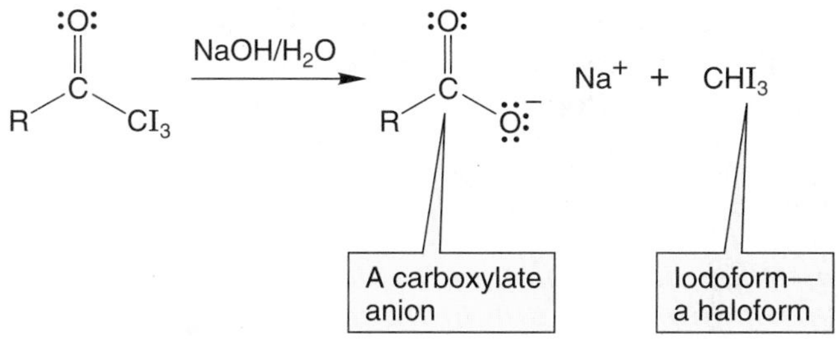

FIGURE 18.35 Trihalocarbonyl compounds react in base to give a molecule of a carboxylate anion and a haloform (trihalomethane).

Haloform formation is a new reaction whose mechanism is well within your powers to figure out. First, ask yourself what bonds must be broken and made in the reaction. Then ask what reactions you know that can make this happen. Almost always there will be a connection between the answers to these two questions. First of all, the formation of a carboxylic acid means that a new oxygen–carbon bond must be made. Similarly, the formation of a haloform, a one-carbon fragment, means that a carbon–carbon bond must have been broken. Those are the bond-making and bond-breaking requirements, and any mechanistic hypothesis must allow for them to take place in a reasonable way. It also seems obvious that the "CI_3" portion of iodoform (HCI_3) will come from the CI_3 present in the starting trihaloketone (Fig. 18.36).

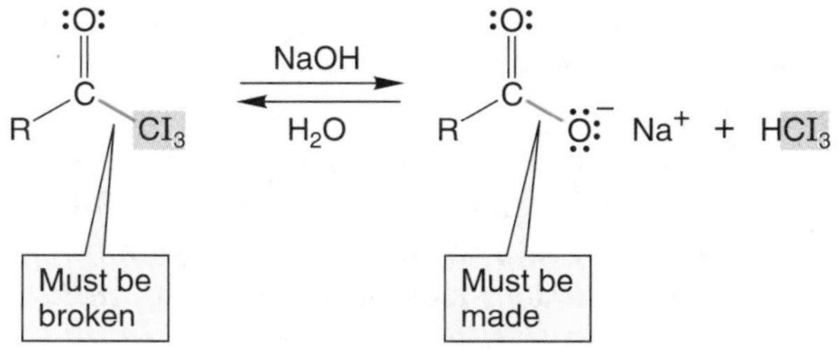

FIGURE 18.36 The bond-making and bond-breaking requirements for this reaction.

Now what's possible? The easiest possible connection would be to make the oxygen–carbon bond at the same time that the carbon–carbon bond is broken. But this requires an S_N2 displacement at an sp^2 hybridized carbon, and that is not a reaction for which there is any precedent (Fig. 18.37).

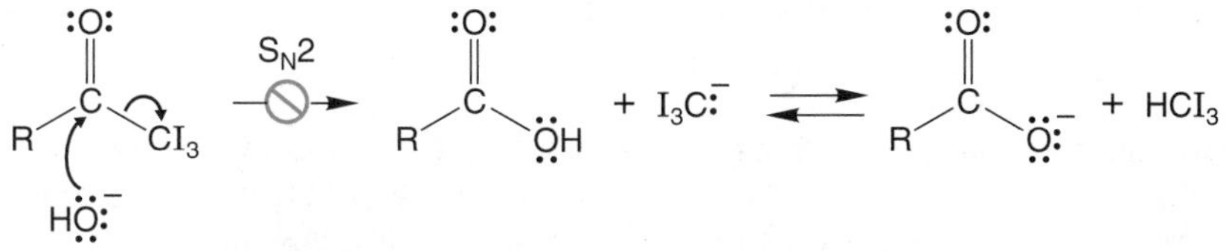

FIGURE 18.37 There is no precedent for this hypothetical S_N2 displacement at an sp^2 carbon.

PROBLEM 18.8

Explain why S_N2 displacement at an sp^2 hybridized atom is difficult.

What other reaction can happen between a carbonyl group and a base? Essentially all nucleophilic reagents add to carbonyl groups. Moreover, the addition reaction is strongly favored by a halogen substitution (Chapter 16, p. 770)! So, the addition of hydroxide to the carbonyl looks like a reaction that is almost certain to occur. Once the addition reaction has taken place, there is an opportunity to generate the acid and the haloform if the triiodomethyl anion can be lost as the carbonyl group re-forms. A proton transfer completes the reaction (Fig. 18.38).

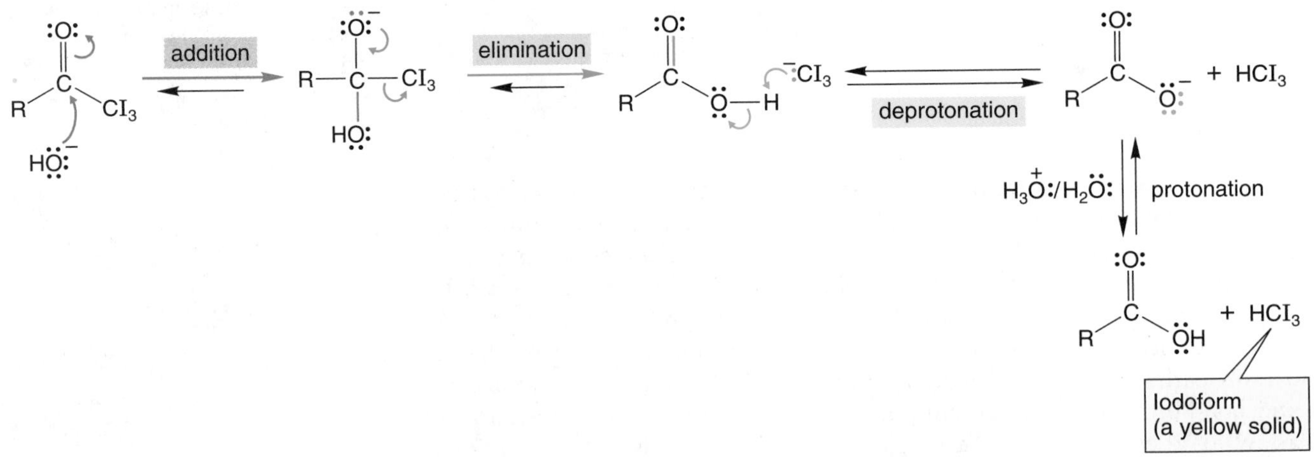

addition

elimination

deprotonation

$H_3\overset{+}{O}$:/$H_2\ddot{O}$:

protonation

$+ HCI_3$

Iodoform (a yellow solid)

FIGURE 18.38 Addition of hydroxide to the carbonyl group leads to a tetrahedral intermediate that can lose triiodomethide anion to generate the carboxylic acid. Transfer of a proton completes the reaction.

Is this a reasonable mechanism? Is the triiodomethyl anion a good enough leaving group to make this step a sensible one? The iodines are strongly electron withdrawing, and that will stabilize the anion. As a check we might look up the pK_a of iodoform to see how easily the molecule is deprotonated to form the anion. The value is about 14, and so iodoform is a relatively strong acid, at least compared to water (pK_a = 15.7). So the loss of $^-CI_3$ in Figure 18.38 does look reasonable (Fig. 18.39).

$$I_3C{-}H \;\; :\ddot{O}H^- \rightleftharpoons I_3C:^- + H{-}\ddot{O}H$$

pK_a ~ 14   pK_a = 15.7

FIGURE 18.39 The pK_a of iodoform is about 14. Iodoform is a strong acid, and the loss of $^-CI_3$ as a leaving group is a reasonable step.

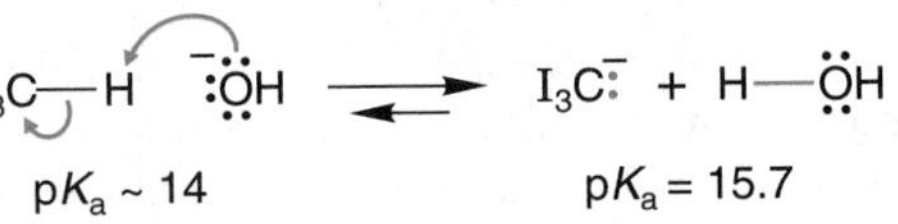

Deprotonation of the carboxylic acid, pK_a ~ 4.5, gives iodoform and the carboxylate anion. A final acidification step gives us back the carboxylic acid. In fact, this reaction serves as a test for methyl ketones. Formation of iodoform, a yellow solid, is a positive test for a molecule containing a methyl group attached to a carbonyl carbon.

PROBLEM 18.9 If we measure the rates of three reactions of the ketone shown in Figure 18.40, exchange of the α-hydrogen for deuterium in D_2O/DO^-, racemization in H_2O/HO^-, and α-bromination using $Br_2/H_2O/HO^-$, we find that they are identical. How can the rates of three such different reactions be the same? Explain, using an Energy versus Reaction progress diagram.

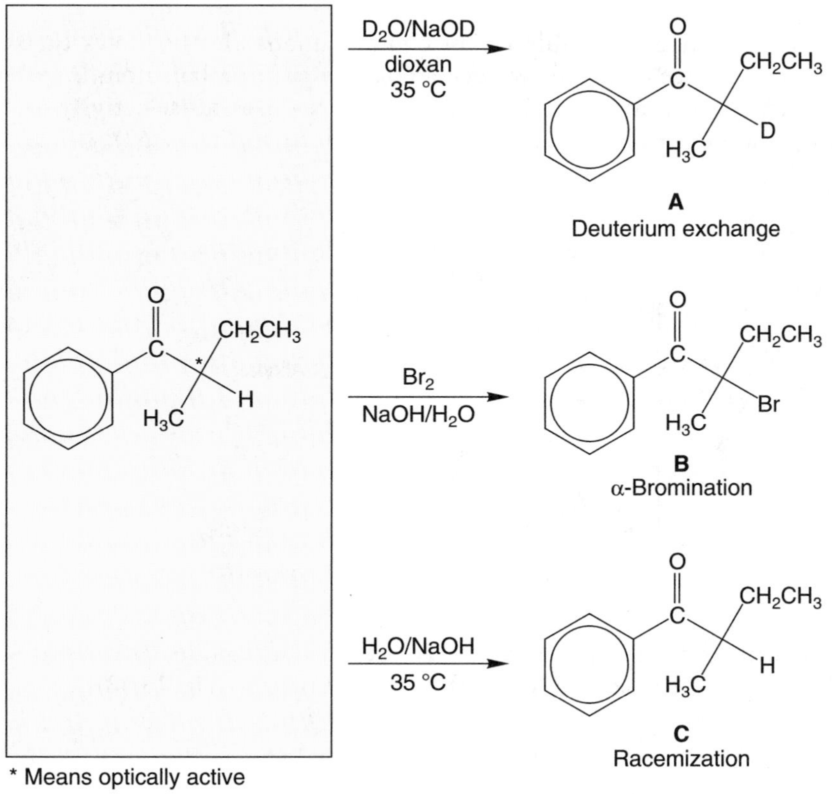

D_2O/NaOD, dioxan, 35 °C → A, Deuterium exchange

Br_2, NaOH/H_2O → B, α-Bromination

H_2O/NaOH, 35 °C → C, Racemization

* Means optically active

FIGURE **18.40**

We have now developed three reactions taking place at the α-position of carbonyl compounds. We will use these as prototypes on which to base further, more complicated processes. The unifying theme is formation of an enolate ion (in base) or an enol (in acid), that can act as a nucleophile. What other reactions might such a nucleophile undergo?

18.2d Alkylation Reactions

In the previous sections, we have seen several examples of enolate anions acting as Lewis bases and nucleophiles. A logical extension would be to attempt to use enolates as nucleophiles in S_N2 displacements. If we could do this reaction, we would have a way to alkylate the α-position of carbonyl compounds, and a new and most useful carbon–carbon bond-forming reaction would result (Fig. 18.41).

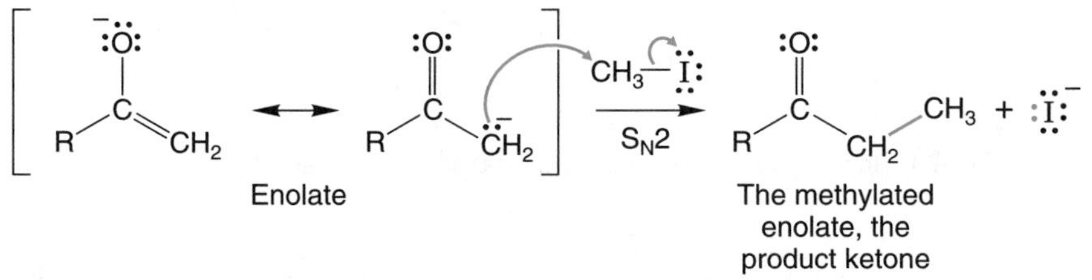

Enolate

CH_3–I, S_N2

The methylated enolate, the product ketone

FIGURE **18.41** If the enolate could act as a nucleophile in the S_N2 reaction, we might have a way of alkylating at the α-position.

We can see some potential difficulties and limitations right away. Alkylation of an enolate is an S_N2 reaction, and that means we could not use a tertiary halide, a species too hindered for participation in the S_N2 reaction.

The resonance formulation of enolate anions clearly shows that the negative charge is shared between an oxygen and a carbon atom. At which atom will alkylation be faster? If there is little or no selectivity in the alkylation reaction, it will surely be of limited use (Fig. 18.42).

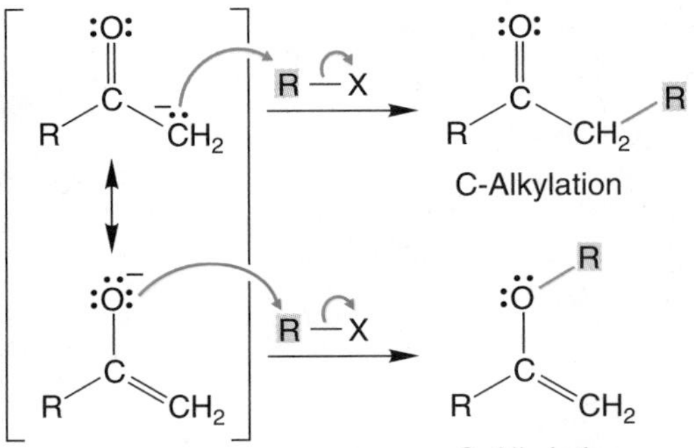

C-Alkylation

O-Alkylation

FIGURE 18.42 In principle, alkylation of the enolate could take place at either carbon or oxygen.

It turns out that most enolates are better nucleophiles at carbon. So, alkylation takes place faster at carbon than at oxygen (Fig. 18.43).

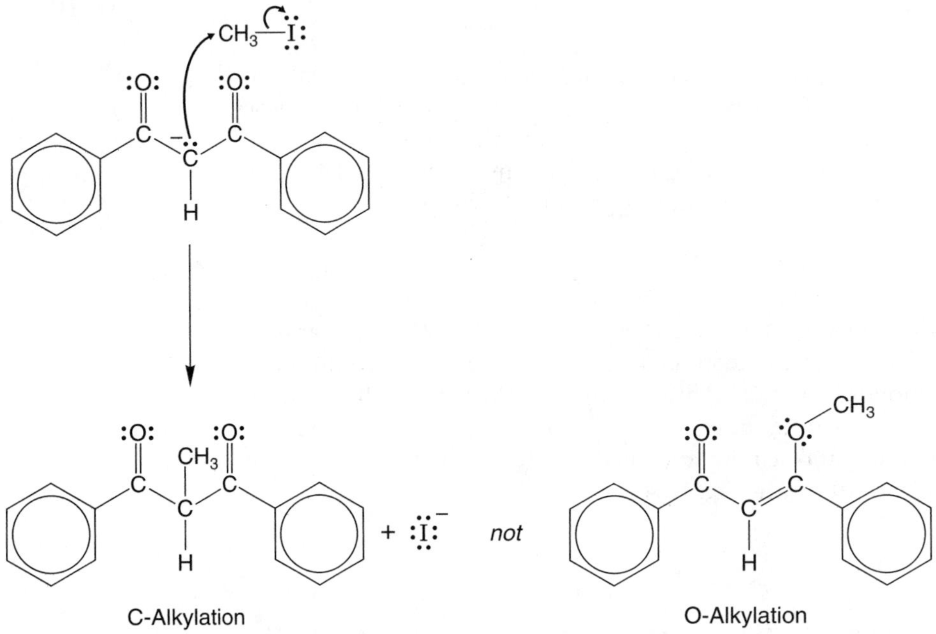

C-Alkylation

not

O-Alkylation

FIGURE 18.43 In practice, alkylation generally takes place at carbon.

Even though alkylation at oxygen is not common, there are still other problems, and this alkylation reaction of simple ketones is not very useful in practice. Consider 2-methylcyclohexanone as an example. Both α-positions are active, and two enolates will be produced (Fig. 18.44).

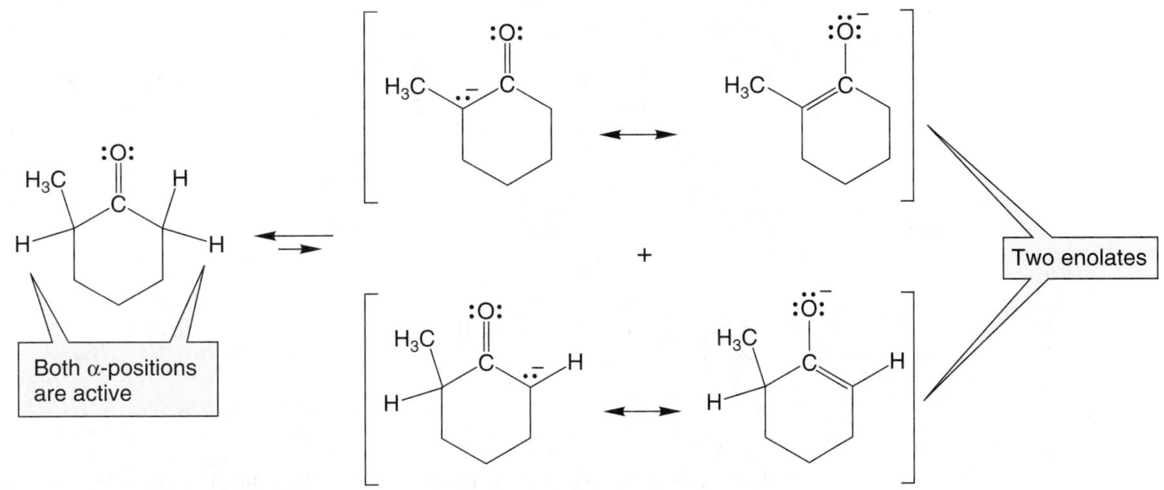

FIGURE 18.44 For many ketones there are at least two possible enolates, and therefore mixtures are obtained in the alkylation reaction.

Moreover, the once-alkylated compounds can undergo further alkylation. Reactions of the initial products compete with the desired single alkylation reaction. Mixtures of products are generally formed, and that is not a very useful situation (Fig. 18.45).

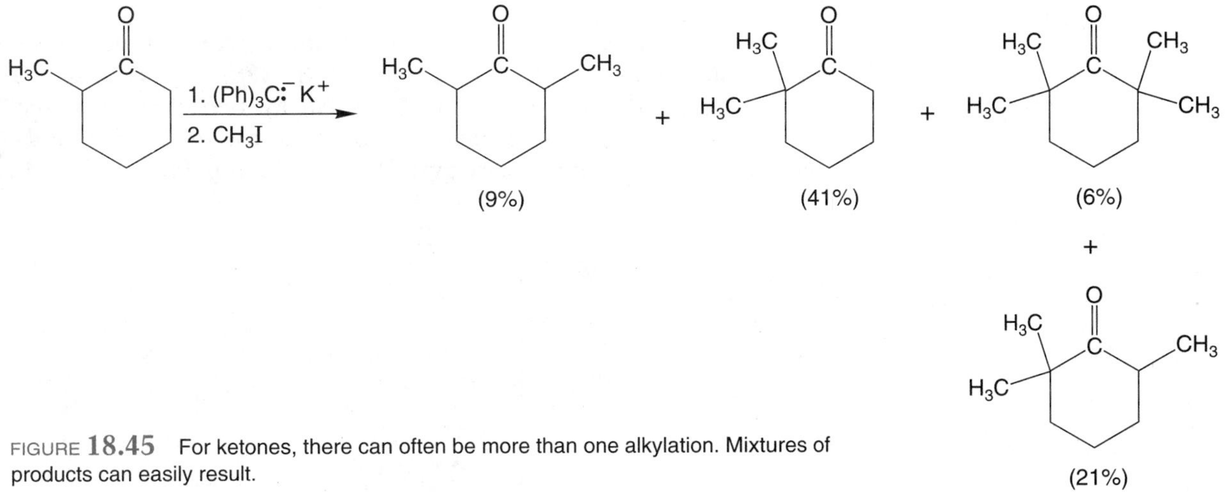

FIGURE 18.45 For ketones, there can often be more than one alkylation. Mixtures of products can easily result.

Aldehydes, which can only contain a single kind of α-hydrogen, are also problematic in this reaction. Not only do we have to deal with this problem of overalkylation, but several side reactions (see Section 18.3) tend to dominate the reaction.

Only if we restrict ourselves to carbonyl compounds containing a single α-hydrogen is some measure of specificity found, and we are now very far indeed from the general alkylation reaction we had hoped for. Strongly basic, but poorly nucleophilic species are used to form the enolate. Sodium hydride is effective, as is **lithium diisopropylamide (LDA)**. Both of these strong bases are only weakly nucleophilic, and addition to the carbonyl group is minimized in favor of deprotonation (Fig. 18.46).

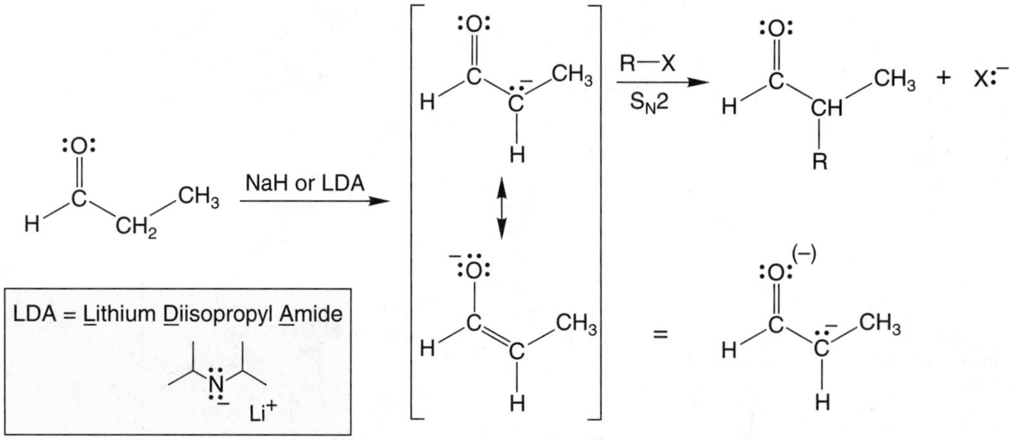
:O:
H
C
CH$_2$
CH$_3$
NaH or LDA
LDA = Lithium Diisopropyl Amide
N
Li$^+$
R—X
S$_N$2
CH
R
+ X:$^-$
(–)

FIGURE **18.46** Strong bases such as LDA or NaH are effective at forming enolates.

Convention Alert!

Note again the use in Figure 18.46 of the shorthand notation for resonance forms. Instead of always drawing out each important resonance form, one summary form is drawn. The positions sharing the charge or electron are indicated with parentheses, (+), (–), or (·).

PROBLEM **18.10** Explain why LDA is a poor nucleophile.

Perhaps not surprisingly, a way around some of these problems has been found, and it uses **enamines** (Chapter 16, p. 786) as substitutes for carbonyl compounds. Enamines can be formed by the addition of secondary amines (not primary or tertiary amines) to carbonyl compounds. Two favorite secondary amines are pyrrolidine and morpholine, shown in Figure 18.47.

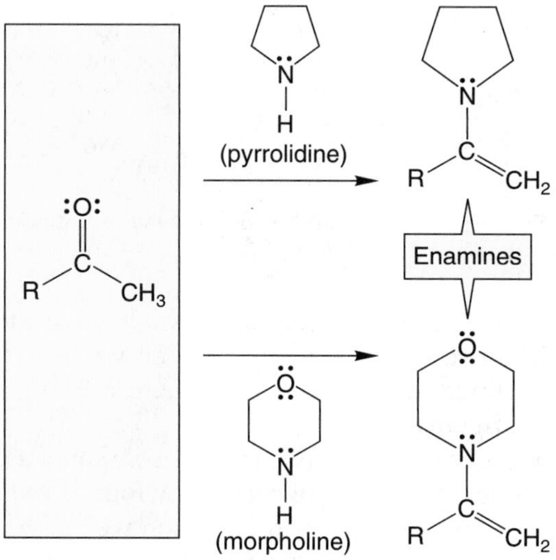
(pyrrolidine)
R
CH$_3$
CH$_2$
Enamines
(morpholine)

FIGURE **18.47** Enamine formation using the secondary amines pyrrolidine and morpholine.

PROBLEM **18.11** Why can't primary or tertiary amines be used to make enamines (Chapter 16, p. 786)?

Enamines are electron-rich and nucleophilic and will attack S_N2-active halides (no tertiary halides!) to form immonium (or imminium) ions. The reaction is ended by adding water to hydrolyze the immonium ions to the corresponding ketones (Fig. 18.48).

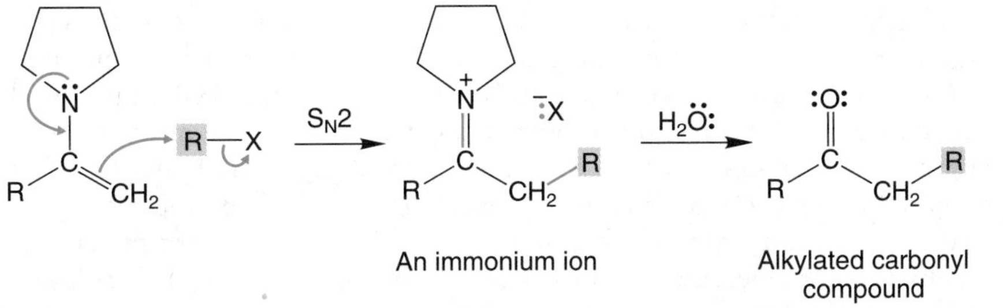

FIGURE **18.48** Alkylation of an enamine initially gives an immonium ion. The immonium ion can be hydrolyzed to give the alkylated carbonyl compound.

PROBLEM **18.12**

Write the mechanism for the hydrolysis of the immonium ion to the ketone. *Hint*: You have seen this reaction before, in reverse, in the mechanism for acid-catalyzed enamine formation (Chapter 16, p. 786).

Figure 18.49 summarizes the procedure: The ketone is first converted into the enamine, the enamine is alkylated, and then the ketone is regenerated by hydrolysis.

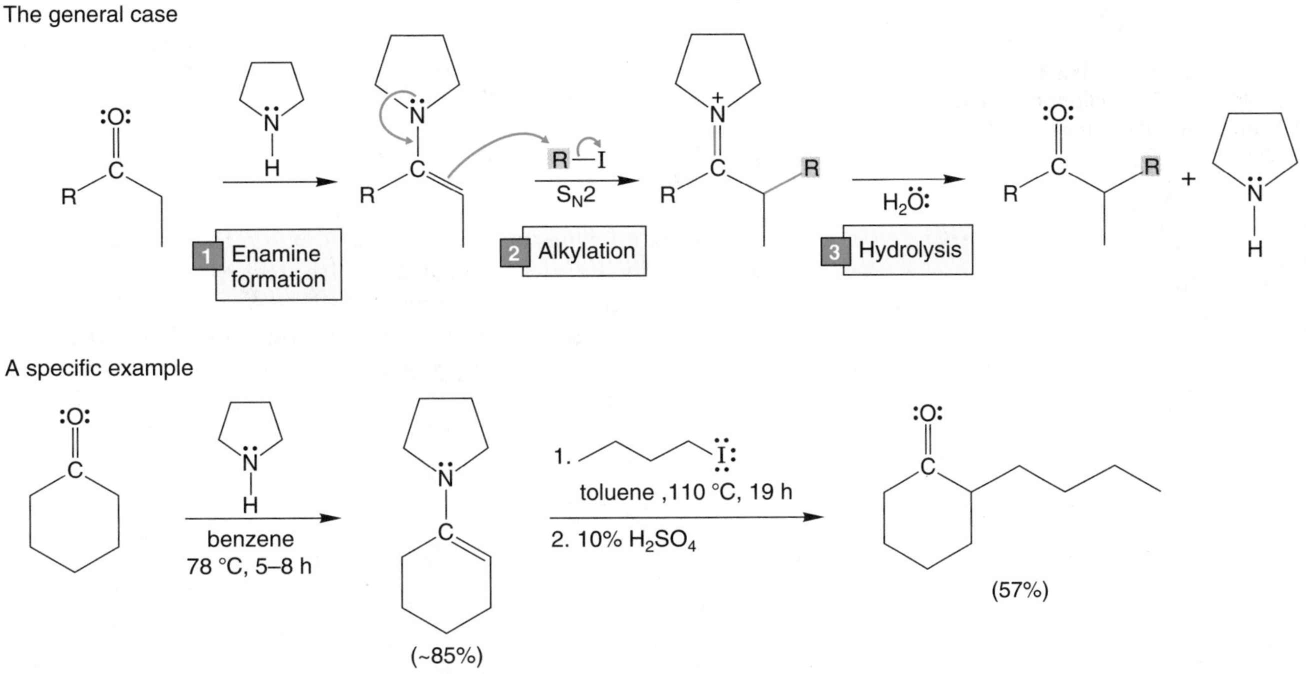

FIGURE **18.49** An overview of the three-step alkylation process using an enamine.

There are still some problems, however. It is not easy to use alkylations of enamines when more than one possible enamine can be formed, and there are occasionally difficulties with overalkylation or substitution on nitrogen.

18.3 CONDENSATION REACTIONS OF CARBONYL COMPOUNDS: THE ALDOL CONDENSATION

So far, we have seen the reactions of enolates and enols with a variety of Brønsted and Lewis acids, for example: D_2O, H_2O, Br_2, Cl_2, I_2, and alkyl halides. What happens if these acids are absent, or, as is the case for water, simply regenerate starting material and do not result in new products? Suppose, for example, we treat acetaldehyde with KOH/H_2O. First of all, the answer to the question of what happens when a base and a carbonyl compound are combined is almost never, "nothing." Nucleophiles add to carbonyl groups, and hydroxide will surely initiate hydrate formation. The hydrate is usually not isolable, but it is formed in equilibrium with the aldehyde. Moreover, enolate formation is also possible and the enolate, as well as the hydrate, will be in equilibrium with the starting aldehyde. Be very careful about "nothing happens" answers. They are almost always wrong (Fig. 18.50).

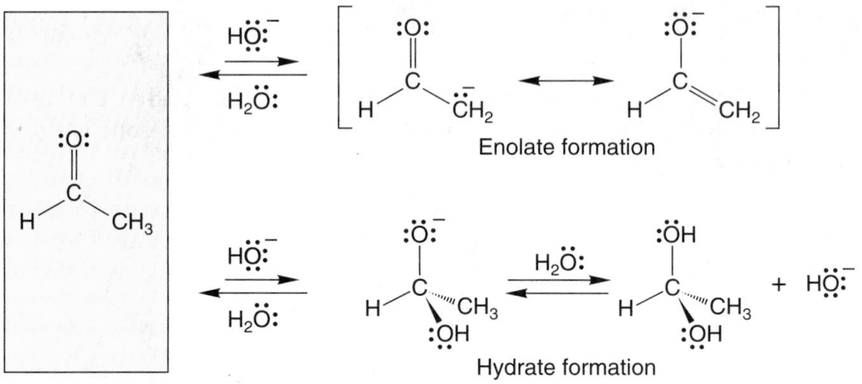

FIGURE 18.50 Two reactions of acetaldehyde with hydroxide ion: addition (hydrate formation) and enolate formation.

Although hydrate formation does not lead directly to an isolable product, formation of the enolate does. Over the last few pages we have seen several examples of the reactions of nucleophilic enolates with Lewis and Brønsted acids. What acids are present in this solution? Water is surely one, but reaction with water simply generates the enol in equilibrium with acetaldehyde (Fig. 18.51).

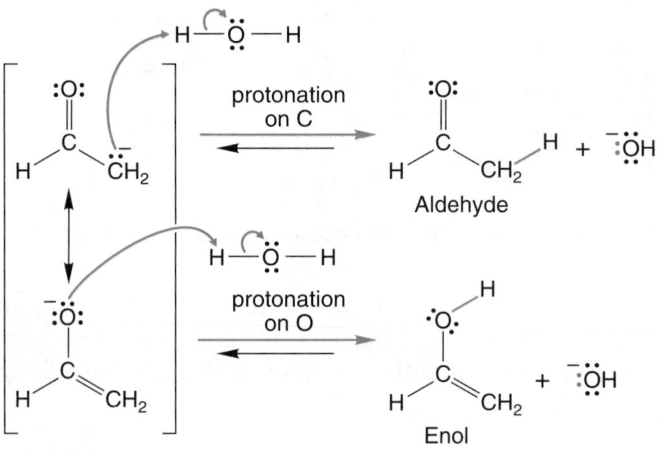

FIGURE 18.51 The enolate can be reprotonated at either carbon or oxygen. Reaction at oxygen usually dominates, but the resulting enol equilibrates with the more stable carbonyl form.

There is another acid present, the carbonyl group of acetaldehyde. Nucleophiles add to carbonyl groups, and the enolate anion is surely a nucleophile. There is no essential difference between the addition of hydroxide to the carbonyl group of acetaldehyde and the addition of the enolate anion (Fig. 18.52).

Nor is there much structural difference between the hydrate formed by protonation of the hydroxide addition product, and the molecule known as "aldol" formed by protonation of the enolate addition product. Note that hydroxide ion, the catalyst, is regenerated in the final step of hydrate formation or of aldol formation (Fig. 18.53).

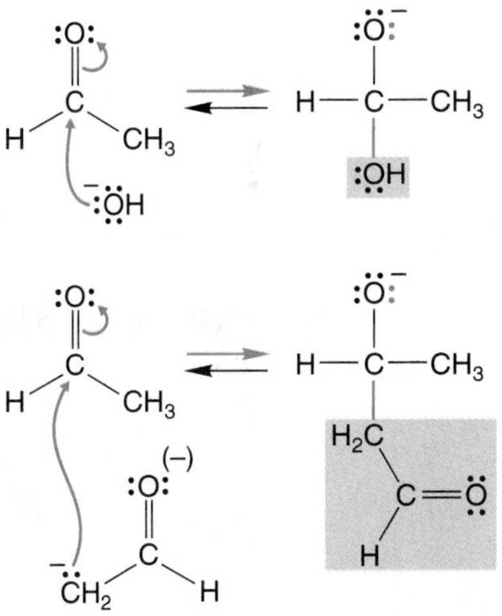

FIGURE 18.52 Addition of hydroxide and the enolate anion to the carbonyl group are simply two examples of the addition reaction of nucleophiles to carbonyl groups.

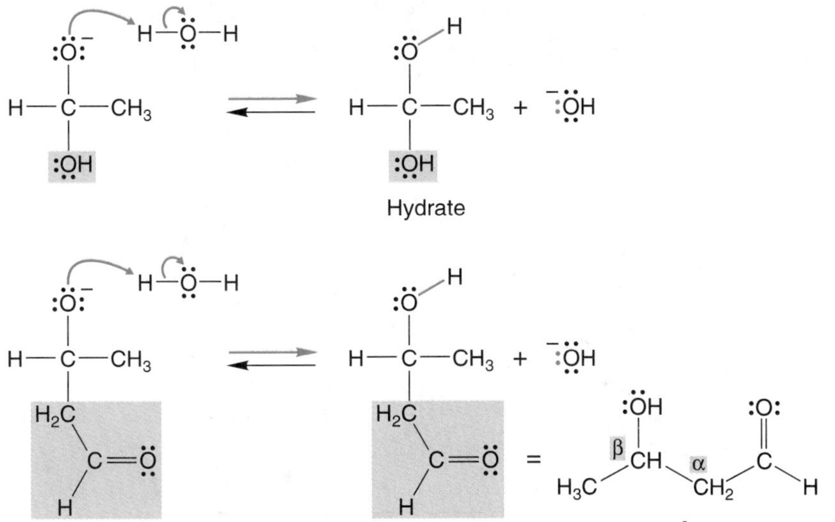

Hydrate

A
Aldol
a β-hydroxy aldehyde

FIGURE 18.53 Protonation of these intermediates gives the hydrate, or, in the enolate case, a compound known as "aldol", a β-hydroxy aldehyde.

But the product itself is surely more complex than the relatively simple hydrate! The conversion of an aldehyde or ketone into a β-hydroxy carbonyl compound is called the **aldol condensation*** after the product of the reaction of acetaldehyde, and is quite general for carbonyl compounds with α-hydrogens.

As usual, this base-catalyzed reaction has its acid-catalyzed counterpart. For the acid-catalyzed reaction the catalyst is not hydroxide, but H_3O^+, and the active ingredient is not the enolate anion, but the enol itself. The first step in the reaction is acid-catalyzed enol formation (Fig. 18.54).

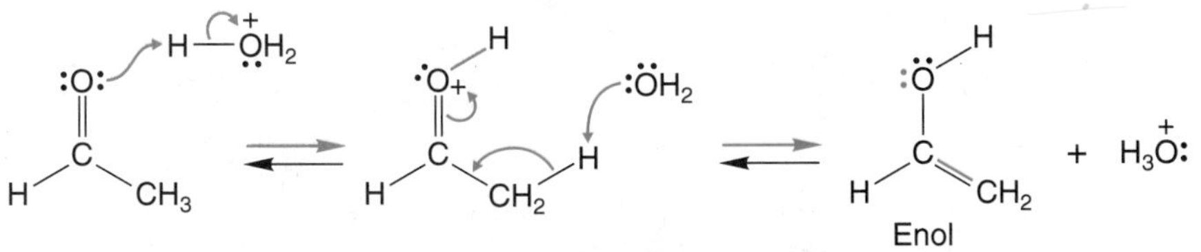

Enol

FIGURE 18.54 The acid-catalyzed aldol condensation begins with enol formation.

*Credit for discovery of the aldol condensation generally goes to Charles Adolphe Wurtz (1817–1884) who coined the word "aldol" in 1872. However, equal billing at least should also go to Alexandr Borodin, the Russian composer—physician—chemist (1833–1887). Borodin was the illegitimate son of Prince Gedianov, and actually spent his formative years as his father's serf. Happily, he was given his freedom and became an eminent chemist who is far better known for the somewhat schmaltzy products of his avocation, music, than for his excellent chemistry.

The enol, though nucleophilic, is not nearly as strong a nucleophile as hydroxide. However, the Lewis acid present is the protonated carbonyl, and it is a far stronger Lewis acid than the carbonyl group itself. The reaction between the enol and the protonated carbonyl group is analogous to the reaction of the enol with H_3O^+. Once again, the final step of the reaction regenerates the catalyst, this time H_3O^+ (Fig. 18.55).

Protonation/deprotonation of the enol to regenerate acetaldehyde

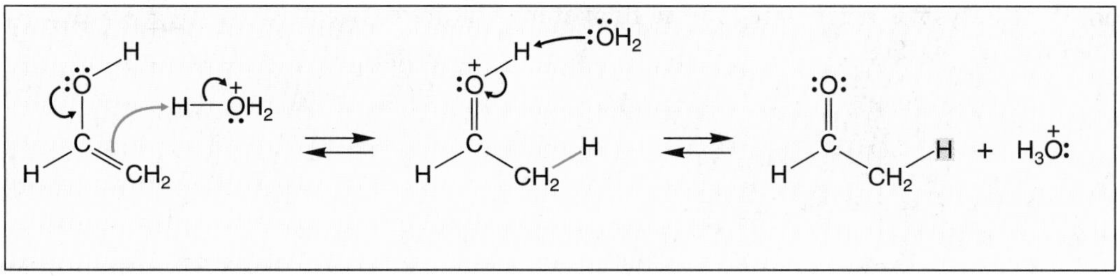

Reaction of the enol with the protonated carbonyl to give aldol

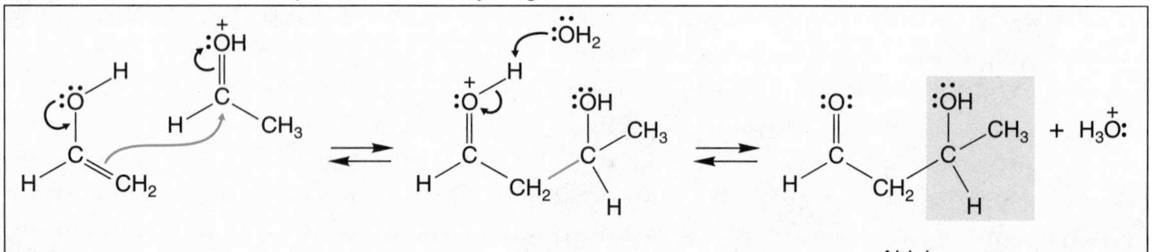

Aldol

FIGURE **18.55** Two reactions of the weakly nucleophilic enol with Lewis acids. In the first case, it is protonated to regenerate acetaldehyde; in the second the enol adds to the strongly Lewis acidic protonated carbonyl group to give aldol.

The acid- and base-catalyzed aldol condensations of acetaldehyde give the same product, the β-hydroxy carbonyl compound called aldol. The aldol condensation is a general synthesis of β-hydroxy ketones and aldehydes.

*PROBLEM **18.13** Write the products of the aldol condensations of the following compounds (Fig. 18.56). Write both acid- and base-catalyzed mechanisms for reactions of (b).

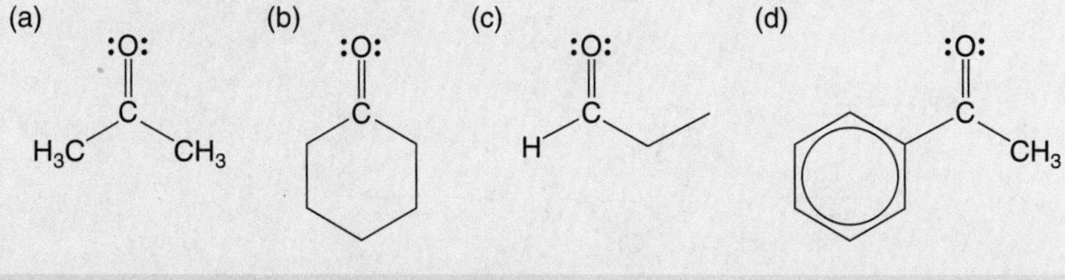

(a) (b) (c) (d)

FIGURE **18.56**

ANSWER (b) As usual, in base the enolate is first formed and then adds to the carbonyl group of another molecule. A final protonation generates the condensation product and regenerates the catalyst, here hydroxide.

ANSWER (CONTINUED)

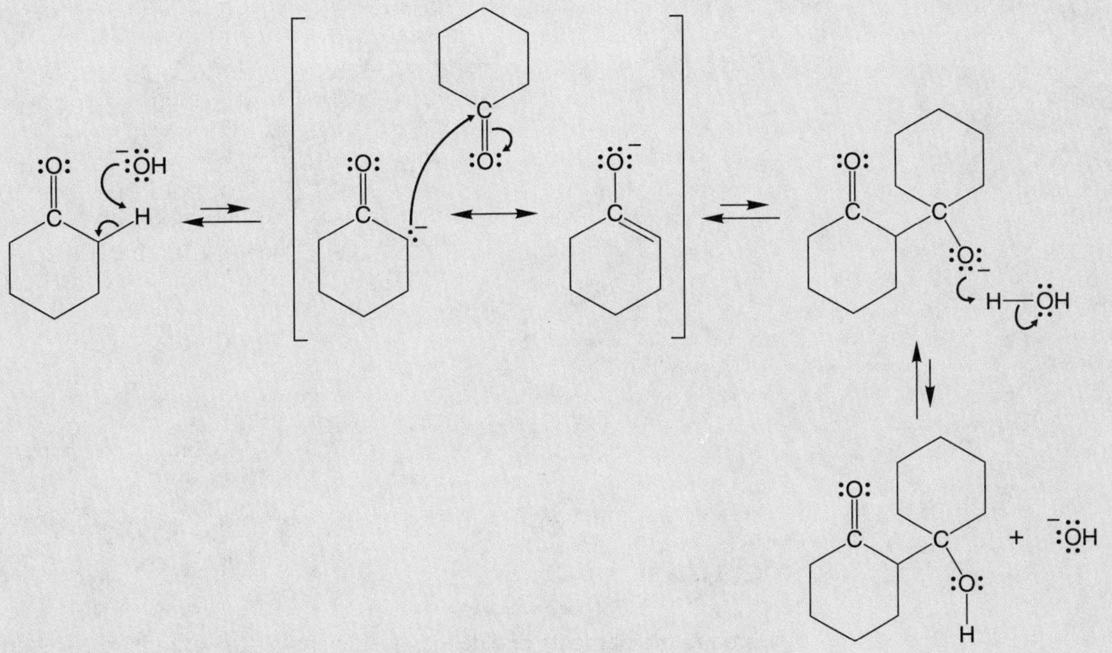

Watch out! In this figure, arrows emanate from one of a pair of resonance forms, which is done for simplicity's sake only. Write the arrow formalism using the other resonance form.

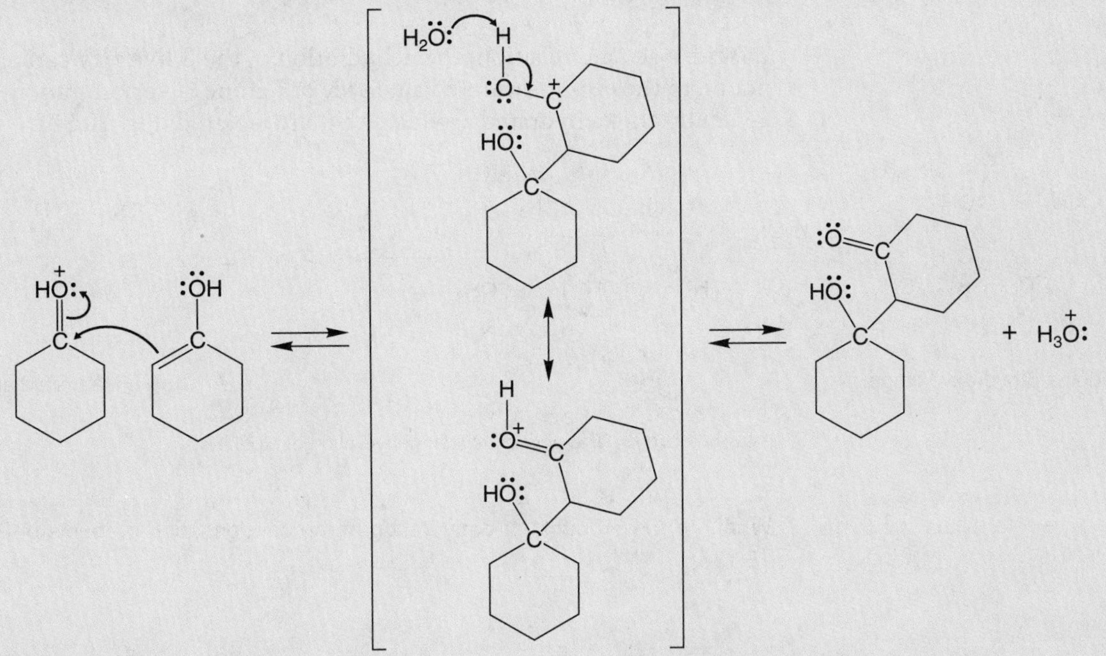

In acid, the protonated carbonyl group reacts with the enol form of the ketone to give a resonance-stabilized intermediate. Deprotonation gives the product and regenerates the acid catalyst, H_3O^+.

ANSWER (CONTINUED)

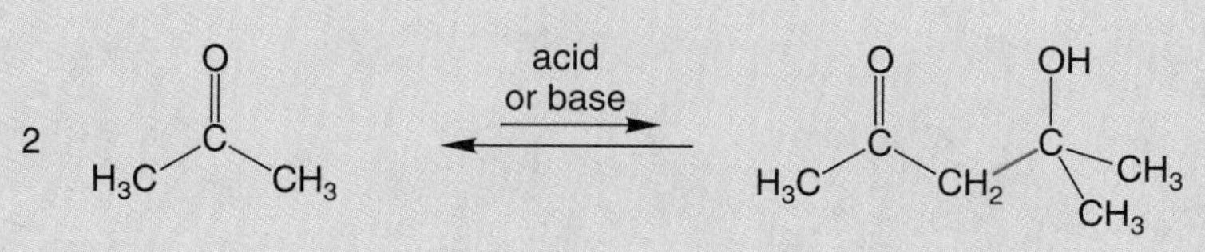

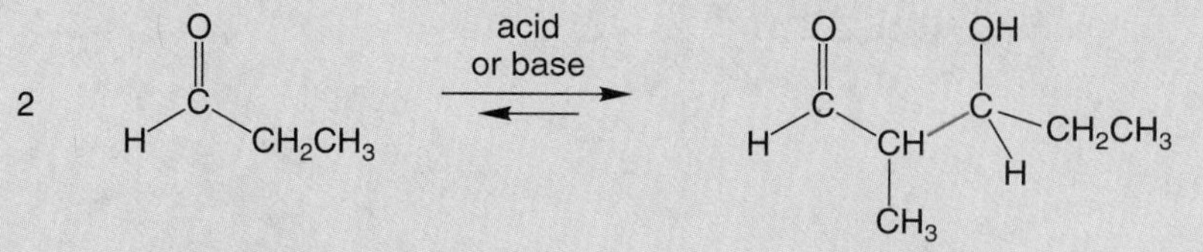

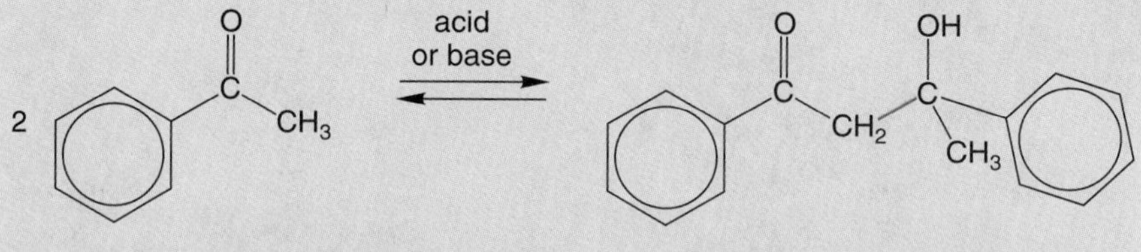

Here are the products of aldol condensations of (a), (c), and (d). The new bond between the two carbonyl compounds is shown as a red line. Practice seeing these reactions in reverse. Ask yourself what compounds could make these β-hydroxy ketones and aldehydes.

In acid, a second reaction, the dehydration of the β-hydroxy carbonyl compound to the α,β-unsaturated aldehyde or ketone is very common, and it is generally the dehydrated products that are isolated (Fig. 18.57).

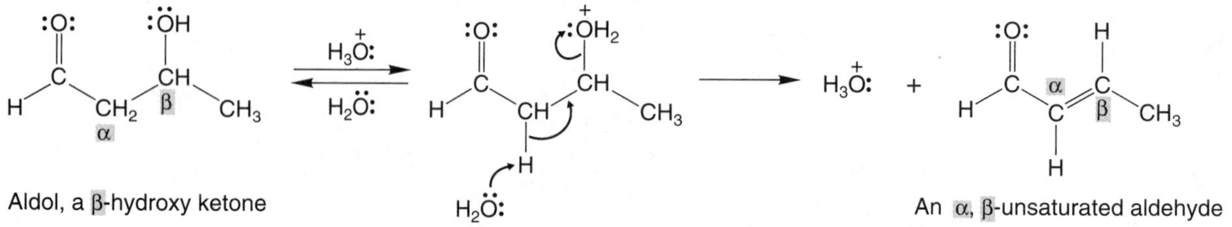

FIGURE 18.57 The acid-catalyzed dehydration of aldol.

PROBLEM 18.14 What are the products of dehydration of the condensation products of Problem 18.13?

*PROBLEM 18.15 Figure 18.57 shows protonation of the hydroxyl oxygen. Is this likely to be faster or slower than protonation of the carbonyl oxygen? Why or why not?

ANSWER Protonation of the hydroxyl oxygen leads to an oxonium ion. Protonation of the carbonyl oxygen leads to a resonance-stabilized, and therefore more stable, cation. The resonance stabilization will be felt in the transition state for protonation, and therefore protonation of the carbonyl oxygen will be faster. Are you getting the idea that one of the generic answers to questions asked in organic chemistry is, "It's resonance-stabilized and therefore more stable."? If you are, you're right.

ANSWER (CONTINUED)

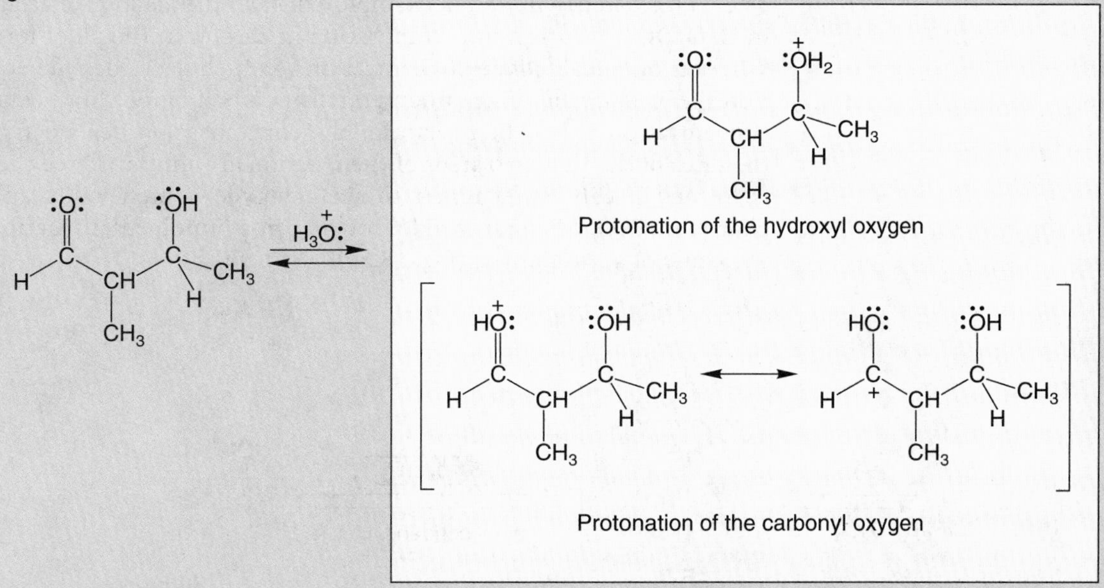

Protonation of the hydroxyl oxygen

Protonation of the carbonyl oxygen

Protonation of the carbonyl oxygen, though faster, is a dead end as far as dehydration is concerned. In order to form the α,β-unsaturated compound, the hydroxyl group must be protonated.

Hydroxide is a poor leaving group. Therefore, dehydration is more difficult in base and the β-hydroxy compounds can often be isolated. Even under basic conditions dehydration can be achieved at relatively high temperature. The mechanism is not a simple E2 reaction, however. Removal of the α-hydrogen is relatively easy, as this gives a resonance-stabilized enolate anion. In a second step, hydroxide is lost and the α,β-unsaturated compound is formed. This kind of elimination mechanism, called the E1cB reaction, was the subject of Section 7.13 (Fig. 18.58).

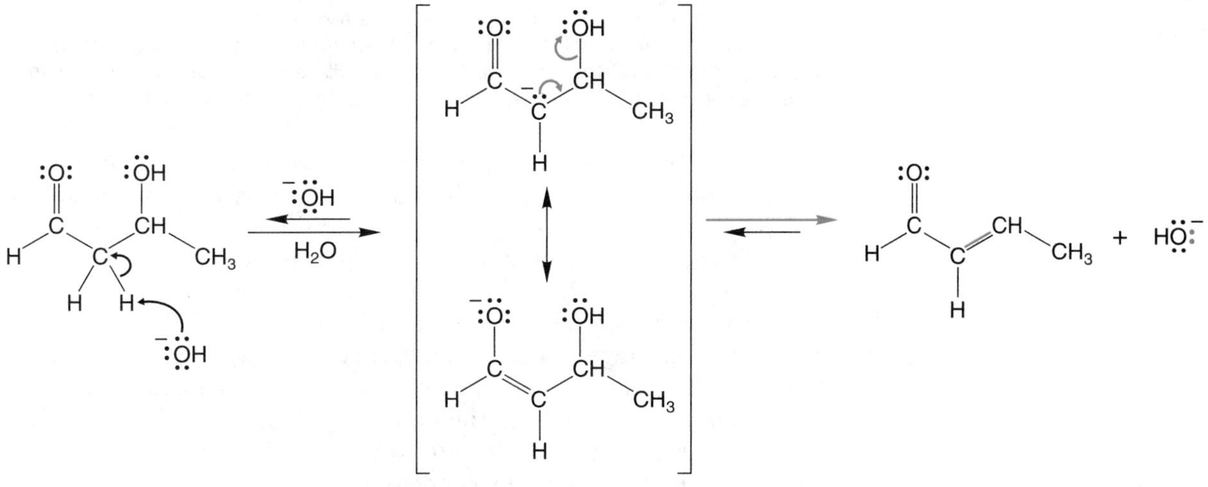

Resonance stabilized enolate

FIGURE **18.58** The base-catalyzed elimination of water from aldol. The reaction mechanism is E1cB.

Like many reactions, the aldol condensation is a series of equilibria. It is not always clear what the slow step in the aldol condensation, the rate-determining step, will be. For example, consider the reaction of acetaldehyde in D_2O/DO^-. Deuterium incorporation at carbon in the aldol formed is a function of acetaldehyde concentration. At high acetaldehyde concentration the product aldol is not deuterated at carbon, but at low acetaldehyde concentration it is. The reason for the concentration dependence is that the rates of the first two steps of the reaction are similar. Once the enolate is formed it can either revert to acetaldehyde, which will result in exchange in D_2O, or go on to aldol product in a bimolecular reaction with a molecule of acetaldehyde (Fig. 18.59).

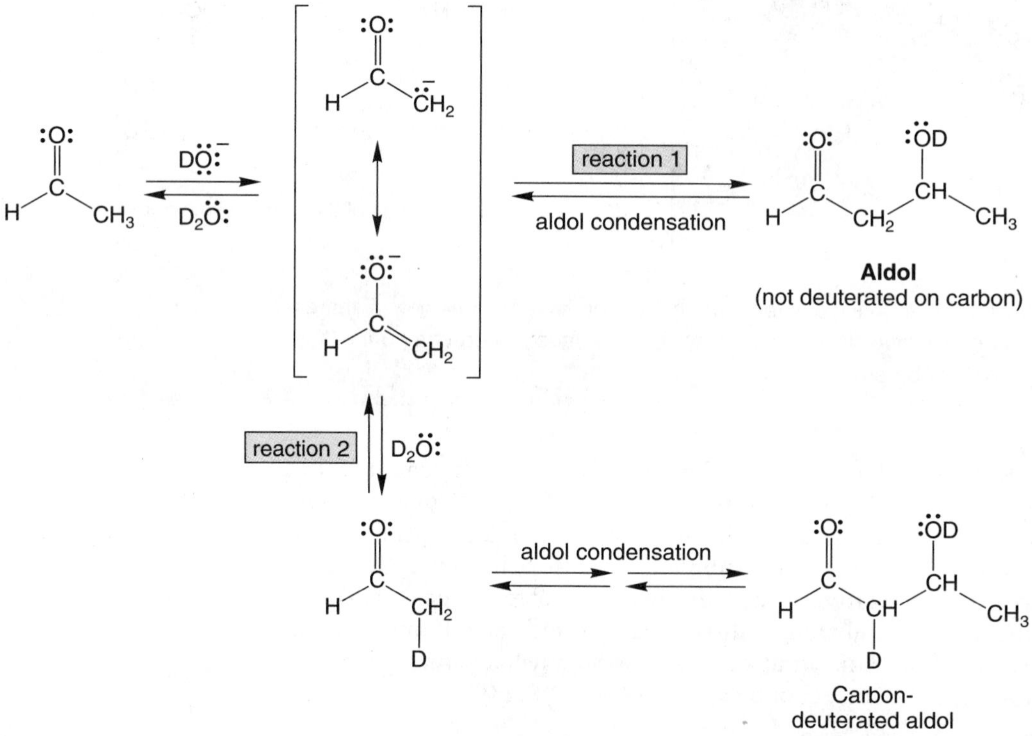

FIGURE 18.59 A high concentration of acetaldehyde results in a relatively rapid aldol condensation and little or no exchange at carbon. If, however, the concentration of acetaldehyde is low, the bimolecular condensation reaction will be slowed and the enolate can revert to acetaldehyde. In D_2O, this leads to exchange of hydrogens attached to carbon.

If the concentration of acetaldehyde is high, the rate of the bimolecular reaction (reaction 1) is faster ($v = k$[enolate][acetaldehyde]), and acetaldehyde is not re-formed from the enolate. Under these conditions, it is the formation of the enolate that is the slow step in the reaction, the rate-determining step (Chapter 8, p. 319). If the concentration of acetaldehyde is low, the enolate is reconverted to acetaldehyde (reaction 2), and in D_2O this results in exchange. When exchanged acetaldehyde goes on to form aldol, the product also contains deuterium attached to carbon. At low acetaldehyde concentrations it is the condensation step itself that is the slower one, and therefore the rate-determining step.

Ketones can undergo the aldol condensation too, and the mechanism is similar to that for the aldol condensation of aldehydes. We will use acetone as an example. In base, the enolate is formed first, and adds to the

Lewis acidic carbonyl compound. Protonation by water yields a molecule once known as diacetone alcohol, 4-hydroxy-4-methyl-2-pentanone (Fig. 18.60).

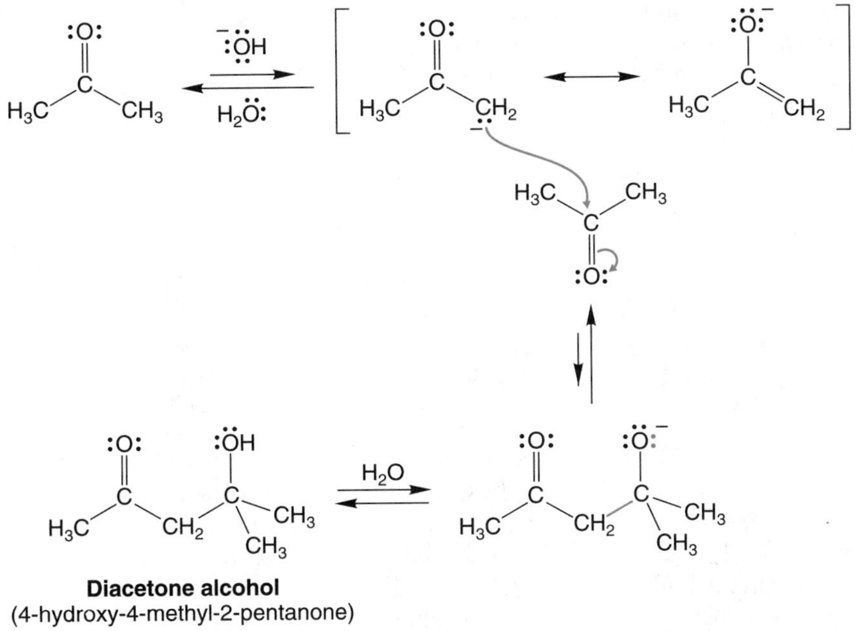

Diacetone alcohol
(4-hydroxy-4-methyl-2-pentanone)

FIGURE 18.60 The base-catalyzed aldol condensation of acetone.

In acid, the enol is the intermediate, and it adds to the very strong Lewis acid, the protonated carbonyl group, to give initially the same product as the base-catalyzed reaction, diacetone alcohol. Dehydration generally follows to give the α,β-unsaturated ketone, 4-methyl-3-penten-2-one, which is still often called mesityl oxide (Fig. 18.61).

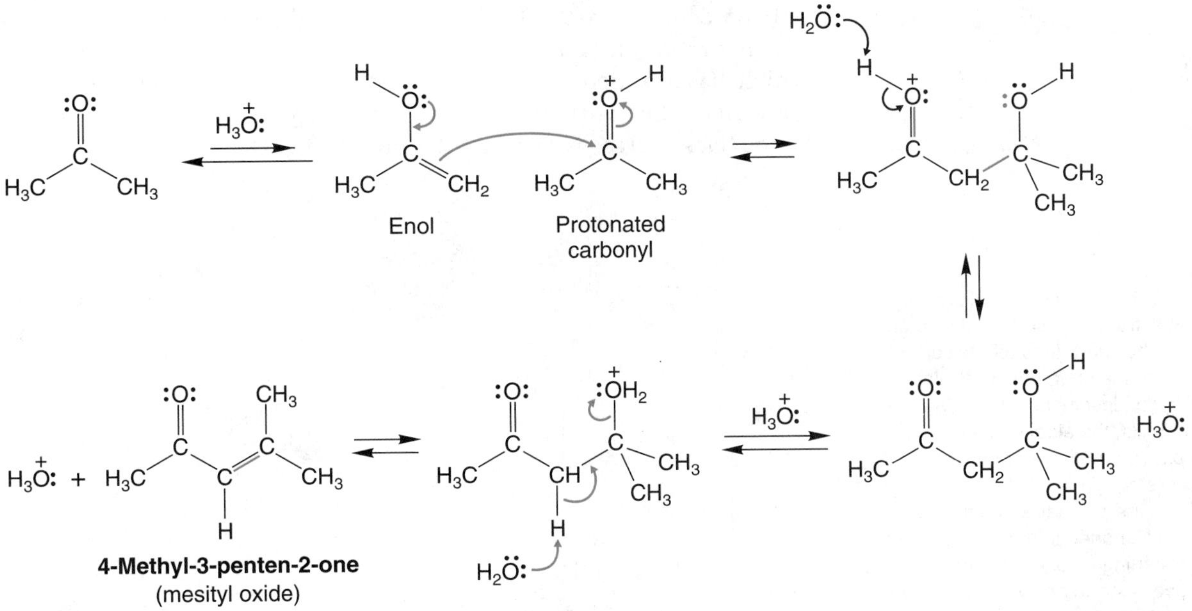

Enol

Protonated carbonyl

4-Methyl-3-penten-2-one
(mesityl oxide)

FIGURE 18.61 The acid-catalyzed aldol condensation of acetone. The first product, diacetone alcohol, is generally dehydrated in acid to give mesityl oxide, 4-methyl-3-penten-2-one.

We have been at pains to point out the similarities between these aldol condensations and other addition reactions of carbonyl compounds.

Like simple additions, such as hydration, the aldol condensations are series of equilibria. They are affected by structural factors in the same way as the simple reactions. Product formation is unfavorable in base-catalyzed aldol reactions of ketones, just as it is in hydration (Fig. 18.62). In acid, dehydration generally occurs and formation of the α,β-unsaturated ketone pulls the equilibrium to the right.

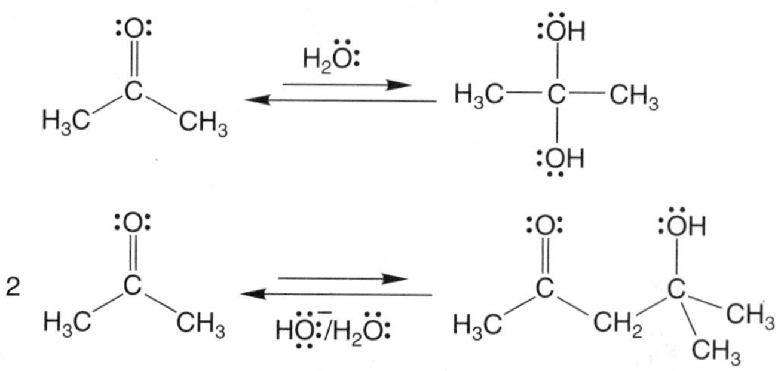

$$H_3C{-}C({=}O){-}CH_3 \overset{H_2O}{\rightleftharpoons} H_3C{-}C(OH)_2{-}CH_3$$

$$2\ H_3C{-}C({=}O){-}CH_3 \underset{HO^-/H_2O}{\rightleftharpoons} H_3C{-}C({=}O){-}CH_2{-}C(OH)(CH_3){-}CH_3$$

FIGURE 18.62 In the hydration of ketones, the hydrate is usually not favored at equilibrium. Similarly, in the aldol condensation of ketones, the product molecule is not usually favored.

If so, how can the aldol condensation of ketones be useful? If equilibrium favors starting material, how can the product be isolated? Ingenious ways have been found to thwart the dictates of thermodynamics. In one, a spectacularly clever piece of apparatus called a Soxhlet extractor is used. The ketone is boiled in a flask separated from the catalyst, usually barium hydroxide, $Ba(OH)_2$, which is insoluble in organic materials. The ketone distills to the condenser, off of which it drips to come in contact with the catalyst. Now the aldol condensation takes place to give an equilibrium mixture of starting ketone (favored) and the product of aldol condensation (unfavored). As the starting ketone continues to boil, and the level of the liquid in the trap rises, the whole solution eventually is siphoned from the chamber into the flask that previously contained only ketone. There is still no catalyst in the flask, and so the aldol product *cannot revert to starting material.* The mixture continues to boil, the remaining ketone, which has a much lower boiling point than the product, continues to come in contact with the catalyst and slowly the concentration of product in the flask increases to a useful amount (Fig. 18.63).

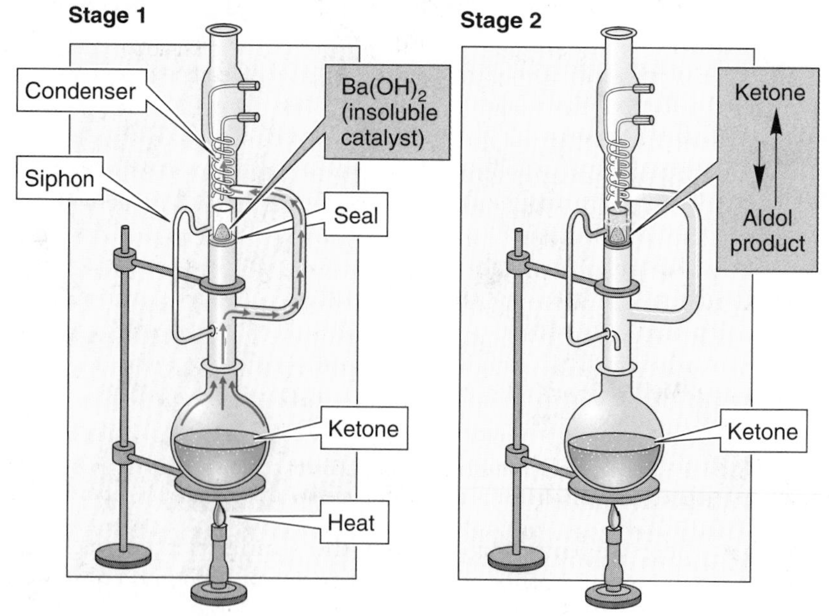

FIGURE 18.63 The operation of a Soxhlet extractor. In stage 1 the ketone begins to boil and drips off the condenser into the chamber containing $Ba(OH)_2$, an insoluble catalyst. As the chamber fills, the aldol condensation takes place, creating a small amount of product (stage 2). In stage 3, the chamber fills to the point at which the contents are siphoned to the flask containing the starting ketone. At stage 4, when the process starts anew, the flask contains starting ketone, a very small amount of product, *and no catalyst.* As the process cycles, more and more ketone is transformed into aldol product. As long as the product never comes in contact with catalyst, it cannot revert to starting material.

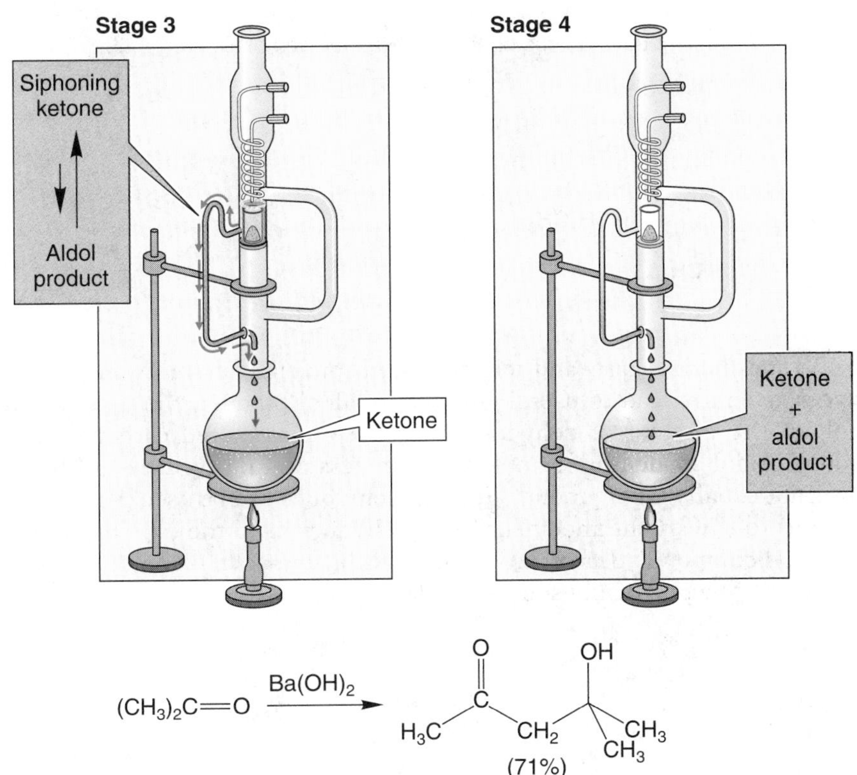

FIGURE 18.63 (CONTINUED)

All β-hydroxy carbonyl compounds are potential products of an aldol condensation. Whenever you see one, your thoughts about synthesis must turn first to the aldol condensation. The same is true for the dehydration products, the α,β-unsaturated carbonyl compounds. It is important to be able to go quickly to the new bond, to find the one formed during the condensation, and to be able to dissect the molecule into its two halves. In doing this operation, you are merely carrying out a retrosynthetic analysis and following the mechanism backward (Fig. 18.64).

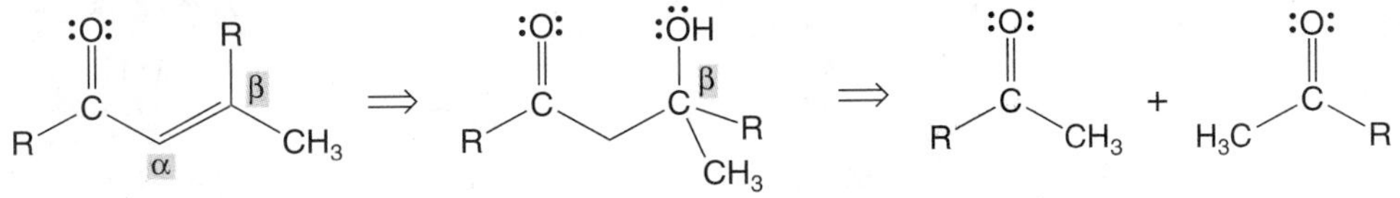

$$R{-}C(=\ddot{O}{:}){-}CH_{\alpha}{=}C_{\beta}(R){-}CH_3 \Rightarrow R{-}C(=\ddot{O}{:}){-}CH_2{-}C_{\beta}(\ddot{O}H)(R){-}CH_3 \Rightarrow R{-}C(=\ddot{O}{:}){-}CH_3 + H_3C{-}C(=\ddot{O}{:}){-}R$$

FIGURE 18.64 A retrosynthetic analysis for the product of an aldol condensation followed by dehydration.

Write mechanisms for the acid- and base-catalyzed reverse aldol condensation of diacetone alcohol (Fig. 18.65). PROBLEM 18.16

$$H_3C{-}C(=\ddot{O}{:}){-}CH_2{-}C(\ddot{O}H)(CH_3){-}CH_3$$

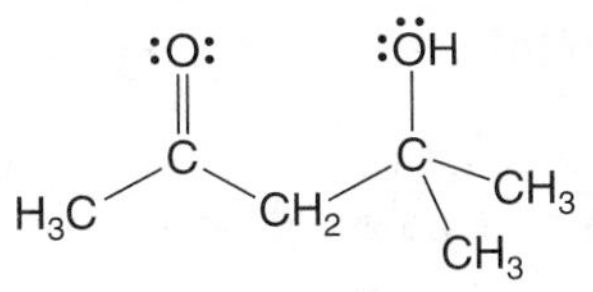

Diacetone alcohol

FIGURE 18.65

PROBLEM 18.17 Perform retrosynthetic analyses on the three molecules of Figure 18.66.

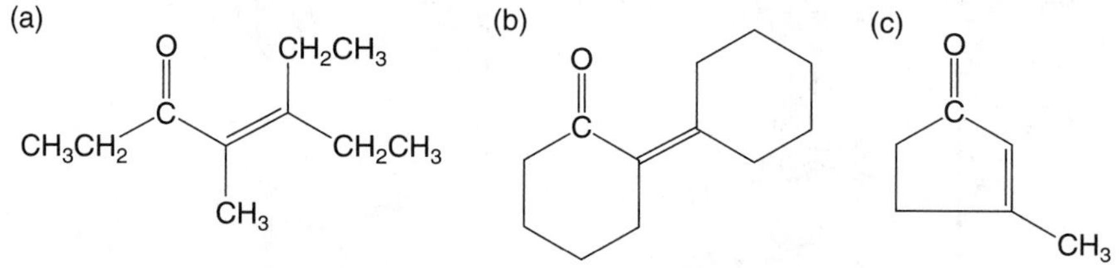

FIGURE 18.66

To summarize: Base- and acid-catalyzed condensation reactions of aldehydes generally lead to useful amounts of aldol product, the β-keto alcohol, or the product of dehydration, the α,β-unsaturated carbonyl compound. Aldol condensations of ketones are less useful because equilibrium does not usually favor product. Special techniques can be used (Soxhlet extractor) to thwart thermodynamics. If dehydration to the α,β-unsaturated carbonyl compound occurs, as it often does in acid-catalyzed aldols, the reaction may be practically successful (Fig. 18.67).

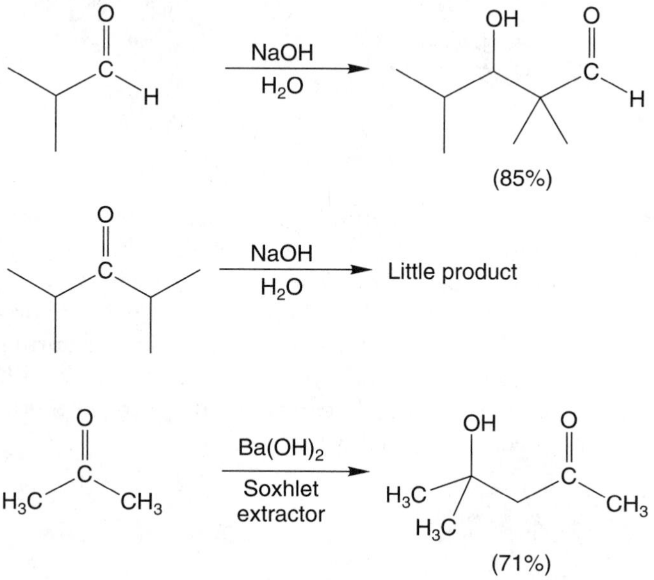

FIGURE 18.67 Although equilibrium generally favors the condensation product in the aldol reaction of aldehydes, the reaction of ketones generally favors starting material. Special techniques must be used to make the reaction of ketones practical.

18.4 REACTIONS RELATED TO THE ALDOL CONDENSATION

In the next few pages, we will use what we know of the simple aldol condensation to understand some related reactions. We will start with reactions that are very little different from simple aldols, and slowly increase the complexity. The connection to the simple, prototypal acid- and base-catalyzed aldols will always be maintained.

18.4a Intramolecular Aldol Condensations

Like most intermolecular reactions, the aldol condensation has an intramolecular version. If a molecule contains both an enolizable hydrogen and a receptor carbonyl group, an intramolecular aldol condensation will be theoretically possible. Particularly favorable are intramolecular condensations that form the relatively strain-free five- and six-membered rings. The products are still β-hydroxy ketones (or the dehydration products, α,β-unsaturated ketones), and the mechanism of the reaction remains the same (Fig. 18.68).

The general base-catalyzed case

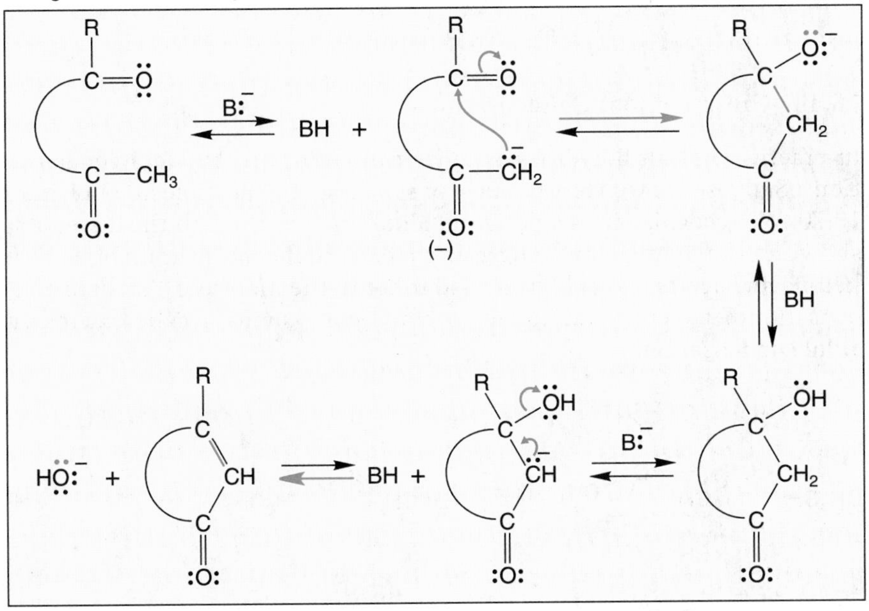

Specific base- and acid-catalyzed examples

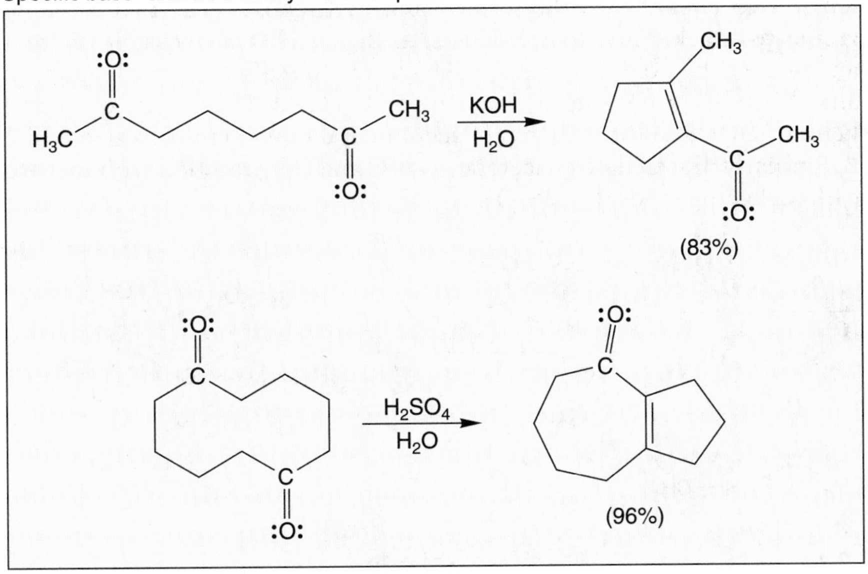

FIGURE 18.68 The mechanism for a base-catalyzed intramolecular aldol condensation.

PROBLEM 18.18

Write the mechanisms for the acid- and base-catalyzed aldol condensations of the molecule at the bottom of Figure 18.68.

PROBLEM **18.19** Write the mechanisms for the acid- and base-catalyzed reverse aldol condensations for the molecule in Figure 18.69.

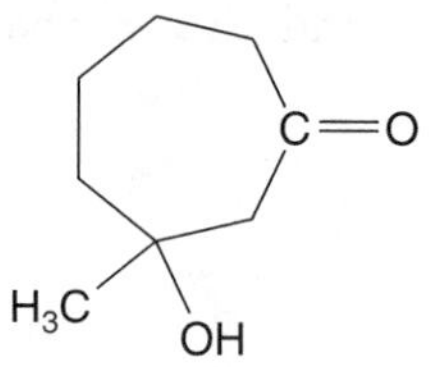

FIGURE **18.69**

18.4b Crossed Aldol Condensations

In practice, not all β-hydroxy ketones can be formed by aldol condensations. Suppose we were set the task of synthesizing the molecule in Figure 18.70. We recognize it as a potential aldol product through the dissection shown in the figure, which suggests that an aldol condensation of diethyl ketone and acetone should be a good route to the molecule. An aldol condensation between two different carbonyl compounds is called a **crossed aldol condensation**.

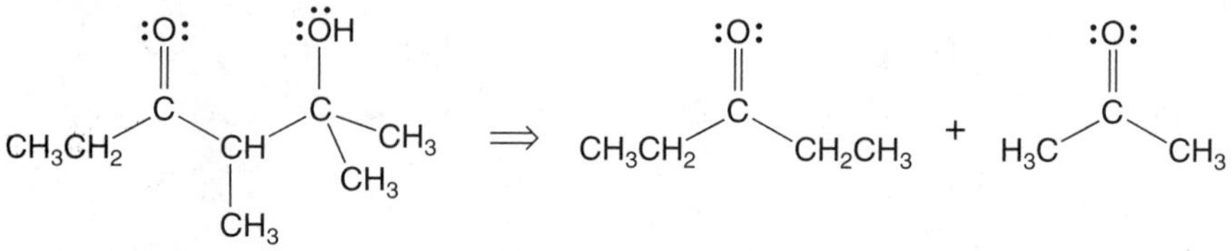

FIGURE **18.70** A retrosynthetic analysis suggests that a condensation between diethyl ketone and acetone should give the β-hydroxy ketone shown.

However, a little thought about the mechanism reveals potential problems. Two enolates can be formed and each enolate has two carbonyl groups to attack. Thus, four products, **A**, **B**, **C**, and **D**, are possible (more if any of the β-hydroxy ketones are dehydrated to α,β-unsaturated ketones), and all are likely to be formed in comparable yield. The enolate of diethyl ketone can add to the carbonyl of acetone or diethyl ketone to give **A** and **B**. Similarly, the enolate of acetone gives **C** and **D** by reaction with the two carbonyl compounds (Fig. 18.71).

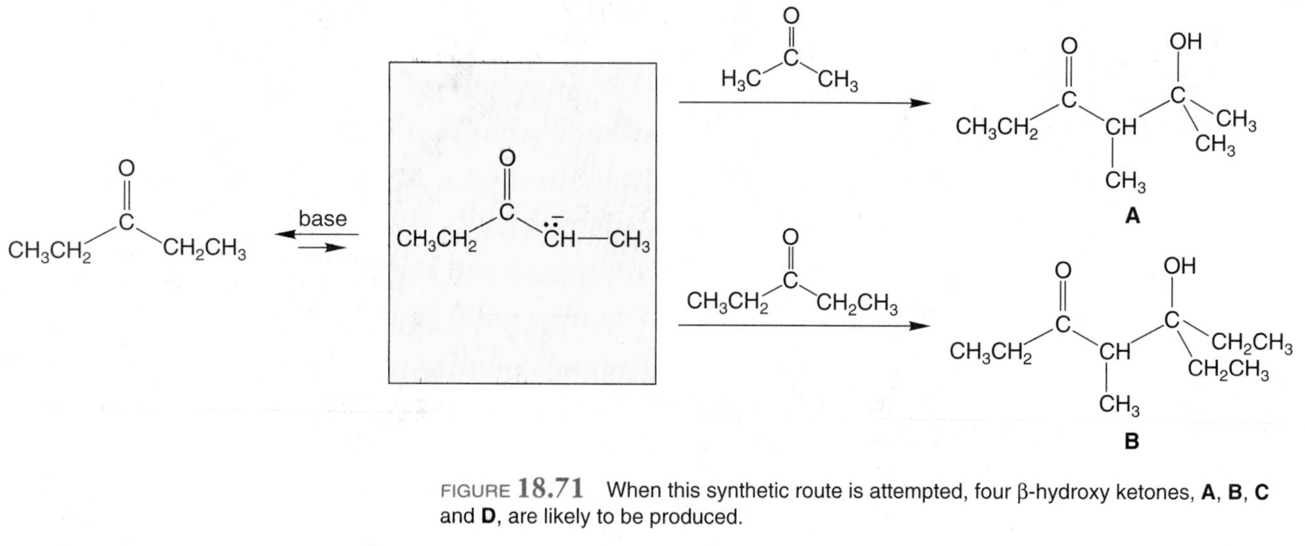

FIGURE **18.71** When this synthetic route is attempted, four β-hydroxy ketones, **A**, **B**, **C** and **D**, are likely to be produced.

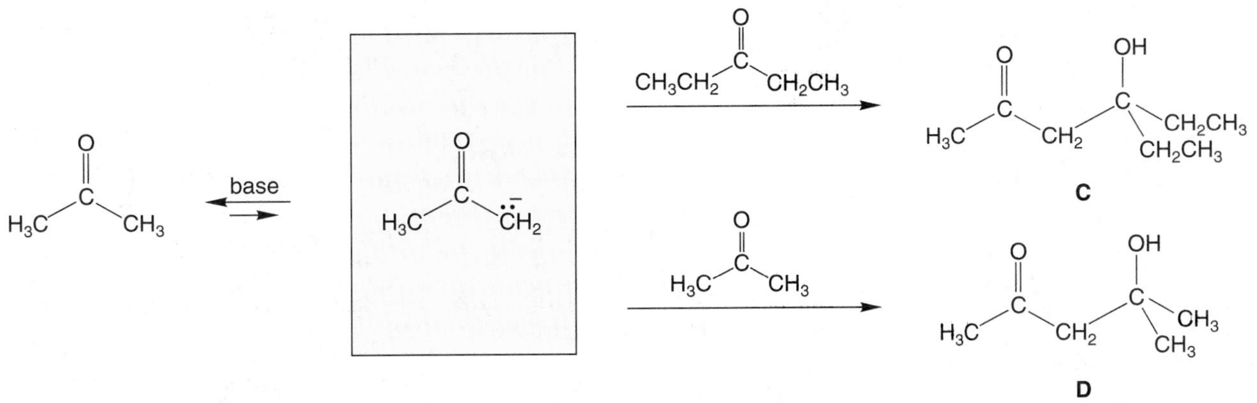

FIGURE 18.71 (CONTINUED)

There are some simple ways to limit the possibilities in a crossed aldol reaction. If one partner has no α-hydrogens, it can function only as an acceptor, and not as the nucleophilic enolate partner in the reaction. For example, we might hope for better luck in a reaction pairing *tert*-butyl methyl ketone and benzaldehyde. In principle, however, there are still two possibilities, as the enolate can add to either ketone or aldehyde to generate two different β-hydroxy ketones (Fig. 18.72).

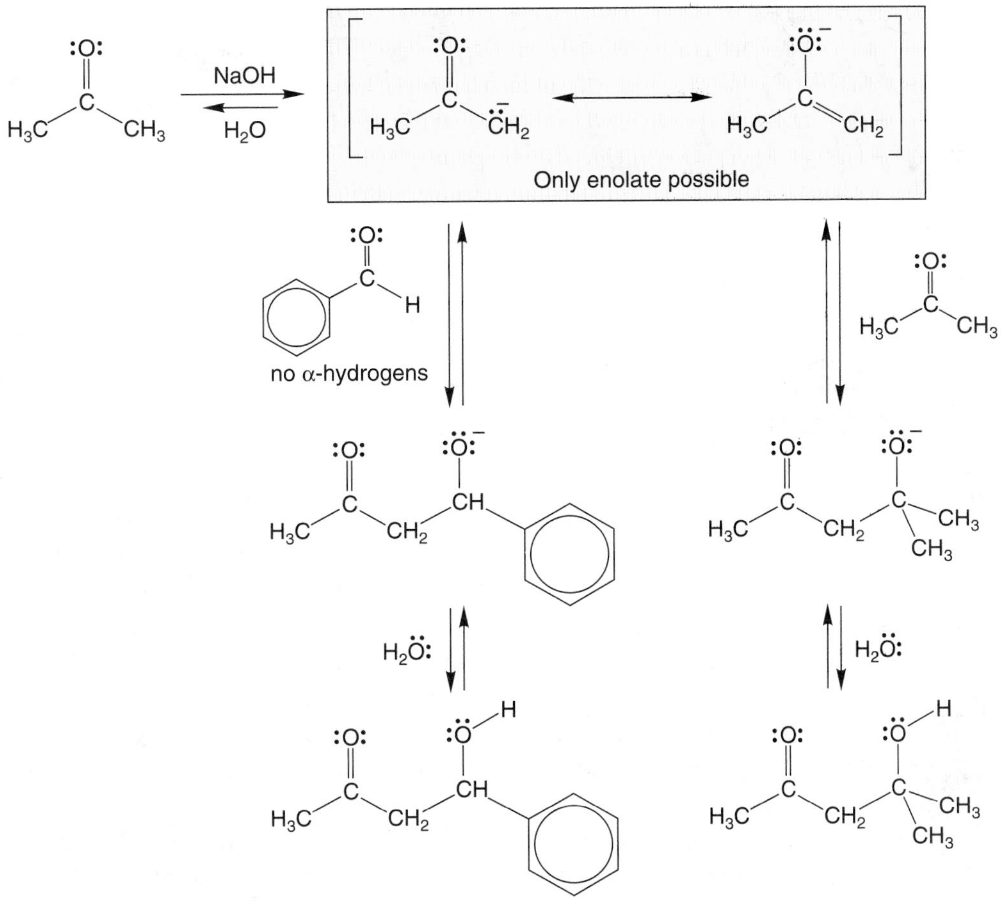

FIGURE 18.72 The crossed aldol reaction of *tert*-butyl methyl ketone and benzaldehyde can give only two products. Benzaldehyde has no α-hydrogens and cannot form an enolate.

When this reaction is run, however, only one product is formed in substantial yield (Fig. 18.73).

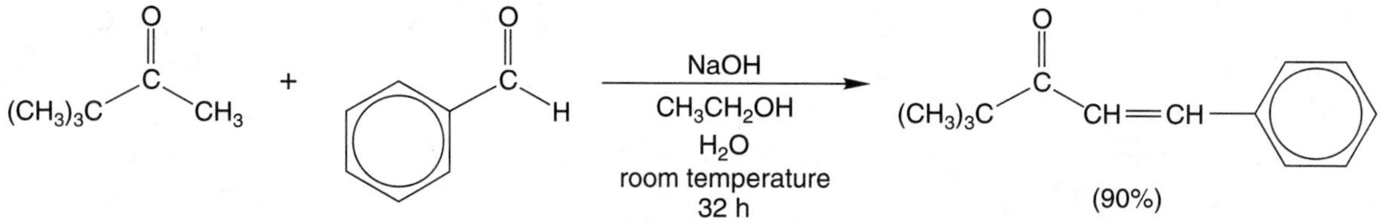

$(CH_3)_3C$–CO–CH_3 + benzaldehyde → ($NaOH$, CH_3CH_2OH, H_2O, room temperature, 32 h) → $(CH_3)_3C$–CO–CH=CH–Ph (90%)

FIGURE 18.73 In fact, there is only one major product in the aldol condensation of *tert*-butyl methyl ketone and benzaldehyde.

There are two reasons that make this crossed aldol successful. First, the rate of addition of the enolate to the carbonyl group of benzaldehyde is much greater than that of addition to *tert*-butyl methyl ketone. The carbonyl groups of aldehydes are more reactive than those of ketones. Second, the equilibrium constant for the addition to an aldehyde carbonyl is more favorable than that for addition to a ketone carbonyl (Chapter 16, p. 768; Fig. 18.74).

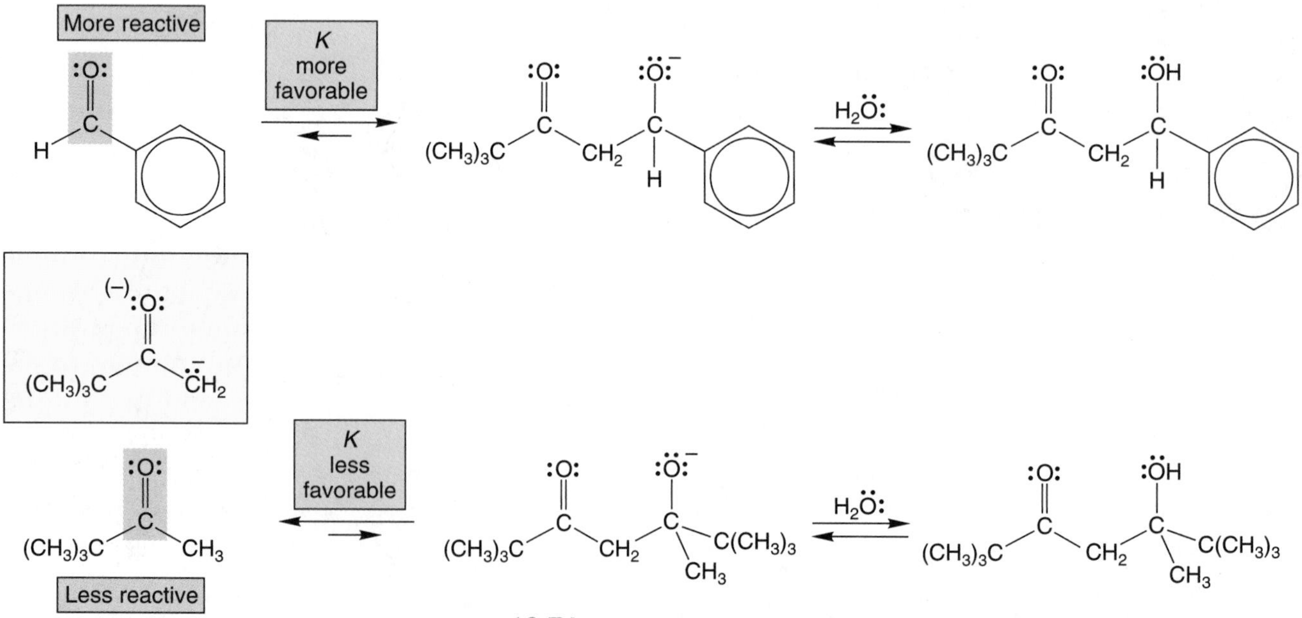

More reactive

K more favorable

$H_2\ddot{O}$:

K less favorable

$H_2\ddot{O}$:

Less reactive

FIGURE 18.74 The carbonyl group of benzaldehyde is more reactive than that of *tert*-butyl methyl ketone, and equilibrium favors the product in the reaction with benzaldehyde, but not for the reaction with *tert*-butyl methyl ketone.

This reaction is important enough to have been given its own name, the **Claisen–Schmidt condensation** (Ludwig Claisen, 1851–1930).

PROBLEM 18.20 There is a reaction between benzaldehyde and base. What is it, and why does it not interfere with the aldol condensation?

Earlier in this chapter we saw lithium diisopropylamide (LDA), a base that was effective in producing enolates without the complications of addition reactions to the carbonyl group. These enolates can be used to do crossed aldol reactions too. Lithium diisopropylamide is first used to create a large amount of enolate to which a carbonyl-containing compound is added. Addition of the enolate followed by protonation gives the aldol product. When there is a choice, LDA forms the more accessible, less substituted enolate. Lithium diisopropylamide is a large, sterically encumbered base and removes a proton from the sterically less hindered position (Fig. 18.75).

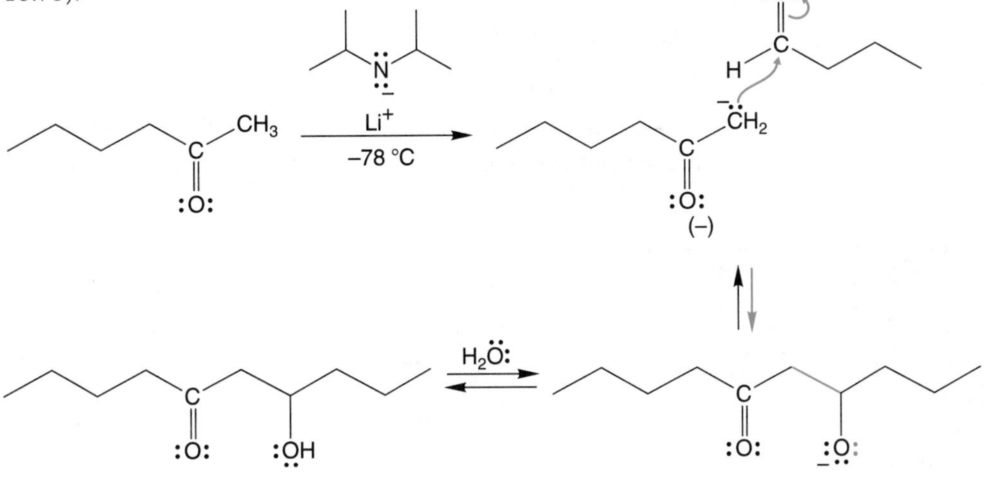

(65%)

FIGURE 18.75 A crossed aldol condensation with LDA as base.

18.4c Knoevenagel Condensations and Related Reactions

Condensation reactions are not limited to simple aldehydes and ketones. Any compound that can yield an anion can attack a carbonyl group to give a condensed product. Diketones and similar compounds are quite strong acids (pK_a = 5–13, Table 18.2) and readily form anions when treated with bases.

TABLE 18.2 Some pK_a Values for Some Dicarbonyl Compounds

Compound	Conjugate Base (enolate)	pK_a
EtO–C(=O)–CH_2–C(=O)–OEt	EtO–C(=O)–$\overset{-}{C}H$–C(=O)–OEt	13.3
NC–CH_2–CN	NC–$\overset{-}{C}H$–CN	11
H_3C–C(=O)–CH_2–C(=O)–OEt	H_3C–C(=O)–$\overset{-}{C}H$–C(=O)–OEt	10.7
H_3C–C(=O)–CH_2–C(=O)–CH_3	H_3C–C(=O)–$\overset{-}{C}H$–C(=O)–CH_3	8.9
Ph–C(=O)–CH_2–C(=O)–CH_3	Ph–C(=O)–$\overset{-}{C}H$–C(=O)–CH_3	8.5
H–C(=O)–CH_2–C(=O)–H	H–C(=O)–$\overset{-}{C}H$–C(=O)–H	5

PROBLEM **18.21** Draw resonance forms for the anions of the first two entries of Table 18.2.

Bases, often amines, can generate significant amounts of enolates from the compounds in Table 18.2, and similar molecules. In the presence of an acceptor aldehyde, condensation occurs to give, ultimately, the α,β-unsaturated diester shown in Figure 18.76. This reaction is called the **Knoevenagel condensation** after Emil Knoevenagel (1865–1921).

The general case

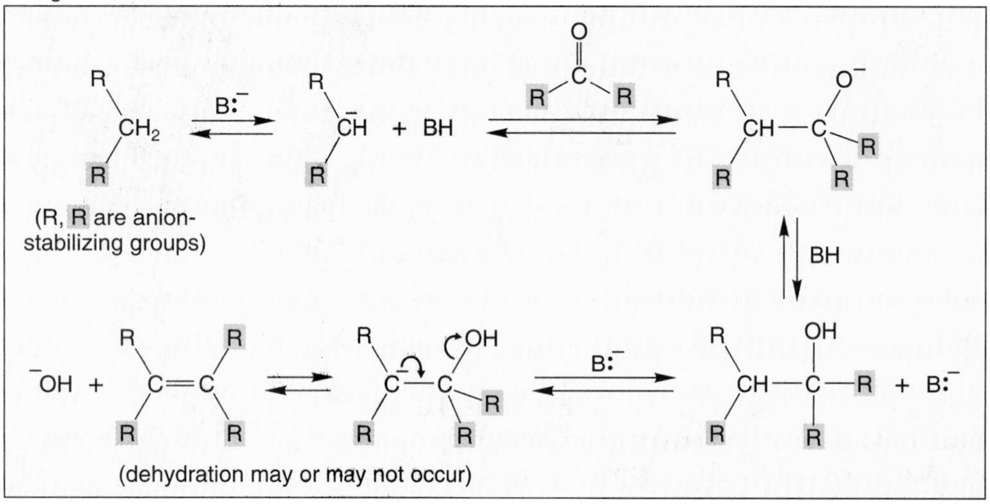

(R, R are anion-stabilizing groups)

BH

B:⁻

(dehydration may or may not occur)

Specific examples

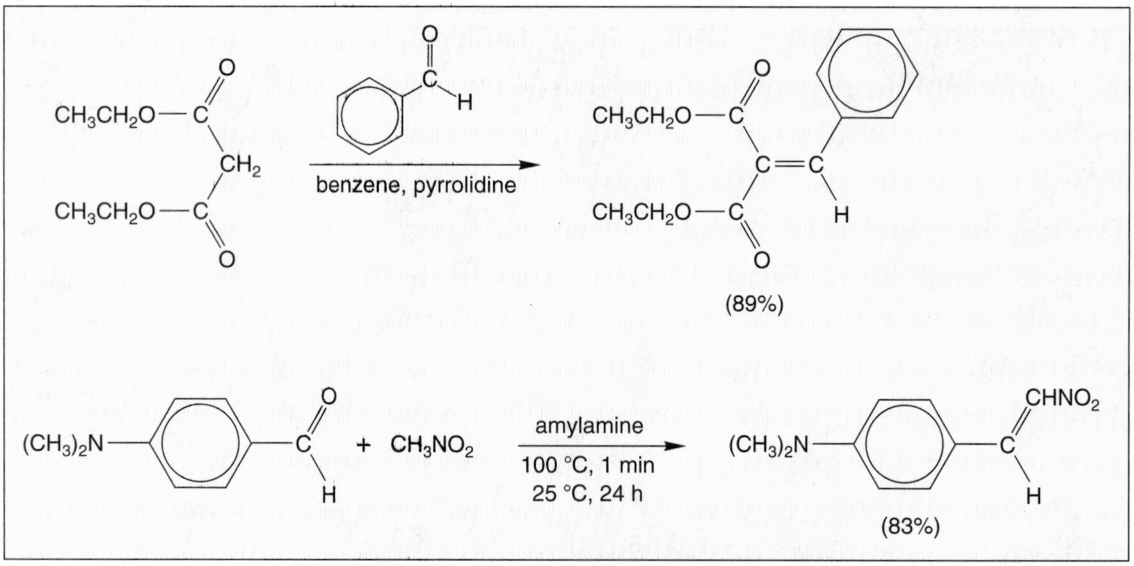

benzene, pyrrolidine

(89%)

amylamine

100 °C, 1 min

25 °C, 24 h

(83%)

FIGURE **18.76** The Knoevenagel condensation.

Many other reactions, some with names, many without, are conceptually similar, in that they involve the addition of a complex anion to a carbonyl group. But, in these other reactions the sources of the nucleophilic anionic partner are not always aldehydes or ketones. For example, cyclopentadiene (pK_a = 16) is a strong enough acid to give an anion when treated with a base, and the cyclopentadienide ion (Chapter 13, p. 000) can add to aldehydes or ketones. The end result is a synthesis of the beautifully colored compounds known as **fulvenes** (Fig. 18.77).

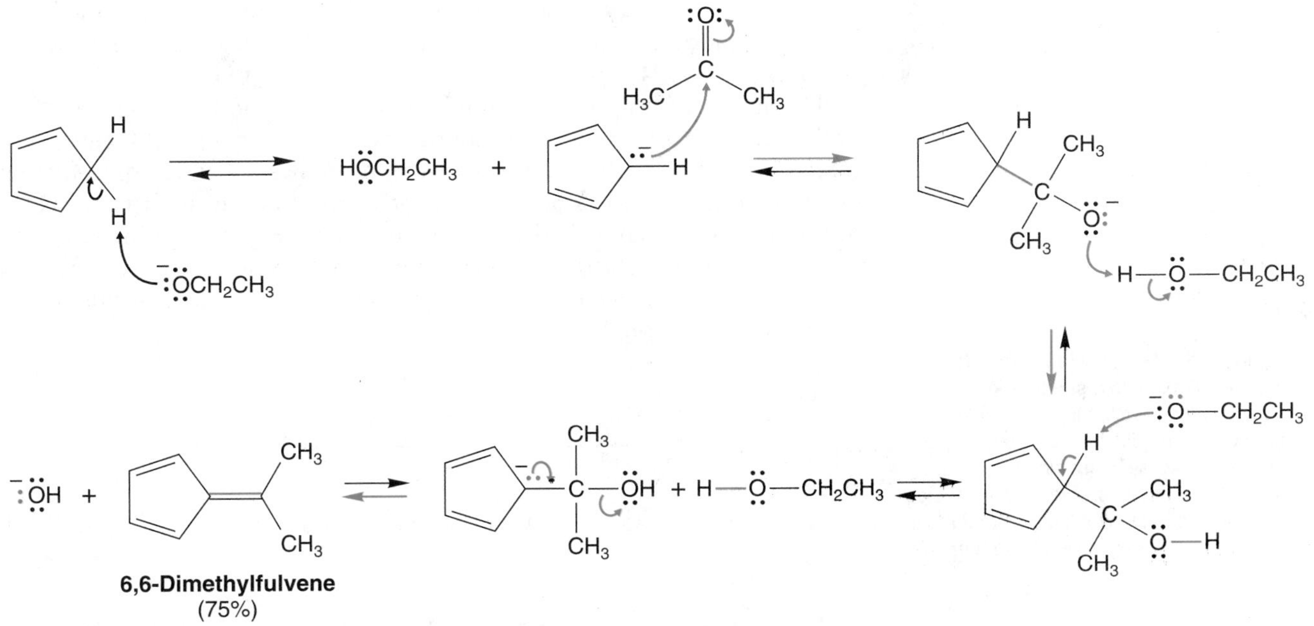

6,6-Dimethylfulvene (75%)

FIGURE 18.77 A condensation reaction leading to fulvenes.

So now your analytical task is more complicated. It is quite easy to learn to dissect β-hydroxy ketones or α,β-unsaturated ketones into their components, but how is one to deal with the less obvious compounds like fulvenes? There is no easy answer to this question. A double bond is always the potential result of a condensation reaction followed by a dehydration. You must learn to look on them that way, and to evaluate the pieces thus created as parts of a possible synthesis (Fig. 18.78).

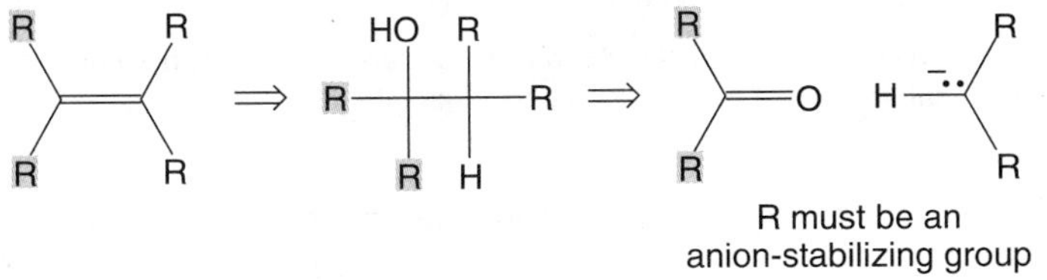

R must be an anion-stabilizing group

FIGURE 18.78 Any double bond is the formal result of a condensation reaction followed by dehydration.

The following problem gives some practice at devising syntheses of molecules formed by condensation reactions.

PROBLEM 18.22

Propose syntheses for the molecules in Figure 18.79.

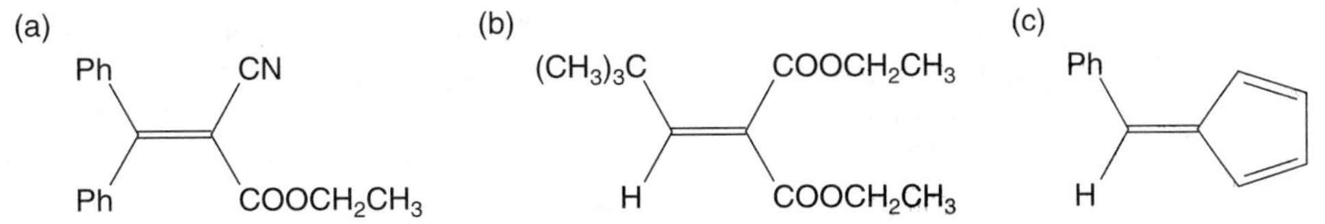

(a) (b) (c)

FIGURE 18.79

18.4d More Related Condensations: The Michael Reaction

We now have a synthetic route to α,β-unsaturated carbonyl compounds, the aldol (and related) condensation reactions. There is some especially important chemistry associated with these compounds, called the **Michael reaction**, after its discoverer, the American chemist Arthur Michael (1853–1942).

Ordinary carbon–carbon double bonds are unreactive toward nucleophiles. In particular, addition is not observed, because it would require formation of an unstabilized anion, the conjugate base of an alkane. Alkanes are extraordinarily weak acids. Their pK_a values are difficult to measure, but numbers as high as 70 have been suggested (Fig. 18.80). In answering problems, an unstabilized carbanion is the functional equivalent of a methyl or primary carbocation. It is a stop sign: A signal that you are almost certainly wrong!

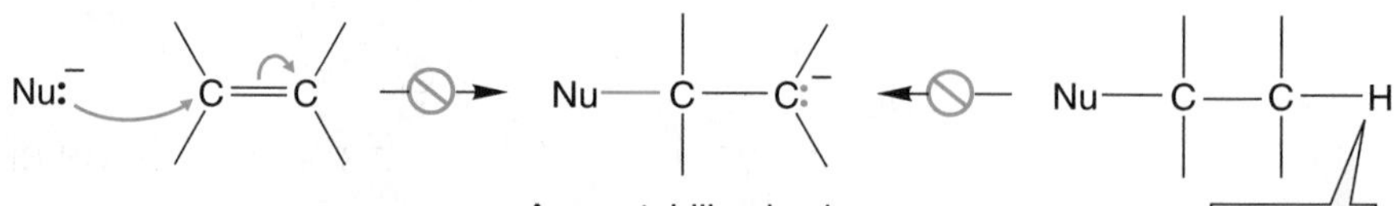

An unstabilized anion

$pK_a \gtrsim 60$

FIGURE 18.80 Addition of a base such as alkoxide to a simple alkene would yield an unstabilized anion. A measure of the difficulty of this reaction can be gained by examining the acidity of the related hydrocarbon. Such species are extraordinarily weak bases and their pK_a values are very difficult to determine.

By contrast, many nucleophiles add easily to the double bonds of α,β-unsaturated carbonyl compounds (Fig. 18.81).

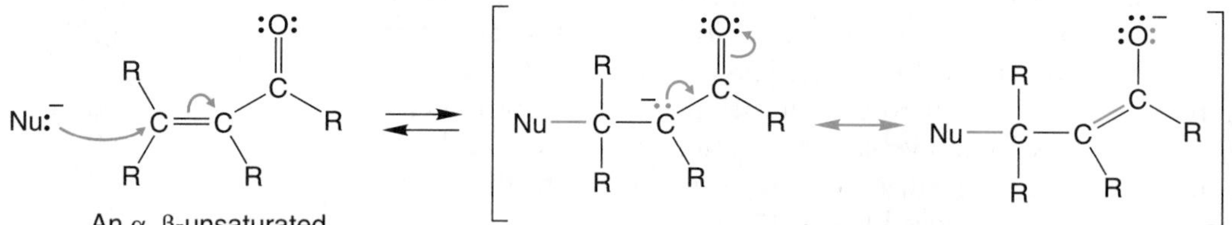

An α, β-unsaturated carbonyl compound

An enolate (resonance stabilized)

FIGURE 18.81 By contrast, additions to α,β-unsaturated carbonyls are common. Notice the resonance stabilization of the resulting enolate anion.

The reason is the initial formation, not of a bare, unstabilized carbanion as was the case in Figure 18.80, but of a nicely resonance-stabilized species. The carbonyl group stabilizes any α-anion, no matter how it is formed. The Michael reaction is another way of generating an enolate.

When the species adding to the α,β-unsaturated compound is itself an enolate, the reaction is properly called the Michael reaction, but more simple related additions are often also lumped under this heading (Fig. 18.82).

There is also an acid-catalyzed version of the Michael reaction, in which 1,4-addition first occurs to the α,β-unsaturated system. The carbonyl group is initially protonated, and then the carbon–carbon double bond is attacked by a nucleophile to give the enol (Fig. 18.83).

The enol equilibrates with the more stable keto compound to produce the final product of addition.

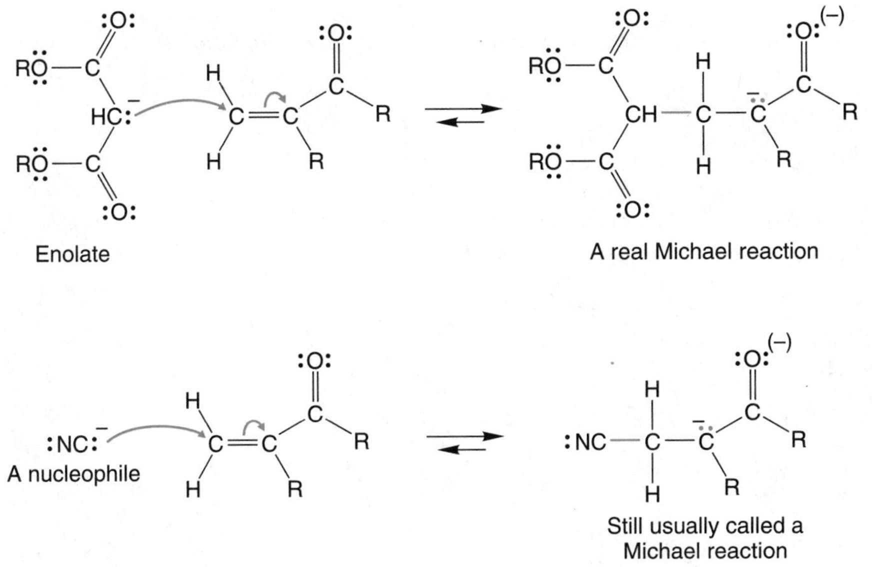

FIGURE 18.82 Two Michael additions.

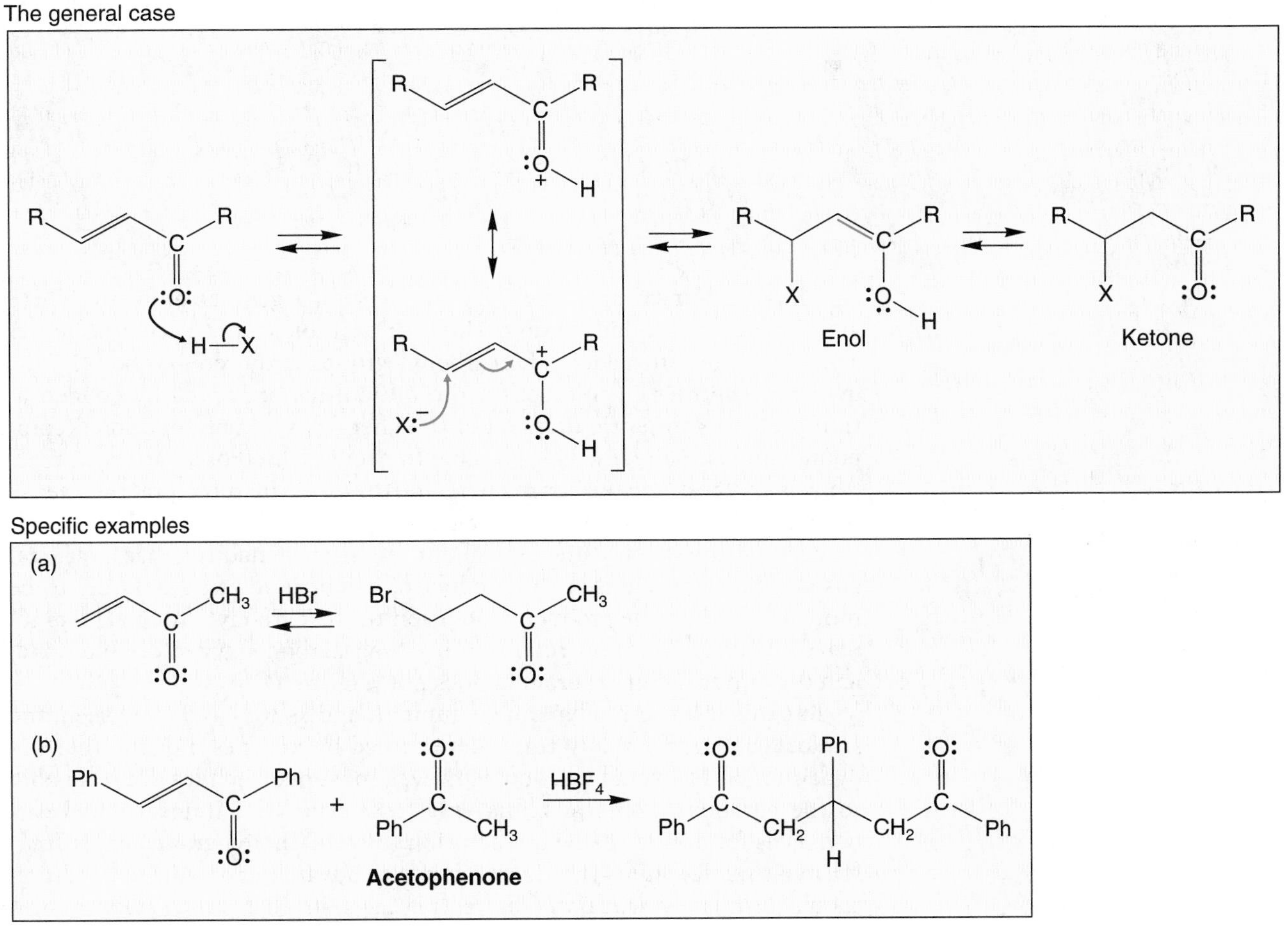

FIGURE 18.83 Two acid-catalyzed Michael reactions.

*PROBLEM 18.23 Write a mechanism for the reaction at the bottom of Figure 18.83.

ANSWER In acid, methyl phenyl ketone (acetophenone) will be in equilibrium with its enol form, and the other ketone will be protonated, thus forming a strong Lewis acid. Addition of the modest nucleophile, the enol, to the strong Lewis acid leads to intermediate **A**, in a standard, acid-catalyzed Michael reaction. Subsequently, **A** is deprotonated and ketonized to give the product. If the formation of the ketone from the enol is either surprising or difficult, *please* go over the mechanism.

Enol

Protonated carbonyl

A

HBF_4

HBF_4

deprotonation

ketonize the enol

Ph =

These α,β-unsaturated compounds present a general problem. There are too many reactions possible. How, for example, can we decide whether a nucleophile will add to the carbonyl group or the double bond? This question captures in microcosm the difficulties presented by condensation problems in particular, and of the chemistry of polyfunctional compounds in general. How is it possible to decide which of a variety of reactions is the best one (either in energy terms or in the more practical sense of finding the route to the end of an exam problem)?

At least the specific question about the Michael addition has a rational answer. First, notice that the product of Michael addition is likely to be more stable than the product of addition to the carbonyl group. The product of Michael addition retains the strong carbon–oxygen double bond, and the simple addition product does not (Fig. 18.84).

But this is a thermodynamic argument, and as we have often seen, the site of reactivity is many times determined by kinetics, not the thermodynamics of the overall reaction. However, most nucleophiles add *reversibly* to the carbonyl group, and this will be as true for α,β-unsaturated carbonyls as for others. So even if addition to the carbonyl group occurs first, for most nucleophiles the thermodynamic advantage of Michael addition should ultimately win out. This notion suggests that strong nucleophiles

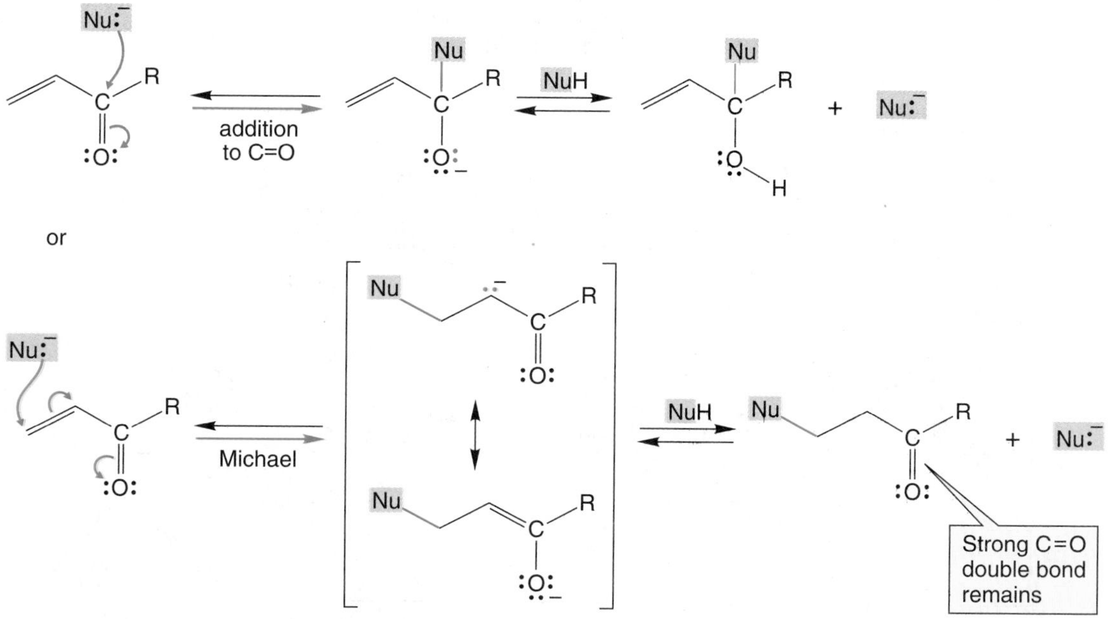

FIGURE 18.84 The two possible reactions of a nucleophile (Nu:⁻) with an α,β-unsaturated carbonyl compound. Michael addition preserves the carbonyl group and is usually favored thermodynamically.

such as hydride (H:⁻) and alkyllithium compounds, which surely add *irreversibly,* might be found to give the products of attack at the carbonyl group, and this is the case (Fig. 18.85).

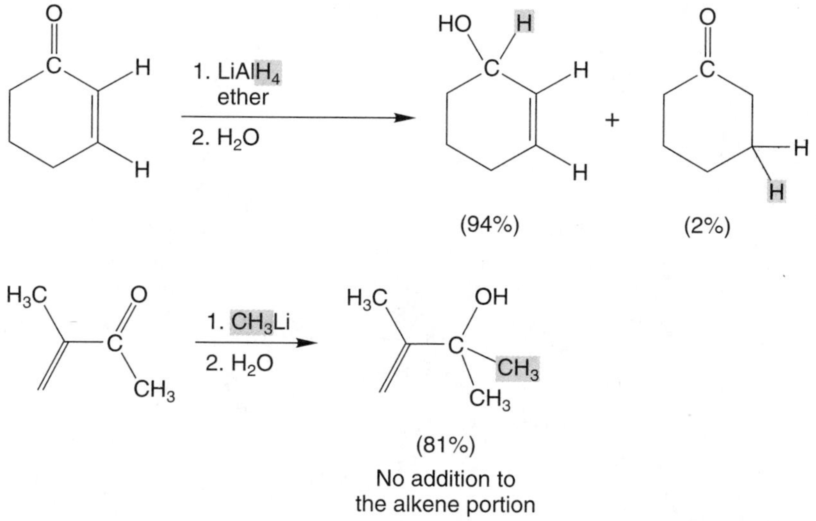

FIGURE 18.85 Two irreversible addition reactions to the carbonyl group of an α,β-unsaturated carbonyl compound.

There are exceptions to the general rule that weak bases add to the double bond and strong bases add to the carbonyl group. Cuprates, for example, give exclusively Michael addition, and Grignard reagents generally give mixtures of 1,2- and 1,4-addition unless they are copper catalyzed, in which case Michael addition is again the rule (Fig. 18.86).

Cuprate additions

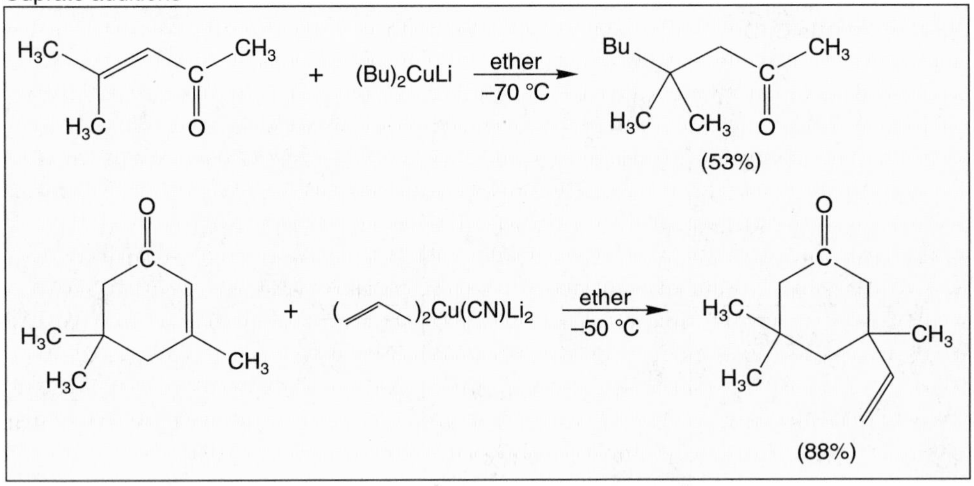

FIGURE 18.86 Cuprates (and often Grignard reagents) add in Michael fashion to α,β-unsaturated carbonyl compounds. Bu = $CH_3CH_2CH_2CH_2$

Calicheamicin

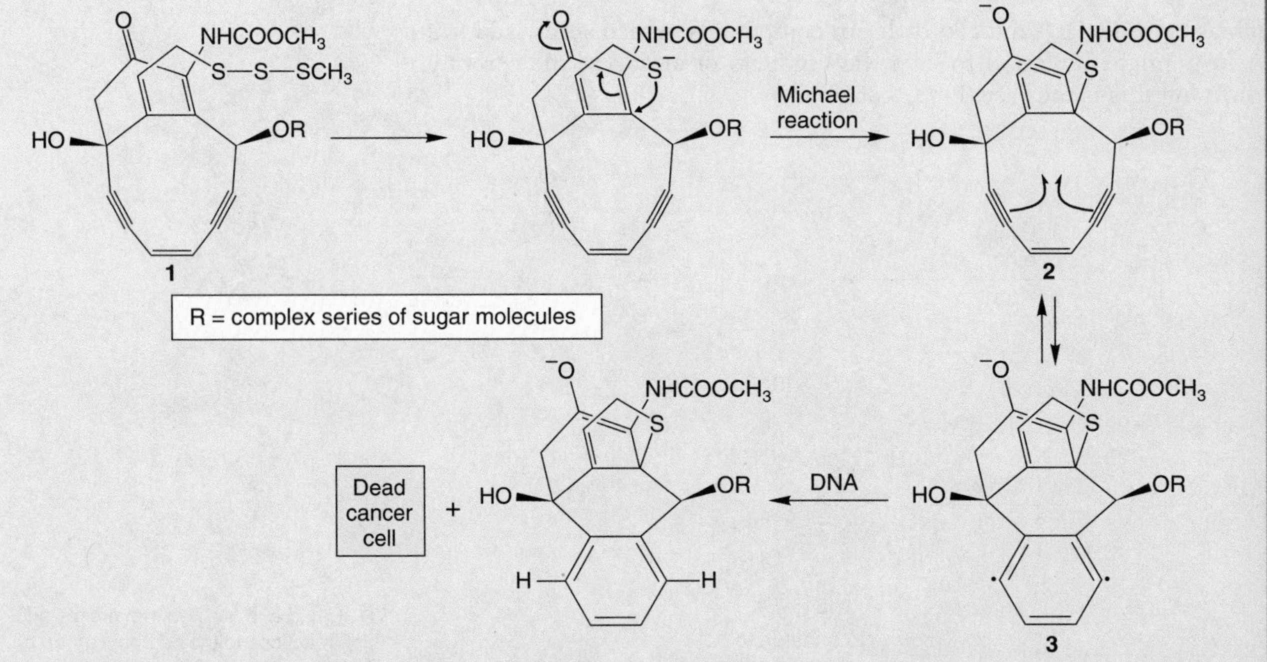

The Michael reaction is critical to the operation of a rather promising anticancer drug, calicheamicin. In the first step in its operation, the trisulfide bond is broken (**1**). The nucleophilic sulfide then adds in Michael fashion to give the enolate (**2**). This addition changes the shape of the molecule, bringing the ends of the two acetylenes closer together. A cyclization occurs to give a diradical (**3**), and this diradical abstracts hydrogen from the cancer cell's DNA, killing it. Calicheamicin depends for its action on the change of shape. Before the Michael reaction, the ends of the two acetylenes are too far apart to cyclize. They are freed to do so when the sulfur adds to the α,β-unsaturated carbonyl.

In this section, we have seen several variations on the aldol theme. In all of these, a single nucleophile adds to a single Lewis acid. The reactions are fundamentally the same, the variations come mainly in the structural differences in the reacting partners. Now we move to a more exciting subject: condensation reactions in sequence.

18.5 PROBLEM SOLVING

18.5a Condensation Reactions in Combination: Magid's Second Rule

Alas, neither Nature nor a typical chemistry teacher shows much mercy, and one is more likely than not to find condensation reactions in combination. None of these reactions is particularly difficult to understand when encountered by itself. Similarly, the synthetic consequences of the reactions are relatively easy to grasp when they are encountered one at a time, but are by no means simple when combinations of these reactions are involved. Condensation problems, occasionally called "Fun in Base" problems (a term that seems to encompass reactions run in acid as well as base; there are no "Fun in Acid" problems) can be very hard. The best way to become proficient is to work lots of them, but there are also some practical hints that may help.

One particularly common sequence is a combination of the Michael and aldol condensations. Indeed, there is a synthetic procedure for constructing six-membered rings, invented by Sir Robert Robinson (1886–1975) and called the **Robinson annulation**, that uses exactly that combination. In it, a ketone is treated with methyl vinyl ketone in a base-catalyzed reaction. A Michael addition is followed by an intramolecular aldol condensation to give the new six-membered ring (Fig. 18.87).

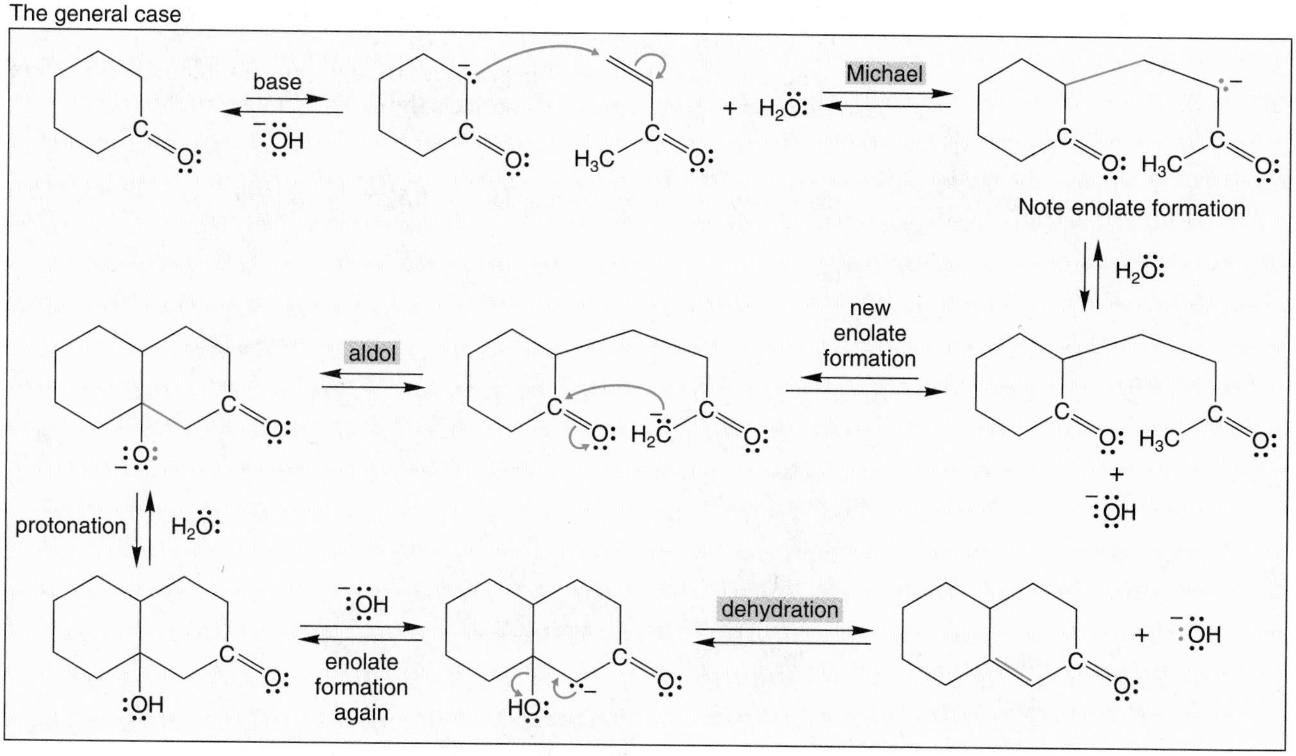

The general case

base
$^-$:ÖH
+ $H_2\ddot{O}$:
Michael
H_3C
Note enolate formation
$H_2\ddot{O}$:
new enolate formation
aldol
H_2C
+
$^-$:ÖH
protonation
$H_2\ddot{O}$:
:ÖH
$^-$:ÖH
enolate formation again
HÖ:
dehydration
+ $^-$:ÖH

FIGURE 18.87 The mechanism of the Robinson annulation.

A specific example

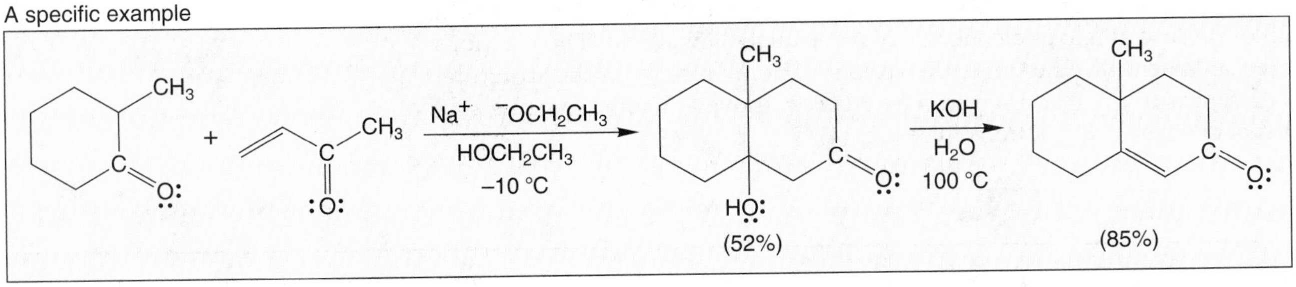

Na⁺ ⁻OCH₂CH₃ / HOCH₂CH₃ / −10 °C → (52%); KOH / H₂O / 100 °C → (85%)

FIGURE 18.87 (CONTINUED)

PROBLEM 18.24 Although this is the way the Robinson annulation is always described mechanistically, there is another (harder) way to write the reaction. Write a mechanism that involves doing the aldol condensation first.

Problem 18.24 brings up a serious practical question. You may well be able to analyze a problem by figuring out the different possible reactions, but choosing among them is more difficult. How do you find the easiest route (there are very often several) to the product and the solution to the problem? Clearly, experience is invaluable and you are by definition short on that commodity. In such a situation, it is wise to be receptive to good advice, and here is a practical hint, from one of the best problem solvers I know, Ronald M. Magid (b. 1938) of the University of Tennessee, in the form of **Magid's Second Rule**,* which says: "Always try the Michael First."

Magid's Second Rule is more than a wisecrack. In fact, there are sound reasons for doing the Michael reaction first. Aldol condensation products of ketones are not generally favored at equilibrium, for example, and doing the Michael reaction makes the subsequent aldol condensation intramolecular, and thus more favorable.

There is no substitute for practice, though, and Problem 18.25 offers some good opportunities to practice solving Fun in Base problems.

PROBLEM 18.25 Provide mechanisms for the reactions shown in Figure 18.88.

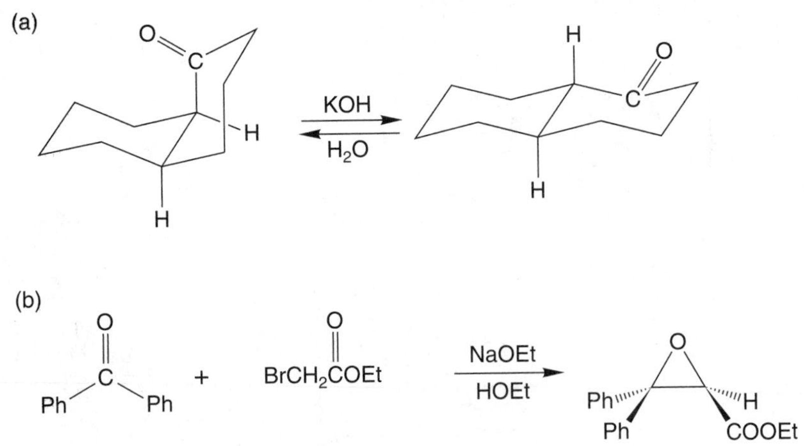

(a) KOH / H_2O

(b) Ph(C=O)Ph + $BrCH_2COEt$ (NaOEt / HOEt) → epoxide (Ph, Ph, H, COOEt)

FIGURE 18.88

*Magid's First Rule is archaic, and we will shortly encounter Magid's Third Rule.

(c) NaOH, H_2O

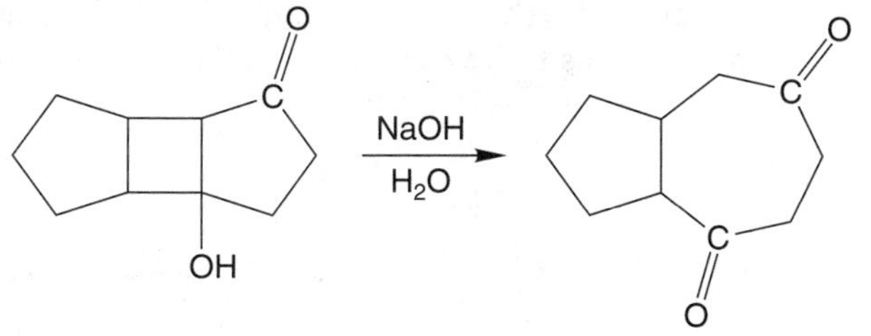

(d) KOH, H_2O

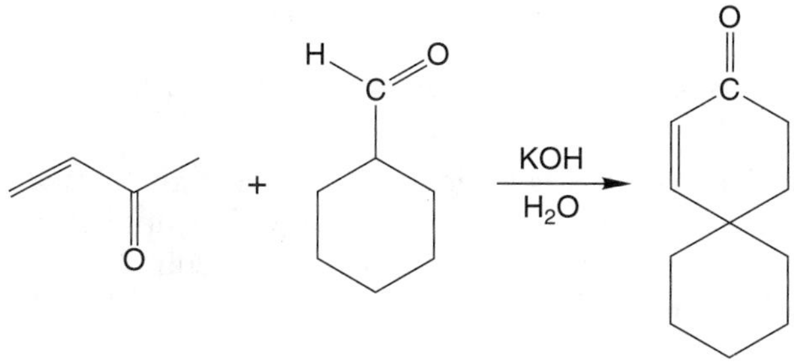

(e) KOH, H_2O

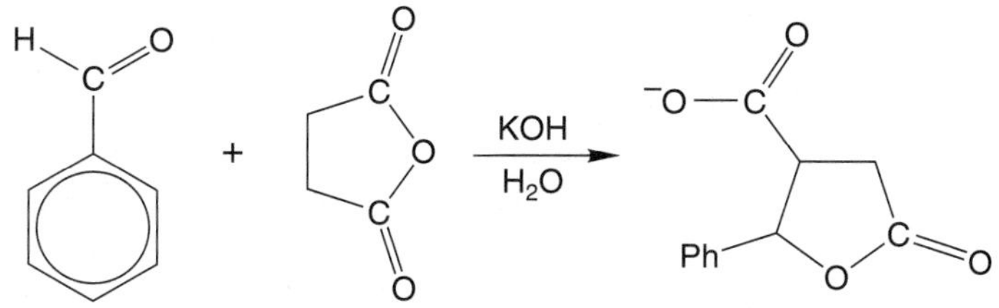

(f) KOH, H_2O

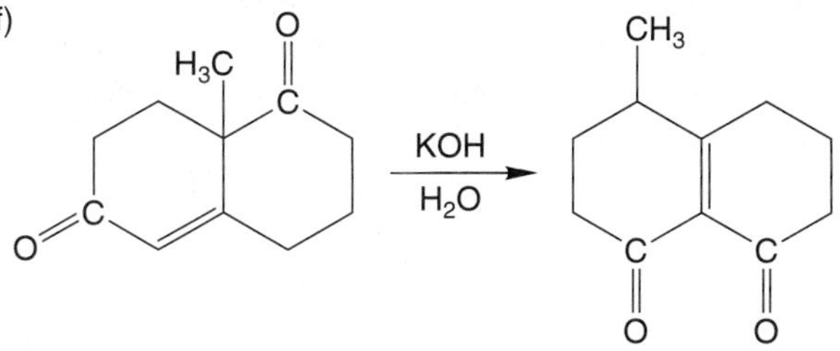

Hint: There is a 10-membered ring involved.

FIGURE 18.88 (CONTINUED)

18.6 CARBONYL COMPOUNDS WITHOUT α-HYDROGENS: MAGID'S THIRD RULE

Throughout this chapter we have concentrated on molecules capable of forming enolates or enols, and the subsequent chemistry has been dependent on these intermediates. It is a legitimate question to ask if there is a base-induced chemistry of carbonyl compounds that do *not* have α-hydrogens, and that cannot form enolate anions.

18.6a The Benzoin Condensation

If a molecule such as benzaldehyde, which has no α-hydrogens, is treated with base, there is no visible rapid reaction. There is a rapid invisible reaction, however, which is addition of the hydroxide to the carbonyl group, followed by protonation to form the hydrate (Fig. 18.89). Of course, no aldol condensation is possible, as there can be no enolate formation.

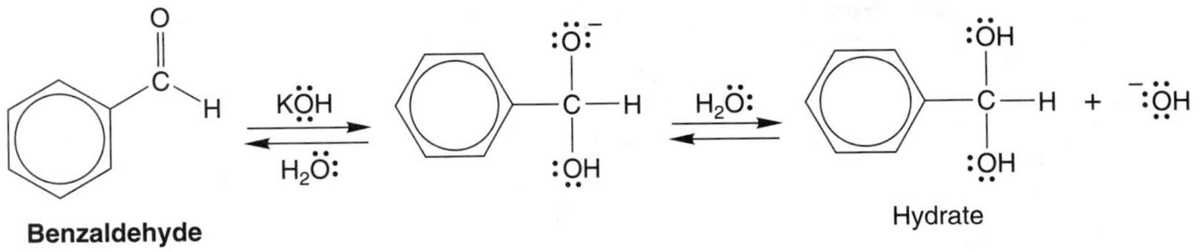

FIGURE 18.89 The base-induced hydration reaction of benzaldehyde.

The situation changes dramatically, however, if the nature of the basic catalyst is changed. If hydroxide is replaced with cyanide, there is a rapid and efficient formation of a molecule named benzoin. The reaction itself is called the **benzoin condensation** (Fig. 18.90).

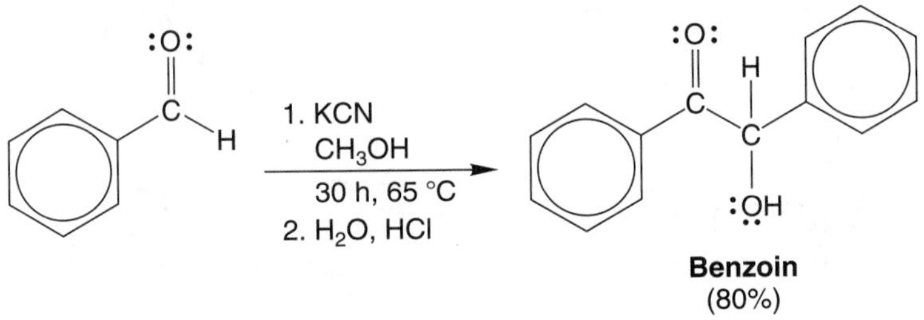

FIGURE 18.90 The benzoin condensation. Cyanide ion is a specific catalyst for this reaction.

If this were phrased as an exercise, it would be an excellent example of what might be called a "misdirection" problem, because it has an obvious (and wrong) answer. Benzoin appears to be the result of removal of the aldehydic hydrogen, followed by a typical aldol-like attack on the carbonyl group of another molecule of benzaldehyde (Fig. 18.91).

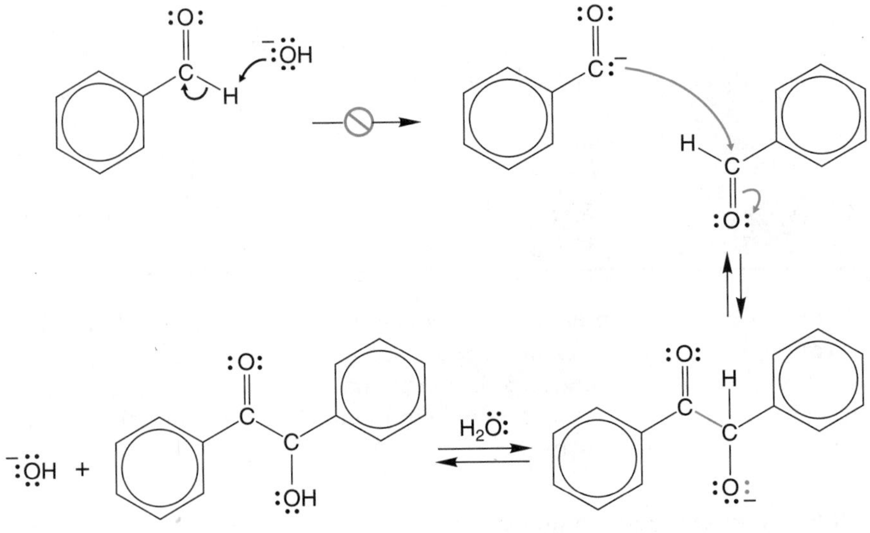

FIGURE 18.91 An "obvious" and quite incorrect mechanism for the benzoin condensation.

No matter how attractive this sequence, it cannot be correct, for there is no way the crucial hydrogen can be directly removed. The resulting anion is not stabilized by resonance and cannot be made. Be sure you are clear on this point. Why is it that an α-anion is resonance stabilized but this one is not (Fig. 18.92)?

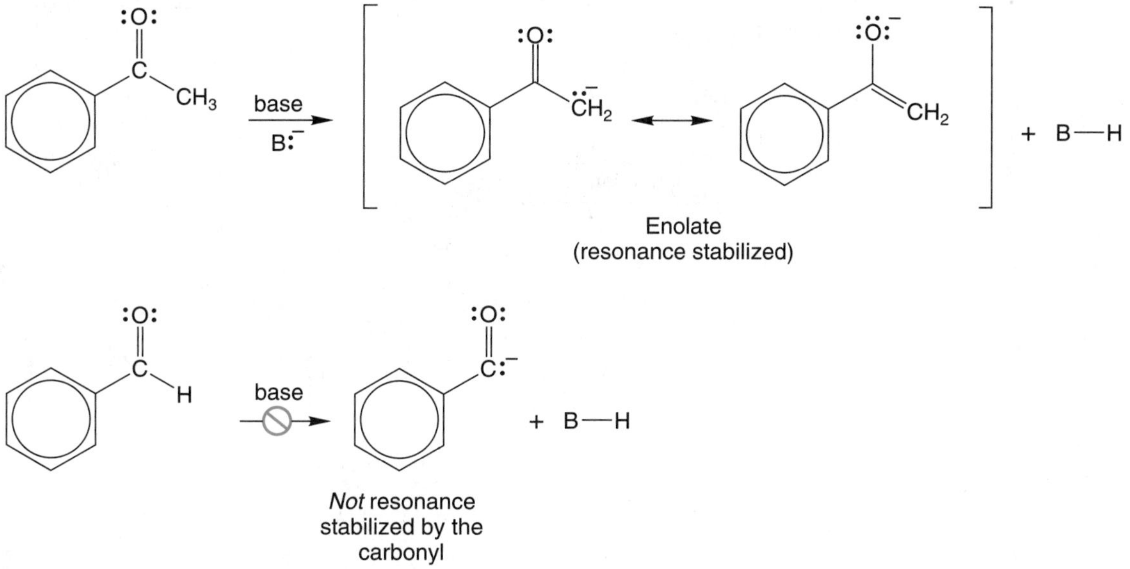

FIGURE 18.92 Removal of an α-hydrogen is quite different from removal of an aldehydic hydrogen. Removal of the aldehydic hydrogen does not lead to a resonance-stabilized anion.

So what is there to do? We must fall back on simple reactions that we know will happen. Nucleophiles add to carbonyl groups, and, in fact, we discussed the attack of cyanide on carbonyl groups at the start of Chapter 16. Cyanide will surely add to the carbonyl group of benzaldehyde to generate the cyanohydrin in Figure 18.93.

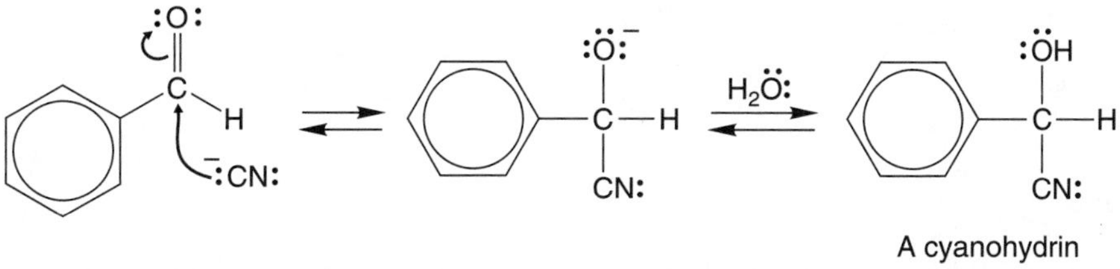

FIGURE 18.93 Cyanohydrin formation from cyanide ion and benzaldehyde.

Now we need a bit of insight to see that the nature of the old aldehydic hydrogen has been fundamentally changed by the addition reaction. It now is acidic because the anion formed by its removal is resonance stabilized by the cyano group (Fig. 18.94).

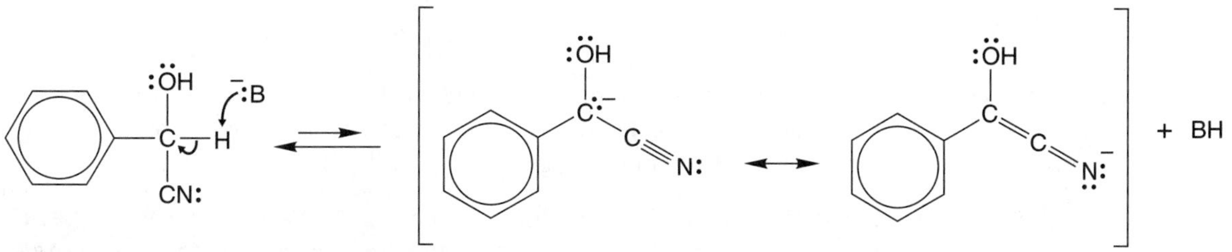

FIGURE 18.94 Removal of the old aldehydic hydrogen from the newly formed cyanohydrin does give a resonance-stabilized anion.

If we can form this anion, it can attack another aldehyde carbonyl to form the link between the two molecules present in benzoin. Two deprotonation–protonation steps lead to loss of cyanide, the formation of the product, benzoin, and the regeneration of the catalyst for the reaction, cyanide (Fig. 18.95).

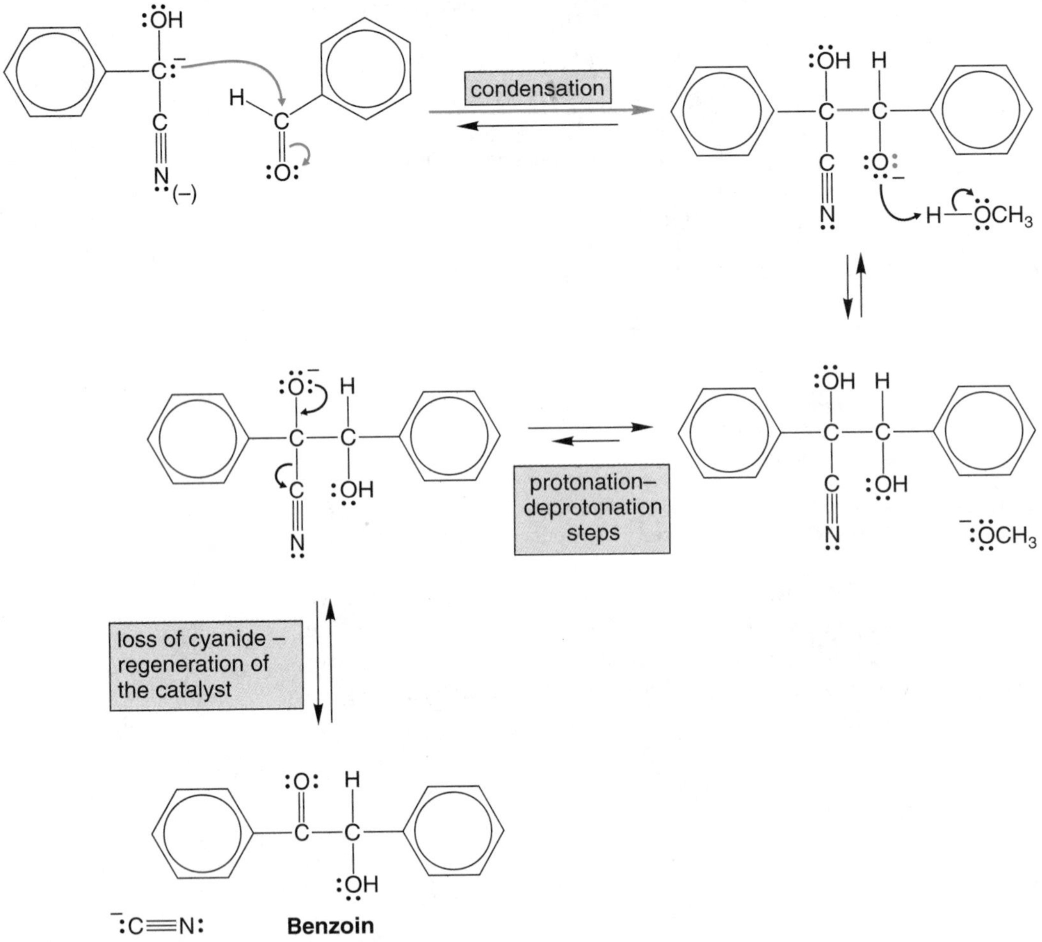

FIGURE 18.95 The mechanism of the benzoin condensation. Note here the use of a convention in which one resonance form is drawn and the other atoms sharing the negative charge are indicated with (–).

So there is some reactivity, even in the absence of acidic α-hydrogens, at least if special catalysts such as cyanide are used.

*PROBLEM 18.26 Show clearly why hydroxide or ethoxide ion is ineffective in promoting the benzoin condensation. These alkoxides do add to the carbonyl group to form compounds much like the cyanohydrin.

ANSWER This problem exemplifies what the text points out. There is something special about cyanide's promotion of the benzoin condensation. Addition of hydroxide and ethoxide takes place, but neither group can stabilize an adjacent pair of

ANSWER (CONTINUED)

electrons as can a cyanide. There is no way to form the anion crucial to the success of the benzoin condensation.

This hydrogen is critical; for the benzoin condensation to succeed, it must be removed to give an anion

R = H or Et

This anion is not stabilized by resonance

This one is

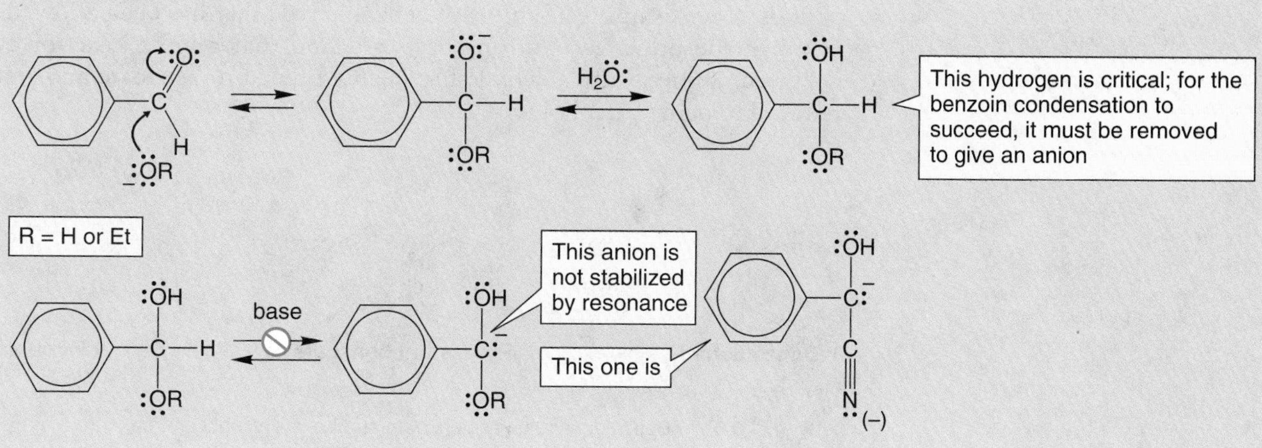

*PROBLEM **18.27**

A few other catalysts will promote benzoin-like condensations. Deprotonated thiazolium ions are examples. Show how the compound in Figure 18.96 can catalyze the condensation of pivaldehyde (*tert*-butylcarboxaldehyde).

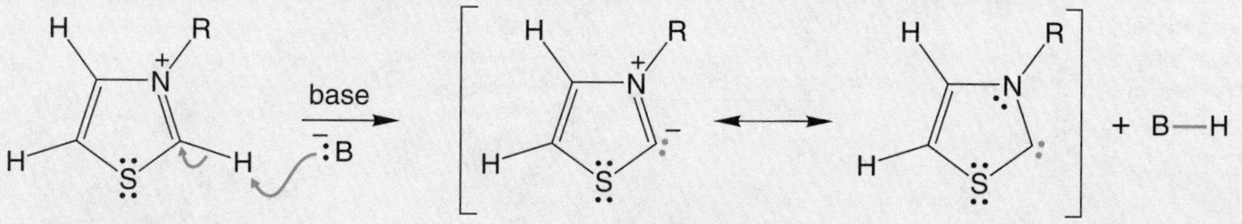

Thiazolium ion

Deprotonated thiazolium ion

FIGURE **18.96**

ANSWER

Like cyanide, this catalyst will stabilize an adjacent anion. The first step in the condensation reaction is addition. Now the critical hydrogen becomes acidic, as an adjacent anion can be stabilized through delocalization after this nucleophile adds.

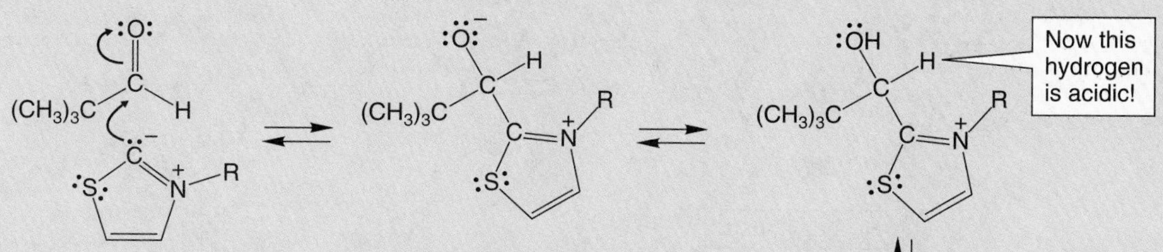

Now this hydrogen is acidic!

This drawing shows how this catalyst can mimic cyanide. You should finish the reaction off with the remaining steps of the benzoin-like condensation. How is the product formed?

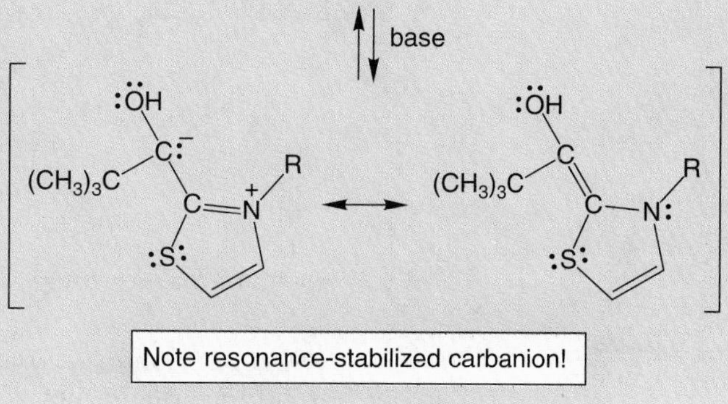

Note resonance-stabilized carbanion!

18.6b The Cannizzaro Reaction

Even hydroxide will give products from non-enolizable carbonyl compounds. If we examine a sample of benzaldehyde that has been allowed to stand in the presence of strong base, we find that a redox reaction has occurred. Some of the aldehyde has been oxidized to benzoate and some reduced to benzyl alcohol (Fig. 18.97).

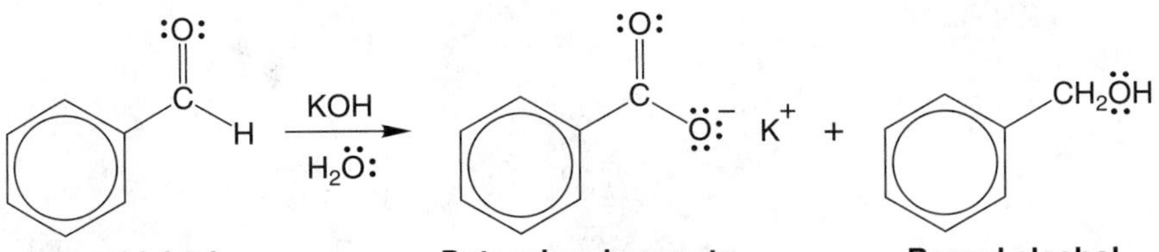

Benzaldehyde —(KOH, H_2O)→ Potassium benzoate (oxidized) + Benzyl alcohol (reduced)

FIGURE 18.97 Benzaldehyde gives benzoate and benzyl alcohol on treatment with hydroxide ion.

This process is known as the **Cannizzaro reaction**, for Stanislao Cannizzaro (1826–1910). If the reaction is carried out in deuterated solvent, no carbon deuterium bonds appear in the alcohol (Fig. 18.98).

FIGURE 18.98 A labeling experiment shows that no solvent deuterium becomes attached to carbon. The hydrogen that reduces the aldehyde must come from another aldehyde, not the solvent.

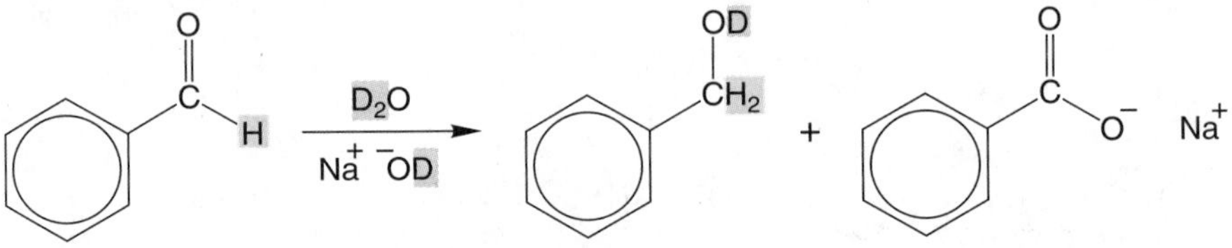

Benzaldehyde is especially prone to hydrate formation, and the first step of the reaction is addition of hydroxide to benzaldehyde. Much of the time this reaction simply reverses to re-form hydroxide and benzaldehyde, but some of the time the intermediate finds a molecule of benzaldehyde in a position to accept a transferred hydride ion (Fig. 18.99).

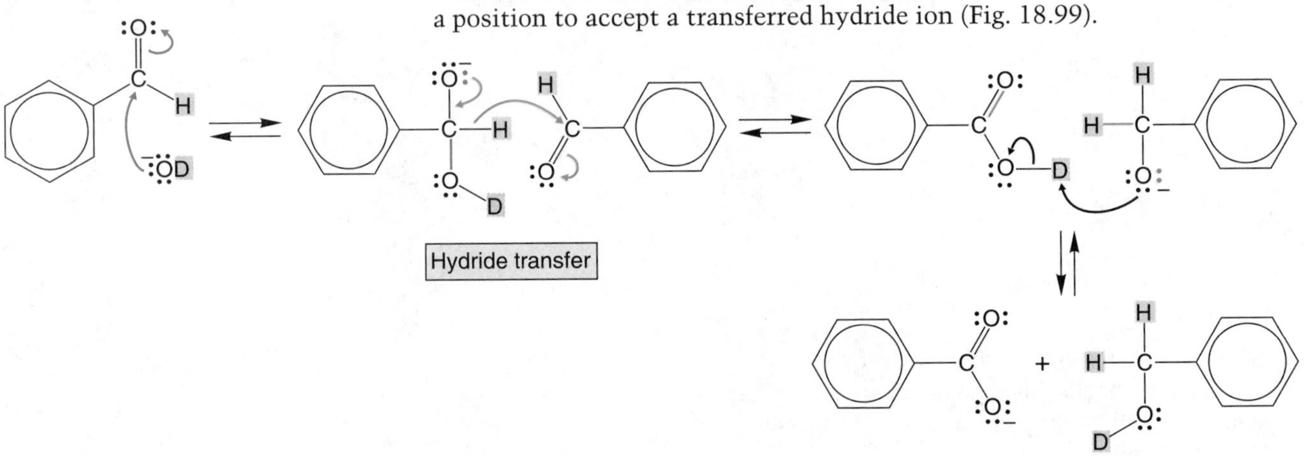

Hydride transfer

FIGURE 18.99 A hydride shift is at the heart of the mechanism of the Cannizzaro reaction.

Do not confuse this step with the expulsion of hydride as a leaving group. Hydride is not a decent leaving group and *cannot* be lost in this way. However, if there is a Lewis acid positioned in exactly the right place to accept the hydride, it can be transferred in a step that simultaneously reduces the receptor aldehyde and oxidizes the donor aldehyde. It is very important to be aware of the requirement for the Lewis acid acceptor of hydride in the Cannizzaro reaction. The hydride must be *transferred,* and in the transition state for the reaction there is a partial bond between the hydride and the accepting carbonyl group. A deprotonation of the acid and protonation of the alkoxide formed from benzaldehyde completes the reaction. Notice how critical the observation about the nonparticipation of the deuterated solvent is. It identifies the source of the reducing hydrogen atom. As it cannot come from solvent (all the "hydrogens" in the solvent are deuteriums) it must come from the aldehyde.

PROBLEM **18.28**

Draw a transition state for the hydride shift reaction.

If there were a way to hold the hydride acceptor in the right position, the Cannizzaro reaction would be much easier, as the requirements for ordering the reactants in just the correct way (entropy) would be automatically satisifed. The two partners in the reaction would not have to find each other. This notion leads to the next reaction in this section, which is largely devoted to hydride shifts.

18.6c The Meerwein–Ponndorf–Verley–Oppenauer Equilibration

In the Meerwein–Ponndorf–Verley–Oppenauer (MPVO) reaction, an aluminum atom is used to clamp together the two halves (oxidation and reduction) of the Cannizzaro reaction. An aluminum alkoxide, typically aluminum isopropoxide, is used as a hydride source to reduce a second carbonyl compound (Fig. 18.100).

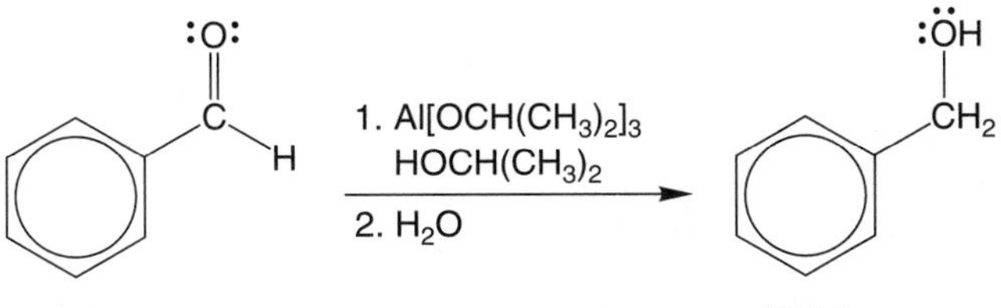

1. $Al[OCH(CH_3)_2]_3$
$HOCH(CH_3)_2$
2. H_2O

(89%)

FIGURE **18.100** The reduction of a benzaldehyde using the Meerwein–Ponndorf–Verley–Oppenauer reaction.

The first step in this reaction is the formation of a bond between the carbonyl oxygen and aluminum. Aluminum is a very strong Lewis acid with respect to oxygen as Lewis base, and this bond forms easily. An analogous reaction is the protonation of the carbonyl oxygen (Fig. 18.101).

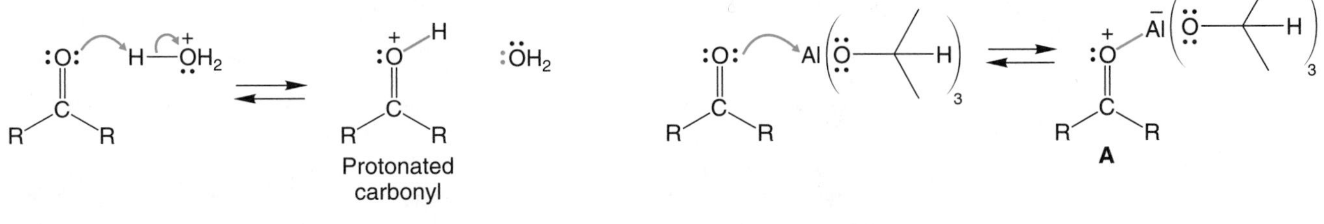

Protonated carbonyl

A

FIGURE **18.101** The formation of a bond between the nucleophile carbonyl oxygen and the Lewis acid aluminum alkoxide to give complex **A**.

This new molecule (**A**) is set up for an intramolecular transfer of hydride, which simultaneously reduces a molecule of the carbonyl compound and oxidizes an isopropoxide group to acetone (Fig. 18.102).

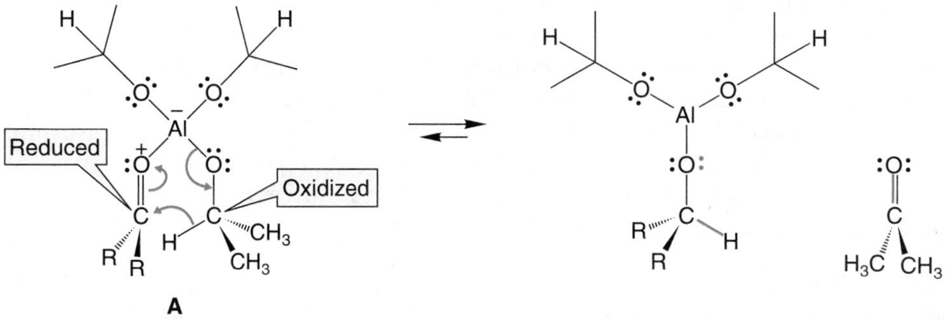

FIGURE 18.102 The crucial intramolecular hydride transfer is facilitated by the aluminum

PROBLEM 18.29 Aluminum alkoxides may look very strange to you. Compare the structure of $Al(OR)_3$ to that of a trivalent boron compound such as $B(OH)_3$.

Repetition leads to a new aluminum oxide that is hydrolyzed at the end of the reaction to give the corresponding alcohol (Fig. 18.103).

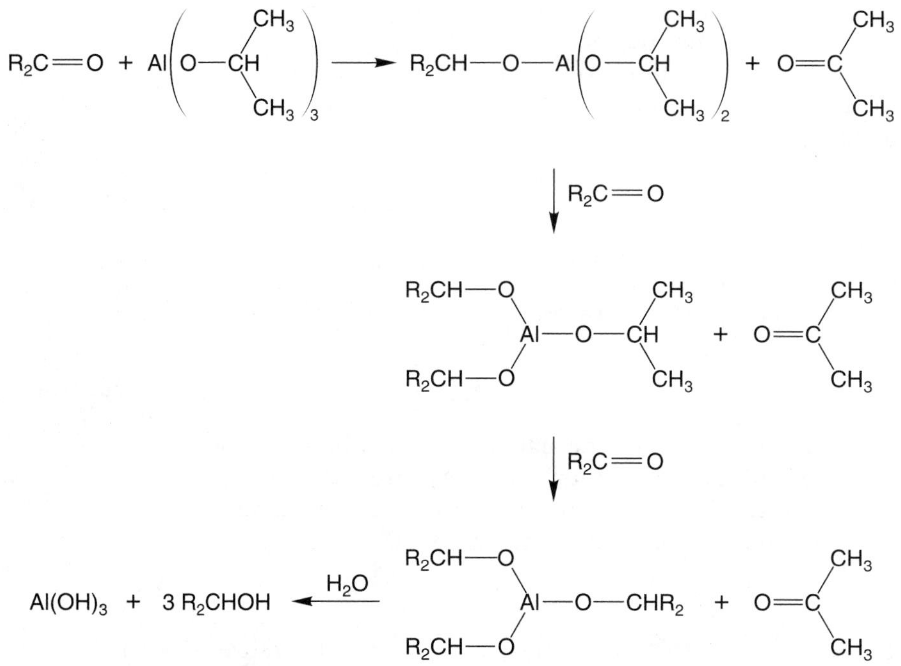

FIGURE 18.103 Repetition and hydrolysis leads to more acetone and the new alcohol.

PROBLEM 18.30 Write a mechanism for the second stage in the reaction, the formation of $(R_2CH\text{–}O)_2Al\text{–}O\text{–}CH(CH_3)_2$.

Write a mechanism for the final hydrolysis reaction of $Al(OR)_3$ to $Al(OH)_3$. PROBLEM 18.31

The effect of the clamp on the rate of the reaction is profound. This reaction, here shown as a reduction, takes place under very mild conditions and with few side reactions.

The reaction can also be used to oxidize alcohols to carbonyl compounds. In this case, a simple ketone such as acetone is used as the acceptor, and the alcohol to be oxidized is used to form the aluminium oxide (Fig. 18.104).

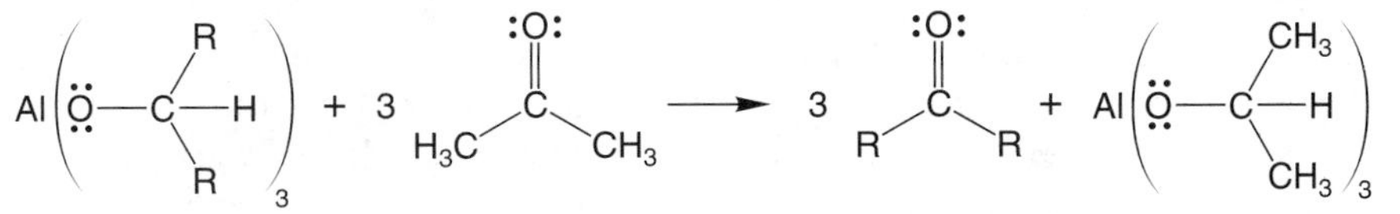

FIGURE 18.104 The Meerwein–Ponndorf–Verley–Oppenauer equilibration can be used as an oxidation reaction.

18.6d The Bartlett–Condon–Schneider Reaction*

Strictly speaking, this reaction does not belong here, as it does not involve a carbonyl compound. However, it surely fits logically in any section devoted to hydride shifts. This reaction shows that hydride shifts can also occur under acidic conditions. We are already quite familiar with an intramolecular version of this reaction. For example, in the addition of hydrogen chloride to 3-methyl-1-butene, the major product results from addition of chloride to a rearranged carbocation formed by an intramolecular shift of hydride (Fig. 18.105).

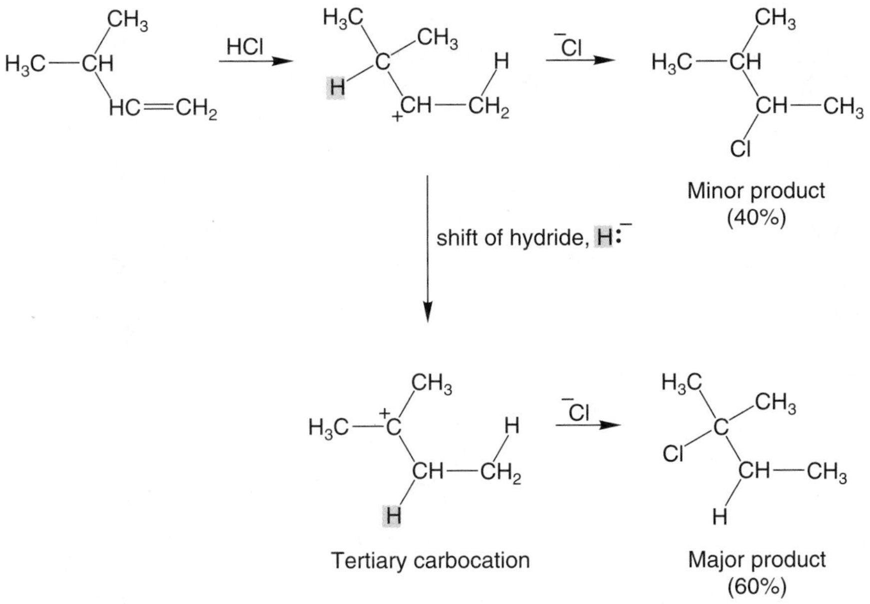

FIGURE 18.105 The major product from the addition of hydrogen chloride to 3-methyl-1-butene is formed by a mechanism involving an intramolecular hydride shift.

*Paul D. Bartlett (b. 1907), Francis E. Condon (b. 1919), Abraham Schneider (b. 1919).

There are intermolecular versions of this reaction and one of them makes an exquisite problem (Fig. 18.106).

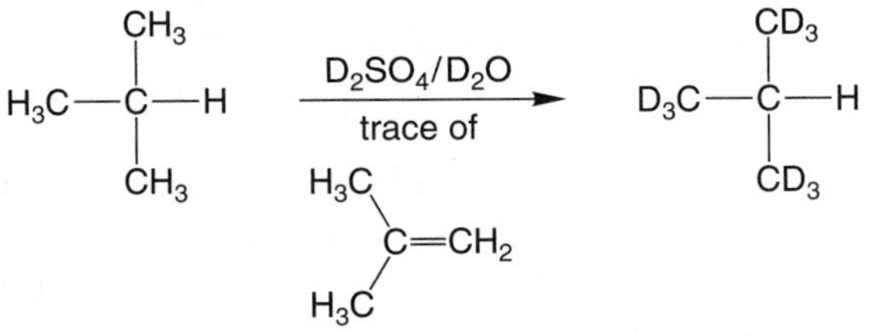

FIGURE 18.106 Isobutane exchanges all the primary hydrogens in D_2SO_4, but the tertiary hydrogen remains H, not D.

The beauty of this problem lies in its ability to misdirect. When one sees acidic conditions and a tertiary hydrogen, one is naturally led to think of the tertiary cation. In turn, that leads one to think of the tertiary position as the center of reactivity. The exchange of all hydrogens *except* for the tertiary one is seemingly contradictory!

The exchange actually requires a small amount of alkene impurity to initiate the reaction. We'll use a molecule of isobutene, but any alkene really will do. Transfer of D^+ takes place to give the more stable tertiary cation in which one deuterium has been incorporated into a methyl group (Fig. 18.107).

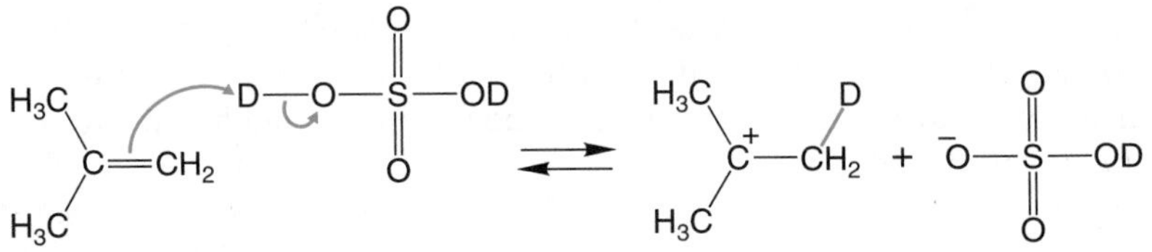

FIGURE 18.107 The reaction starts with the addition of D^+ to isobutene to give the relatively stable tertiary carbocation.

Loss of a proton in either of two ways leads back to an isobutene in which one deuterium is retained (Fig. 18.108).

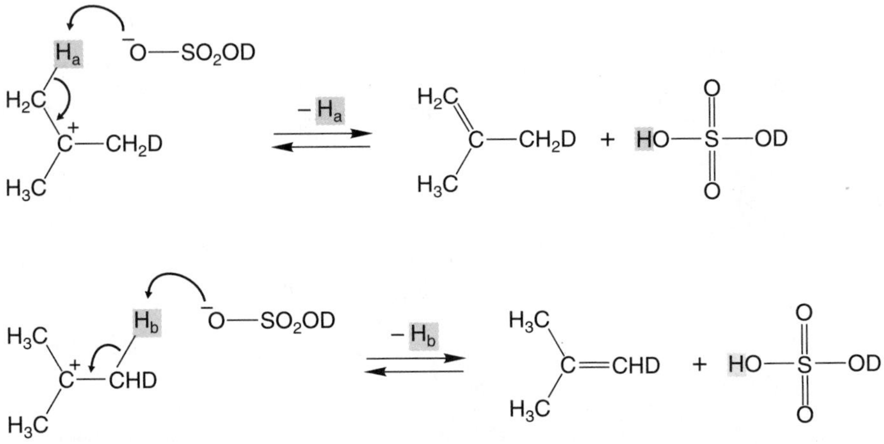

FIGURE 18.108 Loss of a proton (H_a or H_b) leads to deuterated isobutene.

PROBLEM 18.32 What happens if D^+ is removed from the intermediate in Figure 18.108?

Continued reaction with D_2SO_4 leads to further exchange until all nine of the methyl hydrogens have been washed out of the methyl groups by deuterium (Fig. 18.109).

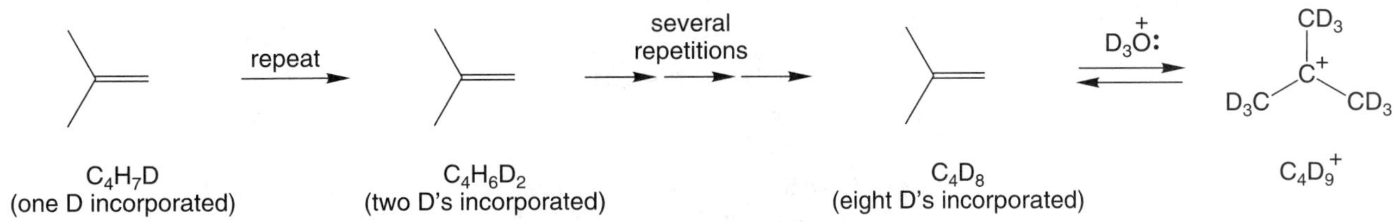

repeat
several repetitions
$\overset{+}{D_3O}$:
CD_3 C^+ D_3C CD_3
C_4H_7D (one D incorporated)
$C_4H_6D_2$ (two D's incorporated)
C_4D_8 (eight D's incorporated)
$C_4D_9^+$

FIGURE **18.109** Repetitions of this process lead to the fully deuterated *tert*-butyl cation.

Now nothing further can happen unless a hydride transfer occurs from isobutane. The hydride can *only* come from the tertiary position because only this transfer will generate a tertiary cation. This reaction opens another molecule for exchange (Fig. 18.110).

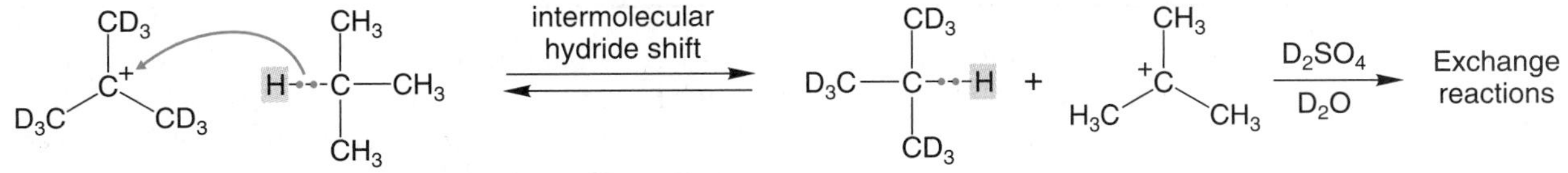

intermolecular hydride shift
D_2SO_4 / D_2O
Exchange reactions

FIGURE **18.110** Hydride transfer can only come from the *tertiary* position of isobutane, because only this transfer leads to a relatively stable tertiary carbocation.

PROBLEM **18.33**

Write a mechanism for the second stage in which a monodeuterioisobutene forms a dideuterioisobutene.

Actually, there is no need to wait for full exchange before the hydride transfer. It can occur at any point in the reaction, and doubtless there is a soup of partially exchanged isobutane molecules and *tert*-butyl cations throughout the reaction. No matter, however; the exchanges will continue and as long as there is excess D_2SO_4, all primary hydrogens, but *never* the tertiary hydrogen, will be exchanged. It's a beautiful problem.

Problems involving hydride transfers tend to be hard, to thwart even excellent problem solvers. That leads to **Magid's Third Rule**, which states, "When all else fails and desperation is setting in, look for the hydride shift." Problem 18.34 provides some chances to practice hydride shifts.

PROBLEM **18.34**

Write mechanisms for the following reactions (Fig. 18.111):

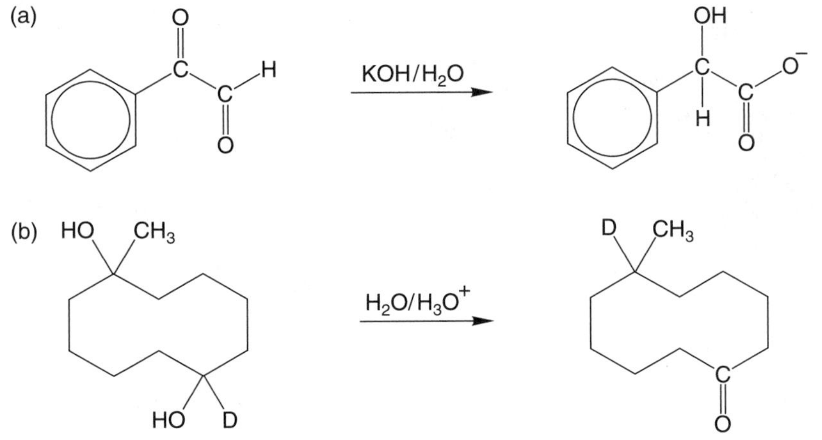

(a) KOH/H_2O

(b) H_2O/H_3O^+

FIGURE **18.111**

(c)

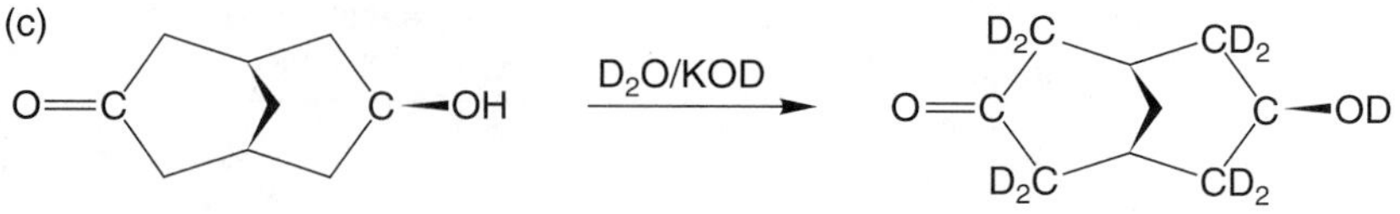

Hint: Draw a good three-dimensional picture of this molecule first!
Hint: The other stereoisomer exchanges only five hydrogens for deuteriums.

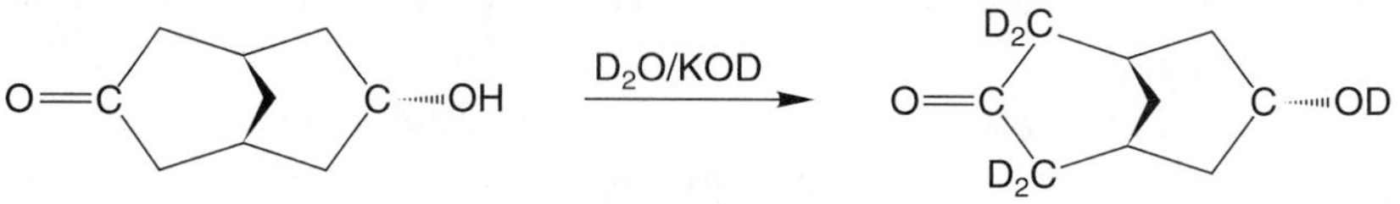

FIGURE 18.111 (CONTINUED)

18.7 THE ALDOL CONDENSATION IN THE REAL WORLD: MODERN SYNTHESIS

In this section, we are going to introduce some of the difficulties of real world chemistry. The detail isn't so important here, although it will be if you go on to become a synthetic chemist! What is important is to get an idea of the magnitude of the problems, and of the ways in which chemists try to solve them. The general principles and relatively simple reactions we have studied (for example, the straightforward aldol condensation) set the stage for a glance at the difficulties encountered when chemists try to do something with these reactions.

In the real world of practical organic synthesis, one rarely needs to do a simple aldol condensation between two identical aldehydes or two identical ketones. Far more commonly, it is necessary to do a crossed aldol between two different aldehydes, two ketones, or an aldehyde and a ketone. As noted earlier (Section 18.4b, p. 912), there are difficulties in doing crossed aldol reactions. Suppose, for example, that we want to condense 2-pentanone with benzaldehyde. Benzaldehyde has no α-hydrogen, so no enolate can be formed from it. Some version of the Claisen–Schmidt reaction (p. 914) seems feasible. But 2-pentanone can form two enolates, and the first problem to solve is the specific formation of one or the other enolate (Fig. 18.112).

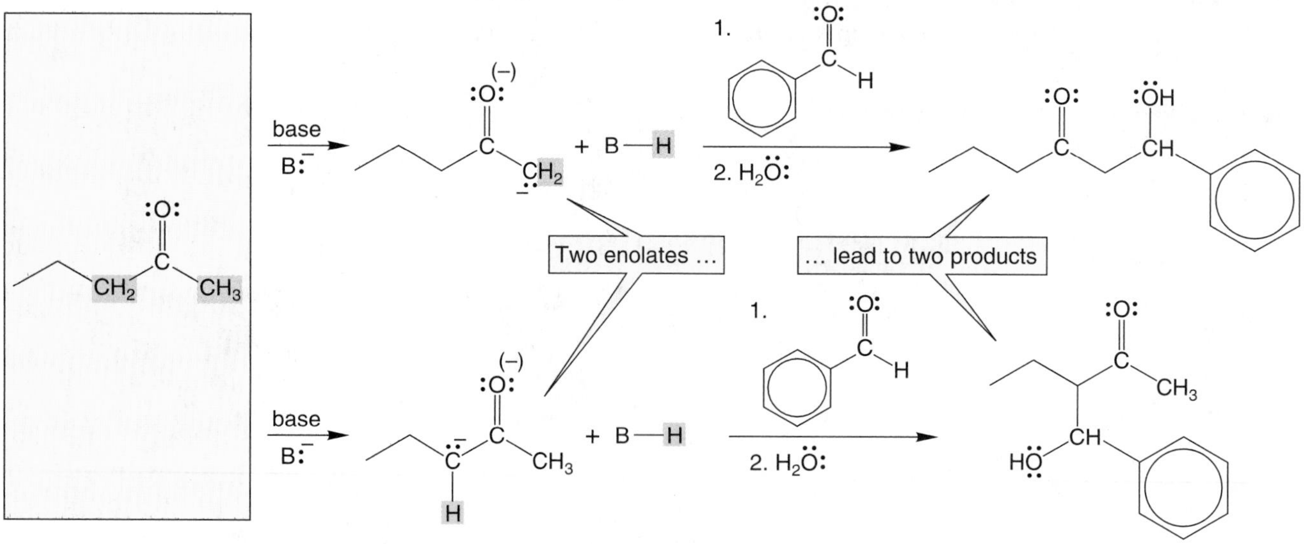

Two enolates ...

... lead to two products

FIGURE 18.112 In principle, an unsymmetrical ketone leads to two enolates, and therefore two products.

The two enolates are quite different in stability; one contains a disubstituted double bond and will be more stable, whereas the other has only a monosubstituted double bond and will be less stable (Fig. 18.113).

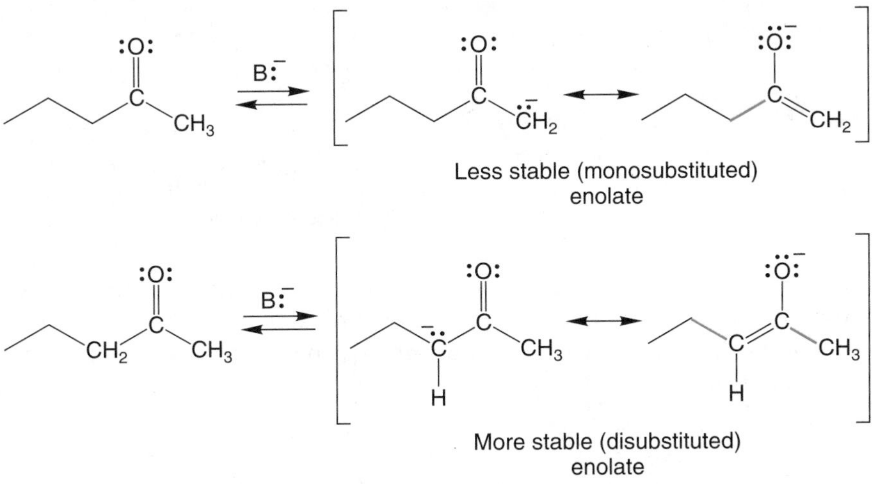

Less stable (monosubstituted) enolate

More stable (disubstituted) enolate

FIGURE **18.113** The two enolates will differ in stability, depending on the number of alkyl groups attached to the double bond.

Effective methods have evolved to form the less stable enolate by taking advantage of the relative ease of access to the less hindered α-hydrogen. Lithium diisopropylamide (p. 897) is one base especially effective at forming these less stable **kinetic enolates**. The key point is that this strong but large base has difficulty in gaining access to the more hindered parts of the carbonyl compound. Removal of the less hindered proton leads to the less stable, or "kinetic" enolate (Fig. 18.114).

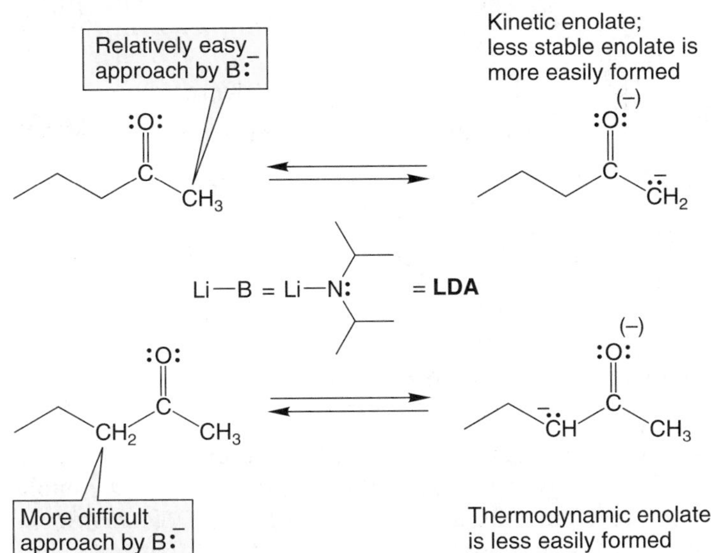
Relatively easy approach by B:⁻
Kinetic enolate; less stable enolate is more easily formed
Li—B = Li—N: = LDA
More difficult approach by B:⁻
Thermodynamic enolate is less easily formed

FIGURE **18.114** Formation of the "thermodynamic" enolate involves making the more stable enolate. The "kinetic" enolate is the one formed more easily (faster). It need not be the more stable enolate.

Other methods allow formation of the more stable, **thermodynamic enols**. In one of these, enol borinates are formed. Variation of the R group allows formation of either the kinetic (less stable, but more easily formed) or thermodynamic (more stable) enol (Fig. 18.115).

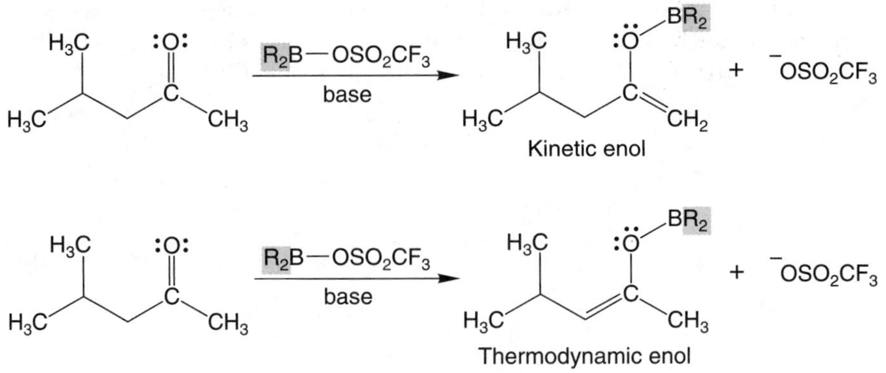

FIGURE 18.115 Methods have been found that favor either the thermodynamic or kinetic enolates and enols.

But our problems are far from over. Aldol condensation of two different carbonyl compounds typically produces a molecule with two new stereogenic carbons. Four new compounds are possible, two diastereomers and their enantiomers (Fig. 18.116).

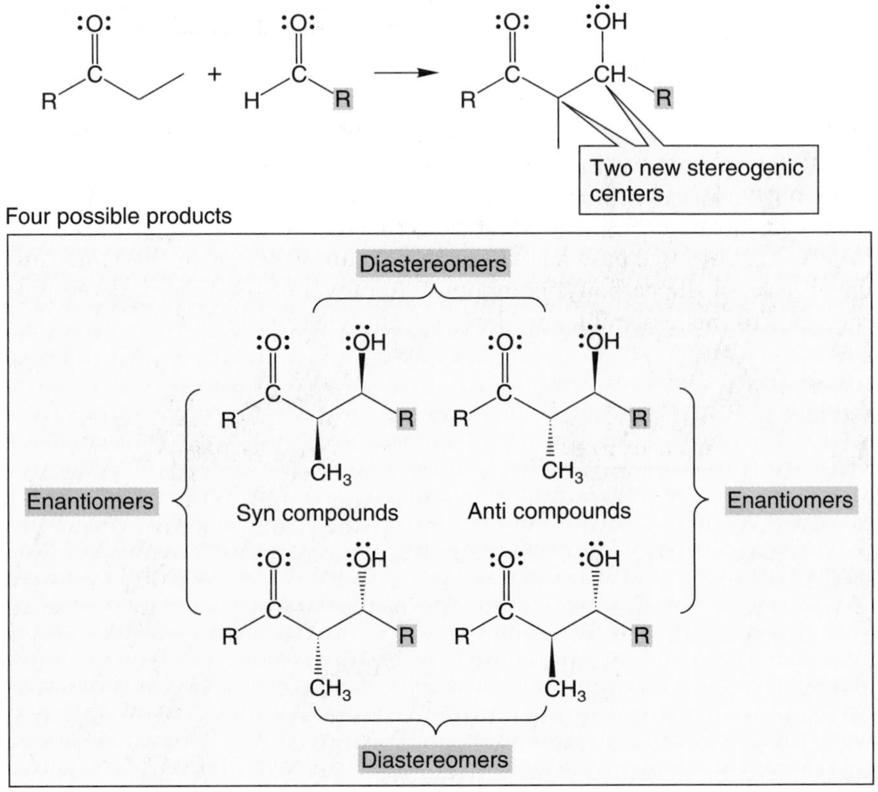

FIGURE 18.116 The typical aldol condensation generates two new stereogenic carbon atoms.

Now we need to ask if it is possible to select for one diastereomer over the other, which leads to an even more detailed look at the enolates involved in the reaction. Enolates exist in (*Z*) and (*E*) forms. It has been found that the (*Z*) enols favor formation of the "syn" diastereomer. As one would expect, ways have been sought (and found) to make the (*Z*) enols selectively. The (*E*) enols lead primarily to the "anti" compounds (Fig. 18.117).

Finally, can one favor one enantiomer over the other? The answer to this question must be "no" unless we use a chiral reagent. In the absence of introduced chirality, a racemic pair of molecules *must* be formed. In

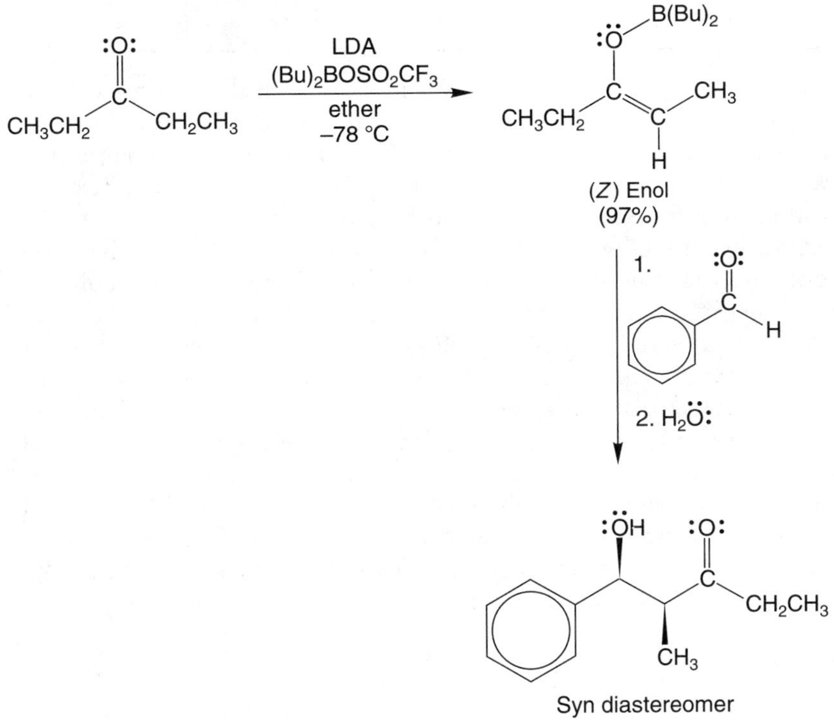

FIGURE 18.117 It is possible to find conditions that favor one diastereomer over the other.

this case, chirality is introduced into one partner in the aldol condensation in such a way that it can be later removed. The aldol condensation now selects for one enantiomer over the other, sometimes quite effectively (Fig. 18.118).

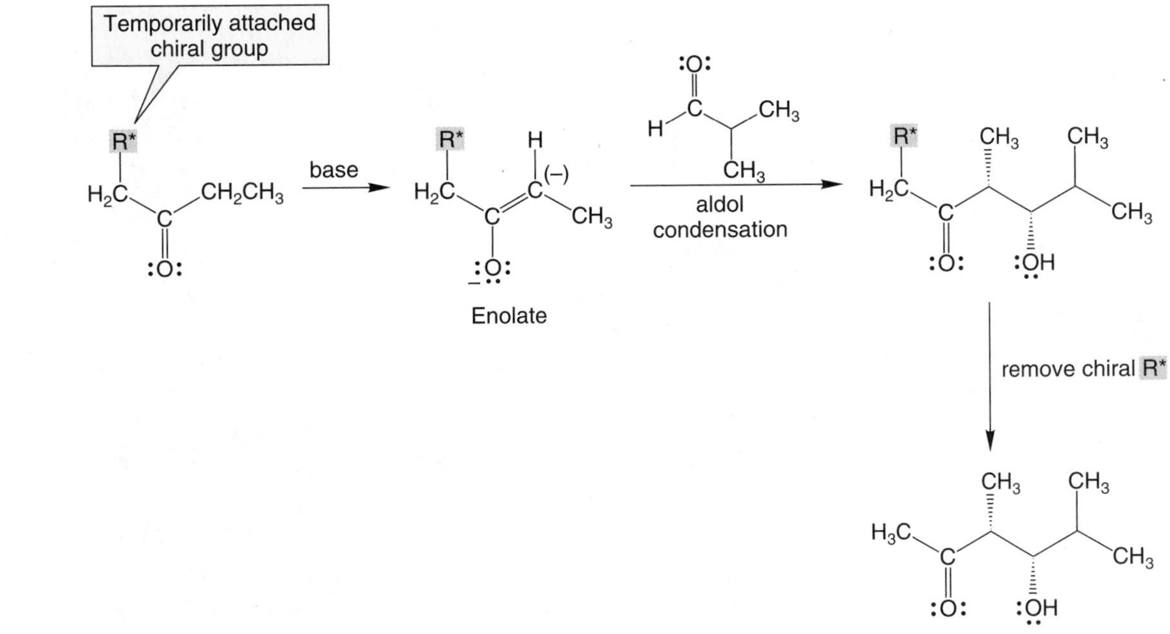

FIGURE 18.118 Use of a removable chiral group allows for the preferential formation of one enantiomer.

18.8 SOMETHING MORE: STATE-OF-THE-ART ORGANIC SYNTHESIS

The previous section describes only a few of the difficulties encountered in just one important synthetic reaction, the aldol condensation. One goal of all synthetic chemists is specificity. How do we do only one reaction? How do we form one, and only one, stereoisomer of the many often possible? Molecules found in nature are sometimes quite spectacularly complicated. Yet we are now able to modify the simple reactions we have been describing so as to be able to make breathtakingly complicated molecules. Here is just one example, from the laboratory of Professor Yoshito Kishi (b. 1937) of Harvard University. Professor Kishi and his co-workers set out to make a molecule called palytoxin, a compound of the formula $C_{129}H_{223}N_3O_{54}$, found in a coral resident in a small tidal pool on the island of Maui in Hawaii. A derivative of this extraordinarily toxic molecule, palytoxin carboxylic acid, is shown in Figure 18.119. Palytoxin is difficult to make, not so much because of its size, but because of its stereochemical complexity and delicacy. It contains no less than 61 stereogenic atoms, each of which must be generated specifically and perfectly in order to produce the real palytoxin.

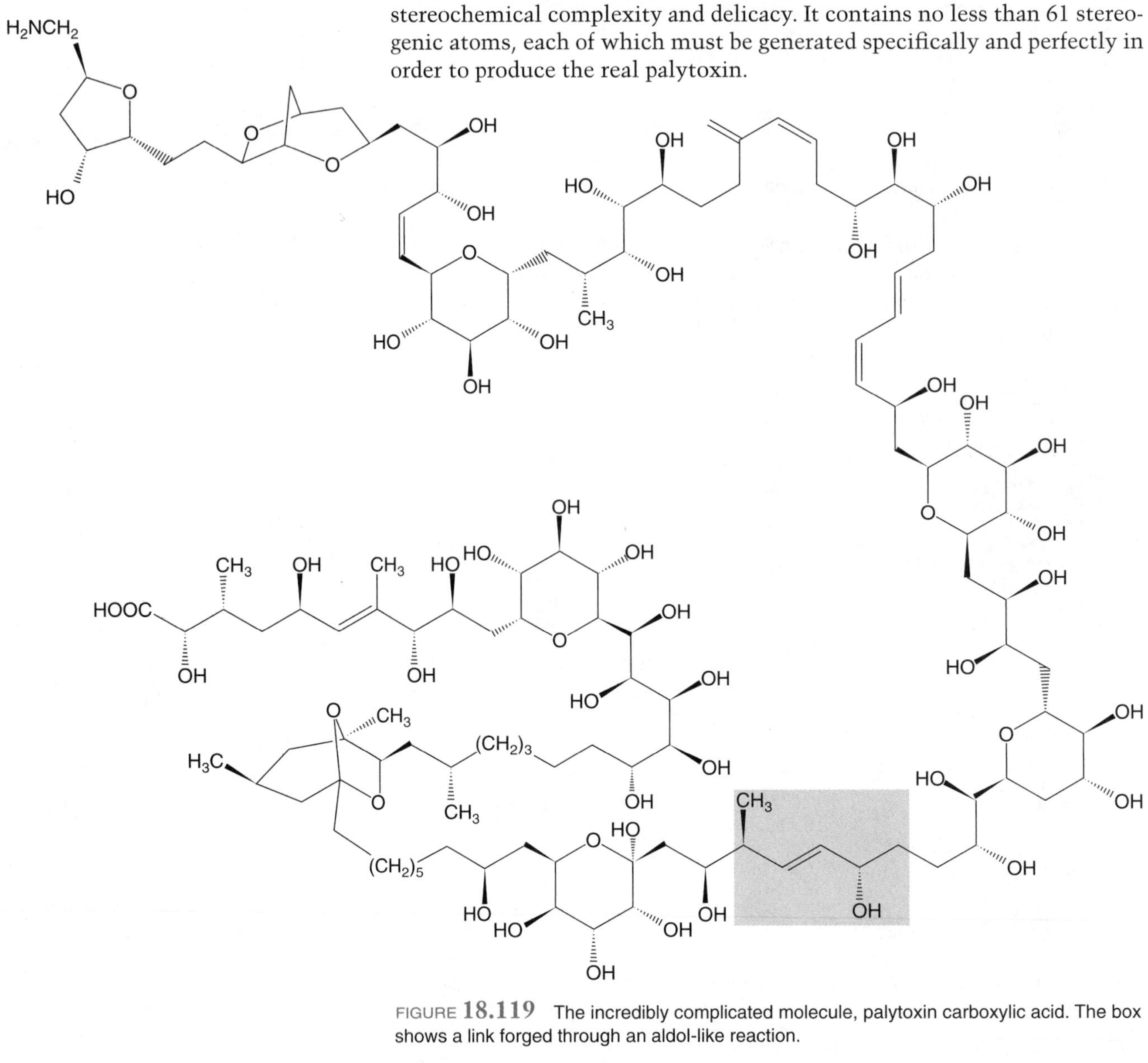

FIGURE **18.119** The incredibly complicated molecule, palytoxin carboxylic acid. The box shows a link forged through an aldol-like reaction.

Several aldol-related reactions were necessary in Kishi's spectacular synthesis. For example, one of the last critical reactions in making palytoxin was a variation of the aldol condensation, a so-called Horner–Emmons reaction, which sewed the molecule together with an α,β-unsaturated carbonyl group that was later reduced (stereospecifically) to give palytoxin. The box shows the site of condensation and subsequent reduction (Fig. 18.120).

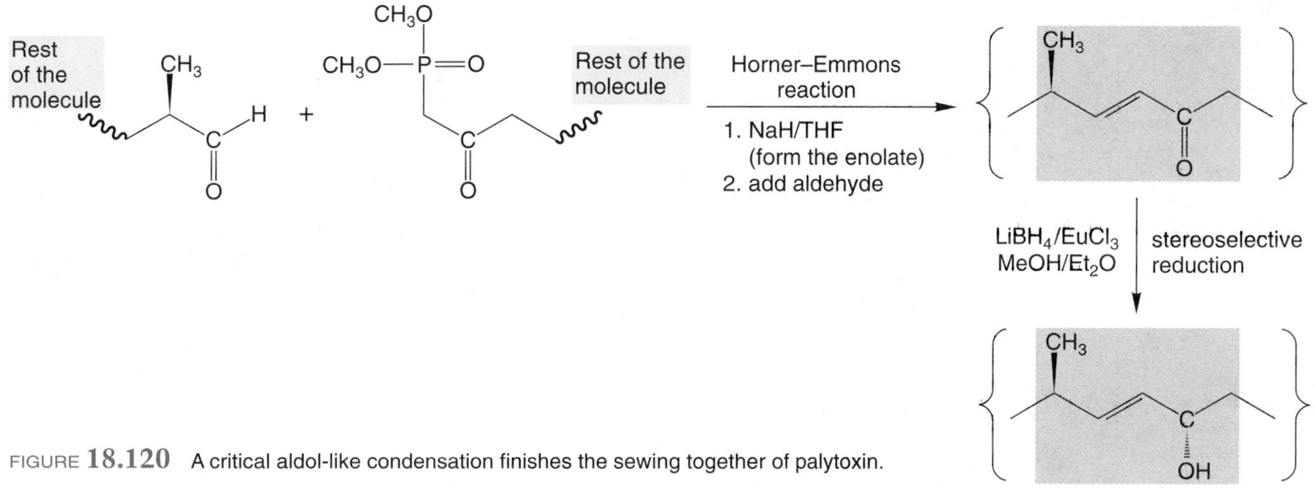

FIGURE **18.120** A critical aldol-like condensation finishes the sewing together of palytoxin.

18.9 SUMMARY

NEW CONCEPTS

It becomes harder and harder to introduce a truly new concept. This chapter deals almost exclusively with the consequences of the acidity of a hydrogen α to a carbonyl group. This concept is really not new, however. You have encountered the idea of resonance stabilization of an enolate before (for example, Problem 9.6), and could surely have dealt with the question of why an α-hydrogen is removable in base, whereas other carbon–hydrogen bonds are not.

In base, most carbonyl compounds containing α-hydrogens equilibrate with an enolate anion. In acid, it is the enol that is formed in equilibrium with the carbonyl compound (Fig. 18.121).

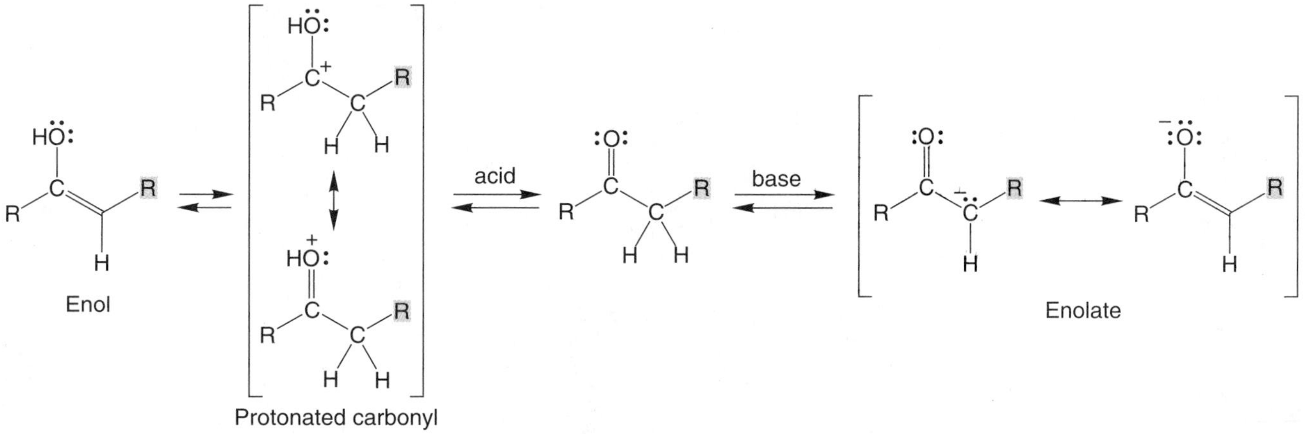

FIGURE **18.121** Enolate formation in base and enol formation in acid are typical reactions of carbonyl compounds bearing α-hydrogens.

The consequences of enol and enolate formation are vitally important for understanding the mechanisms of reactions of aldehydes, ketones, and other carbonyl-containing compounds, as well as for doing synthesis. Enol or enolate formation leads to exchange, halogenation, and alkylation at the α-position, as well as the more complicated condensation reactions. All these reactions involve fundamentally simple Lewis base–Lewis acid chemistry (Fig. 18.122).

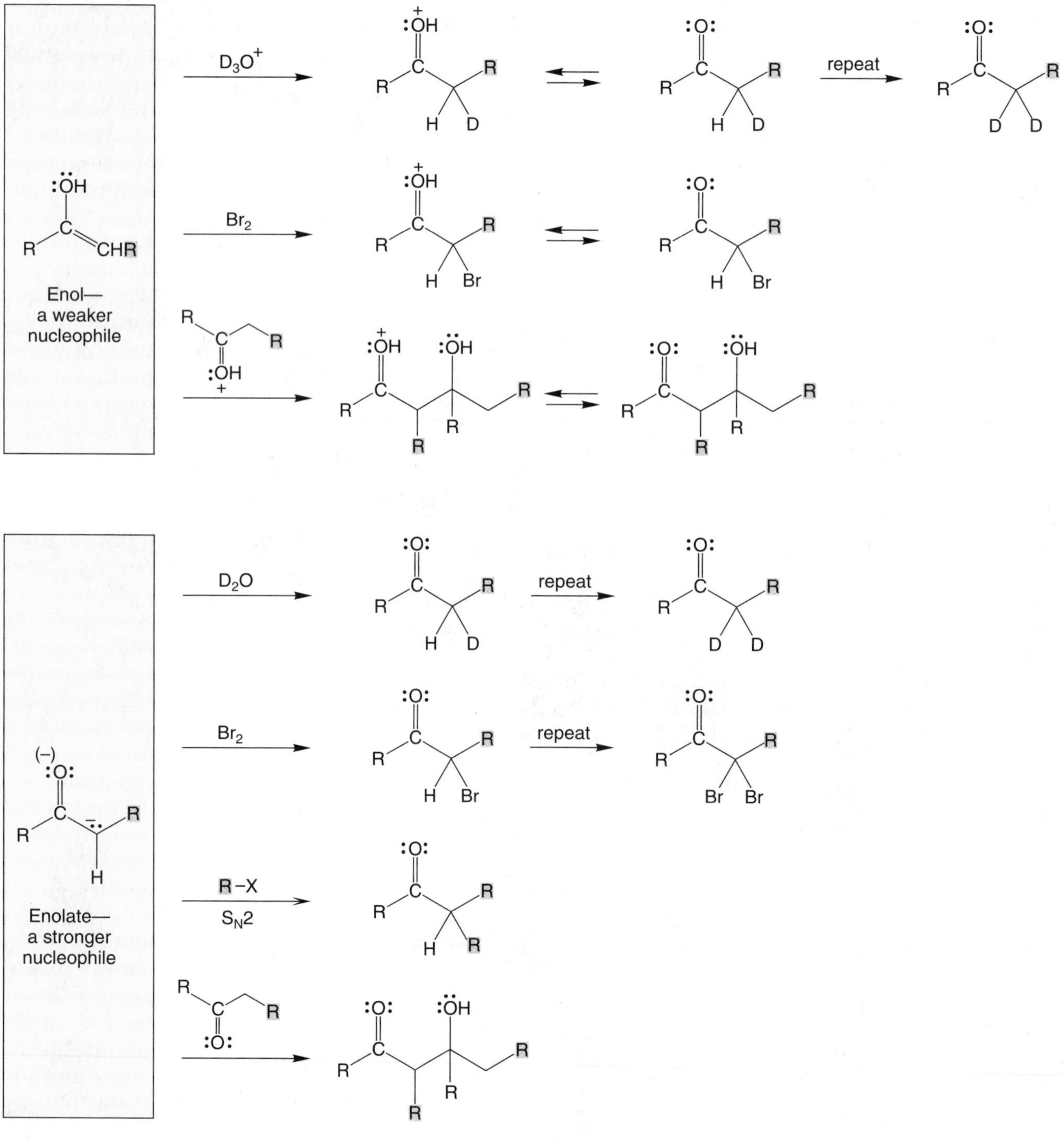

Enol—
a weaker
nucleophile

D_3O^+

repeat

Br_2

Enolate—
a stronger
nucleophile

D_2O

repeat

Br_2

repeat

R–X
S_N2

FIGURE 18.122 Reactions of enolates and enols with various Lewis acids.

This chapter also introduces the notion of forming an enolate indirectly, through Michael addition to an α,β-unsaturated carbonyl group (Fig. 18.82).

The idea of transferring a hydride ion ($H:^-$) is elaborated in Section 18.6. *Remember*: Hydride is not a good leaving group; it cannot be displaced by a nucleophile, but it can sometimes be transferred, providing only that a suitable acceptor Lewis acid is available.

REACTIONS, MECHANISMS, AND TOOLS

The two central reactions in this chapter, from which almost everything else is derived, are enolate formation in base and enol formation in acid (Fig. 18.121).

The enols and enolates are capable of undergoing many reactions, among them α-exchange, racemization, α-halogenation, and α-alkylation. Reaction of the enolate, a strong nucleophile, with the Lewis acidic carbonyl compound, or of the less strongly nucleophilic enol with the powerful Lewis acid, the protonated carbonyl, leads to the aldol condensation. These reactions have been previously summarized in Figure 18.122.

Intramolecular aldols, crossed aldols, and the related Knoevenagel condensations involve similar mechanisms.

The Michael reaction, in which a nucleophile adds to the carbon–carbon double bond of an α,β-unsaturated carbonyl group, is another way of forming an enolate anion (Fig. 18.82).

A variety of reactions involving hydride shifts is described. The Cannizzaro reaction is the most famous of these and involves, as do all hydride shift reactions, a simultaneous oxidation of the hydride donor and reduction of the hydride acceptor (Fig. 18.99).

Three other reactions are worthy of mention. In the benzoin condensation, addition of the basic catalyst, usually cyanide, creates a newly acidic hydrogen that can be removed to initiate a condensation reaction (Figs. 18.93–18.95).

Enamines, formed through reaction of carbonyl groups with secondary amines, can be used to alkylate α-positions. The carbonyl group can be regenerated through hydrolysis of the immonium ion product (Fig. 18.49).

Finally, in a little reaction that leads us to Chapter 19, the idea of addition–elimination reactions of carbonyl groups is introduced. The example in this chapter involves trihalomethyl groups, which can be lost as anions in the elimination phase of the process known as the haloform reaction (Fig. 18.38). In Chapters 19 and 20, we will see just how general and important this process is.

SYNTHESIS

Figure 18.123 summarizes the new synthetic procedures of this chapter.

COMMON ERRORS

Condensation reaction problems can be daunting. At the same time, many chemists agree that they provide a great deal of fun, and almost everyone has his or her favorites. A common error is to set out without a plan. Don't start on a complicated problem without an analysis of what needs to be accomplished. Is the reaction done in acid or base? What connections (bonds) must be made? Are rings opened or closed in the reaction? These are the kinds of questions that should be asked before starting on a problem.

1. Acids

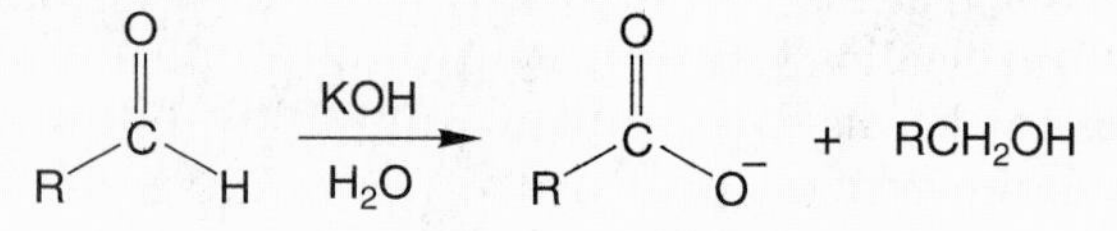

$$RCHO \xrightarrow[H_2O]{KOH} RCOO^- + RCH_2OH$$

The Cannizzaro reaction; R may not have an α-hydrogen

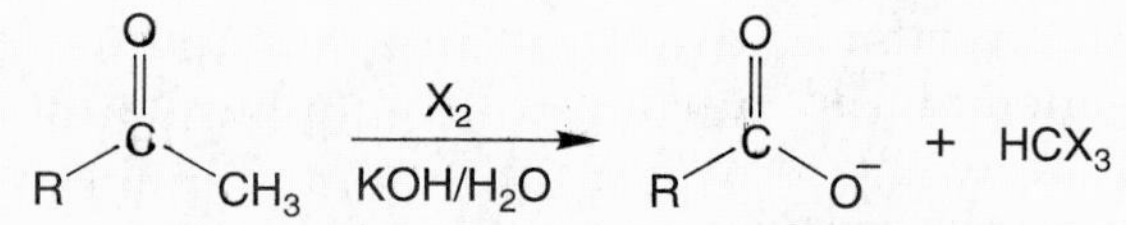

$$RCOCH_3 \xrightarrow[KOH/H_2O]{X_2} RCOO^- + HCX_3$$

The haloform reaction; the haloform (CHX_3) is also formed; works for methyl compounds, X = Cl, Br, and I

2. Alcohols

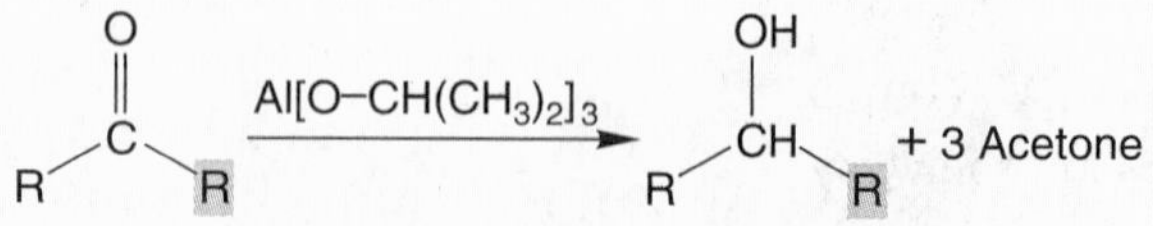

$$RCOR \xrightarrow{Al[O{-}CH(CH_3)_2]_3} RCH(OH)R + 3\ Acetone$$

The MPVO reaction; the mechanism involves complex formation followed by hydride shift

See also the Cannizzaro reaction under "acids," "α-hydroxy ketones," and "β-hydroxy aldehydes and ketones"

3. Alklated aldehydes and ketones

$$RCOCH_2R \xrightarrow{1.\ R_2NH;\ 2.\ R{-}X;\ 3.\ H_2O} RCOCH(R)R$$

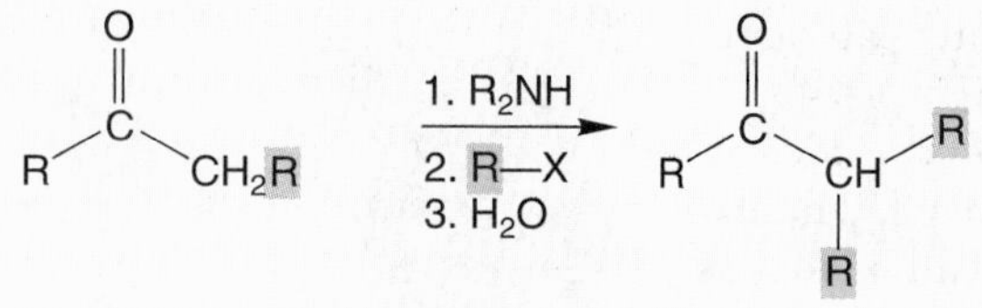

The enamine is an intermediate and R cannot be tertiary

4. Deuterated aldehydes and ketones

$$RCOCH_2R \xrightarrow[D_3O^+]{KOD/D_2O\ \text{or}} RCOCD_2R$$

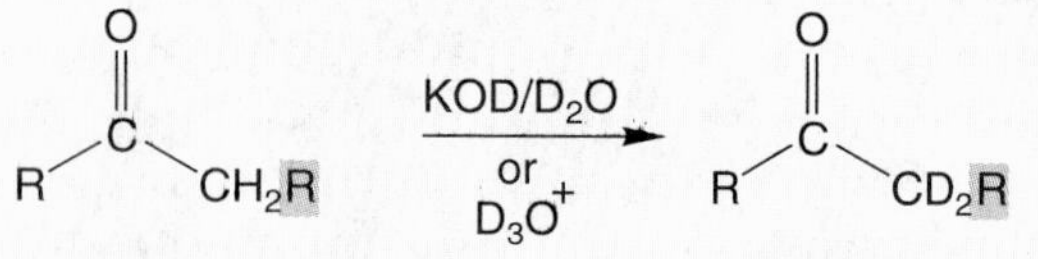

All α-hydrogens will be exchanged in either acid or base

5. α-Halo aldehydes and ketones

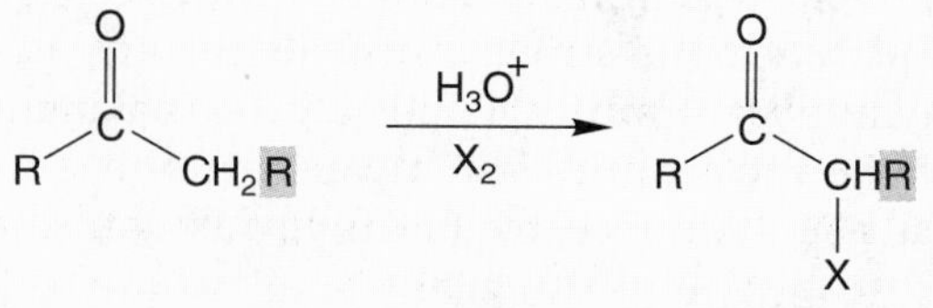

$$RCOCH_2R \xrightarrow[X_2]{H_3O^+} RCOCHR(X)$$

In acid, the reaction stops after one halogenation; X = Cl, Br, and I

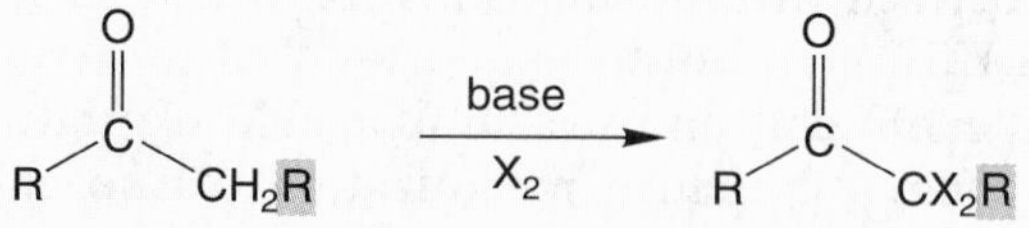

$$RCOCH_2R \xrightarrow[X_2]{base} RCOCX_2R$$

In base, up to three α-hydrogens can be replaced by X; the trihalo carbonyl compounds react further in the haloform reaction

6. Haloforms

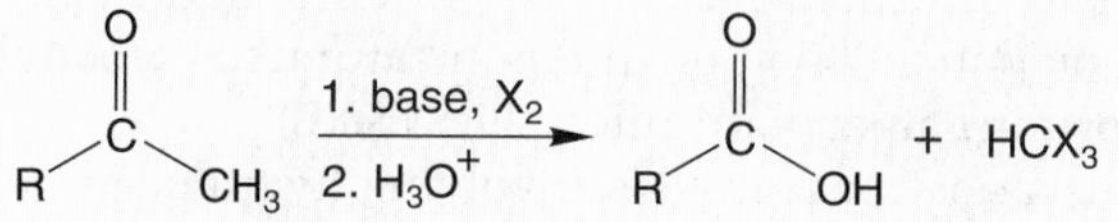

$$RCOCH_3 \xrightarrow{1.\ base,\ X_2;\ 2.\ H_3O^+} RCOOH + HCX_3$$

Starting carbonyl compound must have a methyl group; X = Cl, Br, or I

7. α-Hydroxy ketones

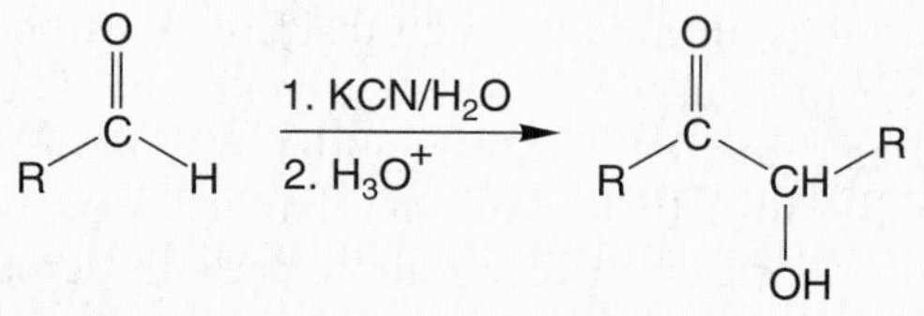

$$RCHO \xrightarrow{1.\ KCN/H_2O;\ 2.\ H_3O^+} RCOCH(OH)R$$

The benzoin condensation, where R must have no α-hydrogens; cyanide or a related catalyst is necessary

8. β-Hydroxy aldehydes and ketones

$$RCOCH_3 \xrightarrow{acid\ or\ base} RCOCH_2C(OH)(R)CH_3$$

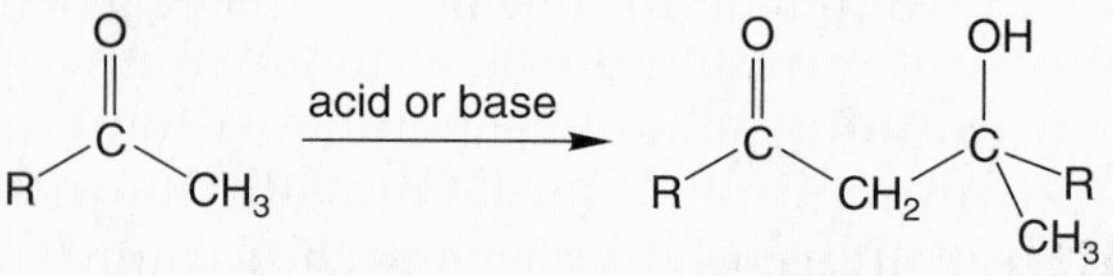

The aldol condensation; dehydration often occurs in acid, and on occasion in base

FIGURE 18.123 New syntheses in Chapter 18.

9. Ketones

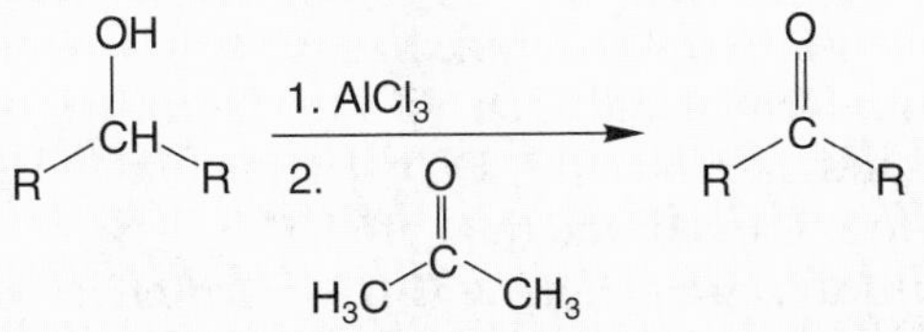

The MPVO reaction; the first step is formation of a new aluminum alkoxide; the mechanism involves complex formation and a hydride shift

10. α, β-Unsaturated aldehydes and ketones

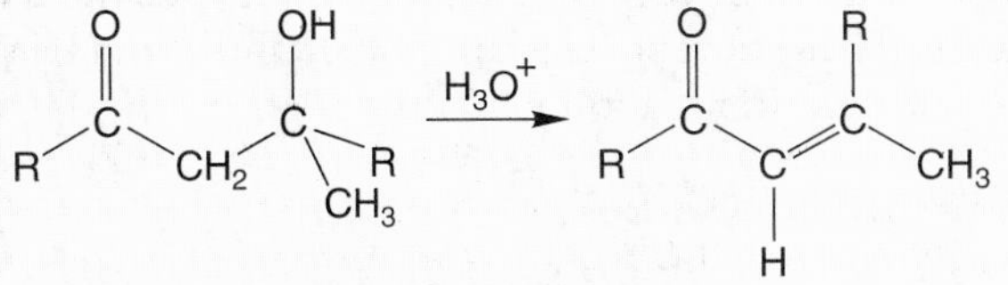

The dehydration of the products of aldol condensations is usually acid-catalyzed

FIGURE **18.123** (CONTINUED)

Two conventions, used widely in this chapter, have the potential for creating misunderstanding. In the less dangerous of the two, charges or dots in parentheses are used to indicate the atoms sharing the charge, electron, or electrons in a molecule best described by more than one electronic description (resonance form).

The second involves drawings in which arrows of an arrow formalism emanate from only one of several resonance forms. Although extraordinarily convenient, this is a dangerous practice, as it carries the inevitable implication that the resonance form used has some individual reality. It does not. What can be done with one resonance form can always be done from all resonance forms. For simplicity, and bookkeeping purposes, we often draw arrow formalisms using only one of several resonance forms. This practice can also be dangerous unless we are very clear about the shortcuts we are taking.

18.10 KEY TERMS

Aldol condensation The acid- or base-catalyzed conversion of a ketone or aldehyde into a β-hydroxy aldehyde or β-hydroxy ketone. In acid, the enol is an intermediate; in base, it is the enolate anion.

Alkylation reaction An aldehyde or ketone is first converted into the enamine by reaction with a secondary amine. The nucleophilic enamine reacts with an alkyl halide to form a new carbon–carbon bond. Hydrolysis regenerates the carbonyl compound and completes the reaction.

Bartlett–Condon–Schneider reaction An acid-catalyzed deuterium exchange reaction of alkanes. In the classic example, a trace of alkene catalyzes the exchange of all nine primary hydrogens of isobutane, but leaves the tertiary position unexchanged.

Benzoin condensation An aldehyde containing no α-hydrogens reacts with cyanide ion to form a cyanohydrin. The old aldehydic hydrogen is now acidic because of the resonance stabilization afforded by the cyano group. Its removal leads to a condensation reaction and the formation of an α-hydroxy ketone.

Cannizzaro reaction The redox reaction of an aldehyde containing no α-hydrogens with hydroxide ion. Addition of hydroxide to the aldehyde is followed by hydride transfer to another aldehyde. Protonation generates a molecule of the carboxylic acid and the alcohol related to the original aldehyde.

Claisen–Schmidt condensation A crossed aldol condensation of an aldehyde without α-hydrogens with a ketone that does have at least one α-hydrogen.

Crossed aldol condensation An aldol condensation between two different carbonyl compounds. This reaction is not very useful unless strategies are employed to limit the number of possible products.

Enamines The compound formed by reaction between an aldehyde or ketone and a secondary amine. These compounds are nucleophilic and useful in alkylation reactions.

$$\mathrm{R{-}C(=CH_2){-}\ddot{N}R_2}$$

Enol A vinyl alcohol. These compounds usually equilibrate with the more stable keto forms.

Enolate The resonance-stabilized anion formed on treatment of an aldehyde or ketone containing an α-hydrogen with base.

Fulvene A 5-methylene-1,3-cyclopentadiene. These are often synthesized through a Knoevenagel condensation using the cyclopentadienide anion and an aldehyde or ketone.

Haloform A compound of the structure HCX_3, where X is F, Cl, Br, or I.

Haloform reaction The conversion of a methyl ketone into a molecule of a carboxylic acid and a molecule of haloform. The trihalo carbonyl compound is formed, base adds to the carbonyl group, and the trihalomethyl anion ($^-{:}CX_3$) is eliminated. The reaction works for X = Cl, Br, or I.

Kinetic enolate The most easily formed enolate. It may or may not be the same as the most stable possible enolate, the thermodynamic enolate.

Knoevenagel condensation Any of a number of condensation reactions related to the crossed aldol condensation. A stabilized anion, often an enolate, is first formed and then adds to the carbonyl group of another molecule. Dehydration often leads to the formation of the final product.

Lithium diisopropylamide (LDA) An effective base for alkylation reactions of carbonyl compounds. This compound is a strong base, but a poor nucleophile, and thus is effective at removing α-hydrogens but not at additions to carbonyl compounds.

Magid's Second Rule "Always try the Michael reaction first."

Magid's Third Rule "In times of desperation and/or despair, when all attempts at solution seem to have failed, try a hydride shift."

Meerwein–Ponndorf–Verley–Oppenauer (MPVO) equilibration A method of oxidizing alcohols to carbonyl compounds, or, equivalently, of reducing carbonyls to alcohols, that uses an aluminum atom to clamp the partners in the reaction together. A hydride shift is involved in the crucial step.

Michael reaction The addition of a nucleophile, often an enolate, to the β-position of a double bond of an α,β-unsaturated carbonyl group.

Robinson annulation A classic method of construction for six-membered rings using a ketone and methyl vinyl ketone. It involves a two-step sequence of a Michael reaction between the enolate of the original ketone and methyl vinyl ketone, followed by an intramolecular aldol condensation.

Thermodynamic enolate The most stable enolate. It may or may not be the same as the most rapidly formed enolate, the "kinetic enolate."

18.11 ADDITIONAL PROBLEMS

PROBLEM **18.35** Write mechanisms for the following acid- and base-catalyzed equilibrations. There is repetition in this drill problem, but being able to do these prototypal reactions quickly and easily is an essential skill. Be certain you can solve this problem easily before going on to more challenging ones.

(a) $$\mathrm{H_3C{-}C(=O){-}CH_3 \rightleftharpoons H_3C{-}C(OH){=}CH_2}$$

(b) $$\mathrm{H_3C{-}C(=NH){-}CH_3 \rightleftharpoons H_3C{-}C(NH_2){=}CH_2}$$

(c) $$\mathrm{H_3C{-}C(=O){-}CH_3 \rightleftharpoons H_3C{-}C(OH)_2{-}CH_3}$$

PROBLEM **18.36** The following compound contains two different types of α-hydrogens. Removal of which α-hydrogen will yield the *more stable* enolate? Which α-hydrogen is likely to be removed *faster*? Explain why these questions need not have the same answer. That is, why might the less stable enolate be formed faster than the more stable enolate and *vice versa*?

$$\mathrm{C_6H_5{-}CH_2{-}C(=O){-}CH_3}$$

PROBLEM **18.37** Which positions in the following molecules will exchange H for D in D_2O/DO^-?

(a) (b) (c) (d) (e)

PROBLEM **18.38** As we saw in Problem 18.37c, methyl vinyl ketone exchanges the three methyl hydrogens for deuterium in D_2O/DO^-. Why doesn't the other formally "α" hydrogen exchange as well?

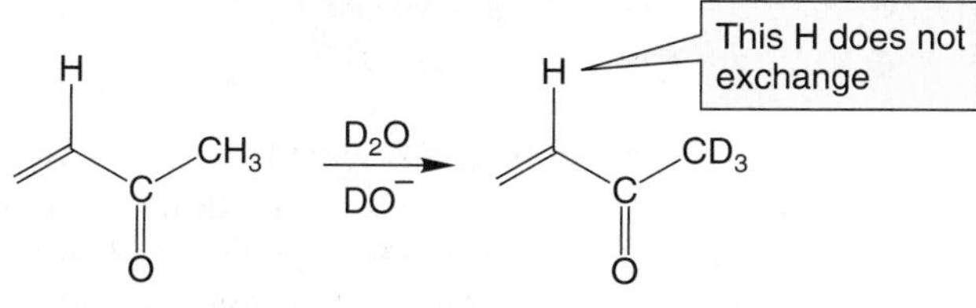

PROBLEM **18.39** Which positions in the following molecules will exchange H for D in D_2O/DO^-?

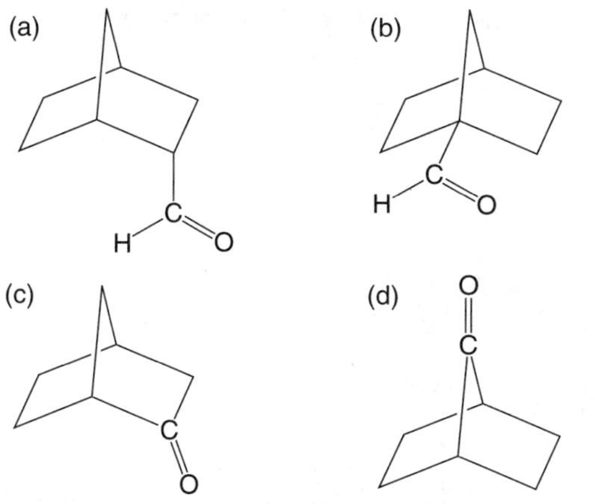

PROBLEM **18.40** The proton-decoupled ^{13}C NMR spectrum of 2,4-pentanedione consists of six lines (δ 24.3, 30.2, 58.2, 100.3, 191.4, and 201.9 ppm), not the "expected" three lines. Explain.

PROBLEM **18.41** Provide simple synthetic routes from cyclohexanone to the following compounds. *Hint*: These questions are all review.

(a) (b) (c) (d) (e)

PROBLEM **18.42** Provide simple synthetic routes from cyclohexanone to the following compounds:

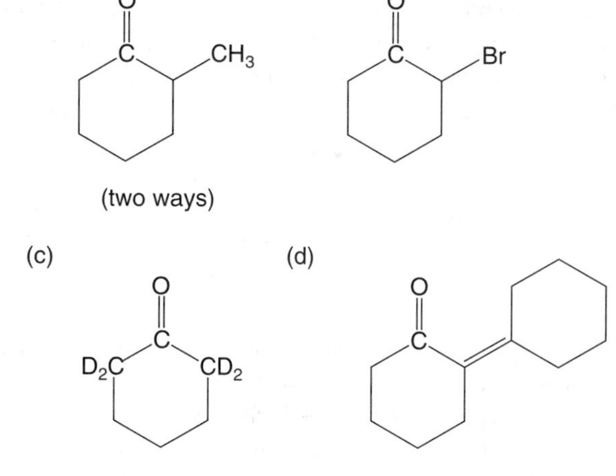

PROBLEM **18.43** Write mechanisms for the following isomerizations:

(a) H_2O/HO^-

(b) H_2O/H_3O^+

PROBLEM **18.44** Perform retrosynthetic analyses on the following molecules, each of which, in principle, can be synthesized by an aldol condensation–dehydration sequence. Which syntheses will be practical? Explain why some of your suggested routes may not work in practice. What are some of the potential problems in each part?

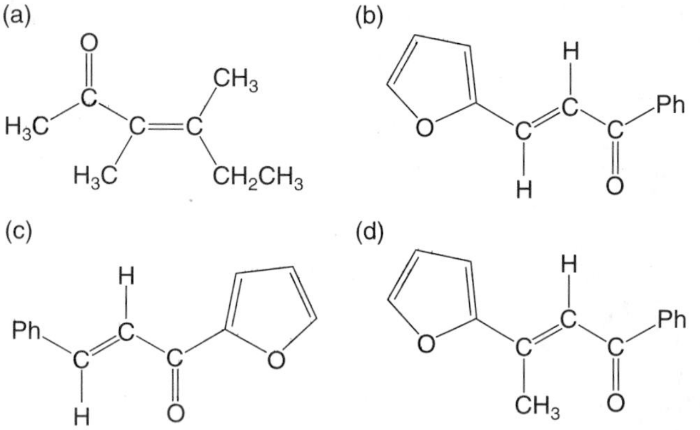

(a) (b) (c) (d)

PROBLEM **18.45** Treatment of compound **1** with an aqueous solution of bromine and sodium hydroxide affords, after acidification, pivalic acid (**2**) and bromoform. Deduce the structure of **1** and write an arrow formalism mechanism for its conversion into **2** and bromoform. Use your mechanism to predict the number of equivalents of bromine and sodium hydroxide necessary for this reaction.

$$\mathbf{1} \xrightarrow[\text{2. } H_2O/H_3O^+]{\text{1. } NaOH/Br_2/H_2O} (CH_3)_3C{-}COOH + CHBr_3$$

2

PROBLEM **18.46** Treatment of ketone **1** with LDA in dimethylformamide (DMF) solvent, followed by addition of methyl iodide leads to two methylated ketones. Write arrow formalisms for their formations and explain why one product is greatly favored over the other.

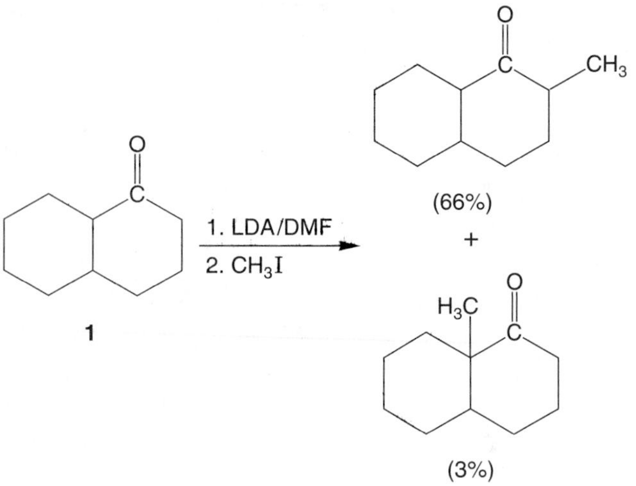

1. LDA/DMF
2. CH_3I

1

(66%) + (3%)

PROBLEM **18.47** As we have seen, alkylation of enolates generally occurs at carbon (see Problem 18.46 for an example). However, this is not always the case. First, use the following estimated bond strengths to explain why treatment of an enolate with trimethylsilyl chloride leads to substitution at oxygen (Si—C~69 kcal/mol; Si—O~109 kcal/mol).

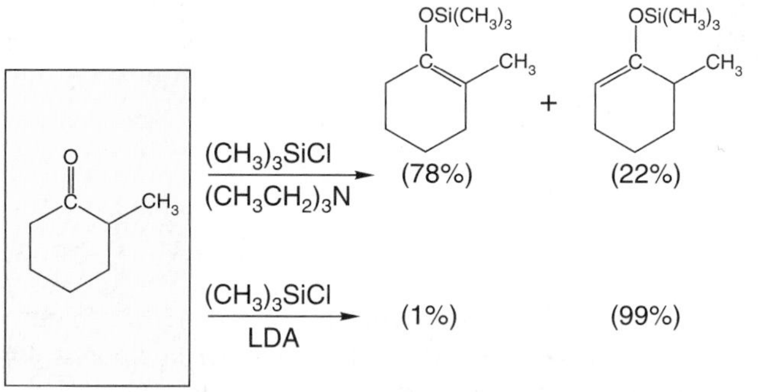

$(CH_3)_3SiCl$, $(CH_3CH_2)_3N$ → (78%) + (22%)

$(CH_3)_3SiCl$, LDA → (1%) (99%)

Second, explain the different regiochemical results under the two reaction conditions shown. *Hint*: Remember that enolate formation with LDA is irreversible.

PROBLEM **18.48** When bromo ketone **1** is treated with potassium *tert*-butoxide in *tert*-butyl alcohol at room temperature, it gives exclusively the 5,5-fused bicyclic ketone **2**. In contrast, when **1** is treated with LDA in tetrahydrofuran (THF) at –72 °C, followed by heating, the product is predominately the 5,7-fused ketone **3**. Write arrow formalism mechanisms for these cycloalkylation reactions and explain why the different reaction conditions favor different products.

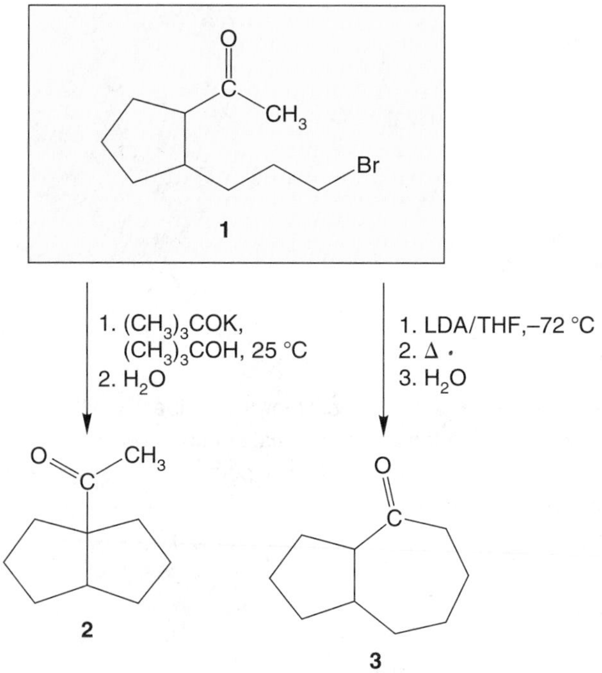

1

1. $(CH_3)_3COK$, $(CH_3)_3COH$, 25 °C
2. H_2O

1. LDA/THF, –72 °C
2. Δ
3. H_2O

2 **3**

PROBLEM **18.49** Condensation of 1,3-diphenylacetone (**1**) and benzil (**2**) in the presence of alcoholic potassium hydroxide affords tetraphenylcyclopentadienone (**3**), a dark purple solid. Write an arrow formalism mechanism for the formation of **3**.

PhH$_2$C–CO–CH$_2$Ph (**1**) + PhCO–COPh (**2**) $\xrightarrow{\text{KOH, CH}_3\text{CH}_2\text{OH}}$ **3** (tetraphenylcyclopentadienone; Ph, Ph, Ph, Ph)

PROBLEM **18.50** Aldehydes, but not ketones, readily react with 5,5-dimethyl-1,3-cyclohexanedione (dimedone, **1**) in the presence of a base (such as piperidine) to give crystalline dimedone derivatives, **2**. These can be converted into octahydroxanthenes, **3**, upon heating in ethyl alcohol with a trace of acid. Provide arrow formalism mechanisms for the formation of **2** from **1** and an aldehyde (RCHO), as well as for the conversion of **2** into **3**.

1 + RCHO $\xrightarrow{\text{H}_2\text{O/EtOH/}\Delta}$ **2** (Piperidine) $\xrightarrow{\Delta,\ \text{H}_3\text{O}^+}$ **3**

PROBLEM **18.51** Write an arrow formalism mechanism for the following reaction (the Darzens condensation).

Cyclohexanone + ClCH$_2$CO–OCH$_2$CH$_3$ $\xrightarrow{(\text{CH}_3)_3\text{COK},\ (\text{CH}_3)_3\text{COH},\ 10\ °\text{C, 1.5 h}}$ epoxy ester, C(=O)OCH$_2$CH$_3$ (89%)

PROBLEM **18.52** Here is a pair of relatively simple Knoevenagel condensations. Write arrow formalism mechanisms for the following conversions:

(a) PhCHO + CH$_3$NO$_2$ $\xrightarrow{\text{1. NaOH/H}_2\text{O, CH}_3\text{OH, }\Delta;\ \text{2. HCl}}$ Ph(H)C=C(H)NO$_2$ (81%)

(b) N-(2-methylphenyl)formamide (N–H, CHO, CH$_3$) $\xrightarrow{(\text{CH}_3)_3\text{COK},\ (\text{CH}_3)_3\text{COH},\ \Delta}$ indole (N–H) (79%)

PROBLEM **18.53** Now here is a pair of slightly more complicated Knoevenagel reactions: Provide arrow formalism mechanisms.

(a) H$_3$C–CHO + H$_3$C–CO–CH$_2$–CO–OCH$_2$CH$_3$ $\xrightarrow{\text{piperidine, CH}_3\text{CH}_2\text{OH, 0 °C}}$ H$_3$C–CO–CH(CO–OCH$_2$CH$_3$)–CH(CH$_3$)–CH(CO–CH$_3$)(CO–OCH$_2$CH$_3$) (>50%)

PROBLEM **18.53** (CONTINUED)

(b) CH_3CH_2CHO + $NCCH_2COOCH_2CH_3$ —(piperidine, CH_3COOH, H_2/Pd)→ $CH_3CH_2CH_2CH(CN)COOCH_2CH_3$

PROBLEM **18.54** Write an arrow formalism mechanism for the conversion of diketone **1** into the important perfumery ingredient *cis*-jasmone (**2**). There is another possible cyclopentenone that could be formed from **1**. What is its structure? Write an arrow formalism mechanism for its formation.

1 —(NaOH, CH_3CH_2OH/H_2O, Δ)→ **2**

PROBLEM **18.55** When benzaldehyde and an excess of acetone react in the presence of aqueous sodium hydroxide, compound **1** is obtained as the major product as well as a small amount of compound **2**. However, when acetone and an excess (at least 2 equiv) of benzaldehyde react under the same conditions, the predominant product is compound **2**. Spectral data for compounds **1** and **2** are summarized below. Deduce structures for compounds **1** and **2** and rationalize their formation.

PhCHO + H_3C–CO–CH_3 —(NaOH, H_2O)→ **1** + **2**

Compound **1**

Mass spectrum: *m/z* = 146 (p, 75%), 145 (50%), 131 (100%),103 (80%)

IR (Nujol); 1667(s), 973 (s), 747 (s), 689 (s) cm^{-1}

^{1}H NMR ($CDCl_3$): δ 2.38 (s 3H), 6.71 (d, *J* = 16 Hz, 1H), 7.30–7.66 (m, 5H), 7.54 (d, *J* = 16 Hz, 1H)

Compound **2**

Mass spectrum: *m/z* = 234 (p, 49%), 233 (44%), 131 (42%), 128 (26%), 103 (90%), 91 (32%), 77 (100%), 51 (49%)

IR (melt): 1651 (s), 984 (s), 762 (s), 670 (s) cm^{-1}

^{1}H NMR ($CDCl_3$); δ 7.10 (d, *J* = 16 Hz, 1H), 7.30–7.70 (m, 5H), 7.78 (d, *J* = 16 Hz, 1H)

PROBLEM **18.56** Write an arrow formalism mechanism for the following transformation:

biphenyl with $CH_2CH(OCH_3)_2$ groups at 2,2′ —(H_3O^+, H_2O)→ dibenzocycloheptene carbaldehyde (CHO)

PROBLEM **18.57** Reaction of 1-morpholinocyclohexene (**1**) and β-nitrostyrene (**2**), followed by hydrolysis, yields the nitro ketone **3**. Write an arrow formalism mechanism for this reaction sequence and be sure to explain the observed regiochemistry.

1 —(1. PhCH=CHNO$_2$ (**2**), CH_3CN; 2. H_2O/H_3O^+)→ **3** (2-(1-phenyl-2-nitroethyl)cyclohexanone)

PROBLEM **18.58** Disobutylaluminum hydride (DIBAL-H), like lithium aluminum hydride ($LiAlH_4$) and sodium borohydride ($NaBH_4$), can reduce aldehydes and ketones to alcohols.

DIBAL-H

R–CHO → 1. DIBAL-H ($[(CH_3)_2CHCH_2]_2AlH$); 2. H_2O → $R—CH_2—OH$

Write a mechanism for this reaction. Interestingly, 1 mol of DIBAL-H can reduce up to 3 mol of aldehyde despite possession of only one apparent hydride equivalent. Propose a mechanism that accounts for the second and third equivalents of hydride in these aldehyde reductions. *Hint*: Remember the Meerwein–Ponndorf–Verley–Oppenauer equilibration.

PROBLEM **18.59** Provide structures for intermediates **A–D**. A mechanistic analysis is not required for this problem, but will almost certainly be of use in developing an answer. *Hint*: You will discover that there are certain structural ambiguities in this problem. The following experimental observation should prove helpful. Mild hydrolysis of compound **B** affords compound **E** ($C_{12}H_{18}O_2$), which is characterized by the following spectra:

IR ($CHCl_3$): 1715 cm^{-1}

^1H NMR ($CDCl_3$): δ 1.55–1.84 (m, 9H), 2.60 (s, 4H), 3.77 (s, 3H), 5.52 (br d, 1H), 6.08 (d, 1H).

1 (cyclopentyl methyl ketone, C(=O)CH_3) + **2** (pyrrolidine, N–H) → H_3O^+ → ($C_{11}H_{19}N$) **A**

A + $H_2C{=}C(CH_2Br)COOCH_3$ (**3**) → benzene → ($C_{16}H_{26}NO_2^+\ Br^-$) **B**

B → $(CH_3CH_2)_3N$, CH_3CN, Δ → ($C_{16}H_{25}NO_2$) **C** → ($C_{16}H_{25}NO_2$) **D** → H_3O^+, H_2O → **4** (spiro ketone with $COOCH_3$)

PROBLEM **18.60** Acetolactate (**2**) is a precursor of the amino acid valine in bacterial cells. Acetolactate is synthesized by the condensation of two molecules of pyruvate (**1**).

2 H_3C–C(=O)–C(=O)–O^- (**1**) + "H^+" ⇌ H_3C–C(=O)–C(CH_3)(OH)–COO^- (**2**) + CO_2

The enzyme that catalyzes this reaction (acetolactate synthetase) requires thiamine pyrophosphate (TPP) as a coenzyme.

TPP: H_3C, N, NH_2, CH_2, $\overset{+}{N}$, CH_3, S, $—CH_2CH_2O—P(=O)(OH)—O—P(=O)(OH)—OH$

II

R–$\overset{+}{N}$ (thiazolium, CH_3, S) –**R**

Suggest a plausible mechanism for the synthesis of acetolactate (**2**). [*Hints*: Review Section 18.6a; particularly Problem 18.27. Note that this condensation is accompanied by a decarboxylation (loss of CO_2) that actually precedes coupling. Finally, you may use the enzyme to do acid- or base-catalyzed reactions as required.]

PROBLEM **18.61** Provide mechanisms for the following changes. *Hint*: Definitely do (a) before (b).

(a)

PhCH=CH–C(=O)–CH=CHPh + NC–CH_2–$COOCH_2CH_3$ → $NaOCH_2CH_3$, $HOCH_2CH_3$ → cyclohexanone with Ph, Ph, CH_3CH_2OOC, CN

PROBLEM **18.61** (CONTINUED)

(b)

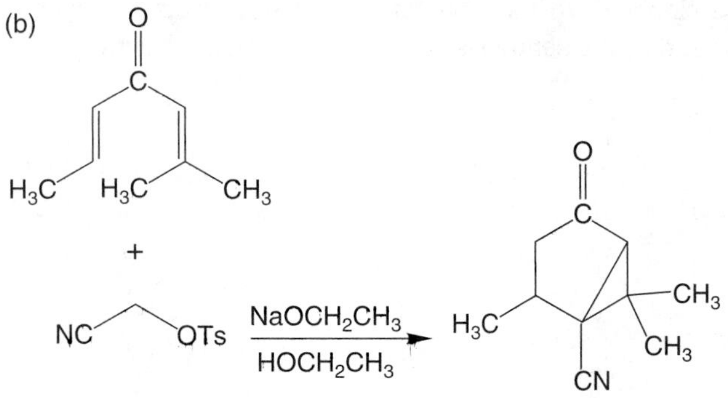

H$_3$C H$_3$C CH$_3$

+

NC OTs $\xrightarrow[\text{HOCH}_2\text{CH}_3]{\text{NaOCH}_2\text{CH}_3}$ H$_3$C CH$_3$ CH$_3$ CN

PROBLEM **18.62** Write a mechanism for the following transformation:

H$_3$C CH$_3$ CH$_2$

CH$_3$CH$_2$COOC COOCH$_2$CH$_3$

NaOCH$_2$CH$_3$ | HOCH$_2$CH$_3$

COOCH$_2$CH$_3$ H$_3$C COOCH$_2$CH$_3$ CH$_3$

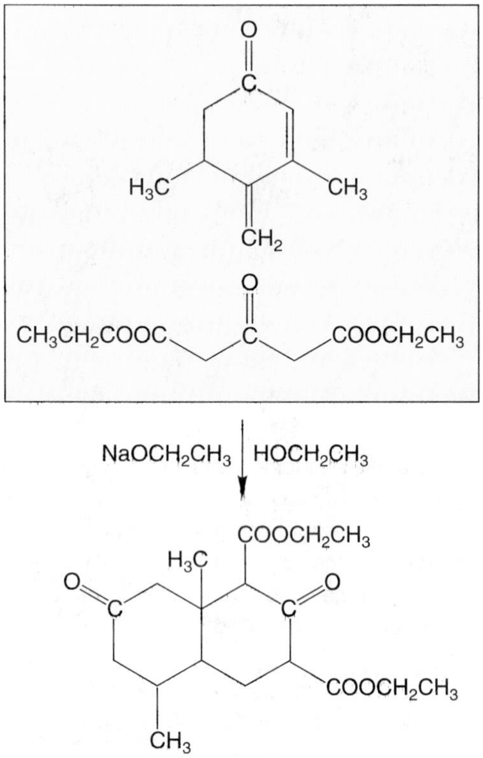

PROBLEM **18.63** Explain the striking observation that methylation of the alcohol **1** leads to a methoxy compound in which the stereochemistry of the oxygen has changed. *Hint*: Note that **1** is a β-keto alcohol.

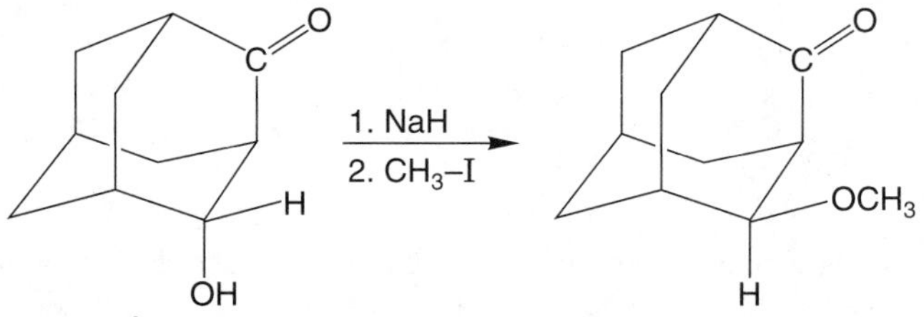

H OH **1** $\xrightarrow[\text{2. CH}_3\text{–I}]{\text{1. NaH}}$ OCH$_3$ H

PROBLEM **18.64** Write a mechanism for the following transformation:

Ph Ph $\xrightarrow[\text{2. H}_2\text{SO}_4]{\text{1. NaOH/H}_2\text{O, 100 °C}}$ Ph—C(OH)(Ph)—COOH

(>90%)

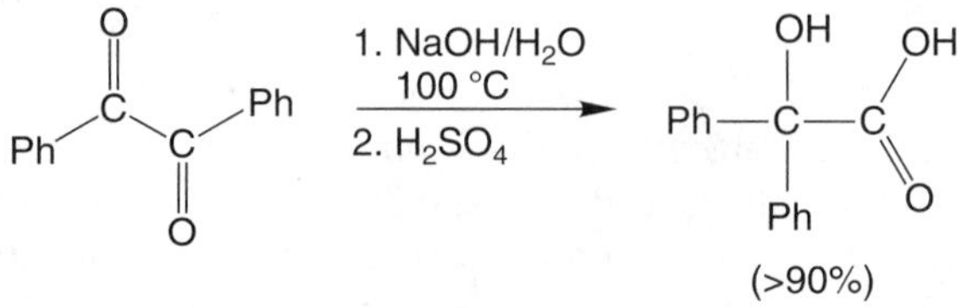

Carboxylic Acids

If I mix CH_2 with NH_4 and boil the atoms in osmotic fog, I should get speckled nitrogen!

—Donald Duck*

He's talking chemical talk!

—Huey

But he knows nothing

—Dewey

About chemicals!

—Louie

In this chapter, we explore the properties and chemistry of carboxylic acids, R–COOH. As with other functional groups, we will begin with a naming system, and then move on to structure. Carboxylic acids have a rich chemistry; they are both acids (hence the name) and bases, and we have seen in Chapters 16 and 18 how the carbonyl group is a locus of reactivity. Although many new reactions will appear, most important will be the emergence of a general reaction mechanism, addition–elimination, that we have only seen in one or two examples. Chapter 20 will extend the exploration of this process. Almost all of this chapter is material you should be able to grasp easily or figure out if it is presented in a reasonable problem. What's a "reasonable" problem? I think problems that ask for extrapolations from known material are generally reasonable, but questions that are open-ended are often not. For example, it is reasonable to ask how the aldol condensation product is formed from acetone, but not reasonable to ask what happens when acetone is treated with base, at least before the aldol condensation is discussed (Fig. 19.1).

:O: KOH / H_2O → :O: :ÖH Mechanism?

:O: KOH / H_2O → ?

FIGURE 19.1 Examples of a reasonable and an unreasonable problem.

The difficulty with the second question is that it has too many answers, all of them correct but incomplete in the absence of all the others.

*This quotation comes from a 1944 issue of *Walt Disney's Comics and Stories*, "Donald Duck the Mad Chemist." It was discovered by Peter P. Gaspar, then a postdoctoral fellow at Caltech, now a professor at Washington University. Those of us who work with CH_2 have still not reproduced Professor Duck's experiment with osmotic fog, but a few of us are still trying.

*PROBLEM **19.1** How many answers can you find to the question, What happens when acetone is treated with hydroxide ion in water?

ANSWER This open-ended problem is hard to answer because so much is possible. Don't worry if you find an extra answer (be sure that it is right, though). In base, the chemistry of carbonyl groups is dominated by addition of nucleophiles to the carbon–oxygen double bond (hydrate formation in this case) and removal of α-hydrogens (enolate formation). Both these reactions are possible for acetone, and these will be the most important processes going on.

addition
protonation
Alkoxide **A**
Hydrate

Once the enolate is formed, further reaction can lead to aldol products. Of course, one is at the mercy of thermodynamics, and for ketones such as acetone the aldol product is not favored. Still, it is a reaction that happens and is a legitimate answer to the problem.

removal of an α-hydrogen
Resonance-stabilized enolate
protonation
Enol

Diacetone alcohol

One could imagine many other answers. The alkoxide **A**, formed as an intermediate in the hydration reaction, could add to a second molecule of acetone, for example.

19.1 NOMENCLATURE AND PROPERTIES OF CARBOXYLIC ACIDS

In the systematic nomenclature, the final "e" of the parent alkane is dropped and the suffix "oic" is added. Diacids are named similarly, except that the suffix is "dioic," and the final "e" is not dropped. Cyclic acids are named as "cycloalkanecarboxylic acids." Table 19.1 shows some common mono- and diacids along with some of their physical properties and common names, many of which are still used.

TABLE 19.1 Carboxylic Acids and Their Properties

Structure	IUPAC Name	Common Name	bp (° C)	mp (° C)	pK_a
HCOOH	Methanoic acid	Formic acid	100.7	8.4	3.75
CH_3COOH	Ethanoic acid	Acetic acid	117.9	16.6	4.75
CH_3CH_2COOH	Propanoic acid	Propionic acid	141	− 20.8	4.87
$CH_3CH_2CH_2COOH$	Butanoic acid	Butyric acid	165.5	− 4.5	4.81
$CH_3CH_2CH_2CH_2COOH$	Pentanoic acid	Valeric acid	186	− 33.8	4.82
$CH_3CH_2CH_2CH_2CH_2COOH$	Hexanoic acid	Caproic acid	205	− 2	4.83
$CH_2{=}CHCOOH$	Propenoic acid	Acrylic acid	141.6	13	4.25
—COOH	Cyclopentanecarboxylic acid		216	− 7	4.91
—COOH	Cyclohexanecarboxylic acid		232	31	4.88
—COOH	Benzenecarboxylic acid	Benzoic acid	249	122	4.19
HOOC−COOH	Ethanedioic acid	Oxalic acid		190	1.23 4.19[a]
HOOC−CH_2−COOH	Propanedioic acid	Malonic acid		136	2.83 5.69[a]
HOOC−$(CH_2)_2$−COOH	Butanedioic acid	Succinic acid		188	4.16 5.61[a]
HOOC−$(CH_2)_3$−COOH	Pentanedioic acid	Glutaric acid	~300	99	4.31 5.41[a]
HOOC−$(CH_2)_4$−COOH	Hexanedioic acid	Adipic acid	>300	156	4.43 5.41[a]
HOOC−CH=CH−COOH					
cis Isomer	*cis*-Butenedioic acid	Maleic acid		140	1.83 6.07[a]
trans Isomer	*trans*-Butenedioic acid	Fumaric acid		~300	3.03 4.44[a]

[a]These are the pK_a values for loss of the second proton.

For substituted acids, the numbering of the longest chain begins with the acid itself, which is given the number "1." In the cyclic compounds as well, it is the "COOH" that gets the first priority in naming. Polyfunctional acids are generally named as acids (Fig. 19.2).

FIGURE 19.2 Some carboxylic acids and their names.

3-Bromobutanoic acid
(3-bromobutyric acid)

3,5-Dichlorohexanoic acid

***m*-Ethylbenzoic acid**
(3-ethylbenzoic acid)

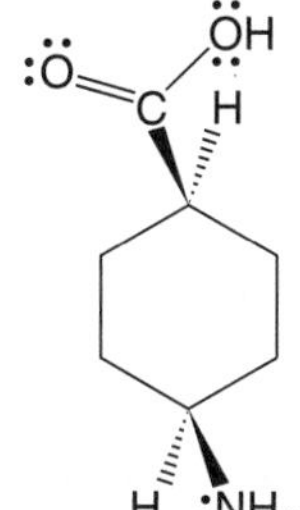

***cis*-4-Aminocyclohexane-carboxylic acid**

PROBLEM 19.2 Name the compounds in Figure 19.3.

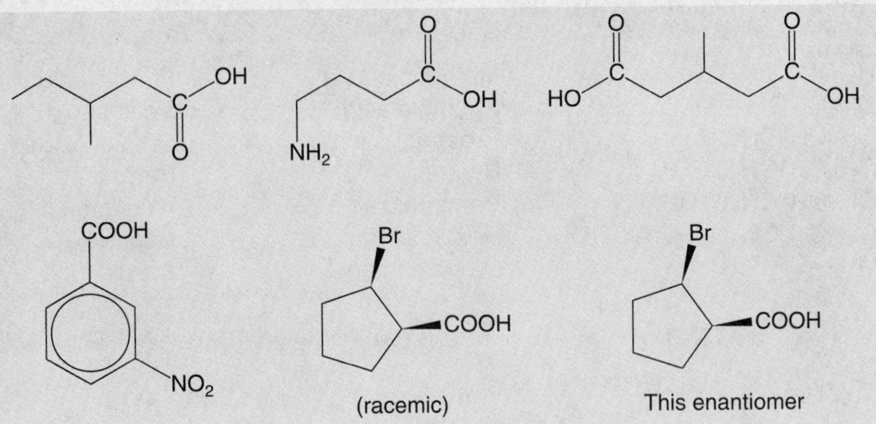

FIGURE 19.3

19.2 STRUCTURE OF CARBOXYLIC ACIDS

Figure 19.4 shows the structure of formic acid and compares it to formaldehyde. The structures of acids are essentially what one would expect by analogy to other carbonyl compounds. The carbonyl carbons are sp^2 hybridized, and the molecules are flat. The strong carbon–oxygen double bonds are appropriately short, about 1.23 Å.

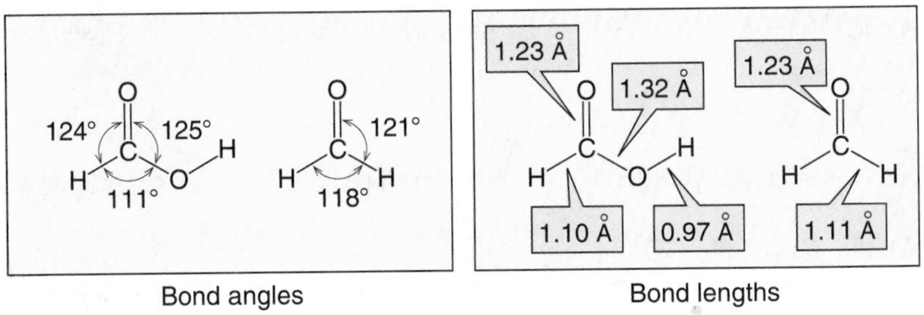

FIGURE 19.4 The structure of formic acid compared to that of formaldehyde.

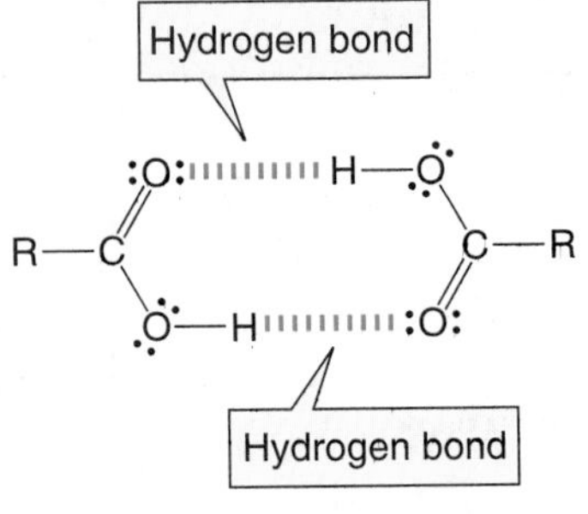

FIGURE 19.5 A dimeric carboxylic acid.

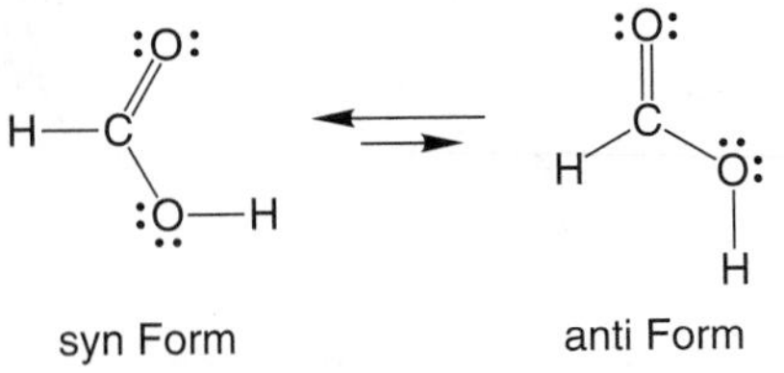

There are two complications with the structural picture. First, in solution the simplest carboxylic acids are largely dimerized. It is possible to form two rather strong hydrogen bonds (~7 kcal/mol, each) in the dimeric forms, and this accounts for the ease of dimer formation. In turn, the ease of dimer formation helps explain the high boiling points (Table 19.1) of carboxylic acids (Fig. 19.5).

Second, there are two energy minima for simple carboxylic acids. The hydroxylic hydrogen can point away (anti) from the carbonyl group, or toward it as in the syn structure of Figure 19.6. These are real energy minima separated in the gas phase by a barrier of about 13 kcal/mol. The syn form for formic acid is more stable by about 6 kcal/mol.

FIGURE 19.6 The two energy minima for a simple carboxylic acid. The syn form is more stable than the anti form.

Explain why the syn form of the carboxylic acid is more stable than the anti form. PROBLEM 19.3

Calculate the equilibrium constant for the equilibrium between syn and anti formic acid at 25 °C. PROBLEM 19.4

Interconversion of the anti and syn forms can occur very easily in water solution. Devise a mechanism not involving simple rotation about a bond. *PROBLEM 19.5

ANSWER

A series of protonations and deprotonations will do the job.

syn

anti

19.3 INFRARED AND NUCLEAR MAGNETIC RESONANCE SPECTRA OF CARBOXYLIC ACIDS

The most prominent features of the IR spectra of carboxylic acids are the strong C=O stretch for the dimeric form at about 1710 cm^{-1} and the strong and very broad band of O–H stretching frequencies centered at about 3100 cm^{-1}. As usual, conjugation shifts the C=O stretch to lower frequency by about 10–20 cm^{-1}. The O–H stretching band is broad for the same reasons that the O–H stretching bands of alcohols are broad (Chapter 17, p. 821). Many hydrogen-bonded dimers and oligomers with different O–H bond strengths are present. Figure 19.7 shows the IR spectrum of acetic acid.

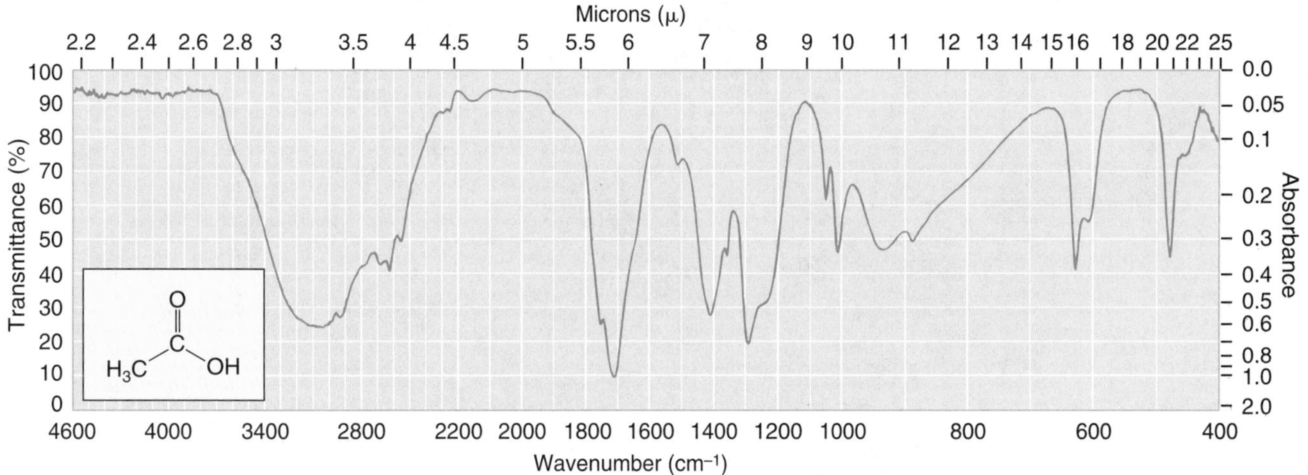

FIGURE 19.7 The IR spectrum of acetic acid.

Hydrogens in the α-position to the carbonyl group of carboxylic acids are deshielded by both the electron-withdrawing nature of the carbonyl group and by local magnetic fields set up by the circulation of electrons in the π bond (Chapter 17, p. 822). They appear in the same region as other hydrogens α to carbonyl groups, δ 2–2.5 ppm.

The carbon of the carboxylic acid group is also strongly deshielded and appears at low field, as do other carbonyl carbons. The position is about 180 ppm, slightly higher field than the chemical shifts of the corresponding carbons of aldehydes and ketones.

19.4 ACIDITY AND BASICITY OF CARBOXYLIC ACIDS

Because carboxylic acids are both acids and bases, we might well expect that a rather diverse chemistry would be found, as with the aldehydes and ketones we saw in Chapters 16 and 18. Carboxylic acids wouldn't be called acids if they were not quite strong Brønsted acids. Table 19.1 gave the pK_a values for some mono- and dicarboxylic acids.

Organic acids (pK_a = 3–5) are much stronger acids than alcohols (pK_a = 15–17) or the related aldehydes or ketones (pK_a = 15–20). Reasons for this increased acidity are not hard to find, but assessing their relative importance is more difficult. One explanation focuses on the formation of a resonance-stabilized **carboxylate anion** after proton loss (Fig. 19.8).

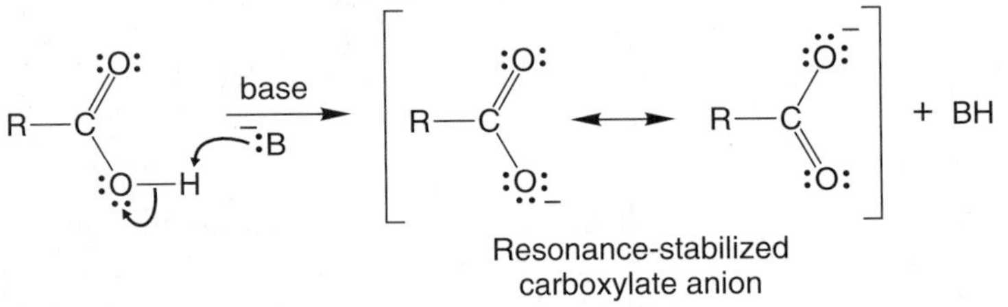

base

Resonance-stabilized carboxylate anion

FIGURE 19.8 The formation of a resonance-stabilized carboxylate anion through removal of a proton from the hydroxyl group of an acid.

Deprotonation of an alcohol gives an oxyanion, an alkoxide, but it is not stabilized by resonance. Loss of a proton from the α-position of other carbonyl compounds does give a resonance-stabilized species, the enolate, but the negative charge is shared between oxygen and carbon, not a pair of highly electronegative oxygen atoms as in the carboxylate anions (Fig. 19.9). It is by no means unreasonable to accept the notion that the formation of the highly stabilized carboxylate is responsible for the high acidity of carboxylic acids.

Carboxylate resonance stabilized

Alkoxide, *not* resonance stabilized

Enolate resonance stabilized

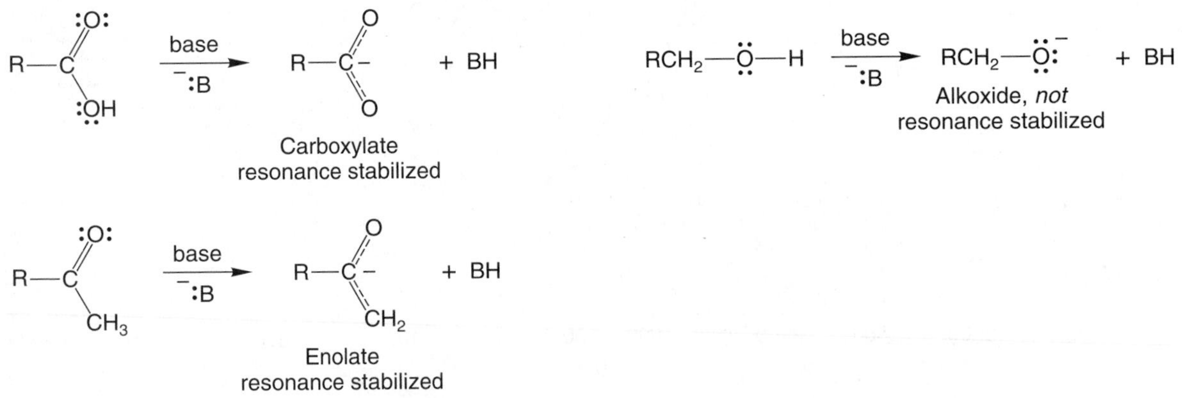

FIGURE 19.9 A comparison of carboxylate, enolate, and alkoxide anions.

Recently, however, this traditional view has been sharply challenged by at least three research groups in both theoretical and experimental arguments. In various ways, the research groups of Andrew Streitwieser (b. 1927) at Berkeley, Darrah Thomas (b. 1932) at Oregon State, and Kenneth Wiberg (b. 1927) at Yale pointed out the importance of the inductive effect of the highly polar carbonyl group. In their view, the increased acidity of acetic acid over ethyl alcohol derives not from resonance stabilization of the product carboxylate anion, but from the electrostatic stabilization afforded the developing negative charge by the adjacent polar carbonyl group in which the carbon bears a partial positive charge. They reckoned that resonance could account for no more than about 15% of the total stabilization (Fig. 19.10).

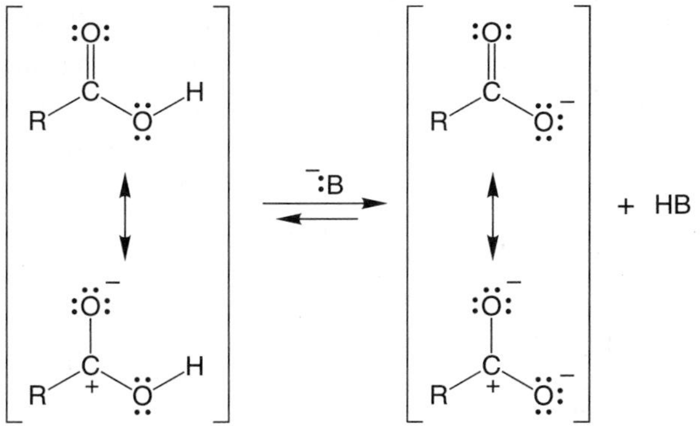

FIGURE 19.10 The highly polar carbonyl group contributes to the stability of the carboxylate anion.

But now we should ask *why* resonance might not be so important. The carboxylate anion is surely delocalized, whereas the alkoxide ion is not. The key point is that the carbonyl group of the acid is *already* so polar that little further delocalization can occur as the anion is formed. The negative charge on the *carbonyl* oxygen isn't developed as the oxygen–hydrogen bond breaks; it is already largely there (Fig. 19.11)!

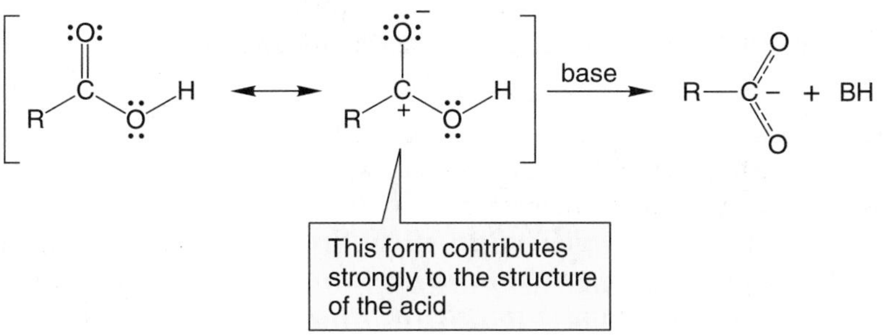

This form contributes strongly to the structure of the acid

FIGURE 19.11 In the resonance picture, charge develops on the carbonyl oxygen as the hydroxyl proton is removed; the new picture points out that the carbonyl oxygen already bears much negative charge before the oxygen–hydrogen bond begins to break.

Both factors surely contribute to the acidity of carboxylic acids; the argument is only over their relative importance.

*PROBLEM 19.6

Explain why fluoroacetic acid has a lower pK_a than acetic acid (2.66 vs. 4.75).

ANSWER

The structure of the anion formed by removal of a proton tells the story. The dipole in the carbon–fluorine bond stabilizes the carboxylate anion **A** and makes removal of the proton relatively easy.

ANSWER (CONTINUED)

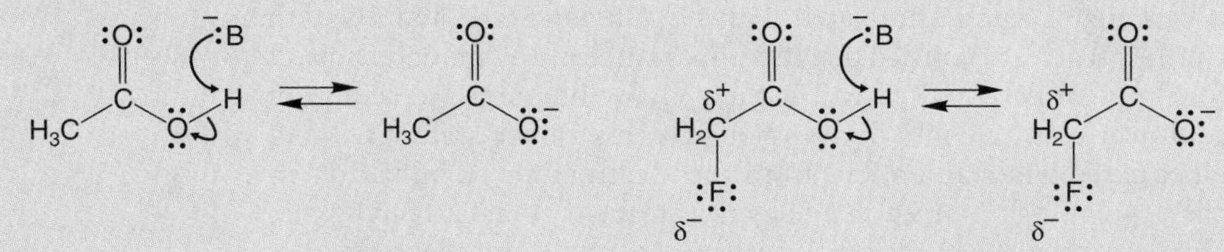

The δ^+ stabilizes the carboxylate anion

A

TABLE 19.2 Acidities of Some Substituted Carboxylic Acids

Acid	pK_a
Acetic	4.75
α-Chloroacetic	2.86
α,α-Dichloroacetic	1.29
α,α,α-Trichloroacetic	0.65
α-Fluoroacetic	2.66
α,α-Difluoroacetic	1.24
α,α,α-Trifluoroacetic	0.23
α-Bromoacetic	2.86
α-Iodoacetic	3.12
α-Nitroacetic	1.68

The stabilization by electron-withdrawing groups pointed out in Problem 19.6 is a general phenomenon. As Table 19.2 shows, acids bearing electron-withdrawing groups are stronger acids than their parent compounds. The further the electron-withdrawing group from the acid, the smaller the effect (Fig. 19.12).

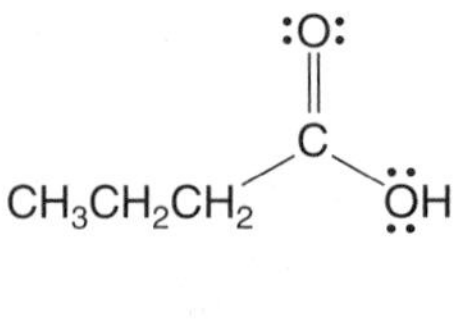

pK_a = 4.81 **Butyric acid**

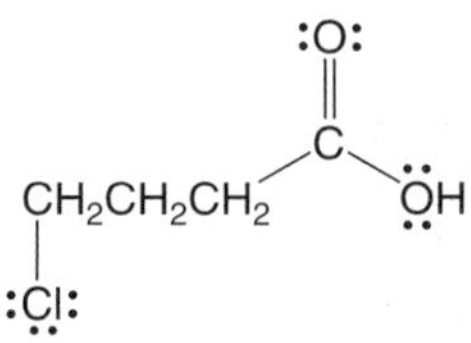

pK_a = 4.52 **4-Chlorobutyric acid**

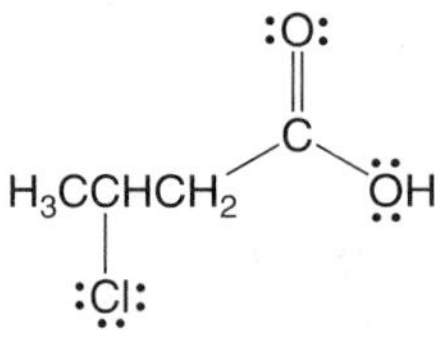

pK_a = 4.06 **3-Chlorobutyric acid**

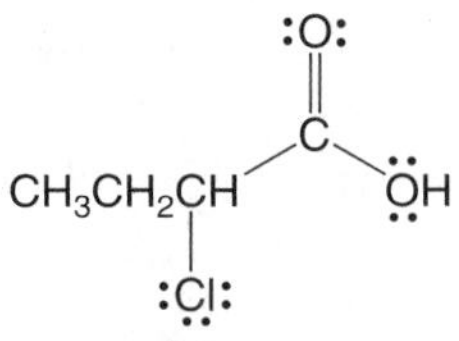

pK_a = 2.84 **2-Chlorobutyric acid**

FIGURE 19.12 The greater the distance (number of bonds) between an electron-withdrawing group, here chlorine, and the point of ionization, the less effect an electron-withdrawing group has.

Carboxylic acids are Lewis acids as well as Brønsted acids. The presence of the carbonyl group ensures that. Remember all the addition reactions of carbonyl groups encountered in Chapters 16 and 18. However, expression of this Lewis acidity is often thwarted because the easiest reaction with base is not addition to the carbonyl group, but removal of the acid's hydroxyl hydrogen to give the carboxylate anion. The anion is far more resistant to addition than an ordinary carbonyl because addition would introduce a *second* negative charge (Fig. 19.13).

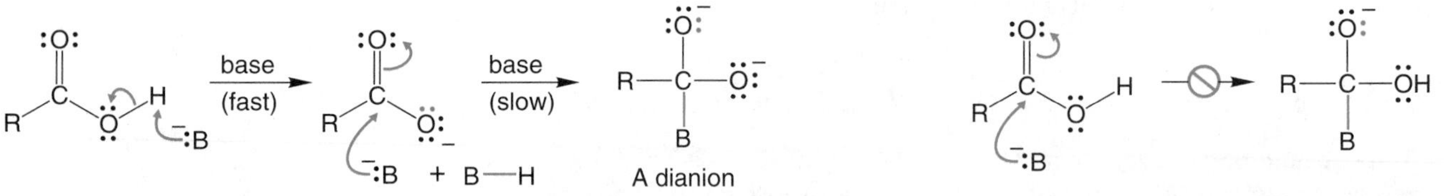

base (fast)

base (slow)

A dianion

FIGURE 19.13 The fastest reaction in base is usually removal of the acidic hydroxyl hydrogen to give the carboxylate anion. This anion is resistant to addition reactions that would introduce a second negative charge.

Nevertheless, some especially strong nucleophiles are able to do this second addition. An example is the organolithium reagent, RLi, and shortly we will see the synthetic consequences of this reaction (Fig. 19.14).

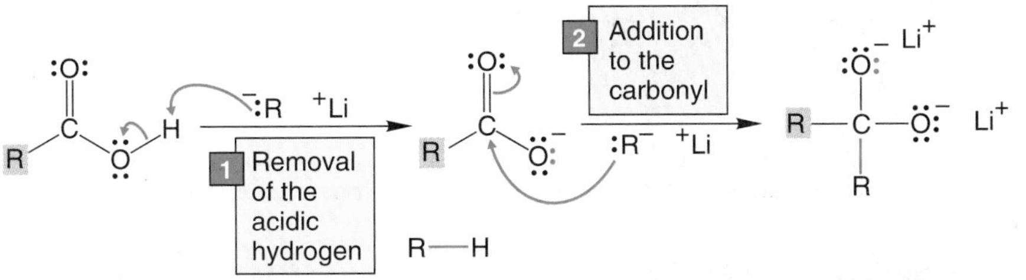

FIGURE 19.14 Some strong nucleophiles can do additions to the carboxylate anion. The alkyllithium reagent is an example.

*PROBLEM 19.7

Anticipate a little. What synthetic use can you see for the reaction in Figure 19.14? *Hint*: What will happen when water is added at the end of the reaction?

ANSWER

When water is added, a hydrate will be formed. You know from Chapter 16 that simple hydrates are generally unstable relative to a carbonyl compound and water.

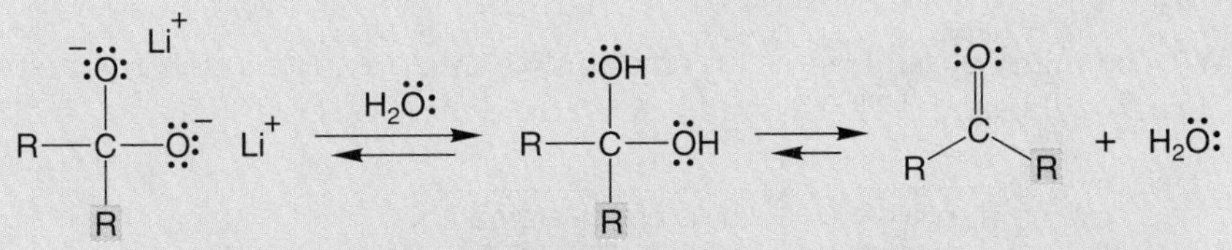

So, this reaction should be a good synthesis of ketones. It is, as we will see later on in the chapter.

Finally, organic acids are Lewis bases, and can react with Lewis and Brønsted acids. The simplest reaction is the protonation of a carboxylic acid. There are two possible sites for protonation, the carbonyl or hydroxyl oxygen (Fig. 19.15). Which will it be?

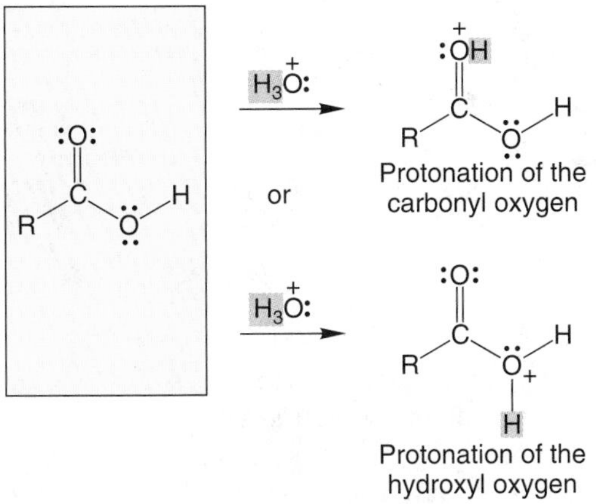

FIGURE 19.15 The two possible sites for protonation of a carboxylic acid.

This question is not hard. The compound in which protonation has taken place at the carbonyl oxygen is resonance stabilized, whereas that in which the hydroxyl oxygen is protonated is not. Moreover, protonation of the hydroxyl oxygen is destabilized by the dipole in the carbon–oxygen double bond that places a partial positive charge on carbon. It will not be energetically favorable to introduce a positive charge adjacent to this already partially positive carbon. The more stable cation, in which the carbonyl oxygen is protonated, will be preferred (Fig. 19.16).

FIGURE 19.16 Delocalization stabilizes the product of protonation of the carbonyl oxygen. The carbon–oxygen dipole destabilizes the product of protonation of the hydroxyl group.

Resonance stabilization of the product of protonation of the carbonyl group

Unstabilized by resonance, destabilized by the C=O dipole

Accordingly, carboxylic acids are more strongly basic at the carbonyl oxygen than at the hydroxyl oxygen. Does this mean that the hydroxyl group is never protonated? Certainly not, but reaction at the more basic site is favored and will be faster.

PROBLEM 19.8 Explain why the thermodynamic stability of the product (the protonated carbonyl in this case) should influence the *rate* (a kinetic parameter) of protonation.

With so many possible sites for reaction we might expect a rich chemistry of carboxylic acids. That would be exactly right (Fig. 19.17).

FIGURE 19.17 The various sites of reactivity for a carboxylic acid.

α-Hydrogen chemistry

Lewis basicity (greater)

Lewis basicity (lesser)

Brønsted acidity

Lewis acidity

Salicylic Acid

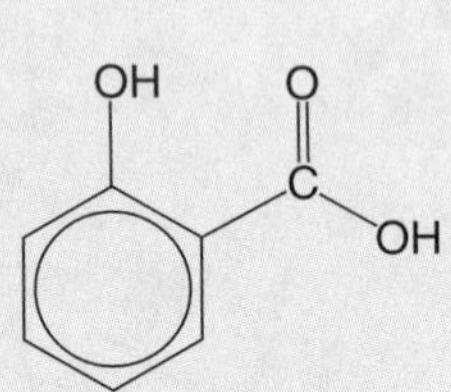

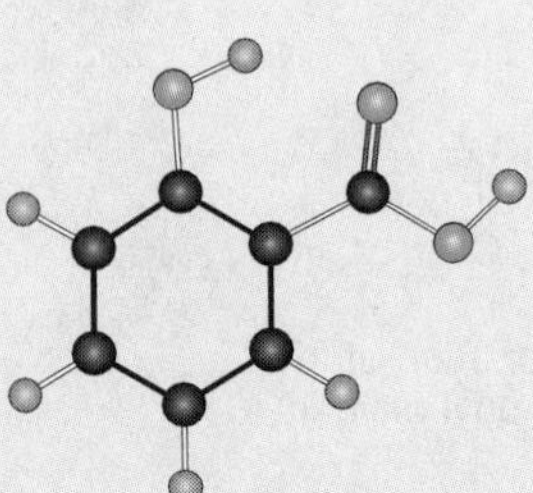

Voodoo lily

The simple aromatic carboxylic acid, salicylic acid, is a plant hormone, and is involved in many of the marvelously complex interactions between plants and animals that makes the study of biology so fascinating. Here's a typical example involving the voodoo lily. The flower of the voodoo lily emits foul odors that attract flies, and the flies are used to transmit pollen from the male reproductive organs to the female organs of another lily. In the late afternoon, salicylic acid triggers the first of two surges of heat, some 10–20 °C above normal. This first surge releases the odor and attracts the flies, which become trapped and coated with pollen. A second heat wave the next morning opens the flower, awakens the pollen-covered flies, which escape till the afternoon when they are attracted to another voodoo lily and deposit the pollen.

Salicylic Acid (Continued)

Salicylic acid is also an analgesic. Indeed, the chewing of willow bark, which contains salicylic acid, has been used to control pain for thousands of years. You are probably most familiar with salicylic acid in its acetylated form, acetyl salicylic acid, or aspirin. Aspirin is a versatile pain killer and has myriad uses. It was synthesized as early as 1853, and its analgesic properties were recognized by a group at Farbenfabriken Baeyer in 1897. It apparently works by inhibiting the production of an enzyme, prostaglandin cyclooxygenase, that catalyzes the synthesis of molecules called prostaglandins. Prostaglandins are active in many ways, one of which is to help transmit pain signals across synapses. No prostaglandins, no signal transmission; no signal transmission, no pain.

19.5 REACTIONS OF CARBOXYLIC ACIDS

19.5a Formation of Esters: Fischer Esterification

An **ester** is a derivative of a carboxylic acid in which the hydrogen of the hydroxyl group has been replaced with an R group. We will examine the nomenclature, properties, and chemistry of esters in detail very soon in Chapter 20, but one of the most important syntheses of these species involves the acid-catalyzed reaction of carboxylic acids with excess alcohol, and must be given here. The reaction is called **Fischer esterification** after the great German chemist Emil Fischer (1852–1919) (Fig. 19.18).

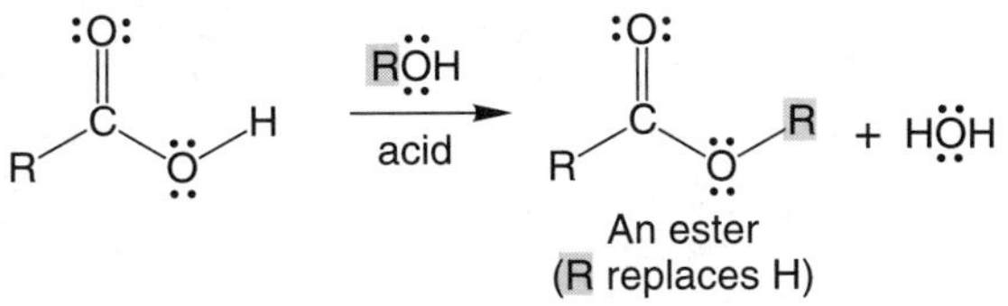

ROH / acid

An ester (R replaces H)

+ HOH

FIGURE 19.18 The acid-catalyzed reaction of a carboxylic acid with an alcohol—Fischer esterification.

We can write the mechanism for this reaction rather easily, as its important steps are quite analogous to reactions of other carbonyl-containing compounds. The first step is surely protonation of the carbonyl group, as it is in the reaction of an aldehyde or ketone in acid (Fig. 19.19).

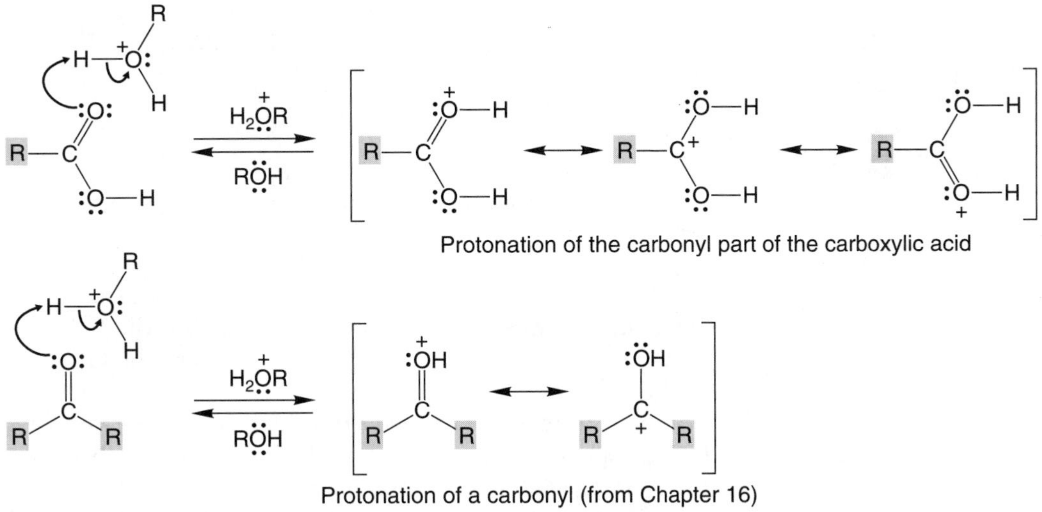

$H_2\overset{+}{O}R$ / ROH

Protonation of the carbonyl part of the carboxylic acid

$H_2\overset{+}{O}R$ / ROH

Protonation of a carbonyl (from Chapter 16)

FIGURE 19.19 The first step in the mechanism is protonation of the carbonyl oxygen to give a resonance-stabilized intermediate.

PROBLEM 19.9 Explain why treatment of acetic acid with $^{18}\overset{+}{O}H_3/^{18}OH_2$ leads to exchange of both oxygen atoms (Fig. 19.20).

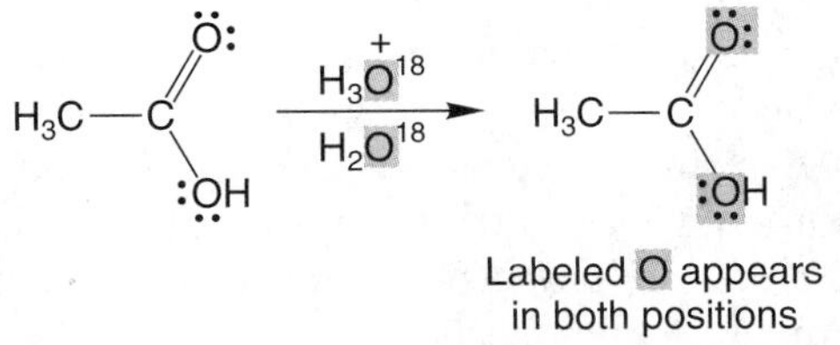

Labeled O appears in both positions

FIGURE 19.20

The second step in Fischer esterification is also completely analogous to aldehyde or ketone chemistry. A molecule of alcohol adds to the protonated carbonyl group. For a ketone, this addition merely gives the hemiacetal; for the acid, a somewhat more complicated intermediate with one more hydroxyl group is formed. In both reactions, however, a planar, sp^2 hybridized species has been converted into a tetrahedral intermediate (Fig. 19.21).

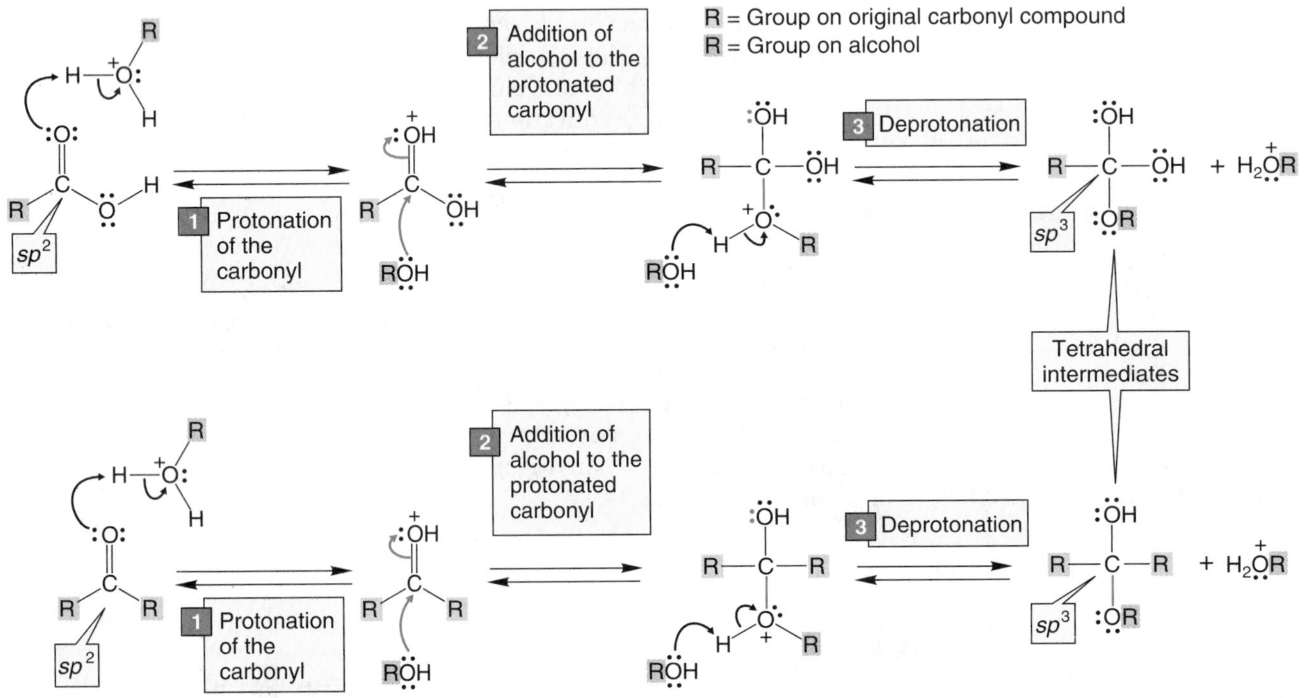

FIGURE 19.21 The second step in Fischer esterification involves addition of the alcohol to the protonated acid to give a tetrahedral intermediate.

Now what can these tetrahedral intermediates do in acid? The hemiacetal has only two options. It can reprotonate the OR group and then revert to the starting ketone, or it can protonate the OH group and proceed to the full acetal (Fig. 19.22).

The intermediate formed from the acid is more versatile, as it has two OH groups. Protonation of the OR simply takes us back along the path to the starting acid. Protonation of one of the OH groups leads to an intermediate that can lose water, deprotonate, and give the ester. In excess alcohol, the reaction is driven in this direction (Fig. 19.23).

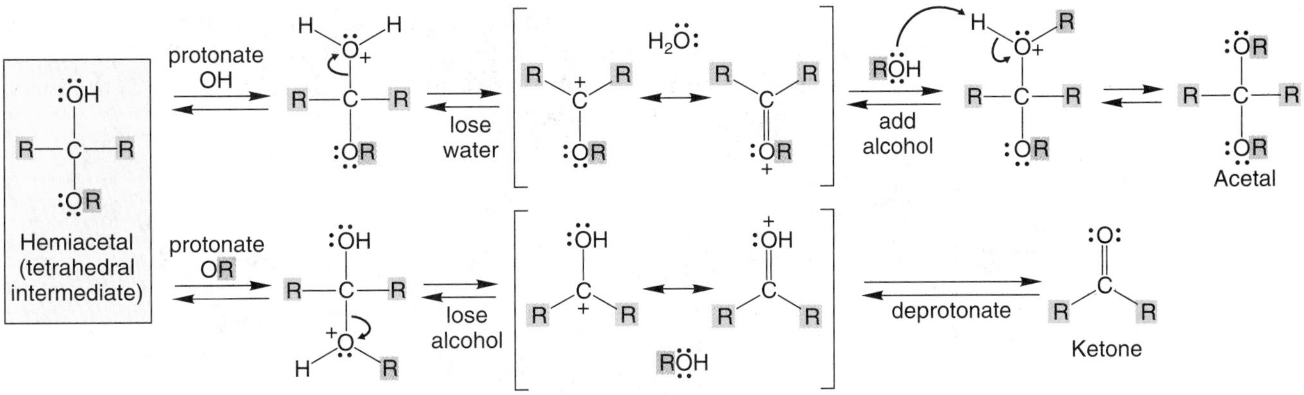

FIGURE 19.22 For a ketone, there are two further possibilities: protonation of the OH leads to water loss and formation of the full acetal, whereas protonation of the OR leads back to the original starting ketone.

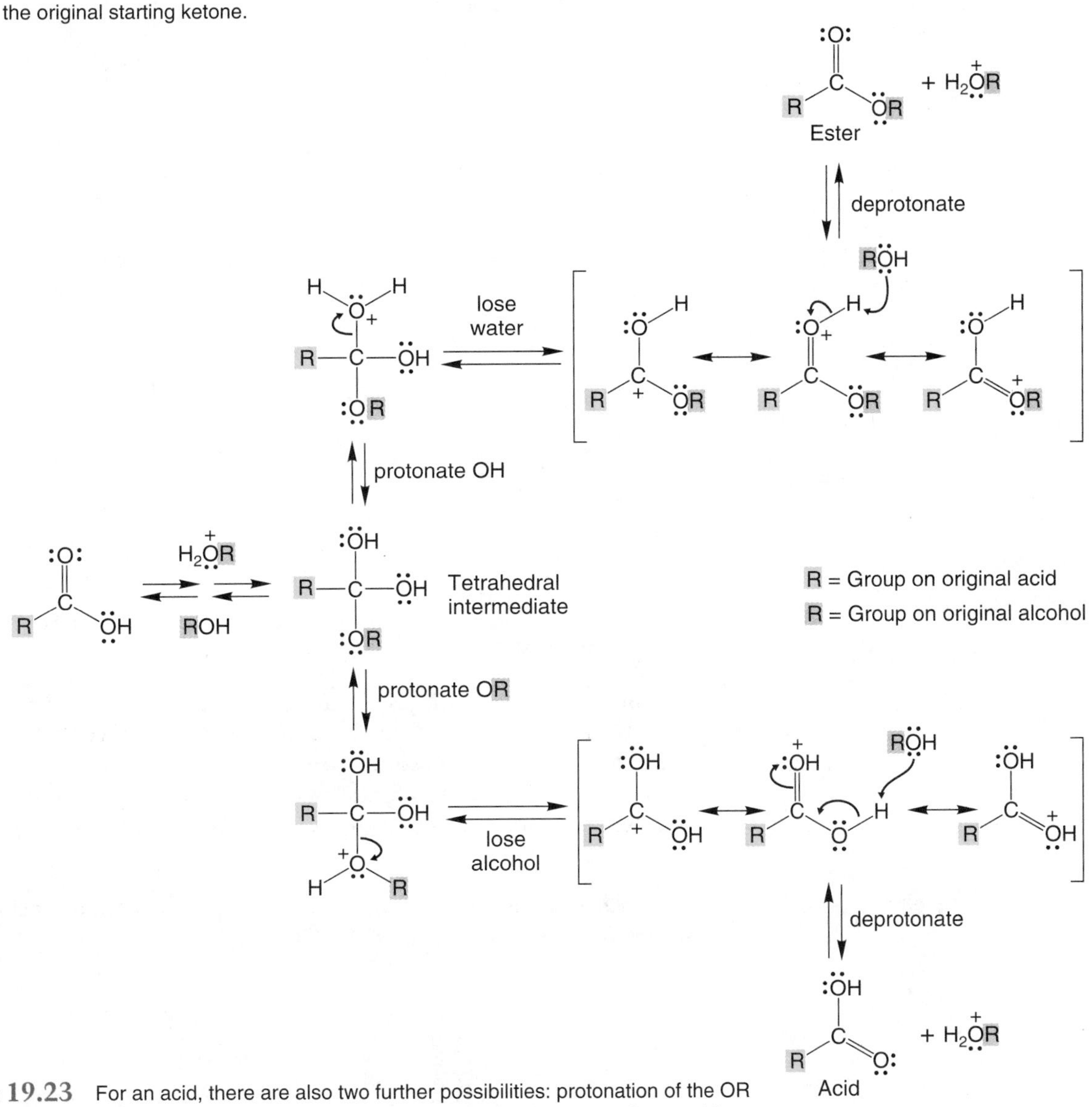

FIGURE 19.23 For an acid, there are also two further possibilities: protonation of the OR leads back to the starting acid, whereas protonation of the OH leads to water loss and formation of the ester.

The mechanism for Fischer esterification is repeated below in Figure 19.24. Note the symmetry of the process—it is symmetrical about the tetrahedral intermediate **A**. Note also the similarity of **B** and **C**, the intermediates on either side of **A**.

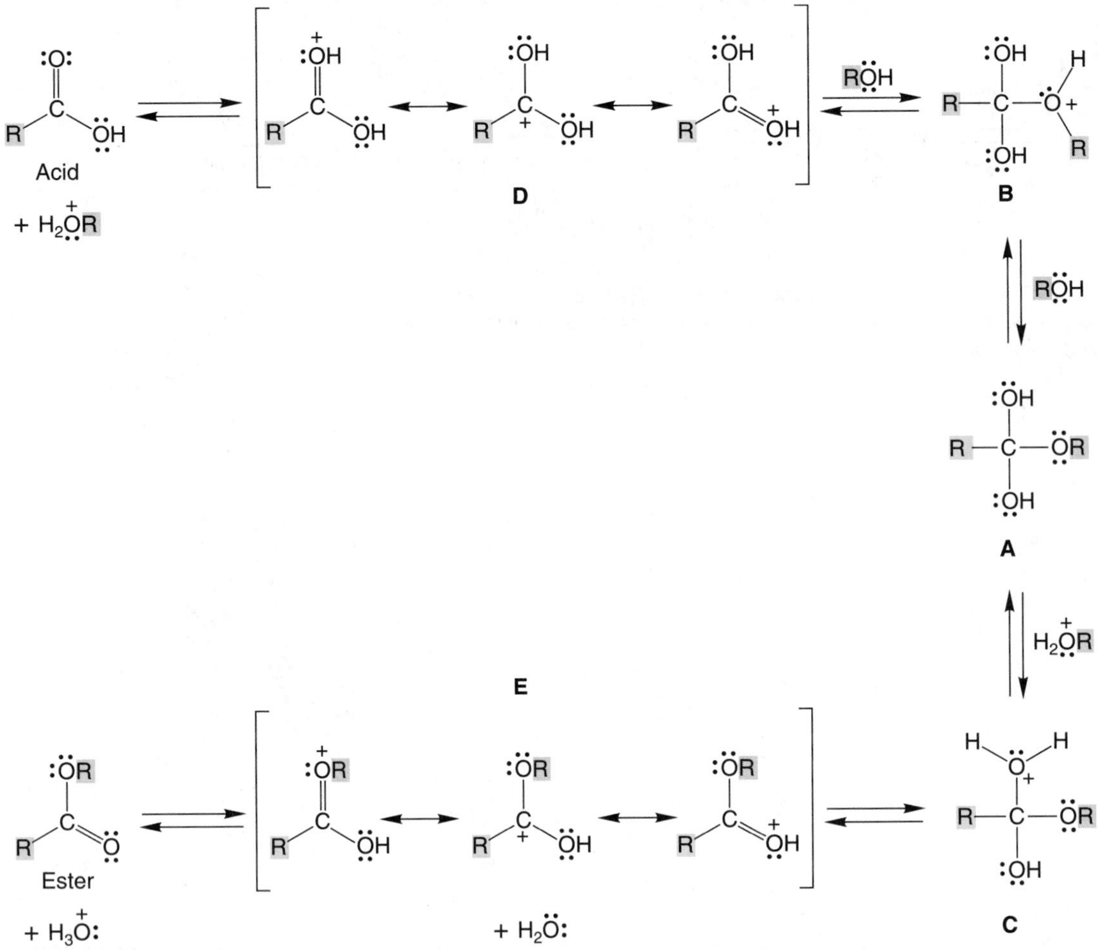

FIGURE 19.24 The full mechanism for Fischer esterification as well as for the reverse reaction, hydrolysis of an ester to a carboxylic acid.

In one of these intermediates, **B**, the OR is protonated. This pathway leads back to the starting acid. In the other, **C**, it is an OH that is protonated. This path leads to the ester.

Flanking **B** and **C** are **D** and **E**, the two resonance-stabilized intermediates resulting from protonation of the acid or ester carbonyl.

PROBLEM 19.10 Write the mechanism for the acid-catalyzed formation of an organic acid from the reaction of an ester with excess water (Fig. 19.25).

FIGURE 19.25

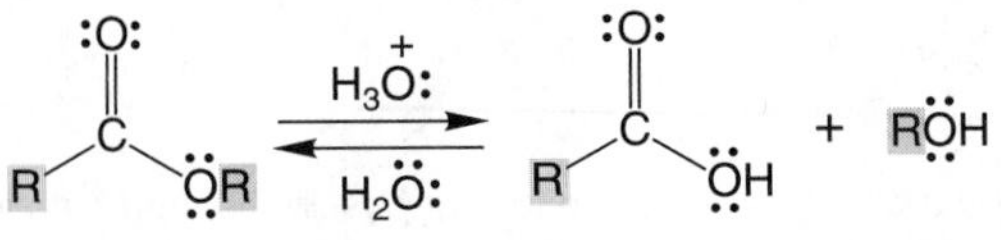

PROBLEM 19.11

Ester formation is not new to you. Give products and mechanisms for the following reactions, which appear in earlier chapters (Fig. 19.26).

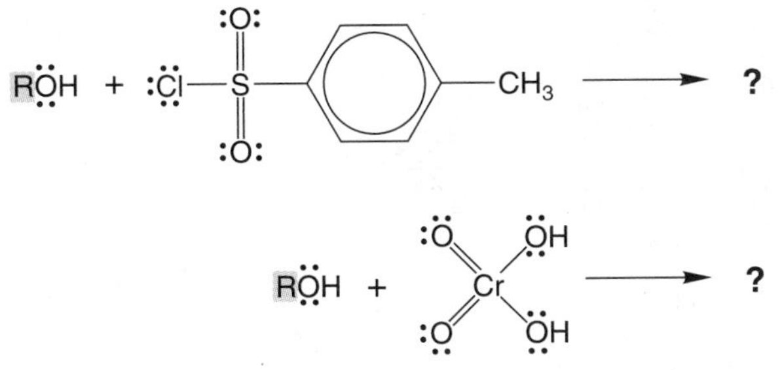

FIGURE 19.26

We have spent a lot of time and space on Fischer esterification and its reverse, acid hydrolysis. We did this because this reaction is prototypal, an anchor for further discussion. As for other anchor reactions (for example, the S_N2 reaction and addition of Br_2 to alkenes), knowing this reaction, quite literally backward and forward, allows you to generalize and to organize the many following reactions that show a similar pattern, although the details may be different. Fischer esterification is definitely a reaction to know well.

Note here also what makes the chemistry of aldehydes and ketones different from that of acids (or esters). In the aldehydes and ketones there is no leaving group present, whereas in an acid (or ester) there is. The chemistries of aldehydes and ketones on the one hand, and acids and esters on the other, diverge only after the addition to the carbonyl. In the acids and esters there is a group that can be lost, providing a route to other molecules; in the aldehydes and ketones there isn't, which means that the acids and esters can undergo an **addition–elimination reaction** that the aldehydes and ketones can't (Fig. 19.27).

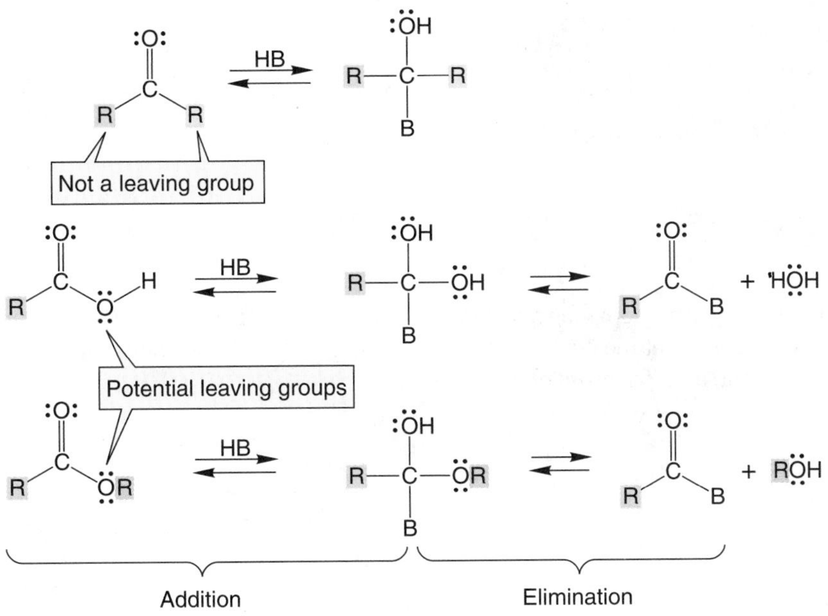

FIGURE 19.27 Unlike aldehydes and ketones, acids and esters bear OH and OR groups that are potential leaving groups.

When Fischer esterification takes place in intramolecular fashion, the result is a cyclic ester, known as a **lactone**. As usual, the relatively unstrained five- and six-membered rings are formed more easily than other sized rings. Figure 19.28 gives an example of lactone formation.*

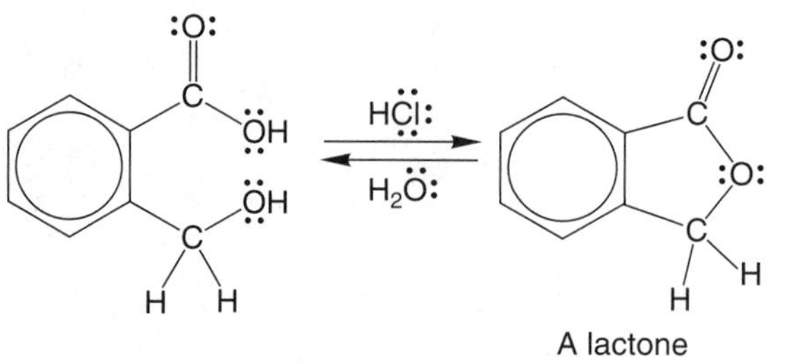

HCl: / H_2O: — A lactone

FIGURE 19.28 Lactone formation involves an intramolecular Fischer esterification.

Lactones are commonly named in reference to the acid containing the same number of carbon atoms. Thus a butanolide has four carbons, a pentanolide has five, and so on. A Greek letter is used to designate the size of the ring. An α-lactone has one carbon joining the oxygen and carbonyl carbon, a β-lactone has two carbons, a γ-lactone three carbons, and so on. In the IUPAC system, lactones are named as "oxacycloalkanones" (Fig. 19.29).

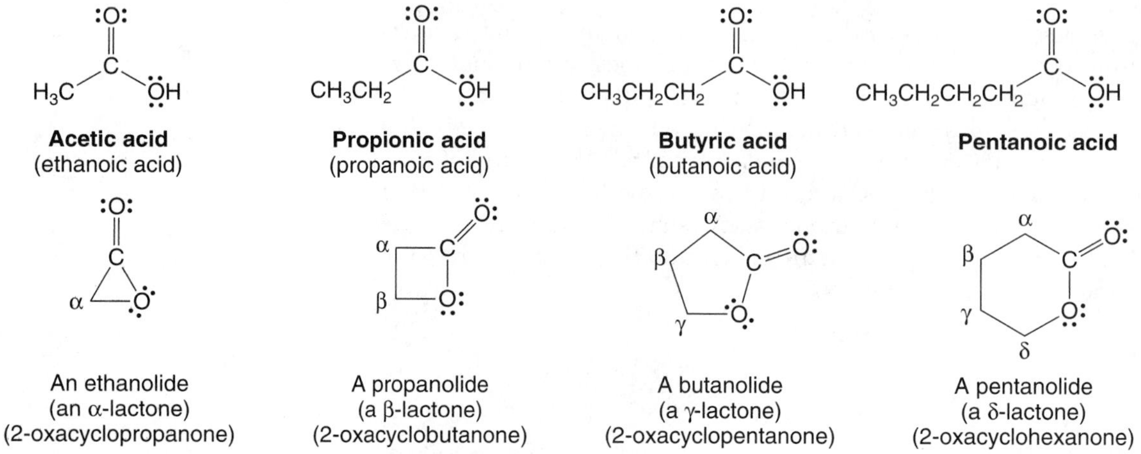

Acetic acid (ethanoic acid) — An ethanolide (an α-lactone) (2-oxacyclopropanone)

Propionic acid (propanoic acid) — A propanolide (a β-lactone) (2-oxacyclobutanone)

Butyric acid (butanoic acid) — A butanolide (a γ-lactone) (2-oxacyclopentanone)

Pentanoic acid — A pentanolide (a δ-lactone) (2-oxacyclohexanone)

FIGURE 19.29 A naming protocol for lactones.

PROBLEM 19.12 Write a detailed mechanism for the reaction in Figure 19.28.

PROBLEM 19.13 Treatment of the compound in Figure 19.30 with I_2/KI and Na_2CO_3 in water leads to a neutral molecule of the formula $C_8H_{11}IO_2$. Give a structure for the product and a mechanism for its formation.

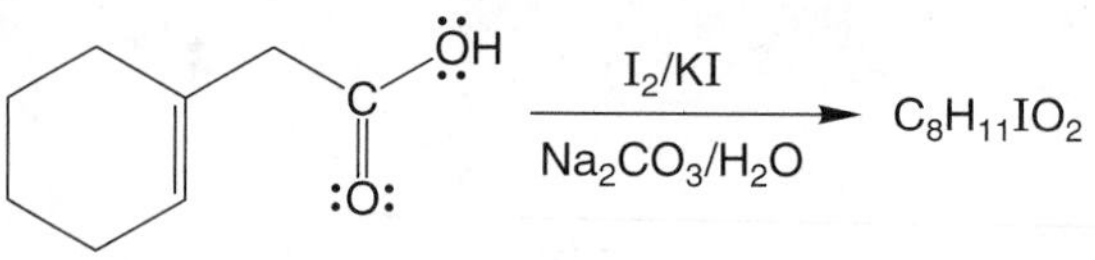

I_2/KI, Na_2CO_3/H_2O → $C_8H_{11}IO_2$

FIGURE 19.30

*Lactones figure in the first of Magid's Rules, which says, "All neutral products are lactones." This statement refers to a kind of problem now out of fashion. The word "neutral" was a clue that a carboxylic *acid* was no longer present. Problem 19.13 gives an example.

But we should wait a moment. Other mechanisms for Fischer esterification can be imagined, and it is not fair to dismiss them without evidence. For example, why not use the protonated alcohol as the Lewis acid rather than the protonated carboxylic acid (Fig. 19.24)? Displacement of water using the acid as nucleophile could give the ester (Fig. 19.31).

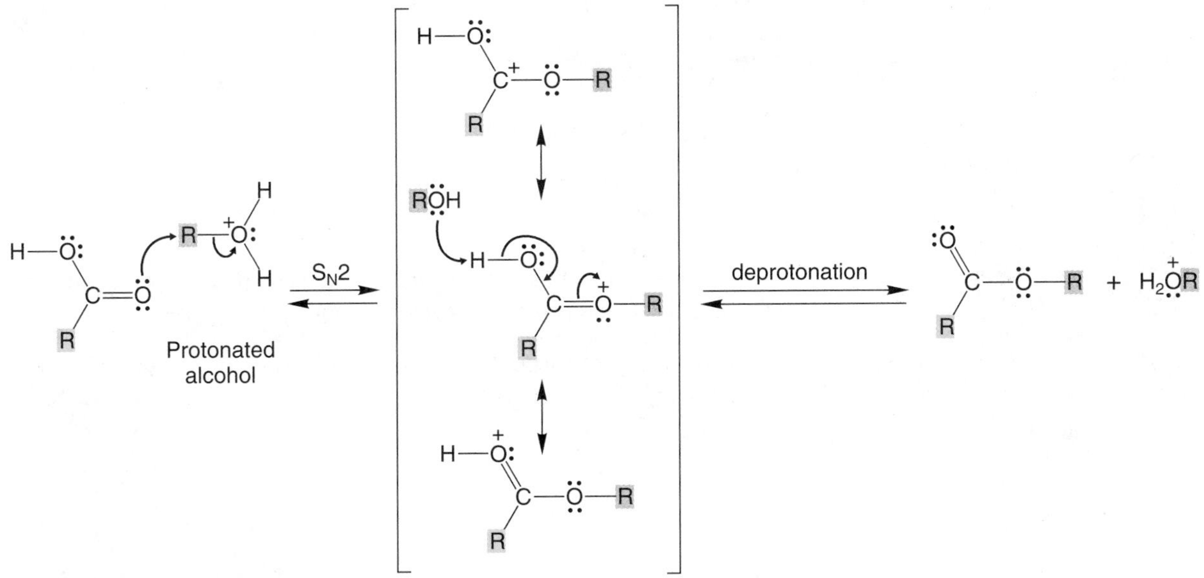

FIGURE **19.31** A different potential mechanism for Fischer esterification. Here the acid is the nucleophile and the protonated alcohol is the Lewis acid. Watch out for conventions! The figure shows proton removal from one resonance form, which is purely for convenience sake—the resonance form has no individual existence.

The key difference between the mechanism for Fischer esterification in Figure 19.24 (which is correct) and this hypothetical one in Figure 19.31 is the site of carbon–oxygen bond breaking. In the first (correct) mechanism, it is a carbon–oxygen bond in the *acid* that is broken; in the new (incorrect) mechanism, it is the carbon–oxygen bond in the *alcohol* that breaks (Fig. 19.32).

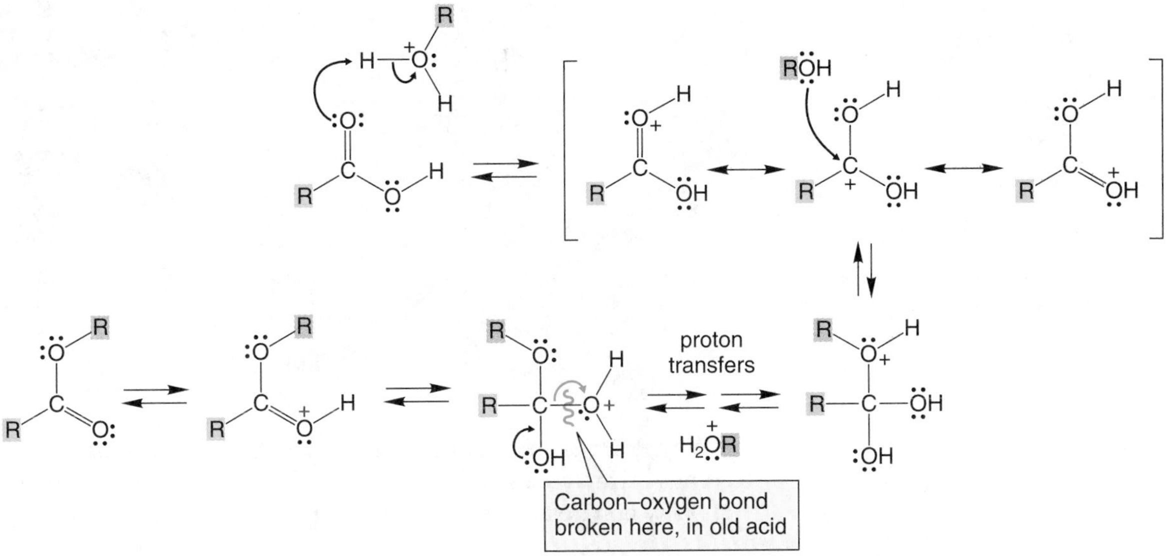

FIGURE **19.32** The two possible mechanisms for Fischer esterification compared (one here, one at the top of the next page). Notice the difference in the carbon–oxygen bonds that are being broken in the two mechanisms.

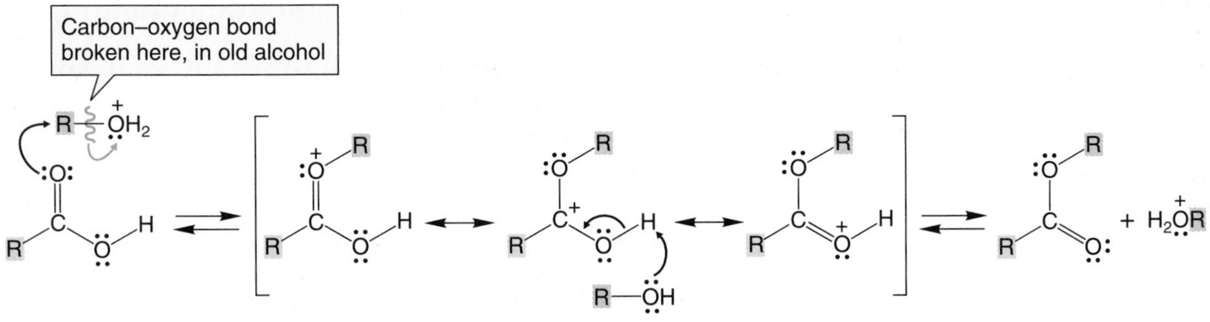

Carbon–oxygen bond broken here, in old alcohol

FIGURE 19.32 (CONTINUED)

*PROBLEM 19.14 Design an experiment to tell these two mechanisms (Fig. 19.32) apart. Assume access to any labeled compounds you may need.

ANSWER Using labeled alcohol will do the trick. If the mechanism involved carbon–oxygen bond breaking in the alcohol (it does not), the use of ^{18}O-labeled alcohol would not lead to ^{18}O incorporation in the product ester.

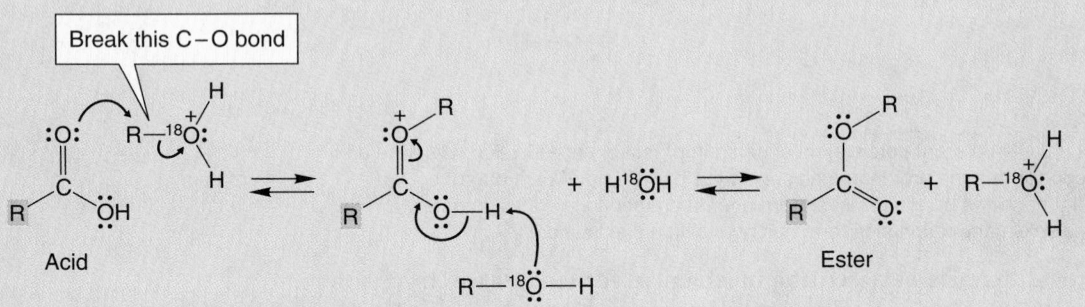

Break this C–O bond

Acid

Ester

However, if the mechanism involves breaking of a carbon–oxygen bond in the original acid (it does) ^{18}O will be incorporated.

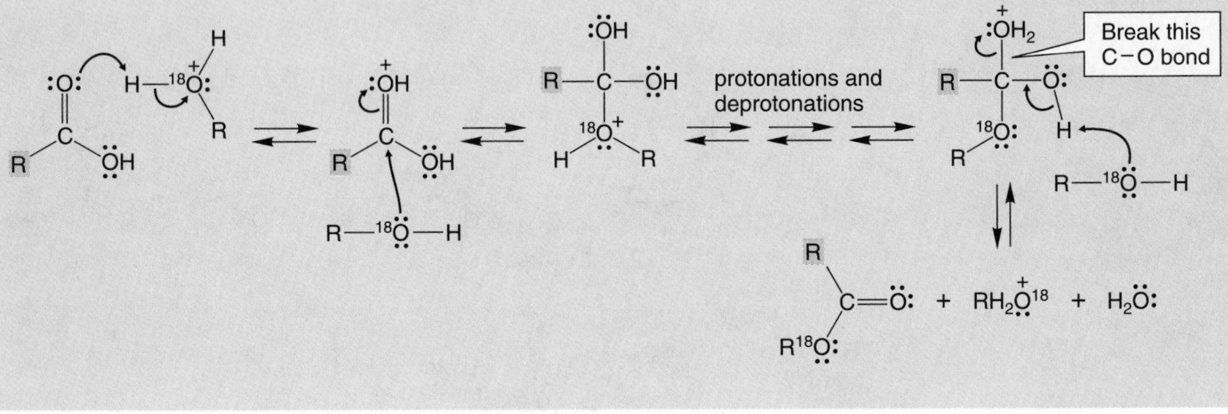

protonations and deprotonations

Break this C–O bond

Why don't esters react further under these conditions? Aldehydes and ketones form acetals when treated with an acid catalyst and excess alcohol. Why don't esters go on to form **ortho esters** (Fig. 19.33)?

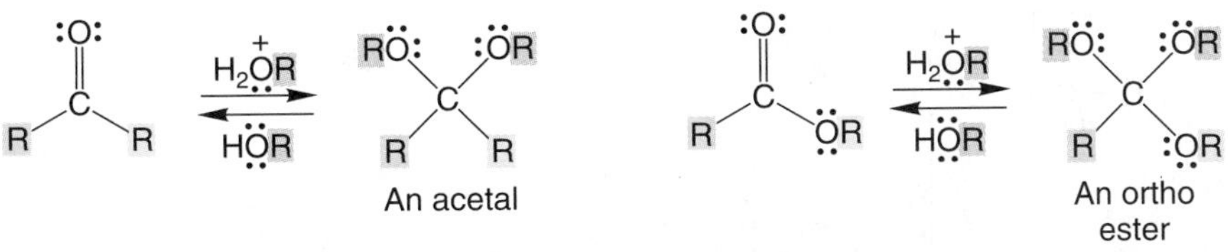

FIGURE 19.33 The conversion of an ester into an ortho ester is analogous to the formation of acetals from aldehydes and ketones.

Figure 19.34 outlines a reasonable mechanism for the conversion of an ester into an ortho ester. Ortho esters are known compounds, and there is nothing fundamentally wrong with the steps outlined in Figure 19.34.

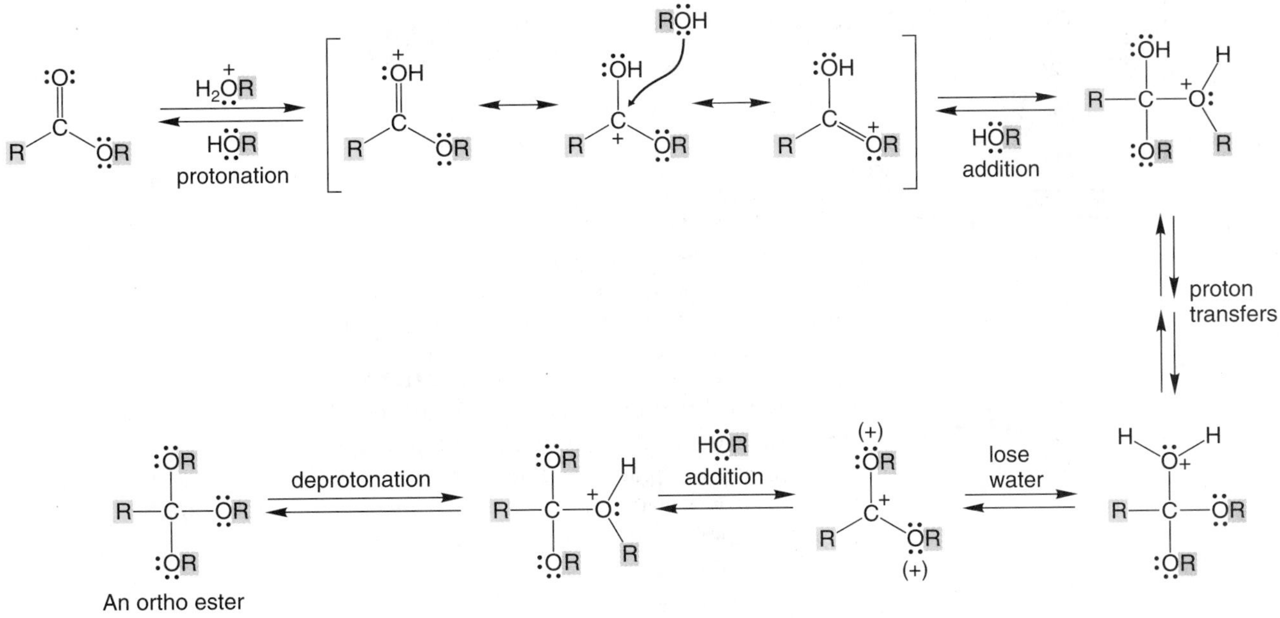

FIGURE 19.34 A reasonable mechanism for the conversion of an ester into an ortho ester. The steps are quite analogous to those for acetal formation.

However, ortho esters cannot be made this way. The problem is one of thermodynamics. The sequence of Figure 19.34 is a series of equilibria, and at equilibrium it is the ester that is greatly favored, not the ortho ester. Note that the starting material in this reaction, the ester, is stabilized by resonance and the product ortho ester is not. The ortho ester is analogous to the thermodynamically unstable intermediate **A** in Figure 19.24. Aldehydes and ketones lack this ester resonance, and are therefore less stable, and more prone to further addition reactions. Figure 19.35 makes this point with a partial Energy versus Reaction progress diagram. The acetal and the ortho ester are roughly equivalent in energy, but the ester lies well below the ketone. The result is that acetals are thermodynamically accessible from ketones, but ortho esters are disfavored relative to esters.

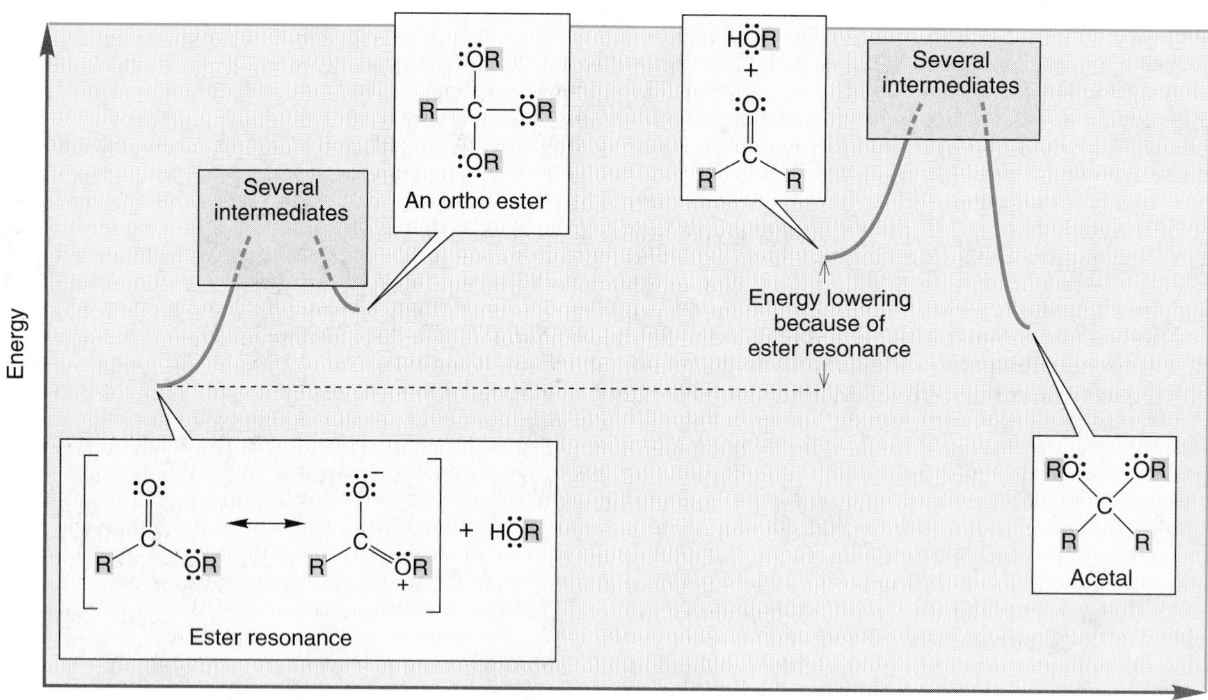

FIGURE 19.35 A truncated Energy versus Reaction progress diagram comparing acetal formation from a ketone with ortho ester formation from an ester.

There are other mechanisms for ester formation, and some of them do not involve breaking a carbon–oxygen bond in the starting acid. Sometimes the carboxylate anion is nucleophilic enough to act as the displacing agent in an S_N2 reaction. The partner in the displacement reaction must be especially reactive. In practice, this means that a primary halide or even more reactive species must be used (Fig. 19.36).

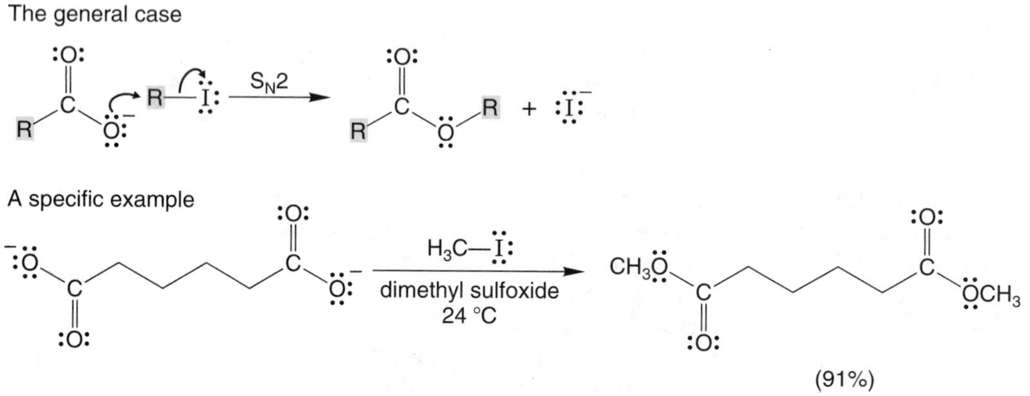

The general case

S_N2

A specific example

$H_3C—I$

dimethyl sulfoxide

24 °C

(91%)

FIGURE 19.36 Carboxylate anions can be used to displace very reactive halides in an ester-forming S_N2 reaction.

Another reaction in which the carboxylate acts as a nucleophile is the ester-forming reaction of acids with diazomethane, CH_2N_2. Diazomethane, though easily made, is quite toxic and a powerful explosive. So, this method is generally used only when small amounts of the methyl esters are needed (Fig. 19.37).

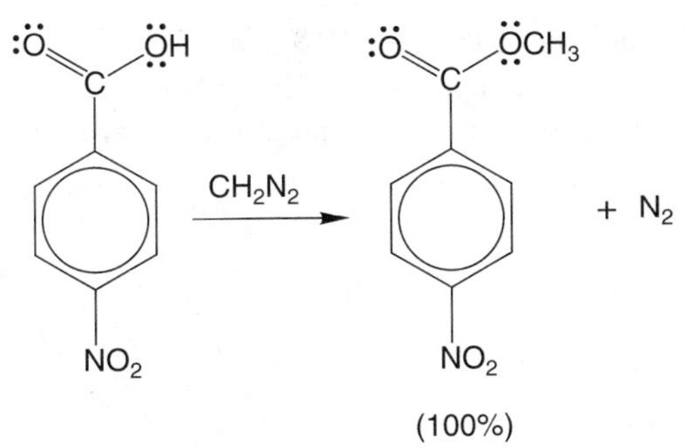

FIGURE 19.37 Diazomethane can be used to make methyl esters from acids in yields as high as 100%.

The first step in this reaction is deprotonation of the acid by the basic carbon of the diazo compound. The result is a carboxylate anion and a diazonium ion, an extraordinarily reactive alkylating agent (Fig. 19.38).

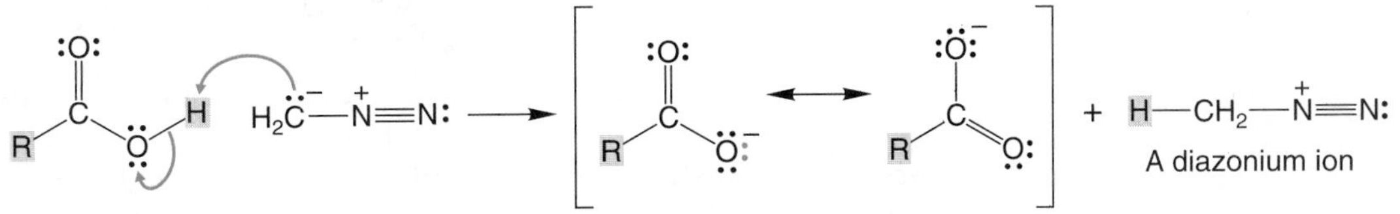

FIGURE 19.38 In the first step, diazomethane acts as a Brønsted base and deprotonates the acid.

Now the alkylation of Figure 19.39 takes place, and nitrogen, a superb leaving group, is displaced to give the methyl ester. In practice, this method of esterification is restricted to the synthesis of methyl esters. The reason is the accessibility of diazomethane and the relative difficulty of making other simple diazo compounds.

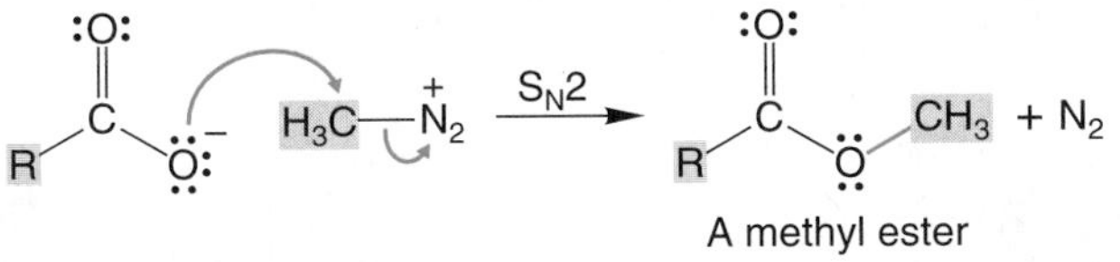

FIGURE 19.39 In the second step, the carboxylate anion displaces nitrogen to give the methyl ester.

19.5b Formation of Amides

Although carboxylic acids do react with amines to form **amides**,* in general,this is not a very useful reaction. The dominant reaction between the basic amine and the carboxylic acid must be proton transfer to give the ammonium salt of the carboxylic acid. Heating of the salts has been used as a source of amides (Fig. 19.40).

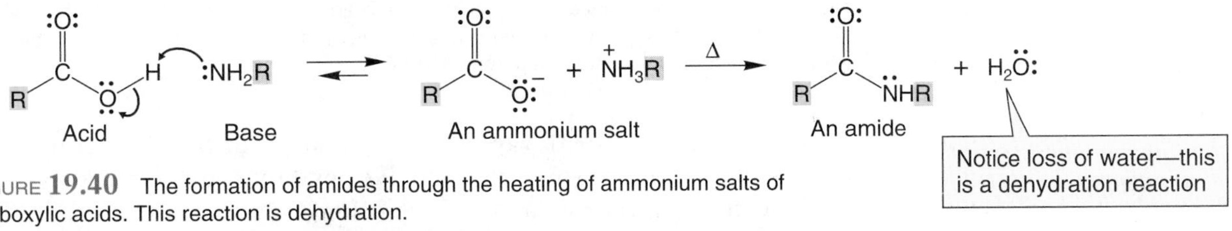

FIGURE 19.40 The formation of amides through the heating of ammonium salts of carboxylic acids. This reaction is dehydration.

Notice loss of water—this is a dehydration reaction

*There is potential for confusion here. There are two kinds of amides, the carbonyl derivative shown here (R–CO–NHR) and the negatively charged ions, $^{-}NR_2$.

Note that the reaction of Figure 19.40 is a dehydration. Water is lost in the reaction. Several dehydrating agents have been developed that greatly faciliate amide formation. In one, **dicyclohexylcarbodiimide** (**DCC**) is used (Fig. 19.41).

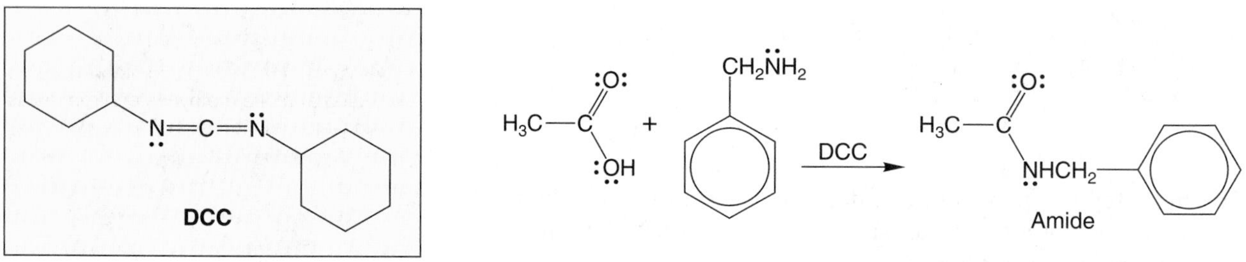

DCC

H_3C—C(=O)OH + $C_6H_5CH_2NH_2$ —DCC→ H_3C—C(=O)$NHCH_2$—C_6H_5

Amide

FIGURE 19.41 The use of DCC, a dehydrating agent, to produce amides.

The strategy is to convert the poor leaving group, OH, into a better one. We used this technique extensively in Chapter 17 (p. 828), when we discussed the transformation of alcohols. Here, the acid first adds to one carbon–nitrogen double bond of DCC, accomplishing the transformation of the leaving group (Fig. 19.42).

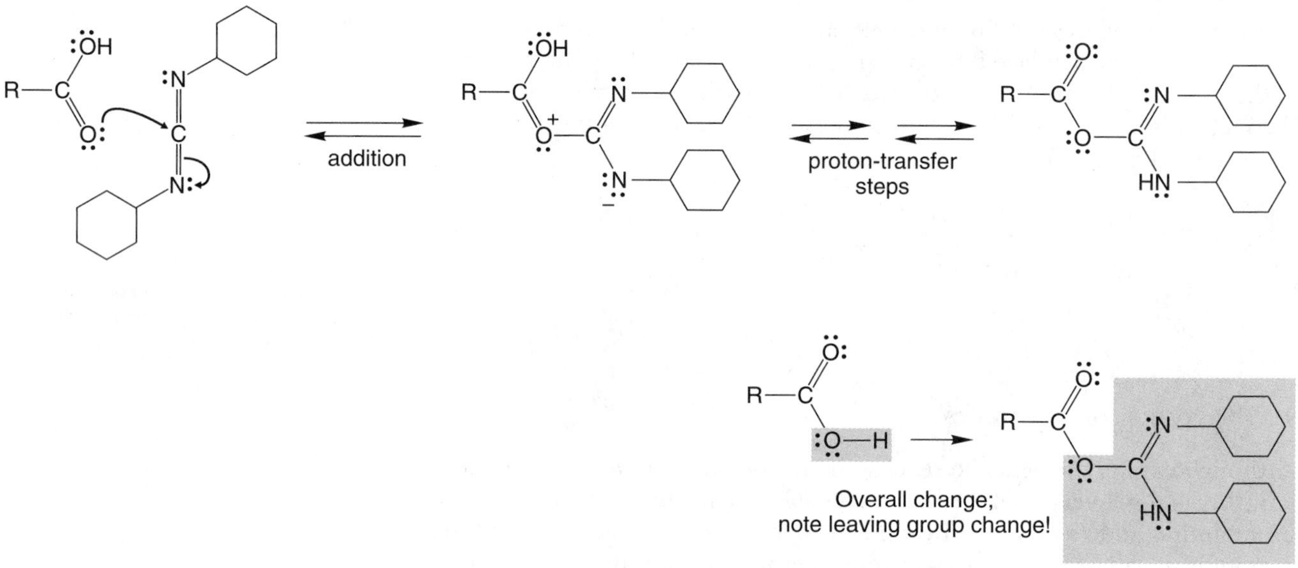

addition

proton-transfer steps

Overall change; note leaving group change!

FIGURE 19.42 The first step in the mechanism is addition of the acid to one carbon–nitrogen double bond. Proton transfers complete the process. The overall transformation involves a change of leaving group from OH to OR.

Two mechanistic pathways are now possible. In the simpler one, the amine adds to the carbon–oxygen double bond to give a tetrahedral intermediate that expels a relatively stable ion to generate the amide (Fig. 19.43).

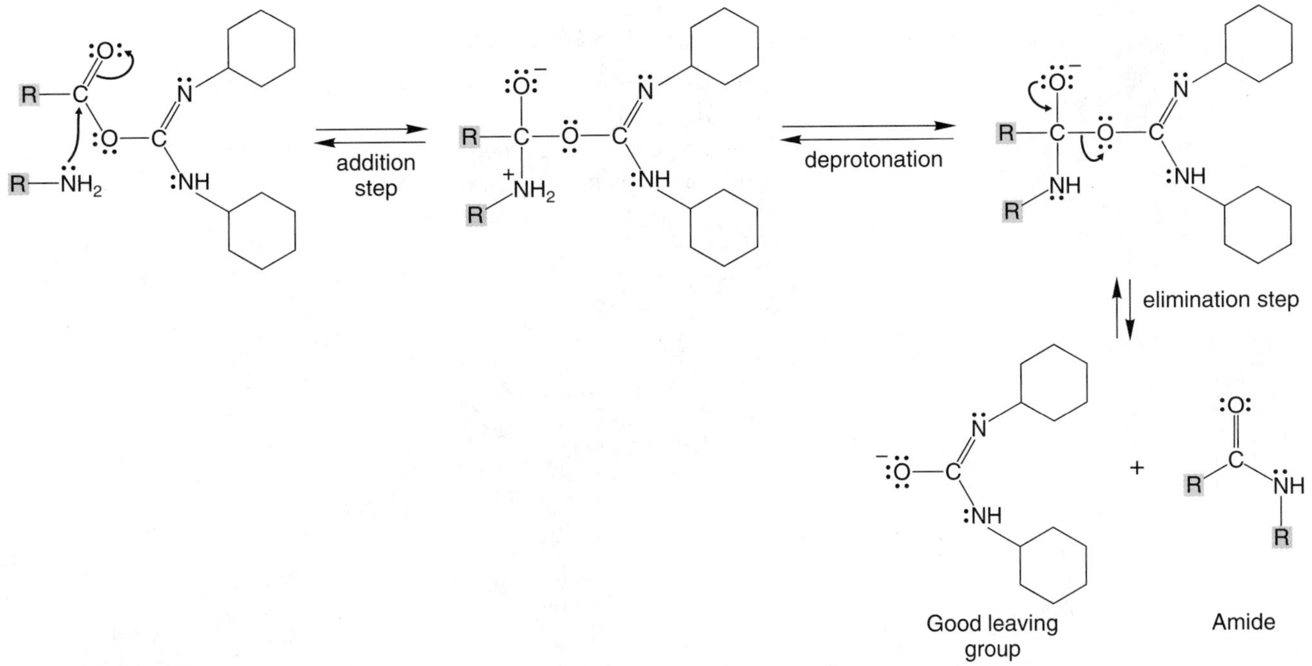

FIGURE 19.43 Addition of the amide to the newly formed intermediate leads to a tetrahedral intermediate that can decompose to give the amide. Can you see why the leaving group is so good in this reaction?

In another, more complicated mechanism, the initially formed intermediate reacts not with the amine, as in Figure 19.43, but with another molecule of acid. The result is an anhydride. The anhydride is the actual reagent that reacts with the amine to give the product amide (Fig. 19.44).

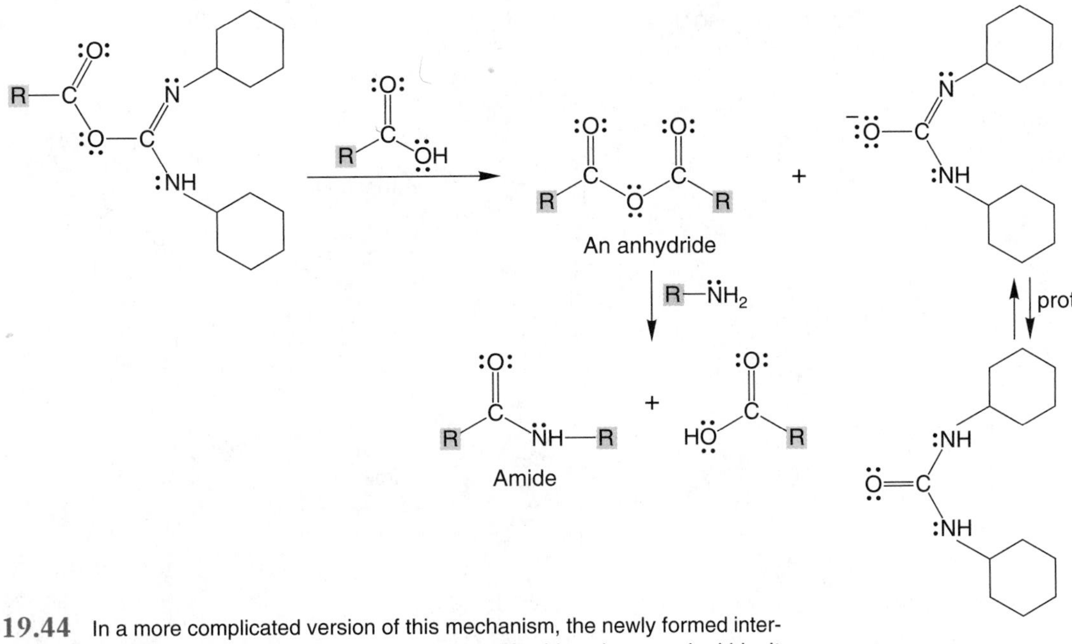

FIGURE 19.44 In a more complicated version of this mechanism, the newly formed intermediate reacts not with the amine but with another molecule of acid to give an anhydride. It is the anhydride that then goes on to react with the amine to give the amide.

*PROBLEM **19.15** Sketch mechanisms for the reactions in Figure 19.44.

ANSWER Both of these steps are standard, addition–elimination reactions. In the first step, the DCC-activated acid (or the protonated acid) is attacked by the carboxylic acid. Loss of the good leaving group leads to the anhydride. Anion **A** is first protonated, and then converted into dicyclohexylurea, DCU.

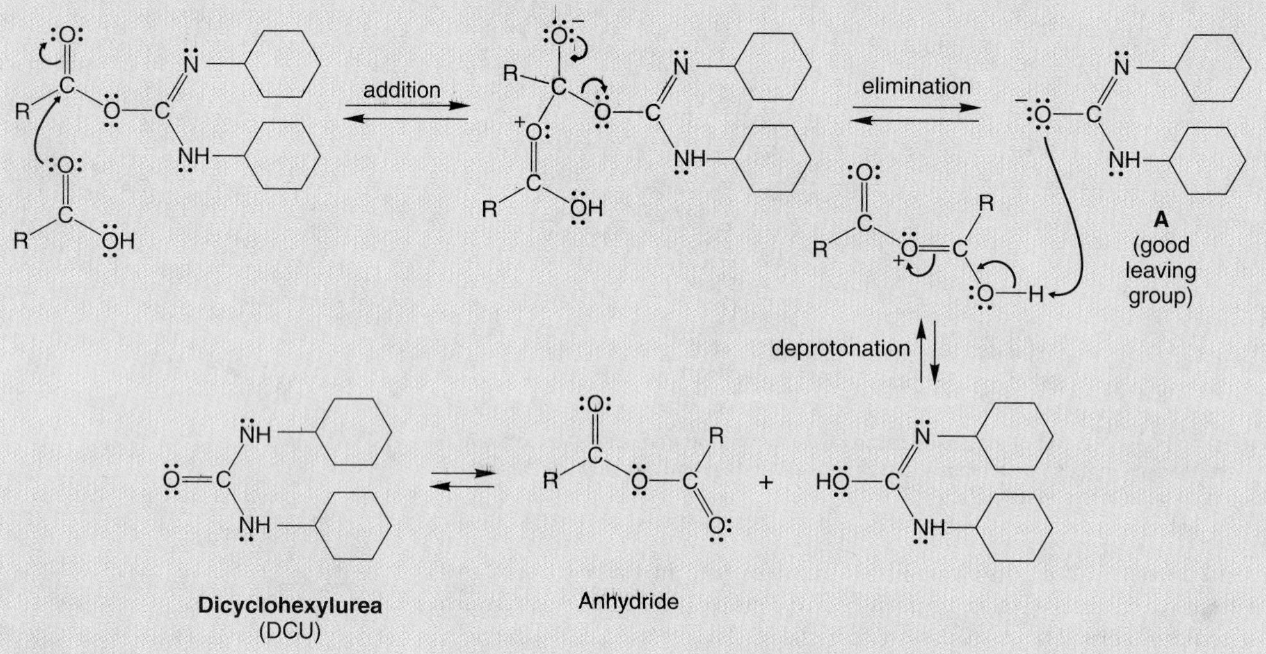

addition

elimination

A (good leaving group)

deprotonation

Dicyclohexylurea (DCU)

Anhydride

In the second step, the amine adds to the anhydride and, after deprotonation on nitrogen and reprotonation on oxygen, a carboxylic acid is lost in the elimination step that completes the reaction.

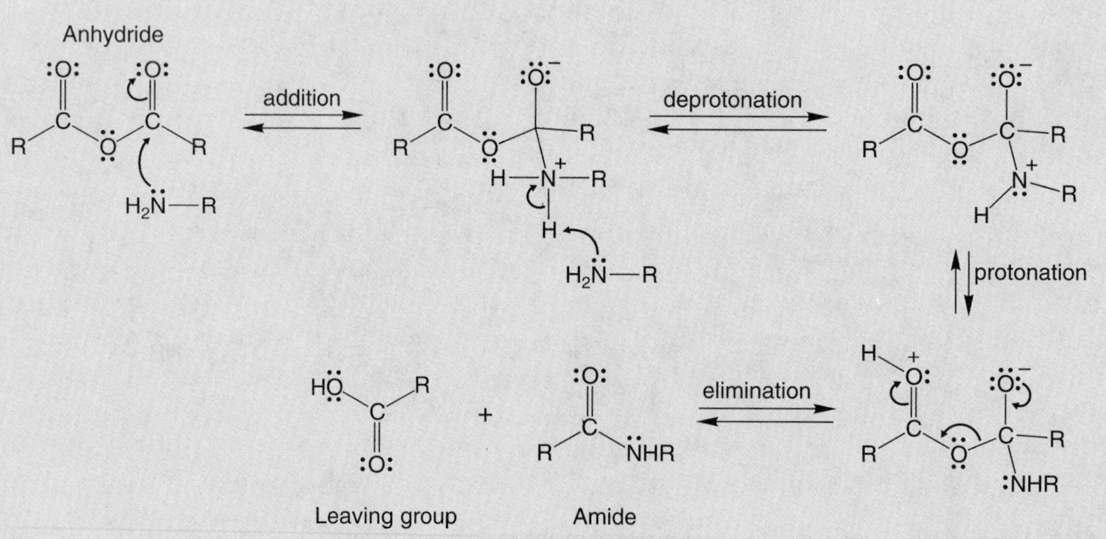

Anhydride

addition

deprotonation

protonation

elimination

Leaving group

Amide

Although this amide synthesis may seem complicated, it is in fact merely a collection of steps closely resembling those in Fischer esterification or acid hydrolysis. We are beginning to see the generality of the addition–elimination process. The addition–elimination reaction will continue to be prominant throughout this and other chapters.

19.5c Formation of Acid Chlorides

Treatment of a carboxylic acid with thionyl chloride or phosphorus pentachloride gives a good yield of the corresponding **acid chloride**. This reaction closely resembles the formation of chlorides by the reaction of alcohols with thionyl chloride (Chapter 17, p. 830). Once again, the key concept in each case is the transformation of a very poor leaving group, OH, into a better one (Fig. 19.45).

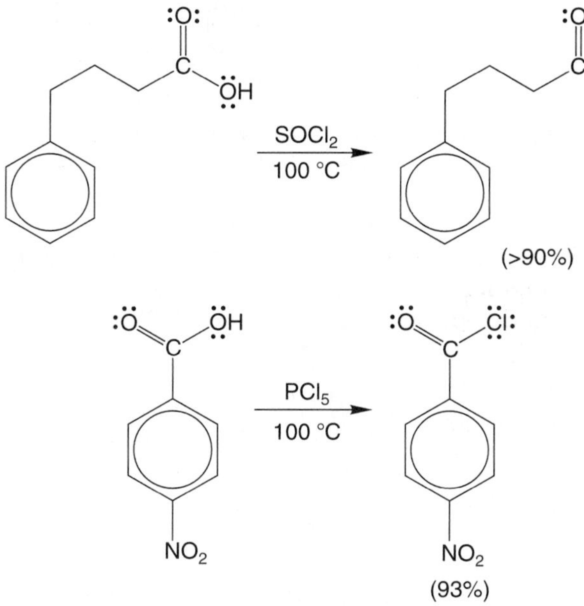

$SOCl_2$, 100 °C (>90%)

PCl_5, 100 °C (93%)

(Recall from Chapter 17: R–ÖH $\xrightarrow{SOCl_2}$ R–Cl:)

FIGURE 19.45 The reactions of alcohols and acids with thionyl chloride ($SOCl_2$) or phosphorus pentachloride (PCl_5) to give chlorides and acid chlorides are similar.

In the first step an ester is formed, in this case an ester of chlorosulfinic acid (Fig. 19.46).

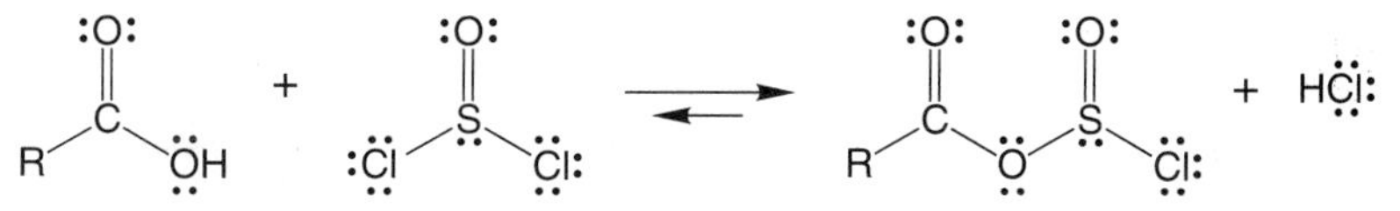

+ HCl

A chlorosulfite ester

FIGURE 19.46 A chlorosulfite ester is formed in the first step of the reaction.

PROBLEM 19.16 Write a mechanism for the formation of the product in Figure 19.46. Watch out for the details! Almost everyone makes a small error the first time on this problem.

The hydrogen chloride generated in the formation of the product next protonates the carbonyl group and the chloride ion adds to the strong Lewis acid. The intermediate so-formed then breaks down into chloride, sulfur dioxide, and the acid chloride. Notice the change in leaving group induced by the overall transformation of OH into OSOCl (Fig. 19.47).

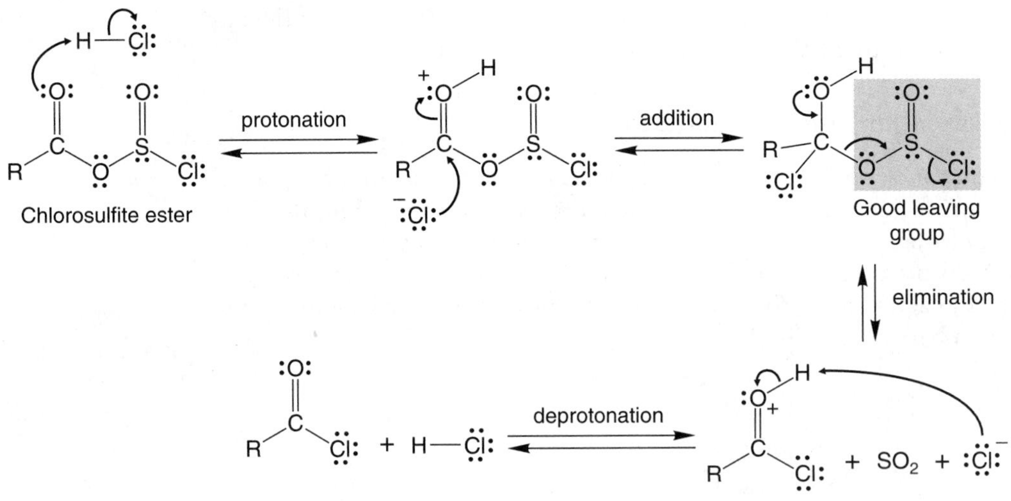

protonation
addition
elimination
deprotonation
Chlorosulfite ester
Good leaving group

FIGURE 19.47 Hydrogen chloride then protonates the carbonyl group, and chloride attacks this strong Lewis acid. In the crucial step, sulfur dioxide and chloride ion are lost as the acid chloride is formed.

Acid chlorides contain a good potential leaving group, the chloride, and the addition–elimination process can lead to all manner of acyl derivatives, R–CO–X, as we will see soon in Chapter 20 (Fig. 19.48).

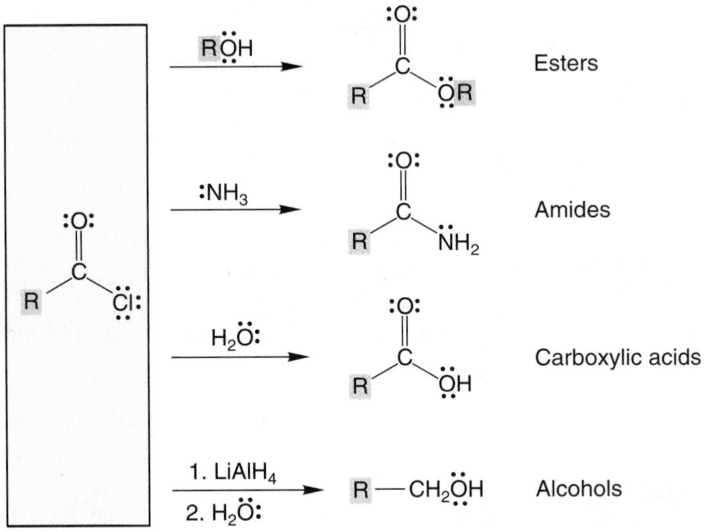

ROH → Esters
:NH_3 → Amides
H_2O: → Carboxylic acids
1. $LiAlH_4$ 2. H_2O: → R—CH_2OH Alcohols

FIGURE 19.48 Some compounds formed from addition–elimination reactions of acid chlorides. See Problem 19.17.

PROBLEM 19.17 Write a mechanism for the last transformation of acid chlorides shown in Figure 19.48.

PROBLEM 19.18 Another excellent synthesis of acid chlorides from carboxylic acids uses phosgene (Cl–CO–Cl) or oxalyl chloride (Cl–CO–CO–Cl) as the reactive agent. Suggest a mechanism for the second transformation in Figure 19.49.

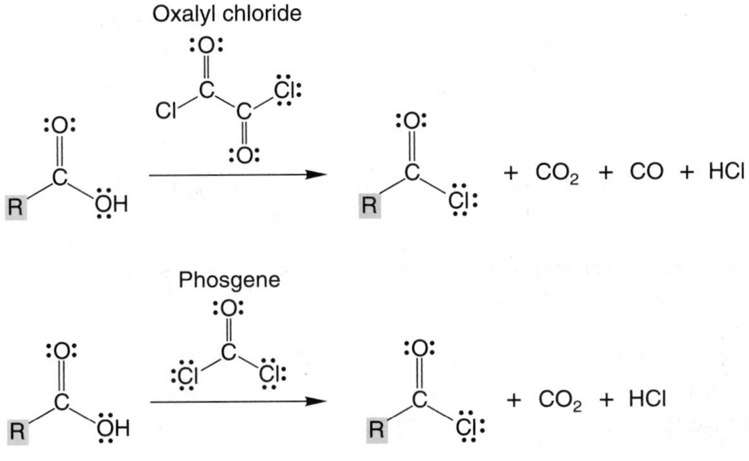

FIGURE 19.49

PROBLEM 19.19

Phosgene (see Problem 19.18) is a most effective poison. Can you guess its mode of action? What happens when phosgene is absorbed by moist lung tissue?

19.5d Anhydride Formation

Anhydrides are, as the name suggests, related to acids by a formal loss of water (Fig. 19.50).

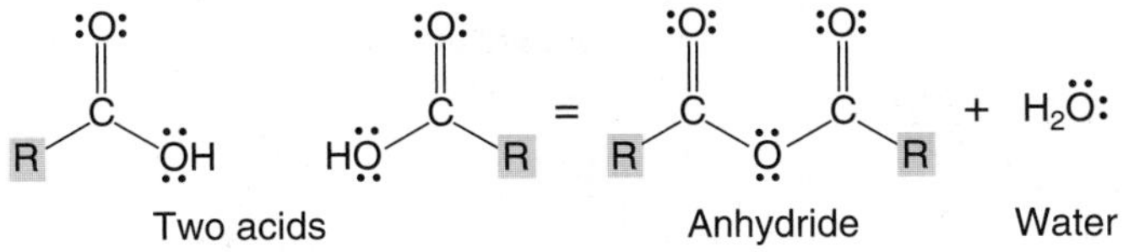

FIGURE 19.50 An anhydride is "two acids less a molecule of water."

Carboxylic acids and their conjugate bases, the carboxylate anions, react with acid halides to give **anhydrides** (Fig. 19.51).

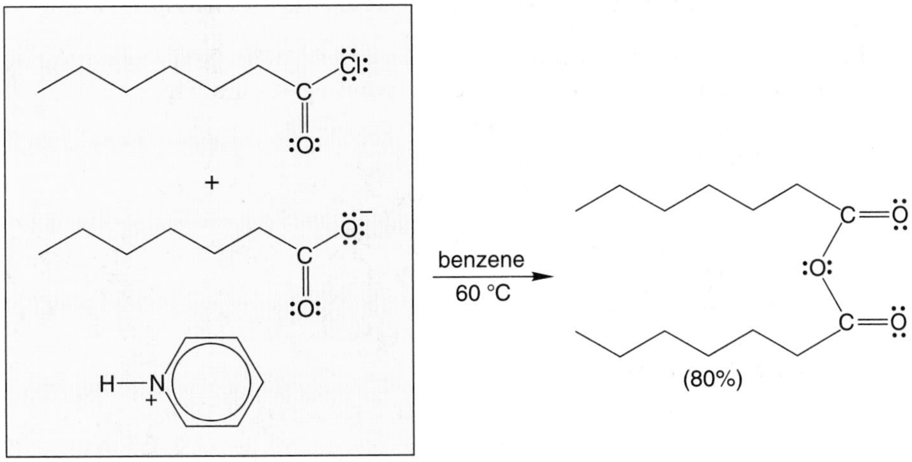

FIGURE 19.51 Anhydride formation through the reaction of a carboxylate salt with an acid chloride.

The mechanisms for these reactions involve addition of the carboxylate anion to the carbonyl group of the acid chloride to give a tetrahedral intermediate that can lose the good leaving group, a chloride ion (Fig. 19.52).

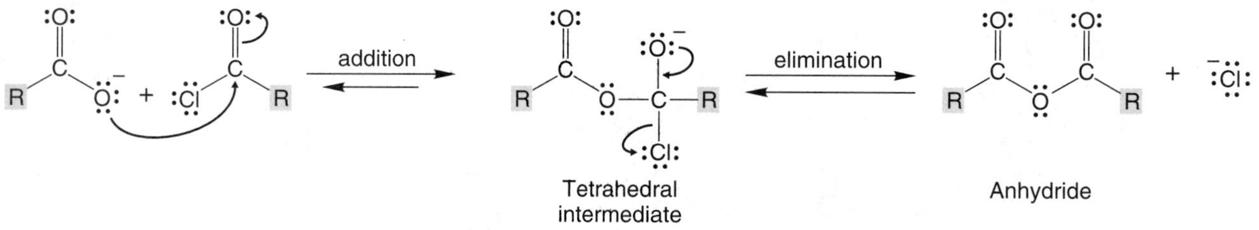

addition

elimination

Tetrahedral intermediate

Anhydride

FIGURE **19.52** The mechanism of anhydride formation involves the generation of a tetrahedral intermediate and loss of chloride ion.

Other **dehydrating agents** such as DCC or P_2O_5 can be used to form anhydrides. Cyclic anhydrides can be formed through reaction with another anhydride, or even by heating of the diacid (Fig. 19.53).

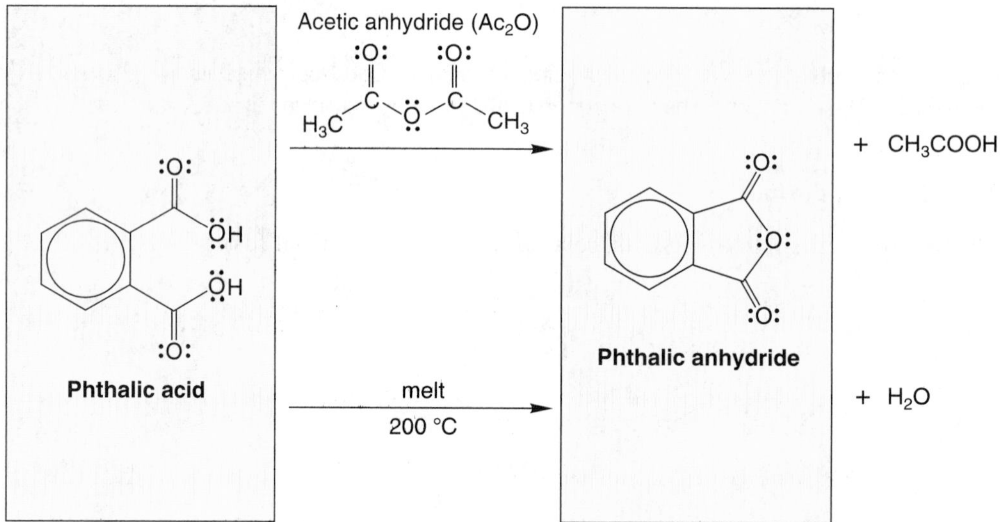

FIGURE **19.53** Two formations of phthalic anhydride, a cyclic anhydride.

19.5e Reactions with Organolithium Reagents and Metal Hydrides

In a useful synthetic reaction, organic acids react with 2 equivalents of an organolithium reagent to give the corresponding ketone (Fig. 19.54).

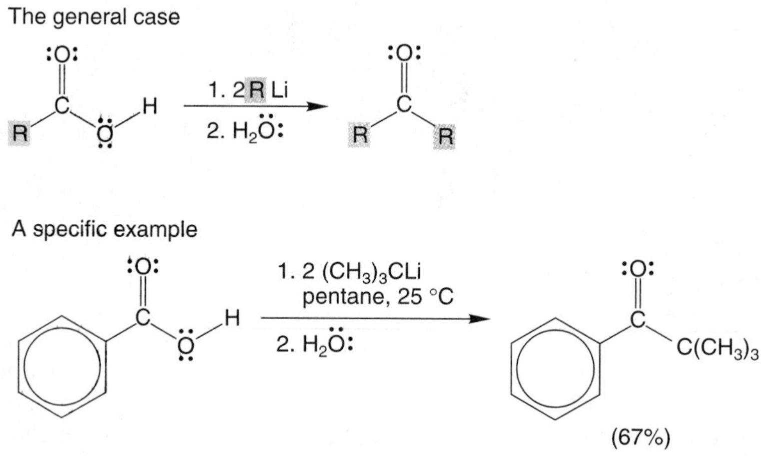

The general case

1. 2R Li
2. H_2O

A specific example

1. 2 $(CH_3)_3CLi$ pentane, 25 °C
2. H_2O

$C(CH_3)_3$

(67%)

FIGURE **19.54** A general ketone synthesis from carboxylic acids and organolithium reagents. Note again the convention that differentiates sequential reactions (1. RLi, 2. H_2O) from "dump 'em all together" reactions (RLi, H_2O).

The first step in the reaction is the formation of the lithium salt of the carboxylic acid. Although this species already bears a negative charge, the organolithium reagent is a strong enough nucleophile to add to it to give the dianion (Fig. 19.55).

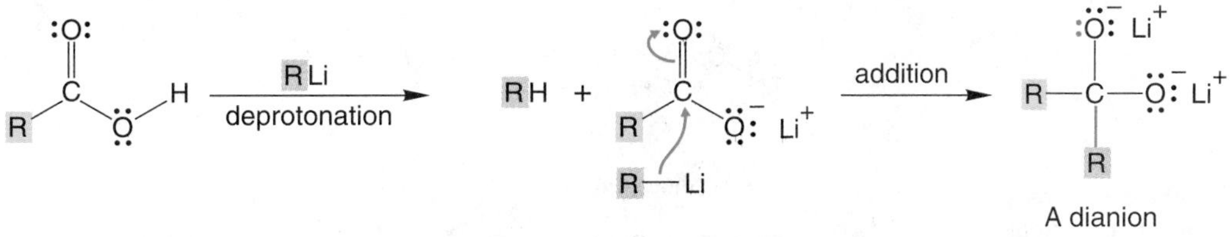

FIGURE 19.55 The first step in the reaction is formation of the carboxylate anion. The organolithium reagent is strong enough to add to the carboxylate anion to give a dianion.

Now what? What can happen to the product dianion in base? This question is one of the rare instances when the answer really is, "nothing." There is no possible leaving group, as neither $R:^-$ nor O^{2-} can be lost. So, the dianion remains in solution until the reaction mixture is hydrolyzed, at which point a hydrate is formed. Hydrates are unstable relative to their ketone precursors (Chapter 16, p. 763), and so the end result is the ketone (Fig. 19.56).

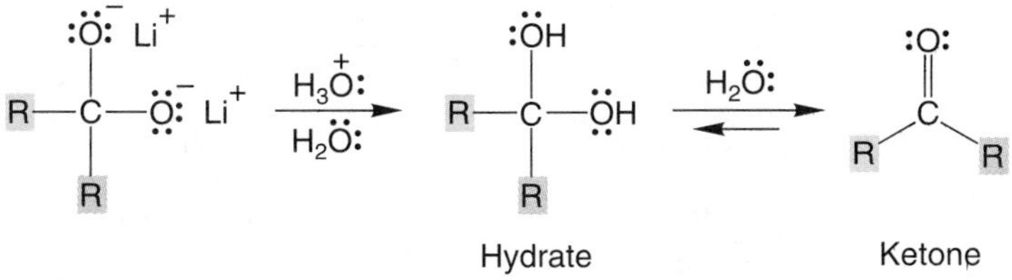

FIGURE 19.56 When the dianion is protonated in the second step, a hydrate is formed. Hydrates are generally unstable compared to the related ketones.

The key to the formation of the ketone is the inability of the dianionic intermediate to expel a leaving group. Compare this nonreaction with a general process in which a leaving group is present. We have already seen many reactions of this type, and in Chapter 20 we will see many more (Fig. 19.57).

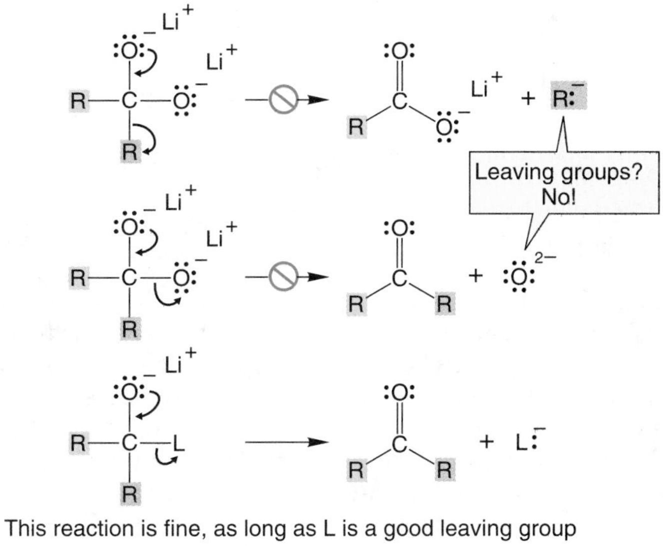

FIGURE 19.57 The dianion is foiled in any attempt to expel a negative charge by the lack of a possible leaving group.

Carboxylic acids are reduced by lithium aluminum hydride to the corresponding primary alcohols (Fig. 19.58).

The general case

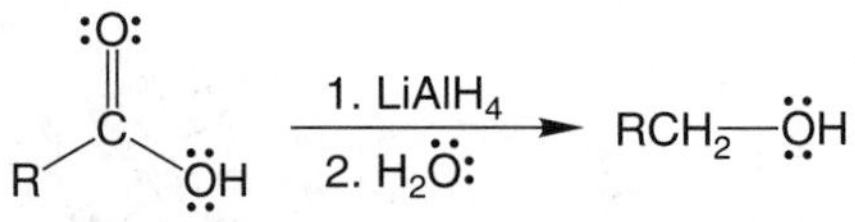

$$RCOOH \xrightarrow[2.\ H_2O]{1.\ LiAlH_4} RCH_2{-}OH$$

A specific example

$$HOOC{-}(CH_2)_8{-}COOH \xrightarrow[2.\ H_2O]{1.\ LiAlH_4} HOCH_2{-}(CH_2)_8{-}CH_2OH\ (97\%)$$

FIGURE 19.58 Carboxylic acids are reduced to primary alcohols by treatment with lithium aluminum hydride, followed by hydrolysis.

There is no unanimity on the details of the reaction mechanism, but the first step must be formation of the carboxylate anion and hydrogen. Anion formation is followed by addition of hydride to the carbonyl group. The result is another dianion. Now, however, a molecule of metal oxide can be lost to generate the aldehyde (Fig. 19.59).

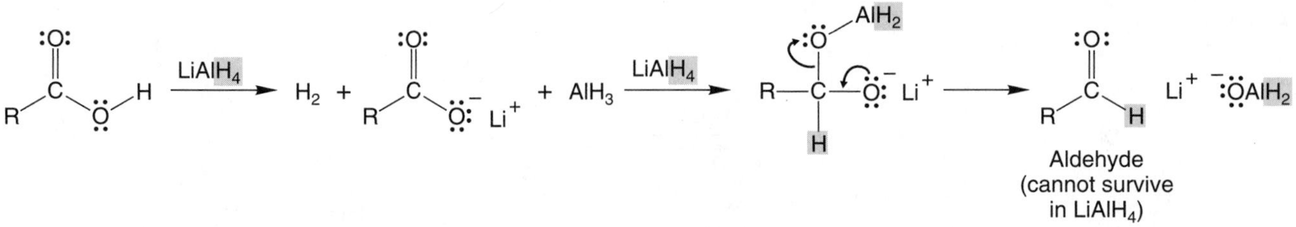

$$RCOOH \xrightarrow{LiAlH_4} H_2 + RCOO^- \ Li^+ + AlH_3 \xrightarrow{LiAlH_4} R{-}CH(O{-}AlH_2)(O^-)\ Li^+ \longrightarrow RCHO + Li^+\ {}^-OAlH_2$$

Aldehyde (cannot survive in $LiAlH_4$)

FIGURE 19.59 Deprotonation to give the carboxylate anion is followed by hydride addition to give a dianion. Loss of $LiOAlH_2$ leads to an intermediate aldehyde.

The aldehyde is formed in the presence of several reducing agents, and cannot survive for long (Chapter 16, p. 793). Another addition of hydride gives the alkoxide, and subsequent hydrolysis forms the alcohol (Fig. 19.60).

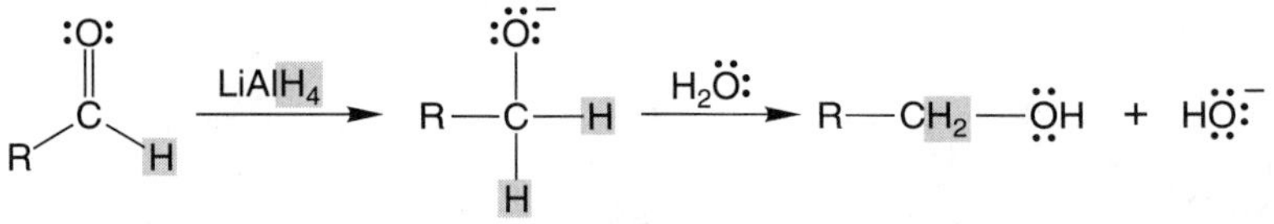

$$RCHO \xrightarrow{LiAlH_4} R{-}CH_2{-}O^- \xrightarrow{H_2O} R{-}CH_2{-}OH + HO^-$$

FIGURE 19.60 The aldehyde cannot survive long in the presence of powerful reducing agents and is attacked by hydride to give the alkoxide. A subsequent hydrolysis step produces the alcohol.

19.5f Decarboxylation

Some carboxylic acids easily lose carbon dioxide. The best examples are β-ketoacids, and 1,3-diacids such as malonic acid (propanedioic acid) in which there is also a carbonyl group β to the acid. The reactions often take place at room temperature or on gentle heating. Note that these reactions provide new syntheses of ketones and acids (Fig. 19.61).

We will defer discussion of the **decarboxylation** of β-ketoacids until we learn how to make such compounds in Chapter 20, but carbonates and malonic acids will be mentioned here.

General reactions

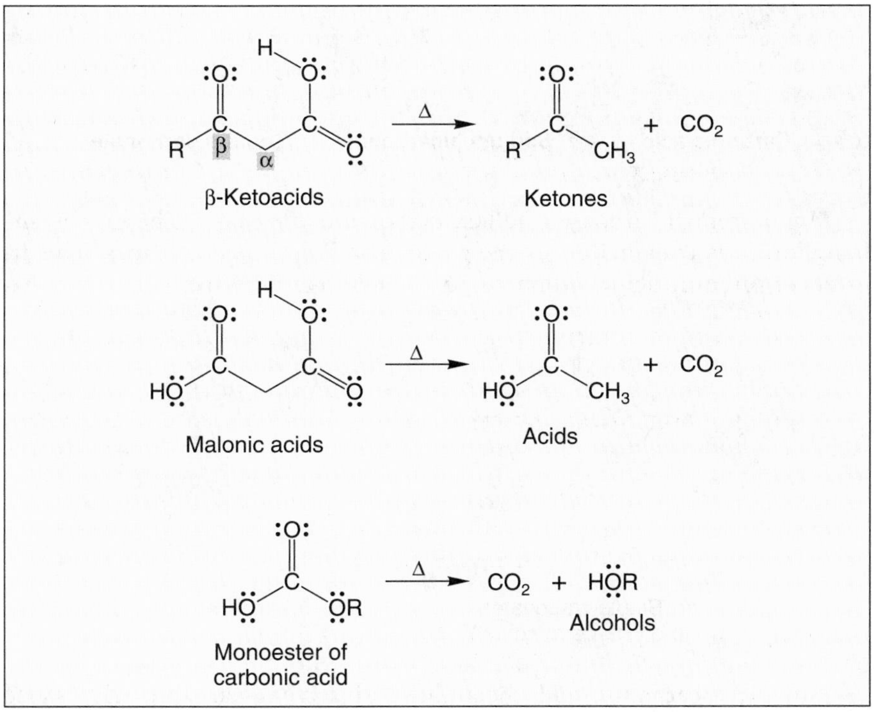

β-Ketoacids → (Δ) Ketones + CO_2

Malonic acids → (Δ) Acids + CO_2

Monoester of carbonic acid → (Δ) CO_2 + HOR Alcohols

Specific examples

~25 °C (> 60%)

~25 °C (> 70%)

30 °C (~100%)

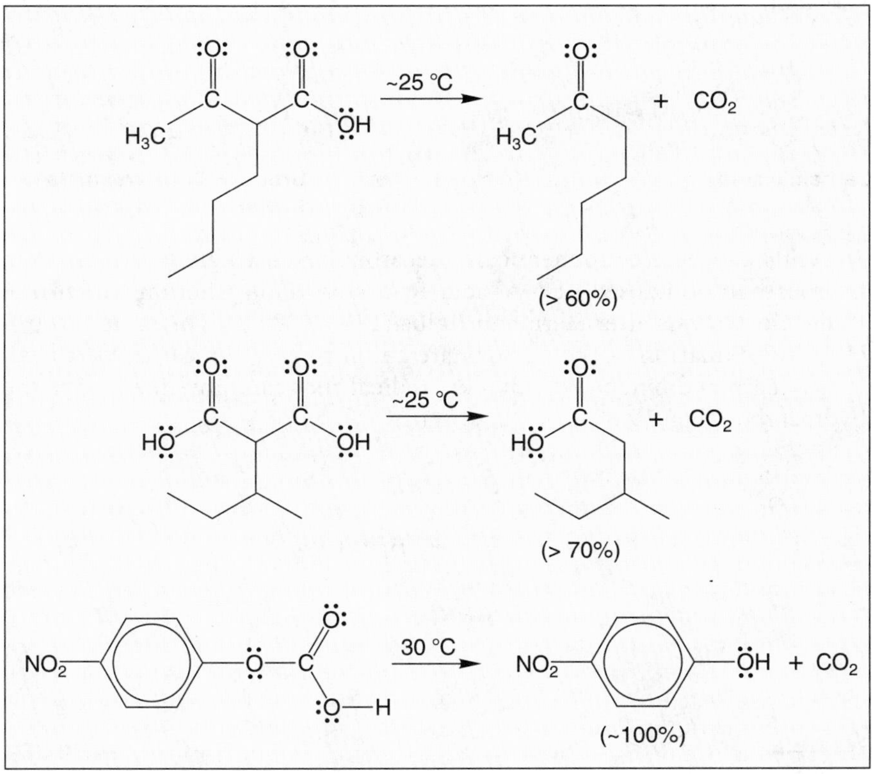

FIGURE **19.61** Three kinds of carboxylic acid that decarboxylate (lose CO_2) easily.

Carbonic acid (H_2CO_3 or HOCOOH) is itself unstable, although its salts are isolable and familiar as **sodium bicarbonate** and **sodium carbonate** (Fig. 19.62).

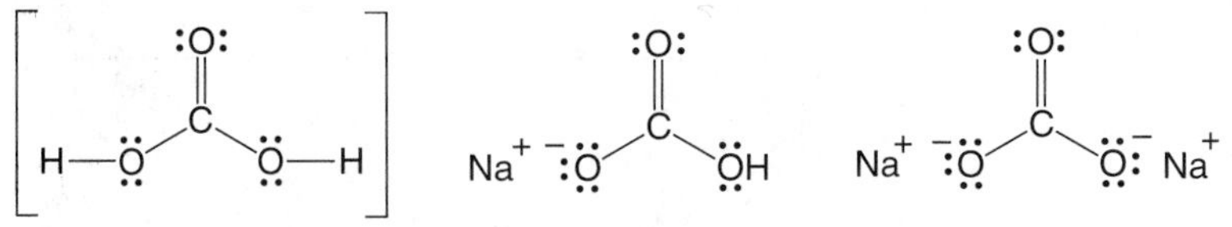

FIGURE 19.62 Carbonic acid is unstable, but sodium bicarbonate and sodium carbonate are stable.

Carbonic acid (unstable) Sodium bicarbonate Sodium carbonate

Monoesters of carbonic acid lose carbon dioxide easily. The loss of carbon dioxide is irreversible, as the gaseous carbon dioxide escapes. The diesters of carbonic acid (carbonates) are stable (Fig. 19.63).

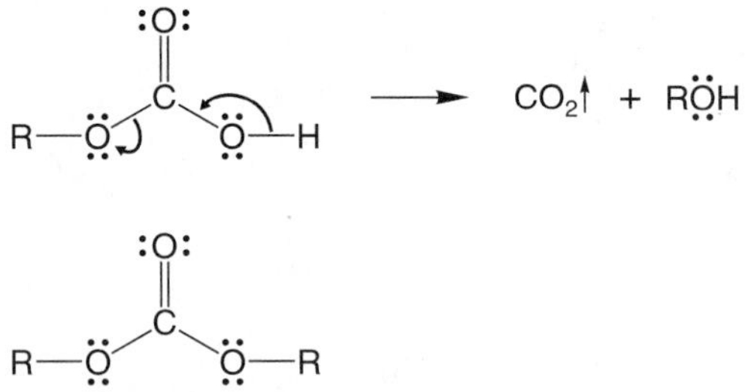

$\longrightarrow CO_2\uparrow + ROH$

Stable carbonate (no H to be removed)

FIGURE 19.63 Monoesters of carbonic acid decarboxylate easily but carbonates do not.

Similarly, **carbamic acid**, the amide derivative of carbonic acid, is also unstable, and loses carbon dioxide to give amines. The diamide, **urea**, and **carbamates**, which cannot lose carbon dioxide, are stable (Fig. 19.64).

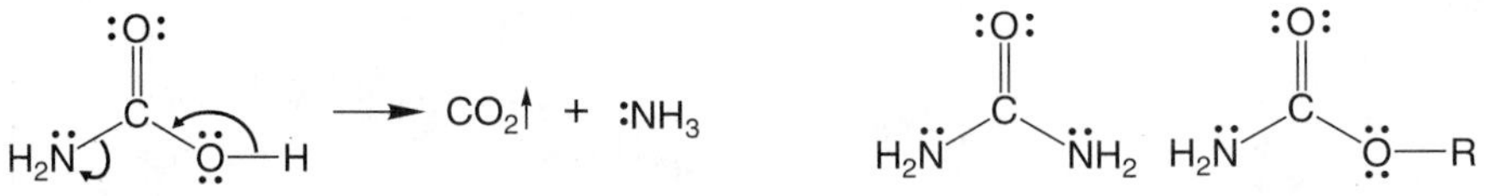

$\longrightarrow CO_2\uparrow + :NH_3$

Carbamic acid Urea (stable) Carbamate (stable)

FIGURE 19.64 Similarly, carbamic acids are unstable, but urea and carbamates are stable.

Acids can be electrochemically decarboxylated to give, ultimately, a hydrocarbon composed of two acid R groups. This reaction, called the **Kolbe electrolysis** after Hermann Kolbe (1818–1884), involves an electrochemical oxidation of the carboxylate anion to give the carboxyl radical. Loss of carbon dioxide gives an alkyl radical that can dimerize to give the hydrocarbon (Fig. 19.65).

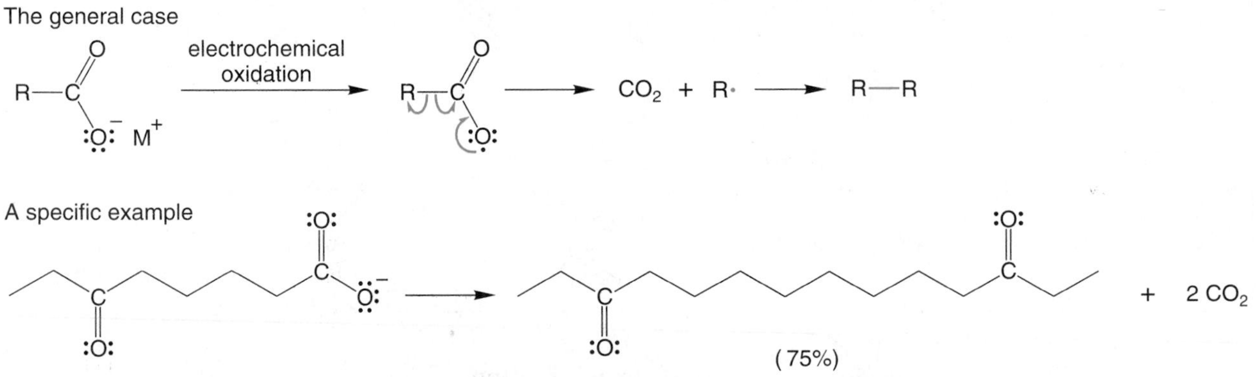

The general case

electrochemical oxidation

$\longrightarrow CO_2 + R\cdot \longrightarrow R{-}R$

A specific example

$+ 2\,CO_2$

(75%)

FIGURE 19.65 The Kolbe electrolysis is a source of molecules made up of coupled R groups of R–COOH.

19.5g Formation of Alkyl Bromides: The Hunsdiecker Reaction

The carboxyl radical just encountered is also the active ingredient in the **Hunsdiecker reaction** in which alkyl bromides are formed from acids (Heinz Hunsdiecker, b. 1904). Typically, the silver salt of the acid is used and allowed to react with bromine. The first step is the formation of the hypobromite. On heating, the weak oxygen–bromine bond breaks homolytically to give the carboxyl radical and a bromine atom. As in the Kolbe reaction, the carboxyl radical loses carbon dioxide to give an alkyl radical. In this case, however, the alkyl radical can carry the chain reaction further by recycling to react with the hypobromite to produce an alkyl bromide and another molecule of carboxyl radical (Fig. 19.66).

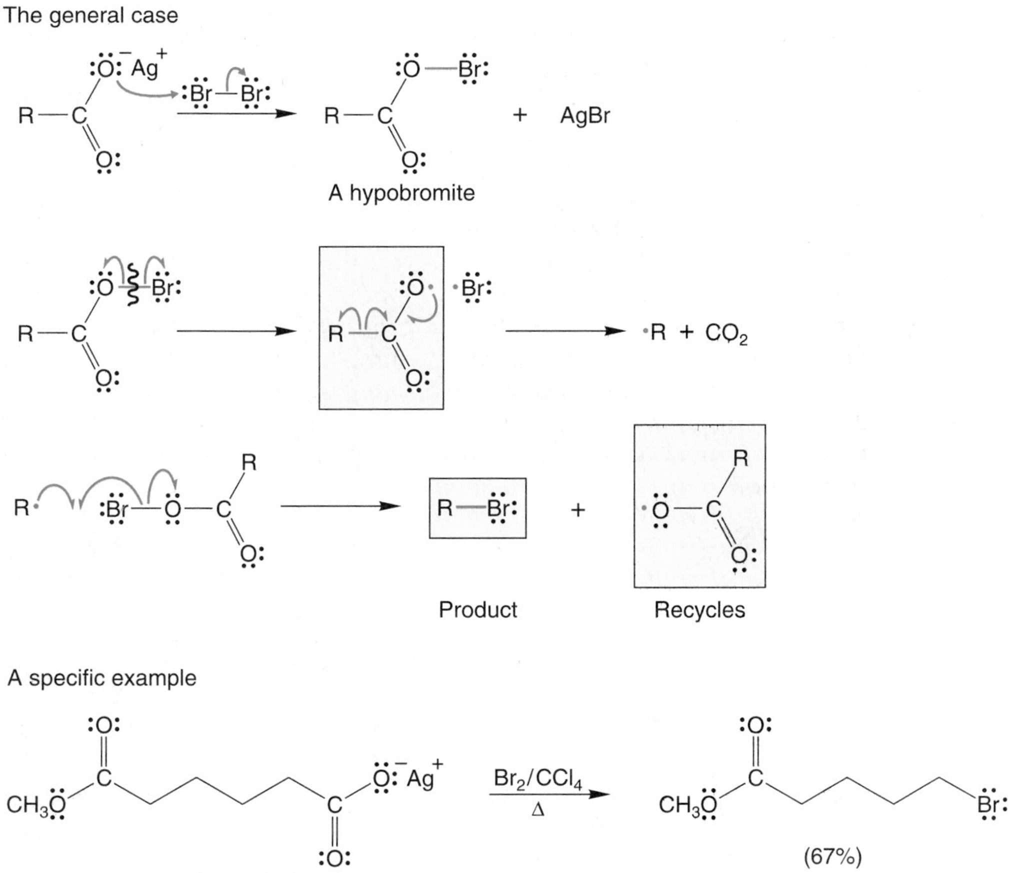

The general case

A hypobromite

Product

Recycles

A specific example

Br_2/CCl_4, Δ

(67%)

FIGURE 19.66 The decarboxylation of the carboxylate radical is also involved in the Hunsdiecker reaction, a synthesis of alkyl bromides from acids. The recycling, chain-carrying carboxyl radical is boxed.

19.6 REACTIVITY OF CARBOXYLIC ACIDS AT THE α-POSITION

19.6a Formation and Reactions of α-Halo Derivatives: Alkylation at the α-Position

As we saw in Chapter 18, in base, aldehydes and ketones can be brominated in the α-position. The enolate anions are the active ingredients (Fig. 19.67).

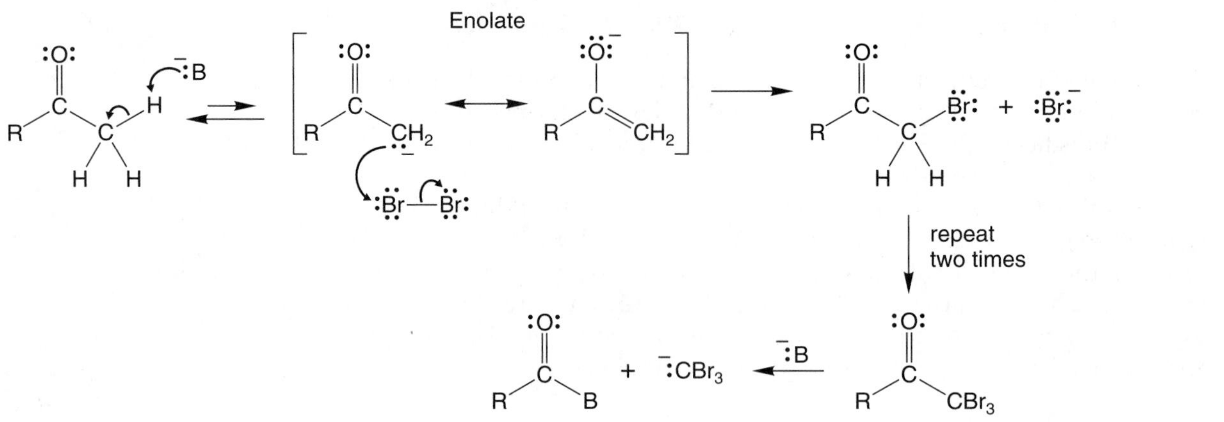

FIGURE 19.67 An enolate is the intermediate in the base-catalyzed α-bromination of aldehydes and ketones.

A similar reaction is possible with carboxylic acids, but two equivalents of strong base are necessary, as this reaction proceeds through a dianion. The first equivalent of base removes the carboxyl hydrogen to give the carboxylate salt. If the base used is a strong *base* and a weak *nucleophile,* a second hydrogen can be removed, this time from the α-position. The typical base used is lithium diisopropylamide (LDA), which is a strongly basic but poorly nucleophilic species because of its steric demands. The dianion can now be brominated or alkylated at the α-position, as long as the alkylating agent is reactive. Primary halides work well in alkylation reactions because the S_N2 reaction is especially facile, but more substituted halides lead mostly to the products of an E2 reaction (Fig. 19.68).

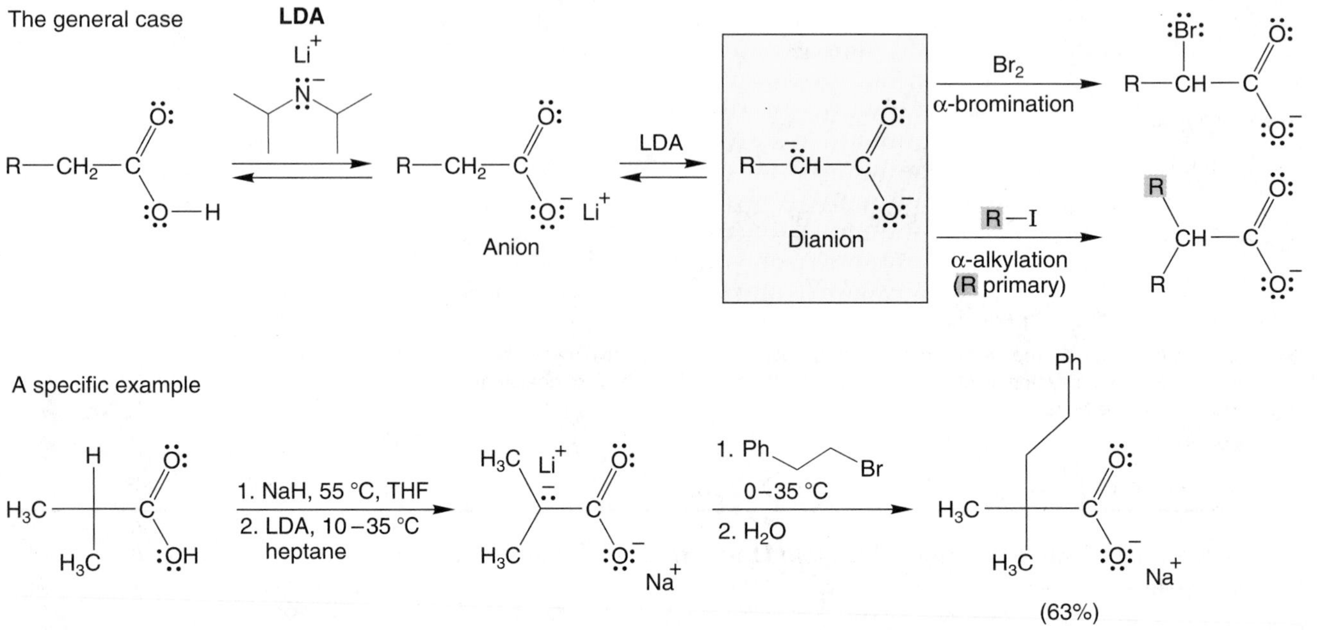

FIGURE 19.68 Dianions of carboxylic acids can function as enolates. They can be brominated or alkylated in the α-position with primary or other reactive halides.

Draw the resonance forms for the dianion. Which is the best representation?

PROBLEM **19.20**

Provide mechanisms for the following reactions that show two of the reactions possible at the α-position of carboxylic acids (Fig. 19.69).

PROBLEM **19.21**

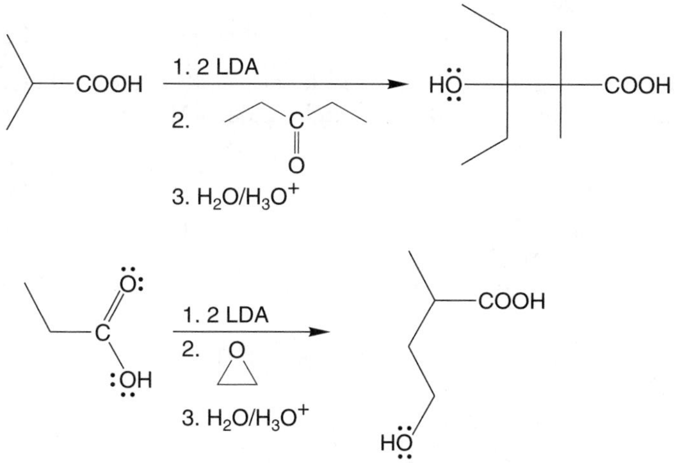

1. 2 LDA
2.
3. H_2O/H_3O^+

1. 2 LDA
2.
3. H_2O/H_3O^+

FIGURE **19.69**

19.6b Another Bromination at the α-Position: The Hell–Volhard–Zelinsky Reaction

Treatment of carboxylic acids with Br_2 and PBr_3, or the equivalent, a mixture of phosphorus and bromine, leads ultimately to formation of the α-bromo acid (Fig. 19.70).

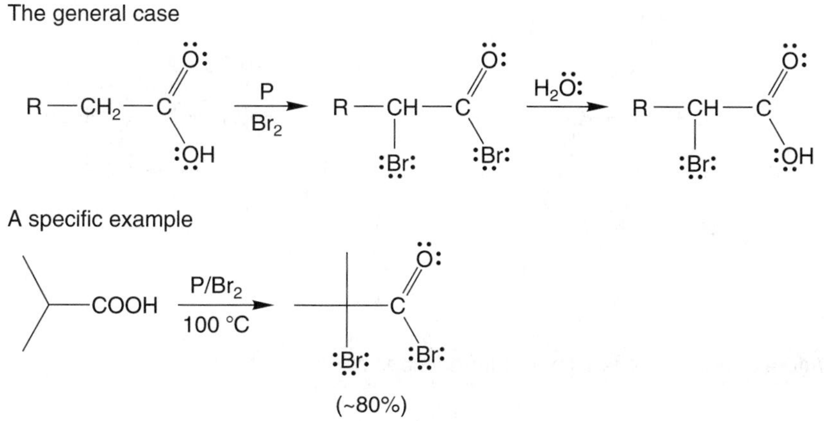

The general case

A specific example

(~80%)

FIGURE **19.70** The Hell–Volhard–Zelinsky reaction.

The process is known as the **Hell–Volhard–Zelinsky reaction** after Carl M. Hell (1849–1926), Jacob Volhard (1834–1910), and Nicolai D. Zelinsky (1861–1953). The first step is formation of the acid bromide through reaction with PBr_3 (p. 977). The acid bromide is in equilibrium with its enol form. Bromination of the enol form with Br_2 gives an isolable α-bromo acid bromide. Recall the bromination of carbonyl compounds (Chapter 18, p. 890), a reaction that also proceeds through the enol form (Fig. 19.71).

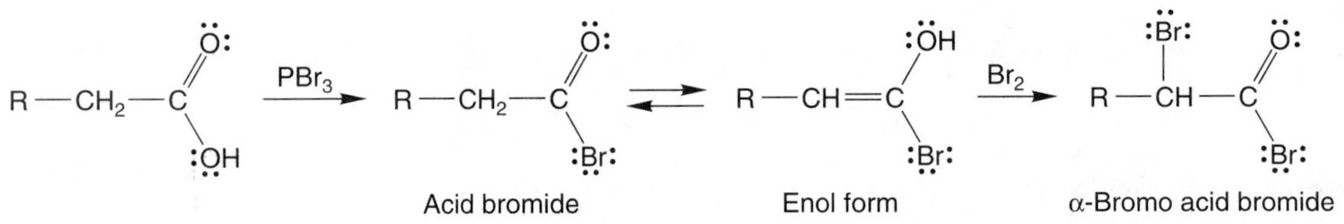

Acid bromide

Enol form

α-Bromo acid bromide

FIGURE 19.71 An intermediate in the Hell–Volhard–Zelinsky reaction is the α-bromo acid bromide. This compound can be isolated if a full equivalent of phosphorus tribromide (PBr_3) is used in the reaction.

An α-bromo acid halide reacts like any acid halide, and can be used to generate the α-bromo acid, ester, and amide, for example (Fig. 19.72).

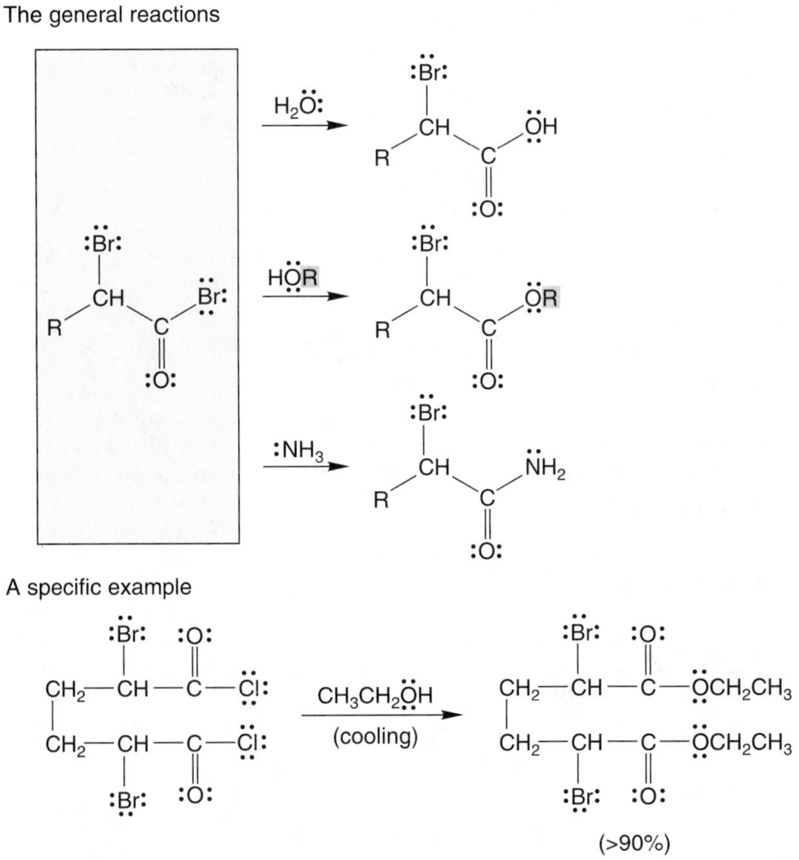

The general reactions

A specific example

(cooling)

(>90%)

FIGURE 19.72 The α-bromo acid bromide can be used to make the related α-bromo acid, ester, and amide.

PROBLEM 19.22 Write a general mechanism for the reactions of Figure 19.72.

If a catalytic amount of PBr_3 is used, and not the full equivalent of the reaction in Figure 19.71, the product of the reaction is the α-bromo acid, not the α-bromo acid bromide. The α-bromo acid bromide is still an intermediate, but only a small amount of it can be made at any one time, as it depends on PBr_3 for its formation. The small amount of α-bromo acid bromide produced is attacked by a molecule of the acid in an anhydride-forming reaction (Section 19.5d, p. 979). The anhydride is opened by bromide to give the carboxylate anion and a new molecule of α-bromo acid bromide that recycles. A final hydrolysis gives the α-bromo acid itself. This reaction surely involves a complicated mechanism (Fig. 19.73).

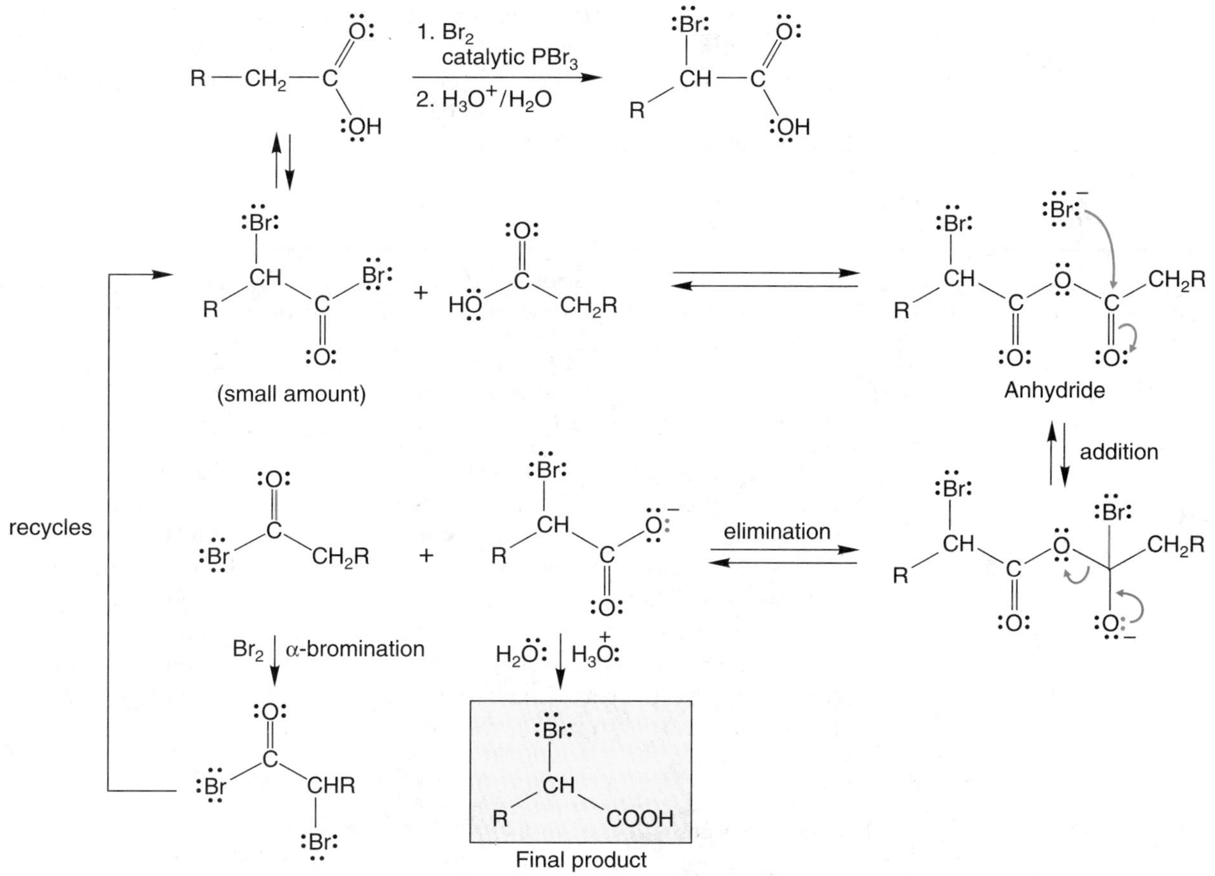

FIGURE 19.73 The complex mechanism of the Hell–Volhard–Zelinsky reaction using a catalytic amount of PBr_3.

Like the α-halo aldehydes and ketones, the α-bromo acids are very reactive in displacement reactions. They serve as sources of many α-substituted acids. A particularly important example is the formation of α-amino acids through reaction with ammonia (Fig. 19.74).

The general case

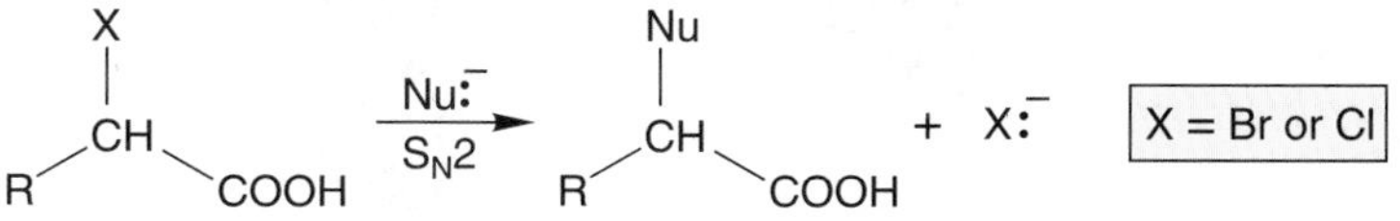

Specific examples

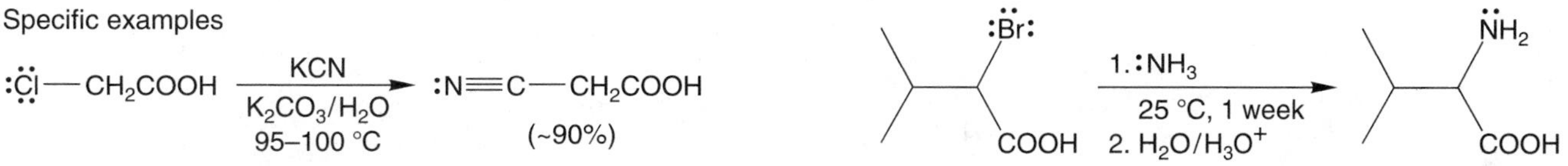

An α-amino acid, valine (48%)

FIGURE 19.74 Some reactions of α-bromo acids.

Phew! That's a lot of material. Many new kinds of compounds have appeared and there may seem to be an overwhelming number of new reactions. Don't be dismayed! Almost all of this section involves versions of the addition–elimination reaction. Keep this in mind and you will be able to figure out what is happening in these reactions relatively easily. It is not necessary, or productive, to try to memorize all this material. What is far more effective is to learn the basic process well, and then practice applying it in different circumstances.

19.7 SYNTHESES OF CARBOXYLIC ACIDS

We already know a number of routes to carboxylic acids, and this section will add an important new one, the reaction of organometallic reagents with carbon dioxide.

19.7a Oxidative Routes

As we know (Chapter 16, p. 795), alcohols and aldehydes can be oxidized to acids. Alkenes can also be oxidatively cleaved to acids by basic potassium permanganate. We also already know that alkenes form 1,2-diols when treated with basic permanganate (Chapter 17, p. 836). The reactions of alkenes with $KMnO_4$ can go further to form two acids. The trick is to use 18-crown-6 (Chapter 17, p. 863) to make $KMnO_4$ soluble in benzene. The crown ether has a great ability to bind potassium ions, and this allows the negatively charged permanganate ion (MnO_4^-) to follow the bound cation, K^+(crown), into solution (Fig. 19.75).

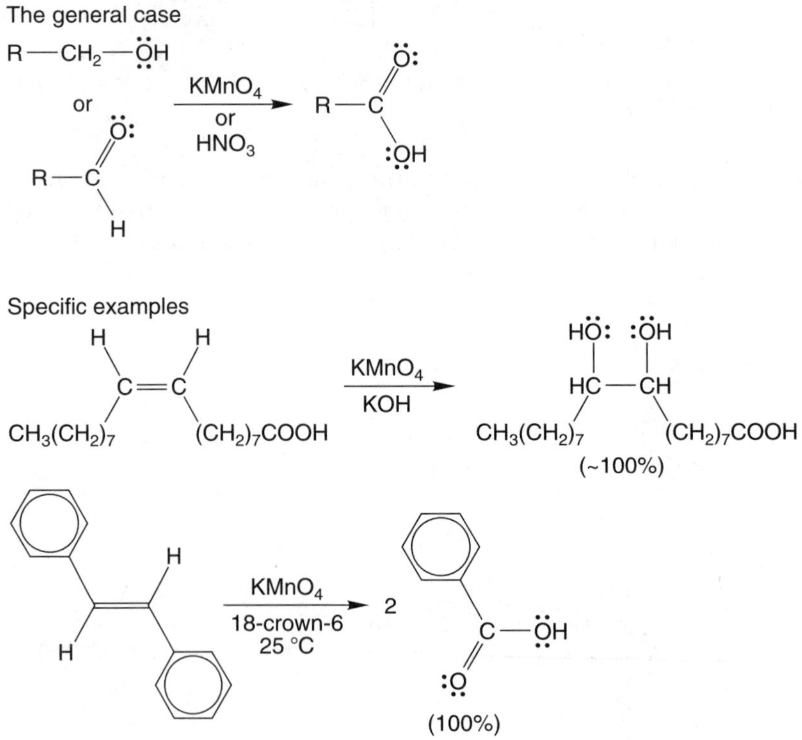

The general case

R—CH_2—OH or R—CHO $\xrightarrow[\text{or } HNO_3]{KMnO_4}$ R—COOH

Specific examples

$CH_3(CH_2)_7$CH=CH$(CH_2)_7$COOH $\xrightarrow[KOH]{KMnO_4}$ $CH_3(CH_2)_7$CH(OH)—CH(OH)$(CH_2)_7$COOH (~100%)

PhCH=CHPh $\xrightarrow[\text{18-crown-6, 25 °C}]{KMnO_4}$ 2 PhCOOH (100%)

FIGURE 19.75 Some oxidative routes to carboxylic acids.

Many alkenes can be ozonized using an oxidative workup to give acids, and the side chains of alkyl aromatic compounds can be oxidized to acid groups (Chapter 13, p. 611; Fig. 19.76).

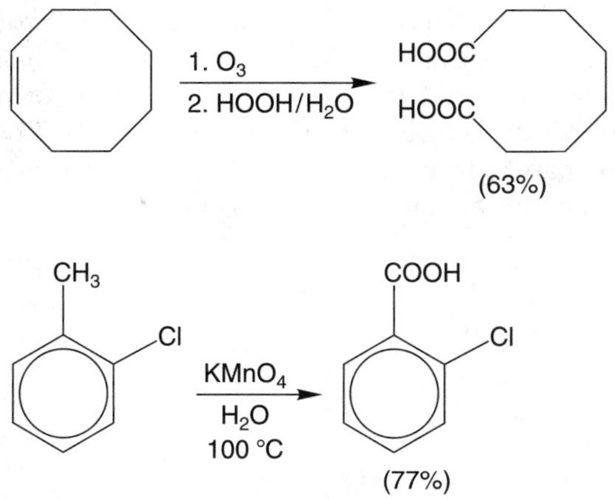

FIGURE 19.76 More oxidative routes to acids.

19.7b Haloform Reaction

The reaction in base of methyl ketones with iodine, bromine, or chlorine gives the acid and the haloform (Chapter 18, p. 890; Fig. 19.77).

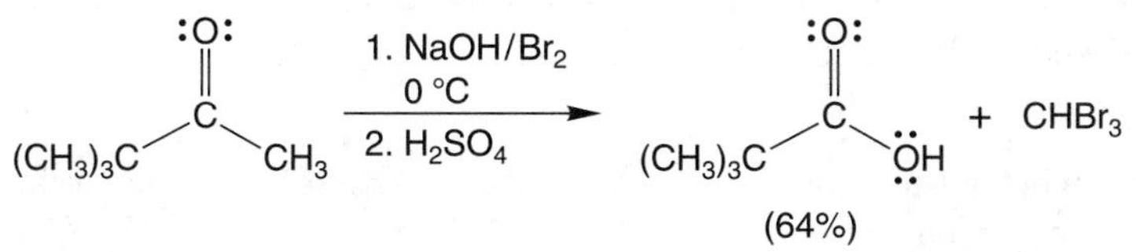

FIGURE 19.77 The haloform reaction is another source of acids.

19.7c Reaction of Organometallic Reagents with Carbon Dioxide

Like any other compound containing a carbonyl group, carbon dioxide reacts with organometallic reagents through an addition reaction. The initial product is a carboxylate salt that is converted into the acid when the reaction mixture is acidified. This reaction is very general and a source of many acids (Fig. 19.78).

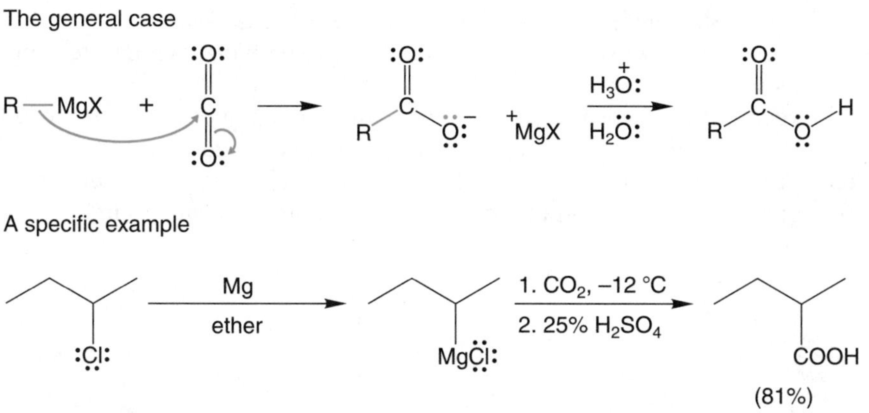

FIGURE 19.78 A general route to carboxylic acids uses the carboxylation of Grignard reagents with carbon dioxide.

19.8 SOMETHING MORE: FATTY ACIDS

We have spent many pages looking at the reactions of carboxylic acids. Now let's see a reaction of molecules containing long hydrocarbon chains and a single acid. The properties of such molecules are strongly influenced by the presence of both the long chain and the acid group. Even-numbered long-chain acids are called **fatty acids** because they are formed by the base-induced **ester hydrolysis (saponification)** of fats (Fig. 19.79). Fatty acids are synthesized in nature from acetic acid, a two-carbon acid, and this accounts for their containing even numbers of carbons.

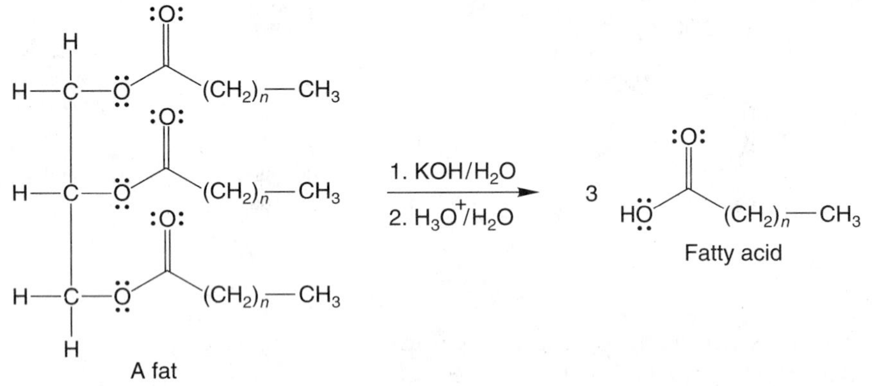

A fat — 1. KOH/H_2O; 2. H_3O^+/H_2O → 3 $HO–C(=O)–(CH_2)_n–CH_3$ (Fatty acid)

$n = 10$	Lauric acid
$n = 12$	Myristic acid
$n = 14$	Palmitic acid
$n = 16$	Stearic acid
$n = 18$	Arachidic acid

FIGURE 19.79 Some fatty acids. These molecules can be made from the hydrolysis (saponification) of fats.

PROBLEM 19.23 Give a mechanism for the hydroxide ion induced hydrolysis (saponification) of a fat to a fatty acid.

The salts of long-chain fatty acids are called **soaps**, and work in the following way. These compounds contain both a highly polar end, the acid salt, and a most nonpolar end, the long-chain hydrocarbon end. The polar ends will be soluble in polar solvents such as water, but the nonpolar hydrocarbon ends will not be. In aqueous solution, the **hydrophobic** (water-hating) hydrocarbon ends cluster together at the inside of a sphere, protected from the aqueous environment by the **hydrophilic** (water-loving) polar acid salt groups forming the interface with the polar water molecules. Such aggregates are called **micelles** (Fig. 19.80).

Similar micellar species are formed from **detergents**, which are long-chain sulfonic acid salts (Fig. 19.81).

Organic dirt (grease) is not water soluble and is not easily removed without the aid of a soap or a detergent. However, when a detergent is used, the grease is dissolved by the hydrocarbon-like internal portion of the micelle and made soluble in water by the polar acid salts outside (Fig. 19.82).

A micelle

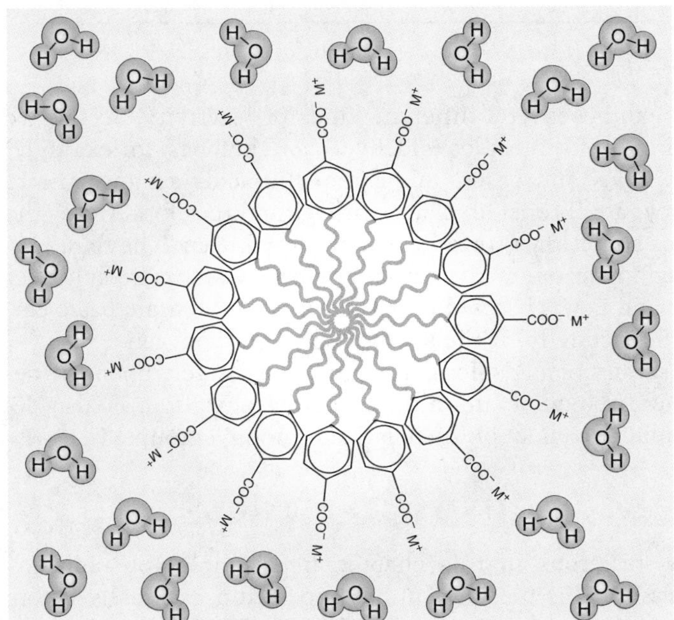

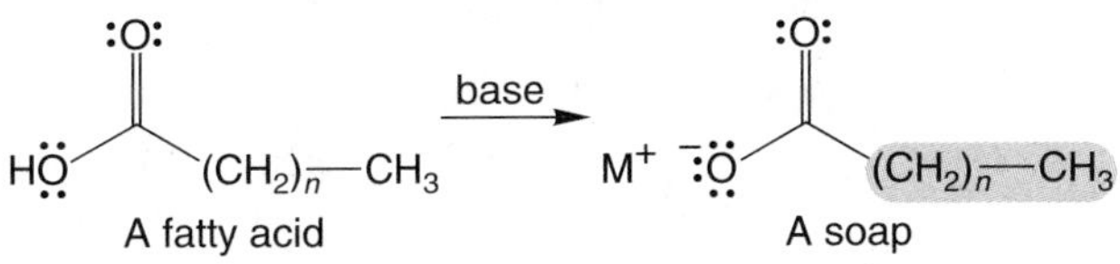

$$\text{HÖ–C(=O)–(CH}_2)_n\text{–CH}_3 \xrightarrow{\text{base}} \text{M}^+\ {}^-\text{:Ö–C(=O)–(CH}_2)_n\text{–CH}_3$$

A fatty acid → A soap

FIGURE **19.80** The formation of a soap from a fatty acid. In polar media, soaps form micelles, with the hydrophobic hydrocarbon chains directed toward the center of a sphere, away from the polar solvent. The surface of the sphere contains the polar carboxylate anions and their positive counterions.

A long-chain hydrocarbon

$$\text{Na}^+\ {}^-\text{:Ö–S(=O)}_2\text{–C}_6\text{H}_4\text{–R}$$

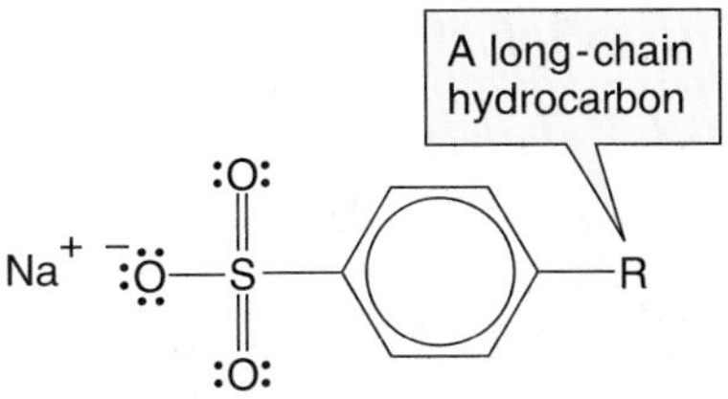

FIGURE **19.81** A synthetic soap, a detergent.

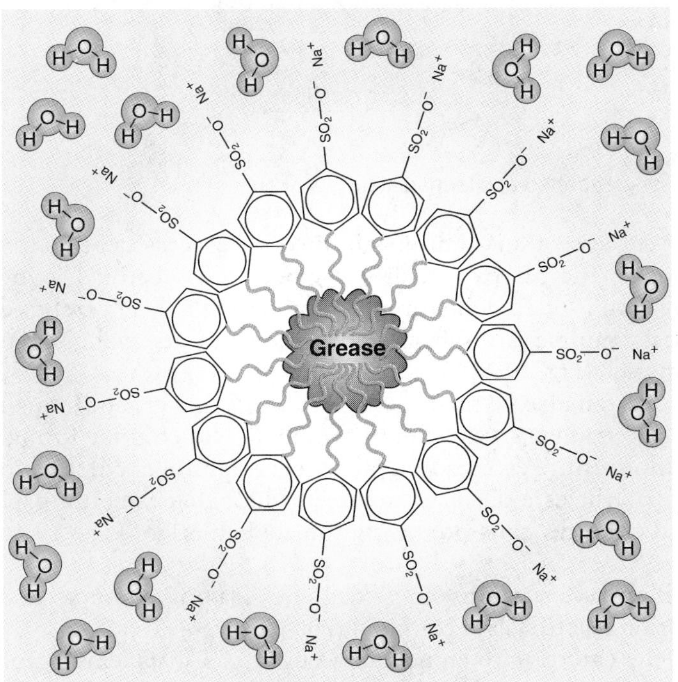

FIGURE **19.82** A micelle of a detergent solubilizing a grease molecule.

19.9 SUMMARY

NEW CONCEPTS

Carboxylic acids exhibit several different kinds of reactivity. So far, we have seen many kinds of molecules (aldehydes and ketones, for example) that can be both acids and bases. But carboxylic acids are even more polyreactive. They are Brønsted acids through proton loss from the hydroxyl group, or sometimes from loss of an α-hydrogen. They are also Lewis acids through reaction at the carbonyl group. Carboxylic acids act as Brønsted bases and Lewis bases through reaction at the more basic carbonyl oxygen or, occasionally, at the less basic hydroxyl oxygen.

The Brønsted acidity of acids, long thought to rest largely upon the delocalization of the carboxylate anion produced by ionization, is also dependent on the inductive effect of the polarized carbonyl group.

REACTIONS, MECHANISMS, AND TOOLS

Most of the new reactions in this chapter are examples of addition–elimination processes. This mechanism has appeared occasionally before but it is emphasized in this chapter for the first time. Carbonyl compounds bearing leaving groups can be attacked by nucleophiles to give tetrahedral intermediates that can then expel the leaving group (Fig. 19.83).

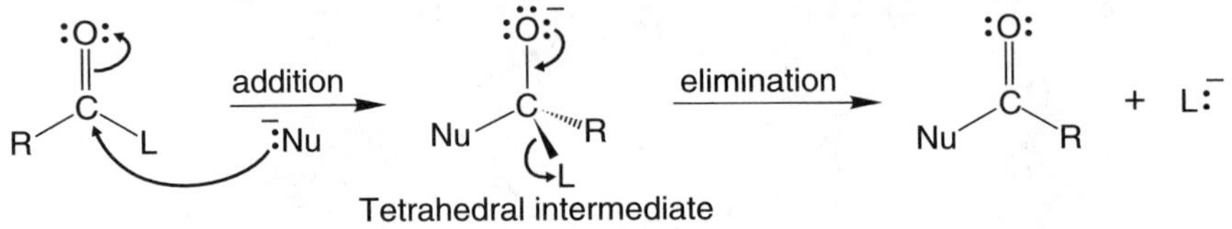

addition
:Nu
Nu
R
L
Tetrahedral intermediate
elimination
Nu
R
+ L:⁻

FIGURE 19.83 The addition–elimination reaction.

For acids, the success of an addition–elimination process depends on the transformation of the carboxyl hydroxyl group into a better leaving group. This conversion is often done through protonation in acid-catalyzed reactions. The best example is Fischer esterification, and its cyclic variation, lactone formation (Fig. 19.28).

The carboxyl OH can also be made into a better leaving group through other chemical conversions. An excellent example is acid chloride formation in which a chlorosulfite ester is first produced. Addition of chloride to the carbonyl group of the ester leads to an intermediate that can eliminate sulfur dioxide and chloride, thus producing the acid chloride (Figs. 19.46 and 19.47).

The acid chlorides contain an excellent leaving group, and addition–elimination reactions are common (Fig. 19.48).

Occasionally, the carboxylate anion can be used as a displacing agent in an S_N2 reaction. In order for this to be successful, the Lewis acid partner in the reaction must be very reactive (Fig. 19.36).

The electrochemical reaction of carboxylate anions leads to hydrocarbons (Kolbe electrolysis). Other reactions involving decarboxylations are briefly mentioned (Fig. 19.61).

SYNTHESES

The many synthetically useful reactions in this chapter are summarized in Figure 19.84.

1. Acid halides

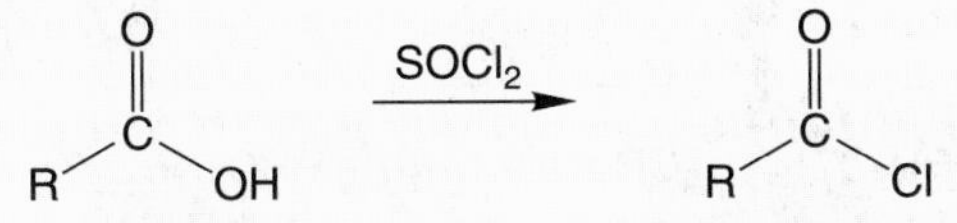

$$R{-}C(=O){-}OH \xrightarrow{SOCl_2} R{-}C(=O){-}Cl$$

A chlorosulfite ester is an intermediate

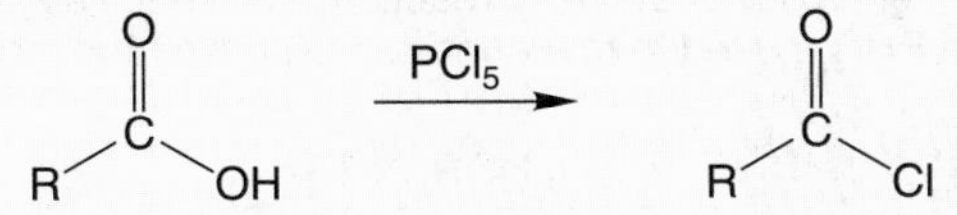

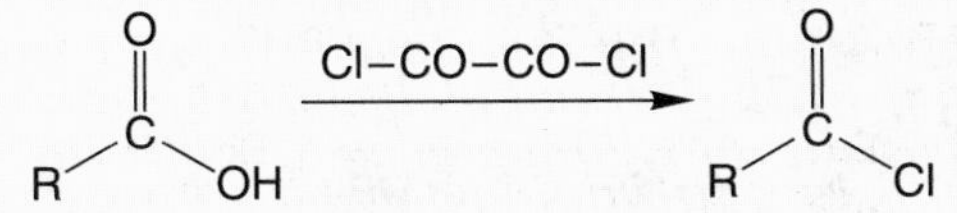

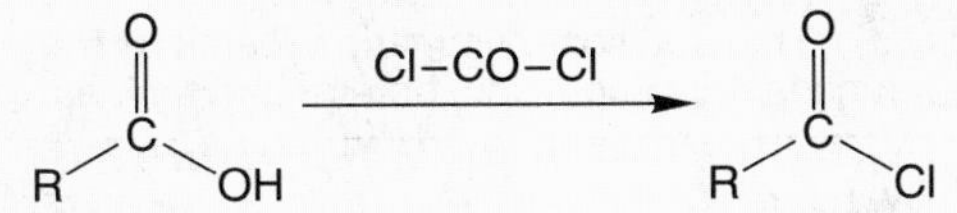

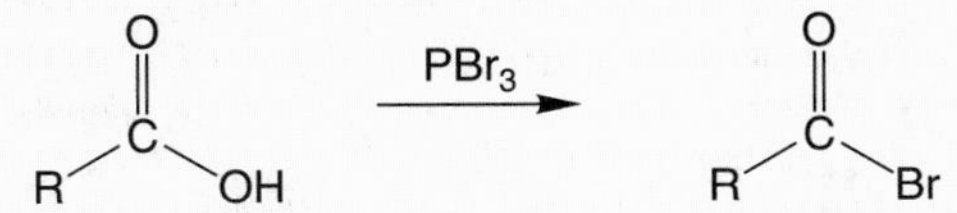

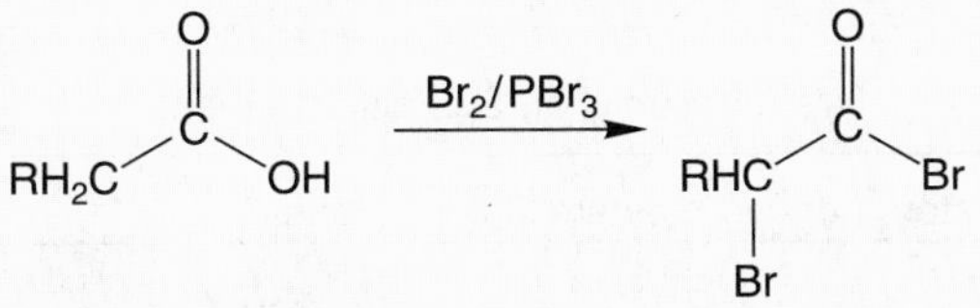

$$R{-}C(=O){-}OH \xrightarrow{PCl_5} R{-}C(=O){-}Cl$$

$$R{-}C(=O){-}OH \xrightarrow{Cl{-}CO{-}CO{-}Cl} R{-}C(=O){-}Cl$$

$$R{-}C(=O){-}OH \xrightarrow{Cl{-}CO{-}Cl} R{-}C(=O){-}Cl$$

$$R{-}C(=O){-}OH \xrightarrow{PBr_3} R{-}C(=O){-}Br$$

$$RH_2C{-}C(=O){-}OH \xrightarrow{Br_2/PBr_3} RHC(Br){-}C(=O){-}Br$$

Hell–Volhard–Zelinsky reaction; a full equivalent of PBr_3 is used to make the α-bromo acid bromide

2. Acids

$$R{-}CH_2{-}OH \xrightarrow{KMnO_4} R{-}C(=O){-}OH$$

Other oxidizing agents also work; the aldehyde is an intermediate

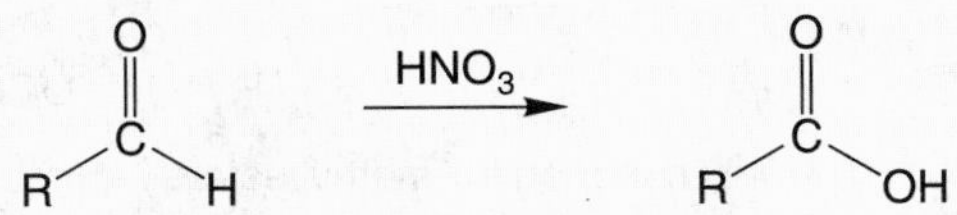

$$R{-}C(=O){-}H \xrightarrow{HNO_3} R{-}C(=O){-}OH$$

Other oxidizing agents work

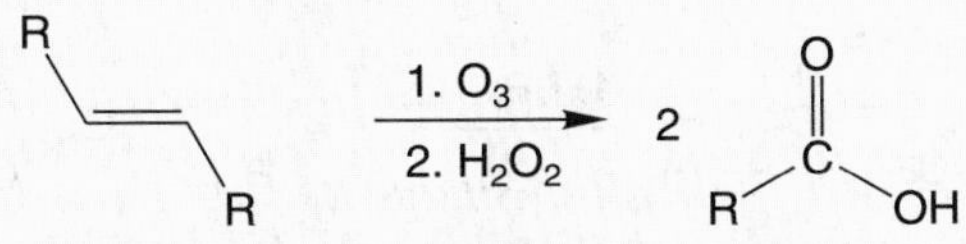

$$RCH{=}CHR \xrightarrow[2.\ H_2O_2]{1.\ O_3} 2\ R{-}C(=O){-}OH$$

Note the oxidative workup; the aldehyde can be obtained by using dimethyl sulfide or H_2/Pd

2. Acids (continued)

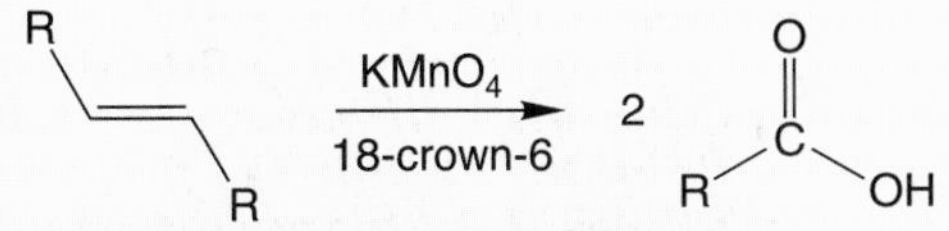

$$RCH{=}CHR \xrightarrow[18\text{-crown-6}]{KMnO_4} 2\ R{-}C(=O){-}OH$$

The crown ether is essential, as it brings the $KMnO_4$ into solution

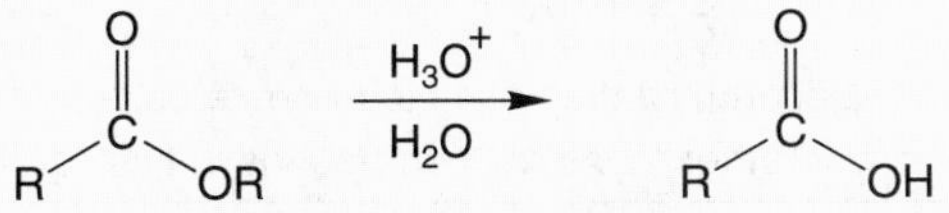

$$R{-}C(=O){-}OR \xrightarrow[H_2O]{H_3O^+} R{-}C(=O){-}OH$$

Acid-catalyzed ester hydrolysis

$$R{-}C(=O){-}OR \xrightarrow[2.\ H_2O/H_3O^+]{1.\ HO^-} R{-}C(=O){-}OH$$

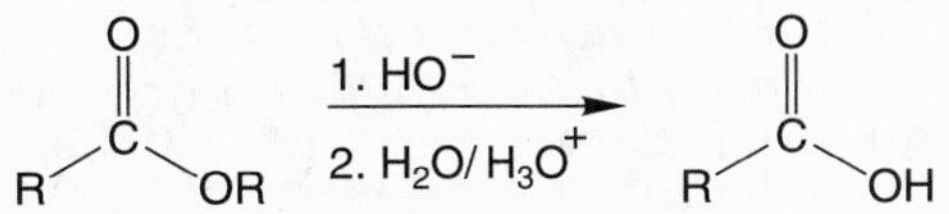

Base-catalyzed ester hydrolysis "saponification"

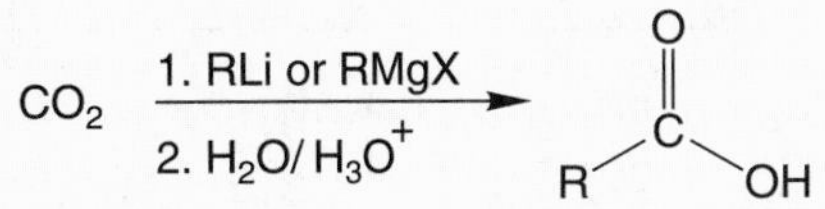

$$CO_2 \xrightarrow[2.\ H_2O/H_3O^+]{1.\ RLi\ \text{or}\ RMgX} R{-}C(=O){-}OH$$

Carboxylation of alkyllithium or Grignard reagents

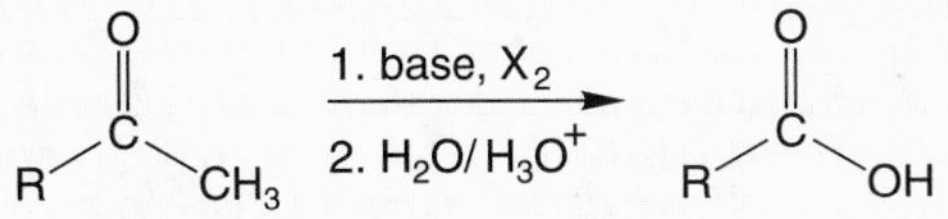

$$R{-}C(=O){-}CH_3 \xrightarrow[2.\ H_2O/H_3O^+]{1.\ \text{base},\ X_2} R{-}C(=O){-}OH$$

This haloform reaction works for methyl ketones and X = I, Br, and Cl. The haloform is the other product

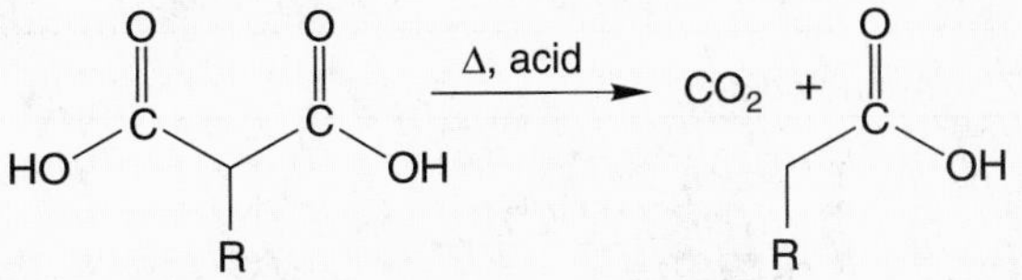

$$HO{-}C(=O){-}CH(R){-}C(=O){-}OH \xrightarrow{\Delta,\ \text{acid}} CO_2 + R{-}CH_2{-}C(=O){-}OH$$

Decarboxylation of malonic acids

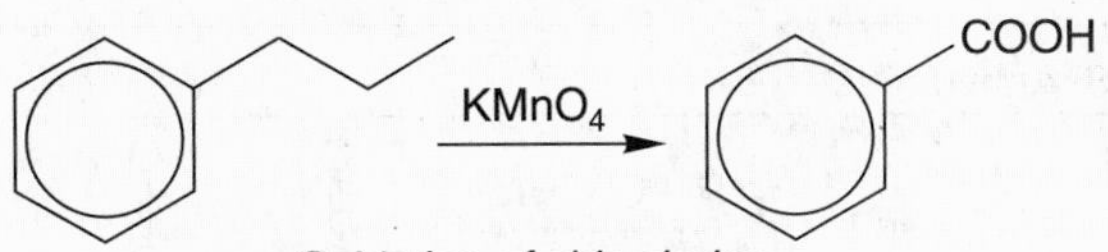

$$C_6H_5{-}CH_2CH_2CH_3 \xrightarrow{KMnO_4} C_6H_5{-}COOH$$

Oxidation of side chains

3. Alcohols

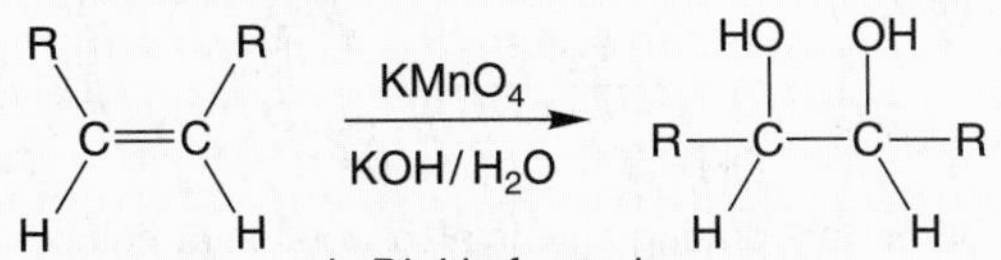

$$RHC{=}CHR \xrightarrow[KOH/H_2O]{KMnO_4} R{-}CH(OH){-}CH(OH){-}R$$

cis Diol is formed

FIGURE 19.84 The synthetic reactions of Chapter 19.

3. Alcohols (continued)

$RCOOH \xrightarrow[2.\ H_2O]{1.\ LiAlH_4} RCH_2OH$

An aluminum alkoxide is a leaving group in this reaction

$ROCOOH \xrightarrow{\Delta} ROH$

Decarboxylation of a bicarbonate ester

See also other "hydroxy" compounds in this figure

4. Alkyl halides

$RCH_2COO^- \ Ag^+ \xrightarrow[\Delta]{Br_2} RCH_2{-}Br + CO_2 + AgBr$

The Hunsdiecker reaction is used almost exclusively for bromides

5. Amides

$RCOOH \xrightarrow[DCC]{NHR_2} RCONR_2$

An addition–elimination process

See also "imides" and "lactams"

6. Amines

$HOCONR_2 \xrightarrow{\Delta} HNR_2 + CO_2$

Decarboxylation of a carbamic acid

7. α-Amino acids

$HOCOCH_2{-}Br \xrightarrow{NHR_2} HOCOCH_2{-}NR_2$

An S_N2 displacement reaction

8. Anhydrides

$RCOCl \xrightarrow{R{-}COOH} RCOOCOR$

This addition–elimination process works even better with the carboxylate anion as nucleophile

$RCOOH \xrightarrow{P_2O_5} RCOOCOR$

Other dehydrating agents also work; intramolecular reactions can often be accomplished by simple heating

9. α-Bromo carboxyl compounds

$RH_2CCOOH \xrightarrow{Br_2/PBr_3} RHC(Br)COBr$

Hell–Volhard–Zelinsky reaction; a full equivalent of PBr_3 is used to make the α-bromo acid bromide

$RHC(Br)COBr \xrightarrow{H_2O,\ HOR,\ NHR_2} RHC(Br)COX$

X = OH, OR, or NR_2

This addition–elimination reaction makes α-bromo acids, esters, and amides

10. α-Cyano acids

$RHC(Br)COOH \xrightarrow{KCN} RHC(CN)COOH$

This reaction is an S_N2 displacement

11. Esters (for cyclic esters see, "lactones")

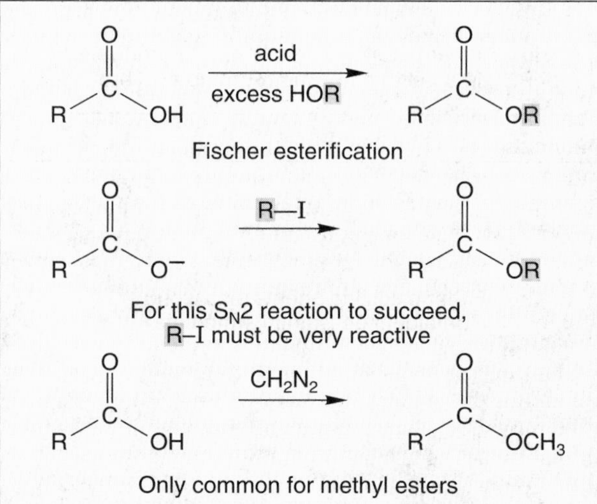
acid
excess HOR
Fischer esterification
R—I
For this S_N2 reaction to succeed, R–I must be very reactive
CH_2N_2
Only common for methyl esters

12. α-Halo acids

$RCH_2COOH \xrightarrow[2.\ H_2O/H_3O^+]{1.\ PBr_3/Br_2} RHC(Br)COOH$

To get the α-bromo acid, a catalytic amount of PBr_3 must be used in the Hell–Volhard–Zelinsky reaction

FIGURE **19.84** (CONTINUED)

12. α-Halo acids (continued)

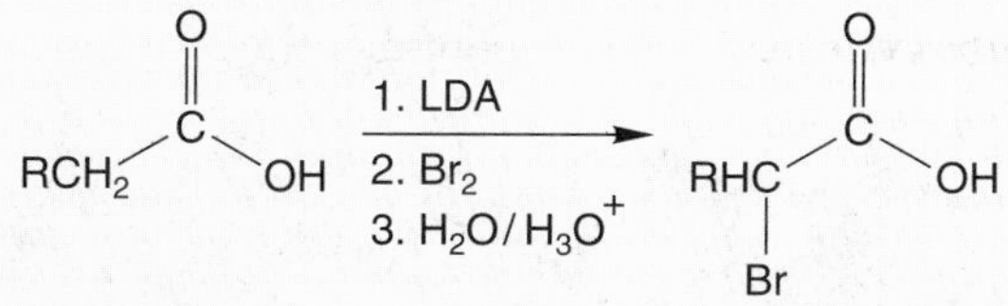

$$RCH_2COOH \xrightarrow[\text{3. } H_2O/H_3O^+]{\text{1. LDA; 2. } Br_2} RHC(Br)COOH$$

The intermediate is the dianion, and two equivalents of the hindered base LDA must be used

13. Hydrocarbons

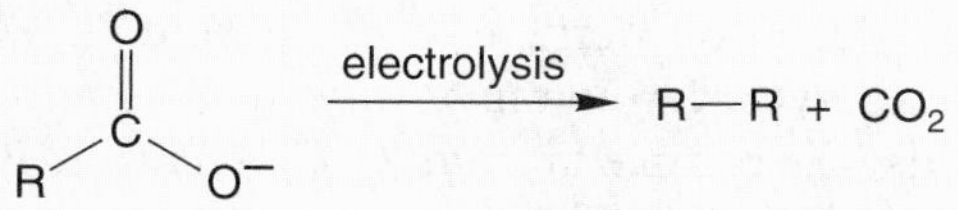

$$RCOO^- \xrightarrow{\text{electrolysis}} R{-}R + CO_2$$

The Kolbe electrolysis, a radical dimerization process

14. α-Hydroxy acids

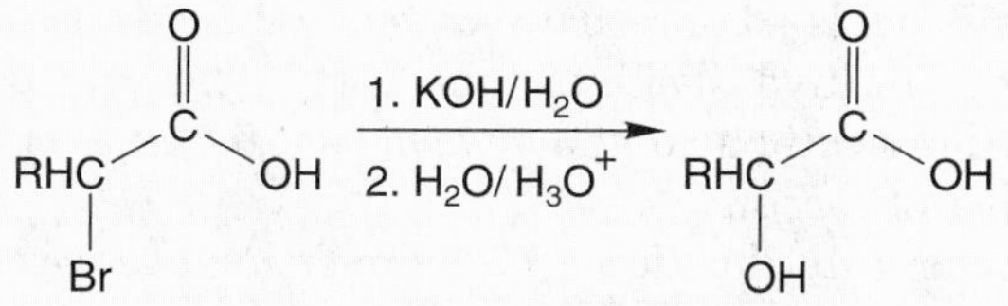

$$RHC(Br)COOH \xrightarrow[\text{2. } H_2O/H_3O^+]{\text{1. KOH/}H_2O} RHC(OH)COOH$$

This reaction is an S_N2 displacement

15. Imides

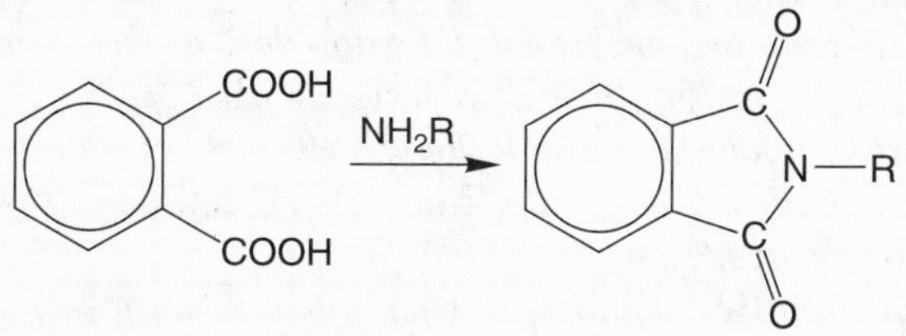

COOH, COOH (phthalic acid) $\xrightarrow{NH_2R}$ N—R (phthalimide)

A double amide formation

16. Ketones

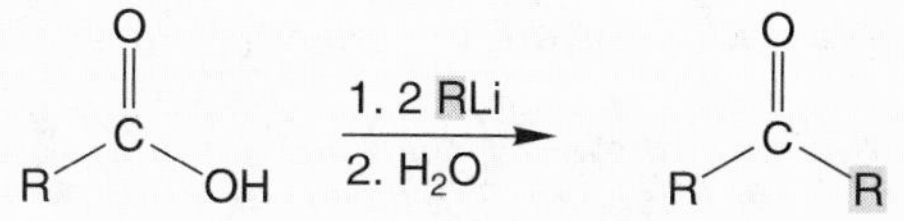

$$RCOOH \xrightarrow[\text{2. } H_2O]{\text{1. 2 RLi}} RCOR$$

The hydrate is an intermediate, 2 mol of RLi are required

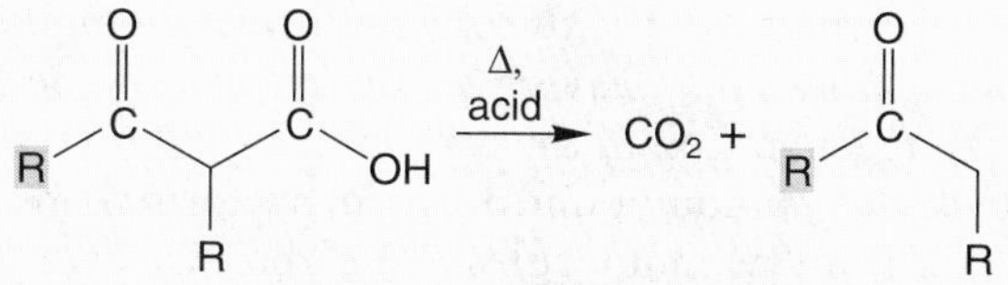

$$RCOCH(R)COOH \xrightarrow{\Delta,\ \text{acid}} CO_2 + RCOCH_2R$$

Decarboxylation of β-keto acids

17. Lactams

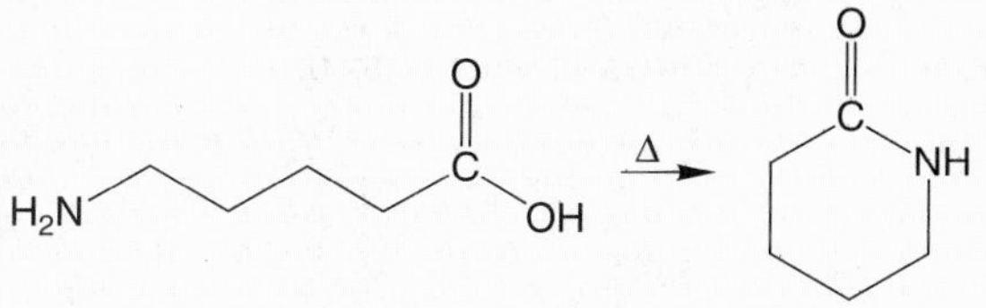

$$H_2N(CH_2)_4COOH \xrightarrow{\Delta} \text{cyclic amide (NH)}$$

This formation of cyclic amides works best for unstrained rings

18. Lactones

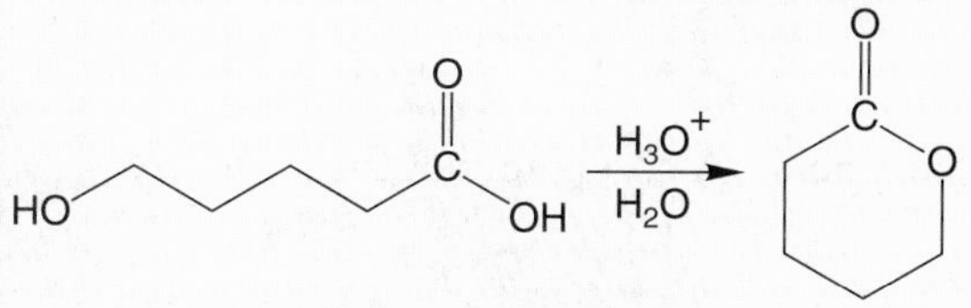

$$HO(CH_2)_4COOH \xrightarrow[H_2O]{H_3O^+} \text{cyclic ester (lactone)}$$

Intramolecular Fischer esterification

FIGURE **19.84** (CONTINUED)

COMMON ERRORS

Addition–elimination reactions are basically simple: The nucleophile adds to the carbonyl, then the carbonyl is reconstituted as a leaving group is displaced. But there is such variety! That's where the difficulty lies—in seeing through all the "spinach" hanging on the nucleophile and hiding the carbonyl to find the simple mechanism at the heart of matters.

Also, there are vast numbers of small steps—proton transfers mostly—in these mechanisms, and it is hard to get them all right.

Finally, there is a small error that all of us seem to make on occasion. This mistake is to add an electrophile to the wrong oxygen of a carboxylic acid. It is the carbonyl oxygen that is the more basic and nucleophilic site, not the hydroxyl oxygen. You can find this error in most textbooks, including, probably, this one.

19.10 KEY TERMS

Acid chloride A compound of the structure

$$R-C(=O)-Cl$$

Addition–elimination reaction A carbonyl group can be attacked to give a tetrahedral intermediate. If the carbon atom of the carbonyl was originally attached to a good leaving group it can now be lost with regeneration of the carbonyl.

Anhydride A compound formed by formal loss of water from two molecules of a carboxylic acid. The structure is

$$R-C(=O)-O-C(=O)-R$$

Amide A compound of the structure

$$R-C(=O)-NR_2$$

Carbamates Esters of carbamic acid. These molecules do not decarboxylate (lose CO_2) easily.

$$RO-C(=O)-NR_2$$

Carbamic acid A compound of the structure

$$HO-C(=O)-NR_2$$

These acids easily decarboxylate to give amines.

Carbonic acid An unstable acid, H_2CO_3.

$$HO-C(=O)-OH$$

Carboxylate anion The resonance-stabilized anion formed on deprotonation of a carboxylic acid.

$$\left[R-C(=O)-O^- \longleftrightarrow R-C(-O^-)=O \right]$$

Carboxylic acid A compound of the structure

$$R-C(=O)-OH$$

Decarboxylation Loss of carbon dioxide, usually from a carboxylic acid.

Detergent A long-chain alkyl sulfonic acid salt.

Dicyclohexylcarbodiimide (DCC) A particularly effective dehydrating agent.

Ester An alkyl derivative of an acid in which the acidic hydrogen is replaced with R′. For carboxylic acids, the change is from R—COOH to R—COOR′.

Ester hydrolysis The conversion of an ester into an acid through treatment with an acid catalyst in excess water. The reverse of Fischer esterification. This reaction also occurs in base, and is called "saponification."

Fatty acids Long-chain carboxylic acids generated by the hydrolysis of fats. Fatty acids are derived from acetic acid and always contain an even number of carbons.

Fischer esterification The conversion of a carboxylic acid into an ester by treatment with an acid catalyst in excess alcohol. The reverse of ester hydrolysis.

Hell–Volhard–Zelinsky reaction The conversion of an acid into either the α-bromo acid or the α-bromo acid bromide through reaction with PBr_3/Br_2.

Hunsdiecker reaction The conversion of silver salts of carboxylic acids into alkyl halides, usually bromides.

Hydrophilic "Water-loving," hence a polar group soluble in water.

Hydrophobic "Water-hating," hence a nonpolar group insoluble in water.

Kolbe electrolysis The electrochemical conversion of carboxylate anions into hydrocarbons. The carboxyl radical produced loses CO_2 to give an alkyl radical that dimerizes.

Lactone A cyclic ester.

Micelle An aggregated group of fatty acid salts (or other molecules) in which a hydrophobic center consisting of the hydrocarbon chains is protected from an aqueous environment by a spherical skin of polar hydrophilic groups.

Ortho esters Compounds of the structure

$$R-C(OR)_3$$

Saponification Base-induced hydrolysis of an ester.

Soap The salt of a fatty acid.

Sodium bicarbonate

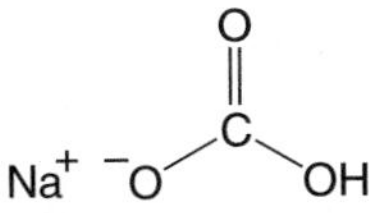

$Na^+ \ ^-O-C(=O)-OH$

Sodium carbonate

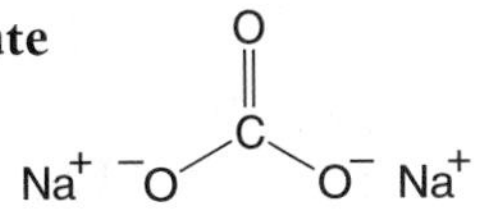

$Na^+ \ ^-O-C(=O)-O^- \ Na^+$

Urea

$H_2N-C(=O)-NH_2$

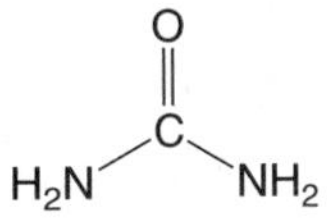

19.11 ADDITIONAL PROBLEMS

PROBLEM **19.24** Write names for the following structures:

(a) cyclopropyl–COOH

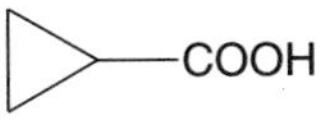

(b) Br, COOH

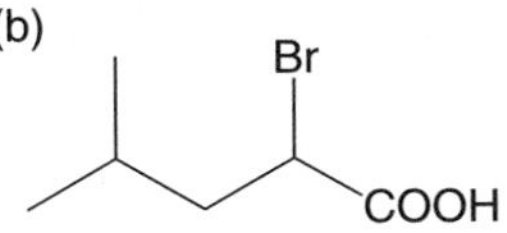

(c) COOH, HO (cyclohexane)

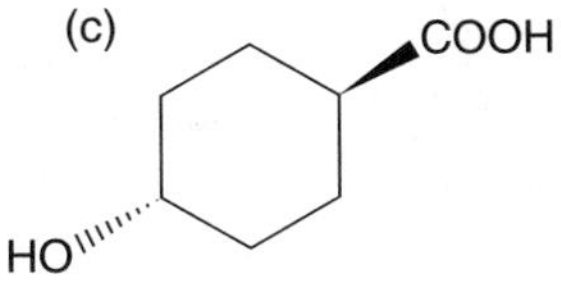

(d) Cl, COOH (benzene ring)

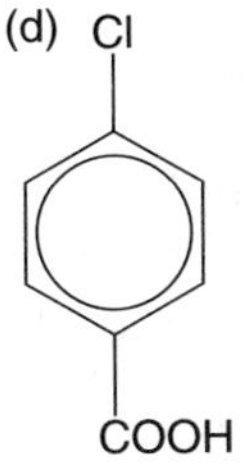

(e) $CH(CH_3)_2$, H_2N, COOH

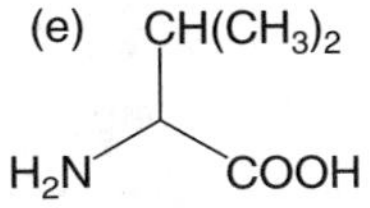

(f) CH_3, C=O

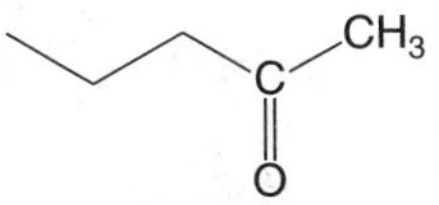

(g) CH_3, OH

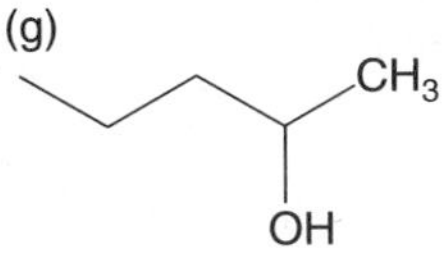

(h) Br, COOH

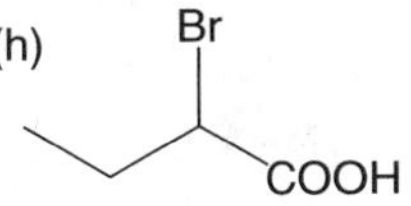

(i) Br, COBr

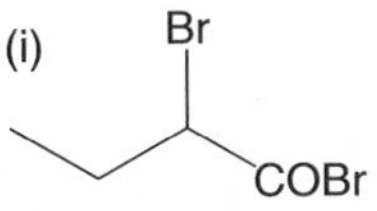

(j) CH_3, COOH

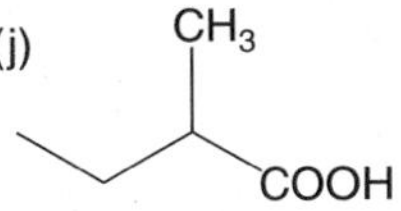

PROBLEM **19.25** Draw structures for the following molecules:

(a) 2-Aminopropanoic acid (alanine)
(b) 2-Hydroxypropanoic acid (lactic acid)
(c) (*Z*)-2-Methyl-2-butenoic acid
(d) 2-Hydroxy-1,2,3-propanetricarboxylic acid (citric acid)
(e) 2-Hydroxybenzoic acid (*o*-hydroxybenzoic acid or salicylic acid)
(f) 2-Oxopropanoic acid (pyruvic acid)

PROBLEM **19.26** Give reagents for converting butyric acid (butanoic acid) into the following compounds:

(a) $C(=O)OCH_3$

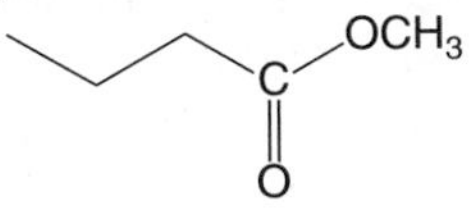

(b) $C(=O)NHCH_3$

(c) $C(=O)Cl$

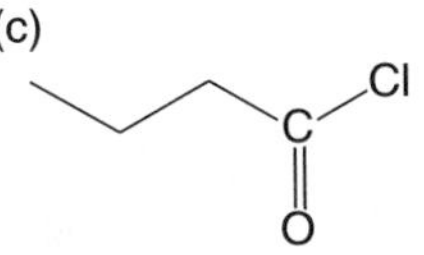

(d) OH

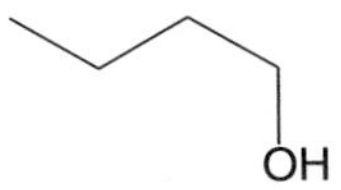

(e) $C(=O)-O-C(=O)$

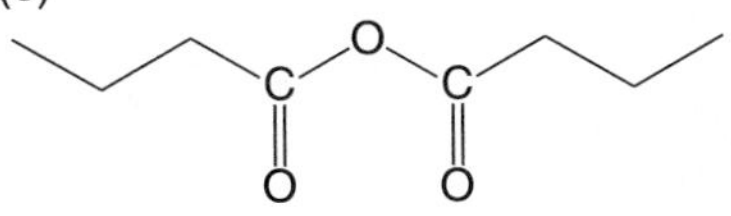

PROBLEM **19.27** Give the major organic products expected in each of the following reactions:

(a) Br, C(=O)OH $\xrightarrow[CH_3OH,\ \Delta]{CH_3\overset{+}{O}H_2}$

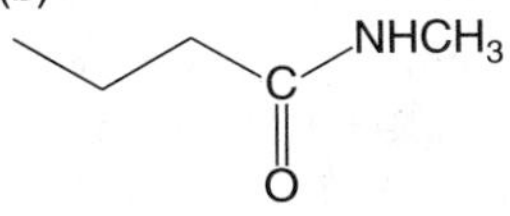

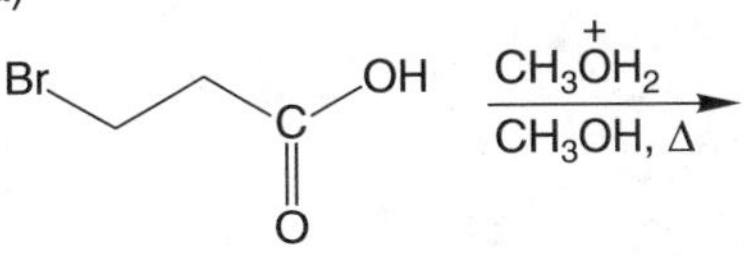

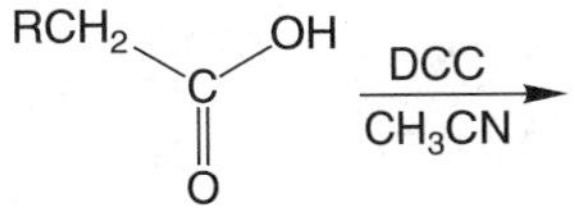

(b) $RCH_2-C(=O)-OH \xrightarrow[CH_3CN]{DCC}$

(c) $C_{13}H_{27}-C(=O)-O$, O, $O-C(=O)-C_{13}H_{27}$, $C_{13}H_{27}$, C=O $\xrightarrow[2.\ H_3\overset{+}{O}/H_2O]{1.\ NaOH/H_2O,\ \Delta}$

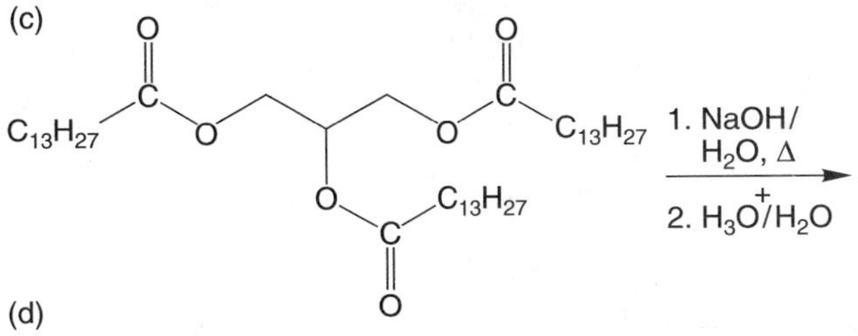

(d) $RCH_2-C(=O)-OH$ + $RRCHNH_2$ $\xrightarrow[CH_3CN]{DCC}$

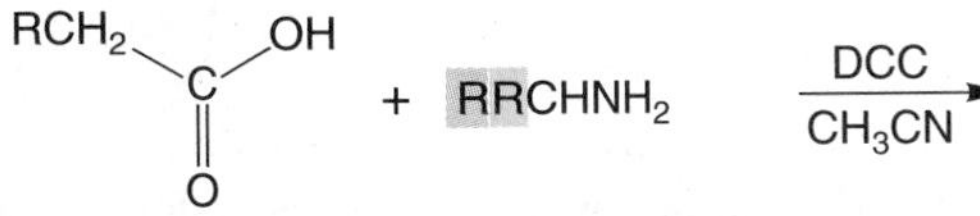

(e) cyclohexyl–C(=O)OH $\xrightarrow[2.\ H_3\overset{+}{O}/H_2O]{1.\ LiH}$

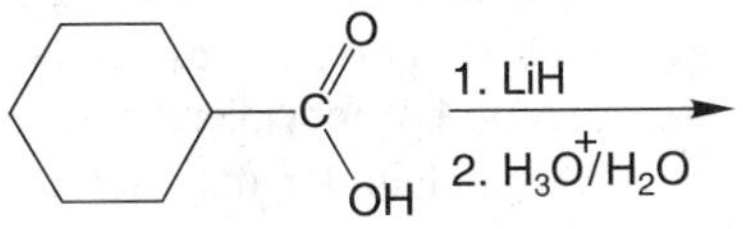

(f) Cyclohexanecarboxylic acid: 1. LiH; 2. CH_3Li; 3. H_2O →

(g) Hexanoic acid: Br_2, PBr_3 (cat.) →

(h) 1-Bromonaphthalene: 1. Mg, ether; 2. CO_2; 3. H_2O/H_3O^+ →

PROBLEM 19.28 Provide brief explanations for the pK_a differences noted below.

(a) HCOOH 3.75; H_3CCOOH 4.75

(b) H_3CCOOH 4.75; NO_2CH_2COOH 1.68

(c) Benzoic acid (C₆H₅–COOH) 4.19; cyclohexanecarboxylic acid 4.88

(d) HOOC–CH=CH–COOH (cis) 1.83; HOOC–CH=CH–COOH (trans) 3.03

(e) HOOC–CH=CH–COO^- (cis) 6.07; HOOC–CH=CH–COO^- (trans) 4.44

PROBLEM 19.29 In Section 19.5b, we saw how dicyclohexylcarbodiimide (DCC) could be used as a dehydrating agent for the formation of amides from carboxylic acids and amines. Another reagent that can be used to good effect is the phosgene derivative *N,N′*-carbonyldiimidazole (CDI). Carboxylic acids and CDI react under mild conditions to form imidazolides, **1**, which then easily react with amines to give amides. Propose mechanisms for the formation of **1** from carboxylic acids and CDI, and for the formation of amides from **1** and amines.

RCOOH + **CDI** —THF, 25 °C→ **1** —RRNH→ RCONR₂

PROBLEM 19.30 In Section 19.5d and Figure 19.53, we saw that one method of forming an anhydride is to allow a dicarboxylic acid to react with another anhydride such as acetic anhydride, Ac_2O. Propose a mechanism for this process.

Phthalic acid + H_3C–CO–O–CO–CH_3 → phthalic anhydride + 2 H_3CCOOH

PROBLEM 19.31 In Problem 19.12, you worked out the mechanism for lactone formation from an hydroxy acid. Here is a related reaction, the formation of a cyclic amide, a lactam. Provide a mechanism.

H_2N–CH₂CH₂CH₂–COOH —acid catalyst, 25 °C, 5 h→ A lactam (97%)

PROBLEM **19.32** There is a compound called the Vilsmeier reagent (we will ask you about its formation in Chapter 20) that reacts with carboxylic acids to give acid chlorides and the molecule dimethylformamide (DMF). Provide a mechanism for this process.

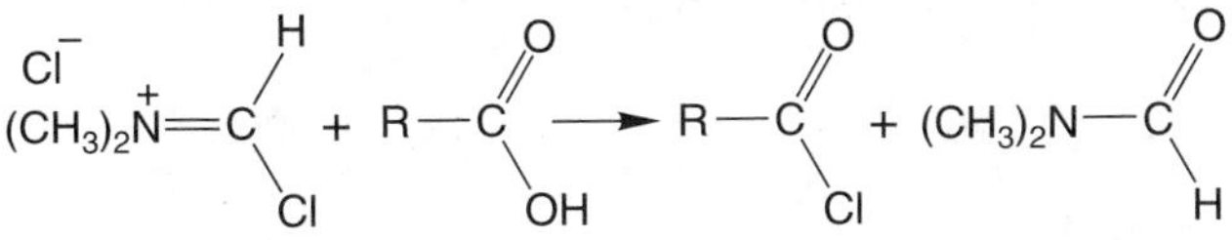

$$\mathrm{Cl^-}\ (CH_3)_2\overset{+}{N}{=}C(H)(Cl) + R{-}C({=}O)OH \longrightarrow R{-}C({=}O)Cl + (CH_3)_2N{-}C({=}O)H$$

Vilsmeier reagent — **DMF**

PROBLEM **19.33** Propose syntheses of the following molecules from the readily available fatty acid, lauric acid (**1**).

$CH_3(CH_2)_{10}COOH$ (**1**) ⟶

(a) $CH_3(CH_2)_{11}Br$

(b) $CH_3(CH_2)_{10}Br$

(c) $CH_3(CH_2)_{11}COOH$

PROBLEM **19.34** Propose an arrow formalism mechanism for the following reaction. *Hint*: See Section 19.5f and Figure 19.65.

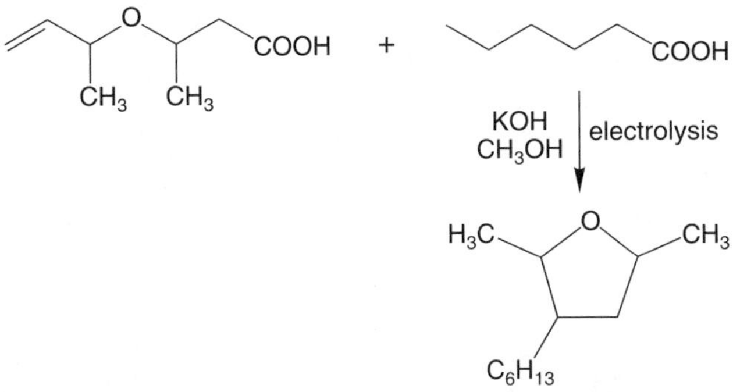

[Structures: O, COOH, CH_3, CH_3 + COOH; KOH, CH_3OH, electrolysis; product: H_3C, O, CH_3, C_6H_{13}]

PROBLEM **19.35** Explain the following observations. Note that the only difference in the starting materials is the stereochemistry of the carbon–carbon double bond.

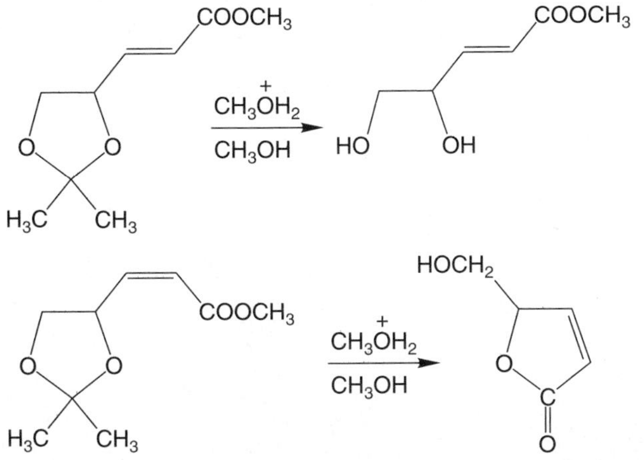

[Structures: $COOCH_3$, O, O, H_3C, CH_3 ⟶ ($CH_3\overset{+}{O}H_2$, CH_3OH) HO, OH, $COOCH_3$]

[Structures: $COOCH_3$, O, O, H_3C, CH_3 ⟶ ($CH_3\overset{+}{O}H_2$, CH_3OH) $HOCH_2$, O, C, O]

PROBLEM **19.36** Here are some somewhat more complicated syntheses. Devise ways to make the compounds shown below from cyclopentanecarboxylic acid.

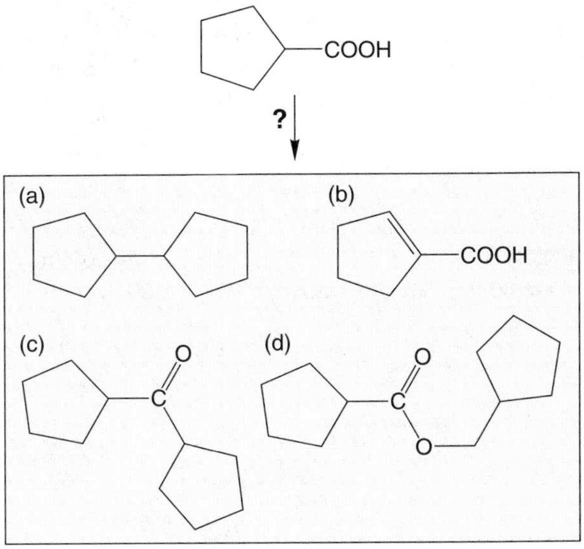

[Structure: cyclopentane–COOH ⟶ ?]

(a) [bicyclopentyl]

(b) [cyclopentene–COOH]

(c) [dicyclopentyl ketone: C=O]

(d) [ester: C(=O)–O–CH_2–cyclopentyl]

PROBLEM **19.37** Estragole (**1**) is a major constituent of tarragon oil. Treatment of **1** with hydrogen bromide in the presence of peroxides affords **2**. Compound **2** reacts with Mg in ether, followed by addition of carbon dioxide and acidification, to give **3**. Spectral data for compound **3** are summarized below. Deduce the structure of compound **3** and propose structures for estragole and compound **2**.

Compound **3**

Mass spectrum: $m/z = 194$ (p)
IR (Nujol): 3330–2500 (br, s), 1696 (s) cm^{-1}
^1H NMR ($CDCl_3$): δ 1.95 (quintet, $J = 7$ Hz, 2H), 2.34 (t, $J = 7$ Hz, 2H), 2.59 (t, $J = 7$ Hz, 2H), 3.74 (s, 3H), 6.75 (d, $J = 8$ Hz, 2H), 7.05 (d, $J = 8$ Hz, 2H), 11.6 (s, 1H)

PROBLEM **19.38** Reaction of cyclopentadiene and maleic anhydride affords compound **1**. Hydrolysis of **1** leads to **2**, which gives an isomeric compound **3** upon treatment with concentrated sulfuric acid. Spectral data for compounds **1–3** are shown below. Propose structures for compounds **1–3** and provide mechanisms for their formations.

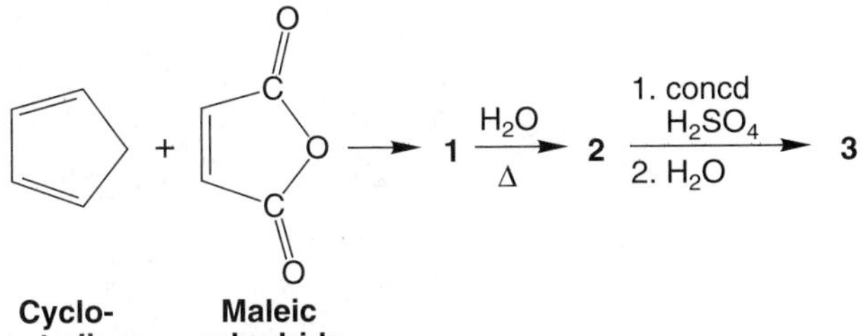

Cyclopentadiene + **Maleic anhydride** ⟶ **1** $\xrightarrow[\Delta]{H_2O}$ **2** $\xrightarrow[\text{2. } H_2O]{\text{1. concd } H_2SO_4}$ **3**

Compound **1**

Mass spectrum: m/z = 164 (p, 3%), 91 (23%), 66 (100%), 65 (22%)
IR (Nujol): 1854 (m) and 1774 (s) cm^{-1}
^1H NMR ($CDCl_3$): δ 1.60–1.70 (m, 1H), 3.40–3.50 (m, 1H), 3.70–3.80 (m, 1H), 6.20–6.30 (m, 1H)

Compound **2**

IR (Nujol): 3450-2500 (br, s), 1710 (s) cm^{-1}
^1H NMR ($CDCl_3$): δ 0.75–0.95 (m, 1H), 2.50–2.60 (m, 1H), 2.70–2.80 (m, 1H), 5.60–5.70 (m, 1H), 11.35 (s, 1H)

Compound **3**

IR (KBr): 3450–2500 (br, s), 1770 (s), 1690 (s) cm^{-1}
^1H NMR ($CDCl_3$): δ 1.50–1.75 (m, 3H), 1.90–2.05 (m, 1H), 2.45–2.55 (m, 1H), 2.65–2.75 (m, 1H), 2.90–3.05 (m, 1H), 3.20–3.30 (m, 1H), 4.75–4.85 (m, 1H), 12.4 (s, 1H)

Derivatives of Carboxylic Acids: Acyl Compounds

Strange shadows from the flames
will grow,
Till things we've never seen seem
familiar ...
—Jerry Garcia* and Robert Hunter*
Terrapin Station

In this chapter, we focus upon things that should seem familiar. Many of the molecules of Chapter 19 will reappear, for example. Acid halides, anhydrides, esters, and amides are all **acyl** compounds of the general structure R–CO–L. These compounds are shown in Figure 20.1 along with the related nitriles (cyanides) and ketenes.

Acids

Esters — Acid chlorides — Acid bromides — Amides — Anhydrides: Acyl compounds

Nitriles (cyanides) — Ketenes: Acyl-like compounds

FIGURE 20.1 Some acyl and related derivatives of carboxylic acids.

*Jerry Garcia (1942–1995) played lead guitar with the Grateful Dead and Robert Hunter (b. 1941) was a frequent songwriting collaborator.

Not only are the structural types of these molecules related, but their chemistries are also closely intertwined—a reaction of one acyl compound is generally a synthesis of another, for example. Luckily, there are unifying properties of these seemingly disparate compounds. The structures of the acyl compounds are determined largely by the interaction between the L group and the carbonyl, and their chemistries are dominated by addition–elimination mechanisms in which the L group is replaced with some other substituent (Chapter 19). Be sure to note that this addition–elimination process is an equilibrium, and can proceed in both the "forward" and "backward" directions. The old leaving group L:⁻ is also a nucleophile (Fig. 20.2).

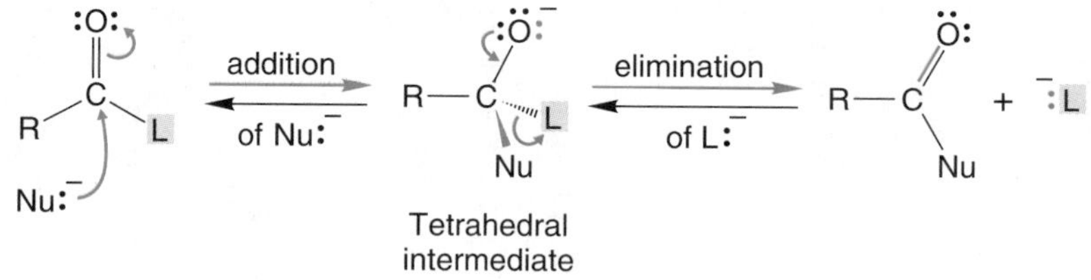

addition of Nu:⁻ — Tetrahedral intermediate — elimination of L:⁻

FIGURE 20.2 The most important mechanism for reactions of acyl compounds is the addition–elimination process. Note the tetrahedral intermediate at the center of this generic mechanism.

Esters, in particular, have a rich chemistry of the enolate anions produced by removal of an α-hydrogen. We will see some new condensation reactions that will greatly expand your ability to do "Fun in Base" problems.

We will first go over some common ground—nomenclature, structure, and spectra—and then go on to examine reactions and syntheses for the different derivatives.

20.1 NOMENCLATURE

All the names of these compounds are based on the parent carboxylic acid. For molecules containing up to five carbons, common names prevail, but after that the IUPAC system takes over. For small **esters** (R–COOR′), the R group is named first and the "ic" of the parent acid is replaced with the suffix "ate." So we have alkyl (methyl, ethyl, etc.) formates, acetates, propionates, and butyrates before we get to the systematic pentanoates. In the systematic naming protocol used for larger esters, the R′ group of the OR′ is again first named and then the suffix "oate" replaces the final "e" of the parent alkane. As usual, the carbonyl group of the acid or ester gets top priority in the naming scheme and substituents are numbered with respect to the carbonyl as position 1 (Fig. 20.3).

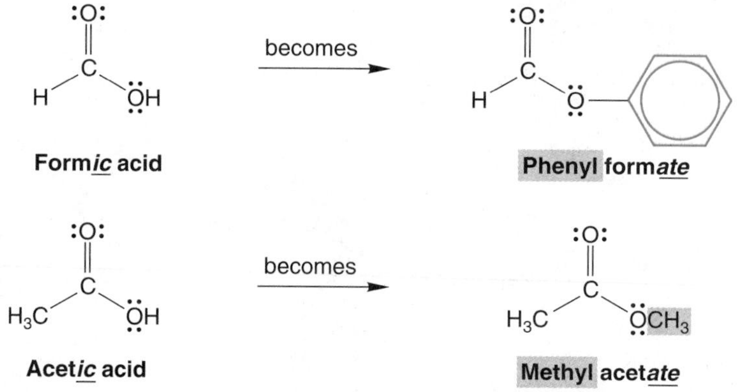

Formic acid becomes Phenyl formate

Acetic acid becomes Methyl acetate

FIGURE 20.3 Some examples of the naming protocol for esters.

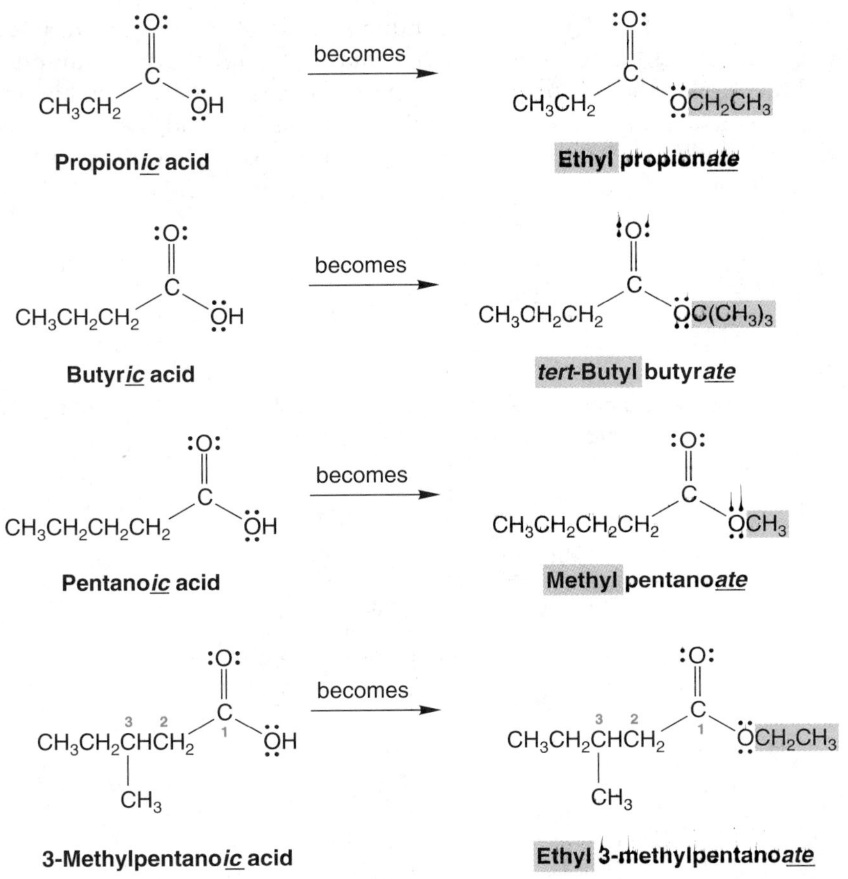

CH_3CH_2COOH → becomes → $CH_3CH_2CO_2CH_2CH_3$

Propion*ic* acid → **Ethyl propion*ate***

$CH_3CH_2CH_2COOH$ → becomes → $CH_3CH_2CH_2CO_2C(CH_3)_3$

Butyr*ic* acid → ***tert*-Butyl butyr*ate***

$CH_3CH_2CH_2CH_2COOH$ → becomes → $CH_3CH_2CH_2CH_2CO_2CH_3$

Pentano*ic* acid → **Methyl pentano*ate***

$CH_3CH_2CH(CH_3)CH_2COOH$ → becomes → $CH_3CH_2CH(CH_3)CH_2CO_2CH_2CH_3$

3-Methylpentano*ic* acid → **Ethyl 3-methylpentano*ate***

FIGURE 20.3 (CONTINUED)

Cyclic esters are called **lactones** (Chapter 19, p. 968). The IUPAC system names these compounds as "oxa" cycloalkanones. The ring size of a lactone can be described by starting at the carbonyl carbon and designating the other carbons in the ring with Greek letters until the oxygen atom is reached. A three-membered ring is an α-lactone, a four-membered ring is a β-lactone, and so on (Fig. 20.4).

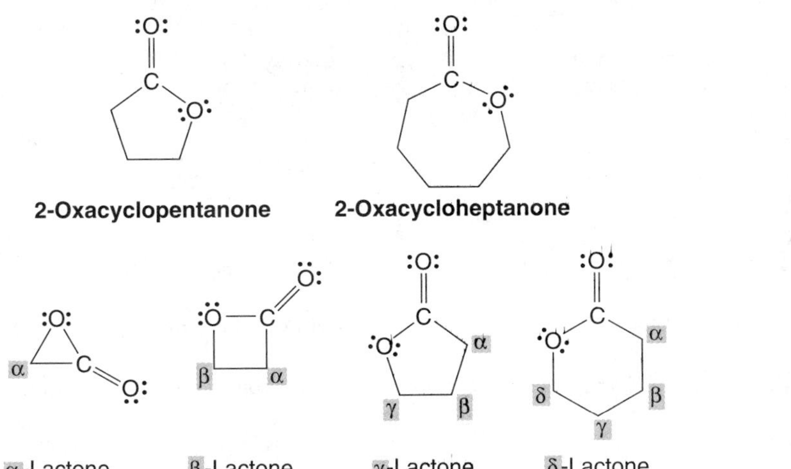

FIGURE 20.4 Cyclic esters are lactones or oxacycloalkanones.

Acid halides are named systematically by replacing the final "e" of the parent alkane with "oyl halide." In practice, the names for the smaller acid halides are based upon the common names for the related acids. Thus, "acetyl chloride," "propionyl bromide," and "butyryl fluoride" would be understood everywhere. Formyl chloride is most unstable and efforts to make it are rewarded only with HCl and CO. The diacid chloride of formic acid (Cl–CO–Cl) is known, however, and this deadly poison is called phosgene (Fig. 20.5).

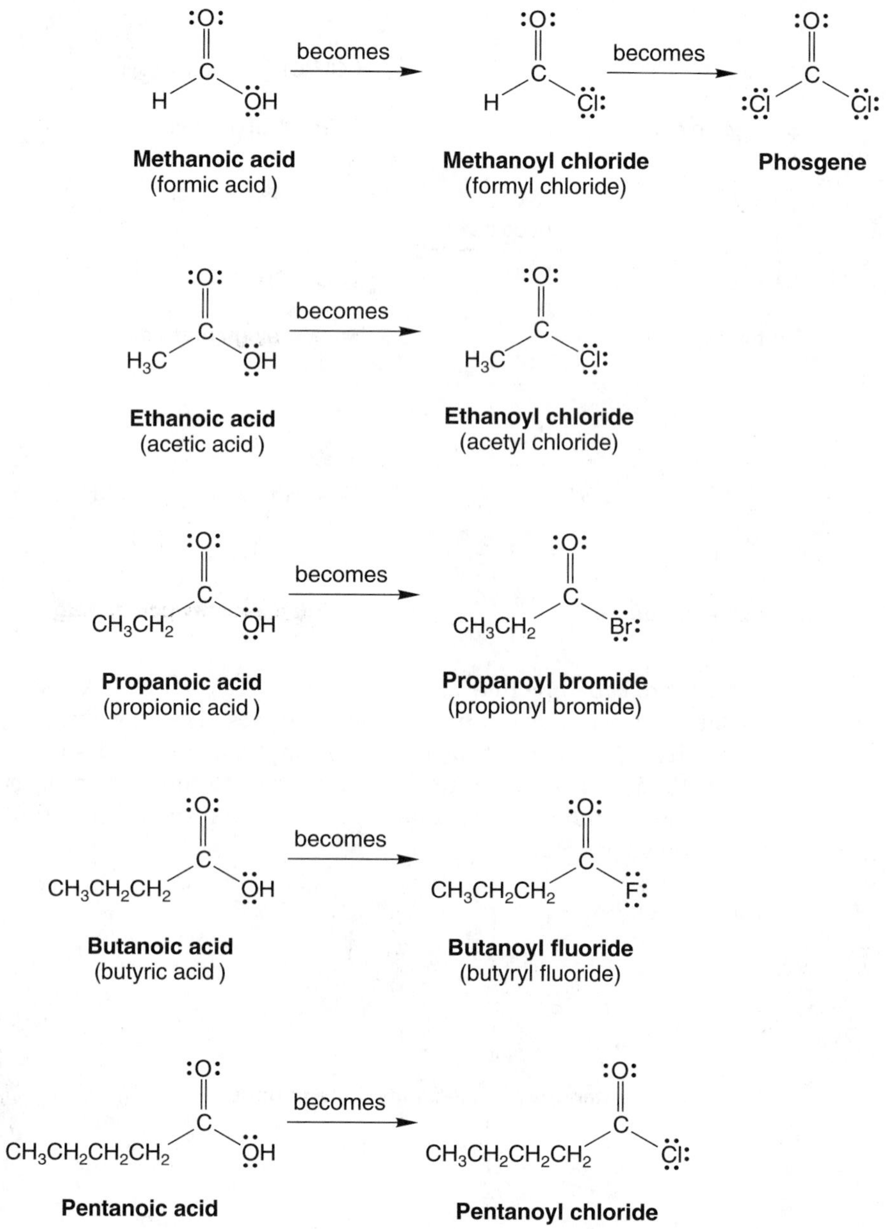

Methanoic acid (formic acid) becomes Methanoyl chloride (formyl chloride) becomes Phosgene

Ethanoic acid (acetic acid) becomes Ethanoyl chloride (acetyl chloride)

Propanoic acid (propionic acid) becomes Propanoyl bromide (propionyl bromide)

Butanoic acid (butyric acid) becomes Butanoyl fluoride (butyryl fluoride)

Pentanoic acid becomes Pentanoyl chloride

FIGURE 20.5 Some examples of the systematic and common naming protocols for acid halides.

Anhydrides are named by reference to the acid from which they are formally derived by loss of water. Both systematic and common names are used in this process (Fig. 20.6).

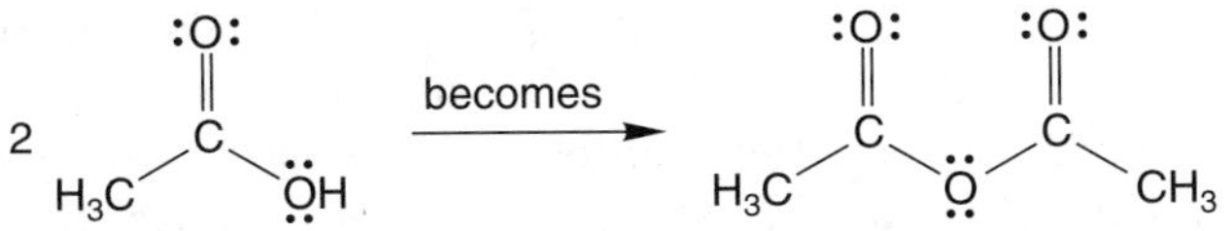

2 H_3C–C(=O:)–ÖH becomes H_3C–C(=O:)–Ö–C(=O:)–CH_3

Ethanoic acid (acetic acid) → **Ethanoic anhydride** (acetic anhydride)

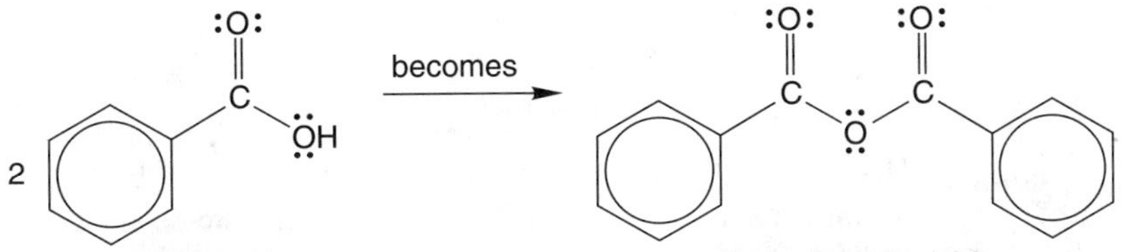

2 **Benzoic acid** becomes **Benzoic anhydride**

FIGURE 20.6 Two examples of the naming protocol for symmetrical anhydrides.

Mixed anhydrides present more problems. Here, both constituent acids must be named alphabetically. Some examples appear in Figure 20.7.

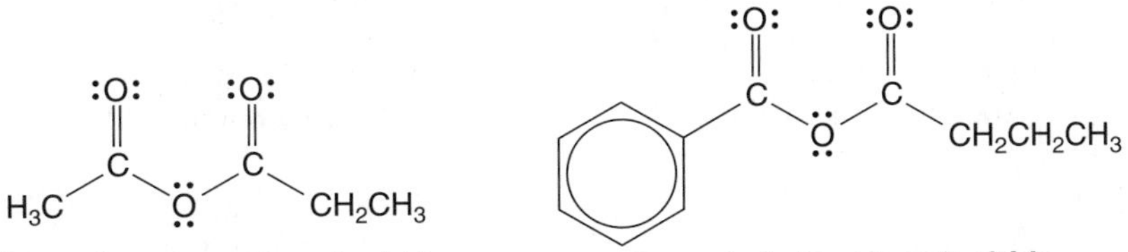

H_3C–C(=O:)–Ö–C(=O:)–CH_2CH_3

Ethanoic propanoic anhydride (acetic propionic anhydride)

C_6H_5–C(=O:)–Ö–C(=O:)–$CH_2CH_2CH_3$

Benzoic butanoic anhydride (benzoic butyric anhydride)

FIGURE 20.7 Two examples of the naming protocol for unsymmetrical anhydrides.

Cyclic anhydrides are named as derivatives of the related carboxylic acids (Fig. 20.8).

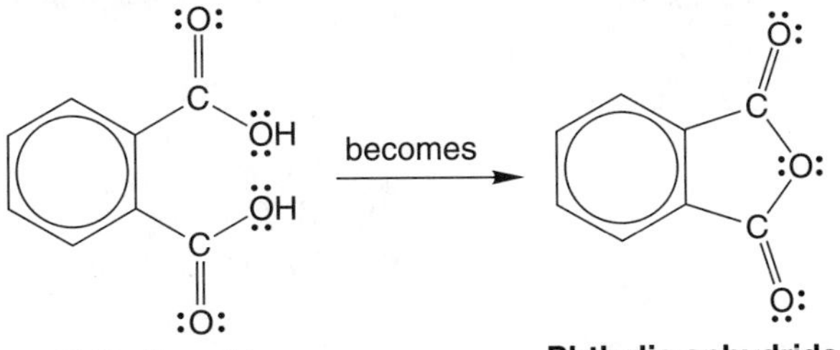

Phthalic acid becomes **Phthalic anhydride**

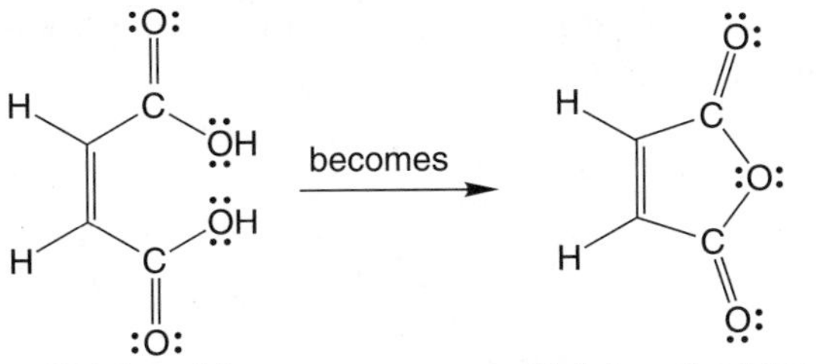

Maleic acid becomes **Maleic anhydride**

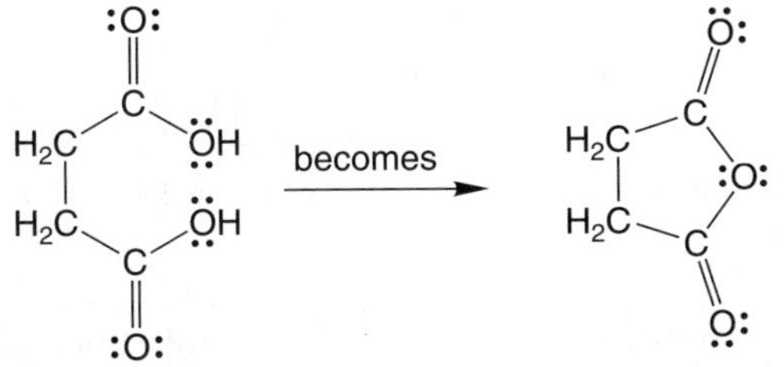

Succinic acid becomes **Succinic anhydride**

FIGURE 20.8 Some cyclic anhydrides.

Amides, the nitrogen counterparts of esters, are systematically named by dropping the final "e" of the parent alkane and adding the suffix "amide." As usual, the smaller members of the class have retained their common names, which are based on the common names for the smaller carboxylic acids. Substitution on nitrogen is indicated with a prefix *N*, as in *N*-methyl, *N,N*-diethyl, and so on (Fig. 20.9).

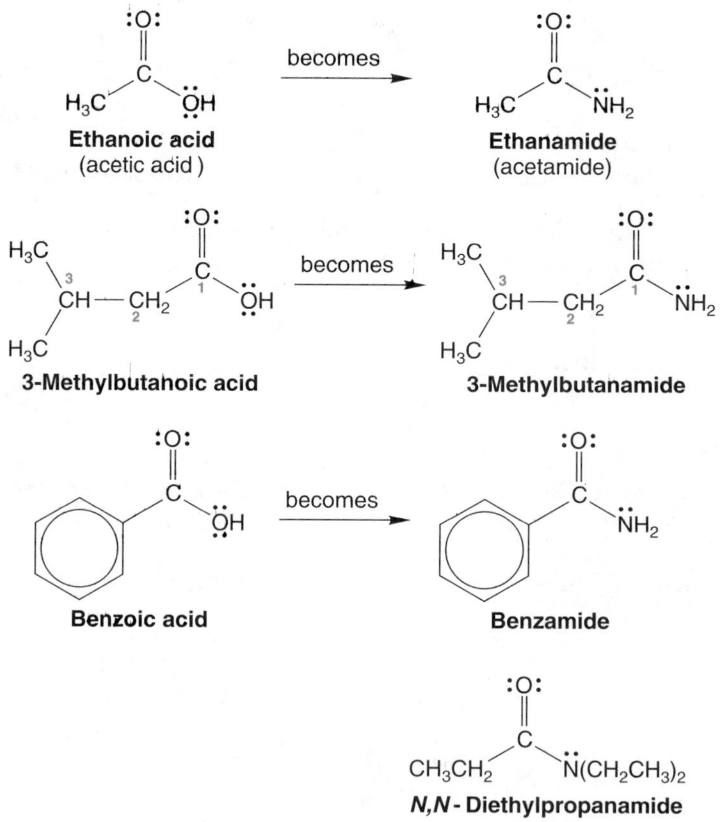

Ethanoic acid (acetic acid) becomes Ethanamide (acetamide)

3-Methylbutanoic acid becomes 3-Methylbutanamide

Benzoic acid becomes Benzamide

N,N-Diethylpropanamide

FIGURE 20.9 Examples of the naming protocol for amides.

Cyclic amides are known as **lactams**, and the naming protocol closely follows that for lactones, the cyclic esters (p. 1005). Lactams are named systematically as "azacycloalkanones." **Imides** are the nitrogen counterparts of anhydrides (Fig. 20.10)

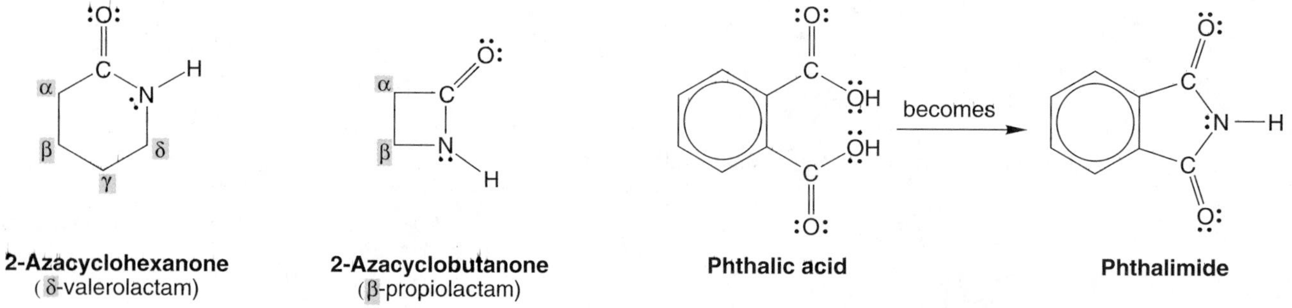

2-Azacyclohexanone (δ-valerolactam)

2-Azacyclobutanone (β-propiolactam)

Phthalic acid becomes Phthalimide

FIGURE 20.10 Cyclic amides are called lactams, and imides are the nitrogen counterparts of anhydrides.

Compounds containing carbon–nitrogen triple bonds, **nitriles**, are also known as **cyanides**. Systematically, these compounds are alkanenitriles.

Notice that the final "e" of the alkane is not dropped. Small nitriles are commonly named as derivatives of the parent carboxylic acid or as cyanides (Fig. 20.11).

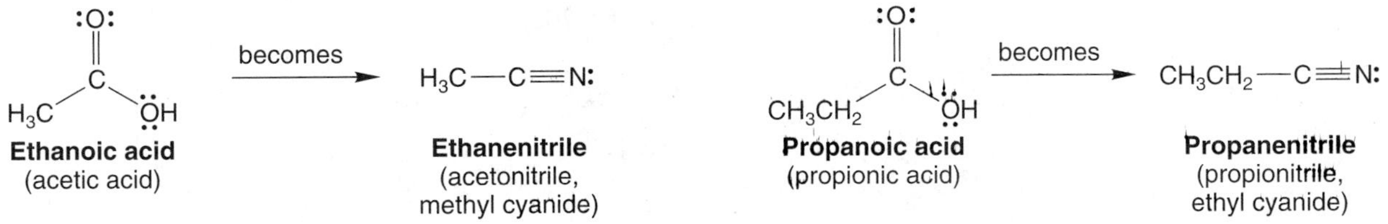

FIGURE **20.11** Some examples of the naming protocol for nitriles, or cyanides.

Ketenes (not to be confused with ketones), are named simply as derivatives of the parent unsubstituted ketene. Recall the allenes, compounds with two directly attached carbon–carbon double bonds (Chapter 12, p. 500; Fig. 20.12).

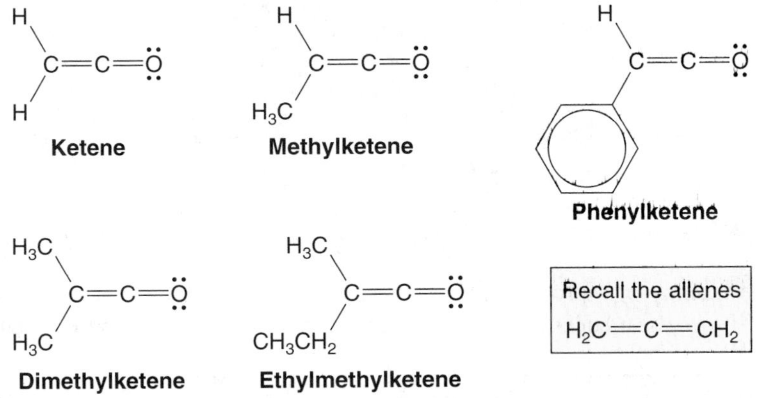

FIGURE **20.12** Ketenes are named as derivatives of the parent unsubstituted ketene.

20.2 PHYSICAL PROPERTIES AND STRUCTURES OF ACYL COMPOUNDS

Acyl compounds are quite polar and have boiling points substantially higher than those of the alkanes from which they are formally derived. Amides, like carboxylic acids, form hydrogen-bonded dimers and oligomers, and are exceptionally high boiling. Table 20.1 gives some physical properties of the acyl compounds related to acetic acid.

TABLE **20.1** Some Properties of Acyl Compounds Related to Acetic Acid.

Formula	Name	bp (°C)	mp (°C)
CH_3COOH	Acetic acid	117.9	16.6
CH_3COOCH_3	Methyl acetate	57	−98.1
$CH_3COOCH_2CH_3$	Ethyl acetate	77	−83.6
CH_3COCl	Acetyl chloride	50.9	−112
CH_3COBr	Acetyl bromide	76	−98
$CH_3CO{-}O{-}COCH_3$	Acetic anhydride	139.6	−73.1
CH_3CONH_2	Acetamide	221.2	82.3
CH_3CN	Acetonitrile	81.6	−45.7
$CH_2{=}C{=}O$	Ketene	−56	−151

Acyl compounds lack the hydroxyl group of carboxylic acids, and so are not strong Brønsted acids. However, hydrogens in the α-position are acidic for the same reasons that α-hydrogens are acidic in aldehydes and ketones. The conjugate bases are resonance stabilized by the adjacent carbonyl or nitrile group (Fig. 20.13).

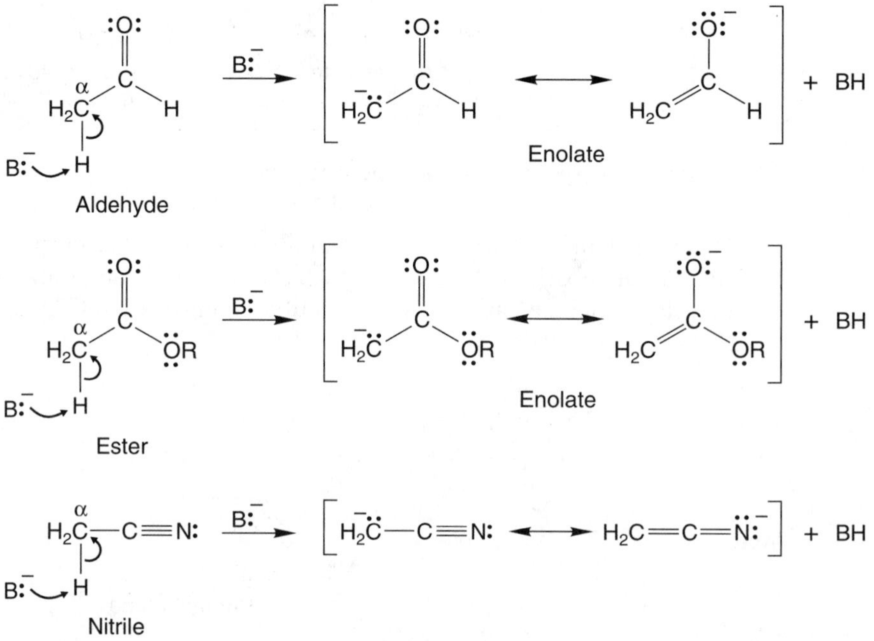

Aldehyde — Enolate — + BH

Ester — Enolate — + BH

Nitrile — + BH

FIGURE **20.13** A hydrogen at the α-position of an acyl compound or nitrile is acidic and can be removed by base to give an enolate, or, in the case of nitriles, a resonance-stabilized species much like an enolate.

The lability of a proton at the α-position varies, however, depending on the ability of the adjacent carbonyl group to provide stabilization (Table 20.2).

TABLE **20.2** The Acidity of Some Acyl Compounds[a]

Compound	Name	pK_a
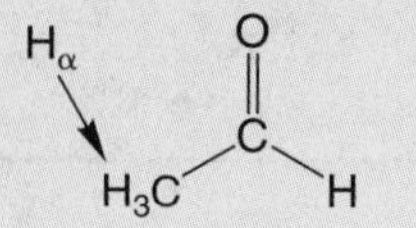 H_3C–C(=O)–H	Acetaldehyde	16.5
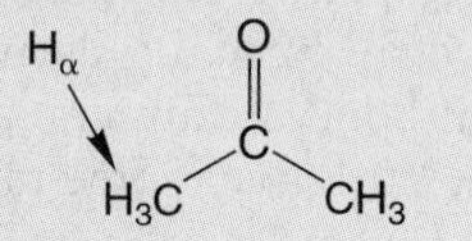 H_3C–C(=O)–CH_3	Acetone	18.9
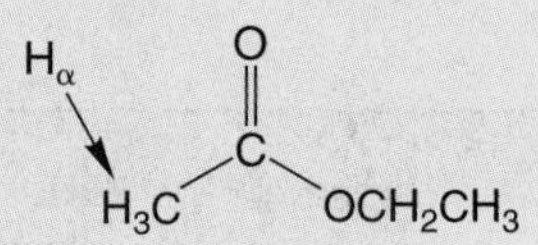 H_3C–C(=O)–OCH_2CH_3	Ethyl acetate	24
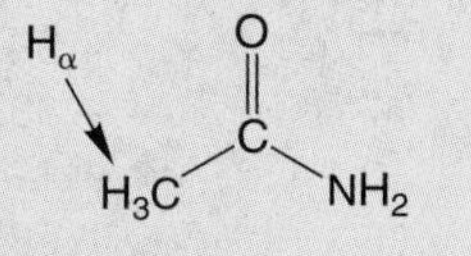 H_3C–C(=O)–NH_2	Acetamide	25
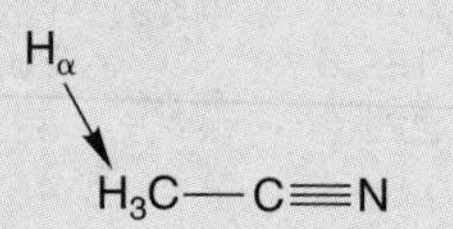 H_3C–C≡N	Acetonitrile	24

[a] The arrow shows the proton lost.

Esters and amides are more stabilized by resonance than are aldehydes and ketones. Therefore their carbonyl groups have less double-bond character, and they are weaker acids at the α-position. Notice the connection between resonance stabilization and acidity. The more resonance stabilized the starting material, the more difficult it will be to remove a proton from it.

Amides are the most resonance stabilized of acyl compounds, as in the polar resonance form it is the relatively electropositive nitrogen atom that bears the positive charge. For this reason these compounds are the weakest acids (Table 20.2; Fig. 20.14).

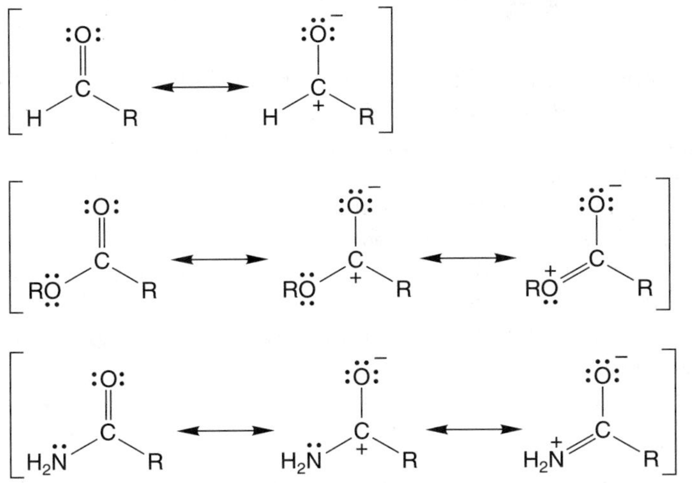

FIGURE **20.14** Esters and amides are more stablized by resonance than aldehydes and ketones—it is harder to remove an α-hydrogen from the R group of these more stable species. They are weaker acids as shown by their higher pK_a values.

PROBLEM **20.1**

Draw structures for the following compounds: cyclopropyl 2-methylbutyrate, 3-chlorobutanoyl chloride, benzoic propionic anhydride, *N,N*-diethyl-4-phenylpentanamide, and ethylpropylketene.

Acyl compounds are also Lewis acids. The carbonyl group provides the locus of Lewis acidity for most of them, but nitriles, even though they have no carbon–oxygen double bond, are still Lewis acids. The electronegative nitrogen strongly polarizes the carbon–nitrogen triple bond, which can react like a carbonyl group in reactions with nucleophiles (Fig. 20.15).

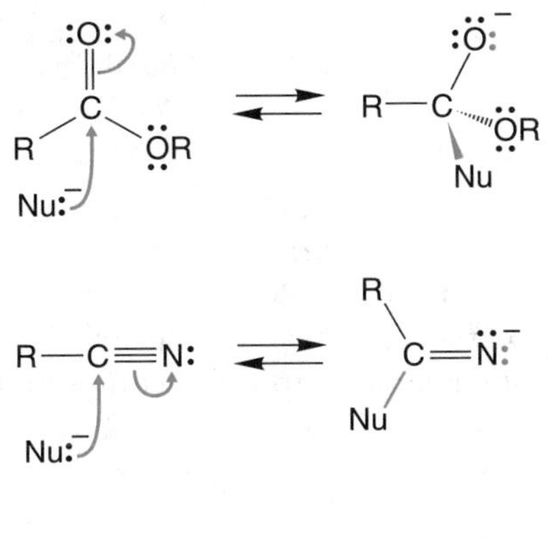

FIGURE **20.15** The carbonyl group of acyl compounds is the center of Lewis acidity and can be attacked by nucleophiles (Nu:⁻) just as are aldehydes and ketones. An analogous reaction occurs with nitriles.

Acyl compounds are Brønsted bases and Lewis bases as well, with the carbonyl oxygen or the nitrile nitrogen the center of basicity (Fig. 20.16).

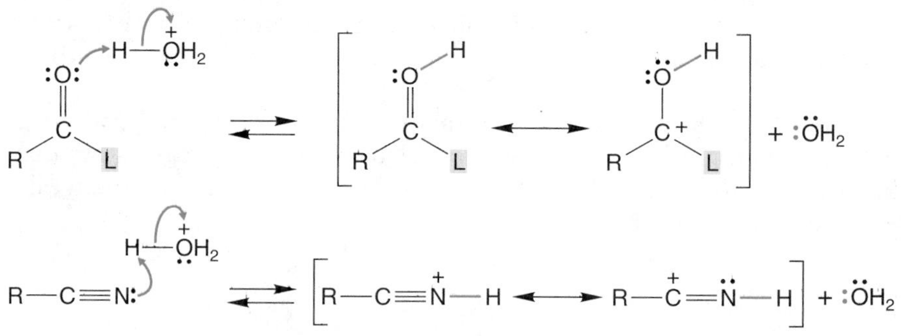

FIGURE **20.16** As these protonation reactions show, acyl compounds and nitriles can act as Brønsted bases.

The carbonyl-containing molecules vary in their base strengths in the following order: amides > esters > acid halides. As we have seen, all these compounds are stabilized by dipolar resonance forms in which the L group bears a positive charge and the carbonyl oxygen a negative charge (Fig. 20.14). Acid chlorides are the least stabilized by this resonance, and therefore have the least basic carbonyl oxygen atoms.

Esters, anhydrides, and acids are similarly stabilized by resonance, but in amides it is a nitrogen atom, not an oxygen atom, that bears the positive charge. Nitrogen is less electronegative than oxygen and therefore bears this positive charge with more grace than oxygen. The result is a more polar carbonyl group and a more basic carbonyl oxygen atom (Fig. 20.17).

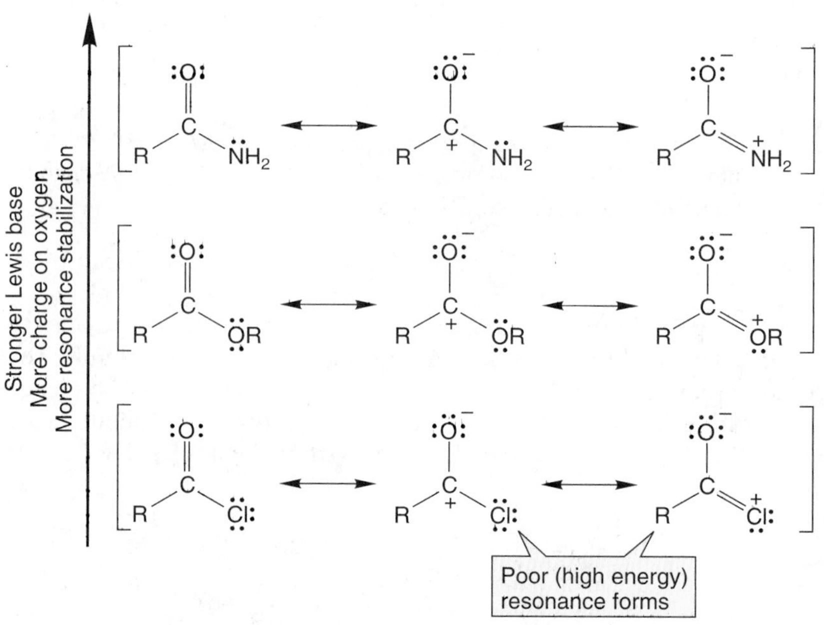

FIGURE **20.17** The carbonyl oxygens of amides are stronger Lewis bases (and nucleophiles) than are those of esters. Acid halide carbonyls are the weakest nucleophiles of all.

Resonance stabilization influences the structures of these compounds as well. Although the carbon–chlorine bond distance is little changed in acid chlorides from that in alkyl chlorides, the carbon–oxygen and carbon–nitrogen bond distances in esters and amides are shortened, reflecting some double-bond character in these compounds (Fig. 20.18).

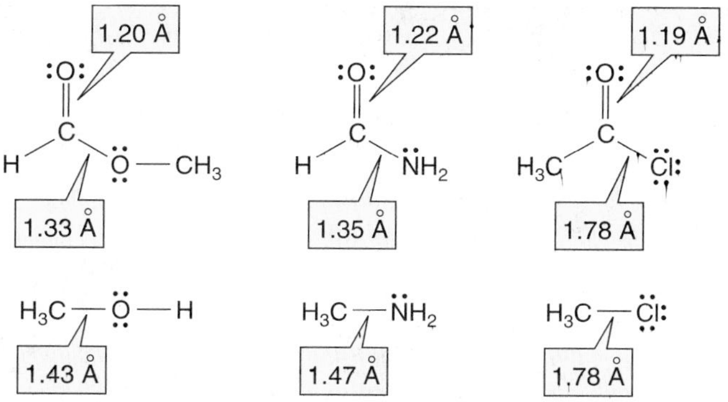

FIGURE **20.18** The structure of some simple acyl compounds. Notice how the carbon–oxygen and carbon–nitrogen single bonds are shortened by their double-bond character. In the much less well resonance stabilized acyl halides there is little double-bond character in the carbon-halide bond and little shortening.

Nitriles (cyanides) are linear molecules with short carbon–nitrogen bond distances. The *sp* hybridization of the carbon and nitrogen atoms ensures this. Figure 20.19 gives orbital pictures of both the σ and π systems, which closely resemble those of the alkynes (Chapter 4, p. 142).

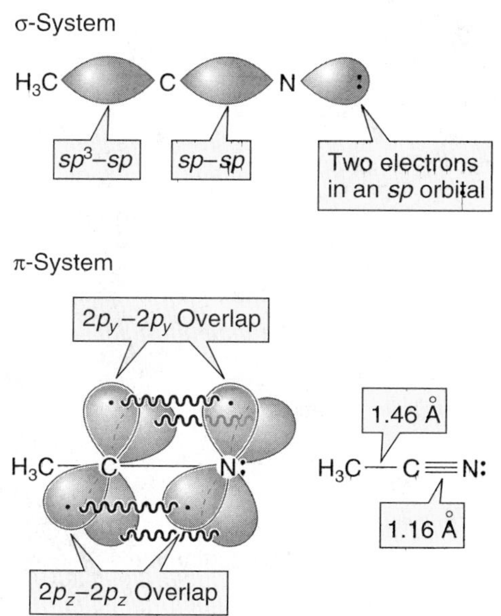

FIGURE **20.19** The σ and π orbital systems of acetonitrile.

20.3 SPECTRAL CHARACTERISTICS

20.3a Infrared Spectra

Characteristic strong carbonyl stretching frequencies are observed for all these compounds (except, of course, for the cyanides, which contain no carbonyl group). The contribution of the dipolar resonance forms can be seen in the position of the C=O stretch. The acid chloride carbonyl is the strongest, with the most double-bond character, so its C=O stretch appears at the highest frequency. Esters, acids, and anhydrides are next. Amides, in which the contribution from the dipolar resonance form is strongest, and the *single*-bond character of the carbonyl group is greatest, have the lowest carbonyl stretching frequencies (Table 20.3).

TABLE 20.3 Infrared Stretching Frequencies of Some Carbonyl Compounds

Compound	Name	Frequency of C=O in CCl_4 (cm^{-1})
$H_2C{=}C{=}O$	Ketene	2151
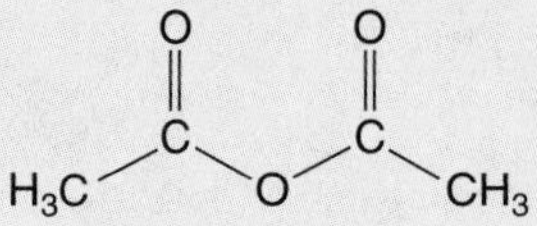 $H_3C{-}C(=O){-}O{-}C(=O){-}CH_3$	Acetic anhydride	1833, 1767
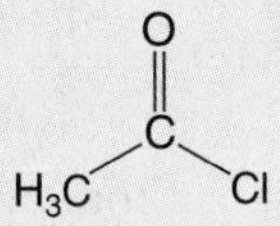 $H_3C{-}C(=O){-}Cl$	Acetyl chloride	1799
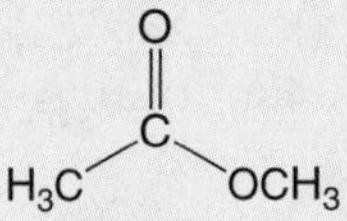 $H_3C{-}C(=O){-}OCH_3$	Methyl acetate	1750
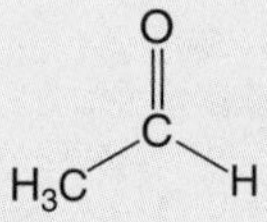 $H_3C{-}C(=O){-}H$	Acetaldehyde	1733
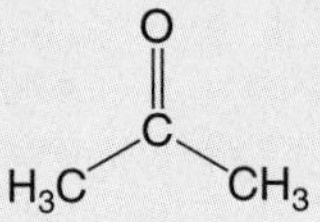 $H_3C{-}C(=O){-}CH_3$	Acetone	1719
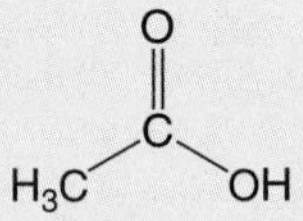 $H_3C{-}C(=O){-}OH$	Acetic acid	1717
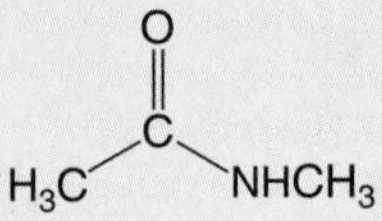 $H_3C{-}C(=O){-}NHCH_3$	*N*-Methyl acetamide	1688

Ketenes have very strong carbonyl groups, as the inner carbon is hybridized *sp*, not the usual sp^2 (Fig. 20.20). The more *s* character in a bond, the stronger it is. Therefore, ketenes absorb in the IR at very high frequency (~ 2150 cm^{-1}).

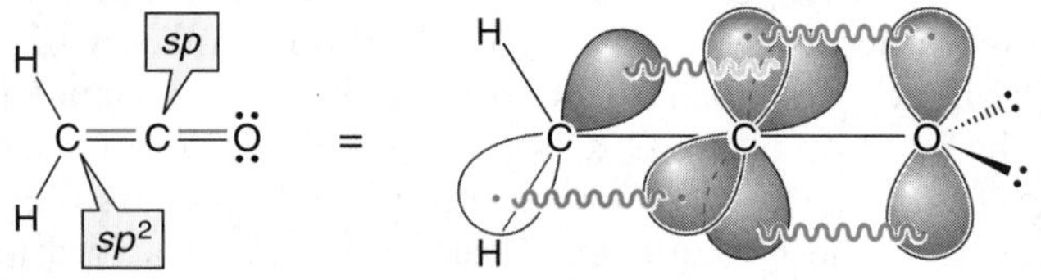

FIGURE 20.20 The bonding scheme in ketene.

In the anhydrides, which contain two carbonyl groups, there are usually two carbonyl stretching frequencies. This doubling is not the result of independent stretching of the separate carbonyls, because symmetrical anhydrides as well as unsymmetrical anhydrides show the two IR bands. Instead, the two bands are caused by the coupled symmetrical and unsymmetrical stretching modes of these compounds (Fig. 20.21). Remember, vibrations of bonds in molecules are no more independent of each other than are the vibrations of attached real springs (Chapter 15, p. 698).

Nitriles show a strong carbon–nitrogen triple-bond stretch at characteristically high frequency (2200–2300 cm^{-1}). Notice that this is quite near the frequency of the related carbon–carbon triple bonds (2100–2300 cm^{-1}).

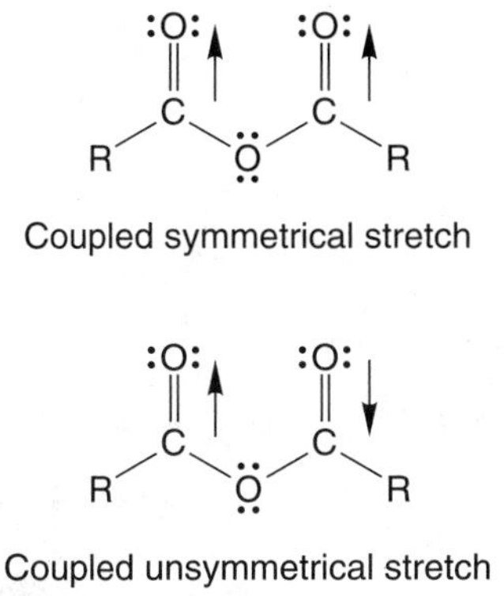

Coupled symmetrical stretch

Coupled unsymmetrical stretch

FIGURE 20.21 The two carbonyl stretching modes of anhydrides.

20.3b Nuclear Magnetic Resonance Spectra

Hydrogens in the α-position are deshielded by the carbonyl or nitrile group, and therefore appear at relatively low field. The carbonyl carbons of acyl compounds, and the nitrile carbons of cyanides bear partial positive charges and appear at relatively low field in the ^{13}C NMR spectra (Table 20.4).

TABLE 20.4 Some NMR Properties of Carbonyl Compounds and CH_3CN (ppm, $CDCl_3$)

Compound	Name	$\delta^{13}C$ (C=O, C≡N)	δ (H_α)
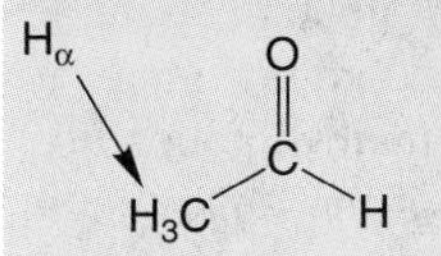 $H_3C{-}C(=O){-}H$	Acetaldehyde	199	2.25
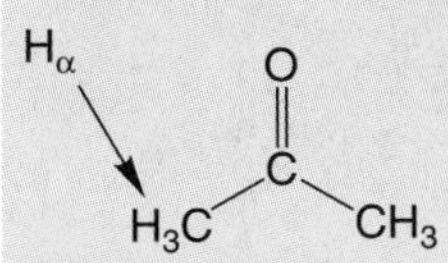 $H_3C{-}C(=O){-}CH_3$	Acetone	205	2.18
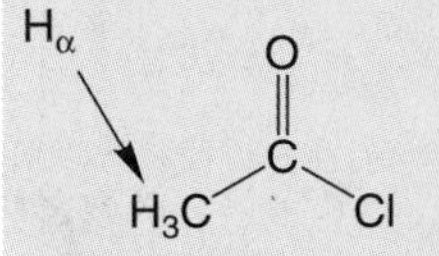 $H_3C{-}C(=O){-}Cl$	Acetyl chloride	169	2.66
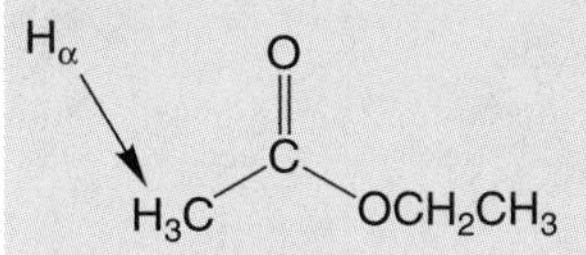 $H_3C{-}C(=O){-}OCH_2CH_3$	Ethyl acetate	169	2.05
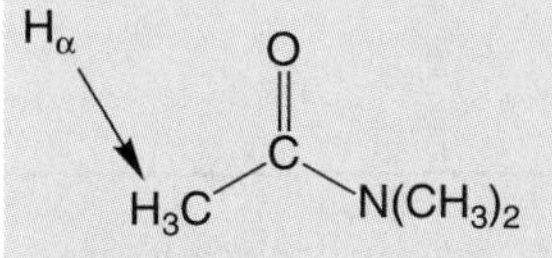 $H_3C{-}C(=O){-}N(CH_3)_2$	*N,N*-Dimethyl acetamide	170	2.09
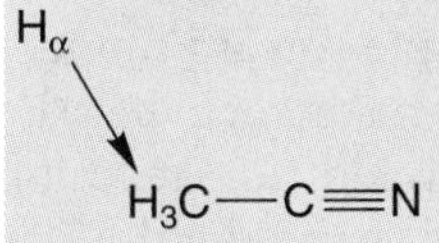 $H_3C{-}C{\equiv}N$	Acetonitrile	117	2.05

*PROBLEM 20.2 In the room temperature ^1H NMR spectrum of dimethylformamide (DMF) (Fig. 20.22), *two* methyl resonances appear. As the temperature is raised they merge into a single signal. Explain.

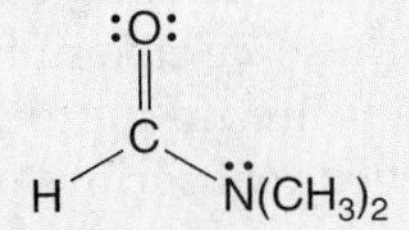

:O:
C
H N(CH$_3$)$_2$

FIGURE 20.22

Dimethylformamide
(DMF)

ANSWER Dimethylformamide is stabilized by resonance, like any ester or amide in which heteroatoms bearing lone pairs of electrons are adjacent to the carbon–oxygen double bond. There is double-bond character to the carbon–carbon bond,

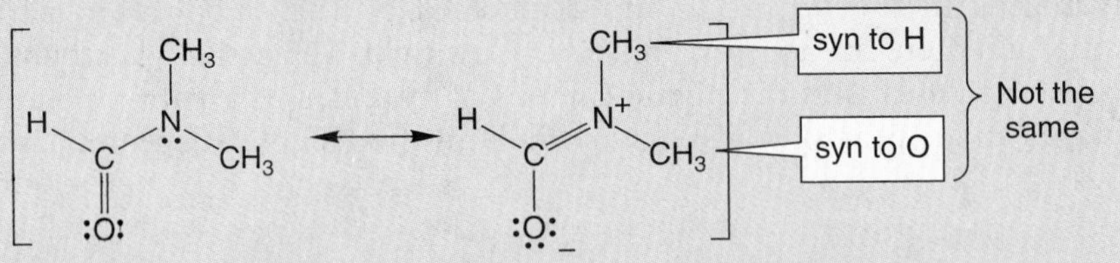

syn to H
syn to O
Not the same

There must be a barrier to rotation about the carbon–carbon bond, *and the two methyl groups are different*! At room temperature there is not enough energy to overcome the barrier to rotation and the two different methyl groups appear at different positions in the ^1H NMR spectrum. At higher temperature there is enough energy to overcome the barrier to rotation and the methyl groups become equivalent. The figure shows both resonance forms of the amide. The carbon–nitrogen bond is not a full double bond, but it does have partial double-bond character.

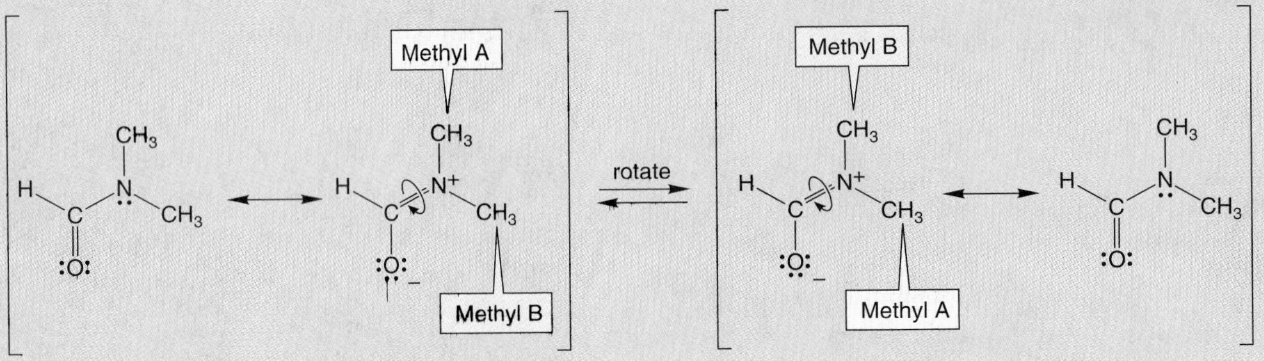

Methyl A
Methyl B
rotate
Methyl B
Methyl A

20.4 REACTIONS OF ACID CHLORIDES

As mentioned before, all acyl compounds participate in the addition–elimination process. Acid chlorides are especially reactive toward nucleophiles. The carbonyl group, being the least stabilized by resonance, has the highest energy and is the most reactive. So, an initial addition reaction with a nucleophile is relatively easy. The chloride atom of acid chlorides is a excellent leaving group, and sits poised, ready to depart once the tetrahedral addition product has been formed. The result is an exceedingly facile and common example of the addition–elimination mechanism (recall Chapter 19, p. 967; Fig. 20.23).

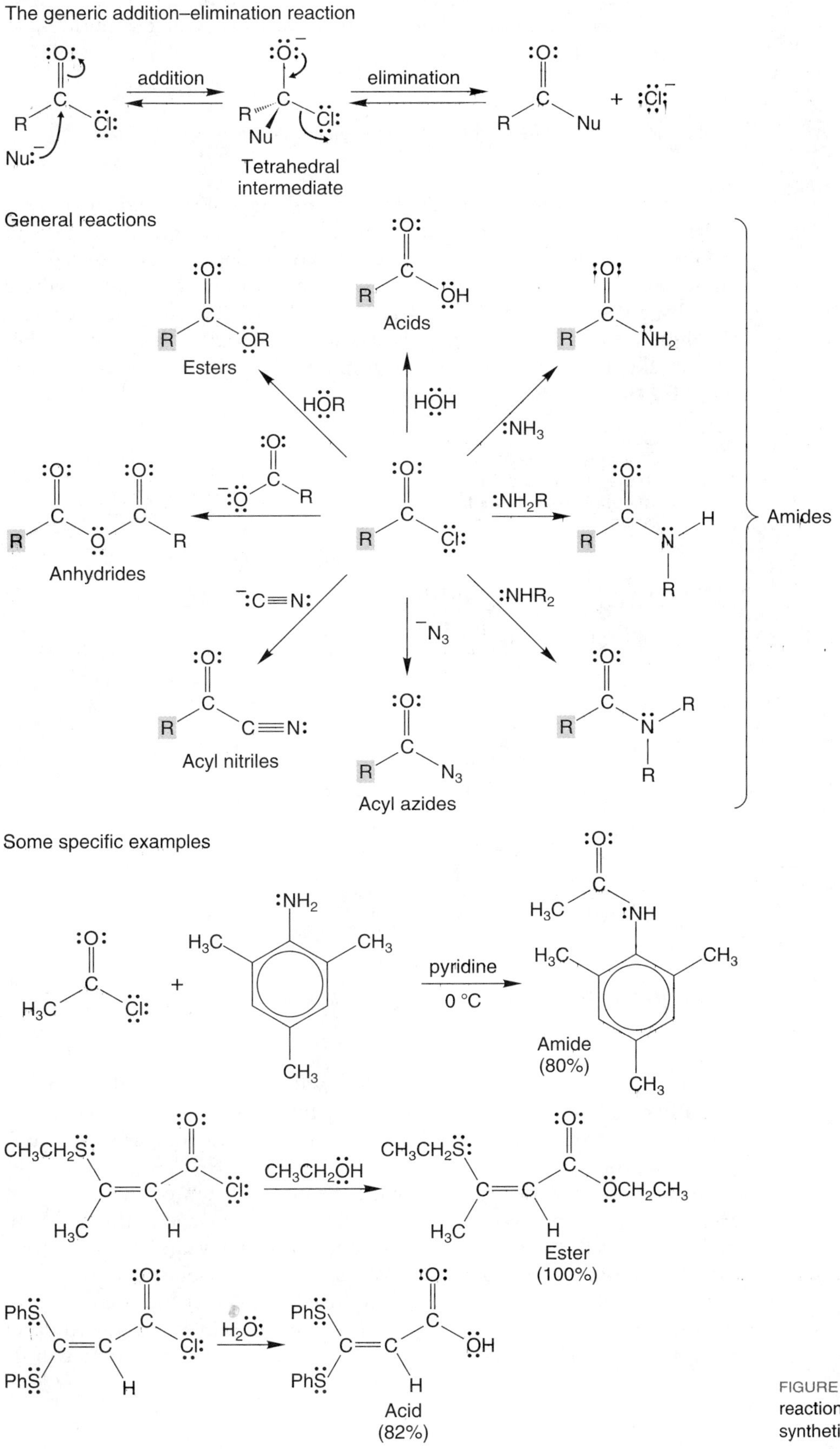

FIGURE 20.23 Addition–elimination reactions of acid chlorides. Note the synthetic potential.

Many nucleophiles are effective in this reaction, and a great many acyl compounds can be made using acid halides, usually acid chlorides, as starting materials.

PROBLEM **20.3** Acid chlorides often smell like hydrochloric acid. Explain.

Of course, strong nucleophiles such as organolithium compounds, Grignard reagents, and the metal hydrides also react rapidly with acid halides. The problem here is one of control of secondary reactions. For example, reduction of an acid chloride with lithium aluminum hydride initially gives an aldehyde. But the newly born aldehyde finds itself in the presence of a reducing agent easily strong enough to reduce it further to the alkoxide, and hence to the primary alcohol after water is added (Fig. 20.24).

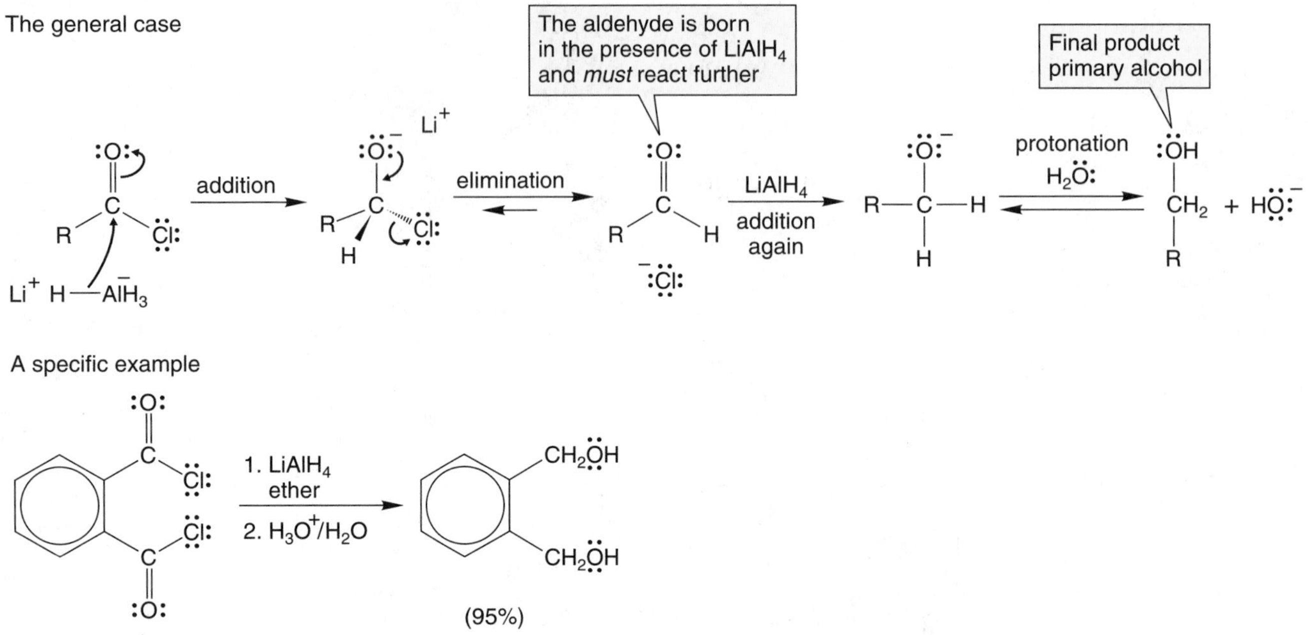

The general case

The aldehyde is born in the presence of $LiAlH_4$ and *must* react further

Final product primary alcohol

addition

elimination

$LiAlH_4$ addition again

protonation H_2O:

A specific example

1. $LiAlH_4$ ether

2. H_3O^+/H_2O

(95%)

FIGURE **20.24** The reaction of an acid chloride with lithium aluminum hydride leads to a primary alcohol

Similarly, reaction of an acid halide with an organometallic reagent initially gives a ketone that usually reacts further to give the tertiary alcohol. Note that this reaction must give an alcohol in which at least two R groups are the same (Fig. 20.25).

All is fine if we are trying to make the alcohols, but often we are not. How to stop the process at the aldehyde or ketone stage is the problem, and several solutions have been found over the years. Less reactive organometallic reagents or metal hydrides allow the isolation of the intermediate aldehydes or ketones. The reactivity of lithium aluminum hydride can be attenuated by replacing some of the hydrogens with other groups. For example, use of lithium aluminum tri-*tert*-butoxyhydride

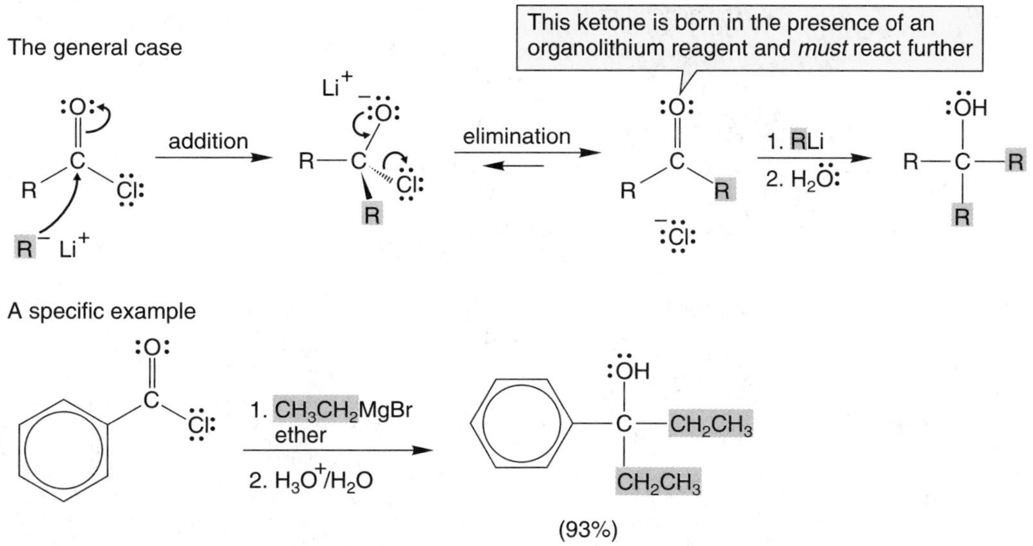

The general case

This ketone is born in the presence of an organolithium reagent and *must* react further

addition

elimination

1. RLi
2. H_2O

A specific example

1. CH_3CH_2MgBr ether
2. H_3O^+/H_2O

(93%)

FIGURE 20.25 The reaction of an acid chloride with an organometallic reagent proceeds all the way to the tertiary alcohol.

allows isolation of the aldehyde. Presumably, the bulky metal hydride is unable to reduce the aldehyde, which is less reactive than the original acid halide (Fig. 20.26).

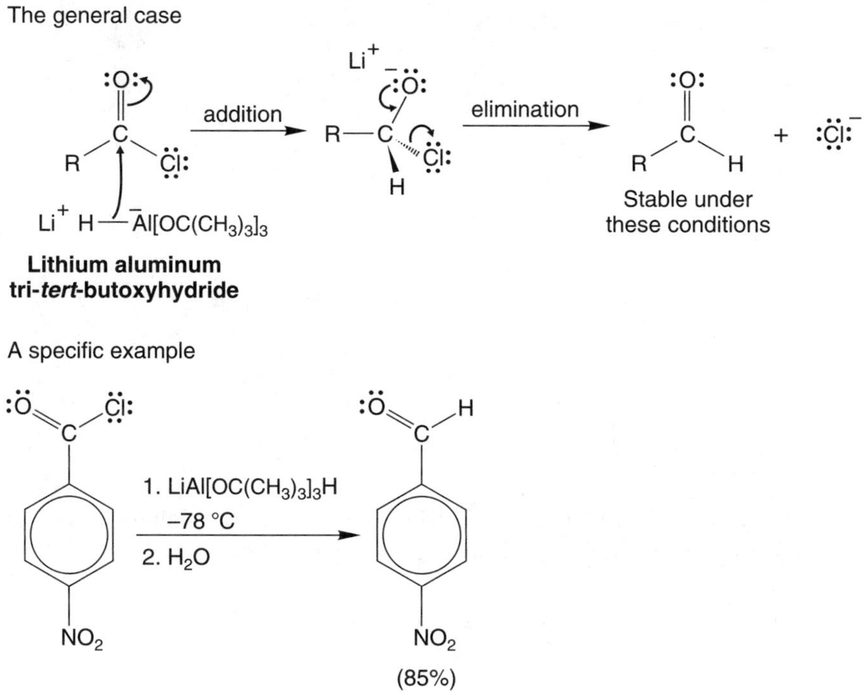

The general case

addition

elimination

Stable under these conditions

Lithium aluminum tri-*tert*-butoxyhydride

A specific example

1. $LiAl[OC(CH_3)_3]_3H$ −78 °C
2. H_2O

(85%)

FIGURE 20.26 Lithium aluminum tri-*tert*-butoxyhydride is not reactive enough to add to the initially formed aldehyde.

Catalytic reduction using a deactivated, or "poisoned" catalyst also allows the isolation of the aldehyde. This reaction is called the **Rosenmund reduction** after Karl W. Rosenmund (1884–1964) (Fig. 20.27).

The general case

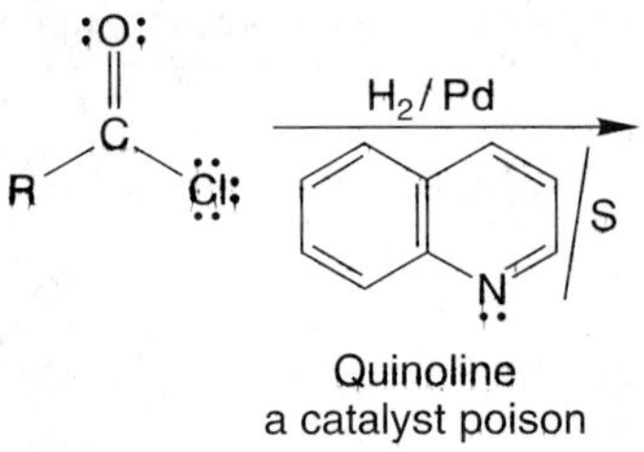

$$RCOCl \xrightarrow{H_2/Pd,\ \text{quinoline}/S} RCHO$$

Quinoline
a catalyst poison

A specific example

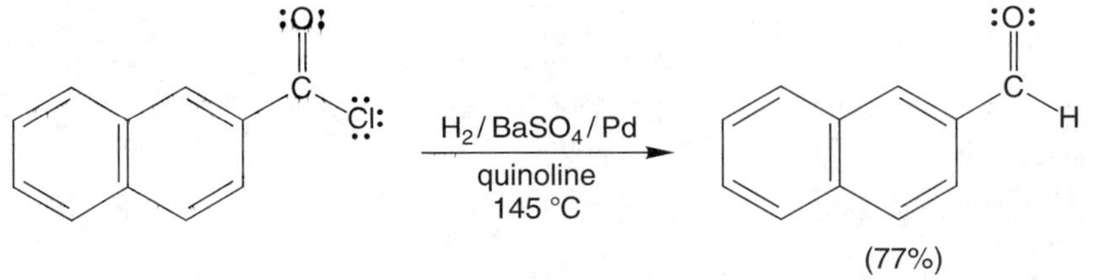

2-naphthoyl chloride $\xrightarrow[\text{quinoline, 145 °C}]{H_2/BaSO_4/Pd}$ 2-naphthaldehyde (77%)

FIGURE 20.27 The Rosenmund reduction uses a deactivated or "poisoned" catalyst so that only the more reactive acid chloride is attacked. The more stable, less reactive aldehyde can be isolated.

Organocuprates react with acid chlorides but are not reactive enough to add to the product ketones (Fig. 20.28).

The general case

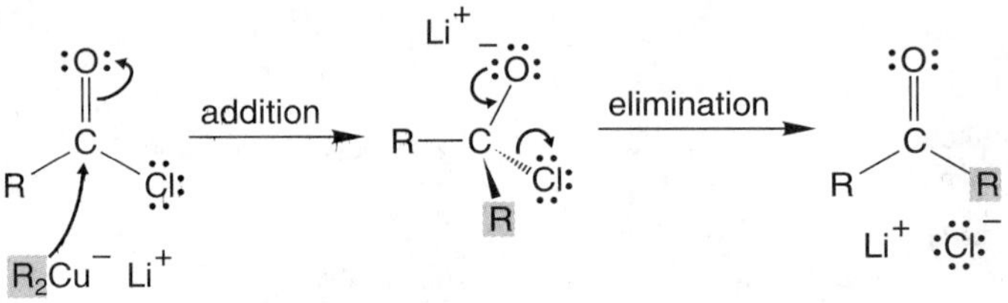

$RCOCl + R_2Cu^- \ Li^+ \xrightarrow{\text{addition}} R{-}C(O^-)(R)(Cl)\ Li^+ \xrightarrow{\text{elimination}} RCOR + Li^+ \ Cl^-$

A specific example

$PhS(CH_3)_3CCuLi + Ph{-}COCl \xrightarrow[2.\ CH_3OH/H_2O]{1.\ -60\ °C} (CH_3)_3C{-}CO{-}Ph$ (85%)

Ph = C_6H_5 (benzene ring)

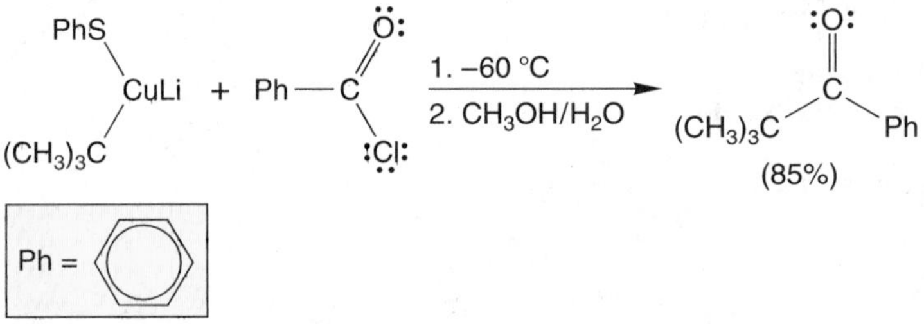

FIGURE 20.28 Cuprates are reactive enough to add R^- to the carbonyl group of the acid chloride, but not reactive enough to attack the product ketones.

Finally, don't forget that acid chlorides are used in the synthesis of aromatic ketones through Friedel–Crafts acylation (Chapter 14, p. 639; Fig. 20.29).

The general case

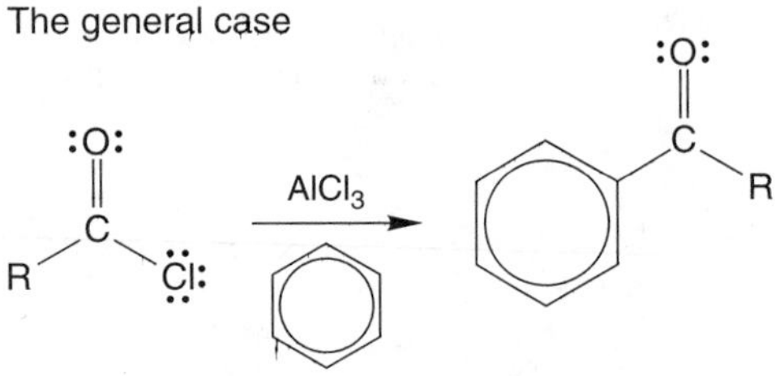

$RCOCl \xrightarrow[\text{benzene}]{AlCl_3} C_6H_5COR$

A specific example

$H_3CCOCl \xrightarrow[\text{benzene}]{AlCl_3} C_6H_5COCH_3$ (97%)

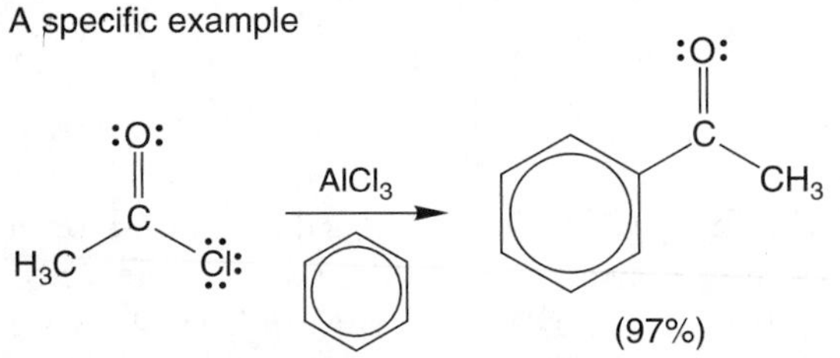

FIGURE 20.29 Friedel–Crafts acylation of benzene

Write a mechanism for the formation of the product of Figure 20.29. PROBLEM **20.4**

20.5 REACTIONS OF ANHYDRIDES

The reactivity of anhydrides is qualitatively similar to that of acid chlorides. A carboxylate salt is the leaving group in a variety of syntheses of acyl derivatives, all of which are examples of the addition–elimination process. One reaction of this kind is the basic hydrolysis of phthalic anhydride to phthalic acid. Here the leaving group is an internal carboxylate anion (Fig. 20.30).

The general case

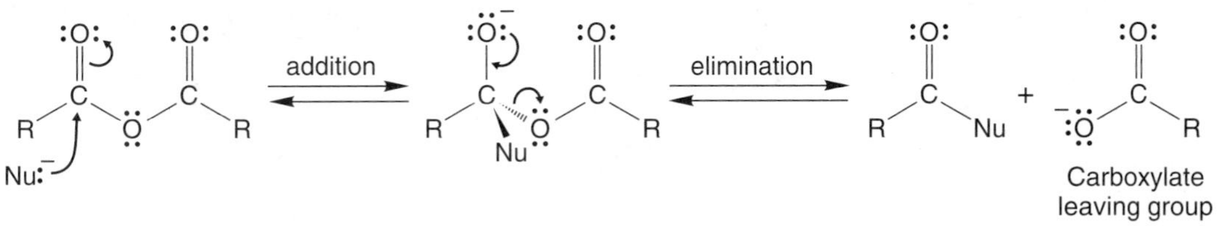

A specific example

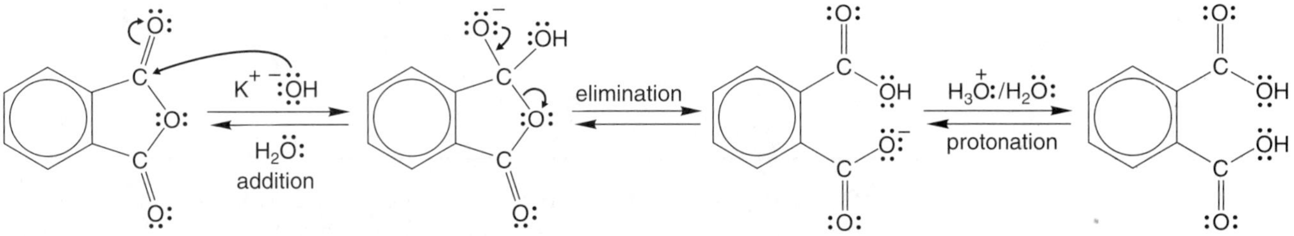

Phthalic anhydride **Phthalic acid**

FIGURE **20.30** Addition–elimination reactions of an anhydride.

Provide a mechanism for the formation of *N*-phenylmaleimide from maleic anhydride and aniline. Be sure you give a structure for the intermediate **A**, and account for the role of the acetic anhydride in the second step (Fig. 20.31). PROBLEM **20.5**

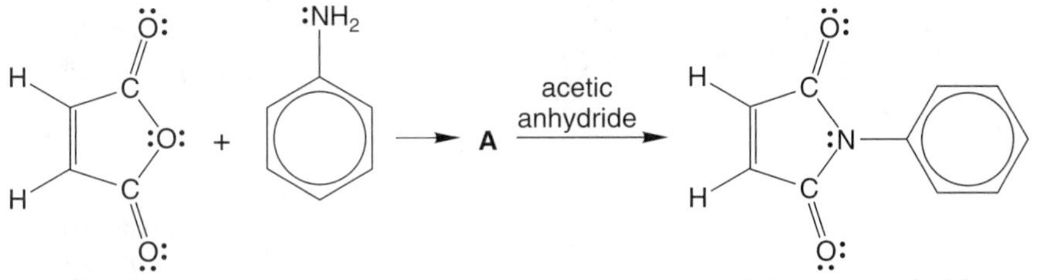

Maleic anhydride **Aniline** ***N*-Phenylmaleimide**

FIGURE **20.31**

Like acid chlorides, anhydrides can be used in the Friedel–Crafts reaction. Once again, a strong Lewis acid catalyst such as aluminum chloride is used to activate the anhydride. The mechanism follows the pattern of the Friedel–Crafts reactions we saw earlier (Chapter 14, p. 635; Fig. 20.32).

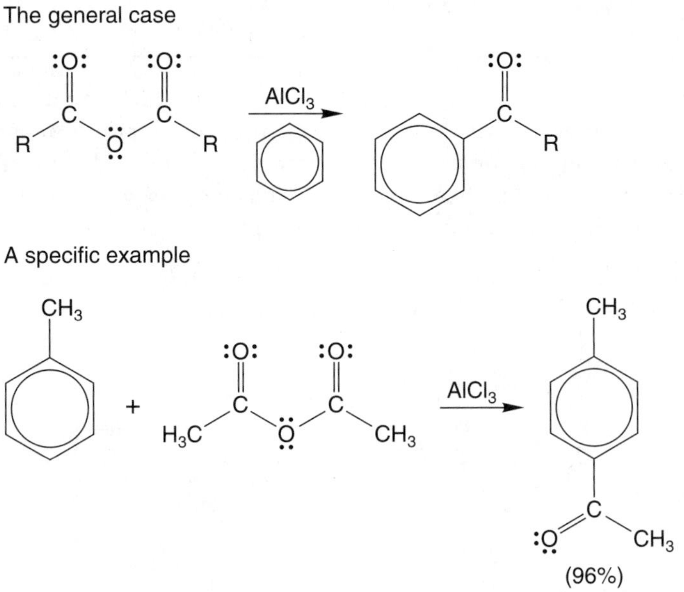

The general case

A specific example

(96%)

FIGURE **20.32** A Friedel–Crafts reaction using an anhydride as the source of the acyl group.

PROBLEM **20.6** Write a mechanism for the reaction of Figure 20.32.

20.6 ADDITION–ELIMINATION REACTIONS OF ESTERS

Esters are less reactive than acid chlorides in addition reactions, but more reactive than amides. Esters can be hydrolyzed to their parent acids under either basic or acid conditions in a process called, logically enough, **ester hydrolysis**. In base, the mechanism is the familiar addition–elimination one: Hydroxide ion attacks the carbonyl group to form a tetrahedral intermediate. Loss of alkoxide then gives the acid, which is rapidly deprotonated to the carboxylate anion in basic solution. Notice that this reaction, given the special name of **saponification**, is not catalytic. The hydroxide ion used up in the reaction is not regenerated at the end. To get the acid itself, a final acidification step is necessary (Fig. 20.33).

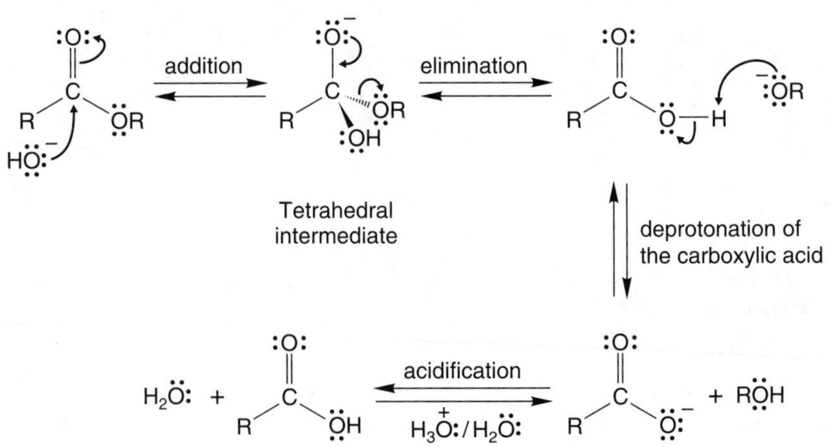

addition

elimination

Tetrahedral intermediate

deprotonation of the carboxylic acid

acidification

H_3O^+/H_2O

FIGURE **20.33** The base-induced ester hydrolysis reaction, saponification.

Esters can also be converted into carboxylic acids in acid. The carbonyl oxygen is first protonated to give a resonance-stabilized cation to which water adds. Water is by no means as nucleophilic as hydroxide, but the protonated carbonyl is a very strong Lewis acid and the overall reaction is favorable (Fig. 20.34).

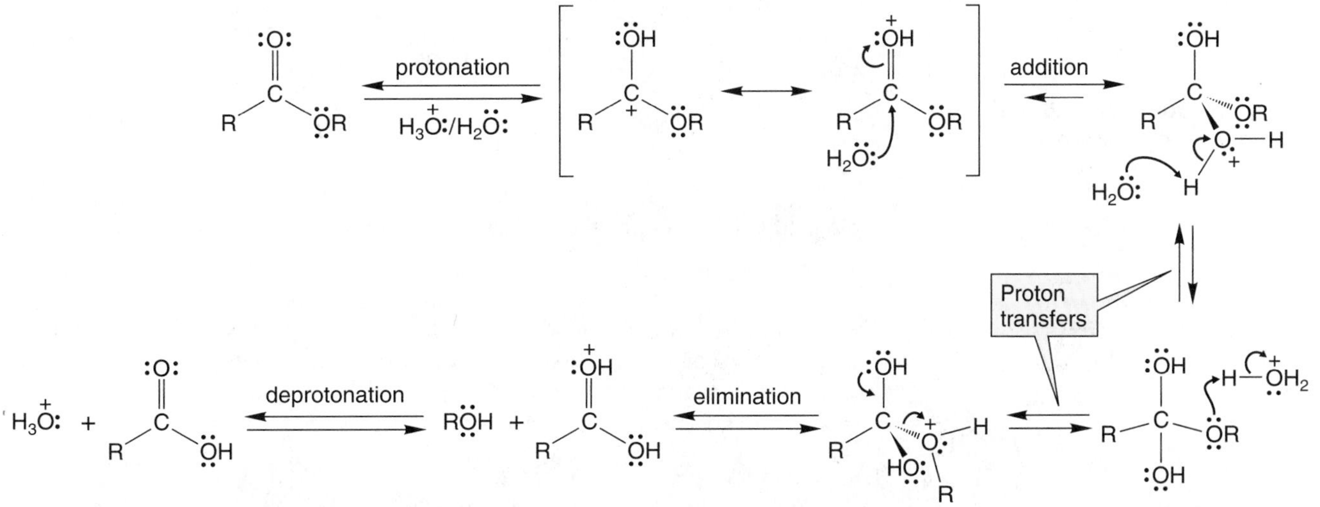

FIGURE 20.34 The acid-catalyzed ester hydrolysis reaction.

The reaction concludes with proton transfers and loss of alcohol. Have you seen this reaction before? You certainly have. It is the exact reverse of the mechanism for Fischer esterification (Chapter 19, p. 963), so there is nothing new here at all. What determines where the overall equilibrium settles out? The structure of the ester is important, but much more important are the reaction conditions. Excess water favors the acid; excess alcohol favors the ester. LeChatelier's principle is at work here, and the chemist has the ability to manipulate this equilibrium to favor either side. In practice, most acids and esters are easily interconvertable (Fig. 20.35).

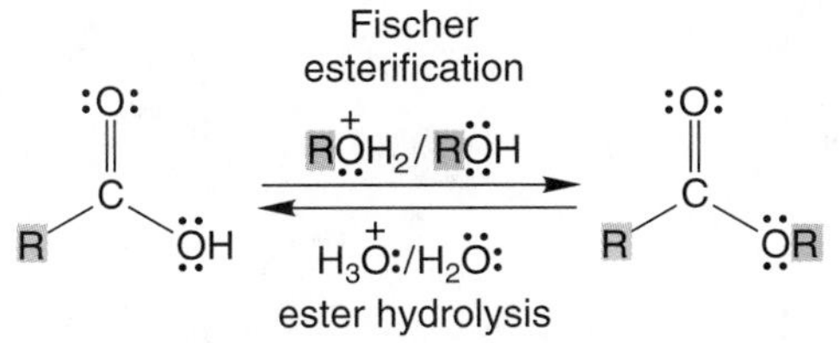

FIGURE 20.35 Acids and esters are usually interconvertable.

PROBLEM 20.7

Where appropriate, draw resonance forms for the intermediate species in Figure 20.35. You will have to write mechanisms for the reactions first.

A very closely related reaction is acid-catalyzed **transesterification**. In this reaction, an ester is treated with an excess of an alcohol and an acid catalyst. The result is replacement of the ester OR group with the alcohol OR group (Fig. 20.36). Like ester hydrolysis, transesterification can be carried out under either acid or basic conditions. The mechanisms are uncomplicated extensions of those you have already seen in the hydrolysis reaction or Fischer esterification.

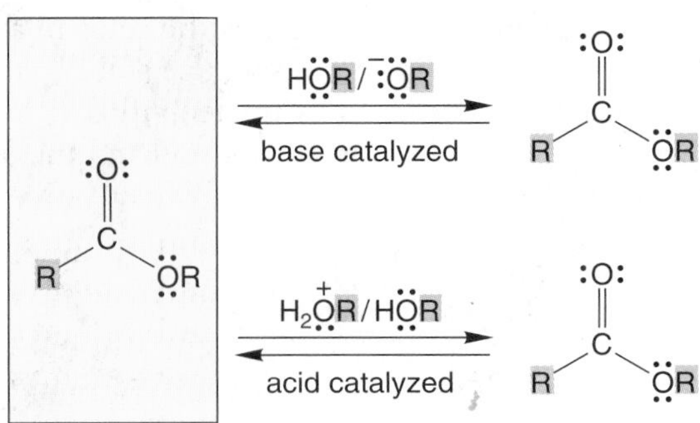

FIGURE 20.36 Base- and acid-catalyzed transesterification.

PROBLEM 20.8 Write mechanisms for acid- and base-induced transesterification.

*PROBLEM 20.9 There is an important difference between the reaction of an ester with hydroxide ($^-$OH) and alkoxide ($^-$OR). The reaction with hydroxide is neither catalytic nor reversible, whereas the reaction with alkoxide is both catalytic and reversible. Analyze the mechanisms of these reactions and explain these observations.

ANSWER Look at the final products of these two reactions. Reaction with hydroxide leads to a carboxylic acid (pK_a ~ 4.5) and an alkoxide ion. These two species *must* react very rapidly to make the more stable carboxylate anion (resonance stabilized) and the alcohol (pK_a ~ 17). The catalyst, hydroxide, is destroyed in this reaction, and the overall process is so thermodynamically favorable that it is irreversible in a practical sense.

$$\mathrm{RCOOR} \xrightleftharpoons{^-:\ddot{O}H} \mathrm{RCO{-}OH} + {}^-:\ddot{O}R \rightleftharpoons \mathrm{RCOO^{-}} + \mathrm{H{-}\ddot{O}R}$$

With an alkoxide, things are different. Now the product is not a carboxylic acid but another ester. There is no proton that can be removed, and there is a new molecule of alkoxide generated. The reaction is approximately thermoneutral.

$$\mathrm{RCOOR} + {}^-:\ddot{O}R \rightleftharpoons \mathrm{RCOOR} + \mathrm{R\ddot{O}:^-}$$

To summarize: There are both acid- and base-catalyzed versions of the reaction of esters with water or alcohols (hydrolysis or transesterification). In the last chapter (Chapter 19, p. 963), we saw the acid-catalyzed version of the conversion of an acid into an ester, called Fischer esterification.

*PROBLEM 20.10 Given that there are acid- and base-catalyzed transesterification reactions, should there not be both acid- and base-catalyzed Fischer esterifications as well? Explain carefully why there is no base-catalyzed version of Fischer esterification (RCOOH + RO$^-$ → RCOOR + HO$^-$).

ANSWER

The reaction of a carboxylic acid with an alkoxide can't proceed by addition–elimination to give an ester because there is another much easier reaction available; that reaction is simple removal of the carboxylic acid hydroxyl proton to give the resonance-stabilized carboxylate anion. They don't call these compounds "acids" for nothing! The lesson in this problem is that you have to "think simple." Look first for "trivial" reactions (loss of the proton) before proceeding on to more complicated processes (addition–elimination).

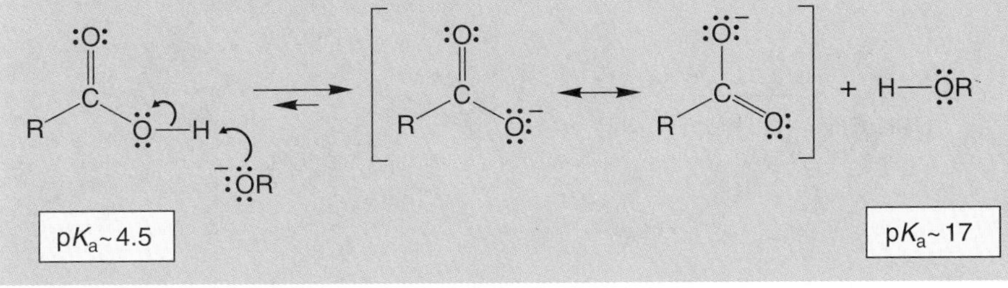

pK_a~4.5

pK_a~17

PROBLEM 20.11

Esters are also able to react with amines to give amides. Write a mechanism for the reaction of methyl acetate (CH_3COOCH_3) and ammonia to form acetamide (CH_3CONH_2).

Organometallic reagents and metal hydrides both react with esters in still another example of the addition–elimination process. Usually, the organometallic reagents react further with the ketones that are the initial products of the reactions. The ketones are not as well stabilized by resonance as are the esters and so are more reactive in the addition reaction (Fig. 20.37).

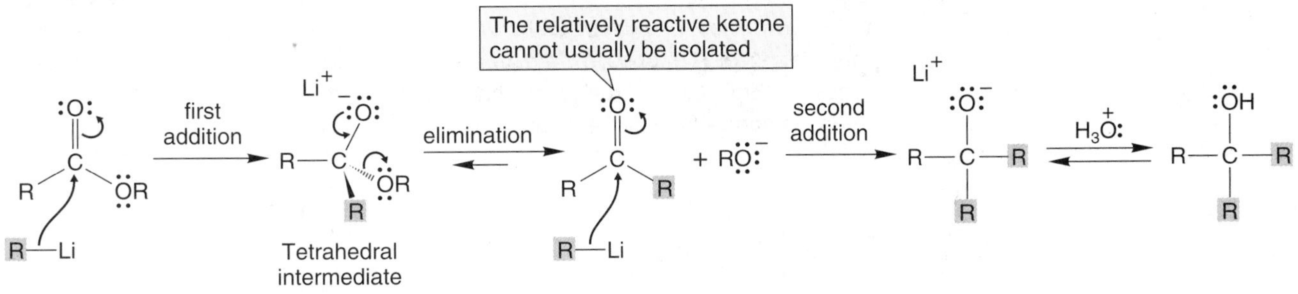

The relatively reactive ketone cannot usually be isolated

first addition

Tetrahedral intermediate

elimination

second addition

FIGURE 20.37 The reaction of esters with organolithium reagents to form complex alcohols.

This reaction is another general synthesis of complex tertiary alcohols. Here are the possibilities. First, tertiary alcohols in which all three R groups are the same can be made if the R group of the organometallic reagent is the same as that of the ester alkyl group. Second, the R of the organometallic reagent can be different from that of the ester. In this case, the tertiary alcohol will contain two different R groups (Fig. 20.38).

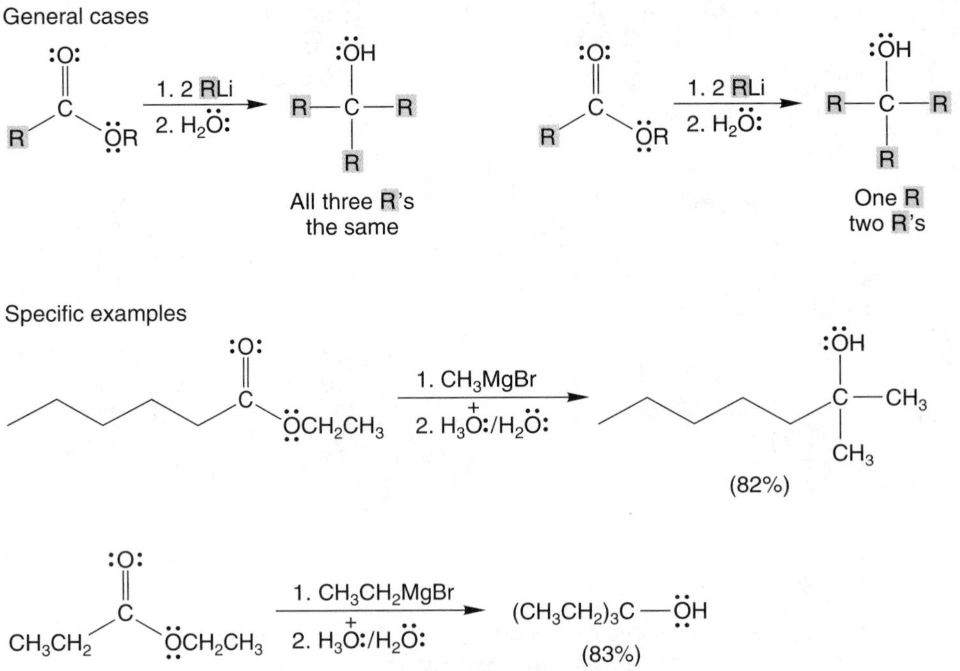

General cases

R–C(=O)–OR →(1. 2 RLi; 2. H_2O) R–C(OH)(R)–R — All three R's the same

R–C(=O)–OR →(1. 2 RLi; 2. H_2O) R–C(OH)(R)–R — One R two R's

Specific examples

$CH_3(CH_2)_4$–C(=O)–OCH_2CH_3 →(1. CH_3MgBr; 2. $H_3\overset{+}{O}$:/H_2O:) $CH_3(CH_2)_4$–C(OH)(CH_3)–CH_3 (82%)

CH_3CH_2–C(=O)–OCH_2CH_3 →(1. CH_3CH_2MgBr; 2. $H_3\overset{+}{O}$:/H_2O:) $(CH_3CH_2)_3C$—OH (83%)

FIGURE 20.38 Tertiary alcohols in which all three R groups are the same, or molecules in which only two R's are the same can be made, but the structural type in which all three R groups are different cannot be made this way.

Because two R groups must come from the organometallic reagent, there is no way to use this ester-based synthesis to make a tertiary alcohol with three different R groups.

PROBLEM 20.12 Dialkyl carbonates can be made from the reaction of phosgene (Chapter 19, p. 978) with alcohols. These carbonates are converted into tertiary alcohols with three identical R groups on reaction with Grignard reagents or organolithium reagents (Fig. 20.39). Provide mechanisms for the formations of carbonates and tertiary alcohols in these reactions.

Cl–C(=O)–Cl →(ROH) RO–C(=O)–OR + HCl →(1. 3 RLi; 2. H_2O) R–C(R)(R)–OH

Phosgene Carbonates Alcohols

FIGURE 20.39

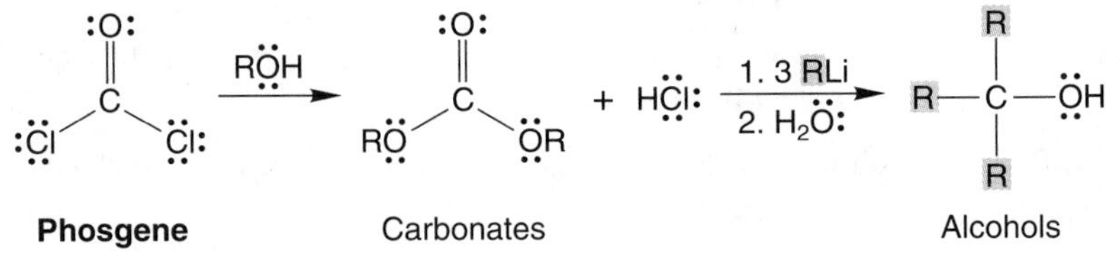

Metal hydride reduction of esters is difficult to stop at the intermediate aldehyde stage, and the usual result is further reduction to the alcohol. Either $LiAlH_4$ or $LiBH_4$ (but not $NaBH_4$, which will not reduce esters) is the reagent of choice (Fig. 20.40).

The general case

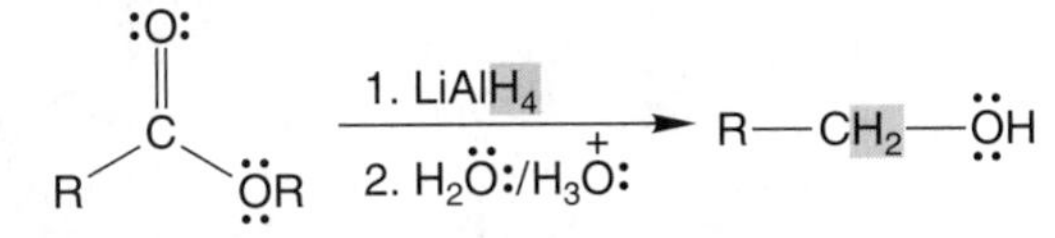

$$RC(=O)\ddot{O}R \xrightarrow[2.\ H_2\ddot{O}:/H_3\overset{+}{O}:]{1.\ LiAlH_4} R{-}CH_2{-}\ddot{O}H$$

Specific examples

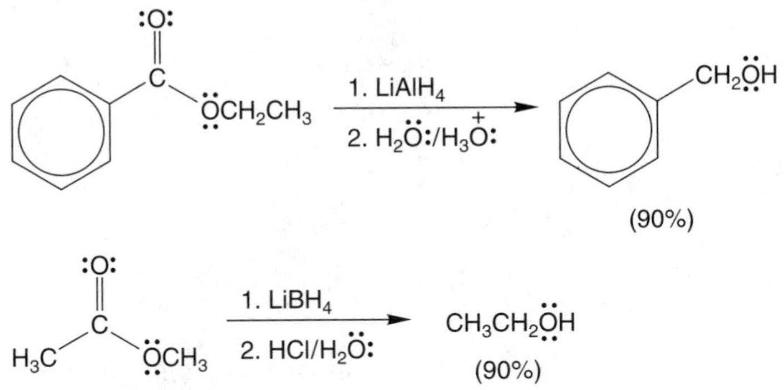

$$C_6H_5C(=O)\ddot{O}CH_2CH_3 \xrightarrow[2.\ H_2\ddot{O}:/H_3\overset{+}{O}:]{1.\ LiAlH_4} C_6H_5CH_2\ddot{O}H \quad (90\%)$$

$$H_3CC(=O)\ddot{O}CH_3 \xrightarrow[2.\ HCl/H_2\ddot{O}:]{1.\ LiBH_4} CH_3CH_2\ddot{O}H \quad (90\%)$$

FIGURE **20.40** The reduction of esters with powerful hydride donors such as lithium aluminum hydride and lithium borohydride leads to primary alkoxides and then, after acidification, to primary alcohols.

As with acid chlorides, other metal hydrides that permit isolation of the intermediate aldehydes have been developed. For example, diisobutylaluminum hydride (DIBAL-H) reduces ketones to alcohols, but can be used to isolate the intermediate aldehydes from esters if care is taken (Fig. 20.41).*

The general case

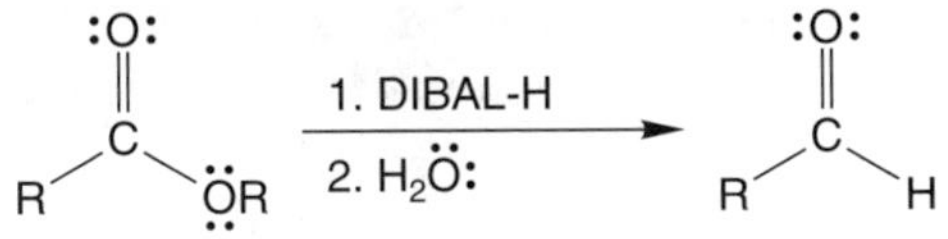

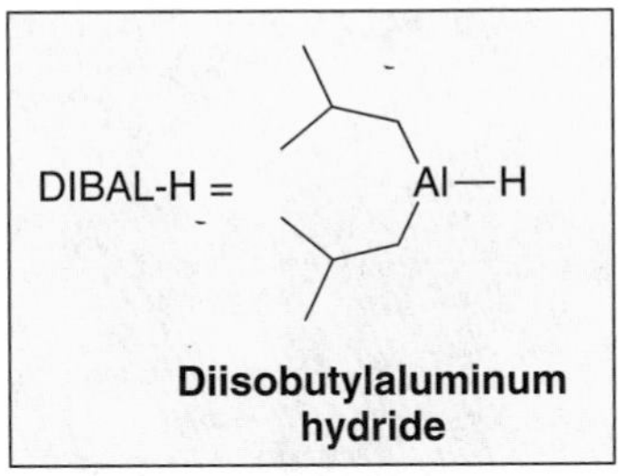

$$RC(=O)\ddot{O}R \xrightarrow[2.\ H_2\ddot{O}:]{1.\ DIBAL\text{-}H} RC(=O)H$$

A specific example

$$CH_3(CH_2)_{10}C(=\ddot{O}:)\ddot{O}CH_2CH_3 \xrightarrow[2.\ H_2\ddot{O}:]{1.\ DIBAL\text{-}H,\ -70\ °C,\ hexane} CH_3(CH_2)_{10}C(=\ddot{O}:)H$$

DIBAL-H = $[(CH_3)_2CHCH_2]_2Al{-}H$

Diisobutylaluminum hydride

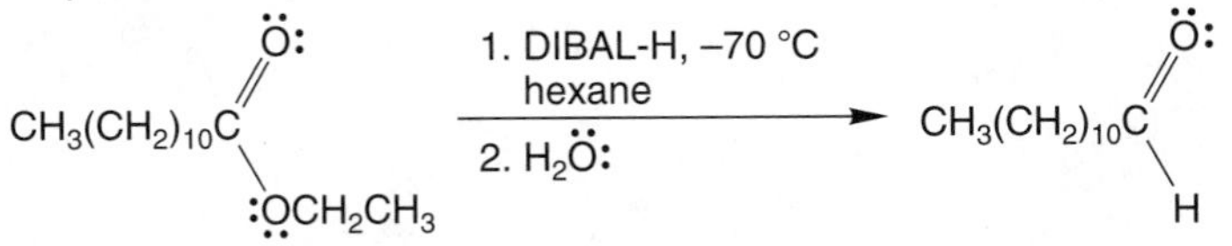

FIGURE **20.41** If DIBAL-H is used as the hydride donor, the reduction of esters can often be stopped at the aldehyde stage.

PROBLEM **20.13** Suggest one reason why DIBAL-H might be slower than $LiAlH_4$ to react with the aldehyde of Figure 20.41.

Esters also have a substantial chemistry involving the related enolate anions. We will develop this subject in detail later in this chapter.

*This reaction is actually complicated and tricky. See Problem 20.53 in the end-of-chapter problems.

Methyl Jasmonates

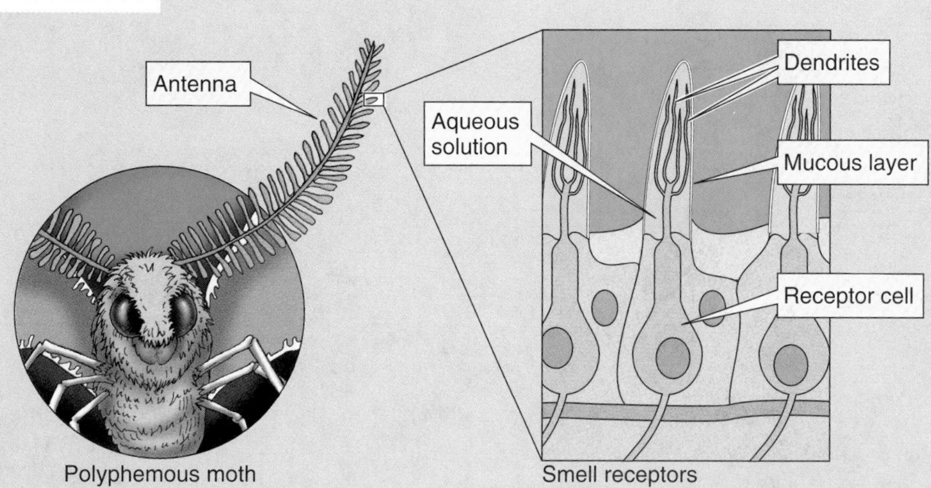

You are perhaps familiar with the magnificent olfactory receptors—antennae—of the insects. These spectacular organs can detect single molecules. We are far less efficient, and it requires millions of molecules to fire the olfactory receptors buried in our nasal passages. Yet our sense of smell is still remarkably delicate and specific. Take, for example, the methyl jasmonates and methyl epijasmonates, the four simple esters shown below. These molecules function as pheromones for both animals and plants. The male oriental fruit moth, *Grapholitha molesta* Busk. responds when a female moth emits one of these compounds. In plants, they act to promote senescence, but do not apparently carry signals to other species. However, it has been cleverly pointed out that humans' perception of the smell of flowers as pleasant is adaptive for the flower, so maybe this judgment is too quick.

The four esters shown are stereoisomers; they differ very little in structure. Yet we respond only to the third structure shown above, 1(*R*),2(*S*)-(*Z*)-methyl epijasmonate. Our sensors are stereochemically sophisticated enough so that only this stereoisomer fits into the molecular receptor and is detected.

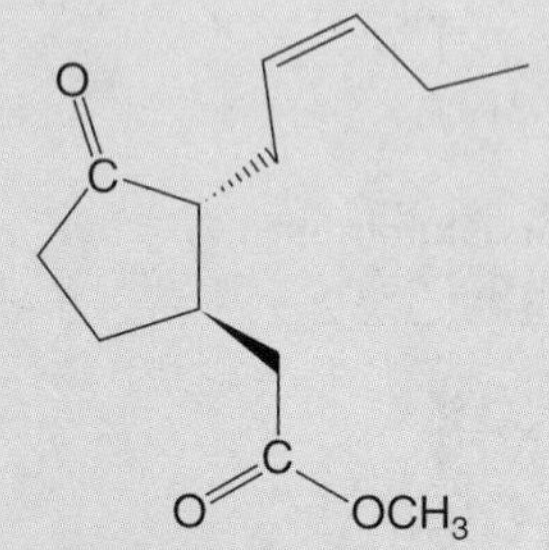

1(*R*),2(*R*)-Z-Methyl jasmonate

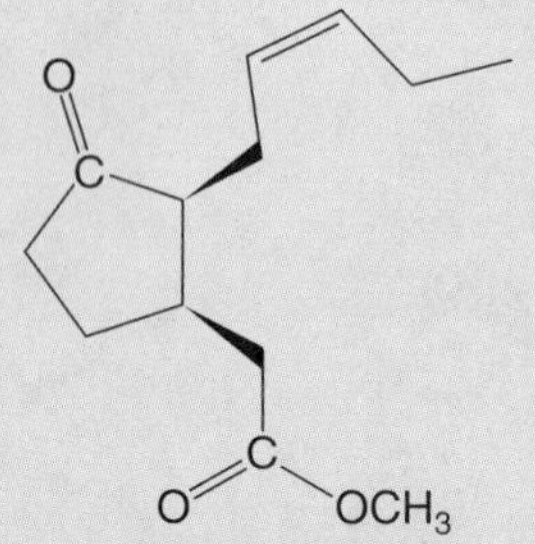

1(*R*),2(*S*)-Z-Methyl epijasmonate

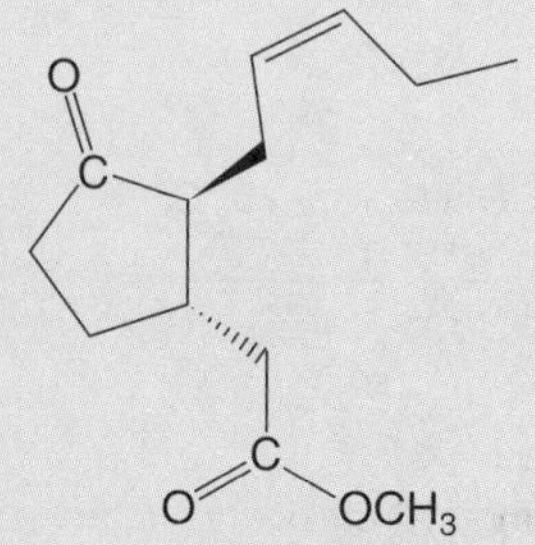

1(*R*),2(*S*)-Z-Methyl epijasmonate

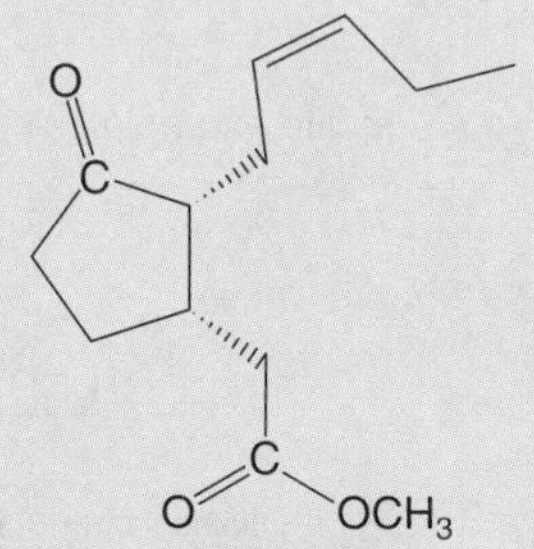

1(*S*),2(*R*)-Z-Methyl jasmonate

20.7 REACTIONS OF AMIDES

These least reactive of the acyl compounds can nevertheless be hydrolyzed in either acid or base. The mechanisms of these reactions are directly related to those of acid- or base-induced ester hydrolysis. In base, the carbonyl is attacked to give tetrahedral intermediate **A**. Loss of amide ion ($^{-}NH_2$) gives the carboxylic acid, and deprotonation of the acid gives the carboxylate salt. A final acidification regenerates the carboxylic acid in isolable form. Overall, it's addition–elimination once again (Fig. 20.42).

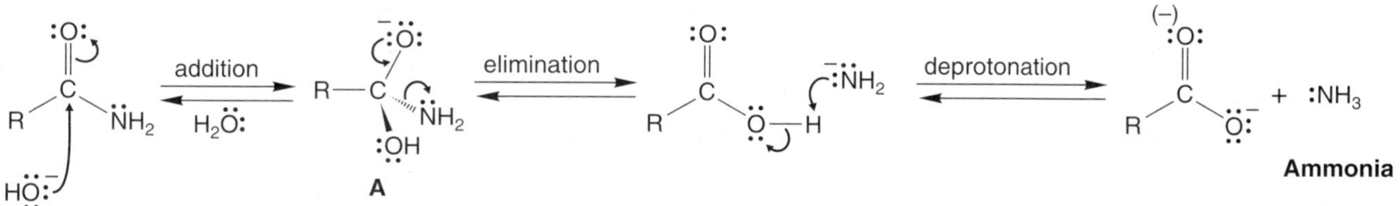

addition, elimination, deprotonation, Ammonia, A

FIGURE 20.42 The base-induced hydrolysis of amides gives carboxylate anions. Acidification gives the acid.

In acid, the carbonyl oxygen is protonated creating a Lewis acid strong enough to react with the relatively weak nucleophile water. The leaving group in the acid-induced reaction is the amine (Fig. 20.43).

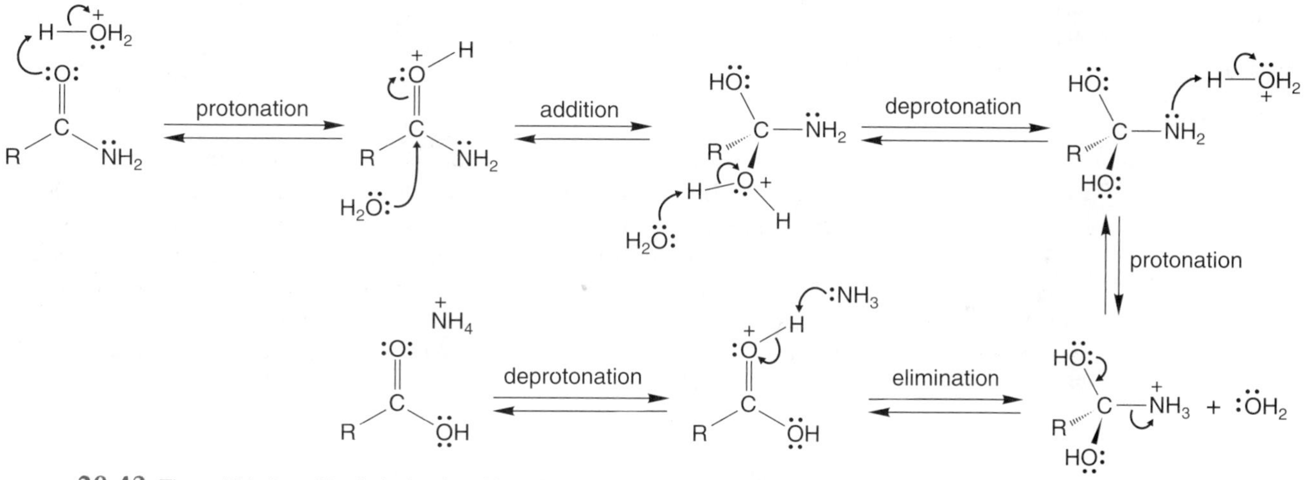

protonation, addition, deprotonation, protonation, elimination, deprotonation

FIGURE 20.43 The acid-induced hydrolysis of amides gives the carboxylic acid directly.

*PROBLEM 20.14

Write resonance forms for the structures in Figure 20.43 where appropriate.

ANSWER

This problem is very similar to Problem 20.7. In this case, it is the protonated amide and protonated acid that are most stabilized by resonance.

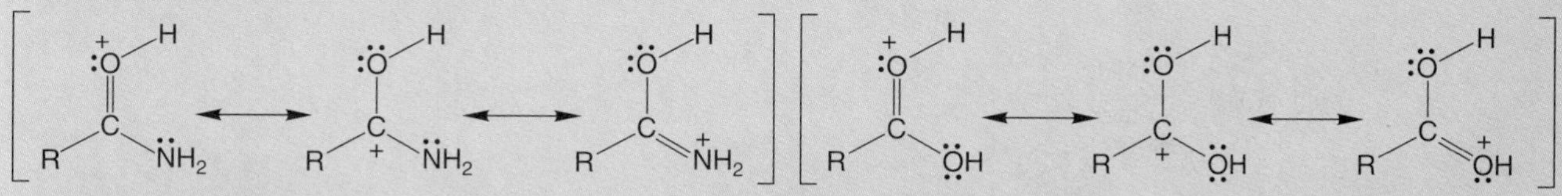

The starting amide and acid are also resonance stabilized, but this kind of stabilization is more important for the charged intermediates. For example, the resonance forms for the amide and acid are charge-separated species.

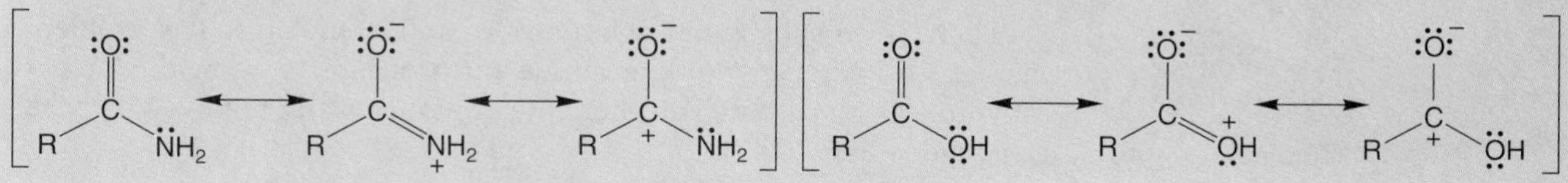

Similar acid- or base-catalyzed reactions of amides with alcohols lead to esters (Fig. 20.44).

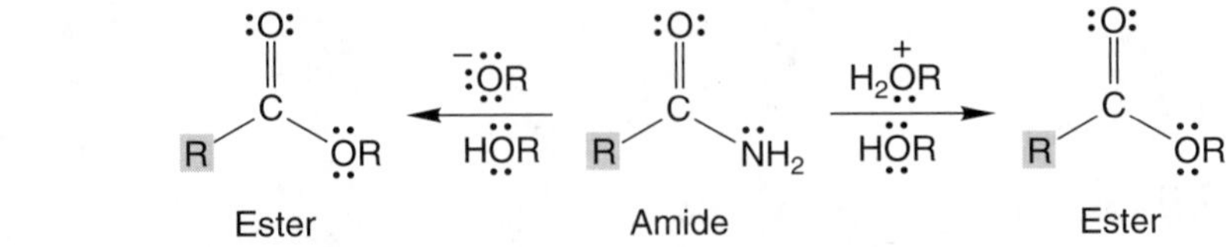

Ester ← (−:ÖR / HÖR) — Amide — ($H_2\overset{+}{O}R$ / HÖR) → Ester

FIGURE 20.44 Amides can be converted into esters in either acid or base.

PROBLEM 20.15 Write detailed mechanisms for the reactions of Figure 20.44.

Metal hydrides reduce amides to amines. Clearly, something new is happening here, as analogy to the reactions of acids, esters, and acid chlorides would lead us to expect alcohols as products (Fig. 20.45).

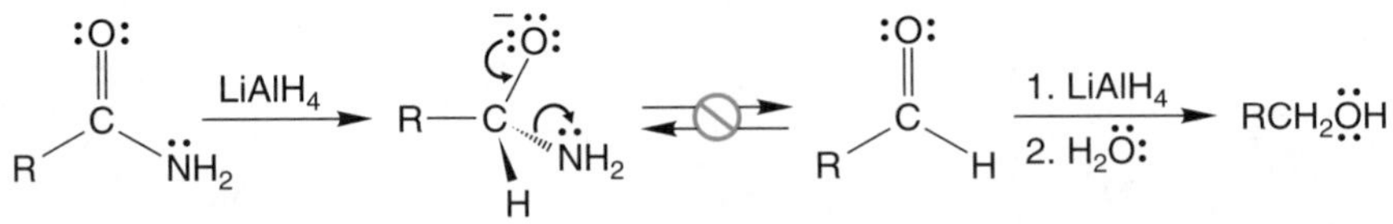

$LiAlH_4$ → R—C(O⁻)(H)—NH_2 ⇄̸ RCHO → 1. $LiAlH_4$ 2. H_2O: → RCH_2OH

FIGURE 20.45 By analogy to what we know of other acyl compounds, we might well expect the following conversion of amides to alcohols to be the result of treatment of an amide with a metal hydride. In fact, this reaction does *not* take place.

The mechanism of this reaction is difficult, and may vary with the number of alkyl groups on nitrogen. However, the first step in this reaction is the same as in the other reductions by metal hydrides; the hydride adds to the carbonyl to give a complex alkoxide. Now, however, the metal oxide is lost with formation of an **immonium ion**. There are several possible ways to describe the formation of the immonium ion, but once it is produced, reduction by the metal hydride to give the amine is sure to be rapid (Fig. 20.46).

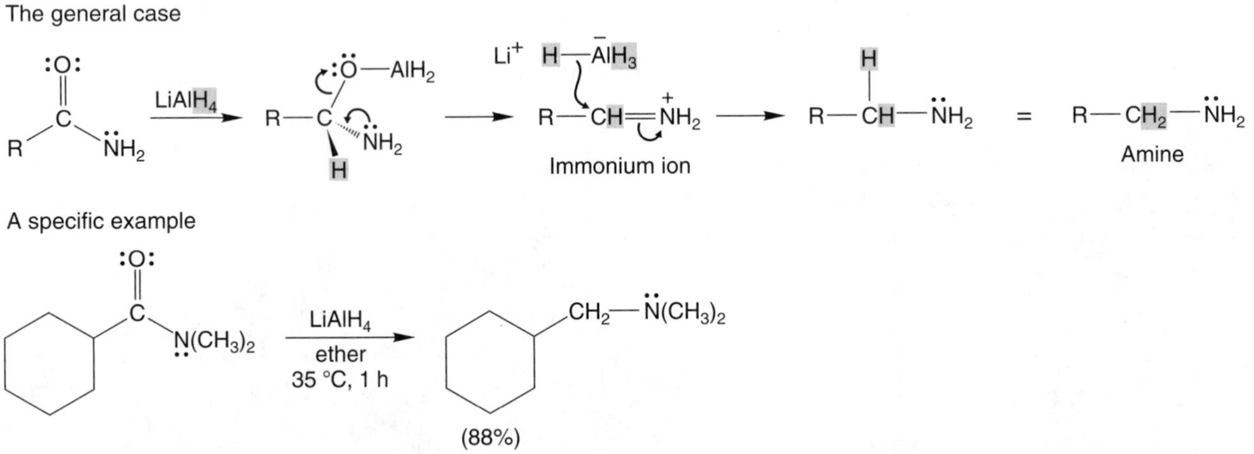

The general case

$LiAlH_4$ → R—C(O—AlH_2)(H)—NH_2 → (Li⁺ H—$\overline{Al}H_3$) R—CH=$\overset{+}{N}H_2$ (Immonium ion) → R—CH(H)—NH_2 = R—CH_2—NH_2 (Amine)

A specific example

Cyclohexyl—C(=O)—$N(CH_3)_2$ → $LiAlH_4$, ether, 35 °C, 1 h → Cyclohexyl—CH_2—$N(CH_3)_2$ (88%)

FIGURE 20.46 Amines are the real products of the reduction of amides with metal hydrides.

There is no substantial chemistry of amide enolates. The problem is that the α-hydrogens on the amide nitrogen are more acidic than the α-hydrogens on carbon, and therefore are more easily removed by base to give anions (Fig. 20.47).

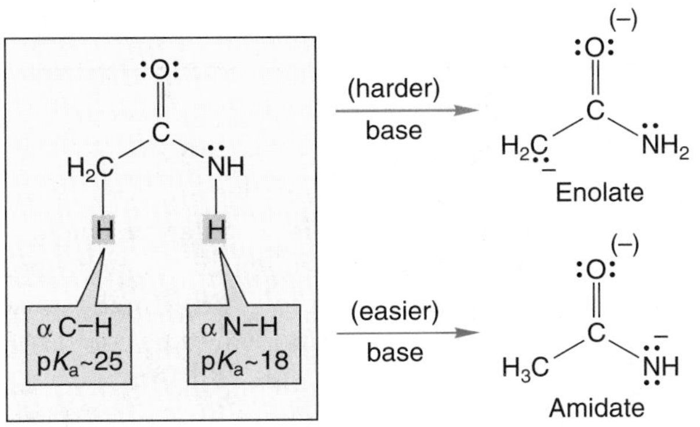

FIGURE 20.47 There are two kinds of α-hydrogen in most amides. It is much easier to remove the proton attached to nitrogen than the proton attached to carbon.

Explain why α-hydrogens on the nitrogen of amides are more acidic than α-hydrogens on carbon (Fig. 20.47).

PROBLEM 20.16

The resulting anions can be alkylated by reactive alkyl halides (Fig. 20.48).

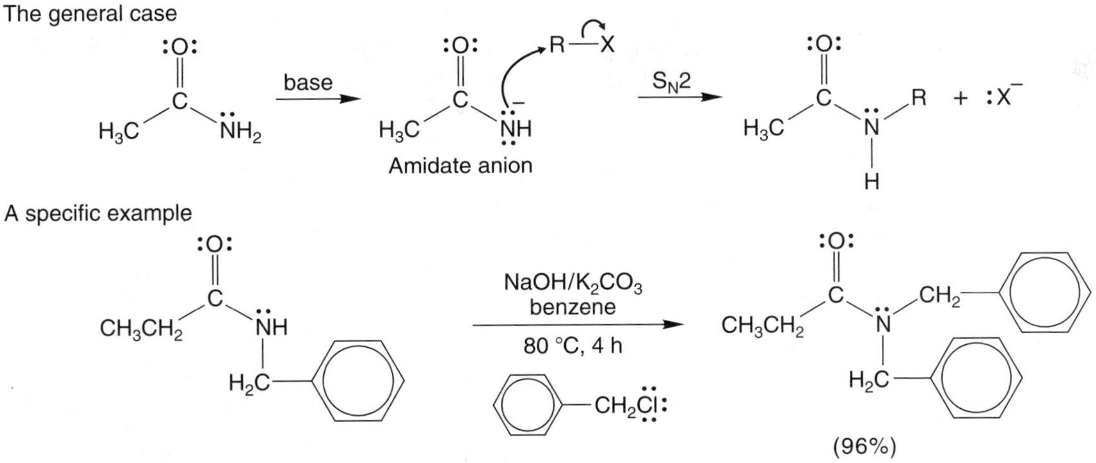

FIGURE 20.48 Alkylation of amidate anions gives substituted amides. This reaction is just an S_N2 substitution in which the amidate acts as nucleophile.

If there are no hydrogens on the nitrogen, it becomes possible to form enolates. Strong, hindered bases such as lithium diisopropylamide (LDA, Chapter 19, p. 986) are generally used (Fig. 20.49).

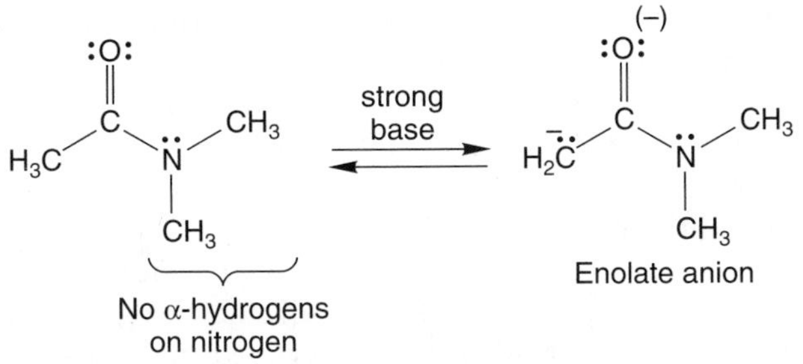

FIGURE 20.49 Amides in which there are no hydrogens attached to nitrogen can form enolates.

20.8 REACTIONS OF NITRILES

Of course nitriles, or cyanides, are not carbonyl compounds. But they are closely related, and an examination of their chemistry naturally falls into this section. Like carbonyl compounds, nitriles are both Lewis acids and Lewis bases. As in carbonyl compounds, a polar resonance form contributes to the structure of nitriles. Consequently, the triple bond is polarized and can act as a Lewis acid at carbon. One might anticipate that nucleophiles would add to nitriles, and that would be correct (Fig. 20.50).

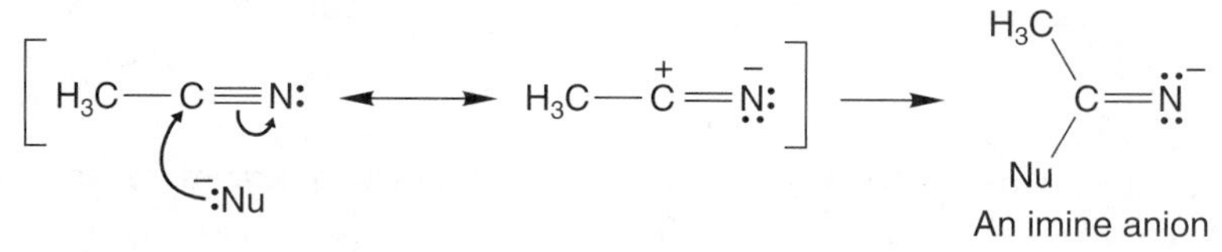

$$\left[H_3C{-}C{\equiv}N{:} \longleftrightarrow H_3C{-}\overset{+}{C}{=}\overset{-}{\ddot{N}}{:} \right] \xrightarrow{:Nu^-} \underset{\text{An imine anion}}{(H_3C)(Nu)C{=}\ddot{N}^{-}}$$

FIGURE 20.50 Nitriles are Lewis acids and nucleophiles add to them to give imine anions. There is a close analogy to carbonyl chemistry here.

The nitrogen atom bears a pair of nonbonding electrons and is therefore the center of Lewis basicity and nucleophilicity. The nitrogen is *sp* hybridized, however, and is a relatively weak Lewis base.

PROBLEM 20.17 Explain why an *sp* hybridized nitrogen should be a weaker base than an sp^2 hybridized nitrogen. What about sp^3 hybridized nitrogens?

Nitriles are also Brønsted acids, as removal of an α-hydrogen gives a resonance-stabilized anion (Fig. 20.51). As a result, the pK_a values of nitriles are similar to those of esters (~24). Be sure to compare the resonance-stabilized species in Figure 20.51 to the more conventional enolates formed from aldehydes, ketones, and esters.

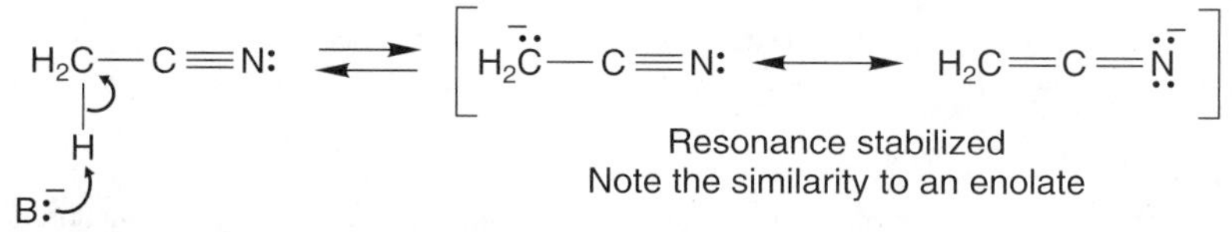

$$H_3C{-}C{\equiv}N{:} \xrightarrow{B:^-} \left[H_2\overset{-}{\ddot{C}}{-}C{\equiv}N{:} \longleftrightarrow H_2C{=}C{=}\overset{-}{\ddot{N}} \right]$$

Resonance stabilized
Note the similarity to an enolate

FIGURE 20.51 In another reaction reminiscent of carbonyl chemistry, an α-hydrogen of a nitrile can be removed in base to give a resonance-stabilized species closely related to an enolate.

These similarities mean that many of the reactions of carbonyl groups are possible for nitriles, and the mechanisms are generally closely related. Students sometimes have trouble with these reactions, however, so it is important not to dismiss them with only a cursory look. For some reason, the analogy between the carbon–oxygen double bond and the carbon–nitrogen triple bond is not always easy to keep in mind.

Both acid- and base-induced hydrolysis of a nitrile gives the amide, but rather severe conditions are required for the reactions, and further hydrolysis of the intermediate amides gives the carboxylic acids (p. 1028). In acid, the first step is protonation of the nitrile nitrogen to give a stong Lewis acid (**A**), which is attacked by water. A series of proton shifts gives an amide that is hydrolyzed to the acid (Fig. 20.52).

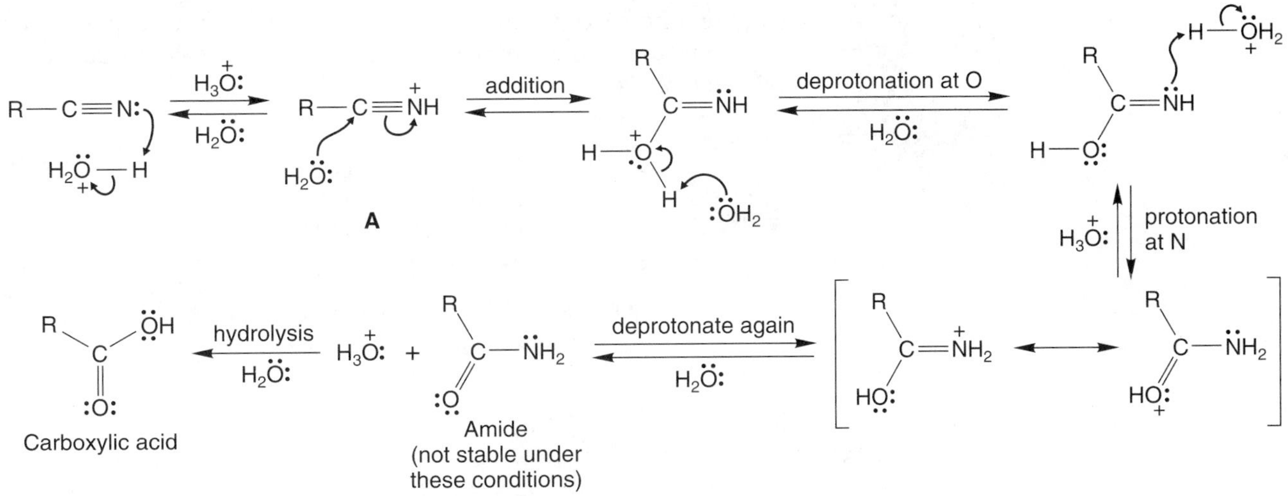

FIGURE 20.52 The acid-induced hydrolysis of a nitrile to a carboxylic acid.

In base, the nucleophilic hydroxide ion adds to the carbon–nitrogen triple bond. Protonation on nitrogen and deprotonation on oxygen leads to the amidate anion, which is protonated to give an intermediate amide. Basic hydrolysis of the intermediate amide leads to the carboxylate salt (Fig. 20.53).

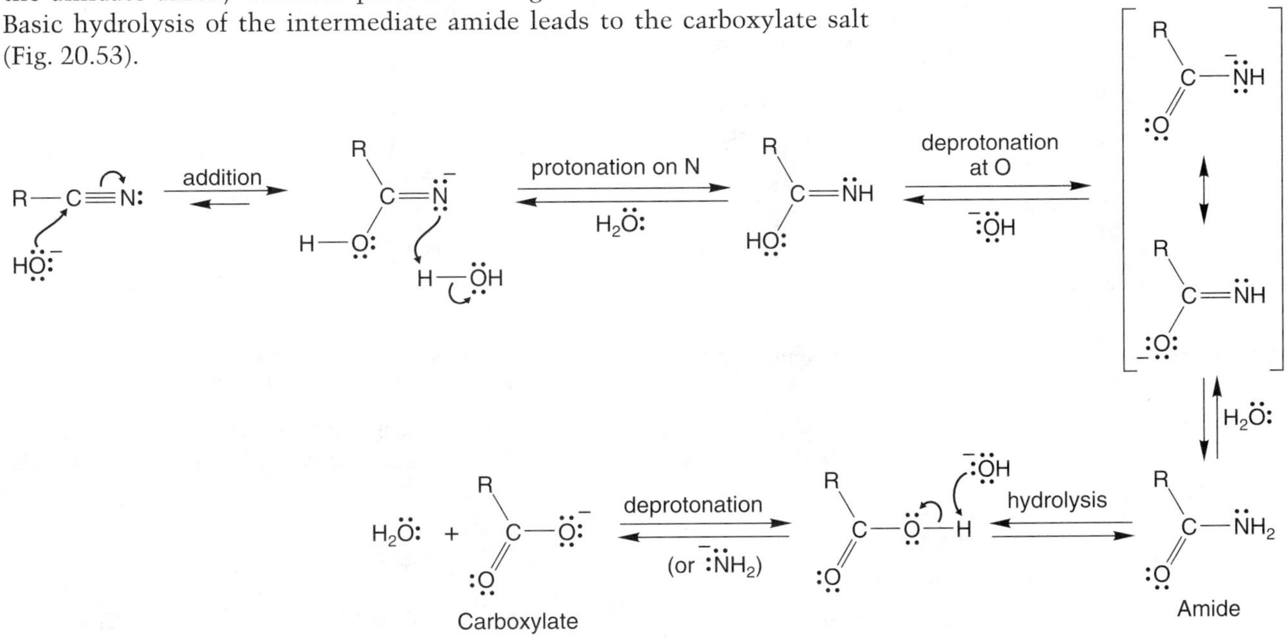

FIGURE 20.53 The base-induced hydrolysis of a nitrile to a carboxylate anion.

Stronger bases also add to nitriles. Organometallic reagents give ketones, although the product before the hydrolysis step is the imine. Addition to the nitrile first takes place in normal fashion to give the imine salt. When the reaction is worked up with water, protonation gives the imine, which in acid undergoes further hydrolysis to the ketone. Note the difference between this reaction and those of alkyllithium or Grignard reagents with acid chlorides or esters, which go all the way to alcohols. With nitriles the reaction can be stopped at the carbonyl stage. With acid chlorides and esters the intermediate carbonyl compounds are born in the presence of the organometallic reagent and must react further. In the reaction with nitriles, there is no leaving group present and the imine salt is stable until hydrolysis (Fig. 20.54).

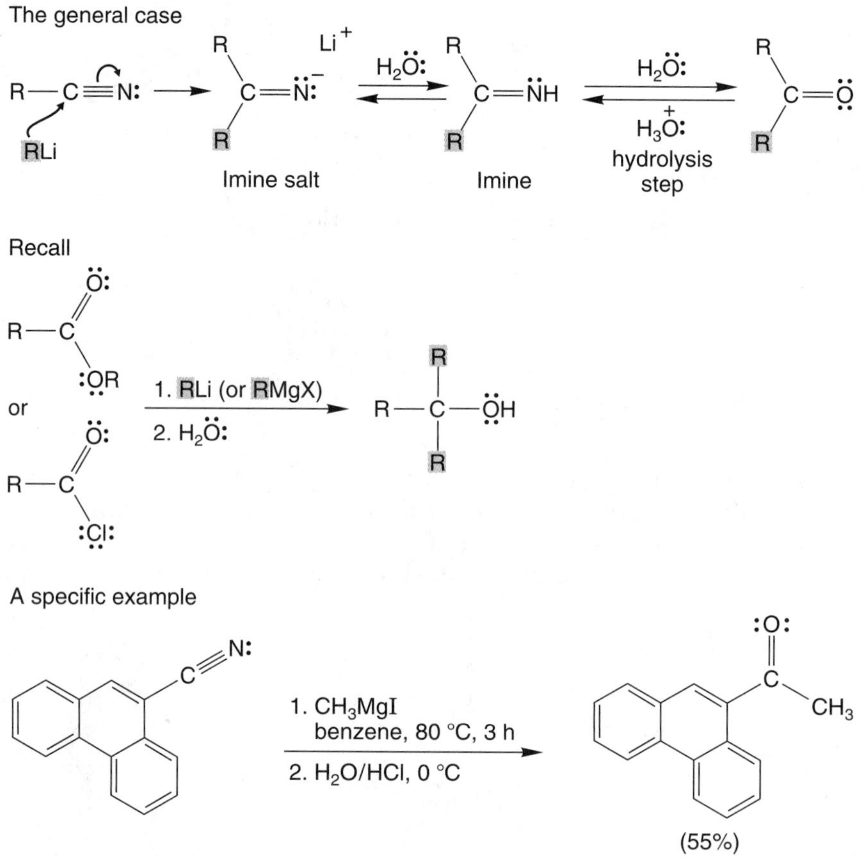

The general case

R—C≡N: + RLi → $R_2C{=}N^{-}$ Li^+ (Imine salt) ⇌ (H_2O) $R_2C{=}NH$ (Imine) ⇌ (H_2O / H_3O^+, hydrolysis step) $R_2C{=}O$

Recall

R—CO—OR or R—CO—Cl → 1. RLi (or RMgX) 2. H_2O → R_3C—OH

A specific example

9-Cyanophenanthrene → 1. CH_3MgI, benzene, 80 °C, 3 h; 2. H_2O/HCl, 0 °C → 9-acetylphenanthrene (55%)

FIGURE 20.54 Reaction of a nitrile with an organometallic reagent gives an imine salt. Acidification first generates the imine and then the ketone.

PROBLEM 20.18 Write a mechanism for the hydrolysis of imines under neutral conditions.

Metal hydrides reduce nitriles to give primary amines. Two equivalents of hydride are delivered, and a final addition of water yields the amine (Fig. 20.55).

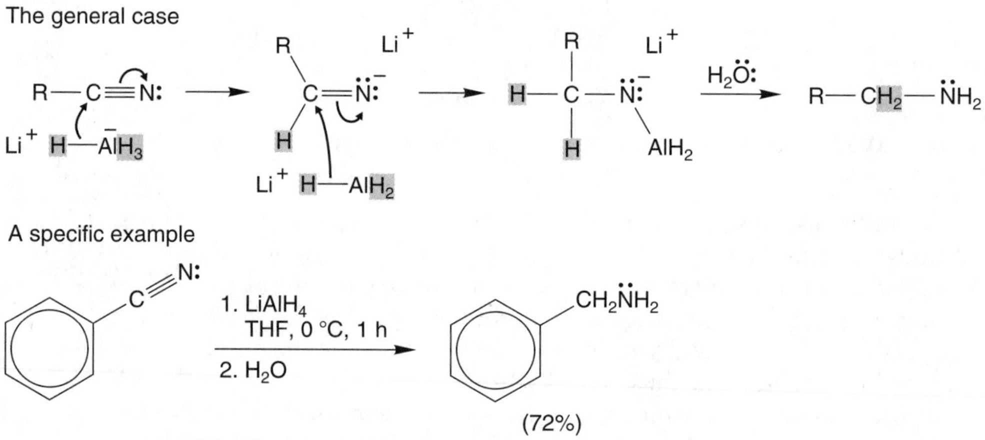

The general case

R—C≡N: + Li^+ H—$\bar{A}lH_3$ → R(H)C=N^- Li^+ + Li^+ H—AlH_2 → H—C(R)(H)—N^-(—AlH_2) Li^+ → (H_2O) R—CH_2—NH_2

A specific example

Benzonitrile → 1. $LiAlH_4$, THF, 0 °C, 1 h; 2. H_2O → $C_6H_5CH_2NH_2$ (72%)

FIGURE 20.55 The metal hydride reduction of nitriles followed by hydrolysis yields primary amines.

Catalytic hydrogenation of nitriles also gives primary amines. Notice that only primary amines can be formed this way (Fig. 20.56).

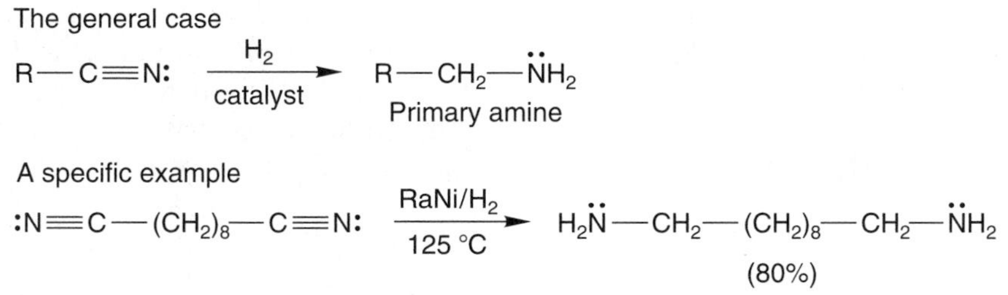

The general case

$$R{-}C{\equiv}N\colon \xrightarrow[\text{catalyst}]{H_2} R{-}CH_2{-}\ddot{N}H_2$$

Primary amine

A specific example

$$\colon N{\equiv}C{-}(CH_2)_8{-}C{\equiv}N\colon \xrightarrow[125\ ^\circ C]{RaNi/H_2} H_2\ddot{N}{-}CH_2{-}(CH_2)_8{-}CH_2{-}\ddot{N}H_2$$

(80%)

FIGURE 20.56 Catalytic hydrogenation of nitriles also gives primary amines.

20.9 REACTIONS OF KETENES

Ketenes are strained compounds and are very reactive even at low temperature. Simple ketenes are so reactive that they must be carefully protected from atmospheric moisture. Most nucleophiles add to the carbonyl group to give acyl derivatives. This addition is another reaction that gives students trouble, but shouldn't. The ketene reacts with nucleophiles like any other carbonyl compound to give an addition product. In this case, the addition product is an enolate, and protonation at oxygen gives an enol, which equilibrates with the more stable carbonyl compound. Protonation at carbon (probably the slower process) gives the carbonyl compound directly. It's all routine addition-to-carbonyl chemistry. Figure 20.57 gives some examples.

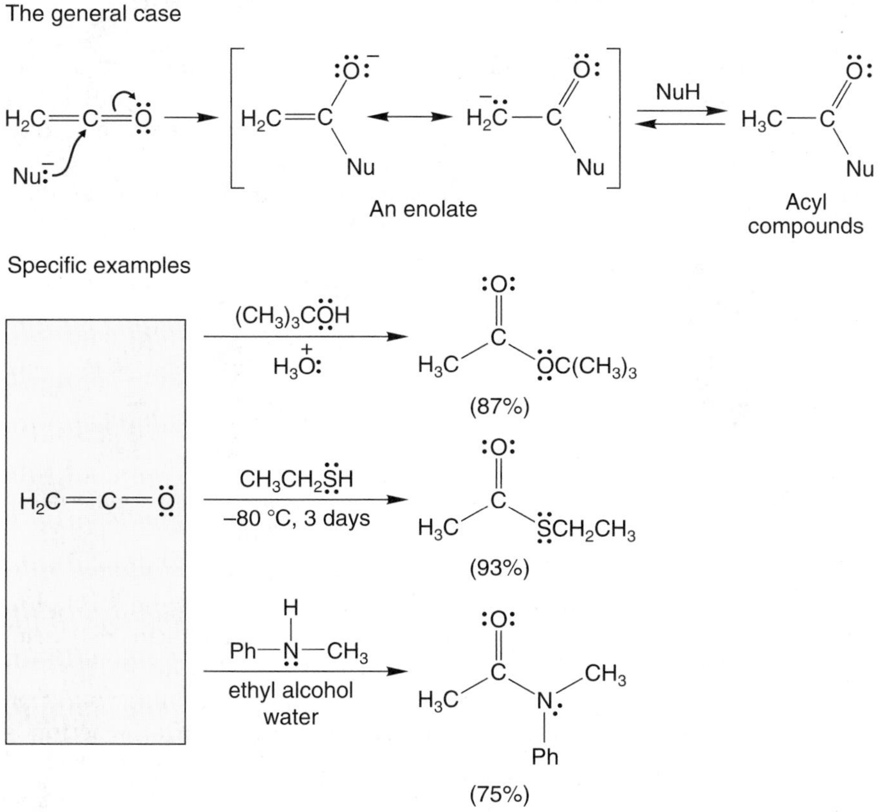

The general case

$$H_2C{=}C{=}\ddot{O} + Nu\colon^- \longrightarrow \left[H_2C{=}C(Nu){-}\ddot{O}\colon^- \longleftrightarrow H_2\ddot{C}^-{-}C(Nu){=}\ddot{O} \right] \underset{}{\overset{NuH}{\rightleftarrows}} H_3C{-}C(Nu){=}\ddot{O}\colon$$

An enolate

Acyl compounds

Specific examples

$$H_2C{=}C{=}\ddot{O} \xrightarrow[H_3\overset{+}{O}\colon]{(CH_3)_3C\ddot{O}H} H_3C{-}C({=}\colon O\colon){-}\ddot{O}C(CH_3)_3 \quad (87\%)$$

$$H_2C{=}C{=}\ddot{O} \xrightarrow[-80\ ^\circ C,\ 3\ \text{days}]{CH_3CH_2\ddot{S}H} H_3C{-}C({=}\colon O\colon){-}\ddot{S}CH_2CH_3 \quad (93\%)$$

$$H_2C{=}C{=}\ddot{O} \xrightarrow[\text{ethyl alcohol, water}]{Ph{-}\ddot{N}(H){-}CH_3} H_3C{-}C({=}\colon O\colon){-}\ddot{N}(CH_3)(Ph) \quad (75\%)$$

FIGURE 20.57 Ketenes react with nucleophiles to give acyl compounds.

20.10 ENOLATE CHEMISTRY OF ESTERS: THE CLAISEN AND RELATED CONDENSATION REACTIONS

Esters are not as acidic as aldehydes and ketones; their pK_a values are about 24 (Table 20.2). Esters are therefore some five pK_a units less acidic than their less well resonance-stabilized relatives. Nevertheless, it is possible for some enolate to be formed in the presence of a base such as an alkoxide ion, even though the reaction must be substantially endothermic (Fig. 20.58).

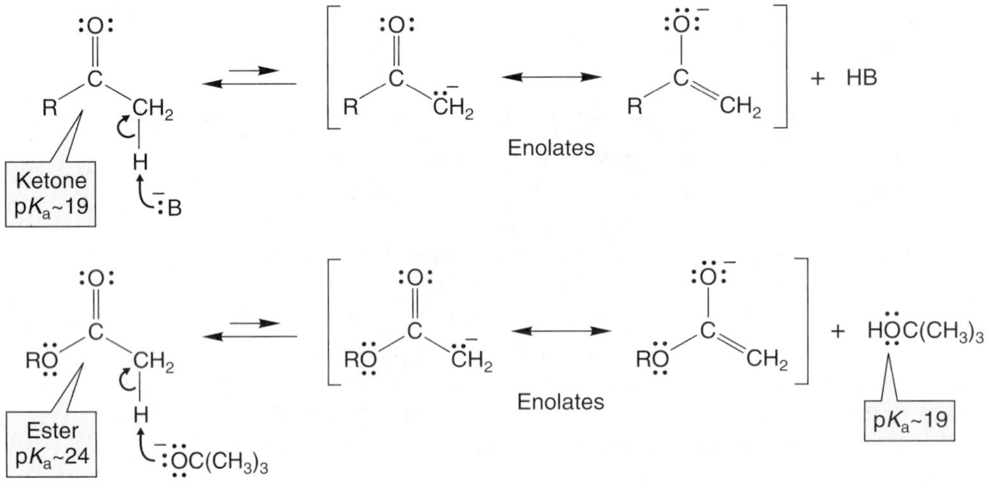

Ketone pK_a~19

Enolates

+ HB

Ester pK_a~24

Enolates

+ $HOC(CH_3)_3$

pK_a~19

FIGURE 20.58 Formation of an enolate anion from an ester. As *tert*-butyl alcohol has a pK_a of about 19, and the ester a pK_a of about 24, this reaction must be highly endothermic.

Why doesn't the alkoxide add to the carbonyl group and start a base-catalyzed transesterification reaction (Section 20.6)? It does! If one is not careful to use the same OR group in the alkoxide and the ester, then mixtures of products are found. As long as the OR groups match, however, the base-catalyzed transesterification doesn't change anything—the product is the same as the starting material (Fig. 20.59).

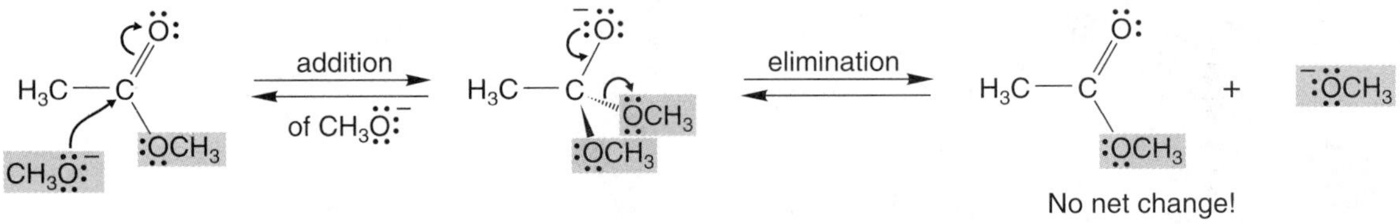

addition

of CH_3O^-

elimination

No net change!

FIGURE 20.59 A base-catalyzed transesterification reaction will generate a new ester unless the OR group of the alkoxide is identical to the OR group of the ester.

What can the ester enolate do? It can react with the acidic alcohol to reprotonate, regenerating the starting ester. However, it can also react *with any other Lewis acid in the system.* The carbonyl group of the ester is just such a Lewis acid, and it can react with the enolate in an addition reaction. Although the product is more complex, the addition reaction is no different from any other addition of a nucleophile to an ester carbonyl (Fig. 20.60).

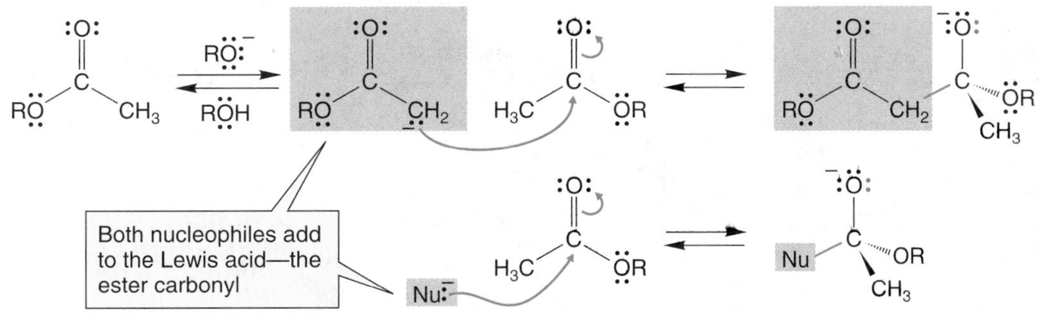

Both nucleophiles add to the Lewis acid—the ester carbonyl

FIGURE 20.60 In the first step of the Claisen condensation, the nucleophilic ester enolate adds to the Lewis acid ester carbonyl.

The next reaction is also analogous to reactions of esters with simpler nucleophiles. The addition product can do two things. It can expel the enolate in a simple reverse of the original addition reaction, or it can lose alkoxide to give a molecule of **acetoacetic ester**, a β-keto ester (Fig. 20.61). This reaction is called the **Claisen condensation** after Ludwig Claisen (1851–1930), who has many reactions named for him. Recall the Claisen–Schmidt condensation, for example (Chapter 18, p. 914).

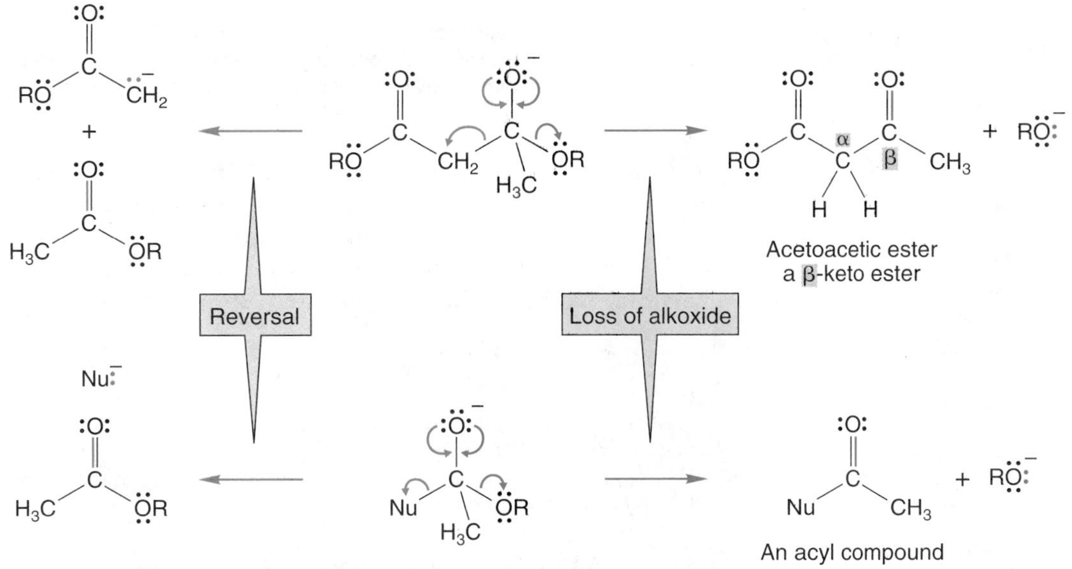

Reversal

Loss of alkoxide

Acetoacetic ester
a β-keto ester

An acyl compound

FIGURE 20.61 In the second step of the Claisen condensation, an alkoxide ion is lost and acetoacetic ester is formed. This process closely resembles the loss of alkoxide to give acyl compounds.

The Claisen condensation bears the same relationship to the aldol condensation that the addition of a nucleophile to an ester does to the addition to an aldehyde or ketone. Whereas the alkoxides formed in the aldol condensation can only revert to starting material or protonate, both ester reactions have the added option of loss of the ester alkoxide. What is added in ester chemistry is the possibilty of the elimination phase of the addition–elimination process (Fig. 20.62).

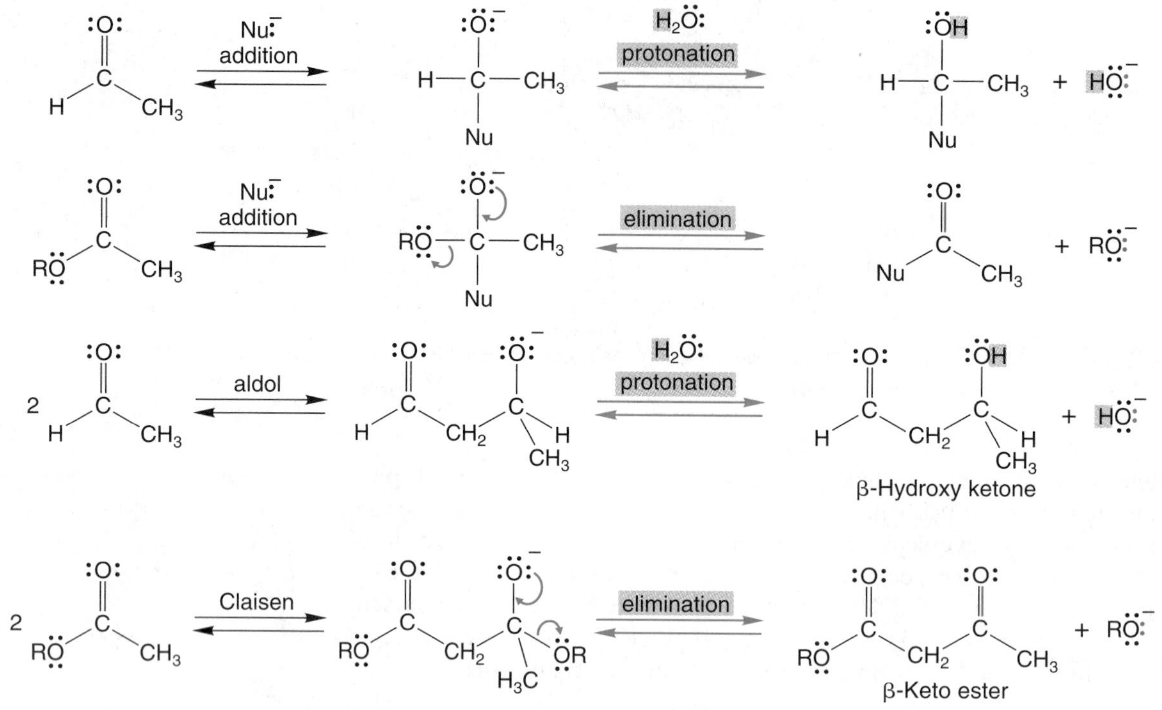

FIGURE 20.62 The aldol condensation is just an example of the addition of a nucleophile to a carbonyl group followed by protonation. The Claisen condensation follows the addition–elimination sequence available to esters because of the presence of a leaving group, OR, within the molecule.

As in aldol condensations of ketones, in the Claisen condensation it is the traditional starting materials that are favored thermodynamically. Esters are stabilized by resonance and there are two ester groups in the starting materials but only one in the product. The result is a thermodynamic favoring of the two separated esters (Fig. 20.63).

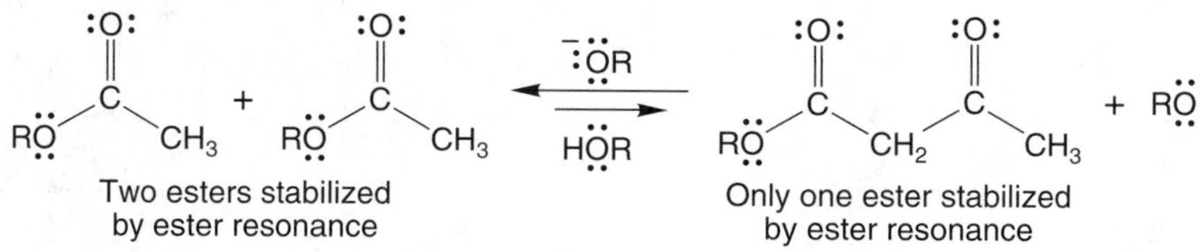

FIGURE 20.63 Thermodynamics favors the pair of esters (starting material) in the Claisen condensation, not the β-keto ester product. The reaction to form the β-keto ester is endothermic.

However, there is an easy way out of this thermodynamic bind. In the last step of the Claisen condensation a molecule of alkoxide catalyst is regenerated. The catalyst is regenerated in the aldol condensation as well, and in the aldol condensation the catalyst goes on to initiate another condensation. In the Claisen condensation, however, there is another, faster reaction possible. The β-keto ester product is by far the strongest acid in the system as it contains hydrogens that are α to *two* carbonyl groups. The anion formed by deprotonation of the β-keto ester is resonance stabilized by *both* carbonyl groups (Fig. 20.64).

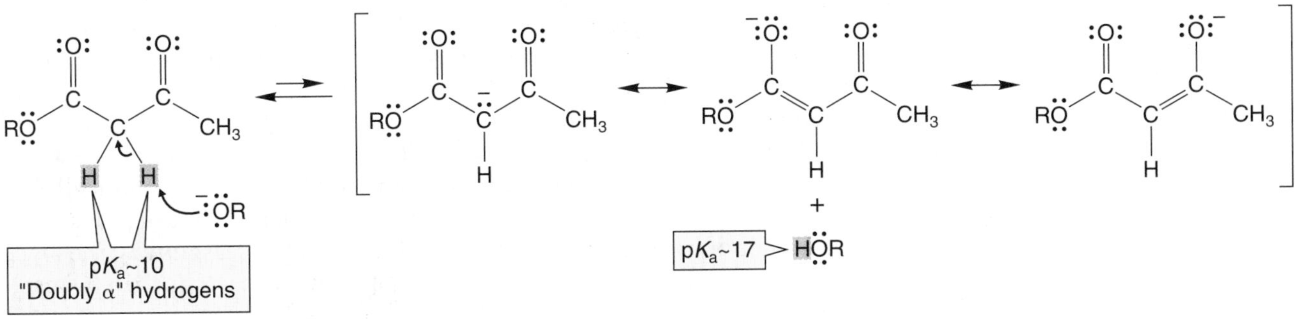

FIGURE 20.64 The removal of a doubly α proton by alkoxide is an exothermic reaction.

The pK_a values of β-keto esters are about 10, and therefore these molecules are about 10^{14} more acidic than the starting esters! By far, the best reaction for the alkoxide regenerated at the end of the condensation step is removal of one of the doubly α hydrogens to give the highly resonance-stabilized salt of the β-keto ester (Fig. 20.65).

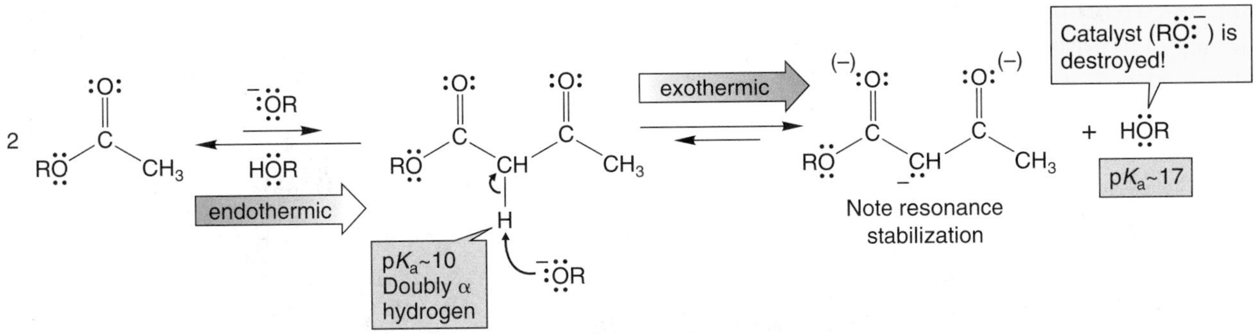

FIGURE 20.65 Alkoxide is destroyed in the last step of the Claisen condensation as a doubly α proton is removed to give a stable anion.

*PROBLEM 20.19

There is another reaction, encountered earlier in this chapter, that also relies for its success on formation of a resonance-stabilized anion. What is it?

ANSWER

The base-induced hydrolysis of an ester to an acid (saponification) generates the resonance-stabilized carboxylate anion. This procedure makes the reaction essentially irreversible, but also demands that the reaction be run with a full equivalent of hydroxide ion.

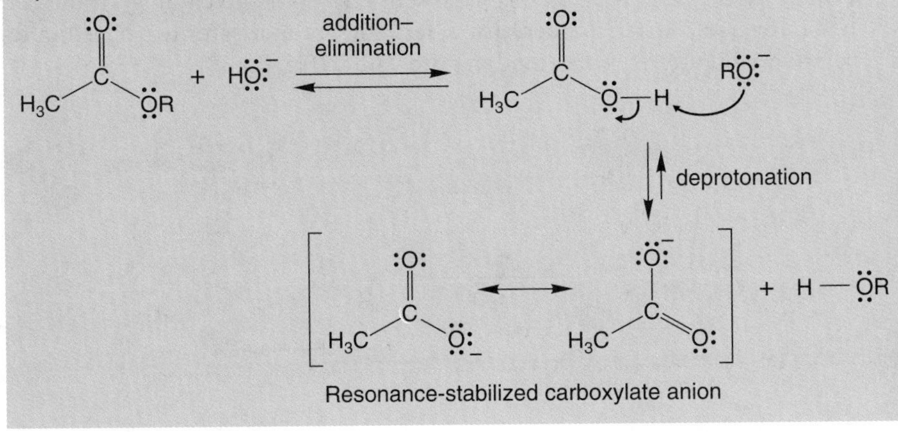

Resonance-stabilized carboxylate anion

If the doubly α proton is removed, all is well thermodynamically; now the product is more stable than the starting material, *but the catalyst for the reaction has been destroyed*! A Claisen condensation using a catalytic amount of base must fail, as it can only generate an amount of product equal to the original amount of base. The simple remedy is to use a full equivalent of alkoxide, not a catalytic amount. A thermodynamically favorable Claisen condensation is not a catalytic reaction, but requires a full equivalent of base. Be certain you are clear as to the reason why the aldol condensation requires only a catalytic amount of base but the Claisen needs a full equivalent (Fig. 20.66).

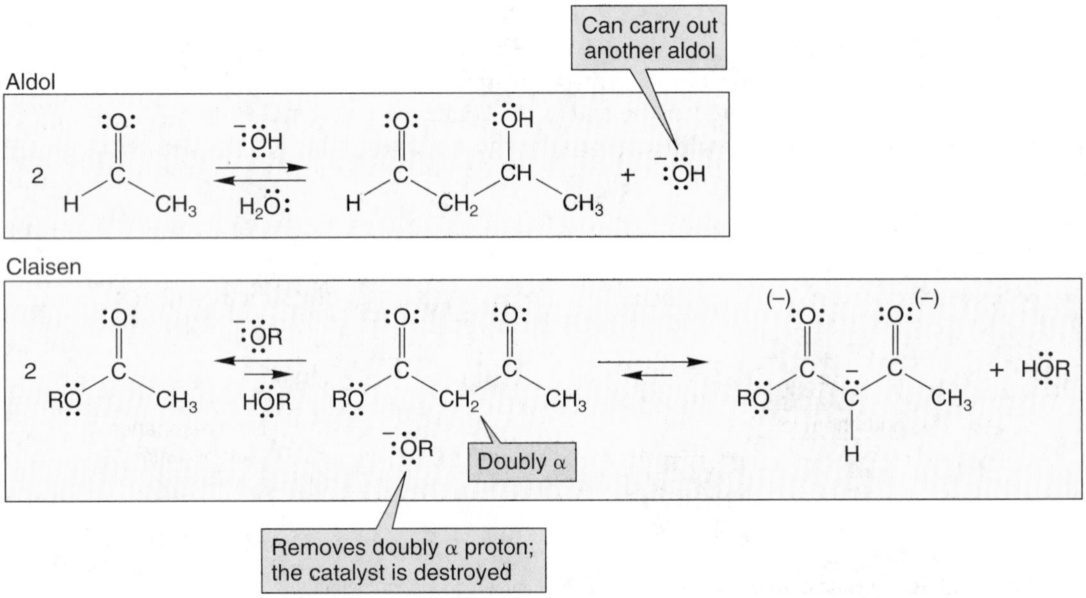

Aldol

Can carry out another aldol

Claisen

Doubly α

Removes doubly α proton; the catalyst is destroyed

FIGURE 20.66 The aldol condensation can be carried out using a catalytic amount of base because the final protonation step regenerates a molecule of hydroxide. In the last step of the Claisen condensation, the catalyst alkoxide ion is destroyed by removal of the doubly α proton.

Acidification of the reaction mixture leads to protonation of the anion to form the β-keto ester. What prevents this product, thermodynamically unstable with respect to the starting ester molecules, from re-forming the esters? There is no basic catalyst present! Without the catalyst, there can be no reverse Claisen condensation, and the product ester is trapped in its relatively high-energy thermodynamic valley (Fig. 20.67).

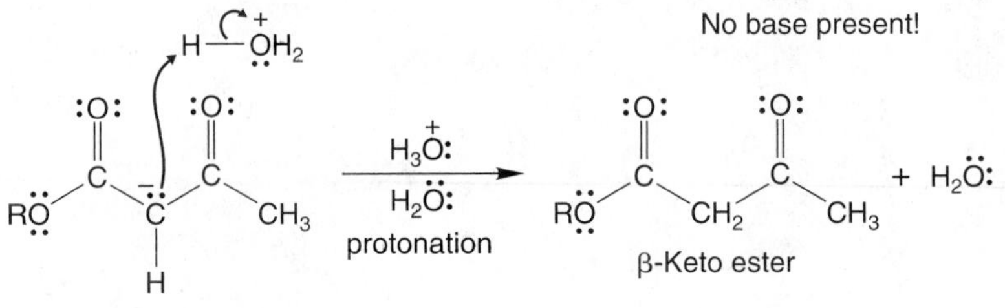

No base present!

protonation

β-Keto ester

FIGURE 20.67 The final neutralization step produces a β-keto ester *in the absence of base.*

Figure 20.68 shows Energy versus Reaction progress diagrams for the aldol and Claisen condensations.

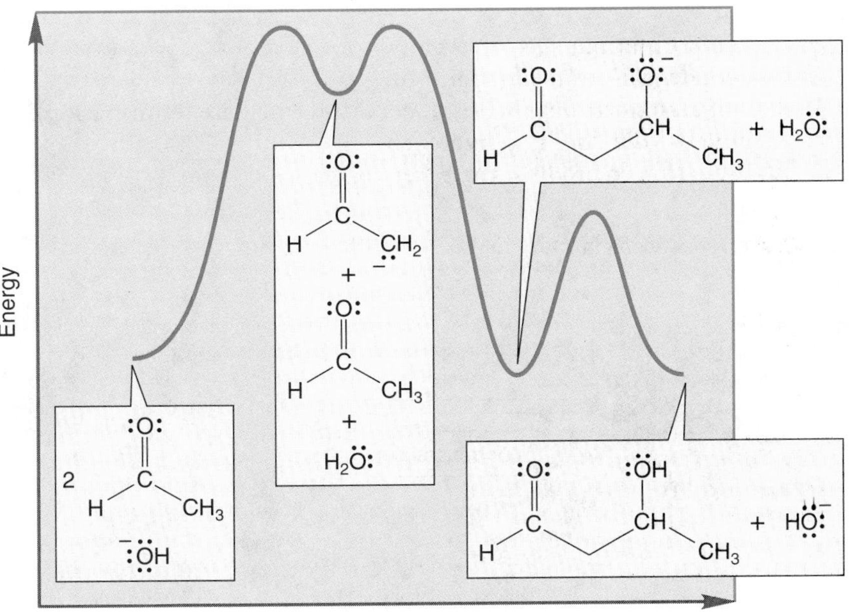
Energy
Reaction progress
The aldol condensation for acetaldehyde is approximately thermoneutral

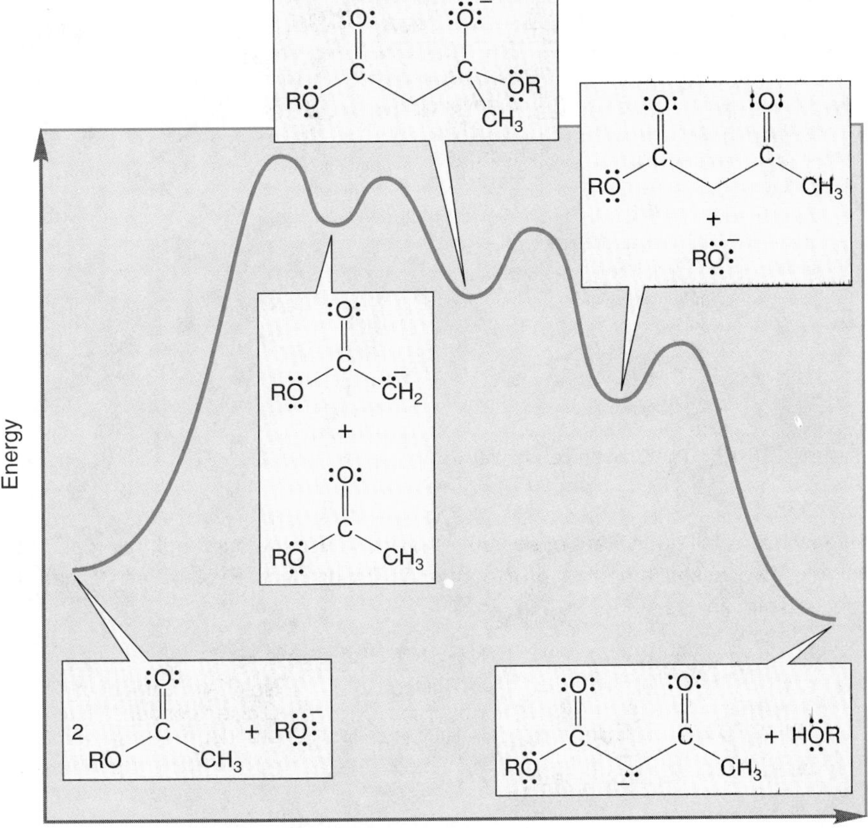
Energy
Reaction progress

FIGURE **20.68** The Energy versus Reaction progress diagrams for the aldol and Claisen condensations.

So, a full equivalent of base makes the Claisen condensation a practical source of acetoacetic esters, β-keto esters. Remember the five steps of the process (Fig. 20.69):

1. Enolate formation.
2. The condensation itself (addition).
3. Loss of alkoxide from the tetrahedral intermediate (elimination).
4. Removal of the doubly α hydrogen.
5. Acidification to produce the β-keto ester.

The Claisen condensation

The general case

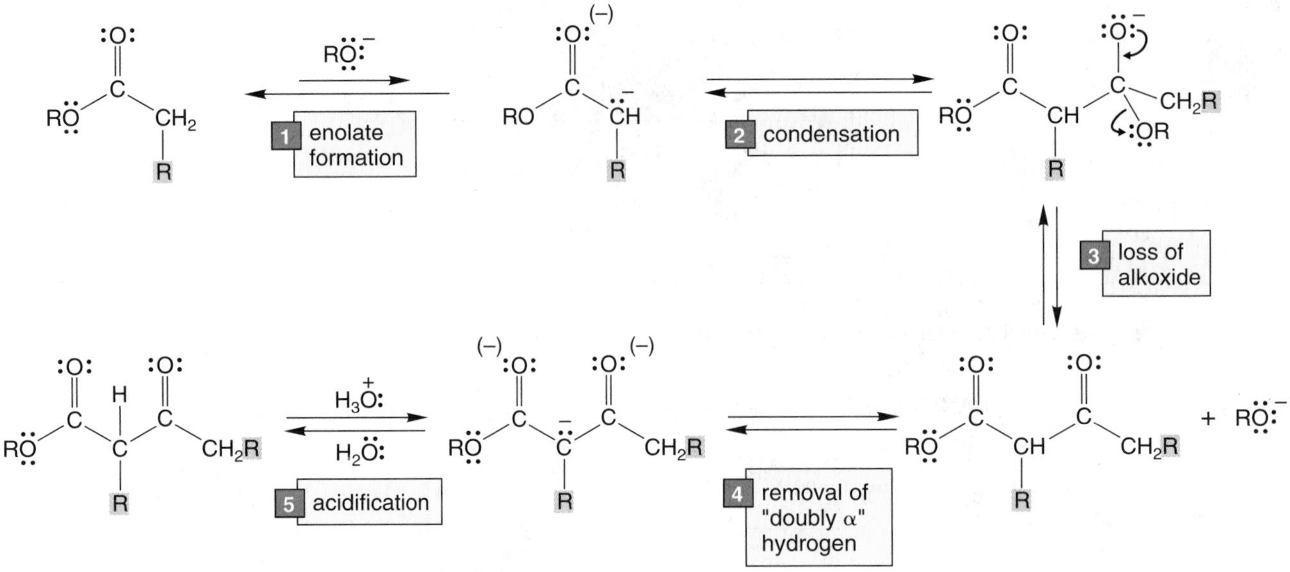

1 enolate formation

2 condensation

3 loss of alkoxide

4 removal of "doubly α" hydrogen

5 acidification

Specific example

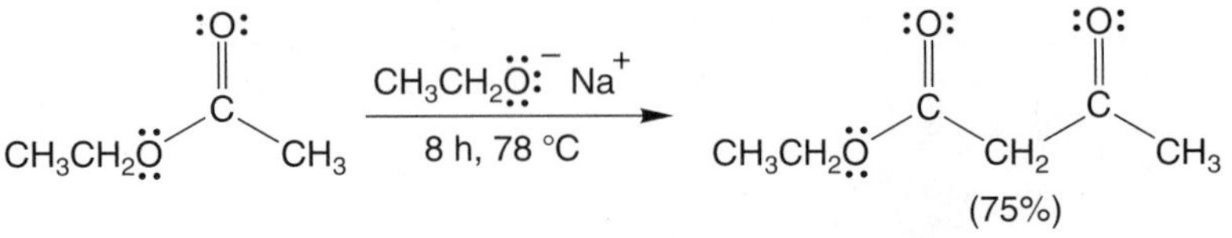

$CH_3CH_2\ddot{O}^- \ Na^+$, 8 h, 78 °C

(75%)

FIGURE 20.69 The Claisen condensation.

*PROBLEM 20.20 Even when a full equivalent of sodium ethoxide is used, ethyl 2-methylpropionate fails to give useful amounts of a condensation product. Explain with a careful mechanistic analysis (Fig. 20.70).

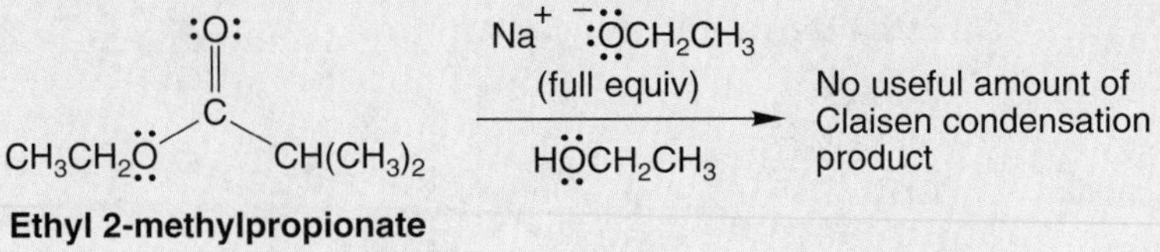

$Na^+ \ ^-\!:\!\ddot{O}CH_2CH_3$ (full equiv), $H\ddot{O}CH_2CH_3$ → No useful amount of Claisen condensation product

Ethyl 2-methylpropionate

FIGURE 20.70

In this classic problem, Claisen condensation leads to product **A** through the usual addition–elimination scheme:

ANSWER

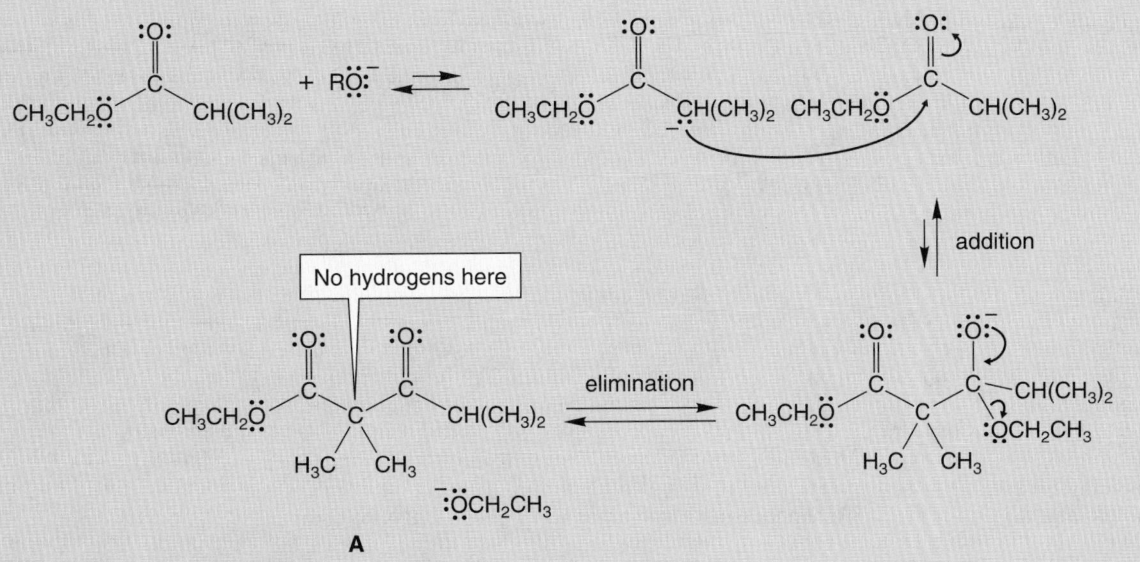

No hydrogens here

addition

elimination

A

However, the product **A** contains no doubly α hydrogens! There can be no formation of the stable, highly resonance-stabilized anion that accounts for the success of the Claisen condensation. Thermodynamics dictates that the starting materials are more stable than the product, and the reaction settles out at a point at which there is little product. The mechanism of the reverse reaction is written for you: Just read the scheme backward.

20.11 VARIATIONS ON THE CLAISEN CONDENSATION

So, that's the Claisen condensation: two esters combine with an equivalent of base to make an acetoacetic ester. Now it's time to look at how changes in structure of the starting materials affect the structure of the product and the mechanism of its formation.

20.1a Cyclic Claisen Condensations: The Dieckmann Condensation

The cyclic version of the Claisen condensation is called the **Dieckmann condensation** after Walter Dieckmann (1869–1925). Five- and six-membered rings are favored, and there is little beside the intramolecular nature of the reaction to distinguish the Dieckmann from the standard Claisen condensation. Like the Claisen, the Dieckmann condensation is thermodynamically unfavorable if only a catalytic amount of base is used. Unless there is sufficient base available to remove the doubly α hydrogen and drive the reaction to a thermodynamically favorable conclusion, the reaction is not synthetically useful. A final acidification liberates the β-keto ester in the absence of base. Figure 20.71 gives an example.

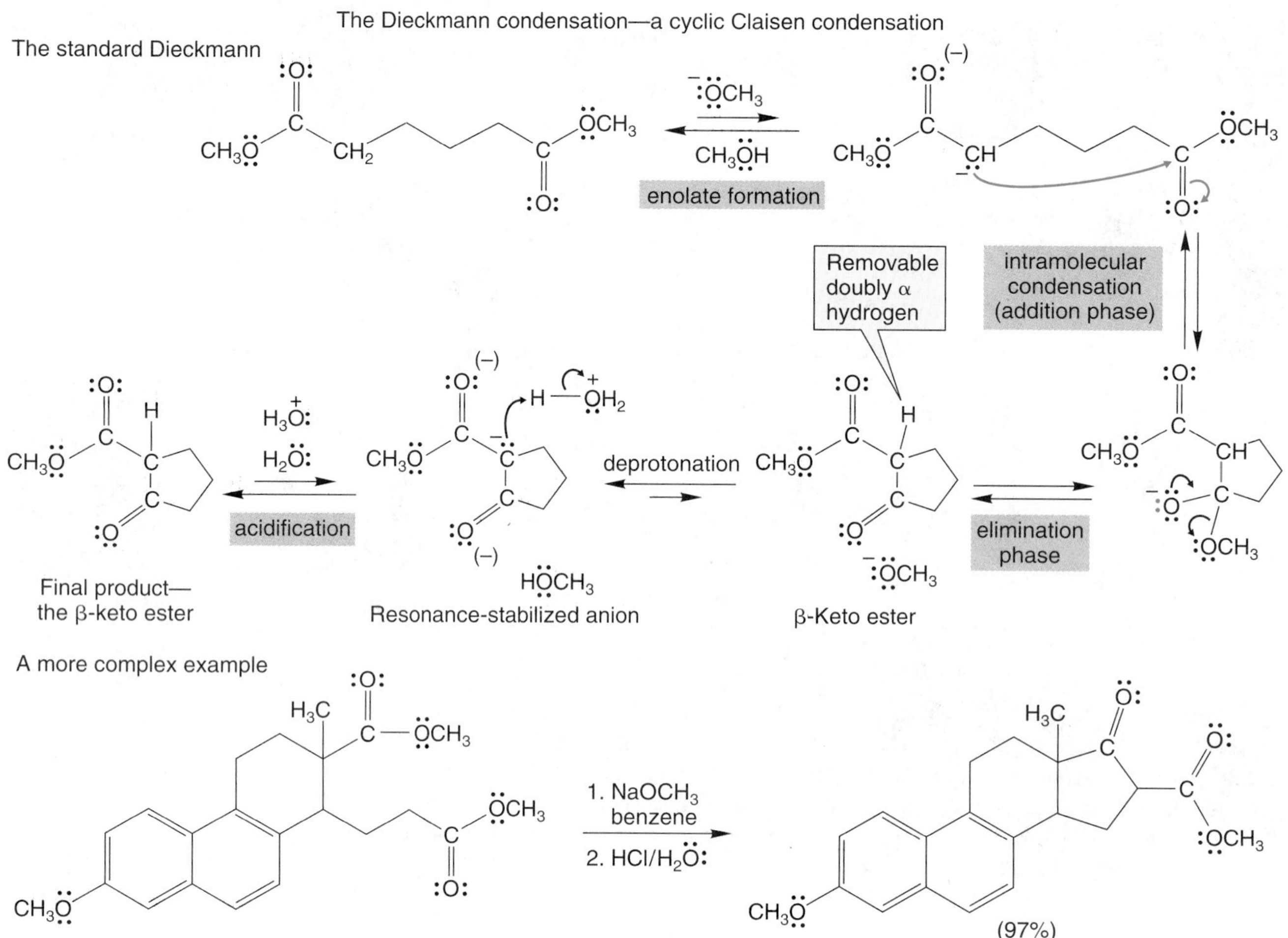

FIGURE 20.71 A Dieckmann condensation is just an intramolecular, or cyclic Claisen condensation.

20.11b Crossed Claisen Condensations

A naively designed crossed Claisen condensation is doomed to the same kind of failure as is the crossed aldol condensation. There are four possible products, and therefore no easy source of the selectivity we need for a practical synthetic reaction (Fig. 20.72).

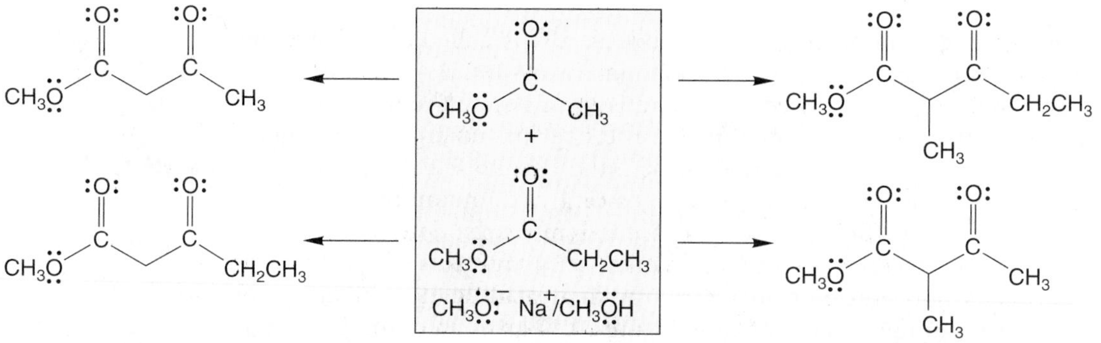

FIGURE 20.72 There are four possible—and likely—products from a crossed Claisen condensation.

PROBLEM 20.21

Show in a general way how these products are produced and write a mechanism for the formation of one of the four products shown in Figure 20.72.

As in the crossed aldol, however, there are some simple steps we can take to improve matters. If one partner has no α-hydrogen, there can be no enolate formation from it, and it can act only as a Lewis acid, never as a nucleophile. It is helpful to add the ester with an α-hydrogen to a mixture of the partner without an α-hydrogen and the base. In this case, the enolate will be very likely to react as it is formed with the large amount of the first ester, and a successful crossed Claisen can be achieved. Benzoates (PhCOOR), formates (HCOOR), and carbonates (ROCOOR) are examples of esters without α-hydrogens and are often used (Fig. 20.73).

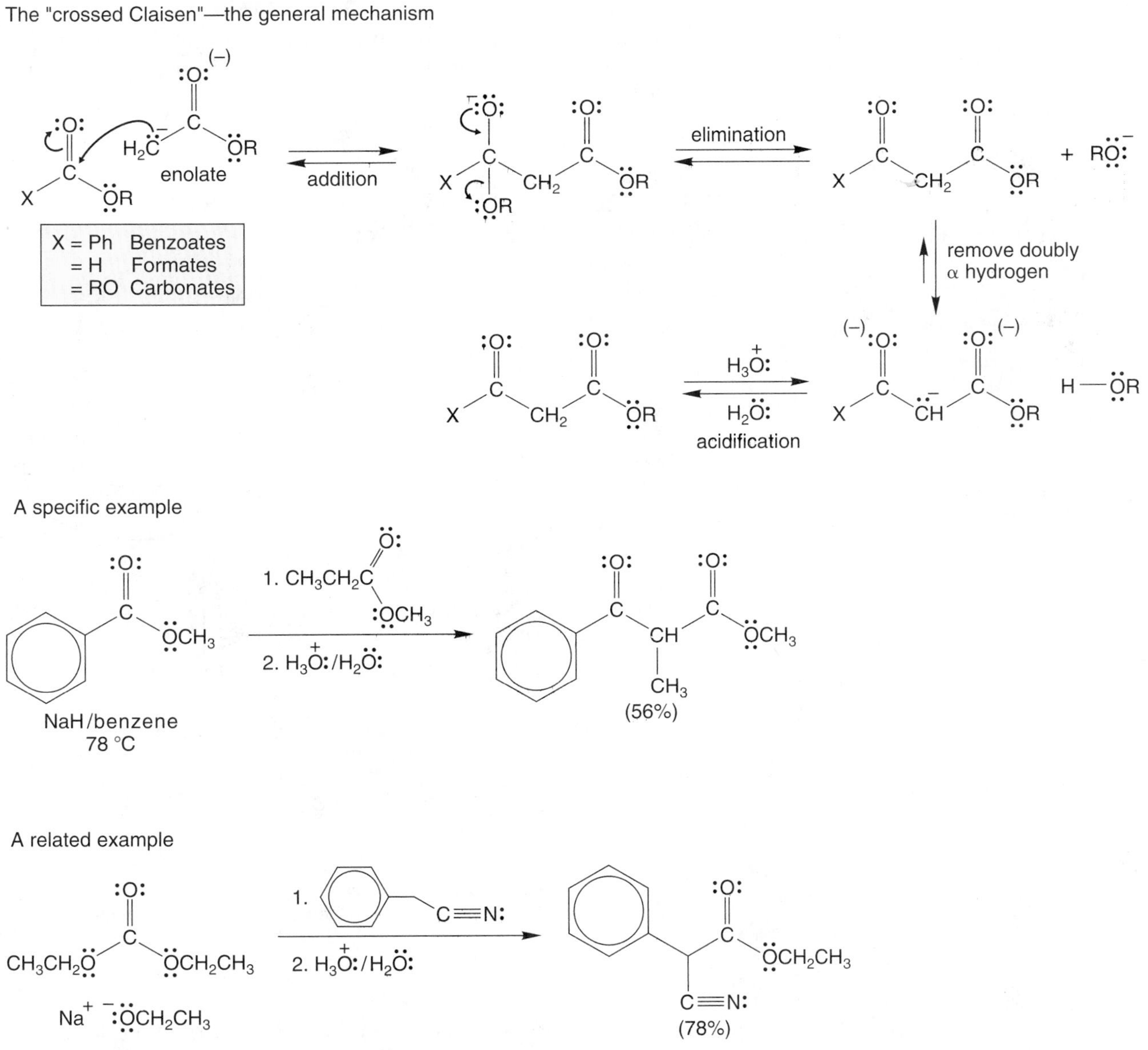

The "crossed Claisen"—the general mechanism

enolate

addition

elimination

+ RO:⁻

X = Ph Benzoates
= H Formates
= RO Carbonates

remove doubly α hydrogen

H_3O^+ / H_2O: acidification

H—OR

A specific example

1. $CH_3CH_2COOCH_3$
2. H_3O^+/H_2O

NaH/benzene
78 °C

(56%)

A related example

1. $PhCH_2C{\equiv}N$
2. H_3O^+/H_2O

$CH_3CH_2OCOOCH_2CH_3$

$Na^+ \; ^-OCH_2CH_3$

(78%)

FIGURE 20.73 Successful crossed Claisen condensations.

20.11c Condensations of Esters with Ketones

A combination aldol-Claisen condensation can occur between an ester and a ketone. It is unlikely that formation of the product of aldol condensation of the ketone will be a problem. Aldol condensations of ketones are reversible, and usually endothermic. So even though the product of the aldol condensation *will* be formed, starting ketone will be favored. If ketone is drained away through reaction with the ester, the reversibly formed aldol product merely serves as a storage point for the ketone. The normal Claisen condensation of two esters is usually not a problem because the ketone is a much stronger acid than the ester (Table 20.2, p. 1010), and formation of the ketone enolate will be much preferred to formation of the ester enolate. The ketone enolate adds to the ester carbonyl, alkoxide is lost, and a β-dicarbonyl compound is formed. If a full equivalent of base is used, the doubly α hydrogen is removed to give a nicely resonance-stabilized anion. The reaction is completed by an acidification step that regenerates the β-dicarbonyl compound in the absence of base. Note that the reaction contains the same five steps as the Claisen condensation itself (Fig. 20.74).

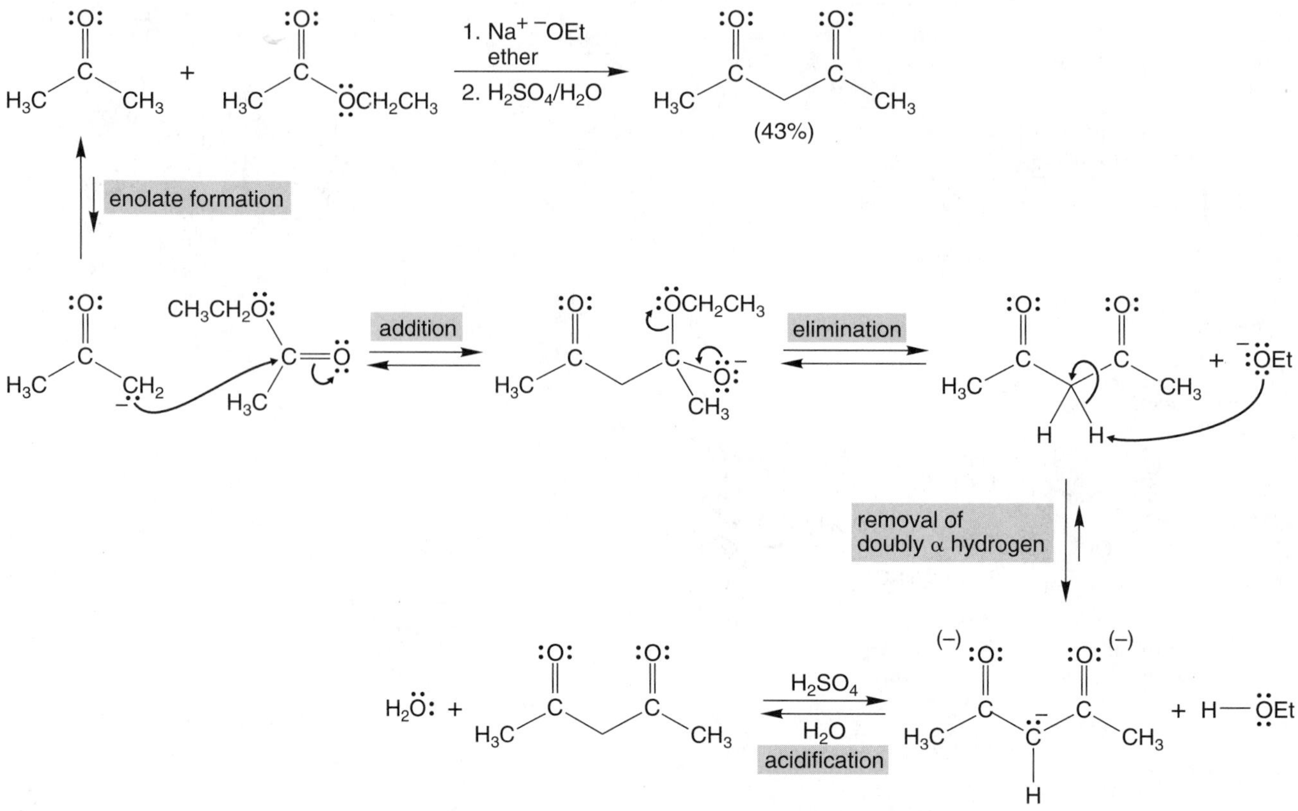

1. $Na^+ \,^-OEt$ ether
2. H_2SO_4/H_2O

(43%)

enolate formation

addition

elimination

removal of doubly α hydrogen

H_2SO_4

H_2O acidification

FIGURE 20.74 The mechanism of a crossed condensation reaction between a ketone and an ester.

PROBLEM 20.22 Why are esters less reactive in addition reactions than aldehydes and ketones?

Frequently, the strong base sodamide ($NaNH_2$), or a dialkylamide such as LDA, is used in this reaction. It is generally the kinetically favored, more accessible hydrogen that is removed when there is a choice of possible enolates (Fig. 20.75).

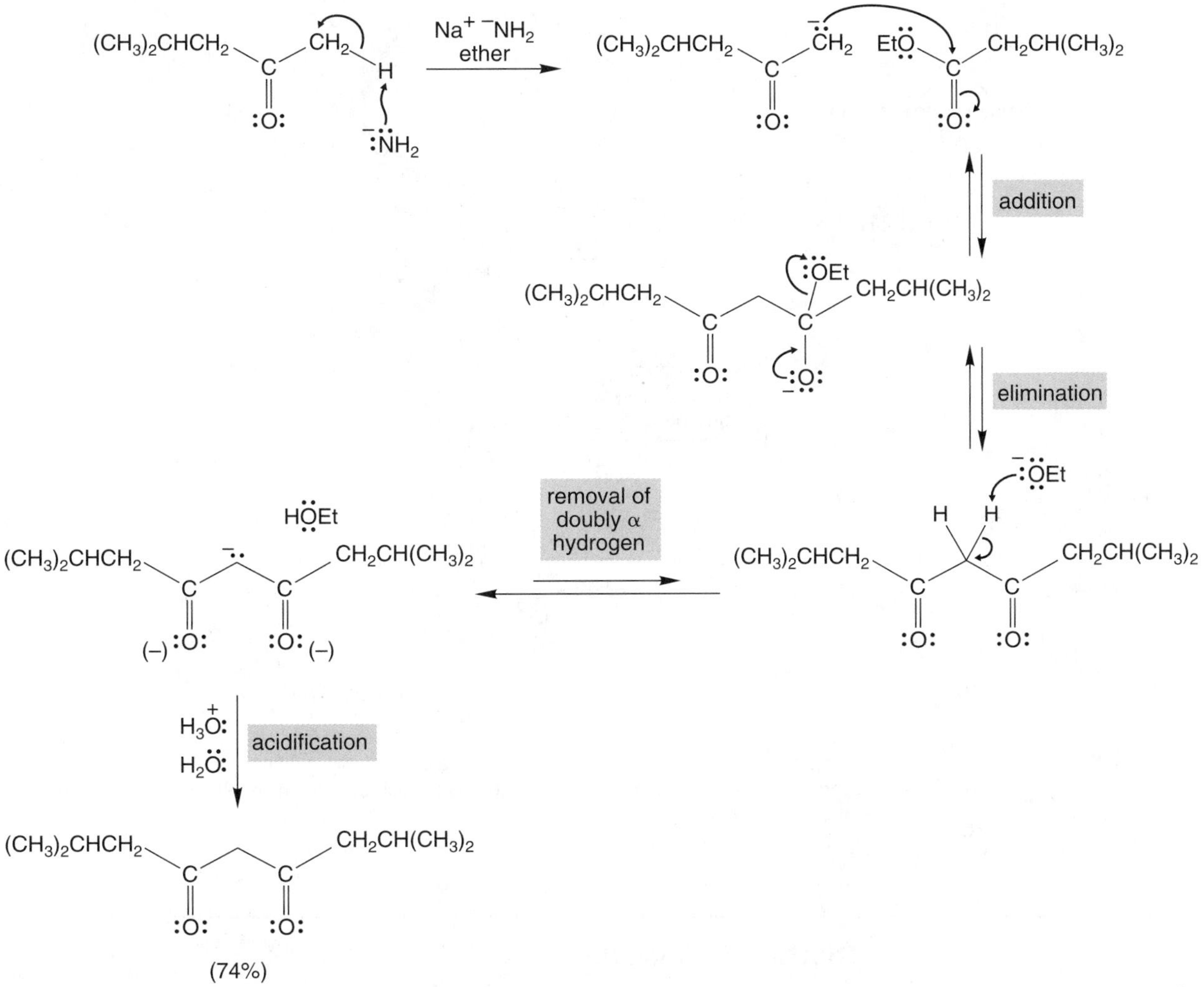

(74%)

FIGURE 20.75 Amide bases are often used in this reaction and lead to the formation of the "kinetic" enolate through removal of the most accessible hydrogen.

20.11d Reverse Claisen Condensations

As we have mentioned, formation of the initial condensation product in the Claisen condensation is usually endothermic. If no doubly α hydrogens exist for salt formation, β-keto esters will revert to their component ester starting materials when treated with base. Ethoxide ion is not a strong enough base to remove the lone singly α hydrogen in the product (recall Problem 20.20). However, if a strong enough base is used, enolates *can* be formed through removal of the singly α hydrogen. The classic example is the Claisen condensation induced by the very strong base, triphenylmethyl sodium (Fig. 20.76).

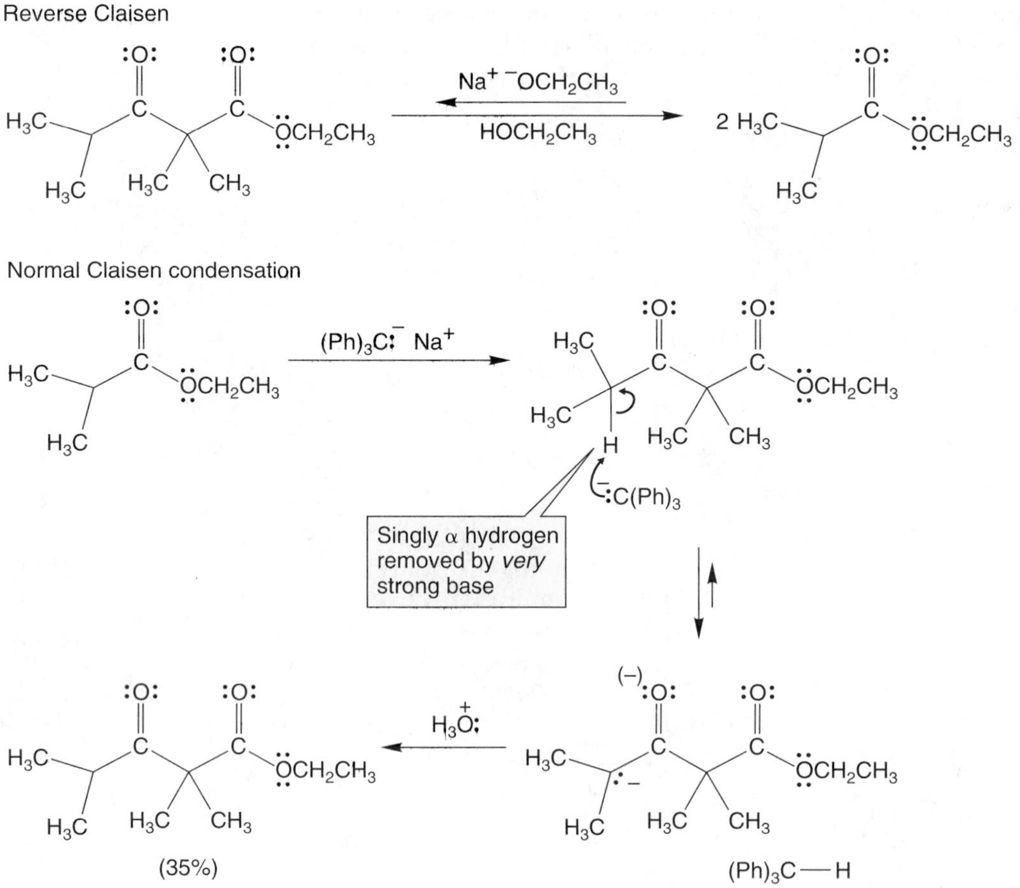

FIGURE 20.76 If no doubly α hydrogens are available, a β-keto ester reacts with base to regenerate a pair of esters. Very strong bases such as triphenylmethyl sodium can remove the singly α hydrogen.

20.12 REACTIONS OF β-KETO ESTERS

20.12a Alkylation Reactions

As we have seen, β-keto esters are quite strong acids. The pK_a of acetoacetic ester itself is about 10 and other β-dicarbonyl compounds, β-cyano carbonyl compounds, and dicyano compounds are similarly acidic (Table 20.5).

The anions formed from these compounds can be alkylated in S_N2 reactions. The once-alkylated compounds still contain one doubly α hydrogen, and therefore remain acidic. A second alkylation reaction is often possible, as long as the requirements of the S_N2 reaction are kept in mind. **Malonic esters** are sufficiently acidic ($pK_a \sim 13$) so that alkylation through the enolate is easy. Both mono- and dialkylation reactions of β-keto esters and related 1,1-diesters are common (Fig. 20.77).

PROBLEM 20.23 Write a mechanism for the base-induced alkylation of 1,1-dicyanomethane (malononitrile) with ethyl iodide.

TABLE 20.5 The Acidity of Some β-Dicarbonyl and Related Compounds

Compound	Name	pK_a
$HC(=O)CH_2C(=O)H$	Malonaldehyde	5
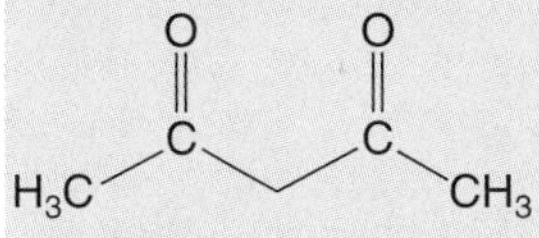 $H_3CC(=O)CH_2C(=O)CH_3$	2,4-Pentandione	9 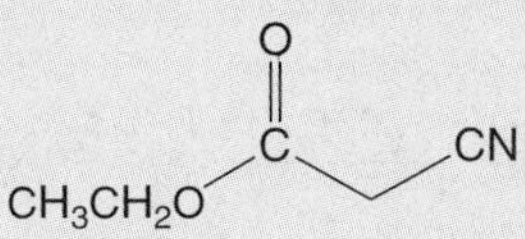
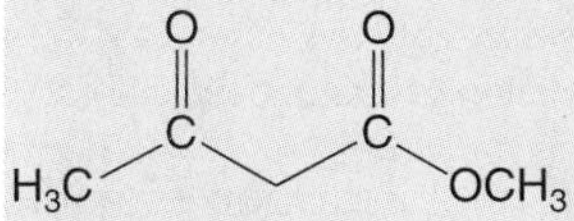 $H_3CC(=O)CH_2C(=O)OCH_3$	Methyl acetoacetate	10 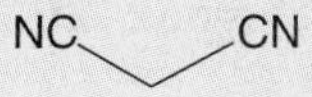
$CH_3CH_2OC(=O)CH_2C(=O)OCH_2CH_3$	Ethyl malonate	13
$CH_3CH_2OC(=O)CH_2CN$	Ethyl cyanoacetate	9
$NCCH_2CN$	Malononitrile	11

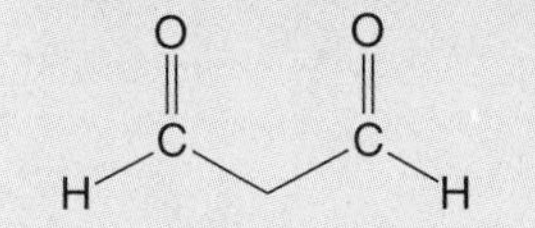

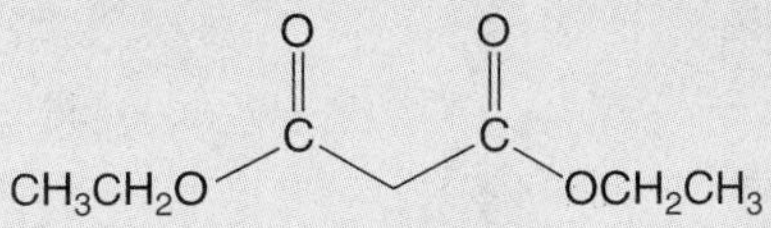

General cases

$$RO{-}C(=O){-}CH_2{-}C(=O){-}CH_3 \xrightarrow[\text{HOR}]{\text{RO}^-} RO{-}C(=O){-}\overset{-}{C}H{-}C(=O){-}CH_3 \xrightarrow[(S_N2)]{\text{R—X}} RO{-}C(=O){-}CH(R){-}C(=O){-}CH_3$$

Once alkylated

$$RO{-}C(=O){-}CH(R){-}C(=O){-}CH_3 \xrightarrow[\text{HOR}]{\text{RO}^-} RO{-}C(=O){-}\overset{-}{C}(R){-}C(=O){-}CH_3 \xrightarrow[(S_N2)]{\text{R—X}} RO{-}C(=O){-}C(R)_2{-}C(=O){-}CH_3$$

Twice alkylated

Specific examples

$$H_3C{-}C(=O){-}CH_2{-}C(=O){-}OEt \xrightarrow[\text{2. } CH_3CH_2CH_2CH_2Br]{\text{1. NaOEt/HOEt, 78 °C}} H_3C{-}C(=O){-}CH(CH_2CH_2CH_2CH_3){-}C(=O){-}OEt \; (70\%)$$

$$(NC)_2CH_2 \xrightarrow[\text{2. } C_6H_5CH_2Cl\text{, 25 °C}]{\text{1. NaH, dimethyl sulfoxide}} [(NC)_2CH{-}CH_2C_6H_5] \xrightarrow{\text{repeat}} (NC)_2C(CH_2C_6H_5)_2 \; (75\%)$$

Not isolated

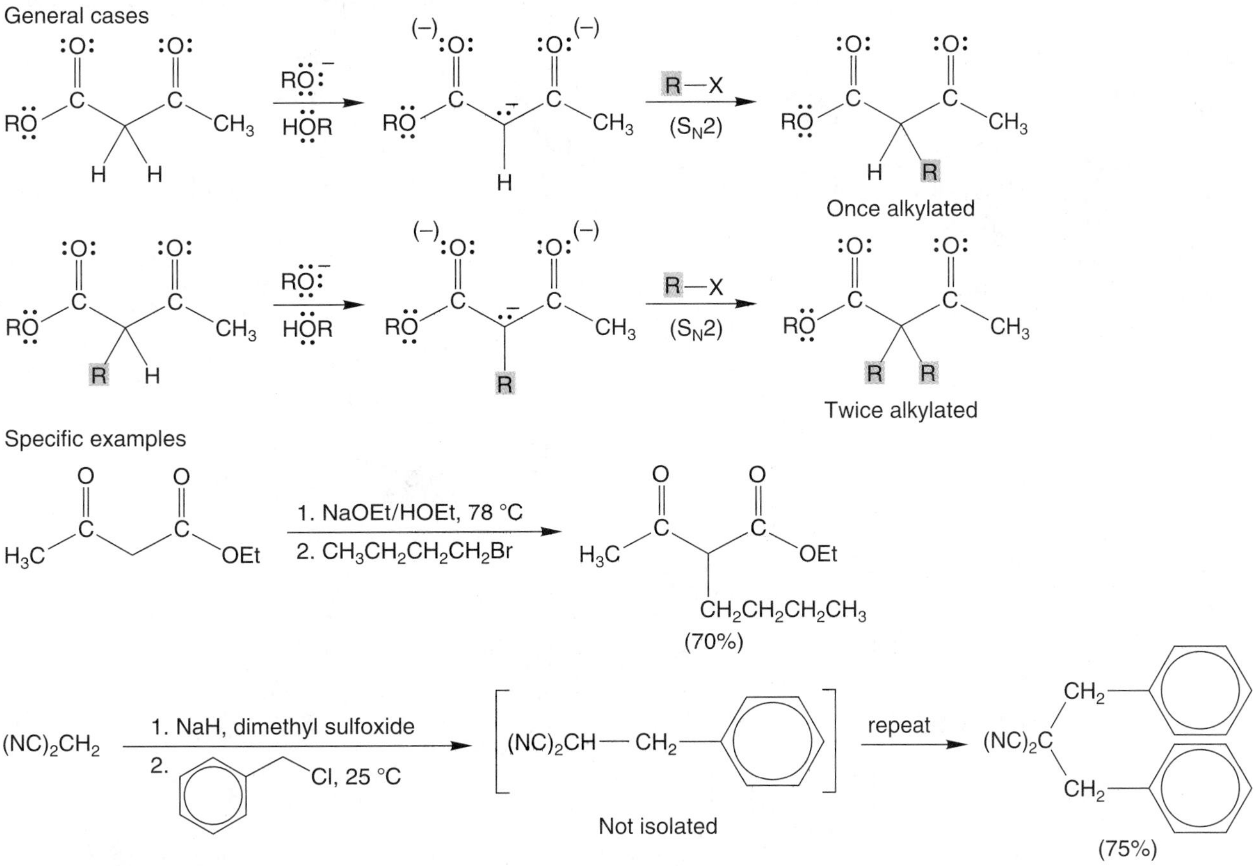

FIGURE 20.77 Alkylations of some ß-dicarbonyl compounds

20.12b Hydrolysis and Decarboxylation of β-Dicarbonyl Compounds

When a β-keto ester such as acetoacetic ester is hydrolyzed in acid, the resulting β-keto acid is most unstable and decarboxylates easily (Fig. 20.78).

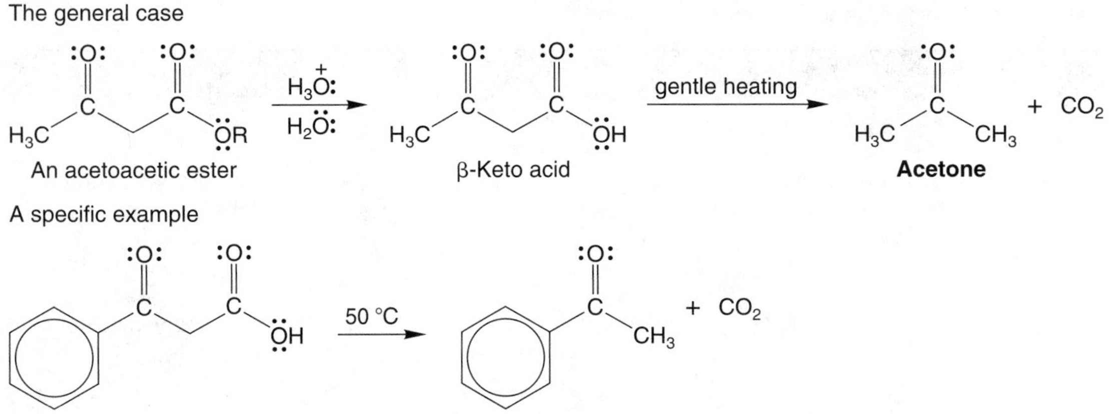

The general case

FIGURE 20.78 The acid-catalyzed hydrolysis and decarboxylation of acetoacetic esters.

The end result is one molecule of acetone and one of carbon dioxide. This reaction is a roundabout method of making this simple ketone. However, the combination of the alkylation reactions of Figure 20.77 followed by acid- or base-hydrolysis and **decarboxylation** is a useful source of complicated ketones (Fig. 20.79).

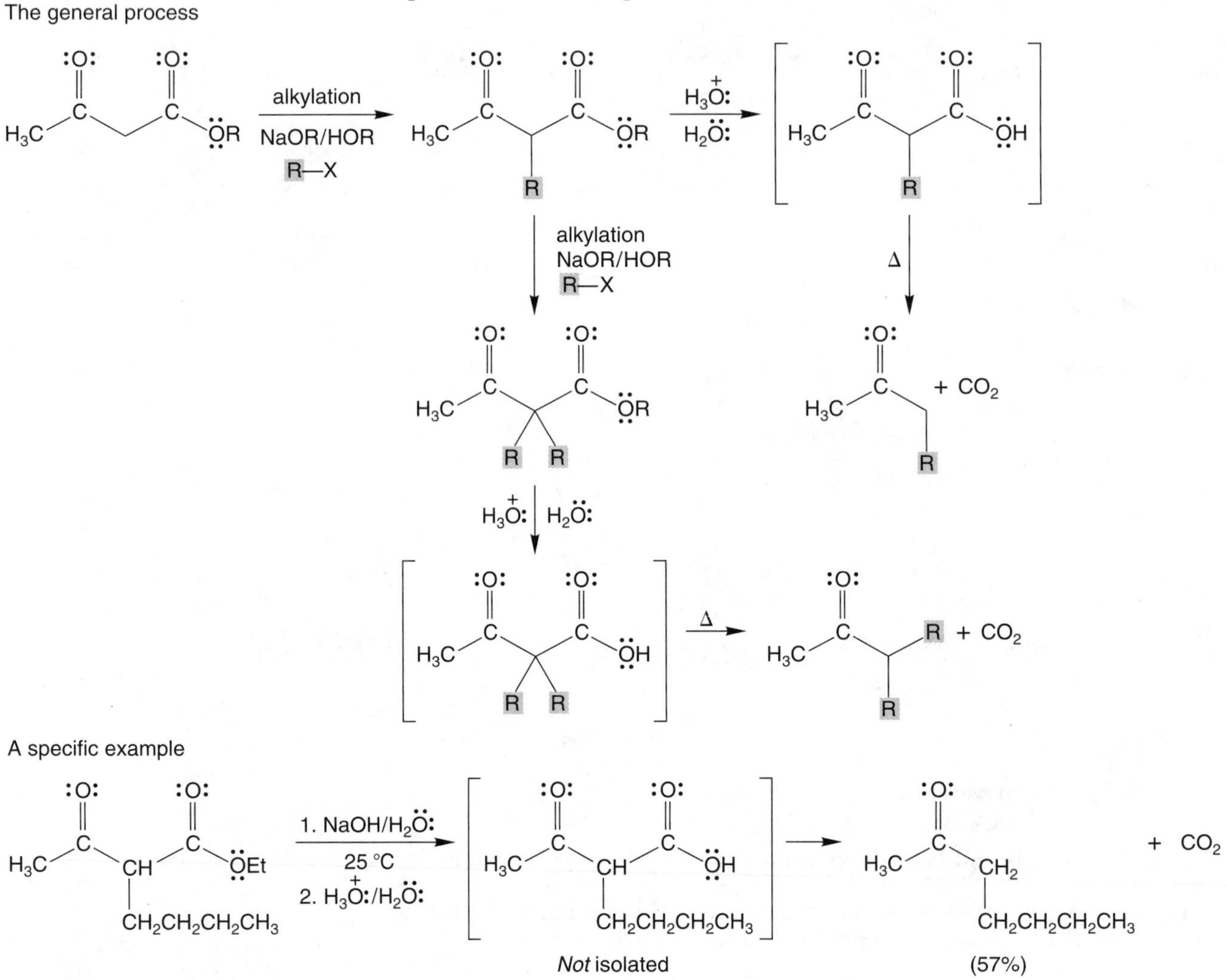

The general process

FIGURE 20.79 The decarboxylation route to substituted ketones.

In the decarboxylation reaction, the carbonyl oxygen acts as base and removes the hydroxyl hydrogen. Carbon dioxide is lost and the enol of the ketone is formed. Normal equilibration with the more stable ketone gives the final product (Fig. 20.80).

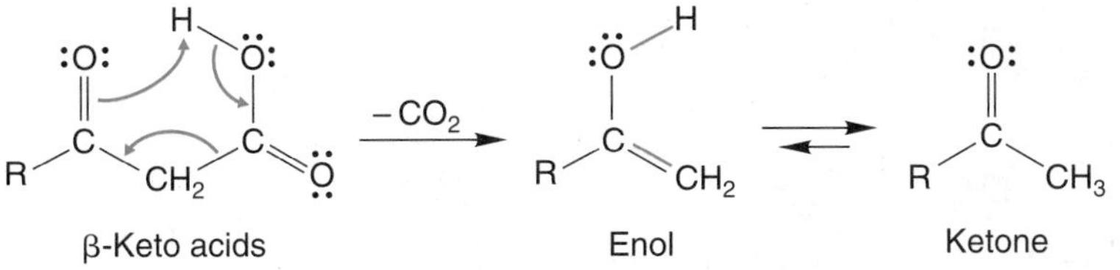

β-Keto acids — $-CO_2$ → Enol ⇌ Ketone

FIGURE 20.80 The mechanism for the decarboxylation of β-keto acids.

Be sure to update your file cards for ketone syntheses with this reaction. Moreover, this is not such a simple reaction to see, as it requires a number of steps. In practice, you must dissect a potential target molecule into its parts to see if it is available through this route. You must identify the R groups to be attached through alkylation, then piece together the original Claisen condensation. It's not always easy.

Some examples are probably useful here, and as usual they will appear as problems.

*PROBLEM 20.24

Show how the following ketones can be made through acetoacetic ester syntheses. Mechanisms are not required (Fig. 20.81).

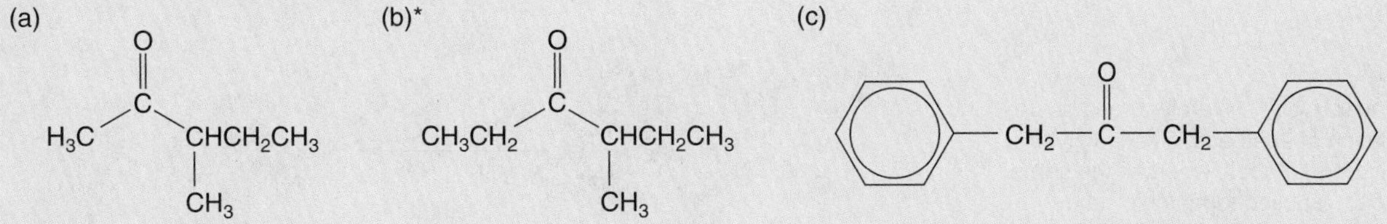

(a) $H_3C-CO-CH(CH_3)CH_2CH_3$ (b)* $CH_3CH_2-CO-CH(CH_3)CH_2CH_3$ (c) $C_6H_5CH_2-CO-CH_2C_6H_5$

FIGURE 20.81

ANSWER

(b) A Claisen condensation using a propionic ester gives a β-keto ester that can be alkylated with ethyl iodide. Hydrolysis and decarboxylation gives the desired product. For clarity, the following answer shows the isolation of the product of the Claisen condensation, but this is not necessary. The anion formed at the end of the Claisen condensation could be alkylated directly.

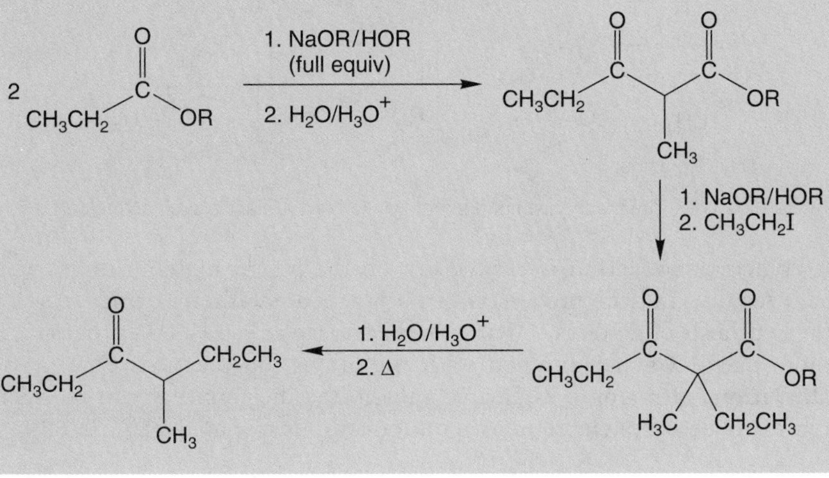

2 CH_3CH_2COOR —(1. NaOR/HOR (full equiv); 2. H_2O/H_3O^+)→ $CH_3CH_2COCH(CH_3)COOR$ —(1. NaOR/HOR; 2. CH_3CH_2I)→ $CH_3CH_2COC(CH_3)(CH_2CH_3)COOR$ —(1. H_2O/H_3O^+; 2. Δ)→ $CH_3CH_2COCH(CH_3)CH_2CH_3$

Diesters such as malonic ester can also be used in this kind of synthetic sequence. The diester is first hydrolyzed in acid or base to the diacid, and then heated to induce loss of carbon dioxide. This process is called the **Malonic ester synthesis**. The end product is acetic acid, not the ketones formed in the related reactions of β-keto esters (Fig. 20.82).

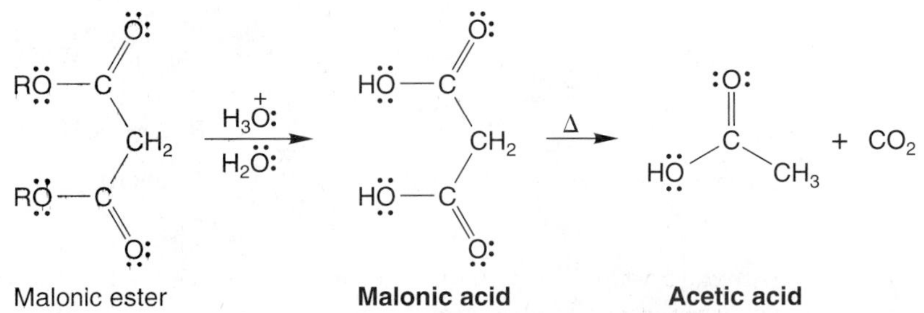

Malonic ester → (H_3O^+, H_2O) → **Malonic acid** → (Δ) → **Acetic acid** + CO_2

FIGURE **20.82** The synthesis of acetic acid from malonic acid.

Note an intuitive point here. The acetoacetic esters, which contain a ketone, are the sources of ketones, and the malonic esters, which contain only ester carbonyls, are the source of acids (Fig. 20.83).

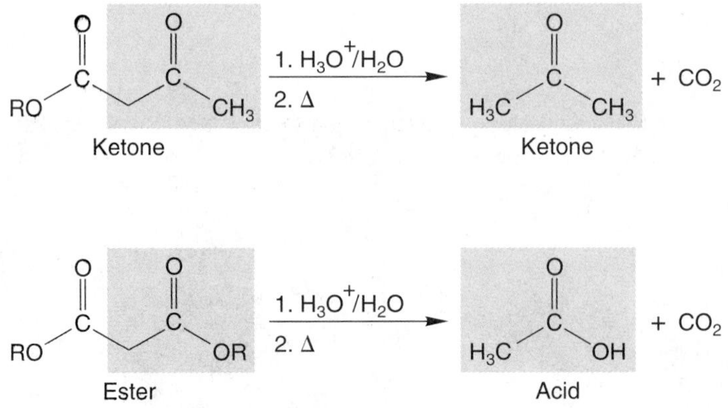

Ketone → (1. H_3O^+/H_2O; 2. Δ) → Ketone + CO_2

Ester → (1. H_3O^+/H_2O; 2. Δ) → Acid + CO_2

FIGURE **20.83** Hydrolysis and decarboxylation of acetoacetic esters gives substituted *ketones*, whereas the same sequence applied to malonic esters gives substituted acetic *acids*.

Here is how the reaction works. The hydrolysis converts the diester into a diacid. These compounds are similar to β-keto acids and will decarboxylate on heating to give the enols of acetic acids. The enols are in equilibrium with the more stable acids, which are the isolated products (Fig. 20.84).

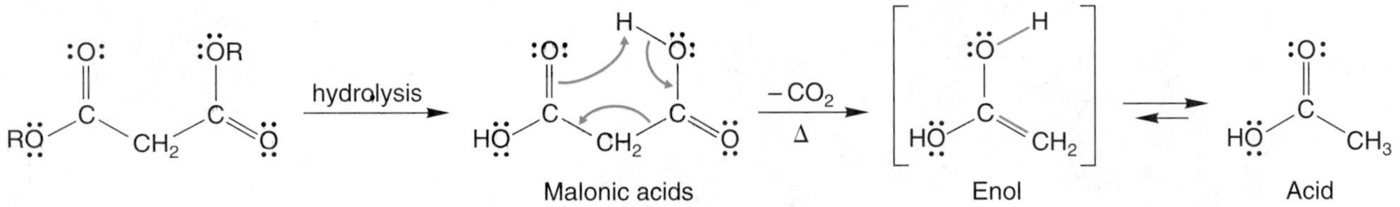

hydrolysis → Malonic acids → ($-CO_2$, Δ) → [Enol] ⇌ Acid

FIGURE **20.84** The mechanism of decarboxylation of 1,1-diacids such as malonic acid.

When this reaction is carried out on the parent malonic ester, acetic acid results, and the process is of no practical synthetic utility. As with the synthesis of ketones starting from acetoacetic esters, this process is far more useful when combined with the alkylation reaction. One or two alkylations of malonic ester give substituted diesters that can be hydrolyzed and decarboxylated to give rather complicated acids (Fig. 20.85).

The general sequence

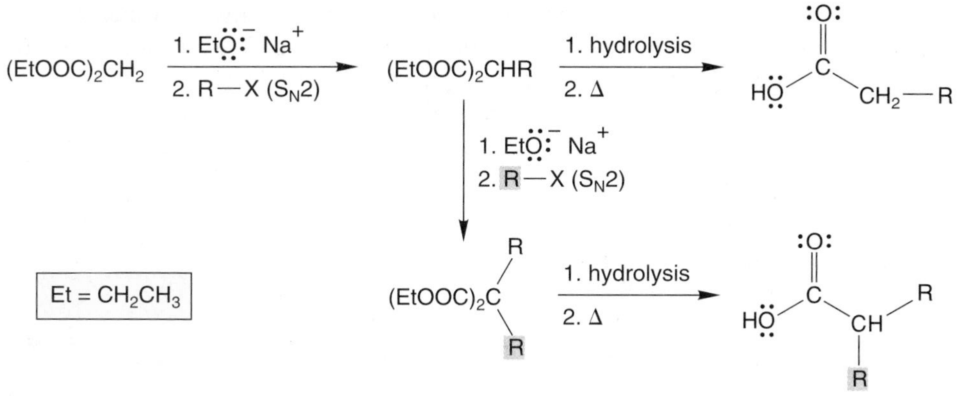

$(EtOOC)_2CH_2$ —[1. $Et\ddot{O}:^- Na^+$; 2. R—X (S_N2)]→ $(EtOOC)_2CHR$ —[1. hydrolysis; 2. Δ]→ HO–C(=O)–CH_2—R

$(EtOOC)_2CHR$ —[1. $Et\ddot{O}:^- Na^+$; 2. R—X (S_N2)]→ $(EtOOC)_2CR_2$ —[1. hydrolysis; 2. Δ]→ HO–C(=O)–CH(R)–R

Et = CH_2CH_3

A specific example

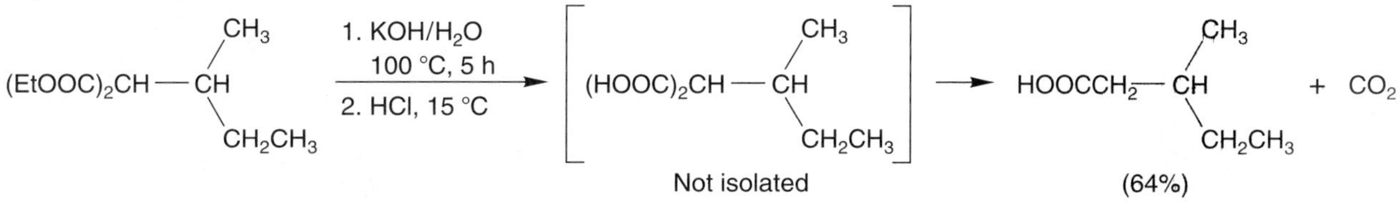

$(EtOOC)_2CH$—$CH(CH_3)(CH_2CH_3)$ —[1. KOH/H_2O, 100 °C, 5 h; 2. HCl, 15 °C]→ [$(HOOC)_2CH$—$CH(CH_3)(CH_2CH_3)$] Not isolated → $HOOCCH_2$—$CH(CH_3)(CH_2CH_3)$ (64%) + CO_2

FIGURE 20.85 The alkylation–decarboxylation sequence for the synthesis of substituted acetic acids from malonic acids.

*PROBLEM 20.25

How can the following molecules be made through a malonic ester synthesis (Fig. 20.86)?

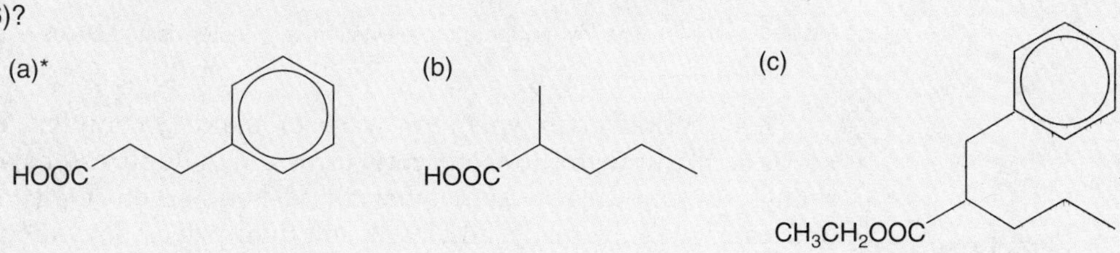

(a)* HOOC

(b) HOOC

(c) CH_3CH_2OOC

FIGURE 20.86

ANSWER

(a) Malonic ester is alkylated, hydrolyzed, and decarboxylated.

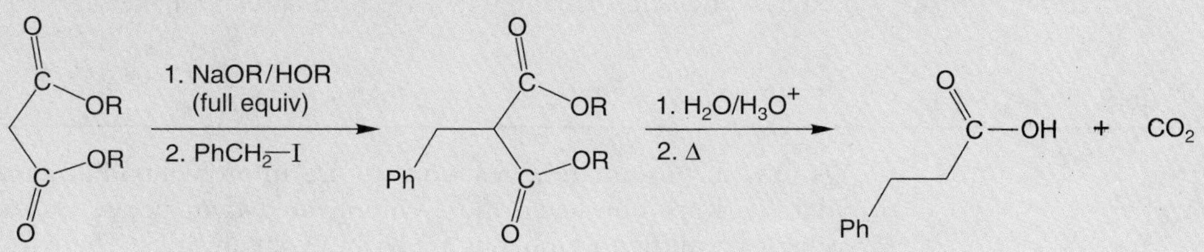

Malonic ester (ROOC–CH_2–COOR) —[1. NaOR/HOR (full equiv); 2. $PhCH_2$–I]→ $PhCH_2CH(COOR)_2$ —[1. H_2O/H_3O^+; 2. Δ]→ $PhCH_2CH_2$–C(=O)–OH + CO_2

Substituted acetic acids are potential sources of esters, acid halides, and any other compound that can be made from a carboxylic acid.

20.12c Direct Alkylation of Esters

It is possible to alkylate esters directly and thereby avoid the extra steps of the malonic ester synthesis. Very strong bases such as dialkyl amides

(LDA, for example) can be used to remove the α-hydrogens of esters. When simple esters are added to these bases at low temperature, an α-hydrogen is removed to give the enolate anion. This anion can then be alkylated by any alkyl halides able to participate in the S_N2 reaction (Fig. 20.87).

The general case

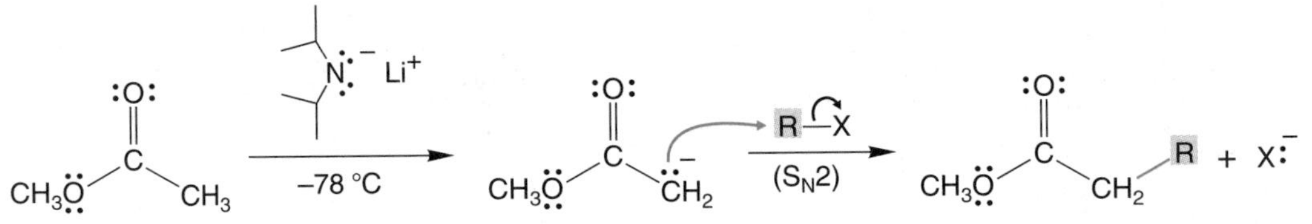

$CH_3\ddot{O}C(=O)CH_3$ —(LDA, Li^+; −78 °C)→ $CH_3\ddot{O}C(=O)\ddot{C}H_2^-$ —(R–X; (S_N2))→ $CH_3\ddot{O}C(=O)CH_2R$ + X^-

A specific example

$CH_3CH_2CH_2C(=O)\ddot{O}CH_3$ —(1. LDA, −78 °C, THF, 30 min; 2. CH_3CH_2Br, −78 °C)→ $CH_3CH_2CH(CH_2CH_3)C(=O)\ddot{O}CH_3$ (92%)

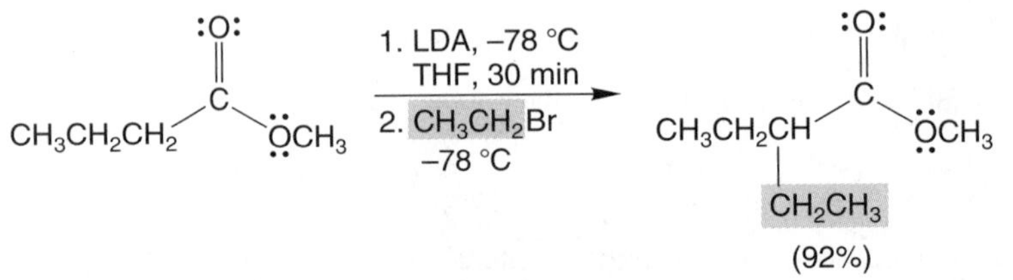

FIGURE 20.87 The low-temperature alkylation of an ester using LDA as base.

This section gives you new ways of making complex ketones and acetic acids through the acetoacetic ester and malonic ester syntheses. Although these routes are somewhat roundabout—acetoacetic ester or malonic ester must be made, alkylated, and finally decarboxylated—in a practical setting they are efficient and solve some of the synthetic problems posed earlier. You might want to revisit our earlier discussion of the difficulties involved in straightforward alkylation reactions of aldehydes, ketones, and acids (Chapter 18, p. 895; Chapter 19, p. 985).

20.13 COMPLICATED "FUN IN BASE" PROBLEMS

We now have some new reactions to use in problems. Put another way, your lives are now even more complicated than before, as even more sadistically difficult problems are within your abilities. There really aren't too many new things to think about. You only have to add the Claisen-like processes that involve the loss of a leaving group from a carbonyl group that has been attacked by an enolate. In other words, you always have to keep the addition–elimination reaction in mind.

But be careful—reactions are driven by thermodynamics, not our concepts of forward and backward. The reverse Claisen is much harder to detect in problems than the forward version, but no less reasonable mechanistically. As always in matters such as these, experience counts a lot, and there is no way to get it except to work a lot of problems.

Provide mechanisms for the reactions shown in Figure 20.88.

PROBLEM 20.26

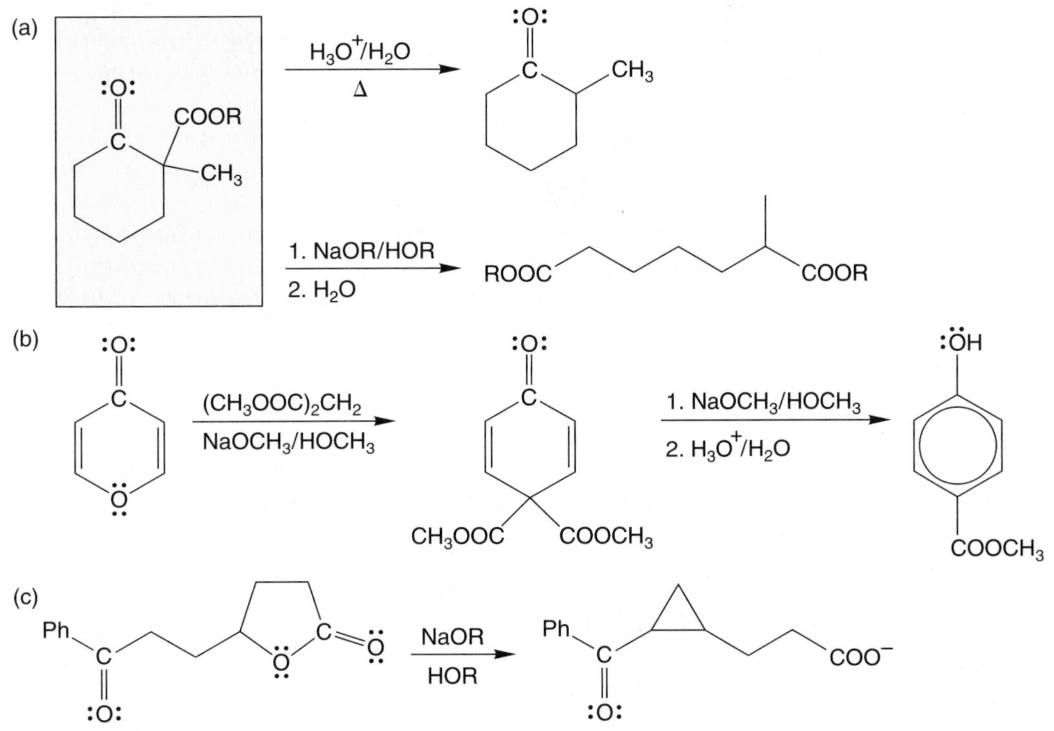

(a) H_3O^+/H_2O, Δ; 1. NaOR/HOR 2. H_2O

COOR, CH_3, CH_3, ROOC, COOR

(b) $(CH_3OOC)_2CH_2$, $NaOCH_3/HOCH_3$; 1. $NaOCH_3/HOCH_3$ 2. H_3O^+/H_2O

CH_3OOC, $COOCH_3$, OH, $COOCH_3$

(c) Ph, C, NaOR, HOR, Ph, C, COO^-

FIGURE 20.88

Provide a mechanism for the following transformation (Fig. 20.89). Be careful. This problem is much harder than it looks at first.

*PROBLEM 20.27

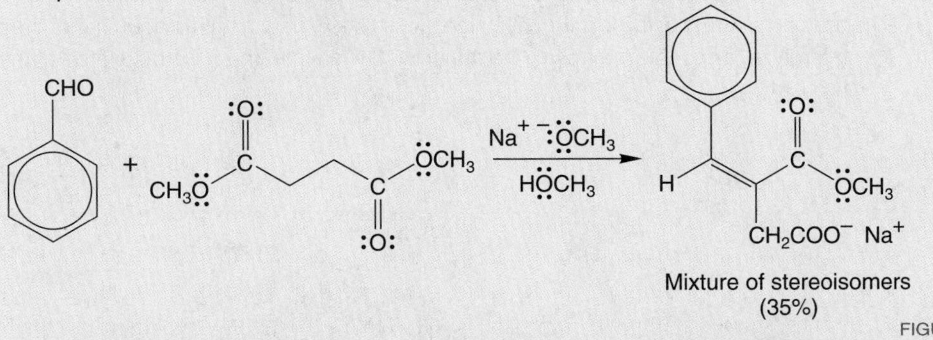

CHO + CH_3O–C(=O)–CH$_2$CH$_2$–C(=O)–OCH_3 $\xrightarrow[HOCH_3]{Na^+\ ^-OCH_3}$ H, C, OCH_3, $CH_2COO^-\ Na^+$

Mixture of stereoisomers (35%)

FIGURE 20.89

ANSWER

This beautiful problem is tantalizing because it involves serious misdirection. It has an "obvious" answer that is wrong. By far the hardest part of this very hard problem is resisting the temptation to take the short, wrong route. The reaction is called the Stobbe condensation.

First of all, it is a member of the class of ester plus ketone reactions we discussed in Section 20.11c. As benzaldehyde has no α-hydrogens, the answer must start with removal of the α-hydrogen of the succinic ester.

ANSWER (CONTINUED)

Dimethyl succinate

Enolate

No enolate possible

The end product clearly is a combination of the ester and benzaldehyde, so we need waste no time with self-condensation of the ester, but turn our attention immediately to a reaction between the two partners. The new bond is simple to find, and easily made through attack by the enolate on the aldehyde carbonyl group. So far, the problem seems simple, and it is.

We are almost there, or so it seems. Now temptation enters. How easy it is to protonate the newly formed alkoxide, lose water in an elimination, and then hydrolyze one ester to give the disarmingly simple monocarboxylate in the product.

eliminate water (E1cB)

Hydrolyze this ester only

ANSWER (CONTINUED)

But wait—why hydrolyze only *that* ester? Why not the other one? And how does the hydrolysis work? The only hydroxide available is that produced in the elimination reaction, and it is hard to believe that it could compete effectively in attacking one of the esters. I see no good answers to these questions. Unless you do, you are forced to conclude that there is something wrong with this answer, no matter how painful it may be to do so. It is necessary to backtrack to the alkoxide and ask what other reaction it might be able to do. There's not much else possible beyond attack of the alkoxide on one of the esters, an intramolecular, base-induced transesterification reaction. There are two esters and therefore two possible reaction paths, (a) and (b). Don't worry that (b) seems awkwardly long. That's an artifact of the two-dimensional representation of these three-dimensional molecules.

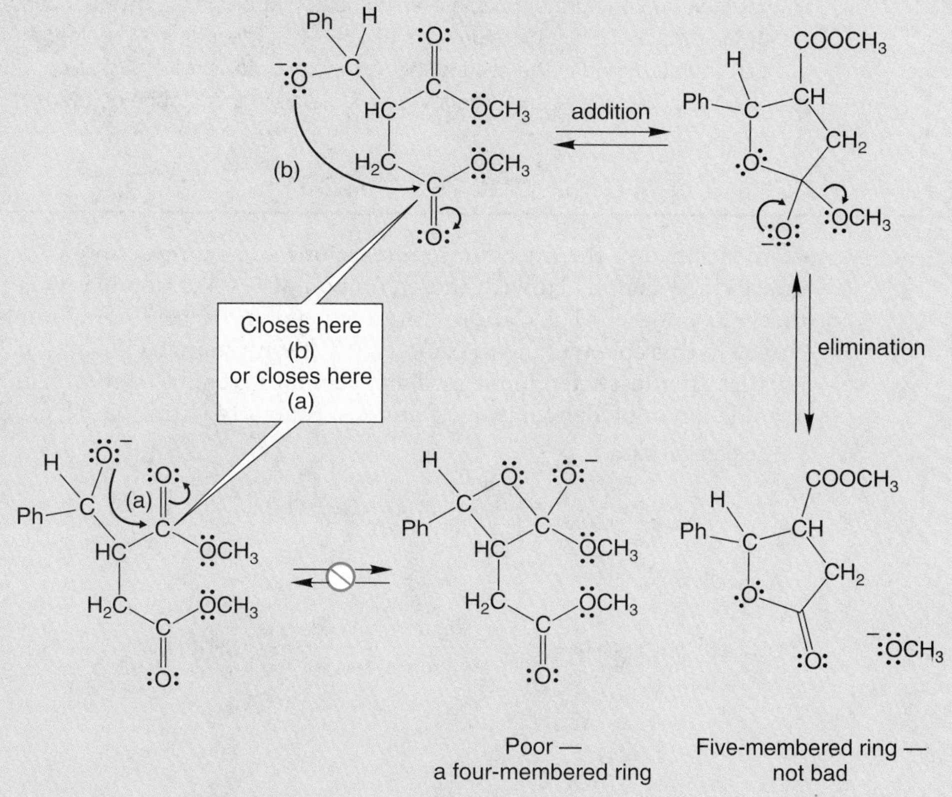

addition

elimination

Closes here (b) or closes here (a)

Poor — a four-membered ring

Five-membered ring — not bad

Path (a) produces a strained four-membered ring and should be disfavored, so concentrate first on the five-membered ring-forming reaction, (b). The initially formed intermediate can lose alkoxide to generate a five-membered γ-lactone. The last step of the reaction is an elimination. The breaking bonds are the carbon–hydrogen and carbon–oxygen bonds shown in the next drawing. Notice that the hydrogen lost is α to an ester, and is therefore relatively easily removed. The leaving group is a rather stable carboxylate anion. The final result is exactly what we want.

ANSWER (CONTINUED)

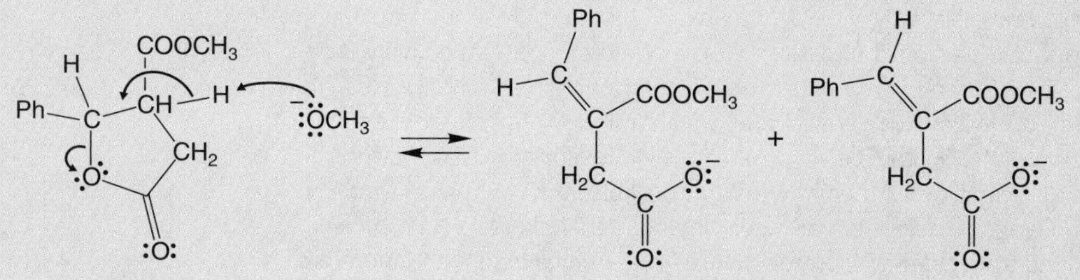

Mixture of stereoisomers

This mechanism *demands* the formation of the CH_2COO^- group found in the product. It is the specificity of this mechanism that gives us confidence in it. Rather than present further puzzles—Why does hydrolysis of one ester occur, but not of the other?—Why any hydrolysis at all?—this mechanism requires that just the carboxylate found in the product be formed and no other. It is also comforting to know that γ-lactones can sometimes be isolated in the Stobbe condensation.

20.14 SYNTHETIC ROUTES TO ACYL COMPOUNDS

As mentioned in the introduction, reactions and syntheses are even more intertwined than usual with these compounds. A reaction of one is a synthesis of another. The various interconversions of acyl compounds discussed in this chapter (and in Chapter 19) are summarized in Section 20.16. Nitriles are most often made by displacement reactions of alkyl halides by cyanide ion or by dehydration of amides, generally with P_2O_5 (Fig. 20.90).

The general case

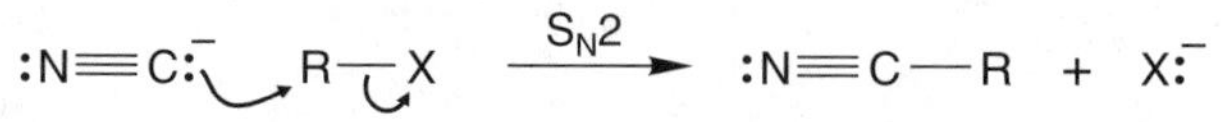

$$:N{\equiv}C:^- \;\; R{-}X \xrightarrow{S_N2} :N{\equiv}C{-}R \; + \; X:^-$$

A specific example

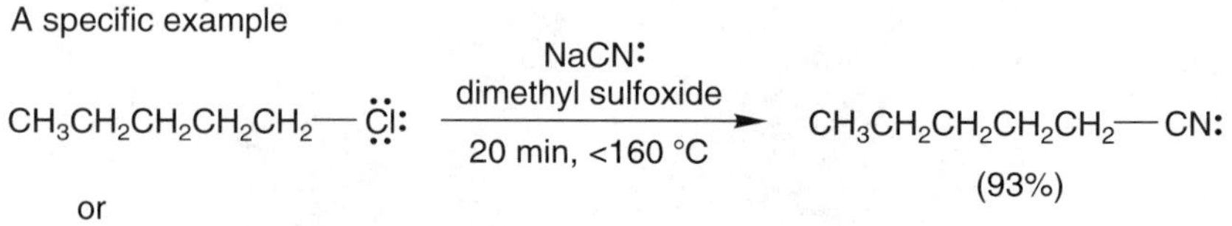

$$CH_3CH_2CH_2CH_2CH_2{-}Cl: \xrightarrow[\text{20 min, <160 °C}]{\text{NaCN: dimethyl sulfoxide}} CH_3CH_2CH_2CH_2CH_2{-}CN: \; (93\%)$$

or

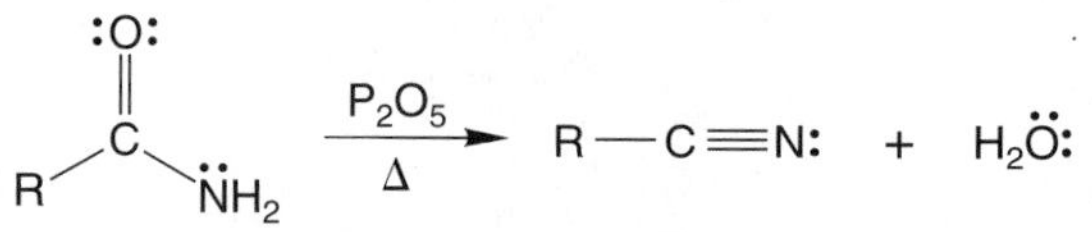

$$RC(=O)NH_2 \xrightarrow[\Delta]{P_2O_5} R{-}C{\equiv}N: \; + \; H_2O:$$

A specific example

$$(CH_3)_2CHC(=O)NH_2 \xrightarrow[\text{210 °C, 9 h}]{P_2O_5} (CH_3)_2CH{-}C{\equiv}N: \; (\sim 75\%)$$

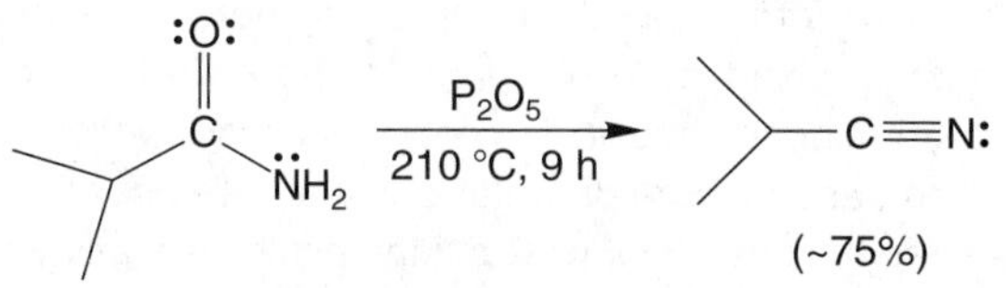

FIGURE 20.90 Nitriles are usually made through S_N2 reactions using cyanide as the nucleophile, or by dehydrations of amides.

Ketenes are available through elimination reactions of acyl chlorides using tertiary amines as bases (Fig. 20.91)

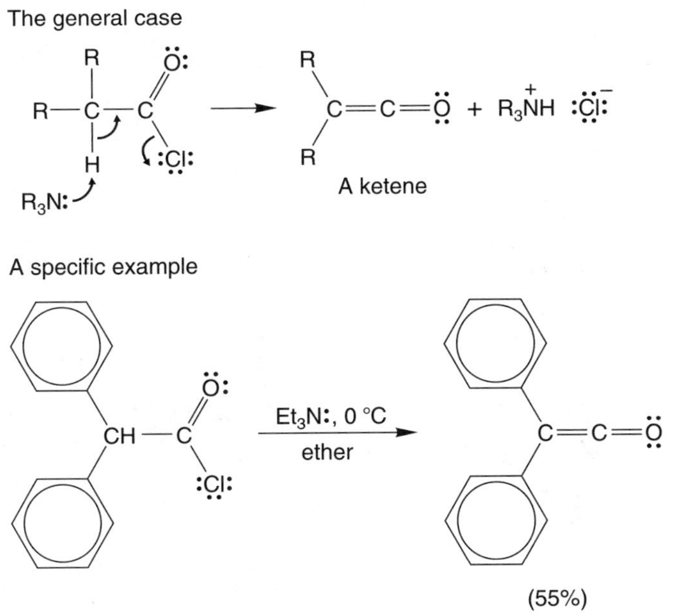

The general case

A ketene

A specific example

Et$_3$N:, 0 °C
ether

(55%)

FIGURE 20.91 One common synthesis of ketenes involves an elimination reaction of acyl halides.

20.15 SOMETHING MORE: THERMAL ELIMINATION REACTIONS OF ESTERS

Esters that contain a hydrogen in the β-position of the OR group give alkenes when heated to high temperatures (> 400 °C) (Fig. 20.92).

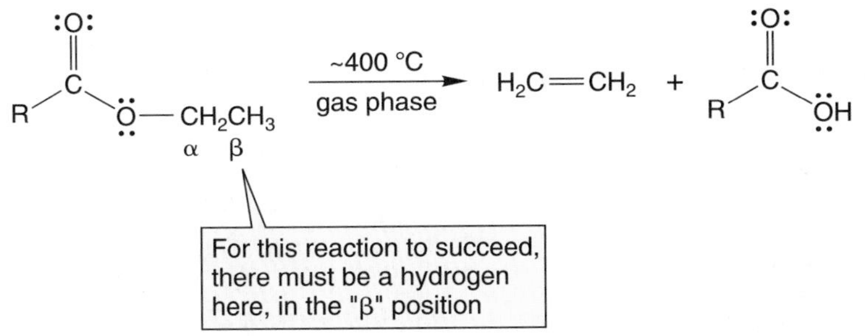

~400 °C
gas phase

For this reaction to succeed, there must be a hydrogen here, in the "β" position

FIGURE 20.92 The thermal elimination reaction of esters containing a hydrogen in the β-position.

In this reaction, the carbonyl group acts as a base, attacking the β-hydrogen intramolecularly. At the same time the ester acts as a leaving group, and acetic acid is lost in a single, concerted step (Fig. 20.93).

In contrast to most E2 reactions, this pyrolytic, intramolecular version proceeds through loss of a syn hydrogen. The reason is simple: Only the syn hydrogen is within reach of the carbonyl group. There is no possibility of doing an anti elimination in this reaction. The elimination reactions of the deuterium-labeled compounds shown in Figure 20.94 provide an example of the stereospecificity of this elimination reaction. In diastereomer **A**, it is only the syn hydrogen that is lost as *trans*-stilbene is formed. By contrast, in **B** it is the deuterium atom that must occupy the syn position in the formation of *trans*-stilbene, and this time it is D, not H, that is lost.

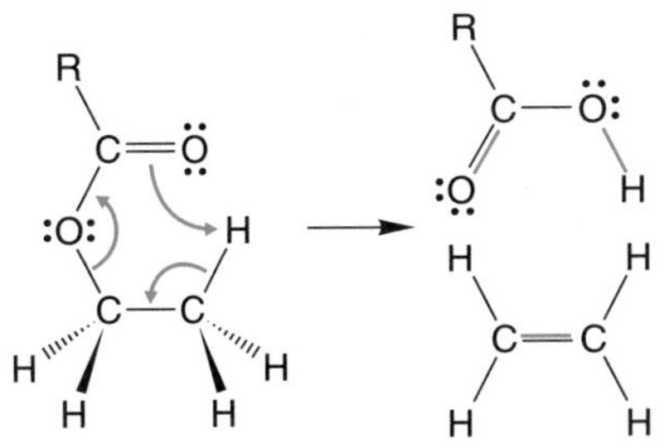

FIGURE 20.93 An arrow formalism description of the thermal elimination reaction of esters.

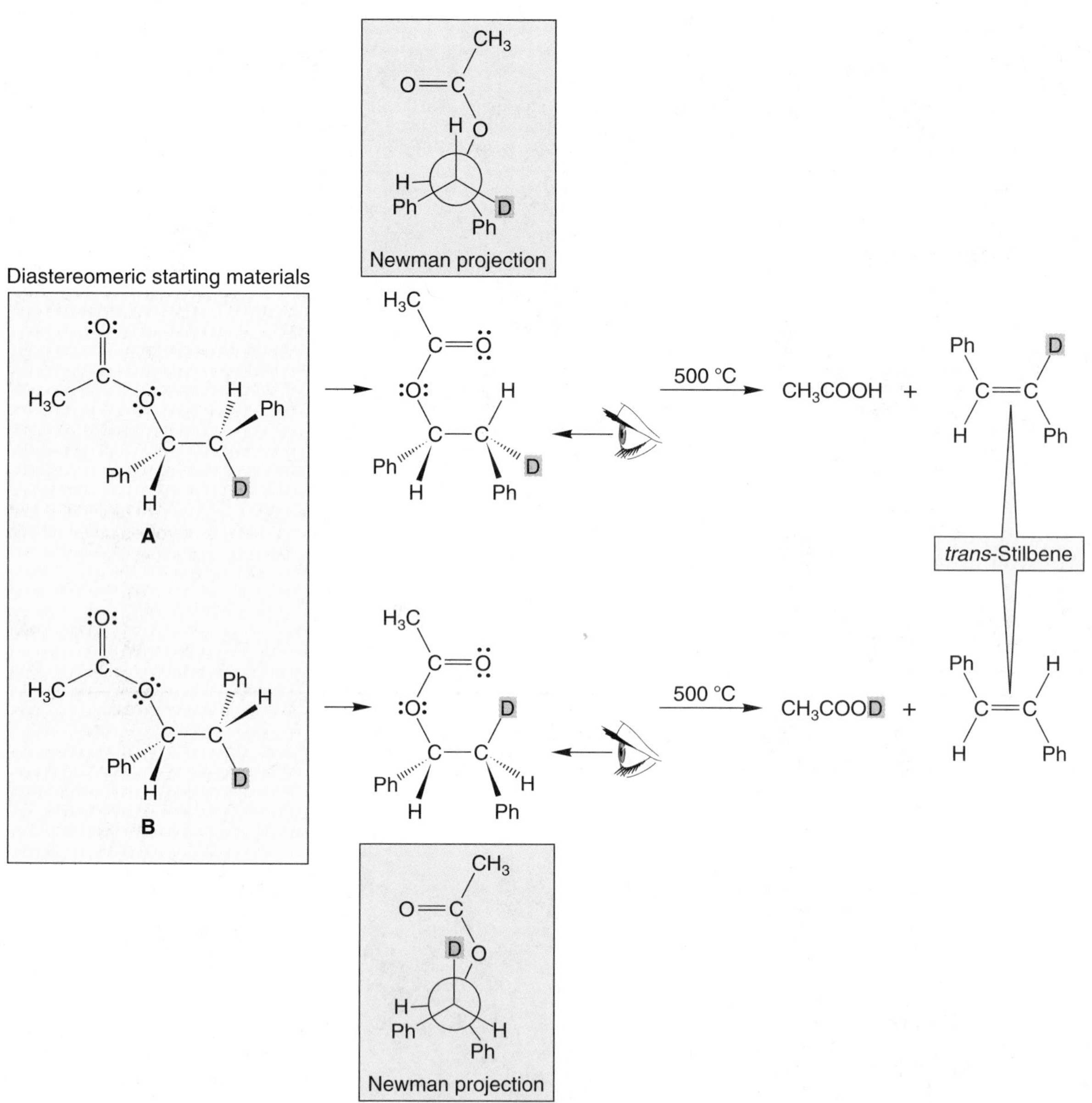

FIGURE 20.94 In these intramolecular elimination reactions, only a syn hydrogen is removed. If the phenyl groups are required to stay as far from each other as possible, syn elimination requires that H be lost from the first stereoisomer (**A**) and D from the second (**B**).

There are other versions of this new elimination process, some of which proceed under much milder conditions than those necessary for ester pyrolysis. For example, the sulfur-containing **xanthate esters** undergo elimination especially easily (Fig. 20.95). Xanthate esters can be prepared as shown in the figure by addition of alkoxide to CS_2 (recall the addition of nucleophiles to carbon dioxide, Chapter 19, p. 991). These intramolecular thermal elimination reactions provide a new route to alkenes, so update your file cards.

PROBLEM 20.28 Write arrow formalisms for the reactions of Figure 20.95.

The general case

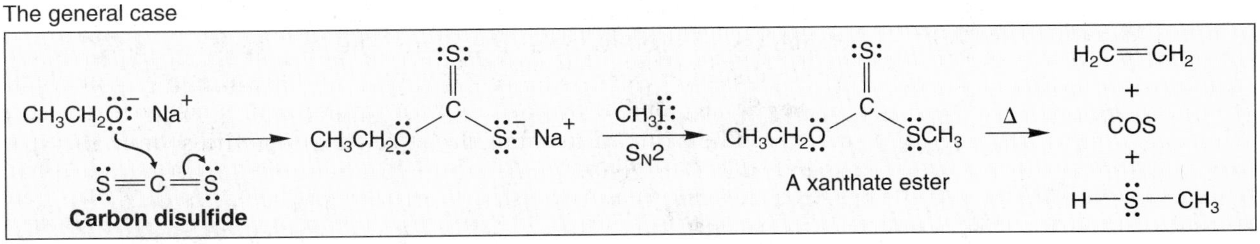

$CH_3CH_2\ddot{O}:^- Na^+$ + S=C=S (Carbon disulfide) → $CH_3CH_2O-C(=S)-S:^- Na^+$ —CH_3I, S_N2→ $CH_3CH_2O-C(=S)-SCH_3$ (A xanthate ester) —Δ→ $H_2C{=}CH_2$ + COS + $H-S-CH_3$

A specific example

4-tert-butylcyclohexanol (C(CH$_3$)$_3$, :OH) —1. NaH, 2. CS_2, 3. CH_3I→ xanthate ester (C(CH$_3$)$_3$; O–C(=S)–S–CH$_3$) —208 °C→ 4-tert-butylcyclohexene (C(CH$_3$)$_3$) (62%)

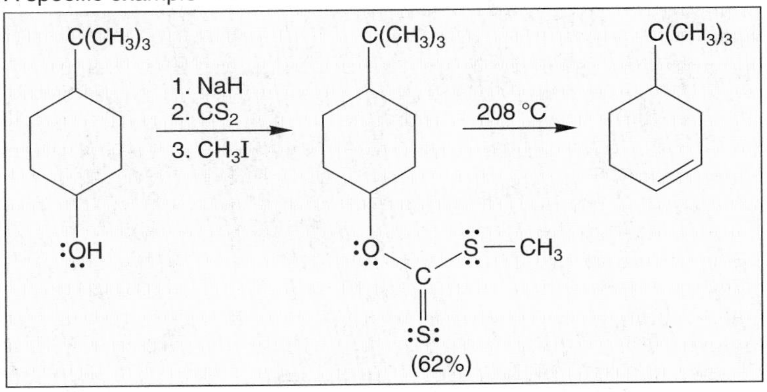

FIGURE 20.95 The formation and thermal fragmentation of a xanthate ester.

PROBLEM 20.29 Provide mechanisms for the following reactions. The first reaction is called the Cope elimination, after Arthur C. Cope (1909–1966) (Fig. 20.96).

(a) Cope elimination

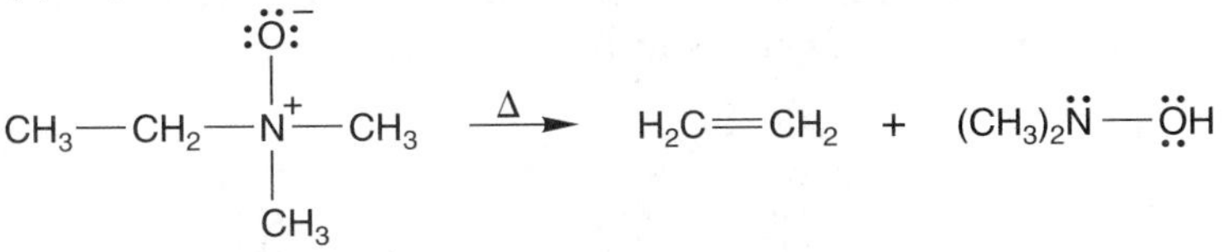

$CH_3-CH_2-N^+(O^-)(CH_3)-CH_3$ —Δ→ $H_2C{=}CH_2$ + $(CH_3)_2N-OH$

(b) Ester pyrolysis

$H_3C-C(=O)-O-(CH_2)_6-O-C(=O)-CH_3$ —575 °C→ CH_3COOH + $CH_2{=}CH-CH_2-CH_2-CH{=}CH_2$ (67%)

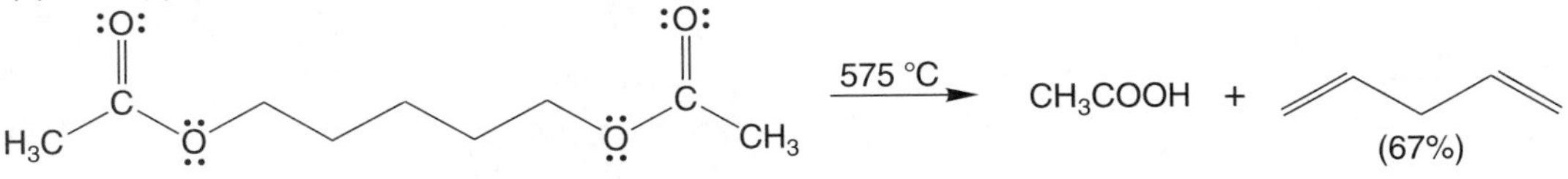

FIGURE 20.96

One of the themes of this and Chapter 19 has been the use of enolates formed from ketones and esters on reaction with strong bases such as LDA. For example, condensation reactions can be directed (p. 1047), and alkylations of simple esters and ketones achieved (p. 1053; Chapter 18, p. 895). Here we combine the alkylation of enolates with a Cope-like elimination reaction to do most useful chemistry. The four- step combination of enolate formation, thioether or selenoether formation, oxidation, and heating can be used to convert saturated ketones and esters into α,β-unsaturated compounds in very high yield (Fig 20.97).

$R-C(=O)-CH_2CH_2R$ → $R-C(=O)-CH{=}CHR$

FIGURE 20.97 The potential transformation of a saturated ketone into an α,β-unsaturated ketone.

The base LDA or a close relative, lithium *N*-cyclohexyl-*N*-isopropylamide, can be used to form an enolate. The enolate reacts in S_N2 fashion with a disulfide (R—S—S—R), or if you can stand the smell, a diselenide (R—Se—Se—R), or Ph—Se—Br to give a thioether or a selenoether. These ethers can be oxidized to **sulfoxides** or **selenoxides** with sodium metaperiodate $NaIO_4$ (Fig. 20.98)

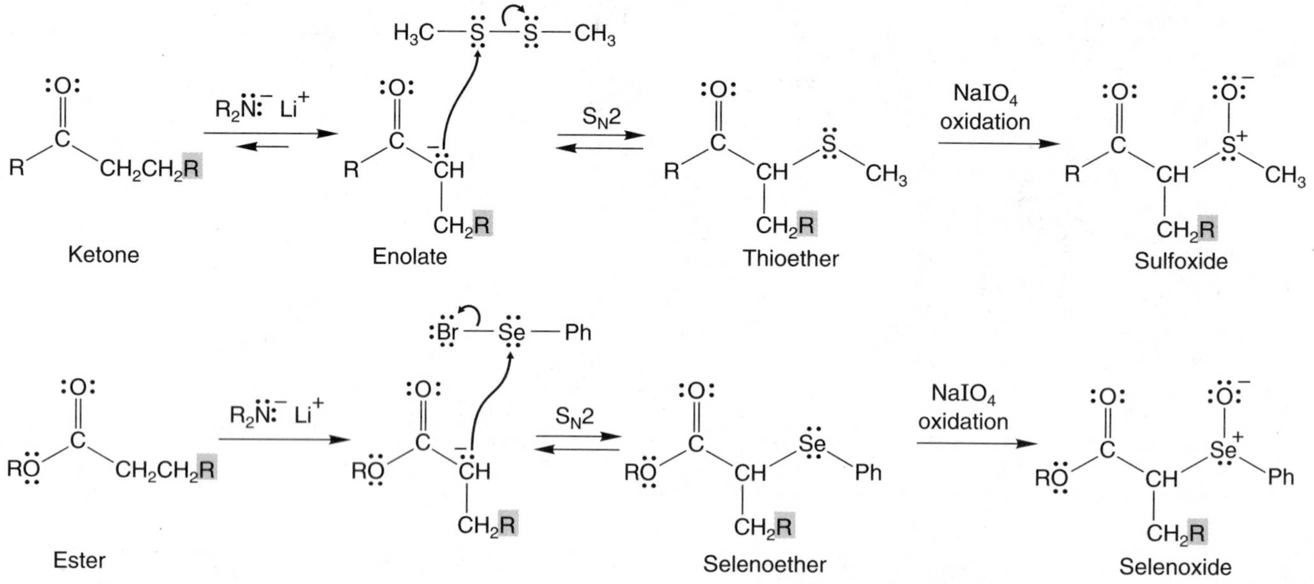

FIGURE 20.98 The formation of sulfoxides and selenoxides from thioethers and selenoethers.

The sulfoxides undergo a Cope-like elimination at about 100 °C, and the selenoxides decompose at about room temperature (Fig. 20.99).

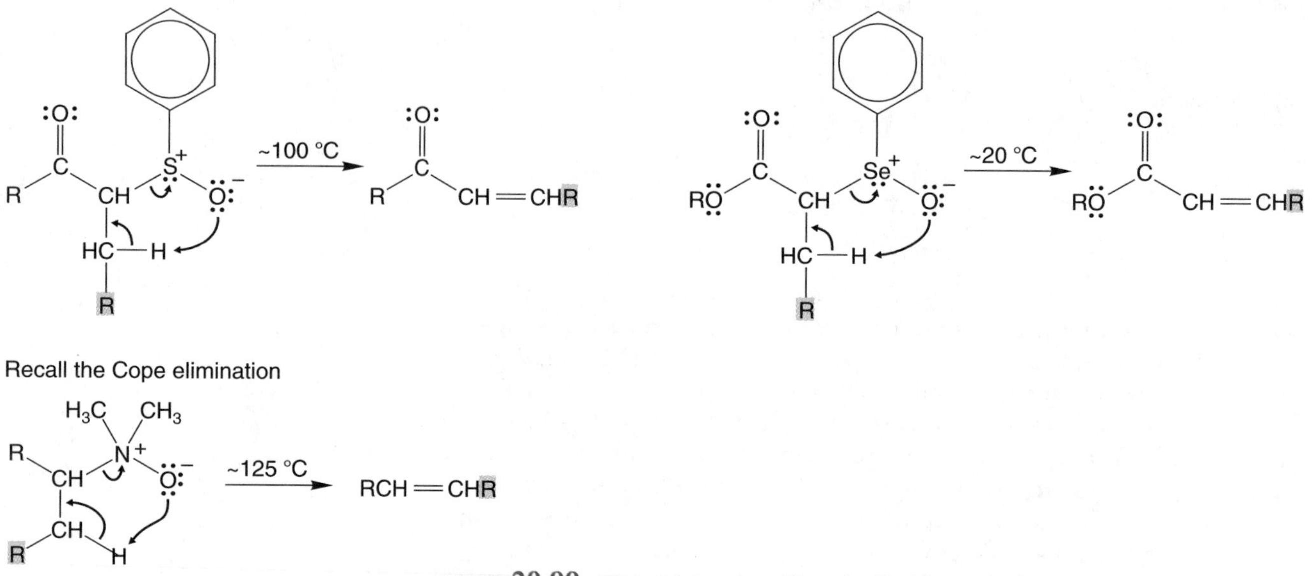

FIGURE 20.99 Thermal decomposition of sulfoxides and selenoxides resembles that of *N*-oxides.

The result is a simple, high-yield introduction of a double bond at the α,β-position of a ketone or ester (Fig. 20.100).

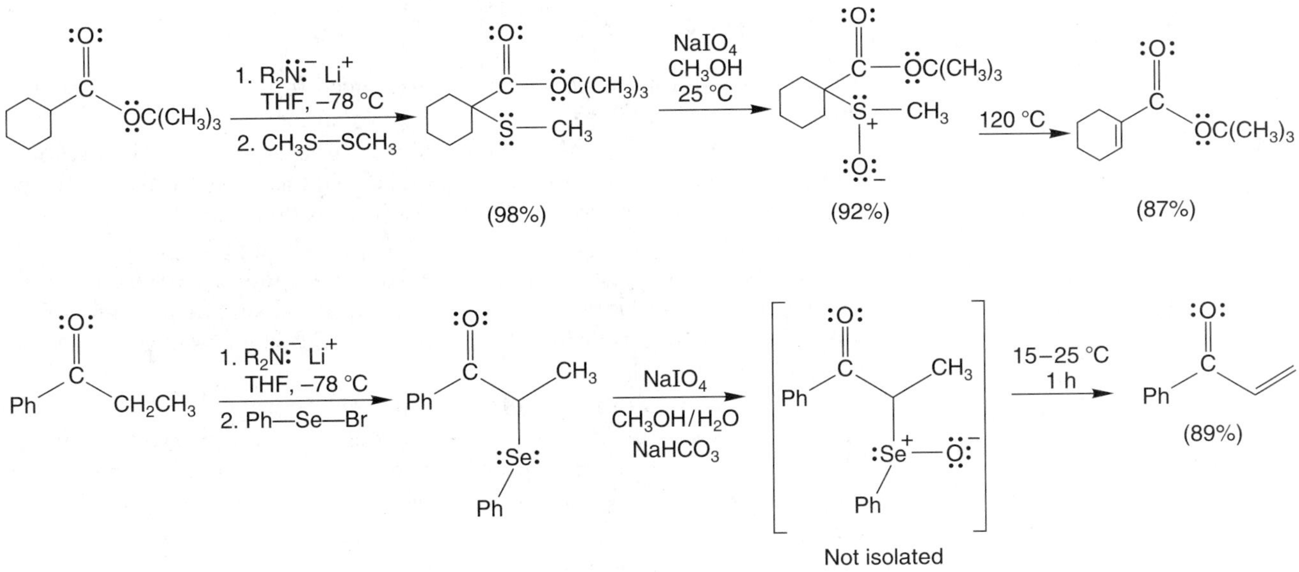

FIGURE **20.100** Formation of alkenes by heating sulfoxides and selenoxides.

20.16 SUMMARY

NEW CONCEPTS

This chapter is filled with detail, but there are no new basic concepts. Most important, although not really new, is the continuation of the exploration of the addition–elimination reaction of carbonyl compounds begun in Chapter 19. When nucleophiles add to carbonyl compounds that do not bear a leaving group (for example, aldehydes or ketones) reversal of the addition or protonation are the only two possible reactions. When there is a leaving group attached to the carbonyl, as there is with most acyl compounds, another option exists, loss of the leaving group (Fig. 20.101).

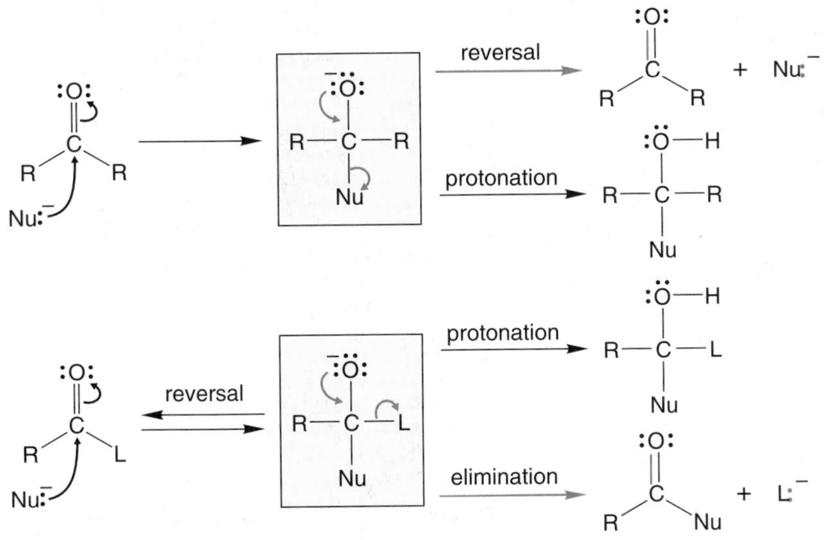

FIGURE **20.101** Carbonyl groups bearing good leaving groups can undergo the addition–elimination reaction; others cannot.

The more stable a carbonyl compound, the less easy will be the endothermic loss of an α-hydrogen to give an enolate. Well resonance-stabilized carbonyl compounds such as esters and amides are weaker acids than aldehydes and ketones.

REACTIONS, MECHANISMS, AND TOOLS

The addition–elimination reaction, introduced in Chapter 19, dominates the reaction mechanisms discussed in this chapter.

The chemistry of nitriles (cyanides) is similar to that of carbonyl compounds. The carbon–nitrogen triple bond can act as an acceptor of nucleophiles in addition reactions, and will provide resonance stabilization to an α-anion, much as will a carbonyl group (Figs. 20.50 and 20.51).

Dicarbonyl compounds, such as 1,1-diesters and β-keto esters (both species are β-dicarbonyl compounds), are quite strong acids and can be alkylated through formation of the anion produced by loss of the doubly α proton (Fig. 20.77).

There is a class of intramolecular thermal elimination reactions that provides a new route to alkenes. Esters, xanthates, amine oxides, sulfoxides, and selenoxides are commonly used in this reaction. The reactions are concerted (one-step), and steric requirements dictate that a syn hydrogen must be lost in the reaction, as the carbonyl group cannot reach a hydrogen in an anti position (Fig. 20.93).

SYNTHESES

Figure 20.102 gives the many important synthetic reactions found in this chapter. A few have been touched upon already in Chapter 19, but most are new, at least in a formal sense. Many of the following reactions share a common (addition–elimination) mechanism.

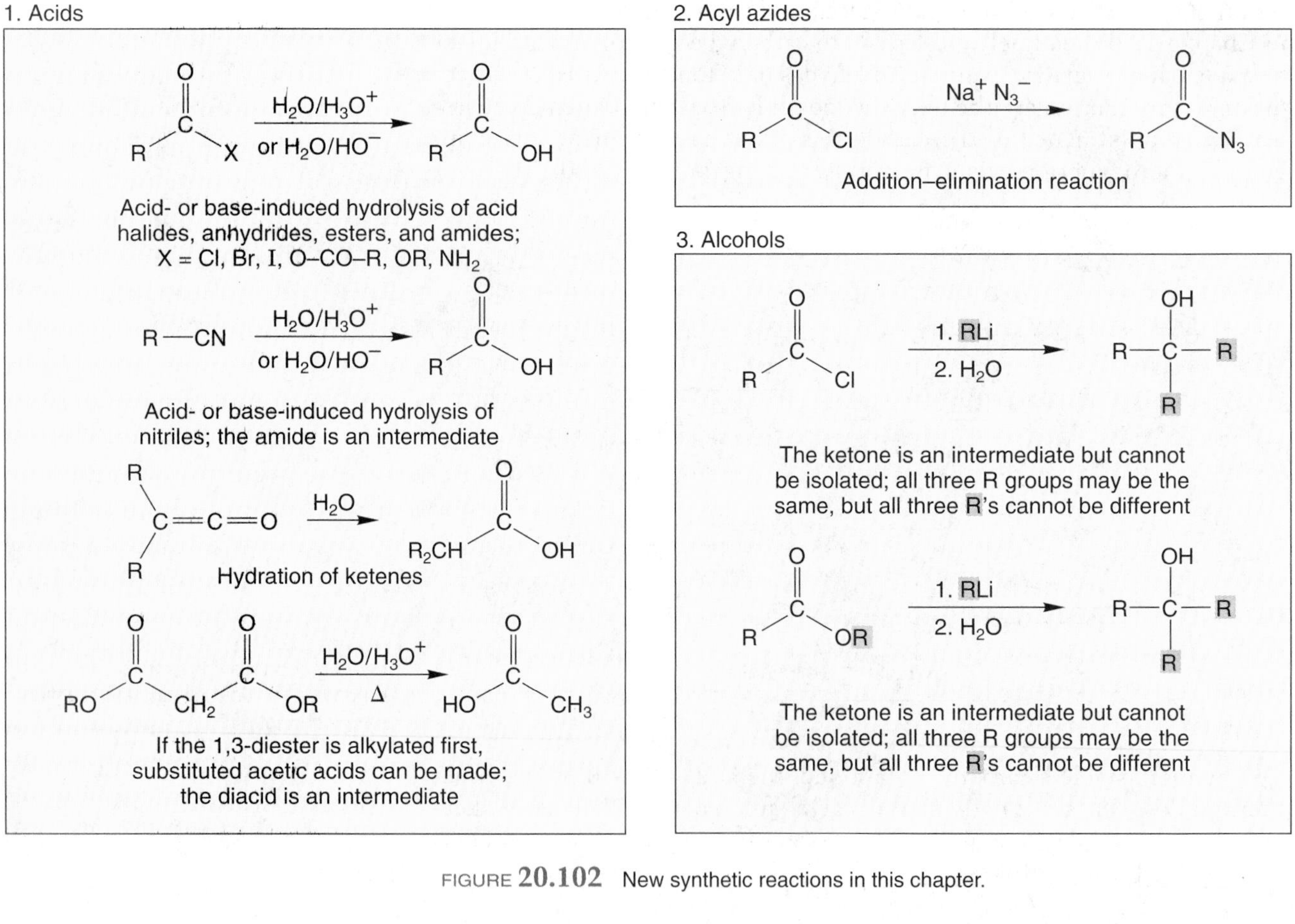

1. Acids

$$\mathrm{RCOX} \xrightarrow[\text{or } H_2O/HO^-]{H_2O/H_3O^+} \mathrm{RCOOH}$$

Acid- or base-induced hydrolysis of acid halides, anhydrides, esters, and amides; X = Cl, Br, I, O–CO–R, OR, NH_2

$$\mathrm{R{-}CN} \xrightarrow[\text{or } H_2O/HO^-]{H_2O/H_3O^+} \mathrm{RCOOH}$$

Acid- or base-induced hydrolysis of nitriles; the amide is an intermediate

$$\mathrm{R_2C{=}C{=}O} \xrightarrow{H_2O} \mathrm{R_2CH{-}COOH}$$

Hydration of ketenes

$$\mathrm{RO{-}CO{-}CH_2{-}CO{-}OR} \xrightarrow[\Delta]{H_2O/H_3O^+} \mathrm{HO{-}CO{-}CH_3}$$

If the 1,3-diester is alkylated first, substituted acetic acids can be made; the diacid is an intermediate

2. Acyl azides

$$\mathrm{RCOCl} \xrightarrow{Na^+\ N_3^-} \mathrm{RCON_3}$$

Addition–elimination reaction

3. Alcohols

$$\mathrm{RCOCl} \xrightarrow[2.\ H_2O]{1.\ RLi} \mathrm{R{-}C(OH)(R)(R)}$$

The ketone is an intermediate but cannot be isolated; all three R groups may be the same, but all three R's cannot be different

$$\mathrm{RCOOR} \xrightarrow[2.\ H_2O]{1.\ RLi} \mathrm{R{-}C(OH)(R)(R)}$$

The ketone is an intermediate but cannot be isolated; all three R groups may be the same, but all three R's cannot be different

FIGURE 20.102 New synthetic reactions in this chapter.

3. Alcohols (continued)

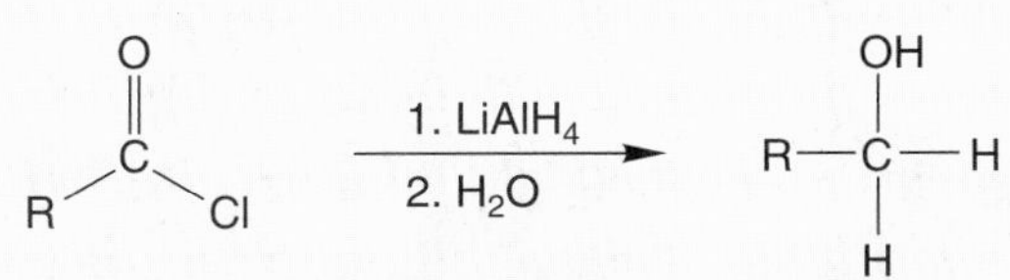

Other metal hydride donors also work; the aldehyde is an intermediate, but cannot be isolated

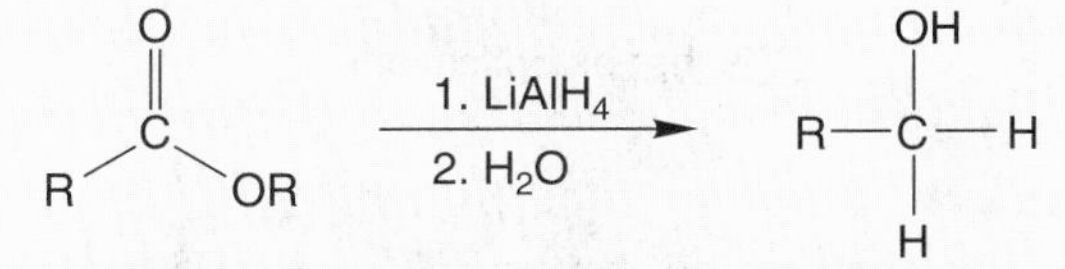

1. $LiAlH_4$
2. H_2O

Other metal hydride donors also work; the aldehyde is an intermediate, but cannot be isolated

1. RLi
2. H_2O

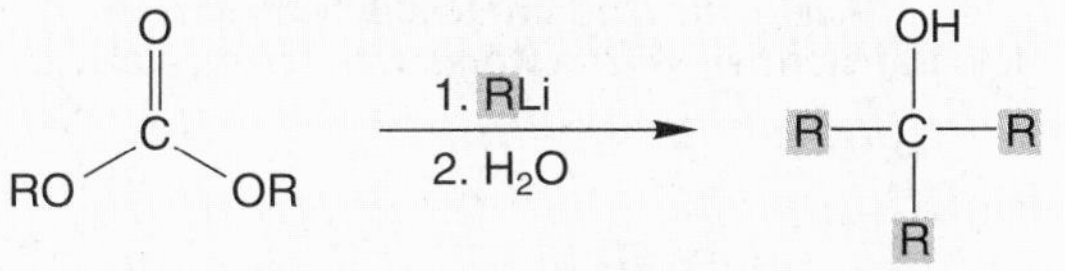

The ester and ketone are intermediates, but cannot be isolated; all three R groups must be the same

4. Aldehydes

1. $LiAlH[OC(CH_3)_3]_3$
2. H_2O

Other encumbered metal hydrides also work

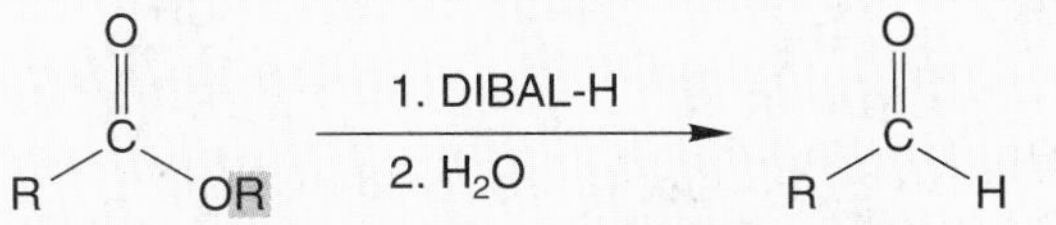

1. DIBAL-H
2. H_2O

Diisobutylaluminum hydride (DIBAL-H), will not reduce the aldehyde if the reaction is carefully run

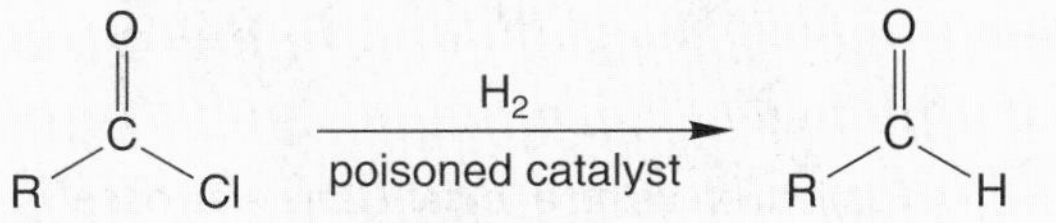

H_2
poisoned catalyst

Rosenmund reduction

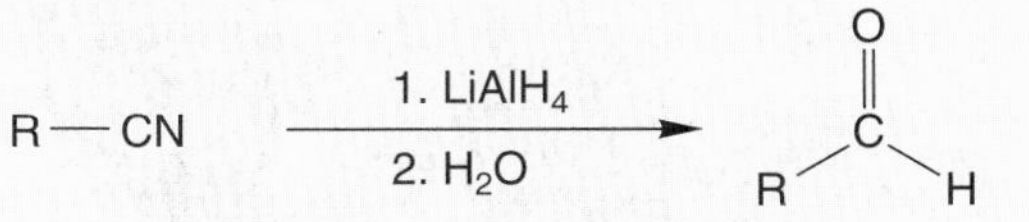

1. $LiAlH_4$
2. H_2O

The imine is the initial product, but is hydrolyzed to the aldehyde

5. Alkenes

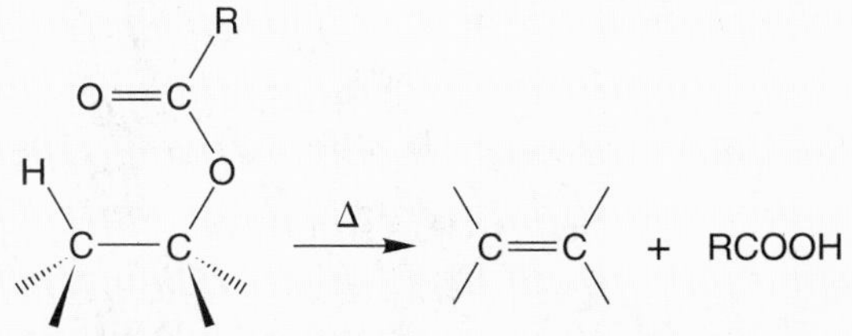

Δ

+ RCOOH

Note that only a syn hydrogen can be removed

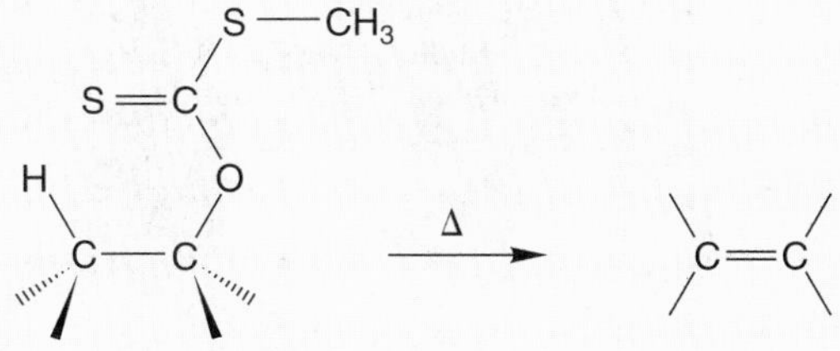

Δ

Xanthate ester pyrolysis requires lower temperature than ester pyrolysis

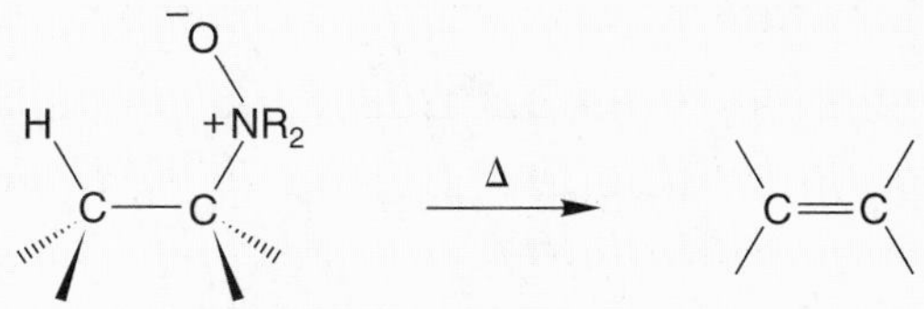

Δ

A Cope elimination

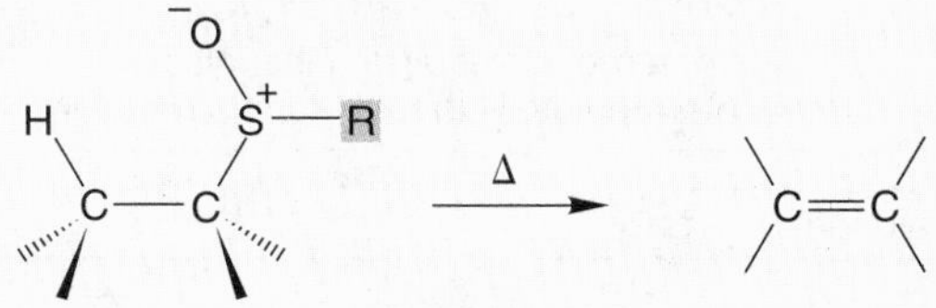

Δ

Sulfoxide decomposition

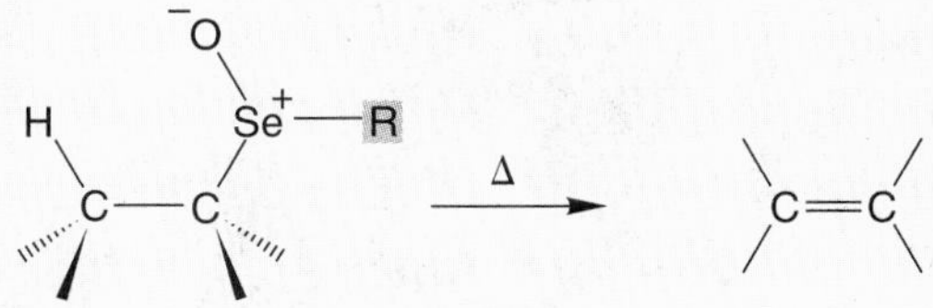

Δ

Selenoxide decomposition

6. Amides

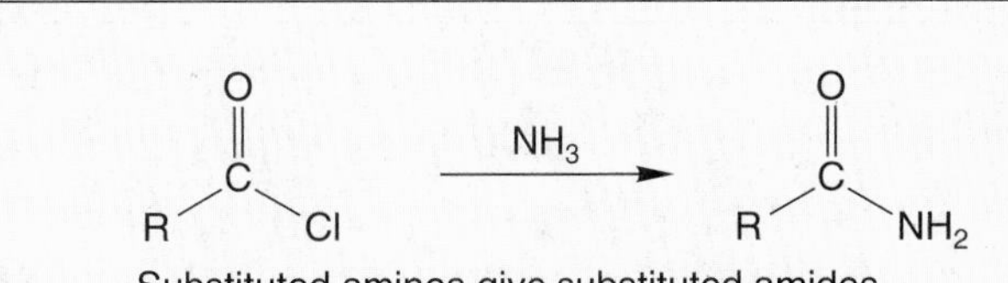

Substituted amines give substituted amides

FIGURE 20.102 (CONTINUED)

6. Amides (continued)

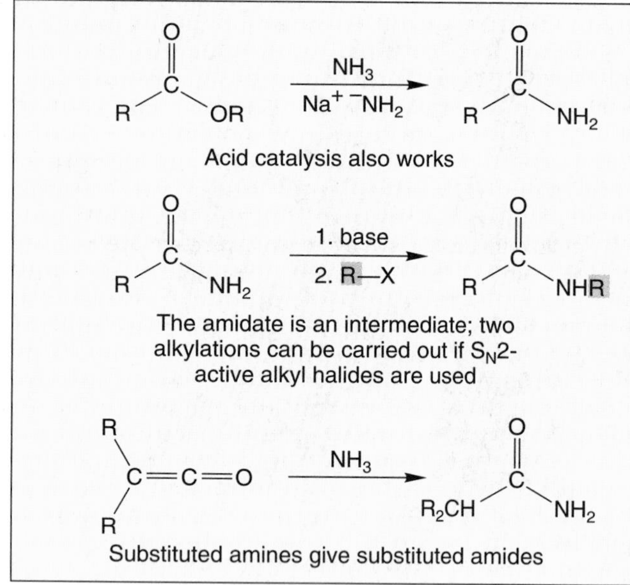

NH3
Na+ −NH2
Acid catalysis also works
1. base
2. R—X
The amidate is an intermediate; two alkylations can be carried out if SN2-active alkyl halides are used
NH3
Substituted amines give substituted amides

7. Amines

$R—CN \xrightarrow{H_2/\text{catalyst}} R—CH_2NH_2$

Catalytic hydrogenation

$RC(=O)NHR \xrightarrow[2.\ H_2O]{1.\ LiAlH_4} R—CH_2NHR$

Loss of a metal oxide from the initially formed intermediate is the key to this reaction

8. Anhydrides

$RC(=O)Cl \xrightarrow{Na^+\ ^-OCOR} RC(=O)OC(=O)R$

$RC(=C=O)R \xrightarrow{Na^+\ ^-OCOR} R_2CHC(=O)OC(=O)R$

9. Carbonates

$ClC(=O)Cl \xrightarrow{ROH} ROC(=O)OR$

The mixed carbonate cannot be made this way

10. Esters

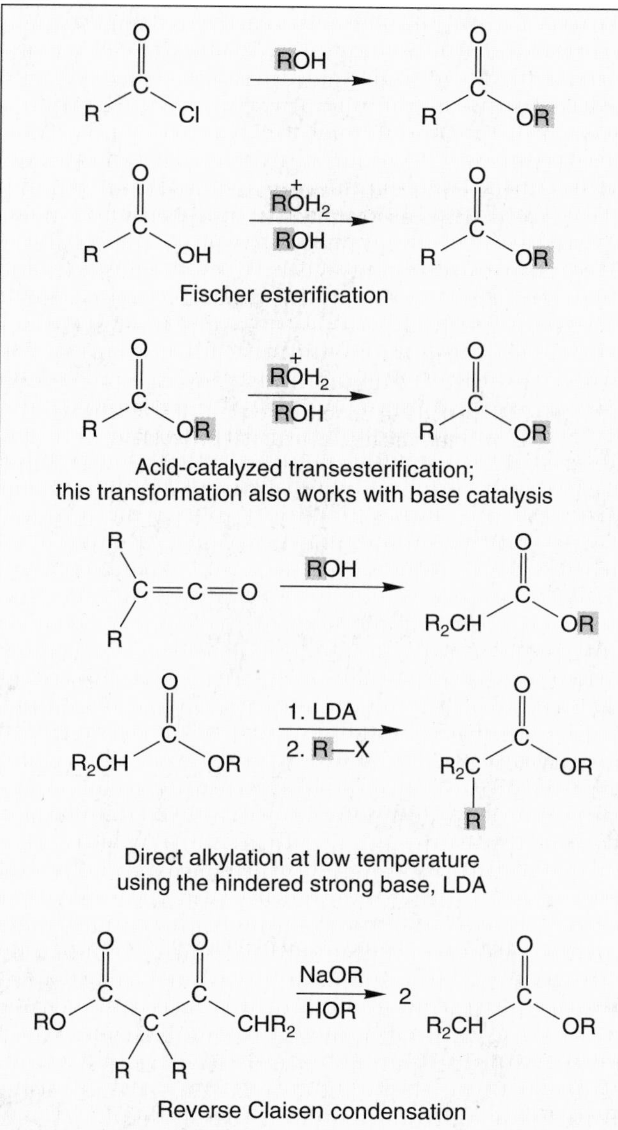

Direct alkylation at low temperature using the hindered strong base, LDA

11. Ketenes

$RCH_2C(=O)Cl \xrightarrow{R_3N} R(H)C=C=O$

Elimination using a tertiary amine as base

12. β-Keto esters

$RCH_2C(=O)OR \xrightarrow[HOR]{NaOR} RCH_2C(=O)CH(R)C(=O)OR$

Claisen condensation

FIGURE 20.102 (CONTINUED)

13. Ketones

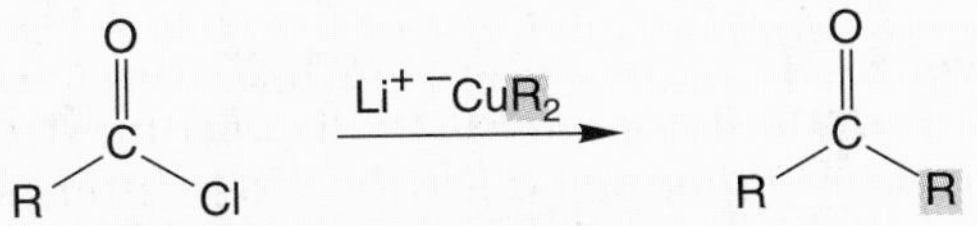

$RCOCl \xrightarrow{Li^+ \; ^-CuR_2} RCOR$

The cuprate will react with the acid chloride but not with the product ketone

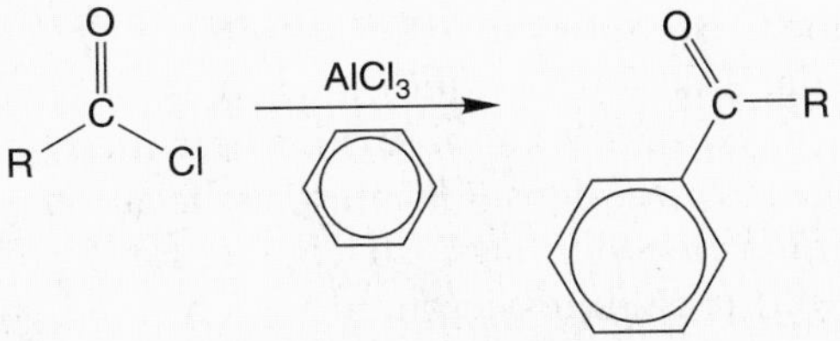

$RCOCl + C_6H_6 \xrightarrow{AlCl_3} C_6H_5COR$

Friedel–Crafts acylation; this reaction works with the anhydride as well

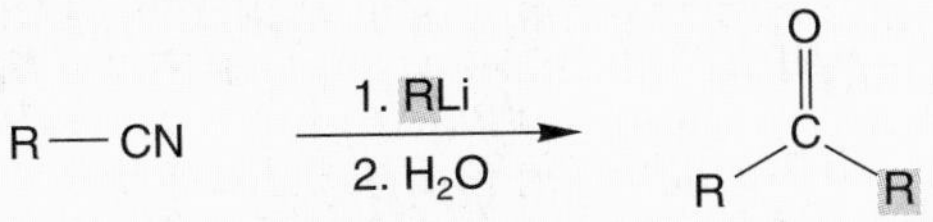

$R{-}CN \xrightarrow[2.\ H_2O]{1.\ RLi} RCOR$

The imine is an intermediate

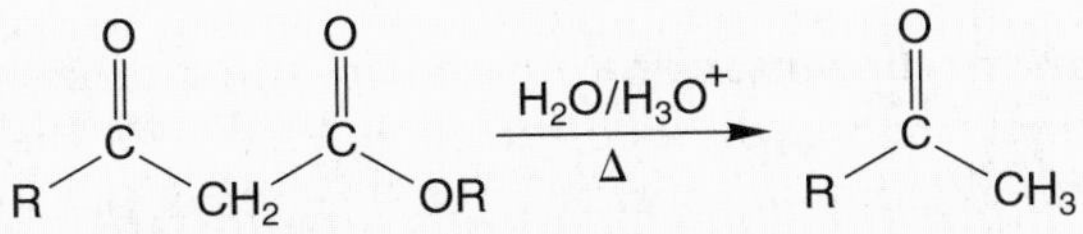

$RCOCH_2COOR \xrightarrow[\Delta]{H_2O/H_3O^+} RCOCH_3$

Hydrolysis gives the β-keto acid which decarboxylates; more complicated ketones can be made by alkylating the β-ketoester first

14. Lactams

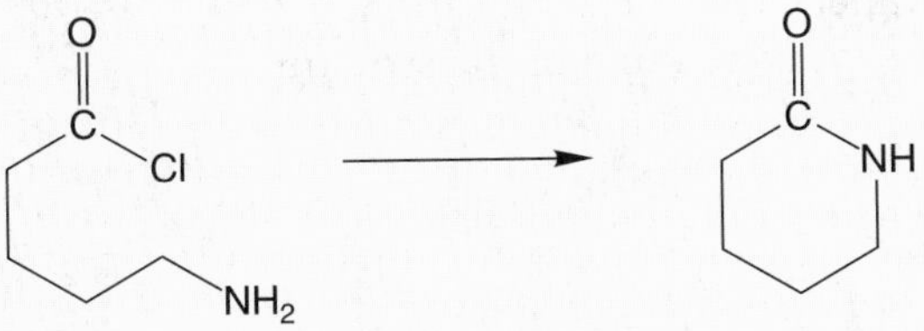

$Cl{-}CO{-}(CH_2)_4{-}NH_2 \longrightarrow$ six-membered lactam (NH)

Amide-forming reactions applied in an intramolecular way will work

15. Lactones

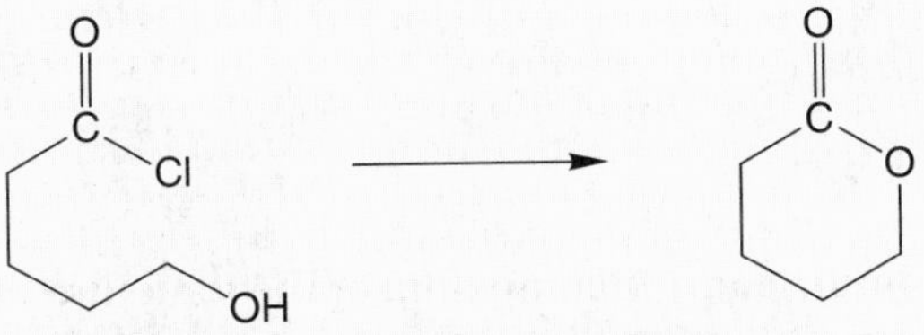

$Cl{-}CO{-}(CH_2)_4{-}OH \longrightarrow$ six-membered lactone (O)

Ester-forming reactions applied in an intramolecular way will work

16. Nitriles

$NC^- + R{-}X \longrightarrow NC{-}R$

S_N2 displacement

$RCONH_2 \xrightarrow[\Delta]{P_2O_5} R{-}CN$

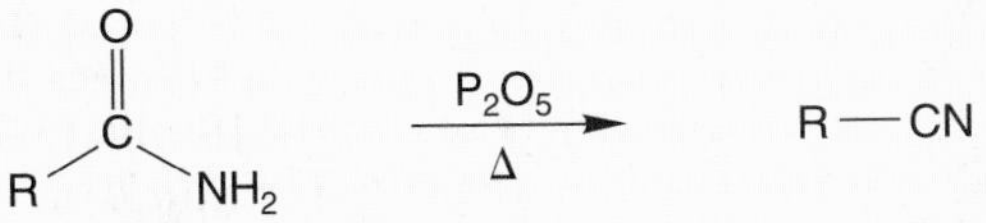

Dehydration of amides

17. Selenoxides

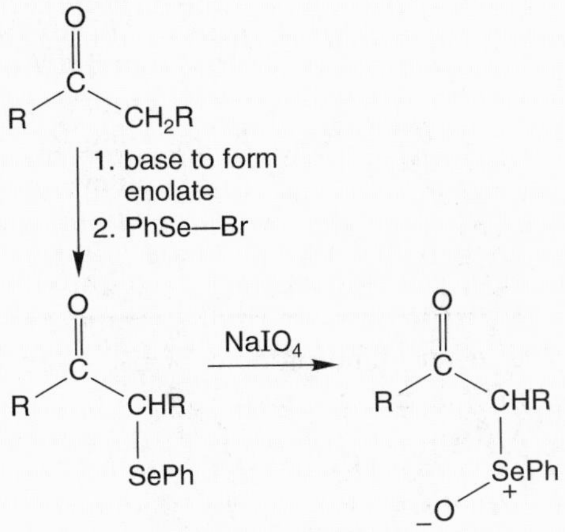

$RCOCH_2R \xrightarrow[2.\ PhSe{-}Br]{1.\ \text{base to form enolate}} RCOCHR(SePh) \xrightarrow{NaIO_4} RCOCHR({}^{+}SePh{-}O^-)$

18. Sulfoxides

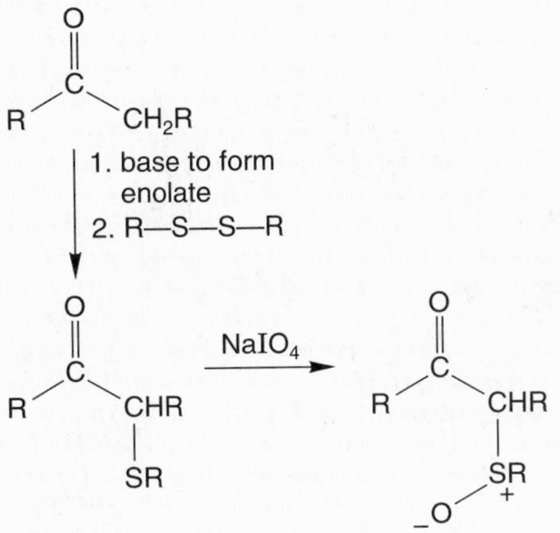

$RCOCH_2R \xrightarrow[2.\ R{-}S{-}S{-}R]{1.\ \text{base to form enolate}} RCOCHR(SR) \xrightarrow{NaIO_4} RCOCHR({}^{+}SR{-}O^-)$

19. Xanthate esters

$RO^- + CS_2 \longrightarrow RO{-}C(=S){-}S^- \xrightarrow[S_N2]{CH_3{-}I} RO{-}C(=S){-}SCH_3$

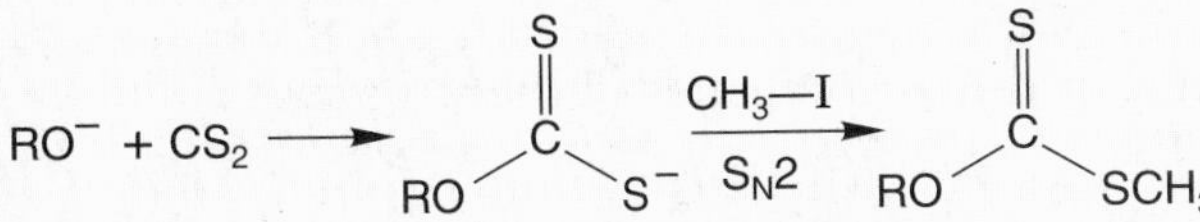

FIGURE **20.102** (CONTINUED)

COMMON ERRORS

Try to analyze problems before starting. Do not be too proud to do the obvious. Use the molecular formula to see what molecules must be combined. In acid (H_3O^+, HCl, etc.), carbocations are the likely intermediates. In base (^-OH, ^-OR, etc.), carbanions that are probably involved. Try to identify what must be accomplished in a problem. Although sometimes it will be hard to plan, and you will have to try all possible intermediates [Problem 20.26(c) is a good example], usually a problem will tell you much of what you must do. Ask yourself questions. What rings must be opened? What rings must be closed? What atoms become incorporated in the product? Pay attention to "stop signs." Primary carbocations are stop signs. Unstabilized carbanions are stop signs. If you are forced to invoke such a species in an answer, it is almost certainly wrong, and you must backtrack. Hard problems become easier if you keep such questions in mind. All of this sounds obvious, but it's not. Thinking a bit about what must be done in a problem saves much time in the long run and avoids the generally hopeless random arrow pushing that can trap you into a wrong answer.

20.17 KEY TERMS

Acetoacetic ester The product of the Claisen condensation of an acetate, such as

H_3C–C(=O)–CH_2–C(=O)–OR

Acid halide A compound of the structure

R–C(=O)–X

R = F, Cl, Br, or I

Acyl compound A compound of the structure

R–C(=O)–X

Amide A compound of the structure

R–C(=O)–NR_2

R can be H

Remember that this term also refers to the ion $^-NH_2$.

Anhydride A compound of the structure

R–C(=O)–O–C(=O)–R'

Claisen condensation A condensation reaction of esters in which an ester enolate adds to the carbonyl group of another ester. The result of this addition–elimination process is a β-keto ester.

Cyanides Compounds of the structure RCN. These compounds are also commonly called nitriles.

Decarboxylation The loss of carbon dioxide, a common reaction of 1,1-diacids and β-keto acids.

Dieckmann condensation An intramolecular, or cyclic, Claisen condensation.

Ester A compound of the structure

R–C(=O)–OR'

Ester hydrolysis The conversion of esters into carboxylic acids in a reaction induced by either acid or base.

Imides The nitrogen counterpart of an anhydride.

R–C(=O)–N(H)–C(=O)–R'

Immonium ion A compound of the structure

R–C(=$^+NR_2$)–R'

Ketene A compound of the structure

R'(R)C=C=O

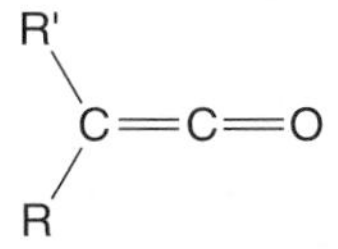

Lactam A cyclic amide.
Lactone A cyclic ester.
Malonic ester The 1,1-diester, ROOC–CH_2–COOR.
Malonic ester synthesis The alkylation of malonic esters to produce substituted 1,1-diesters, followed by hydrolysis and decarboxylation of the malonic acids to give substituted acetic acids.
Nitrile A compound of the structure RCN, also commonly called cyanides.
Rosenmund reduction The reduction of acid chlorides to aldehydes using a poisoned catalyst.
Saponification Base-induced hydrolysis of an ester, usually a fatty acid.
Selenoxide A compound of the structure

R–Se^+(O^-)–R

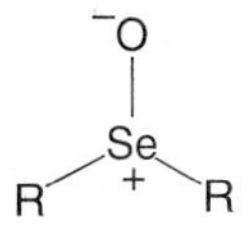

Sulfoxide A compound of the structure

R–S^+(O^-)–R

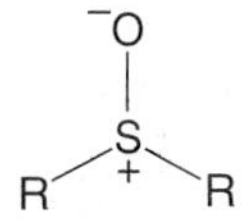

Thermal eliminations A group of intramolecular, syn β-elimination reactions in which an ester or related group acts both as base and leaving group.
Transesterification The formation of one ester from another. The reaction can be catalyzed by either acid or base.
Xanthate ester A compound of the structure

RO–C(=S)–SR

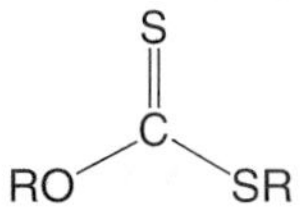

20.18 ADDITIONAL PROBLEMS

Let's warm up with some simple problems, just to be certain that the basic (and acidic) mechanisms of this chapter are under control. Then we can go on to other things. Please be sure you can do the "easy" problems before you try the more difficult ones at the end of this section.

PROBLEM **20.30** Write mechanisms for the following conversions in base:

(a) $(CH_3)_2CH$–C(=O)–Cl $\xrightarrow{H_2NCH_3}$ $(CH_3)_2CH$–C(=O)–$NHCH_3$

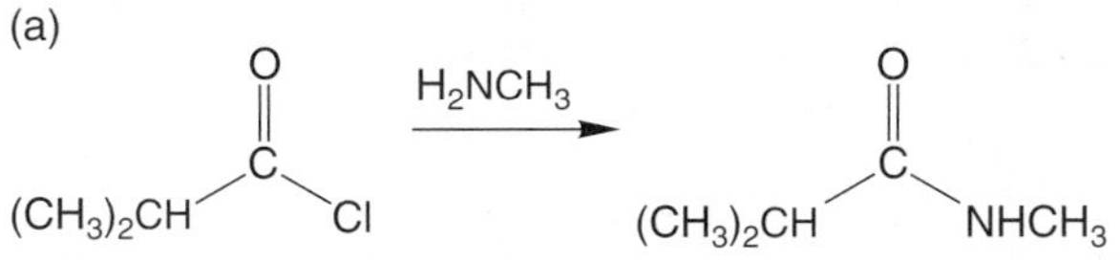

(b) $(CH_3)_2CH$–C(=O)–Cl $\xrightarrow{NaN_3}$ $(CH_3)_2CH$–C(=O)–N_3

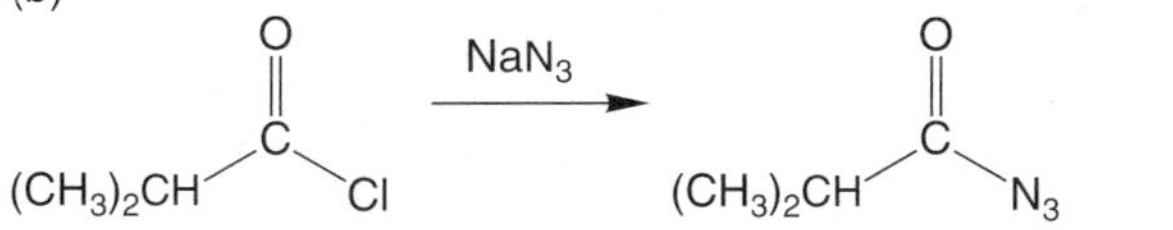

PROBLEM **20.31** Write mechanisms for the following conversions in acid:

(a) $(CH_3)_2CH$–C(=O)–OCH_3 $\xrightarrow[H_3O^+]{H_2O}$ $(CH_3)_2CH$–C(=O)–OH + $HOCH_3$

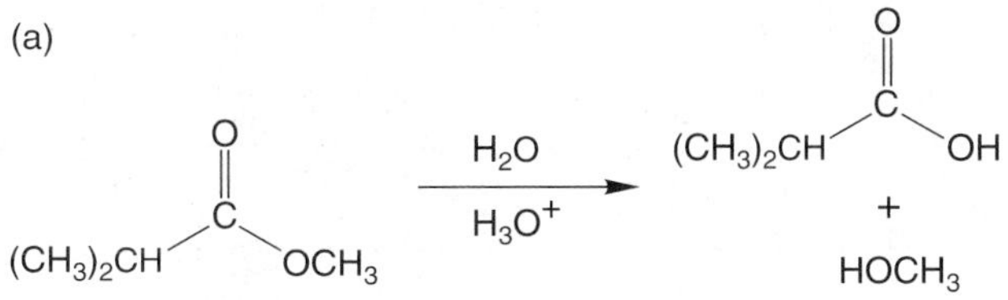

(b) $(CH_3)_2CH$–C(=O)–OCH_3 $\xrightarrow[CD_3\overset{+}{O}H_2]{CD_3OH}$ $(CH_3)_2CH$–C(=O)–OCD_3 + $HOCH_3$

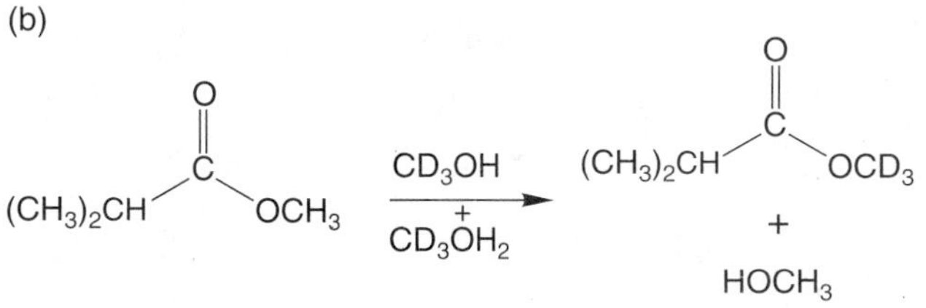

PROBLEM **20.32** Now try a slightly more complicated acid-catalyzed hydrolysis.

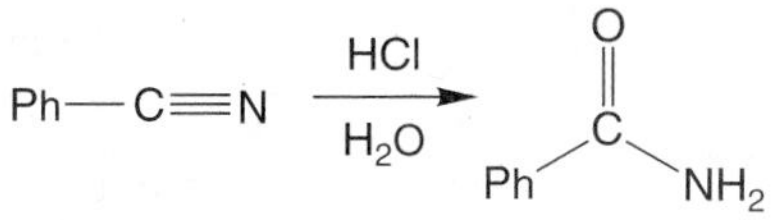

Ph—C≡N $\xrightarrow[H_2O]{HCl}$ Ph–C(=O)–NH_2

PROBLEM **20.33** Write a mechanism for the following Dieckmann condensation:

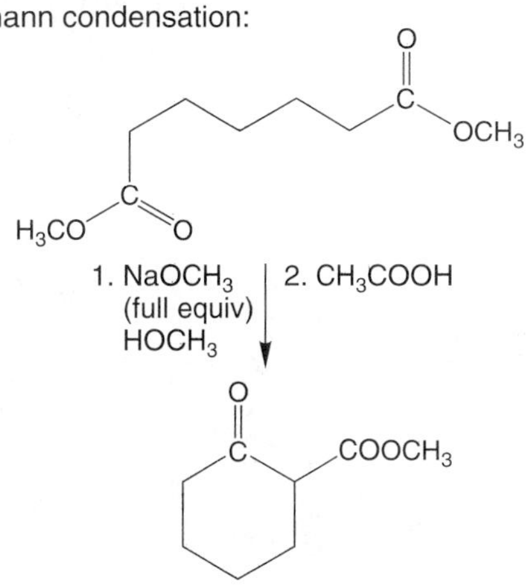

H_3CO–C(=O)–$(CH_2)_5$–C(=O)–OCH_3 $\xrightarrow{\text{1. NaOCH}_3\text{ (full equiv) HOCH}_3\text{; 2. CH}_3\text{COOH}}$ 2-(COOCH$_3$)cyclohexanone

PROBLEM 20.34 Here is a reverse Dieckmann. Write a mechanism.

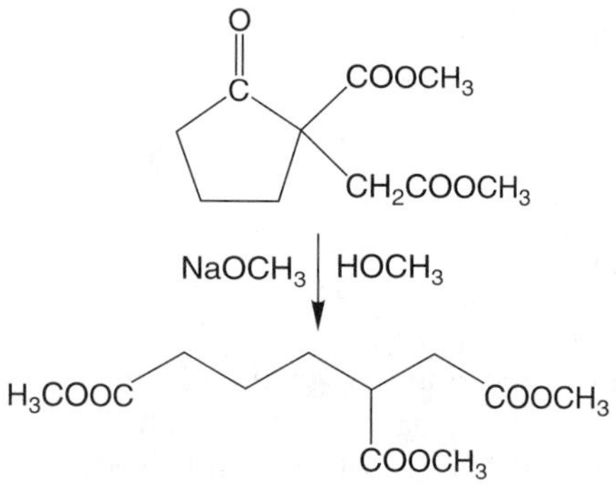

$NaOCH_3$ | $HOCH_3$

H_3COOC $COOCH_3$ $COOCH_3$

PROBLEM 20.35 Write good Lewis structures for the following compounds:

(a) Propionic acid (propanoic acid)
(b) Propionyl chloride (propanoyl chloride)
(c) *N,N*-Diethyl-2-phenylpropanamide
(d) Methyl ketene
(e) Propyl cyclopropanecarboxylate
(f) Phenyl benzoate

PROBLEM 20.36 Give reasonable names for the following molecules:

(a) $COOCH_3$

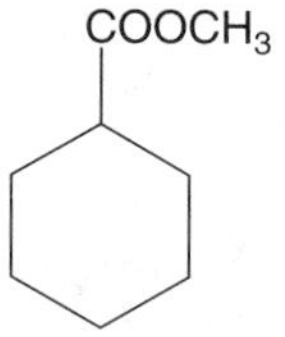

(b) H_3C C(=O) OCH_2CH_3

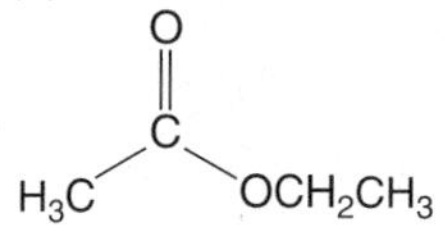

(c) CH_3CH_2 C(=O) OCH_3

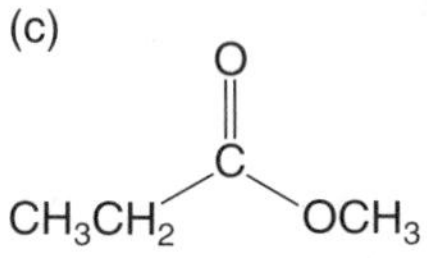

(d)

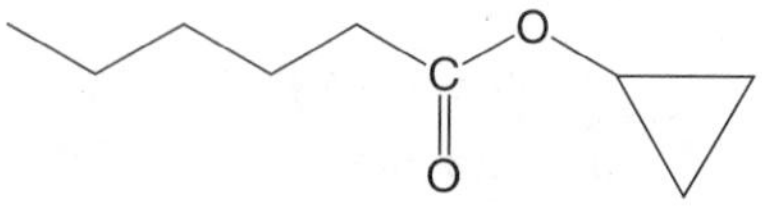

(e) OH

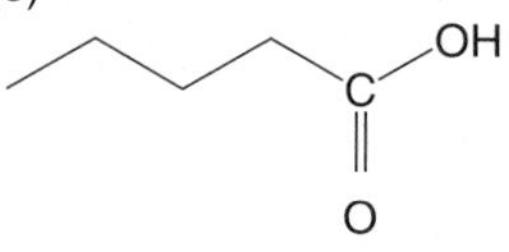

(f) Cl

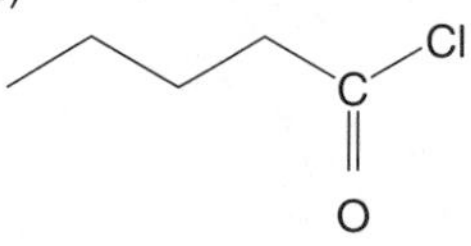

(g) NH_2

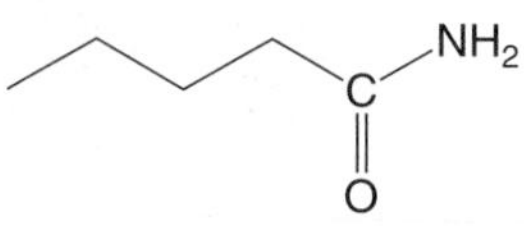

(h) NH_2 Br

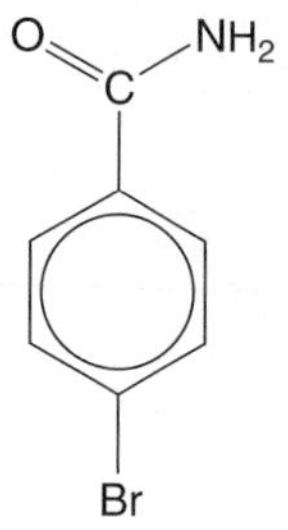

(i) $N(CH_3)_2$

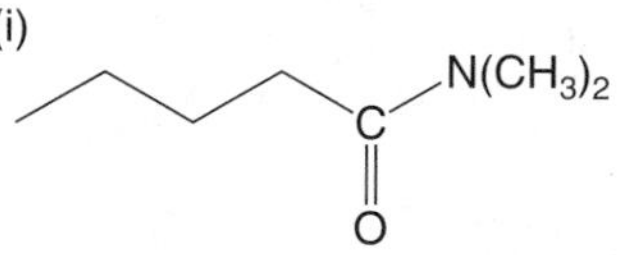

(j) CN

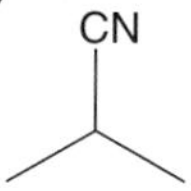

(k) H_3C C=C=O Ph

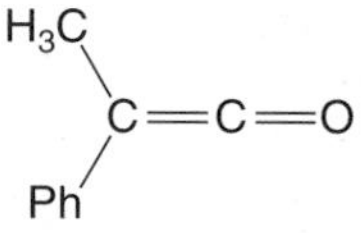

PROBLEM 20.37 Analysis of the IR stretching frequencies of acyl compounds involves an assessment of resonance and inductive effects. This assessment is a somewhat risky business, and it must be admitted that near-circular reasoning is sometimes encountered. Nonetheless, see if you can make sense of the following observations taken from Table 20.3. In each case, it will be profitable to think of all the important resonance forms of the acyl compound.

(a) Acetone absorbs at 1719 cm^{-1}, whereas acetaldehyde absorbs at 1733 cm^{-1}.

(b) Explain why methyl acetate absorbs at higher frequency (1750 cm^{-1}) than acetone (1719 cm^{-1}).

(c) *N*-Methyl acetamide absorbs much lower (1688 cm^{-1}) than methyl acetate (1750 cm^{-1}).

PROBLEM 20.38 Is your explanation in Problem 20.37(b) consistent with the observation that the carbonyl carbon of methyl acetate appears at δ 169 ppm in the ^{13}C NMR spectrum, whereas the carbonyl carbon of acetone is far downfield at δ 205 ppm? Explain.

PROBLEM 20.39 Give the major organic products expected in each of the following reactions or reaction sequences. Mechanisms are not required.

(a) COOH 1. $SOCl_2$ 2. $HN(CH_3)_2$ 3. $LiAlH_4$ 4. H_2O

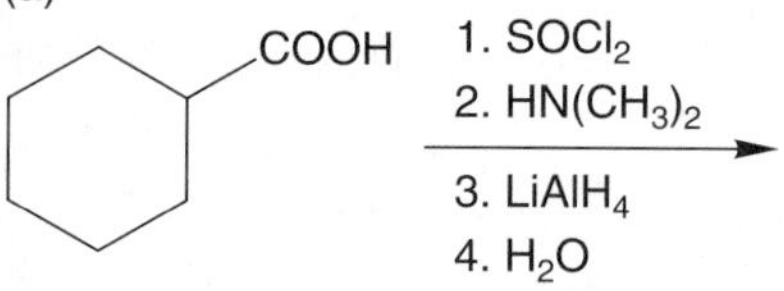

(b) CH_3OH, Δ

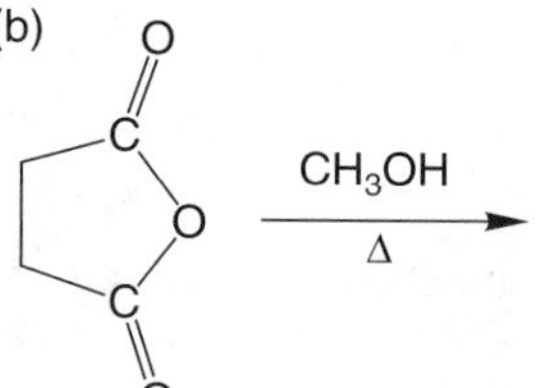

(c) CH_3 CH_3ONa / CH_3OH, Δ

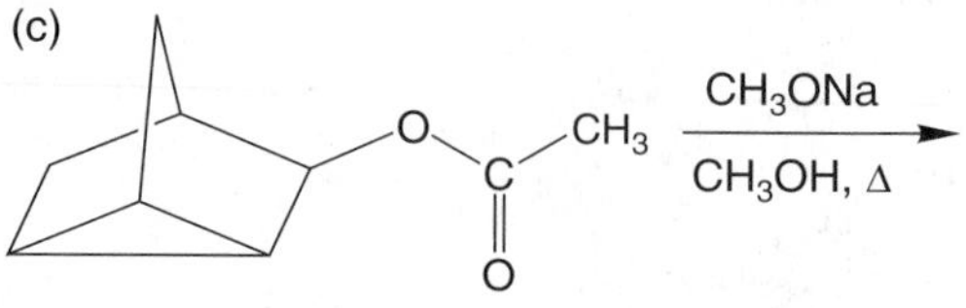

(d) Caprolactam (cyclic amide, C=O, NH in seven-membered ring) $\xrightarrow[H_2O,\ \Delta]{HCl}$

(e) Cyclohexane bearing $COOCH_2CH_3$ and OH (trans)
1. NaH
2. CS_2
3. CH_3I
4. 210–235 °C

(f) Cyclohexyl–$COOCH_2CH_3$
1. LICA, THF, –78 °C
2. I_2
3. $(CH_3CH_2)_3N$

LICA = cyclohexyl–$\bar{N}$–isopropyl Li^+

PROBLEM **20.40** In Section 20.14, we saw that primary amides could be dehydrated to nitriles (cyanides) with P_2O_5. Nitriles can also be prepared by treating aldoximes (**1**) with dehydrating agents such as acetic anhydride. Propose a mechanism for the formation of benzonitrile from the reaction of **1** with acetic anhydride. (How are oximes prepared?)

H(Ph)C=N–OH (**1**) $\xrightarrow[\Delta]{H_3C-CO-O-CO-CH_3}$ Ph—C≡N

PROBLEM **20.41** Reaction of benzonitrile (**1**) and *tert*-butyl alcohol in the presence of concentrated sulfuric acid, followed by treatment with water, gives *N-tert*-butylbenzamide (**2**). Provide an arrow formalism mechanism for this reaction.

Ph—C≡N (**1**) + $(CH_3)_3COH$

1. concd H_2SO_4 2. H_2O

Ph–C(=O)–$NHC(CH_3)_3$ (**2**)

PROBLEM **20.42** Devise syntheses for the following molecules. You may start with benzene, methyl alcohol, sodium methoxide, ethyl alcohol, sodium ethoxide, butyl alcohol (BuOH), phosgene, and propionic acid as your sources of carbon. You may also use any inorganic reagent and solvents as needed. Mechanisms are not required.

(a) Bu_3C—OH

(b) CH_3CH_2–C(=O)–Bu

(c) CH_3CH_2–C(OH)(Bu)–Bu

(d) Ph–C(=O)–CH_2CH_3

(e) CH_3CH_2CN

(f) CH_3CH_2–C(=O)–CH(CH_3)–C(=O)–OCH_2CH_3

(g) CH_3CH_2–C(=O)–CH(CH_3)–Bu

Bu = $CH_3CH_2CH_2CH_2$

PROBLEM **20.43** Propose syntheses for the following target molecules starting from the indicated material. You may use any other reagents that you need. Mechanisms are not required. Use a retrosynthetic analysis in each case.

(a) CH_2=CH–CH_2CH_2–CH(CH($CH_3)_2$)–C(=O)–CH_3 from H_3C–C(=O)–OCH_2CH_3

(b) $CH_3(CH_2)_9$–CH(CH_3)–COOH from $CH_2(COOCH_3)_2$

(c) 2-Benzylcyclopentanone (C(=O) ring with CH_2Ph) from cyclohexene

(d) Cyclobutyl—COOH from $CH_2(COOCH_3)_2$ and $BrCH_2CH_2CH_2Cl$

PROBLEM **20.44** Provide an arrow formalism mechanism for the following reaction:

1. NaOAc, Δ
2. H_2O/H_3O^+

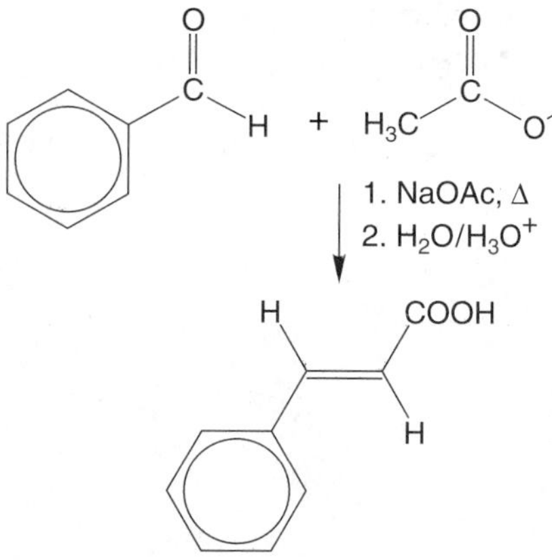

Cinnamic acid

This problem is harder than it looks. Your answer must take into account the fact that compound **B** is an intermediate.

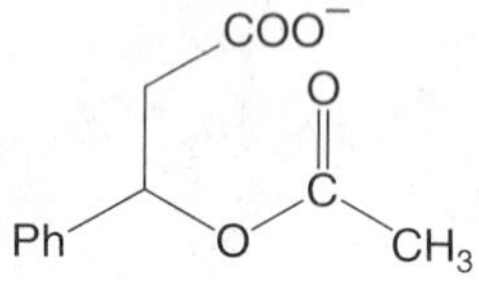

B

PROBLEM **20.45** Here is the problem concerning the formation of the Vilsmeier reagent we promised you in Chapter 19. As we saw in Section 19.5c, acid chlorides can be prepared from carboxylic acids and reagents such as thionyl chloride, phosphorus pentachloride, phosgene, and oxalyl chloride. The conditions for these reactions are sometimes quite severe. It is possible to prepare acid chlorides under appreciably milder conditions by using the Vilsmeier reagent (Problem 19.32). The Vilsmeier reagent is prepared from DMF and oxalyl chloride. Write an arrow formalism mechanism. *Hint*: DMF acts as a nucleophile, but not via the nitrogen atom.

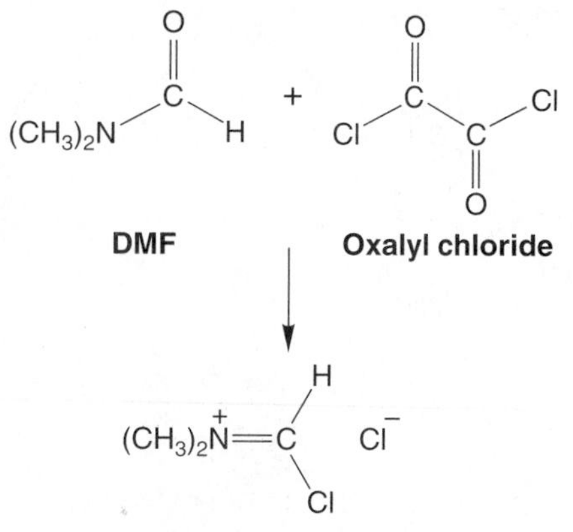

DMF **Oxalyl chloride**

$(CH_3)_2\overset{+}{N}{=}C(H)Cl$ Cl^-

Vilsmeier reagent

PROBLEM **20.46** Provide an arrow formalism for the following reaction:

1. $NaOCH_3/CH_3OH$, Δ
2. H_3COOC–CH=CH–Ph
3. CH_3COOH

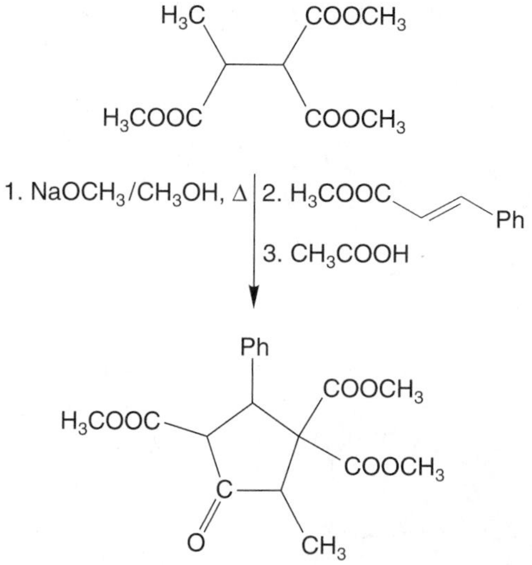

PROBLEM **20.47** Rationalize the following transformation. *Hint*: See Problem 20.26(a).

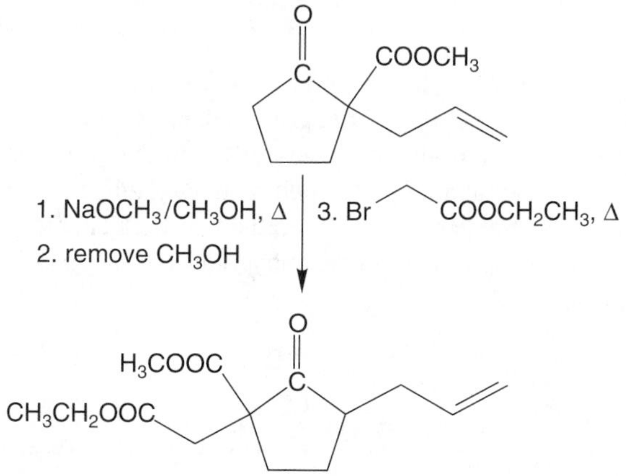

1. $NaOCH_3/CH_3OH$, Δ
2. remove CH_3OH
3. $BrCH_2COOCH_2CH_3$, Δ

PROBLEM **20.48** Provide an arrow formalism mechanism for the following reaction:

+ $CH_2(COOCH_2CH_3)_2$

CH_3CH_2OH, Δ

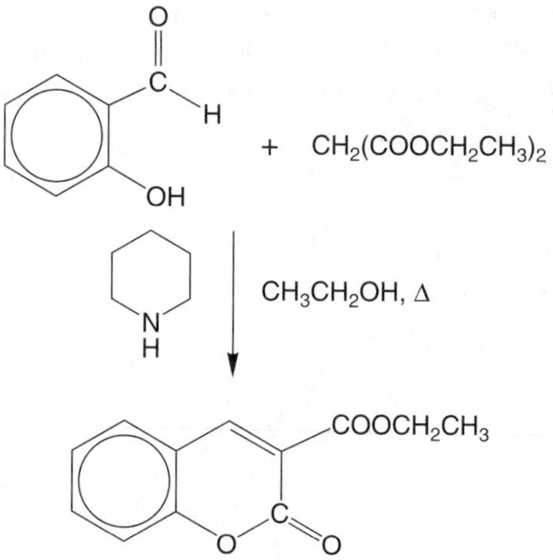

The following two problems should be done together.

PROBLEM 20.49 Provide an arrow formalism for the following reaction:

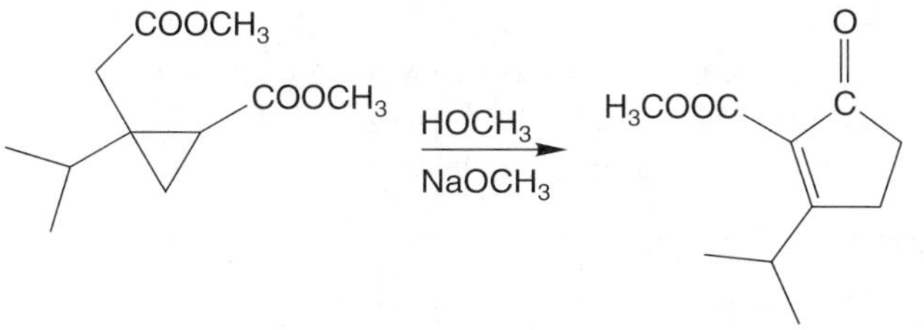

PROBLEM 20.50 Here is an answer to Problem 20.49, which appears in the original paper describing the reaction, as well as at least one compilation of problems. Perhaps it is the answer you came up with. First, criticize this proposal—what makes it unlikely? Next, you can either feel good about finding another mechanism already, or, if this was your answer, set about finding a better mechanism. *Hint*: Find a way to remove strain in the starting material.

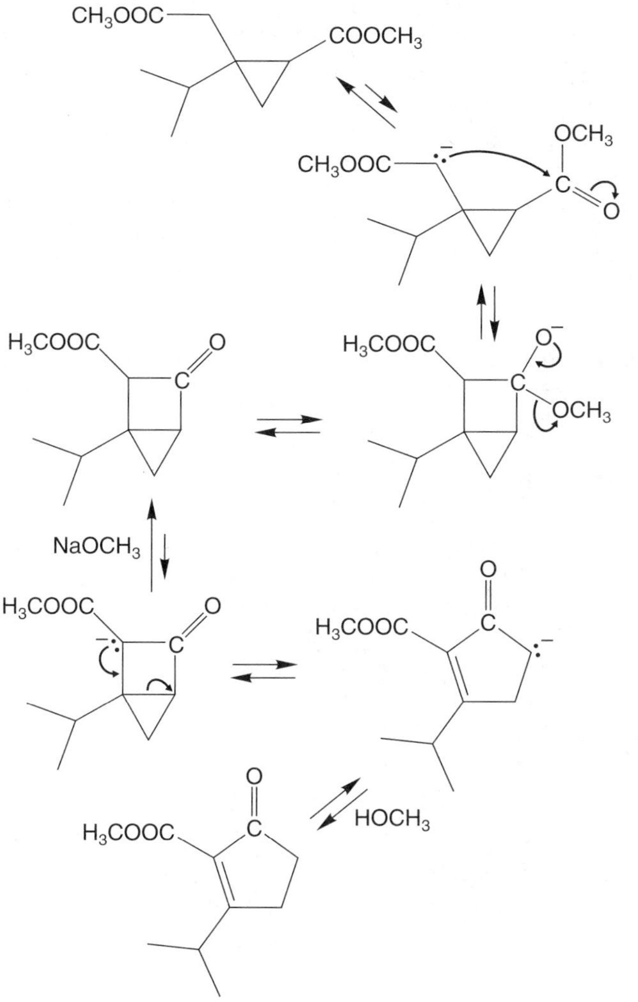

PROBLEM 20.51 Reaction of salicylaldehyde (**1**) and acetic anhydride in the presence of sodium acetate gives compound **2**, a fragrant compound occurring in sweet clover.

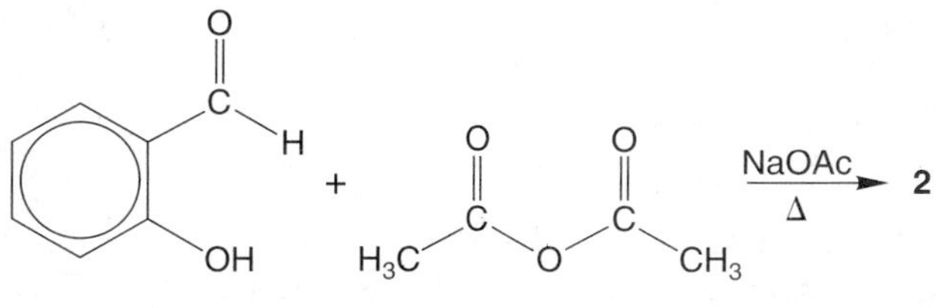

Spectral data for compound **2** are collected below. Deduce the structure of **2**. Naturally, it is an interesting question as to how **2** is formed. This problem turns out to be harder than it looks at first. Nonetheless, you might want to try to formulate a mechanism. However, note that *O*-acetylsalicylaldehyde (**3**) gives only traces of **2** in the absence of acetic anhydride.

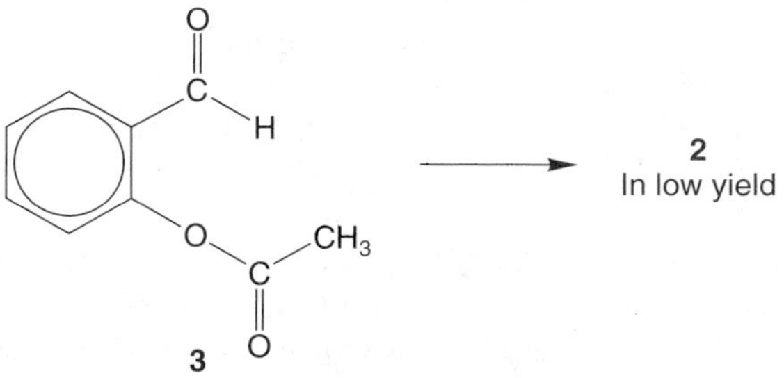

Compound **2**
Mass spectrum: $m/z = 146$ (p, 100%), 118 (80%), 90 (31%), 89 (26%)
IR (Nujol): 1704 cm^{-1} (s)
^1H NMR ($CDCl_3$): δ 6.42 (d, $J = 9$ Hz, 1H), 7.23–7.35 (m, 2H), 7.46–7.56 (m, 2H), 7.72 (d, $J = 9$ Hz, 1H)
^{13}C NMR ($CDCl_3$): δ 116.6 (d), 116.8 (d), 118.8 (s), 124.4, (d), 127.9 (d), 131.7 (d), 143.4 (d), 154.0 (s), 160.6 (s)

PROBLEM 20.52 Provide structures for compounds **A–F**. Spectral data for compounds **E** and **F** are summarized below.

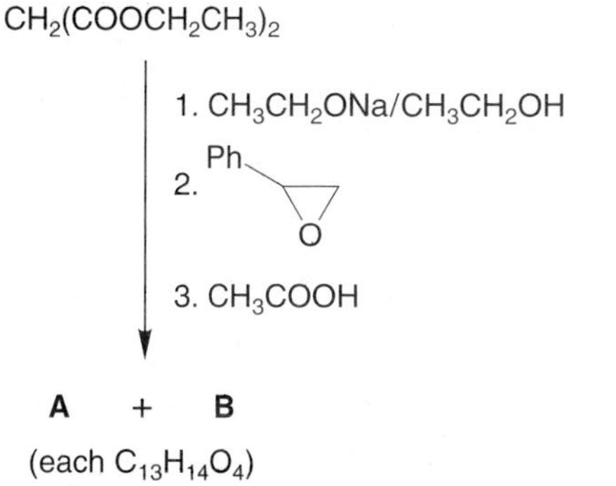

$CH_2(COOCH_2CH_3)_2$
1. CH_3CH_2ONa/CH_3CH_2OH
2. Ph-oxirane
3. CH_3COOH

A + **B**
(each $C_{13}H_{14}O_4$)

PROBLEM **20.52** (CONTINUED)

A + **B** $\xrightarrow[\text{2. } H_2O/H_3O^+]{\text{1. } KOH/H_2O}$ **C** + **D** (each $C_{11}H_{10}O_4$) $\xrightarrow{130\text{–}140\ °C}$ **E** + **F**

Compound **E**
Mass spectrum: *m/z* = 162 (p, 100%), 106 (77%), 104 (51%), 43 (61%)
IR (melt): 1778 cm^{-1} (s)
^1H NMR ($CDCl_3$): δ 2.1–2.15 (m, 1H), 2.5–2.7 (m, 3H), 5.45–5.55 (m, 1H), 7.2–7.4 (m, 5H)
^{13}C NMR ($CDCl_3$): δ 28.9 (t), 30.9 (t), 81.2 (d),125.2 (d), 128.4 (d), 128.7 (d), 139.3 (s), 176.9 (s)

Compound **F**
Mass spectrum: *m/z* = 162 (p, 28%), 104 (100%)
IR (film): 1780 cm^{-1} (s)
^1H NMR ($CDCl_3$): δ 2.67 (dd, *J* = 17.5 Hz, 6.1 Hz,1H), 2.93 (dd, =17.5 Hz, 8.1 Hz, 1H), 3.80 (m, 1H), 4.28 (dd, *J* = 9.0 Hz, 8.1 Hz), 4.67 (dd, *J* = 9.0 Hz, 7.9 Hz, 1H), 7.22–7.41 (m, 5H)
^{13}C NMR ($CDCl_3$: δ 35.4 (t), 40.8 (d), 73.8 (t), 126.5 (d), 127.4 (d), 128.9 (d), 139.2 (s), 176.3 (s)

PROBLEM **20.53** In Section 20.6 we saw that aldehydes are available from the reduction of esters at low temperature (–70 °C) by diisobutylaluminum hydride (DIBAL-H), (see Fig. 20.41). If the reaction mixture is allowed to warm before hydrolysis, the yield of aldehyde drops appreciably. For example, reduction of ethyl butyrate with DIBAL-H at –70 °C followed by hydrolysis at –70 °C, affords a 90% yield of butyraldehyde. However, if the reaction mixture is allowed to warm to 0 °C before hydrolysis, the yield of aldehyde is less than 20%. Among the products under these conditions are ethyl butyrate (recovered starting material) and butyl alcohol ($CH_3CH_2CH_2CH_2OH$), a molecule formed from hydrolysis of $CH_3CH_2CH_2CH_2O\text{–}AlR_2$ (R = isobutyl). Rationalize these observations with arrow formalism mechanisms. *Hint*: Recall the MPVO equilibration and the Cannizzaro reactions, Chapter 18.

Ethyl butyrate ($C_3H_7C(=O)OCH_2CH_3$)

1. DIBAL-H, –70 °C; 2. H_2O → butyraldehyde ($C_3H_7C(=O)H$) (90%)

1. DIBAL-H, –70 °C; 2. warm to 0 °C; 3. H_2O → butyraldehyde (minor) + ethyl butyrate ($C_3H_7C(=O)OCH_2CH_3$) (major)

$[CH_3CH_2CH_2CH_2OAlR_2] \xrightarrow{H_2O} CH_3CH_2CH_2CH_2OH$ (major)

Introduction to the Chemistry of Nitrogen-Containing Compounds: Amines

So, if Sunday you're free,
Why don't you come with me,
And we'll poison the pigeons in the park.

My pulse will be quickenin',
With each drop of strych-i-nin',
We feed to a pigeon,
(It just takes a smidgin),
To poison a pigeon, in the park,
Tra-La.

—Tom Lehrer*

Amines and amides have appeared in some earlier chapters. For example, we saw the chemistry of aromatic amines in Chapter 14, and the chemistry of amides was covered in Chapter 20. Here, amine chemistry is collated and extended. This chapter also provides a necessary introduction to Chapter 26, which discusses the chemistry of some of the many biologically important molecules containing nitrogen. A nitrogen atom provides one of the crucial linking agents in the formation of polyamino acids, called **peptides** or **proteins** (Fig. 21.1).

FIGURE 21.1 A nitrogen-based link is vital in the polymeric molecules known as proteins or peptides.

*Thomas Lehrer (b. 1928) is an American mathematician and songwriter, most of whose work was done in the 1950s and 1960s. Despite the unfortunate fact that there has been little new material for many years, his songs, often sardonic and always very funny, retain their appeal.

Alkaloids are naturally occurring amines that are powerfully active biologically. Alkaloids are among the most useful medicinal agents known. The painkiller morphine and the antimalarial agent quinine are typical examples. At the same time, some of the most destructively addictive substances known are also alkaloids. Heroin and cocaine are examples from our own time, and anyone at all familiar with popular literature knows of the notorious poison strychnine (Fig. 21.2; Chapter 5 Box).*

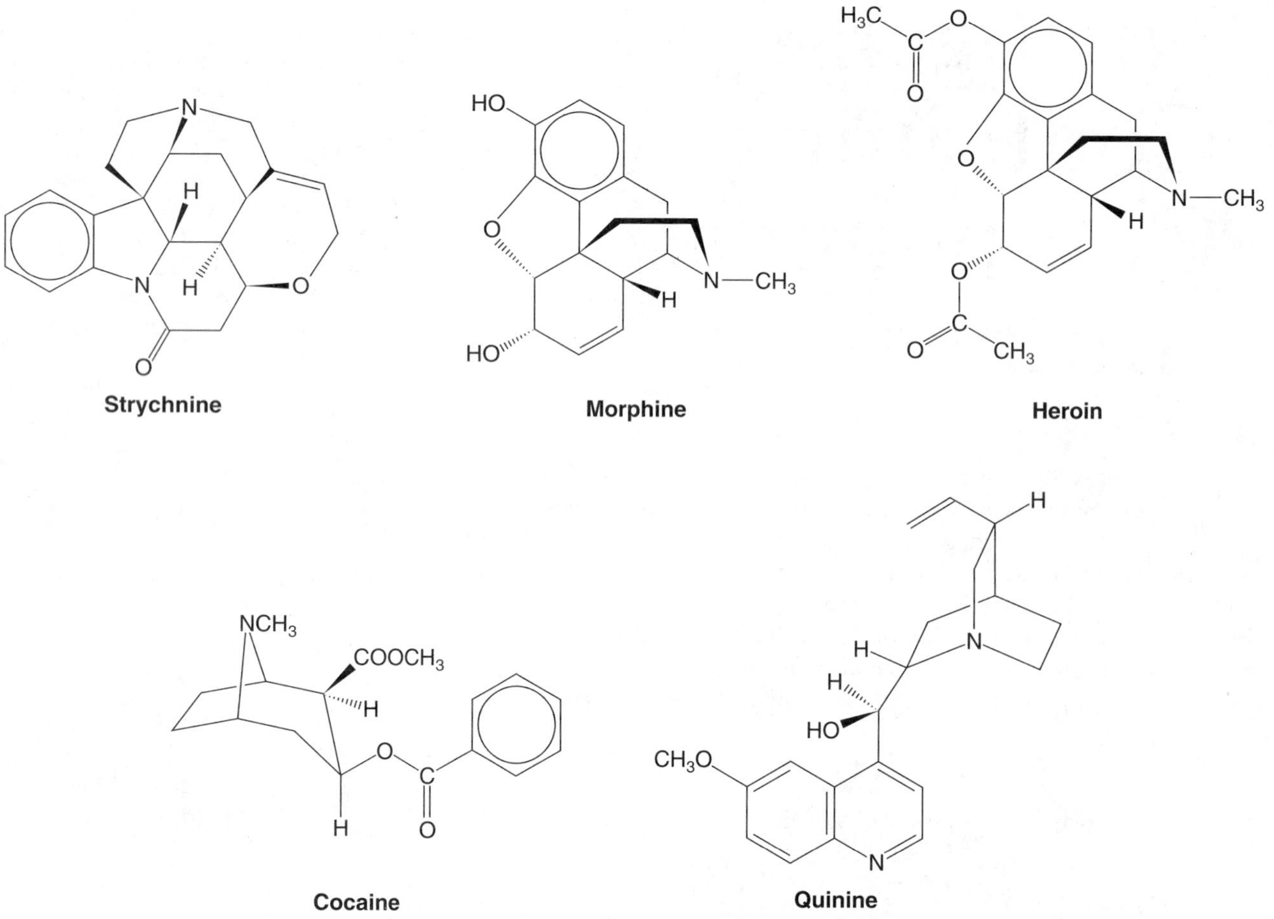

Strychnine

Morphine

Heroin

Cocaine

Quinine

FIGURE **21.2** Some alkaloids—nitrogen-containing molecules with great biological activity.

21.1 NOMENCLATURE

Many systems for naming amines exist, but the chemical world seems to be resisting attempts to bring order out of this mini-chaos by keeping to the old common names. The parent of all amines, $:NH_3$, is called **ammonia**. Successive replacement of the hydrogens leads to **primary, secondary**, and **tertiary amines**. Quaternary nitrogen compounds are positively charged and are called **ammonium ions** (Fig. 21.3).

*It was presumably just this notoriety that led R. B. Woodward to open his report of the synthesis of this molecule, a most remarkable achievement, with the singular exclamation, "Strychnine!"

:NH_3	:NH_2R	:NHR_2	:NR_3	$\overset{+}{N}R_4$
Ammonia	Primary amine	Secondary amine	Tertiary amine	Ammonium ion

FIGURE **21.3** Substituted amines.

Primary amines are commonly named by using the name of the substituent, R, and appending the suffix, amine. Secondary and tertiary amines in which the R groups are all the same are simply named as di- and trialkylamines. Amines bearing different R groups are named by ordering the groups alphabetically. Ethylmethylamine is correct, methylethylamine is not, although both are surely intelligible (Fig. 21.4).

$CH_3\ddot{N}H_2$ **Methylamine**
$CH_3CH_2\ddot{N}H_2$ **Ethylamine**
$PhCH_2\ddot{N}H_2$ **Benzylamine**
$(CH_3)_2\ddot{N}H$ **Dimethylamine**
$(CH_3CH_2)_3N$: **Triethylamine**
$H(CH_3)\ddot{N}CH_2CH_3$ **Ethylmethylamine,** (*not* methylethylamine)
$PhCH_2(CH_3)\ddot{N}CH_2CH_3$ **Benzylethyl-methylamine**
$(CH_3)_3C\ddot{N}H_2$ ***tert*-Butylamine**
$CH_3CH_2CH_2\ddot{N}H_2$ **Propylamine**
Cyclohexylamine
Dicyclopentylamine

FIGURE **21.4** A naming scheme for primary, secondary, and tertiary amines.

The naming system used by the official chemical world names amines analogously to alcohols. If CH_3OH is sometimes known as methan<u>ol</u>, then CH_3NH_2 is methan<u>amine</u>. Figure 21.5 gives some examples.

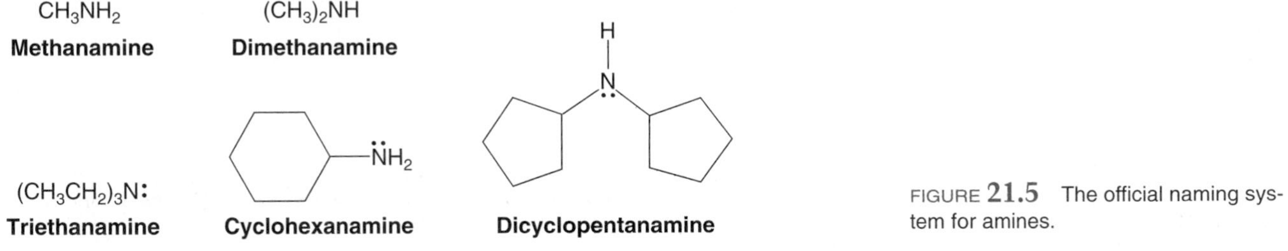

FIGURE **21.5** The official naming system for amines.

Still another method names amines as substituted alkanes. Now CH_3NH_2 becomes aminomethane. Figure 21.6 gives some examples of amines named this way.

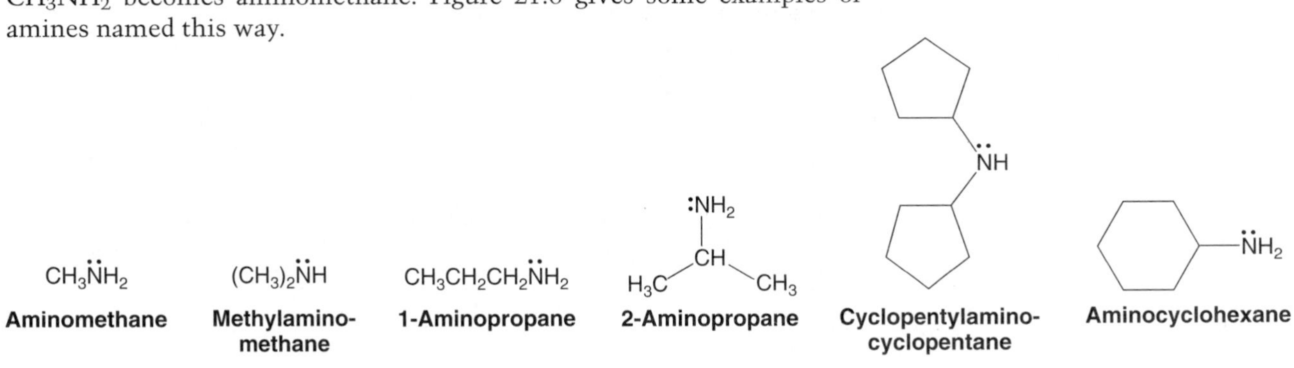

FIGURE **21.6** Still another method for naming amines.

The amino group is just behind the alcohol group in the naming protocol priority system. Figure 21.7 gives some correct and incorrect names of polyfunctional compounds containing amino groups.

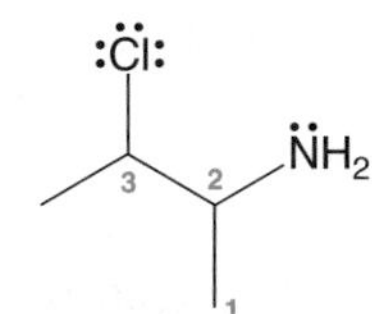

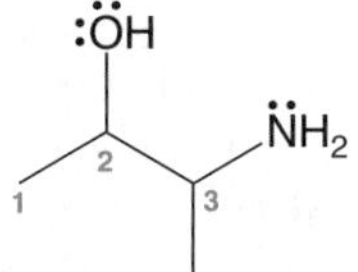

Isobutylamine
1-amino-2-methylpropane

2-Amino-3-chlorobutane
(*not* 3-amino-2-chlorobutane)

3-Amino-2-butanol
(or 3-Amino-2-hydroxybutane
not, 2-amino-3-butanol
or 2-amino-3-hydroxybutane)

3-Bromocyclopentylamine
(*neither* 3-aminobromocyclopentane,
nor, 4-bromoaminocyclopentane)

3-Aminocyclopentanol
(*not*, 3-hydroxycyclopentylamine)

FIGURE 21.7 Some polyfunctional amines and their names.

The names of aromatic amines are based on the parent compound, called benzenamine by the official world and **aniline** (or aminobenzene) by everyone else (Fig. 21.8).

Aniline
(aminobenzene)

***p*-Methylaniline**
(4-methylaniline,
p-toluidine)

***m*-Chloroaniline**
(3-chloroaniline)

***o*-Aminophenol**
(2-aminophenol)

FIGURE 21.8 Some aromatic amines.

Aromatic compounds can include a nitrogen atom in the ring, and these compounds almost inevitably have widely used common names. The parent molecule is **pyridine**, and there is little one can do but learn the names of these compounds (Fig. 21.9).

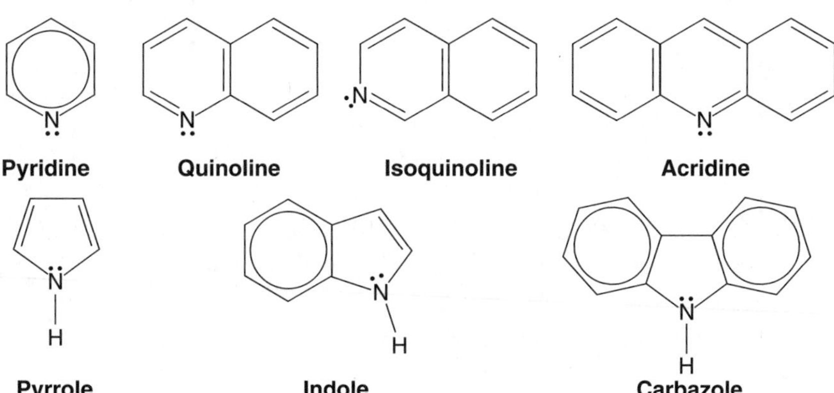

FIGURE 21.9 Some aromatic molecules containing nitrogen atoms in the ring.

The numbering in monocyclic aromatic compounds begins with the nitrogen and proceeds in such a way as to minimize the numbers of any substituents. In polycyclic compounds, the numbering begins adjacent to the bridge and proceeds so as to place the heteroatom at the lowest possible position (Fig. 21.10).

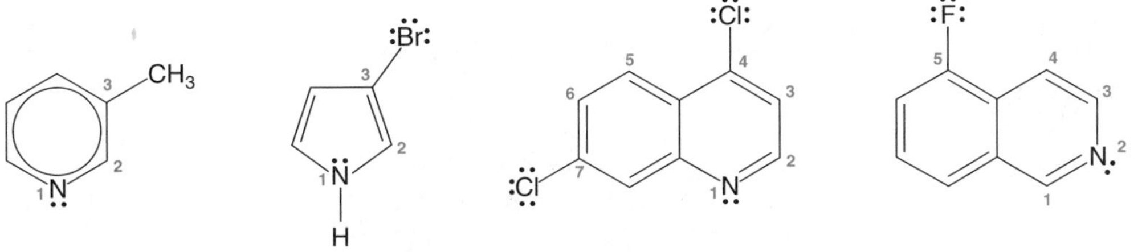

3-Methylpyridine *not* 5-methylpyridine | **3-Bromopyrrole** *not* 4-bromopyrrole | **4,7-Dichloroquinoline** | **5-Fluoroisoquinoline**

FIGURE **21.10** The naming scheme for aromatic compounds with a nitrogen in the ring.

Nonaromatic, cyclic amino compounds are also common in nature, and are almost invariably known by their common names. These compounds are often named as "aza" analogues of an all-carbon system. Thus aziridine becomes azacyclopropane. Methylamine could be called azaethane, although in point of fact this system is almost never used for acyclic compounds. As with the aromatic compounds, unless there is an oxygen atom in the ring, the numbering scheme places the nitrogen at 1 and proceeds so as to minimize the numbering of any substituents. Figure 21.11 gives a few important examples. Once again there is no way out other than to learn these names.

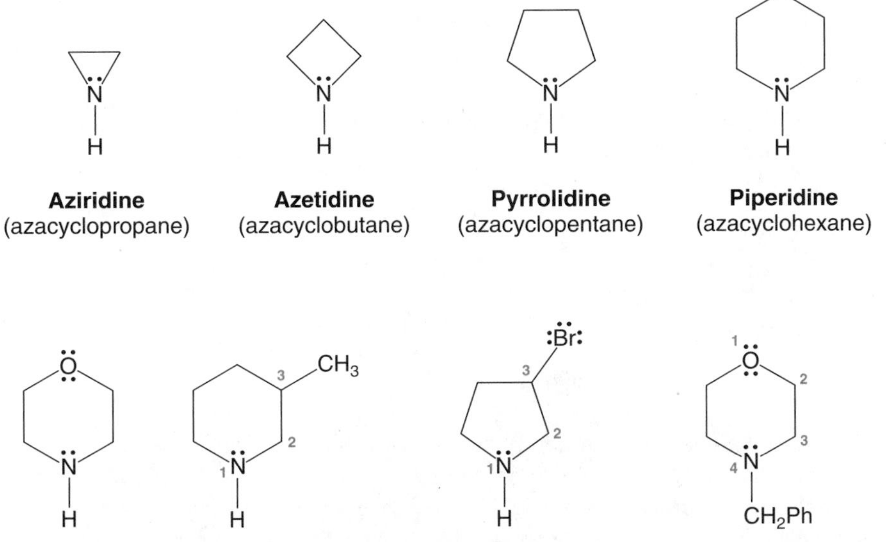

Morpholine | **3-Methylpiperidine** (*not* 5-methylpiperidine) | **3-Bromopyrrolidine** (*not* 4-bromopyrrolidine) | **4-Benzylmorpholine** (*N*-benzylmorpholine)

FIGURE **21.11** Cyclic amines and their names.

Despite this seeming excess of nomenclature, another system must be mentioned. Substituted amines can also be named as *N*-substituted amines. In this system, the substituent on nitrogen is designated with the locating prefix *N*. Some examples are shown in Figure 21.12.

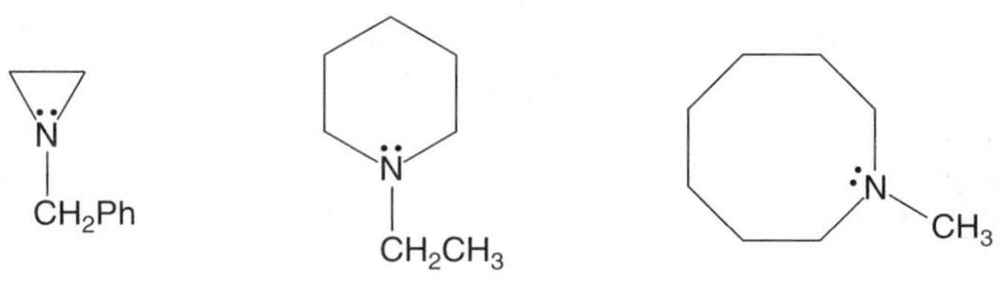

N-Benzylaziridine **N-Ethylpiperidine** **N-Methylazacyclooctane** (heptamethylenemethylamine)

FIGURE **21.12** Still another naming system for amines.

Quaternary ammonium ions are named by alphabetically attaching the names of the substituents. Don't forget to append the name of the negatively charged counterion (Fig. 21.13).

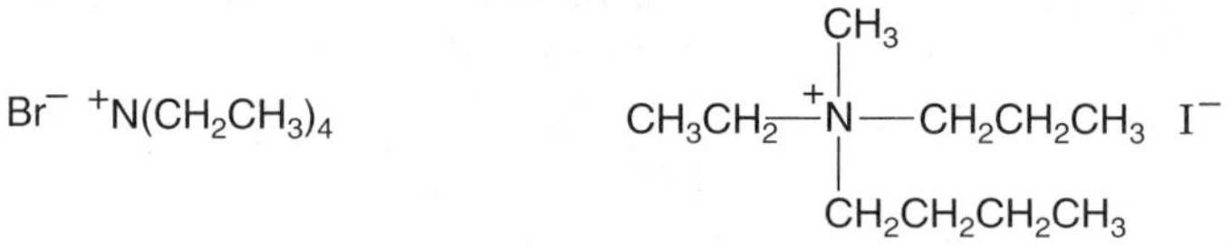

$Br^- \ {}^+N(CH_2CH_3)_4$

Tetraethylammonium bromide

$CH_3CH_2\text{—}{}^+N(CH_3)(CH_2CH_2CH_2CH_3)\text{—}CH_2CH_2CH_3 \ I^-$

Butylethylmethylpropyl ammonium iodide

FIGURE **21.13** Names for two quaternary ammonium ions.

21.2 STRUCTURE AND PHYSICAL PROPERTIES OF AMINES

In simple amines, the carbon–nitrogen bond distance is a little shorter than the corresponding carbon–carbon bond distance in alkanes. This phenomenon is similar to that found in the alcohols and ethers (Fig. 21.14).

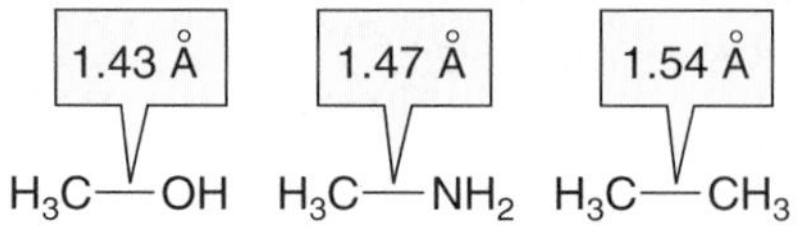

$H_3C\text{—}OH$ (1.43 Å) $H_3C\text{—}NH_2$ (1.47 Å) $H_3C\text{—}CH_3$ (1.54 Å)

FIGURE **21.14** The carbon–nitrogen bond in amines is shorter than a normal carbon–carbon bond.

*PROBLEM **21.1** Why is the carbon–nitrogen bond distance in aniline slightly shorter than that in methylamine (Fig. 21.15)?

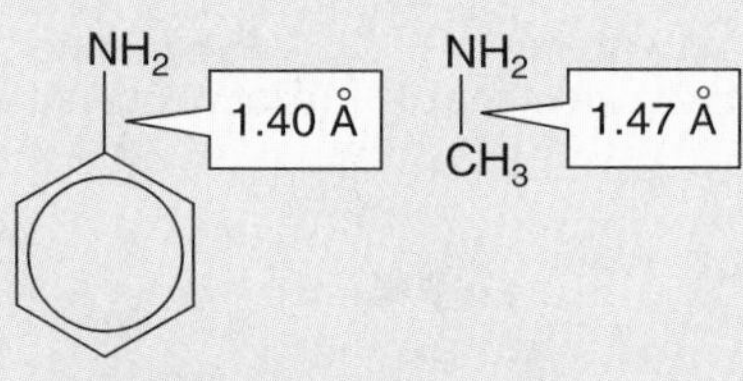

Aniline

FIGURE **21.15**

ANSWER First of all, the carbon–nitrogen bond in aniline is an sp^2–sp^3 bond and will be shorter than the sp^3–sp^3 bond of methylamine. Second, in aniline there is overlap between the orbital containing the two nonbonding electrons on nitrogen and the orbitals of the ring. The resonance forms show this well. The result is double-bond character in the bond attaching the nitrogen to the ring. As double bonds are shorter than single bonds, this leads to a shortening of the carbon–nitrogen bond distance as compared to that in alkylamines, in which there is no resonance effect.

ANSWER (CONTINUED)

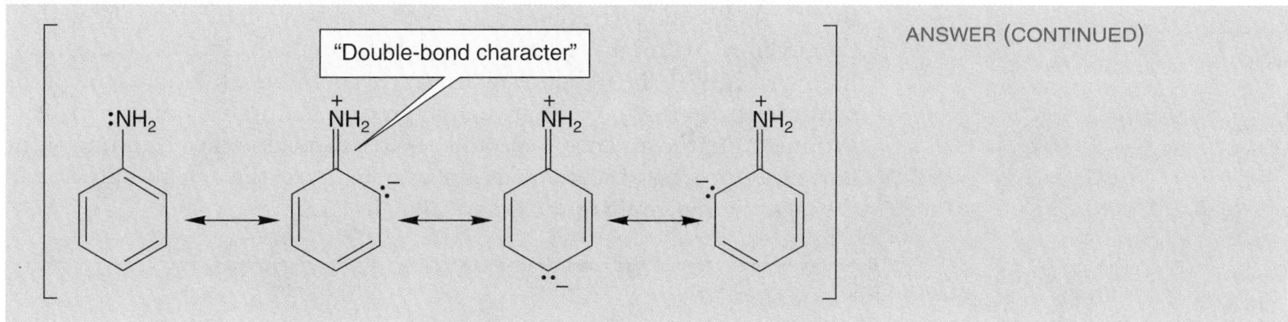

"Double-bond character"
:NH2
+NH2
+NH2
+NH2

Simple amines are hybridized approximately sp^3 at nitrogen and thus are pyramidal. The bond angles are close to the tetrahedral angle of 109.5°, but, of course, cannot be *exactly* the tetrahedral angle.

PROBLEM **21.2**

Why not?

For amines in which all three R groups are identical, there will be only a single R—N—R angle, but for less symmetrical amines, the different R—N—R angles cannot be the same (Fig. 21.16).

107.3° 108.4° 105.9° 112.9°

FIGURE **21.16** Some bond angles in simple amines.

The most important structural feature of amines comes from the presence of the pair of nonbonding electrons acting as the fourth substituent on the pyramid. In alkanes, where there are four substituents, there is nothing more to consider once the pyramid is described. In amines there is, as the pyramid can undergo an "umbrella flip," **amine inversion**, forming a new, mirror-image pyramid (Fig. 21.17).

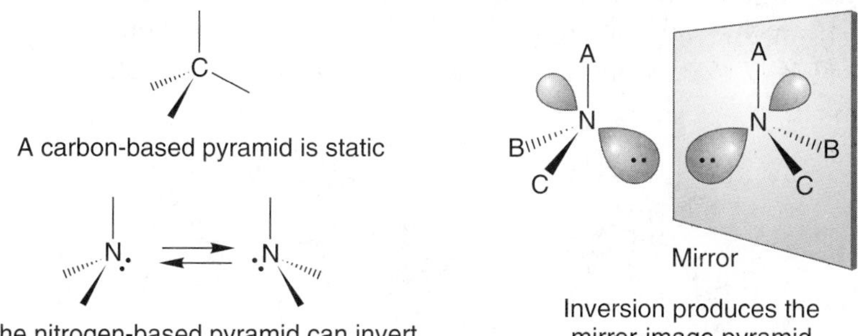

A carbon-based pyramid is static
The nitrogen-based pyramid can invert
Mirror
Inversion produces the mirror-image pyramid

FIGURE **21.17** Amine inversion interconverts enantiomeric amines

This inversion process is sometimes confusing when first encountered, so let's work through it, making comparisons to the alkanes as we go along. At the beginning of the process, the pyramid begins to flatten and the R—N—R bond angles increase. At the halfway point along the path to the inverted pyramid the transition state is reached. Here the molecule is flat, with 120° R—N—R bond angles. At this point, the hybridization of the central nitrogen is sp^2! The inversion process maintains a mirror plane of symmetry. In the alkanes, we could certainly imagine the beginning of the flattening process, but there is no way to force the fourth substituent through the molecule to the other side (Fig. 21.18).

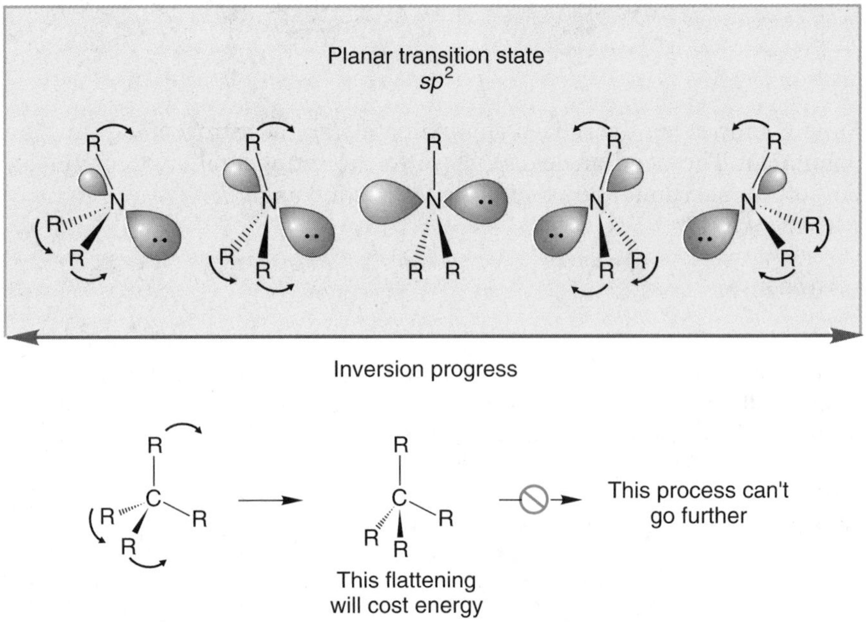

Planar transition state
sp^2

Inversion progress

This process can't go further

This flattening will cost energy

FIGURE **21.18** Amine inversion has a planar transition state.

PROBLEM **21.3** In general terms, what is the hybridization of nitrogen at some point between the starting pyramidal amine and the transition state?

A most important concern is the rate of this inversion process. If it is slow at room temperature, then we should be able to isolate appropriately substituted optically active amines. If inversion is fast, however, racemization will also be fast as the inversion process interconverts enantiomers. In practice, this is what happens. Barriers to inversion in simple amines are very low, about 5–6 kcal/mol, and isolation of one enantiomer, even at very low temperature, is difficult. Inversion is simply too facile (Fig. 21.19).

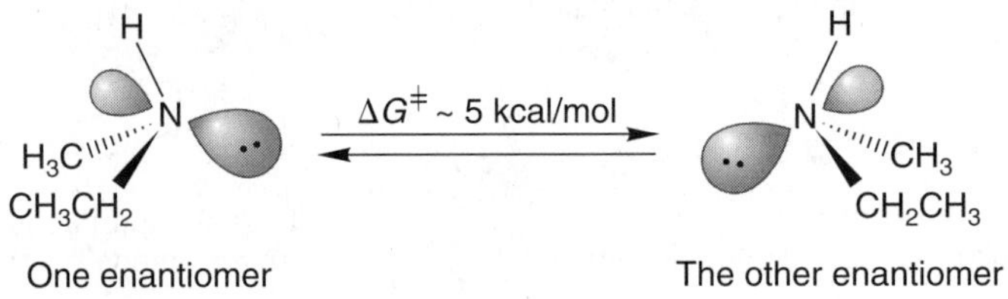

$\Delta G^{\ddagger}$ ~ 5 kcal/mol

One enantiomer

The other enantiomer

FIGURE **21.19** Rapid inversion interconverts enantiomeric amines.

PROBLEM **21.4** In contrast to simple amines, **aziridines** (three-membered rings containing a nitrogen) can often be separated into enantiomers. For example, the activation energy for the inversion of 1,2,2-trimethylaziridine is about 18.5 kcal/mol, much higher than that for simple amines (Fig. 21.20). Explain.

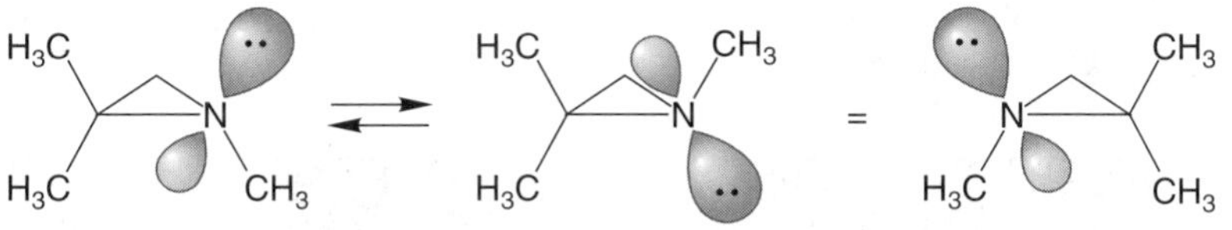

The barrier to this inversion is unusually high, $\Delta G^{\ddagger}$ = 18.5 kcal/mol

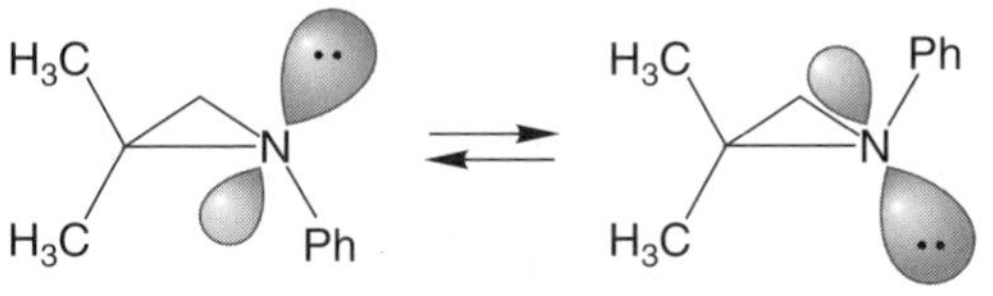

When the methyl on nitrogen is replaced by phenyl, $\Delta G^{\ddagger}$ decreases to 11.2 kcal/mol

FIGURE **21.20**

*PROBLEM **21.5**

If there is a phenyl group attached to the nitrogen atom of the aziridine, the barrier to inversion decreases (Fig. 21.20). Explain.

ANSWER

A subtle, tough question. In the starting aziridine, there is relatively poor overlap between the hybrid orbital on nitrogen containing the nonbonding electrons and the orbitals of the phenyl ring. There is some overlap, but the connection between the unsymmetrical hybrid and the 2*p* orbital on carbon is not ideal. In the transition state, the nitrogen is hybridized sp^2 and now there is excellent overlap, and hence good delocalization of the nonbonding electrons into the benzene ring. The transition state is stabilized, lowered in energy, and becomes easier to reach from starting material.

Starting material

Transition state

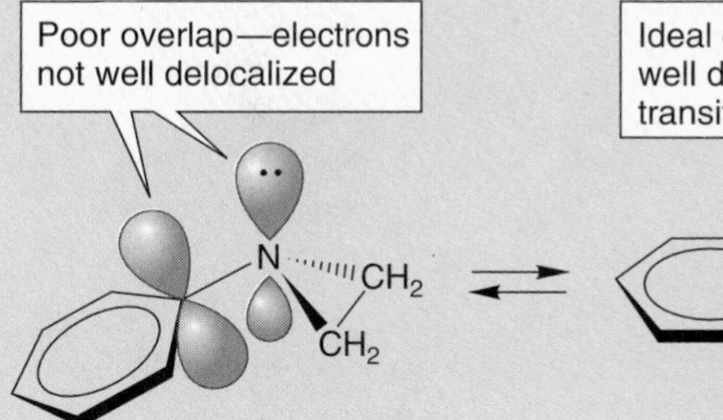
Poor overlap—electrons not well delocalized
N
CH2
CH2

Ideal overlap—electrons well delocalized—transition state stabilized

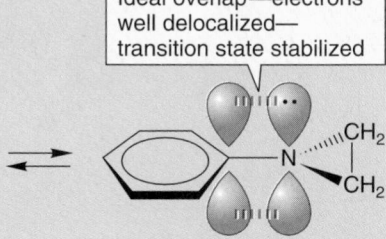

Like alcohols, amines are much higher boiling than hydrocarbons of similar molecular weight, although the effect is not as great as with alcohols (Table 21.1). The reasons for the increased boiling points are the same as for alcohols. Amines are polar compounds and aggregate in solution. Like alcohols, small amines are miscible with water. Boiling requires overcoming the intermolecular attractive forces between the dipoles. Moreover, amines form hydrogen-bonded oligomers in solution, effectively increasing the molecular weight and further increasing the boiling point (Fig. 21.21).

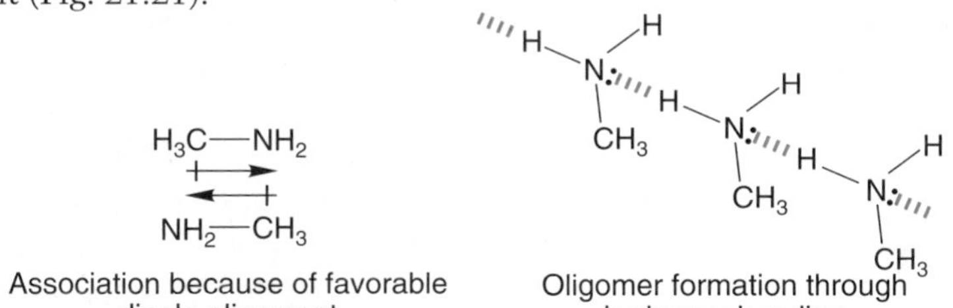

Association because of favorable dipole alignment

Oligomer formation through hydrogen bonding

FIGURE **21.21** The effective molecular weight of amines is increased through association and hydrogen bonding.

When hydrogen bonding is impossible, as with tertiary amines, boiling points decrease (Table 21.1).

TABLE **21.1** Some Physical Properties of Amines and a Few Related Alkanes

Compound	bp (°C)	mp (°C)	Density (g/mL)
NH_3 Ammonia	−33.4	− 77.7	0.77
CH_3NH_2 Methylamine	− 6.3	− 93.5	0.70
CH_3CH_3 Ethane	−88.6	−183.3	0.57
$CH_3CH_2NH_2$ Ethylamine	16.6	− 81	0.68
$CH_3CH_2CH_3$ Propane	−43.1	−189.7	0.59
$(CH_3)_2CHNH_2$ Isopropylamine	32.4	−101	0.69
$CH_3CH_2CH_2CH_2NH_2$ Butylamine	77.8	− 49.1	0.74
$CH_3CH_2CH_2CH_3$ Butane	− 0.5	−138.4	0.58
$(CH_3)_3CNH_2$ *tert*-Butylamine	44.4	− 67.5	0.70
$(CH_3)_2NH$ Dimethylamine	7.4	− 93	0.68
$(CH_3)_3N$ Trimethylamine	2.9	−117.2	0.64
C_6H_5—NH_2 Aniline	184	− 6.3	1.02

It is impossible to discuss the properties of amines without mentioning smell! The smell of amines ranges from the merely fishy to the truly vile.* Indeed, the odor of decomposing fish owes its characteristic unpleasant nature to amines.

*Diamines are even worse. Some indication of how bad smelling they often are can be gained from their names. 1,4-Diaminobutane is called putrescine and 1,5-diaminopentane is called cadaverine.

Many people use lemon when eating fish. This custom is a carryover from the days when it was difficult to preserve fish, and the lemon acted to diminish the unpleasant odor (if not the decomposition). Lemons contain 5–8% citric acid and this contributes to their sour taste. Explain why lemon juice is effective at reducing the odor of fish.

PROBLEM **21.6**

21.3 SPECTROSCOPIC PROPERTIES OF AMINES

21.3a Infrared Spectroscopy

The N—H stretching frequency appears in the same region as that of the oxygen–hydrogen bond, 3100–3500 cm^{-1}. As with the O—H stretch, the peak for primary amines is not sharp but either split into two bands or an envelope of bands arising from a variety of hydrogen-bonded molecules. Secondary amines show single peaks in the same region, presumably because hydrogen bonding is more difficult for steric reasons. The C—N stretch appears in the same region as the C—O stretch, at 1050–1250 cm^{-1}. Figure 21.22 shows the IR spectrum of a typical amine, propylamine.

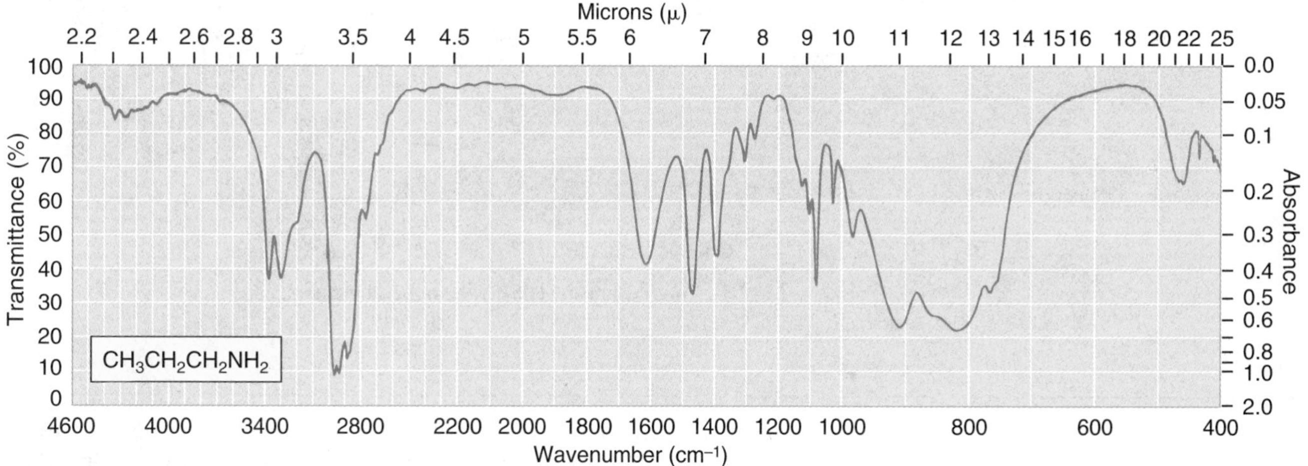

FIGURE **21.22** The IR spectrum of propylamine.

21.3b Nuclear Magnetic Resonance Spectroscopy

Carbons adjacent to the electronegative nitrogen atom are deshielded and appear at relatively low field in the ^{13}C NMR spectra. Hydrogens adjacent to the electron-withdrawing nitrogen atom are also deshielded and appear at relatively low field, δ 2.5–3.0. Like the hydroxyl hydrogens of alcohols, the resonance position of the NH hydrogens of amines is highly dependent on concentration. The carbon atoms of amines behave in a similar way. Those adjacent to the nitrogen appear at relatively low field. The further from the electronegative nitrogen atom, the more "normal" the chemical shift of the carbon.

The NH hydrogens are also subject to rapid exchange. Typically, adjacent hydrogens "feel" only an average NH spin of zero and are not split. Under conditions of no exchange, which are difficult to attain, the usual $n + 1$ splitting rules apply to NH hydrogens. Figure 21.23 shows the ^{13}C and ^{1}H NMR spectra of a typical amine, ethylamine.

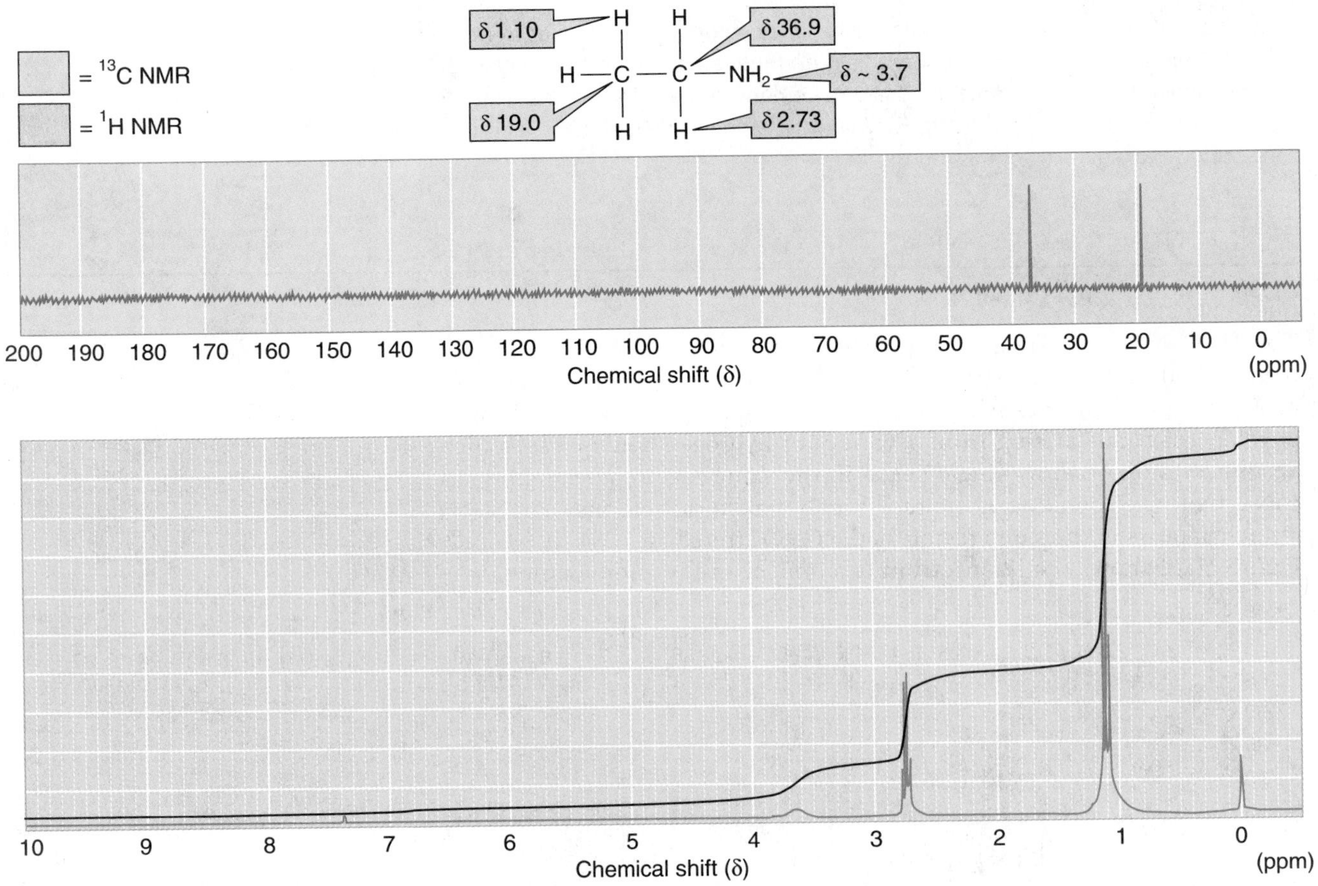

FIGURE 21.23 The ^{13}C and ^{1}H NMR spectra of ethylamine.

21.4 ACID AND BASE PROPERTIES OF AMINES

Like alcohols, amines are both Brønsted bases and nucleophiles. Primary and secondary amines are weak Brønsted acids as well. Brønsted basicity results in the formation of ammonium ions through proton transfer, whereas nucleophilicity is nicely illustrated by displacement by an amine in an S_N2 reaction (Fig. 21.24).

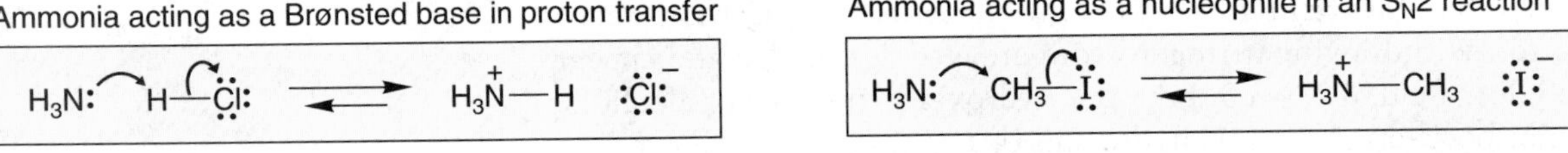
Ammonia acting as a Brønsted base in proton transfer
Ammonia acting as a nucleophile in an S$_N$2 reaction

FIGURE 21.24 Ammonia acting as a Brønsted base and as a nucleophile.

The better competitor the amine is in the proton-transfer reaction, the stronger is its Brønsted basicity. A way of turning this statement around is to focus not on the basicity of the amine, but on the Brønsted acidity of its conjugate acid, the ammonium ion. If the ammonium ion is a strong acid, the related amine must be a weak base. If it is easy to remove a proton from the ammonium ion to give the amine, the amine itself must be a poor competitor in the proton-transfer reaction, which is exactly what we mean by a weak Brønsted base. Strongly basic amines give ammonium ions from which it is difficult to remove a proton, ammonium ions with high pK_a values. So the pK_a of the ammonium ion has come to be a common measure of the basicity of the related amine. Ammonium ions with high pK_a values (weak acids) are related to strongly basic amines, and ammonium ions with low pK_a values (strong acids) are related to weakly basic amines (Fig. 21.25).

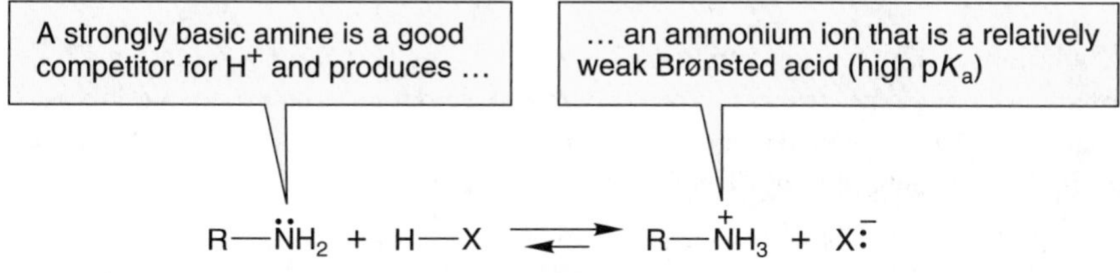
A strongly basic amine is a good competitor for H+ and produces ...
... an ammonium ion that is a relatively weak Brønsted acid (high pKa)
R—NH2 + H—X ⇌ R—NH3+ + X:−
A weakly basic amine is a less effective competitor for H+ and produces an ammonium ion that is a relatively strong Brønsted acid (low pKa).

FIGURE 21.25 Ammonium ions with high pK_a values are related to strongly basic amines, and ammonium ions with low pK_a values are related to weakly basic amines.

Table 21.2 relates the basicity of ammonia and some very simple alkylamines by tabulating the pK_a values of the related ammonium ions.

TABLE 21.2 The pK_a Values for Some Simple Ammonium Ions

Amine	Formula	Ammonium Ion	pK_a (in aqueous solution)
Ammonia	NH_3	$\overset{+}{N}H_4$	9.24
Methylamine	CH_3NH_2	$CH_3\overset{+}{N}H_3$	10.63
Dimethylamine	$(CH_3)_2NH$	$(CH_3)_2\overset{+}{N}H_2$	10.78
Trimethylamine	$(CH_3)_3N$	$(CH_3)_3\overset{+}{N}H$	9.80

The general trend of the first three entries in this table is understandable if we make the analogy between ammonium ions and carbocations. The more substituted a carbocation, the more stable it is. Similarly, the more substituted an ammonium ion, the more stable it is. The more stable the ammonium ion, the less readily it loses a proton, and the higher its pK_a. At least for the first three entries of Table 21.2, increasing substitution of the ammonium ion carries with it a decreasing ease of proton loss.

*PROBLEM 21.7

Remember why increased substitution makes carbocations more stable (Chapter 9, p. 350), then transfer this explanation to ammonium ions. Be careful. This one is a "looks easy, is hard" problem. It's much more subtle than it looks at first glance.

ANSWER Increased substitution stabilizes carbocations through the delocalization of electrons called hyperconjugation (Chapter 9, p. 354).

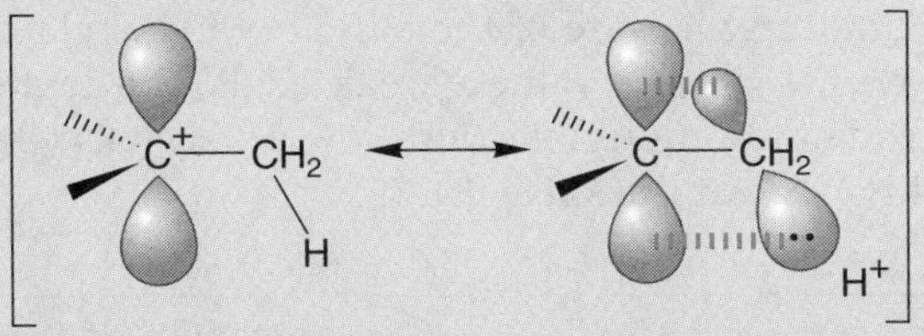

An ammonium ion, though superficially similar, is really quite different from a carbocation. There is no empty orbital on nitrogen, and there can be no overlap with an adjacent filled orbital. Resonance forms can be written, and they will be stabilizing, but there should be a smaller effect than in carbocations.

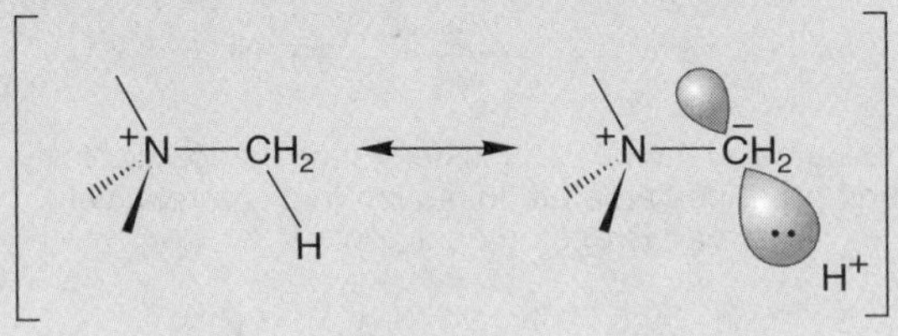

Actually, even the first three entries in Table 21.2 should make you suspicious. Why is there a much bigger change between ammonia and methylamine (1.4 pK_a units) than between methylamine and dimethylamine (only 0.15 pK_a units)? The structural change is the same in each case, the replacement of a hydrogen with a methyl group. When we get to the fourth entry, we find that things have truly gone awry; trimethylammonium ion is a stronger acid than the dimethylammonium ion and the methyl ammonium ion, and almost as strong as the ammonium ion itself, which implies that trimethylamine is a weaker base than dimethylamine or methylamine. There seems to be no continuous trend in these data.

Indeed there isn't, and it takes a look at matters in the gas phase to straighten things out. In the gas phase, the trend *is* regular. The basicity of amines increases with substitution, and the acidity of ammonium ions decreases with substitution (Fig. 21.26).

$^+NH_4 < {}^+NRH_3 < {}^+NR_2H_2 < {}^+NR_3H$

Order of increasing gas-phase pK_a

Lowest pK_a Highest pK_a

FIGURE 21.26 The gas-phase acidity of ammonium ions.

There must be some effect present in solution that disappears in the gas phase, and this effect must be responsible for the irregularities in Table 21.2. The culprit is the solvent itself. Ions in solution are strongly stabilized by solvation, by interaction of the solvent molecules with the ion. These interactions can take the form of electrostatic stabilization, or of partial covalent bond formation as in hydrogen bonding (Fig. 21.27).

Stabilization through dipole–dipole interaction with a polar solvent

Stabilization though hydrogen bonding

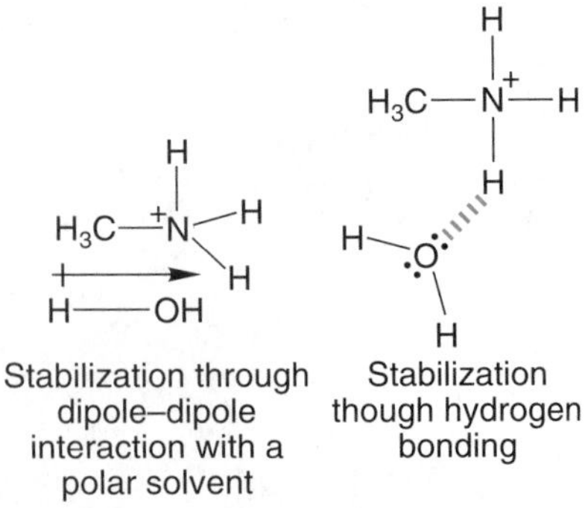

FIGURE 21.27 In solution, amines are stabilized through dipole–dipole interactions and hydrogen bonding. Large groups attached to nitrogen will interfere with these stabilizing interactions.

In solution, an alkyl group has two effects on the stability of the ammonium ion. It stabilizes by helping it to disperse the charge, but it destabilizes the ion by interfering with solvation, by making intermolecular solvation (charge dispersal) more difficult. These two effects operate in different directions! There is little steric interference with solvation when we replace one hydrogen of the ammonium ion with a methyl group, and the pK_a change is more than a full pK_a unit. This change reflects the stabilizing effect of the methyl group on the ammonium ion. However, when a second hydrogen is changed for methyl, the stabilization and interference with solvation nearly balance. The result is practically no change in the overall stability of the ammonium ion. When a third hydrogen is replaced with methyl, the destabilizing effects outweigh the stabilizing forces, and the result is a less stable, more acidic ammonium ion. In the gas phase, where there can be no solvation, only the stabilizing effects remain, and each replacement of a hydrogen with a methyl is stabilizing.

In addition, as we progress along the series, $^+NH_4$, $^+NRH_3$, $^+NR_2H_2$, $^+NR_3H$, $^+NR_4$, there are fewer hydrogens available for hydrogen bonding, and therefore less stabilization of the ammonium ion in solution.

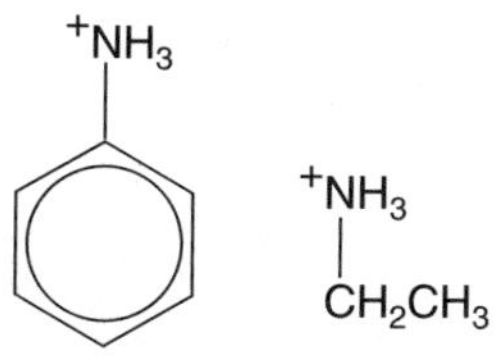

	$^+NH_3$ (on benzene ring)	$^+NH_3$–CH_2CH_3
pK_a	4.6	10.8

FIGURE 21.28

PROBLEM 21.8

Amines can be stabilized by other factors. Explain the pK_a data in Figure 21.28.

It must be admitted that this traditional discussion of base strength of amines in terms of the pK_a of the related ammonium ion is indirect. There is another, less common, but undeniably more direct method. That is to frame the discussion in terms of the pK_b (–log K_b) of the amine (Fig. 21.29).

$$\text{For}\quad RNH_2 + H_2O \overset{K}{\rightleftharpoons} R\overset{+}{N}H_3 + HO^-$$

$$K_b = \frac{[R\overset{+}{N}H_3][HO^-]}{[RNH_2]}$$

FIGURE 21.29 The expression for pK_b.

The higher the pK_b, the weaker the base, just as the higher the pK_a, the weaker the acid. Typical amines have pK_b values of about 4.

Amines are much stronger bases than alcohols. The pK_a value for the ammonium ion is much higher than that for the oxonium ion. It is much harder to deprotonate $^+NH_4$ than $^+OH_3$ (Fig. 21.30).

$^+NH_4$	$^+OH_3$
Ammonium ion	Oxonium ion
pK_a = 9.2	pK_a = −1.7
$^+N(CH_3)_4$ Cl^-	$^+O(CH_3)_3$ BF_4^-
Tetramethylammonium chloride, a stable solid, easily isolable	Trimethyloxonium fluoroborate, not easily isolated, and very reactive

FIGURE 21.30 Amines are much stronger bases than alcohols, and the corresponding ammonium ions are much weaker acids than oxonium ions.

Why is the oxonium ion much more acidic than the ammonium ion? Oxygen is a more electronegative atom than nitrogen and bears the positive charge less well. We can see the results of the increased stability of the ammonium ion in practical terms. Stable ammonium salts are common, but salts of oxonium ions, though known, are rare and usually unstable (Fig. 21.30).

PROBLEM 21.9 The stability of oxonium ions depends on the nature of the negatively charged counterion. Fluoroborate ($^-BF_4$) is an especially favorable counterion. Why?

Primary and secondary amines are Brønsted acids, but only very weak ones. Amines are much weaker acids than alcohols. Removal of a proton from an alcohol gives an alkoxide in which the negative charge is borne by oxygen. An amine forms an **amide*** in which the less electronegative nitrogen carries the minus charge (Fig. 21.31).

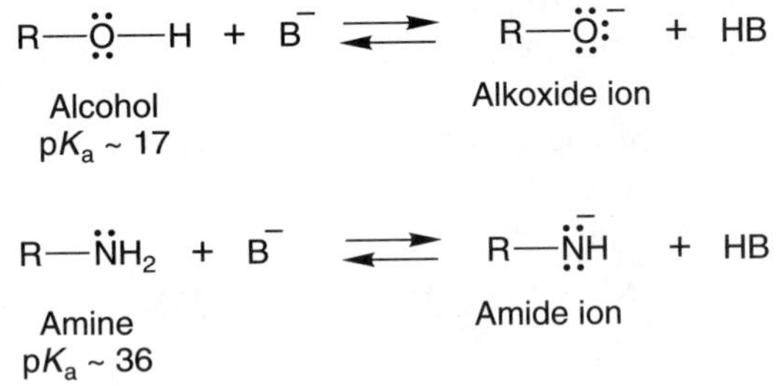

$$R{-}\ddot{\underset{\cdot\cdot}{O}}{-}H + B^- \rightleftharpoons R{-}\ddot{\underset{\cdot\cdot}{O}}{:}^- + HB$$

Alcohol $pK_a \sim 17$ — Alkoxide ion

$$R{-}\ddot{N}H_2 + B^- \rightleftharpoons R{-}\underset{\cdot\cdot}{\overset{-}{N}}H + HB$$

Amine $pK_a \sim 36$ — Amide ion

FIGURE 21.31 Primary and secondary amines are weak Brønsted acids.

It takes a very strong base indeed to remove a proton from an amine to give the amide ion, which is itself a very strong base. Alkyllithium reagents are typically used (Fig. 21.32).

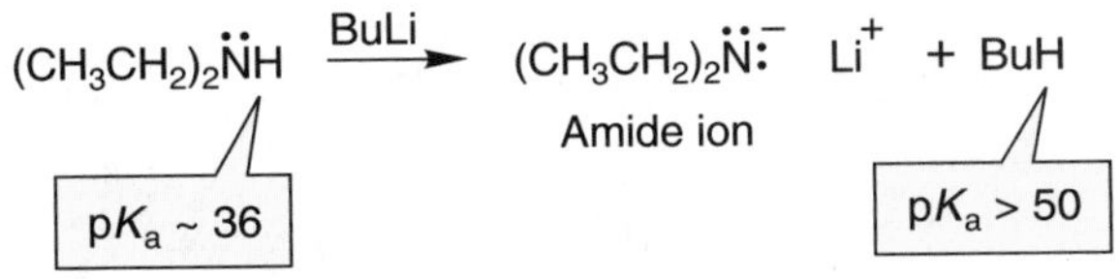

$$(CH_3CH_2)_2\ddot{N}H \xrightarrow{BuLi} (CH_3CH_2)_2\ddot{N}{:}^-\ Li^+ + BuH$$

$pK_a \sim 36$ — Amide ion — $pK_a > 50$

FIGURE 21.32 Very strong bases such as alkyllithium reagents can remove a proton from an amine to give an amide ion. Bu = $CH_2CH_2CH_2CH_3$

So, amines are good bases, but relatively poor acids. Amides, the conjugate bases of amines, are even stronger bases than the amines themselves (Fig. 21.33).

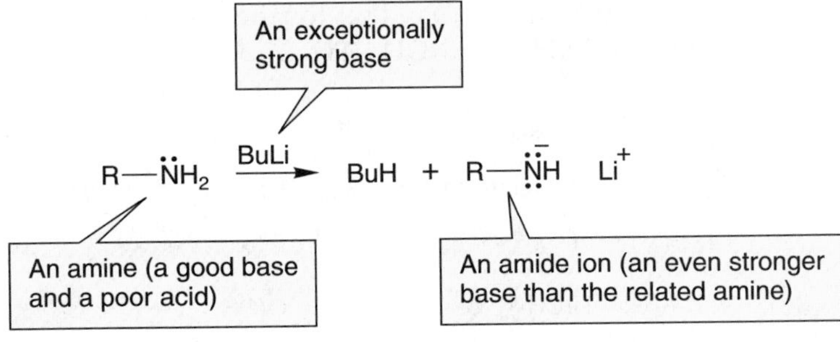

FIGURE 21.33 The amide ion is a far stronger base (and nucleophile) than the parent amine.

21.5 REACTIONS OF AMINES

21.5a Alkylation of Amines

Amines are good nucleophiles and can displace leaving groups in the S_N2 reaction. Of course, the limitations of the S_N2 reaction also apply—there can be no displacements at tertiary carbons. Successive displacement re-

*Remember to keep straight the distinction between the amide ion (RNH^-) and the acyl compound ($R{-}CO{-}NH_2$), which is also called an amide.

actions yield primary, secondary, and tertiary amines. The initial products are alkylammonium ions that are deprotonated by the amine present. These S_N2 reactions produce amines that are also nucleophilic. Further displacement reactions are possible, which can lead to problems. Indeed, the product amine is often a better participant in the S_N2 reaction than the starting material. Therefore it can be difficult to stop the reaction at the desired point, even if an excess of the starting amine is used to minimize overalkylation (Fig. 21.34).

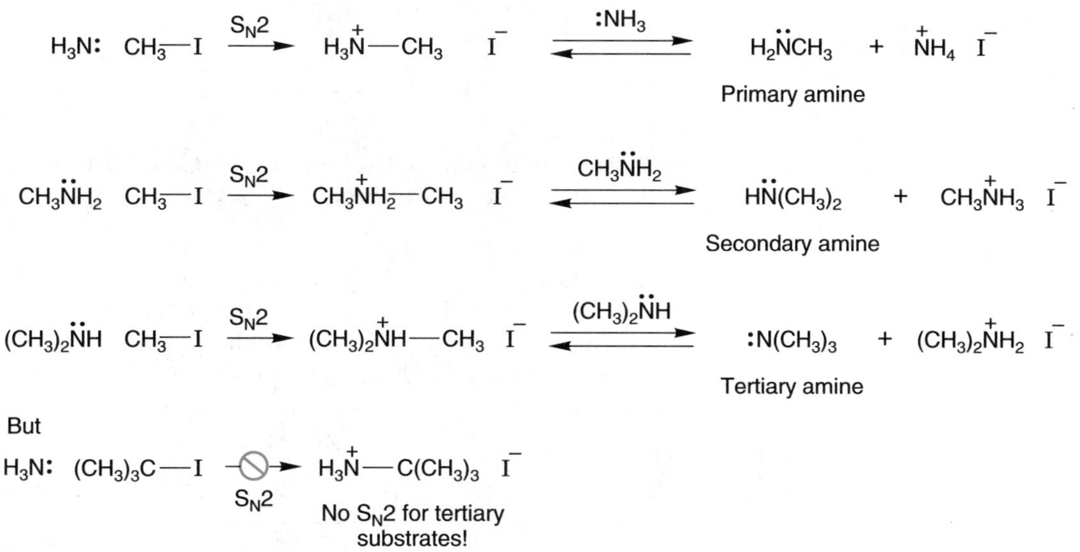

$$H_3N\!: \; CH_3{-}I \xrightarrow{S_N2} H_3\overset{+}{N}{-}CH_3 \;\; I^- \xrightleftharpoons{:NH_3} H_2\ddot{N}CH_3 \;+\; \overset{+}{N}H_4 \; I^-$$

Primary amine

$$CH_3\ddot{N}H_2 \; CH_3{-}I \xrightarrow{S_N2} CH_3\overset{+}{N}H_2{-}CH_3 \;\; I^- \xrightleftharpoons{CH_3\ddot{N}H_2} H\ddot{N}(CH_3)_2 \;+\; CH_3\overset{+}{N}H_3 \; I^-$$

Secondary amine

$$(CH_3)_2\ddot{N}H \; CH_3{-}I \xrightarrow{S_N2} (CH_3)_2\overset{+}{N}H{-}CH_3 \;\; I^- \xrightleftharpoons{(CH_3)_2\ddot{N}H} :N(CH_3)_3 \;+\; (CH_3)_2\overset{+}{N}H_2 \; I^-$$

Tertiary amine

But

$$H_3N\!: \; (CH_3)_3C{-}I \;\underset{S_N2}{\not\longrightarrow}\; H_3\overset{+}{N}{-}C(CH_3)_3 \;\; I^-$$

No S_N2 for tertiary substrates!

FIGURE 21.34 Alkylation reactions of amines.

PROBLEM 21.10

What product do you expect from an attempt to make *tert*-butylamine from the reaction of the amide ion ($^-NH_2$) with *tert*-butyl bromide?

Tertiary amines are also strong nucleophiles, and a final displacement reaction can occur to give a stable quaternary ammonium ion (Fig. 21.35).

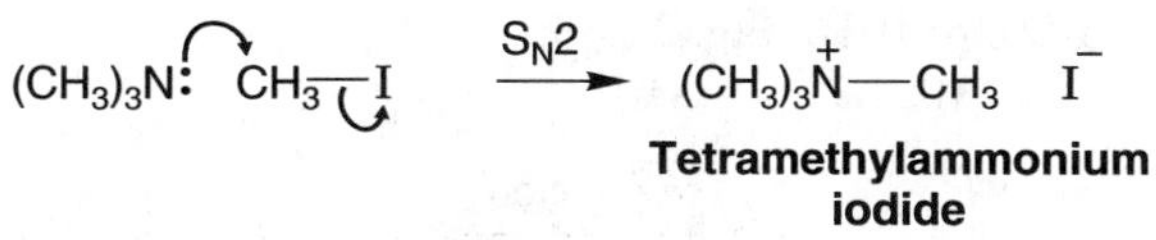

$$(CH_3)_3N\!: \; CH_3{-}I \xrightarrow{S_N2} (CH_3)_3\overset{+}{N}{-}CH_3 \;\; I^-$$

Tetramethylammonium iodide

FIGURE 21.35 The formation of an ammonium ion through alkylation of a tertiary amine.

In practice, the only alkylation procedure without these problems of overalkylation is the full conversion of an amine into the ammonium ion through exhaustive alkylation. As we will soon see in Section 21.6, there are other less direct, but also less problematical ways of making amines.

21.5b Additions of Amines to Carbonyl Groups

Amines are nucleophilic and, like almost all Lewis bases, will add to carbonyl groups. Primary amines react with simple aldehydes and ketones to give imines, and secondary amines yield enamines, as long as a hydrogen at an appropriate position in the carbonyl compound is available for elimination (Chapter 16, p. 780; Fig. 21.36).

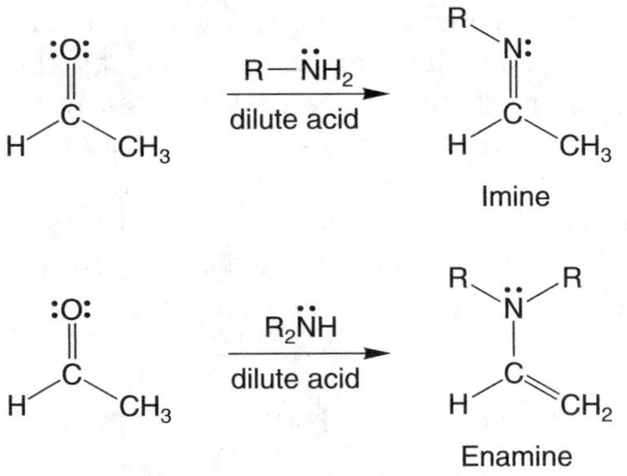

FIGURE 21.36 Imine and enamine formation from reactions of amines with carbonyl compounds.

PROBLEM 21.11 Write mechanisms for the reactions of Figure 21.36. *Hint*: The initial addition reactions are probably not acid catalyzed.

Tertiary amines can undergo the addition process, but not the following reactions that lead to imines or enamines. Equilibrium generally favors the starting materials. Of course ammonium ions, which are not nucleophiles, cannot undergo addition (Fig. 21.37).

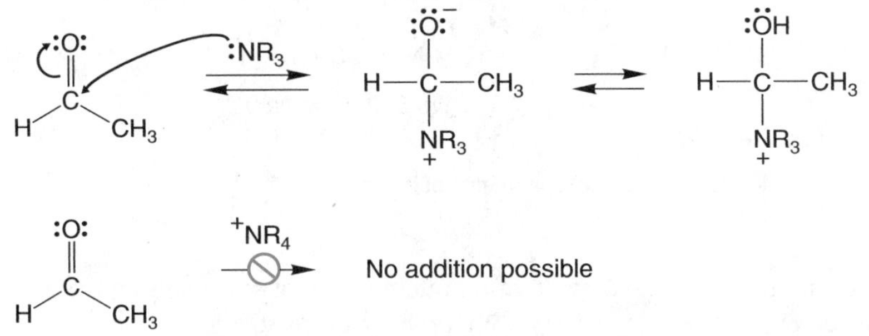

FIGURE 21.37 Tertiary amines can add to carbonyl compounds but no further reaction is possible. Ammonium ions are not nucleophiles and no addition reactions are possible.

If the carbonyl compound contains a leaving group, the initially formed alkoxide can regenerate the carbon–oxygen double bond. The formation of amides from the reaction of amines or amide ions with acid chlorides or esters are simple examples of the addition–elimination process (Chapter 20, p. 1016; Fig. 21.38).

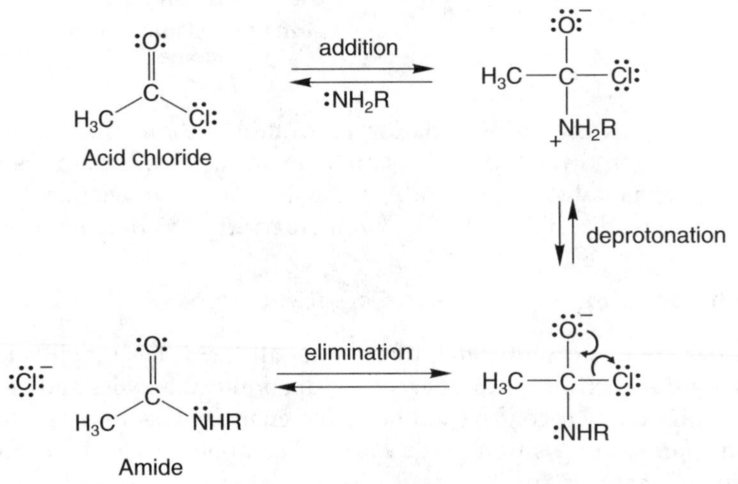

FIGURE 21.38 Addition–elimination reactions of amines and acyl compounds.

21.5c Elimination Reactions of Amines

Although the amide ion is a strong base and thus a poor leaving group, amines are less strong bases and are much better at departing. This situation is analogous to the difference between hydroxide, a notoriously poor leaving group, and water, a good one. The technique of converting hydroxide into water in order to facilitate an elimination reaction is familiar from our discussion of alcohol chemistry (Chapter 17, p. 828). In an analogous process, the alkylation reaction just discussed is first used to convert an amine into an ammonium ion. The ensuing E2 reaction to yield an alkene and a trialkylamine is called the Hofmann elimination (Chapter 7, p. 284) and results in the formation of the less substituted isomer when more than one product is possible (Fig. 21.39).

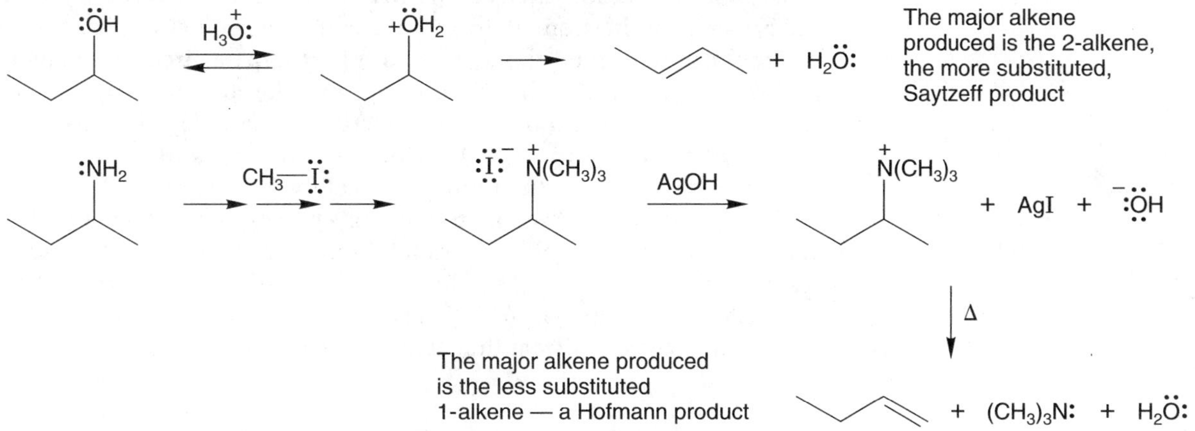

The major alkene produced is the 2-alkene, the more substituted, Saytzeff product

The major alkene produced is the less substituted 1-alkene — a Hofmann product

FIGURE 21.39 Hofmann elimination reactions of amines compared to dehydration reactions.

Provide mechanisms for the reactions of Figure 21.40. PROBLEM 21.12

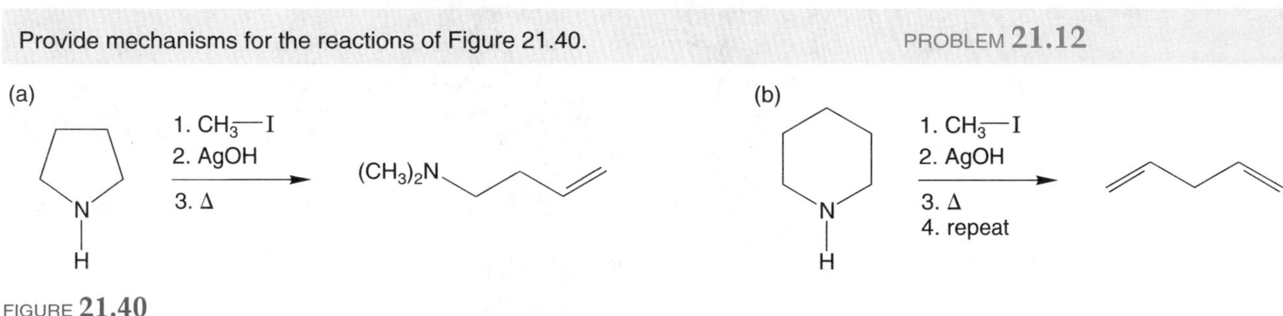

FIGURE 21.40

21.5d Reactions of Aromatic Amines

Some reactions of aromatic amines parallel those of the aliphatic compounds. Alkylation reactions are similar, for example (Fig. 21.41),

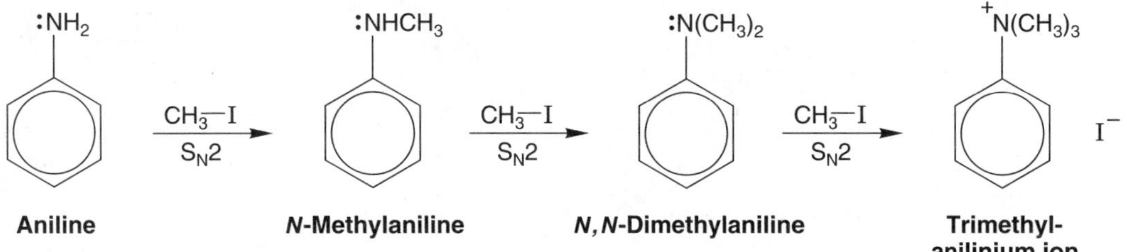

FIGURE 21.41 Alkylation reactions of aniline, a typical aromatic amine.

For other reactions of aliphatic amines there are no counterparts in aromatic chemistry. There is no easy way to do a Hofmann elimination reaction, for example, as benzyne, a most unstable intermediate, would result (Chapter 14, p. 674; Fig. 21.42).

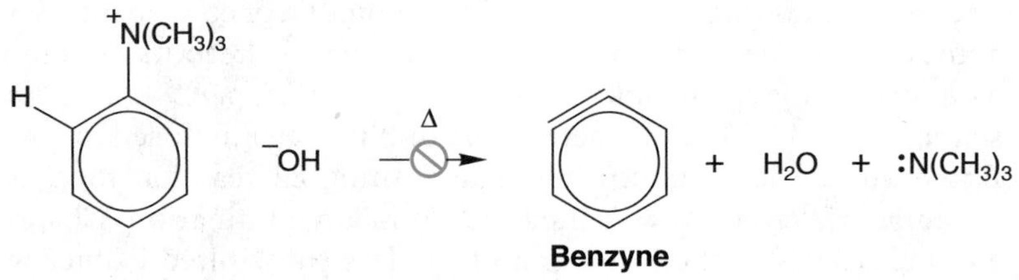

FIGURE 21.42 Hofmann elimination of aromatic amines would lead to dehydrobenzenes, or "benzynes."

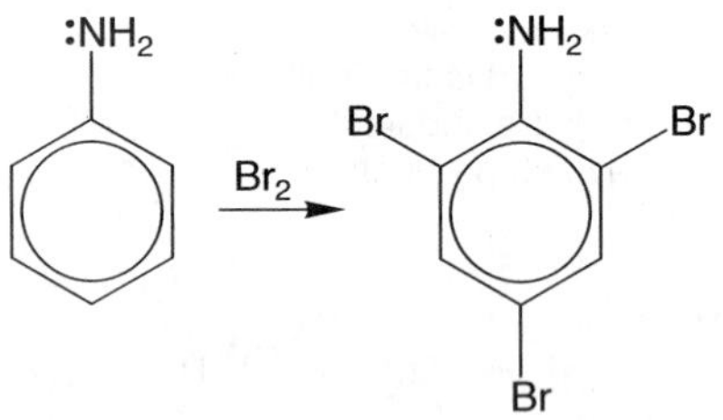

Still other reactions are peculiar to aromatic amines, as they depend on the presence of the aromatic ring. An example is aromatic substitution in which the amine group acts as a powerful ortho/para director (Chapter 14, p. 652). As we saw earlier in Chapter 14, the amino group is such a strongly activating group that overreaction is possible. The classical example is bromination in which aniline is converted into 2,4,6-tribromoaniline on treatment with bromine. No catalyst is necessary (Fig. 21.43).

FIGURE 21.43 Bromination of aniline leads to 2,4,6-tribromoaniline.

The solution to the problem of overalkylation of the ring is to deactivate aniline by acylating the amino group. *N*-Acetylaniline (acetanilide) reacts with nitric acid to give almost exclusively para and ortho substitution (Chapter 14, pp. 667–668). The acetyl group can then be removed by hydrolysis to give *p*-nitroaniline and *o*-nitroaniline (Fig. 21.44).

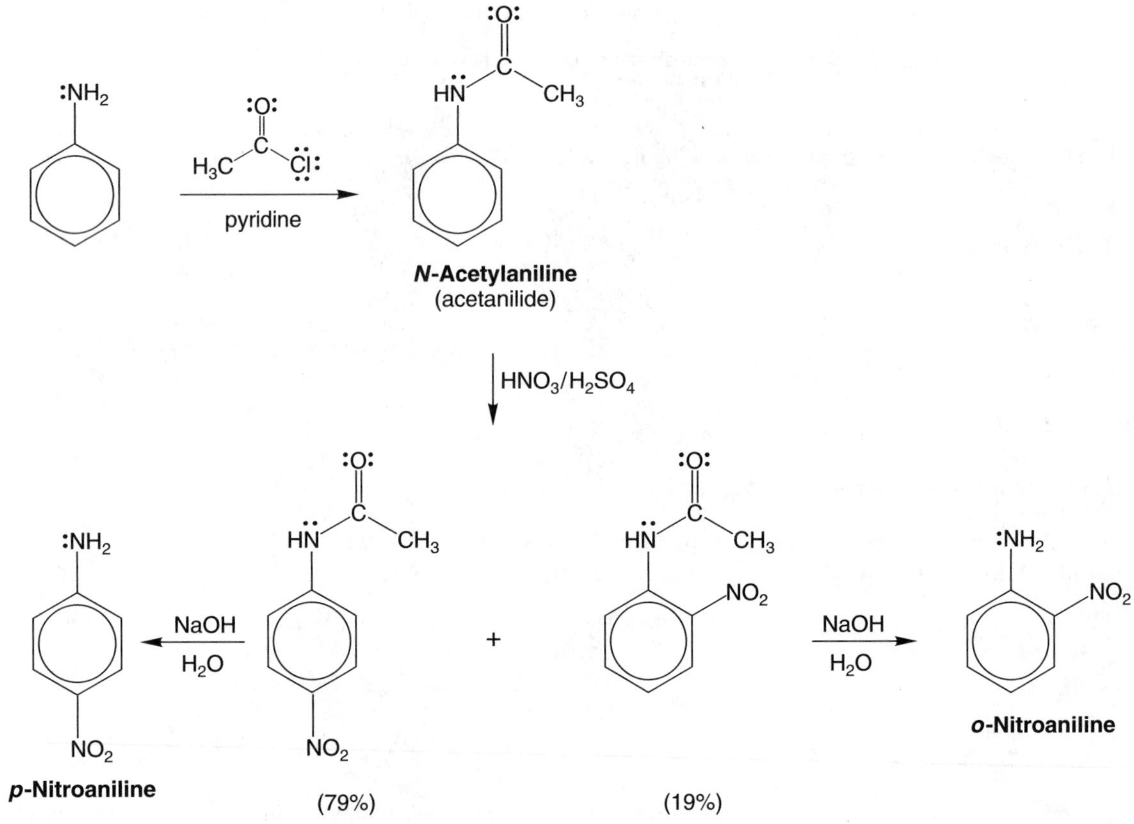

FIGURE 21.44 Acylation deactivates the aromatic compound and allows monosubstitution.

Although the amino group is ortho/para directing, there can be complications in strong acid. Under such conditions the amino group is protonated to give an ammonium ion, a meta director. If there is sufficient acid to induce substantial conversion into the ammonium ion, meta substitution begins to occur. It is important to remember that protonated aniline will undergo further substitution only very slowly. Therefore, the acid strength must be sufficient to convert aniline substantially into its protonated form. Even a small amount of the free amine can lead to dominant formation of the ortho/para products. In practice, the result is often a mixture of ortho, para, and meta substitution (Fig. 21.45).

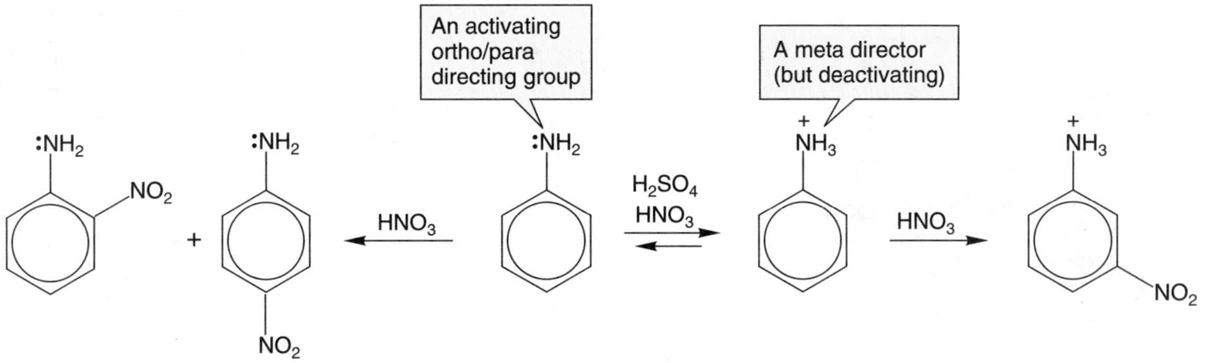

An activating ortho/para directing group

A meta director (but deactivating)

:NH_2 (ortho NO_2) + :NH_2 (para NO_2) ← HNO_3 — :NH_2 (aniline) ⇌ (H_2SO_4, HNO_3) $\overset{+}{N}H_3$ → (HNO_3) $\overset{+}{N}H_3$ (meta NO_2)

FIGURE 21.45 The acid-catalyzed nitration of aniline is complicated. The free amine leads to ortho/para substitution, but the amine group will also be protonated to give an ammonium ion, a meta director.

PROBLEM 21.13 Explain carefully why protonated aniline is a meta director and why its further substitution is slow relative to reactions of benzene (Chapter 14, p. 670).

The solution to the problem lies again in acylation of the amino group prior to reaction. Protonation is now slowed and aromatic substitution occurs primarily in the para position (Fig. 21.46).

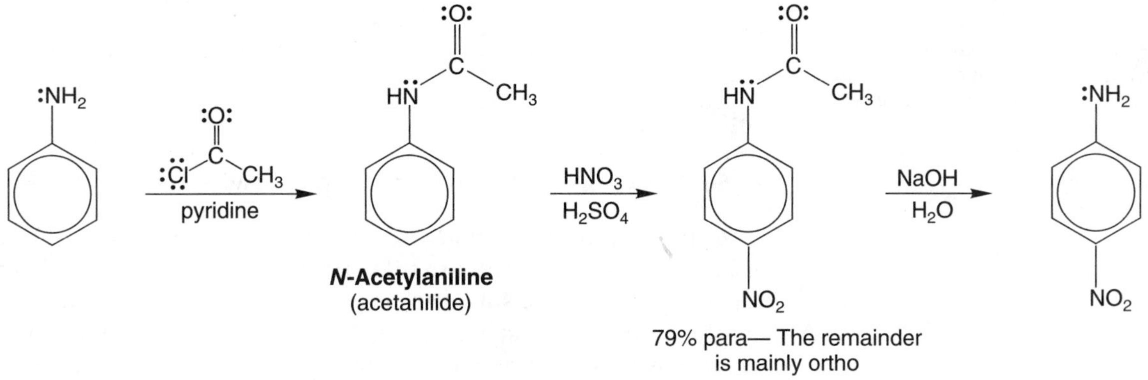

:NH_2 → ($ClC(=O)CH_3$, pyridine) **N-Acetylaniline** (acetanilide) → (HNO_3, H_2SO_4) p-nitroacetanilide → (NaOH, H_2O) p-nitroaniline

79% para— The remainder is mainly ortho

FIGURE 21.46 Acylation reduces the basicity of the amine nitrogen and allows para substitution to dominate.

Reaction of amino aromatics with nitrous acid leads to the formation of diazonium ions. These species are central to the Sandmeyer and related reactions (Chapter 14, p. 646) that can be used to generate all manner of substituted aromatic compounds, such as the compounds shown in Figure 21.47.

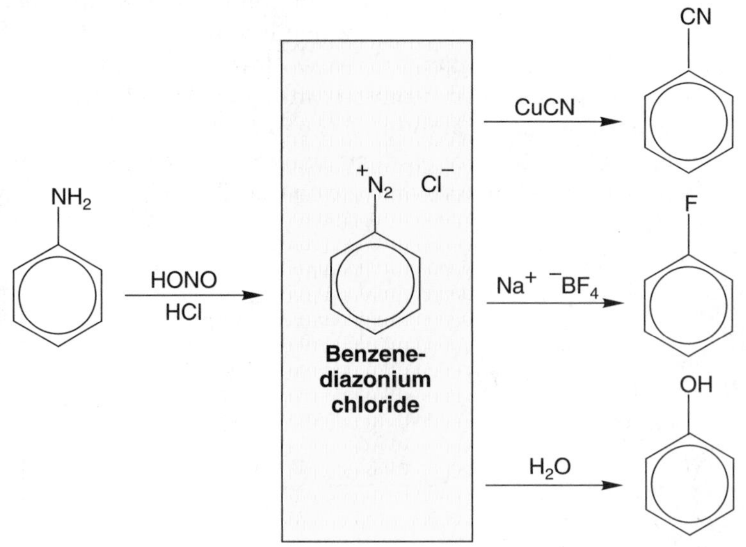

FIGURE 21.47 The formation of an aromatic diazonium ion. Further reactions lead to a variety of substituted benzenes.

The diazonium ion is itself a Lewis acid, although a quite weak one. However, diazonium ions can react with especially activated aromatic compounds to substitute the ring and form **azo compounds** (Fig. 21.48).

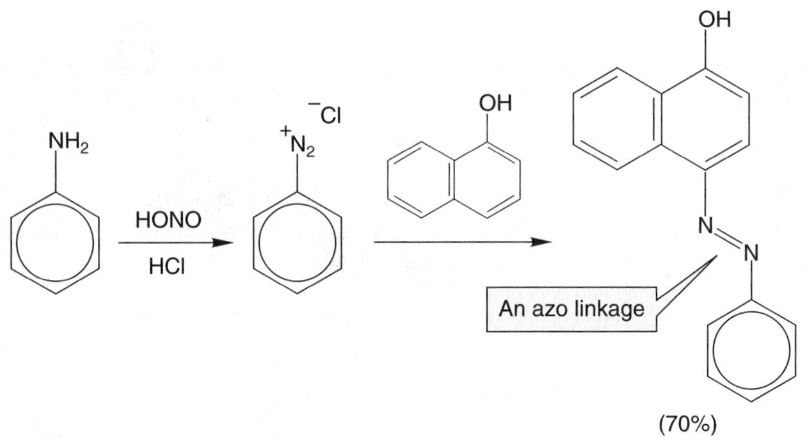

FIGURE 21.48 The diazonium ion can function as a Lewis acid and react with highly activated aromatic compounds.

21.5e Reactions of Nonaromatic Amines with Nitrous Acid

Like aromatic amines, other amines react with nitrous acid. The reaction pattern generally is similar to that of aniline and nitrous acid, but the outcome depends on the substitution pattern of the amine. Like aniline, primary amines form diazonium ions. Although the mechanism is parallel, the stabilities of the products certainly are not. The resonance stabilization of the aryldiazonium ion makes it an accessible, relatively stable species. Simple primary aliphatic diazonium ions are most unstable and generally not isolable, but are used as reactive species as they are generated (Fig. 21.49).

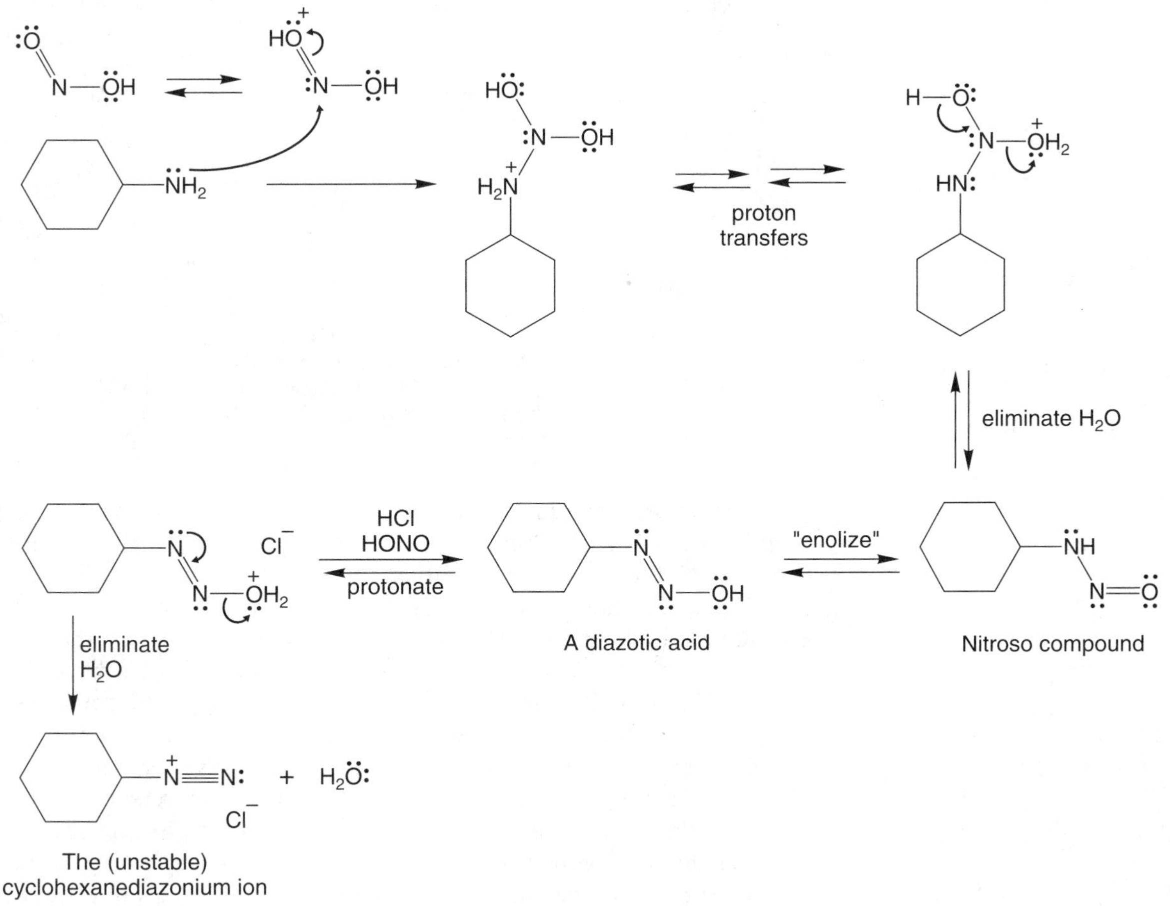

FIGURE **21.49** Primary aliphatic amines react with nitrous acid to give unstable diazonium ions.

PROBLEM **21.14**

Figure 21.49 shows protonated nitrous acid as the active ingredient in the nitrosation reaction. Another mechanism focuses upon dinitrogen trioxide, N_2O_3. Provide a structure for N_2O_3 and a mechanism for its formation. Then show how N_2O_3 could function as a nitrosating agent.

The first step in the reaction is formation of the nitroso compound. Like carbon–oxygen double bonds, which are in equilibrium with their enol forms, the nitroso compound equilibrates with its enol form, a **diazotic acid**. Protonation converts the OH of the diazotic acid into the good leaving group, OH_2, and the diazonium ion can now be formed by elimination of water.

Secondary amines follow a different course. The first step remains formation of the **nitroso compound**, but now no further reaction to give the enol is possible, as no hydrogen remains on the original amine nitrogen atom. The reaction simply stops at this point and the nitrosoamine can be isolated (Fig. 21.50).

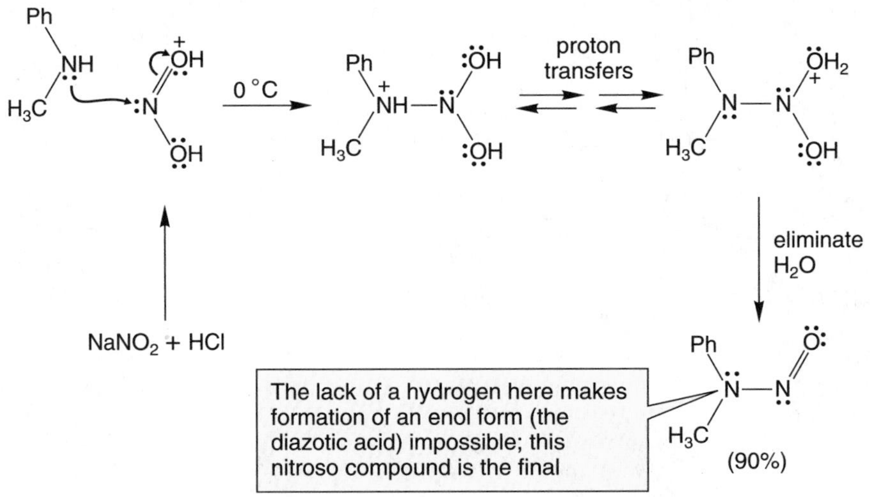

0 °C
proton transfers
eliminate H_2O
$NaNO_2$ + HCl
The lack of a hydrogen here makes formation of an enol form (the diazotic acid) impossible; this nitroso compound is the final
(90%)

FIGURE 21.50 Secondary aliphatic amines react with nitrous acid to give isolable nitroso compounds.

Nitrosoamines are important because many are known to be potent carcinogens. This finding has led to a controversy over the use of sodium nitrite ($NaNO_2$) as a preservative in foods because it might be a nitrosating agent under physiological conditions, or under the conditions in which food is prepared. The classic example is the frying of bacon, which generates nitrosoamines. However, the evidence is not definitive on this point. Many foods (spinach), even our own bodies, produce nitrites.

Aliphatic tertiary amines do not react with nitrous acid to give an isolable product. Surely the initial nitrosation reaction must take place, but there is no pathway onward, no alternative to reversal to starting materials. The initial adduct cannot be stabilized by proton loss because there is no hydrogen attached to nitrogen in a tertiary amine. What differentiates the reaction pathways of primary, secondary, and tertiary amines is the number of hydrogens originally attached to the nitrogen atom (Fig. 21.51).

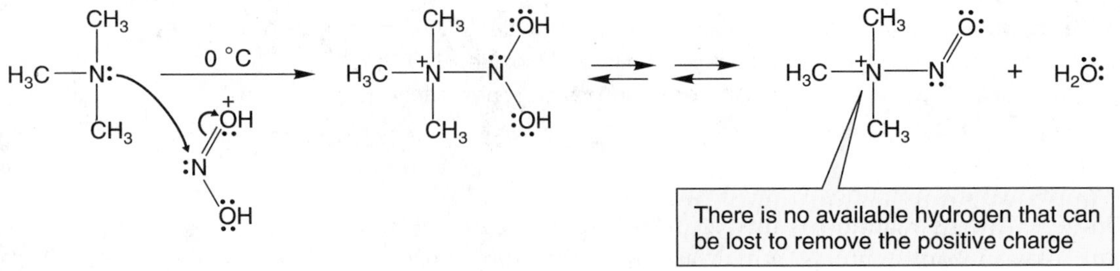

0 °C
There is no available hydrogen that can be lost to remove the positive charge

FIGURE 21.51 Tertiary amines do not give isolable products on reaction with nitrous acid.

Nitrogen is a superb leaving group. Not only is N_2 a very stable species, and therefore easily lost, but displacement of gaseous N_2 is irreversible, unlike so many of the other reactions we have seen. Nitrogen is not nucleophilic, and is not a competitor in the displacement process. So, when primary amines are treated with nitrous acid in the presence of nucleophiles, displacement occurs. Water, alcohols, and acetic acid are typical solvents, and lead to alcohols, ethers, and acetates through displacement of the diazonium ion by the nucleophilic solvent (Fig. 21.52).

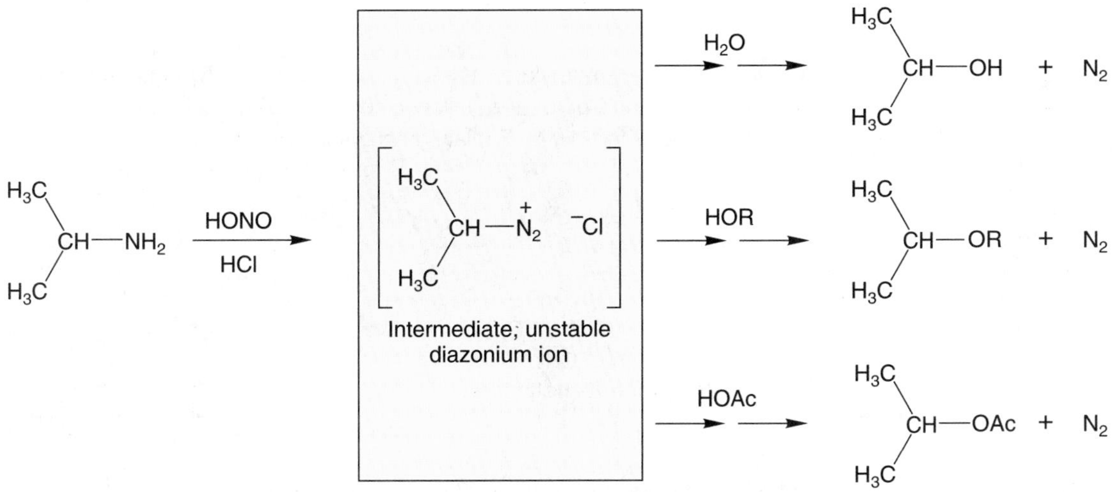

FIGURE 21.52 Nucleophilic solvents can displace nitrogen from an alkyl diazonium ion.

What happens if the nucleophilic displacement of the diazonium ion can be avoided? One useful reaction is the formation of **diazo compounds**. Deprotonation of the diazonium ion leads to a diazo compound that can be extracted into solvent ether, in which it is very soluble. Although diazo compounds are dangerous (poisonous and sometimes explosive), they are also useful sources of carbenes on heating or irradiation with UV light (Chapter 10, p. 411) and excellent esterifying agents for carboxylic acids (Chapter 19, p. 972; Fig. 21.53).

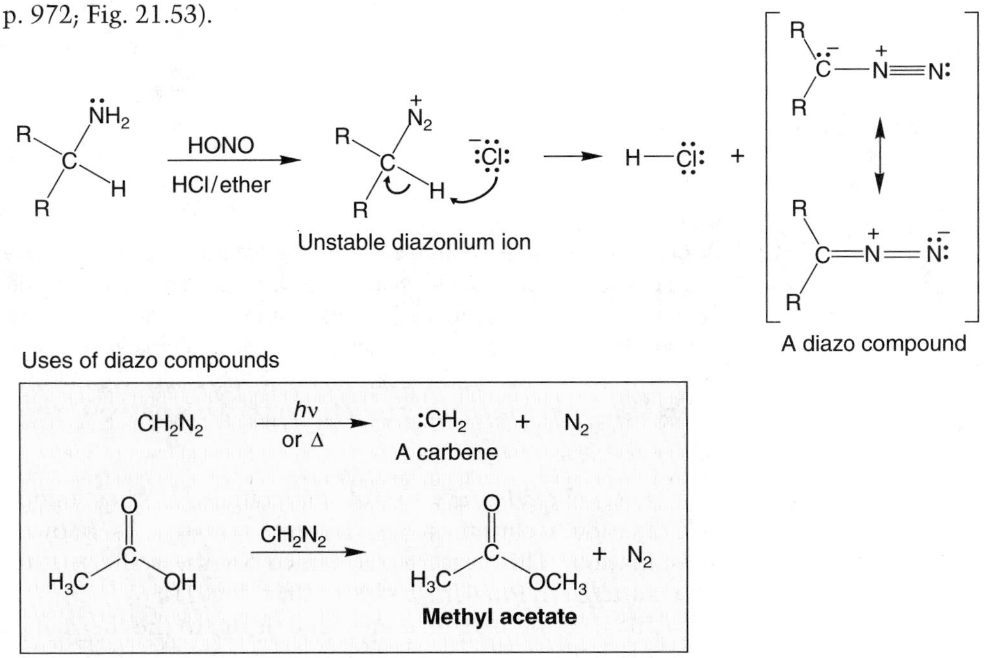

FIGURE 21.53 If nucleophilic displacement can be avoided, a proton can be removed to give a diazo compound.

PROBLEM 21.15 Write a mechanism for the formation of methyl acetate from acetic acid and diazomethane (Fig. 21.53; Chapter 19, p. 972).

PROBLEM 21.16 When *N*-nitroso-*N*-methylurea (NMU) is treated with a two-phase solvent system, H_2O/KOH and ether, the yellow color of diazomethane rapidly appears in the ether layer (Fig. 21.54). Explain mechanistically.

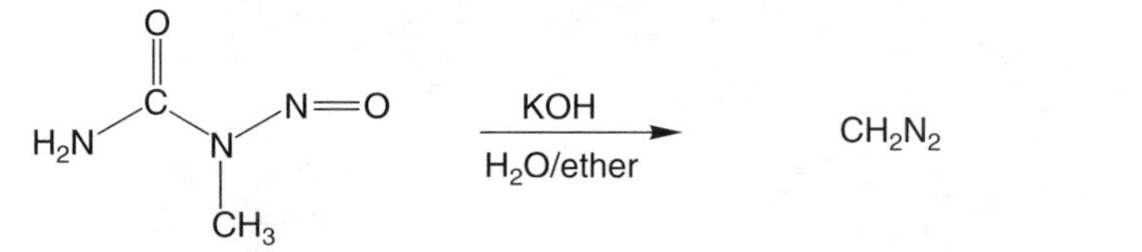

N-Nitroso-N-methylurea (NMU)

Diazomethane (as a yellow solution in ether)

FIGURE 21.54

21.5f *N*-Oxide Formation: The Cope Elimination

Another reaction in which a nitrogen atom acts as a nucleophile is *N*-oxide formation from tertiary amines. Hydroxide ion is displaced from hydrogen peroxide by the tertiary amine to give a hydroxy ammonium ion and hydroxide ion. Hydroxide ion then removes a proton to give the **amine oxide**, or ***N*-oxide** (Fig. 21.55).

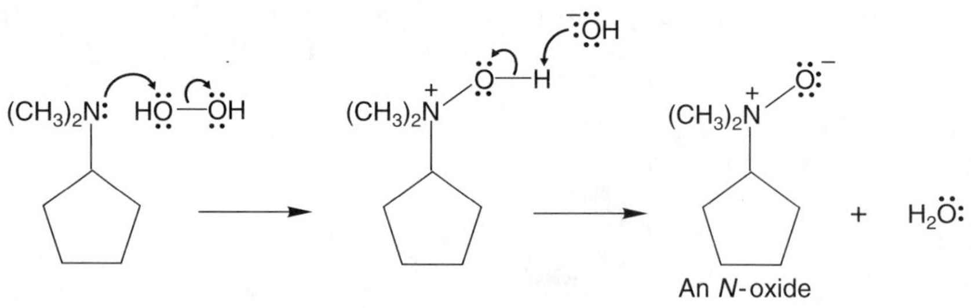

FIGURE 21.55 The mechanism of *N*-oxide formation.

PROBLEM 21.17 Here is a seeming contradiction. Many times the point has been made that hydroxide is a very poor leaving group. Displacements of hydroxide are rare indeed, and we are frequently at pains to transform the hydroxyl group into a better leaving group in order to facilitate an elimination reaction. Yet, *N*-oxide formation involves just such a displacement of hydroxide (Fig. 21.55). Explain why displacement of this normally poor leaving group is possible in this reaction.

These *N*-oxides are useful compounds, as they undergo a thermal elimination reaction to give alkenes in what is known as the **Cope elimination**. This reaction is named for the same Arthur C. Cope we encountered in Problem 20.29 (p. 1061; Fig. 21.56).

This reaction is related mechanistically to the thermal eliminations of acetates (Chapter 20, p. 1059), but the reaction conditions required are less severe. The negatively charged oxygen acts as base in removing a proton from within the same molecule. As in the ester pyrolyses, it is generally only a cis hydrogen that can be reached by the oxygen—the trans hydrogen is too far away. The Cope elimination is a syn elimination (Fig. 21.56).

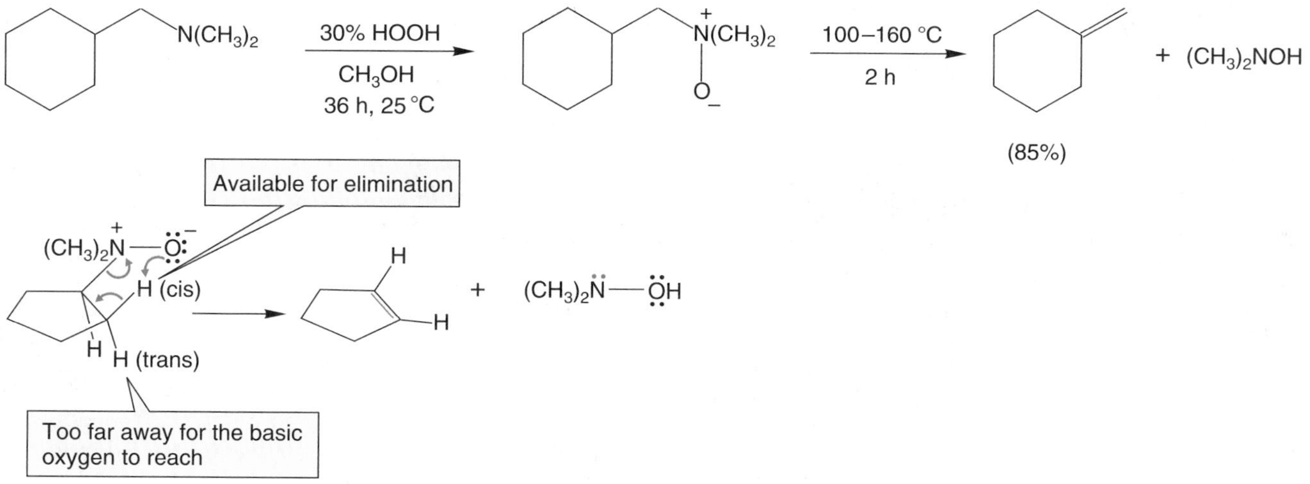

FIGURE 21.56 The Cope elimination reaction is a syn elimination.

Sketch the transition state for the Cope elimination in the second example of Figure 21.56.

PROBLEM 21.18

Design an experiment to test the stereochemistry of the Cope elimination. How would one determine that it is a cis hydrogen that is lost? Assume you can make any labeled compounds you need.

*PROBLEM 21.19

Easy! Just make a specifically deuterated ring compound and see if it is the H or D that is lost. Making such a compound would be hard, but the problem doesn't ask you to do that.

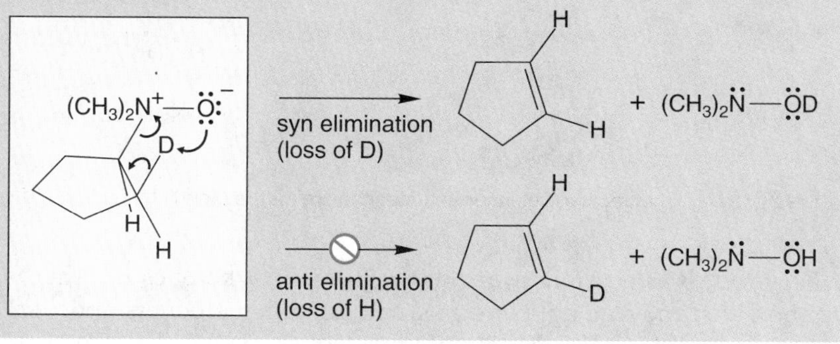

ANSWER

21.5g Amines in Condensation Reactions: The Mannich Reaction

In the acid-catalyzed version of this condensation reaction, named for Carl Mannich (1877–1947), an aldehyde or ketone is heated with an acid catalyst in the presence of formaldehyde and an amine. The initially formed ammonium ion is then treated with base to liberate the final condensation product, the free amine (Fig. 21.57).

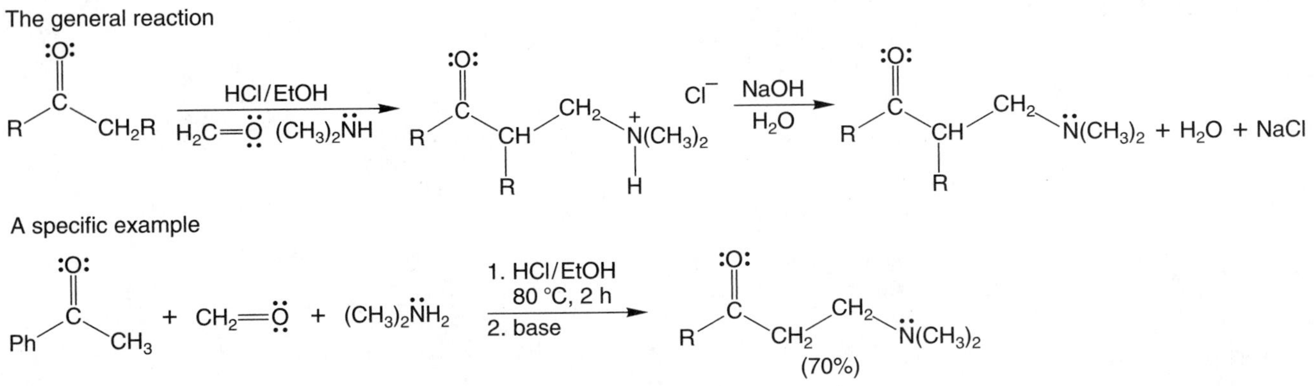

The general reaction

A specific example

FIGURE 21.57 The Mannich reaction.

This reaction presents a nice mechanistic problem because two routes to product seem possible, and it is the simpler (and incorrect) process that is the easier to find. The difficulty is that we are now experienced in working out aldol condensations and a reasonable person would probably start that way. An acid-catalyzed crossed aldol condensation (Chapter 18, p. 912) leads to a β-hydroxy ketone. Displacement of the hydroxyl group (after protonation, of course) would lead to the initial product, the ammonium ion. Treatment with base would surely liberate the free amine (Fig. 21.58).

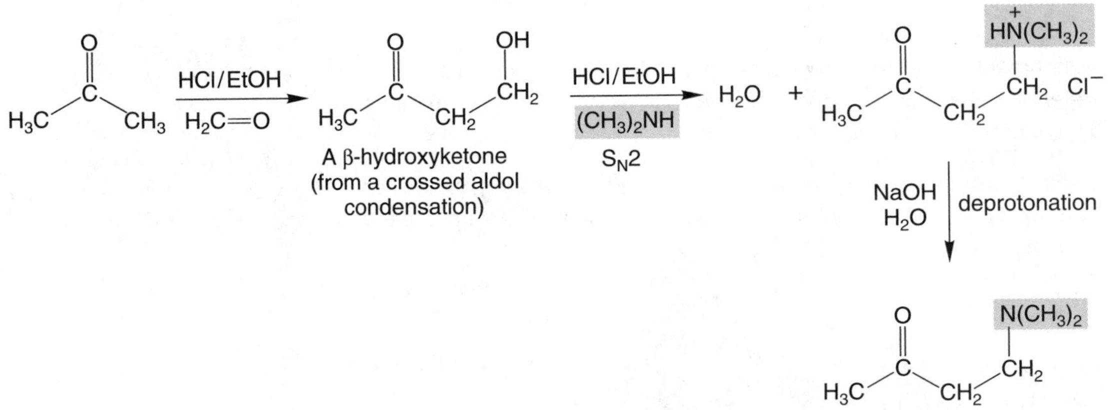

FIGURE 21.58 An attractive, but incorrect, mechanism for the Mannich reaction.

However, it is possible to make the β-hydroxy ketones in other ways and show that they are not converted into Mannich products under the reaction conditions. Sadly, this simple mechanism cannot be correct. What else can happen? Amines are nucleophiles and react rapidly with carbonyl compounds in acid to give immonium ions (Chapter 16, p. 780). It is this immonium ion that is the real participant in the Mannich condensation reaction. Formaldehyde is more reactive than the ketone and the immonium ion will be preferentially formed from this carbonyl compound. In an aldol-like process, the enol of the other carbonyl compound reacts with the immonium ion to give the penultimate product, the ammonium ion. In the second step, the free amine is formed by deprotonation in base (Fig. 21.59).

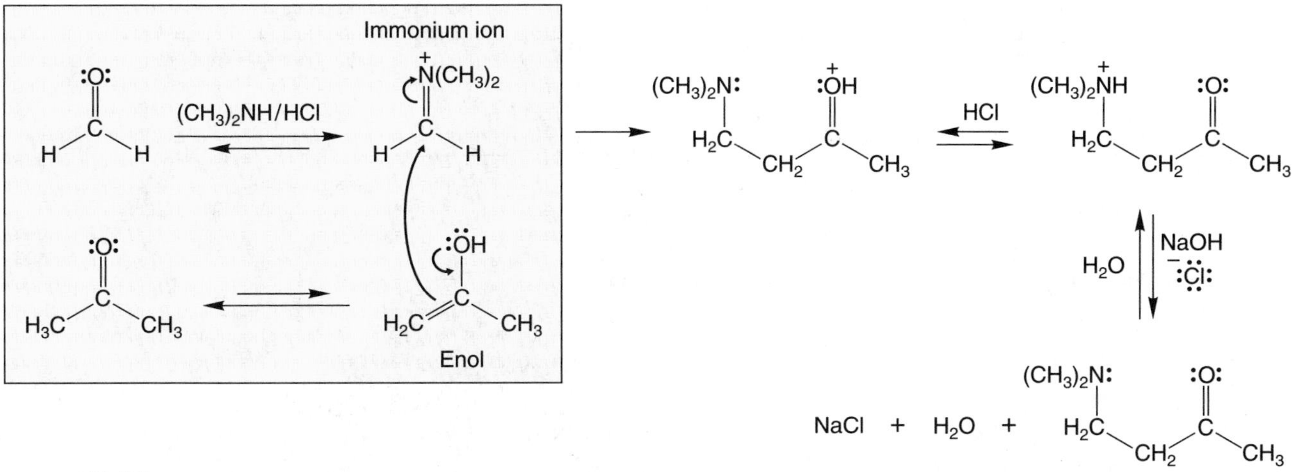

FIGURE 21.59 The correct mechanism of the Mannich reaction.

21.6 SYNTHESES OF AMINES

21.6a Alkylation of Other Amines

A conceptually easy synthesis of amines is the reaction of simpler amines with alkyl halides. Unfortunately, as we saw in Section 21.5, this reaction is not often of great practical use. The product amines are also nucleophiles, and in fact are often stronger nucleophiles than the starting materials. They, too, will undergo alkylation reactions and the result is generally a mixture of alkylated products, even if a large excess of the starting amine can be used to minimize further reaction. Success is only assured if a vast excess of ammonia is used, or the reaction is pushed to its logical conclusion, the complete alkylation to give the ammonium ion (Fig. 21.60).

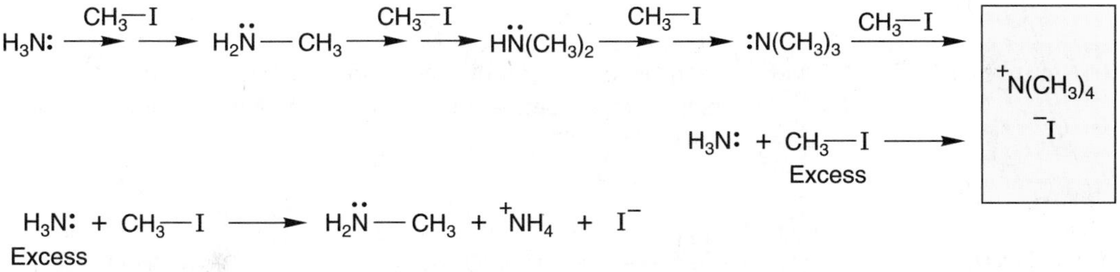

FIGURE 21.60 The synthesis of alkylamines through alkylation reactions (S_N2 reactions) is not very useful. Mono-, di-, and trisubstituted amines are usually formed.

Naturally, procedures to work around this difficulty have evolved. One such device is the **Gabriel amine synthesis**, named for Siegmund Gabriel (1851–1923). This reaction is limited to the synthesis of primary amines, but it does nicely avoid the problem of overalkylation. In this process, phthalic acid (benzene-1,2-dicarboxylic acid) is first converted into the cyclic **imide**, called **phthalimide**, which in turn is used to generate the related anion. This anion is used in an S_N2 reaction with an alkyl halide to generate a substituted phthalimide. This material can be opened in either acid or base to regenerate phthalic acid and either the amine (basic conditions) or the related ammonium ion (acidic conditions). Treatment with acid or base liberates the free amine (Fig. 21.61).

The general procedure

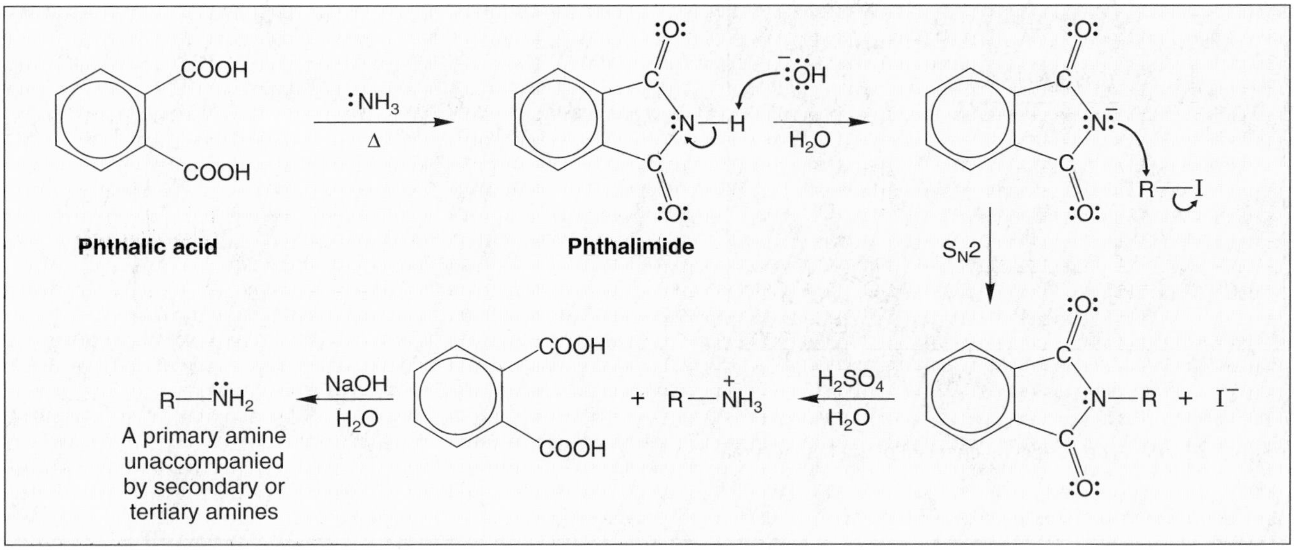

Phthalic acid

$\xrightarrow[\Delta]{:NH_3}$

Phthalimide

$\xrightarrow{H_2O}$ (⁻:ÖH)

S_N2 (R—I)

$\xrightarrow[H_2O]{H_2SO_4}$ + $R—\overset{+}{N}H_3$ + I^-

$\xrightarrow[H_2O]{NaOH}$ R—N̈H₂

A primary amine unaccompanied by secondary or tertiary amines

A specific example

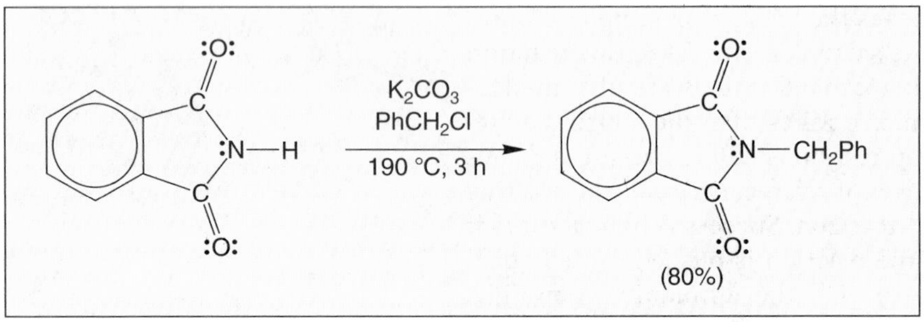

$\xrightarrow[190\ °C,\ 3\ h]{K_2CO_3,\ PhCH_2Cl}$:N—CH_2Ph (80%)

FIGURE **21.61** The Gabriel synthesis of primary amines avoids overalkylation.

PROBLEM **21.20** Other nucleophilic syntheses of amines involve the reaction of simple amines with epoxides or α,β-unsaturated carbonyl compounds. Write mechanisms for the reactions in Figure 21.62.

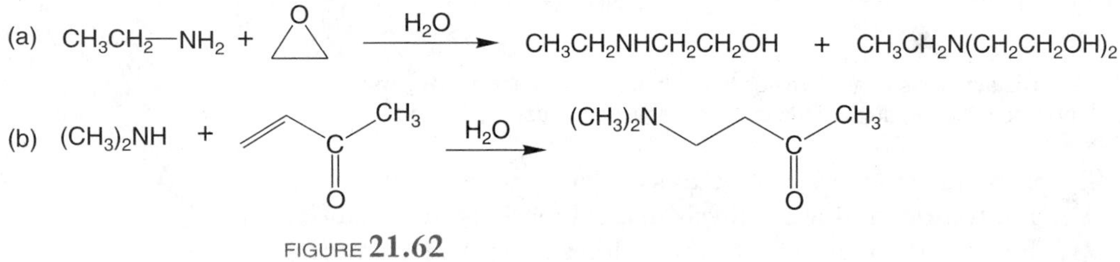

(a) $CH_3CH_2—NH_2$ + (ethylene oxide) $\xrightarrow{H_2O}$ $CH_3CH_2NHCH_2CH_2OH$ + $CH_3CH_2N(CH_2CH_2OH)_2$

(b) $(CH_3)_2NH$ + $CH_2{=}CH{-}C({=}O){-}CH_3$ $\xrightarrow{H_2O}$ $(CH_3)_2NCH_2CH_2C({=}O)CH_3$

FIGURE **21.62**

21.6b Reductive Syntheses of Amines

There is a variety of synthetic procedures designed to generate amines by indirect means in which the crucial step involves a reduction of a masked amine. In one very simple example seen first in Chapter 20 (p. 1035), nitriles (cyanides) are catalytically reduced to give amines. Notice that the common method used to synthesize nitriles, S_N2 displacement on an alkyl halide, necessarily introduces a carbon atom along with the nitrogen that will become the amine. So, this route is limited to the synthesis of primary amines. The reduction step can be carried out using either catalytic hydrogenation or treatment with lithium aluminum hydride (Fig. 21.63).

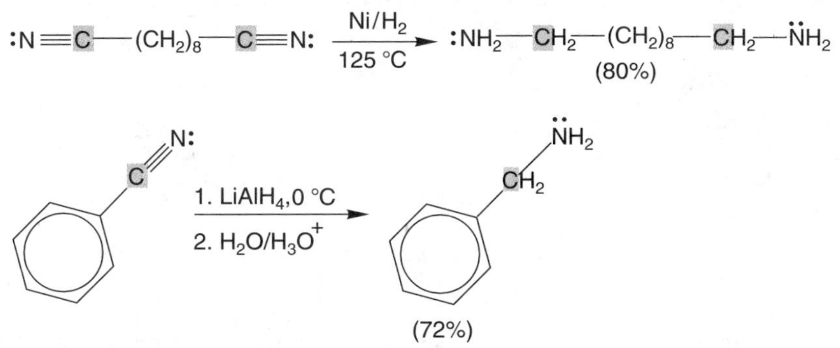

:N≡C—$(CH_2)_8$—C≡N: → (Ni/H_2, 125 °C) :NH_2—CH_2—$(CH_2)_8$—CH_2—NH_2 (80%)

Benzonitrile → (1. $LiAlH_4$, 0 °C; 2. H_2O/H_3O^+) benzylamine (72%)

FIGURE 21.63 The reduction of cyanides to primary amines.

A closely related reaction in which a carbon atom is *not* introduced along with the nitrogen is the reduction of alkyl azides. The azides are generally made through S_N2 displacement of halides by the strongly nucleophilic azide ion, N_3^-. Once again, the reduction step can use either catalytic hydrogenation or lithium aluminum hydride (Fig. 21.64).

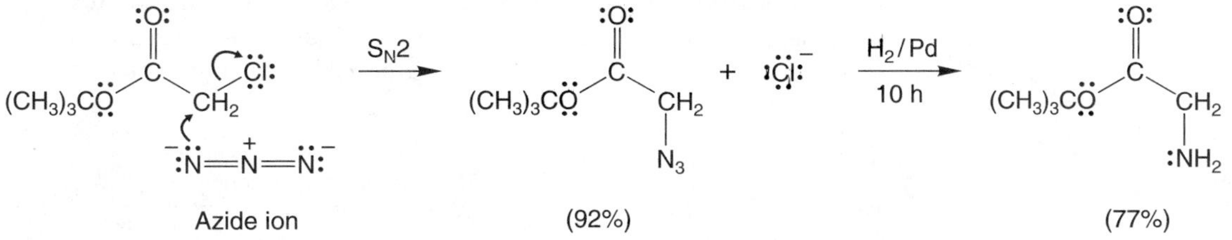

$(CH_3)_3COC(=O)CH_2Cl$ + Azide ion ($^-$:N=N$^+$=N:$^-$) → (S_N2) $(CH_3)_3COC(=O)CH_2N_3$ (92%) + :Cl:$^-$ → (H_2/Pd, 10 h) $(CH_3)_3COC(=O)CH_2NH_2$ (77%)

FIGURE 21.64 The reduction of an azide to an amine.

Another source of amines involves reduction of nitro groups. The nitrite ion ($^-NO_2$) is a nucleophile and can be used to make alkyl nitro compounds. Reduction using either catalytic hydrogenation or Sn/HCl serves to generate the amine. This process should be familiar—we used it to make aniline from nitrobenzene when we studied aromatic chemistry (Chapter 14, p. 644; Fig. 21.65).

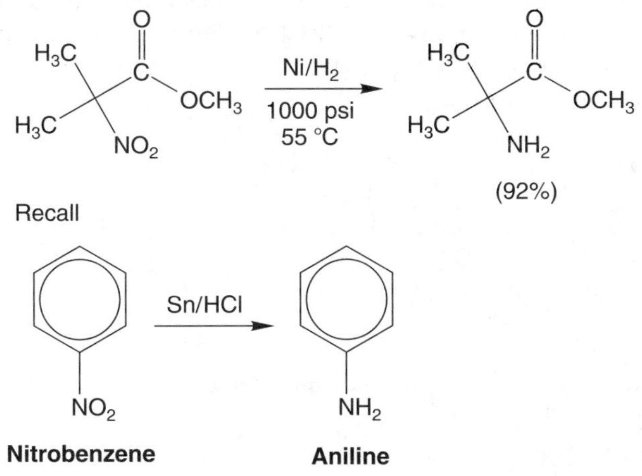

$(H_3C)_2C(NO_2)C(=O)OCH_3$ → (Ni/H_2, 1000 psi, 55 °C) $(H_3C)_2C(NH_2)C(=O)OCH_3$ (92%)

Recall

Nitrobenzene (NO_2) → (Sn/HCl) **Aniline** (NH_2)

FIGURE 21.65 Two related amine syntheses that use reductions.

If carbon–nitrogen triple bonds can be reduced to give primary amines, it would seem that similar reduction of carbon–nitrogen double bonds (imines) should yield primary or secondary amines. So it does, and the general process of imine formation (Chapter 16, p. 780), followed by reduction, is generally called **reductive amination**. Usually, the imine is not isolated, but reduced as it is formed in the reaction (Fig. 21.66).

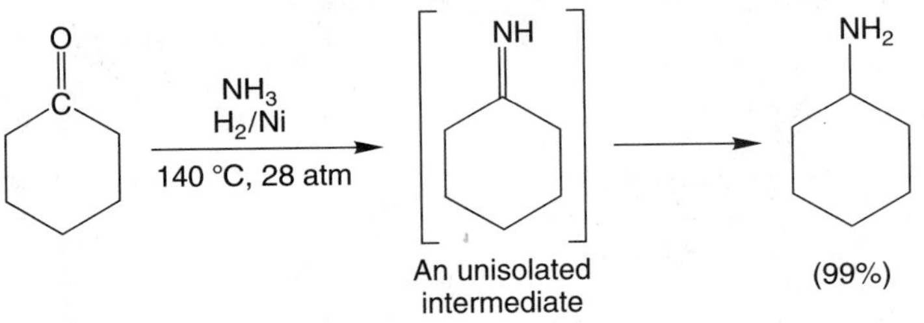

NH_3, H_2/Ni, 140 °C, 28 atm

An unisolated intermediate

(99%)

FIGURE 21.66 Imines can be reduced to amines as they are formed.

A difficulty in reductive aminations is avoiding the reduction of the carbonyl component of the reaction. In order to do this, it is necessary to use a reducing agent that is selective—that will reduce the imine but not the aldehyde or ketone. Sodium cyanoborohydride is often the agent of choice, although catalytic hydrogenation is also effective. The cyano group is strongly electron withdrawing, and diminishes the ability of the borohydride to deliver the hydride ion, H:⁻. The imine is more reactive than the carbon–oxygen double bond, and is more easily reduced. It is not necessary to isolate the intermediate in this process, as the intermediate is generated in the presence of the reducing agent (Fig. 21.67).

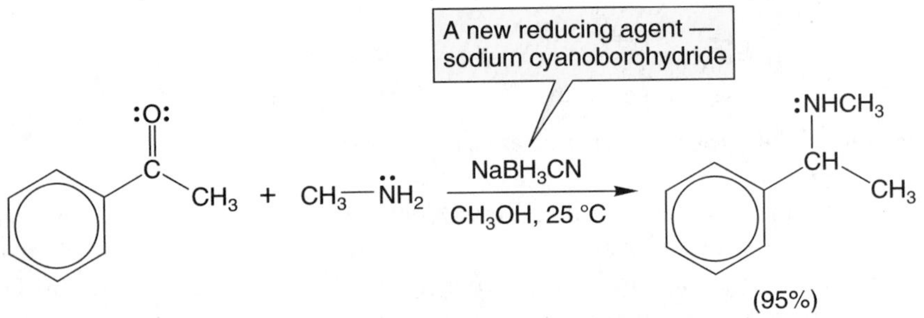

A new reducing agent — sodium cyanoborohydride

CH_3 + CH_3—$\ddot{N}H_2$ → $NaBH_3CN$, CH_3OH, 25 °C → :$NHCH_3$, CH, CH_3

(95%)

FIGURE 21.67 Reduction with sodium cyanoborohydride, a mild (selective) reducing agent.

Substituted imines can also be reduced. Oximes, hydroxyl imines, are typical examples. Here, two outcomes are possible depending on the strength of the reducing agent used. Sodium cyanoborohydride, a mild and selective reducing agent, gives the hydroxylamine, whereas the less selective process of catalytic hydrogenation reduces the oxime all the way to the amine (Fig. 21.68).

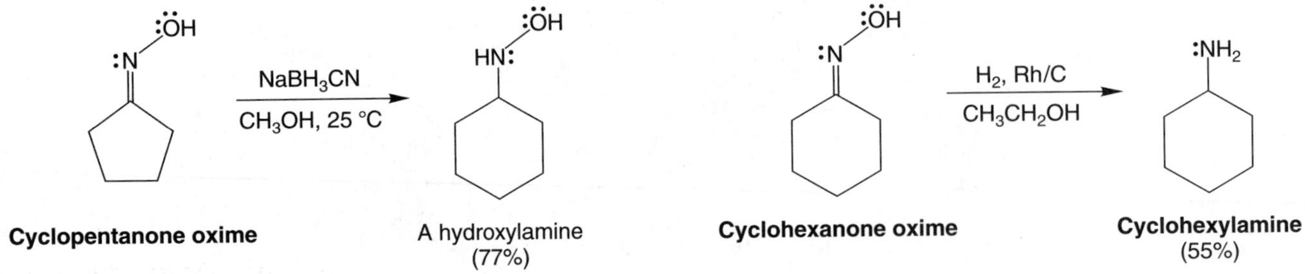

$NaBH_3CN$, CH_3OH, 25 °C

H_2, Rh/C, CH_3CH_2OH

Cyclopentanone oxime

A hydroxylamine (77%)

Cyclohexanone oxime

Cyclohexylamine (55%)

FIGURE 21.68 Oximes can be reduced to hydroxylamines or amines depending on the strength of the reducing agent.

21.6c Amines from Amides

We have already seen two reactions leading to amines from their acyl counterparts, amides. In the simpler of the two reactions, amides are reduced by lithium aluminum hydride to the related amines (Chapter 20, p. 1030). Don't forget that amides can be alkylated—this greatly increases the versatility of this process (Fig. 21.69).

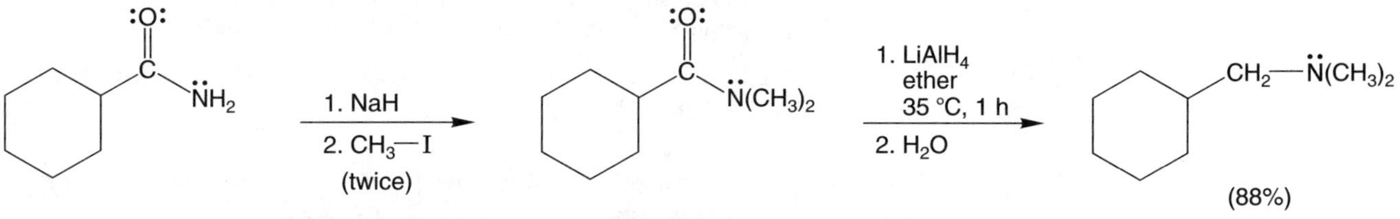

:O:
C
NH2
1. NaH
2. CH3—I
(twice)
:O:
C
N(CH3)2
1. LiAlH4
ether
35 °C, 1 h
2. H2O
CH2—N(CH3)2
(88%)

FIGURE 21.69 The reduction of amides to amines.

A more complex procedure, at least on paper, involves the Hofmann rearrangement of primary amides (Chapter 23, p. 1224). In this example, the first step is the formation of the *N*-chloroamide, which rearranges in base to give an isocyanate. Under aqueous conditions the isocyanate is converted into a carbamic acid that then decarboxylates (Fig. 21.70).

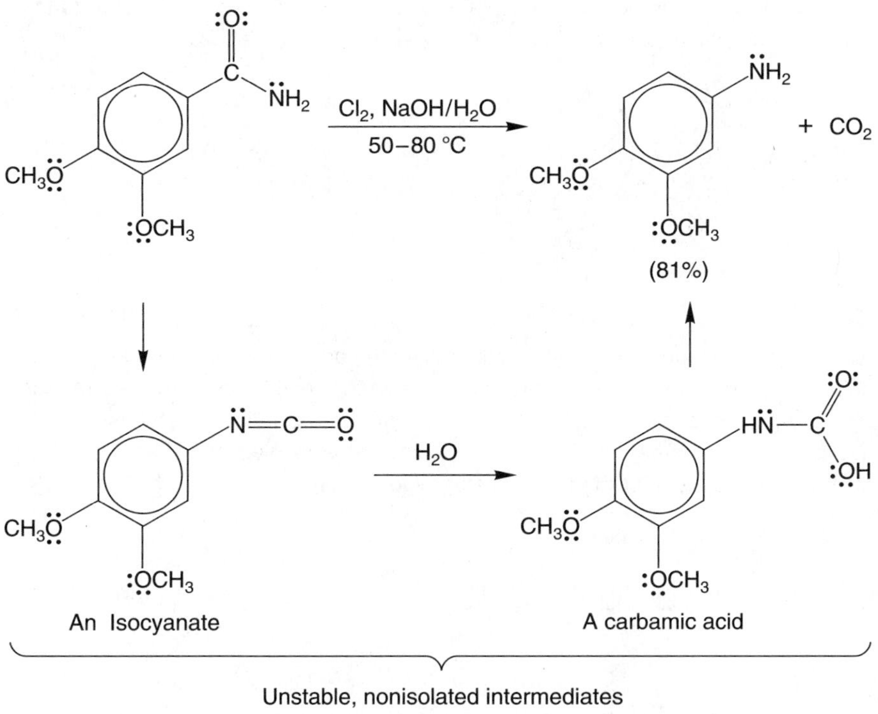

:O:
C
NH2
Cl2, NaOH/H2O
50–80 °C
CH3O
:OCH3
NH2
+ CO2
CH3O
:OCH3
(81%)
N=C=O
H2O
CH3O
:OCH3
An Isocyanate
:O:
HN—C
:OH
CH3O
:OCH3
A carbamic acid
Unstable, nonisolated intermediates

FIGURE 21.70 The Hofmann rearrangement leads to amines through the hydrolysis of an intermediate isocyanate and decarboxylation of the resulting carbamic acid.

In the related Curtius rearrangement (Chapter 23, p. 1222), an acyl azide rearranges to give an isolable isocyanate. Treatment of the isocyanate with water leads to a carbamic acid that decarboxylates, just as it does in the Hofmann rearrangement. The difference is that the Curtius rearrangement is carried out under nonaqueous conditions under which the isocyanate is stable (Fig. 21.71).

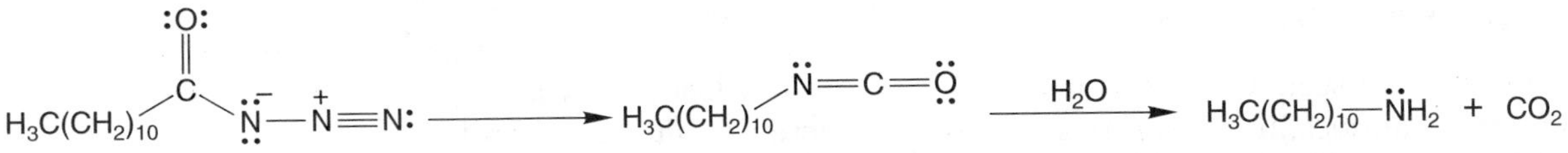

$H_3C(CH_2)_{10}$–C(=O)–N⁻–N⁺≡N: ⟶ $H_3C(CH_2)_{10}$–N=C=O ⟶(H_2O) $H_3C(CH_2)_{10}$–NH_2 + CO_2

This isocyanate is stable— no water is present (83%)

Formed by decarboxylation of an intermediate carbamic acid

FIGURE 21.71 The Curtius rearrangement leads to an isolable isocyanate that can be hydrolyzed in a separate step to give the amine. See Chapter 23 for details.

There is still another variation of this process called the **Schmidt reaction**, after Karl F. Schmidt (1887–1971).

*PROBLEM 21.21 Write a mechanism for the Schmidt reaction. *Hint*: The first step is probably the formation of an acylium ion, $R-\overset{+}{C}{=}O$. Note also that the reagent is HN_3, not NH_3 (Fig. 21.72).

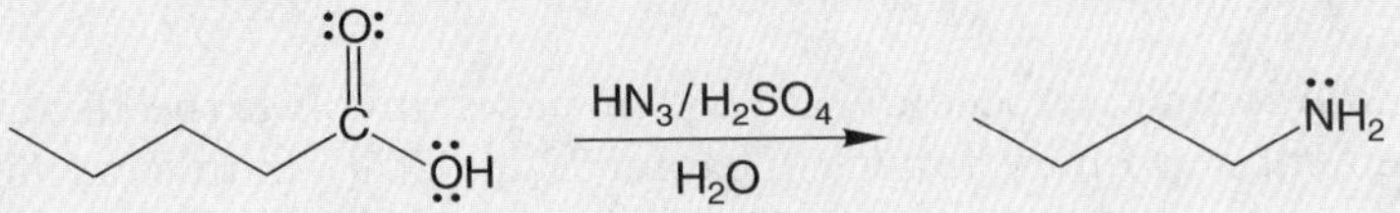

HN_3/H_2SO_4, H_2O; NH_2 (72%)

FIGURE 21.72

ANSWER The hints tell you how to start: Form the acylium ion. Protonation of the hydroxyl oxygen, followed by loss of water, does the trick. Of course the hydroxyl oxygen is not the more basic position in the acid, the carbonyl O is. However, protonation of the carbonyl oxygen leads nowhere—it is a mechanistic dead end.

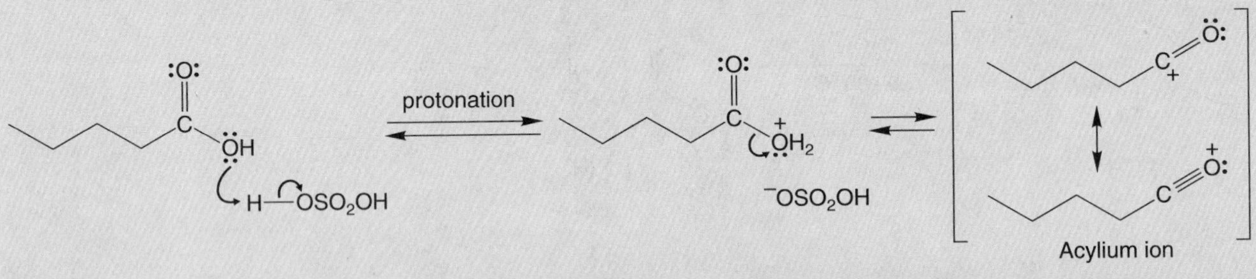

protonation; H–OSO₂OH; $\overset{+}{O}H_2$; ⁻OSO₂OH; Acylium ion

Addition of azide ion to the acylium ion makes the acyl azide, an intermediate capable of rearrangement to an isocyanate. As in the related Hofmann rearrangement, the isocyanate is unstable under aqueous conditions and is rapidly hydrolyzed to a carbamic acid that decarboxylates to give the amine. Mechanisms are not shown for the last two steps (see p. 1107).

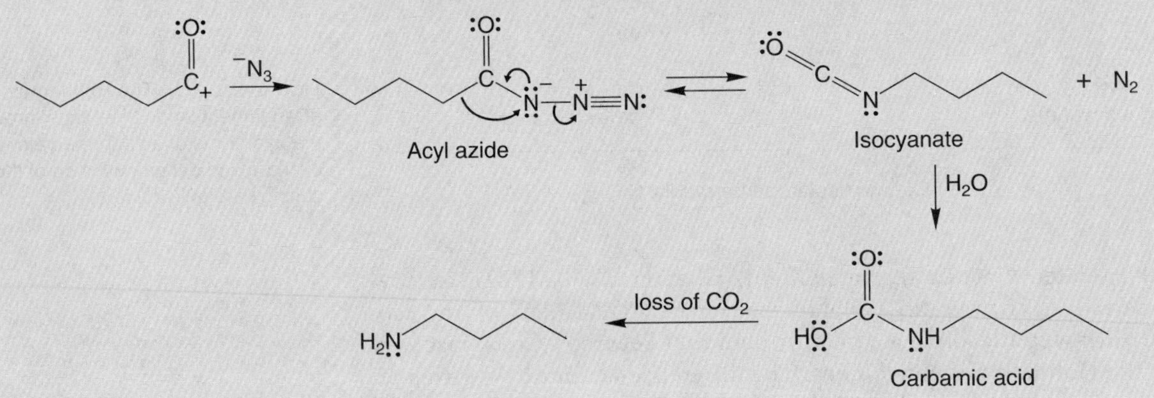

$^-N_3$; Acyl azide; Isocyanate + N_2; H_2O; Carbamic acid; loss of CO_2; H_2N

21.7 ALKALOIDS

The term **alkaloid** refers to any nitrogen-containing compound extracted from plants, although the word is used loosely and some compounds of nonplant origin are also commonly known as alkaloids. These biologically active compounds are basic like all amines, and it is this basicity that led to the name. Presumably, basicity was also important in the relative ease of extraction of alkaloids from the myriad compounds present in an organism as complex as a plant. Extraction of plant mass with acid yields any amines present as water-soluble ammonium ion salts from which they can easily be regenerated. Figure 21.73 shows this process for a simple, but notorious alkaloid known as coniine. Coniine is the active ingredient in hemlock, and it is the physiological activity of coniine that led to the demise of both Socrates and Hamlet's father Claudius.

Alkaloids are very often chiral, and found in Nature as single enantiomers. A classic use of these naturally occurring compounds has been in the resolution of racemic chiral organic acids (Chapter 5, p. 177). A pair of enantiomeric acids will form a pair of diastereomeric ammonium ion salts on reaction with an alkaloid. The diastereomers have different physical properties and can be separated. The optically pure acids can then be regenerated from the salts (Fig. 21.74).

FIGURE 21.73 A scheme for isolation of alkaloids from plant material.

Plant (hemlock)

H_2O/H_3O^+

Ammonium ion (water soluble)

base

(*S*)-Coniine

A racemic mixture of acids

An alkaloid—an optically active amine

A pair of diastereomeric ammonium ion salts; these will have different physical properties and can be separated by crystallization

separation

neutralize

neutralize

Pure (*S*)-acid

Pure (*R*)-acid

FIGURE 21.74 The use of alkaloids in optical resolution.

Many heterocyclic ring types are found in the alkaloids; indeed, many were first encountered in these compounds. Coniine, for example, contains a **piperidine**, a six-membered ring including a single nitrogen atom. Other alkaloids contain **pyrrolidines**, pyridines, indoles, and other nitrogen-containing rings (Fig. 21.75).

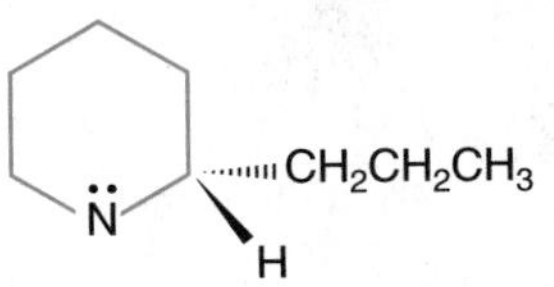

Coniine (piperidine)

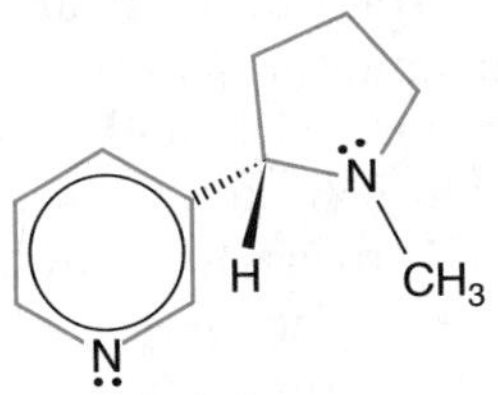

Nicotine (pyrrolidine and pyridine)

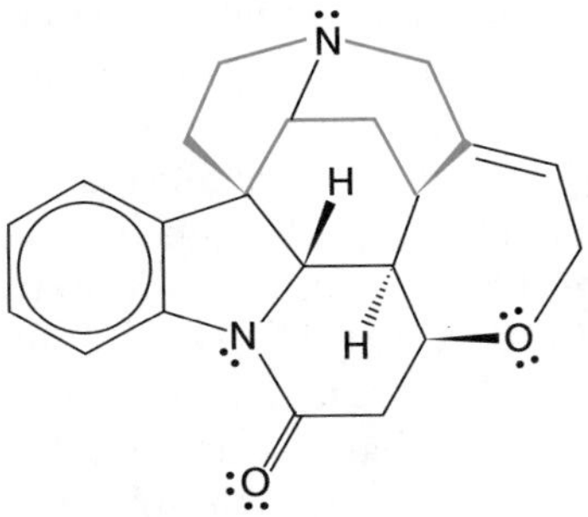

Strychnine piperidine, pyrrolidine, azacyclononane

FIGURE 21.75 Some alkaloids containing various nitrogen heterocycles.

Although the structures of alkaloids are often spectacularly complex, they are still amines, and express the chemistry of amines we have seen throughout this chapter. In the following brief survey of the various structural families of alkaloids, we will see familiar reactions many times.

Morphine, a member of the class of **opium alkaloids** (Fig. 21.76) constitutes about 10% of crude opium, and is an extraordinarily effective painkiller. The body does build up a tolerance for morphine, however, and the large doses eventually necessary can result not only in addiction but in substantial depression of the nervous system.

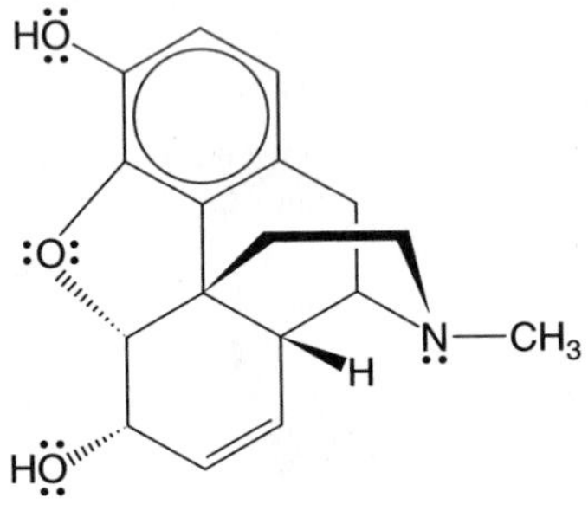

FIGURE 21.76 The structure of morphine.

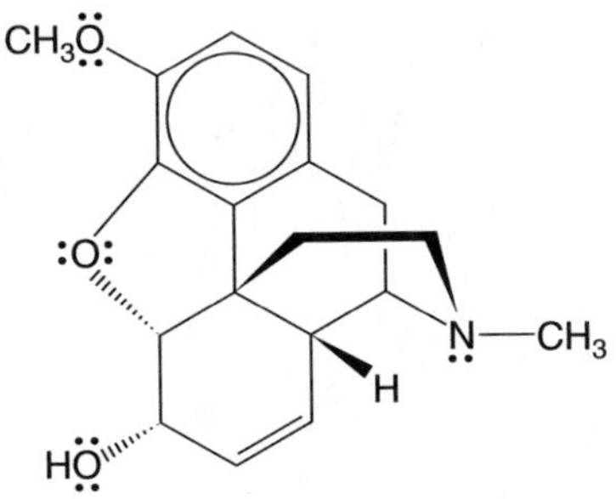

One of the crucial steps in the determination of the general structure of morphine and related alkaloids was the conversion shown in Figure 21.77. The reaction is nothing more than a simple Hofmann elimination, which in this case opens up one of the rings of the polycyclic molecule, simplifying the structure.

Codeine (a close relative of morphine)

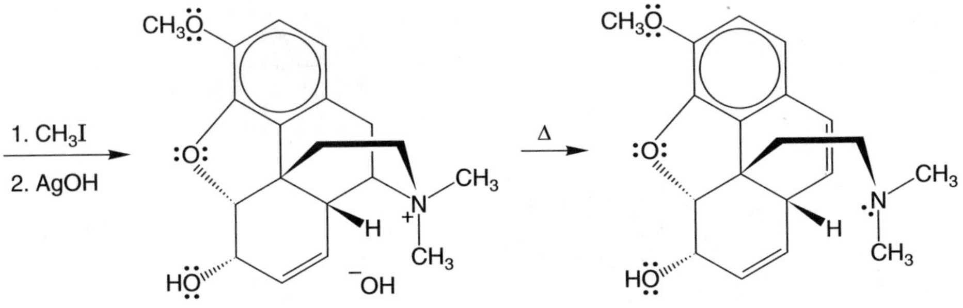

1. CH_3I 2. AgOH

Δ

FIGURE 21.77 A Hofmann elimination applied to codeine.

PROBLEM 21.22 Write arrow formalisms for the reactions of Figure 21.77.

Codeine is another member of this family and is a less addictive and less powerful analgesic than morphine. Heroin, a synthetic derivative of morphine not found in nature is also a close relative (Fig. 21.78).

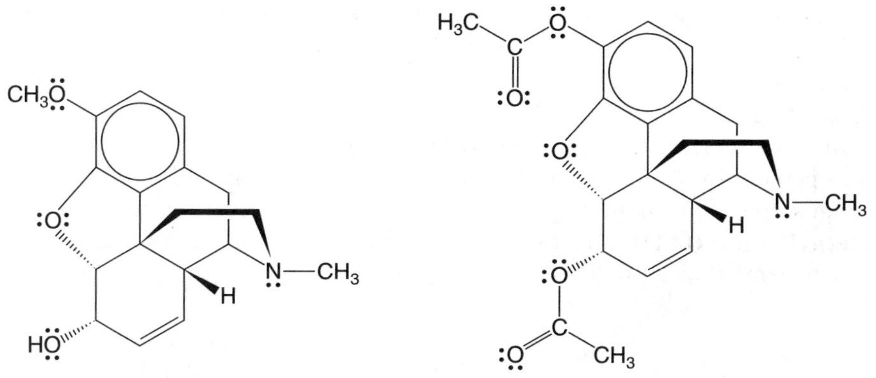

Codeine Heroin

FIGURE 21.78 Codeine and heroin are derivatives of morphine.

PROBLEM 21.23

Unfortunately, morphine is easily converted into heroin, even in a basement lab. Suggest a way to achieve this conversion.

Another important family is the **tropane alkaloids**. These include atropine, found in deadly nightshade, scopolamine, the "truth serum" of too many a detective novel, and cocaine (Fig. 21.79).

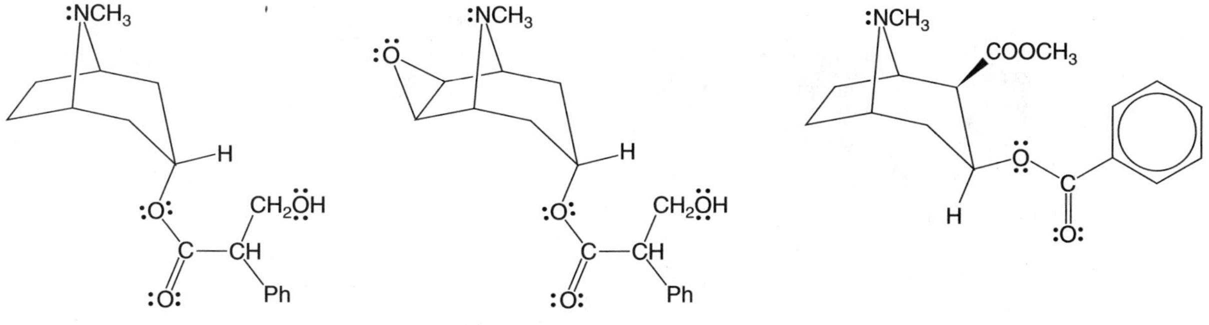

Atropine Scopolamine Cocaine

FIGURE 21.79 Atropine, scopolamine, and cocaine: tropane alkaloids.

PROBLEM 21.24

Cocaine can be hydrolyzed to ecgonine, which can be oxidized with chromium trioxide in acetic acid to give tropinone. Write mechanisms for the reactions shown in Figure 21.80.

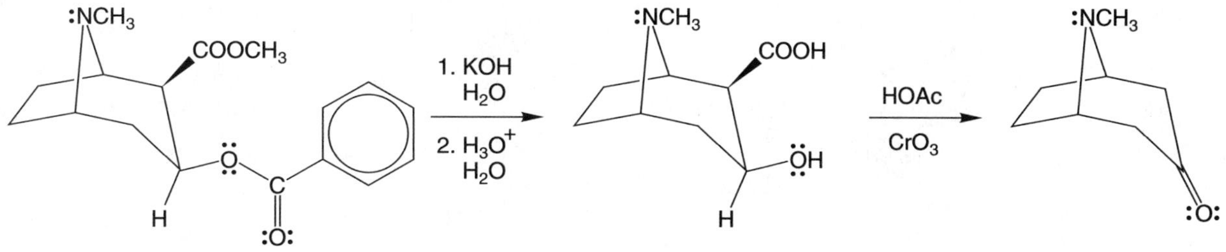

Cocaine Ecgonine Tropinone

FIGURE 21.80

Naturally enough, all **indole alkaloids** contain the aromatic indole system. Examples include a molecule not of plant origin, serotonin, a stimulator of smooth muscle, and important in neurotransmission in the brain. Accordingly, structurally related alkaloids that may interfere with serotonin metabolism cause a wide variety of powerful effects on the brain and perception. These include lysergic acid, which is isolated from the fungus ergot, and is best known as its synthetic derivative, **lysergic acid diethylamide (LSD)**, and psilocin, an alkaloid found in hallucinogenic mushrooms (Fig. 21.81).

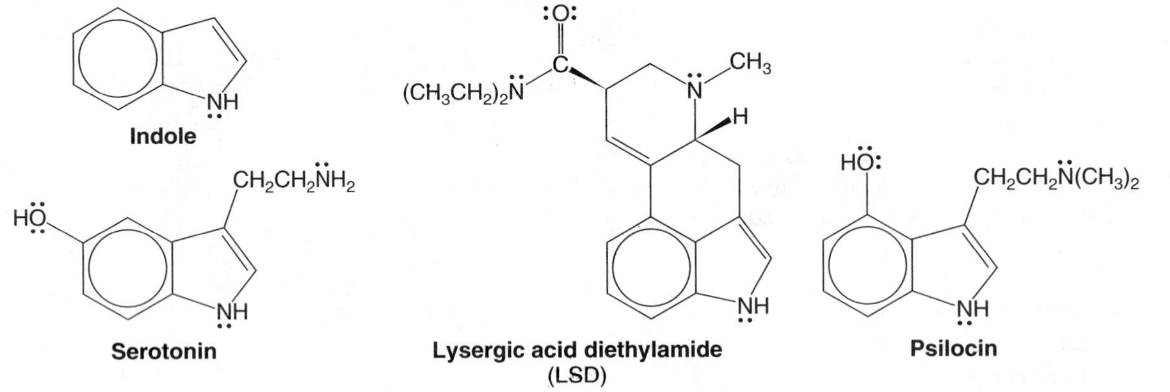

Indole

Serotonin

Lysergic acid diethylamide (LSD)

Psilocin

FIGURE **21.81** Some indole alkaloids.

PROBLEM **21.25** The reactions in Figure 21.82 were important in the synthesis of LSD. Provide structures for the indicated compounds.

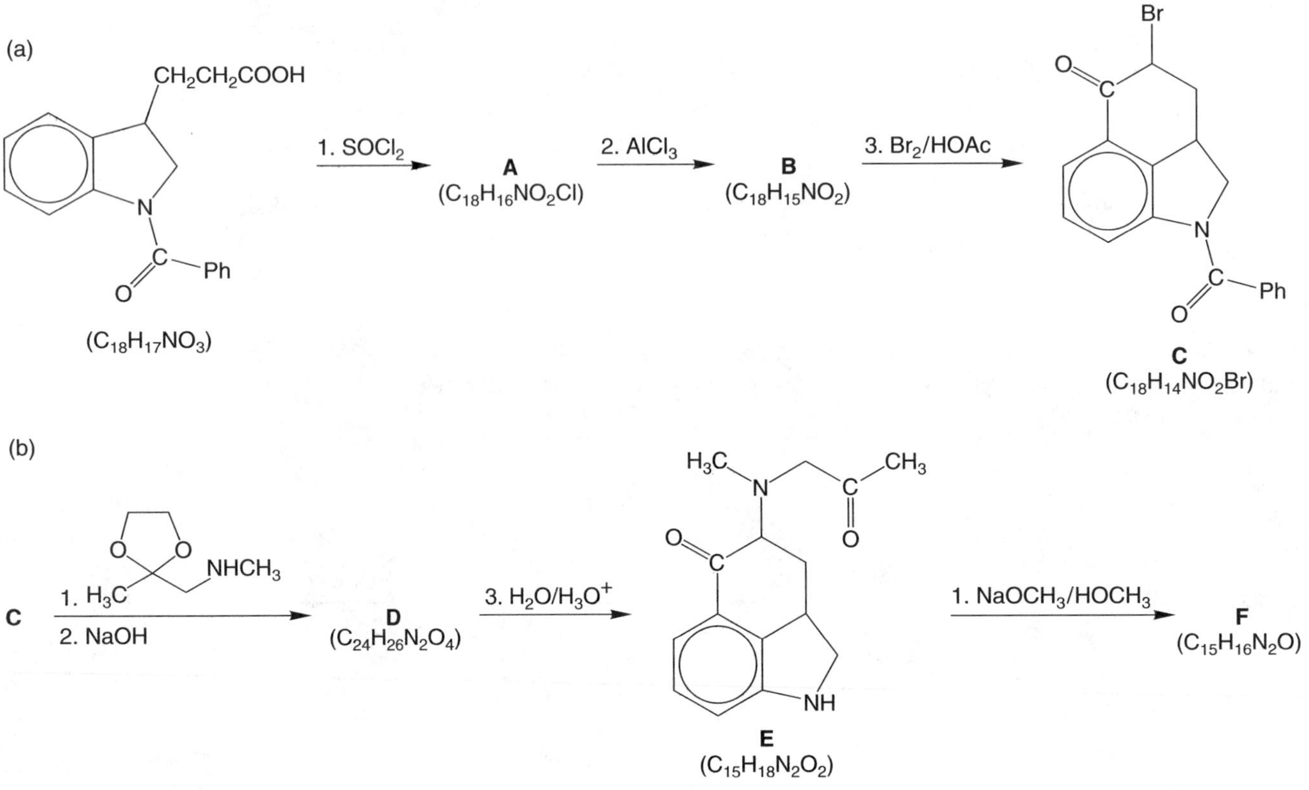

(a)

$(C_{18}H_{17}NO_3)$ → 1. $SOCl_2$ → **A** $(C_{18}H_{16}NO_2Cl)$ → 2. $AlCl_3$ → **B** $(C_{18}H_{15}NO_2)$ → 3. Br_2/HOAc → **C** $(C_{18}H_{14}NO_2Br)$

(b)

C → 1. H_3C(dioxolane)CH_2NHCH_3; 2. NaOH → **D** $(C_{24}H_{26}N_2O_4)$ → 3. H_2O/H_3O^+ → **E** $(C_{15}H_{18}N_2O_2)$ → 1. $NaOCH_3/HOCH_3$ → **F** $(C_{15}H_{16}N_2O)$

FIGURE **21.82**

(c)

F ($C_{15}H_{16}N_2O$) → 2. Ac_2O → G ($C_{17}H_{18}N_2O_2$) → 3. $NaBH_4$ 4. H_2O → H ($C_{17}H_{20}N_2O_2$) → 1. $SOCl_2$ → I ($C_{17}H_{19}N_2OCl$)

(d)

I → 2. NaCN → J ($C_{18}H_{19}N_3O$) → 3. CH_3OH H_2O/H_2SO_4 → K ($C_{17}H_{20}N_2O_2$) → 4. NaOH 5. H_3O^+/H_2O → L ($C_{16}H_{18}N_2O_2$) → several steps → LSD

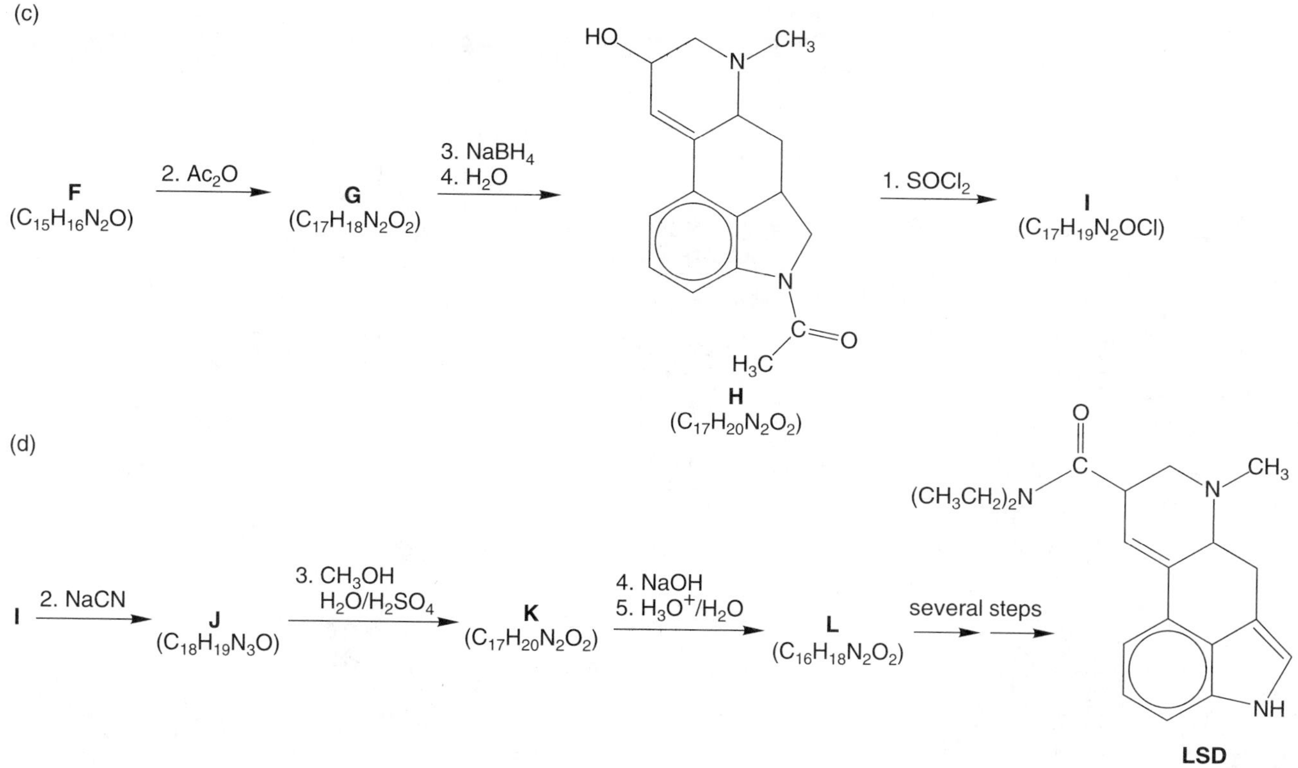

FIGURE **21.82** (CONTINUED)

21.8 SYNTHESIS OF QUININE

Quinine, an alkaloid isolated from its large concentrations in the bark of the cinchona* tree remains one of the most powerful antimalarial agents known (Fig. 21.83).

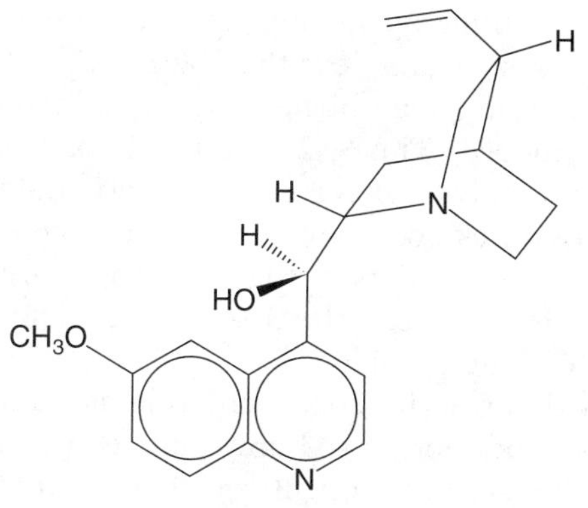

Quinine

FIGURE **21.83** Quinine.

The structure of this complex molecule presents an extraordinary synthetic challenge, and was accomplished in the early 1940s by R. B. Woodward and William von E. Doering. It's worth a look at this synthesis. It was not only an achievement of brilliance, but used many of the reactions we have studied in a mechanistic way.

*Named for the Countess del Chinchon, wife of a seventeenth century Peruvian viceroy, supposedly cured of the vapors by use of the bark.

PROBLEM 21.26 Outline mechanisms for the following transformations leading to quinine (Fig. 21.84):

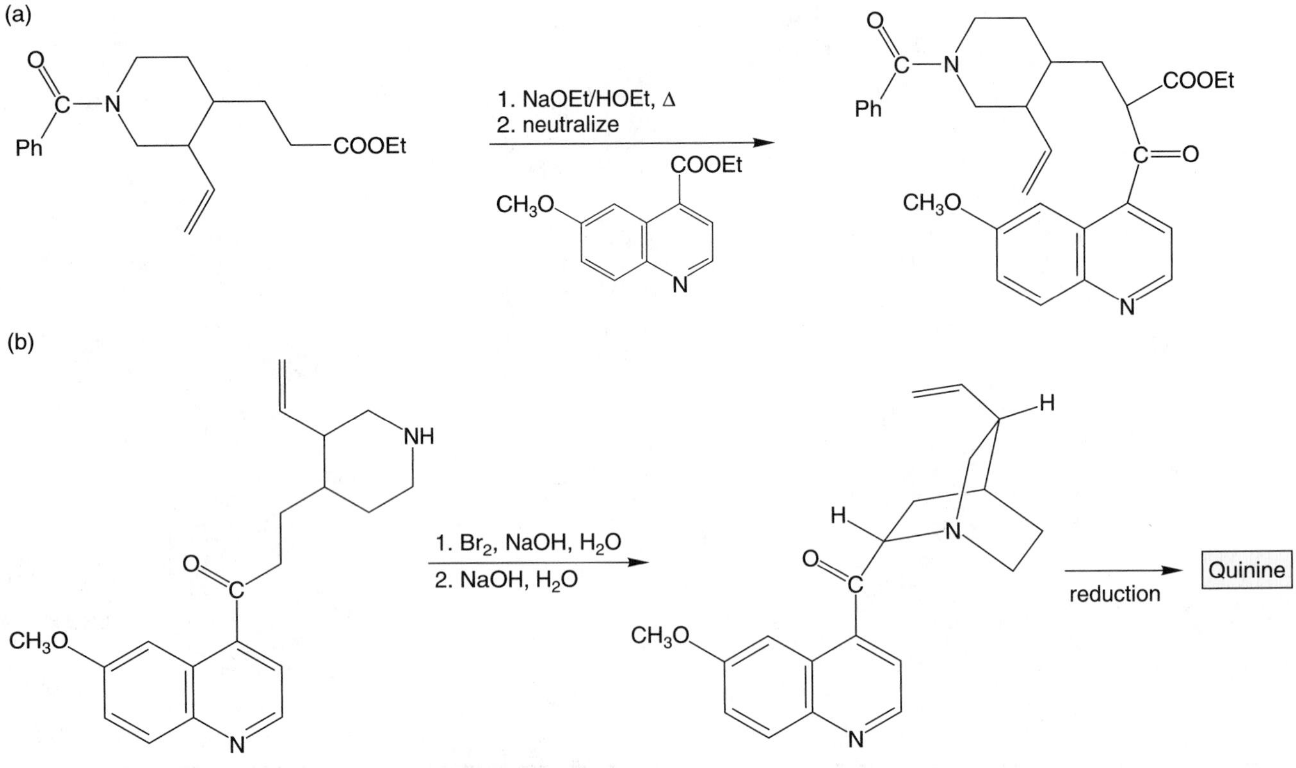

(a)

1. NaOEt/HOEt, Δ
2. neutralize

(b)

1. Br_2, NaOH, H_2O
2. NaOH, H_2O

reduction

Quinine

FIGURE 21.84

21.9 SOMETHING MORE: PHASE-TRANSFER CATALYSIS

Although it is easy to design many chemical reactions on paper, the reality is often different. A classic example is the S_N2 reaction. Displacement of a good leaving group, often a poor nucleophile, by a better nucleophile seems straightforward. Indeed, The S_N2 reaction is one of the cornerstones of our mechanistic analysis of organic chemistry, and we have written such reactions many times over the preceding 20 chapters. Yet, when one looks closely at the constituents of the reaction, practical problems become apparent. Take, for example, the S_N2 displacement of a chloride by the far stronger nucleophile cyanide.

The alkyl halide in this reaction, octyl chloride, is a rather nonpolar liquid. The source of cyanide ion is sodium cyanide, a solid, ionic salt. How are the two reacting species brought together? Throwing $Na^+ \; ^-CN$ into octyl chloride does little good, as cyanide is essentially insoluble in the alkyl halide, and the two species interact only at the surface of the solid (Fig. 21.85).

In practice, the reaction rate under such circumstances is vanishingly low. What about dissolving the ionic sodium cyanide in water and then adding the octyl chloride? This process results in a two-phase system in which the two reacting species encounter each other only at the interface between the two layers (Fig. 21.86). This arrangement may be better, as contact between the reagents is increased, but it is hardly an ideal situation for reaction!

Caffeine and Cyclic-AMP

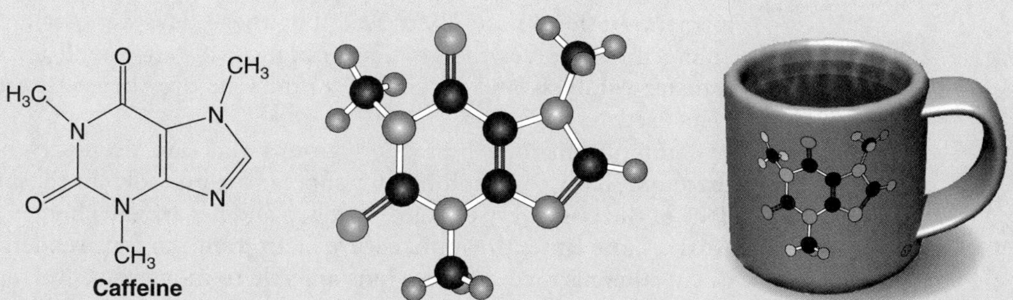

When you drink a cup of coffee late at night while studying your orgo, you ingest about 100 mg of the alkaloid caffeine. As we all know, there is a certain stimulation from this activity. Why? Caffeine inhibits the action of an enzyme, phosphodiesterase, whose job it is to inactivate a molecule called cyclic adenosine monophosphate (AMP). Cyclic-AMP is involved in the formation of glucose in the bloodstream. Deactivation of phosphodiesterase by caffeine frees cyclic-AMP to do its job, more glucose appears, and we feel more energetic.

Caffeine comes from the seeds of *Coffea arabica,* the familar roasted coffee bean. The seeds of *Theobroma cacao* yield a compound that differs from caffeine in that it bears a hydrogen where caffeine had a methyl group. This small change yields *theobromine,* which is the stimulating ingredient in chocolate.

Theobromine

$$Na^{+}\,^{-}CN + CH_3CH_2CH_2CH_2CH_2CH_2CH_2CH_2{-}Cl \xrightarrow{S_N2} Na^{+}\,^{-}Cl + CH_3CH_2CH_2CH_2CH_2CH_2CH_2CH_2{-}CN$$

Liquid octyl chloride

Solid $Na^{+}\,^{-}CN$

FIGURE **21.85** Practical problems in the S_N2 displacement reaction of an alkyl halide.

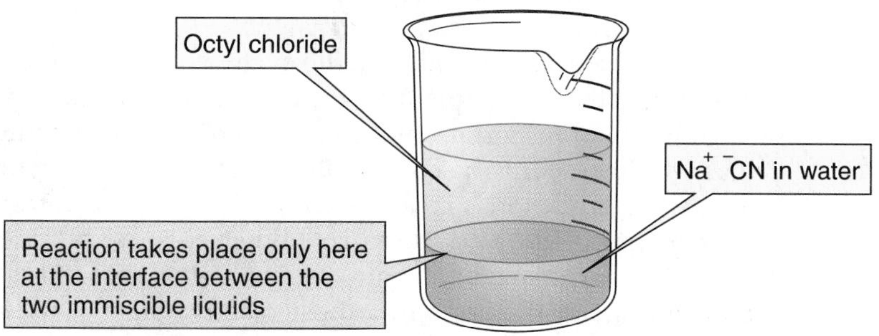

FIGURE **21.86** Contact remains poor even if two liquid phases are used.

Solvents can be found in which both of the reactants are soluble. Alcohols and polar molecules such as tetrahydrofuran (THF) or dimethylformamide (DMF) are often used, but these days another technique has emerged that solves the problem in a most clever way. It is called **phase-transfer catalysis** and alkylammonium ions are the phase-transfer catalysts used.

Ammonium ions bear alkyl groups that are decidedly organic, and therefore likely to be soluble in nonpolar organic solvents despite the presence of the positively charged nitrogen and negatively charged counterion. At the same time, the ionic nature of ammonium ions renders them soluble in aqueous media. These ions are able to move back and forth between the two phases (Fig. 21.87).

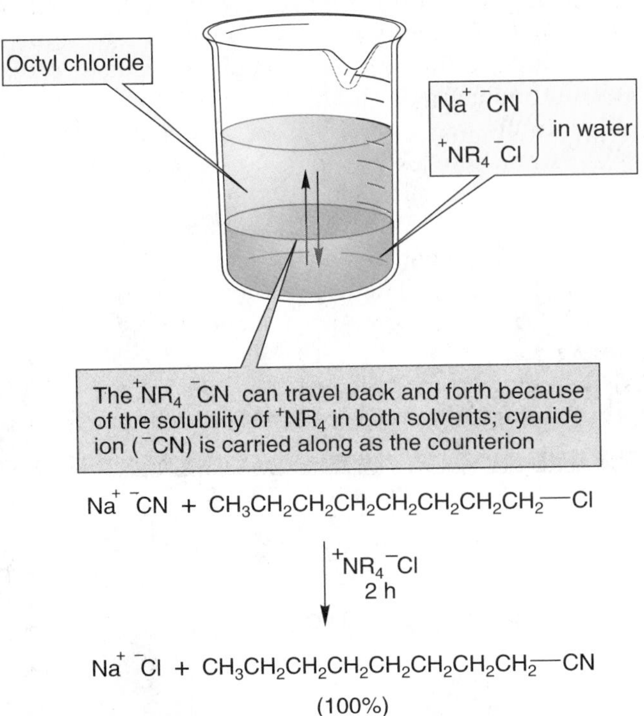
Octyl chloride
$Na^+ \ ^-CN$
$^+NR_4 \ ^-Cl$
in water
The $^+NR_4 \ ^-CN$ can travel back and forth because of the solubility of $^+NR_4$ in both solvents; cyanide ion (^-CN) is carried along as the counterion
$Na^+ \ ^-CN + CH_3CH_2CH_2CH_2CH_2CH_2CH_2CH_2—Cl$
$^+NR_4 \ ^-Cl$
2 h
$Na^+ \ ^-Cl + CH_3CH_2CH_2CH_2CH_2CH_2CH_2CH_2—CN$
(100%)

FIGURE **21.87** The solution is to use phase-transfer catalysis in which ammonium ions are used to transport a reactive species from one phase to the other.

In the example we took up earlier, the S_N2 displacement of chloride by cyanide, a small amount of tetraalkylammonium ion in the two-phase system, octyl chloride/$Na^+ \ ^-CN/H_2O$, can move back and forth between the two phases. As the ammonium ion moves from the aqueous phase into the organic phase it carries with it a negatively charged counterion. If it travels as part of the ammonium ion pair, the cyanide ion can be transported from the aqueous phase in which it is soluble into the organic phase in which it is ordinarily insoluble. No longer do the two reactive partners, the octyl chloride and the cyanide ion, encounter each other only at the interface. One might well expect that the reaction would be facilitated by the transportation of cyanide ion into the organic phase, and this idea would certainly be correct. But there is another factor that helps. Ordinarily, cyanide ion is greatly stabilized through solvation in the polar environment in which it is dissolved. When it is transported into octyl chloride, it arrives naked, shorn of its solvating and stabilizing water molecules, and therefore in a most reactive state. Unsolvated cyanide is far more reactive than solvated cyanide. Under these conditions formation of octyl cyanide is quantitative and complete in 1–2 h.

Thus, there are two effects of the phase-transfer catalyst. One is to bring the partners in the reaction together, and the other is to provide one of the partners in an especially reactive state.

21.10 SUMMARY

NEW CONCEPTS

Several new concepts appear in this chapter, although the trend toward proliferation of detail without the introduction of really new concepts continues.

Amine inversion is new, as are some of the consequences of inversion such as the racemization of potentially chiral amines through "umbrella flipping" (Fig. 21.18).

It is important to keep straight the relationship between the basicity of an amine and the acidity of the related ammonium ion. A strong base competes effectively for a proton, and therefore produces a weakly acidic ammonium ion (relatively high pK_a). By contrast, a weakly basic amine does not compete well for a proton and the related ammonium ion is a stronger acid (lower pK_a) (Fig. 21.25).

The importance of solvation has appeared before [see, for example, the discussion of the acidity of alcohols in Chapter 17 (p. 822), or of phase-transfer catalysis in this chapter], and perhaps it is not surprising that this important, if somewhat neglected subject is raised again in our discussion of amine basicity and ammonium ion acidity. Substitution stabilizes ammonium ions by providing electron-withdrawing groups, but at the same time interferes with solvation, thus destabilizing the ion. The acidity of an ammonium ion depends on both these effects (p. 1086).

Indirect syntheses appear often in this chapter in response to a number of practical problems. Alkylations of amines through simple S_N2 reactions are generally complicated by overalkylation as the initial products also react as nucleophiles in the S_N2 reaction. The Gabriel synthesis of primary amines avoids this problem through the use of phthalimide as a masked amine. Acylation of aromatic amines can be used to avoid overreaction with electrophiles. In a similar vein, a mixture of ortho, para, and meta substitution of aniline can be avoided by acylation of the amino group, which diminishes protonation of the amine nitrogen. See Section 21.6 for details.

REACTIONS, MECHANISMS, AND TOOLS

Two important old mechanisms reappear in this chapter, the displacement reaction of amines acting as nucleophiles in the S_N2 reaction, and the addition of amines to a variety of carbonyl compounds. If a leaving group is present, amides are formed; if not, the product is an imine or enamine (Figs. 21.36 and 21.38).

A few new mechanisms appear as well. Amine inversion through an sp^2 transition state interconverts two enantiomeric forms of amines (Fig. 21.18).

The nitrosation of aliphatic amines leads to different products depending on the number of hydrogens available on the starting amine. Secondary amines give nitroso compounds, whereas primary amines yield diazonium ions in a reaction in which the nitroso compound is an intermediate (Figs. 21.49 and 21.50).

Amines can be converted into *N*-oxides, which on heating undergo an intramolecular elimination reaction to give alkenes (Fig. 21.56).

When an imine is generated in the presence of a reducing agent, amines are produced through a process known as reductive amination (Fig. 21.67).

The Mannich reaction (or Mannich condensation) leads to β-aminocarbonyl compounds in a reaction in which an amine and an aldehyde first react to give the strongly Lewis acidic immonium ion (Fig. 21.59).

The newest method is phase-transfer catalysis. In this process, ammonium ions are used to transport ions between the aqueous and organic phases of a reaction, which facilitates reaction by increasing contact between the reactive species and by providing unsolvated, and therefore unusually reactive reagents.

SYNTHESES

1. Alkenes

^{-}OH $\overset{+}{N}(CH_3)_3$ —Δ→ Major + Minor

Hofmann elimination

$(CH_3)_2N$ —30% H_2O_2 / H_2O→ $(CH_3)_2\overset{+}{N}$–O^- —Δ→

Cope elimination

2. Amides

R–C(=O)–L —RNH_2→ R–C(=O)–NHR   L = Cl or OR

There is a tetrahedral intermediate

3. Amines

$H_3N + CH_3{-}I \xrightarrow{S_N2} H_3\overset{+}{N}{-}CH_3\ I^- \rightleftharpoons H_2NCH_3$ Primary amine

$CH_3NH_2 + CH_3{-}I \xrightarrow{S_N2} CH_3\overset{+}{N}H_2{-}CH_3\ I^- \rightleftharpoons HN(CH_3)_2$ Secondary amine

$(CH_3)_2NH + CH_3{-}I \xrightarrow{S_N2} (CH_3)_2\overset{+}{N}H{-}CH_3\ I^- \rightleftharpoons N(CH_3)_3$ Tertiary amine

Restrictions on S_N2 reaction apply; this also works for aromatic amines; overalkylation is a serious problem

3. Amines (continued)

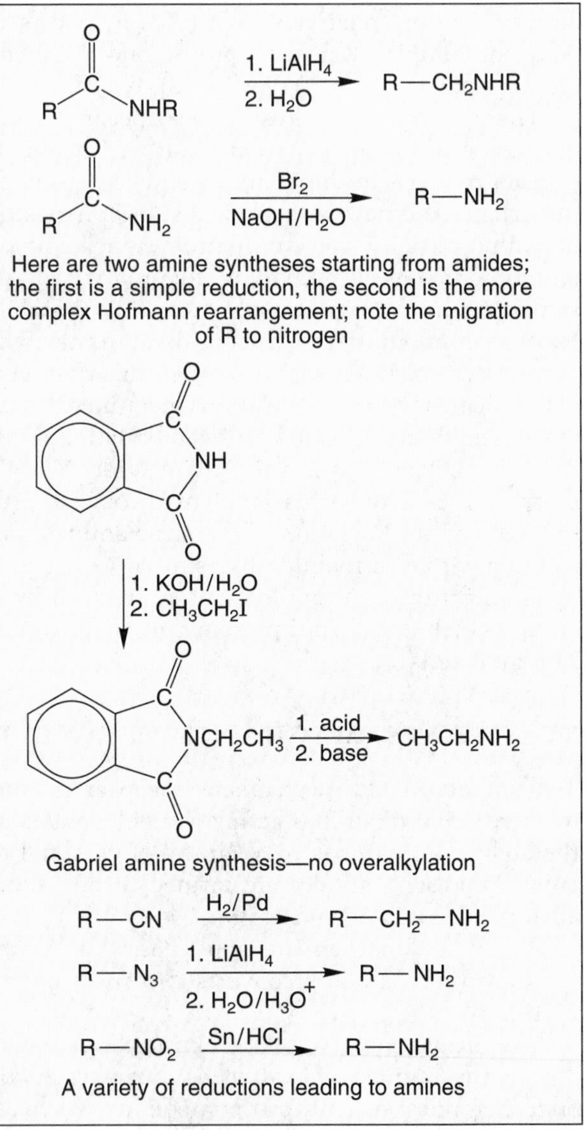

FIGURE 21.88 The syntheses in Chapter 21.

3. Amines (continued)

$$R(H)C{=}N{-}OH \xrightarrow[\text{2. } H_2O/H_3O^+]{\text{1. } LiAlH_4} R{-}CH_2{-}NH_2$$

$$R_2C{=}NH \xrightarrow{H_2/Pd} R_2CH{-}NH_2$$

$$R_2C{=}O + R_2NH \xrightarrow[EtOH]{NaBH_3CN} R_2CH{-}NR_2$$

Reductive amination

$$CH_3CH_2NH_2 + \text{ethylene oxide} \xrightarrow{H_2O} CH_3CH_2NHCH_2CH_2OH + CH_3CH_2N(CH_2CH_2OH)_2$$

Note overalkylation in this epoxide opening

$$(CH_3)_2NH + CH_2{=}CH{-}C({=}O){-}R \xrightarrow{H_2O} (CH_3)_2N{-}CH_2CH_2{-}C({=}O){-}R$$

Michael addition

4. β-Aminoketones (and aldehydes)

$$H_3C{-}C({=}O){-}CH_3 \xrightarrow[\text{2. } H_2O,\ NaOH]{\text{1. } HCl/EtOH/(CH_3)_2NH,\ CH_2{=}O} H_3C{-}C({=}O){-}CH_2{-}CH_2{-}{:}N(CH_3)_2$$

Mannich reaction; there is an intermediate immonium ion; of course this is also an amine synthesis

5. Ammonium ions

$$(CH_3)_3N{:} + CH_3{-}I \xrightarrow{S_N2} (CH_3)_3\overset{+}{N}{-}CH_3\ \ I^-$$

The restrictions of the S_N2 reaction apply

6. Aromatic substitution

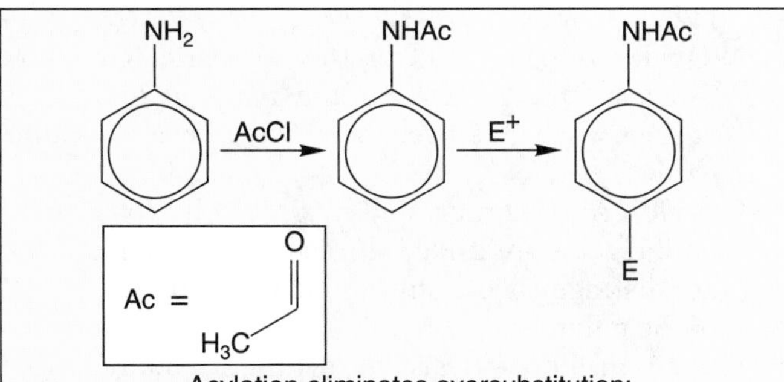

Acylation eliminates oversubstitution; the free amine can be generated by hydrolysis
Acylation also eliminates formation of meta product through reduced protonation of the amine nitrogen

7. Azo compounds

$$C_6H_5{-}N_2^+ + C_6H_5{-}NH_2 \longrightarrow H_2N{-}C_6H_4{-}N{=}N{-}C_6H_5$$

Aromatic substitution using a diazonium ion; the other ring must bear a strongly activating group such as NH_2 or OH

8. Diazo compounds

$$R{-}CH_2NH_2 \xrightarrow[ether]{HONO/HCl} R{-}CHN_2$$

There is an intermediate diazonium ion that can be deprotonated in the absence of nucleophiles

$$RCH_2N(NO){-}C({=}O){-}NH_2 \xrightarrow[ether]{KOH} R{-}CHN_2$$

The ether solvent is necessary to dissolve the dangerous diazo compound

9. Hydroxylamines

$$R_2C{=}N{-}OH \xrightarrow{NaBH_3CN} R_2HC{-}NH{-}OH$$

Reduction with this selective agent does not lead to the amine but stops at the hydroxylamine stage

10. Nitroso compounds

$$(CH_3)_2NH \xrightarrow[HCl]{HONO} (CH_3)_2N{-}N{=}O$$

The nitrosation of secondary amines gives nitroso compounds

11. *N*-Oxides

$$(CH_3CH_2)_3N \xrightarrow[H_2O]{30\%\ HOOH} (CH_3CH_2)_3\overset{+}{N}{-}O^-$$

Peroxide oxidation of tertiary amines gives *N*-oxides (see alkenes)

FIGURE 21.88 (CONTINUED)

COMMON ERRORS

This chapter is quite straightforward. However, there is one point that always seems sticky when first encountered. This point is the relationship between the pK_a of an ammonium ion and the basicity of the related amine. The higher the pK_a of the ammonium ion, the weaker an acid it is, and the stronger the nitrogen–hydrogen bonds. The amine related to an ammonium ion with a high pK_a is a strong base. The lower the pK_a of the ammonium ion, the stronger an acid it is, and the weaker the nitrogen–hydrogen bonds. The amine related to an ammonium ion with a low pK_a is a weak base.

21.11 KEY TERMS

Alkaloids A nitrogen-containing compound, often polycyclic and generally of plant origin. The term is more loosely applied to other naturally occurring amines.

Amide Any of the ions H_2N^-, RHN^-, R_2N^-, $RR'N^-$, or a compound containing the group

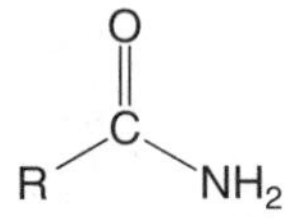

R–C(=O)–NH_2

Amine A compound of the structure R_3N:, where R can be H or another group. Cyclic and aromatic amines are common.

Amine inversion The conversion of one pyramidal form of an amine into the other through a planar, sp^2 hybridized transition state.

Amine oxide A compound of the structure $R_3N^+{-}O^-$. Amine oxides are produced from the reaction of tertiary amines with hydrogen peroxide. See also ***N*-oxide**.

Ammonia, H_3N: The simplest of all amines.

Ammonium ion Tetravalent nitrogen is R_4N^+, R = alkyl, aryl, or H.

Aniline Aminobenzene.

Aziridines Three-membered cyclic amines: azacyclopropanes.

Azo compound A compound of the structure R–N=N–R′.

Cope elimination The thermal formation of alkenes through the pyrolysis of *N*-oxides.

Diazo compounds Compounds of the structure $R_2C{=}N_2$.

Diazotic acid An intermediate in the nitrosation reactions of amines, R–N=N–OH. The enol form of a nitroso compound.

Gabriel amine synthesis A method of forming primary amines without overalkylation to give more substituted amines. Phthalimide is used as a masked amine to introduce the nitrogen atom.

Imide A compound containing the structure O=C–NH–C=O, such as

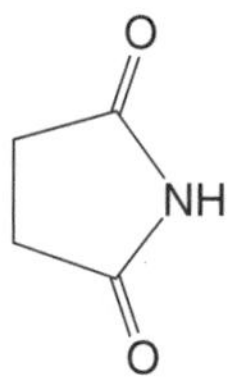

(succinimide structure: O, NH, O)

Indole Any compound containing the following ring system:

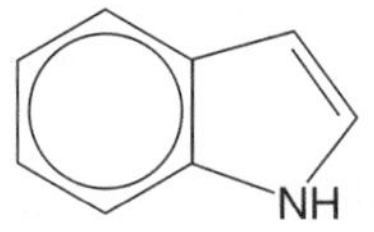

(indole structure: NH)

Indole alkaloids A family of alkaloids, all of which contain the indole ring system.

Lysergic acid diethylamide (LSD) An indole alkaloid of potent hallucinogenic properties.

Morphine An important medicinal alkaloid isolated from the opium poppy.

Nitroso compound A compound containing the N=O group.

***N*-Oxide** A compound of the structure $R_3N^+{-}O^-$. They are produced from the reaction of tertiary amines with hydrogen peroxide. See also **Amine oxide**.

Peptide A polyamino acid in which the constituent amino acids are linked through amide bonds. They are distinguished from proteins only by the length of the polymer.

Phase-transfer catalysis A synthetic method that takes advantage of the ability of alkylammonium ions to transport ions from aqueous to organic media.

Phthalimide The cyclic imide of phthalic acid.

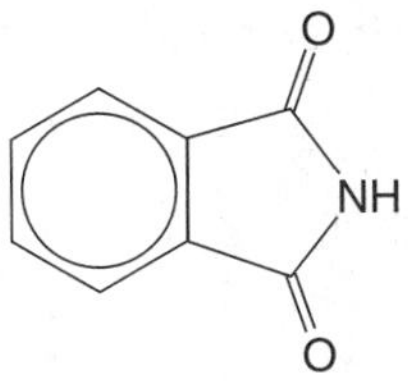

O
NH
O

Piperidine A six-membered ring amine: azacyclohexane.

Primary amines An amine bearing only one R group and two hydrogens.

Protein A polyamino acid in which the constituent amino acids are linked through amide bonds. They are distinguished from peptides only through the length of the polymer.

Pyridine Azabenzene.

Pyrrolidine A saturated five-membered ring amine: azocyclopentane.

Quaternary ammonium ions A compound of the structure $R_4N^{+}\ {}^{-}X$.

Quinine An important antimalarial alkaloid containing the quinoline ring structure.

Reductive amination A synthetic method in which an imine or enamine is generated from a carbonyl compound and an amine only to be reduced in situ to give the substituted amine.

Schmidt reaction The formation of amines through the reaction of carboxylic acids with hydrazoic acid, HN_3. The key step is an intramolecular rearrangement to generate an isocyanate that is then converted into an unstable carbamic acid under the reaction conditions. Decarboxylation gives the amine.

Secondary amine An amine bearing one hydrogen and two R groups.

Tertiary amine An amine bearing no hydrogens and three R groups.

Tropane alkaloids A family of alkaloids, all of which contain the tropane ring system.

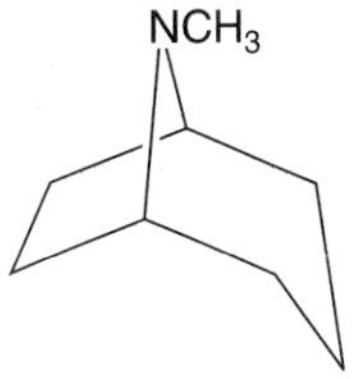

NCH_3

21.12 ADDITIONAL PROBLEMS

PROBLEM **21.27** Name the following compounds (in more than one way, if possible). Identify each as a primary, secondary, or tertiary amine, or as a quaternary ammonium ion.

(a) NH_2

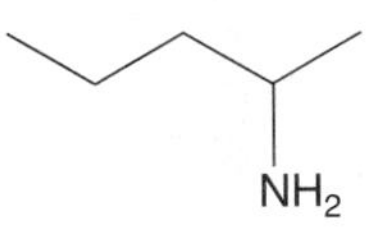

(b) NH, 2

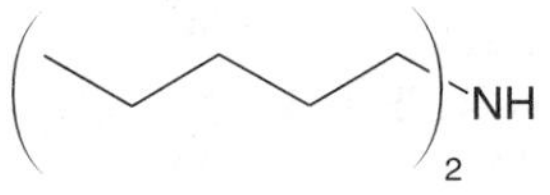

(c) N, 3

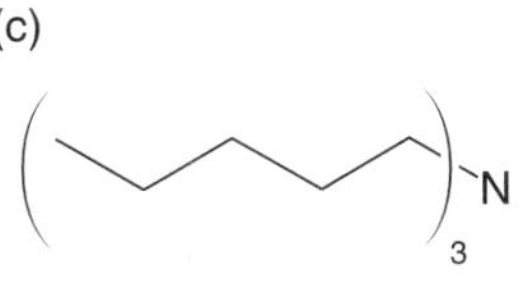

(d) F, NH_2

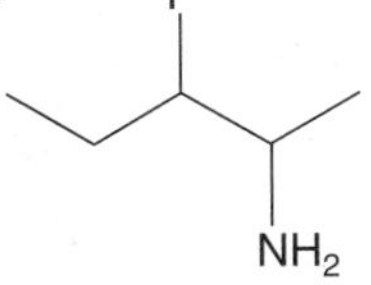

(e) N^{+} I^{-}, 4

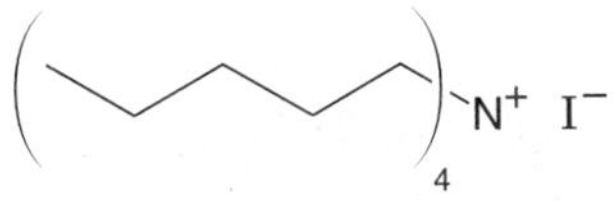

(f) NH_2, NH_2

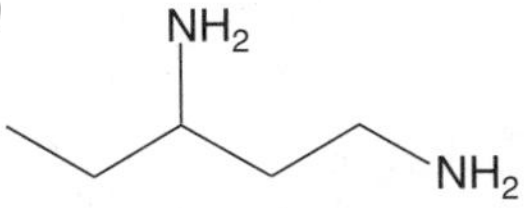

PROBLEM **21.28** Name the following compounds (in more than one way, if possible). Identify each as a primary, secondary, or tertiary amine.

(a) NH_2

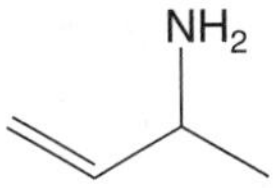

(b) Ph, N, CH_3, CH_2CH_3

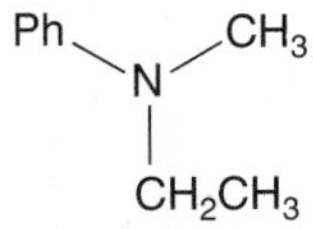

(c) CH_2NH_2

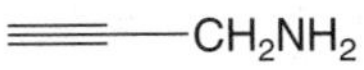

(d) NH_2, OH

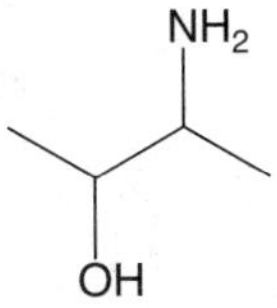

PROBLEM 21.29 Name the following compounds:

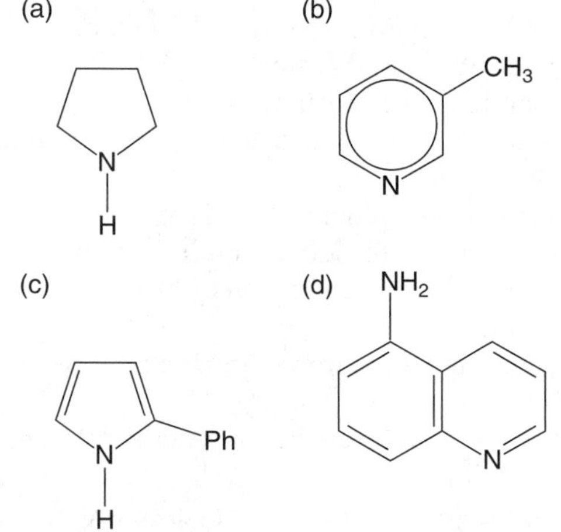

PROBLEM 21.30

(a) Reaction of 1,2-dimethylpyrrolidine (**1**) with ethyl iodide leads to two products, $C_8H_{18}IN$. Explain.

(a)

$$\mathbf{1} \xrightarrow{CH_3CH_2I} \text{Two } C_8H_{18}IN$$

(b) However, when 2-methylpyrrolidine (**2**) undergoes a similar reaction with ethyl iodide, followed by treatment with base, analysis of the product by ^{1}H NMR spectroscopy reveals only one set of signals for $C_7H_{15}N$. Explain.

(b)

$$\mathbf{2} \xrightarrow{CH_3CH_2I} \text{One } C_7H_{15}N$$

PROBLEM 21.31 Explain why compound **1** shows five signals in the ^{13}C NMR spectrum at low temperature but only three at higher temperature.

1

PROBLEM 21.32 Write the expressions for K_a and K_b and show that $pK_a + pK_b = 14$.

PROBLEM 21.33 The pK_b values for a series of simple amines are given below. Explain the relationship between pK_b and base strength, and then rationalize the nonlinear order in the figure.

Amine	$:NH_3$	$:NH_2CH_3$	$:NH(CH_3)_2$	$:N(CH_3)_3$
pK_b	4.76	3.37	3.22	4.20

PROBLEM 21.34 Explain why the hydrogens α to an amino group (R–$\underline{CH_2}$–NH_2) appear at higher field in the ^{1}H NMR spectrum than those α to a hydroxyl group (R–$\underline{CH_2}$–OH).

PROBLEM 21.35 How many signals for the methylene hydrogens (underlined) of $CH_3\underline{CH_2}NHPh$ (**1**) will appear in the ^{1}H NMR spectrum? Analyze **1** exactly as drawn in the figure.

1

In fact, only a single signal appears for the methylene hydrogens of the real compound **1.** Explain.

PROBLEM 21.36 How would you carry out the following conversions of ethylamine and diethylamine? Just give the reagents used in the transformation. Mechanisms are not needed.

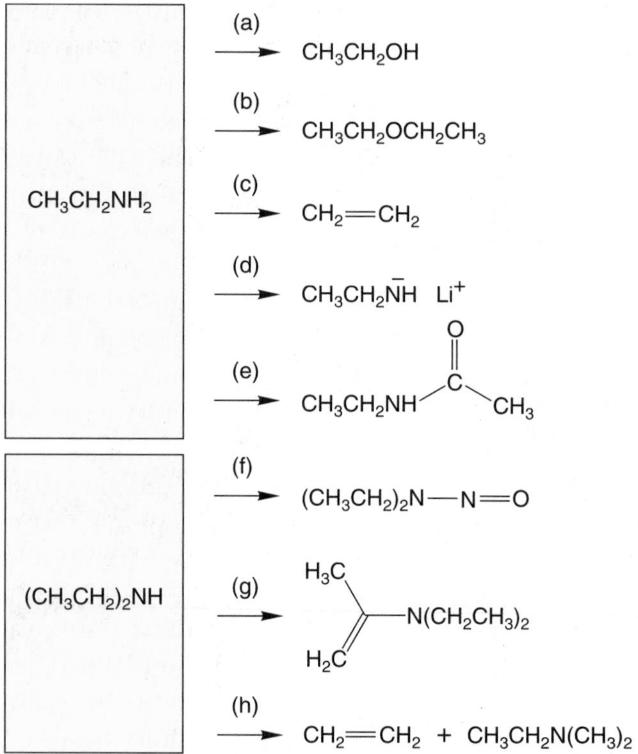

PROBLEM 21.37 How would you convert aniline into each of the following molecules? Mechanisms are not necessary.

(a) 4-bromoaniline (NH_2, Br)
(b) 2,4,6-tribromoaniline (NH_2, Br, Br, Br)
(c) COOH
(d) CH_2NH_2
(e) Br, NO_2

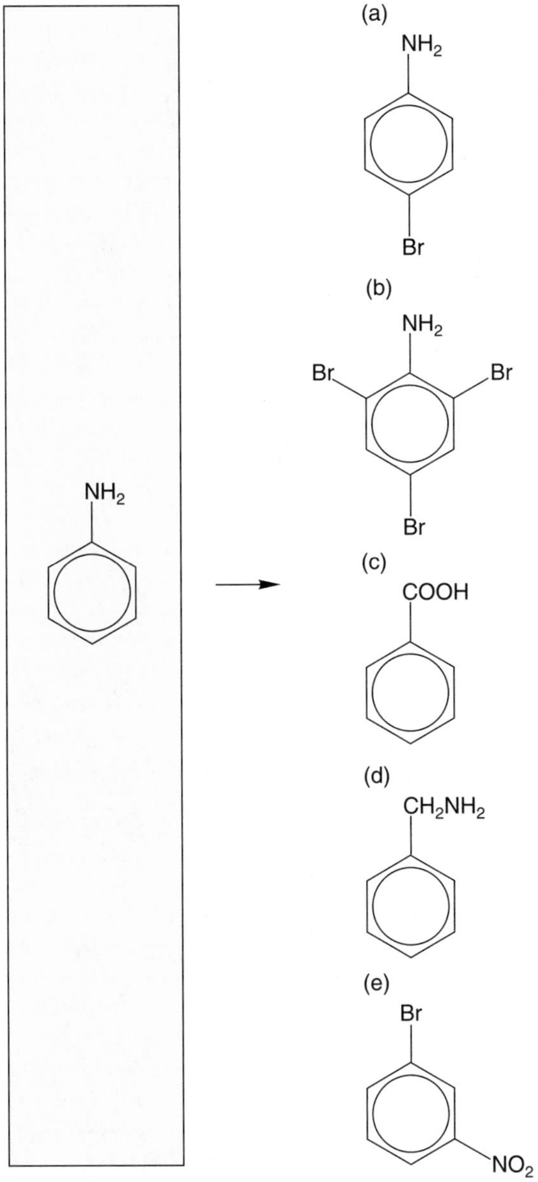

PROBLEM 21.38 Devise a scheme for transforming the amino acid, glycine (H_2NCH_2COOH) into the cyclopropane **1**. You may start with glycine, organic reagents containing no more than one carbon, and the alkene, 2,3-dimethyl-2-butene.

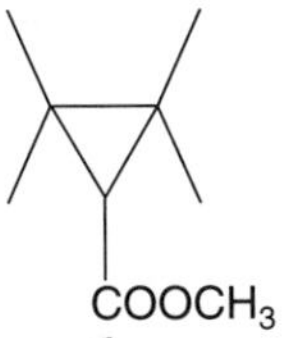

$COOCH_3$
1

PROBLEM 21.39 Provide structures for compounds **A–E**. Mechanisms are not necessary.

COOH → $SOCl_2$, 150 °C → **A** ($C_7H_{11}ClO$) → $(CH_3)_2NH$, benzene → **B** ($C_9H_{17}NO$) → 1. $LiAlH_4$, ether, 35 °C; 2. H_2O → **C** ($C_9H_{19}N$) → 30% H_2O_2, 25 °C → **D** ($C_9H_{19}NO$) → 100 °C, 2 h → **E** (C_7H_{12})

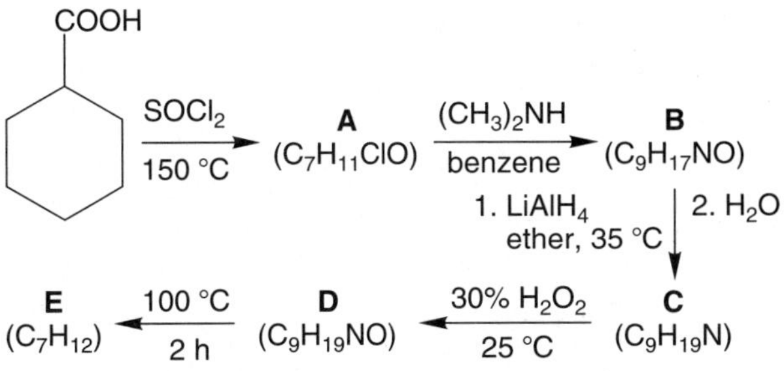

PROBLEM 21.40 Provide a structure for compound **A** and outline a mechanism for its formation. *Hint*: Decide first what reaction this is, then work through the mechanism. *Another hint*: The ^{1}H NMR spectrum is necessary to make a decision about the regiochemistry of this reaction.

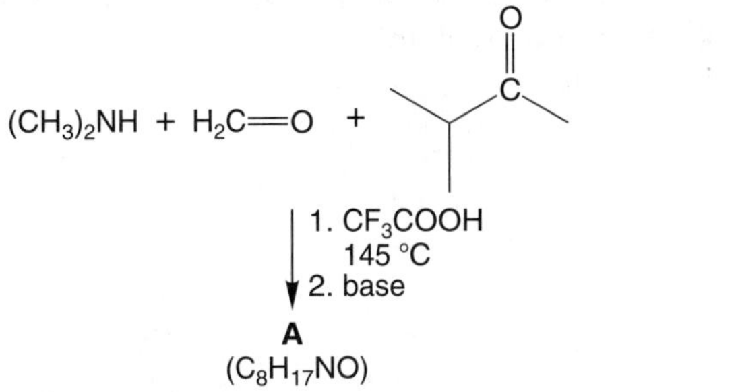

$(CH_3)_2NH$ + $H_2C{=}O$ + (ketone, O, C)

1. CF_3COOH, 145 °C
2. base

A ($C_8H_{17}NO$)

A ^{1}H NMR ($CDCl_3$):
δ: 1.10 (d, J = 7 Hz, 6H), 2.23 (s, 6H), 2.60 (m, 5H)

PROBLEM 21.41 Provide a mechanism for the following transformation of morphine into apomorphine. **Caution**: This problem is very hard.

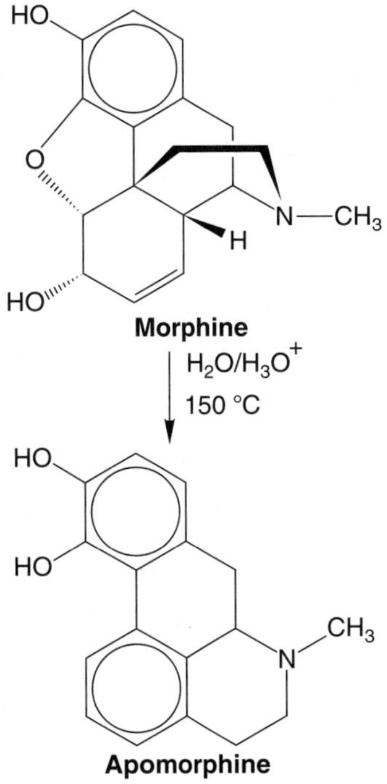

HO, O, N—CH_3, H, HO
Morphine

H_2O/H_3O^+, 150 °C

HO, HO, N, CH_3
Apomorphine

PROBLEM **21.42** Provide arrow formalism mechanisms for the following reactions:

(a)

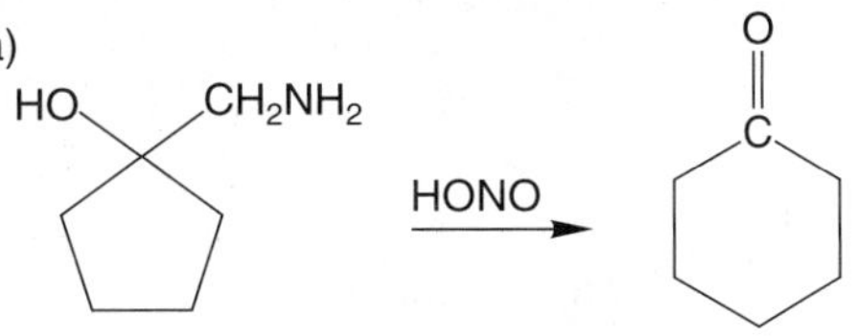

HO, CH_2NH_2 —HONO→ (cyclohexanone)

(b)

OH, CH_2NH_2 —HONO→ (two bicyclic ketones) +

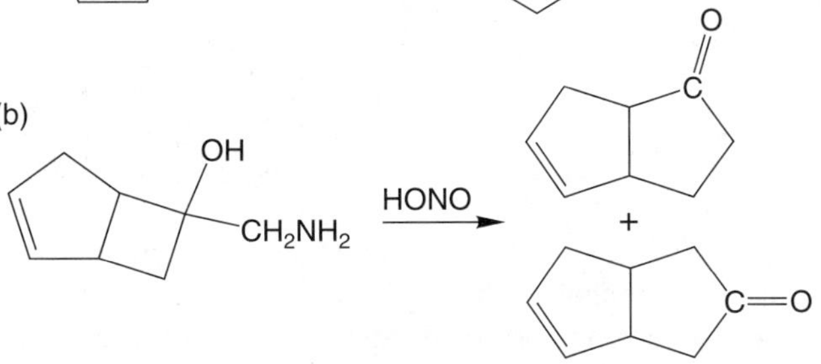

PROBLEM **21.43** Rationalize the different products formed upon treatment of the stereoisomeric aminocyclohexanols **1–4** with nitrous acid.

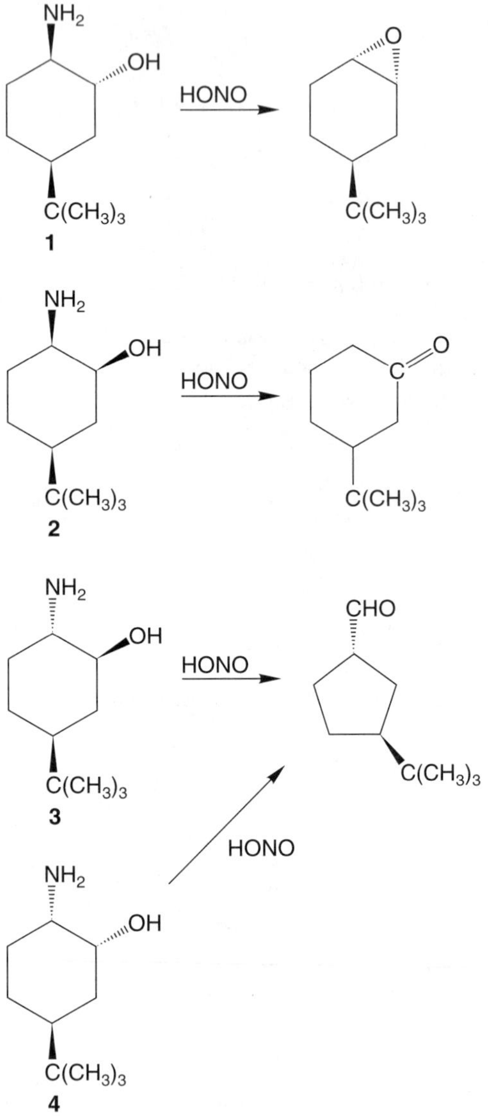

NH_2, OH, $C(CH_3)_3$ **1** —HONO→ $C(CH_3)_3$

NH_2, OH, $C(CH_3)_3$ **2** —HONO→ $C(CH_3)_3$

NH_2, OH, $C(CH_3)_3$ **3** —HONO→ CHO, $C(CH_3)_3$

NH_2, OH, $C(CH_3)_3$ **4** —HONO→

PROBLEM **21.44** As shown in the following figure, deamination of bornylamine (**1**) leads to a variety of rearranged products. Provide mechanisms for the formations of these products.

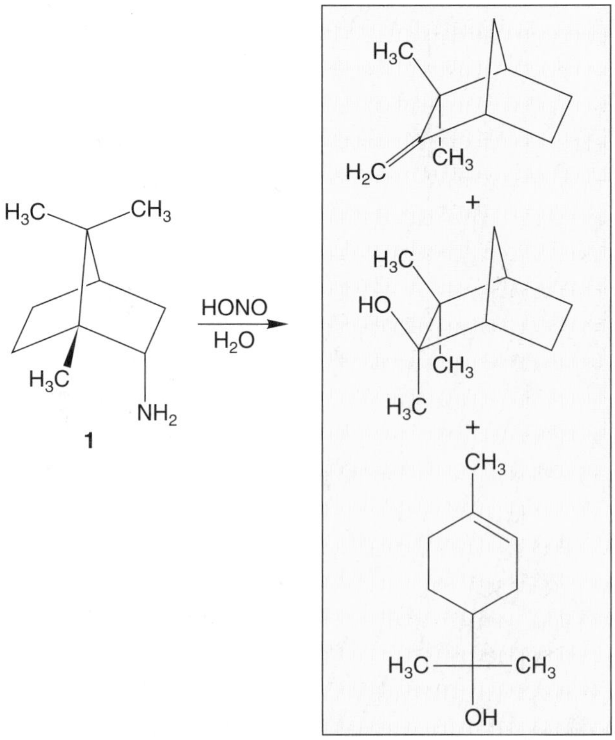

H_3C, CH_3, H_3C, NH_2 **1** —HONO, H_2O→ H_3C, H_2C, CH_3 + H_3C, HO, CH_3, H_3C + CH_3, H_3C, CH_3, OH

PROBLEM **21.45** In the following synthesis of racemic nicotine (**1**), propose structures for intermediates **A–C**. Although mechanisms are not required, a mechanistic analysis will probably prove helpful, as there are several nonisolable intermediates involved.

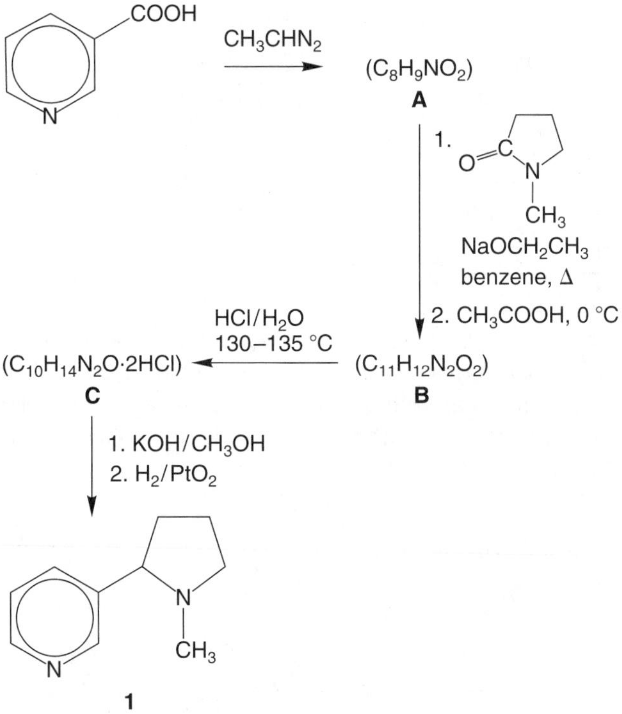

(pyridine-3-COOH) —CH_3CHN_2→ $(C_8H_9NO_2)$ **A**

A —1. (N-methylpyrrolidinone, CH_3), $NaOCH_2CH_3$, benzene, Δ; 2. CH_3COOH, 0 °C→ $(C_{11}H_{12}N_2O_2)$ **B**

B —HCl/H_2O, 130–135 °C→ $(C_{10}H_{14}N_2O{\cdot}2HCl)$ **C**

C —1. KOH/CH_3OH; 2. H_2/PtO_2→ (nicotine, N, CH_3) **1**

PROBLEM **21.46** In 1917, Robinson reported a synthesis (called "von bewundernswert Eleganz" by the German chemist Willstätter) of tropinone (**3**) involving the condensation of succindialdehyde (**1**), methylamine, and 1,3-acetone dicarboxylic acid (**2**). Schöpf subsequently improved the yield of this reaction to about 90%. Propose a plausible mechanism for this reaction. *Hints*: This reaction involves a double condensation of some kind. Acetone can be used in place of **2**, although the yield is reduced.

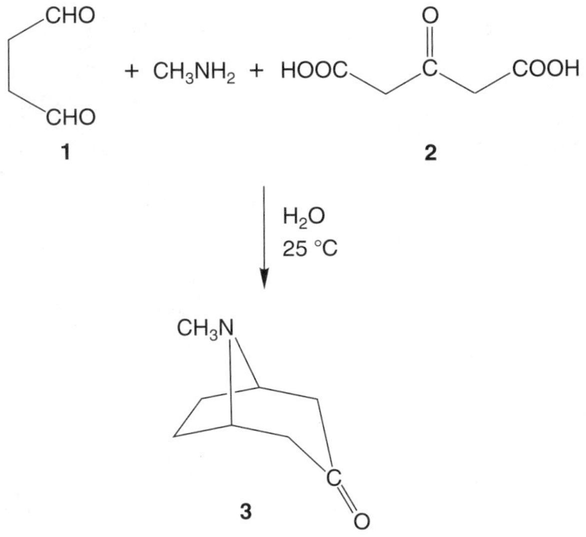

CHO
CHO
1
+ CH_3NH_2 + HOOC C(=O) COOH
2

H_2O
25 °C

CH_3N
C
O
3

Aromatic Transition States: Orbital Symmetry

The fascination of what's difficult
Has dried the sap of my veins, and rent
Spontaneous joy and natural content
Out of my heart.
—William Butler Yeats*
The fascination of what's difficult

In Chapters 12–14, we explored the consequences of conjugation—the sideways overlap of 2*p* orbitals. This study led us to the aromatic compounds, whose great stability can be traced to an especially favorable arrangement of electrons in low-lying bonding molecular orbitals. It can surely be no surprise to find that transition states can also benefit energetically through delocalization, and that most of the effects, including aromaticity, that influence the energies of ground states are important to transition states as well.

In this chapter, we will encounter reactions that were frustratingly difficult for chemists for many, many years. By now, you are used to seeing acid- and base-catalyzed reactions, in which an intermediate is first formed and then produces product with the regeneration of the catalytic agent. Acid-catalyzed additions to alkenes are classic examples, but there are already many other such reactions in your notes. There is a class of reactions that is extraordinarily insensitive to catalysis. Bases, acids, and radicals are largely without effect. Even the presence of solvent seems of little relevance, as the reactions proceed as well in the gas phase as in solution. What is one to make of such reactions? How does one describe a mechanism? We are used to speculating on the structure of an intermediate and then using the postulated picture to predict the structures of the transition states surrounding it. In these uncatalyzed reactions in which starting material and product are separated by a single transition state (called **concerted reactions**), there is very little with which to work in developing a mechanism. Indeed, one may legitimately ask what "mechanism" means in this context. Such processes became known as "no-mechanism" reactions. Some, in which starting material simply rearranges into itself (a **degenerate reaction**) can even be called "no-mechanism, no-reaction, reactions." Figure 22.1 shows an arrow formalism for a typical no-mechanism, no-reaction, reaction.

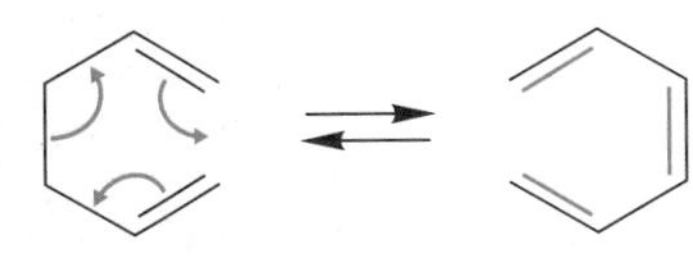

FIGURE 22.1 A single-barrier, one-transition-state, "no-mechanism" reaction.

*William Butler Yeats (1865–1939), was an Irish poet who received the Nobel prize for literature in 1923.

In 1965, R. B. Woodward (1917–1979) and Roald Hoffmann (b. 1937), both then at Harvard, began to publish a series of papers that ventured a mechanistic description of no-mechanism reactions, and gathered a number of these seemingly different processes under the heading of **pericyclic reactions**. Their crucial insight that bonding overlap must be maintained between orbitals during the course of a concerted, pericyclic reaction now seems so simple that you may find it difficult to see why it eluded chemists for so many years. All I can tell you is that simple things are sometimes very hard to see, even for very smart people. **Woodward–Hoffmann theory** had been approached very closely before without the crucial "aha!"; without the lightbulb over the head turning on. Perhaps what was lacking was the ability to combine a knowledge of theory with an awareness of the chemical problem, which is precisely what the brilliant experimentalist Woodward and the young theorist Hoffmann brought to the problem. Ultimately, these papers and their progeny resulted in the Nobel prize of 1981 for Hoffmann [together with Kenichi Fukui (b. 1918) of Kyoto], and would probably have given Woodward his second Nobel prize, had he lived long enough.

The reaction of the chemical community to these papers ranged from enthusiastic admiration to "We knew all this trivial stuff already." I would submit that admiration was by far the more appropriate reaction, and suspect that behind much of the carping was a secret smiting of many foreheads. Woodward–Hoffmann theory is brilliant. It not only answered longstanding and difficult questions, but it had remarkable "legs." Its implications run far into organic and the rest of chemistry, and it makes what are called risky predictions. It is important to separate theory that merely rationalizes—that which explains known phenomena—from that which demands new experiments. The success of Woodward–Hoffmann theory can be partly judged from the flood of experiments it generated. Of course, not all the experiments were incisive, and not all the interpretations were appropriate—the area entered a rococo phase quite early on. Still, the best of the work generated by the early Woodward–Hoffmann papers stands today as an example of the combination of theory and experiment that characterizes the best of our science. In this chapter, we will see electrocyclic reactions in which rings open and close, cycloaddition reactions in which two partners come together to make a new cyclic compound, and sigmatropic shifts in which one part of a molecule flies about, coming to rest at one specific position, and no other. Specificity is the hallmark of all these reactions; atoms move as if in lock step in one direction or another, or move from one place to one specific new place, but no other. The chemical world struggled mightily to understand these mysteriously stereospecific reactions, and finally figured it out in a marvelously simple way. This chapter will show you some wonderful chemistry.

22.1 CONCERTED REACTIONS

Woodward and Hoffmann provided insights into concerted reactions. A reaction is concerted if starting material goes directly to product over a single transition state. Such reactions are "single-barrier" processes. The S_N2 and Diels–Alder reactions (Chapters 7 and 12) are nice examples of concerted reactions you know well (Fig. 22.2).

Of course many reactions involve intermediates, and in these reactions more than one transition state is traversed on the way from starting mate-

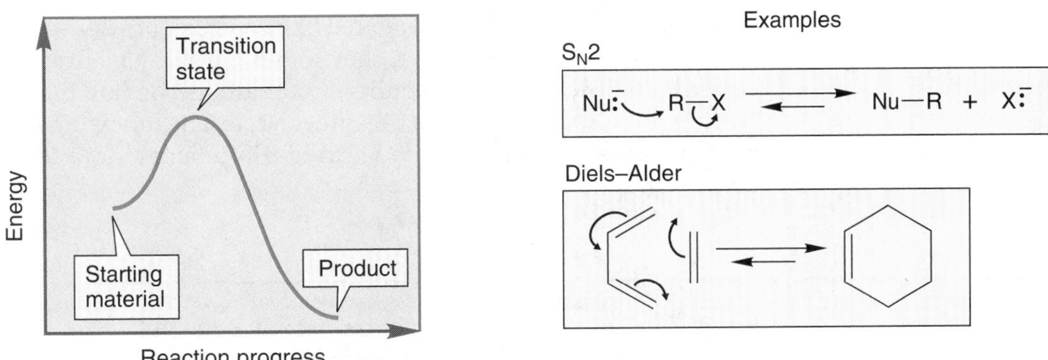

FIGURE 22.2 The S_N2 and Diels–Alder reactions are concerted, single-barrier processes.

rial to product. These are called nonconcerted or stepwise reactions. Typical examples are the S_N1 reaction or the polar addition of HBr to an alkene, two reactions that go through the same intermediate, the *tert*-butyl cation (Chapters 7 and 9; Fig. 22.3).

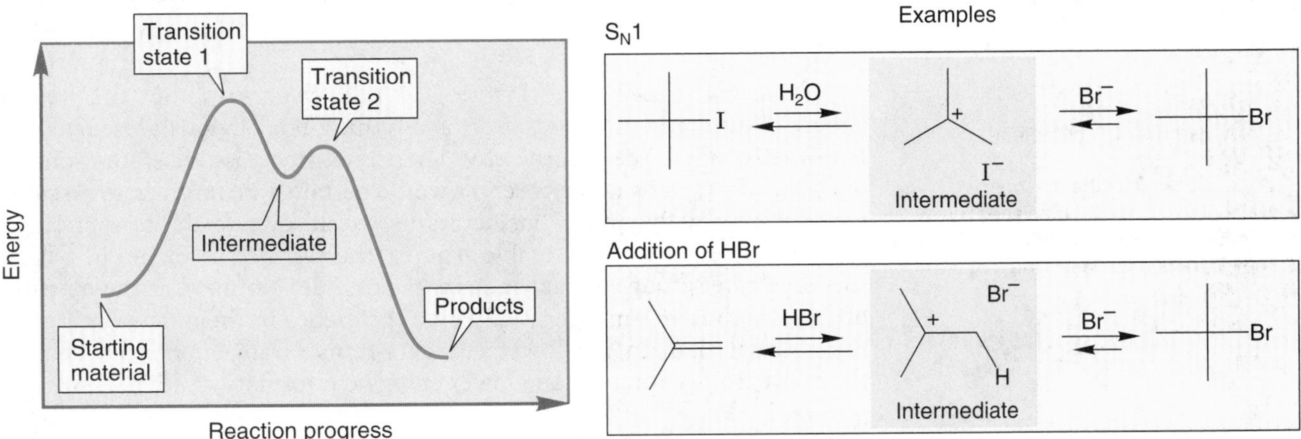

FIGURE 22.3 Two reactions in which starting material and product are separated by a pair of transition states surrounding an intermediate.

It is important to recognize that nonconcerted reactions are made up of series of single-barrier, concerted reactions, each of which could be analyzed independently of whatever other reactions followed or preceded (Fig. 22.4).

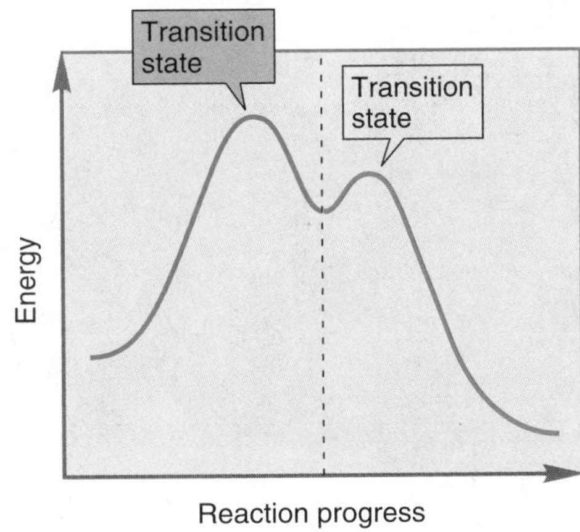

FIGURE 22.4 Any reaction can be separated into a series of single-barrier, concerted processes.

Woodward and Hoffmann offered explanations of why some reactions were concerted and others not; they found a way to analyze reactions by determining the pathways required to maintain bonding interactions between orbital lobes as the reaction progressed. The following sections take up some examples in which Woodward–Hoffmann theory is particularly useful.

22.2 ELECTROCYCLIC REACTIONS

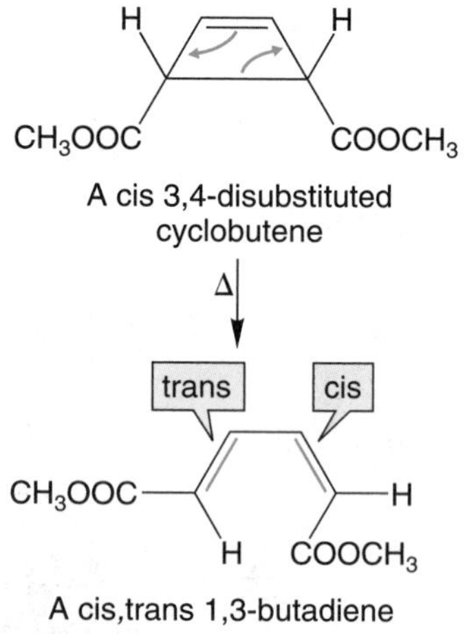

In 1957, a young German chemist named Emanuel Vogel (b. 1927) was finding his way through the labyrinth of the German academic system, working on the experiments necessary for the entrance ticket for a German university: a super Ph.D. degree called the "Habilitätion." His topic included the thermal rearrangements of cyclobutenes. Although cyclobutene itself had been studied by R. Willstätter (1872–1942, Nobel prize in 1915) as early as 1905, and the rearrangement to 1,3-butadiene noted by the American chemist John D. Roberts (b. 1918) in 1949, only Vogel was alert enough to notice the startling stereospecificity of the thermal reaction. For example, the cis 3,4-disubstituted cyclobutene shown in Figure 22.5 rearranges *only* to the cis,trans 1,3-butadiene and no other stereoisomer.

A cis 3,4-disubstituted cyclobutene

A cis,trans 1,3-butadiene

FIGURE 22.5 The thermal opening of a cyclobutene to a 1,3-butadiene. The cis disubstituted cyclobutene rearranges to the cis,trans butadiene. The Δ means heat.

The arrow formalism of Figure 22.5 is easy to write, but the stereochemical outcome of the reaction is anything but obviously predictable. It's worthwhile to see exactly why. There are three possible stereoisomers of the product, and a naive observer would certainly be forgiven for assuming that thermodynamic considerations would dictate the formation of the most energetically favorable isomer, the trans,trans diene, in which the large ester groups are as far apart as possible. Notice that the opening of the cyclobutenes must initially give the dienes in their s-cis conformations rather than in the lower energy s-trans arrangements. Of course, they will rapidly rotate to the lower energy conformations (Fig. 22.6).

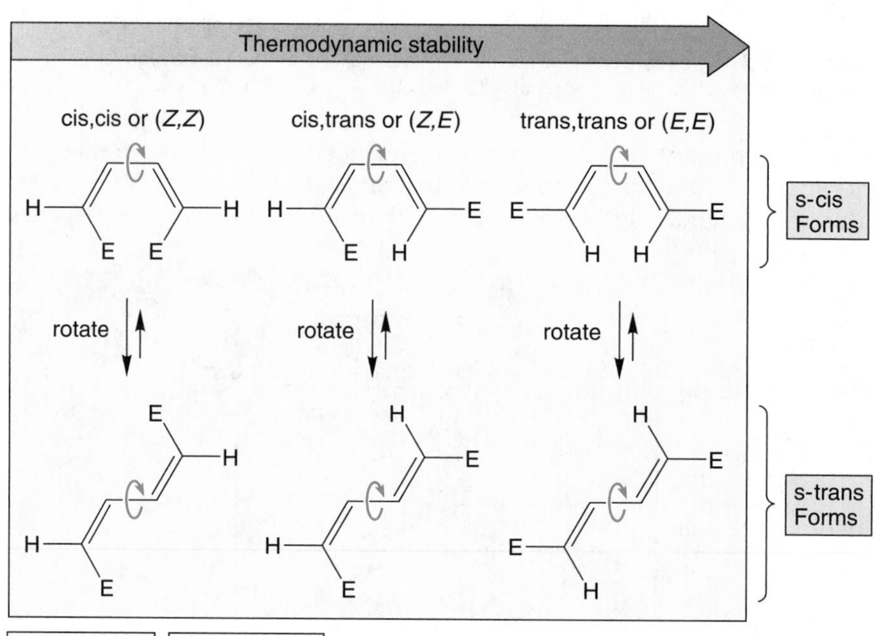

E = $COOCH_3$ *E* = Entgegen

FIGURE 22.6 The three possible stereoisomers of the diene product. But, only one, the cis,trans (*Z*,*E*) diene of intermediate thermodynamic stability, is formed in the reaction.

As Vogel first found, and as others showed later for many other systems, this is not the actual result at all; nor is the reaction stereorandom, as there is only a single stereoisomer found. Vogel well understood that this reaction was rather desperately trying to tell us something. In the intervening years, many similar reactions were found and many a seminar hour was spent in a fruitless search for the explanation. What the reaction had to say to us waited for the papers of Woodward and Hoffmann.

Remarkably, it was discovered that the stereochemical outcome of the reaction was dependent on the energy source. Heat gave one stereochemical result and light another. Figure 22.7 sums up the results of the thermal and photochemical interconversions of cyclobutenes and butadienes. In the thermal reaction, a cis cyclobutene is related to the cis,trans diene. By contrast, in the photochemical process the cis cyclobutene is related to a pair of molecules, the cis,cis diene and the trans,trans diene (Fig. 22.7).

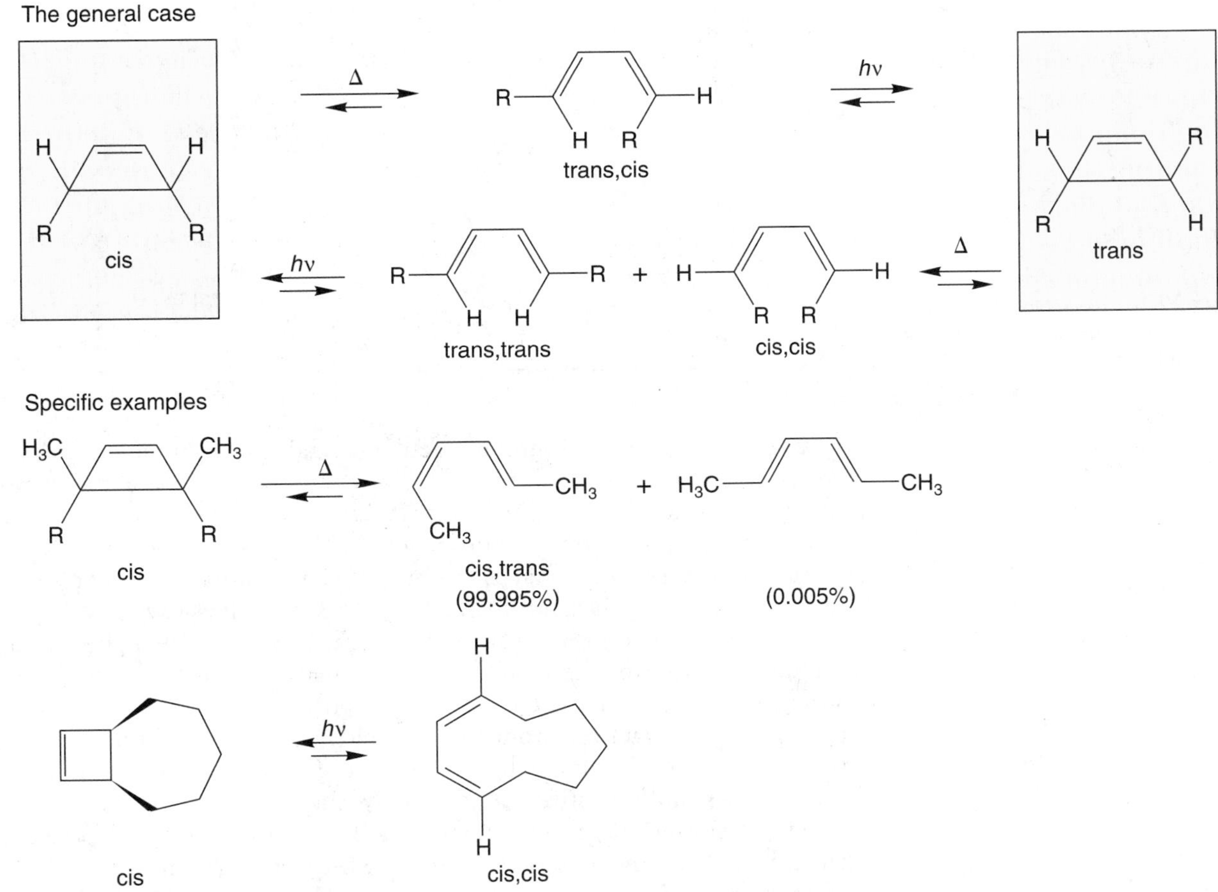

FIGURE 22.7 The stereochemical outcomes of the thermal and photochemical reactions of cyclobutenes and butadienes are different.

Notice that the two reactions of Figure 22.7 run in different directions. The thermal reaction favors the more stable 1,3-butadiene isomer, but the photochemical process uses the energy of the photon to convert the 1,3-butadiene into the less stable, more energetic cyclobutene.

Let's start our analysis by thinking about the thermal opening of cyclobutene to 1,3-butadiene. Here is a general practical hint. It is usually easier to analyze these processes, called **electrocyclic reactions**, by examining *the open-chain partner,* regardless of which way the reaction actually runs. Figure 22.8 shows the molecular orbitals involved in the cyclobutene to butadiene conversion.

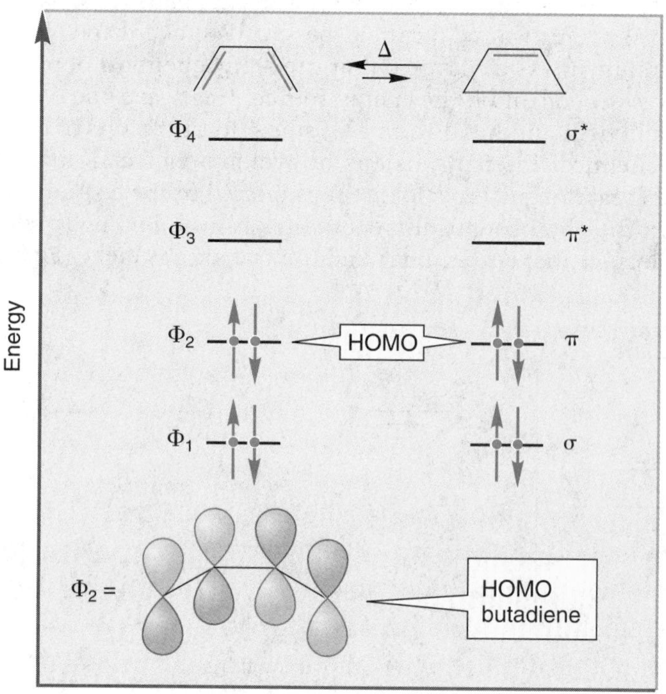

FIGURE 22.8 The molecular orbitals involved in the interconversion of cyclobutene and butadiene.

In this reaction, the four π molecular orbitals of 1,3-butadiene are converted into the σ, σ*, π, and π* orbitals of cyclobutene. Now comes a most important point. *It is assumed that it will be the HOMO of the system that controls the course of this reaction.* This assumption is not ultimately necessary, but it makes working out the results of this and similar reactions very easy. It can be defended in the following way. It is the electrons of highest energy, the valence electrons located in the highest energy orbital, that control the course of atomic reactions, and it should be no surprise that the same is true for molecular reactions. To do a complete analysis of any reaction, atomic or molecular, we really should follow the changes in energy of all electrons. That is difficult or impossible in all but a very few simple reactions. Accordingly, theorists have sought out simplifying assumptions, one of which says that it will be the highest energy electrons, those most loosely held, that will be most important in the reaction. These are the electrons in the Highest Occupied Molecular Orbital, the HOMO. The HOMO of butadiene is Φ_2, and we will assume that this is the controlling molecular orbital. Now ask how Φ_2 must move as 1,3-butadiene becomes cyclobutene. Don't worry that we are analyzing the reaction in the counterthermodynamic sense. If we know about the path in one direction, we know about the path in the other direction as well. In order to create the bond between the end carbons, the *p* orbitals on the end carbons must rotate, and this rotation must be in a fashion that creates a *bond* between the atoms, not an *antibond* (Fig. 22.9).

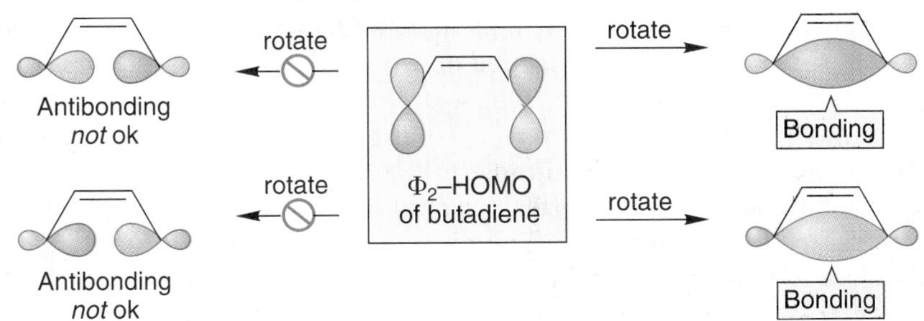

FIGURE 22.9 The π orbitals at the end of the acyclic isomer must rotate so as to create a bond, σ, between the ends of the chain, not an antibond, σ*.

Now look at the symmetry of the end orbitals of Φ_2, the controlling HOMO of butadiene. There are two ways to make a bond: in one the plus (blue) lobes overlap, in the other it is the minus (green) lobes that overlap. In each case we have created a bonding interaction. In this case, the plus/plus overlap requires that the two ends of the molecule each rotate in a clockwise fashion. The minus/minus overlap requires that each end rotate counterclockwise (Fig. 22.10).

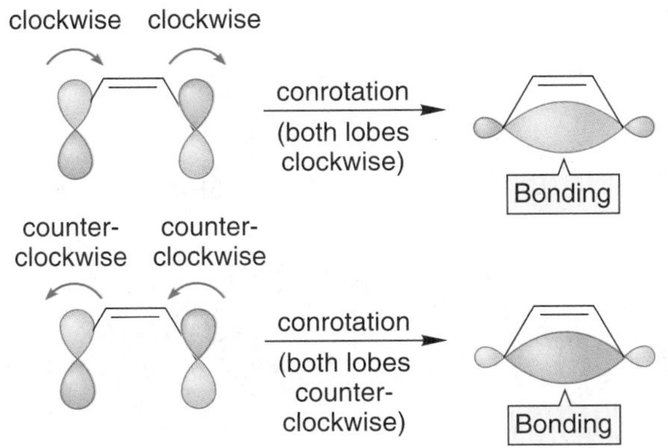

FIGURE 22.10 Formation of a bonding interaction requires what is called conrotation of the ends of the diene. There are two, equivalent conrotatory modes.

These two modes of rotation, in which the two ends rotate in the *same* direction, are called conrotatory motions, and the whole process is known as **conrotation**.

There is another rotatory mode possible, called **disrotation**, in which the two ends of the molecule rotate in a different sense. Figure 22.11 shows the two possible disrotatory motions. In both, an antibond, not a bond, is created between the two end carbons. Disrotation cannot be a favorable process in this case.

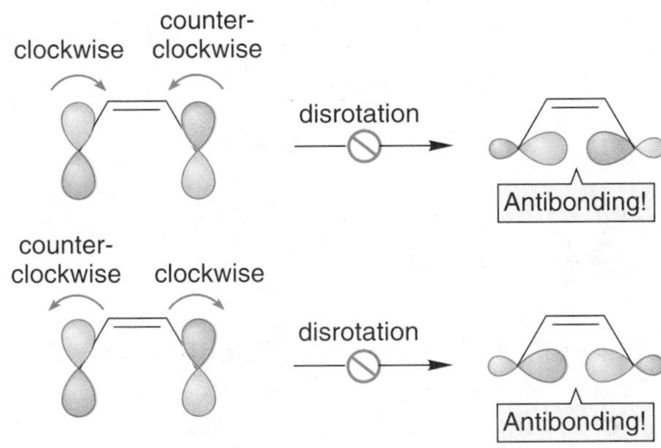

FIGURE 22.11 Disrotation creates an antibonding interaction between the lobes at the ends of the diene. Neither of the two possible disrotatory modes can be a favorable process.

This reaction is as easy to analyze as that. Identify the HOMO, and then see what sort of motion is demanded of the end carbons by the lobes of the molecular orbital. All electrocyclic reactions can be understood in this same simple way. The theory tells us that the thermal interconversion of cyclobutene and 1,3-butadiene must take place in a conrotatory way. For the cyclobutene studied by Vogel, conrotation requires the stereochemical relationship that he observed. The cis 3,4-disubstituted cyclobutene can only open in conrotatory fashion, and conrotation *forces* the formation of the cis,trans diene. Note that there are always two possible conrotatory modes (Fig. 22.12).

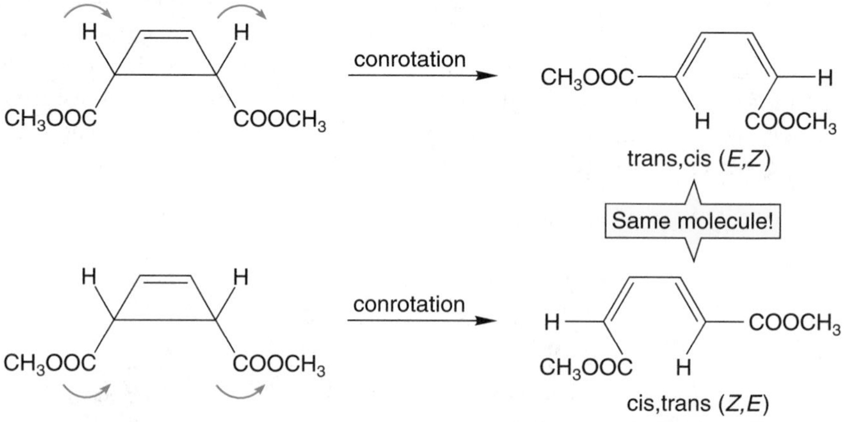

H H conrotation CH_3OOC H CH_3OOC $COOCH_3$ H $COOCH_3$

trans,cis (*E*,*Z*)

Same molecule!

H H conrotation H $COOCH_3$ CH_3OOC $COOCH_3$ CH_3OOC H

cis,trans (*Z*,*E*)

FIGURE 22.12 For the cis disubstituted cyclobutene studied by Vogel, both conrotatory modes demand the formation of the diene with one cis and one trans double bond.

*PROBLEM 22.1 Show through an equivalent analysis that a mixture of the cis,cis and trans,trans dienes must be formed in the thermal opening of trans 3,4-disubstituted cyclobutenes.

ANSWER As the opening is taking place thermally, it must be conrotatory. There are two conrotatory modes for the trans compound. One leads to the cis,cis (*Z*,*Z*) isomer and the other to the trans,trans (*E*,*E*) isomer.

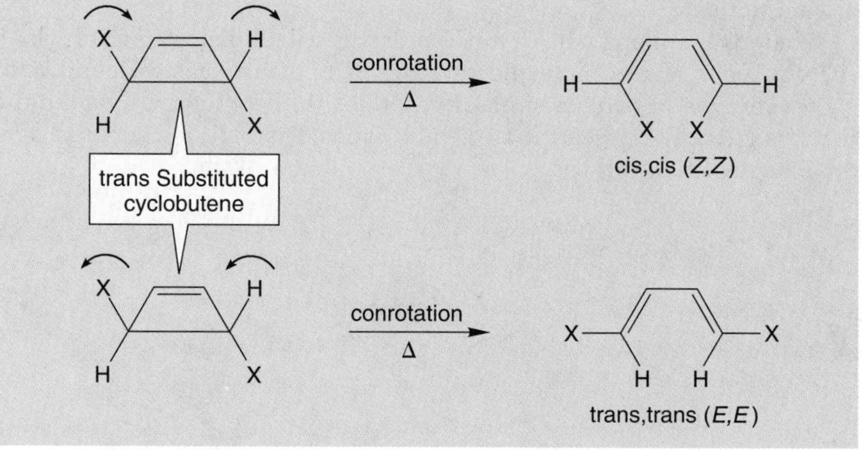

X H conrotation Δ H H H X X X

cis,cis (*Z*,*Z*)

trans Substituted cyclobutene

X H conrotation Δ X X H X H H

trans,trans (*E*,*E*)

Now what about the photochemical reaction? Look again at the molecular orbitals of the open partner, 1,3-butadiene. Recall that absorption of a photon promotes an electron from the HOMO to the LUMO (Chapter 12, p. 514), in this case from Φ_2 to Φ_3, creating a new, "photochemical HOMO" (Fig. 22.13).

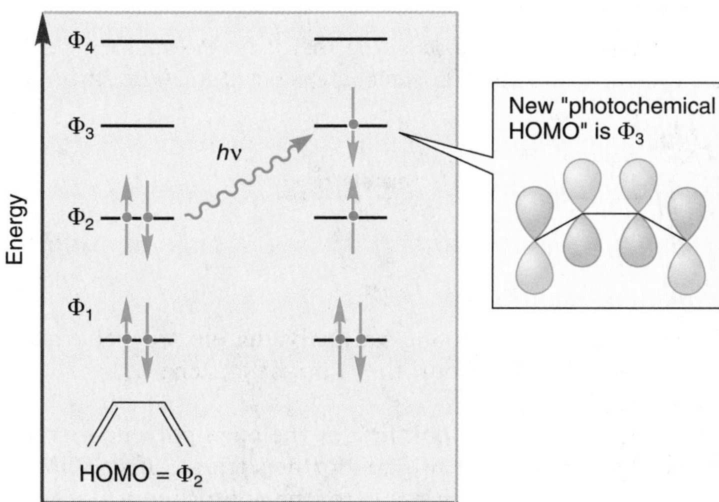

FIGURE 22.13 The HOMO involved in the photochemical reaction of butadiene is Φ_3.

No longer will conrotation in the HOMO, now Φ_3, create a bond between the end carbons. For an orbital of this symmetry, *conrotation creates an antibond.* For the photochemical case, it is disrotation that creates the bonding interaction (Fig. 22.14).

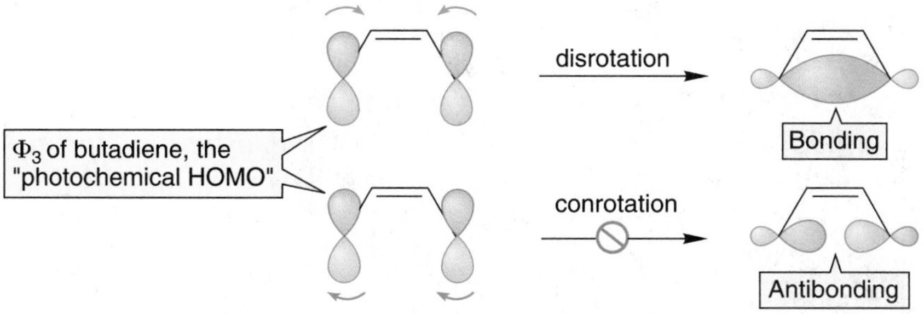

FIGURE 22.14 The new, "photochemical HOMO" (Φ_3) demands disrotation. In this molecular orbital, conrotation produces an antibond.

PROBLEM 22.2 Figure 22.14 shows only one disrotatory and one conrotatory mode. Verify that the other possible disrotation also produces a bond between the two carbons, and that the other possible conrotation produces an antibond.

Two disrotatory modes interconvert the cis 3,4-disubstituted cyclobutene and the cis,cis and trans,trans isomers of the butadiene (Fig. 22.15).

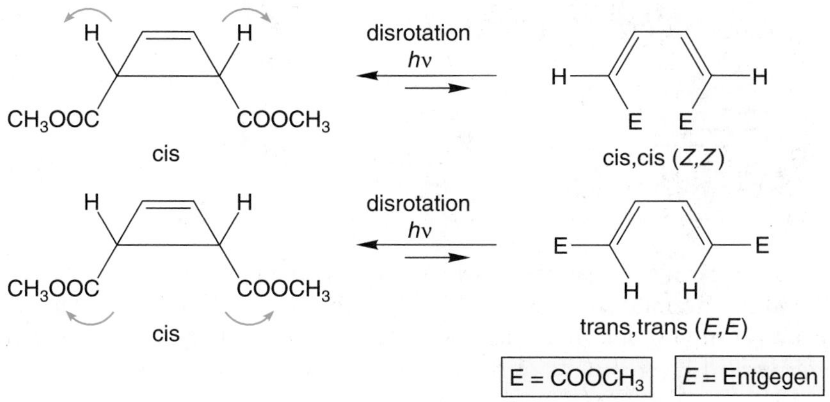

FIGURE 22.15 Disrotation interrelates the cis disubstituted cyclobutene with the cis,cis and trans,trans dienes.

PROBLEM 22.3 Show that disrotation in the photochemical reaction of the trans 3,4-disubstituted cyclobutene leads to the cis,trans isomer of the butadiene (Fig. 22.16).

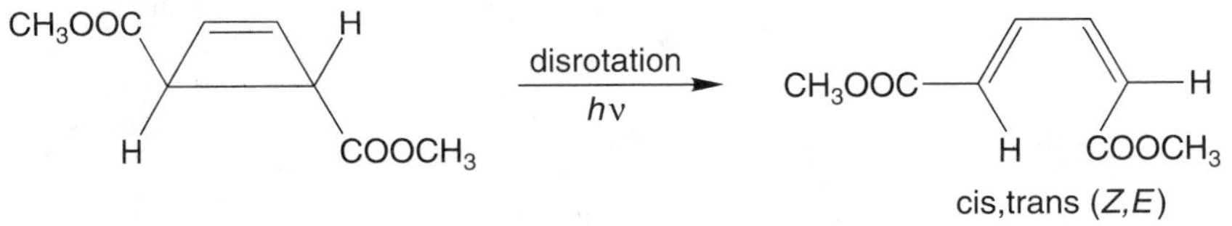

CH_3OOC H H $COOCH_3$ —disrotation, $h\nu$→ CH_3OOC H H $COOCH_3$ cis,trans (*Z*,*E*)

FIGURE 22.16

Let's now use this technique for analyzing electrocyclic reactions to make some predictions about the 1,3,5-hexatriene–1,3-cyclohexadiene system (Fig. 22.17).

As it is always easier to look first at the open polyene system, Figure 22.18 gives the π molecular orbitals for hexatriene. The HOMO for the thermal reaction will be Φ_3, and that for the photochemical reaction, Φ_4.

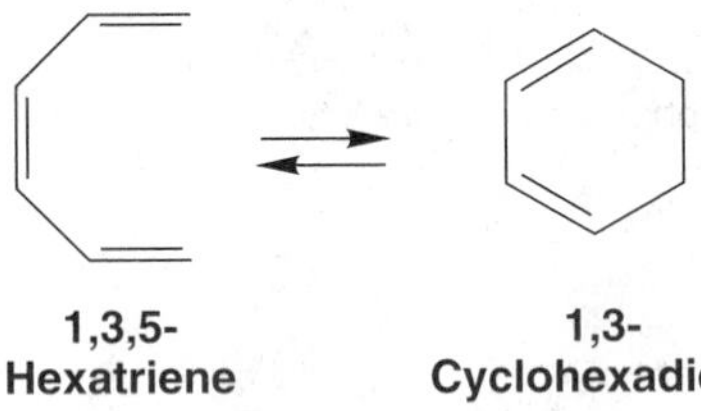

1,3,5-Hexatriene ⇌ **1,3-Cyclohexadiene**

FIGURE 22.17 Another electrocyclic reaction interconverts 1,3,5-hexatriene and 1,3-cyclohexadiene.

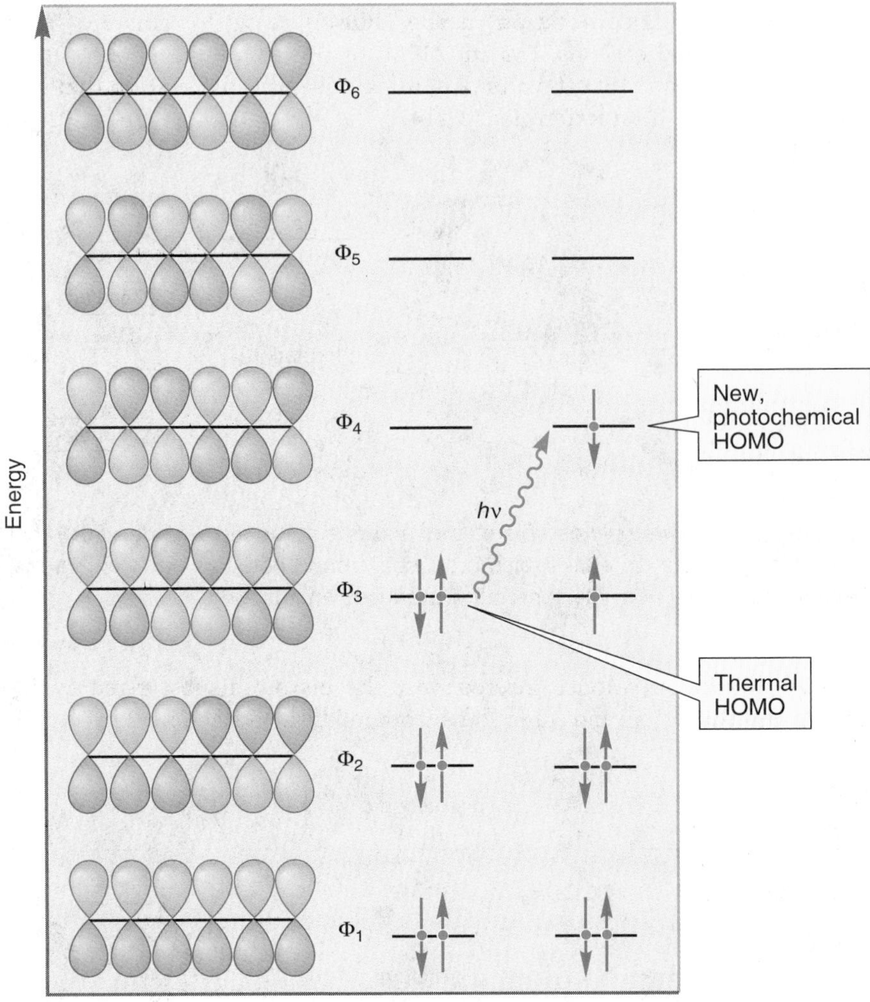

FIGURE 22.18 The π molecular orbitals for hexatriene. The thermal HOMO is Φ_3 and the photochemical HOMO is Φ_4.

As the open triene closes to the cyclohexadiene, the only way to create a bonding interaction between the end carbons is to close in a disrotatory fashion from Φ_3, and in a conrotatory fashion from Φ_4. So, we predict that the thermal interconversion will occur in a disrotatory way, and that the photochemical reaction must involve conrotation (Fig. 22.19).

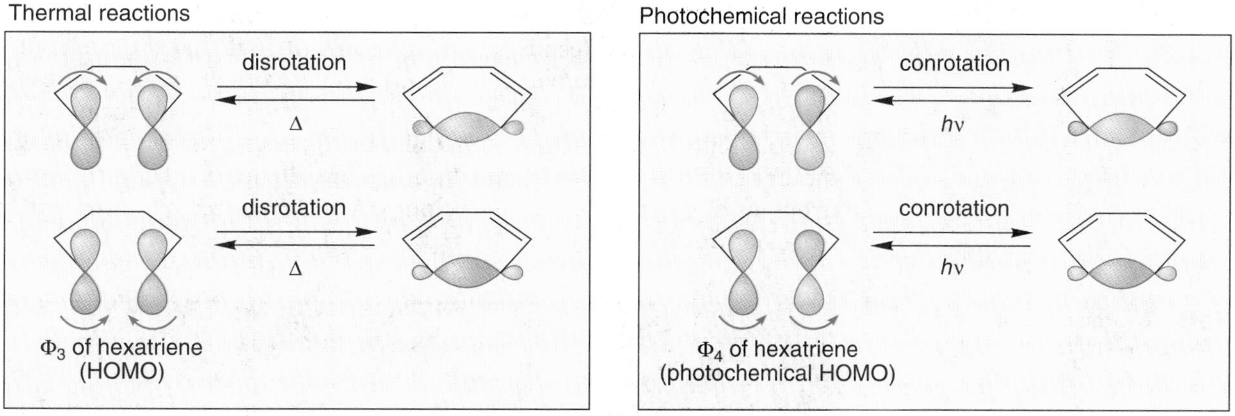

Thermal reactions

Photochemical reactions

disrotation

Δ

disrotation

Δ

Φ_3 of hexatriene (HOMO)

conrotation

$h\nu$

conrotation

$h\nu$

Φ_4 of hexatriene (photochemical HOMO)

FIGURE **22.19** For this system, in order to produce a bonding interaction, the thermal reactions must take place in a disrotatory fashion and the photochemical reactions in a conrotatory way.

To test these predictions we need some labeled molecules. The *trans,cis,trans*-2,4,6-octatriene shown in Figure 22.20 will do fine. A disrotatory thermal closure to *cis*-5,6-dimethylcyclohexa-1,3-diene and a conrotatory photochemical closure to the trans isomer are both known.

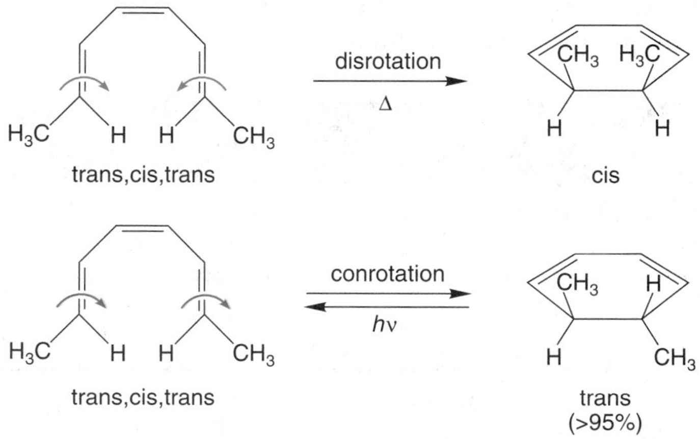

H_3C H H CH_3

trans,cis,trans

disrotation

Δ

CH_3 H_3C

H H

cis

H_3C H H CH_3

trans,cis,trans

conrotation

$h\nu$

CH_3 H

H CH_3

trans (>95%)

FIGURE **22.20** The experimental results bear out the predictions. The thermal reaction is disrotatory and the photochemical reaction is conrotatory.

Figure 22.20 shows the apparent formation of a single enantiomer of *trans*-5,6-dimethylcyclohexa-1,3-diene from an achiral precursor, which cannot be possible. We have not taken the trouble to draw both enantiomers of the racemic mixture of the *trans*-5,6-dimethylcyclohexa-1,3-diene formed in the reaction. This omission leads directly to Problem 22.4.

Convention Alert!

PROBLEM **22.4**

Figure 22.20 shows only one conrotatory and one disrotatory mode. Show that the other disrotation and conrotation give the same stereochemical results.

PROBLEM **22.5**

Why are cyclohexadienes formed only from the trans,cis,trans isomer shown? Why does the trans,trans,trans molecule not give a similar reaction?

PROBLEM **22.6**

Work out the predicted products from the cis,cis,trans and cis,cis,cis isomers of 2,4,6-octatriene for both photochemical and thermal reactions.

Now let's summarize on the way to a generalization of these results. Table 22.1 gives the results so far. It also points out that the cyclobutene–1,3-butadiene interconversion is a four-electron process, whereas the hexatriene–1,3-cyclohexadiene reaction involves six electrons. In the cyclobutene–butadiene reaction, the four electrons in σ and π of the cyclobutene come to occupy Φ_1 and Φ_2 of 1,3-butadiene. Similarly, the six electrons in σ, Φ_1, and Φ_2 of 1,3-cyclohexadiene are interconverted with the six electrons in Φ_1, Φ_2, and Φ_3 of the 1,3,5-hexatriene molecule.

TABLE 22.1 Rotatory Motions in Two Electrocyclic Reactions

Reaction	Number of Electrons	Thermal	Photochemical
Cyclobutene–butadiene	4	Conrotation	Disrotation
Hexatriene–cyclohexadiene	6	Disrotation	Conrotation

It turns out that all $4n$ systems will behave like the four-electron case (thermal reactions conrotatory and photochemical reactions disrotatory) and all $4n + 2$ systems behave like the six-electron case (thermal reactions disrotatory and photochemical reactions conrotatory). So, Table 22.1 can be generalized to Table 22.2, which gives the rules for all electrocyclic reactions. Now, if you can count the number of electrons correctly you can easily predict the stereochemistry of any electrocyclic reaction without even working out the molecular orbitals.

TABLE 22.2 Rotatory Motions in All Electrocyclic Reactions

Reaction	Number of Electrons	Thermal	Photochemical
Cyclobutene–butadiene	4 ($4n$)	Conrotation	Disrotation
Hexatriene–cyclohexadiene	6 ($4n + 2$)	Disrotation	Conrotation
All $4n$ electrocyclic reactions	$4n$	Conrotation	Disrotation
All $4n + 2$ electrocyclic reactions	$4n + 2$	Disrotation	Conrotation

PROBLEM 22.7 Provide mechanisms for the reactions of Figure 22.21.

(a)

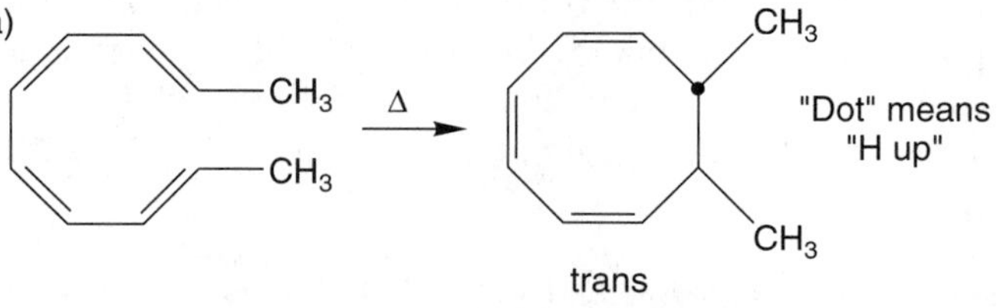

CH_3 CH_3 Δ → CH_3 CH_3 "Dot" means "H up"

trans

(b) Dewar benzene lies perched some 60 kcal/mol above its isomer benzene. It would seem that a simple stretching of the central bond would inevitably and easily give benzene. Yet Dewar benzene certainly does exist and in fact rearranges to benzene only quite slowly. Why?

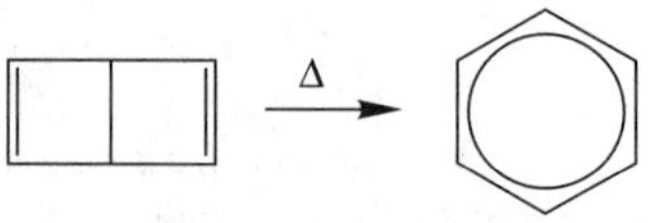

Δ →

FIGURE 22.21

(c)

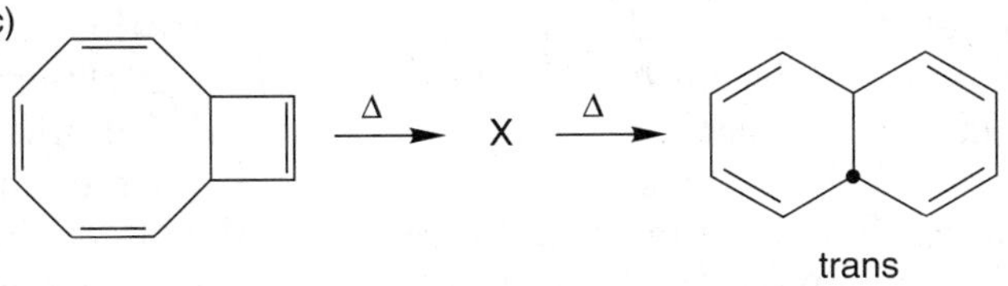

Δ X Δ

trans

(d)

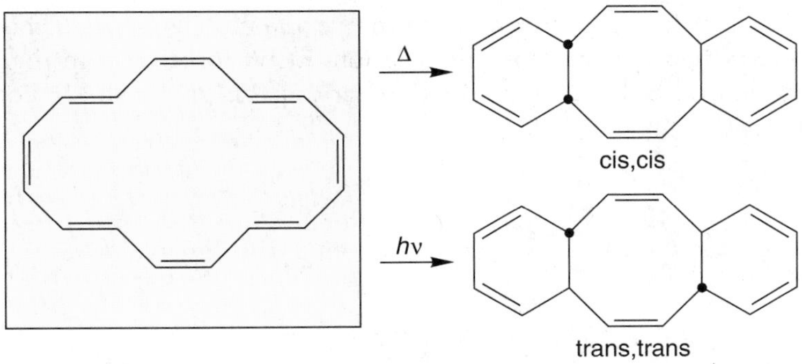

Δ

cis,cis

$h\nu$

trans,trans

FIGURE 22.21 (CONTINUED)

22.3 CYCLOADDITION REACTIONS

Cycloaddition reactions are another kind of orbital symmetry controlled (pericyclic) reaction. We have seen one already, the Diels–Alder reaction, and will use it as our prototype. We found the Diels–Alder cycloaddition to be a thermal process and to take place in a concerted (one-step) fashion. Several stereochemical labeling experiments were described in Chapter 12 (p. 536), all of which showed that the reaction involved neither diradical nor polar intermediates. This stereospecificity is important because orbital symmetry considerations apply only to single-step reactions. Of course, all reactions can be subdivided into series of single-step, single-barrier processes, and each of these steps could be analyzed separately by orbital symmetry. However, it makes no sense to speak of the overall orbital symmetry control of a multistep process.

When the diene and dienophile approach each other in the Diels–Alder reaction, the important orbital interactions are between the HOMOs and LUMOs. As Figure 22.22 shows, there are two possible HOMO–LUMO interactions, $\text{HOMO}_{(\text{diene})}$–$\text{LUMO}_{(\text{dienophile})}$ and $\text{HOMO}_{(\text{dienophile})}$–$\text{LUMO}_{(\text{diene})}$.

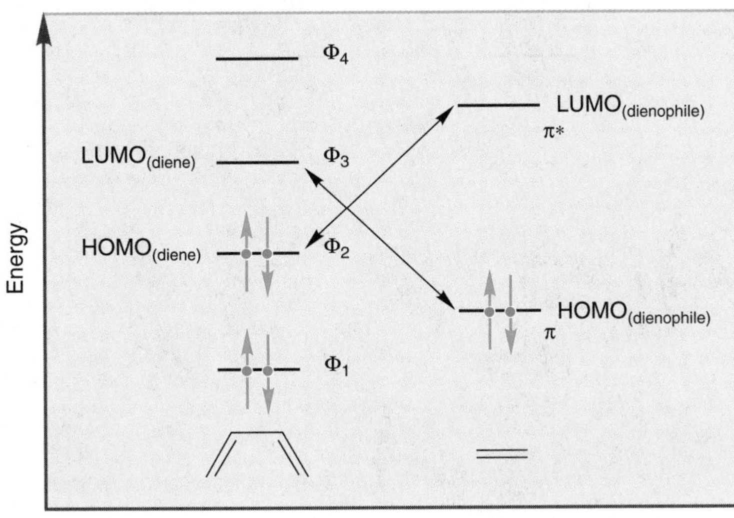

FIGURE 22.22 The two possible HOMO–LUMO interactions in the prototypal Diels–Alder reaction between butadiene and ethylene.

The stronger of these two will control the reaction. In this case, the stronger interaction is between the HOMO and LUMO pair closer in energy. *Remember*: The strength of orbital overlap, and the magnitude of the resulting stabilization, depends on the relative energies of the two orbitals. The closer the two are in energy, the stronger the interaction.

In the simplest, symmetric Diels–Alder reaction between ethylene and 1,3-butadiene, the two HOMO–LUMO interactions are of equal energy and we must look at the orbital symmetry of both. Figure 22.23 shows the symmetry relationships for this reaction. Don't forget that in the Diels–Alder (and related) reactions the two participants approach each other in parallel planes (Chapter 12, p. 536).

FIGURE **22.23** The orbital overlaps in the two HOMO–LUMO interactions of the Diels–Alder reaction. These HOMO–LUMO interactions produce two new bonds.

Both interactions involve bonding overlaps at the points of formation of the two new σ bonds. Accordingly, it's no surprise to find that the thermal Diels–Alder reaction occurs. One says that it is "allowed by orbital symmetry."

What about the photochemical Diels–Alder reaction? This reaction is most uncommon, and this observation might lead to the immediate suspicion that there is something wrong with it. Usually, the absorption of a photon will promote an electron from the HOMO to the LUMO. In this case, the lower energy HOMO–LUMO gap is that in the diene partner. Absorption of light creates a new photochemical HOMO for the diene, Φ_3, and now the HOMO–LUMO interaction with the dienophile partner involves one antibonding overlap. Both new bonds cannot be formed at the same time (Fig. 22.24). So this photochemical Diels–Alder reaction is said to be "forbidden by orbital symmetry."

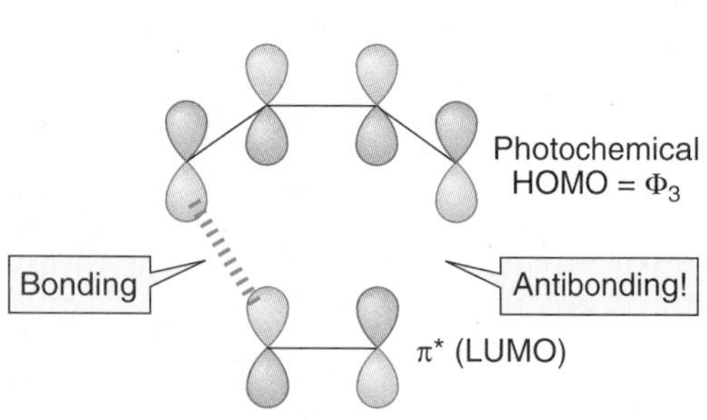

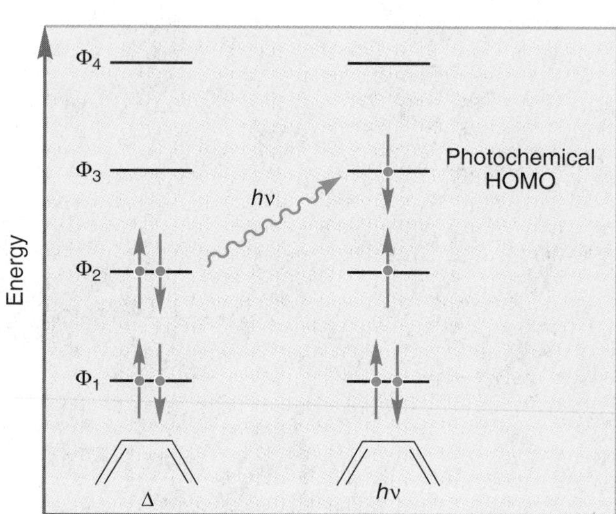

FIGURE **22.24** Absorption of a photon creates a new, photochemical HOMO, Φ_3. Now the HOMO–LUMO interaction involves one antibond. The reaction is forbidden by orbital symmetry.

PROBLEM 22.8

What would happen if the light were used to promote an electron from the HOMO of ethylene, π, to the LUMO of ethylene, π*? Would the Diels–Alder reaction be allowed or forbidden?

Let's now analyze another potential cycloaddition, the simple joining together of a pair of ethylenes to give a cyclobutane (Fig. 22.25).

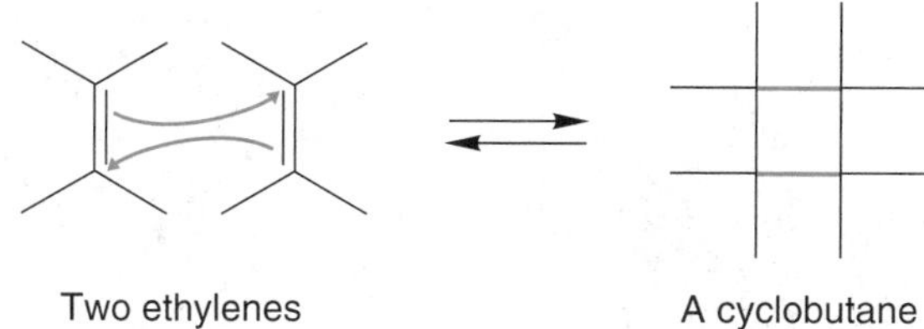

FIGURE 22.25 The hypothetical combination of two alkenes to give a cyclobutane.

PROBLEM 22.9

Will the 2 + 2 reaction of a pair of ethylenes to give a cyclobutane be exothermic or endothermic? Explain.

The HOMO and LUMO are easy to identify, and it is quickly obvious that here, too, the apparently easy bond formation shown by the arrow formalism is thwarted by an antibonding interaction as the two molecules come together (Fig. 22.26).

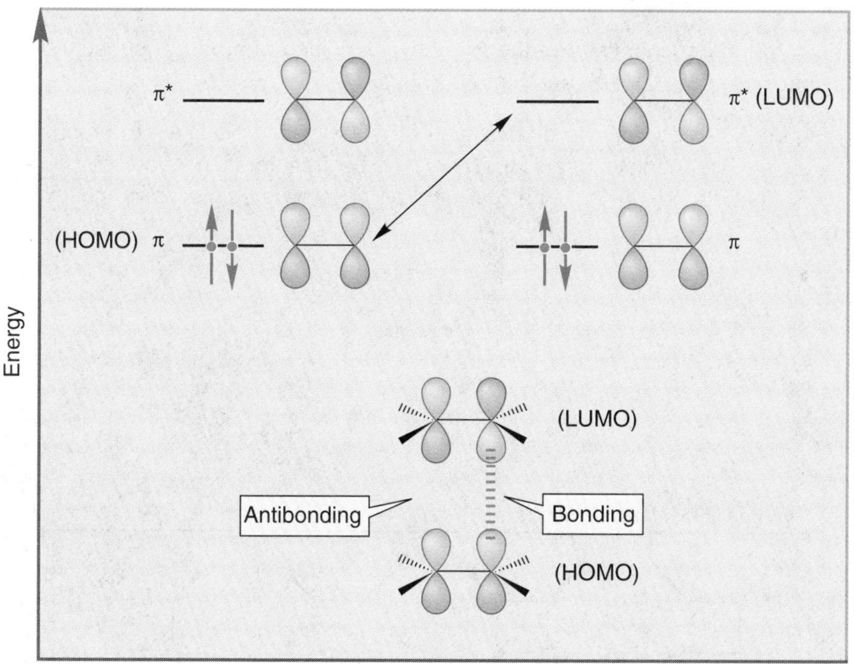

FIGURE 22.26 An analysis of the HOMO and LUMO for the reaction of two alkenes to give a cyclobutane shows that it is forbidden by orbital symmetry. There is an antibonding interaction.

Therefore we might guess that the thermal dimerization of ethylenes would be a rare process and we would be exactly correct. Moreover, it is clear in most of the reactions of this kind that do occur that the two new bonds are formed sequentially, and *not* in a concerted manner (Fig. 22.27).

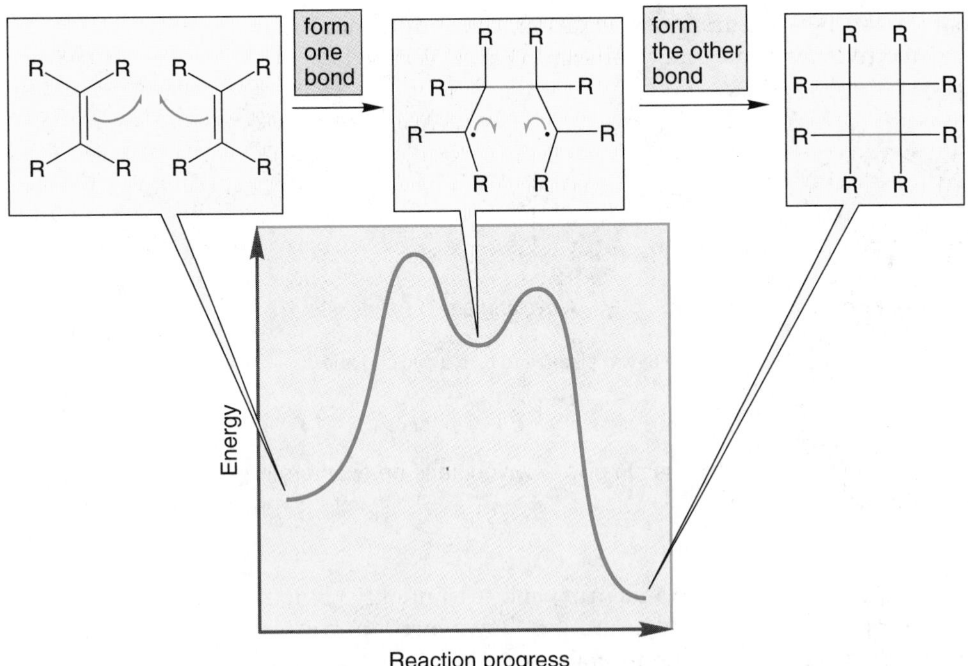

FIGURE 22.27 The known dimerizations of substituted ethylenes can be shown to involve diradical intermediates.

*PROBLEM 22.10 Predict the stereochemistry of the 2 + 2 cycloaddition of a pair of substituted ethylenes, X–CH=CH–X (see Fig. 22.27).

ANSWER As Figure 22.27 shows, the dimerization of two ethylenes is likely to proceed through an intermediate diradical. There will be free rotation about carbon–carbon single bonds in that diradical, and all possible stereoisomers must be formed.

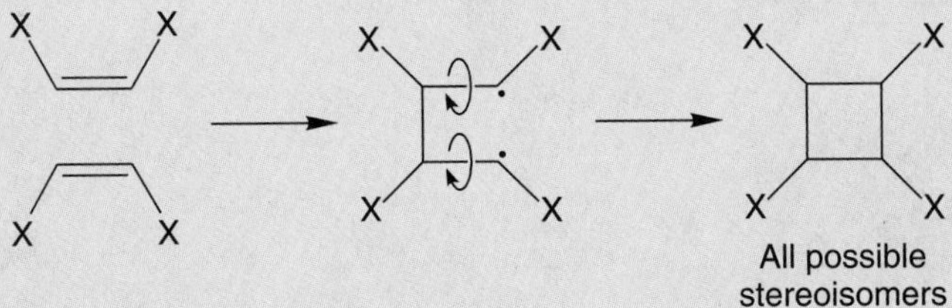

The possible stereoisomers are

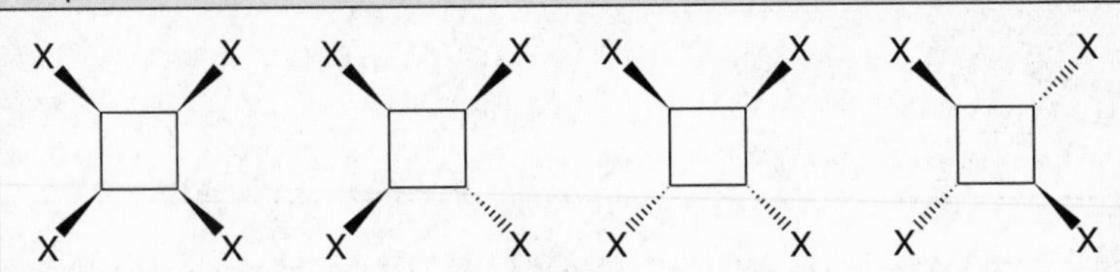

What about the photochemical dimerization of ethylenes? Promotion of a single electron creates a new photochemical HOMO, and now the symmetries are perfect for a HOMO–LUMO interaction involving two bonding overlaps. The photochemical dimerization of two ethylenes seems allowed (Fig. 22.28), and is, in fact, a common reaction.

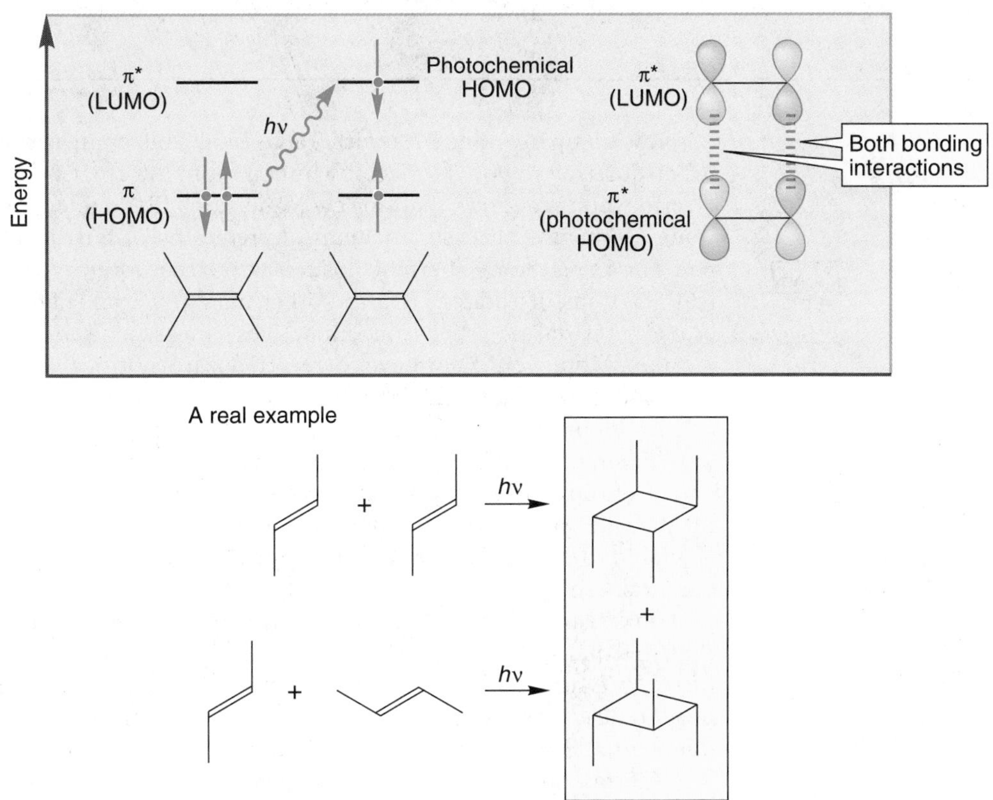

FIGURE 22.28 In contrast to the thermal reaction, the photochemical dimerization of two ethylenes is allowed. For example, the photochemical dimerization of *trans*-2-butene gives the two cyclobutanes shown.

Again, it's time for generalization. Table 22.3 shows what we have so far.

TABLE 22.3 Rules for Two Cycloaddition Reactions

Reaction	Number of Electrons	Thermal	Photochemical
Ethylene + Ethylene ⇄ Cyclobutane	4 (4*n*)	No	Yes
Ethylene + Butadiene ⇄ Cyclohexene (Diels–Alder)	6 (4*n* + 2)	Yes	No

It is possible to expand Table 22.3 to encompass all 4*n* and 4*n* + 2 reactions. Other 4*n* processes will follow the rules for the 2 + 2 dimerization of a pair of ethylenes, and 4*n* + 2 processes will resemble the 4 + 2 cycloaddition we know as the Diels–Alder reaction (Table 22.4).

TABLE 22.4 Rules for All Cycloaddition Reactions

Reaction	Number of Electrons	Thermal	Photochemical
Ethylene + Ethylene ⇄ Cyclobutane	4 (4*n*)	No	Yes
Ethylene + Butadiene ⇄ Cyclohexene (Diels–Alder)	6 (4*n* + 2)	Yes	No
All 4*n* reactions	4*n*	No	Yes
All 4*n* + 2 reactions	4*n* + 2	Yes	No

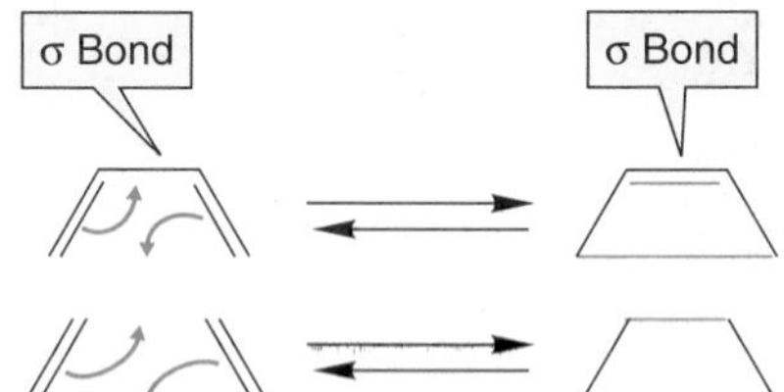

FIGURE 22.29 The equilibration of cyclobutene and 1,3-butadiene and the 2 + 2 dimerization of a pair of ethylenes are very closely related reactions.

Now compare Table 22.2 with Table 22.4. This comparison should be a bit disquieting to you. Notice how much more information is available for electrocyclic reactions than for cycloadditions. For the electrocyclic reactions, we have a clear stereochemical prediction of how things must happen. For 4*n* reactions, thermal closures will be conrotatory and photochemical reactions disrotatory. On the other hand, for cycloadditions, we get a crude "thermal no," "photochemical yes" message; a much less detailed picture. Moreover, electrocyclic reactions are not really different from cycloadditions. Figure 22.29 compares the equilibration of 1,3-butadiene and cyclobutene with the 2 + 2 dimerization of a pair of ethylenes. The only difference is the extra σ bond in butadiene, and this bond is surely not one of the important ones in the reaction—it seems to be just going along for the ride. Why should its presence or absence change the level of detail available to us through an orbital symmetry analysis? It shouldn't, and in fact, it doesn't.

Look again at Figure 22.26, from which we derived the information that the thermal 2 + 2 dimerization of a pair of ethylenes was forbidden. It's only forbidden if we insist on smushing the two ethylenes together head to head exactly as shown in the figure. If we are somewhat more flexible, and allow one ethylene to rotate about its carbon–carbon σ bond as the reaction takes place, the offending antibonding overlap vanishes. What orbital symmetry really says is that for the thermal dimerization of two ethylenes to take place, there must be this rotation (Fig. 22.30).

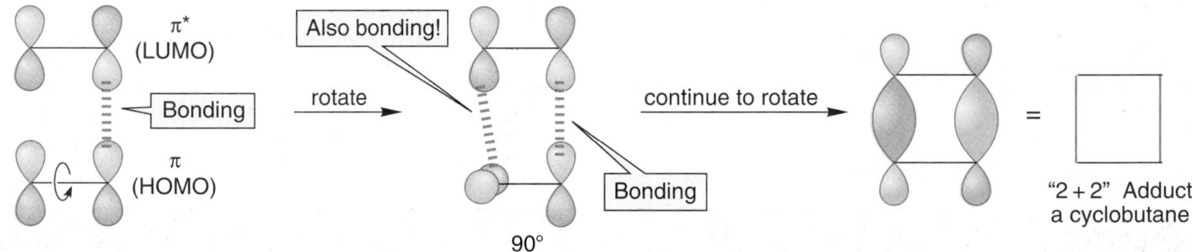

FIGURE 22.30 The thermal 2 + 2 dimerization is allowed if one end of one ethylene can rotate as the reaction occurs.

In practice, this rotation is a very difficult thing to achieve. It is this difficulty that accounts for the absence of the thermal 2 + 2 reaction, not the absolute "no" of Table 22.4. Similarly, the photochemical 4 + 2, Diels–Alder reaction also becomes allowed if we permit one of the partners to undergo a rotation during the reaction (Fig. 22.31).

So we should amend our summary Table 22.4 to produce a new Table 22.5.

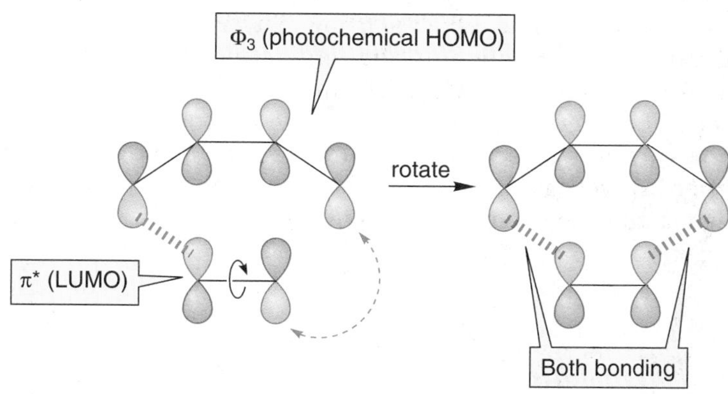

FIGURE 22.31 The photochemical 4 + 2 reaction (the Diels–Alder reaction) is also allowed if a rotation occurs during the cycloaddition.

TABLE 22.5 Rules for All Cycloaddition Reactions

Reaction	Number of Electrons	Thermal	Photochemical
Ethylene + Ethylene ⇄ Cyclobutane	4 (4*n*)	One rotation	No rotation
Ethylene + Butadiene ⇄ Cyclohexene (Diels–Alder)	6 (4*n* + 2)	No rotation	One rotation
All 4*n* reactions	4*n*	One rotation	No rotation
All 4*n* + 2 reactions	4*n* + 2	No rotation	One rotation

Don't forget; these rules do not include any consideration of thermodynamics. The processes allowed with a rotation may well not be very favorable thermodynamically, even though they are permitted electronically.

22.4 SIGMATROPIC SHIFT REACTIONS

When one heats 1,3-pentadiene nothing obvious happens, as the starting material is recovered, apparently unchanged. Finally, at very high temperatures, radical reactions begin as carbon–hydrogen bonds are broken. However, a deuterium-labeling experiment reveals that much is happening long before the energies necessary for the formation of radicals have been reached (Fig. 22.32).

Radicals ($\dot{C}H_2$, •H) ← ΔΔ (high temperature) — 1,3-Pentadiene (CH_3) ⇄ Δ — 1,3-Pentadiene (CH_3) "No reaction"

Labeling experiment

5,5,5-Trideuterio-1,3-pentadiene (CD_3) ⇄ Δ 1,1,5-Trideuterio-1,3-pentadiene (CH_2D, CD_2)

FIGURE 22.32 A labeling experiment reveals the degenerate thermal rearrangement of 1,3-pentadiene.

This kind of process, in which starting material rearranges into itself, is called a degenerate reaction. Isotopically labeled molecules reveal the degenerate reaction and allow a determination of the kinetic activation parameters. The reaction requires an activation energy of 36–38 kcal/mol. Other labels are possible, of course, and substituted molecules will reveal the reaction as well as the more sophisticated isotopic labels. In most cases, the reaction is no longer degenerate, and the equilibrium constant *K* for the reaction can no longer equal 1 (Fig. 22.33).

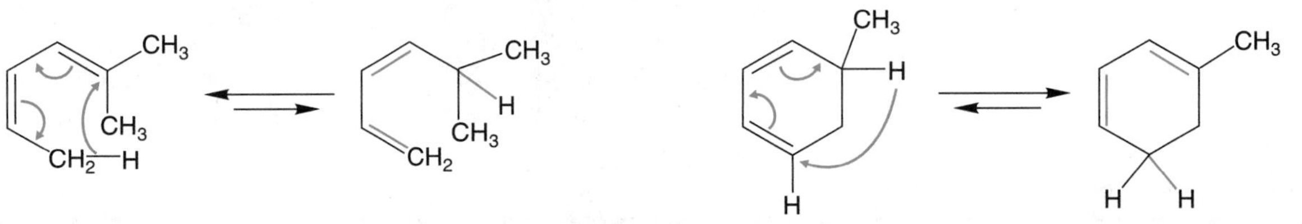

FIGURE 22.33 Substituted molecules also reveal the reaction, but the rearrangement is no longer degenerate, as starting material and product are different.

PROBLEM 22.11 Which of the molecules of the drawing at the right of Figure 22.33 will be favored at equilibrium? Explain your choice.

*PROBLEM 22.12 Estimate the bond dissociation energy (BDE) for the migrating carbon–hydrogen bond. Why is the observed activation energy of 36–38 kcal/mol for the [1,5] shift of hydrogen so much lower than your answer?

ANSWER The carbon–hydrogen bond in methane has a BDE of 104 kcal/mol. However, this is a poor model for the breaking carbon–hydrogen bond in Figure 22.33, because the radical formed from the larger molecule is tertiary, not methyl. The BDE for the tertiary carbon–hydrogen bond in isobutane is about 93 kcal/mol.

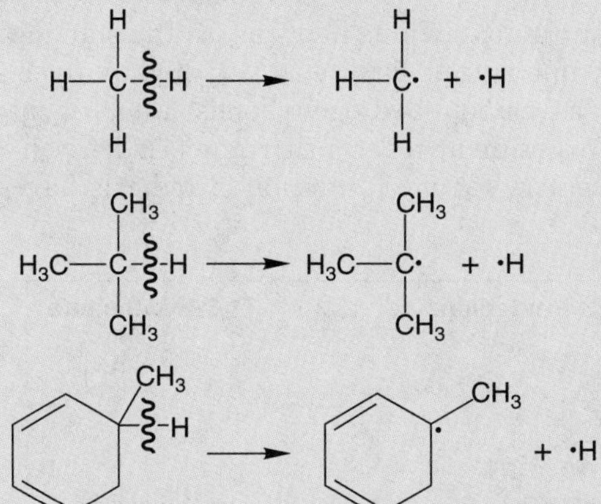

$$CH_4 \longrightarrow \cdot CH_3 + \cdot H \qquad \text{BDE} = 104 \text{ kcal/mol}$$

$$(CH_3)_3C\text{–}H \longrightarrow (CH_3)_3C\cdot + \cdot H \qquad \text{BDE} = 93 \text{ kcal/mol}$$

+ ·H BDE (first guess) = ~93 kcal/mol

But the cyclohexadienyl radical is resonance stabilized and that will further decrease the BDE. If allylic resonance is worth about 12 kcal/mol, we might expect the BDE to be about 93 – 12 = 81 kcal/mol.

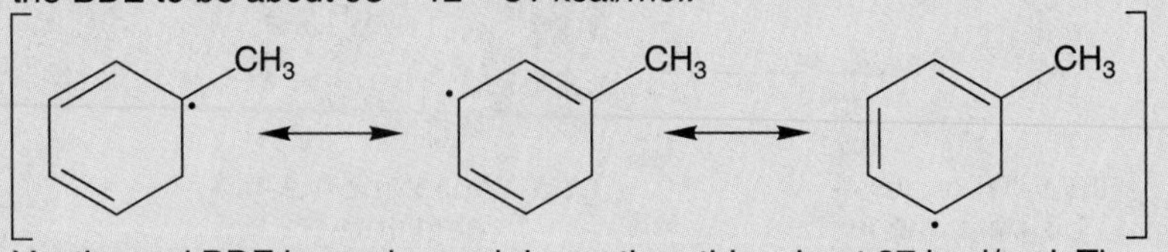

Yet the real BDE is much, much lower than this, about 37 kcal/mol. The reason is that as the carbon–hydrogen bond begins to break, the new carbon–hydrogen

ANSWER (CONTINUED)

bond at C_5 is forming. It is *not* necessary to break the carbon–hydrogen bond fully.

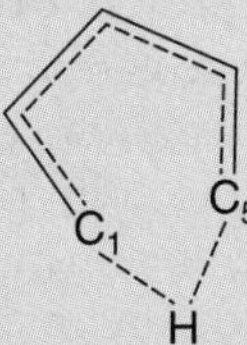

As the C_1—H bond breaks, the new carbon–hydrogen bond at C_5 begins to form; this bond formation decreases the energy required to break the C_1—H bond.

The deuterium labeling experiment reveals that much is happening before the onset of radical reactions. A hydrogen (deuterium, in this case) atom is moving from one position in the molecule to another. This kind of reaction has come to be known as a **sigmatropic shift.*** The starting point of the shift is called atom 1. The rest of the atoms are labeled by counting toward the terminus of the migrating group, and the two numbers denoting the starting and arriving points of the migrating species are enclosed in brackets. In this case, we have a [1,5] shift (Fig. 22.34). Be sure to note that the labeling need not follow the numbering convention used in the naming of the compound.

FIGURE 22.34 In a [1,5] sigmatropic shift of hydrogen, the migrating hydrogen atom moves from carbon 1 to carbon 5.

PROBLEM 22.13

Classify the reactions in Figure 22.35 as [*x*,*y*] sigmatropic shifts.

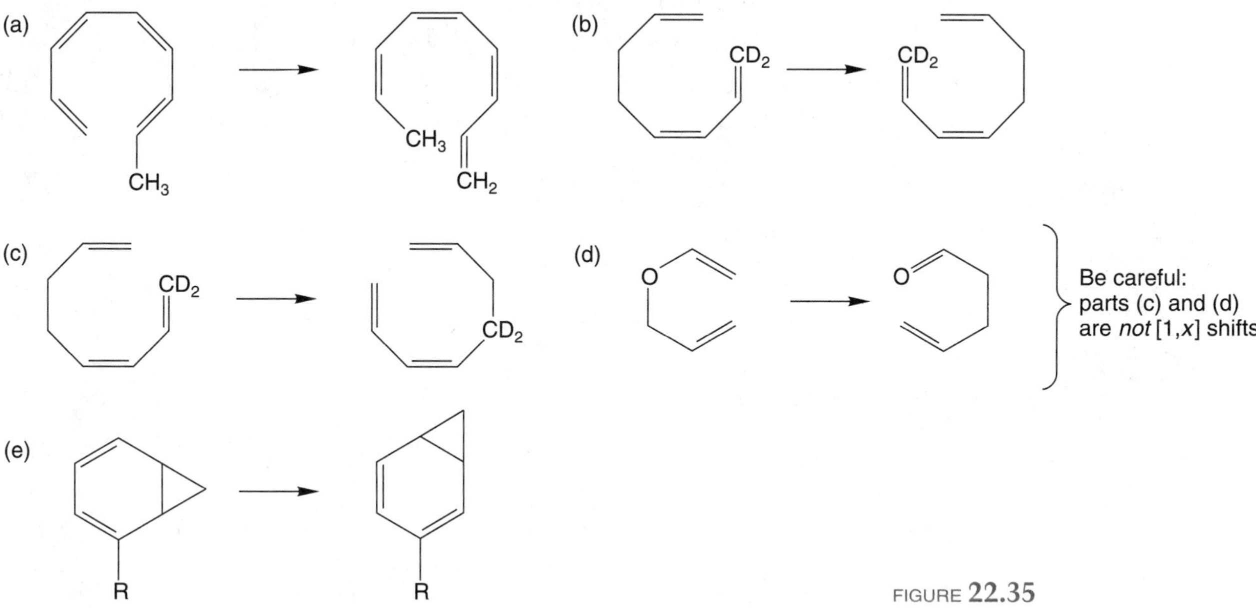

FIGURE 22.35

*It must be admitted that Woodward–Hoffmann theory is filled with jargon. There is nothing to do but learn it.

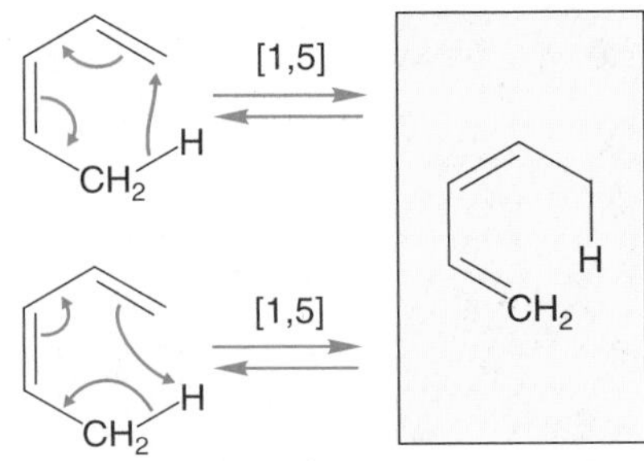

FIGURE 22.36 An arrow formalism description of the [1,5] shift of hydrogen in 1,3-pentadiene.

This reaction responds to attempted mechanistic experiments in true no-mechanism fashion. Neither acids nor bases strongly affect the reaction, and the polarity, or even the presence of solvent, is unimportant. The reaction proceeds quite nicely in the gas phase. It is simple to construct an arrow formalism picture of the reaction, but this does little more than point out the overall change produced by the migration. In these figures, the arrow formalism is even more of a formalism than usual. For example, the arrows of Figure 22.36 could run in either direction, clockwise or counterclockwise. That is not true for a polar reaction in which the convention is to run the arrows from pairs of electrons (the Lewis base) toward the Lewis acid.

It is already possible to identify one curious facet of this reaction, and a little experimentation reveals others. The product of a [1,3] shift is absent. Why, if the hydrogen is willing to travel to the 5-position, does it never stop off at the 3-position? An arrow formalism can easily be written, and it might be reasonably argued that the [1,3] shift, requiring a shorter path than the [1,5] shift, should be easier (Fig. 22.37). Why are [1,3] shifts not observed?

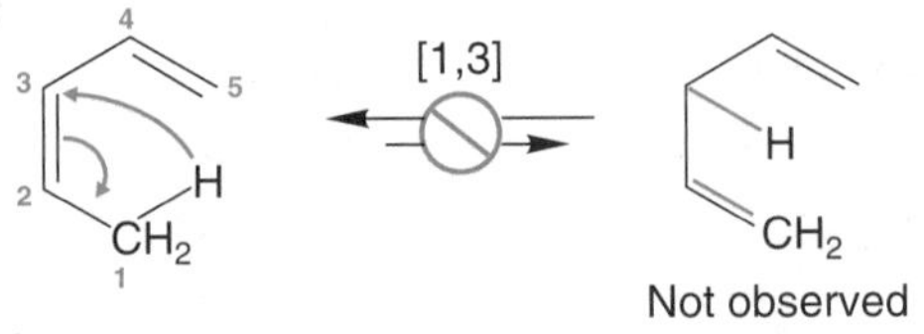

Not observed

FIGURE 22.37 An arrow formalism description for the [1,3] shift of hydrogen that does not occur.

*PROBLEM 22.14 Perhaps it could be argued that the loss of conjugation in the product of the [1,3] shift results in an energetic favoring of the [1,5] process. Design a substituted 1,3-pentadiene that would test this (incorrect) surmise.

ANSWER This problem is hard because you are not used to thinking in the correct way. The hypothesis is that the [1,5] shift is preferred to the [1,3] shift because it preserves the conjugation present in a 1,3-diene. In order to test this surmise, we need to design a molecule in which both the [1,5] and [1,3] shifts preserve the same amount of conjugation. The deuterium label is needed to reveal the [1,5] or [1,3] nature of the reaction.

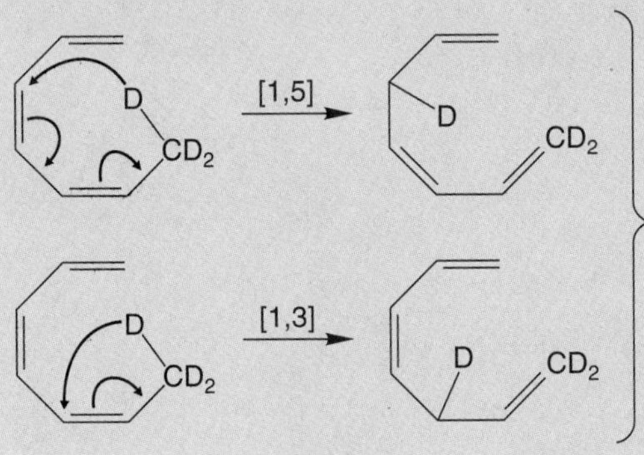

These are the same molecule except for the isotopic label

A second strange aspect of this reaction comes from photochemical experiments. As you already know, it is possible to deliver energy to a molecule in more than one way. One way is to heat up the starting material. However, conjugated molecules absorb light (Chapter 12, p. 514), light quanta contain energy, and this gives us another way to transmit energy to the molecule. Curiously, when 1,3-pentadienes are irradiated, the products of the reaction include the molecules formed through [1,3] shifts, but *not* those of [1,5] shifts. The reactions are sometimes complicated, as many other photochemical reactions take place, but if we just concentrate on the products of shifts, this strange dependence on the method of energy delivery appears. To generalize: Thermal energy induces [1,5] shifts and photochemistry allows [1,3] shifts (Fig. 22.38).

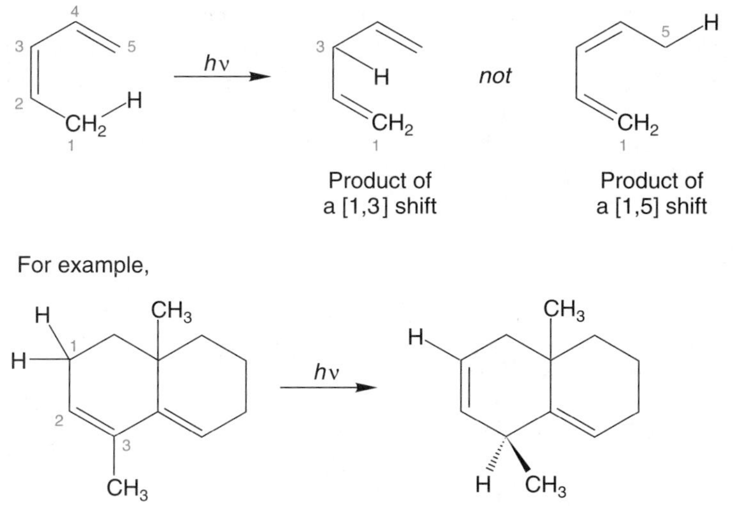

hν

not

Product of a [1,3] shift

Product of a [1,5] shift

For example,

hν

FIGURE 22.38 Photolysis of 1,3-dienes induces [1,3] shifts, but not [1,5] shifts.

So, any mechanism must include an explanation of why thermal shifts are [1,5], whereas photochemically induced shifts are [1,3].

Let's start our analysis by examining the early stages of the reaction. As the reaction begins, a carbon–hydrogen bond of the methyl group begins to stretch. If this stretching were continued to its limit, it would produce two radicals, a hydrogen atom and the pentadienyl radical (Fig. 22.39). We know this does not happen because the products of recombination of these radicals are not found.

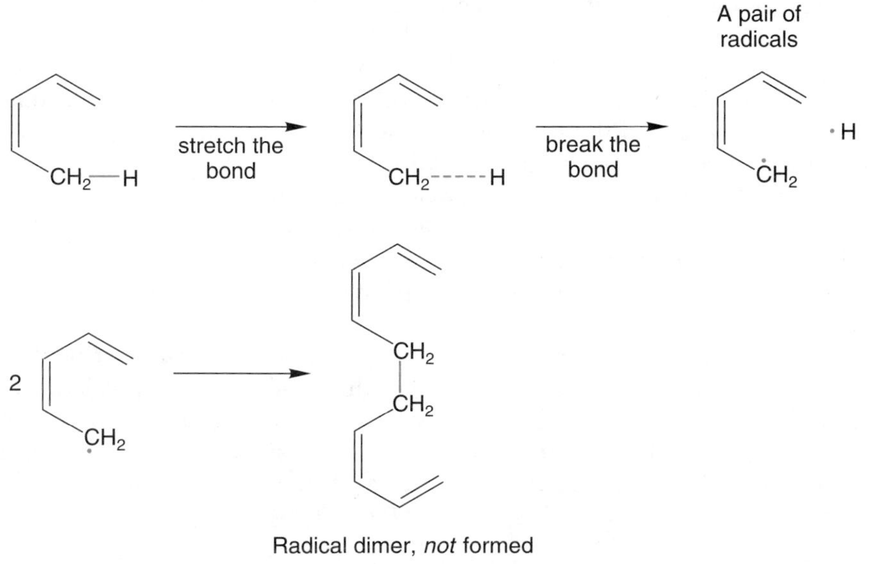

A pair of radicals

stretch the bond

break the bond

Radical dimer, *not* formed

FIGURE 22.39 In the earliest stages of the reaction, a carbon–hydrogen bond begins to stretch. If the carbon–hydrogen bond were to continue this stretching motion, a pair of radicals would result. We know this bond breaking does not take place because the radical dimer is not found.

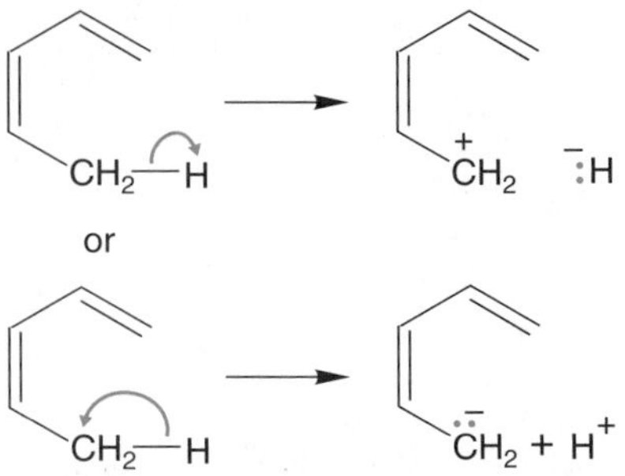

FIGURE 22.40 There are two possible ways to break the carbon–hydrogen bond heterolytically to give a pair of ions. Either way would be strongly dependent on solvent polarity. As there is essentially no dependence of the rate of the reaction on solvent polarity, this mechanistic possibility is excluded.

We should stop right here to question the assumption that the carbon–hydrogen bond is breaking in homolytic (radical) fashion. Why not write a heterolytic cleavage to give a pair of ions rather than radicals (Fig. 22.40)?

The stability of ions depends greatly on solvation. It is inconceivable that a heterolytic bond breaking would not be highly influenced by solvent polarity. As we have just seen, a change in solvent polarity affects the rate of this reaction only very slightly. Only homolytic bond breaking is consistent with the lack of a solvent effect.

As the methyl carbon–hydrogen bond begins to stretch, something must happen that thwarts the formation of the pair of radicals. Perhaps the developing hydrogen atom is intercepted by reattachment at atom 5. In orbital terms, we would say that as the C_1—H bond breaks, the developing 1*s* orbital on hydrogen begins to overlap with the *p* orbital on the carbon at position 5. Now we have a first crude model for the transition state for this sigmatropic shift (Fig. 22.41).

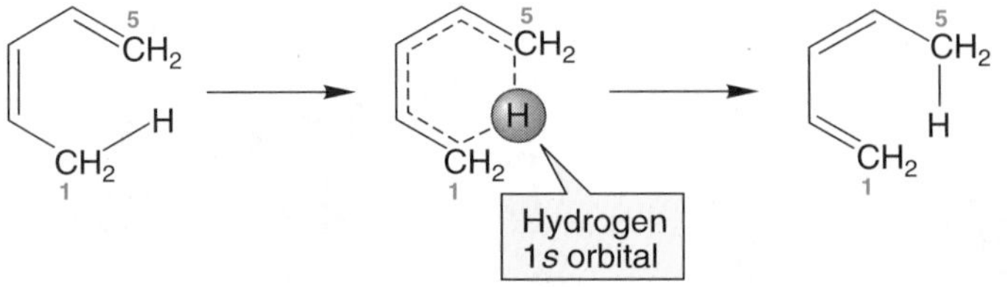

FIGURE 22.41 Radical formation can be thwarted by partial bonding of the migrating hydrogen to the orbitals on carbon atoms 1 and 5 in the transition state. As the C_1—H bond stretches, the hydrogen 1*s* orbital begins to overlap with an orbital on C_5.

This model does describe an intramolecular reaction, but it leaves unanswered the serious questions of mechanism raised before: Why doesn't the migrating hydrogen reattach at the 3-position in thermal reactions? This process does happen in photochemical reactions. We have answered neither the questions of regiochemistry ([1,5] or [1,3]) nor of dependence on the mode of energy input (thermal or photochemical).

Let's look more closely at our transition state model. In essence, it describes a hydrogen atom moving across the π system of a pentadienyl radical (Fig. 22.42).

The hydrogen atom is simple to describe—it merely consists of a proton and a single electron in a 1*s* atomic orbital. The pentadienyl molecular orbital system isn't difficult either. There will be five π molecular orbitals and these can be constructed by taking combinations of five carbon 2*p* orbitals. We assume again that it will be the highest energy electrons, those most loosely held, that will be most important in the reaction. These are the electrons in the HOMO. If we accept this assumption, we can elaborate our picture of the transition state for the reaction to give Figure 22.43.

Notice a very important thing about the transition state outlined in Figure 22.43. In the starting pentadiene, the bond made between the 1*s* orbital of hydrogen and the hybrid orbital on carbon atom 1 must of course be made between lobes of the same sign. As this bond stretches, the signs of the lobes do not change: The 1*s* orbital is always bonded or partially bonded to a lobe of the same symmetry (Fig. 22.44).

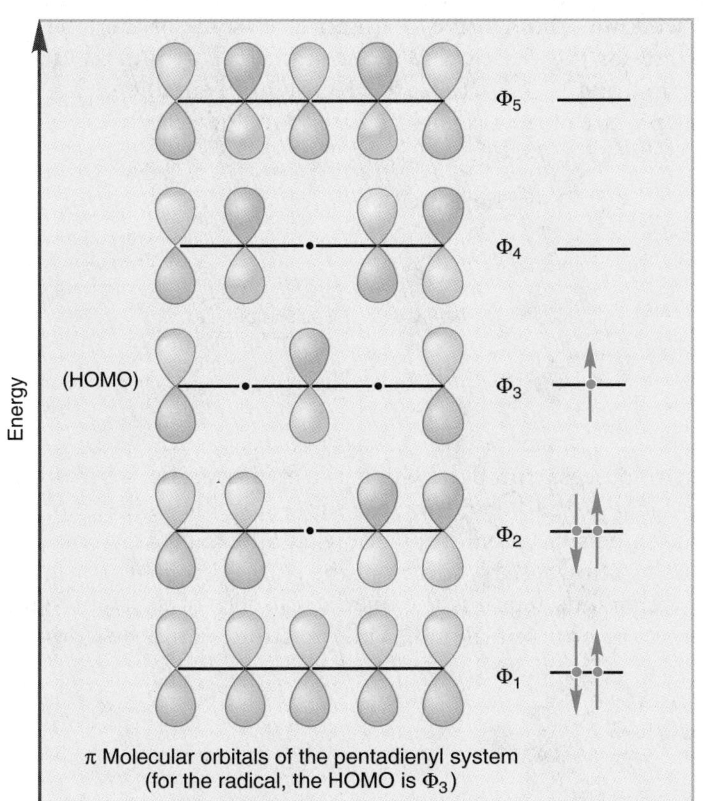

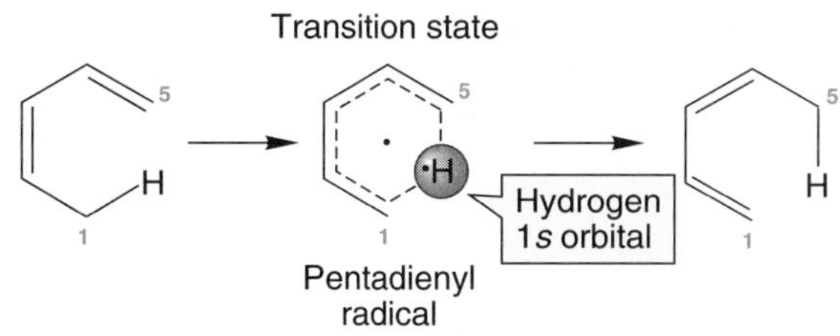

FIGURE 22.42 A model for the transition state shows a hydrogen 1*s* orbital migrating across a pentadienyl radical, for which the π molecular orbitals are shown.

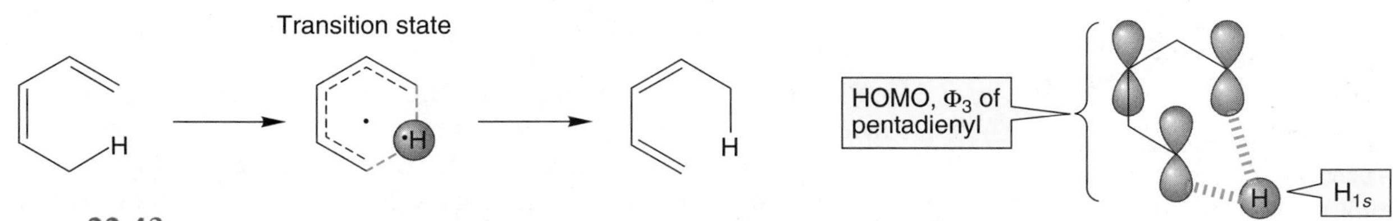

FIGURE 22.43 An elaboration of our model for the transition state, showing a crude outline of the interaction of the hydrogen 1*s* orbital with the lobes of the HOMO for pentadienyl, Φ_3.

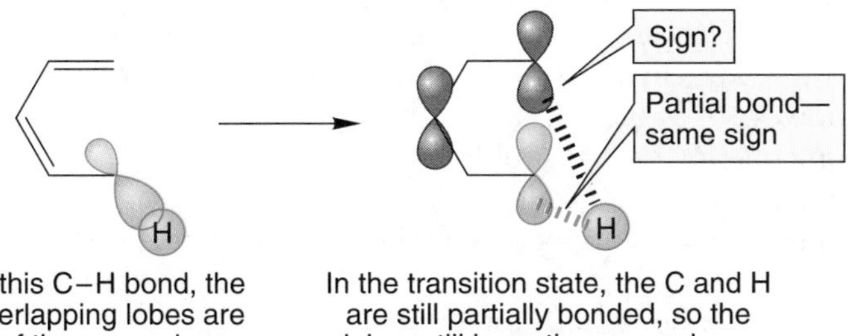

FIGURE 22.44 As the carbon–hydrogen bond lengthens, the phase relationship between the lobes involved in the bond does not change. For maintenance of a bonding relationship, both lobes must be of the same sign.

In this way, we know the relative symmetries of some of the lobes in the transition state for the reaction. We also know the symmetry (from Fig. 22.42, or by figuring it out) of Φ_3, the HOMO of pentadiene. So, we can elaborate the picture of the transition state to include the symmetries of all the orbital lobes (Fig. 22.45).

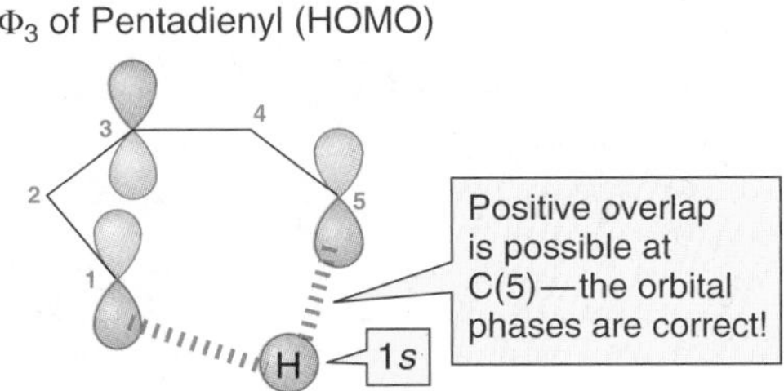
Φ_3 of Pentadienyl (HOMO)
1
2
3
4
5
H
1s
Positive overlap is possible at C(5)—the orbital phases are correct!

FIGURE 22.45 Now we can elaborate our transition state model by filling in the lobes of the HOMO of pentadienyl, Φ_3. The symmetry of the lobes allows the [1,5] shift to take place as shown.

Notice that the process through which the hydrogen detaches from carbon 1 and reattaches to carbon 5 from the same side allows the maintenance of bonding (same sign) lobal interactions at all times. This reaction is "orbital symmetry allowed." Now look at the [1,3] shift. To do the analogous [1,3] shift, rebonding requires the overlap of lobes of different symmetry—the formation of an antibond. This reaction would be "orbital symmetry forbidden" (Fig. 22.46)!

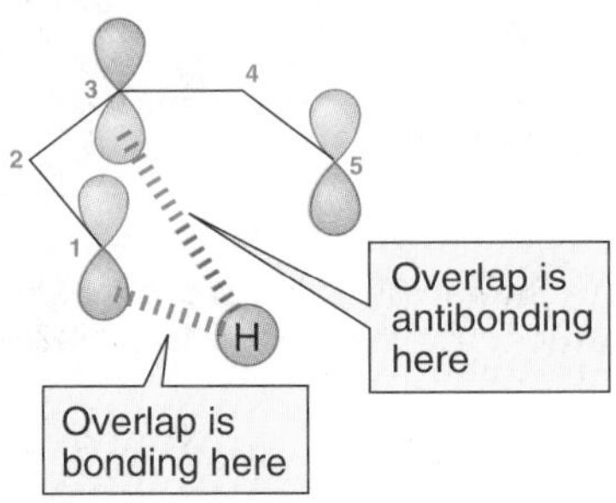
1
2
3
4
5
H
Overlap is antibonding here
Overlap is bonding here

FIGURE 22.46 By contrast, the [1,3] shift involves the formation of an antibond as the hydrogen reattaches to carbon 3.

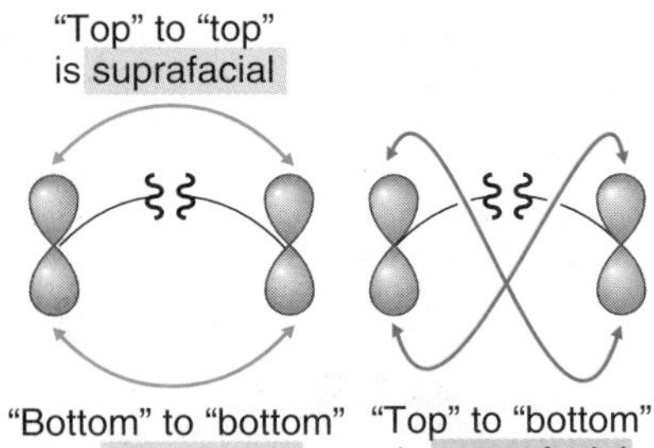
"Top" to "top" is suprafacial
"Bottom" to "bottom" is suprafacial
"Top" to "bottom" is antarafacial

FIGURE 22.47 Suprafacial motion involves bond breaking and bond making on the *same* side of the π system. Antarafacial motion involves bond breaking on one side and bond making on the other.

More important jargon enters here: The motions we have been describing in which the migrating atom leaves and reattaches from the *same* side of the π system are called **suprafacial**. If the migrating atom or group of atoms leaves from one side of the π system and arrives at the other, the process is called **antarafacial** motion (Fig. 22.47). In the [1,5] shift, suprafacial motion is allowed by the symmetries of the orbitals; in [1,3] shift, it is not (Figs. 22.45 and 22.46).

But why not migrate in a different fashion? Why can't the hydrogen depart from one side and reattach from the other? This antarafacial migration would result in an orbital symmetry allowed [1,3] migration (Fig. 22.48). The problem is that the 1*s* orbital is small and cannot effectively span the distance required for an antarafacial migration. There is nothing electronically disallowed or forbidden about the antarafacial [1,3] migration, but there is an insuperable steric problem.

So let's sum up again: The [1,5] shift is intramolecular, and the lack of any dependence on solvent polarity makes intermediate ions or even polar transition states unlikely. In our model of the transition state, we focus on the maintenance of bonding interactions, which explains why [1,5] shifts are possible and [1,3] shifts are exceedingly rare. We have not yet been able to deal with the dependence of the course of this reaction on the mode of energy input—thermal or photochemical.

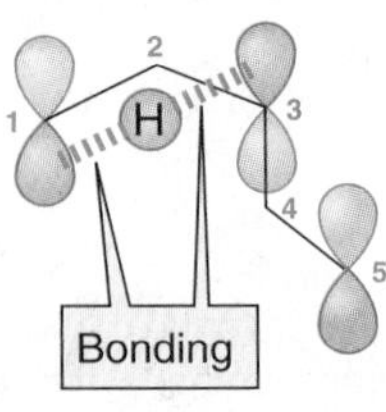
1
2
3
4
5
H
Bonding
Φ_3 of Pentadienyl (HOMO)

FIGURE 22.48 The antarafacial [1,3] shift is allowed by orbital symmetry, but the small size of the 1*s* orbital makes the reaction impossible.

PROBLEM **22.15**

Do you expect the [1,3] shift shown in Figure 22.49 to occur when propene is heated? Explain carefully, using a molecular orbital argument.

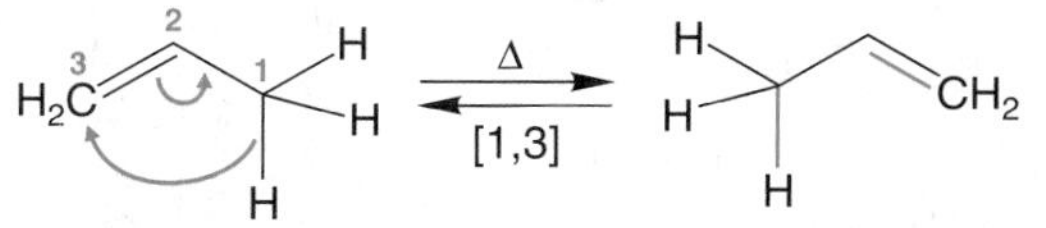

FIGURE **22.49**

Now comes a most important point. Our model not only rationalizes known experimental data, but includes within itself the roots of a critical prediction. The Woodward–Hoffmann explanation predicts—requires—that the [1,5] shift occur in a suprafacial fashion. That is the only way in which bonding interactions can be maintained throughout the migration of a hydrogen atom from carbon 1 to carbon 5. There is *nothing* so far in the available experimental data that allows us to tell whether the shift is suprafacial, antarafacial, or a mixture of both modes.

Let's set about finding an experiment to test the requirement of the theory for suprafacial [1,5] motion. What will we know at the end of the experiment? If we find that the [1,5] shift is indeed strictly suprafacial, the theory will be supported (not proved!), and we will certainly feel better about the mechanistic hypothesis. If we find antarafacial motion, or both suprafacial and antarafacial motions, the theory will be proved (yes, proved) wrong. There is no way our hypothesis can accommodate antarafacial motion; there is an absolute demand for suprafaciality, which nicely illustrates the precarious life of a theory. It can always be disproved by the next experiment, and it can never become free of this state of affairs.

The crucial test was designed by a German chemist, W. R Roth (b. 1930).* It is the first of a series of beautiful experiments you will encounter in this chapter. Roth and his co-workers spent some years in developing a synthesis of the labeled molecule shown in Figure 22.50. Notice that the molecule has the (*R*) configuration at carbon 1, and the (*E*) stereochemistry at the $C_4{-}C_5$ double bond.

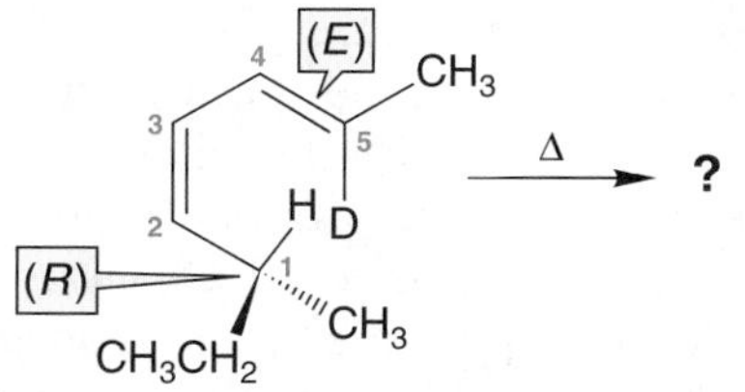

FIGURE **22.50** The 1,3-diene used by Roth and his co-workers to see whether the [1,5] hydrogen shift proceeded in a strictly suprafacial manner.

Now let's work out the possible products from suprafacial and antarafacial [1,5] shifts in this molecule. The situation is a little complicated because there are two forms of this molecule that can undergo the reaction. These are the two conformations (rotational isomers) shown in Figure 22.51. In one, the methyl group lies inside, aimed at the $C_4{-}C_5$ double bond. In the other, it is the ethyl group that is directed toward the $C_4{-}C_5$ double bond. In suprafacial migration of hydrogen in one conformation, the product has the (*S*) configuration at C_5 and the (*E*) stereochemistry at the new double bond located between C_1 and C_2. Another possible suprafacial migration starts from the second conformation. Here it is the opposite stereochemistries that result. The configuration of C_5 is (*R*) and the new double bond is (*Z*) (Fig. 22.51).

*No typo here. There is no period after the R. Roth claims that he was advised by his postdoctoral advisor at Yale, William Doering, to take a middle initial in order not to be lost among the myriad "W. Roth's" in German chemical indexes. Others with common last names, and who were at Yale at the time, rejected similar advice.

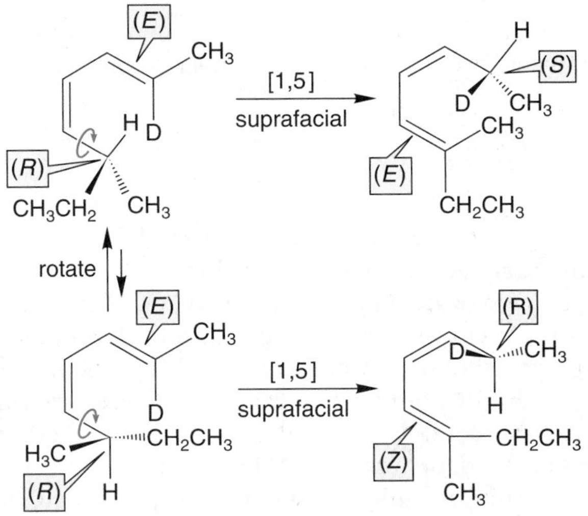

FIGURE 22.51 The products formed by suprafacial migration in the two rotational isomers of Roth's diene.

In antarafacial migrations, the hydrogen leaves from one side and is delivered from the other. Figure 22.52 shows the results of the two possible antarafacial migrations. Note that the products are different from each other and also different from those formed from suprafacial migration.

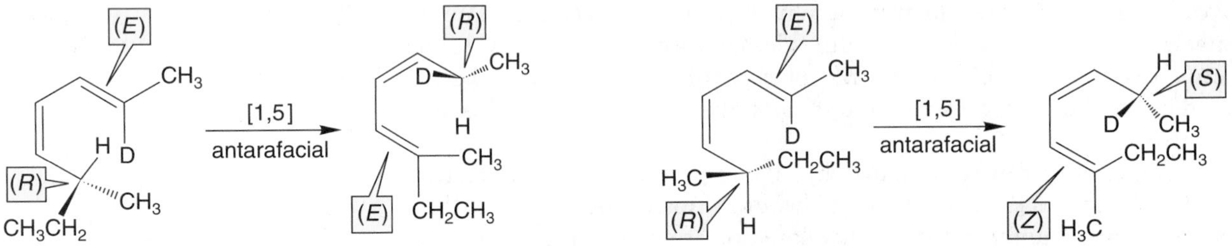

FIGURE 22.52 The possible products of antarafacial migration in Roth's diene.

Roth and his co-workers had the experimental skill necessary to sort out the products and found that only the products of suprafacial migration were formed. The theory is beautifully confirmed by this experiment.

At last we come to the remaining question. Why should the source of energy make a difference in the mode of migration? Remember, the common, thermally induced shifts are [1,5], whereas the relatively rare photochemical shifts are [1,3]. Consider again what happens when a molecule absorbs a photon of light. If you recall Chapter 12, you know that the packet of energy is used to promote an electron from the HOMO to the LUMO. After the absorption of light, there is a new HOMO, what we might call the photochemical HOMO (Fig. 22.13, p. 1135).

The new photochemical HOMO will have different symmetry relationships of the lobes than the old thermal HOMO. Figure 22.53 shows the situation for the simplest [1,3] shift in propene. Migration using Φ_3, the photochemical HOMO, can occur in a suprafacial fashion, and the steric barrier to thermal, necessarily antarafacial [1,3] shifts is gone. By using Φ_3, the hydrogen can migrate in an easy suprafacial manner.

To summarize: Sigmatropic shifts of hydrogen are quite simple to understand. As long as you keep in mind the necessity to form bonds and not antibonds, analysis of the HOMO gives you the positions available for reattachment of a migrating hydrogen. Add to this analysis the idea that hydrogen has only a small 1*s* orbital, and that will generally make any short antarafacial shifts impossible. That is all there is to it. We will now see that shifts of carbon atoms are more versatile and thus more complicated.

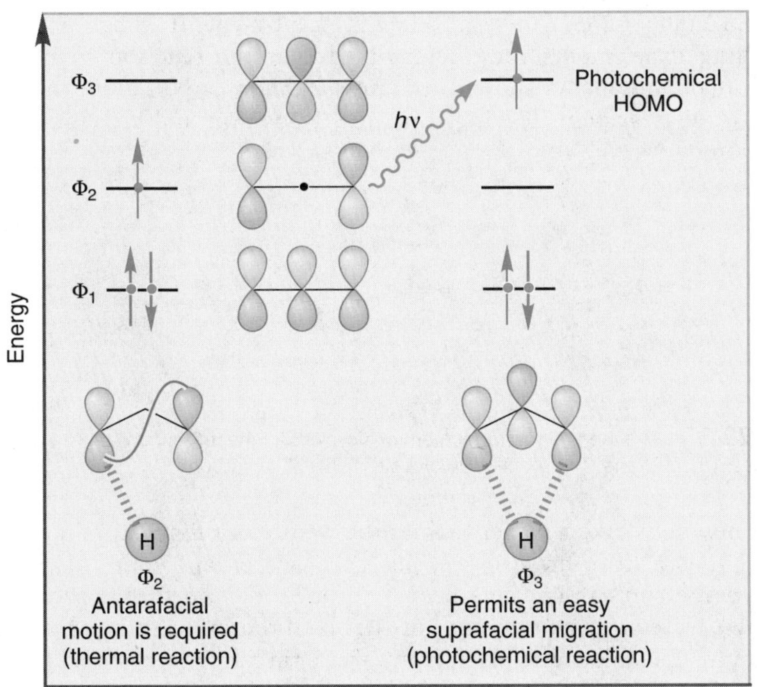

FIGURE 22.53 Suprafacial motion of a hydrogen is possible in Φ_3, the photochemical HOMO for this reaction.

22.5 SIGMATROPIC SHIFTS OF CARBON

There is an important difference between shifts of hydrogen and those of other atoms. Hydrogen is restricted to the use of the small, spherically symmetrical 1*s* orbital in its migrations. We have already seen an instance of how this limits the available motions. Thermal [1,3] shifts of hydrogen are essentially unknown not because there is anything electronically unfavorable about them. There is the requirement that bonding orbital overlap be maintained through antarafacial migration, but it is the steric requirements of this mode of migration that makes the reaction difficult, not anything inherent in antarafacial motion. Carbon and other atoms are not restricted to use of 1*s* or even 2*s* orbitals. They are attached to other atoms through hybrid orbitals and these have two lobes of different sign. The situation is quite different when we stretch a carbon–carbon bond than when we stretch a carbon–hydrogen bond (Fig. 22.54).

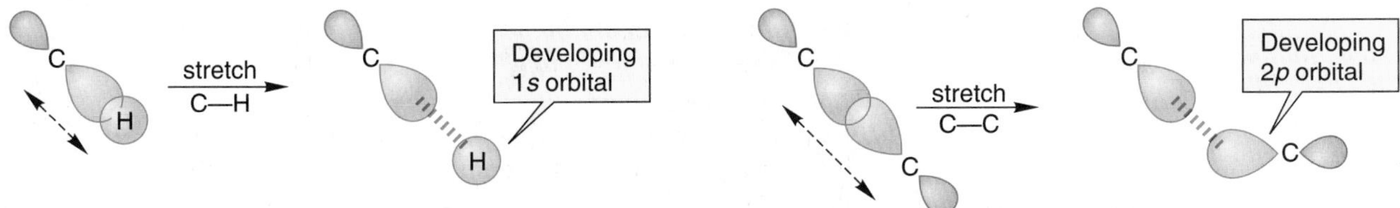

FIGURE 22.54 As a carbon–carbon bond stretches, an orbital with two lobes develops on the migrating atom.

In 1967, Jerome A. Berson (b. 1924) and his co-worker George Nelson worked out the details of the thermal rearrangement of a substituted bicyclo[3.2.0]hept-2-ene to a bicyclo[2.2.1]hept-2-ene. Through a variety of labeling experiments they determined that the reaction involved a [1,3] shift of carbon. As shown in Figure 22.55, a carbon atom migrates across an allyl system, leaving from the 1-position and arriving at the 3-position.

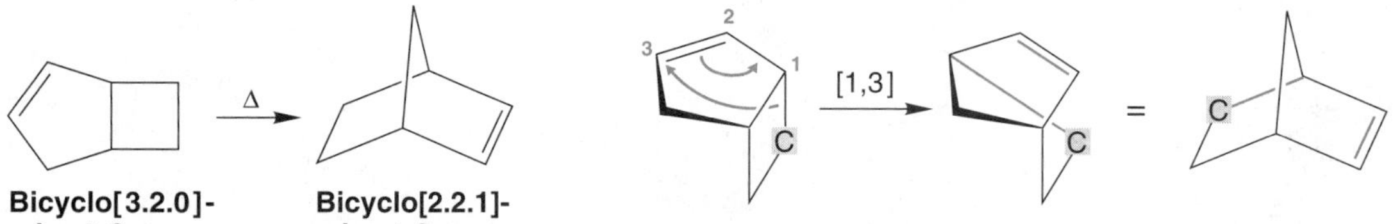

Bicyclo[3.2.0]-hept-2-ene **Bicyclo[2.2.1]-hept-2-ene**

FIGURE 22.55 The thermal rearrangement of bicyclo[3.2.0]hept-2-ene to bicyclo[2.2.1]hept-2-ene. The reaction involves a [1,3] shift of carbon.

But now let's take a closer look at the nature of the reaction, not only because it is a little hard to see, but because perhaps your curiosity is piqued as Berson's must have been. We have just finished a section in which we were at pains to point out that [1,3] thermal shifts of hydrogen faced an insuperable steric barrier, and were, therefore, very rare. Here we do see a thermal [1,3] shift, but of carbon, not hydrogen. What's the difference, and how does this difference make the reaction possible?

Berson and Nelson used a stereolabeling experiment to uncover the details of this reaction. They synthesized the trans-labeled compound shown in Figure 22.56. Now put yourself in the place of a naive observer and predict the stereochemistry of the product. There really are two possibilities. First, the bond might just slide over in what we could call a **least-motion** fashion. In this case, the trans stereochemistry of the starting material would be preserved in the product (Fig. 22.56).

"Least-motion" path

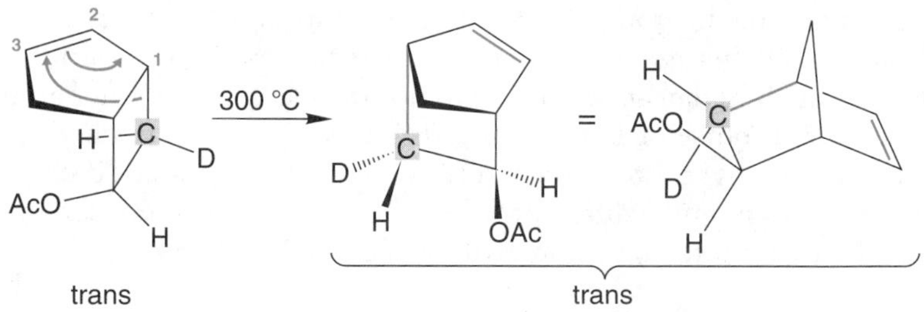

FIGURE 22.56 A least-motion rearrangement preserves the stereochemical relationships present in the starting material.

Second, the bond might break to form a diradical intermediate that could recombine in two ways to give products of both cis and trans stereochemistry (Fig. 22.57).

The experimental result is exactly the naively unpredictable one. The stereochemistry of the product is completely cis. Neither least motion nor the diradical can explain this. Berson and Nelson were scarcely naive observers, however, and had designed their beautiful experiment to be capable of revealing the observed result. Let's use orbital symmetry to recapitulate what must have been their reasoning. As the crucial bond between the allyl system and the migrating carbon stretches, the phase relationship between the two bonded lobes must be maintained. We also know the symmetry of the HOMO of the developing allyl system and can fill in the lobes as in Figure 22.58.

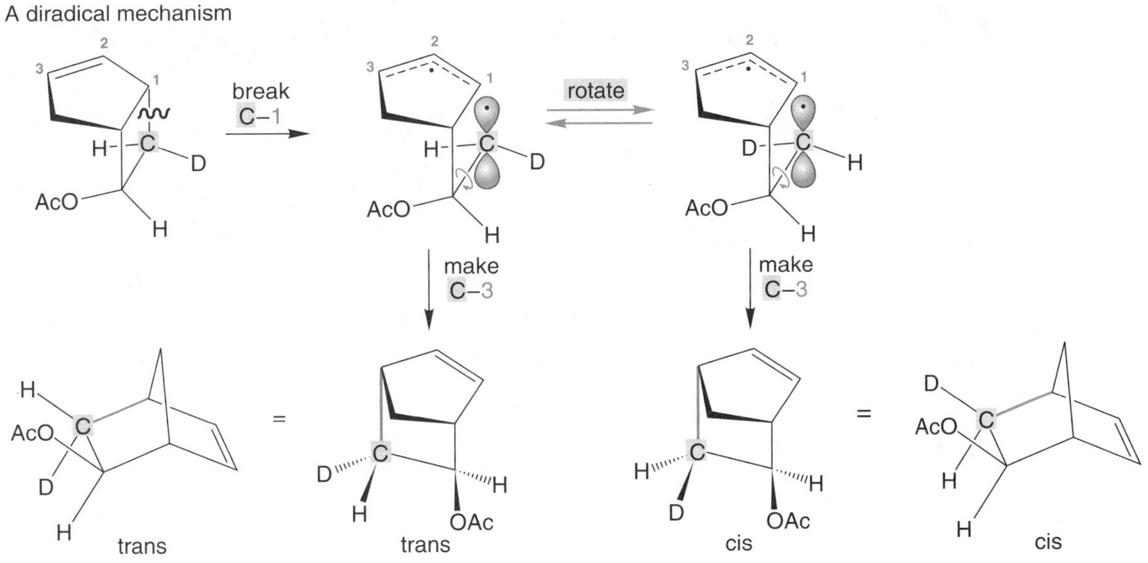

A diradical mechanism

FIGURE 22.57 If a diradical intermediate were involved, both cis and trans products would be formed.

Least motion can now be seen to involve the antibonding interaction that thwarts the suprafacial [1,3] shifts of hydrogen. But now there is an alternative. Migration can occur using the *back* lobe of the migrating carbon, and now a bonding interaction is created. Hydrogen, restricted as it is to the use of a spherically symmetrical 1*s* orbital, lacks this option. The migrating carbon suffers inversion as reattachment takes place to the 3-position of the allyl framework. In order to preserve the bonding overlaps with C_1 and C_3, rotation must occur, and the trans starting material is inevitably converted into cis product (Fig. 22.58).

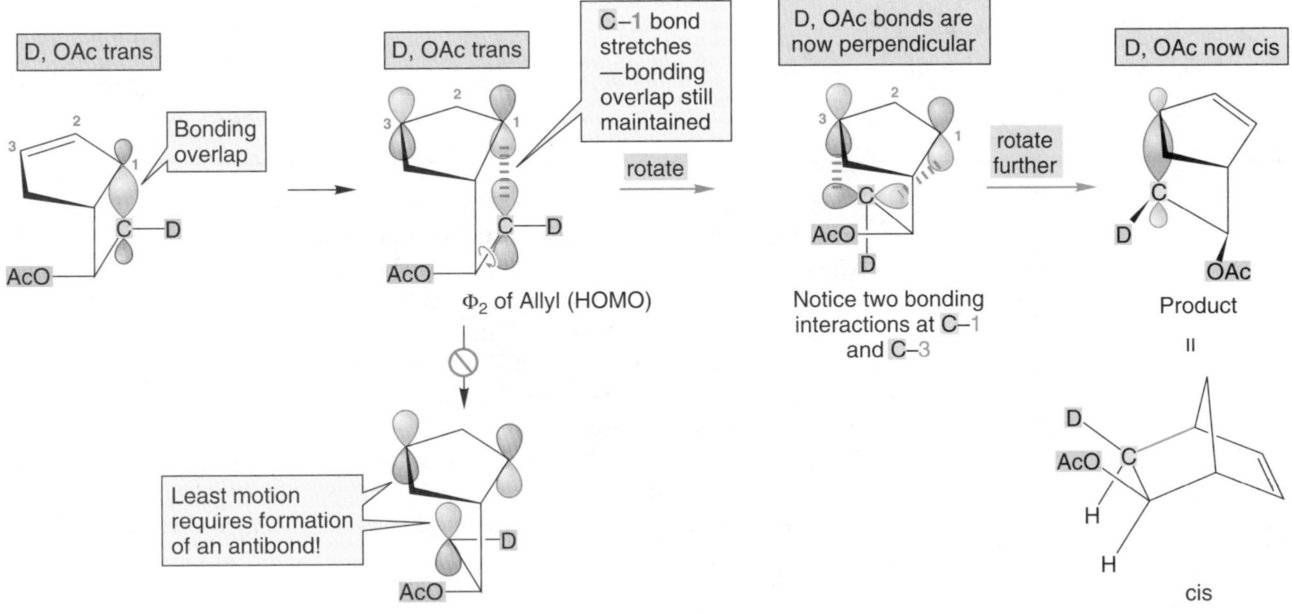

FIGURE 22.58 An analysis of the reaction using orbital symmetry shows how least motion is thwarted by an antibonding interaction. Maintenance of bonding overlap requires that trans starting material become cis product.

PROBLEM **22.16** Provide mechanisms for the reactions shown in Figure 22.59.

(a)

(b)

(c) In molecule **A**, will the X and Y groups remain in their respective positions (X = in and Y = out) as the CXY group walks around the ring, or will the X and Y groups alternate in and out as the reaction takes place? The R group is simply a marker so that the movement of the CXY group is apparent.

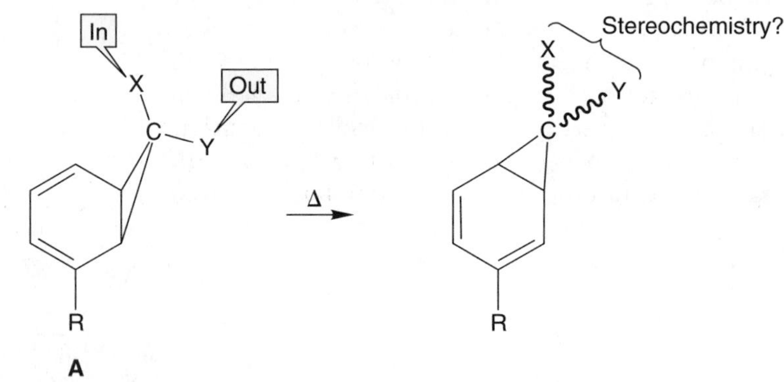

FIGURE **22.59**

22.6 OTHER SIGMATROPIC SHIFTS

Not all shift reactions are [1,*x*] shifts. One very common sigmatropic shift reaction is called the **Cope rearrangement**, and is named for Arthur C. Cope (1909–1966, the same Cope of the Cope elimination, Chapter 20, p. 1061) who spent most of his professional life at MIT.* The Cope rearrangement is a [3,3] shift. This terminology is almost universally confusing, but it is not really so bad. The two parts of the molecule are attached at their respective 1-positions (remember, sigmatropic shift terminology has no relationship to the protocol for naming the molecule). In the reaction, this 1—1 bond breaks and the two parts (allyl groups) are reattached through their respective 3-positions. In its simplest form, the Cope rearrangement is degenerate and must be revealed through a labeling experiment (Fig. 22.60).

*Cope didn't actually discover this reaction; C. Hurd of Northwestern University (b. 1897) apparently did. It was Cope's research group that developed the reaction, however.

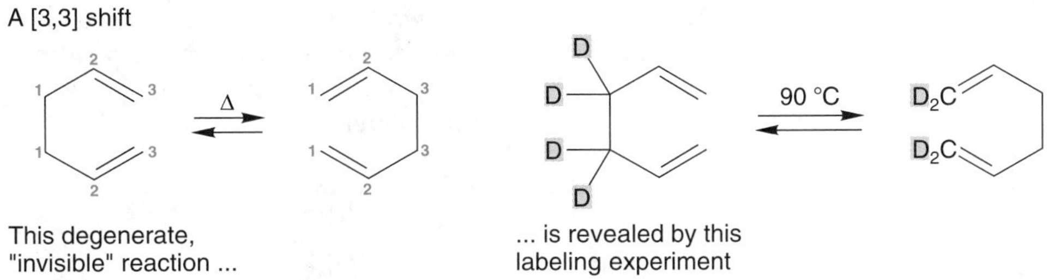

A [3,3] shift

This degenerate, "invisible" reaction ...

... is revealed by this labeling experiment

90 °C

FIGURE 22.60 The degenerate Cope rearrangement, a [3,3] shift, can be revealed by a labeling experiment.

Like the [1,5] shifts we looked at first, the Cope rearrangement is intramolecular, uncatalyzed, and shows no strong dependence on solvent polarity. The activation energy for the Cope rearrangement is about 34 kcal/mol, a value far below what one would estimate for the simple bond breaking into two separated allyl radicals. If the mechanism of the Cope rearrangement did involve two independent allyl radicals, the labeled 1,5-hexadiene would have to give two products, which it does not (Fig. 22.61).

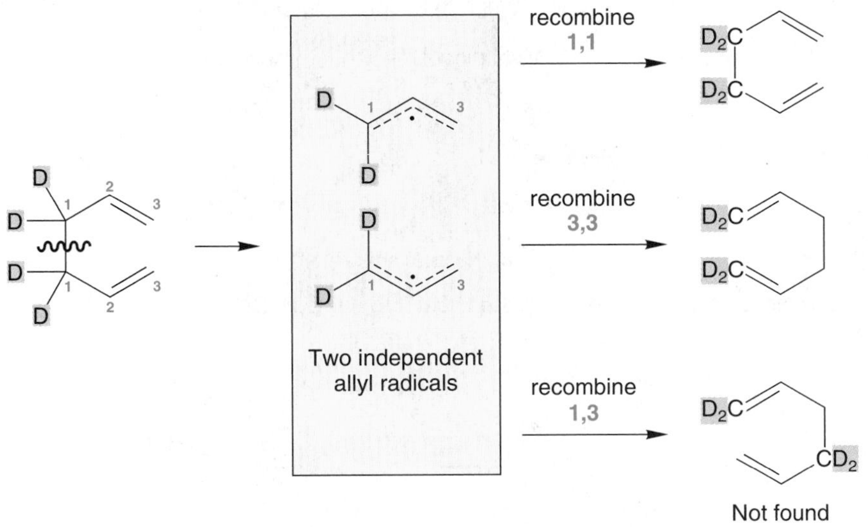

Two independent allyl radicals

recombine 1,1

recombine 3,3

recombine 1,3

Not found

FIGURE 22.61 A labeling experiment shows that the Cope rearrangement does not involve formation of a pair of free allyl radicals.

The Cope rearrangement has a nonpolar transition state, but this time it is not a simple hydrogen atom that migrates but a whole allyl radical. An analysis of the symmetry of the orbitals involved shows why this reaction is a relatively facile thermal process but is not commonly observed on photochemical activation. As we break the $C_1{-}C_1$ bond, the phases of the overlapping lobes must be the same. The HOMO of the allyl radical is Φ_2, and that information allows us to fill in the symmetries of the two allyl radicals making up the transition state (Fig. 22.62).

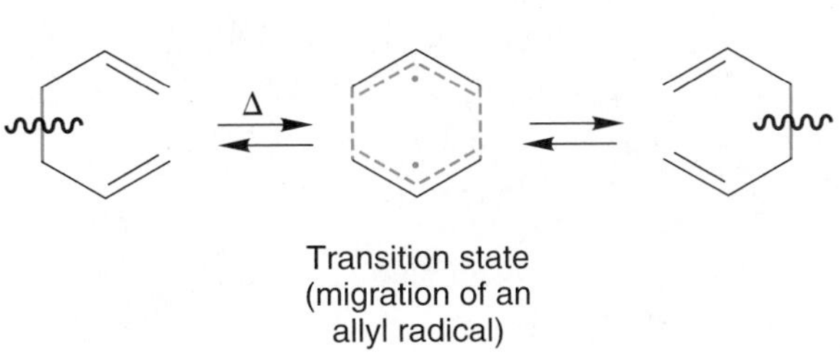

Transition state (migration of an allyl radical)

FIGURE 22.62 The transition state involves the migration of one allyl radical across another.

Reattachment at the two C_3 positions is allowed because the interaction of the two lobes on the two C_3 carbons is bonding. This picture is consistent with the near ubiquity of the Cope rearrangement when 1,5-dienes of all kinds are heated.

Bonding Bonding

Transition state (two partially bonded allyl radicals—Φ_2 is the HOMO for each)

FIGURE 22.63 An orbital symmetry analysis shows that the thermal Cope rearrangement is allowed.

*PROBLEM 22.17 Use orbital symmetry to analyze the photochemical [3,3] shift in 1,5-hexadiene. Be careful! Photons are absorbed *one at a time*. It is most unlikely that a singly excited molecule will live long enough to absorb a second photon.

ANSWER The transition state for the [3,3] shift involves two interacting allyl radicals.

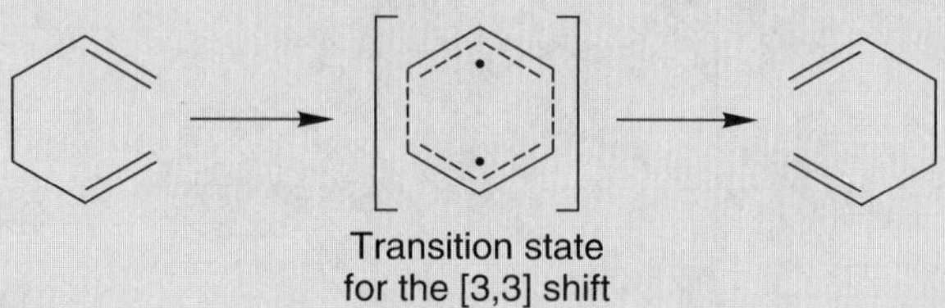

Transition state for the [3,3] shift

If one electron is promoted from the HOMO to the LUMO, the orbital picture will be as follows:

Allyl molecular orbitals

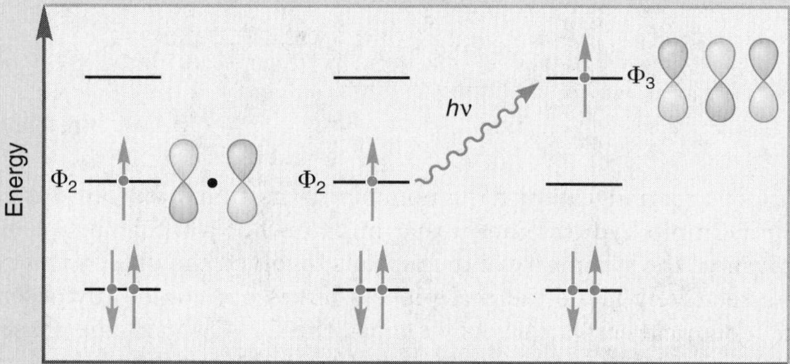

Now an orbital analysis shows that there is no easy path for the photochemical [3,3] shift. One antibonding interaction is always present.

Φ_2

Bonding Antibonding!

Φ_3

If a rotation were energetically feasible; if one end of one of the partners could turn over, a bonding interaction could be achieved where the figure notes the antibonding interaction. However, for the Cope rearrangement this involves a large energy cost.

ANSWER (CONTINUED)

So we now have a general picture of the transition state for the Cope rearrangement. But much more detail is possible if we work a bit at it. How might the two migrating allyl radicals of the transition state arrange themselves in space during the reaction? Two things are certain. First, the two C_1 atoms must start out close together, as they are bonded. Second, the two C_3 positions must also be within bonding distance, as in the transition state there is a partial bond between them (Fig. 22.63).

The question now comes down to determining the position of the two C_2 atoms in the transition state. Two possibilities are shown in Figure 22.64. The two C_2 atoms might be close together as in the boat-like transition state (originally called the **six-center transition state**), or rather far apart, as in the chair, or **four-center transition state**. The figure shows a number of views of these two possibilities, but this is probably a time to get out the models.

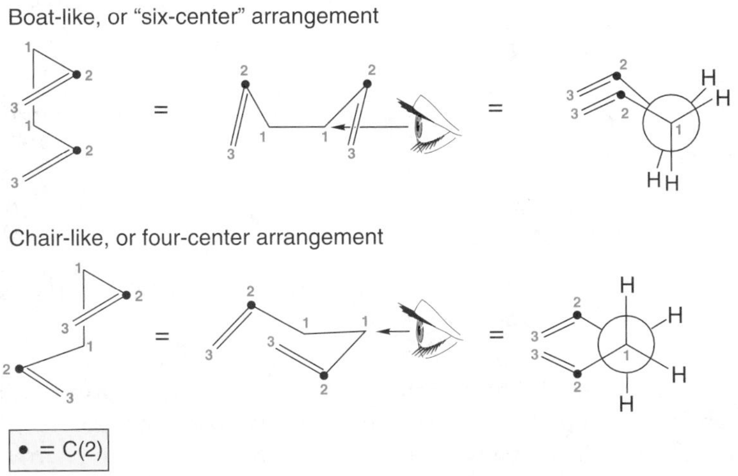

FIGURE 22.64 Several views of the chair-like (four-center) and boat-like (six-center) arrangements for the transition state of the Cope rearrangement.

In 1962, W. von E. Doering and W. R Roth carried out a brilliant experiment that determined which of these two possible transition states was favored. They used a labeled molecule, 3,4-dimethyl-1,5-hexadiene. Notice first that there are three forms of this molecule, a meso, achiral form, and a racemic pair of enantiomers. It all depends on how the methyl labels are attached (Fig. 22.65).

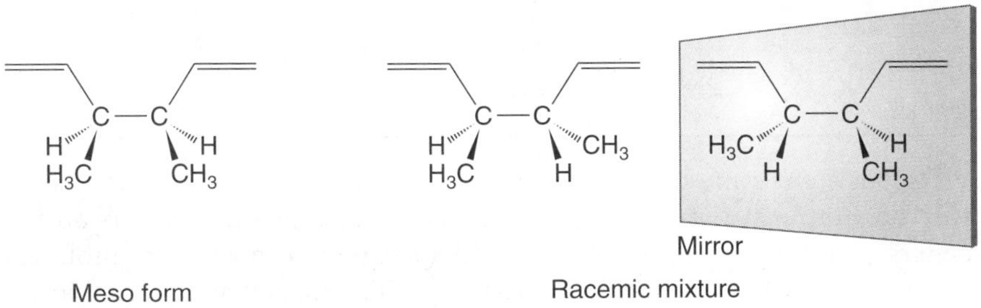

FIGURE 22.65 The meso and chiral forms of 3,4-dimethyl-1,5-hexadiene.

Convention Alert!

In the following figures, we have drawn only one enantiomer of the racemic pair. Be sure to keep in mind that the actual experiment used a 50:50 mixture of the two enantiomers, the racemate (Chapter 5, p. 177).

Both the meso and racemic forms of 3,4-dimethyl-1,5-hexadiene can adopt either the boat (six-center) or chair (four-center) arrangement (Fig. 22.66). The key to Doering and Roth's experiment is that the chair and boat arrangements give different products for both the meso and chiral forms (Fig. 22.66).

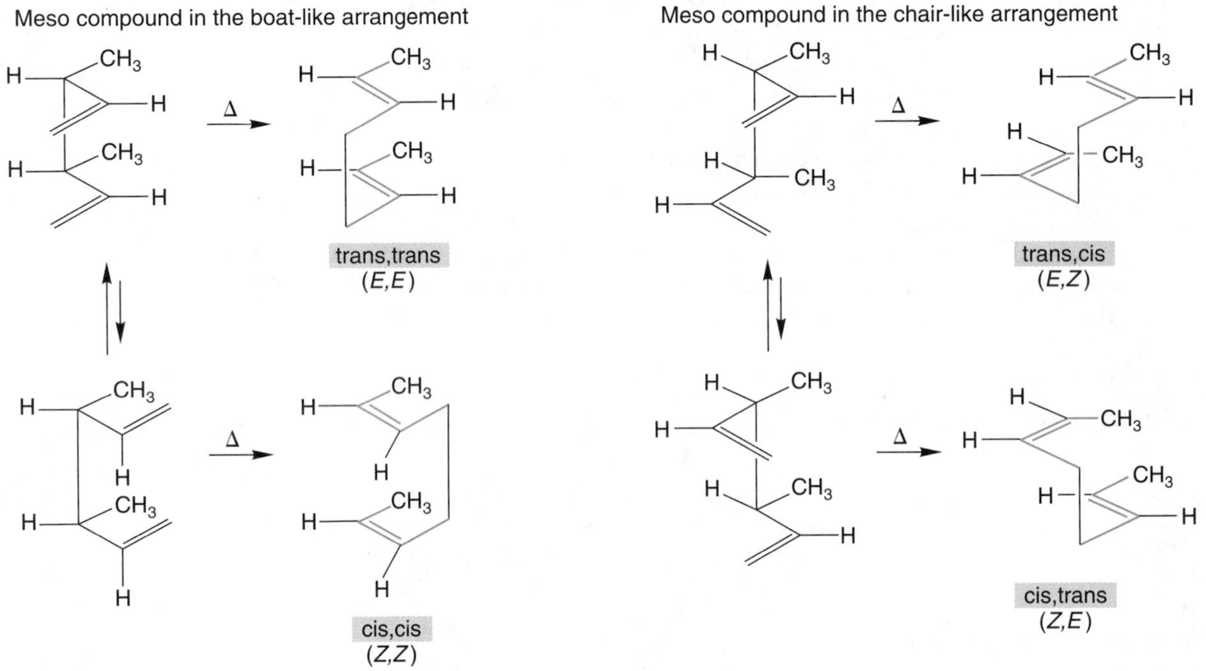

FIGURE 22.66 The products formed from the boat-like (six-center) and chair-like (four-center) arrangements of meso and chiral 3,4-dimethyl-1,5-hexadiene.

The results showed that it is the chair-like four-center form that is favored, and that it is preferred to the boat-like six-center arrangement by at least 6 kcal/mol. Presumably, some of the factors that make chair cyclohexane more stable than the boat form (Chapter 6, p. 197) operate to stabilize the related chair-like transition state.

There are many things that are important about this work. First of all, it is a prototypal example of the way labeling experiments can be used to dig out the details of a reaction mechanism. The second reason is more subtle, but perhaps more important. As we will see in Section 22.7, the study of the Cope rearrangement leads directly to one of the most intellectually creative ideas of the last few years.

22.7 SOMETHING MORE: A MOLECULE WITH A FLUXIONAL STRUCTURE

We spent some time in Section 22.6 working out the details of the structure of the transition state for the Cope rearrangement, which is a most general reaction of 1,5-hexadienes, and appears in almost every molecule incorporating the 1,5-diene substructure. This section will examine the

Chorismate to Prephenate

(*Z*) Double bond
This carbon is (*S*)
(*Z*) Double bond
This carbon would be (*R*)
chorismate mutase
Chorismate
Chair-like transition state for chorismate
Prephenate
If this boat-like transition were used...
...this carbon would be (*R*)

The Cope rearrangement is by no means restricted to the laboratory. Nature uses Cope-like processes as well. For example, in bacteria and plants a critical step in the production of the essential amino acids tyrosine and phenylalanine is the rearrangement of chorismate to prephenate, catalyzed by the enzyme, chorismate mutase. At the heart of this rearrangment is a Cope-like process called the Claisen* rearrangement. The research groups of Jeremy Knowles (b. 1935) at Harvard and Glenn Berchtold (b. 1932) at MIT collaborated on a tritium (^3H = T, a radioactive isotope of hydrogen) labeling experiment, related to that used by Doering and Roth, to determine that the enzymatic reaction also proceeded through a chair-like transition state, not a boat-like one. Notice that the chair-like transition state must take the compound with a (*Z*) double bond to a product in which the stereogenic carbon is (*S*), not (*R*).

*The same Claisen of the Claisen condensation (Chapter 20, p. 1037) and the Claisen–Schmidt reaction (Chapter 18, p. 914)

consequences of the Cope rearrangement and will work toward a molecule whose structure differs in a fundamental way from anything you have seen so far in this book. At room temperature it has no fixed structure; instead it has a **fluxional structure** in which nearest neighbors are constantly changing. Along the way to this molecule there is some extraordinary chemistry and some beautiful insights. This section is rather longer than most Something More sections, but it shows not only clever ideas and beautiful chemistry, but also how much fun this subject can be.

The simple 1,5-hexadiene at the heart of the Cope rearrangement can be elaborated in many ways, and in fact we've already seen one. If we add a central π bond we generate *cis*-1,3,5-hexatriene, a molecule we saw when we considered electrocyclic reactions (p. 1136). The arrow formalism of the Cope rearrangement leads to 1,3-cyclohexadiene (Fig. 22.67).

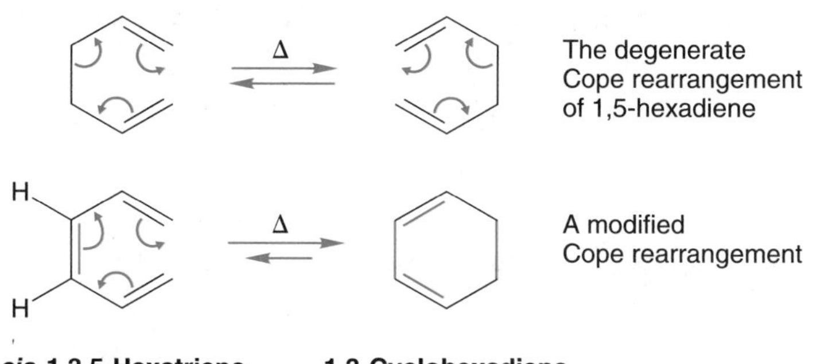

FIGURE 22.67 The electrocyclic ring closure of *cis*-1,3,5-hexatriene to 1,3-cyclohexadiene is a modified Cope rearrangement.

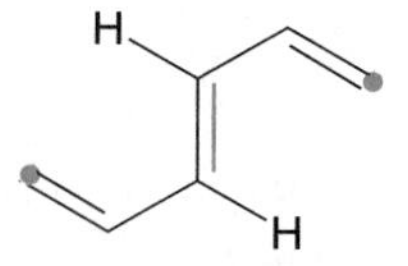

trans-1,3,5-Hexatriene

FIGURE 22.68 *trans*-1,3,5-Hexatriene cannot undergo this rearrangement because the ends of the molecule (dots), where bonding must occur, are too far apart.

But there is a stereochemical problem. This reaction is only possible for the stereoisomer with a cis central double bond. In the trans isomer, the ends of the triene system are too far apart for bond formation (Fig. 22.68).

That's an easy concept, and it appears again in a slightly different modified 1,5-hexadiene, 1,2-divinylcyclopropane. The Cope rearrangement is possible in this molecule and here, too, it is only the cis molecule that is capable of rearrangement. In the trans compound, the ends cannot reach (Fig. 22.69).

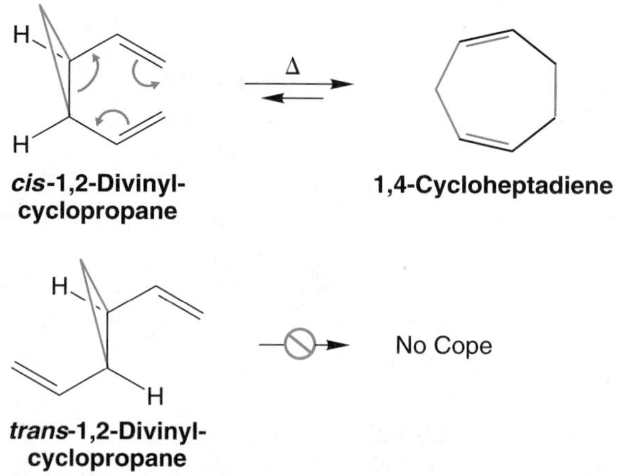

***cis*-1,2-Divinyl-cyclopropane** **1,4-Cycloheptadiene**

***trans*-1,2-Divinyl-cyclopropane** No Cope

FIGURE 22.69 *cis*-1,2-Divinylcyclopropane can undergo the Cope rearrangement, but the trans isomer cannot.

PROBLEM 22.18 This Cope rearrangement is very easy. It is rapid at room temperature. Why should it be so much easier than the other versions we have seen?

There is another stereochemical problem in this reaction that is a bit more subtle. The most favorable arrangement for *cis*-1,2-divinylcyclopropane is shown in Figure 22.70. The two vinyl groups are directed away from the ring, and in particular, away from the syn ring hydrogen. Yet this extended form cannot undergo the Cope rearrangement at all! To do the Cope, a less favored coiled form must be adopted. Now the two vinyl groups point back toward the ring and are closer to the offending syn hydrogen.

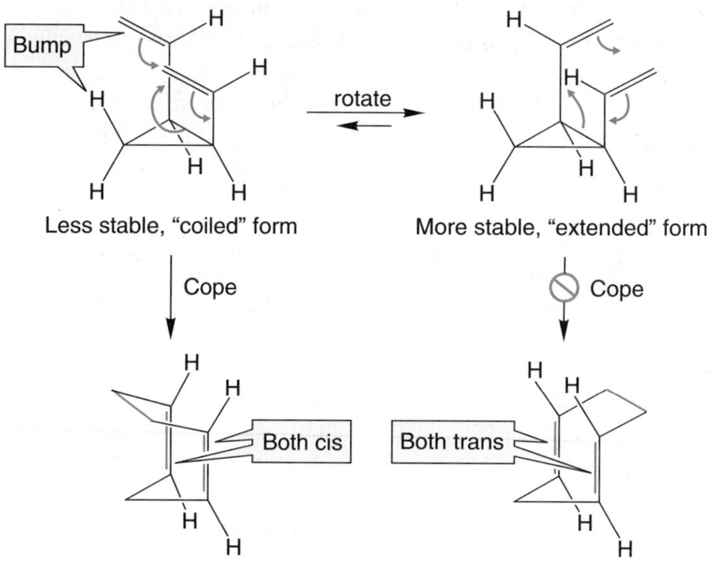

Bump

rotate

Less stable, "coiled" form

More stable, "extended" form

Cope

Cope

Both cis

Both trans

FIGURE 22.70 Only the less stable coiled arrangement of *cis*-1,2-divinylcyclopropane can undergo the Cope rearrangement. The more stable, extended form must produce a hideously unstable *trans,trans*-cyclohepta-1,4-diene.

The problem is that the extended form must produce a 1,4-cycloheptadiene product with two trans double bonds. Although the paper puts up with these trans double bonds without protest, the molecule will not. Remember, trans double bonds in small rings are most unstable. *trans*-Cycloheptene itself is a barely detectable reactive intermediate that can be intercepted only at low temperature. The constraints of the ring lead to severe twisting about the π bond and destabilization. A seven-membered ring containing two trans double bonds is unthinkably unstable.

By contrast, the higher energy coiled form yields a seven-membered ring containing two cis double bonds, and that presents no problem. Be certain you see why the two forms give different stereochemistries. Figure 22.71 shows the Energy versus Reaction progress diagram for the entire reaction.

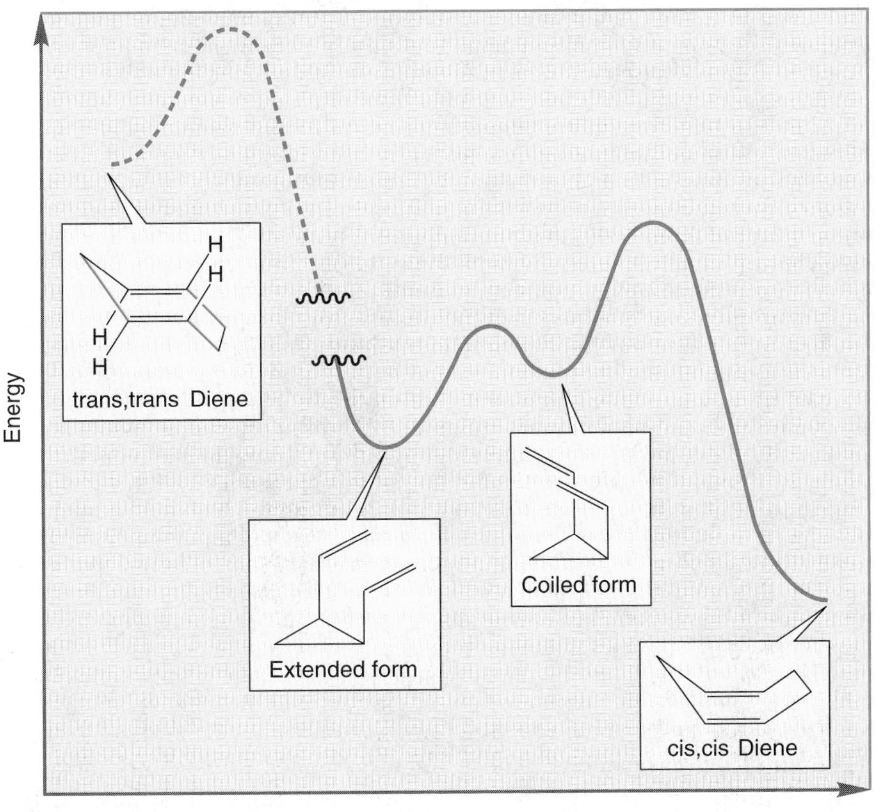

FIGURE 22.71 An Energy versus Reaction progress diagram for the Cope rearrangement of *cis*-1,2-divinylcyclopropane.

Now let's make another addition to the system. Let's add a single methylene group joining the two double bonds to give the molecule named "homotropilidene." We can still draw the arrow formalism, and the Cope rearrangement still proceeds rapidly. As Figure 22.72 shows, we are back to a degenerate Cope rearrangement, as the starting material and product are the same.

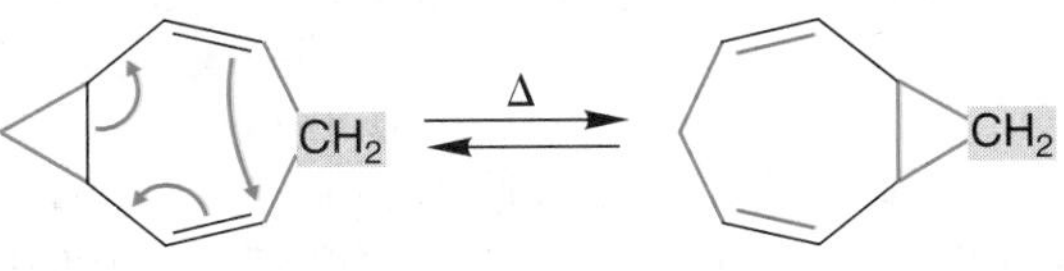

FIGURE 22.72 The degenerate Cope rearrangement of homotropilidene.

PROBLEM 22.19 This rapid Cope led to problems for W. R Roth who first made the molecule. Can you explain in general terms the perplexing ^{1}H NMR spectra for this molecule (Fig. 22.73)? Note that at both low and high temperature the spectrum is sharp, whereas at an intermediate temperature, 20 °C, it consists of two "blobs." You do not have to analyze the spectra in detail—just give the general picture of what must be happening.

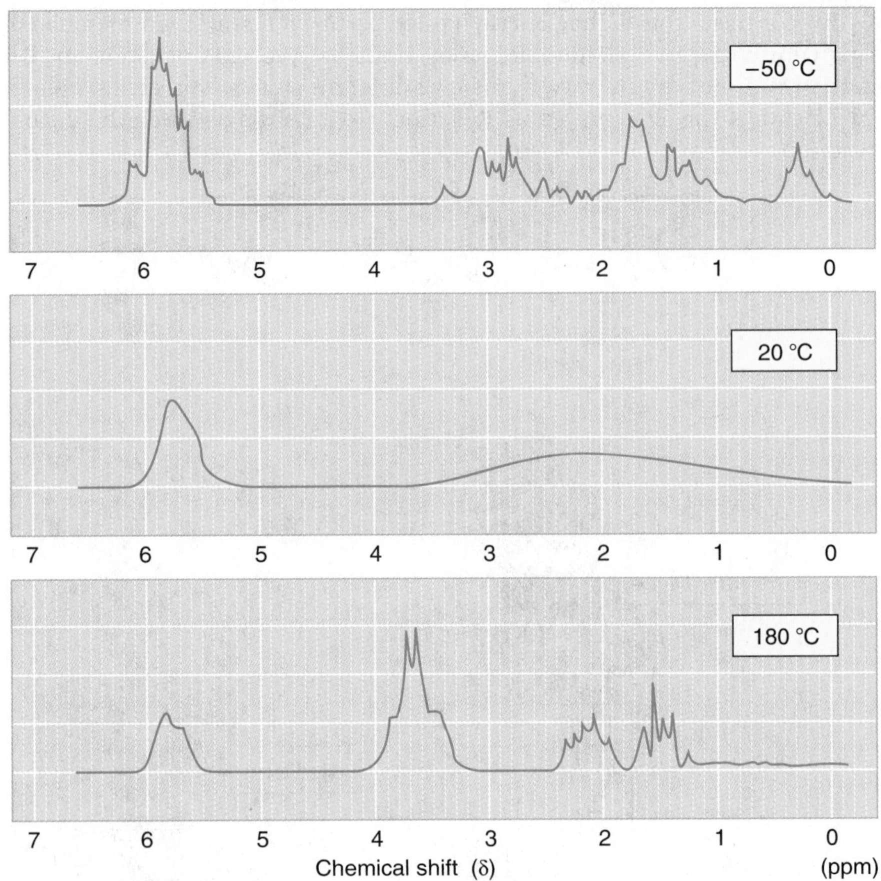

FIGURE 22.73

There are also stereochemical problems for this molecule. Like 1,2-divinylcyclopropane, the most stable extended form of homotropilidene is unable to undergo the Cope rearrangement, and it must be the higher energy coiled arrangement that is actually active in the reaction. Here the destabilization of the coiled arrangement is especially obvious, as the two methylene hydrogens badly oppose each other. As before, the extended form must give an unstable molecule with a pair of trans double bonds incorporated into a seven-membered ring. The less stable coiled arrangement has no such problem, as it can give a product containing two cis double bonds (Fig. 22.74).

Although *cis*-1,2-divinylcyclopropane and homotropilidene undergo the Cope rearrangement quite rapidly even at room temperature and below, one might consider how to make the reaction even faster. For the reaction to occur, an unfavorable equilibrium between the more stable, but unproductive extended form, and the higher energy, but productive coiled form must be overcome. The activation energy for the reaction includes this unfavorable equilibrium (Fig. 22.75).

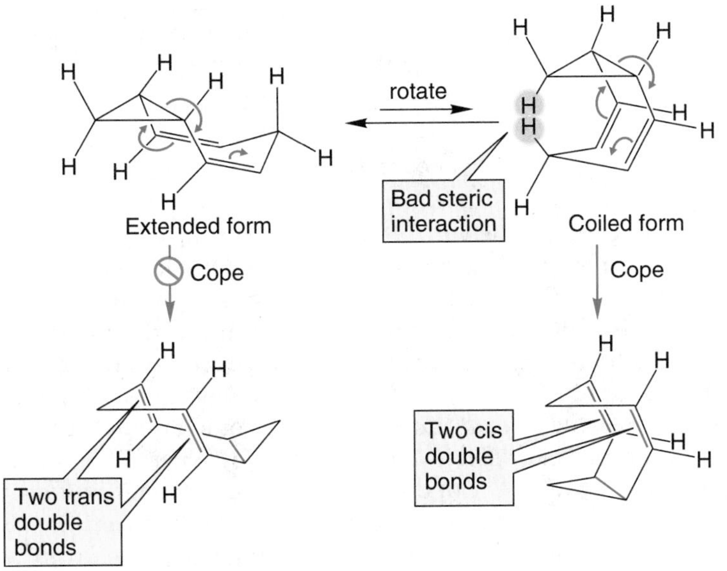

FIGURE 22.74 Only the higher energy coiled form of homotropilidene can give a stable product in the Cope rearrangement. The lower energy extended form must produce a product containing two trans double bonds in a seven-membered ring.

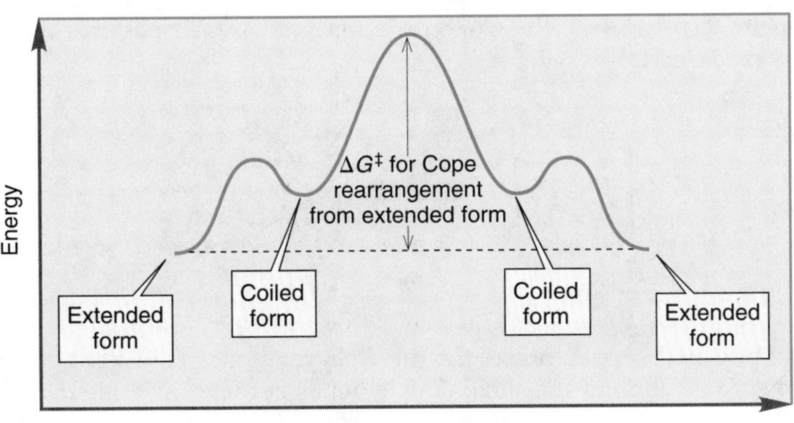

FIGURE 22.75 An Energy versus Reaction progress diagram for the Cope rearrangement of homotropilidene. The activation energy ($\Delta G^{\ddagger}$) for the reaction includes the energy difference between the extended and coiled forms.

If this unfavorable equilibrium could be avoided in some way, and the molecule locked into a coiled arrangement, the activation energy would be lower, and the reaction faster (Fig. 22.76).

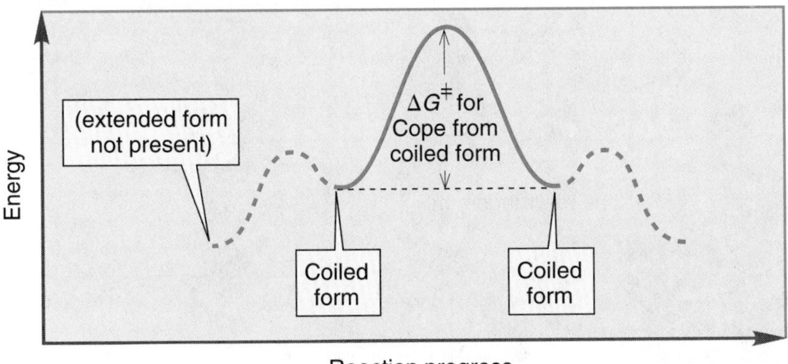

FIGURE 22.76 An Energy versus Reaction progress diagram for the Cope rearrangement of a homotropilidene locked into the coiled arrangement. The activation energy is lower, as it is no longer necessary to fight an unfavorable equilibrium from the extended form, shown here as a phantom (dashes) just to remind you of the previous figure.

Our next variation of the Cope rearrangement includes the clever notion of bridging the two methylene groups of homotropilidene to lock the molecule in a coiled arrangement (Fig. 22.77).

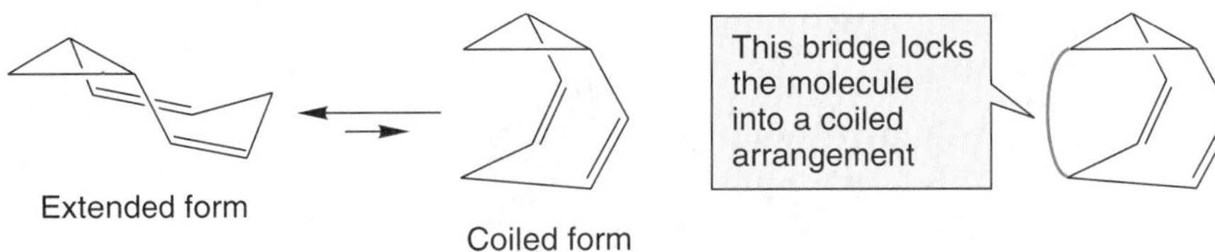

FIGURE 22.77 A homotropilidene locked into the coiled arrangement can be achieved if the two methylene groups are connected with a bridge.

This modification greatly increases the rate of the Cope rearrangement as the activation energy is decreased (Fig. 22.76). No longer is it necessary to fight an unfavorable equilibrium in order for the Cope rearrangement to occur; the coiled form is built into the structure of the molecule itself. Many bridged homotropilidenes are now known, and some are shown in Figure 22.78.

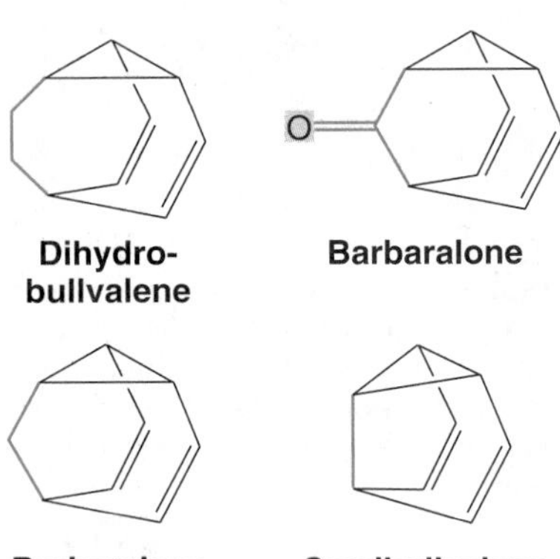

FIGURE 22.78 Four bridged homotropilidenes. All undergo very rapid Cope rearrangements.

Now that's a clever idea, and it has led to some fascinating molecules and some very nice chemistry. But it's only a good idea. What comes next is an extraordinary idea; one of those insights that anyone would envy, and which define through their example what is meant by creativity. Suppose we bridge the two methylene groups with a double bond to give the molecule known as **bullvalene**.* At first, this molecule seems nothing special. As Figure 22.79 shows, the Cope rearrangement is still possible in this highly modified 1,5-hexadiene.

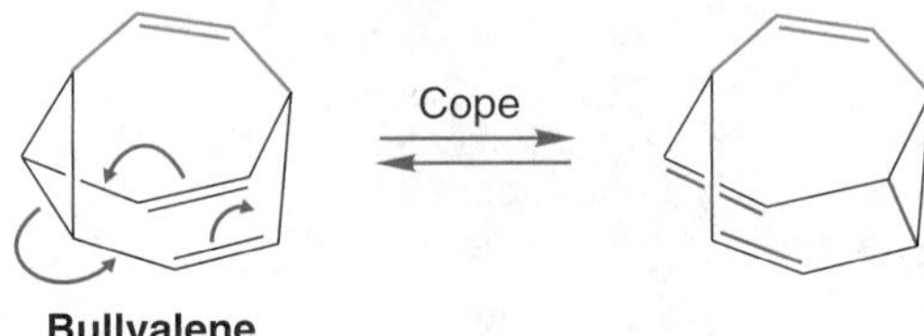

FIGURE 22.79 If the new bridge is a carbon–carbon double bond, the resulting molecule is known as bullvalene.

But now perhaps you notice the threefold rotational axis in bullvalene. Remove the color and the three double bonds become indistinguishable. In fact, there is a Cope rearrangement possible in each of the three faces of the molecule, and every Cope rearrangement is degenerate; each re-forms the same molecule (Fig. 22.80).

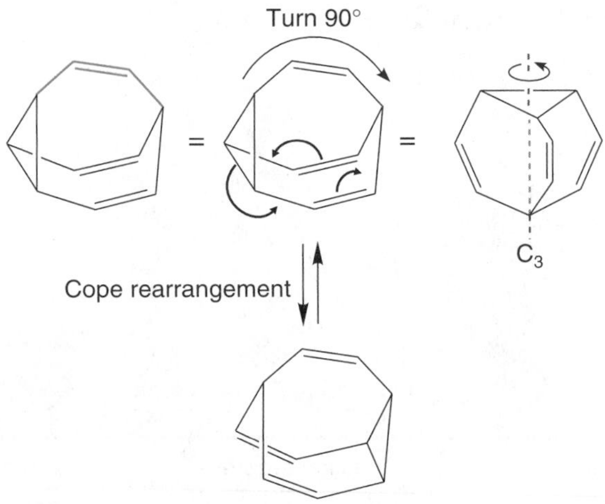

FIGURE 22.80 Bullvalene has a threefold rotational axis: All three double bonds are equivalent. The Cope rearrangement can occur in each of the three faces of the molecule, and is degenerate in every case.

*What an odd name! If you are curious about its origin, see Alex Nickon and Ernest F. Silversmith's nice book, *The Name Game* (Pergamon, New York, 1987) on chemical naming. You won't find the IUPAC system in here, but you will find hundreds of stories about the origins of the names chemists give to their favorite molecules, including bullvalene.

That's quite interesting, but now begin to look at the structural consequences for bullvalene. As Cope rearrangements occur, *the carbons begin to wander apart!* Figure 22.81 only begins the process, but it can be demonstrated that the Cope rearrangement is capable of completely scrambling the 10 CH units of the molecule.* As long as the Cope rearrangement is fast, bullvalene *has no fixed structure;* no carbon has fixed nearest neighbors, but is instead bonded on time average to each of the 9 other methine carbons. It has a fluxional structure (Fig. 22.81).

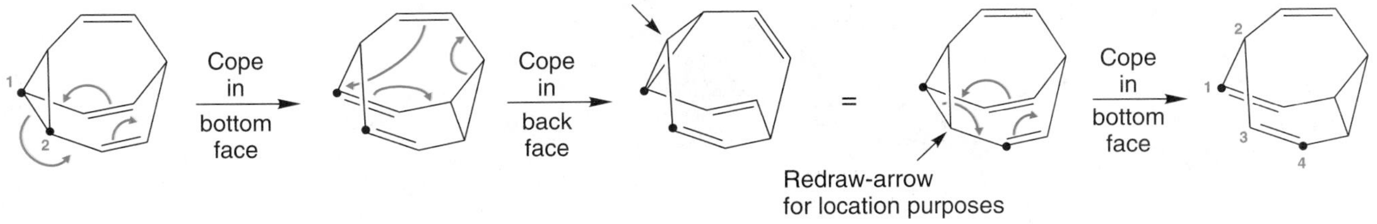

Cope in bottom face → Cope in back face → = Cope in bottom face →

Redraw-arrow for location purposes

FIGURE 22.81 A sequence of Cope rearrangements serves to move neighboring carbon atoms apart.

*PROBLEM 22.20

At somewhat above room temperature the ^1H NMR spectrum of bullvalene consists of a single broad line at δ 4.2 ppm. Explain.

ANSWER

If all the carbon–hydrogen bonds are the same on time average, the NMR spectrum must consist of a single line. We see the averaged spectrum.

*PROBLEM 22.21

Explain why the ketone in Figure 22.82 exchanges *all* its hydrogens for deuteriums when treated with deuterated base.

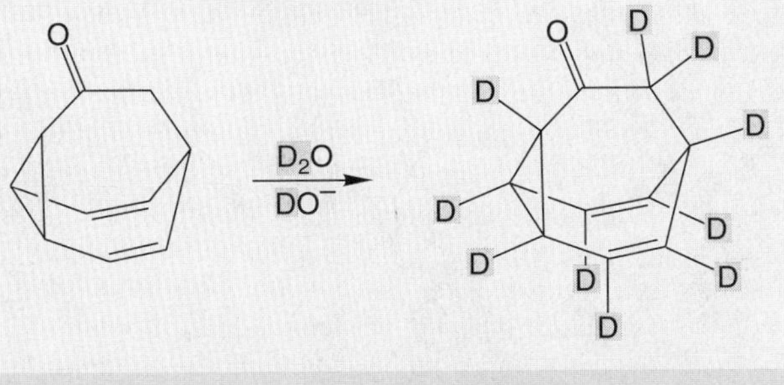

D_2O / DO^- →

FIGURE 22.82

ANSWER

Exchange at the α-position is routine (Chapter 18, p. 888).

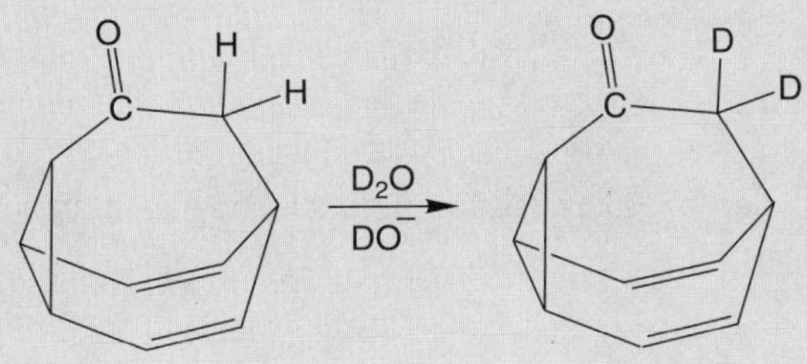

D_2O / DO^- →

*In an early paper, Doering noted that a sequence of 47 Copes sufficed to demonstrate that two adjacent carbons could be exchanged without disturbing the rest of the 10-carbon sequence, and that therefore all 1,209,600 possible isomers could be reached. No claim was made that this represented the minimum path. A later computer-aided effort by two Princeton undergraduates located a sequence of 17 Copes, which could do the same thing.

ANSWER (CONTINUED) However, simple ketones are in equilibrium with their enol forms, and the enol of this molecule is a substituted bullvalene. The Cope rearrangement will make all the carbons equivalent, and thus, on time average, all carbon–hydrogens will come to occupy the exchangeable α-position.

Now this hydrogen can exchange

Cope

D_2O

DO^-

Repetition leads to exchange of all the hydrogens for deuterium.

Naturally, efforts at synthesis began as soon as the prediction of the properties of bullvalene appeared in the chemical literature. Long before the efforts at a rational synthesis were sucessful, the molecule was discovered by Gerhard Schröder (b. 1929), a German chemist then working in Belgium. Schröder was working out the structures of the several dimers of cyclooctatetraene, and in the course of this work irradiated one of the dimers. It conveniently fell apart to benzene and bullvalene (Fig. 22.83). Schröder was alert enough to understand what had happened, and went on to a highly successful career, which included an examination of the properties of bullvalene and substituted versions of this molecule. Doering's predictions were completely confirmed by Schröder's synthesis.

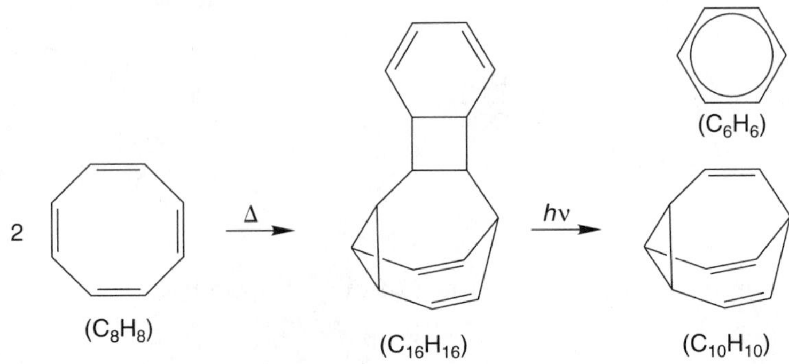

FIGURE 22.83 Schröder's accidental synthesis of bullvalene.

PROBLEM 22.22 Why is the second step of Figure 22.83, the photochemical fragmentation, so easy?

Other neutral, completely fluxional organic molecules have not appeared, although the phenomenon of fluxionality appears to be rather more common in organic cations and organometallic compounds.

Provide mechanisms for the reactions in Figure 22.84. PROBLEM 22.23

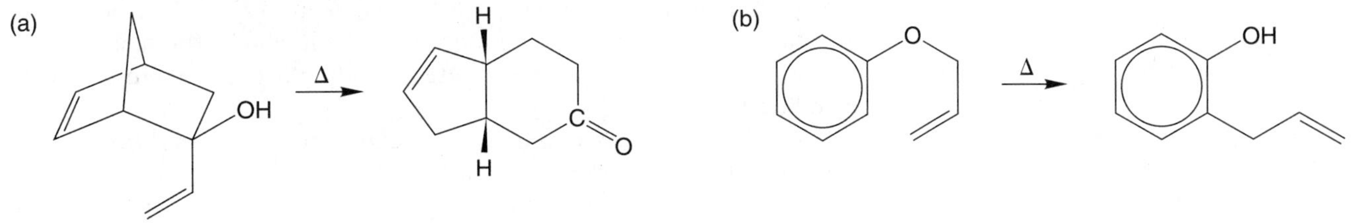

FIGURE 22.84

The Cope reaction is far more than just a chemical curiosity. It is a general reaction of almost all 1,5-hexadienes, and appears almost no matter how the basic hexadiene form is perturbed and disguised. All sorts of chemical "spinach" can be added, often making it quite difficult to find the bare bones of the hexadiene system, but the reaction will persist. Moreover, as we have just seen, digging deep into the Cope reaction, following the intellectual implications of substitution, leads to the marvelous molecule bullvalene, without precedent in organic chemistry, a compound without a fixed structure.

22.8 SUMMARY

NEW CONCEPTS

The central insight of Woodward–Hoffmann theory is that one must keep track of the phase relationships of orbital interactions as a concerted reaction proceeds. If bonding overlap can be maintained throughout, the reaction is allowed; if not, the reaction is forbidden by orbital symmetry. This notion leads to a variety of mechanistic explanations for reactions that can otherwise be baffling. Some of these so-called pericyclic reactions are discussed in the following section.

The notion of a fluxional structure is introduced through the $(CH)_{10}$ molecule bullvalene. In this molecule, a given carbon atom does not have fixed nearest neighbors, but instead is bonded, over time, to each of the other nine carbons in the molecule. In bullvalene, the fluxionality is the result of a rapid degenerate Cope rearrangement.

Many stereochemical labeling experiments are described in this chapter. A careful labeling experiment is able to probe deeply into the details of a reaction mechanism. Examples include W. R Roth's determination of the suprafacial nature of the [1,5] shift of hydrogen and Doering and Roth's experiments on the structure of the transition state for the Cope rearrangement.

REACTIONS, MECHANISMS, AND TOOLS

Certain polyenes and cyclic compounds can be interconverted through a pericyclic process known as an electrocyclic reaction. Examples include the 1,3-butadiene–cyclobutene and 1,3-cyclohexadiene–1,3,5-hexatriene interconversions (Figs. 22.5 and 22.17).

Orbital symmetry considerations dictate that in $4n$-electron reactions the thermal process must use a conrotatory motion, whereas the photochemical reaction must be disrotatory. Just the opposite rules apply for reactions involving $4n + 2$ electrons. The key to analyzing electrocyclic reactions is to look at the way the p orbitals at the end of the open-chain π system must move in order to generate a bonding interaction in the developing σ bond.

Cycloaddition reactions are closely related to electrocyclic reactions. The phases of the lobes in the HOMO and LUMOs must match so that bonding interactions are preserved in the transition state for the reaction. Sometimes a straightforward head-to-head motion is possible in which the two π systems approach each other in parallel planes to produce two bonding interactions. Such reactions are typically easy and are quite common. An example is the Diels–Alder reaction (Fig. 22.23).

In other reactions, the simple approach of HOMO and LUMO involves an antibonding overlap. In such reactions, a rotation is required in order to bring the lobes of the same sign together (Fig. 22.30). Although in principle this rotation maintains bonding interactions, in practice it requires substantial amounts of energy, and such reactions are rare.

Sigmatropic shifts involve the migration of hydrogen or another atom along a π system. Allowed migrations can be either suprafacial (leave and reattach from the same side of the π system) or antarafacial (leave from one side and reattach from the other) depending on the phase of the orbitals involved in the migration. The important point is that in an allowed shift, bonding overlap must be maintained at all times both to the lobe from which the migrating group departs and to that to which it reattaches. Steric considerations can also be important, especially for hydrogen, which must migrate using a small $1s$ orbital (Fig. 22.85).

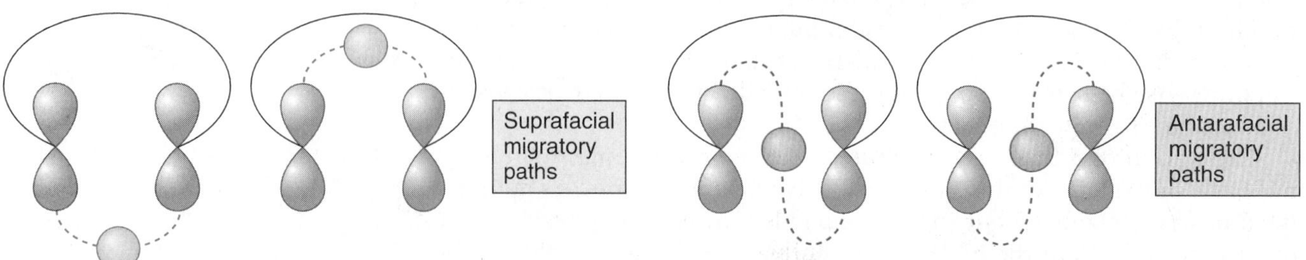

FIGURE 22.85 Suprafacial and antarafacial migratory paths for sigmatropic shifts.

SYNTHESES

The photochemical 2 + 2 dimerization of alkenes (Fig. 22.86).

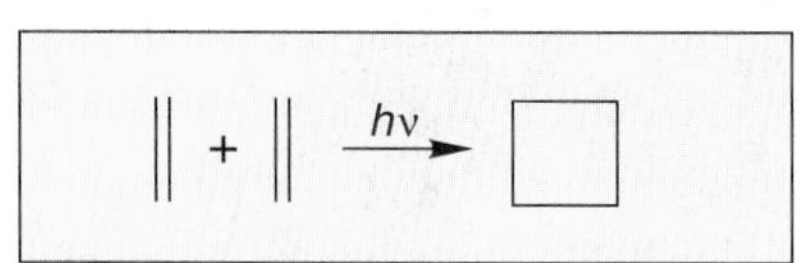

FIGURE 22.86 Synthetic reaction for Chapter 22.

COMMON ERRORS

For many students the big problem in this material is to determine the kind of reaction involved. The way to start a problem in this area is to spend some time and thought identifying what kind of reaction is taking place. Cycloadditions are generally easy to find, although even here confusion exists between these reactions and electrocyclic processes. Sigmatropic shifts are even harder.

Are two molecules coming together to form a new ring? If so, it's a cycloaddition. Is a ring opening or closing in an intramolecular reaction?

That's an electrocyclic process. In sigmatropic shifts, something moves from one point in the molecule to another.

Often it is not easy to identify exactly what has happened in a sigmatropic reaction. When one atom or a group of atoms goes flying from one part of the molecule to another, there is often a rather substantial structural change. The product sometimes doesn't look much like the starting material.

22.9 KEY TERMS

Antarafacial motion Migration of a group from one side of a π system to the other in a sigmatropic shift.

Bullvalene Bullvalene is a molecule with a fluxional structure. Every carbon of this $(CH)_{10}$ compound is bonded on time average to each of the other nine carbons.

Concerted reaction A single-barrier process. In a concerted reaction, starting material is converted into product with no intermediate structures.

Conrotation In a conrotatory process, the end *p* orbitals of a polyene HOMO rotate in the same sense (both clockwise or both counterclockwise).

Cope rearrangement This [3,3] sigmatropic shift converts one 1,5-diene into another.

Cycloaddition reaction A reaction in which two π systems are converted into a ring. The Diels–Alder reaction and the 2 + 2 reaction of a pair of ethylenes to give a cyclobutane are examples.

Degenerate reaction In a degenerate reaction, the starting material and product have the same structure.

Disrotation In a disrotatory process, the end *p* orbitals of a polyene HOMO rotate in opposite senses (one clockwise and the other counterclockwise).

Electrocyclic reaction The interconversion of a polyene and a ring compound. The end *p* orbitals of the polyene rotate so as to form the new σ bond of the ring compound.

Fluxional structure In a molecule with a fluxional structure, a given atom does not have fixed nearest neighbors. Instead, the framework atoms move about over time, each being bonded on time average to all the others.

Four-center transition state The chair-like transition state for the Cope rearrangement in which carbons 1 and 1′ and 3 and 3′ are within bonding distance (eclipsed) but carbons 2 and 2′ are as far apart as possible.

Least-motion mechanism The reaction mechanism involving the minimum translational and rotational motions for the atoms involved.

Pericyclic reaction A reaction in which the maintenance of bonding overlap between the lobes of the orbitals involved in bond making and breaking is controlling. All single-barrier (concerted) reactions can be regarded as pericyclic reactions.

Sigmatropic shift The migration of an atom, under orbital symmetry control, along a π system.

Six-center transition state The boat-like transition state for the Cope rearrangement in which carbons 1 and 1′, 2 and 2′, and 3 and 3′ are all eclipsed.

Suprafacial motion Migration of a group in a sigmatropic shift in which bond breaking and bond making take place on the same side of the π system.

Woodward–Hoffmann theory The notion that one must take account of the phase relationships between orbitals in order to understand reaction mechanisms. In a concerted reaction, bonding relationships must be maintained at all times.

22.10 ADDITIONAL PROBLEMS

PROBLEM **22.24** Analyze the photochemical and thermal [1,7] shifts of hydrogen in 1,3,5-heptatriene. Be sure to look at all possible stereoisomers of the starting material. What kind of [1,7] shift will be allowed thermally? What kind photochemically? How will the starting stereochemistry of the triene affect matters? Top views of the π molecular orbitals of heptatrienyl are given.

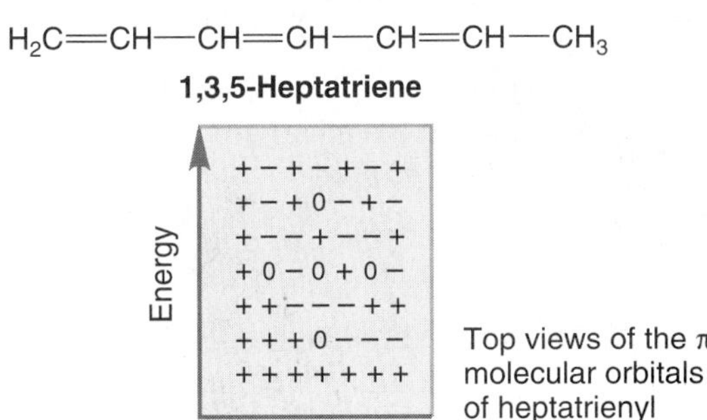

Top views of the π molecular orbitals of heptatrienyl

PROBLEM 22.25 Vitamin D_3 (**3**) is produced in the skin as a result of UV irradiation. It was once believed that 7-dehydrocholesterol (**1**) was converted directly into **3** upon photolysis. It is now recognized that there is an intermediate, previtamin D_3 (**2**), involved in the reaction. This metabolic process formally incorporates two pericyclic reactions, **1** → **2** and **2** → **3**. Identify and analyze the two reactions.

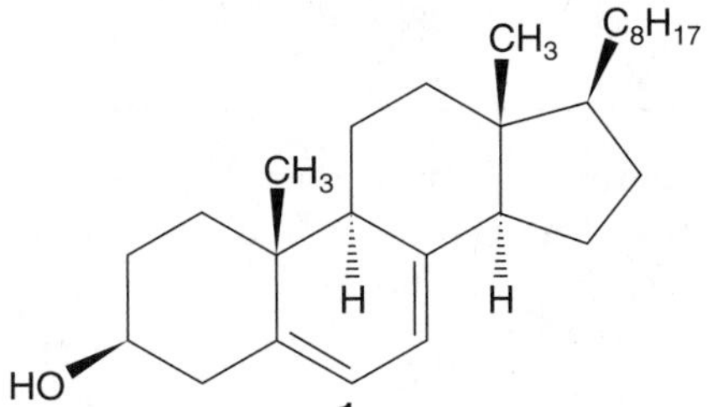

CH3 C8H17 CH3 H H HO **1**

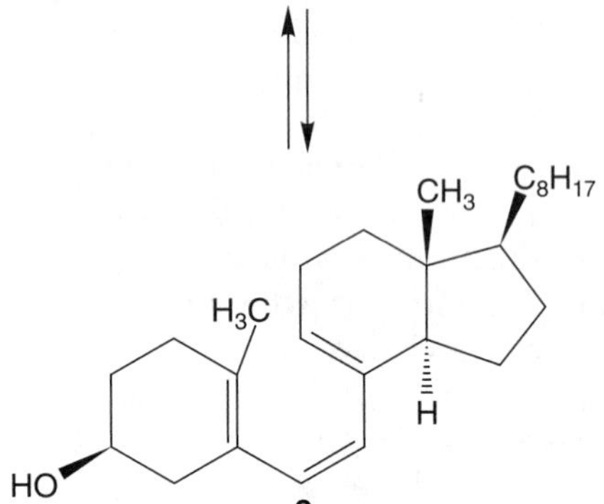

H3C CH3 C8H17 H HO **2**

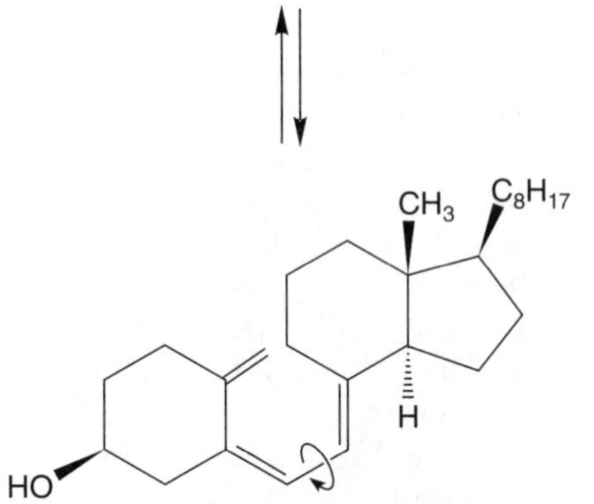

CH3 C8H17 H HO *s-cis*-**3**

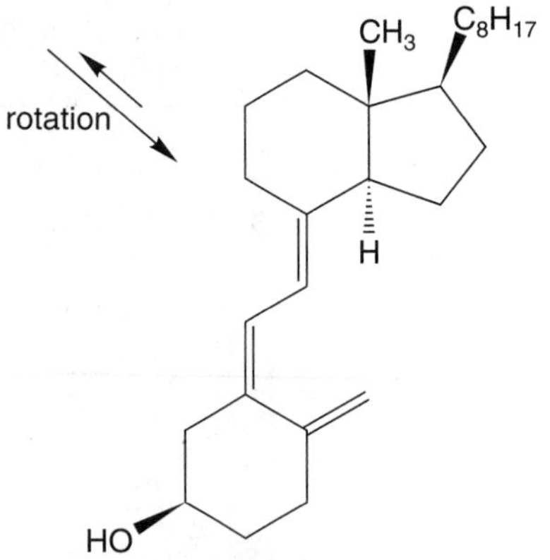

rotation

CH3 C8H17 H HO *s-trans*-**3**

PROBLEM 22.26 The thermally induced interconversion of cyclooctatetraene (**1**) and bicyclo[4.2.0]octa-2,4,7-triene (**2**) can be described with two different sets of arrows. Although these two arrow formalisms may seem to give equivalent results, one of them is, in fact, not a reasonable depiction of the reaction. Explain.

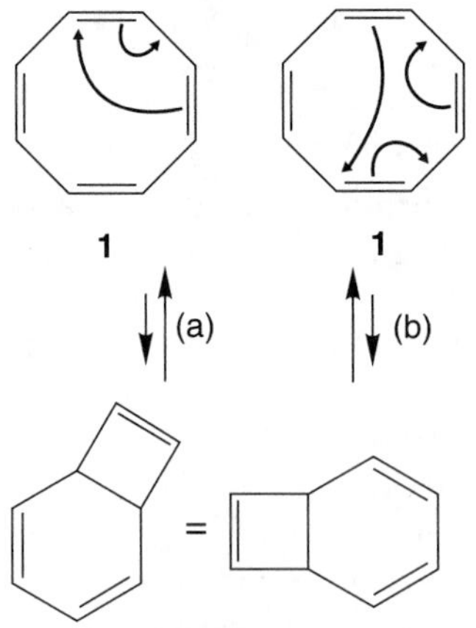

1 **1** (a) (b) **2**

PROBLEM 22.27 Analyze the thermal ring opening to allyl ions of the two cyclopropyl ions shown below. What stereochemistry do you expect in the products? Will the reactions involve conrotation or disrotation? Explain. Do not be put off by the charges! They only serve to allow you to count electrons properly.

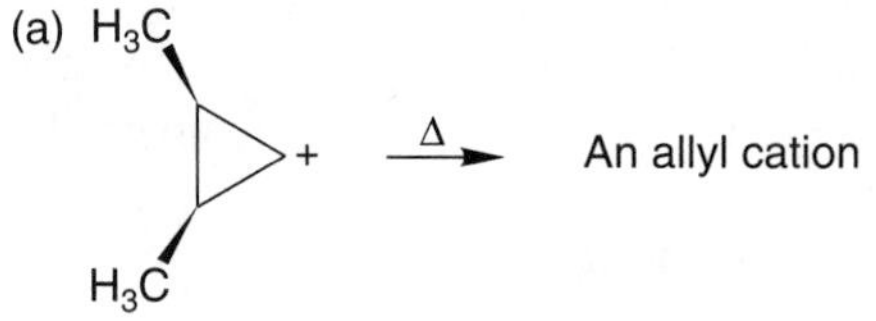

(a) H3C H3C + Δ An allyl cation

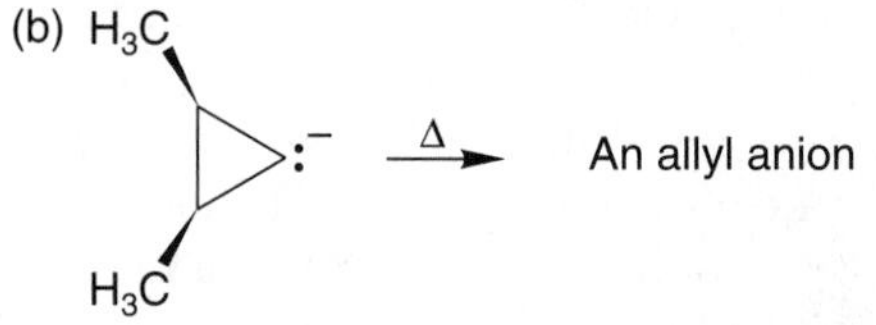

(b) H3C H3C :⁻ Δ An allyl anion

PROBLEM 22.28 Orbital symmetry permits one of the two following photochemical cycloaddition reactions to take place in a concerted fashion. What would be the products, which reaction is the concerted one, and why?

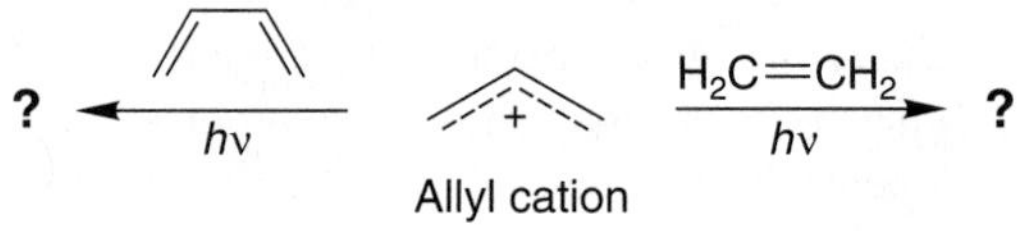

? ← $h\nu$ — Allyl cation — $H_2C{=}CH_2$, $h\nu$ → ?

PROBLEM 22.29 The following two cyclobutenes can be opened to butadienes thermally. However, they react at very different rates. Which is the fast one, which is the slow one, and why?

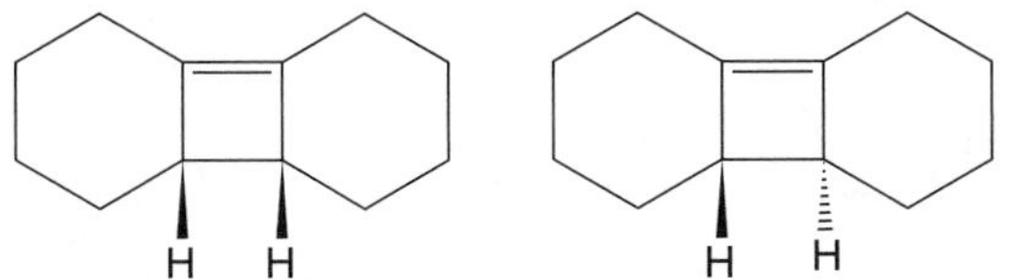

PROBLEM 22.30 These same two cyclobutenes (Problem 22.29) react photochemically to give two monocyclic (one ring) compounds isomeric with starting material. Predict the products, and explain the stereospecificity of the reaction.

PROBLEM 22.31 Benzocyclobutene (**1**) reacts with maleic anhydride when heated at 200 °C to give the single product shown. Write a mechanism for this reaction.

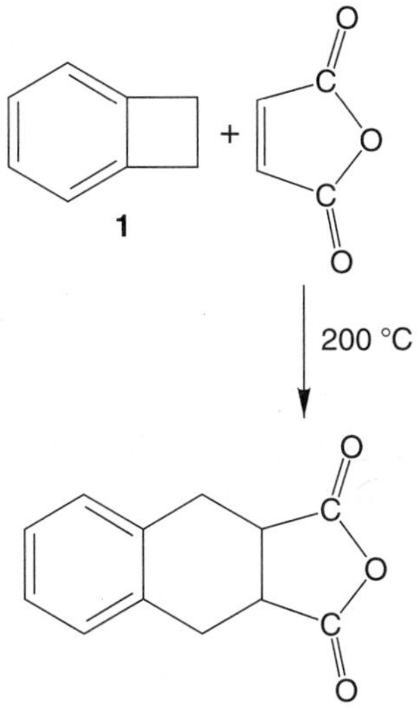

PROBLEM 22.32 Given your answer to Problem 22.31, explain the stereochemistry observed in the following reaction:

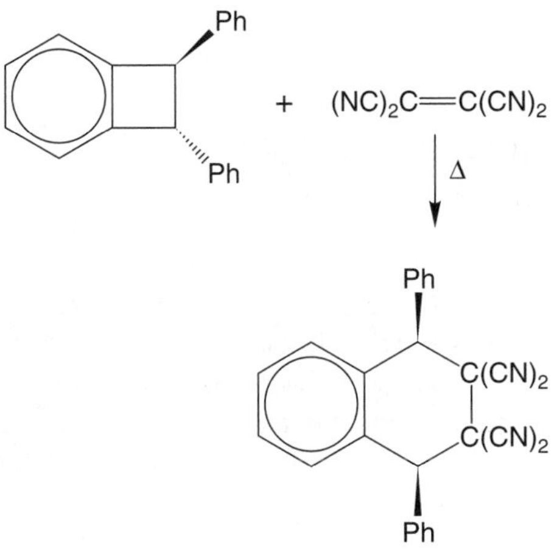

PROBLEM 22.33 Write a mechanism for the following reaction. Note that no similar reaction appears for *cis*-1,3,5-hexatriene. It may be useful to work backward from the product. What molecules must lead to it?

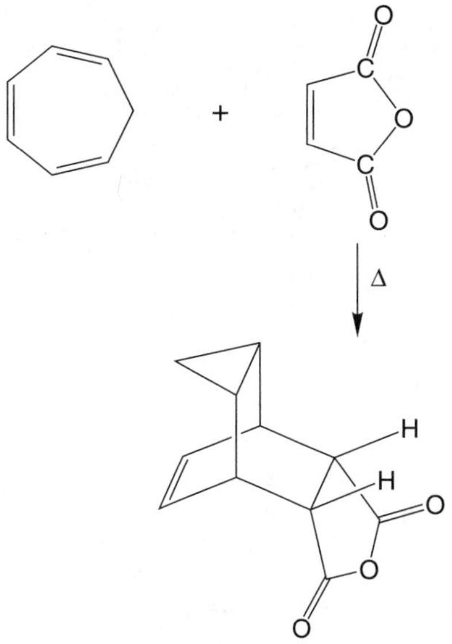

PROBLEM 22.34 Now, here's an interesting variation on the reaction of Problem 22.33. Diels–Alder additions typically take place as shown in Problem 22.33, with the ring-closed partner (**A** in the answer to Problem 22.33) actually doing the addition reaction. However, the molecule shown below is an exception, as Diels–Alder reaction takes place in the normal way to give the product shown. Draw an arrow formalism for the reaction and then explain why **1** reacts differently from other cycloheptatrienes.

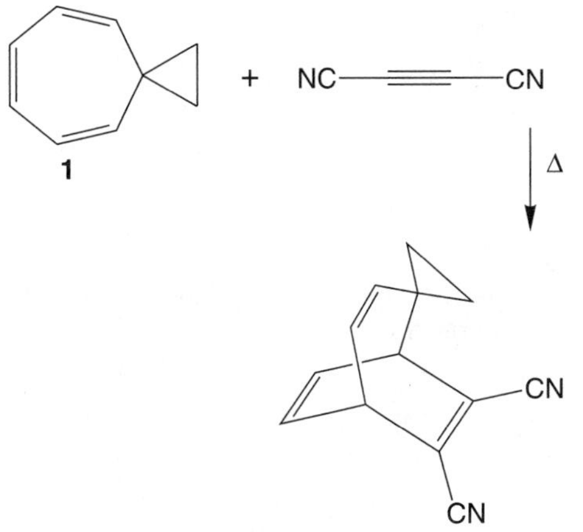

PROBLEM 22.35 Write an arrow formalism mechanism for the following thermal reaction and explain the specific trans stereochemistry. Careful here—note the acid catalyst for this reaction.

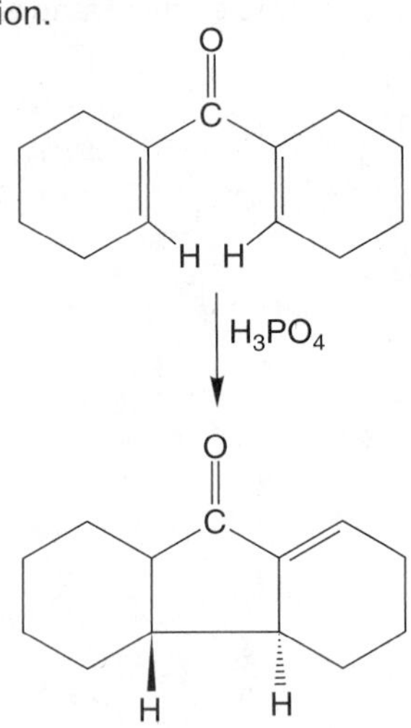

H_3PO_4

PROBLEM 22.36 Write an arrow formalism mechanism for the following reaction and explain the observed trans stereochemistry:

Δ

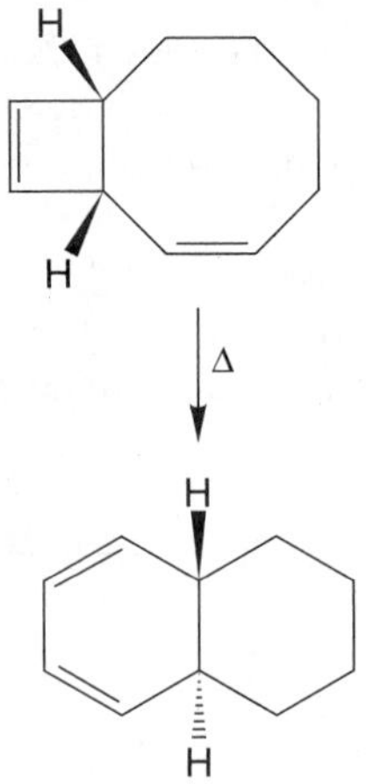

PROBLEM 22.37 Explain the following differences in product stereochemistry:

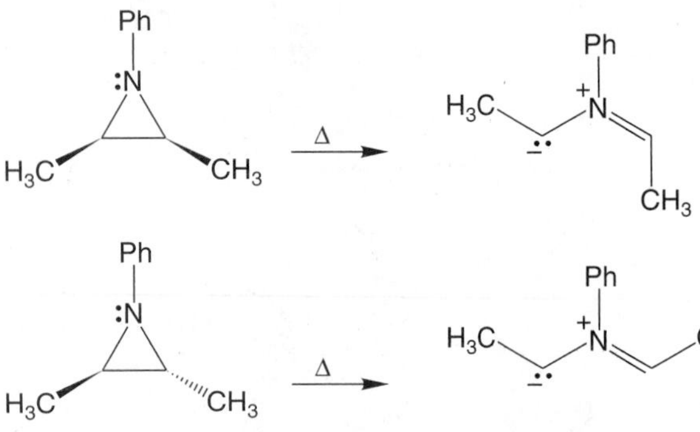

PROBLEM 22.38 Provide mechanisms for the following transformations:

(a) $COOCH_2CH_3$ CH_3CH_2OOC Δ CH_3CH_2OOC CH_3CH_2OOC

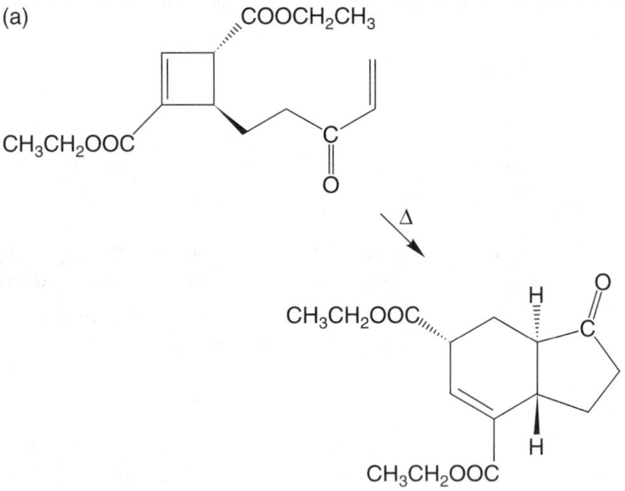

(b) Δ

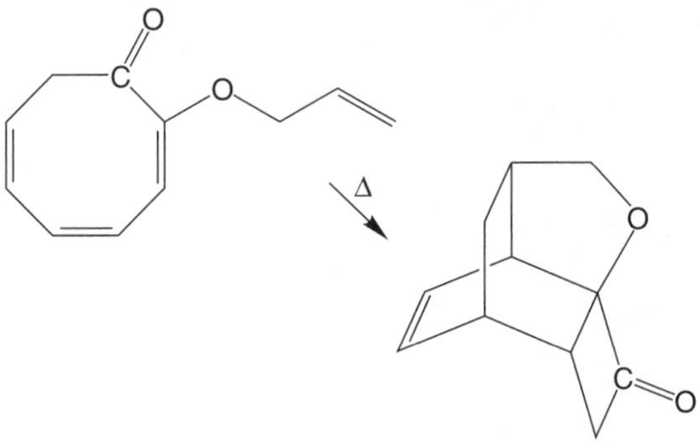

PROBLEM 22.39 Provide mechanisms for the following transformations. Part (a) is harder than it looks (*Hint*: There are two intermediates in which the three-membered ring is gone) and (b) is easier than it looks (*Hint*: What simple photochemical pericyclic reactions do you know?).

(a) Δ

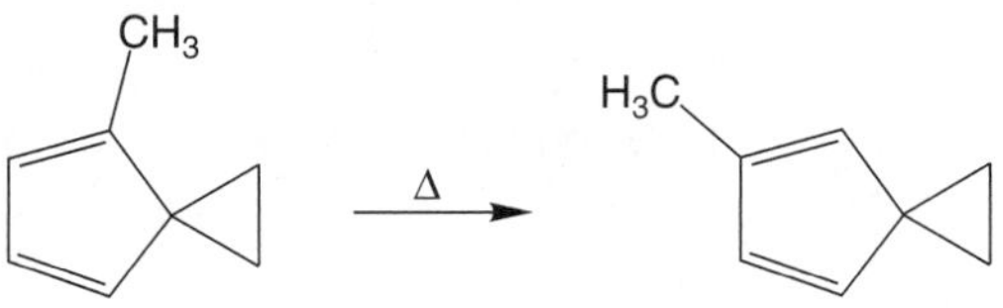

(b) $h\nu$

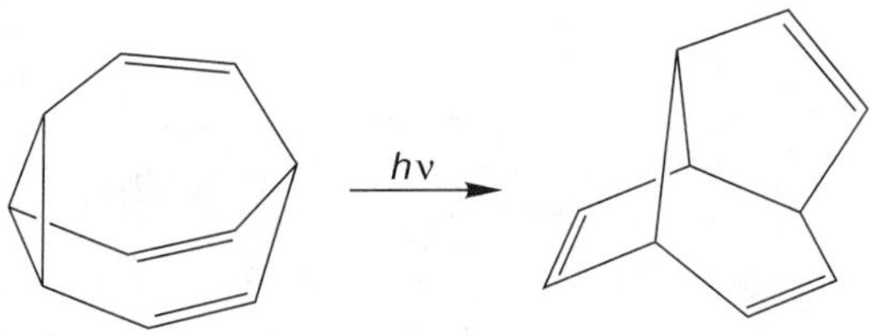

PROBLEM 22.40 In Problem 22.23(b), we saw an example of a [3,3] sigmatropic shift known as the Claisen rearrangement. This reaction involves the thermally induced intramolecular rearrangement of an allyl phenyl ether to an *o*-allylphenol.

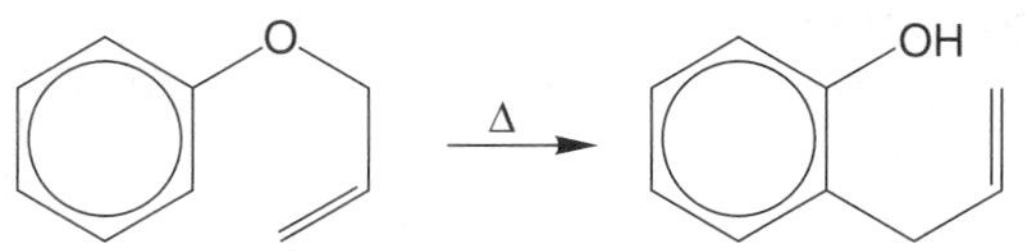

(a) It might be argued that this rearrangement is not a concerted process, but instead involves a pair of radical intermediates. Write such a mechanism and design a labeling experiment that would test this hypothesis.
(b) If both the ortho positions of the allyl phenyl ether are blocked with methyl groups, the product of this rearrangement is the *p*-allylphenol. This version of the Claisen reaction is often called the para-Claisen rearrangement.

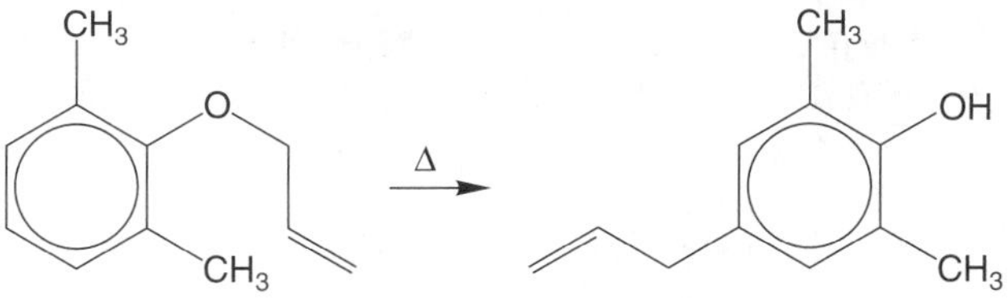

Explain why the rearrangement takes a different course in this case. In addition, would the label you suggested in (a) be useful in this case for distinguishing concerted from nonconcerted (radical) mechanisms?

PROBLEM **22.41** Propose an arrow formalism mechanism for the following reaction. Be sure that your mechanism accommodates the specific labeling results shown by the dot (dot = ^{14}C). Also, be sure that any sigmatropic shifts you propose do not involve impossibly long distances.

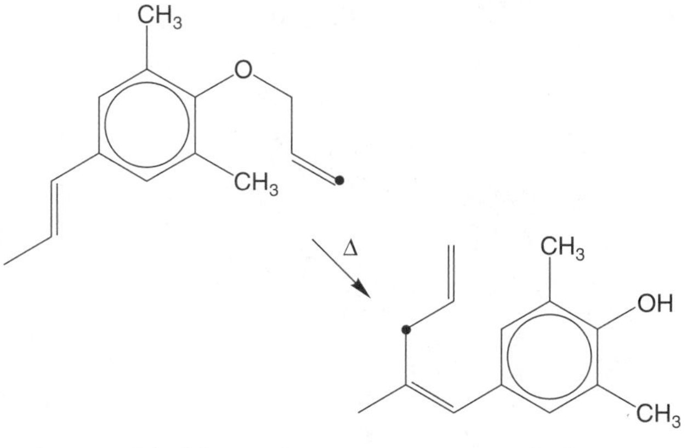

PROBLEM **22.42** We saw in the Box (p. 1163) that chorismate undergoes a [3,3] sigmatropic shift (Claisen rearrangement) to prephenate catalyzed by the enzyme chorismate mutase. This enzyme-catalyzed reaction occurs more than 1 million times faster than the corresponding nonenzymatic reaction. How does the enzyme effect rate enhancement? Both enzymatic and nonenzymatic reactions proceed through a chair-like transition state (see Box). However, the predominant conformer of chorismate in solution, as determined by ^1H NMR spectroscopy, is the pseudodiequatorial conformer **2**, that cannot undergo the reaction.

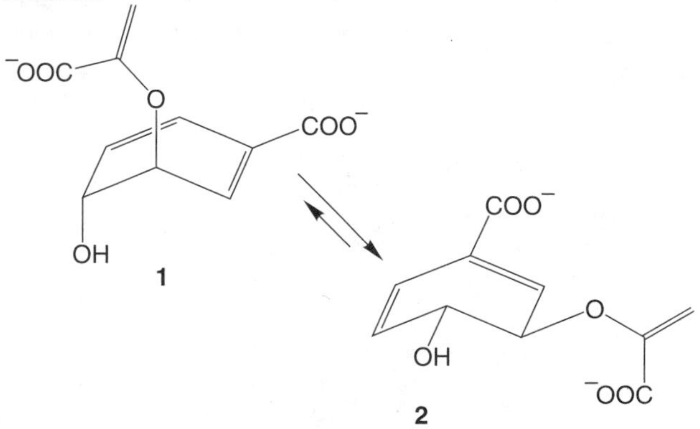

Perhaps the role of the enzyme is to bind the predominant pseudodiequatorial conformer **2**. The enzyme-substrate complex then undergoes a rate-determining conformational change to convert the substrate to the pseudodiaxial form **1**, that rapidly undergoes rearrangement.
(a) Explain why **1** can rearrange to prephenate but **2** cannot.
(b) Suggest a reason why conformer **2** is preferred over conformer **1**. *Hint*: Focus on the OH. What can it do in **2** that it cannot do in **1**?
(c) Draw an appropriate Energy versus Reaction progress diagram that illustrates the situation outlined above. That is, a reaction in which the rate-determining step is the conversion of **2** into **1**. What would the Energy versus Reaction progress diagram look like if the rate-determining step were the conversion of **1** into prephenate?

PROBLEM **22.43** In 1907, Staudinger reported the reaction of diphenylketene (**1**) and benzylideneaniline (**2**) to give β-lactam **3**. Although this reaction is formally a 2 + 2 cycloaddition reaction, there is ample experimental evidence to suggest that the reaction is not concerted, but is instead a two-step process involving a dipolar intermediate.

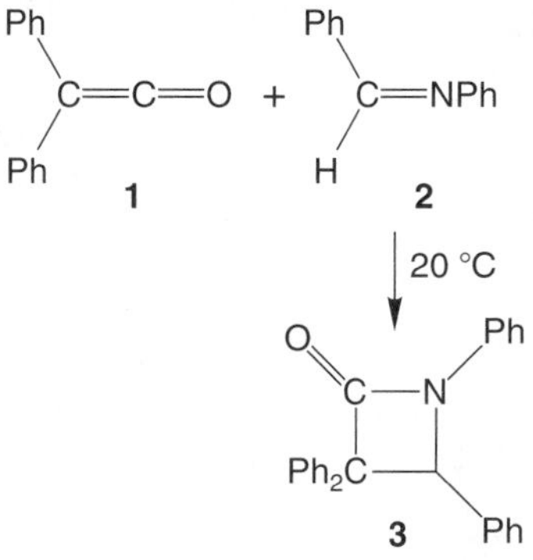

(a) Propose an arrow formalism mechanism for this cycloaddition reaction. Don't forget to include the dipolar intermediate.

PROBLEM **22.43** (CONTINUED)

(b) Although cycloaddition reactions of this type are not concerted, they are often stereospecific, as in the following example:

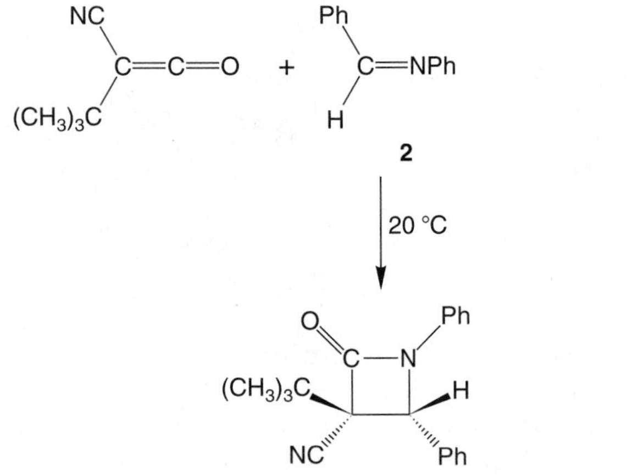

Only isomer formed

Attempt to explain the stereochemistry observed in the reaction above. Start by minimizing the steric interactions in your proposed dipolar intermediate. Then, see if you can find a concerted process that converts your proposed intermediate into the observed β-lactam. *Hint*: A *tert*-butyl group is much bigger than a cyano group.

PROBLEM **22.44** Heating α-phenylallyl acetate (**1**) affords isomeric compound **2**. Spectral data for compound **2** are summarized below. Deduce the structure of **2** and provide an arrow formalism mechanism for this isomerization.

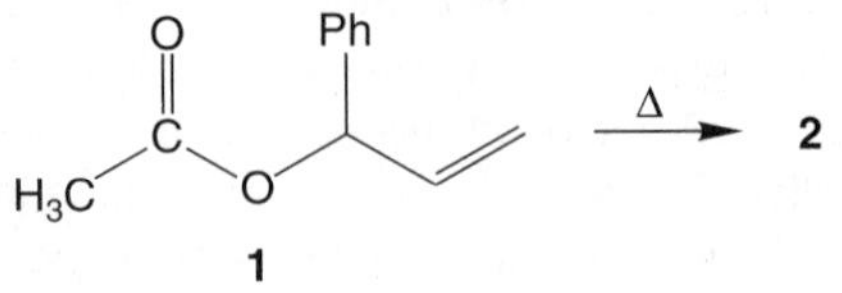

Compound **2**

IR (film): 1740 (s), 1225 (s), 1025 (m), 960 (m), 745 (m), 690 (m) cm^{-1}

^{1}H NMR ($CDCl_3$): δ 2.08 (s, 3H), 4.71 (d, J = 5.5 Hz, 2H), 6.27 (dt, J = 16, 5.5 Hz, 1H), 6.63 (d, J = 16 Hz, 1H), 7.2–7.4 (m, 5H)

^{13}C NMR ($CDCl_3$): δ 20.9 (q), 65.0 (t), 123.1 (d), 126.5 (d), 128.0 (d), 128.5 (d), 134.1 (d), 136.1 (s), 170.6 (s)

PROBLEM **22.45** Reaction of bicyclopropenyl (**1**) and *s*-tetrazine (**2**) gives the semibullvalene (**3**). Provide an arrow formalism mechanism for this transformation. *Hint*: There is at least one nonisolable intermediate involved. Also, explain the relatively simple ^{1}H and ^{13}C NMR spectra of compound **3**, summarized below.

Compound **3**

^{1}H NMR ($CDCl_3$): δ 1.13 (s, 3H), 3.73 (s, 3H), 4.79 (s, 2H)

^{13}C NMR ($CDCl_3$): δ 14.9 (q), 51.4 (q), 60.6 (s), 93.7 (d), 127.2 (s), 164.7 (s)

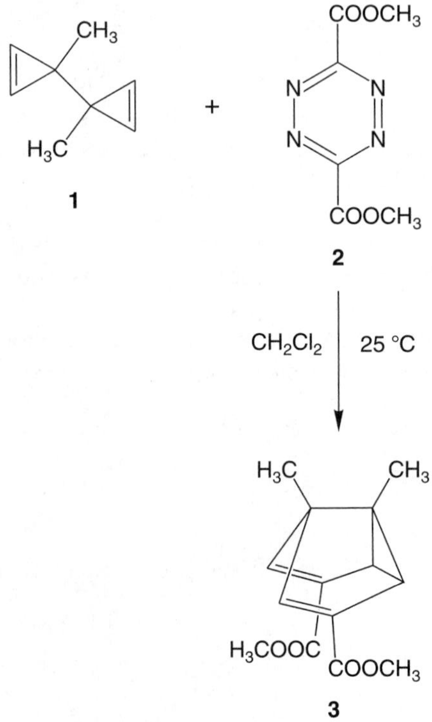

Introduction to Polyfunctional Compounds: Intramolecular Reactions and Neighboring Group Participation

Our race would not have gotten far,
Had we not learned to bluff it out
And look more certain than we are
Of what our motion is about.
—W. H. Auden*

We have now completed our catalog of monofunctional compounds. Here we introduce the beginnings of the chemistry of polyfunctional molecules. This chapter by no means contains an exhaustive treatment. The chemistry of di- and polyfunctional compounds could easily be the subject of an advanced course in organic chemistry, and we will only catch some highlights here. These highlights are important, however, and this chapter is by no means fringe material.

In principle, two functional groups in a single molecule might not interact at all, and the chemistry of the molecule would be the simple combination of the chemistries of the two separate functional groups. This occurrence is rare, as groups in molecules often do interact, even if they are quite remote from each other. Through inductive and resonance effects, one group can affect the chemistry of another group at a distant point in a molecule. Sometimes, groups interact even more strongly, and the chemistry is best described not as a sum of the two individual chemistries, but as something quite new. These are the most interesting examples.

A **neighboring group effect** can change the rates of formation of products as well as the very structures of the molecules produced. Don't be put off by the fancy name. All of intramolecular chemistry involves interactions between neighboring groups. You have already seen lots of intra-

*Wystan Hugh Auden (1907–1975) was one of the most influential British poets of his generation.

molecular chemistry. Figure 23.1 gives a few examples of such reactions, and we will see more as we go along.

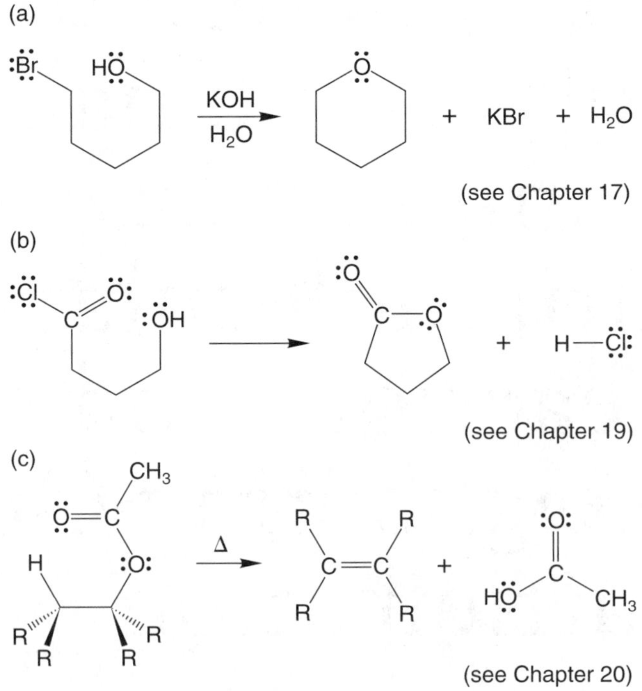

FIGURE 23.1 Some familiar, or at least old, examples of neighboring group interactions.

*PROBLEM 23.1 Write mechanisms for the reactions of Figure 23.1.

ANSWER (a) The first step is certainly loss of the hydroxyl proton to give an alkoxide. Intramolecular S_N2 displacement gives the product.

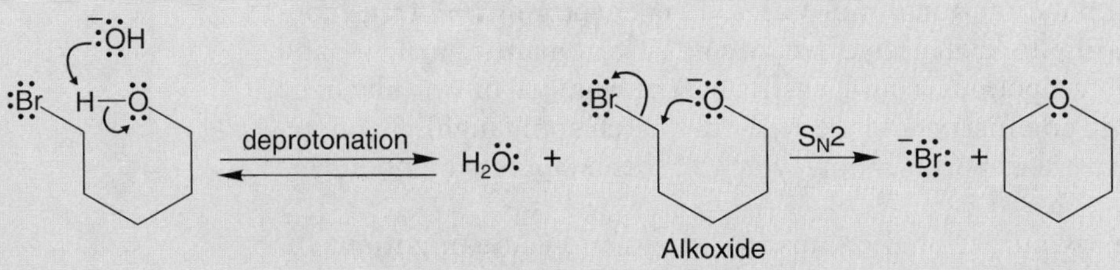

(b) Addition–elimination gives the lactone (a cyclic ester).

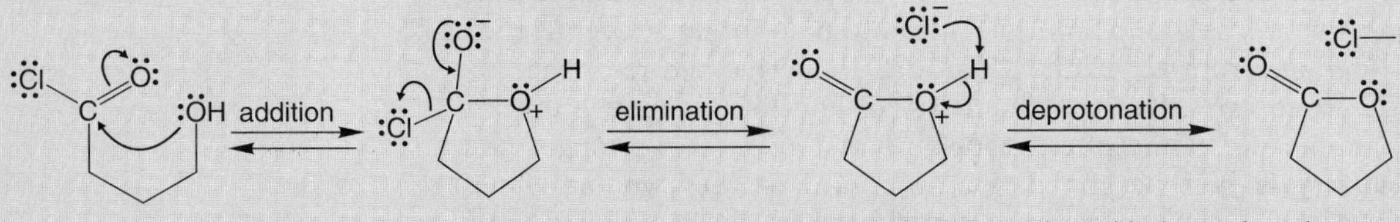

(c) This reaction is an intramolecular elimination with the carbonyl oxygen playing the part of the nucleophile (neighboring group).

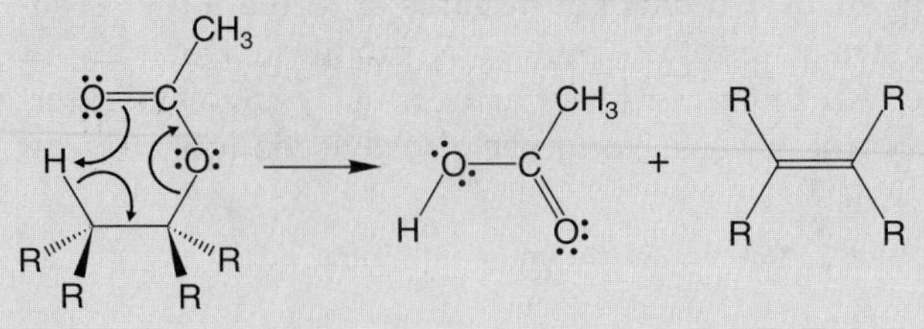

The exploitation of the intuitively obvious notion that groups might interact in a synergistic fashion constituted one of the most fascinating areas of chemical exploration during the middle of the twentieth century, and it continues even today. As we examine this area, we will encounter some sociology as well as chemistry. One of the most heated arguments in the history of organic chemistry centered upon this subject, and it is worth our time to see what we can learn about science from a look at this intensely emotional issue. There will be some opinion here,* but I'll try to keep it labeled as such.

23.1 HETEROATOMS AS NEIGHBORING GROUPS

In its simplest form, the study of neighboring group effects notes that a heteroatom with a pair of nonbonding electrons can act as a nucleophile in an intramolecular displacement reaction, just as external nucleophiles can participate in intermolecular S_N2 reactions. In an example we have seen, oxiranes can be formed from halohydrins in base (Chapter 17, p. 847; Fig. 23.2).

First, base removes the hydroxyl hydrogen to make the alkoxide. Intramolecular S_N2 displacement of the bromide generates the oxirane. Here the alkoxide is the neighboring group, and the whole process couldn't be simpler.

Another example is intramolecular hemiacetal formation. If a relatively unstrained ring can be formed, compounds containing both a carbonyl and a hydroxyl group exist in equilibrium with the cyclic hemiacetal (Chapter 16, p. 775; Fig. 23.3).

Intermolecular S_N2

Intramolecular S_N2 by "neighboring" alkoxide

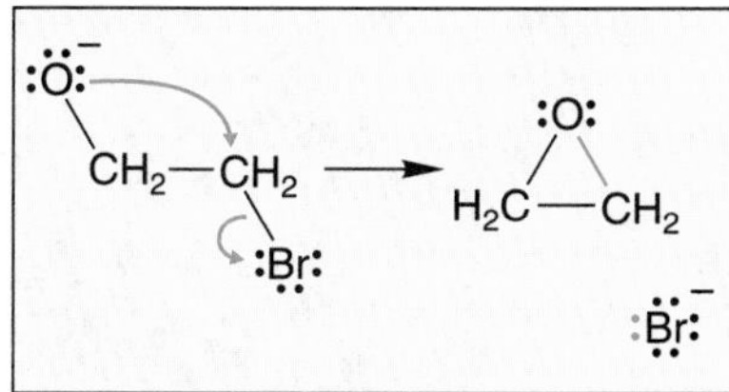

FIGURE 23.2 Formation of an oxirane by treatment of a bromohydrin with hydroxide involves an alkoxide acting as neighboring group.

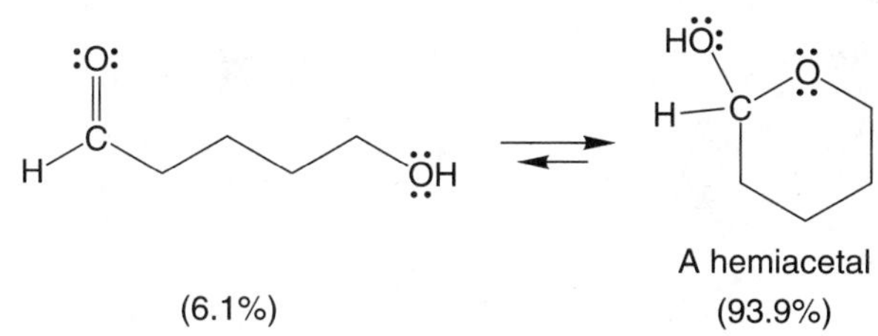

A hemiacetal

(6.1%) (93.9%)

FIGURE 23.3 Cyclic hemiacetal formation involves a hydroxyl group acting as a neighboring group.

PROBLEM 23.2

Write a mechanism for the reaction of Figure 23.3.

Other reactions are more subtle than these straightforward intramolecular processes, but there are always clues to the operation of a neighboring group effect that you can find. For example, consider the reaction of the optically active α-bromocarboxylate in methyl alcohol (Fig. 23.4).

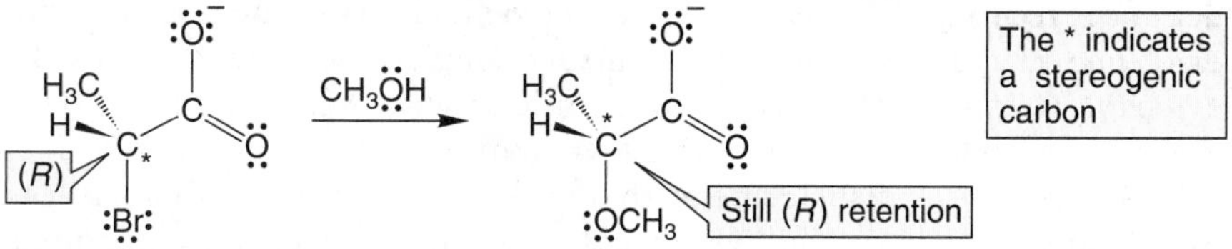

The * indicates a stereogenic carbon

Still (*R*) retention

FIGURE 23.4 Treatment of α-bromopropionate in methyl alcohol. Notice that the reaction involves retention of configuration.

*Actually, everything in this book is, in the limit, nothing but opinion. Pushed far enough, every fact must yield to the uncertainty principle. Your study of science will not lead you to Ultimate Truth.

The product is formed with the stereochemistry opposite to what we might naively expect. An S_N2 displacement of bromide by methyl alcohol or methoxide would give an optically active product, but the stereochemistry would be exactly the reverse of what is observed (Fig. 23.5)!

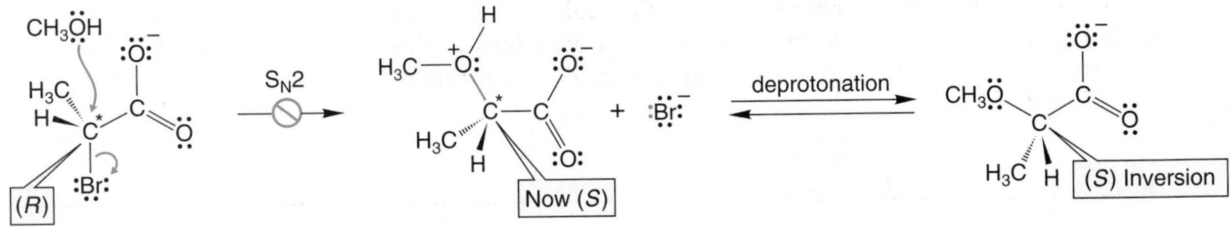

FIGURE 23.5 A straightforward mechanism in which the bromide is displaced by methyl alcohol gives the wrong stereochemical result—inversion.

The observation of a strange stereochemical result is the first kind of clue to the operation of a neighboring group effect. The simple mechanism of Figure 23.5 must be wrong, as it cannot accommodate the observed retention of configuration in the product. In this reaction, the carboxylate salt acts as an internal nucleophile, displacing bromide to give an intermediate called an α-**lactone** (Chapter 20, p. 1005; Fig. 23.6).

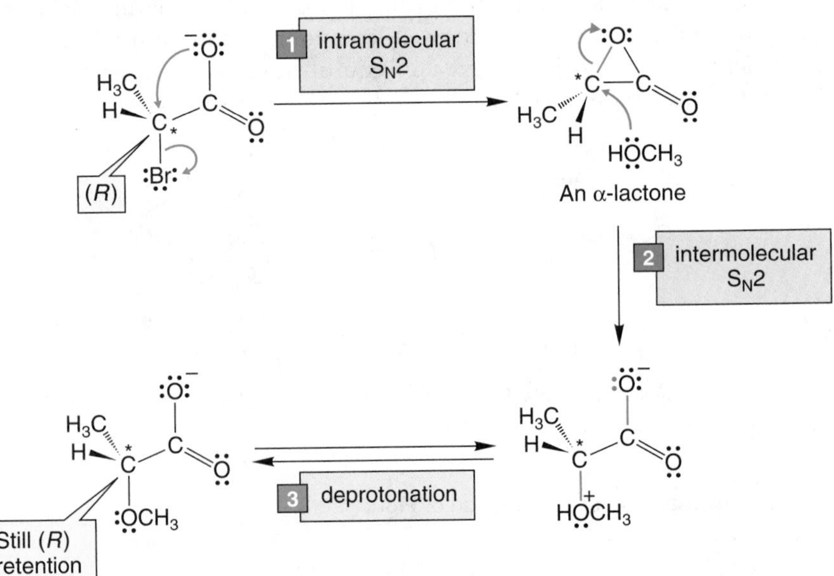

FIGURE 23.6 If the carboxylate anion acts as a neighboring group and displaces the bromide in an intramolecular fashion, methyl alcohol can open the resulting α-lactone. The overall result of this double inversion mechanism is retention.

These smallest cyclic esters cannot be isolated, but they are known as reactive intermediates. In this case, methoxide or methyl alcohol adds to the α-lactone in an intermolecular S_N2 displacement to give the real product of the reaction. Now examine the stereochemistry of the reaction with care. The first displacement is an *intra*molecular S_N2 reaction with the carboxylate anion acting as the nucleophile and the bromide as the leaving group. As with all S_N2 reactions, inversion must be the stereochemical result. A second S_N2 displacement, this time *inter*molecular, occurs as the α-lactone is opened by methyl alcohol. Once again inversion is required. The end result of two inversions (or any even number of inversions) is retention of configuration (Fig. 23.7).

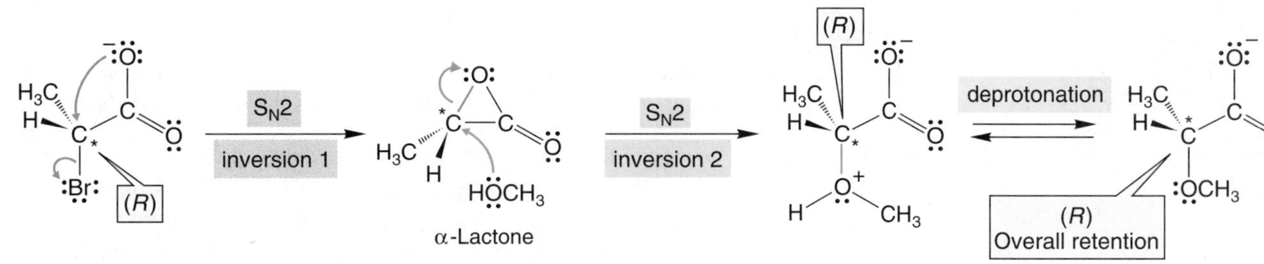

FIGURE 23.7 Net retention is the result of two (or any even number of) inversions.

Remember: Whenever you see a backward stereochemical result (usually retention where inversion would be expected), look for an intramolecular displacement reaction—a neighboring group effect. You will almost always find it.

Oxygen need not be negatively charged to participate in an intramolecular displacement reaction. The reaction of the "brosylate" (*p*-bromobenzenesulfonate) of 4-methoxy-1-butanol in acetic acid (HOAc = CH_3COOH) gives the corresponding acetate (Fig. 23.8).

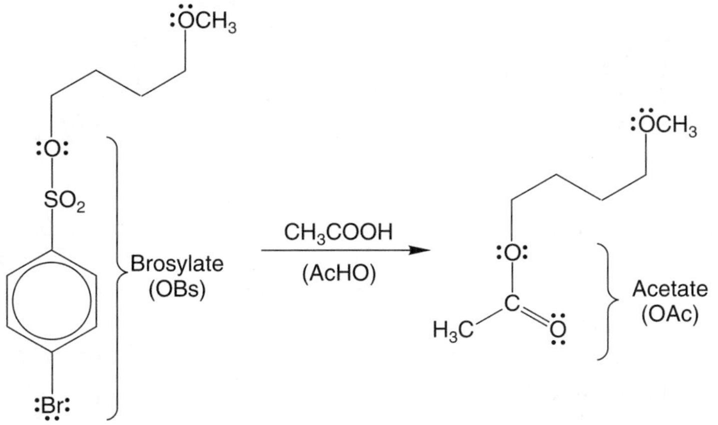

FIGURE 23.8 Acetolysis of the brosylate of 4-methoxy-1-butanol gives the corresponding acetate.

Remember that the acetate group (CH_3COO) is often abbreviated OAc. The acetyl group (Ac) is CH_3CO.

Convention Alert!

Sulfonates are excellent leaving groups and a reasonable first-guess mechanism for the reaction involves intermolecular displacement by acetic acid or acetate (Fig. 23.9).

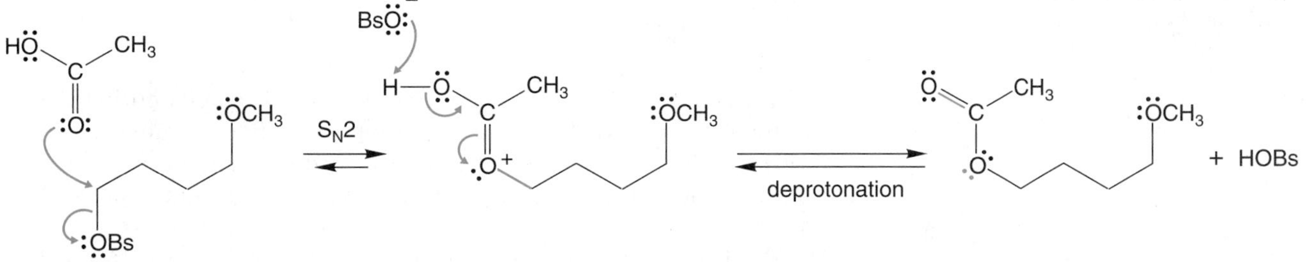

FIGURE 23.9 A simple S_N2 displacement of the brosylate by acetic acid might explain the reaction.

*PROBLEM 23.3

There are two possible nucleophiles for the S_N2 displacement in Figure 23.9, the acetate ion or acetic acid itself. In the displacement using acetic acid, why was the carbonyl oxygen, not the hydroxyl oxygen, used as the displacing agent?

ANSWER

The carbonyl oxygen is the better nucleophile. Recall that protonation of acetic acid goes faster on the carbonyl oxygen (Chapter 19, p. 961). A resonance-stabilized intermediate is formed in this case, but not from protonation of the hydroxyl oxygen. Reaction of a Lewis acid with acetic acid will be similar. In the figure, only the "new" resonance form is shown; in each case there is another form in which the carbon atom bears the positive charge—ordinary carbonyl resonance.

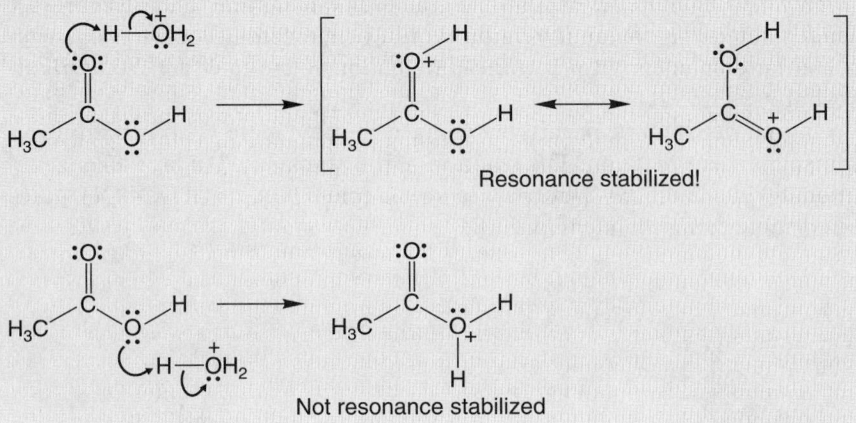

Resonance stabilized!

Not resonance stabilized

When this reaction is run with a similar, but methyl-substituted brosylate, we can see that something more complicated than the simple S_N2 reaction of Figure 23.9 is going on. Our first mechanism can accommodate one of the products but not the other (Fig. 23.10), which is the second clue to neighboring group participation—the observation of an unusual rearrangement reaction.

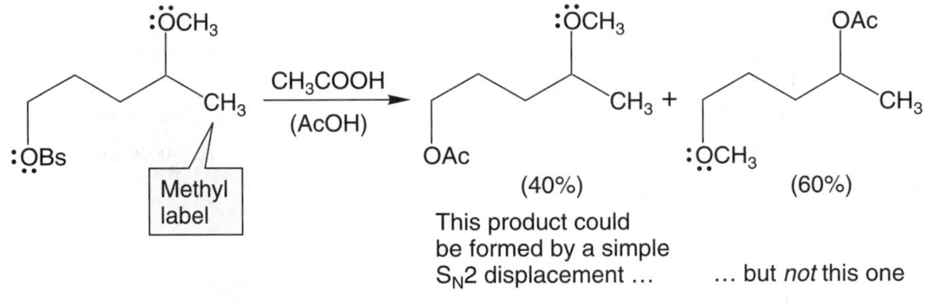

CH_3COOH (AcOH)

Methyl label

(40%) This product could be formed by a simple S_N2 displacement ...

(60%) ... but *not* this one

FIGURE 23.10 Reaction of the methyl-labeled compound shows that the simple mechanism of Figure 23.9 is inadequate.

In this case, it is the ether oxygen that is acting as a neighboring group. The sulfonate is an excellent leaving group and a five-membered ring is relatively unstrained, so the intramolecular displacement of *p*-bromobenzenesulfonate ($^{-}$OBs) to give a cyclic oxonium ion takes place (Fig. 23.11).

The cyclic oxonium ion can be opened by acetic acid in two ways (Fig. 23.11) to give the two observed products. The neutral ether oxygen has acted as a neighboring group to provide an intermediate that can react further in more than one way. The result is an apparent rearrangement.

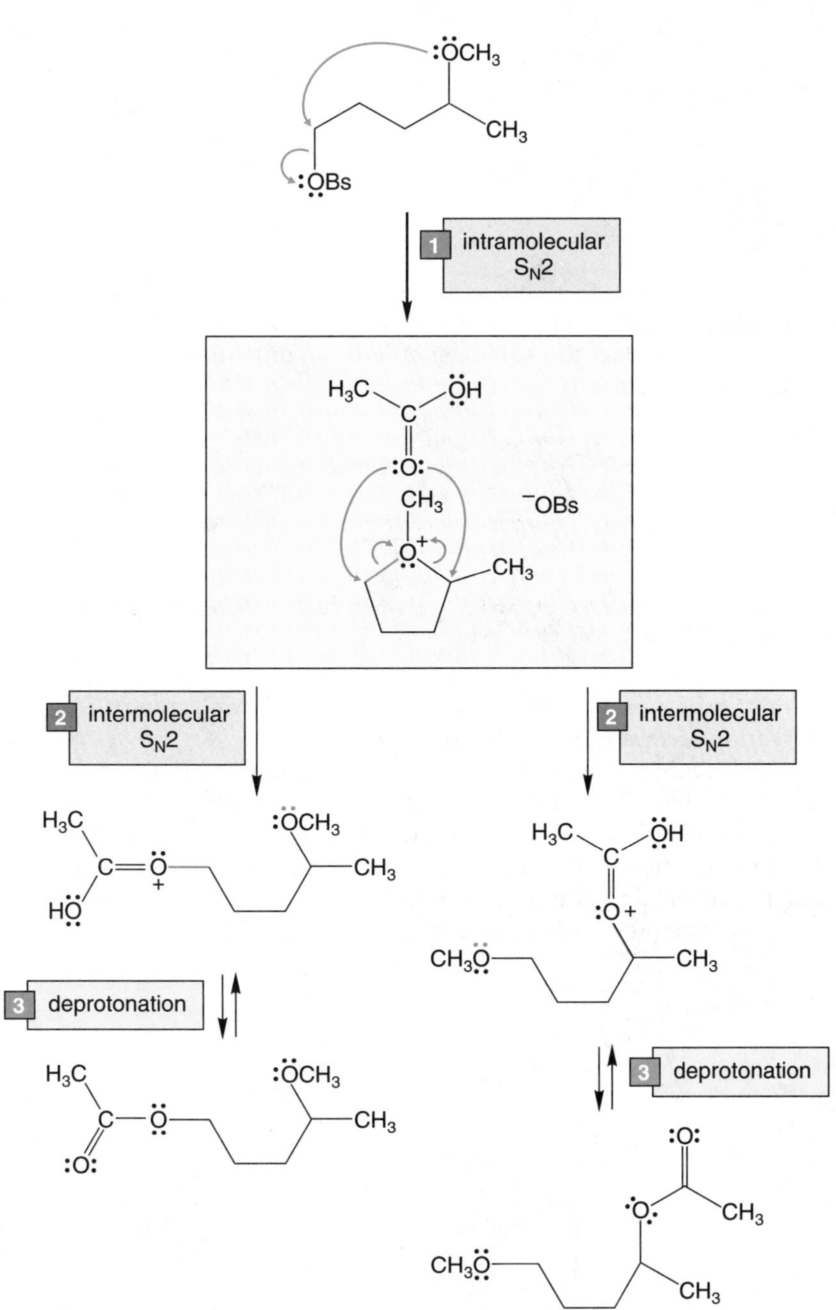

FIGURE 23.11 Here it is the methoxyl oxygen that acts as neighboring group. A cyclic ion is formed that can open in two ways. "Rearrangement" is the result of one mode of opening of the intermediate oxonium ion.

PROBLEM 23.4

There is actually a third product in this reaction, 2-methyltetrahydrofuran (Fig. 23.12). Can you explain the formation of a small amount of this molecule? *Hint*: The intermediate ring is obviously retained in the formation of this product. Think about how this can happen.

Sulfur atoms are often powerful nucleophiles, and the participation of sulfur as a neighboring group is common. Here is just one example. Both hexyl chloride and 2-chloroethyl ethyl sulfide react in water to give the corresponding alcohols (Fig. 23.13). The rate of reaction of the sulfur-containing compound is about 10^3 greater than that of the alkyl chloride, however.

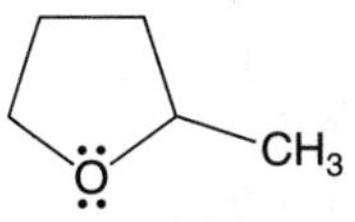

CH_3

2-Methyltetrahydrofuran

FIGURE 23.12

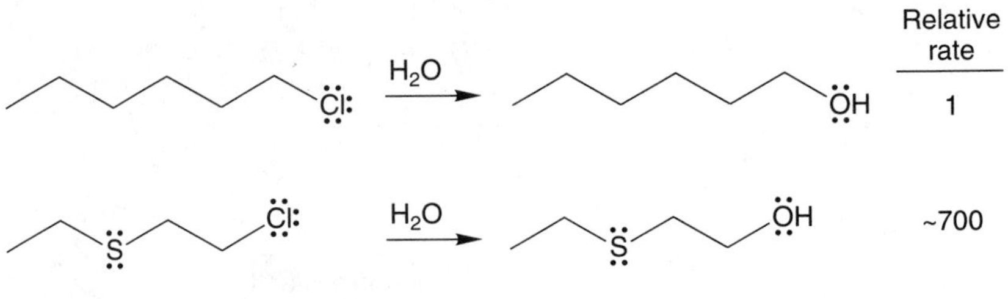

	Relative rate
hexyl chloride →(H_2O) hexanol	1
chlorosulfide →(H_2O) hydroxysulfide	~700

FIGURE 23.13 If a sulfur atom is substituted for the C(3) of hexyl chloride, the formation of the related alcohol is greatly accelerated.

The mechanism of alcohol formation from the primary halide must be a simple S_N2 displacement with water acting as nucleophile (Fig. 23.14). It is difficult to see why the sulfide should be so much faster if the reaction mechanism is analogous.

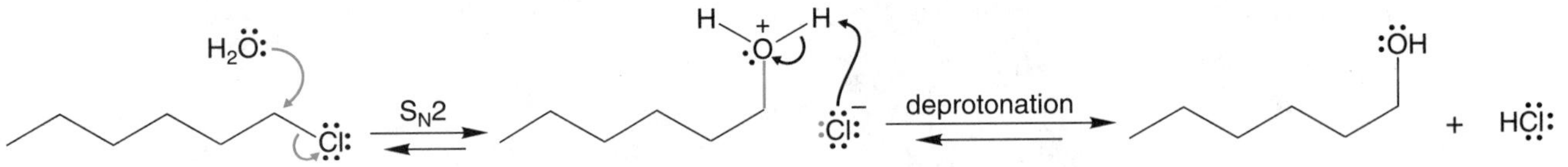

FIGURE 23.14 The reaction of hexyl chloride must be a simple S_N2 displacement of chloride by water, followed by deprotonation.

The third kind of clue to the operation of a neighboring group effect is the observation of an unexpectedly rapid rate.

In the reaction of the sulfide, it is the sulfur atom that displaces the leaving chloride. This intramolecular reaction is much faster than the intermolecular displacement of chloride by water either in the alkyl chloride or the chlorosulfide. The product is an **episulfonium ion**. This intermediate is opened by water in a second, intermolecular S_N2 displacement step to give the product (Fig. 23.15).

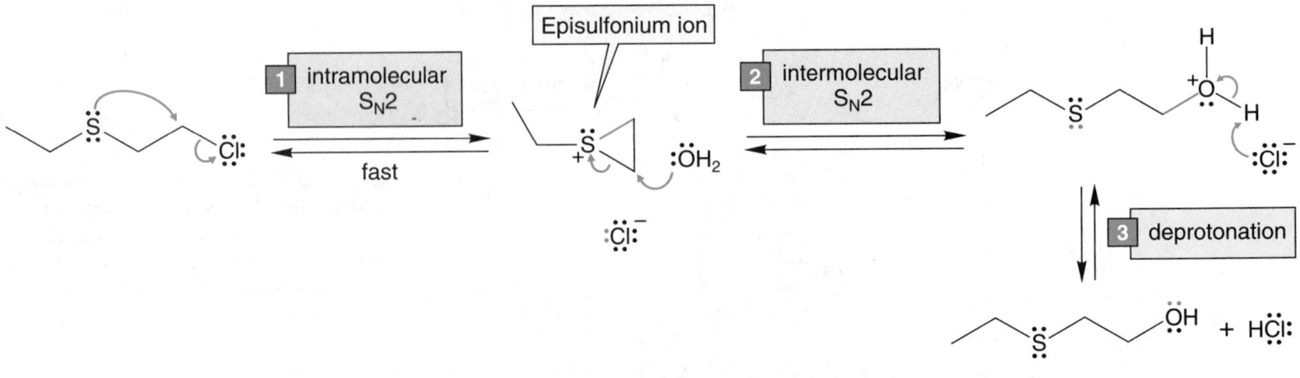

FIGURE 23.15 The mechanism of the sulfur reaction must be different. Here, sulfur can act as a neighboring group, displacing chloride in an intramolecular S_N2 reaction. An episulfonium ion is formed and can be opened in a second, intermolecular S_N2 reaction to give the product.

It is important to be certain you understand the source of the rate increase in the sulfide. Intramolecular reactions often have a large rate advantage over intermolecular reactions as there is no need for the nucleophile and substrate to find each other in solution. Intramolecular assistance in the rate-determining step of a reaction is called **anchimeric assistance**. Once again, it is important not to be put off by the elaborate name. The concept is quite simple: Intramolecular nucleophiles are often more effective displacing agents than intermolecular nucleophiles. What is somewhat more difficult to anticipate is that an intramolecular displacement by a very weak, but perfectly situated nucleophile can often still compete effectively with intermolecular reactions.

PROBLEM 23.5

There are many possible labeling experiments that would support (not prove) the mechanism suggested for the reaction of the chlorosulfide. Design two, one using a methyl group as a marker, and the other using an isotopic label.

Bis(2-chloroethyl)thioether

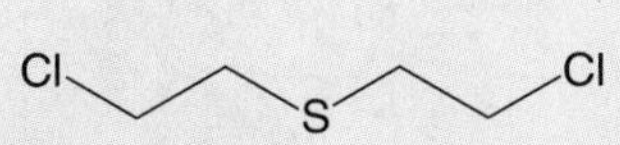

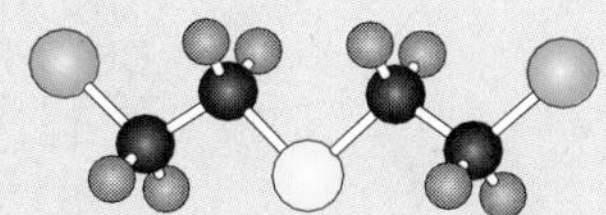

This seemingly routine molecule exposes the dark side of our science; indeed, a review of the history of this compound causes one to reflect on the depths to which human behavior can sink. For this is mustard gas, developed in the early part of this century as a replacement for other less efficient toxic agents for which defenses had been worked out. Mustard gas penetrated the rubber in use in gas masks at the time and thus was more effective. This mustard is lethal in a few minutes at 0.02–0.05% concentration. Less severe exposure nonetheless leads to irreversable cell damage and only delays the inevitable. Introduced by the Germans at Ypres in the first world war in 1917, it was also employed in the China campaign by the Japanese in the second world war. Stories of experiments on prisoners of war are easy to find, if not easy to read about. Nor is there any reason for national pride in not being the first to use this agent, for we in America were the developers of napalm, essentially jellied hydrocarbons, and surely easily defined as a chemical agent.

I leave it to you to consider whether it is morally superior to shoot people, or drop an atomic bomb on them, rather than to dose them with mustard gas or burn them with napalm. Even more complicated for us is the question of the participation by chemists in the development of such agents. Of course, often it is not easy to see the consequences of one's research. But what if one can? Where are the limits?

To summarize: There are three clues that point to the operation of a neighboring group effect. The first is an unusual, often "backward" stereochemical result (clue 1). Generally, this involves retention of configuration when inversion might have been anticipated. Second, the operation of a neighboring group effect can often present the opportunity for a rearrangement, as an intermediate is formed that can react further in more than one way (clue 2). Finally, the anchimeric assistance provided by an intramolecular nucleophile often reveals itself in the form of a rate increase (clue 3).

PROBLEM 23.6 Go back and find examples of these three clues in the preceding reactions.

Now let's look at more heteroatoms with nonbonding electrons that might participate in neighboring group reactions. For example, nitrogen bears a pair of electrons, and so can act as an internal nucleophile. Let's look at the reactions of the two bicyclic amines in Figure 23.16. Treatment of the endo chloride with ethyl alcohol leads to the endo ether, but the exo chloride gives neither the exo nor endo ether under the reaction conditions. Now we'll see why.

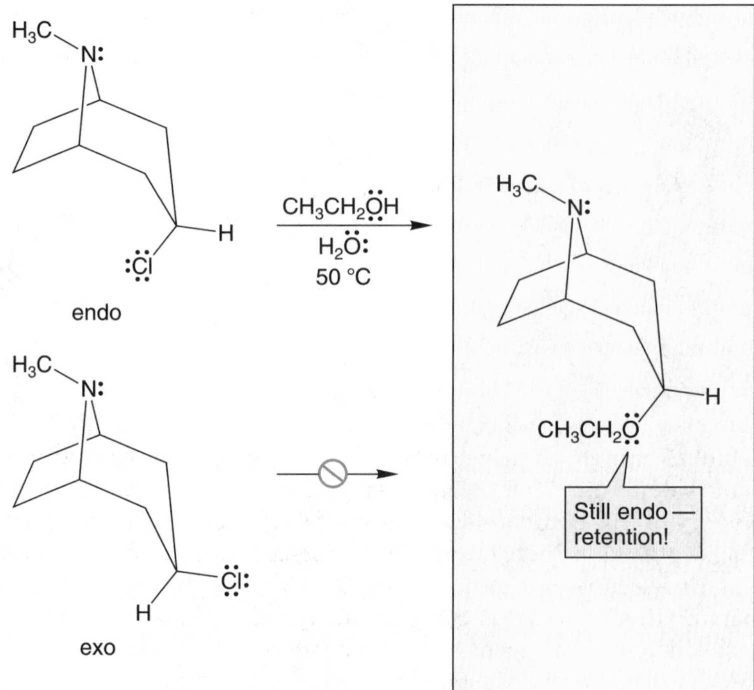

FIGURE 23.16 Treatment of only one isomer of this bicyclic amine gives the related ether. Note the retention of configuration.

Direct displacement of the chloride by ethyl alcohol cannot be the answer, as this potential reaction leads to a product that is not observed, the exo ether. The ether *is* formed, but it is formed with *retention* of stereochemistry, not the *inversion* demanded by the simple S_N2 reaction (Fig. 23.17).

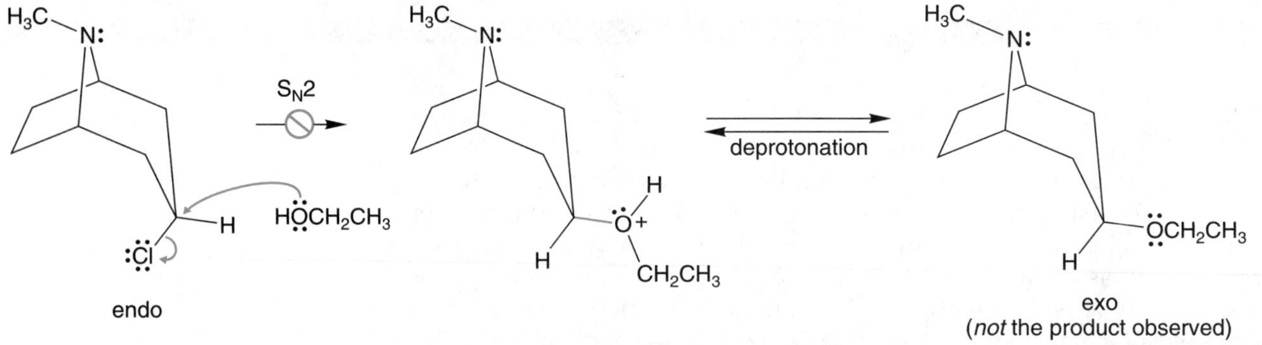

H3C N: H :Cl: endo S_N2 HÖCH2CH3 H3C N: H Ö+ H CH2CH3 deprotonation H3C N: H ÖCH2CH3 exo (*not* the product observed)

FIGURE 23.17 A simple intermolecular S_N2 displacement predicts inversion of configuration.

At least one of the clues to the operation of a neighboring group effect is present in this reaction. The stereochemistry of the product is the opposite to what one would expect from a simple displacement (clue 1).

What's happening? By now you should be alert to the possibility of intramolecular displacement. The nitrogen with its pair of nonbonding electrons is an obvious possible neighboring group, and it is located in the proper position for a rear side displacement of the chloride (Fig. 23.18). The result is a cyclic ammonium ion, a ring containing a positively charged nitrogen. If this ring is opened by ethyl alcohol, the product ether must be formed with retention of stereochemistry. The intramolecular S_N2 displacement by nitrogen is the first inversion reaction, and the second intermolecular displacement, the opening of the ammonium ion by ethyl alcohol, provides the second. Two inversions produce net retention (Fig. 23.18).

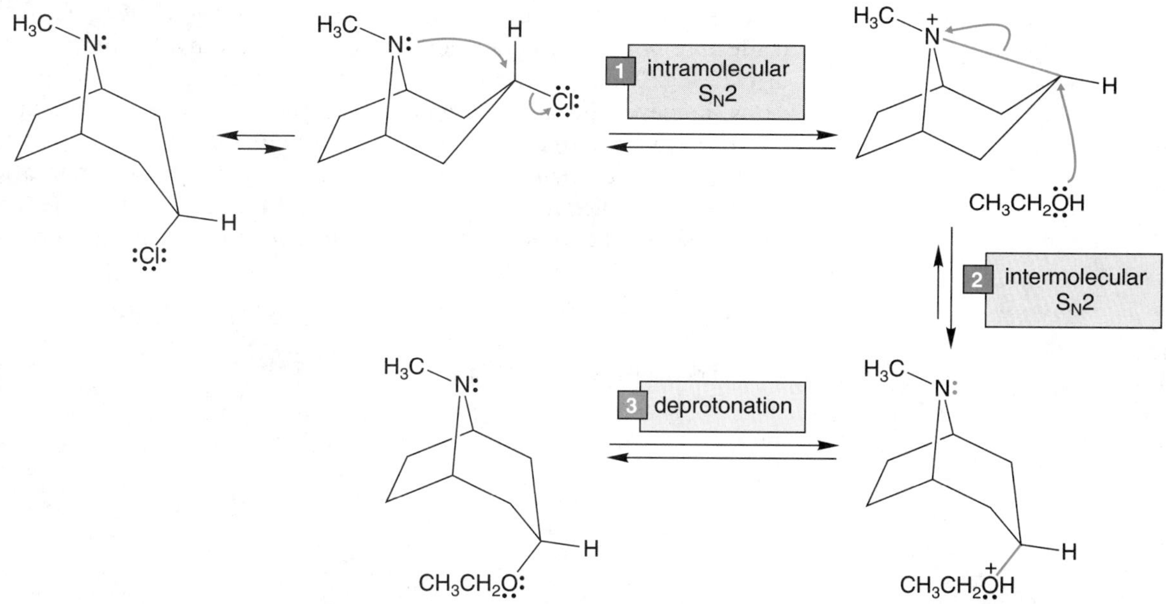

FIGURE 23.18 If nitrogen acts as a neighboring group and displaces the leaving group, a cyclic ammonium ion is formed. The cyclic ion can be opened to give the observed product of retention of configuration.

Why does the exo compound not react in the same way? In this molecule, displacement of the leaving chloride would require a frontside S_N2 reaction, and this is a strongly disfavored process (Chapter 7, p. 244; Fig. 23.19). So, the exo chloride reacts more slowly (and in a different way) from the endo isomer.

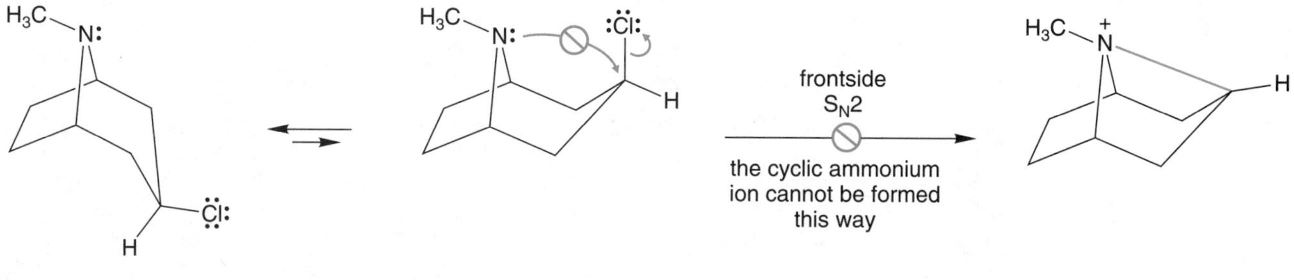

FIGURE 23.19 The exo chloride cannot be displaced by nitrogen, as this reaction would require a frontside S_N2 reaction.

PROBLEM 23.7 The exo chloride gives the product shown in Figure 23.20. Can you suggest a mechanism for this reaction?

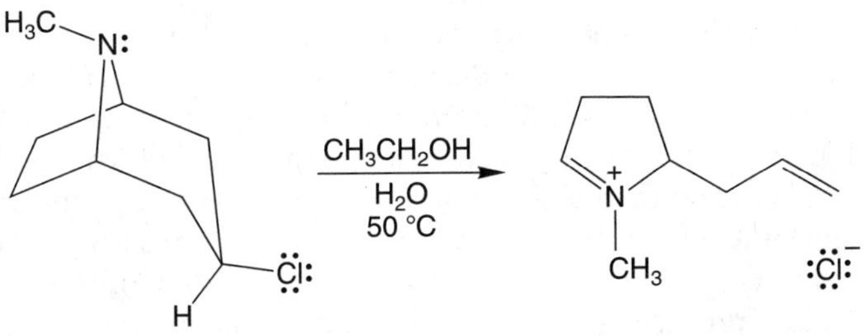
H3C
N:
H
Cl:
CH3CH2OH
H2O
50 °C
N
CH3
:Cl:

FIGURE 23.20

It is often difficult to see these reactions in bicyclic systems. The molecules are not easy to draw, and the extra atoms involved in the cage are distracting. There is nothing *fundamentally* difficult about these reactions, however. In fact, one could argue that once the drawing problems are mastered, the rigidity of the cage is a help. One need not worry about rotations about single bonds, or about finding the proper orientation for the reacting molecules. The stereochemical relationships are defined and preserved by the rigid cages. Here are some problems for practice. It should help to use models.

*PROBLEM 23.8 Write mechanisms for the following reactions (Fig. 23.21):

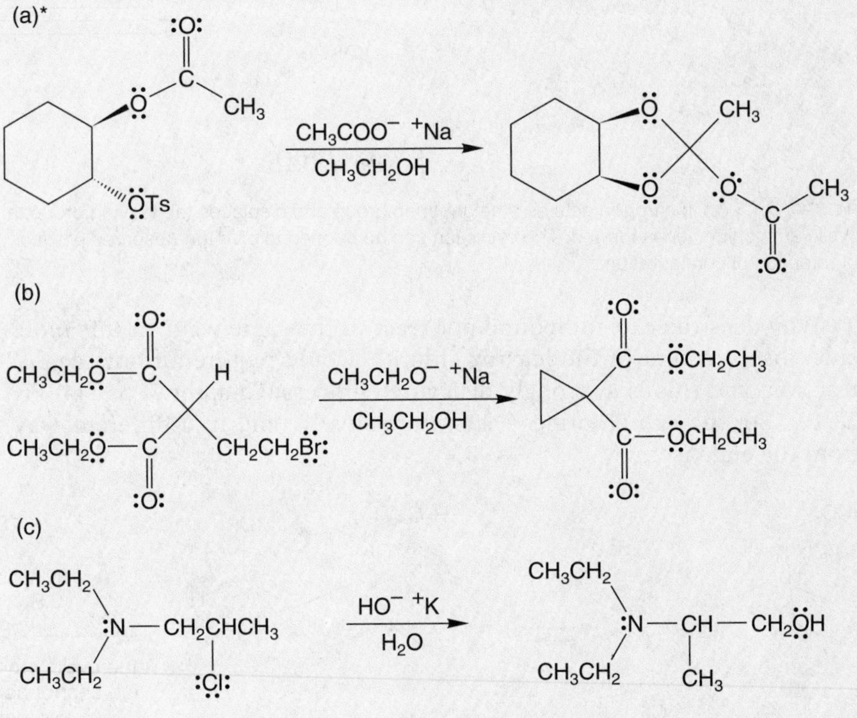
(a)*
:O:
C
O
CH3
OTs
CH3COO⁻ +Na
CH3CH2OH
O
CH3
O
O
CH3
C
:O:
(b)
:O:
CH3CH2O—C
H
CH3CH2O—C
CH2CH2Br:
:O:
CH3CH2O⁻ +Na
CH3CH2OH
:O:
C—OCH2CH3
C—OCH2CH3
:O:
(c)
CH3CH2
:N—CH2CHCH3
CH3CH2
:Cl:
HO⁻ +K
H2O
CH3CH2
:N—CH—CH2OH
CH3CH2
CH3

FIGURE 23.21

ANSWER

Part (a) is the hardest part of this problem. The internal nucleophile is the carbonyl oxygen. Notice that the trans stereochemistry of the starting material allows for a rearside S_N2 displacement of the good leaving group, tosylate. Displacement yields a cyclic cation that is nicely stabilized by resonance. Acetate ion adds to the carbon sharing the positive charge, and gives the product.

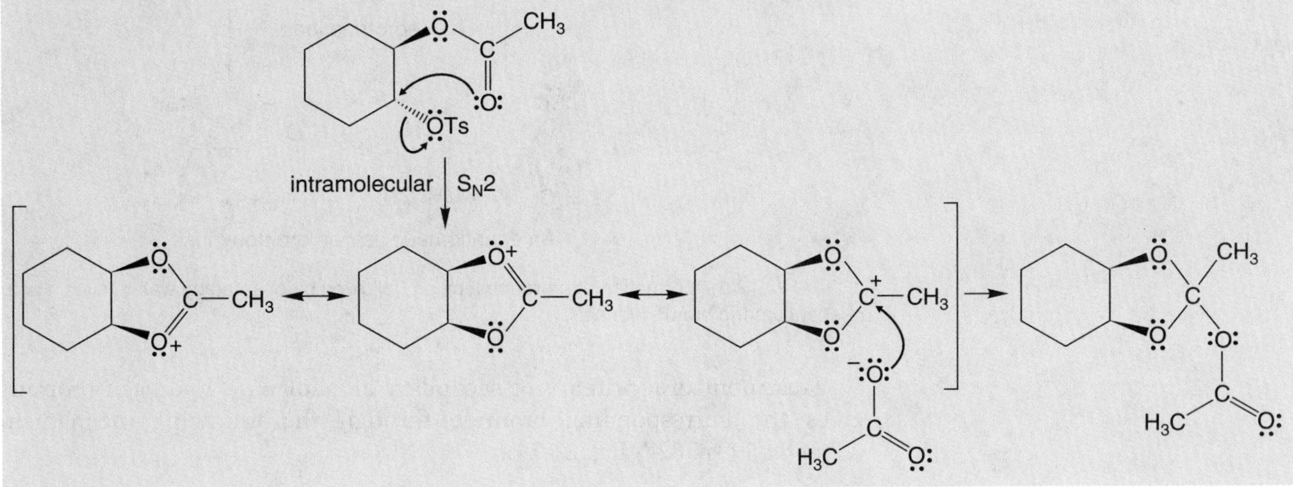

intramolecular S_N2

The halogens are another set of heteroatoms with nonbonding electrons. Neighboring group reactions of halogens are almost entirely restricted to iodine, bromine, and chlorine, and very often reveal themselves through the stereochemistry of the resulting products.

PROBLEM 23.9

What's wrong with fluorine? Why are neighboring group effects involving fluorine very rare?

As an example we will work through the conversion of *cis*-2-butene into 2,3-dibromobutane through the synthetic scheme shown in Figure 23.22.

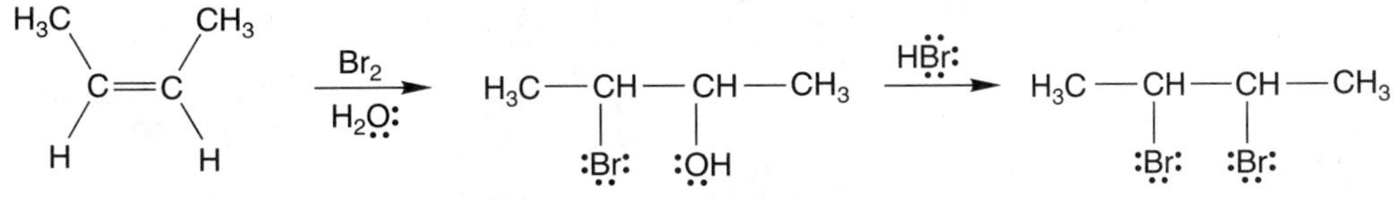

FIGURE 23.22 A possible synthesis of 2,3-dibromobutane (no stereochemistry implied).

PROBLEM 23.10

What is a much easier way of making the 2,3-dibromobutane of Figure 23.22?

In the bromination of *cis*-2-butene, the initial intermediate is the cyclic bromonium ion (Chapter 10, p. 393). It is opened by water to give the racemic pair of enantiomers shown in Figure 23.23.

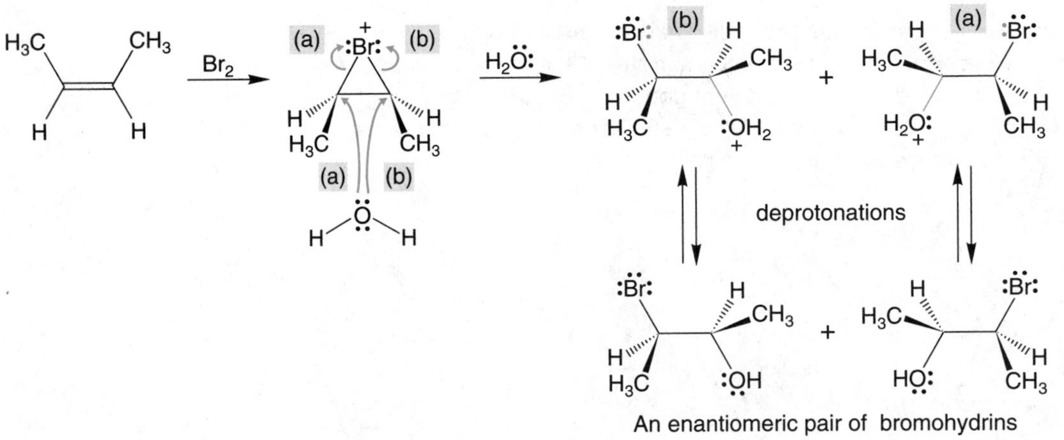

FIGURE 23.23 Formation of a bromonium ion, followed by opening by water, gives a pair of enantiomeric bromohydrins.

Treatment of a primary or secondary alcohol with hydrogen bromide gives the corresponding bromide through the following mechanism (Chapter 17, p. 827; Fig. 23.24).

FIGURE 23.24 The mechanism for formation of a bromide from alcohols and hydrogen bromide.

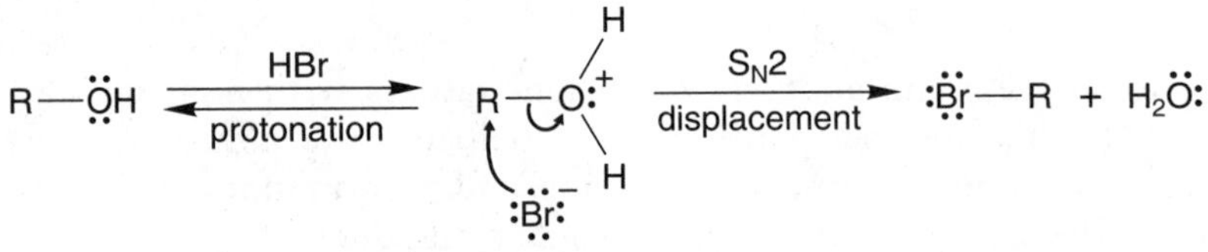

The reaction of Figure 23.24 would give the *meso*-2,3-dibromobutane from the bromohydrins of Figure 23.23. This compound is *not* the product that is really formed. The product *is* 2,3-dibromobutane, but it is the racemic form, not the meso compound that is produced (Fig. 23.25).

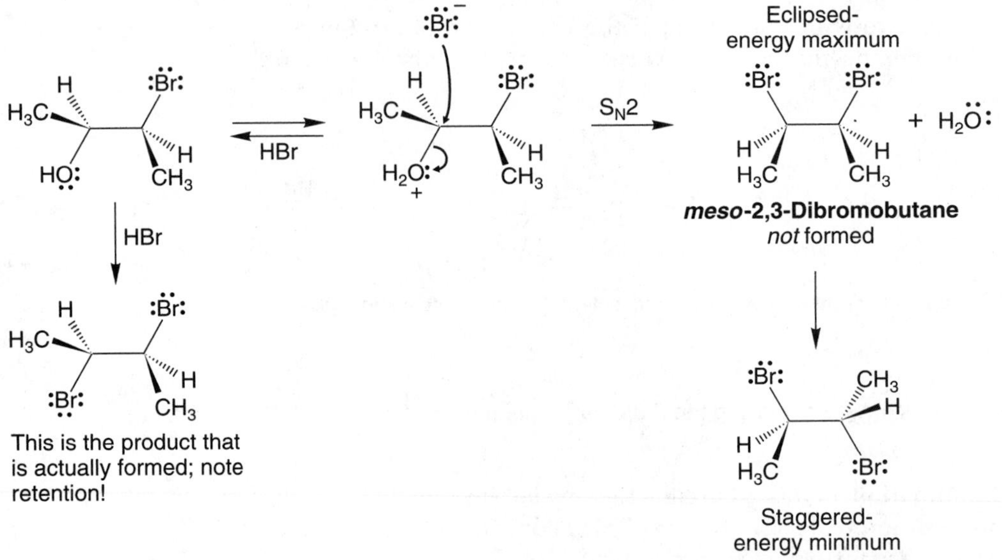

FIGURE 23.25 Reaction of the bromohydrin of Figure 23.23 by S_N2 displacement should give *meso*-2,3-dibromobutane. This compound is *not* the actual product.

Be absolutely certain you see this difference. If you are lost, reread Chapter 5, which works through the differences between meso compounds and racemic mixtures.

Two conventions are being used here. First, the figures often carry through with only one enantiomer of a pair. Be sure to do Problem 23.11. Second, for clarity, eclipsed forms are often drawn. Be sure you remember that these are not really energy minima, but energy maxima.

Convention Alert!

PROBLEM 23.11

Figure 23.25 works the conventional reaction mechanism through with one of the enantiomers of Figure 23.23. You work it through with the other enantiomer.

The formation of the "wrong" stereoisomer is the clue to the operation of a neighboring group effect in this reaction. The product is formed not with the inversion required by the mechanism of Figure 23.24, but with retention (clue 1). There must be a neighboring group effect operating. After the hydroxyl group is protonated, it is displaced not by an external bromide ion, but by the bromine atom lurking within the same molecule. It is most important at this point to keep the stereochemical relationships clear. Displacement must be from the rear, as this is an intramolecular S_N2 reaction (Fig. 23.26).

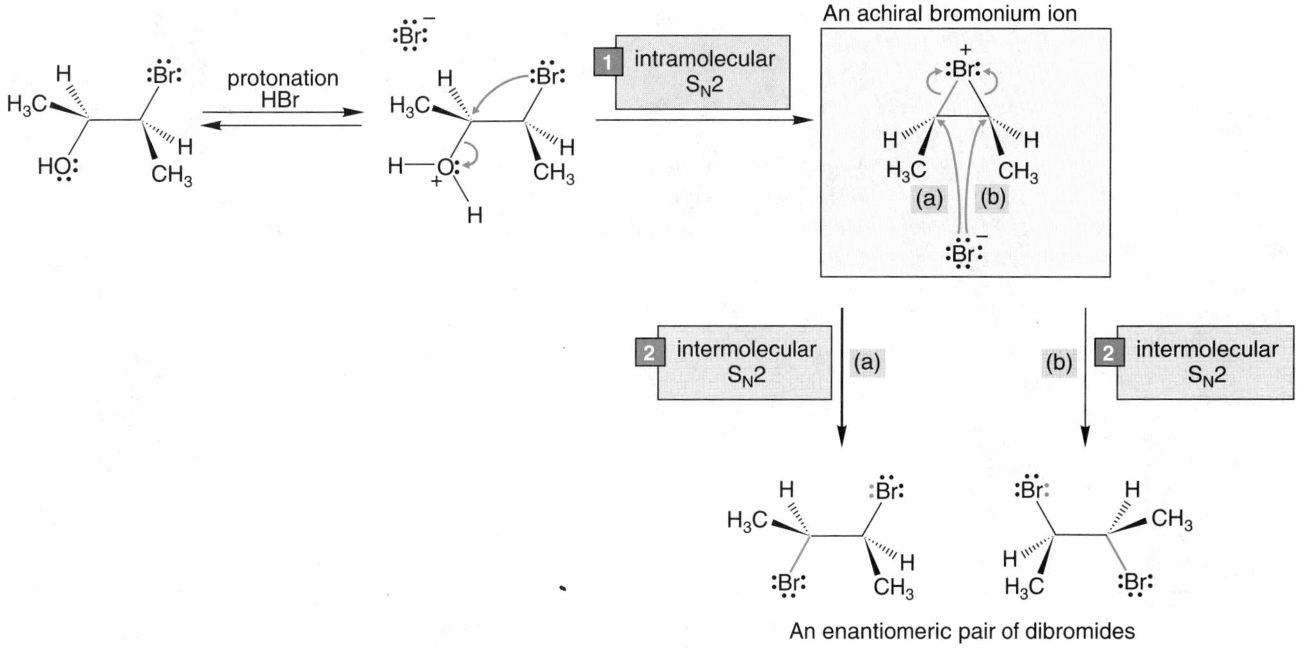

protonation HBr

1 intramolecular S_N2

An achiral bromonium ion

2 intermolecular S_N2 (a)

(b) 2 intermolecular S_N2

An enantiomeric pair of dibromides

FIGURE 23.26 If bromine acts as a neighboring group and displaces water, a bromonium ion is formed. Opening by bromide ion gives the observed product, a pair of enantiomeric dibromides.

Notice that we have formed the same bromonium ion as in Figure 23.23, though by a different route. Like all bromonium ions, this one is subject to opening by nucleophilic attack. In this case, bromide ion can open the cyclic ion at two points (paths a and b) leading to the pair of enantiomers shown in Figure 23.26. The product shows no optical activity not because it is a meso compound, but because it is a racemic mixture of enantiomers.

PROBLEM 23.12 Work through the stereochemical relationships in the following reactions of *trans*-2-butene in great detail. Follow the reactions of all pairs of enantiomers. Contrast them with what you would expect from a reaction not involving a neighboring group effect (Fig. 23.27).

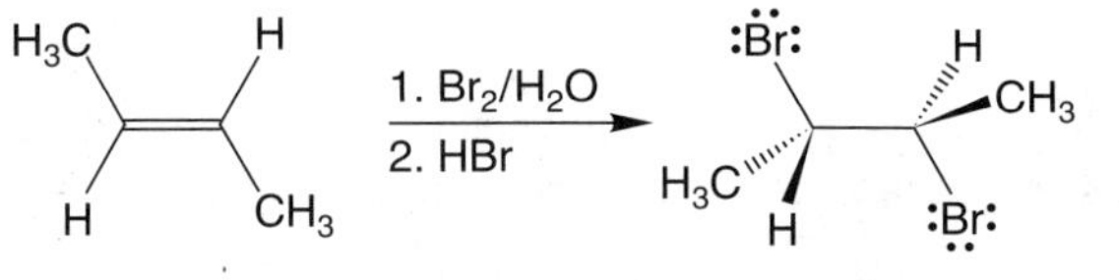

1. Br_2/H_2O
2. HBr

meso-2,3-Dibromobutane

FIGURE 23.27

*PROBLEM 23.13 There is a fundamental difference in the formation of a cyclic ion using a pair of nonbonding electrons or a pair of σ electrons. Carefully draw out the cyclic species for both cases and explain the difference. *Hint*: Count electrons.

ANSWER Problem 23.12 contains many examples of formation of a three-membered ring using a pair of nonbonding electrons. The result is a normal compound in which all bonds are two-electron bonds.

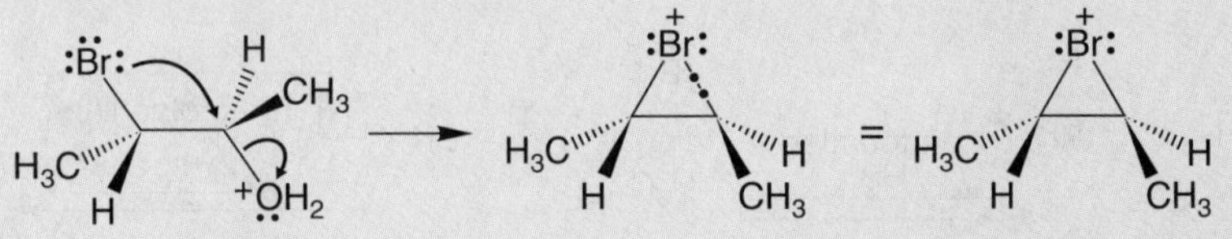

When participation is by a σ bond, everything changes. No longer is it possible to make what we could call a normal three-membered ring; there are not enough electrons. The two electrons originally in the σ bond must serve to bind three nuclei.

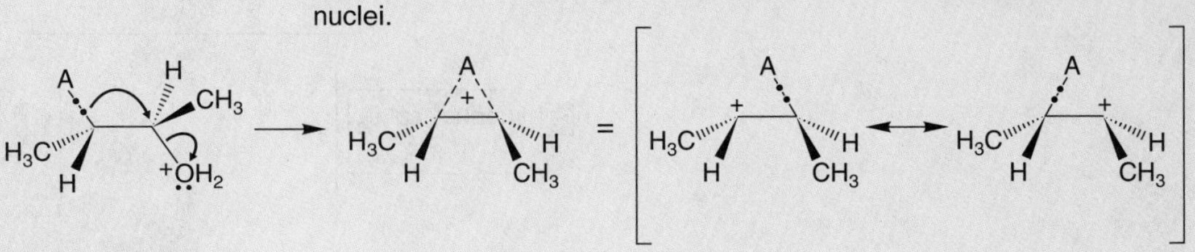

23.2 NEIGHBORING π SYSTEMS

23.2a Aromatic Neighboring Groups

We have progressed from participation reactions of very strong nucleophiles such as alkoxides and sulfur, to those of less powerful displacing agents such as the halogens. Now we examine even weaker nucleophiles, the π systems of aromatic molecules and alkenes. We will finally move on to very weak nucleophiles indeed, and see σ bonds acting as intramolecular displacing agents. The π systems of aromatic rings can act as neighboring groups. The reactions of the 3-phenyl-2-butyl tosylates are good examples. D. J. Cram [b. 1919, awarded the Nobel prize in 1987 for work on crown ethers (Chapter 17, p. 863)] and his co-workers showed that in acetic acid, the optically active 3-phenyl-2-butyl tosylate shown in Figure 23.28 gives a racemic mixture of the corresponding 3-phenyl-2-butyl acetates.

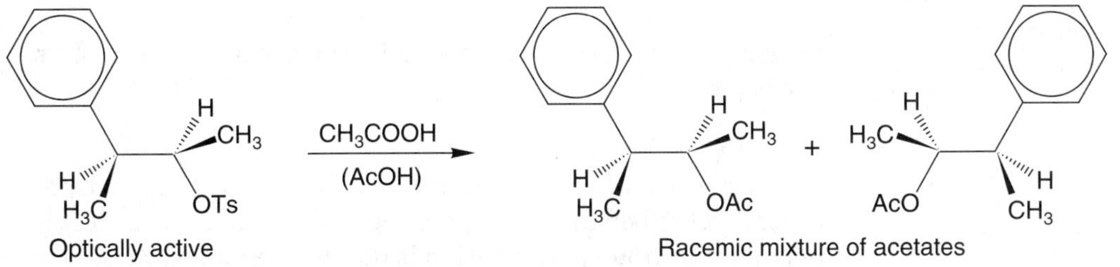

Optically active — Racemic mixture of acetates

FIGURE 23.28 This optically active isomer of 3-phenyl-2-butyl tosylate gives a racemic mixture of the corresponding acetates.

But this result is not what we would expect from a simple analysis of the possible displacement reactions of the starting material by either S_N2 or S_N1 substitution. The S_N2 reaction must give inversion in the substitution step, and so must form the optically active acetate shown in Figure 23.29.

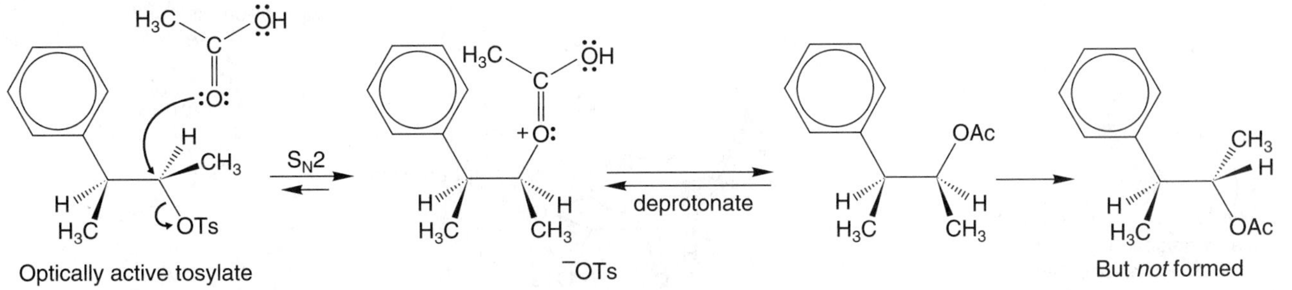

Optically active tosylate — But *not* formed — Optically active acetate

FIGURE 23.29 Simple S_N2 displacement by acetic acid fails to predict the correct product. Note the eclipsed forms drawn in this figure! They are there for clarity only and should be replaced with the staggered forms.

Ionization in an S_N1 process gives an intermediate open carbocation that can be attacked at either face by acetic acid to give the two different optically active products of Figure 23.30.

FIGURE 23.30 An S_N1 ionization also fails to predict the product with the correct stereochemistry. A mixture of two optically active acetates should result.

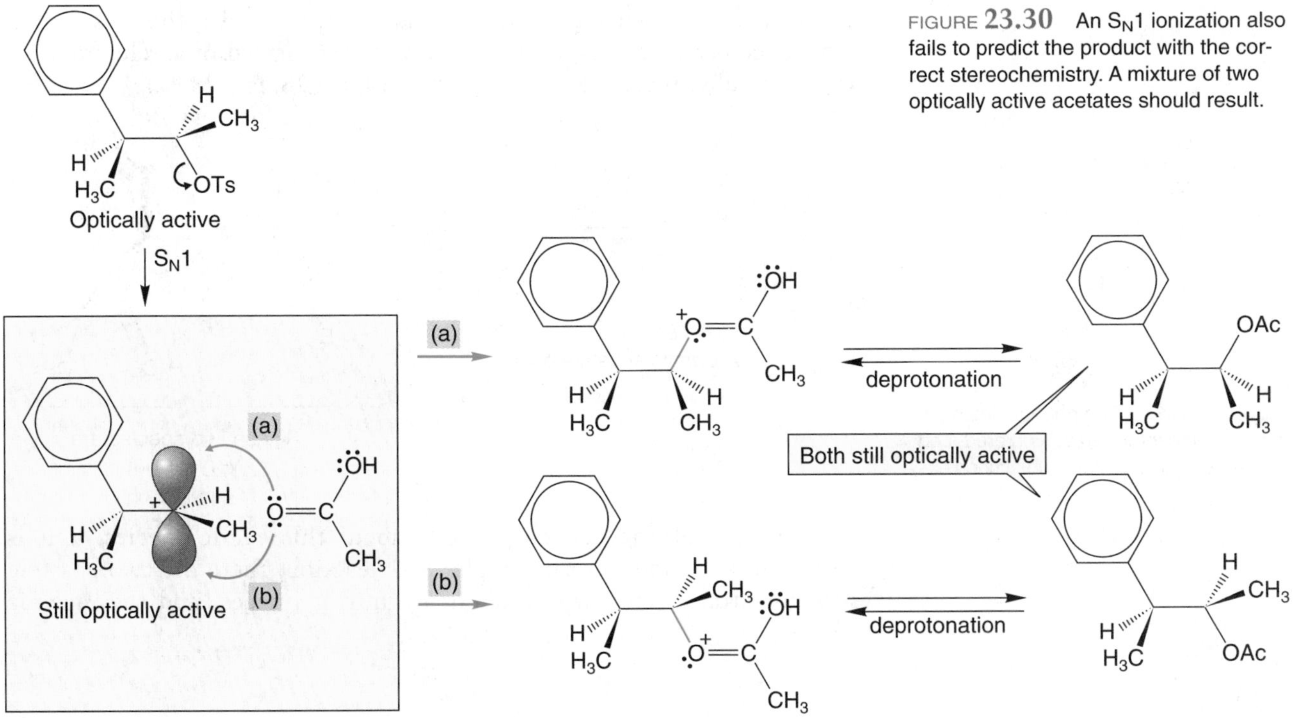

Optically active — Still optically active — Both still optically active

PROBLEM 23.14 What is the stereochemical relationship (names) of the two products at the bottom of Figure 23.30?

As neither of these reactions produces the observed product, some other mechanism must be operating in this reaction. One of the clues for a neighboring group participation is present—an unexpected stereochemical result, clue 1—so let's look around for possible internal nucleophiles.

Aromatic rings act as nucleophiles toward all manner of Lewis acids in the aromatic substitution reaction (Chapter 14), so perhaps we can use the π system of the phenyl ring present in 3-phenyl-2-butyl tosylate as a nucleophile. If the π system displaces the tosylate, it must do so from the rear. The result is the **phenonium ion** shown in Figure 23.31.

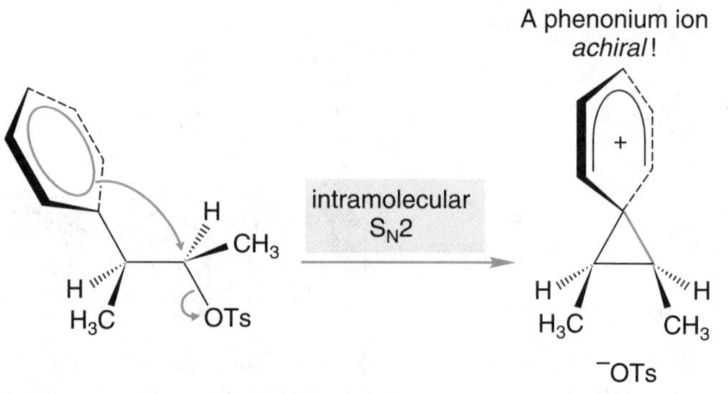

A phenonium ion *achiral*!

intramolecular S_N2

FIGURE 23.31 If the benzene ring acts as a neighboring group, a cyclic phenonium ion results. Notice that the phenonium ion has a plane of symmetry and is achiral.

PROBLEM 23.15 Draw resonance forms for the phenonium ion of Figure 23.31.

Note carefully the geometry of the phenonium ion. The two rings are perpendicular, not coplanar, and the ion is achiral, a meso compound. All chance of optical activity is lost at this point. Note also the close correspondence between this intermediate and that formed in an electrophilic aromatic substitution reaction (Chapter 14, p. 628; Fig. 23.32).

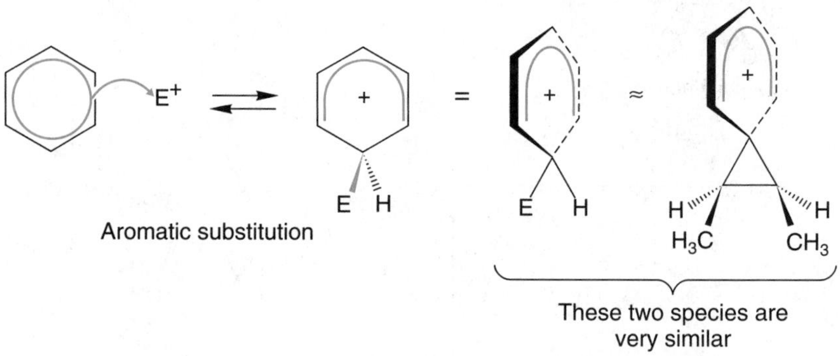

Aromatic substitution

These two species are very similar

FIGURE 23.32 The phenonium ion is similar to the intermediate ion produced in electrophilic aromatic substitution.

There is nothing particularly odd about this cyclic species. If it is opened in a second, intermolecular S_N2 reaction, there are two possible sites for attack. The two products are mirror images, and the racemic mixture of enantiomers of Figure 23.33 is exactly what was found experimentally.

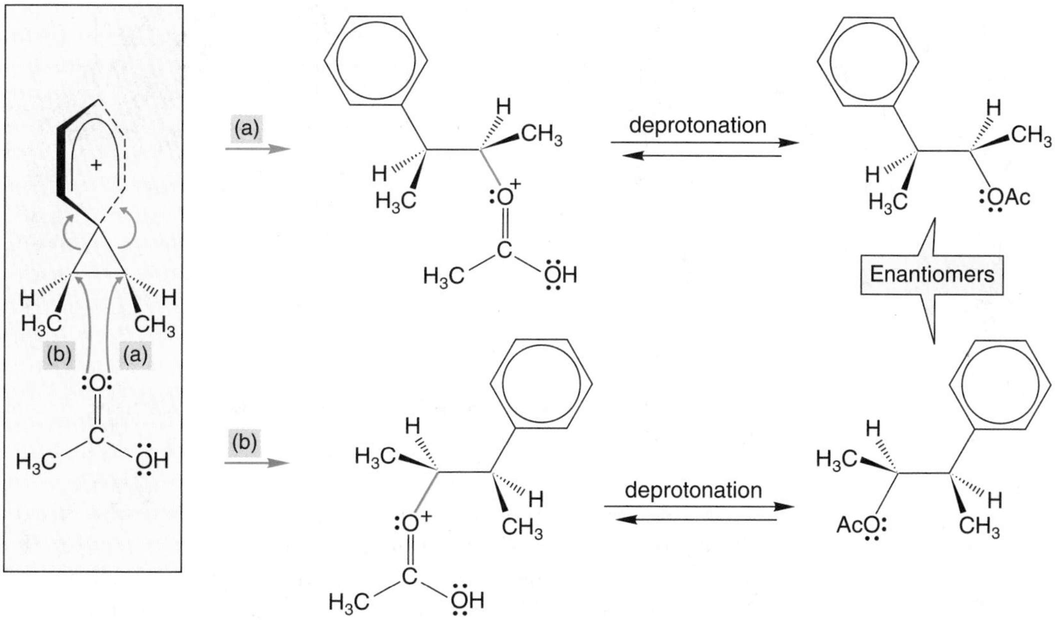

FIGURE 23.33 Opening of the phenonium ion must occur in two equivalent ways (a and b) to give a pair of enantiomers—a racemic mixture.

Work through the reaction of the other possible optically active isomer of 3-phenyl-2-butyl tosylate in acetic acid (Fig. 23.34). This time the product is optically active acetate.

PROBLEM 23.16

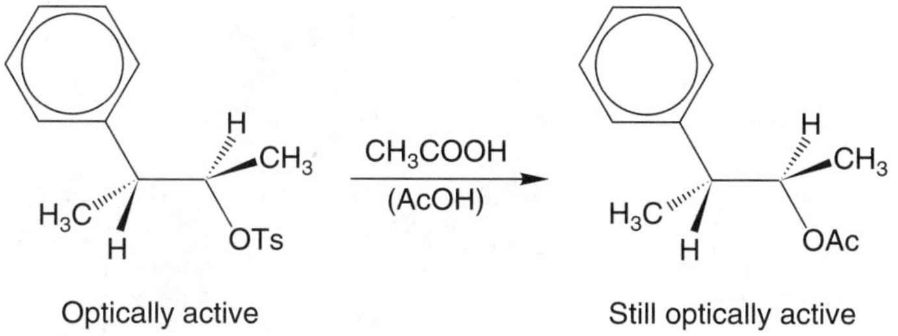

Optically active

Still optically active

FIGURE 23.34

23.2b Carbon–Carbon Double Bonds as Neighboring Groups

Carbon–carbon double bonds are also nucleophiles and can act as neighboring groups in intramolecular displacement reactions. Among the examples you already know are the sequence of intramolecular carbocation additions leading to steroids (Chapter 12, p. 552) and intramolecular carbene additions (Chapter 10, p. 411). In each of these reactions, a carbon–carbon double bond acts as a nucleophile toward an empty carbon 2*p* orbital as Lewis acid (Fig. 23.35).

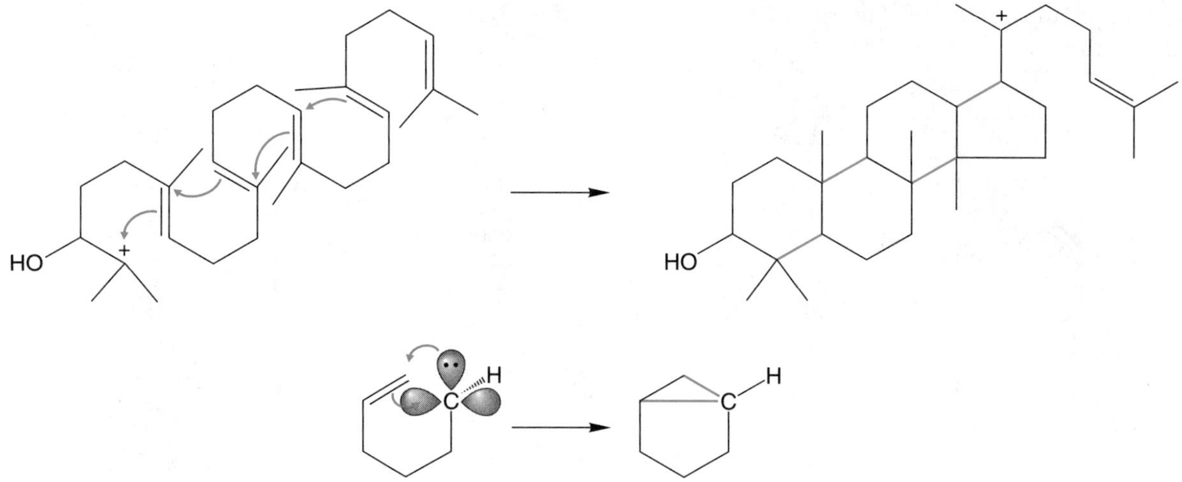

FIGURE 23.35 Two examples of carbon–carbon double bonds acting as intramolecular nucleophiles.

Neighboring carbon–carbon π bonds can influence the rate and control the stereochemistry of a reaction just as a heteroatom can. For example, the tosylate of anti-7-hydroxybicyclo[2.2.1]hept-2-ene (7-hydroxynorbornene)* reacts in acetic acid 10^7 times faster than the corresponding syn compound, and a staggering 10^{11} times faster than the tosylate of the saturated alcohol, 7-hydroxynorbornane. The product is exclusively the anti acetate (Fig. 23.36).

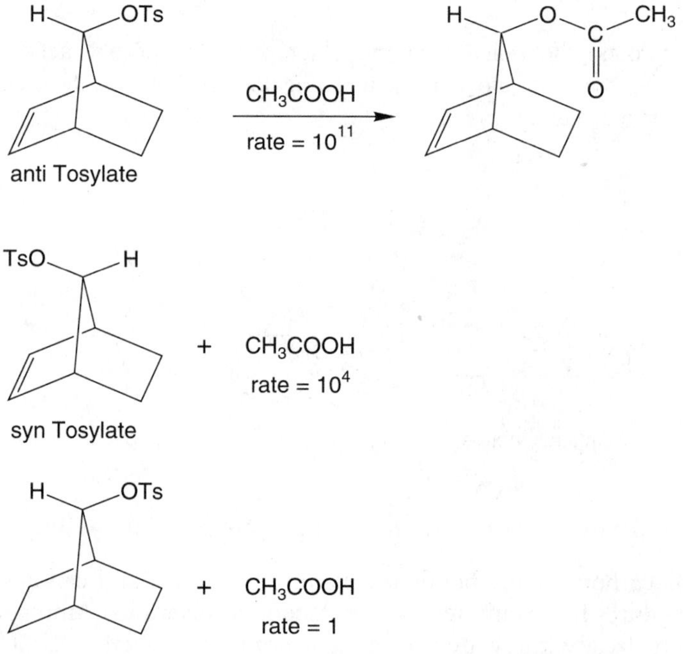

FIGURE 23.36 The anti isomer of 7-norbornenyl tosylate gives the anti acetate in acetic acid. The anti tosylate reacts 10^7 faster than the syn isomer and 10^{11} faster than the saturated compound.

*The bicyclo[2.2.1]heptane cage is commonly known as the **norbornyl system**. This obscure name comes from a natural product, borneol, which has the structure shown in Figure 23.37, and which comes from the camphor tree, common in Borneo. The prefixes syn and anti locate the direction of the OH or OTs group with respect to the double bond and **nor** means "no methyl groups." Thus, 7-hydroxybicyclo[2.2.1]hept-2-ene is 7-hydroxynorborn-2-ene, or, as there is only one possible place for the double bond, 7-hydroxynorbornene.

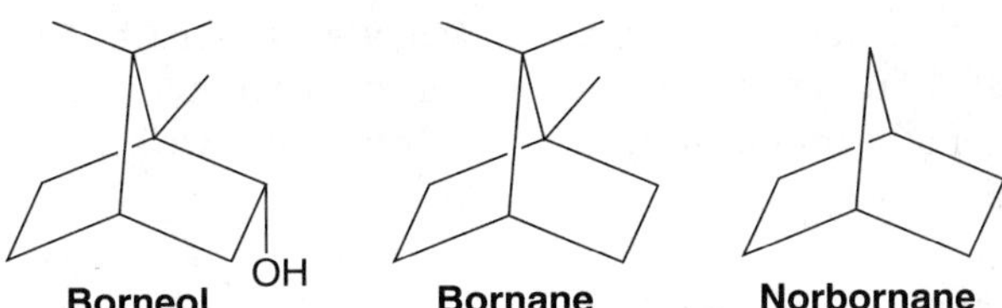

FIGURE 23.37 Norbornane (bicyclo[2.2.1]heptane) is related to the natural product borneol, *endo*-1,7,7-trimethylbicyclo[2.2.1]heptan-2-ol.

PROBLEM 23.17

There is only one possible place for the double bond in the bicyclo[2.2.1]heptyl ring system. Explain why.

A consistent mechanistic explanation notes that only in the anti compound is the π system of the double bond in the correct position to assist in ionization of the tosylate. For the syn double bond to assist in ionization would require a frontside S_N2 displacement, which, as we know, is an unknown process.

The rate enhancement is attributed to anchimeric assistance by the double bond. Intramolecular displacement of the leaving group is faster than the possible intermolecular reactions and is therefore the dominant process. The stereochemistry of the reaction is explained by noting that addition of acetic acid to the intermediate ion is also an S_N2 reaction and must occur from the rear of the departing leaving group. The net result of the two S_N2 reactions (one intramolecular, one intermolecular) is retention (Fig. 23.38).

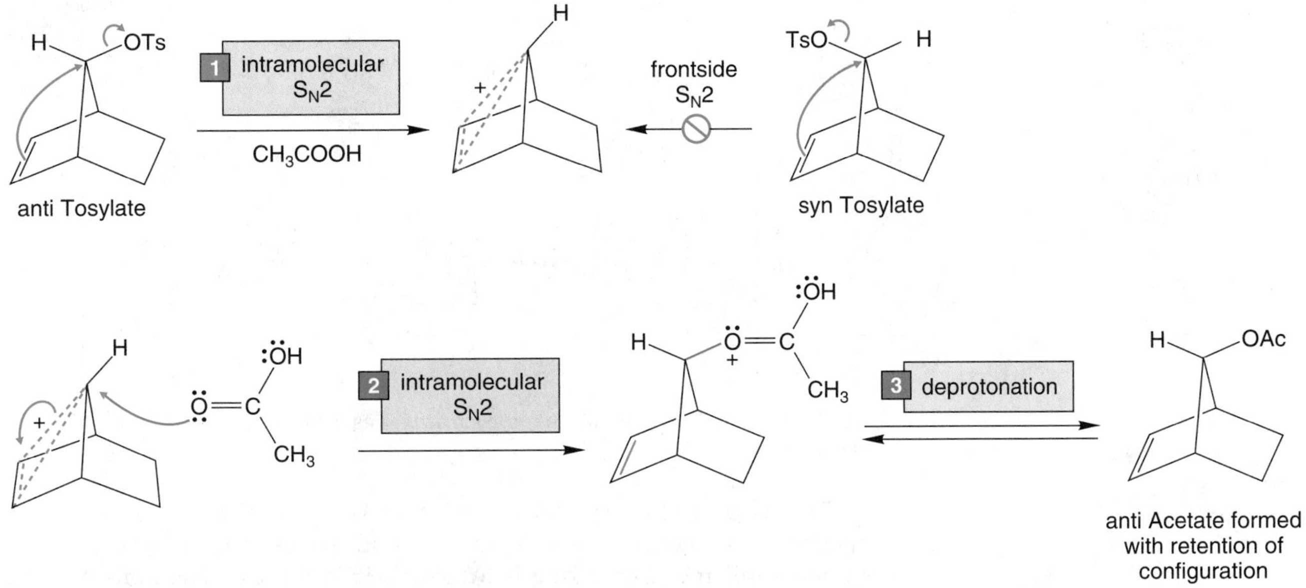

H OTs
anti Tosylate
1 intramolecular S_N2
CH_3COOH
H
+
frontside S_N2
TsO H
syn Tosylate

H
+
:ÖH
Ö=C
CH_3
2 intramolecular S_N2
H
:ÖH
Ö=C
+
CH_3
3 deprotonation
H OAc
anti Acetate formed with retention of configuration

FIGURE 23.38 The carbon–carbon double bond of the anti tosylate, but not of the syn isomer, can react in an intramolecular S_N2 reaction to displace the leaving group. Opening of the cyclic ion must give the anti acetate, the result of net retention.

Think a bit now about the nature of the leaving group in the second, intermolecular S_N2 displacement. Just what is the leaving group, and what do all those dashed bonds in Figure 23.38 mean? To answer these questions well, we must first look closely at the structure of the postulated intermediate in this reaction. When we formed three-membered ring intermediates from heteroatoms, there was always a pair of nonbonding electrons available to do the intramolecular displacement. An example is the formation of a bromonium ion by displacement of water by neighboring bromine (Fig. 23.39).

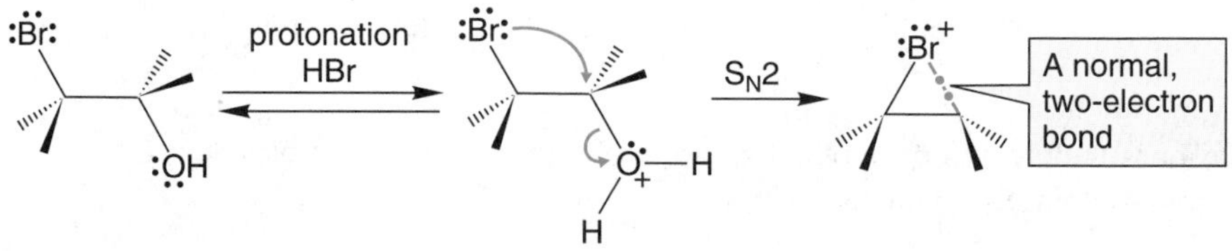

protonation HBr

S_N2

A normal, two-electron bond

FIGURE 23.39 In a bromonium ion, all bonds are normal two-electron bonds.

Contrast the bromonium ion of Figure 23.39 with the species formed by intramolecular displacement of tosylate using the π bond. In the bromonium ion, there are sufficient electrons to form the new σ bond in the three-membered ring. In the all-carbon species there are not. In this cyclic ion, two electrons must suffice to bind three atoms. This is an example of **three-center, two-electron bonding**, and the partial bonds are represented with dashed lines. We have seen such bonds before (Chapter 9, p. 380), and as usual, a resonance formulation shows most clearly how the charge is shared by the three atoms in the ring (Fig. 23.40).

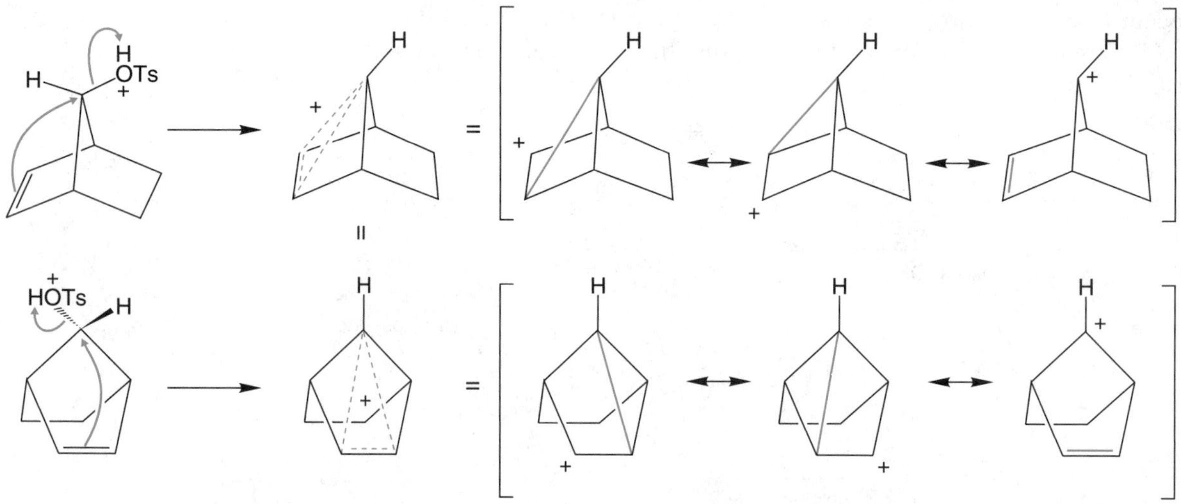

FIGURE 23.40 Two views of the cyclic ion as well as of the resonance forms contributing to the ion.

The two species are *very* different. When the bromonium ion is opened, it is a normal, two-electron σ bond that is broken. When the all-carbon three-membered ring is attacked, it is the two electrons binding the three carbons that act as the leaving group (Fig. 23.41).

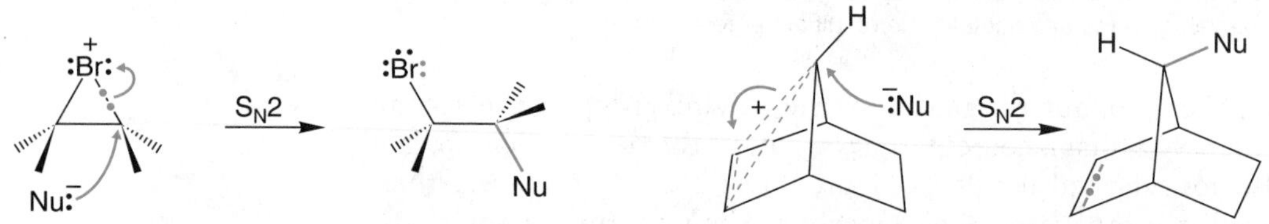

S_N2

S_N2

FIGURE 23.41 When a nucleophile opens a bromonium ion, a pair of electrons is displaced. The same is true for the opening of the bridged, cyclic ion.

The process is still an S_N2 reaction, however, and there should be no reason for softening the requirement for rearside displacement.

Acceptance of the carbon–carbon double bond as a neighboring group requires you to expand your notions of bonding, and to accept the partial bonds making up the dashed three-center, two-electron system as a valid bonding arrangement. There are those who resisted this notion mightily, and although I believe they were ultimately shown to be incorrect, their resistance was anything but foolish. The postulated intermediate in this reaction is not a normal species and should be scrutinized with the greatest care.

PROBLEM 23.18

If three atoms share the positive charge, why is it only the 7-position that is attacked by the nucleophile? Explain the lack of another product. *Hint*: Look at the structure of the product formed by addition at the other position (Fig. 23.42).

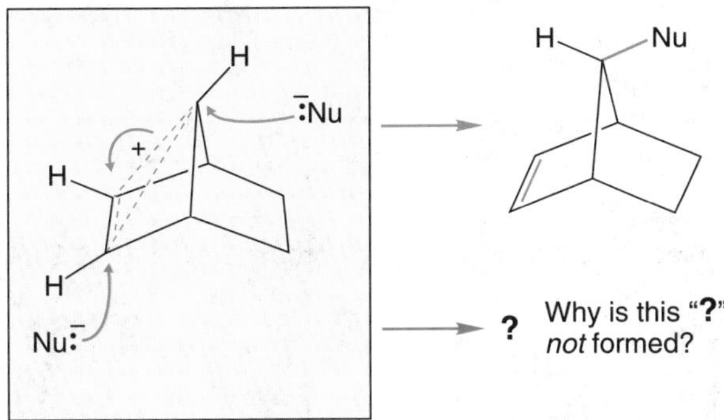

FIGURE 23.42

A molecular orbital approach shows how the two electrons can be stabilized through a symmetrical interaction with an empty carbon 2*p* orbital. The intellectual antecedents for this kind of bonding go back to the cyclic H_3^+ molecule in which two electrons are stabilized in a bonding molecular orbital through a cyclic overlap of hydrogen 1*s* orbitals. We can construct this system by allowing a 1*s* orbital to interact with σ and σ* of H_2 as in Figure 23.43 (recall Problems 2.18 and 2.19, and our discussion of the allyl and cyclopropenyl π systems in Chapter 12).

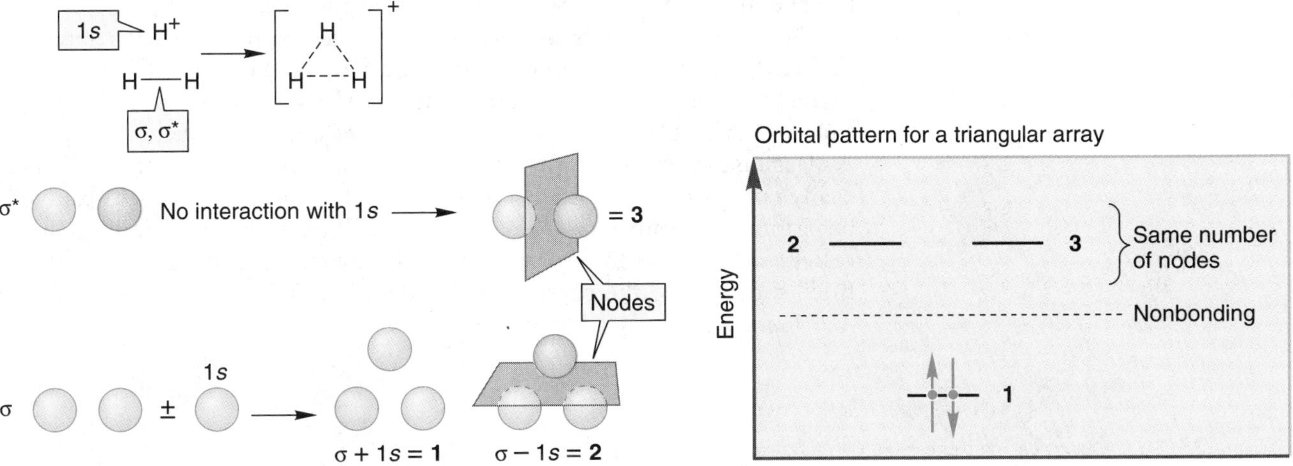

FIGURE 23.43 The construction of the molecular orbitals for cyclic H_3^+.

The result is a system of three molecular orbitals: one bonding, two antibonding. Two, but only two, electrons can be accommodated in the bonding orbital. The same orbital pattern will emerge for any triangular array of orbitals. For example, in Chapter 9 (p. 380) we examined the protonation of ethylene to give a cyclic ion. The cyclic ion of Figure 23.43 is not very different (Fig. 23.44).

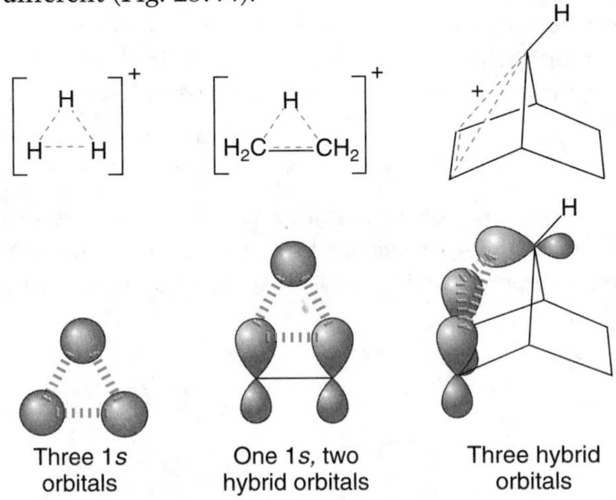

Three 1*s* orbitals — One 1*s*, two hybrid orbitals — Three hybrid orbitals

FIGURE **23.44** The patterns for cyclic ions will be similar. In all cases, two electrons can be accommodated in the low-energy bonding molecular orbital. The antibonding orbitals are both empty.

PROBLEM **23.19** Work out the π molecular orbitals for the cyclopropenium ion (Fig. 23.45).

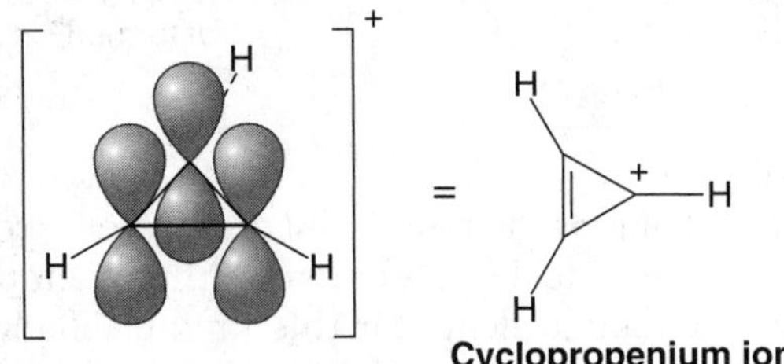

FIGURE **23.45**

Those resisting this notion (the "classicists") suggested that there was no need for this new kind of bonding, and that normal, two-electron models would do as well. Let's follow their argument. Remember as we go that they must explain both a rate acceleration for the anti tosylate as well as the overall retention of stereochemistry in the reaction.

The two compounds in question here, the syn and anti tosylates, are *different*, and there *must* be some rate difference between them. One can scarcely argue with this point, and one is reduced to relying on the magnitude and direction of the rate acceleration as evidence for the bridged ion. A classicist might argue that in cases such as this, the rate increase could easily come from the formation of a normal, unbridged or "classical," carbocationic intermediate. The anti tosylate has the leaving group in an excellent orientation to be attacked by the intramolecular nucleophile, the filled π orbital of the carbon–carbon double bond. The syn tosylate does not (Fig. 23.46).

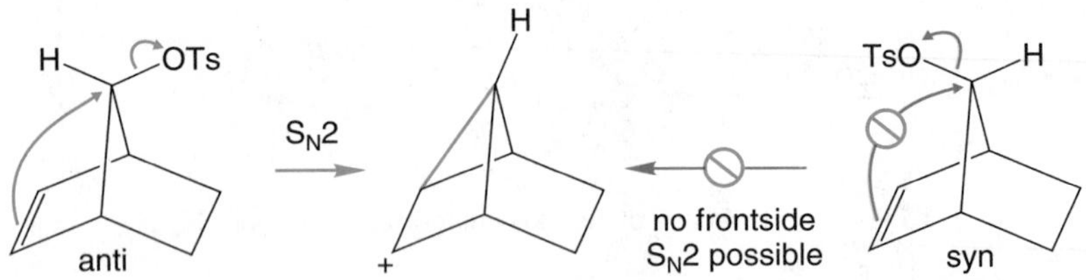

FIGURE **23.46** A "classical" displacement reaction by the π bond of the anti isomer would give an open ion that could not be formed from the syn isomer.

As in the symmetrical displacement to form the bridged ion, the required attack on the leaving group from the rear could occur only in the anti compound. Of course, there are two equivalent intermediates possible (**A** and **B**), and it was suggested that they were in equilibrium. The compound with the charge at the 7-position (**C**) might also be a partner in the equilibrium. Be sure you see the difference between the *three* equilibrating structures of Figure 23.47 and the three resonance forms contributing to the *single* structure of Figures 23.40 and 23.44 (Fig. 23.47).

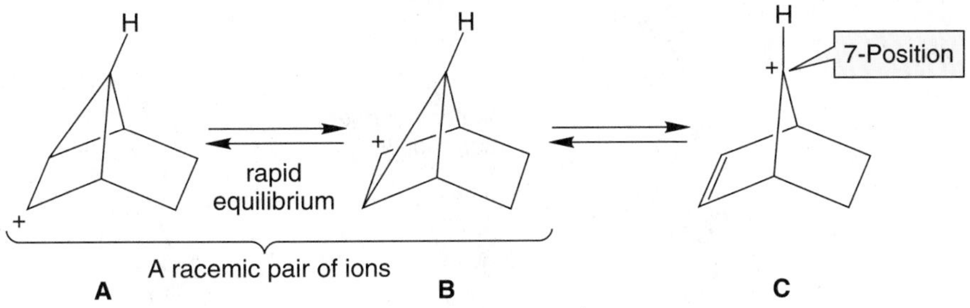

7-Position
rapid equilibrium
A racemic pair of ions
A B C

FIGURE 23.47 The first-formed ion **A** equilibrates with **B** and, perhaps **C**.

Addition of acetic acid to the new ions **A** and **B** must give the anti acetate, as the opening is an S_N2 reaction. Attack on ion **C** is more problematic, as it is difficult to explain the stereospecificity of the reaction. Still, it must be admitted that the syn and anti faces of the ion are different, and in principle, this difference could lead to a stereospecific reaction (Fig. 23.48).

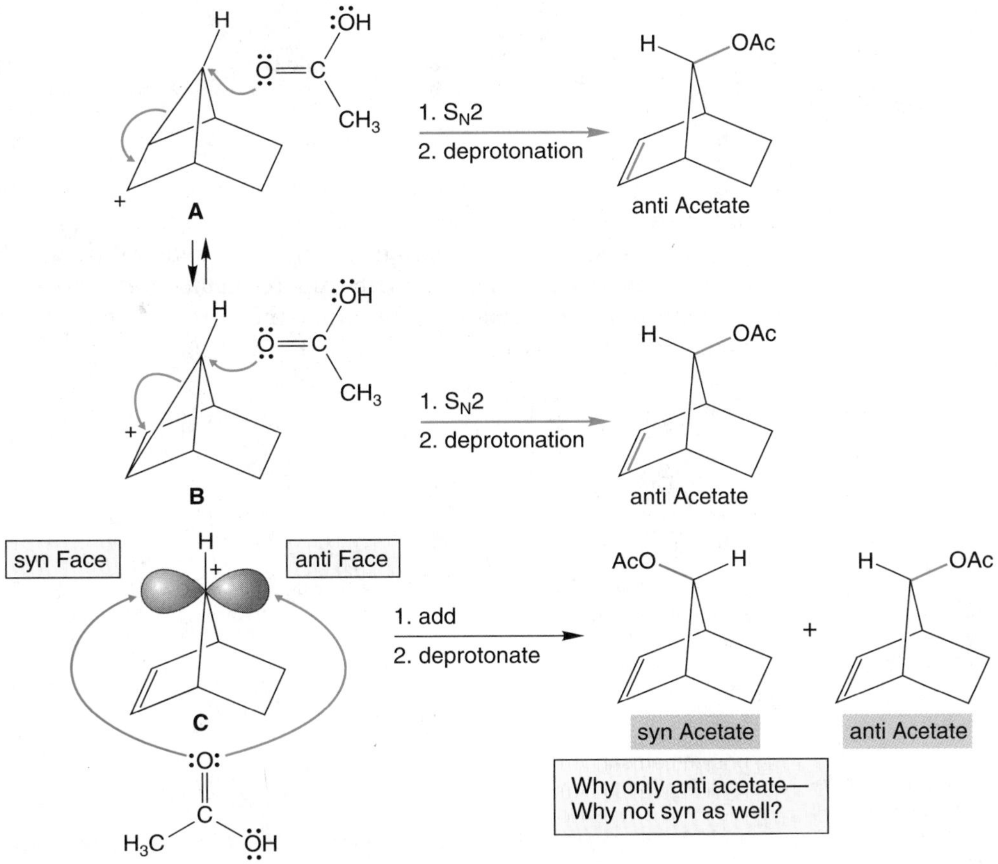

1. S_N2
2. deprotonation
anti Acetate
syn Face
anti Face
1. add
2. deprotonate
syn Acetate
anti Acetate
Why only anti acetate—
Why not syn as well?

FIGURE 23.48 Addition of acetic acid to the equilibrating ions **A** and **B** must lead to anti acetate. Attack at the 7-position (ion **C**) to give only the anti acetate presents some difficulties.

Stop right now and compare the two mechanistic hypotheses. In one, there is a symmetrical displacement of the leaving group to give a resonance-stabilized bridged intermediate. In the other, either the leaving group is displaced to give a carbocationic intermediate containing a three-membered ring (**A** and **B**) in equilibrium with at least one other ion, **C**, or a simple S_N1 ionization to **C** takes place (Fig. 23.49).

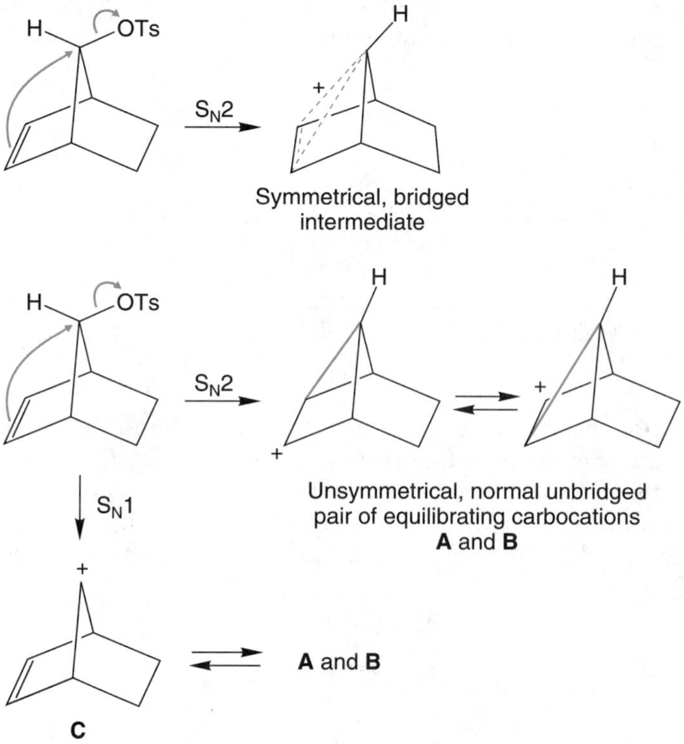

H OTs
S_N2
H
Symmetrical, bridged intermediate

H OTs
S_N2
H H
Unsymmetrical, normal unbridged pair of equilibrating carbocations **A** and **B**
S_N1
A and **B**
C

FIGURE 23.49 A comparison of the symmetrical, bridged intermediate with the open ions **A**, **B**, and **C**.

The whole difference between the two mechanisms is the formulation of the structure of the intermediate! The structures in the two versions of the mechanism for this reaction are only very subtly different. In one, the constituent structures are contributing resonance forms to a single intermediate bridged species; in the other they are equilibrating molecules, each with its own separate existence (Fig. 23.50).

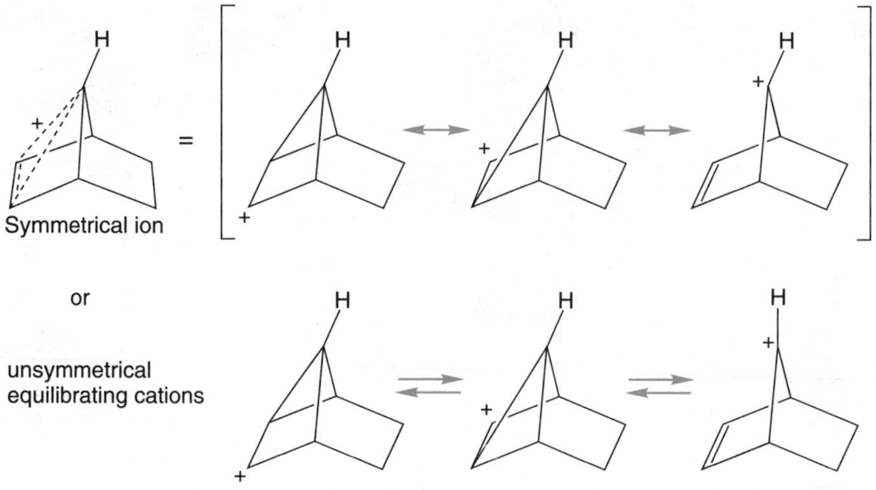

Symmetrical ion

or

unsymmetrical equilibrating cations

FIGURE 23.50 The bridged ion is described with three resonance forms that resemble the open ions of the set of equilibrating structures.

Are we now down to questions of chemical trivia? Are we counting angels on pin heads? Many thought so, but I maintain that they were quite misguided in their criticisms of those engaged in the argument. All questions of mechanism come down to the existence and structures of intermediates, and to the number of barriers separating energy minima.

This argument has been recast many times in related reactions. Not all questions have been resolved, but those that have been have generally involved two kinds of evidence. Sometimes a direct observation of the ion in question was made, usually under **superacid** conditions in which a powerfully acidic, but nonnucleophilic medium is provided in which the ion is stable, at least at low temperature. In other reactions, an experiment like the following one provides evidence for symmetrical participation to produce a bridged intermediate.

The reactions in acetic acid of the monomethyl and dimethyl derivatives of the *p*-nitrobenzoate of *anti*-bicyclo[2.2.1]hepten-7-ol have been examined. The monomethyl compound reacts faster than the parent, as predicted by either mechanistic scheme. A bridged ion would be stabilized because one of its contributing resonance forms is tertiary. Another way of putting this is to note that the methyl group makes the double bond a stronger nucleophile (Fig. 23.51).

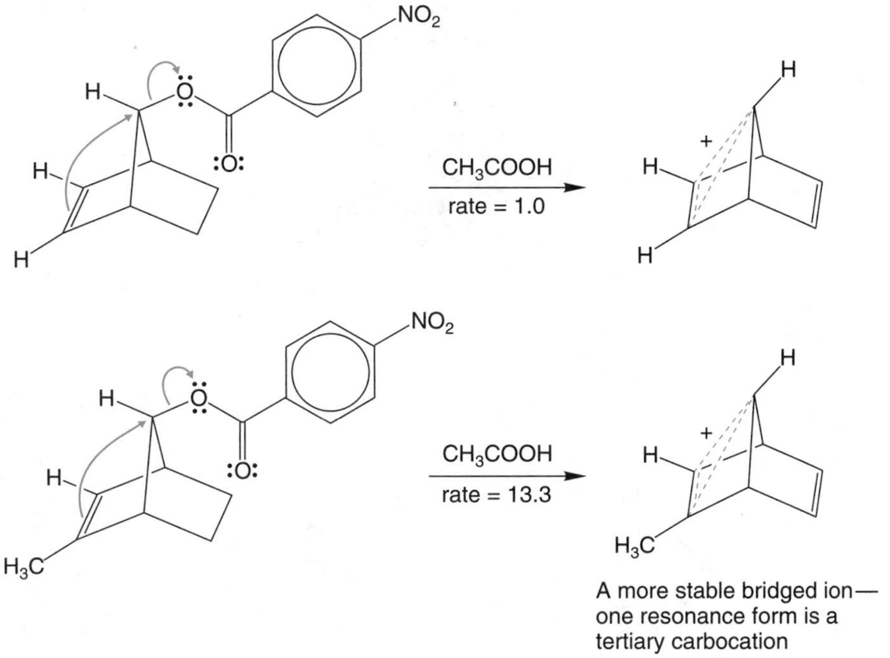

FIGURE 23.51 Formation of a symmetrical, bridged ion from the methyl-substituted molecule should be faster than from the parent.

PROBLEM 23.20

Draw resonance forms for the intermediate of Figure 23.51.

In the more traditional picture of the reaction, a tertiary carbocation can be formed and this would surely accelerate the rate of the reaction relative to the parent, which can produce only a less stable secondary carbocation (Fig. 23.52).

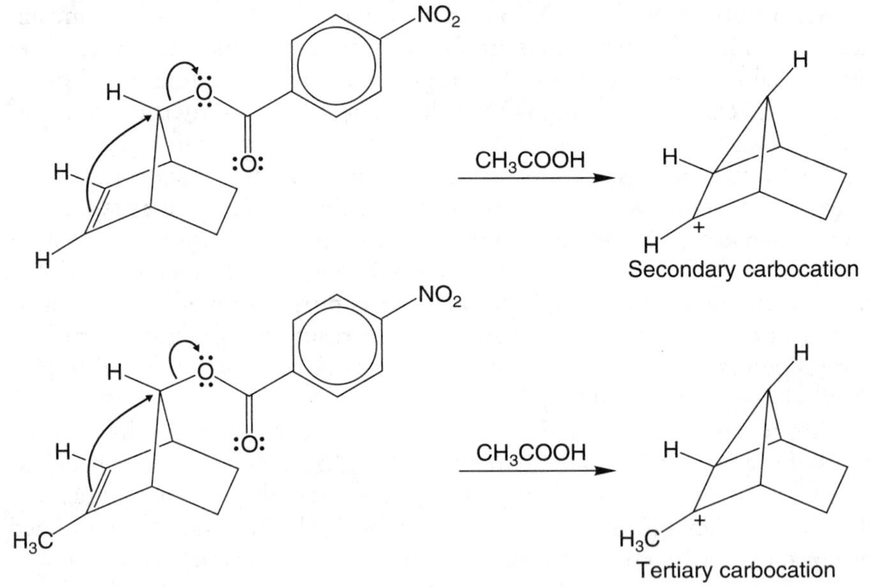

FIGURE 23.52 There should also be a rate increase in the formation of the unsymmetrical, open ion postulated by the classicists.

The observed rate increase of 13.3 is accommodated by both mechanisms.

Now consider the dimethyl compound. If a bridged intermediate is formed through a symmetrical displacement, both methyl groups exert their influence at the same time. If one methyl group induces a rate effect of 13.3, two should give an effect of 13.3 × 13.3 = 177 (Fig. 23.53).

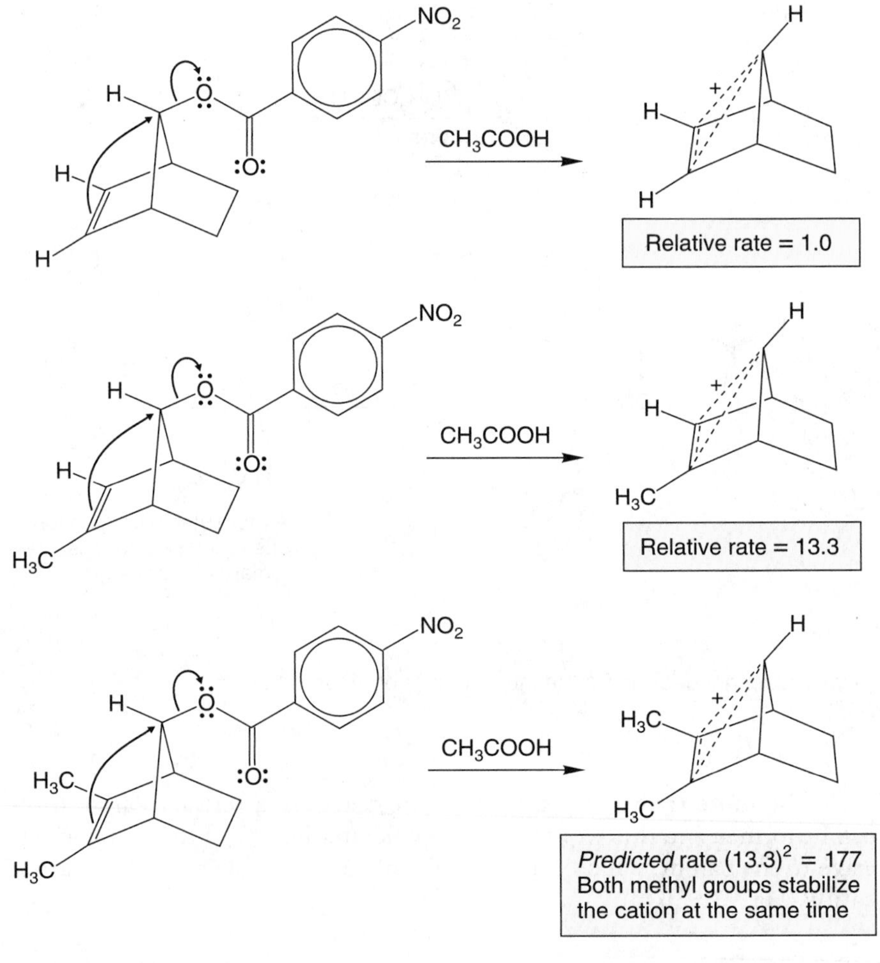

FIGURE 23.53 Reaction of the dimethyl compound to give a bridged ion should be still faster than reaction of the monomethyl compound. The two methyl groups simultaneously stabilize the cation.

On the other hand, if it is the formation of an open tertiary ion that is important, the second methyl increases the possibilities by a factor of 2, *but the two methyl groups never exert their influence at the same time.* The rate change should approximate a factor of 2 (Fig. 23.54).

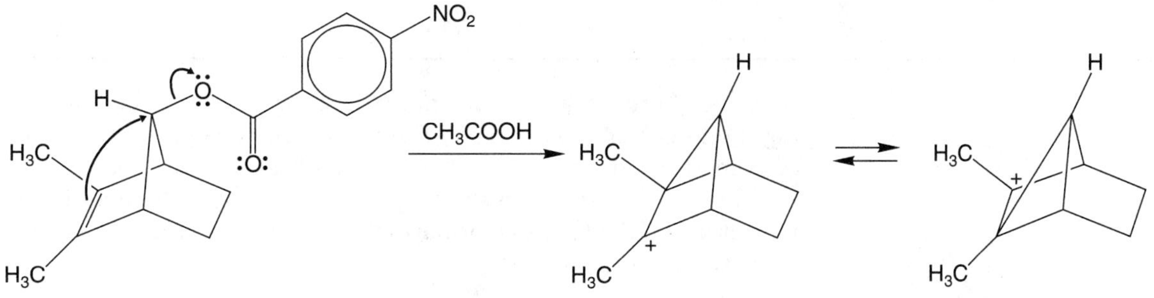

Predicted rate = 2 × 13.3 = 26.6

FIGURE 23.54 In the open, unbridged ions, the two methyl groups can never stabilize the ion at the same time.

The fundamental difference between a mechanism involving a symmetrical bridged ion and the traditional mechanism, which postulates an open, less symmetrical ion, is that in the former case both methyl groups exert their stabilizing influence at the same time. In the pair of equilibrating open ions they cannot do this.

The observed rate for the dimethyl compound is 148, much closer to that expected for the bridged intermediate than the open ion (Fig. 23.55).

Compound	Predicted by Bridged Symmetrical Ion	Predicted by Open Unsymmetrical Ion	Observed
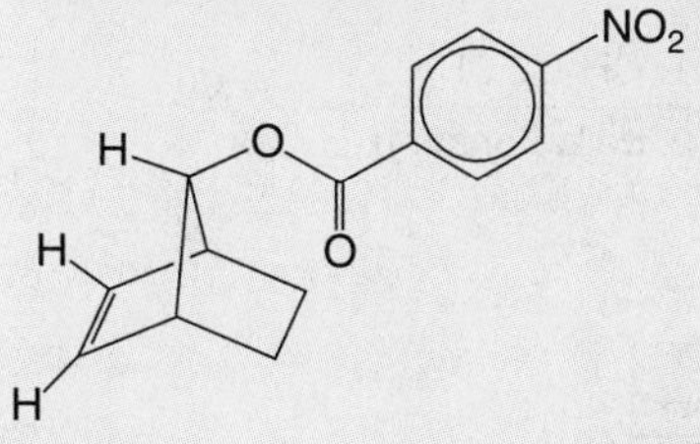	1	1	1
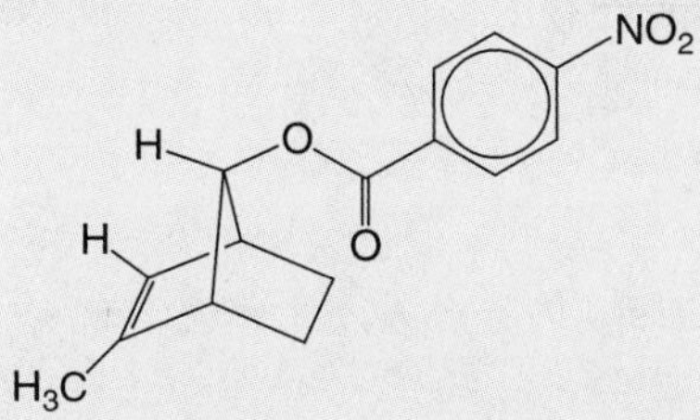	n	n	13.3
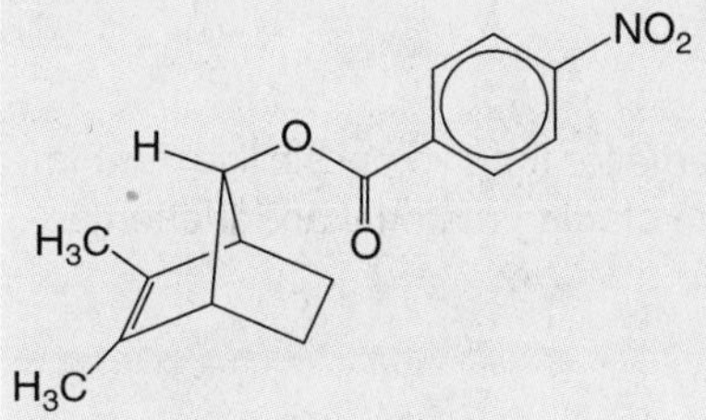	n^2 (177)	$2n$ (26.6)	148

FIGURE 23.55 The experimentally determined rates favor a symmetrical, bridged ion in which both methyl groups act simultaneously.

We have progressed from a discussion of obvious neighboring groups, heteroatoms with nonbonding electrons, to less obvious internal nucleophiles, the filled π orbitals of double bonds. Now we turn to really obscure nucleophiles, filled σ orbitals.

23.3 SINGLE BONDS AS NEIGHBORING GROUPS

Surely, one of the weakest nucleophiles of all would be the electrons occupying the low-energy bonding σ orbitals. Yet there is persuasive evidence that even these tightly held electrons can assist in the ionization of a nearby leaving group. When the tosylate of 3-methyl-2-butanol is treated with acetic acid the result is formation of the tertiary acetate, not the "expected" secondary acetate (Fig. 23.56).

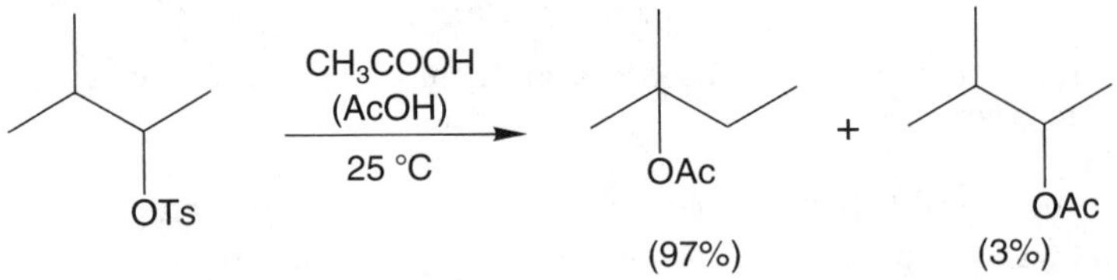

FIGURE 23.56 The reaction of the tosylate of 3-methyl-2-butanol in acetic acid involves a rearrangement.

This result is easily rationalized through a hydride shift to form the relatively stable tertiary carbocation. The interesting mechanistic question is whether the secondary carbocation is an intermediate, or whether the hydride moves *as the leaving group departs.* This example recapitulates the classic mechanistic question: Is there a second intermediate—in this case the secondary carbocation—or is the reaction a concerted formation of the tertiary carbocation (Fig. 23.57)?

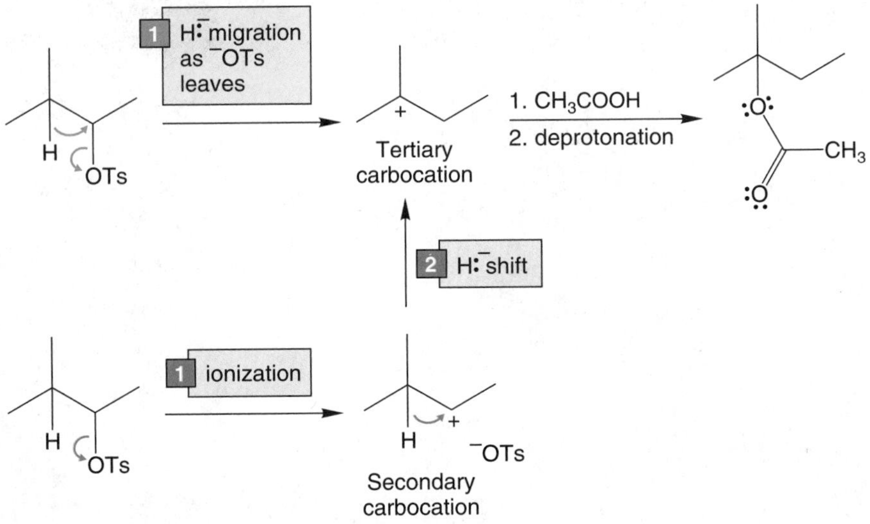

FIGURE 23.57 The mechanistic question involves the timing of the rearrangement. Is a secondary carbocation initially formed, only to be converted into the more stable tertiary carbocation afterward or is movement of the hydride concurrent with ionization? Is formation of the tertiary carbocation a two- or one-step process?

PROBLEM 23.21 Draw Energy versus Reaction progress diagrams for the two possibilities of Figure 23.57. How many energy barriers separate starting material and the tertiary cation in the two mechanisms?

This question has been answered in this system by the observation of an **isotope effect**. The rates of reaction of the unlabeled tosylate and the deuterated tosylate of Figure 23.58 are measured. If the mechanism involves slow formation of the secondary carbocation, followed by a faster hydride shift to give the more stable tertiary carbocation, substitution of deuterium for hydrogen can have no great effect on the rate. Even though the carbon–deuterium bond is stronger than the carbon–hydrogen bond, *it is not broken in the rate-determining step of the reaction.* By contrast, if the hydride (or deuteride) migrates as the leaving group departs, the carbon–hydrogen or carbon–deuterium bond *is* broken in the rate-determining step and a substantial effect on the rate can be expected.

The rate for the unlabeled compound was found to be more than twice as fast as that for the deuterated tosylate, which shows that the carbon–hydrogen or carbon–deuterium bond must be breaking in the rate-determining step of the reaction. This isotope effect ($k_H/k_D = 2.2$) reveals that the hydride must move *as the leaving group departs.* If the slow step in the reaction had involved a simple ionization to a secondary carbocation, *followed* by a hydride shift, only a very small effect of the label could have been observed (precedent suggests $k_H/k_D = 1.1 - 1.3$). The observation of a substantial isotope effect shows that the reaction mechanism must include a neighboring group effect, this time of neighboring hydride!

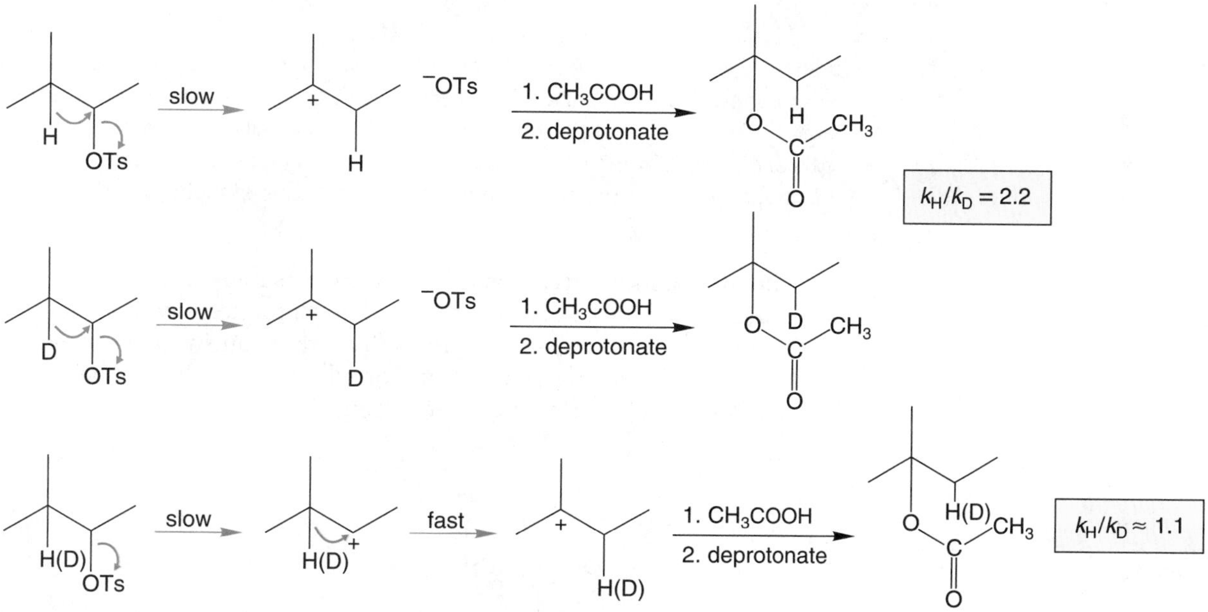

H OTs — slow → + H $^{-}$OTs — 1. CH_3COOH 2. deprotonate → O H CH_3 C O; $k_H/k_D = 2.2$

D OTs — slow → + D $^{-}$OTs — 1. CH_3COOH 2. deprotonate → O D CH_3 C O

H(D) OTs — slow → H(D) + — fast → + H(D) — 1. CH_3COOH 2. deprotonate → O H(D) CH_3 C O; $k_H/k_D \approx 1.1$

FIGURE 23.58 There can only be a substantial isotope effect if breaking the carbon–hydrogen or carbon–deuterium bond occurs in the rate-determining ionization step. The observation of a large isotope effect means that the carbon–hydrogen or carbon–deuterium bond must be breaking in the ionization step of the reaction.

Now comes an even more subtle question. Given that even single bonds can participate in ionization reactions, are cyclic intermediates possible? In Section 9.13 we asked whether the structure for the ethyl cation was a simple open primary carbocation, or whether it was bridged (Fig. 23.59).

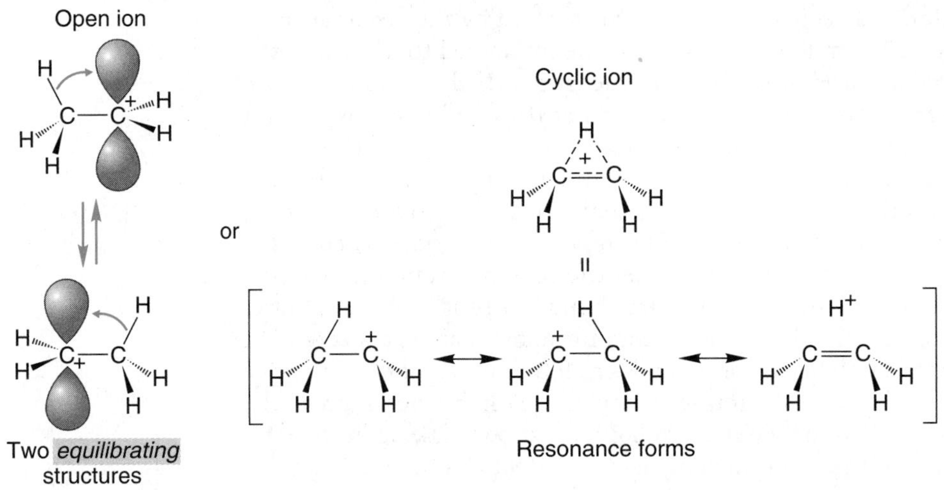

Open ion

or

Cyclic ion

Two *equilibrating* structures

Resonance forms

FIGURE 23.59 Two possibilities for the structure of the ethyl cation.

We could not resolve this question in Chapter 9, but there were indications that the cyclic, bridged ion, at least in the gas phase, might be the more stable structure. This ion is a straightforward example of three-center, two-electron bonding.

PROBLEM 23.22 Although it is hard to see why now, phenonium ions (p. 1094) were originally very controversial. Despite their evident similarity to the intermediates of aromatic substitution (Fig. 23.32), there was fierce opposition to the notion of their existence. Explain why a confusion between neighboring π participation and neighboring σ participation may have led to a misunderstanding.

The most famous argument of modern times in organic chemistry concerned the question of bridged intermediates produced by σ participation, and a fierce one it was—and still is. The reaction in question was the solvolysis (S_N1) reaction of the 2-norbornyl tosylates, the tosylates of *exo*- and *endo*-bicyclo[2.2.1]heptan-2-ol (Fig. 23.60).

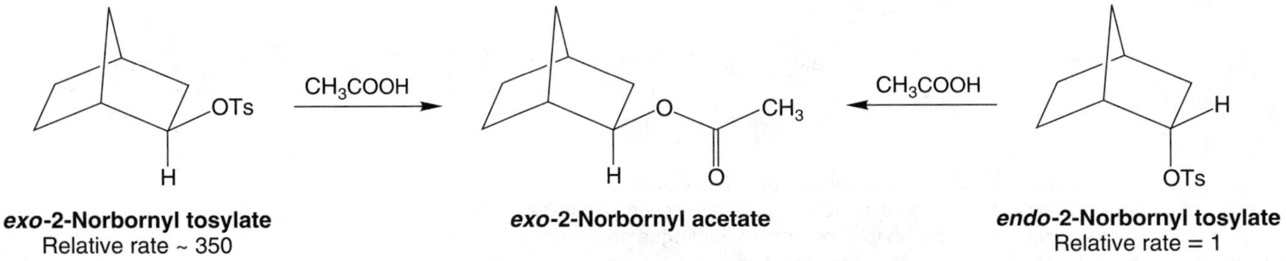

***exo*-2-Norbornyl tosylate**
Relative rate ~ 350

CH_3COOH

***exo*-2-Norbornyl acetate**

CH_3COOH

***endo*-2-Norbornyl tosylate**
Relative rate = 1

FIGURE 23.60 Both *endo*- and *exo*-2-norbornyl tosylate react in acetic acid to give *exo*-2-norbornyl acetate. The reaction of the exo isomer is faster.

The exo compound reacted at a faster rate than the endo compound, with the exact difference depending on reaction conditions. At first sight this would seem to provide evidence for bridging, but the opponents of participation quite rightly pointed out that the rate acceleration was very small (~ 350) and suggested that this effect could be the result of other differences between the exo and endo tosylates. This argument brings up a most important point. Whenever you are given a rate difference, you must ask yourself, Exactly what reaction does this rate difference refer to? In this case, the factor of 350 referred to the formation of acetate product. Is that what we are really interested in? Certainly not. When we speak of participation, we mean participation in *ionization,* ion formation, and that is not necessarily equivalent to product formation. In this case, the ion might be formed much faster than product. So we need a way to measure the rate acceleration in the ionization reaction itself, not the probably slower product-forming reaction (Fig. 23.61).

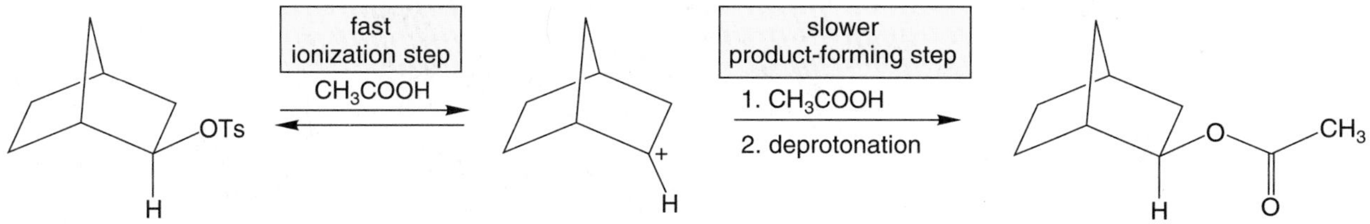

fast ionization step
CH_3COOH
slower product-forming step
1. CH_3COOH
2. deprotonation

FIGURE 23.61 If the rate differential is determined by observing formation of acetate, we may be missing a fast, reversible ionization step.

Perhaps an Energy versus Reaction progress diagram makes the point more clearly. If we measure the rate of appearance of acetate, we miss the faster ionization reaction (Fig. 23.62).

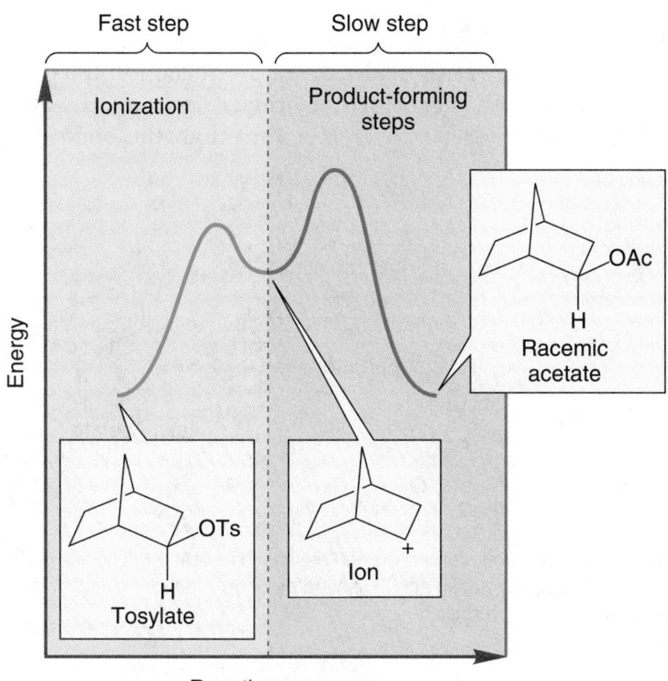

FIGURE 23.62 An Energy versus Reaction progress diagram for this reaction. A rapid ionization precedes a slower formation of acetate.

The stereochemistry of the reaction also points to participation. The product from the optically active exo tosylate is not that of a simple S_N2 reaction, which would be the inverted optically active endo acetate. Instead, it is the product of retention, the exo acetate. Moreover, optically active tosylate gives racemic acetate. We have been alerted to look for just this kind of stereochemical result as evidence of a neighboring group effect (Fig. 23.63)—it's clues 1 and 2.

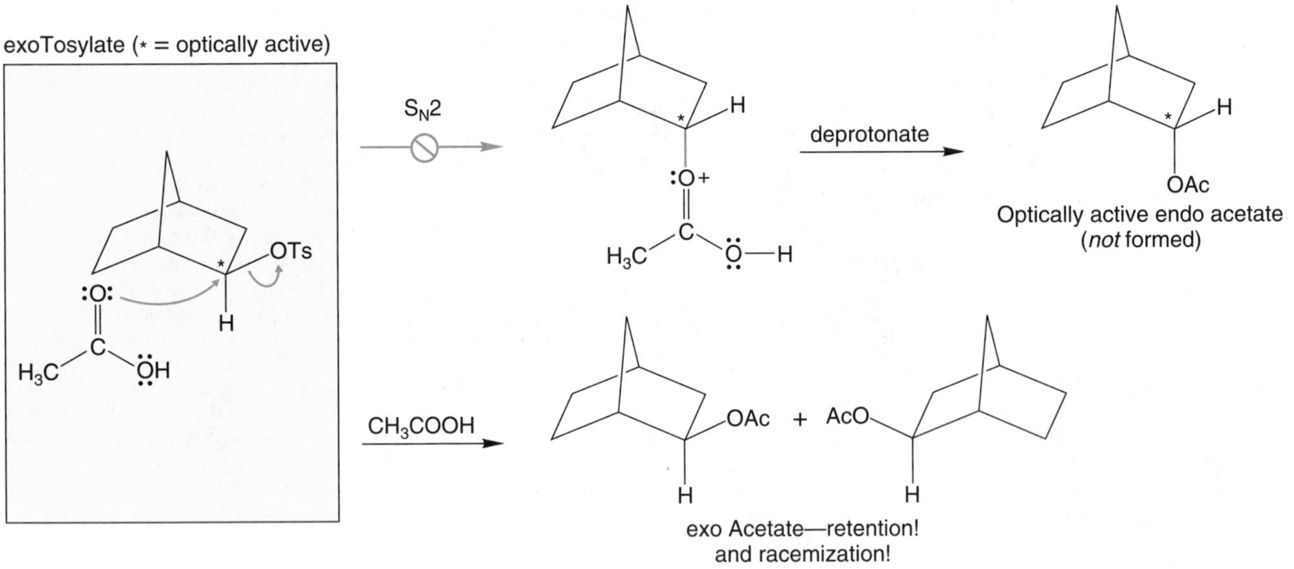

exoTosylate (* = optically active)

S_N2

deprotonate

Optically active endo acetate (*not* formed)

CH_3COOH

exo Acetate—retention! and racemization!

FIGURE 23.63 A simple mechanism involving S_N2 displacement by acetic acid predicts optically active endo acetate, which is not what happens. The product is racemic exo acetate.

Proponents of participation asserted that these pieces of stereochemical data constituted powerful evidence for a bridged ion. First of all, it is only the exo tosylate that can form a bridged ion directly, and the exo compound did indeed react at a greater rate than the endo isomer (Fig. 23.64).

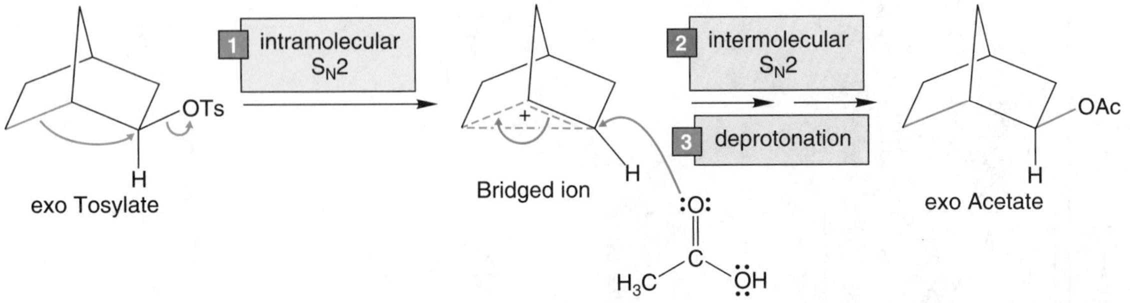

exo Tosylate

1 intramolecular S_N2

Bridged ion

2 intermolecular S_N2

3 deprotonation

exo Acetate

FIGURE 23.64 If the carbon–carbon σ bond assists in ionization, a cyclic, bridged ion can result. Opening of this ion by acetic acid, followed by deprotonation, would give the observed stereoisomer of the product.

PROBLEM 23.23 Draw resonance forms for the bridged ion of Figure 23.64.

It was also quite rightly pointed out that the bridged ion not only explained the observed retention of stereochemistry (exo tosylate gave exo acetate), but demanded a racemic product, as it itself was achiral. The bridged ion doesn't look symmetrical at first, but if you turn it over, it clearly is (Fig. 23.65).

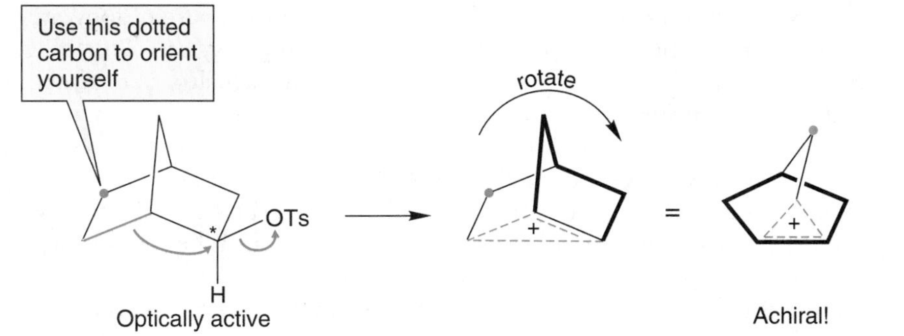

FIGURE 23.65 The observed formation of racemic product (as well as the racemization of recovered starting tosylate) can be explained by the achiral nature of the cyclic ion.

Addition of acetic acid must occur with equal probability at the two equivalent carbons of the bridged ion to give both enantiomers of *exo*-2-norbornyl acetate (Fig. 23.66).

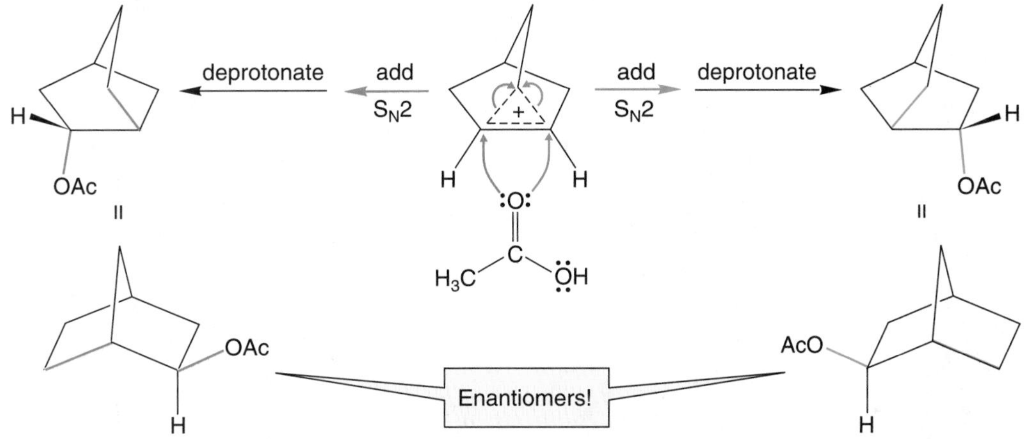

FIGURE 23.66 Addition of acetic acid to the bridged ion must give a pair of enantiomers in equal amounts—a racemic mixture.

This notion also gives a clue as to how to get a better measure of the rate of ionization than simply measuring the rate of acetate formation. *If the initial ionization is reversible,* then starting tosylate will be racemized as soon as the leaving group becomes symmetrically disposed with respect to the cation (Fig. 23.67).

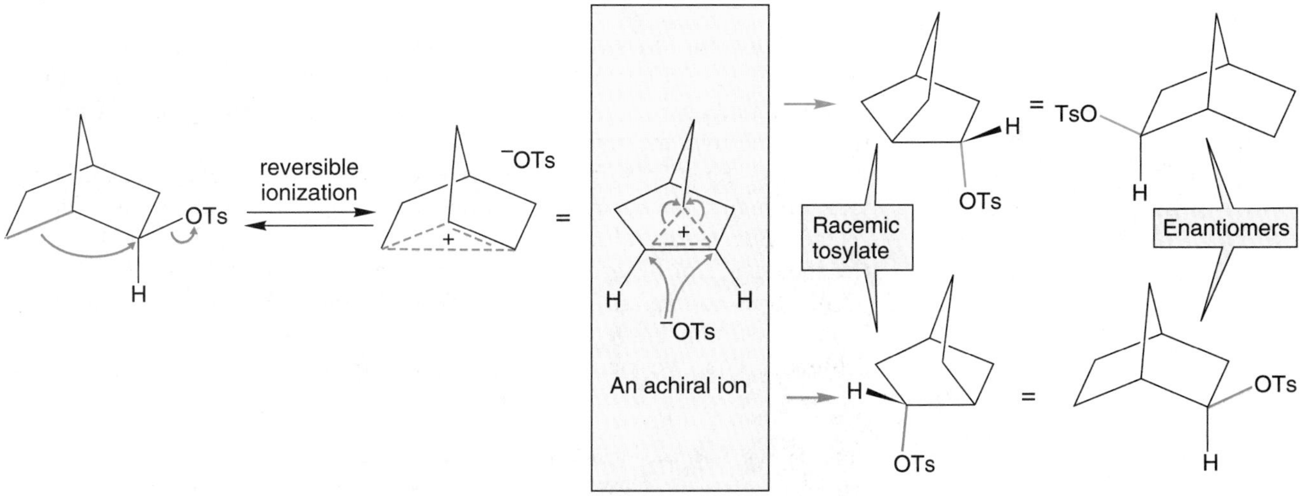

FIGURE 23.67 If the leaving group (tosylate ion) re-adds to the cyclic ion to re-form starting material, racemization results.

So, we should be able to get a better measure of the rate of ionization by determining the rate of racemization of starting tosylate. This clever experiment has now been done and reveals a rate acceleration of about 1500 for the exo tosylate over the endo, a more impressive rate acceleration than the originally reported factor of 350.

Now we have the two traditional kinds of evidence for neighboring group participation: rate acceleration and retention of stereochemistry. It looks like a strong case can be made for participation. What could the opponents of bridging say?

First of all, they pointed out that the rate acceleration was small (they preferred to focus on the original factor of 350, but even 1500 isn't very large), and they were right. Moreover, one could invoke assistance by the σ bond in a simple way, without forming a cyclic ion. That would give some rate acceleration in the exo molecule, as there can be no assistance in the endo isomer (Fig. 23.68).*

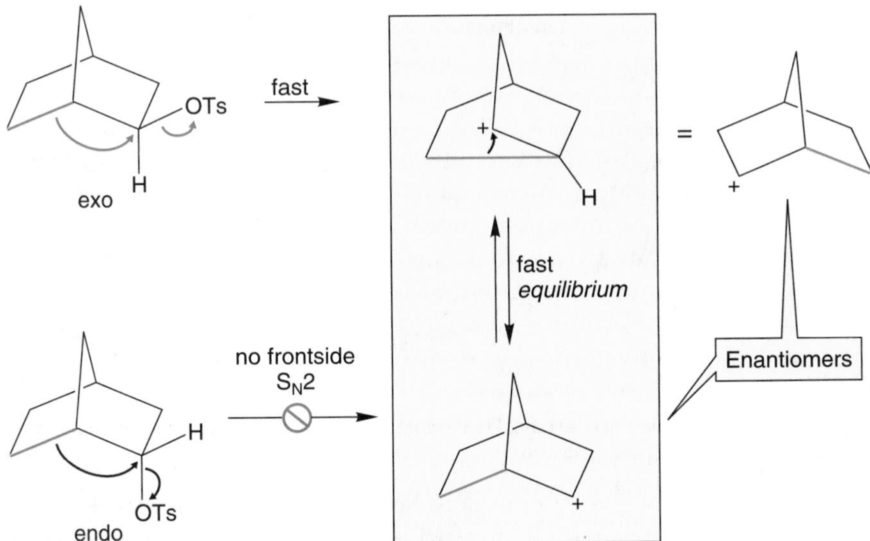

FIGURE 23.68 Formation of two, classical, rapidly equilibrating ions can explain both the fast rate of the exo isomer and the racemization.

But how would such a species explain the stereochemical results: racemization and retention of the exo configuration? Racemization was easy, one had only to permit the initially formed ion to equilibrate rapidly with the equivalent, mirror-image form (Fig. 23.68).

Always keep in mind the difference between the equilibrating open ions of Figure 23.68 and the bridged ion. The bridged ion is a single species. But be careful; the resonance forms of the bridged ion do look like the separate, equilibrating forms postulated by the classicists (Fig. 23.69).

But how to explain retention: the exclusive formation of exo acetate? That was much harder for the classicists, and to my taste, never quite achieved. The two faces of the molecule are different, however, and one

*The classicists actually argued most strongly that the rate of ionization of the endo compound was especially *slow,* and that the exo compound was more or less normal.

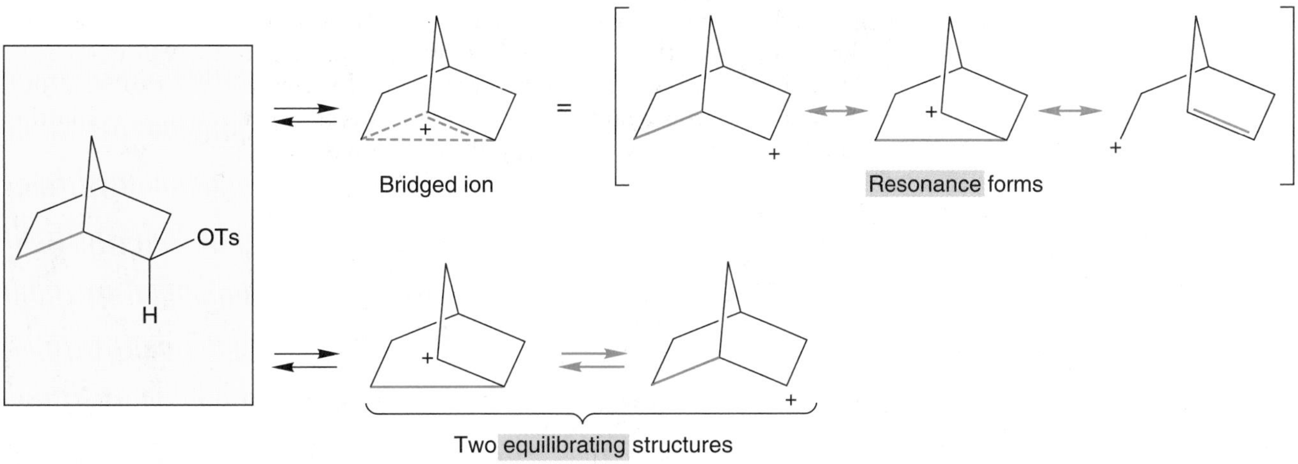

Bridged ion

Resonance forms

Two equilibrating structures

FIGURE 23.69 The two mechanistic proposals compared.

Explain the formation of *exo*-2-norbornyl acetate from the reaction of Figure 23.70.

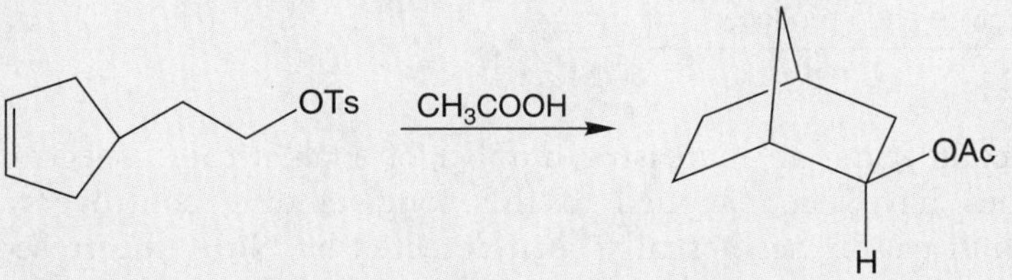

*PROBLEM 23.24

FIGURE 23.70

ANSWER

This reaction is just another way of forming the 2-norbornyl cation. Symmetrical displacement of the leaving group by the double bond of the five-membered ring gives the ion.

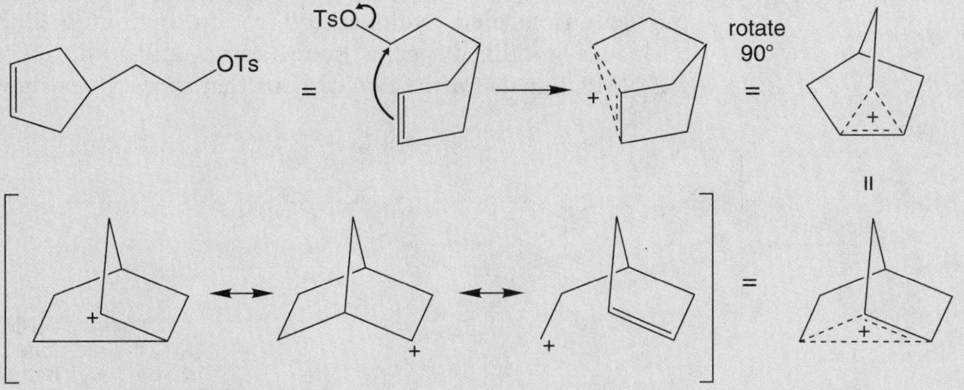

Opening by acetic acid in an intermolecular S_N2 reaction must produce the exo acetate.

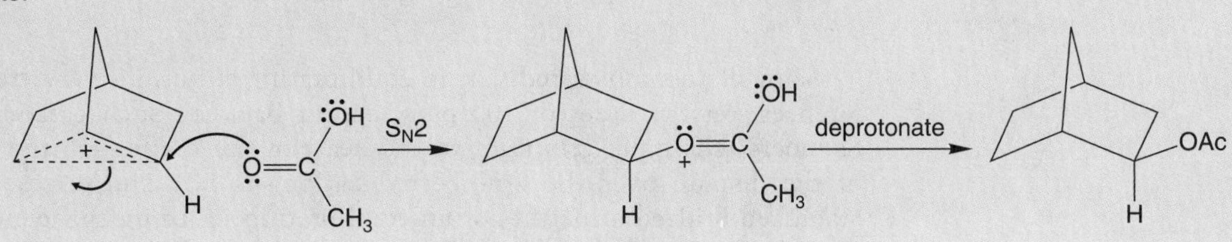

must admit that in principle, there might be sufficient difference between them to permit the exclusive formation of the exo acetate. Much chemistry of the norbornyl system reveals a real preference for exo reactions, and the question again became one of degree. How much exo addition could be tolerated by an open, unbridged ion (Fig. 23.71)?

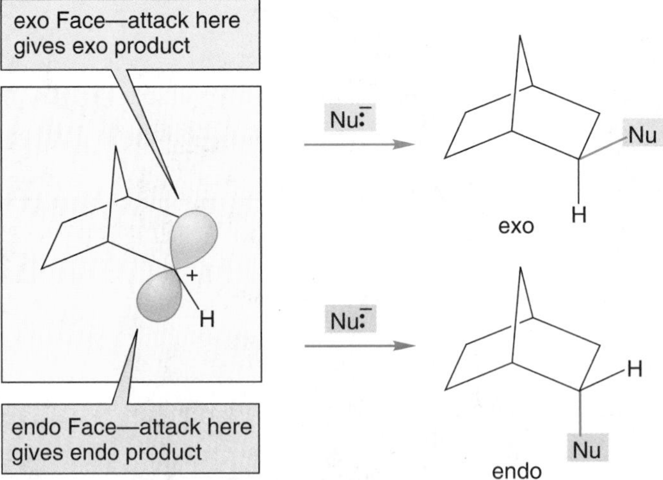

FIGURE 23.71 The exo and endo faces of the open ion are different. Addition of a nucleophile could, in principle, occur from only the exo side.

In recent years, an increasing number of sophisticated spectroscopic techniques have been applied to this long-standing controversy. The 2-norbornyl cation has actually been detected by NMR spectroscopy at very low temperature in a highly polar, but nonnucleophilic solvent. Another spectroscopic technique, pioneered by Martin Saunders of Yale University (b. 1931) focuses on the general differences in the ^{13}C NMR spectra between equilibrating structures and bridged molecules. For example, the deuteriolabeled cyclohexenyl cation in Figure 23.72, certainly a resonance-stabilized species, shows essentially no difference between the ^{13}C NMR signals for the two carbons that share the positive charge.

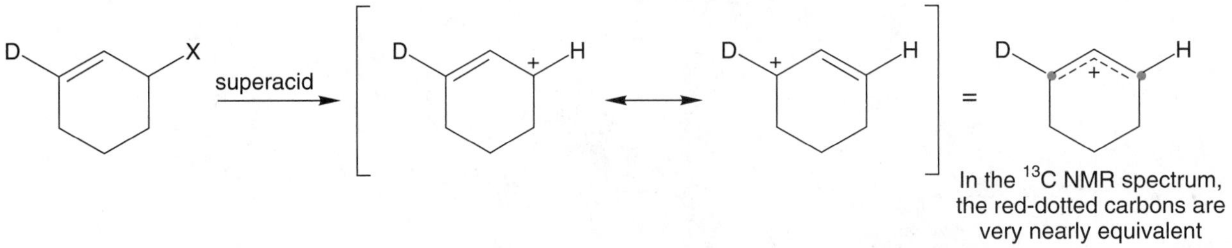

In the ^{13}C NMR spectrum, the red-dotted carbons are very nearly equivalent

FIGURE 23.72 In this resonance-stabilized carbocation, the two carbons sharing the positive charge are not strongly differentiated in the ^{13}C NMR spectrum by deuterium substitution.

Saunders demonstrated that in equilibrating structures very great differences, on the order of 100 ppm, appear between such carbons. The Saunders technique gets right to the heart of the matter—to the center of the dispute over the 2-norbornyl cation—is it a single, resonance-stabilized bridged structure, or an equilibrating set of independent ions? Use of Saunders' technique in this case is complicated by the presence of hydride shifts in the 2-norbornyl cation, but nonetheless the weight of the evidence seems to favor strongly the bridged structure (Fig. 23.73).

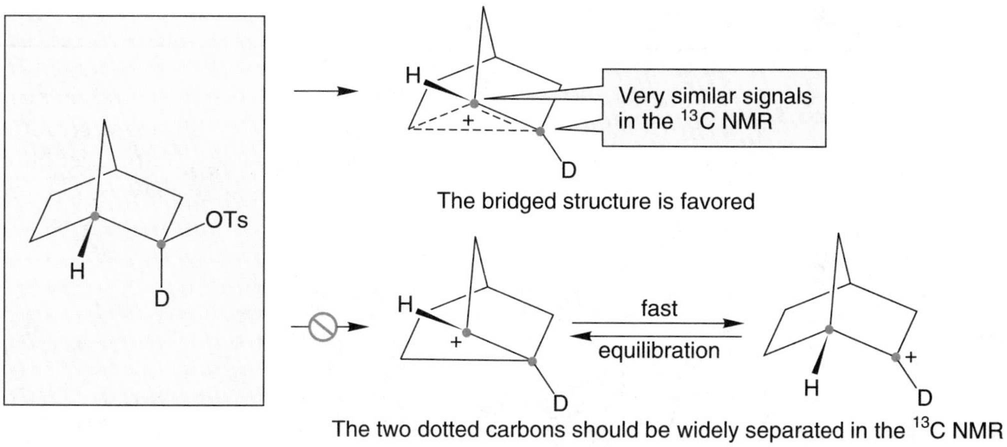

Very similar signals in the ^{13}C NMR

The bridged structure is favored

fast

equilibration

The two dotted carbons should be widely separated in the ^{13}C NMR

FIGURE 23.73 The ^{13}C NMR spectrum shows only a tiny difference between the deuterated and undeuterated positively charged carbons. The 2-norbornyl cation is bridged.

Are all bicyclic cations bridged? For a while it certainly seemed so. At least they were drawn that way during what might be called the rococo phase of this theory. But there is no reason why they should *all* be bridged. Bridging is one form of stabilization, nothing more, and if other forms of stabilization are available, ions will be open, classical species. For example, the classicists were right when they argued that certain tertiary carbocations were unbridged. Tertiary carbocations are more stable than secondary ones, and so there is no need for the special stabilization of three-center, two-electron bridging. The proponents of bridging should not have been upset, but they certainly were (Fig. 23.74).*

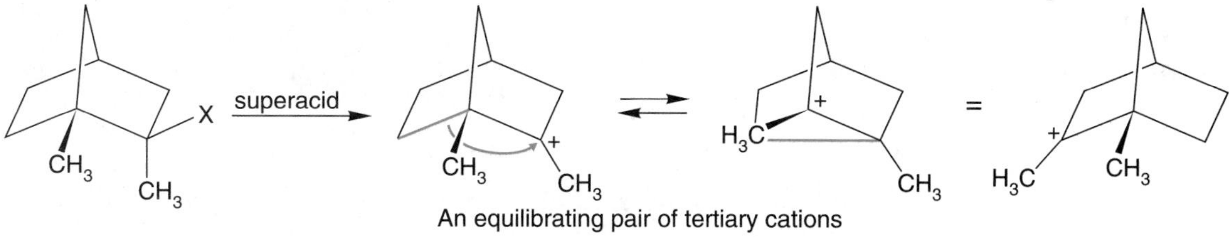

superacid

An equilibrating pair of tertiary cations

FIGURE 23.74 But not all ions are bridged. This related, but tertiary carbocation is not bridged.

*The argument over the nature of the 2-norbornyl cation and related molecules was/is remarkable not only for its longevity, but for its intensity as well. I think the argument became very nasty and both sides definitely lost their cool. Despite the common notion of scientists as Dispassionate Seekers After Truth, we are all-too-prone to the human vices, in this case overdefending our ideas (our intellectual children). In one corner were the proponents of bridging, chief among them Saul Winstein and Donald Cram. In the other was H. C. Brown, who passionately maintained that all could be explained through conventional, unbridged intermediates. This argument was no battle of lightweights! Brown and Cram have since won the Nobel prize (for other work), and Winstein should have been so honored—and would have been, had he lived long enough. The dispute was raucous and featured many a shouting match at meetings and seminars. It is not easy to be insulting in the body of a scientific paper, because the referees (two or more readers who judge whether the work is publishable) will not stand for it. However, it is much easier to be nasty in the footnotes! There are some really juicy footnotes in the literature on this subject. In my view, both sides went overboard. Although there is much virtue in avoiding new, unwarranted explanations, I think the classicists were much too resistant to change. There is nothing inherently crazy about a bridged ion! Three-center, two-electron bonding is a fine way to hold molecules together and is quite common in inorganic chemistry, and even parts of organic chemistry. On the other hand, the bridgers overdefended their ground. Just because some ions are bridged doesn't mean that all are!

23.4 COATES' CATION

Robert M. Coates (b. 1938) of the University of Illinois has found an ion for which there is unequivocal evidence for a bridged structure. The molecule cannot be represented by a rapidly equilibrating set of open ions. The reaction of Coates' molecule in acetic acid is shown in Figure 23.75.

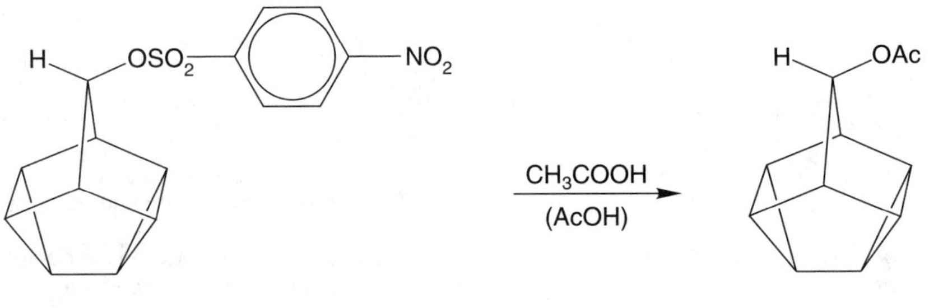

FIGURE 23.75 The solvolysis of this sulfonate in acetic acid gives the related acetate.

At low temperature in superacid, ionization with participation of the cyclopropane σ bond gives a bridged cation that must show three signals in its ^{13}C NMR spectrum (Fig. 23.76).

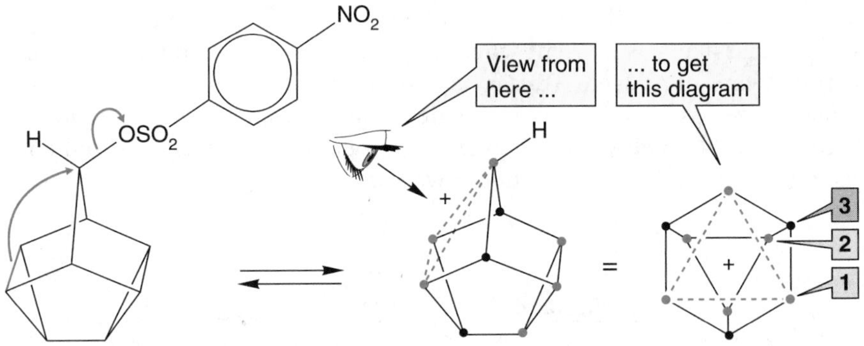

FIGURE 23.76 Participation of the carbon–carbon σ bond to give a symmetrical bridged ion. Notice that three different kinds of carbon exist. The ^{13}C NMR spectrum will consist of three lines.

In this molecule, an equilibrating set of ions must give quite a different ^{13}C NMR spectrum.

*PROBLEM 23.25 How many signals would a rapidly equilibrating "Coates' cation" give in the ^{13}C NMR spectrum (Fig. 23.77)?

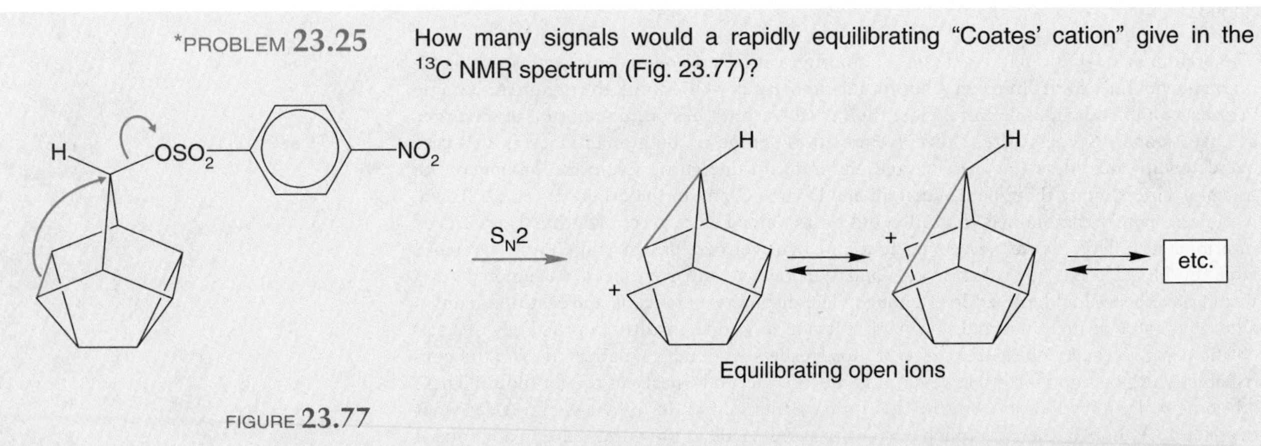

FIGURE 23.77

ANSWER This problem is tough. Let's take the simple Coates' cation and do one rearrangement. What's this ion? Label three carbons with dots, and the two three-

ANSWER (CONTINUED)

membered rings with boldface lines, and then redraw. It's the same cation! It's Coates' cation all over again. The rearrangement is degenerate.

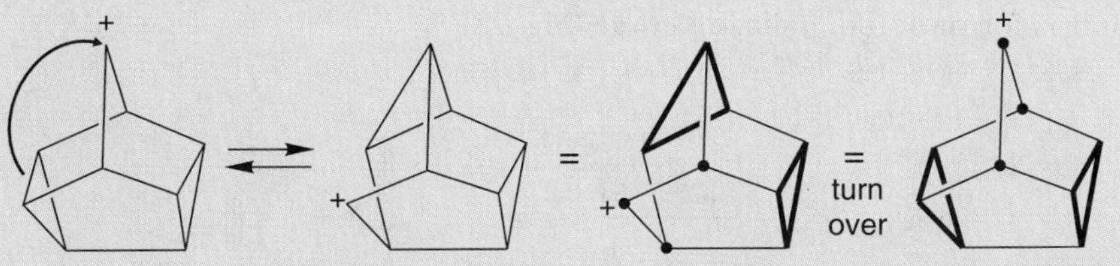

Coates' cation — One rearrangement — Labeled — = turn over — It's the same ion!

In this system, every rearrangement leads to the same ion. In fact, if you do enough rearrangements, all the carbons become equivalent. Start with a Coates' cation in which we have labeled the four equivalent carbons with dots. One rearrangement leads to an ion in which the "top" bridge and the cyclopropyl position have become involved. In fact, we have now shown that all carbons except the bridgehead are equivalent.

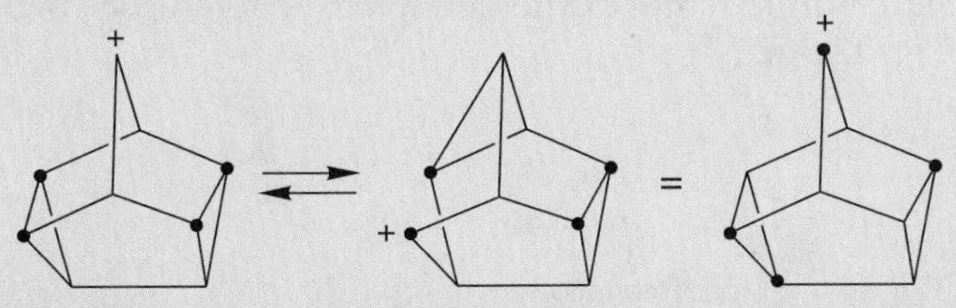

What happens to the bridgehead carbons in this rearrangement process? The labeled ion below answers this question. They equilibrate with the cyclopropyl carbons and thus are equivalent to all the others. All nine carbons are indeed the same, as long as rearrangement is rapid. The ^{13}C NMR spectrum must show only a single signal, not the three that are found experimentally.

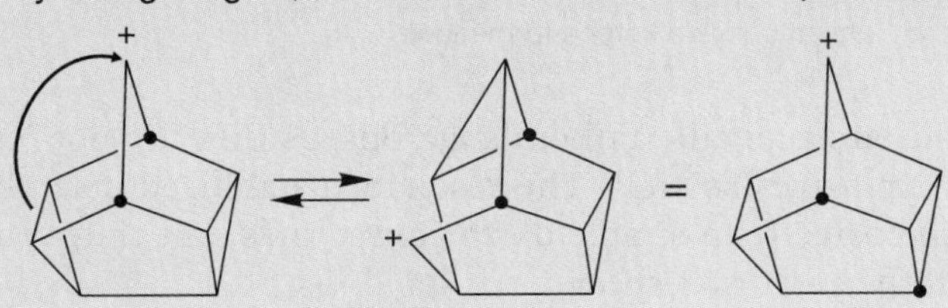

Happily, the sulfonate ionizes to give a stable ion at low temperature in superacid and the case is settled, as the spectrum consists of exactly the three signals required for a bridged structure: no more and no less. Coates' cation is bridged, there is no doubt about it.

23.5 SOMETHING MORE: A FAMILY OF CONCERTED REARRANGEMENTS OF ACYL COMPOUNDS

In this section, we take up a few rearrangement reactions of some new kinds of molecules related to the acyl compounds of Chapter 20. The mechanisms are not always fully worked out, but the reactions are of interest and are often synthetically useful. In each, there is a neighboring group, although it is not traditional to think of these reactions in this way. This section shows how pervasive the concept of assistance by a near-by group is, and introduces a bit of synthetic chemistry as well.

23.5a Diazo Ketones: The Wolff Rearrangement

When an acid chloride is allowed to react with a diazo compound, the result is formation of a **diazo ketone** (Fig. 23.78).

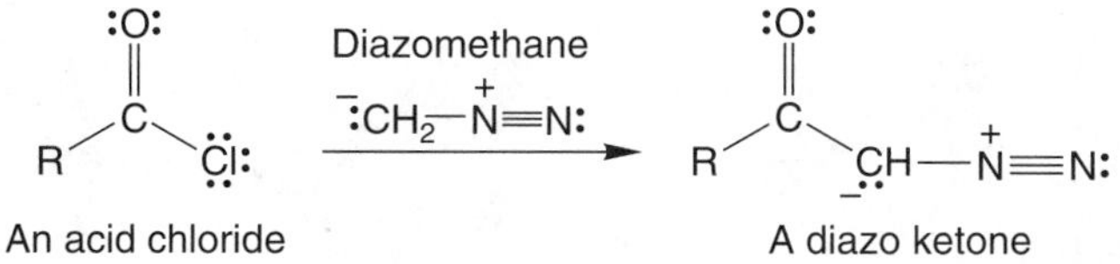

FIGURE 23.78 The reaction of an acid chloride with diazomethane to give a diazo ketone.

The mechanism of this reaction is only slightly more complicated than the usual reaction of nucleophiles with acid chlorides (Chapter 20, p. 1016). Diazo compounds are nucleophiles at carbon and add to the carbonyl group of the reactive acid chloride. Chloride ion is expelled in the elimination phase of this normal addition–elimination process. In the only new step of the reaction, chloride removes the newly acidic hydrogen to give the diazo ketone (Fig. 23.79).

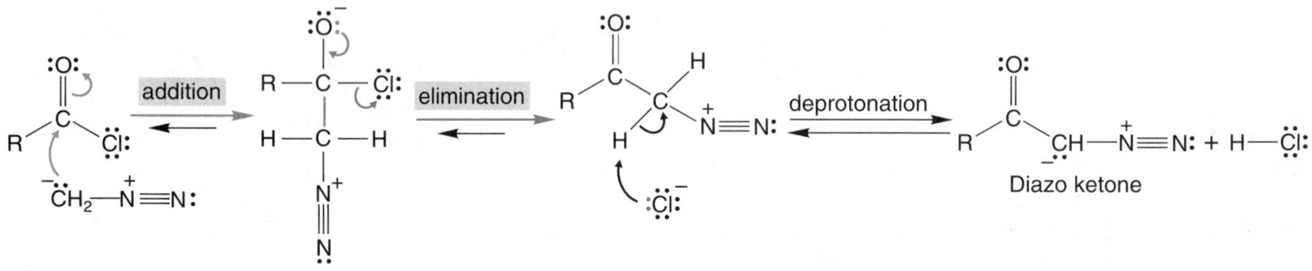

FIGURE 23.79 The mechanism of diazo ketone formation.

Diazo compounds are generally rather dangerous, as they are not only poisonous but are explosives as well. The resonance-stabilized diazo ketones are far less explosive than simple diazo compounds, but they must still be treated with the greatest respect.

PROBLEM 23.26 Draw resonance forms for a diazo ketone.

Like other diazo compounds, diazo ketones are sensitive to both heat and light (Chapter 10, p. 411). Nitrogen is lost and a carbene intermediate, in this case, a ketocarbene, can be formed (Fig. 23.80).

Ketocarbenes are capable of the usual carbene reactions, addition to multiple bonds and insertion into carbon–hydrogen bonds chief among them (Fig. 23.80). In addition, there is a possible intramolecular carbon–carbon insertion reaction, often silver catalyzed, and generally described as migration of the R group. The product of this reaction is a ketene. This reaction is known as the **Wolff rearrangement**, after Ludwig Wolff (1857–1919; remember the Wolff–Kishner reduction, Chapter 14, p. 642? Same Wolff; Fig. 23.81).

The mechanistic complication comes from the notion that it may not be the carbene, but the diazo compound itself that gives the ketene in the Wolff rearrangement. The ketocarbene is not necessarily involved (Fig.

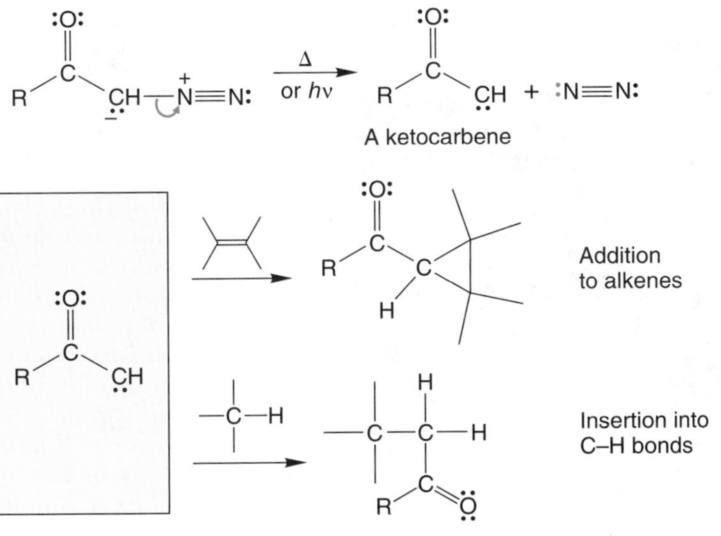

:O: ‖ R–C–$\ddot{C}$H–$\overset{+}{N}$≡N: —Δ or $h\nu$→ R–C(=O)–$\ddot{C}$H + :N≡N:

A ketocarbene

Addition to alkenes

Insertion into C–H bonds

FIGURE 23.80 Diazo ketones decompose to give ketocarbenes on heating or photolysis. The reactions of ketocarbenes include additions to π systems and carbon–hydrogen insertion.

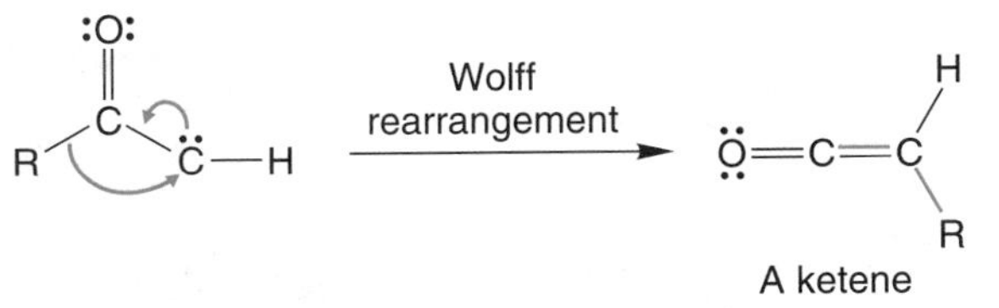

Wolff rearrangement

A ketene

FIGURE 23.81 Intramolecular ketene formation from a ketocarbene.

23.82). The question isn't completely sorted out yet, but the best evidence currently favors at least some direct formation of the ketene from most diazo ketones by a kind of neighboring group effect. This evidence doesn't exclude the reaction of the carbene to give the ketene as well.

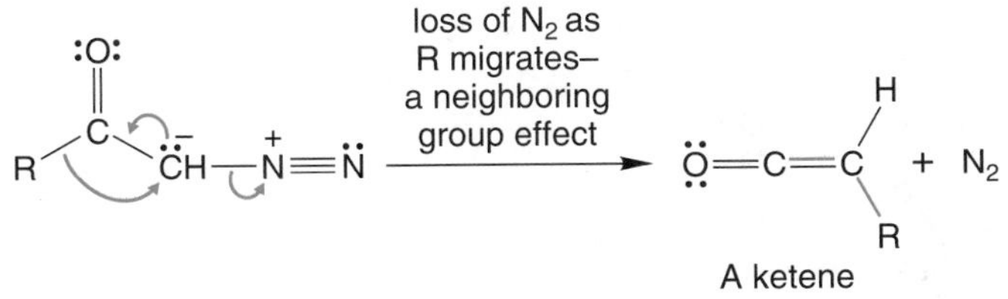

loss of N_2 as R migrates– a neighboring group effect

+ N_2

A ketene

FIGURE 23.82 In the Wolff rearrangement, the C–R bond can break as the nitrogen departs, which is a neighboring group effect.

*PROBLEM 23.27

Much experimentation indicates that only one of the two possible stereoisomers of the diazo ketone can give ketene directly. What are the two stereoisomers of the diazo ketone?

ANSWER

There are syn and anti isomers of diazo ketones.

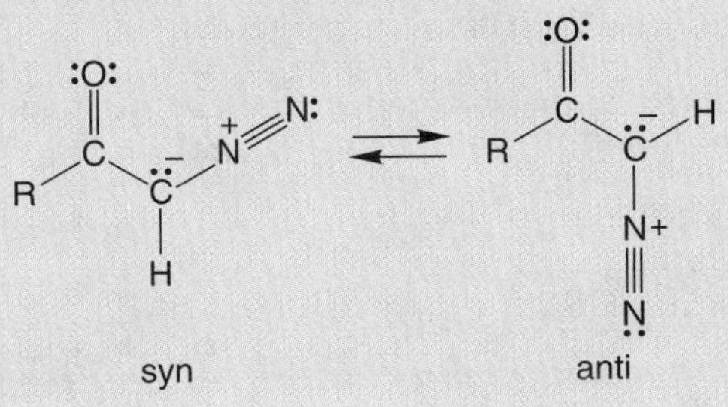

syn ⇌ anti

*PROBLEM 23.28 Explain why only one of these isomers can rearrange easily to a ketene.

ANSWER The syn isomer is set up for migration of R with displacement from the rear in what is really an intramolecular S_N2 reaction. By contrast, the anti isomer would have to do a frontside S_N2 for R to migrate as the nitrogen leaves.

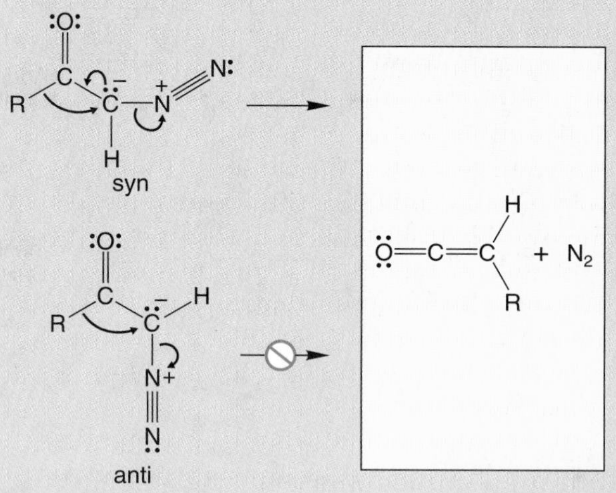

syn

anti

Frontside S_N2—No, No, No

If the mechanistic questions still are not completely resolved, the synthetic applications certainly are. The Wolff rearrangement generates ketenes in the absence of nucleophiles, and therefore they can be isolated in this reaction. If the product ketene is deliberately hydrolyzed to the acid, the overall sequence is called the **Arndt–Eistert reaction** after Fritz Arndt and Bernd Eistert. It constitutes a most useful chain-lengthening reaction of the original acid used to make the acid chloride (Fig. 23.83).

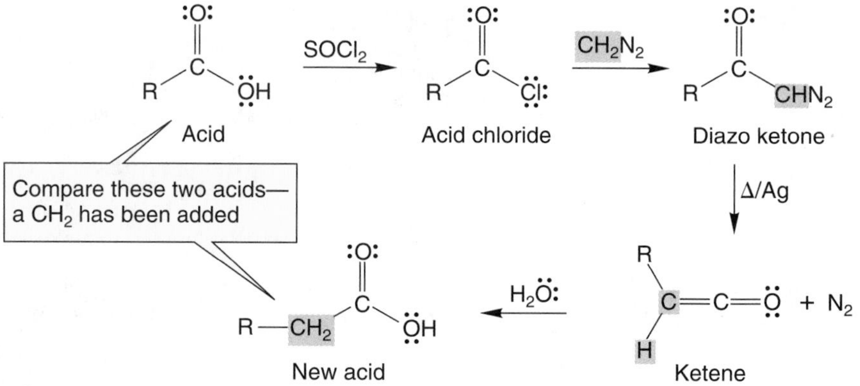

Acid $\xrightarrow{SOCl_2}$ Acid chloride $\xrightarrow{CH_2N_2}$ Diazo ketone $\xrightarrow{\Delta/Ag}$ Ketene $+ N_2$ $\xrightarrow{H_2O}$ New acid

Compare these two acids—a CH_2 has been added

FIGURE 23.83 The Arndt–Eistert synthesis adds a methylene group to a carboxylic acid.

23.5b Acyl Azides: The Curtius Rearrangement

Acyl azides are related to diazo ketones, and can be made by reaction of acid chlorides with the azide ion, $^-N_3$ (Fig. 23.84).

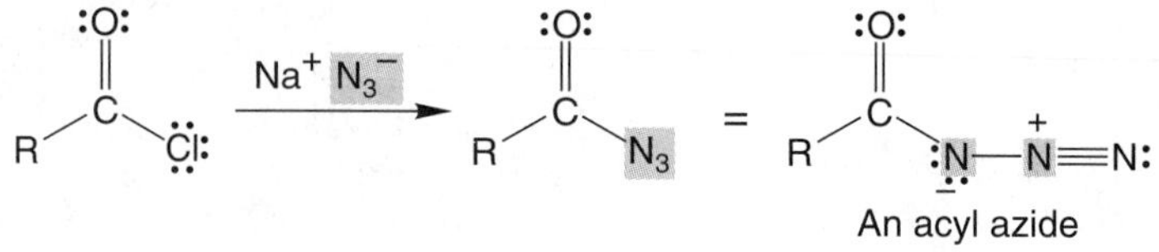

$RCOCl \xrightarrow{Na^+ \; N_3^-} RCON_3$ = An acyl azide

FIGURE 23.84 Acid chlorides react with azide ion ($^-N_3$) to give acyl azides.

Like diazoketones, acyl azides lose nitrogen when heated or irradiated. The end result is an **isocyanate**, formed either by intramolecular rearrangement of a **nitrene** (the nitrogen analogue of a carbene) or directly from the acyl azide in a reaction called the **Curtius rearrangement** after Theodore Curtius (1857–1928) (Fig. 23.85).

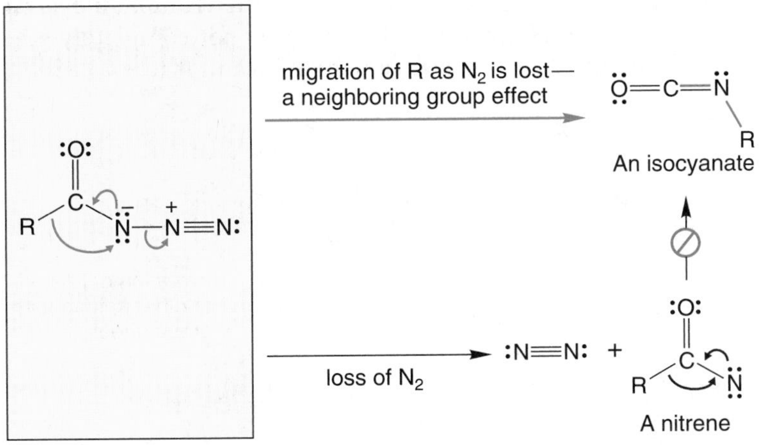

migration of R as N_2 is lost—
a neighboring group effect

An isocyanate

loss of N_2

A nitrene

FIGURE 23.85 The Curtius rearrangement leads to isocyanates. The mechanism probably involves a concerted rearrangement, a neighboring group effect, not a nitrene.

For simple acyl azides, it is known that the isocyanate is formed by direct rearrangement of the acyl compound (a neighboring group effect) and that the nitrene, though also formed from the azide, is not involved in isocyanate production.

PROBLEM 23.29

Can you guess what nitrene chemistry will be like? What products do you anticipate from the reaction of a nitrene with the carbon–carbon double bond? What product would you get from reaction with a carbon–hydrogen bond? Remember that a nitrene is the nitrogen analogue of a carbene (Chapter 10, p. 411).

Isocyanates, like ketenes, are very sensitive to nucleophiles. For example, alcohols add to isocyanates to give carbamate esters (Chapter 19, p. 984; Fig. 23.86).

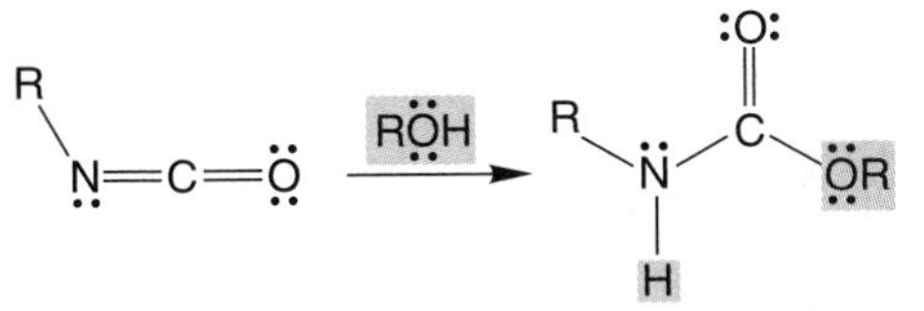

ROH

A carbamate ester

FIGURE 23.86 Isocyanates react with alcohols to give carbamate esters.

PROBLEM 23.30

Write a mechanism for the addition of an alcohol to an isocyanate to give a carbamate ester (Fig. 23.86).

PROBLEM 23.31

Explain the formation of the amine in the following reaction (Fig. 23.87):

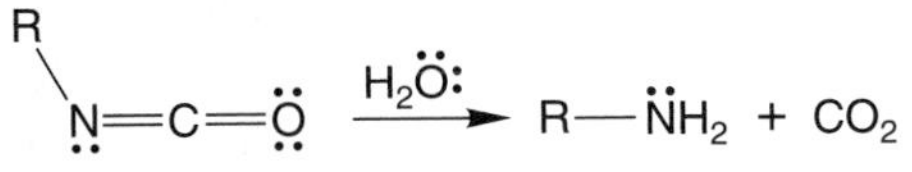

$$R{-}N{=}C{=}O \xrightarrow{H_2O} R{-}NH_2 + CO_2$$

FIGURE 23.87

23.5c Hofmann Rearrangement of Amides

Amides are able to take part in a reaction similar to the Wolff and Curtius rearrangements. This process occurs in two steps. The amide is first brominated to give the *N*-bromoamide (Fig. 23.88). Base then removes the very acidic α-hydrogen and rearrangement to the isocyanate ensues. The question of whether a nitrene is involved in this **Hofmann rearrangement*** (August W. von Hofmann, 1818–1892)† is not completely settled, but I would bet on the direct rearrangement shown in Figure 23.88.

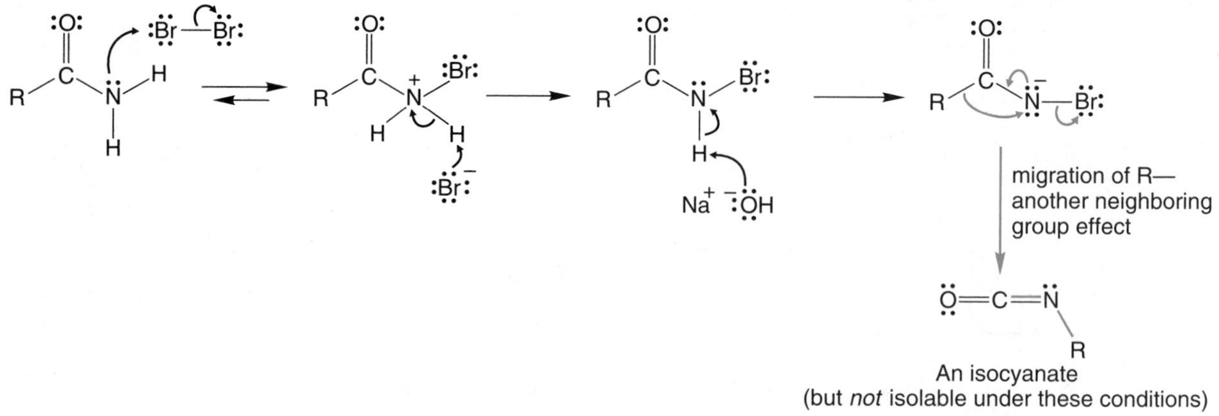

migration of R— another neighboring group effect

An isocyanate (but *not* isolable under these conditions)

FIGURE 23.88 The first stages in the Hofmann rearrangement. An isocyanate is an intermediate, but cannot be isolated.

The isocyanate is not isolable, as it is in the Curtius rearrangement. In this reaction, it is born *in the presence of base* and is hydrolyzed to the carbamic acid, which is unstable and decarboxylates (recall Problem 23.31). The end product is the amine (Fig. 23.89).

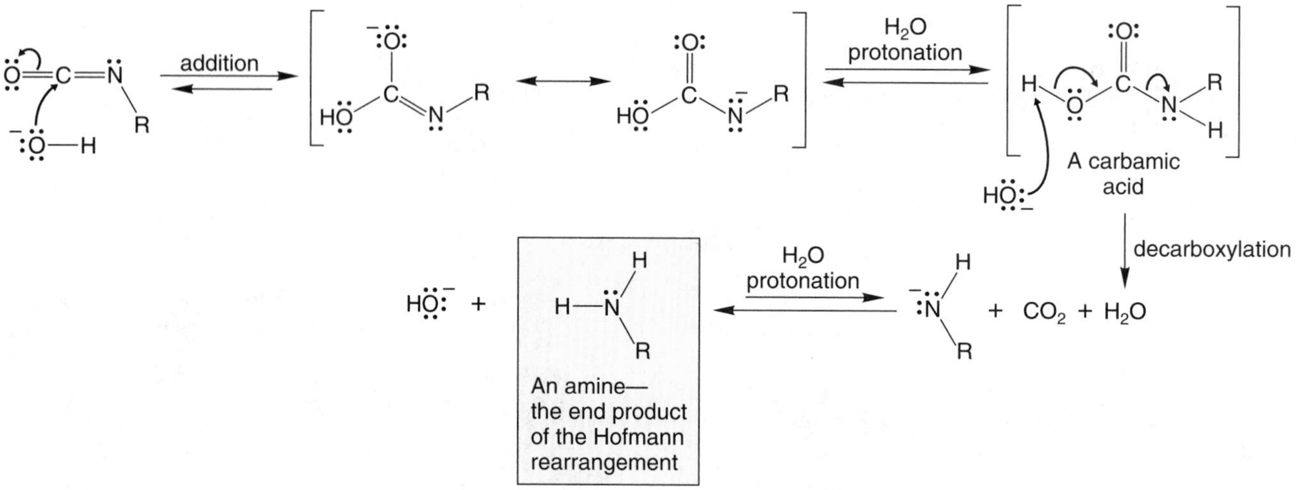

addition

H_2O protonation

A carbamic acid

decarboxylation

H_2O protonation

$+ CO_2 + H_2O$

An amine— the end product of the Hofmann rearrangement

FIGURE 23.89 Under the reaction conditions the isocyanate reacts with hydroxide or water and decarboxylates to give the amine.

*In earlier, more colorful times, this reaction was known as the "Hofmann degradation."

†Do you remember the Hofmann elimination (Chapter 7, p. 284)? It's the same Hofmann.

Remember the essential difference between the Curtius rearrangment and the Hofmann rearrangement. Both involve the formation of isocyanates, but only in the Curtius rearrangement is this intermediate isolable. In the Hofmann rearrangement, base is present and the isocyanate cannot survive. This synthesis of amines is not easy to remember because it involves many steps (Fig. 23.90), thus making it a great favorite of problem writers.

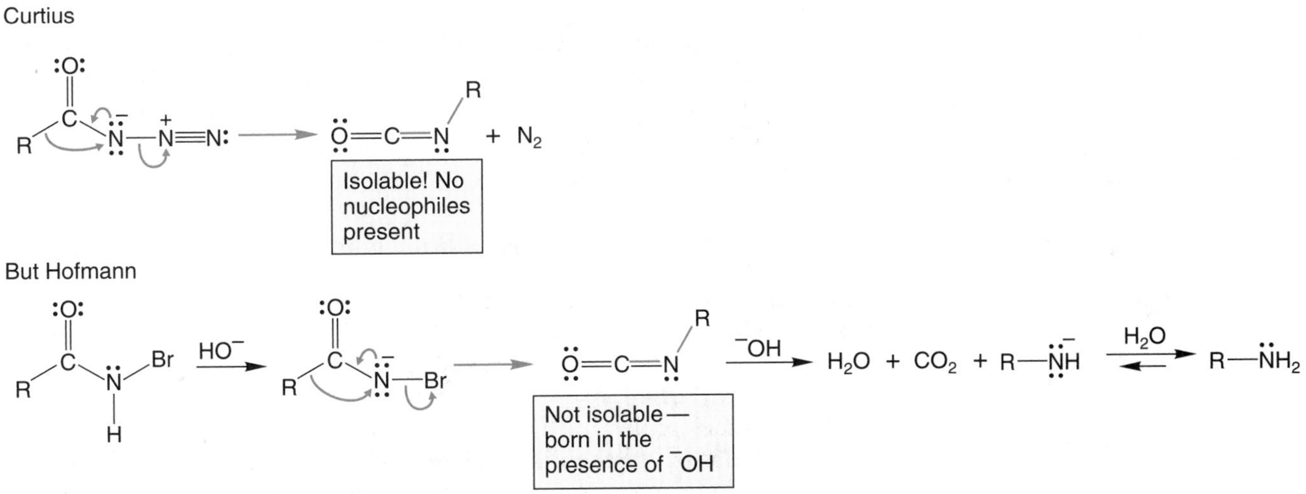

Curtius

Isolable! No nucleophiles present

But Hofmann

Not isolable— born in the presence of ^-OH

FIGURE **23.90** The difference between the Curtius and Hofmann rearrangements. The isocyanate can be isolated only in the Curtius rearrangement, not in the Hofmann.

This section has a set of reactions united by formation of an intermediate containing two cumulated (adjacent) double bonds. Their common ancestor is allene (Chapter 12, p. 500). Are these molecules mere curiosities, chemical oddities of no real importance outside a textbook? Not at all. In 1992, broccoli was shown to contain a sulfur-containing isocyanate, an isothiocyanate, that induces formation of something called a "phase II detoxication enzyme," which is involved in the metabolism of carcinogens (Fig. 23.91). Eat your broccoli and study your orgo.

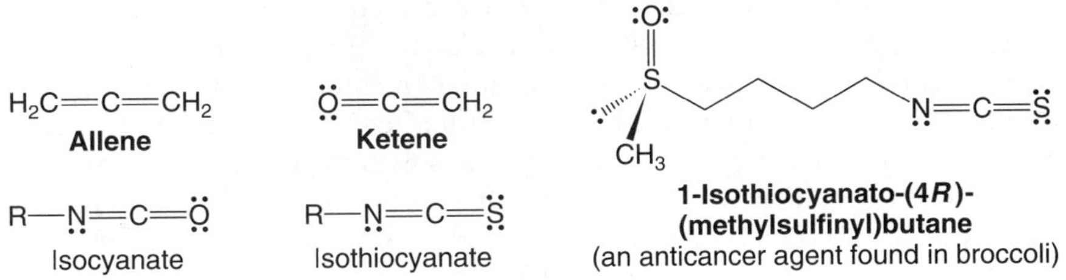

$H_2C{=}C{=}CH_2$ **Allene**

$O{=}C{=}CH_2$ **Ketene**

$R{-}N{=}C{=}O$ Isocyanate

$R{-}N{=}C{=}S$ Isothiocyanate

1-Isothiocyanato-(4*R*)-(methylsulfinyl)butane (an anticancer agent found in broccoli)

FIGURE **23.91** Eat your broccoli!

23.6 SUMMARY

NEW CONCEPTS

One concept dominates this chapter. The chemistry of many organic molecules containing leaving groups is strongly influenced by the presence of an internal nucleophile, a neighboring group. Internal nucleophiles can increase the rate of ionization of such compounds by providing anchimeric assistance in the ionization step. Moreover, the structures of the products derived from the ion produced through intramolecular displacement can be subtly different from those expected of simple intermolecular displacement. The three clues to the operation of a neighboring group effect are (1) an unusual stereochemical result, generally retention of configuration where inversion might have been expected; (2) a rearrangement, often of a label, which is usually the result of formation of a cyclic ion through *intra*molecular displacement of the leaving group; and (3) an unexpectedly fast rate.

The cyclic ions formed by neighboring group displacement are of two kinds. There are normal (often called "classical") species in which all bonds are two-electron bonds, and more complex nonclassical or bridged species in which three-center, two-electron bonding is the rule. Figures 23.39 and 23.40 illustrate the difference.

REACTIONS, MECHANISMS, AND TOOLS

This chapter features intramolecular displacement using a variety of internal nucleophiles. Some, such as the heteroatoms bearing lone pairs of electrons, are easy to understand, and really just recapitulate intermolecular chemistry (Fig. 23.2).

Other nucleophiles, such as the electrons in π or σ bonds, are less obvious extensions of old chemistry. Here, the nucleophiles involved are very weak and there often is not a clear analogy to chemistry you already know. One case in which there is a good connection is formation of phenonium ions through the action of a benzene ring as a neighboring group. The analogy is to the intermolecular reaction of benzenes with electrophiles in the aromatic substitution reaction (Figs. 23.31 and 23.32).

Displacement by carbon–carbon π or σ bonds can lead to bridged ions about which there has been much controversy. It now appears that such intermediates are involved in some ionizations, and that the three-center, two-electron bonding required is an energetically favorable situation, at least for highly electron-deficient systems such as carbocations (Fig. 23.69).

This chapter also introduces a family of intramolecular rearrangements that produces ketenes and isocyanates: molecules with two adjacent double bonds. Further reactions of these compounds depend on the reaction conditions. The Wolff and Curtius rearrangements produce ketenes and isocyanates under reaction conditions to which they are stable—the absence of nucleophiles. Other reactions, such as the Hofmann rearrangement, require nucleophilic solvents. The products are sensitive to nucleophilic attack and cannot survive. They are hydrolyzed to other molecules (Fig. 23.90).

SYNTHESES

In this Chapter, we meet old reactions (S_N1 and S_N2) in intramolecular settings. New synthetically useful reactions appear in the family of intramolecular rearrangements discussed in Section 23.5 (Fig. 23.92).

COMMON ERRORS

The problem with this material is mostly one of recognition. Once it is clear that it is a neighboring group problem, essentially every difficulty is

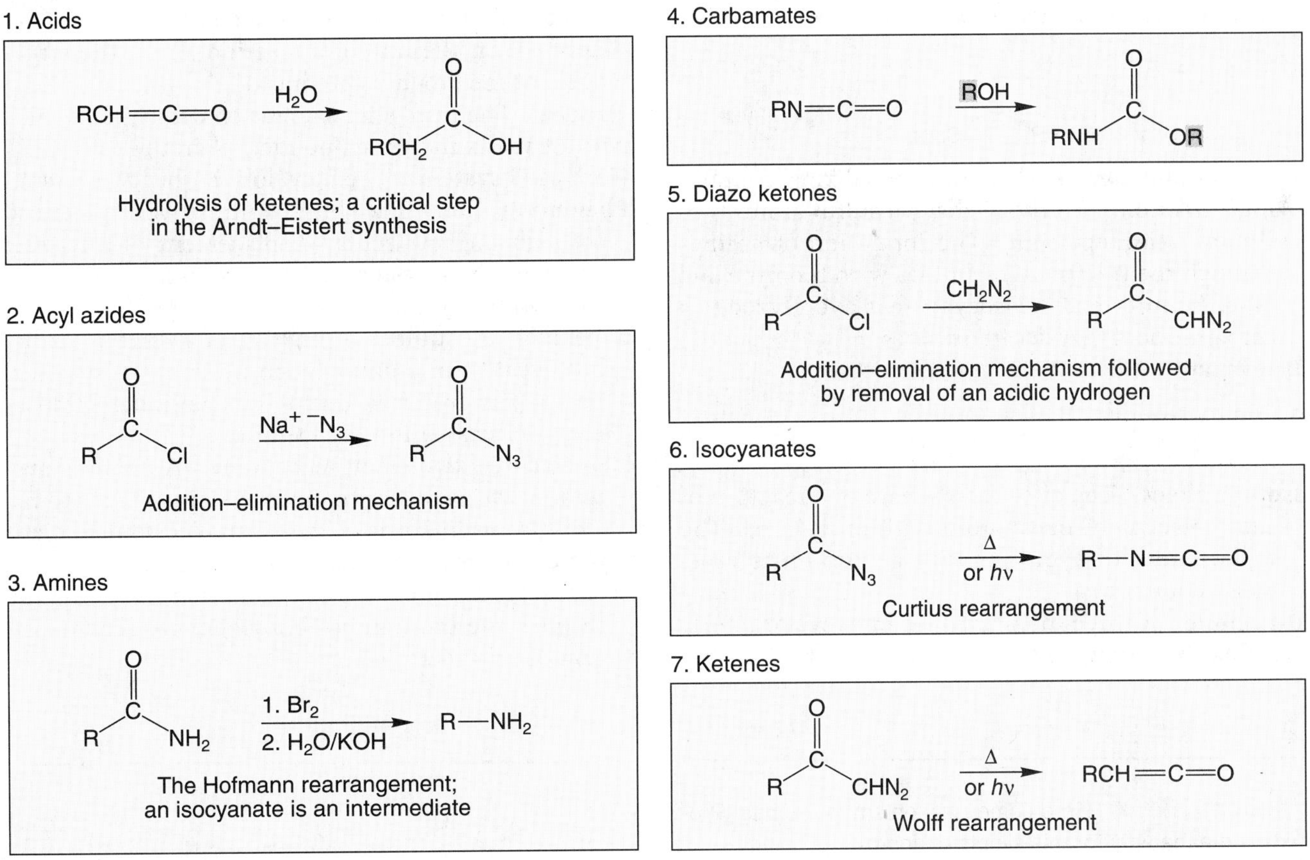

1. Acids

$RCH{=}C{=}O \xrightarrow{H_2O} RCH_2C(=O)OH$

Hydrolysis of ketenes; a critical step in the Arndt–Eistert synthesis

2. Acyl azides

$RC(=O)Cl \xrightarrow{Na^{+}\ {}^{-}N_3} RC(=O)N_3$

Addition–elimination mechanism

3. Amines

$RC(=O)NH_2 \xrightarrow[2.\ H_2O/KOH]{1.\ Br_2} R{-}NH_2$

The Hofmann rearrangement; an isocyanate is an intermediate

4. Carbamates

$RN{=}C{=}O \xrightarrow{ROH} RNHC(=O)OR$

5. Diazo ketones

$RC(=O)Cl \xrightarrow{CH_2N_2} RC(=O)CHN_2$

Addition–elimination mechanism followed by removal of an acidic hydrogen

6. Isocyanates

$RC(=O)N_3 \xrightarrow[\text{or } h\nu]{\Delta} R{-}N{=}C{=}O$

Curtius rearrangement

7. Ketenes

$RC(=O)CHN_2 \xrightarrow[\text{or } h\nu]{\Delta} RCH{=}C{=}O$

Wolff rearrangement

FIGURE 23.92 The synthetic reactions of Chapter 23.

resolved by a simple pair of S_N2 reactions: the first intramolecular, the second intermolecular. Search for the internal nucleophile, and use it to displace the leaving group.

There can be no denying that many of the structures encountered in neighboring group problems are complex and sometimes hard to untangle. A useful technique is to draw the result of the arrow formalism without moving any atoms, then to relax the structure, probably containing long or even bent bonds, to a more realistic picture. Trying to do both steps at once is dangerous.

A minor problem appears in recognizing that in some bridged ions, the web of dashes representing three-center, two-electron bonding can operate as a leaving group. It is easier to see a pair of electrons in a normal carbon-leaving group σ bond (C–L) than it is to recognize the dashes representing partial bonds as a leaving group.

23.7 KEY TERMS

Anchimeric assistance The increase in rate of a reaction that proceeds through intramolecular displacement over that expected of an intermolecular displacement.

Arndt–Eistert reaction The use of the Wolff rearrangement to elongate the chain of a carboxylic acid by one carbon.

Curtius rearrangement The thermal or photochemical decomposition of an acyl azide to give an isocyanate.

Diazo ketone A compound of the structure

$RC(=O)CHN_2$

Episulfonium ion A three-membered ring containing a trivalent, positively charged sulfur atom.

Hofmann rearrangement The formation of amines through the treatment of amides with bromine and base. An intermediate isocyanate is hydrolyzed to a carbamic acid that decarboxylates.

Isocyanate A compound of the structure

$R{-}N{=}C{=}O$

Isotope effect The effect on the rate of a reaction of the replacement of an atom with an isotope. The replacement of hydrogen with deuterium is especially common.

α-Lactone A three-membered lactone (cyclic ester).

Neighboring group effect A general term for the influence of an internal nucleophile (very broadly defined) on the rate of a reaction, or on the structures of the products produced.

Nitrene The nitrogen analogue of a carbene, R–N.

Nor A prefix meaning "no methyl groups."

Norbornyl system The bicyclo[2.2.1]heptyl system.

Phenonium ion The benzenonium ion produced through intramolecular displacement of a leaving group by the π system (not the σ system) of an aromatic ring.

Superacid A number of highly polar, highly acidic, but weakly or nonnucleophilic solvent systems. Superacids are often useful in stabilizing carbocations, at least at low temperature.

Three-center, two-electron bonding A bonding system in which only two electrons bind three atoms. This arrangement is common in electron-deficient molecules such as boranes and carbocations.

Wolff rearrangement The formation of a ketene through the thermal or photochemical decomposition of a diazo ketone.

23.8 ADDITIONAL PROBLEMS

PROBLEM 23.32 Here comes a problem for rabbits. Provide a mechanism for this simple change.

$BrCH_2CH_2CH_2CH_2NH_2$ —(1. H_2O, $(CH_3)_2CHOH$; 2. KOH/H_2O (neutralize))→ pyrrolidine (N–H)

PROBLEM 23.33 Similarly, furan is the product of this next reaction. Provide a mechanism and explain why this reaction is much faster than the similar reaction of 3-chloro-1-propanol, a molecule that gives the corresponding diol as product.

$ClCH_2CH_2CH_2CH_2OH$ —(H_2O)→ tetrahydrofuran (O)

$ClCH_2CH_2CH_2OH$ —(H_2O)→ $HOCH_2CH_2CH_2OH$

PROBLEM 23.34 Now for something a little more interesting. Explain why the regiochemical results of the following two reactions are so different.

Et_2N–CH₂–CH(CH₃)–Cl —(KOH/H_2O)→ Et_2N–CH(CH₃)–CH₂–OH

Bz_2N–CH₂–CH(CH₂CH₃)–Cl —(H_2O)→ Bz_2N–CH₂–CH(CH₂CH₃)–OH

Et = CH_3CH_2
Bz = $PhCH_2$

PROBLEM 23.35 Explain why compound **1** reacts much faster than **2** (140,000 times) and **3** (190,000 times) in acetic acid. Does this information about rates let you decide whether the reaction proceeds through a bridged ion?

OBs OBs

1 **2**

OBs

3

Bs = SO_2–C_6H_4–Br

PROBLEM 23.36 Compound **1** reacts 10^5–10^6 times faster than its cis isomer **2**. The product is **3**. Explain.

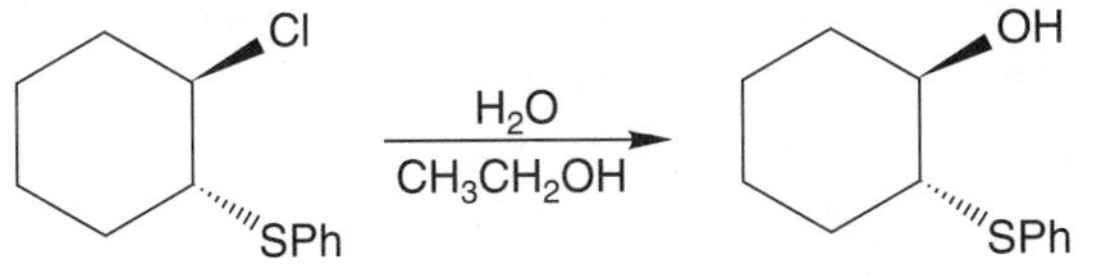

Cl, SPh, **1**; H_2O / CH_3CH_2OH; OH, SPh, **3**; Cl, SPh, **2**

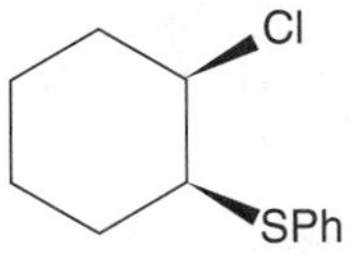

PROBLEM 23.37 When Z = OCH_3, about 93% of the products of solvolysis of **1** can be attributed to neighboring phenyl group participation. When Z = NO_2, essentially none of the products come from neighboring participation by phenyl. Explain.

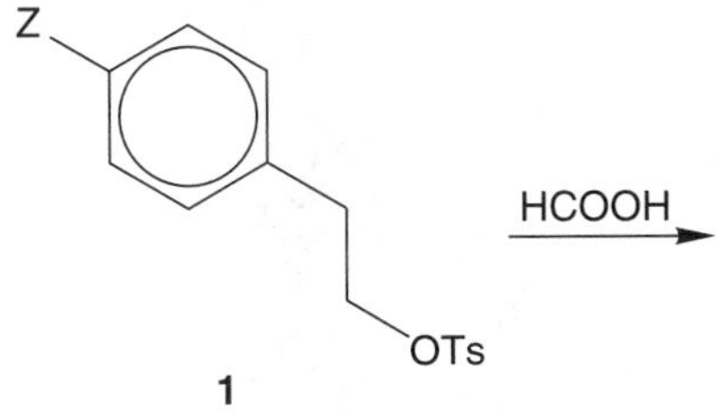

Z, OTs, **1**; HCOOH

PROBLEM 23.38 In Figure 23.36 we neglected to tell you about the product of reaction of *syn*-7-norbornenyl tosylate (**1**) with acetic acid. The product is compound **2**. Provide an arrow formalism mechanism and an explanation for the rate acceleration of 10^4 relative to **3**.

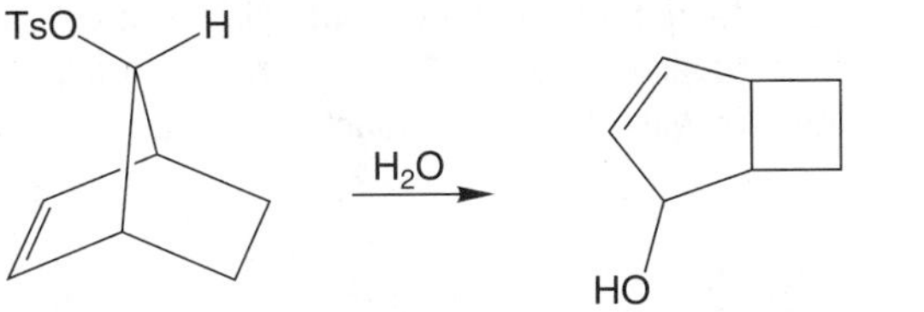

TsO, H, **1**, Rate = 10^4; H_2O; HO, **2** (both stereoisomers)

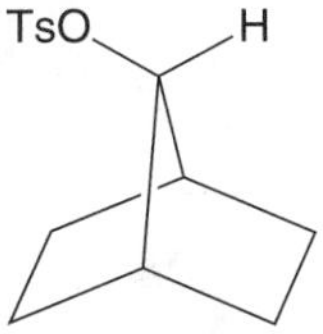

TsO, H, **3**, Rate = 1

PROBLEM 23.39 Provide a mechanism for the following reaction of **1**. Be sure to explain the observed stereochemistry and the fact that the corresponding trans compound (**2**) does not react in this way.

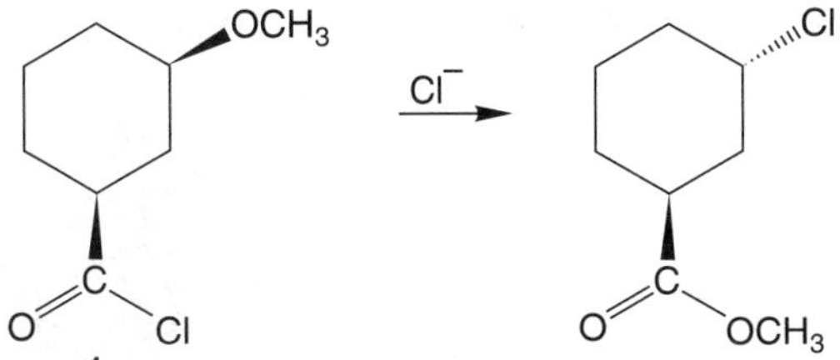

OCH_3, C, O, Cl, **1**; Cl^-; Cl, C, O, OCH_3

OCH_3, C, O, Cl, **2**

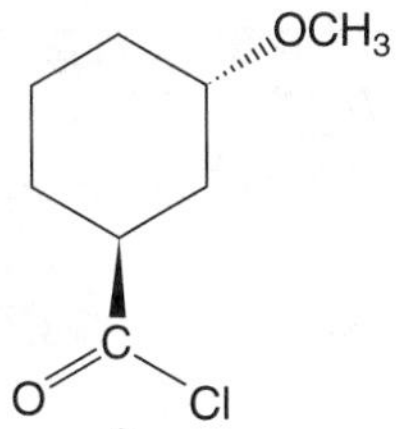

PROBLEM 23.40 Reaction of **1** with bromine leads to a neutral product of the formula $C_{18}H_{17}O_2Br$. Propose a structure for the product and a mechanism for its formation.

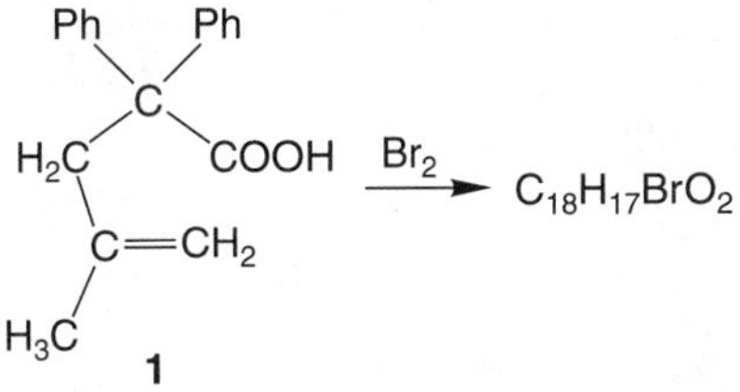

Ph, Ph, C, H_2C, COOH, C=CH_2, H_3C, **1**; Br_2; $C_{18}H_{17}BrO_2$

PROBLEM 23.41 Provide an arrow formalism for the following reaction:

HO—, —$CH_2CH_2CH_2CH_2OBs$; $(CH_3)_3COK$ / $(CH_3)_3COH$; O

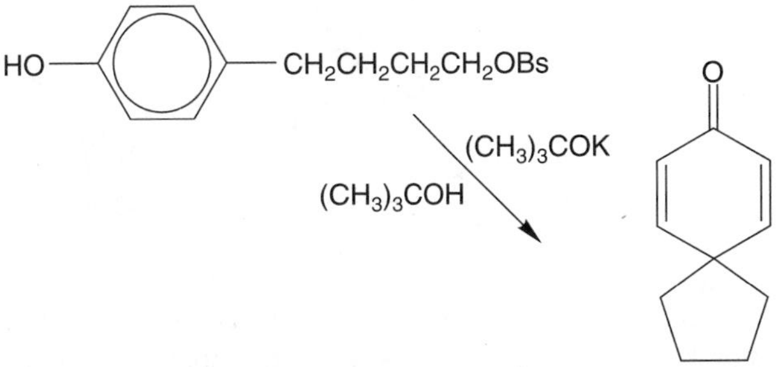

PROBLEM 23.42 Propose a mechanism for the following change. Be sure to explain the observed stereochemistry.

O; H_3O^+ / H_2O; H, OH, OH

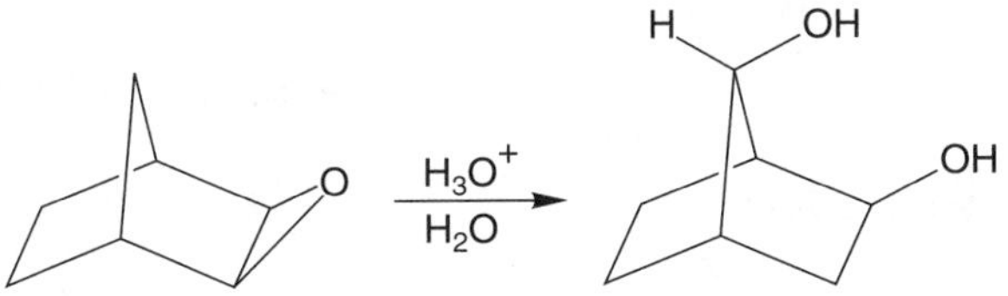

PROBLEM 23.43 What do the relative rate data suggest about the nature of the carbocationic intermediates in the following solvolysis reactions? *Hint*: See Figures 23.53–23.55.

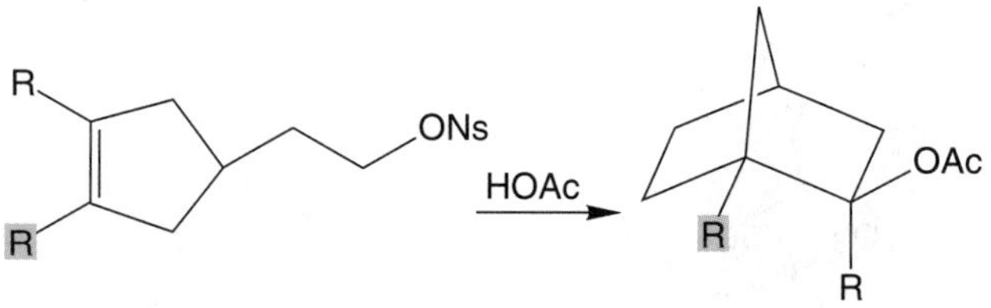

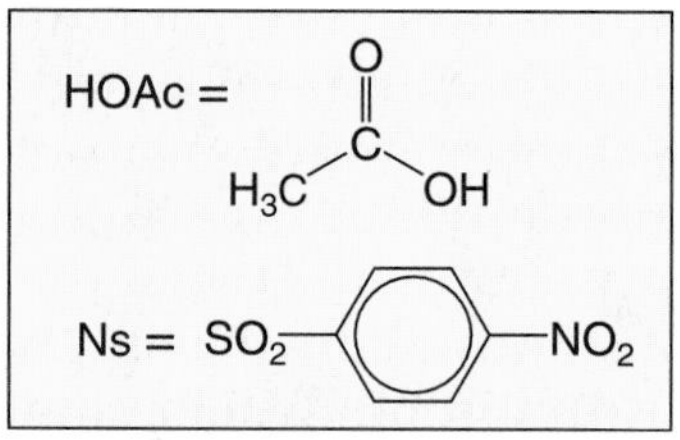

ONs, HOAc, OAc, R, R, R, R

HOAc = $H_3C-C(=O)-OH$

Ns = $SO_2-C_6H_4-NO_2$

R	R	Relative Rate
H	H	1.0
CH_3	H	7.0
CH_3	CH_3	38.5

PROBLEM 23.44 Provide arrow formalism mechanisms for the following reactions:

(a)

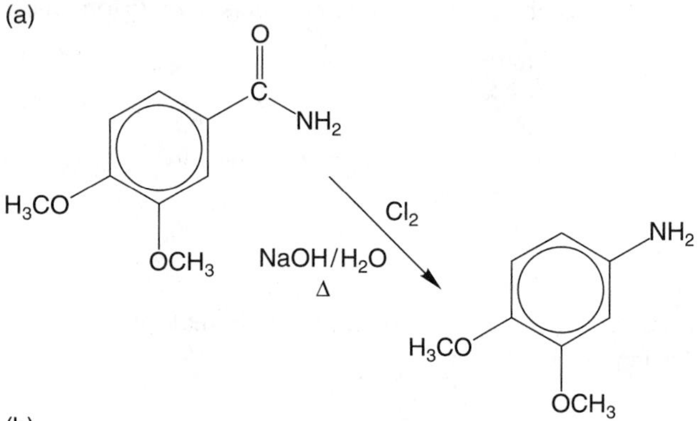

NH_2, H_3CO, OCH_3, Cl_2, $NaOH/H_2O$, Δ, NH_2, H_3CO, OCH_3

(b)

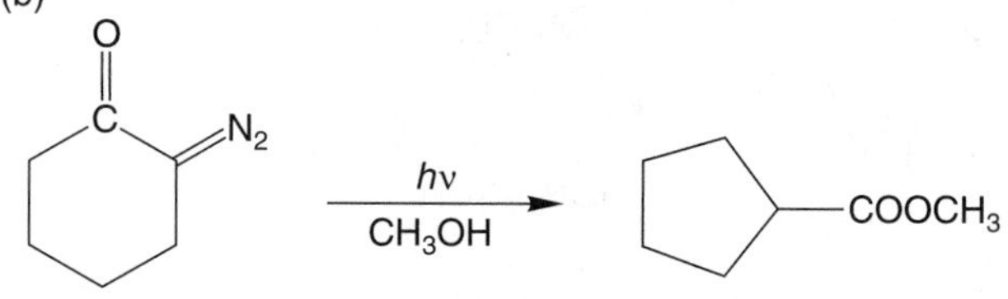

N_2, $h\nu$, CH_3OH, $COOCH_3$

PROBLEM 23.45 Compound **1** reacts about 15 times faster than neopentyl brosylate (**2**). The product is isobutyraldehyde. Provide a mechanism and explain the rate difference.

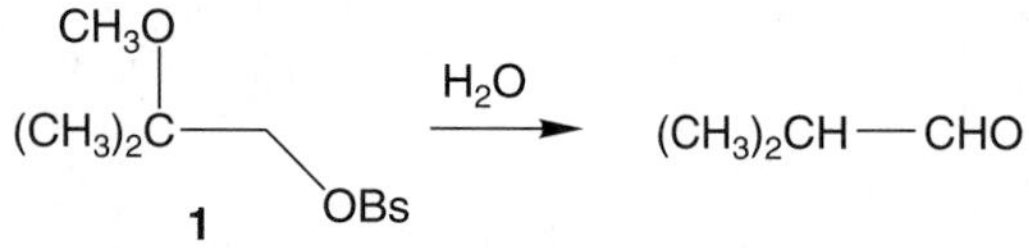

CH_3O, $(CH_3)_2C$, OBs, **1**, Rate = 15, H_2O, $(CH_3)_2CH-CHO$

$(CH_3)_3C$, OBs, **2**, Rate = 1

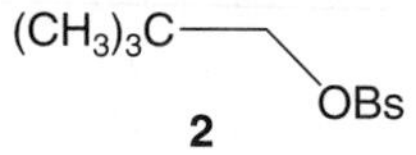

PROBLEM 23.46 Compound **1** is the only product of the reaction of compound **2** under the conditions shown. Provide an arrow formalism mechanism, and deduce the stereochemistry of **2** from the stereochemistry of **1** shown.

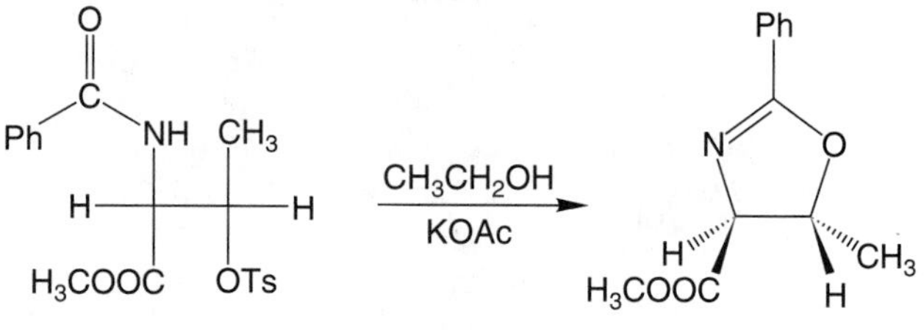

Ph, NH, CH_3, H, H, H_3COOC, OTs, **2**, CH_3CH_2OH, KOAc, Ph, N, O, H, CH_3, H_3COOC, H, **1**

PROBLEM 23.47 Heating β-hydroxyethylamide (**1**) with thionyl chloride affords β-chloroethylamide (**2**). In turn, amide **2** yields oxazoline **3** upon treatment with NaOH.

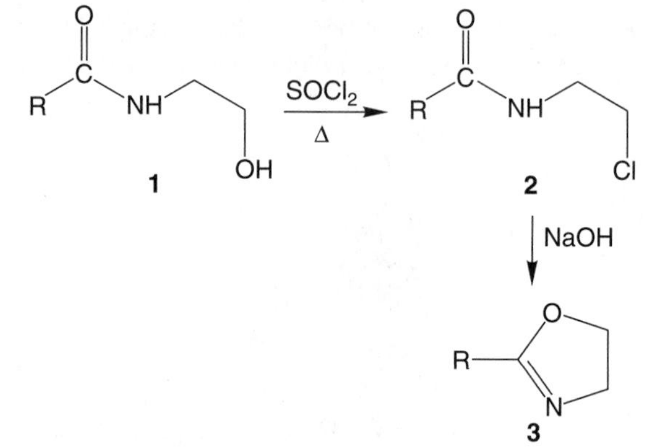

R, NH, OH, **1**, $SOCl_2$, Δ, R, NH, Cl, **2**, NaOH, R, O, N, **3**

(a) Write arrow formalism mechanisms for the formations of **2** and **3**. Things are not quite as simple as they first appear. For example, if **1** is treated with thionyl chloride below 25 °C, compound **4** is obtained instead of **2**. Compound **4** has the same elemental composition as **2**, but unlike **2**, is soluble in water. Upon heating, **4** gives **2**, and compound **4** yields **3** upon treatment with aqueous sodium carbonate.

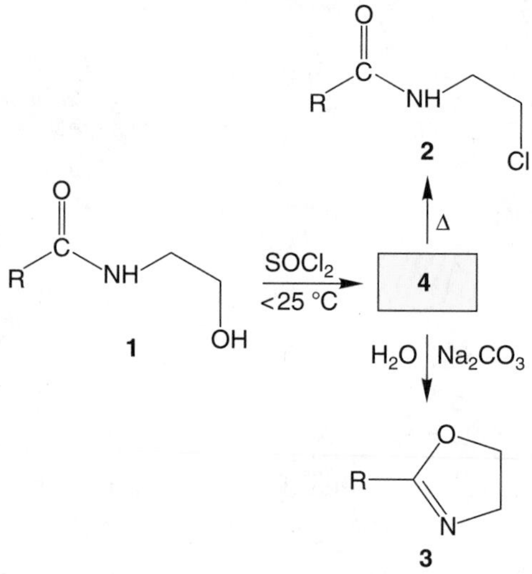

R, NH, Cl, **2**, Δ, R, NH, OH, **1**, $SOCl_2$, <25 °C, **4**, H_2O, Na_2CO_3, R, O, N, **3**

(b) Propose a structure for **4** and show how **4** is formed and converted into **2** and **3**.

PROBLEM 23.48 Outline the mechanistic steps for the following transformation:

1. NaOH/H_2O
2. Br_2
3. HCl (neutralize)

Now explain this more complicated reaction:

1. NaOH/H_2O
2. Br_2
3. HCl (neutralize)

PROBLEM 23.49 The 50 MHz ^{13}C NMR spectrum of the 2-norbornyl cation in $SbF_5/SO_2ClF/SO_2F_2$ solution exhibits an interesting temperature dependence. At −159 °C the spectrum consists of five lines at δ 20.4 [C(5)], 21.2 [C(6)], 36.3 [C(3) and C(7)], 37.7 [C(4)], and 124.5 [C(1) and C(2)]. However, at −80 °C the spectrum consists of three lines at δ 30.8 [C(3), C(5), and C(7)], 37.7 [C(4)], and 91.7 [C(1), C(2), and C(6)] ppm. Explain. *Hint*: Start with the low-temperature spectrum.

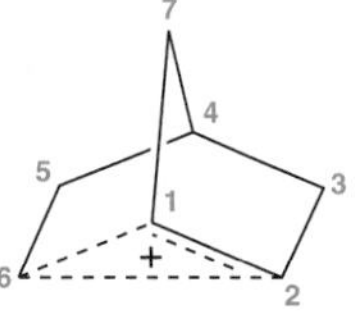

PROBLEM 23.50 γ-Aminobutyrate transaminase (GABA-T) is the key enzyme controlling the metabolism of γ-aminobutyric acid (GABA), a mammalian inhibitory neurotransmitter. In the following synthesis of the hydrochloride salt of the putative GABA-T inhibitor, 3-amino-1,4-cyclohexadiene-1-carboxylic acid (**G**), propose structures for intermediates **A–F**. Mechanisms are not necessary, but may be helpful in some cases.

COOH → (Cl–C(=O)–OEt, EtN(*i*-Pr)$_2$, acetone) → **A** ($C_8H_{10}O_4$)

A → (NaN$_3$, H_2O) → **B** ($C_5H_5N_3O$)

B → (1. toluene, Δ; 2. $(CH_3)_3COH$) → **C** ($C_9H_{15}NO_2$)

C → (O_2N–CH=CH–COOCH$_3$, benzene, 25 °C) → **D** ($C_{13}H_{20}N_2O_6$)

D → (base, THF) → **E** ($C_{13}H_{19}NO_4$)

E → (1. NaOH/H_2O, 55 °C; 2. H_2O/H_3O^+) → **F** ($C_{12}H_{17}NO_4$)

F → (HCl/H_2O, 55 °C) → **G** (COOH, Cl^-, $H_3\overset{+}{N}$)

PROBLEM 23.51 Treatment of phthalimide (**1**) with bromine in aqueous sodium hydroxide, followed by acidification with acetic acid, affords compound **2**. Spectral data for **2** are summarized below. Deduce the structure of **2**, and then propose a mechanism for its formation.

Compound **2**
Mass spectrum: m/z = 137 (p, 59%), 119 (100%), 93 (79%), 92 (59%)
IR (KBr): 3490 (m), 3380 (m), 3300–2400 (br), 1665 (s), 1245 (s), and 765 (m) cm^{-1}
^1H NMR (DMSO-d_6):* δ 6.52 (t, J = 8 Hz, 1H), 6.77 (d, J = 8 Hz, 1H), 7.23 (t, J = 8 Hz, 1H), 7.72 (d, J = 8 Hz, 1H), 8.60 (br s, 3H, washes out with D_2O)
^{13}C NMR (DMSO-d_6): δ 109.5 (s), 114.5 (d), 116.2 (d), 131.1 (d), 133.6 (d), 151.4 (s), 169.5 (s)

*Note that J_{meta} and J_{para} were not observed under these conditions

Polyfunctional Natural Products: Carbohydrates

There exists no better thing in the world than beer to blur class and social differences and to make men equal.

—Emil Fischer*

In the last chapter, we encountered neighboring group effects, the influence of one functional group upon the chemistry of another. In this chapter, we find molecules that bristle with functionality, and for which intramolecular neighboring group effects play most important roles, influencing both structure and reactivity. These are the **sugars**, also called **carbohydrates** or **saccharides**.

Sugars serve as the storage depots for the energy harvested from the sun through photosynthesis by green plants. Moreover, sugars are present in nucleic acids (Chapter 26) and are thus vital in controlling protein synthesis and in transmission of genetic information. More than 100 million tons of sugar is produced yearly, and each American consumes about 64 lb of sugar per year (down from over 100 lb just a few years ago). Sugars are vital to human survival, and it is somewhat surprising that their study took so long to emerge as a glamorous area of research. Perhaps the development of sugar chemistry was delayed by the simple technical difficulties of working with goopy syrups rather than nice, well-behaved solids. A full scale renaissance is underway now, however, and sugar chemistry is very much a hot topic.†

The name carbohydrate comes from the formulas of these compounds, which can be factored into $C_x(H_2O)_y$. These molecules appear to be hydrates of carbon. Of course they are not literal hydrates, in which water molecules cluster around a carbon, but the classic demonstration experiment in which a sugar cube is treated with sulfuric acid does yield quite a

*Emil Fischer (1852–1919) was perhaps the greatest of all organic chemists. He won the Nobel prize of 1902 and made extraordinary contributions to many fields, including carbohydrates and proteins.

†Although commercial sugars are mostly harvested from plants (sugar beets, or sugar cane), there are some marvelously exotic laboratory syntheses. My favorite is that of Philip Shevlin (b. 1939) of Auburn University, who fired bare carbon atoms into water and obtained carbohydrates. Shevlin has suggested that this reaction may well have occurred on the icy surfaces of comets, transforming them into giant orbiting sugar cones.

Saccharin, Cyclamates, and Aspartame

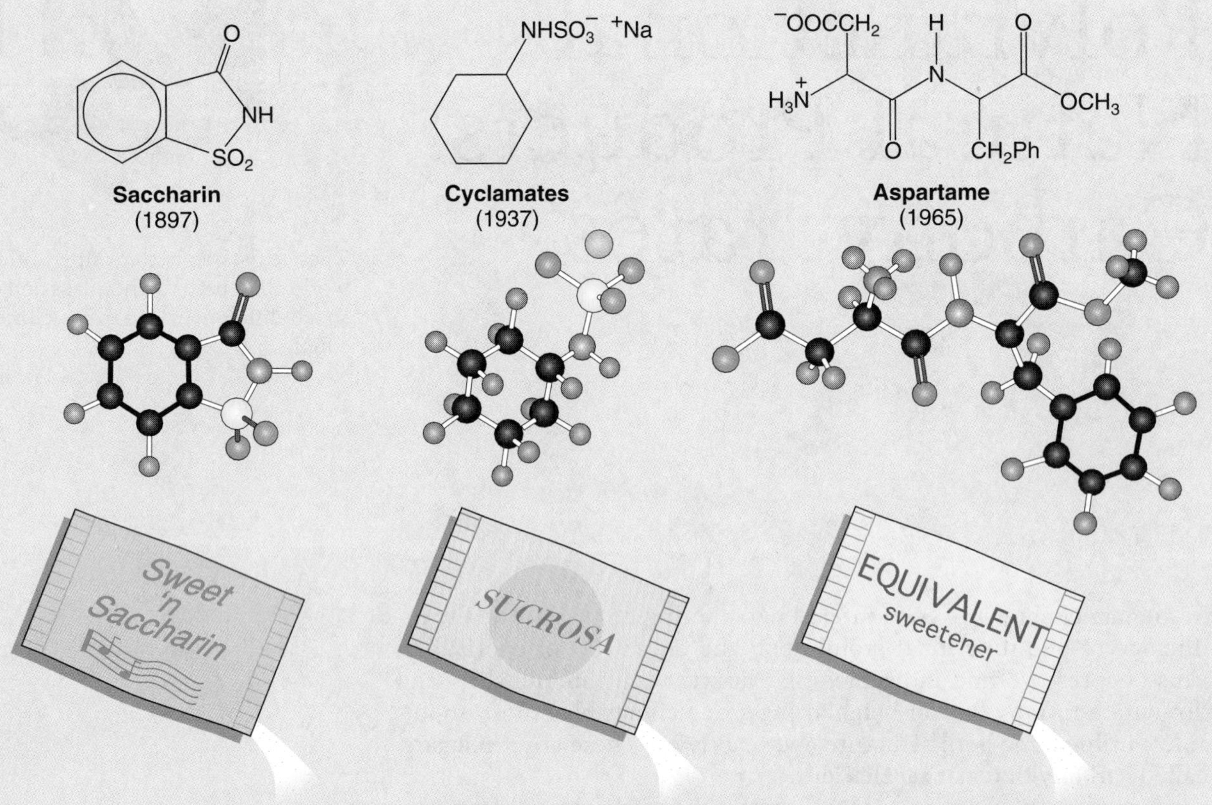

Saccharin (1897)

Cyclamates (1937)

Aspartame (1965)

These three sugar substitutes vary widely in structure, but are united by a single thread; each was found not by a directed search but by accident. Indeed, all three were discovered when a chemist tasted a compound made for other purposes. In fact, cyclamates were discovered because a cigarette smoked in the lab tasted sweet. This observation may seem like courting disaster to you—deliberately ingesting a chemical, not to mention smoking in a chemistry lab—but it was quite common in the old days. Even in my time it was normal to smell—closely and carefully—just about everything one made. I once passed out from smelling the end of an ozonolysis apparatus, for example. After that I discontinued the practice, but whether my sense of smell was destroyed by the ozone or I just became more cautious, I cannot say.

Both saccharin and cyclamates have become less used in this country because tests on animals revealed some possible health dangers. The consensus seems to be that the avoidance of saccharin and the outright banning of cyclamates in the United States is an overreaction.

spectacular sight as exothermic dehydration reactions produce steam and a cone of carbon.*

The sugar chemistry we will see involves only a few new reactions, but the polyfunctionality of these compounds can make for some complications. To be a successful sugar chemist you will have to be able to apply

*DO NOT try this experiment unsupervised! The reaction is very exothermic and sometimes violent. Sulfuric acid can be thrown about, and sulfuric acid in the eyes is no picnic!

old reactions in new settings, and to deal with a good deal of stereochemical complexity as well. It's not a bad idea to review the concepts in Chapter 5 quickly before you start on this chapter. That's a hard thing to do—time is probably pressing, and it's always more interesting to move on to new stuff, but at least reread the summary sections of Chapter 5 and be prepared to look back if things get confusing. Dust off your models. There will be moments when they will be essential.

In this chapter, we will also meet Emil Fischer (1852–1919), surely one of the greatest of organic chemists, and a person of heroic theoretical and experimental achievements. Sugars can be hideously frustrating molecules to work with experimentally. They form syrups and are notoriously difficult to crystallize. These properties were especially important in Emil Fischer's time, as there were no fancy chromatographs and spectrometers in his basement, and separation was routinely achieved through crystallization. One of Fischer's experimental triumphs was the discovery that phenylhydrazone derivatives of many sugars were rather easily crystallized. That may seem a simple point now, but it certainly was not in 1875! One of the great logical triumphs of organic chemistry was the unraveling of the structures of the aldohexoses (we'll see just what this means later, but glucose, a vital source of human energy, is one of the aldohexoses), and Emil Fischer was at the heart of this work for which he won the Nobel prize for chemistry in 1902.

24.1 NOMENCLATURE AND STRUCTURE

Sugars could all be named systematically, but their history and complexity has led to the retention of common names throughout the discipline. There is also a special stereochemical convention, named after Emil Fischer, that is used to simplify representations of these molecules that usually contain several stereogenic carbons.

The simplest sugar is the molecule called **glyceraldehyde**. It contains a single stereogenic carbon and thus can exist in two enantiomeric forms (Fig. 24.1).

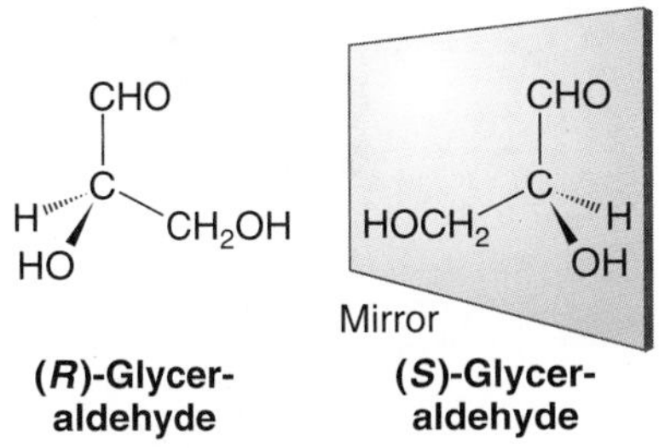

FIGURE **24.1** The two enantiomers of the simplest sugar, glyceraldehyde.

It is fairly easy (I hope) to determine which of these molecules is (*R*) and which is (*S*), so no new convention is really necessary here (Fig. 24.1). Glyceraldehyde is an **aldotriose**. The "tri" tells you it is a three-carbon molecule, and the "aldo" tells you that the molecule contains an aldehyde. The "ose" means that the pattern of an aldehyde (CHO) at one end attached to a primary alcohol (CH_2OH) at the other end through a series of H—C—OH groups (in this case, just one) is present (Fig. 24.2).

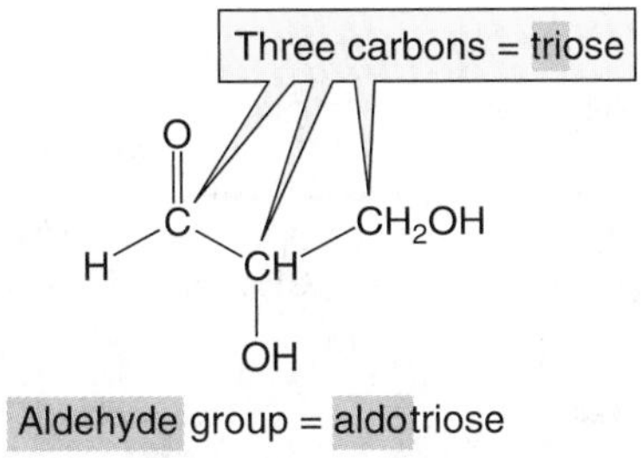

FIGURE **24.2** Glyceraldehyde is an aldotriose because it has a three-carbon backbone and an aldehyde group.

*PROBLEM **24.1** Generalize from the definition of a triose to generate the structures of an aldotetrose and an aldopentose.

ANSWER "Aldo" means that an aldehyde is present. A tetrose will have four carbons and a pentose will have five. All sugars (all ose's) have both a carbonyl group, here an aldehyde, and a primary alcohol (CH_2OH) group attached through a series of H−C−OH groups.

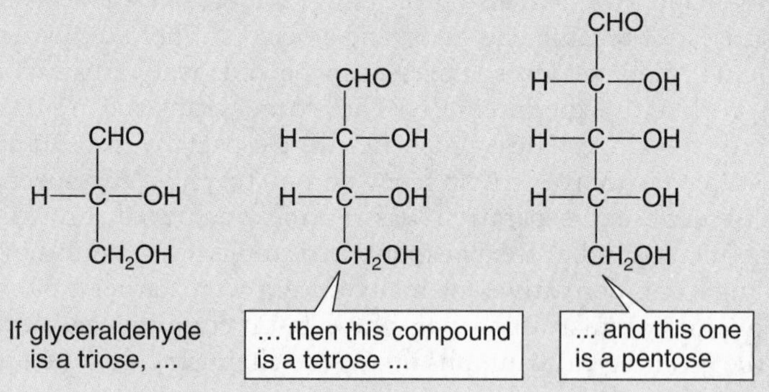

CHO / H—C—OH / CH_2OH — If glyceraldehyde is a triose, ...

CHO / H—C—OH / H—C—OH / CH_2OH — ... then this compound is a tetrose ...

CHO / H—C—OH / H—C—OH / H—C—OH / CH_2OH — ... and this one is a pentose

Convention Alert!

The stereochemical convention called a **Fischer projection** translates the drawings of Figure 24.1 and Problem 24.1 into three-dimensional pictures. Fischer projections do not attempt to represent energy minimum conformations. Instead, they are designed to simplify the specifying of single stereoisomers. An analogy would be the representation of substituted ethanes as eclipsed "sawhorse" conformations (Chapter 3, p. 83). Drawings of these energy maxima are often used because they are easier to visualize than the real energy minima—the staggered conformations. Fischer projections are also eclipsed molecules, and are used because they are simple to draw and to read. We will see soon how to translate the Fischer projections into more accurate representations of the real molecules. In a Fischer projection, all horizontal lines are taken as coming forward, and vertical lines as retreating. The carbons in the frame are not written in. Aldo sugars are almost always written with the aldehyde at the top and the primary alcohol at the bottom (Fig. 24.3).

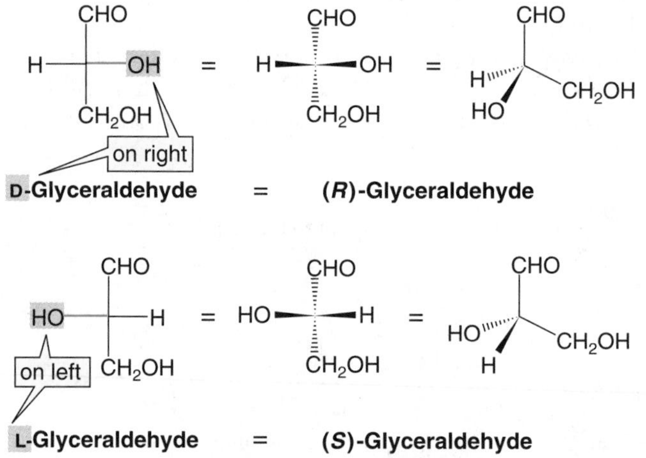

CHO / H—OH / CH_2OH (on right) = H—OH (wedges) = H, HO, CHO, CH_2OH

D-Glyceraldehyde = **(*R*)-Glyceraldehyde**

CHO / HO—H / CH_2OH (on left) = HO—H (wedges) = HO, H, CHO, CH_2OH

L-Glyceraldehyde = **(*S*)-Glyceraldehyde**

FIGURE **24.3** Aldo sugars are written with the aldehyde group at the top and the primary alcohol at the bottom. In this scheme, called a Fischer projection, horizontal bonds are taken as coming toward the viewer and vertical bonds as retreating. If the OH adjacent to the primary alcohol at the bottom is on the right, the sugar is a member of the D series. If it is on the left, it is in the L series.

If the OH group attached to the carbon adjacent to the bottom CH_2OH group, that is the stereogenic carbon most remote from the aldehyde group, is on the right, the molecule is designated as a "D" sugar; if this OH is on the left, the sugar is a member of the "L" series. *Notice that to get the designation right, the sugar must be drawn with the aldehyde at the top and the primary alcohol at the bottom.* There is absolutely no correlation between D and L and the sign of optical rotation (the direction of the rotation of the plane of plane-polarized light). A D sugar can rotate the plane of plane-polarized light either clockwise or counterclockwise. Clockwise rotation is indicated by prefacing the name of the sugar with (+), and counterclockwise rotation with (–). So we might write D-(+)-glyceraldehyde or D-(–)-erythrose. Both sugars are members of the D series, but one is dextrorotatory (rotates the plane of plane-polarized light in a clockwise sense) and the other is levorotatory (rotates the plane of plane-polarized light in a counterclockwise sense). Of course, if the sense of rotation is in one direction for the D stereoisomer, it is the opposite for the enantiomeric L molecule (Fig. 24.4).

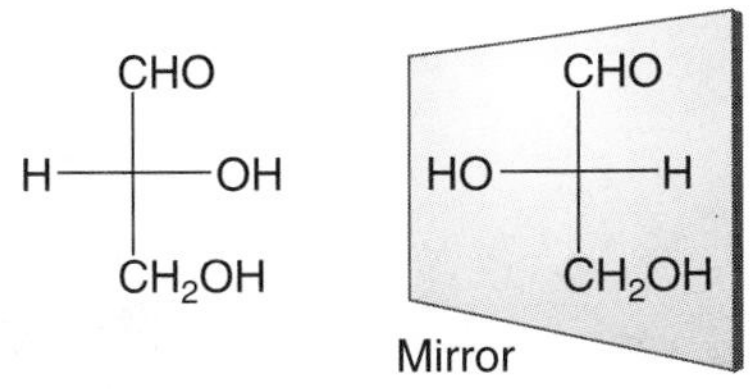

CHO / H—OH / CH_2OH CHO / HO—H / CH_2OH Mirror

D-(+)-Glyceraldehyde   **L-(–)-Glyceraldehyde**

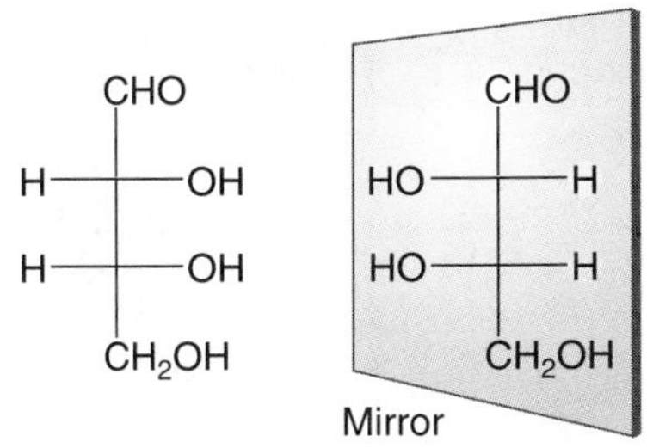

CHO / H—OH / H—OH / CH_2OH CHO / HO—H / HO—H / CH_2OH Mirror

D-(–)-Erythrose   **L-(+)-Erythrose**

FIGURE 24.4 There is no relation between D and L and the sign of optical rotation, (+) and (–).

PROBLEM 24.2

Write three-dimensional representations for the following molecules in their energy-minimum forms. They are shown in Fischer projection in Figure 24.5. Assign each stereogenic carbon of the molecules in Figure 24.5 as either (*R*) or (*S*).

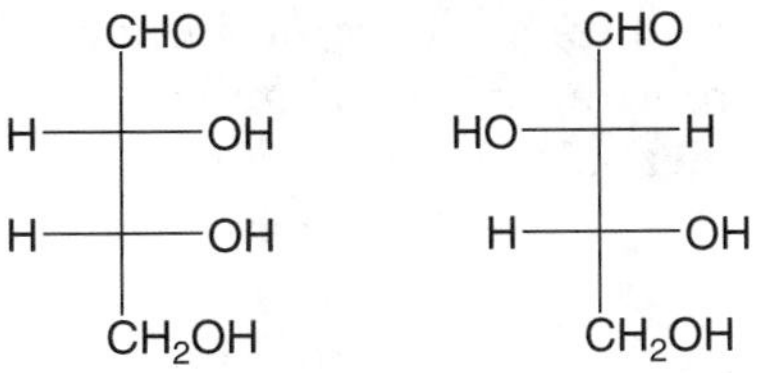

CHO / H—OH / H—OH / CH_2OH CHO / HO—H / H—OH / CH_2OH

FIGURE 24.5

PROBLEM 24.3

Write Fischer projections for the molecules in Figure 24.6.

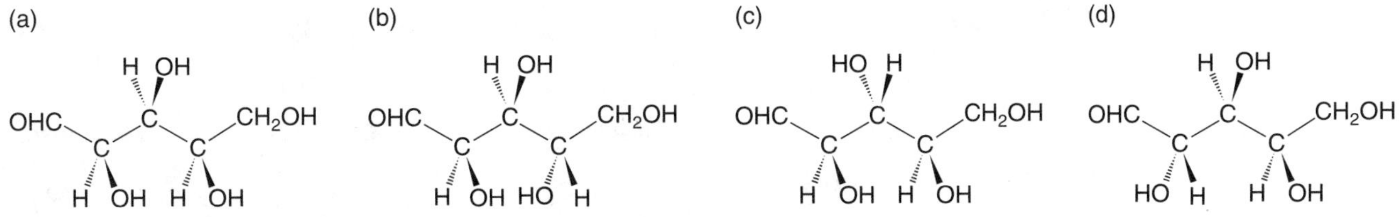

(a) OHC, H OH, C, C, C, CH_2OH, H OH H OH

(b) OHC, H OH, C, C, C, CH_2OH, H OH HO H

(c) OHC, HO H, C, C, C, CH_2OH, H OH H OH

(d) OHC, H OH, C, C, C, CH_2OH, HO H H OH

FIGURE 24.6

There are also **keto sugars**, or **ketoses**, which contain no aldehyde group, but rather a ketone. Figure 24.7 shows 1,3-dihydroxyacetone, a simple ketotriose.

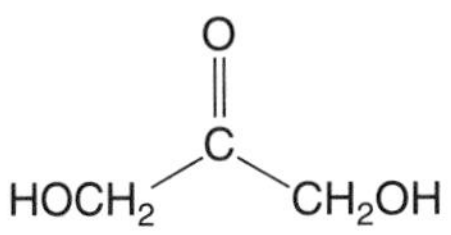

O, C, $HOCH_2$, CH_2OH

FIGURE 24.7 1,3-Dihydroxyacetone, a ketotriose.

Four-carbon sugars are called **tetroses**. There are four aldotetroses, the enantiomeric pairs D- and L-erythrose and D- and L-threose. Remember, for a sugar to be D, the OH next to the bottom CH_2OH must be on the right in a Fischer projection (Fig. 24.8).

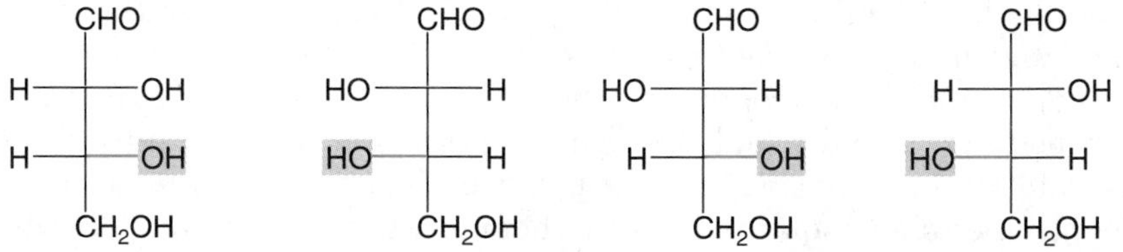

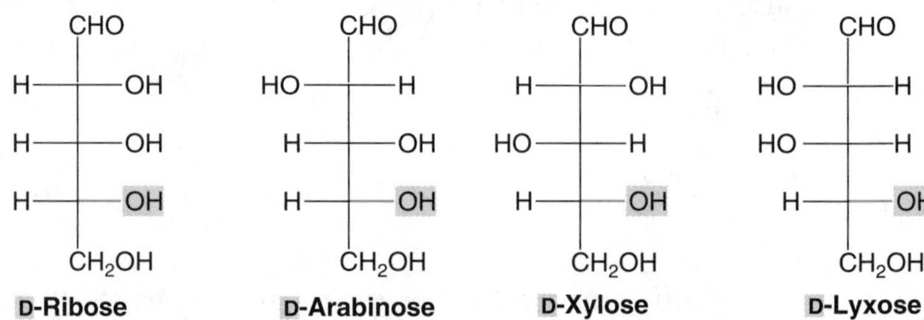

FIGURE 24.8 The four aldotetroses.

Aldopentoses contain three stereogenic carbons, and so there must be a total of eight stereoisomers ($2^3 = 8$). Figure 24.9 shows the four members of the D series of aldopentoses and their names.

FIGURE 24.9 The four D-aldopentoses.

D-Ribose: CHO; H–OH; H–OH; H–OH; CH_2OH
D-Arabinose: CHO; HO–H; H–OH; H–OH; CH_2OH
D-Xylose: CHO; H–OH; HO–H; H–OH; CH_2OH
D-Lyxose: CHO; HO–H; HO–H; H–OH; CH_2OH

In the figures, the internal carbons in the sugar chains are not written in. Watch out for this convention.

Aldohexoses contain four stereogenic carbons, and so there must be a total of $2^4 = 16$ stereoisomers, 8 in the D series and 8 enantiomeric molecules in the L series (Fig. 24.10).

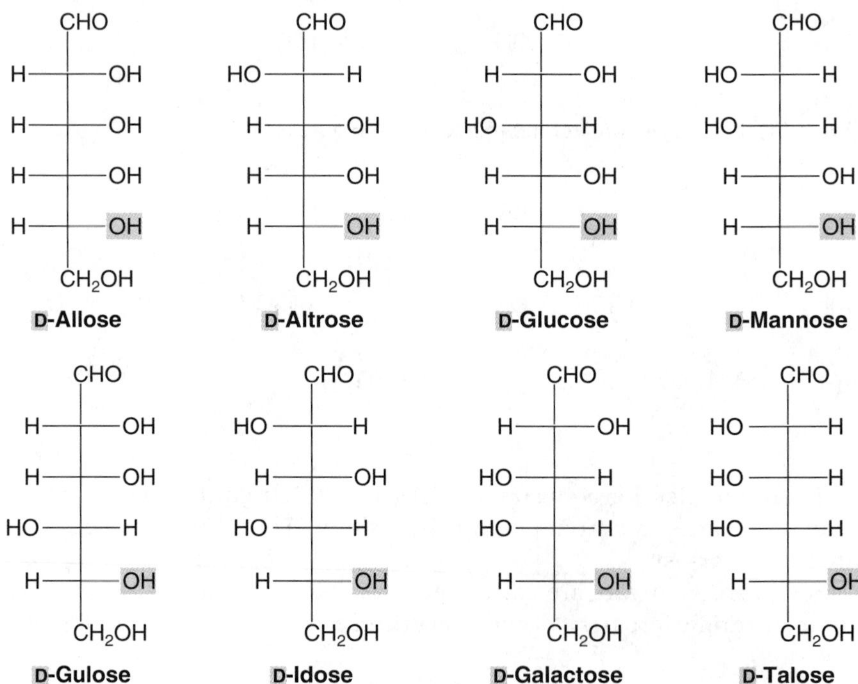

FIGURE 24.10 The eight D-aldohexoses.

Figures 24.9 and 24.10 show only aldohexoses, but there are also many keto sugars among the **pentoses** and **hexoses.** Figure 24.11 shows D-fructose, a common ketohexose.

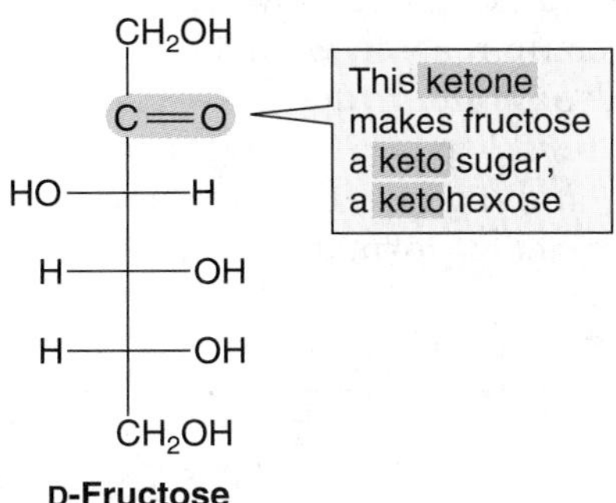

FIGURE **24.11** D-Fructose, a common ketohexose.

PROBLEM 24.4 Draw Fischer projections for the L-aldopentoses.

PROBLEM 24.5 How many D-aldoheptoses are possible?

PROBLEM 24.6 Treatment of D-glucose with sodium borohydride ($NaBH_4$) gives D-glucitol (sorbitol), $C_6H_{14}O_6$. Show the structure of D-glucitol and write a brief mechanism for this simple reaction.

One would expect these aldoses to combine the chemistries of alcohols and aldehydes, and that expectation would be correct. However, the proximity of these functional groups leads to complicated chemistry, and here is an important example. Although reduction with sodium borohydride, followed by hydrolysis, induces the expected change (see Problem 24.6), as the aldehyde is reduced to an alcohol, modern spectroscopic methods such as NMR and IR do *not* reveal the presence of substantial amounts of an aldehyde in the starting hexose (Fig. 24.12). Our first task is to understand this result, which is common in pentose and hexose chemistry. There are many typical reactions of carbonyl groups that take place in the apparent absence of an appropriate NMR or IR signal for the H−C=O functional group.

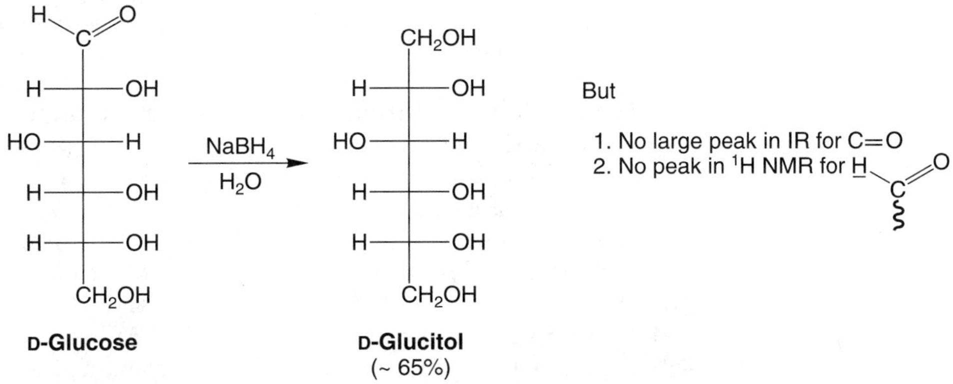

D-Glucose $\xrightarrow[H_2O]{NaBH_4}$ D-Glucitol (~ 65%)

But

1. No large peak in IR for C=O
2. No peak in ^{1}H NMR for H−C=O

FIGURE **24.12** Although reduction with sodium borohydride, followed by hydrolysis, proceeds normally to give an alcohol, neither NMR nor IR spectroscopy reveals large amounts of an aldehyde in the starting material.

You know already (Chapter 16, p. 763) that carbonyl groups react with all manner of nucleophiles. Hydration and hemiacetal formation are typical examples (Fig. 24.13). A hemiacetal could certainly be formed in an intramolecular reaction in which a hydroxyl group at one point in a molecule reacted with a carbonyl group in another. Indeed, 4- and 5-hydroxy aldehydes exist mostly in the cyclic hemiacetal forms (Chapter 23, p. 1181). Figure 24.13 shows a typical reaction of this kind and makes the analogy to both hydration and intermolecular hemiacetal formation.

Hydration

Hemiacetal formation

Intramolecular hemiacetal formation

5-Hydroxyhexanal (6%)

2-Hydroxy-6-methyl-tetrahydropyran A cyclic hemiacetal (94%)

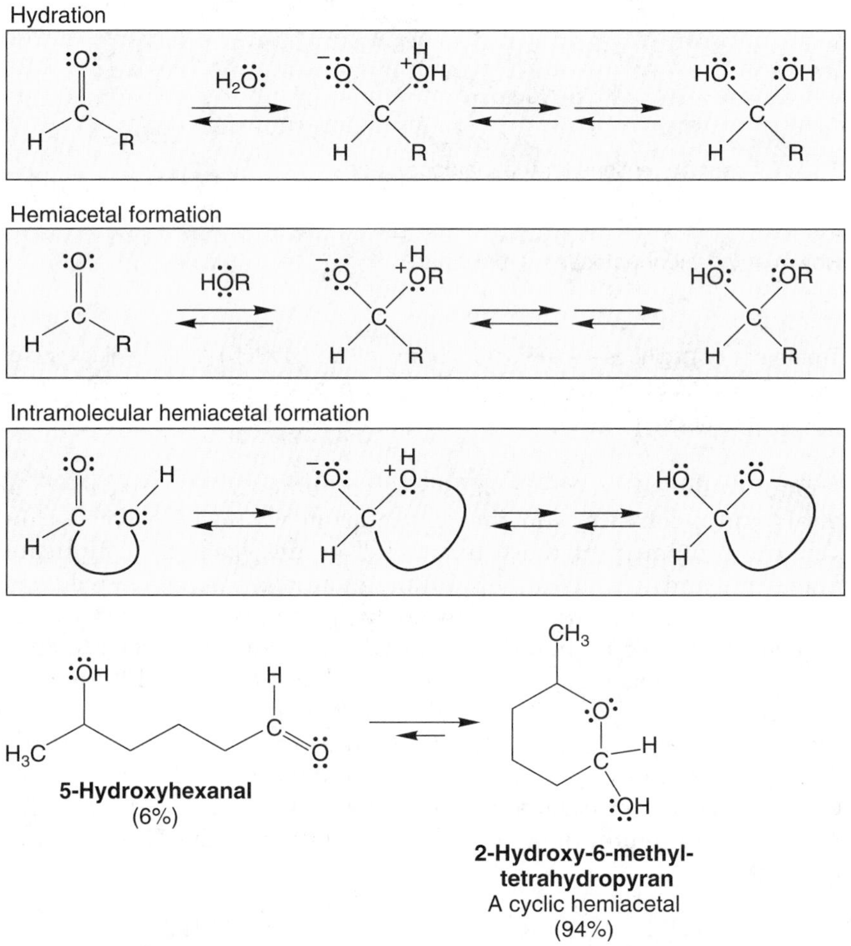

FIGURE **24.13** Intramolecular hemiacetal formation is analogous to hydration and intermolecular hemiacetal formation. Five- and six-membered ring hemiacetals are easily made, and are often more stable than their open forms.

The key to the solution of the "missing carbonyl" problem is that for hexoses, the hemiacetal form is more stable than the open-chain isomer. We can even venture a reasonable guess as to which hydroxyl group participates in the ring-forming reaction. For the cyclic hemiacetal to be more stable than the acyclic hydroxy aldehyde, the ring formed surely must not introduce a lot of strain. That eliminates three of the five OH groups and points to either the OH at C(4), which leads to a five-membered ring, or that at C(5), which would produce a six-membered ring. These cyclic forms are named as **furanoses** (five-membered ring) or **pyranoses** (six-membered ring) by analogy to the five- and six-membered cyclic ethers: tetrahydrofuran and tetrahydropyran (Fig. 24.14).

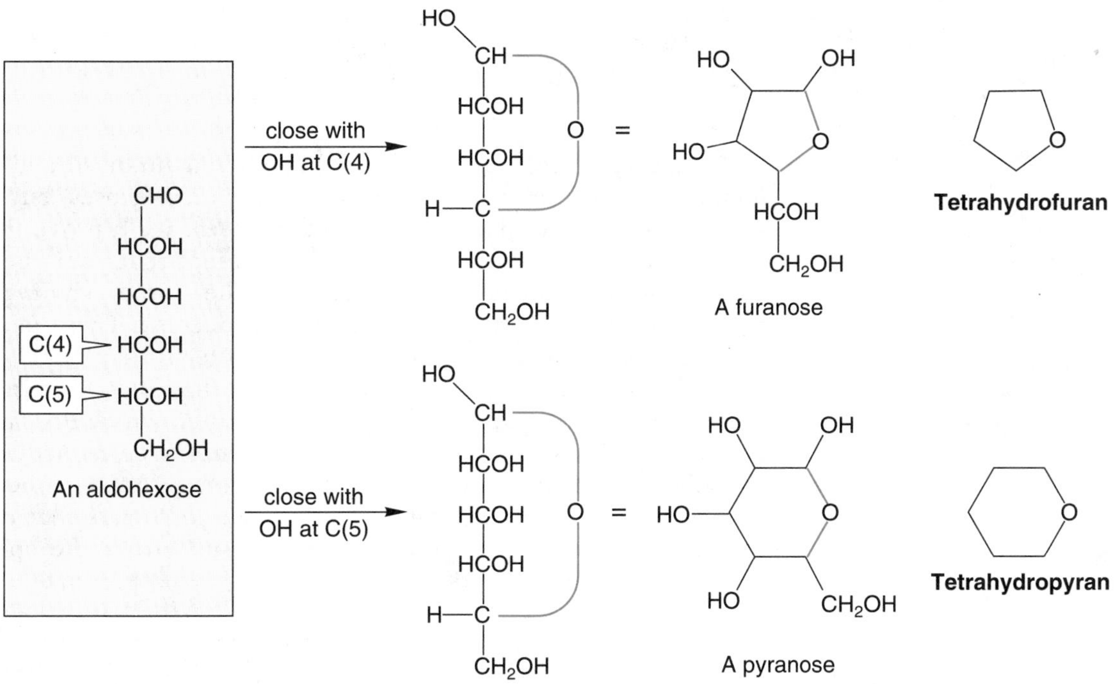

FIGURE **24.14** In an aldohexose, intramolecular hemiacetal formation results in a furanose (five-membered ring) or pyranose (six-membered ring).

Cyclic forms are indicated in the name by combining the simple name of the sugar with "furanose" or "pyranose" to indicate the size of the ring. Figure 24.15 shows D-glucofuranose and D-glucopyranose in Fischer projection.

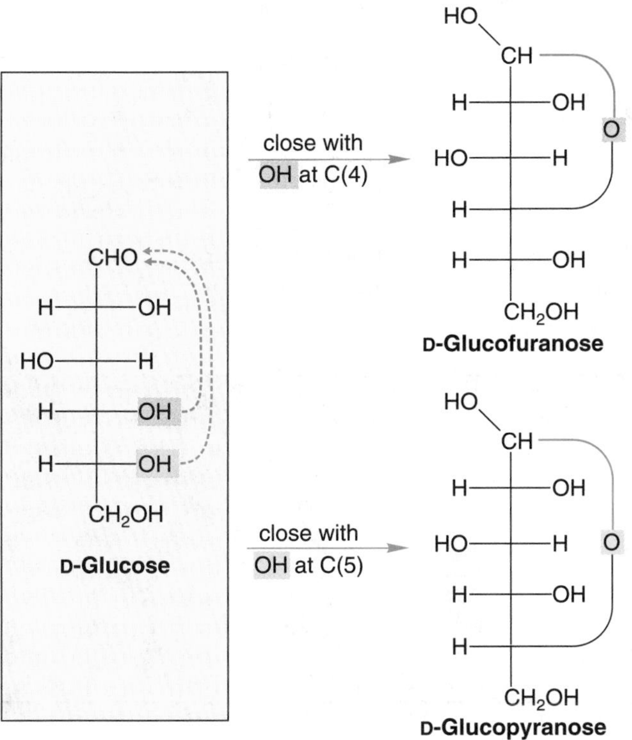

FIGURE **24.15** Fischer projections for D-glucofuranose and D-glucopyranose.

PROBLEM 24.7 Draw Fischer projections for D-mannopyranose, L-gulopyranose, and D-galactofuranose (see Fig. 24.10).

Both cyclic molecules, the five-membered furanose and the six-membered pyranose, are formed from most aldohexoses, but it is the six-membered ring that usually predominates. We'll prove this point a little later in this chapter, but for now accept that your guess was a good one, and let's go on to deal with the question of why typical carbonyl reactions appear for this cyclic hemiacetal. There is a simple answer to this puzzling question. The molecule isn't 100% in the cyclic form, it just exists predominantly in that form. It is in equilibrium with a small amount of the acyclic isomer, which *does* contain a carbonyl group and *does* react with reagents such as sodium borohydride. As the small amount of the reactive partner in the equilibrium is used up, more is generated (Fig. 24.16). Sugars that contain a hemiacetal, and are therefore in equilibrium with some open form, are called **reducing sugars**.

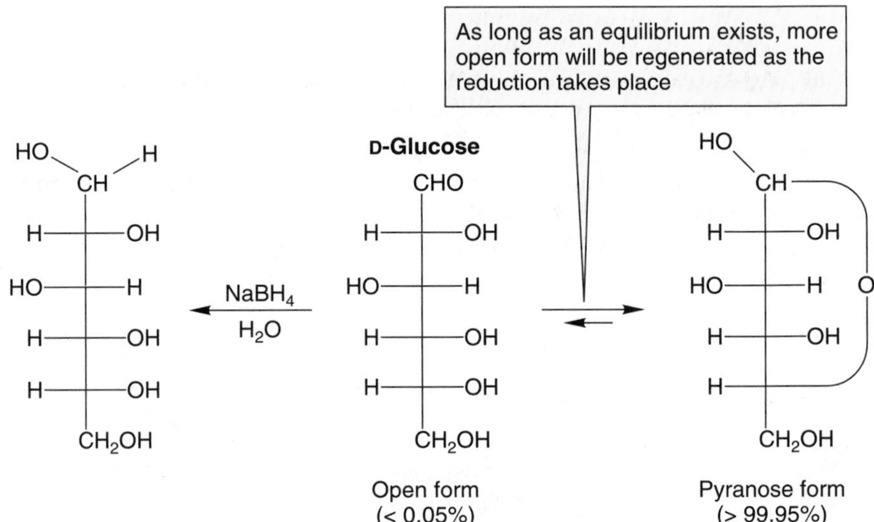

FIGURE 24.16 If two compounds are in equilibrium, irreversible reaction of the minor partner can result in complete conversion into a product. As long as the equilibrium exists, the small amount of the reactive molecule will be replenished as it is used up.

Table 24.1 lists some common carbohydrates and gives the relative amounts found in pyranose and furanose forms. There is very little of the open form present at equilibrium.

TABLE 24.1 Pyranose and Furanose Forms of Aldohexoses at 25 °C

Aldohexose	Pyranose Form	Furanose Form
Allose	92	8
Altrose	70	30
Glucose	~100	<1
Mannose	~100	<1
Gulose	97	3
Idose	75	25
Galactose	93	7
Talose	69	31

However, because there is always a small amount of the reducable open form present, all of the material can eventually be reduced by sodium borohydride. If we return to our spectrometers, crank up the sensitivity, and look very hard for the aldehyde group present in the open-chain form, there it is, in the tiny amount required by our explanation.

Let's turn our attention to the structures of these cyclic, six-membered ring hemiacetals (pyranoses). The first thing to notice is that the intramolecular cyclization reaction creates a new stereogenic carbon that can be either (*R*) or (*S*). These two stereoisomers, called **anomers**, result from addition of the nucleophilic OH to the Lewis acid carbonyl group from the two possible sides. Anomers of aldoses differ only in the stereochemistry at C(1). If the new OH is on the same side in Fischer projection as the oxygen involved in the hemiacetal link, we have the α-anomer. If it is on the opposite side, it is the β-anomer. The stereoisomers are shown in Fischer projection in Figure 24.17, which also shows the schematic "squiggly bond" used to indicate that the new OH group at the anomeric carbon can be on the right or left in Fischer projection.

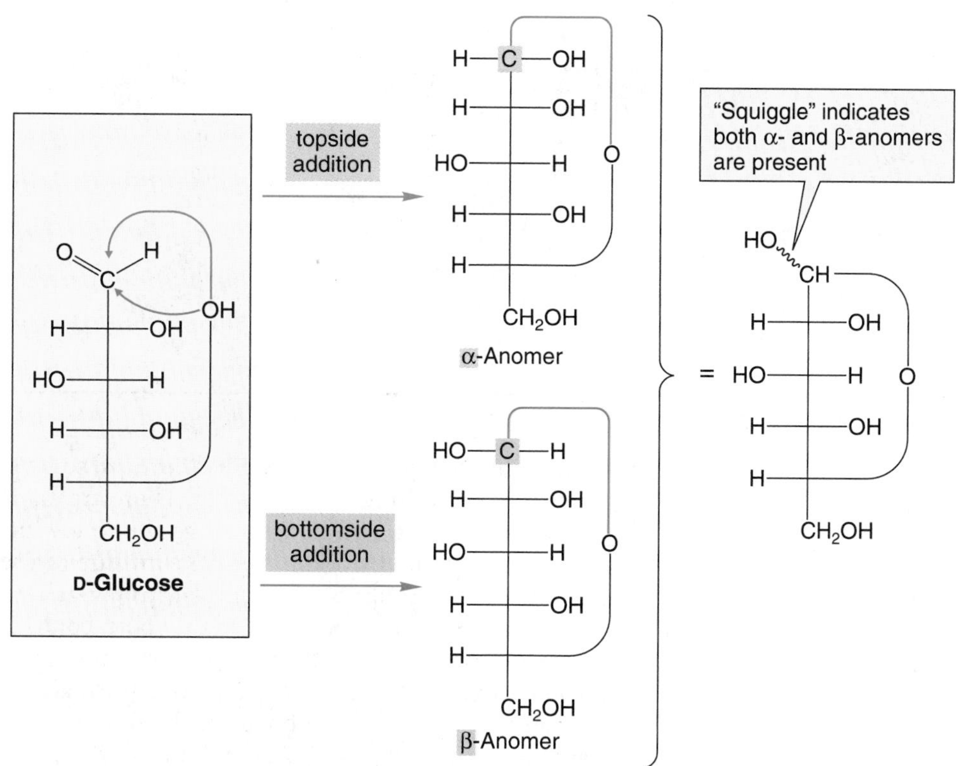

FIGURE 24.17 Intramolecular hemiacetal formation results in two C(1) stereoisomers called anomers.

PROBLEM 24.8 Identify the new stereogenic carbon in the α- and β-anomers of Figure 24.17 as either (*R*) or (*S*).

Convention Alert!

The last few figures have shown molecules with some very peculiar bonds in them. Notice in particular the "around the corner" bonds generally used to show attachment to the ring oxygen. The squiggly bond is used to show that both anomers can be present (Fig. 24.18).

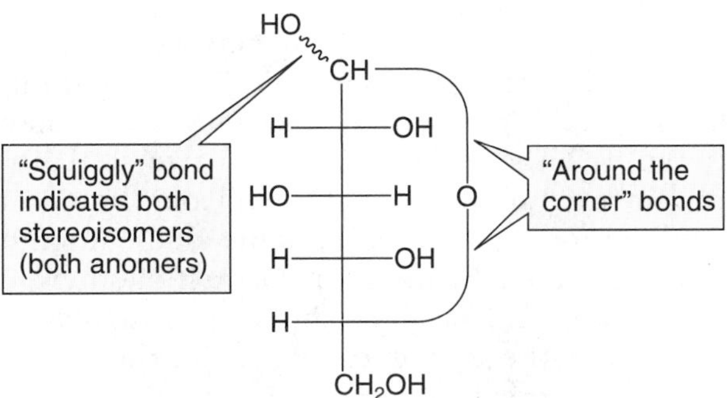

FIGURE **24.18** In these Fischer projections, we have drawn "around the corner" bonds.

Both α- and β-pyranose and furanose forms exist. Table 24.2 shows the relative amounts for the aldohexoses.

TABLE **24.2** α- and β-Pyranose and Furanose Forms of the Aldohexoses at 25 °C

Aldohexose	α-Pyranose Form	β-Pyranose Form	α-Furanose Form	β-Furanose Form
Allose	16	76	3	5
Altrose	27	43	17	13
Glucose	36	64	<1	<1
Mannose	66	34	<1	<1
Gulose	16	81	<1	3
Idose	39	36	11	14
Galactose	29	64	3	4
Talose	37	32	17	14

We know that bonds do not go around corners, so let's try to turn these schematics into something more sensible. There are many ways to do this, and any way you devise will do as well as the following method. Just be certain that your method works! We will continue to use D-glucose as an example and work through a method of arriving at a three-dimensional structure. First, redraw the molecule so that both bonds to the ring oxygen are vertical; so that the ring is retreating from us into the page at both the top and bottom of the molecule. This transformation requires a simple rotation about a carbon–carbon bond. In the α-anomer, the OH at C(1) is on the side opposite the CH_2OH group, and in the β-anomer it is on the same side (Fig. 24.19).

Now tip the molecule over in clockwise fashion to generate a flat six-membered ring. This kind of planar representation is called a **Haworth form**, after its inventor, Walter N. Haworth (1883–1950). Follow the CH_2OH group, and you can see that it is now pointing up. The remainder of the substituent OH groups can now be easily put in by using the CH_2OH as a guide. Figure 24.20 uses the β-anomer as an example.

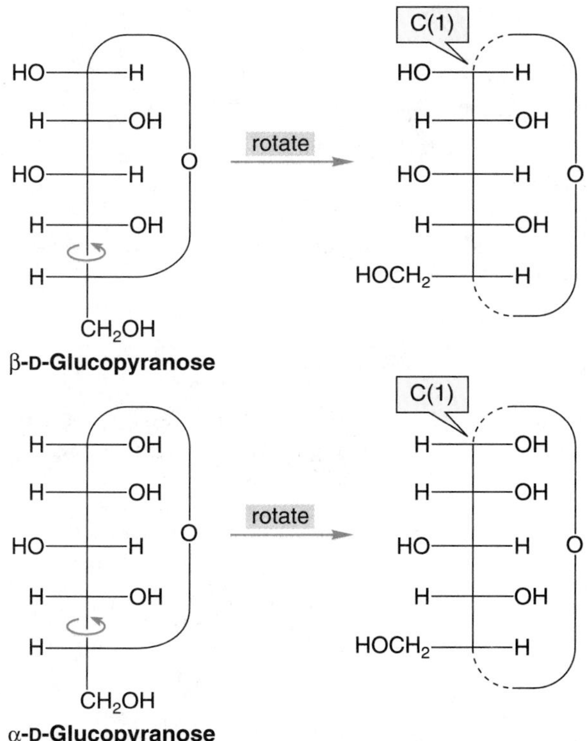

FIGURE **24.19** The first step in creating a three-dimensional drawing is rotation around the indicated carbon–carbon bond. This motion generates a new Fischer projection.

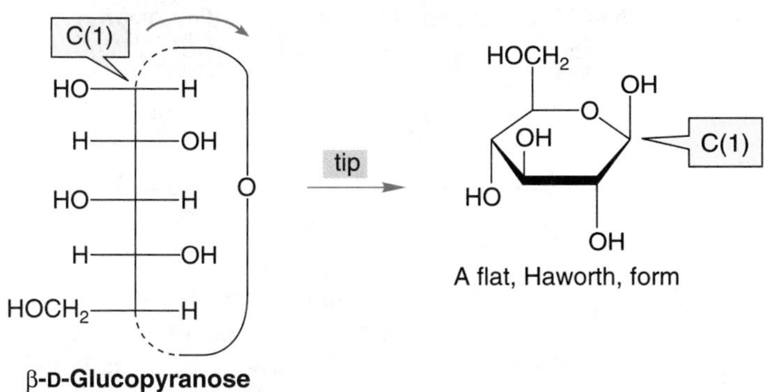

FIGURE **24.20** Next, tip the molecule over in clockwise fashion to produce a flat Haworth form.

PROBLEM **24.9**

Draw the flat, Haworth form of the α-anomer.

Six-membered rings exist in energy minimum chair forms, and are not flat. The final step in this exercise is to allow the flat Haworth form to relax to a chair. There are two chairs possible for any six-membered ring. These are interconverted by the rotations around carbon–carbon bonds we called a ring flip (Chapter 6, p. 204). Be sure to check both forms to see which is lower in energy (Fig. 24.21). Sometimes this will be hard to do, but for β-D-glucopyranose, it is easy. The form with all groups equatorial will be far more stable than that with all groups axial (Chapter 6, p. 207).

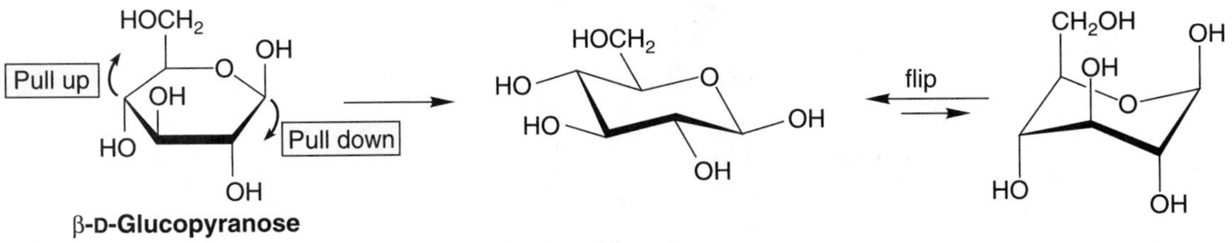
HOCH2
OH
O
Pull up
OH
Pull down
HO
OH
β-D-Glucopyranose
HOCH2
HO
O
HO
OH
OH
flip
CH2OH
OH
OH
O
HO
OH

FIGURE **24.21** Now let the flat, Haworth form relax to a chair. Don't forget that there are always two possible chair forms.

*PROBLEM **24.10** Follow this same procedure for the α-anomer.

ANSWER In the α-anomer, the hydroxyl group at C(1) will be axial, not equatorial as in the β-anomer. Otherwise all the substituent groups remain equatorial.

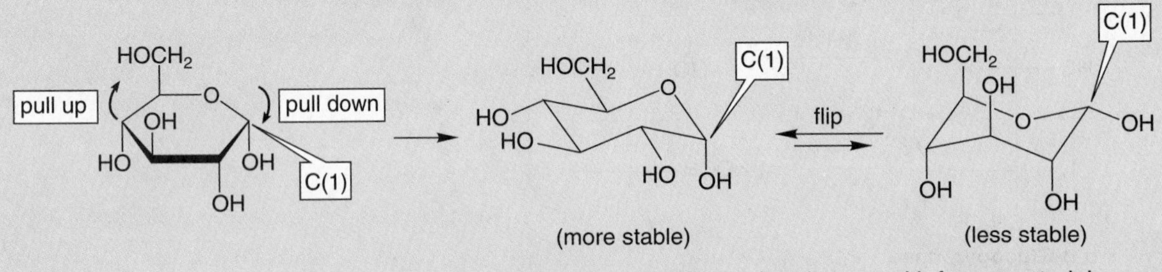
HOCH2
pull up
O
pull down
OH
HO
OH
C(1)
OH
HOCH2
C(1)
O
HO
HO
HO
OH
flip
C(1)
HOCH2
OH
O
OH
OH
OH
(more stable)
(less stable)

Of course, there are two chair forms, but the one with four equatorial groups will be the more stable one.

PROBLEM **24.11** Transform the Fischer projection into a three-dimensional picture of D-mannopyranose.

24.2 REACTIONS OF SUGARS

24.2a Mutarotation of Sugars

The following behavior is typical for aldohexoses and aldopentoses. Pure, crystallized α-D-glucopyranose has a specific rotation of +112°. The pure β-anomer has a specific rotation of +18.7°. However, an aqueous solution of *either* anomer steadily changes rotation until the value of +52.7° is reached. This phenomenon is called **mutarotation** and is common for sugars. Our task is to explain it. The answer comes from the realization that both α- and β-D-glucopyranoses are hemiacetals and exist in solution in equilibrium with a small amount of the open-chain aldohexose. Once the acyclic form is produced, it can re-form the hemiacetal to make *both* the α- and β-pyranose. Over time, an equilibrium mixture is formed and it is that equilibrated pair of anomers that gives the rotation of 52.7° (Fig. 24.22).

24.2b Isomerization of Sugars in Base

In base, sugars rapidly equilibrate with other sugars, which is a more profound change than the equilibration of anomers we saw in the preceding section. Again, let's use D-glucose as an example. As Figure 24.23

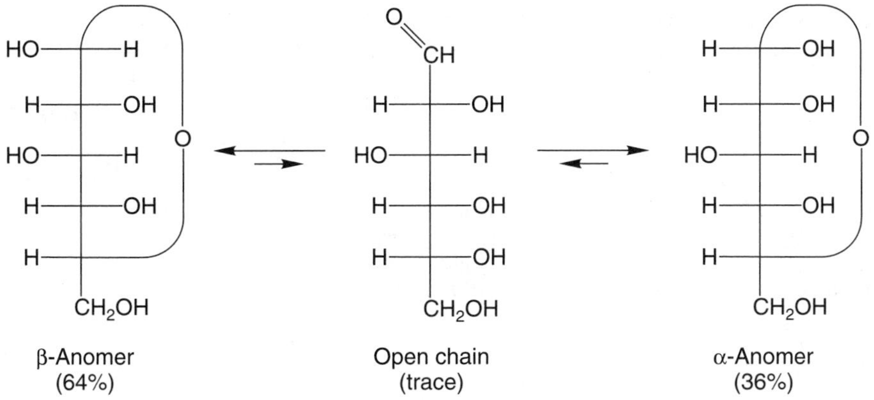

β-Anomer (64%) ⇌ Open chain (trace) ⇌ α-Anomer (36%)

FIGURE 24.22 The α- and β-anomers can equilibrate through the small amount of the open form present at equilibrium.

*PROBLEM 24.12

Write a mechanism for the acid-catalyzed mutarotation of D-glucopyranose.

ANSWER

An ether oxygen of one hemiacetal pyranose form of glucose is protonated. The ring opens to give a resonance-stabilized cation. Reclosure can either regenerate the original pyranose or the other anomer.

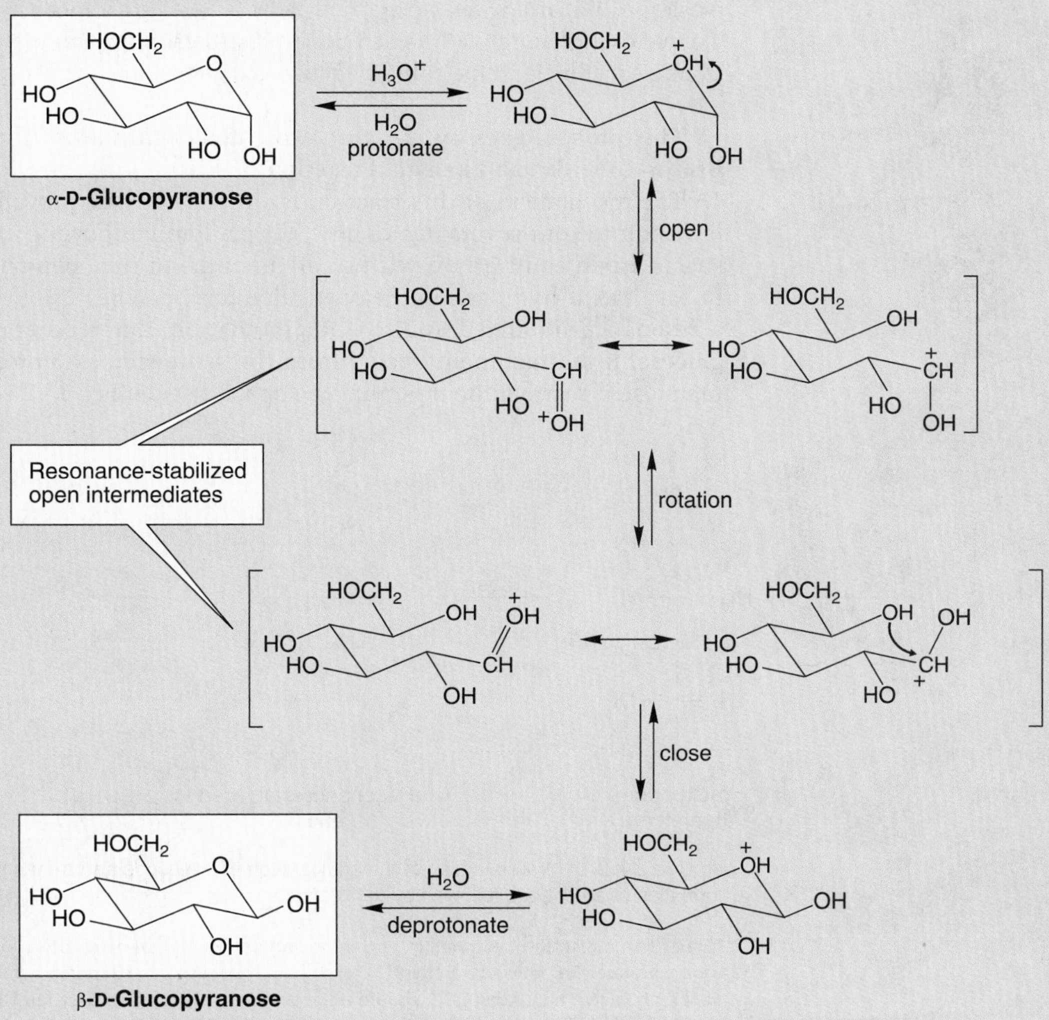

α-D-Glucopyranose

H_3O^+ / H_2O protonate

open

Resonance-stabilized open intermediates

rotation

close

H_2O deprotonate

β-D-Glucopyranose

shows, when treated with base, D-glucose equilibrates with another D-aldohexose, D-mannose, and a keto sugar, D-fructose (Fig. 24.23).

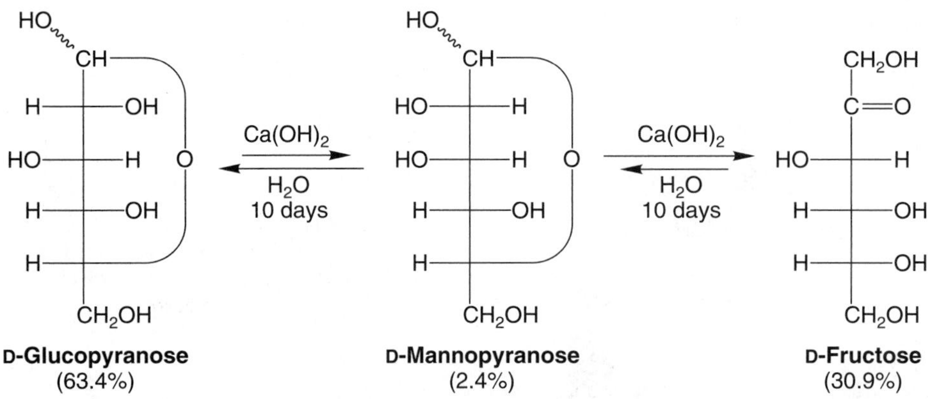

FIGURE **24.23** In base, D-glucose equilibrates with D-mannose and D-fructose, a keto sugar.

PROBLEM 24.13 Although D-fructose is shown in Figure 24.23 in the open chain form, it exists mostly (~ 67%) in a pyranose form and partly (~ 31%) in a furanose form in which the hydroxyl group at C(5) is tied up in hemiacetal formation with the ketone. Draw Fischer projections for the pyranose and furanose forms of D-fructose.

This process goes by the absolutely delightful name of the **Lobry de Bruijn–Alberda van Ekenstein reaction.***

The mechanism of this reaction is simpler than its provenance. As we have pointed out several times now, the predominant cyclic forms of these sugars are in equilibrium with small amounts of the open-chain isomers. In base, the α-hydrogen in the open, aldo form can be removed to produce a resonance-stabilized enolate. Reprotonation can regenerate either D-glucose, if reattachment occurs from the same side as proton loss, or D-mannose if reattachment is from the opposite side (Fig. 24.24).

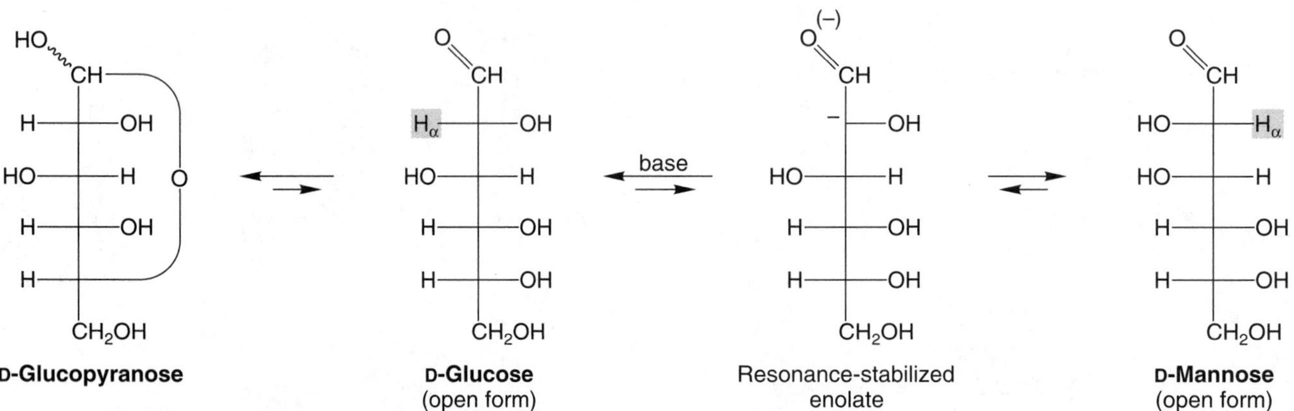

FIGURE **24.24** A mechanism of the equilibration of D-glucose and D-mannose involves formation of an enolate followed by reprotonation.

*This "name reaction" is even better than it sounds. One might be forgiven for assuming that someone meeting these two chemists on a Dutch street in 1880 might say, "Goedemorgen, Lobry; hoe gaat het, Alberda?," should he or she be so presumptuously familiar as to address them by their first names. But that would be wrong! The reaction title, alas, incorporates only parts of their names. The reaction is named for Cornelius Adriaan van Troostenbery Lobry de Bruijn (1857–1904), and the slightly less spectacularly named Willem Alberda van Ekenstein (1858–1907).

PROBLEM 24.14

Make a three-dimensional drawing of the enolate in Figure 24.24 to show how both D-glucose and D-mannose can be formed on reprotonation.

If reprotonation occurs at oxygen, a "double enol" is formed. This compound can re-form a carbon–oxygen double bond in two ways. An aldehyde can be generated, and this simply re-forms the two aldohexoses, D-glucose and D-mannose. If a ketone is produced, it is D-fructose that is the product (Fig. 24.25).

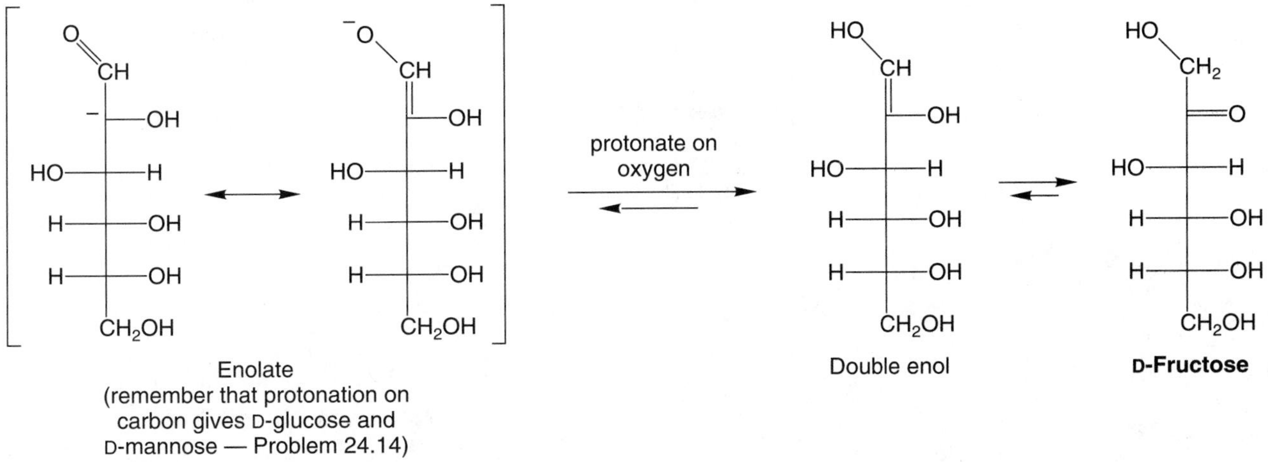

Enolate
(remember that protonation on carbon gives D-glucose and D-mannose — Problem 24.14)

protonate on oxygen

Double enol

D-Fructose

FIGURE 24.25 Protonation on oxygen generates a double enol, which can lead to D-fructose (or the D-aldohexoses, D-glucose, and D-mannose).

24.2c Reduction

We have already seen one aldose reaction, the reduction of an aldohexose with sodium borohydride to give a 1,2,3,4,5,6-hexanehexaol. Other reducing agents also work (Fig. 24.26).

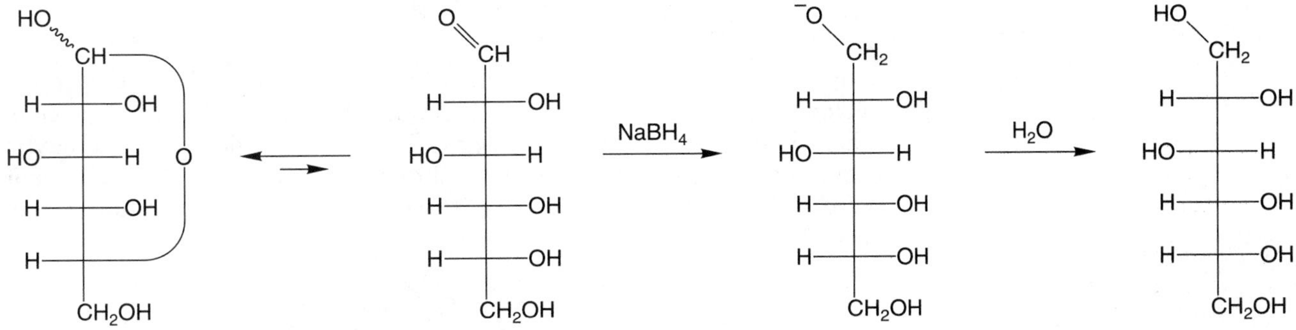

$NaBH_4$

H_2O

FIGURE 24.26 Reduction of D-glucose proceeds through the small amount of the open, aldo form present at equilibrium. As the open form is used up, it is regenerated through equilibration with the pyranose form.

*PROBLEM 24.15

Reduction of D-altrose with sodium borohydride in water gives an optically active molecule, D-altritol. However, the same procedure applied to D-allose gives an optically inactive, meso hexa-alcohol. Explain.

ANSWER

This problem is easy, but critical for the material following in the Fischer proof of the stereochemistry of glucose. The important point is to see that reduction of

ANSWER (CONTINUED) the aldehyde to the alcohol makes the molecule more symmetrical by making the two ends of the chain *both* CH_2OH groups. D-Allose is reduced to a meso compound but reduction of D-altrose gives an alcohol that is still optically active.

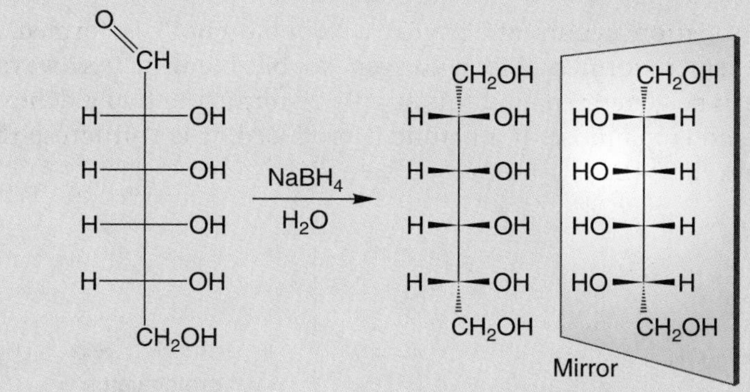

Open form of D-allose
(optically active)

Two superimposable mirror images—
a meso compound
(optically inactive)

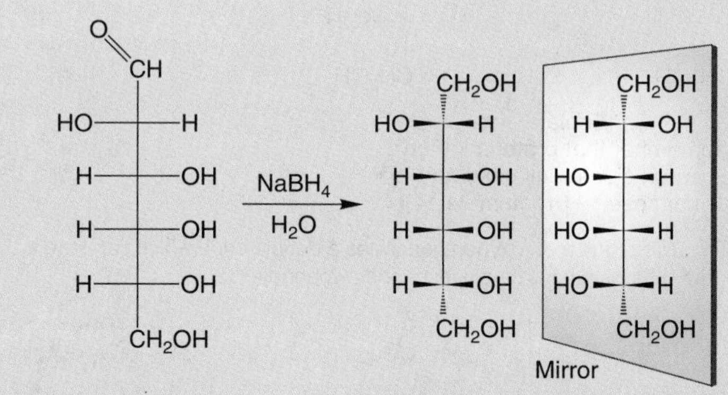

Open form of D-altrose
(optically active)

Two nonsuperimposable mirror images
(this alcohol is still optically active)

24.2d Oxidation

Sugars contain numerous oxidizable groups, and methods have been developed in which one or more of them may be oxidized in the presence of the others. Mild oxidation converts only the aldehyde into a carboxylic acid; the primary and secondary hydroxyls are not touched. The reagent of choice is bromine in water, and the product is called an **aldonic acid** (Fig. 24.27).

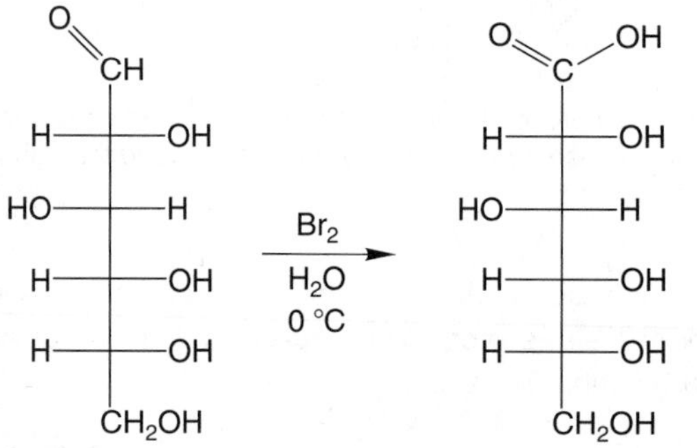

D-Glucose
(open form)

D-Gluconic acid
(~ 96%)

In an aldonic acid, the end groups are still different

FIGURE 24.27 Oxidation of an aldohexose with bromine in water gives an aldonic acid in which the end groups are still different.

PROBLEM **24.16**

Write a mechanism for this oxidation. *Hint for the first step*: What reaction is likely between an aldehyde and water?

PROBLEM **24.17**

Examination of the NMR and IR spectra of typical aldonic acids often shows little evidence of the carboxylic acid group. Explain this odd behavior.

More vigorous oxidation with nitrous or nitric acid creates an **aldaric acid** in which both the aldehyde and primary alcohol have been oxidized to carboxylic acids. Notice that this reaction destroys the difference between the two ends of the molecule. In the aldaric acid, there is an acid group at both the top and bottom of the Fischer projection (Fig. 24.28).

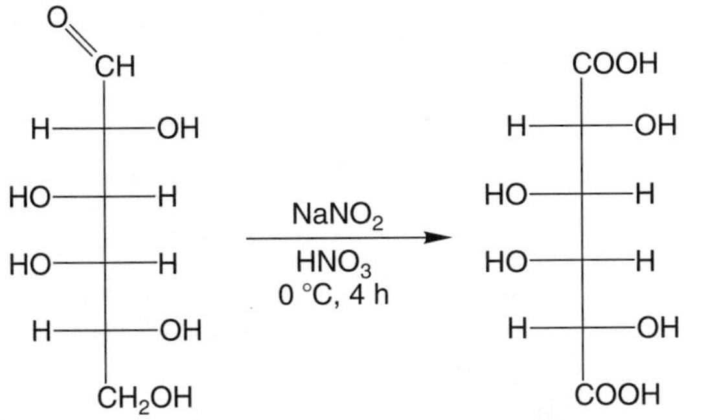

CHO; H–OH; HO–H; HO–H; H–OH; CH_2OH $\xrightarrow[\text{0 °C, 4 h}]{NaNO_2,\ HNO_3}$ COOH; H–OH; HO–H; HO–H; H–OH; COOH

D-Galactose (open form) → **D-Galactaric acid** (89%)

FIGURE **24.28** Oxidation with nitric acid generates an aldaric acid in which the end groups are both carboxylic acids.

In even more vigorous oxidations the secondary hydroxyl groups are attacked, and the six-carbon framework can be ruptured. You already know that vicinal diols (1,2-diols) are cleaved on treatment with periodate (Chapter 17, p. 842). When sugars are treated with periodate, mixtures of products are obtained, as the various 1,2-diols are cleaved. Treatment with periodate can chop the carbon frame into small fragments, and the structures of the fragments can often be used in structure determination.

24.2e Ether and Ester Formation

Alcohols can be made into ethers (Chapter 17, p. 845). One classic method in carbohydrate chemistry involves the treatment of the sugar with methyl iodide and silver oxide (Fig. 24.29).

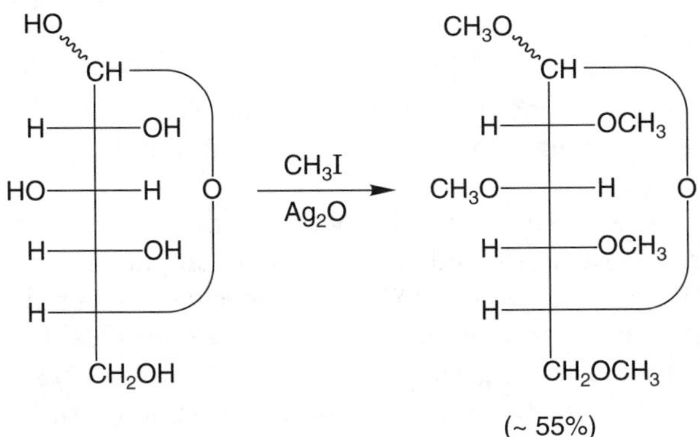

FIGURE **24.29** Treatment of a sugar with methyl iodide and silver oxide leads to methylation at every free hydroxyl group in the molecule.

One of the classic S_N2 reactions, the formation of ethers from alkoxides and alkyl halides, is known as the Williamson ether synthesis (Chapter 17, p. 845). Many variations on this theme are employed. Treatment of an aldohexose with benzyl chloride and potassium hydroxide results in formation of a tetra-ether, monoacetal through a series of Williamson ether syntheses. Be careful to notice the difference between the alkoxide group at C(1) and the others. The former is part of an acetal, whereas the others are simple ethers (Fig. 24.30).

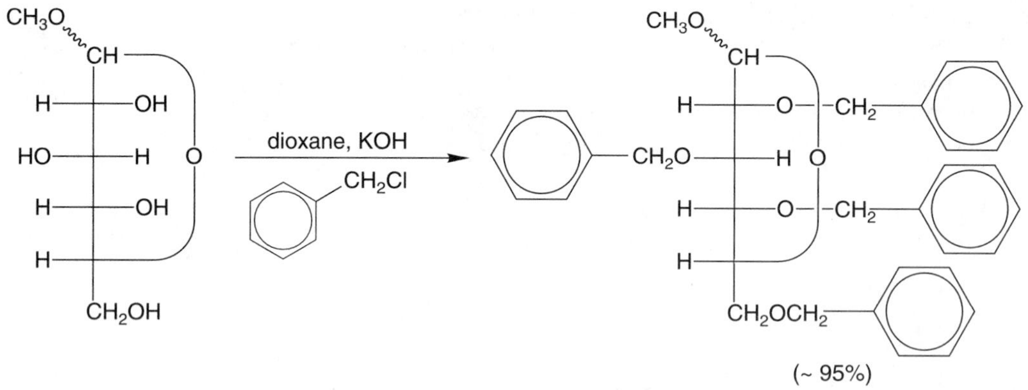

FIGURE **24.30** A similar process can be carried out in base with a Williamson ether synthesis. Notice in this example that neither the existing ether at C(1) nor the pyranose ring connection is disturbed in the benzylation.

Similarly, reaction with acetic anhydride or acetyl chloride will produce a polyacetylated compound (Fig. 24.31).

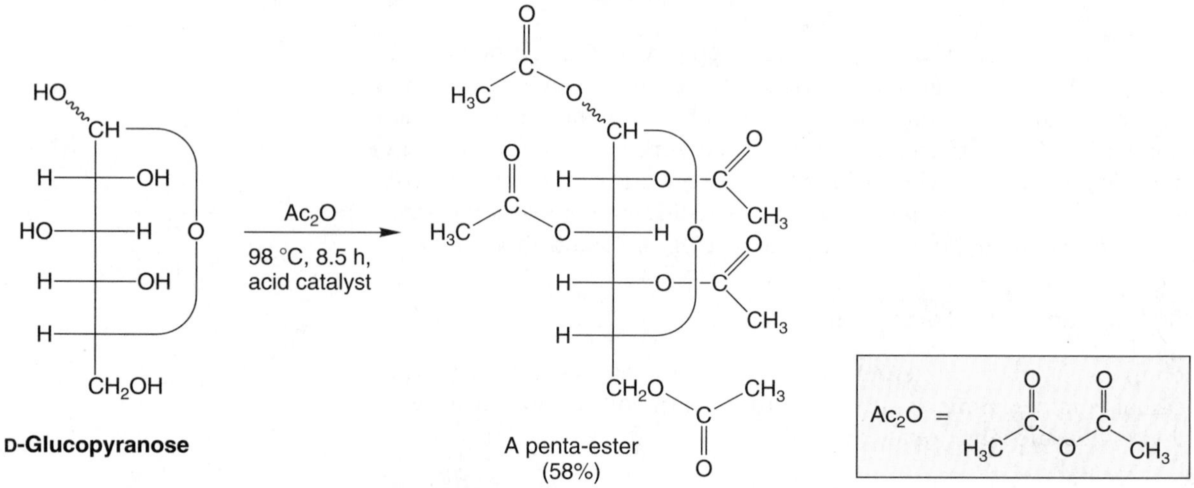

FIGURE **24.31** All the free hydroxyl groups can be esterified with acetic anhydride.

Notice that in both these reactions the hydroxyl group tied up in hemiacetal formation remains unaffected by the reaction. It can form neither an ether nor an ester.

The presence of the hemiacetal function allows for selective ether formation when a sugar is treated with dilute acidic alcohol. The products are full acetals, and their generic name is **glycoside**. Six-membered acetals are **pyranosides** and five-membered acetals are **furanosides** (Fig. 24.32).

Why is it that *only* the acetal hydroxyl group becomes a methoxyl group in this reaction? The first step is protonation of one of the many OH groups. All the OH groups in the molecule are reversibly protonated, but only one protonation leads to a resonance-stabilized cation on loss of water, and it will be this OH that is lost most easily. Addition of alcohol,

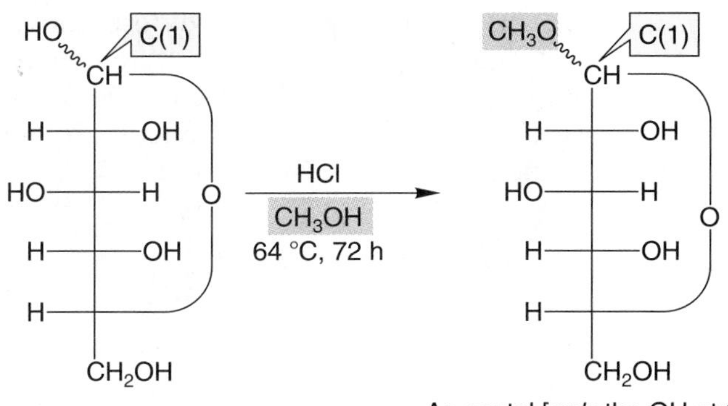

D-Glucopyranose

An acetal [*only* the OH at C(1) is converted into a CH_3O group] (~ 50%)

A "glycoside" (in this case, a "pyranoside")

FIGURE **24.32** Treatment with dilute HCl and alcohol converts only the OH at the anomeric position [C(1)] into an acetal called a glycoside.

followed by deprotonation, gives the acetal in which the hemiacetal OH has been converted into OCH_3 (Fig. 24.33).

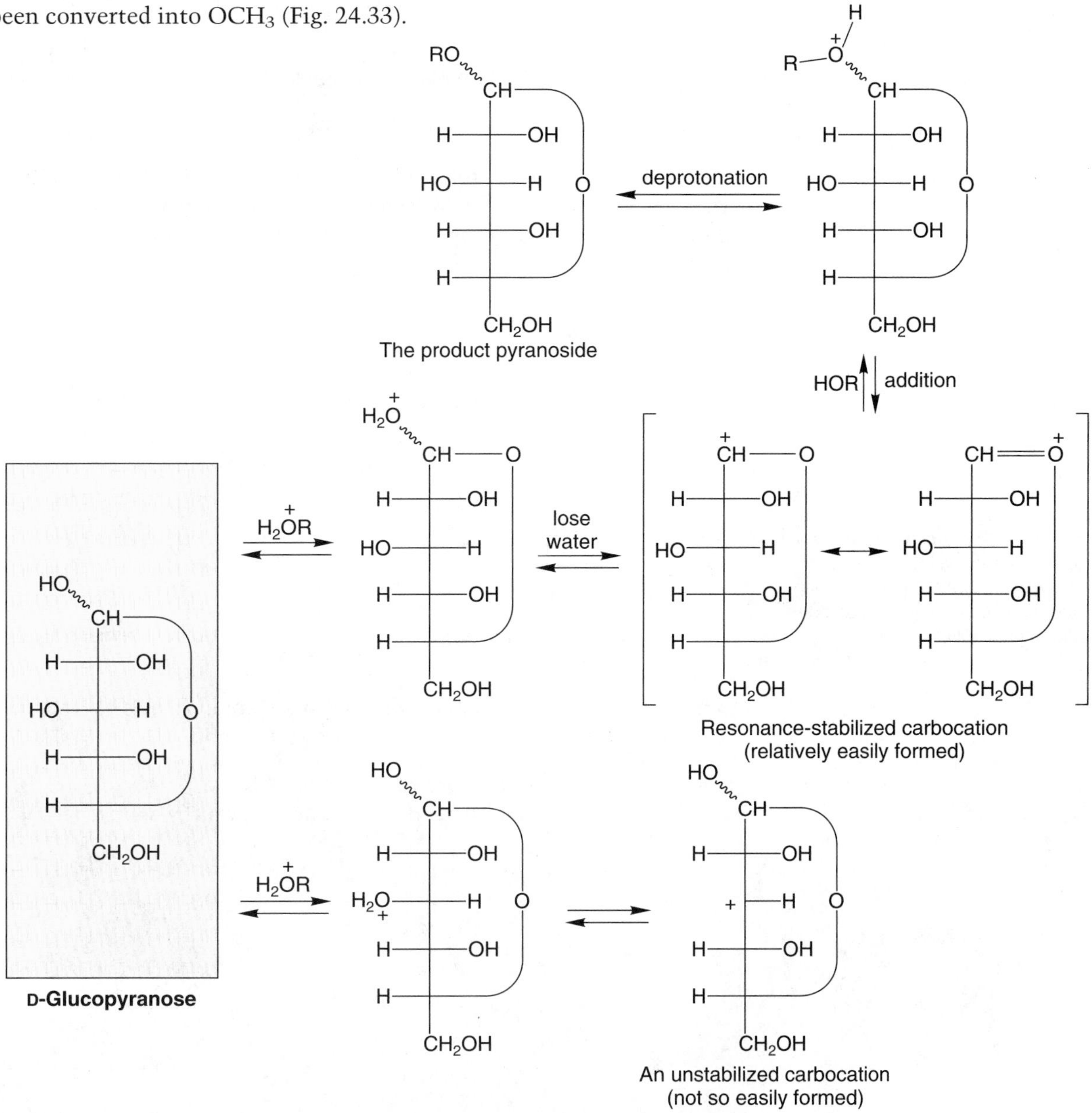

The product pyranoside

deprotonation

HOR addition

$H_2\overset{+}{O}R$

lose water

Resonance-stabilized carbocation (relatively easily formed)

D-Glucopyranose

$H_2\overset{+}{O}R$

An unstabilized carbocation (not so easily formed)

FIGURE **24.33** Although all OH groups can be reversibly protonated, loss of only the anomeric OH leads to a resonance-stabilized cation. Addition of alcohol at this position gives the glycoside.

Specific glycosides are named by replacing the "ose" of the simple sugar's name with "oside" and using the name of the group used to make the ether link. For example, if methyl alcohol were used to form a full acetal with the hemiacetal D-glucopyranose, both methyl α- and methyl β-D-glucopyranosides would be produced (Fig. 24.34).

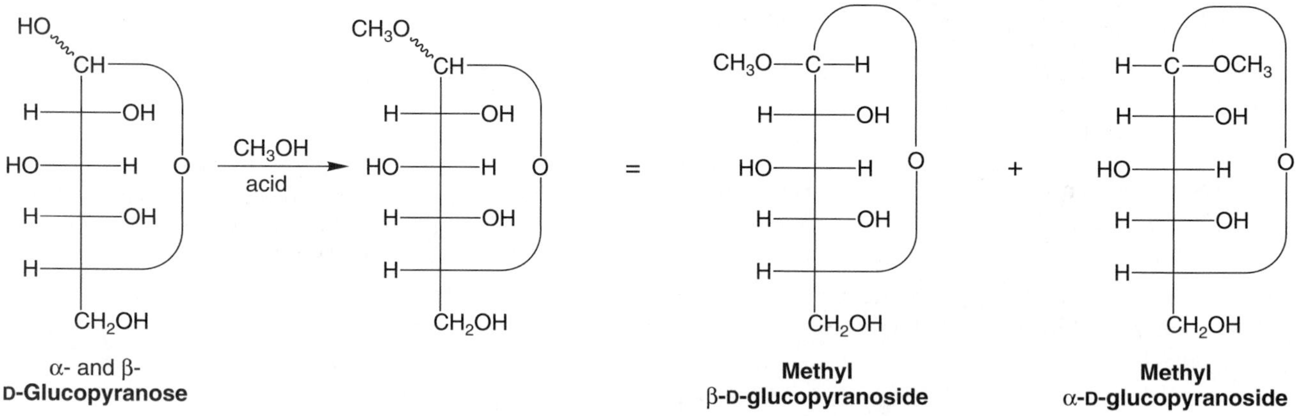

α- and β-
D-Glucopyranose

CH_3OH, acid

Methyl β-D-glucopyranoside

Methyl α-D-glucopyranoside

FIGURE 24.34 Methyl α- and methyl β-D-glucopyranosides.

In a similar vein, the tetra-ether, monoacetal formed in the Williamson ether synthesis can be hydrolyzed in acid to a compound in which only the acetal methoxyl group has been transformed into a hydroxyl group (Fig. 24.35).

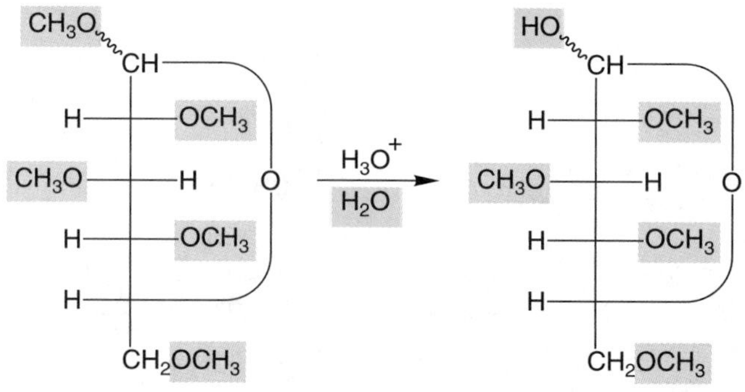

FIGURE 24.35 Hydrolysis of the fully methylated compound leads to a hemiacetal in which only the methoxyl group at the anomeric position [C(1)] has been converted into an OH.

*PROBLEM 24.18 Explain carefully why it is only the acetal methoxyl group that is converted into a hydroxyl group (Fig. 24.35).

ANSWER This kind of reaction takes place by protonation of an ether oxygen, loss of alcohol to give a carbocation, and addition of water followed by deprotonation.

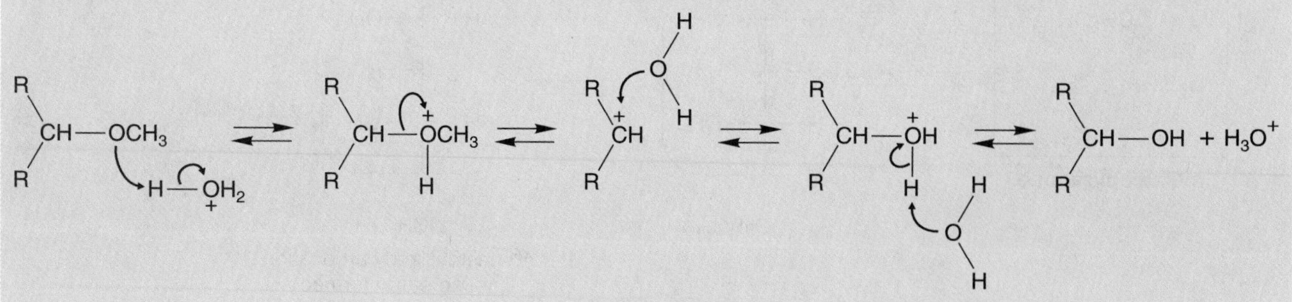

$$R_2CH-OCH_3 + H-\overset{+}{O}H_2 \rightleftarrows R_2CH-\overset{+}{O}(H)CH_3 \rightleftarrows R_2\overset{+}{C}H + H_2O \rightleftarrows R_2CH-\overset{+}{O}H_2 \rightleftarrows R_2CH-OH + H_3O^+$$

ANSWER (CONTINUED)

All of the ether oxygens can be protonated, but only one can lose an alcohol to give a resonance-stabilized carbocation. Addition of water to the cation, followed by deprotonation, leads to the tetra-ether shown in Figure 24.35 (shown in the α-form; both anomers can be formed here).

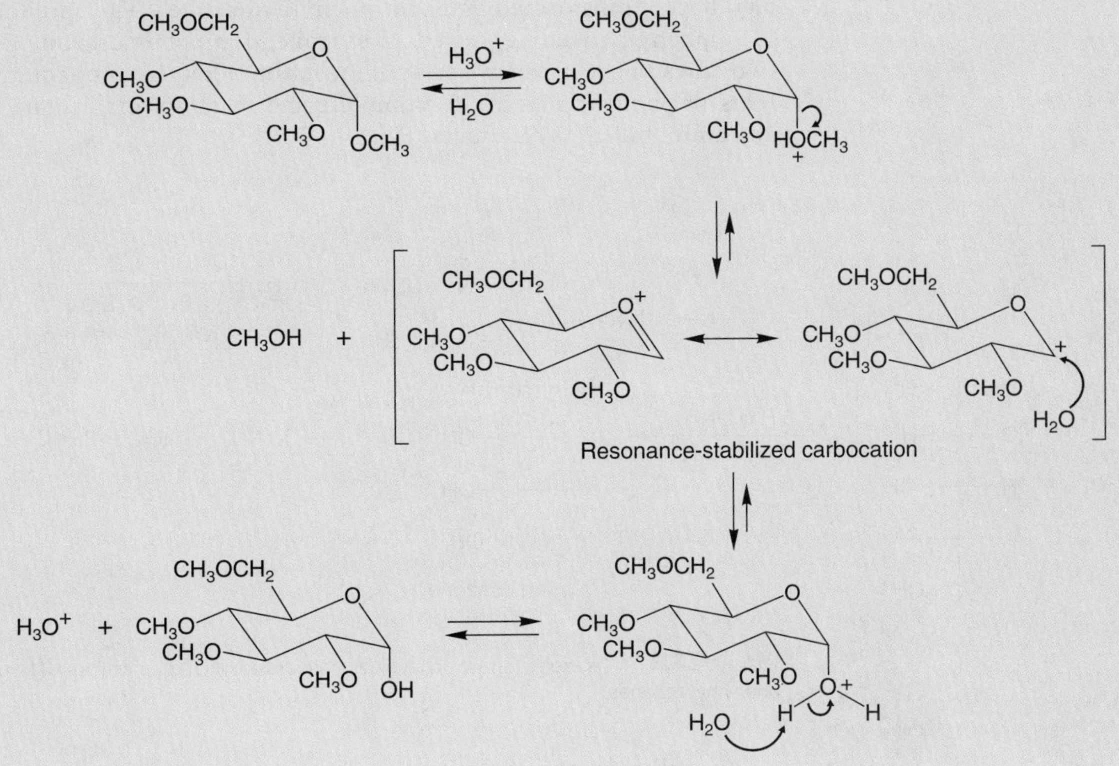

CH_3OCH_2, CH_3O, CH_3O, CH_3O, OCH_3 ⇌ (H_3O^+ / H_2O) CH_3OCH_2, CH_3O, CH_3O, CH_3O, $\overset{+}{H}OCH_3$

CH_3OH + [CH_3OCH_2, CH_3O, CH_3O, CH_3O, O^+ ⟷ CH_3OCH_2, CH_3O, CH_3O, CH_3O, +, H_2O]

Resonance-stabilized carbocation

H_3O^+ + CH_3OCH_2, CH_3O, CH_3O, CH_3O, OH ⇌ CH_3OCH_2, CH_3O, CH_3O, CH_3O, O^+, H, H, H_2O

24.2f Osazone Formation

Treatment of aldo sugars with phenylhydrazine under acidic conditions initiates an odd, and very useful reaction. What would we expect? We know that the cyclic forms of aldohexoses are in equilibrium with the open-chain isomers that contain a free aldehyde group. Aldehydes react with substituted hydrazines to generate hydrazones (Chapter 16, p. 783), and it would be no surprise to see the reaction of Figure 24.36 in which the carbonyl group at C(1) has formed a phenylhydrazone derivative.

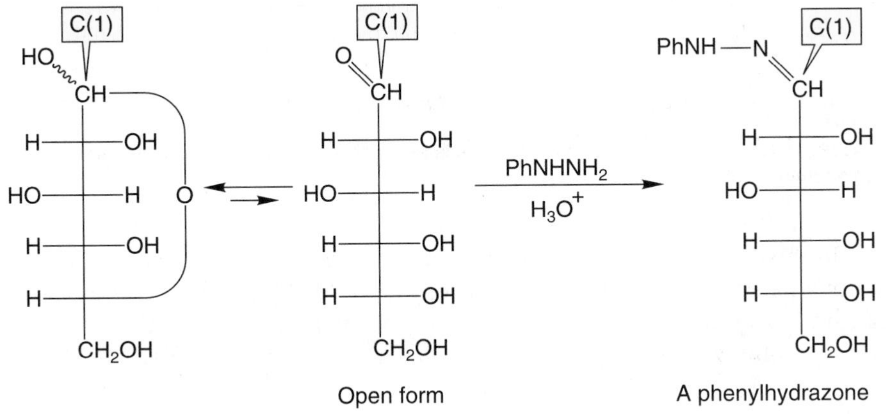

C(1), HO, CH, H, OH, HO, H, O, H, OH, H, CH_2OH ⇌ C(1), O, CH, H, OH, HO, H, H, OH, H, OH, CH_2OH —($PhNHNH_2$ / H_3O^+)→ PhNH—N, C(1), CH, H, OH, HO, H, H, OH, H, OH, CH_2OH

Open form

A phenylhydrazone

FIGURE **24.36** The small amount of free aldehyde present at equilibrium accounts for phenylhydrazone formation at C(1).

PROBLEM **24.19** Write a mechanism for the reaction of Figure 24.36.

What *is* a surprise, is to find that not only C(1) but C(2) as well has been transformed into a phenylhydrazone. The product is a 1,2-diphenylhydrazone, called an **osazone**, or a phenylosazone. Here are two clues to the mechanism of the reaction. First, it takes three equivalents of phenylhydrazine to complete the reaction and second, aniline and ammonia are byproducts (Fig. 24.37).

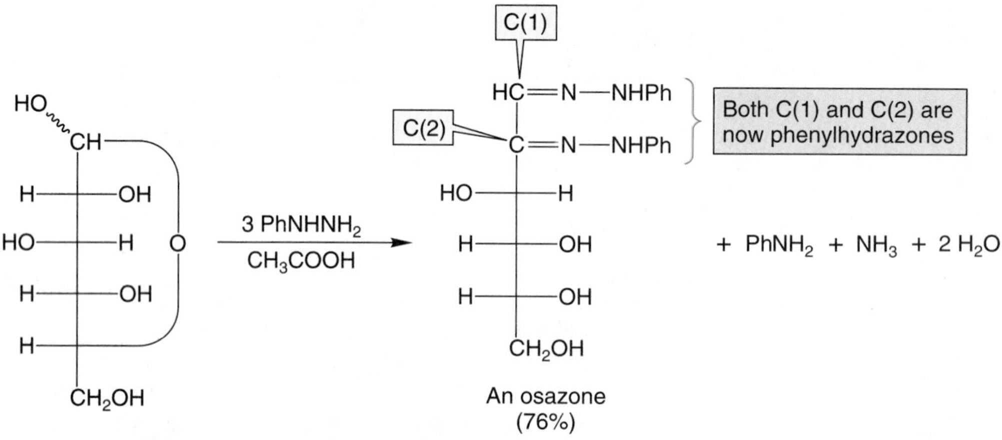

FIGURE **24.37** Osazone formation involves conversion of C(2) as well as C(1) into phenylhydrazones.

We will start our analysis with the reaction at C(1) to form a phenylhydrazone, a substituted imine. Imines (in this case a phenylhydrazone) are in equilibrium with enamines, just as ketones are in equilibrium with enols (Chapter 18, p. 878; Fig. 24.38).

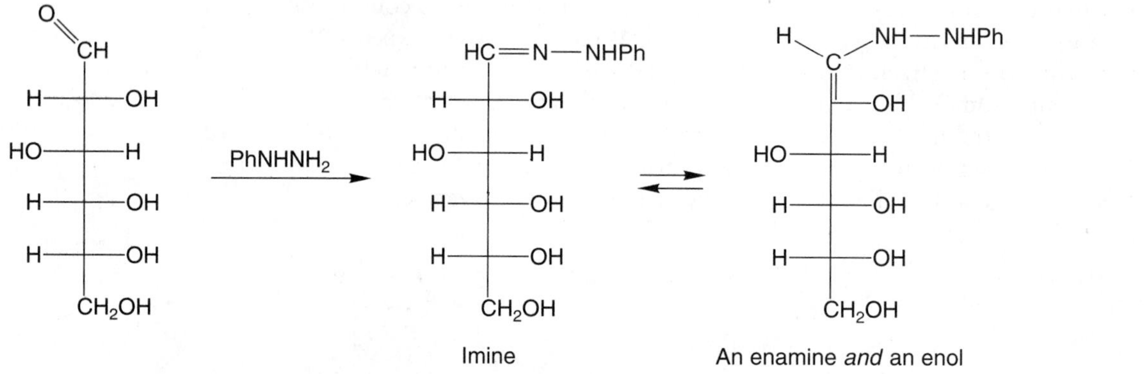

Imine

An enamine *and* an enol

FIGURE **24.38** The C(1) phenylhydrazone is an imine and therefore in equilibrium with an enamine. This enamine is also an enol.

In this case, the enamine is also an enol, and is therefore in equilibrium with a ketone that can react with a second mole of phenylhydrazine. Now aniline is eliminated to give a new imine, which reacts with the third mole of phenylhydrazine to give the final osazone product (Fig. 24.39).

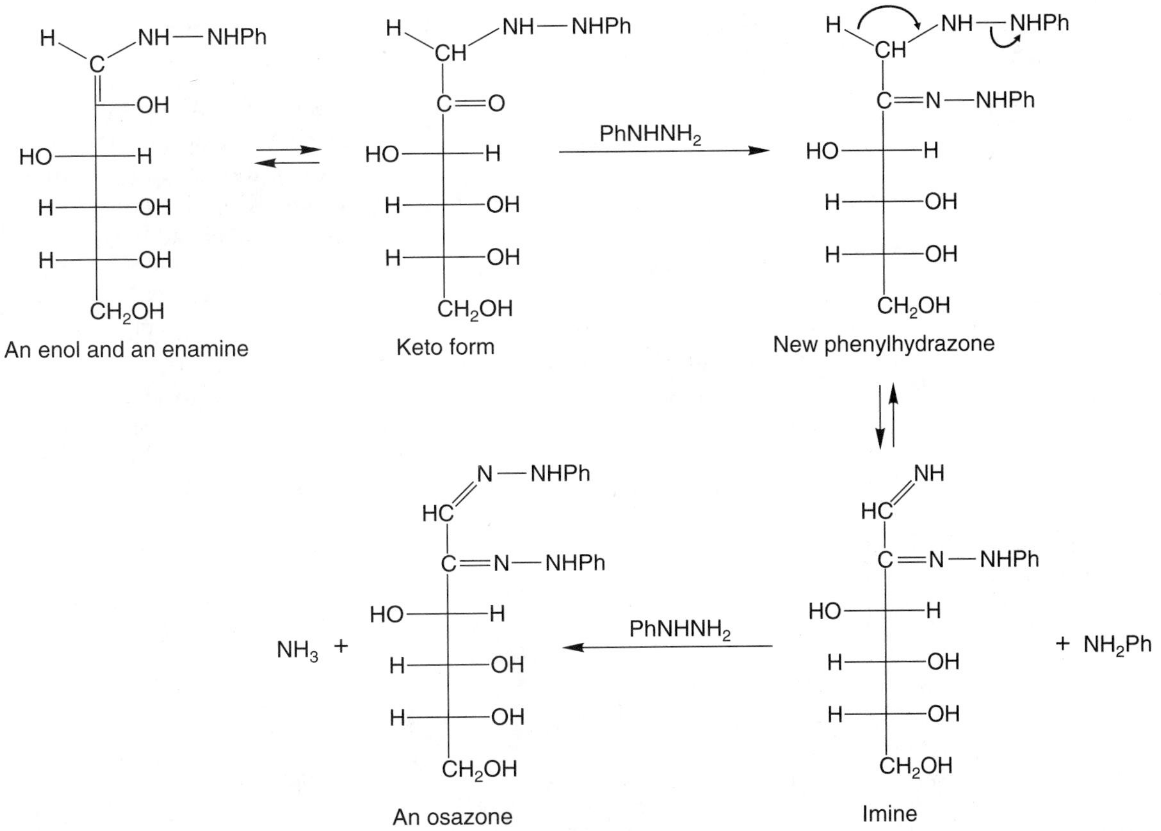

FIGURE 24.39 Reaction of the ketone with phenylhydrazine leads to a new phenylhydrazone that can eliminate aniline to give a new imine. Reaction with a third equivalent of phenylhydrazine leads to the osazone.

Notice that osazone formation destroys the stereogenic carbon at C(2). A given osazone can be formed from *two* sugars, epimeric at C(2). Figure 24.40 makes this point using D-glucose and D-mannose, two aldohexoses that must give the same osazone. When we work through the Fischer proof of the structure of glucose and the other aldohexoses, we will make use of this point a number of times, so remember it.

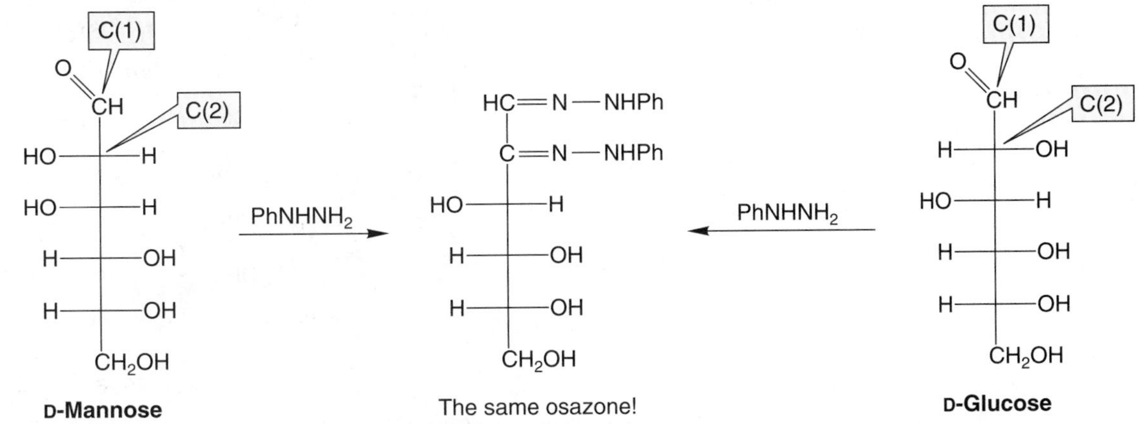

FIGURE 24.40 An osazone can be formed from two different aldo sugars that are stereoisomeric at C(2). The stereochemistry at C(2) is destroyed in the reaction.

24.2g Methods of Lengthening and Shortening Chains in Carbohydrates

Treatment of D-ribose with cyanide, followed by catalytic hydrogenation and hydrolysis, generates *two* new sugars: D-allose and D-altrose. This reaction is an example of a modern variation of the **Kiliani–Fischer synthesis** (Emil Fischer again and Heinrich Kiliani, 1855–1945). In the first step of this reaction, sodium cyanide (NaCN) adds to the carbonyl group to make a cyanohydrin (Chapter 16, p. 772). The key point is that a new stereogenic carbon is generated in this process. The new OH group can be on either the right or left in Fischer projection. The second step of the Kiliani–Fischer synthesis, catalytic reduction, gives a pair of imines that are hydrolyzed under the reaction conditions to aldehydes. The net result is the formation of two new sugars, each one carbon longer than the starting sugar. The new sugars differ *only* in their stereochemistry at the newly formed C(2) [they are C(2) epimers; Fig. 24.41].

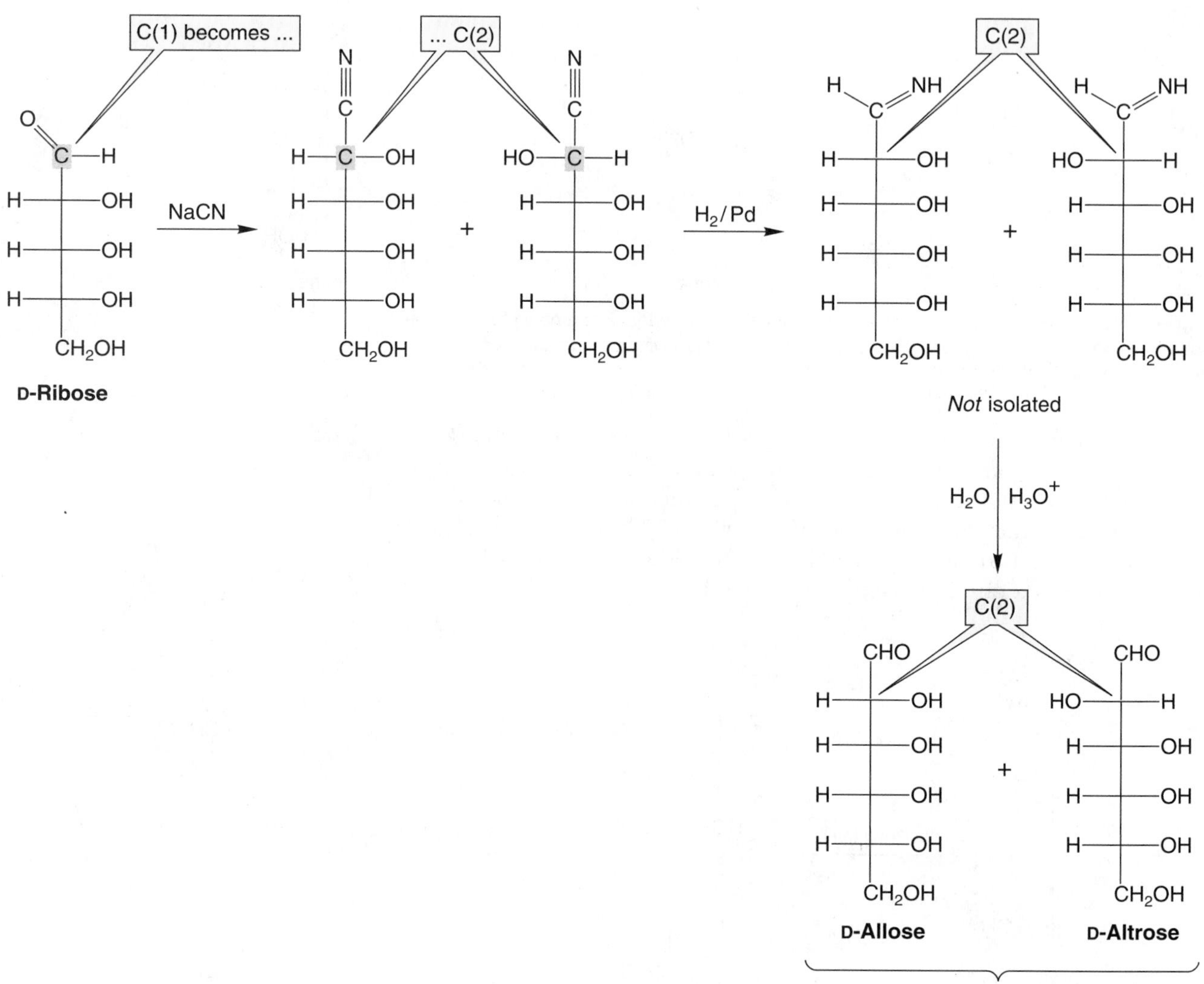

FIGURE 24.41 A modern version of the Kiliani–Fischer synthesis generates two new sugars, each one carbon longer than the starting sugar.

PROBLEM **24.20**

Apply the Kiliani–Fischer synthesis to D-glyceraldehyde. What new sugars are formed? It is not necessary to write mechanisms for the reactions.

PROBLEM **24.21**

The following two sugars are produced by Kiliani–Fischer synthesis from an unknown sugar (Fig. 24.42). What is the structure of that unknown sugar?

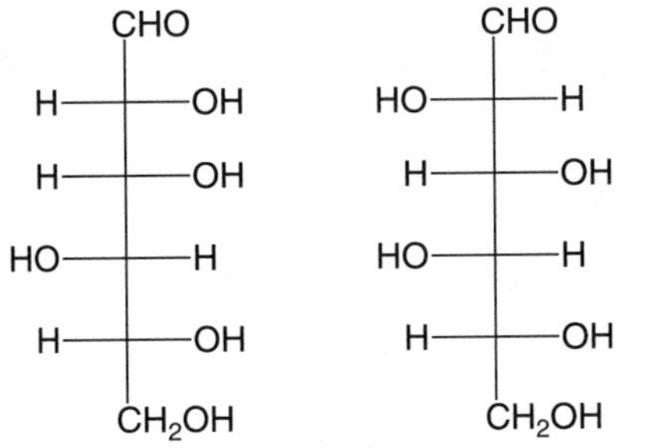

CHO; H—OH; H—OH; HO—H; H—OH; CH_2OH
D-Gulose

CHO; HO—H; H—OH; HO—H; H—OH; CH_2OH
D-Idose

FIGURE **24.42**

The **Ruff degradation** (Otto Ruff, 1871–1939) is a method of shortening the backbone of a sugar by removal of the aldehyde at C(1) and creation of a new aldehyde at the old C(2). In this example, the calcium salt of D-galactonic acid is treated with ferric ion and hydrogen peroxide to give D-lyxose (Fig. 24.43).

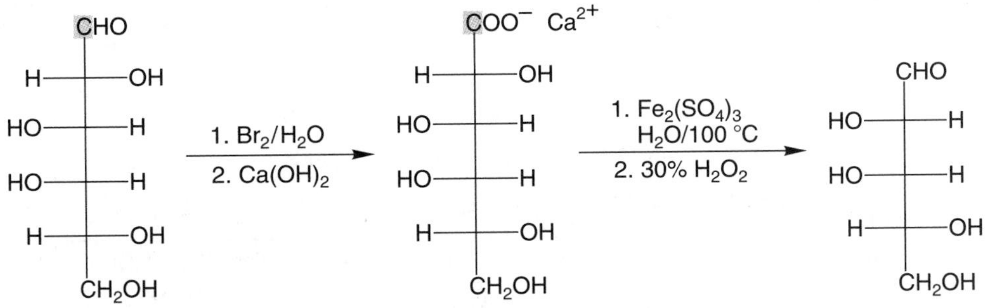

CHO; H—OH; HO—H; HO—H; H—OH; CH_2OH
D-Galactose

1. Br_2/H_2O
2. $Ca(OH)_2$

COO^- Ca^{2+}; H—OH; HO—H; HO—H; H—OH; CH_2OH
Ca Salt of D-galactonic acid

1. $Fe_2(SO_4)_3$ H_2O/100 °C
2. 30% H_2O_2

CHO; HO—H; HO—H; H—OH; CH_2OH
D-Lyxose (41%)

FIGURE **24.43** The Ruff degradation shortens the starting sugar by one carbon. It is the original aldehyde carbon that is lost.

The mechanism of this reaction remains obscure (chemistry is a living science—not everything is fully understood) but one imagines that radicals are involved. Metal ions can act as efficient electron-transfer agents, and a guess at the mechanism might involve formation of the carboxyl radical, decarboxylation, capture of the hydroxyl radical formed, and a final transformation of the hydrate into the aldehyde. Whatever the details of the mechanism, the stereochemistry at the original C(2) is lost, but there can be no changes at the old C(3), C(4), or C(5) (Fig. 24.44).

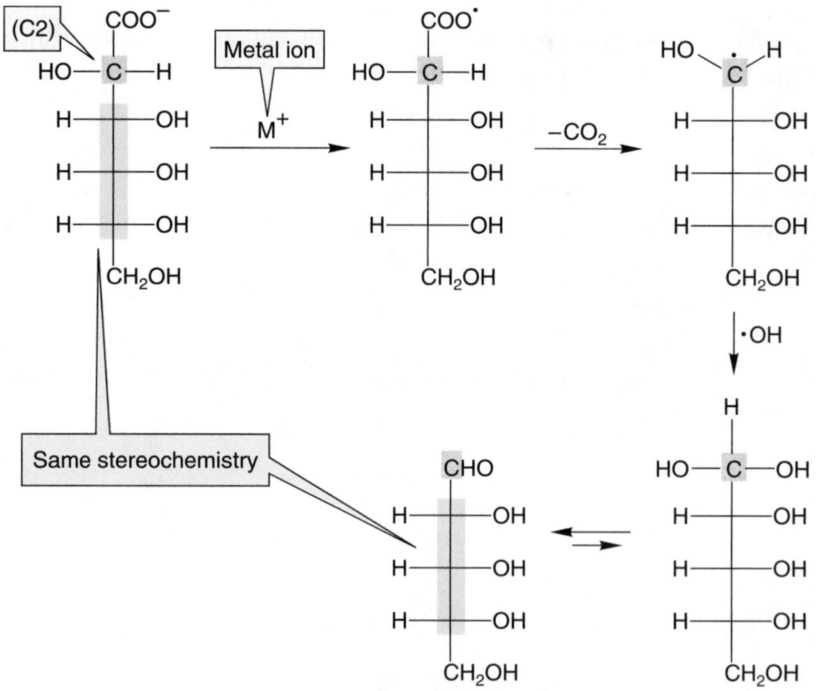

FIGURE 24.44 A possible mechanism for the Ruff degradation.

A similar process, if only in the outcome, is called the **Wohl degradation** (Alfred Wohl, 1863–1933). In a sense, this process reverses the Kiliani–Fischer synthesis. An oxime (a hydroxy imine) is formed first, and then transformed into the nitrile. Treatment with base hydrolyzes five acetoxy groups to hydroxyls. The resulting cyanohydrin then eliminates cyanide and generates the chain-shortened aldehyde (Fig. 24.45).

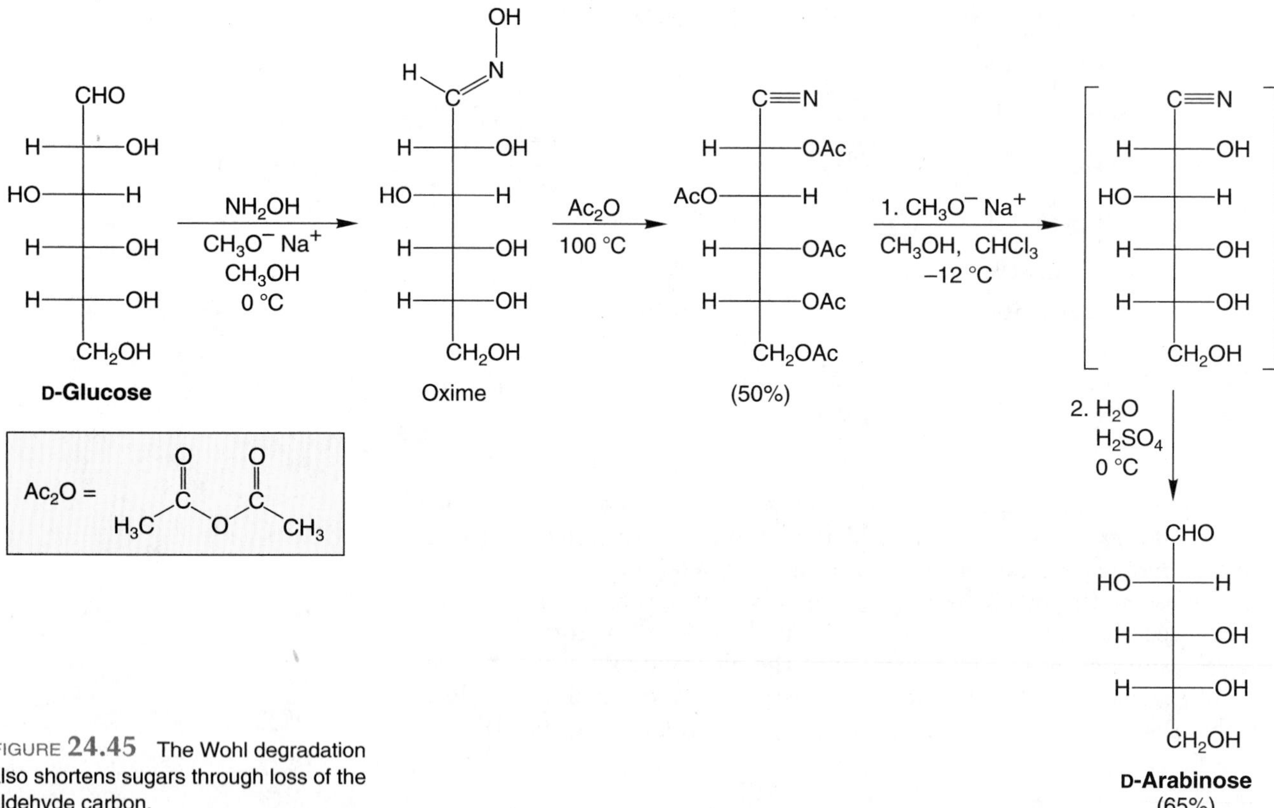

FIGURE 24.45 The Wohl degradation also shortens sugars through loss of the aldehyde carbon.

Both the Ruff and Wohl degradations shorten the chain by one carbon without disturbing the remaining stereogenic carbons.

24.3 THE RING SIZE IN GLUCOSE

We can take advantage of one of these reactions of sugars, combined with an old reaction, to determine the size of the ring in glucose and other aldohexoses. It is really very simple. D-Glucose is first converted into its methyl glycoside by treatment with methyl alcohol in acid. Figure 24.46 shows formation of methyl β-D-glucopyranoside. Now an old reaction is used. Recall that periodate cleaves 1,2-diols to dicarbonyl compounds (Chapter 17, p. 842). Therefore, treatment of methyl β-D-glucopyranoside with HIO_4 leads to the dialdehyde shown in Figure 24.46.

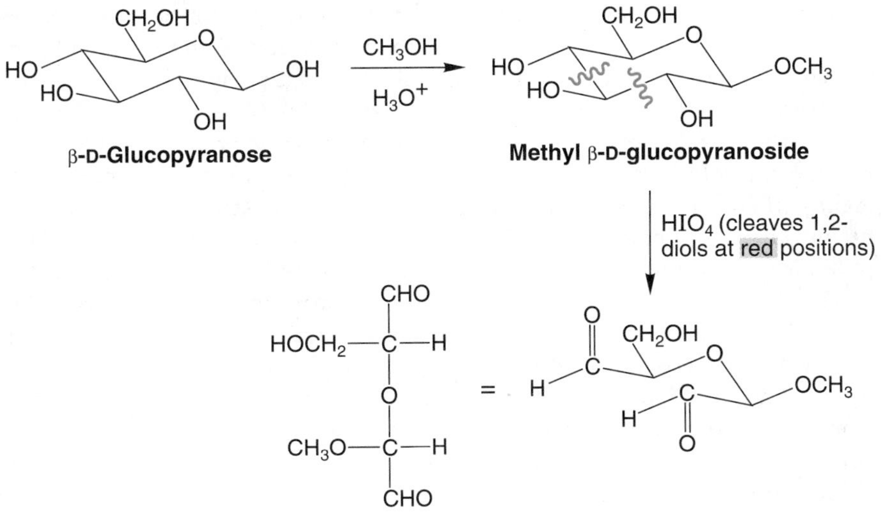

β-D-Glucopyranose

CH_3OH, H_3O^+

Methyl β-D-glucopyranoside

HIO_4 (cleaves 1,2-diols at red positions)

FIGURE 24.46 Formation of methyl β-D-glucopyranose and further oxidation with periodate.

If D-glucose had been mainly in the furanoside form, quite another structure would have been formed. Figure 24.47 shows the reaction of HIO_4 with methyl β-D-glucofuranoside.

β-D-Glucofuranose

CH_3OH, H_3O^+

Methyl β-D-glucofuranoside

HIO_4 (cleaves 1,2-diols at red positions)

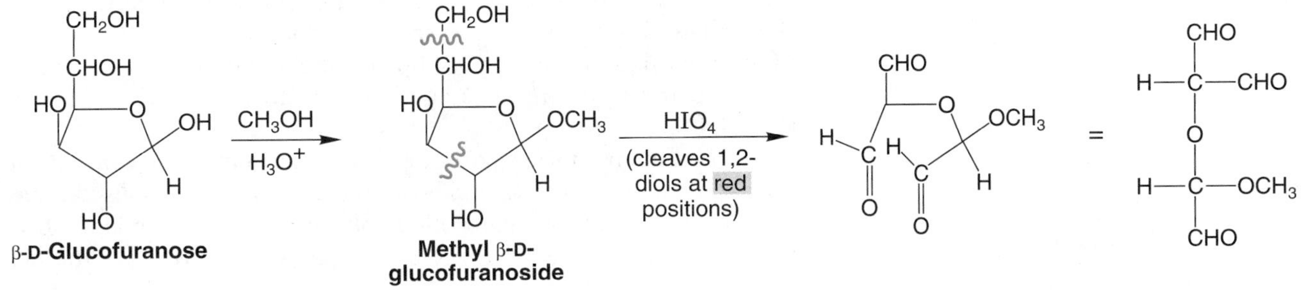

FIGURE 24.47 Formation of methyl β-D-glucofuranose and further oxidation with periodate.

This procedure is one way in which the predominant ring structure in glucose, a six-membered pyranose, was determined.

24.4 THE FISCHER DETERMINATION OF THE STRUCTURE OF D-GLUCOSE (AND THE 15 OTHER ALDOHEXOSES)

In the following pages, we will approximate the reasoning that led Emil Fischer to the structure of D-glucose. The structures of the other 15 aldohexoses can be determined as we go along. As a bonus, we will also be able to derive the structures of the eight aldopentoses. A risk in this kind of discussion, which straightens out the real, highly twisted history of the Fischer determination, is that the work will seem too easy, too straightforward. Fischer's papers appeared in 1891, less than two decades after van't Hoff and LeBel's independent hypothesis of the tetrahedral carbon atom. Determining stereochemistry was no simple matter, and Fischer's work depended critically on a close analysis of the stereochemical relationships of complex molecules. Moreover, the experimental work was extraordinarily difficult and slow.* Fischer's work was monumental, and remains an example of the best we humans are able to accomplish.

Emil Fischer recognized that it was impossible for him to solve one of the central problems in this endeavor, the separation of the D series from the enantiomeric L series. In the 1880s there was no way to determine absolute configuration, which is hardly surprising, as the structural hypothesis of the tetrahedral carbon atom was barely 15 years old at the time. However, Fischer recognized that the 16 aldohexoses must be composed of eight pairs of mirror images, and that if he knew the structures of one set, he automatically could draw the structures of the mirror-image set. The problem was to tell which was which, and this he knew he could not do. So, he did the next best thing, he was arbitrary; one might say that he guessed. The odds aren't bad, 50:50, and there are times when it is best to accept what is possible and not give up because a perfect solution to the problem at hand isn't available. So Fischer simply *defined* one set of eight isomers as the D series, knowing that it might be necessary for history to correct him. In the event, fortune smiled and he made the right guess. I think it is nice that it came out that way. Be sure that you understand that the correctness of Fischer's guess wasn't due to cleverness—his success was not the result of a shrewd guess or the product of a well-developed intuition; there really was no way to know at the time. He was just lucky.

The Fischer proof starts with arabinose, arbitrarily assumed to be the D enantiomer. The first critical observation was that the Kiliani–Fischer synthesis applied to the pentose, D-arabinose, led to a pair of D-aldohexoses (as it must). These were D-glucose and D-mannose. Because the Kiliani–Fischer synthesis must produce a pair of sugars that have different configurations *only* at the newly generated stereogenic carbon, D-glucose and D-mannose must share the following partial structures (Fig. 24.48).

Let's take a moment at the beginning of this section to look carefully at Figure 24.48. What do we know, and how do we know it? Emil Fischer has a flask containing one of the enantiomers of the pentose called arabinose. He really doesn't know whether this enantiomer is D or L, that is, whether

*Here is what Fischer had to say to his mentor, Adolph Bayer, about this matter in 1889. "The investigations on sugars are proceeding very gradually ... . Unfortunately, the experimental difficulties in this group are so great, that a single experiment takes more time in weeks than other classes of compounds take in hours, so only very rarely a student is found who can be used for this work."

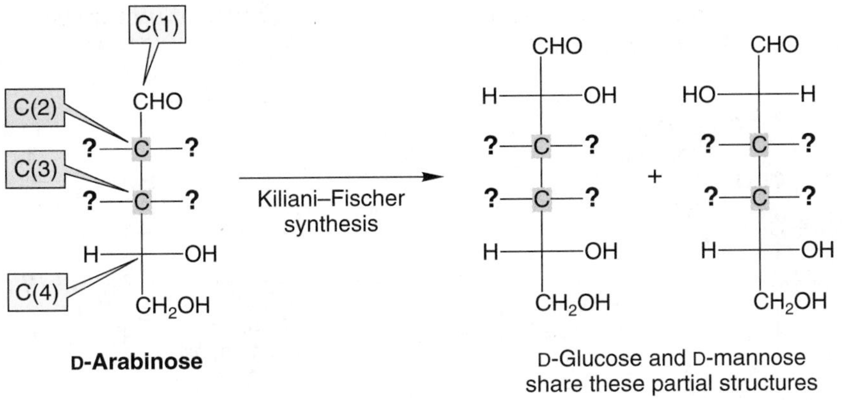

D-Arabinose

D-Glucose and D-mannose share these partial structures

FIGURE 24.48 The first step in Emil Fischer's determination of the structure of glucose. The Kiliani–Fischer synthesis applied to D-arabinose leads to D-glucose and D-mannose, which must share the partial structures shown.

the OH at C(4) of arabinose is on the right or left. He "solves" this problem by guessing, by assuming that the sugar he has in the flask is a member of the D series defined as having the OH at C(4) on the right. We now know that he guessed correctly. He has no idea of the relative arrangement of the OH groups at C(3) and C(2). Kiliani–Fischer synthesis applied to D-arabinose gives D-glucose and D-mannose, one of which has the newly created OH on the carbon adjacent to the aldehyde on the right, the other of which has it on the left.

The second step in the Fischer proof determines the configuration of C(2) in D-arabinose [this carbon becomes C(3) in the two D-aldohexoses: D-glucose and D-mannose]. Fischer observed that the oxidation of D-arabinose with nitric acid gives an *optically active diacid.* The possibilities are shown in Figure 24.49. There are only three possible diacids that can be produced from a D-aldopentose, and two of them are meso. In the other, optically active diacid, the OH at C(2) of the diacid [which corresponds to C(3) in the aldohexoses] is on the left (Fig. 24.49).

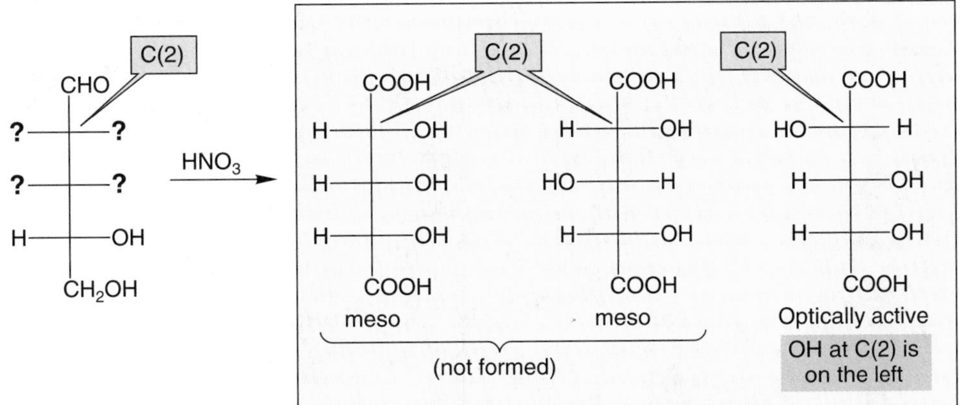

FIGURE 24.49 Oxidation of D-arabinose leads to an optically active diacid, which shows that the OH at C(2) in D-arabinose is on the left.

Now we know a bit more about the partial structures of D-arabinose, D-glucose, and D-mannose. We can specify all but one OH in D-arabinose and in the two structures shared by the two D-aldohexoses. We do not yet know the position of the OH at C(3) of D-arabinose, which becomes C(4) in the D-aldohexoses, and we do not know which partial structure belongs to D-glucose and which to D-mannose (Fig. 24.50).

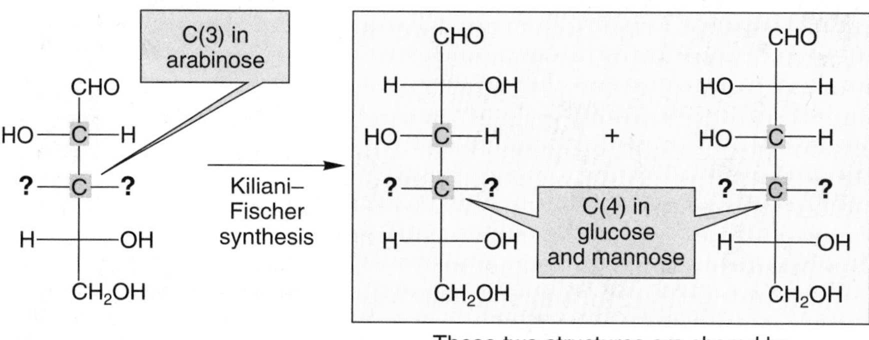

FIGURE 24.50 What we now know about the structures of D-arabinose, D-glucose, and D-mannose. Only the configuration at C(3) of D-arabinose, which becomes C(4) in D-glucose and D-mannose, is left to be determined.

In the third step, the position of the last OH is fixed by the observation that both D-glucose and D-mannose give *optically active diacids* on oxidation with nitric acid. There are two possibilities. The OH at C(4) is either on the left or the right. If it is on the right, both sugars do give an optically active diacid, in agreement with the experimental data (Fig. 24.51).

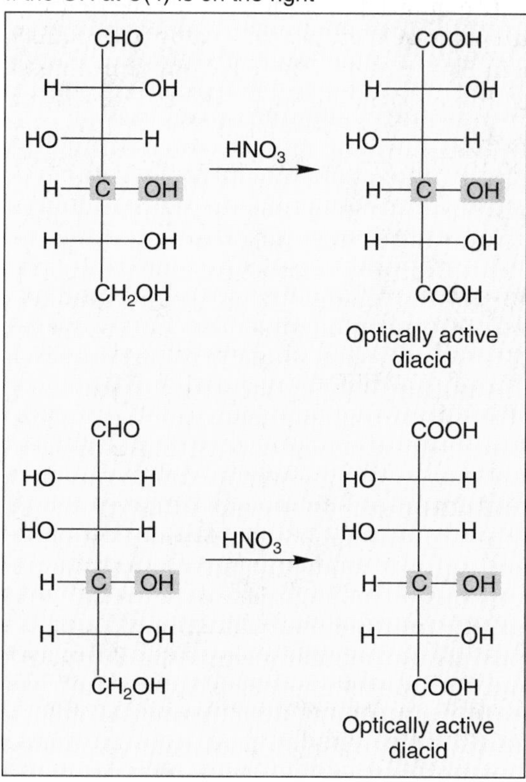

FIGURE 24.51 If the last unknown OH is on the right, two optically active diacids are produced on oxidation with nitric acid. This result matches the experimental results.

On the other hand, if the OH at C(4) is on the left, one of the diacids is meso, not optically active, which does not match the experimental data (Fig. 24.52).

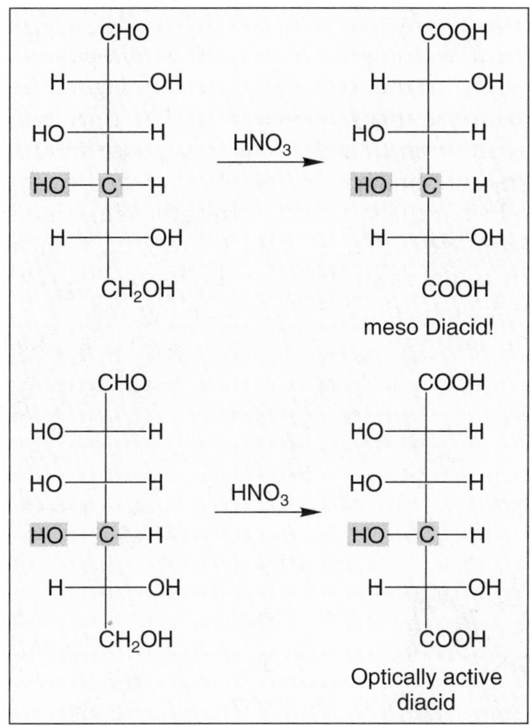

FIGURE 24.52 However, if the unknown OH had been on the left, one possible diacid is meso, not optically active, which does not match the experimental results.

So, the correct position of the OH at C(3) in D-arabinose and at C(4) in the two D-aldohexoses, D-glucose and D-mannose, must be on the right. D-Glucose and D-mannose share the two structures in Figure 24.53. The next problem is to tell which is which.

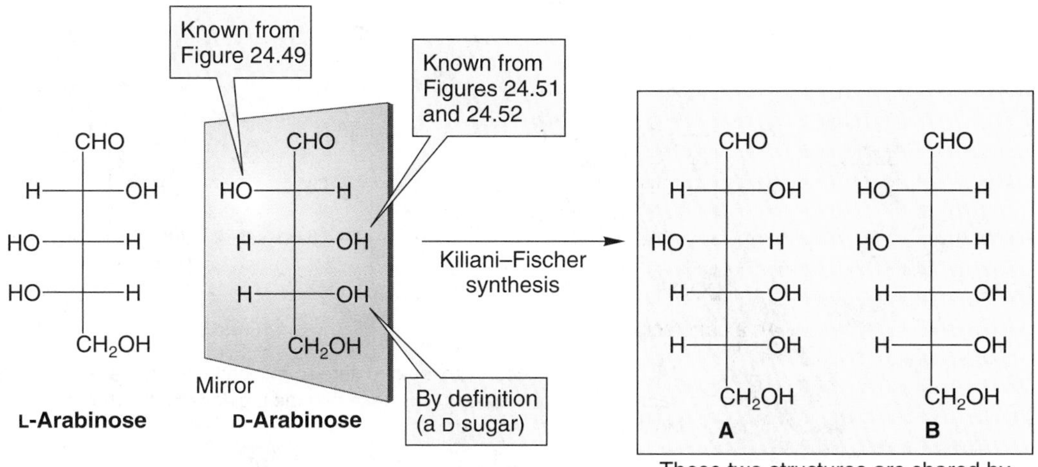

FIGURE 24.53 Now we know the structure of D-arabinose, L-arabinose, and the two structures (**A** and **B**) shared by D-glucose and D-mannose. We do not know which structure belongs to D-glucose and which to D-mannose.

We now know the full structure of D-arabinose as the positions of all three OH groups have been determined. Of course, if we know the structure of the D-enantiomer, we know that of its mirror image, the L-enantiomer (Fig. 24.53).

Now we return to the question of the structures of D-glucose and D-mannose. Which of the two structures in Figure 24.53 is that of D-glucose and which is that of D-mannose? The final necessary observation is that D-glucose *and another sugar,* L-gulose (notice that it's an L sugar!), gave the same optically active diacid when oxidized with nitric acid. We can find what the structure of the other sugar must be in the following way. Treatment with nitric acid destroys the identity of the groups at the ends of the chain, as an aldaric acid is formed (p. 1251). If we simply invert the sugar, placing the CH_2OH at the top and the CHO at the bottom, we will still get an aldaric acid on oxidation (Fig. 24.54).

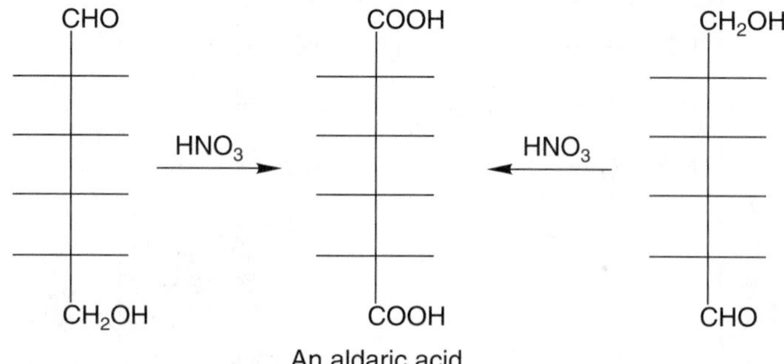

FIGURE 24.54 Oxidation with nitric acid renders the ends of a sugar equivalent. Both the aldehyde end and the primary alcohol end are converted into the same group, a carboxylic acid.

Now there are two possibilities, **A** and **B**. If D-glucose has structure **A** of Figures 24.53 and 24.55, then L-gulose, the sugar that gives the same aldaric acid on oxidation, must have the structure shown at the right in Figure 24.55.

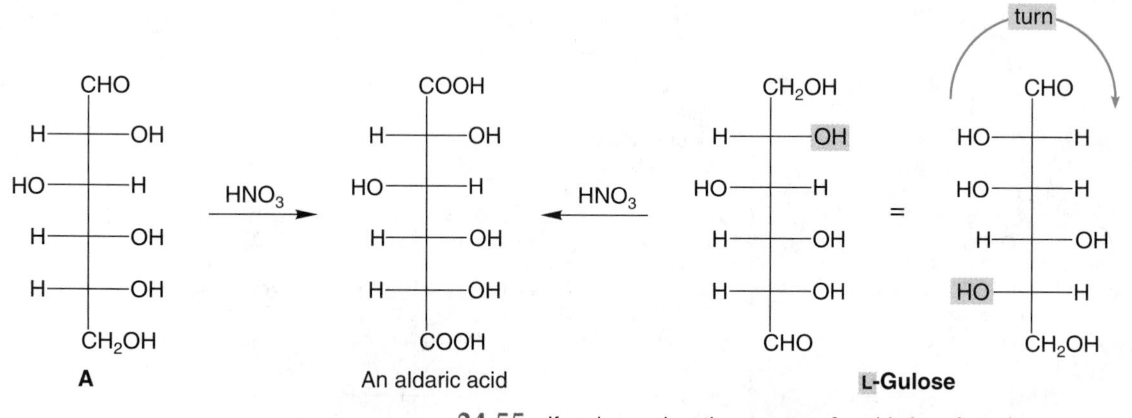

FIGURE 24.55 If D-glucose has the structure **A**, oxidation of another sugar, L-gulose, can give the same aldaric acid. Note in this case that the L-gulose is drawn with the CH_2OH at the top and the CHO at the bottom.

Now let's see what happens if D-glucose has the other possible structure, **B**. Again, we invert the sugar **B** in an attempt to generate another sugar that can give the same aldaric acid. This time, however, the attempt fails, as the "new" sugar is identical with the first. Both are **B** (Fig. 24.56)! There is only one sugar that can give this diacid. There can be no second sugar. D-Glucose must have structure **A**, and the other, **B**, must therefore be D-mannose.

B

HNO_3

An aldaric acid

HNO_3

Still **B**!

turn

FIGURE 24.56 However, there is no sugar other than **B** that can give this aldaric acid. As this fact does not match the experimental results, D-glucose must have structure **A**.

Now let's take stock of what we know, which is rather more than we might think at first. We have the structures of 6 of the 16 possible aldohexoses, as we know D-glucose, D-mannose, and L-gulose and their enantiomers L-glucose, L-mannose, and D-gulose. In addition, we know the structures of D- and L-arabinose (Fig. 24.57).

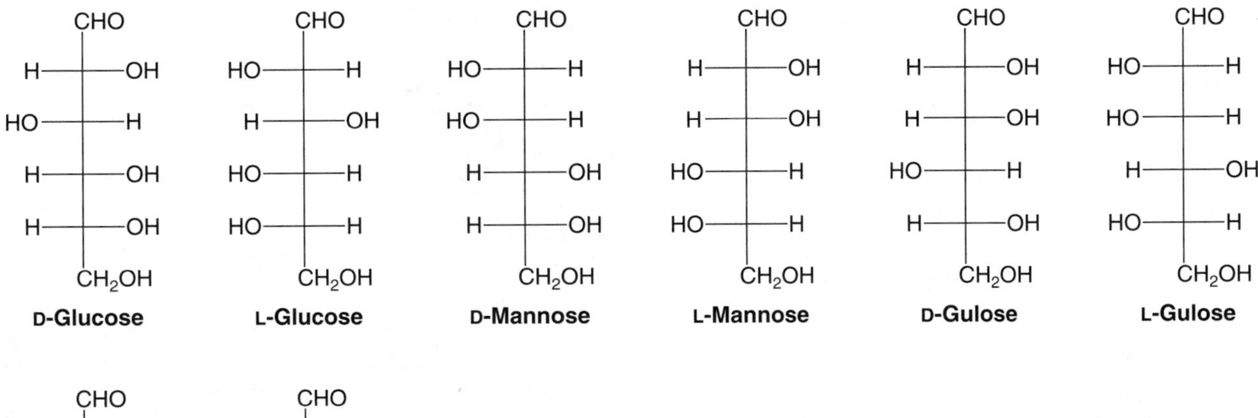

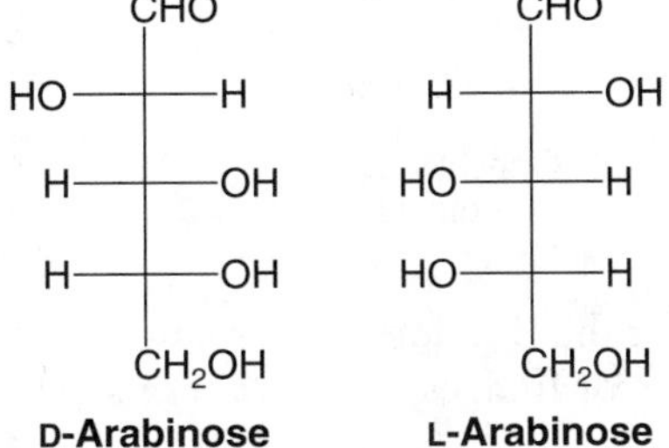

FIGURE 24.57 Now we know the structures of these aldohexoses and aldopentoses.

There are many ways to fill out the structures of the remaining aldopentoses and hexoses. Here is just one possible sequence. First, let's first deal with the three remaining D-aldopentoses, given that we know the structure of D-arabinose. D-Ribose gives the same osazone as D-arabinose (Fig. 24.58), which means that the two sugars must differ only at the configuration at C(2). Therefore D-ribose must have the structure shown in Figure 24.58.

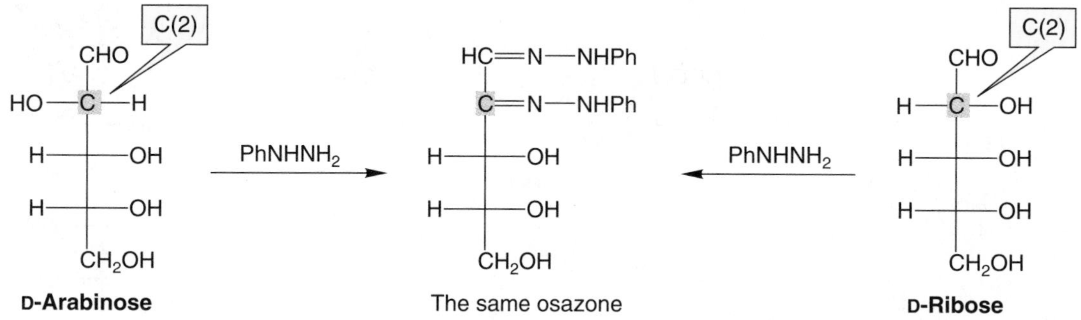

FIGURE 24.58 D-Arabinose and D-ribose give the same osazone, and therefore can differ only at C(2). The structure of D-ribose is therefore known. L-Ribose is simply the mirror image of the D isomer.

There are now only two remaining D-aldopentoses. One of them, D-xylose, gives a meso diacid on oxidation, whereas the other, D-lyxose, gives an optically active diacid. The structures must be as shown in Figure 24.59.

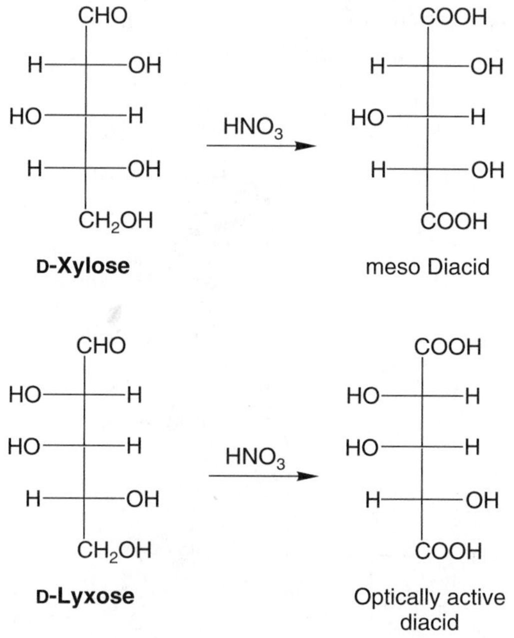

FIGURE 24.59 D-Xylose gives a meso diacid and D-lyxose gives an optically active diacid, which fixes the structures of the remaining D-aldopentoses.

Now we know all four stereoisomers of the D-aldopentoses, and we can construct the L series by drawing the mirror images of the D series. That gives us all eight aldopentoses.

Now let's complete the determination of the structures of the D-aldohexoses. D-Gulose and D-idose give the same osazone and must differ only in their configurations at C(2) (Fig. 24.60).

The Kiliani–Fischer synthesis applied to D-ribose gives D-allose and D-altrose, which must share the two structures of Figure 24.61. The two structures can be assigned by noting that D-allose gives an optically inactive, meso diacid on treatment with nitric acid, whereas D-altrose gives an optically active diacid.

There are now only two remaining D-aldohexoses: D-talose and D-galactose. D-Galactose gives a meso diacid on oxidation with nitric acid, but D-talose gives an optically active diacid (Fig. 24.62).

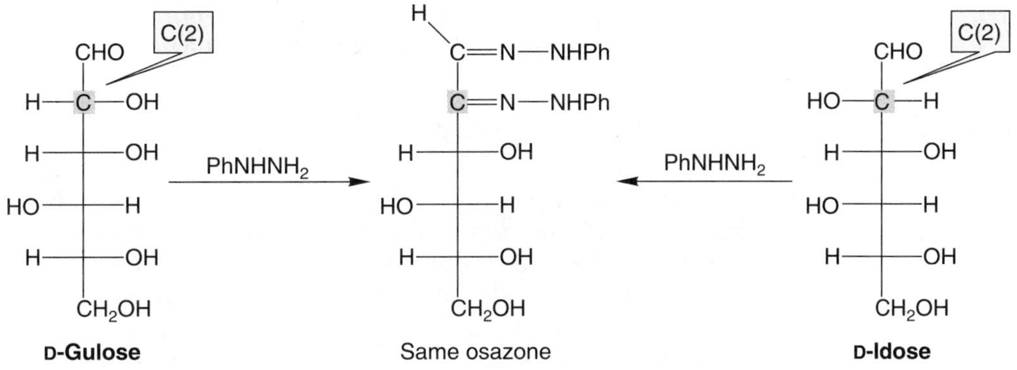

D-Gulose —$PhNHNH_2$→ Same osazone ←$PhNHNH_2$— D-Idose

FIGURE 24.60 D-Gulose and D-idose give the same osazone, which determines the structure of D-idose, and its mirror image L-idose.

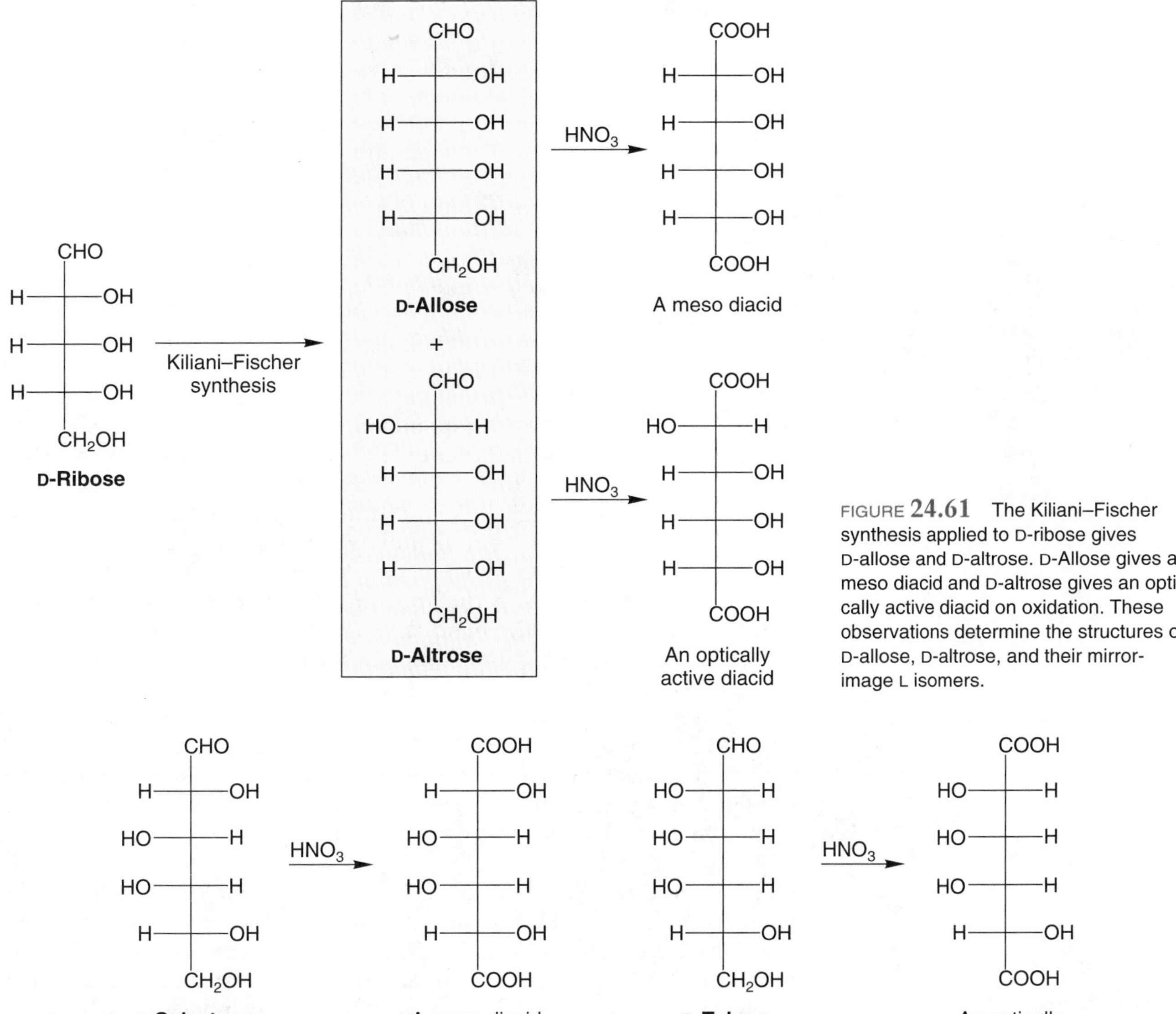

FIGURE 24.61 The Kiliani–Fischer synthesis applied to D-ribose gives D-allose and D-altrose. D-Allose gives a meso diacid and D-altrose gives an optically active diacid on oxidation. These observations determine the structures of D-allose, D-altrose, and their mirror-image L isomers.

FIGURE 24.62 D-Galactose gives a meso diacid and D-talose gives an optically active diacid. Now we know all the structures of all the aldohexoses.

The structures of the eight L isomers are simply the mirror images of those of the D series, and we have at last completed our task.

24.5 SOMETHING MORE: DI- AND POLYSACCHARIDES

A great many sugars containing 12, 18, and other multiples of 6 carbons are known. Some of the common names for sugars actually refer to dimeric species, not the monomeric, relatively simple D-aldohexoses or pentoses. For example, sucrose, lactose, and maltose are all C_{12} sugars. Starch and cellulose are polymers of simple repeating monosaccharide units, less some molecules of water. We will use lactose as an example of how the structures of such molecules can be unraveled.

The first important observation is that (+)-lactose, a disaccharide of the formula $C_{12}H_{22}O_{11}$, can be hydrolyzed in acid to a pair of simple aldohexoses: D-glucose and D-galactose (Fig. 24.63).

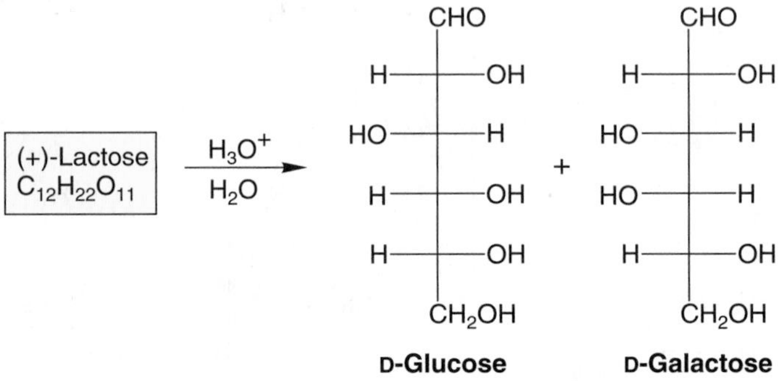

(+)-Lactose $C_{12}H_{22}O_{11}$ $\xrightarrow[H_2O]{H_3O^+}$ D-Glucose + D-Galactose

CHO
H—OH
HO—H
H—OH
H—OH
CH_2OH
D-Glucose

CHO
H—OH
HO—H
HO—H
H—OH
CH_2OH
D-Galactose

FIGURE **24.63** In acid, (+)-lactose is hydrolyzed to D-glucose and D-galactose.

That identifies the two sugars making up (+)-lactose, but this simple experiment gives us another piece of less obvious information. Remember that it is only glycosidic linkages that are cleaved in acid—simple ether connections are not touched. We saw this first in the hydrolysis of *only* the glycosidic methoxyl group of fully methylated D-glucose (p. 1254). D-Glucose and D-galactose must be attached using the hydroxyl group at C(1) of one of the sugars. If the attachment were at any other position, dilute acid would not suffice to break up the dimeric structure (Fig. 24.64).

We now need to know three things about the dimeric structure of lactose (Fig. 24.65).

1. Which sugar is the "attacher" and which is the "attachee"?
2. Which hydroxyl group of one sugar is used to make the linkage to the anomeric position of the other?
3. Are the two sugars linked in an α or β way?

(+)-Lactose can be oxidized by bromine in water to the improbably named lactobionic acid. Bromine in water oxidizes only the aldehydic position of sugars to the acid (p. 1250). In one of the partners in the disaccharide there is no hemiacetal, and therefore no small amount of free aldehyde to be oxidized by bromine/water. Acid hydrolysis of lactobionic

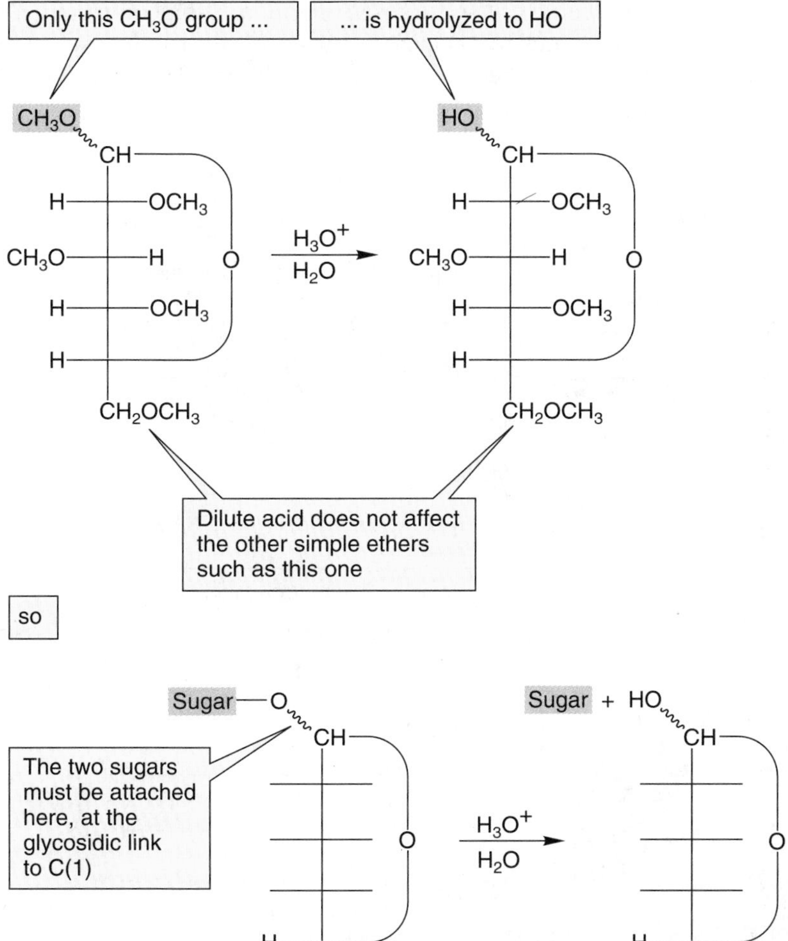

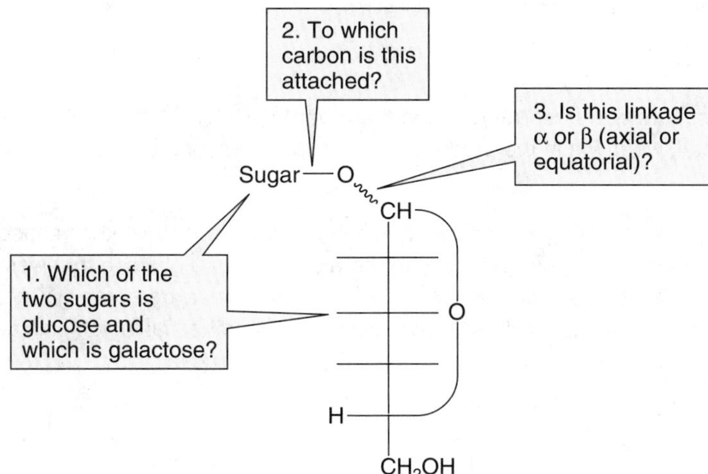

FIGURE 24.64 Hydrolysis in dilute acid means that there must be a glycosidic linkage in (+)-lactose.

FIGURE 24.65 The remaining questions about the structure of (+)-lactose.

acid gives one molecule of D-galactose and one of D-gluconic acid. The remaining aldehyde group in (+)-lactose must belong to glucose—it is the galactose half that is attached to glucose through the OH at C(1) (Fig. 24.64).

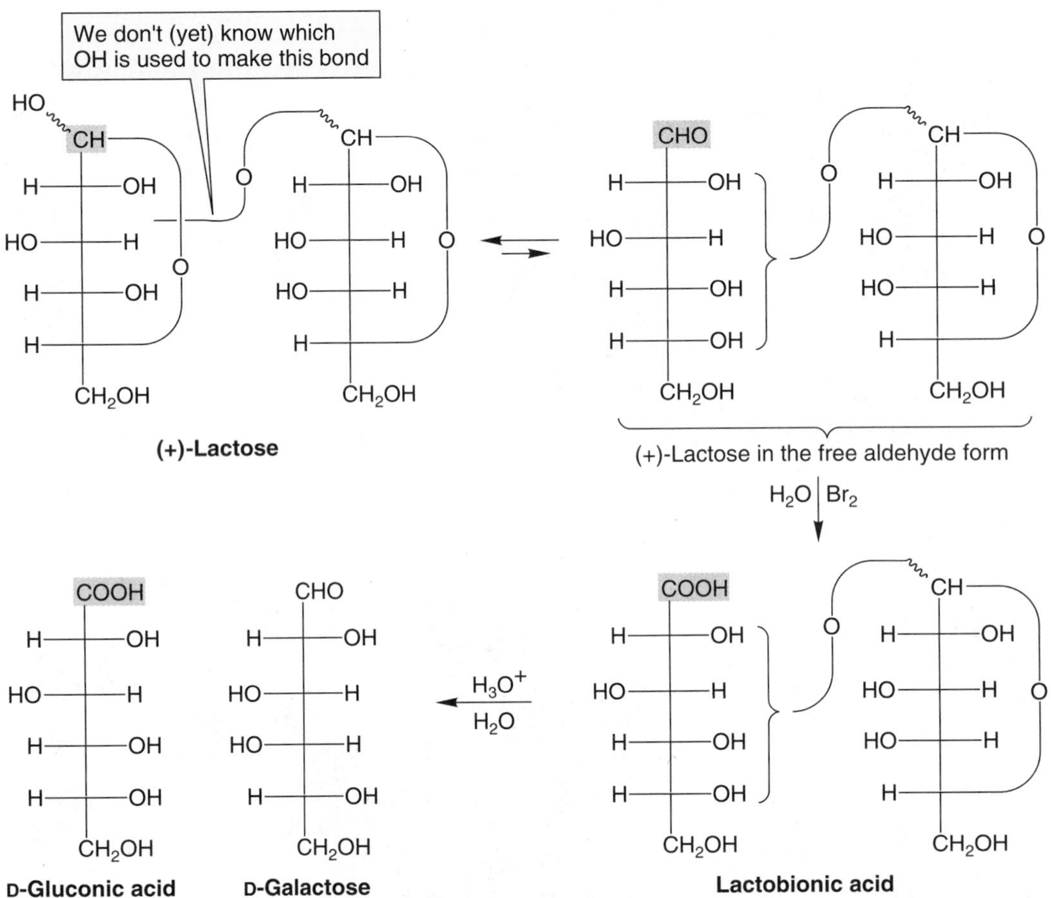

FIGURE 24.66 The acid group in lactobionic acid marks the position of the aldehyde in (+)-lactose. As hydrolysis of lactobionic acid gives a gluconic acid (not a galactonic acid), it is glucose that has the free aldehyde in (+)-lactose.

*PROBLEM 24.22 If the two sugars had been attached in the other way, using C(1) of glucose as the full acetal, what would the products of the experiment shown in Figure 24.66 have been?

ANSWER If the two sugars making up lactose had been attached in the other sense, it would have been the galactose that had the aldehyde, and glucose that was attached through the OH group at C(1). Thus, it would have been the galactose aldehyde group that was oxidized to an acid in this hypothetical lactobionic acid. Hydrolysis would have given D-galactonic acid and D-glucose.

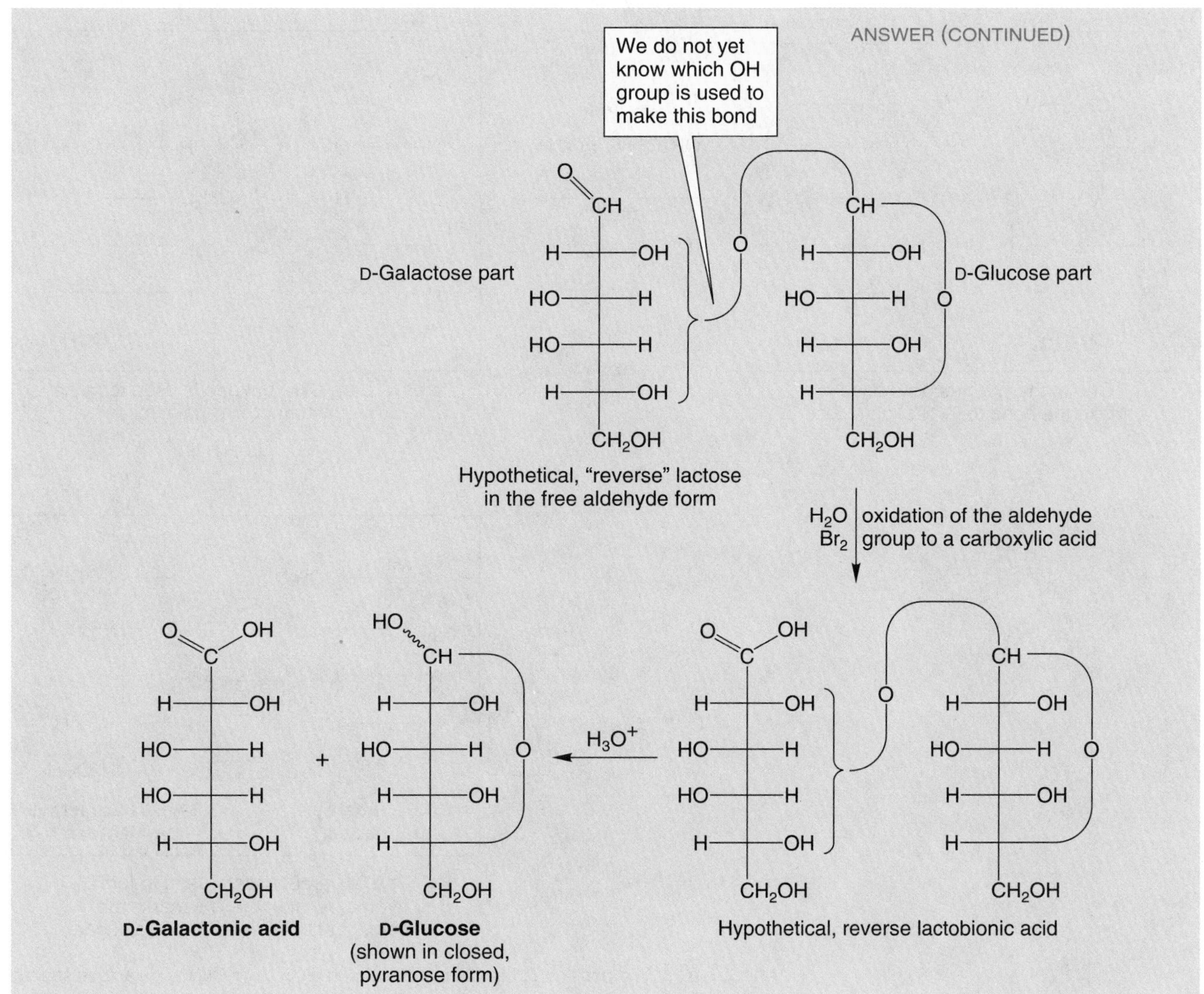

Now, which carbon of glucose is attached to C(1) of galactose? This question is answered by first using a Williamson ether synthesis to methylate all free hydroxyl groups in (+)-lactose. (+)-Lactose is then hydrolyzed in acid to give 2,3,4,6-tetramethylgalactose and 2,3,6-trimethylglucose. Hydrolysis liberates only the OH groups that were tied up in (+)-lactose in hemiacetals or full acetals. It leaves simple ether groups unchanged. It is the OH at C(5) that is used to form the pyranose ring in both sugars, and that accounts for its appearance as an unmethylated OH in the two partially methylated sugars of Figure 24.67. The remaining free OH—at C(4) of glucose—must mark the position attached to C(1) of galactose. It was tied up in an acetal linkage, not as a simple methyl ether (Fig. 24.67).

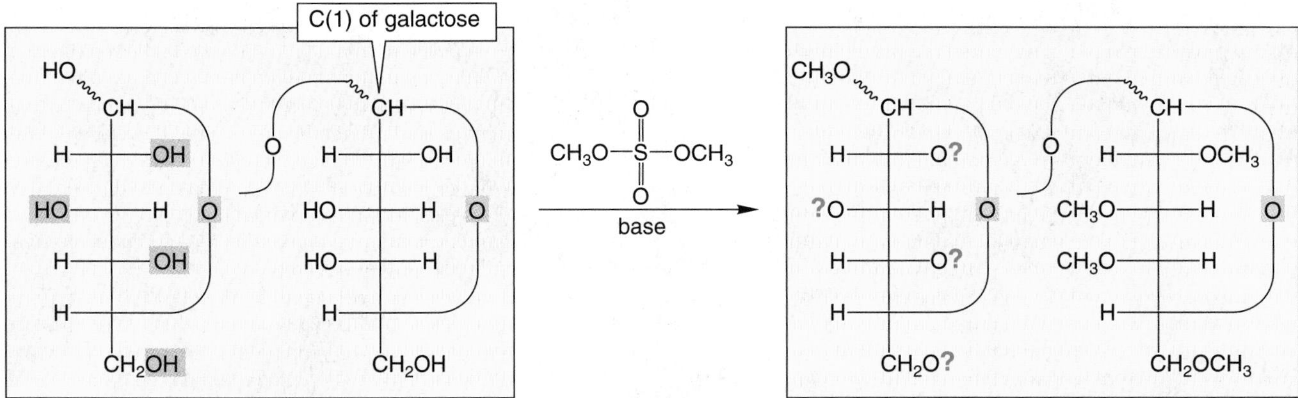
C(1) of galactose
base
One of the OH groups is used to make the bond to C(1) of galactose
Three ?'s are CH3; but one is used to make the attachment to D-galactose
H3O+/H2O

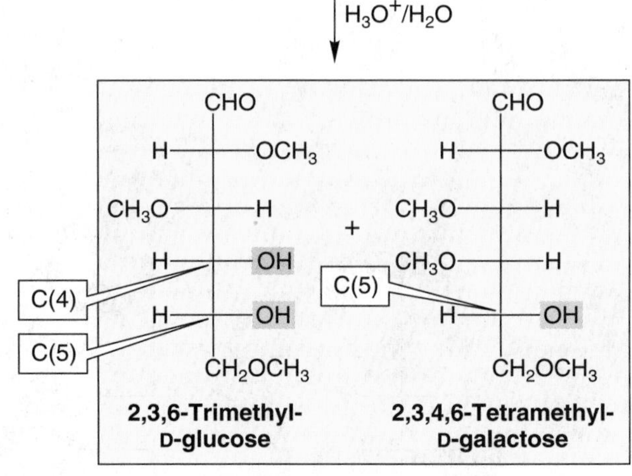
C(4)
C(5)
C(5)
2,3,6-Trimethyl-D-glucose
2,3,4,6-Tetramethyl-D-galactose
The OH groups were used to form pyranose rings; the OH marks the point of attachment of the two sugars

FIGURE 24.67 The position of the OH used to attach glucose to C(1) of galactose can be determined through a series of methylation and hydrolysis experiments.

(+)-Lactose is a β-linked glycoside. There are enzymes such as lactase that have evolved to cleave only β-glycosidic bonds, and to which α-linked disaccharides are inert. (+)-Lactose is cleaved by these enzymes and must be β-linked. Figure 24.68 shows the final Fischer projection. The problem now is to translate this projection into a three-dimensional representation.

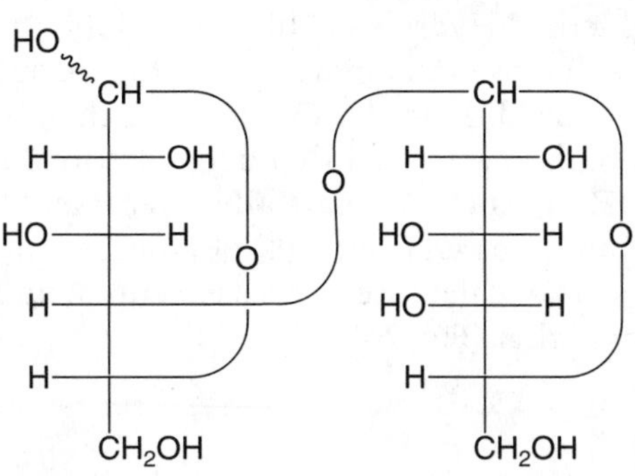

FIGURE 24.68 A Fischer projection for (+)-lactose.

PROBLEM 24.23 Make a good three-dimensional drawing of (+)-lactose.

PROBLEM **24.24**

Cellobiose is a disaccharide cleaved by hydrolysis into two molecules of D-glucose. Treatment of cellobiose with dimethyl sulfate in base leads to an octamethyl derivative. Hydrolysis with acid converts the octamethyl derivative into one molecule of 2,3,6-trimethyl-D-glucose and one of 2,3,4,6-tetramethyl-D-glucose. Cellobiose can be cleaved enzymatically with lactase. Provide a structure for cellobiose.

PROBLEM **24.25**

Maltose is a disaccharide that differs very little from cellobiose. It has the same properties as cellobiose described in Problem 24.24, but is not cleaved by lactase (it is cleaved by another enzyme, maltase). Provide a three-dimensional structure for maltose. *Hint*: How else can the two sugars be linked?

Sometimes two sugars are linked in a way that connects the anomeric carbons [C(1)] of both sugars. These molecules are full acetals and oxidation to an aldonic acid is not possible because there is not even a small amount of free aldehyde. In contrast to other sugars, in which at least a small amount of an oxidizable free aldehyde remains (reducing sugars), these are **nonreducing sugars**. An example is sucrose, common table sugar, in which a molecule of D-glucose and a molecule of D-fructose are connected in this way (Fig. 24.69).

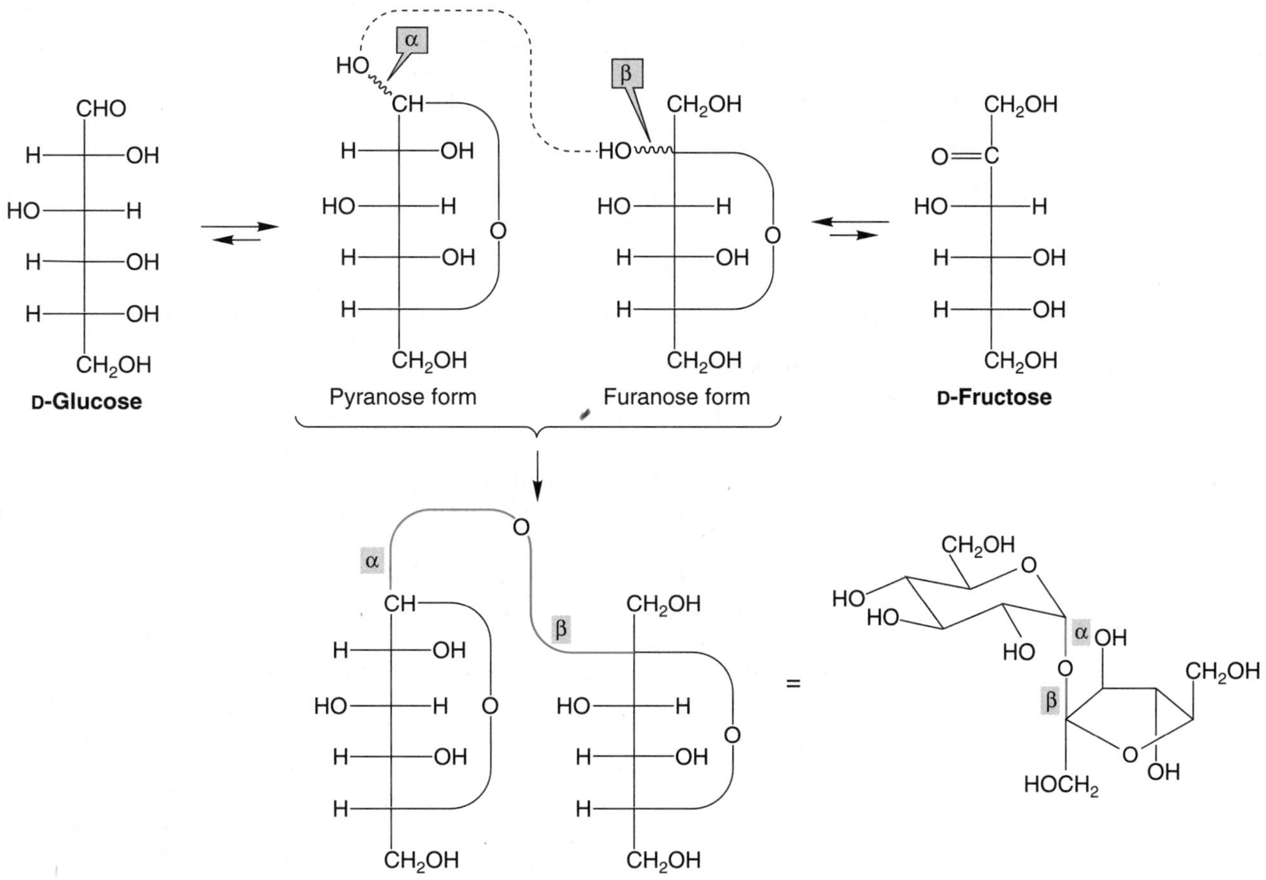

D-Glucose

Pyranose form

Furanose form

D-Fructose

FIGURE **24.69** In nonreducing sugars, there is no free aldehyde group. Attachment must be between both C(1) atoms. Sucrose is an example.

There is also no reason that more than two sugars cannot be connected by glycosidic bonds, and larger polysaccharides are common. The limit of such bonding is a polymer. **Cellulose** is a polymer of glucose in which repeating units are connected linearly by β-glycosidic bonds between C(4) of one sugar and the anomeric C(1) of the other. **Starch** is a different, nonlinear polymer of glucose from which a linear molecule called amylose can be isolated. Amylose has the same structure as cellulose except that the sugars are connected in an α way (Fig. 24.70).

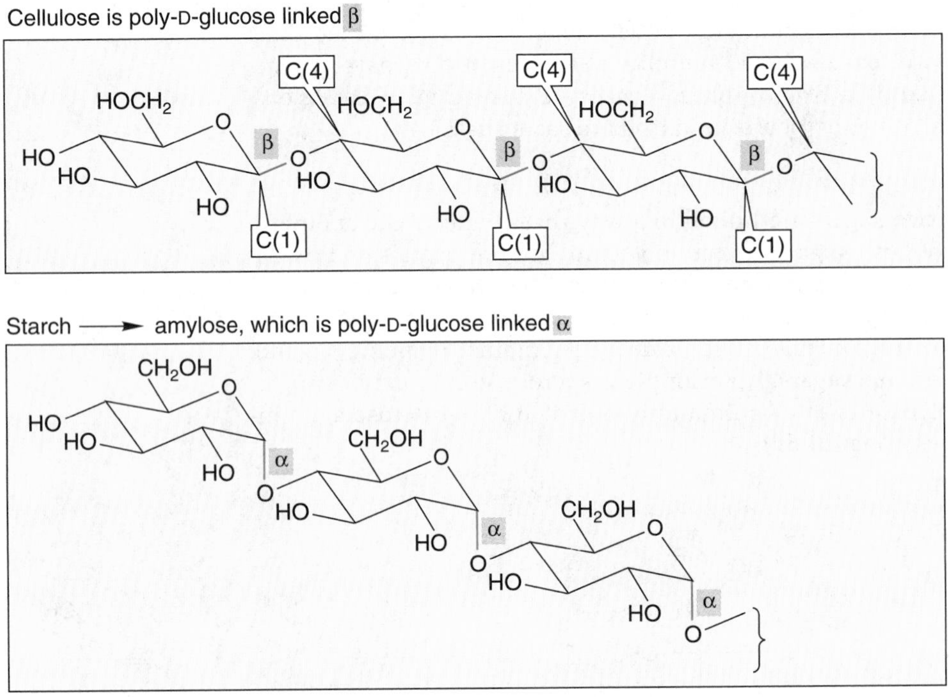

Cellulose is poly-D-glucose linked β

Starch ⟶ amylose, which is poly-D-glucose linked α

FIGURE **24.70** Cellulose and amylose.

24.6 SUMMARY

NEW CONCEPTS

This chapter contains few new concepts, which does not mean that the chemistry is uncomplicated, however. As I have said many times, simple reactions taken in series can become very difficult indeed. The polyfunctionality and stereochemical complexity of sugars make for an intricate chemistry even though there are few really new reactions involved.

The stereochemical complications lead to the simplifying notion of a Fischer projection in which a stylized drawing is used to reduce complexity. The sugar is drawn vertically with the aldehyde group at the top. Horizontal bonds are taken as coming toward you and vertical bonds as retreating. If the OH of the bottom H–C–OH (the one adjacent to the CH_2OH) is on the right, it is a D sugar, if it is on the left, it is an L sugar (Fig. 24.3).

When Emil Fischer worked out the structure of glucose in the late 1800s there was no way to tell absolute configuration. Fischer could not tell the D series from the enantiomeric L series, so he assumed that the OH at C(5) was on the right and worked out the relative configurations of the other groups. As is turned out, he guessed correctly.

A number of times in this chapter we encounter reactions in which a minor partner in a equilibrium reacts to give the product. As long as equilibrium is reestablished, removal of the minor partner can drive the reaction to completion (Fig. 24.16).

REACTIONS, MECHANISMS, AND TOOLS

Most of the reactions in this chapter can be found in the chemistry of carbonyl groups (Chapters 16 and 18) and alcohols (Chapter 17); only a few new reactions appear.

One of these is intramolecular hemiacetal formation. Carbonyl groups react with nucleophiles in the addition reaction (Chapter 16). When the nucleophile is an alcohol, hemiacetals are formed, but generally they are not favored at equilibrium. However, when a relatively strain-free ring can be formed in an intramolecular hemiacetal formation, the cyclic form can be favored. This is the case for aldohexoses (and many other sugars), which exist mainly in the six-membered pyranose forms (Fig. 24.13).

The equilibration of the cyclic hemiacetal with a small amount of the open aldohexose form allows for either the α- or β-anomer to produce an equilibrium mixture of the two anomers in a phenomenon called mutarotation (Fig. 24.22).

The reactivity of the hemiacetal OH allows for a differentiation of this OH from the others in the molecule. For example, protonation of the C(1) hemiacetal OH leads to a molecule that can form a resonance-stabilized carbocation on loss of water. Nucleophiles such as alcohols can add to this carbocation to give the full acetals called glycosides. Only the OH at C(1) can react in this way (Fig. 24.32).

Reaction with three equivalents of phenylhydrazine leads to the 1,2-phenylhydrazones called osazones (Fig. 24.37).

Methods have been developed both to lengthen (Kiliani–Fischer synthesis) and shorten (Ruff and Wohl degradations) the backbone carbon chains of sugars. See the next section for details.

SYNTHESES

Chain lengthening. The Kiliani–Fischer synthesis adds a new C(1) aldehyde and generates a pair of C(2) stereoisomers (Fig. 24.41).

Chain shortening. In the Ruff and Wohl degradations, the original aldehyde group is lost and a new aldehyde created at the old C(2) (Figs. 24.43 and 24.45).

Osazone formation (Fig. 24.37).

Monoacetal and tetra-ether, monoacetal derivatives of carbohydrates (Figs. 24.29 and 24.30).

Aldaric and aldonic acids can be made through oxidation of aldo sugars (Figs. 24.27 and 24.28).

COMMON ERRORS

At the heart of sugar chemistry is the difference between C(1), the anomeric center, and the other carbons. If you lose track of this difference, you lose the ability to follow the labeling studies establishing the ring size of simple sugars and the points of attachment of disaccharides.

It is also sometimes difficult to keep track of the open and closed (hemiacetal) forms of the sugars and sugar derivatives in this chapter. We typically write what is convenient, and this is not always a good representation of structure. For example, open forms of sugars are often written even though most of the material is really tied up in the cyclic hemiacetal form. It is the small amount of the open form that is reactive. This concept can be confusing if you don't have the equilibrium between open and cyclic forms solidly in mind.

24.7 KEY TERMS

Aldaric acid A diacid derived from an aldohexose by oxidation with nitric acid. It has the structure HOOC–$(CHOH)_4$–COOH. In an aldaric acid, the old aldehyde and primary alcohol ends of the sugar have become identical.

Aldohexose A hexose of the structure

$$OCH{-}(CHOH)_4{-}CH_2OH$$

Aldonic acid A monoacid derived from an aldohexose through oxidation with bromine in water. Only the aldehyde group is oxidized to the acid state. It has the structure

$$HOOC{-}(CHOH)_4{-}CH_2OH$$

Aldopentose A pentose of the structure

$$OCH{-}(CHOH)_3{-}CH_2OH$$

Aldotetrose A tetrose of the structure

$$OCH{-}(CHOH)_2{-}CH_2OH$$

Aldotriose A triose of the structure

$$OCH{-}CHOH{-}CH_2OH$$

Anomers For aldoses, these are sugars differing only in the stereochemistry at C(1). They are C(1) stereoisomers.

Carbohydrate A molecule whose formula can be factored into $C_x(H_2O)_y$. A "sugar" or "saccharide."

Cellulose A polymer of glucose in which C(4) of one glucose is linked in β fashion to C(1) of another.

Fischer projection A schematic stereochemical representation of sugars. The aldehyde group is placed at the top and the primary alcohol at the bottom. Horizontal bonds are taken as coming toward the viewer and vertical bonds as retreating. If the OH of the H–C–OH adjacent to the CH_2OH group is on the right, the molecule is a D sugar, if it is on the left, it is an L sugar.

Furanose A sugar containing a five-membered cyclic ether.

Furanoside A furanose in which the anomeric OH has been converted into an acetal.

Glyceraldehyde The aldotriose,

$$OCH{-}CHOH{-}CH_2OH$$

Glycoside A sugar in which the anomeric OH at C(1) has been converted into an OR group.

Haworth form The representation of sugars in which they are shown as planar rings.

Hexose A six-carbon sugar.

Ketose A sugar containing not the usual aldehyde group, but a ketone.

Keto sugar A sugar containing not the usual aldehyde group, but a ketone.

Kiliani–Fischer synthesis A method of lengthening the chain of an aldose by one carbon atom. A pair of sugars, epimeric at the new C(1) is produced.

Lobry de Bruijn–Alberda van Ekenstein reaction The base-catalyzed interconversion of aldo and keto sugars. The key intermediate is the double enol formed by protonation of an enolate.

Mutarotation The interconversion of anomeric sugars in which an equilibrium mixture of α- and β-forms is reached.

Nonreducing sugar A carbohydrate containing no amount of an oxidizable aldehyde group.

Osazone A 1,2-phenylhydrazone formed by treatment of a sugar with three equivalents of phenylhydrazine.

Pentose A five-carbon sugar.

Pyranose A sugar containing a six-membered cyclic ether.

Pyranoside A pyranose in which the anomeric OH at C(1) has been converted into an OR group.

Reducing sugar A sugar containing some amount of an oxidizable free aldehyde group.

Ruff degradation A method for shortening the carbon backbone of a sugar by one carbon. The aldehyde carbon at C(1) is lost and a new aldehyde created at the old C(2).

Saccharide A molecule whose formula can be factored into $C_x(H_2O)_y$. A "sugar" or "carbohydrate."

Starch A nonlinear polymer of glucose containing amylose, an α-linked linear polymer of glucose.

Sugar A molecule whose formula can be factored into $C_x(H_2O)_y$. A "saccharide" or "carbohydrate."

Tetrose A four-carbon sugar.

Wohl degradation A method for shortening the carbon backbone of a sugar by one carbon. The aldehyde carbon at C(1) is lost and a new aldehyde created at the old C(2).

24.8 ADDITIONAL PROBLEMS

PROBLEM 24.26 Draw the following molecules in Fischer projection:

(a) Any D-aldoheptose.
(b) Any L-ketopentose.
(c) A D-aldotetrose that gives a meso diacid on oxidation with dilute nitric acid.
(d) Methyl α-D-galactopyranoside.
(e) The osazone of D-allose.
(f) Phenyl β-D-ribofuranoside.

PROBLEM 24.27 Make perfect three-dimensional drawings of all the pyranose forms of D-altrose. Before you start, decide how many will there be.

PROBLEM 24.28 (a) What sugar or sugars would result from the Ruff degradation applied to D-gulose?
(b) What sugar or sugars would result from the Kiliani–Fischer synthesis applied to D-lyxose?

PROBLEM 24.29 (a) What other sugar, if any, would give the same aldaric acid as D-talose when oxidized with nitric acid?
(b) What other sugar, if any, would give the same aldaric acid as D-xylose when oxidized with nitric acid?
(c) What other sugar, if any, would give the same aldaric acid as D-idose when oxidized with nitric acid?

PROBLEM 24.30 What other sugar would give the same osazone as L-talose?

PROBLEM 24.31 What are the principal organic products when D-lyxose is treated under the following conditions? Mechanisms are not necessary. D-Lyxose is drawn in the open form, but of course it is largely in the cyclic, furanose form. Keep this in mind.

CHO
HO—H
HO—H
H—OH
CH_2OH
D-Lyxose

(a) $NaBH_4$, H_2O
(b) Br_2, H_2O
(c) HNO_3
(d) CH_3I, Ag_2O

→ ?

PROBLEM 24.32 What are the principal organic products when D-ribose is treated under the following conditions? Mechanisms are not necessary. D-Ribose is drawn in the open form, but of course is largely a cyclic furanose.

CHO
H—OH
H—OH
H—OH
CH_2OH
D-Ribose

(a) Ac_2O, H_3O^+
(b) CH_3OH, H_3O^+
(c) $PhNHNH_2$
(d) 1. NaCN
2. H_2/Pd
3. H_2O
(e) 1. NH_2OH, CH_3OH, CH_3ONa
2. Ac_2O, 100 °C
3. CH_3OH, CH_3ONa
4. H_2O, H_3O^+
(f) 1. Br_2/H_2O
2. $Ca(OH)_2$
3. $Fe_2(SO_4)_3$
4. H_2O_2

→ ?

PROBLEM **24.33** Glucose is the most abundant aldohexose. Use a stereochemical analysis to speculate upon the reason for this prevalence.

PROBLEM **24.34** Your task is to distinguish between D-talose and D-galactose. The following sequence is suggested. First, oxidize to aldaric acids. Second, make the methyl esters. Then ... ? Fill in the last step. Oh, we forgot to tell you that at the end of the second step your town's elected officials passed an ordinance forbidding any laboratory work as harmful to their chances for reelection. You are restricted to spectroscopy.

PROBLEM **24.35** Use Table 24.2 (p. 1244) to calculate the energy difference between the α- and β-pyranose forms of D-allose at 25 °C.

PROBLEM **24.36** Calculate the equilibrium ratio of α- and β-D-glucopyranose from the specific rotations of the pure anomers (α = +112°, β = +18.7°; equilibrium value = +52.7°). You might compare your calculation with the values given in Table 24.2.

PROBLEM **24.37** The disaccharide maltose ($C_{12}H_{22}O_{11}$) can be hydrolyzed in acid to two molecules of D-glucose. It is also hydrolyzed by the enzyme maltase, a molecule known to cleave only α-glycosidic linkages. Maltose can be oxidized by bromine in water to maltobionic acid (MBA, $C_{12}H_{22}O_{12}$). When MBA is treated with methyl iodide and base, followed by acid hydrolysis, the products are 2,3,4,6-tetramethyl-D-glucose and 2,3,5,6-tetramethyl-D-gluconic acid. Provide a three-dimensional structure for maltose and explain your reasoning.

PROBLEM **24.38** Draw in Fischer projection the nonreducing disaccharides composed of one molecule of D-allose and one molecule of D-altrose. First determine how many are possible, in principle, and then adopt some scheme for indicating what they are. Don't attempt to draw them all out in three dimensions.

PROBLEM **24.39** Write a mechanism for the acid-induced transformation of a D-aldopentose into furfural.

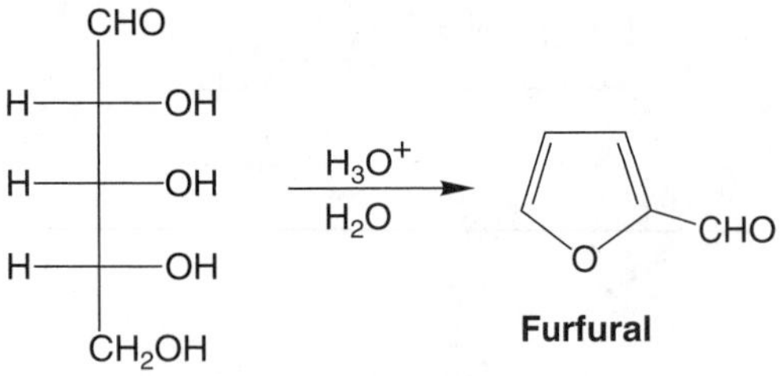

CHO; H—OH; H—OH; H—OH; CH_2OH → (H_3O^+, H_2O) → CHO-substituted furan

Furfural

PROBLEM **24.40** Figures 24.24 and 24.25 give one mechanism for the Lobry de Bruijn–Alberda van Ekenstein reaction. Use the example given, in which D-glucose equilibrates with D-mannose and D-fructose, and provide another mechanism in which a hydride shift occurs.

PROBLEM **24.41** When α-D-xylopyranose is treated with hydrochloric acid in methyl alcohol, the major products are two isomers of methyl D-xylopyranoside.
(a) Draw the starting material and the two major products and explain how the two major products are formed from the single starting material.
(b) There are two minor products formed as well. Both are methyl furanosides. What are these two products and how are they formed?

PROBLEM **24.42** Outline a mechanism for the following reaction:

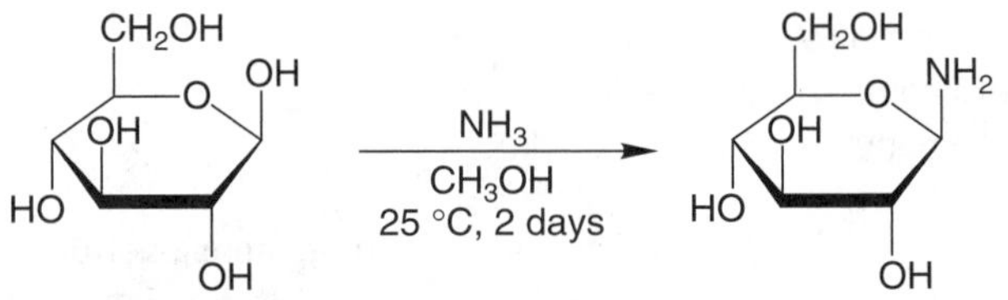

Haworth form of β-D-glucopyranose (CH_2OH, OH, O, OH, HO, OH) → (NH_3; CH_3OH; 25 °C, 2 days) → (CH_2OH, NH_2, O, OH, HO, OH) (40%)

PROBLEM **24.43** Outline a mechanism for the following reaction:

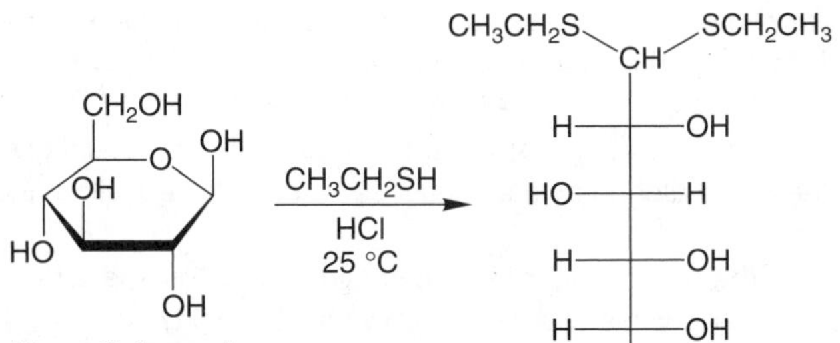

Haworth form of β-D-glucopyranose (CH_2OH, OH, O, OH, HO, OH) → (CH_3CH_2SH; HCl; 25 °C) → CH_3CH_2S–CH–SCH_2CH_3; H—OH; HO—H; H—OH; H—OH; CH_2OH

PROBLEM **24.44** When D-glucose, or any polysaccharide containing D-glucose units such as cellulose, is pyrolyzed with an acid catalyst, a new compound, levoglucosan (**1**), is formed. Write a mechanism.

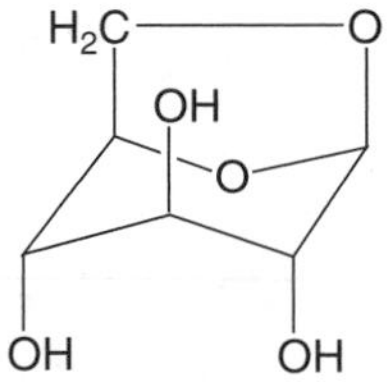

H_2C—O; OH; O; OH; OH

1

PROBLEM **24.45** Further pyrolysis of levoglucosan (**1**), also in the presence of acid, leads to levoglucosenone (**2**).
(a) The original structure assigned to **2** was **3**. Show why this structure is inconsistent with the spectral data for **2**.
(b) Write an arrow formalism mechanism for the transformation of **1** into **2**.
(c) Rationalize the formation of **2** rather than the alternative **4**. Warning! I think this is a hard question. *Hint*: Recall Chapter 23.

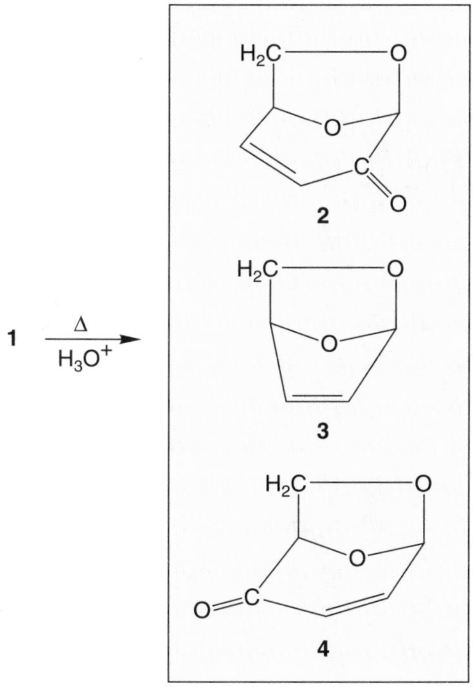

H_2C—O, O, C=O, **2**; H_2C—O, O, **3**; H_2C—O, O, O=C, **4**

1 $\xrightarrow[H_3O^+]{\Delta}$

Spectral data for **2**
UV (λ_{max} in CH_3CH_2OH) = 218 nm, log ε 2.75, 275 nm, log ε 1.5
IR (film): 2990, 2900, 1720 (s), 1700, 1610, 1380, 1100 cm^{-1}
^1H NMR (δ, $CDCl_3$): 3.74 (1H), 3.87 (1H), 5.05 (1H), 5.31 (1H), 6.09 (1H), 7.34 (1H)

PROBLEM **24.46** Treatment of α-D-galactopyranose with acetone and sulfuric acid catalyst leads to a compound $C_{12}H_{20}O_6$ that still contains a primary alcohol. Draw a structure for this compound and explain its formation.

PROBLEM **24.47** Provide structures for **A–D**, and rationalize the following reactions of the starting material, the product formed in Problem 24.46:

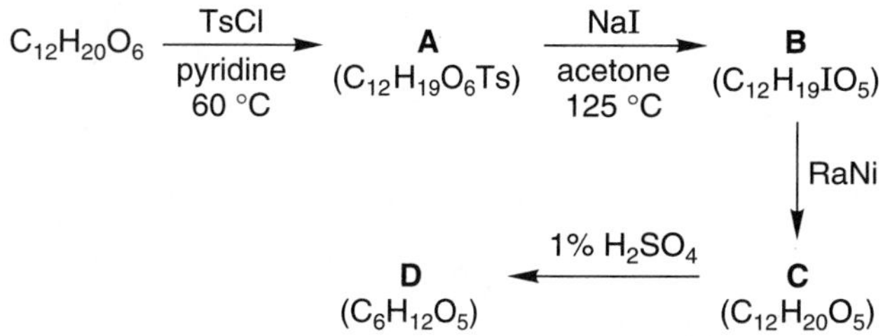

$C_{12}H_{20}O_6$ $\xrightarrow[\text{pyridine, 60 °C}]{TsCl}$ **A** ($C_{12}H_{19}O_6Ts$) $\xrightarrow[\text{acetone, 125 °C}]{NaI}$ **B** ($C_{12}H_{19}IO_5$) $\xrightarrow{RaNi}$ **C** ($C_{12}H_{20}O_5$) $\xrightarrow{1\%\ H_2SO_4}$ **D** ($C_6H_{12}O_5$)

PROBLEM **24.48** In fact, with many carbonyl compounds the primary OH does participate in product formation. For example, many saccharides react with benzaldehyde in acid to give six-membered rings in which the primary alcohol is incorporated.

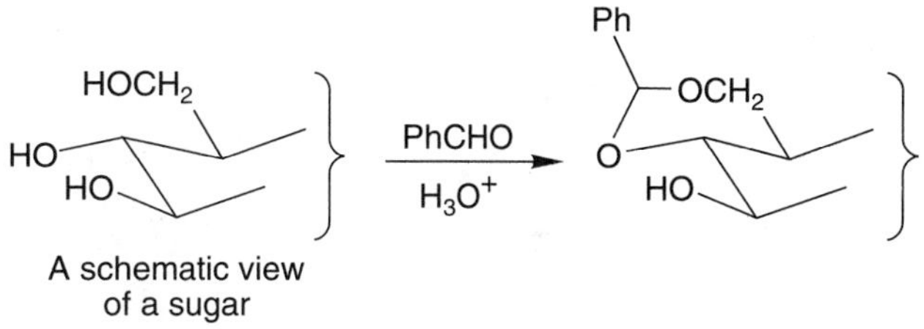

$HOCH_2$, HO, HO $\xrightarrow[H_3O^+]{PhCHO}$ Ph, OCH_2, O, HO

A schematic view of a sugar

However, as shown in Problem 24.46, acetone leads to five-membered rings, not six-membered rings shown above.

First, draw the detailed three-dimensional structures of the six-membered rings formed by reaction between benzaldehyde and acetone with the partial sugar shown above, then explain this odd observation.

Introduction to the Chemistry of Heterocyclic Molecules

I love the old questions. Ah, the old questions, the old answers, there's nothing like them!
—Samuel Beckett*
Endgame

By definition, a **heterocyclic compound** is a ring compound containing at least one non-carbon atom in the ring. Vast numbers of such compounds are possible, and we have already seen many in earlier chapters. Some have been quite normal compounds [the cyclic ethers of Chapter 17 (p. 846) or the cyclic amines of Chapter 21 (p. 1079)], others were less conventional [the halonium ions of Chapter 10 (p. 395), or the heterocations of Chapter 23 (p. 1186) would be good examples], and a few could legitimately qualify as bizarre [the bridged protonium ion of Chapters 9 and 23 (pp. 380 and 1201)].

Heterocycles are worth our attention for many reasons, chief among them their prevalence among biologically active molecules. Many drugs are heterocycles. Pharmacological activity requires chemical reaction and reaction requires functionality. Hydrocarbons are simply too unreactive to be likely candidates for pharmaceutical agents, and it should be no surprise that many specialists in heterocyclic chemistry are employed in drug companies.

By now we are well equipped to understand and even anticipate much of the chemistry of heterocycles, and to appreciate many of the synthetic methods worked out to construct these compounds. We are becoming hard pressed near the end of this book to find truly new chemistry. Details will be different, but new underlying principles will appear only rarely. It's often the details that make chemistry interesting, however, and heterocyclic chemistry makes a splendid reservoir from which to fish out some fascinating reactions and molecules. We will concentrate on the common heterocyclic molecules in which N, O, and S are contained in the ring, but other, more exotic heteroatoms will also make an occasional appearance, and we will spend a small amount of time on molecules containing more than one heteroatom.

*Samuel Barclay Beckett (1906–1989) was an Irish playwright and poet who received the Nobel prize for literature in 1969. His work has been described as "desolate, terminal, obsessional, irradiated with flashes of last-ditch black humor."

25.1 THREE-MEMBERED HETEROCYCLES: OXIRANES (EPOXIDES), THIIRANES (EPISULFIDES), AND AZIRIDINES

Oxiranes have made a substantial appearance already (Chapter 10, Chapter 17) and aziridines have received passing mention (Chapter 21, p. 1082). Here we will collect some old reactions and introduce a few new ones.

25.1a Synthetic Methods

Oxiranes are often made by direct oxidation of alkenes with peracids (Chapter 10, p. 405; Fig. 25.1).

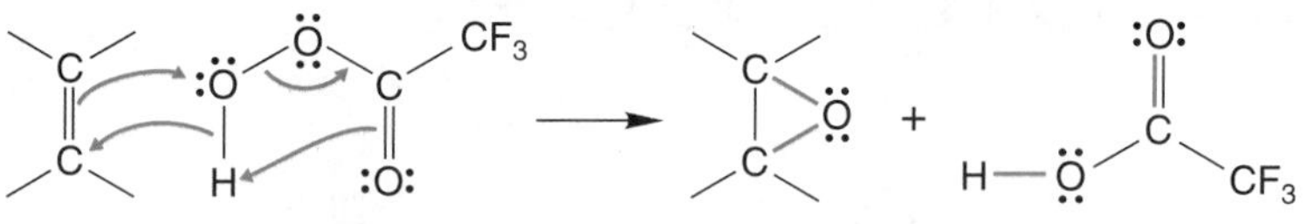

FIGURE 25.1 Oxirane synthesis using a peracid.

As the reaction depends on the alkene acting as a nucleophile, electron-deficient alkenes are either poor performers in this reaction or entirely unreactive. For some electron-poor alkenes, a different reaction in which the oxygen is introduced in a nucleophilic way is effective. Peroxide ion attacks in Michael fashion (Chapter 18, p. 918) to generate an anion that can displace hydroxide (!) intramolecularly (Fig. 25.2).

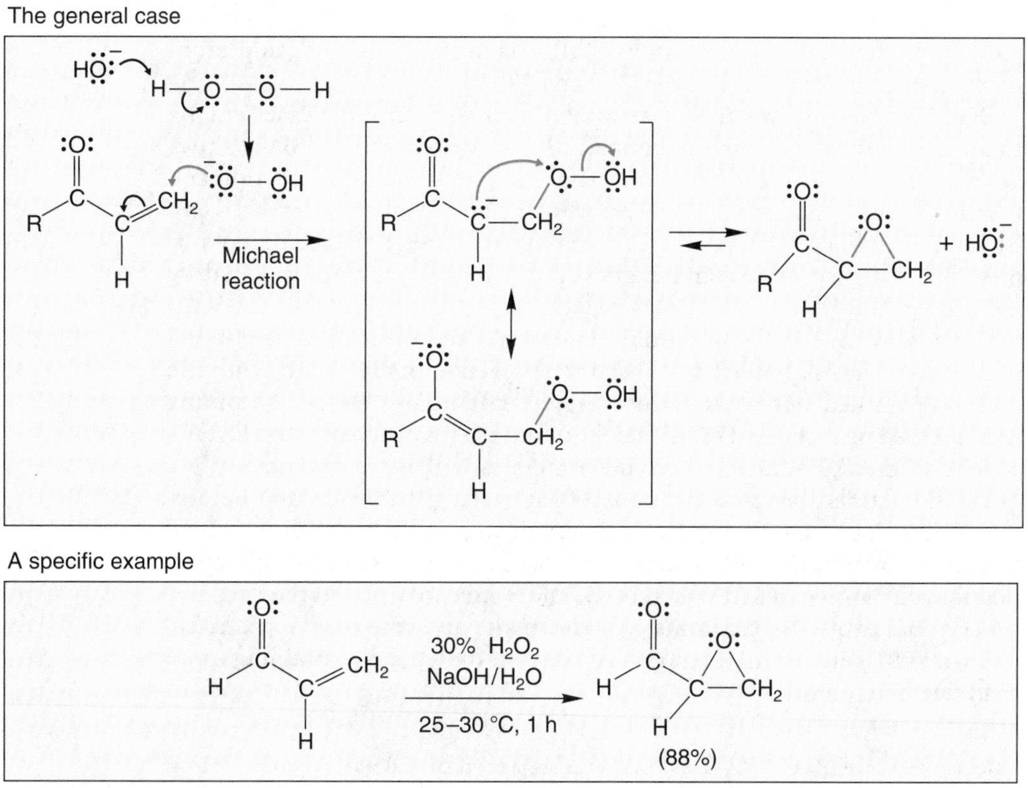

The general case

Michael reaction

A specific example

30% H_2O_2
NaOH/H_2O
25–30 °C, 1 h
(88%)

FIGURE 25.2 A base-catalyzed epoxidation reaction.

PROBLEM 25.1

The difference in nucleophilicity between differently substituted alkenes can lead to selectivity in the epoxidation reaction. Rationalize the position of faster epoxidation in the 1,4-cyclohexadiene of Figure 25.3 and explain your reasoning.

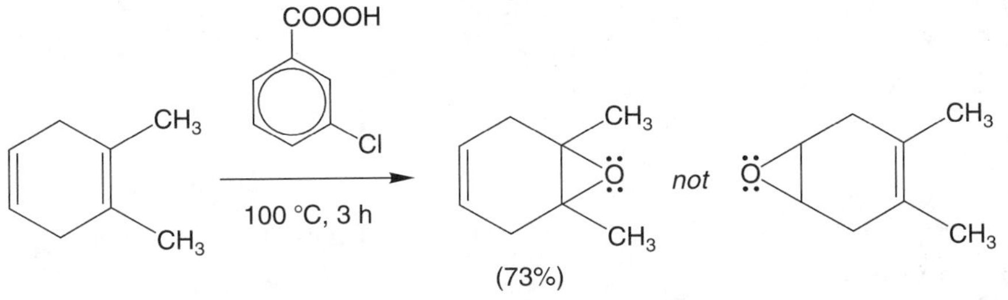

COOOH
CH3
Cl
100 °C, 3 h
CH3
CH3
not
CH3
CH3
CH3
(73%)

FIGURE 25.3

Another classic oxirane synthesis takes advantage of the easy intramolecular S_N2 reaction that occurs when vicinal halohydrins are treated with base (Chapter 17, p. 847; Fig. 25.4).

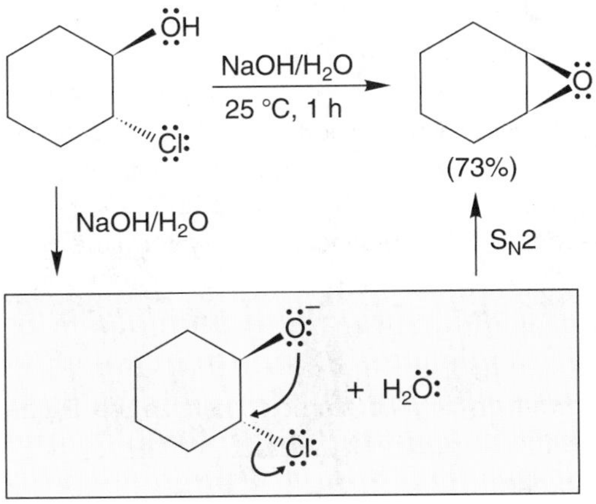

OH
$NaOH/H_2O$
25 °C, 1 h
Cl:
(73%)
$NaOH/H_2O$
S_N2
+ H_2O:

FIGURE 25.4 A synthesis of oxiranes from vicinal halohydrins.

PROBLEM 25.2

Provide a synthesis of the halohydrin of Figure 25.4 starting from cyclohexene. If you forget, see Chapter 10 (p. 398).

Still another intramolecular displacement occurs in the Darzens condensation, named after Georges Darzens (1867–1954). In this condensation reaction, an α-halo ester is treated with base in the presence of an aldehyde or ketone (Fig. 25.5).

The first step is removal of an α-hydrogen to give an enolate. The presence of the halogen atom assists in formation of the haloester enolate. Addition of the enolate to the carbonyl group of the ketone generates the same kind of intermediate active in Figure 25.4, and unsurprisingly, intramolecular S_N2 displacement of halide gives the oxirane.

Thiiranes (episulfides) are the sulfur analogues of oxiranes, but cannot often be made in closely related ways. Direct reaction is not feasible, for example. One reasonably general synthesis of these compounds takes advantage of the relative ease of access to oxiranes and uses the three-membered cyclic ether as starting material. The reagent of choice is a metal thiocyanate, $M^+ {}^-SCN$. Sulfur is an excellent nucleophile (Chapter 7, p. 257; Chapter 23, p. 1186), and begins the reaction with a typical base-

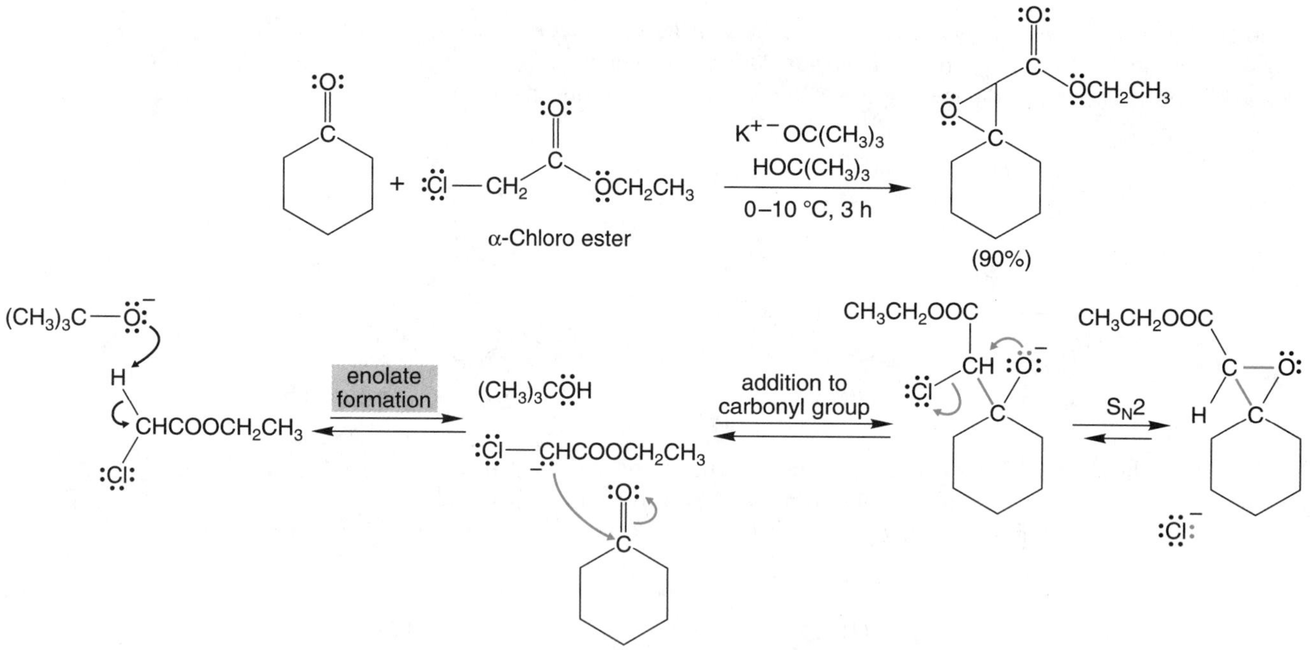

FIGURE 25.5 The Darzens condensation.

PROBLEM 25.3 Why does the presence of the chlorine atom help form the enolate shown in Figure 25.5?

*PROBLEM 25.4 Account for the formation of both stereoisomers in the Darzens condensation shown in Figure 25.6.

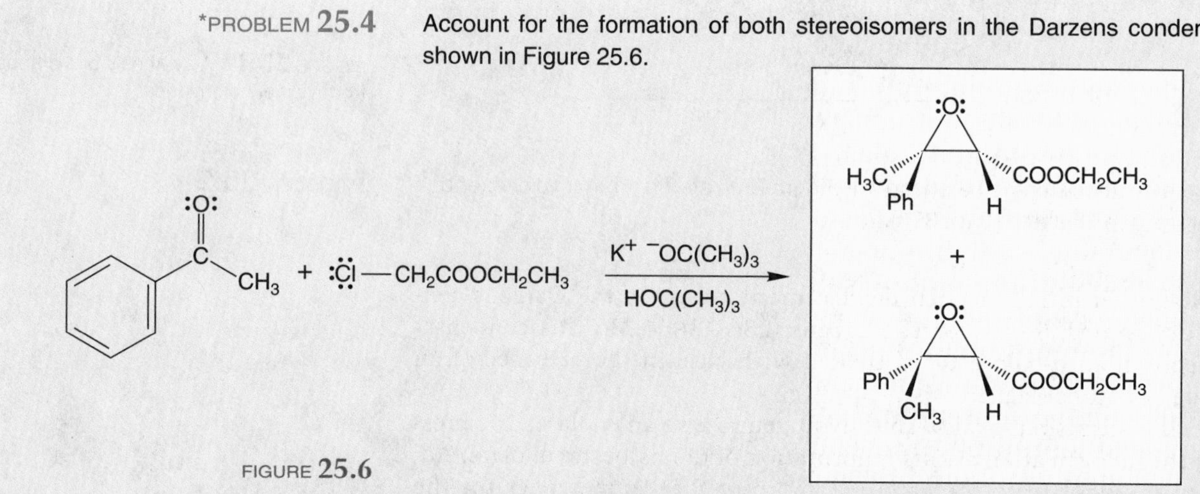

FIGURE 25.6

ANSWER The enolate is first formed and then adds to the Lewis acid carbonyl group

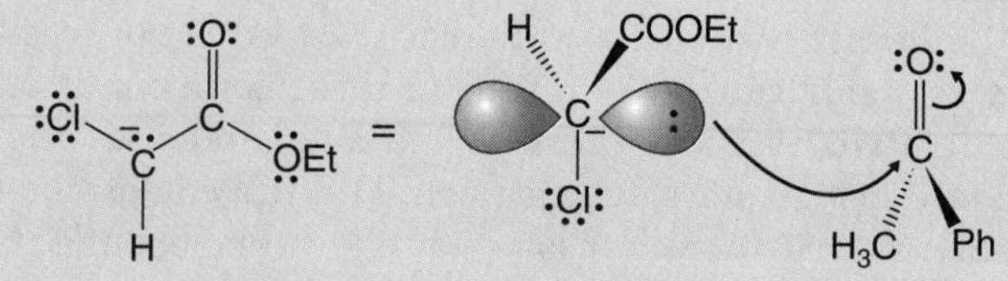

ANSWER (CONTINUED)

But this is not the only way addition can take place. There are two possible orientations for addition, which lead to two diastereomeric alkoxides.

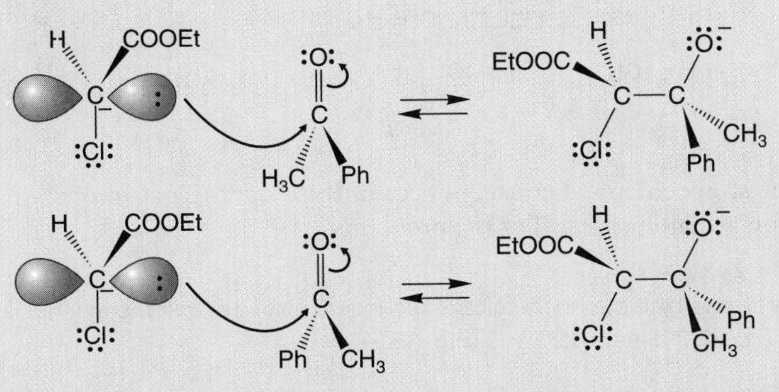

In the final step of the Darzens reaction, the alkoxides undergo an intramolecular S_N2 reaction in which the alkoxide displaces chloride (from the rear) to give the two epoxides.

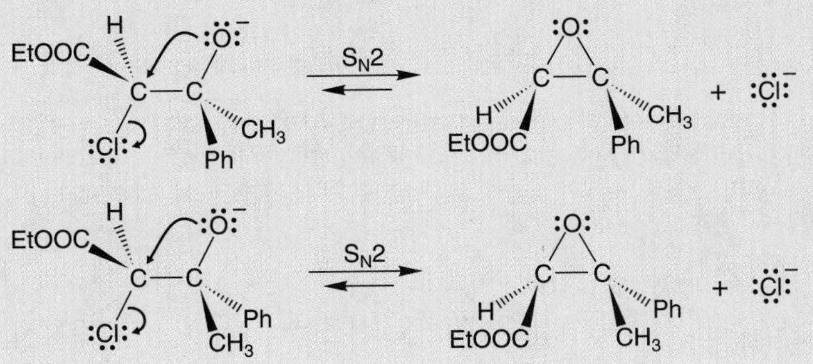

induced opening of the oxirane. The resulting oxide can then add to the cyanide group (Chapter 20, p. 1032) to generate an intermediate five-membered anion (Fig. 25.7).

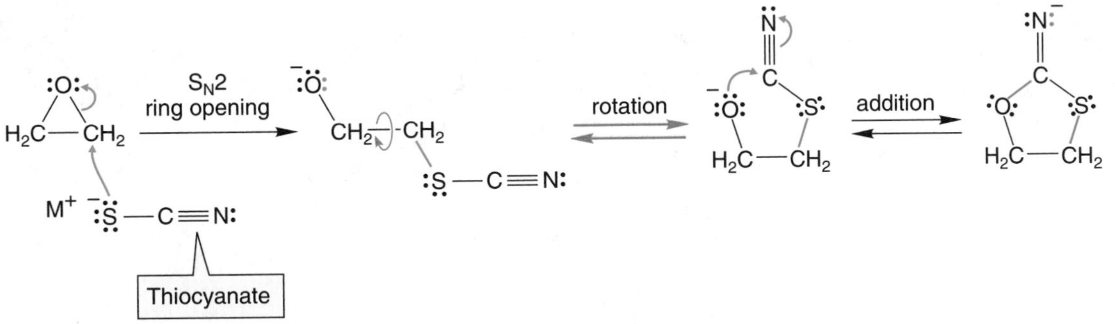

FIGURE 25.7 The first stages of a general synthesis of thiiranes.

The options available to this anion are reversal to the oxyanion, or opening in the other direction to give the sulfur anion. In the second option, the excellent sulfur nucleophile displaces cyanate ($^-$OCN) to give the thiirane (Fig. 25.8).

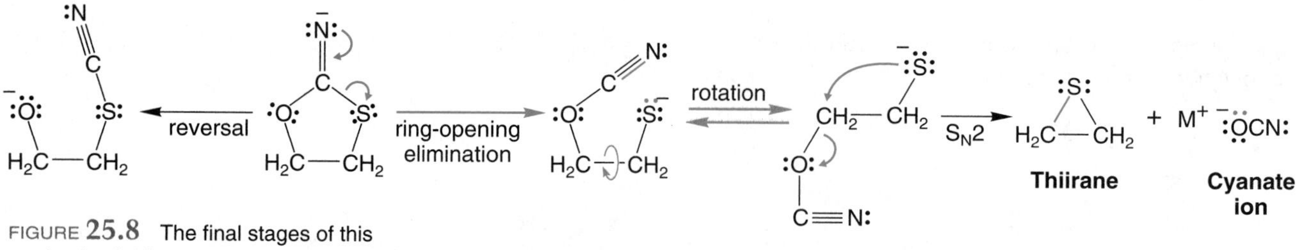

FIGURE 25.8 The final stages of this synthesis of thiiranes.

The stereochemical consequences of this "open–close–reopen–reclose" reaction are interesting. Try the problems.

*PROBLEM 25.5 Follow through the stereochemical consequences of this thiirane synthesis starting with *cis*-2,3-dimethyloxirane (Fig. 25.9).

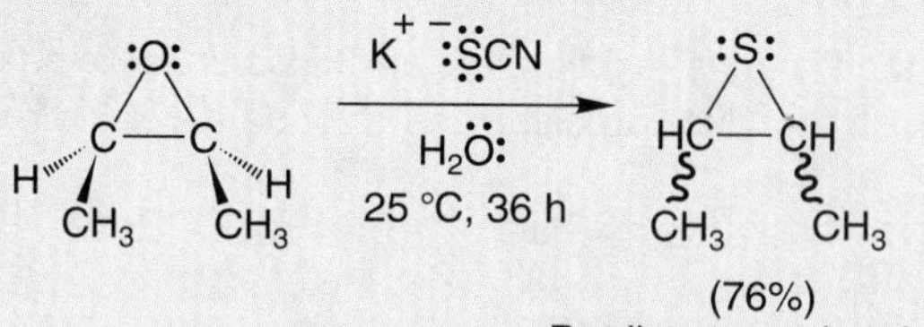

FIGURE 25.9

ANSWER If you follow the sequence of ring openings, rotations, and ring closings of Figures 25.7 and 25.8, keeping in mind the requirements for S_N2 reactions to take place with inversion, you can't go wrong. Stereochemistry is retained in this reaction.

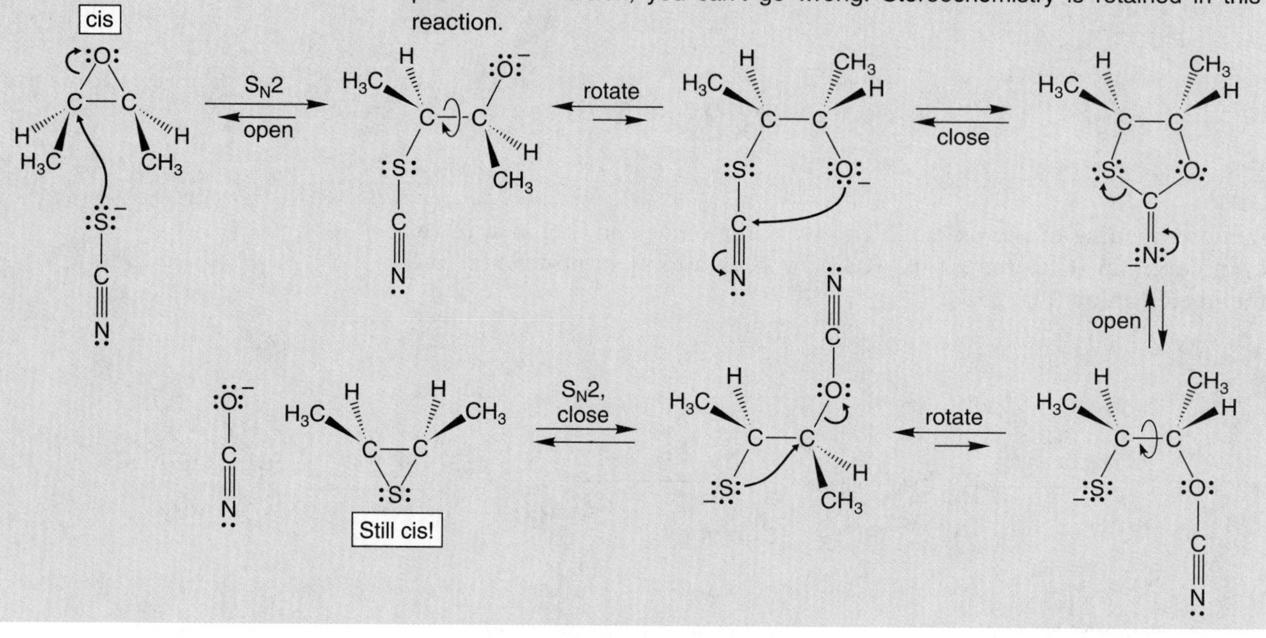

*PROBLEM 25.6 Why does this reaction sequence fail to produce a thiirane when norbornene epoxide is used as starting material (Fig. 25.10)?

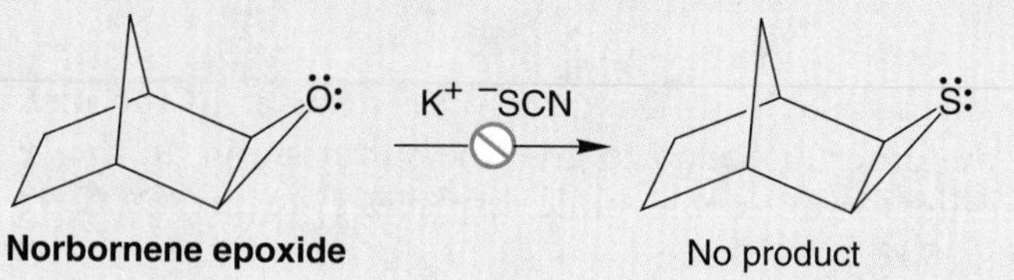

FIGURE 25.10

ANSWER

The rigid frame of norbornane epoxide prevents the rotation necessary to produce the five-membered ring central to this reaction. The initial addition is possible, but the opened intermediate can only revert to starting material.

Norbornene epoxide

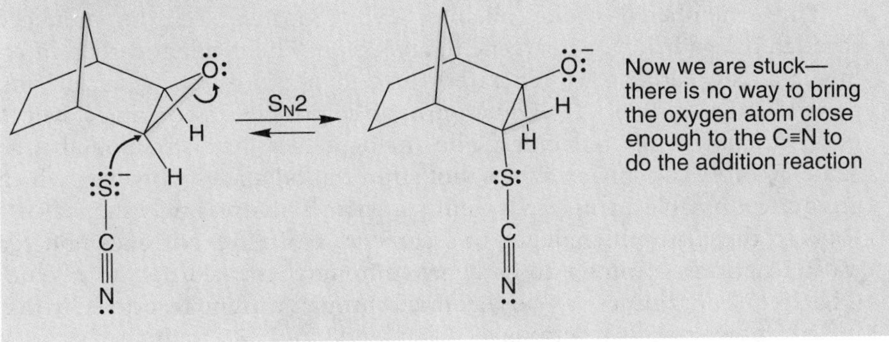

S_N2

Now we are stuck—there is no way to bring the oxygen atom close enough to the C≡N to do the addition reaction

Three-membered rings containing nitrogen, **aziridines**, can be made by conventional reactions, which are quite analogous to processes used to make oxiranes, as well as new reactions using new reagents.

Amines bearing leaving groups in the β-position can often be converted into aziridines. β-Haloamines are classic examples, and vicinal hydroxylamines (2-amino alcohols) can be converted into isolable sulfates that close to aziridines in acid (Fig. 25.11). These reactions are similar to the formation of **epoxides** from vicinal bromohydrins (Chapter 17, p. 847) and to the last step in our thiirane synthesis (Fig 25.8).

The general case

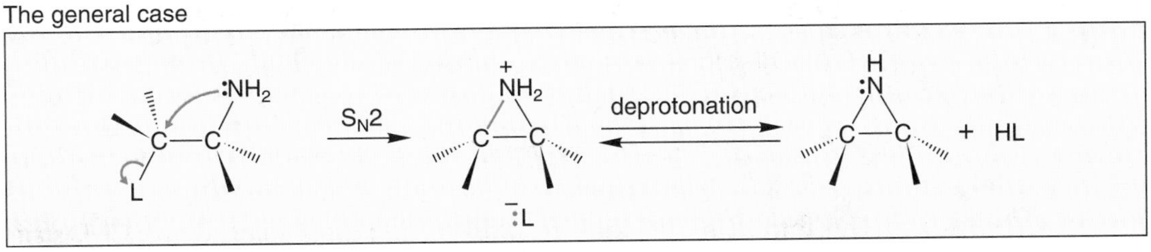

S_N2 ; deprotonation ; + HL

Specific examples

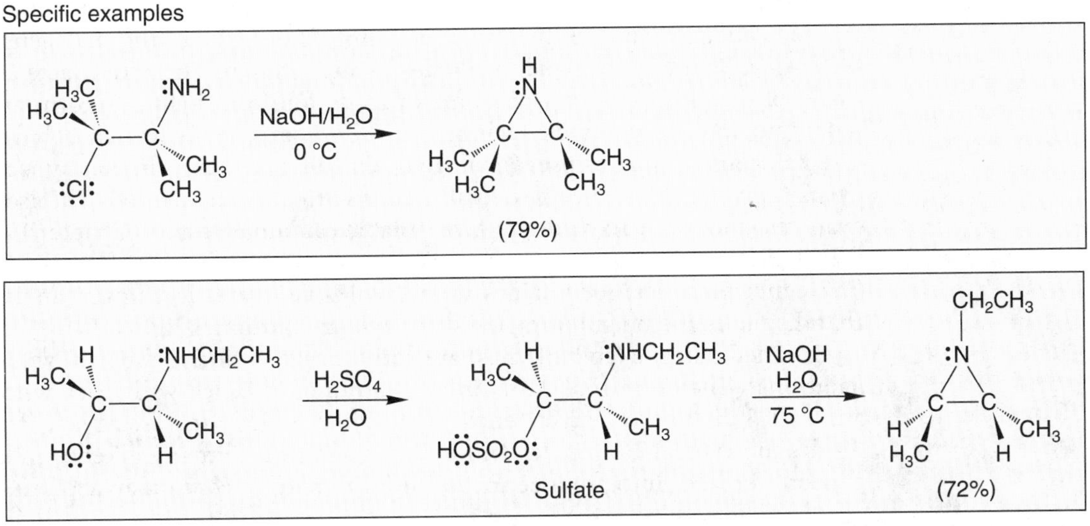

NaOH/H_2O, 0 °C → (79%)

H_2SO_4, H_2O → Sulfate → NaOH, H_2O, 75 °C → (72%)

FIGURE 25.11 Aziridines can be made from 2-haloamines or 2-amino alcohols.

PROBLEM 25.7 Why do you suppose the initial conversion into the sulfate shown in the specific example of Figure 25.11 is important? Why not simply heat the starting amino alcohol?

Three-membered cyclic alkanes, cyclopropanes, can be produced through the addition of a carbene to an alkene. The divalent carbon intermediate, the carbene, is generated through photolysis of a diazo compound (Chapter 10, p. 411). This synthetic sequence can be carried over to nitrogen-containing molecules. The analogue of a diazo compound is an **azide**. Photolysis of many azides (not simple alkyl azides, however, which are quite unstable compounds, and sometimes distressingly explosive*) leads to the nitrogen analogue of a carbene, a **nitrene**. Nitrenes undergo most reactions common to carbenes, among them addition to alkenes (Fig. 25.12). Aziridines can be generated through nitrene reactions by taking advantage of this reaction (see Chapter 23, p. 1222, for a discussion of the nonintermediacy of nitrenes in the Curtius rearrangement).

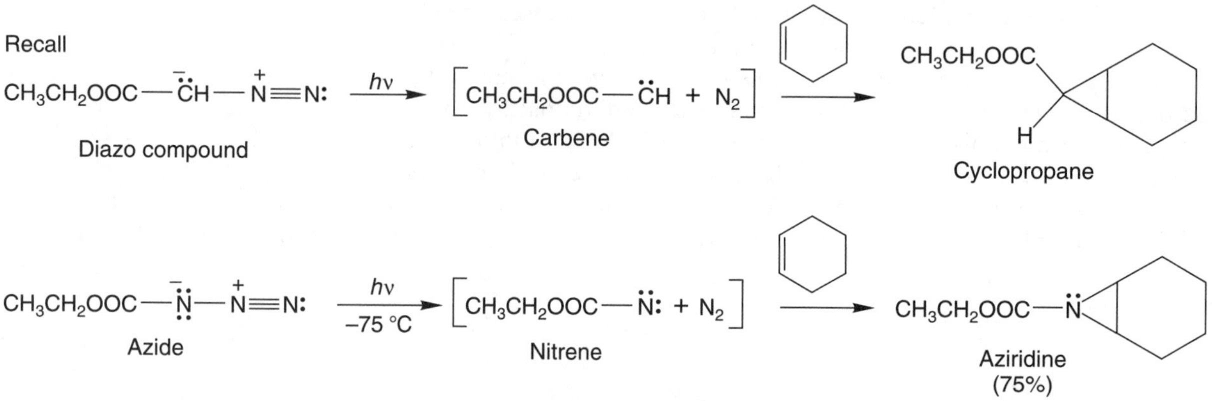

Recall

$CH_3CH_2OOC—\overset{-}{\ddot{C}}H—\overset{+}{N}≡N:$ $\xrightarrow{h\nu}$ $[CH_3CH_2OOC—\ddot{C}H + N_2]$ → Cyclopropane

Diazo compound — Carbene — CH_3CH_2OOC, H — Cyclopropane

$CH_3CH_2OOC—\overset{-}{\ddot{N}}—\overset{+}{N}≡N:$ $\xrightarrow[-75\ °C]{h\nu}$ $[CH_3CH_2OOC—\ddot{N}: + N_2]$ → $CH_3CH_2OOC—\ddot{N}$ (aziridine)

Azide — Nitrene — Aziridine (75%)

FIGURE 25.12 Aziridine formation through addition of a nitrene to an alkene.

The formidable looking reagents iodine isocyanate (I–N=C=O) and iodine azide (I–N_3) probably are not familiar to you. They become tractable if their analogy to the halogens is pointed out. Like the halogens, I–N=C=O and I–N_3 can add to alkenes in anti fashion. The alkene acts as nucleophile and displaces isocyanate or azide ion from iodine to form an iodonium ion. Compounds such as I–N=C=O and I–N_3, which behave like halogens, are called **pseudohalogens**. Opening of the cyclic ion by isocyanate gives the product of overall anti addition to the alkene. The isocyanates react with alkoxides to give isolable carbamate esters (Chapter 19, p. 984; Fig. 25.13).

Reduction of the iodo azides with $LiAlH_4$ generates a nitrogen anion that displaces the neighboring iodide to give an aziridine (Fig. 25.14).

Strained alkenes (and alkynes) are active reagents in 1,3-dipolar cycloadditions (Chapter 10, p. 416). Even unstrained alkenes will react with active 1,3-dipoles or, given enough time, with less reactive reagents. Azides are typical 1,3-dipoles and undergo addition to many alkenes in a concerted reaction to produce **triazolines**. Triazolines contain an azo link-

* Phenyl azide, for example, is the active ingredient in airbags.

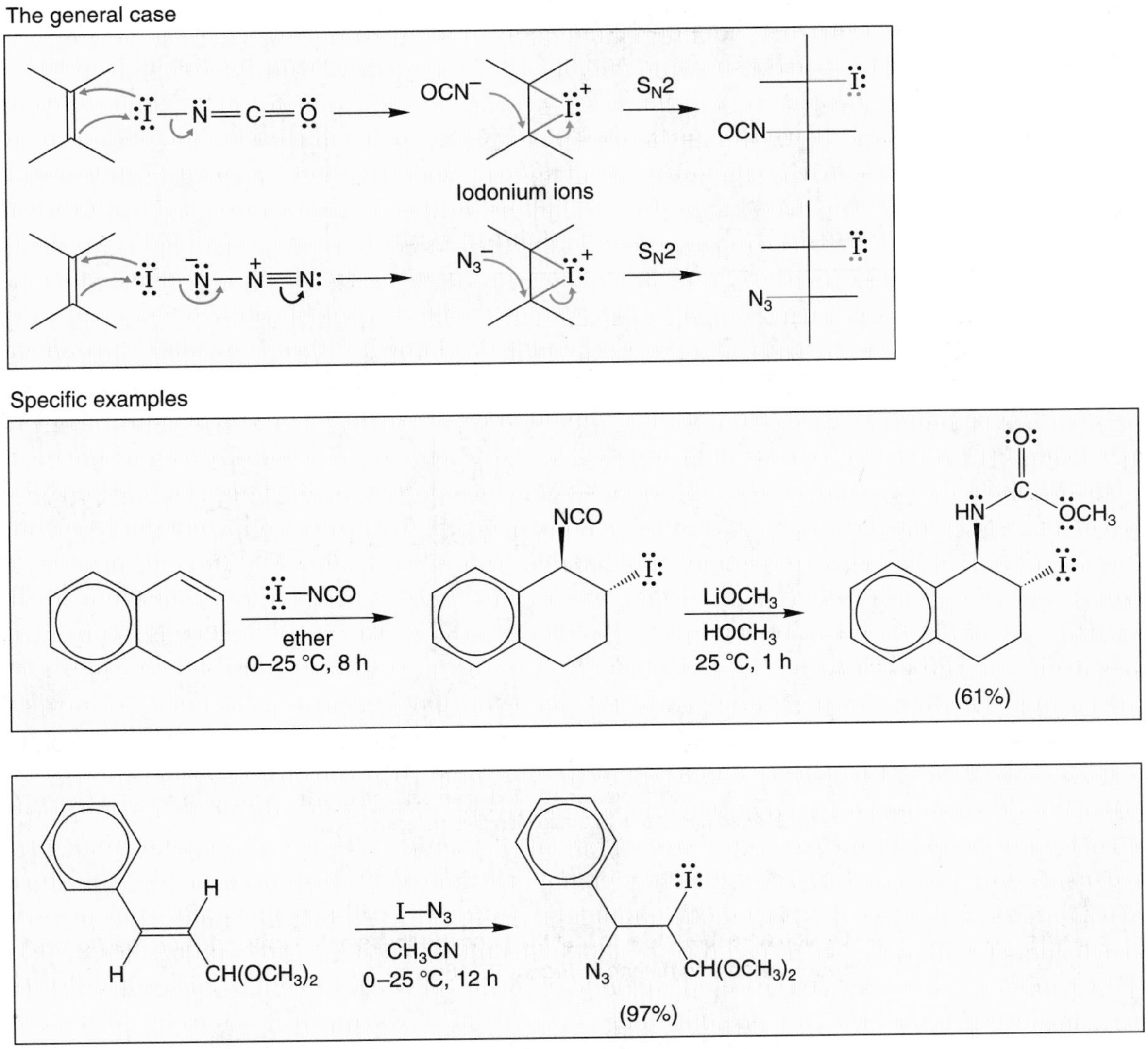

The general case

OCN⁻ Iodonium ions S_N2 OCN

N_3^- S_N2 N_3

Specific examples

:I—NCO ether 0–25 °C, 8 h NCO LiOCH$_3$ HOCH$_3$ 25 °C, 1 h HN OCH$_3$ (61%)

I—N$_3$ CH$_3$CN 0–25 °C, 12 h CH(OCH$_3$)$_2$ N$_3$ (97%)

FIGURE **25.13** Overall anti addition of iodine isocyanate and iodine azide to alkenes.

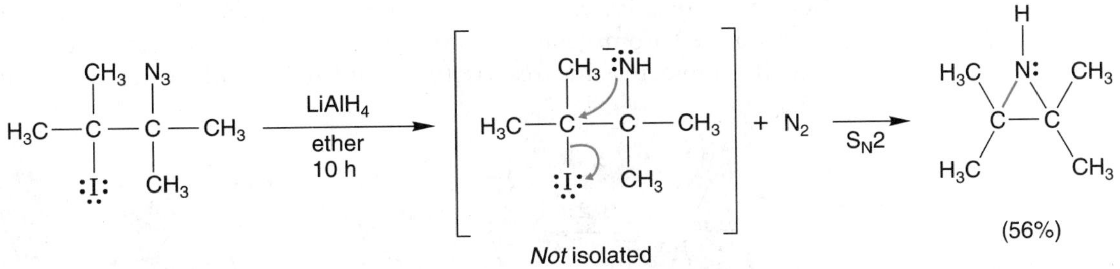

LiAlH$_4$ ether 10 h + N$_2$ S_N2 (56%)

Not isolated

FIGURE **25.14** Reduction of the iodo azide with LiAlH$_4$ leads to an aziridine through intramolecular displacement of iodine.

age (N=N), and thermal or photochemical decomposition of such compounds irreversibly generates nitrogen and produces the aziridine (Fig. 25.15).

The general case

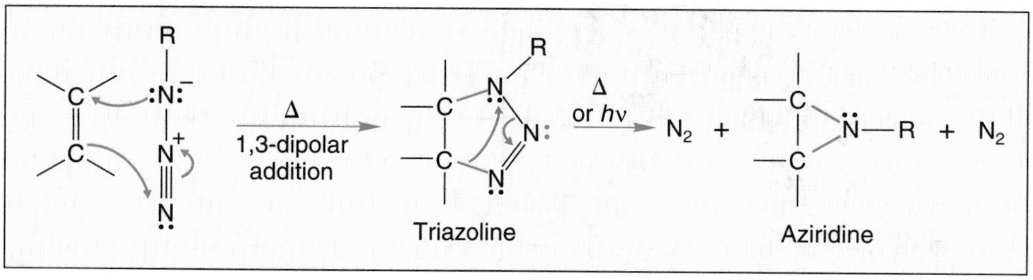

A specific example

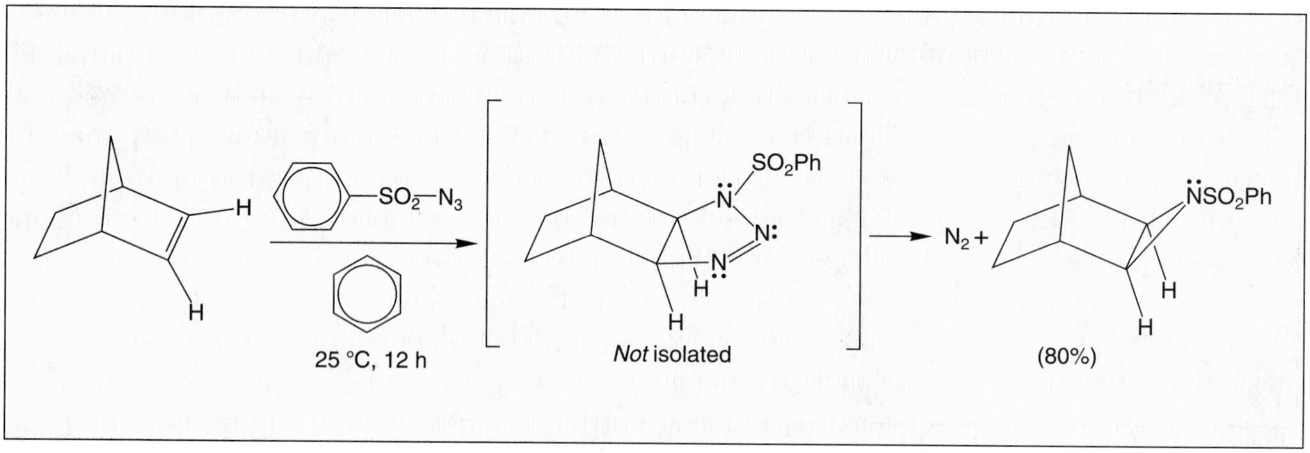

FIGURE 25.15 A 1,3-dipolar addition of an azide to give a triazoline. Thermal or photochemical decompositions of triazolines give aziridines.

PROBLEM 25.8 Design an experiment to test the concertedness of the cycloaddition reaction between azide and alkene in Figure 25.15.

25.1b Reactions of Three-Membered Heterocyclic Compounds

Three-membered rings contain both severe angle strain and torsional strain (Chapter 6, p. 192). The result, even for cyclopropane, is a substantial chemical reactivity. In heterocyclic three-membered rings, the presence of a heteroatom generally provides a relatively good leaving group within the molecule and reactivity toward nucleophiles increases (Fig. 25.16).

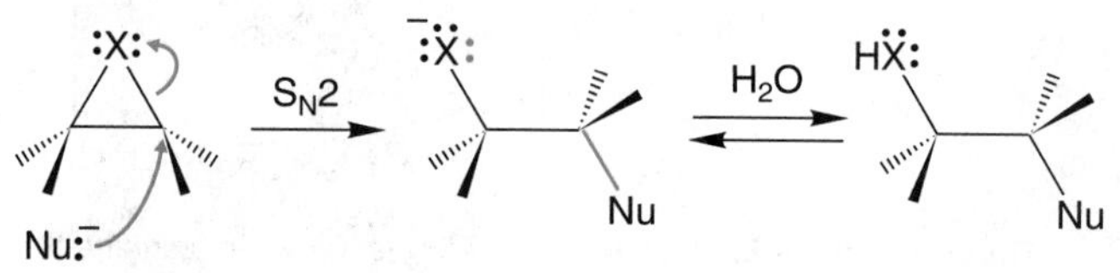

FIGURE 25.16 Nucleophilic opening of heterocyclic three-membered rings.

We have already seen the many acid- and base-induced ring-opening reactions of oxiranes when nucleophiles are present (Chapter 10, p. 408; Chapter 17, p. 853; Fig. 25.17).

Acid-induced oxirane opening

HCl

(45%) (55%)

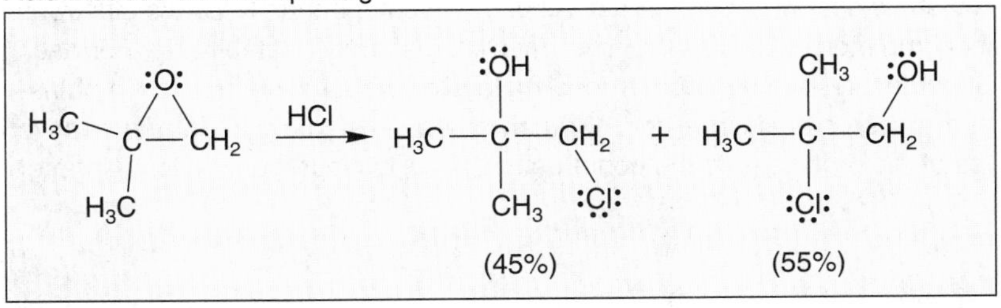

Base-induced oxirane opening

150 °C

(90%)

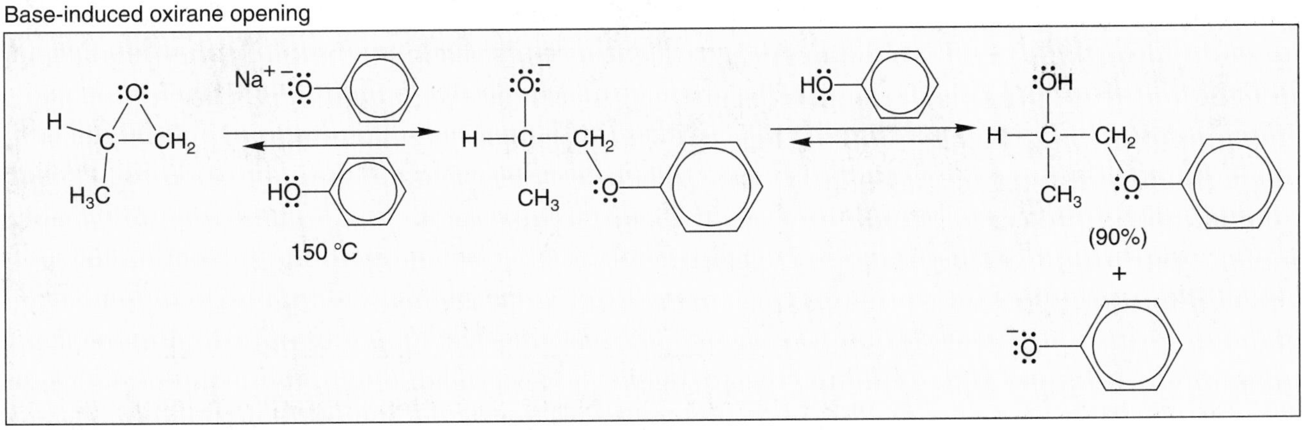

FIGURE 25.17 Typical acid- and base-induced openings of oxiranes.

Both thiiranes and aziridines readily undergo ring-opening reactions, and one example for each species is shown in Figure 25.18.

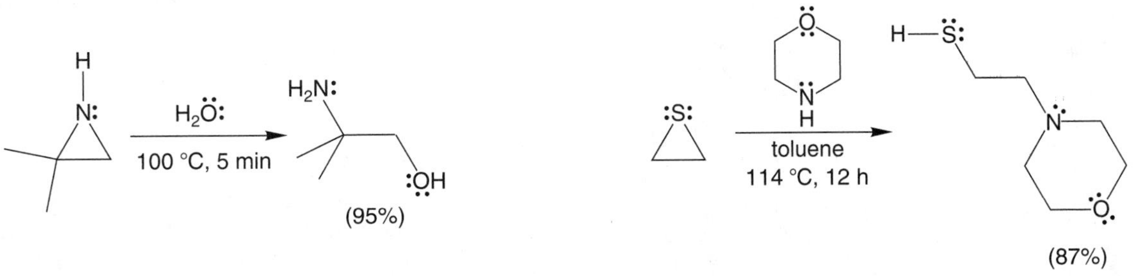

100 °C, 5 min

(95%)

toluene
114 °C, 12 h

(87%)

FIGURE 25.18 Ring-opening reactions of aziridines and thiiranes.

*PROBLEM 25.9

Although heterocyclopropanes are well known, heterocyclopropenes are not. Some have been implicated as reaction intermediates, but it's fair to say that they are clearly much less stable than their more saturated analogues. Explain the increased instability of the compounds shown in Figure 25.19. *Hint*: The reason is not *only* that they are more strained, although they surely are. Think "molecular orbitals."

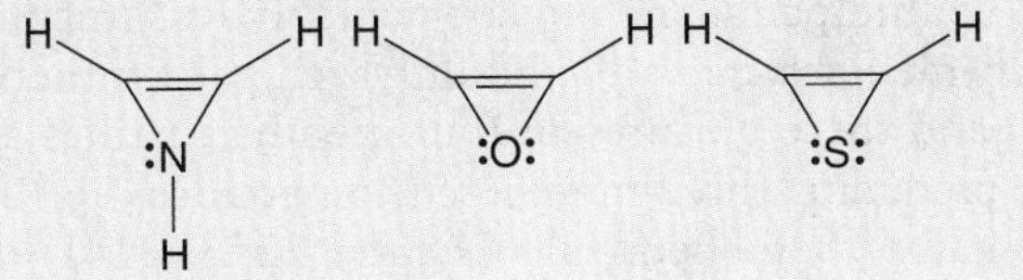

FIGURE 25.19

ANSWER (CONTINUED)

The problem is not only one of strain. All of these compounds will have electrons in antibonding molecular orbitals, and that will be sharply destabilizing. For each of these molecules we can construct a molecular orbital system from combinations of three 2*p* orbitals in a ring. We can use a Frost circle to get the relative energies of the orbitals (see the construction of the cyclopropenium ion, Chapter 13, p. 588).

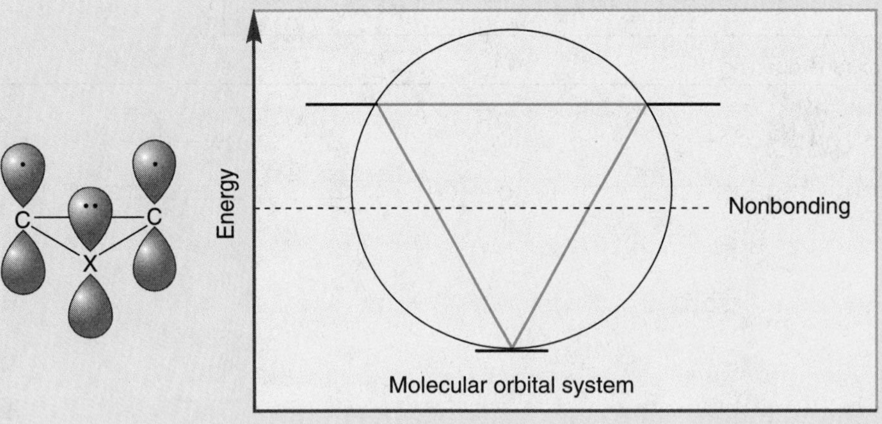

There are four π electrons for each of these species, and there will always be two electrons in antibonding orbitals.

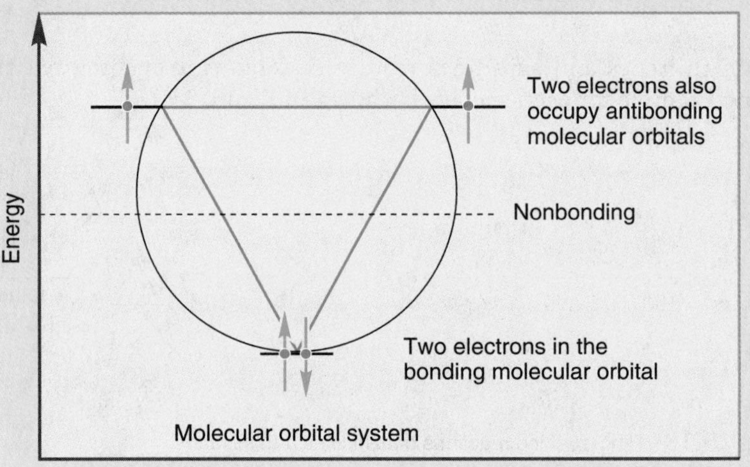

25.2 FOUR-MEMBERED HETEROCYCLES: OXETANES, THIETANES, AND AZETIDINES

Oxetanes, **thietanes**, and **azetidines** are less easily made than their three-membered counterparts, and are considerably less important. Intramolecular displacements constitute the only general synthetic route to these molecules, and these closures to four-membered rings are more difficult than those producing three-membered compounds (Fig. 25.20).

Occasionally photochemical (Chapter 22, p. 1141) or thermal 2 + 2 cycloaddition reactions are useful in making small samples of oxetanes and other four-membered heterocycles (Fig. 25.21).

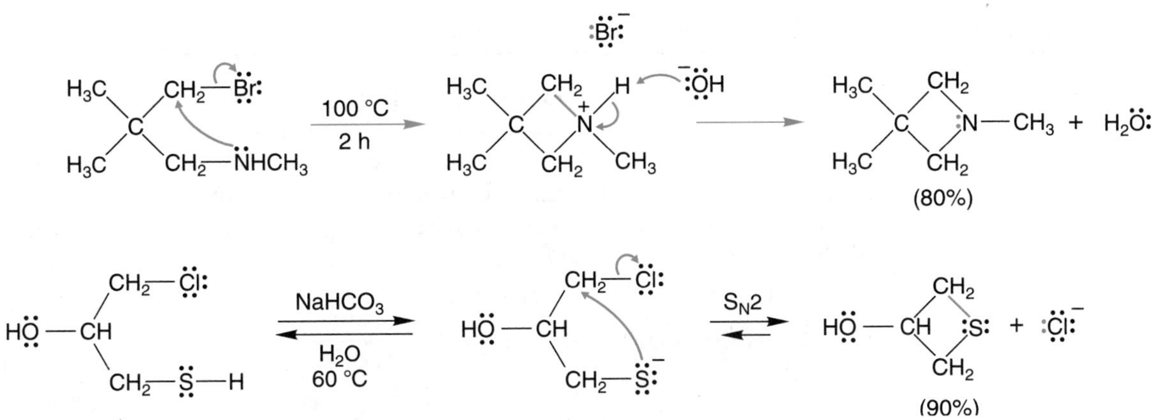

FIGURE 25.20 Some ring-closing reactions leading to four-membered heterocycles.

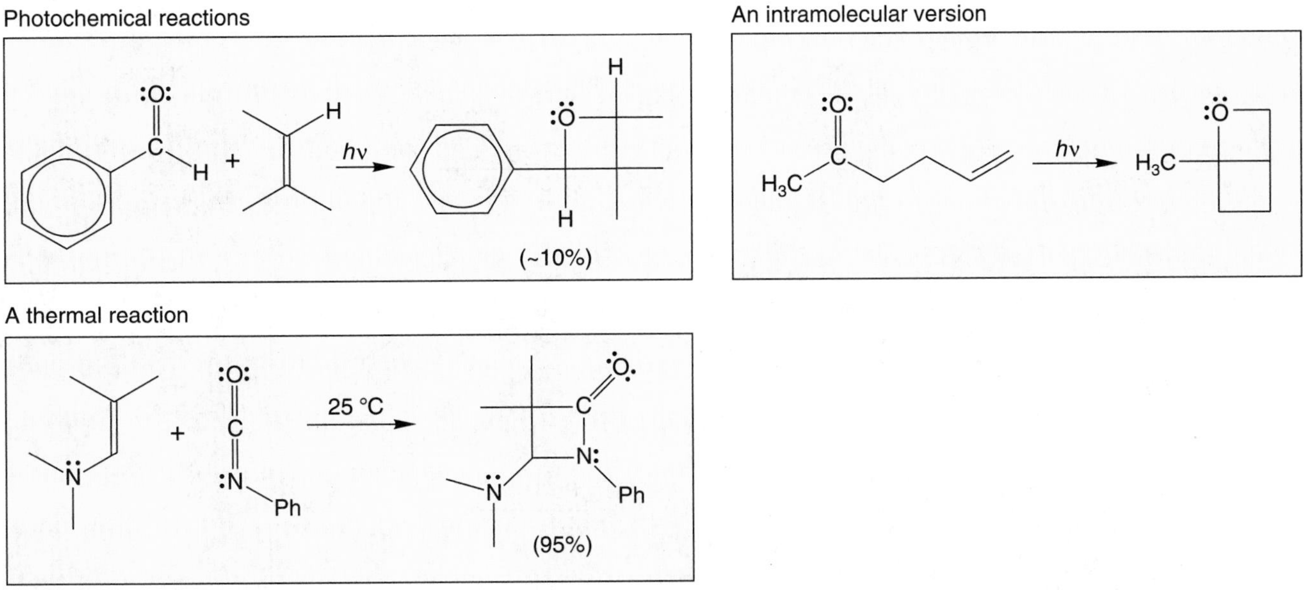

FIGURE 25.21 Photochemical and thermal 2 + 2 cycloaddition reactions lead to oxetanes and other four-membered rings.

These four-membered rings still contain considerable strain and often undergo ring-opening reactions in either acid or base (Fig. 25.22).

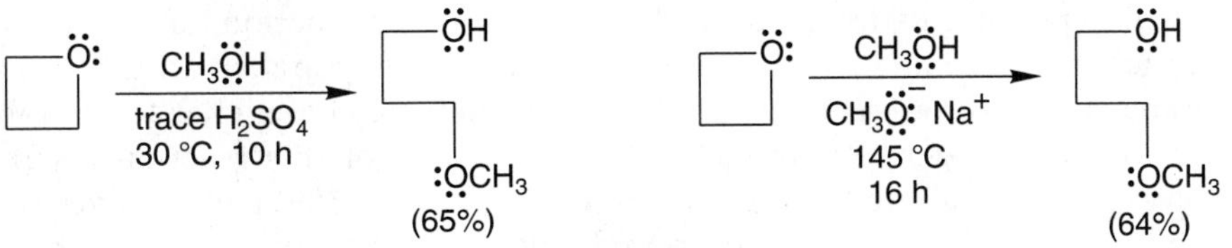

FIGURE 25.22 Ring-opening reactions of four-membered heterocycles.

Write mechanisms for the reactions of Figure 25.22. PROBLEM 25.10

25.3 FIVE- AND SIX-MEMBERED HETEROCYCLES

In these molecules ring strain is not important, and such compounds are approximately as unreactive as their acyclic cousins.

Synthetic methods for making these cyclic heterocycles are as normal as their chemistry. The most common routes are intramolecular and intermolecular double displacement reactions, and hydrogenations of the corresponding unsaturated heterocycles (Fig. 25.23).

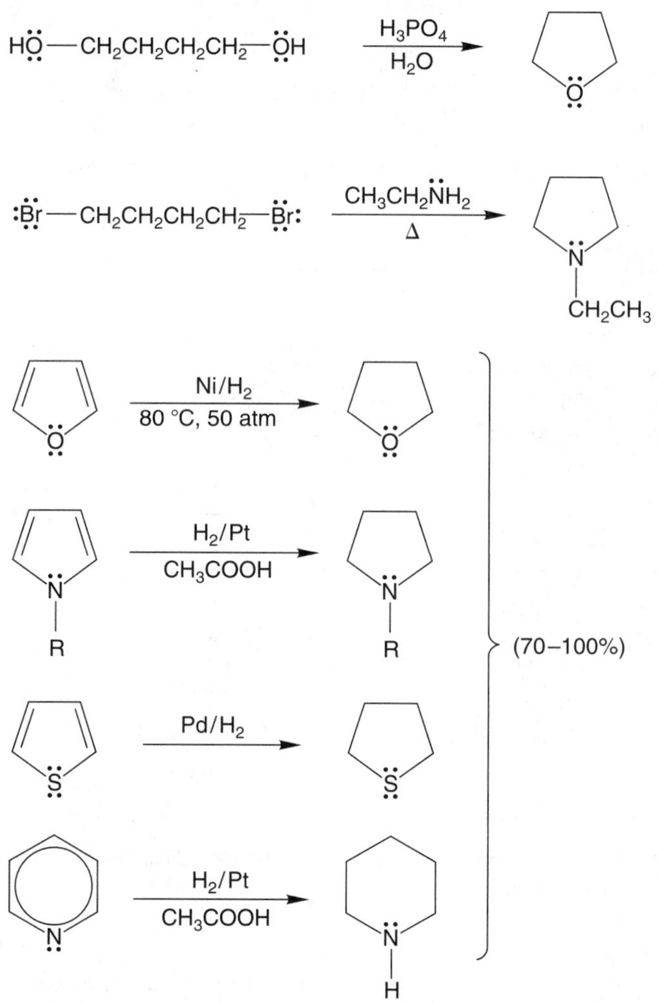

$HO-CH_2CH_2CH_2CH_2-OH$ $\xrightarrow[H_2O]{H_3PO_4}$

$Br-CH_2CH_2CH_2CH_2-Br$ $\xrightarrow[\Delta]{CH_3CH_2NH_2}$ N-CH_2CH_3

$\xrightarrow[80\ °C,\ 50\ atm]{Ni/H_2}$

$\xrightarrow[CH_3COOH]{H_2/Pt}$ R R

$\xrightarrow{Pd/H_2}$

$\xrightarrow[CH_3COOH]{H_2/Pt}$ N H

(70–100%)

FIGURE 25.23 Syntheses of five- and six-membered heterocyclic compounds.

There are important specialized uses of some of these compounds. For example, as we saw in Chapter 17 (p. 861), sulfides usually can be desulfurized to produce hydrocarbons. This hydrogenation reaction takes place at a nickel surface and uses a reagent known as **Raney nickel**, after Murray Raney (1885–1966) who invented this fine dispersion of nickel containing absorbed hydrogen. Presumably, the mechanism of this reaction resembles that of catalytic hydrogenation (Chapter 10, p. 389) and involves the absorption of both the hydrogen and sulfide on the active surface of the metal. Radicals are formed that then abstract hydrogen from absorbed hydrogen (H_2). Sulfur can be removed in this way in the presence of most functional groups, and Raney nickel is a most versatile reagent (Fig. 25.24).

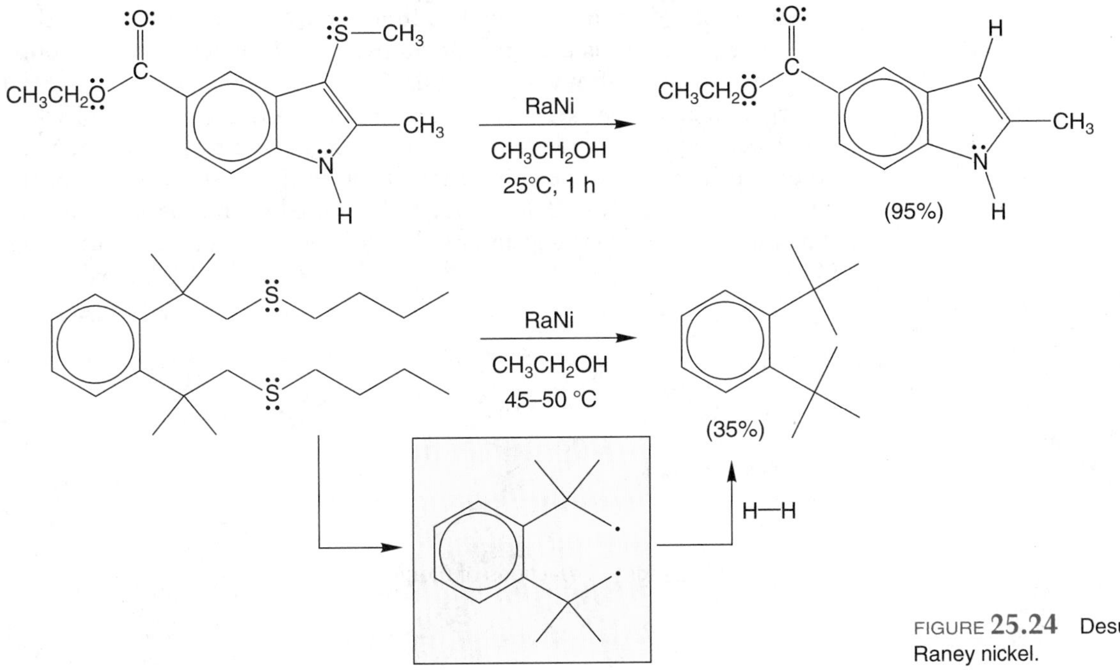

FIGURE 25.24 Desulfurization using Raney nickel.

When the reaction is applied to cyclic sulfides, mixtures of acyclic and cyclic hydrocarbons are obtained through double abstraction or ring closure of an intermediate diradical (Fig. 25.25).

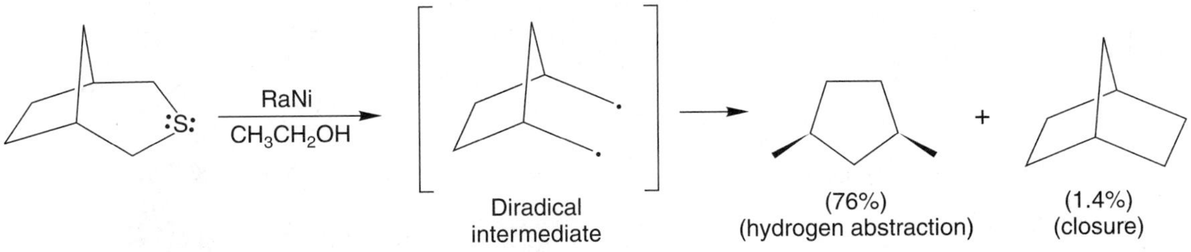

FIGURE 25.25 Desulfurization of a cyclic hydrocarbon leads to cyclic and acyclic products.

The reaction is especially useful when thioacetals are desulfurized. These can be made from the corresponding carbonyl compound and a dithiol (Chapter 17, p. 861). Treatment with Raney nickel leads to the hydrocarbon. The overall procedure is an effective synthetic method for converting a carbon–oxygen double bond into a methylene group without the use of strong acid, as is required in the Clemmensen reaction, or strong base, as is used in the Wolff–Kishner (Chapter 14, p. 642) procedure (Fig. 25.26).

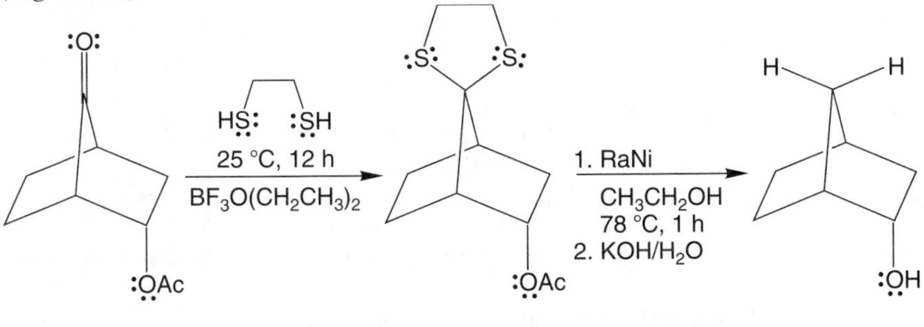

FIGURE 25.26 Desulfurization of thioacetals leads to hydrocarbons.

Heterocyclic compounds can also function as protecting groups. Alcohols can be protected as their tetrahydropyran or **THP derivatives** through treatment with dihydropyran in acid. The THP derivatives are nothing more than acetals, and are stable in base but revert to the alcohol when treated with aqueous acid. As with most reactions of this kind, the position of equilibrium can be shifted by adjusting the concentrations of the reacting molecules. The THP derivative is formed in an excess of dihydropyran and decomposed to regenerate the alcohol in an excess of water (Fig. 25.27).

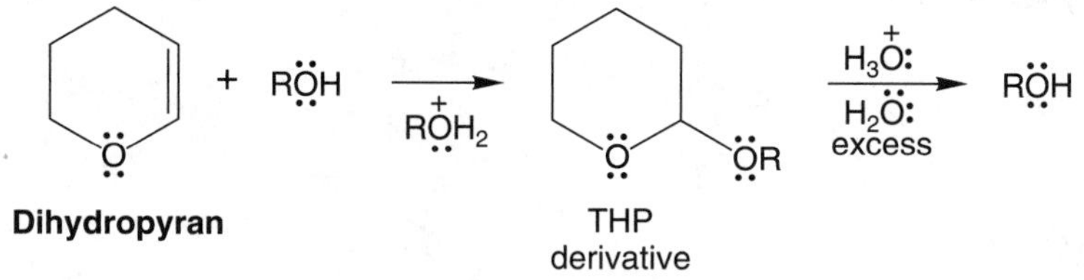

FIGURE 25.27 Alcohols can be protected as THP derivatives, then regenerated.

PROBLEM 25.11 Write mechanisms for the reactions of Figure 25.27.

25.4 NONAROMATIC HETEROCYCLES CONTAINING MORE THAN ONE HETEROATOM

Not every combination of two or three heteroatoms in a ring is known, but many are, and Figure 25.28 gives some examples.

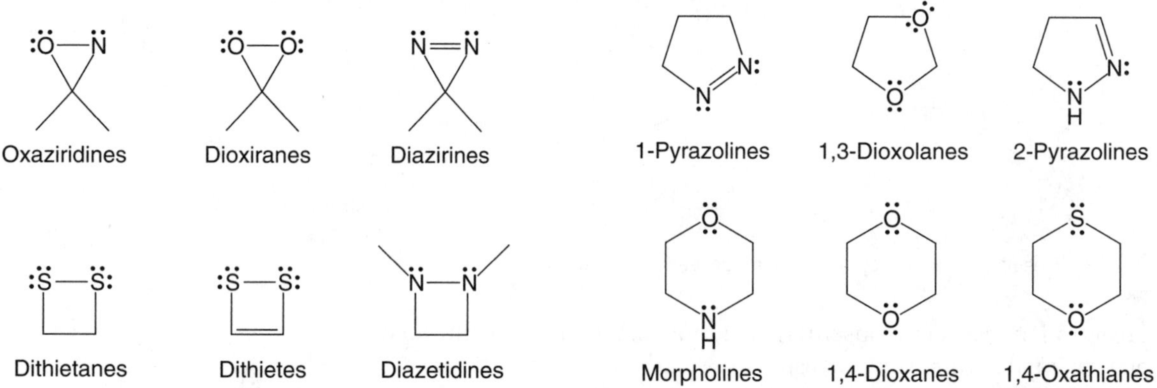

FIGURE 25.28 Some heterocyclic compounds containing more than one heteroatom.

A few we have met already. 1,3-Dipolar cycloadditions of diazomethane yield 1-pyrazolines, and morpholine is a favorite of the enamine makers (Chapter 18, p. 898; Fig. 25.29).

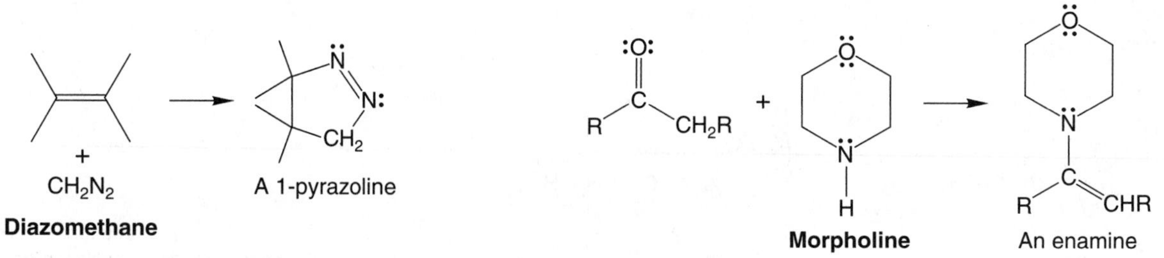

FIGURE 25.29 Two such compounds we have already encountered.

Others have exotic, if often fascinating uses. Robert W. Murray (b. 1928) has shown **dioxiranes** to be powerful oxidizing agents for all sorts of organic compounds, and **diazirines** are a less dangerous (but not completely safe!) source of carbenes than diazo compounds (Fig. 25.30).

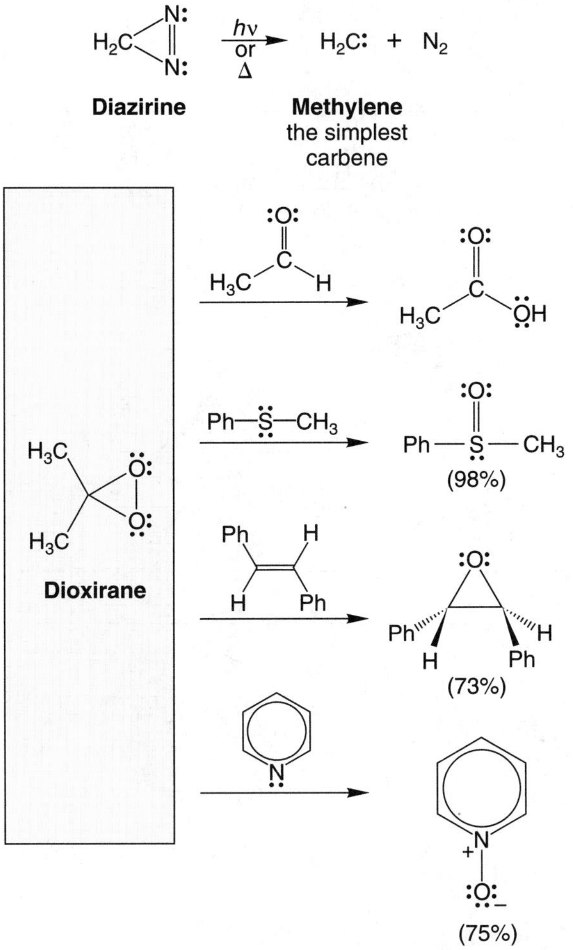

FIGURE 25.30 Some uses of diazirines and dioxiranes.

One of the most practically useful of these molecules is 1,3-dithiane, (1,3-dithiacyclohexane). This compound is a rather strong Brønsted acid (pK_a = 31.1) and can be easily deprotonated with alkyllithium reagents (Fig. 25.31).

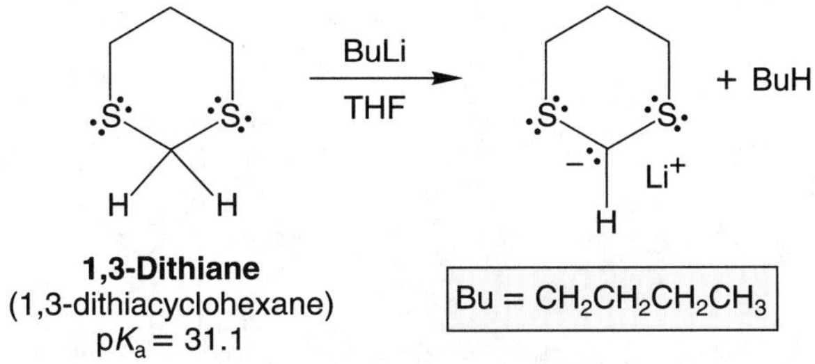

FIGURE 25.31 Deprotonation of 1,3-dithiacyclohexane with butyllithium.

PROBLEM 25.12

Explain how the sulfur atoms operate to increase the acidity of the adjacent hydrogens in 1,3-dithiane (1,3-dithiacyclohexane).

The resulting anion is a powerful nucleophile and can undergo both displacement reactions (subject, of course, to the usual restrictions of the S_N2 reaction) and additions to carbonyl compounds. The products of these reactions are thioacetals and can be hydrolyzed, usually by mercury salts, to aldehydes or ketones. The dithiacyclohexane functions as a masked carbonyl group! Here is a new synthesis of aldehydes and ketones (Fig. 25.32).

The general case

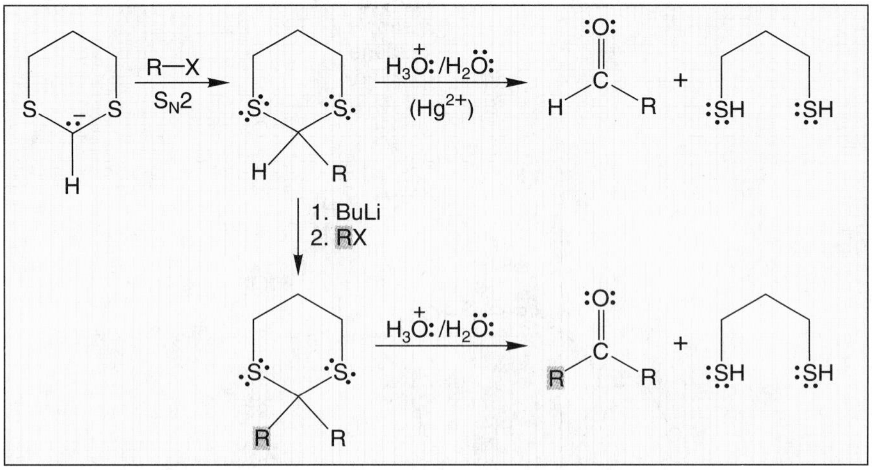

Specific examples

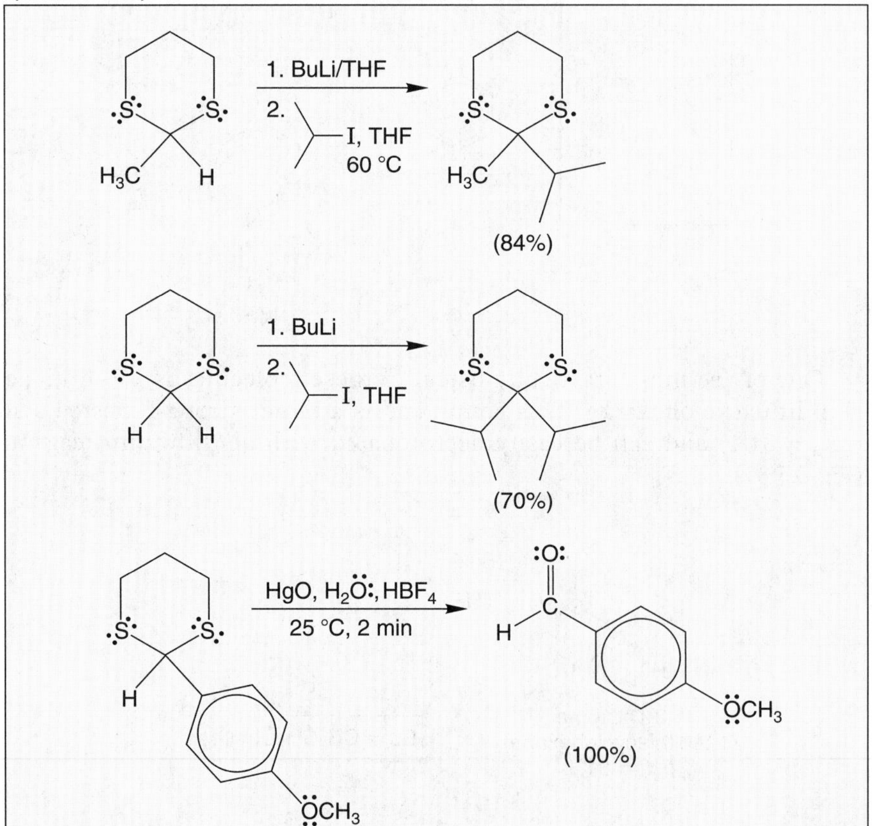

FIGURE 25.32 A synthesis of carbonyl compounds using alkylation of 1,3-dithiane (1,3-dithiacyclohexane).

*PROBLEM 25.13

Suggest a synthetic route to alkanes starting from 1,3-dithiane.

ANSWER

Easy (and effective in practice). First alkylate the dithiane, then desulfurize with Raney nickel.

1,3-dithiane $\xrightarrow{\text{BuLi}}$ anion + R–I $\xrightarrow{S_N2}$ 2-R-dithiane $\xrightarrow{\text{repeat}}$ 2,2-R₂-dithiane $\xrightarrow{\text{Raney Ni}}$ RCH_2R

25.5 NITROGEN-CONTAINING AROMATIC HETEROCYCLIC COMPOUNDS

This section describes **pyridine** and **pyrrole**, six- and five-membered aromatic heterocyclic compounds (Chapter 13, p. 596; Fig. 25.33).

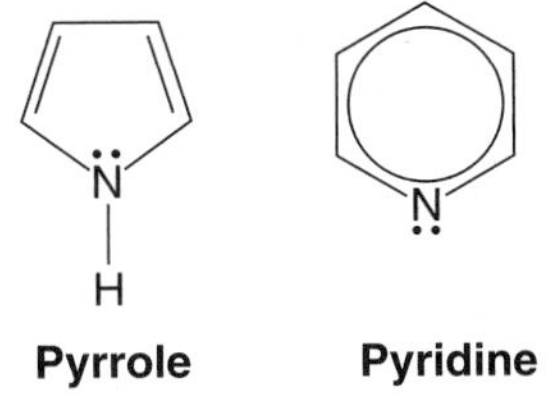

FIGURE 25.33 Pyrrole and pyridine, two aromatic heterocycles.

25.5a Structure

Pyridine and pyrrole provide a nice contrast because of the different ways in which the lone-pair electrons on the nitrogen atom behave. The structure of pyridine closely resembles that of benzene. The lone pair lies in the plane of the σ system, acting like one of the carbon–hydrogen bonds in benzene. By contrast, in pyrrole the pair of nonbonding electrons makes up part of the aromatic sextet as it is incorporated into the π system (Fig. 25.34).

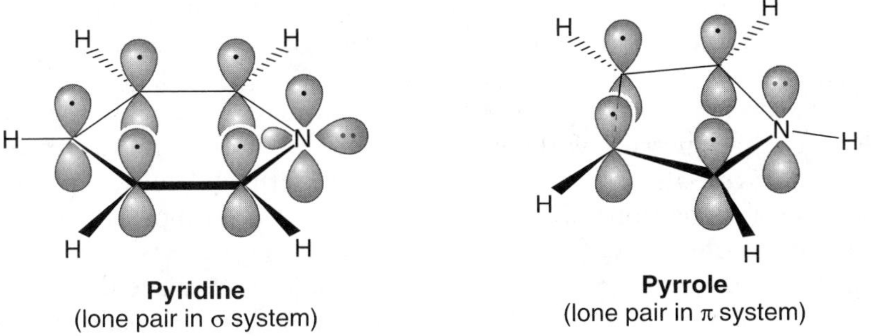

FIGURE 25.34 Orbital pictures of pyrrole and pyridine. Notice the different positions of the lone-pair electrons.

These molecules can be described from either a molecular orbital point of view, or through resonance forms. Figure 25.35 shows both representations.

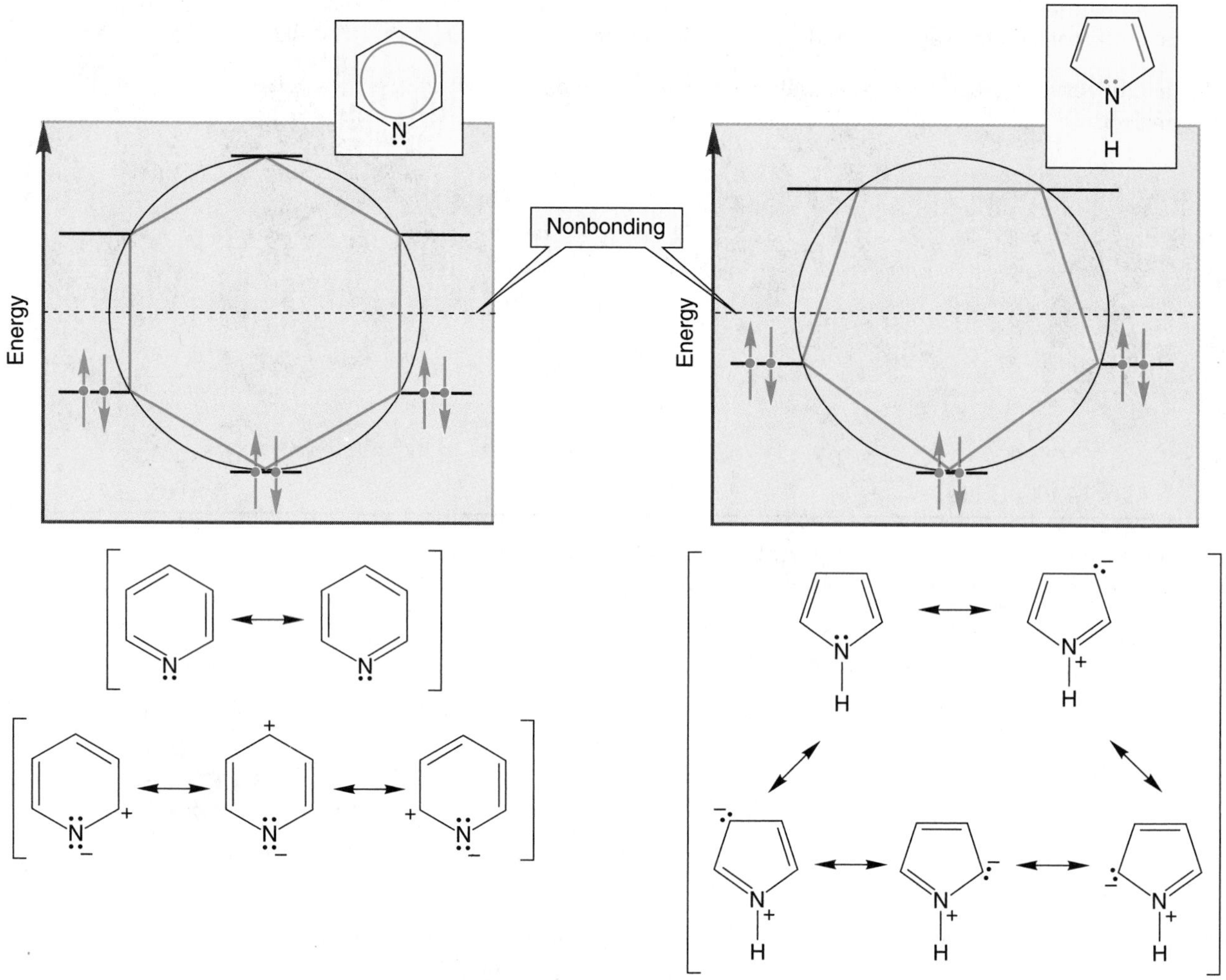

FIGURE 25.35 Molecular orbital and resonance descriptions of pyrrole and pyridine.

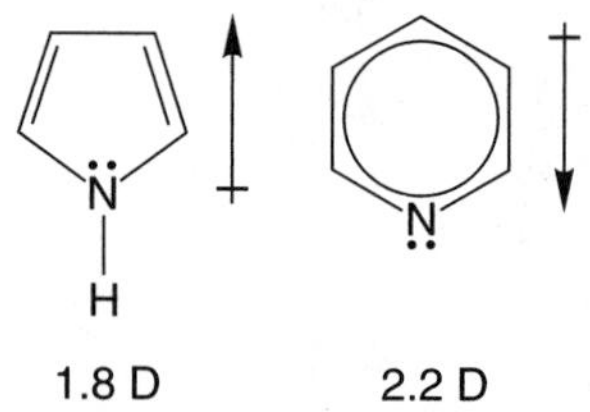

FIGURE 25.36 The dipole moments of pyrrole and pyridine are in different directions.

Note that polar resonance structures are especially important for pyrrole and that the direction of the dipole moment is different in the two aromatic molecules. In pyridine, the electronegative nitrogen is the negative end of the dipole, but in pyrrole nitrogen is the positive end (Fig. 25.36).

How do we know these molecules are aromatic? Probably the best answers to that question will come later when we look at reactivity, but information can also be gained from a look at the ^{1}H NMR spectra. For pyridine, the positions of the ring hydrogens on the ring seem well below what we would expect for a simple polyalkene (Fig. 25.37). As in benzene, the ring hydrogens find themselves in the region of space deshielded by the induced magnetic field (B_i) set up by the ring current, and thus come into resonance at low field. The presence of a ring current implies aromaticity.

Although this argument is probably right, it can be criticized. What other effect could lead to an especially low-field resonance position? Look at the polar resonance forms in Figure 25.35. In pyridine, the ring carbons are electron deficient, and this will lead to deshielding and low-field

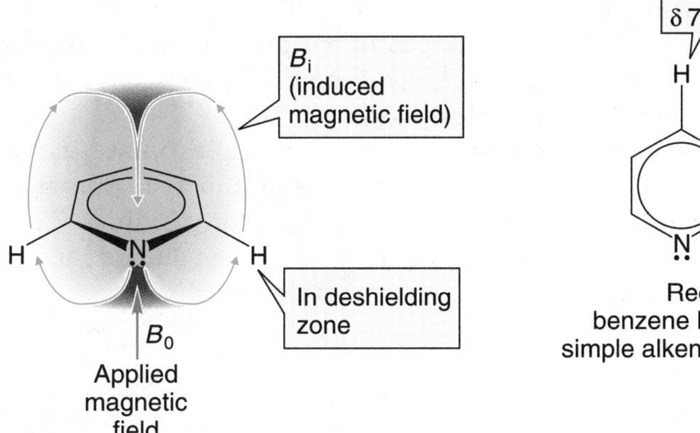

FIGURE 25.37 The ring hydrogens of pyridine are in a deshielded region, and therefore appear downfield in the ^{1}H NMR spectrum.

signals. The NMR shifts seem large enough to imply the presence of a ring current, and thus aromaticity, but it must be admitted that the argument isn't air tight.

PROBLEM 25.14

The carbon-bound hydrogens of pyrrole do not appear at especially low field in the ^{1}H NMR spectrum (δ 6.05 and 6.62 ppm). Does this mean that pyrrole is not aromatic? Comment in light of the previous discussion of pyridine.

PROBLEM 25.15

At room temperature the pyridinophane shown in Figure 25.38 shows a signal in its ^{1}H NMR spectrum at about δ −0.2 ppm. What relevance does this observation have to the question of the aromaticity of pyridine? *Hint*: See Chapter 15 (p. 719).

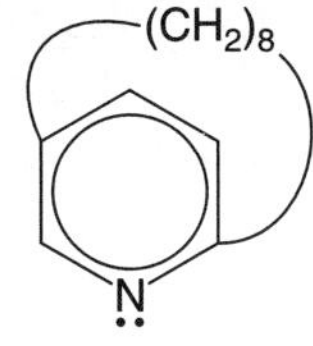

FIGURE 25.38

25.5b Acid and Base Properties

Pyridine can be protonated in acid, but pyridine is a weaker base than a simple cyclic secondary amine such as piperidine. The pK_a values of the conjugate acid ammonium ions show this clearly, as the pyridinium ion is a much stronger acid (pK_a = 5.2) than the ammonium ion formed from a simple secondary amine (pK_a ~ 11). Alternatively, we might make this argument in terms of pK_b values. The pK_b of pyridine is 8.8, whereas that for a simple secondary amine is about 3. So, the simple amine is a far stronger base than pyridine (Fig. 25.39).

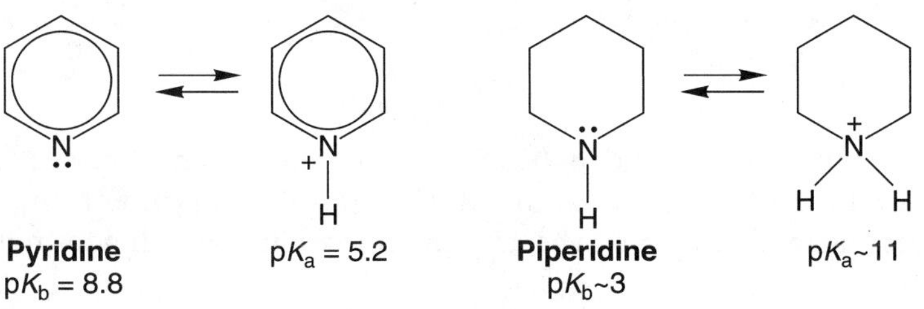

FIGURE 25.39 Pyridine is a weaker base than the similar, but nonaromatic piperidine.

The reason for the decreased basicity of pyridine is that the lone-pair electrons of the sp^2 hybridized nitrogen of pyridine are held more tightly than those of the approximately sp^3 nitrogen of a simple secondary amine like piperidine, and therefore are less basic.

Pyrrole is even less basic than pyridine. One can estimate the pK_a of the conjugate acid at about –4! When pyrrole is protonated, aromaticity is lost as the aromatic sextet is disrupted (*not* true for pyridine). The lone pair of electrons is tied up in the new bond to hydrogen and no longer contributes to the completion of the aromatic sextet (Fig. 25.40)!

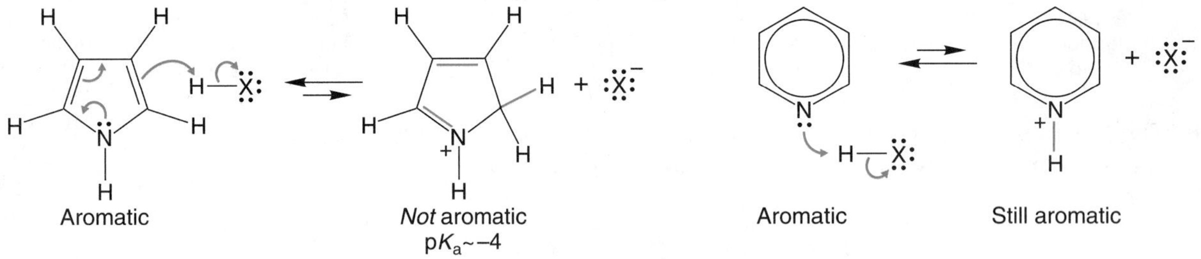

Aromatic — *Not* aromatic pK_a~–4 — Aromatic — Still aromatic

FIGURE 25.40 Pyrrole is a very weak base. Protonation destroys aromaticity.

PROBLEM 25.16 Notice in Figure 25.40 that pyrrole is protonated on carbon, not nitrogen. Explain why.

For an organic compound, pyrrole is a rather strong Brønsted acid, pK_a ~ 23; but it is much weaker than cyclopentadiene, pK_a = 16. This difference may seem curious at first, because removal of a proton from pyrrole leaves a negative charge on a relatively electronegative nitrogen, whereas cyclopentadiene gives a carbon anion, generally a much less happy event, (Fig. 25.41).

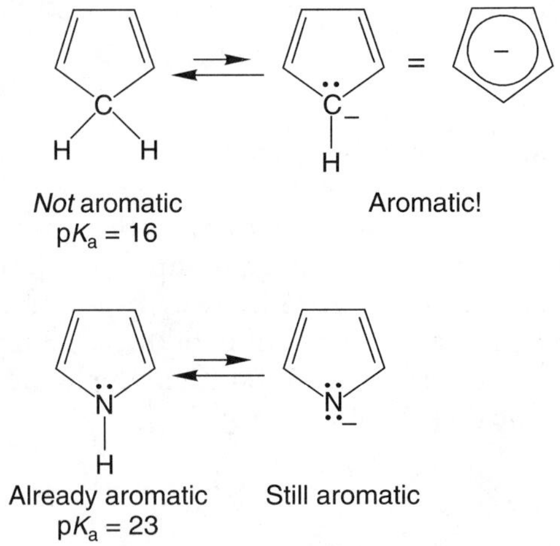

Not aromatic pK_a = 16 — Aromatic!

Already aromatic pK_a = 23 — Still aromatic

FIGURE 25.41 Pyrrole is a relatively weak acid compared to cyclopentadiene.

However, cyclopentadiene *becomes* aromatic when a proton is removed, and pyrrole is *already* aromatic. Removal of a proton from cyclopentadiene is made easier by the gain in aromaticity, which does not happen in pyrrole.

25.5c Pyridine as Nucleophile

Like other amines, pyridine can be alkylated in an S_N2 reaction in which the lone-pair electrons on nitrogen displace a leaving group (Chapter 21, p. 1090). Treatment of pyridine with a primary or secondary alkyl iodide leads to alkyl **pyridinium ions**. In a reaction reminiscent of the formation of *N*-oxides by oxidation of tertiary amines (Chapter 21, p. 1100), pyridine gives **pyridine *N*-oxide** when treated with hydrogen peroxide (Fig. 25.42).

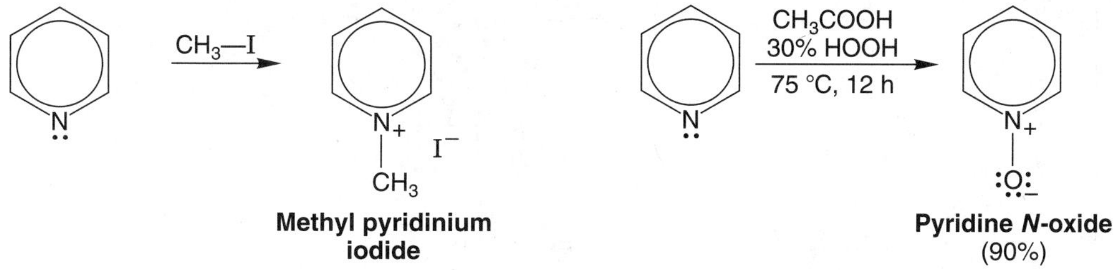

$CH_3—I$; I^- ; CH_3

Methyl pyridinium iodide

CH_3COOH ; 30% HOOH ; 75 °C, 12 h

Pyridine *N*-oxide (90%)

FIGURE 25.42 Pyridine can act as a nucleophile in the formation of pyridinium salts and pyridine *N*-oxides.

25.5d Electrophilic Aromatic Substitution of Pyridine and Pyrrole: Resonance Energy of Heteroaromatic Compounds

The classic reaction of aromatic compounds is electrophilic substitution (Chapter 14). Both pyridine and pyrrole undergo this reaction, which is directly analogous to the substitution reactions of benzene itself, and occurs for the same reason. Delocalization leads to the substantial energy lowering known as resonance energy. We reached a quantitative determination of the resonance energy for benzene through measurements of the heats of hydrogenation (or combustion) for the model compounds cyclohexene and cyclohexadiene. We then used these data to estimate the heat of hydrogenation of the *hypothetical* molecule cyclohexatriene. Finally, we compared our estimate with the experimental value determined by hydrogenating the *real* molecule benzene (Chapter 13, p. 575). Table 25.1 gives similarly determined values for a variety of heteroaromatic compounds.

The heteroatom introduces a complication in aromatic substitution not present in benzene. Benzene has six equivalent positions, but substitution could occur at two positions in pyrrole and three in pyridine (Fig. 25.43).

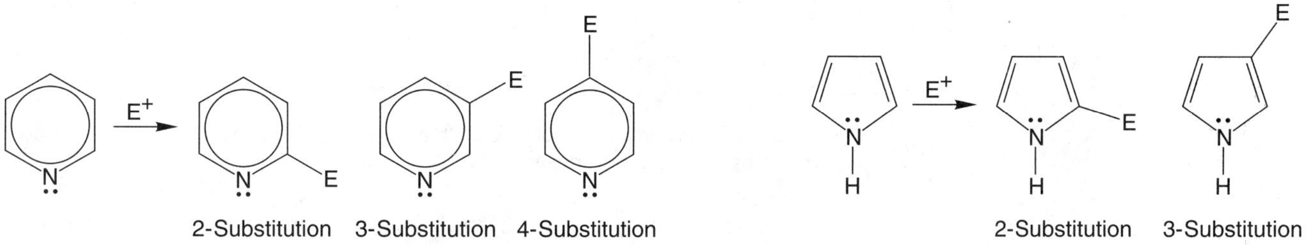

TABLE 25.1 Resonance Energies of Some Heteroaromatic Compounds and Benzene

Compound	Resonance Energy (kcal/mol)
Benzene	36
Thiophene	29
Pyridine	23
Pyrrole	21
Furan	16

E^+ — 2-Substitution 3-Substitution 4-Substitution

E^+ — 2-Substitution 3-Substitution

FIGURE 25.43 Possible positions for electrophilic aromatic substitution in pyridine and pyrrole.

As for benzene, electrophilic attack on pyridine gives an intermediate in which three atoms share the positive charge. Substitution ortho or para to the nitrogen in pyridine places a partial positive charge on the relatively electronegative nitrogen atom. Substitution at the meta, or 3-position does not, and is accordingly favored (Fig. 25.44).

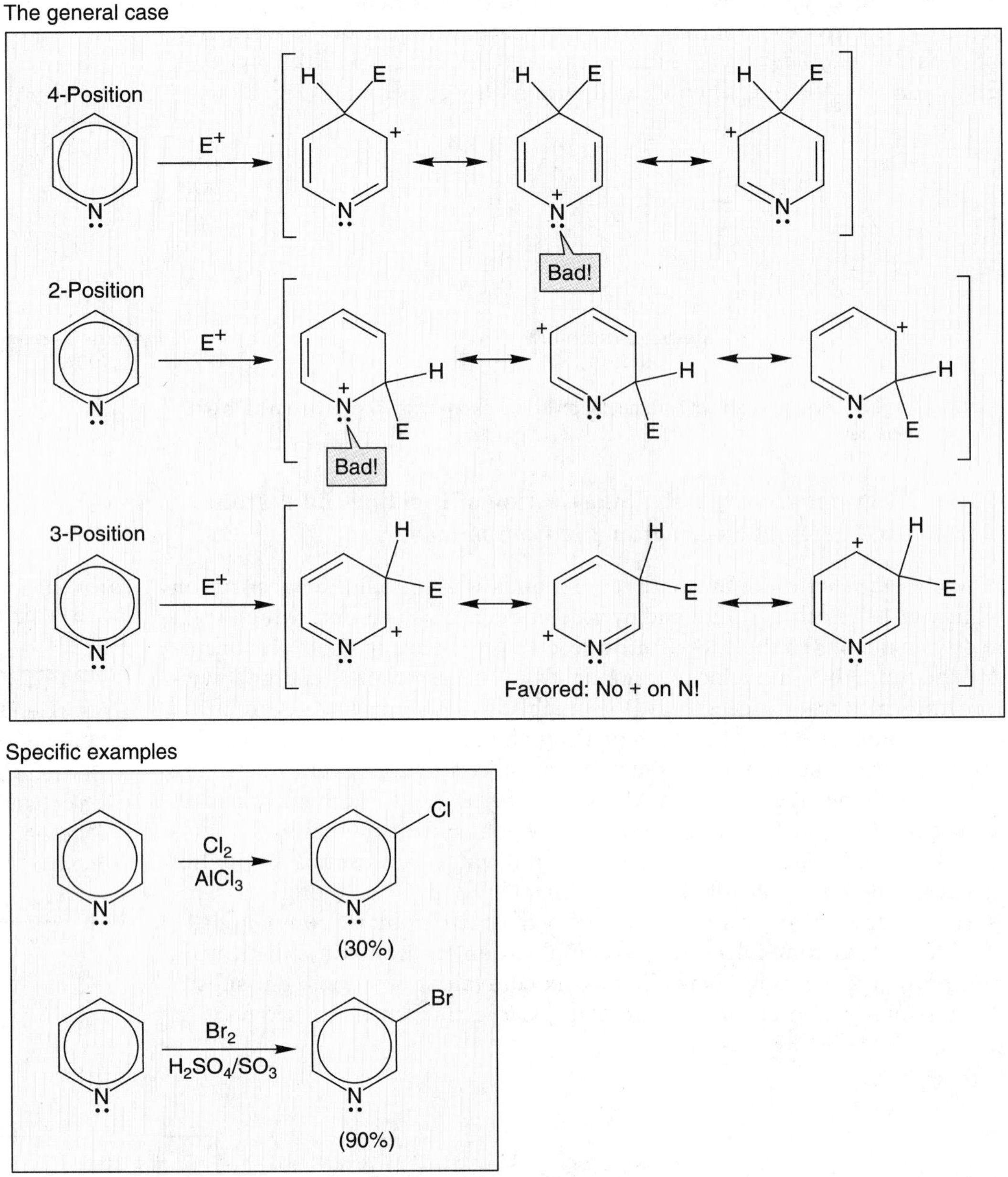

The general case

4-Position

E^+

Bad!

2-Position

E^+

Bad!

3-Position

E^+

Favored: No + on N!

Specific examples

Cl_2 / $AlCl_3$ (30%)

Br_2 / H_2SO_4/SO_3 (90%)

FIGURE 25.44 In pyridine, substitution at the 3-position is favored, although the reaction is slow compared to that of benzene.

This effect is only accentuated if the nitrogen is protonated, as it very often is under the strongly acid conditions necessary to carry out substitution on pyridine. Therefore it is not surprising that substitution of pyridine takes place more slowly than analogous reactions of benzene. Pyridine itself substitutes about as rapidly as nitrobenzene, 10^6 more slowly than benzene itself. The pyridinium ion substitutes an impressive 10^{18} times more slowly than benzene.

A similar analysis serves to locate the more favorable position for substitution of pyrrole. Addition at either carbon 2 or 3 allows the nitrogen to help bear the positive charge. However, in the intermediate produced by substitution at carbon 2, two ring carbons share the charge, whereas substitution at carbon 3 leads to an intermediate in which only one carbon shares the charge. Addition to the 2-position is favored, although the formation of mixtures is commonly observed (Fig. 25.45).

General cases

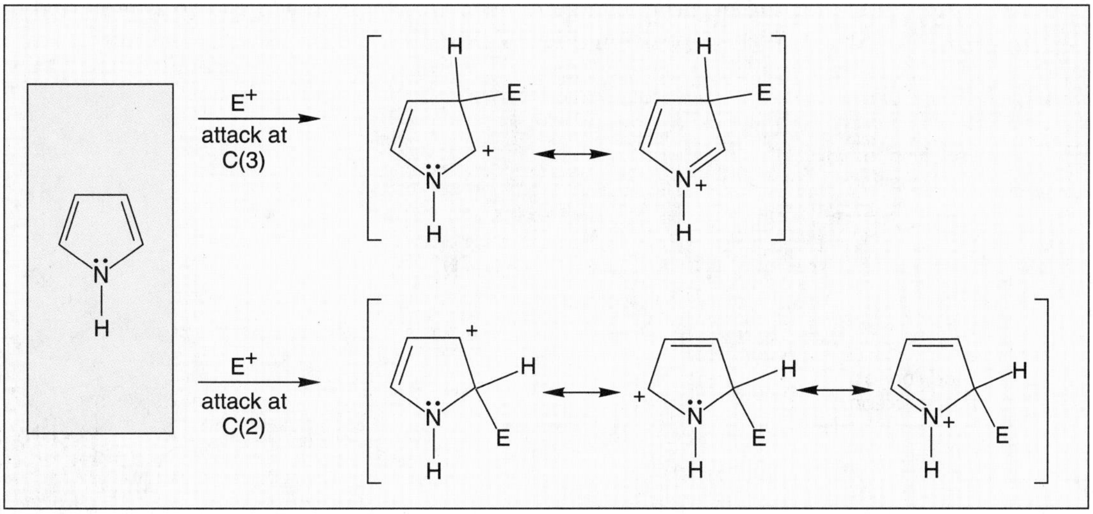

E^+ attack at C(3)

E^+ attack at C(2)

A specific example

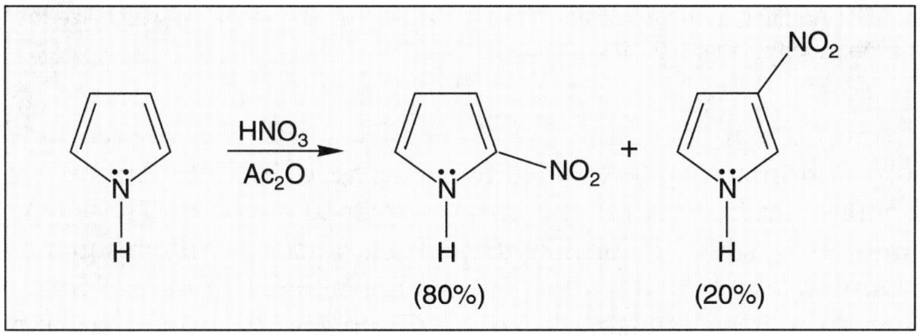

HNO_3, Ac_2O (80%) (20%)

FIGURE 25.45 In pyrrole, substitution at the 2-position is favored.

Pyrrole is more reactive than benzene or pyridine, as pyrrole is less well stabilized by resonance than the six-membered ring molecules (Table 25.1).

25.5e Nucleophilic Aromatic Substitution of Pyridine and Pyrrole

Pyridine is more active in nucleophilic aromatic substitution than benzene. Not only is pyridine less well resonance stabilized than benzene (Table 25.1), but the electronegative nitrogen atom provides an attractive place for an incoming negative charge (Fig. 25.46).

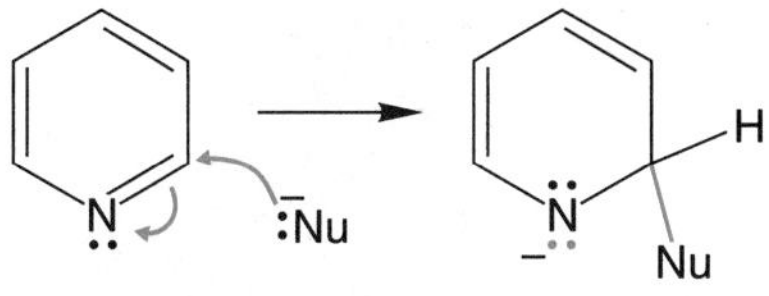

The classic example of nucleophilic aromatic substitution of pyridine is the **Chichibabin reaction**, named for Alexei E. Chichibabin (1871–1945), in which pyridine is converted into 2-aminopyridine through treatment with potassium amide in liquid ammonia (Fig. 25.47).

FIGURE 25.46 Nucleophilic attack on pyridine.

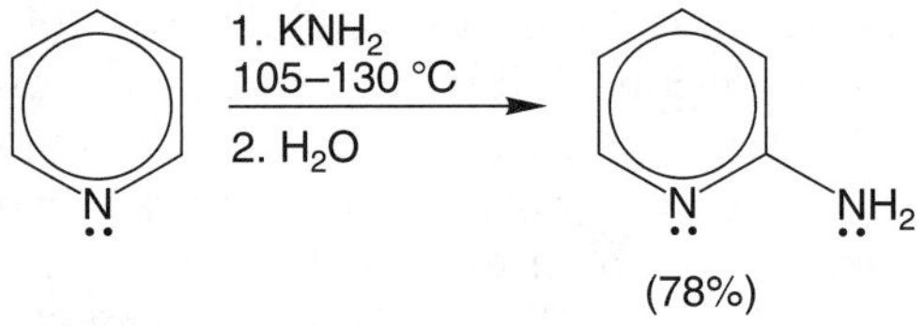

FIGURE 25.47 The Chichibabin reaction.

The first step of the mechanism is conventional, as the strongly basic amide ion adds to pyridine. Addition at the 2- or 4-position makes it possible for the negative charge to be shared by nitrogen and so is preferred to attack at the 3-position (Fig. 25.48).

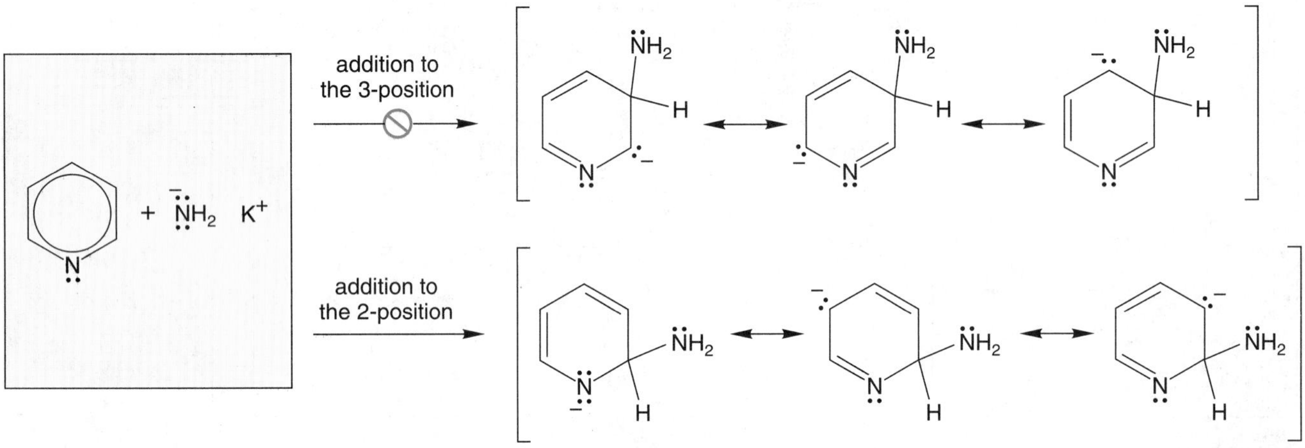

FIGURE 25.48 Addition to the 3-position does not allow the nitrogen to help stabilize the negative charge. Addition to the 2- or 4-position does.

This conventional beginning is followed by a rather unconventional conclusion, as aromaticity is regained through loss of hydride. Aminopyridine and potassium hydride are the initially formed products. These react rapidly and irreversibly to form hydrogen gas and a nicely resonance-stabilized amide ion. A final hydrolysis step regenerates aminopyridine. It is presumably the irreversible nature of hydrogen gas formation that helps drive this reaction to completion (Fig. 25.49).

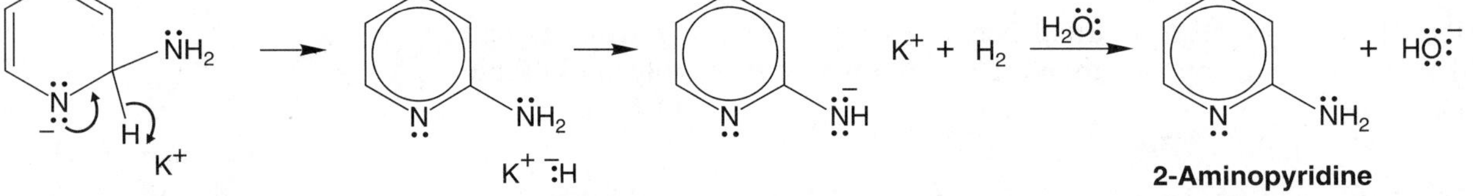

FIGURE 25.49 Loss of hydride is followed by formation of H_2. Hydrolysis regenerates 2-aminopyridine.

Other strong nucleophiles such as organolithium reagents also form substituted pyridines. In these reactions, hydrogen gas cannot be formed until the final hydrolysis step (Fig. 25.50).

Nucleophilic substitution provides a means of converting pyridines bearing a leaving group at the 2- or 4-position into differently substituted molecules. Addition to the position bearing the leaving group (called **ipso**

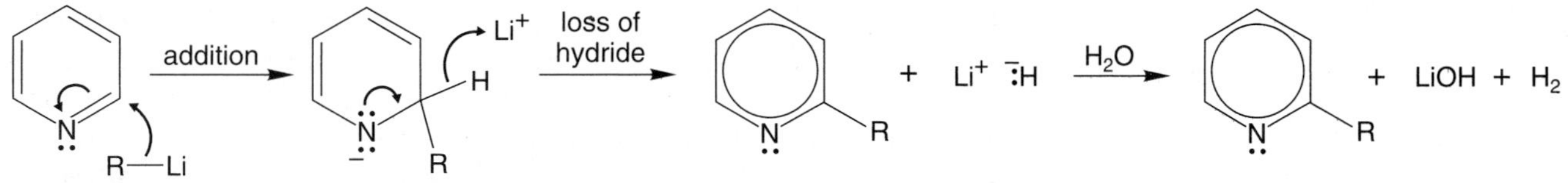

FIGURE 25.50 Addition of alkyllithium reagents also leads to substituted pyridines.

attack) leads to an intermediate that can either revert to starting material or eliminate the leaving group to give a new compound (Fig. 25.51).

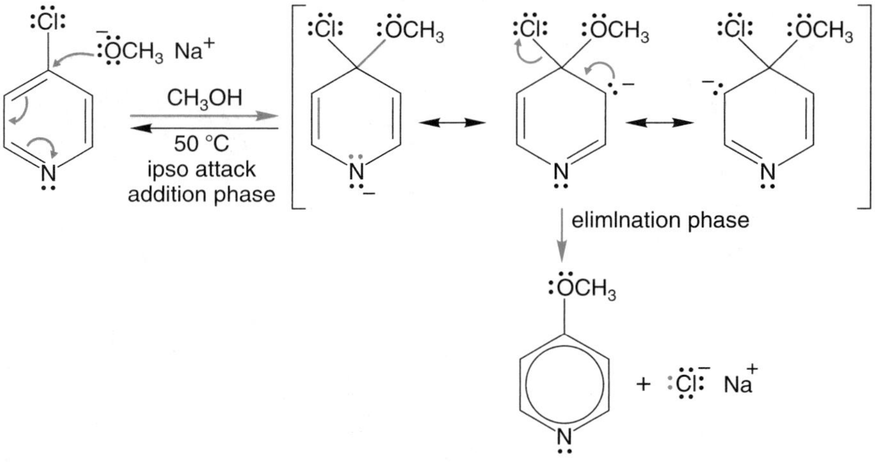

FIGURE 25.51 Substitution of a pyridine through ipso attack. A leaving group is displaced through an addition–elimination reaction.

*PROBLEM 25.17

Explain why this procedure works only for pyridines substituted at the 2- or 4-position and fails for 3-substituted pyridines.

ANSWER

The key to the relatively easy nucleophilic additions to pyridines is the placement of the negative charge on nitrogen. If the leaving group is attached to the 3-position, ipso addition of a nucleophile does not place the negative charge on the electronegative nitrogen atom. Only if the leaving group is at the 2- or 4-position does this happen.

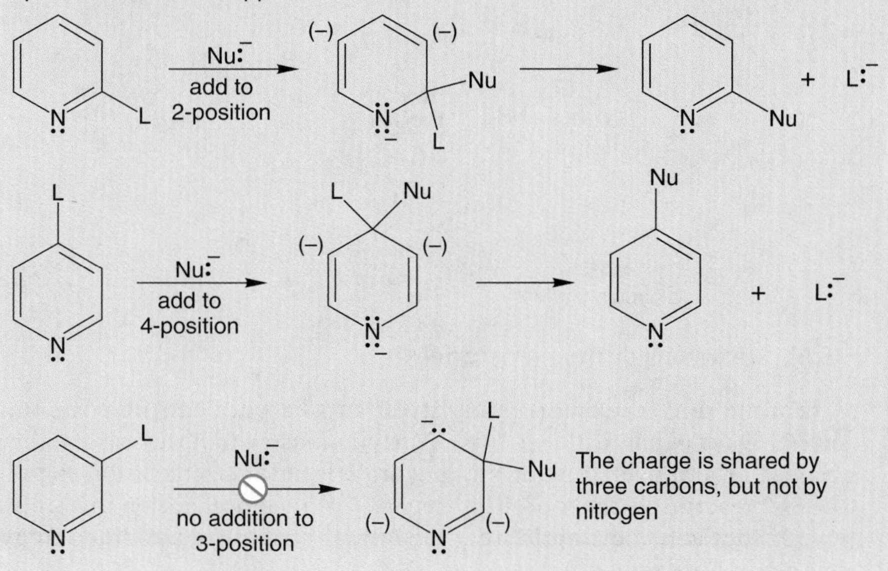

The charge is shared by three carbons, but not by nitrogen

The *N*-oxides mentioned in Section 25.5c are especially active in nucleophilic aromatic sustitution because the incoming nucleophile annihilates the positive charge on nitrogen as it adds. Addition destroys aromaticity, but at least for the *N*-oxides, loss of a leaving group regenerates the aromatic sextet. This reaction is generally carried out in the presence of an acid chloride (here, SO_2Cl_2), in order to produce the necessary good leaving group (Fig. 25.52).

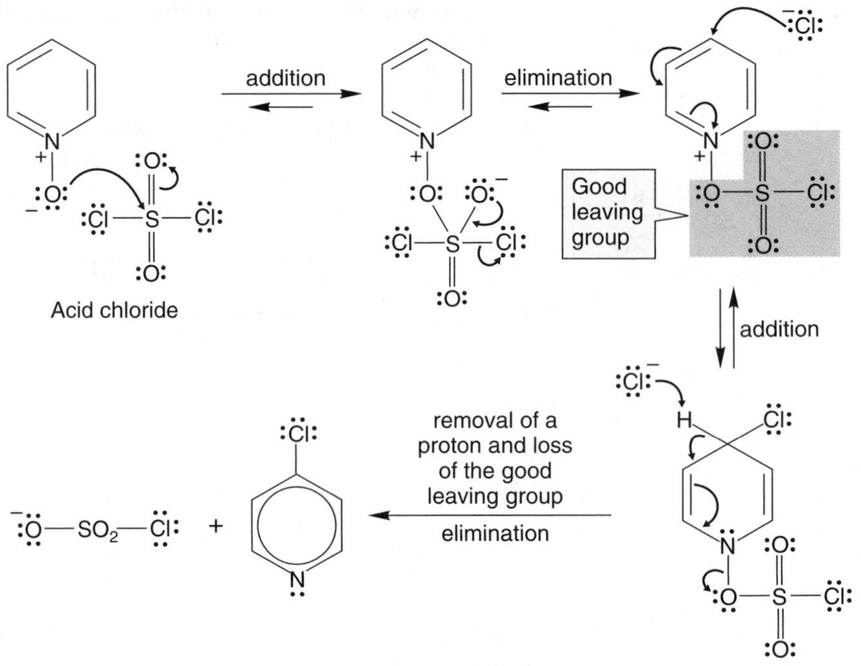

addition

elimination

Acid chloride

Good leaving group

addition

removal of a proton and loss of the good leaving group

elimination

FIGURE 25.52 Pyridine *N*-oxides are relatively easily attacked by nucleophiles. Loss of a leaving group leads to substituted pyridines.

In Chapter 16 (p. 802), we saw a similar reaction, the transfer of hydride to the "para" position of the pyridinium ion, NAD^+, the oxidized form of nicotinamide adenine dinucleotide, in biological oxidation (Fig. 25.53).

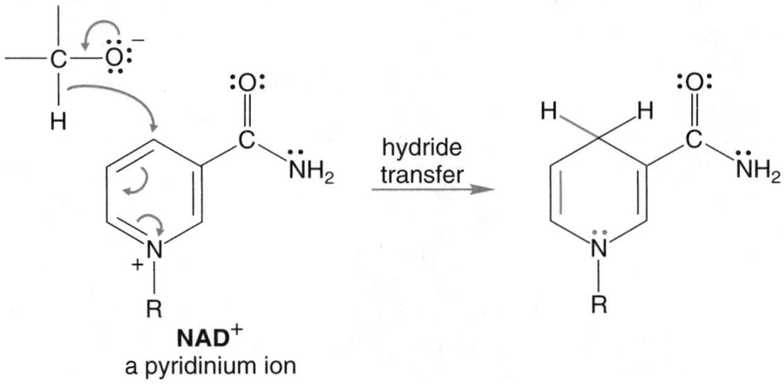

hydride transfer

NAD^+
a pyridinium ion

FIGURE 25.53 Reduction of NAD^+ through hydride transfer.

25.5f Reactions at the α-Position

A benzene ring accelerates reactivity at an adjacent benzylic position in several ways (Chapter 13, p. 606). Benzylic ions or radicals are stabilized through orbital overlap with the ring orbitals, and the transition state for the S_N2 reaction at the α-position benefits from delocalization in a similar way. It should be no surprise that the orbital systems of pyridine and pyrrole act in the same way.

Provide mechanisms for the reactions in Figure 25.54.

PROBLEM 25.18

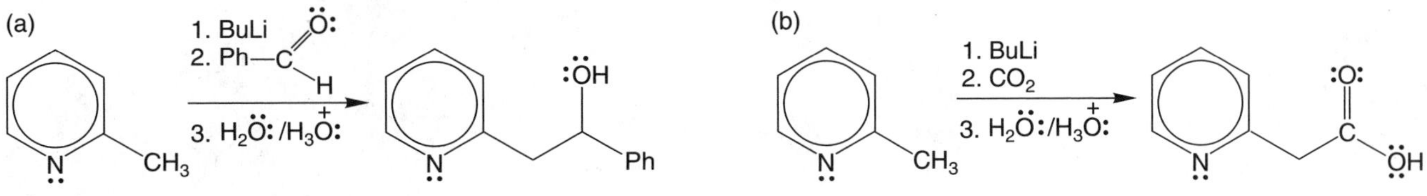
(a)
1. BuLi
2. Ph—C(=O:)H
3. H₂O:/H₃O:⁺
CH₃
:OH
Ph
(b)
1. BuLi
2. CO₂
3. H₂O:/H₃O:⁺
CH₃
:O:
OH

FIGURE 25.54

25.5g Syntheses of Pyridines and Pyrroles

Both pyridine and pyrrole, as well as their simple alkyl derivatives, can be isolated from natural sources, generally by distillation of coal tar. Pyrroles can be made through reaction of appropriate diones with a source of nitrogen in a process involving a series of addition and elimination reactions (Fig. 25.55).

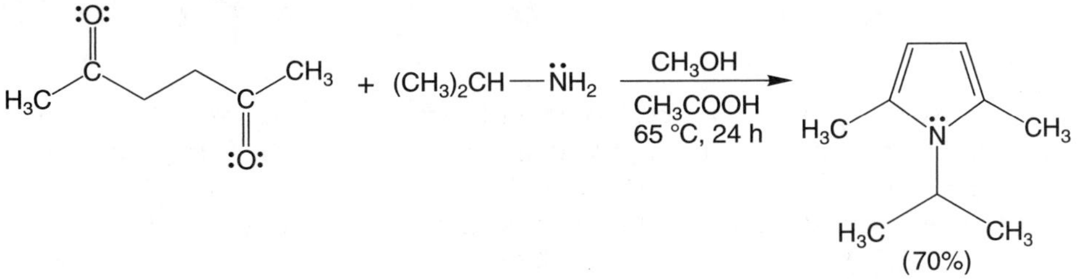
:O:
H₃C
CH₃
:O:
+ (CH₃)₂CH—NH₂
CH₃OH
CH₃COOH
65 °C, 24 h
H₃C
CH₃
H₃C
CH₃
(70%)

FIGURE 25.55 A typical synthesis of a pyrrole.

Write a mechanism for the reaction of Figure 25.55, called the Knorr–Paal synthesis after Ludwig Knorr (1859–1921) and Karl Paal (1860–1935).

*PROBLEM 25.19

ANSWER

Addition of the amine to the protonated carbonyl group gives a carbinolamine. Loss of water can give either an imine or, if the proton is lost from carbon as shown, an enamine (Chapter 16, p. 780).

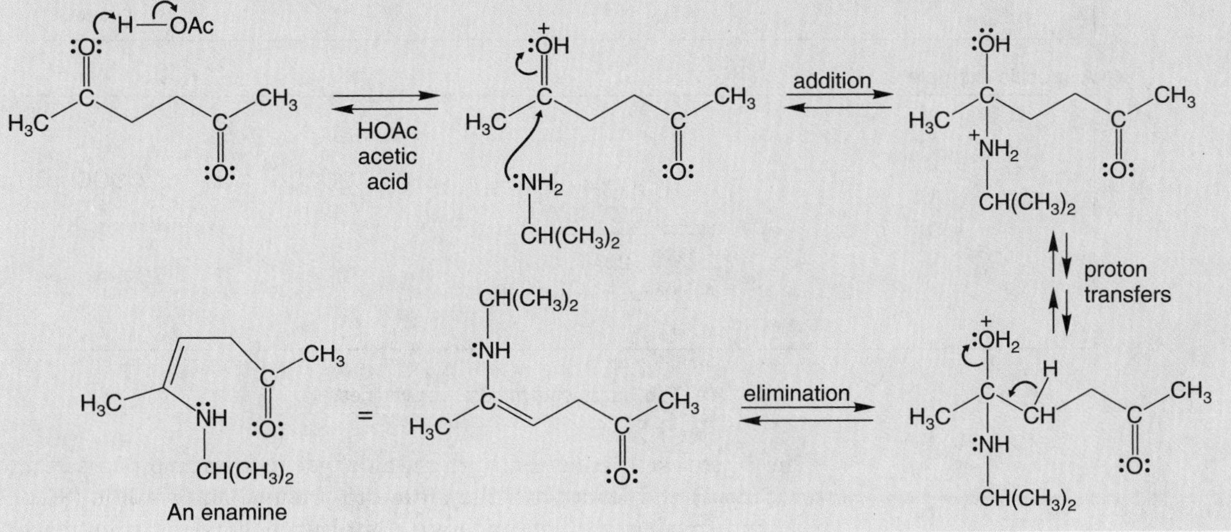
H—OAc
HOAc
acetic
acid
:NH₂
CH(CH₃)₂
addition
proton
transfers
elimination
:NH
CH(CH₃)₂
An enamine

ANSWER (CONTINUED) Now the process is repeated in an intramolecular sense with the nitrogen of the enamine adding to the protonated carbonyl in the same molecule. Another loss of water and the pyrrole is formed.

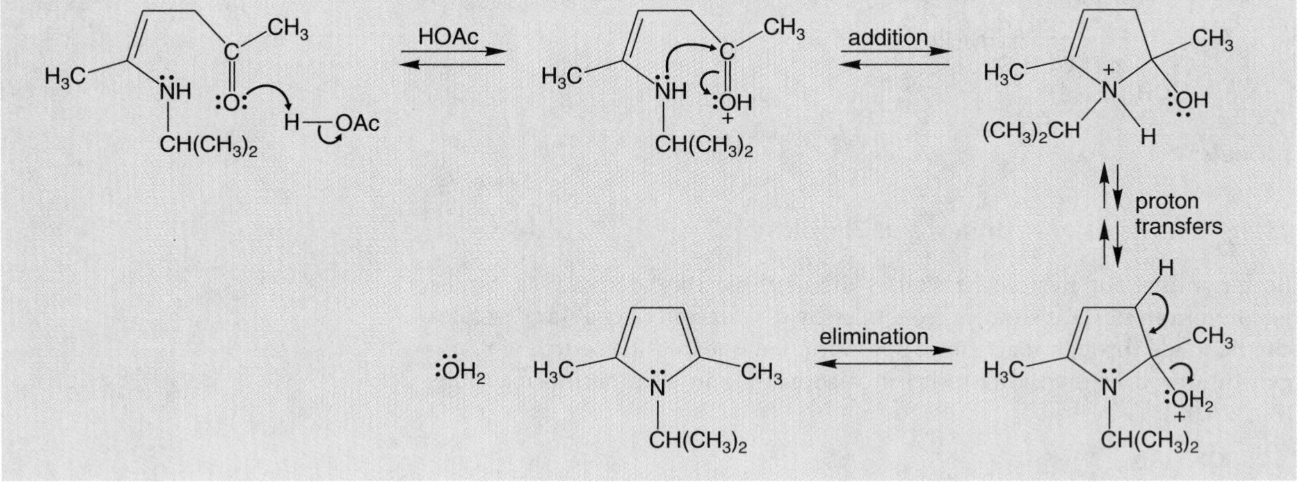
HOAc
addition
proton transfers
elimination
H—OAc
CH(CH3)2
(CH3)2CH

Most pyridine syntheses start with pyridine itself, but there is a general method for making substituted pyridines called the **Hantzsch synthesis** (Arthur Rudolph Hantzsch, 1857–1935). In this reaction, a β-keto ester and an aldehyde provide the carbons, and ammonia provides the nitrogen (Fig. 25.56).

The general case

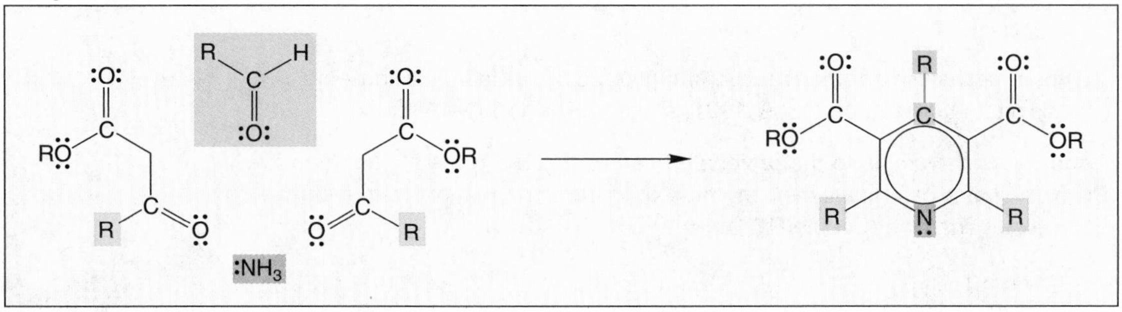
:NH3

A specific example

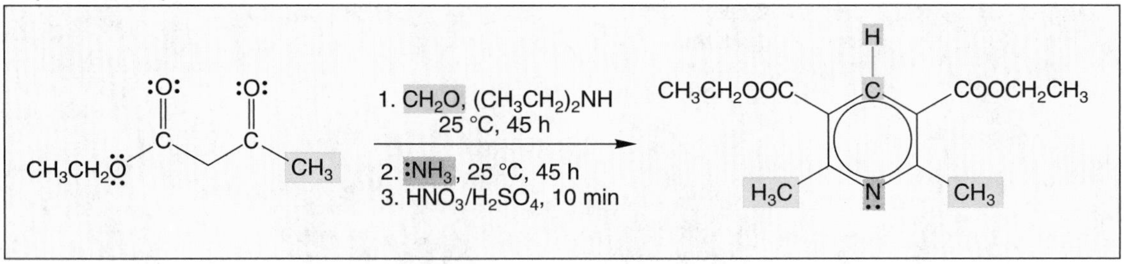
1. CH2O, (CH3CH2)2NH
25 °C, 45 h
2. :NH3, 25 °C, 45 h
3. HNO3/H2SO4, 10 min
CH3CH2OOC
COOCH2CH3

FIGURE **25.56** The Hantzsch synthesis of pyridines.

The β-keto esters condense with the aldehyde to form an α,β-unsaturated ester (Step 1) and react with the amine to produce an enamine (Step 2). These intermediates then undergo a multistep condensation reaction

(Step 3). The product is a dihydropyridine, an isolable intermediate (Fig. 25.57).

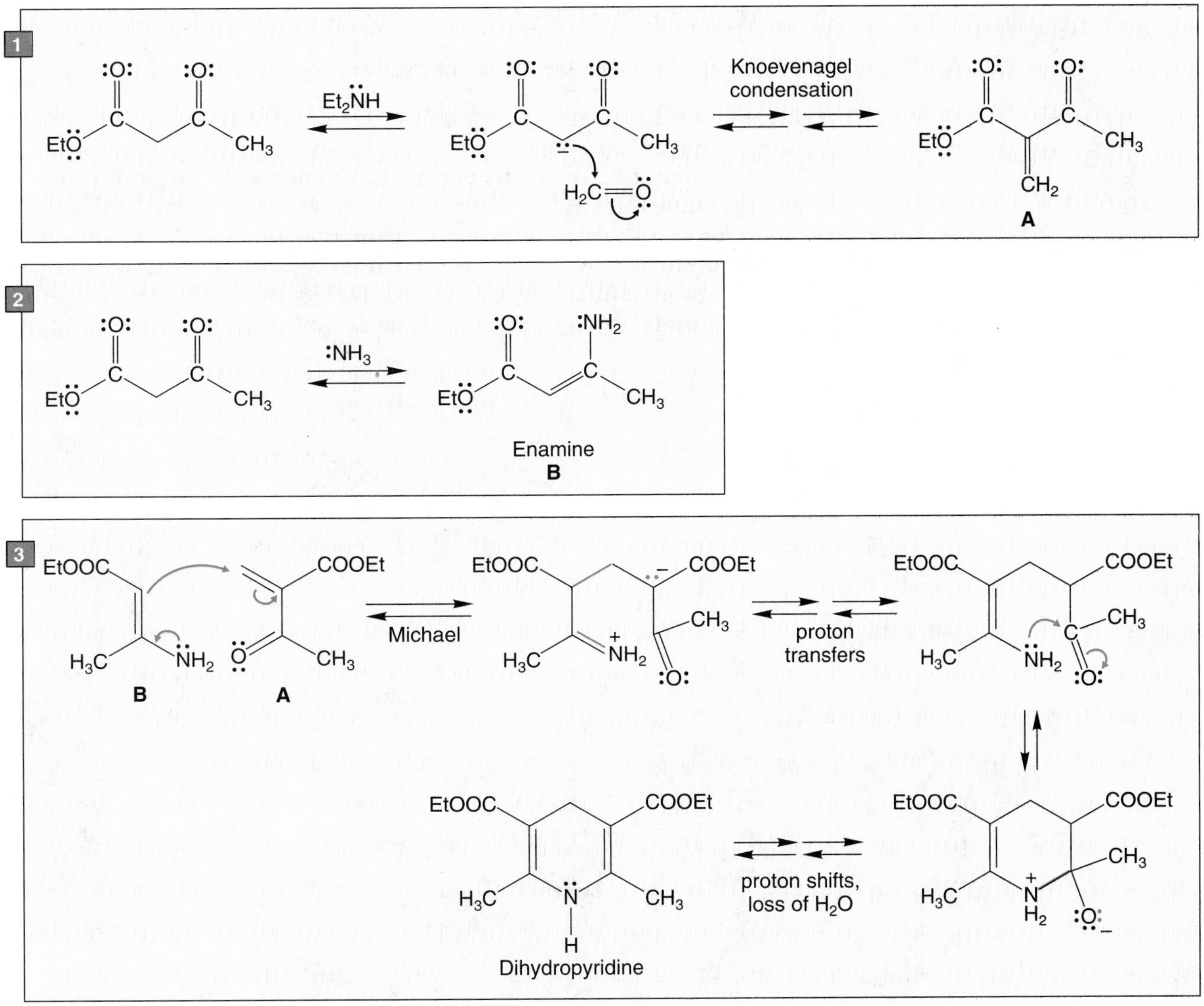

1 Et_2NH Knoevenagel condensation A

2 $:NH_3$ Enamine B

3 Michael B A proton transfers proton shifts, loss of H_2O Dihydropyridine

FIGURE 25.57 The mechanisms of the first steps in the Hantzsch synthesis.

Oxidation completes the reaction by aromatizing the system. The diesters formed in the classic Hantszch synthesis can be decarboxylated by basic hydrolysis, followed by heating, to give the more simply substituted pyridines (Fig. 25.58).

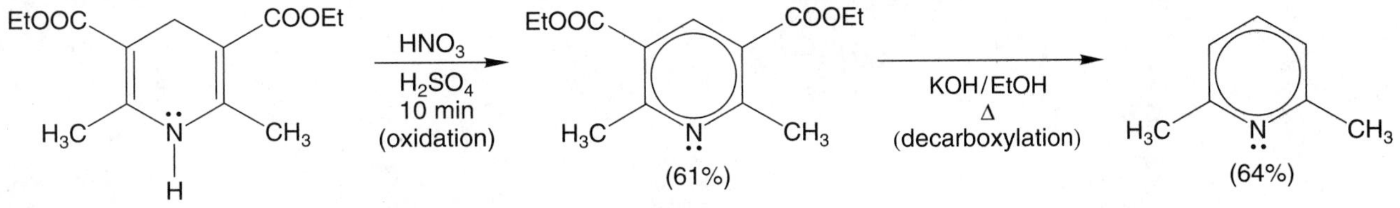

HNO_3, H_2SO_4, 10 min (oxidation) (61%) KOH/EtOH Δ (decarboxylation) (64%)

FIGURE 25.58 The final steps include oxidation and, sometimes, decarboxylation.

PROBLEM 25.20 Outline a Hantzsch synthesis for 2,4,6-trimethylpyridine. No mechanism is necessary and you may start from inorganic materials and any molecule containing no more than four carbons.

25.5h Diels–Alder Reactions of Pyrrole

Pyrroles are 1,3-dienes and one might suspect that the Diels–Alder reaction (Chapter 12, p. 536; Chapter 22, p. 1139) would occur. This suspicion is correct, but the aromaticity of pyrrole introduces complications. Generally speaking, there is always an energy cost to be paid when aromaticity is lost, as in a Diels–Alder reaction of an aromatic pyrrole to give a nonaromatic product. Accordingly, such reactions are often slow, and aromaticity is sometimes regained after loss through a retro Diels–Alder reaction. With those hints, you should be able to tackle Problem 25.21.

***PROBLEM 25.21** Write a mechanism for the following reaction (Fig. 25.59).

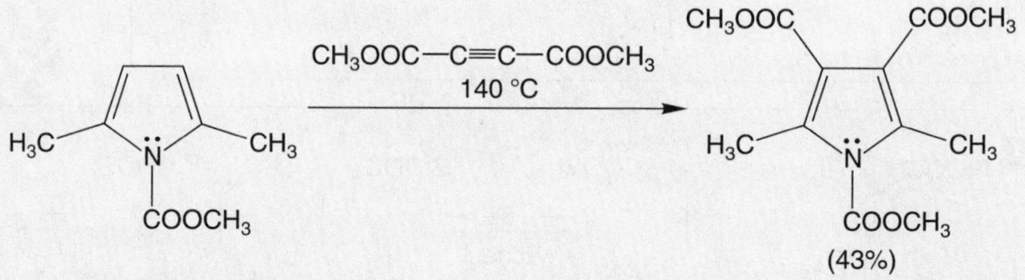

$CH_3OOC{-}C{\equiv}C{-}COOCH_3$, 140 °C

(43%)

FIGURE 25.59

ANSWER All those hints surely point directly at a Diels–Alder reaction. In this case, that leads to intermediate **A**.

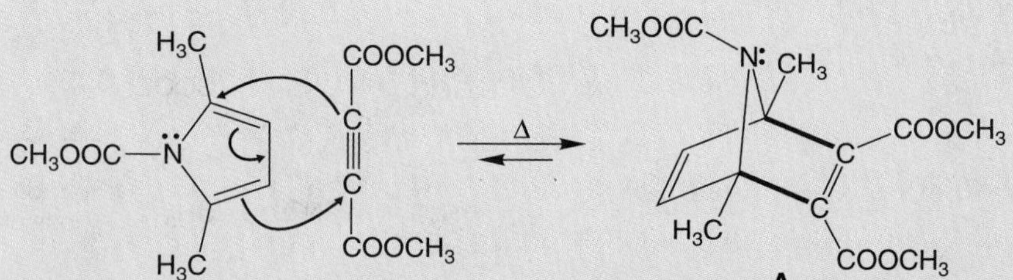

A
(new bonds shown in boldface)

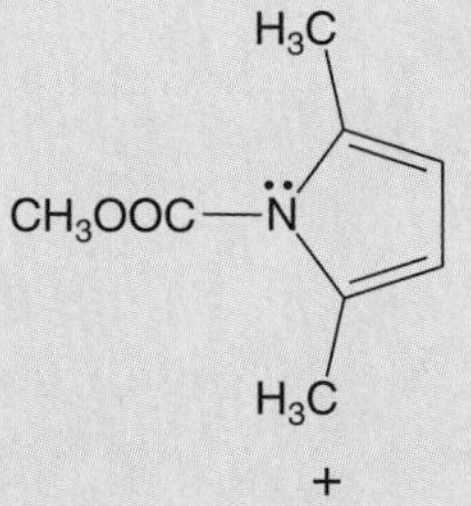

Now, with luck, you can see the symmetry of the azanorbornadiene, **A**. As Diels–Alder reactions are reversible, there are two possible routes for **A** to take. In one, the reaction simply reverses to the starting materials. In the other, a reverse Diels–Alder reaction proceeds in the other direction, and acetylene is lost to give the target molecule. One imagines that it is the irreversible loss of the gas acetylene that drives the reaction toward the product shown. The arrows show the product-forming pathway.

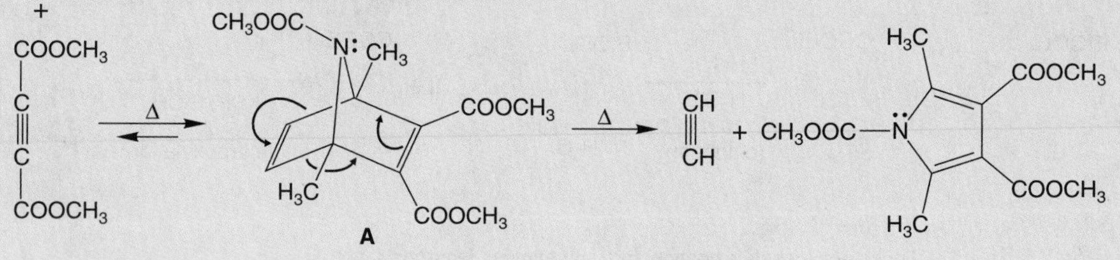

A

25.6 OTHER AROMATIC HETEROCYCLES: FURAN AND THIOPHENE

25.6a Structure

Both **furan** and **thiophene** have two pairs of electrons on the heteroatom and so combine the structural features of pyrrole and pyridine. One pair of electrons is in the six-electron π system and the other lies in the plane of the ring (Fig. 25.60).

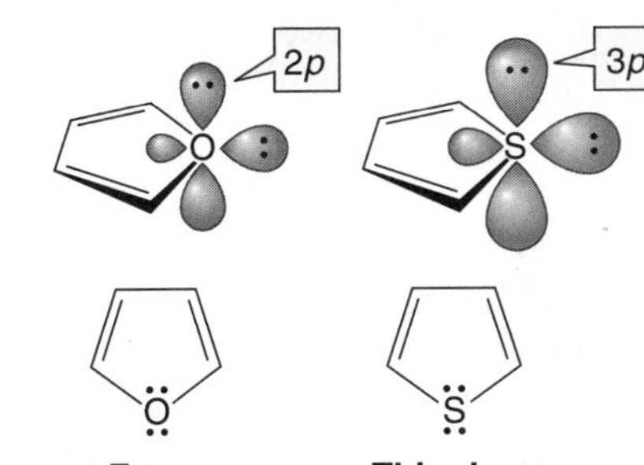

FIGURE 25.60 The structures of furan and thiophene.

25.6b Reactions of Furan and Thiophene: Electrophilic Aromatic Substitution

These compounds are very reactive toward electrophilic reagents, E^+. As in pyrrole, aromatic substitution is directed first to the 2-position, although mixtures are sometimes obtained in which some 3-substitution appears (Fig. 25.61).

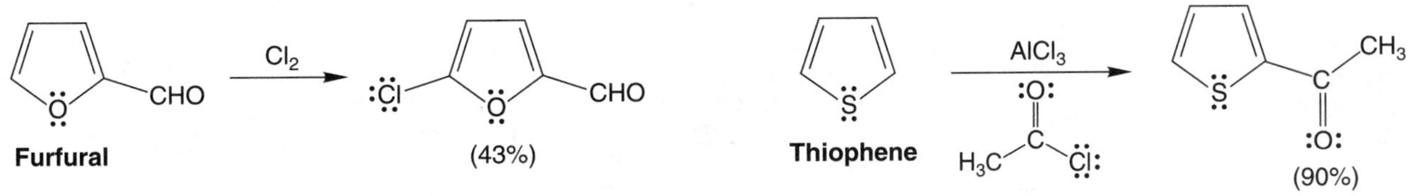

Furfural — Cl_2 → (43%)

Thiophene — $AlCl_3$, H_3C–C(=O)–Cl → (90%)

FIGURE 25.61 Electrophilic substitution reactions of furan and thiophene.

Further substitution depends on the directing effect of the first substituent.

PROBLEM 25.22 Explain why 2-substitution is preferred to 3-substitution in furan.

PROBLEM 25.23 There is another reasonable mechanism for aromatic substitution of furan. Suggest one. *Hint*: Think simple! What are the possible reactions between electrophiles and double bonds (Chapters 9 and 10)? Use the chlorination of furan as an example.

PROBLEM 25.24 There is a difficulty with the answer to Problem 25.23. Find what is wrong (or unlikely) and suggest an alternative mechanism. *Hint*: Recall what you know of the stereoelectronic requirements of the E2 reaction (Chapter 7, p. 276).

25.6c Reactions that Do Not Preserve Aromaticity: Diels–Alder and Other Additions

Furan will undergo addition reactions in which aromaticity is not regained more easily than benzene. There is less delocalization energy in furan than benzene (Table 25.1) and preservation of aromaticity is less important. For example, addition reactions are often accompanied by acid-catalyzed ring openings to 1,4-dicarbonyl compounds. The mechanism of this reaction involves protonation to give a resonance-stabilized cation, followed by addition of water. This reaction is nothing more than a hydration of an alkene, a reaction first encountered in Chapter 9 (p. 358). In this case hydration produces a hemiacetal, a molecule that is reactive in acid (Chapter 16, p. 774). A second protonation leads to ring opening and the formation of the dicarbonyl compound. This process is incorporated into a very nice synthesis of the plant constituent, *cis*-jasmone (Fig. 25.62).

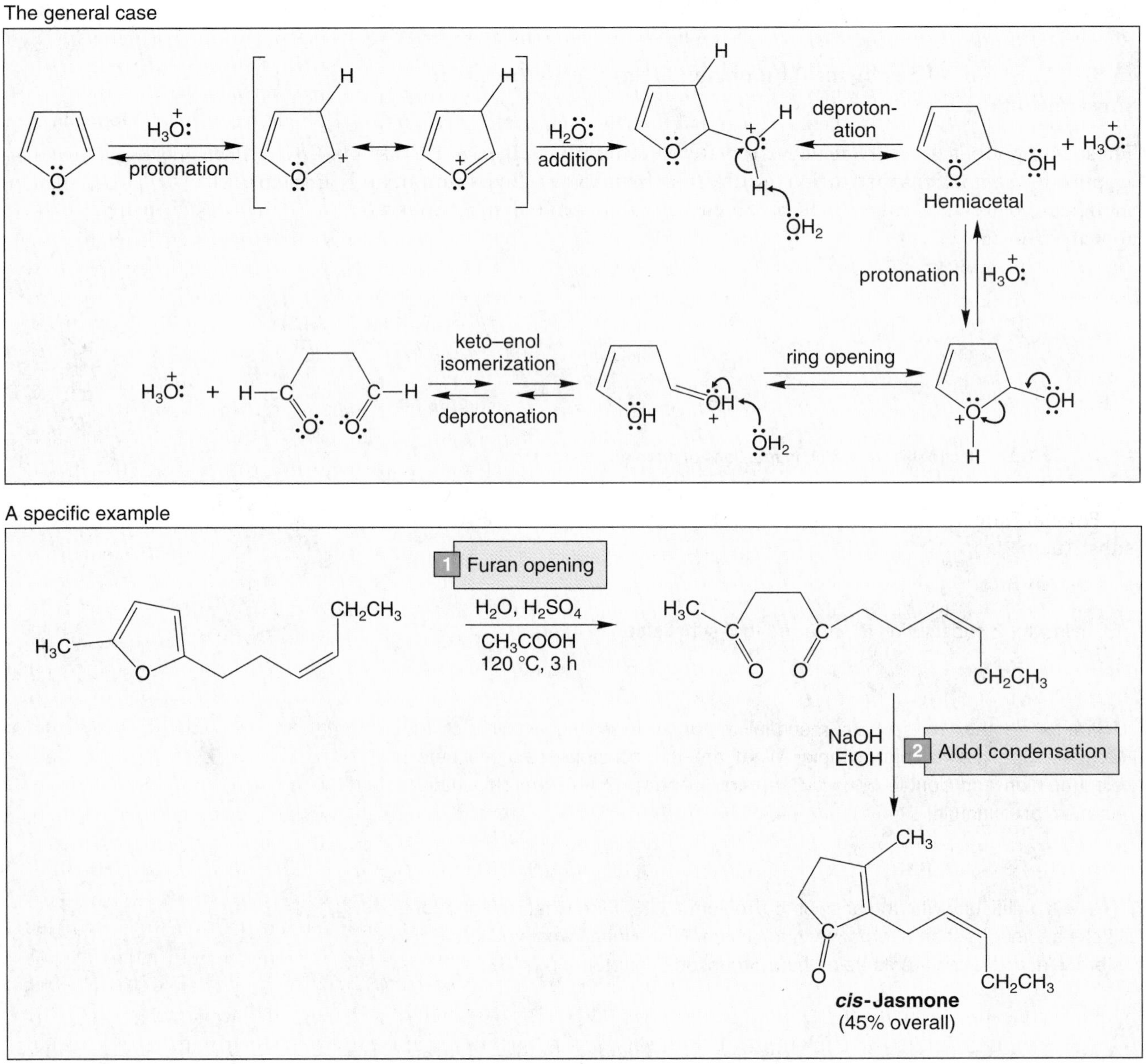

The general case

protonation; $H_3\overset{+}{O}$:; $H_2\ddot{O}$: addition; deprotonation; Hemiacetal; + $H_3\overset{+}{O}$:; protonation; ring opening; keto–enol isomerization; deprotonation; $H_3\overset{+}{O}$: + H—C ... C—H

A specific example

1 Furan opening: H_2O, H_2SO_4, CH_3COOH, 120 °C, 3 h

2 Aldol condensation: NaOH, EtOH

***cis*-Jasmone** (45% overall)

FIGURE 25.62 The mechanism of the acid-catalyzed ring opening of a furan.

Diels–Alder reactions are common for furans, which make excellent trapping agents for reactive π systems (Fig. 25.63).

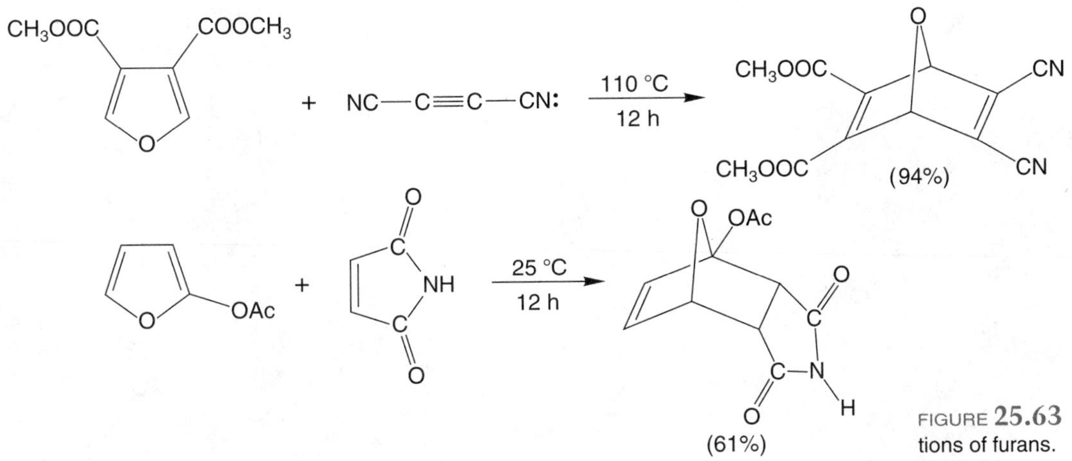

FIGURE 25.63 Two Diels–Alder reactions of furans.

Thiophenes can be desulfurized (Section 25.3, p. 1296) with Raney nickel, which also reduces any available carbon–carbon π bonds (Fig. 25.64).

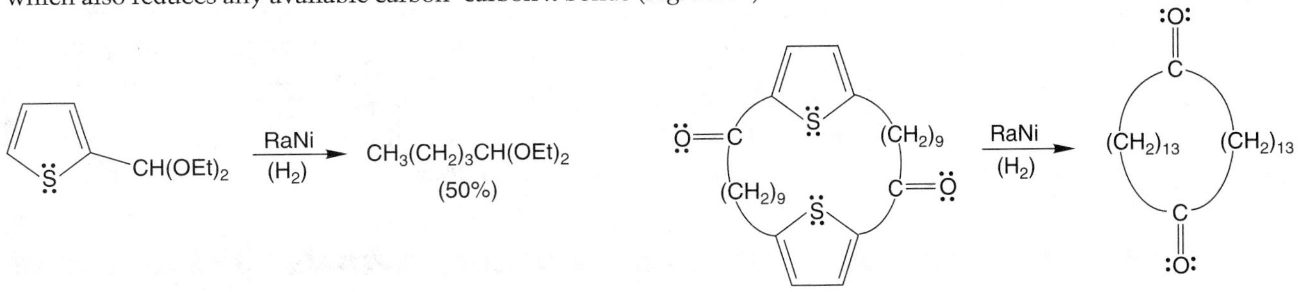

FIGURE 25.64 Thiophenes can be desulfurized with Raney nickel. Carbon–carbon double bonds are also reduced.

25.6d Syntheses of Furans and Thiophenes

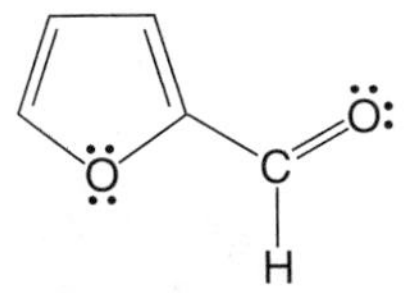

Like the nitrogen-containing five- and six-membered rings, simple furans are often isolated from natural products. Carbohydrates are the prime source, and acid treatment of oat husks is a substantial source of furfural, furan-2-carboxaldehyde (Fig. 25.65).

Another excellent source of these molecules is the formal reversal of the reaction of Figure 25.62. When a 1,4-dicarbonyl compound is treated with dehydrating agents such as polyphosphoric acid, P_2O_5, P_4S_7, or P_2S_5, ring closure and aromatization occur (Fig. 25.66).

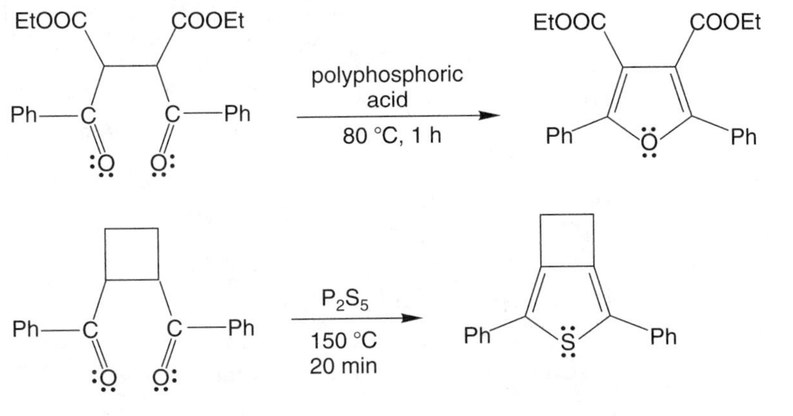

FIGURE 25.65 Furfural.

FIGURE 25.66 Heterocycle synthesis from 1,4-diketones.

2-*sec*-Butyldihydrothiazole and Dehydro-*exo*-brevicomin

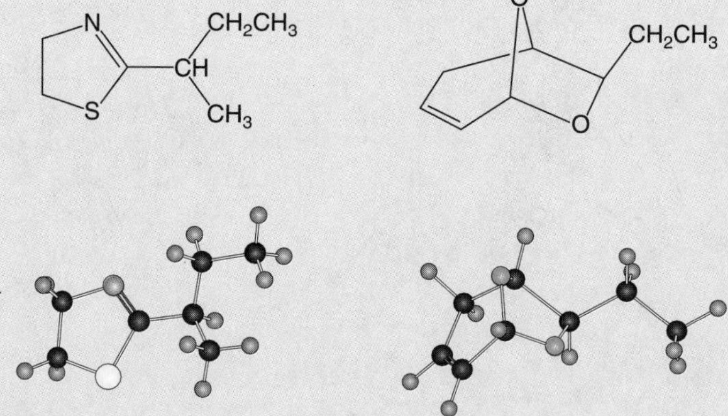

Were you to encounter these seemingly unexciting heterocyclic compounds, there would be no great effect on your personality (probably). But if you were a male mouse, the situation would change drastically. You would be turned into a snarling little gray ball of aggression, ready and willing to bite any other male mouse in sight. For these molecules are chemical signals that provoke aggression in mice. Interestingly, both compounds must be present for fighting behavior to appear. Neither molecule by itself is very active. It is not clear why this is true, but it is a commonly encountered phenomenon in pheromone chemistry.

25.7 AROMATIC HETEROCYCLES CONTAINING MORE THAN ONE HETEROATOM

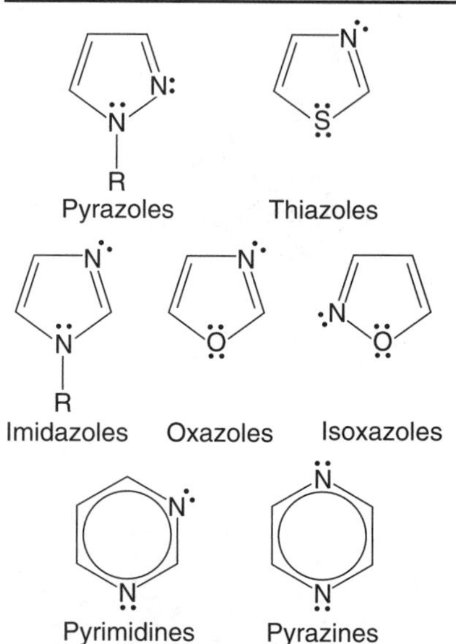

FIGURE 25.67 Some aromatic heterocycles containing two heteroatoms.

There are many such molecules, and Figure 25.67 shows only a few simple examples.

We will see five important heterocyclic compounds in Chapter 26 when we study nucleotides, combinations of sugars and nitrogen heterocycles linked by phosphates to form the polymers called nucleic acids (Fig. 25.68).

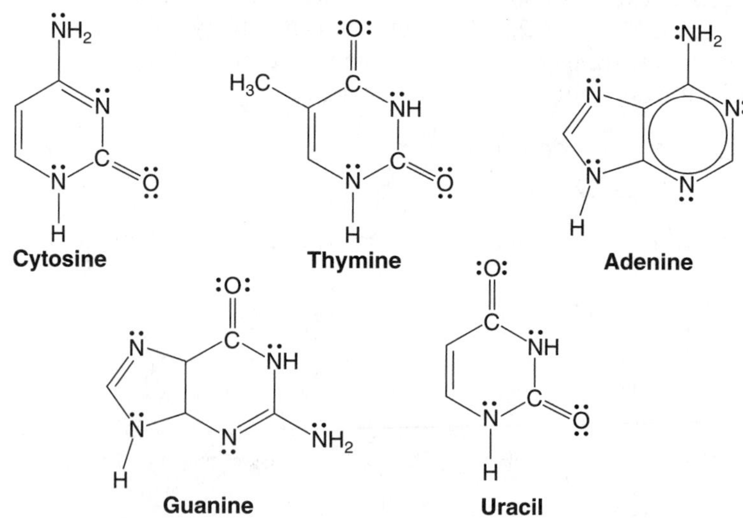

FIGURE 25.68 Five biologically important heterocyclic compounds containing two nitrogen atoms in one ring.

25.8 TWO-RING HETEROCYCLES

25.8a Indole, Benzofuran, and Benzothiophene

Indole, **benzofuran**, and **benzothiophene** are the benzo-fused counterparts of the simple heterocycles pyrrole, furan, and thiophene, and are related to the two-ring, all-carbon aromatic compound naphthalene (Fig. 25.69).

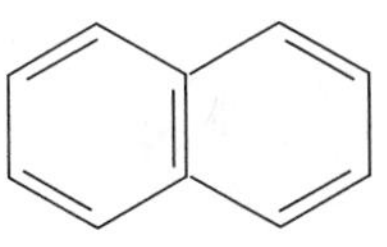

Naphthalene

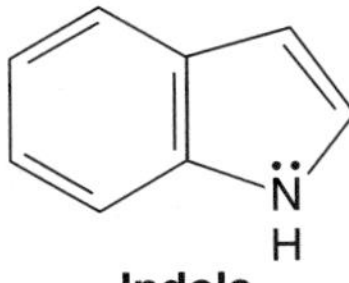

Indole

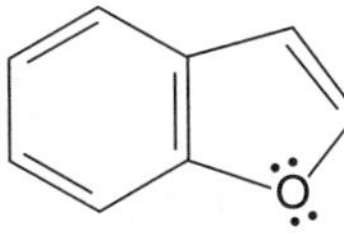

Benzofuran

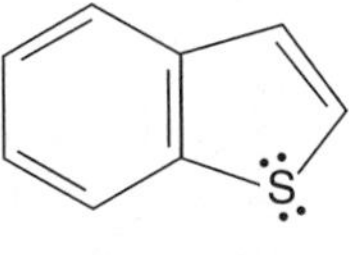

Benzothiophene

FIGURE 25.69 Naphthalene and three related bicyclic heteroaromatic compounds.

They are important chiefly because of the large variety of natural products containing these ring systems (see the indole alkaloids, Chapter 21, p. 1112). Some other examples of natural products containing simple and complex five-membered heterocyclic rings are given in Figure 25.70.

Griseofulvin (an antibiotic)

Khellin

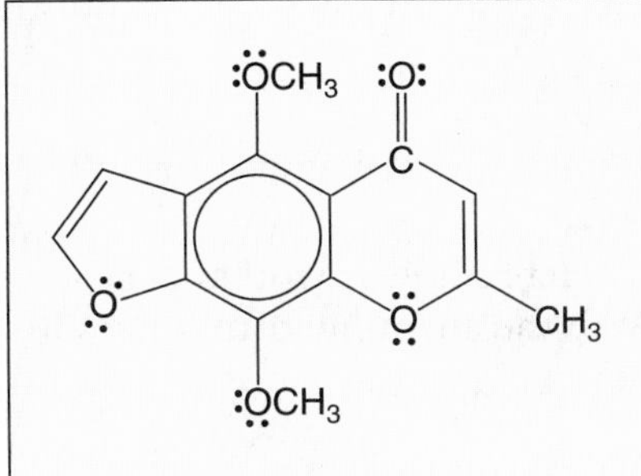

Biotin
(a growth factor present in every cell)

Reserpine (antihypertensive)

FIGURE 25.70 Some natural products containing five-membered heterocyclic rings.

25.8b Indole Syntheses

Two general and important syntheses of indoles are especially worth mentioning. One, the **Fischer indole synthesis** (yes, it's the same Emil Fischer—Chapter 24, p. 1235), is acid catalyzed and the other, the **Reissert synthesis** takes place in strong base. The Fischer synthesis incorporates much fascinating chemistry from many of the areas we have studied, including additions and eliminations (Chapters 7, 9, and 10), carbonyl chemistry (Chapter 16), and even pericyclic reactions (Chapter 22). In the Fischer indole synthesis, an aromatic phenylhydrazone is heated in acid (Fig. 25.71).

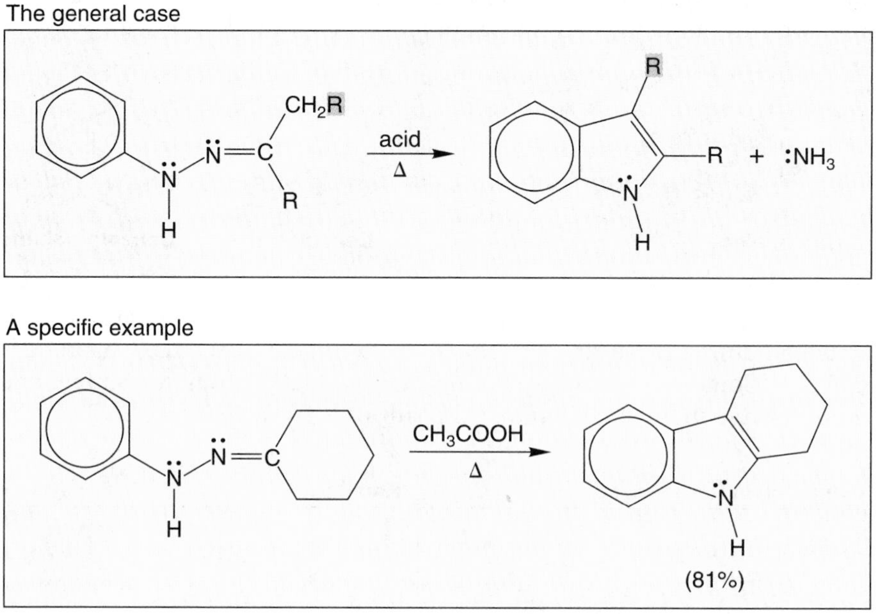

The general case

A specific example

FIGURE 25.71 The Fischer indole synthesis.

Here's how it works. In acid, equilibration of the hydrazone (an imine) with the enamine form occurs first (Chapter 16, p. 780; Fig. 25.72).

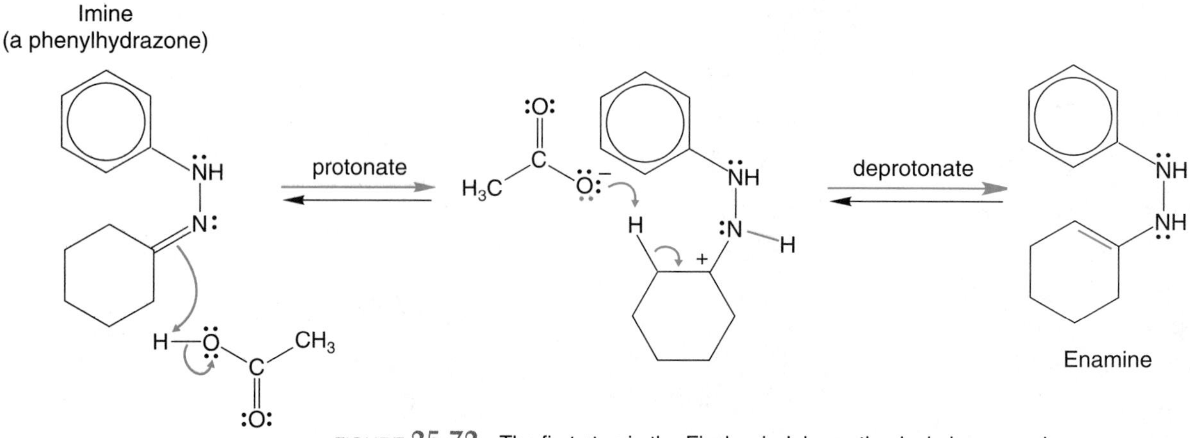

FIGURE 25.72 The first step in the Fischer indole synthesis, imine–enamine interconversion.

The molecule is now set up for a Cope rearrangement (Chapter 22, p. 1158), even though this destroys aromaticity in a benzene ring. Protonation and deprotonation reactions regenerate the benzene ring, and produce an immonium ion that can be attacked by the nearby amine to create the indole skeleton. A final sequence of "little steps"—deprotonation of one nitrogen, protonation of the other, and elimination of ammonia—leads to the indole itself (Fig. 25.73).

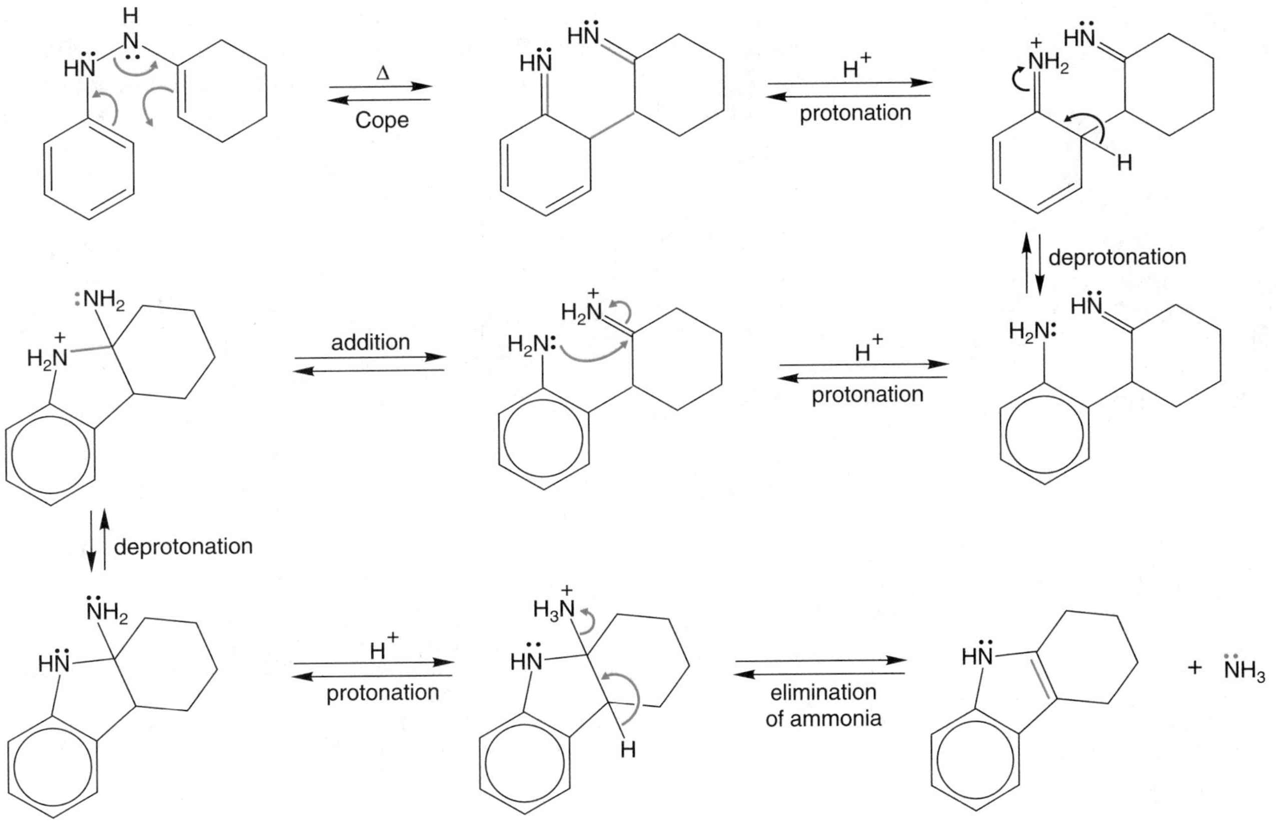

FIGURE 25.73 The remaining steps in the Fischer indole synthesis.

In the Reissert synthesis, an alkyl nitrobenzene reacts with an ester of oxalic acid to give a condensed product in which the alkyl group has been incorporated into the five-membered ring of the indole (Fig. 25.74).

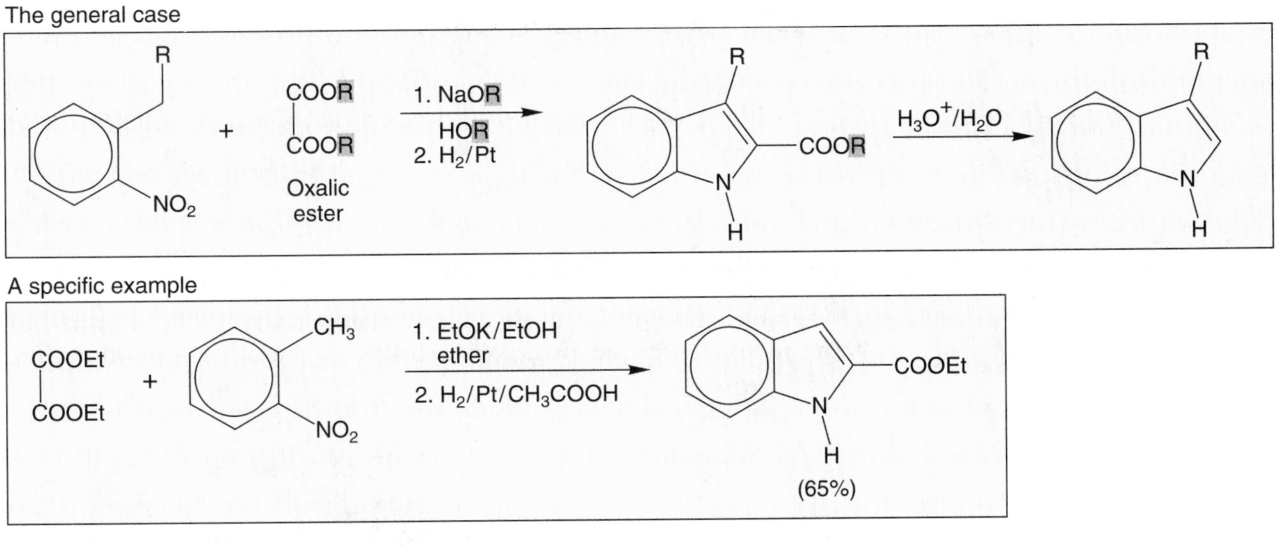

FIGURE 25.74 The Reissert indole synthesis.

The first step of the mechanism is a simple reaction of a carbanion with an ester—an addition–elimination process (Chapter 20, p. 1022). Reduction then converts the nitro group into an amine, which undergoes another addition–elimination reaction, and an indole is the result (Fig. 25.75).

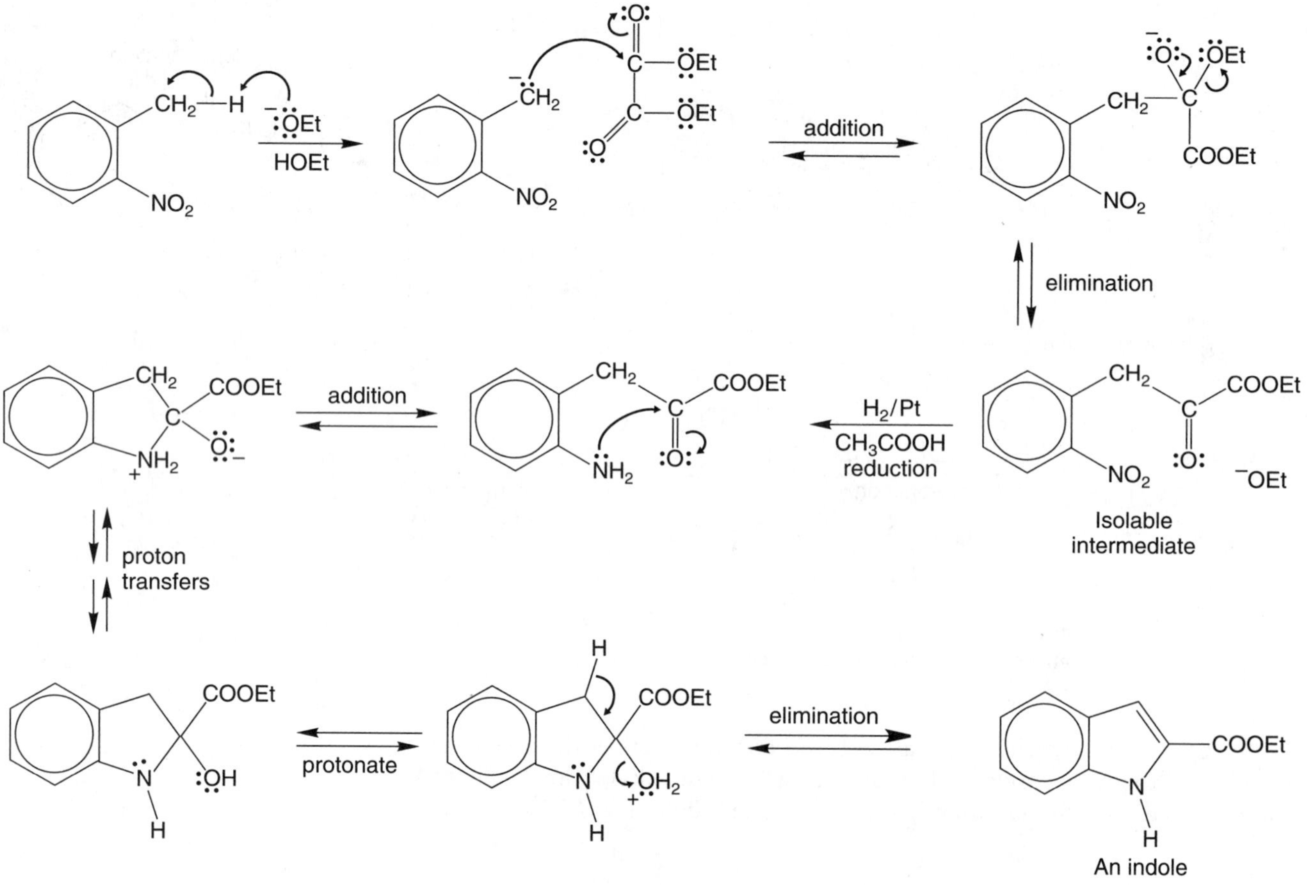

FIGURE 25.75 The mechanism of the Reissert indole synthesis.

PROBLEM 25.25 Explain carefully why the nitro group is essential to the success of the first step of the Reissert synthesis.

25.8c Quinoline and Isoquinoline

These molecules are also made up of two aromatic rings, but in this case both are six-membered rings: one a benzene, the other a pyridine (Fig. 25.76).

FIGURE 25.76 Quinoline and isoquinoline.

Like naphthalene (Chapter 13, p. 599) both **quinoline** and **isoquinoline** undergo aromatic substitution at an α-position in the all-carbon ring. As the pyridine ring is strongly deactivated, substitution in the benzene half of the molecule is preferred (Fig. 25.77).

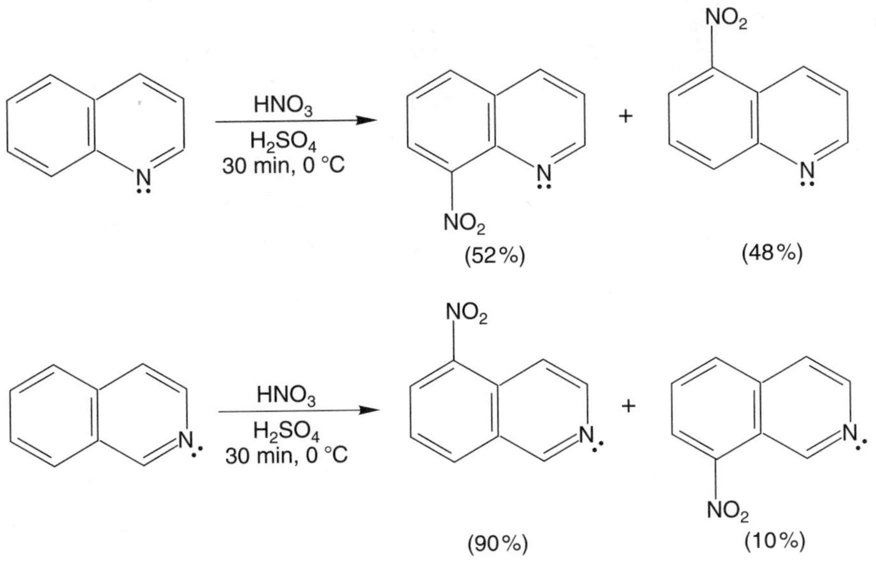

FIGURE 25.77 Electrophilic substitution of quinoline and isoquinoline. The α-position of the benzenoid ring is attacked.

By contrast, nucleophilic substitution (Section 25.5e, p. 1307) is favored in the pyridine ring (Fig. 25.78).

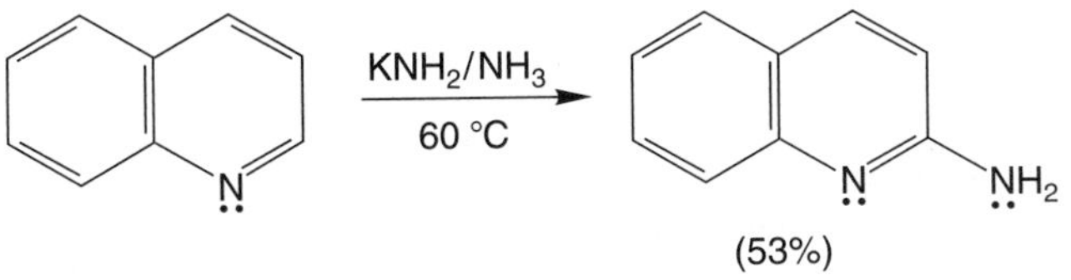

FIGURE 25.78 Nucleophilic substitution of quinoline occurs in the pyridine ring.

There are several syntheses of quinolines and isoquinolines, and many of them are composed of series of familiar reaction mechanisms. For example, the **Skraup synthesis** (Zdenko Hans Skraup, 1850–1910) of quinoline involves the treatment of aniline with glycerol (1,2,3-trihydroxypropane) in acid containing an oxidizing agent (Fig. 25.79).

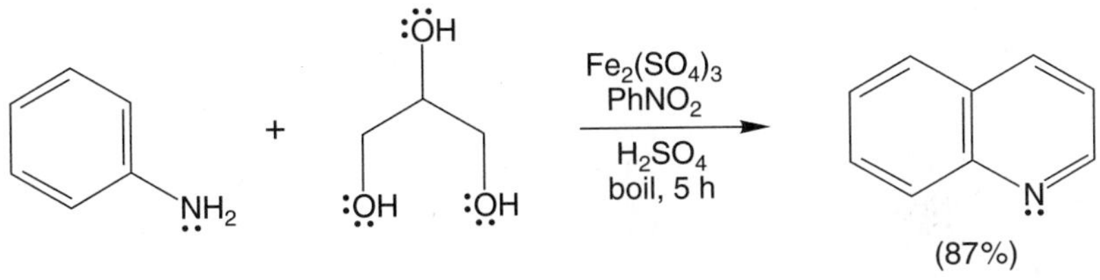

FIGURE 25.79 The Skraup synthesis of quinolines.

The first step is the acid-catalyzed conversion of glycerol into acrolein. Indeed, variations of the Skraup synthesis use preformed acroleins instead of glycerols. The reaction sequence then proceeds though a Michael addition (Chapter 18, p. 918) and an intramolecular aromatic substitution (Chapter 14), followed by a final oxidation (Fig. 25.80).

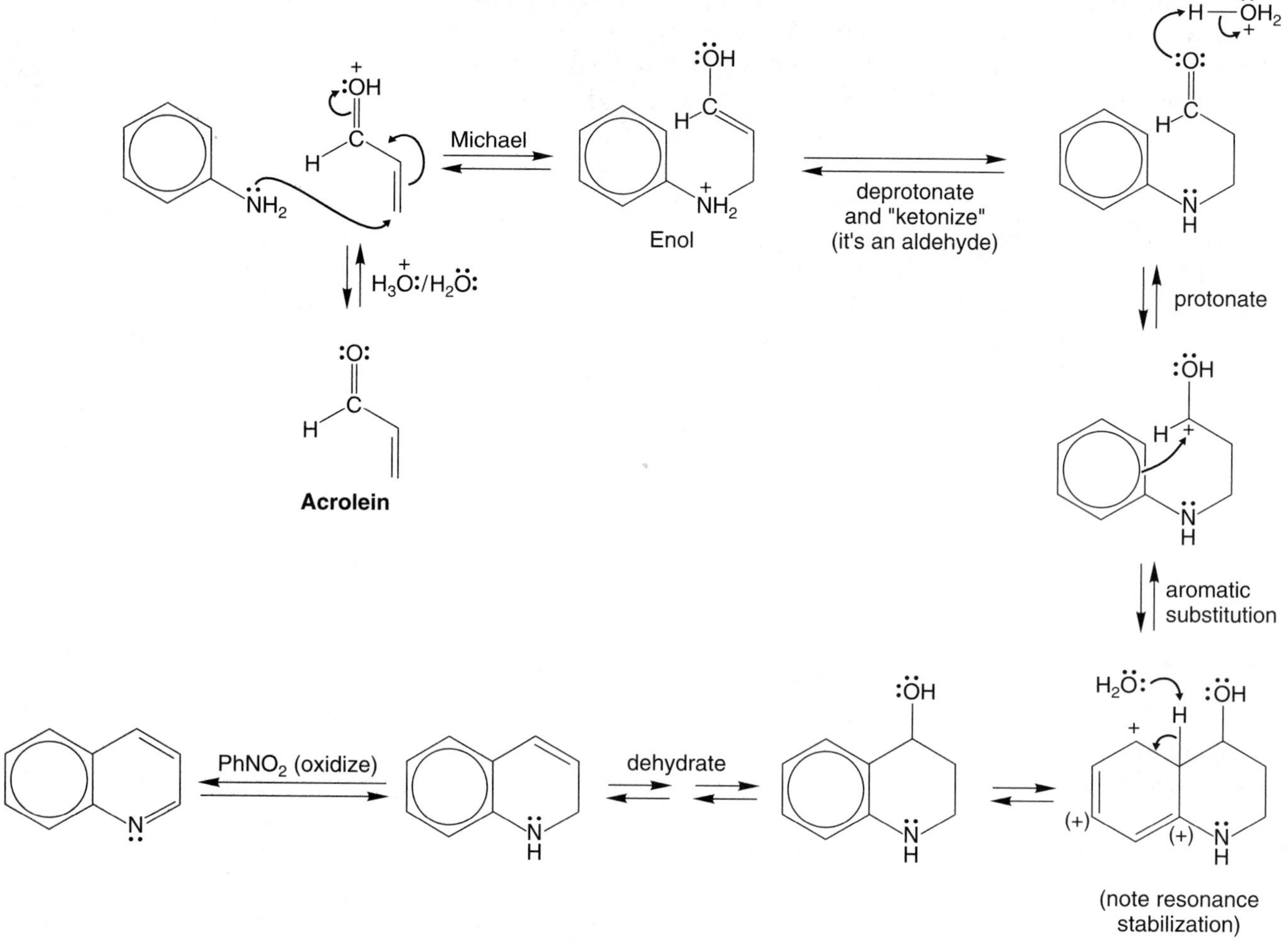

Michael

Enol

deprotonate and "ketonize" (it's an aldehyde)

$H_3\overset{+}{O}$:/H_2O:

Acrolein

protonate

aromatic substitution

PhNO$_2$ (oxidize)

dehydrate

(note resonance stabilization)

FIGURE 25.80 The mechanism of the Skraup synthesis.

PROBLEM 25.26 Write a mechanism for the acid-catalyzed transformation of glycerol into acrolein.

PROBLEM 25.27 Provide mechanisms for the following reactions leading to quinolines (Fig. 25.81).

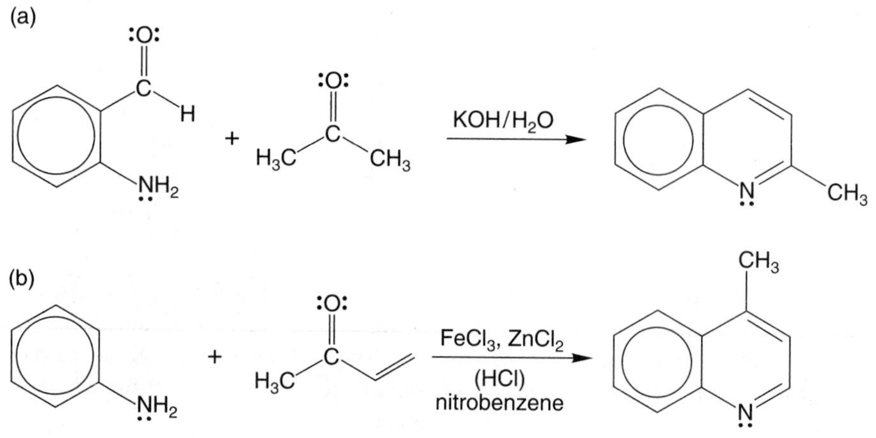

(a)

KOH/H_2O

(b)

$FeCl_3$, $ZnCl_2$ (HCl) nitrobenzene

FIGURE 25.81

25.9 SOMETHING MORE: PENICILLIN AND RELATED ANTIBIOTICS

Cyclic esters are lactones and cyclic amides are lactams. The size of the ring is indicated by a Greek letter as shown in Figure 25.82.

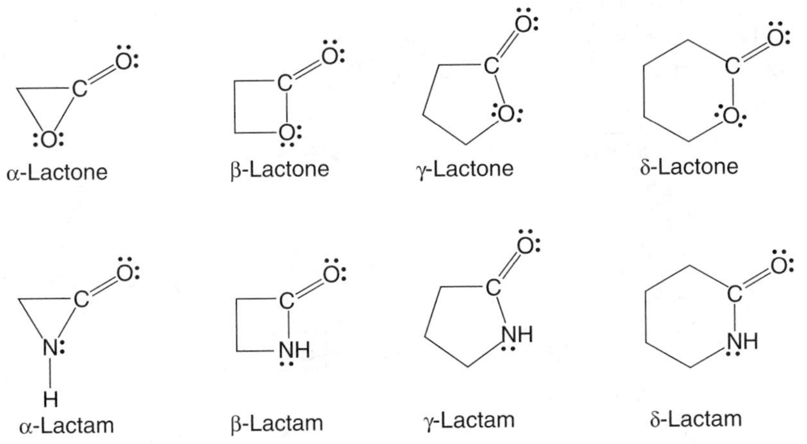

FIGURE 25.82 Small ring lactones and lactams.

Attachment of a carbonyl group to a small cyclic ether or amine to produce the lactone or lactam introduces further angle strain as the sp^2 carbon of the carbonyl group "prefers" 120° bond angles, and the small ring cannot approximate this value (Fig. 25.83).

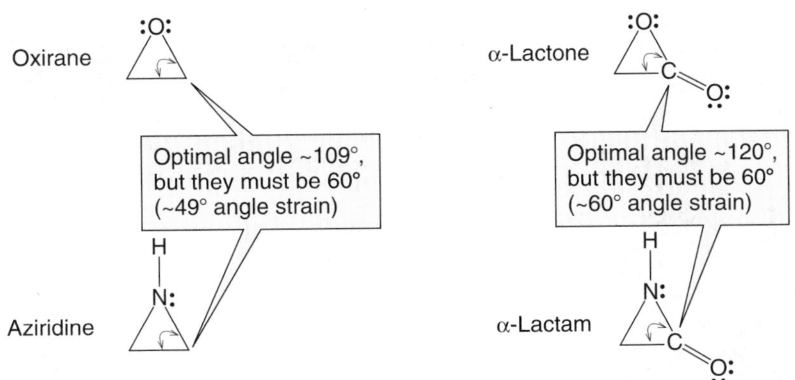
Oxirane
α-Lactone
Optimal angle ~109°, but they must be 60° (~49° angle strain)
Optimal angle ~120°, but they must be 60° (~60° angle strain)
Aziridine
α-Lactam

FIGURE 25.83 Angle strain is even greater in α-lactones and α-lactams than it is in oxiranes and aziridines.

Accordingly, small ring lactones and lactams are highly reactive. We have encountered a three-membered lactone (an α-lactone) before in our section on neighboring group effects (Chapter 23, p. 1181; Fig. 25.84).

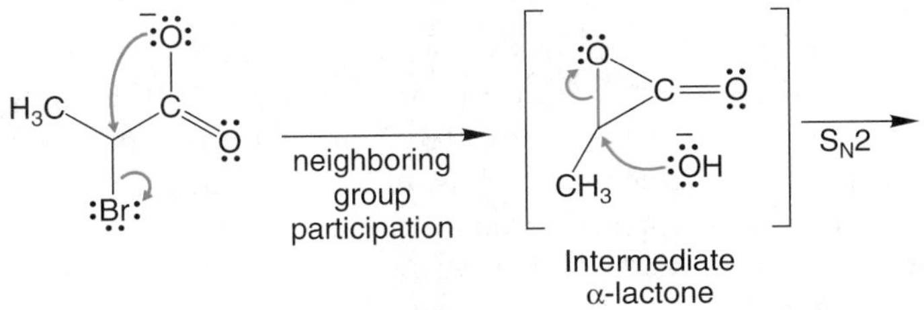

FIGURE 25.84 Neighboring group participation to give an intermediate α-lactone.

The α-lactone cannot be isolated because of the strain, but is captured through hydrolysis as it is formed in the reaction.

Four-membered rings are more stable, and can be isolated, although they, too, are quite reactive. Both the four-membered cyclic esters (β-lactones) and cyclic amides (β-lactams) can be isolated, although these species are still quite susceptible to base-induced ring-opening reactions. Not only is strain relieved in the *overall* reaction, but the first step, attack on the carbonyl group to produce a tetrahedral intermediate, relieves strain as well, as the sp^2 carbon of the carbonyl group is converted into the approximately sp^3 carbon of the tetrahedral intermediate (Fig. 25.85).

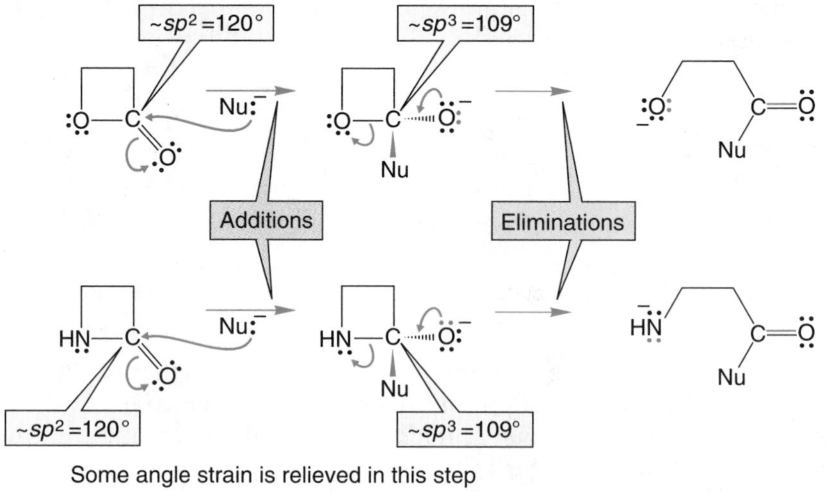

FIGURE 25.85 Nucleophilic ring-opening reactions of four-membered (β) lactones and lactams: addition–elimination reactions.

Two closely related and extraordinarily important classes of antibiotic, the **penicillins** and the **cephalosporins** (Fig. 25.86), contain the β-lactam structural unit.

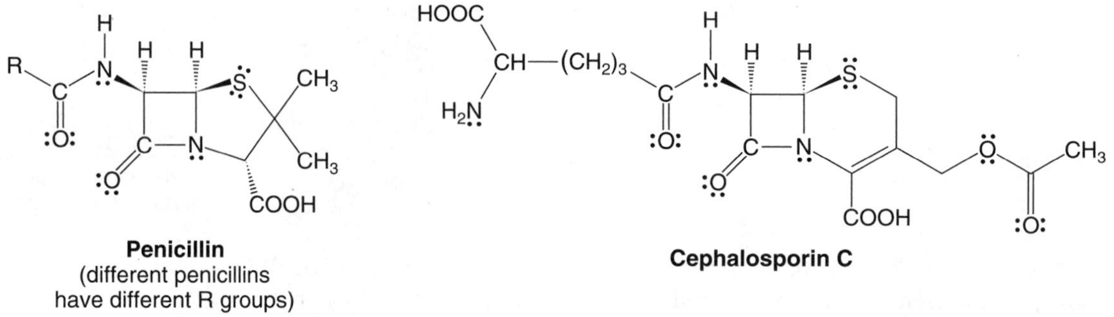

FIGURE 25.86 A penicillin and a cephalosporin.

Biological activity relates directly to the ease of ring opening of their β-lactam rings. Bacterial cells differ from mammalian cells in that they have a structurally solid cell wall constructed mainly from a polymer called peptidoglycan. Peptidoglycan consists of chains built from disaccharides constructed from *N*-acetyl glucosamine (an amino sugar) and the closely related muramic acid. The muramic acid is attached to a polypeptide (an oligomer of α-amino acids; see Chapter 26 for more details) containing 7–10 amino acid units. Chains are cross-linked through other amino acids (Fig. 25.87).

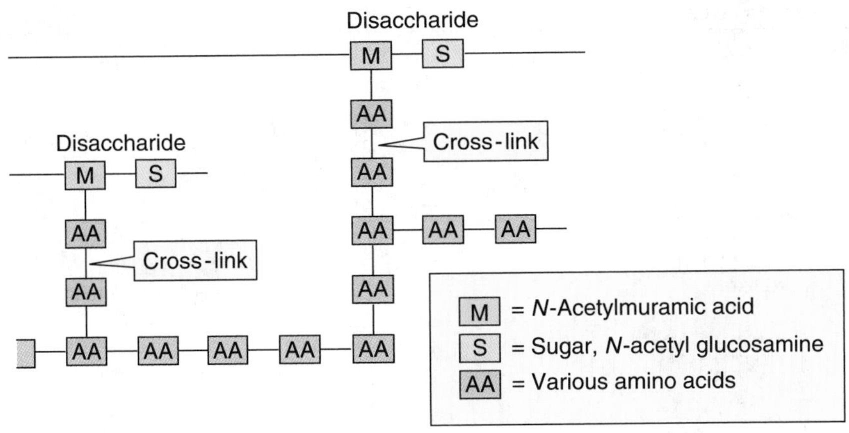

FIGURE 25.87 A schematic picture of the bacterial cell wall made from the polymer, peptidoglycan.

These cross-linking polypeptide chains are constructed by a series of amide bond linkages (Fig. 25.88).

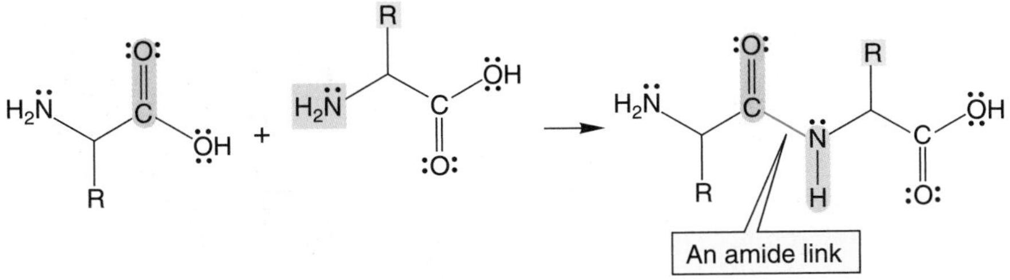

FIGURE 25.88 An amide bond linking two amino acids.

This cross-linking creates a flexible, porous, yet very strong material from which much of the bacterial cell wall is constructed. Penicillin interferes with the cross-linking process and the bacterium is unable to grow a normal, strong cell wall. Cell deformation or lysis (destruction) occurs, the offending organism dies, and you get well.

It is of course vital that an antibiotic agent be more toxic to its target than it is to its host. It should kill the "bug" but not you. Mammalian cells have flexible membranes rather than relatively rigid cell walls. Thus, the source of the specificity so important in antibiotic action becomes apparent. Interference with cell wall construction will have no effect on a mammal but will be lethal to a bacterium.

The question now becomes, "How does penicillin interfere with cell wall construction?" The key to penicillin's action is the high reactivity of the β-lactam link. Cross-linking in cell wall construction is catalyzed by the enzyme transpeptidase. Basically, the enzyme acts as a facilitator of amide formation between two chains made up of amino acids. This kind of thing can be carried out with small molecules as well. Suppose we were set the task of joining the two amino acid derivatives in Figure 25.89.

The reaction of Figure 25.89 won't succeed because the acid group contains no decent leaving group and is prone to proton-transfer reactions, not amide formation. However, if we were to transform the acid group into an acid chloride, the reaction could proceed easily. Now chloride can function as a good leaving group, and there is no proton transfer possible. Recall the long discussion of the addition–elimination process in Chapter 20 (Fig. 25.90).

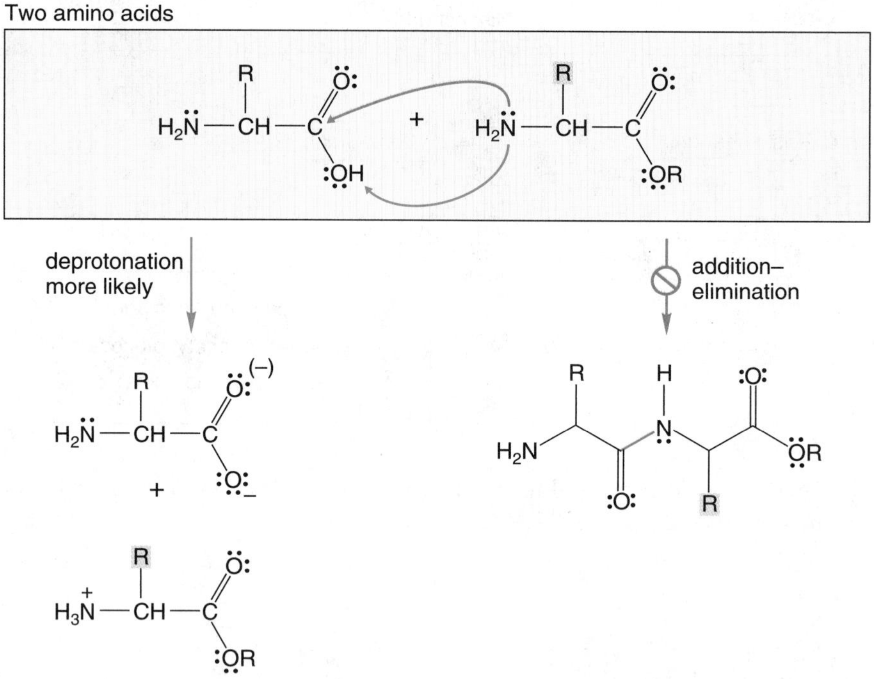

FIGURE 25.89 Amide formation between two unmodified amino acids is unlikely. Acid–base chemistry will dominate.

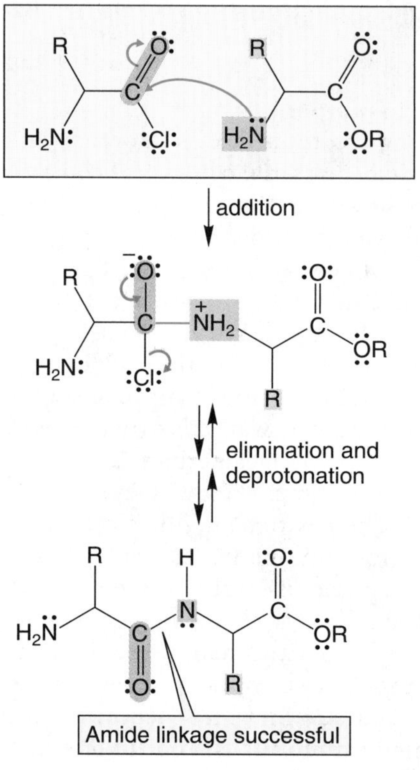

FIGURE 25.90 Amide formation can be facilitated through use of an acid chloride rather than an acid.

Transpeptidase functions as a biological counterpart of the chloride in Figure 25.90. It first reacts at an "active site" (a nucleophilic position somewhere on transpeptidase) to form an acyl derivative that can be attacked by an amino group to regenerate the enzyme and link the two chains (Fig. 25.91).

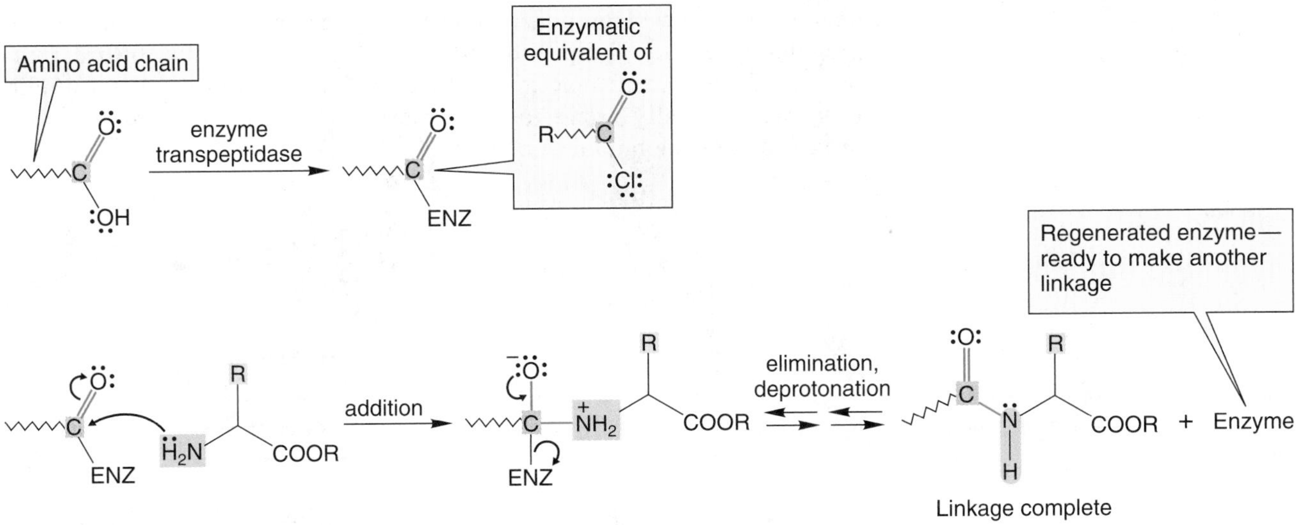

FIGURE **25.91** The enzyme transpeptidase plays the part of the chloride by first forming an acyl derivative of the amino acid chain. A new amino acid can now add and extend the chain.

Penicillin contains a most reactive amide group, the β-lactam, that reacts with the active site of transpeptidase to inactivate it (Fig. 25.92), and to stop cross-linking in cell wall construction.

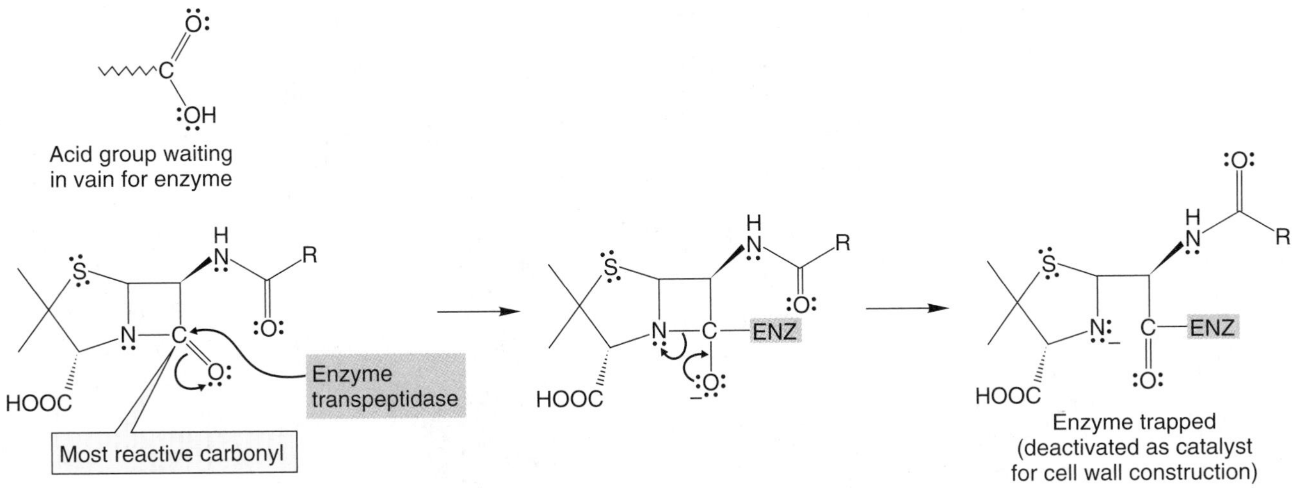

FIGURE **25.92** Penicillin intercepts and deactivates the enzyme through its highly reactive lactam ring.

25.10 SUMMARY

NEW CONCEPTS

Few new concepts appear in this chapter, which consists mostly of applications of ideas you already know. There are a few almost-new ideas, however. First, remember that the angle and torsional strain present in three-membered rings leads to high reactivity. We have already seen this general phenomenon in cyclopropane, but the addition of a heteroatom in the ring provides a leaving group and further increases reactivity toward ring-opening displacement reactions. Three- and four-membered heterocyclic rings are especially prone to this process, which neither simple cyclopropanes nor cyclobutanes undergo (Fig. 25.93).

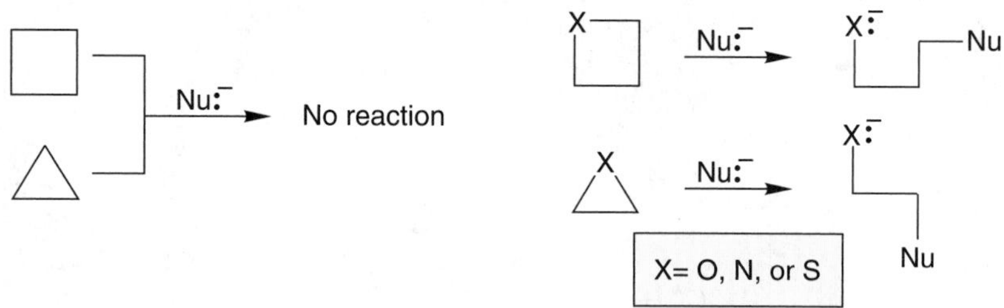

FIGURE 25.93 Heterocyclic small rings are prone to nucleophilic ring openings.

In a related vein, the presence of a heteroatom alters the reactivity of aromatic heterocycles. Although benzene and pyridine both undergo electrophilic substitution, only the heterocycle is usually subject to nucleophilic attack. The presence of the heteroatom is important to this difference (Figs. 25.47 and 25.78).

Finally, the concept of protecting groups arises again. We have seen this concept before in a few places, notably in the reversible conversion of aldehydes and ketones into acetals. In this chapter, we see alcohols being stored as THP derivatives, and carbonyl compounds protected as thioacetals and thioketals (Fig. 25.94).

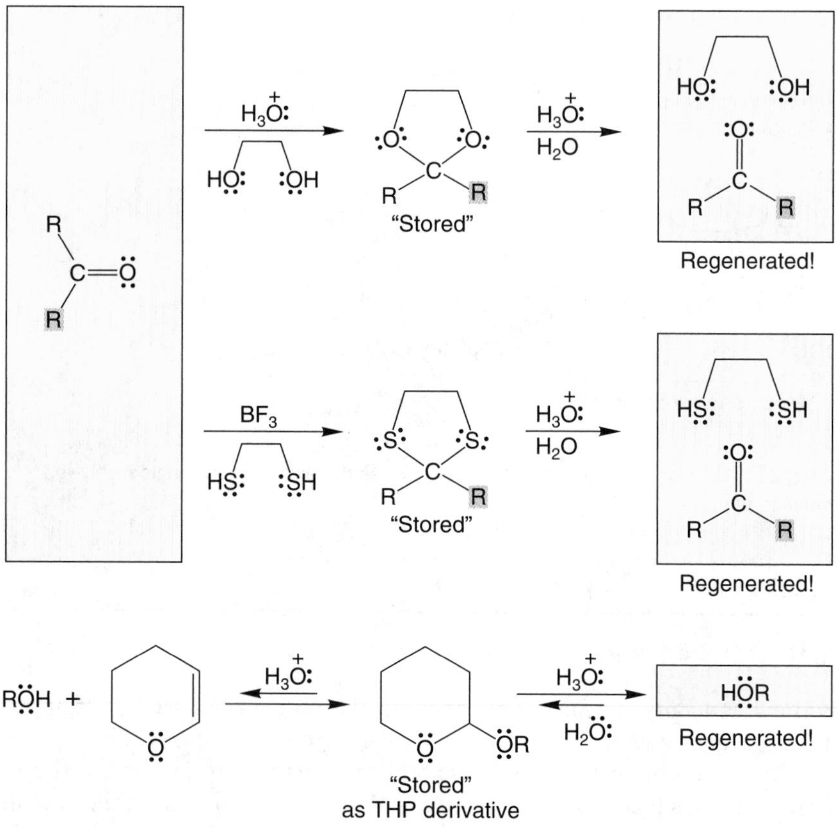

FIGURE 25.94 Some protecting groups in Chapter 25.

REACTIONS, MECHANISMS, AND TOOLS

This chapter includes a large number of reactions designed to lead to small rings. Many of these incorporate a ring-closing displacement as the crucial step, and many seemingly different reactions can be grouped under this heading. Most direct are the intramolecular S_N2 displacements leading to heterocycles of many ring sizes (Figs. 25.11, 25.20, and 25.23).

Moreover, the Darzens reaction, the addition of peroxidate anion to electron-deficient alkenes, the addition of such pseudohalogens as I—NCO or I—N_3 to alkenes, and the conversion of oxiranes to thiiranes with metal thiocyanates are all reactions leading to similar intermediates—nucleophiles poised for ring closure.

Other ring-forming reactions are different. The addition of nitrenes to alkenes leads to aziridines. So does the less direct, two-step process that uses 1,3-dipolar cycloaddition of azides to generate triazolines, followed by thermal or photochemical decomposition (Figs. 25.12 and 25.15).

Just as intramolecular S_N2 ring closures are the most general reactions leading to heterocyclic ring compounds, so the intermolecular ring-opening S_N2 reaction is the chief method for opening heterocycles (Fig. 25.93).

Both electrophilic and nucleophilic additions to aromatic heterocycles are important (Figs. 25.44 and 25.47).

Five-membered heterocyclic dienes, especially furans, are able to undergo the Diels–Alder reaction (Fig. 25.63).

Sulfides and thiophenes are desulfurized by Raney nickel to give hydrocarbons (Fig. 25.64).

Several other synthetic procedures arise in this chapter (Skraup, Reissert, and Fischer syntheses, for example). See the next section for details.

SYNTHESES

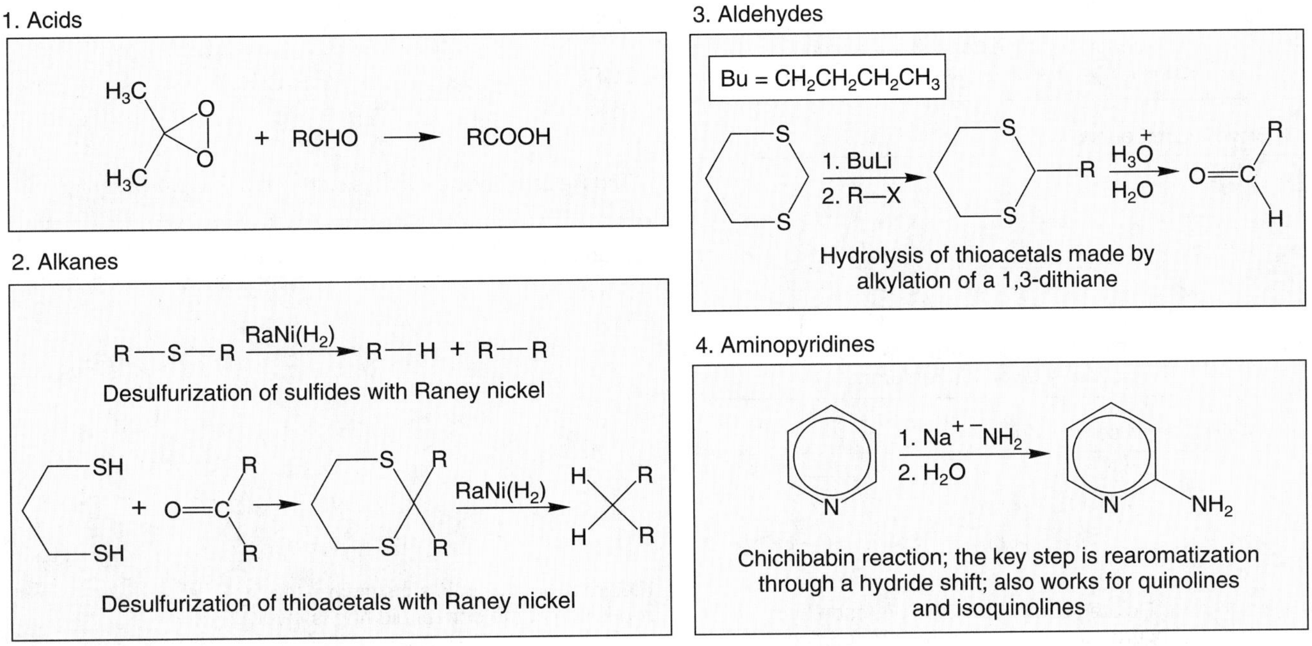

1. Acids

$(CH_3)_2C$(dioxirane) + RCHO ⟶ RCOOH

2. Alkanes

R—S—R $\xrightarrow{RaNi(H_2)}$ R—H + R—R

Desulfurization of sulfides with Raney nickel

HS–CH₂CH₂CH₂–SH + O=CR₂ ⟶ 2,2-dialkyl-1,3-dithiane $\xrightarrow{RaNi(H_2)}$ H_2CR_2

Desulfurization of thioacetals with Raney nickel

3. Aldehydes

Bu = $CH_2CH_2CH_2CH_3$

1,3-dithiane $\xrightarrow{\text{1. BuLi; 2. R—X}}$ 2-R-1,3-dithiane $\xrightarrow{H_3O^+,\ H_2O}$ O=C(R)H

Hydrolysis of thioacetals made by alkylation of a 1,3-dithiane

4. Aminopyridines

pyridine $\xrightarrow{\text{1. } Na^+ \ ^-NH_2\text{; 2. } H_2O}$ 2-aminopyridine (NH_2)

Chichibabin reaction; the key step is rearomatization through a hydride shift; also works for quinolines and isoquinolines

FIGURE 25.95 New synthetic reactions of Chapter 25.

5. Azetidines

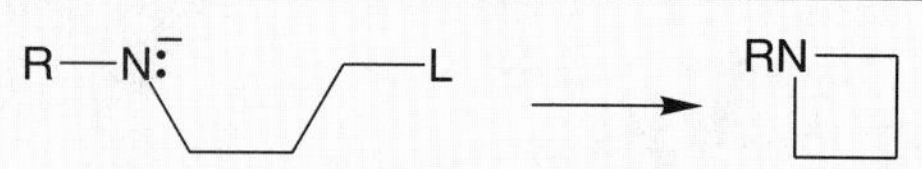

Various intramolecular displacement reactions of different charge types are possible

6. Aziridines

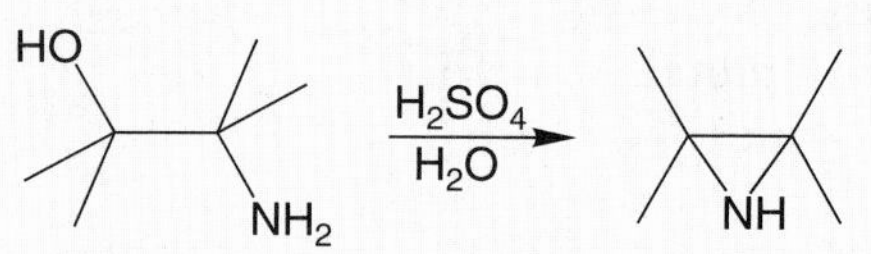

Intramolecular displacement

1. I—NCO
2. ROH
3. KOH
4. Δ

N
H

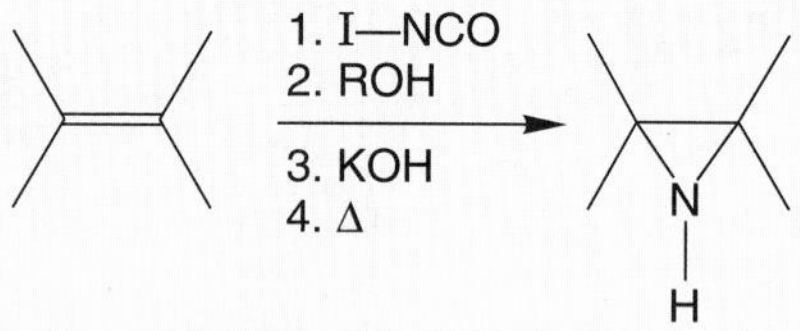

trans Addition of I–NCO is followed by other transformations

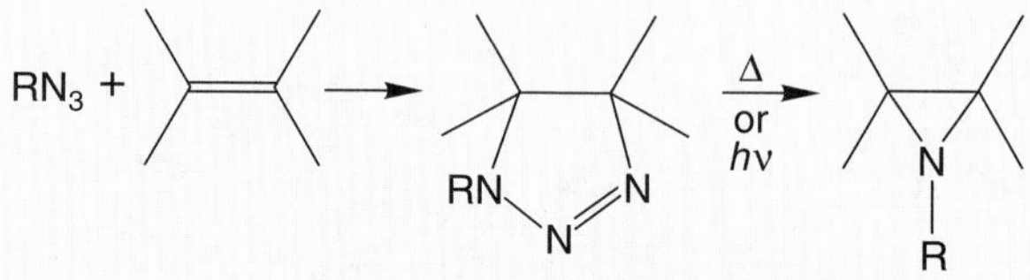

RN_3 +

Δ or $h\nu$

RN N N

N R

1,3-Dipolar cycloaddition followed by loss of N_2

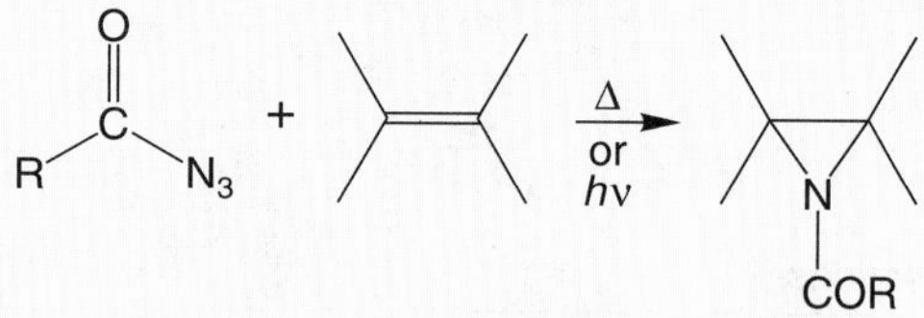

O C R N_3 +

Δ or $h\nu$

N COR

An addition of a nitrene; the reaction does not work well for simple alkyl azides

7. Bicyclic compounds

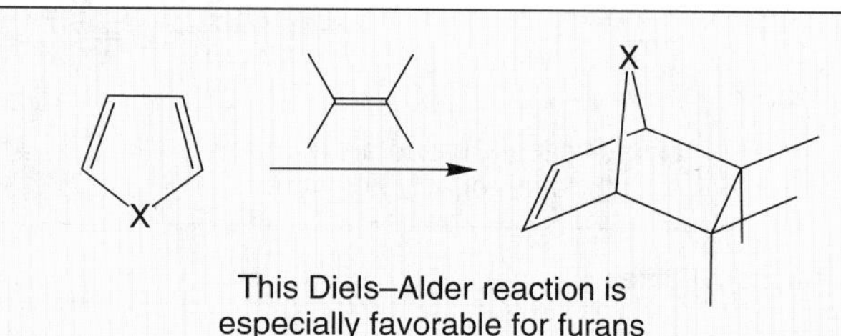

This Diels–Alder reaction is especially favorable for furans

8. 1,4-Dicarbonyl compounds

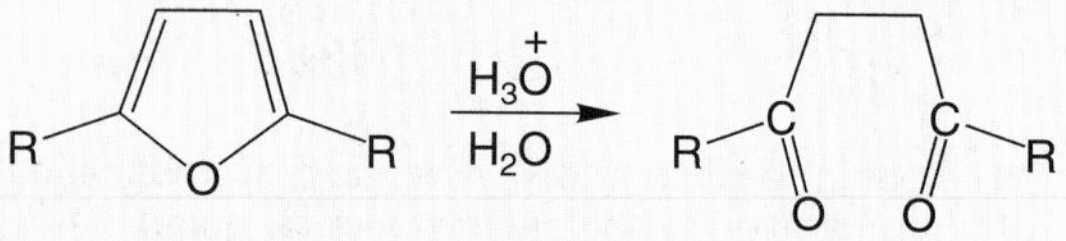

R O R

$H_3\overset{+}{O}$
H_2O

R C C R
O O

Related compounds can be made by hydrolysis of other five-membered rings

9. THP derivatives

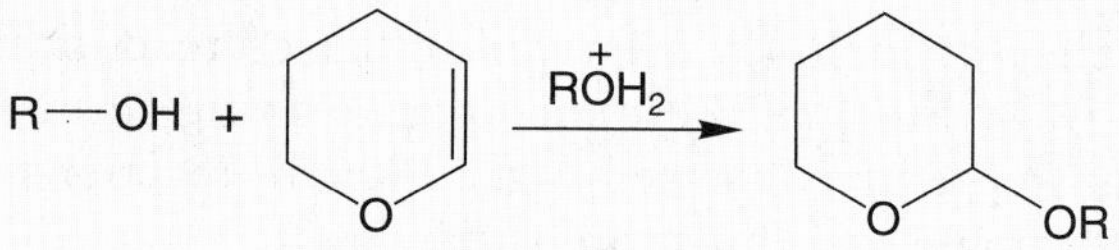

R—OH + O

$R\overset{+}{O}H_2$

O OR

The THP derivative is an acetal; hydrolysis will regenerate the alcohol; THPs are protecting groups for alcohols

10. Furans

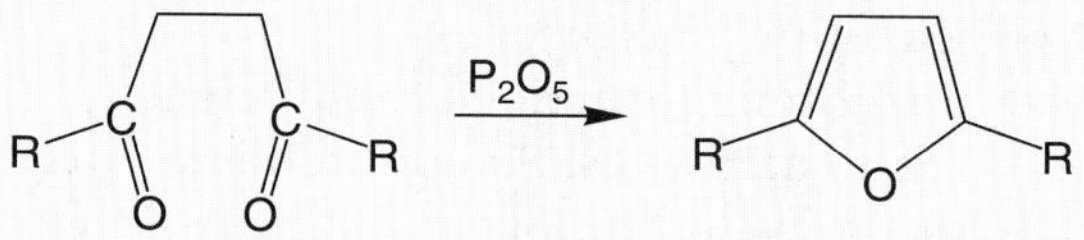

R C C R
O O

P_2O_5

R O R

See item 8, this reaction is the reverse process

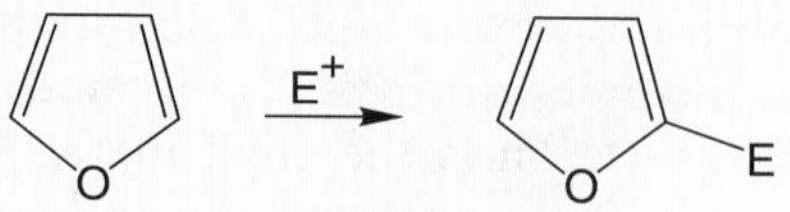

O

E^+

O E

Electrophilic aromatic substitution; reaction at the 2-position is favored

11. Indoles

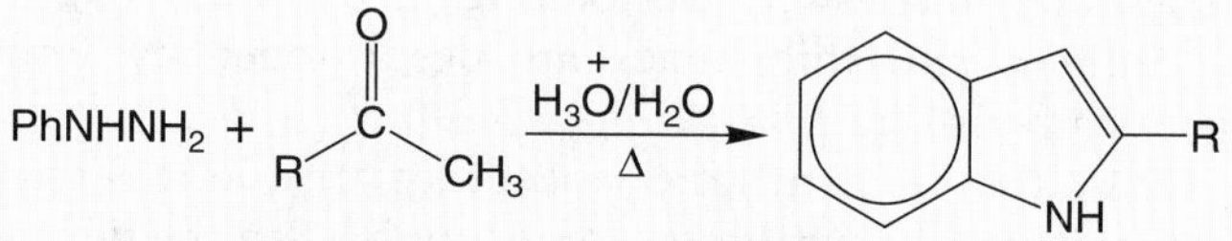

$PhNHNH_2$ + O C R CH_3

$H_3\overset{+}{O}/H_2O$
Δ

R NH

The Fischer indole synthesis; the carbonyl compound must have an α-hydrogen

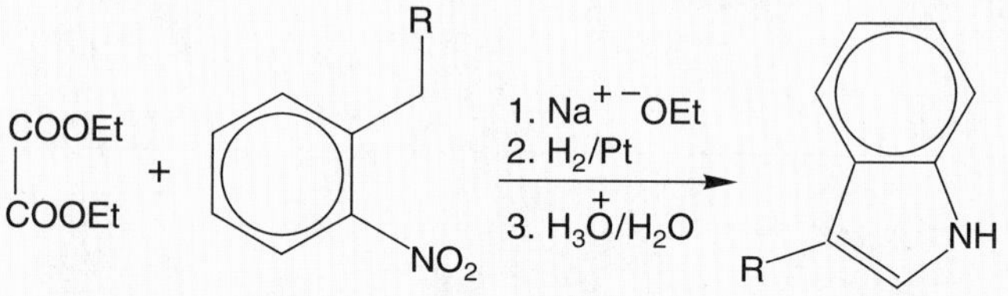

COOEt
COOEt
+ R NO_2

1. $Na^+\ ^-OEt$
2. H_2/Pt
3. $H_3\overset{+}{O}/H_2O$

NH R

The Reissert indole synthesis; the nitro group is essential

12. Isoquinolines

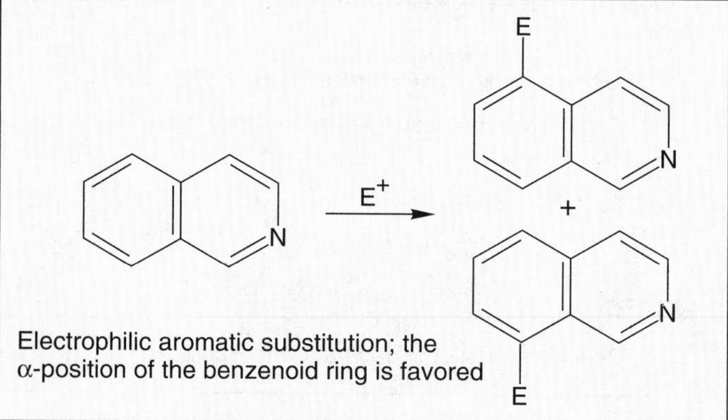

Electrophilic aromatic substitution; the α-position of the benzenoid ring is favored

FIGURE **25.95** (CONTINUED)

13. Ketones (see also 1,4-dicarbonyl compounds)

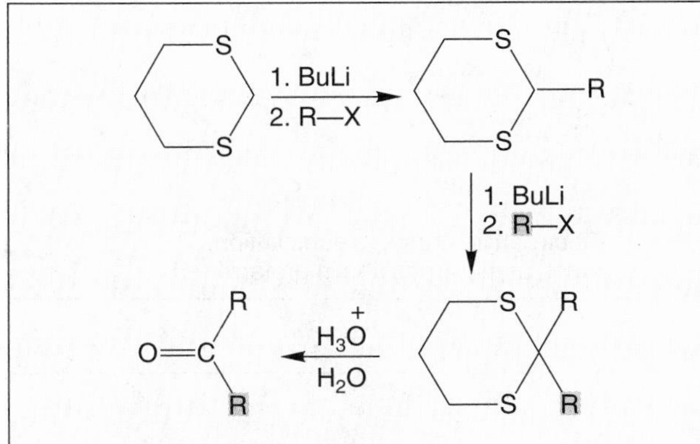

14. Methylene (the parent carbene)

Works for other carbenes as well; carbenes are not stable molecules and react as they are generated

15. Nitrenes

Nitrenes are not stable molecules and react as they are generated

16. Oxetanes

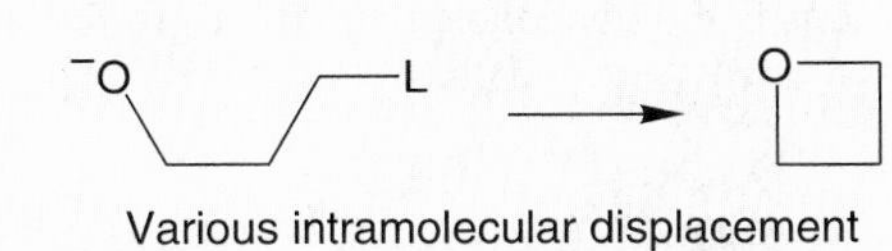

Various intramolecular displacement reactions of different charge types are possible

A photochemical 2 + 2 reaction

17. Oxiranes

Direct, cis epoxidation, slow if the alkene is substituted with electron-withdrawing groups

17. Oxiranes (continued)

Michael addition followed by intramolecular displacement of hydroxide; works only for alkenes substituted with electron-withdrawing groups

Deprotonation followed by intramolecular displacement

Enolate formation, attack on the carbonyl and intramolecular displacement; the Darzens condensation

18. Pyridines

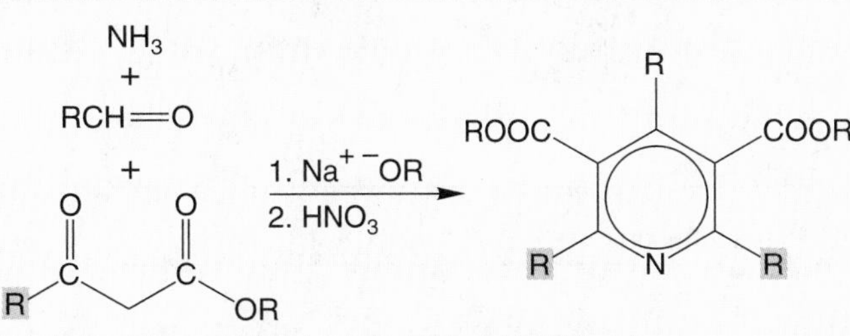

Hantzsch pyridine synthesis; a dihydropyridine is an intermediate

Decarboxylation of the Hantzsch products

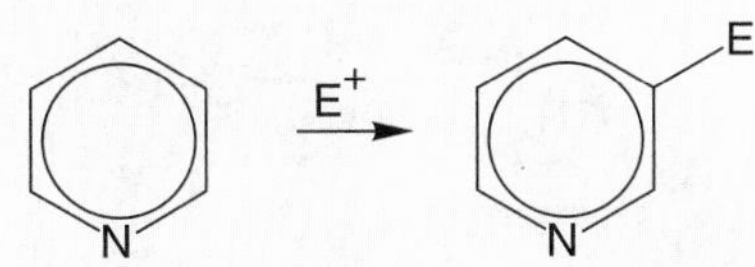

Electrophilic aromatic substitution; the 3-position is favored

FIGURE **25.95** (CONTINUED)

18. Pyridines (continued)

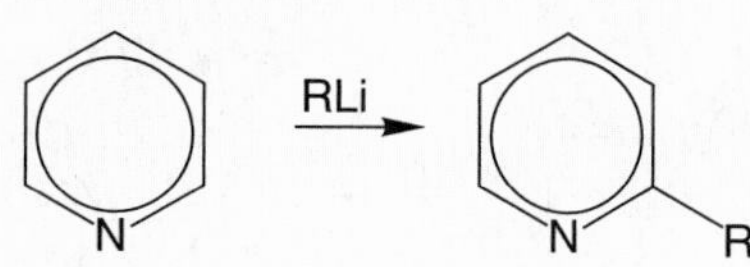

Nucleophilic aromatic substitution; a hydride shift is the key step; see the Chichibabin reaction for a related process

1. $Na^+ \, ^-OR$, HOR

Nucleophilic aromatic substitution through ipso attack

19. Pyridine *N*-oxides

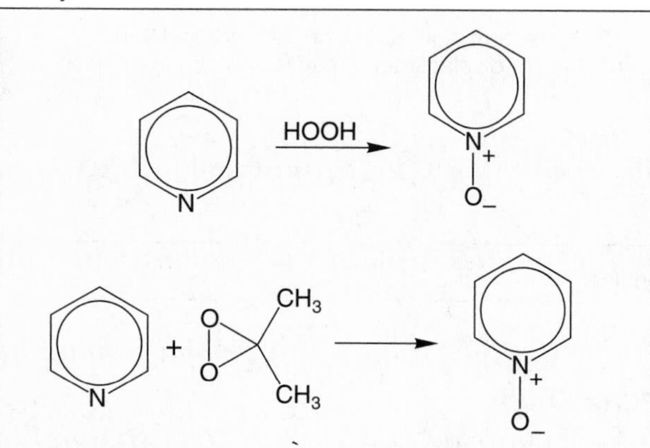

20. Pyridinium ions

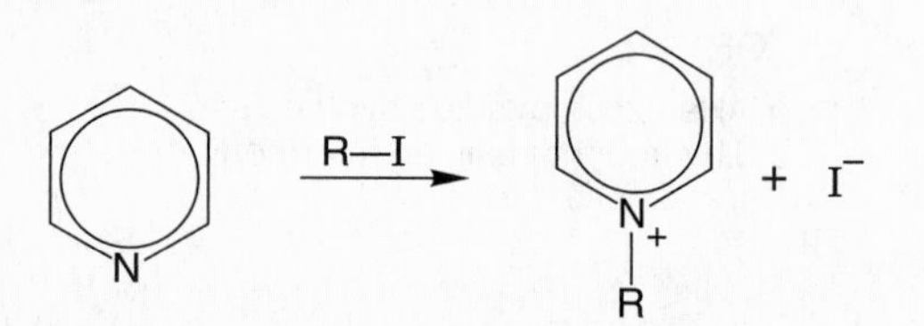

Pyridine acts as nucleophile in this S_N2 reaction; the "R–I" must not be tertiary

21. Pyrroles

R—NH_2, HOAc

Knorr–Paal synthesis

21. Pyrroles (continued)

Electrophilic aromatic substitution; attack at the 2-position is preferred

22. Quinolines

H_2SO_4, $PhNO_2$

The Skraup synthesis; a dihydroquinoline is an intermediate

Electrophilic aromatic substitution; the two α-positions in the benzenoid ring are favored

1. $Na^+ \, ^-NH_2$ 2. H_2O/H_3O^+

Nucleophilic aromatic substitution—a Chichibabin reaction

23. Sulfoxides

FIGURE 25.95 (CONTINUED)

24. Thietanes

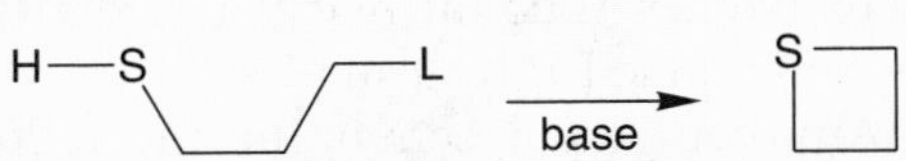

Various intramolecular displacement reactions of different charge types are possible

25. Thiiranes

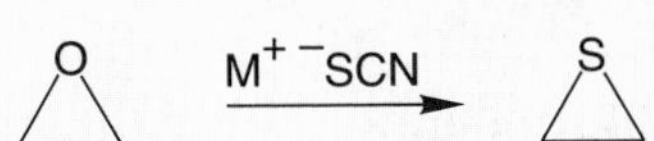

This mechanism involves a sequence of ring openings and closings; the original stereochemistry of the oxirane is preserved

26. Thiophenes

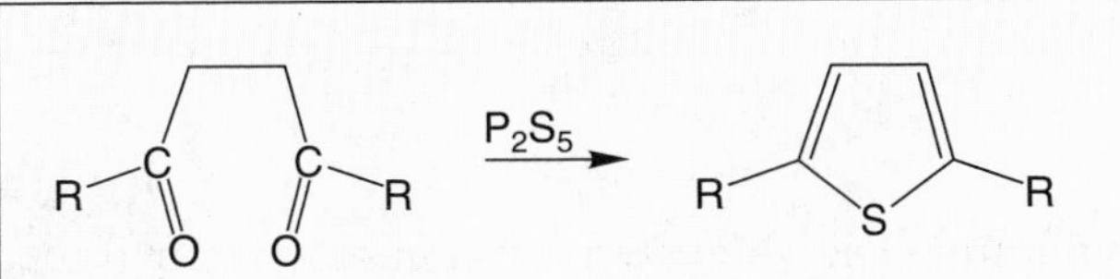

26. Thiophenes (continued)

E^+

Electrophilic aromatic substitution; reaction at the 2-position is favored

27. Triazolines

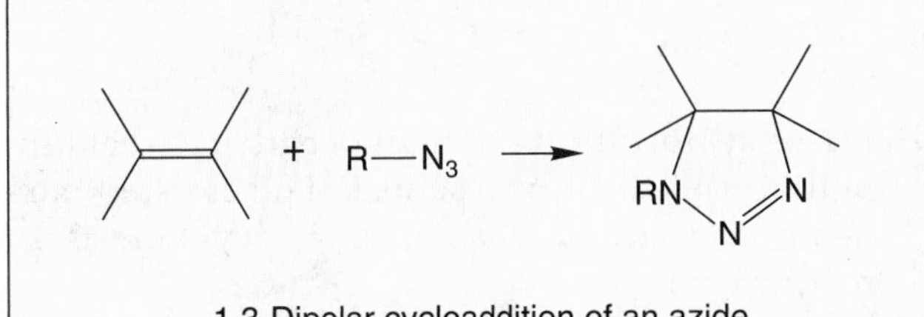

1,3-Dipolar cycloaddition of an azide

FIGURE **25.95** (CONTINUED)

COMMON ERRORS

Once again, the problems come mainly from the wealth of detail. There is also a balancing act to be mastered—the combination of ring chemistry (strain and geometrical effects) and heteroatom chemistry (electronegativity and nucleophilicity).

In heterocyclic chemistry, there are also many "open the ring—close the ring" problems. The trick is to see the nature of the problem *before* trying to solve it, which sounds simple but in reality is not. It is all too tempting to push ahead without doing the analysis that allows you to see what you need to do.

25.11 KEY TERMS

Azetidine A saturated four-membered ring containing one nitrogen atom.

Azides Compounds of the structure

$$R{-}\ddot{\underset{\cdot\cdot}{N}}^{-}{-}\overset{+}{N}{\equiv}N:$$

Aziridine A saturated three-membered ring containing one nitrogen atom.

Benzofuran A compound incorporating the structure

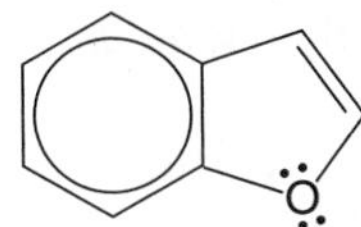

Benzothiophene A compound incorporating the structure

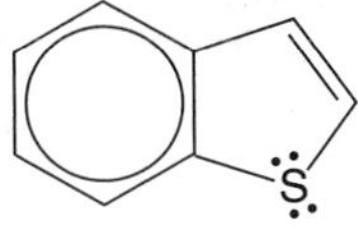

Cephalosporin A powerful antibiotic related to penicillin incorporating a β-lactam ring fused to a six-membered ring containing sulfur.

Chichibabin reaction Nucleophilic addition of an amide ion to pyridine (or a related heteroaromatic compound) leading to an aminopyridine. A key step involves hydride transfer.

Diazirine A three-membered ring containing both a double bond and two nitrogen atoms.

Dioxirane A three-membered ring containing two oxygen atoms. A versatile oxidizing agent.

Episulfide A saturated three-membered ring containing a single sulfur atom. See **Thiirane**.

Epoxide A saturated three-membered ring containing a single oxygen atom. See **Oxirane**.

Fischer indole synthesis A synthesis of indoles starting from phenylhydrazones.

Furan Any compound containing the following five-membered aromatic heterocyclic ring:

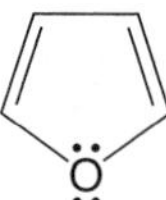

Hantzsch synthesis A synthesis of pyridines that uses multiple condensation reactions of β-ketoesters, ammonia, and an aldehyde. A dihydropyridine is an intermediate.

Heterocyclic compound A ring compound containing at least one heteroatom in the ring.

Indole Any compound containing the following ring system:

Ipso attack Addition to an aromatic ring at a position already occupied by a non-hydrogen substituent.

Isoquinoline Any compound containing the following ring system:

Nitrene A reactive intermediate containing a neutral, monovalent nitrogen atom. The nitrogen counterpart of a carbene.

Oxetane A saturated, four-membered ring containing a single oxygen atom.

Oxirane A saturated three-membered ring containing a single oxygen atom. See **Epoxide**.

Penicillin A powerful antibiotic. The key feature is the reactive β-lactam ring.

Pseudohalogen Any of a variety of X—Y compounds that add to alkenes as do the halogens. Examples in this chapter include I—NCO and $I{-}N_3$.

Pyridine Any compound containing the following "azabenzene" ring system:

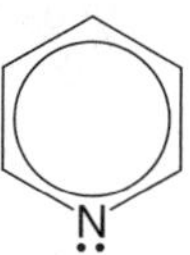

Pyridine-*N*-oxide A pyridine in which the nitrogen has been oxidized to the oxide stage.

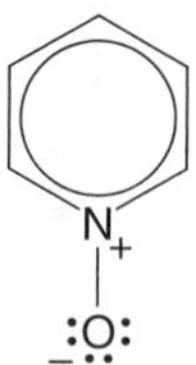

Pyridinium ion A quaternary ammonium ion formed from a pyridine.

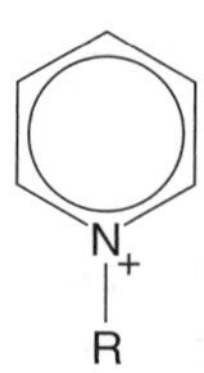

Pyrrole Any compound containing the following ring structure:

Quinoline Any compound containing the following ring structure:

Raney nickel A good reducing agent composed of finely divided nickel on which hydrogen has been adsorbed.

Reissert synthesis A synthesis of indoles starting from a nitroalkylbenzene and an ester of oxalic acid.

Skraup synthesis A synthesis of quinolines starting from aniline and glycerol (1,2,3-trihydroxypropane).

Thietane A saturated four-membered ring containing a single sulfur atom.

Thiirane A saturated three-membered ring containing a single, unoxidized sulfur atom. See **Episulfide**.

Thiophene Any compound containing the following five-membered aromatic heterocyclic ring:

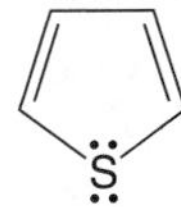

THP derivatives Protecting groups for alcohols of the structure

O
OR

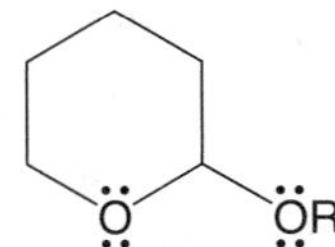

Triazoline Any compound containing the following ring system, usually formed by the 1,3-dipolar cycloaddition of an azide to an alkene:

R
N
N
N

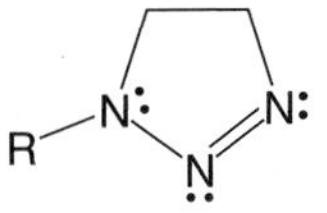

25.12 ADDITIONAL PROBLEMS

PROBLEM **25.28** Design an aromatic heterocyclic compound that

(a) Contains one boron atom.

(b) Contains one boron and one oxygen atom in a six-membered ring.

(c) Contains one oxygen and one nitrogen atom in a five-membered ring.

(d) Contains two nitrogen atoms in a five-membered ring.

(e) Contains three boron and three nitrogen atoms in a six-membered ring.

PROBLEM **25.29** Try to construct an aromatic heterocycle in which there is an oxygen and a nitrogen in a neutral six-membered ring.

PROBLEM **25.30** Carefully explain why azirines of structure **1** have never been isolated, whereas the isomeric azirines of structure **2** are well known.

R1
R2
N
R3

1

R1
R3
R2
N

2

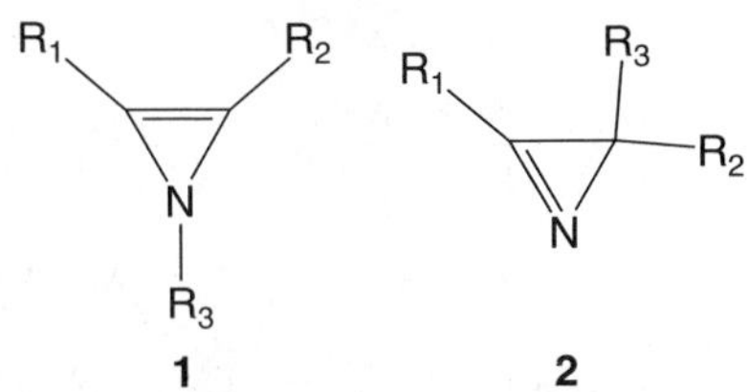

PROBLEM **25.31** Write the principal organic products of the following reactions. Mechanisms are not necessary.

(a)

1. CF3COOOH
2. KOH/H2O

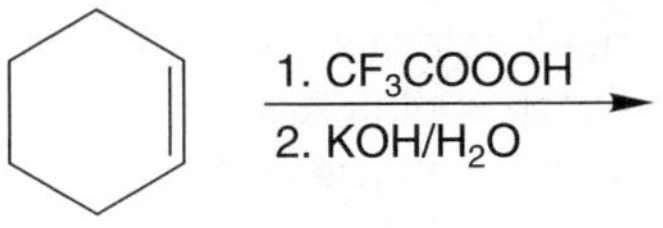

(b)

H2O2
NaOH
O

(c)

Ph
O
K+ −SCN
H2O

(d)

NH2
Cl
H2O
NaOH

(e)

O
C
N3
OCH3
hν

(f)

H3C
H3C
S
PhNH2
Δ

(g)

O
Ph
CH3O−
CH3OH

(h)

S
S
Ra Ni

(i)

CH3CH2OH
O
+
H3O

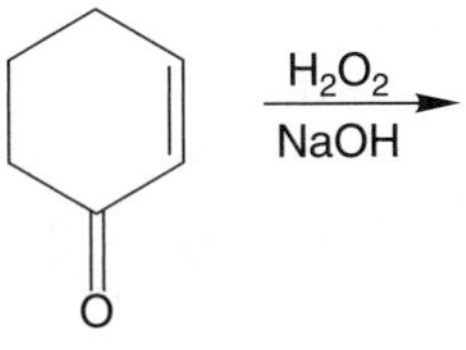

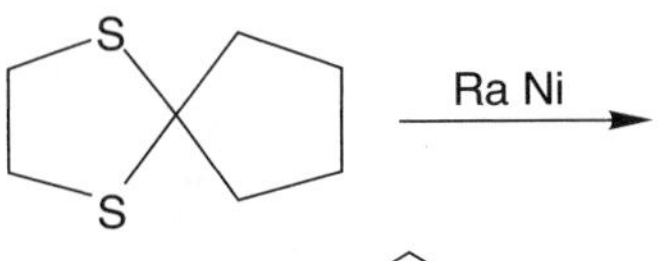

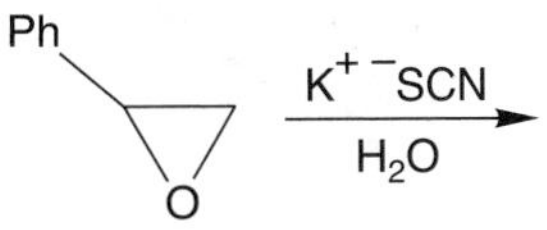

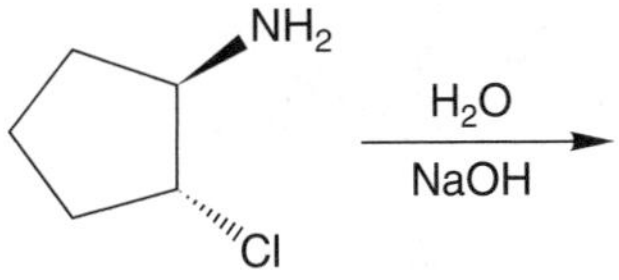

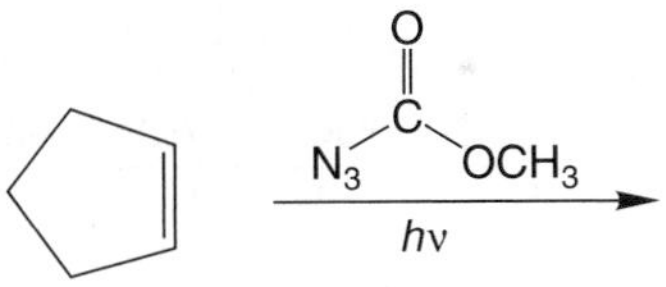

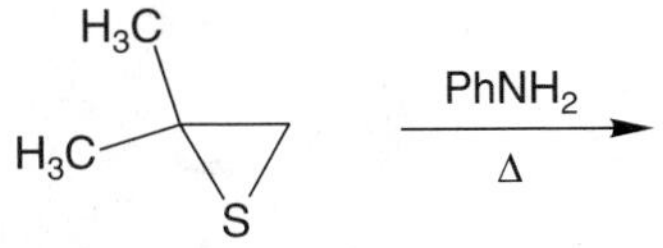

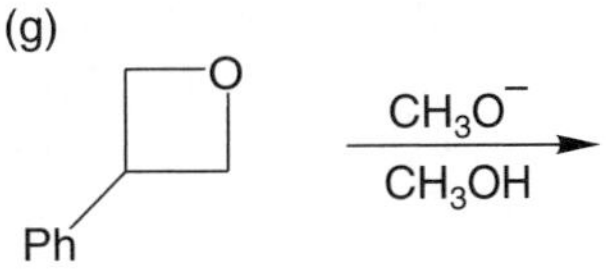

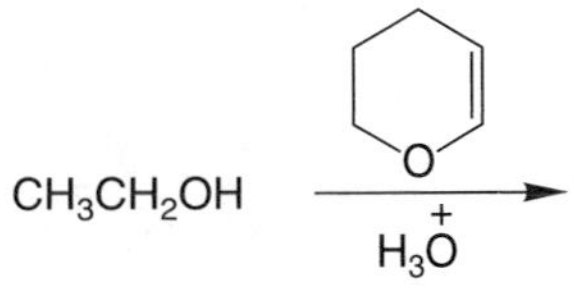

PROBLEM 25.31 (CONTINUED)

(j) [dimethyldioxirane] $\xrightarrow{Ph—CH=CH_2}$

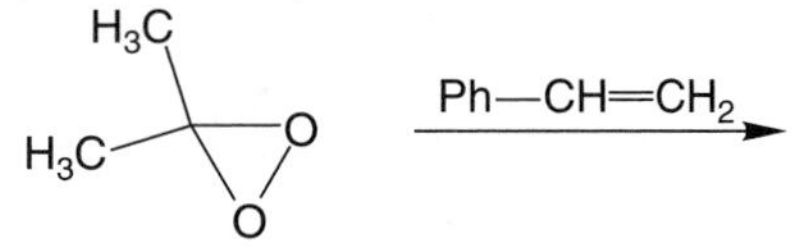

(k) [pyridine] $\xrightarrow[2.\ H_2O]{1.\ KNH_2/NH_3}$

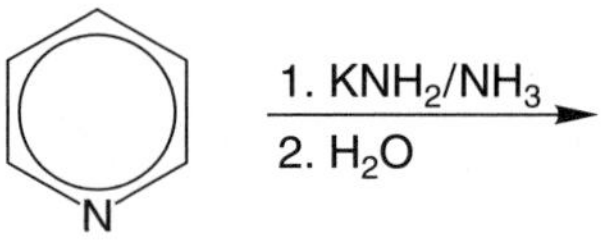

(l) [2-chloropyridine] $\xrightarrow[CH_3OH]{CH_3O^-}$

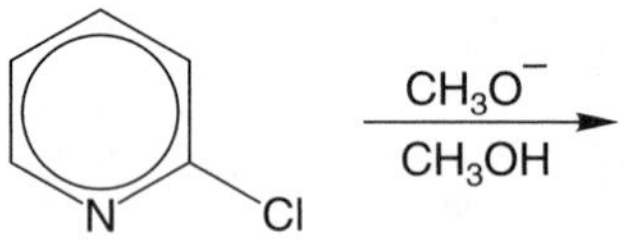

(m) [3-chloropyridine] $\xrightarrow[CH_3OH]{CH_3O^-}$

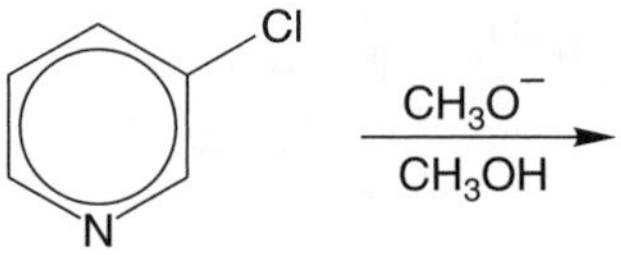

(n) [furan] $\xrightarrow[\Delta]{CH_3OOC—C\equiv C—COOCH_3}$

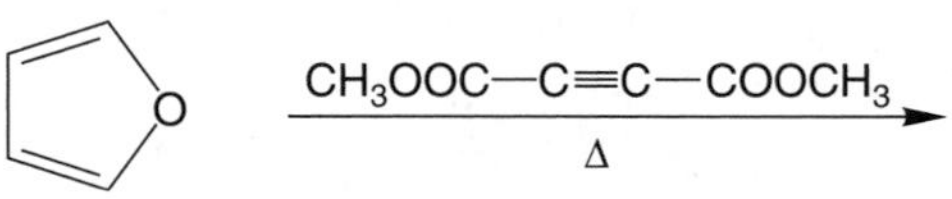

(o) [2,5-diphenylfuran] $\xrightarrow{H_2O/H_3O^+}$

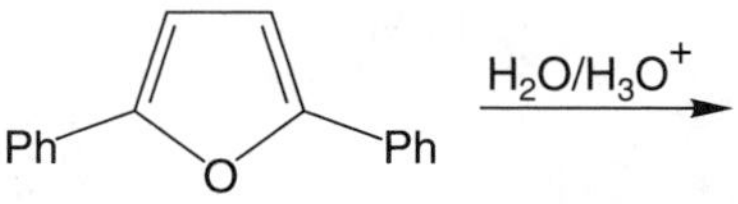

(p) Ph–CO–CH₂CH₂–CO–Ph $\xrightarrow[\Delta]{P_2S_5}$

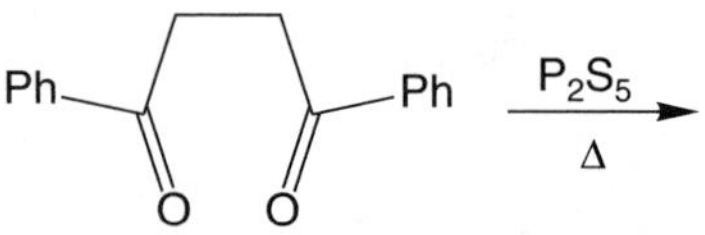

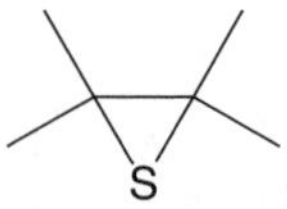

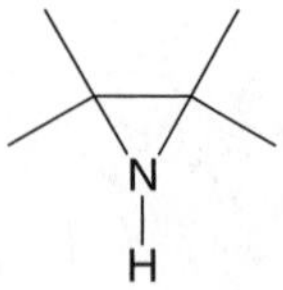

PROBLEM 25.32 How would you convert 2,3-dimethyl-2-butene into the following compounds?

(a) [tetramethyloxirane] (b) [tetramethylthiirane] (c) [tetramethylaziridine] (d) [tetramethylcyclopropane]

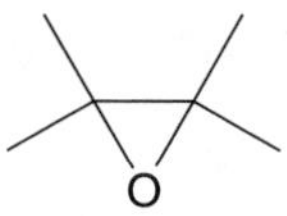

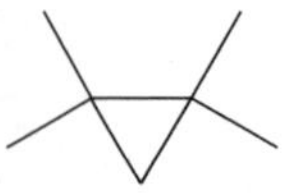

PROBLEM 25.33 The dipole moment of THF is 1.7 D, but that of furan is only 0.7 D. Explain.

PROBLEM 25.34 Predict which nitrogen of imidazole (**1**) will be protonated faster.

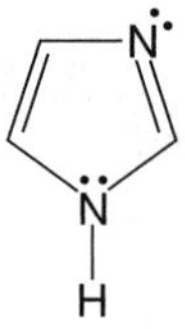

[imidazole, **1**]

PROBLEM 25.35 Write an arrow formalism mechanism for the following change:

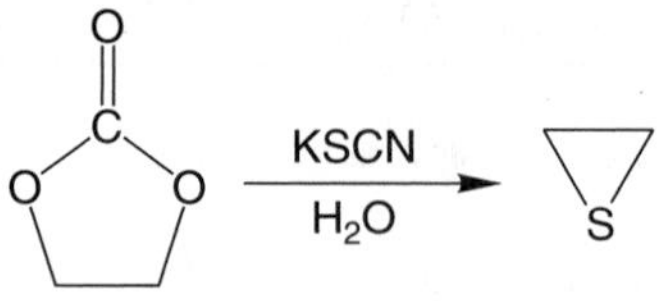

[ethylene carbonate] $\xrightarrow[H_2O]{KSCN}$ [thiirane]

PROBLEM 25.36 Outline how you would carry out the following conversions. Mechanisms are not necessary. Each part requires the initial synthesis of an intermediate.

(a) [thiophene] ⟶ $CH_3CH_2CH_2CH_2$–C(=O)–CH_3

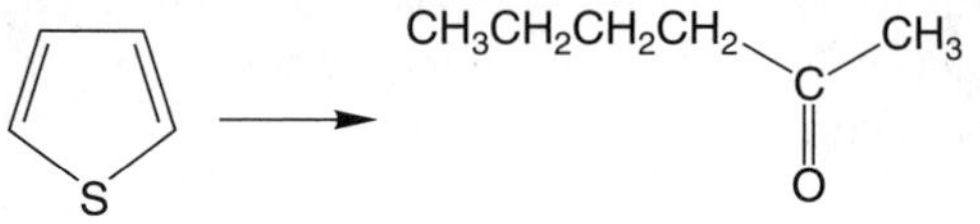

(b) [aniline, NH_2] ⟶ [2-aminoquinoline, NH_2]

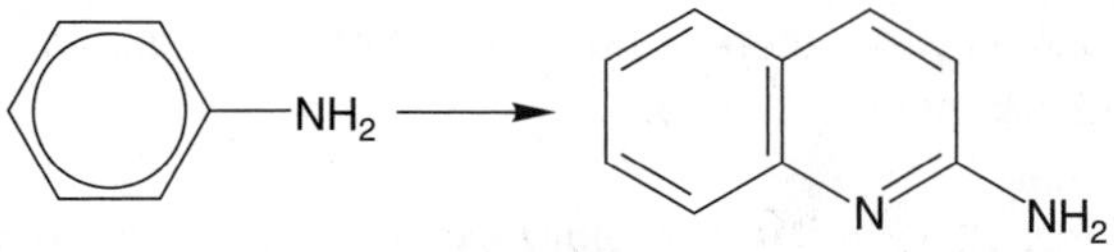

PROBLEM 25.37 Write structures for compounds **A**, **B**, and **C**.

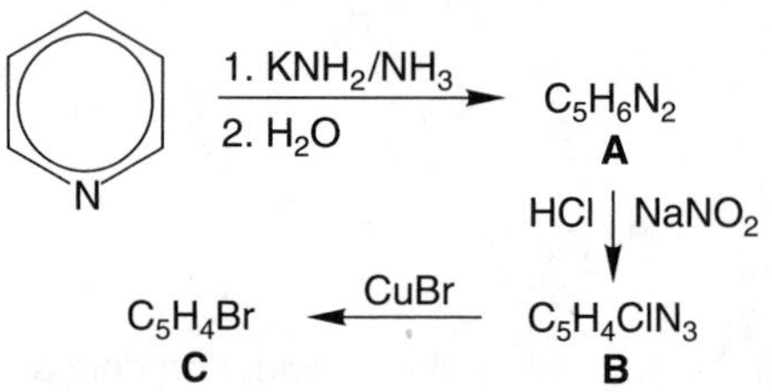

[pyridine] $\xrightarrow[2.\ H_2O]{1.\ KNH_2/NH_3}$ $C_5H_6N_2$ **A** $\xrightarrow{HCl\ |\ NaNO_2}$ $C_5H_4ClN_3$ **B** $\xrightarrow{CuBr}$ C_5H_4Br **C**

PROBLEM 25.38

(a) Outline Fischer indole syntheses for **1** and **2**.
(b) Outline a Reissert synthesis for **3**.
(d) Outline a Skraup synthesis for **4**.
(e) Outline a Hantzsch synthesis for **5**.

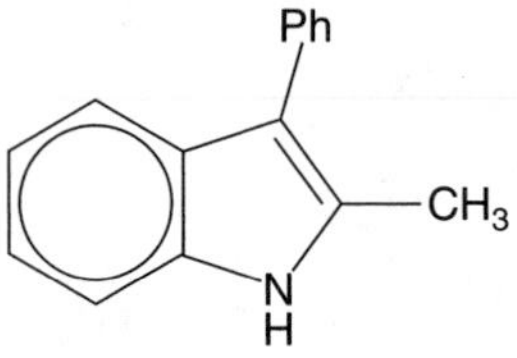

[3-phenyl-2-methylindole, **1**] [5-methyl-3-phenylindole, **2**]

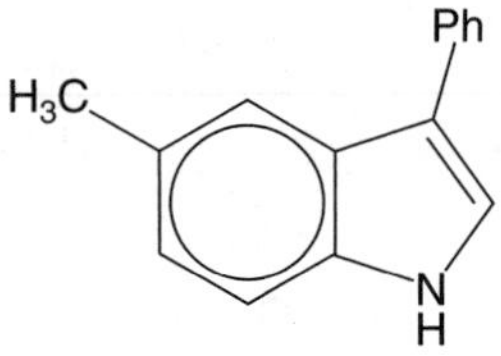

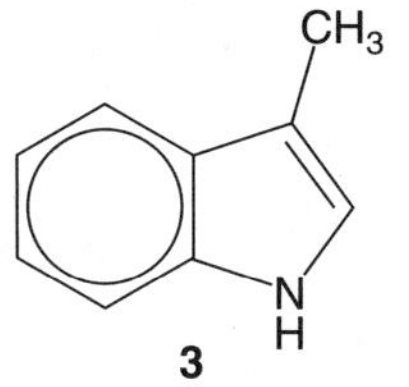
CH3
N
H

3

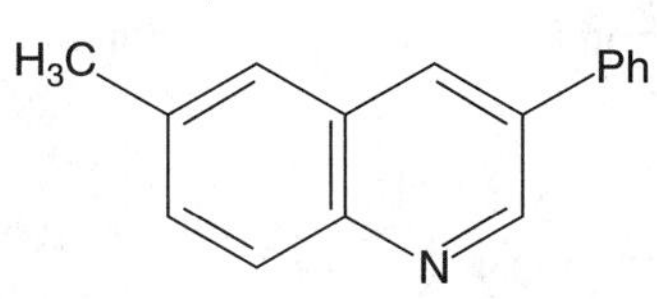
H3C
Ph
N

4

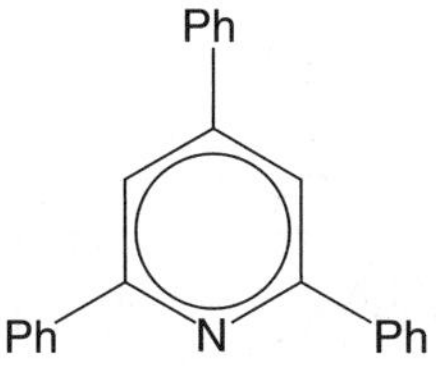
Ph
Ph
N
Ph

5

PROBLEM 25.39 Supply structures for compounds **A–D**. Mechanisms are not necessary, and there are intermediates involved that are not isolable under normal conditions. Pay attention to stereochemistry.

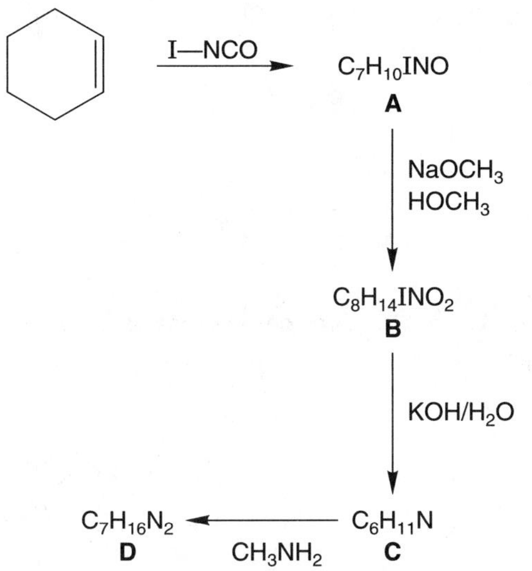
I—NCO
C7H10INO
A
NaOCH3
HOCH3
C8H14INO2
B
KOH/H2O
C6H11N
C
CH3NH2
C7H16N2
D

PROBLEM 25.40 Propose a mechanism for the following reaction:

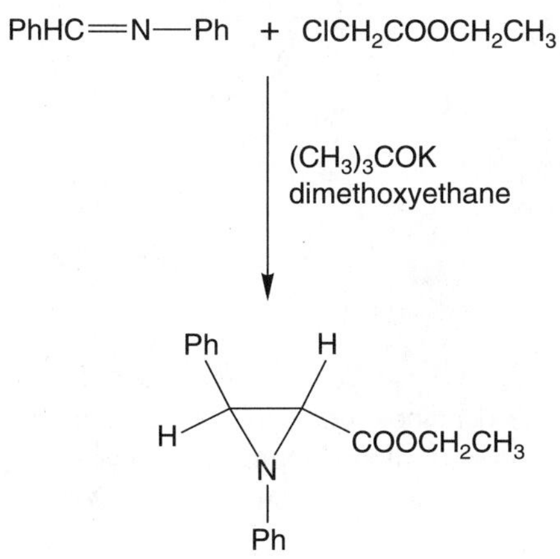
PhHC=N—Ph + ClCH2COOCH2CH3
(CH3)3COK
dimethoxyethane
Ph
H
H
COOCH2CH3
N
Ph

PROBLEM 25.41 Although simple alkyl and aryl nitriles are poor dienophiles, nitriles substituted with electron-withdrawing groups are more reactive. Provide mechanisms for the formations of the two products in the following reaction:

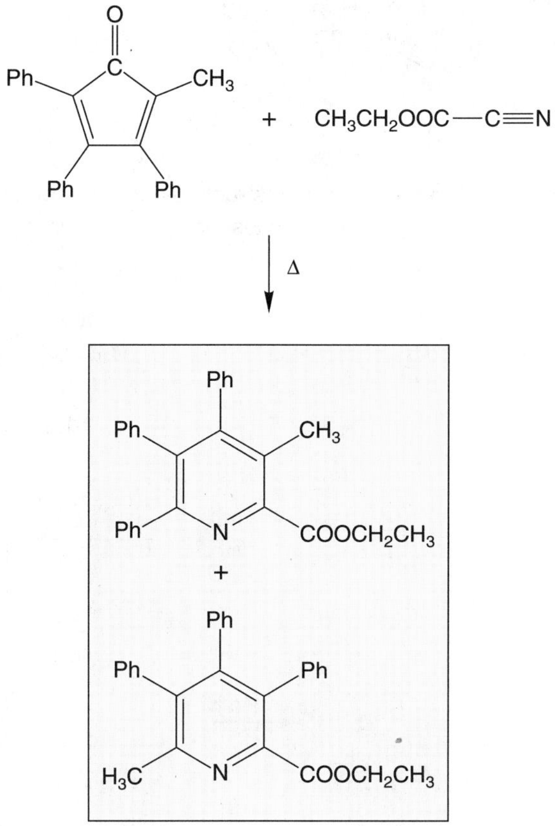
O
C
Ph
CH3
Ph
Ph
+ CH3CH2OOC—C≡N
Δ
Ph
Ph
CH3
Ph
N
COOCH2CH3
+
Ph
Ph
Ph
H3C
N
COOCH2CH3

PROBLEM 25.42 Reaction of 1,3-diphenylisobenzofuran (**1**) and vinylene carbonate (**2**) gives compound **3**. Acid hydrolysis of **3** affords **4**. Propose a structure for **3** and write a mechanism for its formation and hydrolysis to **4**.

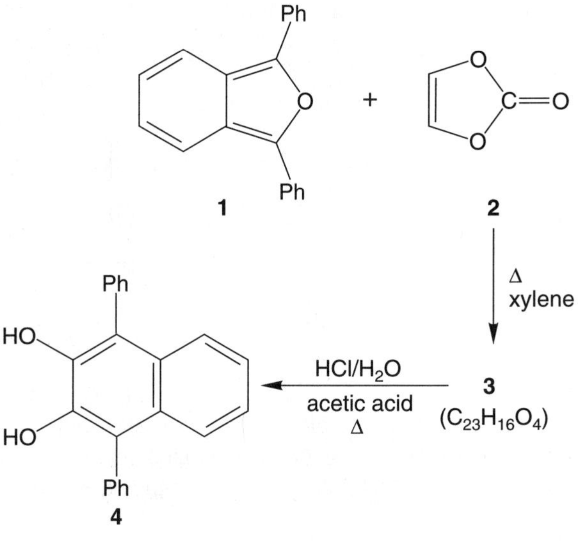
Ph
O
Ph
1
+
O
C=O
O
2
Δ
xylene
3
(C23H16O4)
HCl/H2O
acetic acid
Δ
Ph
HO
HO
Ph
4

PROBLEM 25.43 Write an arrow formalism mechanism for the following reaction:

1. Ph–C(=O)–Cl
2. KCN/H_2O

N, CN

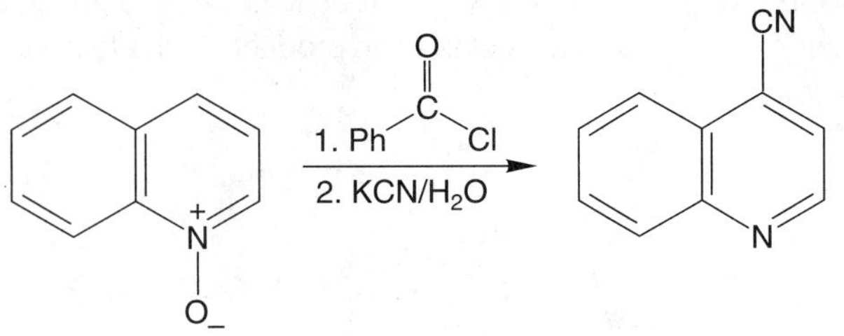

PROBLEM 25.44 Provide structures for intermediates **A–D**. Mechanisms are not necessary.

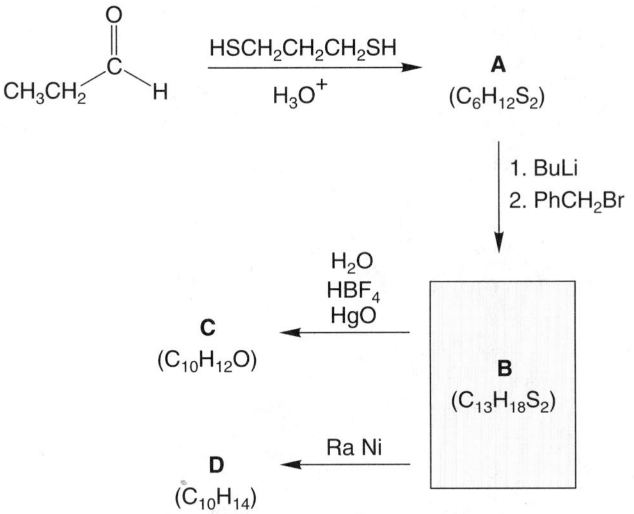

CH_3CH_2–C(=O)–H → ($HSCH_2CH_2CH_2SH$, H_3O^+) → **A** ($C_6H_{12}S_2$)

1. BuLi
2. $PhCH_2Br$

B ($C_{13}H_{18}S_2$)

H_2O, HBF_4, HgO → **C** ($C_{10}H_{12}O$)

Ra Ni → **D** ($C_{10}H_{14}$)

PROBLEM 25.45 In Section 25.1a, we saw that 2-amino alcohols can be converted into aziridines if the hydroxyl group is first changed into a better leaving group (see Problem 25.7). A related approach to aziridines might involve treating a 2-amino alcohol with thionyl chloride in the presence of base.

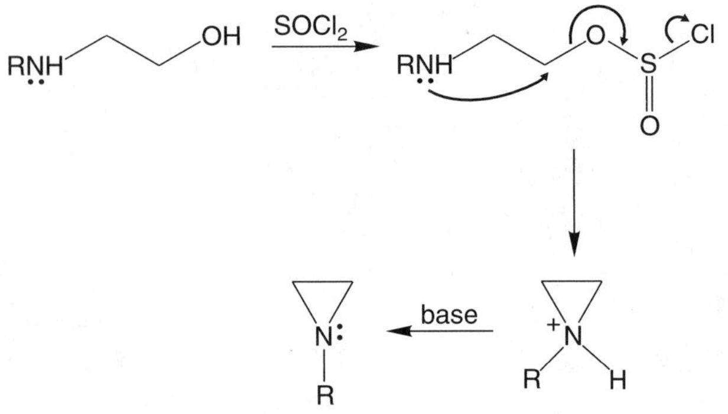

RNH–CH$_2$CH$_2$–OH → ($SOCl_2$) → RNH–CH$_2$CH$_2$–O–S(=O)–Cl → aziridinium (R, H) → (base) → aziridine N–R

However, when 2-amino alcohol **1** was treated with thionyl chloride in the presence of triethylamine, isomeric compounds **3** and **4** ($C_{12}H_{17}NO_2S$) were obtained rather than the anticipated **2**. Propose structures for isomers **3** and **4** and explain their formation.

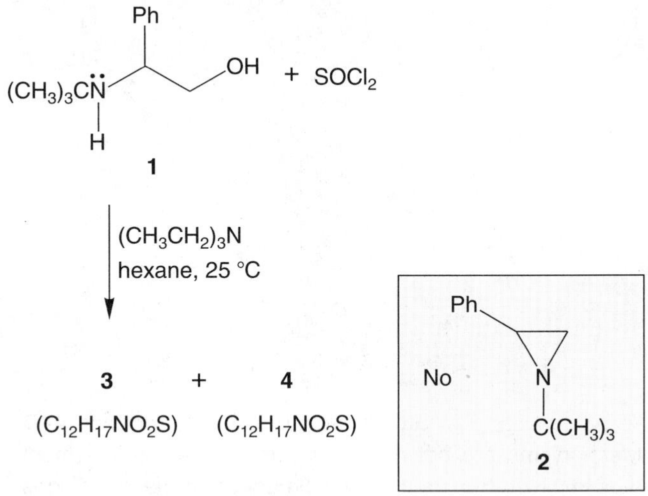

$(CH_3)_3CNH$–CH$_2$–CH(Ph)–CH$_2$OH + $SOCl_2$

1

$(CH_3CH_2)_3N$
hexane, 25 °C

3 + **4**
($C_{12}H_{17}NO_2S$) ($C_{12}H_{17}NO_2S$)

No: Ph, N, $C(CH_3)_3$ **2**

PROBLEM 25.46 Provide a mechanism for the following reaction. *Hint*: What is :CHCl? What do such species do? See Chapter 10.

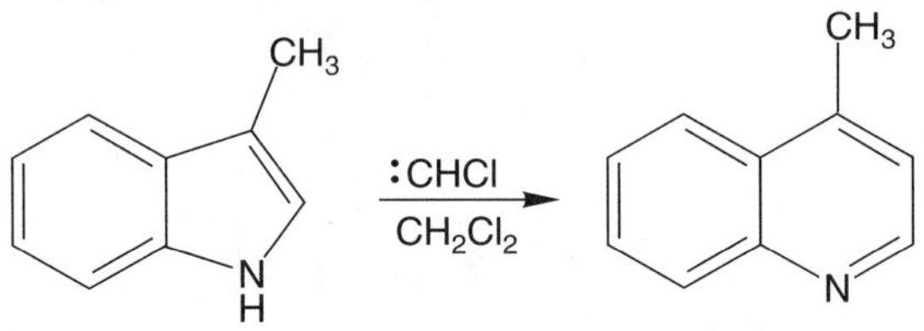

CH_3, N–H → (:CHCl, CH_2Cl_2) → CH_3, N

PROBLEM 25.47 Provide a mechanism for the following reaction:

H_3C–C(=O)–CH$_2$CH$_2$–C(=O)–CH_3 + cyclopentyl–NH_2

CH_3COOH
Δ

H_3C, N, CH_3

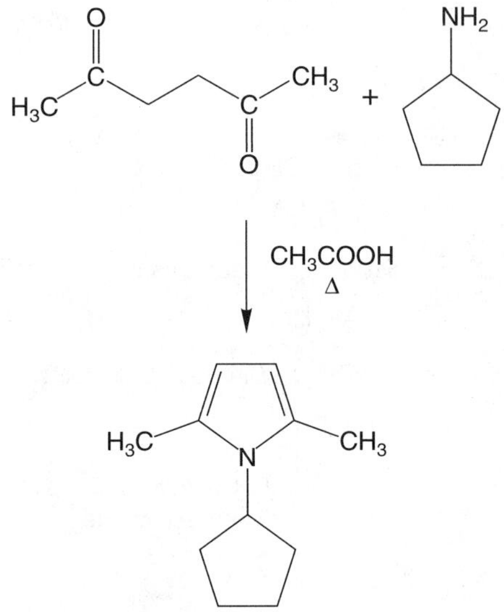

PROBLEM 25.48 In Section 25.4 (see Fig. 25.32), we saw how 1,3-dithianes function as masked carbonyl groups or as α-anion equivalents. Other heterocycles can be used for similar purposes. For example, 2-oxazoline **1**, a mole-

cule readily available from acetic acid, affords the lithium salt **2** upon treatment with butyllithium. Alkylation with a variety of alkyl halides leads to the substituted oxazoline **3**. This process can be repeated to yield dialkyl oxazoline **4**. Acid hydrolysis of either **3** or **4** gives α-alkylcarboxylic acid **5**, or α,α-dialkylcarboxylic acid **6**, respectively.

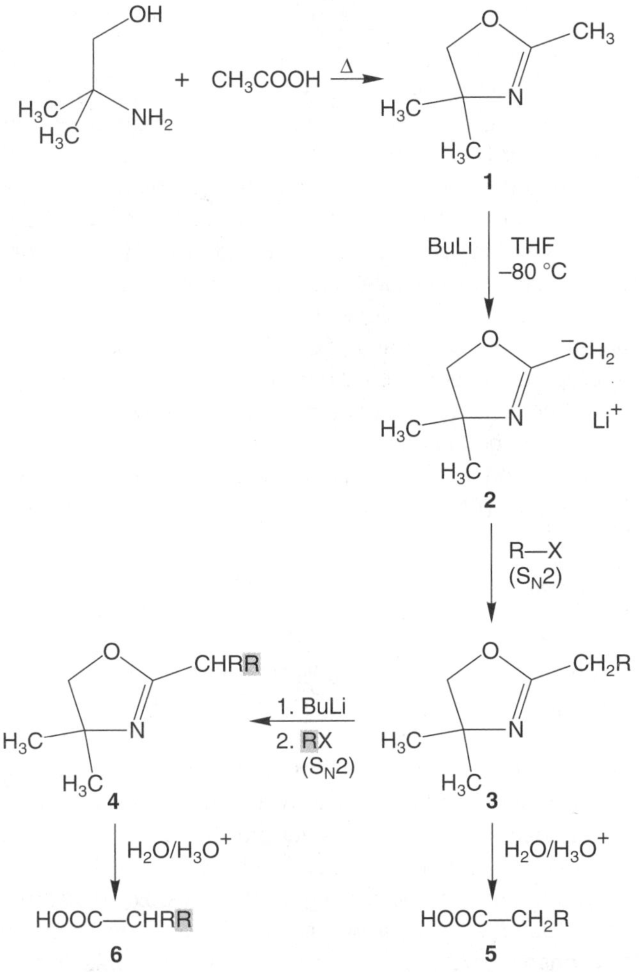

(a) Explain why the hydrogens of 2-oxazolines **1** and **3** are so acidic.
(b) Propose a mechanism for the acid hydrolysis of **3** to carboxylic acid **5**.
(c) What other synthetic method that we have already studied could be used to prepare carboxylic acids such as **5** and **6**?

PROBLEM **25.49** Isoquinoline (**1**) undergoes the Chichibabin amination reaction when treated with potassium amide in liquid ammonia to yield 1-aminoisoquinoline (**2**). When the reaction mixture containing **1**, potassium amide, and liquid ammonia was examined at −10 °C by ^{1}H NMR spectroscopy, the following signals were observed (solvent hydrogens and NH_2 hydrogens not reported):

δ 4.87 (d, *J* = 5.5 Hz, 1H), 5.34 (t, *J* = 7.0 Hz, 1H; signal collapses to a singlet when amide ion is present in excess), 6.35–7.3 (br m, 5H)

Rationalize this low-temperature ^{1}H NMR spectrum. *Hint*: For reference sake, the ^{1}H NMR spectrum of isoquinoline itself (**1**) is summarized below.

δ 7.3–8.2 (m, H_4, H_5, H_6, H_7, H_8), 8.61 (d, *J* = 6 Hz, H_3), 9.33 (s, H_1).

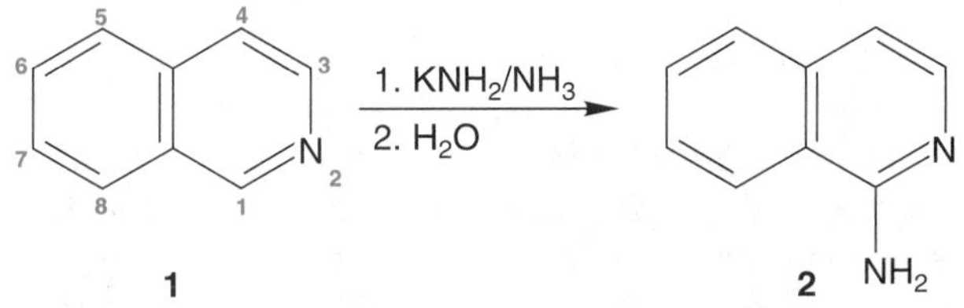

PROBLEM **25.50** In Section 25.5g, we examined the Hantzsch synthesis of pyridines (see Figs. 25.56–25.58). A conceptually related synthesis of pyridines involves the condensation of an α,β-unsaturated ketone (**1**) with a methyl ketone (**2**). This reaction gives a pyrilium salt (**3**), which then reacts with ammonia to produce a 2,4,6-trisubstituted pyridine.

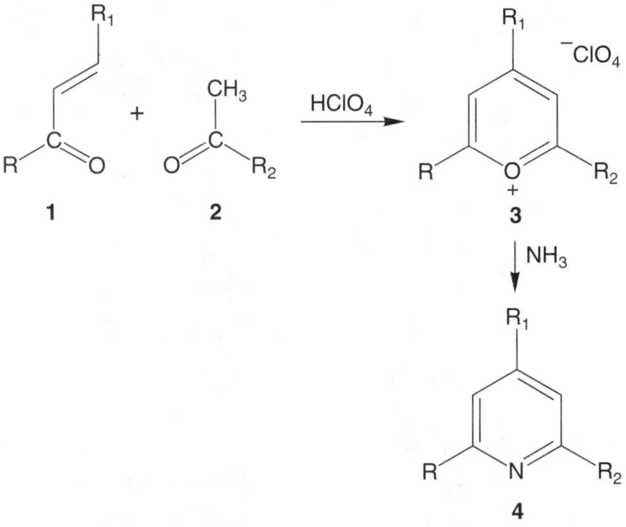

Propose mechanisms for the formation of pyrilium salt **3** from the condensation of **1** and **2**, and for the formation of the pyridine **4** from **3** and ammonia. *Hints*: The formation of **3** occurs best when 1.5–2.0 equivalents of **1** are used. The ketone **5** is a byproduct of the condensation; how can it be formed? Think "redox".

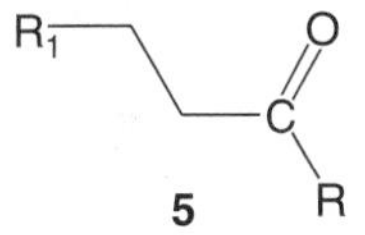

PROBLEM 25.51

(a) Reaction of pyridine *N*-oxide (**1**) and phenyl isocyanate (**2**) gives 2-anilinopyridine (**3**). Propose a mechanism for this reaction. *Hint*: Pyridine *N*-oxide can be viewed as a 1,3-dipole.

+ Ph—N═C═O (2) → DMF, 110 °C → NHPh (3)

1

(b) When 3,5-lutidine *N*-oxide (**4**) and phenyl isocyanate are allowed to react under comparable conditions, a 1:1 adduct **5** ($C_{14}H_{14}N_2O_2$) can be isolated. Spectral data for **5** are summarized below. Adduct **5** yields 2-anilino-3,5-lutidine (**6**) upon heating alone at 160 °C or on treatment with alcoholic KOH. Reduction of **5** with sodium borohydride in ethyl alcohol gives compound **7**, $C_{14}H_{16}N_2O_2$. Compound **7** displays an N—H stretch and a carbonyl stretch in its IR spectrum. Propose structures for 1:1 adduct **5** and compound **7**.

H_3C, CH_3 (4) + Ph—N═C═O (2) → DMF, 110 °C → 5

5 → 160 °C or KOH/CH_3CH_2OH, Δ → H_3C, CH_3, NHPh (6)

5 → $NaBH_4$ / CH_3CH_2OH → $C_{14}H_{16}N_2O_2$ (7)

Data for **5**

IR (KBr): 1725 cm^{-1}

^1H NMR: δ 1.60 (s, 3H), 1.93 (d, *J* = 1.8 Hz, 3H), 5.50 (d, *J* = 2.3 Hz, 1H), 5.92 (dq, 1H), 7.2–7.9 (m, 6H)

(c) Finally, 3,5-dibromopyridine *N*-oxide (**8**) reacts with phenyl isocyanate (**2**) again under comparable conditions, to yield the bromooxazolopyridine **9**. Modify your mechanism from (a) to account for the formation of 1:1 adduct **5** and oxazolopyridine **9**.

Ph—N═C═O (2) + Br, Br (8) → DMF, 110 °C → Br, O, C═O, N, N, Ph (9)

PROBLEM 25.52 In the good old days before IR and NMR spectroscopy, you might have been privileged to take a laboratory course called qualitative organic. In this course, a laboratory instructor would supply a series of unknowns, which you would attempt to identify. Typically, you would perform some "wet chemistry" on your unknown, including tests of solubility, qualitative elemental analysis, and functional group analysis. Ultimately, you would have to prepare one or two solid derivatives of your unknown so as to be able to compare melting points with those of known compounds in the literature. For example, if your unknown had been benzoylacetone (**1**), you might have attempted to prepare the oxime derivative **X**, as shown below.

Ph, C(═O), C(═O), CH_3 (1) → $HONH_2$/HCl, sodium acetate, H_2O/CH_3CH_2OH → X

You would have been gratified to obtain a colorless solid with a melting point of about 68 °C, after recrystallization. Upon checking your trusty, ever-present Table of Organic Derivatives of Ketones, however, you would have been dismayed to find that the melting point of this oxime was not listed. Bummer! Further investigation would have revealed the reason for the omission; your oxime (**X**) was not an oxime. Compound **X** has a molecular formula of $C_{10}H_9NO$, and shows neither an O—H nor C═O stretching band in the IR. Propose a structure for **X** and suggest a mechanism for its formation. Based on the available information, there are at least two viable structures for **X**. Either is acceptable.

PROBLEM 25.53 There is another way to form an oxime in Problem 25.52 other than by adding to one of the carbonyl groups. Your job is to find it. *Hints*: β-Diketones exist largely in their enol forms in solution. Remember the Michael reaction.

Introduction to Amino Acids and Polyamino Acids (Peptides and Proteins)

When I returned from the physical shock of Nagasaki ... , I tried to persuade my colleagues in governments and the United Nations that Nagasaki should be preserved exactly as it was then. I wanted all future conferences on disarmament ... to be held in that ashy, clinical sea of rubble. I still think as I did then, that only in this forbidding context could statesmen make realistic judgements of the problems which they handle on our behalf.

Alas, my official colleagues thought nothing of my scheme. On the contrary, they pointed out to me that delegates would be uncomfortable in Nagasaki.

—Jacob Bronowski*

For many people, fascination with science depends on the relationship between the material they study and biological processes. This connection seems natural as no one can deny the excitement of today's incredible pace of discovery in biological chemistry and molecular biology. It's quite impossible to be a decent molecular biologist without being first a chemist. Organic chemistry is particularly, if not uniquely, important. Molecular biology is organic chemistry (and inorganic and physical chemistry) applied to molecules of a scale not generally encountered in traditional, "small molecule" organic chemistry.†

Size, and the attendant complexity of structure, generates genuinely new chemistry. We have already seen how important molecular shape is in small molecules, and now, in molecules of biological size, stereo-

* Jacob Bronowski (1908–1974) was a mathematician, poet, and playwright. He was the author and compelling narrator of the brilliant television series, *The Ascent of Man.* Don't miss it if it is replayed.

† Scales change depending on your point of view. To a theoretician, a molecule with 10–20 non-hydrogen "heavy atoms" is often too large to be easily approached. Daunting problems are presented, if not of intellectual content, certainly of cost. Even with the decrease in price of computer time, calculations on molecules of "biological" size often simply cost too much. To a molecular biologist, a molecule of only 10 heavy atoms would be laughably tiny —hardly visible on the scale of proteins and enzymes.

chemistry becomes more complex and even more vital to understanding mechanism. To a great extent, biological action depends on shape. Much of the immense architecture of amino acid polymers serves "only" to bring a reactive group to an appropriate point on a substrate molecule, which is itself held by the enzyme in just the right position for reaction.

This chapter can only begin to introduce this subject and attempt to show you how important organic chemistry is to it. The choice of subjects is certainly arbitrary—there is a vast number of possible topics. This chapter is held together, somewhat tenuously it must be admitted, by nitrogen atoms. Nitrogen atoms are always components of the alkaloids we saw in Chapter 21, and of amino acids, the subjects of this chapter.

26.1 AMINO ACIDS

26.1a Nomenclature

Of course there is a systematic nomenclature for amino acids, but as we have so often seen, the real world eschews the system and goes right on using familiar old common names. The generic term "amino acid" itself is shorthand for α-**amino acid**. It is the α that locates the amino group on the carbon adjacent to the carboxylate carbon, and distinguishes the α-amino acids from the isomeric, but less important β and γ compounds (Fig. 26.1).

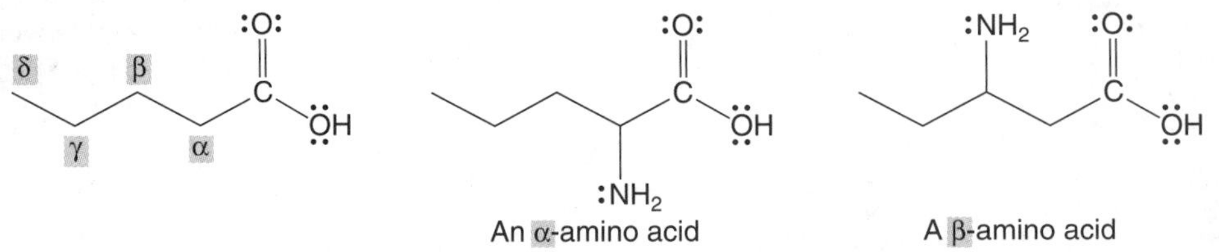

δ β γ α :O: C ÖH

:O: C ÖH :NH₂ An α-amino acid

:NH₂ :O: C ÖH A β-amino acid

FIGURE 26.1 The amino group can be located on the hydrocarbon chain with a Greek letter, α, β, γ, or δ.

In another naming system, we use a "2" instead of the "α" and name the α-amino acids as members of a series of 2-amino-substituted carboxylic acids (Fig. 26.2).

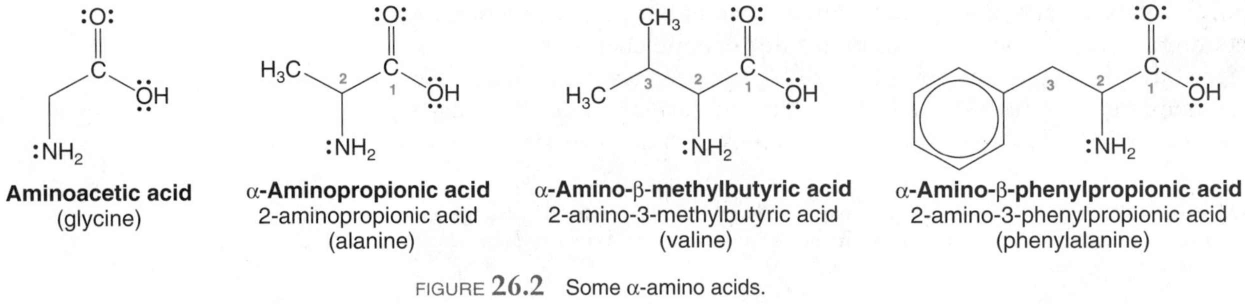

Aminoacetic acid
(glycine)

α-Aminopropionic acid
2-aminopropionic acid
(alanine)

α-Amino-β-methylbutyric acid
2-amino-3-methylbutyric acid
(valine)

α-Amino-β-phenylpropionic acid
2-amino-3-phenylpropionic acid
(phenylalanine)

FIGURE 26.2 Some α-amino acids.

In practice, the systematic name is rarely used, and common names are retained for most of these molecules. Biochemists tend to view amino acids as substituted α-aminoacetic acids. Note that as long as R doesn't equal H, the amino acids contain a stereogenic carbon atom (Fig. 26.3).

In addition, three-letter codes have been developed for the amino acids and are very commonly used, especially when describing the polymeric collections of amino acids known as poly**peptides** or in especially large examples, as **proteins**. The boundary between the smaller peptides, or "polypeptides," and the larger proteins is neither fixed nor always clear, and the terms are sometimes used interchangeably. More recently, an even shorter set of one-letter abbreviations has come into use. Table 26.1 collects the 20 α-amino acids, gives their common names and abbreviations, and collates some of their physical properties. Notice that R groups of many kinds appear, from simple hydrocarbon chains to more complicated acidic and basic groups.

Why pick out this particular set of 20 from the vast number conceivable? These are the 20 amino acids encountered in peptides and proteins. Your body can synthesize one-half of these 20 from starting materials provided by the food you eat. One might think that a reasonable way for your insides to proceed would be to break down ingested amino acid polymers to their constituent amino acids, and then use these directly. But this is not how you do it. Instead, amino acids are metabolized (broken down chemically) into much simpler molecules from which the amino acids you need are resynthesized. But even your body is not a perfect synthetic chemist—unless you are a superior mutant, there are 10 of these 20 amino acids it is unable to make. These you must ingest directly, and if your diet fails to include one or more of these **essential amino acids** you will not last long. The essential amino acids are marked with an asterisk in Table 26.1.

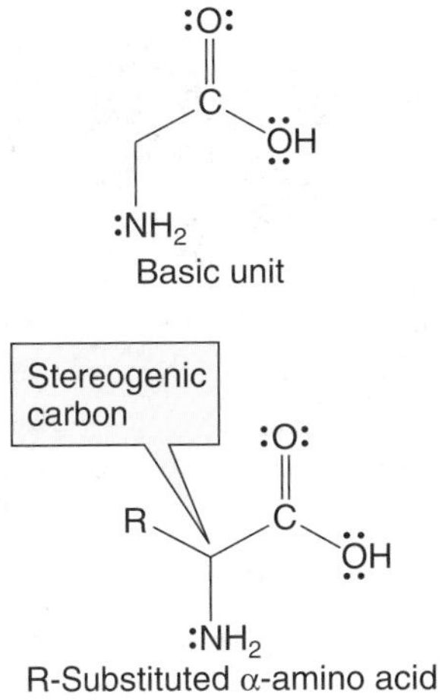

Basic unit

Stereogenic carbon

R-Substituted α-amino acid

FIGURE 26.3 An α-amino acid is just an "R-substituted" 2-amino acetic acid.

26.1b Structure of Amino Acids

All but 1 of the 20 common amino acids in Table 26.1 is a primary amine. The only exception is proline, an amino acid containing a secondary amine (Fig. 26.4). All biologically important amino acids but the achiral glycine and one other are found in the (*S*) configuration. Many (*R*) amino acids are known in nature, but most play no important role in human biochemistry. Why? That's a good question and no certain answer exists. Did the first resolution produce an (*S*) amino acid, which then determined the sense of chirality of all the amino acids formed thereafter? Is the preference for (*S*) no more than the result of some primeval accident? No one knows. If it is, we may one day encounter (*R*) civilizations somewhere in the universe, provided we are sufficiently lucky and clever to survive long enough.

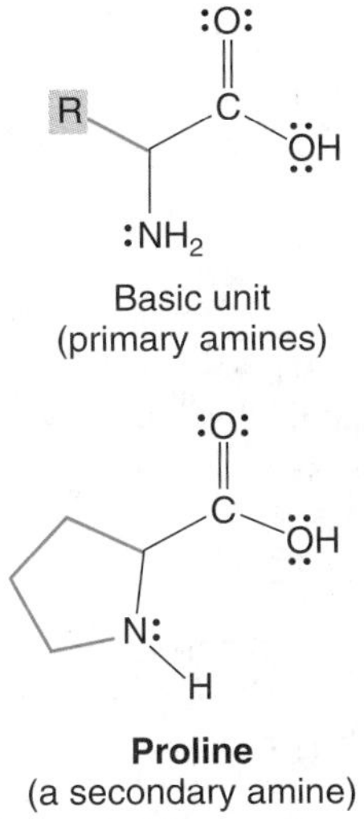

Basic unit (primary amines)

Proline (a secondary amine)

FIGURE 26.4 Of the 20 common α-amino acids, only proline is not a primary amine.

*PROBLEM 26.1

Draw tetrahedral three-dimensional structures for (*R*) and (*S*) valine and aspartic acid.

ANSWER

Draw the tetrahedron, assign priorities (Chapter 5, p. 160), and draw the arrow 1→2→3. If it runs clockwise, the molecule is (*R*); if counterclockwise, it is (*S*).

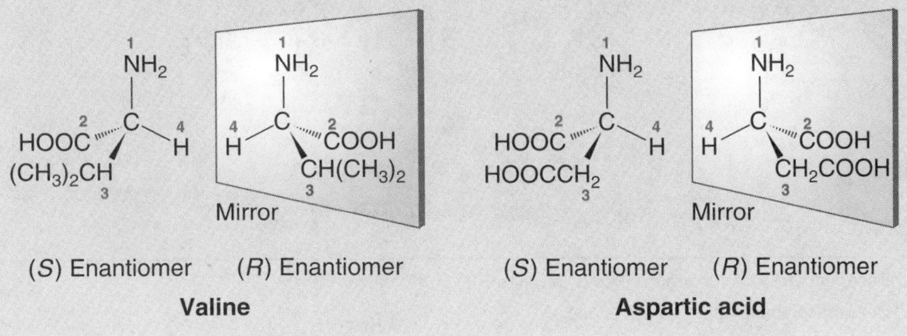

Mirror

(*S*) Enantiomer (*R*) Enantiomer

Valine

Mirror

(*S*) Enantiomer (*R*) Enantiomer

Aspartic acid

TABLE **26.1** Properties of 20 Common Amino Acids

R	Common[a] Name	Abbreviations	mp[b] (°C) L-Form	pK_a (COOH)	pK_a ($^+NH_3$)	pK_a (Other)	Isoelectric Point pH
Simple alkane groups							
H	Glycine	Gly, G	229 (dec)	2.3	9.6		6.0
CH_3	Alanine	Ala, A	297 (dec)	2.3	9.7		6.0
$CH(CH_3)_2$	Valine*	Val, V	315 (dec)	2.3	9.7		6.0
$CH_2CH(CH_3)_2$	Leucine*	Leu, L	337 (dec)	2.3	9.6		6.0
$CHCH_3(CH_2CH_3)$	Isoleucine*	Ile, I	284 (dec)	2.3	9.6		6.0
Aromatic groups							
H_2C– (benzene ring)	Phenylalanine*	Phe, F	283 (dec)	1.8	9.2		5.9
H_2C– (benzene ring) –OH	Tyrosine	Tyr, Y	316 (dec)	2.2	9.1	10.1	5.7
H_2C– (imidazole ring: N, N–H)	Histidine*	His, H	288 (dec)	1.8	9.0	6.0	7.6
H_2C– (indole ring: N–H)	Tryptophan*	Trp, W	290 (dec)	2.4	9.4		5.9
Groups containing an OH (see also Tyr)							
CH_2OH	Serine	Ser, S	228 (dec)	2.2	9.2		5.7
$CH(CH_3)OH$	Threonine*	Thr, T	226 (dec)	2.2	9.2		5.6
Groups containing S							
CH_2SH	Cysteine	Cys, C		1.7	10.8	8.3	5.0
$CH_2CH_2SCH_3$	Methionine*	Met, M	283 (dec)	2.3	9.2		5.8
Groups containing a carbonyl group							
CH_2COOH	Aspartic Acid	Asp, D	270 (dec)	1.9	9.6	3.7	2.9
CH_2CH_2COOH	Glutamic Acid	Glu, E	211 (dec)	2.2	9.7	4.3	3.2
CH_2CONH_2	Asparagine	Asn, N	214 (dec)	2.0	8.8		5.4
$CH_2CH_2CONH_2$	Glutamine	Gln, Q	185 (dec)	2.2	9.1		5.7
Groups containing an amino group (see also Trp and His)							
$(CH_2)_4NH_2$	Lysine*	Lys, K	224 (dec)	2.2	8.9	10.3	9.7
$(CH_2)_3NH{-}C{=}NH(NH_2)$	Arginine*	Arg, R	207 (dec)	2.2	9.1	12.5	10.8
A secondary amine							
(pyrrolidine ring: N–H) COOH	Proline	Pro, P	221 (dec)	2.0	10.6		6.1

[a] Essential amino acids are marked by an asterisk (*).

[b] Decomposition is abbreviated as dec.

Amino acids are usually not described as (*S*) or (*R*). Instead, the Fischer projection system is used (Chapter 24, p. 1236). The acid group is drawn at the top, and, as with the sugars, vertical bonds are retreating from you and horizontal bonds are coming toward you. The position, left or right, of the amino group determines whether the amino acid is D (right) or L (left). As shown in Figure 26.5, a simple (*S*) amino acid is also L in the terminology used for carbohydrates.

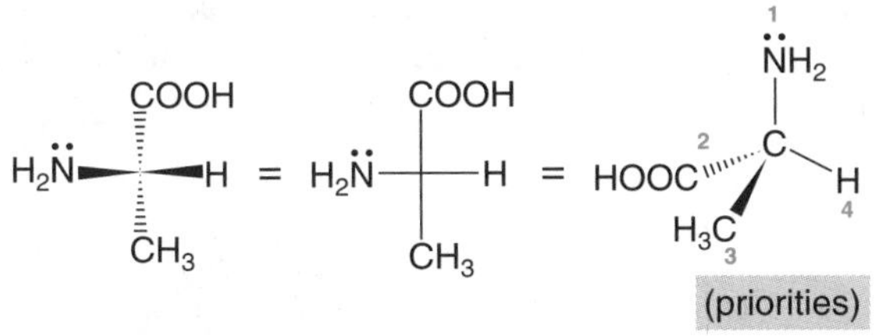

(priorities)

Fischer projections of L-alanine = 2-(*S*)-alanine

FIGURE 26.5 An (*S*) amino acid is L in the Fischer terminology.

*PROBLEM 26.2 Draw the L-isomers of valine and cysteine in Fischer projection.

ANSWER Remember the convention for Fischer projections. Horizontal bonds are coming toward you and vertical bonds are retreating from you.

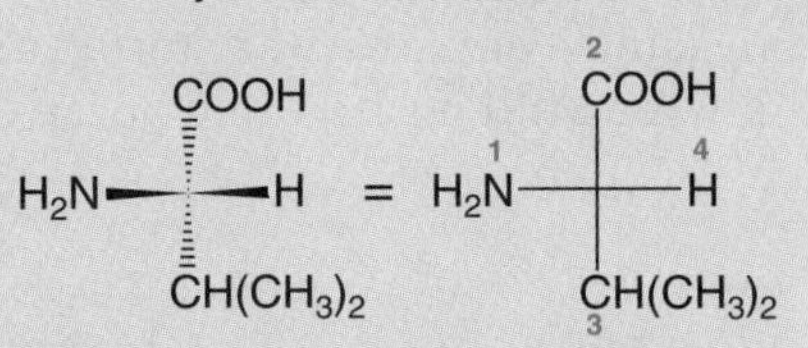

L-Valine is the (*S*) enantiomer

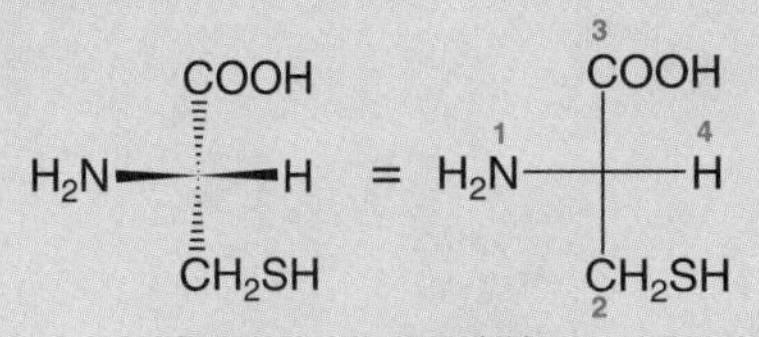

L-Cysteine is the (*R*) enantiomer

Canavanine

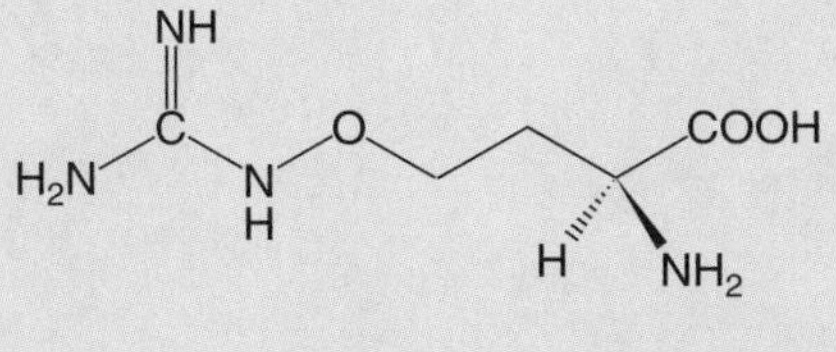

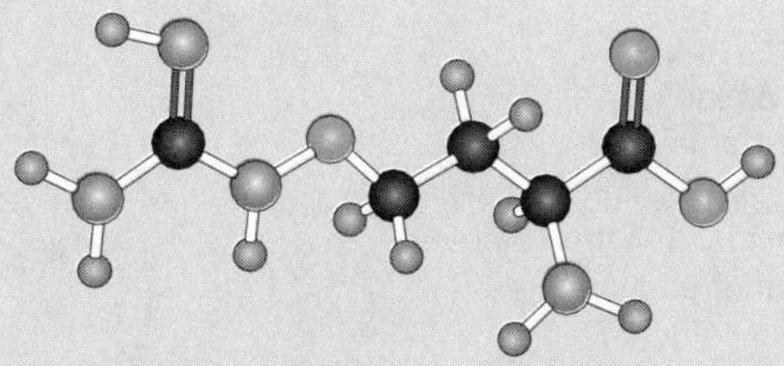

We tend to think of plants as passive defenseless creatures, growing quietly at the mercy of herbivorous predators. But this seems odd; shouldn't evolutionary pressure lead to defense mechanisms in which the poor plant finds a way to prevent animals from eating it? Indeed it has, and plants are full of all sorts of toxins evolved to discourage predators. Of course, this leads to evolutionary pressure on the predator population, and sometimes one species will work its way around the plant's defenses. The signal that this has happened is the presence of a single predator. All others have been discouraged, but one has found a way to defeat the plant's defense system. Take, for example, *Dioclea megacarpa,* a vine whose only predator is the beetle *Caryedes brasiliensis.* This plant uses canavanine in place of the closely related arginine (replace the O in canavanine with CH_2 and you get arginine). In most herbivores, replacement of arginine with canavanine leads to incorrect protein folding and arrested development. The single successful beetle has developed an enzyme that can tell the difference between arginine and canavinine, as well as an enzyme specially designed to destroy canavanine. So, *C. brasiliensis* happily munches away on *Dioclea* while all other beetles stay away.

As you can see from Table 26.1, amino acids are high-melting solids. Does this not seem curious? Common amino acids have molecular weights between 90 and 150, and this does not seem sufficient to produce solids. This conundrum leads directly to Section 26.1c.

26.1c Acid–Base Properties of Amino Acids

Why are amino acids solids? To answer this question, let's ask another. What reaction do you expect between the generic amine and the generic acid of Figure 26.6?

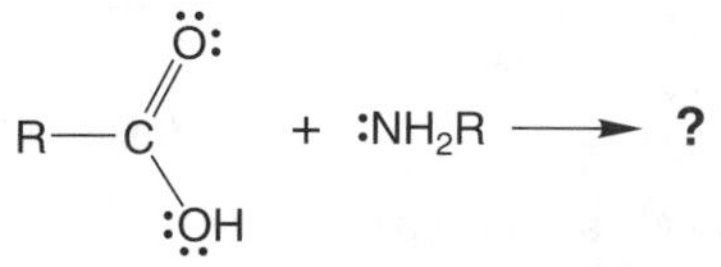

FIGURE 26.6 What reaction must occur between these two molecules?

A simple organic acid has a pK_a of about 4.5, and an ammonium ion has a pK_a of about 8–10. An amine and an acid must react to give the ammonium salt of Figure 26.7, as the carboxylic acid is a much stronger acid than an ammonium ion. Exactly the same reaction must take place within a single amino acid to form the "inner salt" or **zwitterion**, a molecule in which full positive and negative charges coexist (Fig. 26.7).

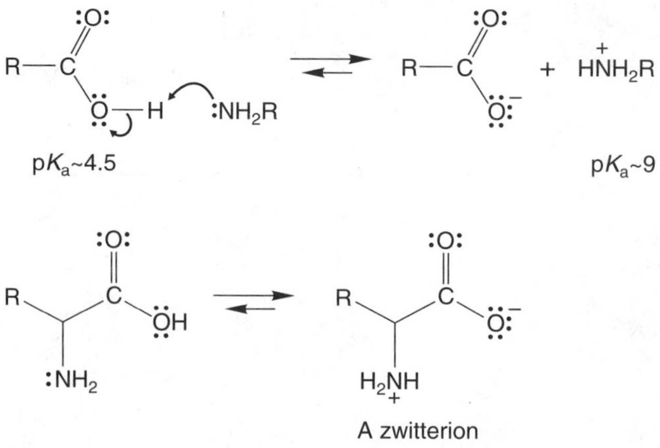

FIGURE 26.7 An amine and a carboxylic acid must undergo acid–base chemistry to give an ammonium salt of the acid. In amino acids, an intramolecular version of the same reaction gives a zwitterion.

Amino acids are ionic compounds under normal conditions, and that accounts for their high melting points. Other properties are in accord with this notion. They are miscible with water, but quite insoluble in typical nonpolar organic solvents. They have high dipole moments. Of course, the charge state of an amino acid depends on the pH of the medium in which it finds itself. Under neutral, or nearly neutral conditions they will exist as the zwitterions of Figure 26.7. If we lower the pH, eventually the carboxylate anion will protonate, and the amino acid will exist primarily in the ammonium ion form. Conversely, if we raise the pH, eventually the ammonium ion site will be deprotonated, and the molecule will exist as the carboxylate anion (Fig. 26.8).

Amino acids can be described by a series of pK_a values. The first refers to the deprotonation of the acid function, and, as you can see from Table 26.1, is in the range 1.7–2.6 for the 20 commonly encountered molecules.

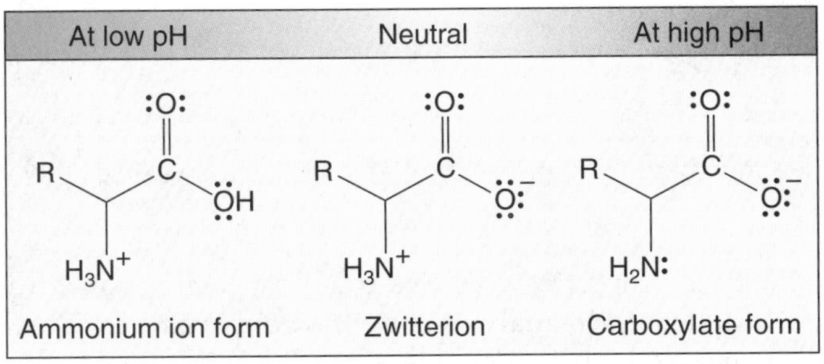

FIGURE 26.8 At low pH, the zwitterion will protonate to give a net positively charged ion. At high pH, a carboxylate anion, net negatively charged, will be formed.

PROBLEM 26.3

The pK_a for a simple organic acid is 4–5. Why are amino acids so much more acidic?

The second pK_a value is for deprotonation of the ammonium ion ($^+NH_3CHRCOOH$), and is in the range 9–10.8 (Table 26.1).

The pH at which the concentration of the ammonium ion ($^+NH_3CHRCOOH$) equals that of the carboxylate ion ($NH_2CHRCOO^-$) is called the **isoelectric point**.

PROBLEM 26.4

Calculate the pH of the isoelectric point for alanine (pK_a = 2.3 for COOH and pK_a = 9.7 for $^+NH_3$). *Hint*: Write expressions for the K_a for each of the two equilibria involved, then multiply them. Remember that at the isoelectric point the concentrations of $^+NH_3{-}CHR{-}COOH$ and $H_2N{-}CHR{-}COO^-$ are equal.

At this pH the amino acid is primarily in the zwitterionic form, and the number of positively charged species exactly equals the number of negatively charged molecules.

As you can see from Table 26.1, many amino acids bear acidic or basic side chains, which complicates the picture somewhat, and can make the determination of the isoelectric point a bit more difficult. These side groups are also involved in acid–base equilibria, and there will be pK_a values associated with them. For example, arginine will be doubly protonated at very low pH (Fig. 26.9).

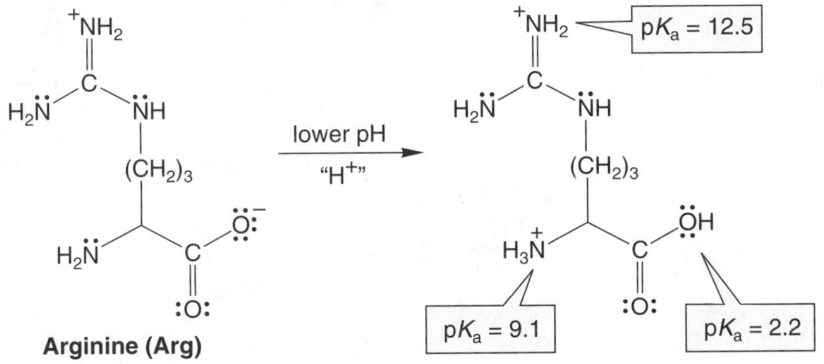

FIGURE 26.9 Arginine is twice protonated at low pH.

The most acidic group is the carboxylic acid, pK_a = 2.2. Second is the ammonium ion, pK_a = 9.1. The guanidinium ion has a pK_a of 12.5 and will be the last position deprotonated as the pH is raised.

PROBLEM **26.5** In the guanidino group of arginine, where is the site of initial protonation?

PROBLEM **26.6** Why is the five-membered ring (an imidazole) in histidine so acidic (Table 26.1)?

These acid–base properties give rise to an especially powerful separation method widely used in analysis of amino acid mixtures. This method is called **electrophoresis**. It works this way. A mixture of amino acids is placed on moistened paper. At a given pH, different amino acids will exist as different mixtures of positive and negative ions. When a current is applied to the paper, the positively charged ions will migrate to the cathode (negative pole) and the negatively charged ions to the anode (positive pole) at rates depending on the pH and the strength of the field, which are controllable variables, and the size and net electric charge of the amino acid, which are individual properties of the individual amino acids. The net result is a separation, as the amino acids migrate at different rates. For example, at pH = 5.5, the amino acids Lys, Phe, and Glu exist in the forms shown in Figure 26.10.

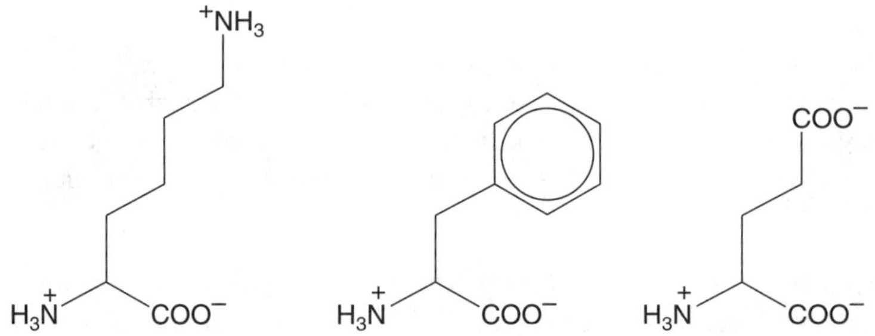

Lys (cation = 1^+) **Phe** (neutral) **Glu** (anion = 1^-)

FIGURE **26.10** The charge states of Lys, Phe, and Glu at pH 5.5.

When a current is applied, Phe will not move, Lys will migrate to the cathode and Glu to the anode (Fig. 26.11).

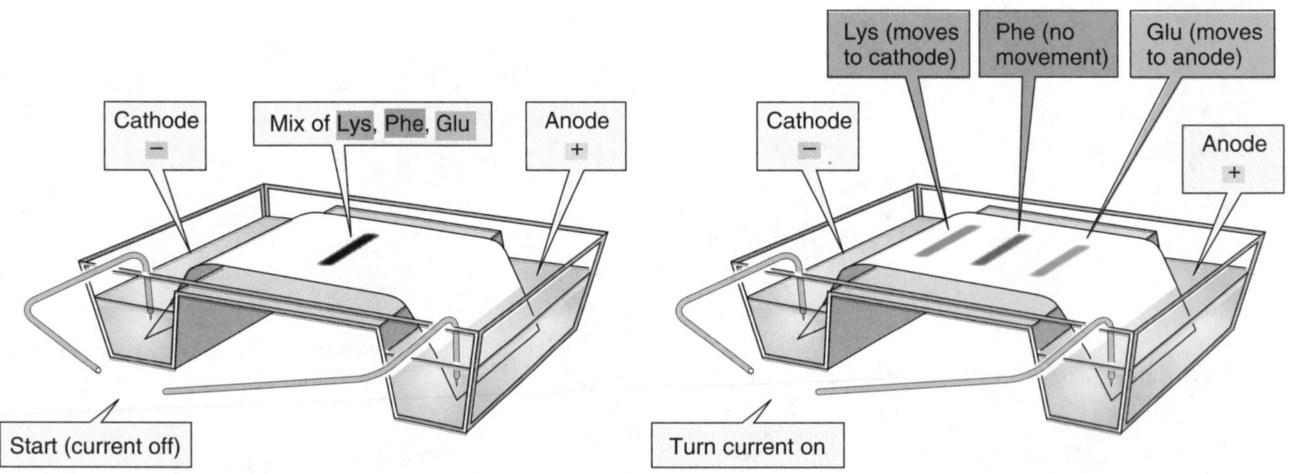

FIGURE **26.11** Separation of three amino acids by electrophoresis.

26.1d Syntheses of Amino Acids

One very simple method is to alkylate ammonia using the S_N2 reaction of ammonia with an α-halo acid (Fig. 26.12).

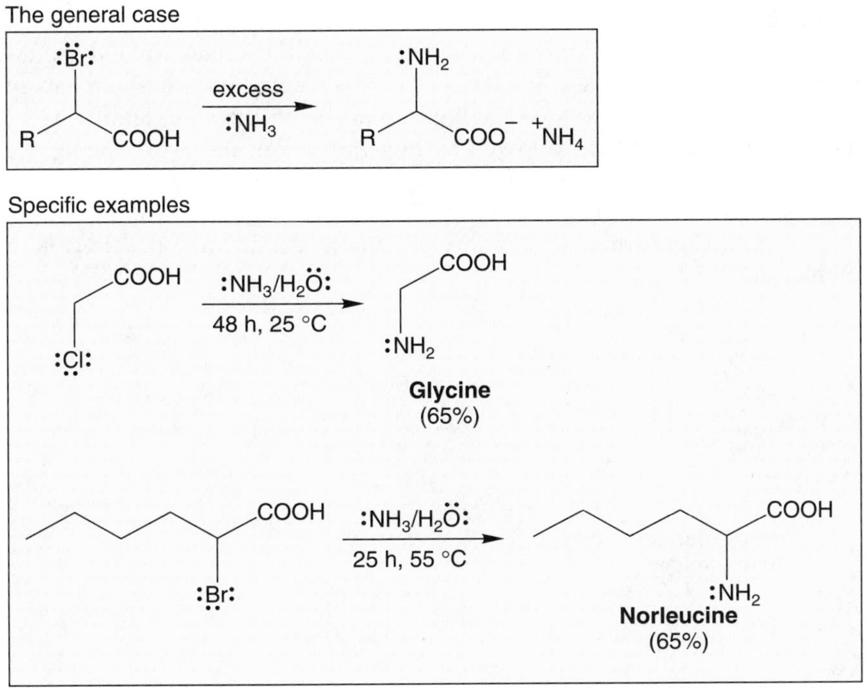

The general case

Specific examples

Glycine (65%)

Norleucine (65%)

FIGURE **26.12** Simple alkylation of an α-halo acid will give an α-amino acid.

In Chapter 21 (p. 1103), a closely related alkylation method appeared as a means of producing alkylamines. A problem for this synthetic procedure was the ease of overalkylation. The simple alkylation reaction is not generally a practical method for producing alkylamines. It is rather more useful in making α-amino acids. The α-bromo acids are easily available, and a vast excess of ammonia can be used to minimize overalkylation. Moreover, the first formed α-amino acids are not especially effective competitors in the S_N2 reaction. They are sterically congested, and less basic than simple amines. This technique is an effective way of making both "natural" and "unnatural" amino acids.

PROBLEM **26.7**

Devise simple syntheses of the following α-bromo acids starting from alcohols containing no more than four carbon atoms and inorganic reagents of your choice (Fig. 26.13) (see Section 19.6b, p. 987).

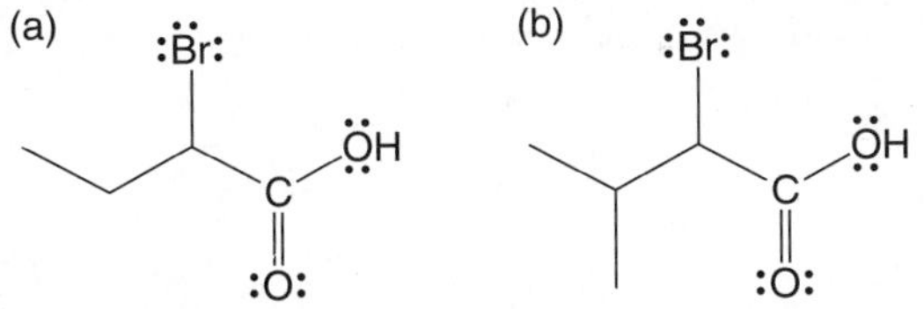

(a) (b)

FIGURE **26.13**

PROBLEM **26.8**

Why are α-bromo acids (or their carboxylate salts) more easily alkylated than simple alkyl bromides in the S_N2 reaction (Chapter 12, p. 533)?

As seen in amine chemistry, a successful variation of the simple alkylation synthesis involves a protected, disguised amine, and is called the Gabriel synthesis, or, sometimes, the **Gabriel malonic ester synthesis**. In this procedure, bromomalonic ester is first treated with potassium phthalimide in an S_N2 displacement of bromide ion. The nitrogen atom destined to become the amine nitrogen of the amino acid is introduced at this point, and rendered nonnucleophilic by the flanking carbonyl groups. In the next step, the malonic ester product is treated with base and an active alkyl halide to introduce the R group of the amino acid. Finally, treatment with acid or base, followed by neutralization, uncovers the amine. The phthalimide group is hydrolyzed to phthalic acid, and the malonic ester is hydrolyzed to a malonic acid. Gentle heating induces decarboxylation to give the amino acid. There are many variations on this general theme (Fig. 26.14).

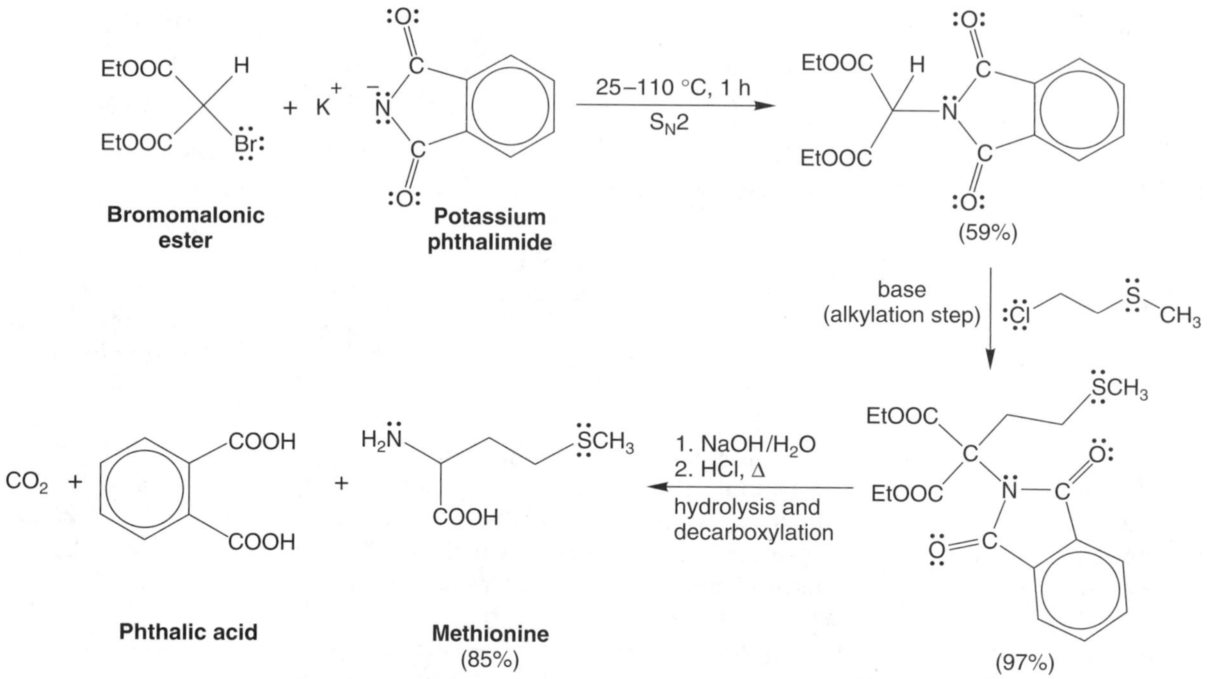

EtOOC H EtOOC Br: + K^+ Potassium phthalimide → 25–110 °C, 1 h, S_N2 → (59%)

Bromomalonic ester

base (alkylation step) $:\ddot{Cl}$–CH_2CH_2–$\ddot{S}$–CH_3

CO_2 + Phthalic acid (COOH, COOH) + Methionine (85%) ($H_2\ddot{N}$, COOH, $\ddot{S}CH_3$) ← 1. $NaOH/H_2O$ 2. HCl, Δ; hydrolysis and decarboxylation ← (97%)

FIGURE **26.14** The Gabriel malonic ester synthesis.

PROBLEM **26.9** Provide detailed mechanisms for the reactions at the bottom of Figure 26.14.

*PROBLEM **26.10** In a simpler version of this reaction, acetamidomalonic ester is used (Fig. 26.15). Sketch out the steps in this related amino acid synthesis.

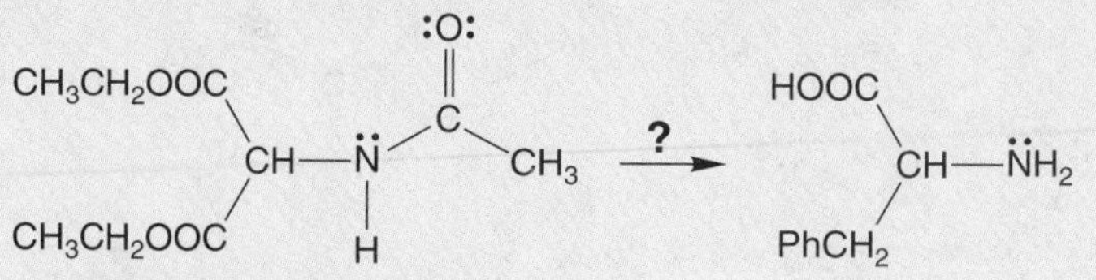

CH_3CH_2OOC, CH_3CH_2OOC, CH—$\ddot{N}$(H)—C(=$\ddot{O}$)—CH_3 → ? → HOOC, PhCH$_2$, CH—$\ddot{N}H_2$

FIGURE **26.15** **Acetamidomalonic ester** **Phenylalanine**

ANSWER

It is basically the same process as that of Figure 26.14 and Problem 26.9. Advantage is first taken of the acidity of the doubly α hydrogen to introduce an R group.

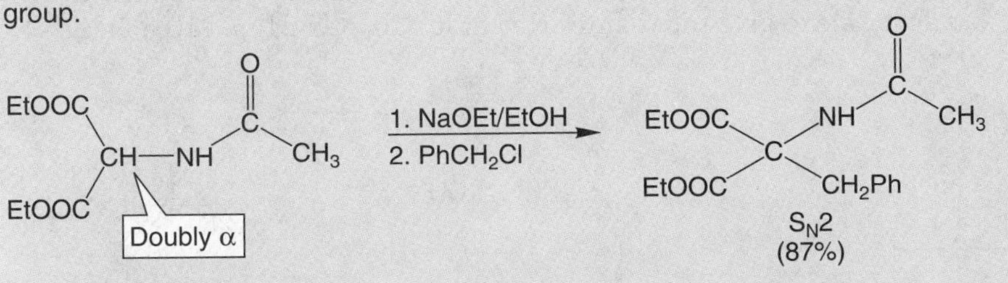

EtOOC / CH—NH—C(=O)—CH_3 / EtOOC (Doubly α) → 1. NaOEt/EtOH 2. $PhCH_2Cl$ → $(EtOOC)_2C(NH—C(=O)—CH_3)(CH_2Ph)$ S_N2 (87%)

Now acid (or base) is used to convert the esters into carboxylic acids, and to hydrolyze the amide. Heating decarboxylates the malonic acid and liberates the amino acid.

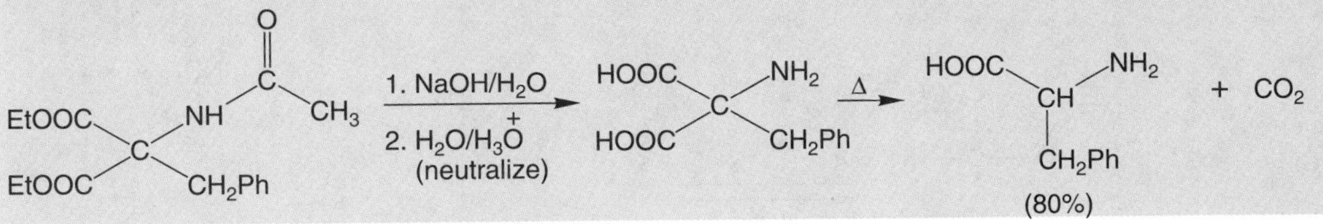

$(EtOOC)_2C(NH—C(=O)—CH_3)(CH_2Ph)$ → 1. $NaOH/H_2O$ 2. $H_2O/H_3\overset{+}{O}$ (neutralize) → $(HOOC)_2C(NH_2)(CH_2Ph)$ →Δ→ HOOC—CH(NH_2)—CH_2Ph (80%) + CO_2

In the **Strecker synthesis** (Adolph Strecker, 1822–1871), a cyanide is used as the source of the acid portion of an amino acid. An aldehyde is treated with ammonia, the ultimate source of the amino group, and cyanide ion. In the first step of the reaction, a carbinolamine is formed (Chapter 16, p. 780) and water is lost to give an imine. The imine is then attacked by cyanide to give an α-amino cyanide. Acid hydrolysis of the nitrile gives the acid (Chapter 20, p. 1032; Fig. 26.16).

The general case

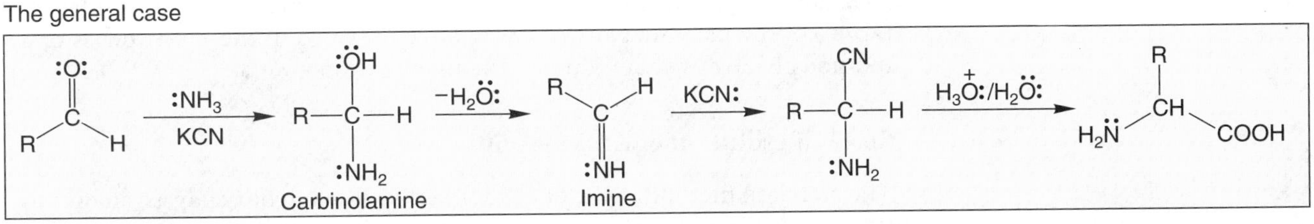

R—C(=O:)—H →$:NH_3$, KCN→ R—C(ÖH)(H)—$:NH_2$ (Carbinolamine) →$-H_2\ddot{O}:$→ R(H)C=:NH (Imine) →KCN:→ R—C(CN)(H)—$:NH_2$ →$H_3\overset{+}{\ddot{O}}:/H_2\ddot{O}:$→ $H_2\ddot{N}$—CH(R)—COOH

A specific example

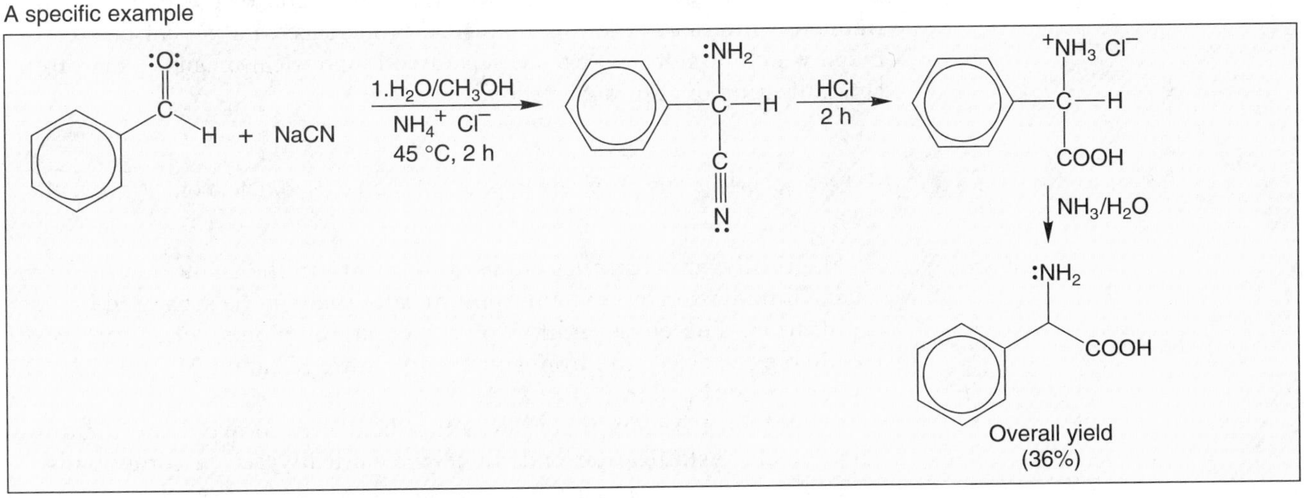

Ph—C(=O:)—H + NaCN →1. H_2O/CH_3OH, $NH_4^+\ Cl^-$, 45 °C, 2 h→ Ph—C(:NH_2)(H)—C≡N: →HCl, 2 h→ Ph—C($^+NH_3\ Cl^-$)(H)—COOH →NH_3/H_2O→ Ph—CH(:NH_2)—COOH

Overall yield (36%)

FIGURE **26.16** The Strecker amino acid synthesis.

The first step in this reaction, imine formation, is analogous to the first step in the Mannich condensation in which the role of the amine is also to produce a reactive imine. This imine can next be attacked by the base cyanide to give the α-amino cyanide (Chapter 21, p. 1101; Fig. 26.17).

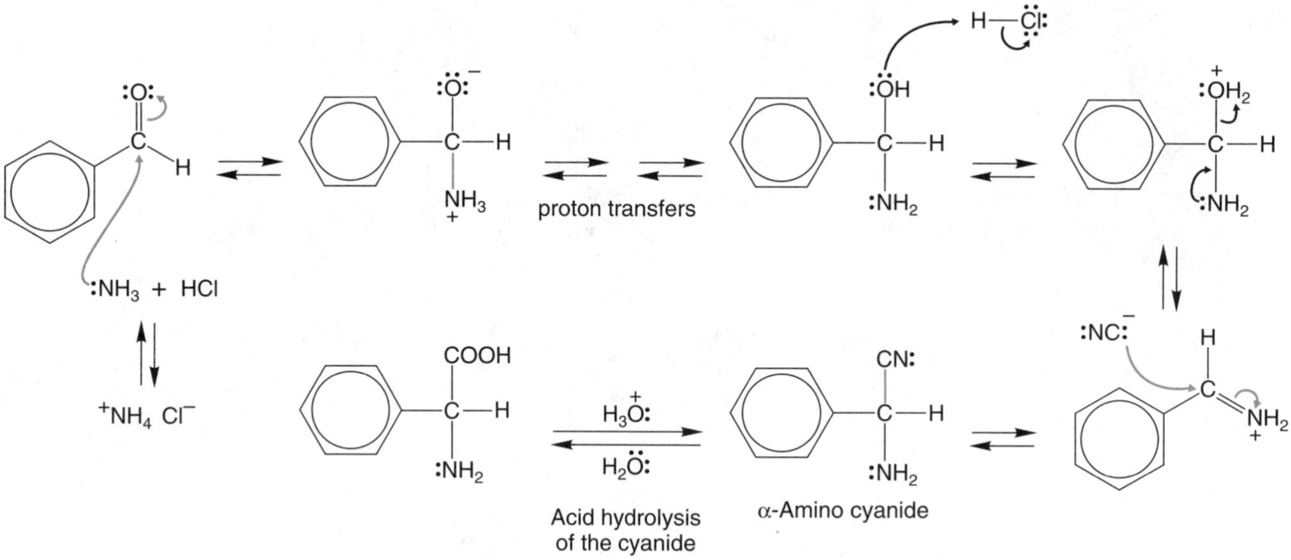

:NH$_3$ + HCl

$^+NH_4$ Cl^-

proton transfers

Acid hydrolysis of the cyanide

α-Amino cyanide

FIGURE 26.17 The mechanism of the Strecker synthesis.

As we have seen, amino acids in nature are almost all encountered in optically active (*S*), or L forms (save glycine, which is achiral). None of the syntheses outlined in this section automatically produces the natural (*S*) enantiomer specifically. So we either need a method for resolution (separation) of the racemic pairs of enantiomers formed in these syntheses or a method of selectively producing the (*S*) enantiomers.

26.1e Resolution of Amino Acids

The standard method is the one we first discussed in our chapter on chirality (Chapter 5, p. 177) and involves transformation of the pair of enantiomers, which have identical physical properties, to a pair of diastereomers, which do not, and can be separated by physical means (most often crystallization or chromatography).

PROBLEM 26.11 There actually is one physical property not shared by enantiomers. What is it?

Optically active amines are used to transform the enantiomeric amino acids into diastereomeric ammonium salts that can be separated by crystallization. The chiral amines of choice are alkaloids, which are easily available optically pure from sources in nature (Chapter 21, p. 1109). The process is outlined in Figure 26.18.

However, it seems that there must be a way to avoid the tedium of fractional crystallization and make the optically active compounds directly. After all, Nature does this somehow, and we should be clever enough to find out how this happens and imitate the process. One method

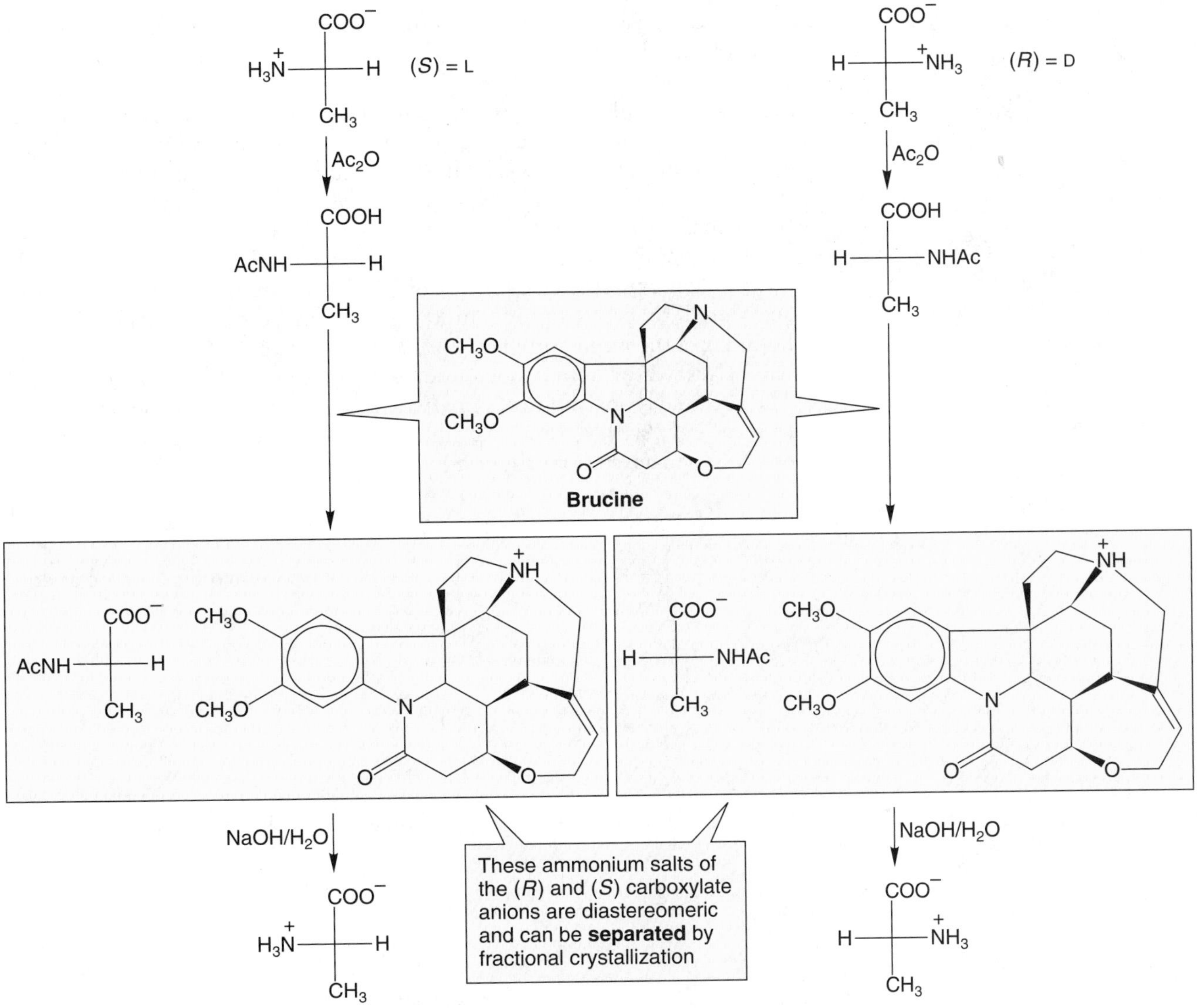

FIGURE 26.18 Separation, or resolution, of a pair of enantiomeric amino acids.

nature uses to make optically active amino acids involves a reductive amination (Chapter 21, p. 1106). Ammonia and an α-keto acid react in the presence of an enzyme and a source of hydride, usually NADH (Chapter 16, p. 802), to produce an optically active chiral amino acid (Fig. 26.19).

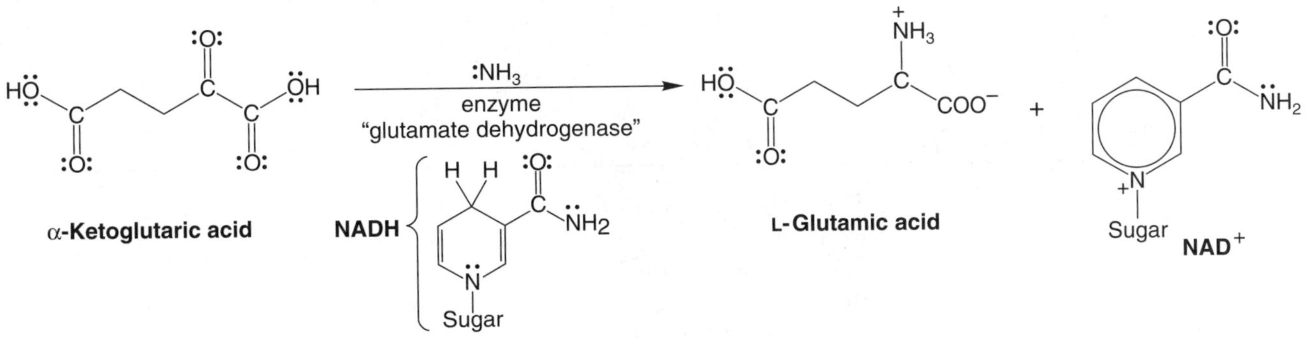

FIGURE 26.19 Synthesis of an optically active L-amino acid through reductive amination.

There are two basic imitative procedures. In one, actual microorganisms are used in fermentation reactions that can directly produce some L-amino acids.

In a more subtle imitative (biomimetic) procedure, enzymes are used to react with one enantiomer of a racemic pair of mirror-image molecules. As you might guess, it is usually the L isomer that is "eaten" by the enzyme, because enzymes have evolved in an environment of L isomers. So it is usually the D isomers that are ignored by enzymes. In an extraordinarily clever procedure, this preference for natural, L, enantiomers can be used to achieve a **kinetic resolution**. The mixture of enantiomers is first acetylated to create amide linkages. An enzyme, hog-kidney acylase, hydrolyzes the amide linkages *of only L-amino acids.* So, treatment of the pair of acetylated amino acids leads to formation of the free L-amino acid. The acylated D enantiomer is unaffected by the enzyme, which has evolved to react only with natural L-amino acids. The free L-amino acid can then easily be separated from the residual D acetylated material (Fig. 26.20).

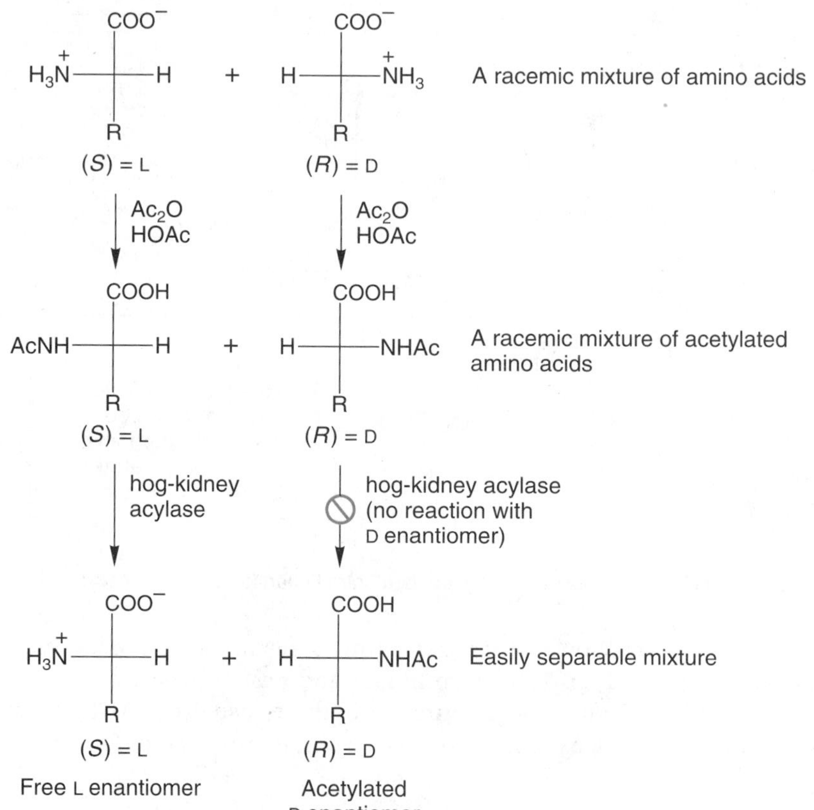

FIGURE **26.20** A kinetic resolution of a pair of enantiomeric amino acids.

26.2 REACTIONS OF AMINO ACIDS

The simple expectation that the chemical reactions of the amino acids would be a combination of acid (Chapters 19 and 20) and amine (Chapter 21) chemistry would be right. Indeed, we have already seen some of this in the acid–base chemistry described in Section 26.1c. A few more important and useful examples follow.

26.2a Acylation Reactions

The amino group of amino acids can be acylated using all manner of **acylating agents**. Typical examples of effective acylating agents are acetic anhydride, benzoyl chloride, and acetyl chloride (Fig. 26.21).

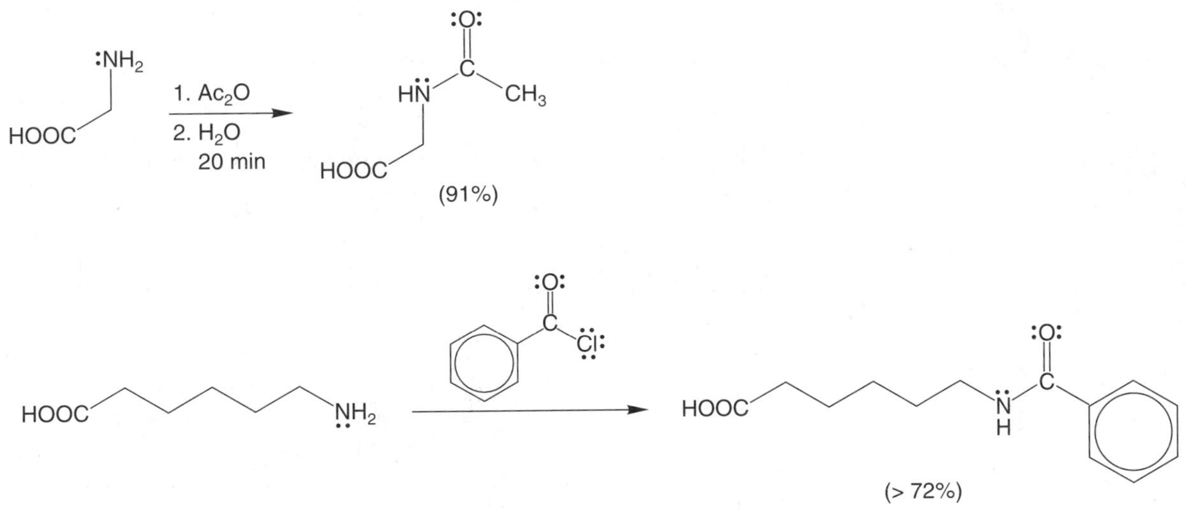

FIGURE 26.21 The amino group of an amino acid can be acylated.

26.2b Esterification Reactions

The acid function of amino acids can be esterified using typical esterification reactions. Fischer esterification is the simplest process (Fig. 26.22).

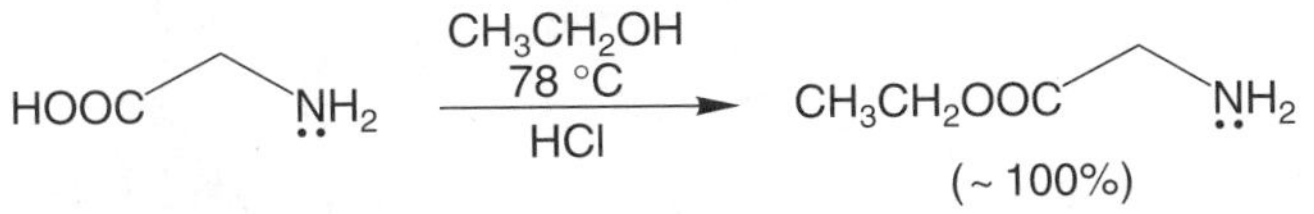

FIGURE 26.22 Fischer esterification of the acid group of an amino acid.

PROBLEM 26.12

Write a mechanism for the acylation reactions of Figure 26.21.

PROBLEM 26.13

If amino acids are typically in their zwitterionic forms, how are acylation and esterification reactions, which depend on the presence of free amine or acid groups, possible?

26.2c Reaction with Ninhydrin

Amino acids react with **ninhydrin** to form a deep purple molecule. It is this purple color that is used to detect amino acids. Although it may seem curious at first, all amino acids except proline, regardless of the identity of the R group, form the *same* purple molecule on reaction with ninhydrin. How can this be?

Ninhydrin is the hydrate of indan-1,2,3-trione, and like all hydrates, is in equilibrium with the related carbonyl compound (Fig. 26.23).

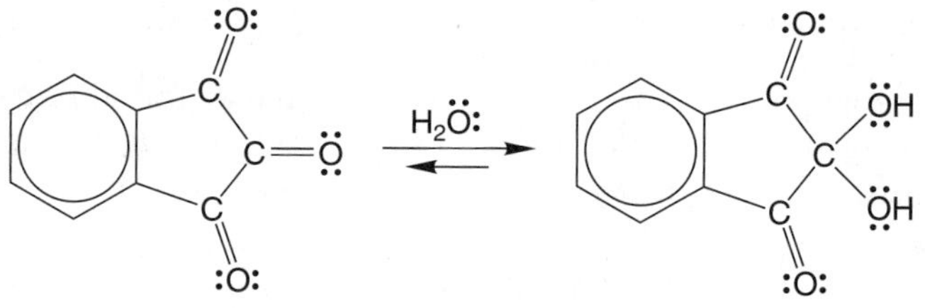

Indan-1,2,3-trione

Ninhydrin

FIGURE 26.23 Ninhydrin is the hydrate of indan-1,2,3-trione.

PROBLEM 26.14 Why is the trione of Figure 26.23 hydrated at the middle, not the side carbonyl group?

The amino acid first reacts with the trione to form an imine through the small amount of the free amine present at equilibrium. We have seen this kind of reaction many times (Chapter 16, p. 780; Chapter 21, p. 1091). Decarboxylation produces a second imine that is hydrolyzed under the aqueous conditions to give an amine. It is in this hydrolysis step that the R group of the original amino acid is lost. Still another imine, the third in this sequence, is then formed through reaction of the amine with another molecule of ninhydrin (Fig. 26.24). The extensive conjugation in the product leads to a very low energy absorption in the visible region (Chapter 12, p. 514), and the evident purple color.

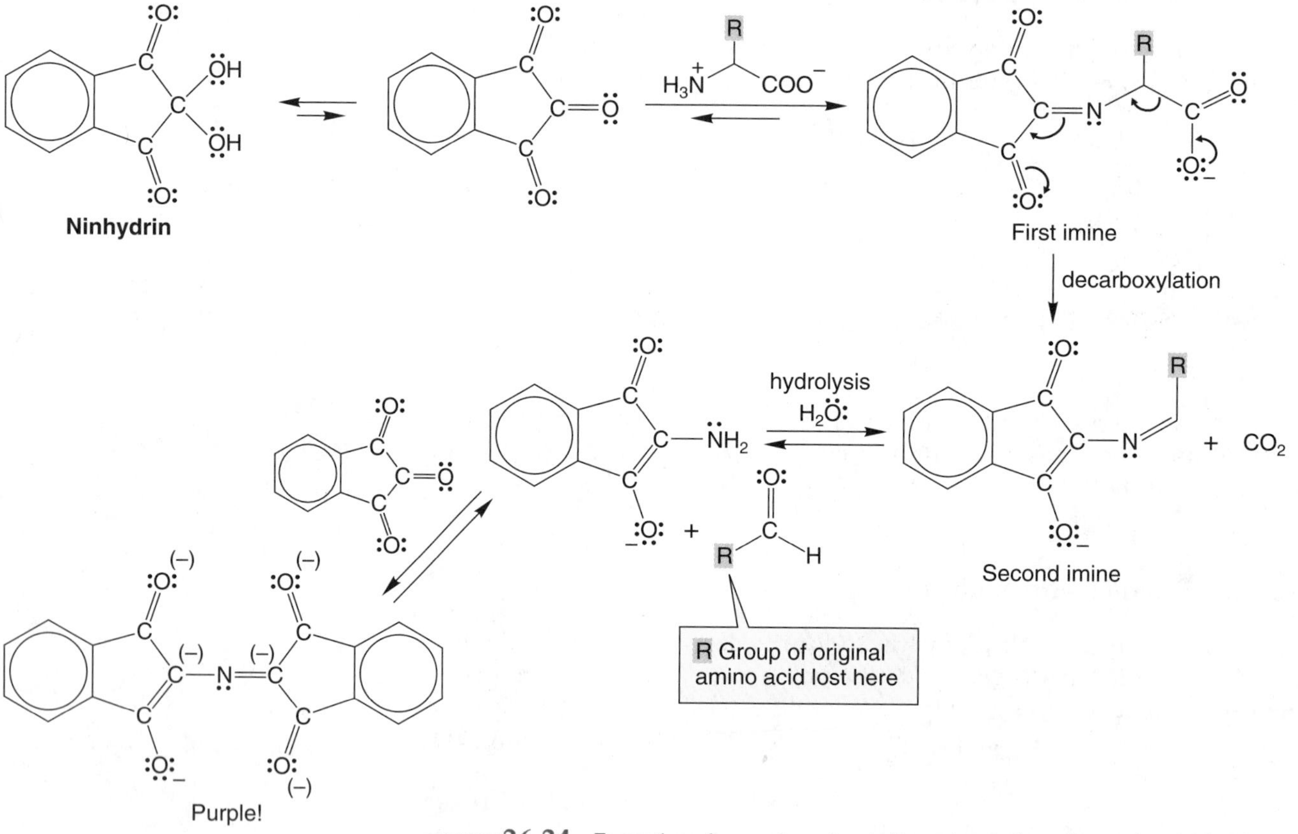

Ninhydrin

First imine

decarboxylation

hydrolysis

Second imine

R Group of original amino acid lost here

Purple!

FIGURE 26.24 Formation of a purple molecule through reaction of an amino acid with ninhydrin.

By far the most important reaction of amino acids is their polymerization to peptides and proteins through the formation of new amide bonds (Fig. 26.25). The following sections will be devoted to this reaction and some of its consequences.

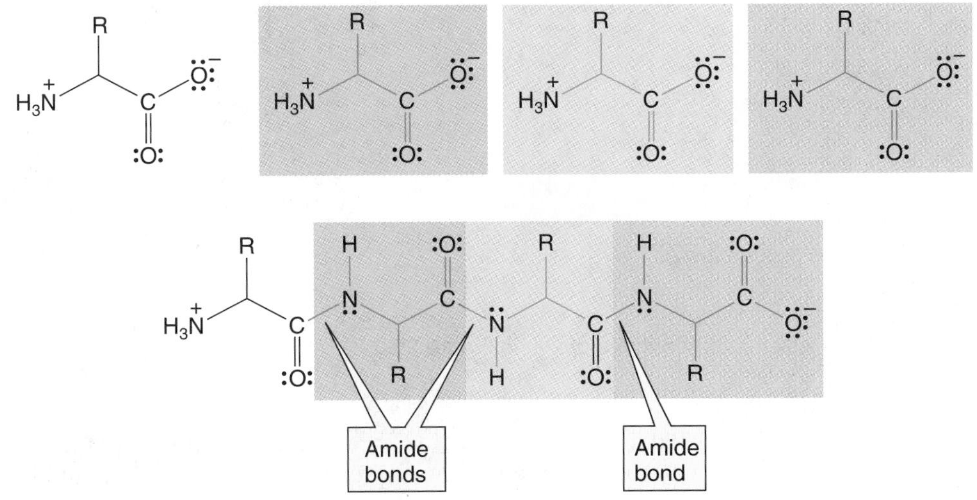

FIGURE **26.25** Polypeptides are polyamino acids linked through amide bonds.

26.3 PEPTIDE CHEMISTRY

26.3a Nomenclature and Structure

Figure 26.26 shows the schematic construction of a three amino acid peptide (a tripeptide) made from alanine, serine, and valine. Don't confuse this picture with real synthetic procedures—we will get to them soon enough. Peptides are always written and named with the **amino terminus** on the left, with the structure, and name, proceeding toward the **carboxy terminus** on the right. So the tripeptide in Figure 26.26 is alanylserinylvaline, or Ala·Ser·Val, or A·S·V.

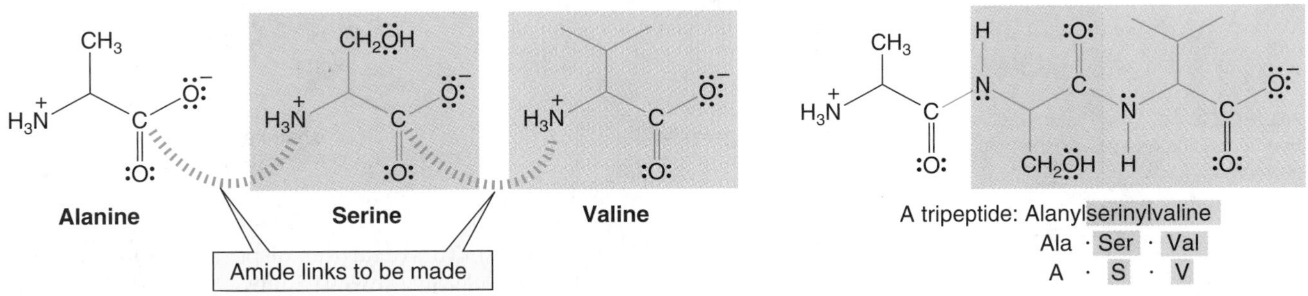

FIGURE **26.26** The tripeptide, alanylserinylvaline, Ala·Ser·Val. The amino acid terminus starts the name, and the carboxy terminus ends it.

Write structures for the peptides named in the yellow box in Figure 26.27. PROBLEM **26.15**

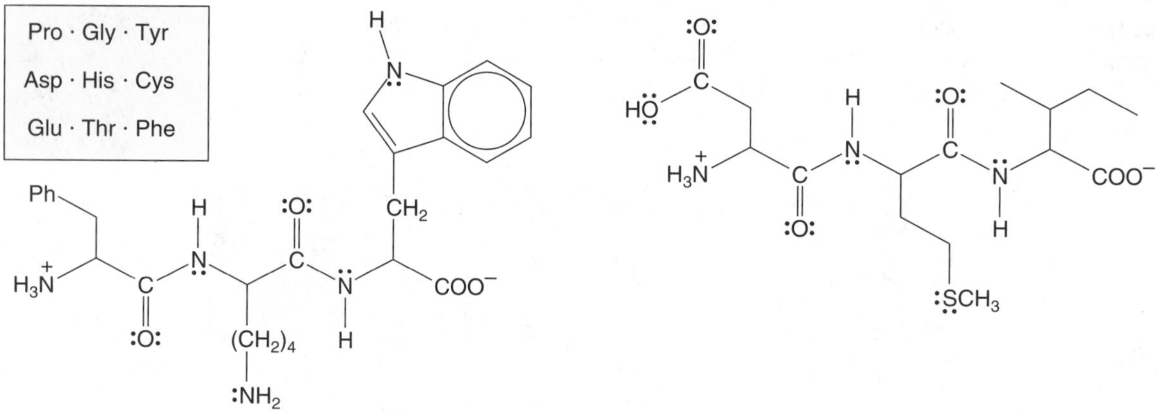

Pro · Gly · Tyr

Asp · His · Cys

Glu · Thr · Phe

FIGURE 26.27

PROBLEM 26.16 Name the peptides drawn in Figure 26.27.

The sequence of amino acids in a peptide or protein constitutes its **primary structure**. Often these complex series of amino acids are linked in another, intermolecular way. Connections between chains are commonly formed through **disulfide bridges**. Cysteine is usually the amino acid involved in these bridges, which are made through oxidation of the thiol groups (Fig. 26.28). For the chemical oxidation of thiols to disulfides, see Chapter 17, page 861.

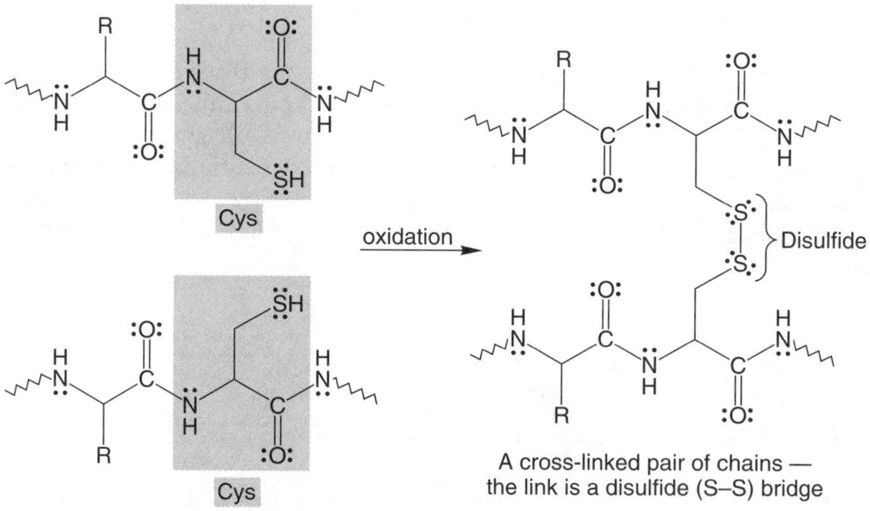

FIGURE 26.28 Formation of disulfide bonds can link one polyamino acid chain to another.

A cross-linked pair of chains — the link is a disulfide (S–S) bridge

There are other important structural features of polyamino acids. The amide groups prefer planarity and the internal hydrogen bonds between the NH hydrogens and carboxy carbonyl groups, along with the structural constraints imposed by the disulfide links, lead to regions of order in most large polypeptides.

*PROBLEM 26.17 Explain carefully why amides prefer planarity.

Amides are stabilized by resonance.

This resonance stabilization is maximized when the $2p$ orbitals on N, C, and O overlap as much as possible, which requires sp^2 hybridization and planarity.

ANSWER

Especially common are right-handed, or α-**helical** regions, and repeating folded sections known as **β-pleated sheets**. Both of these arrangements allow regular patterns of hydrogen bonding to develop, thereby imposing order on the formally linear sequence of amino acids. In the α-helical regions the hydrogen bonds are between the coils of the helix, whereas in the pleated sheets they hold lengths of the chains in roughly parallel lines (Fig. 26.29). Other regions consist of completely disordered, **random coil**, series of amino acids. The combination of these structural features determines the folding pattern, or **secondary structure** of the polypeptide.

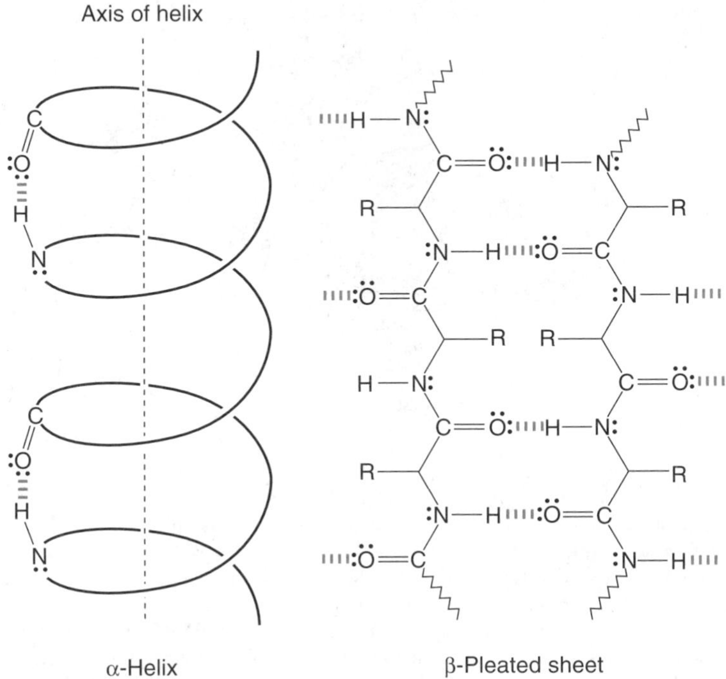

FIGURE 26.29 Two examples of ordered secondary structure of polypeptides, the α-helix and β-pleated sheet.

There are even higher structural orders for these molecules, which is not an obvious point. It might have been that polypeptides were best described as ordered regions of secondary structure (α-helix or β-pleated

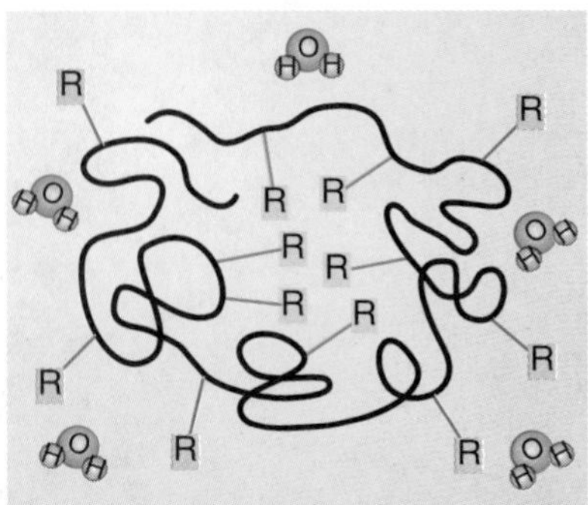

FIGURE 26.30 A globular protein, ordered so as to put the nonpolar side chains in the inside of the "glob" and the polar side chains outside in the polar solvent medium.

sheets) connected by sections of random coil. In such a case, a protein would have no fixed overall structure beyond the sections defined by secondary structure. But this has been shown not to be the case. X-ray structure determinations of crystallized proteins have revealed these species to have *specific*, three-dimensional structures.* This **tertiary structure** of the protein is determined partly by hydrogen bonding, but also by van der Waals and electrostatic forces. These giant molecules are composed of a carbon backbone to which side chains, or R groups, are attached. These R groups are either hydrocarbon-like, nonpolar groups such as methyl, isopropyl, or benzyl, or highly polar amino, hydroxyl, or sulfhydryl groups. In the natural environment of a protein, water, it should be no surprise to find that the protein adjusts its shape so that the polar, **hydrophilic** groups are aimed outward toward the polar solvent, whereas the nonpolar, "greasy" **hydrophobic** hydrocarbon portions cluster inside the molecule, maximally protected from the hostile aqueous environment (Fig. 26.30).

Many proteins adopt these globular shapes, which maximize interior hydrophilic and exterior hydrophobic interactions, but others are fibrous, taking the form of a superhelix composed of rope-like coils of α-helices. Figure 26.31 shows a typical polyamino acid, the enzyme lactate dehydrogenase which incorporates five α-helices and six β-sheets.

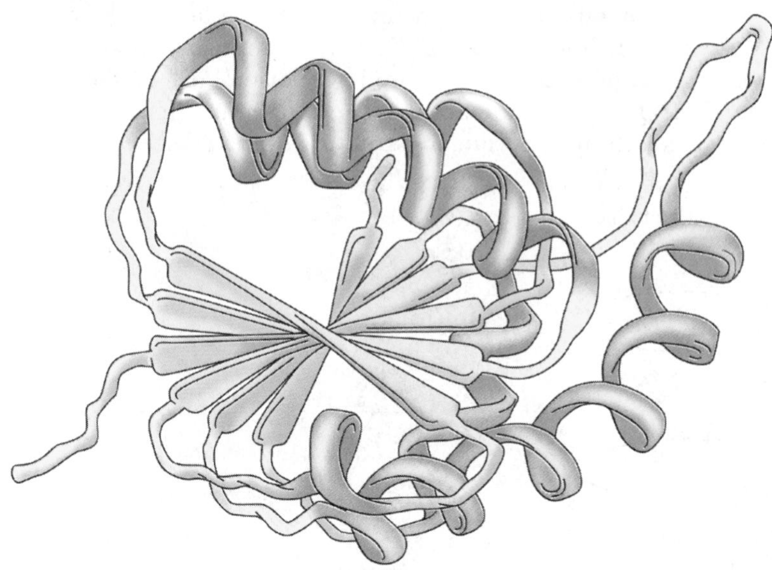

FIGURE 26.31 Lactate dehydrogenase.

* It is a general problem that until very recently the structures of proteins could only be determined by X-ray diffraction studies on crystallized materials. Work was limited by the availability of crystals, and crystallization of these molecules is usually difficult. Successful formation of X-ray quality crystals of a protein is cause for celebration (and publication) even before the work of structure determination begins. But how do we know that the structure of the molecule in solution is the same as that in the solid state crystal? Remember, biological activity is tied intimately to the detailed structure of these molecules—might we not be led astray by a structure that owed its shape to crystal packing forces in the solid state, and which was quite different in solution? Indeed we might. These days, it is becoming possible to use very high field NMR spectrometers, along with increasingly sophisticated NMR pulse techniques, to derive structures directly in solution. Remember that high-field spectrometers are anything but cheap; the cost is about \$$10^3$ per megahertz of your tax dollars.

This tertiary structure of a protein is extraordinarily important for it determines the shape of these huge molecules and biological activity depends intimately on these shapes. Evolution has produced proteins in which pockets, or bays, exist into which certain molecules, or substrates, fit and can be bound in the proper orientation for reaction. The reaction could be complicated, or as simple as the hydrolysis of an ester. In other molecules, substrates are bound only for transportation purposes. These **active sites** are a direct consequence of the primary structure that determines the secondary and tertiary structures. Figure 26.32 gives a schematic representation of this process.

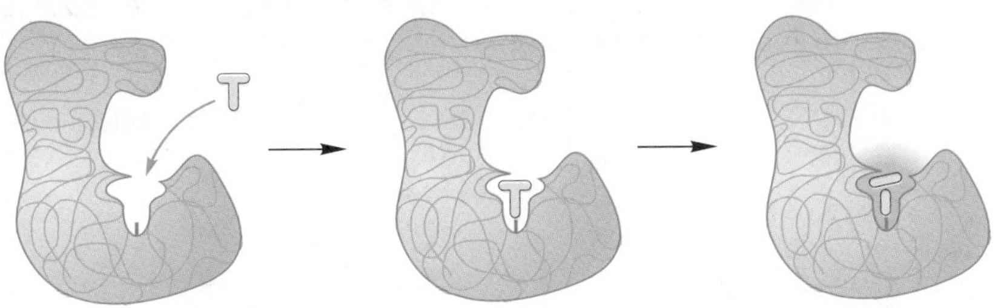

FIGURE 26.32 A binding site in a protein captures (binds) a small molecule for transport, or reaction with an appropriately located reactive site.

How do we know that the *same* tertiary structure is adopted by every protein molecule of a given amino acid sequence or primary structure? One clue is that synthetic proteins have the same biological activity as the natural materials. If we reproduce the primary sequence of a protein correctly, the proper biological activity appears. As this biological activity depends *directly* on the details of shape, the molecule must be folding properly into the appropriate tertiary structure.

Another clue comes from experiments in which the tertiary structure of a protein is deliberately disrupted in a reversible way, which is called **denaturing** the protein. Many ways of denaturing a protein are irreversible. When an egg is cooked, the thermal energy supplied denatures the proteins in the egg white in an irreversible way. Once the egg is fried, there is no unfrying it. Similar chemistry can be induced by pH change. The curdling of milk products is an example. But some denaturing reactions are reversible. For example, disulfide bonds can be broken through treatment with thiols. This disruption of the secondary structure transforms the ordered sections of the protein into the random coil arrangements. Similar reversible disruptions of order can be achieved through treatment with urea (in a reaction whose precise mechanism remains unknown), or, sometimes, through heating (Fig. 26.33).

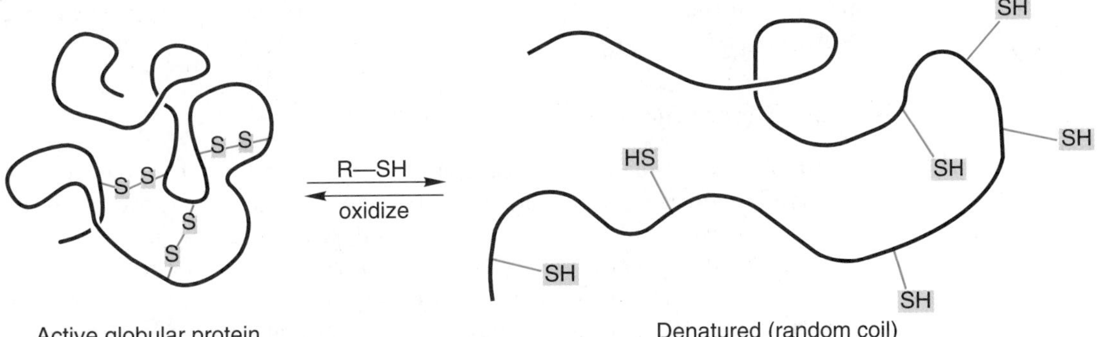

FIGURE 26.33 One method of denaturing proteins, of destroying the secondary and tertiary structure, is to break the disulfide bonds. If air oxidization re-forms the disulfides, the higher order structures are regenerated and biological activity is reestablished.

If these denatured proteins are allowed to stand, the thiol groups are reoxidized by air to the appropriate disulfide linkages and the original tertiary structure with its biological activity returns!

There is even higher order to some proteins. Some proteins are collections of smaller polyamino acid units held together by intramolecular attractions—van der Waals and electrostatic forces—into super structures with **quaternary structure**. The classic example of quaternary structure is **hemoglobin**, the protein responsible for oxygen absorption and transport in humans. To accomplish these functions it incorporates four heme units, each consisting of an iron atom surrounded by heterocyclic rings positioned so as to stabilize the metal. Figure 26.34 shows the heme unit in blue, before and after oxygen absorption.

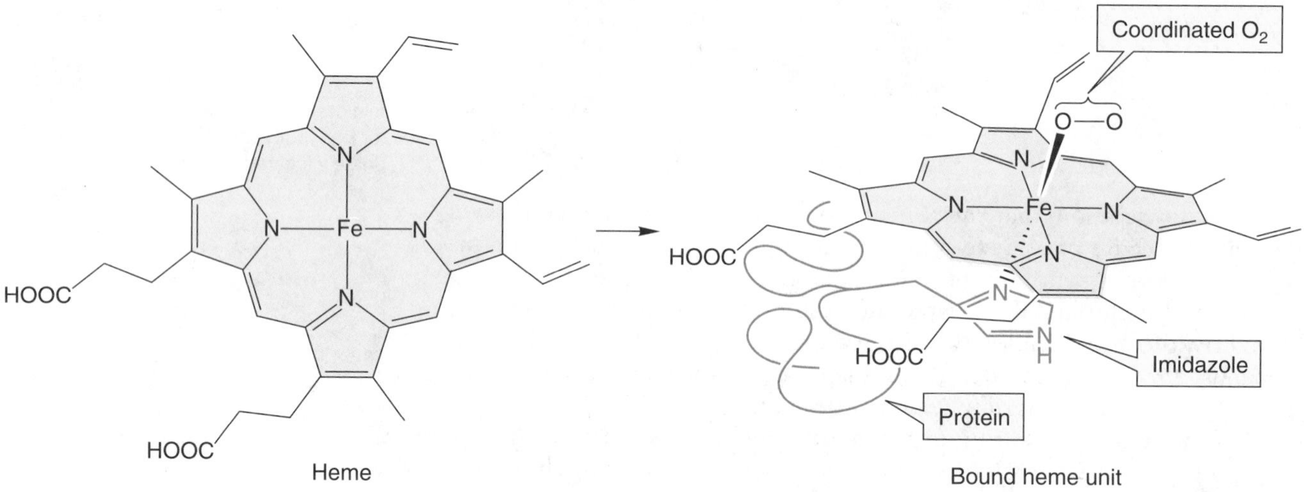

Heme

Bound heme unit

FIGURE **26.34** The unoxidized and oxidized heme structures of hemoglobin.

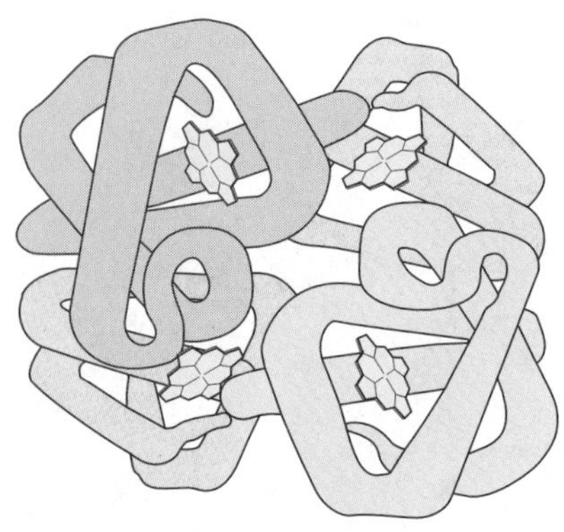

FIGURE **26.35** The quaternary structure of hemoglobin.

The protein surrounding the heme is called globin, and in each of the subunits of globin an imidazole is poised to hold the heme in position. There are four subunits in hemoglobin, two each of two slightly different peptide chains. These are held together by electrostatic and van der Waals forces as well as hydrogen bonding, and take the shape of a giant tetrahedron (Fig. 26.35).

Given all this general information about structure, we still face two difficult problems. First, how do we determine the primary sequence of an unknown protein, and second, once we know that structure, how can we synthesize it?

26.3b Determination of Protein Structure

If suitable crystals can be obtained, and if the X-ray diffraction pattern can be solved in a reasonable amount of time, the structure of the protein is apparent. However, this is not yet a general procedure. Crystallization of these molecules remains difficult, and although the determination of structure of small molecules by X-ray is now routine and rapid, the determination of the structures of proteins remains difficult and time consuming.

In principle, we can get an overall picture of the composition of a protein by first destroying all disulfide links in the molecule, and then hydrolyzing all the peptide bonds. *Remember*: These peptide bonds are amides and simple acid hydrolysis should convert them into acids (Chapter 20,

p. 1028). This procedure regenerates all the amino acids of which the molecule is built. Figure 26.36 shows a hypothetical example. Of course, this technique does nothing to reveal primary, or higher structures.

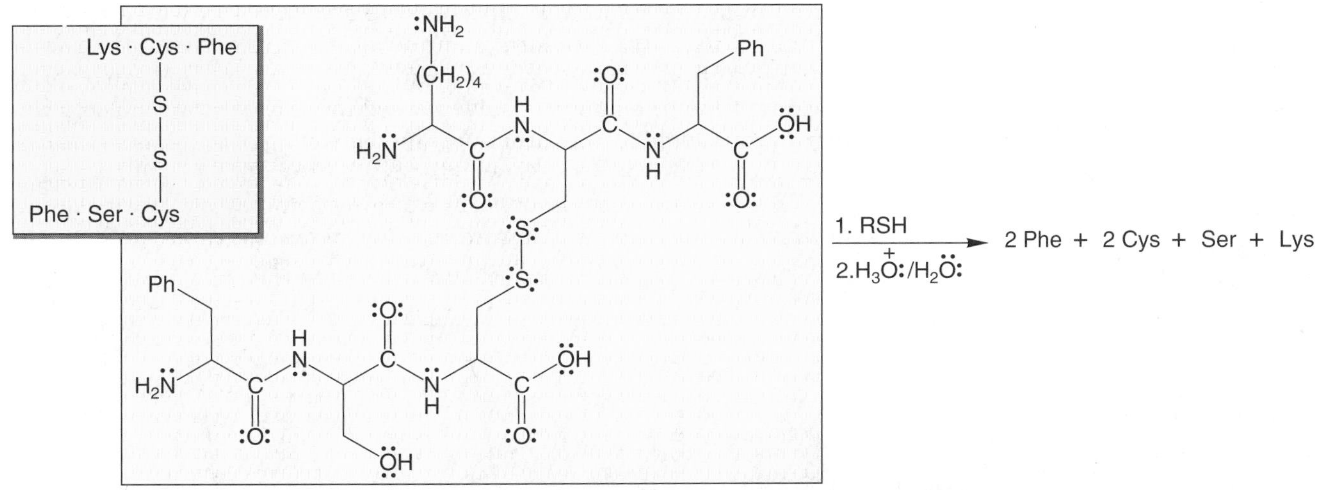

Lys · Cys · Phe
S
S
Phe · Ser · Cys

$$\text{Hexapeptide} \xrightarrow[\text{2. } H_3O^+/H_2O]{\text{1. RSH}} \text{2 Phe + 2 Cys + Ser + Lys}$$

FIGURE 26.36 Cleavage of a hexapeptide to its constituent amino acids. The disulfide bonds are first broken to produce two fragments, which are hydrolyzed to break the amide bonds.

In practice, the pieces formed when the disulfide bonds are broken are first separated and then hydrolyzed. This separation is no trivial task, but several methods are now available. Electrophoresis is a technique we saw as useful for separating simple amino acids (p. 1350) and it works for polyamino acids as well. Various kinds of chromatography are also effective. In **gel-filtration chromatography** the mixture is passed over a column of polymer beads which, on the molecular level, contain holes into which the smaller peptide fragments fit more easily than the larger pieces. Accordingly, the larger fragments pass more rapidly down the column, and a separation is achieved. **Ion-exchange chromatography** uses electrostatic attractions to hold more highly charged pieces on the column longer than more nearly neutral molecules. As in any chromatographic technique, the more tightly held molecules move more slowly than those less tightly held by the medium (Fig. 26.37).

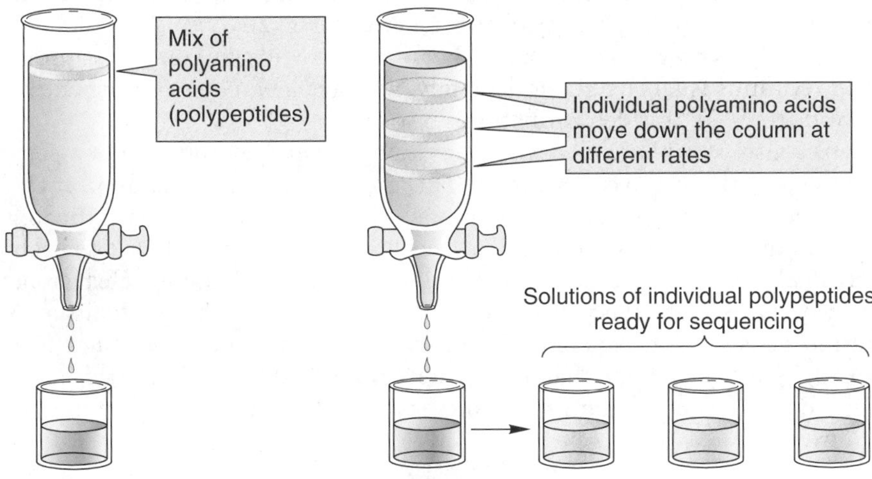
Mix of polyamino acids (polypeptides)
Individual polyamino acids move down the column at different rates
Solutions of individual polypeptides ready for sequencing

FIGURE 26.37 A schematic description of chromatographic separation of polyamino acids.

Once the peptide pieces are separated, each can be hydrolyzed to its constituent amino acids. Now we need methods for determining how much of each amino acid is present. Again chromatography is used. The amino acids are separated on an ion-exchange column. As each amino acid is eluted from the column, it emerges into a chamber in which it reacts with ninhydrin (p. 1357) to form a purple dye. The intensity of the purple color is proportional to the amount of amino acid present, and can be plotted against time or amount of solvent used for elution from the column. This procedure produces a **chromatogram** consisting of a series of peaks of various sizes. How do we tell which peak corresponds to which amino acid? This identification is done by running samples of known amino acids through the column, determining their **retention times**, and matching these with those of the unknown amino acids (Fig. 26.38).

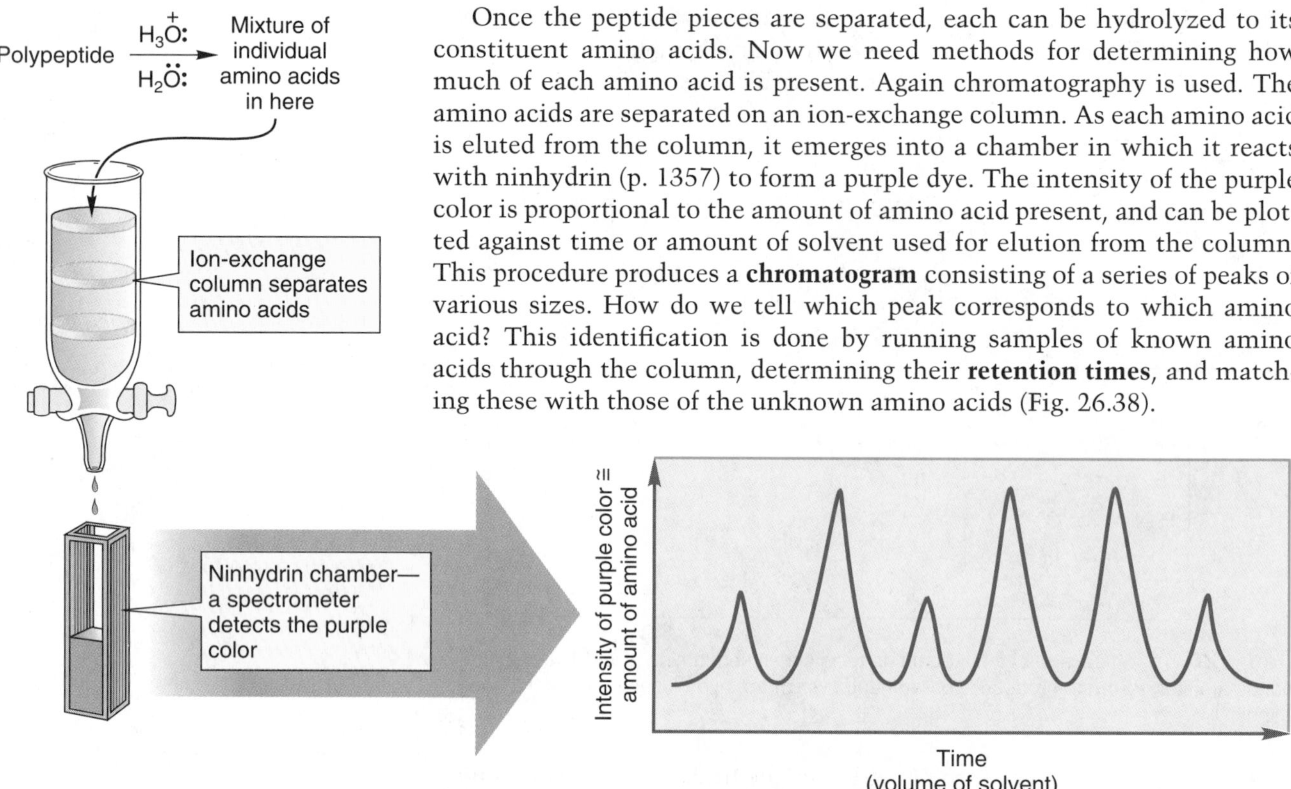

FIGURE 26.38 Polypeptides can be hydrolyzed to constituent amino acids, which can be separated by chromatography. Reaction of each amino acid with ninhydrin gives a purple color that can be quantitatively analyzed spectroscopically. The intensity of the purple color is proportional to the amount of amino acid formed.

But this technique only gives us the coarsest picture of the structure of the protein: We know the constituent amino acids and their relative amounts, but we have no idea of their order, of the primary structure. An important task is to find a better way to get the actual sequence of amino acids in a protein. The first step is again to destroy the disulfide links and separate the constituent polypeptides. A simple method of determining the amino terminus is called the **Sanger degradation**, after Frederick Sanger (b. 1918). The peptide is allowed to react with 2,4-dinitrofluorobenzene. An adduct is formed that labels the terminal amino group with a 2,4-dinitrobenzene group (Fig. 26.39).

Hydrolysis now generates a number of amino acids, but only the amino acid terminus is labeled! Unfortunately, we have had to destroy the entire peptide to make this determination!

It is also possible to discover the amino acid at the other end of a peptide chain, the carboxy terminus. The method of choice uses an enzymatic reaction in which one of a number of carboxypeptidases specifically cleaves the terminal amino acid at the carboxy end of the chain. As each terminal amino acid reacts, a new amino acid is revealed to be cleaved in turn by more carboxypeptidase. Careful monitoring of the production of amino acids as a function of time can give a good idea of the sequence (Fig. 26.40). Still, it is clear that things will get messy as long peptides are turned into a soup containing an increasingly complex mixture of amino acids.

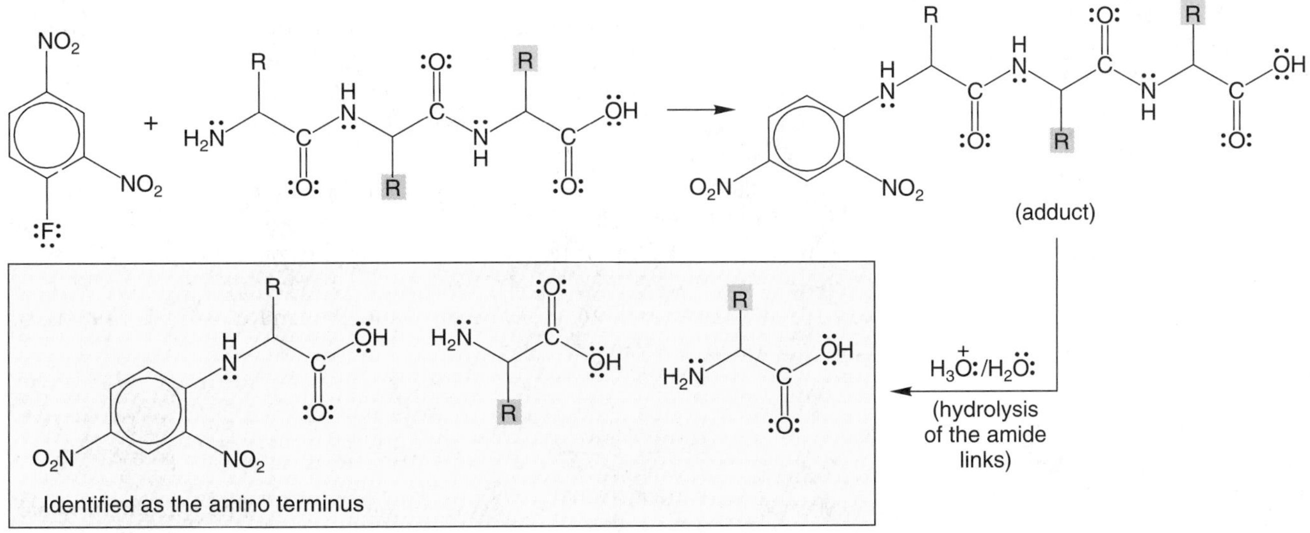

FIGURE 26.39 The amino acid at the amino terminus can be identified by reaction with 2,4-dinitrofluorobenzene, followed by hydrolysis. Only the amino acid at the amino end of the chain will be labeled.

PROBLEM 26.18 What is the mechanism of adduct formation in Figure 26.39? What is the function of the nitro groups (Chapter 14, p. 671)?

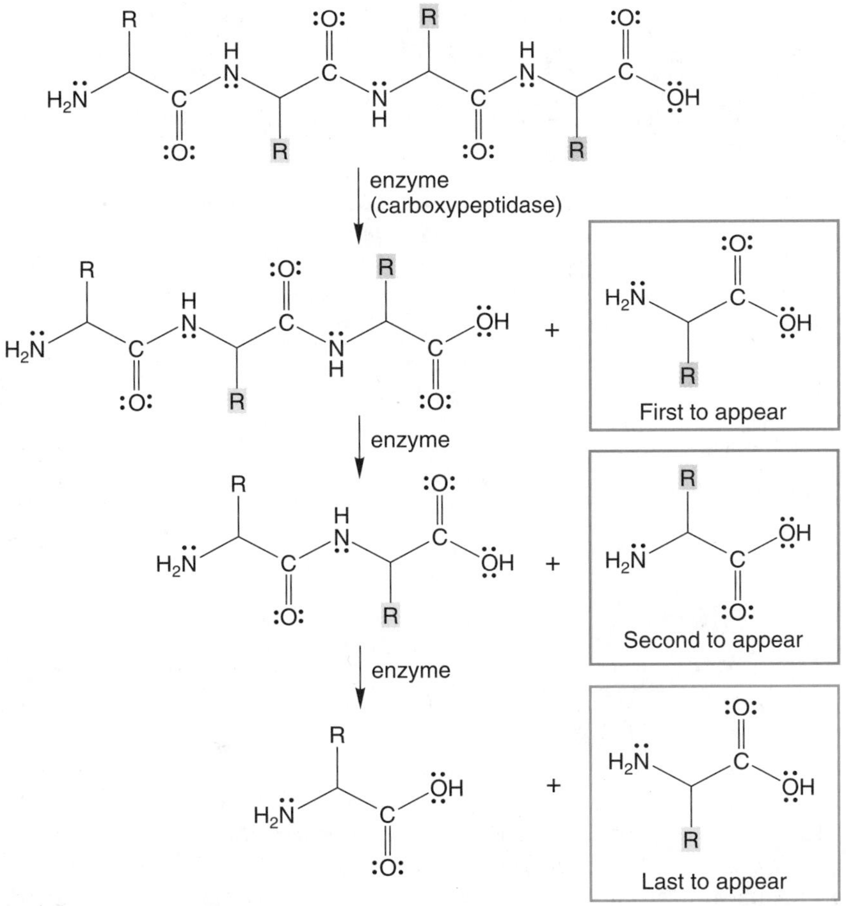

FIGURE 26.40 Carboxypeptidase is an enzyme that cleaves only at the carboxy end of a polypeptide.

A much better method would be one that included no overall hydrolysis step and that could be more controlled than the cleavage reactions set in motion by treatment with a carboxypeptidase. We need a method that unzips the protein from one end or the other, revealing at each step the terminal amino acid.

Methods have now been developed that allow just such a step-by-step sequencing of peptides. One effective way is called the **Edman degradation**, after Pehr Edman (b. 1916), and uses the molecule **phenylisothiocyanate**, Ph–N=C=S (Fig. 26.41). Isocyanates have appeared before (Chapter 23, p. 1224; Fig. 26.41), and an isothiocyanate is just the sulfur analogue of an isocyanate.

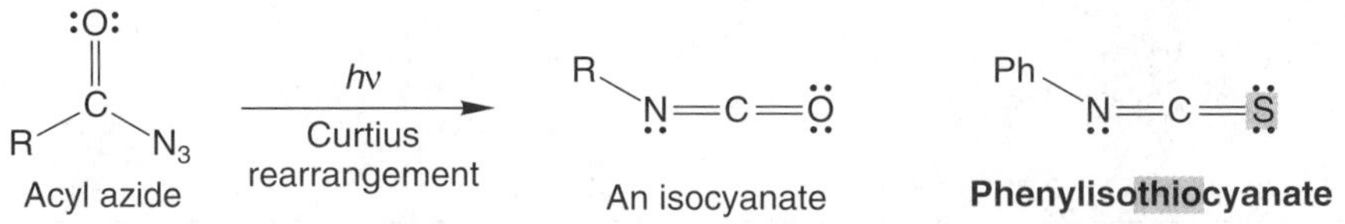

FIGURE 26.41 An isothiocyanate is just the sulfur version of an isocyanate, the product of a Curtius rearrangement.

PROBLEM 26.19 What is the mechanism of the isocyanate formation in Figure 26.41 (Chapter 23, p. 1225)?

Like isocyanates, the isothiocyanates react rapidly with nucleophiles. Reactions of isocyanates with ammonia or amines yield **ureas**. Iso<u>thio</u>cyanates give a **<u>thio</u>urea** as the first formed intermediate (Fig. 26.42).

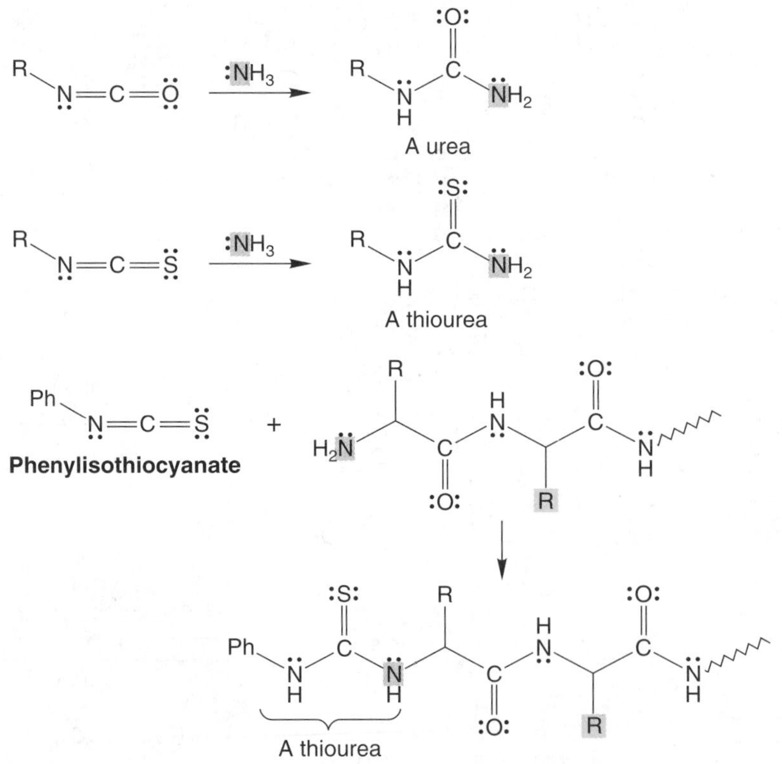

FIGURE 26.42 Isothiocyanates react with amines to give thioureas. Phenylisothiocyanate can be used to label the amino terminus of a polypeptide.

The thiourea can be cleaved in acid *without breaking the other amide bonds in the peptide.* As we saw in Chapter 23 (p. 1185), sulfur is an excellent neighboring group, and the introduction of the thiourea has the effect of placing a sulfur atom in just the right position to aid in cleavage of the terminal amino acid. The adjacent carbonyl is protonated, and then it is the neighboring sulfur that adds to it, triggering the cleavage reaction to give a molecule called a thiazolinone (Fig. 26.43).

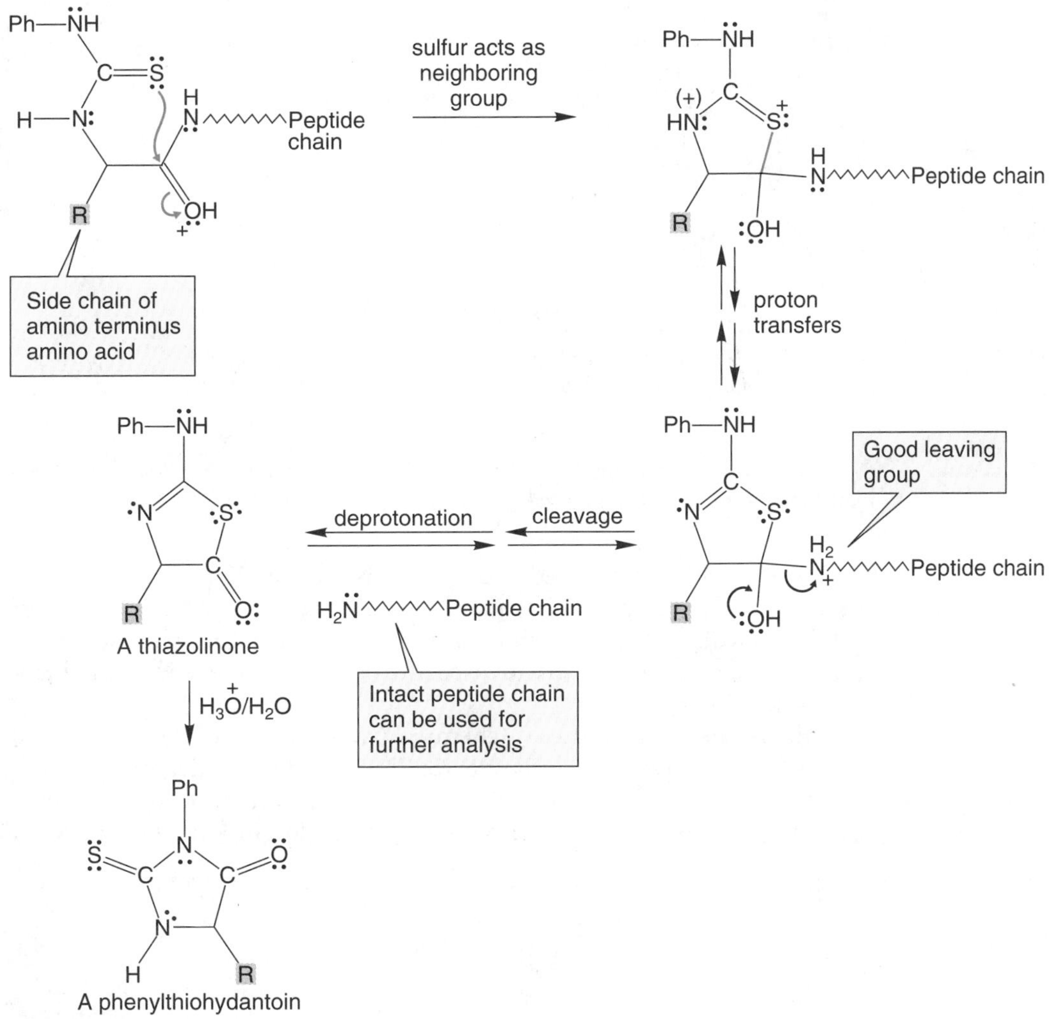

FIGURE **26.43** The Edman procedure for sequencing polyamino acids uses phenylisothiocyanate. The amino terminus is identified by the structure of the phenylthiohydantoin formed. Notice that this method does not destroy the peptide chain as does the Sanger procedure.

Further acid treatment converts the thiazolinone into a phenylthiohydantoin containing the R group of the terminal amino acid, which can now be identified (Fig. 26.43). This process is indirect, as the initial product of reaction is not the phenylthiohydantoin, but a thiazolinone. The critical point is that the rest of the peptide remains intact and the Edman procedure can now be repeated, in principle, as many times as are needed to sequence the entire protein (Fig. 26.44).

But the practical world intrudes, and even the Edman procedure, good as it is, becomes ineffective for polypeptides longer than about 20–30 amino acids. Impurities build up at each stage in the sequence of determinations and eventually the reaction mixture becomes too complex to yield unequivocal results in phenylthiohydantoin formation. Accordingly,

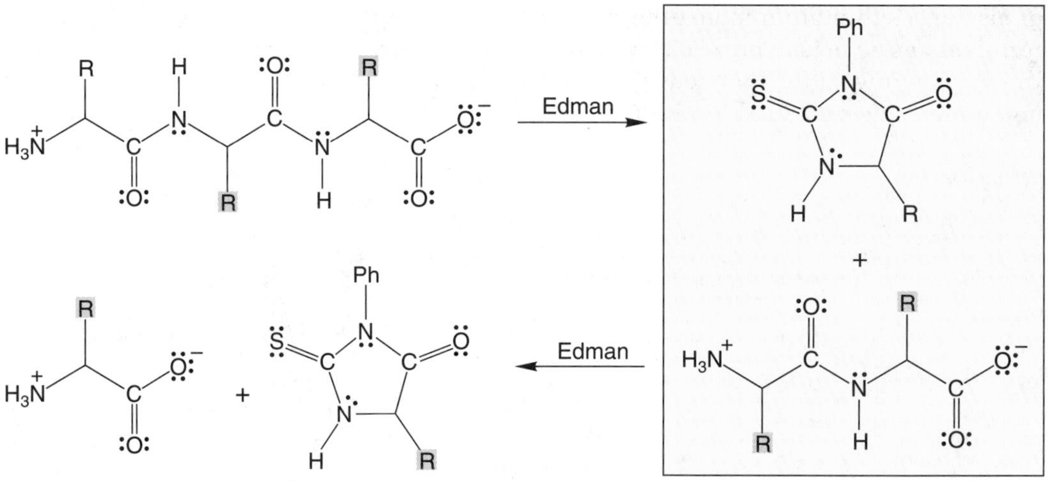

FIGURE 26.44 The structures of polyamino acids can be determined through repeated applications of the Edman procedure, which is one method of sequencing.

PROBLEM 26.20 Provide a mechanism for the formation of the phenylthiohydantoin from the thiazolinone (Fig. 26.43).

methods have been developed to cut proteins into smaller pieces at known positions in the amino acid sequence. These methods take advantage of both biological reactions in which enzymes break peptides only at certain amino acids, as well as a very few specific cleavages induced by small laboratory reagents.

Perhaps the best of these "chemical" (enzymes are chemicals, too) cleavages uses **cyanogen bromide (BrCN)** and takes advantage of the nucleophilicity of sulfur and the neighboring group effect to induce cleavage at the carboxy side of each methionine residue in a polypeptide (Fig. 26.45).

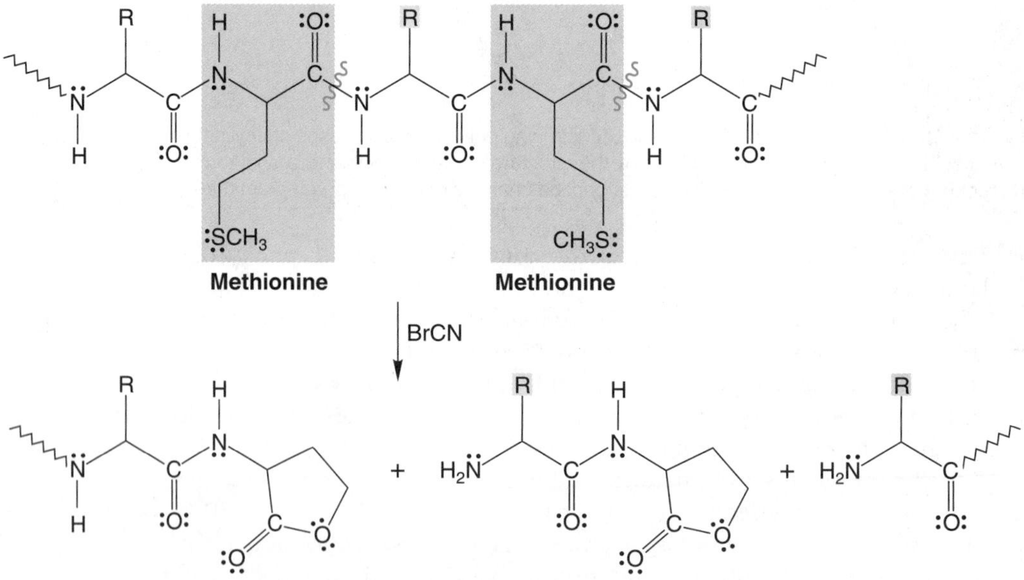

FIGURE 26.45 The carboxy end of each methionine is cleaved on treatment of a polypeptide with cyanogen bromide (BrCN).

The sulfur atom of methionine first displaces bromide from cyanogen bromide to form a sulfonium ion. Attack by the proximate carbonyl group of methionine displaces the excellent leaving group, methylthiocyanate, and forms a five-membered ring called a homoserine lactone. Hydrolysis steps follow to produce the end products, two smaller peptides (Fig. 26.46).

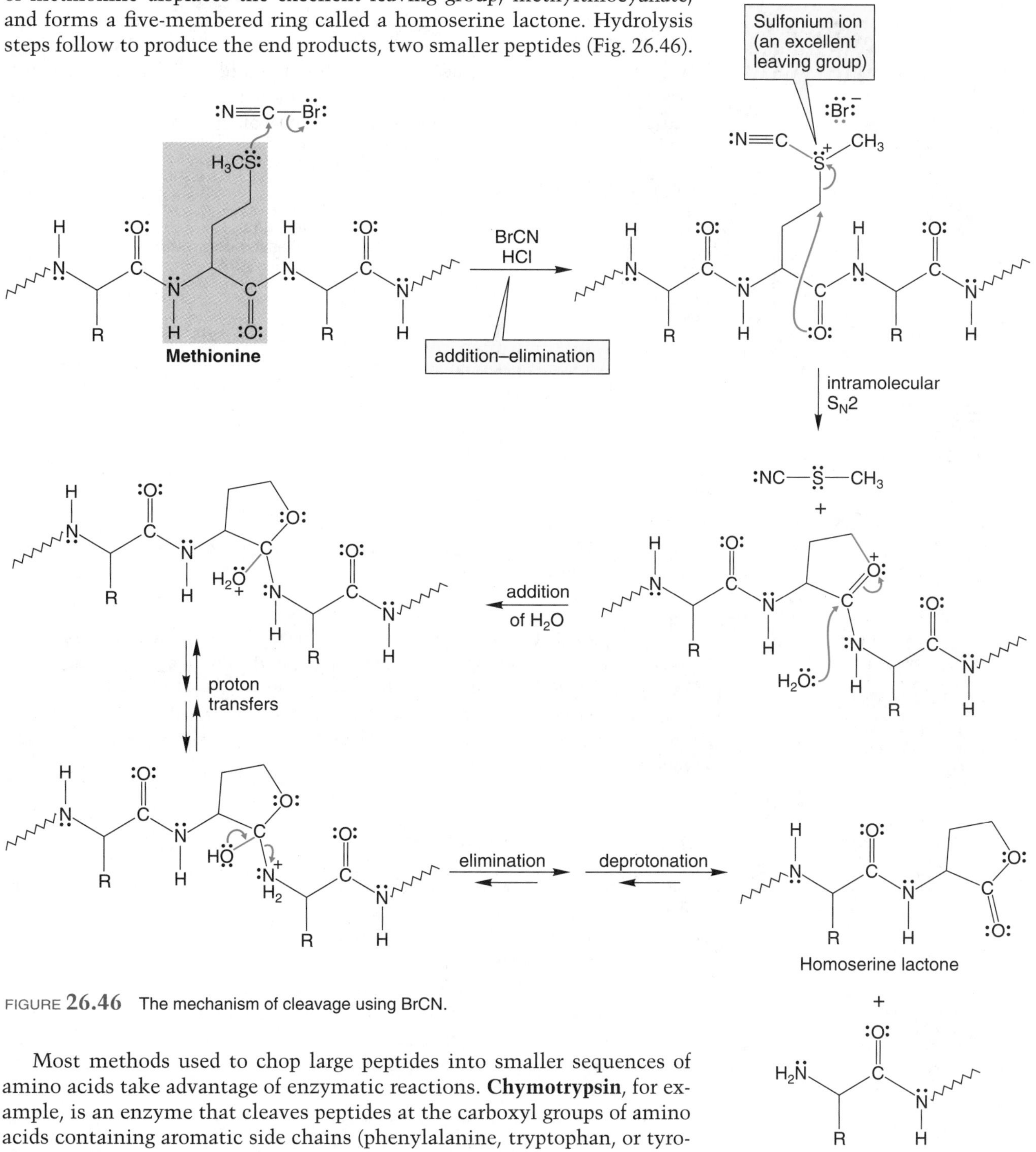

FIGURE 26.46 The mechanism of cleavage using BrCN.

Most methods used to chop large peptides into smaller sequences of amino acids take advantage of enzymatic reactions. **Chymotrypsin**, for example, is an enzyme that cleaves peptides at the carboxyl groups of amino acids containing aromatic side chains (phenylalanine, tryptophan, or tyrosine). **Trypsin** cleaves only at the carboxy end of lysine and arginine. There are several other examples of this kind of specificity. Long peptide chains can be degraded into shorter molecules that can be effectively sequenced by the Edman procedure. A little detective work suffices to put together the entire sequence. Let's look at a simple example and then do a real problem.

Consider the decapeptide in Figure 26.47. We can determine the amino acid terminus through the Sanger procedure to be Phe. Next, we use the enzyme trypsin to cleave the molecule into three fragments. As we know the peptide must start with Phe, the first four amino acids are Phe·Tyr·Trp·Lys. The only other cleavage point is at Arg, so the Asp·Ile·Arg fragment must be the middle piece. If this tripeptide were at the end, with Arg as the carboxy terminus, no cleavage would be possible. The sequence must be Phe·Tyr·Trp·Lys·Asp·Ile·Arg·Glu·Leu·Val.

Phe·Tyr·Trp·Lys·Asp·Ile·Arg·Glu·Leu·Val

Sanger

O_2N NO_2 N H COOH Ph

Phe must be the amino terminus

trypsin (cleaves only after Lys or Arg)

Must start with Phe → Phe·Tyr·Trp·Lys

\+

Contains a cleavage point (Arg) must be middle fragment → Asp·Ile·Arg

\+

Glu·Leu·Val

Can be sequenced using Edman

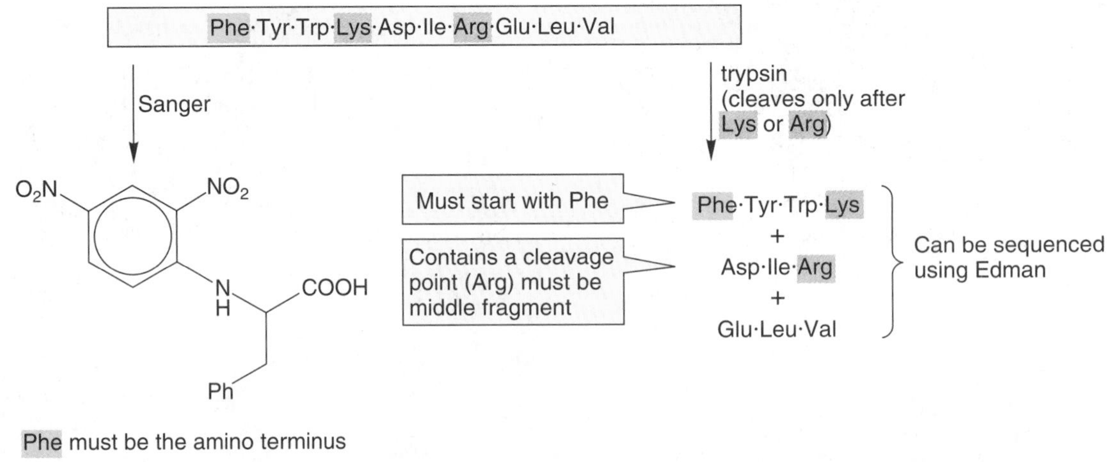

FIGURE 26.47 An example of peptide sequencing using Sanger and Edman procedures as well as enzymatic cleavage reactions.

PROBLEM 26.21 The nonapeptide bradykinin can be completely hydrolyzed in acid to give three molecules of Pro, two molecules each of Arg and Phe, and single molecules of Ser and Gly. Treatment with chymotrypsin gives the pentapeptide Arg·Pro·Pro·Gly·Phe, the tripeptide Ser·Pro·Phe, and Arg. End-group analysis shows that both the amino and carboxy termini are the same. Provide the sequence for bradykinin.

26.3c Synthesis of Peptides

Now that we know how to sequence proteins, how about the reverse process—how can we put amino acids together in any sequence we want? How can we synthesize proteins? It is simply a matter of forming amide bonds in the proper order, and this might seem an easy task. However, two difficulties surface as soon as we begin to think hard about the problem. First, there's a lot of work to do in order to make even a small protein. Imagine that we only want to create the nine amide linkages of the decapeptide in Figure 26.47. Even if we can make each amide bond in 95% yield, we are in trouble. A series of nine reactions each proceeding in 95% yield, which probably seems quite good to anyone who has spent some time in an organic lab, gives an overall yield of only $(0.95)^9 = 63\%$. Obviously, to get anywhere in the real world of proteins, in which chain lengths of hundreds of amino acids are common, we are going to have to do much better than this.

Second, there is a problem of specificity. Suppose we want to make the trivial dipeptide Ala·Leu. Even if we can avoid simple acid–base chemistry, random reaction between these two amino acids will give us at least the four possible dimeric molecules, and in practice, other larger peptides will be produced as well (Fig. 26.48).

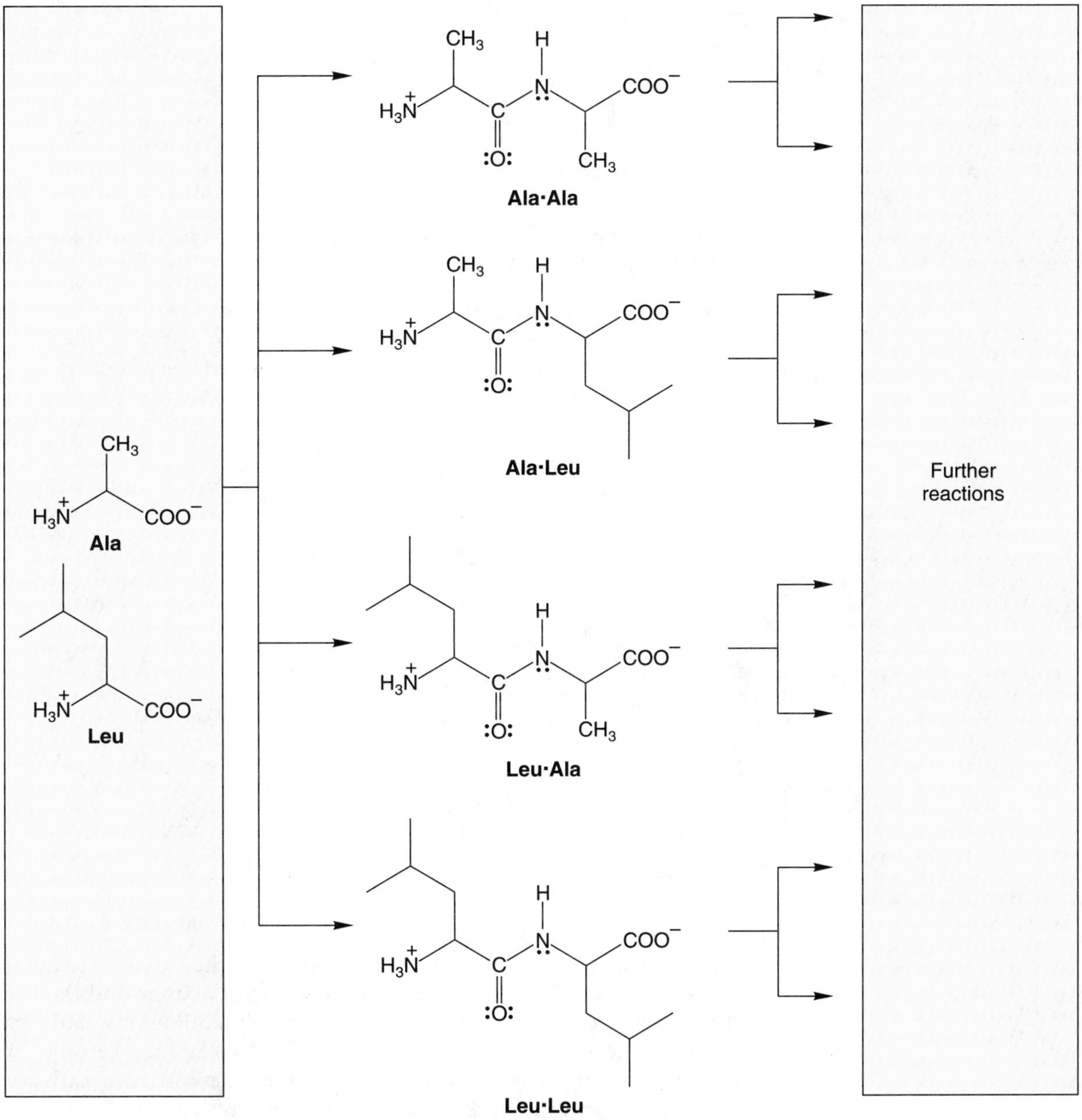

FIGURE 26.48 The possible combinations of two unprotected amino acids.

One simple strategy for avoiding random reaction would be to activate the carbonyl group of alanine by transforming the acid group of alanine into the acid chloride, and then treating it with leucine (Fig. 26.49).

However, even this is hopeless! What is to prevent the original amino group of alanine from reacting with the acid chloride made from alanine? Nothing. A mixture is certain to result (Fig. 26.50). Remember, we ultimately will have to make dozens, perhaps hundreds of amide bonds. We

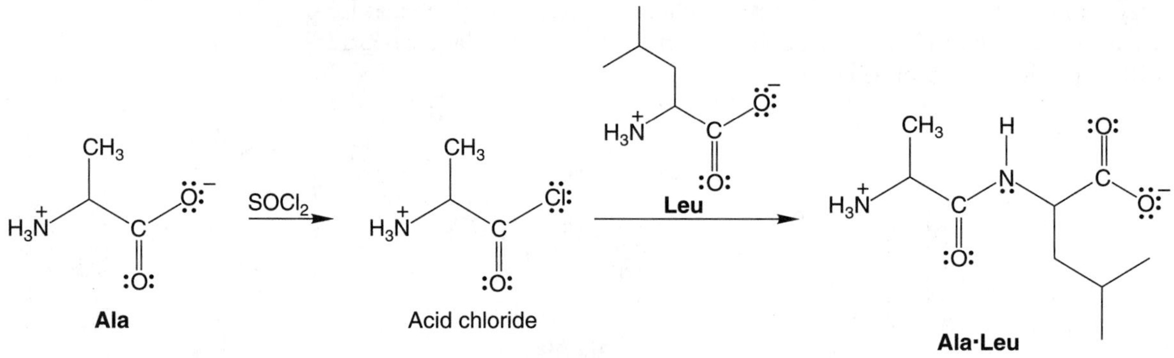

FIGURE 26.49 Some specificity might be obtained by allowing the acid chloride of one amino acid (activated) to react with the other amino acid.

can tolerate no mediocre yields if we hope to make useful amounts of product. We need to find most efficient ways of making the amide bond specifically at the point we want it, and of avoiding side reactions.

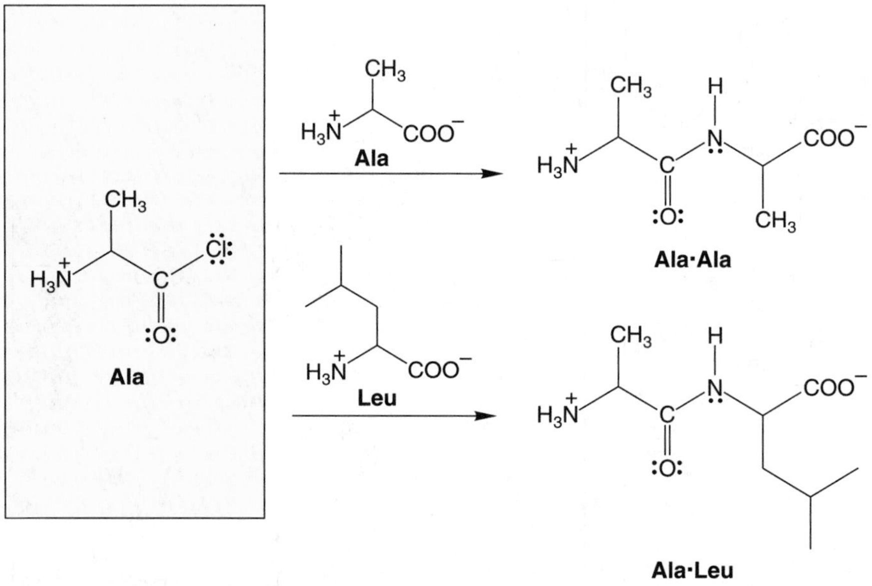

FIGURE 26.50 Here, too, more than one product is inevitable.

In this trivial case, we need not only to activate the carboxyl group of alanine, but to block reaction at the amino group of alanine and the carboxyl group of leucine. Three positions must be modified (the carboxyl groups of alanine and leucine and the amino group of alanine). Left free for reaction are the two positions we hope to join, the activated carboxyl group of alanine and the unmodified amino group of leucine. Remember also that eventually the blocking, or protecting, groups will have to be removed (Fig. 26.51).

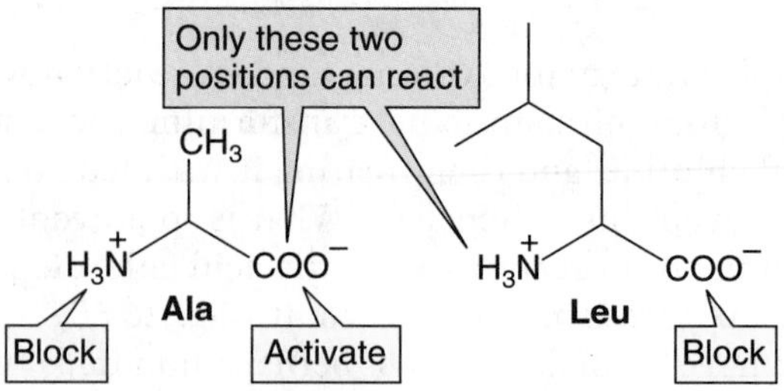

FIGURE 26.51 In any successful strategy, we must activate the carboxy end and block the amino end of one amino acid while blocking reaction at the carboxy end of the other amino acid.

Amino groups can be deactivated by converting them into carbamates through simple addition–elimination reactions (Chapter 20, p. 1022). Two popular methods involve the transformation of the free amine into a carbamate by reaction with di-*tert*-butyl dicarbonate or benzyl chloroformate (benzyloxycarbonyl chloride). Biochemists are even more addicted to acronyms than are organic chemists, and these protecting groups are called **tBoc** and **Cbz**, respectively (Fig. 26.52).

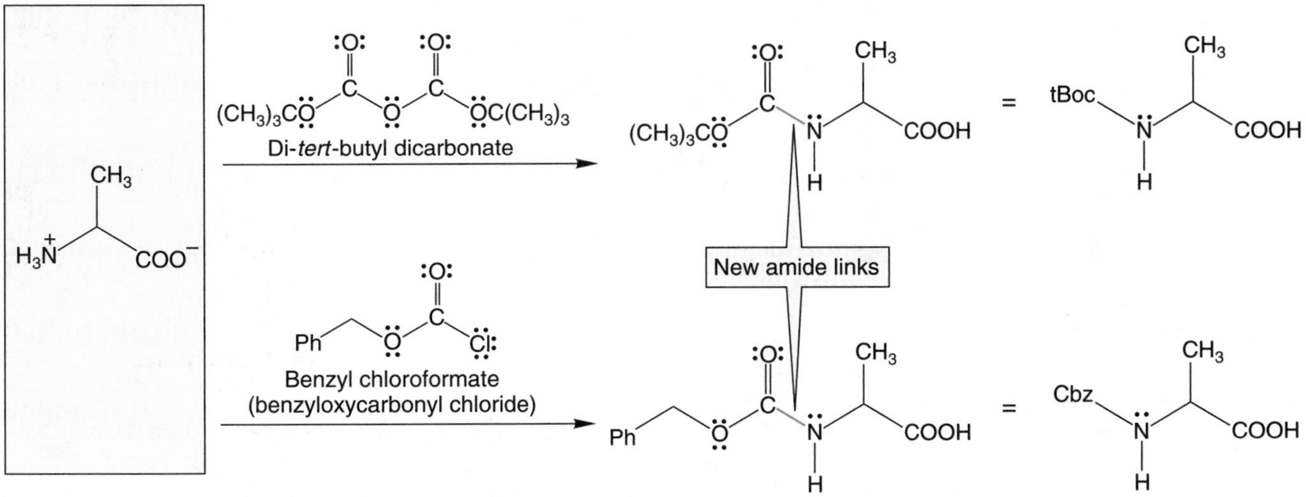

FIGURE 26.52 Two protecting, or blocking groups for amine ends of amino acids.

So now we have two ways of deactivating one amino group, in this case the amino group of alanine, so it cannot participate in the formation of the peptide amide bond. Now we need to deactivate the carboxy end of the other amino acid, leucine. This deactivation can be accomplished by simply transforming the acid into an ester (Fig. 26.53).

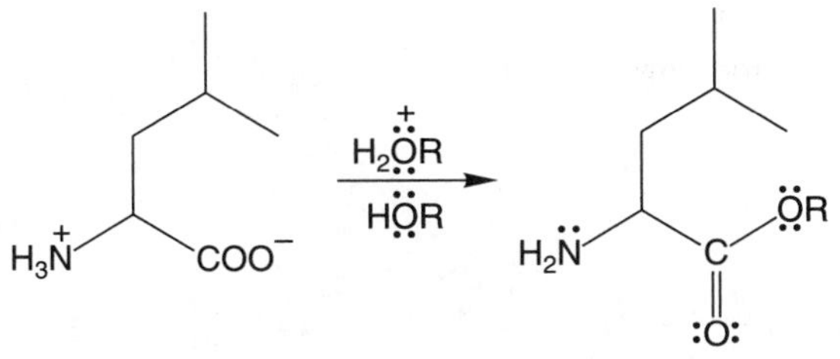

FIGURE 26.53 A carboxy end can be protected as a simple ester.

Now we need to do two more things. First, we still have to do the actual joining of the unprotected amino group of leucine to the unprotected carboxyl group of alanine, and we need to do this very efficiently and under the mildest possible conditions. Second, we need to deprotect the tied-up amino and carboxyl groups after the amide linkage has been formed.

Dicyclohexylcarbodiimide (DCC) is the reagent of choice for peptide bond making. Formation of the amide bond is a dehydration reaction (water is the other product), and DCC is a powerful dehydrating agent. At the end of the reaction it appears as dicyclohexylurea (DCU), the hydrated form of DCC (Fig. 26.54).

What is the mechanism of this coupling reaction? Remember that other cumulated double bonds as in ketenes (Chapter 20, p. 1035), isocyanates (Chapter 23, p. 1223), and isothiocyanates (Chapter 26, p. 1368)

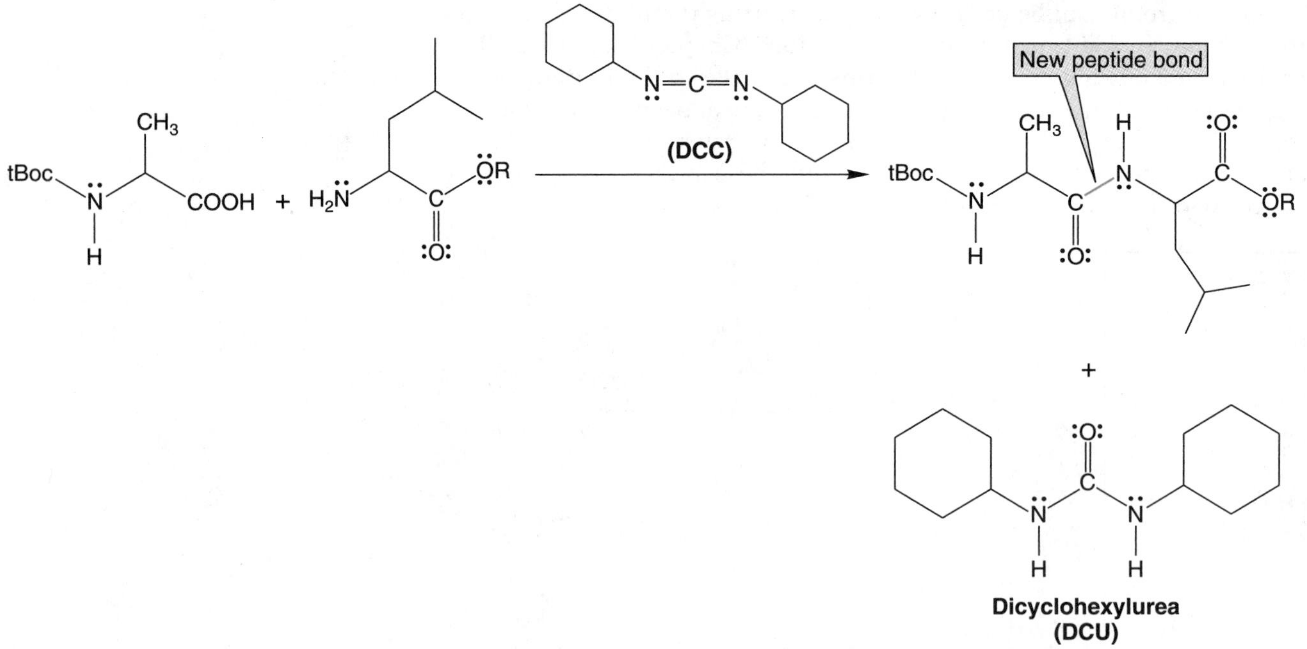

FIGURE 26.54 Coupling can be achieved using dicyclohexylcarbodiimide (DCC).

all react rapidly with nucleophiles. The related DCC is no exception, and it is attacked by amines to give molecules called guanidines. Carboxylates are nucleophiles too, and will also add to DCC (Fig. 26.55).

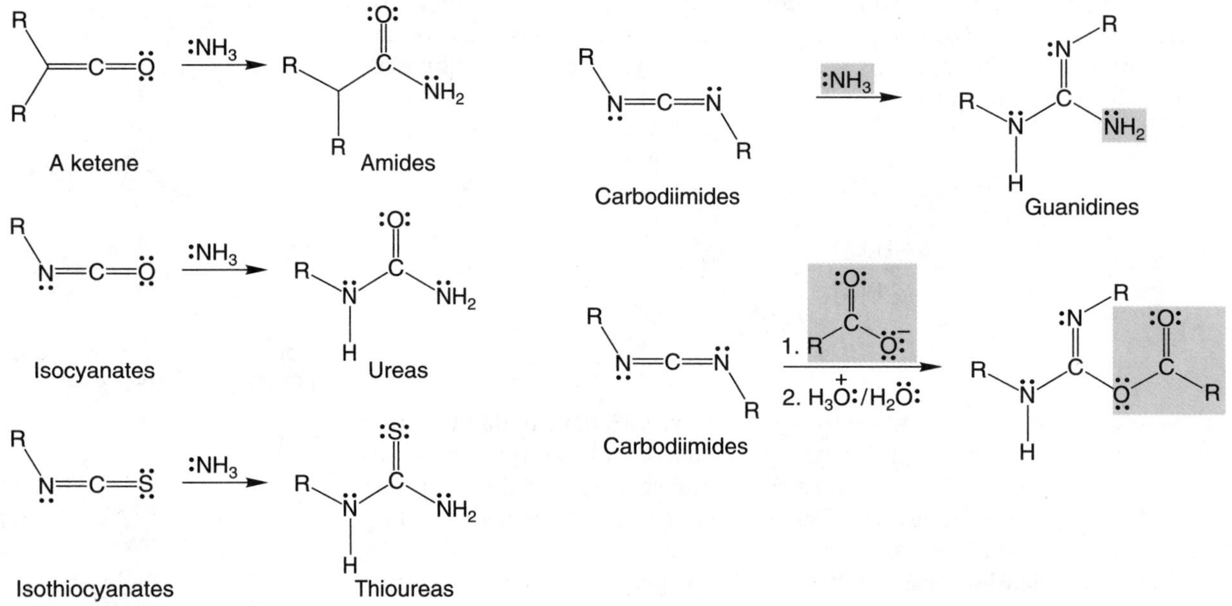

FIGURE 26.55 Cumulated double bonds react with all manner of nucleophiles, including amines and carboxylate anions.

The unprotected carboxylic acid group of alanine adds to DCC to give an intermediate that can react with a second carboxylic acid to give the anhydride. The amino acid, leucine, then adds through an addition–

elimination mechanism to give the dipeptide. Note that the dipeptide is still protected in two places (Fig. 26.56).

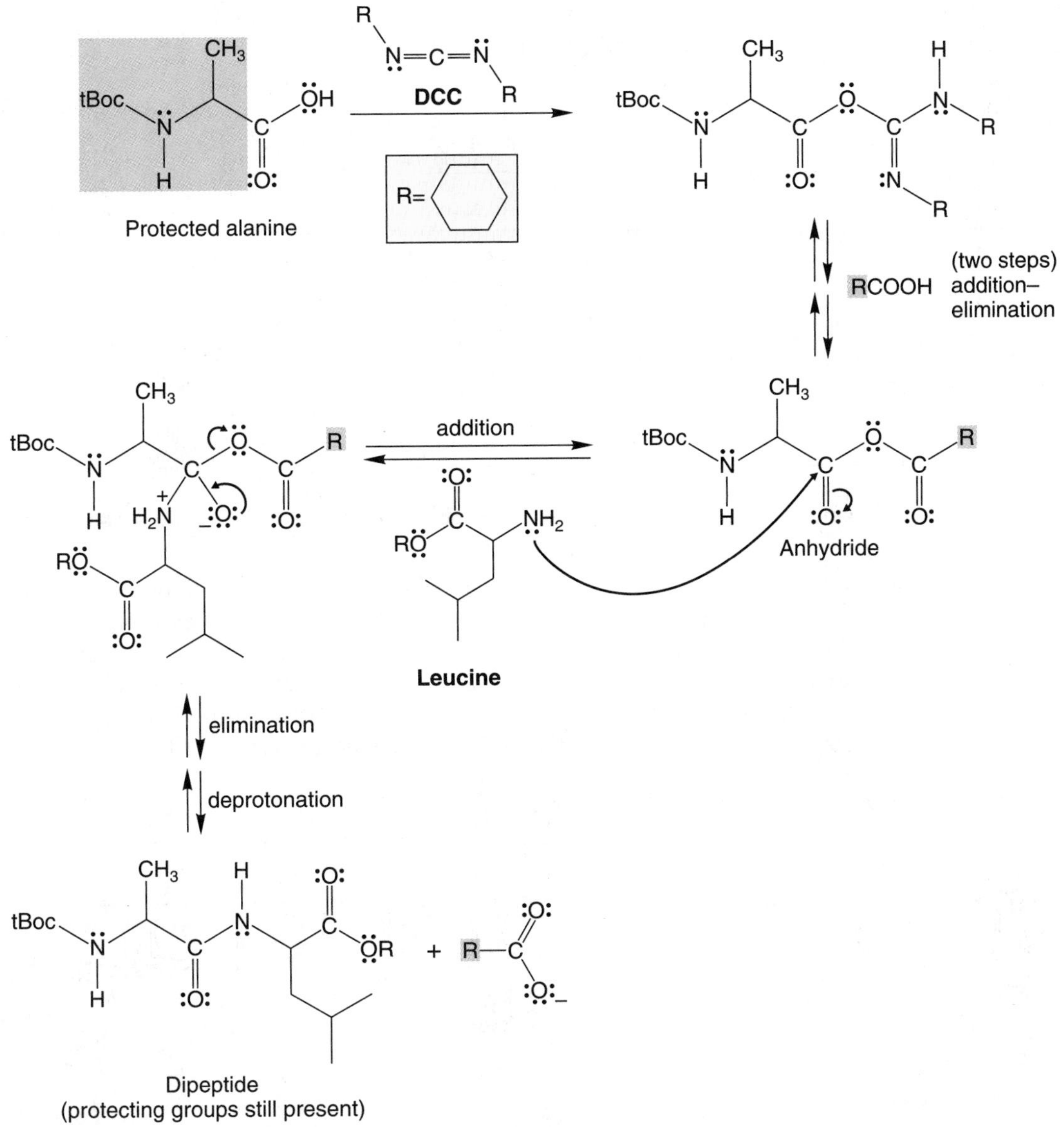

FIGURE 26.56 The mechanism of amino acid coupling using DCC.

Now all we have to do is remove these protecting groups and, at last, we will have made our dipeptide, Ala·Leu. The Cbz blocking group is easily removed by catalytic hydrogenation, and tBoc by very mild acid hydrolysis, which does not cleave amide bonds. The acid can be regenerated by treatment of the ester with base. The ester will hydrolyze faster than the amide (Fig. 26.57).

What about the problem of yield? Remember, this process, with all its protection and deprotection steps, must be repeated many, many times in the synthesis of a large peptide or protein. These steps are efficient—they proceed in high yield as essentially no side pathways are followed. The problem of tedious repetition has been solved in an old fashioned way—by

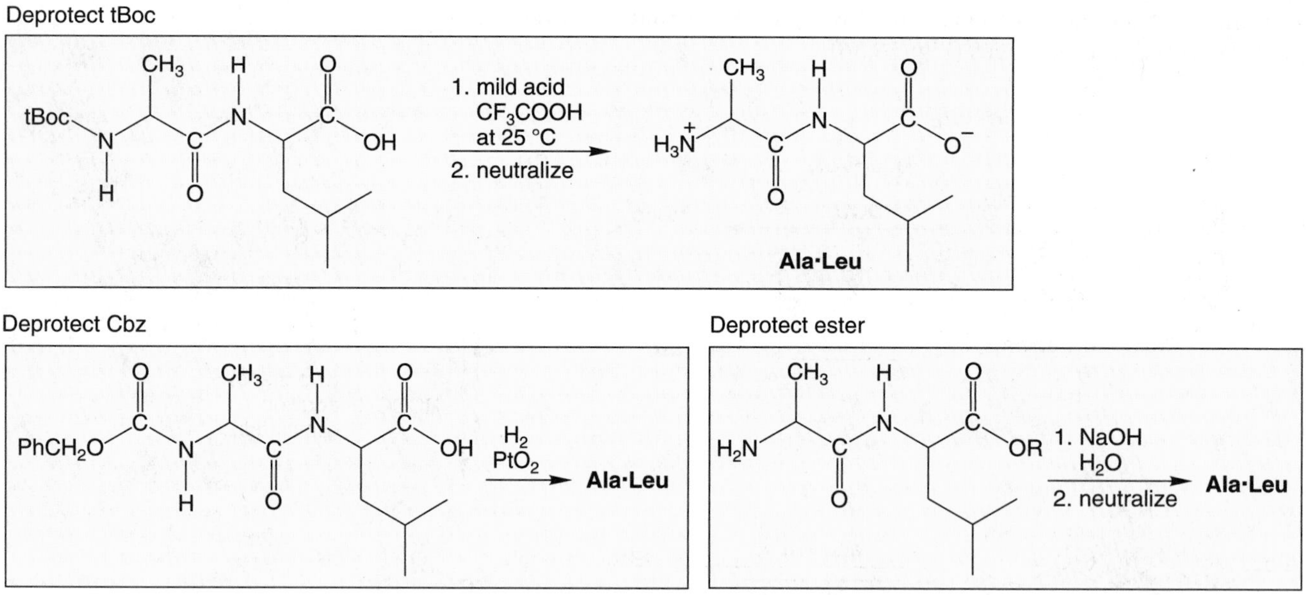

Deprotect tBoc

tBoc–NH–CH(CH_3)–C(=O)–NH–CH(isobutyl)–COOH → 1. mild acid CF_3COOH at 25 °C; 2. neutralize → $H_3\overset{+}{N}$–CH(CH_3)–C(=O)–NH–CH(isobutyl)–COO^-

Ala·Leu

Deprotect Cbz

$PhCH_2O$–C(=O)–NH–CH(CH_3)–C(=O)–NH–CH(isobutyl)–COOH → H_2, PtO_2 → **Ala·Leu**

Deprotect ester

H_2N–CH(CH_3)–C(=O)–NH–CH(isobutyl)–C(=O)OR → 1. NaOH, H_2O; 2. neutralize → **Ala·Leu**

FIGURE 26.57 Methods for removing the protecting groups used in our scheme.

automation. Polypeptides can be synthesized by a machine invented by R. Bruce Merrifield (b. 1921). In Merrifield's procedure, the carboxy end of a tBoc-protected amino acid is first anchored to a polymeric material constructed of polystyrene in which some phenyl rings have been substituted with chloromethyl groups. The attachment of the amino acid involves a displacement of chlorine through an S_N2 reaction (Fig. 26.58).

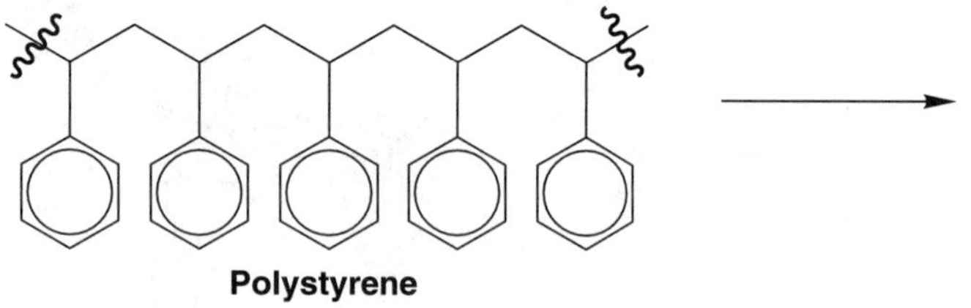

Polystyrene

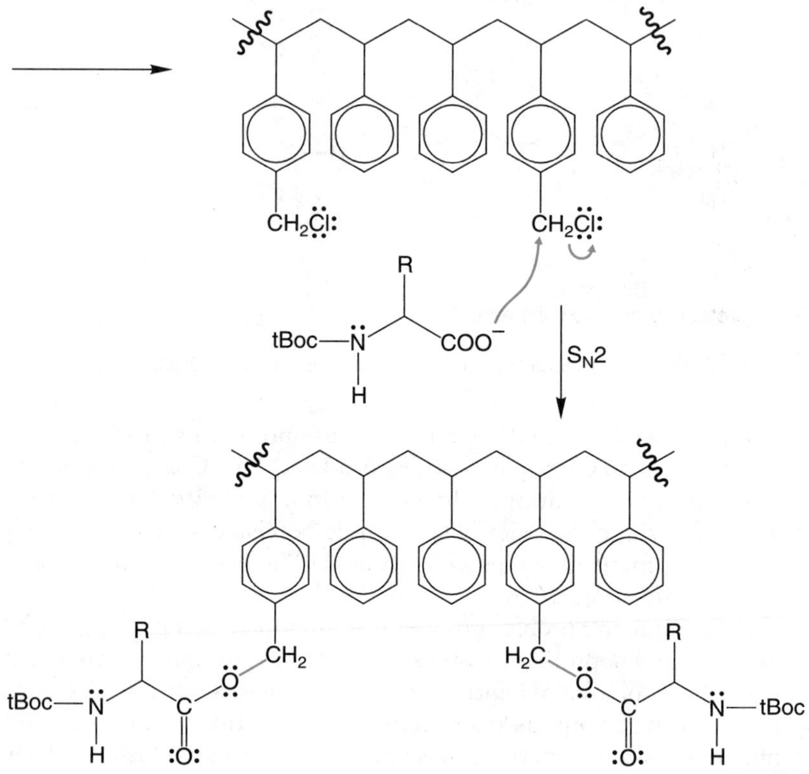

$CH_2\ddot{Cl}$: $CH_2\ddot{Cl}$:

tBoc–N(H)–CH(R)–COO^-

S_N2

tBoc–N(H)–CH(R)–C(=O)–O–CH_2 H_2C–O–C(=O)–CH(R)–N(H)–tBoc

FIGURE 26.58 Polystyrene can be chloromethylated in a Friedel–Crafts procedure, and the chlorines replaced through S_N2 reaction with the unprotected carboxy end of an amino acid. This technique anchors an amino acid to the polymer chain.

PROBLEM 26.22

What is the function of the benzene ring of polystyrene? Could any ring-containing polymer have been used? Would cyclohexane have done as well, for example?

The tBoc group is then removed with mild acid hydrolysis, and a new, tBoc-protected amino acid is attached to the bound amino acid through DCC-mediated coupling (Fig. 26.59).

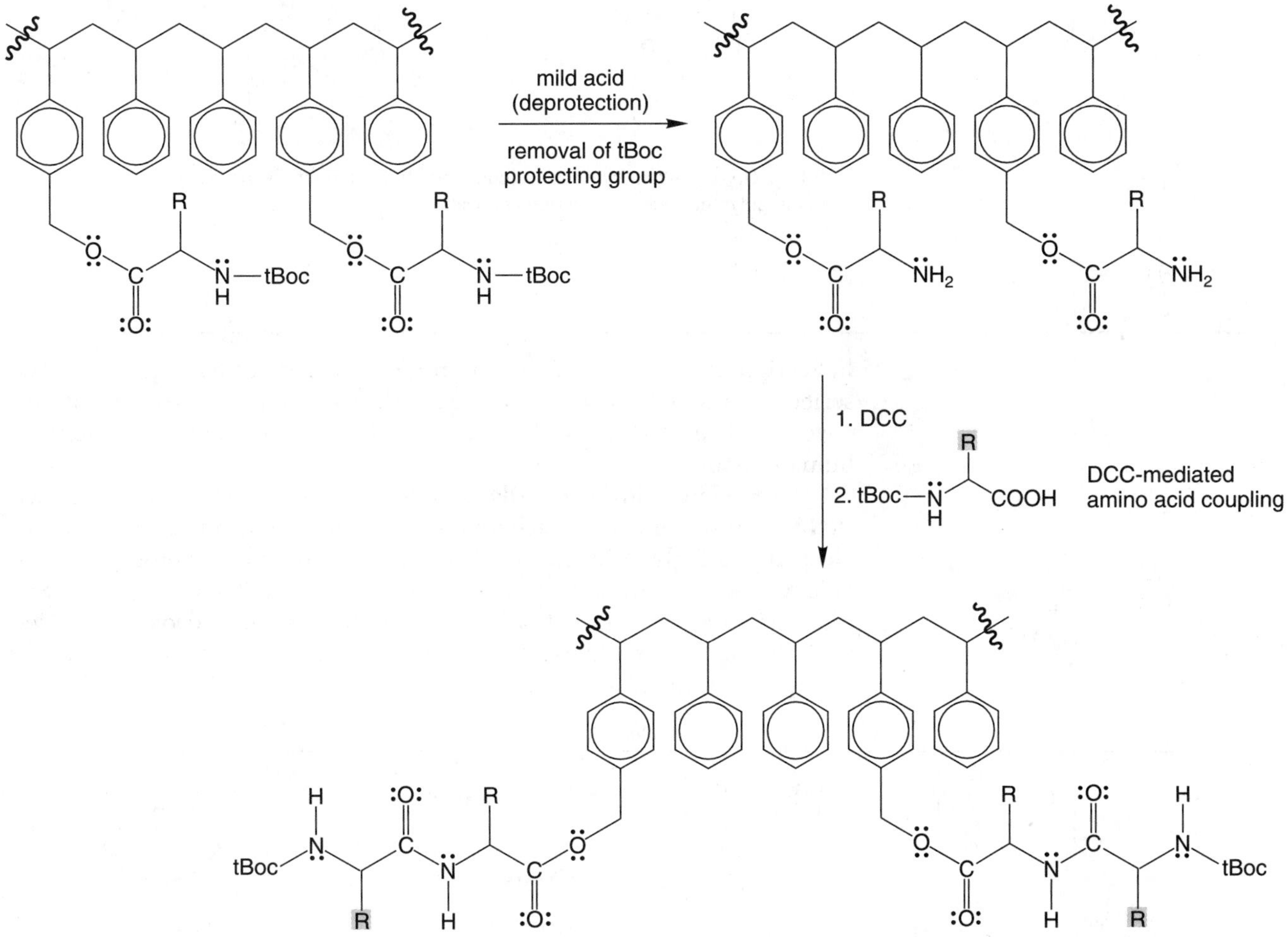

FIGURE 26.59 The steps of amino acid synthesis can now be carried out on this immobilized amino acid.

This procedure can be repeated as many times as necessary. The growing chain cannot escape, as it is securely bound to the polystyrene; reagents can be added and byproducts can be washed away after each reaction. The final step is detachment of the peptide from the polymeric resin, usually through reaction with hydrogen fluoride (HF) or another acid (Fig. 26.60).

The first relatively primitive homemade Merrifield machine managed to produce a chain of 125 amino acids in an overall 17% yield, a staggering accomplishment. Merrifield was quite rightly rewarded with the Nobel prize for 1984.

FIGURE 26.60 When synthesis is complete, the polypeptide can be detached from the polymer backbone by reaction with HF.

26.4 SOMETHING MORE: NUCLEOSIDES, NUCLEOTIDES, AND NUCLEIC ACIDS

In Section 26.3c, the synthesis of proteins carried out by humans was described. In the natural world* quite another technique is used, and our uncovering of the outlines of how it works is one of the great discoveries of human history.

The **nucleic acids, deoxyribonucleic acid (DNA)**, and **ribonucleic acid (RNA)**, are polymers of **nucleotides** (phosphorylated nucleosides). A **nucleoside** is a β-glycoside formed between a sugar and a heterocyclic molecule (a **base**). In a ribonucleoside, the sugar ribose (Chapter 24, p. 1268) is attached to a base through a β-glycosidic linkage. In a deoxyribonucleoside, the sugar deoxyribose replaces ribose. 5-Phosphorylated nucleosides are called nucleotides (Fig. 26.61).

Sugars

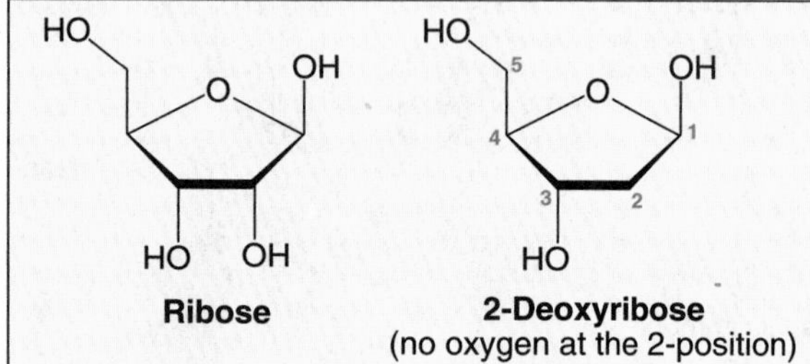

Nucleosides

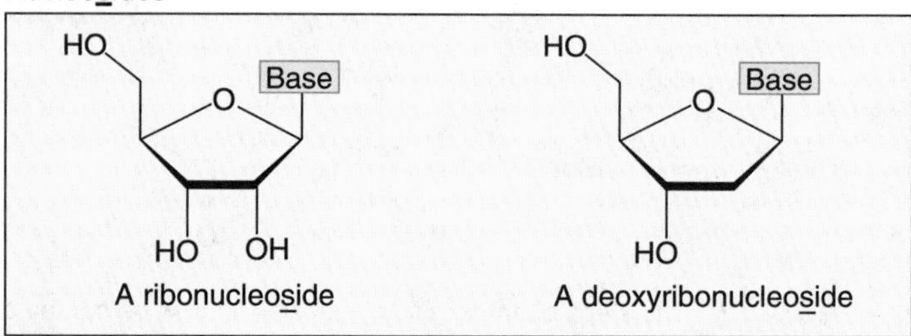

Nucleotides

FIGURE 26.61 The sugars ribose and deoxyribose, and the corresponding nucleosides and nucleotides, phosphorylated nucleosides.

* It's an artificial distinction, almost as foolish as labels describing natural as opposed to artificial vitamin C. The Merrifield procedure is no less natural than what we are about to describe. It's not unnatural, it's just human-mediated.

In the polymers (the nucleic acids DNA and RNA), the sugars, ribose in RNA and deoxyribose in DNA, are linked to each other through phosphoric acid groups attached to the 5′ position of one sugar and the 3′ position of the other. Each sugar still carries a heterocyclic base attached to the 1′ carbon. We saw these heterocyclic bases first in (Chapter 25, p. 1318; Fig. 26.62).

One of the striking features of this replication machine, for that is exactly what we have begun to describe, is the economy with which it operates. So far, we have phosphoric acid and two sugars, ribose and deoxyribose, held together in polymeric fashion. One might have been pardoned for imagining that all manner of bases would have been attached to this backbone, and that the diversity of the world of living things would come from the introduction of detail in this way. But that would be wrong: There are only five bases attached to C(1) in DNA and RNA, and, in fact, three of them (cytosine, adenine, and guanine), are common to *both* kinds of nucleic acid and the other two (thymine and uracil) differ only in the presence or absence of a single methyl group. These bases are written in shorthand by using a single letter, C, A, G, T, or U (Fig. 26.63). The introduction of detailed information must come in another way.

Notice the correspondence between these polymers and the other polymeric species we have seen, the proteins. In the polyamino acids, the

FIGURE 26.62 Nucleic acids (DNA and RNA) are polymeric collections of nucleotides.

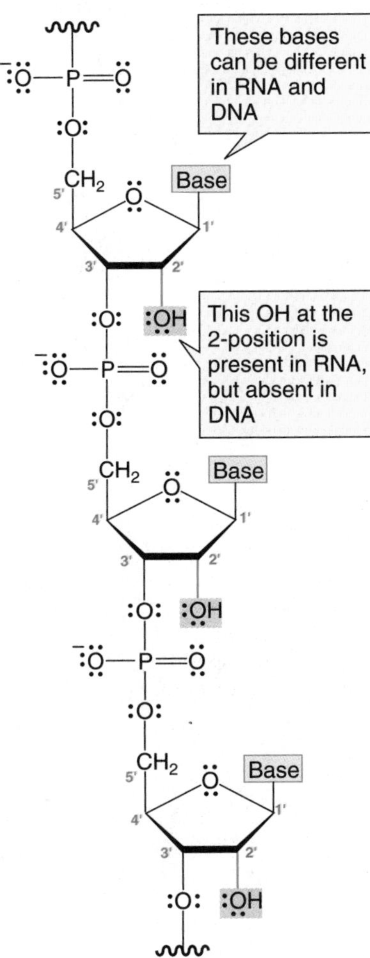
These bases can be different in RNA and DNA
Base
This OH at the 2-position is present in RNA, but absent in DNA
Base
Base

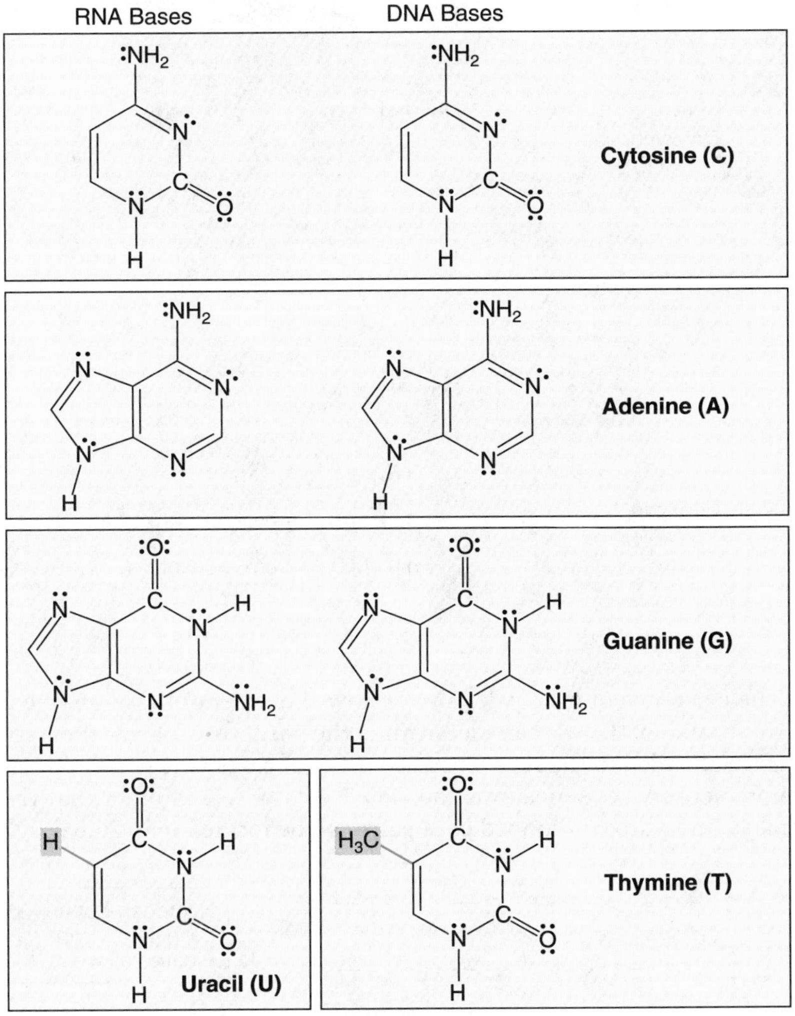

RNA Bases | DNA Bases

Cytosine (C)

Adenine (A)

Guanine (G)

Uracil (U)

Thymine (T)

FIGURE 26.63 The five bases present in RNA and DNA. Three are common to both kinds of nucleic acid.

units are linked by amide bonds. In DNA and RNA they are attached by phosphates. The information in proteins is carried by the side chains, the R groups whose steric and electronic demands determine the higher order structures—secondary, tertiary, and quaternary—of these molecules. In DNA and RNA, it is the bases that perform the function of the R groups of proteins. There are only 20 common R groups, but there are even fewer bases—a mere four for each nucleic acid. Nevertheless, higher ordered structures are also found in the nucleic acids. Our next task is to see how this comes about.

A crucial observation was made by Erwin Chargaff (b. 1905) who noticed in 1950 that in many different DNAs the amounts of the bases adenine (A) and thymine (T) were always equal, as were the amounts of guanine (G) and cytosine (C). The four bases of DNA seemed to be found in pairs. This observation led to the reasonable notion that the members of the pairs must be associated in some way, and to the detailed picture of hydrogen-bonded dimeric structures of **base pairs** in which A was always associated with T and G with C. These combinations are the result of especially facile hydrogen bonding; the molecules fit well together (Fig. 26.64).

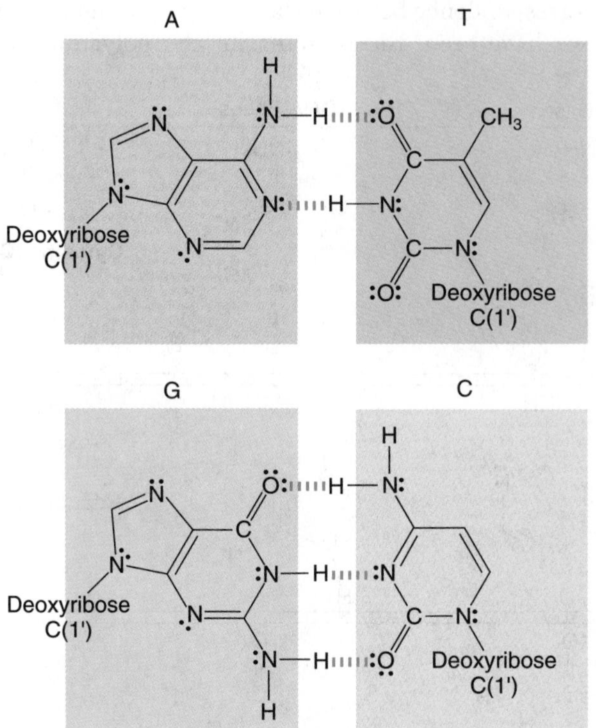

FIGURE 26.64 Adenine-thymine and guanine-cytosine form base pairs through effective hydrogen-bond formation.

The existence of base pairs shows how the order of bases on one polymeric strand of DNA can determine the structure of another strand, hydrogen bonded to the first. For a sequence of bases C C A T G C T A, the complementary sequence must be G G T A C G A T in order that the optimal set of hydrogen-bonded base pairs can be formed (Fig. 26.65).

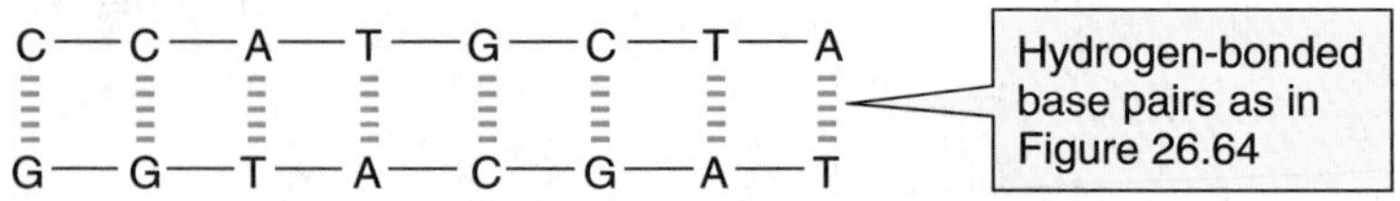

C—C—A—T—G—C—T—A

G—G—T—A—C—G—A—T

Hydrogen-bonded base pairs as in Figure 26.64

FIGURE 26.65 The sequence of nucleotides on one polymer determines the sequence in another through formation of hydrogen-bonded base pairs, C-G and A-T.

This notion was also instrumental in the realization by James D. Watson (b. 1928) and Francis H. C. Crick (b. 1916), with immense help from X-ray diffraction patterns determined by Rosalind Franklin (1920–1958) and Maurice Wilkins (b. 1916), that the structure of DNA was that of a double-stranded helix in which the two DNA chains were held together in the helix by A-T and G-C hydrogen-bonded base pairs. The diameter of the helix is about 20 Å, and there are 10 residues per turn of the helix. Each turn is about 34 Å long (Fig. 26.66).

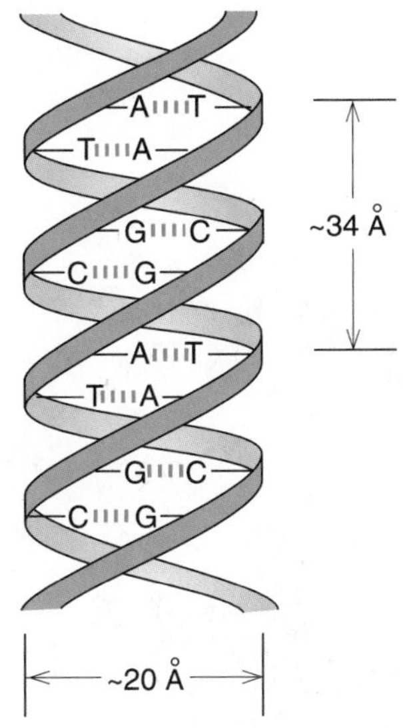

FIGURE 26.66 The famous double helix is held together by hydrogen bonds between base pairs.

Now we still have two outstanding problems: (1) Where does the density of information come from? How is the seemingly inadequate pool of five bases translated into the wealth of information that must be transferred in the synthesis of proteins containing specific sequences made from the 20 amino acids? (2) What is the actual mechanism of the information transferal? The second of these two questions is probably the easier to tackle.

The replication of DNA involves partial unwinding of the helix to yield two templates for the development of new DNA chains. Note that it is the order of bases in each strand that determines the growth of the next strand. Where there is a T, an A must appear, where there is a G, a C must attach, and so on. *Each strand of the original DNA must induce the formation of an exact duplicate of the strand to which it was originally attached* (Fig. 26.67).

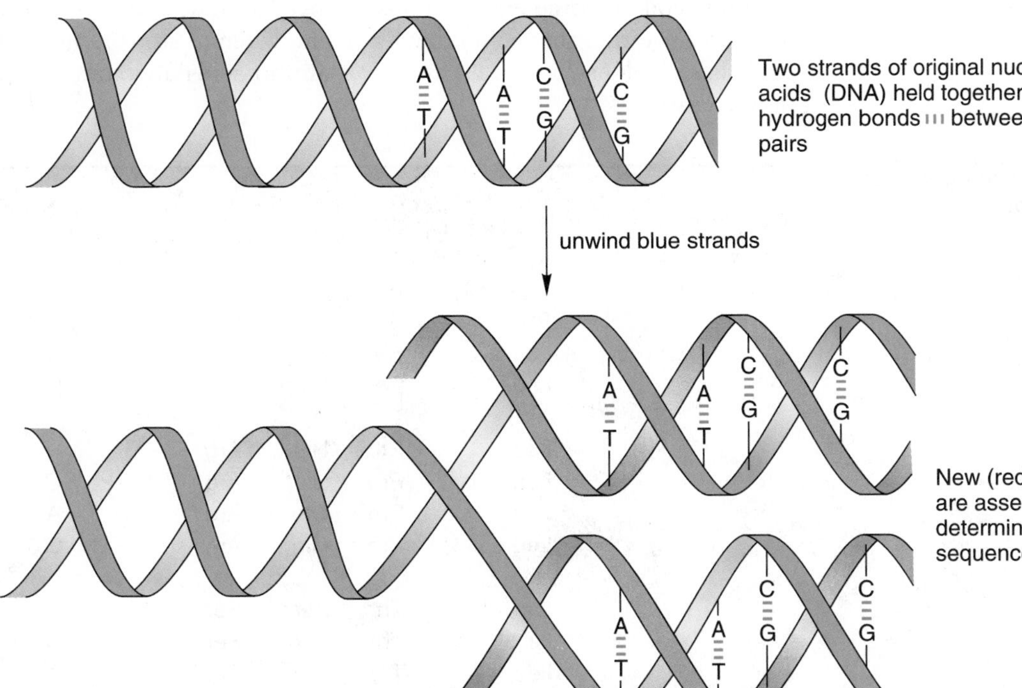

FIGURE 26.67 The replication machine works by first unraveling to form single-stranded sections. Assembly of new polymers is determined by the original base sequence. Where there is an A, a T must be added; a G requires a C.

Protein synthesis starts with a similar, but not quite identical process. A DNA strand can also act as a template for the production of an RNA polymer. The differences are that in RNA it is the sugar ribose that helps construct the sugar–phosphate backbone, not deoxyribose, and that the base uracil replaces thymine. The RNA produced is called **messenger RNA (mRNA)** (Fig. 26.68).

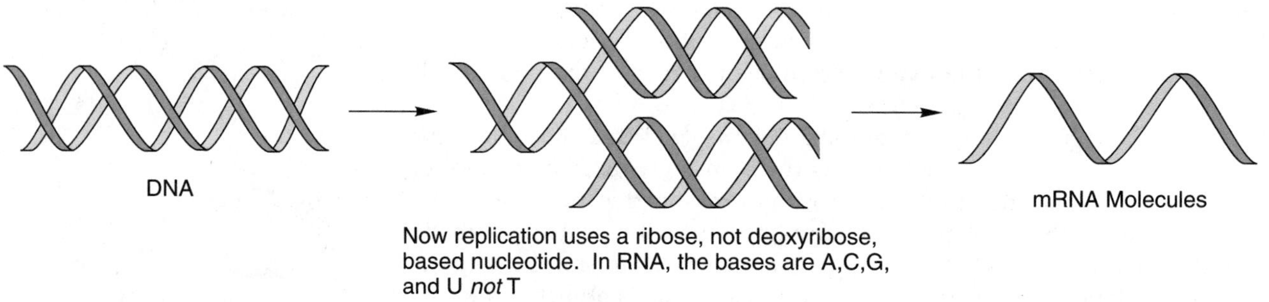

DNA

mRNA Molecules

Now replication uses a ribose, not deoxyribose, based nucleotide. In RNA, the bases are A,C,G, and U *not* T

FIGURE 26.68 Synthesis of mRNA.

This new RNA molecule acts to direct protein synthesis in the following spectacularly simple way: Synthesis of each of the 20 common amino acids is directed by one or more sequences of three bases, called **codons**. How many sequences of three bases can be constructed from the available four RNA bases? The answer is $4^3 = 64$, more than enough to code for the 20 common amino acids. Indeed, some amino acids are coded by more than one codon. Some extra codons are needed to deal with the problems of when to start amino acid production and when to stop. Table 26.2 gives three-base codons for the 20 common amino acids and some frequently used commands.

TABLE 26.2 Codons for Amino Acids and Simple Commands

First Position (5′)	Second Position				Third Position (3′)
	U	C	A	G	
U	Phe	Ser	Tyr	Cys	U
	Phe	Ser	Tyr	Cys	C
	Leu	Ser	Stop	Stop	A
	Leu	Ser	Stop	Trp	G
C	Leu	Pro	His	Arg	U
	Leu	Pro	His	Arg	C
	Leu	Pro	Gln	Arg	A
	Leu	Pro	Gln	Arg	G
A	Ile	Thr	Asn	Ser	U
	Ile	Thr	Asn	Ser	C
	Ile	Thr	Lys	Arg	A
	Met	Thr	Lys	Arg	G
	Start	Thr	Lys	Arg	G
G	Val	Ala	Asp	Gly	U
	Val	Ala	Asp	Gly	C
	Val	Ala	Glu	Gly	A
	Val	Ala	Glu	Gly	G

As an example of how this process works, let's look at a strand of DNA with the sequence: T A C C G A A G C A C G A T T. Base pairing must produce a mRNA fragment: A U G G C U U C G U G C U A A (remember that in RNA the base U replaces T). A look at Table 26.2 shows that this translates into the following message: "Start - add Ala - add Ser - add Cys - Stop."

*PROBLEM 26.23

Why would 20 amino acids be coded for by a total of 64 three-base sequences? Why is there so much overlap? Consider the alternative method of having two-base sequences coding for amino acid production. How many possible two-base sequences are there for a total of four bases?

ANSWER

There are only 2^4 possible combinations of four objects taken two at a time. So there are only 16 things that could be coded by two-base sequences of four different bases, which is not sufficient to code for the 20 common amino acids. Three-base codons turns out to be the most economical way to code for a system of 20 amino acids.

One further problem remains. How does mRNA direct amino acid synthesis? Other RNAs are used, called transfer RNA (tRNAs), which are relatively small and designed to interact with an enzyme, aminoacyl tRNA synthetase, to acquire one specific amino acid and carry it to the mRNA where it is added to the growing chain at the position of the correct codon.

26.5 SUMMARY

NEW CONCEPTS

By far the major new concept in this chapter is that the assembly of α-amino acids into long polymeric chains leads to structures of higher order. The secondary structure of these peptides or proteins involves the regions of order within the polymer. Most often these are in the α-helical or β-pleated sheet arrangements, which are separated from each other by disordered sections of the chains, random coils. Disulfide bonds as well as electrostatic forces, van der Waals forces, and hydrogen bonding twist these molecules into shapes characteristic of individual proteins, the tertiary structure. Secondary and tertiary structure can be destroyed, sometimes only temporarily, by any of a number of denaturing processes. Finally, intermolecular forces can hold a number of these molecules together to form super molecules. We call this the quaternary structure.

These higher ordered structures are affected by the identities of the R groups on the constituent amino acids. It is the electronic and steric properties of the R groups that generate the particular secondary, tertiary, and quaternary structures of proteins.

These new structures require new analytical and synthetic techniques. Methods for determining the sequence of peptides and proteins involve physical techniques such as X-ray diffraction, as well as chemical techniques for revealing the terminal amino acids of the chains, and even for sequencing entire polymers.

Other biopolymers, the nucleic acids, are constructed not of amino acids, but of nucleotides, sugars bearing heterocyclic base groups and joined through phosphoric acid linkages. These molecules also have higher order, the most famous being the double-helical arrangement of DNA. Most remarkable of all is the ability of these molecules to carry the genetic code, to uncoil and direct the assembly not only of replicas of themselves, but also, through the mediation of RNA, of the polyamino acids called proteins. In the nucleic acids, the information is not carried in a diverse supply of R groups as it is in peptides and proteins, but in a small number of base groups. Hydrogen bonding in base pairs allows for the replication reactions, and three-base units called codons direct the assembly of amino acids into proteins.

REACTIONS, MECHANISMS, AND TOOLS

Reactions important in determining the sequence of a polymer of amino acids include several techniques for end-group analysis. There are enzymes (proteins themselves) that will cleave the amino acid marking the carboxy terminus of the polymer. The amino terminus can be found through the Sanger reaction.

In principle, an entire sequence can be unraveled using the Edman degradation in which phenylisothiocyanate is used in a procedure that cleaves the amino acid at the amino terminus and labels it as a phenylthiohydantoin.

Other agents are available for inducing specific cleavages in the polymers, which can yield smaller fragments for sequencing through the Edman procedure. These include various enzymes and small chemical reagents such as cyanogen bromide.

Synthesis of these polymeric molecules requires nothing more than the sequential formation of a large number of amide bonds. The DCC coupling procedure is effective, but it requires that the amino acid monomers be supplied in a form in which the ends to be attached are activated, or at least available, and other reactive points blocked, or protected (Fig. 26.69).

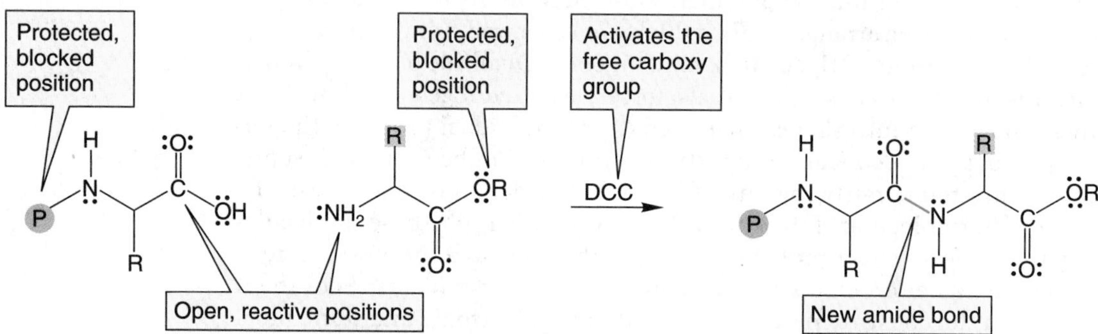

FIGURE 26.69 The DCC-mediated amino acid coupling. Protecting and activating groups must be added in appropriate places and removed after the reaction is over.

In turn, we need sources of the amino acids themselves. As with simple amines, a straightforward S_N2 alkylation procedure is usually inadequate. The best known syntheses of amino acids are the Gabriel and Strecker procedures.

Naturally occurring amino acids are optically active (except for glycine, which is achiral) and occur in the L form. Resolution can be achieved through the classical technique of diastereomer formation followed by crystallization, or through other methods. The most clever of these involves a kinetic resolution in which advantage is taken of the ability of enzymes to attack only the naturally occurring (*S*) enantiomers, while the (*R*) isomer is left alone (Fig. 26.20).

SYNTHESES

1. Amino acids

NH_3

Simple alkylation of ammonia using an α-bromo acid; overalkylation is a problem

1. (phthalimide anion)
2. R—X, base
3. $H_3O^+/H_2O/\Delta$

Gabriel synthesis. The R—X must be S_N2 active

R—CHO
1. KCN/NH_3
2. H_2O/H_3O^+

Strecker sythesis

2. Optically active amino acids

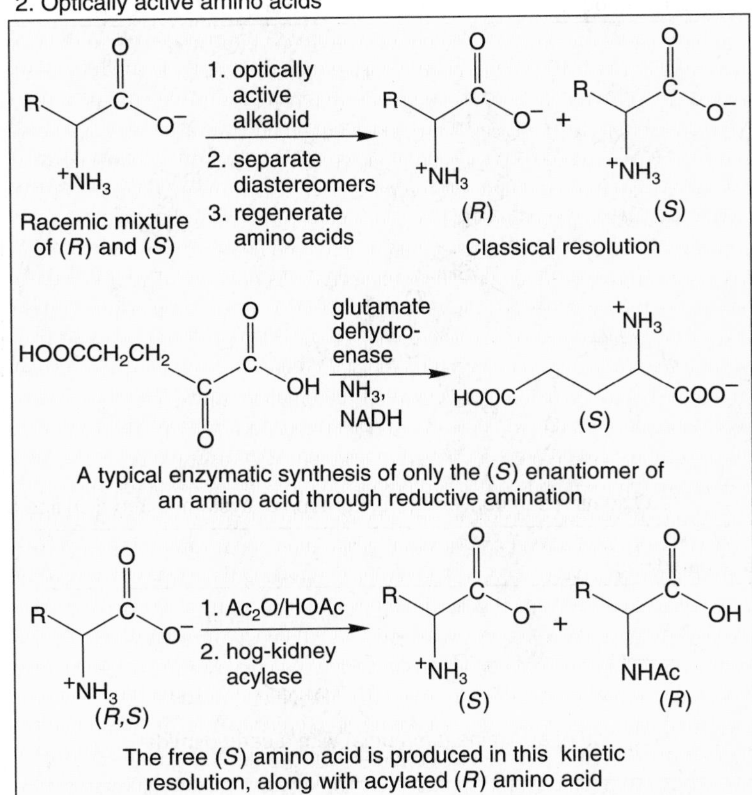

3. Protected amino acids

$(CH_3)_3CO-C(=O)-O-C(=O)-OC(CH_3)_3$

A tBoc-protected amino acid

$PhCH_2O-C(=O)-Cl$

A Cbz-protected amino acid

FIGURE 26.70 The new synthetic reactions of Chapter 26.

3. Protected amino acids (continued)

P–NH–CH(R)–C(=O)OH + H_2N–CH(R)–C(=O)OR → (1. DCC; 2. removal of protecting groups) → $H_3\overset{+}{N}$–CH(R)–C(=O)–NH–CH(R)–COO^-

P = tBoc or Cbz

DCC-Mediated coupling; notice that the starting amino acids are prepared for reaction with protecting groups that are removed in later steps

4. Disulfide bridges

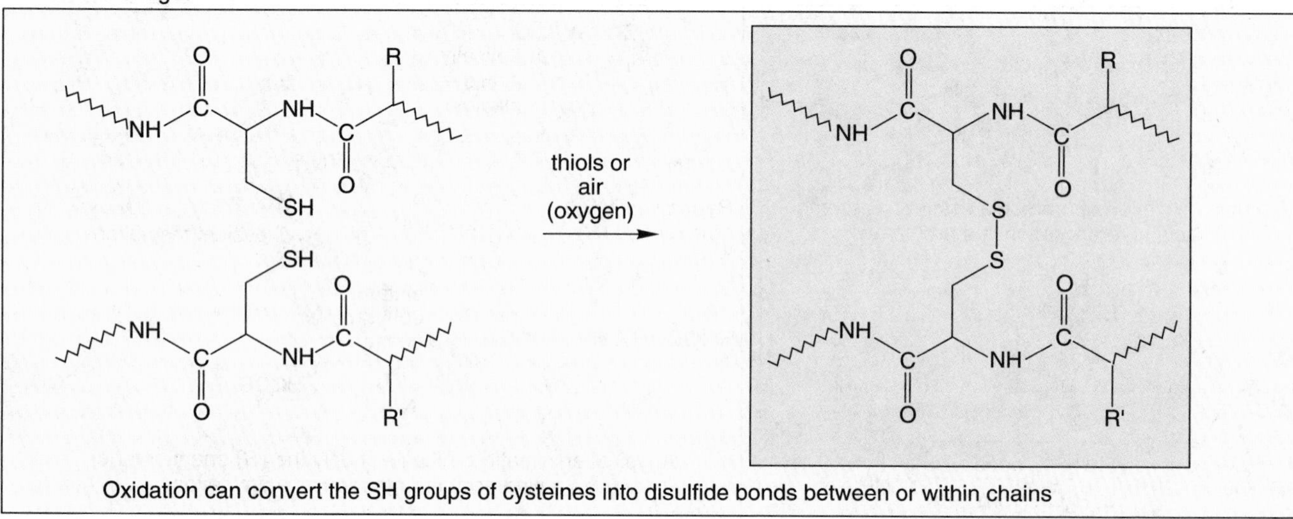

Oxidation can convert the SH groups of cysteines into disulfide bonds between or within chains

5. Guanidines

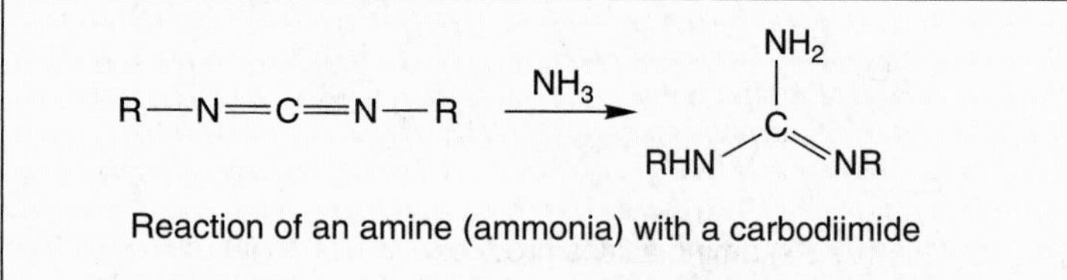

Reaction of an amine (ammonia) with a carbodiimide

6. Thiohydantoin

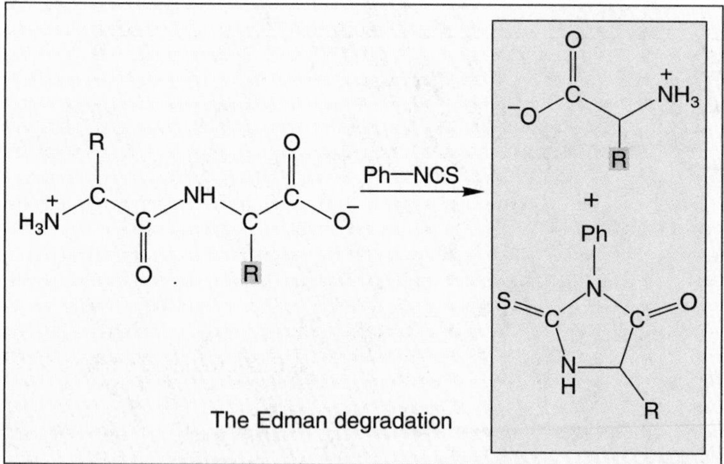

The Edman degradation

7. Thioureas

R–N=C=S $\xrightarrow{RNH_2}$ RHN–C(=S)–NHR

Addition of an amine to an isothiocyanate

8. Ureas

R–N=C=O $\xrightarrow{RNH_2}$ RHN–C(=O)–NHR

Addition of an amine to an isocyanate

R–N=C=N–R $\xrightarrow{H_2O}$ RHN–C(=O)–NHR

Hydration of a carbodiimide

FIGURE **26.70** (CONTINUED)

26.6 KEY TERMS

Active site The region in a protein in which a substrate molecule is bound and in which chemical reaction often takes place.

Acylating agent Any of a number of reagents capable of transferring an acyl group (RC=O) to a nucleophilic site, usually N or O. Examples are acetyl chloride and acetic anhydide.

α-Amino acids 2-Amino acetic acids, the monomeric constituents of the polymeric peptides and proteins.

Amino terminus The amino acid at the free amine end of the peptide polymer.

Base One of the five heterocyclic molecules, adenine (A), thymine (T), uracil (U), cytosine (C), and guanine (G), attached to the 1′ position of the sugar (ribose or deoxyribose) in a nucleotide or nucleoside.

Base pair A hydrogen-bonded pair of bases, always A-T in DNA or A-U in RNA, and C-G in both DNA and RNA.

tBOC A protecting group for the amino end of an amino acid that works by transforming the amine into a less basic carbamate.

Carboxy terminus The amino acid at the free carboxylic acid end of the peptide polymer.

Cbz A protecting group for the amino end of an amino acid that works by transforming the amine into a less basic carbamate.

Chromatogram A plot of amount of component versus time for any chromatographic technique.

Chymotrypsin An enzyme that cleaves sequences of amino acids after any amino acid containing an aromatic side chain.

Cyanogen bromide (BrCN) A reagent able to cleave peptide chains after a methionine residue.

Codon A three-base sequence in a polynucleotide that directs the addition of a particular amino acid to a growing chain of amino acids. Some codons also give directions: "start assembly" "stop assembly," and so on.

Deoxyribonucleic acid (DNA) A polymer of nucleotides made up of deoxyribose units connected by phosphoric acid links. Each sugar is attached to one of the bases, A, T, G, or C.

Denaturing The destruction of the higher order structures of a protein, sometimes reversible, sometimes not.

Dicyclohexylcarbodiimide (DCC) A dehydrating agent effective in the coupling of amino acids through the formation of amide bonds.

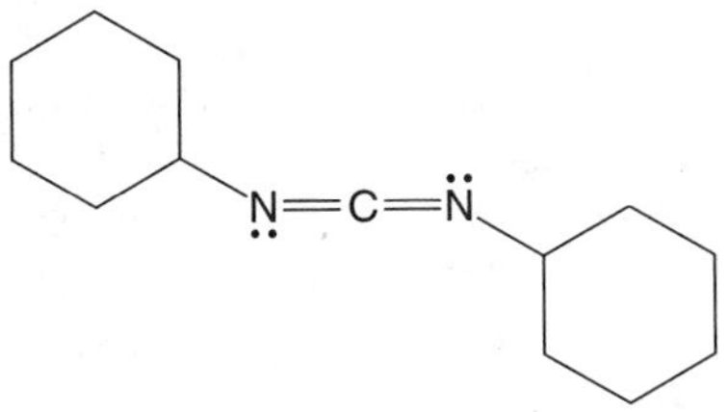

Disulfide bridges The attachment of amino acids through sulfur–sulfur bonds formed from the oxidation of cysteine CH_2SH side chains. Disulfide bridges can be formed within a single peptide or between two peptides.

Edman degradation The phenylisothiocyanate-induced cleavage of the amino acid at the amino terminus of a peptide. Successive applications of the Edman technique can determine the sequence of a peptide.

Electrophoresis A technique for separating amino acids or chains of amino acids that takes advantage of the different charge states of different amino acids (or their polymers) at a given pH.

Essential amino acid Any of the 10 amino acids that cannot be synthesized by humans and must be ingested directly.

Gabriel malonic ester synthesis A synthesis of amino acids that uses phthalimide as a source of the amine nitrogen. Overalkylation is avoided by decreasing the nucleophilicity of the nitrogen in this way.

Gel-filtration chromatography A chromatographic technique that relies on polymeric beads containing molecule-sized holes. Molecules that fit easily into the holes pass more slowly down the column than larger molecules, which fit less well into the holes.

α-Helix A right-handed coiled form adopted by many proteins in their secondary structures.

Hemoglobin The protein in human blood responsible for oxygen transport.

Hydrophilic group Literally a "water-loving" group, that is, a polar side chain or group.

Hydrophobic group Literally, a "water-hating" group, that is, a nonpolar side chain or group.

Ion exchange chromatography A separation technique that relies on differences in the affinity of molecules for a charged substrate.

Isoelectric point The pH at which the amount of an amino acid present as the ammonium ion exactly equals the amount present as the carboxylate anion.

Kinetic resolution A technique for separating a pair of enantiomers based on the selective transformation of one of them, often by an enzyme.

Messenger RNA (mRNA) A polynucleotide produced from DNA whose base sequence codes for amino acid assembly.

Ninhydrin The hydrate of indan-1,2,3-trione, a molecule that reacts with amino acids to give the purple dye used in quantitative analysis of amino acids.

Nucleic acid The polynucleotides DNA and RNA.

Nucleoside A sugar, either ribose or deoxyribose, bonded to a heterocyclic base at its 1′ position.

Nucleotide A phosphorylated nucleoside; one of the monomers of which DNA and RNA are composed.

Peptide A polymer of α-amino acids linked through amide bonds.

Phenylisothiocyanate The active ingredient, PhNCS, in the Edman degradation that converts the amino terminus of a peptide into a phenylthiohydantoin.

Pleated sheet One of the common forms of secondary peptide structure in which hydrogen bonding holds chains of amino acids roughly in parallel lines.

Primary structure The sequence of amino acids in a protein.

Protein A polymer of α-amino acids, generally larger than a peptide.

Quaternary structure A self-assembled aggregate of two or more protein units.

Random coil Disordered portions of a chain of amino acids.

Retention time The time necessary for elution of a component of a mixture in chromatography.

Ribonucleic acid (RNA) A polymer of nucleotides made up of ribose units connected by phosphoric acid links. Each sugar is attached to one of the bases, A, U, G, or C.

Sanger degradation A method for determining the amino acid at the amino terminus of a chain; using 2,4-dinitrofluorobenzene. Unfortunately, the Sanger degradation hydrolyzes, and thus destroys, the entire peptide.

Secondary structure Ordered regions of a protein chain. The two most common types of secondary structure are the α-helix and the β-pleated sheet.

Strecker synthesis A synthesis of amino acids using an aldehyde, cyanide, and ammonia, followed by hydrolysis.

Tertiary structure The structure of a protein induced by its folding pattern.

Thiourea A compound of the structure

:S:
C
RHN NHR

Trypsin An enzyme capable of cleaving a peptide after lysine or arginine.

Urea A compound of the structure

:O:
C
RHN NHR

Zwitterion A dipolar species in which full plus and minus charges coexist within the same molecule.

26.7 ADDITIONAL PROBLEMS

PROBLEM **26.24**

(a) Amides protonate on oxygen rather than nitrogen. Why?

(b) Amides (pK_b ~ 14) are weaker bases than amines (pK_b ~ 5). Why?

PROBLEM **26.25** Write structures for the following tripeptides and name them:

(a) Ala·Ser·Cys (b) Met·Phe·Pro (c) Val·Asp·His

PROBLEM **26.26** Name the following dipeptides:

(a) H_3C, OH, H, $H_3\overset{+}{N}$, N, COO^-, C, O, $CH_2CH(CH_3)_2$

(b) $NH_2CH_2CH_2CH_2CH_2$, H, $H_3\overset{+}{N}$, N, COO^-, C, O, H_2C, OH

PROBLEM **26.27** Draw (*S*)-serine and (*S*)-proline as tetrahedra.

PROBLEM **26.28** Write Fischer projections for L-serine and L-proline.

PROBLEM **26.29** Give the structures for the amino acids His, Arg, and Phe at pH = 3 and 12.

PROBLEM **26.30** Toward which electrode (anode or cathode) will the amino acids Asp and Lys migrate at pH = 7?

PROBLEM **26.31** All of the amino acids listed in Table 26.1, except the achiral glycine, have the L configuration. Of the 19 amino acids with the L configuration, 18 are (*S*) amino acids. Which amino acid has an (*R*) configuration? Carefully explain why the configuration of this L amino acid is (*R*) rather than (*S*).

PROBLEM **26.32** In Section 26.2c, we saw that all the amino acids in Table 26.1, except proline, react with ninhydrin to form the same purple compound. In contrast, proline reacts with ninhydrin to give a yellow compound thought to have structure **1**. Propose a mechanism for the formation of **1**.

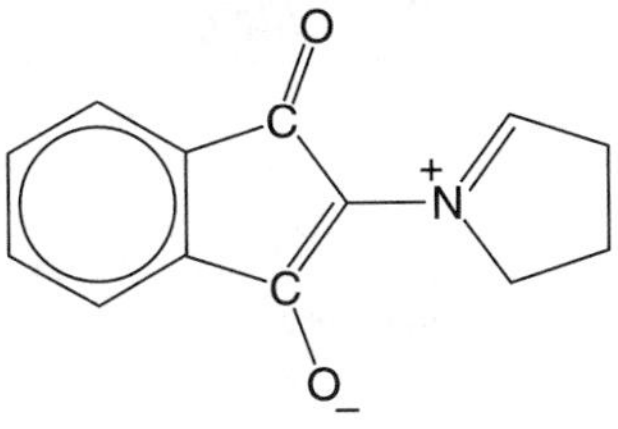

1

PROBLEM **26.33**

(a) A peptide of unknown structure is treated with the Sanger reagent, 2,4-dinitrofluorobenzene, followed by acid hydrolysis. The result is formation of a compound containing signals integrating for eight aromatic hydrogens in its ^{1}H NMR spectrum. What information can you derive about the structure of this peptide?

(b) A second peptide, similarly treated, gives a compound in which a single hydrogen appears in the high resolution ^{1}H NMR spectrum as an eight-line signal in the region δ 3.5 ppm. What information can you derive about the structure of this peptide?

PROBLEM **26.34** A tripeptide of unknown structure is hydrolyzed in acid to one molecule of Ala, one of Cys, and one of Met. The tripeptide reacts with phenyl isothiocyanate, followed by treatment with acid, to give a phenylthiohydantoin that shows only a three-hydrogen doublet above δ 3 ppm in the ^{1}H NMR spectrum. The tripeptide is not cleaved by BrCN. Deduce the structure from these data.

PROBLEM **26.35** A dodecapeptide **1** is hydrolyzed in acid to give 2 Arg, Asp, Cys, His, 2 Leu, Lys, 2 Phe, and 2 Val. The Edman procedure yields a phenylthiohydantoin that shows only a doublet integrating for 6H and a multiplet integrating for 1H at δ > 4.0 ppm in the ^{1}H NMR spectrum. Treatment of **1** with trypsin leads to Val·His·Phe·Leu·Arg, Asp·Cys·Leu·Phe·Lys, and Val·Arg. Treatment of **1** with chymotrypsin leads to Val·His·Phe, Leu·Arg·Asp·Cys·Leu·Phe, and Lys·Val·Arg. Deduce the structure of **1** and explain your reasoning.

PROBLEM **26.36** In Section 26.3, we saw that the tBoc group can be used to protect (or block) the amino group of an amino acid during peptide synthesis. This group can be introduced by allowing the amino acid to react with di-*tert*-butyl dicarbonate (**1**). Once the amide linkage has been formed, the tBoc protecting group can be removed by mild acid hydrolysis. Both of these reactions are illustrated below. Provide mechanisms for the introduction of the tBoc group and its removal. *Hint*: One of the products of the hydrolysis is isobutene.

(a) $H_3\overset{+}{N}CH(R)COO^-$ + $(H_3C)_3CO\text{–}C(=O)\text{–}O\text{–}C(=O)\text{–}OC(CH_3)_3$ (**1**)

1. NaOH, $(CH_3)_3COH/H_2O$
2. neutralize

$(H_3C)_3CO\text{–}C(=O)\text{–}NH\text{–}CH(R)\text{–}COOH$

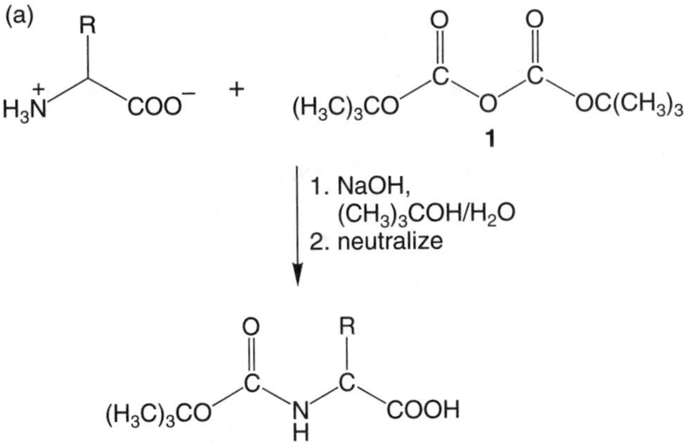

(b) $(H_3C)_3CO\text{–}C(=O)\text{–}NH\text{–}CH(R)\text{–}C(=O)\text{–}$NH—Peptide

1. CF_3COOH
2. $(CH_3CH_2)_3N$

$H_2N\text{–}CH(R)\text{–}C(=O)\text{–}$NH—Peptide + CO_2 + $(CH_3)_2C{=}CH_2$

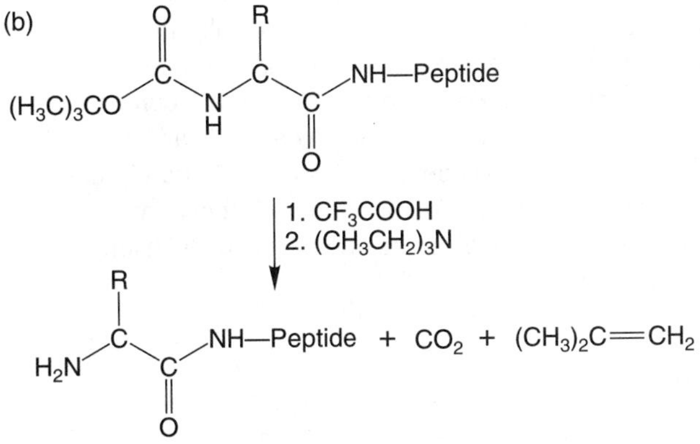

PROBLEM **26.37** In Section 26.3c (Fig. 26.56), we saw that the acylating agent in DCC-mediated peptide bond formation is the anhydride. There is another intermediate that could act as the acylating agent. What is it, and how can it function as the acylating agent? *Hint*: This intermediate appears in Figure 26.56.

PROBLEM 26.38 Reaction of *N*-phenylalanine (**1**) with nitrous acid affords *N*-nitroso amino acid (**2**), which upon treatment with acetic anhydride yields "sydnone" (**3**). Sydnones belong to a class of compounds known as mesoionic compounds. These cannot be satisfactorily represented by Lewis structures not involving charge separation. The name sydnone derives from the University of Sydney where the first examples were prepared in 1935. Reaction of **3** and dimethyl acetylenedicarboxylate (DMAD) gives pyrazole (**4**). Propose mechanisms for the formations of **2**, **3**, and **4**. *Hint*: For the formation of **4**, it may be helpful to consider the other possible Lewis structures for sydnone **3**.

Ph, NH, H_3C, COOH, **1** → $NaNO_2$/HCl, H_2O → Ph, N=O, N, H_3C, COOH, **2** → Ac_2O → Ph, $\overset{+}{N}$=N, O, H_3C, O^-, **3** → $H_3COOCC{\equiv}CCOOCH_3$ DMAD, xylene, 120 °C → H_3COOC, $COOCH_3$, N, N, CH_3, Ph, **4**

PROBLEM 26.39 In Problem 26.12, you were asked to write a mechanism for the acylation of glycine. The mechanism presented in the answer is actually incomplete, at least when the reaction is run in the presence of excess acetic anhydride. For example, what is the purpose of the water in the second step? There is actually an intermediate involved in this reaction, the 2-oxazolin-5-one (**1**), colloquially referred to as an azlactone. Propose mechanisms for the formation of azlactone (**1**) and its hydrolysis to *N*-acetylglycine.

NH_2, COOH → H_3C–C(=O)–O–C(=O)–CH_3 → CH_3, N, O, C, O, **1** → H_2O → O, C, HN, CH_3, COOH

PROBLEM 26.40 In the preceding problem, we saw that the azlactone (**1**) is an intermediate in the acetylation of glycine in the presence of excess acetic anhydride. The generation of azlactone (**1**) in this reaction lies at the heart of one of the oldest known amino acid syntheses. For example, if glycine (or *N*-acetylglycine) is treated with acetic anhydride in the presence of benzaldehyde and sodium acetate, the benzylidene azlactone (**2**) is formed. Azlactone (**2**) can then be converted into racemic phenylalanine (**3**) by the following sequence. Propose a mechanism for the formation of (**2**). *Hint*: Azlactone (**1**) is still an intermediate in this reaction.

NH_2, COOH → H_3C–C(=O)–O–C(=O)–CH_3, PhCHO, $NaOCOCH_3$ (NaOAc) → CH_3, N, O, Ph, C, H, O, **2** → H_2O, Δ → O, C, HN, CH_3, Ph, COOH, H → H_2/PtO_2 → O, C, HN, CH_3, Ph, COOH → 1. HCl/H_2O/Δ 2. neutralize → $\overset{+}{N}H_3$, $PhCH_2$, COO^-, **3**

PROBLEM 26.41 When optically active *N*-methyl-L-alanine (**1**) is heated with an excess of acetic anhydride, followed by treatment with water, the resulting *N*-acetyl-*N*-methylalanine (**2**) is racemic. If the acetylation reaction is run in the presence of dimethyl acetylenedicarboxylate (DMAD), the pyrrole **3** can be isolated in good yield. Account for the racemization of **2** and the formation of **3**. *Hint*: There is a common intermediate involved in these two reactions. Also, see Problem 26.38.

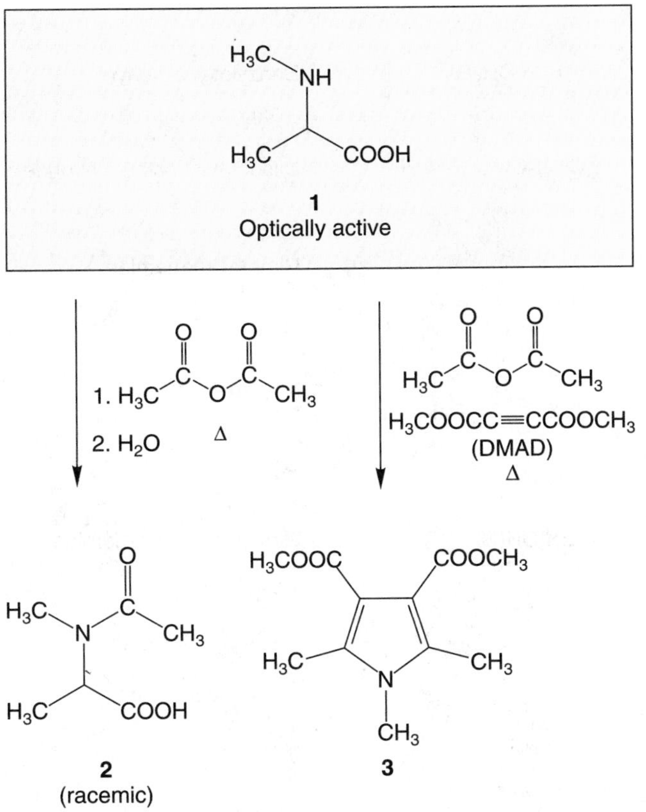

H_3C NH H_3C COOH

1
Optically active

1. H_3C–C(=O)–O–C(=O)–CH_3 Δ
2. H_2O

H_3C–C(=O)–O–C(=O)–CH_3
$H_3COOCC{\equiv}CCOOCH_3$
(DMAD)
Δ

H_3C N C(=O) CH_3 H_3C COOH

2
(racemic)

H_3COOC $COOCH_3$ H_3C N CH_3 CH_3

3

PROBLEM 26.42
(a) What mRNA sequence does the following DNA sequence produce? T A C G G G T T T A T C?
(b) What message does that mRNA sequence produce?

Some of the Commonly Used Abbreviations in this Book

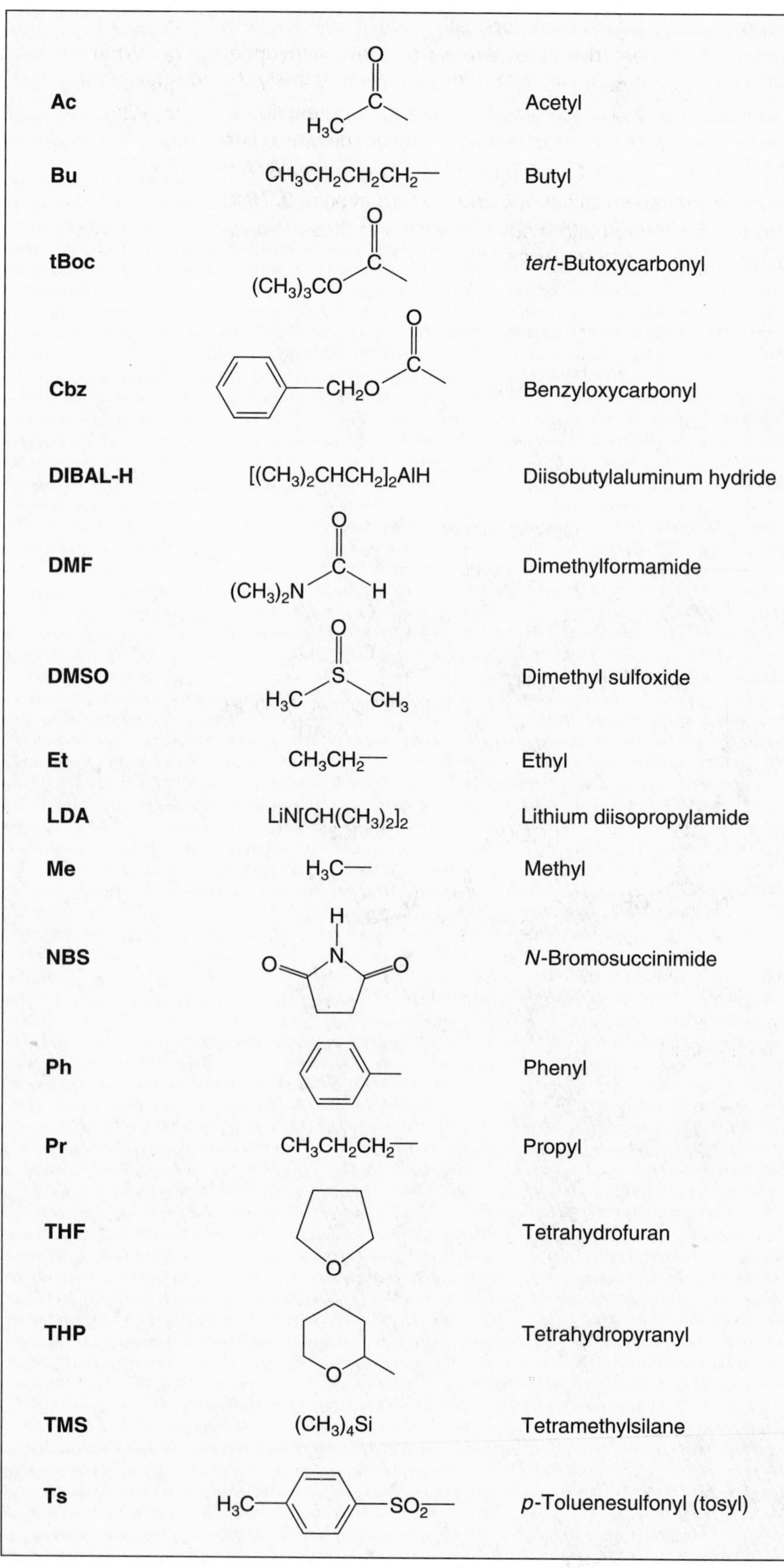

Ac	$H_3C-C(=O)-$	Acetyl
Bu	$CH_3CH_2CH_2CH_2-$	Butyl
tBoc	$(CH_3)_3CO-C(=O)-$	*tert*-Butoxycarbonyl
Cbz	$C_6H_5-CH_2O-C(=O)-$	Benzyloxycarbonyl
DIBAL-H	$[(CH_3)_2CHCH_2]_2AlH$	Diisobutylaluminum hydride
DMF	$(CH_3)_2N-C(=O)-H$	Dimethylformamide
DMSO	$H_3C-S(=O)-CH_3$	Dimethyl sulfoxide
Et	CH_3CH_2-	Ethyl
LDA	$LiN[CH(CH_3)_2]_2$	Lithium diisopropylamide
Me	H_3C-	Methyl
NBS	H, N, O, O	*N*-Bromosuccinimide
Ph		Phenyl
Pr	$CH_3CH_2CH_2-$	Propyl
THF	O	Tetrahydrofuran
THP	O	Tetrahydropyranyl
TMS	$(CH_3)_4Si$	Tetramethylsilane
Ts	$H_3C-C_6H_4-SO_2-$	*p*-Toluenesulfonyl (tosyl)

Index

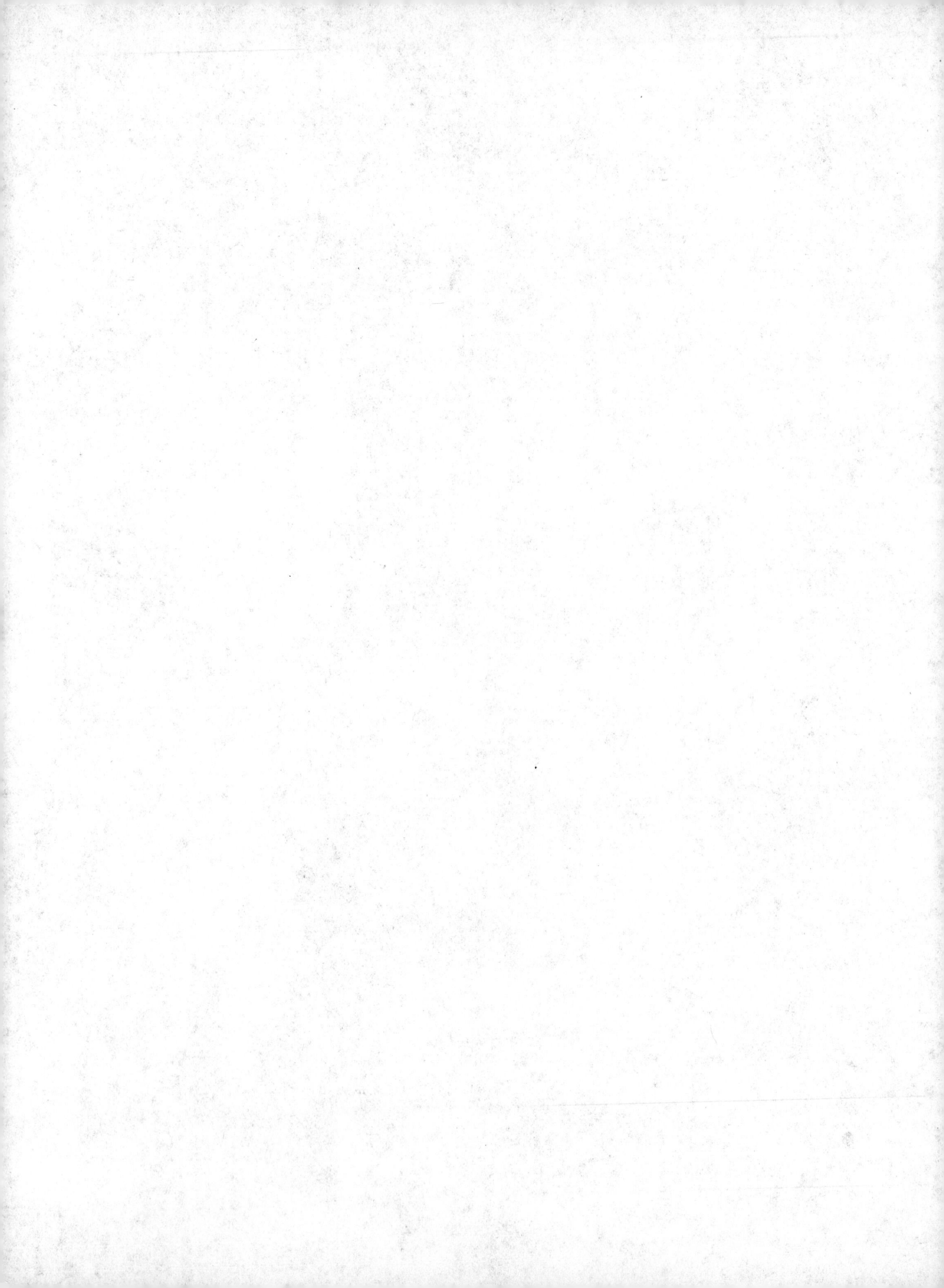

PERIODIC TABLE OF THE ELEMENTS

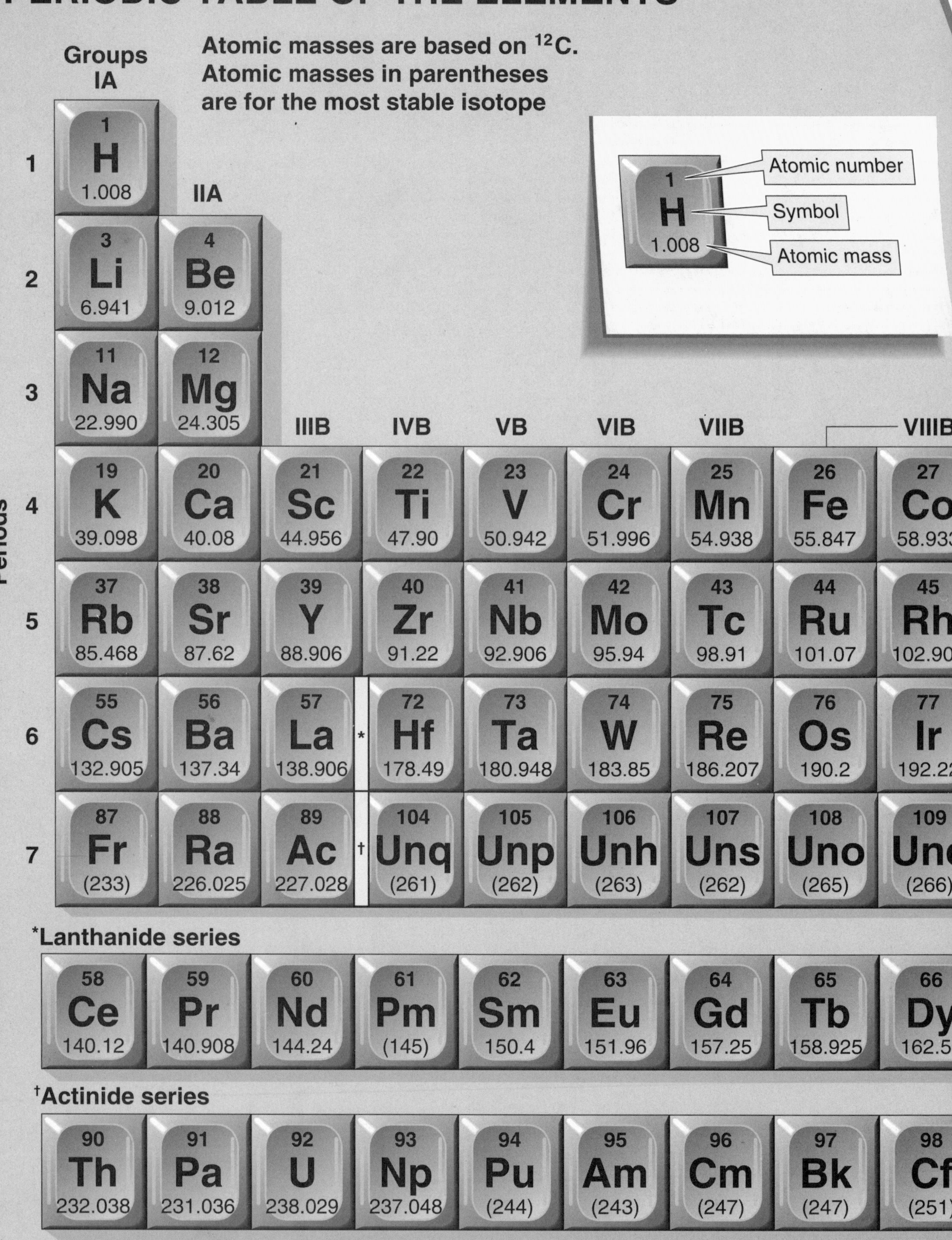

Groups IA
IIA
Atomic masses are based on ^{12}C.
Atomic masses in parentheses are for the most stable isotope
1 H 1.008
Atomic number
Symbol
Atomic mass
IIIB
IVB
VB
VIB
VIIB
VIIIB
Periods
1
1 H 1.008
2
3 Li 6.941
4 Be 9.012
3
11 Na 22.990
12 Mg 24.305
4
19 K 39.098
20 Ca 40.08
21 Sc 44.956
22 Ti 47.90
23 V 50.942
24 Cr 51.996
25 Mn 54.938
26 Fe 55.847
27 Co 58.933
5
37 Rb 85.468
38 Sr 87.62
39 Y 88.906
40 Zr 91.22
41 Nb 92.906
42 Mo 95.94
43 Tc 98.91
44 Ru 101.07
45 Rh 102.90
6
55 Cs 132.905
56 Ba 137.34
57 La 138.906
*
72 Hf 178.49
73 Ta 180.948
74 W 183.85
75 Re 186.207
76 Os 190.2
77 Ir 192.22
7
87 Fr (233)
88 Ra 226.025
89 Ac 227.028
†
104 Unq (261)
105 Unp (262)
106 Unh (263)
107 Uns (262)
108 Uno (265)
109 Une (266)
*Lanthanide series
58 Ce 140.12
59 Pr 140.908
60 Nd 144.24
61 Pm (145)
62 Sm 150.4
63 Eu 151.96
64 Gd 157.25
65 Tb 158.925
66 Dy 162.50
†Actinide series
90 Th 232.038
91 Pa 231.036
92 U 238.029
93 Np 237.048
94 Pu (244)
95 Am (243)
96 Cm (247)
97 Bk (247)
98 Cf (251)